RAND McNALLY

M000118509

GOODE'S

ATLAS OF Physical Geography

to accompany titles published by
John Wiley & Sons, Inc.

Howard Veregin, Ph.D., Editor

Editorial Advisory Board

Byron Augustin, D.A., Texas State University-San Marcos
Joshua Comenetz, Ph.D., University of Florida
Francis Galgano, Ph.D., United States Military Academy
Sallie A. Marston, Ph.D., University of Arizona
Virginia Thompson, Ph.D., Towson University

Abridgement of
21ST Edition

John Wiley & Sons, Inc. and **RAND McNALLY**

Working together to bring you the best in geography education

Few publishers can claim as rich a history as John Wiley & Sons, Inc. (publishers since 1807) and Rand McNally & Company (publishers since 1856). Even fewer can claim as long-standing a commitment to geographic education.

Wiley's partnership with the geographic community began at the very beginning of the 20th century with the publication of textbooks on surveying. Rand McNally's partnership began even earlier, with the publication of the first Rand McNally maps in 1872. Since then, both companies have worked in parallel to help students visualize spatial relationships and appreciate the earth's dynamic landscapes and diverse cultures.

Now these two publishers have combined their efforts to bring you this new atlas, which represents the very best in educational resources for geography.

Based on the 21st edition of the *Goode's World Atlas*, the *Goode's Atlas of Physical Geography* features:

- An emphasis on map accuracy and legibility, and the mixture of maps of different types and scales to facilitate interpretation of geographic phenomena.
- World, continental, and regional population density maps, which have been created using LandScan, a digital population database developed using satellite and computer-mapping technology.
- Graphs accompanying many of the maps, to show important statistical information, trends over time, and relationships between variables.
- Maps and graphs that have been updated, based on the most current available data in accordance with the high standards and quality that have always been a defining feature of the *Goode's World Atlas*.

Wiley and Rand McNally are currently offering seven new course-specific atlases, which can be packaged with any of Wiley's best-selling textbooks, or sold separately as stand-alones. These atlases include:

Rand McNally Goode's Atlas of Political Geography	0-471-70694-9
Rand McNally Goode's Atlas of Latin America	0-471-70697-3
Rand McNally Goode's Atlas of North America	0-471-70696-5
Rand McNally Goode's Atlas of Asia	0-471-70699-X
Rand McNally Goode's Atlas of Urban Geography	0-471-70695-7
Rand McNally Goode's Atlas of Physical Geography	0-471-70693-0
Rand McNally Goode's Atlas of Human Geography	0-471-70692-2

This book was set by GGS Book Services and printed and bound by Walsworth Press. The cover was printed by Phoenix Color.

To order books or for customer service please, call 1-800-CALL WILEY (225-5945).

ISBN 0471-70693-0

Printed in the United States

10 9 8 7 6 5 4 3 2 1

Table of Contents

Tables and Indexes

Introduction

Basic Earth Properties

The subject matter of **geography** includes people, landforms, climate, and all the other physical and human phenomena that make up the earth's environments and give unique character to different places. Geographers construct maps to visualize the **spatial distributions** of these phenomena: that is, how the phenomena vary over geographic space. Maps help geographers understand and explain phenomena and their interactions.

To better understand how maps portray geographic distributions, it is helpful to have an understanding of the basic properties of the earth.

The earth is essentially **spherical** in shape. Two basic reference points — the **North and South Poles** — mark the locations of the earth's axis of rotation. Equidistant between the two poles and encircling the earth is the **equator**. The equator divides the earth into two halves, called the **northern and southern hemispheres**. (See the figures to the right.)

Latitude and longitude are used to identify the locations of features on the earth's surface. They are measured in degrees, minutes and seconds. There are 60 minutes in a degree and 60 seconds in a minute. Latitude is the angle north or south of the equator. The symbols °, ', and " represent degrees, minutes and seconds, respectively. The N means north of the equator. For latitudes south of the equator, S is used. For example, the Rand McNally head office in Skokie, Illinois, is located at 42°1'51" N. The minimum latitude of 0° occurs at the equator. The maximum latitudes of 90° N and 90° S occur at the North and South Poles.

A **line of latitude** is a line connecting all points on the earth having the same latitude. Lines of latitude are also called **parallels**, as they run parallel to each other. Two parallels of special importance are the **Tropic of Cancer** and the **Tropic of Capricorn**, at approximately 23°30' N and S respectively. This angle coincides with the inclination of the earth's axis relative to its orbital plane around the sun. These tropics are the lines of latitude where the noon sun is directly overhead on the solstices. (See figure on page 66.) Two other important parallels are the **Arctic Circle** and the **Antarctic Circle**, at approximately 66°30' N and S respectively. These lines mark the most northerly and southerly points at which the sun can be seen on the solstices.

While latitude measures locations in a north-south direction, longitude measures them east-west. Longitude is the angle east or west of the **Prime Meridian**. A **meridian** is a line of longitude, a straight line extending from the North Pole to the South Pole. The Prime Meridian is the meridian passing through the Royal Observatory in Greenwich, England. For this reason the Prime Meridian is sometimes referred to as the **Greenwich Meridian**. This location for the Prime Meridian was adopted at the International Meridian Conference in Washington, D.C., in 1884.

Like latitude, longitude is measured in degrees, minutes, and seconds. For example, the Rand McNally head office is located at 87°43'6" W. The qualifiers E and W indicate whether a location is east or west of the Greenwich Meridian. Longitude ranges from 0° at Greenwich to 180° E or W. The meridian at 180° E is the same as the meridian at 180° W. This meridian, together with the Greenwich Meridian, divides the earth into **eastern and western hemispheres**.

Any circle that divides the earth into equal hemispheres is called a **great circle**. The equator is an example. The shortest distance between any two points on the earth is along a great circle. Other circles, including all other lines of latitude, are called **small circles**. Small circles divide the earth into two unequal pieces.

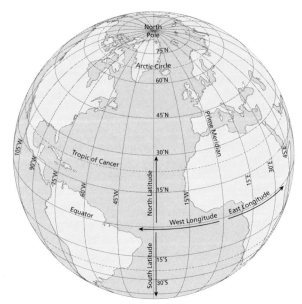

View of earth centered on 30° N, 30° W

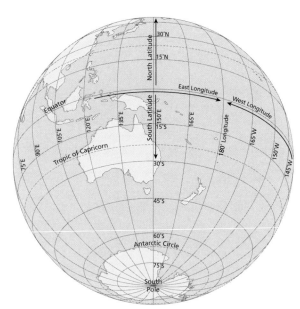

View of earth centered on 30° S, 150° E

The Geographic Grid

The grid of lines of latitude and longitude is known as the **geographic grid**. The following are some important characteristics of the grid.

All lines of longitude are equal in length and meet at the North and South Poles. These lines are called meridians.

All lines of latitude are parallel and equally spaced along meridians. These lines are called parallels.

The length of parallels increases with distance from the poles. For example, the length of the parallel at 60° latitude is one-half the length of the equator.

Meridians get closer together with increasing distance from the equator, and finally converge at the poles.

Parallels and meridians meet at right angles.

Map Scale

To use maps effectively it is important to have a basic understanding of map scale.

Map scale is defined as the ratio of distance on the map to distance on the earth's surface. For example, if a map shows two towns as separated by a distance of 1 inch, and these towns are actually 1 mile apart, then the scale of the map is 1 inch to 1 mile.

The statement "1 inch to 1 mile" is called a **verbal scale**. Verbal scales are simple and intuitive, but a drawback is that they are tied to the specific set of map and real-world units in the numerator and denominator of the ratio. This makes it difficult to compare the scales of different maps.

A more flexible way of expressing scale is as a **representative fraction**. In this case, both the numerator and denominator are converted to the same unit of measurement. For example, since there are 63,360 inches in a mile, the verbal scale "1 inch to 1 mile" can be expressed as the representative fraction 1:63,360. This means that 1 inch on the map represents 63,360 inches on the earth's surface. The advantage of the representative fraction is that it applies to any linear unit of measurement, including inches, feet, miles, meters, and kilometers.

Map scale can also be represented in graphical form. Many maps contain a **graphic scale** (or **bar scale**) showing real-world units such as miles or kilometers. The bar scale is usually subdivided to allow easy calculation of distance on the map.

Map scale has a significant effect on the amount of detail that can be portrayed on a map. This concept is illustrated here using a series of maps of the Washington, D.C., area. (See the figures to the right.) The scales of these maps range from 1:40,000,000 (top map) to 1:4,000,000 (center map) to 1:25,000 (bottom map). The top map has the **smallest scale** of the three maps, and the bottom map has the **largest scale**.

Note that as scale increases, the area of the earth's surface covered by the map decreases. The smallest-scale map covers thousands of square miles, while the largest-scale map covers only a few square miles within the city of Washington. This means that a given feature on the earth's surface will appear larger as map scale increases. On the smallest-scale map, Washington is represented by a small dot. As scale increases the dot becomes an orange shape representing the built-up area of Washington. At the largest scale Washington is so large that only a portion of it fits on the map.

Because small-scale maps cover such a large area, only the largest and most important features can be shown, such as large cities, major rivers and lakes, and international boundaries. In contrast, large-scale maps contain relatively small features, such as city streets, buildings, parks, and monuments.

Small-scale maps depict features in a more simplified manner than large-scale maps. As map scale decreases, the shapes of rivers and other features must be simplified to allow them to be depicted at a highly reduced size. This simplification process is known as **map generalization**.

Maps in *Goode's Atlas of Physical Geography* have a wide range of scales. The smallest scales are used for the world thematic map series, where scales range from approximately 1:200,000,000 to 1:75,000,000. Reference map scales range from a minimum of 1:100,000,000 for world maps to a maximum of 1:1,000,000 for city maps. Most reference maps are regional views with a scale of 1:4,000,000.

1:40,000,000 scale

1:4,000,000 scale

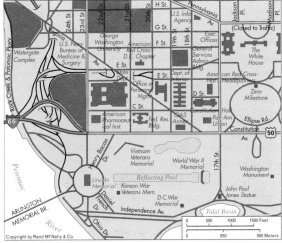

1:25,000 scale

Map Projections

Map projections influence the appearance of features on the map and the ability to interpret geographic phenomena.

A **map projection** is a geometric representation of the earth's surface on a flat or plane surface. Since the earth's surface is curved, a map projection is needed to produce any flat map, whether a page in this atlas or a computer-generated map of driving directions on www.randmcnally.com. Hundreds of projections have been developed since the dawn of mapmaking. A limitation of all projections is that they distort some geometric properties of the earth, such as shape, area, distance, or direction. However, certain properties are preserved on some projections.

If shape is preserved, the projection is called **conformal**. On conformal projections the shapes of features agree with the shapes these features have on the earth. A limitation of conformal projections is that they necessarily distort area, sometimes severely.

Equal-area projections preserve area. On equal area projections the areas of features correspond to their areas on the earth. To achieve this effect, equal-area projections distort shape.

Some projections preserve neither shape nor area, but instead balance shape and area distortion to create an aesthetically-pleasing result. These are often referred to as **compromise** projections.

Distance is preserved on **equidistant** projections, but this can only be achieved selectively, such as along specific meridians or parallels. No projection correctly preserves distance in all directions at all locations. As a result, the stated scale of a map may be accurate for only a limited set of locations. This problem is especially acute for small-scale maps covering large areas.

The projection selected for a particular map depends on the relative importance of different types of distortion, which often depends on the purpose of the map. For example, world maps showing phenomena that vary with area, such as population density or the distribution of agricultural crops, often use an equal-area projection to give an accurate depiction of the importance of each region.

Map projections are created using mathematical procedures. To illustrate the general principles of projections without using mathematics, we can view a projection as the geometric transfer of information from a globe to a flat projection surface, such as a sheet of paper. If we allow the paper to be rolled in different ways, we can derive three basic types of map projections: **cylindrical, conic,** and **azimuthal.** (See the figures to the right.)

For cylindrical projections, the sheet of paper is rolled into a tube and wrapped around the globe so that it is **tangent** (touching) along the equator. Information from the globe is transferred to the tube, and the tube is then unrolled to produce the final flat map.

Conic projections use a cone rather than a cylinder. The figure shows the cone tangent to the earth along a line of latitude with the apex of the cone over the pole. The line of tangency is called the **standard parallel** of the projection.

Azimuthal projections use a flat projection surface that is tangent to the globe at a single point, such as one of the poles.

The figures show the **normal orientation** of each type of surface relative to the globe. The **transverse orientation** is produced when the surface is rotated 90 degrees from normal. For azimuthal projections this orientation is usually called **equatorial** rather than transverse. An **oblique orientation** is created if the projection surface is oriented at an angle between normal and transverse. In general, map distortion increases with distance away from the point or line of tangency. This is why the normal orientations of the cylindrical, conic, and azimuthal projections are often used for mapping equatorial, mid-latitude, and polar regions, respectively.

The projection surface model is a visual tool useful for illustrating how information from the globe can be projected to the map. However, each of the three projection surfaces actually represents scores of individual projections. There are, for example, many projections with the term "cylindrical" in the name, each of which has the same basic rectangular shape, but different spacings of parallels and meridians. The projection surface model does not account for the numerous mathematical details that differentiate one cylindrical, conic, or azimuthal projection from another.

Cylindrical Projection

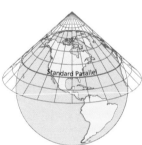

Conic Projection

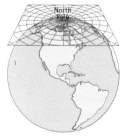
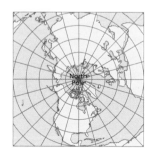

Azimuthal Projection

Map Projections Used in *Goode's Atlas of Physical Geography*

Of the hundreds of projections that have been developed, only a fraction are in everyday use. The main projections used in *Goode's Atlas of Physical Geography* are described below.

Simple Conic

Type: Conic   **Conformal:** No   **Equal-area:** No

Notes: Shape and area distortion on the Simple Conic projection are relatively low, even though the projection is neither conformal nor equal-area. The origins of the Simple Conic can be traced back nearly two thousand years, with the modern form of the projection dating to the 18th century.

Uses in *Goode's Atlas of Physical Geography*: Larger-scale reference maps of North America, Europe, Asia, and other regions.

Simple Conic Projection

Lambert Conformal Conic

Type: Conic   **Conformal:** Yes   **Equal-area:** No

Notes: On the Lambert Conformal Conic projection, spacing between parallels increases with distance away from the standard parallel, which allows the property of shape to be preserved. The projection is named after Johann Lambert, an 18th century mathematician who developed some of the most important projections in use today. It became widely used in the United States in the 20th century following its adoption for many statewide mapping programs.

Uses in *Goode's Atlas of Physical Geography*: Thematic maps of the United States and Canada, and reference maps of parts of Asia.

Lambert Conformal Conic Projection

Albers Equal-Area Conic

Type: Conic   **Conformal:** No   **Equal-area:** Yes

Notes: On the Albers Equal-Area Conic projection, spacing between parallels decreases with distance away from the standard parallel, which allows the property of area to be preserved. The projection is named after Heinrich Albers, who developed it in 1805. It became widely used in the 20th century, when the United States Coast and Geodetic Survey made it a standard for equal area maps of the United States.

Uses in *Goode's Atlas of Physical Geography*: Thematic maps of North America and Asia.

Albers Equal-Area Conic Projection

Polyconic

Type: Conic   **Conformal:** No   **Equal-area:** No

Notes: The term polyconic — literally "many-cones" — refers to the fact that this projection is an assemblage of different cones, each tangent at a different line of latitude. In contrast to many other conic projections, parallels are not concentric, and meridians are curved rather than straight. The Polyconic was first proposed by Ferdinand Hassler, who became Head of the United States Survey of the Coast (later renamed the Coast and Geodetic Survey) in 1807. The United States Geological Survey used this projection exclusively for large-scale topographic maps until the mid-20th century.

Uses in *Goode's Atlas of Physical Geography*: Reference maps of North America and Asia.

Polyconic Projection

Lambert Azimuthal Equal-Area

Type: Azimuthal   **Conformal:** No   **Equal-area:** Yes

Notes: This projection (another named after Johann Lambert) is useful for mapping large regions, as area is correctly preserved while shape distortion is relatively low. All orientations — polar, equatorial, and oblique — are common.

Uses in *Goode's Atlas of Physical Geography*: Thematic and reference maps of North and South America, Asia, Africa, Australia, and polar regions.

Lambert Azimuthal Equal-Area Projection

Miller Cylindrical

Type: Cylindrical   Conformal: No   Equal-area: No

Notes: This projection is useful for showing the entire earth in a simple rectangular form. However, polar areas exhibit significant exaggeration of area, a problem common to many cylindrical projections. The projection is named after Osborn Miller, Director of the American Geographical Society, who developed it in 1942 as a compromise projection that is neither conformal nor equal-area.

Uses in *Goode's Atlas of Physical Geography*: World climate and time zone maps.

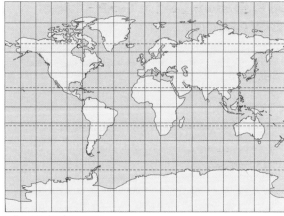
Miller Cylindrical Projection

Sinusoidal

Type: Pseudocylindrical   Conformal: No   Equal-area: Yes

Notes: The straight, evenly spaced parallels on this projection resemble the parallels on cylindrical projections. Unlike cylindrical projections, however, meridians are curved and converge at the poles. This causes significant shape distortion in polar regions. The Sinusoidal is the oldest-known pseudocylindrical projection, dating to the 16th century.

Uses in *Goode's Atlas of Physical Geography*: Reference maps of equatorial regions.

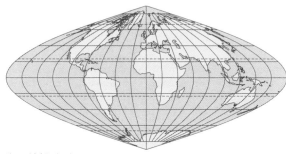
Sinusoidal Projection

Mollweide

Type: Pseudocylindrical   Conformal: No   Equal-area: Yes

Notes: The Mollweide (or Homolographic) projection resembles the Sinusoidal but has less shape distortion in polar areas due to its elliptical (or oval) form. One of several pseudocylindrical projections developed in the 19th century, it is named after Karl Mollweide, an astronomer and mathematician.

Uses in *Goode's Atlas of Physical Geography*: Oceanic reference maps.

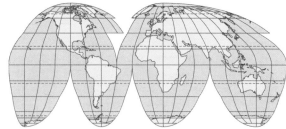
Mollweide Projection

Goode's Interrupted Homolosine

Type: Pseudocylindrical   Conformal: No   Equal-area: Yes

Notes: This projection is a fusion of the Sinusoidal between 40°44'N and S, and the Mollweide between these parallels and the poles. The unique appearance of the projection is due to the introduction of discontinuities in oceanic regions, the goal of which is to reduce distortion for continental landmasses. A condensed version of the projection also exists in which the Atlantic Ocean is compressed in an east-west direction. This modification helps maximize the scale of the map on the page. The Interrupted Homolosine projection is named after J. Paul Goode of the University of Chicago, who developed it in 1923. Goode was an advocate of interrupted projections and, as editor of *Goode's School Atlas*, promoted their use in education.

Uses in *Goode's Atlas of Physical Geography*: Small-scale world thematic and reference maps. Both condensed and non-condensed forms are used. An uninterrupted example is used for the Pacific Ocean map.

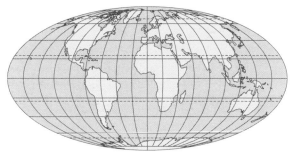
Goode's Interrupted Homolosine Projection

Robinson

Type: Pseudocylindrical   Conformal: No   Equal-area: No

Notes: This projection resembles the Mollweide except that polar regions are flattened and stretched out. While it is neither conformal nor equal-area, both shape and area distortion are relatively low. The projection was developed in 1963 by Arthur Robinson of the University of Wisconsin, at the request of Rand McNally.

Uses in *Goode's Atlas of Physical Geography*: World maps where the interrupted nature of Goode's Homolosine would be inappropriate, such as the World Oceanic Environments map.

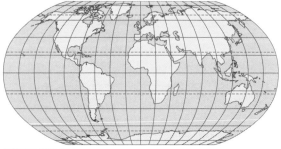
Robinson Projection

Thematic Maps in *Goode's Atlas of Physical Geography*

Thematic maps depict a single "theme" such as population density, agricultural productivity, or annual precipitation. The selected theme is presented on a base of locational information, such as coastlines, country boundaries, and major drainage features. The primary purpose of a thematic map is to convey an impression of the overall geographic distribution of the theme. It is usually not the intent of the map to provide exact numerical values. To obtain such information, the graphs and tables accompanying the map should be used.

Goode's Atlas of Physical Geography contains many different types of thematic maps. The characteristics of each are summarized below.

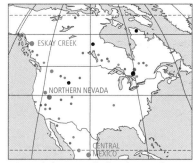

Point symbol map: Detail of Precious Metals

Point Symbol Maps

Point symbol maps are perhaps the simplest type of thematic map. They show features that occur at discrete locations. Examples include earthquakes, nuclear power plants, and minerals-producing areas. The Precious Metals map is an example of a point symbol map showing the locations of areas producing gold, silver, and platinum. A different color is used for each type of metal, while symbol size indicates relative importance.

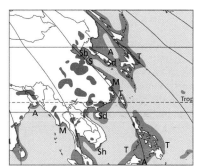

Area symbol map: Detail of Tobacco and Fisheries

Area Symbol Maps

Area symbol maps are useful for delineating regions of interest on the earth's surface. For example, the Tobacco and Fisheries map shows major tobacco-producing regions in one color and important fishing areas in another. On some area symbol maps, different shadings or colors are used to differentiate between major and minor areas.

Dot Maps

Dot maps show a distribution using a pattern of dots, where each dot represents a certain quantity or amount. For example, on the Sugar map, each dot represents 20,000 metric tons of sugar produced. Different dot colors are used to distinguish cane sugar from beet sugar. Dot maps are an effective way of representing the variable density of geographic phenomena over the earth's surface. This type of map is used extensively in *Goode's Atlas of Physical Geography* to show the distribution of agricultural commodities.

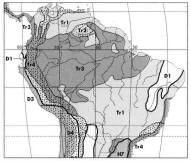

Dot map: Detail of Sugar

Area Class Maps

On area class maps, the earth's surface is divided into areas based on different classes or categories of a particular geographic phenomenon. For example, the Ecoregions map differentiates natural landscape categories, such as Tundra, Savanna, and Prairie. Other examples of area class maps in *Goode's Atlas of Physical Geography* include Landforms, Climatic Regions, Natural Vegetation, Soils, Agricultural Areas, Languages and Religions.

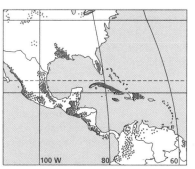

Area class map: Detail of Ecoregions

Isoline Maps

Isoline maps are used to portray quantities that vary smoothly over the surface of the earth. These maps are frequently used for climatic variables such as precipitation and temperature, but a variety of other quantities — from crop yield to population density — can also be treated in this way.

An isoline is a line on the map that joins locations with the same value. For example, the Summer (May to October) Precipitation map contains isolines at 5, 10, 20, and 40 inches. On this map, any 10-inch isoline separates areas that have less than 10 inches of precipitation from areas that have more than 10 inches. Note that the areas between isolines are given different colors to assist in map interpretation.

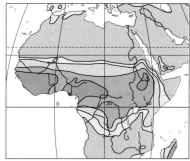

Isoline map: Detail of Precipitation

Proportional Symbol Maps

Proportional symbol maps portray numerical quantities, such as the total population of each state, the total value of agricultural goods produced in different regions, or the amount of hydroelectricity generated in different countries. The symbols on these maps — usually circles — are drawn such that the size of each is proportional to the value at that location. For example the Exports map shows the value of goods exported by each country in the world, in millions of U.S. dollars.

Proportional symbols are frequently subdivided based on the percentage of individual components making up the total. The Exports map uses wedges of different color to show the percentages of various types of exports, such as manufactured articles and raw materials.

Flow Line Maps

Flow line maps show flows between locations. Usually, the thickness of the flow lines is proportional to flow volume. Flows may be physical commodities like petroleum, or less tangible quantities like information. The flow lines on the Mineral Fuels map represent movement of petroleum measured in billions of U.S. dollars. Note that the locations of flow lines may not represent actual physical routes.

Choropleth Maps

Choropleth maps apply distinctive colors to predefined areas, such as counties or states, to represent different quantities in each area. The quantities shown are usually rates, percentages, or densities. For example, the Birth Rate map shows the annual number of births per one thousand people for each country.

Digital Images

Some maps are actually digital images, analogous to the pictures captured by digital cameras. These maps are created from a very fine grid of cells called **pixels**, each of which is assigned a color that corresponds to a specific value or range of values. The population density maps in this atlas are examples of this type. The effect is much like an isoline map, but the isolines themselves are not shown and the resulting geographic patterns are more subtle and variable. This approach is increasingly being used to map environmental phenomena observable from remote sensing systems.

Cartograms

Cartograms deliberately distort map shapes to achieve specific effects. On **area cartograms**, the size of each area, such as a country, is made proportional to its population. Countries with large populations are therefore drawn larger than countries with smaller populations, regardless of the actual size of these countries on the earth.

The world cartogram series in this atlas depicts each country as a rectangle. This is a departure from cartograms in earlier editions of the atlas, which attempted to preserve some of the salient shape characteristics for each country. The advantage of the rectangle method is that it is easier to compare the area of countries when their shapes are consistent.

The cartogram series incorporates choropleth shading on top of the rectangular cartogram base. In this way map readers can make inferences about the relationship between population and another thematic variable, such as HIV-infection rates.

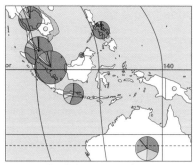
Proportional symbol map: Detail of Exports

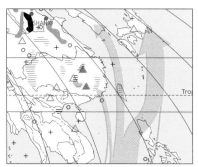
Flow line map: Detail of Mineral Fuels

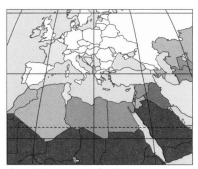
Choropleth map: Detail of Birth Rate

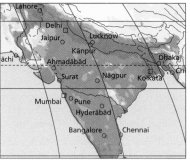
Digital image map: Detail of Population Density

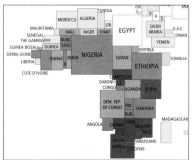
Cartogram: Detail of HIV Infection

Map Legend

Political Boundaries

Political maps	Physical maps	
┅┅┅	▬▬▬	International (Demarcated, Undemarcated, and Administrative)
━‧━‧	▬‧▬‧	Disputed de facto
▬ ▬	▬ ▬	Indefinite or Undefined
━‧‧━	▬‧‧▬	Secondary, State, Provincial, etc.

Parks, Indian Reservations

City Limits

Urbanized Areas

Transportation

Political maps	Physical maps	
━━━	━━━	Railroads
┄┄┄	┄┄┄	Railroad Ferries
	━━━	Major Roads
	━━━	Minor Roads
	┈┈┈	Caravan Routes
	✈	Airports

Cultural Features

~~⌐✓ Dams

┄┄┄┄ Pipelines

▲ Points of Interest

∴ Ruins

Populated Places

⊙	1,000,000 and over
◎	250,000 to 1,000,000
⊙	100,000 to 250,000
•	25,000 to 100,000
∘	Under 25,000
▫	Neighborhoods, Sections of Cities
TŌKYŌ	National Capitals
Boise	Secondary Capitals

Note: On maps at 1:20,000,000 and smaller, symbols do not follow the population classification shown above. Some other maps use a slightly different classification, which is shown in a separate legend in the map margin. On all maps, type size indicates the relative importance of the city.

Land Features

△ Peaks, Spot Heights

≍ Passes

Sand

Contours

Elevation

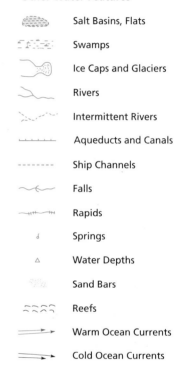

Meters		Feet
3050		10 000
1525		5000
610		2000
305		1000
152.5		500
0	Sea Level	0
152.5		500 Below
1525		5000 Sea Level
3050		10 000
6100		20 000

Lakes and Reservoirs

Fresh Water

Fresh Water: Intermittent

Salt Water

Salt Water: Intermittent

Other Water Features

Salt Basins, Flats

Swamps

Ice Caps and Glaciers

Rivers

Intermittent Rivers

Aqueducts and Canals

Ship Channels

Falls

Rapids

Springs

△ Water Depths

Sand Bars

Reefs

⟶ Warm Ocean Currents

⟶ Cold Ocean Currents

The legend above shows the symbols used for the political and physical reference maps in *Goode's Atlas of Physical Geography*.

To portray relative areas correctly, uniform map scales have been used wherever possible:

Continents – 1:40,000,000
Countries and regions – between 1:4,000,000 and 1:20,000,000
World, polar areas and oceans – between 1:50,000,000 and 1:100,000,000
Urbanized areas – 1:1,000,000

Elevations on the maps are shown using a combination of shaded relief and hypsometric tints. Shaded relief (or hill-shading) gives a three-dimensional impression of the landscape, while hypsometric tints show elevation ranges in different colors.

The choice of names for mapped features is complicated by the fact that a variety of languages and alphabets are used throughout the world. A local-names policy is used in *Goode's Atlas of Physical Geography* for populated places and local physical features. For some major features, an English form of the name is used with the local name given below in parentheses. Examples include Moscow (Moskva), Vienna (Wien) and Naples (Napoli). In countries where more than one official language is used, names are given in the dominant local language. For large physical features spanning international borders, the conventional English form of the name is used. In cases where a non-Roman alphabet is used, names have been transliterated according to accepted practice.

Selected features are also listed in the Index (pp. 217-322), which includes a pronunciation guide. A list of foreign geographic terms is provided in the Glossary (p. 214).

2

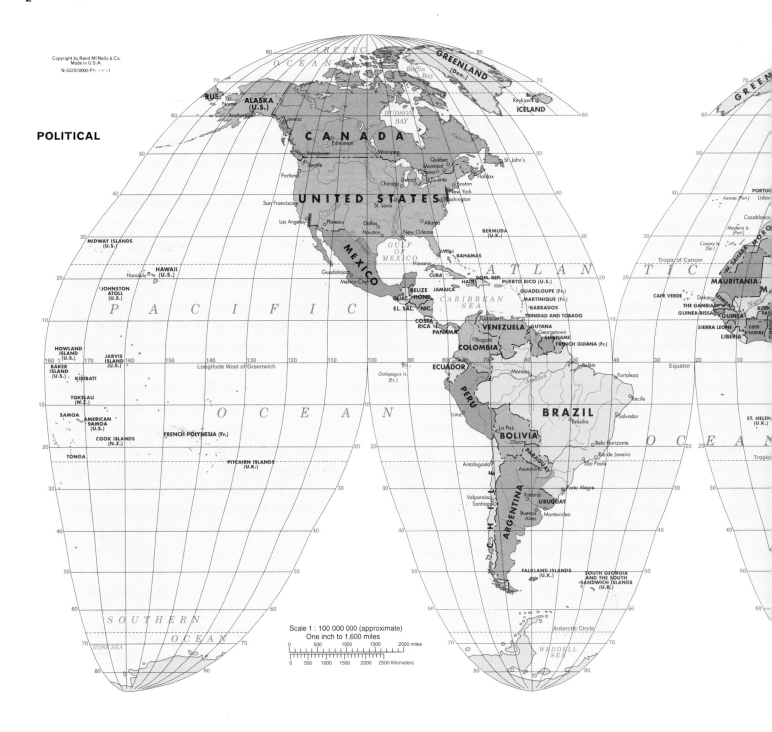

Copyright by Rand McNally & Co.
Made in U.S.A.
N-GDS10000-P1- -1- -1

POLITICAL

ARCTIC OCEAN

GREENLAND (Den.)

Baffin Bay

RUS. Nome
ALASKA (U.S.)
Anchorage
Juneau

Reykjavik
ICELAND

C A N A D A
Edmonton
Vancouver
Winnipeg
Seattle
Portland
Québec
Montréal
Ottawa
St. John's
Chicago Detroit Toronto
Halifax
Boston
San Francisco
UNITED STATES
St. Louis
New York
Washington

Los Angeles
Phoenix
Dallas
Atlanta
Houston
New Orleans
BERMUDA (U.K.)

HUDSON BAY

A T L A N T I C

MIDWAY ISLANDS (U.S.)

HAWAII (U.S.)
Honolulu

JOHNSTON ATOLL (U.S.)

GULF OF MEXICO
MEXICO
Guadalajara
Mexico City
Havana
BAHAMAS
CUBA
HAITI
DOM. REP.
PUERTO RICO (U.S.)
GUADELOUPE (Fr.)
MARTINIQUE (Fr.)
BARBADOS
TRINIDAD AND TOBAGO

Tropic of Cancer

W. SAHARA MOR.
MAURITANIA

CAPE VERDE
THE GAMBIA
GUINEA-BISSAU
SIERRA LEONE
LIBERIA

Azores (Port.)
Lisbo
Casablanca
Madeira Is. (Port.)
Canary Is. (Sp.)
Dakar
SENEGAL
GUINEA
COTE D'IVOIRE
BURK FAS
Niger

PORTU
PORTUG

GREEN

P A C I F I C

BELIZE
GUAT. HOND.
EL SAL. NIC.
COSTA RICA
PANAMA
JAMAICA
CARIBBEAN SEA
Caracas
VENEZUELA
GUYANA
Georgetown
SURINAME
FRENCH GUIANA (Fr.)
COLOMBIA
Bogotá

HOWLAND ISLAND (U.S.)
BAKER ISLAND (U.S.)
JARVIS ISLAND (U.S.)
KIRIBATI

TOKELAU (N.Z.)
SAMOA
AMERICAN SAMOA (U.S.)
COOK ISLANDS (N.Z.)
TONGA

FRENCH POLYNESIA (Fr.)

PITCAIRN ISLANDS (U.K.)

Longitude West of Greenwich

O C E A N

Galapagos Is. (Ec.)
ECUADOR
Quito
PERU
Lima
BOLIVIA
La Paz
Sucre
Manaus
Amazon
Belém
Fortaleza
Recife
BRAZIL
Brasília
Salvador
Belo Horizonte
PARAGUAY
Asunción
Rio de Janeiro
São Paulo
Porto Alegre
URUGUAY
Buenos Aires
Montevideo
ARGENTINA
Antofagasta
Valparaíso
Santiago
Rosario

Equator

ST. HELENA (U.K.)

O C E A N
Tropic

SOUTHERN OCEAN
ROSS SEA

FALKLAND ISLANDS (U.K.)
SOUTH GEORGIA AND THE SOUTH SANDWICH ISLANDS (U.K.)

Antarctic Circle

WEDDELL SEA

Scale 1 : 100 000 000 (approximate)
One inch to 1,600 miles
0   500   1000   1500   2000 miles
0   500   1000   1500   2000   2500 Kilometers

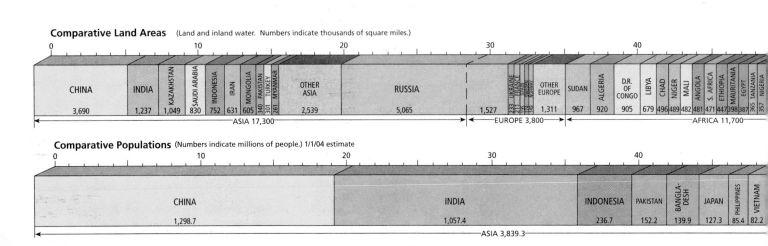

Comparative Land Areas (Land and inland water. Numbers indicate thousands of square miles.)

| CHINA 3,690 | INDIA 1,237 | KAZAKHSTAN 1,049 | SAUDI ARABIA 830 | INDONESIA 752 | IRAN 631 | MONGOLIA 605 | PAKISTAN 340 | TURKEY 301 | MYANMAR 261 | OTHER ASIA 2,539 | RUSSIA 5,065 | 1,527 | UKRAINE 233 | FRANCE 211 | SPAIN | SWEDEN | GERMANY | NORWAY | OTHER EUROPE 1,311 | SUDAN 967 | ALGERIA 920 | D.R. OF CONGO 905 | LIBYA 679 | CHAD 496 | NIGER 489 | MALI 482 | ANGOLA 481 | S. AFRICA 471 | ETHIOPIA 447 | MAURITANIA 398 | EGYPT 387 | TANZANIA 365 | NIGERIA 357 |

ASIA 17,300 — EUROPE 3,800 — AFRICA 11,700

Comparative Populations (Numbers indicate millions of people.) 1/1/04 estimate

| CHINA 1,298.7 | INDIA 1,057.4 | INDONESIA 236.7 | PAKISTAN 152.2 | BANGLADESH 139.9 | JAPAN 127.3 | PHILIPPINES 85.4 | VIETNAM 82.2 |

ASIA 3,839.3

ARCTIC OCEAN

SVALBARD (Nor.)

JAN MAYEN (Nor.)

ICELAND   Arctic Circle

NORWAY   ALASKA (U.S.)

Rjvik   FAROE ISLANDS (Den.)

R U S S I A

UNITED KINGDOM   FINLAND   SWEDEN   Arkhangelsk

DEN.   St. Petersburg   Ob   Yenisey   Lena

NETH.   London   Stockholm   EST.   Moscow   Novosibirsk   SEA OF OKHOTSK   BERING SEA

GERMANY   POLAND   BELARUS   Volga   Irkutsk   Magadan

FRANCE   SWITZ.   UKRAINE   Kiev   KAZAKHSTAN   MONGOLIA   Ulan Bator   Vladivostok   SEA OF JAPAN

SLOV.   AUS.   HUNG.   ROM.   MOLD.   Ürümqi   Harbin   NORTH KOREA   JAPAN

ITALY   CRO.   BUL.   Black Sea   Caspian Sea   UZBEKISTAN   Tashkent   KYRG.   Shenyang   SOUTH KOREA   Seoul   Tōkyō

SPAIN   GREECE   Istanbul   Ankara   GEO.   ARM. AZER.   TURKMENISTAN   TAJIK.   Beijing   C H I N A   Ōsaka

Rome   TURKEY   SYRIA   Tehran   AFGHANISTAN   Nanjing   Shanghai

Algiers   TUNISIA   MEDITERRANEAN SEA   CYPRUS   LEB.   IRAQ   Baghdad   IRAN   PAKISTAN   New Delhi   NEPAL   BHU.   Chongqing   Wuhan

Tripoli   ISRAEL   JORDAN   KUWAIT   Karachi   Ganges   BNGL.   Guangzhou   TAIWAN   WAKE ISLAND (U.S.)

ALGERIA   LIBYA   EGYPT   SAUDI ARABIA   QATAR   U.A.E.   Kolkata   MYANMAR   Hanoi   Hong Kong   NORTHERN MARIANA ISLANDS (U.S.)

Riyadh   OMAN   I N D I A   LAOS   NIGER   CHAD   SUDAN   Mecca   YEMEN   ARABIAN SEA   Mumbai   BAY OF BENGAL   Rangoon   THAILAND   VIETNAM   Manila   GUAM (U.S.)

NIGERIA   Lagos   CENTRAL AFRICAN REPUBLIC   Addis Ababa   DJIBOUTI   Hyderabad   Chennai   CAMBODIA   Ho Chi Minh City   PHILIPPINES

EQUATORIAL GUINEA   CAMEROON   Congo   ETHIOPIA   SOMALIA   Columbo   SRI LANKA   MALDIVES   MALAYSIA   BRUNEI   PALAU   FED. STATES OF MICRONESIA   MARSHALL ISLANDS

GABON   CONGO   UGANDA   KENYA   Mogadishu   Kuala Lumpur   SINGAPORE   BORNEO

Brazzaville   DEM. REP. OF THE CONGO   RWANDA   Nairobi   INDIAN   Longitude East of Greenwich   SUMATRA   I N D O N E S I A   NAURU   KIRIBATI

Kinshasa   BURUNDI   Mombasa   Jakarta   NEW GUINEA   PAPUA NEW GUINEA

Luanda   TANZANIA   Dar Es Salaam   SEYCHELLES   O C E A N   Surabaya   EAST TIMOR   SOLOMON ISLANDS   TUVALU

ANGOLA   ZAMBIA   COMOROS   CHRISTMAS ISLAND (Austl.)   CORAL SEA   VANUATU   FIJI

Capricorn   NAMIBIA   BOTSWANA   ZIMBABWE   MADAGASCAR   Antananarivo   COCOS ISLANDS (Austl.)   Darwin   Suva

MOZAMBIQUE   Pretoria   MAURITIUS   REUNION (Fr.)   NEW CALEDONIA (Fr.)

Johannesburg   Maputo   SWAZILAND   A U S T R A L I A   Brisbane

SOUTH AFRICA   LESOTHO   Durban   Perth   Sydney   Canberra   Auckland

Cape Town   Adelaide   NEW ZEALAND   Wellington

Melbourne

S O U T H E R N   O C E A N

The Antarctic territorial claims of Argentina, Australia, Chile, France, New Zealand, Norway, and the United Kingdom are not recognized by other nations. Antarctica is administered under the provisions of the Antarctic Treaty of 1959.

A N T A R C T I C A

Goode's Homolosine Equal Area Projection

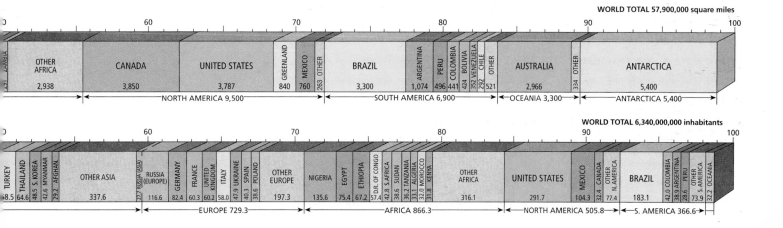

WORLD TOTAL 57,900,000 square miles

ZAMBIA	OTHER AFRICA	CANADA	UNITED STATES	GREENLAND	MEXICO	OTHER	BRAZIL	ARGENTINA	PERU	COLOMBIA	BOLIVIA	VENEZUELA	CHILE	OTHER	AUSTRALIA	OTHER	ANTARCTICA
291	2,938	3,850	3,787	840	760	263	3,300	1,074	496	441	424	352	292	521	2,966	334	5,400

NORTH AMERICA 9,500 — SOUTH AMERICA 6,900 — OCEANIA 3,300 — ANTARCTICA 5,400

WORLD TOTAL 6,340,000,000 inhabitants

TURKEY	THAILAND	S. KOREA	MYANMAR	AFGHAN	OTHER ASIA	RUSSIA (ASIA)	RUSSIA (EUROPE)	GERMANY	FRANCE	UNITED KINGDOM	ITALY	UKRAINE	SPAIN	POLAND	OTHER EUROPE	NIGERIA	EGYPT	ETHIOPIA	D.R. OF CONGO	S. AFRICA	SUDAN	TANZANIA	MOROCCO	KENYA	OTHER AFRICA	UNITED STATES	MEXICO	CANADA	OTHER N. AMERICA	BRAZIL	COLOMBIA	ARGENTINA	PERU	OTHER S. AMERICA	OCEANIA
68.5	64.6	48.5	42.6	29.2	337.6	27.7	116.6	82.4	60.3	60.2	58.0	47.9	40.3	38.6	197.3	135.6	75.4	67.2	57.4	42.8	38.6	36.2	33.1	32.0	31.8	316.1	291.7	104.3	32.4	77.4	183.1	42.0	38.9	28.6	73.9	32.2

EUROPE 729.3 — AFRICA 866.3 — NORTH AMERICA 505.8 — S. AMERICA 366.6

4

PHYSICAL

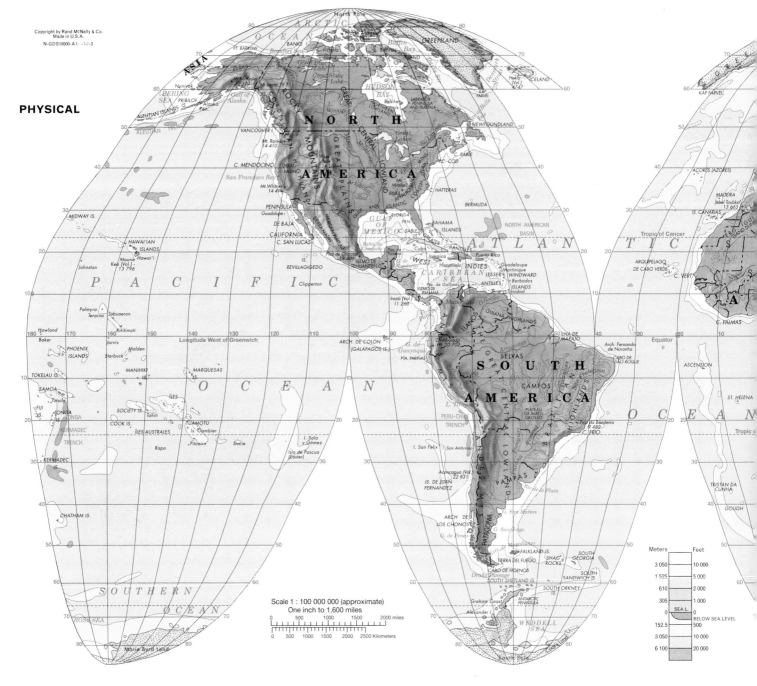

Scale 1 : 100 000 000 (approximate)
One inch to 1,600 miles

Meters	Feet
3 050	10 000
1 525	5 000
610	2 000
305	1 000
SEA L	
152.5	BELOW SEA LEVEL 500
3 050	10 000
6 100	20 000

Land Elevations in Profile

Ocean Depths in Profile

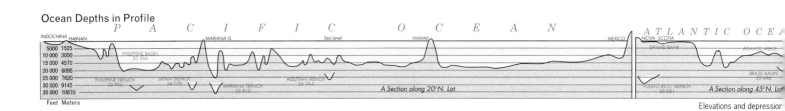

Elevations and depression

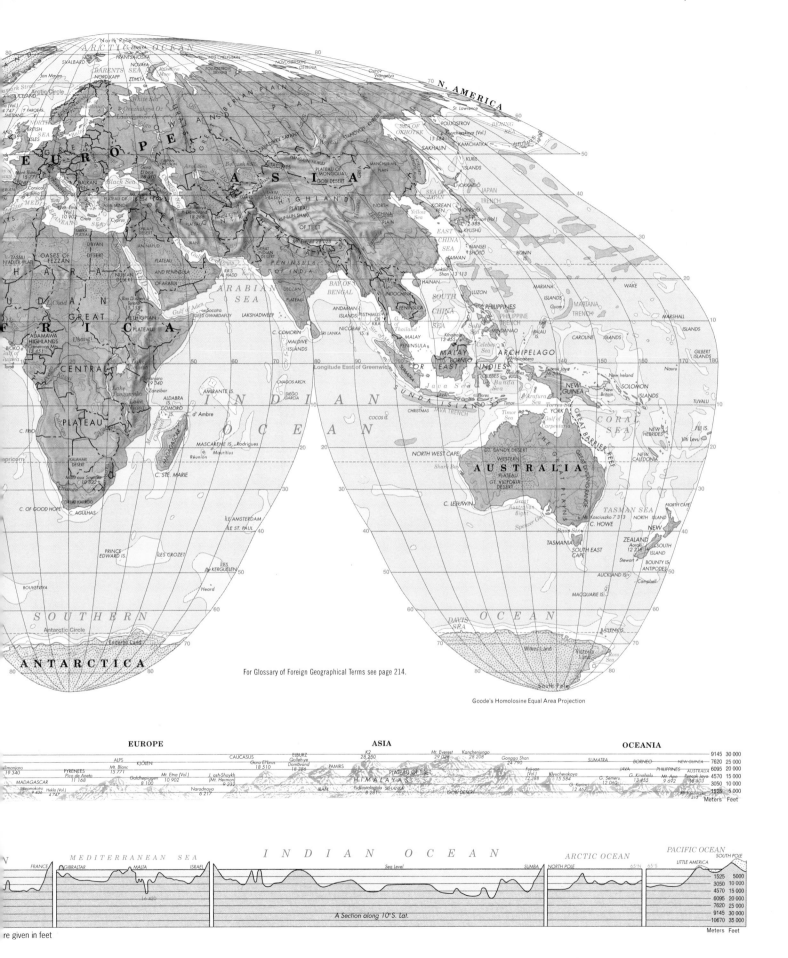

For Glossary of Foreign Geographical Terms see page 214.

Goode's Homolosine Equal Area Projection

A Section along 10°S. Lat.

re given in feet

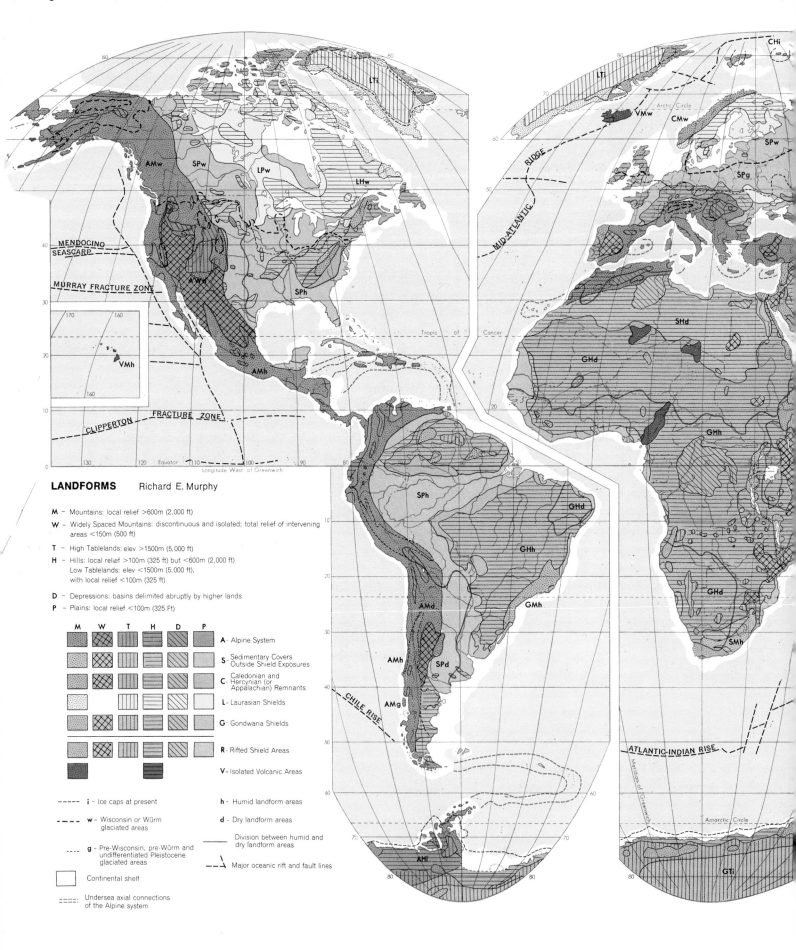

6

LANDFORMS   Richard E. Murphy

M – Mountains: local relief >600m (2,000 ft)

W – Widely Spaced Mountains: discontinuous and isolated; total relief of intervening areas <150m (500 ft)

T – High Tablelands: elev >1500m (5,000 ft)

H – Hills: local relief >100m (325 ft) but <600m (2,000 ft)
Low Tablelands: elev <1500m (5,000 ft), with local relief <100m (325 ft)

D – Depressions: basins delimited abruptly by higher lands

P – Plains: local relief <100m (325 Ft)

M	W	T	H	D	P	
						A- Alpine System
						S- Sedimentary Covers Outside Shield Exposures
						C- Caledonian and Hercynian (or Appalachian) Remnants
						L- Laurasian Shields
						G- Gondwana Shields
						R- Rifted Shield Areas
						V- Isolated Volcanic Areas

- - - - - **i** - Ice caps at present

- - - - **w** - Wisconsin or Würm glaciated areas

- - - - **g** - Pre-Wisconsin, pre-Würm and undifferentiated Pleistocene glaciated areas

h - Humid landform areas

d - Dry landform areas

——— Division between humid and dry landform areas

- - - Major oceanic rift and fault lines

☐ Continental shelf

===== Undersea axial connections of the Alpine system

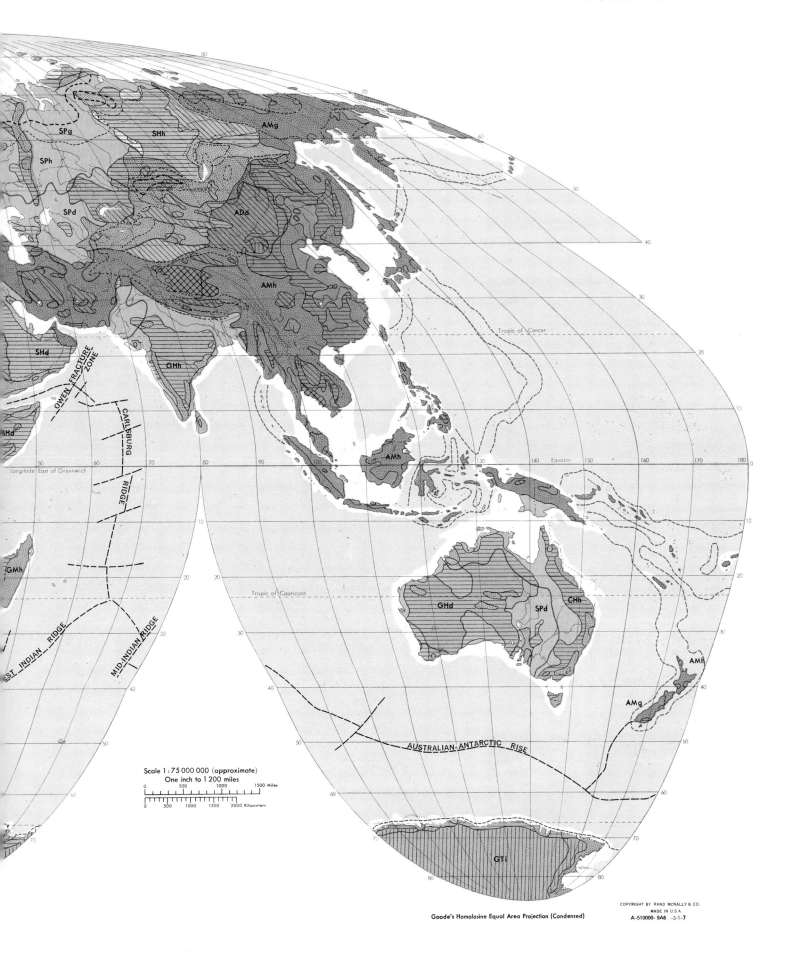

SPg

SPh

SHh

AMg

SPd

ADd

AMh

SHd

OWEN FRACTURE ZONE

CARLSBURG RIDGE

GHh

SHd

GMh

WEST INDIAN RIDGE

MID-INDIAN RIDGE

AMh

Longitude East of Greenwich

Tropic of Cancer

Equator

Tropic of Capricorn

GHd

SPd

CHh

AMh

AMg

AUSTRALIAN-ANTARCTIC RISE

GTi

Scale 1 : 75 000 000 (approximate)
One inch to 1 200 miles

0   500   1000   1500 Miles

0   500   1000   1500   2000 Kilometers

Goode's Homolosine Equal Area Projection (Condensed)

CONTINENTAL DRIFT

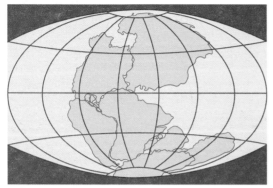

225 million years ago the supercontinent of Pangaea exists and Panthalassa forms the ancestral ocean. Tethys Sea separates Eurasia and Africa.

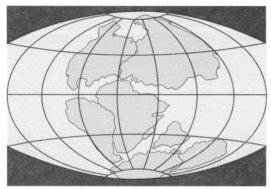

180 million years ago Pangaea splits, Laurasia drifts north. Gondwanaland breaks into South America/Africa, India, and Australia/Antarctica.

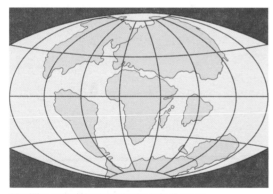

65 million years ago ocean basins take shape as South America and India move from Africa and the Tethys Sea closes to form the Mediterranean Sea.

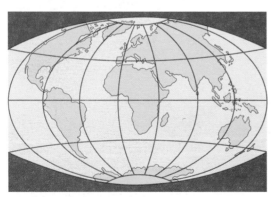

The present day: India has merged with Asia, Australia is free of Antarctica, and North America is free of Eurasia.

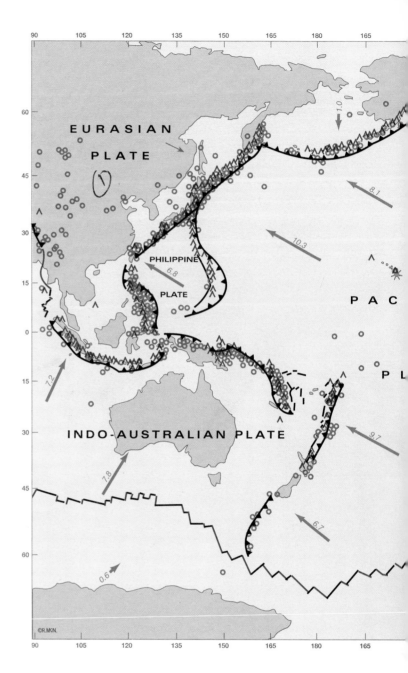

PLATE TECTONICS

Types of plate boundaries

— **Divergent:** magma emerges from the earth's mantle at the mid-ocean ridges forming new crust and forcing the plates to spread apart at the ridges.

▲▲▲ **Convergent:** plates collide at subduction zones where the denser plate is forced back into the earth's mantle forming deep ocean trenches.

— **Transform:** plates slide past one another producing faults and fracture zones.

Other map symbols

→ Direction of plate movement

6.7 → Length of arrow is proportional to the amount of plate movement (number indicates centimeters of movement per year)

○ Earthquake of magnitude 7.5 and above (from 10 A.D. to the present)

∧ Volcano (eruption since 1900)

✳ Selected hot spots

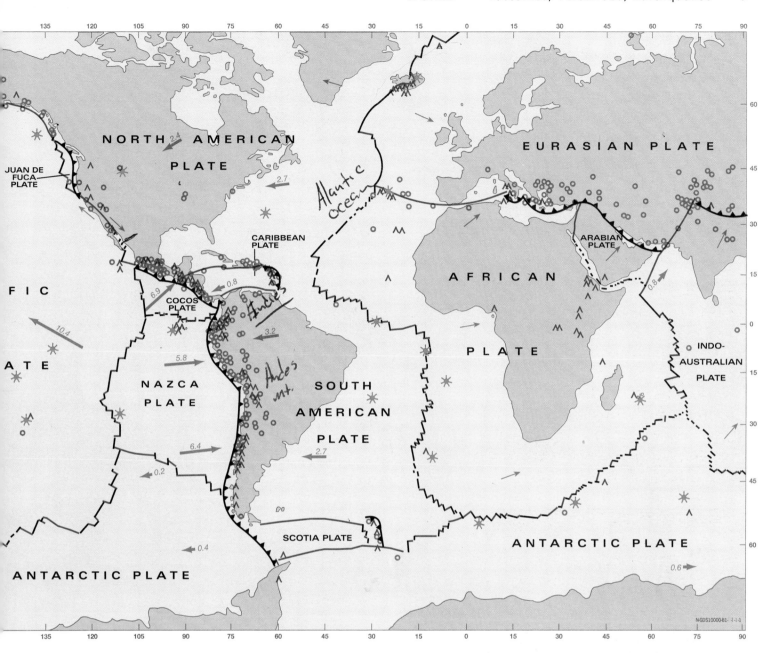

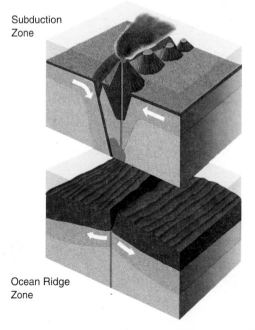

Subduction
Zone

Ocean Ridge
Zone

The plate tectonic theory describes the movement of the earth's surface and subsurface and explains why surface features are where they are.

Stated concisely, the theory presumes the lithosphere - the outside crust and uppermost mantle of the earth - is divided into about a dozen major rigid plates and several smaller platelets that move relative to one another. The position and names of the plates are shown on the map above.

The motor that drives the plates is found deep in the mantle. The theory states that because of temperature differences in the mantle, slow convection currents circulate there. Where two molten currents converge and move upward, they separate, causing the crustal plates to bulge and move apart in mid-ocean regions. Transverse fractures disrupt these broad regions. Lava wells up at these points to cause volcanic activity and to form ridges. The plates grow larger by accretion along these mid-ocean ridges, cause vast regions of the crust to move apart, and force the plates to collide with one another. As the plates do so, they are destroyed at subduction zones, where the plates are consumed downward, back into the earth's mantle, forming deep ocean trenches. The diagrams to the right illustrate the processes.

Most of the earth's volcanic and seismic activities

occur where plates slide past each other at transform boundaries or collide along subduction zones. The friction and heat caused by the grinding motion of the subducted plates causes rock to liquify and rise to the surface as volcanoes and eventually form vast mountain ranges. Strong and deep earthquakes are common here.

Volcanoes and earthquakes also occur at random locations around the earth known as "hot spots". Hot rock from deep in the mantle rises to the surface creating some of the earth's tallest mountains. As the lithospheric plates move slowly over these stationary plumes of magma, island chains (such as the Hawaiian Islands) are formed.

The overall result of tectonic movement is that the crustal plates move slowly and inexorably as relatively rigid entitles, carrying the continents along with them. The history of this continental drifting is illustrated in the four maps to the left. It began with a single landmass called the supercontinent of Pangaea and the ancestral sea, the Panthalassa Ocean. Pangaea first split into a northern landmass called Laurasia and a southern block called Gondwanaland and subsequently into the continents we map today. The map of the future will be significantly different as the continents continue to drift.

Scale 1:72 000 000 at 40° latitude.

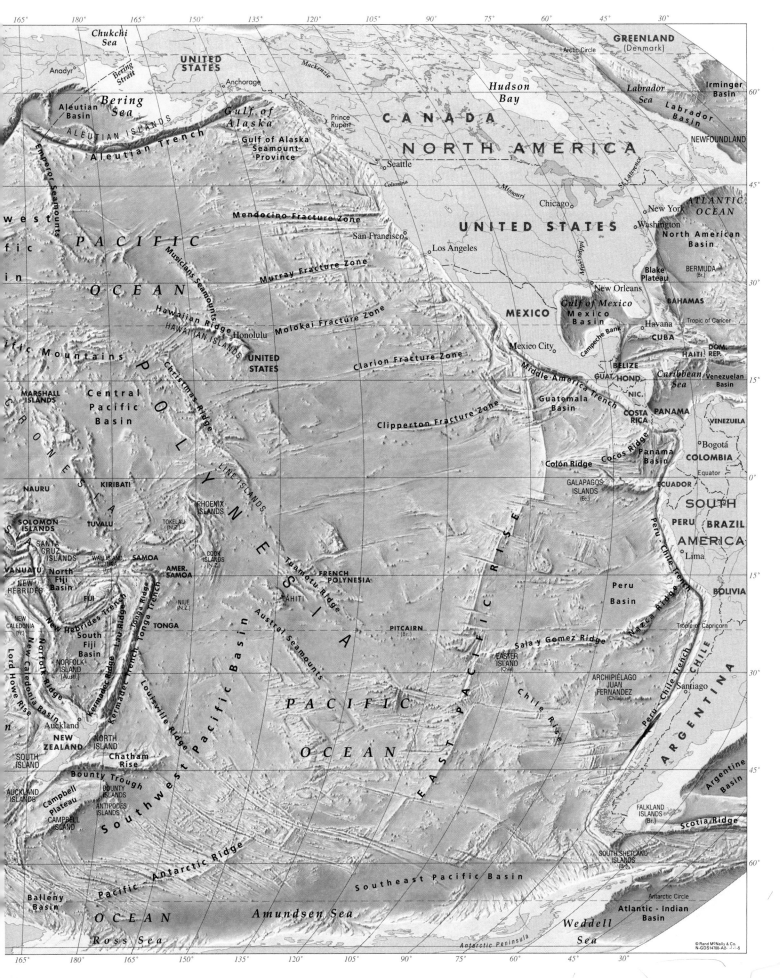

ROBINSON PROJECTION

Scale 1:72 000 000 at 40° latitude. ROBINSON PROJECTION

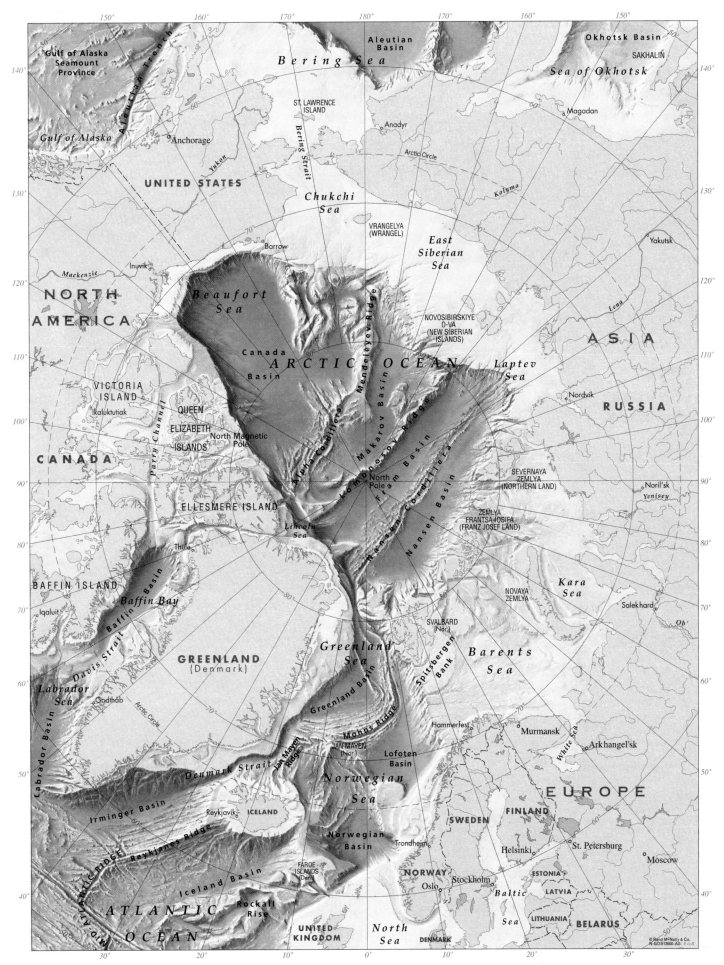

Scale 1:30 000 000. LAMBERT AZIMUTHAL EQUAL AREA PROJECTION

14

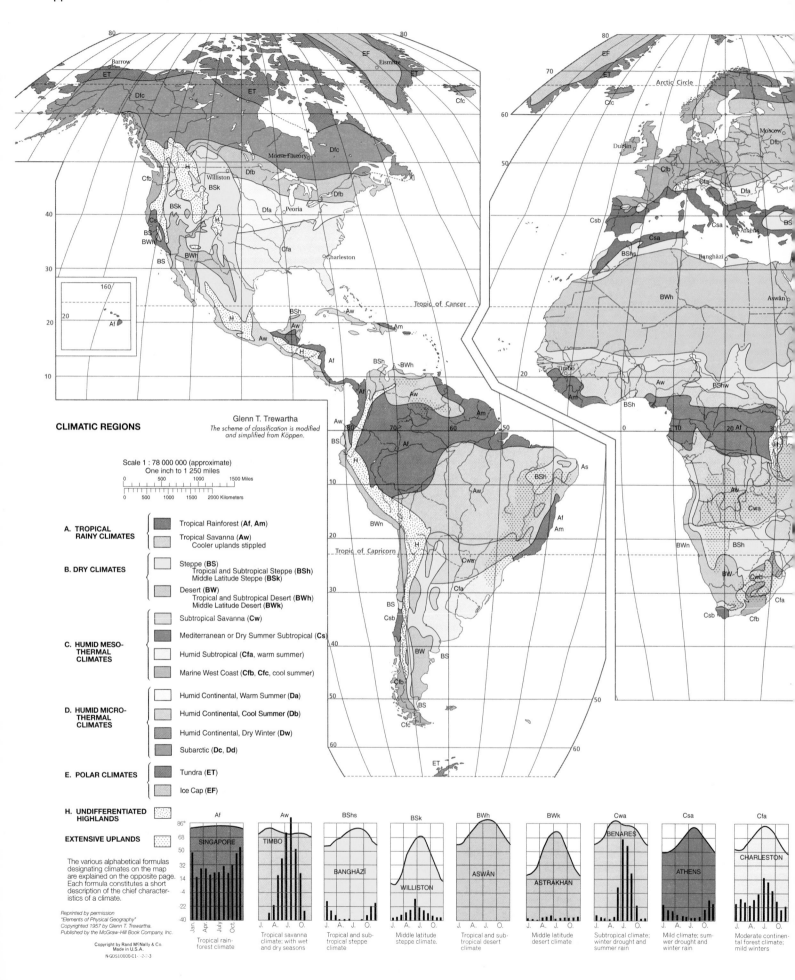

CLIMATIC REGIONS

Glenn T. Trewartha
The scheme of classification is modified and simplified from Köppen.

Scale 1 : 78 000 000 (approximate)
One inch to 1 250 miles

| | | | |
| 0 | 500 | 1000 | 1500 Miles |

| | | | |
| 0 | 500 | 1000 | 1500 | 2000 Kilometers |

A. TROPICAL RAINY CLIMATES
- Tropical Rainforest (**Af, Am**)
- Tropical Savanna (**Aw**)
  Cooler uplands stippled

B. DRY CLIMATES
- Steppe (**BS**)
  Tropical and Subtropical Steppe (**BSh**)
  Middle Latitude Steppe (**BSk**)
- Desert (**BW**)
  Tropical and Subtropical Desert (**BWh**)
  Middle Latitude Desert (**BWk**)

C. HUMID MESO-THERMAL CLIMATES
- Subtropical Savanna (**Cw**)
- Mediterranean or Dry Summer Subtropical (**Cs**)
- Humid Subtropical (**Cfa**, warm summer)
- Marine West Coast (**Cfb, Cfc**, cool summer)

D. HUMID MICRO-THERMAL CLIMATES
- Humid Continental, Warm Summer (**Da**)
- Humid Continental, Cool Summer (**Db**)
- Humid Continental, Dry Winter (**Dw**)
- Subarctic (**Dc, Dd**)

E. POLAR CLIMATES
- Tundra (**ET**)
- Ice Cap (**EF**)

H. UNDIFFERENTIATED HIGHLANDS

EXTENSIVE UPLANDS

The various alphabetical formulas designating climates on the map are explained on the opposite page. Each formula constitutes a short description of the chief characteristics of a climate.

Reprinted by permission
"Elements of Physical Geography"
Copyrighted 1957 by Glenn T. Trewartha.
Published by the McGraw-Hill Book Company, Inc.

Af	Aw	BShs	BSk	BWh	BWk	Cwa	Csa	Cfa
SINGAPORE	TIMBO	BANGHĀZĪ	WILLISTON	ASWĀN	ASTRAKHĀN	BENARES	ATHENS	CHARLESTON
Tropical rain-forest climate	Tropical savanna climate; with wet and dry seasons	Tropical and sub-tropical steppe climate	Middle latitude steppe climate.	Tropical and sub-tropical desert climate	Middle latitude desert climate	Subtropical climate; winter drought and summer rain	Mild climate; sum-wer drought and winter rain	Moderate continental forest climate; mild winters

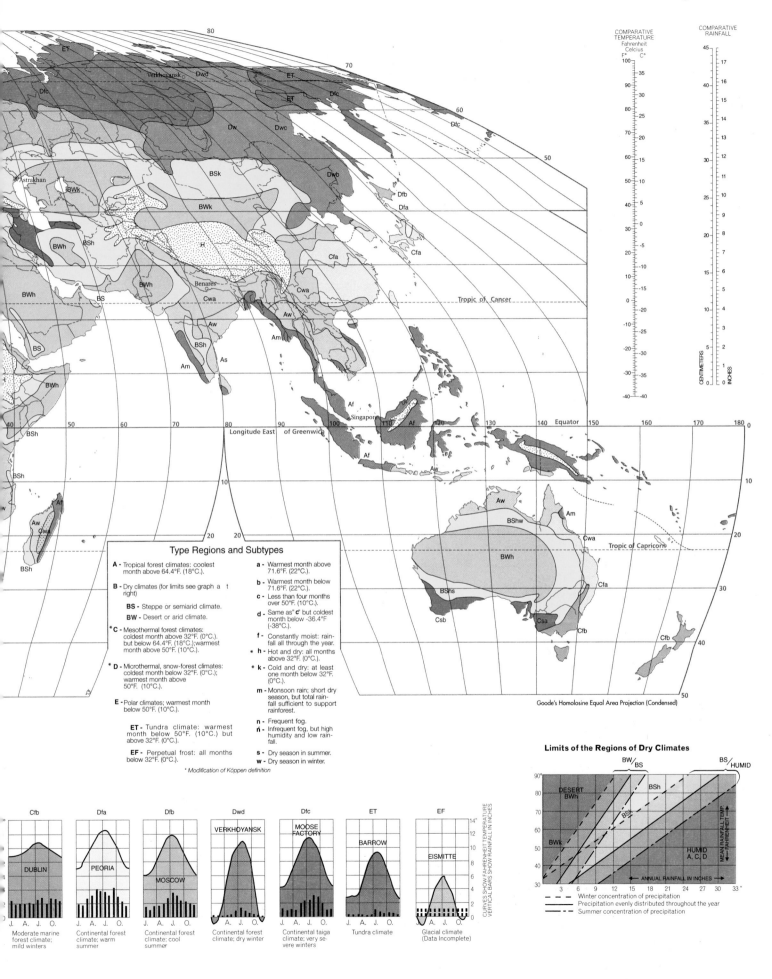

Type Regions and Subtypes

A - Tropical forest climates: coolest month above 64.4°F. (18°C.).

B - Dry climates (for limits see graph at right)

　　BS - Steppe or semiarid climate.

　　BW - Desert or arid climate.

*C - Mesothermal forest climates: coldest month above 32°F. (0°C.). but below 64.4°F. (18°C.);warmest month above 50°F. (10°C.).

*D - Microthermal, snow-forest climates: coldest month below 32°F. (0°C.); warmest month above 50°F. (10°C.).

E - Polar climates; warmest month below 50°F. (10°C.).

　　ET - Tundra climate: warmest month below 50°F. (10°C.) but above 32°F. (0°C.).

　　EF - Perpetual frost: all months below 32°F. (0°C.).

* Modification of Köppen definition

a - Warmest month above 71.6°F. (22°C.).

b - Warmest month below 71.6°F. (22°C.).

c - Less than four months over 50°F. (10°C.).

d - Same as "**c**" but coldest month below -36.4°F (-38°C.).

f - Constantly moist: rainfall all through the year.

* h - Hot and dry: all months above 32°F. (0°C.).

* k - Cold and dry: at least one month below 32°F. (0°C.).

m - Monsoon rain; short dry season, but total rainfall sufficient to support rainforest.

n - Frequent fog.

ń - Infrequent fog, but high humidity and low rainfall.

s - Dry season in summer.

w - Dry season in winter.

COMPARATIVE TEMPERATURE Fahrenheit Celcius

COMPARATIVE RAINFALL

Goode's Homolosine Equal Area Projection (Condensed)

Limits of the Regions of Dry Climates

DESERT BWh

BWk

BSh

BSk

HUMID A, C, D

MEAN RAINFALL TEMP. FAHRENHEIT

ANNUAL RAINFALL IN INCHES

- - - -　Winter concentration of precipitation
──────　Precipitation evenly distributed throughout the year
- · - ·　Summer concentration of precipitation

CURVES SHOW FAHRENHEIT TEMPERATURE
VERTICAL BARS SHOW RAINFALL IN INCHES

Cfb
DUBLIN
J. A. J. O.
Moderate marine forest climate; mild winters

Dfa
PEORIA
J. A. J. O.
Continental forest climate; warm summer

Dfb
MOSCOW
J. A. J. O.
Continental forest climate; cool summer

Dwd
VERKHOYANSK
A. J. O.
Continental forest climate; dry winter

Dfc
MOOSE FACTORY
J. A. J. O.
Continental taiga climate; very severe winters

ET
BARROW
J. A. J. O.
Tundra climate

EF
EISMITTE
A. J. O.
Glacial climate (Data Incomplete)

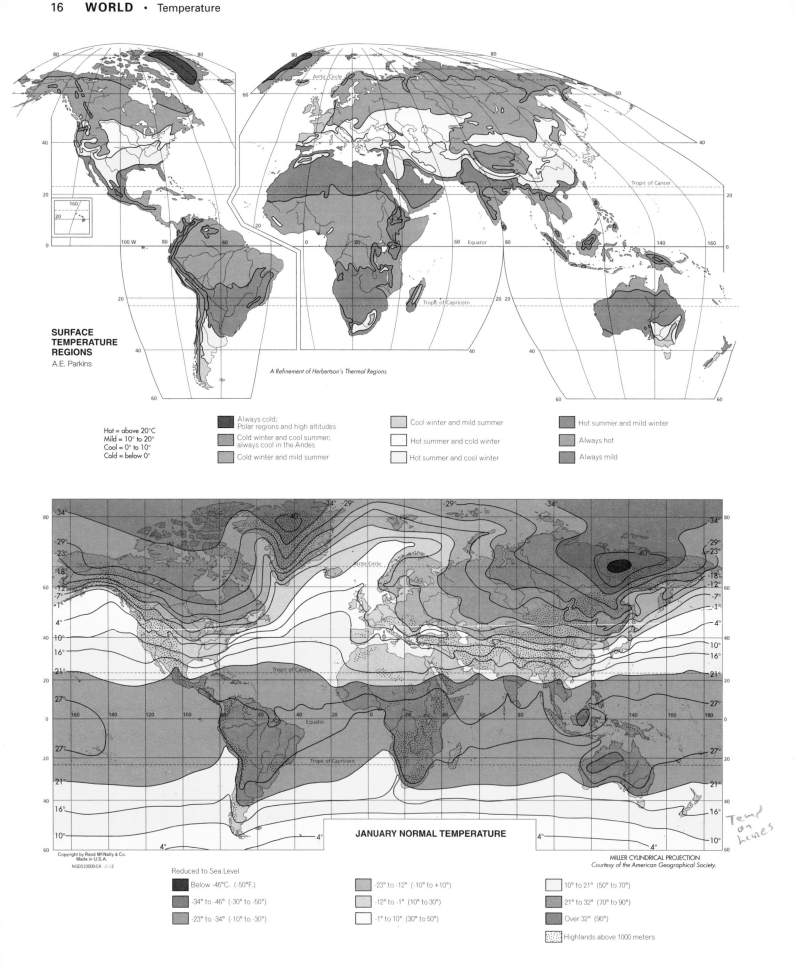

SURFACE TEMPERATURE REGIONS

A.E. Parkins

A Refinement of Herbertson's Thermal Regions

Hot = above 20°C
Mild = 10° to 20°
Cool = 0° to 10°
Cold = below 0°

- Always cold; Polar regions and high altitudes
- Cold winter and cool summer; always cool in the Andes
- Cold winter and mild summer
- Cool winter and mild summer
- Hot summer and cold winter
- Hot summer and cool winter
- Hot summer and mild winter
- Always hot
- Always mild

JANUARY NORMAL TEMPERATURE

Copyright by Rand McNally & Co.
Made in U.S.A.
N-GDS10000-C4-·2-2-2

MILLER CYLINDRICAL PROJECTION
Courtesy of the American Geographical Society.

Reduced to Sea Level

- Below -46°C. (-50°F.)
- -34° to -46° (-30° to -50°)
- -23° to -34° (-10° to -30°)
- -23° to -12° (-10° to +10°)
- -12° to -1° (10° to 30°)
- -1° to 10° (30° to 50°)
- 10° to 21° (50° to 70°)
- 21° to 32° (70° to 90°)
- Over 32° (90°)
- Highlands above 1000 meters

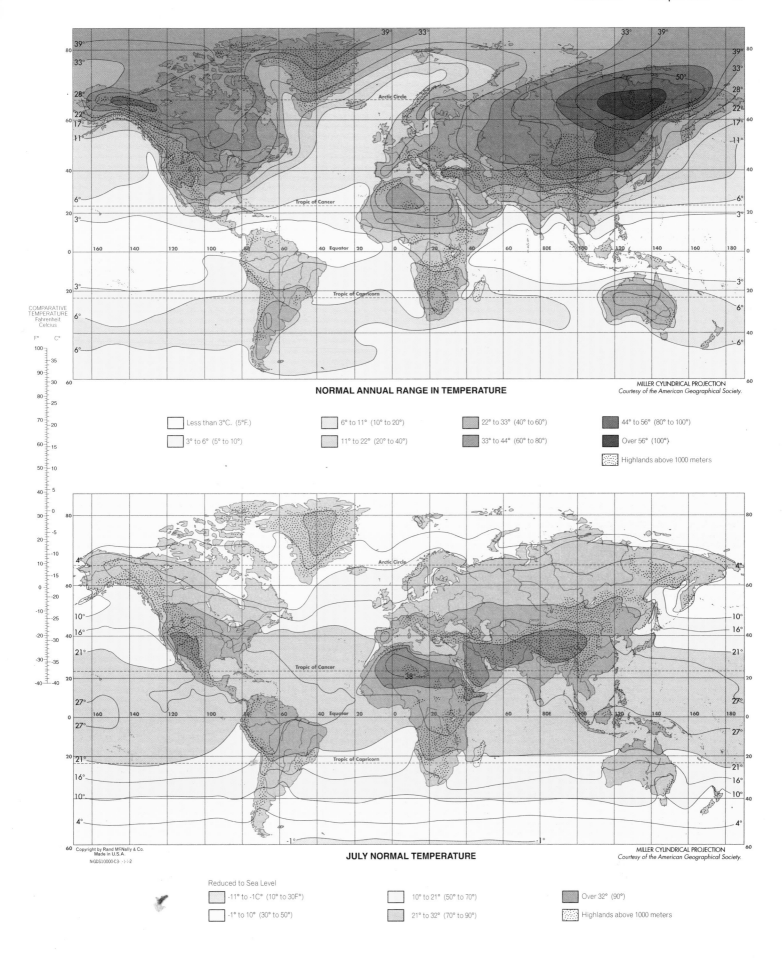

COMPARATIVE
TEMPERATURE
Fahrenheit
Celcius

F°   C°

NORMAL ANNUAL RANGE IN TEMPERATURE

MILLER CYLINDRICAL PROJECTION
Courtesy of the American Geographical Society.

Less than 3°C. (5°F.)   6° to 11° (10° to 20°)   22° to 33° (40° to 60°)   44° to 56° (80° to 100°)

3° to 6° (5° to 10°)   11° to 22° (20° to 40°)   33° to 44° (60° to 80°)   Over 56° (100°)

Highlands above 1000 meters

JULY NORMAL TEMPERATURE

MILLER CYLINDRICAL PROJECTION
Courtesy of the American Geographical Society.

Reduced to Sea Level

-11° to -1C° (10° to 30F°)   10° to 21° (50° to 70°)   Over 32° (90°)

-1° to 10° (30° to 50°)   21° to 32° (70° to 90°)   Highlands above 1000 meters

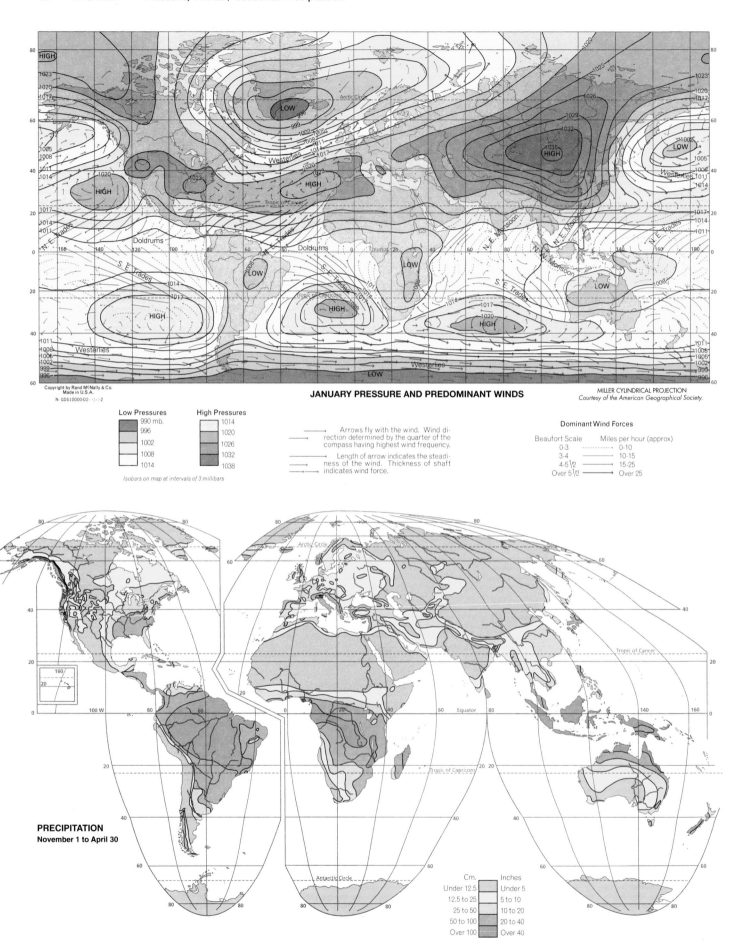

JANUARY PRESSURE AND PREDOMINANT WINDS

Copyright by Rand McNally & Co.
Made in U.S.A.
N- GDS10000-D2- -1-:-2

MILLER CYLINDRICAL PROJECTION
Courtesy of the American Geographical Society.

Low Pressures	High Pressures
990 mb.	1014
996	1020
1002	1026
1008	1032
1014	1038

Isobars on map at intervals of 3 millibars

Arrows fly with the wind. Wind direction determined by the quarter of the compass having highest wind frequency.

Length of arrow indicates the steadiness of the wind. Thickness of shaft indicates wind force.

Dominant Wind Forces

Beaufort Scale	Miles per hour (approx)
0-3	0-10
3-4	10-15
4-5½	15-25
Over 5½	Over 25

PRECIPITATION
November 1 to April 30

Cm.	Inches
Under 12.5	Under 5
12.5 to 25	5 to 10
25 to 50	10 to 20
50 to 100	20 to 40
Over 100	Over 40

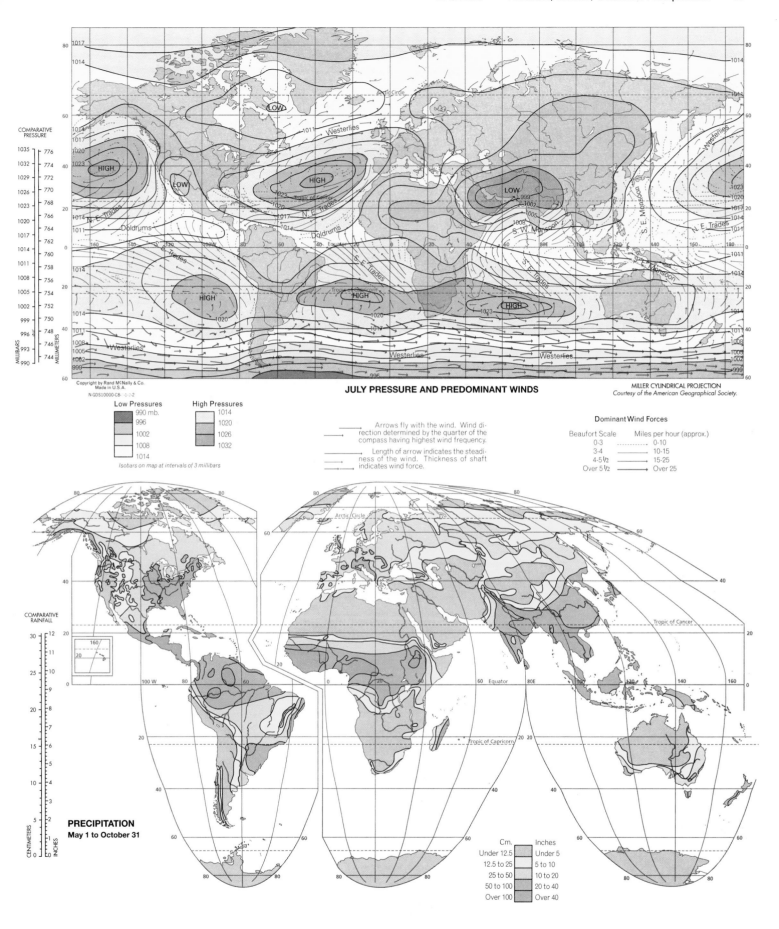

JULY PRESSURE AND PREDOMINANT WINDS

MILLER CYLINDRICAL PROJECTION
Courtesy of the American Geographical Society.

COMPARATIVE PRESSURE

MILLIBARS	MILLIMETERS
1035	776
1032	774
1029	772
1026	770
1023	768
1020	766
1017	764
1014	762
1011	760
1008	758
1005	756
1002	754
999	752
996	750
993	748
	746
990	744

Copyright by Rand McNally & Co.
Made in U.S.A.
N-GDS10000-C8- -1-2-2

Low Pressures

	mb.
	990
	996
	1002
	1008
	1014

High Pressures

	1014
	1020
	1026
	1032

Isobars on map at intervals of 3 millibars

Arrows fly with the wind. Wind direction determined by the quarter of the compass having highest wind frequency.

Length of arrow indicates the steadiness of the wind. Thickness of shaft indicates wind force.

Dominant Wind Forces

Beaufort Scale	Miles per hour (approx.)
0-3	0-10
3-4	10-15
4-5½	15-25
Over 5½	Over 25

PRECIPITATION
May 1 to October 31

COMPARATIVE RAINFALL

CENTIMETERS	INCHES
30	12
	11
25	10
	9
20	8
	7
15	6
	5
10	4
	3
5	2
	1
0	0

Cm.	Inches
Under 12.5	Under 5
12.5 to 25	5 to 10
25 to 50	10 to 20
50 to 100	20 to 40
Over 100	Over 40

20

ANNUAL
PRECIPITATON
AND OCEAN
CURRENTS

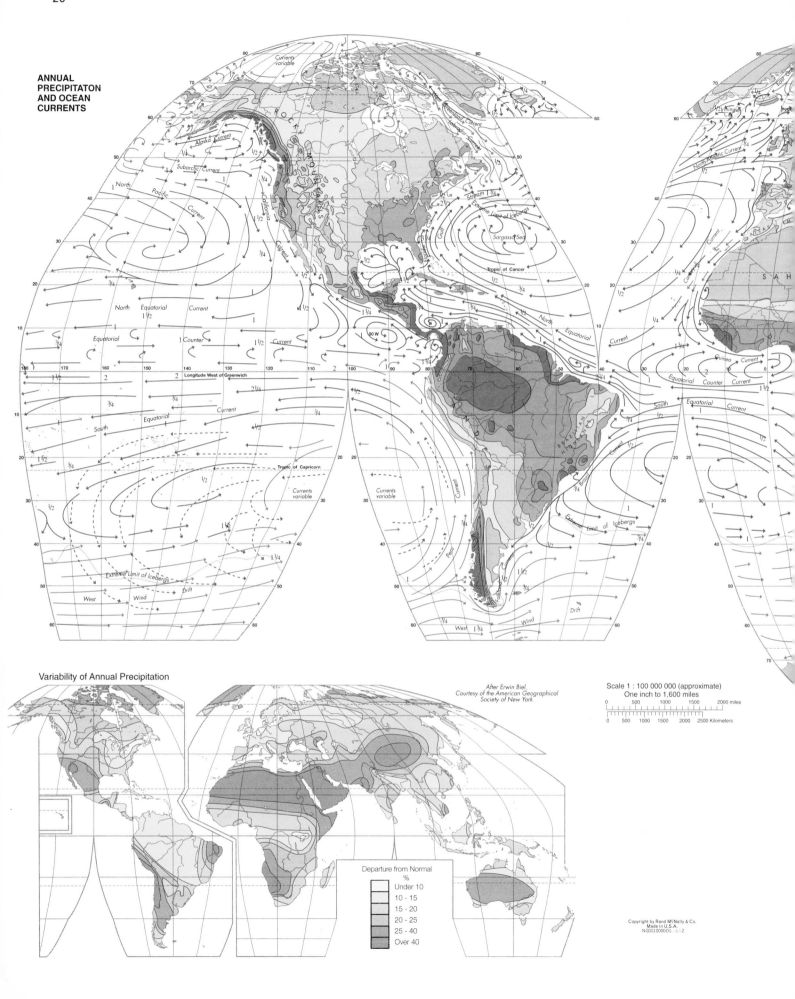

Variability of Annual Precipitation

After Erwin Biel,
Courtesy of the American Geographical
Society of New York

Scale 1 : 100 000 000 (approximate)
One inch to 1,600 miles

0   500   1000   1500   2000 miles

0   500   1000   1500   2000   2500 Kilometers

Departure from Normal
%
Under 10
10 - 15
15 - 20
20 - 25
25 - 40
Over 40

Copyright by Rand McNally & Co.
Made in U.S.A.
NGDS10000-D1- -1-1-2

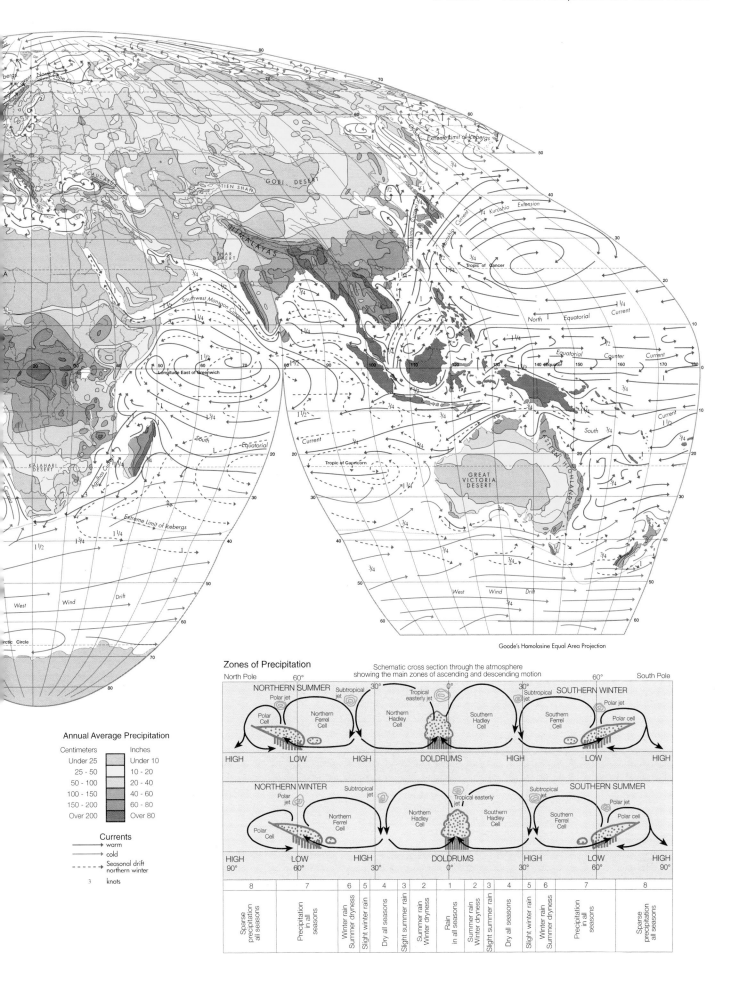

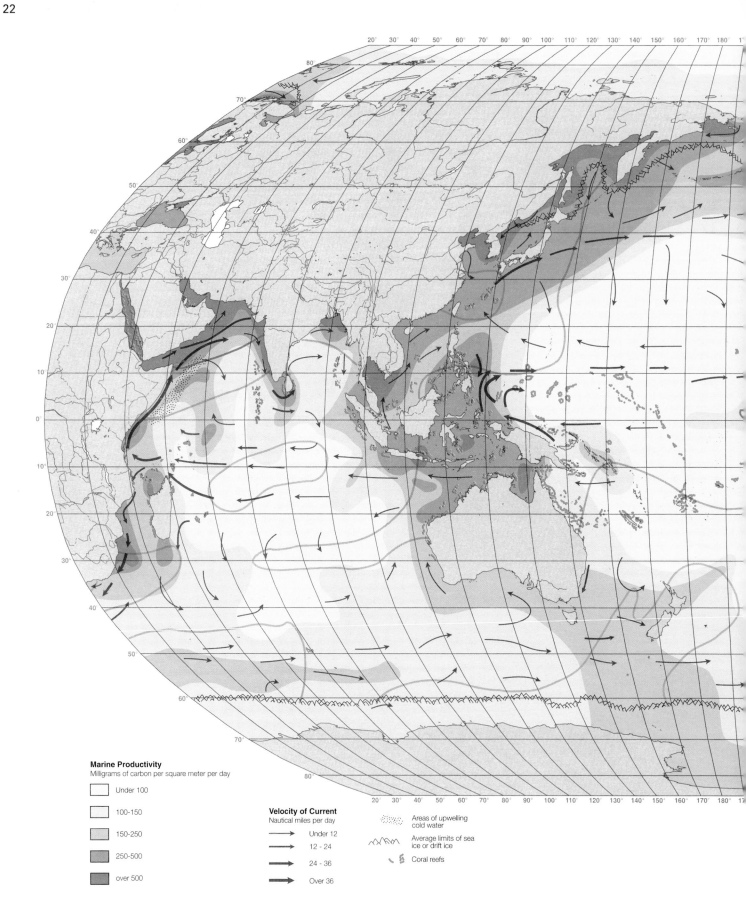

Marine Productivity
Milligrams of carbon per square meter per day

- Under 100
- 100-150
- 150-250
- 250-500
- over 500

Velocity of Current
Nautical miles per day

- → Under 12
- → 12 - 24
- → 24 - 36
- → Over 36

Areas of upwelling
cold water

Average limits of sea
ice or drift ice

Coral reefs

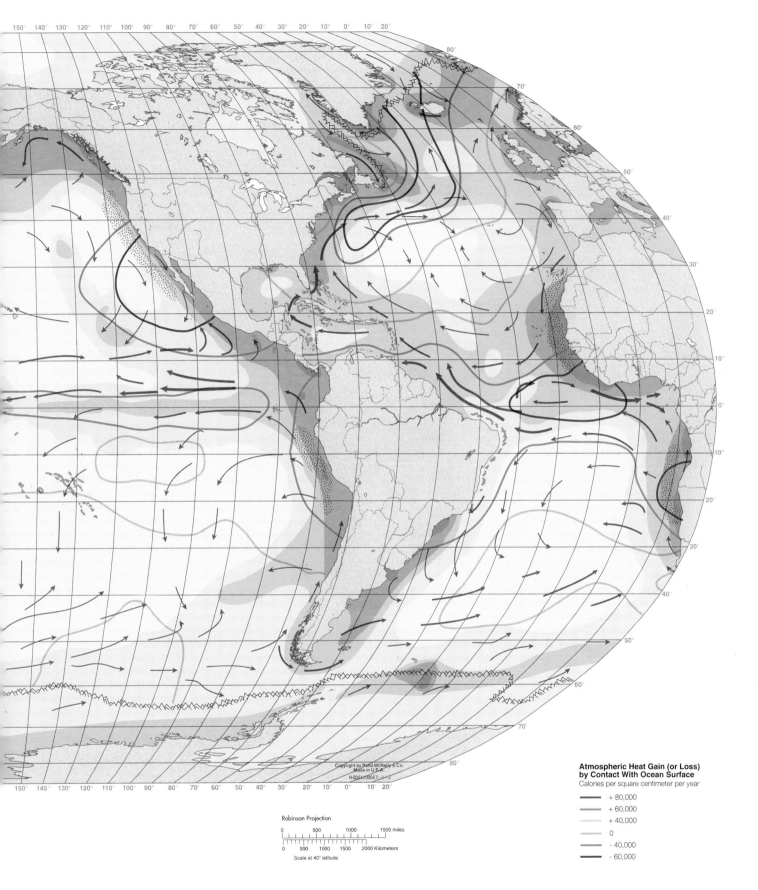

Copyright by Rand McNally & Co.
Made in U.S.A.
NGDS11000 E7...1-2

Robinson Projection

0   500   1000   1500 miles

0   500   1000   1500   2000 Kilometers

Scale at 40° latitude

Atmospheric Heat Gain (or Loss)
by Contact With Ocean Surface
Calories per square centimeter per year

+ 80,000
+ 60,000
+ 40,000
0
- 40,000
- 60,000

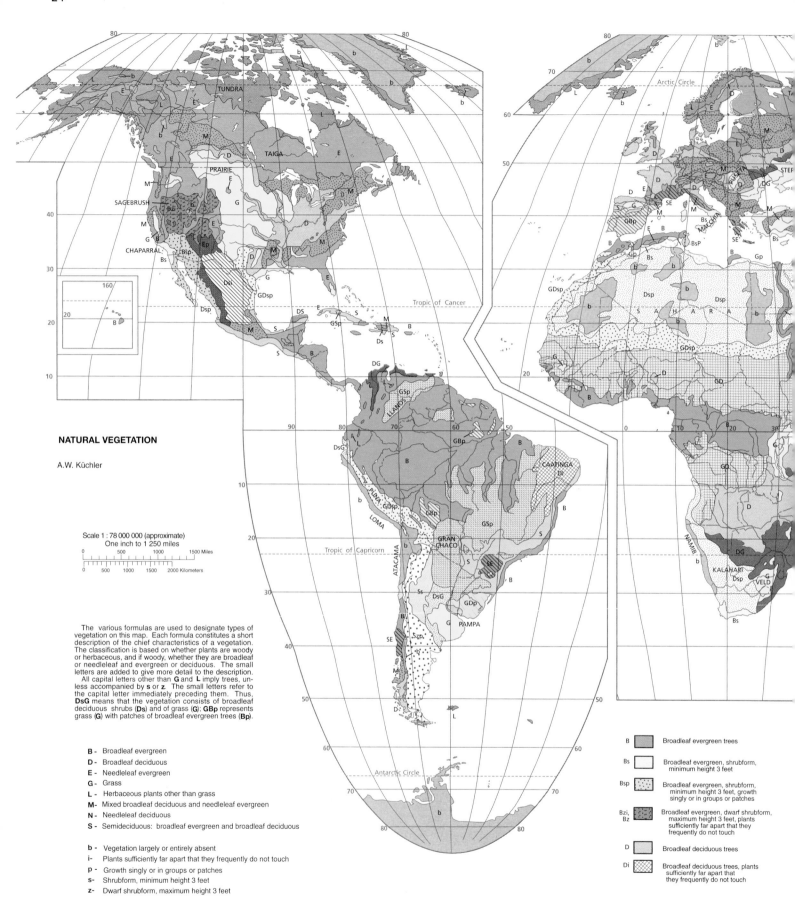

NATURAL VEGETATION

A.W. Küchler

Scale 1 : 78 000 000 (approximate)
One inch to 1 250 miles

0   500   1000   1500 Miles

0   500   1000   1500   2000 Kilometers

The various formulas are used to designate types of
vegetation on this map. Each formula constitutes a short
description of the chief characteristics of a vegetation.
The classification is based on whether plants are woody
or herbaceous, and if woody, whether they are broadleaf
or needleleaf and evergreen or deciduous. The small
letters are added to give more detail to the description.
   All capital letters other than **G** and **L** imply trees, un-
less accompanied by **s** or **z**. The small letters refer to
the capital letter immediately preceding them. Thus,
DsG means that the vegetation consists of broadleaf
deciduous shrubs (**Ds**) and of grass (**G**); **GBp** represents
grass (**G**) with patches of broadleaf evergreen trees (**Bp**).

B - Broadleaf evergreen
D - Broadleaf deciduous
E - Needleleaf evergreen
G - Grass
L - Herbaceous plants other than grass
M - Mixed broadleaf deciduous and needleleaf evergreen
N - Needleleaf deciduous
S - Semideciduous: broadleaf evergreen and broadleaf deciduous

b - Vegetation largely or entirely absent
i - Plants sufficiently far apart that they frequently do not touch
p - Growth singly or in groups or patches
s - Shrubform, minimum height 3 feet
z - Dwarf shrubform, maximum height 3 feet

B -		Broadleaf evergreen trees
Bs		Broadleaf evergreen, shrubform, minimum height 3 feet
Bsp		Broadleaf evergreen, shrubform, minimum height 3 feet, growth singly or in groups or patches
Bzi, Bz		Broadleaf evergreen, dwarf shrubform, maximum height 3 feet, plants sufficiently far apart that they frequently do not touch
D		Broadleaf deciduous trees
Di		Broadleaf deciduous trees, plants sufficiently far apart that they frequently do not touch

TUNDRA

TAIGA

GOBI

TAKLA MAKAN

TERAI

Tropic of Cancer

Longitude East of Greenwich

Equator

Tropic of Capricorn

MALLEE

Goode's Homolosine Equal Area Projection (Condensed)

Ds		Broadleaf deciduous, shrubform, minimum height 3 feet
Dsi		Broadleaf deciduous, shrubform, minimum height 3 feet, plants sufficiently far apart that they frequently do not touch
Dsp		Broadleaf deciduous, shrubform, minimum height 3 feet, growth singly or in groups or patches
Dzp		Broadleaf deciduous, dwarf shrubform, maximum height 3 feet, growth singly or in groups or patches
DsG		Broadleaf deciduous, shrubform, minimum height 3 feet Grass and other herbaceous plants
DG		Broadleaf deciduous trees Grass and other herbaceous plants
DBs		Broadleaf deciduous trees Broadleaf evergreen, shrubform, minimum height 3 feet

E		Needleleaf evergreen trees
Ep		Needleleaf evergreen trees, growth singly or in groups or patches
G		Grass and other herbaceous plants
Gp		Grass and other herbaceous plants, growth singly or in groups or patches
GBp		Grass and other herbaceous plants Broadleaf evergreen trees, growth singly or in groups or patches
GD		Grass and other herbaceous plants Broadleaf deciduous trees
GDp		Grass and other herbaceous plants Broadleaf deciduous trees, growth singly or in groups or patches

GDsp		Grass and other herbaceous plants Broadleaf deciduous, shrubform, minimum height 3 feet, growth singly or in groups or patches
GSp		Grass and other herbaceous plants Semideciduous: broadleaf evergreen and broadleaf deciduous trees, growth singly or in groups or patches
L		Herbaceous plants other than grass
M		Mixed: broadleaf deciduous and needleleaf evergreen trees
N		Needleleaf deciduous trees
ND		Needleleaf deciduous trees Broadleaf deciduous trees

S		Semideciduous: broadleaf evergreen and broadleaf deciduous trees
Ss		Semideciduous: broadleaf evergreen and broadleaf deciduous, shrubform, minimum height 3 feet
SsG		Semideciduous: broadleaf evergreen and broadleaf deciduous, shrubform, minimum height 3 feet Grass and other herbaceous plants
Szp		Semideciduous: broadleaf evergeen and broadleaf deciduous, dwarf shrub-form, maximum height 3 feet, growth singly or in groups or patches
SE		Semideciduous: broadleaf evergreen and broadleaf deciduous trees Needleleaf evergreen trees
b		Vegetation largely or entirely absent

26

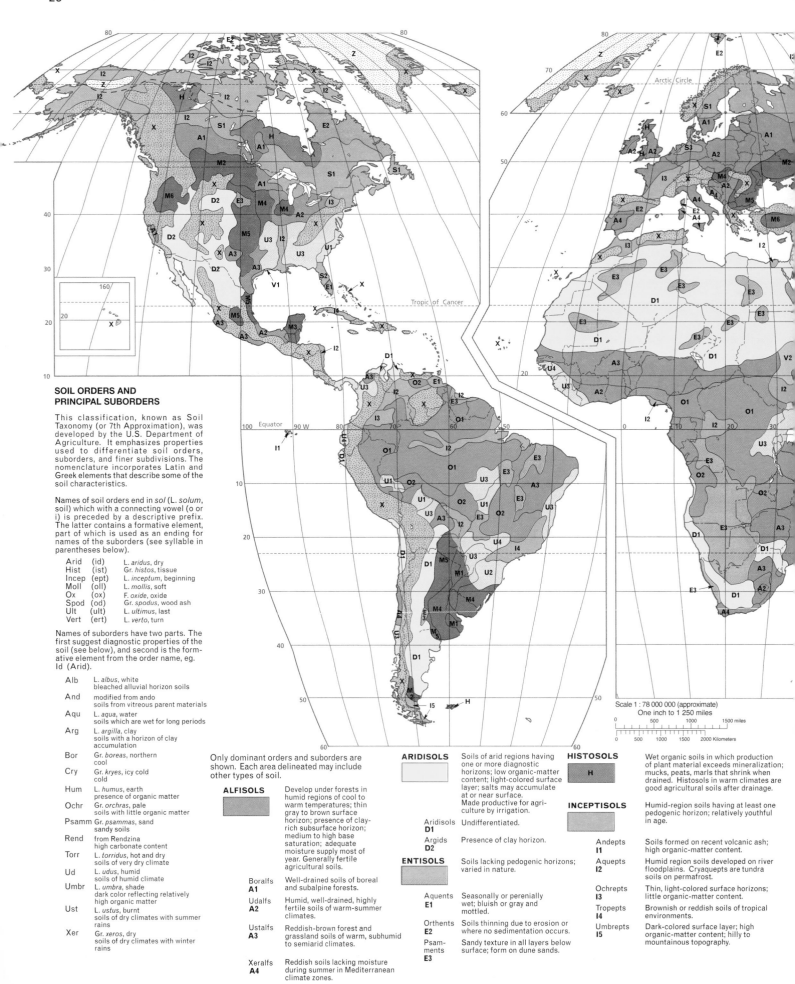

SOIL ORDERS AND PRINCIPAL SUBORDERS

This classification, known as Soil Taxonomy (or 7th Approximation), was developed by the U.S. Department of Agriculture. It emphasizes properties used to differentiate soil orders, suborders, and finer subdivisions. The nomenclature incorporates Latin and Greek elements that describe some of the soil characteristics.

Names of soil orders end in *sol* (L. *solum*, soil) which with a connecting vowel (o or i) is preceded by a descriptive prefix. The latter contains a formative element, part of which is used as an ending for names of the suborders (see syllable in parentheses below).

Arid	(id)	L. *aridus*, dry
Hist	(ist)	Gr. *histos*, tissue
Incep	(ept)	L. *inceptum*, beginning
Moll	(oll)	L. *mollis*, soft
Ox	(ox)	F. *oxide*, oxide
Spod	(od)	Gr. *spodus*, wood ash
Ult	(ult)	L. *ultimus*, last
Vert	(ert)	L. *verto*, turn

Names of suborders have two parts. The first suggest diagnostic properties of the soil (see below), and second is the formative element from the order name, eg. Id (Arid).

Alb	L. *albus*, white bleached alluvial horizon soils
And	modified from ando soils from vitreous parent materials
Aqu	L. *aqua*, water soils which are wet for long periods
Arg	L. *argilla*, clay soils with a horizon of clay accumulation
Bor	Gr. *boreas*, northern cool
Cry	Gr. *kryes*, icy cold cold
Hum	L. *humus*, earth presence of organic matter
Ochr	Gr. *orchras*, pale soils with little organic matter
Psamm	Gr. *psammas*, sand sandy soils
Rend	from Rendzina high carbonate content
Torr	L. *torridus*, hot and dry soils of very dry climate
Ud	L. *udus*, humid soils of humid climate
Umbr	L. *umbra*, shade dark color reflecting relatively high organic matter
Ust	L. *ustus*, burnt soils of dry climates with summer rains
Xer	Gr. *xeros*, dry soils of dry climates with winter rains

Only dominant orders and suborders are shown. Each area delineated may include other types of soil.

ALFISOLS
Develop under forests in humid regions of cool to warm temperatures; thin gray to brown surface horizon; presence of clay-rich subsurface horizon; medium to high base saturation; adequate moisture supply most of year. Generally fertile agricultural soils.

Boralfs A1 — Well-drained soils of boreal and subalpine forests.
Udalfs A2 — Humid, well-drained, highly fertile soils of warm-summer climates.
Ustalfs A3 — Reddish-brown forest and grassland soils of warm, subhumid to semiarid climates.
Xeralfs A4 — Reddish soils lacking moisture during summer in Mediterranean climate zones.

ARIDISOLS
Soils of arid regions having one or more diagnostic horizons; low organic-matter content; light-colored surface layer; salts may accumulate at or near surface. Made productive for agriculture by irrigation.

Aridisols D1 — Undifferentiated.
Argids D2 — Presence of clay horizon.

ENTISOLS
Soils lacking pedogenic horizons; varied in nature.

Aquents E1 — Seasonally or perenially wet; bluish or gray and mottled.
Orthents E2 — Soils thinning due to erosion or where no sedimentation occurs.
Psamments E3 — Sandy texture in all layers below surface; form on dune sands.

HISTOSOLS
Wet organic soils in which production of plant material exceeds mineralization; mucks, peats, marls that shrink when drained. Histosols in warm climates are good agricultural soils after drainage.

INCEPTISOLS
Humid-region soils having at least one pedogenic horizon; relatively youthful in age.

Andepts I1 — Soils formed on recent volcanic ash; high organic-matter content.
Aquepts I2 — Humid region soils developed on river floodplains. Cryaquepts are tundra soils on permafrost.
Ochrepts I3 — Thin, light-colored surface horizons; little organic-matter content.
Tropepts I4 — Brownish or reddish soils of tropical environments.
Umbrepts I5 — Dark-colored surface layer; high organic-matter content; hilly to mountainous topography.

Scale 1 : 78 000 000 (approximate)
One inch to 1 250 miles

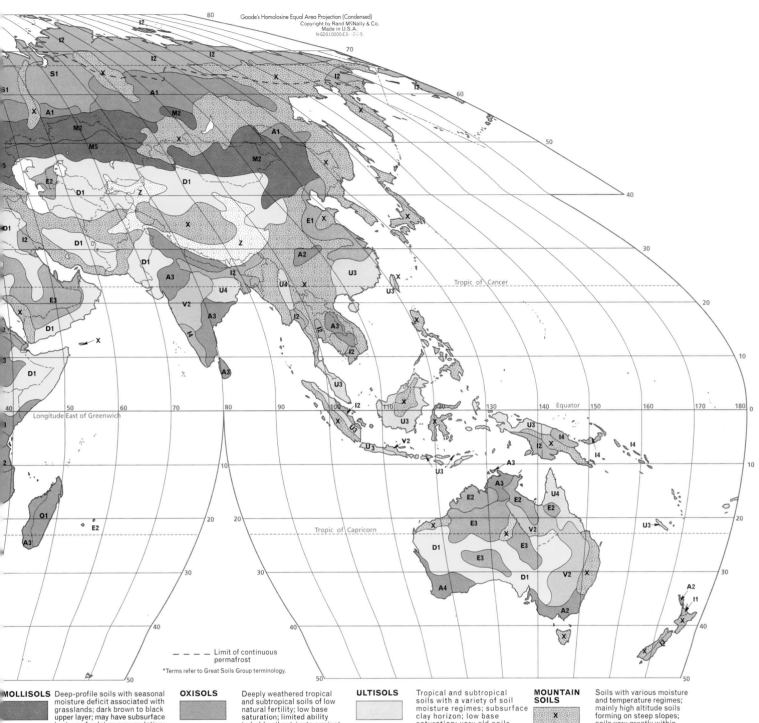

Goode's Homolosine Equal Area Projection (Condensed)
Copyright by Rand McNally & Co.
Made in U.S.A.
N-GDS10000-E3 -2-5

Longitude East of Greenwich

Tropic of Cancer

Equator

Tropic of Capricorn

– – – – Limit of continuous permafrost

*Terms refer to Great Soils Group terminology.

MOLLISOLS Deep-profile soils with seasonal moisture deficit associated with grasslands; dark brown to black upper layer; may have subsurface horizon of calcium accumulation; high base saturation. Very productive for grain crops.

Albolls **M1**	Soils with a grayish subsurface horizon over clay layer and a fluctuating water table.
Borolls **M2**	Well-drained, fertile grassland soils of cool summers and cold winters.
Rendolls **M3**	Formed on calcareous limestones.
Udolls **M4**	Freely drained soils of humid regions with warm summers; excellent agricultural soils.
Ustolls **M5**	Fertile agricultural soils of subhumid climates.
Xerolls **M6**	Pronounced soil-moisture deficit during high-sun season; associated with Mediterranean climates.

OXISOLS Deeply weathered tropical and subtropical soils of low natural fertility; low base saturation; limited ability to hold soil nutrients against leaching; presence of plinthite (laterite) layers. Generally unsuited to large-scale agricultural production.

Orthox **O1**	Hot and nearly always moist; associated with tropical rainforests.
Ustox **O2**	Hot to warm forest and savanna soils with a drier season of low soil-moisture availability.

SPODOSOLS Soils of moist climates ranging from subtropical to cold conditions; include a spodic subsurface horizon incorporating active organic matter beneath a light-colored, leached, sandy horizon. Generally marginal for agriculture.

Spodosols **S1**	Undifferentiated, mostly in high latitudes.
Aquods **S2**	Seasonally wet developed on sandy parent material.
Humods **S3**	Considerable organic matter present in subsurface horizon.
Orthods **S4**	Subsurface accumulations of iron, aluminum, and organic matter.

ULTISOLS Tropical and subtropical soils with a variety of soil moisture regimes; subsurface clay horizon; low base saturation; very old soils characterized by long weathering of clay minerals; low ability to hold nutrients against leaching. Often marginal for agriculture.

Aquults **U1**	Seasonally wet with mottled, gray subsurface horizon.
Humults **U2**	Dark soils with high organic-matter content, warm temperatures.
Udults **U3**	Low organic-matter content and temperate to hot conditions.
Ustults **U4**	Seasonally dry, warm to hot conditions.

VERTISOLS Dark tropical and subtropical soils developed on heavy clays; deep shrinkage cracks appear during dry season which become filled with loose surface materials that absorb moisture and swell during wet season. Generally fertile and well suited to crop production.

Uderts **V1**	Generally moist with limited period for shrinkage cracks to develop.
Usterts **V2**	Over three months of shrinkage-crack formation.

MOUNTAIN SOILS Soils with various moisture and temperature regimes; mainly high altitude soils forming on steep slopes; soils vary greatly within a short distance.

X

Z Areas with little or no soils.

APPROXIMATE CORRELATION WITH OTHER SOIL CLASSIFICATION SYSTEMS

Soil Taxonomy	Great Soil Groups (former U.S. system)	Canadian system
Udalfs	Gray-brown Podzolic	Luvisolic Gray-Brown
Ustalfs	Reddish Chestnut; Red and Yellow Podzolic	
Aridisols	Desert and Reddish Desert Solonetz, Solonchak	
Entisols	Lithosols	Regosolic
Histosols	Bog	Organic
Inceptisol		Brunisolic
Orthents	Lithosols	
Aquepts	Humic Gley	Gleysolic
Cryaquept	Tundra	Cryosolic
Boralfs		Luvisolic Gray; Solonetzic
Borolls	Chernozem Chestnut Brown	Chernozemic, Solonetzic
Rendolls	Rendzina	
Udolls	Prairie	
Ustolls	Brown	
Oxisols	Latosols	
Humod		Humic Podzolic
Orthods	Podzols	Podzolic
Udults	Red and Yellow Podzolic Reddish Brown Lateritic	
Vertisols	Rendzina	

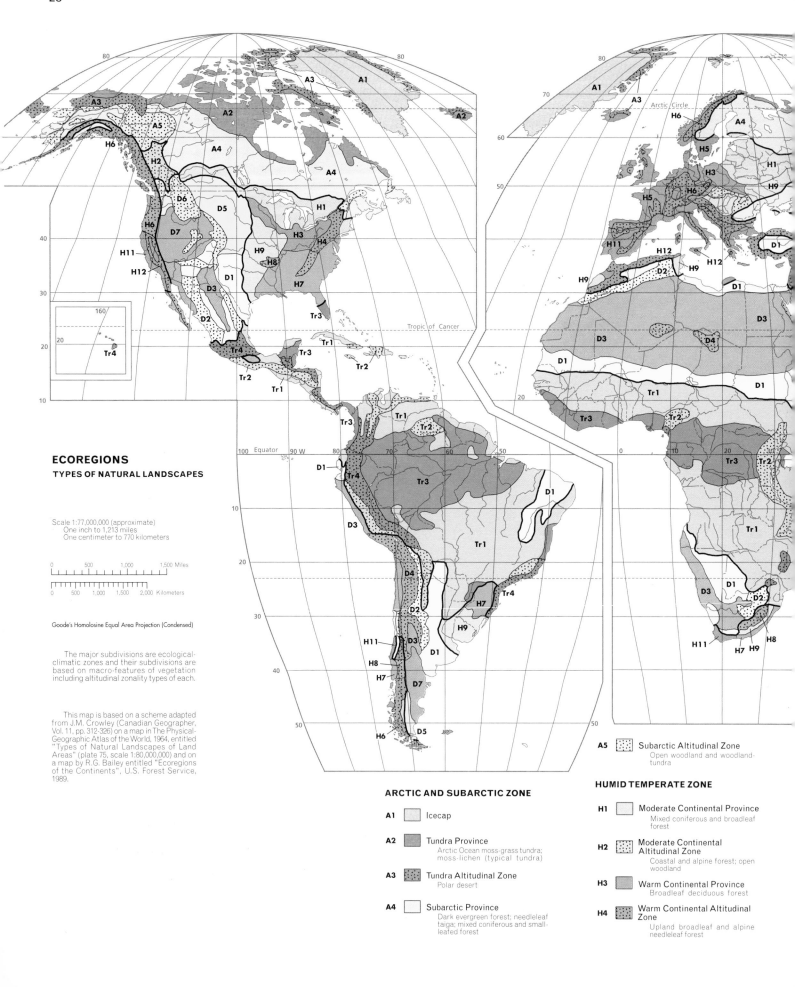

28

ECOREGIONS

TYPES OF NATURAL LANDSCAPES

Scale 1:77,000,000 (approximate)
One inch to 1,213 miles
One centimeter to 770 kilometers

```
0        500      1,000         1,500 Miles
├──┼──┼──┼──┼──┼──┼──┼──┤

0      500   1,000   1,500    2,000 Kilometers
```

Goode's Homolosine Equal Area Projection (Condensed)

The major subdivisions are ecological-
climatic zones and their subdivisions are
based on macro-features of vegetation
including altitudinal zonality types of each.

This map is based on a scheme adapted
from J.M. Crowley (Canadian Geographer,
Vol. 11, pp. 312-326) on a map in The Physical-
Geographic Atlas of the World, 1964, entitled
"Types of Natural Landscapes of Land
Areas" (plate 75, scale 1:80,000,000) and on
a map by R.G. Bailey entitled "Ecoregions
of the Continents", U.S. Forest Service,
1989.

A5 [symbol] Subarctic Altitudinal Zone
Open woodland and woodland-
tundra

ARCTIC AND SUBARCTIC ZONE

A1 [symbol] Icecap

A2 [symbol] Tundra Province
Arctic Ocean moss-grass tundra;
moss-lichen (typical tundra)

A3 [symbol] Tundra Altitudinal Zone
Polar desert

A4 [symbol] Subarctic Province
Dark evergreen forest; needleleaf
taiga; mixed coniferous and small-
leafed forest

HUMID TEMPERATE ZONE

H1 [symbol] Moderate Continental Province
Mixed coniferous and broadleaf
forest

H2 [symbol] Moderate Continental
Altitudinal Zone
Coastal and alpine forest; open
woodland

H3 [symbol] Warm Continental Province
Broadleaf deciduous forest

H4 [symbol] Warm Continental Altitudinal
Zone
Upland broadleaf and alpine
needleleaf forest

Copyright by Rand McNally & Co.
Made in U.S.A.
N-GDS10000-E5- -1-2-5

H5 Marine Province
Lowland, west-coastal humid forest

H6 Marine Altitudinal Zone
Humid coastal and alpine coniferous forest

H7 Humid Subtropical Province
Broadleaf evergreen and broadleaf deciduous forest

H8 Humid Subtropical Altitudinal Zone
Upland, subtropical broadleaf forest

H9 Prairie Province

H10 Prairie Altitudinal Zone
Upland mixed prairie and woodland

H11 Mediterranean Province
Sclerophyll woodland, shrub, and steppe

H12 Mediterranean Altitudinal Zone
Upland shrub and steppe

DRY AND DESERT ZONE

D1 Tropical/Subtropical Steppe Province
Dry steppe, desert shrub, semi-desert savanna

D2 Tropical/Subtropical Steppe Altitudinal Zone
Upland steppe and desert shrub

D3 Tropical/Subtropical Desert Province
Hot, lowland desert at subtropical and coastal locations

D4 Tropical/Subtropical Desert Altitudinal Zone
Desert shrub

D5 Temperate Steppe Province
Medium to short steppe grassland

D6 Temperate Steppe Altitudinal Zone
Alpine meadow and coniferous woodland

D7 Temperate Desert Province
Midlatitude rainshadow desert

D8 Temperate Desert Altitudinal Zone
Extreme continental desert-steppe

HUMID TROPICAL ZONE

Tr1 Savanna Province
Seasonally dry forest, open woodland, tall grass

Tr2 Savanna Altitudinal Zone
Open woodland-steppe

Tr3 Rainforest Province
Constantly humid, broadleaf evergreen forest

Tr4 Rainforest Altitudinal Zone
Broadleaf evergreen and subtropical deciduous forest

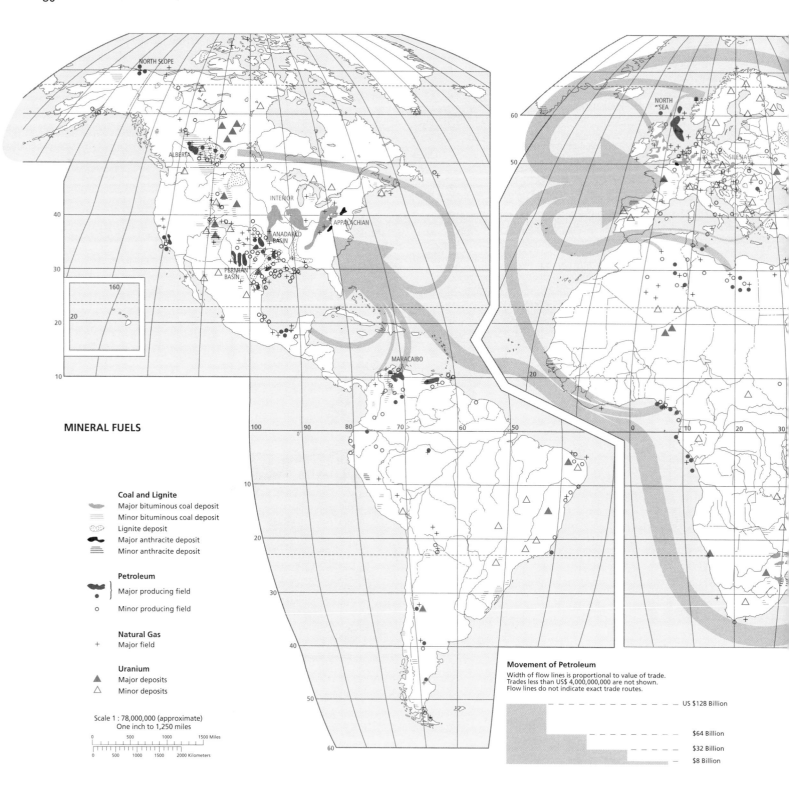

MINERAL FUELS

Coal and Lignite
- Major bituminous coal deposit
- Minor bituminous coal deposit
- Lignite deposit
- Major anthracite deposit
- Minor anthracite deposit

Petroleum
- } Major producing field
- o Minor producing field

Natural Gas
- + Major field

Uranium
- ▲ Major deposits
- △ Minor deposits

Scale 1 : 78,000,000 (approximate)
One inch to 1,250 miles

0   500   1000   1500 Miles
0   500   1000   1500   2000 Kilometers

Map labels: NORTH SLOPE, ALBERTA, INTERIOR, ANADARKO BASIN, APPALACHIAN, PERMIAN BASIN, MARACAIBO, NORTH SEA, SILESIA

Movement of Petroleum
Width of flow lines is proportional to value of trade.
Trades less than US$ 4,000,000,000 are not shown.
Flow lines do not indicate exact trade routes.

- - - - US $128 Billion
$64 Billion
- - - - $32 Billion
$8 Billion

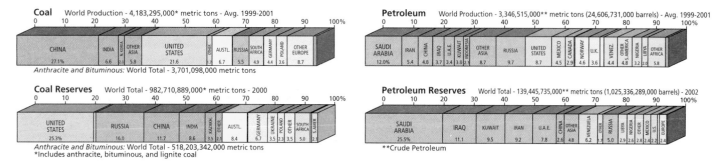

Coal   World Production - 4,183,295,000* metric tons - Avg. 1999-2001

0	10	20	30	40	50	60	70	80	90	100%

CHINA 27.1% | INDIA 6.6 | N. KOREA 2.6 | OTHER ASIA 5.8 | UNITED STATES 21.6 | OTHER 3.7 | AUSTL. 6.7 | RUSSIA 5.5 | SOUTH AFRICA 4.9 | GERMANY 4.4 | POLAND 3.6 | OTHER EUROPE 8.7

Anthracite and Bituminous: World Total - 3,701,098,000 metric tons

Coal Reserves   World Total - 982,710,889,000* metric tons - 2000

0	10	20	30	40	50	60	70	80	90	100%

UNITED STATES 25.3% | RUSSIA 16.0 | CHINA 11.7 | INDIA 8.6 | KAZAKH. 3.5 | OTHER 2.1 | AUSTL. 8.4 | GERMANY 6.7 | UKRAINE 3.5 | POLAND 2.3 | OTHER 5.0 | SOUTH AFRICA 5.0 | S. AMER. 2.1

Anthracite and Bituminous: World Total - 518,203,342,000 metric tons
*Includes anthracite, bituminous, and lignite coal

Petroleum   World Production - 3,346,515,000** metric tons (24,606,731,000 barrels) - Avg. 1999-2001

0	10	20	30	40	50	60	70	80	90	100%

SAUDI ARABIA 12.0% | IRAN 5.4 | CHINA 4.8 | IRAQ 3.7 | U.A.E. 3.4 | KUWAIT 3.0 | INDONESIA 2.1 | OTHER ASIA 8.7 | RUSSIA 9.7 | UNITED STATES 8.7 | MEXICO 4.5 | CANADA 2.9 | NORWAY 4.6 | U.K. 3.6 | VENEZ. 4.4 | OTHER S. AMERICA 4.8 | NIGERIA 3.2 | OTHER AFRICA 5.8

Petroleum Reserves   World Total - 139,445,735,000** metric tons (1,025,336,289,000 barrels) - 2002

0	10	20	30	40	50	60	70	80	90	100%

SAUDI ARABIA 25.5% | IRAQ 11.1 | KUWAIT 9.5 | IRAN 9.2 | U.A.E. 7.8 | CHINA 2.6 | OTHER ASIA 4.8 | VENEZUELA 6.2 | OTHER 1.5 | RUSSIA 5.0 | LIBYA 2.9 | NIGERIA 2.6 | MEXICO 2.8 | U.S. 2.2 | EUROPE 2.4

**Crude Petroleum

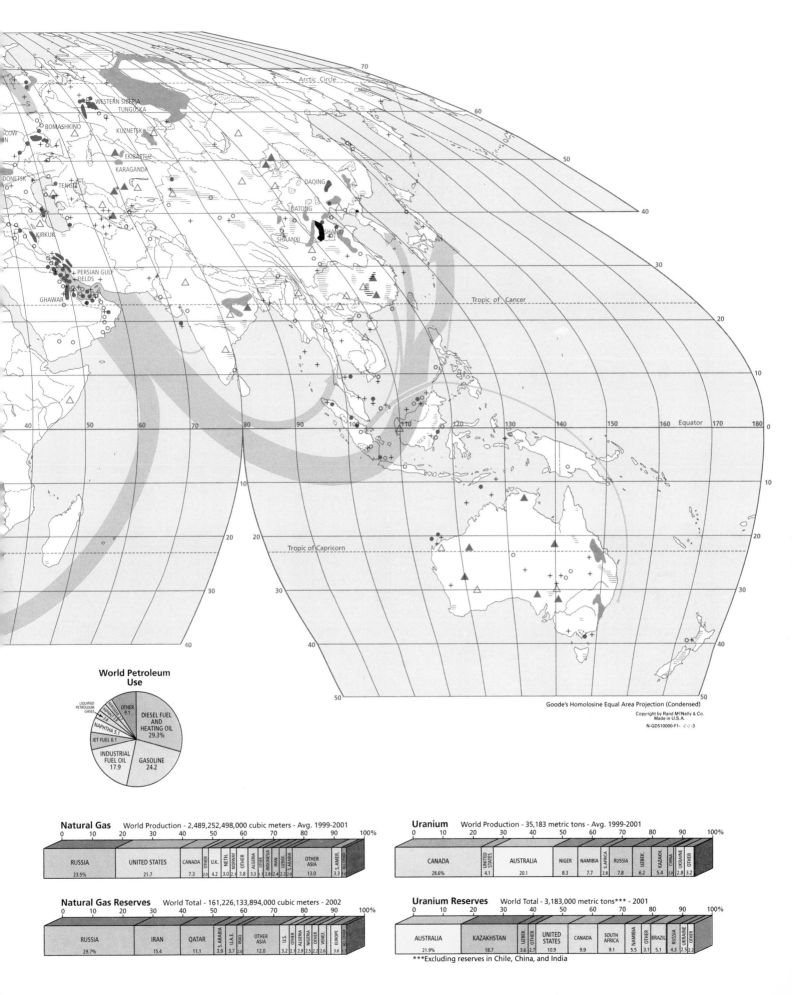

World Petroleum Use

- LIQUIFIED PETROLEUM GASES
- NAPHTHA 5.1
- JET FUEL 6.1
- KEROSENE
- ASPHALT
- OTHER 9.1
- DIESEL FUEL AND HEATING OIL 29.3%
- GASOLINE 24.2
- INDUSTRIAL FUEL OIL 17.9

Goode's Homolosine Equal Area Projection (Condensed)
Copyright by Rand McNally & Co.
Made in U.S.A.
N-GDS10000-F1- -3

Natural Gas World Production - 2,489,252,498,000 cubic meters - Avg. 1999-2001

RUSSIA	UNITED STATES	CANADA	OTHER	U.K.	NETH.	NORWAY	OTHER	ALGERIA	IRAN	INDONESIA	UZBEK.	S. ARABIA	OTHER ASIA	S. AMER.	ALL OTHER
23.5%	21.7	7.3	2.0	4.2	3.0	2.1	3.8	3.3	2.4	2.8	2.4	2.3	13.0	3.3	

Natural Gas Reserves World Total - 161,226,133,894,000 cubic meters - 2002

RUSSIA	IRAN	QATAR	S. ARABIA	U.A.E.	IRAQ	OTHER ASIA	U.S.	OTHER	ALGERIA	NIGERIA	OTHER	VENEZ.	EUROPE	ALL OTHER
29.7%	15.4	11.1	3.9	3.7	2.0	12.0	3.2	2.1	2.9	2.5	2.2	2.2	3.6	

Uranium World Production - 35,183 metric tons - Avg. 1999-2001

CANADA	UNITED STATES	AUSTRALIA	NIGER	NAMIBIA	S. AFRICA	RUSSIA	UZBEK.	KAZAKH.	CHINA	UKRAINE	OTHER
28.6%	4.1	20.1	8.3	7.7	2.8	7.8	6.2	5.4	2.8	3.2	

Uranium Reserves World Total - 3,183,000 metric tons*** - 2001

AUSTRALIA	KAZAKHSTAN	UZBEK.	OTHER	UNITED STATES	CANADA	SOUTH AFRICA	NAMIBIA	OTHER	BRAZIL	RUSSIA	UKRAINE	OTHER
21.9%	18.7	3.6	2.7	10.9	9.9	9.1	5.5	3.1	5.1	4.3	2.5	2.2

***Excluding reserves in Chile, China, and India

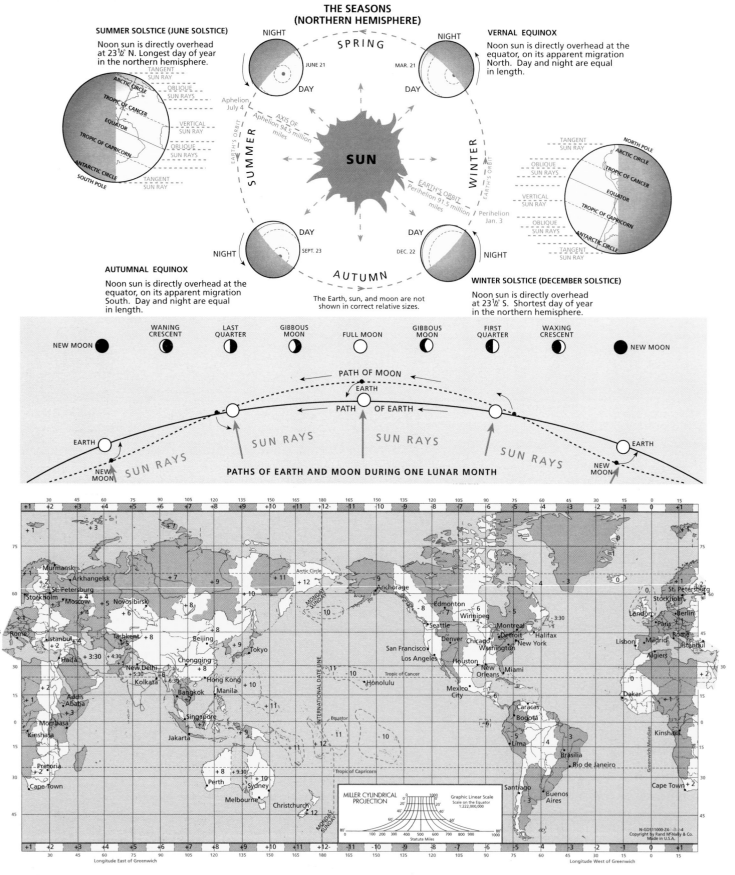

THE SEASONS
(NORTHERN HEMISPHERE)

SUMMER SOLSTICE (JUNE SOLSTICE)
Noon sun is directly overhead at 23½° N. Longest day of year in the northern hemisphere.

VERNAL EQUINOX
Noon sun is directly overhead at the equator, on its apparent migration North. Day and night are equal in length.

AUTUMNAL EQUINOX
Noon sun is directly overhead at the equator, on its apparent migration South. Day and night are equal in length.

The Earth, sun, and moon are not shown in correct relative sizes.

WINTER SOLSTICE (DECEMBER SOLSTICE)
Noon sun is directly overhead at 23½° S. Shortest day of year in the northern hemisphere.

PATHS OF EARTH AND MOON DURING ONE LUNAR MONTH

MILLER CYLINDRICAL PROJECTION
Graphic Linear Scale
Scale on the Equator
1:222,000,000

Longitude East of Greenwich

Longitude West of Greenwich

Time Zones

The surface of the earth is divided into 24 time zones. Each zone represents 15° of longitude or one hour of time. The time of the initial, or zero, zone is based on the Greenwich Meridian and extends eastward and westward for a distance of 7½° of longitude. Each of the zones is designated by a number representing the hours (+ or -) by which its standard time differs from Greenwich mean time. These standard time zones are indicated by bands of orange and yellow. Areas which have a fractional deviation from standard time are shown in an intermediate color. The irregularities in the zones and the fractional deviations are due to political and economic factors.

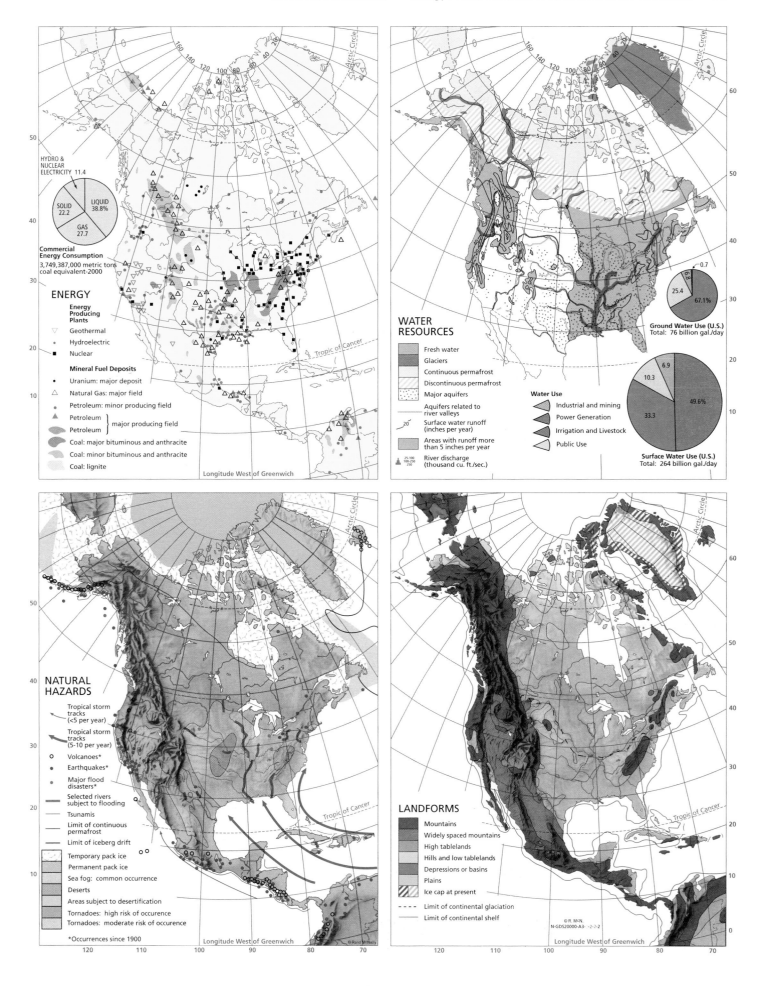

ENERGY

Energy Producing Plants
- ▽ Geothermal
- • Hydroelectric
- ■ Nuclear

Mineral Fuel Deposits
- • Uranium: major deposit
- △ Natural Gas: major field
- • Petroleum: minor producing field
- ▲ Petroleum } major producing field
- ▬ Petroleum
- Coal: major bituminous and anthracite
- Coal: minor bituminous and anthracite
- Coal: lignite

HYDRO & NUCLEAR ELECTRICITY 11.4

LIQUID 38.8%
SOLID 22.2
GAS 27.7

Commercial Energy Consumption
3,749,387,000 metric tons coal equivalent-2000

Longitude West of Greenwich

WATER RESOURCES

- Fresh water
- Glaciers
- Continuous permafrost
- Discontinuous permafrost
- Major aquifers
- Aquifers related to river valleys
- 20 Surface water runoff (inches per year)
- Areas with runoff more than 5 inches per year
- River discharge (thousand cu. ft./sec.) 25-100 100-250 250

Ground Water Use (U.S.)
Total: 76 billion gal./day

0.7
6.8
25.4
67.1%

Water Use
- ◁ Industrial and mining
- ◁ Power Generation
- ◁ Irrigation and Livestock
- ◁ Public Use

Surface Water Use (U.S.)
Total: 264 billion gal./day

6.9
10.3
33.3
49.6%

NATURAL HAZARDS

- ⟶ Tropical storm tracks (<5 per year)
- ⟹ Tropical storm tracks (5-10 per year)
- ○ Volcanoes*
- • Earthquakes*
- • Major flood disasters*
- ▬ Selected rivers subject to flooding
- Tsunamis
- Limit of continuous permafrost
- Limit of iceberg drift
- Temporary pack ice
- Permanent pack ice
- Sea fog: common occurrence
- Deserts
- Areas subject to desertification
- Tornadoes: high risk of occurence
- Tornadoes: moderate risk of occurence

*Occurrences since 1900

Longitude West of Greenwich

© Rand McNally

LANDFORMS

- Mountains
- Widely spaced mountains
- High tablelands
- Hills and low tablelands
- Depressions or basins
- Plains
- Ice cap at present
- --- Limit of continental glaciation
- — Limit of continental shelf

© R. McN.
N-GD520000-A3- -2-2-2

Longitude West of Greenwich

Tropic of Cancer

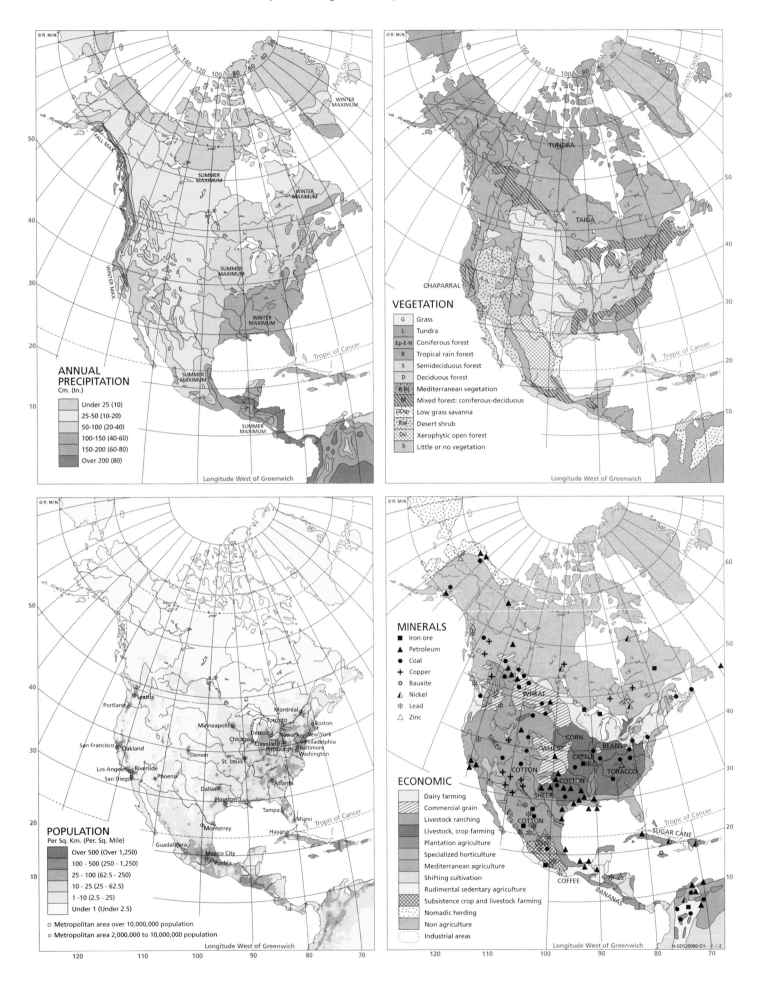

ANNUAL PRECIPITATION
Cm. (In.)

- Under 25 (10)
- 25-50 (10-20)
- 50-100 (20-40)
- 100-150 (40-60)
- 150-200 (60-80)
- Over 200 (80)

VEGETATION

G	Grass
L	Tundra
Ep-E-N	Coniferous forest
B	Tropical rain forest
S	Semideciduous forest
D	Deciduous forest
a-Bs	Mediterranean vegetation
	Mixed forest: coniferous-deciduous
GDsp	Low grass savanna
Bsp	Desert shrub
Os	Xerophytic open forest
b	Little or no vegetation

POPULATION
Per Sq. Km. (Per. Sq. Mile)

- Over 500 (Over 1,250)
- 100 - 500 (250 - 1,250)
- 25 - 100 (62.5 - 250)
- 10 - 25 (25 - 62.5)
- 1 - 10 (2.5 - 25)
- Under 1 (Under 2.5)

□ Metropolitan area over 10,000,000 population

o Metropolitan area 2,000,000 to 10,000,000 population

MINERALS

- ■ Iron ore
- ▲ Petroleum
- ● Coal
- + Copper
- ○ Bauxite
- ◭ Nickel
- ✳ Lead
- △ Zinc

ECONOMIC

- Dairy farming
- Commercial grain
- Livestock ranching
- Livestock, crop farming
- Plantation agriculture
- Specialized horticulture
- Mediterranean agriculture
- Shifting cultivation
- Rudimental sedentary agriculture
- Subsistence crop and livestock farming
- Nomadic herding
- Non agriculture
- Industrial areas

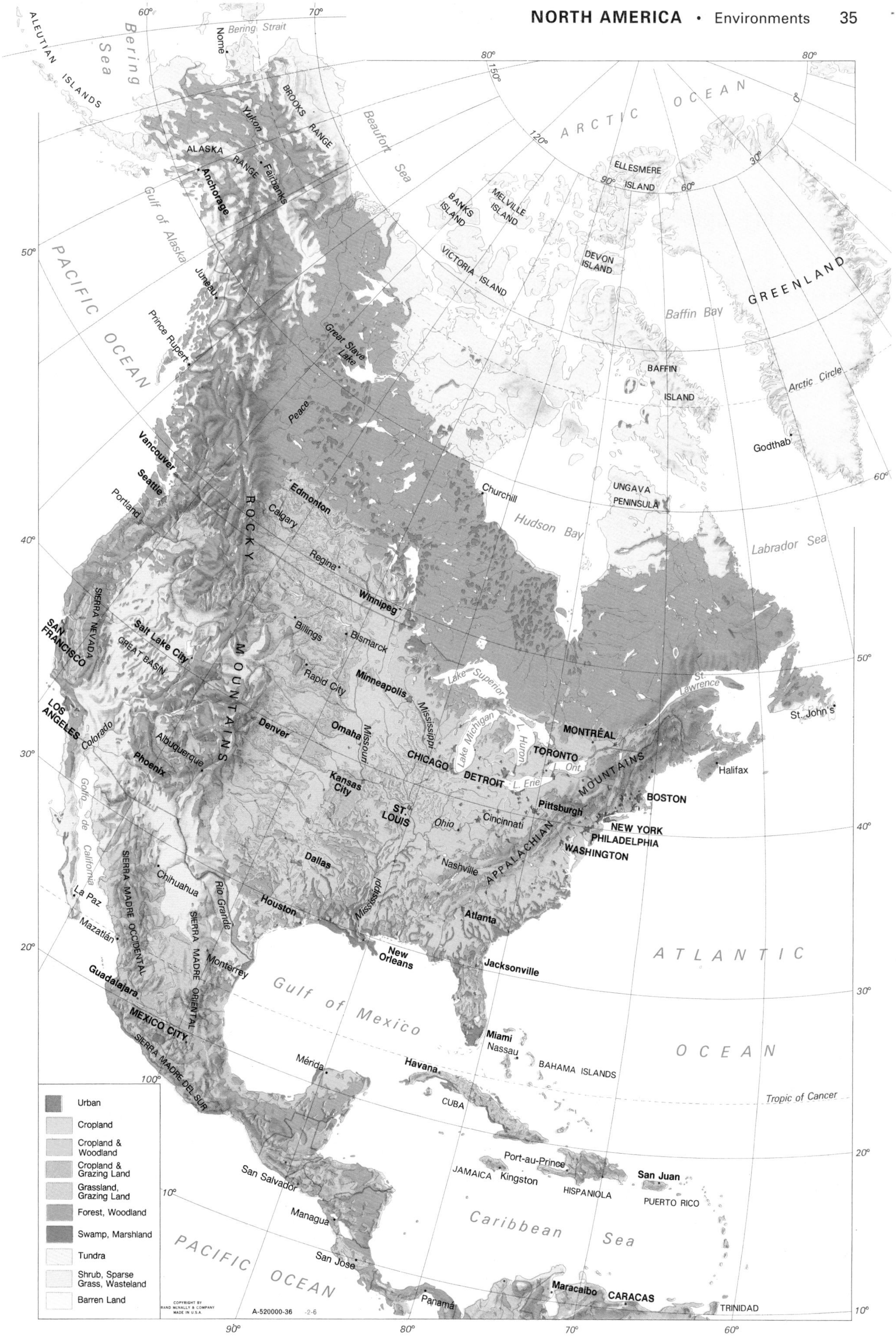
Urban
Cropland
Cropland & Woodland
Cropland & Grazing Land
Grassland, Grazing Land
Forest, Woodland
Swamp, Marshland
Tundra
Shrub, Sparse Grass, Wasteland
Barren Land
ARCTIC OCEAN
PACIFIC OCEAN
ATLANTIC OCEAN
Bering Sea
Bering Strait
ALEUTIAN ISLANDS
Beaufort Sea
Gulf of Alaska
Baffin Bay
Hudson Bay
Labrador Sea
Gulf of Mexico
Caribbean Sea
Golfo de California
Arctic Circle
Tropic of Cancer
ELLESMERE ISLAND
BANKS ISLAND
MELVILLE ISLAND
VICTORIA ISLAND
DEVON ISLAND
BAFFIN ISLAND
GREENLAND
UNGAVA PENINSULA
BROOKS RANGE
ALASKA RANGE
ROCKY MOUNTAINS
SIERRA NEVADA
GREAT BASIN
APPALACHIAN MOUNTAINS
SIERRA MADRE OCCIDENTAL
SIERRA MADRE ORIENTAL
SIERRA MADRE DEL SUR
Yukon
Peace
Great Slave Lake
Colorado
Rio Grande
Missouri
Mississippi
Ohio
St. Lawrence
Lake Superior
Lake Michigan
L. Huron
L. Erie
L. Ont.
Nome
Fairbanks
Anchorage
Juneau
Prince Rupert
Vancouver
Seattle
Portland
Edmonton
Calgary
Regina
Winnipeg
Churchill
Godthab
St. John's
Halifax
SAN FRANCISCO
Salt Lake City
LOS ANGELES
Phoenix
Albuquerque
Denver
Billings
Bismarck
Rapid City
Minneapolis
Omaha
Kansas City
ST. LOUIS
CHICAGO
DETROIT
TORONTO
MONTRÉAL
BOSTON
NEW YORK
PHILADELPHIA
WASHINGTON
Pittsburgh
Cincinnati
Nashville
Atlanta
Jacksonville
Miami
Nassau
BAHAMA ISLANDS
Havana
CUBA
Dallas
Houston
New Orleans
Chihuahua
Monterrey
La Paz
Mazatlán
Guadalajara
MEXICO CITY
Mérida
JAMAICA
Kingston
Port-au-Prince
HISPANIOLA
San Juan
PUERTO RICO
San Salvador
Managua
San José
Panamá
Maracaibo
CARACAS
TRINIDAD
COPYRIGHT BY RAND McNALLY & COMPANY MADE IN U.S.A.
A-520000-36 -2-6
0 100 200 400 600 800 Miles
0 150 300 600 900 1200 Kilometers

Scale 1:36,000,000; one inch to 570 miles. Lambert Azimuthal Equal-Area Projection

PHYSIOGRAPHIC DIVISIONS

1 Pacific Mountain System
2 Intermontane Plateaus
3 Rocky Mountain System
4 Interior Plains
5 Ozark-Ouachita Highlands
6 Gulf-Atlantic Plain
7 Appalachian Highlands
8 Laurentian Upland (Canadian Shield)
9 Hudson Bay Lowland

0 25 50 75 100   200   300   400   500 Miles

0   50   100   200   400   600   800 Kilometers

Scale 1: 12 000 000; One inch to 190 miles.   POLYCONIC PROJECTION

PHYSIOGRAPHY
BY
ERWIN RAISZ

LITHOLOGY AND STRUCTURE

Unconsolidated deposits: alluvium, sands, playa deposits, etc.

Essentially horizontal sedimentary rocks; many partially unconsolidated.

Slightly to moderately tilted, older sedimentary rocks.

Steeply folded or faulted, sedimentary rocks

Volcanics; largely lava flows.

Metamorphic and intrusive igneous rocks; structure complex.

— — — Limits of continental glaciation.

LANDFORMS

PLATEAUS   BASIN RANGES

HILLS   VOLCANO AND LAVA

MOUNTAINS   SAND

MESAS   SINKS

CUESTAS   MORAINES

FOLDED   DRUMLINS
MOUNTAINS

A-520500-9A6 -3-3-7
Copyright by Rand McNally & Co.
Made in U.S.A.

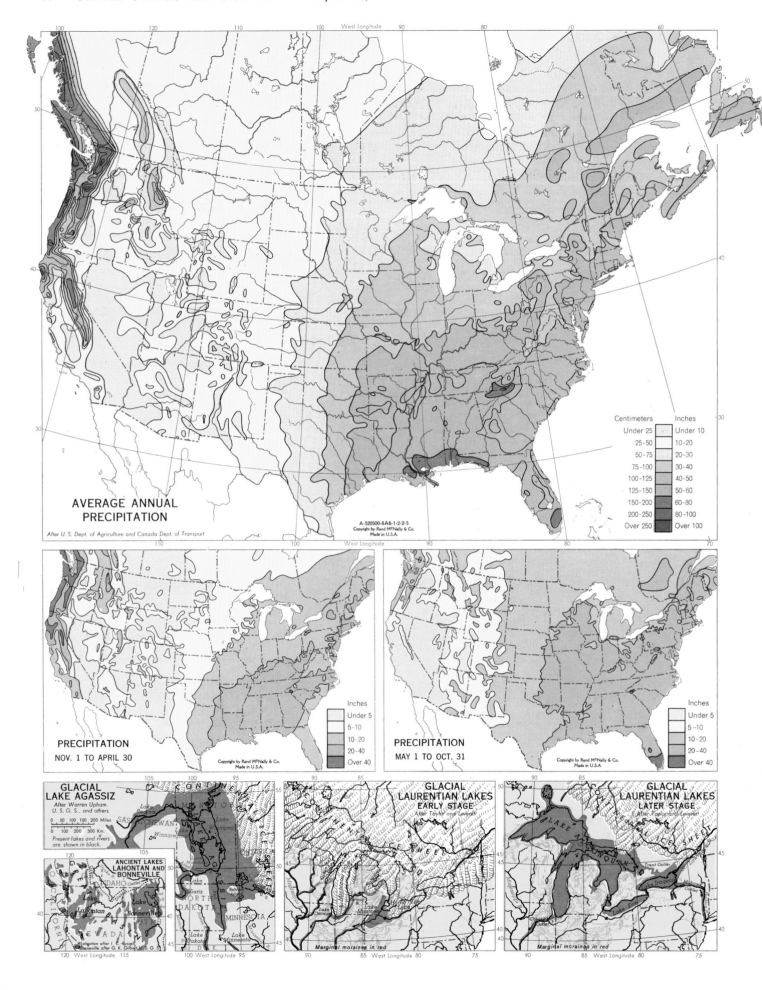

AVERAGE ANNUAL PRECIPITATION

After U. S. Dept. of Agriculture and Canada Dept. of Transport

A-520500-6A6-1-2-2-5
Copyright by Rand M°Nally & Co.
Made in U.S.A.

Centimeters	Inches
Under 25	Under 10
25-50	10-20
50-75	20-30
75-100	30-40
100-125	40-50
125-150	50-60
150-200	60-80
200-250	80-100
Over 250	Over 100

PRECIPITATION
NOV. 1 TO APRIL 30

Inches
Under 5
5-10
10-20
20-40
Over 40

PRECIPITATION
MAY 1 TO OCT. 31

Copyright by Rand M°Nally & Co.
Made in U.S.A.

Inches
Under 5
5-10
10-20
20-40
Over 40

GLACIAL LAKE AGASSIZ
After Warren Upham,
U. S. G. S., and others

0   50 100 150 200 Miles
0 100 200 300 Km.

Present lakes and rivers
are shown in black.

ANCIENT LAKES LAHONTAN AND BONNEVILLE

Lahontan after I. C. Russell
Bonneville after G. K. Gilbert, U. S. G. S.

GLACIAL LAURENTIAN LAKES EARLY STAGE
After Taylor and Leverett

Marginal moraines in red

GLACIAL LAURENTIAN LAKES LATER STAGE
After Taylor and Leverett

Marginal moraines in red

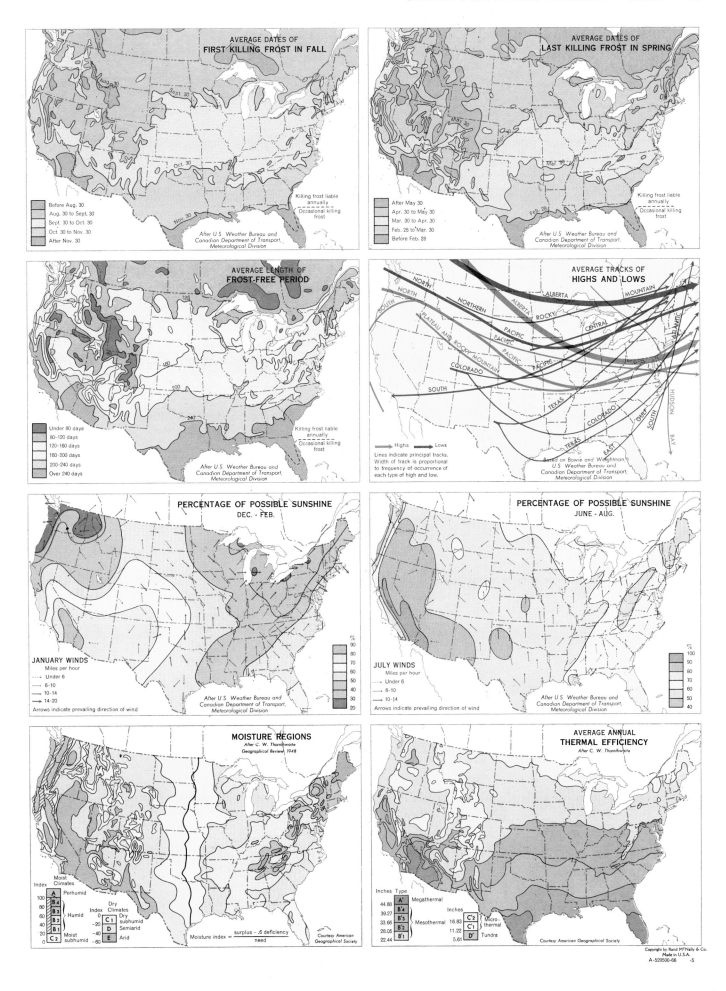

AVERAGE DATES OF
FIRST KILLING FROST IN FALL

Before Aug. 30
Aug. 30 to Sept. 30
Sept. 30 to Oct. 30
Oct. 30 to Nov. 30
After Nov. 30

Killing frost liable
annually
Occasional killing
frost

After U.S. Weather Bureau and
Canadian Department of Transport,
Meteorological Division

AVERAGE DATES OF
LAST KILLING FROST IN SPRING

After May 30
Apr. 30 to May 30
Mar. 30 to Apr. 30
Feb. 28 to Mar. 30
Before Feb. 28

Killing frost liable
annually
Occasional killing
frost

After U.S. Weather Bureau and
Canadian Department of Transport,
Meteorological Division

AVERAGE LENGTH OF
FROST-FREE PERIOD

Under 80 days
80-120 days
120-160 days
160-200 days
200-240 days
Over 240 days

Killing frost liable
annually
Occasional killing
frost

After U.S. Weather Bureau and
Canadian Department of Transport,
Meteorological Division

AVERAGE TRACKS OF
HIGHS AND LOWS

→ Highs   → Lows
Lines indicate principal tracks.
Width of track is proportional
to frequency of occurrence of
each type of high and low.

Based on Bowie and Weightman,
U.S. Weather Bureau and
Canadian Department of Transport,
Meteorological Division

PERCENTAGE OF POSSIBLE SUNSHINE
DEC. - FEB.

JANUARY WINDS
Miles per hour
--→ Under 6
--→ 6-10
→ 10-14
→ 14-20
Arrows indicate prevailing direction of wind

%
90
80
70
60
50
40
30
20

After U.S. Weather Bureau and
Canadian Department of Transport,
Meteorological Division

PERCENTAGE OF POSSIBLE SUNSHINE
JUNE - AUG.

JULY WINDS
Miles per hour
--→ Under 6
--→ 6-10
→ 10-14
Arrows indicate prevailing direction of wind

%
100
90
80
70
60
50
40

After U.S. Weather Bureau and
Canadian Department of Transport,
Meteorological Division

MOISTURE REGIONS
After C. W. Thornthwaite
Geographical Review, 1948

Index | Moist Climates
100 | A | Perhumid
80 | B4
60 | B3 | Humid
40 | B2
20 | B1 | Moist
0 | C2 | subhumid

Index | Dry Climates
0 | C1 | Dry subhumid
-20 | D | Semiarid
-40 | E | Arid
-60

Moisture index = surplus - .6 deficiency / need

Courtesy American
Geographical Society

AVERAGE ANNUAL
THERMAL EFFICIENCY
After C. W. Thornthwaite

Inches | Type
44.88 | A' | Megathermal
39.27 | B'4
33.66 | B'3 | Mesothermal
28.05 | B'2
22.44 | B'1

Inches
16.83 | C'2 | Microthermal
11.22 | C'1
5.61 | D' | Tundra

Courtesy American Geographical Society

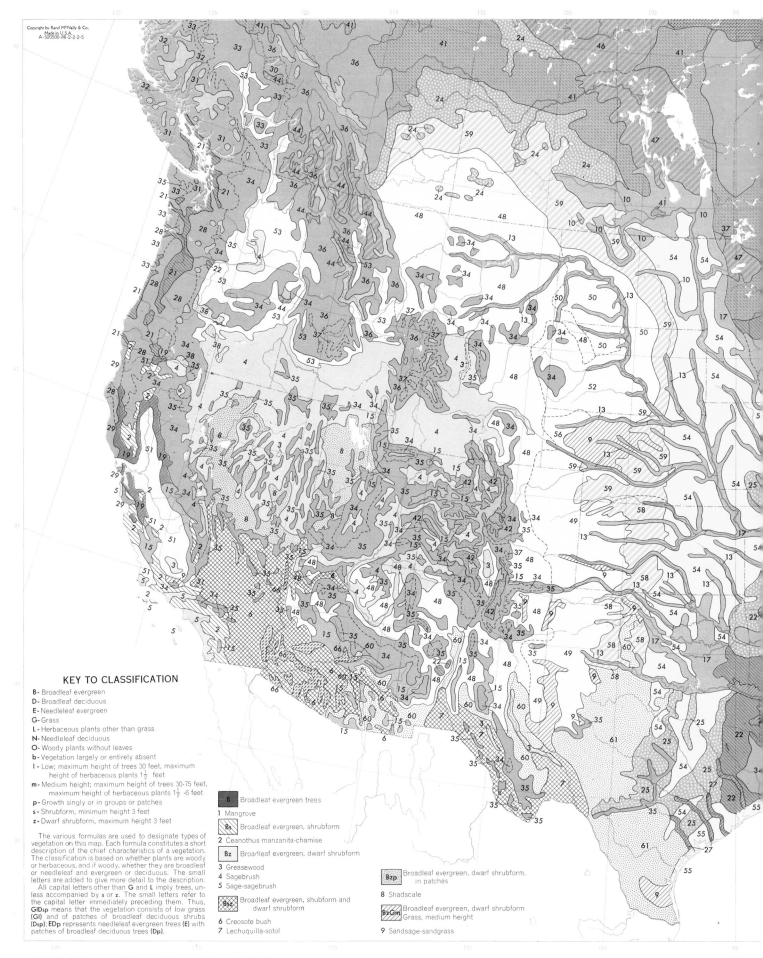

Copyright by Rand McNally & Co.
Made in U.S.A.
A-520500-86-2-2-2-5

KEY TO CLASSIFICATION

B- Broadleaf evergreen
D- Broadleaf deciduous
E- Needleleaf evergreen
G- Grass
L- Herbaceous plants other than grass
N- Needleleaf deciduous
O- Woody plants without leaves
b- Vegetation largely or entirely absent
l- Low; maximum height of trees 30 feet, maximum
   height of herbaceous plants 1½ feet
m- Medium height; maximum height of trees 30-75 feet,
   maximum height of herbaceous plants 1½ -6 feet
p- Growth singly or in groups or patches
s- Shrubform, minimum height 3 feet
z- Dwarf shrubform, maximum height 3 feet

The various formulas are used to designate types of
vegetation on this map. Each formula constitutes a short
description of the chief characteristics of a vegetation.
The classification is based on whether plants are woody
or herbaceous, and if woody, whether they are broadleaf
or needleleaf and evergreen or deciduous. The small
letters are added to give more detail to the description.
All capital letters other than G and L imply trees, un-
less accompanied by s or z. The small letters refer to
the capital letter immediately preceding them. Thus,
GlDsp means that the vegetation consists of low grass
(Gl) and of patches of broadleaf deciduous shrubs
(Dsp); EDp represents needleleaf evergreen trees (E) with
patches of broadleaf deciduous trees (Dp).

B Broadleaf evergreen trees	
1 Mangrove	
Bs Broadleaf evergreen, shrubform	
2 Ceanothus manzanita-chamise	
Bz Broadleaf evergreen, dwarf shrubform	
3 Greasewood	
4 Sagebrush	
5 Sage-sagebrush	
Bsz Broadleaf evergreen, shrubform and dwarf shrubform	**Bzp** Broadleaf evergreen, dwarf shrubform, in patches
6 Creosote bush	8 Shadscale
7 Lechuquilla-sotol	**BzGm** Broadleaf evergreen, dwarf shrubform Grass, medium height
	9 Sandsage-sandgrass

0 25 50 75 100   200   300   400   500 Miles

0 50 100   400   600   800 Kilometers

Scale 1 : 14 000 000;   One inch to 220 miles

NATURAL VEGETATION

BY A. W. KÜCHLER

Based on "A Physiognomic Classification of Vegetation"
Annals of the Assoc. of American Geographers, Vol. 39, September, 1949

AMBERT CONFORMAL CONIC PROJECTION

D Broadleaf deciduous trees

10 Aspen-oak
11 Beech-maple
12 Beech-tulip tree-maple-basswood
13 Cottonwood-willow
14 Maple-basswood
15 Oak
16 Oak-ash-maple
17 Oak-hickory
18 Oak-tulip tree

DB Broadleaf deciduous trees
     Broadleaf evergreen trees

19 Oak-madrone

DE Broadleaf deciduous trees
     Needleleaf evergreen trees

20 Maple-yellow birch-hemlock-pine
21 Oak-Douglas fir
22 Oak-pine
23 Maple-beech-hemlock

D Gmp Broadleaf deciduous trees
         Grass, medium height, in patches

24 Aspen-needle grass-wheat grass
25 Oak-hickory-bluestem

DN Broadleaf deciduous trees
     Needleleaf deciduous trees

26 Bay trees-bald cypress
27 Tupelo-gum-bald cypress

E Needleleaf evergreen trees

28 Douglas fir
29 Douglas fir-redwood
30 Hemlock-arbor vitae
31 Hemlock-arbor vitae-Douglas fir
32 Hemlock-arbor vitae-fir
33 Hemlock-spruce
34 Pine
35 Pine-juniper
36 Pine-spruce
37 Spruce-fir

Esp Needleleaf evergreen, shrubform,
       in patches

38 Juniper

EDp Needleleaf evergreen trees
       Broadleaf deciduous trees, in patches

39 Douglas fir-pine-aspen
40 Pine-spruce-birch
41 Spruce-aspen
42 Spruce-fir-aspen
43 Spruce-poplar-birch

EN Needleleaf evergreen trees
      Needleleaf deciduous trees

44 Hemlock-arbor vitae-Douglas fir-larch
45 Pine-bald cypress
46 Pine-spruce-larch
47 Spruce-larch

Gl Grass, low

48 Grama grass
49 Grama grass-buffalo grass
50 Grama grass-needle grass
51 Needle grass-blue grass
52 Wheat grass
53 Wheat grass-blue grass

Gm Grass, medium height

54 Bluestem
55 Broom grass-water grass
56 Marsh grass
57 Saw grass

Gml Grass, medium and low height

58 Bluestem-bunch grass
59 Needle grass-wheat grass

Gl Grass, low
Dsp Broadleaf deciduous, shrubform, in patches

60 Bunch grass-oak

Gm Grass, medium height
Dsp Broadleaf deciduous, shrubform, in patches

61 Mesquite grass-mesquite

L Herbaceous plants other than grass

62 Lichens, etc.

LEp Herbaceous plants other than grass
       Needleleaf evergreen trees, in patches

63 Lichens-spruce

LEp Herbaceous plants other than grass
Np Needleleaf evergreen trees, in patches
        Needleleaf deciduous trees, in patches

64 Lichens-spruce-larch

N Needleleaf deciduous trees

65 Bald cypress

Op Woody plants without leaves, in patches

66 Palo verde-cacti-ocotillo

b Vegetation largely or entirely absent

42

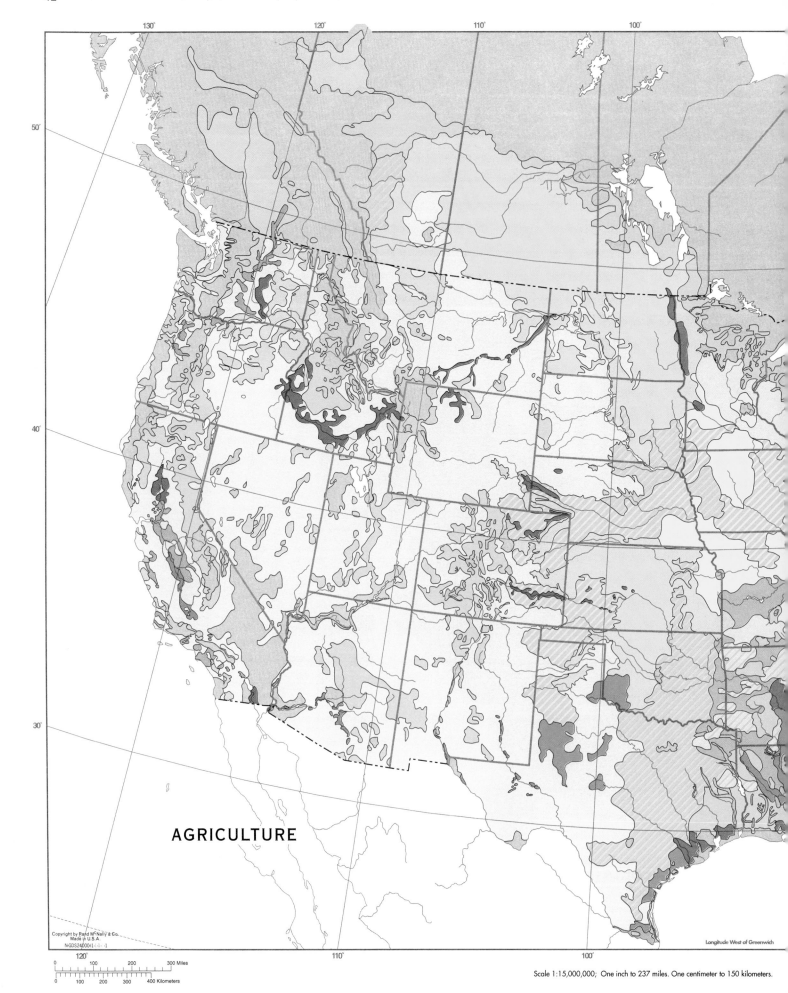

AGRICULTURE

Longitude West of Greenwich

0 100 200 300 Miles

0 100 200 300 400 Kilometers

Scale 1:15,000,000; One inch to 237 miles. One centimeter to 150 kilometers.

Dairying

Fruits and Vegetables

Wheat, Barley, and Oilseeds

Cash Corn and Soybeans

Tobacco

Cotton

Livestock and Feed Grains: Beef

Livestock and Feed Grains: Hogs

Livestock and Feed Grains: Poultry

Livestock and Feed Grains: Mixed

Specialty Crops (Peanuts, Potatoes, Rice, Sugar)

Western Livestock Ranching

Western Feedlots

Agriculture and Forestry

Non-Agricultural Areas

ALBERS CONIC PROJECTION

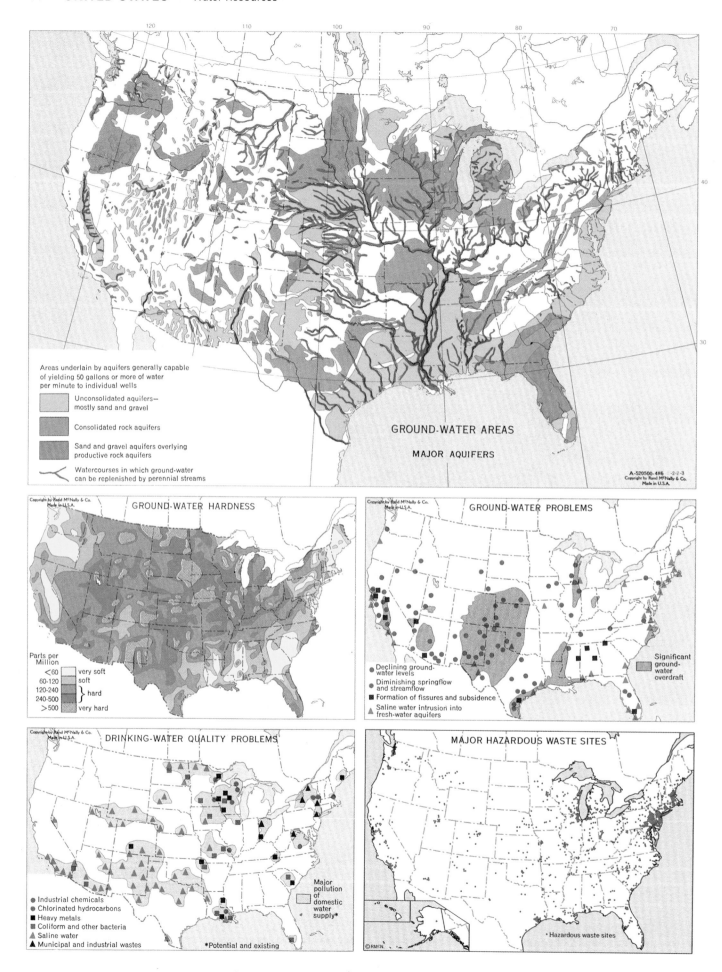

GROUND-WATER AREAS

MAJOR AQUIFERS

Areas underlain by aquifers generally capable
of yielding 50 gallons or more of water
per minute to individual wells

Unconsolidated aquifers—
mostly sand and gravel

Consolidated rock aquifers

Sand and gravel aquifers overlying
productive rock aquifers

Watercourses in which ground-water
can be replenished by perennial streams

A-520500-4H6 -2-2-3
Copyright by Rand McNally & Co.
Made in U.S.A.

GROUND-WATER HARDNESS

Parts per
Million
<60   very soft
60-120   soft
120-240 } hard
240-500
>500   very hard

GROUND-WATER PROBLEMS

● Declining ground-
  water levels
● Diminishing springflow
  and streamflow
■ Formation of fissures and subsidence
▲ Saline water intrusion into
  fresh-water aquifers

Significant
ground-
water
overdraft

DRINKING-WATER QUALITY PROBLEMS

● Industrial chemicals
● Chlorinated hydrocarbons
■ Heavy metals
■ Coliform and other bacteria
▲ Saline water
▲ Municipal and industrial wastes

Major
pollution
of
domestic
water
supply*

*Potential and existing

MAJOR HAZARDOUS WASTE SITES

• Hazardous waste sites

©R.M.N.

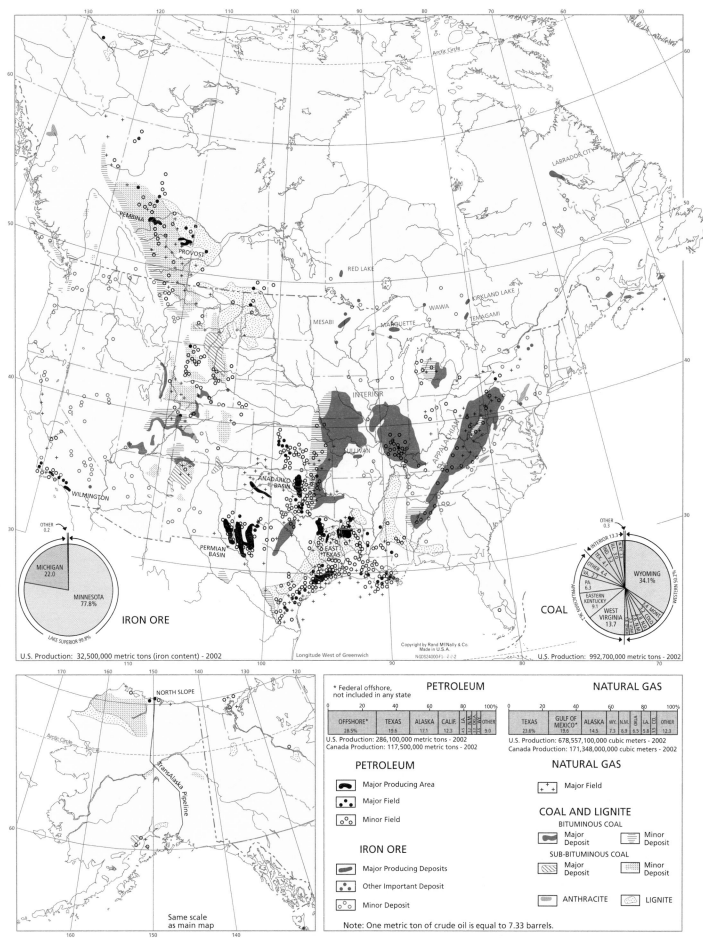

IRON ORE

OTHER
0.2

MICHIGAN
22.0

MINNESOTA
77.8%

LAKE SUPERIOR 99.8%

U.S. Production: 32,500,000 metric tons (iron content) - 2002

COAL

OTHER
0.3

INTERIOR 13.3

MO.
IND. 3.0
ILL.

OTHER 4.4

TEX. 4.1

VA. 2.7

PA.
6.3

EASTERN
KENTUCKY
9.1

WEST
VIRGINIA
13.7

WYOMING
34.1%

WESTERN 50.2%

APPALACHIAN 36.2

U.S. Production: 992,700,000 metric tons - 2002

PEMBINA

PROVOST

RED LAKE

LABRADOR CITY

MESABI

MARQUETTE

WAWA

KIRKLAND LAKE

TEMAGAMI

INTERIOR

SULLIVAN

APPALACHIAN

WILMINGTON

ANADARKO
BASIN

PERMIAN
BASIN

EAST
TEXAS

Copyright by Rand McNally & Co.
Made in U.S.A.
NGDS24000-F1- 2-2-2

Longitude West of Greenwich

NORTH SLOPE

Arctic Circle

TransAlaska Pipeline

Same scale
as main map

* Federal offshore,
not included in any state

PETROLEUM

OFFSHORE* 28.5%	TEXAS 19.6	ALASKA 17.1	CALIF. 12.3	LA. 4.3	N.M. 3.2	OKLA. 2.6	WY.	OTHER 9.0

U.S. Production: 286,100,000 metric tons - 2002
Canada Production: 117,500,000 metric tons - 2002

NATURAL GAS

TEXAS 23.6%	GULF OF MEXICO* 19.6	ALASKA 14.5	WY. 7.3	N.M. 6.9	OKLA. 6.5	LA. 5.8	CO. 3.5	OTHER 12.3

U.S. Production: 678,557,100,000 cubic meters - 2002
Canada Production: 171,348,000,000 cubic meters - 2002

PETROLEUM

🦴 Major Producing Area

Major Field

Minor Field

IRON ORE

Major Producing Deposits

Other Important Deposit

Minor Deposit

NATURAL GAS

+ + Major Field

COAL AND LIGNITE

BITUMINOUS COAL

Major Deposit   Minor Deposit

SUB-BITUMINOUS COAL

Major Deposit   Minor Deposit

ANTHRACITE   LIGNITE

Note: One metric ton of crude oil is equal to 7.33 barrels.

Scale 1:29,000,000; One inch to 457 miles.   ALBERS CONIC PROJECTION

Scale 1:12,000,000.   One inch to 190 miles.   Albers Conic Projection
                     One centimeter to 120 kilometers.

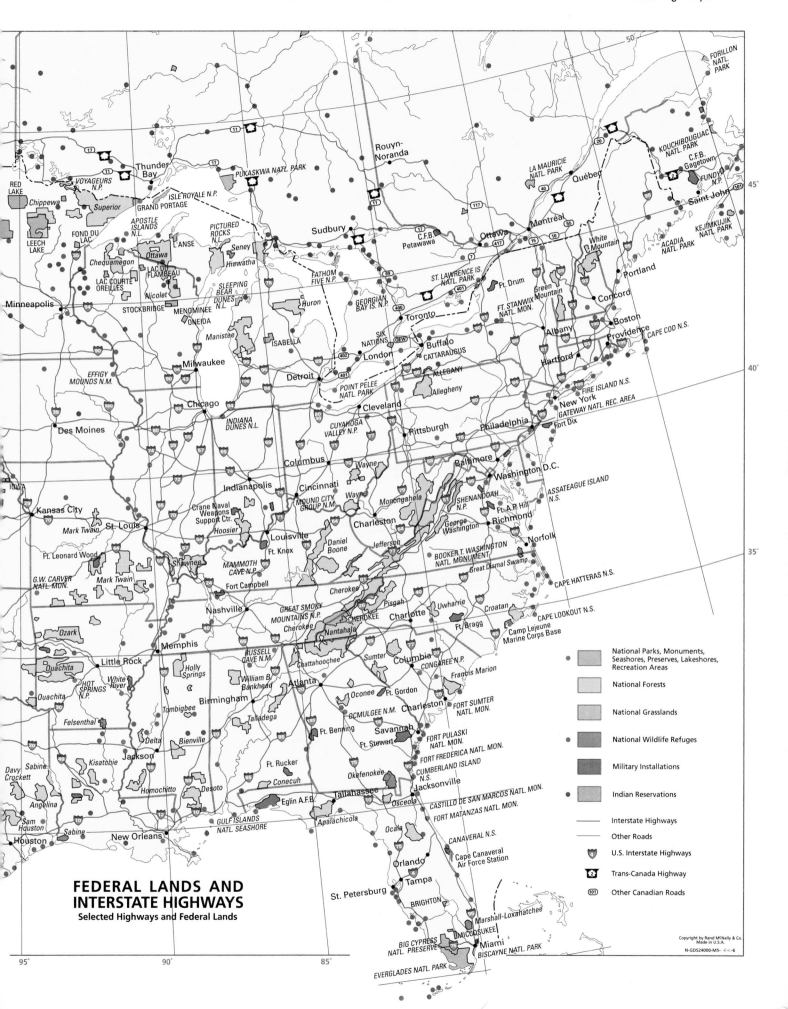

FEDERAL LANDS AND INTERSTATE HIGHWAYS
Selected Highways and Federal Lands

Legend:

National Parks, Monuments, Seashores, Preserves, Lakeshores, Recreation Areas

National Forests

National Grasslands

National Wildlife Refuges

Military Installations

Indian Reservations

Interstate Highways

Other Roads

U.S. Interstate Highways

Trans-Canada Highway

Other Canadian Roads

Copyright by Rand McNally & Co.
Made in U.S.A.

N-GDS24000-M5- -6

ASIA
RUSSIA
UNITED KINGDOM
IRELAND
GREENLAND (Denmark)
ICELAND
Reykjavík

North Pole
North Magnetic Pole

ARCTIC OCEAN

ALASKA
BROOKS RANGE
ALASKA RANGE
Fairbanks
Anchorage
Nome
Seward
Sitka

Whitehorse
Dawson
KLONDIKE REGION

Mt. Logan
Mt. McKinley

Bering Strait
BERING SEA
ALEUTIAN ISLANDS
ALEUTIAN TROUGH

Gulf of Alaska
PRIBILOF ISLANDS
KODIAK ISLAND
Bristol Bay

QUEEN CHARLOTTE ISLANDS
Prince Rupert
VANCOUVER ISLAND

C A N A D A

Edmonton
Calgary
Regina
Winnipeg
Vancouver
Seattle
Spokane
Portland
Butte

Great Bear Lake
Great Slave Lake
Athabasca Lake
Reindeer Lake
Lake Winnipeg
Lake of the Woods
Churchill
Ft. Simpson

HUDSON BAY
James Bay
Southampton
Foxe Basin
BAFFIN ISLAND
VICTORIA ISLAND
BANKS ISLAND
QUEEN ELIZABETH ISLANDS
Viscount Melville Sound
Resolute

Hudson Strait
Ungava Bay
LABRADOR
UNGAVA PEN.
CAPE CHIDLEY

Baffin Bay
Davis Strait
Denmark Strait
Thule
Etah
Upernavik
Godhavn
Godthåb
Angmagssalik
KAP FARVEL
KAP YORK
DISKO

Lincoln Sea
McKinley Sea
GREENLAND SEA
JAN MAYEN (Nor.)
SHETLAND IS. (Br.)
FAROE IS. (Den.)
North Sea

U N I T E D   S T A T E S

ROCKY MOUNTAINS
CASCADE RANGE
COAST RANGES
SIERRA NEVADA
GREAT BASIN
GREAT PLAINS
APPALACHIAN MTS.
Mt. Shasta 14,162
Mt. Whitney 14,494
Pikes Peak 14,110

San Francisco
Oakland
Los Angeles
Salt Lake City
Denver
Wichita
Kansas City
Omaha
Minneapolis
St. Paul
Milwaukee
Duluth
Fargo
CHICAGO
DETROIT
Cleveland
Buffalo
Pittsburgh
Cincinnati
St. Louis
Memphis
Birmingham
Atlanta
Mobile
New Orleans
Houston
San Antonio
Galveston
Dallas
Fort Worth
El Paso
NEW YORK
PHILADELPHIA
Baltimore
Washington
Richmond
Norfolk
Boston
Buffalo
Toronto
Ottawa
MONTRÉAL
Québec
Saint John
Halifax
St. John's
Savannah
Jacksonville
Miami

Columbia River
Missouri River
Arkansas
Red
Rio Grande
Colorado
Yellowstone
LAURENTIAN HIGHLANDS
NOVA SCOTIA
CAPE BRETON ISLAND
NEWFOUNDLAND
CAPE RACE
Gulf of St. Lawrence
CAPE SABLE
CAPE COD
Chesapeake Bay
CAPE HATTERAS
CAPE MENDOCINO

L. Superior
L. Michigan
L. Huron
L. Erie
L. Ontario
Lake Nipigon

ATLANTIC OCEAN
BERMUDA (Br.)

M E X I C O

SIERRA MADRE OCCIDENTAL
SIERRA MADRE ORIENTAL
BAJA CALIFORNIA
Golfo de California
CABO SAN LUCAS
GUADALUPE (Mex.)
ISLAS REVILLAGIGEDO (Mex.)
Tropic of Cancer

MÉXICO CITY
Guadalajara
Tampico
Veracruz
Pico de Orizaba 18,406 (Vol.)
Popocatépetl 17,887 (Vol.)
Bahía de Campeche
YUCATÁN PEN.

GULF OF MEXICO
Straits of Florida
Yucatán Channel
HAVANA
CUBA
BAHAMAS
SAN SALVADOR
JAMAICA
Kingston
HAITI
DOM. REP.
Port-au-Prince
Santo Domingo
San Juan
PUERTO RICO (U.S.A.)
PUERTO RICO TRENCH
GUADELOUPE (Fr.)
MARTINIQUE (Fr.)
BARBADOS
TRINIDAD AND TOBAGO
WEST INDIES
Windward Pass.

CARIBBEAN SEA

BELIZE
GUATEMALA
HONDURAS
EL SALVADOR
NICARAGUA
COSTA RICA
PANAMA
ISTMO DE PANAMÁ
G. de Panamá
Golfo de Honduras
CENTRAL AMERICA
PTA. DE GALLINAS

ISLA DEL COCO (Costa Rica)
ISLA DE MALPELO (Colombia)

S O U T H   A M E R I C A
Bogotá
Caracas
Rio Orinoco
Equator
Quito

PACIFIC OCEAN

40,000 SQ MI AREA

0 300 600
Miles

0 200 400 600 800 1000 Miles
0 400 800 1200 1600 Kilometers

A-520000-26 I-5-5-18
COPYRIGHT BY
RAND McNALLY & COMPANY
MADE IN U.S.A.

Longitude West of Greenwich

Scale 1:40 000 000; one inch to 630 miles. Lambert's Azimuthal Equal Area Projection
Elevations and depressions are given in feet

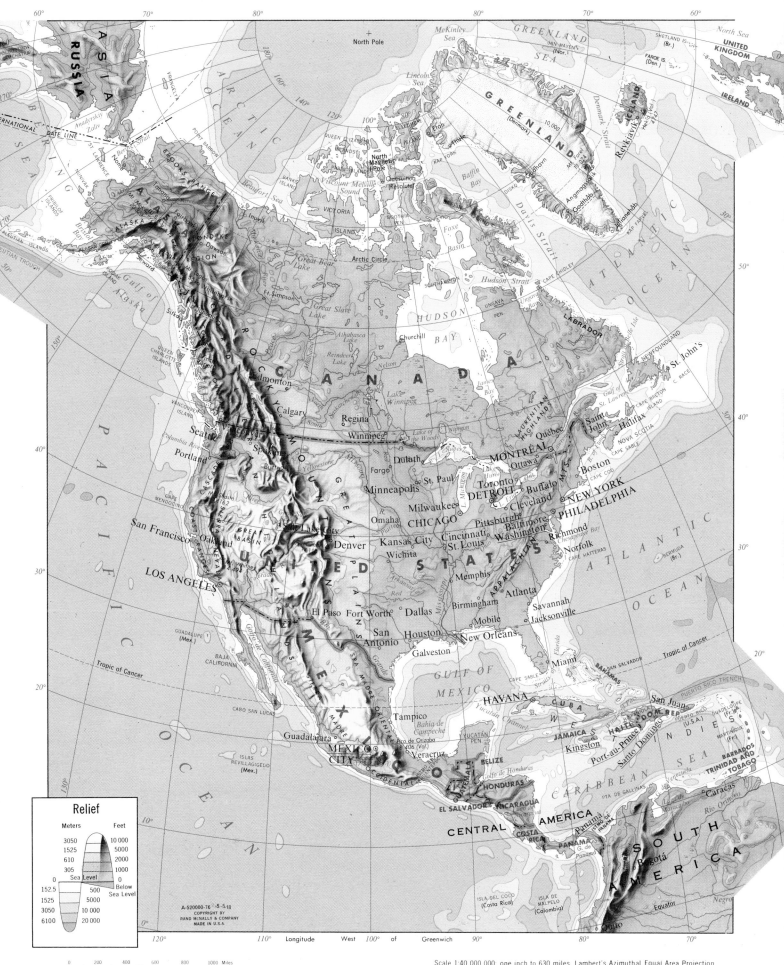

RUSSIA

ASIA

North Pole

GREENLAND SEA

United Kingdom

Ireland

GREENLAND (Denmark)

ICELAND

ARCTIC OCEAN

CANADA

HUDSON BAY

LABRADOR

NEWFOUNDLAND

ATLANTIC OCEAN

ALASKA

ROCKY MOUNTAINS

PACIFIC OCEAN

UNITED STATES

Seattle
Portland
San Francisco
Oakland
LOS ANGELES

Edmonton
Calgary
Regina
Winnipeg
Minneapolis
Omaha
Denver
Wichita
Kansas City

Duluth
St. Paul
Fargo
Milwaukee
CHICAGO
DETROIT
Cleveland
Pittsburgh
Cincinnati
St. Louis
Memphis
Birmingham

Toronto
Ottawa
MONTRÉAL
Québec
Buffalo
NEW YORK
PHILADELPHIA
Baltimore
Washington
Richmond
Norfolk

Boston
Halifax
Nova Scotia
St. John's

Atlanta
Savannah
Jacksonville
Mobile
New Orleans
Houston
San Antonio
Galveston
El Paso   Fort Worth   Dallas

Miami

BAHAMAS

Tropic of Cancer

GULF OF MEXICO

BAJA CALIFORNIA

MEXICO CITY
Guadalajara
Tampico
Veracruz

HAVANA   CUBA

HAITI   DOM. REP.
JAMAICA   Kingston
Port-au-Prince   Santo Domingo
San Juan
PUERTO RICO (U.S.A.)

WEST INDIES

CARIBBEAN SEA

BARBADOS
TRINIDAD AND TOBAGO

BELIZE
GUATEMALA
HONDURAS
EL SALVADOR   NICARAGUA
COSTA RICA
PANAMA

CENTRAL AMERICA

Caracas

SOUTH AMERICA

Bogotá

Equator

Relief

Meters	Feet
3050	10 000
1525	5000
610	2000
305	1000
0	Sea Level 0
	Below
152.5	Sea Level 500
1525	5000
3050	10 000
6100	20 000

0   200   400   600   800   1000   Miles

0   400   800   1200   1600   Kilometers

120°   110°   Longitude   West   100°   of   Greenwich   90°   80°   70°

Scale 1:40 000 000, one inch to 630 miles. Lambert's Azimuthal Equal Area Projection
Elevations and depressions are given in feet

50

Scale 1: 12 000 000; one inch to 190 miles. Conic Projection
Elevations and depressions are given in feet

QUEBEC

NEWFOUNDLAND AND LABRADOR

Same scale as main map

Longitude West of Greenwich

ATLANTIC OCEAN

HUDSON BAY

All islands within bays and straits lie within Nunavut.

ONTARIO

QUEBEC

NEWFOUNDLAND AND LABRADOR

MONTRÉAL

TORONTO

MICHIGAN

WISCONSIN

NEW YORK

PENNSYLVANIA

MAINE

NEW BRUNSWICK

NOVA SCOTIA

ATLANTIC OCEAN

A-520200-26 · -10-9-23
COPYRIGHT BY
RAND McNALLY & COMPANY
MADE IN U.S.A.

40,000 SQ MI
AREA
0   100   200
Miles

0 25 50 75 100   200   300   400   500 Miles
0   100   200   400   600   800 Kilometers

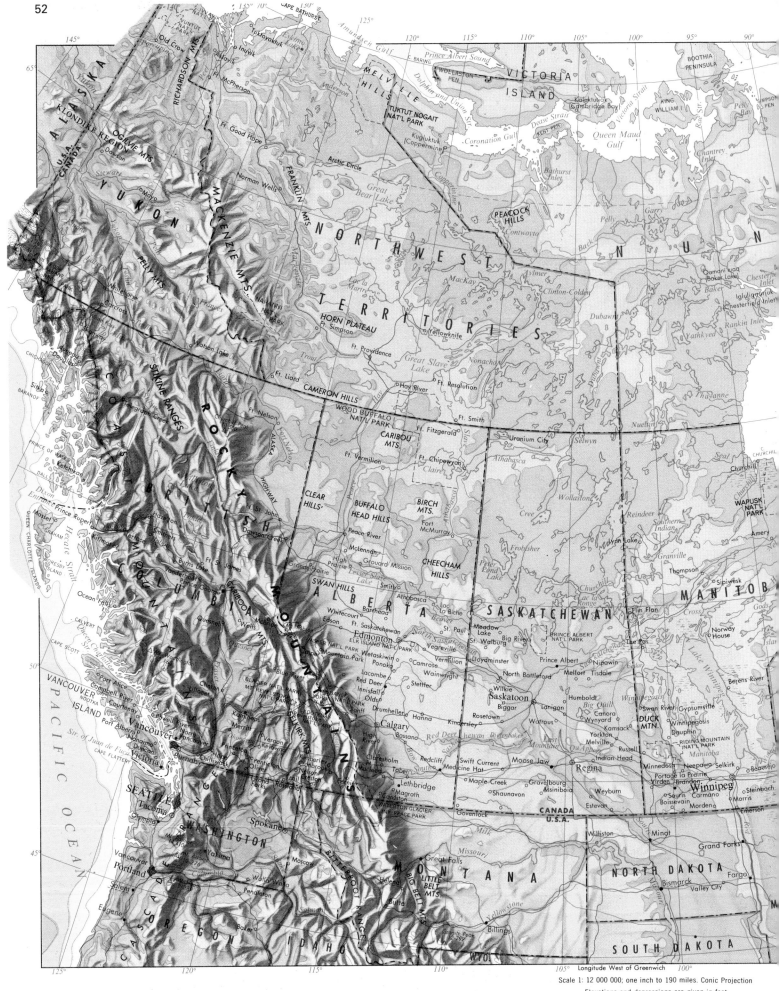

a

Longitude West of Greenwich

QUEBEC

Same scale as
main map

NEWFOUNDLAND
AND
LABRADOR

CAPE BAULD

Gulf of
St. Lawrence

LONG RANGE MTS.

GROS MORNE
NAT'L PARK
Deer Lake
Corner Brook
Stephenville
St.
George s

Windsor
Grand Falls

Twillingate
Gander
Bonavista

TERRA NOVA
NAT'L PARK

NEWFOUNDLAND
Trinity

CAPE RAY
C. ST. GEORGE

Channel-Port-aux-Basques

St. John's

CAPE NORTH

Grand Bank
Burin

CAPE BRETON
ISLAND

ST. PIERRE AND MIQUELON (Fr.)

ATLANTIC OCEAN

MELVILLE
PENINSULA

Foxe
Basin

BAFFIN
ISLAND

PRINCE
CHARLES
ISLAND

BAFFIN ISLAND
NAT'L PARK

Pangnirtung

Cumberland Sound

C. MERCY

Arctic Circle

Iqaluit

HALL
PEN.

EVERETT
MTS.

RESOLUTION

Foxe
Channel

FOXE
PEN.

SOUTHAMPTON
ISLAND

BELL
PEN.

Fisher Strait

COATS

MANSEL

NOTTINGHAM
ISLAND

SALISBURY

DE
NOUVELLE-
FRANCE

Hudson
Strait

C. HOPES
ADVANCE

AKPATOK

KILLINIQ I.

TORNGAT
MTS.

Hebron

Nain

HUDSON
BAY

Ottawa
Islands

PENINSULE
D'UNGAVA

Ungava
Bay

Kuujjuaq

Hopedale
Makkovik

NEWFOUNDLAND

All islands within bays and straits
lie within Nunavut.

Povungnituk

Minto

Lac
Bienville

BELCHER
ISLANDS

Happy
Valley-Goose Bay

Rigolet

MEALY MTS.

AND

Cartwright

Battle Harbour

C. Henrietta Maria

Grande de la Baleine

Schefferville

LABRADOR

Anthony

PTE. LOUIS-XIV

James
Bay

Chisasibi

AKIMISKI

Ft. Severn

Ft. Albany

Moosonee

La Grande

Eastmain

Nichicun

MTS.
OTISH

LONG RANGE MTS.

GROS MORNE
NAT'L PARK

Corner Brook
Stephenville
St. George

ONTARIO

Winisk

Severn

St.
Joseph

Trout

Red Lake

Lac Seul

Sioux Lookout

nora
Dryden

Armstrong Sta.

Geraldton
Longlac

Nipigon

Nakina

Hearst
Cochrane
Kapuskasing
Iroquois Falls

QUEBEC

Misteassini

Chibougamau

Dolbeau
Alma

Kenogami
Arvida

Roberval
Chambord

St. Félicien
Jonquière

Lac Mistassini

La Sarre
Amos

Rouyn
Malartic

Senneterre

La Tuque

Parent

Chicoutimi

La Baie
St-Paul

ILE D'ANTICOSTI

Clarke City
Sept-Iles

Betsiamites

Rimouski

Baie-Comeau

St. Lawrence River

Natashquan

Mingan

CHIC-CHOCS

Cap-Chat
Matane

MTS.

Gaspé

Chandler

PEN. DE GASPE
New Carlisle

ILES DE LA
MADELEINE

Rainy
River

Lake of the Woods

Thunder Bay

Nipigon

Lake
Superior

Marathon

PUKASKWA
NAT'L PARK

MICHIPICOTEN I.

Timmins
Kirkland Lake

Cobalt

Ville-Marie
Temiscaming

Chapleau

Sudbury

Sturgeon Falls
North
Bay

MONTREAL

Shawinigan
Trois-
Rivieres

Joliette
Sorel

Drummondville

Granby

Quebec

Levis

Montmagny

La Pocatière

Rivière-du-Loup

Edmundston

Grand
Mere

Victoriaville

Sherbrooke

NEW
BRUNSWICK

CANADA
U.S.A.

Campbellton
Bathurst

Chatham
Newcastle

Richibucto

Moncton

PRINCE EDWARD
P.E.I.
Charlottetown

Summerside

NOVA
SCOTIA

Antigonish
Amherst

New Waterford
Sydney Mines
Glace Bay
Sydney

CAPE BRETON
HIGHLANDS
NAT'L PARK

Thessalon
Blind River
Espanola

Sault Ste. Marie

MANITOULIN

Georgian
Bay

Parry Sound

Huntsville
Bracebridge

Renfrew

Pembroke

Ottawa
Hull

Bancroft
Smiths Falls

Perth

Ogdensburg

Brockville
Kingston

Alexandria
Bay

Cornwall

Valleyfield

VERMONT

NEW
HAMPSHIRE

MAINE

FUNDY
NAT'L PARK

Fredericton
Woodstock

Saint John

St. Stephen
St. Andrews

St. George

KEJIMKUJIK
NAT'L PARK

Kentville

Yarmouth

Digby

Lunenburg
Bridgewater
Liverpool

Dartmouth
Halifax

Shelburne

CAPE SABLE

Duluth
Superior

Marquette

Escanaba

MICHIGAN

Sault Ste. Marie

Blind River

Espanola

Sudbury

Owen Sound
Kincardine

Wiarton

Midland
Barrie

Orillia

Peterborough
Lindsay
Cobourg

Trenton

Whitby
Oshawa

Coburg

Lake
Ontario

Rochester

Concord

Hartford

MASS.
CONN.

Boston

CAPE COD

R.I. Providence

ATLANTIC
OCEAN

MINNESOTA

St. Paul

WISCONSIN

Green Bay

Lake Michigan

Saginaw Bay

Owen Sound

London

St.
Catharines

St. Thomas

Sarnia

Chatham

Windsor

Leamington

Lake Erie

TORONTO
Kitchener
Hamilton
Niagara
Falls

BUFFALO

NEW YORK

Albany

N.J.

NEW YORK

Madison

Grand
Rapids

MILWAUKEE

Flint
Lansing

DETROIT

CHICAGO

ILLINOIS

Toledo

OHIO

PENNSYLVANIA

Scranton

Port
Huron

Relief

Meters   Feet

3050   10 000
1525   5000
610   2000
305   1000
152.5   500

Sea Level

152.5   500
1525   5000
3050   10 000

A-520200-76 -10. -23
COPYRIGHT BY
RAND McNALLY & COMPANY
MADE IN U.S.A.

0 25 50 75 100   200   300   400   500 Miles

0   100   200   400   600   800 Kilometers

134°   132°   130°   128°   126°   124°

WILLISTON Lake

OMINECA MOUNTAINS

SKEENA MOUNTAINS

PRINCE OF WALES ISLAND

MT. REID 4592

REVILLAGIGEDO ISLAND

Klawock
Hydaburg
Copper Mtn 3916
Ketchikan
Metlakatla
DALL ISLAND
ANNETTE ISLAND

Stedini Pk 8750

Mt. Thomlinson 8050

HAZELTON MOUNTAINS

Alice Arm

Hazelton

Smithers

Takla Lake

Tchentlo Lake

McLeod Lake

UNITED STATES
CANADA

DUNDAS ISLAND

CHATHAM SOUND

Terrace

BULKLEY RANGES

Hawson Pk 9050

Burns Lake

Stuart Lake

Fort St. James

NECHAKO

Babine Lake

CAPE KNOX

Dixon Entrance

54°

Masset

Massell Inlet

QUEEN CHARLOTTE

GRAHAM ISLAND

PORCHER ISLAND

Prince Rupert

Kitimat

Morice Lake

Endako

Vanderhoof

Hecate

BANKS ISLAND

Harley Bay

COAST

KITIMAT RANGES

BRITISH

Ootsa Lake

Tahtsa Lake

Michel Pk 7396

Whitesail Lake

Eutsuk Lake

Nechako Reservoir

PLATEAU

KENNEY DAM

NECHAKO RANGE

Tetachuck Lake

CHARLOTTE

MORESBY ISLAND

Mount Kermode 3550

Skidegate Inlet

Strait

FLTY ISLAND

ESTEVAN GROUP

PRINCESS ROYAL ISLAND

West Road

West Road

ISLANDS

RANGES

ARISTAZABAL ISLAND

Mt. Parry 3450

ROYAL ISLAND

Dean

COLUM

FRASER

52°

CAPE ST. JAMES

MOUNTAINS

Ocean Falls

Bella Coola

Charlotte Lake

Redstone

Bella Bella

Monarch Mtn 1570

Queen

Charlotte

Sound

CALVERT ISLAND

Namu

Rivers Inlet

PACIFIC

Razorback Mtn 10432

Silverthrone Mtn 9700

Mt. Tatlow 10058

PLATEAU

Chilko Lake

CAPE CAUTION

Mt. Waddington 13163

Mt. Queen Bess 10291

Good Hope Mtn 10157

Monmouth Mtn 10480

Bull Harbour

Queen Charlotte Strait

CAPE SCOTT

Port Hardy

Simood Sound

RANGES

Mt. Gilbert 10207

Bralorne

50°

PACIFIC

Quatsino Sound

Port Alice

CAPE COOK

VANCOUVER

Victoria Pk 7095

Bloedel

Kelsey Bay

Campbell River

Wedge Mtn 9484

Mt. Garibaldi 8787

Squamish

OCEAN

VANCOUVER

ISLAND

Nootka Sound

Golden Hinde 7297

Courtenay

Comox

Vanandan

Powell River

North Vancouver

Vancouver
Burnaby
New Westminster

Tofino

Port Alberni

Nanaimo

ISLAND

RANGES

Mt. Whymper 5056

PACIFIC RIM NATIONAL PARK

Barkley Sound

CAPE BEALE

Ladysmith

Lake Cowichan

Duncan

Cowichan Lake

48°

CAPE FLATTERY

Strait of Juan de Fuca

Esquimalt

Oak Bay

Victoria

OLYMPIC NATIONAL PARK

Port Angeles

OLYMPIC NATIONAL PARK

Port Townsend

Continued on pages 72-73

Longitude West of Greenwich

Scale 1:4 000 000; one inch to 64 miles. Conic Projection

Elevations and depressions are given in feet.

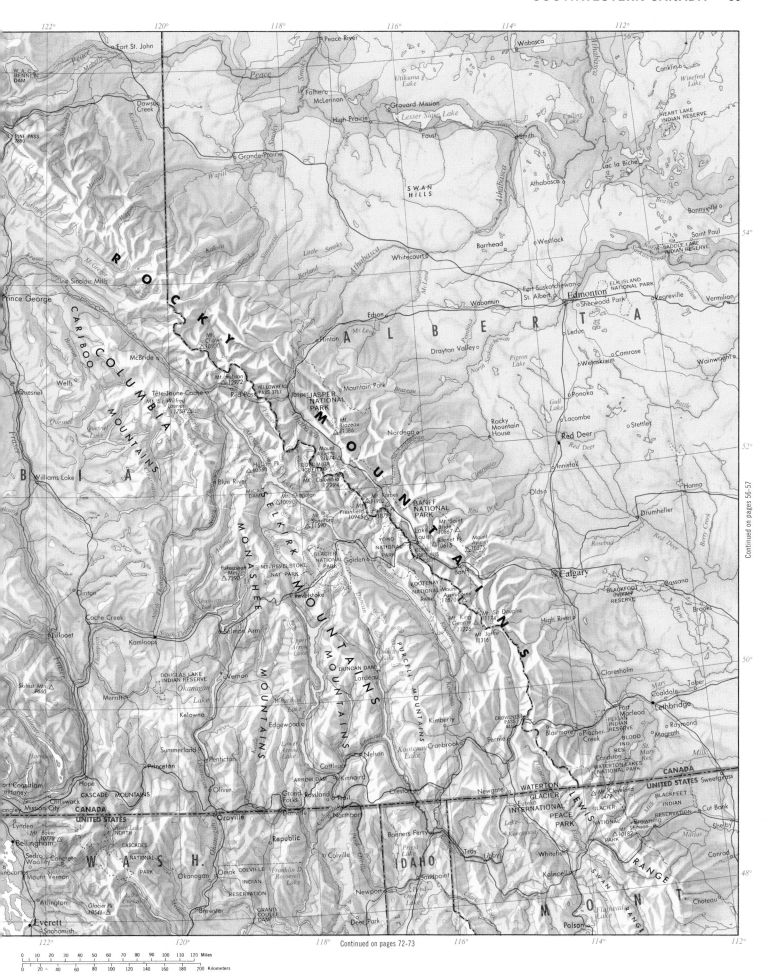

Continued on pages 56-57

Continued on pages 72-73

0 10 20 30 40 50 60 70 80 90 100 110 120 Miles

0 20 40 60 80 100 120 140 160 180 200 Kilometers

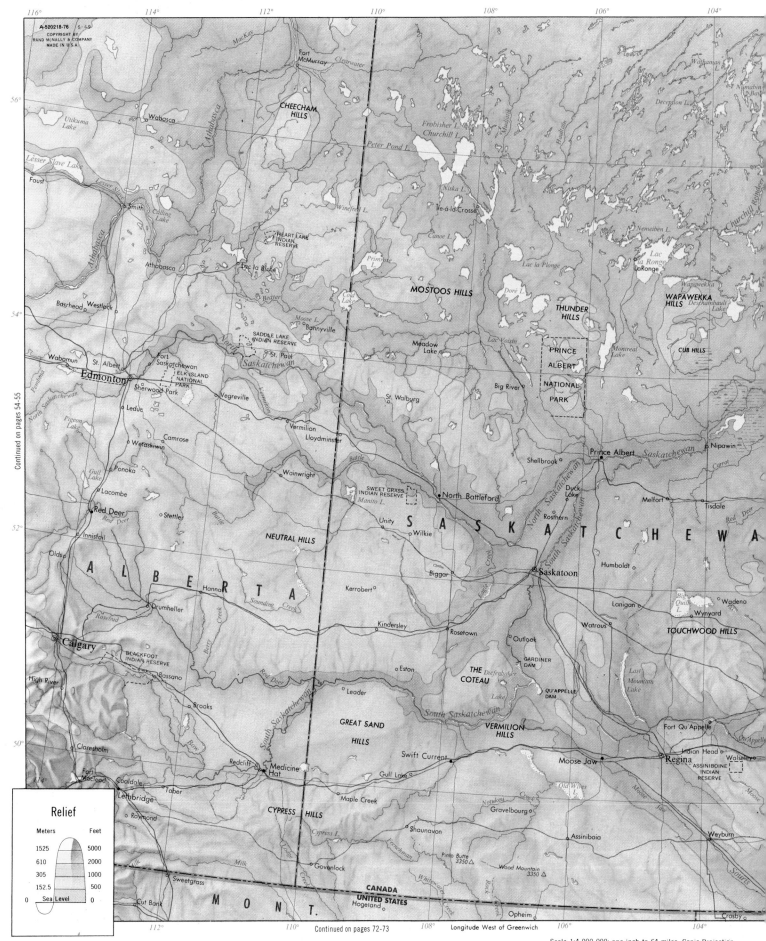

A-520218-76 5-49
COPYRIGHT BY
RAND McNALLY & COMPANY
MADE IN U.S.A.

MacKay

Fort
McMurray   Clearwater

Wabasca   CHEECHAM
                                  HILLS   Frobisher L.
Utikuma   Churchill L.
Lake
                                          Peter Pond L.

Lesser Slave Lake   Niska L.
Faust
                                  HEART LAKE   Ile-à-la-Crosse
                                  INDIAN   Winefred L.
         Smith   Calling   Canoe L.
                  Lake   RESERVE   Lac la Plonge
                                                                Primrose   Lac
                                                                 L.   la Ronge
         Athabasca   Lac la Biche   MOSTOOS HILLS   Doré L.   LaRonge
Barrhead   Westlock   WAPAWEKKA
                                          Beaver   Cold   HILLS   Deschambault
Wabamun   Saddle Lake   Moose L.   Lake   THUNDER   Lake
         St. Albert   North   INDIAN RESERVE Bonnyville   Meadow   HILLS
Edmonton   Saskatchewan   St. Paul   Lake   PRINCE
         Sherwood Park Fort   Lac Vaisin   ALBERT   CUB HILLS
                        Saskatchewan ELK ISLAND   NATIONAL
         Leduc   NATIONAL   St. Walburg   Big River   PARK
Pigeon   PARK   Vegreville
Lake   Vermilion   Lloydminster   Prince Albert   Saskatchewan Nipawin
         Wetaskiwin Camrose   Battle   Shellbrook
Gull   Duck   Rosthern   Melfort   Tisdale
Lake   Ponoka   North Battleford   Lake   Red Deer
         Lacombe   SWEET GRASS
Red Deer   INDIAN RESERVE   S A S K A T C H E W A
         Stettler   Battle   Manito L.   Unity   Wilkie   Humboldt
Innisfail
Olds   NEUTRAL HILLS   Biggar   Saskatoon   South Saskatchewan
                                                                                                                      Lanigan   Big Wadena
A L B E R T A   Hanna   Sounding   Creek   Kerrobert   Watrous   Quill Wynyard
                                                                                                                              L.
         Drumheller   Berry   Creek   Kindersley   Outlook   GARDINER   TOUCHWOOD HILLS
Calgary   Rosebud   Rosetown   DAM
         BLACKFOOT   THE   Diefenbaker   Last
         INDIAN RESERVE   Eston   COTEAU   Lake   QU'APPELLE Mountain
High River   Bassano   DAM   Lake
                                  Red Deer   Leader   South Saskatchewan   VERMILION
         Brooks   HILLS   Fort Qu'Appelle
                                                                                                                           Qu'Appelle
                  South Saskatchewan   GREAT SAND   Indian Head
Claresholm   HILLS   Moose Jaw   Regina Wolseley
         Redcliff Medicine   Swift Current   Gull Lake   ASSINIBOINE
Fort   Hat   INDIAN
Macleod Coaldale   Old Wives   RESERVE
Lethbridge   Taber   Maple Creek   Notukeu   L.   Moose
         Raymond   CYPRESS HILLS   Gravelbourg   Weyburn
                                  Cypress L.   Shaunavon   Assiniboia
                                                            Pinto Butte
                              Milk   3350 △   Wood Mountain   Souris
         Sweetgrass   Govenlock   3350 △
Cut Bark   CANADA   Crosby
         M O N T.   UNITED STATES
                                               Hogeland
                                                          Opheim

Relief
Meters   Feet
1525   5000
610   2000
305   1000
152.5   500
0   Sea Level   0

Continued on pages 54-55

Continued on pages 72-73

Longitude West of Greenwich

Scale 1:4 000 000; one inch to 64 miles. Conic Projection
Elevations and depressions are given in feet.

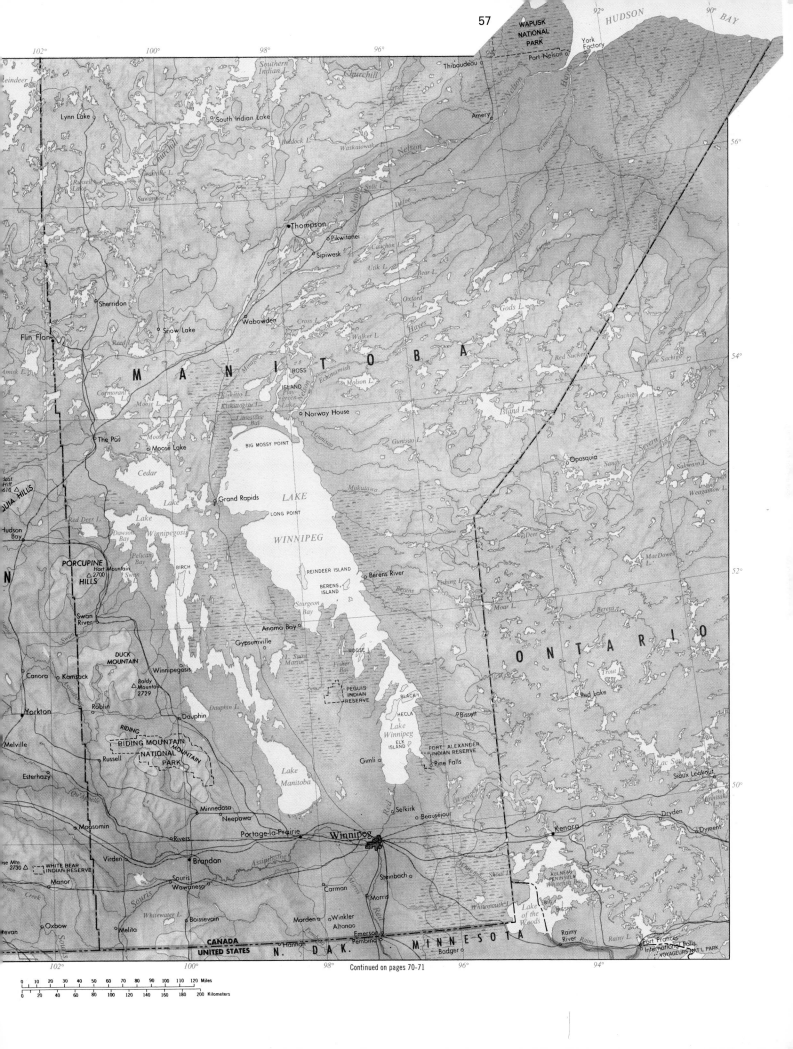

HUDSON BAY

WAPUSK NATIONAL PARK

York Factory

M A N I T O B A

O N T A R I O

Lynn Lake

South Indian Lake

Thompson

Flin Flon

Sherridon

Snow Lake

Wabowden

The Pas

Moose Lake

Grand Rapids

LAKE WINNIPEG

Norway House

PORCUPINE HILLS

Swan River

Canora

Kamsack

Yorkton

Roblin

Dauphin

Melville

Esterhazy

Russell

RIDING MOUNTAIN NATIONAL PARK

DUCK MOUNTAIN

Winnipegosis

Gypsumville

Anama Bay

Berens River

PEGUIS INDIAN RESERVE

HECLA I.

Bissett

Lake Winnipeg

FORT ALEXANDER INDIAN RESERVE

Gimli

Pine Falls

Minnedosa

Neepawa

Moosomin

Rivers

Portage-la-Prairie

Winnipeg

Selkirk

Beauséjour

Kenora

Dryden

Virden

Brandon

Souris

Wawanesa

Manor

Steinbach

Carman

Morris

Oxbow

Melita

Boissevain

Morden

Winkler

Altona

Emerson

Pembina

CANADA
UNITED STATES

N. DAK.

MINNESOTA

Lake of the Woods

International Falls

Fort Frances

Continued on pages 70-71

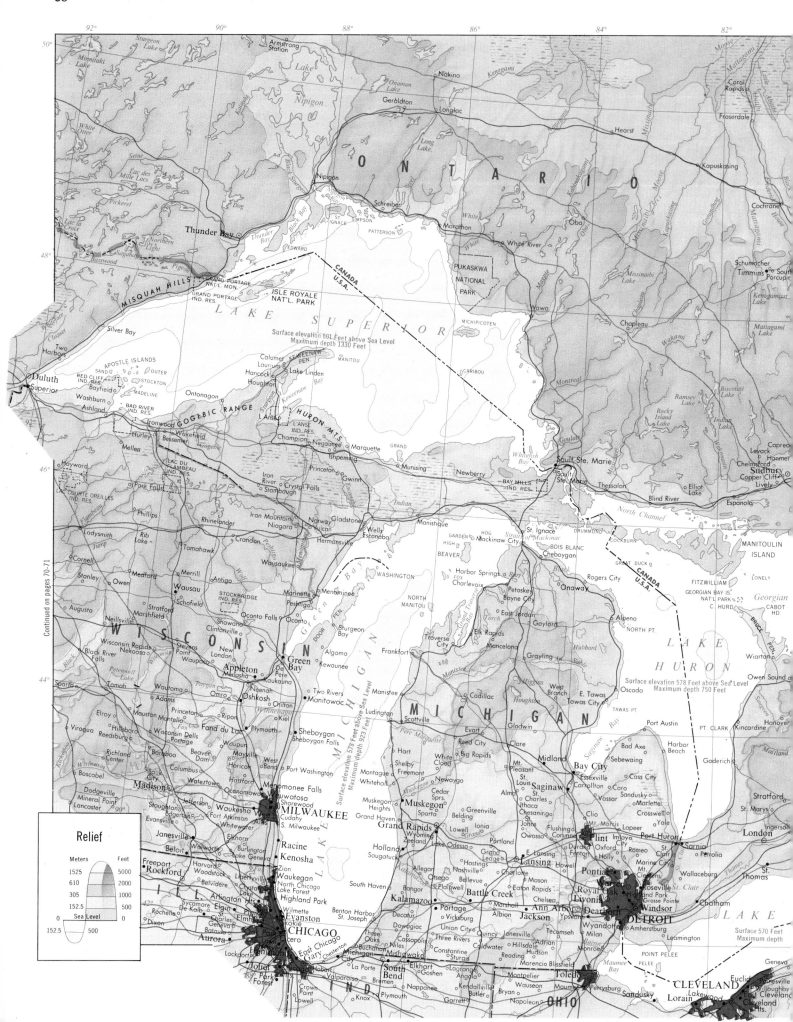

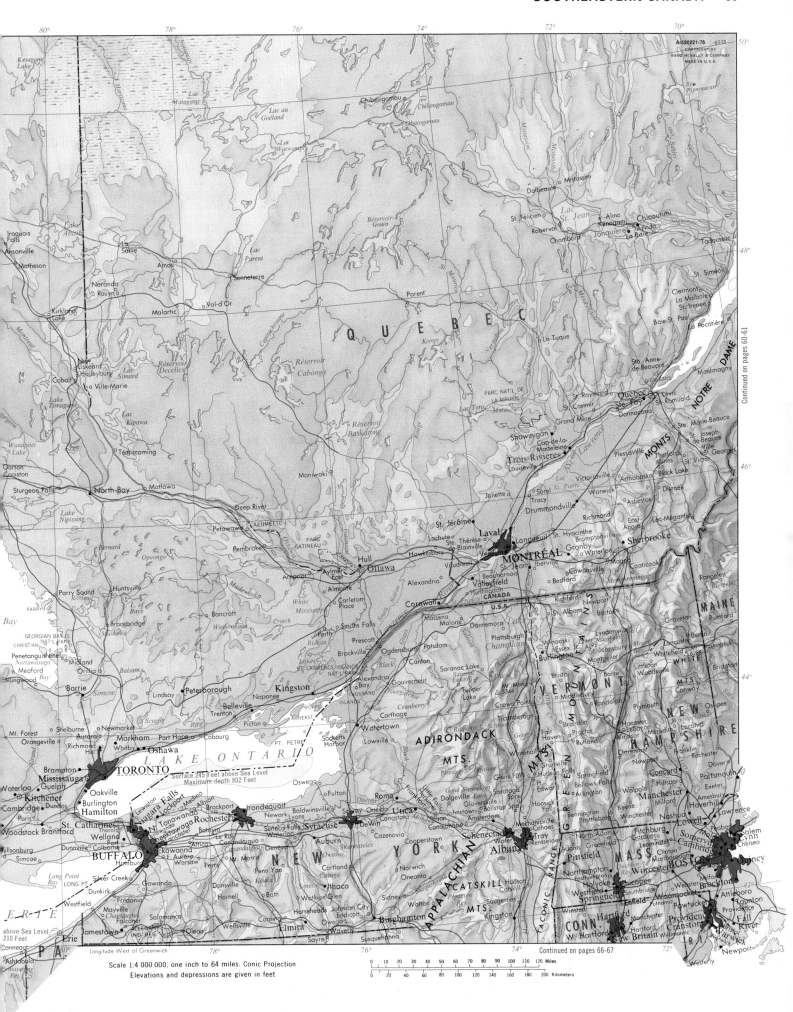

Scale 1:4 000 000; one inch to 64 miles. Conic Projection
Elevations and depressions are given in feet

Continued on pages 60-61

Continued on pages 66-67

60

Continued on pages 58-59

Continued on pages 66-67

Scale 1:4 000 000; one inch to 64 miles. Conic Projection
Elevations and depressions are given in feet.

Longitude West of Greenwich

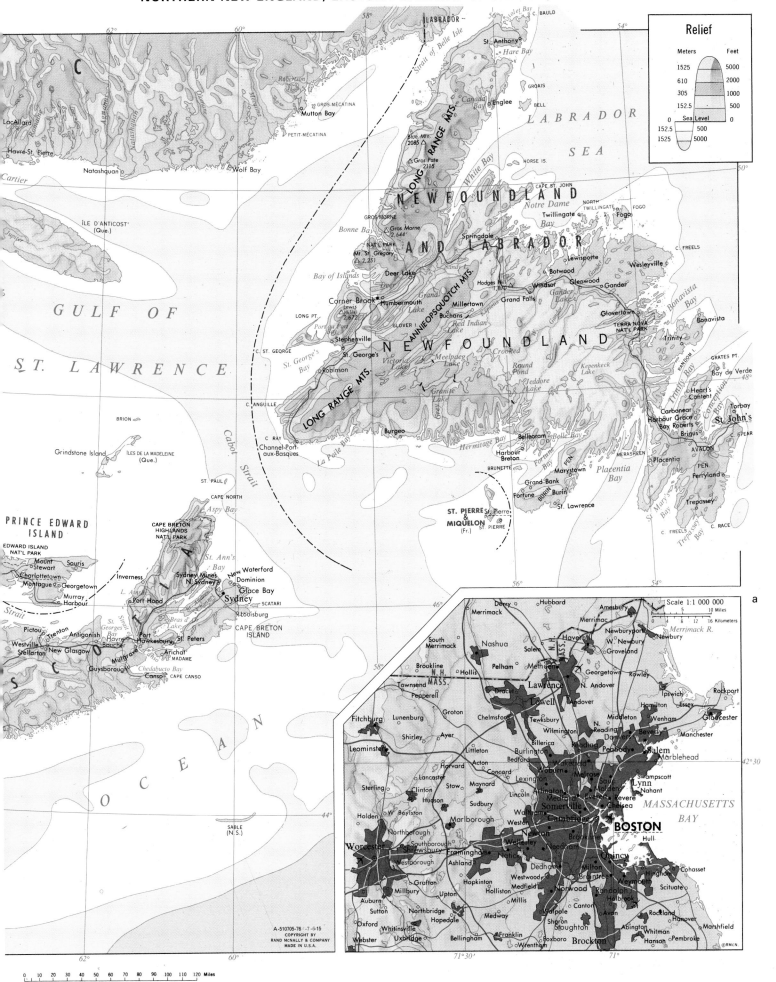

Relief

Meters	Feet
1525	5000
610	2000
305	1000
152.5	500
0	Sea Level	0
152.5	500
1525	5000

LABRADOR

C. BAULD
St. Anthony
Hare Bay
GROAIS
BELL
Canada Bay
Englee

LABRADOR SEA

Strait of Belle Isle
Blue Mtn. 2085
Gros Pate 2115
HORSE IS.

NEWFOUNDLAND

CAPE ST. JOHN
NORTH TWILLINGATE
Notre Dame Bay
Twillingate
Fogo
FOGO

AND LABRADOR

GROS MORNE
Gros Morne 2,644
NAT'L PARK
Springdale
Botwood
Lewisporte
Wesleyville

Mt. St. Gregory 2,251
Bonne Bay
Deer Lake
Hodges Hill 1,870
Windsor
Glenwood
Gander
Gander Lake

Bay of Islands
Corner Brook
Lewis Hills 2,672
Humbermouth
Humber
Grand Lake
Millertown
Buchans
Grand Falls
Glovertown
TERRA NOVA NAT'L PARK
Bonavista Bay
Bonavista

LONG PT.
GLOVER I.
ANNIEOPSQUOTCH MTS.
Red Indian Lake
Trinity

NEWFOUNDLAND

Port au Port Bay
Stephenville
Victoria Lake
Meelpaeg Lake
Crooked Lake
Round Pond
Kepenkeck Lake
Jeddore Lake
GRATES PT.
Bay de Verde
Heart's Content

C. ST. GEORGE
St. George's Bay
St. George's
Robinson
Granite Lake

RANDOM I.
TRINITY BAY
Carbonear
Harbour Grace
Bay Roberts
Brigus
St. John's
C. SPEAR

C. ANGUILLE

LONG RANGE MTS.

Burgeo

La Poile Bay
Hermitage Bay

MERASHEEN
AVALON PEN.
Placentia
Ferryland

CAPE NORTH
Aspy Bay
St. Ann's Bay

C. RAY
Channel-Port-aux-Basques

Belleoram
Belle Bay
Harbour Breton
FORTUNE BAY
BURIN PEN.
BRUNETTE

Grand Bank
Fortune
Marystown
Burin
St. Lawrence

Placentia Bay

St. Mary's Bay
Trepassey
C. RACE
C. FREELS
Trepassey Bay

GULF OF ST. LAWRENCE

C
Natashquan
LacAllard
Havre-St. Pierre
Mutton Bay
PETIT-MÉCATINA
GROS-MÉCATINA
Wolf Bay
Natashquan

Cartier

ÎLE D'ANTICOST (Que.)
Jupiter

BRION
Grindstone Island
Îles de la Madeleine (Que.)

ST. PAUL
Cabot Strait

PRINCE EDWARD ISLAND
EDWARD ISLAND NAT'L PARK
Mount Stewart
Souris
Charlottetown
Montague
Georgetown
Murray Harbour

CAPE BRETON HIGHLANDS NAT'L PARK

Inverness
Sydney Mines
N. Sydney
New Waterford
Dominion
Glace Bay
Port Hood
L. Ainslie
Sydney
SCATARI

Pictou
Trenton
Antigonish
Havre Boucher
Port Hawkesbury
St. Georges Bay
Bras d'Or Lake
Arichat
MADAME
St. Peters
Louisburg
CAPE BRETON ISLAND

Westville
Stellarton
New Glasgow
Mulgrave
Guysborough
Canso
CAPE CANSO
Chedabucto Bay

SCOTIA

ST. PIERRE & MIQUELON (Fr.)
St. Pierre
St. Pierre

O C E A N

SABLE (N.S.)

A-510705-76-7-6-15
COPYRIGHT BY
RAND McNALLY & COMPANY
MADE IN U.S.A.

Inset map (a):

Scale 1:1 000 000

Derry
Hubbard
Amesbury
Merrimack
Merrimac
Haverhill
Newburyport
W. Newbury
Newbury
Merrimack R.

South Merrimack
Nashua
Salem
Methuen
Georgetown
Rowley
Ipswich

Brookline
N.H.
Hollis
Pelham
Lawrence
N. Andover
Rockport

Townsend
MASS.
Pepperell
Dracut
Lowell
Andover
Hamilton
Essex

Fitchburg
Lunenburg
Groton
Chelmsford
Tewksbury
Wilmington
N. Reading
Middleton
Wenham
Gloucester

Leominster
Shirley
Ayer
Billerica
Reading
Danvers
Beverly
Salem

Littleton
Burlington
Bedford
Woburn
Peabody
Manchester
Marblehead

Sterling
Lancaster
Harvard
Acton
Concord
Lexington
Melrose
Swampscott
Lynn
Nahant

Holden
W. Boylston
Clinton
Hudson
Stow
Maynard
Sudbury
Lincoln
Arlington
Everett
Revere
Chelsea

Northborough
Marlborough
Weston
Waltham
Cambridge
Somerville
MASSACHUSETTS BAY

Worcester
Shrewsbury
Southborough
Westborough
Framingham
Ashland
Natick
Wellesley
Newton
Dedham
Needham
Brookline
BOSTON
Hull

Grafton
Millbury
Upton
Hopkinton
Holliston
Medfield
Millis
Norwood
Milton
Braintree
Weymouth
Hingham
Scituate
Cohasset
Quincy

Auburn
Sutton
Northbridge
Hopedale
Medway
Walpole
Canton
Avon
Randolph
Holbrook
Rockland
Hanover
Marshfield

Oxford
Whitinsville
Uxbridge
Bellingham
Franklin
Foxboro
Sharon
Stoughton
Abington
Whitman
Hanson
Pembroke

Webster
Wrentham
Brockton

a

0 10 20 30 40 50 60 70 80 90 100 110 120 Miles
0 20 40 60 80 100 120 140 160 180 200 Kilometers

a — MONTRÉAL

Laurentides, L'Épiphanie, St-Sulpice, L'Assomption, ST. JÉRÔME, Ste. Anne-des-Plaines, Mascouche, Repentigny, Charlemagne, Verchères, St. Canut, St. Janvier, Terrebonne, Ste. Scholastique, Bois-des-Filion, Dalesville, Brownsburg, Lachute, St. Philippe-d'Argenteuil, St. Augustin, Deux-Montagnes, Ste. Thérèse-de-Blainville, Rosemère, PTE-AUX-TREMBLES, Varennes, St. Benoit, St. Eustache, St. André-Est, St. Placide, MONTRÉAL NORD, ANJOU, ST. LÉONARD, Boucherville, LAVAL, Mont-Royal, Ste. Fortune, St. Joseph-du-Lac, Deux-Montagnes, ST. LAURENT, OUTREMONT, MONTRÉAL, LONGUEUIL, St. Bruno, Hudson Hts., Como-Est, Lac des Deux Montagnes, Westmount, VERDUN, St. Lambert, ST. HUBERT, Rigaud, Hudson, Greenfield Park, Pte. Claire, Dorval, Beaconsfield, LACHINE, Brossard, Chambly, St. Lazare-de-Vaudreuil, Vaudreuil, LA SALLE, La Prairie, Ste. Justine-de-Newton, Dorion-Vaudreuil, Î. Perrot, St. Louis, Lac St. Louis, Caughnawaga, St. Constant, Delson, St. Philippe-de-Laprairie, L'Acadie, St. Clet, Léry, Pte-des-Cascades, Les Cèdres, Maple Grove, Châteauguay, Mercier, St. Dominique, Coteau-du-Lac, Melocheville, Beauharnois, St. Isidore-de-Laprairie, St. Édouard-de-Napierville, Coteau-Landing, St. Timothée, St. Rémi, Rivière-Beaudette, VALLEYFIELD, Canal de Beauharnois, Ste. Martine, St. Michel-de-Napierville, Napierville, St. Anicet, St. Stanislas-de-Kostka, Ste. Barbe, Howick, Aubrey, Barrington, St. Valentin, Ormstown

Lake St. Francis, St. Louis-de-Gonzague

Copyright by Rand McNally & Co.

b — QUÉBEC

St. Féréol, Île AUX GRUES, Beaupré, St-Joachim-de-Montmorency, Cap-St. Ignace, Ste. Anne-de-Beaupré, Stoneham, Lac-Beauport, Château-Richer, St. François, St. Pierre-Montmagny, MONTMAGNY, Valcartier-Village, L'Ange-Gardien, Ste. Famille, Berthier, François-Montmagny, St. Charles, St. Jean, CHARLESBOURG, Beauport, Ste. Pétronille, St. Pierre-d'Orléans, St. Michel, Villier, LORETTEVILLE, QUÉBEC, St. Laurent-d'Orléans, St. Raphaël, Ancienne-Lorette, Lauzon, La Durantaye, Ste. Euphémie, STE. FOY, Lévis, St. David, St. Charles, Armagh, Neuville, Cap-Rouge, St. Romuald-d'Etchemin, St. Gervais, St. Nérée, St. Augustin-de-Québec, Charny, Carrier, St. Nicolas, St. Jean-Chrysostome, St. Henri, St. Philémon, Rédempteur, Breakeyville, St. Antoine-de-Tilly, St. Étienne-de-Lauzon, Honfleur, St. Lazare, Ste. Euphémie, St. Apollinaire, Ste. Anselme, Buckland, St. Lambert-de-Lévis, Ste. Claire, St. Damien-de-Buckland, St. Isidore-Dorchester

Copyright by Rand McNally & Co.

c — OTTAWA

Alcove, Montebello, Wakefield, Papineauville, PARC DE LA GATINEAU, Perkins, Thurso, Plaisance, Chelsea, Buckingham, Alfred, Angers, Masson, Rockland, Wendover, Plantagenet, Templeton, Gatineau, Cumberland, Curran, Pointe-Gatineau, HULL, Eastview Park, Orleans, Bourget, St. Isidore-de-Prescott, Aylmer, VANIER, Navan, Deschênes, OTTAWA, Ramsayville, Vars, Limoges, Casselman, Maxville, Bells Corners, Leitrim, Embrun, Stittsville, Russell, Crysler, Moose Creek, Richmond, Metcalfe, Manotick, Vernon, Morewood, Finch, Monkland, Osgoode, Avonmore, N. Gower, Newington

Copyright by Rand McNally & Co.

d — TORONTO

Orangeville, Nobleton, King City, Caledon, Bolton, RICHMOND HILL, MARKHAM, Alton, Inglewood, Vaughan, Hillsburgh, Erin, Bramalea, Snelgrove, BRAMPTON, Georgetown, Acton, Norval, Rockwood, GUELPH, Streetsville, MISSISSAUGA, TORONTO, Milton, Port Credit, OAKVILLE, LAKE ONTARIO, Sheffield, Freelton, Waterdown, BURLINGTON, St. George, Lynden, Dundas, Niagara-on-the-Lake, Youngstown, Hamilton Hbr., HAMILTON, Winona, Welland Canal, BRANTFORD, Cainsville, Mt. Hope, Stoney Creek, Grimsby, Lincoln, ST. CATHARINES, Lewiston, Thorold, NEW YORK

Copyright by Rand McNally & Co.

e — CALGARY

Ghost Lake, Bow, Balzac, Kathryn, Keoma, STONY IND. RES., Cochrane, McDonald L., Morley, Delacour, Dalroy, Conrich, Lyalta, CALGARY, Chestermere L., Bragg Creek, SARCEE IND. RES., Shepard, Langdon, Priddis, Indus, Bragg Creek, Lloyd L., Dalemead, Elbow, Jumpingpound, Fish Cr.

Copyright by Rand McNally & Co.

f — WINNIPEG

Delta Beach, Argyle, Stonewall, Warren, Reaburn, Marquette, Grosse Isle, Stony Mountain, Lockport, Poplar Point, Meadows, Gonor, High Bluff, St. Eustache, Pigeon Lake, PORTAGE LA PRAIRIE, Fortier, Rosser, Gordon, Birds Hill, Newton, Oakville, Elie, St. François Xavier, WINNIPEG, Dacotah, Springstein, Prairie Grove, Grande Pointe, Fannystelle, Oak Bluff, Culross, Starbuck, Sanford, La Salle, St. Adolphe

RELIEF

Meters		Feet
3 050		10 000
1 525		5 000
610		2 000
305		1 000
152.5		500
	Sea Level	0
152.5		500

A-520055-76 -7 -83

Copyright by Rand McNally & Co.

g — EDMONTON

ALEXANDER IND. RES., Morinville, Bruderheim, Cardiff, Rivière Qui Barre, Carbondale, Duagh, Fort Saskatchewan, Josephburg, Calahoo, Namao, Villeneuve, St. Albert, Oliver, Cannell, Bremner, EDMONTON, Clover Bar, Sherwood Park, ELK ISLAND NAT'L PARK, Stony Plain, Spruce Grove, STONY PLAIN IND. RES., Ardrossan, Uncas, N. Cooking Lake, Devon, Hercules, Ellerslie, Cooking Lake, Looma, Ministik L., Nisku, Beaumont, Buford, Calmar, Leduc, New Sarepta, Saunders L., Miquelon L.

Copyright by Rand McNally & Co.

Scale 1:1 000 000; One inch to 16 miles.
Elevations and depressions are given in feet.

Miles: 0 2 4 6 8 10 12 14 16 18 20 22 24
Kilometers: 0 4 8 12 16 20 24 28 32 36 40

64

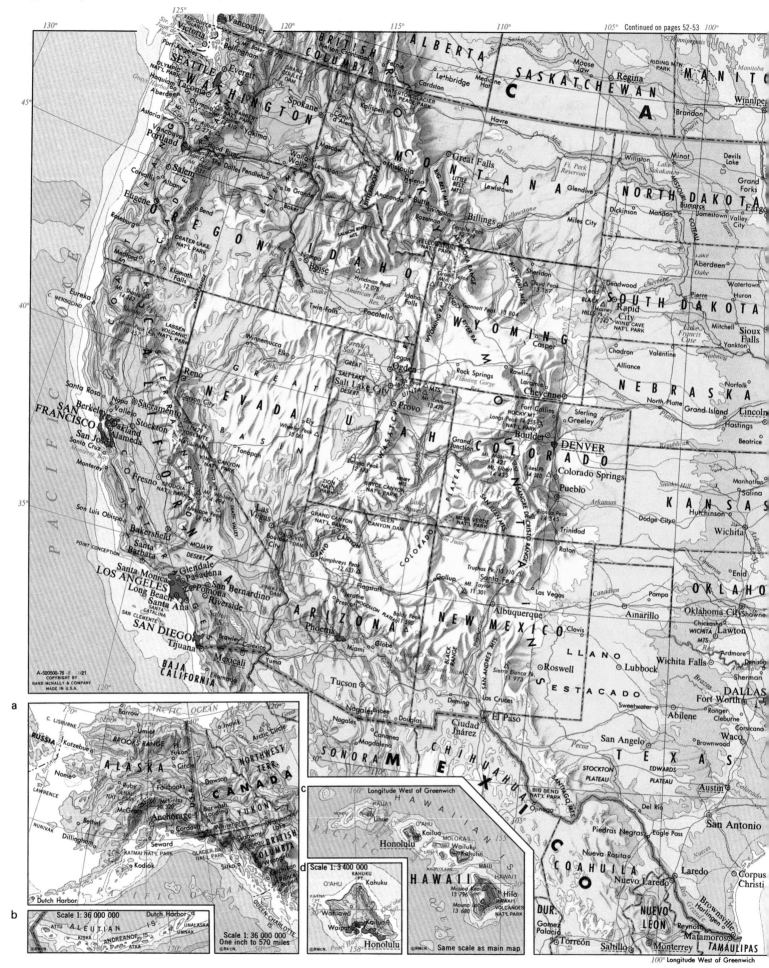

Continued on pages 52-53

Scale 1:12 000 000; one inch to 190 miles. Polyconic Projection
Elevations and depressions are given in feet

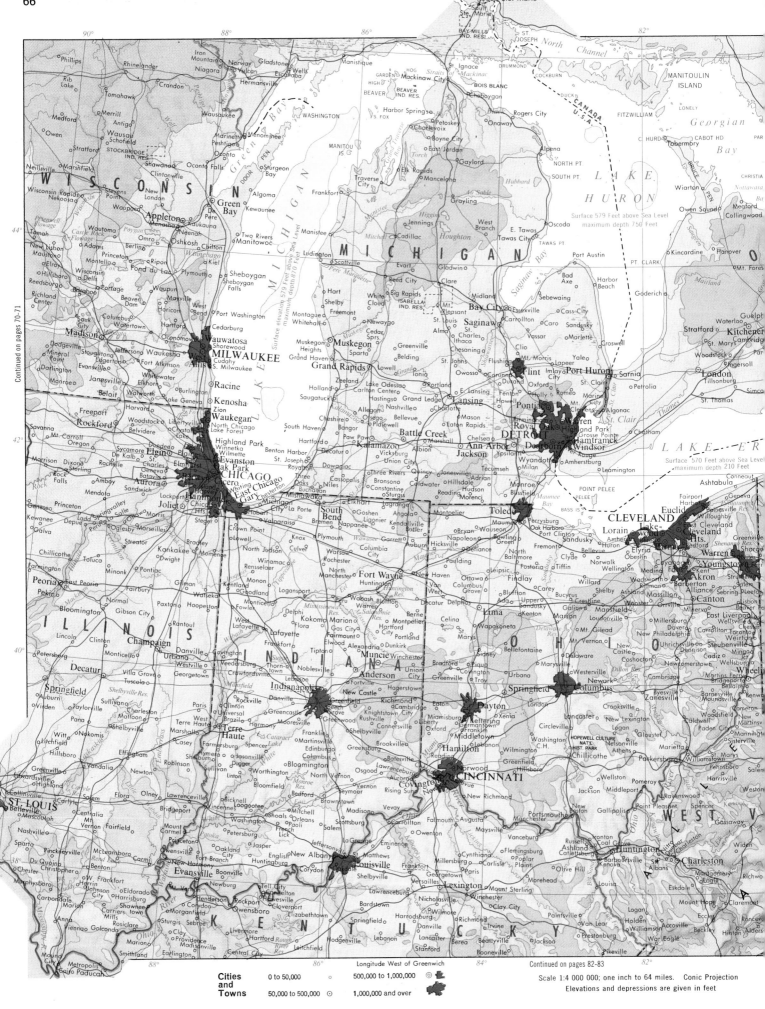

Continued on pages 70-71

Continued on pages 82-83

Cities and Towns		
0 to 50,000		500,000 to 1,000,000
50,000 to 500,000 ⊙		1,000,000 and over

Longitude West of Greenwich

Scale 1:4 000 000; one inch to 64 miles. Conic Projection
Elevations and depressions are given in feet

Continued on pages 58-59

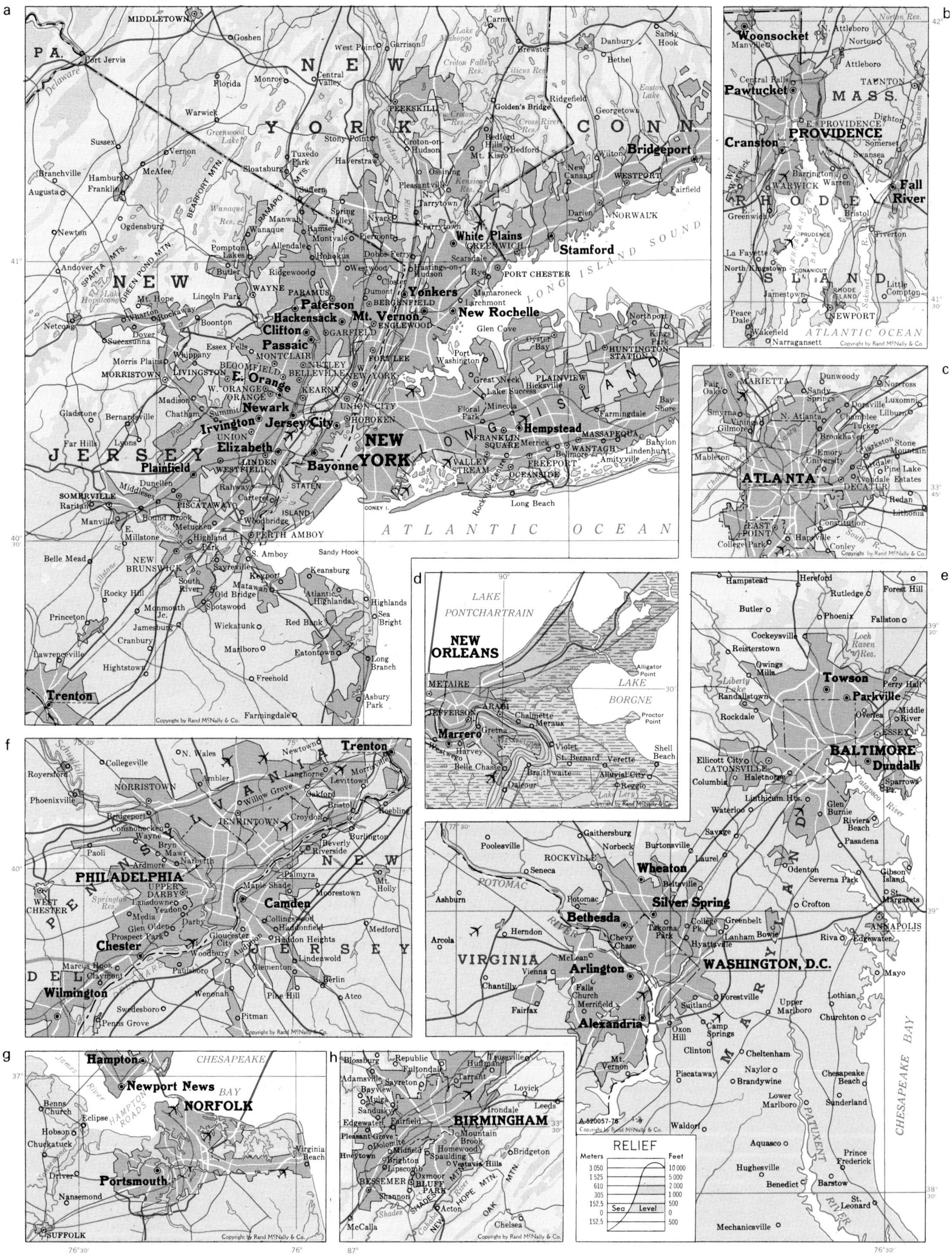

a

PA.
MIDDLETOWN
Goshen
Port Jervis
Florida
Monroe
Central Valley
West Point
Garrison
Carmel
Danbury
Sandy Hook
Bethel

NEW YORK
Warwick
Sussex
Vernon
McAfee
Tuxedo Park
Sloatsburg
Stony Point
Haverstraw
PEEKSKILL
Croton Falls Res.
Golden's Bridge
Ridgefield
Georgetown
Brewster

Branchville
Hamburg
Franklin
Augusta
Newton
Andover
Ogdensburg
SPARTA MTS.
GREEN POND MTN.
Mt. Hope
Netcong
Dover
Succasunna
Morris Plains
MORRISTOWN
Gladstone
Bernardsville
Madison
Chatham
Summit
Far Hills
Lyons
SOMERVILLE
Raritan
Manville
Belle Mead
Rocky Hill
Princeton
Lawrenceville
Hightstown
Trenton

Suffern
Ramsey
Montvale
Allendale
Hohokus
Ridgewood
WAYNE
PARAMUS
BERGENFIELD
Dumont
Closter
Westwood
Hillsdale

Spring Valley
Nyack
Piermont
Dobbs Ferry
Hastings-on-Hudson
Tarrytown
White Plains
GREENWICH
Scarsdale
Rye
PORT CHESTER
Stamford

Yonkers
Mt. Vernon
New Rochelle
Larchmont
Mamaroneck
Glen Cove
Oyster Bay
Port Washington
Great Neck
Hicksville
PLAINVIEW
Lake Success
Mineola
Farmingdale
HUNTINGTON STATION
Northport
Kings Park

Paterson
Hackensack
Clifton
Passaic
MONTCLAIR
NUTLEY
BELLEVILLE
BLOOMFIELD
E. Orange
W. ORANGE
ORANGE
KEARNY
FORT LEE
ENGLEWOOD
GARFIELD

Newark
Irvington
Jersey City
HOBOKEN
UNION CITY
NEW YORK
NEW YORK

Elizabeth
Bayonne
STATEN ISLAND
Hempstead
FRANKLIN SQUARE
MASSAPEQUA
Babylon
Amityville
Lindenhurst
Bellmore
Merrick
Freeport
Wantagh
OCEANSIDE
VALLEY STREAM
LONG ISLAND
Long Beach

Plainfield
LINDEN
WESTFIELD
Rahway
Carteret
Woodbridge
PERTH AMBOY
Metuchen
Highland Park
NEW BRUNSWICK
PISCATAWAY
Middlesex
Bound Brook
E. Millstone

ATLANTIC OCEAN
CONEY I.
Sandy Hook
Keansburg
Keyport
Matawan
Old Bridge
Spotswood
Sayreville
South River
S. Amboy
Atlantic Highlands
Highlands
Sea Bright
Red Bank
Wickatunk
Marlboro
Eatontown
Long Branch
Asbury Park
Freehold
Cranbury
Jamesburg
Monmouth Jc.
Farmingdale

b
N. Attleboro
Norton
Woonsocket
Manville
Attleboro
Central Falls
MASS.
TAUNTON
Pawtucket
Dighton
E. PROVIDENCE
Somerset
Swansea
PROVIDENCE
Cranston
Barrington
Warren
Bristol
Fall River
WARWICK
E. Greenwich
RHODE
Tiverton
La Fayette
North Kingstown
CONANICUT
Jamestown
PRUDENCE
RHODE ISLAND
Little Compton
Peace Dale
Wakefield
Newport
Narragansett
ATLANTIC OCEAN

c
MARIETTA
Dunwoody
Norcross
Fair Oaks
Sandy Springs
Luxomni
Smyrna
Vinings
Gilmore
N. Atlanta
Roswell
Doraville
Chamblee
Lilburn
Tucker
Mableton
Emory University
Scottdale
Clarkston
Stone Mountain
Pine Lake
ATLANTA
DECATUR
Avondale Estates
Redan
Lithonia
Mableton
Constitution
EAST POINT
Hapville
Conley
College Park

d
LAKE PONTCHARTRAIN
NEW ORLEANS
METAIRE
Alligator Point
LAKE BORGNE
JEFFERSON
ARABI
Chalmette
Meraux
Proctor Point
Marrero
Gretna
Westwego
Harvey
Violet
St. Bernard
Verette
Shell Beach
Belle Chasse
Braithwaite
Oakuer
Alluvial City
Reggio
Lake Lery

e
Hampstead
Hereford
Forest Hill
Butler
Rutledge
Phoenix
Fallston
Cockeysville
Loch Raven Res.
Reisterstown
Owings Mills
Towson
Perry Hall
Randallstown
Parkville
Overlea
Middle River
Rockdale
ESSEX
Ellicott City
CATONSVILLE
Linthicum Hts.
Glen Burnie
Riviera Beach
Columbia
Halethorpe
BALTIMORE
Dundalk
Waterloo
Sparrows Pt.
Savage
Gibson Island
St. Margarets
Odenton
Severna Park
ANNAPOLIS
Riva
Edgewater
Mayo

f
Collegeville
N. Wales
Newtown
Trenton
Royersford
NORRISTOWN
Ambler
Langhorne
Levittown
Morrisville
Phoenixville
Bridgeport
Conshohocken
Willow Grove
Oakford
Bristol
Wayne
Bryn Mawr
Jenkintown
Croydon
Roebling
Paoli
Narberth
Beverly
Burlington
Ardmore
Riverside
PHILADELPHIA
UPPER DARBY
Palmyra
Maple Shade
Mt. Holly
Lansdowne
Yeadon
Camden
Moorestown
WEST CHESTER
Springfield Res.
Media
Glen Olden
Darby
Collingswood
Haddonfield
Medford
Chester
Prospect Park
Haddon Heights
Lindenwold
Marcus Hook
Woodbury
Clementon
Pine Hill
Atco
Claymont
Paulsboro
Berlin
Wilmington
Swedesboro
Wenonah
Pitman
Penns Grove

g
Hampton
CHESAPEAKE
Newport News
BAY
NORFOLK
Benns Church
Eclipse
HAMPTON ROADS
Hobson
Chuckatuck
Virginia Beach
Driver
Portsmouth
Nansemond
SUFFOLK

h
Blossburg
Republic
Trussville
Adamsville
Sayreton
Fultondale
Huffman
Bayview
Mulga
Tarrant
Loxiey
Sandusky
Irondale
Leeds
Edgewater
Fairfield
BIRMINGHAM
Pleasant Grove
Mountain Brook
Dolomite
Homewood
Bridgeton
Hueytown
Midfield
Spaulding
Brighton
Vestavia Hills
BESSEMER
Oxmoor
BLUFF PARK
Shannon
Acton
Chelsea
McCalla
Shades

WASHINGTON, D.C.
Gaithersburg
Burtonsville
Poolesville
Norbeck
Laurel
ROCKVILLE
Seneca
Beltsville
Wheaton
Potomac
POTOMAC
Ashburn
Odenton
Severna Park
Crofton
Herndon
Silver Spring
College Pk.
Greenbelt
Lanham Bowie
Bethesda
Takoma Pk.
Hyattsville
Arcola
Chevy Chase
McLean
Arlington
Vienna
Falls Church
Merrifield
Chantilly
Alexandria
Suitland
Forestville
Upper Marlboro
Lothian
Fairfax
Camp Springs
Churchton
Oxon Hill
Clinton
Cheltenham
VIRGINIA
Mt. Vernon
Naylor
Brandywine
Chesapeake Beach
Piscataway
Lower Marlboro
Sunderland
Waldorf
Aquasco
Prince Frederick
Hughesville
Benedict
Barstow
Mechanicsville
St. Leonard
CHESAPEAKE BAY
PATUXENT RIVER
MARYLAND

RELIEF
Meters		Feet
3 050		10 000
1 525		5 000
610		2 000
305		1 000
152.5		500
0	Sea Level	0
152.5		500

A-520057-76
Copyright by Rand McNally & Co.

Scale 1:1 000 000; One inch to 16 miles.
Elevations and depressions are given in feet.

Miles 0 2 4 6 8 10 12 14 16 18 20 22 24
Kilometers 0 4 8 12 16 20 24 28 32 36 40

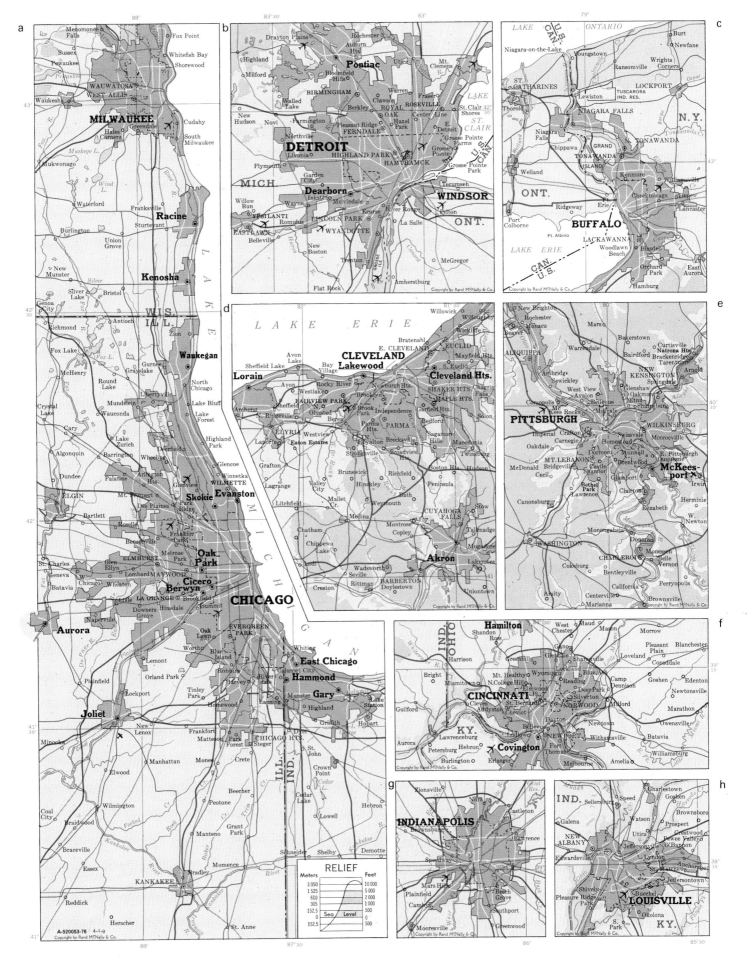

a

Menomonee Falls
Sussex
Pewaukee
Fox Point
Whitefish Bay
Shorewood
WAUWATOSA
WEST ALLIS
Waukesha
MILWAUKEE
Greendale
Hales Corners
Cudahy
South Milwaukee
Muskego L.
Wind L.
Mukwonago
Waterford
Franksville
Burlington
Sturtevant
Racine
Union Grove
New Munster
Silver Lake
Bristol
Genoa City
Kenosha

b

Drayton Plains
Highland
Auburn Hts.
Rochester
Utica
Mt. Clemens
Milford
Pontiac
Bloomfield Hills
Walled Lake
BIRMINGHAM
Berkley
Clawson
Warren
Fraser
ROSEVILLE
New Hudson
Novi
Farmington
Northville
Pleasant Ridge
FERNDALE
ROYAL OAK
Hazel Park
Center Line
E. Detroit
St. Clair Shores
LAKE ST. CLAIR
Plymouth
Livonia
DETROIT
HIGHLAND PARK
HAMTRAMCK
Grosse Pointe Farms
Grosse Pointe
Grosse Pointe Park
Garden City
Dearborn
Inkster
Melvindale
WINDSOR
Tecumseh
Willow Run
YPSILANTI
Wayne
LINCOLN PARK
Ecorse
River Rouge
La Salle
ONT.
EASTLAWN
Belleville
WYANDOTTE
New Boston
Trenton
Amherstburg
McGregor
Flat Rock
MICH.

c

LAKE ONTARIO
Niagara-on-the-Lake
Burt
Newfane
Youngstown
Ransomville
Wrights Corners
ST. CATHARINES
Lewiston
LOCKPORT
TUSCARORA IND. RES.
Thorold
NIAGARA FALLS
N.Y.
Niagara Falls
Chippawa
GRAND ISLAND
N. TONAWANDA
Welland
TONAWANDA
Kenmore
Williamsville
ONT.
Ridgeway
Erie
Cheektowaga
Depew
Port Colborne
Lancaster
Pt. Albino
BUFFALO
LACKAWANNA
Woodlawn Beach
CAN.-U.S.
Orchard Park
Blasdell
East Aurora
LAKE ERIE
Hamburg

d

LAKE ERIE
Willowick
Avon Lake
Bratenahl
E. CLEVELAND
Wickliffe
Willoughby
Sheffield Lake
Bay Village
CLEVELAND
EUCLID
Mayfield Hts.
Lorain
Avon
Rocky River
Lakewood
Newburgh Hts.
S. Euclid
Cleveland Hts.
Amherst
Sheffield
FAIRVIEW PARK
Brooklyn
SHAKER HTS.
Chagrin Falls
N. Ridgeville
Olmsted
Brook Park
Independence
MAPLE HTS.
ELYRIA
Westview
Berea
Parma Hts.
Garfield Hts.
Lanofee
Eaton Estates
PARMA
Bedford
Solon
Grafton
Royalton
Breckville
Sagamore Hills
Brunswick
Broadview
Macedonia
Lagrange
Valley City
Hinckley
Richfield
Boston Hts.
Hudson
Twinsburg
Litchfield
Mallet Cr.
Weymouth
Bath
Stow
Chatham
Medina
Montrose
Copley
CUYAHOGA FALLS
Talmadge
Chippewa Lake
Lodi
Wadsworth
Akron
Lakemore
Creston
Seville
Rittman
BARBERTON
Doylestown
Uniontown

e

New Brighton
Rochester
Monaca
Mars
Bakerstown
Curtisville
Natrona Hts.
Beaver
Warrendale
Brackenridge
Tarentum
ALIQUIPPA
Bairdford
NEW KENSINGTON
Ambridge
Sewickley
West View
Glenshaw
Springdale
Arnold
Coraopolis
Avalon
Bellevue
Oakmont
Sharpsburg
Mc Kees Rocks
Etna
Imperial
Crafton
Millvale
Oakdale
Carnegie
PITTSBURGH
Swissvale
WILKINSBURG
Monroeville
McDonald
Dormont
Homestead
E. Pittsburgh
Cecil
Bridgeville
MT. LEBANON
Castle Shannon
Munhall
Duquesne
McKees-port
Irwin
Canonsburg
Bethel Park
Lawrence
Clairton
Glassport
Hermine
Elizabeth
W. Newton
Monongahela
Donora
WASHINGTON
Mogadore
Charleroi
Monessen
Belle Vernon
Cokeburg
Bentleyville
Amity
Centerville
California
Perryopolis
Marianna
Brownsville

f

Hamilton
West Chester
Maud
Morrow
Shandon
Ross
Mason
IND.
OHIO
Harrison
Greenhills
Gano
Glendale
Sharonville
Loveland
Pleasant Plain
Blanchester
Bright
Miamitown
Mt. Healthy
Wyoming
Blue Ash
Cozaddale
Mt. Healthy
N.College Hill
Lockland
Camp Dennison
Goshen
Edenton
Elmwood
Reading
Newtonsville
CINCINNATI
St. Bernard
Deer Park
Silverton
Milford
Cleves
Addyston
NORWOOD
Marathon
Cheviot
Dayton
Guilford
Lawrenceburg
Ludlow
Bellevue
NEWPORT
Newtown
Owensville
Aurora
Petersburg
Hebron
Covington
Fort Thomas
Williamsburg
Burlington
Erlanger
Melbourne
Amelia
KY.

g

Zionsville
Crist Res.
Nora
Sellersburg
Castleton
INDIANAPOLIS
Brownsburg
Lawrence
Speedway
Mars Hills
Beech Grove
Plainfield
Camby
Southport
Mooresville
Greenwood

h

IND.
Charlestown
Goshen
Speed
Sellersburg
Galena
Watson
Brownsboro
Utica
Prospect
NEW ALBANY
Jeffersonville
Crestwood
Pewee Valley
O'Bannon
Edwardsville
Lyndon
Anchorage
St. Matthews
Jeffersontown
Shively
Buechel
Pleasure Ridge
LOUISVILLE
Park
Okolona
KY.

RELIEF

Meters		Feet
3 050		10 000
1 525		5 000
610		2 000
305		1 000
152.5		500
0	Sea Level	0
152.5		500

Miles: 0 2 4 6 8 10 12 14 16 18 20 22 24
Kilometers: 0 4 8 12 16 20 24 28 32 36 40

Scale 1:1 000 000; One inch to 16 miles.
Elevations and depressions are given in feet.

A-520053-76 4-4-9
Copyright by Rand McNally & Co.

Continued on pages 56-57

Continued on pages 72-73

Continued on pages 78-79

Longitude West of Greenwich

A-511005-76-6-9-8-15
COPYRIGHT BY
RAND McNALLY & COMPANY
MADE IN U.S.A.

Cities and Towns

0 to 50,000 ∘	500,000 to 1,000,000 ◉
50,000 to 500,000 ⊙	1,000,000 and over

Scale 1:4 000 000; one inch to 64 miles. Conic Projection
Elevations and depressions are given in feet

DENVER

Continued on pages 66-67

Continued on pages 78-79

Relief

Meters	Feet
1525	5000
610	2000
305	1000
152.5	500
0 Sea Level	0
152.5	500

LAKE SUPERIOR
Surface elev. 600 Feet above Sea Level
Maximum depth 1333 Feet

LAKE MICHIGAN
Surface elevation 579 Feet above Sea Level
Maximum depth 870 Feet

Continued on pages 54-55

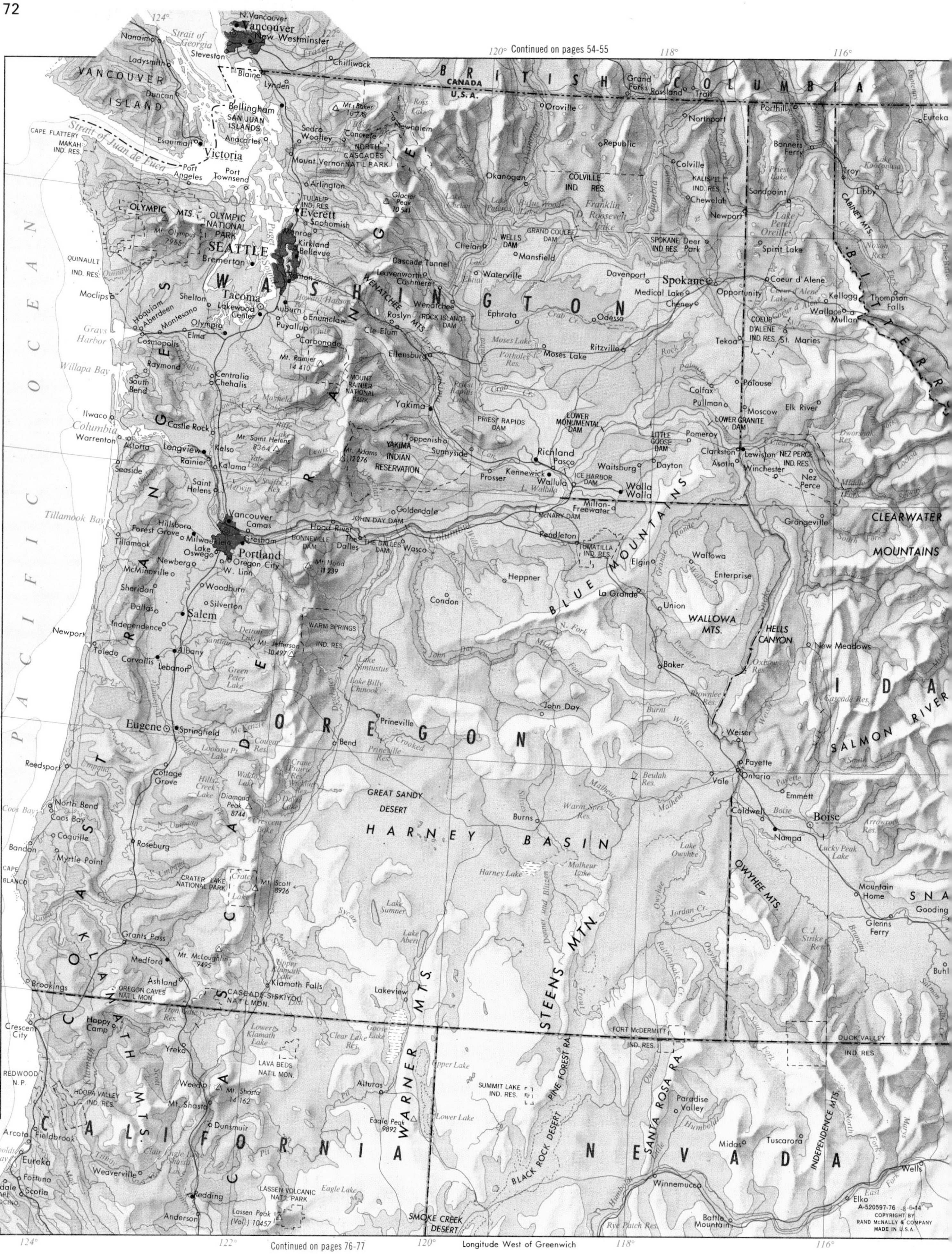

BRITISH COLUMBIA
CANADA
U.S.A.

VANCOUVER ISLAND

WASHINGTON

OREGON

CALIFORNIA

NEVADA

IDA

PACIFIC OCEAN

Strait of Georgia
Strait of Juan de Fuca

Vancouver
New Westminster
Victoria
Bellingham
Seattle
Tacoma
Everett
Olympia
Spokane
Portland
Salem
Eugene
Boise

CASCADE RANGE
COAST RANGE
KLAMATH MTS.
BLUE MOUNTAINS
WALLOWA MTS.
CLEARWATER MOUNTAINS
STEENS MTN.
OWYHEE MTS.
WARNER MTS.
SANTA ROSA RA.
INDEPENDENCE MTS.

OLYMPIC NATIONAL PARK
MOUNT RAINIER NATIONAL PARK
NORTH CASCADES NAT'L PARK
CRATER LAKE NATIONAL PARK
REDWOOD N.P.

GREAT SANDY DESERT
HARNEY BASIN
BLACK ROCK DESERT
SMOKE CREEK DESERT

Continued on pages 76-77   Longitude West of Greenwich

A-520597-76
COPYRIGHT BY
RAND McNALLY & COMPANY
MADE IN U.S.A.

Scale 1: 4,000,000; one inch to 64 miles. Conic Projection
Elevations and depressions are given in feet

114° 112° Continued on pages 56-57 110° 108° 106°

ALBERTA SASKATCHEWAN
CANADA
U.S.A.

WATERTON-GLACIER
INTERNATIONAL
PEACE PARK

BLACKFEET IND. RES.
Cut Bank
Sunburst
Shelby
Browning
Valier
Conrad
Whitefish
Kalispell
Polson
FLATHEAD INDIAN RESERVATION
NATIONAL BISON RANGE
Ronan
Missoula
Lolo
Stevensville
Hamilton
Philipsburg
Anaconda
Deer Lodge
Helena
East Helena
Townsend
Walkerville
Butte
Three Forks
Twin Bridges
BIG HOLE NAT'L BATTLEFIELD
PIONEER MTS.
Homer Youngs Peak 10 621
Dillon
Salmon
LEMHI RANGE
BEAVERHEAD MTS.
LOST RIVER RA.
Borah Pk. 12 662
Mackay
Boulder Peak 10 981
Hyndman Peak 12 009
Arco
CRATERS OF THE MOON NAT'L MON.
St. Anthony
Ashton
Rexburg
Rigby
Idaho Falls
Shelley
SNAKE RIVER PLAIN
Bailey
Shoshone
Jerome
Rupert
Twin Falls
Burley
Oakley
Blackfoot
FORT HALL IND. RES.
Pocatello
American Falls Res.
American Falls
Soda Springs
Lava Hot Spgs.
Meade Peak 9957
Afton
Montpelier
Malad City
Preston
Lewiston
Richmond
Smithfield
Logan
Providence
Garland
Wellsville
Brigham
Huntsville
Morgan
Ogden
Farmington
Bountiful
Salt Lake City
Murray
Midvale
Tooele
Park City
Heber City
Wendover
GREAT SALT LAKE DESERT
Great Salt Lake
Lucin
UTAH
WASATCH RANGE
BEAR RIVER RANGE
Bear Lake
Kemmerer
Evanston
UINTA MTS.
Kings Peak 13 528
Mt. Emmons 13 440
UINTAH AND OURAY IND. RES.
DINOSAUR NAT'L MON.
Vernal
COLO.
UINTAH RANGE

MONTANA
Great Falls
Belt
LITTLE BELT MTS.
Neihart
White Sulphur Spgs.
Choteau
Fort Benton
Winifred
Lewistown
Winnett
Harlowton
Roundup
CRAZY MTS.
Big Timber
Bozeman
Livingston
Columbus
Laurel
Billings
Huntley
Hardin
Forsyth
Colstrip
Red Lodge
Bear Creek
Granite Peak 12 799
Gardiner
Electric Peak 10 992
ABSAROKA RANGE
Mammoth Hot Springs
Mt. Washburn 10 243
YELLOWSTONE NATIONAL PARK
7733 ft above sea level
Shoshone Lake
Jackson Lake
GRAND TETON NAT'L PARK
Grand Teton 13 770
Fremont Peak 13 745
Gannett Peak 13 804
WIND RIVER RANGE
WYOMING RANGE
WIND RIVER IND. RES.
WIND RIVER
Riverton
Lander
Shoshoni
Thermopolis
Gebo
Worland
Ten Sleep
Greybull
Basin
Cody
Powell
Lovell
BIGHORN MOUNTAINS
Sheridan
Buffalo
Kaycee
Midwest
Powder River
Casper
Douglas
Orin
Glenrock
WYOMING
GREAT DIVIDE BASIN
Superior
Rawlins
Wheatland
Hanna
Seminoe Res.
Pathfinder Res.
Alcova Res.
Granger
Green River
Rock Springs
Flaming Gorge Res.
Fontenelle Res.
Craig
Steamboat Spgs.
Oak Creek
PARK RANGE

Sunburst
Chinook
Havre
Harlem
Malta
Hogeland
Opheim
Scobey
Plentywood
Grenora
Williston
FORT PECK IND. RES.
Glasgow
Ft. Peck
Wolf Point
Poplar
Sidney
Brockway
Glendive
Beach
Terry
Miles City
Baker
Marmarth
Forsyth
N. DAK.
Fort Peck Lake
Medicine Lake
Missouri
CROW IND. RES.
Crow Agency
LITTLE BIGHORN BATTLEFIELD NAT'L MON.
Lame Deer
NORTHERN CHEYENNE IND. RES.
DEVILS TOWER NAT'L MON.
Syndance
Moorcroft
Gillette

FT. BELKNAP IND. RES.
ROCKY BOYS IND. RES.

Morgan

Relief

Meters		Feet
3050		10000
1525		5000
610		2000
305		1000
152.5		500
Sea Level		0
1525		500

0 20 40 60 80 100 120 Miles
0 20 40 60 80 100 120 140 160 180 200 Kilometers

Continued on pages 76-77
Continued on pages 70-71

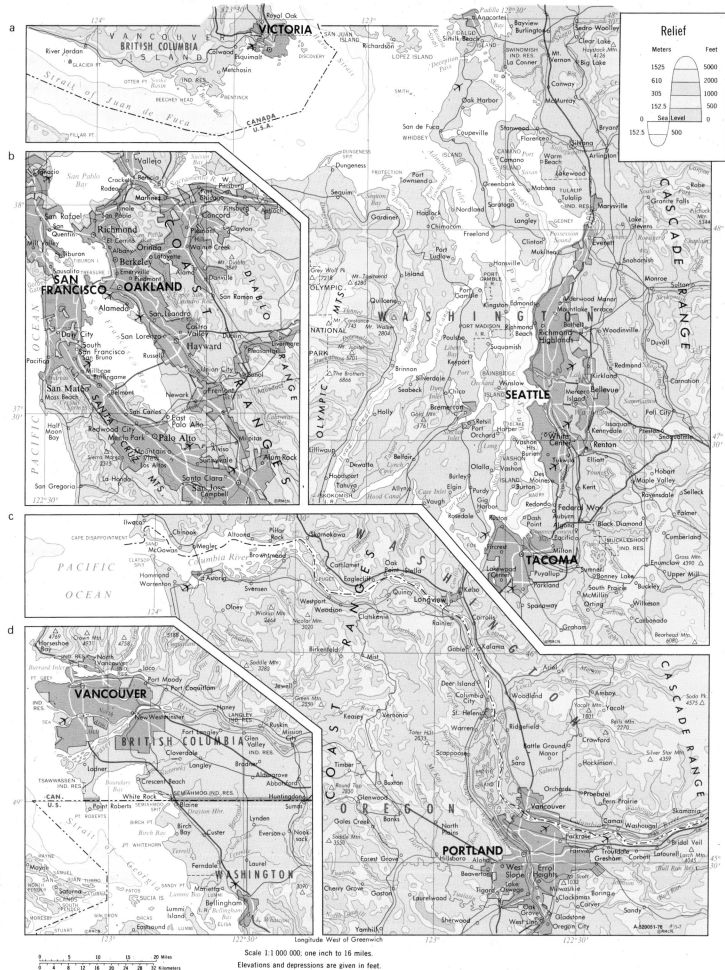

Relief

Meters	Feet
1525	5000
610	2000
305	1000
152.5	500
0 Sea Level	0
152.5	500

Scale 1:1 000 000; one inch to 16 miles.
Elevations and depressions are given in feet.

Longitude West of Greenwich

0 5 10 15 20 Miles
0 4 8 12 16 20 24 28 32 Kilometers

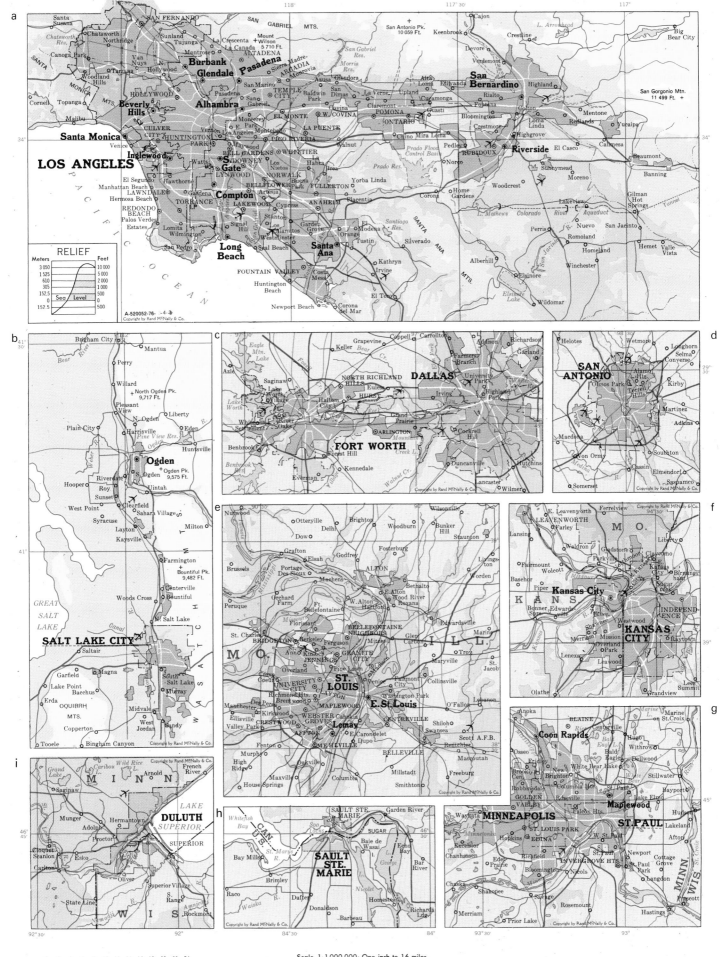

a

Santa Susana • Santa Susana
Chatsworth
Chatsworth Res.
Northridge
Canoga Park
Woodland Hills
Tarzana
Cornell
Topanga
Malibu
SAN FERNANDO
Sunland
Tujunga
Van Nuys
Hollywood
Burbank
Glendale
Pasadena
SAN GABRIEL MTS.
La Crescenta
La Canada
ALTADENA
Mount Wilson 5 710 Ft.
Sierra Madre
ARCADIA
Monrovia
Azusa
Glendora
Alta Loma
San Antonio Pk. 10 059 Ft. +
Keenbrook
Cajon
Crestline
L. Arrowhead
Devore
Verdemont
Big Bear City

HOLLYWOOD
Beverly Hills
Alhambra
San Marino
TEMPLE CITY
Baldwin Park
La Verne
Upland
Cucamonga
Claremont
Pomona
ONTARIO
Guasti
San Bernardino
Rialto
Fontana
Colton
Bloomington
Loma Linda
Highland
Redlands
Mentone
Yucaipa
San Gorgonio Mtn. 11 499 Ft. +

Santa Monica
Venice
CULVER CITY
HUNTINGTON PARK
Vernon
E. Los Angeles
Montebello
PICO RIVERA
W. COVINA
EL MONTE
Monterey Park
San Gabriel
Chino
Mira Loma
Pedley
RUBIDOUX
Riverside
El Casco
Calimesa
Beaumont
Banning

Los Angeles
Inglewood
Watts
S. Gate
Downey
Maywood
BELL GARDENS
WHITTIER
NORWALK
Habra
Brea
Yorba Linda
Buena Park
Corona
Home Gardens
Woodcrest
Sunnymead
Moreno
Gilman Hot Springs
Tunnel

El Segundo
Manhattan Beach
LAWNDALE
Hermosa Beach
Gardena
Hawthorne
Compton
LYNWOOD
BELLFLOWER
Cypress
FULLERTON
Placentia
ANAHEIM
Lakeview
Perris
Nuevo
San Jacinto
Romoland
Hemet
Valle Vista
Winchester

REDONDO BEACH
Palos Verdes Estates
San Pedro
TORRANCE
LAKEWOOD
Stanton
Garden Grove
Westminster
Los Alamitos
Seal Beach
Orange
Tustin
El Modena
Santiago Res.
Silverado
SANTA ANA MTS.
Colorado River Aqueduct
L. Mathews
Elsinore
Alberhill
Homeland
Wildomar

Lomita
Wilmington
Signal Hill
Long Beach
Santa Ana
Kathryn
Irvine
El Toro
Corona del Mar
Bel Mar
Elsinore Lake

FOUNTAIN VALLEY
Huntington Beach
Costa Mesa
Newport Beach

PACIFIC OCEAN
SANTA MONICA MTS.

RELIEF
Meters — Feet
3 050 — 10 000
1 525 — 5 000
610 — 2 000
305 — 1 000
152.5 — 500
0 — Sea Level
152.5 — 500

A-520052-76- -4-3
Copyright by Rand McNally & Co.

b
41°30'
Brigham City
Mantua
Bear River
Perry
Willard
North Ogden Pk. 9,717 Ft. +
Liberty
Plain City
Pleasant View
N. Ogden
Eden
Harrisville
Pine View Res.
Huntsville
Ogden
Ogden Pk. 9,575 Ft. +
Hooper
Riverdale
S. Ogden
Roy
Sunset
Uintah
West Point
Clearfield
Sahara Village
Milton
Syracuse
Layton
Kaysville
41°
Farmington
Bountiful Pk. 9,482 Ft. +
Centerville
Woods Cross
Bountiful
GREAT SALT LAKE
Canal
N. Salt Lake
SALT LAKE CITY
Saltair
Garfield
Magna
Lake Point
Bacchus
South Salt Lake
Murray
Erda
OQUIRRH MTS.
Midvale
West Jordan
Sandy
Copperton
Tooele
Bingham Canyon
WASATCH
Weber R.
Jordan R.
Copyright by Rand McNally & Co.

c
32°30'
Eagle Mtn. Lake
Grapevine
Coppell
Carrollton
Addison
Richardson
Wetmore
97°
Keller
Bear Cr.
Farmers Branch
Garland
Azle
Saginaw
NORTH RICHLAND HILLS
Euless
University Park
Highland Park
DALLAS
Lake Worth
Haltom City
HURST
Irving
White Settlement
River Oaks
Grand Prairie
FORT WORTH
ARLINGTON
Cockrell Hill
Benbrook
Mountain Creek L.
Forest Hill
Duncanville
Hutchins
32°
Benbrook Res.
Kennedale
Everman
Lancaster
Wilmer
Walnut Cr.
Copyright by Rand McNally & Co.

d
98°30'
Helotes
SAN ANTONIO
Alamo Hts.
Terrell Hills
Kirby
Olmos Park
29°30'
Selma
Converse
Longhorn
Martinez
Macdona
Von Ormy
Adkins
Cassin
Elmendort
Somerset
Southton
Saspamco
Medina R.
Copyright by Rand McNally & Co.

e
90°30'
Nutwood
Otterville
Delhi
Brighton
Woodburn
Wilsonville
90°
Dow
Grafton
Godfrey
Fosterburg
Bunker Hill
Staunton
Livingston
Brussels
Elsah
ALTON
Bethalto
Worden
Portage Des Sioux
Machens
W. Alton
E. Alton
Wood River
Roxana
Peruque
Orchard Farm
Ft. Bellefontaine
Hartford
Edwardsville
St. Charles
Florissant
BELLEFONTAINE NEIGHBORS
Glen Carbon
Marine
Bridgeton
Berkeley
Ferguson
Mitchell
Troy
MO.
JENNINGS
GRANITE CITY
Maryville
Creve Coeur
Kinloch
Pine Lawn
Venice
St. Jacob
UNIVERSITY CITY
Richmond Hts.
Clayton
ST. LOUIS
Fairmont Washington Park
Collinsville
Lebanon
Manchester
Des Peres
Brentwood
Maplewood
E. ST. LOUIS
O'Fallon
Ellisville
Kirkwood
WEBSTER GROVES
Cahokia
Shiloh
Valley Park
CRESTWOOD
AFFTON
Lemay
Carondelet
Dupo
Swansea
Scott A.F.B.
Rentchler
BELLEVILLE
Fenton
MEHLVILLE
Murphy
Oakville
Mascoutah
38°
High Ridge
Columbia
Freeburg
Maxville
Millstadt
House Springs
Smithton
Copyright by Rand McNally & Co.

f
94°30'
E. Leavenworth
Ferrelview
Copyright by Rand McNally & Co.
LEAVENWORTH
Farley
Liberty
Lansing
Gladstone
Claycomo
MO.
Fairmount
Wolcott
Parkville
Kansas City
Birmingham
Basehor
Piper
Bonner Springs
Edwardsville
Kansas City
Sugar Cr.
INDEPENDENCE
KANS.
Merriam
Mission
Westwood
KANSAS CITY
Raytown
Lenexa
Overland Park
Leawood
Lees Summit
Olathe
Grandview
39°

g
93°30'
Anoka
BLAINE
Marine on St. Croix
Marine
Osseo
Coon Rapids
Centerville
Hugo
Withrow
Fridley
New Brighton
White Bear Lake
Dellwood
Brooklyn Park
Robbinsdale
Columbia Hts.
Roseville
GOLDEN VALLEY
Bald Eagle
Stillwater
45°
Wayzata
MINNEAPOLIS
St. Louis Park
St. Paul
Falcon Hts.
Lake Elmo
Maplewood
Hopkins
EDINA
W. St. Paul
ST. PAUL
Minnetonka
Excelsior
Chanhassen
Richfield
INVER GROVE HTS.
St. Paul Park
Eden Prairie
Bloomington
Nicols
Newport
Cottage Grove
Chaska
Shakopee
Savage
Rosemount
Prior Lake
Merriam
Hastings
MINN.
WIS.
St. Croix R.
Copyright by Rand McNally & Co.

i
92°30'
Grand Lake
Caribou
Wild Rice L.
French River
MINN.
92°
Saginaw
Arnold
Munger
Hermantown
DULUTH
LAKE SUPERIOR
46°45'
Adolph
Proctor
SUPERIOR
Cloquet
Scanlon
Esko
Superior Village
S. Range
Carlton
Oliver
State Line
Rockmont
WIS.

h
Whitefish Bay
SAULT STE. MARIE
Garden River
CAN.
U.S.
Soo Locks
SUGAR
46°30'
St. Marys R.
Bay Mills
Baie de Wasai
Echo Bay
SAULT STE. MARIE
L. George
Bar River
Brimley
Raco
Nicolet
Dafter
Waiska R.
84°30'
Donaldson
Homestead
Richards Ldg.
Barbeau
84°
Copyright by Rand McNally & Co.

Scale 1:1 000 000; One inch to 16 miles.
Elevations and depressions are given in feet.

Miles
0 2 4 6 8 10 12 14 16 18 20 22 24
Kilometers
0 4 8 12 16 20 24 28 32 36 40

76

Continued on pages 72-73

Scale 1:4 000 000; one inch to 64 miles. Conic Projection
Elevations and depressions are given in feet

Longitude West of Greenwich

a

SAN DIEGO

Scale 1:1 000 000

A-520599-76 -8-a22
COPYRIGHT BY
RAND McNALLY & COMPANY
MADE IN U.S.A.
®RMCN.

Relief

Meters	Feet
3050	10000
1525	5000
610	2000
305	1000
152.5	500
0	Sea Level
152.5	500 Below Sea Level
1525	5000
3050	10000

Continued on pages 78-79
Continued on pages 80-81

114° 110° 108°

38° 36° 34° 32°

112°

GREAT SALT LAKE DESERT

Great Salt Lake

Salt Lake City
Murray · Park City
Tooele · Midvale · Heber City
West Jordan
Lehi
American Fork
Orem · Provo
Springville
Spanish Fork
Payson
Eureka
Nephi
Fairview
Mount Pleasant
Moroni
Ephraim
Manti
Delta
Gunnison
Salina
Fillmore
Richfield
Monroe
Milford
Delano Pk. 12 169
Beaver
Parowan
Panguitch
Escalante
Cedar City
CEDAR BREAKS NATL. MON.
BRYCE CANYON NATL. PARK
GRAND STAIRCASE-ESCALANTE NATL. MON.
ZION NATL. PARK
Hurricane
Saint George
Kanab
Pioche
Caliente
Little Salt Lake

WASATCH PLAT.
TIMPANOGOS CAVE N.M.
UTAH
UINTAH AND OURAY IND. RES.
Vernal
Roosevelt
Duchesne
WEST TAVAPUTS PLATEAU
Helper
Price
Hiawatha
Sunnyside
EAST TAVAPUTS PLATEAU
Castle Dale
Green River
CAPITOL REEF NATL. PARK
Mt. Ellen 11 522
HENRY MTS.
Abajo Pk. 11 360
Blanding
NATURAL BRIDGES NATL. MON.
GLEN CANYON NATL. RECR. AREA
RAINBOW BRIDGE NATL. MON.
Mexican Hat
Lake Powell
Page
INSCRIPTION HOUSE RUIN
KEET SEEL RUIN
BETATAKIN RUIN
NAVAJO NATL. MON.
BLACK MESA
NAVAJO HOPI JOINT USE AREA
CANYON DE CHELLY NATL. MON.

UINTAH AND OURAY IND. RES.
West
White
Meeker
Oak Creek
Bond
Rifle
Glenwood Springs
ROCKY
Leadville
Mt. Massive 14 421
Aspen
Mt. Elbert 14 433
Castle Pk. 14 265
Plata Pk. 14 361
Mt. Harvard 14 420
Buena Vista
Cripple Creek
Canon City
Grand Junction
COLORADO NATL. MON.
Fruita
Delta
Paonia
Crested Butte
Gunnison
Salida
Montrose
BLACK CANYON OF THE GUNNISON NATL. PARK
Morrow Point Res.
Blue Mesa Res.
COLORADO
UNCOMPAHGRE PLATEAU
Mt. Sneffels 14 150
Ouray
Uncompahgre Pk. 14 309
SAN JUAN MTS.
Telluride
Silverton
Summit Peak 13 300
Durango
Pagosa Springs
Del Norte
Monte Vista
Alamosa
Blanca Pk. 14 345
GREAT SAND DUNES N.M.
Saguache
SANGRE DE CRISTO MTS.
Antonito
CANYONLANDS NATL. PARK
La Sal
Monticello
CANYONS OF THE ANCIENTS NATL. MON.
HOVENWEEP NATL. MON.
MESA VERDE NATL. PARK
Cortez
Bluff
SOUTHERN UTE INDIAN RES.
UTE MTN. IND. RES.
AZTEC RUINS NATL. MON.
Aztec
Farmington
JICARILLA APACHE
El Vado Res.
Wheeler Pk. 13 161
Taos
INDIAN
Abiquiu Res.
RESERVATION
SANTA CLARA IND. RES.
Truchas Pk. 13 101
Los Alamos
JEMEZ IND. RES.
BANDELIER NATL. MON.
Santa Fe
SANTO DOMINGO IND. RES.
SAN FELIPE IND. RES.
Galisteo
ZIA IND. RES.
CHACO CANYON NATL. MON.
CHACO CULTURE NATL. HIST. PARK
Gallup
Bernalillo
SANDIA IND. RES.
CANONCITO IND. RES.
Albuquerque
LAGUNA IND. RES.
Mt. Taylor 11 301
ZUNI
ZUNI MTS.
Sandees
ACOMA IND. RES.
LAGUNA IND. RES.
ISLETA IND. RES.
EL MORRO NATL. MON.
IND. RES.
Belen
ALAMO IND. RES.
NEW MEXICO
PETRIFIED FOREST NATL. PARK
Saint Johns
SALINAS NATL. MON.
Magdalena
Socorro
San Marcial
Carrizozo
Elephant Butte
Truth or Consequences
Sierra Blanca Peak 11 973
MESCALERO APACHE IND. RES.
BLACK RANGE
Tularosa
Alamogordo
SAN ANDRES MTS.
WHITE SANDS NATL. MON.
Caballo Res.
Silver City
Bayard
GILA CLIFF DWELLINGS NATL. MON.
Glenwood
Morenci
Clifton
Safford
PELONCILLO MTS.
Lordsburg
Deming
Las Cruces
Mesilla
FLORIDA MTS.
Columbus
Playas Lake
Franklin Mtn. 7 192
EL PASO
TEXAS
Ciudad Juárez
Yleta
CHIHUAHUA

SNAKE RA.
GREAT BASIN NATL. PARK
Wheeler Peak 13 061
GOSHUTE IND. RES.
Sevier Lake

COLORADO PLATEAU
COLORADO PLATEAUS
Moab
Mt. Peale 12 721
ARCHES NATL. PARK
GLEN CANYON NATL. RECR. AREA

Mt. Bangs 8 012
UINKARET PLATEAU
PIPE SPRING NATL. MON.
KAIBAB IND. RES.
GRAND CANYON-PARASHANT NATL. MON.
KANAB PLATEAU
SHIVWITS PLATEAU
Lake Mead
LAKE MEAD NATL. RECR. AREA
KAIBAB PLATEAU
GRAND CANYON NATIONAL PARK
MARBLE CANYON
HAVASUPAI IND. RES.
Grand Canyon
HUALAPAI IND. RES.
Chloride
Kingman
Oatman
HUALAPAI MTS.
Topock
Lake Havasu
Lake Havasu City
PARKER DAM
Bill Williams
COLORADO RIVER IND. RES.
Quartzsite
Yuma

NAVAJO INDIAN RESERVATION
PAINTED DESERT
Moenkopi
HOPI INDIAN RESERVATION
Moenkopi
COCONINO PLATEAU
WUPATKI NATL. MON.
SUNSET CRATER N.M.
Humphreys Pk. 12 633
Ash Fork
Williams
Flagstaff
WALNUT CANYON NATL. MON.
Winslow
Holbrook
CHUSKA MTS.
NAVAJO INDIAN RESERVATION
APACHE
INDIAN

AGUA FRIA NATL. MON.
Clarkdale
Jerome
TUZIGOOT N.M.
MONTEZUMA CASTLE NATL. MON.
Prescott
MOGOLLON RIM
ARIZONA
Wickenburg
Theodore Roosevelt Lake
THEODORE ROOSEVELT DAM
TONTO NATL. MON.
SALT RIVER IND. RES.
Glendale
Phoenix
Tempe
Mesa
Miami
Globe
Superior
GILA RIVER IND. RES.
Florence
CASA GRANDE RUINS NATL. MON.
Gila Bend
Casa Grande
Painted Rock Res.
IRONWOOD FOREST NATL. MON.
Ajo
ORGAN PIPE CACTUS N.M.
TOHONO O'ODHAM INDIAN RESERVATION
San Manuel
Tucson
SAGUARO NATL. PARK
SAN XAVIER IND. RES.
Benson
TUMACACORI NATL. MON.
Nogales
Fort Huachuca
Bisbee
Lowell
Pirtleville
Douglas
Tombstone
Willcox
Willcox Playa Lake
CHIRICAHUA NATL. MON.
McNary
Springerville
Mt. Ord 11 357
Boldy Peak 11 403
FORT APACHE INDIAN RESERVATION
Maverick
SAN CARLOS INDIAN RESERVATION
San Carlos Lake
Hayden
GRAN DESIERTO
SONORA
USA MEXICO

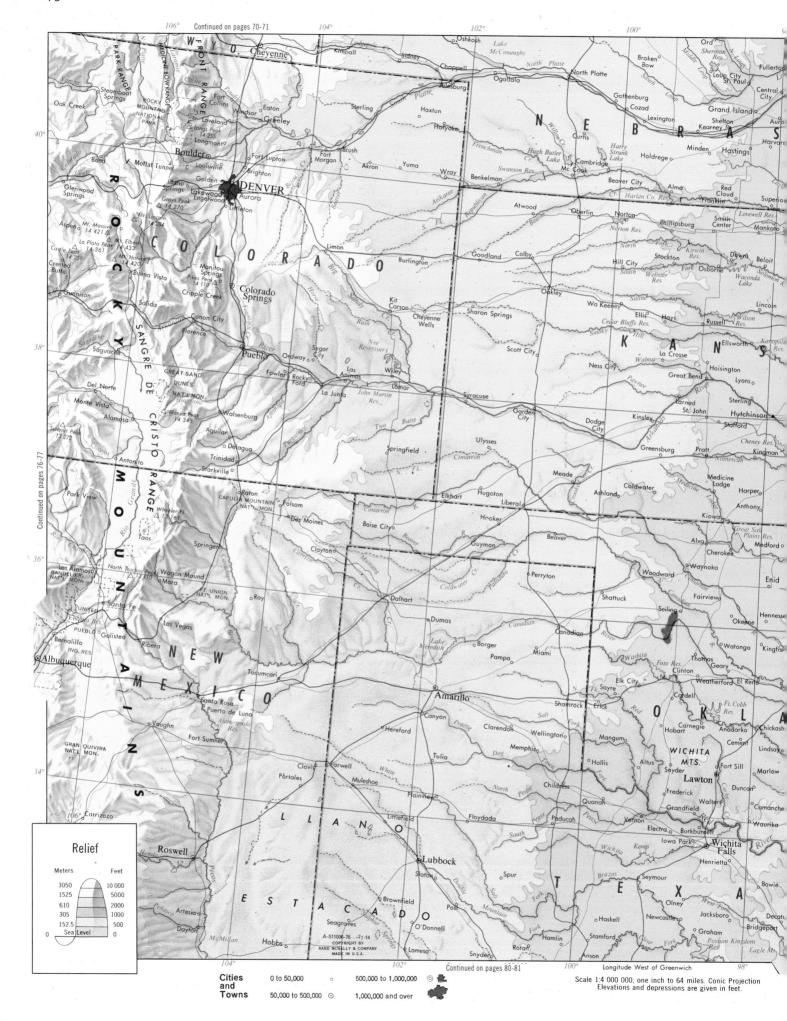

Continued on pages 70-71

106°   104°   102°   100°

W Y O.

Cheyenne

FRONT RANGE
PARK RANGE
MEDICINE BOW RANGE
ROCKY MOUNTAIN NATIONAL PARK
Steamboat Springs
Yampa
Oak Creek
Bond
Glenwood Springs
Aspen
Mt. Massive 14 421
Mt. Lincoln 14 284
La Plata Peak 14 361
Mt. Elbert 14 433
Castle Peak 14 259
Mt. Harvard 14 420
Crested Butte
Gunnison
Buena Vista
Pikes Peak 14 110
Manitou Springs
Cripple Creek
Salida
Saguache
Monte Vista
Del Norte
Alamosa
Summit Peak 13 272
Blanca Peak 14 345
Great Sand Dunes NAT'L MON.
Huerfano
Walsenburg
Aguilar
Delagua
Trinidad
Starkville
Park View
Antonito
Rio Grande
Chama
Taos
North Truchas Peak 13 110
Wheeler Pk. 13 161
Wagon Mound
Mora
Springer
FT. UNION NAT'L MON.
Roy
Los Alamos
BANDELIER NAT'L MON.
JEMEZ IND. RES.
Santa Fe
Galisteo
Bernalillo
PUEBLO IND. RES.
Albuquerque
Las Vegas
Ribera
Tucumcari

Longs Peak 14 255
Grays Peak 14 270
Fort Collins
Windsor
Loveland
Longmont
Boulder
Louisville
Golden
Idaho Springs
Moffat Tunnel
Eaton
Greeley
Brighton
DENVER
Lakewood
Englewood
Aurora
Littleton
Fort Lupton
Fort Morgan
Brush
Akron
Yuma
Limon
Burlington
Ordway
Sugar City
Las Animas
Rocky Ford
Fowler
La Junta
Lamar
Wiley
John Martin Res.
Two Butte
Springfield
Elkhart
Clayton
Folsom
Des Moines
Boise City
CAPULIN MOUNTAIN NAT'L MON.
Raton

ROCKY MOUNTAINS

S A N G R E   D E   C R I S T O   R A N G E

C O L O R A D O

40°

38°

36°

34°

N E W   M E X I C O

Carrizozo
GRAN QUIVIRA NAT'L MON.
Vaughn
Fort Sumner
Santa Rosa
Puerto de Luna
Alamogordo Res.
Clovis
Portales
Farwell
Muleshoe

N E B R A S K A

Oshkosh
Lake McConaughy
Kimball
Sidney
Chappell
North Platte
Broken Bow
Ord
Sherman Res.
Loup City
St. Paul
Grand Island
Fullerton
Central City
Aurora
Harvard
Hastings
Shelton
Kearney
Minden
Gothenburg
Cozad
Lexington
Ogallala
Julesburg
Sterling
Haxtun
Holyoke
Wray
Benkelman
Curtis
Hugh Butler Lake
Harry Strunk Lake
Cambridge
Mc Cook
Holdrege
Alma
Red Cloud
Franklin
Superior
Beaver City
Harlan Co. Res.

K A N S A S

Atwood
Oberlin
Norton
Phillipsburg
Smith Center
Mankato
Lovewell Res.
Goodland
Colby
Hill City
Stockton
Osborne
Downs
Beloit
Kit Carson
Cheyenne Wells
Sharon Springs
Scott City
Oakley
Wa Keeney
Ellis
Hays
Russell
Lincoln
Ness City
Great Bend
La Crosse
Hoisington
Lyons
Ellsworth
Syracuse
Garden City
Larned
St. John
Sterling
Hutchinson
Dodge City
Kinsley
Pratt
Greensburg
Stafford
Cheney Res.
Kingman
Ulysses
Cimarron
Meade
Ashland
Coldwater
Medicine Lodge
Harper
Anthony
Kiowa
Hugoton
Liberal
Hooker
Guymon
Beaver
Great Salt Plains Res.
Alva
Cherokee
Medford
Waynoka
Woodward
Fairview
Enid
Perryton
Shattuck
Seiling
Okeene
Hennessey
Kingfisher
Watonga
Thomas
Geary
Clinton
Weatherford
El Reno
Cordell
Elk City
Sayre
Erick
Shamrock
Wellington
Mangum
Hobart
Carnegie
Anadarko
Cement
Chickasha
Lindsay
Hollis
Altus
Snyder
Fort Sill
Lawton
WICHITA MTS.
Marlow
Duncan
Frederick
Walters
Comanche
Grandfield
Vernon
Electra
Burkburnett
Iowa Park
Wichita Falls
Henrietta
Bowie
Decatur
Bridgeport
Possum Kingdom Res.
Eagle Mt.

O K L A H O M A

T E X A S

Dalhart
Dumas
Canadian
Borger
Pampa
Miami
Canadian
Lake Meredith
Amarillo
Canyon
Hereford
Clarendon
Memphis
Tulia
Childress
Quanah
Paducah
Floydada
Plainview
Littlefield
Lubbock
Slaton
Spur
Post
Brownfield
Seagraves
O'Donnell
Lamesa
Snyder
Rotan
Hamlin
Stamford
Anson
Seymour
Graham
Olney
Jacksboro
Newcastle

L L A N O   E S T A C A D O

Roswell
Artesia
Dayton
McMillan
Hobbs
Pecos
Hondo

Continued on pages 76-77

106°   104°   102°   100°   98°

Longitude West of Greenwich

Continued on pages 80-81

Relief

Meters		Feet
3050		10 000
1525		5000
610		2000
305		1000
152.5		500
0	Sea Level	0

A-511000-76- -27-14
COPYRIGHT BY
RAND McNALLY & COMPANY
MADE IN U.S.A.

Cities
and
Towns

0 to 50,000   o   500,000 to 1,000,000   ◎
50,000 to 500,000   ⊙   1,000,000 and over

Scale 1:4 000 000; one inch to 64 miles. Conic Projection
Elevations and depressions are given in feet.

Continued on pages 70-71 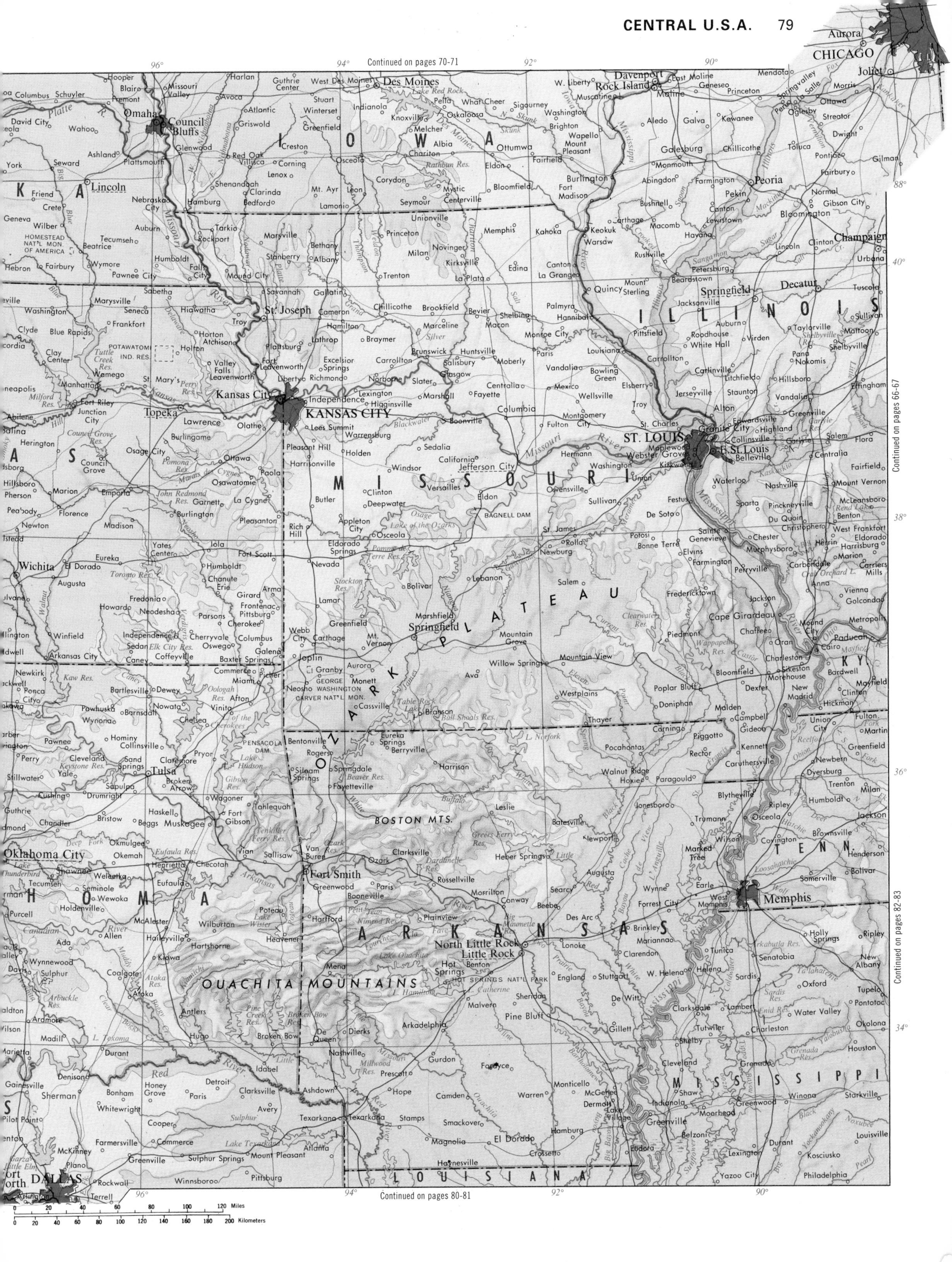
Continued on pages 66-67
Continued on pages 82-83
Continued on pages 80-81

Aurora
CHICAGO
Joliet

96° 94° 92° 90° 88°
40°
38°
36°
34°

IOWA

ILLINOIS

MISSOURI

KANSAS

OKLAHOMA

ARKANSAS

TENN.

MISSISSIPPI

KY.

LOUISIANA

OZARK PLATEAU

BOSTON MTS.

OUACHITA MOUNTAINS

Omaha
Council Bluffs
Des Moines
Davenport
Rock Island
East Moline
Moline
Lincoln
Topeka
Kansas City
KANSAS CITY
St. Joseph
ST. LOUIS
E. St. Louis
Springfield
Decatur
Champaign
Peoria
Bloomington
Wichita
Tulsa
Oklahoma City
Springfield
Fort Smith
North Little Rock
Little Rock
Hot Springs
Memphis
West Memphis
Cape Girardeau
Cairo
Paducah
Jefferson City
Columbia
North DALLAS

HOMESTEAD NAT'L MON. OF AMERICA
POTAWATOMI IND. RES.
GEORGE WASHINGTON CARVER NAT'L MON.
HOT SPRINGS NAT'L PARK
BAGNELL DAM
PENSACOLA DAM

Lake of the Ozarks
Bull Shoals Res.
Table Rock Lake
L. Norfork
Lake Texoma

Missouri R.
Mississippi River
Arkansas River
Red River

0 20 40 60 80 100 120 Miles
0 20 40 60 80 100 120 140 160 180 200 Kilometers

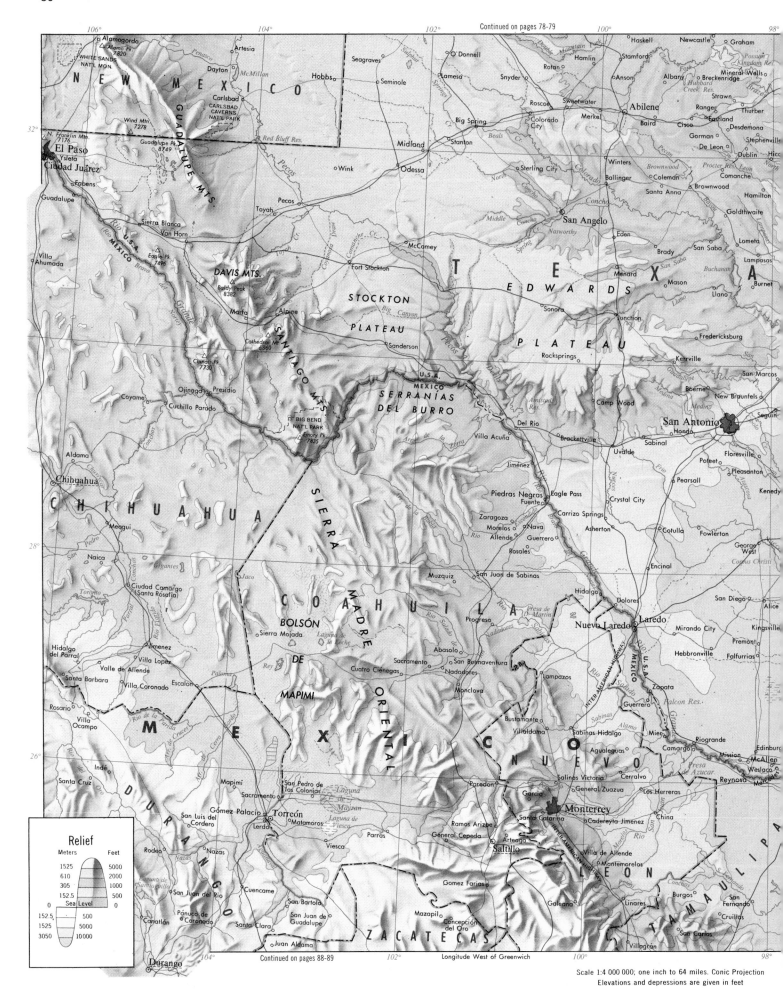

80

Continued on pages 78-79

NEW MEXICO

Alamogordo
White Sands Nat'l Mon.
Alamo Pk. 7820
Artesia
Dayton
McMillan
Hobbs
O'Donnell
Seagraves
Seminole
Lamesa
Snyder
Rotan
Hamlin
Stamford
Haskell
Newcastle
Graham

Wind Mtn. 7278
Carlsbad
Carlsbad Caverns Nat'l Park
Red Bluff Res.

N. Franklin Mtn. 7176
Guadalupe Pk. 8749
Roscoe
Sweetwater
Colorado City
Big Spring
Stanton
Merkel
Baird
Abilene
Ranger
Cisco
Eastland
Desdemona
Thurber
Dublin
Hico

El Paso
Ysleta
Ciudad Juárez
Fabens
Guadalupe
Villa Ahumada

GUADALUPE MTS.

Toyah
Pecos
Wink
Odessa
Midland
Sterling City
Gorman
De Leon
Winters
Ballinger
Coleman
Santa Anna
Brownwood
Comanche
Hamilton
Goldthwaite
Stephenville

Sierra Blanca
Van Horn
Eagle Pk. 7496

Sonora
Eden
Brady
San Saba
Lometa
Lampasas
Buchanan

DAVIS MTS.
Baldy Peak 8382
Marfa
Alpine

San Angelo
Menard
Mason
Llano
Burnet

Fort Stockton
McCamey

TEXAS
EDWARDS
PLATEAU

Cathedral Mt. 6860
SANTIAGO MTS.
Chinati Pk. 7730

STOCKTON
PLATEAU
Big Canyon
Sanderson
Rockspring
Junction
Kerrville
Fredericksburg

Ojinaga
Presidio
Coyame
Cuchillo Parado

U.S.A.
MEXICO
SERRANIAS
DEL BURRO

Del Rio
Camp Wood
Boerne
San Marcos
New Braunfels

CHIHUAHUA
Chihuahua
Aldama
Meoqui

BIG BEND NAT'L PARK
Emory Pk. 7835

Villa Acuña
Jiménez

Brackettville
Uvalde
Sabinal
San Antonio
Hondo
Seguin
Floresville
Poteet
Pleasanton
Kenedy

Piedras Negras
Fuente
Eagle Pass
Crystal City
Pearsall

Zaragoza
Morelos
Nava
Allende
Guerrero
Rosales

Carrizo Springs
Asherton
Cotulla
Fowlerton

SIERRA

Naica
Gigantes
Jaco

Muzquiz
San Juan de Sabinas

Encinal
George West
Corpus Christi

Ciudad Camargo (Santa Rosalia)

COAHUILA
MADRE

Hidalgo
Dolores
San Diego
Alice

Jimenez
Villa Lopez
Sierra Mojada
BOLSÓN
Laguna de la Leche

Progreso
Nadadores
Presa de D. Martin
Nuevo Laredo
Laredo
Mirando City
Kingsville
Premont

Hidalgo del Parral
Santa Barbara
Valle de Allende
Villa Coronado
Escalon

DE
ORIENTAL
Abasolo
Sacramento
San Buenaventura
Nadadores
Cuatro Ciénegas
Monclova

Lampazos
Bustamante
Villaldama
Sabinas Hidalgo

Guerrero
Zapata
Mier
Riogrande
Camargo

Hebbronville
Falfurrias

Rosario
Villa Ocampo
Indé

MEXICO
MAPIMI
Rey

Aguaguas

NUEVO

Mission
McAllen
Weslaco
Reynosa

Santa Cruz
Mapimi
Sacramento
San Pedro de las Colonias
Laguna de Mayran

Paredon
Salinas Victoria
General Zuazua
Cerralvo
Los Herreras
China

Gómez Palacio
Lerdo
Torreón
Matamoros
Laguna de Viesca
Viesca
Parras

Garcia
Ramos Arizpe
Santa Catarina
Monterrey
Cadereyta Jimenez

Edinburg

DURANGO
San Luis del Cordero
Rodeo
Nazas

General Cepeda
Arteaga
Saltillo
Villa de Allende
Montemorelos

LEON

Burgos
San Fernando

Santa Cruz
San Juan del Rio
Cuencame
San Bartolo
Gomez Farias
Galeana
Linares

TAMAULIPAS

Pánuco de Coronado
Canatlán
Santa Clara
San Juan de Guadalupe
Mazapil
Concepción del Oro

Durango
Juan Aldama
ZACATECAS
Villagran
San Carlos
Cruillas

Continued on pages 88-89

Longitude West of Greenwich

Relief
Meters | Feet
1525 | 5000
610 | 2000
305 | 1000
152.5 | 500
0 Sea Level 0
152.5 | 500
1525 | 5000
3050 | 10000

Scale 1:4 000 000; one inch to 64 miles. Conic Projection
Elevations and depressions are given in feet

Continued on pages 78-79

Continued on pages 82-83

GULF OF MEXICO

HOUSTON

GALVESTON BAY

EAST BAY

BOLIVAR PENINSULA

GALVESTON ISLAND

GULF OF MEXICO

Scale 1:1 000 000

0   5   10 Miles

0   4   8   12   16 Kilometers

a

0   20   40   60   80   100   120 Miles

0   20   40   60   80   100   120   160   180   200 Kilometers

Cities and Towns	0 to 50,000	500,000 to 1,000,000
	50,000 to 500,000	1,000,000 and over

82

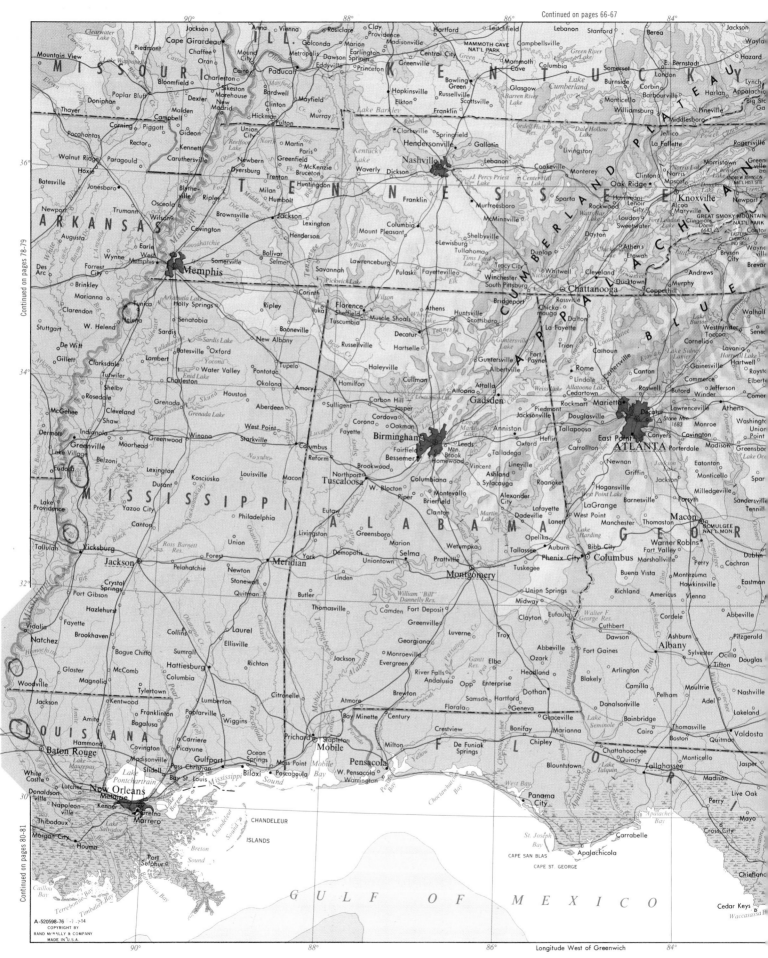

Continued on pages 66-67

Continued on pages 78-79

Continued on pages 80-81

Scale 1:4 000 000; one inch to 64 miles. Conic Projection
Elevations and depressions are given in feet

Longitude West of Greenwich

A-520598-76 -7 -7-14
COPYRIGHT BY
RAND McNALLY & COMPANY
MADE IN U.S.A.

Relief

Meters		Feet
1525		5000
610		2000
305		1000
152.5		500
0	Sea Level	0
152.5		500
1525		5000

Same scale as main map

a

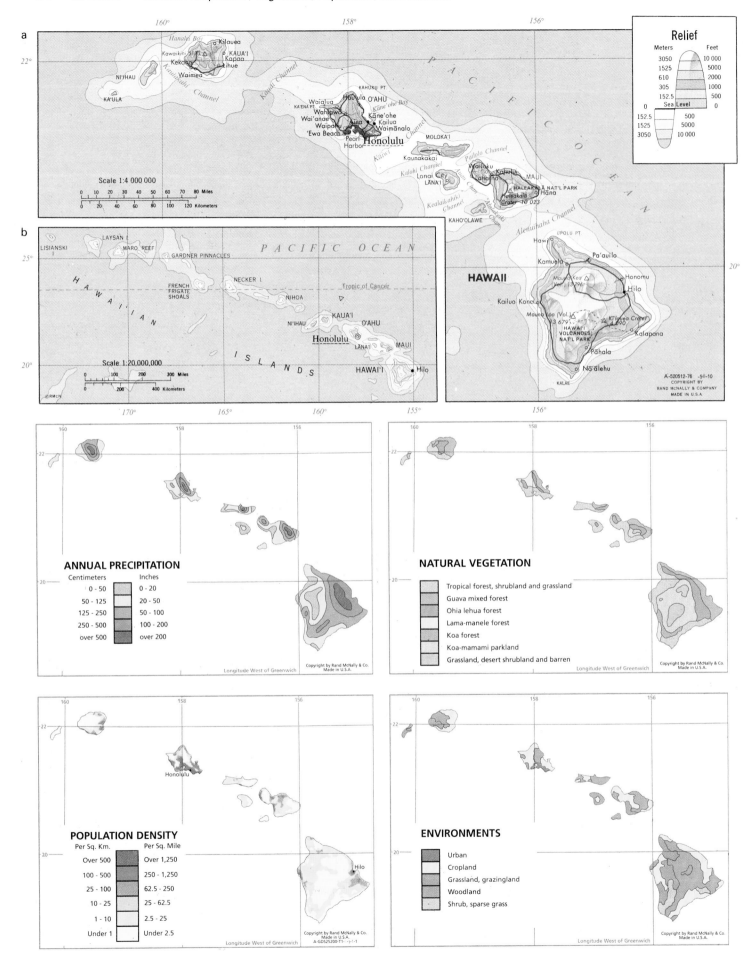

ANNUAL PRECIPITATION

Centimeters	Inches
0 - 50	0 - 20
50 - 125	20 - 50
125 - 250	50 - 100
250 - 500	100 - 200
over 500	over 200

NATURAL VEGETATION

Tropical forest, shrubland and grassland
Guava mixed forest
Ohia lehua forest
Lama-manele forest
Koa forest
Koa-mamami parkland
Grassland, desert shrubland and barren

POPULATION DENSITY

Per Sq. Km.	Per Sq. Mile
Over 500	Over 1,250
100 - 500	250 - 1,250
25 - 100	62.5 - 250
10 - 25	25 - 62.5
1 - 10	2.5 - 25
Under 1	Under 2.5

ENVIRONMENTS

Urban
Cropland
Grassland, grazingland
Woodland
Shrub, sparse grass

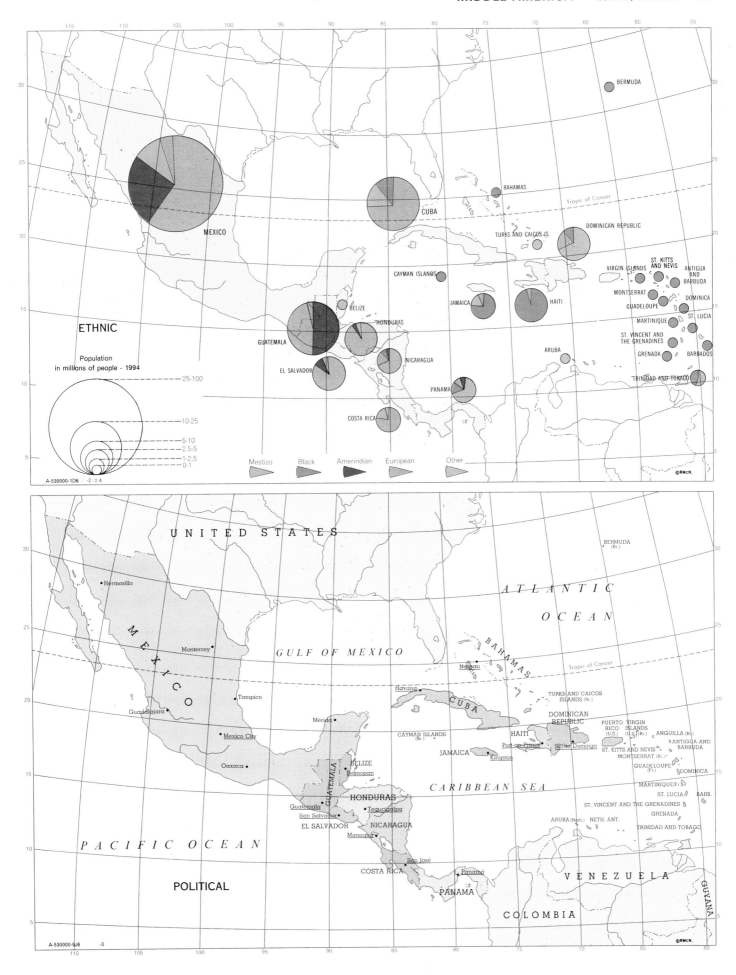

ETHNIC

Population
in millions of people - 1994

25-100

10-25

5-10

2.5-5

1-2.5

0-1

Mestizo   Black   Amerindian   European   Other

A-530000-1D6   -2 -2 -4

MEXICO

CUBA

BAHAMAS

Tropic of Cancer

TURKS AND CAICOS IS.

DOMINICAN REPUBLIC

CAYMAN ISLANDS

VIRGIN ISLANDS

ST. KITTS
AND NEVIS

ANTIGUA
AND
BARBUDA

JAMAICA

HAITI

MONTSERRAT

DOMINICA

GUADELOUPE

ST. LUCIA

BELIZE

HONDURAS

MARTINIQUE

GUATEMALA

ST. VINCENT AND
THE GRENADINES

ARUBA

GRENADA

BARBADOS

EL SALVADOR

NICARAGUA

TRINIDAD AND TOBAGO

COSTA RICA

PANAMA

BERMUDA

©RMCN.

UNITED STATES

BERMUDA
(Br.)

ATLANTIC

OCEAN

• Hermosillo

M
E
X
I
C
O

GULF OF MEXICO

BAHAMAS

Monterrey •

Tropic of Cancer

Nassau

• Tampico

TURKS AND CAICOS
ISLANDS (Br.)

Guadalajara •

Havana

CUBA

DOMINICAN
REPUBLIC

PUERTO
RICO
(U.S.)

VIRGIN
ISLANDS
(U.S.) (Br.)

ANGUILLA (Br.)

Mérida •

CAYMAN ISLANDS
(Br.)

HAITI

Santo Domingo

ANTIGUA AND
BARBUDA

Mexico City •

Port-au-Prince

ST. KITTS AND NEVIS

JAMAICA

MONTSERRAT (Br.)

Oaxaca •

BELIZE

Kingston

GUADELOUPE
(Fr.)

DOMINICA

Belmopan

CARIBBEAN SEA

MARTINIQUE(Fr.)

G
U
A
T
E
M
A
L
A

HONDURAS

ST. LUCIA

BARB.

Guatemala •

• Tegucigalpa

ST. VINCENT AND THE GRENADINES

San Salvador •

GRENADA

EL SALVADOR

NICARAGUA

ARUBA (Neth.)

NETH. ANT.

TRINIDAD AND TOBAGO

Managua •

P A C I F I C   O C E A N

COSTA RICA

San José •

V E N E Z U E L A

POLITICAL

• Panamá

G
U
Y
A
N
A

PANAMA

COLOMBIA

A-530000-SJ6   -3

©RMCN.

a

Scale 1:16 000 000; one inch to 250 miles. Polyconic Projection
Elevations and depressions are given in feet

Continued on pages 100-101

Continued on pages 80-81

106°   104°   102°   100°

24°

S I E R R A

DURANGO

San Dimas

Durango

Miguel Auza

Juan Aldama

Ascension

Hidalgo   Pilon

Jiménez   Abasolo

Aramberri

N U E V O

Padilla

El Salto

Nombre de Dios

Nieves

Gruñidora

L E O N

Zaragoza

Ciudad Victoria

Soto la Marina

Pánuco

Siqueros

Concordia

El Salto

Muleros

Sombrerete

10 100

Río Grande

Vanegas

Cedral

La Paz

Doctor Arroyo

Peña Nevada
13 300

Miquihuana

Jaumave

TAMAULIPA

Mier y
Noriega

Llera

S I E R R A

A Z U L

Pueblo
Nuevo

SIERRA DE NAYARIT

Mezquital

Chalchihuites

11 700

Catorce

Matehuala

La Paz

Xicotencatl

Aldama

Villa Unión

Santa María
de Ocotán

C. Pimol

Z A C A T E C A S

Fresnillo

Charcas

Venado

Moctezuma

S A N   L U I S

Guadalcázar

Ocampo

Magiscatzin

Gonzáles

Ciudad Mante

Rosario

Escuinapa

SINALOA

Valparaíso

Calera

Victor Rosales

Morelos

Ramos

Cerritos

Ciudad del Maiz

El Ebano

Panuco

Huejuquilla el Alto

Ciudad
García

Zacatecas

Troncoso

Zaragoza

P O T O S I

Tamuín

Huajicori

Acaponeta

Mezquitic

Huejúcar

Villanueva

Rincón
de Romos

Luis Moya

Ojocaliente

Salinas

Bocas

Peotillos

Soledad Díez Gutiérrez

Pozos

Ciudad Fernández

Ojo Caliente

Sta. María del Río

Pastora

Riverde

Rayón

Cárdenas

Ciudad de Valles

San Felipe

Monte
Escobedo

Sta. María
de los Angeles

Tepezala

Villa García

San Luis
Potosí

Zaragoza

Villa de
Reyes

San Diego
de la Unión

Villa Pedro
Montoya

Lagunillas

Arroyo Seco

Xilitla

General Pedro
Antonio Santos

Tabasco

Calvillo

AGUASCALIENTES

Villa
Hidalgo

Encarnación
de Díaz

San Felipe

San Luis
de la Paz

Jalpan

Tamazunchale

Tantoyu

Platón
Sánche

Tuxpan

NAYARIT

San Blas

Jalisco

Tepic

Ruiz

Bolaños

Chimaltitán

J (Tlaltenango)
Sánchez Román

Jalpa

Jiménez del Téul

M E X I C A N A

Villa
Hidalgo

Teocaltiche

San Juan
de los Lagos

Lagos de
Moreno

León

Dolores Hidalgo

G U A N A J U A T O

San José
Iturbide

Colón

Cadereyta

Jalpan

Zimapan

Jacala

Calnali

Huitzilingo

Huejutla

Chicontep

Sta. María del Oro

San Pedro
Lagunillas

Compostela

SIERRA
DE VALLEJO

Ahuacatlán
Ixtlán
del Río

Juchipila

García de la Cadena

Moyahua

Nochistlán

Mexticacán

Yahualica

Jalostotitlán

Unión de
San Antonio

San Miguel
el Alto

Tepatitlán de Morelos

San Francisco
del Rincón

La Luz

Silao

Guanajuato

San Miguel
de Allende

Comonfort

Querétaro

Cayetano Rubio

Q U E R E T A R O

Tolimán

Zacualtipan

Metztitlán

HIDALGO

Huichapan

Ixmiquilpan

Tasquillo

PUNTA
DE MITA

Bahía de
Banderas

CABO
CORRIENTES

O C C I D E N T A L

Puerto Vallarta

Talpa de Allende

Mascota

Ameca

Ahualulco

S. Martín
Hidalgo

Hostotipaquillo

Etzatlán

Tequila

Zapopan

Tlaquepaque

Tala

Tonalá

Cuquío

Arandas

Atotonilco el Alto
Ayo
el Chico

Ciudad Manuel
Doblado

Juventino
Rosas

Irapuato

Apaseo

San Juan
del Río

Tequisquiapan

Tecozautla

Actopan

Mineral del
Chica

Atotonilco el Grar

Huauchinang

Zapotlanejo

Cuerámaro

Salamanca

Cortazar

Celaya

Huasca

Mixquiahuala

Mineral del
Monte

Tulancingo

Cuautepec

Guadalajara

Tepeji del
Río

Zumpango

Tezontepec

Chignahuapar

20°

SA. DEL CUALE

PTA. FARALLON

Cihuatlán

Tomatlan

COLIMA

Autlán

El Grullo

Cocula

Teocaltitlán
de Corona

Jocotepec

Ayutla

Zacoalco
de Torres

Tecolotlán

Tamazula
de Gordiano

Sayula

Atoyac

Tuxcueca

Jiquilpan
de Juárez

La Barca

Chapala

Lago de Chapala

Ocotlán

Jamay

Tlajomulco
de Zúñiga

Degollado

Yurécuaro

Ayo

La Piedra

Cabadas

Angamacutiro

Penjamillo

Valle de
Santiago

Jaral del Progreso

Salvatierra

Puruándiro

Moroleón

Acámbaro

Maravatío

El Oro

Ixtlahuaca
de Rayón

Atlacomulco

Jilotepec

Polotitlán

Tula

Apan

Otumba

Tepeapulco

Teoloyucan

Tepexpan

Texcoco

Zumpango

TLAXC

Tlaxcala

Zacate

Apizaco

Sta. Ana

Calpulalpan

Huamantla

Puebla

Nevado de Colima
13 911
Colima
12 620

Ciudad Guzmán

Zapotiltic

Tuxpan

Tecalitlán

Venustiano
Carranza

Tapalpa

Tizapán

Cotija
de la Paz

Los Reyes

Zamora

Purépero

Villa Morelos

Cuitzeo

Laguna
de Cuitzeo

Zinapécuaro

Tlalpujahua

Temascalcingo

Toluca

COLIMA

Comala

Villa de
Alvarez

Minatitlán

Tonila

Quitupan

Tingüindín

Paracho

Chilchota

Cherán

Zacapú

Quiroga

Lago de
Pátzcuaro

Morelia

Ciudad
Hidalgo

Angangueo

Villa Escalante

Pátzcuaro

Tzitzio

Zitácuaro

Almoloya

Ixtapan
de la Sal

Valle de Bravo

Tenango

Metepec

Tlalpan

Tianguistengo

MÉXICO
CITY

Azcapotzalco

Coyoacán

DISTRITO FEDERAL

Iztaccihuatl
17 159

Coatepec

Texmelucan

Cholula

Pueb

18°

Manzanillo

Cuyutlán

Tecomán

M I C H O A C A N

Tepalcatepec

Apatzingán
de la Constitución

Tancítaro
12 661

Tancítaro
Nueva Italia
de Ruiz

Uruapan

Paricutín

Parícutin
9 186

Acuitzio del Canje

Capulhuac

G. A.
MADERO

Tlapan

Tepepan

Tenango

Tepoztlán

Amecameca

Huejotzingo

Popocatépetl
17 930

Volcán
Iztaccíhuatl

MORELOS

Atlixco

Huaquechula

Huatlatlauc

Coalcomán de Matamoros

Aguililla

S I E R R A   D E   C O A L C O M A N

Presa de
Infiernillo

Tumbiscatío

Churumuco

Huetamo de Núñez

Cuzamala
de Pinzón

Zirándaro

Ciudad Altamirano

Coyuca de Catalán

Tejupilco
de Hidalgo

Tenancingo

Sultepec

Zacualpan

Taxco
de Alarcón

Cuernavaca

Coatetelco

Jojutla

Cuautla

Tepalcingo

Axochiapan

Tecomatlán

Chiautla

Acatlán
de Osorio

Tulcingo

PUNTA TEJUPAN

Coahuayutla

Coyuca de Catalán

Ajuchitlán del Progreso

Tlapehuala

Apipilulco

Iguala

Huitzuco

Tepecoacuilco
de Trujano

Cuetzalá
del Progreso

Río Balsas

Olinalá

Huamixtitlán

Apango

Atliaca

Zitlala

Chilapa

Tlapa

Tlacotepec

Silocayoapc

Juxtlahuac

Zapotitlán

La Unión

Petatlán

S I E R R A

G U E R R E R O

Tixtla de Guerrero

Chilpancingo
de los Bravo

Mochitlán

Hueycatenango

M A D R E

Tecpan
de Galeana

Atoyac de Alvarez

San Jerónimo
de Juárez

San Marcos

Cuautepec

Azoyú

Tecoanapa

Ometepec

Acapulco

Coyuca de Benítez

Laguna
Papagayo

Malinaltepec

Cozoyoapan

TS. Pedr

Amusg

PTA MALDONADO

Pinotepa
Nacional

16°

Puerto Miniso

P A C I F I C   O C E A N

104°   102°   Longitude West of Greenwich   100°   9

Relief

Meters		Feet
3050		10 000
1525		5000
610		2000
305		1000
152.5		500
Sea Level		0
152.5		500
1525		5000
3050		10 000

Cities and Towns

0 to 50,000 ○

50,000 to 500,000 ⊙

500,000 to 1,000,000 ◎

1,000,000 and over

Scale 1:4 000 000; one inch to 64 miles. Conic Projection
Elevations and depressions are given in feet

a

Inset map (Mexico City area, Scale 1:1 000 000)

MÉXICO

Morelos o
Cuautitlán
Nicolás Romero
Cahuacán
Tutitlán
Coacalco
Tecamac
Teotihuacán o
Acolman
Chiconautla
Tepexpan
▲ Pyramids of Teotihuacán
Otumba
Apan
HIDALGO

San Bartolo o
Ixtlahuaca
Atizapán
Tlalnepantla o
Tepetlaoxtoc
Calpulalpan
TLAXCALA

Jiquipilco
Cerro La Catedral 13 000 △
Mazatla
San Jerónimo
Texcoco
Nanacamilpa

Temoaya
Atzcapotzalco
Naucalpan de Juárez
Gustavo A. Madero
Lago de Texcoco (Dry Lake)
Coatlinchán
Chicoloapan

Río Lerma
Mimiapan
Chimalpa
MEXICO CITY
Iztacalco
Nezahualcóyotl
Los Reyes
Río Frío
HY.

Toluca
Huixquilucan
Cuajimalpa
Villa Obregón Contreras
Iztapalapa
Ixtapaluca
Ayotla
INTER-AMERICAN
Texmelucan

Lerma
Coyoacán
Tláhuac
PUEBLA

Capultitlán
Metepec
Mexicalcingo
Cerro Muneco 12 655 △
San Andrés
Tlálpan
Xochimilco
Chalco

DISTRITO
Ajusco
Tecómitl
Tlalmanalco

Cerro Ajusco 12 850 △
Topilejo
Milpa Alta
Iztaccíhuatl 17 343 △

Nevado de Toluca 14 409 △
Almoloya
Oxtotepec
Tenango

FEDERAL
Coatepec
Amecameca

Tenango
Tres Cumbres
Ozumba
Volcán Popocatépetl 17 887 △ 19°

Huitzilac
Tepoztlán
Tlalnepantla

Scale 1:1 000 000
0 5 10 Miles
0 4 8 12 16 Kilometers
©RMcN

MORELOS
Tepoztlán
Tlayacapan
Cuernavaca

Main map

GULF OF MEXICO

Laguna Almagre
Tropic of Cancer

PTA. JEREZ

Laguna de San Andres

Altamira
Ciudad Madero
Tampico
Villa Cuauhtémoc
Tampico Alto

Laguna Tamiahua
CABO ROJO
ARRECIFE BLANQUILLA
ISLA DE LOBOS

zuluama
Tancoco
Tamiahua
Alamo
Túxpan
ARRECIFE TANQUIJO
ARRECIFE TÚXPAN

Tihuatlán
Poza Rica
Tecolutla
Gutiérrez Zamora
apalapa
Furbero
Coyutla
Nautla

Coxquihui
Cuetzalan del Progreso
Tlapacoyan
Misantla
Vega de Alatorre

acatlán
Atempan
Jalacingo
Altotonga
Naolinco
PUNTA ZEMPOALA

capoaxtla
Teziutlán
Las Vigas
Antigua Veracruz

Libres
Perote
Xalapa
△14 048
Coatepec
Veracruz

amantla
Teocelo
ARRECIFE CABEZA

San Juan Ixtenco
Huatusco
Coscomatepec

Ciudad Serdán △8 406
Pico de Orizaba (Vol.)
Medellín

peaca
Orizaba
Córdoba
Tlalixcoyan
Alvarado

Acatzingo de Hidalgo
Heroica Nogales
Omealca
Cotaxtla

Atoyatempan
Maltrata
Tierra Blanca
San Martín (Vol.) △6000
PTA ZAPOTITLÁN

Tlacotepec
Tehuacán
Tlacotalpan
San Andrés Tuxtla
Catemaco

San Gabriel Chilac
Ajalpan
Zoquitlán
Santiago Tuxtla

Petlalcingo
Zinacatepec
Huatla de Jiménez
Ojitlán (S. Lucas)
Pajapan
Coatzacoalcos (Puerto México)

S. Miguel
Teotitlán del Camino (San Felipe)
Jalapa de Díaz
Tuxtepec
Tesecheacan
Soteapan

Chazumba
Tepelmeme
Cosamaloápan
Chacaltianguis
Jaltipan
Cosoleacaque

Huajuapan de León
Coixtlahuaca
Cuicatlán
Playa Vicente
San Juan Evangelista
Minatitlán
Texistepec

amazulapan del Progreso
Tejúpan (Santiago)
Acayucan
Sayula

utla de Guerrero
Chalcatongo
San Mateo (Etlatongo)
Nochixtlán (Asunción)
Talea de Castro (San Miguel)
Jesús Carranza
Pueblo Viejo

Teposcolula
an Pedro y San Pablo
Yosonú (Sta. Catarina)
Zaachila
Ixtlán de Juárez
Villa Alta (San Ildefonso)
ISTMO

Tlaxiaco
Sta. María Asunción
Oaxaca
Tlacolula de Matamoros
Hidalgo Yalalag
Zacatepec (Santiago)
DE

Itundujia Sta. Cruz
Ocotlán de Morelos
Zimatlán de Alvarez
Táviche
Zempoaltepetl 11 142
Mazatlán (San Juan)
Guichicovi (San Juan)

Sola de Vega (S. Miguel)
Ejutla de Crespo
TEHUANTEPEC
Ixtepec (Asunción)
Unión Hidalgo

Huazolotitlán (Sta. María)
amilpec
Miahuatlán
Las Vacas
Jalapa del Márques
Ixtaltepec (Sto. Domingo)
Juchitán de Zaragoza
Tapanatepec

Loxicha (Sta. Catarina)
Pluma Hidalgo
Tehuantepec Sto. Domingo
Laguna Superior
Laguna Inferior
Ixhuatán (San Francisco)
Arriaga

Salina Cruz
Mar Muerto
Tonalá

SIERRA DE OAXACA
OAXACA DEL SUR
Pochutla (San Pedro)
Puerto Ángel
Golfo de Tehuantepec

BAHÍA DE CAMPECHE

YUCATÁN
Sisal
Hunucmá
Maxcanú
Halachá
Calkini
Dzitbalché
Hecelchakán

Lerma
Campeche
Seybaplaya
Champotón
Pustunich

Sabancuy
Chicbul
CAMPECHE
Mamantel

ISLA DEL CARMEN
Laguna de Términos
San Pedro
Ciudad del Carmen
Arroyo Caribe

PUNTA FONTERA
Paraíso
Frontera
Palizada
Allende
Comalcalco
Jalpa
Cunduacán
Jonuta

Cárdenas
Villahermosa
San Carlos
Balancán
MEXICO
GUATEMALA

Huimanguillo
Tacotalpa
Emiliano Zapata
Teapa
Pichucalco
Palenque
Tenosique

Chapultenango
Yajalón
Tecpatán
Pantepec
Simojovel
Bachajón
Ococingo

Compañalá
Jitotol
MESETA DE AGUA ESCONDIDA

Berriozábal
Tuxtla Gutiérrez 9400
Bohóm
Cancuc
Oxchuc
San Cristóbal de las Casas

Ocozocoautla
Chiapa de Corzo
Acala
Amatenango
Las Rosas

Cintalapa
Las Cruces
Suchiapa
Teopisca
Comitán

Zanatepec
Venustiano Carranza
Socoltenango

Villa Flores
La Concordia
Trinitaria
CHIAPAS

SIERRA MADRE
CORD. DE CHIAPAS
Cuauhtemoc
SA. CUCHUMATANES
GUATEMALA

Mapastepec
Jacaltenango
Pijijiapan
San Miguel

0 20 40 60 80 100 120 Miles
0 20 40 60 80 100 120 140 160 180 200 Kilometers

90

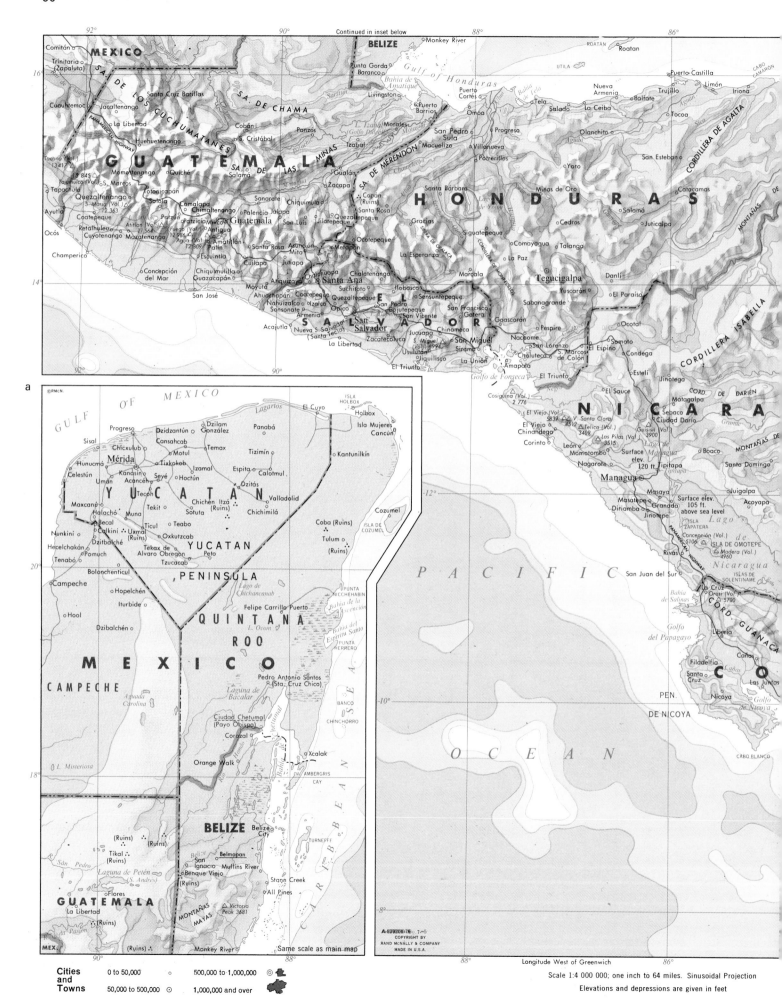

Longitude West of Greenwich

Scale 1:4 000 000; one inch to 64 miles. Sinusoidal Projection

Elevations and depressions are given in feet

Cities and Towns

0 to 50,000	○	500,000 to 1,000,000
50,000 to 500,000	⊙	1,000,000 and over

Same scale as main map

ANGUILLA
(Br.)
ST. MARTIN
(Neth. and Fr.)

ST. BARTHÉLEMY
(Fr.)

18°

16°

SABA
(Neth.)

Codrington   BARBUDA

ST. EUSTATIUS
(Neth.)   ST. KITTS
Mt. Misery
3792   Basseterre
Charlestown   Nevis Peak   **ST. KITTS AND NEVIS**
3596   St. Johns
NEVIS   Boggy Peak   **ANTIGUA
AND
BARBUDA**
1319

REDONDA

MONTSERRAT
(Br.)
Plymouth   Chaves Pk.
3000

Relief

Meters	Feet
3050	10 000
1525	5000
610	2000
305	1000
152.5	500
Sea Level	
152.5	500
1525	5000
3050	10 000

b

PUNTA PATUCA

COLÓN

Cabo Gracias a Dios

CAYOS
MISKITO

POINTE DE
LA GRANDE VIGIE
GRANDE TERRE
Ste. Rose   Le Moule   DÉSIRADE
(Fr.)
Pointe-à-Pitre   Ste. Anne   PETITE TERRE
(Fr.)
BASSE TERRE
Soufrière   **GUADELOUPE**
4813   Capesterre   (Fr.)
Basse Terre   MARIE GALANTE
Grand Bourg   (Fr.)
LES SAINTES IS.

14°

LEEWARD   IS.

60°

Puerto Cabezas

Lone Star   Laguna Caratasca

Huaunta   Laguna Huaunta

Prinzapolca

C A R I B B E A N

Portsmouth   Morne Diablotins
4747
St. Joseph   **DOMINICA**
Roseau

Dominica Channel

HUAPÍ

ISLA DE PROVIDENCIA
(Colombia)

S E A

Laguna
las Perlas

Rama

Bluefields

SAN ANDRÉS
(Colombia)
CAYOS DE ESE

LITTLE CORN

GREAT CORN
(Nicaragua)

CAYOS DE ALBUQUERQUE
(Colombia)

Mt. Pelée (Vol.)   Trinité
4583   Pitons du Carbet
St. Pierre   3960
Fort-de-France   Le François
Le Marin   **MARTINIQUE**
(Fr.)
POINTE D'ENFER

St. Lucia Channel

Castries
Morne Gimie   **ST. LUCIA**
3117
Soufrière

16°

ISLA DE LA CIERVO

Bahía
de San Juan
del Norte

PUNTA MICO

San Carlos

San Juan del Norte
(Greytown)

Soufrière   NORTH POINT
4048   **BARBADOS**
ST. VINCENT   1115
Kingstown   **AND THE**   Mt. Hillaby
GRENADINES   Bathsheba
BEQUIA   Bridgetown
SOUTH POINT
MUSTIQUE

CANOUAN

14°

W I N D W A R D   I S.

12°

COLÓN

COSTA DE MOSQUITOS

ATLANTIC OCEAN

CARIBBEAN SEA

St. Vincent Passage

THE GRENADINES

CARRIACOU

©RMcN

80°

Mt. St. Catherine
2757   Grenville
St.
George's   **GRENADA**

78°

Same scale as main map

12°

San Ramón   Guápiles   Caño
San José   Heredia   Matina
Esparta   Alajuela   Irazú   Turrialba
Puntarenas   Cartago   Vol.   Limón
San José   Paraíso   11 260

Parrita
R I C A
Quepos   PUNTA CAHUITA

San Isidro
Cerro Chirripó
12 530
Buenos Aires   Cerro Kámuk
11 696
Cerro Echandi   **T A L A M A N C A**
10 394
Boquete
Chiriquí Grande
Volcán Barú
11 401

Bahía
de Coronada

Puerto Cortés

ISLA DE CAÑO

PENÍNSULA
Puerto Jiménez
DE OSA

CABO MATAPALO

Golfito

La Cuesta

David
Horconcitos

Puerto Armuelles

PUNTA BURICA

Bahía Charco
de Azul

ISLA COIBA

Concepción

C. de Santa
Catalina
5249   **SERRANÍA
DE TABASARÁ**

C. Negro 4429

Remedios
Las Palmas   Santiago
Soná

Río de Jesús

Chitré
Las Santos

**PENÍNSULA
DE AZUERO**

Bahía
ISLA CEBACO

ISLA JICARÓN

PUNTA MARIATO

Bocas del Toro
Bahía de Almirante
Almirante   PUNTA CHIRIQUÍ
Laguna
de Chiriquí   ESCUDO
DE VERAGUAS

Golfo
de los Mosquitos

PUNTA MANZANILLO   Nombre   El
de Dios   Porvenir   PUNTA SAN BLAS
Portobelo   Mandinga   Golfo de San Blas
Colón
Gatúnillo   Silver City   C. Brewster
Lago   North Gamboa   3018   Chepo
Gatún   Balboa Heights
Balboa   **Panamá**
Chorrera   Bahía de Panamá
Bejuco
PUNTA CHAME

10°

CORD. DE SAN
BLAS

P A N A M Á

SERRANÍA DEL DARIÉN

CABO
TIBURÓN

ARCHIPIÉLAGO
DE LAS PERLAS

San Miguel
ISLA
DEL REY

ISLA DE SAN JOSÉ

PUNTA GARACHINÉ

Penonomé
Natá   Antón
Aguadulce
Río Hato

Golfo
de Parita

Golfo de Panamá

PUNTA MALA

La Palma

Bahía
San Miguel

Garachiné   El Real

COLOMBIA

ISTMO DE PANAMÁ

84°   82°   80°   78°

0   20   40   60   80   100   120 Miles
0   20   40   60   80   100   120   140   160   180   200 Kilometers

FLORIDA

Sanibel

Naples

Big Cypress
Swamp

SEMINOLE
IND. RES.

Delray Beach

Fort Lauderdale

Dania

CAPE ROMANO

Everglades

TEN THOUSAND
ISLANDS

EVERGLADES

MIAMI

Miami Beach

NATIONAL PARK

Homestead

Biscayne
Bay

Whitewater
Bay

KEY
LARGO

CAPE SABLE

Florida Bay

PINE IS.

FLORIDA KEYS

MARQUESAS
KEYS

Key
West

DRY TORTUGAS

G   U   L   F

O   F

M   E   X   I   C   O

Straits   of   Florida

LITTLE BAHAMA
BANK

GREAT SALE
CAY

SETTLEMENT PT.

West
End

Freeport

GRAND BAHAMA

PINDER POINT

Carrion
Crow Harbor

Whale Cay

Marsh
Harbour

GREAT
ABACO

LITTLE
ABACO

MORES

Pelican
Harbor

ELBOW C

GREAT
ABACO

Cherokee
Sound

Northwest Providence Channel

GREAT ISAAC

BROTHERS

LITTLE ISAAC

GORDA CAY

Cornwall

Cross Harbor

NORTH BIMINI

SOUTH BIMINI

Barnett Harbor

N. CAT CAY

Dollar Harbor

RIDING
ROCKS

ORANGE CAY

GREAT STIRRUP
CAY

GREAT
HARBOR CAY

BERRY

ISLANDS

FRAZIERS HOG CAY

WHALE CAY

BONDS CAY

SOUTHWEST
PT.

BRIDGE C

ROYAL

CURRE

JOULTER'S CAYS

Nicolls Town

WILLIAMS

Nassau

SIMMS PT.

PARADISE

NEW PROVIDENCE

Staniard Creek

Ship Channel Cay

HIGHBORNE CAY

SHROUD CAY

SALVADOR PT.
North Bight

Middle Bight

South Bight

GREEN CAY

BOOBY ROCKS

ANDROS ISLAND

Turner Sound

DOG ROCKS

NORTH ELBOW CAYS

CAY SAL

DAMAS
CAYS

CAY SAL
BANK

ANGUILLA
CAYS

Santaren Channel

HURRICANE FLATS

Nicholas Channel

Old Bahama Channel

TONGUE OF THE OCEAN

SNAP PT.

CURLY CUT CAYS

Tropic of Cancer

Bahía Honda

HAVANA

CIUDAD DE
LA HABANA

Marianao

Regla

Guanabacoa

Santa Lucía

Guanajay

San Antonio de los Baños

Artemisa

Pan de Guajaibón
2532

Candelaria

ARCHIPIÉLAGO
DE LOS COLORADOS

Consolación del Sur

SIERRA

DE LOS

ÓRGANOS

PINAR

DEL RÍO

VUELTABAJO

Mantua

Guane

Bahía de
Guadiana

San Juan
y Martínez

Pinar del Río

Los Palacios

Bejucal

Güines

Güira de
Melena

Union de Reyes

Batabanó

Alacranes

Matanzas

Cárdenas

Bahía
de Cárdenas

Corralillo

Colón

MATANZAS

Jovellanos

Pedro
Betancourt

Jagüey
Grande

Martí

Quemado
de Güines

Sagua la
Grande

Santo
Domingo

Esperanza

Santa
Clara

VILLA CLARA

Remedios

Camajuaní

Zulueta

Bahía Matanzas

Bahía de Santa Clara

ARCHIPIÉLAGO DE SABANA

CAYO BLANCOS

FRAGOSO

CAYO SANTA MARÍA

Bahía Buena
Vista

CAYO COCO

CAYO LOBOS

CAYO
CRUZ

CAYO GUAJABA

CAYO SABINAL

Bahía
Perro's
TIBURGUANO

Laguna
de Leche

PINAR   DEL   RÍO

PEN.
DE GUANAHACABIBES

CABO
FRANCÉS

CABO CORRIENTES

Ensenada
de Cortés

CAYOS DE SAN FELIPE

CAYOS DE LOS INDIOS

Nueva Gerona

Santa Fé

ISLA DE LA
JUVENTUD

ISLAS DE MANGLES

PTA. FRANCÉS

ARCHIPIÉLAGO
DE LOS CANARREOS

CAYO ROSARIO

CAYO CANTILES

CABO PEPE

CAYO LARGO

GOLFO DE BATABANÓ

PUNTA GORDA

CAYOS DE
JUAN LUIS

PENÍNSULA DE ZAPATA

Ensenada
de la Broa

Ensenada de

Ciénaga de Zapata

BANCO JARDINES

H A B A N A

Hanábana

Aguada

Rodas

Lajas

Cruces

Palmira

C

CIENFUEGOS

Cienfuegos

Bahía de
Cienfuegos

Pico San Juan

SIERRA DE
TRINIDAD

Trinidad

Casilda

Banco
Cienfuegos

Golfo de
Cazones

Bahía
Cochinos

BANCO
XAGUA

Florida

Santo
Spíritus

SANCTI
SPÍRITUS

Sancti
Spíritus

Jatibónico

Tunas de Zaza

Ícaro

Morón

CIEGO DE
ÁVILA

Ciego de Ávila

Fomento

CAMAGÜEY

Minas

Santa Lucía

Camagüey

Nuevitas

Bahía de Nuevit

Puerto Padre

LAS
TUNAS

Victoria de
las Tunas

CAYOS
ANA MARÍA

CAYOS CINCO BALAS

CAYOS
DE LAS DOCE LEGUAS

Canal de Caballones

LABERINTO DE LAS
DOCE LEGUAS

San Pedro

Nipe

Guayabal

Santa Cruz
del Sur

GOLFO DE
GUACANAYABO

Manzanillo

Campechuela

GRANMA

Niquero

SIERRA

Pico Ojo del Toro
1748

Pico Turquino 1974

CABO CRUZ

C   A   R   I   B   B   E   A   N

LITTLE CAYMAN

CAYMAN BRAC

(Br.)

CAYMAN   ISLANDS

George Town

GRAND CAYMAN

S   E   A

Montego Bay

Lucea

Falmouth

St. Ann's Bay

GALINA PT

Port Ma

JAMAICA

SOUTH NEGRIL PT.

Savanna la Mar

Mt. Denham
325

Annotto Bay

Black River

GT. PEDRO BLUFF

May Pen

Spanish Tov

Kingston

Port

PORTLAND PT.

PORTLAND PT.

Portland
Bight

Relief

Meters	Feet
3050	10 000
1525	5000
610	2000
305	1000
152.5	500
0 Sea Level	0
152.5	500
1525	5000
3050	10 000
6100	20 000

**Cities
and
Towns**

0 to 50,000

50,000 to 500,000

500,000 to 1,000,000

1,000,000 and over

Longitude West of Greenwich

Scale 1:4 000 000; one inch to 64 miles. Conic Projection

Elevations and depressions are given in feet.

GULF OF MEXICO

Cojimar
Playa de Guanabo

Guanabacoa
Regla
Campo Florido
San Francisco de Paula
Playa de Santa Fé
Baracoa
Marianao
Cotorro
Cuatro Caminos
Arroya Arena
Calabazar
Managua
Bauta
Rancho Bayeros
San José de las Lajas
Caimito del Guayabal
Santiago de las Vegas
La Sabina
Bejucal
Ceiba del Agua
Buenaventura
San Antonio de los Baños
San Antonio de las Vegas
L. de Ariguanabo
△ 950

ATLANTIC

James Pt.
Governor's Harbour
Palmetto Pt.
ELEUTHERA
Arpum Bay
Well
Rock Sound
Eleuthera Pt.
San Salvador

Arthur's Town
Northeast Pt.
Little San Salvador
CAT

Exuma Sound
Old Bight
Columbus Pt.
SAN SALVADOR
(WATLING)
(Columbus, Oct. 12, 1492)
Southwest Pt.

OCEAN

Great Guana Cay
Hawks Nest Pt.
Conception

Darby
Lee Stocking
Cape Sta. Maria
Rum Cay
Rolleville
Great Exuma
George Town
Little Exuma
Hog Cay
LONG
Clarence Town
Tropic of Cancer

Jumento Cays
Water Cay
Samana or Atwood Cay
Flamingo Cay
Cap Verde
Bird Rock
Crooked
Jamaica Cay
Northeast Pt.
Seal Cays
Fortune
Plana or Flat Cays
Man of War Channel
Diana Bank
The Bight of Acklins
Fish Cay
Nurse Cay
Acklins
Cochinos Banks
Raccoon Cay
Abraham's Bay
Mayaguana Passage
Great Ragged
Salina Pt.
Castle
Mayaguana
Columbus Bank
Mira Por Vos Islets
Cay Verde
Mira Por Vos Pass.
Caicos Passage
North Caicos
Providenciales
Grand Caicos
Cape Comete
East Caicos
Cay Sta. Domingo
Hogsty Reef
West Caicos
Caicos Is. (Br.)
South Caicos
Grand Turk
Caicos Bank
Turks Is. (Br.)
Brown Bank
Little Inagua
West Sand Spit
Salt Cay
Ambergris Cays
Northeast Pt.
Palmetto Pt.
Ocean Bight
Seal Cays
Mouchoir Passage
Turks I. Passage
Man of War Bay
The Lake
Great Inagua
Mouchoir Bank
Silver Bank
Gibara
Cabo Lucrecia
Matthew Town
South Bay
Banes
Antilla
Bahia de Nipe
Holguin
Silver Bank Passage
Navidad Bank
Olguin
Mayari
Sagua de Tánamo
Cauto
Cuchillas de Toa
Baracoa
Guantanamo
Sa. de Purial
Santiago de Cuba
Alto Songo
Punta Maisi
Ile de la Tortue
Estra
San Luis
Caney
Bahia de Ovando
Cabo Isabela
Maestra
Piedra
Santiago de Cuba
Caimanera
Yateras
Canal de la Tortue
Monte Cristi
Puerto Plata
Naval Station (U.S.A.)
Port de Paix
Cordillera Septentrional
Cabo Francés Viejo
Cap Haitien
Le Borgne
Guayubin
Gasper Hernandez
Bahia de Guantánamo
Cap St. Nicolas
Le Môle
Limbé
Fort Liberté
Dajabon
Santiago
Moca
Nagua
Cabo Samaná
Pte. Plateforme
Grande Riviere du Nord
Ouanaminthe
Rodriguez
Santiago de los
Cabelleros
Salcedo
San Francisco
Sánchez
Bahia Escocesa
Windward Passage
Gonaives
Valliere
Hinche
La Vega
Riva
Samaná
Cabo San Rafael
Golfe des Gonaives
St. Michel de l'Atalaye
DOMINICAN
Sabana de la Mar
Bani
St. Marc
Bonhamme
Jarabacoa
Cotui
Cordillera Oriental
Jérémie
Ile Grande Cayemite
Canal de Saint-Marc
Mirebalais
Lascahobas
Cordillera Central
Hato Mayor
Seibo
Bayaguana
Los Llanos
Point Ouest
Ile de la Gonave
San Juan
Sierra de Neiba
Azua
Higüey
Cap Dame Marie
Petit Goave
Port-au-Prince
Petionville
Neiba
San Cristobal
La Romana
Anse d'Hainault
Canal du Sud
Léogane
Sierra de Neiba
Catalina
Massif de la Hotte
Mirogoane
Massif de la Selle
Duverge
Barahona
Santo Domingo
S. Pedro de Macoris
Pico de Macaya
Anse à Veau
Jacmel
Belle-Anse
Bahoruco
Pta. Palenque
Saona
Coteaux
Tiburon
Les Cayes
Ile à Vache
Enriquillo
Navassa (U.S.A.)
Roche à Bateau
Pointe à Gravois
Cabo Falso
Port Antonio
HISPANIOLA
HAITI
REPUBLIC
Morant Pt.
Oviedo
Cabo Beata
Beata
Alto Velo

0 10 20 30 40 50 60 70 80 90 100 110 120 Miles
0 20 40 60 80 100 120 140 160 180 200 Kilometers

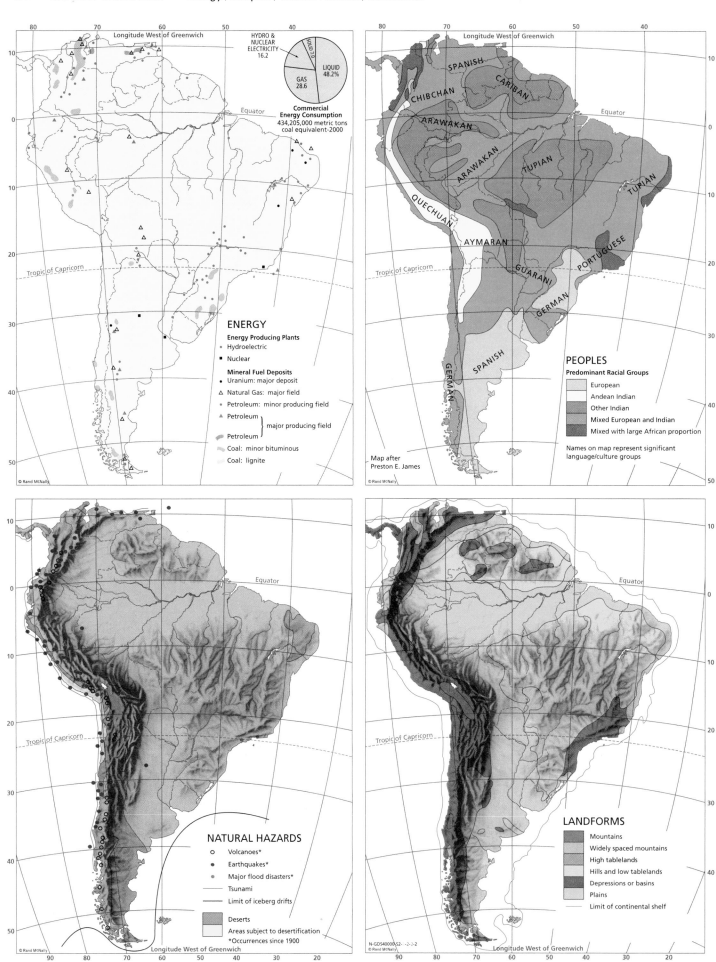

ENERGY

Energy Producing Plants
- Hydroelectric
- Nuclear

Mineral Fuel Deposits
- Uranium: major deposit
- Natural Gas: major field
- Petroleum: minor producing field
- Petroleum } major producing field
- Petroleum }
- Coal: minor bituminous
- Coal: lignite

HYDRO & NUCLEAR ELECTRICITY 16.2

SOLID 7.0

LIQUID 48.2%

GAS 28.6

Commercial Energy Consumption
434,205,000 metric tons coal equivalent-2000

© Rand McNally

PEOPLES

Predominant Racial Groups
- European
- Andean Indian
- Other Indian
- Mixed European and Indian
- Mixed with large African proportion

Names on map represent significant language/culture groups

Map after Preston E. James

© Rand McNally

CHIBCHAN
SPANISH
CARIBAN
ARAWAKAN
ARAWAKAN
QUECHUAN
TUPIAN
TUPIAN
AYMARAN
GUARANI
PORTUGUESE
GERMAN
SPANISH
GERMAN

NATURAL HAZARDS
- Volcanoes*
- Earthquakes*
- Major flood disasters*
- Tsunami
- Limit of iceberg drifts
- Deserts
- Areas subject to desertification
*Occurrences since 1900

© Rand McNally

LANDFORMS
- Mountains
- Widely spaced mountains
- High tablelands
- Hills and low tablelands
- Depressions or basins
- Plains
- Limit of continental shelf

N-GDS40000-S2- --2-2-2
© Rand McNally

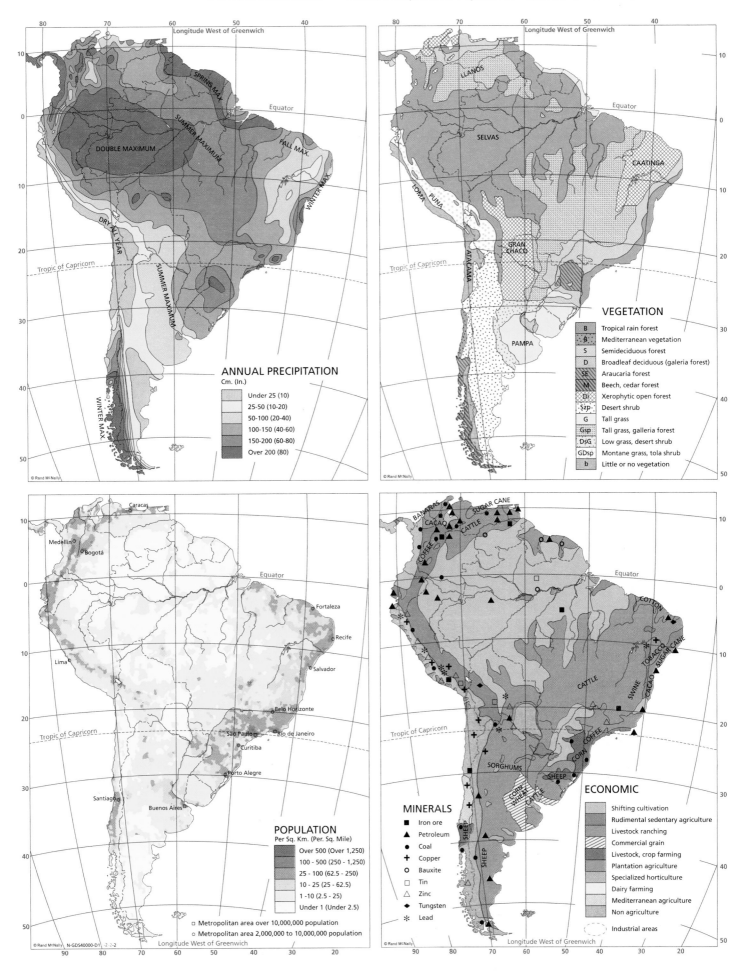

ANNUAL PRECIPITATION
Cm. (In.)

- Under 25 (10)
- 25-50 (10-20)
- 50-100 (20-40)
- 100-150 (40-60)
- 150-200 (60-80)
- Over 200 (80)

VEGETATION

B	Tropical rain forest
B̃	Mediterranean vegetation
S	Semideciduous forest
D	Broadleaf deciduous (galeria forest)
SE	Araucaria forest
M	Beech, cedar forest
Di	Xerophytic open forest
Szp	Desert shrub
G	Tall grass
Gsp	Tall grass, galleria forest
DsG	Low grass, desert shrub
GDsp	Montane grass, tola shrub
b	Little or no vegetation

POPULATION
Per Sq. Km. (Per. Sq. Mile)

- Over 500 (Over 1,250)
- 100 - 500 (250 - 1,250)
- 25 - 100 (62.5 - 250)
- 10 - 25 (25 - 62.5)
- 1 - 10 (2.5 - 25)
- Under 1 (Under 2.5)

□ Metropolitan area over 10,000,000 population
○ Metropolitan area 2,000,000 to 10,000,000 population

MINERALS

- ■ Iron ore
- ▲ Petroleum
- ● Coal
- ✛ Copper
- ○ Bauxite
- □ Tin
- △ Zinc
- ◆ Tungsten
- ✳ Lead

ECONOMIC

- Shifting cultivation
- Rudimental sedentary agriculture
- Livestock ranching
- Commercial grain
- Livestock, crop farming
- Plantation agriculture
- Specialized horticulture
- Dairy farming
- Mediterranean agriculture
- Non agriculture
- Industrial areas

Scale 1:40 000 000; one inch to 630 miles. Lambert's Azimuthal, Equal Area Projection
Elevations and depressions are given in feet

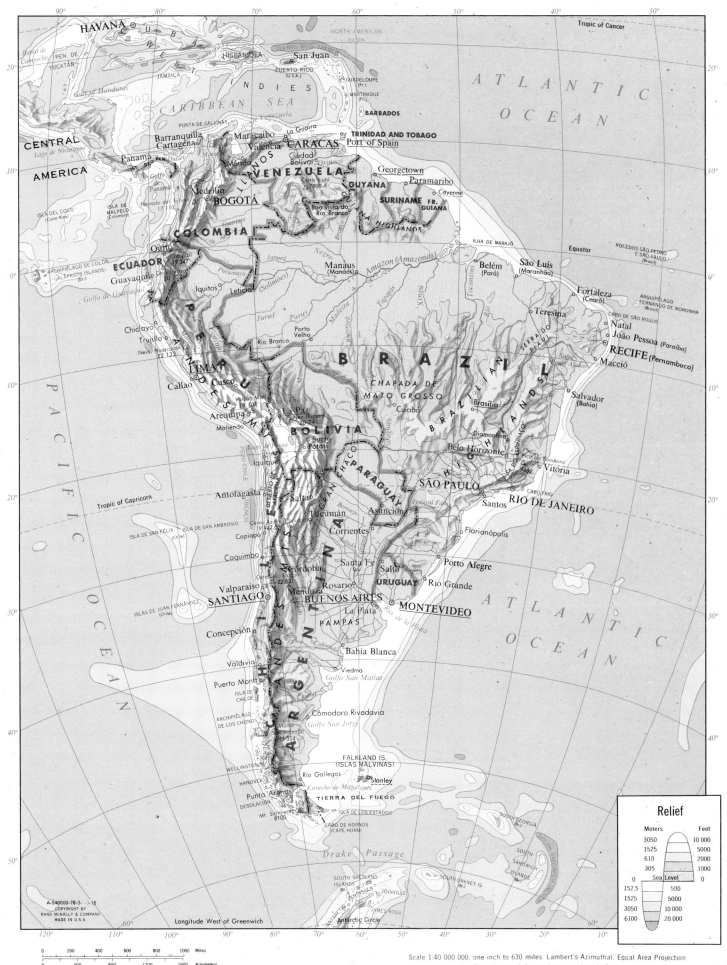

Scale 1:40 000 000, one inch to 630 miles. Lambert's Azimuthal, Equal Area Projection
Elevations and depressions are given in feet

Urban
Cropland
Cropland & Woodland
Cropland & Grazing Land
Grassland, Grazing Land
Forest, Woodland
Swamp, Marshland
Shrub, Sparse Grass, Wasteland
Barren Land

Scale 1:36,000,000; one inch to 570 miles Lambert Azimuthal Equal-Area Projection

0   100   200   400   600   800 Miles
0   150   300   600   900   1200 Kilometers

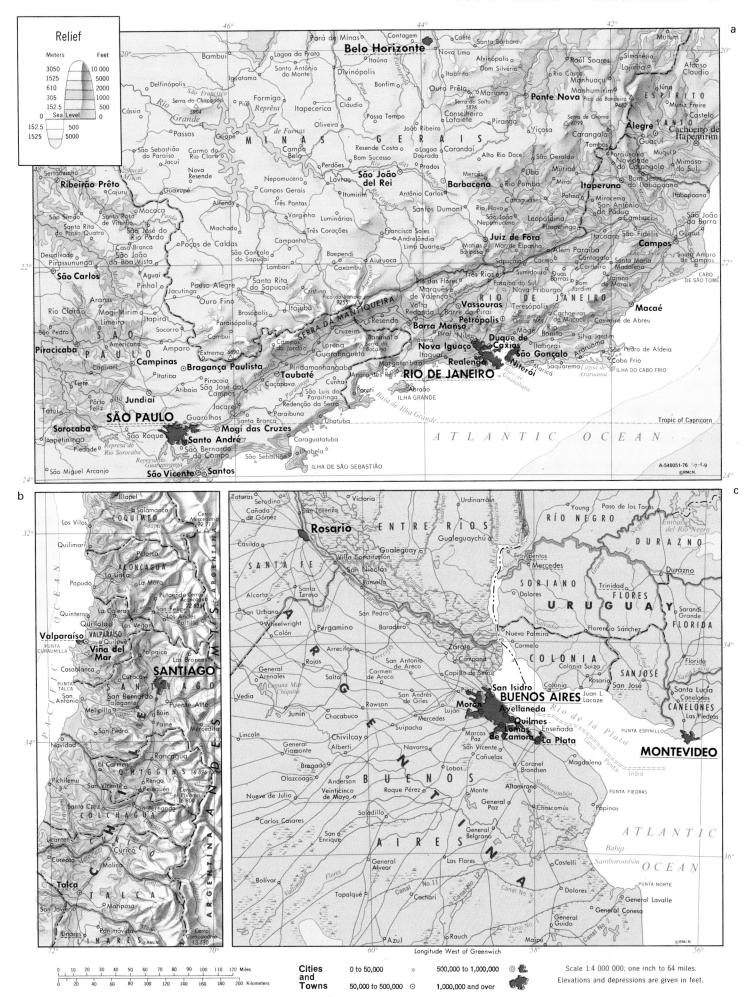

a

Relief

Meters	Feet
3050	10 000
1525	5000
610	2000
305	1000
152.5	500
Sea Level	0
152.5	500
1525	5000

Belo Horizonte

Pará de Minas · Contagem · Caeté · Santa Bárbara · Mutum

Bambuí · Lagoa da Prata · Nova Lima · Alvinópolis · Simonésia · Lajinha

Lagoa da Prata · Itaúna · Dom Silvério · Raul Soares · Manhuaçu · Afonso Cláudio

Delfinópolis · Santo Antônio do Monte · Divinópolis · Bonfim · Ouro Prêto · Mariana · **Ponte Nova** · Rio Casca · Manhumirim · Iúna · ESPÍRITO

Formiga · Cláudio · Passa Tempo · Conselheiro Lafaiete · Piranga · Serra do Grama 5896 · Pico da Bandeira 9482 · Muniz Freire · Castelo · SANTO

Iguatama · Itapecerica · João Ribeiro · 5904 · Serra de Canastra · Alto Rio Doce · Carangaia · Viçosa · 6099 · **Alegre** · Cachoeiro de Itapemirim

Rio Grande · Serra de Chapadão · Piuí · Resende Costa · Lagoa Dourada · Carandaí · São Geraldo · Tombos · Porciúncula · Navidade de Carangola · Mimoso do Sul

Ribeirão Prêto · de Furnas · Campo Belo · Perdões · Bom Sucesso · Prados · Mercês · Ubá · Muriaé · Mirai · **Itaperuna** · San Antônio de Pádua · São João da Barra

Sertãozinho · São Sebastião do Paraíso · Jacuí · Represa · São João del Rei · Antônio Carlos · Santos Dumont · Rio Pomba · Leopoldina · Pirapetinga · Itaocara · Cambuci · São Fidélis · Guarus

Santa Rosa de Viterbo · Nova Resende · Lavras · Itumirim · Rio Novo · Nepomuceno · Palma · Miracema · Santa Amaro de Campos

Descalvado · Pirassununga · Campos Gerais · Três Pontas · Luminárias · Andrelândia · Lima Duarte · Mar de Espanha · Alem Paraíba · Cantagalo · Santa Maria Madalena

São Carlos · Poços de Caldas · São Gonçalo do Sapucaí · Baependi · Caxambu · Sapucaia · Sumidouro · Três Rios · Duas Barras · Cordeiro · CABO DE SÃO TOMÉ

Aguaí · Pinhal · Machado · Campanha · Aiuruoca · **Juiz de Fora** · Carmo · Santa Maria · Itabapoana

Casa Branca · Poços · Lambari · Santa Rita do Sapucaí · Cristina · Marquês de Valença · Nova Friburgo · Bom Jardim · Duas Barras

Piracicaba · Limeira · Araras · Mogi-Mirim · São Simão · Amparo · Extrema 6890 · Pico da Itatiaia 9255 · Pataiba do Sul · Teresópolis · de Macacu · Casimiro de Abreu · **Macaé**

Rio Claro · Itapira · Socorro · Pouso Alegre · Itajubá · Resende · Volta Redonda · **Vassouras** · Cachoeiras · Rio Bonito · Silva Jardim

Campinas · Americana · São Sebastião do Rio Pardo · Santa Rita do Sapucaí · Serra da Mantiqueira · Cruzeiro · Barra do Piraí · **Petrópolis** · Magé · São Pedro de Aldeia

Bragança Paulista · Capivari · Itatiba · Serra da Bocaina · Lorena · Banana · **Barra Mansa** · Pirai · Nilópolis · Itaboraí · Cabo Frio

São Pedro · Tietê · Piracaia · Itu · **Taubaté** · Pindamonhangaba · Guaratinguetá · **Nova Iguaçu** · Itaguaí · **Duque de Caxias** · **São Gonçalo** · Araruama · Lagoa de Araruama · ILHA DO CABO FRIO

Sorocaba · Jundiaí · São José dos Campos · Cunha · Angra dos Reis · **Realengo** · **Niterói** · Maricá · Saquarema

São Paulo · Atibaia · Jacareí · Redenção da Serra · Parati · **RIO DE JANEIRO** · Baía de Guanabara

Itapetininga · Piedade · Guarulhos · Santa Branca · Paraibuna · Ubatuba · Abraão · ILHA GRANDE · Tropic of Capricorn

Sorocaba · São Roque · Caraguatatuba · Baía de Ilha Grande

São Vicente · **Santos** · **Mogi das Cruzes** · **Santo André** · São Bernardo do Campo · São Sebastião · Ilhabela · **ATLANTIC OCEAN**

São Miguel Arcanjo · Represa do Rio Sorocaba · Represa de Guarapiranga · ILHA DE SÃO SEBASTIÃO

A-540051-76 -7-4-9 ©RMcN

b

ILLAPEL · Salamanca · Cerro Mercedario 22 211

Los Vilos · COQUIMBO · Petorca · ACONCAGUA · Cerro Aconcagua 22 831

Quilimari · La Ligua · La Mora · Putaendo · Cerro Portillo · CORDILLERA

Papudo · Quintero · La Calera · Quillota · San Felipe · Los Andes · Portillo

Valparaíso · VALPARAÍSO · Quilpué · Las Vegas

PUNTA CURRUMILLA · **Viña del Mar** · Polpaico · Casablanca · Curacaví · Los Bronces

PUNTA TALCA · San Antonio · **SANTIAGO** · San Bernardo · Puente Alto

Navidad · Melipilla · Talagante · Buin · Paine · Mercedita

El Carmen · San Pedro

Pichilemu · Rancagua · Rengo · O'HIGGINS 16 896

San Vicente · Peleguén · Cerro el Palomo 16 800

Santa Cruz · San Fernando

Licanten · COLCHAGUA · Molina

Curepto · Curicó

Talca · TALCA · Mariposa

San Javier · LINARES · Cerro Campanario 13 130

c

Totoras · Serodino · Victoria · Urdinarráin · Young · Paso de los Toros · RÍO NEGRO

Cañada de Gómez · San Lorenzo · ENTRE RIOS · Embalse del Río Negro

Rosario · Gualeguay · Gualeguaychú · Fray Bentos · DURAZNO

Casilda · SANTA FE · Villa Constitución · Mercedes · SORIANO · Durazno

Alcorta · San Nicolás · Dolores · Trinidad · URUGUAY · FLORES · Sarandí Grande

San Urbana · Santa Teresa · Ramallo · Nueva Palmira · FLORIDA

Wheelwright · San Pedro · Baradero · Florencio Sánchez · Florida

Vedia · Colón · Arrecifes · Zárate · COLONIA · SAN JOSÉ · Santa Lucia · CANELONES

General Arenales · Rojas · San Antonio de Areco · Campana · Colonia Suiza · Rosario · San José · Las Piedras

Junín · Chacabuco · Carmen de Areco · Capilla de Señor · **San Isidro** · Colonia · Juan L. Lacaze

Lincoln · Rawson · San Andrés de Giles · Pilar · **BUENOS AIRES** · **Morón** · Luján · **Avellaneda** · **MONTEVIDEO**

General Viamonte · Chivilcoy · Suipacha · Mercedes · Marcos Paz · **Quilmes** · Ensenada · PUNTA ESPINILLO

Nueve de Julio · Alberti · Navarro · San Vicente · Cañuelas · **Lomas de Zamora** · **La Plata** · Río de la Plata

Veinticinco de Mayo · Roque Pérez · Lobos · Monte · Coronel Brandsen · Magdalena · PUNTA PIEDRAS

Olazcoaga · Saladillo · Altamirano · Chascomús

Carlos Casares · BUENOS · General Belgrano · Papinas

San Enrique · General Alvear · Las Flores · General Paz · Bahía Samborombón · ATLANTIC OCEAN

Bolívar · AIRES · Tapalqué · Cachari · Castelli · PUNTA NORTE

Azul · Rauch · Maipú · Dolores · General Lavalle · General Conesa · General Guido

Longitude West of Greenwich ©RMcN

| 0 10 20 30 40 50 60 70 80 90 100 110 120 Miles |
| 0 20 40 60 80 100 120 140 160 180 200 Kilometers |

Cities and Towns

| | 0 to 50,000 | ○ | 500,000 to 1,000,000 | ◎ |
| | 50,000 to 500,000 | ⊙ | 1,000,000 and over | |

Scale 1:4 000 000; one inch to 64 miles.
Elevations and depressions are given in feet.

EL SALVADOR
NICARAGUA
León   Managua   Bluefields
San Juan del Sur   San Juan del Norte (Greytown)

CARIBBEAN SEA

Continued on pages 86-87

ARUBA (Neth.)   CURAÇAO (Neth.)   BONAIRE   ISLAS LOS ROQUES
PTA DE GALLINAS   PENÍNSULA DE GUAJIRA
Willemstad
Ríohacha   Punto Fijo   Golfo de Venezuela   Puerto Cabello   I. ORCHILA   ISLA DE MARGARITA
Irazú (Vol.) 11,260   Limón
Bocas del Toro   Golfo de los Mosquitos   Santa Marta   Coro   Maracay   La Asunción   Porlamar
Barranquilla   Maracaibo   La Guaira   Maiquetía
Puntarenas   Cartagena   Ciénaga   Tucacas   Barcelona   ISLA DE TORTUGA
San José   Colón   El Carmen   Calamar   Cabimas   San Felipe   Valencia   CARACAS   Cumaná
PANAMA   Soledad   Plato   Altagracia   Los Teques   La Victoria   Maturín
Golfo Dulce   Panamá   Sincelejo   Sincé   Mompós   El Banco   Lago de   Barquisimeto   Ocumare del Tuy   Aragua   Tucupita
Golfo de Chiriquí   Golfo de Panamá   Lorica   Magangué   Ocaña   Maracaibo   Trujillo   San Carlos   Cumaná   Ciudad Guayana
COIBA   Montería   Cereté   Pico Bolívar 16,427   Mérida   Barinas   VENEZUELA
PENÍNSULA DE AZUERO   Turbo   Puerto Wilches   Pamplona   Guanare   Calabozo   El Tigre   El Pao
David   Ituango   Barrancabermeja   Cúcuta   Valle de la Pascua   Ciudad Bolívar
Tarumá   Socorro   San Cristóbal   La Grita   Calabozo
Urrao   Antioquia   Bello   Bucaramanga   Málaga   Arauca   Cerro Lluri 800
Quibdó   Aguada   La Dorada   Alto Ritacuva 18,022   San Fernando de Apure   SIERRA MAIGUALIDA
Andagoya   Sonsón   Chiquinquirá   Duitama   Sogamoso   Arauca
MEDELLÍN   Honda   Zipaquirá   Tunja   Miraflores   LLANOS
Manizales   Ibagué   Girardot   BOGOTÁ   Orocué   Puerto de Nutrias
Pereira   Armenia   Espinal   Villavicencio   San Fernando de Atabapo
Buenaventura   Cali   Palmira   Purificación   Meta   SERRANÍA PARIMA
Bahía de Buenaventura   Chaparral   Salto de Tequendama   Maroa   SIERRA CURUPIRA

PACIFIC OCEAN

COLOMBIA
Popayán   Neiva   Campoalegre   Inírida
ISLA DE MALPELO (Colombia)   Bolívar   Garzón   Calamar   MESA DE YAMBÍ   Vaupés   Içana
Tumaco   Pitalito   Florencia   Apaporis
Barbacoas   La Cruz   Galeras (Vol.) 13,992   Caquetá   Uaupés   Negro
Esmeraldas   Túquerres   Pasto   Putumayo   Içá
Equator   Ipiales   Tulcán
PINTA   MARCHENA   Otavalo   Ibarra   Cayambe   Japurá
GENOVESA   Quito   Cotopaxi 19,347   Archidona   Napa   São Paulo de Olivença   Tefé
ISLA SAN SALVADOR   ARCHIPIÉLAGO   Bahía de Caráquez   Chone   Latacunga   Fonte Boa   Coari
ISABELA   SANTA CRUZ   DE COLÓN   Manta   ECUADOR   Tigre   Içá (Solimões)
SAN CRISTOBAL   (GALÁPAGOS ISLANDS)   Portoviejo   Ambato   Baños   Chimborazo 20,702   Marañón
(Ecuador)   Jipijapa   Guaranda   Riobamba   Pastaza   AMAZONAS
Babahoyo   Alausí   SELVAS
Guayaquil   Iquitos   Coari
Golfo de Guayaquil   Cuenca   Azogues   Moyobamba   Yurimaguas   Juruá
Machala   Sigsig   Chachapoyas   Lamas   Eirunepé   Lábrea   Huma
Santa Rosa   Loja   Ferreñafe   Tarapoto   Purus
Tumbes   CERROS DE CANCHYUAYA
PTA PARIÑAS   Talara   Sullana   CORDILLERA AZUL   ACRE   B
Paita   Chulucanas   Chiclayo   Cajamarca   Cruzeiro do Sul   Porto Velho
Piura   Lambayeque   Puerto Eten   GRAN PAJONAL
LOBOS DE TIERRA   Pacasmayo   Chepén   Huamachuco   Porto Acre   RONDÔ
Castilla   Puerto Chicama   Cajabamba   Rio Branco   Villa Bella   MASSIÇO DE PAC
PTA AGUJA   Trujillo   Salaverry   Tingo María   Puerto Bermúdez   Cobija   Guajará Mirim
Chimbote   Huánuco   Riberalta   Guaporé
Huaraz   Nudo de Pasco   Cerro de Pasco   Rogoaguado   Magdalena
Huacho   Tarma   GRAN PAJONAL
ISLAS CHINCHAS   Huaral   La Oroya   Puerto Maldonado   MASSIÇO DE PAC
Callao   LIMA   Jauja   Reyes
Chorrillos   Huancayo   Machu Picchu   Trinidad
Huancavelica   Ayacucho   Cusco
Cañete   Abancay   Sicuani   Ayaviri
Chincha Alta   Cotabambas   Pucallpa   Magdalena
Bahía de Pisco   Pisco   Ica   Coracora   Juliaca   Ayata
PTA CARRETAS   Puquio   Nev. Huama 21,066
Nudo Coropuna 21,698   Volcán Misti 19,101   Puno   Ayata   Guaqui   La Paz
Arequipa   Lago Titicaca   Achacachi   BOLIVIA
Camaná   Miraflores   Nev. Illimani 20,741   Viacha   Cochabamba
Mollendo   Moquegua   ALTIPLANO   Oruro   Funata   Tarata   Valle Grande   Portachu
Ilo   Corocoro   Uncia   Santa Cruz
Tacna   Nev. Sajama 21,391   Huanuni   Colquechaca
Arica   Salar de Coipasa   Sucre   Logunillas
Iquique   PUNA DE ATACAMA   Huanchaca   Potosí   Monteagudo   Villa
Pisagua   Uyuni   Pulacayo   San Lucas
Tocopilla   Challapata   Tupiza   Villazón
Chuquicamata   Ollagüe   Salar de Uyuni   Yacuiba
Calama   Tocopilla   Pedro de Valdivia   PUNA DE ATACAMA   JUJUY   ARGENTINA
Mejillones   Antofagasta   SALTA

A-549100-76- 11-10-22
COPYRIGHT BY
RAND MCNALLY & COMPANY
MADE IN U.S.A.

Tropic of Capricorn

Pavarandocito   Alto de Tres Morros 11,155   Ituango   Valdivia   Segovia
Dabeiba   Paramillo 12,990   Yarumal   Anorí   Amalfi   Remedios
San Andrés   Yolombó
Cañasgordas   ANTIOQUIA   Cisneros   Puerto Berrío
Alto Musinga 12,631   Sabanas   Santa Rosa   Sopetrán   Barbosa   San Roque
Urrao   Maceo Jaramelo 9,186   Antioquia   Itagüí   Bello   San Rafael
Arza   Concordia   Rionegro   San Carlos
Bebará   Titiribí   Envigado   Caldas   La Ceja   San Luis
Neguá   Andes   Aguadas   Cerro de las Paredes 10,991   Puerto Niño
Quibdó   Cerro Caramanta 12,795   Sonsón   Puerto Salgar
MEDELLÍN
CHOCÓ   Fredonia   Pensilvania   La Dorada   Honda
Certeguí   Salamina   Manzanares   Victoria
Tadó   Riosucio   Neira   Fresno   Mariquita
Istmina   Cerro Tamaná 13,780   Anserma   Apía   CALDAS   Villeta   Zipaquirá
El Cajón   Ansermanuevo   RISARALDA   Santa Rosa de Cabal   Líbano   Gachetá
Cerro Torrá 12,721   Manizales   Armero   Guasca
Sipí   Nevado del Ruiz 17,717   Facatativá   CUNDINAMARCA   Junín
Cartago   Quimbaya   Pereira   Ambalema   La Calera
Roldanillo   Zarzal   Finlandia   Nevado del Tolima 17,110   La Mesa   Fontibón   Fómeque
Cerro Torrá 13,944   Armenia   Ibagué   Tocaima   BOGOTÁ   Quetame   Restrepo
Sevilla   Caicedonia   QUINDÍO   Pico de Chili 17,894   Girardot   Fusagasugá   Villavicencio
VALLE DEL CAUCA   Buga   Calarcá   Rovira   Espinal   Pico de Mendanueva 13,120
Darién   Guacarí   San Antonio   Ortega   Guamo   Acacias
Restrepo   Cerrito   Coyaima   Purificación   Prado   Cerro el Nevado 14,961
Pradera   Chaparral   Natagaima   Dolores   San Martín
Cali   Palmira   Ataco   Alpujarra   Colombia   META
Jamundí   Miranda   Baraya   Villavieja   San Juan
Florida   Guacarí   Puerto Tejada   Toribío   Tello   San Antonio
Buenos Aires   Santander   Nevado de Huila 18,865   Neiva   HUILA   Palermo
CORDILLERA CENTRAL   ANDES   CORDILLERA ORIENTAL

Scale 1:4,000,000
0 10 20 30 40 Miles
0 10 20 30 40 50 60 Kilometers
© R.M.C.N.

Cities and Towns   0 to 50,000   500,000 to 1,000,000
                    50,000 to 500,000   1,000,000 and over

Scale 1:16,000,000, one inch to 250 miles. Sinusoidal Projection
Elevations and depressions are given in feet

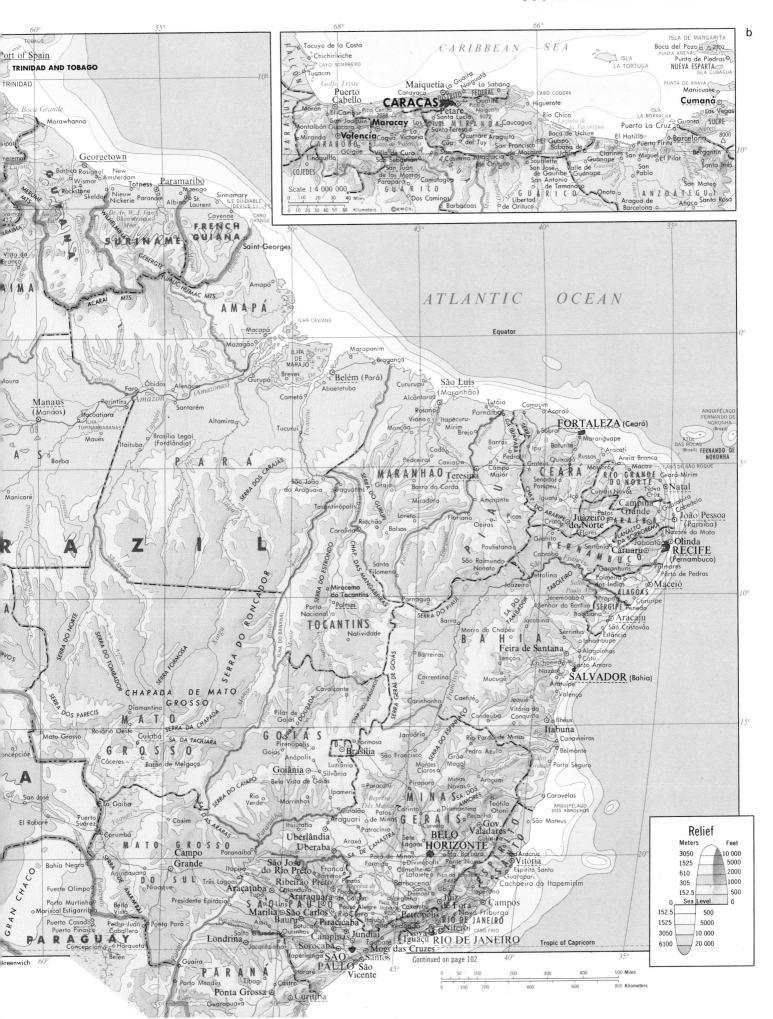

b

CARIBBEAN SEA

ISLA DE MARGARITA
Boca del Pozo △ 2002
PUNTA ARENAS
Punta de Piedras
NUEVA ESPARTA
ISLA CUBAGUA

PUNTA DE ARAYA

Tocuyo de la Costa
Chichiriviche
Cayo Sombrero
Tucacas

Golfo Triste

Maiquetía La Guaira Naiguatá La Sabana
Carayaca Macuto
Puerto
Cabello Pico Naiguatá
CARACAS
Morón El Cambur Petare
San Joaquín ○ Los Teques
Montalbán Guacara La
Maracay Victoria
Miranda San José
Valencia Villa de Cura
Tinaquillo Güigüe

Cumaná
Manicuare Las Vegas
SUCRE
Guanta △ 8000
Puerto La Cruz Barcelona Bergantín
El Hatillo Puerto Píritu
Clarines San Miguel El Pilar Santa Inés
Guanape San Pablo
San Mateo
Aragua de Anaco Santa Rosa
Barcelona

CARACAS
ESPÍRITU
FEDERAL
Río Chico
Higuerote
Caucagua
Santa Teresa
Boca de Uchire
Cúa Sabana
del Tuy San Francisco
de Macaira
Ocumare Araguita
del Tuy
CABO CODERA
Laguna de la Tacarigua
M I R A N D A

San Juan
de los Morros
Parapara
Camatagua
Dos Caminos

Barbacoas
de Orituco

ARAGUA
Casimiro
Altagracia
de Orituco
Sombrero
El Guapo
San José
de Gauribe
San Antonio
de Tamanaco

G U Á R I C O

Libertad
de Orituco

Scale 1:4 000 000
0 10 20 30 40 Miles
0 10 20 30 40 50 60 Kilometers
©R.M.C.N.

Port of Spain
TRINIDAD AND TOBAGO
TRINIDAD

Boca Grande
Morawhanna

Georgetown
Bartica Rosignol New
Wismar Amsterdam
Rockstone Skeldon
Nieuw
Nickerie Paranam
Dr. Ir. W.J. Van
Blommesteinmeer

MERUME
MTS.

ACARAÍ MTS.

Totness Paramaribo
Albina Moengo
St.
Laurent Sinnamary
ÍLE DU-DIABLE
(DEVIL'S I.)
**FRENCH
GUIANA**
Cayenne
CABO
ORANGE
Saint-Georges

SURINAME

GEBERGTE

KUMUC-HUMAC MTS.

A T L A N T I C   O C E A N

Amapá

ILHA CAVIANA

Equator 0°

Manaus
(Manáos)
Itacoatiara
ILHA
TUPINAMBARANAS
Maués
Itaituba
Borba
Manicoré

Óbidos
Alenquer
Faro
Parintins
Santarém
Amazon
Amazonas

Brasília
Legal
(Fordlândia)

Belém (Pará)
Breves Abaetetuba
ILHA
DE
MARAJÓ
Cametá

Altamira
Tucuruí

P A R Á

SERRA DOS CARAJÁS

São João
do Araguaia
Araguatins

Tocantinópolis
Riachão
Carolina

SERRA DO GURUPI

São Luís
(Maranhão)
Alcântara
Rosário
Viana
Monção
Codó
Caxias
Pedreiras

Cururupu
Bragança
Marapanim

M A R A N H Ã O
Teresina
Barra do Corda
Miradora
Loreto
Balsas

Grajaú

Tutóia
Parnaíba
Camocim
Acaraú
FORTALEZA (Ceará)
Maranguape
Sobral
Ipu Baturité
Aracati
Barras Russas Areia Branca
Quixadá Macau
Crateús Mossoró
**RIO GRANDE
DO NORTE**
C E A R Á Ceará-Mirim
Nova Natal
Campina Cruz
Grande
Guarabira
João Pessoa
(Paraíba)
Nazaré da Mata
Olinda
RECIFE
(Pernambuco)

Viana
Mirim
Brejo
Itapecuru-
Mirim
Monção

Pedro II
Senador
Pompeu

Picos
Oeiras
Paulistana

Amarante
Floriano
Santa
Filomena

São Raimundo
Nonato

P I A U Í

Petrolina
Juazeiro

Iguatu
Patos
Crato
Juazeiro
do Norte
Flores
Granito
Cabrobó
Sertânia
Caruaru
P E R N A M B U C O
Palmeira
dos Índios
Garanhuns
Palmares
Pôrto de Pedras
Maceió
A L A G O A S
Penedo
S E R G I P E
Coruripe
Propriá
Itabaiana
Aracaju
Estância
São Cristóvão
Inhambupe

ARQUIPÉLAGO
FERNANDO DE
NORONHA
(Brazil)
ATOL
DAS ROCAS
(Brazil)
FERNANDO DE
NORONHA

5°

10°

B R A Z I L

Moura

Vista do
Branco

Miracema
do Tocantins
Palmas
Porto
Nacional

SERRA DO RONCADOR

SERRA DO ESTRONDO

CHAP. DAS MANGABEIRAS

ILHA DO BANANAL

T O C A N T I N S
Natividade

SERRA GERAL DE GOIÁS

Barreiras

Correntina

Carinhanha

B A H I A
Morro do Chapéu
Jacobina
Serrinha
Senhor do Bonfim
Feira de Santana
Cachoeira
Santo Amaro
Nazaré
Valença

Barra

SERRA DO
TABADOR

Catú
Alagoinhas

SALVADOR (Bahia)

Januária

SERRA DO ESPINHAÇO

Lençóis
Mucugê
Caetité
Condeúba
Vitória da
Conquista
Jequié

Ilhéus
Itabuna
Canavieiras
Belmonte
Pôrto Seguro

15°

SERRA DO NORTE
SERRA DO TOMBADOR
Serra Formosa
SERRA DO CATAPO

CHAPADA DE MATO
GROSSO
SERRA DOS PARECIS

Diamantino

M A T O

G R O S S O

Rosário Oeste
Mato Grosso
Cáceres
Cuiabá
Barão de Melgaço

SA. DA TAQUARA
SERRA DA CHAPADA

G O I Á S
Pirenópolis
Goiás
Anápolis
Goiânia
Luziânia
Bela Vista de Goiás
Silvânia

Formosa
D. F.
Brasília
São Francisco
Paracatu
Unaí

Cavalcante

Pilar de
Goiás

Barra do
Garças

Rio Pardo de Minas
Pedra Azul
Grão
Mogol
Montes
Claros
Pirapora
Minas
Novas
Araçuaí

SA. DOS
AIMORÉS
ARQUIPÉLAGO
DOS ABROLHOS

Caravelas

San José

El Roboré

Puerto
Suárez
Corumbá

La Gaiba

Coxim
Rio
Verde
Morrinhos

M A T O   G R O S S O

Bahía Negra
Fuerte Olimpo

Mariscal Estigarribia

Puerto Casado

Pedro Juan
Caballero

Campo
Grande
Bela
Vista

Aquidauana
Nioaque

Rosário Oeste

D O   S U L

Coxim

Três Lagoas

Araçatuba

Tupã
Marília
Assis
Presidente Epitácio
Presidente Prudente

Bauru

Catalão
Ituiutaba

Araguari
Patrocínio
Uberlândia
Uberaba
SA. DE CANASTRA
Araxá
Barretos
Pará de Minas
Ribeirão Prêto
Franca
Batatais
Pico da
Bandeira
9482
Poços de
Caldas
São José
do Rio Prêto
São Carlos
Sorocaba

M I N A S
G E R A I S
Curvelo
Sete
Lagoas
**BELO
HORIZONTE**
Sta. Bárbara
Ponte Nova
Conselheiro
Lafaiete
Barbacena
Juiz
de Fora
Petrópolis
Nova Friburgo
Teresópolis
Niterói
RIO DE JANEIRO

Gov.
Valadares
Colatina
Teófilo
Otoni
Diamantina
São Mateus

Aracruz
Vitória
Espírito Santo
Guarapari
Cachoeiro do Itapemirim

E. SANTO

Campos
CABO FRIO

20°

P A R A G U A Y

Porto Murtinho
Gran Chaco
Concepción

Puerto Pinasco

San José

Bella
Vista

Londrina

Caarapó
Ponta Porã

Iguaçu
P A R A N Á
Guaíra
Porto Mendes
Tibagi
Ponta Grossa
Guarapuava
Curitiba
Castro

Piracicaba
Campinas
Jundiaí
Sorocaba
Itararé
**SÃO
PAULO** São
Vicente
Santos
Mogi das Cruzes
Taubaté

Tropic of Capricorn

25°

Continued on page 102

0 50 100 200 300 400 500 Miles
0 100 200 300 400 600 800 Kilometers

Relief

Meters	Feet
3050	10 000
1525	5000
610	2000
305	1000
152.5	500
Sea Level	0
152.5	500
1525	5000
3050	10 000
6100	20 000

Continued on pages 100-101

BOLIVIA

PARAGUAY

GRAN CHACO

CHACO

BRAZIL

MATO GROSSO DO SUL

MINAS GERAIS

BELO HORIZONTE

SÃO PAULO

PARANÁ

SANTA CATARINA

RIO GRANDE DO SUL

PORTO ALEGRE

URUGUAY

MONTEVIDEO

ARGENTINA

BUENOS AIRES

LA PAMPA

RÍO NEGRO

CHUBUT

SANTA CRUZ

PATAGONIA

TIERRA DEL FUEGO

PACIFIC OCEAN

ATLANTIC OCEAN

RÍO DE LA PLATA

Golfo San Matías

Golfo San Jorge

PENÍNSULA VALDÉS

FALKLAND IS. (ISLAS MALVINAS) (Br.) (Claimed by Argentina)

Stanley

CABO DE HORNOS (CAPE HORN)

Relief

Meters	Feet
3050	10 000
1525	5000
610	2000
305	1000
152.5	500
0 Sea Level	Sea Level 0
152.5	500 Below
1525	5000 Sea Level
3050	10 000
6100	20 000

Longitude West of Greenwich

A-549200-76 -11-11-14

COPYRIGHT BY RAND McNALLY & COMPANY MADE IN U.S.A.

Scale 1:16 000 000; one inch to 250 miles. Sinusoidal Projection
Elevations and depressions are given in feet

0 50 100 200 300 400 500 Miles
0 100 200 400 600 800 Kilometers

a

BUENOS AIRES

RIO DE LA PLATA

Scale 1:1 000 000
0 4 8 12 16 Kilometers
0 10 Miles
© RMCN.

b

SERRA DAS ARARAS

RIO DE JANEIRO

Petrópolis

Baía de Guanabara

ATLANTIC OCEAN

Scale 1:1 000 000
0 4 8 12 16 Kilometers
0 10 Miles
© RMCN.

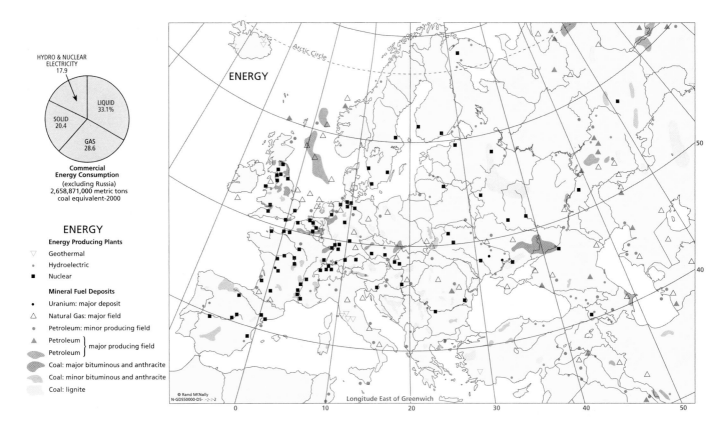

ENERGY

HYDRO & NUCLEAR
ELECTRICITY
17.9

LIQUID
33.1%

SOLID
20.4

GAS
28.6

**Commercial
Energy Consumption**
(excluding Russia)
2,658,871,000 metric tons
coal equivalent-2000

ENERGY

Energy Producing Plants

▽ Geothermal

• Hydroelectric

■ Nuclear

Mineral Fuel Deposits

• Uranium: major deposit

△ Natural Gas: major field

• Petroleum: minor producing field

▲ Petroleum } major producing field
⬟ Petroleum }

Coal: major bituminous and anthracite

Coal: minor bituminous and anthracite

Coal: lignite

© Rand McNally
N-GDS50000-D5- -2-/-2

Longitude East of Greenwich

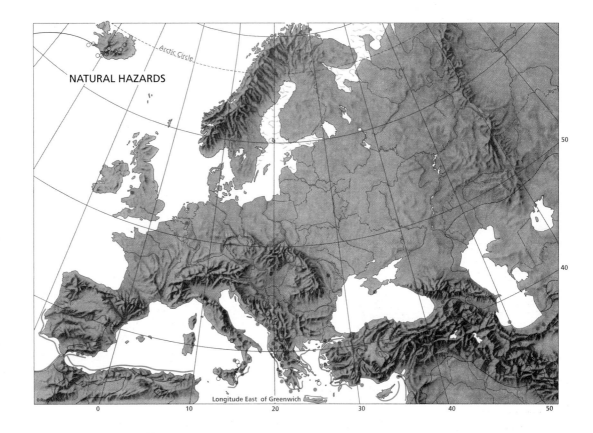

NATURAL HAZARDS

NATURAL HAZARDS

○ Volcanoes*

• Earthquakes*

• Major flood disasters*

—— Tsunamis

—— Limit of iceburg drift

Temporary pack ice

Areas subject to desertification

*Occurrences since 1900

Longitude East of Greenwich

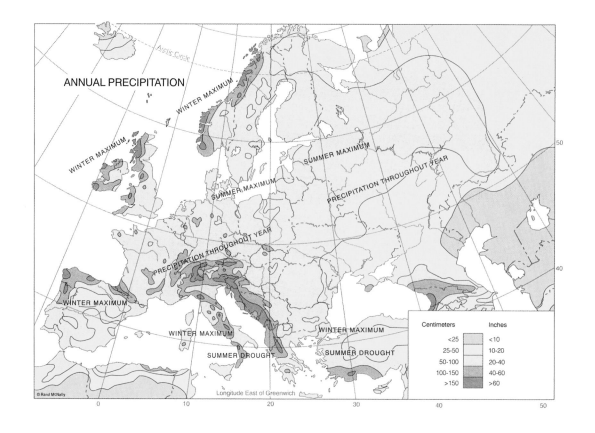

ANNUAL PRECIPITATION

WINTER MAXIMUM

WINTER MAXIMUM

SUMMER MAXIMUM

SUMMER MAXIMUM

PRECIPITATION THROUGHOUT YEAR

PRECIPITATION THROUGHOUT YEAR

WINTER MAXIMUM

WINTER MAXIMUM

WINTER MAXIMUM

SUMMER DROUGHT

SUMMER DROUGHT

Longitude East of Greenwich

© Rand McNally

Centimeters		Inches
<25		<10
25-50		10-20
50-100		20-40
100-150		40-60
>150		>60

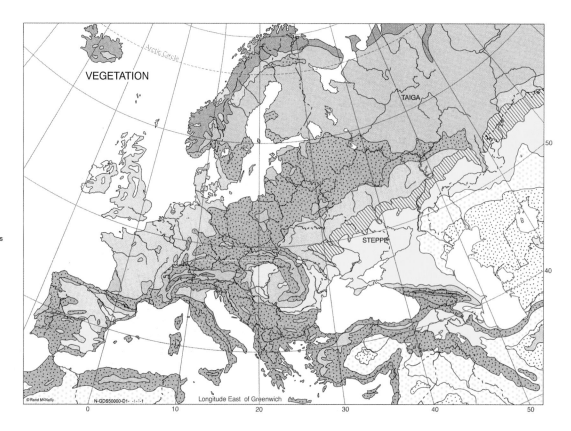

VEGETATION

TAIGA

STEPPE

© Rand McNally   N-GDS50000-D1- -1-1-1   Longitude East of Greenwich

VEGETATION

E	Coniferous forest
B,Bs	Mediterranean vegetation
M	Mixed forest: coniferous-deciduous
S	Semi-deciduous forest
D	Deciduous forest
DG	Wooded steppe
G	Grass (steppe)
Gp	Short grass
Dsh	Desert shrub
L	Heath and moor
L	Alpine vegetation, tundra
b	Little or no vegetation

For explanation of letters in boxes,
see Natural Vegetation Map
by A. W. Kuchler, p. 24

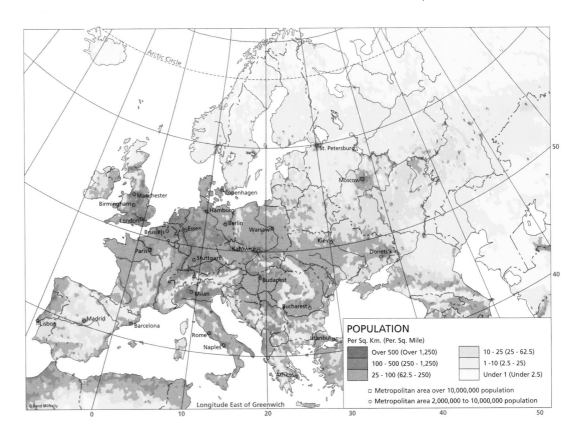

POPULATION

Per Sq. Km. (Per. Sq. Mile)

Over 500 (Over 1,250)	10 - 25 (25 - 62.5)
100 - 500 (250 - 1,250)	1 -10 (2.5 - 25)
25 - 100 (62.5 - 250)	Under 1 (Under 2.5)

□ Metropolitan area over 10,000,000 population

○ Metropolitan area 2,000,000 to 10,000,000 population

Longitude East of Greenwich

© Rand McNally

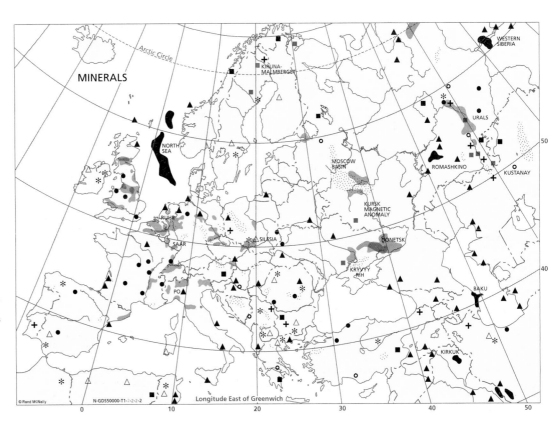

MINERALS

- Industrial areas
- Major coal deposits
- Major petroleum deposits
- Lignite deposits
- ▲ Minor petroleum deposits
- ● Minor coal deposits
- ■ Major iron ore
- ■ Minor iron ore
- ✳ Lead
- ○ Bauxite
- △ Zinc
- ✛ Copper

© Rand McNally   N-GD550000-T1-3-2-2-2   Longitude East of Greenwich

106

Urban

Cropland

Cropland & Woodland

Cropland & Grazing Land

Grassland, Grazing Land

Forest, Woodland

Swamp, Marshland

Tundra

Shrub, Sparse Grass;
Wasteland (pattern)

Barren Land

Oasis

20°   10°   0°   10°   20°   30°

Reykjavik

Narvik   Murmansk

ATLANTIC

Trondheim   Ume

Gulf of Bothnia

60°

Bergen   Oslo   Helsinki   ST. PETERSBURG

Tallinn

OCEAN

Glasgow   Göteborg   Stockholm

Belfast   Riga

North Sea

50°   MANCHESTER   Copenhagen   Baltic Sea   Vilnius

Dublin   Kaliningrad   Minsk

Amsterdam   Hamburg   Elbe   BERLIN

LONDON   Leipzig   Oder   Warsaw   Pripet

Antwerp   Essen

Brest   Frankfurt   Prague   Kraków   L'viv

PARIS   Seine   Strasbourg   CARPATHIANS   Dniester

Loire   Rhine   Danube   VIENNA   Tisza

Munich   BUDAPEST

A Coruña   Zürich   ALPS

Bordeaux   Lyon   MILAN   Zagreb

40°   Bilbao   Garonne   Rhône   Venice   Sava   Belgrade

Duero   PYRENEES   Genoa   BUCHAREST

MADRID   Ebro   Marseille   CORSICA   Adriatic Sea   Danube   Sofia

Lisbon   BARCELONA   ROME   Tirane

Sevilla   SARDINIA   Naples

ISLAS BALEARES   Tyrrhenian Sea   Athens   Aegean Sea

Tanger   Palermo

Algiers   SICILY

Oran   Tunis

Casablanca   ATLAS   MOUNTAINS   MALTA   Mediterranean Sea   CRETE

Longitude West of Greenwich   0°   Longitude East of Greenwich   10°   20°

Scale 1: 16,000,000; one inch to 250 miles. Conic Projection

0   50   100   200   300   400   500 Miles

0   100   200   400   600   800 Kilometers

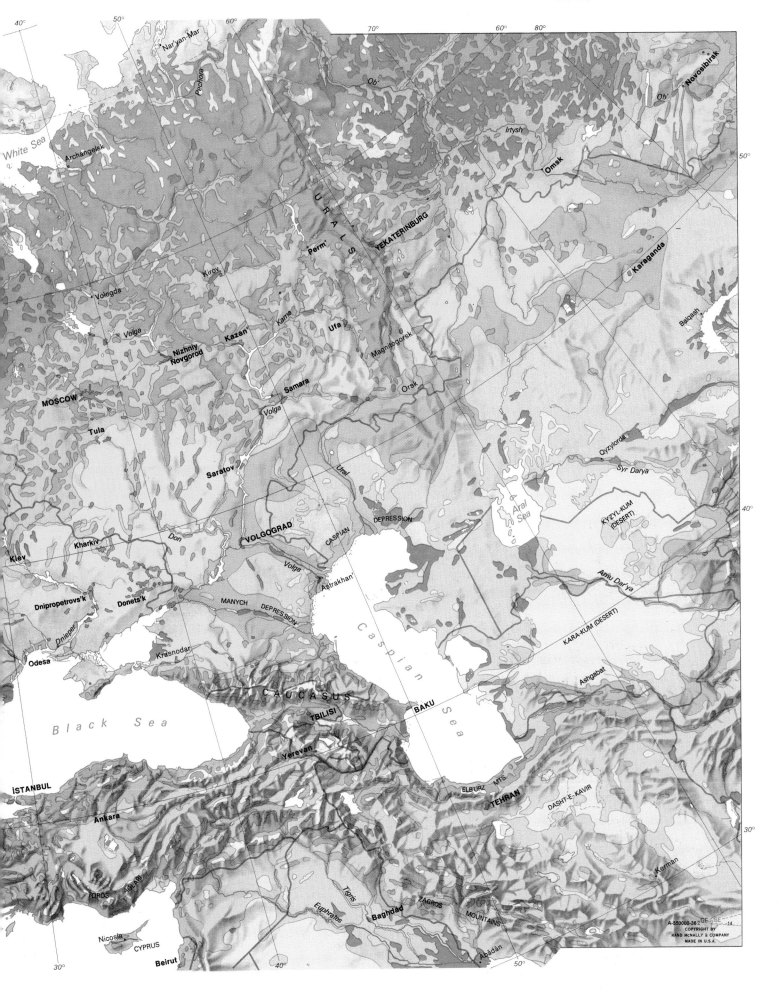

White Sea

Nar'yan-Mar
Pechora
Archangelsk
Ob'
Irtysh
Omsk
Novosibirsk
Ob'
URALS
Perm'
YEKATERINBURG
Karaganda
Kirov
Vologda
Kama
Ufa
Volga
Kazan'
Magnitogorsk
Balqash
Nizhniy Novgorod
Samara
Orsk
MOSCOW
Volga
Tula
Ural
Qyzylorda
Syr Darya
Saratov
Aral Sea
DEPRESSION
KYZYL-KUM (DESERT)
Kharkiv
CASPIAN
Don
VOLGOGRAD
Amu Dar'ya
Kiev
Volga
Dnipropetrovs'k
Astrakhan'
KARA-KUM (DESERT)
Donets'k
MANYCH   DEPRESSION
Dnieper
Krasnodar
Caspian Sea
Ashgabat
Odesa
CAUCASUS
TBILISI
BAKU
Black   Sea
Yerevan
İSTANBUL
ELBURZ   MTS.
DASHT-E-KAVIR
Ankara
TEHRAN
TOROS
TAURUS
Nicosia
Kerman
CYPRUS
Tigris
ZAGROS
Euphrates
Baghdad
MOUNTAINS
Beirut
Ābādān

A-550000-362
COPYRIGHT BY
RAND McNALLY & COMPANY
MADE IN U.S.A.

Scale 1:16 000 000; one inch to 250 miles. Conic Projection
Elevations and depressions are given in feet.

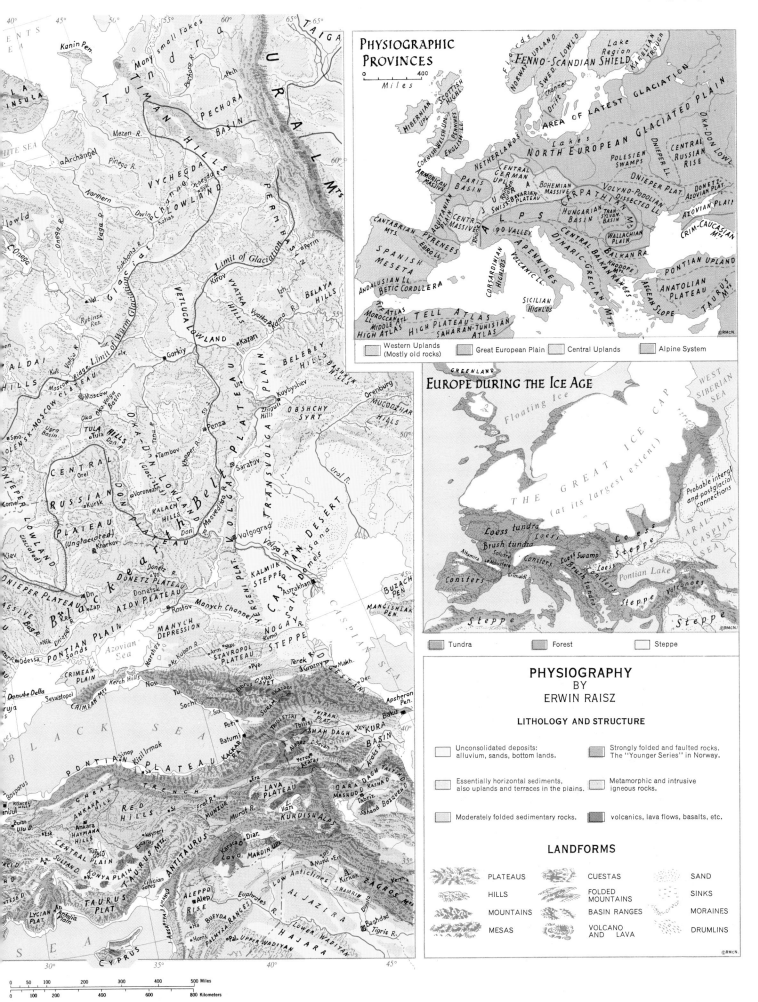

PHYSIOGRAPHIC PROVINCES

EUROPE DURING THE ICE AGE

PHYSIOGRAPHY
BY
ERWIN RAISZ

40,000 SQ MI
AREA

0   100   200
Miles

Scale 1: 16 000 000; one inch to 250 miles. Conic Projection

Elevations and depressions are given in feet

Longitude West of Greenwich   0   Longitude East of Greenwich

0   50   100   200   300   400   500 Miles

0   100   200   400   600   800 Kilometers

Relief

Meters	Feet
3050	10 000
1525	5000
610	2000
305	1000
152.5	500
0	Sea Level	0
	Below Sea Level
152.5	500
1525	5000
3050	10 000

Continued on pages 184–185

Longitude West of Greenwich   Longitude East of Greenwich

Scale 1: 16 000 000; one inch to 250 miles. Conic Projection

Elevations and depressions are given in feet

| 0 | 50 | 100 | 200 | 300 | 400 | 500 Miles |

| 0 | 100 | 200 | 400 | 600 | 800 Kilometers |

Continued on pages 140-141

Continued on pages 154-155

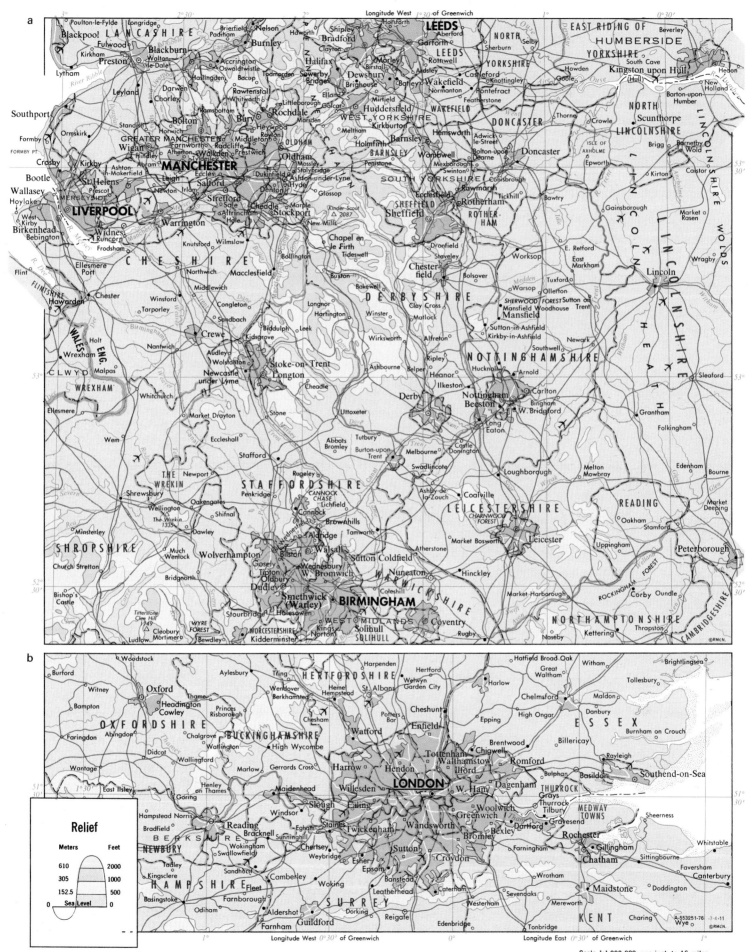

Scale 1:1 000 000; one inch to 16 miles.
Elevations and depressions are given in feet.

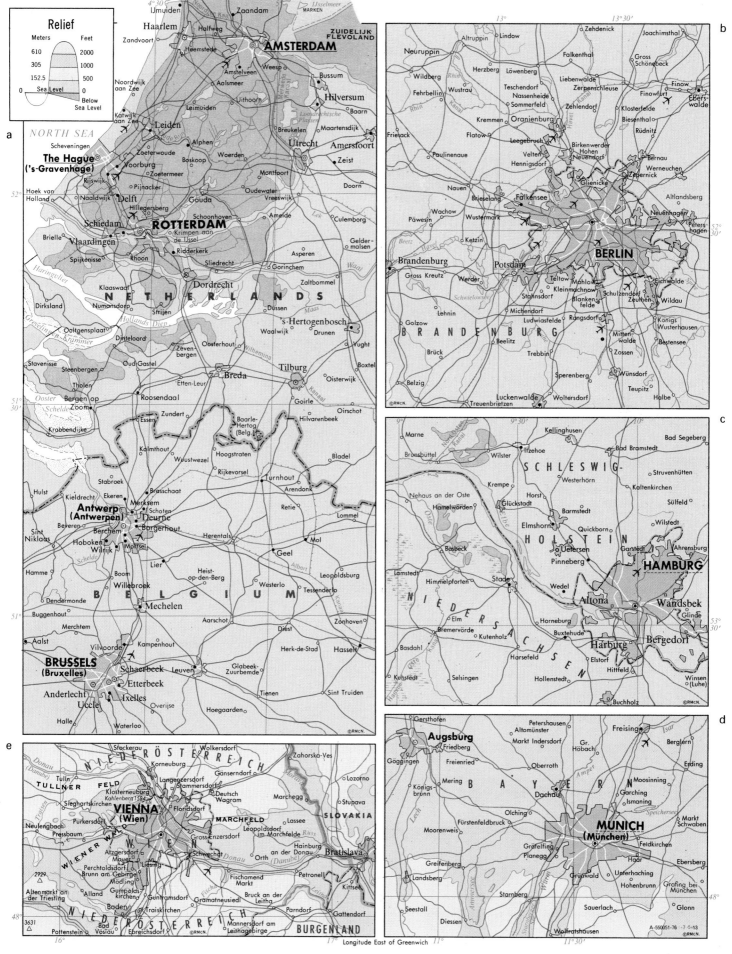

Relief

Meters	Feet
610	2000
305	1000
152.5	500
0 Sea Level	0 Below Sea Level

a

NORTH SEA

IJmuiden · Zaandam · MARKEN · *IJsselmeer*

Haarlem · Halfweg · ZUIDELIJK FLEVOLAND

Zandvoort · Heemstede · **AMSTERDAM**

Amstelveen · Aalsmeer · Weesp · Bussum

Noordwijk aan Zee · Uithoorn · Hilversum

Leimuiden · *Loosdrechtse Plassen* · Baarn

Katwijk aan Zee · Alphen · Breukelen · Maartensdijk

Leiden · Woerden · **Utrecht** · Amersfoort

Scheveningen · Zoeterwoude · Boskoop · Zeist

The Hague ('s-Gravenhage) · Voorburg · Zoetermeer · Montfoort · Doorn

Rijswijk · Pijnacker · Oudewater · Vreeswijk

52° · Hoek van Holland · Naaldwijk · **Delft** · Gouda · Ameide · Culemborg

Hillegersberg · Schoonhoven · *Lek*

Schiedam · **ROTTERDAM** · Krimpen aan de IJssel · Gelder-malsen

Brielle · Vlaardingen · Rhoon · Ridderkerk · Asperen · Gorinchem

Spijkenisse · Sliedrecht · *Waal*

Haringvliet · Klaaswaal · **Dordrecht** · Zaltbommel

Dirksland · Numansdorp · Strijen · Dussen · *Maas*

Grevelingen-Krammer · Ooltgensplaat · *Hollands Diep* · s-Hertogenbosch

Stavenisse · Steenbergen · Oud-Gastel · Zeven-bergen · Oosterhout · Waalwijk · Drunen · Vught

Tholen · **NETHERLANDS** · Etten-Leur · **Breda** · **Tilburg** · Boxtel

51° 30' · *Ooster Schelde* · Bergen op Zoom · Roosendaal · Goirle · Oisterwijk

Krabbendijke · Zundert · Baarle-Hertog (Belg.) · Hilvarenbeek · Oirschot

Essen · Hoogstraten · Bladel

Hulst · Kalmthout · Wuustwezel · Rijkevorsel · Turnhout · Arendonk

Stabroek · Brasschaat · Retie · Lommel

Kieldrecht · Ekeren · Merksem · Schoten · **Antwerp (Antwerpen)** · Deurne · Herentals · Mol · Geel

Beveren · Berchem · Borgerhout · Leopoldsburg

Sint Niklaas · Hoboken · Wilrijk · Mortsel · Lier

Hamme · Boom · Heist-op-den-Berg · Westerlo · Tessenderlo

Willebroek · **B E L G I U M** · Zonhoven

51° · Buggenhout · Dendermonde · **Mechelen** · Aarschot · Diest · Herk-de-Stad · Hasselt

Aalst · Merchtem · Vilvoorde · Kampenhout · Sint Truiden

BRUSSELS (Bruxelles) · Schaerbeek · Leuven · Glabbeek-Zuurbemde

Anderlecht · Etterbeek · Tienen

Uccle · Ixelles · Overijse · Hoegaarden

Halle · Waterloo

b

13° · 13° 30'

Altruppin · Lindow · Zehdenick · Joachimsthal

Neuruppin · Herzberg · Löwenberg · Falkenthal · Gross Schönebeck

Wildberg · Wustrau · Teschendorf · Liebenwalde · Zerpenschleuse · Finow · Ebers-walde

Fehrbellin · Nassenheide · Zehlendorf · Finowfurt

Rhin · Sommerfeld · Klosterfelde · Biesenthal · Rüdnitz

Friesack · Kremmen · **Oranienburg** · Birkenwerder · Hohen Neuendorf · Bernau

Flatow · Leegebruch · Werneuchen

Paulinenaue · Velten · Hennigsdorf · Zepernick

Nauen · Brieselang · **Falkensee** · Glienicke · Altlandsberg

Wachow · Wustermark · Neuenhagen · Peters-hagen

52° 30' · Päwesin · *Havel* · **BERLIN**

Ketzin · Teltow · Mahlow · Eichwalde

Brandenburg · **Potsdam** · Stahnsdorf · Kleinmachnow · Blanken-felde · Schulzendorf · Zeuthen · Wildau

Gross Kreutz · Werder · Michendorf · Rangsdorf · Königs Wusterhausen

Golzow · Lehnin · Ludwigsfelde · Mitten-walde · Bestensee

B R A N D E N B U R G · Beelitz · Zossen

Brück · Trebbin · Sperenberg · Wünsdorf · Teupitz

Belzig · Luckenwalde · Woltersdorf · Halbe

®RMcN · Treuenbrietzen

c

9° · 9° 30' · 10°

Marne · Kellinghusen · Bad Segeberg

Bronsbüttel · Wilster · Itzehoe · Bad Bramstedt · Struvenhütten

Nord-Ostsee-Kanal · **SCHLESWIG-** · Westerhörn · Kaltenkirchen

Nehaus an der Oste · Krempe · Horst · Barmstedt · Sülfeld

Hamelwörden · Glückstadt · Elmshorn · Quickborn · Wilstedt

Basbeck · *Oste* · **H O L S T E I N** · Garstedt · Ahrensburg

Lamstedt · Uetersen · Pinneberg · **HAMBURG**

Himmelpforten · Stade · *Elbe* · Wedel · Altona · Wandsbek · Glinde

Bremervörde · Kutenholz · Horneburg · Buxtehude · Bergedorf

Basdahl · **N I E D E R S A C H S E N** · Harburg

Kuhstedt · Harsefeld · Elstorf · Hittfeld · Winsen (Luhe)

Selsingen · Hollenstedt · Buchholz

53° 30'

d

11° · Gersthofen · Petershausen · Freising

Augsburg · Friedberg · Altomünster · Markt Indersdorf · Gr. Höbach · Berglern

Göggingen · Freienried · Oberroth · Erding

Königs-brunn · Mering · **B A Y E R N** · Dachau · Moosinning · Garching · Ismaning

Olching · Speichersee · Markt Schwaben

Fürstenfeldbruck · **MUNICH (München)** · Feldkirchen

Moorenweis · Gräfelfing · Planegg · Haar · Ebersberg

Greifenberg · Grünwald · Unterhaching · Grafing bei München

Landsberg · Hohenbrunn

Starnberg · Sauerlach · Glonn

Seestall · Diessen · Wolfratshausen

A-550051-76 -7-5·13 · ®RMcN

48°

e

Donau (Danube) · Stockerau · Wolkersdorf · Zahorska-Ves

NIEDERÖSTERREICH · Korneuburg · Gänserndorf

Tulln · Langenzersdorf · Stammersdorf · Deutsch Wagram · Lozorno

TULLNER FELD · Klosterneuburg · Kahlenberg 1584 · Floridsdorf · Marchegg · Stupava

Sieghartskirchen · **VIENNA (Wien)** · Gross-Enzersdorf · **SLOVAKIA**

Neulengbach · Purkersdorf · **MARCHFELD** · Lassee

Pressbaum · **W I E N E R W** · Leopoldsdorf im Marchfelde · Hainburg an der Donau · **Bratislava**

2929 · Atzgersdorf · Mauer · Liesing · Schwechat · Orth (Danube) · Fischamend Markt · Petronell · Kittsee

Perchtoldsdorf · Brunn am Gebirge · *Fischa* · Bruck an der Leitha

Mödling · Gumpolds-kirchen · Guntramsdorf · Gramatneusiedl · Parndorf · Gattendorf

Altenmarkt an der Triesting · Alland · Baden · Traiskirchen · *Leitha* · Mannersdorf am Leithagebirge

3631 · **N I E D E R Ö S T E R R E I C H** · **BURGENLAND**

Pottenstein · Bad Vöslau · Ebreichsdorf

16° · 17° · Longitude East of Greenwich · 11° · 11° 30'

Scale 1:1 000 000; one inch to 16 miles.
Elevations and depressions are given in feet.

0 5 10 15 20 Miles
0 4 8 12 16 20 24 28 32 Kilometers

Continued on pages 136-137

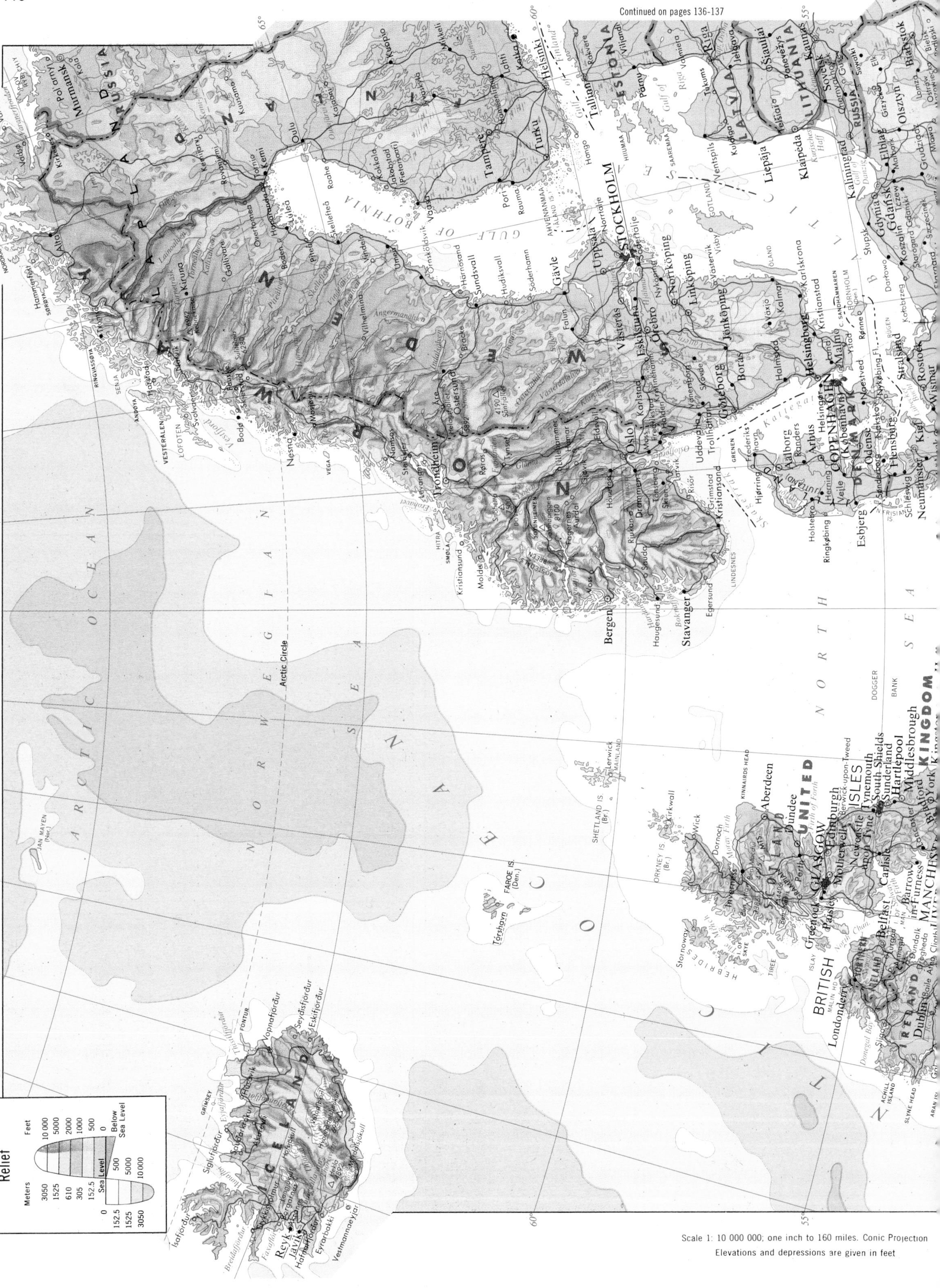

BELARUS

RUSSIA

Murmansk
Kola
Pol arnyj

LAPLAND

FINLAND

Helsinki

ESTONIA
Tallinn

LATVIA
Riga

LITHUANIA
Kaunas

RUSSIA
Kaliningrad

Gulf of Finland

Oulu

GULF OF BOTHNIA

Tampere
Turku

Gävle

STOCKHOLM
Uppsala
Södertälje
Norrköping
Linköping

Kiruna

SWEDEN

Östersund

Trollhättan
Göteborg
Borås
Halmstad
Helsingborg
Malmö

COPENHAGEN

DENMARK

Odense

Esbjerg

Oslo

NORWAY

Drammen
Kristiansand

Bergen

Stavanger

Trondheim

Namsos

Bodø

LOFOTEN
VESTERÅLEN

Narvik

NORWEGIAN SEA

Molde
Kristiansund

ARCTIC OCEAN

Arctic Circle

JAN MAYEN

NORTH SEA

DOGGER BANK

Aberdeen
Dundee
Edinburgh
GLASGOW
Newcastle upon Tyne
Sunderland
Middlesbrough
UNITED KINGDOM

SCOTLAND

SHETLAND IS. (Br.)
Lerwick

ORKNEY IS. (Br.)
Kirkwall

FAROE IS. (Den.)
Tórshavn

Stornoway

HEBRIDES

BRITISH ISLES

MANCHESTER
York
Bradford

Belfast
NORTHERN IRELAND

IRELAND
Dublin
Londonderry

ICELAND

Reykjavík

ATLANTIC OCEAN

Relief

Meters	Feet
3050	10 000
1525	5000
610	2000
305	1000
152.5	500
0	0
Sea Level	Below Sea Level
	Sea Level
152.5	500
1525	5000
3050	10 000

Scale 1: 10 000 000; one inch to 160 miles. Conic Projection
Elevations and depressions are given in feet

118

Continued on pages 116-117

Relief

Meters		Feet
3050		10000
1525		5000
610		2000
305		1000
152.5		500
0	Sea Level	0
		Below
		Sea Level
152.5		500
1525		5000
3050		10000

A-556300-76 '78-'86
COPYRIGHT BY
RAND M^{c}NALLY & COMPANY
MADE IN U.S.A.

Longitude West of Greenwich 0° Longitude East of Greenwich

Scale 1:10 000 000; one inch to 160 miles. Bonne's Projection
Elevations and depressions are given in feet

Continued on pages 136-137

The Turkish Republic of Northern Cyprus
unilaterally declared its independence
on Nov. 15, 1983.

Areas occupied by Israel since 1967.

BLACK SEA

SEA OF AZOV

AEGEAN SEA

MEDITERRANEAN SEA

IONIAN SEA

RED SEA

DEAD SEA

POLAND
SLOVAKIA
HUNGARY
UKRAINE
MOLDOVA
ROMANIA
CROATIA
BOSNIA AND HERZEGOVINA
SERBIA AND MONTENEGRO
BULGARIA
MACEDONIA
ALBANIA
GREECE
TURKEY
ASIA MINOR
RUSSIA
GEORGIA
CYPRUS
SYRIA
LEBANON
ISRAEL
JORDAN
EGYPT
SAUDI ARABIA
LIBYA

KATOWICE
BUDAPEST
BUCHAREST
BELGRADE (Beograd)
Sofia (Sofiya)
ISTANBUL
Ankara (Angora)
ATHENS (Athína)
Nicosia
Beirut
Damascus (Dimashq)
Tel Aviv-Yafo
Jerusalem
Amman
ALEXANDRIA (Al Iskandariyah)
CAIRO (Al Qahirah)
Banghazi

BARQAH (CYRENAICA)
LIBYAN PLATEAU
MUNKHAFAD AL QATTARAH
LIBYAN DESERT
SINAI PEN.

0   50 100 150 200 250 300 Miles
0   100   200   300   400   500 Kilometers

a

Same scale as main map

ATLANTIC

SHETLAND
ISLANDS
(Br.)

St. Magnus Bay

YELL

MAINLAND

Lerwick

FOULA

OCEAN

FAIR
ISLAND

WESTRAY
ROUSAY
N. RONALDSAY
SANDAY
STRONSAY
Kirkwall
MAINLAND
ORKNEY
ISLANDS
(Br.)
HOY
S. RONALDSAY
Pentland Firth
Thurso
DUNCANSBY HD.
SCOTLAND

©RMCN.

Relief

Meters		Feet
610		2000
305		1000
152.5		500
0	Sea Level	0
152.5		500 Below
1525		5000 Sea Level

A-559700-76--9.-7-17
COPYRIGHT BY
RAND McNALLY & COMPANY
MADE IN U.S.A.

Scale 1: 4 000 000; one inch to 64 miles. Conic Projection
Elevations and depressions are given in feet

Continued on pages 122-123

Continued on pages 124-125

Continued on pages 126-127

Longitude East of Greenwich

| 0 | 10 | 20 | 30 | 40 | 50 | 60 | 70 | 80 | 90 | 100 | 110 | 120 Miles |

| 0 | 20 | 40 | 60 | 80 | 100 | 120 | 140 | 160 | 180 | 200 Kilometers |

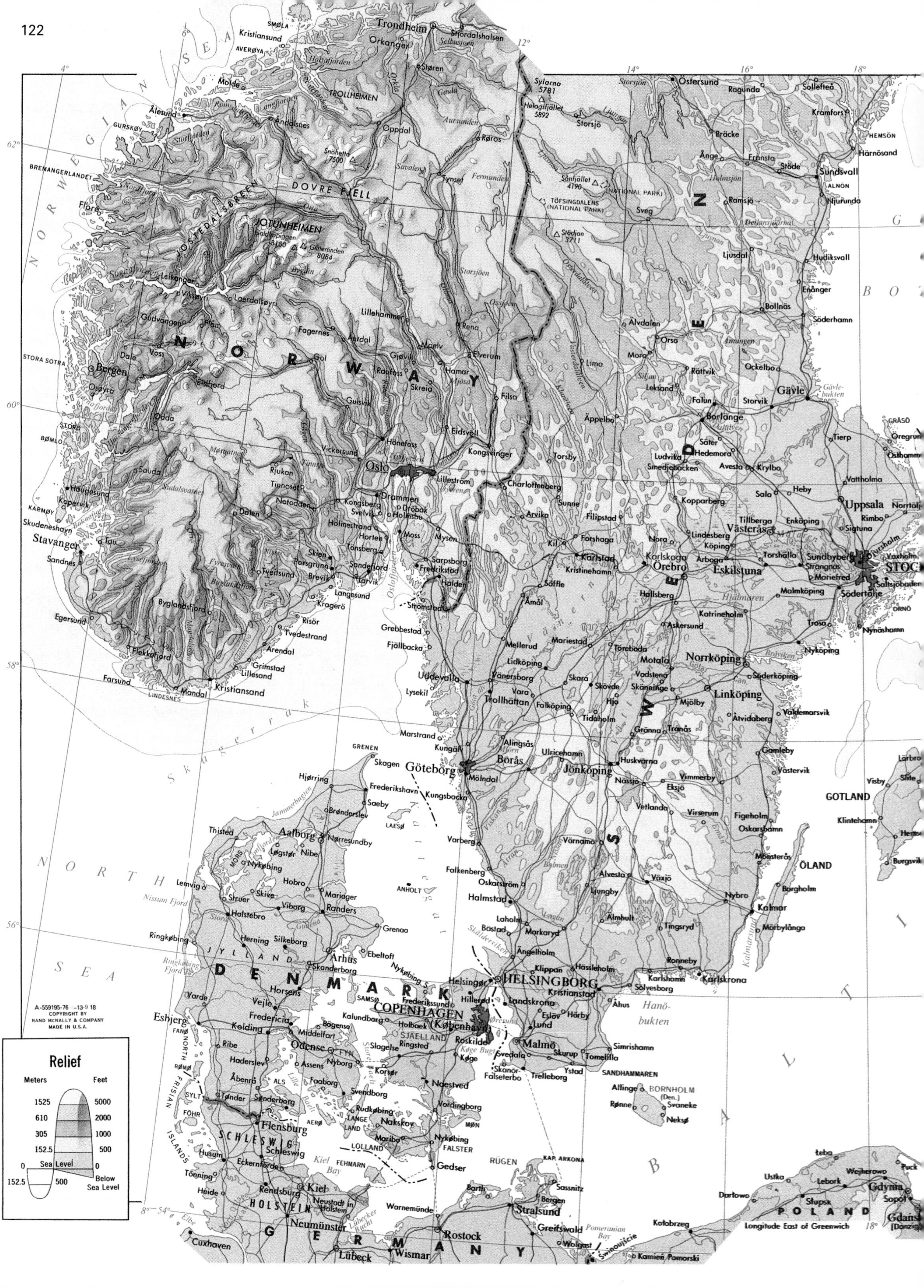

NORWEGIAN SEA

NORTH SEA

Skagerrak

Kattegat

BALTIC SEA

NORWAY

SWEDEN

DENMARK

GERMANY

POLAND

GOTLAND

ÖLAND

BORNHOLM (Den.)

JYLLAND

HOLSTEIN

SCHLESWIG

Trondheim
Stjørdalshalsen
Orkanger
Støren
Oppdal
Røros
Molde
Ålesund
Åndalsnes
Kristiansund
Smøla
Averøya
Sylarna 5781
Helagsfjället 5892
Storsjön
Östersund
Ragunda
Sollefteå
Kramfors
Härnösand
Sundsvall
Bräcke
Ånge
Fränsta
Stöde
Alnön
Njurunda
Ramsjö
Hudiksvall
Ljusdal
Enånger
Bollnäs
Söderhamn
Snøhetta 7500
Dovre Fjell
Jotunheimen
Galdhøpiggen 8100
Glittertinden 8084
Tynset
Fermunden
Sönfjället 4790
Töfsingdalens (National Park)
Sveg
Stödjan 3711
Leikanger
Vik
Lærdalsøyri
Fagernes
Lillehammer
Rena
Alvdalen
Orsa
Lima
Mora
Siljan
Rättvik
Ockelbo
Falun
Borlänge
Gävle
Gävlebukten
Gudvangen
Flåm
Voss
Dale
Bergen
Osøyra
Aurdal
Gjøvik
Raufoss
Mjøsa
Hamar
Skreia
Elverum
Filsa
Appelbo
Ludvika
Säter
Hedemora
Avesta
Krylbo
Gol
Gulsvik
Eidsvoll
Hønefoss
Kongsvinger
Torsby
Smedjebacken
Kopparberg
Sala
Heby
Uppsala
Sigtuna
Odda
Sauda
Rjukan
Tinnoset
Oslo
Lillestrøm
Charlottenberg
Arvika
Sunne
Filipstad
Nora
Lindesberg
Köping
Torshälla
Sundbyberg
Mariefred
STOCKHOLM
Haugesund
Kopervik
Karmøy
Skudeneshavn
Notodden
Kongsberg
Svelvik
Drammen
Drøbak
Holmsbu
Kil
Forshaga
Karlstad
Karlskoga
Arboga
Eskilstuna
Strängnäs
Södertälje
Saltsjöbaden
Stavanger
Sandnes
Tau
Dalen
Skien
Porsgrunn
Brevik
Larvik
Sandefjord
Tønsberg
Horten
Holmestrand
Moss
Mysen
Sarpsborg
Fredrikstad
Halden
Säffle
Åmål
Kristinehamn
Örebro
Hallsberg
Malmköping
Katrineholm
Trosa
Nynäshamn
Egersund
Flekkefjord
Byglandsfjord
Risør
Tvedestrand
Arendal
Grimstad
Lillesand
Farsund
Mandal
Kristiansand
Strömstad
Grebbestad
Fjällbacka
Uddevalla
Lysekil
Mellerud
Vänersborg
Lidköping
Mariestad
Skara
Skövde
Töreboda
Motala
Vadstena
Askersund
Norrköping
Söderköping
Linköping
Åtvidaberg
Valdemarsvik
Gamleby
Västervik
Marstrand
Kungälv
Göteborg
Mölndal
Alingsås
Borås
Ulricehamn
Jönköping
Huskvarna
Gränna
Tranås
Mjölby
Eksjö
Vimmerby
Vetlanda
Virserum
Figeholm
Oskarshamn
Mönsterås
Visby
Klintehamn
Skagen
Hjørring
Frederikshavn
Saeby
Brønderslev
Kungsbacka
Varberg
Falkenberg
Nässjö
Värnamo
Alvesta
Ljungby
Växjö
Nybro
Kalmar
Borgholm
Thisted
Aalborg
Nørresundby
Øgstør
Nibe
Nykøbing
Hobro
Mariager
Randers
Oskarström
Halmstad
Laholm
Markaryd
Älmhult
Tingsryd
Ronneby
Mörbylånga
Lemvig
Struer
Skive
Viborg
Grenaa
Båstad
Ängelholm
Hässleholm
Karlshamn
Karlskrona
Sölvesborg
Åhus
Hanöbukten
Ringkøbing
Herning
Silkeborg
Århus
Ebeltoft
Skanderborg
Nykøbing
Klippan
Kristianstad
Varde
Horsens
Helsingør
Helsingborg
Landskrona
Eslöv
Hörby
Simrishamn
Esbjerg
Kolding
Fredericia
Vejle
Middelfart
Bogense
Kalundborg
Holbaek
Hillerød
København
COPENHAGEN
Roskilde
Lund
Malmö
Svedala
Skurup
Tomelilla
Ribe
Haderslev
Assens
Odense
Nyborg
Slagelse
Ringsted
Køge
Skanör
Falsterbo
Trelleborg
Ystad
Åbenrå
Tønder
Sønderborg
Svendborg
Rudkøbing
Nakskov
Maribo
Nykøbing
Næstved
Vordingborg
Gedser
Rønne
Svaneke
Neksø
Flensburg
Husum
Eckernförde
Schleswig
Rendsburg
Kiel
Neumünster
Lübeck
Cuxhaven
Wismar
Rostock
Warnemünde
Stralsund
Greifswald
Sassnitz
Barth
Kołobrzeg
Darłowo
Słupsk
Ustka
Lębork
Wejherowo
Gdynia
Sopot
Gdańsk

Relief
Meters / Feet
1525 / 5000
610 / 2000
305 / 1000
152.5 / 500
0 Sea Level 0
152.5 / 500 Below Sea Level

A-559195-76 -13-9 18
COPYRIGHT BY
RAND MCNALLY & COMPANY
MADE IN U.S.A.

Longitude East of Greenwich

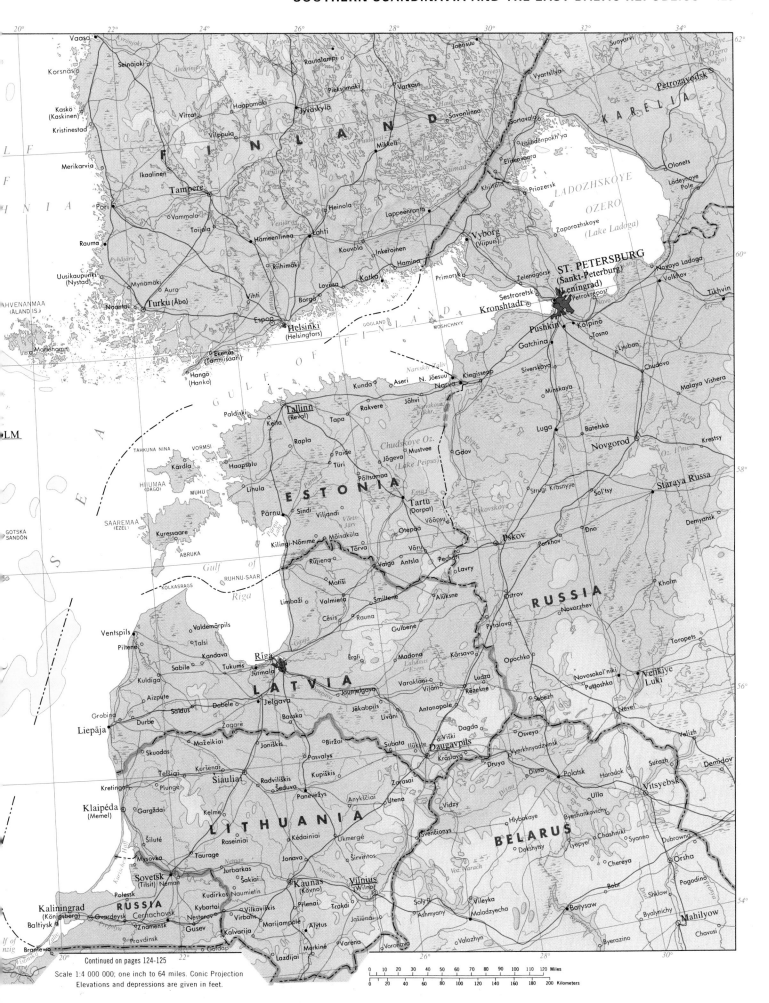

Continued on pages 124-125

Scale 1:4 000 000; one inch to 64 miles. Conic Projection
Elevations and depressions are given in feet.

Continued on pages 122-123

Continued on pages 126-127

Continued on pages 130-131

Longitude East of Greenwich

Scale 1:4 000 000; one inch to 64 miles. Conic Projection
Elevations and depressions are given in feet.

COPYRIGHT BY RAND McNALLY & COMPANY
MADE IN U.S.A.

Continued on pages 122-123

Continued on pages 132-133

Relief

Meters	Feet
3050	10 000
1525	5000
610	2000
305	1000
152.5	500
0 Sea Level	0
	Below Sea Level

RUSSIA

LITHUANIA

BELARUS

POLAND

UKRAINE

SLOVAKIA

HUNGARY

ROMANIA

MOLDOVA

TRANSYLVANIA

CARPATHIAN MOUNTAINS

GALICIA

MASURIA

SEA

Gulf of Danzig

WARSAW (Warszawa)

BUDAPEST

KATOWICE

Kraków

Kaunas (Kovno)

Vilnius

Minsk

0 10 20 30 40 50 60 70 80 90 100 110 120 Miles
0 20 40 60 80 100 120 140 160 180 200 Kilometers

126

Continued on pages 120-121

Relief

Meters	Feet
3050	10 000
1525	5000
610	2000
305	1000
152.5	500
Sea Level	0
152.5	500
1525	5000

UNITED KINGDOM

Exeter
Honiton
Dorchester
Southampton Portsmouth
Hove
Dover
Folkestone
Calais
Dunkerque
Roeselare Ieper
Gent Aalst Mechelen
Leuven
Anderlecht BRUSSELS
BELGIUM

Launceston
Exmouth
Poole
Bournemouth
Cowes
Chichester
Brighton
Worthing
Lewes
Hastings
Bexhill
Eastbourne
Boulogne-sur-Mer
St. Omer
Armentières
Lille
Tourcoing Roubaix
Kortrijk
Nivelles
Namur

Torquay (Torbay)
Dartmouth
Weymouth
Newport
ISLE OF WIGHT
Ryde
Portsmouth
Étaples
Berck
Bruay-en-Artois
Béthune
Douai
Valenciennes
Mons
Charleroi
Maubeuge
Dinant

Plymouth
START POINT
Arras
Denain
Crécy-en-Ponthieu
Hautmont
Cambrai
Givet
ARDEN

CHANNEL
Abbeville
St. Valéry-sur-Somme
PICARDIE
Bohain-en-Vermandois
Fourmies
Hirson
Nouzonville

ENGLISH
Dieppe
Le Tréport
Amiens
Albert
Péronne
St. Quentin
Corbie
Guise
Rethel
Charleville-Mézières
Sedan

C. DE LA HAGUE
ALDERNEY
Fécamp
Neufchâtel-en-Bray
Montdidier
Roye
Chauny
Laon
Vouziers
ARGONNE
Verdun

GUERNSEY
St. Peter Port
SARK
CHANNEL ISLANDS (Br.)
PTE. DE BARFLEUR
Cherbourg
Baie de la Seine
Bolbec
Yvetot
Le Havre
Trouville
Honfleur
Pont-Audemer
Rouen
Beauvais
Compiègne
Soissons
Reims
Châlons-sur-Marne
Bar-le-Duc

JERSEY
St. Helier
Valognes
Carentan
Bayeux
Caen
Lisieux
Elbeuf
Louviers
Vernon
Gisors
Méru
Creil
Pontoise
St-Denis
Meaux
Epernay
CHAMPAGNE
Vitry-le-François
Arcis-sur-Aube
Joinville
St. Dizier

St. Pol-de-Léon
Morlaix
Guingamp
St. Brieuc
St. Malo
Dinard
Dinan
Granville
Condé
Flers
Argentan
L'Aigle
Évreux
St. Germain-en-Laye
Argenteuil
St-Denis
Clichy
PARIS
Troyes
Chaumont
Langres

I. D'OUESSANT
Brest
Landerneau
MTS. D'ARRÉE
Carhaix-Plouguer
Pontivy
Lamballe
Avranches
COLLINES DE NORMANDIE
Dreux
Rambouillet
Versailles
Boulogne-Billancourt
Melun
Corbeil-Essonnes
Romilly-sur-Seine
Aube
Sens
Montbard

PTE. DU RAZ
Douarnenez
Audierne
Pont-l'Abbé
Quimper
BRETAGNE
Montfort
Vitré
Fougères
Alençon
Nogent-le-Rotrou
Chartres
Étampes
Fontainebleau
Nemours
Pithiviers
Montargis
Joigny
Auxerre
Avallon
MORVAN
Dijon
CÔTE D'OR
PLATEAU

Concarneau
ÎLES DE GLÉNAN
Lorient
ÎLE DE GROIX
Hennebont
Vannes
Ploërmel
Rennes
Laval
Le Mans
Château-Gontier
Sablé-sur-Sarthe
La Flèche
Vendôme
Châteaudun
Orléans
Clamecy
Beaune

Quiberon
BELLE-ÎLE
St. Nazaire
Redon
Châteaubriant
Angers
Trélazé
Tours
Amboise
Blois
SOLOGNE
Romorantin-Lanthenay
Cosne-sur-Loire
Gien
Briare
Autun
Le Creusot
Chalon-sur-Saône
Montceau

ÎLE DE NOIRMOUTIER
Pornic
Nantes
Cholet
Chemillé
Saumur
Chinon
Loches
FRANCE
Vierzon
Mehun-sur-Yèvre
Bourges
Nevers
Paray-le-Monial
Digoin
Cluny
Mâcon
Bourg-en-Bresse

ÎLE D'YEU
L. de Grand Lieu
Thouars
Bressuire
Loudun
Descartes
Issoudun
St. Florent-sur-Cher
Moulins
Commentry

La Roche-sur-Yon
Fontenay-le-Comte
Parthenay
HAUTEURS DE GÂTINE
Châtellerault
Poitiers
Le Blanc
Argenton-sur-Creuse
Châteauroux
St. Amand-Mont Rond
Montluçon
Commentry

Les Sables-d'Olonne
Luçon
Montmorillon
Guéret
Vichy
Roanne
Villefranche

Pertuis Breton
ÎLE DE RÉ
La Rochelle
Surgères
Rochefort
Ruffec
Confolens
Aubusson
MTS DU FOREZ
Thiers
Tarare
Lyon

BAY OF BISCAY
ÎLE D'OLÉRON
Marennes
St. Jean-d'Angely
St. Junien
Limoges
Riom
Clermont-Ferrand
Villeurbanne

La Tremblade
Saintes
Cognac
Barbezieux
Jonzac
Angoulême
St. Yrieix-la-Perche
PLATEAUX DU LIMOUSIN
Ussel
Bort-les-Orgues
Puy de Sancy 6185
Issoire
Ambert
Montbrison
St. Chamond
Rive-de-Gier
Vienne

Royan
Blaye-et-Ste. Luce
Périgueux
Brive-la-Gaillarde
Tulle
Argentat
AUVERGNE
Murat
Plomb du Cant. 6076
St. Flour
Le Puy
Mt. Mézenc 5751
Annonay
Yssingeaux
St. Étienne
Firminy
Romans

Étang de Carcans
Blanquefort
Mérignac
Pessac
Bègles
Bordeaux
Libourne
Bergerac
Dordogne
Sarlat-la-Canéda
Aurillac
MASSIF CENTRAL
Langogne
Privas
Aubenas
Le Teil

Arcachon
Bassin d'Arcachon
La Teste-de-Buch
Étang de Cazaux
LANDES
La Réole
Langon
Marmande
Tonneins
Lot
Figeac
Decazeville
Mende
Montélimar
Valréas

Labouheyre
Étang de Biscarosse
Villeneuve-sur-Lot
Cahors
Aubin
Rodez
Villefranche-de-Rouergue
Millau
CÉVENNES
Bagnols-sur-Cèze
Orange

Mont-de-Marsan
Agen
Nérac
Condom
Castelsarrasin
Moissac
Carmaux
Gaillac
Albi
St. Affrique
Lodève
Vigan
Alès
La Grand-Combe
Avignon
Beaucaire
Nîmes
Tarascon

GASCOGNE
Dax
Adour
Aire-sur-l'Adour
Verdun
Montauban
Gaillac
St. Afrique
Bédarieux
Montpellier
Lunel
Miramas
Arles

Biarritz
Bayonne
Salies-de-Béarn
Orthez
Auch
Toulouse
Muret
Castres
Baziège
Castelnaudary
Pézenas
Béziers
Sète
Agde
Golfe du Lion
Martigues

Irún
St. Jean-de-Luz
Pau
Tarbes
St. Gaudens
Carcassonne
Limoux
Narbonne

Roncesvalles
Oloron-Ste. Marie
Lourdes
Bagnères-de-Bigorre
St. Girons
Foix
Tarascon
Quillan
Sigean
Rivesaltes
Perpignan

Pamplona
Jaca
PYRÉNÉES
Bagnères-de-Luchon
Mt. Perdido 11007
Pico de Aneto 11168
ANDORRA
Ax-les-Thermes
Prades
Céret
Port Vendres
C. DE CREUS

Tafalla
Boltaña
SPAIN
Andorra
MED

Continued on pages 128-129

Scale 1:4 000 000; one inch to 64 miles. Conic Projection
Elevations and depressions are given in feet

A-550900-76 3-7-6-14
COPYRIGHT BY
RAND MCNALLY & COMPANY
MADE IN U.S.A.

a

Miramas
St. Chamas
Équilles
Aix-en-Provence
Istres
Berre-l'Étang
Rognac
Gardanne
Simiane
Étang de Berre
Marignane
St. Victoret
Allauch
Port-de-Bouc
Lavéra
Martigues
Châteauneuf
L'Estaque
Carry-le-Rouet
La Penne-sur-Huveaune
Marseille
La Couronne
Carro
Sausset-les-Pins
COTE DE LA GINESTE 1075
Mazargue
La Madrague
Golfe du Lion

MEDITERRANEAN SEA

Scale 1:1 000 000
0   5   10 Miles
0 4 8 12 16 Kilometers

©RMCN.

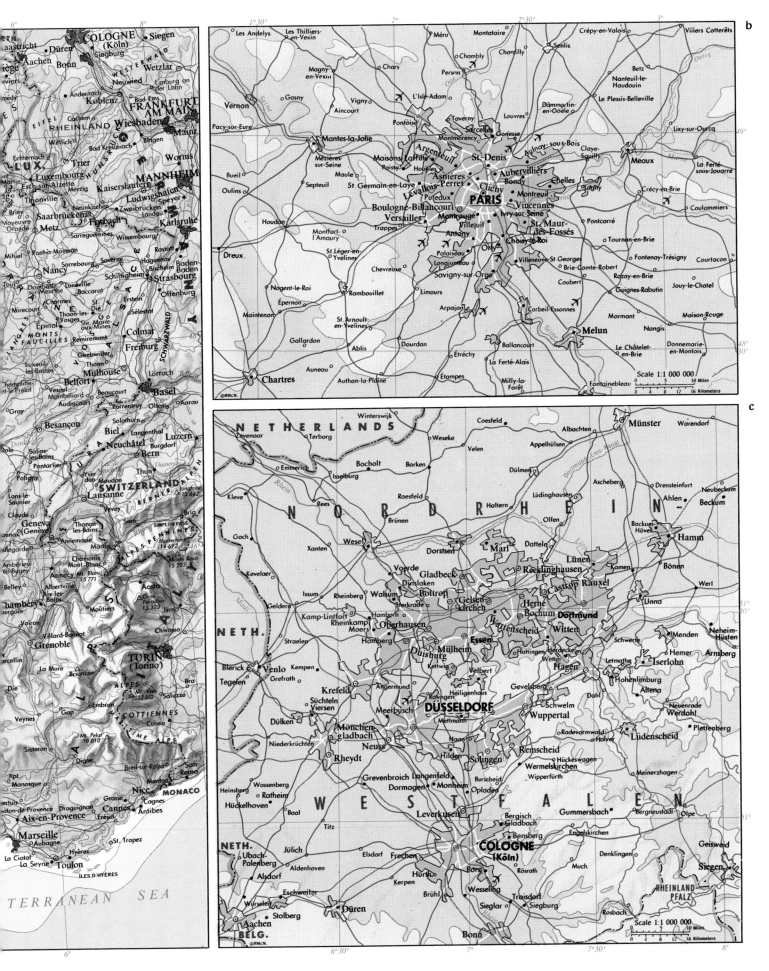

b

c

Map b (Paris region)

Les Andelys
Les Thilliers-en-Vexin
Montataire
Crépy-en-Valois
Viliers Cotterêts

Méru
Chambly
Chantilly
Senlis
Betz

Magny-en-Vexin
Chars
Persan
Nanteuil-le-Haudouin

Vernon
Gasny
Vigny
L'Isle-Adam
Louvres
Dammartin-en-Goële
Le Plessis-Belleville

Pacy-sur-Eure
Aincourt
Pontoise
Taverny
Goussainville
Lixy-sur-Ourcq

Mantes-la-Jolie
Sarcelles
Claye-Souilly
Meaux

Mézières-sur-Seine
Maisons-Laffitte
Montmorency
Aulnay-sous-Bois
La Ferté-sous-Jouarre

Bueil
Poissy
Argenteuil
St-Denis
Maule
Asnières
Aubervilliers
Bondy
Chelles
Lagny
Crécy-en-Brie

Septeuil
St. Germain-en-Laye
Levallois-Perret
Clichy
Montreuil
Vincennes

Oulins
Houdan
PARIS
Ivry-sur-Seine
St. Maur-des-Fossés
Coulommiers

Boulogne-Billancourt
Puteaux
Versailles
Montrouge
Choisy-le-Roi
Pontcarré

Montfort-l'Amaury
Trappes
Villejuif
Antony
Orly
Tournan-en-Brie

Dreux
St-Léger-en-Yvelines
Longjumeau
Villeneuve-St Georges
Fontenay-Trésigny
Courtacon

Chevreuse
Savigny-sur-Orge
Brie-Comte-Robert
Rozay-en-Brie
Jouy-le-Chatel

Nogent-le-Roi
Limours
Arpajon
Coubert
Guignes-Rabutin

Maintenon
Épernon
Rambouillet
Corbeil-Essonnes
Mormant
Maison-Rouge

St. Arnoult-en-Yvelines
Dourdan
Melun
Nangis
Donnemarie-en-Montois

Gallardon
Auneau
Ballancourt
Le Châtelet-en-Brie

Ablis
Étréchy
La Ferté-Alais

Chartres
Authon-la-Plaine
Étampes
Milly-la-Forêt
Fontainebleau

Scale 1:1 000 000
0   4   8   12   16 Kilometers
0   5   10 Miles

Map c (Düsseldorf / Cologne region)

Winterswijk
Coesfeld
Albachten
Münster
Warendorf

NETHERLANDS
Zevenaar
Terborg
Weseke
Appelhülsen

Emmerich
Bocholt
Borken
Velen
Dülmen
Ascheberg
Drensteinfurt
Neubeckum

Kleve
Rees
Isselburg
Lüdinghausen
Ahlen
Beckum

N O R D R H E I N -
Raesfeld
Halterh
Olfen
Bockum-Hövel
Hamm

Goch
Xanten
Wesel
Dorstsen
Marl
Lünen
Bönen
Werl

Kevelaer
Voerde
Gladbeck
Recklinghausen
Kamen

Issum
Dinslaken
Bottrop
Castrop Rauxel
Unna

Geldern
Rheinberg
Walsum
Gelsenkirchen
Herne
Bochum
Dortmund

Kamp-Lintfort
Sterkrade
Wattenscheid
Witten
Schwerte
Menden
Neheim-Hüsten

Straelen
Rheinkamp
Oberhausen
Essen
Herdecke
Hemer
Arnsberg

NETH.
Moers
Homberg
Mülheim
Hattingen
Wetter
Hagen
Iserlohn

Blerick
Kempen
Duisburg
Kettwig
Velbert
Gevelsberg
Letmathe

Venlo
Grefrath
Angermund
Heiligenhaus
Schwelm
Hohenlimburg
Altena

Tegelen
Krefeld
Ratingen
Gevelsberg
Dahl
Neuenrade
Werdohl

Süchteln
Viersen
Meerbusch
DÜSSELDORF
Mettmann
Wuppertal
Lüdenscheid

Dülken
Haan
Radevormwald
Halver
Plettenberg

Mönchengladbach
Neuss
Hilden
Solingen
Remscheid
Hückeswagen
Meinerzhagen

Niederkrüchten
Rheydt
Grevenbroich
Langenfeld
Burscheid
Wermelskirchen
Wipperfürth

Wassenberg
Dormagen
Monheim
Opladen
Bergisch Gladbach
Gummersbach
Bergneustadt
Olpe

Heinsberg
Ratheim
Leverkusen
Bensberg
Engelskirchen

Hückelhoven
Baal
Titz
W E S T F A L E N

NETH.
Jülich
Elsdorf
Frechen
COLOGNE (Köln)
Denklingen

Ubach-Palenberg
Aldenhoven
Kerpen
Hürth
Porz
Rösrath
Much
Siegen

Wurselen
Eschweiler
Düren
Brühl
Wesseling
Troisdorf
RHEINLAND PFALZ

Aachen
Stolberg
Sieglar
Siegburg

BELG.
Bonn
Rosbach

Scale 1:1 000 000
0   4   8   12   16 Kilometers
0   5   10 Miles

Map a (left column — Rhine / Alps)

COLOGNE (Köln)
Siegen

Aachen
Bonn
Siegburg

Maastricht
Düren
WESTERWALD

Andernach
Neuwied
Limburg an der Lahn
Wetzlar

Koblenz
Bad Ems

EIFEL
Cochem
RHEINLAND
Wiesbaden
FRANKFURT AM MAIN

Wittlich
Bad Kreuznach
Bingen
Mainz

Echternach
Trier
Worms

LUX.
Luxembourg
HUNSRÜCK
MANNHEIM

Esch-sur-Alzette
Merzig
Kaiserslautern
Ludwigshafen
Speyer

Thionville
Saarbrücken
Forbach
Landau
Karlsruhe

Briey
Neunkirchen
Zweibrücken

Metz
Sarreguemines
Wissembourg

Sarrebourg
Saverne
Hagenau
Baden-Baden

Pont-à-Mousson
Bischeim
Strasbourg

Nancy
Schiltigheim
Offenburg

Lunéville
Baccarat
Erstein
Sélestat

Charmes
St. Dié
Ste. Marie-aux-Mines

Mirecourt
Thaon-les-Vosges
Colmar
Freiburg

Épinal
Remiremont
Guebwiller

MONTS FAUCILLES
SCHWARZWALD

Luxeuil-les-Bains
Belfort
Mulhouse
Lörrach

Vesoul
Montbéliard
Beaucourt

Gray
Audincourt
Basel
Aarau

Besançon
Biel
Langenthal
Olten

Salins-les-Bains
Neuchâtel
Luzern

Dole
Pontarlier
Burgdorf
Bern

Poligny
JURA
SWITZERLAND
Thun

Lons-le-Saunier
Yverdon
Moudon
Jungfrau 13 642

St. Claude
Lausanne
BERNER ALPEN
Brig

Geneva (Genève)
Vevey
Sion
SIMPLON PASS

Thonon-les-Bains
Martigny
ALPES PENNINES
Matterhorn 14 692

Annemasse
Chamonix Mont-Blanc 15 771
Monte Rosa 15 203

Amberieu-en-Bugey
Mont-Blanc 15 781
Aosta

Belley
Albertville
Gran Paradiso 13 323
Ivrea

Chambéry
Moûtiers
Chivasso

Aix-les-Bains
Villard-Bonnot

Voiron
Grenoble
TURIN (Torino)

La Mure
Briançon
ALPES
Bra

Veynes
Embrun
Mt. Viso 12 602
Saluzzo
COTTIENNES

Gap
Cuneo

Sisteron
Digne
Mt. Pelat 10 010
MARITIME ALPS

Apt
Manosque
Breil-sur-Roya
San Remo

Draguignan
Menton
Cagnes

Grasse
Nice
MONACO

Aix-en-Provence
Cannes
Antibes

Marseille
Aubagne
Fréjus

La Ciotat
Hyères
St. Tropez

La Seyne
Toulon
ÎLES D'HYÈRES

MEDITERRANEAN SEA

0 10 20 30 40 50 60 70 80 90 100 110 120 Miles
0   20   40   60   80   100 120 140 160 180 200 Kilometers

128

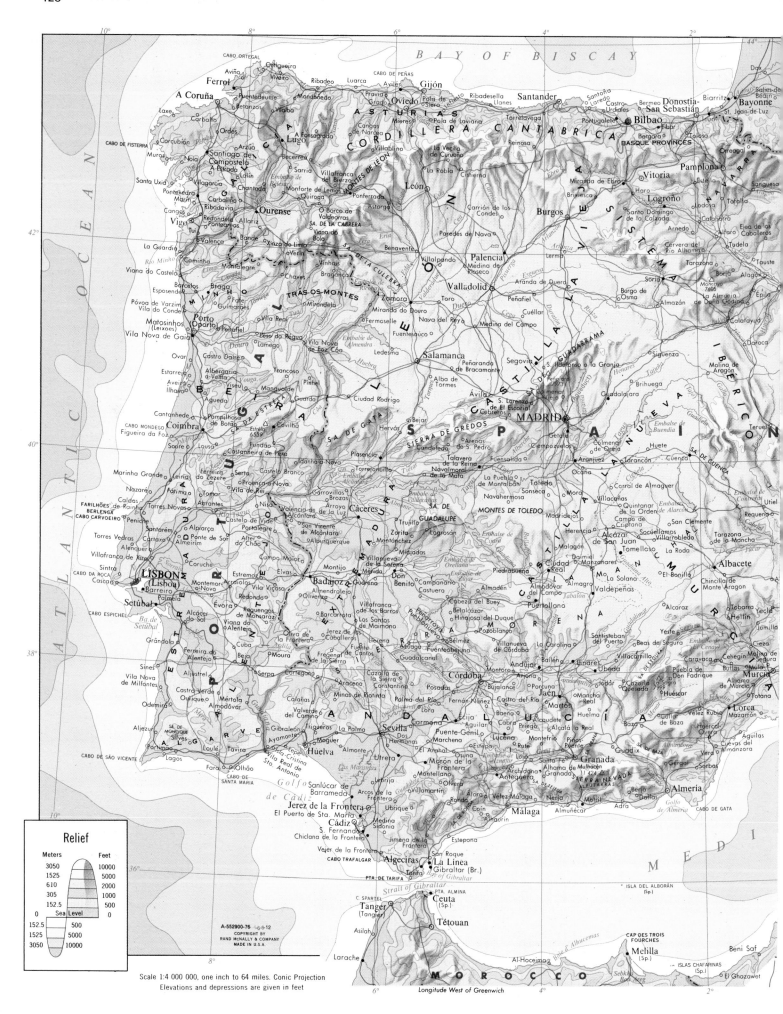

BAY OF BISCAY

CABO ORTEGAL
Ortigueira
Aviño
Ferrol   Ribadeo Luarca CABO DE PEÑAS
A Coruña Puentedeume Viveiro   Avilés Gijón
          Betanzos Mondoñedo   Provia Grado Oviedo Pola de   Ribadesella Santander Santoña
Laxe Carballo   A Fonsagrada Cangas Siero Llanes   Laredo
CABO DE FISTERRA Corcubión Ordés Arzúa Lugo de Nardea de Laviana Torrelavega Reinosa Bermeo Donostia- Biarritz
        Muros Noia A Estrada Becerreá Sarria   La Vecilla Cistierna   Urdiales San Sebastián St. Jean-de-Luz Bayonne
        Santiago de Chantada Villafranca de Curueño   Portugalete Bilbao Eibar
        Compostela   del Bierzo La Robla   Miranda de Ebro Bergara BASQUE PROVINCES Irún Orreaga
        Santa Uxía Villagarcía Monforte de Lemos Ponferrada León   Briviesca   Vitoria Pamplona
        Pontevedra Marín Quiroga Astorga   Carrión de los Burgos   Haro Logroño
        Cangas Ribadavia O Barco de   Condes   Santo Domingo Estella Sangüesa
        Vigo Redondela Valdeorras   Benavente   de la Calzada Arnedo Calahorra
        Tui Ponteareas SA. DE LA CABRERA Villalpando Palencia   Lerma Burgo de Soria Cervera del Ejea de los
La Guardia Valença Bande Vila do Paredes de Nava   Medina de   Osma   Río Alhama Caballeros
        Caminha Xinzo de Limia Bolo   Rioseco Valladolid Aranda de Duero   Almazán Tarazona Tauste
Viana do Castelo Montalegre Verín   Zamora Toro   Peñafiel   Borja Alagón
Esposende Barcelos Chaves Vinhais Bragança Miranda do Douro Nava del Rey Medina del Campo   Monzón Epila
Póvoa de Varzim Braga Fafe Mirandela   Fermoselle Fuentesaúco   La Almunia Daroca
Vila do Conde Guimarães   Vila Real Vila Nova Ledesma Salamanca Segovia SA. DE GUADARRAMA Sigüenza de Doña Godina
Matosinhos Fâo   de Foz Côa   Peñaranda   S. Ildefonso o la Granja Guadalajara Molina de
(Leixões) Porto Peñafiel Lamego   Huebra de Bracamonte Ávila Cebreros Brihuega Aragón
Vila Nova de Gaia Oporto Amarante   Alba de   S. Lorenzo Colmenar Henares IBÉRICO
        Ovar Castro Daire Peso da Regua Tormes de El Escorial Guadix   Teruel
Estarreja Albergaria Viseu Trancoso Pinhel   Hervás SIERRA DE GREDOS MADRID Embalse de
Aveiro a-Velha Mangualde Guarda   Béjar   Getafe Buendia
Ílhavo Águeda   Castelheira de Pêra   Arenas Candeleda Ciempozuelos
CABO MONDEGO Coimbra Pampilhosa Estrela Covilhã   Plasencia de S. Pedro   Colmenar Huete
Figueira da Foz do Botão 6539 Fundão   Navalmoral Talavera de Oreja SA. DE CUENCA
        Soure Lousã   Idanha-a-Nova Torrecillas de la Mata de la Reina Aranjuez Tarancón Cuenca
Marinha Grande Serta Castelo Proença-a-Nova   La Puebla Toledo Ocaña
Nazaré Leiria Ferreira Branco Vila de Rei   de Montalbán Sonseca Corral de Almaguer
FARILHÕES Fátima do Zêzere   Garrovillas Cáceres Navahermosa Mora Villacañas Embalse de
BERLENGA Caldas Tomar Niso Arroyo de Brozas   MONTES DE TOLEDO Quintanar Alcaraz Utiel
CABO CARVOEIRO da Rainha Torres Novas Abrantes Valencia de la Luz SA. DE Madridejos de la Orden
Peniche Santarém Alpiarça Castelo de Vide de Alcántara GUADALUPE Herencia Campo de San Clemente
Torres Vedras Cartaxo Ponte de Portalegre   Trujillo   Criptana Socuéllamos Requena
Villafranca de Xira Almeirim Sor Alter San Vicente Logrosán   Madrigalejo Alcázar Villarrobledo Tarazona
        Alenquer do Chão de Alcántara Zorita   Piedrabuena de San Juan La Roda de la Mancha
Sintra Coruche   Campo Maior Montánchez   Daimiel Manzanares El Bonillo Albacete
CABO DA ROCA LISBON   Elvas Mérida Don Villanueva Ciudad Almagro La Solana Chinchilla de
Cascais (Lisboa) Montemor-o- Arronches Badajoz Benito de la Serena Real   Monte Aragón
        Barreiro Novo Vila Viçosa Guareña   Almodóvar Valdepeñas
Setúbal Palmela Redondo   Campanario   Cabeza del Campo   Yecla
CABO ESPICHEL Ba. de Alcácer Évora Reguengos de Almendralejo Castuera del Buey Puertollano Alcaraz Tobarra Hellín
        Setúbal do Sal Monsaraz Villafranca Hinojosa del Duque   Jabalón Caravaca Molina de
Grândola Viana do Oliva de de los Barros   Belálcazar   Almadén Almadén   de Segura
        Alentejo la Frontera Los Santos Peñarroya-   Fuenteovejuna   Cehegín Segura
Ferreira do Cuba Jerez de de Maimona Pueblonuevo   Pozoblanco Santisteban Beas de Segura Cieza Murcia
Alentejo Beja Caballeros   Azuaga Villanueva La Carolina del Puerto   Puebla de Mula
Sines Aljustrel Moura Fuente de Llerena Guadalcanal de Córdoba   Villacarrillo Don Fadrique Alhama de Murcia
Vila Nova Serpa Cantos Fregenal   Montoro Bailén Linares Úbeda Cazorla Lorca
de Milfontes Castro Verde Corfegana de la Sierra   Andújar   7999 Hércal- Mazarrón
        Almodóvar Minas de Riotinto   Posadas Bujalance   Huéscar Overa
Odemira Ourique Valverde Aracena Cazalla de   Córdoba Porcuna Jaén Cabra Bazas Vélez Rubio Cuevas del
        del Camino Calañas la Sierra   Fernán-Núñez Castro del Río Mancha de Baza   Almanzora Aguilas
Aljezur SA. DE MONCHIQUE Silves Constantina   Palma del Río   Real Martos   Vera Gérgal
        Loulé   ANDALUCÍA Gibraleón Trigueros La Palma Écija Puente-Genil Aguilar Huelma Cambil Guadix Sorbas
Portimão Dos Tavira   Ayamonte Moguer Almonte Sevilla   Estepa Montefrío Iznalloz SA. NEVADA Almería
Lagos Faro Isla Vila Real de Huelva Dos Marchena Lucena Rute   Granada Mulhacén CABO DE GATA
        CABO DE Cristina Sto. Antonio Utrera Hermanas Morón de Arahal   Santa Fe 11 424 Berja Almería
        SANTA MARIA   Lebrija la Frontera Montellano Antequera Alhama de Granada ALPUJARRAS Dalías Golfo de
Golfo   Sanlúcar de Arcos de la Villamartín Ronda Granada   Adra Almería
de Cádiz   Barrameda Frontera Olvera Coín Vélez-Málaga Nerja Motril Almuñécar
        Jerez de la Frontera Ubrique   Álora   Málaga
        El Puerto de Sta. María Medina Alhaurín
        Cádiz Sidonia Jimena de la
        S. Fernando Frontera Estepona
        Chiclana de la Frontera Vejer de la San Roque
CABO TRAFALGAR Frontera Algeciras La Línea Gibraltar (Br.) MEDI
PTA. DE TARIFA   Tarifa Bay of Gibraltar ISLA DEL ALBORÁN (Sp.)
        Strait of Gibraltar
C. SPARTEL PTA. ALMINA Ceuta (Sp.) CAP DES TROIS FOURCHES
Tanger (Tangier) Tétouan
Asilah   Melilla (Sp.)
                Al-Hoceima ISLAS CHAFARINAS (Sp.) Beni Saf
Larache MOROCCO   El Ghazawet

OCEAN ATLANTIC

GALICIA ASTURIAS CORDILLERA CANTABRICA NAVARRA
MINHO TRÁS-OS-MONTES LEÓN CASTILLA LA VIEJA SISTEMA
BEIRA DOURO SA. DA ESTRELA SPAIN CASTILLA LA NUEVA MURCIA
ESTREMADURA ALTA EXTREMADURA SA. MORENA ANDALUCÍA
ALGARVE BAIXO ALENTEJO

Relief
Meters   Feet
3050   10000
1525   5000
610   2000
305   1000
152.5   500
0   Sea Level   0
152.5   500
1525   5000
3050   10000

A-552900-76
COPYRIGHT BY
RAND McNALLY & COMPANY
MADE IN U.S.A.

Scale 1:4 000 000, one inch to 64 miles. Conic Projection
Elevations and depressions are given in feet

Longitude West of Greenwich

Scale 1:4 000 000; one inch to 64 miles. Conic Projection
Elevations and depressions are given in feet

Continued on pages 124-125

Continued on pages 126-127

Relief

Meters		Feet
3050		10 000
1525		5000
610		2000
305		1000
152.5		500
Sea Level		0
152.5		500
1525		5000
3050		10 000

HUNGARY

Szeged Makó Arad
Nádlac
Baja Bácsalmás
Pécs Mohács Kanizsa Timişoara Hunedoara
Subotica Senta
Bačka Topola Ada
Sombor Bečej Kikinda Jimbolia Lugoj Hateg
Apatin Vrbas Srbobran
Čurug BANAT Caransebeş
Osijek Bač Palanka Novi Sad
Vukovar Bačka Palanka Reşita Anina
Vinkovci Srem. Karlovci Vršac Rešiţa
Brod Danube Ruma Pančevo Bela Crkva
Bosanski Šamac Sremska Mitrovica Zemun (Semlin) Orşova Drobeta-Turnu Severin
Brčka Belgrade (Beograd) IRON GATE Strehaia
Gradačac Požarevac
Gračanica Smederevo (Semendria)
Tuzla Smed. Palanka
Maglaj Loznica Negotin
Zvornik Bregovo
AND Valjevo Svilajnac Vidin
Zenica Vares Gornji Milanovac Kragujevac Zaječar Calafat Lom
Sarajevo Čačak Paraćin Gramada
Višegrad Kraljevo Trstenik Berkovitsa
Rogatica Užice Kruševac Aleksinac Niš Pirot
Foča Pribaj Nova Varoš Raška Prokuplje
Kalinovik Prijepolje Novi Pazar Kuršumlija Vlasotince
Čainice Sjenica Bijelo Polje KOSOVO Leskovac
Nevesinje Pljevlja Kos. Mitrovica Vranje
Gacko Priboj Priština
Bileća CRNA GORA (MONTENEGRO) Peć Gnjilane Kriva Palanka
Trebinje Nikšić Gusinje Đakovica Kumanovo Kratovo
Dubrovnik (Ragusa) Podgorica Prizren ŠAR PLANINA Skopje
Kotor Cetinje Tetovo Kočani
Virpazar Shkodër (Scutari) Kičevo Titov Veles Štip
Bar Kukës Gostivar Radoviš
Ulcin (Dulcigno) Lesh (Alessio) Debar (Dibra) MACEDONIA Strumica
Krujë Kruševo Prilep
Durrës Tiranë Ohrid Bitola (Monastir) Kilkis
Kavajë Edessa
Peqin Elbasan Flórina Giannitsá
Fier Berat Kastoria Náoussa Thessaloniki
Vlorë (Valona) Korçë Kozáni Véroia
Gjirokastër Grevená Katerini

ROMANIA

CARPATII MERIDIONALI (TRANSYLVANIAN ALPS)
Alba Iulia Sfântu Gheorghe
Mureş Makó Sebeş Braşov
Deva Orăştie Sibiu Sinaia Slănic Buzău
Târgu Jiu Petroşani Câmpulung Câmpina
Curtea de Argeş Târgovişte Ploieşti Urziceni
Piteşti Mizil
Slatina BUCHAREST (Bucureşti)
Drăgăşani VALACHIA
Roşiori de Vede Giurgiu Ruse (Russe) Vetovo Dobrich
Craiova Balş Slatina Alexandria Tutrakan Balchik
Caracal Turnu Măgurele Zimnicea Razgrad Novi Pazar
Corabia Svishtov Popovo Shumen Provadiya
Bǎileşti Orekhovo Nikopol Byala Smyadovo Varna (Stalin)
Kozloduy Pleven Türgovishte
Selanovtsi Lukovit Gorna Oryakhovitsa Veliko Lyaskovets
Vǎlchedrǎm Knezha Sevlievo Türnovo
Byala Slatina Lovech Gabrovo STARA PLANINA (BALKAN MTS.) Sliven
Borovan Ugürchin Teteven Sliven Straldzha Karnobat
Vratsa Botevgrad Kazanlŭk R. Sliven
BULGARIA Karlovo Stara Zagora Yambol Burgas
Trǔn Breznik Sofia (Sofiya) Panagyurishte Nova Zagora
Pernik Ikhtiman Golyamo Konare Chirpan Topolovgrad ISTRANCA DAĞLARI
Radomir Samokov Vetren Plovdiv (Philippopolis) Sozopol
Kyustendil Blagoevgrad (Gorna Dzhumaya) Pazardzhik Pazardzhik Asenovgrad Kharmanli Akhtopol Malko Türnovo
Musala Peshtera Khaskovo Edirne (Adrianople) Babaeski
Razlog RHODOPE Svilengrad Kürdzhali Lyuleburgas
Bansko Zlatograd Dhidhimótikhon Babaeski
Petrich MTS. Komotiní THRACE Tekirdağ (Rodosto)
Sidírokastro Dráma Xánthi Keşan Malkara
Sérres Kavála Alexandroúpolis (Dédéagatch) Enez

THÁSOS
Theologos GALLIPOLI PENINSULA Lâpseki
SAMOTHRÁKI Eceabat (Maidos) Biga
GÖKÇEADA Çanakkale Troy Ruins
Çanakkale Boğazı (Dardanelles) Kumkale Ezine Bayramiç
LIMNOS Bozcaada
Mýrina BOZCA ADA Edremit
LÉSVOS Ayvalik
Mytilíni Agiásos Edremit Körfezi
Polichnítos Plomári

AEGEAN SEA

Strait of Otranto
SAZANIT
Brindisi
Lecce
Manduria Nardò Galatina Otranto
Gallipoli Maglie
Taranto
C. S. MARIA DI LEUCA

GREECE
Thermaïkós Kólpos
Kólpos Kassándras Kólpos Agíon Óros Athos
CHALKIDIKÍ Karyés
Mt. Ólimbos (Ólympos) 2570 Livádi
Siátista Melívoia
Tyrnavos Lárisa
Kónitsa THESSALÍA Vólos
Ioánnina Tríkala Pagasitikós Kólpos
Métsovon Kardítsa VÓREIOI SPORÁDES
Farsala (Pharsalus) Almyrós SKIATHOS SKÓPELOS SKÝROS
PINDOS OROS ÓROS ÓRTHRYS
Préveza Ágrafa
Astakós Lamía
Agrínio Amfissa Vóreios Evvoïkós Kólpos Kými
Delphi Ruins Leivádia Khalkída
Galaxídi Thíva (Thebes) EVVOIA

IONIAN ISLANDS
Kérkyra KÉRKYRA
PAXOÍ
OTHONOÍ
LEFKÁDA Lefkáda
ITHÁKI Aitolikó Mesolóngi Aígio
Pátra (Patras) Korinthiakós Kólpos
Argóstoli KEFALLONÍA Patraïkós Kólpos Athens (Athína)
ZÁKYNTHOS Zákynthos Amaliáda Mégara Peiraiás (Piraeus)
Pýrgos Olympia Trípoli Kórinthos (Corinth) Salamína
Kyparissiakós Kólpos Argos Náfplio Saronikós Kólpos Aígina
Kyparissía PELOPÓNNISOS AÍGINA
Filiátra Spárti (Sparta)
Messíni Kalamáta Gýtheio
Gargaliáni Messiniakós Kólpos Neápoli
AKRA TAÍNARO AKRA MALÉAS

TURKEY
Uzunköprü Hayrabolu
Sarköy
Gelibolu (Gallipoli)

DODEKÁNISOS (DODECANESE)
ÁGIOS EFSTRÁTIOS
PSARÁ Chíos CHIOS Çeşme Urla
ÁNDROS Ándros TÍNOS MYKÓNOS IKARIA
KÉA SÝROS Ermoúpoli NÁXOS DONOÚSSA
KÍKLÁDES PÁROS NÁXOS AMORGÓS
SÉRIFOS Sérifos PÁROS
SÍFNOS ÍOS SIKINOS
KÍMOLOS MÍLOS ÍOS

18° Longitude East of Greenwich 20° 24° 26°

0 10 20 30 40 50 60 70 80 90 100 110 120 Miles
0 20 40 60 80 100 120 140 160 180 200 Kilometers

Relief

Feet	Meters
5000	1525
2000	610
1000	305
500	152.5
0 Sea Level	0
500	152.5

Continued on pages 122-123

Cities and Towns

0 to 50,000 ○
50,000 to 500,000 ◉
500,000 to 1,000,000 ◎
1,000,000 and over

Scale 1:4 000 000; one inch to 64 miles. Conic Projection
Elevations and depressions are given in feet

Scale 1:20 000 000; one inch to 315 miles
Lambert's Azimuthal, Equal Area Projectio
Elevations and depressions are given in fe

Relief

Meters		Feet
3050		10 000
1525		5000
610		2000
305		1000
152.5		500
0	Sea Level	0
152.5		Below
1525		5000	Sea Level
3050		10 000

VRANGELYA
(WRANGEL)

CHUKOTSKOYE NAGOR'YE

M. SHELAGSKIY

AYON

M. Billings

KORYAKSKIY KHREBET

Arctic Circle

Nizhne-Kolymsk

Srednee-Kolymsk

KHREBET GYDAN (KOLYMSKIY)

Anadyr

Penzhino

Markovo

A R C T I C   O C E A N

SEVERNAYA ZEMLYA
(NORTHERN LAND)

P-OV
GORY
TAYMYR
BYRRANGA

M. CHELYUSKIN

BOL'SHOY
BEGICHEV

DE-LONGA

NOVOSIBIRSKIYE O-VA
(NEW SIBERIAN ISLANDS)

NOVAYA SIBIR

FADDEYA

KOTEL'NYY

MALYY LYAKHOVSKIY
LYAKHOVSKIYE

M. SVYATOY NOS

M. BUOR-KHAYA

E A S T   S I B E R I A N   S E A

M. MEDVEZH'I

M. TAIGONOS

KAMCHATKA

Verkhne-Kamchatsk

Petropavlovsk-Kamchatskiy

Ust'-Bol'sheretsk

L A P T E V   S E A

Taymyr

Khatangskiy Zaliv

Nordvik

Taymyr

Ust'-Olenek

Tiksi

Zashiversk

Zyryanka

Omolon

Grizhiga

Ola

Magadan

M. ALEVINA

S E A   O F   O K H O T S K

SAKHALIN

P-OV
GORY
PUTORANA

Noril'sk

Turukhansk

rka

Baykit

Khatanga

Bulun

Abyy

Verkhoyansk

Gora Chen
10 771

Omyakon

Nel'kan

Okhotsk

Aleksandrovsk

Poronaysk

Ust'-Bol'sheretsk

Nizhnyaya Tunguska

Tura

Suntar

Vilyuysk

YAKUTSK

Amga

Ust'-Maya

Ayan

Chumikan

Udskaya Guba

SHANTAR

Nikolayevsk-na-Amure

Okha

M. VELIKAYA

M. TERPENIYA

Uglegorsk

Podkamennaya Tunguska

Yartsevo

G. Polkan
3543

Yeniseysk

Baykit

Muknuya

Peleduy

Vitim

Olekminsk

Aldan

Tommot

Aldanskaya

Venyu

DZHUGDZHUR KHREBET

Komsomol'sk
na-Amure

KHREBET BUREINSKIY

Nikolayevsk

Kholmsk

Yuzhno-Sakhalinsk

Korsakov

G. Golets-Purpula
3177

PATOM
PLATEAU

Bodaybo

Golets-Skalistyy
9186

STANOVOY KHREBET

Tyndinskiy

Zeya

Svobodnyy

Belogorsk

Ust' Tyrma

Bureya

Birobidzhan

Khabarovsk

SIKHOTE ALIN'

Sovetskaya Gavan'

Tatar Strait

HOKKAIDO

Sapporo

Otaru

Wakkanai

TSK

Krasnoyarsk

Bogotol

Kansk

Tayshet

Bratsk

Ilimsk

Nizhne-Angarsk

Kachuga

Barguzin

Chita

Nerchinsk

Nerchinskiy
Zavod

Blagoveshchensk

Skovorodino

Nenjiang

Goukou

LESSER KHINGAN
RANGE

Boli

Suifenhe

Spassk-Dal'niy

Dal'nerechensk

Arsen'yev

USSURIYSKIY KHREBET

O'ga

Esashi

znetski

Balakhta

Nizhneudinsk

G. Piramida
10 801

Tulun

Cheremkhovo

Angarsk

Irkutsk

Zhigalova

Lake Baykal
Surface elev. 1535 ft.
above sea level

Ulan-Ude

BAYKAL'SKIY KHREBET

Sretensk

Nerchinskiy

NERCHINSKIY KHREBET

Aginskoye

Borzya

KHINGAN

Hailun

Suihua

Harbin

Mudanjiang

Artem

Pamzhsan

Nakhodka

Vladivostok

Minusinsk

Abakan

Kyzyl

TANNU-OLA

SAYAN
KHREBET

Kutulik

Munku Sardyk
11 457

Kyren

Petrovsk-Zabaykal'skiy

Goradok

Kyakhta

YABLONOVYY KHREBET

Aksha

VITIM KHREBET

Onon

Kerulen

GREATER KHINGAN
RANGE

Wenquan

Tao'an

Fuyu

Qiqihar

Jarud Qi

Shuangliao

Dunhua

Hunchun

MANCHURIA

Jilin

Najin

Chongjin

S E A   O F   J A P A N

HONSHU

Ust'-
Usu

Hissary
Nuur

Har Us Nuur

HANGAYN NURUU
KHANGAI MTS.

Hovd

Uliastay

Tsast Bogd
13 419

Selenge

Ulan Bator
(Ulaanbaatar)

Ondorhaan

M O N G O L I A

Sayr-Usa

G O B I   O R   S H A M O
(D E S E R T)

CHANGCHUN

SHENYANG

FUSHUN

Chifeng

Weichang

Chengde

NORTH
KOREA

P'yongyang

J A P A N

Kanazawa

KYOTO

KOBE

Tottori

Matsue

OSAKA

Hami

Zhangjiakou

Fengzhen

BEIJING

TIANJIN

Baoding

Lüshun

Dalian

Kaesong

Andong

SHANDONG BANDAO

Y E L L O W   S E A

Bo Hai

Korea Bay

SEOUL

SOUTH
KOREA

Taegu

PUSAN

Hiroshima

Okayama

Kochi

K O R E A

R U S S I A

S I B E R I A

C H I N A

0   100 200 300 400 500 600 Miles
0 200 400 600 800 1000 Kilometers

136

Cities and Towns

0 to 50,000	○	500,000 to 1,000,000	◎
50,000 to 500,000	⊙	1,000,000 and over	

Relief

Feet	Meters
10000	3050
5000	1525
2000	610
1000	305
500	152.5
0 Sea Level	Sea Level
Below Sea Level	
500	152.5
5000	1525
10000	3050

0   50   100   150   200   250   300 Miles

0   100   200   300   400   500 Kilometers

Continued on pages 116-117

Scale 1:10 000 000; one inch to 160 miles. Conic Projection
Elevations and depressions are given in feet.

Continued on pages 118-119

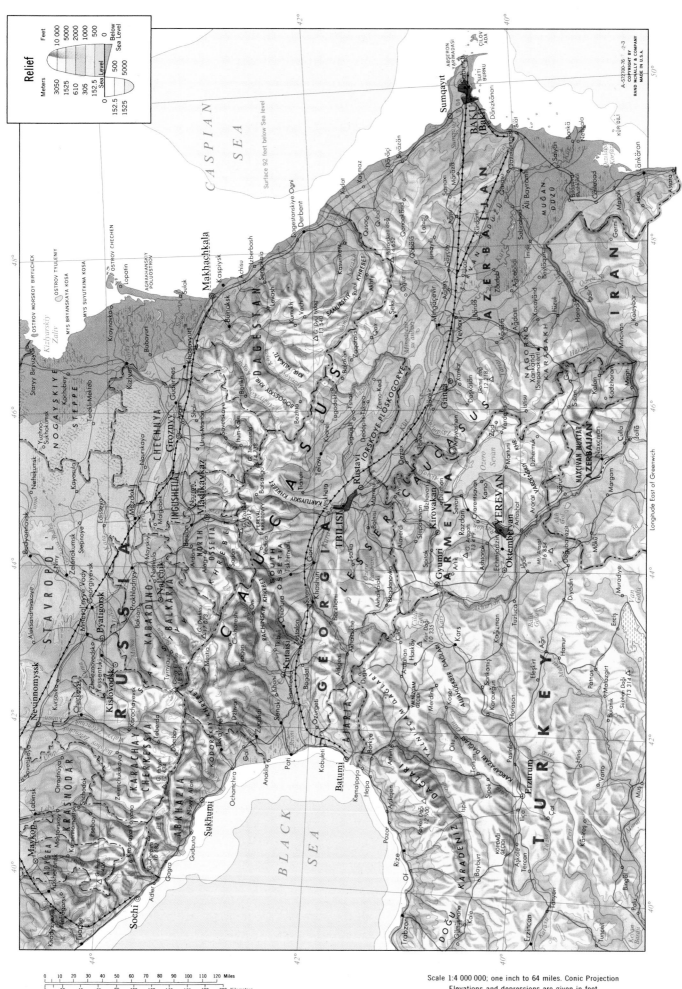

Relief

Meters	Feet
3050	10 000
1525	5000
610	2000
305	1000
152.5	500
0	Sea Level
	Below
	Sea Level

Sea Level

| 152.5 | 500 |
| 1525 | 5000 |

CASPIAN SEA

Surface 92 feet below Sea Level

BLACK SEA

Scale 1:4 000 000; one inch to 64 miles. Conic Projection
Elevations and depressions are given in feet

0 10 20 30 40 50 60 70 80 90 100 110 120 Miles
0 20 40 60 80 100 120 140 160 180 200 Kilometers

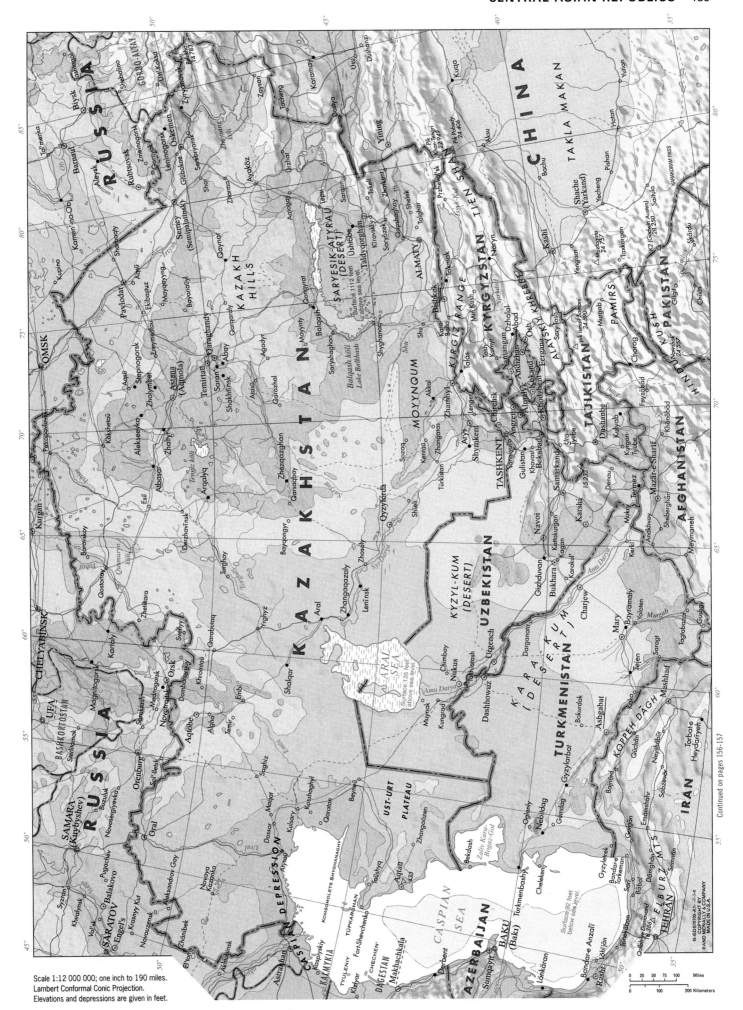

Continued on pages 156-157

Scale 1:12 000 000; one inch to 190 miles.
Lambert Conformal Conic Projection.
Elevations and depressions are given in feet.

0   25 50 75 100   Miles
0   100   200 Kilometers

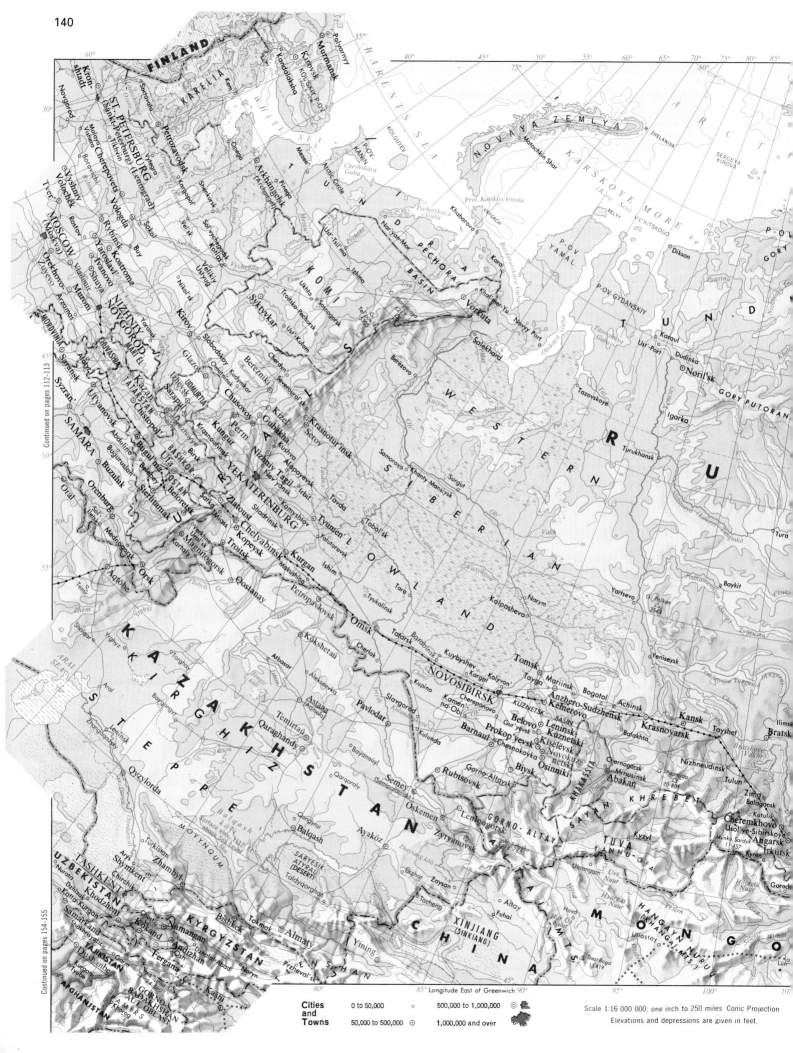

Relief

Meters	Feet
3050	10 000
1525	5000
610	2000
305	1000
152.5	500
0	Sea Level
152.5	500
1525	5000
3050	10000

Continued on pages 160-161

0 50 100 200 300 400 500 Miles
0 100 200 400 600 800 Kilometers

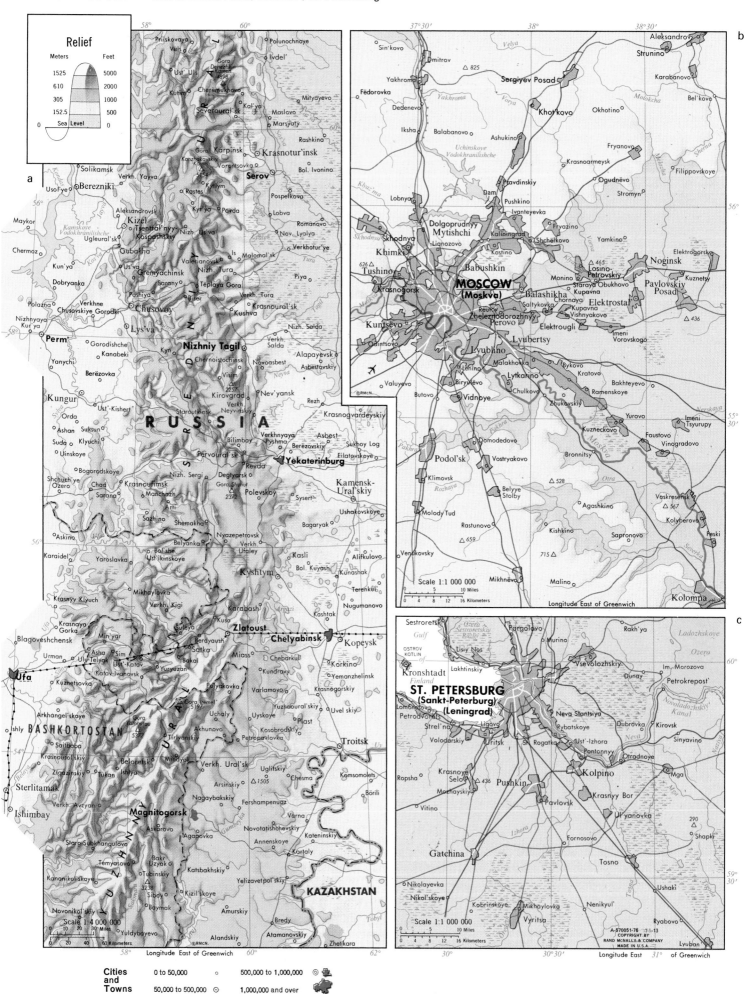

Relief

Meters		Feet
1525		5000
610		2000
305		1000
152.5		500
0	Sea Level	0

a

Scale 1:4 000 000

Longitude East of Greenwich

b

Scale 1:1 000 000

Longitude East of Greenwich

c

Scale 1:1 000 000

A-570051-76 7-5-13
COPYRIGHT BY
RAND McNALLY & COMPANY
MADE IN U.S.A.

Longitude East of Greenwich

Cities and Towns

| 0 to 50,000 | ○ | 500,000 to 1,000,000 | ◎ |
| 50,000 to 500,000 | ⊙ | 1,000,000 and over | |

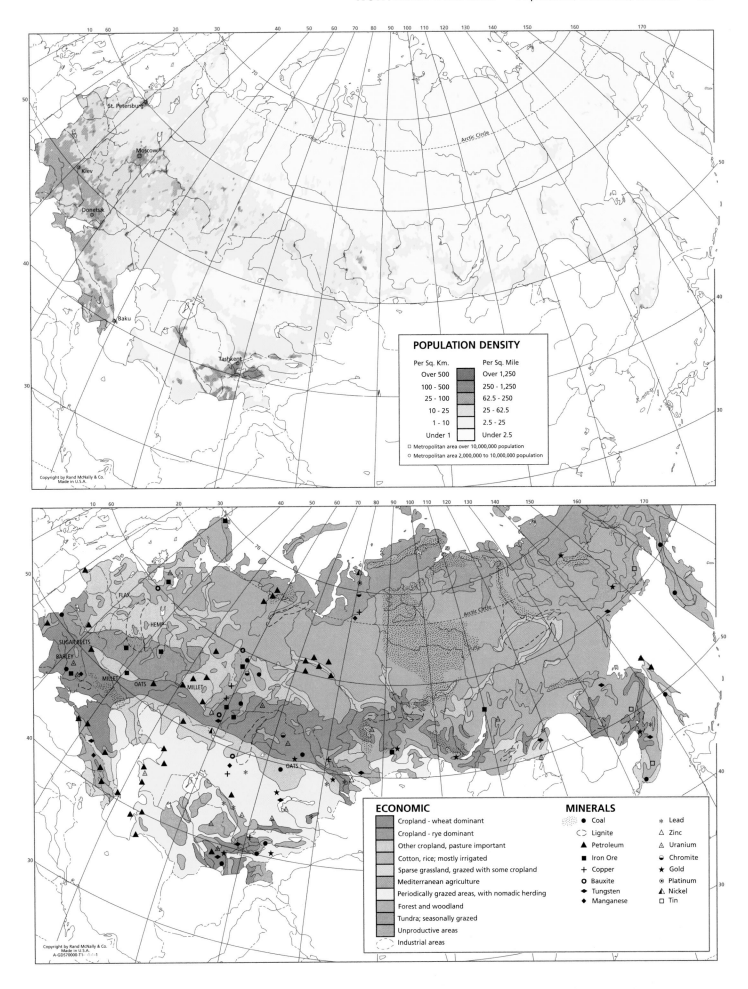

POPULATION DENSITY

Per Sq. Km.	Per Sq. Mile
Over 500	Over 1,250
100 - 500	250 - 1,250
25 - 100	62.5 - 250
10 - 25	25 - 62.5
1 - 10	2.5 - 25
Under 1	Under 2.5

□ Metropolitan area over 10,000,000 population
○ Metropolitan area 2,000,000 to 10,000,000 population

Copyright by Rand McNally & Co.
Made in U.S.A.

St. Petersburg
Moscow
Kiev
Donetsk
Baku
Tashkent

ECONOMIC

- Cropland - wheat dominant
- Cropland - rye dominant
- Other cropland, pasture important
- Cotton, rice; mostly irrigated
- Sparse grassland, grazed with some cropland
- Mediterranean agriculture
- Periodically grazed areas, with nomadic herding
- Forest and woodland
- Tundra; seasonally grazed
- Unproductive areas
- Industrial areas

MINERALS

●	Coal	✳	Lead
◖	Lignite	△	Zinc
▲	Petroleum	△	Uranium
■	Iron Ore	◖	Chromite
+	Copper	★	Gold
○	Bauxite	◉	Platinum
◆	Tungsten	▲	Nickel
◆	Manganese	□	Tin

FLAX
HEMP
SUGAR BEETS
BARLEY
MILLET
OATS
MILLET
OATS

Copyright by Rand McNally & Co.
Made in U.S.A.
A-GDS70000-T1-·-·-1

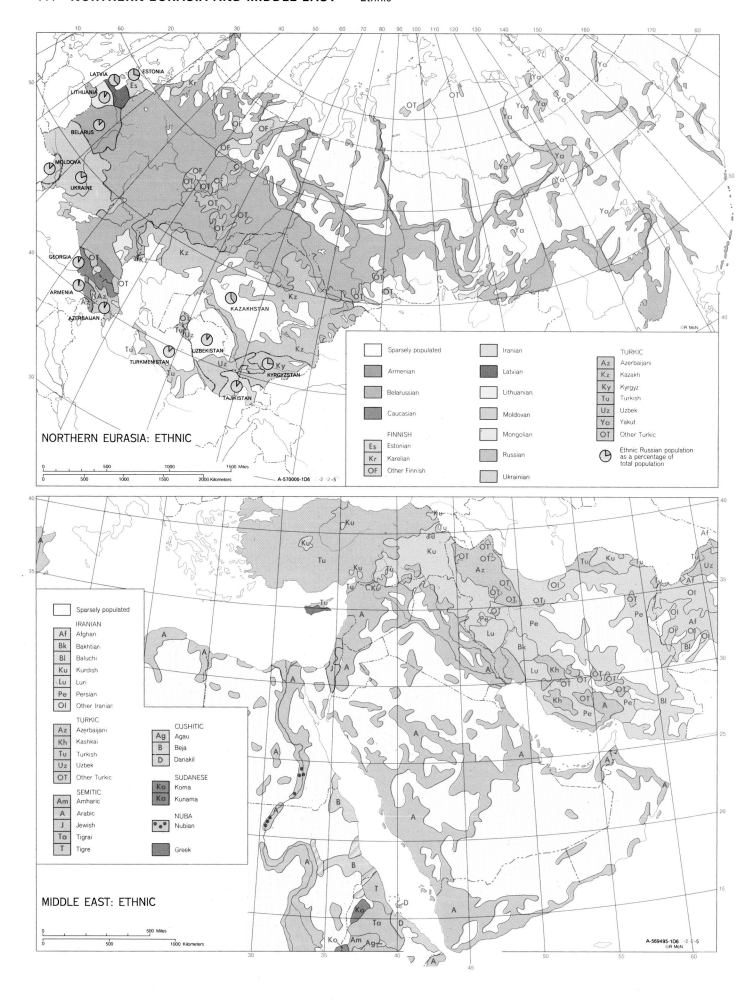

NORTHERN EURASIA: ETHNIC

LATVIA
ESTONIA
Es
LITHUANIA
BELARUS
MOLDOVA
UKRAINE
GEORGIA
ARMENIA
Az
AZERBAIJAN
TURKMENISTAN
UZBEKISTAN
KAZAKHSTAN
KYRGYZSTAN
TAJIKISTAN
Kr
OF
OT
Kz
Tu
Uz
Ky
Ya

0 500 1000 1500 Miles
0 500 1000 1500 2000 Kilometers
A-570000-1D6 -2 -2 -5

☐ Sparsely populated	☐ Iranian	**TURKIC**
☐ Armenian	☐ Latvian	Az Azerbaijani
☐ Belarussian	☐ Lithuanian	Kz Kazakh
☐ Caucasian	☐ Moldovan	Ky Kyrgyz
FINNISH	☐ Mongolian	Tu Turkish
Es Estonian	☐ Russian	Uz Uzbek
Kr Karelian	☐ Ukrainian	Ya Yakut
OF Other Finnish		OT Other Turkic

◔ Ethnic Russian population
as a percentage of
total population

MIDDLE EAST: ETHNIC

☐ Sparsely populated	
IRANIAN	
Af Afghan	
Bk Bakhtiari	
Bl Baluchi	
Ku Kurdish	
Lu Luri	
Pe Persian	
OI Other Iranian	
TURKIC	**CUSHITIC**
Az Azerbaijani	Ag Agau
Kh Kashkai	B Beja
Tu Turkish	D Danakil
Uz Uzbek	**SUDANESE**
OT Other Turkic	Ko Koma
SEMITIC	Ka Kunama
Am Amharic	**NUBA**
A Arabic	Nubian
J Jewish	☐ Greek
Ta Tigrai	
T Tigre	

0 500 Miles
0 500 1000 Kilometers
A-569495-1D6 -2 -1 -5
©R McN

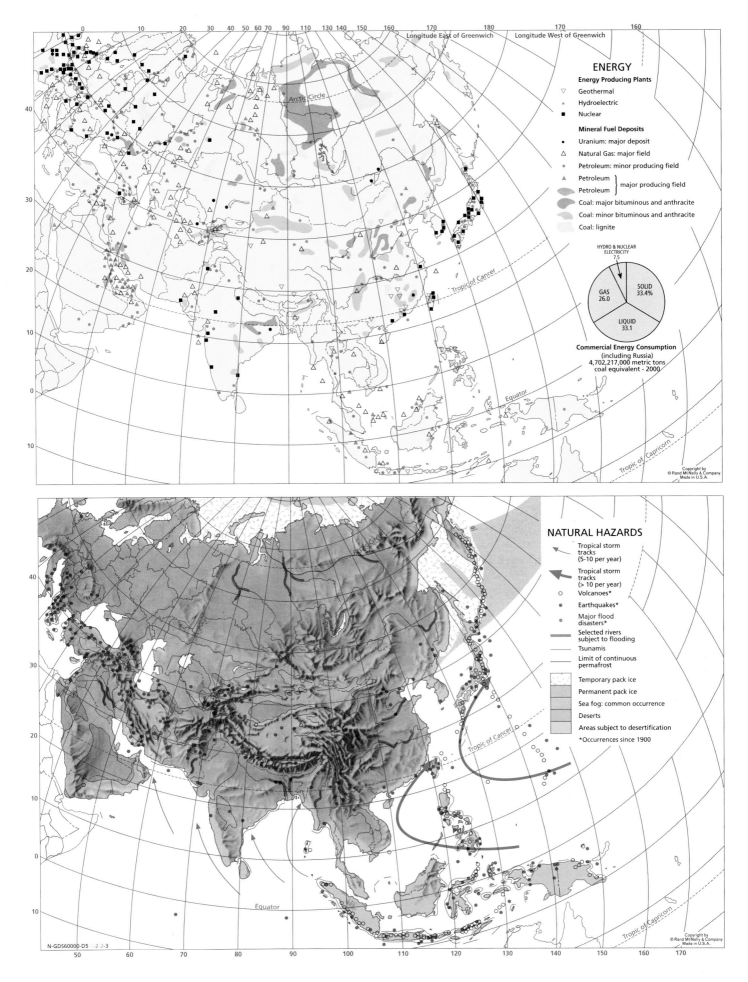

ENERGY

Energy Producing Plants

▽ Geothermal
· Hydroelectric
■ Nuclear

Mineral Fuel Deposits

· Uranium: major deposit
△ Natural Gas: major field
· Petroleum: minor producing field
▲ Petroleum }
  Petroleum } major producing field
Coal: major bituminous and anthracite
Coal: minor bituminous and anthracite
Coal: lignite

HYDRO & NUCLEAR
ELECTRICITY
7.5

GAS
26.0

SOLID
33.4%

LIQUID
33.1

Commercial Energy Consumption
(including Russia)
4,702,217,000 metric tons
coal equivalent - 2000

Longitude East of Greenwich   Longitude West of Greenwich

Copyright by
© Rand McNally & Company
Made in U.S.A.

NATURAL HAZARDS

↖ Tropical storm
  tracks
  (5-10 per year)
◄ Tropical storm
  tracks
  (> 10 per year)
○ Volcanoes*
· Earthquakes*
· Major flood
  disasters*
— Selected rivers
  subject to flooding
— Tsunamis
— Limit of continuous
  permafrost
Temporary pack ice
Permanent pack ice
Sea fog: common occurrence
Deserts
Areas subject to desertification

*Occurrences since 1900

Copyright by
© Rand McNally & Company
Made in U.S.A.

N-GDS60000-D5 -2-2-3

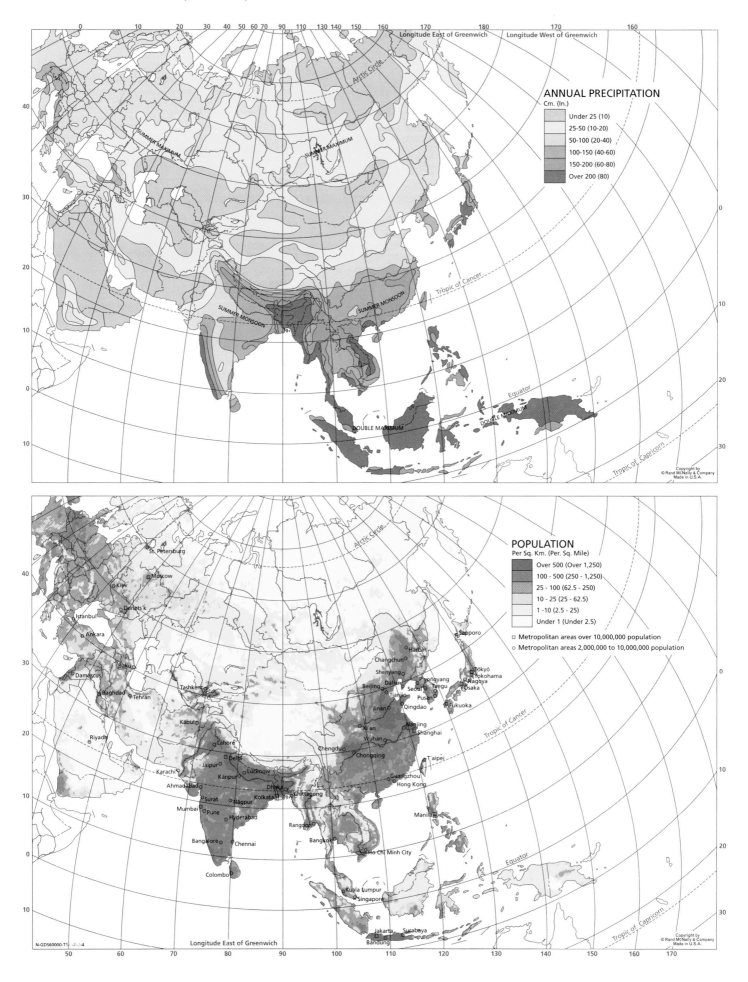

ANNUAL PRECIPITATION
Cm. (In.)

- Under 25 (10)
- 25-50 (10-20)
- 50-100 (20-40)
- 100-150 (40-60)
- 150-200 (60-80)
- Over 200 (80)

POPULATION
Per Sq. Km. (Per. Sq. Mile)

- Over 500 (Over 1,250)
- 100 - 500 (250 - 1,250)
- 25 - 100 (62.5 - 250)
- 10 - 25 (25 - 62.5)
- 1 -10 (2.5 - 25)
- Under 1 (Under 2.5)

□ Metropolitan areas over 10,000,000 population
○ Metropolitan areas 2,000,000 to 10,000,000 population

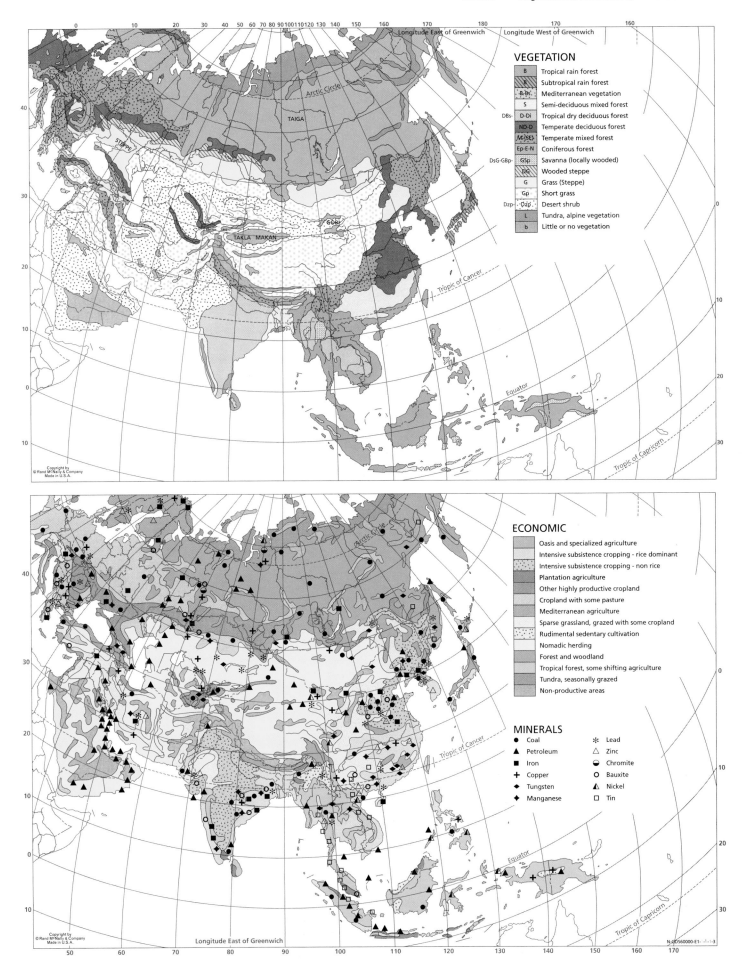

VEGETATION

B	Tropical rain forest
R	Subtropical rain forest
B-Bs	Mediterranean vegetation
S	Semi-deciduous mixed forest
DBs- D-Di	Tropical dry deciduous forest
ND-D	Temperate deciduous forest
M-(SE)	Temperate mixed forest
Ep-E-N	Coniferous forest
DsG-GBp- GSp	Savanna (locally wooded)
DG	Wooded steppe
G	Grass (Steppe)
Gp	Short grass
Dzp- Dzp	Desert shrub
L	Tundra, alpine vegetation
b	Little or no vegetation

ECONOMIC

	Oasis and specialized agriculture
	Intensive subsistence cropping - rice dominant
	Intensive subsistence cropping - non rice
	Plantation agriculture
	Other highly productive cropland
	Cropland with some pasture
	Mediterranean agriculture
	Sparse grassland, grazed with some cropland
	Rudimental sedentary cultivation
	Nomadic herding
	Forest and woodland
	Tropical forest, some shifting agriculture
	Tundra, seasonally grazed
	Non-productive areas

MINERALS

●	Coal	✳	Lead
▲	Petroleum	△	Zinc
■	Iron	◖	Chromite
✛	Copper	○	Bauxite
◆	Tungsten	▲	Nickel
◆	Manganese	☐	Tin

N-GOS60000-E1- -1- -3

Scale 1:36,000,000; one inch to 570 miles. Lambert Azimuthal Equal-Area Projection

ASIA · Environments / Political

POLITICAL

POLITICAL

Map labels — Southeast Asia / Indian Ocean inset:

MINDANAO, Cebu, Manado, MANADO, Celebes Sea, CELEBES, Ujung Pandang, Java Sea, JAVA, JAKARTA, SINGAPORE, Kuching, BORNEO, Kota Kinabalu, South China Sea, HO CHI MINH CITY, BANGKOK, Singapore, SUMATRA, Medan, Rangoon, Andaman Sea, Salween, Bay of Bengal, CHENNAI (Madras), SRI LANKA, Colombo, Kozhikode, WESTERN GHATS, EASTERN GHATS, MUMBAI (Bombay), Arabian Sea, INDIAN OCEAN, Equator, Aden, Gulf of Aden, Berbera

Scale: 0 100 200 400 600 800 Miles / 0 150 300 600 900 1200 Kilometers

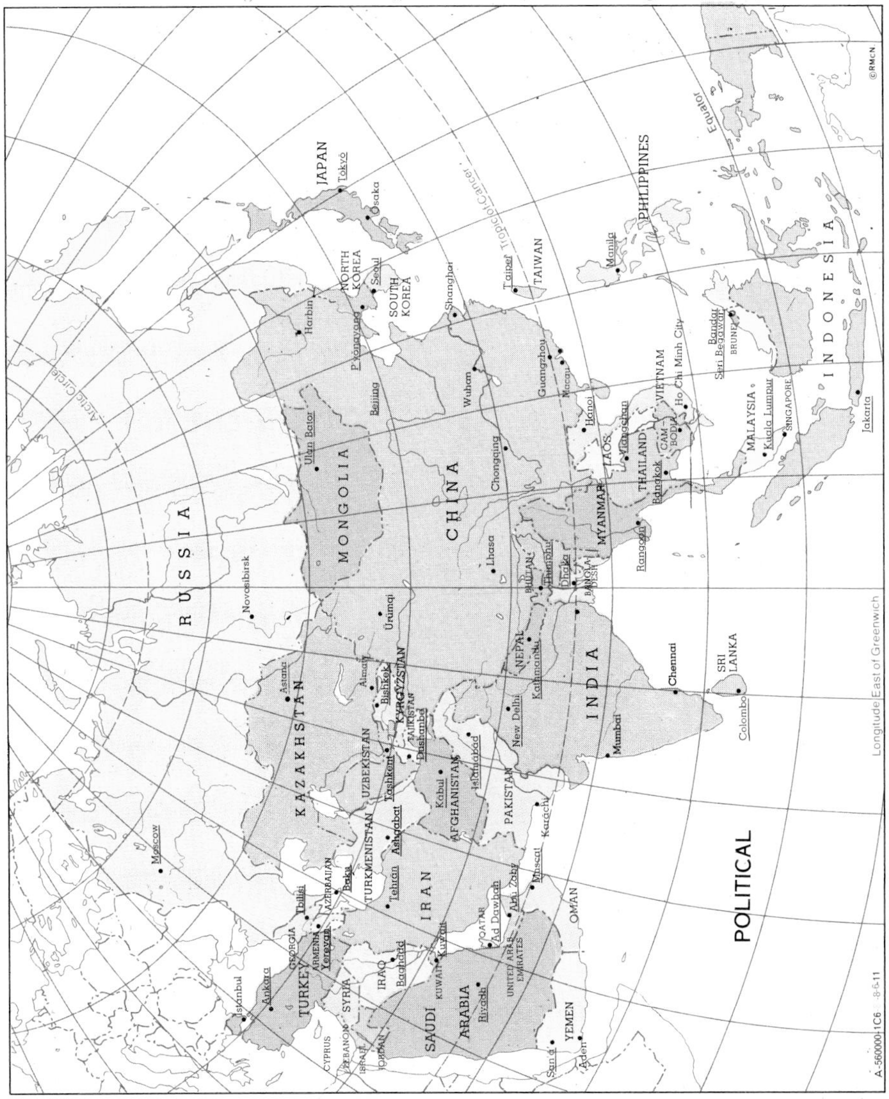

A.560000-36 -4 I-521 COPYRIGHT BY RAND McNALLY & COMPANY MADE IN U.S.A.

Map labels — Asia Political (upper right):

JAPAN, Tokyo, Osaka, NORTH KOREA, Pyongyang, SOUTH KOREA, Seoul, Harbin, Shenyang, Beijing, Wuhan, Shanghai, Chongqing, Guangzhou, Chengdu, CHINA, MONGOLIA, Ulan Bator, RUSSIA, Novosibirsk, Ürümqi, Almaty, Moscow, Astana, KAZAKHSTAN, Bishkek, KYRGYZSTAN, UZBEKISTAN, Tashkent, TAJIKISTAN, Dushanbe, TURKMENISTAN, Ashgabat, Tehran, IRAN, Baghdad, IRAQ, KUWAIT, Kuwait, Riyadh, SAUDI ARABIA, YEMEN, San'a, Aden, OMAN, Muscat, UNITED ARAB EMIRATES, QATAR, Ad Dawhah, BAHRAIN, Abu Zaby, Kabul, AFGHANISTAN, PAKISTAN, Karachi, Islamabad, New Delhi, INDIA, Mumbai, Chennai, SRI LANKA, Colombo, NEPAL, Kathmandu, BHUTAN, BANGLADESH, Dhaka, MYANMAR, Rangoon, Lhasa, THAILAND, Bangkok, LAOS, VIETNAM, Hanoi, Ho Chi Minh City, CAMBODIA, MALAYSIA, Kuala Lumpur, SINGAPORE, BRUNEI, PHILIPPINES, Manila, TAIWAN, INDONESIA, Jakarta, GEORGIA, Tbilisi, ARMENIA, Yerevan, AZERBAIJAN, Baku, TURKEY, Istanbul, Ankara, CYPRUS, LEBANON, SYRIA, ISRAEL, JORDAN, Arctic Circle, Tropic of Cancer, Equator, Longitude East of Greenwich

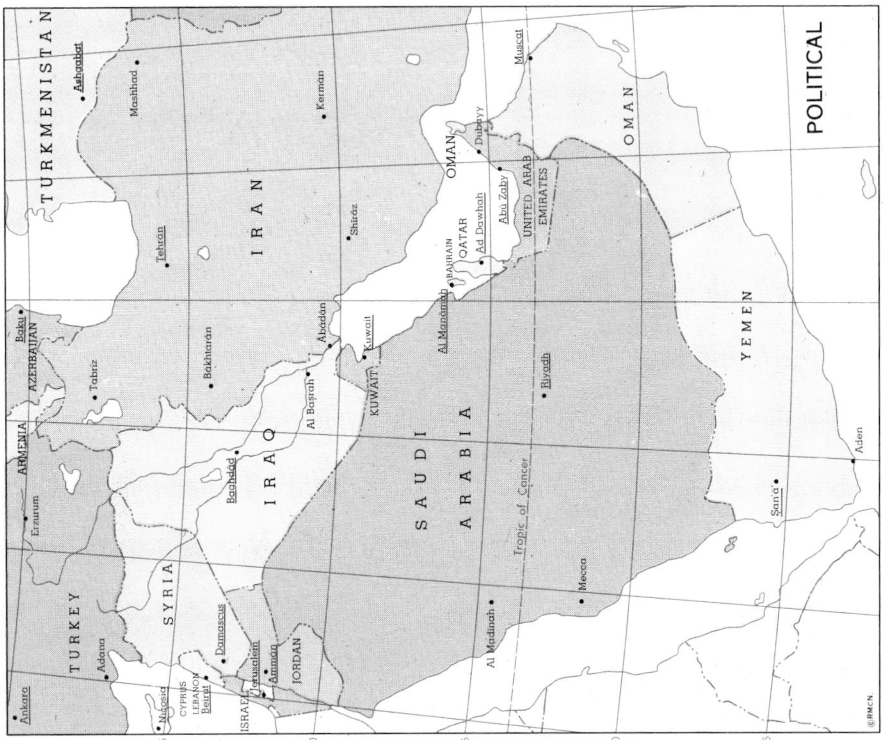

A-560001-IC6   9-8-11
©RM.CN

POLITICAL

Map labels — Middle East Political (lower right):

TURKMENISTAN, Ashgabat, Mashhad, IRAN, Tehran, Kerman, Shiraz, Muscat, OMAN, Dubayy, Abu Zaby, UNITED ARAB EMIRATES, QATAR, Ad Dawhah, BAHRAIN, Al Manamah, Abadan, KUWAIT, Kuwait, Al Basrah, IRAQ, Baghdad, Al Mubarraz, Riyadh, SAUDI ARABIA, YEMEN, Aden, San'a, Mecca, Al Madinah, Tropic of Cancer, JORDAN, Amman, ISRAEL, Jerusalem, LEBANON, Beirut, CYPRUS, Nicosia, SYRIA, Damascus, TURKEY, Adana, Ankara, Erzurum, ARMENIA, AZERBAIJAN, Baku, Tabriz, Bakhtaran

©RM.CN

Scale 1:40 000 000; one inch to 630 miles. Lambert's Azimuthal, Equal Area Projection
Elevations and depressions are given in feet

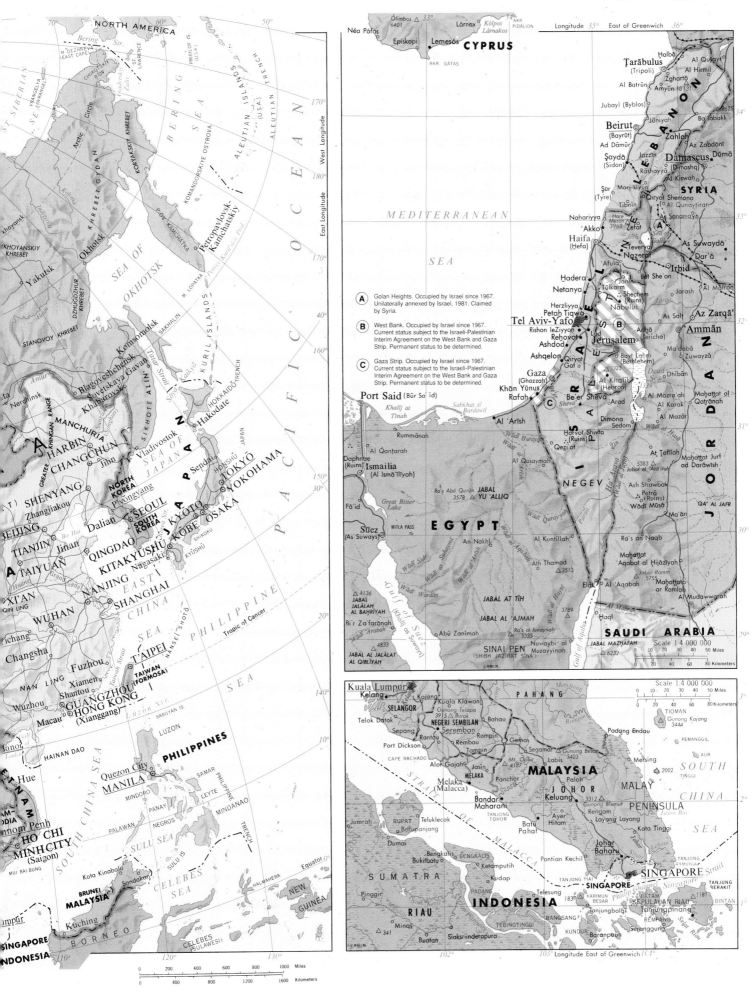

Scale 1:40 000 000; one inch to 630 miles. Lambert's Azimuthal, Equal Area Projection
Elevations and depressions are given in feet

Relief

Meters	Feet
3050	10 000
1525	5000
610	2000
305	1000
0 Sea Level	0 Sea Level
152.5	500 Below Sea Level
1525	5000
3050	10 000
6100	20 000

A-519695-76 1-24 2946
COPYRIGHT BY
RAND McNALLY & COMPANY
MADE IN U.S.A.

Continued on page 183

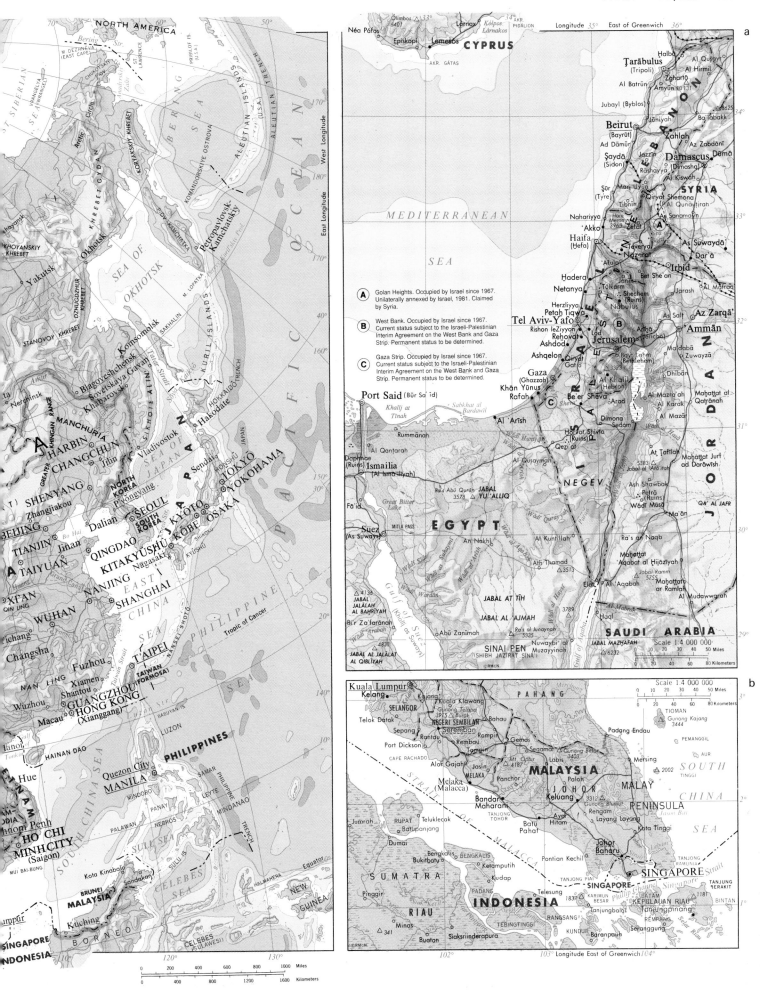

BLACK SEA

İstanbul Boğazı (Bosporus)
Marmara Denizi
İstanbul
(Ruins)
Mytilini
İzmir
Bergama
Bursa
Zonguldak
Kastamonu
Sinop
Kütahya
Eskişehir
Çankırı
Merzifon
Samsun
Ankara
Kırşehir
Yozgat
Çorum
Tokat
Giresun
Trabzon
Kayseri
Sivas
Erzincan
Erzurum 16,854
Konya
Kahramanmaraş
Malatya
Elazığ
Diyarbakır
Siverek
Cizre
TOROS DAGLARI
Adana
Tarsus
Şanlıurfa
İskenderun
Hatay
Gaziantep
Al Mawşil
Nineveh

TURKEY
KURDISTAN

CAUCASUS
Vladikavkaz
Kutaisi
Batumi
Poti
Makhachkala
Grozny
GEORGIA
Tbilisi
Gyumri
Kars
Yerevan
ARMENIA
AZER.
AZERBAIJAN
Gänca BAKU (Baki)
Khvoy
Ardabil
Tabriz
Orūmiyeh
Van Gölü
Van

RUSSIA
Fort-Shevchenko
Aqtaū
CASPIAN SEA
Surface 92 feet below Sea Level

UST-URT PLATEAU

KAZAK
ARAL SEA
135
UZBEKISTAN
Kungrad
Chimbay
Nukus
KYZYL-KUM (DESERT)
Khiva
Turtkul

KARA-KUM (DESERT)
TURKESTAN
Turkmenbashy
Nebitdag
Chekishler
Charjew
Bukhara

MEDITERRANEAN SEA
CYPRUS
Nicosia
Tarābulus (Tripoli)
Al Lādhiqīyah (Latakia)
Hims
Hamāh
Beirut
Saydā (Sidon)
LEBANON
Areas occupied by Israel since 1967
Rashid
Damietta
Tel Aviv-Yafo
ISRAEL
Haifa
Jerusalem
Gaza
Port Said
ALEXANDRIA (Al Iskandarīyah)
CAIRO (Al Qāhirah)
Suez (As Suways)
SINAI

Aleppo
Dayr az Zawr
SYRIA
Damascus (Dimashq)
As Suwaydā
Palmyra (Ruins)
Abū Kamāl
Tikrīt
Al Mawşil
As Sulaymānīyah
Arbīl
Karkūk
Tabrīz
Orūmiyeh
Zanjān
Bandar-e Anzali
Rasht
Bandar-e Torkeman
Bojnūrd
Gorgān
Gonbad-e Kāvūs
ELBURZ MTS.
Qazvīn
Sanandaj
Hamadān
Bakhtarān
Qom
TEHRAN
Dāmghān
Semnān
Neyshābūr
Mashhad
Meymaneh
HERAT

KOPPEH DAGH
Ashgabat
Kara-Kum Canal
Mary
Gushgy

Jerusalem
Amman
JORDAN
Ma'ān
Al 'Aqabah
PEN Elat
Jabal Katrina 8668
Jabal Shāyib 7175
EGYPT
Būr Safājah
Al Quşayr
RA'S BANAS

SYRIAN DESERT
Ar Ramādī
BAGHDAD
Karbalā
An Najaf
Babylon (Ruins)
Al Turayf
Badanah
Al Jawf
Sakākah
Rafhā
An Nāşirīyah
Al Başrah
KUWAIT
Kuwait (Al Kuwayt)
Al Qaysūmah
IRAQ
Ahvaz
Khorramshahr
Abādān
Bandar-e Khomeyni
Shūshtar
Dezfūl
Masjed Soleymān
Qomsheh
Eşfahān
Shīrāz
Kāzerūn
Borāzjān
Bandar-e Büshehr
Jahrom
Lār
Kermān
Rafsanjān
Yazd
Bāfq
ZAGROS MTS.
PLATEAU OF IRAN
DASHT-E KAVIR DESERT
Daryācheh-ye Namak
Kāshān
Arāk
Borūjerd
Qayen
Birjand
Ferdows
Bajestān
IRAN
DASHT-E LUT (DESERT)
Namakzār-e Shāhdād
Zāhedān
Rīgān
Bampūr
Khāsh
Hāmūn-i Māshkel
CHAGAI HILLS
BA

AFGHAN
Herāt
Farāh
RIG

SAUDI ARABIA
An Nafūd
Taymā
Jabal Shammar
Hā'il
Khaybar
Buraydah
Unayzah
Sudair
Ash Shaqrā
NAJD
AD DAHNĀ
Al Qaţīf
Az Zahrān (Dhahran)
Ad Dammām
BAHRAIN
Al Manāmah
QATAR
Ad Dawhah
Abū Zaby
UNITED ARAB EMIRATES
Dubayy
Ajman
Al Buraymī
JABAL AL AKHDAR 9957 Jabal ash Shām
Muscat
Matrah
RA'S AL HADD
Şūr
GULF OF OMAN
Bandar-e Lengeh
Bandar-e Abbas
Qeshm
STRAIT OF HORMUZ
Jāsk
Bandar Beheshtī
Gwādar

RA'S AT TANNŪRAH
Al Jubayl
Al Hufūf
AL HASĀ
Riyadh (Ar Riyāḍ)
Ad Dilam
AL AFLAJ
AD DAWĀSIR
AL ARABIA
JABAL TUWAYQ
AR RUB' AL KHĀLĪ
OMAN
RA'S AL MADRAKAH
Al Maşīrah

RED SEA
Yanbu
Al Madīnah (Medina)
AL HIJĀZ
Jiddah
Mecca (Makkah)
Aţ Ţā'if
Al Khurmah
Mubarraz
NAFŪD
AD DAHY
Wādī ad Dawāsir
Al Lidām
ASĪR
Abhā
Najran
Qal'at Bishah
Jabal Ibrāhīm
Jabal Şabir

ERITREA
Kassalā
Sebderat
Keren
Akordat
Barentū
'Adī Ugrī
Asmera
Mitsiwa (Massawa)
DAHLAK ARCH.
KAMARĀN
Mersa Fatma
DENAKIL
Ed
Beylul
Tadjoura
DJIBOUTI
Aysha
Seylac
Berbera
Zeila
ETHIOPIA

SUDAN
Būr Sūdān
Sawākin
Tawkar
Al Qunfudhah
Erba 7274

JAZĀ'IR FARASAN
Qīzān
Abū 'Arish
Ṣa'dah
Al Luḥayyah
Ḥajar Shu'ayb 12,008
San'ā'
Al Ḥudaydah
Jabal Radā 10,729
Al Mukhā (Mocha)
Madīnat ash Sha'b
Aden ('Adan)
YEMEN
RAMLAT AS SAB'ATAYN
Shibām
Tarīm
Say'ūn
Al Hawtah
Shuqrah
Ash Shiḥr
Al Mukallā
HADRAMAWT
Mirbāṭ
Sayhūt
RA'S FARTAK
KHŪRYĀN-MŪRYĀN (Oman)
RA'S AL HADD

GULF OF ADEN
Caluula
SUQUTRA (SOCOTRA) (Yemen)
Hadībū
GEES GWARDAFUY
Lass Qoray
SOMALIA

ARAB SEA

Continued on pages 184-185

ADMINISTR. BDY.

Tropic of Cancer

Longitude East of Greenwich

Scale 1:16 000 000; one inch to 250 miles. Polyconic Projection
Elevations and depressions are given in feet

A-569400-76   -04-23-43
COPYRIGHT BY
RAND McNALLY & COMPANY
MADE IN U.S.A.

Relief

Meters		Feet
3050		10 000
1525		5000
610		2000
305		1000
152.5		500
0	Sea Level	0
152.5		Below Sea Level
1525		500
3050		5000
		10 000

Continued on pages 140-141

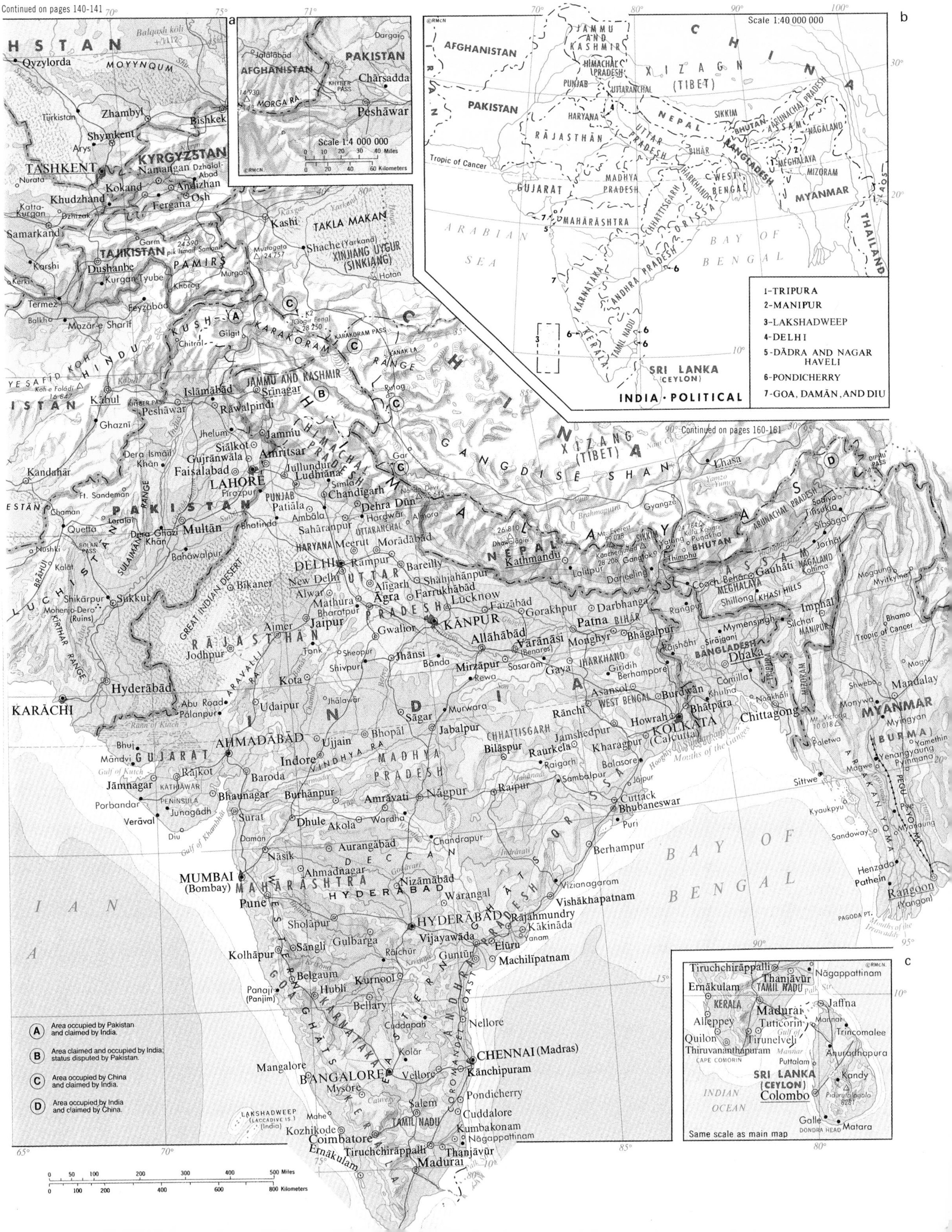

a

Scale 1:4 000 000

AFGHANISTAN

PAKISTAN
Chārsadda
Peshāwar
KHYBER PASS
MORGA RA.
Jalālābād
Dargai

0 10 20 30 40 Miles
0 20 40 60 Kilometers

b

Scale 1:40 000 000

AFGHANISTAN
PAKISTAN
IRAN
CHINA
XIZANG (TIBET)

JAMMU AND KASHMIR
HIMACHAL PRADESH
PUNJAB
UTTARANCHAL
HARYANA
NEPAL
SIKKIM
BHUTAN
ARUNACHAL PRADESH
ASSAM
NAGALAND
MEGHALAYA
MANIPUR
MIZORAM
BANGLADESH
MYANMAR
THAILAND
RĀJASTHĀN
UTTAR PRADESH
BIHĀR
JHARKHAND
WEST BENGAL
GUJARAT
MADHYA PRADESH
CHHATTISGARH
ORISSA
MAHĀRASHTRA
ANDHRA PRADESH
KARNATAKA
KERALA
TAMIL NADU
ARABIAN SEA
BAY OF BENGAL
Tropic of Cancer
SRI LANKA (CEYLON)

INDIA · POLITICAL

1-TRIPURA
2-MANIPUR
3-LAKSHADWEEP
4-DELHI
5-DĀDRA AND NAGAR HAVELI
6-PONDICHERRY
7-GOA, DAMĀN, AND DIU

KAZAKHSTAN
Qyzylorda
MOYYNQUM
Balqash köli
Türkistan
Zhambyl
Bishkek
Shymkent
Arys
KYRGYZSTAN
Nurata
Namangan Dzhalol-Abad
Kokand
Andizhan
Osh
TASHKENT
Khudzhand
Fergana
Samarkand
Karshi
TAJIKISTAN
Dushanbe
Kurgan-Tyube
PAMIRS
Kerki
Termez
Balkh
Mazār-e Sharif
KUSH
HINDU
Feyzabad
Khorog
Murgab
Kashi
TAKLA MAKAN
Shache (Yarkand)
XINJIANG UYGUR (SINKIANG)
Hotan

AFGHANISTAN
Kābul
Peshāwar
KHYBER PASS
Ghaznī
Kandahār
Chaman
Quetta
BALUCHISTAN
SULAIMAN RANGE
KIRTHAR RANGE
PAKISTAN
Islāmābād
Rāwalpindi
Srīnagar
JAMMU AND KASHMIR
KARAKORAM RANGE
K2 28 250
KARAKORAM PASS
Gilgit
Chitral
HIMACHAL PRADESH
Jammu
Siālkot
Gujrānwāla
Amritsar
LAHORE
Faisalabad
Jullundur
Ludhiāna
Simla
Chandigarh
Dehra Dūn
PUNJAB
Patiāla
Fīrozpur
Ambāla
Saharanpur
Bhatinda
HARYANA
Meerut
Morādābād
DELHI
New Delhi
Rampur
Bareilly
UTTAR PRADESH
Alīgarh
Farrukhābād
Lucknow
Mathura
Āgra
Bharatpur
KĀNPUR
Faizābād
Gorakhpur
Allāhābād
Vārānasi (Benares)
Mirzāpur
NEPAL
Kathmandu
Lalitpur
Mt. Everest 29 028
Darjeeling
SIKKIM
Gangtok
BHUTAN
Thimphu
Cooch Behar
Rangpur
ASSAM
Gauhāti
Shillong
KHASI HILLS
MEGHALAYA
Silchar
Imphāl
MANIPUR
Mymensingh
BANGLADESH
Dhaka
Comilla
Noākhāli
Chittagong
MYANMAR (BURMA)
Mandalay
Monywa
Myingyan
Bhamo
Myitkyinā
Moggaung

Multan
Dera Ghāzi Khān
Bahāwalpur
Shikārpur
Sukkur
Mohenjo-Daro (Ruins)
Kalāt
Hyderābād
KARACHI
Bhuj
Mãndvi
GUJARAT
Jāmnagar
Rājkot
KATHIĀWĀR PENINSULA
Porbandar
Junāgadh
Verāval
Diu
Bhaunagar
Gulf of Kutch
Gulf of Khambat
AHMADĀBĀD
Ahmadnagar
Ujjain
Indore
Baroda
Surat
Dhule
Nāsik
Aurangābād
Daman

GREAT INDIAN DESERT
Bīkaner
Jodhpur
RĀJASTHĀN
ARAVALLI RANGE
Ajmer
Jaipur
Alwar
Gwalior
Tonk
Kota
Jhalawar
Sheopur
Shivpuri
Jhānsi
Sāgar
Bhopāl
VINDHYA RANGE
MADHYA PRADESH
Jabalpur
Bilāspur
CHHATTISGARH
Nāgpur
Amrāvati
Akola
Wardha
Raipur
Raigarh
Sambalpur
ORISSA
Cuttack
Bhubaneswar
Puri
Berhampur
Chandrapur
DECCAN

Udaipur
Pālanpur
Abu Road

GANGES
Patna
BIHĀR
Darbhanga
Bhāgalpur
Monghyr
Sāsarām
Gaya
Giridih
Rānchī
JHARKHAND
Asansol
Burdwān
Bhātpāra
Howrah
KOLKATA (Calcutta)
WEST BENGAL
Khulna
Kharagpur
Jamshedpur
Raurkela
Balasore
Mouths of the Ganges
Hooghly

INDIA

MAHĀRASHTRA
MUMBAI (Bombay)
Pune
Sholāpur
Sāngli
Kolhāpur
Belgaum
Hubli
Panaji (Panjim)
GOA
KARNATAKA
WESTERN GHĀTS
Bellary
Gulbarga
Gadag
Rāichūr
Kurnool
Mangalore
BANGALORE
Mysore
Kolār
Vellore
CHENNAI (Madras)
Kānchipuram
Pondicherry
Cuddalore
Salem
Coimbatore
Tiruchchirāppalli
Thanjāvūr
Kozhikode
Madurai
TAMIL NADU
KERALA
LAKSHADWEEP (LACCADIVE IS.) (India)
Mahe

Nizāmābād
HYDERĀBĀD
Warangal
Vijayawāda
Rājahmundry
Kākināda
Eluru
ANDHRA PRADESH
Guntur
Machilipatnam
Nellore
Cuddapah
Vishākhapatnam
Vizianagaram
EASTERN GHĀTS
CORROMANDEL COAST

ARABIAN SEA
IRANIAN

BAY OF BENGAL
PEGU YOMA
ARAKAN YOMA
Sittwe
Kyaukpyu
Sandoway
Henzada
Pathein
Rangoon (Yangon)
PAGODA PT.
Mouths of the Irrawaddy
Pye (Prome)
Pyinmana
Yenangyaung
Magwe
Minbu
Yamethin
Sagu
Pyu

MYANMAR (BURMA)
Tropic of Cancer

D
DIPHU PASS
ARUNACHAL PRADESH
Sadiya
Tinsukia
Sibsāgar
Jorhat
NAGALAND
Kohima
Mt. Victoria 10 018

A - Area occupied by Pakistan and claimed by India.
B - Area claimed and occupied by India, status disputed by Pakistan.
C - Area occupied by China and claimed by India.
D - Area occupied by India and claimed by China.

0 50 100 200 300 400 500 Miles
0 100 200 400 600 800 Kilometers

c

Tiruchchirāppalli
Thanjāvūr
Nāgappattinam
Ernākulam
TAMIL NADU
KERALA
Madurai
Jaffna
Alleppey
Tuticorin
Tirunelveli
Mannar
Trincomalee
Quilon
Thiruvananthapuram
CAPE COMORIN
Puttalam
Anurādhapura
SRI LANKA (CEYLON)
Colombo
Kandy
Pidurutalagala 8281
INDIAN OCEAN
Galle
Matara
DONDRA HEAD
Gulf of Mannar
Palk Strait

Same scale as main map

Continued on pages 160-161

XIZANG (TIBET)
Lhasa
Gyangzê
DISE SHAN
HIMALAYA
Brahmaputra
CHINA

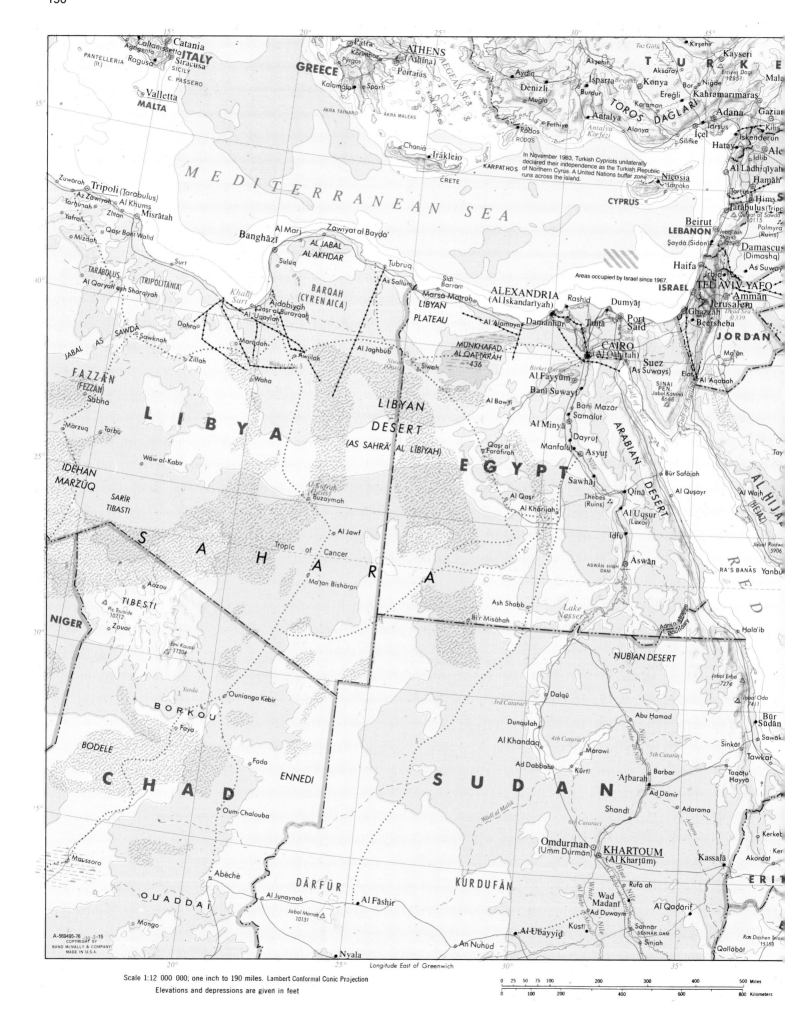

Scale 1:12 000 000; one inch to 190 miles. Lambert Conformal Conic Projection

Elevations and depressions are given in feet

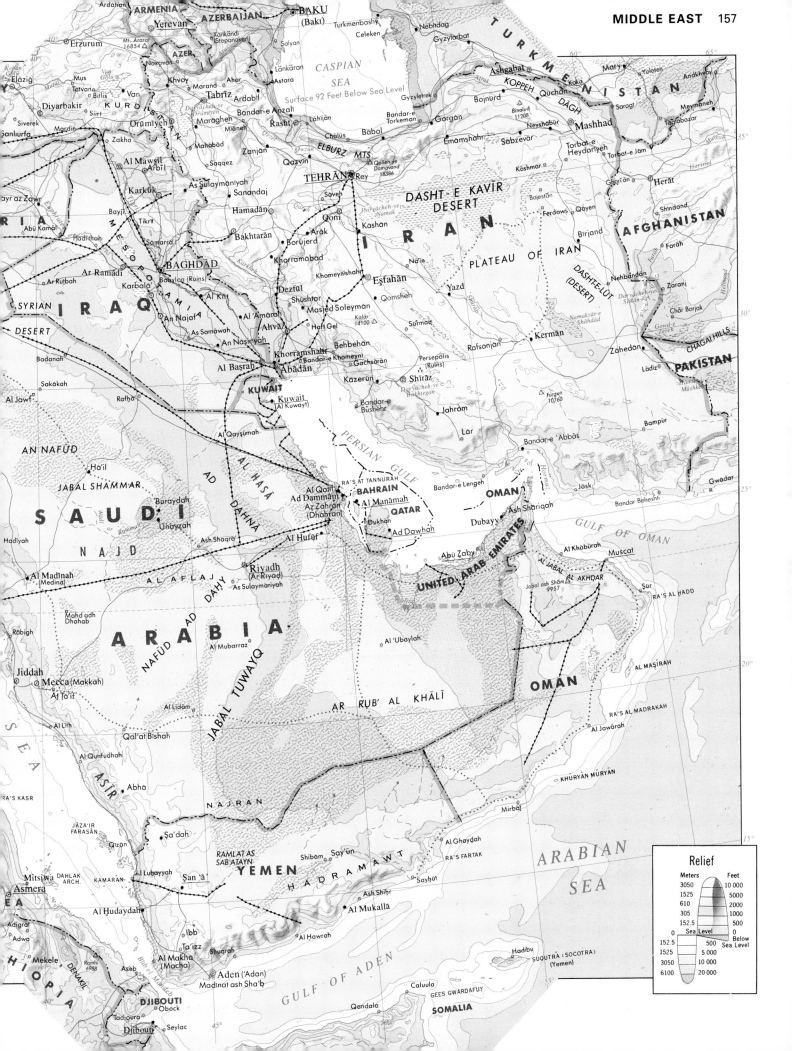

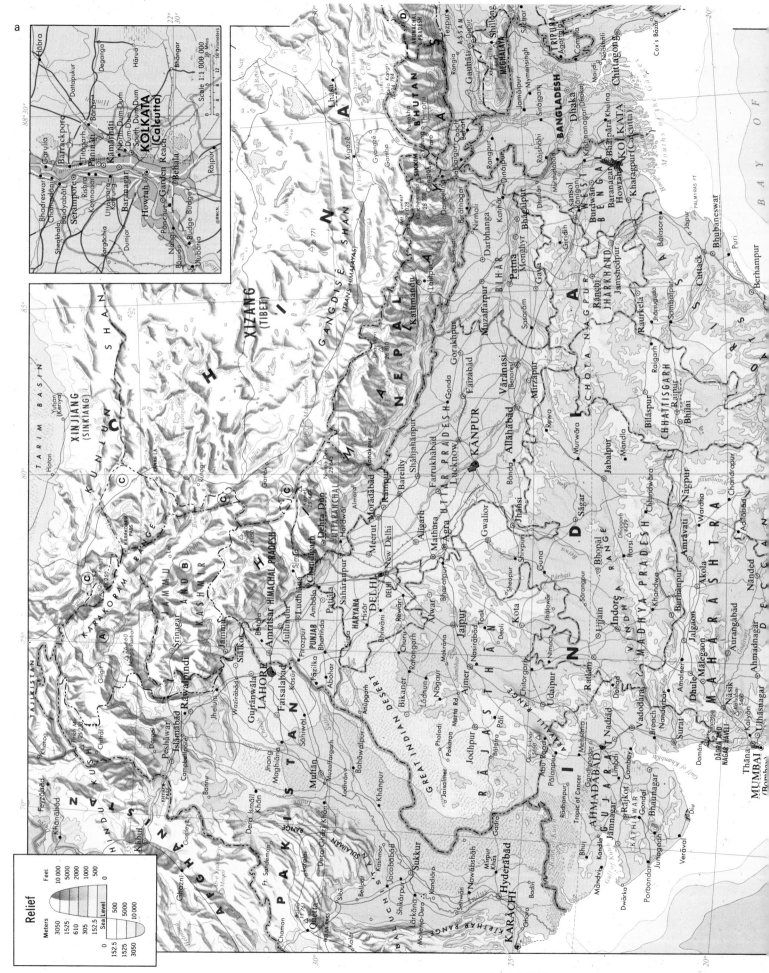

a

KOLKATA
[Calcutta]

Scale 1:1 000 000

Relief

Meters	Feet	
3050	10 000	
1525	5000	
610	2000	
305	1000	
152.5	500	
0	Sea Level	
152.5		
1525	5000	
3050	10 000	

Scale 1:10 000 000; one inch to 160 miles. Lambert Conformal Conic Projection
Elevations and depressions are given in feet

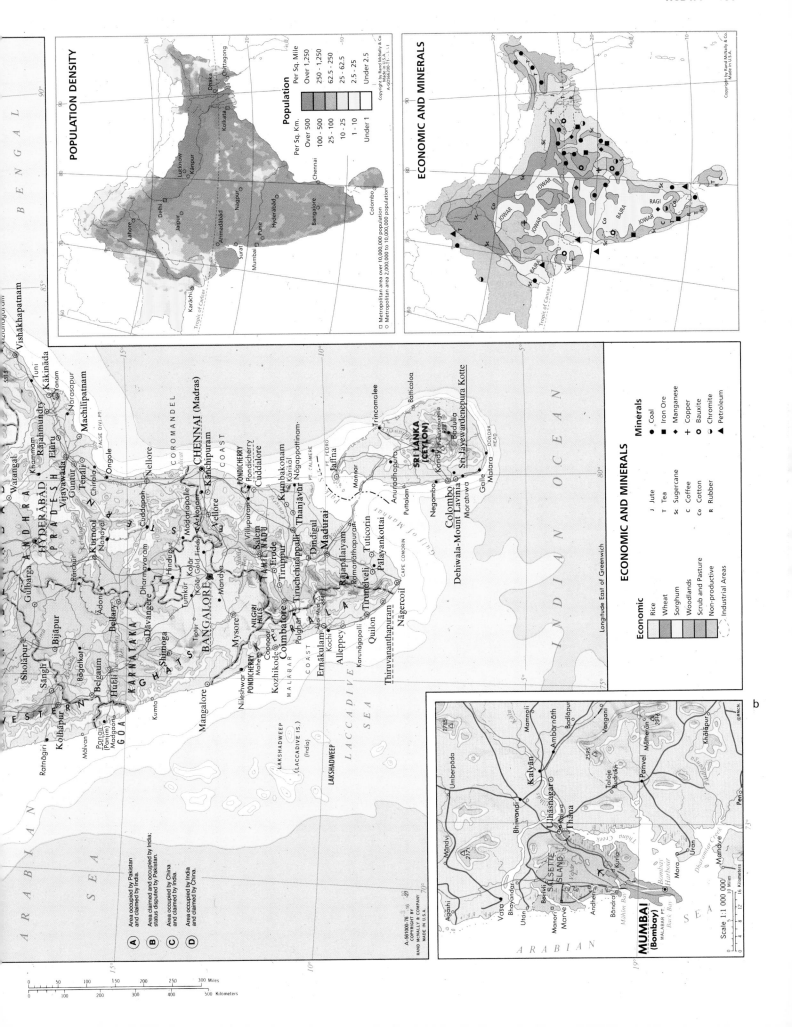

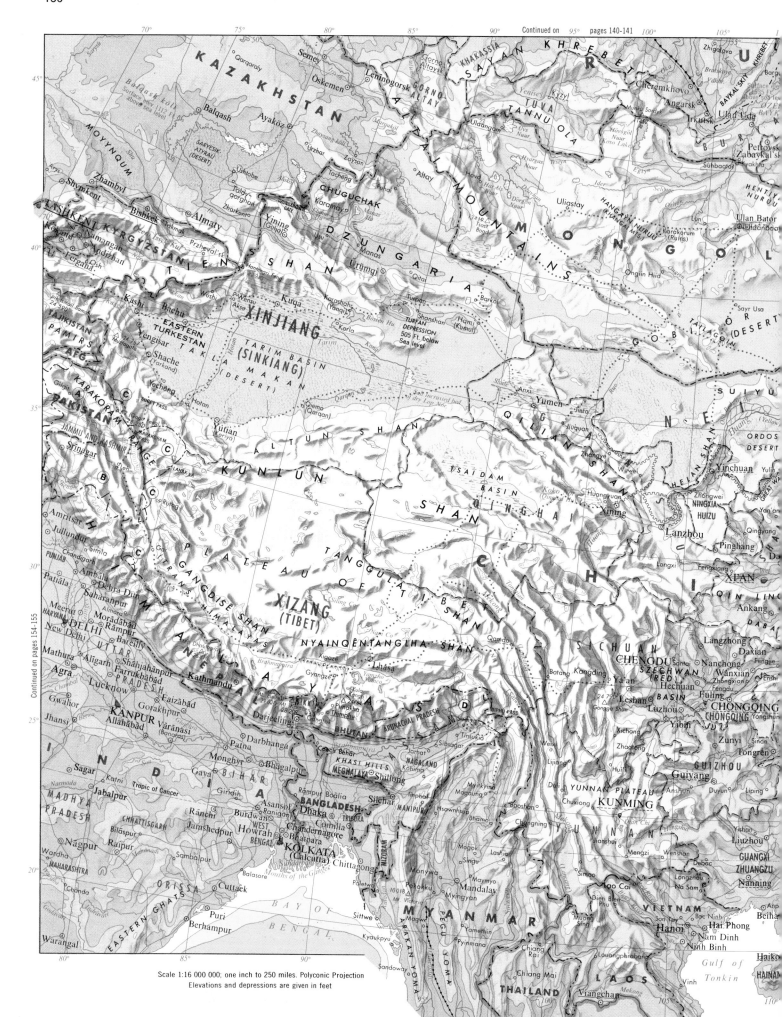

Continued on pages 140-141

Continued on pages 154-155

Scale 1:16 000 000; one inch to 250 miles. Polyconic Projection
Elevations and depressions are given in feet

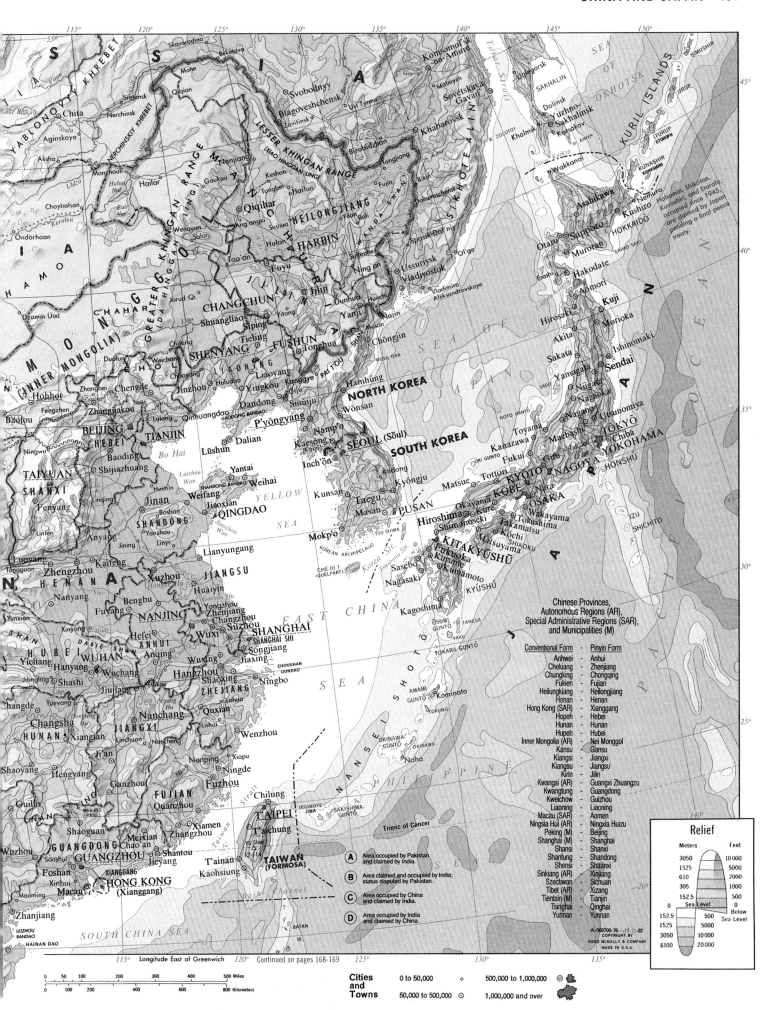

Chinese Provinces,
Autonomous Regions (AR),
Special Administrative Regions (SAR),
and Municipalities (M)

Conventional Form	-	Pinyin Form
Anhwei	-	Anhui
Chekiang	-	Zhejiang
Chungking	-	Chongqing
Fukien	-	Fujian
Heilungkiang	-	Heilongjiang
Honan	-	Henan
Hong Kong (SAR)	-	Xianggang
Hopeh	-	Hebei
Hunan	-	Hunan
Hupeh	-	Hubei
Inner Mongolia (AR)	-	Nei Monggol
Kansu	-	Gansu
Kiangsi	-	Jiangxi
Kiangsu	-	Jiangsu
Kirin	-	Jilin
Kwangsi (AR)	-	Guangxi Zhuangzu
Kwangtung	-	Guangdong
Kweichow	-	Guizhou
Liaoning	-	Liaoning
Macau (SAR)	-	Aomen
Ningsia Hui (AR)	-	Ningxia Huizu
Peking (M)	-	Beijing
Shanghai (M)	-	Shanghai
Shansi	-	Shanxi
Shantung	-	Shandong
Shensi	-	Shaanxi
Sinkiang (AR)	-	Xinjiang
Szechwan	-	Sichuan
Tibet (AR)	-	Xizang
Tientsin (M)	-	Tianjin
Tsinghai	-	Qinghai
Yunnan	-	Yunnan

(A) Area occupied by Pakistan and claimed by India.

(B) Area claimed and occupied by India; status disputed by Pakistan.

(C) Area occupied by China and claimed by India.

(D) Area occupied by India and claimed by China.

Relief

Meters		Feet
3050		10 000
1525		5000
610		2000
305		1000
152.5		500
0	Sea Level	0
152.5		500 Below Sea Level
1525		5000
3050		10 000
6100		20 000

A-569700-76--17-10-82
COPYRIGHT BY
RAND McNALLY & COMPANY
MADE IN U.S.A.

Longitude East of Greenwich

0 50 100 200 300 400 500 Miles
0 100 200 400 600 800 Kilometers

| Cities and Towns | 0 to 50,000 | ○ | 500,000 to 1,000,000 | ◎ |
| | 50,000 to 500,000 | ⊙ | 1,000,000 and over | |

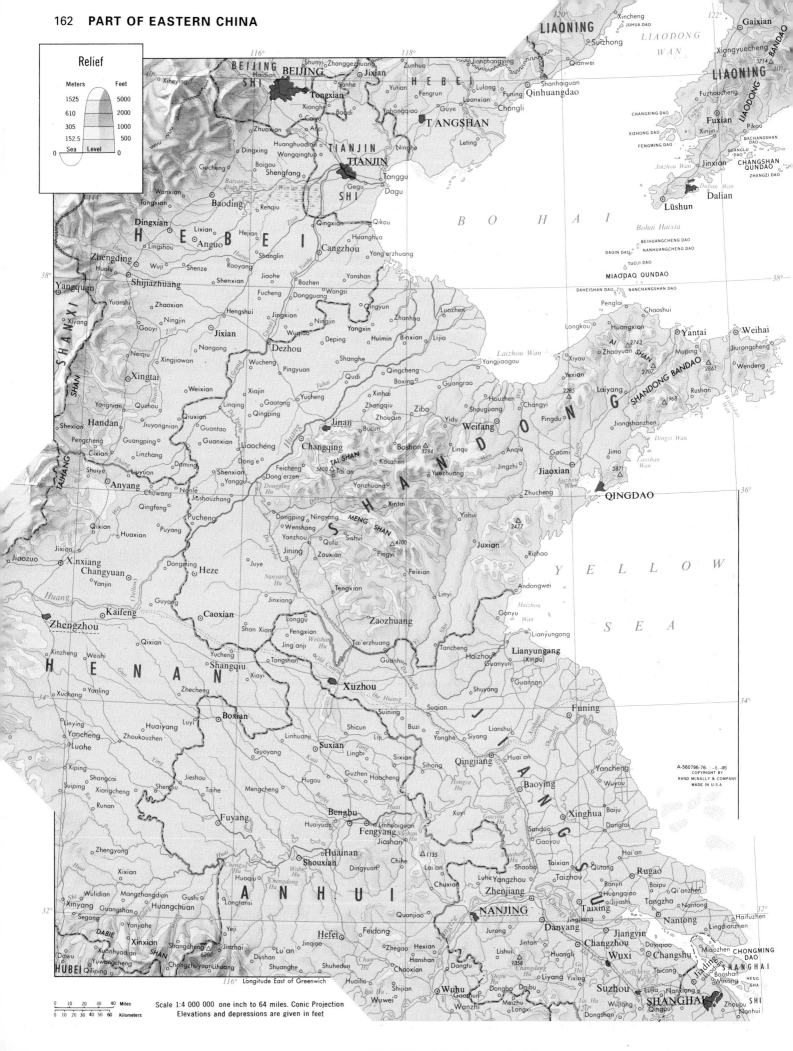

Relief

Meters	Feet
1525	5000
610	2000
305	1000
152.5	500
0 Sea Level	0

LIAONING

LIAODONG WAN

Xincheng JUHUA DAO Gaixian

Suizhong

Qianwei

Xiangyuecheng 3714

LIAODONG BANDAO

LIAONING

Fuzhoucheng

Fuxian

CHANGXING DAO

XIZHONG DAO

FENGMING DAO

Xinjin Pikou DACHANGSHAN DAO

GUANGLU DAO CHANGSHAN QUNDAO

Jinxian ZHANGZI DAO

Jinzhou Wan

Dalian Wan

Lüshun Dalian

BEIJING SHI Beijing BEIJING

Xiheying Haidian Shunyi Zhanggezhuang Jixian Zunhua Jianchangying

Tongxian Sanhe Yutian Fengrun Lulong Shanhaiguan Qinhuangdao

Xianghe Baodi Yahongqiao Güye Changli

HEBEI

Zhuoxian Caiyu Anci Luanxian Funing

Huanghuadian TANGSHAN Leting

Dingxing Wangqingtuo Ninghe

TIANJIN TIANJIN

Baigou Shengfang SHI Tanggu

Gucheng Balyang Dian Wen an Wa Gegu Dagu

Wanxian Tangxian Renqiu Qingxian Qikou

Baoding Qingxian

BOHAI

Dingxian Lixian Hejian Huanghua

Anguo Yang erzhuang

HEBEI Shanglin Cangzhou Bohai Haixia

Zhengding Wuji Raoyang Yanshan BEIHUANGCHENG DAO

Huolu Shenze Jiaohe Bozhen DAQIN DAO NANHUANGCHENG DAO

Yangquan Shijiazhuang Shenxian Dongguang TUOJI DAO

SHANXI Zhaoxian Hengshui Jingxian Ningjin MIAODAO QUNDAO

Yuanshi Fucheng Wuqiao Zhanhua Luozhen DAHEISHAN DAO NANCHANGSHAN DAO

Xiyang Ningjin Nangong Deping Huimin Binxian Lijia Penglai Chaoshui

Gaoyi Jixian Dezhou Yanshan Longkou Huangxian Yantai Weihai

Zhao xian Wucheng Shanghe Qingcheng Boxing Yangjiaogou AI 2743 SHAN Muping Jiurongcheng

Xingtai Weixian Xiajin Yucheng Pingyuan Qingcheng Zhaoyuan Xiyou 2702 Wendeng

Yongnian Quzhou Linqing Gaotang Qudi SHANDONG Laiyang 2861

Handan Qingping Zhangqiu Guangrao Houzhen 2285 Rushan

Shexian Jiuyongnian Qiuxian Guantao Dong e Jinan Zhoucun Zibo Yidu Changyi 1968 Dingzi Wan

Pengcheng Guangping Guanxian Liaocheng Changqing Bucun Weifang Pingdu Jimo

Cixian Linzhang Daming Shenxian Boshan 3284 Linqu Gaomi Laoshan Wan

Anyang Chuwang Nonle Yanggu TAI SHAN Kouzhen Anqiu Jiaoxian 387?

Qingfeng 5600 Tai an Yanzhuang Jingzhi Jimo

Qixian Shuiye Liuyuan Dong erzen Zhucheng Jiaozhou Wan

Jixian Huaxian Pucheng Shenxian Yanzhuang QINGDAO

Jiaozuo Xinxiang Puyang Dongming Juye Dongping Hu Xintai 2427

Changyuan Heze Guyang Ningyang MENG SHAN Yishui Juxian

Yanjin Dongping Wenshang 4100 Rizhao

Kaifeng Caoxian Yanzhou Qufu Pingyi

Zhengzhou Shan Xian Longgu Jining Zouxian Feixian Linyi Andongwei

Qixian Fengxian Weishan Tai erzhuang Tancheng Haizhou Wan

Xinzheng Weishi Yucheng Jing anji Hu Zaozhuang Guanhu Haizhou Lianyungang

HENAN Shangqiu Tongshan Guanhu Da Yunhe Lianyungang (Xinpu)

Xuchang Zhecheng Xiayi Xuzhou Suining Sugian Guannan

Yanling Boxian Suining Shicun Liji Yanghe Shuyang SEA

Linying Luyi Linhuanji Buzi Siyang Guannan

Yancheng Huaiyang Guoyang Suxian Lingbi Sixian Funing

Luohe Zhoukouzhen Jieshou Guzhen Haocheng Sihong Qingjiang Huai an

Xiping Shangcai Taihe Mengcheng Hugou Lianshui Baoying

Suiping Xiangcheng Bengbu Beifei Xuyi Yancheng Baiju

Runan ANHUI Huaiyuan Huai Gaoyou Xinghua Dongtai

Zhengyang Fuyang Linhuaiguan Nishan Hu Shaobo Hu Taixian Qutang Rugao

Xixian Shouxian Fengyang Jiashan Hai an Baipu Qi anzhen

Xinyang Huoqiu Huainan Dingyuan Chihe Shaobo Taizhou Banjin Tangzha

Wulidian Mangzhangdian Chengxi Hu Wabie Lai an Luhe Yangzhou Huangqiao Jiajiash Nantong

Gushan Longtansi Chengdong Hu Quanjiao Chuxian Zhenjiang Jingjiang Haifuzhen

DABIE Yejii Shuanghe Dangtu NANJING Taixing Lingdianzhen

Xinyang Guangshan Xinxian Shangcheng Jinzhai Lu an Jingiao Jurong Danyang Nantong

Segang Yanjiahe 6200 Changzhuoyan Liheuang Feidong Lishui Jiangyin CHONGMING DAO

HUBEI Yuwangcheng SHAN Changfeng Hefei Hexian Hanshan Dangtu 1358 Changzhou Changshu SHANGHAI SHI

Qiliping Huaihe Zhegao Shijiu Chaoxian Wuxi Jiading Baoshan

116° Longitude East of Greenwich Wuhu Gaoshun Meizhou Suzhou HENG Wusong

Huaiba Langxi Tai Hu Kunshan SHANGHAI

Wanzhi Dongshan Wujiang Qingpu Zhoupu

Scale 1:4 000 000 one inch to 64 miles. Conic Projection
Elevations and depressions are given in feet

A-560796-76- -6-40
COPYRIGHT BY
RAND McNALLY & COMPANY
MADE IN U.S.A.

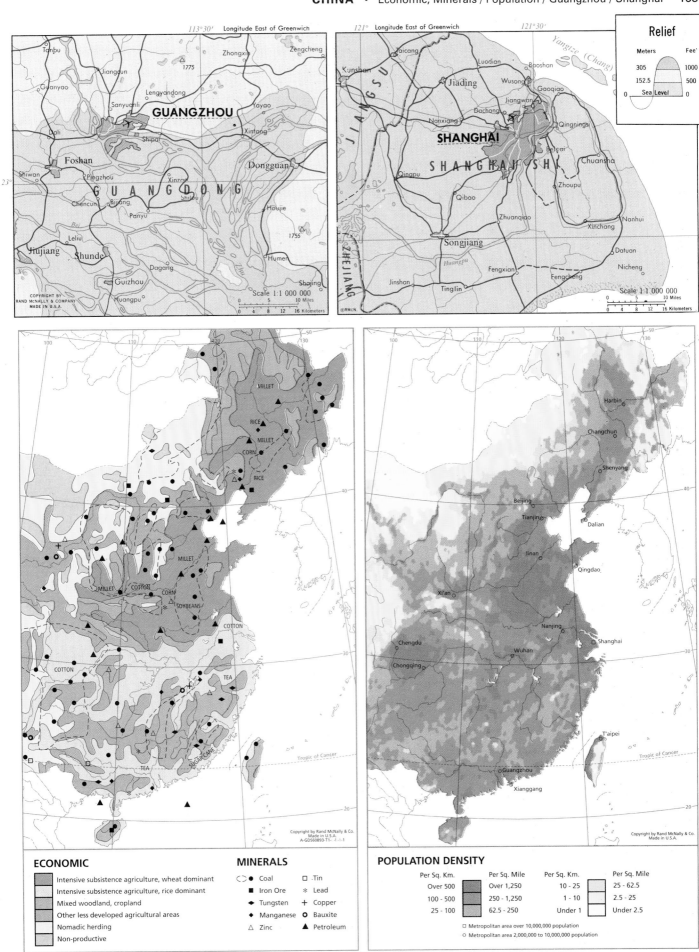

Scale 1:1 000 000

0   5   10 Miles
0 4 8 12 16 Kilometers

Scale 1:1 000 000

0   5   10 Miles
0 4 8 12 16 Kilometers

Relief

Meters		Feet
305		1000
152.5		500
0	Sea Level	0

Copyright by Rand McNally & Co.
Made in U.S.A.
A-GDS60893-T1- -1- -1- -1

Copyright by Rand McNally & Co.
Made in U.S.A.

ECONOMIC

- Intensive subsistence agriculture, wheat dominant
- Intensive subsistence agriculture, rice dominant
- Mixed woodland, cropland
- Other less developed agricultural areas
- Nomadic herding
- Non-productive

MINERALS

- ⊂● Coal
- ■ Iron Ore
- ◆ Tungsten
- ◆ Manganese
- △ Zinc
- □ Tin
- ✳ Lead
- + Copper
- ○ Bauxite
- ▲ Petroleum

POPULATION DENSITY

Per Sq. Km.	Per Sq. Mile	Per Sq. Km.	Per Sq. Mile
Over 500	Over 1,250	10 - 25	25 - 62.5
100 - 500	250 - 1,250	1 - 10	2.5 - 25
25 - 100	62.5 - 250	Under 1	Under 2.5

□ Metropolitan area over 10,000,000 population
○ Metropolitan area 2,000,000 to 10,000,000 population

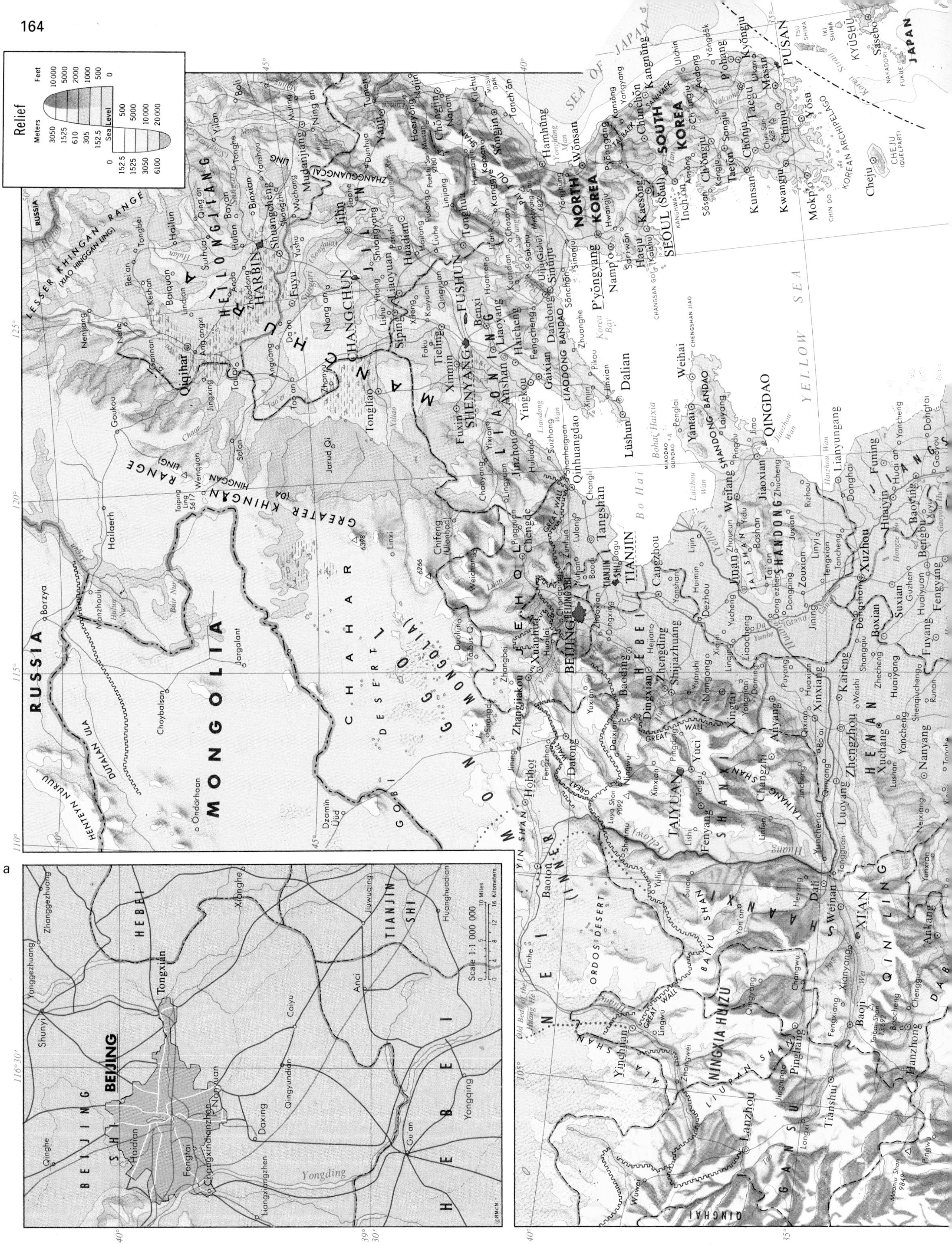

Relief

Feet
10 000
5000
2000
1000
500
0

Meters		Sea Level
3050		500
1525		5000
610		20 000
305		
152.5		
0	152.5	
	1525	
	3050	
	6100	

SEA OF JAPAN

RUSSIA

LESSER KHINGAN RANGE (XIAO HINGGAN LING)

HEILONGJIANG

Qiqihar

HARBIN

CHANGCHUN

JILIN

GREATER KHINGAN RANGE (DA HINGGAN LING)

CHAHAR

MONGOLIA

GOBI DESERT

HENTIYN NURUU

DUTALAN ULA

Choybalsan

Öndörhaan

LIAONING

SHENYANG

FUSHUN

Benxi

Anshan

LIAODONG BANDAO

Dalian

Lüshun

NORTH KOREA

P'yongyang

Namp'o

SOUTH KOREA

SEOUL (Soul)

Inch'on

Taejon

Taegu

PUSAN

Kwangju

Mokp'o

Cheju

CHEJU (QUELPART)

KOREAN ARCHIPELAGO

JAPAN

KYŪSHŪ

Sasebo

YELLOW SEA

Bo Hai

QINGDAO

SHANDONG

SHANDONG BANDAO

Yantai

Weihai

Weifang

Jinan

TIANJIN

TIANJIN SHI

BEIJING

BEIJING SHI

HEBEI

Tangshan

Shijiazhuang

Baoding

Chengde

GREAT WALL

SHANXI

TAIYUAN

Datong

Hohhot

Baotou

INNER MONGOLIA

NEI MONGGOL

ORDOS DESERT

YIN SHAN

TAIHANG SHAN

HENAN

Zhengzhou

Luoyang

Kaifeng

Xuzhou

SHAANXI

XI'AN

Xianyang

Baoji

QIN LING

NINGXIA HUIZU

Yinchuan

LIUPAN SHAN

GANSU

Lanzhou

Tianshui

QINGHAI

Old Bed of the Huang He

GREAT WALL

GREAT WALL

Huang (Yellow)

Inset map (a):

HEBEI

BEIJING SHI

BEIJING

TIANJIN SHI

Tongxian

Shunyi

Haidian

Fengtai

Daxing

Yongding

Scale 1:1 000 000

0 to 10 Miles

0 4 8 12 16 Kilometers

Cities and Towns

0 to 50,000	500,000 to 1,000,000
50,000 to 500,000	1,000,000 and over

Scale 1:10 000 000; one inch to 160 miles. Lambert Conformal Conic Projection
Elevations and depressions are given in feet

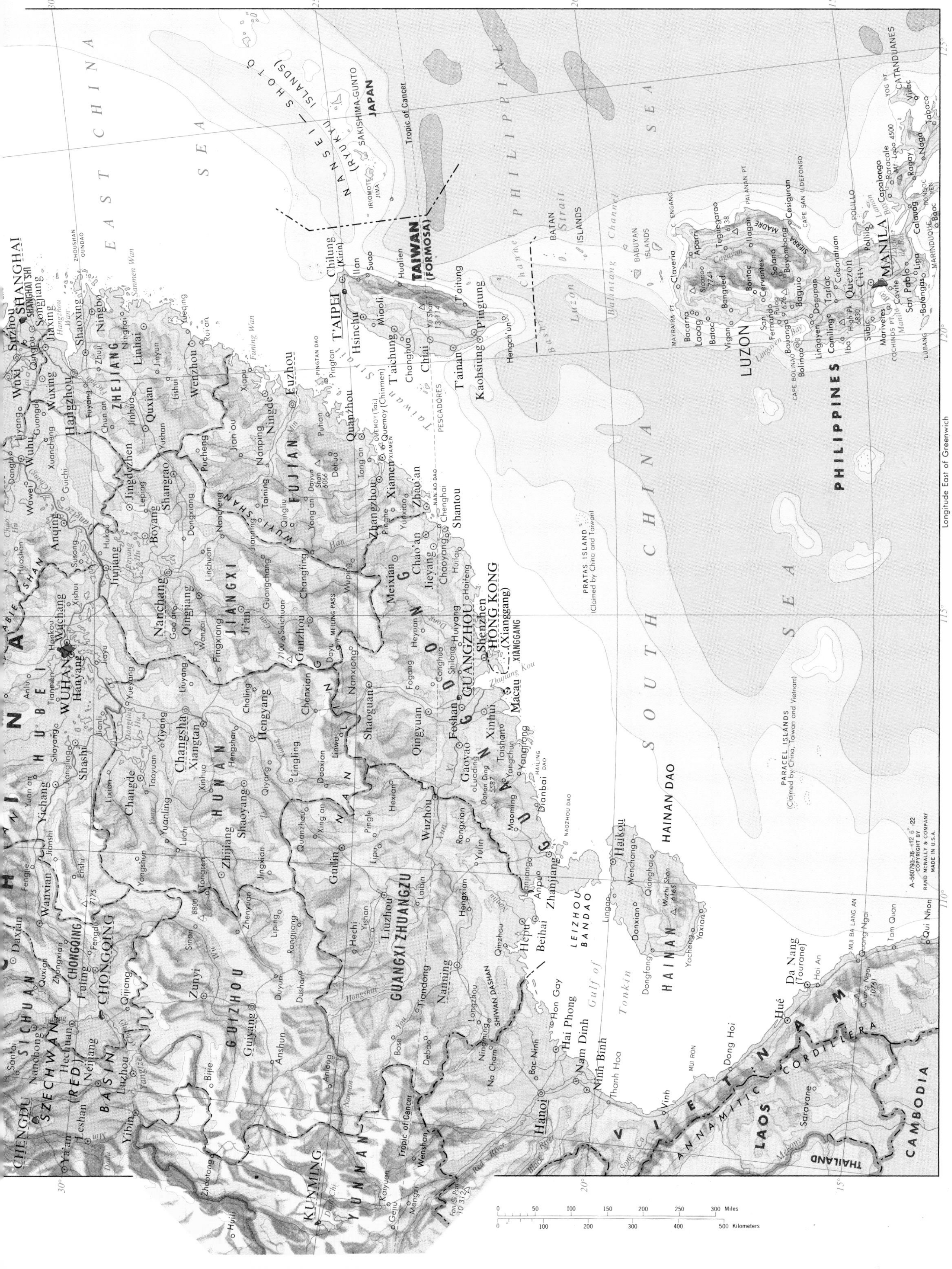

EAST CHINA SEA

SOUTH CHINA SEA

PHILIPPINE SEA

JAPAN

NANSEI SHOTO (RYUKYU ISLANDS)

SAKISHIMA-GUNTO

IRIOMOTE-JIMA

Tropic of Cancer

TAIWAN (FORMOSA)

T'AIPEI

Chilung (Kirin)
Ilan
Suao
Hualien
Hsinchu
Miaoli
Taichung
Changhua
Chiai
T'ai-nan
Yu Shan 13114
Taitung
Kaohsiung
Pingtung
Hengch'un

PESCADORES

Quemoy (Taj)
Quemoy (Chinmen)
NAN-AO DAO

PINGTAN DAO
Pingtan

Taiwan Strait

BATAN ISLANDS
Batan
BABUYAN ISLANDS
BALINTANG CHANNEL
Bashi Channel
Luzon Strait

LUZON

PHILIPPINES

MANILA
Quezon City
Cavite
Manila Bay
Baguio
Aparri
Tuguegarao
Laoag
Vigan
Bacarra
San Fernando
Dagupan
Tarlac
Cabanatuan
San Pablo
Batangas
Lipa
Lubang
Mindoro
MARINDUQUE
CATANDUANES
Naga
Legaspi

CAPE ENGAÑO
PALANAN PT
MT. LABO 4500
Capalonga

MANILA

SHANGHAI
Songjiang
Suzhou
Jiaxing
Wuxing
Hangzhou
Shaoxing
Ningbo
ZHOUSHAN QUNDAO
Sanmen Wan
Linhai
Wenzhou
Ningde
Fuzhou

ZHEJIANG

Huzhou
Yixing
Wuxi
Wuhu
Xuancheng
Fuyang
Jinhua
Quxian
Jian'ou
Nanping
Putian
Quanzhou

FUJIAN

WU YI SHAN

Jiangshan
Pucheng
Jianyang
Yong'an
Tong'an
XIAMEN
Xiamen
Zhangzhou
Pinghe
Zhao'an
Chenghai
Shantou

GUANGDONG

Meixian
Jieyang
Chaoyang
Haifeng
Huiyang
Huizhou
Heyuan
Shilong
Shenzhen
HONG KONG (Xianggang)
Macau
Guangzhou
Foshan
Xinhui
Taishan
Gaoyao
Zhaoqing
Yangjiang
HAILING DAO
Maoming

JIANGXI

Nanchang
Qingjiang
Ji'an
Ganzhou
Pingxiang
Gao'an
Fengxin
Linchuan
Guangchang
Changting
Ruijin

Boyang
Shangrao
Dongxiang
Leping
Jingdezhen

POYANG HU
MEILING PASS

HUNAN

Changsha
Xiangtan
Zhuzhou
Hengyang
Shaoyang
Lingling
Daoxian
Chenxian
Chaling
Leiyang
Lianyuan
Yiyang
Changde

Yueyang
DONGTING HU

HUBEI

WUHAN
Wuchang
Hanyang
Hankou
Huangshi
Xianning
Yichang
Shashi
Anlu

ANHUI

HENAN

GUANGXI ZHUANGZU

Guilin
Liuzhou
Nanning
Wuzhou
Yulin
Beihai
Hepu
Qinzhou
Hengxian
Laibin
Guiping
Pingnan
Bose
Baise
Longzhou

HAINAN

HAINAN DAO
Haikou
Qiongshan
Wenchang
Danxian
Wuzhi Shan 6165
Yaxian
Dongfang

LEIZHOU BANDAO
Zhanjiang
NAOZHOU DAO

GULF OF TONKIN

GUIZHOU

Guiyang
Zunyi
Anshun
Duyun
Dushan

SICHUAN

SZECHWAN

Chengdu
Chongqing
CHONGQING
Neijiang
Luzhou
Yibin
Fuling
Wanxian
Fengjie
Daxian
Nanchong
Quxian

RED BASIN

YUNNAN

KUNMING
Gejiu
Mengzi

VIETNAM

Hanoi
Hai Phong
Hon Gay
Nam Dinh
Ninh Binh
Thanh Hoa
Vinh
Dong Hoi
Hué
Da Nang (Tourane)
Qui Nhon
Nha Trang

ANNAMITIC CORDILLERA

LAOS

THAILAND

CAMBODIA

PRATAS ISLAND
(Claimed by China and Taiwan)

PARACEL ISLANDS
(Claimed by China, Taiwan and Vietnam)

Longitude East of Greenwich

A-560793-76-112 6-22
COPYRIGHT BY
RAND McNALLY & COMPANY
MADE IN U.S.A.

Tropic of Cancer

| 0 | 50 | 100 | 150 | 200 | 250 | 300 Miles |

| 0 | 100 | 200 | 300 | 400 | 500 Kilometers |

Continued on pages 164-165

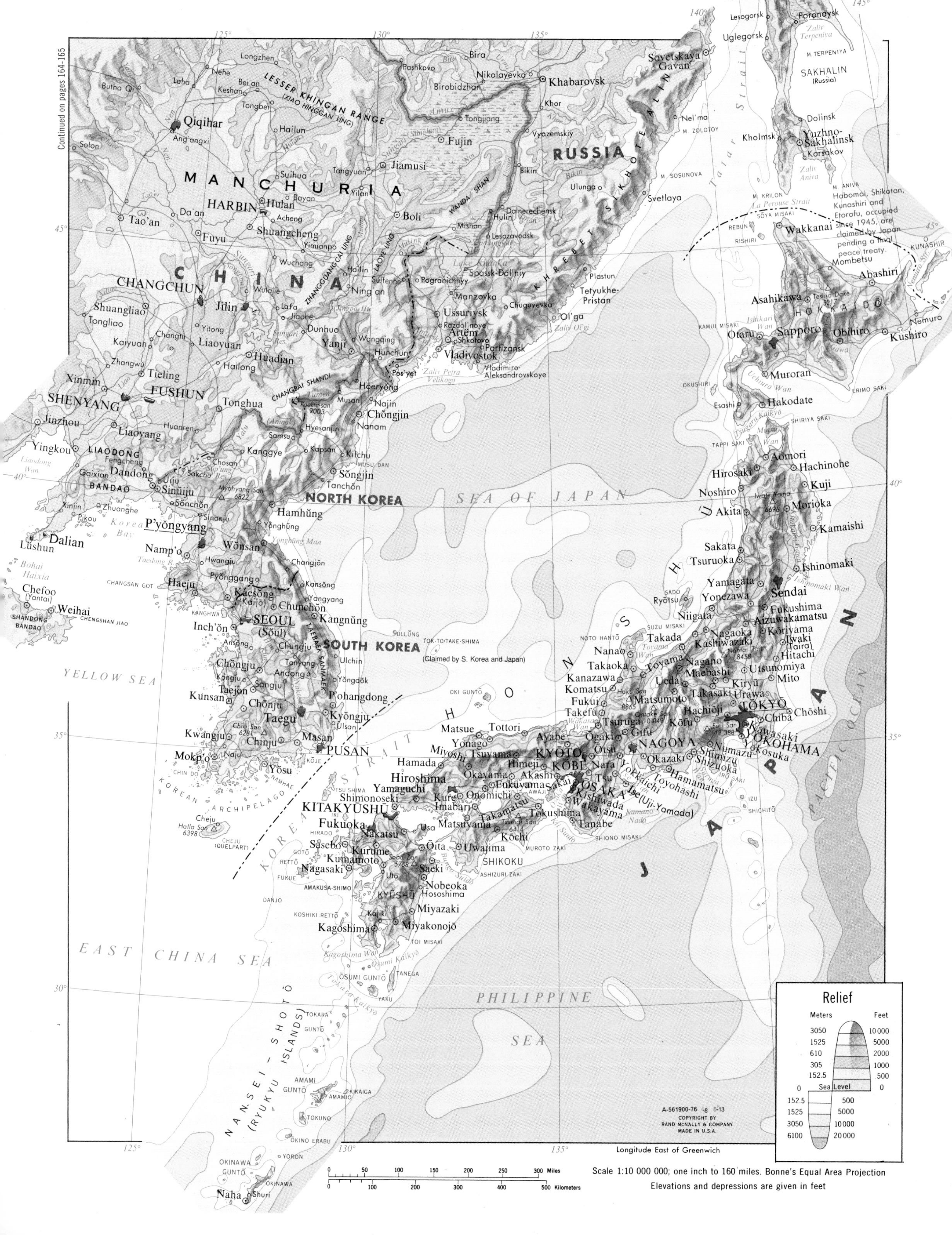

MANCHURIA

HARBIN

CHINA

RUSSIA

LESSER KHINGAN RANGE (XIAO HINGGAN LING)

SAKHALIN (Russia)

Qiqihar

Longzhen · Nehe · Bei'an · Laha · Keshan · Tongbei · Hailun · Suihua · Bayan · Acheng · Shuangcheng · Yilan · Boli · Tangyuan · Jiamusi · Fujin · Nikolayevka · Bira · Birobidzhan · Pashkovo · Khabarovsk · Khor · Vyazemskiy · Bikin · Lesozavodsk · Dalnerechensk

Tao'an · Da'an · Fuyu · Ang'angxi · Solon

CHANGCHUN · Shuangliao · Tongliao · Yitong · Lafa · Jiaohe · Dunhua · Wangqing · Yanji · Hunchun · Suifenhe · Pogranichnyy · Manzovka · Spassk-Dal'niy · Mishan · Chuguyevka · Plastun · Tetyukhe-Pristan

Kaiyuan · Liaoyuan · Changtu · Huadian · Hailong

Zhangwu · Tieling · Huadian

Xinmin · SHENYANG · FUSHUN · Huanren · Tonghua · CHANGBAI SHANDI

Jinzhou · Liaoyang · Kaita · Baekdusan 9003

LIAODONG · Fengcheng · Dandong · Sinŭiju · Sŏnch'ŏn

Gaixian · BANDAO · Zhuanghe · Uiju · Sakchu · Chosan

Yingkou · Xinjin · Pikou · Sinanju · Anju · P'yŏngyang

Lüshun · Dalian · Korea Bay

Bohai Haixia

Chefoo (Yantai) · Weihai · SHANDONG BANDAO · Chengshan Jiao

YELLOW SEA

Kanggye · Kapsan · Hyesanjin · Musan · Hoeryŏng · Najin · Ch'ŏngjin · Nanam · Kilchu · Tanch'ŏn · Sŏngjin

NORTH KOREA

Myohyang San 6822

Hamhŭng · Yŏnghŭng · Wŏnsan · Changjŏn · Kansŏng

Namp'o · Hwangju · P'yŏnggang · Haeju · Kaesŏng (Kaijo) · Ch'unch'ŏn · Kangnŭng

Inch'ŏn · SEOUL (Sŏul) · Chungju · Wŏnju · Ulchin · Yŏngdŏk

Anyang · Taebaek Sanmaek

SOUTH KOREA

Ch'ŏngju · Sangju · Andong · P'ohangdong

Kongju · Taejŏn · Kyŏngju · Ulsan

Kunsan · Chŏnju · Taegu · Masan

Kwangju · Chinju · PUSAN

Mokp'o · Naju · Yŏsu

Chin Do · Cheju · Halla San 6398 · CHEJU (QUELPART)

KOREAN ARCHIPELAGO · KOREA STRAIT

SEA OF JAPAN

HOKKAIDO

Wakkanai · Asahikawa · Otaru · Sapporo · Obihiro · Kushiro · Abashiri · Muroran · Hakodate · Esashi

Habomai, Shikotan, Kunashiri and Etorofu, occupied since 1945, are claimed by Japan pending a final peace treaty.

Aomori · Hirosaki · Hachinohe · Noshiro · Kuji · Akita · Morioka · Kamaishi · Sakata · Ishinomaki · Tsuruoka · Yamagata · Sendai · Yonezawa · Fukushima · Niigata · Aizuwakamatsu · Kōriyama · Nagaoka · Iwaki (Taira)

HONSHU

Ryōtsu · SADO · Takada · Nanao · Toyama · Nagano · Maebashi · Utsunomiya · Mito · Kanazawa · Ueda · Kiryū · Komatsu · Matsumoto · Takasaki · Urawa · Fukui · Takefu · Kōfu · Hachiōji · TOKYO · Chōshi · Tsuruga · Gifu · Ogaki · Nagoya · YOKOHAMA · Chiba · Fuji San 12388 · Kawasaki · Yokosuka

Matsue · Tottori · Ayabe · Ōtsu · KYOTO · Numazu · Shizuoka

Yonago · Tsuyama · Himeji · KOBE · NARA · Tsu · Yokkaichi · Toyohashi · Hamamatsu

Hamada · Okayama · Akashi · OSAKA · Ise (Uji-Yamada) · Kishiwada · Wakayama

Yamaguchi · Hiroshima · Fukuyama · Onomichi · Kure · Imabari · Kōchi · Tokushima · Tanabe

Shimonoseki · KITAKYŪSHŪ · Matsuyama · Takamatsu

Fukuoka · Nakatsu · Usa · Ōita · Uwajima · SHIKOKU

Sasebo · Kurume · Kumamoto · Uto · Saeki · Nobeoka · Hososhima

Nagasaki · KYŪSHŪ · Miyazaki · Miyakonojō

Kagoshima · Kagoshima Wan

TOK-TO/TAKE-SHIMA (Claimed by S. Korea and Japan)

ULLŬNG

OKI GUNTŌ

JAPAN

PACIFIC OCEAN

EAST CHINA SEA

PHILIPPINE SEA

NANSEI-SHOTŌ (RYUKYU ISLANDS)

AMAMI GUNTŌ · TOKARA GUNTŌ

OKINAWA GUNTŌ · Naha · Shuri

A-561900-76

COPYRIGHT BY RAND McNALLY & COMPANY MADE IN U.S.A.

Longitude East of Greenwich

Relief		
Meters		Feet
3050		10 000
1525		5000
610		2000
305		1000
152.5		500
0	Sea Level	0
152.5		500
1525		5000
3050		10 000
6100		20 000

0 50 100 150 200 250 300 Miles
0 100 200 300 400 500 Kilometers

Scale 1:10 000 000; one inch to 160 miles. Bonne's Equal Area Projection
Elevations and depressions are given in feet

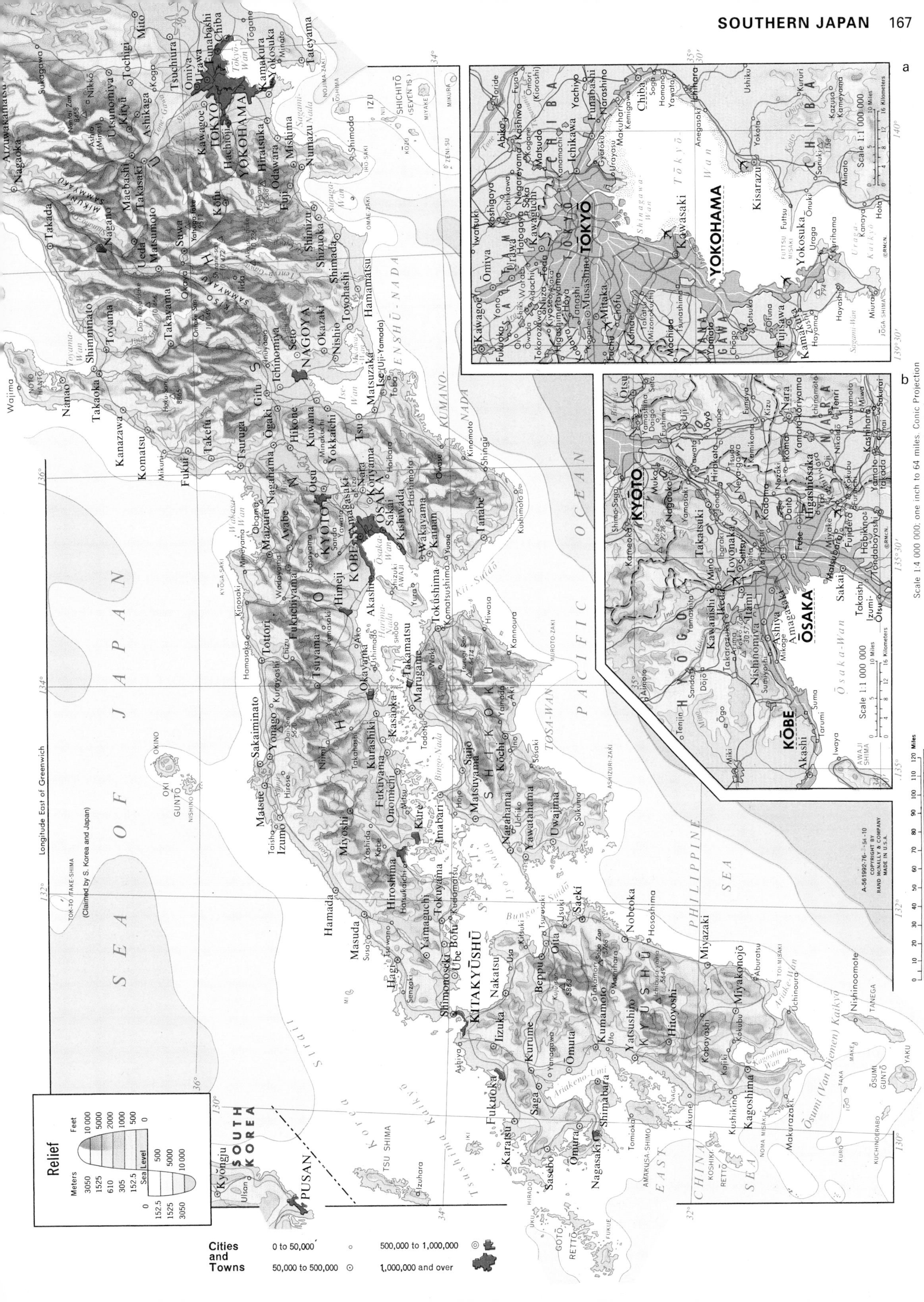

a

b

Scale 1:4 000 000; one inch to 64 miles. Conic Projection
Elevations and depressions are given in feet.

Scale 1:1 000 000

TŌKYŌ

YOKOHAMA

KYŌTO

ŌSAKA

KŌBE

Relief

Meters	Feet
3050	10 000
1525	5000
610	2000
305	1000
152.5	500
0	Sea Level
152.5	500
1525	5000
3050	10000

SEA OF JAPAN

PACIFIC OCEAN

PHILIPPINE SEA

EAST CHINA SEA

SOUTH KOREA

PUSAN

TOKYO
YOKOHAMA
NAGOYA
KYOTO
OSAKA
KOBE
KITAKYŪSHŪ

HONSHŪ
SHIKOKU
KYŪSHŪ

Longitude East of Greenwich

TOK-TO (TAKE-SHIMA)
(Claimed by S. Korea and Japan)

Cities and Towns

0 to 50,000 ○
50,000 to 500,000 ⊙
500,000 to 1,000,000 ◎
1,000,000 and over

A-561992-76----54-10
COPYRIGHT BY
RAND McNALLY & COMPANY
MADE IN U.S.A.

CHINA

Wuzhou · Jieyang Chao'an Tropic of Cancer Shantou
Foshan Xinhui GUANGZHOU
Macau HONG KONG
(Aomen) (Xianggang)

T'ainan · 13 114
Kaohsiung TAIWAN

Maoming
Zhanjiang
Beihai
LEIZHOU BANDAO
Wuzhi Shan 6125
Haikou
HAINAN DAO

Monywa · Maymyo
Pakokku Mandalay
Myingyan Myinmu
MYANMAR
(BURMA)
Paletwa
Sittwe
Kyaukpyu
RAMREE ISLAND
CHEDUBA ISLAND
Sandoway
Pyinmana
Toungoo
Prome (Pyè)

LAOS
Muong Sing
Lang Son
Hanoi
Hai Phong
Ninh Binh
Nam Dinh
Thanh Hoa

Chiang Rai
Louangphrabang
Chau Bia 9249
Chiang Mai
Vianchan

Henzada
Rangoon (Yangon)
Pathein
Bago
Mouths of the Irrawaddy
Mawlamyine
Gulf of Martaban

Uttaradit
Udon Thani
Vinh
Dong Hoi
GULF OF TONKIN
VIETNAM

Tak
Phitsanulok
Khon Kaen
Savannakhet
Hue
Da Nang

THAILAND
Nakhon Sawan
Ubon Ratchathani
Quang Ngai

Ye
Dawei
Nakhon Si Ayutthaya
Nakhon Ratchasima
An Nhon
Qui Nhon

PARACEL ISLANDS
(Claimed by China, Taiwan and Vietnam)

Prachin Buri
BANGKOK (Krung Thep)
Angkor (Ruins)
Siem Reap
Stoeng Trêng
Nha Trang
MUI KE GA

Chanthaburi
Battambang
Kâmpóng Thum
Krâchéh

Mergui
CAMBODIA (KAMPUCHEA)
Tenasserim
Phnom Penh
Kâmpóng Saôm
Chau-phu
Kâmpot
Loc Ninh
Bien Hoa
Phan Thiet

Gulf of Thailand
Long Xuyen
HO CHI MINH CITY (Saigon)

DAO PHU QUOC
Bac Lieu

SOUTH

MANILA

Laoag
Vigan
Aparri
Tuguega
LUZON
San Fernando
Baguio
Lingayen
Tarlac
Olongapo Quezon City
POLIL ISLAN
Lipa Batangas
Lubang Islands
MINDORO

CHINA

CON SON

SPRATLY (Claimed by China, Malaysia, Philippines, Taiwan and Vietnam)

SEA

Calamian Group
Culion
CUYO IS
Iloilo
PANAY
Bacolo
NEGROS

Surat Thani
Nakhon Si Thammarat
PALAWAN
Puerto Princesa

Phuket
Kantang
MALAY
Songkhla
Pattani
BALABAC ISLAND
Balabac Strait
PULAU BANGGI
CAGAYAN SULU
Zamboanga
SULU SEA

Hat Yai
Alor Setar
PENINSULA
Kota Baharu
Kudat
Kota Kinabalu 13 455
Gunong Kinabalu
Sandakan
JOLO ISLAND
Jolo
SULU ARCHIPELAGO

Banda Aceh
Sabang
George Town (Pinang)
KEDAH
PULAU LABUAN
SABAH
TAWITAWI GROUP
SIBUTU ISLAND

Idi
Ta'ping
Ipoh
7174 G.G. Tahan
Bandar Seri Begawan
BRUNEI
Bukit Pagon 6070
CELEB

Langsa
MALAYSIA
Miri

Medan
Belawan
Tarakan

Pematangsiantar
NATUNA BESAR
Binjai
KEPULAUAN BUNGURAN UTARA
MALAYSIA
Bintulu
SEA

PULAU TOBA
Danau Toba
KEPULAUAN BANYAK
Sibolga
Kuala Lumpur
Kelang
Melaka (Malacca)
KEPULAUAN ANAMBAS
SARAWAK
UPPER KAPUAS MTS.
IRAN MTS.
Kuching
PEGUNUNGAN MULLER

PULAU SIMEULUE
Batu Pahat
Johor Baharu
SINGAPORE
SINGAPORE
KEPULAUAN RIAU

PULAU NIAS
Bengkalis
Pakanbaru
KEPULAUAN TAMBELAN
Pontianak
B O R N E O
Samarinda
KEPULAUA

PULAU PINI
Equator
KEPULAUAN BATU
SUMATRA (SUMATERA)
Bukittinggi
Sawahlunto
LINGA
KEPULAUAN KARIMATA (Karimata Strait)
Bukit Raya 7474
SCHWANER
Donggala

Gorontalo

PULAU TANAHMASA
Indragiri
Jambi
Muntok
Ketapang
K A L I M A N T A N
Balikpapan

PULAU TANAHBALA
PULAU SIBERUT
Padang
Gunung Kerinci 12 467
Pangkalpinang
BANGKA
Sukadana PEG
CELEBES (SULAWESI)
335
Bulu Rantekombolo

MENTAWAI
PULAU SIPURA
PULAU PAGAI UTARA
KEPULAUAN
Palembang
Tanjungpandan
TG PUTING
Kotabaru
Majene
Parepare

PULAU PAGAI SELATAN
Lampu 10 365
BELITUNG
Banjarmasin
Martapura
PULAU LAUT

Bengkulu
PEGUNUNGAN BARISAN
G R E A T E R
I N D O N E
S U N D A
Ujungpandang (Makasar)
PULAU KABAENA

PULAU ENGGANG
TG PUTING
TG SELATAN
KEPULAUAN LAUT KECIL
Bonthain
PULAU SELAYAR

Bandar Lampung
KEPULAUAN KARIMUNJAWA
LAUT JAWA (JAVA SEA)
MASALEMBO-BESAR
I S L A N D S
LAUT FLORES (FLORES SEA)
PULAU KALAO

Serang
PULAU BAWEAN
JAKARTA
Cirebon
Semarang
Bangkalan
MADURA
KEPULAUAN KANGEAN
FLORES

Bogor
Sukabumi
BANDUNG
Surakarta
SURABAYA
Pasuruan
BALI
LAUT SA
(SAVU SE

Yogyakarta
Malang
Semeru 12 060
Banyuwangi
Denpasar
G. Rinjani
Agung 12 224
Mataram
LOMBOK
Sumbawa Besar
SUMBAWA
Raba
Waingapu
SUMBA

JAVA (JAWA)
JAVA
LESSER
SUNDA
ISLANDS
PULAU SAWU

SOUTH

INDIAN

OCEAN

JAVA TRENCH
24 442

CHRISTMAS ISLAND (Austl.)

NICOBAR
NORTH ANDAMAN
MIDDLE ANDAMAN
SOUTH ANDAMAN
Port Blair
LITTLE ANDAMAN

Andaman Sea

MERGUI ARCHIPELAGO

ISTHMUS OF KRA

NICOBAR ISLANDS (India)
GREAT NICOBAR

Relief

Meters		Feet
3050		10 000
1525		5000
610		2000
305		1000
152.5		500
	Sea Level	
152.5		500
1525		5000
3050		10 000
6100		20 000

95°   100°   105°   Longitude East of Greenwich   110°   115°   120°

Scale 1:16 000 000; one inch to 250 miles. Polyconic Projection
Elevations and depressions are given in feet

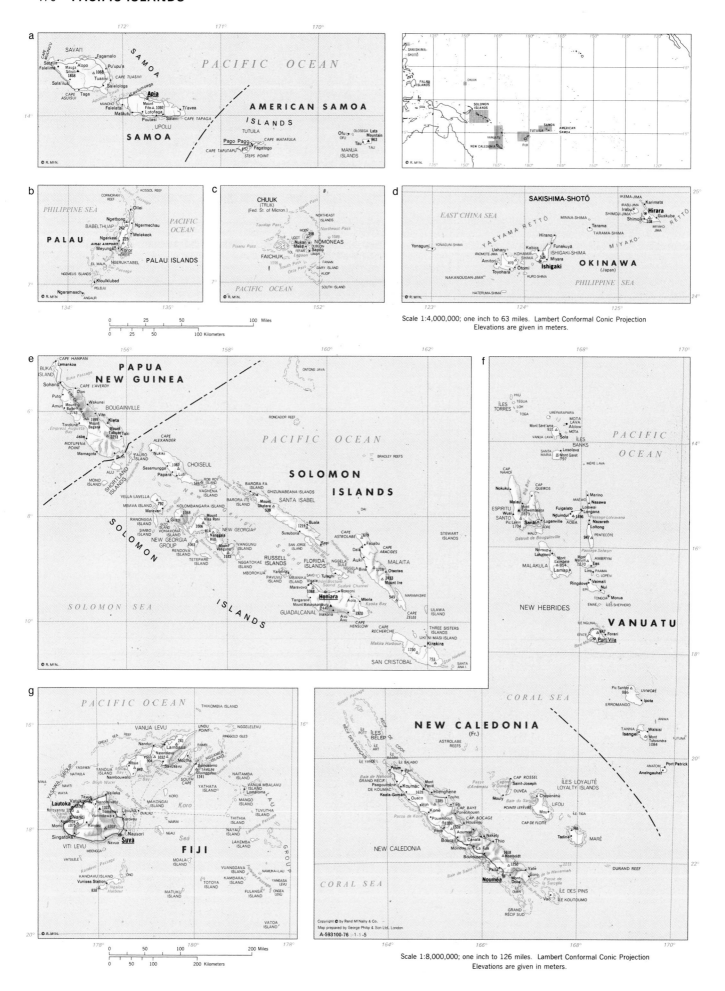

Scale 1:4,000,000; one inch to 63 miles. Lambert Conformal Conic Projection
Elevations are given in meters.

Scale 1:8,000,000; one inch to 126 miles. Lambert Conformal Conic Projection
Elevations are given in meters.

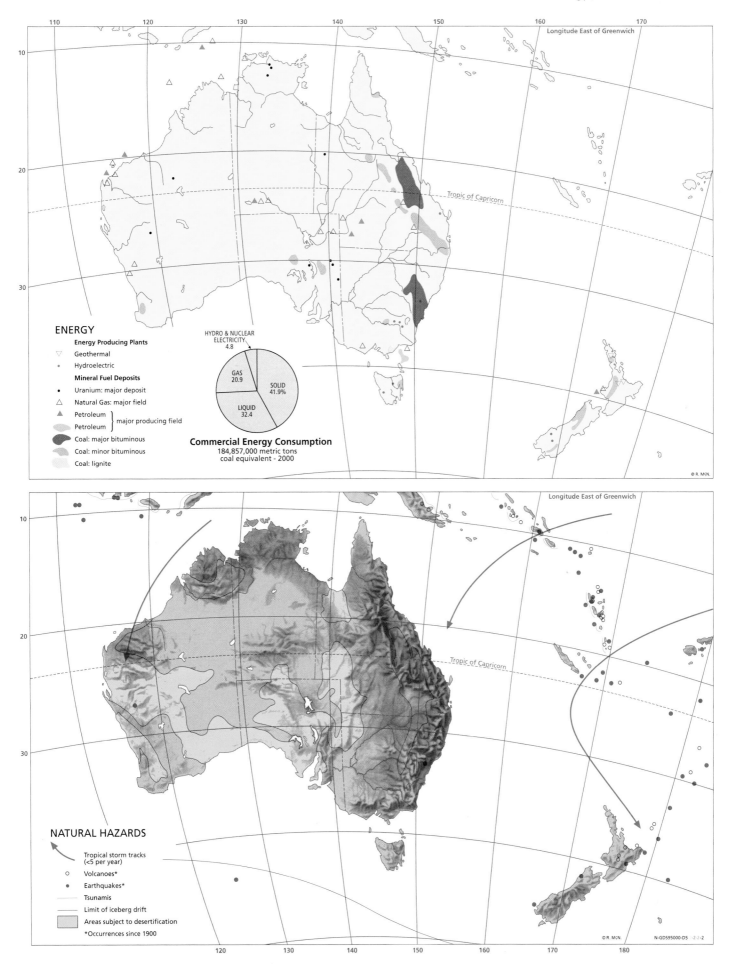

ENERGY

Energy Producing Plants

▽ Geothermal

• Hydroelectric

Mineral Fuel Deposits

• Uranium: major deposit

△ Natural Gas: major field

▲ Petroleum }
Petroleum } major producing field

Coal: major bituminous

Coal: minor bituminous

Coal: lignite

HYDRO & NUCLEAR
ELECTRICITY
4.8

GAS
20.9

SOLID
41.9%

LIQUID
32.4

Commercial Energy Consumption
184,857,000 metric tons
coal equivalent - 2000

Longitude East of Greenwich

Tropic of Capricorn

© R. McN.

NATURAL HAZARDS

⟶ Tropical storm tracks
(<5 per year)

○ Volcanoes*

• Earthquakes*

— Tsunamis

— Limit of iceberg drift

▨ Areas subject to desertification

*Occurrences since 1900

Longitude East of Greenwich

Tropic of Capricorn

© R. McN.   N-GD595000-DS -2-2-2

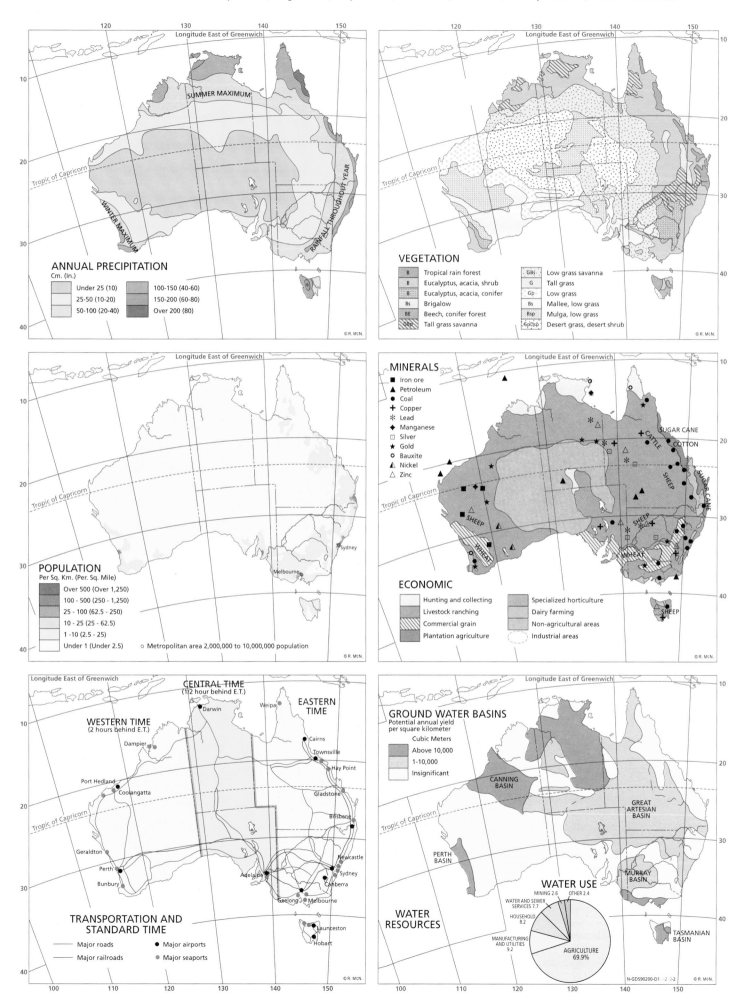

ANNUAL PRECIPITATION
Cm. (In.)

Under 25 (10)	100-150 (40-60)
25-50 (10-20)	150-200 (60-80)
50-100 (20-40)	Over 200 (80)

SUMMER MAXIMUM

WINTER MAXIMUM

RAINFALL THROUGHOUT YEAR

Tropic of Capricorn

VEGETATION

B	Tropical rain forest	GBs	Low grass savanna
B	Eucalyptus, acacia, shrub	G	Tall grass
B	Eucalyptus, acacia, conifer	Gp	Low grass
Bs	Brigalow	Bs	Mallee, low grass
BE	Beech, conifer forest	Bsp	Mulga, low grass
GBp	Tall grass savanna	GpDsp	Desert grass, desert shrub

POPULATION
Per Sq. Km. (Per. Sq. Mile)

Over 500 (Over 1,250)	
100 - 500 (250 - 1,250)	
25 - 100 (62.5 - 250)	
10 - 25 (25 - 62.5)	
1 -10 (2.5 - 25)	
Under 1 (Under 2.5)	o Metropolitan area 2,000,000 to 10,000,000 population

Sydney

Melbourne

MINERALS

- ■ Iron ore
- ▲ Petroleum
- ● Coal
- + Copper
- ✳ Lead
- ◆ Manganese
- ◻ Silver
- ★ Gold
- ○ Bauxite
- ◣ Nickel
- △ Zinc

SUGAR CANE

CATTLE

COTTON

SHEEP

SUGAR CANE

SHEEP

WHEAT

SHEEP

WHEAT

SHEEP

ECONOMIC

Hunting and collecting	Specialized horticulture
Livestock ranching	Dairy farming
Commercial grain	Non-agricultural areas
Plantation agriculture	Industrial areas

TRANSPORTATION AND STANDARD TIME

CENTRAL TIME
(1/2 hour behind E.T.)

EASTERN TIME

WESTERN TIME
(2 hours behind E.T.)

Darwin

Weipa

Cairns

Dampier

Townsville

Hay Point

Port Hedland

Coolangatta

Gladstone

Geraldton

Brisbane

Perth

Newcastle

Bunbury

Sydney

Adelaide

Canberra

Geelong

Melbourne

Launceston

Hobart

—— Major roads	● Major airports
—— Major railroads	● Major seaports

GROUND WATER BASINS
Potential annual yield per square kilometer

Cubic Meters

Above 10,000	
1-10,000	
Insignificant	

CANNING BASIN

GREAT ARTESIAN BASIN

PERTH BASIN

MURRAY BASIN

WATER RESOURCES

WATER USE

MINING 2.6   OTHER 2.4

WATER AND SEWER SERVICES 7.7

HOUSEHOLD 8.2

MANUFACTURING AND UTILITIES 9.2

AGRICULTURE 69.9%

TASMANIAN BASIN

N-GD590200-D1

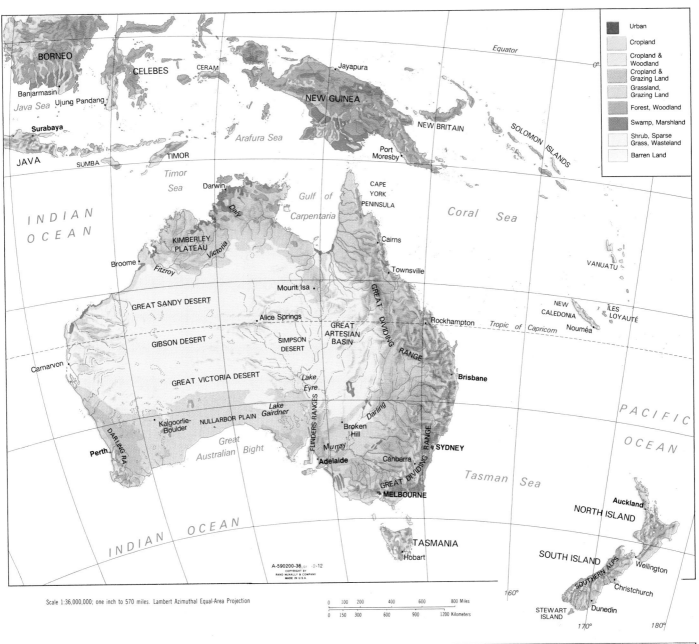

Urban
Cropland
Cropland & Woodland
Cropland & Grazing Land
Grassland, Grazing Land
Forest, Woodland
Swamp, Marshland
Shrub, Sparse Grass, Wasteland
Barren Land

BORNEO
CELEBES
CERAM
Banjarmasin
Ujung Pandang
Java Sea
Surabaya
JAVA
SUMBA
TIMOR
Timor Sea
Jayapura
NEW GUINEA
NEW BRITAIN
Port Moresby
SOLOMON ISLANDS
Equator
0°

Arafura Sea

Darwin
Daly
Gulf of Carpentaria
CAPE YORK PENINSULA
Coral Sea

INDIAN OCEAN
KIMBERLEY PLATEAU
Victoria
Broome
Fitzroy
Cairns
Townsville
VANUATU
NEW CALEDONIA
ÎLES LOYAUTÉ
Nouméa

GREAT SANDY DESERT
Mount Isa
Alice Springs
GREAT ARTESIAN BASIN
GREAT DIVIDING RANGE
Rockhampton
Tropic of Capricorn

GIBSON DESERT
SIMPSON DESERT
Carnarvon
GREAT VICTORIA DESERT
Lake Eyre
Brisbane
PACIFIC OCEAN

FLINDERS RANGES
Lake Gairdner
Darling
Kalgoorlie-Boulder
NULLARBOR PLAIN
Broken Hill
SYDNEY
Murray
Canberra
Perth
DARLING RA.
Great Australian Bight
Adelaide
GREAT DIVIDING RANGE
Tasman Sea
MELBOURNE

INDIAN OCEAN
TASMANIA
Hobart
Auckland
NORTH ISLAND
SOUTH ISLAND
SOUTHERN ALPS
Wellington
Christchurch
STEWART ISLAND
Dunedin

160°
170°
180°

A-590200-36 GT -2-12
COPYRIGHT BY
RAND McNALLY & COMPANY
MADE IN U.S.A.

Scale 1:36,000,000; one inch to 570 miles. Lambert Azimuthal Equal-Area Projection

0 100 200 400 600 800 Miles
0 150 300 600 900 1200 Kilometers

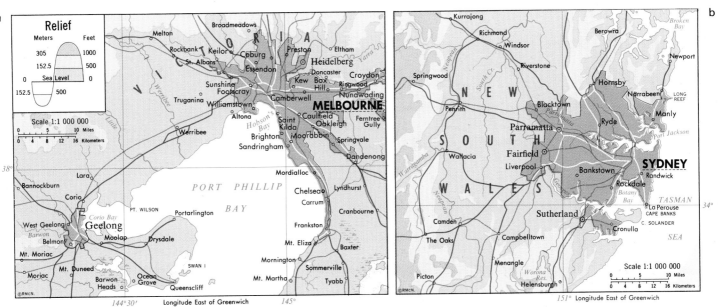

a

Relief

Meters		Feet
305		1000
152.5		500
0	Sea Level	0
152.5		500

Scale 1:1 000 000

0 5 10 Miles
0 4 8 12 16 Kilometers

VICTORIA
Melton
Broadmeadows
Rockbank
Keilor
St. Albans
Coburg
Preston
Eltham
Essendon
Heidelberg
Sunshine
Doncaster
Croydon
Footscray
Kew
Box Hill
Ringwood
Truganina
Williamstown
Camberwell
Nunawading
MELBOURNE
Altona
Saint Kilda
Caulfield
Oakleigh
Ferntree Gully
Werribee
Brighton
Moorabbin
Springvale
Sandringham
Dandenong
Lara
Bannockburn
Mordialloc
Corio
PORT PHILLIP BAY
Chelsea
Lyndhurst
Corio Bay
Carrum
PT. WILSON
Portarlington
Cranbourne
West Geelong
Geelong
Frankston
Barwon
Belmont
Moolap
Drysdale
Mt. Moriac
Mt. Duneed
Ocean Grove
Mt. Eliza
Moriac
Barwon Heads
Queenscliff
Mornington
Sommerville
SWAN I.
Mt. Martha
Baxter
Tyabb

38°
®RMCN.
144°30'   Longitude East of Greenwich   145°

b

Kurrajong
Richmond
Windsor
Berowra
Broken Bay
Newport
Springwood
NEW
Riverstone
Hornsby
Narrabeen
LONG REEF
Penrith
Blacktown
Manly
SOUTH
Parramatta
Ryde
Port Jackson
Wallacia
Fairfield
Liverpool
Bankstown
SYDNEY
WALES
Randwick
Rockdale
Botany Bay
La Perouse
CAPE BANKS
TASMAN
Camden
Sutherland
Cronulla
C. SOLANDER
The Oaks
Campbelltown
SEA
Menangle
Worona Res.
Picton
Helensburgh

Scale 1:1 000 000

0 5 10 Miles
0 4 8 12 16 Kilometers

®RMCN.
151°   Longitude East of Greenwich

34°

174

Continued on pages 168-169

INDONESIA

Pasuruan

G. Semeru 12 060
G. Mahameru 10 932
Rinjani 12 224
G. Raung

Singaraja
LOMBOK
Rabo
SUMBAWA
Sumbawa Besar
SUMBA
Waingapu
SUMBA

FLORES
SAVU SEA
SAWU
ROTI
Kupang
LOMBLEN PANTAR
ALOR
TIMOR
Dili
EAST TIMOR

SELARU
ARAFURA SEA
TANJUNG VALS

SELATA

SUNDA ISLANDS

TIMOR SEA

CAPE LONDONDERRY
Joseph Bonaparte Gulf
Queens Chan.
Anson Bay

C. VAN DIEMEN
CROKER
BATHURST
MELVILLE
Van Diemen Gulf
Clarence Str.
Darwin
WESSEL IS.
CAPE ARNHEM
Blue Mud Bay
GROOTE EYLANDT
Limmen Bight
SIR EDWARD PELLEW GROUP
WELLESLEY IS.
GULF OF CARPENTARIA

ARNHEM LAND
Pine Creek
Katherine

SUNDA TRENCH

INDIAN

OCEAN

CAPE LEVEQUE
BUCCANEER
ARCH.
King Sd.
Sunday Str.
Collier Bay
Wyndham
Mt. Hann 2800
KING LEOPOLD RANGES
DAMPIER
Derby
BROOME LAND
GEIKIE RANGE
Fitzroy Crossing
Halls Creek
Fitzroy
Ord
Sturt Cr.

Roebuck Bay
Broome
LaGrange
EIGHTY MILE BEACH

Birdum
Victoria River Downs
Daly Waters
Newcastle Waters
Victoria
Roper
Woods L.

NORTHERN

Borroloola

Burketown

LARREY POINT

Tanami

Tennant Creek

TERRITORY

Dobbyn
Camooweal
Mount Isa
Malbon
Duchess
Dajarra

QU

DAMPIER ARCH.
RIPON
Port Hedland
DeGrey
Roebourne
MONTE BELLO IS.
BARROW
NORTH WEST CAPE
Exmouth Gulf
Onslow
Millstream
HAMERSLEY RANGE
Mt. Bruce 4052
Marble Bar
Nullagine
GREAT SANDY DESERT
Mackay

RANGES
Mt. Ziel 4955
MACDONNELL
Barrow Creek
Arltunga
Alice Springs
JAMES RANGE
SIMPSON

POINT CLOATES
Tropic of Capricorn
CAPE FARQUHAR
Geographe Chan.
Carnarvon
Gascoyne
Peak Hill
Nabberu
Jiggalong
Disappointment
WESTERN
Macdonald
Gibson DESERT
Uluru (Ayers Rock)
Amadeus
Charlotte Waters
MUSGRAVE RANGES
Mt. Woodroffe 4724
EVERARD RANGES
Finke
DESERT
Birdsville
Diamantina
A
B

BERNIER
DORRE
Shark Bay
Nannine
Carnegie
Wells
Gillen
The Alberga
Oodnadatta
Cooper

DIRK HARTOG
STEEP POINT
Meekatharra
Cue
Sandstone
Austin
Mount Magnet
Laverton
Carey
Yeo
Eyre 39
William Creek
Marree
Gregory
SOUTH AUSTRALIA

HOUTMAN ROCKS
Ajana
Northampton
Geraldton
Dongara
Mingenew
Moore
AUSTRALIA
Ballard
Barlee
Menzies
Kalgoorlie-Boulder
Coolgardie
GREAT VICTORIA DESERT
Rawlinna
Oldea Station
Hughes
NULLARBOR PLAIN
Penong
Ceduna
POINT FOWLER
Farina
Pimba
Woomera
STUART RANGE
Torrens
Parachilna
FLINDERS RANGES
FLINDE

Pithara
Milling
Moora
Lake Brown
Southern Cross
SWANLAND
Cowan
Norseman
Dundas
Eyre
Eucla
Everard
Whyalla
Port Augusta
Port Pirie
Gladstone
EYRE PENINSULA
Peterborough
Moonta
Wallaroo
Port Wakefield
Gawler
Murray

Perth
Fremantle
DARLING RANGE
Northam
York
Narrogin
Collie
Bunbury
Busselton
CAPE NATURALISTE
CAPE LEEUWIN
Normalup
Albany
PT. D'ENTRECASTEAUX
WEST CAPE HOWE
King George Sd.
Salmon Gums
Ravensthorpe
Esperance
Hopetoun
ARCHIPELAGO OF THE RECHERCHE
GREAT AUSTRALIAN BIGHT
Port Lincoln
Gulf St. Vincent
Spencer Gulf
KANGAROO
Encounter Bay
Adelaide
Murray Bridge
Nardcoorte
Kingston
CAPE JAFFA
Mt. Gambier

INDIAN OCEAN

A-590200-76 7-5-18
COPYRIGHT BY
RAND McNALLY & COMPANY
MADE IN U.S.A.

Relief

Meters		Feet
3050		10 000
1525		5000
610		2000
305		1000
152.5		500
0	Sea Level	0
152.5		500
		Below Sea Level
1525		5000
3050		10 000
6100		20 000

Longitude 115° East of Greenwich

Scale 1:16 000 000; one inch to 250 miles. Lambert's Azimuthal, Equal Area Projection
Elevations and depressions are given in feet

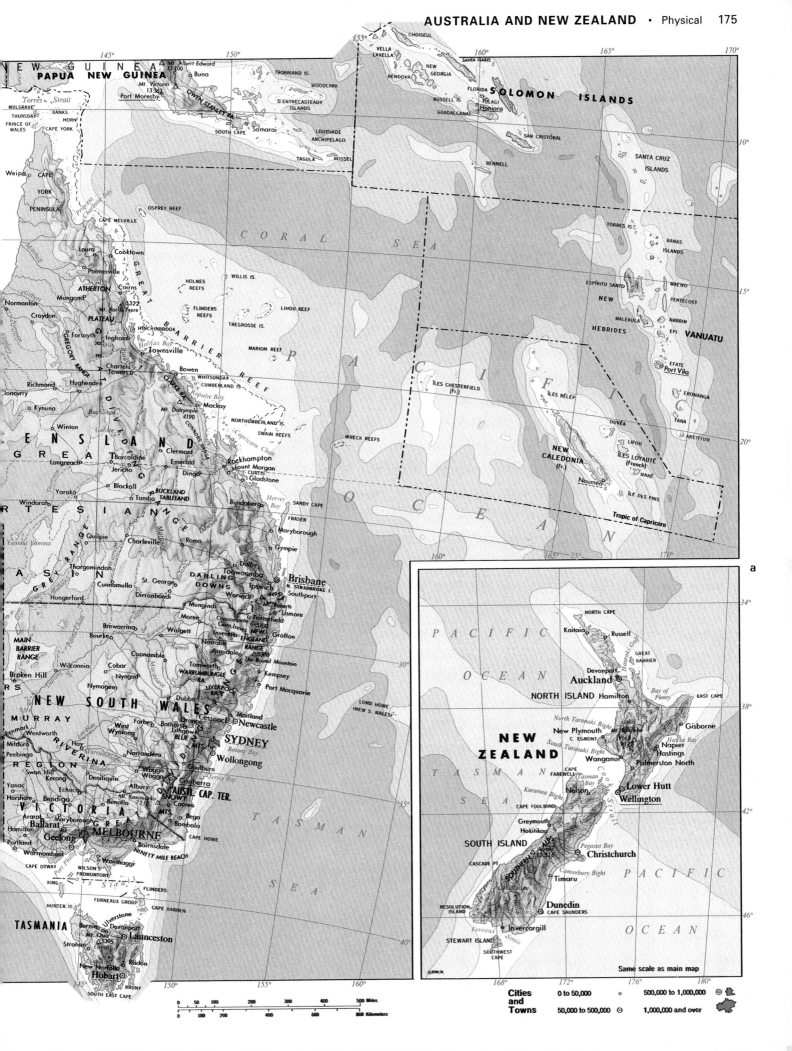

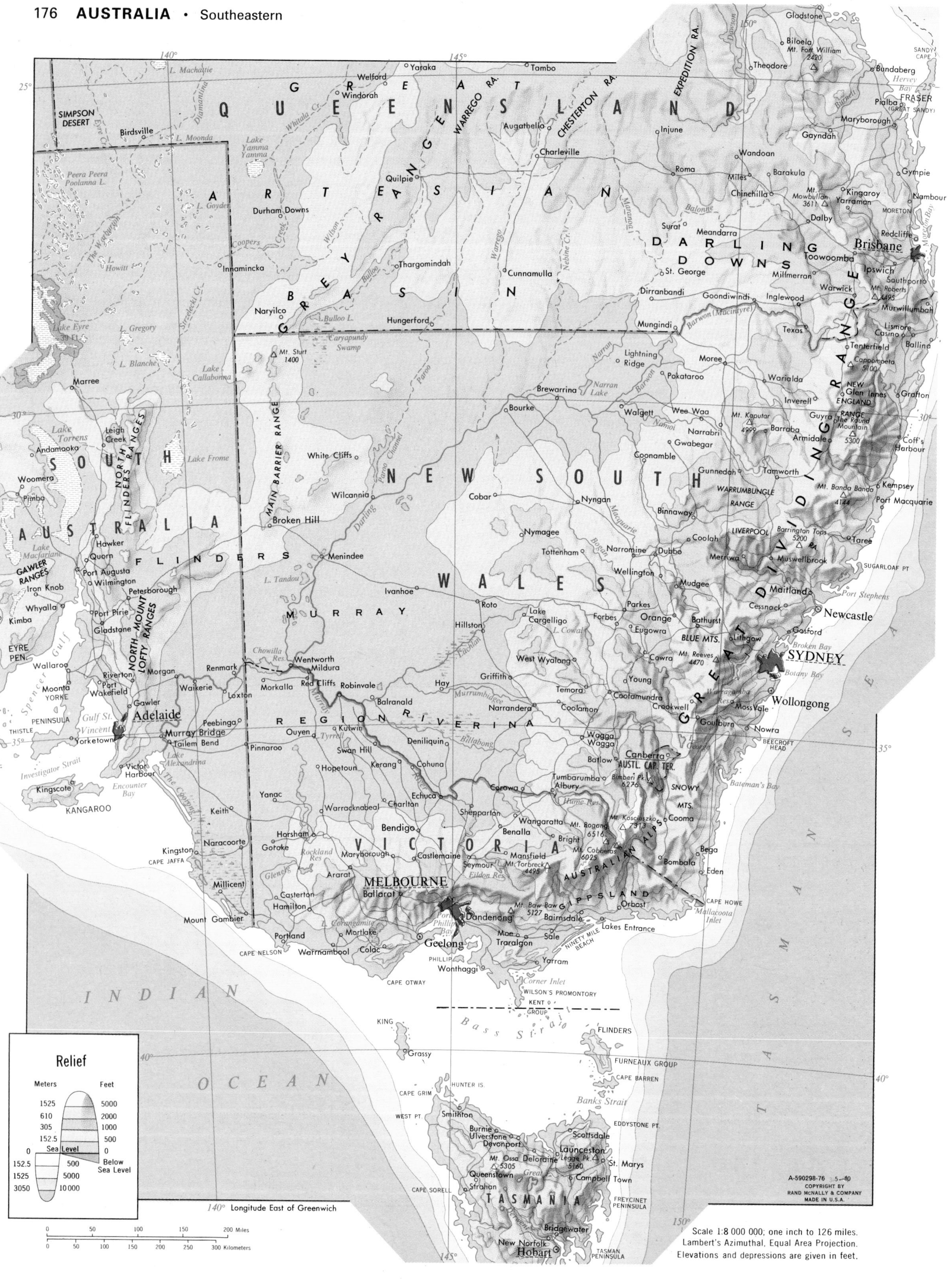

SIMPSON DESERT

QUEENSLAND

GREAT DIAMANTINA

GREAT ARTESIAN BASIN

GREY RANGE

WARREGO RA.

CHESTERTON RA.

EXPEDITION RA.

Gladstone
Biloela
Mt. Fort William 2420
Theodore
Bundaberg
SANDY CAPE
HERVEY BAY (GREAT SANDY)
FRASER

Welford
Yaraka
Tambo
Windorah
Augathella
Injune
Wandoan
Barakula
Maryborough
Gayndah
Gympie
Nambour
MORETON

Durham Downs
Quilpie
Charleville
Roma
Miles
Chinchilla
Dalby
Kingaroy
Yarraman
Mt. Mowbullan 3611
Redcliffe

Birdsville
L. Machattie
L. Yamma Yamma
Lake Yamma Yamma
L. Goyder

Innamincka
Thargomindah
Surat
Meandarra
St. George
Millmerran
Warwick
Toowoomba
Ipswich
Southport
Mt. Roberts 4495
Brisbane

DARLING DOWNS

Naryilco
Hungerford
Cunnamulla
Dirranbandi
Goondiwindi
Inglewood
Texas
Murwillumbah

L. Eyre
L. Gregory
L. Blanche
Lake Callabonna

Mt. Sturt 1400
Bulloo L.
Caryapundy Swamp
Mungindi
Barwon (Macintyre)
Lismore
Casino
Ballina

Marree
Lightning Ridge
Moree
Pokataroo
Wee Waa
Walgett
Narrabri
Warialda
Inverell
Glen Innes
NEW ENGLAND RANGE
The Round Mountain
Grafton
Coff's Harbour

SOUTH AUSTRALIA

FLINDERS RANGES
NORTH FLINDERS RANGES

Andamooka
Leigh Creek
Lake Torrens
Lake Frome
White Cliffs
Brewarrina
Bourke
Narran Lake
Paroo
Darling
Namoi
Gunnedah
Tamworth
Mt. Kaputar 4999
Barraba
Guyra
Armidale 5300
Mt. Banda Banda 4144
Kempsey
Port Macquarie

Woomera
Pimba
Hawker
Quorn
GAWLER RANGES
Iron Knob
Whyalla
Kimba
FLINDERS
Peterborough
MURRAY
Wilcannia
Cobar
Nymagee
Nyngan
Coonamble
WARRUMBUNGLE RANGE
Mt. Banda Banda
LIVERPOOL RANGE
Barrington Tops 5200
Merriwa
Muswellbrook
Taree
SUGARLOAF PT.

NEW SOUTH WALES

MAIN BARRIER RANGE

Broken Hill
Menindee
L. Tandou
Ivanhoe
Roto
Tottenham
Narromine
Dubbo
Wellington
Mudgee
Cessnock
Maitland
Port Stephens

Port Augusta
Wilmington
Port Pirie
Gladstone
NORTH MOUNT LOFTY RANGES
Morgan
Waikerie
Renmark
Chowilla Res.
Wentworth
Mildura
Red Cliffs
Morkalla
Robinvale
Balranald
Hay
Hillston
Griffith
Lake Cargelligo
L. Cowal
Forbes
Parkes
Orange
Eugowra
Bathurst
BLUE MTS.
Mt. Reeves 4470
Lithgow
Gosford
Broken Bay
Newcastle

EYRE PEN.
Wallaroo
Moonta
Riverton
Port Wakefield
Gawler
YORKE PENINSULA
Adelaide
Murray Bridge
Peebinga
Ouyen
Kulwin
West Wyalong
Young
Cowra
Cootamundra
Crookwell
Goulburn
Nowra
Sydney
Botany Bay
Wollongong
MossVale
BEECROFT HEAD

THISTLE
Yorketown
Kingscote
Victor Harbour
Encounter Bay
Tailem Bend
Pinnaroo
Swan Hill
Deniliquin
Murrumbidgee
Narrandera
Coolamon
Wagga Wagga
Temora
Bateman's Bay

KANGAROO
Investigator Strait
L. Alexandrina
The Coorong
Yanac
Kerang
Cohuna
Batlow
Canberra
AUSTL. CAP. TER.
SNOWY MTS.
Cooma

REGION
RIVERINA
Billabong
Corowa
Albury
Tumbarumba
Bimberi Pk. 6276
Bega
Eden
CAPE HOWE

Kingston
Naracoorte
Keith
Hopetoun
Warracknabeal
Charlton
Echuca
Shepparton
Wangaratta
Benalla
Bright
Mt. Bogong 6516
Mt. Kosciuszko 7313
AUSTRALIAN ALPS
Bombala
Orbost
Mallacoota Inlet

CAPE JAFFA
Millicent
Horsham
Gotoke
Maryborough
Castlemaine
Bendigo
Mansfield
Mt. Cobberas 6025
Mt. Torbreck 4495
GIPPSLAND

Mount Gambier
Casterton
Ararat
Seymour
Eildon Res.
Mt. Baw Baw 5127
Bairnsdale
Sale
Lakes Entrance

VICTORIA

Hamilton
Melbourne
Ballarat
Dandenong
Moe
Traralgon
NINETY MILE BEACH

Portland
Warrnambool
Colac
Mortlake
Port Phillip Bay
Geelong
Yarram
PHILLIP I.
Wonthaggi

CAPE NELSON
L. Corangamite
CAPE OTWAY
Corner Inlet
WILSON'S PROMONTORY
KENT GROUP

INDIAN OCEAN

KING
Grassy
Bass Strait
FLINDERS
FURNEAUX GROUP
CAPE BARREN
Banks Strait

CAPE GRIM
HUNTER IS.
WEST PT.
Smithton
Burnie
Ulverstone
Devonport
Scottsdale
EDDYSTONE PT.
Mt. Ossa 5305
Deloraine
Launceston
Legge Pk. 5160
St. Marys

TASMANIA

Queenstown
Strahan
Campbell Town
FREYCINET PENINSULA

CAPE SOREL
Bridgewater
New Norfolk
Hobart
TASMAN PENINSULA

TASMAN SEA

Relief

Meters		Feet
1525		5000
610		2000
305		1000
152.5		500
0	Sea Level	0
152.5		500 Below Sea Level
1525		5000
3050		10 000

140° Longitude East of Greenwich

0   50   100   150   200 Miles
0   50 100 150 200 250 300 Kilometers

A-590298-76 5-40
COPYRIGHT BY
RAND McNALLY & COMPANY
MADE IN U.S.A.

Scale 1:8 000 000; one inch to 126 miles.
Lambert's Azimuthal, Equal Area Projection.
Elevations and depressions are given in feet.

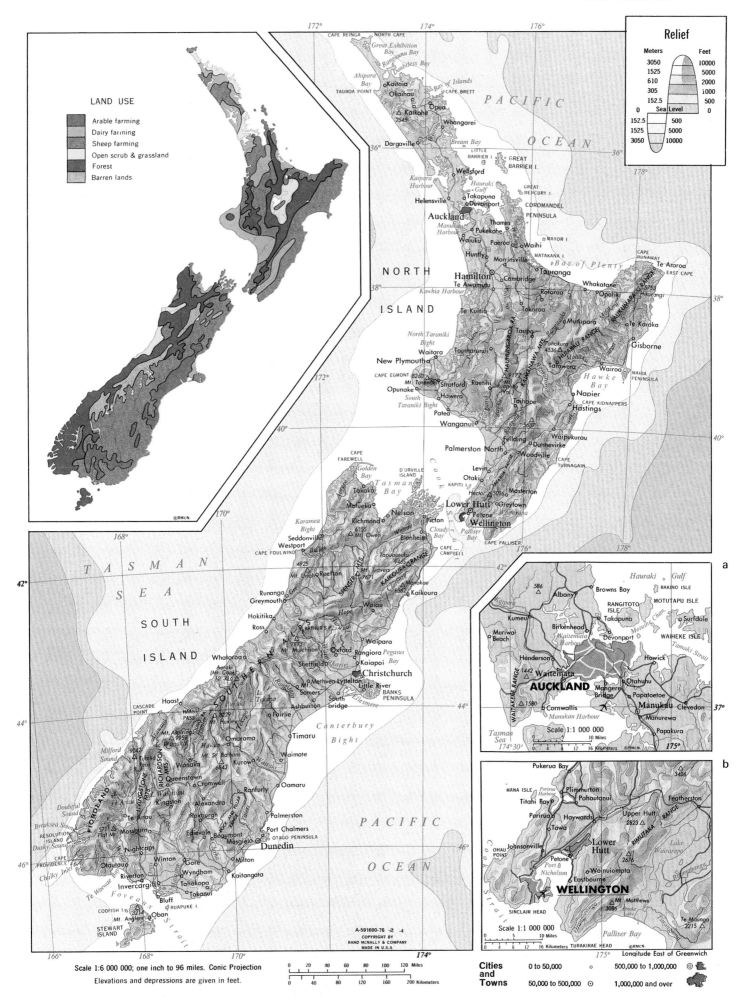

Relief

Meters	Feet
3050	10000
1525	5000
610	2000
305	1000
152.5	500
0 Sea Level	0
152.5	500
1525	5000
3050	10000

LAND USE

- Arable farming
- Dairy farming
- Sheep farming
- Open scrub & grassland
- Forest
- Barren lands

©RMcN.

PACIFIC OCEAN

NORTH ISLAND

TASMAN SEA

SOUTH ISLAND

PACIFIC OCEAN

a

AUCKLAND

Scale 1:1 000 000

b

WELLINGTON

Scale 1:1 000 000

Longitude East of Greenwich

A-591600-76 -2 -4
COPYRIGHT BY
RAND McNALLY & COMPANY
MADE IN U.S.A.

Scale 1:6 000 000; one inch to 96 miles. Conic Projection

Elevations and depressions are given in feet.

Cities and Towns

0 to 50,000	o	500,000 to 1,000,000	◉
50,000 to 500,000	⊙	1,000,000 and over	

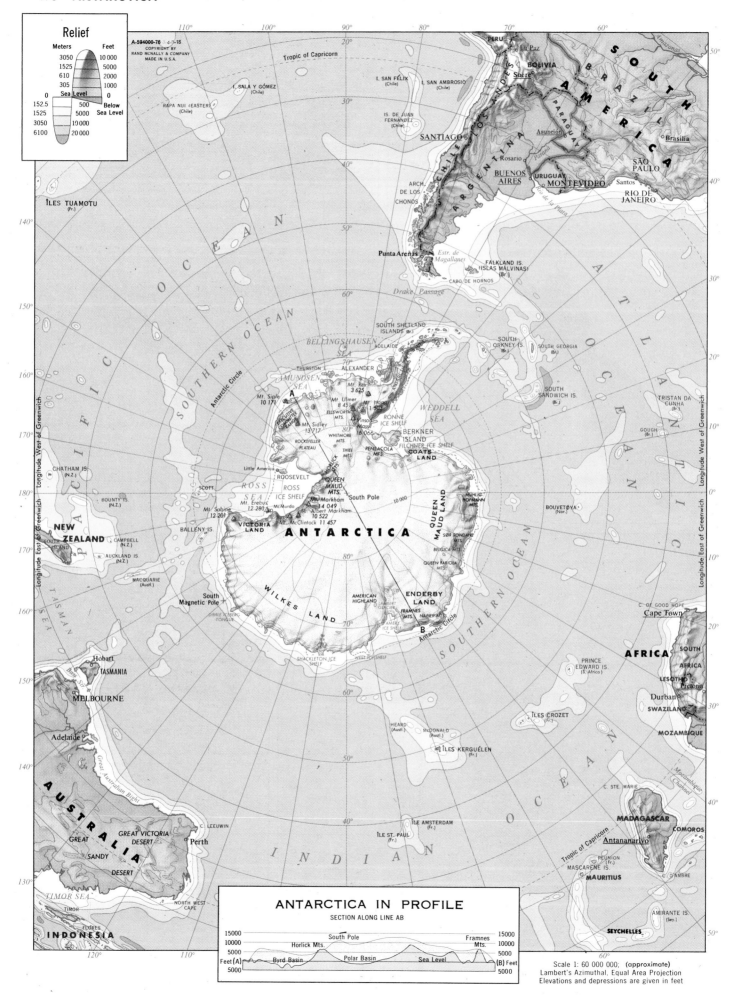

Relief

Meters		Feet
3050		10 000
1525		5000
610		2000
305		1000
0	Sea Level	0
152.5		500 Below
1525		5000 Sea Level
3050		19 000
6100		20 000

A-594000-76 4-7-18
COPYRIGHT BY
RAND McNALLY & COMPANY
MADE IN U.S.A.

ANTARCTICA IN PROFILE
SECTION ALONG LINE AB

15000	South Pole		15000
10000		Framnes	10000
5000	Horlick Mts.	Mts.	5000
Feet (A)	Byrd Basin Polar Basin Sea Level		(B) Feet
5000			5000

Scale 1: 60 000 000; (approximate)
Lambert's Azimuthal, Equal Area Projection
Elevations and depressions are given in feet

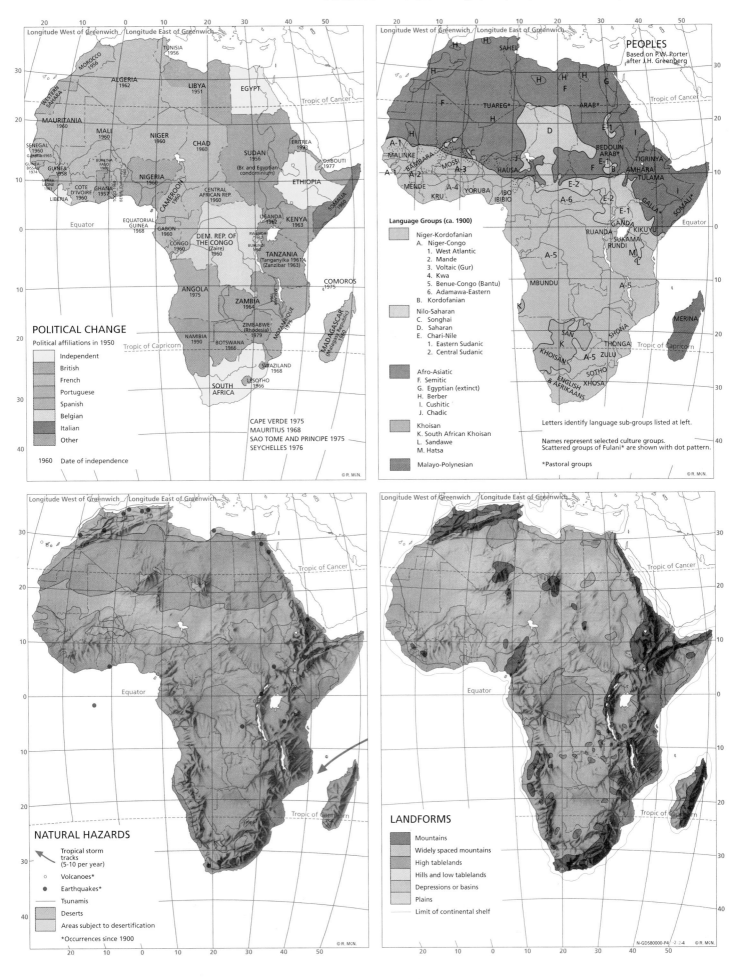

POLITICAL CHANGE

Political affiliations in 1950

- Independent
- British
- French
- Portuguese
- Spanish
- Belgian
- Italian
- Other

1960   Date of independence

CAPE VERDE 1975
MAURITIUS 1968
SAO TOME AND PRINCIPE 1975
SEYCHELLES 1976

© R. McN.

PEOPLES
Based on P.W. Porter
after J.H. Greenberg

Language Groups (ca. 1900)

Niger-Kordofanian
  A. Niger-Congo
    1. West Atlantic
    2. Mande
    3. Voltaic (Gur)
    4. Kwa
    5. Benue-Congo (Bantu)
    6. Adamawa-Eastern
  B. Kordofanian

Nilo-Saharan
  C. Songhai
  D. Saharan
  E. Chari-Nile
    1. Eastern Sudanic
    2. Central Sudanic

Afro-Asiatic
  F. Semitic
  G. Egyptian (extinct)
  H. Berber
  I. Cushitic
  J. Chadic

Khoisan
  K. South African Khoisan
  L. Sandawe
  M. Hatsa

Malayo-Polynesian

Letters identify language sub-groups listed at left.

Names represent selected culture groups.
Scattered groups of Fulani* are shown with dot pattern.

*Pastoral groups

© R. McN.

NATURAL HAZARDS

→   Tropical storm tracks (5-10 per year)
○   Volcanoes*
●   Earthquakes*
   Tsunamis

- Deserts
- Areas subject to desertification

*Occurrences since 1900

© R. McN.

LANDFORMS

- Mountains
- Widely spaced mountains
- High tablelands
- Hills and low tablelands
- Depressions or basins
- Plains
- Limit of continental shelf

N-GD580000-P4 -2-2-4   © R. McN.

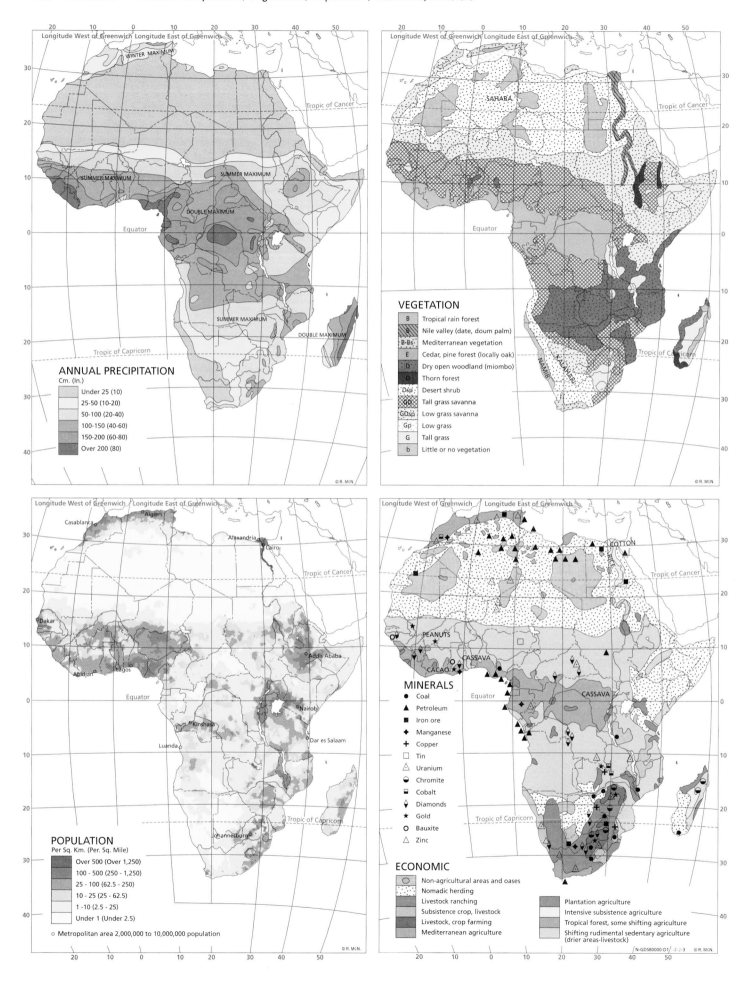

ANNUAL PRECIPITATION

WINTER MAXIMUM

SUMMER MAXIMUM

DOUBLE MAXIMUM

Equator

SUMMER MAXIMUM

SUMMER MAXIMUM

DOUBLE MAXIMUM

Tropic of Capricorn

ANNUAL PRECIPITATION
Cm. (In.)

- Under 25 (10)
- 25-50 (10-20)
- 50-100 (20-40)
- 100-150 (40-60)
- 150-200 (60-80)
- Over 200 (80)

VEGETATION

B	Tropical rain forest
B	Nile valley (date, doum palm)
B-Bs	Mediterranean vegetation
E	Cedar, pine forest (locally oak)
D	Dry open woodland (miombo)
D	Thorn forest
Dsp	Desert shrub
GD	Tall grass savanna
GDsp	Low grass savanna
Gp	Low grass
G	Tall grass
b	Little or no vegetation

SAHARA

NAMIB KALAHARI

POPULATION
Per Sq. Km. (Per. Sq. Mile)

- Over 500 (Over 1,250)
- 100 - 500 (250 - 1,250)
- 25 - 100 (62.5 - 250)
- 10 - 25 (25 - 62.5)
- 1 - 10 (2.5 - 25)
- Under 1 (Under 2.5)

o Metropolitan area 2,000,000 to 10,000,000 population

Casablanca
Algiers
Alexandria
Cairo
Dakar
Addis Ababa
Abidjan
Lagos
Nairobi
Kinshasa
Luanda
Dar es Salaam
Johannesburg

MINERALS

- ● Coal
- ▲ Petroleum
- ■ Iron ore
- ◆ Manganese
- ＋ Copper
- □ Tin
- △ Uranium
- ◓ Chromite
- ▭ Cobalt
- ◇ Diamonds
- ★ Gold
- ○ Bauxite
- △ Zinc

COTTON
RICE
PEANUTS
CASSAVA
CACAO
CASSAVA

ECONOMIC

- Non-agricultural areas and oases
- Nomadic herding
- Livestock ranching
- Subsistence crop, livestock
- Livestock, crop farming
- Mediterranean agriculture
- Plantation agriculture
- Intensive subsistence agriculture
- Tropical forest, some shifting agriculture
- Shifting rudimental sedentary agriculture (drier areas-livestock)

N-GDS80000-D1 / -2.2-3   © R. McN.

MADRID

CORSICA
ROME
SARDINIA
SICILY
MALTA

İSTANBUL

Black Sea

BAKU

Caspian Sea

ATLANTIC
OCEAN

Athens

TEHRAN

CRETE
CYPRUS

Beirut

Baghdad

Algiers
Tunis

Casablanca

ATLAS MOUNTAINS

Tripoli

Banghāzī

SYRIAN
DESERT

Tigris

Euphrates

Alexandria

CANARY ISLANDS

GRAND ERG OCCIDENTAL

GRAND ERG ORIENTAL

CAIRO

AN NAFŪD

El Aaíun

Nile

ARABIAN DESERT

Red Sea

Riyadh

Tropic of Cancer

S   A   H   A   R   A

LIBYAN DESERT

Lake Nasser

Mecca

Tamenghest

AHAGGAR

NUBIAN DESERT

ADRAR
DES IFÔGHAS

TIBESTI

Nile

Tombouctou

ENNEDI

S   U   D   A   N

Khartoum

Asmera

Dakar

Niger

Bamako

Kano

Lake Chad

N'Djamena

Al-Fāshir

White Nile

Blue Nile

DANAKIL

Aden
Gulf of Aden

Berbera

Freetown

Niger

Lake Volta

Addis Ababa

Mountain Nile

Abidjan

Lagos

Yaoundé

Bangui

Uele

Gulf of Guinea

Ubangi

Congo

Kisangani

Mogadishu

Equator

Lake
Victoria

Nairobi

INDIAN

OCEAN

Congo

Kasai

Kinshasa

Luanda

Lake
Tanganyika

Dar es Salaam

ATLANTIC   OCEAN

Lubumbashi

COMORO ISLANDS

Lake Nyasa

Lusaka

Blantyre

Moçambique

Zambezi

Harare

Mozambique Channel

MADAGASCAR

Antananarivo

NAMIB DESERT

Windhoek

KALAHARI
DESERT

Limpopo

Tropic of Capricorn

Johannesburg

Durban

Orange

Orange

INDIAN   OCEAN

Cape
Town

	Urban
	Cropland
	Cropland & Woodland
	Cropland & Grazing Land
	Grassland, Grazing Land
	Forest, Woodland
	Swamp, Marshland
	Shrub, Sparse Grass, Wasteland
	Barren Land
	Oasis

A-580000-36   -2 3-13
COPYRIGHT BY
RAND McNALLY & COMPANY
MADE IN U.S.A.

Scale 1:36,000,000; one inch to 570 miles.   Lambert Azimuthal Equal-Area Projection

0 100 200   400   600   800 Miles
0 150 300   600   900   1200 Kilometers

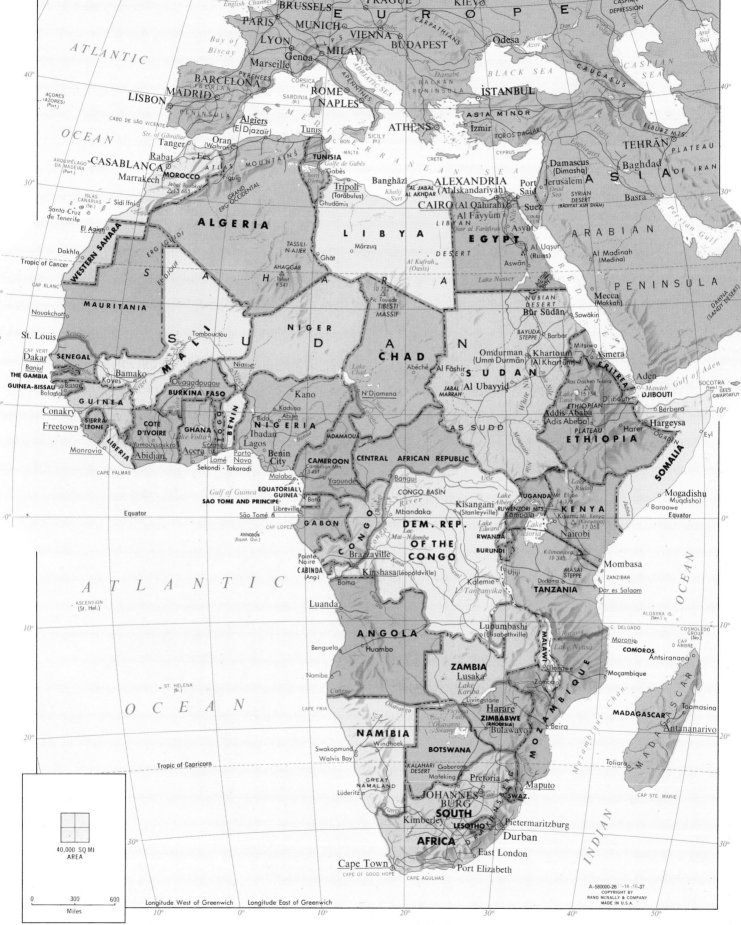

Scale 1:40 000 000; one inch to 630 miles. Lambert's Azimuthal, Equal Area Projection
Elevations and depressions are given in feet.

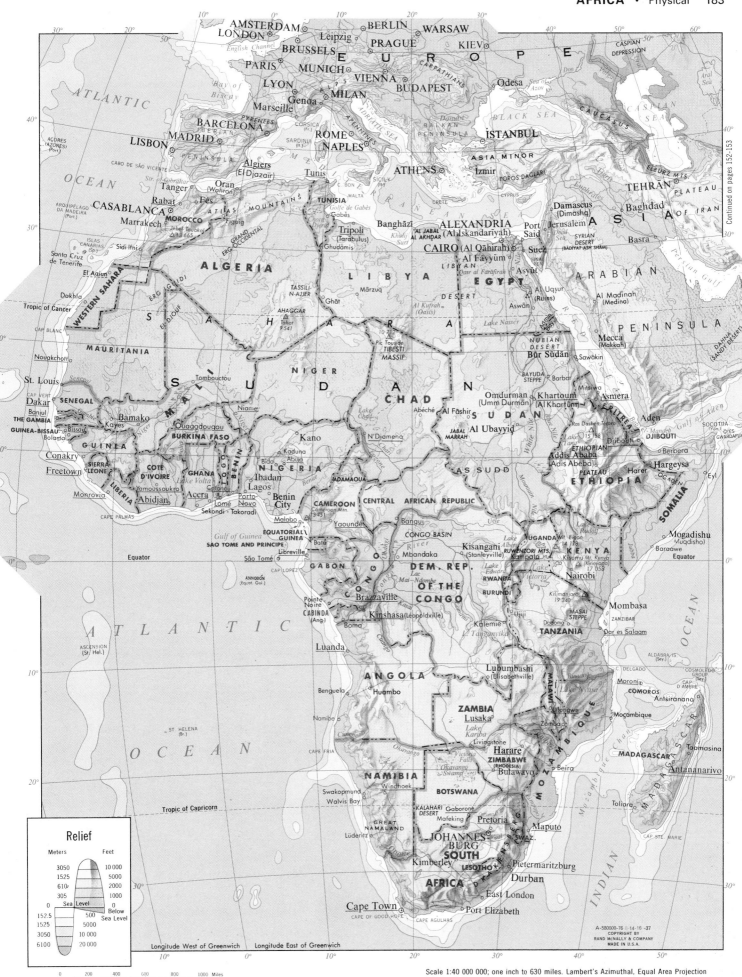

Continued on pages 152-153

Relief

Meters		Feet
3050		10 000
1525		5000
610		2000
305		1000
0	Sea Level	0
		Below Sea Level
152.5		500
1525		5000
3050		10 000
6100		20 000

Longitude West of Greenwich   Longitude East of Greenwich

0   200   400   600   800   1000 Miles

0   400   800   1200   1600 Kilometers

A-580000-76 14-16 -37
COPYRIGHT BY
RAND MCNALLY & COMPANY
MADE IN U.S.A.

Scale 1:40 000 000; one inch to 630 miles. Lambert's Azimuthal, Equal Area Projection
Elevations and depressions are given in feet.

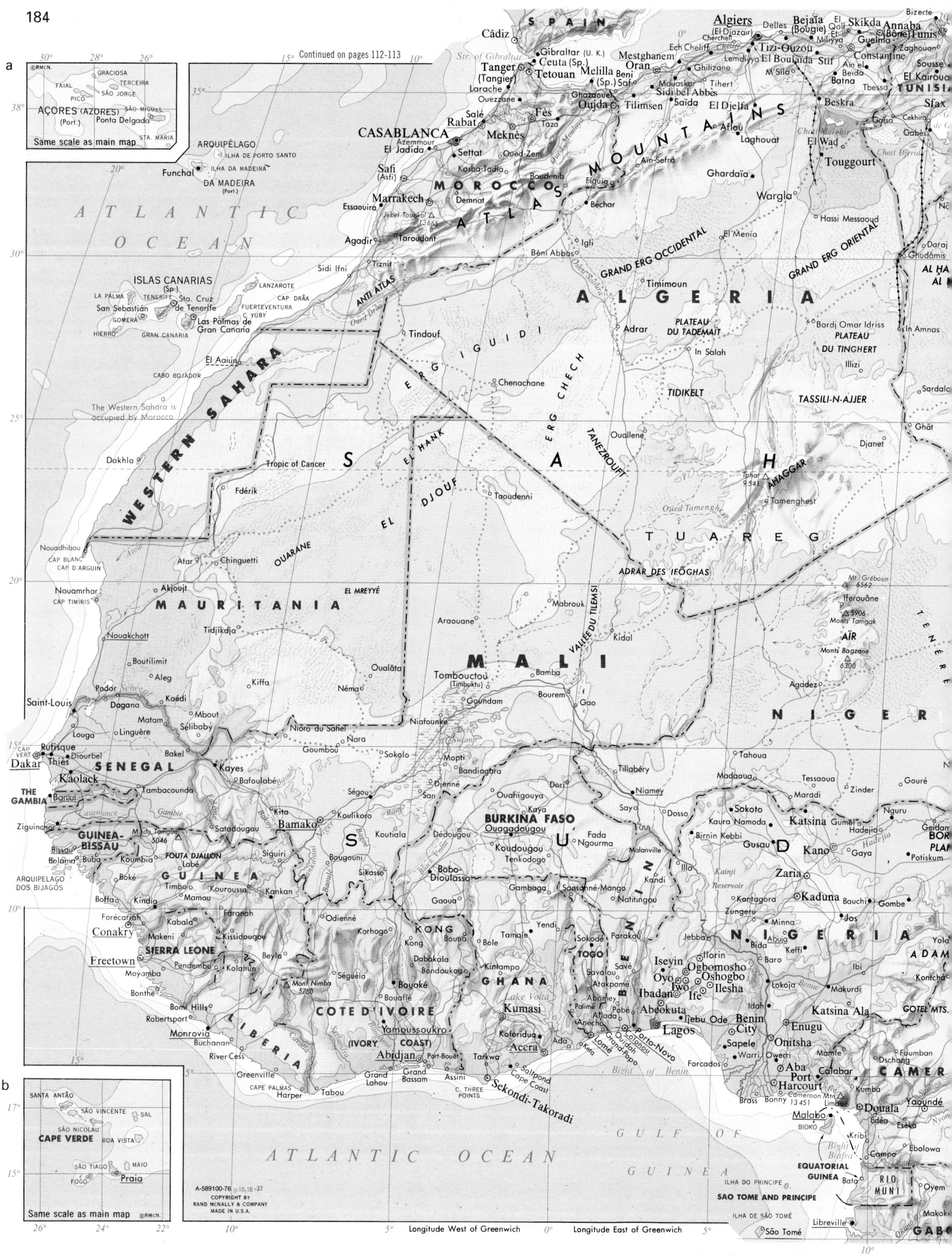

184

SPAIN
Cádiz
Str. of Gibraltar
Gibraltar (U.K.)
Ceuta (Sp.)
Tanger (Tangier)
Tetouan
Larache
Salé
Rabat
CASABLANCA
El Jadida
Azemmour
Settat
Safi (Asfi)
Marrakech
Essaouira
Agadir
Taroudant
Jebel Toubkal 13665
Sidi Ifni
Tiznit
ANTI ATLAS
Algiers (El Djazaïr)
Delles
Cherchell
Mestghanem
Oran
Sidi bel Abbès
Saïda
Ech Cheliff
Beni Saf
Ghazaouet
Tilimsen
Oujda
Fès
Taza
Meknès
Oued Zem
Kosba-Tadla
Demnat
Boudenib
Béchar
Figuig
Béni Abbès
Igli
Ghardaïa
Laghouat
El Menia
GRAND ERG OCCIDENTAL
Timimoun
Adrar
Chenachane
PLATEAU DU TADEMAÏT
In Salah
GRAND ERG ORIENTAL
ERG CHECH
ERG IGUIDI
ALGERIA
Bejaïa (Bougie)
Skikda
Annaba (Bône)
Tunis
Constantine
Sétif
Batna
Tébessa
TUNISIA
Sfax
Gabès
Tozeur
El Oued
Touggourt
Wargla
Hassi Messaoud
Bordj Omar Idriss
In Amnas
PLATEAU DU TINGHERT
Illizi
TASSILI-N-AJJER
Ghat
Djanet
Sardalas

ATLANTIC OCEAN
AÇORES (AZORES) (Port.)
GRACIOSA
TERCEIRA
SÃO JORGE
FAIAL PICO
SÃO MIGUEL
Ponta Delgada
STA. MARIA

ARQUIPÉLAGO DE PORTO SANTO
ILHA DA MADEIRA
DA MADEIRA (Port.)
Funchal

ISLAS CANARIAS (Sp.)
LA PALMA
TENERIFE
Sta. Cruz de Tenerife
LANZAROTE
FUERTEVENTURA
C. YUBY
San Sebastián
GOMERA
HIERRO
GRAN CANARIA
Las Palmas de Gran Canaria
CAP DRÂA
Oued Drâa

CABO BOJADOR
El Aaiún
WESTERN SAHARA
The Western Sahara is occupied by Morocco
Dakhla
Tindouf
Tropic of Cancer
Fdérik
EL HANK
EL DJOUF
Taoudenni
S A H A R A
TANEZROUFT
AHAGGAR
Tahat 9541
Tamenghest
Oued Tamenghest
TUAREG
ADRAR DES IFÔGHAS
Mt Gréboun 6562
Iférouâne
Monts Tamgak 5906
Monts Bagzane 6300
AÏR
Agadez
TÉNÉRÉ

Nouadhibou
CAP BLANC
CAP D'ARGUIN
Atar
Chinguetti
OUARANE
EL MREYYÉ
Mabrouk
Araouane
Kidal
Nouamrhar
CAP TIMIRIS
MAURITANIA
Akjoujt
Tidjikdja
VALLÉE DU TILEMSI
Nouakchott
Boutilimit
Ouâlâta
Tombouctou (Timbuktu)
Bamba
Gao
Bourem
Goundam
NIGER
Saint-Louis
Dagana
Kaédi
Mbout
Sélibaby
Kiffa
Néma
Niafunké
Tahoua
Madaoua
Louga
Linguère
Matam
Nioro du Sahel
Nara
Goumbou
Sokolo
Mopti
Bandiagara
Niamey
Tillabéry
Say
Dosso
Sokoto
Tessaoua
Zinder
Gouré
Maradi
Nguru

CAP VERT
Rufisque
Dakar
Thiès
Diourbel
SENEGAL
Kaolack
THE GAMBIA
Banjul
Ziguinchor
GUINEA-BISSAU
Bissau
Bolama
Buba
ARQUIPÉLAGO DOS BIJAGÓS
Boké
Boffa
Kindia
Conakry
Freetown
SIERRA LEONE
Moyamba
Bonthe
Bomi Hills
Robertsport
Monrovia
Buchanan
River Cess
Greenville
CAPE PALMAS
Harper
Tabou
Kayes
Bafoulabé
Kita
Bakel
Satadougou
M'du Tamgué 5046
FOUTA DJALLON
Labé
Timbo
Mamou
Kouroussa
Kankan
Siguiri
GUINEA
Faranah
Kabala
Makeni
Kissidougou
Beyla
Kolahun
Pendembu
LIBERIA
Mont Nimba 5760
Séguéla
Ségou
San
Djenné
Bamako
Koulikoro
Koutiala
Dédougou
Bougouni
Sikasso
Bobo-Dioulasso
Gaoua
Gambaga
Sansanné-Mango
Natitingou
Odienné
Korhogo
KONG
Kong
Dabakala
Bondoukou
Bouna
Bole
Kintampo
Yendi
Tamale
BURKINA FASO
Ouagadougou
Koudougou
Tenkodogo
Kaya
Ouahigouya
Dori
Fada Ngourma
Malanville
Kandi
Illo
Kontagora
Zungeru
Minna
Birnin Kebbi
Gusau
Katsina
Kano
Zaria
Kaduna
Hadejia
Gumel
NIGERIA
Jos
Bauchi
Gombe
Bida
Baro
Keffi
ABUJA
Ibi
Makurdi
Katsina Ala
CÔTE D'IVOIRE (IVORY COAST)
Bouaké
Bouaflé
Yamoussoukro
Kumasi
Koforidua
GHANA
Tarkwa
Sekondi-Takoradi
Accra
Ada
Lake Volta
TOGO
Sokodé
Atakpamé
Savalou
Abomey
Savé
Palimé
Anécho
Lomé
BENIN
Porto-Novo
Grand-Popo
Cotonou
Abeokuta
Lagos
Ijebu Ode
Benin City
Sapele
Warri
Forcados
Owerri
Aba
Port Harcourt
Brass
Bonny
Calabar
CAMEROON
Iseyin
Oyo
Ogbomosho
Oshogbo
Ilesha
Iwo
Ife
Ibadan
Ilorin
Lokoja
Idah
Enugu
Onitsha
Mamfe
Kumba
Dschang
Foumban
Bafoussam
GOTEL MTS.
ADAMAWA
Kontcha
Cameroon Mtn. 13451
Limbe
Douala
Malabo
BIOKO
Yaoundé
Edéa
Eséka
Kribi
EQUATORIAL GUINEA
SÃO TOMÉ AND PRINCIPE
ILHA DO PRINCIPE
Bata
RIO MUNI
Oyem
Ebolowa
Campo
ILHA DE SÃO TOMÉ
São Tomé
Libreville
GABON
GULF OF GUINEA
Bight of Biafra
Bight of Benin

CAPE VERDE
SANTA ANTÃO
SÃO VICENTE
SAL
SÃO NICOLAU
BOA VISTA
SÃO TIAGO
MAIO
FOGO
Praia
Same scale as main map

ATLANTIC OCEAN

A-589100-76
COPYRIGHT BY
RAND McNALLY & COMPANY
MADE IN U.S.A.

Longitude West of Greenwich
Longitude East of Greenwich

Scale 1:16 000 000; one inch to 250 miles. Sinusoidal Projection
Elevations and depressions are given in feet

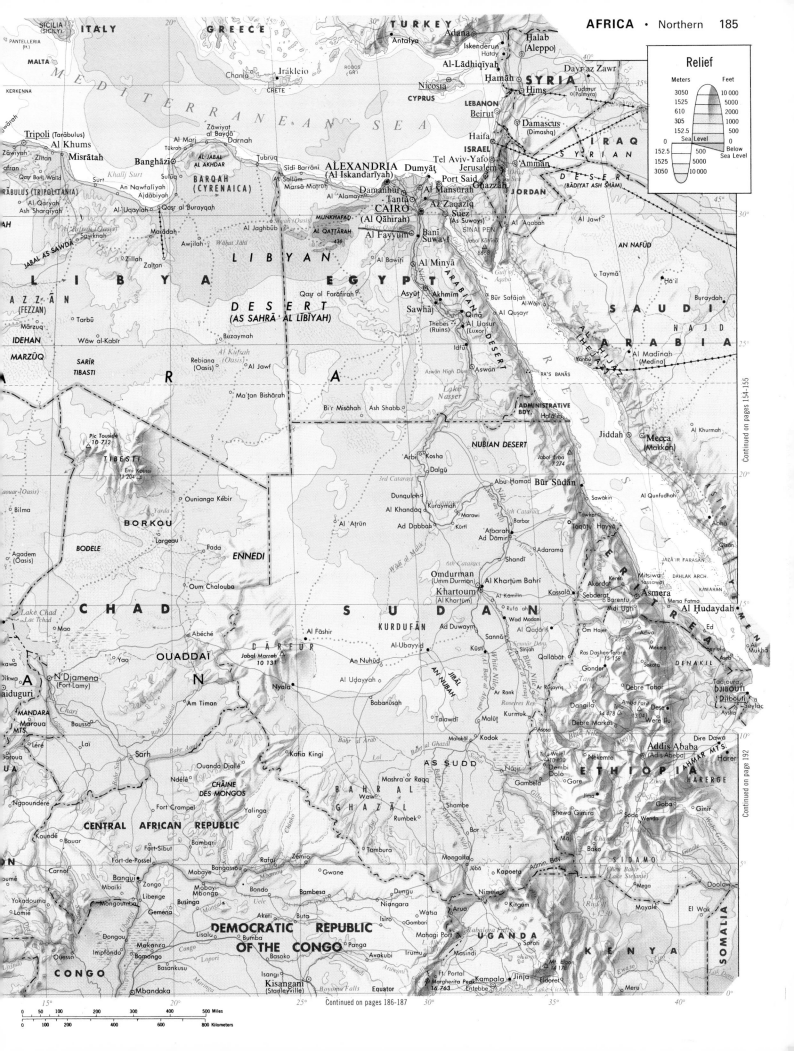

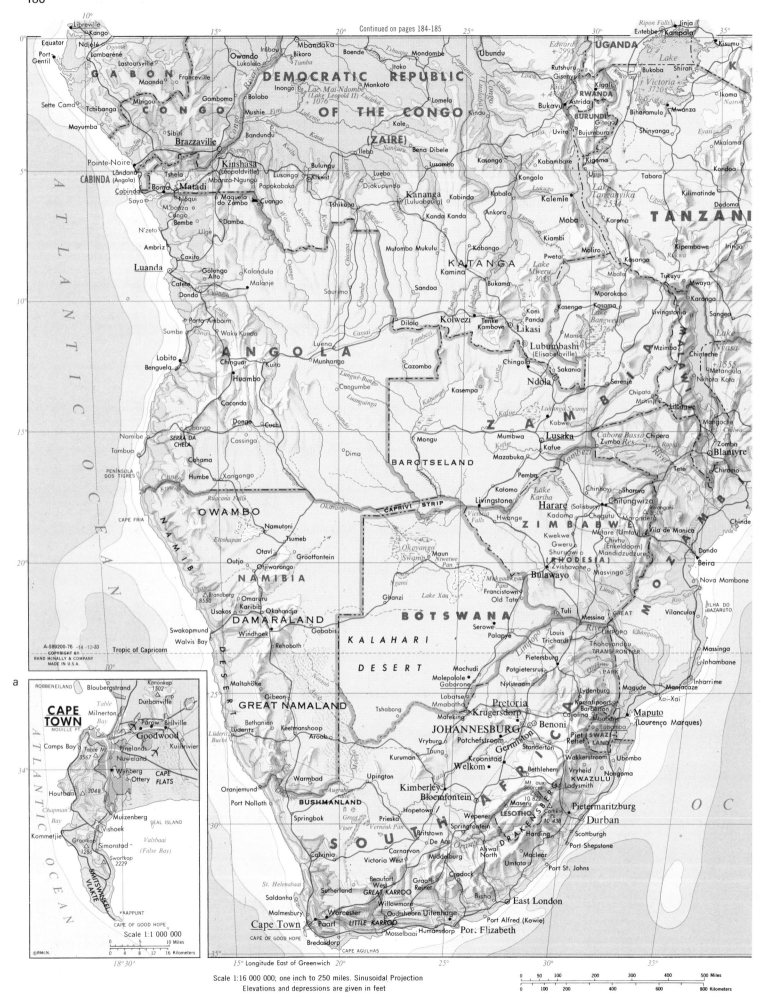

186

Continued on pages 184-185

10° 15° 20° 25° 30° 35°

Ripon Falls Jinja
Entebbe Kampala
UGANDA Kisumu
Edward Lake
+ 2995
Rutshuru Bukoba Shirati K
Giseny Victoria Ikoma Eyasi Natron
Libreville RWANDA +3720 Mwanza
Kango Kigali BURUNDI Biharamulo
Ndjolé Ogooué Mbandaka Mondombe Ubundu Bukavu Gitega Astrida Mwanza

GABON Lambaréné DEMOCRATIC REPUBLIC Uvira Bujumbura Shinyanga
Lastoursville Irébou Bikoro Boende Kindu Kigoma Kondoa
Owando Lukolela Itoko TANZANIA
Franceville Lac Mai-Ndombe Monkoto Ujiji Tabora Dodoma
Moanda (Lake Leopold II) Lomela Kabambare Kilimatinde
Sette Cama CONGO Gamboma Inongo Mushie +1076 Lukenie OF THE CONGO Kole Kongolo Lake Tanganyika Kipembawe Iringa
Tchibanga Bolobo Fimi Kasai Lusambo Kasongo + 2534 Karema
Mayumba Sibiti Bandundu Lomami Bena Dibele Kalemie Mkalama

Pointe-Noire Brazzaville (ZAIRE) Ilebo Sankuru Kabalo Moba Kasanga Tukuyu
Stanley Kwilu Luebo Kananga Kabinda Ankoro Mporokoso Mwaya Karonga
Landana Kinshasa Popokabaka Kikwit (Luluabourg) Kanda Kanda Kiambi Lake Mweru Kasama Livingstonia
CABINDA (Angola) (Léopoldville) Lusanga Kwenge Tshikapa + 3055 Molino Sengea
Cabinda Boma Matadi Mbanza-Ngungu Kasai Mutombo Mukulu Kabongo Bukama Pweto Lake Bangweulu + 3764 Mzimba

ATLANTIC OCEAN

Scale 1:16 000 000; one inch to 250 miles. Sinusoidal Projection
Elevations and depressions are given in feet

A-589200-76 -14-12-33
COPYRIGHT BY
RAND McNALLY & COMPANY
MADE IN U.S.A.

a

CAPE TOWN

ROBBENEILAND Bloubergstrand Kanonkop 1502
Table Bay Durbanville
MOUILLE PT. Milnerton Bellville
Camps Bay Parow
Table Mt. 3567 Goodwood Kuilsrivier
Pinelands Nuweland CAPE FLATS
Wynberg Ottery
Houtbaai 3048 Muizenberg
Chapman's Bay Vishoek SEAL ISLAND
Kommetjie Simonstad Valsbaai (False Bay)
Grootkop 1280 Swartkop 2229
SMITSWINKEL VLAKTE
KAAPPUNT
CAPE OF GOOD HOPE

ATLANTIC OCEAN

Scale 1:1 000 000
0 5 10 Miles
0 4 8 16 Kilometers
®RMCN.

18°30' 15° Longitude East of Greenwich 20° 25° 30° 35°

0 50 100 200 300 400 500 Miles
0 100 200 400 600 800 Kilometers

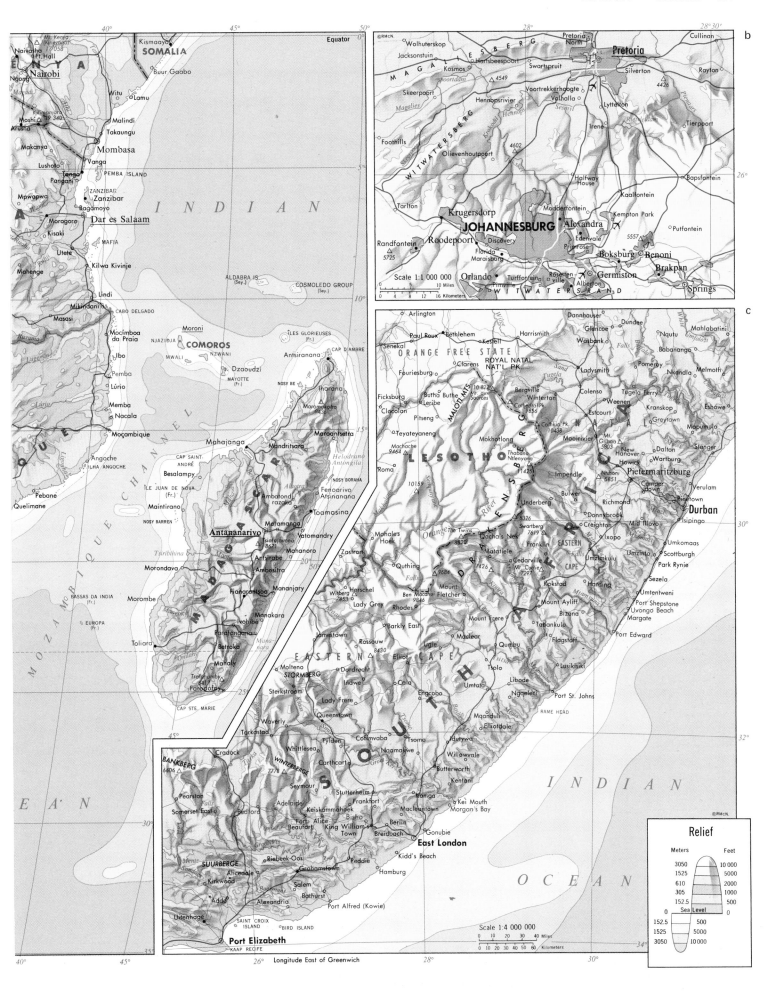

b

KENYA
Naivasha
Mt. Kenya
(Kirinyaga)
17,058
Fr. Hall
Ngong
Nairobi
Kismaayo
SOMALIA
Buur Gaabo
Equator
Kilimanjaro
19,340
Moshi
Arusha
Witu
Lamu
Makanya
Lushoto
Tanga
Pangani
Malindi
Takaungu
Mpwapwa
Vanga
Mombasa
PEMBA ISLAND
ZANZIBAR
Zanzibar
Morogoro
Bagamoyo
Dar es Salaam
Kisaki
Utete
MAFIA
INDIAN
Mahenge
Kilwa Kivinje
Lindi
Mikindani
CABO DELGADO
Masasi
Moçimboa
da Praia
Moroni
ILES GLORIEUSES
(Fr.)
CAP D'AMBRE
NJAZIDJA
COMOROS
Antsiranana
ALDABRA IS
(Sey.)
COSMOLEDO GROUP
(Sey.)
MWALI
NZWANI
Dzaoudzi
MAYOTTE
(Fr.)
NOSY BE
Iharana

Relief

Meters	Feet
3050	10 000
1525	5000
610	2000
305	1000
152.5	500
Sea Level	0
152.5	500
1525	5000
3050	10 000

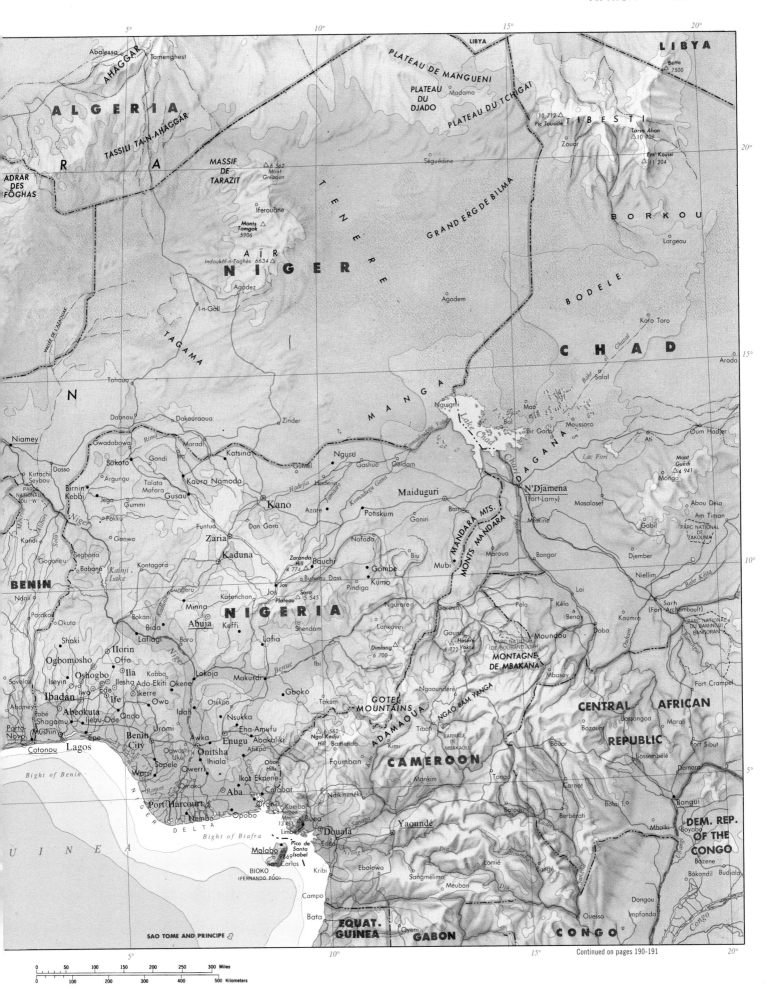

ALGERIA

AHAGGAR
Abalessa
Tamenghest

TASSILI TA-N-AHAGGAR

ADRAR
DES
FOGHAS

MASSIF
DE
TARAZIT
△ 6 562
Mont
Grébaun

Iferouâne

Monts
Tamgak
5906 △

NIGER

AÏR
Indoukâl-n-Taghès 6634 △
Agadez

I-n-Gall

TAGAMA

TENERE

GRAND ERG DE BILMA

LIBYA

PLATEAU DE MANGUENI

PLATEAU
DU
DJADO
Madama

PLATEAU DU TCHIGAÏ

10 712 △
Pic Toussidé
Zouar

TIBESTI

LIBYA

Bette
△ 7500

Tarso Ahon
△ 10 909

Emi Koussi
△ 11 204

BORKOU

Largeau

Séguédine

Agadem

BODELE

Koro Toro

CHAD

Arada

Tahoua

N

Niamey

Dabnou

Dakouraoua

Zinder

MANGA

DAGANA

Nguigmi

Lake Chad

Mao

Bol

Bir Gara

Moussoro

Salal
Bahr el Ghazal

Ati

Oum Hadjer

Mont
Guédi
△ 4 941
Mongo

Lac Fitri

Abou Deia

Am Timan

Gabil

Djember

PARC NATIONAL
DE
ZAKOUMA

Gwadabawa
Kirtachi
Seybou
Dosso
Sokoto
Gandi
Isa

Maradi

Katsina

Gumel

Nguru

Gashua

Geidam

Komadugu Yobé

N'Djamena
(Fort-Lamy)

Masalasef

PARC
NATIONAUX
DU "W"
Birnin
Kebbi
Argungu
Jega
Talata
Mafara
Gummi
Gusau

Kaura Namoda

Hadejia Hadejia
Jama'are

Komadugu Gana

Maiduguri

Bama

Mekine

Niger

Fokku

Fakai

Dan Gora

Funtua

Kano

Azare

Potiskum

Goniri

Mandara Mts.

Monts Mandara

Maroua

Bongor

Niellim

Bahr Kéita

Kandi

Ganwo

Zaria

Nafada

Biu

Mubi

Lai

Sarh
(Fort-Archambault)

Gogonou

Kontagora

Kaduna

Zaranda
Hill
4 774 △

Bauchi

Gombe

Kumo

Pindiga

Gombe

Kélo

Koumra

Doba

PARC NATIONAL
DU BAMINGUI
BANGORAN

BENIN

Ndali

Segbana
Babana

Kainji
Lake

Zungeru

Jos Jos
Plateau

Sara
△ 5 545

Ngurore

Garoua

Pala

Benoy

Fort Crampel

Parakou

Okuta

Bokani

Bida

Minna

Kafenchan

Shendam

Lankoviri

Goundi
△ 722 Yakra

Moundou

Mbasay

Bozoum

Marali

Shaki

Lafiagi

Baro

Keffi

Abuja

Lafia

Dimlang △
6 700

PARC NATIONAL
DE BOUBA NDJIDAH

MONTAGNE
DE MBAKANA

Bouar

Bossangoa

CENTRAL   AFRICAN

Ilorin

Offa

NIGERIA

Ibi

Takúm

GOTEL
MOUNTAINS

Ngaoundéré

NGAO BAM YANGA

Fort Sibut

Ogbomosho
Oshogbo
Iseyin
Oyo
Iwo
Ila
Ede
Ilesha
Ikerre
Ado-Ekiti
Owo
Kabba
Okene

Lokoja

Makurdi

Benue

Gboko

ADAMAOUA

Tibati

Kimi

Bossembélé

Ibadan

Ife

Otukpa

Ngol Kedju
Hill
△ 6 562

Bamenda

Kim

BARRAGE
DE
MBAKAOU

Batouri

Berbérati

REPUBLIC

Damara

Savalou

Abomey
Bobé
Shagamu

Abeokuta

Ondo

Idah

Nsukka

Eha-Amufu

Bamenda

Foumban

CAMEROON

Tongo

Bangui

Porto-
Novo

Ibadan
Mushin

Ijebu-Ode

Awka
Ogwashi
Uku

Enugu

Abakaliki

Afikpo

Mankim

Mbaïki

Boyabo

Cotonou

Lagos

Benin
City

Onitsha
Ihiala

Oban
Hills

Ndikiniméki

DEM. REP.

Warri

Sapele

Owerri

Ikot Ekpene

Calabar

Yaoundé

Carnot

Bayanga

OF THE

Bight of Benin

Ramos

Port Harcourt

Nemba

Opobo

Oron

Kumba

Cameroon
Mts
△ 13 451
Buea
Limbe

Douala

Edéa

Sanaga

Batouri I.

Bazène

CONGO

Aba

GUINEA

DELTA

Bight of Biafra

Malabo
San Carlos
△ 9869
Pico de
Santa
Isabel

Kribi

Eséka

Nyong

Lomié

Bangé

Bökondil

Budjala

BIOKO
(FERNANDO PÓO)

Ebolowa

Sangmélima

Dja

Dja

Ouesso

Impfondo

Dongou

Campo

Meuban

Congo

SAO TOME AND PRINCIPE

Bata

EQUAT.
GUINEA

Oyem

GABON

CONGO

Continued on pages 190-191

0   50   100   150   200   250   300 Miles

0   100   200   300   400   500 Kilometers

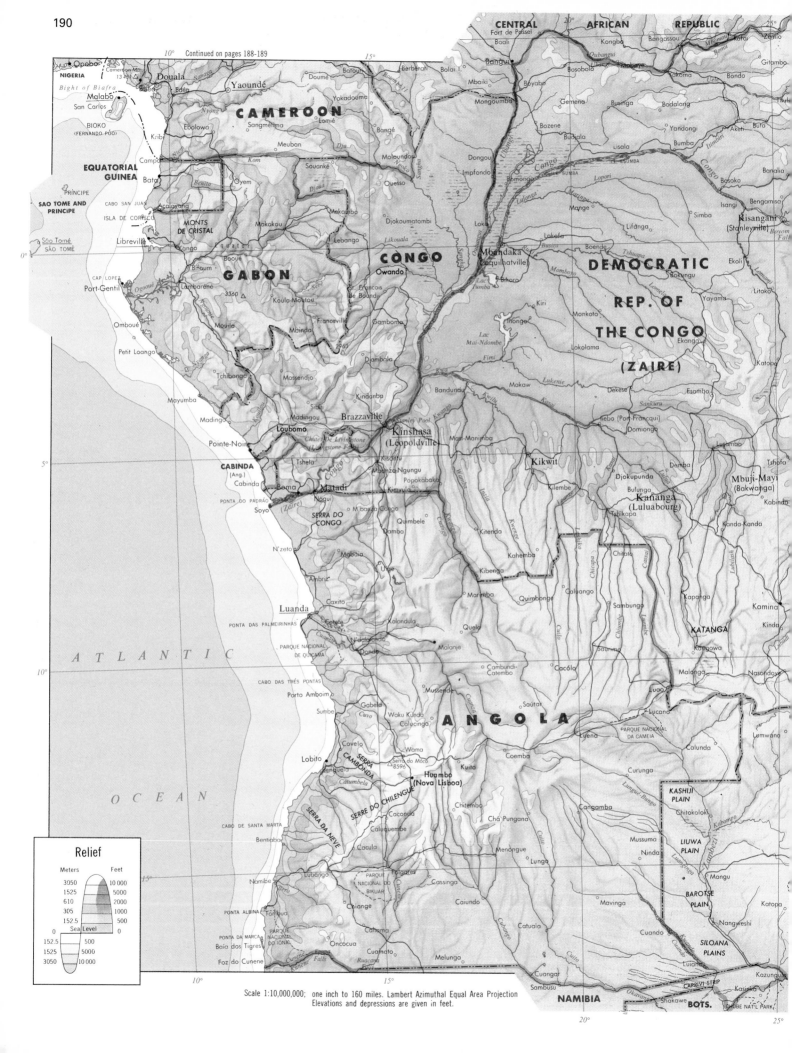

Continued on pages 188-189

CENTRAL AFRICAN REPUBLIC
Fort de Possel
Baali
Kongbo
Bangassou
Rafai
Zemio

NIGERIA
Opobo
Cameroon Mtn. 13 451
Douala
Yaoundé
Doumé
Batouri
Berbérati
Bolai I.
Bangui
Boyabo
Bosobolo
Gemena
Businga
Bodalang
Bondo
Bumba
Buta
Aket
Titule

Bight of Biafra
Malabo
San Carlos
BIOKO
(FERNANDO-PÓO)
Bata
Kribi
Edéa
Ebolowa
Sangmelima
Lomié
Yokadouma
Moloundou
Ncolo
Ouesso
Mongoumba
Dongou
Impfondo
Bomongo
Budjala
Lisala
Isangi
Simba
Bengamisa
Kisangani
(Stanleyville)

CAMEROON
Nyong
Meuban
Dja
Souanké
Sangha
Loka
Mbandaka
(Coquilhatville)
Lomako
Boende
Lifanga
Basoko
Bovoma Fall

EQUATORIAL GUINEA
PRÍNCIPE
CABO SAN JUAN
ISLA DE CORISCO
MONTS DE CRISTAL
Oyem
Makokou
Mekambo
Djokoumatombi
Likouala
CONGO
Tshuapa
Bokungu
Ekoli
Litoko

SAO TOME AND PRINCIPE
São Tomé
SÃO TOMÉ
Libreville
Kango
Booué
Lebango
Owando
Mbandaka
Lac Tumba
Bikoro
DEMOCRATIC REP. OF THE CONGO (ZAIRE)
Yayama
Katopa

CAP LOPEZ
Port-Gentil
Bifoum
Lambaréné
3360
Koula-Moutou
Franceville
St. François de Boundji
Gamboma
Kiri
Monkoto
Lokolama
Lac Mai-Ndombe
Ikela (Port-Francqui)
Domiongo

GABON
Mouila
Mbinda
Djambala
Fimi
Makaw
Lukenie
Dekese
Esombo

Omboué
Petit Loango
Sibiti
Kindanba
Bandundu
Kwa
Kwango
Sankuru
Kwilu
Lusambo
Tshofa

Tchibanga
Mossendjo
Madingou
Brazzaville
Stanley Pool
Mosi-Manimba
Kikwit
Djokupunda
Damba
Mbuji-Mayi
(Bakwanga)

Mayumba
Madingo
Loubomo
Chutes de Livingstone
(Livingstone Falls)
Kinshasa
(Léopoldville)
Kilembe
Kananga
(Luluabourg)
Tshikapa
Kabinda

Pointe-Noire
Tshela
Kisantu
Mbanza-Ngungu
Popokabaka
Wamba
Kwenge
Kitenda
Chikapa
Chitato
Kanda-Kanda

CABINDA (Ang.)
Cabinda
Boma
Matadi
Kimyula
Kahemba
Kibenga
Lulua
Lubilash

PONTA DO PADRÃO
Soyo
Náqui
M'banza Congo
Quimbele
Kibunga
Kaluango
Kapanga
Kamina

SERRA DO CONGO
N'zeto
Mbaia
Uíge
Marimba
Quimbonge
Sambungo
KATANGA
Kamba

ATLANTIC
Ambriz
Caxito
Kalandula
Quela
Cullo
Malanga
Nasondoye

Luanda
PONTA DAS PALMEIRINHAS
N'dalatando
Dondo
Malanje
Cacóla
Luau
Lucano
Lomwana

PARQUE NACIONAL DE QUICAMA
Cambundi-Catembo
Cuango
Lungue-Bungo
KASHIJI PLAIN

CABO DAS TRÊS PONTAS
Porto Amboim
Mussende
ANGOLA
Parque Nacional da Cameia
Chitokoloki

Sumbe
Gabela
Waku Kundu
Calucinga
Wama
Kuito
Coemba
Luena
Curunga
Cangamba
LIUWA PLAIN

OCEAN
Lobito
Benguela
Covelo
SERRA CAMBONDA
Serra do Môco 8596
Huambo
(Nova Lisboa)
Chitembo
Chá Pungana
BAROTSE PLAIN

Catumbela
Cubal
SERRE DO CHILENGUE
Caconda
Calucquembe
Chitembo
Mussuma
Ninda
Mongu
Katopa

CABO DE SANTA MARTA
Bentiaba
SERRA DA NEVE
Cacula
Menongue
Lunga
Lungoebungo
SILOANA PLAINS
Nangweshi

Relief
Meters   Feet
3050   10 000
1525   5000
610   2000
305   1000
152.5   500
0   Sea Level 0
152.5   500
1525   5000
3050   10 000

Namibe
Lubango
Parque Nacional do Bikuar
Folgares
Cassinga
Caiundo
Mavinga
Katopa

PONTA ALBINA
Tombua
Chiange
Cahama
Caundo
Catuala
Cuando
Luiana
Kazungu

PONTA DA MARCA
Baía dos Tigres
Parque Nacional do Iona
Oncocua
Cuamato
Melunga
CAPRIVI STRIP
Kasika

Foz do Cunene
Ruacana Falls
NAMIBIA
Sambusu
Shakawe
CHOBE NAT'L PARK
BOTS.

Scale 1:10,000,000; one inch to 160 miles. Lambert Azimuthal Equal Area Projection
Elevations and depressions are given in feet.

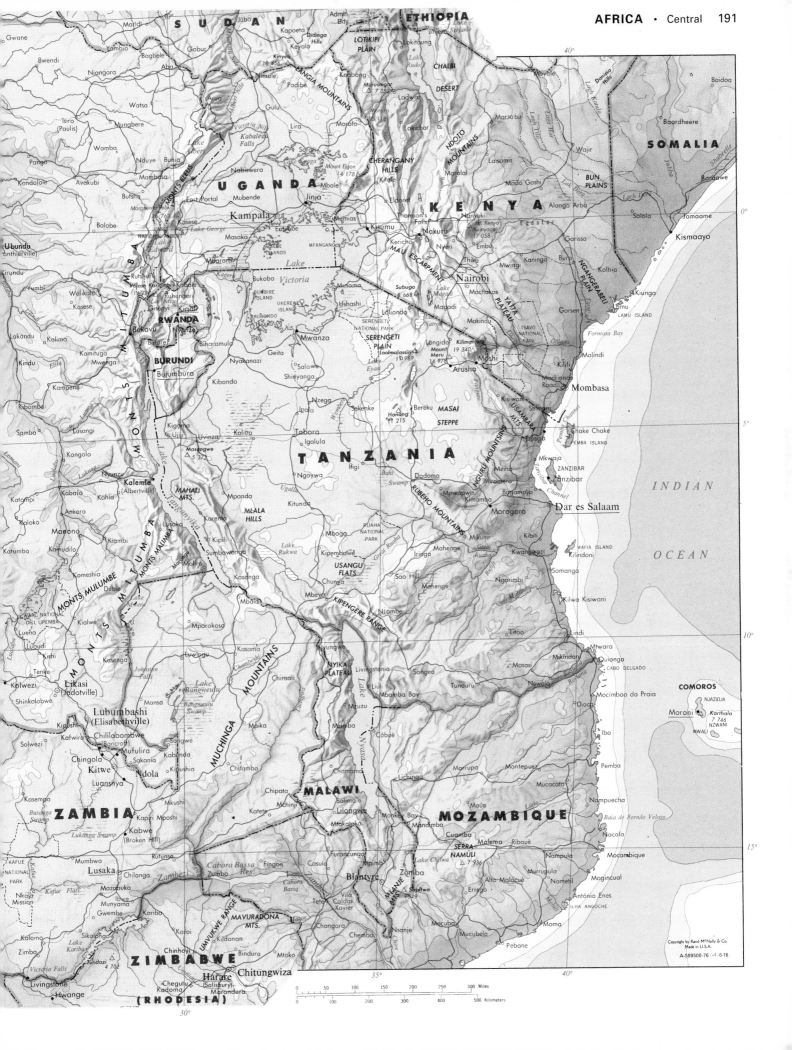

SUDAN
ETHIOPIA
SOMALIA
UGANDA
KENYA
RWANDA
BURUNDI
TANZANIA
ZAMBIA
MALAWI
MOZAMBIQUE
ZIMBABWE
(RHODESIA)

INDIAN
OCEAN

COMOROS

Nairobi
Mombasa
ZANZIBAR
Zanzibar
Dar es Salaam
Kampala
Moroni

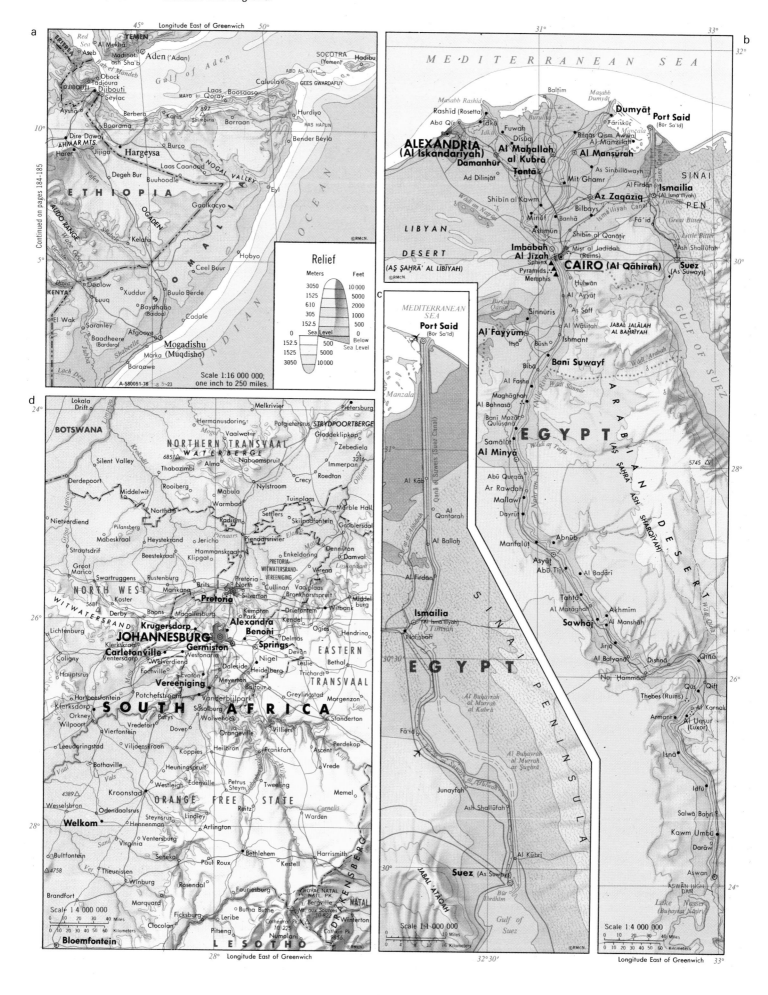

a

Red Sea
YEMEN
Al Mukha
Aden ('Adan)
Madinat ash Sha'b
SOCOTRA (Yemen) Hadibu
GEES GWARDAFUY
Gulf of Aden
ABD AL-KURI
Calula
Obock
Tadjoura
Djibouti
Seylac
DJIBOUTI
Aysha
Berbera
Karin
MAYD Is.
Shimbiris 897
Boosaaso
Borraan
RAS HAFUN
Hurdiyo
Dire Dawa
AHMAR MTS.
Jijiga
Harer
Hargeysa
Degeh Bur
Laas Caanood
Buuhoodle
NOGAL VALLEY
Bender Beyla
ETHIOPIA
Gaalkacyo
Eyl
OGADEN
AUDO RANGE
Kelafo
Ceel Buur
Hobyo
KENYA
Doolow
Xuddur
Buulo Berde
El Wak
Luuq
Baydhabo (Baidoa)
Cadale
Saranley
Afgooye
Marka (Merca)
Mogadishu (Muqdisho)
Baadheere (Bardera)
Baraawe
INDIAN OCEAN

Relief

Meters	Feet
3050	10 000
1525	5000
610	2000
305	1000
152.5	500
0 Sea Level	0
152.5	500 Below Sea Level
1525	5000
3050	10 000

Scale 1:16 000 000; one inch to 250 miles.
A-580051-76 -8 5-23

Continued on pages 184-185

b

MEDITERRANEAN SEA
Baltîm
Maşabb Dumyât
Dumyât
Port Said (Bûr Sa'îd)
Maşabb Rashîd
Rashîd (Rosetta)
Abû Qîr
Idkû
Fuwah
Fâriskûr
Manzala
Disûq
Bilqas Qism Awwal Al Manzilah
ALEXANDRIA (Al Iskandarîyah)
Al Mahallah al Kubrâ
Al Manşûrah
Damanhûr
Tantâ
As Sinbillâwayn
SINAI PEN.
Ad Dilinjât
Mît Ghamr
Al Firdân
Ismailia (Al Ismâ'îlîyah)
Shibîn al Kawm
Az Zaqâzîq
Minûf
Bilbays
Ismâ'îlîyah Canal
Fâ'id
Great Bitter
LIBYAN
Banhâ
Ashmûn
Shibîn al Qanâtir
Little Bitter
DESERT
(AŞ ŞAHRÂ' AL LÎBÎYAH)
Imbâbah
Al Jîzah
Mişr al Jadîdah (Ruins)
Ash Shallûfah
CAIRO (Al Qâhirah)
Suez (As Suways)
Pyramids
Memphis
Hulwân
Al 'Ayyât
Birkat Qârûn
As Şaff
Sinnûris
JABAL JALÂLAH AL BAHRÎYAH
Al Fayyûm
Al Wâsitah
Itsâ
Bûsh
Ishmant
Biba
Banî Suwayf
Maghâghah
Al Bahnasâ
Wâdî 'Arabah
Banî Mazâr
Qulûşnâ
EGYPT (AŞ ŞAHRÂ' ASH SHARQÎYAH)
Samâlût
Al Minyâ
5745
Abû Qurqâş
Ar Rawdah
Mallawî
Dayrût
Abnûb
Manfalût
Asyût
Abû Tîj
Al Badârî
Tahtâ
Al Marâghah
Akhmîm
Sawhâj
Al Manshâh
Jirjâ
Al Balyanâ
Dishnâ
Qinâ
Qûş
Qift
Thebes (Ruins)
Armant
Al Karnak
Al Uqsur (Luxor)
Isnâ
Idfû
Salwâ Bahrî
Kawm Umbû
Darâw
Aswân
ASWÂN HIGH DAM
Lake Nasser (Buhayrat Nâşir)
ARABIAN DESERT
GULF OF SUEZ

c

MEDITERRANEAN SEA
Port Said (Bûr Sa'îd)
Manzala
Qanâ el Suweis (Suez Canal)
Al Kâb
Al Qantarah
Al Ballah
Al Firdân
SINAI PENINSULA
Ismailia (Al Ismâ'îlîyah)
Timsah
Nafîshah
EGYPT
Al Buhayrah al Murrah al Kubrâ
Al Buhayrah al Murrah aş Şughrâ
Fâ'id
Qanâ es Suweis (Suez Canal)
Junayfah
Ash Shallûfah
Al Kûbrî
Suez (As Suways)
JABAL 'ATÂQAH
Bûr Ibrâhîm
Gulf of Suez

Scale 1:4 000 000
0 10 20 30 40 Miles
0 10 20 30 40 50 60 Kilometers

d

Lokala Drift
Melkrivier
Pietersburg
BOTSWANA
Hermanusdoring
Potgietersrus STRYDPOORTBERGE
Mogol
Vaalwater
Gladdeklipkop
NORTHERN TRANSVAAL
WATERBERGE
6851
Alma
Naboomspruit
Zebediela
Silent Valley
Thabazimbi
Immerpan
3216
Derdepoort
Rooiberg
Crecy
Roedtan
Middelwit
Mabula
Nylstroom
Nietverdiend
Pilansberg
Warmbad
Tuinplaas
Marble Hall
Northam
Radium
Settlers
Skilpadfontein
Groblersdal
Mabeskraal
Heystekrand
Jericho
Pienaars
Enkeldoring
Dennilton
Straatsdrif
Beesteekraal
Klipgat
Hammanskraal
PRETORIA-WITWATERSRAND-VEREENIGING
Verena
Damval
Loskopdam
Groot Marico
Swartruggens
Rustenburg
Pretoria North
Cullinan
Vaalplaas
Bronkhorstspruit
NORTH WEST
Marikana
Brits
Silverton
Middelburg
Koster
5681
Magaliesburg
Kempton Park
Driefontein
Kendal
Witbank
Lichtenburg
Derby
Krugersdorp
Alexandra
Ogies
Hendrina
Coligny
Klerksdorp
JOHANNESBURG
Benoni
Delmas
EASTERN
Carletonville
Germiston
Springs
Westonaria
Nigel
Leslie
Bethal
Hauptrus
Fochville
Welverdiend
Daleside
Devon
TRANSVAAL
Vereeniging
Evaton
Heidelberg
Trichardt
Meyerton
Balfour
Morgenzon
Hartbeesfontein
Potchefstroom
Vanderbijlpark
Greylingstad
SOUTH AFRICA
Sasolburg
Wolwehoek
Standerton
Klerksdorp
Parys
Dover
Villiers
Orkney
Vredefort
Orangeville
Wilpoort
Vierfontein
Heilbron
Frankfort
Perdekop
Leeudoringstad
Viljoenskroon
Vrede
Bothaville
Westleigh
Edenville
Petrus Steyn
Tweeling
Memel
4389
Kroonstad
Lindley
Reitz
Cornelis
Wesselsbron
Odendaalsrus
Steynsrus
Warden
Welkom
Hennenman
Arlington
ORANGE FREE STATE
Virginia
Ventersburg
4758
Winburg
Senekal
Bethlehem
Harrismith
Theunissen
Paul Roux
Kestell
Brandfort
Rosendal
Fouriesburg
ROYAL NATAL NATL. PK.
Aux Sources 10 822
Winterton
Marquard
Ficksburg
Leribe
Clocolan
Butha Buthe
Cathedral Pk. 10 225
Cathkin Pk. 9856
Bloemfontein
Numalani
LESOTHO
NATAL
DRAKENSBERG

Scale 1:4 000 000
0 10 20 30 40 Miles
0 20 40 60 Kilometers

28° Longitude East of Greenwich

Longitude East of Greenwich

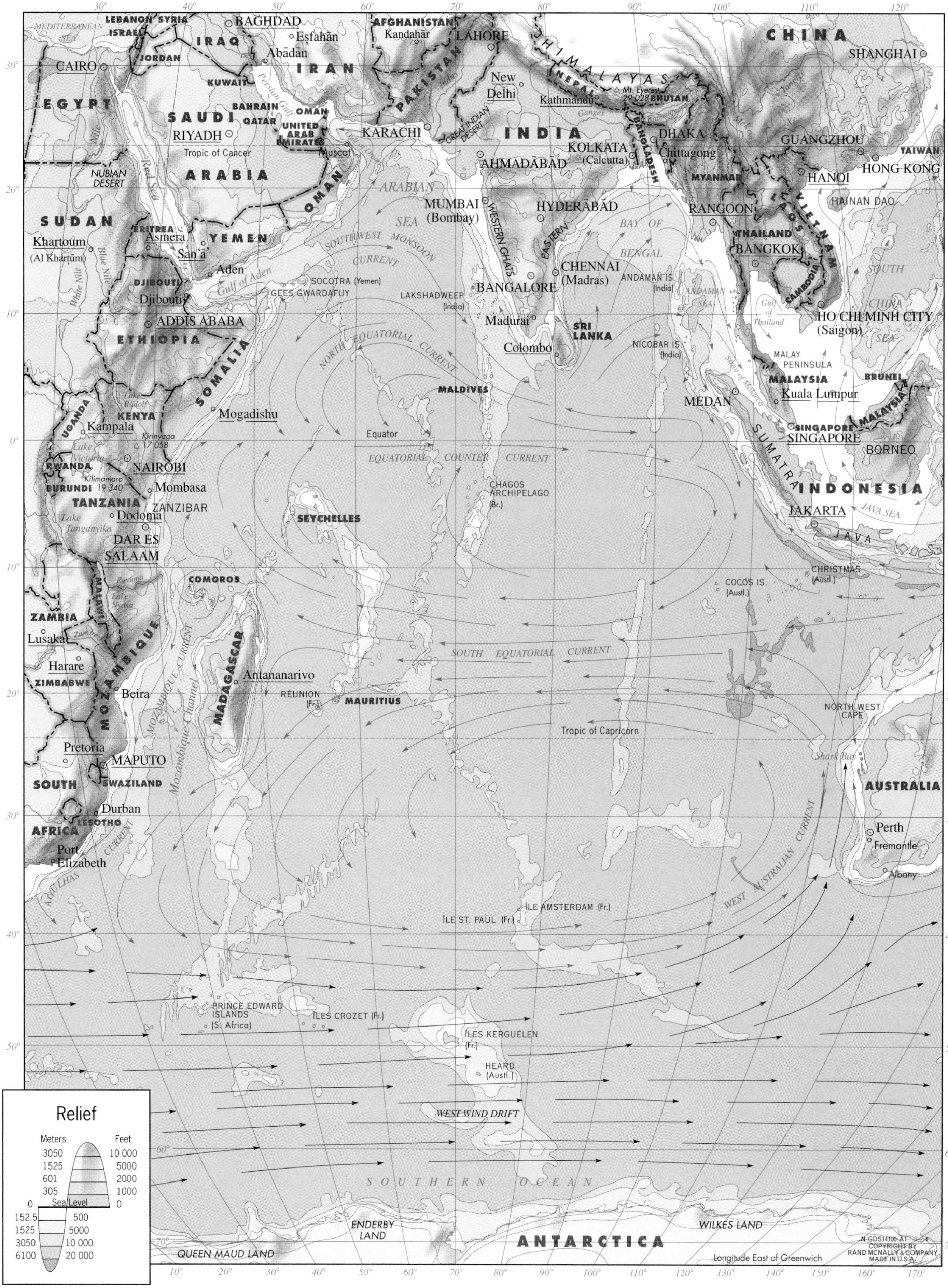

LEBANON SYRIA
ISRAEL
JORDAN
MEDITERRANEAN SEA
CAIRO
EGYPT
NUBIAN DESERT
SUDAN
Khartoum (Al Khartūm)
White Nile
Blue Nile
ERITREA
Asmera
San'a
YEMEN
Aden
Gulf of Aden
DJIBOUTI
Djibouti
ADDIS ABABA
ETHIOPIA
SOMALIA
GEES GWARDAFUY
SOCOTRA (Yemen)

BAGHDAD
Eşfahān
Abādān
IRAQ
IRAN
KUWAIT
BAHRAIN
QATAR
OMAN
UNITED ARAB EMIRATES
SAUDI
RIYADH
ARABIA
Tropic of Cancer
Muscat
OMAN
Persian Gulf
Red Sea

AFGHANISTAN
Kandahār
LAHORE
PAKISTAN
HIMALAYAS
New Delhi
Mt. Everest 29,028
NEPAL
BHUTAN
Kathmandu
Ganges
GREAT INDIAN DESERT
KARACHI
INDIA
AHMADĀBAD
MUMBAI (Bombay)
WESTERN GHATS
HYDERABAD
EASTERN GHATS
KOLKATA (Calcutta)
DHAKA
BANGLADESH
Chittagong
ARABIAN SEA
BANGALORE
CHENNAI (Madras)
BAY OF BENGAL
MADURAI
LAKSHADWEEP (India)
SRI LANKA
Colombo
MALDIVES

CHINA
SHANGHAI
GUANGZHOU
HONG KONG
TAIWAN
HANOI
HAINAN DAO
MYANMAR
RANGOON
THAILAND
BANGKOK
LAOS
VIETNAM
CAMBODIA
HO CHI MINH CITY (Saigon)
ANDAMAN IS. (India)
ANDAMAN SEA
NICOBAR IS. (India)
Gulf of Thailand
SOUTH CHINA SEA
MALAY PENINSULA
MALAYSIA
Kuala Lumpur
BRUNEI
MALAYSIA
SINGAPORE
BORNEO
Strait of Malacca
MEDAN
SUMATRA
INDONESIA
JAKARTA
JAVA
JAVA SEA

SOUTHWEST MONSOON CURRENT
NORTH EQUATORIAL CURRENT
Equator
EQUATORIAL COUNTER CURRENT
CHAGOS ARCHIPELAGO (Br.)
SEYCHELLES
COMOROS
SOUTH EQUATORIAL CURRENT
COCOS IS. (Austl.)
CHRISTMAS (Austl.)

Mogadishu
UGANDA
Kampala
Lake Victoria
Kirinyaga 17,058
KENYA
NAIROBI
RWANDA
BURUNDI
Kilimanjaro 19,340
Mombasa
TANZANIA
ZANZIBAR
Dodoma
DAR ES SALAAM
Lake Tanganyika
MALAWI
Lake Nyasa
ZAMBIA
Lusaka
Zambezi
Harare
ZIMBABWE
MOZAMBIQUE
Beira
MOZAMBIQUE CURRENT
MADAGASCAR
Antananarivo
RÉUNION (Fr.)
MAURITIUS
Mozambique Channel
Tropic of Capricorn

Pretoria
MAPUTO
SWAZILAND
SOUTH AFRICA
LESOTHO
Durban
Port Elizabeth
AGULHAS CURRENT

NORTHWEST CAPE
Shark Bay
AUSTRALIA
Perth
Fremantle
Albany
WEST AUSTRALIAN CURRENT

ÎLE AMSTERDAM (Fr.)
ÎLE ST. PAUL (Fr.)
PRINCE EDWARD ISLANDS (S. Africa)
ÎLES CROZET (Fr.)
ÎLES KERGUELEN (Fr.)
HEARD (Austl.)
WEST WIND DRIFT

SOUTHERN OCEAN
ENDERBY LAND
WILKES LAND
ANTARCTICA
QUEEN MAUD LAND
Longitude East of Greenwich

N-GDS14100-AT
COPYRIGHT BY
RAND McNALLY & COMPANY
MADE IN U.S.A.

Relief

Meters	Feet
3050	10 000
1525	5000
601	2000
305	1000
0 Sea Level	0
152.5	500
1525	5000
3050	10 000
6100	20 000

→ Warm ocean currents
→ Cold ocean currents

Scale 1:50 000 000; one inch to 790 miles. Mollweide Projection
Elevations and depressions are given in feet

0 200 400 600 800 1000 Miles
0 400 800 1200 1600 Kilometers

Warm ocean currents
Cold ocean currents

Scale 1:50 000 000; one inch to 800 miles. Goode's Homolosine Equal Area Projection
Elevations and depressions are given in feet

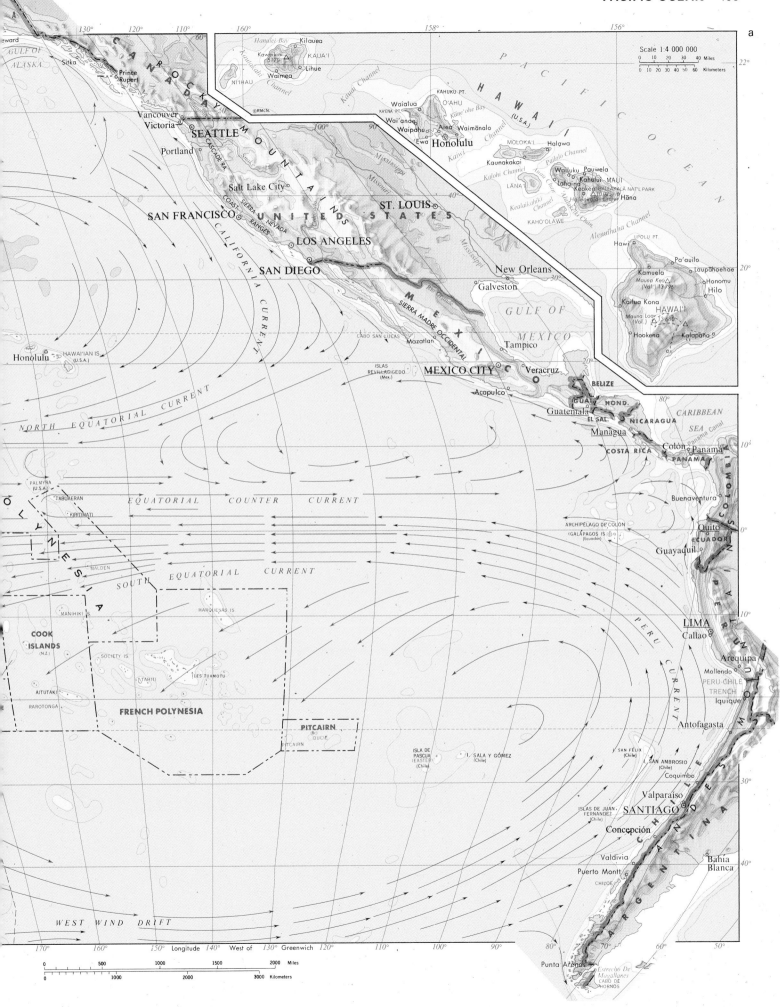

a

Scale 1:4 000 000

0 10 20 30 40 Miles
0 10 20 30 40 50 60 Kilometers

HAWAI'I
(U.S.A.)

PACIFIC OCEAN

Handlei Bay · Kilauea
Kawaikini
5170 △ KAUA'I
NI'IHAU · Waimea · Lihue
Kauai Channel
©RMCN.

KAHUKU PT.
Waialua · O'AHU
Wai'anae · Kane'ohe Bay
Waipahu Aiea Waimānalo
Ewa Honolulu
Kaiwi Channel

MOLOKA'I · Halawa
Kaunakakai
Kalohi Channel Pailolo Channel
LĀNA'I Wailuku · Pauwela
Lahaina · Kahului MAUI
Keokea PALEAKALA NAT'L PARK
Kealaikahiki △10025 Hāna
Channel Halaeakala Crater
KAHO'OLAWE Alenuihaha Channel

UPOLU PT.
Hawi · Pa'auilo
Kamuela · Laupahoehoe
Mauna Kea △ Honomu
(Vol.) 13 796 Hilo
HAWAI'I
Kailua Kona
Mauna Loa △13 680
(Vol.) △
Hookena Kalapana
Kalapana

GULF OF ALASKA
Seward
Sitka
CANADA
ROCKY MOUNTAINS
Prince Rupert
Vancouver
Victoria
SEATTLE
Portland
CASCADE RA.
Salt Lake City
SIERRA NEVADA
Snake
COAST RANGES
SAN FRANCISCO
UNITED STATES
ST. LOUIS
Mississippi
Missouri
LOS ANGELES
SAN DIEGO
CALIFORNIA CURRENT
Rio Grande
SIERRA MADRE OCCIDENTAL
M E X I C O
New Orleans
Galveston
GULF OF MEXICO
CABO SAN LUCAS
Mazatlan
Tampico
ISLAS REVILLAGIGEDO
(Mex.)
MEXICO CITY
Veracruz
Honolulu
HAWAI'IAN IS.
(U.S.A.)
Acapulco
BELIZE
GUAT.
Guatemala
EL SAL. HOND.
NICARAGUA
CARIBBEAN SEA
Managua
COSTA RICA
Colón · Panama
PANAMA
Panama Canal
NORTH EQUATORIAL CURRENT
PALMYRA
(U.S.A.)
TABUAERAN
KIRITIMATI
EQUATORIAL COUNTER CURRENT
ARCHIPÉLAGO DE COLÓN
(GALÁPAGOS IS.)
(Ecuador)
Buenaventura
COLOMBIA
Quito
ECUADOR
Guayaquil
MALDEN
SOUTH EQUATORIAL CURRENT
P O L Y N E S I A
MANIHIKI IS.
MARQUESAS IS.
LIMA
Callao
PERU
COOK ISLANDS
(N.Z.)
SOCIETY IS.
ÎLES TUAMOTU
PERU CURRENT
Arequipa
AITUTAKI
TAHITI
Mollendo
RAROTONGA
FRENCH POLYNESIA
PERU-CHILE TRENCH
Iquique
PITCAIRN
(Br.)
DUCIE
PITCAIRN
Antofagasta
ISLA DE PASCUA
(EASTER)
(Chile)
I. SALA Y GÓMEZ
(Chile)
I. SAN FÉLIX
(Chile)
I. SAN AMBROSIO
(Chile)
Coquimbo
Valparaíso
ISLAS DE JUAN FERNANDEZ
(Chile)
SANTIAGO
CHILE
ANDES
ARGENTINA
Concepción
Valdivia
Bahia Blanca
Puerto Montt
CHILOE
WEST WIND DRIFT
Punta Arenas
Estrecho De Magallanes
CABO DE HORNOS

Longitude West of Greenwich

0 500 1000 1500 2000 Miles
0 1000 2000 3000 Kilometers

196

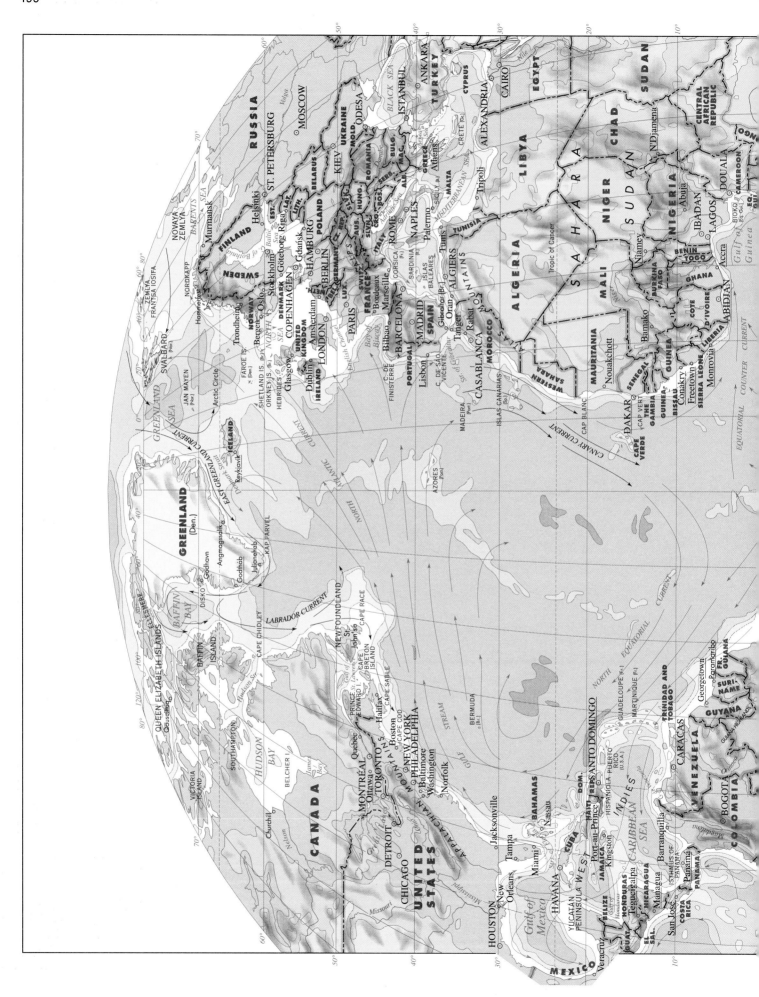

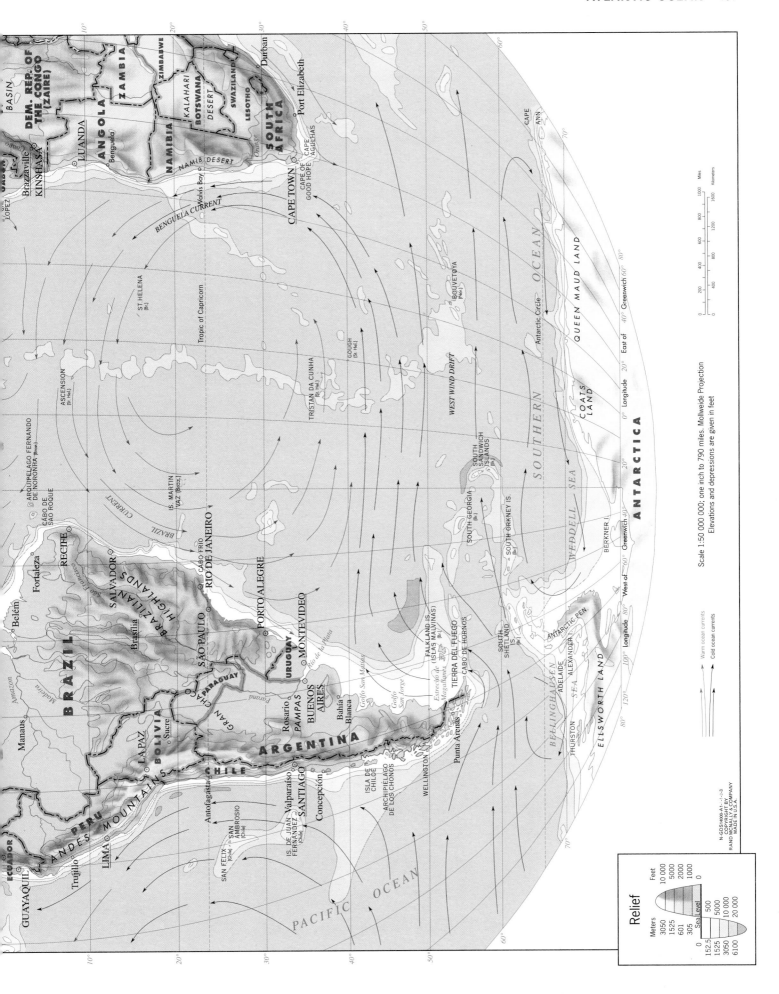

DEM. REP. OF THE CONGO (ZAIRE)
KINSHASA
Brazzaville
LOPEZ
GABON
BASIN
LUANDA
ANGOLA
Benguela
ZAMBIA
ZIMBABWE
NAMIBIA
BOTSWANA
KALAHARI DESERT
SWAZILAND
LESOTHO
SOUTH AFRICA
Durban
Port Elizabeth
NAMIB DESERT
Walvis Bay
Orange
CAPE AGULHAS
CAPE OF GOOD HOPE
CAPE TOWN
CAPE ANN

Benguela Current

ST. HELENA
[Br.]

Tropic of Capricorn

ASCENSION
[St. Hel.]

ARQUIPÉLAGO FERNANDO DE NORONHA [Braz.]
CABO DE SÃO ROQUE
RECIFE
Fortaleza
Belém
Manaus
Amazon
Madeira
BRAZIL
BRAZILIAN HIGHLANDS
Brasília
São Francisco
SALVADOR
SÃO PAULO
RIO DE JANEIRO
CABO FRIO
I.S. MARTIN VAZ [Braz.]

BRAZIL CURRENT

TRISTAN DA CUNHA
[St. Hel.]

GOUGH
[St. Hel.]

BOUVETØYA
[Nor.]

SOUTHERN   OCEAN

Antarctic Circle

East of   40° Greenwich   60°   80°

QUEEN MAUD LAND

COATS LAND

WEST WIND DRIFT

SOUTH SANDWICH ISLANDS
[Br.]

SOUTH GEORGIA
[Br.]

SOUTH ORKNEY IS.
[Br.]

WEDDELL SEA

BERKNER I.

ANTARCTICA

20°   0° Longitude   20°

PORTO ALEGRE
MONTEVIDEO
URUGUAY
Rio de la Plata
Rosario
BUENOS AIRES
PAMPAS
Bahía Blanca
PARAGUAY
GRAN CHACO
Paraná
LA PAZ
BOLIVIA
Sucre
ARGENTINA
CHILE
PERU
LIMA
Trujillo
ANDES MOUNTAINS
ECUADOR
GUAYAQUIL
Antofagasta
SAN FÉLIX [Chile]
SAN AMBROSIO [Chile]
IS. DE JUAN FERNÁNDEZ [Chile]
Valparaíso
SANTIAGO
Concepción
ISLA DE CHILOÉ
ARCHIPIÉLAGO DE LOS CHONOS
WELLINGTON
Punta Arenas
Golfo San Matías
Golfo San Jorge
Estrecho de Magallanes
TIERRA DEL FUEGO
CABO DE HORNOS
FALKLAND IS. (ISLAS MALVINAS)
[Br.]
SOUTH SHETLAND IS.
[Br.]
ANTARCTIC PEN.
ALEXANDER I.
Adelaide
BELLINGHAUSEN SEA
THURSTON
ELLSWORTH LAND

West of   60° Greenwich   40°
100° Longitude   80°
120°
70°

PACIFIC   OCEAN

Scale 1:50 000 000; one inch to 790 miles. Mollweide Projection
Elevations and depressions are given in feet

Miles
Kilometers
0   200   400   600   800   1000
0   400   800   1200   1600

Warm ocean currents
Cold ocean currents

N-GDS1400C-A1--1:---3
COPYRIGHT BY
RAND MCNALLY & COMPANY
MADE IN U.S.A.

Relief
Meters   Feet
3050   10 000
1525   5000
601   2000
305   1000
Sea Level   Sea Level
152.5   500
1525   5000
3050   10 000
6100   20 000

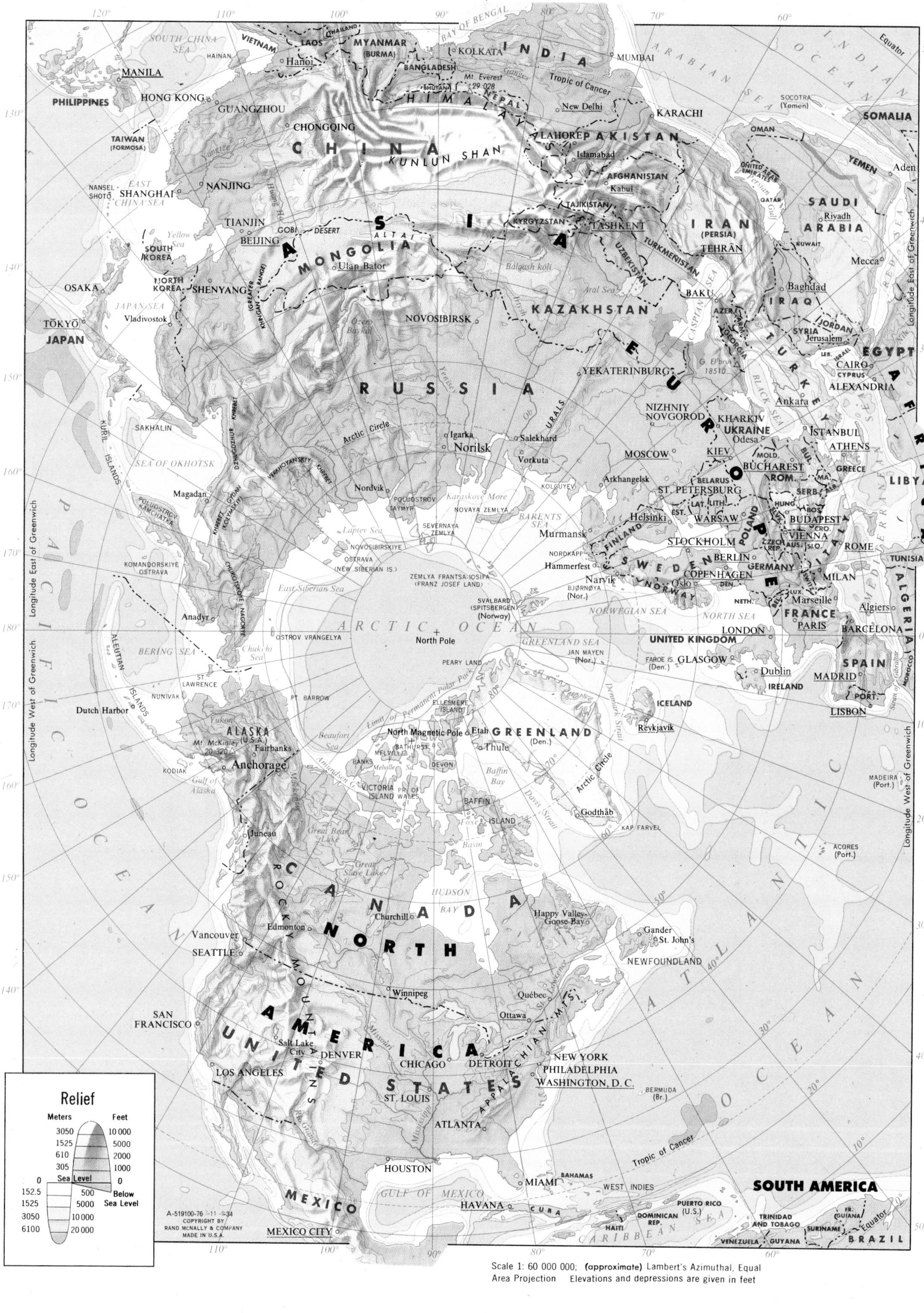

SOUTH CHINA SEA
VIETNAM LAOS THAILAND MYANMAR (BURMA) KOLKATA INDIA BAY OF BENGAL ARABIAN SEA INDIAN OCEAN Equator
MANILA Hanoi HAINAN BANGLADESH Mt. Everest 29,028 Tropic of Cancer MUMBAI SOCOTRA (Yemen) SOMALIA
PHILIPPINES HONG KONG GUANGZHOU BHUTAN NEPAL New Delhi KARACHI OMAN YEMEN Aden
TAIWAN (FORMOSA) CHONGQING HIMALAYA LAHORE PAKISTAN Islamabad UNITED ARAB EMIRATES SAUDI
NANSEI SHOTO SHANGHAI NANJING CHINA KUNLUN SHAN AFGHANISTAN Kabul QATAR Riyadh ARABIA
EAST CHINA SEA TIANJIN GOBI DESERT ALTAI KYRGYZSTAN TASHKENT IRAN (PERSIA) Baghdad Mecca
SOUTH KOREA BEIJING A S I A MONGOLIA Ulan Bator TAJIKISTAN UZBEKISTAN TURKMENISTAN TEHRĀN BAKU KUWAIT
OSAKA NORTH KOREA SHENYANG GREATER KHINGAN RANGE Aral Sea KAZAKHSTAN AZER. IRAQ JORDAN
JAPAN SEA Vladivostok Ozero Baykal NOVOSIBIRSK Balqash koli CASPIAN SEA G. El'brus 18510 SYRIA Jerusalem EGYPT
TŌKYŌ JAPAN YEKATERINBURG EUROPE GEORGIA BLACK SEA Ankara TURKEY CYPRUS ALEXANDRIA
SAKHALIN R U S S I A URALS NIZHNIY NOVGOROD KHARKIV UKRAINE ISTANBUL ATHENS AFRICA
SEA OF OKHOTSK Arctic Circle Igarka Salekhard Ob' MOSCOW KIEV Odesa BUCHAREST ROM. GREECE LIBYA
Magadan Norilsk Vorkuta Arkhangelsk ST. PETERSBURG BELARUS LAT. LITH. BUDAPEST VIENNA ROME
POLUOSTROV KAMCHATKA Nordvik KOLGUYEV Murmansk Helsinki WARSAW POLAND CZECH REP. SLO. MILAN TUNISIA
KOMANDORSKIYE OSTRAVA TAYMYR NOVAYA ZEMLYA BARENTS SEA Nordkapp STOCKHOLM BERLIN GERMANY
Anadyr Laptev Sea NOVOSIBIRSKIYE OSTRAVA (NEW SIBERIAN IS.) SEVERNAYA ZEMLYA Hammerfest SWEDEN NORWAY Oslo COPENHAGEN DEN. LUX. Marseille Algiers ALGERIA
BERING SEA Ostrov Vrangelya East Siberian Sea ZEMLYA FRANTSA-IOSIFA (FRANZ JOSEF LAND) SVALBARD (SPITSBERGEN) (Norway) BJØRNØYA (Nor.) Narvik NORWEGIAN SEA NORTH SEA FRANCE PARIS BARCELONA
ALEUTIAN ISLANDS Chukchi Sea ARCTIC OCEAN North Pole JAN MAYEN (Nor.) UNITED KINGDOM LONDON SPAIN
ST. LAWRENCE NUNIVAK PT. BARROW North Magnetic Pole GREENLAND SEA FAROE IS. GLASGOW MADRID
Dutch Harbor Beaufort Sea BATHURST MELVILLE North Magnetic Pole Permanent Polar Pack ICELAND Dublin IRELAND LISBON PORT.
ALASKA (U.S.A.) Mt. McKinley 20,320 Fairbanks Anchorage Amundsen G. BANKS VICTORIA ISLAND PR. OF WALES DEVON ELLESMERE ISLAND PEARY LAND Limit of Permanent Polar Pack Etah Thule GREENLAND (Den.) Reykjavik Denmark Strait MADEIRA (Port.)
KODIAK Juneau Great Bear Lake Great Slave Lake BAFFIN ISLAND Baffin Bay Godthåb Davis Strait KAP FARVEL ACORES (Port.)
Vancouver Edmonton C A N A D A FOXE BASIN HUDSON BAY Churchill Happy Valley-Goose Bay Gander St. John's Arctic Circle
SEATTLE NORTH Winnipeg Québec NEWFOUNDLAND
SAN FRANCISCO A M E R I C A Ottawa APPALACHIAN MTS.
ROCKY MOUNTAINS UNITED STATES Salt Lake City DENVER CHICAGO DETROIT NEW YORK PHILADELPHIA WASHINGTON, D.C. BERMUDA (Br.)
LOS ANGELES ST. LOUIS ATLANTA
HOUSTON BAHAMAS Tropic of Cancer WEST INDIES PUERTO RICO (U.S.) TRINIDAD AND TOBAGO FR. GUIANA
MEXICO GULF OF MEXICO MIAMI HAVANA CUBA DOMINICAN REP. HAITI SURINAME SOUTH AMERICA
MEXICO CITY CARIBBEAN SEA VENEZUELA GUYANA BRAZIL Equator

PACIFIC OCEAN
ATLANTIC OCEAN
Longitude East of Greenwich
Longitude West of Greenwich

Relief
Meters   Feet
3050   10 000
1525   5000
610   2000
305   1000
0   Sea Level
152.5   500
          Below Sea Level
1525   5000
3050   10 000
6100   20 000

A-519100-76-11-9-34
COPYRIGHT BY
RAND MCNALLY & COMPANY
MADE IN U.S.A.

Scale 1: 60 000 000; (approximate) Lambert's Azimuthal, Equal
Area Projection   Elevations and depressions are given in feet

WORLD POLITICAL INFORMATION TABLE

This table gives the area, population, population density, political status, capital, and predominant languages for every country in the world. The political units listed are categorized by political status in the form of government column of the table, as follows: A—independent countries; B—internally independent political entities which are under the protection of another country in matters of defense and foreign affairs; C—colonies and other dependent political units; and D—the major administrative subdivisions

of Australia, Canada, China, the United Kingdom, and the United States. For comparison, the table also includes the continents and the world. All footnotes appear at the end of the table.

The populations are estimates for January 1, 2004, made by Rand McNally on the basis of official data, United States Census Bureau estimates, and other available information. Area figures include inland water.

REGION OR POLITICAL DIVISION	Area Sq. Mi.	Est. Pop. 1/1/04	Pop. Per Sq. Mi.	Form of Government and Ruling Power	Capital	Predominant Languages	International Organizations
Afars and Issas see Djibouti							
Afghanistan	251,773	29,205,000	116	Transitional ... A	Kābul	Dari, Pashto, Uzbek, Turkmen	UN
Africa	11,700,000	866,305,000	74				
Alabama	52,419	4,515,000	86	State (U.S.) ... D	Montgomery	English	
Alaska	663,267	650,000	1.0	State (U.S.) ... D	Juneau	English, indigenous	
Albania	11,100	3,535,000	318	Republic ... A	Tiranë	Albanian, Greek	NATO/PP, UN
Alberta	255,541	3,215,000	13	Province (Canada) ... D	Edmonton	English	
Algeria	919,595	33,090,000	36	Republic ... A	Algiers (El Djazaïr)	Arabic, Berber dialects, French	AL, AU, OPEC, UN
American Samoa	77	58,000	753	Unincorporated territory (U.S.) ... C	Pago Pago	Samoan, English	
Andorra	181	70,000	387	Parliamentary co-principality (Spanish and French) ... B	Andorra	Catalan, Spanish (Castilian), French, Portuguese	UN
Angola	481,354	10,875,000	23	Republic ... A	Luanda	Portuguese, indigenous	AU, COMESA, UN
Anguilla	37	13,000	351	Overseas territory (U.K.) ... C	The Valley	English	
Anhui	53,668	61,215,000	1,141	Province (China) ... D	Hefei	Chinese (Mandarin)	
Antarctica	5,400,000	(¹)					
Antigua and Barbuda	171	68,000	398	Parliamentary state ... A	St. John's	English, local dialects	OAS, UN
Aomen (Macau)	6.9	445,000	64,493	Special administrative region (China) ... D	Macau (Aomen)	Chinese (Cantonese), Portuguese	
Argentina	1,073,519	38,945,000	36	Republic ... A	Buenos Aires	Spanish, English, Italian, German, French	MERCOSUR, OAS, UN
Arizona	113,998	5,600,000	49	State (U.S.) ... D	Phoenix	English	
Arkansas	53,179	2,735,000	51	State (U.S.) ... D	Little Rock	English	
Armenia	11,506	3,325,000	289	Republic ... A	Yerevan	Armenian, Russian	CIS, NATO/PP, UN
Aruba	75	71,000	947	Self-governing territory (Netherlands protection) ... B	Oranjestad	Dutch, Papiamento, English, Spanish	
Ascension	34	1,000	29	Dependency (St. Helena) ... C	Georgetown	English	
Asia	17,300,000	3,839,320,000	222				
Australia	2,969,910	19,825,000	6.7	Federal parliamentary state ... A	Canberra	English, indigenous	ANZUS, UN
Australian Capital Territory	911	325,000	357	Territory (Australia) ... D	Canberra	English	
Austria	32,378	8,170,000	252	Federal republic ... A	Vienna (Wien)	German	EU, NATO/PP, UN
Azerbaijan	33,437	7,850,000	235	Republic ... A	Baku (Bakı)	Azeri, Russian, Armenian	CIS, NATO/PP, UN
Bahamas	5,382	300,000	56	Parliamentary state ... A	Nassau	English, Creole	OAS, UN
Bahrain	267	675,000	2,528	Monarchy ... A	Al Manāmah	Arabic, English, Persian, Urdu	AL, UN
Bangladesh	55,598	139,875,000	2,516	Republic ... A	Dkaha (Dacca)	Bangla, English	UN
Barbados	166	280,000	1,687	Parliamentary state ... A	Bridgetown	English	OAS, UN
Beijing (Peking)	6,487	14,135,000	2,179	Autonomous city (China) ... D	Beijing (Peking)	Chinese (Mandarin)	
Belarus	80,155	10,315,000	129	Republic ... A	Minsk	Belarussian, Russian	CIS, NATO/PP, UN
Belau see Palau							
Belgium	11,787	10,340,000	877	Constitutional monarchy ... A	Brussels (Bruxelles)	Dutch (Flemish), French, German	EU, NATO, UN
Belize	8,867	270,000	30	Parliamentary state ... A	Belmopan	English, Spanish, Mayan, Garifuna, Creole	OAS, UN
Benin	43,484	7,145,000	164	Republic ... A	Porto-Novo and Cotonou	French, Fon, Yoruba, indigenous	AU, UN
Bermuda	21	65,000	3,095	Overseas territory (U.K. protection) ... B	Hamilton	English, Portuguese	
Bhutan	17,954	2,160,000	120	Monarchy (Indian protection) ... B	Thimphu	Dzongkha, Tibetan and Nepalese dialects	UN
Bolivia	424,165	8,655,000	20	Republic ... A	La Paz and Sucre	Aymara, Quechua, Spanish	OAS, UN
Bosnia and Herzegovina	19,767	4,000,000	202	Republic ... A	Sarajevo	Bosnian, Serbian, Croatian	UN
Botswana	224,607	1,570,000	7.0	Republic ... A	Gaborone	English, Tswana	AU, UN
Brazil	3,300,172	183,080,000	55	Federal republic ... A	Brasília	Portuguese, Spanish, English, French	MERCOSUR, OAS, UN
British Columbia	364,764	4,245,000	12	Province (Canada) ... D	Victoria	English	
British Indian Ocean Territory	23	(¹)		Overseas territory (U.K.) ... C		English	
British Virgin Islands	58	22,000	379	Overseas territory (U.K.) ... C	Road Town	English	
Brunei	2,226	360,000	162	Monarchy ... A	Bandar Seri Begawan	Malay, English, Chinese	ASEAN, UN
Bulgaria	42,855	7,550,000	176	Republic ... A	Sofia (Sofiya)	Bulgarian, Turkish	NATO, UN
Burkina Faso	105,869	13,400,000	127	Republic ... A	Ouagadougou	French, indigenous	AU, UN
Burma see Myanmar							
Burundi	10,745	6,165,000	574	Republic ... A	Bujumbura	French, Kirundi, Swahili	AU, COMESA, UN
California	163,696	35,590,000	217	State (U.S.) ... D	Sacramento	English	
Cambodia	69,898	13,245,000	189	Constitutional monarchy ... A	Phnom Penh (Phnum Pénh)	Khmer, French, English	ASEAN, UN
Cameroon	183,568	15,905,000	87	Republic ... A	Yaoundé	English, French, indigenous	AU, UN
Canada	3,855,103	32,360,000	8.4	Federal parliamentary state ... A	Ottawa	English, French, other	NAFTA, NATO, OAS, UN
Cape Verde	1,557	415,000	267	Republic ... A	Praia	Portuguese, Crioulo	AU, UN
Cayman Islands	102	43,000	422	Overseas territory (U.K.) ... C	George Town	English	
Central African Republic	240,536	3,715,000	15	Republic ... A	Bangui	French, Sango, indigenous	AU, UN
Ceylon see Sri Lanka							
Chad	495,755	9,395,000	19	Republic ... A	N'Djamena	Arabic, French, indigenous	AU, UN
Channel Islands	75	155,000	2,067	Two crown dependencies (U.K. protection) ...		English, French	
Chile	291,930	15,745,000	54	Republic ... A	Santiago	Spanish	OAS, UN
China (excl. Taiwan)	3,690,045	1,298,720,000	352	Socialist republic ... A	Beijing (Peking)	Chinese dialects	UN
Chongqing	31,815	31,600,000	993	Autonomous city (China) ... D	Chongqing (Chungking)	Chinese (Mandarin)	
Christmas Island	52	400	7.7	External territory (Australia) ... C	Settlement	English, Chinese, Malay	
Cocos (Keeling) Islands	5.4	600	111	External territory (Australia) ... C	West Island	English, Cocos-Malay	
Colombia	439,737	41,985,000	95	Republic ... A	Bogotá	Spanish	OAS, UN
Colorado	104,094	4,565,000	44	State (U.S.) ... D	Denver	English	
Comoros (excl. Mayotte)	863	640,000	742	Republic ... A	Moroni	Arabic, French, Shikomoro	AL, AU, COMESA, UN
Congo	132,047	2,975,000	23	Republic ... A	Brazzaville	French, Lingala, Monokutuba, indigenous	AU, UN
Congo, Democratic Republic of the (Zaire)	905,446	57,445,000	63	Republic ... A	Kinshasa	French, Lingala, indigenous	AU, COMESA, UN
Connecticut	5,543	3,495,000	631	State (U.S.) ... D	Hartford	English	

REGION OR POLITICAL DIVISION	Area Sq. Mi.	Est. Pop. 1/1/04	Pop. Per Sq. Mi.	Form of Government and Ruling Power	Capital	Predominant Languages	International Organizations
Cook Islands	91	21,000	231	Self-governing territory (New Zealand protection) . . . B	Avarua	English, Maori	
Costa Rica	19,730	3,925,000	199	Republic . . . A	San José	Spanish, English	OAS, UN
Cote d'Ivoire (Ivory Coast)	124,504	17,145,000	138	Republic . . . A	Abidjan and Yamoussoukro	French, Dioula and other indigenous	AU, UN
Croatia	21,829	4,430,000	203	Republic . . . A	Zagreb	Croatian	NATO/PP, UN
Cuba	42,804	11,290,000	264	Socialist republic . . . A	Havana (La Habana)	Spanish	OAS, UN
Cyprus	3,572	775,000	217	Republic . . . A	Nicosia	Greek, Turkish, English	EU, UN
Czech Republic	30,450	10,250,000	337	Republic . . . A	Prague (Praha)	Czech	EU, NATO, UN
Delaware	2,489	820,000	329	State (U.S.) . . . D	Dover	English	
Denmark	16,640	5,405,000	325	Constitutional monarchy . . . A	Copenhagen (København)	Danish	EU, NATO, UN
District of Columbia	68	565,000	8,309	Federal district (U.S.) . . . D	Washington	English	
Djibouti	8,958	460,000	51	Republic . . . A	Djibouti	French, Arabic, Somali, Afar	AL, AU, COMESA, UN
Dominica	290	69,000	238	Republic . . . A	Roseau	English, French	OAS, UN
Dominican Republic	18,730	8,775,000	468	Republic . . . A	Santo Domingo	Spanish	OAS, UN
East Timor	5,743	1,010,000	176	Republic . . . A	Dili	Portuguese, Tetum, Bahasa Indonesia (Malay), English	UN
Ecuador	109,484	13,840,000	126	Republic . . . A	Quito	Spanish, Quechua, indigenous	OAS, UN
Egypt	386,662	75,420,000	195	Republic . . . A	Cairo (Al Qāhirah)	Arabic	AL, AU, CAEU, COMESA, UN
Ellice Islands see Tuvalu							
El Salvador	8,124	6,530,000	804	Republic . . . A	San Salvador	Spanish, Nahua	OAS, UN
England	50,356	50,360,000	1,000	Administrative division (U.K.) . . . D	London	English	
Equatorial Guinea	10,831	515,000	48	Republic . . . A	Malabo	French, Spanish, indigenous, English	AU, UN
Eritrea	45,406	4,390,000	97	Republic . . . A	Asmera	Afar, Arabic, Tigre, Kunama, Tigrinya, other	AU, COMESA, UN
Estonia	17,462	1,405,000	80	Republic . . . A	Tallinn	Estonian, Russian, Ukrainian, Finnish, other	EU, NATO, UN
Ethiopia	426,373	67,210,000	158	Federal republic . . . A	Addis Ababa (Adis Abeba)	Amharic, Tigrinya, Orominga, Guaraginga, Somali, Arabic	AU, COMESA, UN
Europe	3,800,000	729,330,000	192				
Falkland Islands (²)	4,700	3,000	0.6	Overseas territory (U.K.) . . . C	Stanley	English	
Faroe Islands	540	47,000	87	Self-governing territory (Danish protection) . . . B	Tórshavn	Danish, Faroese	
Fiji	7,056	875,000	124	Republic . . . A	Suva	English, Fijian, Hindustani	UN
Finland	130,559	5,210,000	40	Republic . . . A	Helsinki (Helsingfors)	Finnish, Swedish, Sami, Russian	EU, NATO/PP, UN
Florida	65,755	17,070,000	260	State (U.S.) . . . D	Tallahassee	English	
France (excl. Overseas Departments)	208,482	60,305,000	289	Republic . . . A	Paris	French	EU, NATO, UN
French Guiana	32,253	190,000	5.9	Overseas department (France) . . . C	Cayenne	French	
French Polynesia	1,544	265,000	172	Overseas territory (France) . . . C	Papeete	French, Tahitian	
Fujian	46,332	35,495,000	766	Province (China) . . . D	Fuzhou	Chinese dialects	
Gabon	103,347	1,340,000	13	Republic . . . A	Libreville	French, Fang, indigenous	AU, UN
Gambia, The	4,127	1,525,000	370	Republic . . . A	Banjul	English, Malinke, Wolof, Fula, indigenous	AU, UN
Gansu	173,746	26,200,000	151	Province (China) . . . D	Lanzhou	Chinese (Mandarin), Mongolian, Tibetan dialects	
Gaza Strip	139	1,300,000	9,353	Israeli territory with limited self-government . . .		Arabic, Hebrew	(⁴)
Georgia	59,425	8,710,000	147	State (U.S.) . . . D	Atlanta	English	
Georgia	26,911	4,920,000	183	Republic . . . A	Tbilisi	Georgian, Russian, Armenian, Azeri, other	NATO/PP, UN
Germany	137,847	82,415,000	598	Federal republic . . . A	Berlin	German	EU, NATO, UN
Ghana	92,098	20,615,000	224	Republic . . . A	Accra	English, Akan and other indigenous	AU, UN
Gibraltar (¹)	2.3	28,000	12,174	Overseas territory (U.K.) . . . C	Gibraltar	English, Spanish, Italian, Portuguese	
Gilbert Islands see Kiribati							
Golan Heights	454	37,000	81	Occupied by Israel . . .		Arabic, Hebrew	
Great Britain see United Kingdom							
Greece	50,949	10,635,000	209	Republic . . . A	Athens (Athína)	Greek, English, French	EU, NATO, UN
Greenland	836,331	56,000	0.07	Self-governing territory (Danish protection) . . . B	Godthåb (Nuuk)	Danish, Greenlandic, English	
Grenada	133	89,000	669	Parliamentary state . . . A	St. George's	English, French	OAS, UN
Guadeloupe (incl. Dependencies)	687	440,000	640	Overseas department (France) . . . C	Basse-Terre	French, Creole	
Guam	212	165,000	778	Unincorporated territory (U.S.) . . . C	Hagåtña (Agana)	English, Chamorro, Japanese	
Guangdong	68,649	88,375,000	1,287	Province (China) . . . D	Guangzhou (Canton)	Chinese dialects, Miao-Yao	
Guangxi Zhuangzu	91,236	45,905,000	503	Autonomous region (China) . . . D	Nanning	Chinese dialects, Thai, Miao-Yao	
Guatemala	42,042	14,095,000	335	Republic . . . A	Guatemala	Spanish, indigenous	OAS, UN
Guernsey (incl. Dependencies)	30	65,000	2,167	Crown dependency (U.K. protection) . . . B	St. Peter Port	English, French	
Guinea	94,926	9,135,000	96	Republic . . . A	Conakry	French, indigenous	AU, UN
Guinea-Bissau	13,948	1,375,000	99	Republic . . . A	Bissau	Portuguese, Crioulo, indigenous	AU, UN
Guizhou	65,637	36,045,000	549	Province (China) . . . D	Guiyang	Chinese (Mandarin), Thai, Miao-Yao	
Guyana	83,000	705,000	8.5	Republic . . . A	Georgetown	English, indigenous, Creole, Hindi, Urdu	OAS, UN
Hainan	13,205	8,050,000	610	Province (China) . . . D	Haikou	Chinese, Min, Tai	
Haiti	10,714	7,590,000	708	Republic . . . A	Port-au-Prince	Creole, French	OAS, UN
Hawaii	10,931	1,260,000	115	State (U.S.) . . . D	Honolulu	English, Hawaiian, Japanese	
Hebei	73,359	68,965,000	940	Province (China) . . . D	Shijiazhuang	Chinese (Mandarin)	
Heilongjiang	181,082	37,725,000	208	Province (China) . . . D	Harbin	Chinese dialects, Mongolian, Tungus	
Henan	64,479	94,655,000	1,468	Province (China) . . . D	Zhengzhou	Chinese (Mandarin)	
Holland see Netherlands							
Honduras	43,277	6,745,000	156	Republic . . . A	Tegucigalpa	Spanish, indigenous	OAS, UN
Hubei	72,356	61,645,000	852	Province (China) . . . D	Wuhan	Chinese dialects	
Hunan	81,082	65,855,000	812	Province (China) . . . D	Changsha	Chinese dialects, Miao-Yao	
Hungary	35,919	10,045,000	280	Republic . . . A	Budapest	Hungarian	EU, NATO, UN
Iceland	39,769	280,000	7.0	Republic . . . A	Reykjavik	Icelandic, English, other	EFTA, NATO, UN
Idaho	83,570	1,370,000	16	State (U.S.) . . . D	Boise	English	
Illinois	57,914	12,690,000	219	State (U.S.) . . . D	Springfield	English	
India (incl. part of Jammu and Kashmir)	1,222,510	1,057,415,000	865	Federal republic . . . A	New Delhi	English, Hindi, Telugu, Bengali, indigenous	UN
Indiana	36,418	6,215,000	171	State (U.S.) . . . D	Indianapolis	English	
Indonesia	735,310	236,680,000	322	Republic . . . A	Jakarta	Bahasa Indonesia (Malay), English, Dutch, indigenous	ASEAN, OPEC, UN
Iowa	56,272	2,955,000	53	State (U.S.) . . . D	Des Moines	English	
Iran	636,372	68,650,000	108	Islamic republic . . . A	Tehrān	Persian, Turkish dialects, Kurdish, other	OPEC, UN
Iraq	169,235	25,025,000	148	Republic . . . A	Baghdād	Arabic, Kurdish, Assyrian, Armenian	AL, CAEU, OPEC, UN
Ireland	27,133	3,945,000	145	Republic . . . A	Dublin (Baile Átha Cliath)	English, Irish Gaelic	EU, NATO/PP, UN
Isle of Man	221	74,000	335	Crown dependency (U.K. protection) . . . B	Douglas	English, Manx Gaelic	

REGION OR POLITICAL DIVISION	Area Sq. Mi.	Est. Pop. 1/1/04	Pop. Per Sq. Mi.	Form of Government and Ruling Power	Capital	Predominant Languages	International Organizations
Israel (excl. Occupied Areas)	8,019	6,160,000	768	Republic ... A	Jerusalem (Yerushalayim)	Hebrew, Arabic	UN
Italy	116,342	58,030,000	499	Republic ... A	Rome (Roma)	Italian, German, French, Slovene	EU, NATO, UN
Ivory Coast see Cote d'Ivoire							
Jamaica	4,244	2,705,000	637	Parliamentary state ... A	Kingston	English, Creole	OAS, UN
Japan	145,850	127,285,000	873	Constitutional monarchy ... A	Tōkyō	Japanese	UN
Jersey	45	90,000	2,000	Crown dependency (U.K. protection) ... B	St. Helier	English, French	
Jiangsu	39,614	76,065,000	1,920	Province (China) ... D	Nanjing (Nanking)	Chinese dialects	
Jiangxi	64,325	42,335,000	658	Province (China) ... D	Nanchang	Chinese dialects	
Jilin	72,201	27,895,000	386	Province (China) ... D	Changchun	Chinese (Mandarin), Mongolian, Korean	
Jordan	34,495	5,535,000	160	Constitutional monarchy ... A	'Ammān	Arabic	AL, CAEU, UN
Kansas	82,277	2,730,000	33	State (U.S.) ... D	Topeka	English	
Kazakhstan	1,049,156	16,780,000	16	Republic ... A	Astana (Aqmola)	Kazakh, Russian	CIS, NATO/PP, UN
Kentucky	40,409	4,130,000	102	State (U.S.) ... D	Frankfort	English	
Kenya	224,961	31,840,000	142	Republic ... A	Nairobi	English, Swahili, indigenous	AU, COMESA, UN
Kiribati	313	100,000	319	Republic ... A	Bairiki	English, I-Kiribati	UN
Korea, North	46,540	22,585,000	485	Socialist republic ... A	P'yŏngyang	Korean	UN
Korea, South	38,328	48,450,000	1,264	Republic ... A	Seoul (Sŏul)	Korean	UN
Kuwait	6,880	2,220,000	323	Constitutional monarchy ... A	Kuwait (Al Kuwayt)	Arabic, English	AL, CAEU, OPEC, UN
Kyrgyzstan	77,182	4,930,000	64	Republic ... A	Bishkek	Kirghiz, Russian	CIS, NATO/PP, UN
Laos	91,429	5,995,000	66	Socialist republic ... A	Viangchan (Vientiane)	Lao, French, English	ASEAN, UN
Latvia	24,942	2,340,000	94	Republic ... A	Riga	Latvian, Lithuanian, Russian, other	EU, NATO, UN
Lebanon	4,016	3,755,000	935	Republic ... A	Beirut (Bayrūt)	Arabic, French, Armenian, English	AL, UN
Lesotho	11,720	1,865,000	159	Constitutional monarchy ... A	Maseru	English, Sesotho, Zulu, Xhosa	AU, UN
Liaoning	56,255	43,340,000	770	Province (China) ... D	Shenyang (Mukden)	Chinese (Mandarin), Mongolian	
Liberia	43,000	3,345,000	78	Republic ... A	Monrovia	English, indigenous	AU, UN
Libya	679,362	5,565,000	8.2	Socialist republic ... A	Tripoli (Ṭarābulus)	Arabic	AL, AU, CAEU, OPEC, UN
Liechtenstein	62	33,000	532	Constitutional monarchy ... A	Vaduz	German	EFTA, UN
Lithuania	25,213	3,590,000	142	Republic ... A	Vilnius	Lithuanian, Polish, Russian	EU, NATO, UN
Louisiana	51,840	4,510,000	87	State (U.S.) ... D	Baton Rouge	English	
Luxembourg	999	460,000	460	Constitutional monarchy ... A	Luxembourg	French, Luxembourgish, German	EU, NATO, UN
Macedonia	9,928	2,065,000	208	Republic ... A	Skopje	Macedonian, Albanian, other	NATO/PP, UN
Madagascar	226,658	17,235,000	76	Republic ... A	Antananarivo	French, Malagasy	AU, COMESA, UN
Maine	35,385	1,310,000	37	State (U.S.) ... D	Augusta	English	
Malawi	45,747	11,780,000	258	Republic ... A	Lilongwe	Chichewa, English, indigenous	AU, COMESA, UN
Malaysia	127,320	23,310,000	183	Federal constitutional monarchy ... A	Kuala Lumpur and Putrajaya (¹)	Bahasa Melayu, Chinese dialects, English, other	ASEAN, UN
Maldives	115	335,000	2,913	Republic ... A	Male'	Dhivehi	UN
Mali	478,841	11,790,000	25	Republic ... A	Bamako	French, Bambara, indigenous	AU, UN
Malta	122	400,000	3,279	Republic ... A	Valletta	English, Maltese	EU, UN
Manitoba	250,116	1,190,000	4.8	Province (Canada) ... D	Winnipeg	English	
Marshall Islands	70	57,000	814	Republic (U.S. protection) ... B	Majuro (island)	English, indigenous, Japanese	UN
Martinique	425	430,000	1,012	Overseas department (France) ... C	Fort-de-France	French, Creole	
Maryland	12,407	5,525,000	445	State (U.S.) ... D	Annapolis	English	
Massachusetts	10,555	6,455,000	612	State (U.S.) ... D	Boston	English	
Mauritania	397,956	2,955,000	7.4	Republic ... A	Nouakchott	Arabic, Wolof, Pular, Soninke, French	AL, AU, CAEU, UN
Mauritius (incl. Dependencies)	788	1,215,000	1,542	Republic ... A	Port Louis	English, French, Creole, other	AU, COMESA, UN
Mayotte (⁶)	144	180,000	1,250	Departmental collectivity (France) ... C	Mamoutzou	French, Swahili (Mahorian)	
Mexico	758,452	104,340,000	138	Federal republic ... A	Mexico City (Ciudad de México)	Spanish, indigenous	NAFTA, OAS, UN
Michigan	96,716	10,110,000	105	State (U.S.) ... D	Lansing	English	
Micronesia, Federated States of	271	110,000	406	Republic (U.S. protection) ... B	Palikir	English, indigenous	UN
Midway Islands	2.0	(¹)		Unincorporated territory (U.S.) ... C		English	
Minnesota	86,939	5,075,000	58	State (U.S.) ... D	St. Paul	English	
Mississippi	48,430	2,890,000	60	State (U.S.) ... D	Jackson	English	
Missouri	69,704	5,720,000	82	State (U.S.) ... D	Jefferson City	English	
Moldova	13,070	4,440,000	340	Republic ... A	Chişinău (Kishinev)	Romanian (Moldovan), Russian, Gagauz	CIS, NATO/PP, UN
Monaco	0.8	32,000	40,000	Constitutional monarchy ... A	Monaco	French, English, Italian, Monegasque	UN
Mongolia	604,829	2,730,000	4.5	Republic ... A	Ulan Bator (Ulaanbaatar)	Khalkha Mongol, Turkish dialects, Russian	UN
Montana	4,095	920,000	225	State (U.S.) ... D	Helena	English	
Montserrat	39	9,000	231	Overseas territory (U.K.) ... C	Plymouth	English	
Morocco (excl. Western Sahara)	172,414	31,950,000	185	Constitutional monarchy ... A	Rabat	Arabic, Berber dialects, French	AL, UN
Mozambique	309,496	18,695,000	60	Republic ... A	Maputo	Portuguese, indigenous	AU, UN
Myanmar (Burma)	261,228	42,620,000	163	Provisional military government ... A	Rangoon (Yangon)	Burmese, indigenous	ASEAN, UN
Namibia	317,818	1,940,000	6.1	Republic ... A	Windhoek	English, Afrikaans, German, indigenous	AU, COMESA, UN
Nauru	8.1	13,000	1,605	Republic ... A	Yaren District	Nauruan, English	UN
Nebraska	77,354	1,745,000	23	State (U.S.) ... D	Lincoln	English	
Nei Mongol (Inner Mongolia)	456,759	24,295,000	53	Autonomous region (China) ... D	Hohhot	Mongolian	
Nepal	56,827	26,770,000	471	Constitutional monarchy ... A	Kathmandu	Nepali, indigenous	UN
Netherlands	16,164	16,270,000	1,007	Constitutional monarchy ... A	Amsterdam and The Hague ('s-Gravenhage)	Dutch, Frisian	EU, NATO, UN
Netherlands Antilles	309	215,000	696	Self-governing territory (Netherlands protection) ... B	Willemstad	Dutch, Papiamento, English, Spanish	
Nevada	110,561	2,250,000	20	State (U.S.) ... D	Carson City	English	
New Brunswick	28,150	770,000	27	Province (Canada) ... D	Fredericton	English, French	
New Caledonia	7,172	210,000	29	Territorial collectivity (France) ... C	Nouméa	French, indigenous	
Newfoundland and Labrador	156,453	535,000	3.4	Province (Canada) ... D	St. John's	English	
New Hampshire	9,350	1,290,000	138	State (U.S.) ... D	Concord	English	
New Hebrides see Vanuatu							
New Jersey	8,721	8,665,000	994	State (U.S.) ... D	Trenton	English	
New Mexico	121,590	1,880,000	15	State (U.S.) ... D	Santa Fe	English, Spanish	
New South Wales	309,129	6,665,000	22	State (Australia) ... D	Sydney	English	
New York	54,556	19,245,000	353	State (U.S.) ... D	Albany	English	
New Zealand	104,454	3,975,000	38	Parliamentary state ... A	Wellington	English, Maori	ANZUS, UN
Nicaragua	50,054	5,180,000	103	Republic ... A	Managua	Spanish, English, indigenous	OAS, UN
Niger	489,192	11,210,000	23	Republic ... A	Niamey	French, Hausa, Djerma, indigenous	AU, UN
Nigeria	356,669	135,570,000	380	Transitional military government ... A	Abuja	English, Hausa, Fulani, Yoruba, Ibo, indigenous	AU, OPEC, UN
Ningxia Huizu	25,637	5,745,000	224	Autonomous region (China) ... D	Yinchuan	Chinese (Mandarin)	
Niue	100	2,000	20	Self-governing territory (New Zealand protection) ... B	Alofi	Niuean, English	
Norfolk Island	14	2,000	143	External territory (Australia) ... C	Kingston	English, Norfolk	

REGION OR POLITICAL DIVISION	Area Sq. Mi.	Est. Pop. 1/1/04	Pop. Per Sq. Mi.	Form of Government and Ruling Power	Capital	Predominant Languages	International Organizations
North America	9,500,000	505,780,000	53				
North Carolina	53,819	8,430,000	157	State (U.S.) ... D	Raleigh	English	
North Dakota	70,700	635,000	9.0	State (U.S.) ... D	Bismarck	English	
Northern Ireland	5,242	1,725,000	329	Administrative division (U.K.) ... D	Belfast	English	
Northern Mariana Islands	179	77,000	430	Commonwealth (U.S. protection) ... B	Saipan (island)	English, Chamorro, Carolinian	
Northern Territory	520,902	200,000	0.4	Territory (Australia) ... D	Darwin	English, indigenous	
Northwest Territories	519,735	43,000	0.08	Territory (Canada) ... D	Yellowknife	English, indigenous	
Norway (incl. Svalbard and Jan Mayen)	125,050	4,565,000	37	Constitutional monarchy ... A	Oslo	Norwegian, Sami, Finnish	EFTA, NATO, UN
Nova Scotia	21,345	965,000	45	Province (Canada) ... D	Halifax	English	
Nunavut	808,185	30,000	0.04	Territory (Canada) ... D	Iqaluit	English, indigenous	
Oceania (incl. Australia)	3,300,000	32,170,000	9.7				
Ohio	44,825	11,470,000	256	State (U.S.) ... D	Columbus	English	
Oklahoma	69,898	3,520,000	50	State (U.S.) ... D	Oklahoma City	English	
Oman	119,499	2,855,000	24	Monarchy ... A	Muscat (Masqat)	Arabic, English, Baluchi, Urdu, Indian dialects	AL, UN
Ontario	415,599	12,495,000	30	Province (Canada) ... D	Toronto	English	
Oregon	98,381	3,570,000	36	State (U.S.) ... D	Salem	English	
Pakistan (incl. part of Jammu and Kashmir)	339,732	152,210,000	448	Federal Islamic republic ... A	Islāmābād	English, Urdu, Punjabi, Sindhi, Pashto, other	UN
Palau (Belau)	188	20,000	106	Republic (U.S. protection) ... B	Koror and Melekeok (²)	Angaur, English, Japanese, Palauan, Sonsorolese, Tobi	UN
Panama	29,157	2,980,000	102	Republic ... A	Panamá	Spanish, English	OAS, UN
Papua New Guinea	178,704	5,360,000	30	Parliamentary state ... A	Port Moresby	English, Motu, Pidgin, indigenous	UN
Paraguay	157,048	6,115,000	39	Republic ... A	Asunción	Guarani, Spanish	MERCOSUR, OAS, UN
Pennsylvania	46,055	12,400,000	269	State (U.S.) ... D	Harrisburg	English	
Peru	496,225	28,640,000	58	Republic ... A	Lima	Quechua, Spanish, Aymara	OAS, UN
Philippines	115,831	85,430,000	738	Republic ... A	Manila	English, Filipino, indigenous	ASEAN, UN
Pitcairn Islands (incl. Dependencies)	19	100	5.3	Overseas territory (U.K.) ... C	Adamstown	English, Pitcairnese	
Poland	120,728	38,625,000	320	Republic ... A	Warsaw (Warszawa)	Polish	EU, NATO, UN
Portugal	35,516	10,110,000	285	Republic ... A	Lisbon (Lisboa)	Portuguese, Mirandese	EU, NATO, UN
Prince Edward Island	2,185	140,000	64	Province (Canada) ... D	Charlottetown	English	
Puerto Rico	3,515	3,890,000	1,107	Commonwealth (U.S. protection) ... B	San Juan	Spanish, English	
Qatar	4,412	830,000	188	Monarchy ... A	Ad Dawḥah (Doha)	Arabic	AL, OPEC, UN
Qinghai	277,994	5,295,000	19	Province (China) ... D	Xining	Tibetan dialects, Mongolian, Turkish dialects, Chinese (Mandarin)	
Quebec	595,391	7,675,000	13	Province (Canada) ... D	Québec	French, English	
Queensland	668,208	3,785,000	5.7	State (Australia) ... D	Brisbane	English	
Reunion	969	760,000	784	Overseas department (France) ... C	Saint-Denis	French, Creole	
Rhode Island	1,545	1,080,000	699	State (U.S.) ... D	Providence	English	
Rhodesia see Zimbabwe							
Romania	91,699	22,370,000	244	Republic ... A	Bucharest (Bucureşti)	Romanian, Hungarian, German	NATO, UN
Russia	6,592,849	144,310,000	22	Federal republic ... A	Moscow (Moskva)	Russian, other	CIS, NATO/PP, UN
Rwanda	10,169	7,880,000	775	Republic ... A	Kigali	English, French, Kinyarwanda, Kiswahili	AU, COMESA, UN
St. Helena (incl. Dependencies)	121	7,500	62	Overseas territory (U.K.) ... C	Jamestown	English	
St. Kitts and Nevis	101	39,000	386	Parliamentary state ... A	Basseterre	English	OAS, UN
St. Lucia	238	165,000	693	Parliamentary state ... A	Castries	English, French	OAS, UN
St. Pierre and Miquelon	93	7,000	75	Territorial collectivity (France) ... C	Saint-Pierre	French	
St. Vincent and the Grenadines	150	115,000	767	Parliamentary state ... A	Kingstown	English, French	OAS, UN
Samoa	1,093	180,000	165	Constitutional monarchy ... A	Apia	English, Samoan	UN
San Marino	24	28,000	1,167	Republic ... A	San Marino	Italian	UN
Sao Tome and Principe	372	180,000	484	Republic ... A	São Tomé	Portuguese	AU, UN
Saskatchewan	251,366	1,025,000	4.1	Province (Canada) ... D	Regina	English	
Saudi Arabia	830,000	24,690,000	30	Monarchy ... A	Riyadh (Ar Riyāḑ)	Arabic	AL, OPEC, UN
Scotland	30,167	5,135,000	170	Administrative division (U.K.) ... D	Edinburgh	English, Scots Gaelic	
Senegal	75,951	10,715,000	141	Republic ... A	Dakar	French, Wolof and other indigenous	AU, UN
Serbia and Montenegro (Yugoslavia)	39,449	10,660,000	270	Republic ... A	Belgrade (Beograd)	Serbian, Albanian	UN
Seychelles	176	81,000	460	Republic ... A	Victoria	English, French, Creole	AU, COMESA, UN
Shaanxi	79,151	36,865,000	466	Province (China) ... D	Xi'an (Sian)	Chinese (Mandarin)	
Shandong	59,074	92,845,000	1,572	Province (China) ... D	Jinan	Chinese (Mandarin)	
Shanghai	2,394	17,120,000	7,151	Autonomous city (China) ... D	Shanghai	Chinese (Wu)	
Shanxi	60,232	33,715,000	560	Province (China) ... D	Taiyuan	Chinese (Mandarin)	
Sichuan	188,263	85,175,000	452	Province (China) ... D	Chengdu	Chinese (Mandarin), Tibetan dialects, Miao-Yao	
Sierra Leone	27,699	5,815,000	210	Republic ... A	Freetown	English, Krio, Mende, Temne, indigenous	AU, UN
Singapore	264	4,685,000	17,746	Republic ... A	Singapore	Chinese (Mandarin), English, Malay, Tamil	ASEAN, UN
Slovakia	18,924	5,420,000	286	Republic ... A	Bratislava	Slovak, Hungarian	EU, NATO, UN
Slovenia	7,821	1,935,000	247	Republic ... A	Ljubljana	Slovenian, Croatian, Serbian	EU, NATO, UN
Solomon Islands	10,954	515,000	47	Parliamentary state ... A	Honiara	English, indigenous	UN
Somalia	246,201	8,165,000	33	Transitional ... A	Mogadishu (Muqdisho)	Arabic, Somali, English, Italian	AL, AU, CAEU, UN
South Africa	470,693	42,770,000	91	Republic ... A	Pretoria, Cape Town, and Bloemfontein	Afrikaans, English, Xhosa, Zulu, other indigenous	AU, UN
South America	6,900,000	366,600,000	53				
South Australia	379,724	1,525,000	4.0	State (Australia) ... D	Adelaide	English	
South Carolina	32,020	4,160,000	130	State (U.S.) ... D	Columbia	English	
South Dakota	77,117	765,000	9.9	State (U.S.) ... D	Pierre	English	
South Georgia and the South Sandwich Islands (²)	1,450	(¹)	.....	Overseas territory (U.K.) ... C		English	
South West Africa see Namibia							
Spain	194,885	40,250,000	207	Constitutional monarchy ... A	Madrid	Spanish (Castilian), Catalan, Galician, Basque	EU, NATO, UN
Spanish North Africa (³)	12	140,000	11,667	Five possessions (Spain) ... C		Spanish, Arabic, Berber dialects	
Spanish Sahara see Western Sahara							
Sri Lanka	25,332	19,825,000	783	Socialist republic ... A	Colombo and Sri Jayewardenepura Kotte	English, Sinhala, Tamil	UN
Sudan	967,500	38,630,000	40	Provisional military government ... A	Khartoum (Al Kharṭūm)	Arabic, Nubian, and other indigenous, English	AL, AU, CAEU, COMESA, UN
Suriname	63,037	435,000	6.9	Republic ... A	Paramaribo	Dutch, Sranan Tongo, English, Hindustani, Javanese	OAS, UN

REGION OR POLITICAL DIVISION	Area Sq. Mi.	Est. Pop. 1/1/04	Pop. Per Sq. Mi.	Form of Government and Ruling Power	Capital	Predominant Languages	International Organizations
Swaziland	6,704	1,165,000	174	Monarchy ... A	Mbabane and Lobamba	English, siSwati	AU, COMESA, UN
Sweden	173,732	8,980,000	52	Constitutional monarchy ... A	Stockholm	Swedish, Sami, Finnish	EU, NATO/PP, UN
Switzerland	15,943	7,430,000	466	Federal republic ... A	Bern (Berne)	German, French, Italian, Romansch	EFTA, NATO/PP, UN
Syria	71,498	17,800,000	249	Republic ... A	Damascus (Dimashq)	Arabic, Kurdish, Armenian, Aramaic, Circassian	AL, CAEU, UN
Taiwan	13,901	22,675,000	1,631	Republic ... A	T'aipei	Chinese (Mandarin), Taiwanese (Min), Hakka	
Tajikistan	55,251	6,935,000	126	Republic ... A	Dushanbe	Tajik, Russian	CIS, NATO/PP, UN
Tanzania	364,900	36,230,000	99	Republic ... A	Dar es Salaam and Dodoma	English, Swahili, indigenous	AU, UN
Tasmania	26,409	475,000	18	State (Australia) ... D	Hobart	English	
Tennessee	42,143	5,860,000	139	State (U.S.) ... D	Nashville	English	
Texas	268,581	22,185,000	83	State (U.S.) ... D	Austin	English, Spanish	
Thailand	198,115	64,570,000	326	Constitutional monarchy ... A	Bangkok (Krung Thep)	Thai, indigenous	ASEAN, UN
Tianjin (Tientsin)	4,363	10,235,000	2,346	Autonomous city (China) ... D	Tianjin (Tientsin)	Chinese (Mandarin)	
Togo	21,925	5,495,000	251	Republic ... A	Lomé	French, Ewe, Mina, Kabye, Dagomba	AU, UN
Tokelau	4.6	1,500	326	Island territory (New Zealand) ... C		English, Tokelauan	
Tonga	251	110,000	438	Constitutional monarchy ... A	Nuku'alofa	Tongan, English	UN
Trinidad and Tobago	1,980	1,100,000	556	Republic ... A	Port of Spain	English, Hindi, French, Spanish, Chinese	OAS, UN
Tristan da Cunha	40	300	7.5	Dependency (St. Helena) ... C	Edinburgh	English	
Tunisia	63,170	9,980,000	158	Republic ... A	Tunis	Arabic, French	AL, AU, UN
Turkey	302,541	68,505,000	226	Republic ... A	Ankara	Turkish, Kurdish, Arabic, Armenian, Greek	NATO, UN
Turkmenistan	188,457	4,820,000	26	Republic ... A	Ashgabat (Ashkhabad)	Turkmen, Russian, Uzbek	CIS, NATO/PP, UN
Turks and Caicos Islands	166	20,000	120	Overseas territory (U.K.) ... C	Grand Turk	English	
Tuvalu	10	11,000	1,100	Parliamentary state ... A	Funafuti	Tuvaluan, English, Samoan, I-Kiribati	UN
Uganda	93,065	26,010,000	279	Republic ... A	Kampala	English, Luganda, Swahili, indigenous, Arabic	AU, COMESA, UN
Ukraine	233,090	47,890,000	205	Republic ... A	Kiev (Kyïv)	Ukrainian, Russian, Romanian, Polish, Hungarian	CIS, NATO/PP, UN
United Arab Emirates	32,278	2,505,000	78	Federation of monarchs ... A	Abū Ẓaby (Abu Dhabi)	Arabic, Persian, English, Hindi, Urdu	AL, CAEU, OPEC, UN
United Kingdom	93,788	60,185,000	642	Constitutional monarchy ... A	London	English, Welsh, Scots Gaelic	EU, NATO, UN
United States	3,794,083	291,680,000	77	Federal republic ... A	Washington	English, Spanish	ANZUS, NAFTA, NATO, OAS, UN
Upper Volta see Burkina Faso							
Uruguay	67,574	3,425,000	51	Republic ... A	Montevideo	Spanish	MERCOSUR, OAS, UN
Utah	84,899	2,360,000	28	State (U.S.) ... D	Salt Lake City	English	
Uzbekistan	172,742	26,195,000	152	Republic ... A	Tashkent (Toshkent)	Uzbek, Russian, Tajik	CIS, NATO/PP, UN
Vanuatu	4,707	200,000	42	Republic ... A	Port Vila	Bislama, English, French	UN
Vatican City	0.2	900	4,500	Ecclesiastical state ... A	Vatican City	Italian, Latin, French, other	
Venezuela	352,145	24,835,000	71	Federal republic ... A	Caracas	Spanish, indigenous	OAS, OPEC, UN
Vermont	9,614	620,000	64	State (U.S.) ... D	Montpelier	English	
Victoria	87,807	4,905,000	56	State (Australia) ... D	Melbourne	English	
Vietnam	128,066	82,150,000	641	Socialist republic ... A	Hanoi	Vietnamese, English, French, Chinese, Khmer, indigenous	ASEAN, UN
Virginia	42,774	7,410,000	173	State (U.S.) ... D	Richmond	English	
Virgin Islands (U.S.)	134	110,000	821	Unincorporated territory (U.S.) ... C	Charlotte Amalie	English, Spanish, Creole	
Wake Island	3.0	(¹)		Unincorporated territory (U.S.) ... C		English	
Wales	8,023	2,965,000	370	Administrative division (U.K.) ... D	Cardiff	English, Welsh Gaelic	
Wallis and Futuna	99	16,000	162	Overseas territory (France) ... C	Mata-Utu	French, Wallisian	
Washington	71,300	6,150,000	86	State (U.S.) ... D	Olympia	English	
West Bank (incl. Jericho and East Jerusalem)	2,263	2,275,000	1,005	Israeli territory with limited self-government		Arabic, Hebrew	(⁴)
Western Australia	976,792	1,945,000	2.0	State (Australia) ... D	Perth	English	
Western Sahara	102,703	265,000	2.6	Occupied by Morocco ... C		Arabic	
West Virginia	24,230	1,815,000	75	State (U.S.) ... D	Charleston	English	
Wisconsin	65,498	5,490,000	84	State (U.S.) ... D	Madison	English	
Wyoming	97,814	505,000	5.2	State (U.S.) ... D	Cheyenne	English	
Xianggang (Hong Kong)	425	7,440,000	17,506	Special administrative region (China) ... D	Hong Kong (Xianggang)	Chinese (Cantonese), English	
Xinjiang Uygur (Sinkiang)	617,764	19,685,000	32	Autonomous region (China) ... D	Ürümqi	Turkish dialects, Mongolian, Tungus, English	
Xizang (Tibet)	471,045	2,680,000	5.7	Autonomous region (China) ... D	Lhasa	Tibetan dialects	
Yemen	203,850	19,680,000	97	Republic ... A	Şan'ā' (Sanaa)	Arabic	AL, CAEU, UN
Yugoslavia see Serbia and Montenegro							
Yukon Territory	186,272	32,000	0.2	Territory (Canada) ... D	Whitehorse	English, Inuktitut, indigenous	
Yunnan	152,124	43,850,000	288	Province (China) ... D	Kunming	Chinese (Mandarin), Tibetan dialects, Khmer, Miao-Yao	
Zaire see Congo, Democratic Republic of the							
Zambia	290,586	10,385,000	36	Republic ... A	Lusaka	English, indigenous	AU, COMESA, UN
Zhejiang	39,305	47,830,000	1,217	Province (China) ... D	Hangzhou	Chinese dialects	
Zimbabwe	150,873	12,630,000	84	Republic ... A	Harare (Salisbury)	English, indigenous	AU, COMESA, UN
WORLD	57,900,000	6,339,505,000	109				

... None, or not applicable
(1) No permanent population
(2) Claimed by Argentina
(3) Claimed by Spain
(4) The Palestinian Liberation Organization (PLO) is a member of AL and CAEU
(5) Future capital
(6) Claimed by Comoros
(7) Comprises Ceuta, Melilla, and several small islands

AL	Arab League (League of Arab States)
ANZUS	Australia-New Zealand-U.S. Security Treaty
ASEAN	Association of Southeast Asian Nations
AU	African Union
CAEU	Council of Arab Unity
CIS	Commonwealth of Independent States
COMESA	Common Market for Eastern and Southern Africa
EFTA	European Free Trade Association
EU	European Union
MERCOSUR	Southern Common Market
NAFTA	North American Free Trade Agreement
NATO	North Atlantic Treaty Organization
NATO/PP	NATO-Partnership for Peace Program
OAS	Organization of American States
OPEC	Organization of Petroleum Exporting Countries

WORLD DEMOGRAPHIC TABLE

CONTINENT/Country	Population Estimate 2004	Pop. Per Sq. Mile 2004	Percent Urban[1] 2001	Crude Birth Rate per 1,000[2] 2003	Crude Death Rate per 1,000[2] 2003	Natural Increase Percent[2] 2003	Fertility Rate (Children born/Woman)[3] 2003	Infant Mortality Rate per 1,000[3] 2003	Median Age[2] 2002	Life Expectancy Male[2] 2003	Life Expectancy Female[2] 2003
NORTH AMERICA											
Bahamas	300,000	56	64.7	19	9	1.0%	2	26	27	62	69
Belize	270,000	30	48.1	30	6	2.4%	4	27	19	65	70
Canada	32,360,000	8	78.9	11	8	0.3%	2	5	38	76	83
Costa Rica	3,925,000	199	59.5	19	4	1.5%	2	11	25	74	79
Cuba	11,290,000	264	75.5	12	7	0.5%	2	7	35	75	79
Dominica	69,000	238	71.4	17	7	1.0%	2	15	28	71	77
Dominican Republic	8,775,000	468	66.0	24	7	1.7%	3	34	24	66	70
El Salvador	6,530,000	804	61.5	28	6	2.2%	3	27	21	67	74
Guatemala	14,095,000	335	39.9	35	7	2.8%	5	38	18	64	66
Haiti	7,590,000	708	36.3	34	13	2.1%	5	76	18	50	53
Honduras	6,745,000	156	53.7	32	6	2.5%	4	30	19	65	68
Jamaica	2,705,000	637	56.6	17	5	1.2%	2	13	27	74	78
Mexico	104,340,000	138	74.6	22	5	1.7%	3	22	24	72	78
Nicaragua	5,180,000	103	56.5	26	5	2.2%	3	31	20	68	72
Panama	2,980,000	102	56.5	21	6	1.5%	3	21	26	70	75
St. Lucia	165,000	693	38.0	21	5	1.6%	2	14	24	70	77
Trinidad and Tobago	1,100,000	556	74.5	13	9	0.4%	2	25	30	67	72
United States	291,680,000	77	77.4	14	8	0.6%	2	7	36	74	80
SOUTH AMERICA											
Argentina	38,945,000	36	88.3	17	8	1.0%	2	16	29	72	79
Bolivia	8,655,000	20	62.9	26	8	1.8%	3	56	21	62	67
Brazil	183,080,000	55	81.7	18	6	1.2%	2	32	27	67	75
Chile	15,745,000	54	86.1	16	6	1.0%	2	9	30	73	80
Colombia	41,985,000	95	75.5	22	6	1.6%	3	22	26	67	75
Ecuador	13,840,000	126	63.4	25	5	2.0%	3	32	23	69	75
Guyana	705,000	9	36.7	18	9	0.9%	2	38	26	61	66
Paraguay	6,115,000	39	56.7	30	5	2.6%	4	28	21	72	77
Peru	28,640,000	58	73.1	23	6	1.7%	3	37	24	68	73
Suriname	435,000	7	74.8	19	7	1.3%	2	25	26	67	72
Uruguay	3,425,000	51	92.1	17	9	0.8%	2	14	32	73	79
Venezuela	24,835,000	71	87.2	20	5	1.5%	2	24	25	71	77
EUROPE											
Albania	3,535,000	318	42.9	15	5	1.0%	2	23	27	74	80
Austria	8,170,000	252	67.4	9	9	0%	1	5	39	76	82
Belarus	10,315,000	129	69.6	10	14	-0.4%	1	14	37	63	75
Belgium	10,340,000	877	97.4	11	10	0.1%	2	5	40	75	82
Bosnia and Herzegovina	4,000,000	202	43.4	13	8	0.4%	2	23	36	70	75
Bulgaria	7,550,000	176	67.4	10	14	-0.5%	1	22	41	68	75
Croatia	4,430,000	203	58.1	13	11	0.2%	2	7	39	71	78
Czech Republic	10,250,000	337	74.5	9	11	-0.1%	1	4	38	72	79
Denmark	5,405,000	325	85.1	12	11	0.1%	2	5	39	75	80
Estonia	1,405,000	80	69.4	9	13	-0.4%	1	12	38	64	77
Finland	5,210,000	40	58.5	11	10	0.1%	2	4	40	75	82
France	60,305,000	289	75.5	13	9	0.3%	2	4	38	76	83
Germany	82,415,000	598	87.7	9	10	-0.2%	1	4	41	75	82
Greece	10,635,000	209	60.3	10	10	0%	1	6	40	76	81
Hungary	10,045,000	280	64.8	10	13	-0.3%	1	9	38	68	77
Iceland	280,000	7	92.7	14	7	0.7%	2	4	34	78	82
Ireland	3,945,000	145	59.3	14	8	0.6%	2	6	33	75	80
Italy	58,030,000	499	67.1	9	10	-0.1%	1	6	41	76	83
Latvia	2,340,000	94	59.8	9	15	-0.6%	1	15	39	63	75
Lithuania	3,590,000	142	68.6	10	13	-0.2%	1	14	37	64	76
Luxembourg	460,000	460	91.9	12	8	0.4%	2	5	38	75	82
Macedonia	2,065,000	208	59.4	13	8	0.5%	2	12	33	72	77
Moldova	4,440,000	340	41.4	14	13	0.2%	2	42	32	61	69
Netherlands	16,270,000	1,007	89.6	12	9	0.3%	2	5	39	76	81
Norway	4,565,000	37	75.0	12	10	0.3%	2	4	38	77	82
Poland	38,625,000	320	62.5	10	10	0.1%	1	9	36	70	78
Portugal	10,110,000	285	65.8	11	10	0.1%	1	6	38	73	80
Romania	22,370,000	244	55.2	11	12	-0.1%	1	28	35	67	75
Serbia and Montenegro	10,660,000	270	51.7	13	11	0.2%	2	17	36	71	77
Slovakia	5,420,000	286	57.6	10	10	0.1%	1	8	35	70	78
Slovenia	1,935,000	247	49.1	9	10	-0.1%	1	4	39	72	80
Spain	40,250,000	207	77.8	10	9	0.1%	1	5	39	76	83
Sweden	8,980,000	52	83.3	11	10	0%	2	3	40	78	83
Switzerland	7,430,000	466	67.3	10	8	0.1%	1	4	40	77	83
Ukraine	47,890,000	205	68.0	10	16	-0.7%	1	21	38	61	72
United Kingdom	60,185,000	642	89.5	11	10	0.1%	2	5	38	76	81
Russia	144,310,000	22	72.9	10	14	-0.4%	1	20	38	62	73
ASIA											
Afghanistan	29,205,000	116	22.3	41	17	2.3%	6	142	19	48	46
Armenia	3,325,000	289	67.2	13	10	0.2%	2	41	32	62	71
Azerbaijan	7,850,000	235	51.8	19	10	1.0%	2	82	27	59	68
Bahrain	675,000	2,528	92.5	19	4	1.5%	3	19	29	71	76
Bangladesh	139,875,000	2,516	25.6	30	9	2.1%	3	66	21	61	61
Brunei	360,000	162	72.8	20	3	1.6%	2	14	26	72	77
Cambodia	13,245,000	189	17.5	27	9	1.8%	4	76	19	55	60
China	1,298,720,000	352	37.1	13	7	0.6%	2	25	32	70	74
Cyprus	775,000	217	70.2	13	8	0.5%	2	8	34	75	80
East Timor	1,010,000	176	7.5	28	6	2.1%	4	50	20	63	68
Georgia	4,920,000	183	56.5	12	15	-0.3%	2	51	35	61	68
India	1,057,415,000	865	27.9	23	8	1.5%	3	60	24	63	64
Indonesia	236,680,000	322	42.1	21	6	1.5%	3	38	26	67	71
Iran	68,650,000	108	64.7	17	6	1.2%	2	44	23	68	71
Iraq	25,025,000	148	67.4	34	6	2.8%	5	55	19	67	69
Israel	6,160,000	768	91.8	19	6	1.2%	3	7	29	77	81
Japan	127,285,000	873	78.9	10	9	0.1%	1	3	42	78	84
Jordan	5,535,000	160	78.7	24	3	2.1%	3	19	22	75	81
Kazakhstan	16,780,000	16	55.8	18	11	0.8%	2	59	28	58	69
Korea, North	22,585,000	485	60.5	18	7	1.1%	2	26	31	68	74
Korea, South	48,450,000	1,264	82.5	13	6	0.7%	2	7	33	72	79
Kuwait	2,220,000	323	96.1	22	2	1.9%	3	11	26	76	78

CONTINENT/Country	Population Estimate 2004	Pop. Per Sq. Mile 2004	Percent Urban[1] 2001	Crude Birth Rate per 1,000[2] 2003	Crude Death Rate per 1,000[2] 2003	Natural Increase Percent[2] 2003	Fertility Rate (Children born/Woman)[3] 2003	Infant Mortality Rate per 1,000[3] 2003	Median Age[2] 2002	Life Expectancy Male[2] 2003	Life Expectancy Female[2] 2003
Kyrgyzstan	4,930,000	64	34.3	26	9	1.7%	3	75	23	59	68
Laos	5,995,000	66	19.7	37	12	2.5%	5	89	19	52	56
Lebanon	3,755,000	935	90.1	20	6	1.3%	2	26	26	70	75
Malaysia	23,310,000	183	58.1	24	5	1.9%	3	19	24	69	75
Mongolia	2,730,000	5	56.6	21	7	1.4%	2	57	24	62	66
Myanmar	42,620,000	163	28.1	19	12	0.7%	2	70	25	54	58
Nepal	26,770,000	471	12.2	32	10	2.3%	4	71	20	59	59
Oman	2,855,000	24	76.5	37	4	3.4%	6	21	19	70	75
Pakistan	152,210,000	448	33.4	30	9	2.1%	4	77	20	61	63
Philippines	85,430,000	738	59.4	26	6	2.1%	3	25	22	66	72
Qatar	830,000	188	92.9	16	4	1.1%	3	20	31	71	76
Saudi Arabia	24,690,000	30	86.7	37	6	3.1%	6	48	19	67	71
Singapore	4,685,000	17,746	100.0	13	4	0.8%	1	4	35	77	84
Sri Lanka	19,825,000	783	23.1	16	6	1.0%	2	15	29	70	75
Syria	17,800,000	249	51.8	30	5	2.5%	4	32	20	68	71
Taiwan	22,675,000	1,631	(5)	13	6	0.7%	2	7	33	74	80
Tajikistan	6,935,000	126	27.7	33	8	2.4%	4	113	19	61	68
Thailand	64,570,000	326	20.0	16	7	1.0%	2	22	30	69	74
Turkey	68,505,000	226	66.2	18	6	1.2%	2	44	27	69	74
Turkmenistan	4,820,000	26	44.9	28	9	1.9%	4	73	21	58	65
United Arab Emirates	2,505,000	78	87.2	18	4	1.4%	3	16	28	72	77
Uzbekistan	26,195,000	152	36.6	26	8	1.8%	3	72	22	61	68
Vietnam	82,150,000	641	24.5	20	6	1.3%	2	31	25	68	73
Yemen	19,680,000	97	25.0	43	9	3.4%	7	65	16	59	63
AFRICA											
Algeria	33,090,000	36	57.7	22	5	1.7%	3	38	23	69	72
Angola	10,875,000	23	34.9	46	26	2.0%	6	194	18	36	38
Benin	7,145,000	164	43.0	43	14	3.0%	6	87	16	50	52
Botswana	1,570,000	7	49.4	26	31	-0.6%	3	67	19	32	32
Burkina Faso	13,400,000	127	16.9	45	19	2.6%	6	100	17	43	46
Burundi	6,165,000	574	9.3	40	18	2.2%	6	72	16	43	44
Cameroon	15,905,000	87	49.7	35	15	2.0%	5	70	18	47	49
Cape Verde	415,000	267	63.5	27	7	2.0%	4	51	19	67	73
Central African Republic	3,715,000	15	41.7	36	20	1.6%	5	93	18	40	43
Chad	9,395,000	19	24.1	47	16	3.1%	6	96	16	47	50
Comoros	640,000	742	33.8	39	9	3.0%	5	80	19	59	64
Congo	2,975,000	23	66.1	29	14	1.5%	4	95	20	49	51
Congo, Democratic Republic of the	57,445,000	63	30.7	45	15	3.0%	7	97	16	47	51
Cote d'Ivoire	17,145,000	138	44.0	40	18	2.2%	6	98	17	40	45
Djibouti	460,000	51	84.2	41	19	2.1%	6	107	18	42	44
Egypt	75,420,000	195	42.7	24	5	1.9%	3	35	23	68	73
Equatorial Guinea	515,000	48	49.3	37	13	2.4%	5	89	19	53	57
Eritrea	4,390,000	97	19.1	39	13	2.6%	6	76	18	51	55
Ethiopia	67,210,000	158	15.9	40	20	2.0%	6	103	17	40	42
Gabon	1,340,000	13	82.3	37	11	2.5%	5	55	19	55	59
Gambia, The	1,525,000	370	31.3	41	12	2.8%	6	75	17	52	56
Ghana	20,615,000	224	36.4	26	11	1.5%	3	53	20	56	57
Guinea	9,135,000	96	27.9	43	16	2.7%	6	93	18	48	51
Guinea-Bissau	1,375,000	99	32.3	38	17	2.2%	5	110	19	45	49
Kenya	31,840,000	142	34.4	29	16	1.3%	3	63	18	45	45
Lesotho	1,865,000	159	28.8	27	25	0.3%	4	86	20	37	37
Liberia	3,345,000	78	45.5	45	18	2.7%	6	132	18	47	49
Libya	5,565,000	8	88.0	27	3	2.4%	3	27	22	74	78
Madagascar	17,235,000	76	30.1	42	12	3.0%	6	80	17	54	59
Malawi	11,780,000	258	15.1	45	23	2.2%	6	105	16	38	38
Mali	11,790,000	25	30.9	48	19	2.9%	7	119	16	45	46
Mauritania	2,955,000	7	59.1	42	13	2.9%	6	74	17	50	54
Mauritius	1,215,000	1,542	41.6	16	7	0.9%	2	16	30	68	76
Morocco	31,950,000	185	56.1	23	6	1.7%	3	45	23	68	72
Mozambique	18,695,000	60	33.3	37	23	1.4%	5	138	19	39	37
Namibia	1,940,000	6	31.4	34	19	1.5%	5	68	18	44	41
Niger	11,210,000	23	21.1	50	22	2.8%	7	124	16	42	42
Nigeria	135,570,000	380	44.9	39	14	2.5%	5	71	18	51	51
Rwanda	7,880,000	775	6.3	40	22	1.8%	6	103	18	39	40
Sao Tome and Principe	180,000	484	47.7	42	7	3.5%	6	46	16	65	68
Senegal	10,715,000	141	48.2	36	11	2.5%	5	58	18	55	58
Sierra Leone	5,815,000	210	37.3	44	21	2.3%	6	147	18	40	45
Somalia	8,165,000	33	27.9	46	18	2.9%	7	120	18	46	49
South Africa	42,770,000	91	57.7	19	18	0%	2	61	25	47	47
Sudan	38,630,000	40	37.1	36	10	2.7%	5	66	18	57	59
Swaziland	1,165,000	174	26.7	29	21	0.8%	4	67	19	41	38
Tanzania	36,230,000	99	33.3	40	17	2.2%	5	104	18	43	46
Togo	5,495,000	251	33.9	35	12	2.4%	5	69	17	51	55
Tunisia	9,980,000	158	66.2	17	5	1.2%	2	27	26	73	76
Uganda	26,010,000	279	14.5	47	17	3.0%	7	88	15	43	46
Zambia	10,385,000	36	39.8	40	24	1.5%	5	99	17	35	35
Zimbabwe	12,630,000	84	36.0	30	22	0.8%	4	66	19	40	38
OCEANIA											
Australia	19,825,000	7	91.2	13	7	0.5%	2	5	36	77	83
Fiji	875,000	124	50.2	23	6	1.7%	3	13	24	66	71
Kiribati	100,000	319	38.6	31	9	2.3%	4	51	20	58	64
Micronesia, Federated States of	110,000	406	28.6	26	5	2.1%	4	32	19[4]	67	71
New Zealand	3,975,000	38	85.9	14	8	0.7%	2	6	33	75	81
Papua New Guinea	5,360,000	30	17.6	31	8	2.3%	4	55	21	62	66
Samoa	180,000	165	22.3	15	6	0.9%	3	30	24	67	73
Solomon Islands	515,000	47	20.2	32	4	2.8%	4	23	18	70	75
Tonga	110,000	438	33.0	25	6	1.9%	3	13	20	66	71
Vanuatu	200,000	42	22.1	24	8	1.6%	3	58	22	60	63

This table presents data for most independent nations having an area greater than 200 square miles

(1) Source: United Nations World Urbanization Prospects
(2) Source: United States Census Bureau International Database
(3) Source: United States Central Intelligence Agency World Factbook
(4) 2000 Census preliminary count from www.fsmgov.org/info/people.html
(5) Data for Taiwan is included with China

WORLD AGRICULTURE TABLE

CONTINENT/Country	Total Area Sq. Miles	Agricultural Area 2001 Cropland Area[1] Sq. Miles	Cropland Area[1] %	Pasture Area[1] Sq. Miles	Pasture Area[1] %	Average Production 1999-2001 Wheat[1] 1,000 metric tons	Rice[1] 1,000 metric tons	Corn[1] 1,000 metric tons	Average 1999-2001 Cattle[1] 1,000	Pigs[1] 1,000	Sheep[1] 1,000
NORTH AMERICA											
Bahamas	5,382	46	0.9%	8	0.1%	-	-	-	1	5	6
Belize	8,867	402	4.5%	193	2.2%	-	12	36	52	25	4
Canada	3,855,103	177,144	4.6%	111,970	2.9%	24,676	-	8,168	13,340	12,970	819
Costa Rica	19,730	2,027	10.3%	9,035	45.8%	-	267	20	1,358	438	3
Cuba	42,804	17,239	40.3%	8,494	19.8%	-	342	207	4,305	2,600	310
Dominica	290	77	26.6%	8	2.7%	-	-	-	13	5	8
Dominican Republic	18,730	6,162	32.9%	8,108	43.3%	-	615	30	2,026	548	106
El Salvador	8,124	3,514	43.2%	3,066	37.7%	-	47	605	1,190	195	5
Guatemala	42,042	7,355	17.5%	10,046	23.9%	9	46	1,057	2,500	1,417	270
Haiti	10,714	4,247	39.6%	1,892	17.7%	-	111	211	1,390	934	147
Honduras	43,277	5,514	12.7%	5,822	13.5%	1	9	509	1,737	474	14
Jamaica	4,244	1,097	25.8%	884	20.8%	-	-	2	400	180	1
Mexico	758,452	105,406	13.9%	308,882	40.7%	3,263	324	18,466	30,428	16,112	6,048
Nicaragua	50,054	8,382	16.7%	18,591	37.1%	-	234	374	2,008	402	4
Panama	29,157	2,683	9.2%	5,927	20.3%	-	237	71	1,348	279	-
St. Lucia	238	69	29.2%	8	3.2%	-	-	-	12	15	13
Trinidad and Tobago	1,980	471	23.8%	42	2.1%	-	13	5	36	41	12
United States	3,794,083	684,401	18.0%	903,479	23.8%	58,862	9,222	244,296	98,197	60,229	7,071
SOUTH AMERICA											
Argentina	1,073,519	135,136	12.6%	548,265	51.1%	15,642	1,140	15,217	49,299	4,200	13,588
Bolivia	424,165	11,973	2.8%	130,618	30.8%	121	281	607	6,715	2,786	8,743
Brazil	3,300,172	256,623	7.8%	760,621	23.0%	2,461	10,998	35,119	170,295	30,608	14,728
Chile	291,930	8,880	3.0%	49,942	17.1%	1,490	113	685	4,117	2,395	4,153
Colombia	439,737	16,405	3.7%	161,391	36.7%	37	2,262	1,128	25,274	2,726	2,247
Ecuador	109,484	11,525	10.5%	19,653	18.0%	19	1,340	483	5,261	2,654	2,214
Guyana	83,000	1,969	2.4%	4,749	5.7%	-	560	3	220	20	130
Paraguay	157,048	12,008	7.6%	83,784	53.3%	256	112	804	9,758	2,633	402
Peru	496,225	16,255	3.3%	104,634	21.1%	180	1,963	1,205	4,936	2,795	14,414
Suriname	63,037	259	0.4%	81	0.1%	-	178	-	128	22	8
Uruguay	67,574	5,174	7.7%	52,290	77.4%	284	1,189	190	10,446	375	13,257
Venezuela	352,145	13,158	3.7%	70,425	20.0%	1	696	1,547	14,620	5,555	780
EUROPE											
Albania	11,100	2,699	24.3%	1,699	15.3%	298	-	203	719	96	1,929
Austria	32,378	5,676	17.5%	7,413	22.9%	1,412	-	1,774	2,166	3,556	357
Belarus	80,155	24,151	30.1%	11,564	14.4%	903	-	13	4,411	3,565	96
Belgium	11,787	3,344[2]	26.2%[2]	2,618[2]	20.5%[2]	1,535	-	420	3,165	7,462	150
Bosnia and Herzegovina	19,767	3,243	16.4%	4,633	23.4%	289	-	656	448	345	645
Bulgaria	42,855	17,900	41.8%	6,236	14.6%	3,071	8	1,137	664	1,459	2,536
Croatia	21,829	6,124	28.1%	6,035	27.6%	852	-	1,958	435	1,276	519
Czech Republic	30,450	12,788	42.0%	3,730	12.2%	4,196	-	324	1,604	3,761	87
Denmark	16,640	8,880	53.4%	1,452	8.7%	4,683	-	-	1,887	12,052	147
Estonia	17,462	2,691	15.4%	745	4.3%	123	-	-	276	304	29
Finland	130,559	8,490	6.5%	77	0.1%	427	-	-	1,060	1,303	101
France	208,482	75,618	36.3%	38,788	18.6%	35,327	110	15,928	20,377	14,693	9,754
Germany	137,847	46,409	33.7%	19,355	14.0%	21,358	-	3,362	14,723	26,021	2,746
Greece	50,949	14,873	29.2%	17,954	35.2%	2,111	153	2,007	584	925	8,977
Hungary	35,919	18,548	51.6%	4,097	11.4%	3,843	9	6,664	845	5,216	991
Iceland	39,769	27	0.1%	8,780	22.1%	-	-	-	72	44	477
Ireland	27,133	4,050	14.9%	12,934	47.7%	688	-	-	6,613	1,765	5,311
Italy	116,342	42,379	36.4%	16,907	14.5%	7,239	1,310	10,222	7,167	8,356	11,000
Latvia	24,942	7,220	28.9%	2,355	9.4%	410	-	-	393	407	28
Lithuania	25,213	11,541	45.8%	1,923	7.6%	1,062	-	-	856	984	14
Luxembourg	999	[3]	[3]	[3]	[3]	-	-	2	134	-	-
Macedonia	9,928	2,363	23.8%	2,432	24.5%	308	20	135	267	209	1,285
Moldova	13,070	8,398	64.3%	1,483	11.3%	902	-	1,096	423	646	929
Netherlands	16,164	3,622	22.4%	3,834	23.7%	995	-	148	4,108	13,253	1,335
Norway	125,050	3,398	2.7%	625	0.5%	265	-	-	1,017	414	2,342
Poland	120,728	55,267	45.8%	15,745	13.0%	8,946	-	962	6,124	17,588	366
Portugal	35,516	10,444	29.4%	5,548	15.6%	295	146	907	1,415	2,346	4,337
Romania	91,699	38,305	41.8%	19,039	20.8%	5,610	3	8,317	3,021	5,946	8,062
Serbia and Montenegro	39,449	14,394	36.5%	7,197	18.2%	2,207	-	5,013	1,550	4,012	1,853
Slovakia	18,924	6,085	32.2%	3,375	17.8%	1,445	-	612	671	1,548	344
Slovenia	7,821	784	10.0%	1,185	15.2%	153	-	283	473	585	80
Spain	194,885	69,298	35.6%	44,209	22.7%	5,785	844	4,208	6,140	22,079	24,185
Sweden	173,732	10,413	6.0%	1,726	1.0%	2,135	-	-	1,683	1,975	440
Switzerland	15,943	1,683	10.6%	4,417	27.7%	535	-	214	1,603	1,499	421
Ukraine	233,090	129,321	55.5%	30,541	13.1%	15,043	74	3,075	10,591	9,270	1,074
United Kingdom	93,788	22,019	23.5%	43,440	46.3%	14,380	-	-	11,052	6,537	41,205
Russia	6,592,849	485,400	7.4%	351,905	5.3%	37,455	509	1,133	27,936	17,076	12,954
ASIA											
Afghanistan	251,773	31,097	12.4%	115,831	46.0%	1,821	205	172	2,600	-	12,762
Armenia	11,506	2,162	18.8%	3,089	26.8%	211	-	9	478	75	515
Azerbaijan	33,437	7,471	22.3%	10,039	30.0%	1,172	19	107	1,965	21	5,321
Bahrain	267	23	8.7%	15	5.8%	-	-	-	12	-	17
Bangladesh	55,598	32,761	58.9%	2,317	4.2%	1,807	36,909	8	23,817	-	1,128
Brunei	2,226	27	1.2%	23	1.0%	-	-	-	2	6	2
Cambodia	69,898	14,699	21.0%	5,792	8.3%	-	4,035	146	2,896	2,079	-
China	3,690,045	599,520[4]	16.2%[4]	1,544,412[4]	41.9%[4]	102,463[4]	189,840[4]	116,240[4]	104,179[4]	440,384[4]	130,536[4]
Cyprus	3,572	436	12.2%	15	0.4%	12	-	-	55	419	240
East Timor	5,743	309	5.4%	579	10.1%	-	33	93	173	300	36
Georgia	26,911	4,104	15.3%	7,490	27.8%	207	-	358	1,117	433	541
India	1,222,510	655,987	53.7%	42,124	3.4%	72,140	132,818	12,285	217,773	17,000	57,900
Indonesia	735,310	129,730	17.6%	43,155	5.9%	-	50,953	9,409	11,370	6,098	7,316
Iran	636,372	63,892	10.0%	169,885	26.7%	8,740	2,103	1,113	8,273	-	53,900
Iraq	169,235	23,514	13.9%	15,444	9.1%	667	110	73	1,342	-	6,770
Israel	8,019	1,637	20.4%	548	6.8%	94	-	73	393	138	373
Japan	145,850	18,510	12.7%	1,564	1.1%	657	11,551	-	4,592	9,823	11
Jordan	34,495	1,544	4.5%	2,865	8.3%	12	-	13	66	-	1,900
Kazakhstan	1,049,156	83,672	8.0%	714,667	68.1%	10,938	225	256	4,021	984	8,785
Korea, North	46,540	10,811	23.2%	193	0.4%	88	2,031	1,253	575	3,076	186
Korea, South	38,328	7,293	19.0%	208	0.5%	4	7,204	67	2,191	8,266	1
Kuwait	6,880	58	0.8%	525	7.6%	-	-	-	19	-	543

| | Agricultural Area 2001 | | | | | Average Production 1999-2001 | | | Average 1999-2001 | | |
| | Total Area Sq. Miles | Cropland Area[1] Sq. Miles | Cropland Area[1] % | Pasture Area[1] Sq. Miles | Pasture Area[1] % | Wheat[1] 1,000 metric tons | Rice[1] 1,000 metric tons | Corn[1] 1,000 metric tons | Cattle[1] 1,000 | Pigs[1] 1,000 | Sheep[1] 1,000 |
CONTINENT/Country											
Kyrgyzstan	77,182	5,664	7.3%	35,873	46.5%	1,113	17	363	942	98	3,101
Laos	91,429	3,699	4.0%	3,390	3.7%	-	2,213	108	1,106	1,390	-
Lebanon	4,016	1,208	30.1%	62	1.5%	60	-	4	76	63	354
Malaysia	127,320	29,286	23.0%	1,100	0.9%	-	2,170	63	744	1,943	167
Mongolia	604,829	4,633	0.8%	499,230	82.5%	148	-	-	2,997	17	14,587
Myanmar	261,228	41,023	15.7%	1,212	0.5%	105	20,683	413	10,974	3,923	390
Nepal	56,827	12,324	21.7%	6,784	11.9%	1,143	4,137	1,528	7,012	872	852
Oman	119,499	313	0.3%	3,861	3.2%	1	-	-	299	-	342
Pakistan	339,732	85,560	25.2%	19,305	5.7%	19,319	6,920	1,653	22,007	-	24,067
Philippines	115,831	41,120	35.5%	4,942	4.3%	-	12,377	4,540	2,467	10,724	30
Qatar	4,412	81	1.8%	193	4.4%	-	-	1	15	-	214
Saudi Arabia	830,000	14,649	1.8%	656,373	79.1%	1,871	-	5	304	-	7,848
Singapore	264	4	1.5%	-	0.0%	-	-	-	-	190	-
Sri Lanka	25,332	7,378	29.1%	1,699	6.7%	-	2,804	30	1,580	71	12
Syria	71,498	21,043	29.4%	31,942	44.7%	3,514	-	196	933	-	13,288
Taiwan	13,901	(5)	(5)	(5)	(5)	(5)	(5)	(5)	(5)	(5)	(5)
Tajikistan	55,251	4,093	7.4%	13,514	24.5%	375	67	38	1,045	1	1,481
Thailand	198,115	70,657	35.7%	3,089	1.6%	1	25,578	4,405	4,973	6,539	40
Turkey	302,541	101,757	33.6%	47,792	15.8%	19,341	350	2,266	10,949	4	29,394
Turkmenistan	188,457	7,008	3.7%	118,533	62.9%	1,472	33	9	863	46	5,750
United Arab Emirates	32,278	919	2.8%	1,178	3.6%	-	-	-	94	-	504
Uzbekistan	172,742	18,649	10.8%	88,031	51.0%	3,637	219	133	5,279	83	7,980
Vietnam	128,066	32,579	25.4%	2,479	1.9%	-	31,964	1,961	4,029	20,273	-
Yemen	203,850	6,158	3.0%	62,027	30.4%	145	-	48	1,320	-	4,758
AFRICA											
Algeria	919,595	31,861	3.5%	122,780	13.4%	1,414	-	1	1,667	6	19,000
Angola	481,354	12,741	2.6%	208,495	43.3%	4	16	417	3,995	800	345
Benin	43,484	8,745	20.1%	2,124	4.9%	-	46	740	1,486	463	650
Botswana	224,607	1,440	0.6%	98,842	44.0%	1	-	8	2,035	6	347
Burkina Faso	105,869	15,444	14.6%	23,166	21.9%	-	102	500	4,767	621	6,722
Burundi	10,745	4,865	45.3%	3,610	33.6%	7	57	124	321	67	215
Cameroon	183,568	27,645	15.1%	7,722	4.2%	-	69	759	5,761	1,232	3,734
Cape Verde	1,557	158	10.2%	97	6.2%	-	-	27	22	195	9
Central African Republic	240,536	7,799	3.2%	12,066	5.0%	-	23	101	3,096	669	218
Chad	495,755	14,016	2.8%	173,746	35.0%	3	114	88	5,852	22	2,374
Comoros	863	510	59.1%	58	6.7%	-	17	4	51	-	21
Congo	132,047	849	0.6%	38,610	29.2%	-	1	6	87	46	102
Congo, Democratic Republic of the	905,446	30,425	3.4%	57,915	6.4%	9	338	1,184	823	1,050	925
Cote d'Ivoire	124,504	28,958	23.3%	50,193	40.3%	-	1,217	693	1,398	333	1,439
Djibouti	8,958	4	0.0%	5,019	56.0%	-	-	-	269	-	465
Egypt	386,662	12,888	3.3%	-	0.0%	6,388	5,681	6,487	3,583	29	4,510
Equatorial Guinea	10,831	888	8.2%	402	3.7%	-	-	-	5	6	37
Eritrea	45,406	1,942	4.3%	26,900	59.2%	32	-	13	2,150	-	1,570
Ethiopia	426,373	44,255	10.4%	77,220	18.1%	1,340	-	2,938	35,025	25	22,333
Gabon	103,347	1,911	1.8%	18,012	17.4%	-	1	26	36	213	197
Gambia, The	4,127	985	23.9%	1,772	42.9%	-	28	24	350	12	115
Ghana	92,098	22,780	24.7%	32,240	35.0%	-	244	988	1,297	327	2,715
Guinea	94,926	5,888	6.2%	41,313	43.5%	-	830	96	2,576	93	824
Guinea-Bissau	13,948	2,116	15.2%	4,170	29.9%	-	95	26	509	347	283
Kenya	224,961	19,923	8.9%	82,240	36.6%	184	58	2,419	13,229	311	7,000
Lesotho	11,720	1,290	11.0%	7,722	65.9%	39	-	128	547	63	839
Liberia	43,000	2,317	5.4%	7,722	18.0%	128	188	-	36	127	210
Libya	679,362	8,301	1.2%	51,352	7.6%	128	-	-	207	-	5,100
Madagascar	226,658	13,707	6.0%	92,664	40.9%	9	2,412	175	10,339	1,267	793
Malawi	45,747	9,035	19.7%	7,143	15.6%	2	86	2,190	741	450	110
Mali	478,841	18,147	3.8%	115,831	24.2%	8	801	378	6,594	72	6,282
Mauritania	397,956	1,931	0.5%	151,545	38.1%	-	65	7	1,470	-	7,437
Mauritius	788	409	51.9%	27	3.4%	-	-	-	27	12	10
Morocco	172,414	37,529	21.8%	81,081	47.0%	2,284	33	95	2,629	8	17,059
Mozambique	309,496	16,351	5.3%	169,885	54.9%	1	168	1,136	1,317	179	125
Namibia	317,818	3,166	1.0%	146,719	46.2%	4	-	26	2,436	21	2,330
Niger	489,192	17,375	3.6%	46,332	9.5%	10	66	5	2,217	39	4,386
Nigeria	356,669	120,464	33.8%	151,352	42.4%	75	3,109	4,734	19,677	5,000	20,833
Rwanda	10,169	5,019	49.4%	2,124	20.9%	6	13	66	766	172	264
Sao Tome and Principe	372	205	55.0%	4	1.0%	-	-	2	4	2	3
Senegal	75,951	9,653	12.7%	21,815	28.7%	-	229	84	3,076	263	4,619
Sierra Leone	27,699	2,178	7.9%	8,494	30.7%	-	215	9	413	52	365
Somalia	246,201	4,135	1.7%	166,024	67.4%	1	2	188	5,133	4	13,100
South Africa	470,693	60,664	12.9%	324,048	68.8%	2,200	3	9,147	13,594	1,542	28,677
Sudan	967,500	64,298	6.6%	452,434	46.8%	230	8	48	37,081	-	45,980
Swaziland	6,704	734	10.9%	4,633	69.1%	-	-	94	613	32	27
Tanzania	364,900	19,112	5.2%	135,136	37.0%	87	509	2,567	17,350	449	3,513
Togo	21,925	10,154	46.3%	3,861	17.6%	-	69	480	277	287	1,528
Tunisia	63,170	18,954	30.0%	15,792	25.0%	1,111	-	-	760	6	6,862
Uganda	93,065	27,799	29.9%	19,738	21.2%	12	106	1,108	5,977	1,540	1,065
Zambia	290,586	20,386	7.0%	115,831	39.9%	80	11	768	2,709	324	137
Zimbabwe	150,873	12,934	8.6%	66,410	44.0%	282	-	1,698	5,840	494	602
OCEANIA											
Australia	2,969,910	195,368	6.6%	1,563,327	52.6%	23,654	1,417	363	27,645	2,607	116,736
Fiji	7,056	1,100	15.6%	676	9.6%	-	16	1	335	139	7
Kiribati	313	151	48.1%	-	0.0%	-	-	-	-	10	-
Micronesia, Federated States of	271	139	51.3%	42	15.7%	-	-	-	14	32	-
New Zealand	104,454	13,019	12.5%	53,525	51.2%	337	-	185	9,025	364	45,114
Papua New Guinea	178,704	3,320	1.9%	676	0.4%	-	1	7	87	1,583	6
Samoa	1,093	498	45.6%	8	0.7%	-	-	-	28	179	-
Solomon Islands	10,954	286	2.6%	154	1.4%	-	5	-	11	63	-
Tonga	251	185	73.8%	15	6.2%	-	-	-	11	81	-
Vanuatu	4,707	463	9.8%	162	3.4%	-	-	1	151	62	-

This table presents data for most independent nations having an area greater than 200 square miles
- Zero, insignificant, or not available
(1) Source: United Nations Food and Agriculture Organization
(2) Includes data for Luxembourg
(3) Data for Luxembourg is included with Belgium
(4) Includes data for Taiwan
(5) Data for Taiwan is included with China

WORLD ECONOMIC TABLE

CONTINENT/Country	GDP 2002: Total GDP[1]	GDP 2002: GDP Per Capita[1]	Trade: Value of Exports[1]	Trade: Value of Imports[1]	Commercial Energy Production Avg. 2000[2]: Total (1,000 Metric Tons of Coal Equiv.)	Solid %	Liquid %	Gas %	Hydro & Nuclear %	Average Production 1999-2001 in Metric Tons: Coal[3]	Petroleum[3]	Iron Ore[4]	Bauxite[4]
NORTH AMERICA													
Bahamas	$4,590,000,000	$17,000	$560,700,000	$1,860,000,000	-	-	-	-	-	-	-	-	-
Belize	$1,280,000,000	$4,900	$290,000,000	$430,000,000	12	-	-	-	100%	-	-	-	-
Canada	$934,100,000,000	$29,400	$260,500,000,000	$229,000,000,000	507,218	10%	33%	43%	14%	70,711,084	97,834,913	20,527,000	-
Costa Rica	$32,000,000,000	$8,500	$5,100,000,000	$6,400,000,000	1,937	-	-	-	100%	-	-	-	-
Cuba	$30,690,000,000	$2,300	$1,800,000,000	$4,800,000,000	4,626	-	83%	17%	-	-	2,134,520	-	-
Dominica	$380,000,000	$5,400	$50,000,000	$135,000,000	4	-	-	-	100%	-	-	-	-
Dominican Republic	$53,780,000,000	$6,100	$5,300,000,000	$8,700,000,000	115	-	-	-	100%	-	-	-	-
El Salvador	$29,410,000,000	$4,700	$3,000,000,000	$4,900,000,000	1,110	-	-	-	100%	-	-	-	-
Guatemala	$53,200,000,000	$3,700	$2,700,000,000	$5,600,000,000	1,822	-	81%	1%	18%	-	1,076,526	9,000	-
Haiti	$10,600,000,000	$1,700	$298,000,000	$1,140,000,000	33	-	-	-	100%	-	-	-	-
Honduras	$16,290,000,000	$2,600	$1,300,000,000	$2,700,000,000	347	-	-	-	100%	-	-	-	-
Jamaica	$10,080,000,000	$3,900	$1,400,000,000	$3,100,000,000	18	-	-	-	100%	-	-	-	11,728,000
Mexico	$924,400,000,000	$9,000	$158,400,000,000	$168,400,000,000	340,594	1%	79%	16%	4%	11,097,943	150,165,451	6,860,000	-
Nicaragua	$11,160,000,000	$2,500	$637,000,000	$1,700,000,000	706	-	-	-	100%	-	-	-	-
Panama	$18,060,000,000	$6,000	$5,800,000,000	$6,700,000,000	418	-	-	-	100%	-	-	-	-
St. Lucia	$866,000,000	$5,400	$68,300,000	$319,400,000	-	-	-	-	-	-	-	-	-
Trinidad and Tobago	$11,070,000,000	$9,500	$4,200,000,000	$3,800,000,000	22,768	-	39%	61%	-	-	5,964,991	-	-
United States	$10,450,000,000,000	$37,600	$733,900,000,000	$1,194,100,000,000	2,342,228	33%	22%	30%	14%	996,498,186	289,640,487	35,178,000	-
SOUTH AMERICA													
Argentina	$403,800,000,000	$10,200	$25,300,000,000	$9,000,000,000	118,739	-	50%	45%	5%	260,299	38,783,798	-	-
Bolivia	$21,150,000,000	$2,500	$1,300,000,000	$1,600,000,000	7,732	-	33%	64%	3%	-	1,599,401	-	-
Brazil	$1,376,000,000,000	$7,600	$59,400,000,000	$46,200,000,000	143,640	3%	63%	6%	28%	4,446,477	61,155,586	124,667,000	13,654,000
Chile	$156,100,000,000	$10,000	$17,800,000,000	$15,600,000,000	6,180	6%	11%	45%	38%	475,484	349,201	5,523,000	-
Colombia	$251,600,000,000	$6,500	$12,900,000,000	$12,500,000,000	99,513	36%	52%	9%	4%	38,112,136	34,896,672	348,000	-
Ecuador	$42,650,000,000	$3,100	$4,900,000,000	$6,000,000,000	32,171	-	94%	3%	3%	-	19,520,185	-	-
Guyana	$2,628,000,000	$4,000	$500,000,000	$575,000,000	1	-	-	-	100%	-	-	-	2,272,000
Paraguay	$25,190,000,000	$4,200	$2,000,000,000	$2,400,000,000	6,577	-	-	-	100%	-	-	-	-
Peru	$138,800,000,000	$4,800	$7,600,000,000	$7,300,000,000	10,933	-	73%	9%	18%	52,297	4,932,561	2,701,000	-
Suriname	$1,469,000,000	$3,500	$445,000,000	$300,000,000	1,022	-	84%	-	16%	-	496,400	-	3,946,000
Uruguay	$26,820,000,000	$7,800	$2,100,000,000	$1,870,000,000	867	-	-	-	100%	-	-	-	-
Venezuela	$131,700,000,000	$5,500	$28,600,000,000	$18,800,000,000	311,899	3%	81%	14%	2%	7,482,998	146,621,238	10,497,000	4,309,000
EUROPE													
Albania	$15,690,000,000	$4,500	$340,000,000	$1,500,000,000	1,089	1%	42%	2%	55%	32,666	284,321	-	-
Austria	$227,700,000,000	$27,700	$70,000,000,000	$74,000,000,000	9,611	5%	15%	24%	56%	1,197,660	921,120	525,000	-
Belarus	$90,190,000,000	$8,200	$7,700,000,000	$8,800,000,000	3,644	18%	73%	9%	-	-	1,830,872	-	-
Belgium	$299,700,000,000	$29,000	$162,000,000,000	$152,000,000,000	18,451	2%	-	-	98%	318,998	-	-	-
Bosnia and Herzegovina	$7,300,000,000	$1,900	$1,150,000,000	$2,800,000,000	6,553	90%	-	-	10%	8,414,623	-	50,000	75,000
Bulgaria	$49,230,000,000	$6,600	$5,300,000,000	$6,900,000,000	13,500	46%	-	-	53%	28,841,963	37,048	310,000	-
Croatia	$43,120,000,000	$8,800	$4,900,000,000	$10,700,000,000	4,962	-	42%	43%	15%	5,104	1,191,360	-	-
Czech Republic	$157,100,000,000	$15,300	$40,800,000,000	$43,200,000,000	39,843	85%	1%	1%	14%	63,466,671	283,097	-	-
Denmark	$155,300,000,000	$29,000	$56,300,000,000	$47,900,000,000	36,502	-	70%	29%	2%	-	16,701,163	-	-
Estonia	$15,520,000,000	$10,900	$3,400,000,000	$4,400,000,000	3,892	100%	-	-	-	-	-	-	-
Finland	$133,800,000,000	$26,200	$40,100,000,000	$31,800,000,000	11,933	15%	-	-	85%	-	-	-	-
France	$1,558,000,000,000	$25,700	$307,800,000,000	$303,700,000,000	175,306	2%	4%	1%	93%	3,616,981	1,446,228	12,000	-
Germany	$2,160,000,000,000	$26,600	$608,000,000,000	$487,300,000,000	181,697	47%	2%	13%	38%	204,685,080	3,044,206	5,000	-
Greece	$203,300,000,000	$19,000	$12,600,000,000	$31,400,000,000	12,988	92%	3%	1%	4%	64,503,999	166,807	583,000	1,975,000
Hungary	$134,000,000,000	$13,300	$31,400,000,000	$33,900,000,000	16,319	25%	19%	24%	32%	14,796,257	1,301,710	-	994,000
Iceland	$8,444,000,000	$25,000	$2,300,000,000	$2,100,000,000	1,638	-	-	-	100%	-	-	-	-
Ireland	$113,700,000,000	$30,500	$86,600,000,000	$48,600,000,000	3,232	47%	-	47%	6%	-	-	-	-
Italy	$1,455,000,000,000	$25,000	$259,200,000,000	$238,200,000,000	40,332	-	16%	54%	30%	47,666	4,144,278	-	-
Latvia	$20,990,000,000	$8,300	$2,300,000,000	$3,900,000,000	369	6%	-	-	94%	-	-	-	-
Lithuania	$30,080,000,000	$8,400	$5,400,000,000	$6,800,000,000	3,677	-	12%	-	87%	-	251,824	-	-
Luxembourg	$21,940,000,000	$44,000	$10,100,000,000	$13,250,000,000	113	-	-	-	100%	-	-	-	-
Macedonia	$10,570,000,000	$5,000	$1,100,000,000	$1,900,000,000	3,038	95%	-	-	5%	7,463,628	-	9,000	-
Moldova	$11,510,000,000	$2,500	$590,000,000	$980,000,000	7	-	-	-	100%	-	-	-	-
Netherlands	$437,800,000,000	$26,900	$243,300,000,000	$201,100,000,000	87,974	-	4%	94%	2%	-	1,437,293	-	-
Norway	$149,100,000,000	$31,800	$68,200,000,000	$37,300,000,000	324,396	-	72%	22%	5%	847,996	154,419,533	355,000	-
Poland	$373,200,000,000	$9,500	$32,400,000,000	$43,400,000,000	108,277	94%	1%	5%	-	164,737,813	645,072	-	-
Portugal	$195,200,000,000	$18,000	$25,900,000,000	$39,000,000,000	1,560	-	-	-	100%	-	-	6,000	-
Romania	$169,300,000,000	$7,400	$13,700,000,000	$16,700,000,000	37,598	19%	24%	46%	10%	27,392,191	6,038,110	24,000	-
Serbia and Montenegro	$23,150,000,000	$2,370	$2,400,000,000	$6,300,000,000	14,188	74%	8%	8%	10%	34,480,488	810,787	10,000	580,000
Slovakia	$67,340,000,000	$12,200	$12,900,000,000	$15,400,000,000	8,813	17%	1%	2%	79%	3,606,648	48,134	200,000	-
Slovenia	$37,060,000,000	$18,000	$10,300,000,000	$11,100,000,000	3,644	38%	-	-	62%	4,391,644	991	-	-
Spain	$850,700,000,000	$20,700	$122,200,000,000	$156,600,000,000	40,444	28%	2%	1%	68%	23,479,212	296,665	-	-
Sweden	$230,700,000,000	$25,400	$80,600,000,000	$68,600,000,000	31,413	1%	-	-	99%	-	-	12,114,000	-
Switzerland	$233,400,000,000	$31,700	$100,300,000,000	$94,400,000,000	14,710	-	-	-	100%	-	-	-	-
Ukraine	$218,000,000,000	$4,500	$18,100,000,000	$18,000,000,000	118,973	50%	5%	20%	25%	81,998,575	3,747,936	28,933,000	-
United Kingdom	$1,528,000,000,000	$25,300	$286,300,000,000	$330,100,000,000	397,906	7%	47%	38%	8%	32,758,497	119,820,635	1,000	-
Russia	$1,409,000,000,000	$9,300	$104,600,000,000	$60,700,000,000	1,412,286	10%	33%	52%	5%	253,376,954	324,436,632	48,300,000	3,983,000
ASIA													
Afghanistan	$19,000,000,000	$700	$1,200,000,000	$1,300,000,000	195	1%	-	79%	20%	1,000	-	-	-
Armenia	$12,130,000,000	$3,800	$525,000,000	$991,000,000	901	-	-	-	100%	-	-	-	-
Azerbaijan	$28,610,000,000	$3,500	$2,000,000,000	$1,800,000,000	27,748	-	72%	27%	1%	-	14,183,985	-	-
Bahrain	$9,910,000,000	$14,000	$5,800,000,000	$4,200,000,000	14,442	-	22%	78%	-	-	1,827,397	-	-
Bangladesh	$238,200,000,000	$1,700	$6,200,000,000	$8,500,000,000	11,713	-	-	99%	1%	-	120,476	-	-
Brunei	$6,500,000,000	$18,600	$3,000,000,000	$1,400,000,000	27,922	-	49%	51%	-	-	9,435,323	-	-
Cambodia	$20,420,000,000	$1,500	$1,380,000,000	$1,730,000,000	10	-	-	-	100%	-	-	-	-
China	$5,989,000,000,000	$4,400	$658,260,000,000	$618,930,000,000	1,023,314[5]	70%[5]	23%[5]	4%[5]	3%[5]	1,251,423,183	161,226,848	72,967,000	9,000,000
Cyprus	$9,400,000,000	$15,000	$1,030,000,000	$3,900,000,000	-	-	-	-	-	-	-	-	-
East Timor	$440,000,000	$500	$8,000,000	$237,000,000	-	-	-	-	-	-	-	-	-
Georgia	$16,050,000,000	$3,100	$515,000,000	$750,000,000	963	1%	16%	8%	75%	10,000	102,258	-	-
India	$2,664,000,000,000	$2,540	$44,500,000,000	$53,800,000,000	367,807	73%	14%	8%	4%	304,842,421	32,123,682	48,080,000	7,554,000
Indonesia	$714,200,000,000	$3,100	$52,300,000,000	$32,100,000,000	279,695	27%	45%	26%	2%	79,664,587	70,565,213	282,000	1,168,000
Iran	$458,300,000,000	$7,000	$24,800,000,000	$21,800,000,000	350,729	-	77%	23%	-	1,376,993	181,632,777	5,367,000	136,000
Iraq	$58,000,000,000	$2,400	$13,000,000,000	$7,800,000,000	186,519	-	97%	3%	-	-	124,281,583	-	-
Israel	$117,400,000,000	$19,000	$28,100,000,000	$30,800,000,000	334	94%	2%	4%	1%	-	5,957	-	-
Japan	$3,651,000,000,000	$28,000	$383,800,000,000	$292,100,000,000	142,731	2%	1%	2%	95%	3,286,983	351,650	1,000	-
Jordan	$22,630,000,000	$4,300	$2,500,000,000	$4,400,000,000	316	-	1%	97%	2%	-	1,986	-	-
Kazakhstan	$120,000,000,000	$6,300	$10,300,000,000	$9,600,000,000	113,390	40%	45%	14%	1%	70,311,969	30,508,827	7,467,000	3,668,000
Korea, North	$22,260,000,000	$1,000	$842,000,000	$1,314,000,000	65,932	96%	-	-	4%	94,174,845	-	3,000,000	-
Korea, South	$941,500,000,000	$19,400	$162,600,000,000	$148,400,000,000	43,892	6%	-	-	94%	4,054,646	-	175,000	-
Kuwait	$36,850,000,000	$15,000	$16,000,000,000	$7,300,000,000	161,322	-	92%	8%	-	-	98,844,823	-	-

	GDP 2002		Trade		Commercial Energy Production Avg. 2000[2]					Average Production 1999-2001 in Metric Tons			
CONTINENT/Country	Total GDP[1]	GDP Per Capita[1]	Value of Exports[1]	Value of Imports[1]	Total (1,000 Metric Tons of Coal Equiv.)	Solid %	Liquid %	Gas %	Hydro & Nuclear %	Coal[3]	Petroleum[3]	Iron Ore[4]	Bauxite[4]
Kyrgyzstan	$13,880,000,000	$2,800	$488,000,000	$587,000,000	2,026	9%	5%	2%	83%	423,664	91,503	-	-
Laos	$10,400,000,000	$1,700	$345,000,000	$555,000,000	146	1%	-	-	99%	1,000	-	-	-
Lebanon	$17,610,000,000	$5,400	$1,000,000,000	$6,000,000,000	55	-	-	-	100%	-	-	-	-
Malaysia	$198,400,000,000	$9,300	$95,200,000,000	$76,800,000,000	110,069	-	41%	58%	1%	314,332	33,792,132	208,000	137,000
Mongolia	$5,060,000,000	$1,840	$501,000,000	$659,000,000	2,212	100%	-	-	-	5,099,640	-	-	-
Myanmar	$73,690,000,000	$1,660	$2,700,000,000	$2,500,000,000	9,297	3%	6%	88%	2%	358,331	587,374	-	-
Nepal	$37,320,000,000	$1,400	$720,000,000	$1,600,000,000	172	10%	-	-	90%	9,667	-	-	-
Oman	$22,400,000,000	$8,300	$10,600,000,000	$5,500,000,000	74,376	-	92%	8%	-	-	46,989,489	-	-
Pakistan	$295,300,000,000	$2,100	$9,800,000,000	$11,100,000,000	33,773	6%	12%	74%	7%	3,247,391	2,768,108	-	10,000
Philippines	$379,700,000,000	$4,200	$35,100,000,000	$33,500,000,000	16,244	6%	-	-	94%	1,306,993	173,128	-	-
Qatar	$15,910,000,000	$21,500	$10,900,000,000	$3,900,000,000	92,237	-	57%	43%	-	-	35,018,538	-	-
Saudi Arabia	$268,900,000,000	$10,500	$71,000,000,000	$39,500,000,000	736,996	-	91%	9%	-	-	401,559,222	-	-
Singapore	$112,400,000,000	$24,000	$127,000,000,000	$113,000,000,000	-	-	-	-	-	-	-	-	-
Sri Lanka	$73,700,000,000	$3,700	$4,600,000,000	$5,400,000,000	394	-	-	-	100%	-	-	-	-
Syria	$63,480,000,000	$3,500	$6,200,000,000	$4,900,000,000	47,898	-	83%	15%	2%	-	26,119,029	-	-
Taiwan	$406,000,000,000	$18,000	$130,000,000,000	$113,000,000,000	(6)	(6)	(6)	(6)	(6)	58,284	38,686	-	-
Tajikistan	$8,476,000,000	$1,250	$710,000,000	$830,000,000	1,790	-	1%	3%	95%	20,667	16,613	-	-
Thailand	$445,800,000,000	$6,900	$67,700,000,000	$58,100,000,000	44,127	25%	24%	50%	2%	18,551,756	5,080,720	20,000	-
Turkey	$489,700,000,000	$7,000	$35,100,000,000	$50,800,000,000	28,167	69%	14%	3%	14%	65,334,995	2,642,106	2,300,000	303,000
Turkmenistan	$31,340,000,000	$5,500	$2,970,000,000	$2,250,000,000	71,764	-	15%	85%	-	-	7,139,688	-	-
United Arab Emirates	$53,970,000,000	$22,000	$44,900,000,000	$30,800,000,000	199,656	-	83%	17%	-	-	112,737,023	-	-
Uzbekistan	$66,060,000,000	$2,500	$2,800,000,000	$2,500,000,000	85,806	1%	13%	85%	1%	2,736,319	4,419,300	-	-
Vietnam	$183,800,000,000	$2,250	$16,500,000,000	$16,800,000,000	39,300	30%	59%	5%	7%	9,688,950	15,926,911	-	-
Yemen	$15,070,000,000	$840	$3,400,000,000	$2,900,000,000	30,622	-	100%	-	-	-	21,304,264	-	-
AFRICA													
Algeria	$173,800,000,000	$5,300	$19,500,000,000	$10,600,000,000	222,648	-	47%	53%	-	24,000	61,651,110	757,000	-
Angola	$18,360,000,000	$1,600	$8,600,000,000	$4,100,000,000	53,315	-	98%	1%	-	-	36,961,745	-	-
Benin	$7,380,000,000	$1,070	$207,000,000	$479,000,000	69	-	100%	-	-	-	39,547	-	-
Botswana	$13,480,000,000	$9,500	$2,400,000,000	$1,900,000,000	(7)	(7)	(7)	(7)	(7)	956,767	-	-	-
Burkina Faso	$14,510,000,000	$1,080	$250,000,000	$525,000,000	15	-	-	-	100%	-	-	-	-
Burundi	$3,146,000,000	$600	$26,000,000	$135,000,000	21	29%	-	-	71%	-	-	-	-
Cameroon	$26,840,000,000	$1,700	$1,900,000,000	$1,700,000,000	10,722	-	96%	-	4%	1,000	4,326,440	-	-
Cape Verde	$600,000,000	$1,400	$30,000,000	$220,000,000	-	-	-	-	-	-	-	-	-
Central African Republic	$4,296,000,000	$1,300	$134,000,000	$102,000,000	10	-	-	-	100%	-	-	-	-
Chad	$9,297,000,000	$1,100	$197,000,000	$570,000,000	-	-	-	-	-	-	-	-	-
Comoros	$441,000,000	$720	$16,300,000	$39,800,000	-	-	-	-	-	-	-	-	-
Congo	$2,500,000,000	$900	$2,400,000,000	$73,000,000	19,097	-	99%	1%	-	-	13,651,000	-	-
Congo, Democratic Republic of the	$34,000,000,000	$610	$1,200,000,000	$890,000,000	2,630	4%	71%	-	25%	96,000	1,194,669	-	-
Cote d'Ivoire	$24,030,000,000	$1,500	$4,400,000,000	$2,500,000,000	4,439	-	50%	45%	5%	-	620,450	-	-
Djibouti	$619,000,000	$1,300	$70,000,000	$255,000,000	-	-	-	-	-	-	-	-	-
Egypt	$289,800,000,000	$3,900	$7,000,000,000	$15,200,000,000	86,315	-	65%	32%	2%	-	38,024,058	1,283,000	-
Equatorial Guinea	$1,270,000,000	$2,700	$2,500,000,000	$562,000,000	7,531	-	100%	-	-	-	7,461,521	-	-
Eritrea	$3,300,000,000	$740	$20,000,000	$500,000,000	-	-	-	-	-	-	-	-	-
Ethiopia	$48,530,000,000	$750	$433,000,000	$1,630,000,000	211	-	-	-	100%	-	-	-	-
Gabon	$8,354,000,000	$5,700	$2,600,000,000	$1,100,000,000	23,273	-	95%	5%	-	-	15,674,359	-	-
Gambia, The	$2,582,000,000	$1,800	$138,000,000	$225,000,000	-	-	-	-	-	-	-	-	-
Ghana	$41,250,000,000	$2,100	$2,200,000,000	$2,800,000,000	830	-	2%	-	98%	-	330,933	-	525,000
Guinea	$18,690,000,000	$2,000	$835,000,000	$670,000,000	25	-	-	-	100%	-	-	-	15,663,000
Guinea-Bissau	$901,400,000	$800	$71,000,000	$59,000,000	-	-	-	-	-	-	-	-	-
Kenya	$32,890,000,000	$1,020	$2,100,000,000	$3,000,000,000	642	-	-	-	100%	-	-	-	-
Lesotho	$5,106,000,000	$2,700	$422,000,000	$738,000,000	(7)	(7)	(7)	(7)	(7)	-	-	-	-
Liberia	$3,116,000,000	$1,100	$110,000,000	$165,000,000	24	-	-	-	100%	-	-	-	-
Libya	$33,360,000,000	$7,600	$11,800,000,000	$6,300,000,000	103,205	-	92%	8%	-	-	67,767,436	-	-
Madagascar	$12,590,000,000	$760	$700,000,000	$985,000,000	64	-	-	-	100%	-	-	-	-
Malawi	$6,811,000,000	$670	$435,000,000	$505,000,000	107	-	-	-	100%	-	-	-	-
Mali	$9,775,000,000	$860	$680,000,000	$630,000,000	29	-	-	-	100%	-	-	-	-
Mauritania	$4,891,000,000	$1,900	$355,000,000	$360,000,000	4	-	-	-	100%	-	-	7,492,000	-
Mauritius	$12,150,000,000	$11,000	$1,600,000,000	$1,800,000,000	12	-	-	-	100%	-	-	-	-
Morocco	$121,800,000,000	$3,900	$7,500,000,000	$10,400,000,000	201	14%	9%	33%	43%	61,000	15,223	4,000	-
Mozambique	$19,520,000,000	$1,000	$680,000,000	$1,180,000,000	874	2%	-	-	98%	18,667	-	-	8,000
Namibia	$13,150,000,000	$6,900	$1,210,000,000	$1,380,000,000	(7)	(7)	(7)	(7)	(7)	-	-	-	-
Niger	$8,713,000,000	$830	$293,000,000	$368,000,000	175	100%	-	-	-	151,666	-	-	-
Nigeria	$112,500,000,000	$875	$17,300,000,000	$13,600,000,000	172,641	-	90%	10%	-	61,000	108,397,478	-	-
Rwanda	$8,920,000,000	$1,200	$68,000,000	$253,000,000	20	-	-	-	100%	-	-	-	-
Sao Tome and Principe	$200,000,000	$1,200	$5,500,000	$24,800,000	1	-	-	-	100%	-	-	-	-
Senegal	$15,640,000,000	$1,500	$1,150,000,000	$1,460,000,000	1	-	-	100%	-	-	-	-	-
Sierra Leone	$2,826,000,000	$580	$35,000,000	$190,000,000	(7)	(7)	(7)	(7)	(7)	-	-	-	-
Somalia	$4,270,000,000	$550	$126,000,000	$343,000,000	-	-	-	-	-	-	-	-	-
South Africa	$427,700,000,000	$10,000	$31,800,000,000	$26,600,000,000	245,195[8]	92%[8]	5%[8]	1%[8]	2%[8]	224,286,505	1,277,485	20,751,000	-
Sudan	$52,900,000,000	$1,420	$1,800,000,000	$1,500,000,000	13,436	-	99%	-	1%	-	7,679,837	-	-
Swaziland	$5,542,000,000	$4,400	$820,000,000	$938,000,000	(7)	(7)	(7)	(7)	(7)	288,665	-	-	-
Tanzania	$20,420,000,000	$630	$863,000,000	$1,670,000,000	343	23%	-	-	77%	5,000	-	-	-
Togo	$7,594,000,000	$1,500	$449,000,000	$561,000,000	-	-	-	-	-	-	-	-	-
Tunisia	$67,130,000,000	$6,500	$6,800,000,000	$8,700,000,000	8,065	-	66%	34%	-	-	3,826,400	105,000	-
Uganda	$30,490,000,000	$1,260	$476,000,000	$1,140,000,000	193	-	-	-	100%	-	-	3,000	-
Zambia	$8,240,000,000	$890	$709,000,000	$1,123,000,000	1,117	15%	-	-	85%	192,358	-	-	-
Zimbabwe	$26,070,000,000	$2,400	$1,570,000,000	$1,739,000,000	4,801	92%	-	-	8%	4,508,643	-	237,000	-
OCEANIA													
Australia	$525,500,000,000	$27,000	$66,300,000,000	$68,000,000,000	331,923	71%	14%	14%	1%	307,176,075	31,728,994	104,014,000	51,834,000
Fiji	$4,822,000,000	$5,500	$442,000,000	$642,000,000	53	-	-	-	100%	-	-	-	-
Kiribati	$79,000,000	$840	$6,000,000	$44,000,000	-	-	-	-	-	-	-	-	-
Micronesia, Federated States of	$277,000,000	$2,000	$22,000,000	$149,000,000	-	-	-	-	-	-	-	-	-
New Zealand	$78,400,000,000	$20,200	$15,000,000,000	$12,500,000,000	19,812	14%	13%	40%	33%	3,452,315	1,839,394	660,000	-
Papua New Guinea	$10,860,000,000	$2,300	$1,800,000,000	$1,100,000,000	5,864	-	96%	2%	2%	-	3,874,601	-	-
Samoa	$1,000,000,000	$5,600	$15,500,000	$130,100,000	3	-	-	-	100%	-	-	-	-
Solomon Islands	$800,000,000	$1,700	$47,000,000	$82,000,000	-	-	-	-	-	-	-	-	-
Tonga	$236,000,000	$2,200	$8,900,000	$70,000,000	-	-	-	-	-	-	-	-	-
Vanuatu	$563,000,000	$2,900	$22,000,000	$93,000,000	-	-	-	-	-	-	-	-	-

This table presents data for most independent nations having an area greater than 200 square miles
- Zero, insignificant, or not available
(1) Source: United States Central Intelligence Agency World Factbook
(2) Source: United Nations Energy Statistics Yearbook
(3) Source: United States Energy Information Administration International Energy Annual
(4) Source: United States Geological Survey Minerals Yearbook
(5) Includes data for Taiwan
(6) Data for Taiwan is included with China
(7) Data for countries in the South Africa Customs Union are included with South Africa
(8) Includes data for countries in the South Africa Customs Union

WORLD ENVIRONMENT TABLE

CONTINENT/Country	Total Area Sq. Miles	Protected Area 2002[1,2] Sq. Miles	%	Endangered Species 2003[3] Mammal	Bird	Reptile	Amphib.	Fish	Invrt.	Forest Cover[4] Sq. Miles 2000	Percent Change 1990-2000
NORTH AMERICA											
Bahamas	5,382	-	-	5	4	6	0	15	1	3,251	-
Belize	8,867	3,999	45.1%	5	2	4	0	17	1	5,205	-20.9%
Canada	3,855,103	427,916	11.1%	16	8	2	1	25	11	944,294	-
Costa Rica	19,730	4,538	23.0%	13	13	7	1	13	9	7,598	-7.4%
Cuba	42,804	29,578	69.1%	11	18	7	0	23	3	9,066	13.4%
Dominica	290	-	-	1	3	4	0	11	0	178	-8.0%
Dominican Republic	18,730	9,721	51.9%	5	15	10	1	10	2	5,313	-
El Salvador	8,124	33	0.4%	2	0	4	0	5	1	467	-37.3%
Guatemala	42,042	8,408	20.0%	7	6	8	0	14	8	11,004	-15.9%
Haiti	10,714	43	0.4%	4	14	8	1	12	2	340	-44.3%
Honduras	43,277	2,770	6.4%	10	5	6	0	14	2	20,784	-9.9%
Jamaica	4,244	3,590	84.6%	5	12	8	4	12	5	1,255	-14.2%
Mexico	758,452	77,362	10.2%	72	40	18	4	106	41	213,148	-10.3%
Nicaragua	50,054	8,910	17.8%	6	5	7	0	17	2	12,656	-26.3%
Panama	29,157	6,327	21.7%	17	16	7	0	17	2	11,104	-15.3%
St. Lucia	238	-	-	2	5	6	0	10	0	35	-35.7%
Trinidad and Tobago	1,980	119	6.0%	1	1	5	0	15	0	1,000	-7.8%
United States	3,794,083	982,668	25.9%	39	56	27	25	155	557	872,563	1.7%
SOUTH AMERICA											
Argentina	1,073,519	70,852	6.6%	32	39	5	5	9	10	133,777	-7.6%
Bolivia	424,165	56,838	13.4%	25	28	2	1	0	1	204,897	-2.9%
Brazil	3,300,172	221,112	6.7%	74	113	22	6	33	34	2,100,028	-4.1%
Chile	291,930	55,175	18.9%	21	22	0	3	9	0	59,985	-1.3%
Colombia	439,737	44,853	10.2%	39	78	14	0	23	0	191,510	-3.7%
Ecuador	109,484	20,036	18.3%	34	62	10	0	11	48	40,761	-11.5%
Guyana	83,000	249	0.3%	13	2	6	0	13	1	65,170	-2.8%
Paraguay	157,048	5,497	3.5%	10	26	2	0	0	0	90,240	-5.0%
Peru	496,225	30,270	6.1%	46	76	6	1	8	2	251,796	-4.0%
Suriname	63,037	3,089	4.9%	12	1	6	0	12	0	54,491	-
Uruguay	67,574	203	0.3%	6	11	3	0	8	1	4,988	63.3%
Venezuela	352,145	224,669	63.8%	26	24	13	0	19	1	191,144	-4.2%
EUROPE											
Albania	11,100	422	3.8%	3	3	4	0	16	4	3,826	-7.3%
Austria	32,378	10,685	33.0%	7	3	0	0	7	44	15,004	2.0%
Belarus	80,155	5,050	6.3%	7	3	0	0	0	5	36,301	37.5%
Belgium	11,787	-	-	11	2	0	0	7	11	2,811	-1.8%
Bosnia and Herzegovina	19,767	99	0.5%	10	3	1	1	10	10	8,776	-
Bulgaria	42,855	1,928	4.5%	14	10	2	0	10	9	14,247	5.9%
Croatia	21,829	1,637	7.5%	9	4	1	1	26	11	6,884	1.1%
Czech Republic	30,450	4,902	16.1%	8	2	0	0	7	19	10,162	0.2%
Denmark	16,640	5,658	34.0%	5	1	0	0	7	11	1,757	2.2%
Estonia	17,462	2,061	11.8%	5	3	0	0	1	4	7,954	6.5%
Finland	130,559	12,142	9.3%	4	3	0	0	1	10	84,691	0.4%
France	208,482	27,728	13.3%	18	5	3	2	15	65	59,232	4.2%
Germany	137,847	43,973	31.9%	11	5	0	0	12	31	41,467	-
Greece	50,949	1,834	3.6%	13	7	6	1	26	11	13,896	9.1%
Hungary	35,919	2,514	7.0%	9	8	1	0	8	25	7,104	4.1%
Iceland	39,769	3,897	9.8%	7	0	0	0	8	0	120	24.0%
Ireland	27,133	461	1.7%	6	1	0	0	6	3	2,544	34.8%
Italy	116,342	9,191	7.9%	14	5	4	4	16	58	38,622	3.0%
Latvia	24,942	3,342	13.4%	5	3	0	0	3	8	11,286	4.5%
Lithuania	25,213	2,597	10.3%	6	4	0	0	3	5	7,699	2.5%
Luxembourg	999	-	-	3	1	0	0	0	4	-	-
Macedonia	9,928	705	7.1%	11	3	2	0	4	5	3,498	-
Moldova	13,070	183	1.4%	6	5	1	0	9	5	1,255	2.2%
Netherlands	16,164	2,295	14.2%	10	4	0	0	7	7	1,448	2.7%
Norway	125,050	8,503	6.8%	10	2	0	0	7	9	34,240	3.6%
Poland	120,728	14,970	12.4%	14	4	0	0	3	15	34,931	2.0%
Portugal	35,516	2,344	6.6%	17	7	0	1	19	82	14,154	18.4%
Romania	91,699	4,310	4.7%	17	8	2	0	10	22	24,896	2.3%
Serbia and Montenegro	39,449	1,302	3.3%	12	5	1	0	19	19	11,147	-0.5%
Slovakia	18,924	4,315	22.8%	9	4	1	0	8	19	8,405	9.0%
Slovenia	7,821	469	6.0%	9	1	0	1	15	42	4,274	2.0%
Spain	194,885	16,565	8.5%	24	7	7	3	23	63	55,483	6.4%
Sweden	173,732	15,810	9.1%	6	2	0	0	6	13	104,765	-
Switzerland	15,943	4,783	30.0%	5	2	0	0	4	30	4,629	3.7%
Ukraine	233,090	9,091	3.9%	16	8	2	0	11	14	37,004	3.3%
United Kingdom	93,788	19,602	20.9%	12	2	0	0	11	10	10,788	6.5%
Russia	6,592,849	514,242	7.8%	45	38	6	0	18	30	3,287,242	0.2%
ASIA											
Afghanistan	251,773	755	0.3%	13	11	1	1	0	1	5,216	-
Armenia	11,506	874	7.6%	11	4	5	0	1	7	1,355	13.6%
Azerbaijan	33,437	2,040	6.1%	13	8	5	0	5	6	4,224	13.5%
Bahrain	267	-	-	1	6	0	0	6	0	-	-
Bangladesh	55,598	445	0.8%	22	23	20	0	8	0	5,151	14.1%
Brunei	2,226	-	-	11	14	4	0	6	0	1,707	-2.2%
Cambodia	69,898	12,931	18.5%	24	19	10	0	11	0	36,043	-5.7%
China	3,690,045	287,824	7.8%	81	75	31	1	46	4	631,200	12.4%
Cyprus	3,572	-	-	3	3	3	0	6	0	664	44.5%
East Timor	5,743	-	-	0	6	0	0	2	0	1,958	-6.3%
Georgia	26,911	619	2.3%	13	3	7	1	6	10	11,537	-
India	1,222,510	63,571	5.2%	86	72	25	3	27	23	247,542	0.6%
Indonesia	735,310	151,474	20.6%	147	114	28	0	91	31	405,353	-11.1%
Iran	636,372	30,546	4.8%	22	13	8	2	14	3	28,182	-
Iraq	169,235	-	-	11	11	2	0	3	2	3,085	-
Israel	8,019	1,267	15.8%	15	12	4	0	10	10	510	61.0%
Japan	145,850	9,918	6.8%	37	35	11	10	27	45	92,977	0.1%
Jordan	34,495	1,173	3.4%	9	8	1	0	5	3	332	-
Kazakhstan	1,049,156	28,327	2.7%	17	15	2	1	7	4	46,904	24.5%
Korea, North	46,540	1,210	2.6%	13	19	0	0	5	1	31,699	-
Korea, South	38,328	2,645	6.9%	13	25	0	0	7	0	24,124	-0.8%

CONTINENT/Country	Total Area Sq. Miles	Protected Area 2002[1,2] Sq. Miles	%	Mammal	Bird	Endangered Species 2003[3] Reptile	Amphib.	Fish	Invrt.	Forest Cover[4] Sq. Miles 2000	Percent Change 1990-2000
Kuwait	6,880	103	1.5%	1	7	1	0	6	0	19	66.7%
Kyrgyzstan	77,182	2,779	3.6%	7	4	2	0	0	3	3,873	29.4%
Laos	91,429	11,429	12.5%	31	20	11	0	6	0	48,498	-4.0%
Lebanon	4,016	20	0.5%	6	7	1	0	8	1	139	-2.7%
Malaysia	127,320	7,257	5.7%	50	37	21	0	34	3	74,487	52.4%
Mongolia	604,829	69,555	11.5%	14	16	0	0	1	3	41,101	-5.3%
Myanmar	261,228	784	0.3%	39	35	20	0	7	2	132,892	-13.1%
Nepal	56,827	5,058	8.9%	29	25	6	0	0	1	15,058	-16.7%
Oman	119,499	16,730	14.0%	11	10	4	0	17	1	4	-
Pakistan	339,732	16,647	4.9%	17	17	9	0	14	0	9,116	-14.3%
Philippines	115,831	6,602	5.7%	50	67	8	23	48	19	22,351	-13.3%
Qatar	4,412	-	-	0	6	1	0	4	0	-	-
Saudi Arabia	830,000	317,890	38.3%	9	15	2	0	8	1	5,807	-
Singapore	264	13	4.9%	3	7	3	0	12	1	8	-
Sri Lanka	25,332	3,420	13.5%	22	14	8	0	22	2	7,490	-15.2%
Syria	71,498	-	-	4	8	3	0	8	3	1,780	-
Taiwan	13,901	-	-	12	21	8	0	23	0	-	-
Tajikistan	55,251	2,321	4.2%	9	7	1	0	3	2	1,544	5.3%
Thailand	198,115	27,538	13.9%	37	37	19	0	35	1	56,996	-7.1%
Turkey	302,541	4,841	1.6%	17	11	12	3	29	13	39,479	2.2%
Turkmenistan	188,457	7,915	4.2%	13	6	2	0	8	5	14,498	-
United Arab Emirates	32,278	-	-	4	8	1	0	6	0	1,239	32.1%
Uzbekistan	172,742	3,455	2.0%	9	9	2	0	4	1	7,602	2.4%
Vietnam	128,066	4,738	3.7%	42	37	24	1	22	0	37,911	5.5%
Yemen	203,850	-	-	6	12	2	0	10	2	1,734	-17.0%
AFRICA											
Algeria	919,595	45,980	5.0%	13	6	2	0	9	12	8,282	14.2%
Angola	481,354	31,769	6.6%	19	15	4	0	8	6	269,329	-1.7%
Benin	43,484	4,957	11.4%	9	2	1	0	7	0	10,232	-20.9%
Botswana	224,607	41,552	18.5%	7	7	0	0	0	0	47,981	-8.7%
Burkina Faso	105,869	12,175	11.5%	7	2	1	0	0	0	27,371	-2.1%
Burundi	10,745	612	5.7%	6	7	0	0	0	3	363	-61.0%
Cameroon	183,568	8,261	4.5%	38	15	1	1	34	4	92,116	-8.5%
Cape Verde	1,557	-	-	3	2	0	0	13	0	328	142.9%
Central African Republic	240,536	20,927	8.7%	14	3	1	0	0	0	88,444	-1.3%
Chad	495,755	45,114	9.1%	15	5	1	0	0	1	49,004	-6.0%
Comoros	863	-	-	2	9	2	0	3	4	31	-33.3%
Congo	132,047	6,602	5.0%	15	3	1	0	9	1	85,174	-0.8%
Congo, Democratic Republic of the	905,446	58,854	6.5%	40	28	2	0	9	45	522,037	-3.8%
Cote d'Ivoire	124,504	7,470	6.0%	19	12	2	1	10	1	27,479	-27.1%
Djibouti	8,958	-	-	5	5	0	0	9	0	23	-
Egypt	386,662	37,506	9.7%	13	7	6	0	13	1	278	38.5%
Equatorial Guinea	10,831	-	-	16	5	2	1	7	2	6,765	-5.7%
Eritrea	45,406	1,952	4.3%	12	7	6	0	8	0	6,120	-3.3%
Ethiopia	426,373	72,057	16.9%	35	16	1	0	0	4	17,734	-8.1%
Gabon	103,347	723	0.7%	14	5	1	0	11	1	84,271	-0.5%
Gambia, The	4,127	95	2.3%	3	2	1	0	10	0	1,857	10.3%
Ghana	92,098	5,157	5.6%	14	8	2	0	7	0	24,460	-15.9%
Guinea	94,926	664	0.7%	12	10	1	1	7	3	26,753	-4.8%
Guinea-Bissau	13,948	-	-	3	0	1	0	9	1	8,444	-9.0%
Kenya	224,961	17,997	8.0%	50	24	5	0	27	15	66,008	-5.2%
Lesotho	11,720	23	0.2%	6	7	0	0	1	1	54	-
Liberia	43,000	731	1.7%	16	11	2	0	7	2	13,440	-17.9%
Libya	679,362	679	0.1%	8	1	3	0	8	0	1,382	15.1%
Madagascar	226,658	9,746	4.3%	50	27	18	2	25	32	45,278	-9.1%
Malawi	45,747	5,124	11.2%	8	11	0	0	0	8	9,892	-21.6%
Mali	478,841	17,717	3.7%	13	4	1	0	1	0	50,911	-7.0%
Mauritania	397,956	6,765	1.7%	10	2	2	0	10	1	1,224	-23.6%
Mauritius	788	-	-	3	9	4	0	7	32	62	-5.9%
Morocco	172,414	1,207	0.7%	16	9	2	0	10	8	11,680	-0.4%
Mozambique	309,496	25,998	8.4%	15	16	5	0	19	7	118,151	-2.0%
Namibia	317,818	43,223	13.6%	14	11	3	1	11	1	31,043	-8.4%
Niger	489,192	37,668	7.7%	11	3	0	0	0	1	5,127	-31.7%
Nigeria	356,669	11,770	3.3%	27	9	2	0	11	1	52,189	-22.8%
Rwanda	10,169	630	6.2%	8	9	0	0	0	2	1,185	-32.8%
Sao Tome and Principe	372	-	-	3	9	1	0	6	2	104	-
Senegal	75,951	8,810	11.6%	12	4	6	0	17	0	23,958	-6.8%
Sierra Leone	27,699	582	2.1%	12	10	3	0	7	4	4,073	-25.5%
Somalia	246,201	1,970	0.8%	19	10	2	0	16	1	29,016	-9.3%
South Africa	470,693	25,888	5.5%	36	28	19	9	47	113	34,429	-0.9%
Sudan	967,500	50,310	5.2%	22	6	2	0	7	1	237,943	-13.5%
Swaziland	6,704	-	-	5	5	0	0	0	0	2,015	12.5%
Tanzania	364,900	108,740	29.8%	41	33	5	0	26	47	149,850	-2.3%
Togo	21,925	1,732	7.9%	9	0	2	0	7	0	1,969	-29.1%
Tunisia	63,170	190	0.3%	11	5	3	0	8	5	1,969	2.2%
Uganda	93,065	22,894	24.6%	20	13	0	0	27	10	16,178	-17.9%
Zambia	290,586	92,697	31.9%	11	11	0	0	0	6	120,641	-21.4%
Zimbabwe	150,873	18,256	12.1%	11	10	0	0	0	2	73,514	-14.4%
OCEANIA											
Australia	2,969,910	397,968	13.4%	63	35	38	35	74	282	596,678	-1.8%
Fiji	7,056	78	1.1%	5	13	6	1	8	2	3,147	-2.0%
Kiribati	313	-	-	0	4	1	0	4	1	108	-
Micronesia, Federated States of	271	-	-	6	5	2	0	6	4	58	-37.5%
New Zealand	104,454	30,918	29.6%	8	63	11	1	16	13	30,680	5.2%
Papua New Guinea	178,704	4,110	2.3%	58	32	9	0	31	12	118,151	-3.6%
Samoa	1,093	-	-	3	8	1	0	4	1	405	-19.2%
Solomon Islands	10,954	33	0.3%	20	23	4	0	4	6	9,792	-1.7%
Tonga	251	-	-	2	3	2	0	3	2	15	-
Vanuatu	4,707	-	-	5	8	2	0	4	0	1,726	1.4%

This table presents data for most independent nations having an area greater than 200 square miles
- Zero, insignificant, or not available
(1) Source: World Resources Institute, 2003. Earth Trends: The Environmental Information Portal. Available at http://earthtrends.wri.org. Washington D. C. World Resources Institute
(2) Source: United Nations Environment Programme - World Conservation Monitoring Centre (UNEP-WCMC); World Database on Protected Areas
(3) Source: International Union of Conservation of Nature and Natural Resources; IUCN 2003 Red List of Threatened Species <www.redlist.org>
(4) Source: United Nations Food and Agriculture Organization; Global Forest Resources Assessment 2000

WORLD COMPARISONS

General Information

Equatorial diameter of the earth, 7,926.38 miles.
Polar diameter of the earth, 7,899.80 miles.
Mean diameter of the earth, 7,917.52 miles.
Equatorial circumference of the earth, 24,901.46 miles.
Polar circumference of the earth, 24,855.34 miles.
Mean distance from the earth to the sun, 93,020,000 miles.
Mean distance from the earth to the moon, 238,857 miles.
Total area of the earth, 197,000,000 sq. miles.

Highest elevation on the earth's surface, Mt. Everest, Asia, 29,028 ft.
Lowest elevation on the earth's land surface, shores of the Dead Sea, Asia, 1,339 ft. below sea level.
Greatest known depth of the ocean, southwest of Guam, Pacific Ocean, 35,810 ft.
Total land area of the earth (incl. inland water and Antarctica), 57,900,000 sq. miles.

Area of Africa, 11,700,000 sq. miles.
Area of Antarctica, 5,400,000 sq. miles.
Area of Asia, 17,300,000 sq. miles.
Area of Europe, 3,800,000 sq. miles.
Area of North America, 9,500,000 sq. miles.
Area of Oceania (incl. Australia) 3,300,000 sq. miles.
Area of South America, 6,900,000 sq. miles.
Population of the earth (est. 1/1/04), 6,339,505,000.

Principal Islands and Their Areas

ISLAND	Area (Sq. Mi.)
Baffin I., Canada	195,928
Banks I., Canada	27,038
Borneo (Kalimantan), Asia	287,300
Bougainville, Papua New Guinea	3,591
Cape Breton I., Canada	3,981
Celebes (Sulawesi), Indonesia	73,057
Ceram (Seram), Indonesia	7,191
Corsica, France	3,367
Crete, Greece	3,189
Cuba, N. America	42,780
Cyprus, Asia	3,572
Devon I., Canada	21,331
Ellesmere I., Canada	75,767
Flores, Indonesia	5,502
Great Britain, U.K.	88,795
Greenland, N. America	840,000
Guadalcanal, Solomon Is.	2,060
Hainan Dao, China	13,127
Hawaii, U.S.	4,028
Hispaniola, N. America	29,300
Hokkaidō, Japan	32,245
Honshū, Japan	89,176
Iceland, Europe	39,769
Ireland, Europe	32,587
Jamaica, N. America	4,247
Java (Jawa), Indonesia	51,038
Kodiak I., U.S.	3,670
Kyūshū, Japan	17,129
Lyete, Philippines	2,785
Long Island, U.S.	1,377
Luzon, Philippines	40,420
Madagascar, Africa	226,642
Melville I., Canada	16,274
Mindanao, Philippines	36,537
Mindoro, Philippines	3,759
Negros, Philippines	4,907
New Britain, Papua New Guinea	14,093
New Caledonia, Oceania	6,252
Newfoundland, Canada	42,031
New Guinea, Asia-Oceania	308,882
New Ireland, Papua New Guinea	3,475
North East Land, Norway	6,350
North I., New Zealand	44,333
Novaya Zemlya, Russia	31,892
Palawan, Philippines	4,550
Panay, Philippines	4,446
Prince of Wales I., Canada	12,872
Puerto Rico, N. America	3,514
Sakhalin, Russia	29,498
Samar, Philippines	5,050
Sardinia, Italy	9,301
Shikoku, Japan	7,258
Sicily, Italy	9,926
Somerset I., Canada	9,570
Southampton I., Canada	15,913
South I., New Zealand	57,708
Spitsbergen, Norway	15,260
Sri Lanka, Asia	24,942
Sumatra (Sumatera), Indonesia	182,860
Taiwan, Asia	13,900
Tasmania, Australia	26,178
Tierra del Fuego, S. America	18,600
Timor, Asia	5,743
Vancouver I., Canada	12,079
Victoria I., Canada	83,897
Vrangelya (Wrangel), Russia	2,819

Principal Lakes, Oceans, Seas, and Their Areas

LAKE Country	Area (Sq. Mi.)
Arabian Sea	1,492,000
Aral Sea, Kazakhstan-Uzbekistan	13,000
Arctic Ocean	5,400,000
Athabasca, L., Canada	3,064
Atlantic Ocean	29,600,000
Balqash köli (L. Balkhash), Kazakhstan	7,027
Baltic Sea, Europe	163,000
Baykal, Ozero (L. Baikal), Russia	12,162
Bering Sea, Asia-N.A.	876,000
Black Sea, Europe-Asia	178,000
Caribbean Sea, N.A.-S.A.	1,063,000
Caspian Sea, Asia-Europe	144,402
Chad, L., Cameroon-Chad-Nigeria	595
Erie, L., Canada-U.S.	9,910
Eyre, L., Australia	3,668
Gairdner, L., Australia	1,076
Great Bear Lake, Canada	12,096
Great Salt Lake, U.S.	1,700
Great Slave Lake, Canada	11,030
Hudson Bay, Canada	475,000
Huron, L., Canada-U.S.	23,000
Indian Ocean	26,500,000
Japan, Sea of, Asia	389,000
Koko Nor (Qinghai Hu), China	1,722
Ladozhskoye Ozero (L. Ladoga), Russia	7,002
Manitoba, L., Canada	1,785
Mediterranean Sea, Europe-Africa-Asia	967,000
Mexico, Gulf of, N. America	596,000
Michigan, L., U.S.	22,300
Nicaragua, Lago de, Nicaragua	3,147
North Sea, Europe	222,000
Nyasa, L., Malawi-Mozambique-Tanzania	11,120
Onezhskoye Ozero (L. Onega), Russia	3,819
Ontario, L., Canada-U.S.	7,340
Pacific Ocean	60,100,000
Red Sea, Africa-Asia	169,000
Rudolf, L., Ethiopia-Kenya	2,471
Southern Ocean	7,800,000
Superior, L., Canada-U.S.	31,700
Tanganyika. L., Africa	12,355
Titicaca, Lago, Bolivia-Peru	3,232
Torrens, L., Australia	1,076
Vänern (L.), Sweden	2,181
Van Gölü (L.), Turkey	1,434
Victoria, L., Kenya-Tanzania-Uganda	26,564
Winnipeg, L., Canada	9,416
Winnipegosis, L., Canada	2,075
Yellow Sea, China-Korea	480,000

Principal Mountains and Their Heights

MOUNTAIN Country	Elev. (Ft.)
Aconcagua, Cerro, Argentina	22,831
Annapurna, Nepal	26,504
Aoraki, New Zealand	12,316
Api, Nepal	23,399
Apo, Philippines	9,692
Ararat, Mt., Turkey	16,854
Barú, Volcán, Panama	11,401
Bangueta, Mt., Papua New Guinea	13,520
Belukha, Mt., Kazakhstan-Russia	14,783
Bia, Phou, Laos	9,249
Blanc, Mont (Monte Bianco), France-Italy	15,771
Blanca Pk., Colorado, U.S.	14,345
Bolívar, Pico, Venezuela	16,427
Bonete, Cerro, Argentina	22,546
Borah Pk., Idaho, U.S.	12,662
Boundary Pk., Nevada, U.S.	13,140
Cameroon Mtn., Cameroon	13,451
Carrauntoohil, Ireland	3,406
Chaltel, Cerro (Monte Fitzroy), Argentina-Chile	10,958
Chimborazo, Ecuador	20,702
Chirripó, Cerro, Costa Rica	12,530
Colima, Nevado de, Mexico	13,911
Cotopaxi, Ecuador	19,347
Cristóbal Colón, Pico, Colombia	19,029
Damāvand, Qolleh-ye, Iran	18,386
Dhawalāgiri, Nepal	26,810
Duarte, Pico, Dominican Rep.	10,417
Dufourspitze (Monte Rosa), Italy-Switzerland	15,203
Elbert, Mt., Colorado, U.S.	14,433
El'brus, Gora, Russia	18,510
Elgon, Mt., Kenya-Uganda	14,178
Erciyeş, Dağı, Turkey	12,848
Etna, Mt., Italy	10,902
Everest, Mt., China-Nepal	29,028
Fairweather, Mt., Alaska-Canada	15,300
Folādī, Koh-e, Afghanistan	16,847
Foraker, Mt., Alaska, U.S.	17,400
Fuji San, Japan	12,388
Galdhøpiggen, Norway	8,100
Gannett Pk., Wyoming, U.S.	13,804
Gasherbrum, China-Pakistan	26,470
Gerlachovský štít, Slovakia	8,711
Giluwe, Mt., Papua New Guinea	14,331
Gongga Shan, China	24,790
Grand Teton, Wyoming, U.S.	13,770
Grossglockner, Austria	12,457
Hadūr Shu'ayb, Yemen	12,008
Haleakalā Crater, Hawaii, U.S.	10,023
Hekla, Iceland	4,892
Hood, Mt., Oregon, U.S.	11,239
Huascarán, Nevado, Peru	22,133
Huila, Nevado de, Colombia	18,865
Hvannadalshnúkur, Iceland	6,952
Illampu, Nevado, Bolivia	21,066
Illimani, Nevado, Bolivia	20,741
Ismail Samani, pik, Tajikistan	24,590
Iztaccíhuatl, Mexico	17,159
Jaya, Puncak, Indonesia	16,503
Jungfrau, Switzerland	13,642
K2 (Qogir Feng), China-Pakistan	28,250
Kāmet, China-India	25,447
Kānchenjunga, India-Nepal	28,208
Kātrīnā, Jabal, Egypt	8,668
Kebnekaise, Sweden	6,926
Kenya, Mt. (Kirinyaga), Kenya	17,058
Kerinci, Gunung, Indonesia	12,467
Kilimanjaro, Tanzania	19,340
Kinabalu, Gunong, Malaysia	13,455
Klyuchevskaya, Russia	15,584
Kosciuszko, Mt., Australia	7,313
Koussi, Emi, Chad	11,204
Kula Kangri, Bhutan	24,784
La Selle, Massif de, Haiti	8,793
Lassen Pk., California, U.S.	10,457
Llullaillaco, Volcán, Argentina-Chile	22,110
Logan, Mt., Canada	19,551
Longs Pk., Colorado, U.S.	14,255
Makālu, China-Nepal	27,825
Margherita Peak, Dem. Rep. of the Congo-Uganda	16,763
Markham, Mt., Antarctica	14,049
Maromokotro, Madagascar	9,436
Massive, Mt., Colorado, U.S.	14,421
Matterhorn, Italy-Switzerland	14,692
Mauna Kea, Hawaii, U.S.	13,796
Mauna Loa, Hawaii, U.S.	13,679
Mayon Volcano, Philippines	8,077
McKinley, Mt., Alaska, U.S.	20,320
Meron, Hare, Israel	3,963
Meru, Mt., Tanzania	14,978
Misti, Volcán, Peru	19,101
Mitchell, Mt., North Carolina, U.S.	6,684
Môco, Serra do, Angola	8,596
Moldoveanu, Romania	8,346
Mulhacén, Spain	11,424
Musala, Bulgaria	9,596
Muztag, China	25,338
Muztagata, China	24,757
Namjagbarwa Feng, China	25,446
Nanda Devi, India	25,645
Nanga Parbat, Pakistan	26,660
Narodnaya, Gora, Russia	6,217
Nevis, Ben, United Kingdom	4,406
Ojos del Salado, Nevado, Argentina-Chile	22,615
Ólimbos, Cyprus	6,401
Ólympos, Greece	9,570
Olympus, Mt., Washington, U.S.	7,965
Orizaba, Pico de, Mexico	18,406
Paektu San, North Korea-China	9,003
Paricutín, Mexico	9,186
Parnassós, Greece	8,061
Pelée, Montagne, Martinique	4,583
Pidurutalagala, Sri Lanka	8,281
Pikes Pk., Colorado, U.S.	14,110
Pobedy, pik, China-Kyrgyzstan	24,406
Popocatépetl, Volcán, Mexico	17,930
Pulog, Mt., Philippines	9,626
Rainier, Mt., Washington, U.S.	14,410
Ramm, Jabal, Jordan	5,755
Ras Dashen Terara, Ethiopia	15,158
Rinjani, Gunung, Indonesia	12,224
Robson, Mt., Canada	12,972
Roraima, Mt., Brazil-Guyana-Venezuela	9,432
Ruapehu, Mt., New Zealand	9,177
St. Elias, Mt., Alaska, U.S.-Canada	18,008
Sajama, Nevado, Bolivia	21,391
Semeru, Gunung, Indonesia	12,060
Shām, Jabal ash, Oman	9,957
Shasta, Mt., California, U.S.	14,162
Snowdon, United Kingdom	3,560
Tahat, Algeria	9,541
Tajumulco, Guatemala	13,845
Taranaki, Mt., New Zealand	8,260
Tirich Mīr, Pakistan	25,230
Tomanivi (Victoria), Fiji	4,341
Toubkal, Jebel, Morocco	13,665
Triglav, Slovenia	9,396
Trikora, Puncak, Indonesia	15,584
Tupungato, Cerro, Argentina-Chile	21,555
Turquino, Pico, Cuba	6,470
Uluru (Ayers Rock), Australia	2,844
Uncompahgre Pk, Colorado, U.S.	14,309
Vesuvio (Vesuvius), Italy	4,190
Victoria, Mt., Papua New Guinea	13,238
Vinson Massif, Antarctica	16,066
Waddington, Mt., Canada	13,163
Washington, Mt., New Hampshire, U.S.	6,288
Whitney, Mt., California, U.S.	14,494
Wilhelm, Mt., Papua New Guinea	14,793
Wrangell, Mt., Alaska, U.S.	14,163
Xixabangma Feng (Gosainthan), China	26,286
Yü Shan, Taiwan	13,114
Zugspitze, Austria-Germany	9,718

Principal Rivers and Their Lengths

RIVER Continent	Length (Mi.)
Albany, N. America	610
Aldan, Asia	1,412
Amazonas-Ucayali, S. America	4,000
Amu Darya, Asia	1,578
Amur, Asia	1,752
Araguaia, S. America	1,367
Arkansas, N. America	1,460
Atchafalaya, N. America	1,420
Athabasca, N. America	765
Brahmaputra, Asia	1,770
Brazos, N. America	1,280
Canadian, N. America	906
Churchill, N. America	1,000
Colorado, N. America (U.S.-Mexico)	1,450
Colorado, N. America (Texas)	862
Columbia, N. America	1,240
Congo (Zaïre), Africa	2,715
Danube, Europe	1,777
Darling, Australia	864
Dnieper (Dnipro), Europe	1,367
Don, Europe	1,162
Elbe, Europe	690
Essequibo, S. America	603
Euphrates, Asia	1,510
Fraser, N. America	851
Ganges, Asia	1,864
Gila, N. America	649
Godāvari, Asia	932
Huang (Yellow), Asia	2,902
Indigirka, Asia	1,072
Indus, Asia	1,118
Irrawaddy, Asia	1,300
Juruá, S. America	1,250
Kama, Europe	1,122
Kasai, Africa	1,338
Kolyma, Asia	1,323
Lena, Asia	2,734
Limpopo, Africa	1,100
Loire, Europe	634
Mackenzie, N. America	2,635
Madeira, S. America	2,013
Magdalena, S. America	951
Marañón, S. America	1,000
Mekong, Asia	2,796
Meuse, Europe	575
Mississippi, N. America	2,340
Mississippi-Missouri, N. America	3,710
Missouri, N. America	2,540
Murray-Darling, Australia	2,169
Negro, S. America	1,305
Nelson, N. America	1,600
Niger, Africa	2,585
Nile, Africa	4,132
Ob', Asia	2,268
Oder, Europe	565
Ohio, N. America	1,310
Oka, Europe	932
Orange, Africa	1,300
Orinoco, S. America	1,703
Ottawa, N. America	790
Paraguay, S. America	1,610
Parnaíba, S. America	901
Peace, N. America	1,195
Pechora, Europe	1,125
Pecos, N. America	926
Pilcomayo, S. America	1,550
Plata-Paraná, S. America	2,920
Platte, N. America	990
Purús, S. America	1,860
Red, N. America	1,290
Rhine, Europe	820
Rhône, Europe	503
Rio Grande, N. America	1,900
Roosevelt, S. America	950
St. Lawrence, N. America	1,900
Salado, S. America	870
Salween (Nu), Asia	1,750
São Francisco, S. America	1,740
Saskatchewan-Bow, N. America	1,205
Severnaya Dvina (Northern Dvina), Europe	462
Snake, N. America	1,040
Sungari (Songhua), Asia	1,140
Syr Darya, Asia	1,370
Tagus, Europe	625
Tarim, Asia	1,328
Tennessee, N. America	886
Tigris, Asia	1,180
Tisa, Europe	607
Tocantins, S. America	1,640
Ucayali, S. America	1,220
Ural, Asia	1,509
Uruguay, S. America	1,025
Verkhnyaya Tunguska (Angara), Asia	1,105
Vilyuy, Asia	1,647
Volga, Europe	2,082
Volta, Africa	994
Wisła (Vistula), Europe	630
Xiang, Asia	930
Xingu, S. America	1,230
Yangtze (Chang), Asia	3,915
Yellowstone, N. America	692
Yenisey, Asia	2,169
Yukon, N. America	1,980
Zambezi, Africa	1,653

PRINCIPAL CITIES OF THE WORLD

Abidjan, Cote d'Ivoire	1,929,079
Abū Ẓaby (Abu Dhabi), United Arab Emirates	242,975
Accra, Ghana (1,390,000)	949,113
Addis Ababa, Ethiopia	2,424,000
Ahmadābād, India (4,519,278)	3,515,361
Aleppo (Ḥalab), Syria (1,640,000)	1,591,400
Alexandria (Al Iskandarīyah), Egypt (3,350,000)	3,339,076
Algiers (El Djazaïr), Algeria (2,547,983)	1,507,241
Al Jīzah (Giza), Egypt (*Al Qāhirah)	2,221,817
Almaty, Kazakhstan (1,190,000)	1,129,356
'Ammān, Jordan (1,500,000)	1,147,447
Amsterdam, Netherlands (1,121,303)	727,053
Ankara, Turkey (3,294,220)	2,984,099
Antananarivo, Madagascar	1,250,000
Antwerp (Antwerpen), Belgium (1,135,000)	453,030
Ashgabat (Ashkhabad), Turkmenistan	557,600
Asmera, Eritrea	358,100
Astana (Aqmola), Kazakhstan (319,324)	312,965
Asunción, Paraguay (700,000)	546,637
Athens (Athína), Greece (3,150,000)	772,072
Atlanta, Georgia, U.S. (4,112,198)	416,474
Auckland, New Zealand (1,074,510)	367,737
Baghdād, Iraq	3,841,268
Baku (Bakı), Azerbaijan (2,020,000)	1,792,300
Bamako, Mali	658,275
Bandung, Indonesia	5,919,400
Bangalore, India (5,686,844)	4,292,223
Banghāzī, Libya	800,000
Bangkok (Krung Thep), Thailand (7,060,000)	5,620,591
Bangui, Central African Republic	451,690
Barcelona, Spain (4,000,000)	1,496,266
Beijing, China (7,320,000)	6,690,000
Beirut (Bayrūt), Lebanon (1,675,000)	509,000
Belfast, N. Ireland, U.K. (730,000)	297,300
Belgrade (Beograd), Serbia and Montenegro	1,594,483
Belo Horizonte, Brazil (4,055,000)	1,366,301
Berlin, Germany (4,220,000)	3,386,667
Birmingham, England, U.K. (2,705,000)	1,020,589
Bishkek, Kyrgyzstan	753,400
Bogotá, Colombia	6,422,198
Bonn, Germany (600,000)	301,048
Boston, Massachusetts, U.S. (5,819,100)	589,141
Brasília, Brazil	1,947,133
Bratislava, Slovakia	451,395
Brazzaville, Congo	693,712
Brisbane, Australia (1,627,535)	888,449
Brussels (Bruxelles), Belgium (2,390,000)	133,845
Bucharest (București), Romania (2,300,000)	2,016,131
Budapest, Hungary (2,450,000)	1,825,153
Buenos Aires, Argentina (11,000,000)	2,960,976
Cairo (Al Qāhirah), Egypt (9,300,000)	6,800,992
Calgary, Alberta, Canada (951,395)	878,866
Cali, Colombia	2,128,920
Canberra, Australia (342,798)	311,518
Cape Town, South Africa (1,900,000)	854,616
Caracas, Venezuela (4,000,000)	1,822,465
Cardiff, Wales, U.K. (645,000)	315,040
Casablanca, Morocco (3,400,000)	3,022,000
Changchun, China	2,470,000
Chelyabinsk, Russia (1,320,000)	1,086,300
Chengdu, China	2,760,000
Chennai (Madras), India (6,424,624)	4,216,268
Chicago, Illinois, U.S. (9,157,540)	2,896,016
Chișinău (Kishinev), Moldova (746,500)	658,300
Chittagong, Bangladesh (2,342,662)	1,566,070
Chongqing, China	3,870,000
Cincinnati, Ohio, U.S. (1,979,202)	331,285
Cleveland, Ohio, U.S. (2,945,831)	478,403
Cologne (Köln), Germany (1,830,000)	962,507
Colombo, Sri Lanka (2,050,000)	615,000
Conakry, Guinea	950,000
Copenhagen (København), Denmark (2,030,000)	499,148
Córdoba, Argentina (1,260,000)	1,179,067
Cotonou, Benin	650,660
Curitiba, Brazil (2,595,000)	1,586,848
Dakar, Senegal (1,976,533)	879,703
Dalian, China	2,400,000
Dallas, Texas, U.S. (5,221,801)	1,188,580
Damascus (Dimashq), Syria (2,230,000)	1,549,932
Dar es Salaam, Tanzania	1,360,850
Delhi, India (12,791,458)	9,817,439
Denver, Colorado, U.S. (2,581,506)	554,636
Detroit, Michigan, U.S. (5,456,428)	951,270
Dhaka (Dacca), Bangladesh (6,537,308)	3,637,892
Djibouti, Djibouti	329,337
Dnipropetrovs'k, Ukraine (1,590,000)	1,108,682
Donets'k, Ukraine (2,090,000)	1,050,369
Douala, Cameroon	712,251
Dublin (Baile Átha Cliath), Ireland (1,175,000)	481,854
Durban, South Africa (1,740,000)	669,242
Dushanbe, Tajikistan (700,000)	528,600
Düsseldorf, Germany (1,200,000)	568,855
Edinburgh, Scotland, U.K. (640,000)	448,850
Edmonton, Alberta, Canada (937,845)	666,104
Eşfahān, Iran (1,525,000)	1,266,072
Essen, Germany (5,040,000)	599,515
Fortaleza, Brazil (2,780,000)	788,956
Frankfurt am Main, Germany (1,960,000)	643,821
Fukuoka, Japan (2,000,000)	1,341,489
Geneva (Génève), Switzerland (450,592)	172,598
Glasgow, Scotland, U.K. (1,870,000)	616,430
Goiânia, Brazil	1,075,761
Guadalajara, Mexico (3,669,021)	1,646,183
Guangzhou (Canton), China	3,750,000
Guatemala, Guatemala (1,500,000)	1,006,954
Guayaquil, Ecuador	2,117,553
Halifax, Nova Scotia, Canada (359,183)	119,300
Hamburg, Germany (2,460,000)	1,704,735
Hannover, Germany (1,015,000)	514,718
Hanoi, Vietnam (1,275,000)	1,073,760
Harare, Zimbabwe (1,470,000)	1,189,103
Harbin, China	3,120,000
Havana (La Habana), Cuba (2,285,000)	2,189,716
Helsinki, Finland (939,697)	548,720
Hiroshima, Japan (1,600,000)	1,126,282
Ho Chi Minh City (Saigon), Vietnam (3,300,000)	3,015,743
Hong Kong (Xianggang), China (4,770,000)	1,250,993
Honolulu, Hawaii, U.S. (876,156)	371,657
Houston, Texas, U.S. (4,669,571)	1,953,631
Hyderābād, India (5,533,640)	3,449,878
Ibadan, Nigeria	1,144,000
Islāmābād, Pakistan (*Rāwalpindi)	529,180
İstanbul, Turkey (8,506,026)	8,260,438
İzmir, Turkey (2,554,363)	2,081,556
Jaipur, India	2,324,319
Jakarta, Indonesia (10,200,000)	9,373,900
Jerusalem (Yerushalayim), Israel (685,000)	633,700
Jiddah, Saudi Arabia	1,450,000
Jinan, China	2,150,000
Johannesburg, South Africa (4,000,000)	752,349
Kābul, Afghanistan	1,424,400
Kampala, Uganda	773,463
Kānpur, India (2,690,486)	2,540,069
Kaohsiung, Taiwan (1,845,000)	1,468,586
Karāchi, Pakistan	9,339,023
Katowice, Poland (2,755,000)	343,158
Kharkiv, Ukraine (1,990,000)	1,494,235
Khartoum (Al Kharṭūm), Sudan (1,450,000)	947,483
Kiev (Kyïv), Ukraine (3,250,000)	2,589,541
Kingston, Jamaica (830,000)	516,500
Kinshasa, Dem. Rep. of the Congo	3,000,000
Kitakyūshū, Japan (1,550,000)	1,011,491
Kolkata (Calcutta), India (13,216,546)	4,580,544
Kuala Lumpur, Malaysia (2,500,000)	1,297,526
Kuwait (Al Kuwayt), Kuwait (1,126,000)	28,859
Lagos, Nigeria (3,800,000)	1,213,000
Lahore, Pakistan	5,143,495
La Paz, Bolivia (1,487,854)	792,611
Libreville, Gabon (418,616)	362,386
Lilongwe, Malawi	435,964
Lima, Peru (6,321,173)	340,422
Lisbon (Lisboa), Portugal (2,350,000)	563,210
Liverpool, England, U.K. (1,515,000)	467,995
Ljubljana, Slovenia	263,832
Lomé, Togo	450,000
London, England, U.K. (12,000,000)	7,074,265
Los Angeles, California, U.S. (16,373,645)	3,694,820
Luanda, Angola	1,459,900
Lucknow, India (2,266,933)	2,207,340
Lusaka, Zambia	1,269,848
Lyon, France (1,648,216)	445,452
Madrid, Spain (4,690,000)	2,882,860
Managua, Nicaragua	864,201
Manaus, Brazil	1,394,724
Manchester, England, U.K. (2,760,000)	430,818
Manila, Philippines (11,200,000)	1,654,761
Mannheim, Germany (1,525,000)	307,730
Maputo, Mozambique	966,837
Maracaibo, Venezuela	1,249,670
Marseille, France (1,516,340)	798,430
Mashhad, Iran	1,887,405
Mecca (Makkah), Saudi Arabia	630,000
Medan, Indonesia	1,988,200
Medellín, Colombia (2,290,000)	1,885,001
Melbourne, Australia (3,366,542)	67,784
Mexico City (Ciudad de México), Mexico (17,786,983)	8,605,239
Miami, Florida, U.S. (3,876,380)	362,470
Milan (Milano), Italy (3,790,000)	1,305,591
Milwaukee, Wisconsin, U.S. (1,689,572)	596,974
Minneapolis, Minnesota, U.S. (2,968,806)	382,618
Minsk, Belarus (1,680,567)	1,677,137
Mogadishu (Muqdisho), Somalia	600,000
Monrovia, Liberia	465,000
Monterrey, Mexico (3,236,604)	1,110,909
Montevideo, Uruguay (1,650,000)	1,303,182
Montréal, Quebec, Canada (3,426,350)	1,039,534
Moscow (Moskva), Russia (12,850,000)	8,389,700
Mumbai (Bombay), India (16,368,084)	11,914,398
Munich (München), Germany (1,930,000)	1,194,560
Nagoya, Japan (5,250,000)	2,171,378
Nāgpur, India (2,122,965)	2,051,320
Nairobi, Kenya	2,143,254
Nanjing, China	2,490,000
Naples (Napoli), Italy (3,150,000)	1,046,987
N'Djamena, Chad	546,572
Newcastle upon Tyne, England, U.K. (1,350,000)	282,338
New Delhi, India (*Delhi)	294,783
New York, New York, U.S. (21,199,865)	8,008,278
Niamey, Niger	392,165
Nizhniy Novgorod, Russia (1,950,000)	1,364,900
Nouakchott, Mauritania	393,325
Novosibirsk, Russia (1,505,000)	1,402,400
Nürnberg, Germany (1,065,000)	486,628
Odesa, Ukraine (1,150,000)	1,002,246
Omsk, Russia (1,190,000)	1,157,600
Ōsaka, Japan (17,050,000)	2,598,589
Oslo, Norway (773,498)	504,040
Ottawa, Ontario, Canada (1,063,664)	774,072
Ouagadougou, Burkina Faso	634,479
Palembang, Indonesia	1,415,500
Panamá, Panama (995,000)	415,964
Paris, France (11,174,743)	2,125,246
Patna, India (1,707,429)	1,376,950
Perm', Russia (1,110,000)	1,017,100
Perth, Australia (1,244,320)	10,195
Philadelphia, Pennsylvania, U.S. (6,188,463)	1,517,550
Phnom Penh (Phnum Pénh), Cambodia	570,155
Phoenix, Arizona, U.S. (3,251,876)	1,321,045
Port Moresby, Papua New Guinea	246,664
Port-au-Prince, Haiti (1,425,594)	990,558
Portland, Oregon, U.S. (2,265,223)	529,121
Porto, Portugal (1,230,000)	273,060
Porto Alegre, Brazil (3,375,000)	1,304,998
Prague (Praha), Czech Republic (1,328,000)	1,193,270
Pretoria, South Africa (1,100,000)	692,348
Pune, India (3,755,525)	2,540,069
Pusan, South Korea	3,814,325
P'yŏngyang, North Korea	2,741,260
Qingdao, China	2,300,000
Québec, Quebec, Canada (682,757)	169,076
Quezon City, Philippines (*Manila)	1,989,419
Quito, Ecuador	1,615,809
Rabat, Morocco (1,200,000)	717,000
Rangoon (Yangon), Myanmar (2,800,000)	2,705,039
Recife, Brazil (3,160,000)	1,421,993
Regina, Saskatchewan, Canada (192,800)	178,225
Reykjavík, Iceland (166,015)	107,684
Rīga, Latvia (1,000,000)	792,508
Rio de Janeiro, Brazil (10,465,000)	5,851,914
Riyadh (Ar Riyāḍ), Saudi Arabia	1,800,000
Rome (Roma), Italy (3,235,000)	2,649,765
Rosario, Argentina (1,190,000)	894,645
Rostov-na-Donu, Russia (1,160,000)	1,017,300
Rotterdam, Netherlands (1,089,979)	539,000
Sacramento, California, U.S. (1,796,857)	407,018
St. Louis, Missouri, U.S. (2,603,607)	348,189
St. Petersburg (Leningrad), Russia (6,000,000)	4,728,200
Salvador, Brazil (2,855,000)	2,439,823
Samara, Russia (1,450,000)	1,168,000
San Diego, California, U.S. (2,813,833)	1,223,400
San Francisco, California, U.S. (7,039,362)	776,733
San José, Costa Rica (996,194)	309,672
San Juan, Puerto Rico (1,967,627)	421,958
San Salvador, El Salvador (1,908,921)	473,372
Santiago, Chile	4,788,543
Santo Domingo, Dominican Republic	2,677,056
São Paulo, Brazil (17,380,000)	9,713,692
Sapporo, Japan (2,000,000)	1,822,300
Sarajevo, Bosnia and Herzegovina	367,703
Saratov, Russia (1,135,000)	881,000
Seattle, Washington, U.S. (3,554,760)	563,374
Seoul (Sŏul), South Korea (15,850,000)	10,231,217
Shanghai, China (11,010,000)	8,930,000
Shenyang (Mukden), China	4,050,000
Singapore, Singapore (4,400,000)	4,017,700
Skopje, Macedonia	440,577
Sofia (Sofiya), Bulgaria (1,189,794)	1,138,629
Stockholm, Sweden (1,643,366)	743,703
Stuttgart, Germany (2,020,000)	582,443
Surabaya, Indonesia	2,801,300
Sūrat, India (2,811,466)	2,433,787
Sydney, Australia (3,741,290)	11,115
T'aipei, Taiwan (6,200,000)	2,640,322
Tallinn, Estonia	403,981
Tashkent (Toshkent), Uzbekistan (2,325,000)	2,142,700
Tbilisi, Georgia (1,460,000)	1,279,000
Tegucigalpa, Honduras	576,661
Tehrān, Iran (8,800,000)	6,758,845
Tel Aviv-Yafo, Israel (1,890,000)	348,100
Tianjin (Tientsin), China	5,000,000
Tiranë, Albania	244,153
Tōkyō, Japan (30,300,000)	8,130,408
Toronto, Ontario, Canada (4,682,897)	2,481,494
Tripoli (Ṭarābulus), Libya	1,500,000
Tunis, Tunisia (1,300,000)	702,330
Turin (Torino), Italy (1,550,000)	921,485
Ufa, Russia (1,110,000)	1,088,900
Ulan Bator (Ulaanbaatar), Mongolia	672,882
Ürümqi, China	1,130,000
València, Spain (1,340,000)	739,014
Vancouver, British Columbia, Canada (1,986,965)	545,671
Viangchan (Vientiane), Laos	464,000
Vienna (Wien), Austria (1,950,000)	1,609,631
Vilnius, Lithuania	578,334
Volgograd (Stalingrad), Russia (1,358,000)	1,000,000
Warsaw (Warszawa), Poland (2,300,000)	1,615,369
Washington, D.C., U.S. (7,608,070)	572,059
Wellington, New Zealand (346,500)	167,400
Winnipeg, Manitoba, Canada (671,274)	619,544
Wuhan, China	3,870,000
Xi'an, China	2,410,000
Yekaterinburg, Russia (1,530,000)	1,272,900
Yerevan, Armenia (1,315,000)	1,249,202
Yokohama, Japan (*Tōkyō)	3,426,506
Zagreb, Croatia	867,865
Zürich, Switzerland (932,681)	337,553

Metropolitan area populations are shown in parentheses.
* City is located within the metropolitan area of another city; for example, Yokohama, Japan is located in the Tōkyō metropolitan area.

GLOSSARY OF FOREIGN GEOGRAPHICAL TERMS

Annam . . . Annamese
Arab . . . Arabic
Bantu . . . Bantu
Bur . . . Burmese
Camb . . . Cambodian
Celt . . . Celtic
Chn . . . Chinese
Czech . . . Czech
Dan . . . Danish
Du . . . Dutch
Fin . . . Finnish
Fr . . . French
Ger . . . German
Gr . . . Greek
Hung . . . Hungarian
Ice . . . Icelandic
India . . . India
Indian . . . American Indian
Indon . . . Indonesian
It . . . Italian
Jap . . . Japanese
Kor . . . Korean
Mal . . . Malayan
Mong . . . Mongolian
Nor . . . Norwegian
Per . . . Persian
Pol . . . Polish
Port . . . Portuguese
Rom . . . Romanian
Rus . . . Russian
Siam . . . Siamese
So. Slav . . . Southern Slavonic
Sp . . . Spanish
Swe . . . Swedish
Tib . . . Tibetan
Tur . . . Turkish
Yugo . . . Yugoslav

å, Nor., Swe . . . brook, river
aa, Dan., Nor . . . brook
aas, Dan., Nor . . . ridge
āb, Per . . . water, river
abad, India, Per . . . town, city
ada, Tur . . . island
adrar, Berber . . . mountain
air, Indon . . . stream
akrotírion, Gr . . . cape
älf, Swe . . . river
alp, Ger . . . mountain
altipiano, It . . . plateau
alto, Sp . . . height
archipel, Fr . . . archipelago
archipiélago, Sp . . . archipelago
arquipélago, Port . . . archipelago
arroyo, Sp . . . brook, stream
ås, Nor., Swe . . . ridge
austral, Sp . . . southern
baai, Du . . . bay
bab, Arab . . . gate, port
bach, Ger . . . brook, stream
backe, Swe . . . hill
bad, Ger . . . bath, spa
bahía, Sp . . . bay, gulf
bahr, Arab . . . river, sea, lake
baia, It . . . bay, gulf
baía, Port . . . bay
baie, Fr . . . bay, gulf
bajo, Sp . . . depression
bak, Indon . . . stream
bakke, Dan., Nor . . . hill
balkan, Tur . . . mountain range
bana, Jap . . . point, cape
banco, Sp . . . bank
bandar, Mal., Per. . . . town, port, harbor
bang, Siam . . . village
bassin, Fr . . . basin
batang, Indon., Mal . . . river
ben, Celt . . . mountain, summit
bender, Arab . . . harbor, port
bereg, Rus . . . coast, shore
berg, Du., Ger., Nor., Swe. . . . mountain, hill
bir, Arab . . . well
birkat, Arab . . . lake, pond, pool
bit, Arab . . . house
bjaerg, Dan., Nor . . . mountain
bocche, It . . . mouth
boğazı, Tur . . . strait
bois, Fr . . . forest, wood
boloto, Rus . . . marsh
bolsón, Sp. . . . flat-floored desert valley
boreal, Sp . . . northern
borg, Dan., Nor., Swe . . . castle, town
borgo, It . . . town, suburb
bosch, Du . . . forest, wood
bouche, Fr . . . river mouth
bourg, Fr . . . town, borough
bro, Dan., Nor., Swe . . . bridge
brücke, Ger . . . bridge
bucht, Ger . . . bay, bight
bugt, Dan., Nor., Swe . . . bay, gulf
bulu, Indon . . . mountain
burg, Du., Ger . . . castle, town
buri, Siam . . . town
burun, burnu, Tur . . . cape
by, Dan., Nor., Swe . . . village
caatinga, Port. (Brazil) . . . open brushland
cabezo, Sp . . . summit
cabo, Port., Sp . . . cape
campo, It., Port., Sp . . . plain, field
campos, Port. (Brazil) . . . plains
cañon, Sp . . . canyon
cap, Fr . . . cape
capo, It . . . cape
casa, It., Port., Sp . . . house
castello, It., Port . . . castle, fort
castillo, Sp . . . castle
càte, Fr . . . hill
çay, Tur . . . stream, river
cayo, Sp . . . rock, shoal, islet
cerro, Sp . . . mountain, hill
champ, Fr . . . field
chang, Chn . . . village, middle
château, Fr . . . castle
chen, Chn . . . market town
chiang, Chn . . . river
chott, Arab . . . salt lake
chou, Chn. capital of district; island
chu, Tib . . . water, stream
cidade, Port . . . town, city
cima, Sp . . . summit, peak
città, It . . . town, city
ciudad, Sp . . . town, city
cochilha, Port . . . ridge
col, Fr . . . pass
colina, Sp . . . hill
cordillera, Sp . . . mountain chain
costa, It., Port., Sp . . . coast
côte, Fr . . . coast
cuchilla, Sp . . . mountain ridge
dağ, Tur . . . mountain(s)
dake, Jap . . . peak, summit
dal, Dan., Du., Nor., Swe . . . valley
dan, Kor . . . point, cape
danau, Indon . . . lake
dar, Arab . . . house, abode, country
darya, Per . . . river, sea
dasht, Per . . . plain, desert
deniz, Tur . . . sea
désert, Fr . . . desert
deserto, It . . . desert
desierto, Sp . . . desert
détroit, Fr . . . strait
dijk, Du . . . dam, dike
djebel, Arab . . . mountain
do, Kor . . . island
dorf, Ger . . . village
dorp, Du . . . village
duin, Du . . . dune
dzong, Tib. . . . fort, administrative capital
eau, Fr . . . water
ecuador, Sp . . . equator
eiland, Du . . . island
elv, Dan., Nor . . . river, stream
embalse, Sp . . . reservoir
erg, Arab . . . dune, sandy desert
est, Fr., It . . . east
estado, Sp . . . state
este, Port., Sp . . . east
estrecho, Sp . . . strait
étang, Fr . . . pond, lake
état, Fr . . . state
eyjar, Ice . . . islands
feld, Ger . . . field, plain
festung, Ger . . . fortress
fiume, It . . . river
fjäll, Swe . . . mountain
fjärd, Swe . . . bay, inlet
fjeld, Nor . . . mountain, hill
fjord, Dan., Nor . . . fiord, inlet
fjördur, Ice . . . fiord, inlet
fleuve, Fr . . . river
flod, Dan., Swe . . . river
flói, Ice . . . bay, marshland
fluss, Ger . . . river
foce, It . . . river mouth
fontein, Du . . . a spring
forêt, Fr . . . forest
fors, Swe . . . waterfall
forst, Ger . . . forest
fos, Dan., Nor . . . waterfall
fu, Chn . . . town, residence
fuente, Sp . . . spring, fountain
fuerte, Sp . . . fort
furt, Ger . . . ford
gang, Kor . . . stream, river
gangri, Tib . . . mountain
gat, Dan., Nor . . . channel
gave, Fr . . . stream
gawa, Jap . . . river
gebergte, Du . . . mountain range
gebiet, Ger . . . district, territory
gebirge, Ger . . . mountains
ghat, India . . . pass, mountain range
gobi, Mong . . . desert
gol, Mong . . . river
göl, gölü, Tur . . . lake
golf, Du., Ger . . . gulf, bay
golfe, Fr . . . gulf, bay
golfo, It., Port., Sp . . . gulf, bay
gomba, gompa, Tib . . . monastery
gora, Rus., So. Slav . . . mountain
góra, Pol . . . mountain
gorod, Rus . . . town
grad, Rus., So. Slav . . . town
guba, Rus . . . bay, gulf
gundung, Indon . . . mountain
guntō, Jap . . . archipelago
gunung, Mal . . . mountain
haf, Swe . . . sea, ocean
hafen, Ger . . . port, harbor
haff, Ger . . . gulf, inland sea
hai, Chn . . . sea, lake
hama, Jap . . . beach, shore
hamada, Arab . . . rocky plateau
hamn, Swe . . . harbor
hāmūn, Per . . . swampy lake, plain
hantō, Jap . . . peninsula
hassi, Arab . . . well, spring
haus, Ger . . . house
haut, Fr . . . summit, top
hav, Dan., Nor . . . sea, ocean
havn, Dan., Nor . . . harbor, port
havre, Fr . . . harbor, port
háza, Hung . . . house, dwelling of
heim, Ger . . . hamlet, home
hem, Swe . . . hamlet, home
higashi, Jap . . . east
hisar, Tur . . . fortress
hissar, Arab . . . fort
ho, Chn . . . river
hoek, Du . . . cape
hof, Ger . . . court, farmhouse
höfn, Ice . . . harbor
hoku, Jap . . . north
holm, Dan., Nor., Swe . . . island
hora, Czech . . . mountain
horn, Ger . . . peak
hoved, Dan., Nor . . . cape
hsien, Chn . . . district, district capital
hu, Chn . . . lake
hügel, Ger . . . hill
huk, Dan., Swe . . . point
hus, Dan., Nor., Swe . . . house
île, Fr . . . island
ilha, Port . . . island
indsö, Dan., Nor . . . lake
insel, Ger . . . island
insjö, Swe . . . lake
irmak, irmagi, Tur . . . river
isla, Sp . . . island
isola, It . . . island
istmo, It., Sp . . . isthmus
järvi, jaur, Fin . . . lake
jebel, Arab . . . mountain
jima, Jap . . . island
jökel, Nor . . . glacier
joki, Fin . . . river
jökull, Ice . . . glacier
kaap, Du . . . cape
kai, Jap . . . bay, gulf, sea
kaikyō, Jap . . . channel, strait
kalat, Per . . . castle, fortress
kale, Tur . . . fort
kali, Mal . . . creek, river
kand, Per . . . village
kang, Chn . . . mountain ridge; village
kap, Dan., Ger . . . cape
kapp, Nor., Swe . . . cape
kasr, Arab . . . fort, castle
kawa, Jap . . . river
kefr, Arab . . . village
kei, Jap . . . creek, river
ken, Jap . . . prefecture
khor, Arab . . . bay, inlet
khrebet, Rus . . . mountain range
kiang, Chn . . . large river
king, Chn . . . capital city, town
kita, Jap . . . north
ko, Jap . . . lake
köbstad, Dan . . . market-town
kol, Mong . . . lake
kólpos, Gr . . . gulf
kong, Chn . . . river
kopf, Ger . . . head, summit, peak
köpstad, Swe . . . market-town
körfezi, Tur . . . gulf
kosa, Rus . . . spit
kou, Chn . . . river mouth
köy, Tur . . . village
kraal, Du. (Africa) . . . native village
ksar, Arab . . . fortified village
kuala, Mal . . . bay, river mouth
kuh, Per . . . mountain
kum, Tur . . . sand
kuppe, Ger . . . summit
küste, Ger . . . coast
kyo, Jap . . . town, capital
la, Tib . . . mountain pass
labuan, Mal . . . anchorage, port
lac, Fr . . . lake
lago, It., Port., Sp . . . lake
lagoa, Port . . . lake, marsh
laguna, It., Port., Sp . . . lagoon, lake
lahti, Fin . . . bay, gulf
län, Swe . . . county
landsby, Dan., Nor . . . village
liehtao, Chn . . . archipelago
liman, Tur . . . bay, port
ling, Chn . . . pass, ridge, mountain
llanos, Sp . . . plains
loch, Celt. (Scotland) . . . lake, bay
loma, Sp . . . long, low hill
lough, Celt. (Ireland) . . . lake, bay
machi, Jap . . . town
man, Kor . . . bay
mar, Port., Sp . . . sea
mare, It., Rom . . . sea
marisma, Sp . . . marsh, swamp
mark, Ger . . . boundary, limit
massif, Fr . . . block of mountains
mato, Port . . . forest, thicket
me, Siam . . . river
meer, Du., Ger . . . lake, sea
mer, Fr . . . sea
mesa, Sp . . . flat-topped mountain
meseta, Sp . . . plateau
mina, Port., Sp . . . mine
minami, Jap . . . south
minato, Jap . . . harbor, haven
misaki, Jap . . . cape, headland
mont, Fr . . . mount, mountain
montagna, It . . . mountain
montagne, Fr . . . mountain
montaña, Sp . . . mountain
monte, It., Port., Sp. . . . mount, mountain
more, Rus., So. Slav . . . sea
morro, Port., Sp . . . hill, bluff
mühle, Ger . . . mill
mund, Ger . . . mouth, opening
mündung, Ger . . . river mouth
mura, Jap . . . township
myit, Bur . . . river
mys, Rus . . . cape
nada, Jap . . . sea
nadi, India . . . river, creek
naes, Dan., Nor . . . cape
nafud, Arab . . . desert of sand dunes
nagar, India . . . town, city
nahr, Arab . . . river
nam, Siam . . . river, water
nan, Chn., Jap . . . south
näs, Nor., Swe . . . cape
nez, Fr . . . point, cape
nishi, nisi, Jap . . . west
njarga, Fin . . . peninsula
nong, Siam . . . marsh
noord, Du . . . north
nor, Mong . . . lake
nord, Dan., Fr., Ger., It., Nor., Swe . . . north
norte, Port., Sp . . . north
nos, Rus . . . cape
nyasa, Bantu . . . lake
ö, Dan., Nor., Swe . . . island
occidental, Sp . . . western
ocna, Rom . . . salt mine
odde, Dan., Nor . . . point, cape
oeste, Port., Sp . . . west
oka, Jap . . . hill
oost, Du . . . east
oriental, Sp . . . eastern
óros, Gr . . . mountain
ost, Ger., Swe . . . east
öster, Dan., Nor., Swe . . . eastern
ostrov, Rus . . . island
oued, Arab . . . river, stream
ouest, Fr . . . west
ozero, Rus . . . lake
pää, Fin . . . mountain
padang, Mal . . . plain, field
pampas, Sp. (Argentina) . . . grassy plains
pará, Indian (Brazil) . . . river
pas, Fr . . . channel, passage
paso, Sp . . . mountain pass, passage
passo, It., Port. . . . mountain pass, passage, strait
patam, India . . . city, town
pei, Chn . . . north
pélagos, Gr . . . open sea
pegunungan, Indon . . . mountains
peña, Sp . . . rock
peresheyek, Rus . . . isthmus
pertuis, Fr . . . strait
peski, Rus . . . desert
pic, Fr . . . mountain peak
pico, Port., Sp . . . mountain peak
piedra, Sp . . . stone, rock
ping, Chn . . . plain, flat
planalto, Port . . . plateau
planina, Yugo . . . mountains
playa, Sp . . . shore, beach
pnom, Camb . . . mountain
pointe, Fr . . . point
polder, Du., Ger . . . reclaimed marsh
polje, So. Slav . . . plain, field
poluostrov, Rus . . . peninsula
pont, Fr . . . bridge
ponta, Port . . . point, headland
ponte, It., Port . . . bridge
pore, India . . . city, town
porthmós, Gr . . . strait
porto, It., Port . . . port, harbor
potamós, Gr . . . river
p'ov, Rus . . . peninsula
prado, Sp . . . field, meadow
presqu'île, Fr . . . peninsula
proliv, Rus . . . strait
pu, Chn . . . commercial village
pueblo, Sp . . . town, village
puerto, Sp . . . port, harbor
pulau, Indon . . . island
punkt, Ger . . . point
punt, Du . . . point
punta, It., Sp . . . point
pur, India . . . city, town
puy, Fr . . . peak
qal'a, qal'at, Arab . . . fort, village
qasr, Arab . . . fort, castle
rann, India . . . wasteland
ra's, Arab . . . cape, head
reka, Rus., So. Slav . . . river
reprêsa, Port . . . reservoir
rettō, Jap . . . island chain
ría, Sp . . . estuary
ribeira, Port . . . stream
riberão, Port . . . river
rio, It., Port . . . stream, river
río, Sp . . . river
rivière, Fr . . . river
roca, Sp . . . rock
rt, Yugo . . . cape
rūd, Per . . . river
saari, Fin . . . island
sable, Fr . . . sand
sahara, Arab . . . desert, plain
saki, Jap . . . cape
sal, Sp . . . salt
salar, Sp . . . salt flat, salt lake
salto, Sp . . . waterfall
san, Jap., Kor . . . mountain, hill
sat, satul, Rom . . . village
schloss, Ger . . . castle
sebkha, Arab . . . salt marsh
see, Ger . . . lake, sea
şehir, Tur . . . town, city
selat, Indon . . . stream
selvas, Port. (Brazil) . . . tropical rain forests
seno, Sp . . . bay
serra, Port . . . mountain chain
serranía, Sp . . . mountain ridge
seto, Jap . . . strait
severnaya, Rus . . . northern
shahr, Per . . . town, city
shan, Chn . . . mountain, hill, island
shatt, Arab . . . river
shi, Jap . . . city
shima, Jap . . . island
shōtō, Jap . . . archipelago
si, Chn . . . west, western
sierra, Sp . . . mountain range
sjö, Nor., Swe . . . lake, sea
sö, Dan., Nor . . . lake, sea
söder, södra, Swe . . . south
song, Annam . . . river
sopka, Rus . . . peak, volcano
source, Fr . . . a spring
spitze, Ger . . . summit, point
staat, Ger . . . state
stad, Dan., Du., Nor., Swe. . . . city, town
stadt, Ger . . . city, town
stato, It . . . state
step', Rus . . . treeless plain, steppe
straat, Du . . . strait
strand, Dan., Du., Ger., Nor., Swe . . . shore, beach
stretto, It . . . strait
strom, Ger . . . river, stream
ström, Dan., Nor., Swe. . . . stream, river
stroom, Du . . . stream, river
su, suyu, Tur . . . water, river
sud, Fr., Sp . . . south
süd, Ger . . . south
suidō, Jap . . . channel
sul, Port . . . south
sund, Dan., Nor., Swe . . . sound
sungai, sungei, Indon., Mal . . . river
sur, Sp . . . south
syd, Dan., Nor., Swe . . . south
tafelland, Ger . . . plateau
take, Jap . . . peak, summit
tal, Ger . . . valley
tanjung, tanjong, Mal . . . cape
tao, Chn . . . island
târg, târgul, Rom . . . market, town
tell, Arab . . . hill
teluk, Indon . . . bay, gulf
terra, It . . . land
terre, Fr . . . earth, land
thal, Ger . . . valley
tierra, Sp . . . earth, land
tō, Jap . . . east; island
tonle, Camb . . . river, lake
top, Du . . . peak
torp, Swe . . . hamlet, cottage
tsangpo, Tib . . . river
tsi, Chn . . . village, borough
tso, Tib . . . lake
tsu, Jap . . . harbor, port
tundra, Rus . . . treeless arctic plains
tung, Chn . . . east
tuz, Tur . . . salt
udde, Swe . . . cape
ufer, Ger . . . shore, riverbank
ujung, Indon . . . point, cape
umi, Jap . . . sea, gulf
ura, Jap . . . bay, coast, creek
ust'ye, Rus . . . river mouth
valle, It., Port., Sp . . . valley
vallée, Fr . . . valley
valli, It . . . lake
vár, Hung . . . fortress
város, Hung . . . town
varoš, So. Slav . . . town
veld, Du . . . open plain, field
verkh, Rus . . . top, summit
ves, Czech . . . village
vest, Dan., Nor., Swe . . . west
vik, Swe . . . cove, bay
vila, Port . . . town
villa, Sp . . . town
villar, Sp . . . village, hamlet
ville, Fr . . . town, city
vostok, Rus . . . east
wad, wādī, Arab. . . . intermittent stream
wald, Ger . . . forest, woodland
wan, Chn., Jap . . . bay, gulf
weiler, Ger . . . hamlet, village
westersch, Du . . . western
wüste, Ger . . . desert
yama, Jap . . . mountain
yarimada, Tur . . . peninsula
yug, Rus . . . south
zaki, Jap . . . cape
zaliv, Rus . . . bay, gulf
zapad, Rus . . . west
zee, Du . . . sea
zemlya, Rus . . . land
zuid, Du . . . south

ABBREVIATIONS OF GEOGRAPHICAL NAMES AND TERMS

Afg.	Afghanistan	Grc.	Greece	Ok., U.S.	Oklahoma, U.S.
Afr.	Africa	Gren.	Grenada	Or., U.S.	Oregon, U.S.
Ak., U.S.	Alaska, U.S.	Grnld.	Greenland	p.	Pass
Al., U.S.	Alabama, U.S.	Guad.	Guadeloupe	Pa., U.S.	Pennsylvania, U.S.
Alb.	Albania	Guat.	Guatemala	Pak.	Pakistan
Alg.	Algeria	Guern.	Guernsey	Pan.	Panama
Am. Sam.	American Samoa	Gui.	Guinea	Pap. N. Gui.	Papua New Guinea
And.	Andorra	Gui.-B.	Guinea-Bissau	Para.	Paraguay
Ang.	Angola	Guy.	Guyana	pen.	Peninsula
Ant.	Antarctica			Phil.	Philippines
Antig.	Antigua and Barbuda	Hi., U.S.	Hawaii, U.S.	Pit.	Pitcairn
aq.	Aqueduct	hist.	Historic Site, Ruins	pl.	Plain, Flat
Ar., U.S.	Arkansas, U.S.	hist. reg.	Historic Region	plat.	Plateau, Highland
Arg.	Argentina	Hond.	Honduras	Pol.	Poland
Arm.	Armenia	Hung.	Hungary	Port.	Portugal
arpt.	Airport			P.R.	Puerto Rico
Aus.	Austria	i.	Island	prov.	Province, Region
Austl.	Australia	Ia., U.S.	Iowa, U.S.	pt. of i.	Point of Interest
Az., U.S.	Arizona, U.S.	ice	Ice Feature, Glacier		
Azer.	Azerbaijan	Ice.	Iceland	r.	River, Creek
		Id., U.S.	Idaho, U.S.	Reu.	Reunion
b.	Bay, Gulf, Inlet, Lagoon	Il., U.S.	Illinois, U.S.	rec.	Recreational Site, Park
Bah.	Bahamas	In., U.S.	Indiana, U.S.	reg.	Physical Region
Bahr.	Bahrain	Indon.	Indonesia	rel.	Religious Institution
Barb.	Barbados	I. of Man	Isle of Man	res.	Reservoir
Bdi.	Burundi	I.R.	Indian Reservation	rf.	Reef, Shoal
Bel.	Belgium	Ire.	Ireland	R.I., U.S.	Rhode Island, U.S.
Bela.	Belarus	is.	Islands	Rom.	Romania
Ber.	Bermuda	Isr.	Israel	Rw.	Rwanda
Bhu.	Bhutan	isth.	Isthmus		
bk.	Undersea Bank			S.A.	South America
bldg.	Building	Jam.	Jamaica	S. Afr.	South Africa
Blg.	Bulgaria	Jord.	Jordan	Sau. Ar.	Saudi Arabia
Bngl.	Bangladesh			S.C., U.S.	South Carolina, U.S.
Bol.	Bolivia	Kaz.	Kazakhstan	sci.	Scientific Station
Bos.	Bosnia and Herzegovina	Kir.	Kiribati	Scot., U.K.	Scotland, U.K.
Bots.	Botswana	Kor., N.	Korea, North	S.D., U.S.	South Dakota, U.S.
Braz.	Brazil	Kor., S.	Korea, South	sea feat.	Undersea Feature
Bru.	Brunei	Ks., U.S.	Kansas, U.S.	Sen.	Senegal
Br. Vir. Is.	British Virgin Islands	Kuw.	Kuwait	Serb.	Serbia and Montenegro
bt.	Bight	Ky., U.S.	Kentucky, U.S.	Sey.	Seychelles
Burkina	Burkina Faso	Kyrg.	Kyrgyzstan	S. Geor.	South Georgia
				Sing.	Singapore
c.	Cape, Point	l.	Lake, Pond	S.L.	Sierra Leone
Ca., U.S.	California, U.S.	La., U.S.	Louisiana, U.S.	Slvk.	Slovakia
Cam.	Cameroon	Lat.	Latvia	Slvn.	Slovenia
Camb.	Cambodia	Leb.	Lebanon	S. Mar.	San Marino
can.	Canal	Leso.	Lesotho	Sol. Is.	Solomon Islands
Can.	Canada	Lib.	Liberia	Som.	Somalia
C.A.R.	Central African Republic	Liech.	Liechtenstein	Sp. N. Afr.	Spanish North Africa
Cay. Is.	Cayman Islands	Lith.	Lithuania	Sri L	Sri Lanka
C. Iv.	Cote d'Ivoire	Lux.	Luxembourg	St. Hel.	St. Helena
clf.	Cliff, Escarpment			St. K./N.	St. Kitts and Nevis
co.	County, Parish	Ma., U.S.	Massachusetts, U.S.	St. Luc.	St. Lucia
Co., U.S.	Colorado, U.S.	Mac.	Macedonia	St. P./M.	St. Pierre and Miquelon
Col.	Colombia	Madag.	Madagascar	strt.	Strait, Channel, Sound
Com.	Comoros	Malay.	Malaysia	S. Tom./P.	Sao Tome and Principe
cont.	Continent	Mald.	Maldives	St. Vin.	St. Vincent and
Cook Is.	Cook Islands	Marsh. Is.	Marshall Islands		the Grenadines
C.R.	Costa Rica	Mart.	Martinique	Sur.	Suriname
Cro.	Croatia	Maur.	Mauritania	Sval.	Svalbard
cst.	Coast, Beach	May.	Mayotte	sw.	Swamp, Marsh
Ct., U.S.	Connecticut, U.S.	Md., U.S.	Maryland, U.S.	Swaz.	Swaziland
C.V.	Cape Verde	Me., U.S.	Maine, U.S.	Swe.	Sweden
Cyp.	Cyprus	Mex.	Mexico	Switz.	Switzerland
Czech Rep.	Czech Republic	Mi., U.S.	Michigan, U.S.		
		Micron.	Micronesia,	Tai.	Taiwan
d.	Delta		Federated States of	Taj.	Tajikistan
D.C., U.S.	District of	Mn., U.S.	Minnesota, U.S.	Tan.	Tanzania
	Columbia, U.S.	Mo., U.S.	Missouri, U.S.	T./C. Is.	Turks and Caicos Islands
De., U.S.	Delaware, U.S.	Mol.	Moldova	ter.	Territory
Den.	Denmark	Mong.	Mongolia	Thai.	Thailand
dep.	Dependency, Colony	Monts.	Montserrat	Tn., U.S.	Tennessee, U.S.
depr.	Depression	Mor.	Morocco	trans.	Transportation Facility
dept.	Department, District	Moz.	Mozambique	Trin.	Trinidad and Tobago
des.	Desert	Ms., U.S.	Mississippi, U.S.	Tun.	Tunisia
Dji.	Djibouti	Mt., U.S.	Montana, U.S.	Tur.	Turkey
Dom.	Dominica	mth.	River Mouth or Channel	Turkmen.	Turkmenistan
Dom. Rep.	Dominican Republic	mtn.	Mountain	Tx., U.S.	Texas, U.S.
D.R.C.	Democratic Republic	mts.	Mountains		
	of the Congo	Mwi.	Malawi	U.A.E.	United Arab Emirates
		Mya.	Myanmar	Ug.	Uganda
Ec.	Ecuador			U.K.	United Kingdom
educ.	Educational Facility	N.A.	North America	Ukr.	Ukraine
El Sal.	El Salvador	N.C., U.S.	North Carolina, U.S.	Ur.	Uruguay
Eng., U.K.	England, U.K.	N. Cal.	New Caledonia	U.S.	United States
Eq. Gui.	Equatorial Guinea	N.D., U.S.	North Dakota, U.S.	Ut., U.S.	Utah, U.S.
Erit.	Eritrea	Ne., U.S.	Nebraska, U.S.	Uzb.	Uzbekistan
Est.	Estonia	neigh.	Neighborhood		
est.	Estuary	Neth.	Netherlands	Va., U.S.	Virginia, U.S.
Eth.	Ethiopia	Neth. Ant.	Netherlands Antilles	val.	Valley, Watercourse
E. Timor	East Timor	N.H., U.S.	New Hampshire, U.S.	Ven.	Venezuela
Eur.	Europe	Nic.	Nicaragua	Viet.	Vietnam
		Nig.	Nigeria	V.I.U.S.	Virgin Islands (U.S.)
Falk. Is.	Falkland Islands	N. Ire., U.K.	Northern	vol.	Volcano
Far. Is.	Faroe Islands		Ireland, U.K.	Vt., U.S.	Vermont, U.S.
Fin.	Finland	N.J., U.S.	New Jersey, U.S.		
fj.	Fjord	N.M., U.S.	New Mexico, U.S.	Wa., U.S.	Washington, U.S.
Fl., U.S.	Florida, U.S.	N. Mar. Is.	Northern	W.B.	West Bank
for.	Forest, Moor		Mariana Islands	Wi., U.S.	Wisconsin, U.S.
Fr.	France	Nmb.	Namibia	W. Sah.	Western Sahara
Fr. Gu.	French Guiana	Nor.	Norway	wtfl.	Waterfall
Fr. Poly.	French Polynesia	Nv., U.S.	Nevada, U.S.	W.V., U.S.	West Virginia, U.S.
		N.Y., U.S.	New York, U.S.	Wy., U.S.	Wyoming, U.S.
Ga., U.S.	Georgia, U.S.	N.Z.	New Zealand		
Gam.	The Gambia				
Gaza	Gaza Strip	o.	Ocean	Zam.	Zambia
Geor.	Georgia	Oc.	Oceania	Zimb.	Zimbabwe
Ger.	Germany	Oh., U.S.	Ohio, U.S.		

PRONUNCIATION OF GEOGRAPHICAL NAMES

Key to the Sound Values of Letters and Symbols Used in the Index to Indicate Pronunciation

ă-ăt; băttle
a-fin*a*l; appe*a*l
ā-rāte; elāte
â-senâte; inanimâte
ä-ärm; cälm
à-àsk; bàth
a-sof*a*; m*a*rine (short neutral or indeterminate sound)
â-fâre; prepâre
ch-choose; church
dh-as th in other; either
ē-bē; ēve
ê-êvent; crêate
ĕ-bĕt; ĕnd
e-rec*e*nt (short neutral or indeterminate sound)
ẽ-cratẽr; cindẽr
g-gō; gāme
gh-guttural g
ĭ-bĭt; wĭll
i-(short neutral or indeterminate sound)
ī-rīde; bīte
κ-gutteral k as *ch* in German *ich*
ng-sing
ŋ-baŋk; liŋger
ɴ-indicates nasalized
ŏ-nŏd; ŏdd
o-c*o*mmit; c*o*nnect
ō-ōld; bōld
ô-ôbey; hôtel
ô-ôrder; nôrth
oi-boil
o͞o-fo͞od; ro͞ot
ȯ-as oo in foot; wood
ou-out; thou
s-soft; so; sane
sh-dish; finish
th-thin; thick
ū-pūre; cūre
û-ûnite; ûsûrp
û-ûrn; fûr
ŭ-stŭd; ŭp
u-circ*u*s; s*u*bmit
ü-as in French *tu*
zh-as *z* in a*z*ure
'-indeterminate vowel sound

In many cases the spelling of foreign geographical names does not even remotely indicate the pronunciation to an American, i.e., Słupsk in Poland is pronounced swȯpsk; Jujuy in Argentina is pronounced ho͞oͤhwē', La Spezia in Italy is lä-spē'zyä.

This condition is hardly surprising, however, when we consider that in our own language Worcester, Massachusetts, is pronounced wȯs'tẽr; Sioux City, Iowa, so͞o sĭ'tĭ; Schuylkill Haven, Pennsylvania, sko͞ol'kĭl hā-vĕn; Poughkeepsie, New York, pŏ-kĭp'sĕ.

The indication of pronunciation of geographic names presents several peculiar problems:

1. Many foreign tongues use sounds that are not present in the English language and which an American cannot normally articulate. Thus, though the nearest English equivalent sound has been indicated, only approximate results are possible.

2. There are several dialects in each foreign tongue which cause variation in the local pronunciation of names. This also occurs in identical names in the various divisions of a great language group, as the Slavic or the Latin.

3. Within the United States there are marked differences in pronunciation, not only of local geographic names, but also of common words, indicating that the sound and tone values for letters as well as the placing of the emphasis vary considerably from one part of the country to another.

4. A number of different letters and diacritical combinations could be used to indicate essentially the same or approximate pronunciations.

Some variation in pronunciation other than that indicated in this index may be encountered, but such a difference does not necessarily indicate that either is in error, and in many cases it is a matter of individual choice as to which is preferred. In fact, an exact indication of pronunciation of many foreign names using English letters and diacritical marks is extremely difficult and sometimes impossible.

PRONOUNCING INDEX

This universal index includes in a single alphabetical list approximately 30,000 names of features that appear on the reference maps. Each name is followed by a page reference and geographical coordinates.

Abbreviation and Capitalization Abbreviations of names on the maps have been standardized as much as possible. Names that are abbreviated on the maps are generally spelled out in full in the index. Periods are used after all abbreviations regardless of local practice. The abbreviation "St." is used only for "Saint". "Sankt" and other forms of this term are spelled out.

Most initial letters of names are capitalized, except for a few Dutch names, such as "s-Gravenhage". Capitalization of noninitial words in a name generally follows local practice.

Alphabetization Names are alphabetized in the order of the letters of the English alphabet. Spanish *ll* and *ch*, for example, are not treated as direct letters. Furthermore, diacritical marks are disregarded in alphabetization — German or Scandinavian *ä* or *ö* are treated as *a* or *o*.

The names of physical features may appear inverted, since they are always alphabetized under the proper, not the generic, part of the name, thus: "Gibraltar, Strait of". Otherwise every entry, whether consisting of one word or more, is alphabetized as a single continuous entity. "Lakeland", for example, appears after "La Crosse" and before "La Salle". Names beginning with articles (Le Harve, Den Helder, Al Manāmah, Ad Dawhah) are not inverted.

In the case of identical names, towns are listed first, then political divisions, then physical features.

Generic Terms Except for cities, the names of all features are followed by terms that represent broad classes of features, for example, Mississippi, r. or Alabama, state. A list of all abbreviations used in the index is on page 215.

Country names and the names of features that extend beyond the boundaries of one county are followed by the name of the continent in which each is located. Country designations follow the names of all other places in the index. The locations of places in the United States and the United Kingdom are further defined by abbreviations that include the state or political division in which each is located.

Pronunciations Pronunciations are included for most names listed. An explanation of the pronunciation system used appears on page 215.

Page References and Geographical Coordinates The geographical coordinates and page references are found in the last columns of each entry.

If a page contains several maps or insets, a lowercase letter identifies the specific map or inset.

Latitude and longitude coordinates for point features, such as cities and mountain peaks, indicate the location of the symbols. For extensive areal features, such as countries or mountain ranges, or linear features, such as canals and rivers, locations are given for the position of the type as it appears on the map.

PLACE (Pronunciation)	PAGE	LAT.	LONG.
A			
Aachen, Ger. (ä´kĕn)	117	50°46′N	6°07′E
Aalborg, Den. (ôl´bôr)	110	57°02′N	9°55′E
Aalen, Ger. (ä´lĕn)	124	48°49′N	10°08′E
Aalsmeer, Neth.	115a	52°16′N	4°44′E
Aalst, Bel.	121	50°58′N	4°00′E
Aarau, Switz. (ä´rou)	117	47°22′N	8°03′E
Aarschot, Bel.	115a	50°59′N	4°51′E
Aba, D.R.C.	191	3°52′N	30°14′E
Aba, Nig.	184	5°06′N	7°21′E
Ābādān, Iran (ä-bǔ-dän´)	154	30°15′N	48°30′E
Abaetetuba, Braz. (ä´bắĕ-tĕ-tōō´bȧ)	101	1°44′S	48°45′W
Abajo Peak, mtn., Ut., U.S. (ä-bä´hō)	77	37°51′N	109°28′W
Abakaliki, Nig.	189	6°21′N	8°06′E
Abakan, Russia (ŭ-bá-kän´)	135	53°43′N	91°28′E
Abakan, r., Russia	140	53°00′N	91°06′E
Abancay, Peru (ä-bän-kä´ĕ)	100	13°44′S	72°46′W
Abashiri, Japan (ä-bä-shē´rē)	166	44°00′N	144°13′E
Abasolo, Mex. (ä-bä-sō´lô)	88	24°05′N	98°24′W
Abasolo, Mex. (ä-bä-sō´lô)	80	27°13′N	101°25′W
Abaya, Lake, l., Eth. (ä-bä´yä)	185	6°24′N	38°22′E
'Abbāsah, Tur'at al, can., Egypt	192d	30°45′N	32°15′E
Abbeville, Fr. (ȧb-vēl´)	117	50°08′N	1°49′E
Abbeville, Al., U.S. (ăb´ē-vĭl)	82	31°35′N	85°15′W
Abbeville, Ga., U.S. (ăb´ē-vĭl)	82	31°53′N	83°23′W
Abbeville, La., U.S.	81	29°59′N	92°07′W
Abbeville, S.C., U.S.	83	34°09′N	82°25′W
Abbiategrasso, Italy (äb-byä´tä-gräs´sō)	130	45°23′N	8°52′E
Abbots Bromley, Eng., U.K. (ăb´ŭts brŭm´lē)	114a	52°49′N	1°52′W
Abbotsford, Can. (ăb´ŭts-fĕrd)	74d	49°03′N	122°17′W
'Abd al Kūrī, i., Yemen (äbd-ĕl-kó´rē)	192a	12°12′N	51°00′E
Abdulino, Russia (äb-dò-lē´nō)	136	53°42′N	53°40′E
Abengourou, C. Iv.	188	6°44′N	3°29′W
Abeokuta, Nig. (ä-bä-ô-kōō´tä)	184	7°10′N	3°26′E
Abercorn see Mbala, Zam.	186	8°55′S	31°22′E
Aberdare, Wales, U.K. (ăb-ĕr-dâr´)	120	51°45′N	3°35′W
Aberdeen, Scot., U.K. (ăb-ĕr-dēn´)	110	57°10′N	2°05′W
Aberdeen, Ms., U.S.	82	33°49′N	88°33′W
Aberdeen, S.D., U.S. (ăb-ĕr-dēn´)	64	45°28′N	98°29′W
Aberdeen, Wa., U.S.	64	47°00′N	123°48′W
Aberford, Eng., U.K. (ăb´ĕr-fĕrd)	114a	53°49′N	1°21′W
Abergavenny, Wales, U.K. (ăb´ĕr-gȧ-vĕn´ĭ)	120	51°45′N	3°05′W
Abert, Lake, l., Or., U.S. (ā´bĕrt)	72	42°39′N	120°24′W
Aberystwyth, Wales, U.K. (ă-bĕr-ĭst´wĭth)	120	52°25′N	4°04′W
Abidjan, C. Iv. (ä-bēd-zhäɴ´)	184	5°19′N	4°02′W
Abiko, Japan (ä-bē´kō)	167a	35°53′N	140°01′E
Abilene, Ks., U.S. (ăb´ĭ-lēn)	79	38°54′N	97°12′W
Abilene, Tx., U.S.	64	32°25′N	99°45′W
Abingdon, Eng., U.K.	114b	51°38′N	1°17′W
Abingdon, Il., U.S. (ăb´ĭng-dŭn)	71	40°48′N	90°21′W
Abingdon, Va., U.S.	83	36°42′N	81°57′W
Abington, Ma., U.S. (ăb´ĭng-tŭn)	61a	42°07′N	70°57′W
Abiquiu Reservoir, res., N.M., U.S.	77	36°26′N	106°42′W
Abitibi, l., Can. (ăb-ĭ-tĭb´ĭ)	53	48°27′N	80°20′W
Abitibi, r., Can.	53	49°30′N	81°10′W
Abkhazia, state, Geor.	137	43°10′N	40°45′E
Ablis, Fr. (ä-blē´)	127b	48°31′N	1°50′E
Abnūb, Egypt (äb-nōōb´)	192b	27°18′N	31°11′E
Åbo see Turku, Fin.	110	60°28′N	22°12′E
Abohar, India	158	30°12′N	74°13′E
Aboisso, C. Iv.	184	5°28′N	3°12′W
Abomey, Benin (ab-ô-mā´)	184	7°11′N	1°59′E
Abony, Hung. (ŏ´bô-ny´)	125	47°12′N	20°00′E
Abou Deïa, Chad	189	11°27′N	19°17′E
Abra, r., Phil. (ä´brä)	169a	17°16′N	120°38′E
Abraão, Braz. (äbrä-oun´)	99a	23°10′S	44°10′W
Abraham's Bay, b., Bah.	93	22°20′N	73°50′W
Abram, Eng., U.K. (ā´brăm)	114a	53°31′N	2°36′W
Abrantes, Port. (ä-brän´tĕs)	128	39°28′N	8°13′W
Abrolhos, Arquipélago dos, is., Braz.	101	17°58′S	38°40′W
Abruka, i., Est. (ä-brô´kä)	123	58°09′N	22°30′E
Abruzzi e Molise, hist. reg., Italy	130	42°10′N	13°55′E
Absaroka Range, mts., U.S. (ȧb-sä-rō-kä)	64	44°50′N	109°47′W
Abşeron Yarımadası, pen., Azer.	137	40°20′N	50°30′E
Abū Arīsh, Sau. Ar. (ä-bōō á-rēsh´)	154	16°48′N	43°00′E
Abu Dhabi see Abū Ẓaby, U.A.E.	154	24°15′N	54°28′E
Abū Ḥamad, Sudan (ä´bōō hä´-mĕd)	185	19°37′N	33°21′E
Abuja, Nig.	184	9°12′N	7°11′E
Abū Kamāl, Syria	154	34°45′N	40°46′E
Abunã, r., S.A.	100	10°25′S	67°00′W
Abū Qīr, Egypt (ä´bōō kēr´)	192b	31°18′N	30°06′E
Abū Qurūn, Ra's, mtn., Egypt	153a	30°22′N	33°32′E
Aburatsu, Japan (ä´bó-rät´sōō)	167	31°33′N	131°20′E
Abu Road, India (á´bōō)	155	24°38′N	72°45′E
Abū Tīj, Egypt	192b	27°03′N	31°19′E
Abū Ẓaby, U.A.E.	154	24°15′N	54°28′E
Abū Ẓanīmah, Egypt	153a	29°03′N	33°08′E
Abyy, Russia	135	68°24′N	134°00′E
Acacias, Col. (ä-kä´sēäs)	100a	3°59′N	73°44′W
Acadia National Park, rec., Me., U.S. (ȧ-kā´dĭ-ȧ)	65	44°19′N	68°01′W
Acajutla, El Sal. (ä-kä-hōōt´lä)	90	13°37′N	89°50′W
Acala, Mex. (ä-kä´lä)	89	16°38′N	92°49′W
Acalayong, Eq. Gui.	190	1°05′N	9°40′E
Acámbaro, Mex. (ä-käm´bä-rō)	88	20°03′N	100°42′W
Acancéh, Mex. (ä-kän-sĕ´)	90a	20°50′N	89°27′W
Acapetlahuaya, Mex. (ä-kä-pĕt´lä-hwä´yä)	88	18°24′N	100°04′W
Acaponeta, Mex. (ä-kä-pô-nā´tä)	88	22°31′N	105°25′W
Acaponeta, r., Mex. (ä-kä-pô-nā´tä)	88	22°47′N	105°23′W
Acapulco, Mex. (ä-kä-pōōl´kō)	86	16°49′N	99°57′W
Acaraí Mountains, mts., S.A.	101	1°30′N	57°40′W
Acarigua, Ven. (äkä-rē´gwä)	100	9°29′N	69°11′W
Acatlán de Osorio, Mex. (ä-kät-län´dä ô-sō´rē-ō)	88	18°11′N	98°04′W
Acatzingo de Hidalgo, Mex.	89	18°58′N	97°47′W
Acayucan, Mex. (ä-kä-yōō´kän)	89	17°56′N	94°55′W
Accoville, W.V., U.S. (ăk´kô-vĭl)	66	37°45′N	81°50′W
Accra, Ghana (ä´krä)	184	5°33′N	0°13′W
Accrington, Eng., U.K. (ăk´rĭng-tŭn)	114a	53°45′N	2°22′W
Acerra, Italy (ä-chĕ´r-rä)	129c	40°42′N	14°22′E
Achacachi, Bol. (ä-chä-kä´chĕ)	100	16°11′S	68°32′W
Achelóos, r., Grc.	131	38°45′N	21°26′E
Achill Island, i., Ire. (ȧ-chĭl´)	116	53°55′N	10°05′W
Achinsk, Russia (ȧ-chênsk´)	140	56°13′N	90°32′E
Acireale, Italy (ä-chē-rā-ä´lä)	130	37°37′N	15°12′E
Acklins, i., Bah. (äk´lĭns)	87	22°30′N	73°55′W
Acklins, The Bight of, b., Bah. (äk´lĭns)	93	22°35′N	74°20′W
Acolmán, Mex. (ä-kôl-má´n)	89a	19°38′N	98°56′W
Acoma Indian Reservation, I.R., N.M., U.S.	77	34°52′N	107°40′W
Aconcagua, prov., Chile (ä-kòn-kä´gwä)	99b	32°20′S	71°00′W
Aconcagua, r., Chile (ä-kòn-kä´gwä)	99b	32°43′S	70°53′W
Aconcagua, Cerro, mtn., Arg. (ä-kòn-kä´gwä)	102	32°38′S	70°00′W
Açores (Azores), is., Port.	183	37°44′N	29°25′W
A Coruña, Spain	110	43°20′N	8°20′W
Acoyapa, Nic. (ä-kô-yä´pä)	90	11°54′N	85°11′W
Acqui, Italy (äk´kwē)	130	44°41′N	8°22′W
Acre, state, Braz. (ä´krä)	100	8°40′S	70°45′W
Acre, r., S.A.	100	10°33′S	68°34′W
Acton, Can. (ăk´tŭn)	62d	43°38′N	80°02′W
Acton, Al., U.S. (ăk´tŭn)	68h	33°21′N	86°49′W
Acton, Ma., U.S. (ăk´tŭn)	61a	42°29′N	71°26′W
Actopan, Mex. (äk-tô-pän´)	88	20°16′N	98°57′W
Actopán, Mex. (äk-tô´pän´)	89	19°25′N	96°31′W
Acuitzio del Canje, Mex. (ä-kwēt´zĕ-ô dĕl kän´há)	88	19°28′N	101°21′W
Acul, Baie de l', b., Haiti (ä-kōōl´)	93	19°55′N	72°20′W
Ada, Mn., U.S. (ā´dȧ)	70	47°17′N	96°32′W
Ada, Oh., U.S. (ā´dȧ)	66	40°45′N	83°45′W
Ada, Ok., U.S. (ā´dȧ)	79	34°45′N	96°43′W
Ada, Serb. (ä´dä)	131	45°48′N	20°06′E
Adachi, Japan	167a	35°50′N	39°36′E
Adak, Ak., U.S. (ä-dăk´)	63a	56°50′N	176°48′W
Adak, i., Ak., U.S. (ä-dăk´)	63a	51°40′N	176°28′W
Adak Strait, strt., Ak., U.S. (ä-dăk´)	63a	51°42′N	177°16′W
Adamaoua, mts., Afr.	184	6°30′N	11°50′E
Adams, Ma., U.S. (ăd´ămz)	67	42°35′N	73°10′W
Adams, Wi., U.S. (ăd´ămz)	71	43°55′N	89°48′W

ăt; fīnăl; rāte; senâte; ärm; ȧsk; sofȧ; fâre; ch-choose; dh-as th in other; bē; ēvent; bĕt; recĕnt; cratĕr; g-gō; gh-guttural g; bĭt; ĭ-short neutral; rīde; к-guttural k as ch in German ich;

PLACE (Pronunciation)	PAGE	LAT.	LONG.
Adams, r., Can. (ăd′ămz)	55	51°30′N	119°20′W
Adams, Mount, mtn., Wa., U.S. (ăd′ămz)	64	46°15′N	121°19′W
Adamsville, Al., U.S. (ăd′ămz-vĭl)	68h	33°36′N	86°57′W
Adana, Tur. (ä′dä-nä)	154	37°05′N	35°20′E
Adapazarı, Tur. (ä-dä-pä-zä′rē)	119	40°45′N	30°20′E
Adarama, Sudan (ä-dä-rä′mä)	185	17°11′N	34°56′E
Adda, r., Italy (äd′dä)	130	45°43′N	9°31′E
Ad Dabbah, Sudan	185	18°04′N	30°58′E
Ad Dahnā, des., Sau. Ar.	154	26°05′N	47°15′E
Ad-Dāmir, Sudan (ad-dä′mēr)	185	17°38′N	33°57′E
Ad Dammām, Sau. Ar.	154	26°27′N	49°59′E
Ad Dāmūr, Leb.	153a	33°44′N	35°27′E
Ad Dawhah, Qatar	154	25°02′N	51°28′E
Ad Dilam, Sau. Ar.	154	23°47′N	47°03′E
Ad Dilinjāt, Egypt	192b	30°48′N	30°32′E
Addis Ababa, Eth.	185	9°00′N	38°44′E
Addison, Tx., U.S. (ă′dĭ-sŭn)	75c	32°58′N	96°50′W
Addo, S. Afr. (ădō)	187c	33°33′S	25°43′E
Ad Duwaym, Sudan (ad-dò-âm′)	185	13°56′N	32°22′E
Addyston, Oh., U.S. (ăd′ē-stŭn)	69f	39°09′N	84°42′W
Adel, Ga., U.S. (ā-dĕl′)	82	31°08′N	83°55′W
Adelaide, Austl. (ăd′ē-lād)	174	34°46′S	139°08′E
Adelaide, S. Afr. (ăd-ĕl′ād)	187c	32°41′S	26°07′E
Adelaide Island, i., Ant. (ăd′ē-lād)	178	67°15′S	68°40′W
Aden (′Adan), Yemen (ä′dĕn)	154	12°48′N	45°00′E
Aden, Gulf of, b.,	154	11°45′N	45°45′E
Adi, Pulau, i., Indon. (ä′dē)	169	4°25′S	133°52′E
Adige, r., Italy (ä′dē-jā)	118	46°38′N	10°43′E
Adigrat, Eth.	157	14°17′N	39°28′E
Adilābād, India (ŭ-dĭl-ä-băd′)	158	19°47′N	78°30′E
Adirondack Mountains, mts., N.Y., U.S. (ăd-ĭ-rŏn′dăk)	65	43°45′N	74°40′W
Adis Abeba see Addis Ababa, Eth.	185	9°00′N	38°44′E
Adi Ugri, Erit. (ä-dē ōō′grē)	185	14°54′N	38°52′E
Adjud, Rom. (äd′zhòd)	125	46°05′N	27°12′E
Adkins, Tx., U.S.	75d	29°22′N	98°18′W
Admiralty, i., Ak., U.S. (ăd′mĭ-răl-tē)	63	57°50′N	133°50′W
Admiralty Inlet, Wa., U.S. (ăd′mĭ-răl-tē)	74a	48°10′N	122°45′W
Admiralty Island National Monument, rec., Ak., U.S. (ăd′mĭ-răl-tē)	63	57°50′N	137°30′W
Admiralty Islands, is., Pap. N. Gui. (ăd′mĭ-răl-tē)	169	1°40′S	146°45′E
Ado-Ekiti, Nig.	189	7°38′N	5°12′E
Adolph, Mn., U.S. (ā′dolf)	75h	46°47′N	92°17′W
Ādoni, India	159	15°42′N	77°18′E
Adour, r., Fr. (ä-dōōr′)	117	43°43′N	0°38′W
Adra, Spain (ä′drä)	128	36°45′N	3°02′W
Adrano, Italy (ä-drä′nō)	130	37°42′N	14°52′E
Adrar, Alg.	184	27°53′N	0°15′W
Adria, Italy (ä′drē-ä)	130	45°03′N	12°01′E
Adrian, Mi., U.S. (ā′drĭ-ăn)	66	41°55′N	84°00′W
Adrian, Mn., U.S. (ā′drĭ-ăn)	70	43°39′N	95°56′W
Adrianople see Edirne, Tur.	110	41°41′N	26°35′E
Adriatic Sea, sea, Eur.	112	43°30′N	14°27′E
Adwa, Eth.	185	14°02′N	38°58′E
Adwick-le-Street, Eng., U.K. (ăd′wĭk-lē-strēt′)	114a	53°35′N	1°11′W
Adycha, r., Russia (ä′dĭ-chà)	141	66°11′N	136°45′E
Adygea, prov., Russia	136	45°00′N	40°00′E
Adz′va, r., Russia (ädz′và)	136	67°00′N	59°20′E
Aegean Sea, sea, (ē-jē′ăn)	112	39°04′N	24°56′E
A Estrada, Spain	128	42°42′N	8°29′W
Affton, Mo., U.S.	75e	38°33′N	90°20′W
Afghanistan, nation, Asia (ăf-găn-ĭ-stăn′)	154	33°00′N	63°00′E
Afgooye, Som. (äf-gô′ĭ)	192a	2°08′N	45°08′E
Afikpo, Nig.	189	5°53′N	7°56′E
Aflou, Alg. (ä-flōō′)	184	33°59′N	2°04′E
Afognak, i., Ak., U.S. (ä-fŏg-năk′)	63	58°28′N	151°35′W
A Fonsagrada, Spain	128	43°08′N	7°07′W
Afonso Claudio, Braz. (äl-fōn′sô-klou′dēō)	99a	20°05′S	41°05′W
Afragola, Italy (ä-frä′gō-lä)	129c	40°40′N	14°19′E
Africa, cont.	183	10°00′N	22°00′E
Afton, Mn., U.S. (ăf′tŭn)	75g	44°54′N	92°47′W
Afton, Ok., U.S. (ăf′tŭn)	79	36°42′N	94°56′W
Afton, Wy., U.S. (ăf′tŭn)	73	42°42′N	110°52′W
′Afula, Isr. (ä-fōō′là)	153a	32°36′N	35°17′E
Afyon, Tur. (ä-fē-ōn)	154	38°45′N	30°20′E
Agadem, Niger (ä′gà-dĕm)	185	16°50′N	13°17′E
Agadez, Niger (ä′gà-dĕs)	184	16°58′N	7°59′E
Agadir, Mor. (ä-gà-dēr)	184	30°30′N	9°37′W
Agalta, Cordillera de, mts., Hond. (kôr-dēl-yĕ′rä-dĕ-ä-gä′l-tä)	90	15°15′N	85°42′W
Agapovka, Russia (ä-gä-pôv′kä)	142a	53°18′N	59°10′E
Agartala, India	158	23°53′N	91°22′E
Agāshi, India	159b	19°28′N	72°46′E
Agashkino, Russia (à-gäsh′kĭ-nô)	142b	55°18′N	38°13′E
Agattu, i., Ak., U.S. (à′gä-tōō)	63a	52°14′N	173°40′E
Agboville, C. Iv.	188	5°56′N	4°13′W
Ağdam, Azer. (äg′däm)	137	40°00′N	47°00′E
Agde, Fr. (ägd)	126	43°19′N	3°30′E
Agen, Fr. (à-zhäN′)	117	44°13′N	0°31′E
Agiásos, Grc.	131	39°06′N	26°25′E
Aginskoye, Russia (ä-hĭn′skô-yĕ)	135	51°15′N	113°15′E
Ágios Efstrátios, i., Grc.	119	39°30′N	24°58′E
Agíou Órous, Kólpos, b., Grc.	131	40°15′N	24°00′E
Agno, Phil. (äg′nō)	169a	16°07′N	119°49′E
Agno, r., Phil.	169a	15°42′N	120°28′E
Agnone, Italy (än-yō′nä)	130	41°49′N	14°23′E
Agogo, Ghana	188	6°47′N	1°04′W
Agra, India (ä′grä)	155	27°18′N	78°00′E
Ağrı, Tur.	137	39°50′N	43°10′E
Agri, r., Italy (ä′grē)	130	40°15′N	16°21′E
Agrínio, Grc.	119	38°38′N	21°06′E
Agua, vol., Guat. (ä′gwä)	90	14°28′N	90°43′W
Agua Blanca, Río, r., Mex. (rē′ō-ä-gwä-blä′n-kä)	88	21°46′N	102°54′W
Agua Brava, Laguna de, l., Mex.	88	22°04′N	105°40′W
Agua Caliente Indian Reservation, I.R., Ca., U.S. (ä′gwä kal-yĕn′tä)	76	33°50′N	116°24′W
Aguada, Cuba (ä-gwä′dä)	92	22°25′N	80°50′W
Aguada, l., Mex. (ä-gwä′dä)	90a	18°46′N	89°40′W
Aguadas, Col. (ä-gwä′däs)	100	5°37′N	75°27′W
Aguadilla, P.R. (ä-gwä-dēl′yä)	87b	18°27′N	67°10′W
Aguadulce, Pan. (ä-gwä-dōōl′sā)	91	8°15′N	80°33′W
Agua Escondida, Meseta de, plat., Mex.	89	16°54′N	91°35′W
Agua Fria, r., Az., U.S. (ä′gwä frē-ä)	77	33°43′N	112°22′W
Agua Fria National Monument, rec., Az., U.S.	77	34°13′N	112°03′W
Aguai, Braz. (ägwä-ē′)	99a	22°04′S	46°57′W
Agualeguas, Mex. (ä-gwä-lā′gwäs)	80	26°19′N	99°33′W
Aguán, r., Hond. (ä-gwà′n)	90	15°22′N	87°00′W
Aguanaval, r., Mex. (à-guä-nä-väl′)	80	25°12′N	103°28′W
Aguanus, r., Can. (à-gwä′nŭs)	61	50°45′N	62°03′W
Aguascalientes, Mex. (ä′gwäs-käl-yĕn′täs)	86	21°52′N	102°17′W
Aguascalientes, state, Mex. (ä′gwäs-käl-yĕn′täs)	88	22°00′N	102°18′W
Águeda, Port. (ä-gwā′dä)	128	40°36′N	8°26′W
Agueda, r., Eur. (ä-gē-dä)	128	40°50′N	6°44′W
Aguelhok, Mali	188	19°28′N	0°52′E
Aguilar, Spain	128	37°32′N	4°39′W
Aguilar, Co., U.S. (ä-gē-lär′)	78	37°24′N	104°38′W
Aguilas, Spain (ä-gē-läs)	118	37°26′N	1°35′W
Aguililla, Mex. (ä-gē-lēl-yä)	88	18°44′N	102°44′W
Aguililla, r., Mex. (ä-gē-lēl-yä)	88	18°30′N	102°48′W
Aguja, Punta, c., Peru (pŭn′tà à-gōō′ hä)	100	6°00′S	81°15′W
Agulhas, Cape, c., S. Afr. (ä-gōōl′yäs)	186	34°47′S	20°00′E
Agusan, r., Phil. (ä-gōō′sän)	169	8°12′N	126°07′E
Ahaggar, mts., Alg. (à-hà-gär′)	184	23°14′N	6°00′E
Ahar, Iran	157	38°28′N	47°04′E
Ahlen, Ger. (ä′lĕn)	124	51°45′N	7°52′E
Ahmadābād, India (ŭ-mĕd-ä-băd′)	155	23°04′N	72°38′E
Ahmadnagar, India (ä′mûd-nŭ-gûr)	155	19°09′N	74°45′E
Ahmar Mountains, mts., Eth.	185	9°22′N	42°00′E
Ahoskie, N.C., U.S. (ā-hŏs′kē)	83	36°15′N	77°00′W
Ahrensburg, Ger. (ä′rĕns-bòrg)	115c	53°40′N	10°14′E
Ahrweiler, Ger. (är′vī-lĕr)	124	50°34′N	7°05′E
Ähtärinjärvi, l., Fin.	123	62°46′N	24°25′E
Ahuacatlán, Mex. (ä-wä-kät-län′)	88	21°05′N	104°28′W
Ahuachapán, El Sal. (ä-wä-chä-pän′)	90	13°57′N	89°53′W
Ahualulco, Mex. (ä-wä-lōōl′kō)	88	20°43′N	103°57′W
Ahuatempan, Mex. (ä-wä-tĕm-pän)	88	18°11′N	98°02′W
Åhus, Swe. (ô′hòs)	122	55°56′N	14°19′E
Ahvāz, Iran	154	31°15′N	48°54′E
Ahvenanmaa (Åland), is., Fin. (ä′vĕ-nàn-mô) (ô′lănd)	116	60°36′N	19°55′E
′Aiea, Hi., U.S.	84a	21°18′N	157°52′W
Aígina, Grc.	131	37°43′N	23°35′E
Aígina, i., Grc.	131	37°43′N	23°35′E
Aígio, Grc.	131	38°13′N	22°04′E
Aiken, S.C., U.S. (ā′kĕn)	83	33°32′N	81°43′W
Aimorès, Serra dos, mts., Braz. (sĕ′r-rä-dôs-ī-mō-rĕ′s)	101	17°40′S	42°38′W
Aimoto, Japan (ī-mô-tō)	167b	34°59′N	135°09′E
Aincourt, Fr. (ăN-kōō′r)	127b	49°04′N	1°47′E
Aïn el Beïda, Alg.	184	35°57′N	7°25′E
Ainsworth, Ne., U.S. (ānz′wûrth)	70	42°32′N	99°51′W
Aïn Témouchent, Alg. (ä′ĕntĕ-mōō-shäN′)	118	35°20′N	1°23′W
Aïn Wessara, Alg. (ĕn ōō-sà-rà)	129	35°25′N	2°50′E
Aipe, Col. (ī′pĕ)	100a	3°13′N	75°15′W
Aïr, mts., Niger	184	18°00′N	8°30′E
Aire, r., Eng., U.K.	114a	53°42′N	1°00′W
Aire-sur-l'Adour, Fr. (âr)	126	43°42′N	0°17′W
Airhitam, Selat, strt., Indon.	153b	0°58′N	102°38′E
Ai Shan, mts., China (āī′shän)	162	37°27′N	120°35′E
Aisne, r., Fr. (ĕn)	117	49°28′N	3°32′E
Aitape, Pap. N. Gui. (ä-ē-tä′pā)	169	3°00′S	142°10′E
Aitkin, Mn., U.S. (āt′kĭn)	71	46°32′N	93°43′W
Aitolikó, Grc.	131	38°27′N	21°21′E
Aitos, Blg. (ä-ē′tōs)	131	42°42′N	27°17′E
Aitutaki, i., Cook Is. (ī-tōō-tä′kē)	195	19°00′S	162°00′W
Aiud, Rom. (ä′ē-òd)	119	46°19′N	23°40′E
Aiuruoca, Braz. (äē′ōō-rōōô′-kà)	99a	21°57′S	44°36′W
Aiuruoca, r., Braz.	99a	22°11′S	44°35′W
Aix-en-Provence, Fr. (ĕks-prô-văNs)	117	43°32′N	5°27′E
Aix-les-Bains, Fr. (ĕks′-lā-băN′)	127	45°42′N	5°56′E
Aizpute, Lat. (ä′ēz-pōō-tĕ)	123	56°44′N	21°37′E
Aizuwakamatsu, Japan	166	37°27′N	139°51′E
Ajaccio, Fr. (ä-yät′chō)	110	41°55′N	8°42′E
Ajalpan, Mex. (ä-häl′pän)	89	18°21′N	97°14′W
Ajana, Austl. (ăj-än′ĕr)	174	28°00′S	114°45′E
Ajaria, state, Geor.	138	41°40′N	42°00′E
Ajdābiyah, Libya	185	30°56′N	20°16′E
Ajjer, Tassili-n-, plat., Alg.	184	25°40′N	6°57′E
Ajmah, Jabal al, mts., Egypt	153a	29°12′N	34°03′E
Ajman, U.A.E.	154	25°15′N	54°30′E
Ajmer, India (ŭj-mēr′)	155	26°26′N	74°42′E
Ajo, Az., U.S. (ä′hō)	77	32°20′N	112°55′W
Ajuchitlán del Progreso, Mex. (ä-hōō-chet-län)	88	18°11′N	100°32′W
Ajusco, Mex. (ä-hōō′s-kō)	89a	19°13′N	99°12′W
Ajusco, Cerro, mtn., Mex. (sĕ′r-rô-ä-hōō′s-kō)	89a	19°12′N	99°16′W
Akaishi-dake, mtn., Japan (ä-kī-shē dä′kä)	167	35°30′N	138°00′E
Akashi, Japan (ä′kä-shē)	166	34°38′N	134°59′E
Aketi, D.R.C. (ä-kå-tē)	185	2°44′N	23°46′E
Akhaltsikhe, Geor. (äkä′l-tsĭ-kĕ)	137	41°40′N	42°50′E
Akhdar, Al Jabal al, mts., Libya	185	32°00′N	22°00′E
Akhdar, Al Jabal al, mts., Oman	154	23°30′N	56°43′W
Akhisar, Tur. (äk-hĭs-sär′)	119	38°58′N	27°58′E
Akhtarskaya, Bukhta, b., Russia (bōōk′tä àk-tär′skà-yà)	133	45°53′N	38°22′E
Akhtopol, Blg. (äk′tô-pōl)	131	42°08′N	27°54′E
Akhunovo, Russia (ä-kû′nô-vô)	142a	54°13′N	59°36′E
Aki, Japan (ä′kē)	167	33°31′N	133°51′E
Akiak, Ak., U.S. (äk′yăk)	63	61°00′N	161°02′W
Akimiski, i., Can. (ä-kĭ-mĭ′skĭ)	53	52°54′N	80°22′W
Akita, Japan (ä′kē-tä)	161	39°40′N	140°12′E
Akjoujt, Maur.	184	19°45′N	14°23′W
′Akko, Isr.	153a	32°56′N	35°05′E
Aklavik, Can. (äk′lä-vĭk)	50	68°28′N	135°26′W
Aklé′Âouâna, dunes, Afr.	188	18°07′N	6°00′W
Ako, Japan (ä′kô)	167	34°44′N	134°22′E
Akola, India (à-kō′lä)	155	20°47′N	77°00′E
Akordat, Erit.	185	15°34′N	37°54′E
Akpatok, i., Can. (ăk′pà-tŏk)	53	60°30′N	67°10′W
Akranes, Ice.	116	64°18′N	21°40′W
Akron, Co., U.S. (ăk′rŭn)	78	40°09′N	103°14′W
Akron, Oh., U.S. (ăk′rŭn)	65	41°05′N	81°30′W
Aksaray, Tur. (äk-sà-rī′)	119	38°30′N	34°05′E
Akşehır, Tur. (äk′shà-hēr)	119	38°20′N	31°20′E
Akşehır Gölü, l., Tur. (äk′shà-hēr)	154	38°40′N	31°30′E
Aksha, Russia (äk′shà)	135	50°28′N	113°00′E
Aksu, China (ä-kŭ-sōō)	160	41°29′N	80°15′E
Akune, Japan (ä′kò-nå)	167	32°03′N	130°16′E
Akureyri, Ice. (ä-kò-rå′rē)	116	65°39′N	18°01′W
Akutan, i., Ak., U.S. (ä-kōō-tän′)	63a	53°58′N	169°54′W
Akwatia, Ghana	188	6°04′N	0°49′W
Alabama, state, U.S. (ăl-à-băm′à)	65	32°50′N	87°30′W
Alabama, r., Al., U.S. (ăl-à-băm′à)	65	31°20′N	87°39′W
Alabat, i., Phil. (ä-lä-bät′)	169a	14°14′N	122°05′E
Alacam, Tur. (ä-lä-chäm′)	137	41°30′N	35°40′E
Alacant, Spain	118	38°20′N	0°30′W
Alacranes, Cuba (ä-lä-krä′nås)	92	22°45′N	81°35′W
Al Aflaj, des., Sau. Ar.	154	24°00′N	44°47′E
Alagôas, state, Braz. (ä-lä-gō′äzh)	101	9°50′S	36°33′W
Alagoinhas, Braz. (ä-lä-gō-ēn-yäzh)	101	12°13′S	38°12′W
Alagón, Spain (ä-lä-gōn′)	128	41°46′N	1°07′W
Alagón, r., Spain (ä-lä-gōn′)	128	39°53′N	6°42′W
Alahuatán, r., Mex. (ä-lä-wä-tà′n)	88	18°30′N	100°00′W
Alajuela, C.R. (ä-lä-hwä′lä)	91	10°01′N	84°14′W
Alajuela, Lago, l., Pan. (ä-lä-hwä′lä)	86a	9°15′N	79°34′W
Alaköl, l., Kaz.	139	45°45′N	81°13′E
′Alalakeiki Channel, strt., Hi., U.S. (ä-lä-lä-kā′kē)	84a	20°40′N	156°30′W
Al ′Alamayn, Egypt	185	30°53′N	28°52′E
Al ′Amārah, Iraq	157	31°50′N	47°09′E
Alameda, Ca., U.S. (ăl-à-mā′dà)	64	37°46′N	122°15′W
Alameda, r., Ca., U.S. (ăl-à-mā′dà)	74b	37°36′N	122°02′W
Alaminos, Phil. (ä-lä-mē′nôs)	169a	16°09′N	119°58′E
Al ′Amirīyah, Egypt	119	31°01′N	29°52′E
Alamo, Mex. (ä′lä-mô)	89	20°55′N	97°41′W
Alamo, Ca., U.S. (ă′là-mō)	74b	37°51′N	122°02′W
Alamo, Nv., U.S. (ă′là-mō)	76	37°22′N	115°10′W
Alamo, r., Mex. (ä′lä-mô)	80	26°33′N	99°35′W
Alamogordo, N.M., U.S. (ăl-à-mō-gôr′dō)	77	32°55′N	106°00′W
Alamo Heights, Tx., U.S. (ă′là-mō)	75d	29°28′N	98°27′W
Alamo Indian Reservation, I.R., N.M., U.S.	77	34°30′N	107°30′W
Alamo Peak, mtn., N.M., U.S. (ă′là-mō pēk)	80	32°50′N	105°55′W
Alamosa, Co., U.S. (ăl-à-mō′sà)	77	37°25′N	105°50′W
Åland see Ahvenanmaa, is., Fin.	116	60°36′N	19°55′E
Alandskiy, Russia (ä-länt′skĭ)	142a	52°14′N	59°48′E
Alanga Arba, Kenya	191	0°07′N	40°25′E
Alanya, Tur.	119	36°40′N	32°10′E
Alaotra, l., Madag. (ä-lä-ō′trà)	187	17°15′S	48°17′E
Alapayevsk, Russia (ä-lä-pä′yĕfsk)	134	57°50′N	61°35′E
Al ′Aqabah, Jord.	154	29°32′N	35°00′E
Alaquines, Mex. (ä-là-kē′nås)	88	22°07′N	99°35′W
Al ′Arīsh, Egypt (ä-rēsh′)	153a	31°08′N	33°48′E
Alaska, state, U.S. (à-lăs′kà)	64a	64°00′N	150°00′W
Alaska, Gulf of, b., Ak., U.S. (à-lăs′kà)	63	57°42′N	147°40′W
Alaska Highway, Ak., U.S. (à-lăs′kà)	63	63°00′N	142°00′W
Alaska Peninsula, pen., Ak., U.S. (à-lăs′kà)	63	55°50′N	162°10′W
Alaska Range, mts., Ak., U.S. (à-lăs′kà)	63	62°00′N	152°18′W
Al ′Atrūn, Sudan	185	18°13′N	26°44′E
Alatyr′, Russia (ä′lä-tür)	134	54°55′N	46°30′E
Alazani, r., Asia	138	41°05′N	46°40′E
Alba, Italy (äl′bä)	130	44°41′N	8°02′E
Albacete, Spain (äl-bä-thā′tā)	118	39°00′N	1°49′W
Albachten, Ger. (äl-bä′κ-tĕn)	127c	51°55′N	7°31′E
Alba de Tormes, Spain (äl-bà dā tôr′mās)	128	40°48′N	5°28′W
Alba Iulia, Rom. (äl-bä yōō′lyà)	119	46°05′N	23°32′E
Albani, Colli, hills, Italy	129d	41°46′N	12°45′E
Albania, nation, Eur. (äl-bā′nĭ-à)	110	41°45′N	20°00′E
Albano, Lago, l., Italy (lä′-gō äl-bä′nō)	129d	41°45′N	12°44′E
Albano Laziale, Italy (äl-bä′nō lät-zē-ä′lä)	130	41°44′N	12°43′E
Albany, Austl. (ôl′bà-nĭ)	174	35°00′S	118°00′E
Albany, Ca., U.S. (ôl′bà-nĭ)	74b	37°54′N	122°18′W
Albany, Ga., U.S. (ôl′bà-nĭ)	65	31°35′N	84°10′W
Albany, Mo., U.S. (ôl′bà-nĭ)	79	40°14′N	94°18′W
Albany, N.Y., U.S. (ôl′bà-nĭ)	65	42°40′N	73°50′W
Albany, Or., U.S. (ôl′bà-nĭ)	64	44°38′N	123°06′W
Albany, r., Can. (ôl′bà-nĭ)	53	51°45′N	83°30′W
Al Başrah, Iraq	154	30°35′N	47°59′E

ăt; finăl; rāte; senåte; ärm; åsk; sofà; fâre; ch-choose; dh-as th in other; bē; ėvent; bĕt; recĕnt; cratẽr; g-gō; gh-guttural g; bĭt; ĭ-short neutral; rīde; κ-guttural k as ch in German ich;

PRONOUNCING INDEX

Al Batrūn – Almonte 219

PLACE (Pronunciation)	PAGE	LAT.	LONG.
Al Batrūn, Leb. (äl-bä-tro͞on′)	153a	34°16′N	35°39′E
Albemarle, N.C., U.S. (ăl′bē-märl)	83	35°24′N	80°36′W
Albemarle Sound, strt., N.C., U.S. (ăl′bē-märl)	65	36°00′N	76°17′W
Albenga, Italy (äl-bĕŋ′gä)	130	44°04′N	8°13′E
Alberche, r., Spain (äl-bĕr′chå)	128	40°08′N	4°19′W
Alberga, The, r., Austl. (ăl-bŭr′gȧ)	174	27°15′S	135°00′E
Albergaria-a-Velha, Port.	128	40°47′N	8°31′W
Alberhill, Ca., U.S. (ăl′bẽr-hĭl)	75a	33°43′N	117°23′W
Albert, Fr. (ȧl-bâr′)	126	50°00′N	2°49′E
Albert, l., Afr. (ăl′bẽrt) (ȧl-bâr′)	185	1°50′N	30°40′E
Albert, Parc National, rec., D.R.C.	191	0°05′N	29°30′E
Alberta, prov., Can. (ăl-bûr′tȧ)	50	54°33′N	117°10′W
Alberta, Mount, mtn., Can. (ăl-bûr′tȧ)	55	52°18′N	117°28′W
Albert Edward, Mount, mtn., Pap. N. Gui. (ăl′bẽrt ĕd′wẽrd)	169	8°25′S	147°25′E
Alberti, Arg. (äl-bĕ′r-tē)	99c	35°01′S	60°16′W
Albert Kanaal, can., Bel.	115a	51°07′N	5°07′E
Albert Lea, Mn., U.S. (ăl′bẽrt lē′)	71	43°38′N	93°24′W
Albert Nile, r., Ug.	191	3°25′N	31°35′E
Alberton, Can. (ăl′bẽr-tŭn)	60	46°49′N	64°04′W
Alberton, S. Afr.	187b	26°16′S	28°08′E
Albertville *see* Kalemie, D.R.C.	186	5°56′S	29°12′E
Albertville, Fr. (ȧl-bêr-vēl′)	127	45°42′N	6°25′E
Albertville, Al., U.S. (ăl′bẽrt-vĭl)	82	34°15′N	86°10′W
Albi, Fr. (äl-bē′)	117	43°54′N	2°07′E
Albia, Ia., U.S. (ăl-bĭ-ȧ)	71	41°01′N	92°44′W
Albina, Sur. (äl-bē′nä)	101	5°30′N	54°33′W
Albina, Ponta, c., Ang.	190	15°51′S	11°44′E
Albino, Point, c., Can. (äl-bē′nō)	69c	42°50′N	79°05′W
Albion, Mi., U.S. (ăl′bĭ-ŭn)	66	42°15′N	84°50′W
Albion, Ne., U.S. (ăl′bĭ-ŭn)	70	41°42′N	98°00′W
Albion, N.Y., U.S. (ăl′bĭ-ŭn)	67	43°15′N	78°10′W
Alboran, Isla del, i., Spain (ē′s-lä-dĕl-äl-bō-rä′n)	112	35°58′N	3°02′W
Albuquerque, N.M., U.S. (ăl-bu̇-kûr′kē)	64	35°05′N	106°40′W
Albuquerque, Cayos de, is., Col.	91	12°12′N	81°24′W
Alburquerque, Spain (äl-bo͞or-kĕr′kä)	128	39°13′N	6°58′W
Albury, Austl. (ôl′bẽr-ē)	175	36°00′S	147°00′E
Alcabideche, Port. (äl-kȧ-bē-dā′chå)	129b	38°43′N	9°24′W
Alcácer do Sal, Port. (äb′ĭ-lēn)	128	38°24′N	8°33′W
Alcalá de Henares, Spain (äl-kä-lä′ dā ā-na′räs)	129a	40°29′N	3°22′W
Alcalá la Real, Spain (äl-kä-lä′lä rā-äl′)	128	37°27′N	3°57′W
Alcamo, Italy (äl′kȧ-mō)	130	37°58′N	13°03′E
Alcanadre, r., Spain (äl-kä-nä′drå)	129	41°41′N	0°18′W
Alcanar, Spain (äl-kä-när′)	129	40°35′N	0°27′E
Alcañiz, Spain (äl-kän-yēth′)	118	41°03′N	0°08′W
Alcântara, Braz. (äl-kän′tȧ-rȧ)	101	2°17′S	44°29′W
Alcaraz, Spain (äl-kä-räth′)	128	38°39′N	2°28′W
Alcaudete, Spain (äb′ĭng-dŭn)	128	37°38′N	4°05′W
Alcázar de San Juan, Spain (äl-kä′thär dā sän hwän′)	118	39°22′N	3°12′W
Alcira, Spain (ä-thē′rä)	129	39°09′N	0°26′W
Alcoa, Tn., U.S. (ăl-kō′ȧ)	82	35°45′N	84°00′W
Alcobendas, Spain (äl-kō-bĕn′dås)	129a	40°32′N	3°39′W
Alcochete, Port. (äl-kō-chā′ta)	129b	38°45′N	8°58′W
Alcoi, Spain	118	38°42′N	0°30′W
Alcorcón, Spain	129a	40°22′N	3°50′W
Alcorta, Arg. (ȧl-kôr′tä)	99c	33°32′S	61°08′W
Alcova Reservoir, res., Wy., U.S. (ăl-kō′vȧ)	73	42°31′N	106°33′W
Alcove, Can. (ăl-kōv′)	62c	45°41′N	75°55′W
Alcúdia, Badia d', b., Spain	129	39°48′N	3°20′E
Aldabra Islands, is., Sey. (äl-dä′brä)	187	9°16′S	46°17′E
Aldama, Mex. (äl-dä′mä)	88	22°54′N	98°04′W
Aldama, Mex. (äl-dä′mä)	80	28°50′N	105°54′W
Aldan, Russia	135	58°46′N	125°19′E
Aldan, r., Russia	135	63°00′N	134°00′E
Aldan Plateau, plat., Russia	141	57°42′N	130°28′E
Aldanskaya, Russia	135	61°52′N	135°29′E
Aldenhoven, Ger. (äl′dĕn-hō′vĕn)	127c	50°54′N	6°18′E
Aldergrove, Can. (ôl′dẽr-grōv)	74d	49°03′N	122°28′W
Alderney, i., Guern. (ôl′dẽr-nĭ)	126	49°43′N	2°11′W
Aldershot, Eng., U.K. (ôl′dẽr-shŏt)	120	51°14′N	0°46′W
Alderson, W.V., U.S. (ôl-dẽr-sŭn)	66	37°40′N	80°40′W
Alderwood Manor, Wa., U.S. (ôl′dẽr-wo͝od măn′ŏr)	74a	47°49′N	122°18′W
Aldridge-Brownhills, Eng., U.K.	114a	52°38′N	1°55′W
Aledo, Il., U.S. (ȧ-le′dō)	79	41°12′N	90°47′W
Aleg, Maur.	184	17°03′N	13°55′W
Alegre, Braz. (ȧlĕ′grĕ)	99a	20°41′S	41°32′W
Alegre, r., Braz. (ȧlĕ′grĕ)	102b	22°22′S	43°34′W
Alegrete, Braz. (ä-lå-grā′tä)	102	29°46′S	55°44′W
Aleksandrov, Russia (ä-lyĕk-sän′ drôf)	136	56°24′N	38°45′E
Aleksandrovsk, Russia (ä-lyĕk-sän′drôfsk)	142a	59°11′N	57°36′E
Aleksandrovsk, Russia (ä-lyĕk-sän′drôfsk)	135	51°02′N	142°21′E
Aleksandrów Kujawski, Pol. (ä-lĕk-säh′dro͞ov ko͞o-yav′skē)	125	52°54′N	18°45′E
Alekseyevka, Russia (ä-lyĕk-sä-yĕf′kȧ)	133	50°39′N	38°40′E
Aleksin, Russia (äb′ĭng-tŭn)	132	54°31′N	37°07′E
Aleksinac, Serb. (ȧ-lyĕk-sē-nȧk′)	131	43°33′N	21°42′E
Alemán, Presa, res., Mex. (prä′sä-lĕ-mȧ′n)	89	18°20′N	96°35′W
Alem Paraíba, Braz. (ä-lĕ′m-pȧ-räē′bȧ)	99a	21°54′S	42°40′W
Alençon, Fr. (ȧ-län-sôN′)	117	48°26′N	0°08′E
Alenquer, Braz. (ä-lĕŋ-kĕr′)	101	1°58′S	54°44′W
Alenquer, Port. (ä-lĕŋ-kĕr′)	128	39°04′N	9°01′W
Alentejo, hist. reg., Port. (ä-lĕŋ-tä′zhō)	128	38°05′N	7°45′W

PLACE (Pronunciation)	PAGE	LAT.	LONG.
Alenuihaha Channel, strt., Hi., U.S. (ä′lȧ-no͞o-ē-hä′hä)	84a	20°20′N	156°05′W
Aleppo, Syria (ȧ-lĕp-ō)	154	36°10′N	37°18′E
Alès, Fr. (ä-lĕs′)	117	44°07′N	4°06′E
Alessandria, Italy (ä-lĕs-sän′drē-ä)	118	44°53′N	8°35′E
Ålesund, Nor. (ô′lĕ-so͞on′)	122	62°28′N	6°14′E
Aleutian Islands, is., Ak., U.S. (ȧ-lu′shăn)	64b	52°40′N	177°30′W
Aleutian Trench, deep,	63a	50°40′N	177°10′E
Alevina, Mys, c., Russia	135	58°49′N	151°44′E
Alexander Archipelago, is., Ak., U.S. (ăl-ĕg-zăn′dẽr)	63	57°05′N	138°10′W
Alexander City, Al., U.S.	82	32°55′N	85°55′W
Alexander Indian Reserve, I.R., Can.	62g	53°47′N	114°00′W
Alexander Island, i., Ant.	178	71°00′S	71°00′W
Alexandra, S. Afr. (ȧl-ex-än′drȧ)	192c	26°07′S	28°07′E
Alexandria, Austl. (ăl-ĕg-zăn′drĭ-ȧ)	174	19°00′S	136°56′E
Alexandria, Can. (ăl-ĕg-zăn′drĭ-ȧ)	59	45°50′N	74°35′W
Alexandria, Egypt (ăl-ĕg-zăn′drĭ-ȧ)	185	31°12′N	29°58′E
Alexandria, Rom. (ăl-ĕg-zăn′drĭ-ȧ)	131	43°55′N	25°21′E
Alexandria, S. Afr. (ăl-ĕx-än-drĭ-ȧ)	187c	33°40′S	26°26′E
Alexandria, In., U.S. (ăl-ĕg-zăn′drĭ-ȧ)	66	40°20′N	85°20′W
Alexandria, La., U.S. (ăl-ĕg-zăn′drĭ-ȧ)	65	31°18′N	92°28′W
Alexandria, Mn., U.S. (ăl-ĕg-zăn′drĭ-ȧ)	70	45°53′N	95°23′W
Alexandria, S.D., U.S. (ăl-ĕg-zăn′drĭ-ȧ)	70	43°39′N	97°45′W
Alexandria, Va., U.S. (ăl-ĕg-zăn′drĭ-ȧ)	65	38°50′N	77°05′W
Alexandria Bay, N.Y., U.S. (ăl-ĕg-zăn′drĭ-ȧ)	67	44°20′N	75°55′W
Alexandroúpoli, Grc.	119	40°41′N	25°51′E
Alfaro, Spain (äl-färō)	128	42°08′N	1°43′W
Al-Fāshir, Sudan (ăl-fä′shẽr)	185	13°38′N	25°21′E
Al Fashn, Egypt	192b	28°47′N	30°53′E
Al Fayyūm, Egypt	185	29°14′N	30°48′E
Alfeiós, r., Grc.	131	37°33′N	21°50′E
Alfenas, Braz. (äl-fĕ′nȧs)	99a	21°26′S	45°55′W
Al Firdān, Egypt (äl-fer-dän′)	192b	30°43′N	32°20′E
Alfred, Can. (ăl′frĕd)	62c	45°34′N	74°52′W
Alfreton, Eng., U.K. (ăl′fẽr-tŭn)	114a	53°06′N	1°23′W
Algarve, hist. reg., Port. (äl-gär′vĕ)	128	37°15′N	8°12′W
Algeciras, Spain (äl-hä-thē′räs)	128	36°08′N	5°25′W
Algeria, nation, Afr. (ăl-gē′rĭ-ȧ)	184	28°45′N	1°00′E
Algete, Spain (äl-hā′tä)	129a	40°36′N	3°30′W
Al Ghaydah, Yemen	157	16°12′N	52°15′E
Alghero, Italy (äl-gâ′rō)	118	40°32′N	8°22′E
Algiers, Alg. (ăl-jērs)	184	36°51′N	2°56′E
Algoa, Tx., U.S. (ăl-gō′ȧ)	81a	29°24′N	95°11′W
Algoma, Wa., U.S.	74a	47°17′N	122°15′W
Algoma, Wi., U.S.	71	44°38′N	87°29′W
Algona, Ia., U.S.	71	43°04′N	94°11′W
Algonac, Mi., U.S. (ăl′gô-năk)	66	42°35′N	82°30′W
Algonquin, Il., U.S. (ăl-gŏŋ′kwĭn)	69a	42°10′N	88°17′W
Algonquin Provincial Park, rec., Can.	65	45°50′N	78°20′W
Alhama de Granada, Spain (äl-hä′mä-dĕ-grä-nä′dä)	128	37°00′N	3°59′W
Alhama de Murcia, Spain	128	37°50′N	1°24′W
Alhambra, Ca., U.S. (ăl-hăm′brȧ)	75a	34°05′N	118°08′W
Al Ḩammān, Egypt	119	30°46′N	29°42′E
Alhandra, Port. (äl-yän′drȧ)	129b	38°55′N	9°01′W
Alhaurín, Spain (ä-lou-rēn′)	128	36°40′N	4°40′W
Al Ḩawrah, Yemen	157	13°49′N	47°37′E
Al Ḩawṭah, Yemen	154	15°58′N	48°26′E
Al Ḩijāz, reg., Sau. Ar.	154	23°45′N	39°08′E
Al Hirmil, Leb.	153a	34°23′N	36°22′E
Alhos Vedros, Port. (äl′yōs′vā′drōs)	129b	38°39′N	9°02′W
Alhucemas, Baie d', b., Afr.	128	35°18′N	3°50′W
Al Ḩudaydah, Yemen	154	14°43′N	43°03′E
Al Hufūf, Sau. Ar.	154	25°15′N	49°43′E
Al Ḩulwān, Egypt (äl-hĕl′wän)	192b	29°51′N	31°20′E
Aliákmonas, r., Grc.	119	40°26′N	22°17′E
Äli Bayramli, Azer.	138	39°56′N	48°56′E
Alibori, r., Benin	189	11°40′N	2°55′E
Alice, S. Afr. (ăl-ĭs)	187c	32°47′S	26°51′E
Alice, Tx., U.S. (ăl′ĭs)	80	27°45′N	98°04′W
Alice, Punta, c., Italy (ä-lē′chĕ)	131	39°23′N	17°10′E
Alice Arm, Can.	54	55°29′N	129°29′W
Alicedale, S. Afr. (ăl′ĭs-dāl)	187c	33°18′S	26°04′E
Alice Springs, Austl. (ăl′ĭs)	174	23°38′S	133°56′E
Alicudi, i., Italy (ä-lē-ko͞o′dē)	130	38°34′N	14°21′E
Alifkulovo, Russia (ä-lĭf-ku̇′lô-vô)	142a	55°57′N	62°06′E
Alīgarh, India (ä-lē-gŭr′)	155	27°58′N	78°08′E
Alingsås, Swe. (ȧ′lĭŋ-sôs)	122	57°57′N	12°30′E
Aliquippa, Pa., U.S. (ăl-ĭ-kwĭp′ȧ)	69e	40°37′N	80°15′W
Al Iskandarīyah *see* Alexandria, Egypt	192b	31°12′N	29°58′E
Aliwal North, S. Afr. (ä-lē-wäl′)	186	31°09′S	28°26′E
Al Jafr, Qa'al, pl., Jord.	153a	30°15′N	36°24′E
Al Jaghbūb, Libya	185	29°46′N	24°32′E
Al Jawārah, Oman	157	18°55′N	57°17′E
Al Jawf, Libya	185	24°14′N	23°15′E
Al Jawf, Sau. Ar.	154	29°45′N	39°30′E
Aljezur, Port. (äl-zhā-zo͞or′)	128	37°18′N	8°52′W
Al Jīzah, Egypt	192b	30°01′N	31°12′E
Al Jubayl, Sau. Ar.	154	27°01′N	49°40′E
Al Jufrah, oasis, Libya	185	29°30′N	15°16′E
Al Junaynah, Sudan	156	13°27′N	22°27′E
Aljustrel, Port. (äl-zho͞o-strĕl′)	128	37°44′N	8°23′W
Al Kāb, Egypt	192d	30°56′N	32°19′E
Al Kāmilīn, Sudan (käm-lēn′)	185	15°09′N	33°06′E
Al Karak, Jord. (kĕ-räk′)	153a	31°11′N	35°42′E
Al Karnak, Egypt (kär′nak)	192b	25°42′N	32°43′E
Al Khābūrah, Oman	154	23°45′N	57°30′E
Al Khalīl, W.B.	153a	31°31′N	35°07′E
Al Khandaq, Sudan (kän-däk′)	185	18°38′N	30°29′E
Al Khārijah, Egypt	156	25°26′N	30°33′E

PLACE (Pronunciation)	PAGE	LAT.	LONG.
Al Khums, Libya	185	32°35′N	14°10′E
Al Khurmah, Sau. Ar.	154	21°37′N	41°44′E
Al Kiswah, Syria	153a	33°31′N	36°13′E
Alkmaar, Neth. (älk-mär′)	121	52°39′N	4°42′E
Al Kufrah, oasis, Libya	185	24°45′N	22°45′E
Al Kuntillah, Egypt	153a	29°59′N	34°42′E
Al Kūt, Iraq	157	32°30′N	45°49′E
Al Kuwayt, Kuw. (äl-ko͞o-wit)	154	29°04′N	47°59′E
Al Lādhiqīyah, Syria	154	35°32′N	35°51′E
Allagash, r., Me., U.S. (ăl′ȧ-găsh)	60	46°50′N	69°24′W
Allāhābād, India (ŭl-ŭ-hä-bäd′)	155	25°32′N	81°53′E
All American Canal, can., Ca., U.S. (âl ȧ-mĕr′ĭ-kăn)	76	32°43′N	115°12′W
Alland, Aus.	115e	48°04′N	16°05′E
Allariz, Spain (äl-yä-rēth′)	118	42°10′N	7°48′W
Allatoona Lake, res., Ga., U.S. (ăl′ȧ-to͞on′ȧ)	82	34°05′N	84°57′W
Allauch, Fr. (ä-lĕ′ō)	126a	43°21′N	5°30′E
Allaykha, Russia (ä-lī′kȧ)	135	70°32′N	148°53′E
Allegan, Mi., U.S. (ăl′ē-găn)	66	42°30′N	85°55′W
Allegany Indian Reservation, I.R., N.Y., U.S. (ăl-ē-gā′nĭ)	67	42°05′N	78°55′W
Allegheny, r., Pa., U.S. (ăl-ē-gā′nĭ)	67	41°10′N	79°20′W
Allegheny Front, mtn., U.S. (ăl-ē-gā′nĭ)	66	38°12′N	80°03′W
Allegheny Mountains, mts., U.S. (ăl-ē-gā′nĭ)	65	37°35′N	81°55′W
Allegheny Plateau, plat., U.S. (ăl-ē-gā′nĭ)	66	39°00′N	81°15′W
Allegheny Reservoir, res., U.S. (ăl-ē-gā′nĭ)	67	41°50′N	78°55′W
Allen, Ok., U.S. (ăl′ĕn)	79	34°51′N	96°26′W
Allen, Lough, l., Ire. (lŏk ăl′ĕn)	120	54°07′N	8°09′W
Allendale, N.J., U.S. (ăl′ĕn-dāl)	68a	41°02′N	74°08′W
Allendale, S.C., U.S. (ăl′ĕn-dāl)	83	33°00′N	81°19′W
Allende, Mex. (äl-yĕn′då)	89	18°23′N	92°49′W
Allende, Mex.	80	28°20′N	100°50′W
Allentown, Pa., U.S. (ălĕn-toun)	65	40°35′N	75°30′W
Alleppey, India (ȧ-lĕp′ē)	159	9°33′N	76°22′E
Aller, r., Ger. (äl′ẽr)	124	52°43′N	9°50′E
Alliance, Ne., U.S. (ȧ-lī′ăns)	64	42°06′N	102°53′W
Alliance, Oh., U.S. (ȧ-lī′ăns)	66	40°55′N	81°10′W
Al Lidām, Sau. Ar.	154	20°45′N	44°12′E
Allier, r., Fr. (ȧ-lyā′)	126	46°43′N	3°03′E
Alligator Point, c., La., U.S. (ăl′ĭ-gā-tẽr)	68d	30°57′N	89°41′W
Allinge, Den. (äl′ĭŋ-ĕ)	122	55°16′N	14°48′E
Al Līth, Sau. Ar.	157	20°09′N	40°16′E
All Pines, Belize (ôl pīnz)	90a	16°55′N	88°15′W
Al Luḩayyah, Yemen	154	15°58′N	42°48′E
Alluvial City, La., U.S.	68d	29°51′N	89°42′W
Allyn, Wa., U.S. (ăl′ĭn)	74a	47°23′N	122°51′W
Alma, Can. (ăl′mȧ)	60	45°36′N	64°59′W
Alma, Can.	51	48°29′N	71°42′W
Alma, S. Afr.	192c	24°30′S	28°05′E
Alma, Ga., U.S.	83	31°33′N	82°31′W
Alma, Mi., U.S.	66	43°25′N	84°40′W
Alma, Ne., U.S.	78	40°08′N	99°21′W
Alma, Wi., U.S.	71	44°21′N	91°57′W
Alma-Ata *see* Almaty, Kaz.	139	43°19′N	77°08′E
Almada, Port. (äl-mä′dä)	129b	38°40′N	9°09′W
Almadén, Spain (äl-mȧ-dhān)	128	38°47′N	4°50′W
Al Madīnah, Sau. Ar.	154	24°26′N	39°42′E
Al Mafraq, Jord.	153a	32°21′N	36°13′E
Almagre, Laguna, l., Mex. (lä-gô′nä-ȧl-mä′grĕ)	89	23°48′N	97°45′W
Almagro, Spain (äl-mä′grō)	128	38°52′N	3°41′W
Al Maḩallah al Kubrā, Egypt	192b	30°58′N	31°10′E
Al Manāmah, Bahr.	154	26°01′N	50°33′E
Almanor, Lake, l., Ca., U.S. (ăl-măn′ôr)	76	40°11′N	121°20′W
Almansa, Spain (äl-män′sä)	128	38°52′N	1°09′W
Al Manshāh, Egypt	192b	26°31′N	31°46′E
Almansor, r., Port. (äl-män-sôr)	128	38°41′N	8°27′W
Al Manşūrah, Egypt	185	31°02′N	31°25′E
Al Manzilah, Egypt (män′za-lȧ)	192b	31°09′N	32°05′E
Almanzora, r., Spain (äl-män-thō′rä)	128	37°20′N	2°25′W
Al Marāghah, Egypt	192b	26°41′N	31°35′E
Almargem do Bispo, Port. (äl-mär-zhĕN)	129b	38°51′N	9°16′W
Al-Marj, Libya	185	32°44′N	21°08′E
Al Maşīrah, i., Oman	154	20°43′N	58°58′E
Almaty (Alma-Ata), Kaz.	139	43°19′N	77°08′E
Almaty, val., Sau. Ar.	153a	29°16′N	35°12′E
Al Mawşil, Iraq	154	36°00′N	42°53′E
Almazán, Spain (äl-mä-thän′)	128	41°30′N	2°33′W
Al Mazār, Jord.	153a	31°04′N	35°41′E
Al Mazra'ah, Jord.	153a	31°17′N	35°33′E
Almeirim, Port. (äl-māī-rēN′)	128	39°13′N	8°31′W
Almelo, Neth. (äl′mĕ-lō)	121	52°20′N	6°42′E
Almendra, Embalse de, res., Spain	128	41°15′N	6°10′W
Almendralejo, Spain (äl-män-drä-lā′hō)	128	38°43′N	6°24′W
Almería, Spain (äl-mā-rē′ä)	110	36°52′N	2°28′W
Almería, Golfo de, b., Spain (gôl-fô-dĕ-äl-māī-reN′)	128	36°45′N	2°26′W
Älmhult, Swe. (ĕlm′ho͞olt)	122	56°35′N	14°08′E
Almina, Punta, c., Mor. (äl-mē′nä)	128	35°58′N	5°17′W
Al Minyā, Egypt	185	28°06′N	30°45′E
Almirante, Pan. (äl-mē-rän′tä)	91	9°18′N	82°24′W
Almirante, Bahía de, b., Pan.	91	9°22′N	82°07′W
Almodóvar del Campo, Spain (äl-mō-dhō′vär)	128	38°43′N	4°10′W
Almoloya, Mex. (äl-mō-lō′yä)	88	19°32′N	99°44′W
Almoloya, Mex. (äl-mō-lō′yä)	89a	19°11′N	99°28′W
Almonte, Can. (ăl-mŏn′tē)	59	45°15′N	76°15′W
Almonte, Spain (äl-mōn′tä)	128	37°16′N	6°32′W

220 Almonte – Angeles PRONOUNCING INDEX

PLACE (Pronunciation)	PAGE	LAT.	LONG.
Almonte, r., Spain (äl-mōn′tä)	128	39°35′N	5°50′W
Almora, India	155	29°20′N	79°40′E
Al Mubarraz, Sau. Ar.	154	22°31′N	46°27′E
Al Mudawwarah, Jord.	153a	29°20′N	36°01′E
Al Mukhā (Mocha), Yemen	154	13°11′N	43°20′E
Almuñécar, Spain (äl-mōōn-yā′kär)	128	36°44′N	3°43′W
Almyrós, Grc.	131	39°13′N	22°47′E
Alnön, i., Swe.	122	62°20′N	17°39′E
Aloha, Or., U.S. (ä′lô-hä)	74c	45°29′N	122°52′W
Alor, Pulau, i., Indon. (ä′lôr)	169	8°07′S	125°00′E
Álora, Spain (ä′lô-rä)	128	36°49′N	4°42′W
Alor Gajah, Malay.	153b	2°23′N	102°13′E
Alor Setar, Malay. (ä′lôr stär)	168	6°10′N	100°16′E
Alouette, r., Can. (ä-lōō-ĕt′)	74d	49°16′N	122°32′W
Alpena, Mi., U.S. (ăl-pē′nȧ)	65	45°05′N	83°30′W
Alpes Cotiennes, mts., Eur.	127	44°46′N	7°02′E
Alphen, Neth.	115a	52°07′N	4°38′E
Alpiarça, Port. (äl-pyär′sȧ)	128	39°38′N	8°37′W
Alpine, Tx., U.S. (ăl′pīn)	80	30°21′N	103°41′W
Alps, mts., Eur. (ălps)	112	46°18′N	8°42′E
Alpujarra, Col. (ál-pōō-ĸá′rä)	100a	3°23′N	74°56′W
Al Qaḍārif, Sudan	185	14°03′N	35°11′E
Al Qāhirah *see* Cairo, Egypt	185	30°00′N	31°17′E
Al Qanṭarah, Egypt	192d	30°51′N	32°20′E
Al Qaryah Ash Sharqīyah, Libya	185	30°36′N	13°13′E
Al Qaṣr, Egypt	156	25°42′N	28°53′E
Al Qaṭīf, Sau. Ar.	154	26°30′N	50°00′E
Al Qayṣūmah, Sau. Ar.	154	28°15′N	46°20′E
Al Qunaytirah, Syria	153a	33°09′N	35°49′E
Al Qunfudhah, Sau. Ar.	154	19°08′N	41°05′E
Al Quṣaymah, Egypt	153a	30°40′N	34°23′E
Al Quṣayr, Egypt	185	26°14′N	34°11′E
Al Qusayr, Syria	153a	34°32′N	36°33′E
Als, i., Den. (äls)	122	55°06′N	9°40′E
Alsace, hist. reg., Fr. (äl-sȧ′s)	127	48°25′N	7°24′E
Altadena, Ca., U.S. (ăl-tä-dē′nä)	75a	34°12′N	118°08′W
Alta Gracia, Arg. (äl′tä grä′sē-a)	102	31°41′S	64°19′W
Altagracia, Ven.	100	10°42′N	71°34′W
Altagracia de Orituco, Ven.	101b	9°53′N	66°22′W
Altai Mountains, mts., Asia (äl′tī′)	160	49°11′N	87°15′E
Alta Loma, Ca., U.S. (ăl′tä lō′mä)	75a	34°07′N	117°35′W
Alta Loma, Tx., U.S. (ăl′tȧ lō-mȧ)	81a	29°22′N	95°05′W
Altamaha, r., Ga., U.S. (ôl-tá-má-hô′)	83	31°50′N	82°00′W
Altamira, Braz. (äl-tä-mē′rä)	101	3°13′S	52°14′W
Altamira, Mex.	89	22°25′N	97°55′W
Altamirano, Arg. (äl-tä-mē-rä′nō)	102	35°26′S	58°12′W
Altamura, Italy (äl-tä-mōō′rä)	119	40°40′N	16°35′E
Altavista, Va., U.S. (ăl-tä-vĭs′tȧ)	83	37°08′N	79°14′W
Altay, China (äl-tā)	160	47°52′N	86°50′E
Altenburg, Ger. (äl-tĕn-bŏŏrgh)	124	50°59′N	12°27′E
Altenmarkt an der Triesting, Aus.	115e	48°02′N	16°00′E
Alter do Chão, Port. (äl-tĕr′dô shäɴ′ôɴ)	128	39°13′N	7°38′W
Altiplano, pl., Bol. (äl-tē-plá′nō)	100	18°38′S	68°20′W
Altlandsberg, Ger. (ält länts′bĕrgh)	115b	52°34′N	13°44′E
Alto, La., U.S. (ăl′tō)	81	32°21′N	91°52′W
Alto Marañón, r., Peru (ȧl′tô-mä-rän-yō′n)	100	8°18′S	77°13′W
Altomünster, Ger. (äl′tō-mün′stĕr)	115d	48°24′N	11°16′E
Alton, Can. (ôl′tŭn)	62d	43°52′N	80°05′W
Alton, Il., U.S. (ôl′tŭn)	65	38°53′N	90°11′W
Altona, Austl.	173a	37°52′S	144°50′E
Altona, Can.	57	49°06′N	97°33′W
Altona, Ger. (äl′tō-nä)	115c	53°33′N	9°54′E
Altoona, Al., U.S. (ăl-tōō′nȧ)	82	34°01′N	86°15′W
Altoona, Pa., U.S. (ăl-tōō′nȧ)	65	40°25′N	78°25′W
Altoona, Wa., U.S. (ăl-tōō′nȧ)	74c	46°16′N	123°39′W
Alto Rio Doce, Braz. (äl′tô-rē′ô-dō′sĕ)	99a	21°02′S	43°23′W
Alto Songo, Cuba (äl-fō-sôn′gō)	93	20°10′N	75°45′W
Altotonga, Mex. (äl-tô-tôn′gä)	89	19°44′N	97°13′W
Alto Velo, i., Dom. Rep. (äl-tô-vĕ′lō)	93	17°30′N	71°35′W
Altrincham, Eng., U.K. (ôl′trĭng-ăm)	114a	53°18′N	2°21′W
Altruppin, Ger. (ält rōō′pēn)	115b	52°56′N	12°50′E
Altun Shan, mts., China (äl-tòn shän)	160	36°58′N	85°09′E
Alturas, Ca., U.S. (ăl-tōō′rás)	72	41°29′N	120°33′W
Altus, Ok., U.S. (ăl′tŭs)	78	34°38′N	99°20′W
Al 'Ubaylah, Sau. Ar.	157	21°59′N	50°57′E
Al-Uḍayyah, Sudan	185	12°06′N	28°16′E
Alūksne, Lat. (ä′lòks-nĕ)	136	57°24′N	27°04′E
Alumette Island, i., Can. (á-lü-mĕt′)	59	45°50′N	77°00′W
Alum Rock, Ca., U.S.	74b	37°23′N	121°50′W
Al 'Uqaylah, Libya	185	30°15′N	19°07′E
Al Uqṣur, Egypt	185	25°38′N	32°59′E
Alushta, Ukr. (ä′lshò-tȧ)	133	44°39′N	34°23′E
Alva, Ok., U.S. (ăl′vȧ)	78	36°46′N	98°41′W
Alvarado, Mex. (äl-vä-rä′dhō)	89	18°48′N	95°45′W
Alvarado, Laguna de, l., Mex. (lä-gò′nä-dĕ-äl-vä-rá′dô)	89	18°44′N	95°45′W
Älvdalen, Swe. (ĕlv′dä-lĕn)	122	61°14′N	14°04′E
Alverca, Port. (al-vĕr′kȧ)	129b	38°53′N	9°02′W
Alvesta, Swe. (äl-vĕs′tä)	122	56°55′N	14°29′E
Alvin, Tx., U.S. (ăl′vĭn)	81a	29°25′N	95°14′W
Alvinópolis, Braz. (äl-vēnō′pō-lēs)	99a	20°07′S	43°03′W
Alviso, Ca., U.S. (ăl-vī′sō)	74b	37°26′N	121°59′W
Al Wajh, Sau. Ar.	154	26°15′N	36°32′E
Alwar, India (ŭl′wŭr)	155	27°39′N	76°39′E
Al Wāsiṭah, Egypt	192b	29°21′N	31°15′E
Alytus, Lith. (ä′lē-tòs)	123	54°25′N	24°05′E
Amacuzac, r., Mex. (ä-mä-kōō-zák)	88	18°00′N	99°03′W
Amadeus, l., Austl. (ăm-ȧ-dē′ŭs)	174	24°30′S	131°25′E
Amadjuak, l., Can. (ä-mädj′wäk)	53	64°50′N	69°20′W
Amadora, Port.	129b	38°45′N	9°14′W
Amagasaki, Japan (ä′mä-gä-sä′kê)	167	34°43′N	135°25′E
Amakusa-Shimo, i., Japan (ämä-kōō′sä shē-mō)	166	32°24′N	129°35′E
Åmål, Swe. (ô′môl)	122	59°05′N	12°40′E
Amalfi, Col. (ä′má′l-fē)	100a	6°55′N	75°04′W
Amalfi, Italy (ä-mä′l-fē)	129c	40°23′N	14°36′E
Amaliáda, Grc.	131	37°48′N	21°23′E
Amalner, India	158	21°07′N	75°06′E
Amambai, Serra de, mts., S.A.	101	20°06′S	57°08′W
Amami, i., Japan	161	28°10′N	129°55′E
Amapala, Hond. (ä-mä-pä′lä)	90	13°16′N	87°39′W
Amarante, Braz. (ä-mä-rän′tä)	101	6°17′S	42°43′W
Amargosa, r., Ca., U.S. (ȧ′mär-gō′sȧ)	76	35°55′N	116°45′W
Amarillo, Tx., U.S. (ăm-ȧ-rĭl′ō)	64	35°14′N	101°49′W
Amaro, Mount, mtn., Italy (ä-mä′rō)	118	42°07′N	14°07′E
Amasya, Tur. (ä-mä′sē-ȧ)	119	40°40′N	35°50′E
Amatenango, Mex. (ä-mä-tä-naŋ′gō)	89	16°30′N	92°29′W
Amatignak, i., Ak., U.S. (ä-mä′tē-näk)	63a	51°12′N	178°30′W
Amatique, Bahía de, b., N.A. (bä-ē′ä-dĕ-ä-mä-tē′kä)	90	15°58′N	88°50′W
Amatitlán, Guat. (ä-mä-tē-tlän′)	90	14°27′N	90°39′W
Amatlán de Cañas, Mex. (ä-mät-län′dā kän-yäs)	88	20°50′N	104°22′W
Amazon (Amazonas) (Solimões), r., S.A.	101	2°03′S	53°18′W
Amazonas, state, Braz. (ä-mä-thō′näs)	100	4°15′S	64°30′W
Ambāla, India (ŭm-bä′lŭ)	155	30°31′N	76°48′E
Ambalema, Col. (äm-bä-lā′mä)	100	4°47′N	74°45′W
Ambarchik, Russia (ŭm-bär′chĭk)	135	69°39′N	162°18′E
Ambarnāth, India	159b	19°12′N	73°10′E
Ambato, Ec. (äm-bä′tō)	100	1°15′S	78°30′W
Ambatondrazaka, Madag.	187	17°58′S	48°43′E
Amberg, Ger. (äm′bĕrgh)	124	49°26′N	11°51′E
Ambergris Cay, i., Belize (ăm′bẽr-grēs kāz)	90a	18°04′N	87°43′W
Ambergris Cays, is., T./C. Is.	93	21°20′N	71°40′W
Ambérieu-en-Bugey, Fr. (äɴ-bā-rē-u′)	127	45°57′N	5°21′E
Ambert, Fr. (äɴ-bĕr′)	126	45°32′N	3°41′E
Ambil Island, i., Phil. (äm′bēl)	169a	13°51′N	120°25′E
Ambler, Pa., U.S. (ăm′blẽr)	68f	40°09′N	75°13′W
Amboise, Fr. (äɴ-bwäz′)	126	47°25′N	0°56′E
Ambon, Indon.	169	3°45′S	128°17′E
Ambon, Pulau, i., Indon.	169	4°50′S	128°45′E
Ambositra, Madag. (äm-bô-sē′trä)	187	20°31′S	47°28′E
Amboy, Il., U.S. (ăm′boi)	66	41°41′N	89°15′W
Amboy, Wa., U.S. (ăm′boi)	74c	45°55′N	122°27′W
Ambre, Cap d', c., Madag.	187	12°06′S	49°15′E
Ambridge, Pa., U.S. (ăm′brĭdj)	69e	40°36′N	80°13′W
Ambrim, i., Vanuatu	175	16°25′S	168°15′E
Ambriz, Ang.	186	7°50′S	13°06′E
Amchitka, i., Ak., U.S. (ăm-chĭt′kȧ)	63a	51°25′N	178°10′E
Amchitka Passage, strt., Ak., U.S. (ăm-chĭt′kȧ)	63a	51°30′N	179°36′W
Amealco, Mex. (ä-mā-äl′kō)	88	20°12′N	100°08′W
Ameca, Mex. (ä-mĕ′kä)	86	20°34′N	104°02′W
Amecameca, Mex. (ä-må-kä-mä′kä)	88	19°06′N	98°46′W
Ameide, Neth.	115a	51°57′N	4°57′E
Ameland, i., Neth.	121	53°29′N	5°54′E
Amelia, Oh., U.S. (ȧ-mēl′yȧ)	69f	39°01′N	84°12′W
American, South Fork, r., Ca., U.S. (ȧ-mĕr′ĭ-kăn)	76	38°43′N	120°45′W
Americana, Braz. (ä-mĕ-rē-ká′ná)	99a	22°46′S	47°19′W
American Falls, Id., U.S. (ȧ-mĕr′ĭ-kăn-fâls′)	73	42°45′N	112°53′W
American Falls Reservoir, res., Id., U.S. (ȧ-mĕr′ĭ-kăn-fâls′)	64	42°56′N	113°18′W
American Fork, Ut., U.S.	77	40°20′N	111°50′W
American Highland, plat., Ant.	178	72°00′S	79°00′E
American Samoa, dep., Oc.	2	14°20′S	170°00′W
Americus, Ga., U.S. (ȧ-mĕr′ĭ-kŭs)	65	32°04′N	84°15′W
Amersfoort, Neth. (ä′mẽrz-fōrt)	115a	52°08′N	5°23′E
Amery, Can. (ā′mẽr-ê)	51	56°34′N	94°03′W
Amery, Wi., U.S.	71	45°19′N	92°24′W
Ames, Ia., U.S. (āmz)	71	42°00′N	93°36′W
Amesbury, Ma., U.S. (āmz′bẽr-ê)	61a	42°51′N	70°56′W
Amfissa, Grc. (àm-fĭ′sȧ)	131	38°32′N	22°26′E
Amga, Russia (ŭm-gä′)	135	61°08′N	132°09′E
Amga, r., Russia	141	61°41′N	133°11′E
Amgun', r., Russia	141	52°30′N	138°00′E
Amherst, Can. (ăm′hẽrst)	51	45°49′N	64°14′W
Amherst, Oh., U.S.	69d	41°24′N	82°13′W
Amherst, i., Can. (ăm′hẽrst)	59	44°08′N	76°45′W
Amiens, Fr. (ä-myăɴ′)	117	49°54′N	2°18′E
Amirante Islands, is., Sey.	5	6°02′S	52°30′E
Amisk Lake, l., Can.	57	54°35′N	102°13′W
Amistad Reservoir, res., N.A.	80	29°20′N	101°00′W
Amite, La., U.S. (ă-mēt′)	81	30°43′N	90°32′W
Amite, r., La., U.S.	81	30°45′N	90°48′W
Amity, Pa., U.S. (ăm′ĭ-tĭ)	69e	40°02′N	80°11′W
Amityville, N.Y., U.S. (ăm′ĭ-tĭ-vĭl)	68a	40°41′N	73°24′W
Amlia, i., Ak., U.S. (á′mlēä)	63a	52°00′N	173°28′W
'Ammān, Jord. (äm′mán)	154	31°57′N	35°57′E
Ammersee, l., Ger. (äm′mĕr)	115d	48°00′N	11°08′E
Amnicon, r., Wi., U.S. (ăm′nė-kŏn)	75h	46°35′N	91°56′W
Amorgós, i., Grc. (ä-môr′gōs)	119	36°47′N	25°47′E
Amory, Ms., U.S. (ămô-rē)	82	33°58′N	88°27′W
Amos, Can. (ā′mŭs)	51	48°31′N	78°04′W
Amoy *see* Xiamen, China	161	24°30′N	118°10′E
Amparo, Braz. (äm-pá′-rô)	99a	22°43′S	46°44′W
Amper, r., Ger. (äm′pĕr)	115d	48°18′N	11°32′E
Amposta, Spain (äm-pōs′tä)	129	40°42′N	0°34′E
Amqui, Can.	60	48°28′N	67°28′W
Amrāvati, India	155	20°58′N	77°47′E
Amritsar, India (ŭm-rĭt′sŭr)	155	31°43′N	74°52′E
Amstelveen, Neth.	115a	52°18′N	4°51′E
Amsterdam, Neth. (äm-stĕr-däm′)	110	52°21′N	4°52′E
Amsterdam, N.Y., U.S. (ăm′stẽr-dăm)	67	42°55′N	74°10′W
Amsterdam, Île, i., Afr.	178	37°52′S	77°32′E
Amstetten, Aus. (äm′stĕt-ĕn)	124	48°09′N	14°53′E
Am Timan, Chad (äm′tē-män′)	185	11°18′N	20°30′E
Amu Darya, r., Asia (ä-mò-dä′rēä)	134	38°30′N	64°00′E
Amukta Passage, strt., Ak., U.S. (ä-mōōk′tȧ)	63a	52°30′N	172°00′W
Amundsen Gulf, b., Can. (ä′mŭn-sĕn-gŭlf′)	52	70°17′N	123°28′W
Amundsen Sea, sea, Ant. (ä′mŭn-sĕn-sē′)	178	72°00′S	110°00′W
Amungen, l., Swe.	122	61°07′N	16°00′E
Amur, r., Asia	135	49°00′N	136°00′E
Amurskiy, Russia (ä-mûr′skĭ)	142a	52°35′N	59°36′E
Amurskiy, Zaliv, b., Russia (zä′lĭf ä-môr′skĭ)	166	43°20′N	131°40′E
Amusgos, Mex.	88	16°39′N	98°09′W
Amuyao, Mount, mtn., Phil. (ä-mōō-yä′ō)	169a	17°04′N	121°09′E
Amvrakikos Kólpos, b., Grc.	131	39°00′N	21°00′E
Amyun, Leb.	153a	34°18′N	35°48′E
Anabar, r., Russia (ăn-ă-bär′)	141	71°15′N	113°00′E
Anaco, Ven. (ä-ná′kô)	101b	9°29′N	64°27′W
Anaconda, Mt., U.S. (ăn-ȧ-kŏn′dȧ)	64	46°07′N	112°55′W
Anacortes, Wa., U.S. (ăn-ȧ-kôr′tĕz)	74a	48°30′N	122°37′W
Anadarko, Ok., U.S. (ăn-á-där′kō)	78	35°05′N	98°14′W
Anadyr', Russia (ŭ-nȧ-dĭr′)	135	64°47′N	177°01′E
Anadyr, r., Russia	141	65°30′N	172°45′E
Anadyrskiy Zaliv, b., Russia	134	64°10′N	178°00′W
Anaheim, Ca., U.S. (ăn′á-hīm)	75a	33°50′N	117°55′W
Anahuac, Tx., U.S. (ä-nä′wäk)	81a	29°46′N	94°41′W
Ānai Mudi, mtn., India	159	10°10′N	77°00′E
Anama Bay, Can.	57	51°56′N	98°05′W
Ana María, Cayos, is., Cuba	92	21°25′N	78°50′W
Anambas, Kepulauan, is., Indon. (ä-näm-bäs)	168	2°41′N	106°38′E
Anamosa, Ia., U.S. (ăn-ȧ-mō′sȧ)	71	42°06′N	91°18′W
Anan'ïv, Ukr.	137	47°43′N	29°59′E
Anapa, Russia (á-nä′pä)	137	44°54′N	37°19′E
Anápolis, Braz. (ä-nä′pō-lēs)	101	16°17′S	48°47′W
Añatuya, Arg. (à-nyä-tōō′yà)	102	28°22′S	62°45′W
Anchieta, Braz. (àn-chyĕ′tä)	102b	22°49′S	43°24′W
Ancholme, r., Eng., U.K. (ăn′chŭm)	114a	53°28′N	0°27′W
Anchorage, Ak., U.S. (ăŋ′kẽr-āj)	64a	61°12′N	149°48′W
Anchorage, Ky., U.S.	69h	38°16′N	85°32′W
Anci, China (än-tsŭ)	162	39°31′N	116°41′E
Ancienne-Lorette, Can. (äɴ-syĕn′ lō-rĕt′)	62b	46°48′N	71°21′W
Ancón, Pan. (ăŋ-kōn′)	86a	8°55′N	79°32′W
Ancona, Italy (än-kō′nä)	110	43°37′N	13°32′E
Ancud, Chile (äŋ-kōōdh′)	102	41°52′S	73°45′W
Ancud, Golfo de, b., Chile (gôl-fô-dĕ-äŋ-kōōdh′)	102	41°15′S	73°00′W
Anda, China	164	46°20′N	125°20′E
Åndalsnes, Nor.	122	62°33′N	7°46′E
Andalucia, hist. reg., Spain (än-dä-lōō-sē′ä)	128	37°35′N	5°40′W
Andalusia, Al., U.S. (ăn-dȧ-lōō′zhĭȧ)	82	31°19′N	86°19′W
Andaman Islands, is., India (ăn-dá-măn′)	168	11°38′N	92°17′E
Andaman Sea, sea, Asia	168	12°44′N	95°45′E
Andarax, r., Spain	128	37°00′N	2°40′W
Anderlecht, Bel. (än′dẽr-lĕkt)	115a	50°49′N	4°16′E
Andernach, Ger. (än′dẽr-näk)	124	50°25′N	7°23′E
Anderson, Arg. (á′n-dĕr-sōn)	99c	35°15′S	60°15′W
Anderson, Ca., U.S. (ăn′dẽr-sŭn)	72	40°28′N	122°19′W
Anderson, In., U.S. (ăn′dẽr-sŭn)	66	40°05′N	85°50′W
Anderson, S.C., U.S. (ăn′dẽr-sŭn)	65	34°30′N	82°40′W
Anderson, r., Can. (ăn′dẽr-sŭn)	52	68°32′N	125°12′W
Andes Mountains, mts., S.A. (ăn′dēz) (än′dās)	97	13°00′S	75°00′W
Andheri, neigh., India	159b	19°08′N	72°50′E
Andhra Pradesh, state, India	155	16°00′N	79°00′E
Andikýthira, i., Grc.	119	35°50′N	23°20′E
Andizhan, Uzb. (än-dē-zhän′)	139	40°45′N	72°22′E
Andong, Kor., S. (än′dŭng′)	161	36°31′N	128°42′E
Andongwei, China (än-dôŋ-wā)	162	35°08′N	119°19′E
Andorra, And. (än-dôr′rä)	129	42°38′N	1°30′E
Andorra, nation, Eur. (än-dôr′rä)	110	42°30′N	2°00′E
Andover, Ma., U.S. (ăn′dô-vẽr)	61a	42°39′N	71°08′W
Andover, N.J., U.S. (ăn′dô-vẽr)	68a	40°59′N	74°45′W
Andøya, i., Nor. (änd-ûê)	116	69°12′N	14°58′E
Andreanof Islands, is., Ak., U.S. (än-drå-ä′nôf-ĭ′ăndz)	64b	51°10′N	177°00′W
Andrelândia, Braz. (än-drĕ-lä′n-dyä)	99a	21°45′S	44°18′W
Andrew Johnson National Historic Site, rec., Tn., U.S. (ăn′drōō jŏn′sŭn)	83	36°15′N	82°55′W
Andrews, N.C., U.S. (ăn′drōōz)	82	35°12′N	83°48′W
Andrews, S.C., U.S. (ăn′drōōz)	83	33°25′N	79°32′W
Andria, Italy (än′drē-ä)	119	41°17′N	15°55′E
Andros, Grc. (án′dhròs)	131	37°50′N	24°54′E
Ándros, i., Grc. (án′dròs)	119	37°59′N	24°55′E
Androscoggin, r., Me., U.S. (ăn-drŭs-kŏg′ĭn)	60	44°25′N	70°45′W
Andros Island, i., Bah. (ăn′drŏs)	87	24°30′N	78°00′W
Anefis i-n-Darane, Mali	188	18°03′N	0°36′E
Anegasaki, Japan (ä′nā-gä-sä′kē)	167a	35°29′N	140°02′E
Aneityum, i., Vanuatu (ä-nā-ē′tē-ŭm)	175	20°15′S	169°49′E
Aneta, N.D., U.S. (ă-nē′tȧ)	70	47°41′N	97°57′W
Aneto, Pico de, mtn., Spain (pē′kō-dĕ-ä-nĕ′tô)	112	42°35′N	0°38′E
Angamacutiro, Mex. (äŋ′gä-mä-kōō-tē′rô)	88	20°08′N	101°44′W
Angangueo, Mex. (än-gäŋ′gwā-ō)	88	19°36′N	100°18′W
Ang'angxi, China (äŋ-äŋ-shyē)	161	47°05′N	123°58′E
Angarsk, Russia	135	52°48′N	104°15′E
Ånge, Swe. (ông′å)	122	62°31′N	15°39′E
Ángel, Salto, wtfl., Ven. (säl′tō-á′n-hĕl)	100	5°44′N	62°27′W
Ángel de la Guarda, i., Mex. (á′n-hĕl-dĕ-lä-gwä′r-dä)	86	29°30′N	113°00′W
Angeles, Phil. (än′hå-lās)	169a	15°09′N	120°35′E

PRONOUNCING INDEX Ängelholm – Argentino 221

PLACE (Pronunciation)	PAGE	LAT.	LONG.
Ängelholm, Swe. (ĕng′ĕl-hôlm)	122	56°14′N	12°50′E
Angelina, r., Tx., U.S. (ăn-jê lē′nȧ)	81	31°30′N	94°53′W
Angels Camp, Ca., U.S. (ăn′jĕls kămp′)	76	38°03′N	120°33′W
Ångermanälven, r., Swe.	116	64°10′N	17°30′E
Angermund, Ger. (än′ngĕr-münd)	127c	51°20′N	6°47′E
Angermünde, Ger. (äng′ĕr-mûn-dĕ)	124	53°02′N	14°00′E
Angers, Can. (äN-zhā′)	62c	45°31′N	75°29′W
Angers, Fr.	126	47°29′N	0°36′W
Angkor, hist., Camb. (äng′kôr)	168	13°52′N	103°50′E
Anglesey, i., Wales, U.K. (ăŋ′g'l-sê)	120	53°35′N	4°28′W
Angleton, Tx., U.S. (aŋ′g'l-tŭn)	81a	29°10′N	95°25′W
Angmagssalik, Grnld. (aŋ-má′sȧ-lĭk)	49	65°40′N	37°40′W
Angoche, Ilha, i., Moz. (ê′lä-ăn-gō′chȧ)	187	16°20′S	40°00′E
Angol, Chile (aŋ-gōl′)	102	37°47′S	72°43′W
Angola, In., U.S. (ăŋ-gō′lȧ)	66	41°35′N	85°00′W
Angola, nation, Afr. (ăŋ-gō′lȧ)	186	14°15′S	16°00′E
Angora *see* Ankara, Tur.	154	39°55′N	32°50′E
Angoulême, Fr. (äŋ′gōō-lâm′)	126	45°40′N	0°09′E
Angra dos Reis, Braz. (aŋ′grä dōs rā′ēs)	99a	23°01′S	44°17′W
Angri, Italy (ä′n-grē)	129c	40°30′N	14°35′E
Anguang, China (än-gůäŋ)	164	45°28′N	123°42′E
Anguilla, dep., N.A.	87	18°15′N	62°54′W
Anguilla Cays, is., Bah. (ăŋ-gwĭl′ȧ)	92	23°30′N	79°35′W
Anguille, Cape, c., Can. (kăp′-äŋ-gē′yē)	61	47°55′N	59°25′W
Anguo, China (än-gwô)	162	38°27′N	115°19′E
Anholt, i., Den. (än′hŏlt)	122	56°43′N	11°34′E
Anhui, prov., China (än-hwä)	161	31°30′N	117°15′E
Aniak, Ak., U.S. (ä-nyà′k)	63	61°32′N	159°35′W
Aniakchak National Monument, rec., Ak., U.S.	63	56°50′N	157°50′W
Animas, r., Co., U.S. (ä′nê-más)	77	37°03′N	107°50′W
Anina, Rom. (ä-nē′nä)	131	45°03′N	21°50′E
Anita, Pa., U.S. (á-nē′ȧ)	67	41°05′N	79°00′W
Aniva, Mys, c., Russia (mĭs á-nē′vá)	166	46°08′N	143°13′E
Aniva, Zaliv, b., Russia (zä′lĭf á-nē′vá)	166	46°30′N	143°00′E
Anjou, Can.	62a	45°37′N	73°33′W
Ankang, China (än-käŋ)	160	32°38′N	109°10′E
Ankara, Tur. (än′ká-rá)	154	39°55′N	32°50′E
Anklam, Ger. (än′kläm)	124	53°52′N	13°43′E
Ankoro, D.R.C. (ăŋ-kō′rō)	186	6°45′S	26°57′E
Anloga, Ghana	188	5°47′N	0°50′E
Anlong, China (än-loŋ)	165	25°01′N	105°32′E
Anlu, China (än′lōō′)	165	31°18′N	113°40′E
Ann, Cape, c., Ma., U.S. (kăp′ăn′)	67	42°40′N	70°40′W
Anna, Russia (än′ä)	133	51°31′N	40°27′E
Anna, Il., U.S. (ăn′ȧ)	79	37°28′N	89°15′W
Annaba, Alg.	184	36°57′N	7°39′E
Annaberg-Bucholz, Ger. (än′ä-bĕrgh)	124	50°35′N	13°02′E
An Nafūd, des., Sau. Ar.	154	28°30′N	40°30′E
An Najaf, Iraq (än nä-jăf′)	154	32°00′N	44°25′E
An Nakhl, Egypt	153a	29°55′N	33°45′E
Annamese Cordillera, mts., Asia	168	17°34′N	105°38′E
Annapolis, Md., U.S. (ă-năp′ô-lĭs)	65	39°00′N	76°25′W
Annapolis Royal, Can.	60	44°45′N	65°31′W
Ann Arbor, Mi., U.S. (ăn är′bēr)	65	42°15′N	83°45′W
An Nāşirīyah, Iraq	154	31°08′N	46°15′E
An Nawfalīyah, Libya	185	30°57′N	17°38′E
Annecy, Fr. (ån sē′)	127	45°54′N	6°07′E
Annemasse, Fr. (än′mäs′)	127	46°09′N	6°13′E
Annette Island, i., Ak., U.S.	54	55°13′N	131°30′W
An Nhon, Viet.	168	13°55′N	109°00′E
Annieopsquotch Mountains, mts., Can.	61	48°37′N	57°17′W
Anniston, Al., U.S. (ăn′ĭs-tŭn)	65	33°39′N	85°47′W
Annobón, i., Eq. Gui.	183	2°00′S	3°30′E
Annonay, Fr. (ån′ĭs-tsiŭn)	126	45°16′N	4°36′E
Annotto Bay, Jam. (ăn-nō′tō)	92	18°15′N	76°45′W
An Nuhūd, Sudan	185	12°39′N	28°18′E
Anoka, Mn., U.S. (å-nō′ká)	75g	45°12′N	93°24′W
Anori, Col. (á-nō′rē)	100a	7°01′N	75°09′W
Áno Viánnos, Grc.	130a	35°02′N	25°26′E
Anpu, China (än-pōō)	160	21°28′N	110°00′E
Anqiu, China (än-chyô)	162	36°26′N	119°12′E
Ansbach, Ger. (äns′bäk)	124	49°18′N	10°35′E
Anse à Veau, Haiti (äNs′ ä-vō′)	93	18°30′N	73°25′W
Anse d'Hainault, Haiti (äNs′dĕnō)	93	18°30′N	74°25′W
Anserma, Col. (á′n-sĕ′r-mä)	100a	5°13′N	75°47′W
Ansermanuevo, Col. (á′n-sĕ′r-mä-nwĕ′vō)	100a	4°47′N	75°59′W
Anshan, China	164	41°00′N	123°00′E
Anshun, China (än-shōōn′)	160	26°12′N	105°50′E
Anson, Tx., U.S. (ăn′sŭn)	80	32°45′N	99°52′W
Anson Bay, b., Austl.	174	13°10′S	130°00′E
Ansŏng, Kor., S. (än′sŭng′)	166	37°00′N	127°12′E
Ansongo, Mali	188	15°40′N	0°30′E
Ansonia, Ct., U.S. (ăn-sō′nĭ-ȧ)	67	41°20′N	73°05′W
Antalya, Tur. (än-tä′lê-ä) (ä-dä′lê-ä)	119	37°00′N	30°50′E
Antalya Körfezi, b., Tur.	119	36°40′N	31°20′E
Antananarivo, Madag.	187	18°51′S	47°40′E
Antarctica, cont.	178	80°15′S	127°00′E
Antarctic Peninsula, pen., Ant.	178	70°00′S	65°00′W
Antelope Creek, r., Wy., U.S. (ăn′tē-lōp)	73	43°29′N	105°42′W
Antequera, Spain (än-tĕ-kĕ′rä)	118	37°01′N	4°34′W
Anthony, Ks., U.S. (ăn′thô-nê)	78	37°08′N	98°01′W
Anthony Peak, mtn., Ca., U.S.	76	39°51′N	122°58′W
Anti Atlas, mts., Mor.	184	28°45′N	9°30′W
Antibes, Fr. (äN-tēb′)	127	43°36′N	7°12′E
Anticosti, Île d', i., Can. (ăn-tĭ-kŏs′tē)	53	49°30′N	62°00′W
Antigo, Wi., U.S. (ăn′tĭ-gō)	71	45°09′N	89°11′W
Antigonish, Can. (ăn-tĭ-gô-nēsh′)	61	45°35′N	61°55′W
Antigua, Guat. (än-tē′gwä)	86	14°32′N	90°43′W
Antigua, r., Mex.	89	19°16′N	96°36′W
Antigua and Barbuda, nation, N.A.	87	17°15′N	61°15′W
Antigua Veracruz, Mex. (än-tē′gwä vā-rä-krōōz′)	89	19°18′N	96°17′W
Antilla, Cuba (än-tē′lyä)	93	20°50′N	75°50′W
Antioch, Ca., U.S. (ăn′tĭ-ŏk)	74b	38°00′N	121°48′W
Antioch, Il., U.S.	69a	42°29′N	88°06′W
Antioch, Ne., U.S.	70	42°05′N	102°36′W
Antioquia, Col. (än-tê-ō′kēä)	100	6°34′N	75°49′W
Antioquia, dept., Col.	100a	6°48′N	75°42′W
Antlers, Ok., U.S. (ănt′lērz)	79	34°14′N	95°38′W
Antofagasta, Chile (än-tô-fä-gäs′tä)	102	23°32′S	70°21′W
Antofalla, Salar de, pl., Arg. (sä-lär′de án′tō-fä′lä)	102	26°00′S	67°52′W
Antón, Pan. (än-tōn′)	87	8°24′N	80°15′W
Antongila, Helodrano, b., Madag.	187	16°15′S	50°15′E
Antônio Carlos, Braz. (än-tō′nĕô-kà′r-lôs)	99a	21°19′S	43°45′W
António Enes, Moz. (ăn-to′nyō ĕn′ĕs)	187	16°14′S	39°58′E
Antonito, Co., U.S. (ăn-tô-nē′tō)	78	37°04′N	106°01′W
Antonopole, Lat. (än′tô-nô-pō lyĕ)	123	56°19′N	27°11′E
Antony, Fr.	127b	48°45′N	2°18′E
Antsirabe, Madag. (änt-sê-rä′bä)	187	19°49′S	47°16′E
Antsiranana, Madag.	187	12°18′S	49°16′E
Antsla, Est. (änt′slȧ)	123	57°49′N	26°29′E
Antuco, vol., S.A. (än-tōō′kō)	102	37°30′S	72°30′W
Antwerp, Bel.	110	51°13′N	4°24′E
Antwerpen *see* Antwerp, Bel.	110	51°13′N	4°24′E
Anūpgarh, India (ŭ-nŏp′gŭr)	158	29°22′N	73°20′E
Anuradhapura, Sri L. (ŭ-nōō′rä-dŭ-pōō′rŭ)	159	8°24′N	80°25′E
Anxi, China (än-shyē)	160	40°36′N	95°49′E
Anyang, China (än′yäng)	161	36°05′N	114°22′E
Anykščiai, Lith. (aniksh-chä′ē)	123	55°34′N	25°04′E
Anzhero-Sudzhensk, Russia (än′zhâ-rô-sòd′zhĕnsk)	134	56°08′N	86°08′E
Anzio, Italy (änt′zê-ō)	130	41°28′N	12°39′E
Anzoátegui, dept., Ven. (än-zòá′tĕ-gê)	101b	9°38′N	64°45′W
Aoba, i., Vanuatu	170f	15°25′S	167°50′E
Aomori, Japan (äô-mō′rē)	161	40°45′N	140°52′E
Aoraki (Cook, Mount), mtn., N.Z.	175a	43°27′S	170°13′E
Aosta, Italy (ä-ôs′tä)	130	45°45′N	7°20′E
Aouk, Bahr, r., Afr. (ä-ōk′)	185	9°30′N	20°45′E
Aoukâr, reg., Maur.	188	18°00′N	9°40′W
Apalachicola, Fl., U.S. (ăp-ȧ-lăch-ĭ-kō′lȧ)	82	29°43′N	84°59′W
Apalachicola, r., Fl., U.S. (ăpȧ-lăch′ĭ-cōlā)	65	30°11′N	85°00′W
Apan, Mex. (ä-pä′n)	88	19°43′N	98°27′W
Apango, Mex. (ä-päŋ′gō)	88	17°41′N	99°22′W
Apaporis, r., S.A. (ä-pä-pô′rĭs)	100	0°48′N	72°32′W
Aparri, Phil. (ä-pär′rē)	168	18°15′N	121°40′E
Apasco, Mex. (ä-pà′s-kō)	88	20°33′N	100°43′W
Apatin, Serb. (ŏ′pō-tĭn)	131	45°40′N	19°00′E
Apatzingán de la Constitución, Mex.	88	19°07′N	102°21′W
Apeldoorn, Neth. (ä′pĕl-dōōrn)	117	52°14′N	5°55′E
Apennines *see* Appennino, mts., Italy	112	43°48′N	11°06′E
Apía, Col. (á-pē′ä)	100a	5°07′N	75°58′W
Apia, Samoa	170a	13°50′S	171°44′W
Apipilulco, Mex. (ä-pĭ-pĭ-lōōl′kō)	88	18°09′N	99°40′W
Apishapa, r., Co., U.S. (ăp-ĭ-shä′pȧ)	78	37°40′N	104°08′W
Apizaco, Mex. (ä-pê-zä′kō)	88	19°18′N	98°11′W
Apo, Mount, mtn., Phil. (ä′pō)	169	6°56′N	125°05′E
Apopka, Fl., U.S. (ä-pŏp′kȧ)	83a	28°37′N	81°30′W
Apopka, Lake, l., Fl., U.S.	83a	28°38′N	81°50′W
Apostle Islands, is., Wi., U.S. (ȧ-pŏs′l)	71	47°05′N	90°55′W
Appalachia, Va., U.S. (ăpȧ-lăch′ĭ-ȧ)	83	36°54′N	82°49′W
Appalachian Mountains, mts., N.A. (ăp-ȧ-lăch′ĭ-ăn)	65	37°20′N	82°00′W
Appelbo, Swe. (ĕp-ĕl-bōō)	122	60°30′N	14°02′E
Appelhülsen, Ger. (ä′pĕl-hül′sĕn)	127c	51°55′N	7°26′E
Appennino, mts., Italy (äp-pĕn-nē′nô)	112	43°48′N	11°06′E
Appleton, Mn., U.S. (ăp′l-tŭn)	70	45°10′N	96°01′W
Appleton, Wi., U.S.	65	44°14′N	88°27′W
Appleton City, Mo., U.S.	79	38°10′N	94°02′W
Appomattox, r., Va., U.S. (ăp-ô-măt′ŭks)	83	37°22′N	78°09′W
Aprília, Italy (ȧ-prē′lyȧ)	130	41°36′N	12°40′E
Apsheronsk, Russia	138	44°28′N	39°44′E
Apt, Fr. (äpt)	127	43°54′N	5°19′E
Apure, r., Ven. (ä-pōō′rä)	100	8°08′N	68°46′W
Apurimac, r., Peru (ä-pōō-rê-mäk′)	100	11°39′S	73°48′W
Aqaba, Gulf of, b. (ä′kä-bȧ)	154	28°30′N	34°40′E
Aqabah, Wādī al, r., Egypt	153a	29°48′N	34°05′E
Aqmola *see* Astana, Kaz.	139	51°10′N	71°43′E
Aqtaū, Kaz.	139	43°35′N	51°05′E
Aqtöbe, Kaz.	139	50°20′N	57°00′E
Aquasco, Md., U.S. (á′gwă′scô)	68e	38°35′N	76°44′W
Aquidauana, Braz. (ä-kē-däwä′nä)	101	20°24′S	55°46′W
Aquin, Haiti (ä-kăn′)	93	18°20′N	73°25′W
Ara, r., Japan (ä-rä)	167a	35°40′N	139°52′E
Arab, Baḩr al, r., Sudan	185	9°46′N	26°52′E
'Arabah, Wādī, val., Egypt	192b	29°02′N	32°10′E
Arabats'ka Strilka (Tongue of Arabat), spit, Ukr.	133	45°50′N	35°05′E
Arabi, La., U.S.	68d	29°58′N	90°01′W
Arabian Desert, des., Egypt (á-rā′bĭ-ăn)	185	27°06′N	32°49′E
Arabian Sea, sea, (ȧ-rā′bĭ-ăn)	152	16°00′N	65°15′E
Aracaju, Braz. (ä-rä-kä-zhōō′)	101	11°00′S	37°01′W
Aracati, Braz. (ä-rä′kä-tē′)	101	4°31′S	37°41′W
Araçatuba, Braz. (ä-rä-sá-tōō′bä)	101	21°14′S	50°19′W
Aracena, Spain	128	37°53′N	6°34′W
Arachthos, r., Grc. (är′äk-thôs)	131	39°10′N	21°05′E
Aracruz, Braz. (ä-rä-krōō′s)	101	19°58′S	40°11′W
'Arad, Isr.	153a	31°20′N	35°15′E
Arad, Rom. (ŏ′rŏd)	119	46°10′N	21°18′E
Arafura Sea, sea, (ä-rä-fōō′rä)	169	8°40′S	130°00′E
Aragats, Gora, mtn., Arm.	138	40°32′N	44°14′E
Aragon, hist. reg., Spain (ä-rä-gōn′)	129	40°55′N	0°45′W
Aragón, r., Spain	128	42°35′N	1°10′W
Aragua, dept., Ven. (ä-rä′gwä)	101b	10°00′N	67°05′W
Aragua de Barcelona, Ven.	100	9°29′N	64°48′W
Araguaía, r., Braz. (ä-rä-gwä′yä)	101	8°37′S	49°43′W
Araguari, Braz. (ä-rä-gwä′rē)	101	18°43′S	48°03′W
Araguatins, Braz. (ä-rä-gwä-tēns)	101	5°41′S	48°04′W
Aragüita, Ven. (ärä-gwē′tà)	101b	10°13′N	66°28′W
Araj, oasis, Egypt (ä-räj′)	119	29°05′N	26°51′E
Arāk, Iran	154	34°08′N	49°57′E
Arakan Yoma, mts., Mya. (ŭ-rŭ-kŭn′yō′mà)	155	19°51′N	94°13′E
Aral, Kaz.	139	46°47′N	62°00′E
Aral Sea, sea, Asia	134	45°17′N	60°02′E
Aralsor köli, l., Kaz. (ä-räl′sôr′)	137	49°00′N	48°20′E
Aramberri, Mex. (ä-rám-bĕr-rē′)	88	24°05′N	99°47′W
Arana, Sierra, mts., Spain	128	37°17′N	3°28′W
Aranda de Duero, Spain (ä-rän′dä dā dwā′rō)	128	41°43′N	3°45′W
Arandas, Mex. (ä-rän′däs)	88	20°43′N	102°18′W
Aran Island, i., Ire. (ăr′ăn)	120	54°58′N	8°33′W
Aran Islands, is., Ire.	116	53°04′N	9°59′W
Aranjuez, Spain (ä-rän-hwāth′)	118	40°02′N	3°24′W
Aransas Pass, Tx., U.S. (ȧ-răn′sȧs pȧs)	81	27°55′N	97°09′W
Araouane, Mali	184	18°54′N	3°33′W
Arapkir, Tur. (ä-räp-kēr′)	119	39°00′N	38°10′E
Araraquara, Braz. (ä-rä-rä-kwá′rä)	101	21°47′S	48°08′W
Araras, Braz. (ä-rá′räs)	99a	22°21′S	47°22′W
Araras, Serra das, mts., Braz. (sĕ′r-rä-däs-ä-rá′räs)	101	18°03′S	53°23′W
Araras, Serra das, mts., Braz. (sĕ′r-rä-däs-ä-rä′räs)	102	23°30′S	53°00′W
Araras, Serra das, mts., Braz.	102b	22°24′S	43°15′W
Ararat, Austl. (ăr′ărăt)	175	37°17′S	142°56′E
Ararat, Mount, mtn., Tur.	154	39°50′N	44°20′E
Arari, l., Braz. (ä-rá′rē)	101	0°30′S	48°50′W
Araripe, Chapada do, hills, Braz. (shä-pä′dä-dō-ä-rä-rē′pē)	101	5°55′S	40°42′W
Araruama, Braz. (ä-rä-rōō-ä′mä)	99a	22°53′S	42°19′W
Araruama, Lagoa de, l., Braz.	99a	23°00′S	42°15′W
Aras, r., Asia (ä-räs)	154	39°15′N	47°10′E
Aratuípe, Braz. (ä-rä-tōō-ē′pē)	101	13°12′S	38°58′W
Arauca, Col. (ä-rou′kä)	100	6°56′N	70°45′W
Arauca, r., S.A.	100	7°13′N	68°43′W
Aravalli Range, mts., India (ä-rä′vŭ-lê)	155	24°15′N	72°40′E
Araya, Punta de, c., Ven. (pŭn′tà-dĕ-ä-rä′yà)	101b	10°40′N	64°15′W
Arayat, Phil. (ä-rä′yät)	169a	15°10′N	120°44′E
'Arbi, Sudan	185	20°36′N	29°57′E
Arbīl, Iraq	154	36°10′N	44°00′E
Arboga, Swe. (är-bō′gä)	122	59°26′N	15°50′E
Arborea, Italy (är-bō-rĕ′ä)	130	39°50′N	8°36′E
Arbroath, Scot., U.K. (är-brōth′)	120	56°36′N	2°25′W
Arcachon, Fr. (är-kä-shôN′)	117	44°39′N	1°12′W
Arcachon, Bassin d', Fr. (bä-sēN′ där-kä-shôN′)	126	44°42′N	1°50′W
Arcadia, Ca., U.S. (är-kā′dĭ-ȧ)	75a	34°08′N	118°02′W
Arcadia, Fl., U.S.	83a	27°12′N	81°51′W
Arcadia, La., U.S.	81	32°33′N	92°56′W
Arcadia, Wi., U.S.	71	44°15′N	91°30′W
Arcata, Ca., U.S. (är-kȧ′tȧ)	72	40°54′N	124°05′W
Arc Dome Mountain, mtn., Nv., U.S. (ärk dōm)	76	38°51′N	117°21′W
Arcelia, Mex. (är-sā′lê-ä)	88	18°19′N	100°14′W
Archbald, Pa., U.S. (ärch′bôld)	67	41°30′N	75°35′W
Arches National Park, rec., Ut., U.S. (är′chĕs)	77	38°45′N	109°35′W
Archidona, Ec. (är-chê-dō′nä)	100	1°01′S	77°49′W
Archidona, Spain (är-chē-dô′nä)	128	37°08′N	4°24′W
Arcis-sur-Aube, Fr. (är-sēs′sûr-ōb′)	126	48°31′N	4°04′E
Arco, Id., U.S. (är′kô)	73	43°39′N	113°15′W
Arcola, Tx., U.S.	81a	29°30′N	95°28′W
Arcola, Va., U.S. (är′cōlȧ)	68e	38°57′N	77°32′W
Arcos de la Frontera, Spain (är′kōs-dĕ-lä-frôn-tĕ′rä)	128	36°44′N	5°48′W
Arctic Ocean, o.	198	85°00′N	170°00′E
Arda, r., Blg. (är′dä)	131	41°36′N	25°18′E
Ardabīl, Iran	154	38°15′N	48°00′E
Ardahan, Tur. (är-dá-hän′)	137	41°10′N	42°40′E
Ardatov, Russia (är-dá-tôf′)	136	54°58′N	46°10′E
Ardennes, mts., Eur. (är-dĕn′)	117	50°01′N	5°12′E
Ardila, r., Eur. (är-dē′lȧ)	128	38°10′N	7°15′W
Ardmore, Ok., U.S. (ärd′mōr)	64	34°10′N	97°08′W
Ardmore, Pa., U.S.	68f	40°01′N	75°18′W
Ardrossan, Can. (är-dros′ăn)	62g	53°33′N	113°08′W
Ardsley, Eng., U.K. (ärdz′lê)	114a	53°43′N	1°33′W
Åre, Swe.	116	63°12′N	13°12′E
Arecibo, P.R. (ä-rå-sē′bō)	87b	18°28′N	66°45′W
Areia Branca, Braz. (ä-rĕ′yä-brá′n-kä)	101	4°58′S	37°02′W
Arena, Point, c., Ca., U.S. (ä-rā′nȧ)	76	38°57′N	123°40′W
Arenas, Punta, c., Ven. (pòn′tä-rĕ′näs)	101b	10°57′N	64°24′W
Arenas de San Pedro, Spain	128	40°12′N	5°04′W
Arendal, Nor. (ä′rĕn-däl)	122	58°29′N	8°44′E
Arendonk, Bel.	115a	51°19′N	5°07′E
Arequipa, Peru (ä-rå-kē′pä)	100	16°27′S	71°30′W
Arezzo, Italy (ä-rĕt′sō)	118	43°28′N	11°54′E
Arga, r., Spain (är′gä)	128	42°35′N	1°55′W
Arganda, Spain (är-gän′dä)	129a	40°18′N	3°27′W
Argazi, l., Russia (är′gä-zĭ)	142a	55°24′N	60°37′E
Argazi, r., Russia	142a	55°33′N	57°30′E
Argentan, Fr. (ár-zhäN-täN′)	126	48°45′N	0°01′W
Argentat, Fr. (ár-zhäN-tä′)	126	45°07′N	1°57′E
Argenteuil, Fr. (är-zhäN-tû′y′)	126	48°56′N	2°15′E
Argentina, nation, S.A. (är-jĕn-tē′nȧ)	102	35°30′S	67°00′W
Argentino, l., Arg. (är-kĕn-tē′nō)	102	50°15′S	72°45′W

222 Argenton-sur-Creuse – Athy PRONOUNCING INDEX

PLACE (Pronunciation)	PAGE	LAT.	LONG.
Argenton-sur-Creuse, Fr. (är-zhäɴ´tôɴ-sür-krôs)	126	46°34'N	1°28'E
Argolikos Kólpos, b., Grc.	131	37°20'N	23°00'E
Argonne, mts., Fr. (ä´r-gôn)	127	49°21'N	5°54'E
Argos, Grc. (är´gŏs)	131	37°38'N	22°45'E
Argostóli, Grc.	131	38°10'N	20°30'E
Arguello, Point, c., Ca., U.S. (är-gwāl´yō)	76	34°35'N	120°40'W
Arguin, Cap d', c., Maur.	184	20°28'N	17°46'W
Argun', r., Asia (ȧr-gōōn´)	135	50°00'N	119°00'E
Argungu, Nig.	189	12°45'N	4°31'E
Argyle, Can. (är´gīl)	62f	50°11'N	97°27'W
Argyle, Mn., U.S.	70	48°21'N	96°48'W
Århus, Den. (ôr´hōōs)	116	56°09'N	10°10'E
Ariakeno-Umi, b., Japan (ä-rē´ä-ĸä´nō ōō´nē)	167	33°03'N	130°18'E
Ariake-Wan, b., Japan (ä´rē-ä´kå wän)	167	31°19'N	131°15'E
Ariano, Italy (ä-rē-ä´nō)	130	41°09'N	15°11'E
Ariari, r., Col. (ä-ryȧ´rē)	100a	3°34'N	73°42'W
Aribinda, Burkina	188	14°14'N	0°52'W
Arica, Chile (ä-rē´kä)	100	18°34'S	70°14'W
Arichat, Can. (ä-rĭ-shät´)	61	45°31'N	61°01'W
Ariège, r., Fr. (ȧ-rē-ĕzh´)	126	43°26'N	1°29'E
Ariel, Wa., U.S. (ā´rĭ-ĕl)	74c	45°57'N	122°34'W
Arieș, r., Rom.	125	46°25'N	23°15'E
Ariguanabo, Lago de, l., Cuba (lä´gô-dĕ-ä-rē-gwä-nä´bō)	93a	22°52'N	82°33'W
Arikaree, r., Co., U.S. (ă-rĭ-kȧ-rē´)	78	39°51'N	102°18'W
Arima, Japan (ä´rē-mä´)	167b	34°48'N	135°16'E
Aringay, Phil. (ä-rĭŋ-gä´ē)	169a	16°25'N	120°20'E
Arinos, r., Braz. (ä-rē´nōzsh)	101	12°09'S	56°49'W
Aripuanã, r., Braz. (ȧ-rē-pwän´yȧ)	101	7°06'S	60°29'W
'Arīsh, Wādī al, r., Egypt (ȧ-rēsh´)	153a	30°36'N	34°07'E
Aristazabal Island, i., Can.	54	52°30'N	129°20'W
Arizona, state, U.S. (ăr-ĭ-zō´nȧ)	64	34°00'N	113°00'W
Arjona, Spain (är-hō´nä)	128	37°58'N	4°03'W
Arka, r., Russia	141	60°45'N	142°30'E
Arkabutla Lake, res., Ms., U.S. (är-kȧ-bŭt´lā)	82	34°48'N	90°00'W
Arkadelphia, Ar., U.S. (är-kȧ-dĕl´fĭ-ȧ)	79	34°06'N	93°05'W
Arkansas, state, U.S. (är´kăn-sô) (är-kăn´sȧs)	65	34°50'N	93°40'W
Arkansas, r., U.S.	64	37°30'N	97°00'W
Arkansas City, Ks., U.S.	79	37°04'N	97°02'W
Arkhangelsk (Archangel), Russia (ȧr-ĸän´gĕlsk)	134	64°30'N	40°25'E
Arkhangel'skoye, Russia (är-kän-gĕl´skô-yĕ)	142a	54°25'N	56°48'E
Arklow, Ire. (ärk´lō)	120	52°47'N	6°10'W
Arkonam, India (ār-kō-nām´)	159	13°05'N	79°43'E
Arlanza, r., Spain (är-län-thä´)	128	42°08'N	3°45'W
Arlanzón, r., Spain (är-län-thōn´)	128	42°12'N	3°58'W
Arlberg Tunnel, trans., Aus. (ärl´bĕrgh)	124	47°05'N	10°15'E
Arles, Fr. (ärl)	126	43°42'N	4°38'E
Arlington, S. Afr.	192c	28°02'S	27°52'E
Arlington, Ga., U.S. (är´lĭng-tun´)	82	31°25'N	84°42'W
Arlington, Ma., U.S.	61a	42°26'N	71°13'W
Arlington, S.D., U.S. (är´lĕng-tŭn)	70	44°23'N	97°09'W
Arlington, Tx., U.S. (ȧr´lĭng-tŭn)	75c	32°44'N	97°07'W
Arlington, Va., U.S.	68e	38°55'N	77°10'W
Arlington, Vt., U.S.	67	43°05'N	73°05'W
Arlington, Wa., U.S.	74a	48°11'N	122°08'W
Arlington Heights, Il., U.S. (är´lĕng-tŭn-hī´ts)	69a	42°05'N	87°59'W
Arltunga, Austl. (ärl-tòŋ´gȧ)	174	23°19'S	134°45'E
Arma, Ks., U.S. (är´mȧ)	79	37°34'N	94°43'W
Armagh, Can. (är-mä´) (är-mäĸ´)	62b	46°45'N	70°36'W
Armagh, N. Ire., U.K.	116	54°21'N	6°25'W
Armant, Egypt (är-mänt´)	192b	25°37'N	32°32'E
Armaro, Col. (är-mȧ´rō)	100a	4°58'N	74°54'W
Armavir, Russia (ȧr-mȧ-vĭr´)	134	45°00'N	41°00'E
Armenia, Col. (ȧr-mĕ´nēȧ)	100	4°33'N	75°40'W
Armenia, El Sal. (är-mā´nē-ä)	90	13°44'N	89°31'W
Armenia, nation, Asia	134	41°00'N	44°39'E
Armentières, Fr. (ȧr-mäɴ-tyâr´)	126	50°43'N	2°53'E
Armeria, Río de, r., Mex. (rē´ō-dĕ-är-må-rē´ä)	88	19°36'N	104°10'W
Armherstburg, Can. (ärm´hĕrst-bōōrgh)	58	42°06'N	83°06'W
Armians'k, Ukr.	133	46°06'N	33°42'E
Armidale, Austl. (är´mĭ-dāl)	175	30°27'S	151°50'E
Armour, S.D., U.S. (är´mẽr)	70	43°18'N	98°21'W
Armstrong Station, Can. (ärm´strŏng)	51	50°21'N	89°00'W
Arnedo, Spain (är-nā´dō)	128	42°12'N	2°03'W
Arnhem, Neth. (ärn´hĕm)	117	51°58'N	5°56'E
Arnhem, Cape, c., Austl.	174	12°15'S	137°00'E
Arnhem Land, reg., Austl. (ärn´hĕm-lănd)	174	13°15'S	133°00'E
Arno, r., Italy (ä´r-nō)	118	43°30'N	11°00'E
Arnold, Eng., U.K. (är´nŭld)	114a	53°00'N	1°08'W
Arnold, Mn., U.S. (ȧr´nŭld)	75h	46°53'N	92°06'W
Arnold, Pa., U.S.	69e	40°35'N	79°45'W
Arnprior, Can. (ärn-prī´ẽr)	59	45°25'N	76°20'W
Arnsberg, Ger. (ärns´bĕrgh)	127c	51°25'N	8°02'E
Arnstadt, Ger. (ärn´shtät)	124	50°51'N	10°57'E
Aroab, Nmb. (är´ō-ȧb)	186	25°40'S	19°45'E
Aroostook, r., Me., U.S. (ȧ-rōs´tòk)	60	46°44'N	68°15'W
Aroroy, Phil. (ä-rō-rō´ē)	169a	12°30'N	123°24'E
Arpajon, Fr. (ăr-pă-jō´n)	127b	48°35'N	2°15'E
Arpoador, Ponta do, c., Braz. (pō´n-tä-dō-är´pōä-dō´r)	102b	22°59'S	43°11'W
Arraiolos, Port. (är-rī-ō´lōzh)	128	38°47'N	7°59'W
Ar Ramādī, Iraq	154	33°26'N	43°19'E
Arran, Island of, Scot., U.K. 7	120	55°25'N	5°25'W
Ar Rank, Sudan	185	11°45'N	32°53'E
Arras, Fr. (ȧ-räs´)	117	50°21'N	2°40'E
Ar Rawḍah, Egypt	192b	27°47'N	30°52'E
Arrecifes, Arg. (är-rå-sē´fās)	99c	34°03'S	60°05'W
Arrecifes, r., Arg.	99c	34°07'S	59°50'W
Arrée, Monts d', mts., Fr. (är-rā´)	126	48°27'N	4°00'W
Arriaga, Mex. (är-rēä´gä)	89	16°15'N	93°54'W
Arrone, r., Italy	129d	41°57'N	12°17'E
Arrow Creek, r., Mt., U.S. (ăr´ō)	73	47°29'N	109°53'W
Arrowhead, Lake, l., Ca., U.S. (lāk ăr´ōhĕd)	75a	34°17'N	117°13'W
Arrowrock Reservoir, res., Id., U.S. (ăr´ō-rŏk)	72	43°40'N	115°30'W
Arroya Arena, Cuba (är-rō´yä-rē´nä)	93a	23°01'N	82°30'W
Arroyo de la Luz, Spain (är-rō´yō-dĕ-lä-lōō´z)	128	39°39'N	6°46'W
Arroyo Seco, Mex. (är-rō´yō sā´kō)	88	21°31'N	99°44'W
Ar Rub' al Khālī, des., Asia	154	20°00'N	51°00'E
Ar Ruṭbah, Iraq	157	33°02'N	40°17'E
Arsen'yev, Russia	135	44°13'N	133°32'E
Arsinskiy, Russia (är-sĭn´skĭ)	142a	53°46'N	59°54'E
Árta, Grc. (är´tā)	119	39°08'N	21°02'E
Arteaga, Mex. (är-tā-ä´gä)	80	25°28'N	100°50'W
Artëm, Russia (ȧr-tyôm´)	135	43°28'N	132°29'E
Artemisa, Cuba (är-tå-mē´sä)	92	22°50'N	82°45'W
Artemivs'k, Ukr.	137	48°37'N	38°00'E
Artesia, N.M., U.S. (är-tē´sĭ-ȧ)	78	32°44'N	104°23'W
Arthabaska, Can.	59	46°03'N	71°54'W
Arthur's Town, Bah.	93	24°40'N	75°40'W
Arti, Russia (är´tĭ)	142a	56°20'N	58°38'E
Artibonite, r., N.A. (är-tē-bō-nē´tā)	93	19°00'N	72°25'W
Aru, Kepulauan, is., Indon.	169	6°20'S	133°00'E
Arua, Ug. (ä´rōō-ä)	185	3°01'N	30°55'E
Aruba, dep., N.A. (ä-rōō´bȧ)	87	12°29'N	70°00'W
Arunachal Pradesh, state, India	155	27°35'N	92°56'E
Arusha, Tan. (ȧ-rōō´shä)	186	3°22'S	36°41'E
Arvida, Can.	51	48°26'N	71°11'W
Arvika, Swe. (är-vē´kȧ)	122	59°41'N	12°35'E
Arzamas, Russia (är-zä-mäs´)	136	55°20'N	43°52'E
Arziw, Alg.	118	35°50'N	0°20'W
Arzúa, Spain	128	42°54'N	8°19'W
Aš, Czech Rep. (äsh´)	124	50°12'N	12°13'E
Asahi-Gawa, r., Japan (ȧ-sä´hē-gä´wä)	167	35°01'N	133°40'E
Asahikawa, Japan	161	43°50'N	142°09'E
Asaka, Japan (ä-sä´kä)	167a	35°47'N	139°36'E
Asansol, India	155	23°45'N	86°58'E
Asbest, Russia (äs-bĕst´)	136	57°02'N	61°28'E
Asbestos, Can. (ăs-bĕs´tōs)	59	45°49'N	71°52'W
Asbestovskiy, Russia	142a	57°46'N	61°23'E
Asbury Park, N.J., U.S. (ăz´bĕr-ĭ)	68a	40°13'N	74°01'W
Ascención, Bahía de la, b., Mex.	90a	19°39'N	87°30'W
Ascensión, Mex. (äs-sĕn-sē-ōn´)	88	24°21'N	99°54'W
Ascension, i., St. Hel. (ȧ-sĕn´shŭn)	183	8°00'S	13°00'W
Ascent, S. Afr. (ăs-ĕnt´)	192c	27°14'S	29°06'E
Aschaffenburg, Ger. (ä-shäf´ĕn-bòrgh)	124	49°58'N	9°12'E
Ascheberg, Ger. (ä´shĕ-bĕrg)	127c	51°47'N	7°38'E
Aschersleben, Ger. (äsh´ẽrs-lā-bĕn)	124	51°46'N	11°28'E
Ascoli Piceno, Italy (äs´kō-lēpē-chā´nō)	130	42°50'N	13°55'E
Aseb, Erit.	185	12°52'N	43°39'E
Asenovgrad, Blg.	131	42°00'N	24°49'E
Aseri, Est. (ȧ´sĕ-rĭ)	123	59°26'N	26°58'E
Asha, Russia (ä´shä)	142a	55°01'N	57°17'E
Ashabula, l., N.D., U.S. (ăsh´ȧ-bū-lă)	70	47°07'N	97°51'W
Ashan, Russia (ä´shän)	142a	57°08'N	56°25'E
Ashbourne, Eng., U.K. (ăsh´bŭrn)	114a	53°01'N	1°44'W
Ashburn, Ga., U.S. (ăsh´bŭrn)	82	31°42'N	83°42'W
Ashburn, Va., U.S.	68e	39°02'N	77°30'W
Ashburton, r., Austl. (ăsh´bûr-tŭn)	174	22°30'S	115°30'E
Ashby-de-la-Zouch, Eng., U.K. (ăsh´bĭ-dē-lȧ zōōsh´)	114a	52°44'N	1°23'W
Ashdod, Isr.	153a	31°46'N	34°39'E
Ashdown, Ar., U.S. (ăsh´doun)	79	33°41'N	94°07'W
Asheboro, N.C., U.S. (ăsh´bŭr-ō)	83	35°41'N	79°50'W
Asherton, Tx., U.S. (ăsh´ẽr-tŭn)	80	28°26'N	99°45'W
Asheville, N.C., U.S. (ăsh´vĭl)	65	35°35'N	82°35'W
Ash Fork, Az., U.S.	77	35°13'N	112°29'W
Ashgabat, Turkmen.	139	37°57'N	58°23'E
Ashikaga, Japan (ä´shē-kä´gȧ)	167	36°22'N	139°26'E
Ashiya, Japan (ä´shē-yä´)	167	33°54'N	130°40'E
Ashiya, Japan	167b	34°44'N	135°18'E
Ashizuri-Zaki, c., Japan (ä-shē-zò-rē zä-kē)	166	32°43'N	133°04'E
Ashland, Al., U.S. (ăsh´lȧnd)	82	33°15'N	85°50'W
Ashland, Ks., U.S.	78	37°11'N	99°46'W
Ashland, Ky., U.S.	66	38°25'N	82°40'W
Ashland, Ma., U.S.	61a	42°16'N	71°28'W
Ashland, Me., U.S.	60	46°37'N	68°26'W
Ashland, Ne., U.S.	70	41°02'N	96°23'W
Ashland, Oh., U.S.	66	40°50'N	82°15'W
Ashland, Or., U.S.	72	42°12'N	122°42'W
Ashland, Pa., U.S.	67	40°45'N	76°20'W
Ashland, Wi., U.S.	65	46°34'N	90°55'W
Ashley, N.D., U.S. (ăsh´lē)	70	46°03'N	99°23'W
Ashley, Pa., U.S.	67	41°15'N	75°55'W
Ashmūn, Egypt (ȧsh-mōōn´)	192b	30°19'N	30°57'E
Ashmyany, Bela.	123	54°27'N	25°55'E
Ashqelon, Isr. (äsh´kĕ-lōn)	153a	31°40'N	34°36'E
Ash Shabb, Egypt (shĕb)	185	22°34'N	29°52'E
Ash Shallūfah, Egypt (shäl´lò-fȧ)	192b	30°09'N	32°33'E
Ash Shaqrā', Sau. Ar.	154	25°10'N	45°08'E
Ash Shāriqah, U.A.E.	157	25°22'N	55°23'E
Ash Shawbak, Jord.	153a	30°31'N	35°35'E
Ash Shiḥr, Yemen	154	14°45'N	49°32'E
Ashtabula, Oh., U.S. (ăsh-tȧ-bū´lȧ)	65	41°55'N	80°50'W
Ashton, Id., U.S. (ăsh´tŭn)	73	44°04'N	111°28'W
Ashton-in-Makerfield, Eng., U.K. (ăsh´tŭn-ĭn-māk´ẽr-fēld)	114a	53°29'N	2°39'W
Ashton-under-Lyne, Eng., U.K. (ăsh´tŭn-ŭn-dẽr-līn´)	114a	53°29'N	2°04'W
Ashuanipi, l., Can. (ăsh-wȧ-nĭp´ĭ)	53	52°40'N	67°42'W
Ashukino, Russia (ȧ-shōō´kinô)	142b	56°10'N	37°57'E
Asia, cont.	152	50°00'N	100°00'E
Asia Minor, reg., Tur. (ā´zhȧ)	113	38°18'N	31°18'E
Asientos, Mex. (ä-sē-ĕn´tōs)	88	22°13'N	102°05'W
Asilah, Mor.	128	35°30'N	6°05'W
Asinara, i., Italy	130	41°02'N	8°22'E
Asinara, Golfo dell', b., Italy (gōl´fō-dĕl-ä-sē-nä´rä)	130	40°58'N	8°28'E
Asīr, reg., Sau. Ar. (ä-sēr´)	154	19°30'N	42°00'E
Askarovo, Russia (äs-kä-rô´vō)	142a	53°21'N	58°32'E
Askersund, Swe. (äs´kĕr-sònd)	122	58°43'N	14°53'E
Askino, Russia (äs´kĭ-nô)	142a	56°06'N	56°29'E
Asmara *see* Asmera, Erit.	184	15°17'N	38°56'E
Asmera, Erit. (äs-mä´rä)	185	15°17'N	38°56'E
Asnieres, Fr. (ä-nyâr´)	127b	48°55'N	2°18'E
Asosa, Eth.	185	10°13'N	34°28'E
Asotin, Wa., U.S. (ȧ-sō´tĭn)	72	46°19'N	117°01'W
Aspen, Co., U.S. (ăs´pĕn)	77	39°15'N	106°55'W
Asperen, Neth.	115a	51°52'N	5°07'E
Aspy Bay, b., Can. (ăs´pē)	61	46°55'N	60°25'W
Aş Şaff, Egypt	192b	29°33'N	31°23'E
As Sallūm, Egypt	185	31°35'N	25°05'E
As Salt, Jord.	153a	32°02'N	35°44'E
Assam, state, India (ăs-săm´)	155	26°00'N	91°00'E
As Samāwah, Iraq	157	31°18'N	45°17'E
Assens, Den. (äs´sĕns)	122	55°16'N	9°54'E
As Sinbillāwayn, Egypt	192b	30°53'N	31°37'E
Assini, C. Iv. (ȧ-sē-nē´)	184	4°52'N	3°16'W
Assiniboia, Can.	50	49°38'N	105°59'W
Assiniboine, r., Can. (ă-sĭn´ĭ-boin)	57	50°03'N	97°57'W
Assiniboine, Mount, mtn., Can.	55	50°52'N	115°39'W
Assis, Braz. (ä-sē´s)	101	22°39'S	50°21'W
Assisi, Italy	118	43°04'N	12°37'E
As-Sudd, reg., Sudan	185	8°45'N	30°45'E
As Sulaymānīyah, Iraq	154	35°47'N	45°23'E
As Sulaymānīyah, Sau. Ar.	157	24°09'N	46°19'E
As Suwaydā', Syria	154	32°41'N	36°41'E
Astakós, Grc. (äs´tä-kōs)	131	38°42'N	21°00'E
Astana (Aqmola), Kaz.	139	51°10'N	71°43'E
Astara, Azer.	137	38°30'N	48°50'E
Asti, Italy (äs´tē)	118	44°54'N	8°12'E
Astorga, Spain (äs-tôr´gä)	128	42°28'N	6°03'W
Astoria, Or., U.S. (ăs-tō´rĭ-ȧ)	64	46°11'N	123°51'W
Astrakhan', Russia (äs-trȧ-kän´)	134	46°15'N	48°00'E
Astrida, Rw. (äs-trē´dȧ)	186	2°37'S	29°48'E
Asturias, hist. reg., Spain (äs-tōō´ryäs)	128	43°21'N	6°00'W
Astypalaia, i., Grc.	119	36°31'N	26°19'E
Asunción *see* Ixtaltepec, Mex.	89	16°33'N	95°04'W
Asunción *see* Nochistlán, Mex.	88	21°23'N	102°52'W
Asunción, Para. (ä-sōōn-syōn´)	102	25°25'S	57°30'W
Asunción Mita, Guat. (ä-sōōn-syŏ´n-mē´tä)	90	14°19'N	89°43'W
Aswān, Egypt (ä-swän´)	185	24°05'N	32°57'E
Aswān High Dam, dam, Egypt	185	23°58'N	32°53'E
Atacama, Desierto de, des., Chile (dĕ-syĕ´r-tō-dĕ-ä-tä-kȧ´mä)	97	23°50'S	69°00'W
Atacama, Puna de, plat., Bol. (pōō´nä-dĕ-ä-tä-kȧ´mä)	100	21°35'S	66°58'W
Atacama, Puna de, reg., Chile (pōō´nä-dĕ-ȧtä-kȧ´mä)	102	23°15'S	68°45'W
Atacama, Salar de, l., Chile (sȧ-lȧr´dĕ-ȧtä-kȧ´mä)	102	23°38'S	68°15'W
Ataco, Col. (ä-tȧ´kō)	100a	3°36'N	75°22'W
Atacora, Chaîne de l', mts., Benin	188	10°15'N	1°15'E
Atā 'itah, Jabal al, mtn., Jord.	153a	30°48'N	35°19'E
Atamanovskiy, Russia (ä-tä-mä´nôv-skĭ)	142a	52°15'N	60°47'E
'Atāqah, Jabal, mts., Egypt	192d	29°59'N	32°20'E
Atar, Maur. (ä-tär´)	184	20°45'N	13°16'W
Atascadero, Ca., U.S. (ăt-ăs-kȧ-dâ´rō)	76	35°29'N	120°40'W
Atascosa, r., Tx., U.S. (ăt-ăs-kō´sȧ)	80	28°50'N	98°17'W
Atauro, Ilha de, i., E. Timor (dĕ-ä-tä´ōō-rō)	169	8°20'S	126°15'E
Atbara, r., Afr.	185	17°14'N	34°27'E
'Aṭbarah, Sudan (ät´bȧ-rä)	185	17°45'N	33°15'E
Atbasar, Kaz. (ät´bä-sär´)	139	51°42'N	68°28'E
Atchafalaya, r., La., U.S.	81	30°53'N	91°51'W
Atchafalaya Bay, b., La., U.S. (ăch-ȧ-fȧ-lī´ȧ)	81	29°25'N	91°30'W
Atchison, Ks., U.S. (ăch´ĭ-sŭn)	65	39°33'N	95°08'W
Atco, N.J., U.S. (ăt´kō)	68f	39°46'N	74°53'W
Atempan, Mex. (ä-tĕm-pȧ´n)	89	19°49'N	97°25'W
Atenguillo, r., Mex. (ä-tĕn-gē´l-yō)	88	20°18'N	104°35'W
Athabasca, Can. (ăth-ȧ-băs´kȧ)	50	54°43'N	113°17'W
Athabasca, l., Can.	52	59°04'N	109°10'W
Athabasca, r., Can.	52	57°30'N	112°00'W
Athens (Athína), Grc.	131	38°00'N	23°38'E
Athens, Al., U.S. (ăth´ĕnz)	82	34°47'N	86°58'W
Athens, Ga., U.S.	65	33°55'N	83°24'W
Athens, Oh., U.S.	66	39°20'N	82°10'W
Athens, Pa., U.S.	67	42°00'N	76°30'W
Athens, Tn., U.S.	82	35°26'N	84°36'W
Athens, Tx., U.S.	81	32°13'N	95°51'W
Atherstone, Eng., U.K. (ăth´ẽr-stŭn)	114a	52°34'N	1°33'W
Atherton, Eng., U.K. (ăth´ẽr-tŭn)	114a	53°32'N	2°29'W
Atherton Plateau, plat., Austl. (ădh-ĕr-tŏn)	175	17°00'S	144°30'E
Athi, r., Kenya (ä´tē)	187	2°43'S	38°30'E
Athína *see* Athens, Grc.	110	38°00'N	23°38'E
Athlone, Ire. (ăth-lōn´)	116	53°24'N	7°30'W
Áthos, mtn., Grc. (ăth´ŏs)	131	40°10'N	24°15'E
Ath Thamad, Egypt	153a	29°41'N	34°17'E
Athy, Ire. (ȧ-thī)	120	52°59'N	7°08'W

ăt; finăl; rāte; senāte; ärm; åsk; sofȧ; fâre; ch-choose; dh-as th in other; bē; ēvent; bĕt; recĕnt; cratẽr; g-gō; gh-guttural g; bĭt; ĭ-short neutral; rīde; ĸ-guttural k as ch in German ich;

PLACE (Pronunciation)	PAGE	LAT.	LONG.
Ati, Chad	189	13°13'N	18°20'E
Atibaia, Braz. (ä-tē-bá'yȧ)	99a	23°08'S	46°32'W
Atikonak, l., Can.	53	52°34'N	63°49'W
Atimonan, Phil. (ä-tē-mō'nän)	169a	13°59'N	121°56'E
Atiquizaya, El Sal. (ä'tē-kē-zä'yä)	90	14°00'N	89°42'W
Atitlan, vol., Guat. (ä-tē-tlän')	90	14°35'N	91°11'W
Atitlan, Lago, l., Guat. (ä-tē-tlän')	90	14°38'N	91°23'W
Atizapán, Mex. (ä'tē-zà-pän')	89a	19°33'N	99°16'W
Atka, Ak., U.S. (ăt'kȧ)	63a	52°18'N	174°18'W
Atka, i., Ak., U.S.	64b	51°58'N	174°30'W
Atkarsk, Russia (åt-kärsk')	137	51°50'N	45°00'E
Atkinson, Ne., U.S. (ăt'kĭn-sŭn)	70	42°32'N	98°58'W
Atlanta, Ga., U.S. (ăt-lăn'tȧ)	65	33°45'N	84°23'W
Atlanta, Tx., U.S.	79	33°09'N	94°09'W
Atlantic, Ia., U.S. (ăt-lăn'tĭk)	71	41°23'N	94°58'W
Atlantic, N.C., U.S.	83	34°54'N	76°20'W
Atlantic City, N.J., U.S.	65	39°20'N	74°30'W
Atlantic Highlands, N.J., U.S.	68a	40°25'N	74°04'W
Atlantic Ocean, o.	4	5°00'S	25°00'W
Atlas Mountains, mts., Afr. (ăt'lăs)	184	31°22'N	4°57'W
Atliaca, Mex. (ät-lē-ä'kä)	88	17°38'N	99°24'W
Atlin, l., Can. (ăt'lĭn)	52	59°34'N	133°20'W
Atlixco, Mex. (ät-lēz'kō)	88	18°52'N	98°27'W
Atmore, Al., U.S. (ăt'mōr)	82	31°01'N	87°31'W
Atoka, Ok., U.S. (á-tō'kȧ)	79	34°23'N	96°07'W
Atoka Reservoir, res., Ok., U.S.	79	34°30'N	96°05'W
Atotonilco el Alto, Mex.	88	20°35'N	102°32'W
Atotonilco el Grande, Mex.	88	20°17'N	98°41'W
Atoui, r., Afr. (á-tōō-ē')	184	21°00'N	15°32'W
Atoyac, Mex. (ä-tô-yäk')	88	20°01'N	103°28'W
Atoyac, r., Mex.	88	18°35'N	98°16'W
Atoyac, r., Mex.	89	16°27'N	97°28'W
Atoyac de Alvarez, Mex. (ä-tô-yäk'dā äl'vä-rāz)	88	17°13'N	100°29'W
Atoyatempan, Mex. (ä-tō'yá-tĕm-pän')	89	18°47'N	97°54'W
Atrak, r., Asia	154	37°45'N	56°30'E
Åtran, r., Swe.	122	57°02'N	12°43'E
Atrato, Río, r., Col. (rē'ō-ä-trä'tō)	100	7°15'N	77°18'W
Aţ Ţafilah, Jord. (tä-fē'la)	153a	30°50'N	35°36'E
Aţ Ţā'if, Sau. Ar.	154	21°03'N	41°00'E
Attalla, Al., U.S. (ă-tăl'yȧ)	82	34°01'N	86°05'W
Attawapiskat, r., Can. (ăt'ȧ-wȧ-pĭs'kăt)	53	52°31'N	86°22'W
Attersee, l., Aus.	124	47°57'N	13°25'E
Attica, N.Y., U.S. (ăt'ĭ-kȧ)	67	42°55'N	78°15'W
Attleboro, Ma., U.S. (ăt''l-bŭr-ô)	68b	41°56'N	71°15'W
Attow, Ben, mtn., Scot., U.K. (bĕn ăt'tô)	120	57°15'N	5°25'W
Attoyac Bay, Tx., U.S. (ă-toi'yăk)	81	31°45'N	94°23'W
Attu, i., Ak., U.S. (ăt-tōō')	64b	53°08'N	173°18'E
Aţ Ţūr, Egypt	119	28°09'N	33°47'E
Aţ Ţurayf, Sau. Ar.	154	31°32'N	38°30'E
Åtvidaberg, Swe. (ôt-vē'dá-bĕrgh)	122	58°12'N	15°55'E
Atwood, Ks., U.S. (ăt'wŏod)	78	39°48'N	101°06'W
Atyraū, Kaz.	139	47°10'N	51°50'E
Atzcapotzalco, Mex. (ät'zkä-pô-tzäl'kō)	88	19°29'N	99°11'W
Atzgersdorf, Aus.	115e	48°10'N	16°17'E
Auau Channel, strt., Hi., U.S. (ä'ŏ-ä'ŏ)	84a	20°55'N	156°50'W
Aubagne, Fr. (ō-bän'y')	127	43°18'N	5°34'E
Aube, r., Fr. (ōb)	126	48°42'N	3°49'E
Aubenas, Fr. (ōb-nä')	126	44°37'N	4°22'E
Aubervilliers, Fr. (ō-bĕr-vē-yā')	127b	48°54'N	2°23'E
Aubin, Fr. (ō-băN')	126	44°29'N	2°12'E
Aubrey, Can. (ô-brē')	62a	45°08'N	73°47'W
Auburn, Al., U.S. (ô'bŭrn)	82	32°35'N	85°26'W
Auburn, Ca., U.S.	76	38°52'N	121°05'W
Auburn, Il., U.S.	79	39°36'N	89°46'W
Auburn, In., U.S.	66	41°20'N	85°05'W
Auburn, Ma., U.S.	61a	42°11'N	71°51'W
Auburn, Me., U.S.	65	44°04'N	70°24'W
Auburn, Ne., U.S.	79	40°23'N	95°50'W
Auburn, N.Y., U.S.	67	42°55'N	76°35'W
Auburn, Wa., U.S.	74a	47°18'N	122°14'W
Auburn Heights, Mi., U.S.	69b	42°37'N	83°13'W
Aubusson, Fr. (ō-bü-sôN')	126	45°57'N	2°10'E
Auch, Fr. (ōsh)	117	43°38'N	0°35'E
Aucilla, r., Fl., U.S. (ô-sĭl'ȧ)	82	30°15'N	83°55'W
Auckland, N.Z. (ôk'lănd)	175a	36°53'S	174°45'E
Auckland Islands, is., N.Z.	3	50°30'S	166°30'E
Aude, r., Fr. (ōd)	126	42°55'N	2°08'E
Audierne, Fr. (ō-dyĕrn')	126	48°02'N	4°31'W
Audincourt, Fr. (ō-dăN-kōōr')	127	47°30'N	6°49'E
Audley, Eng., U.K. (ôd'lĭ)	114a	53°03'N	2°18'W
Audo Range, mts., Eth.	192a	6°58'N	41°18'E
Audubon, Ia., U.S. (ô'dŏŏ-bŏn)	71	41°43'N	94°57'W
Audubon, N.J., U.S.	68f	39°54'N	75°04'W
Aue, Ger. (ou'ē)	124	50°35'N	12°44'E
Augathella, Austl. (ôr'gȧ'thĕ-lȧ)	176	25°49'S	146°40'E
Augrabiesvalle, wtfl., S. Afr.	186	28°30'S	20°00'E
Augsburg, Ger. (ouks'bŏŏrgh)	117	48°23'N	10°55'E
Augusta, Ar., U.S. (ô-gŭs'tȧ)	79	35°16'N	91°21'W
Augusta, Ga., U.S.	65	33°26'N	82°00'W
Augusta, Ks., U.S.	79	37°41'N	96°58'W
Augusta, Ky., U.S.	66	38°45'N	84°00'W
Augusta, Me., U.S.	65	44°19'N	69°42'W
Augusta, N.J., U.S.	68a	41°07'N	74°44'W
Augusta, Wi., U.S.	71	44°40'N	91°09'W
Augustow, Pol. (ou-gŏŏs'tŏŏf)	125	53°52'N	23°00'E
Auki, Sol. Is.	170e	8°46'S	160°42'E
Aulnay-sous-Bois, Fr. (ō-nĕ'sōō-bwä')	127b	48°56'N	2°30'E
Aulne, r., Fr. (ōn)	126	48°08'N	3°53'W
Auneau, Fr. (ō-nēō')	127b	48°28'N	1°45'E
Auob, r., Afr. (ä'wŏb)	186	25°00'S	19°00'E
Aur, i., Malay.	153b	2°27'N	104°51'E
Aura, Fin.	123	60°38'N	22°32'E
Aurangābād, India (ou-rŭŋ-gä-bäd')	155	19°56'N	75°19'E
Aurdal, Nor. (äŭr-däl)	116	60°54'N	9°24'E
Aurès, Massif de l', mts., Alg.	118	35°16'N	5°53'E
Aurillac, Fr. (ō-rē-yák')	117	44°57'N	2°27'E
Aurora, Can.	59	43°59'N	79°25'W
Aurora, Co., U.S.	78	39°44'N	104°50'W
Aurora, Il., U.S. (ô-rō'rȧ)	65	41°45'N	88°18'W
Aurora, In., U.S.	69f	39°04'N	84°55'W
Aurora, Mn., U.S.	71	47°31'N	92°17'W
Aurora, Mo., U.S.	79	36°58'N	93°42'W
Aurora, Ne., U.S.	78	40°54'N	98°01'W
Aursunden, l., Nor. (äûr-sûndĕn)	122	62°42'N	11°10'E
Au Sable, r., Mi., U.S. (ô-sā'b'l)	66	44°40'N	84°25'W
Ausable, r., N.Y., U.S.	67	44°25'N	73°50'W
Austin, Mn., U.S. (ôs'tĭn)	71	43°40'N	92°58'W
Austin, Nv., U.S.	76	39°30'N	117°05'W
Austin, Tx., U.S.	64	30°15'N	97°42'W
Austin, l., Austl.	174	27°45'S	117°30'E
Austin Bayou, Tx., U.S. (ôs'tĭn bī-ōō')	81a	29°17'N	95°21'W
Australia, nation, Oc.	174	25°00'S	135°00'E
Australian Alps, mts., Austl.	176	37°10'S	147°55'E
Australian Capital Territory, ter., Austl. (ôs-trā'lĭ-ăn)	175	35°30'S	148°40'E
Austria, nation, Eur. (ôs'trĭ-ȧ)	110	47°15'N	11°53'E
Authon-la-Plaine, Fr. (ō-tôN-lä-plĕ'n)	127b	48°27'N	1°58'E
Autlán, Mex. (ä-ōōt-län')	86	19°47'N	104°24'W
Autun, Fr. (ō-tŭN')	126	46°58'N	4°14'E
Auvergne, mts., Fr. (ō-vĕrn'y')	126	45°12'N	2°31'E
Auxerre, Fr. (ō-sâr')	117	47°48'N	3°32'E
Ava, Mo., U.S. (ā'vȧ)	79	36°56'N	92°40'W
Avakubi, D.R.C. (ä-vá-kōō'bē)	185	1°20'N	27°34'E
Avallon, Fr. (á-vá-lôN')	126	47°30'N	3°58'E
Avalon, Ca., U.S.	76	33°21'N	118°22'W
Avalon, Pa., U.S. (ăv'ȧ-lŏn)	69e	40°31'N	80°05'W
Aveiro, Port. (ä-vā'rō)	118	40°38'N	8°38'W
Avelar, Braz. (ä'vē-lá'r)	102b	22°20'S	43°25'W
Avellaneda, Arg. (ä-vĕl-yä-nā'dhä)	102	34°40'S	58°23'W
Avellino, Italy (ä-vĕl-lē'nō)	130	40°40'N	14°46'E
Averøya, i., Nor. (ävĕr-ûê)	122	63°40'N	7°16'E
Aversa, Italy (ä-vĕr'sä)	130	40°58'N	14°13'E
Avery, Tx., U.S. (ā'vēr-ĭ)	79	33°34'N	94°46'W
Avesta, Swe. (ä-vĕs'tä)	122	60°16'N	16°09'E
Aveyron, r., Fr. (ä-vâ-rôN)	126	44°07'N	1°45'E
Avezzano, Italy (ä-vât-sä'nō)	130	42°03'N	13°27'E
Avigliano, Italy (ä-vēl-yä'nō)	130	40°45'N	15°44'E
Avignon, Fr. (ä-vē-nyôN')	117	43°55'N	4°50'E
Ávila, Spain (ä-vē-lä)	128	40°39'N	4°42'W
Avilés, Spain (ä-vē-lās')	118	43°33'N	5°55'W
Aviño, Spain	128	43°36'N	8°05'W
Avoca, Ia., U.S. (ȧ-vō'kȧ)	79	41°29'N	95°16'W
Avon, Ct., U.S. (ā'vŏn)	67	41°40'N	72°50'W
Avon, Ma., U.S. (ā'vŏn)	61a	42°08'N	71°03'W
Avon, Oh., U.S.	69d	41°27'N	82°02'W
Avon, r., Eng., U.K. (ā'vŭn)	120	52°05'N	1°55'W
Avondale, Ga., U.S.	68c	33°47'N	84°16'W
Avon Lake, Oh., U.S.	69d	41°31'N	82°01'W
Avonmore, Can. (ă'vōN-mōr)	62c	45°11'N	74°58'W
Avon Park, Fl., U.S. (ā'vŏn pärk')	83a	27°35'N	81°29'W
Avranches, Fr. (á-vränsh')	126	48°43'N	1°34'W
Awaji-Shima, i., Japan	166	34°32'N	135°02'E
Awe, Loch, l., Scot., U.K. (lŏk ôr)	120	56°22'N	5°04'W
Awjilah, Libya	185	29°07'N	21°21'E
Ax-les-Thermes, Fr. (äks'lā tĕrm')	126	42°43'N	1°50'E
Axochiapan, Mex. (äks-ō-chyä'pän)	88	18°29'N	98°49'W
Ay, r., Russia	136	55°55'N	57°55'E
Ayabe, Japan (ä'yä-bē)	166	35°16'N	135°17'E
Ayachi, Arin', mtn., Mor.	118	32°29'N	4°57'W
Ayacucho, Arg. (ä-yä-kōō'chō)	102	37°05'S	58°30'W
Ayacucho, Peru	100	13°12'S	74°03'W
Ayaköz, Kaz.	139	48°00'N	80°12'E
Ayamonte, Spain (ä-yä-mô'n-tē)	118	37°14'N	7°28'W
Ayan, Russia (á-yän')	135	56°26'N	138°18'E
Ayata, Bol. (ä-yä'tä)	100	15°17'S	68°43'W
Ayaviri, Peru (ä-yä-vē'rē)	100	14°46'S	70°38'W
Aydar, r., Eur. (ī-där')	133	49°15'N	38°48'E
Ayden, N.C., U.S. (ā'dĕn)	83	35°27'N	77°25'W
Aydın, Tur. (āīy-dĕn)	154	37°40'N	27°40'E
Ayer, Ma., U.S. (âr)	61a	42°33'N	71°36'W
Ayer Hitam, Malay.	153b	1°55'N	103°11'E
Ayers Rock *see* Uluru, mtn., Austl.	174	25°23'S	131°05'E
Aylesbury, Eng., U.K. (ālz'bēr-ĭ)	120	51°47'N	0°49'W
Aylmer, l., Can. (āl'mēr)	52	64°27'N	108°22'W
Aylmer, Mount, mtn., Can.	55	51°19'N	115°26'W
Aylmer East, Can. (āl'mēr)	59	45°24'N	75°50'W
Ayo el Chico, Mex. (ä'yō el chē'kō)	88	20°31'N	102°21'W
Ayon, i., Russia (ī-ôn')	135	69°50'N	168°40'E
Ayorou, Niger	188	14°44'N	0°55'E
Ayotla, Mex. (ä-yōt'lä)	89a	19°18'N	98°55'W
Ayoun el Atrous, Maur.	188	16°40'N	9°37'W
Ayr, Scot., U.K. (âr)	120	55°27'N	4°40'W
Aysha, Eth.	185	10°48'N	42°32'E
Ayutla, Guat. (ä-yōōt'lä)	90	14°44'N	92°11'W
Ayutla, Mex.	88	16°50'N	99°16'W
Ayutla, Mex.	88	20°09'N	104°20'W
Ayvalık, Tur. (āīy-wä-lĭk)	119	39°19'N	26°40'E
Azaouad, reg., Mali	188	18°00'N	3°20'W
Azaouak, Vallée de l', val., Afr.	189	15°50'N	3°10'E
Azare, Nig.	189	11°40'N	10°11'E
Azemmour, Mor. (á-zē-mōōr')	184	33°20'N	8°21'W
Azerbaijan, nation, Asia	134	40°30'N	47°30'E
Azle, Tx., U.S. (āz'lē)	75c	35°54'N	97°33'W
Azogues, Ec. (ä-sō'gās)	100	2°47'S	78°45'W
Azores *see* Açores, is., Port.	183	37°44'N	29°25'W
Azov, Russia (á-zôf') (ä-zôf)	137	47°07'N	39°19'E
Azov, Sea of, sea, Eur.	134	46°00'N	36°20'E
Aztec, N.M., U.S. (ăz'tĕk)	77	36°40'N	108°00'W
Aztec Ruins National Monument, rec., N.M., U.S.	77	36°50'N	108°00'W
Azua, Dom. Rep. (ä'swä)	93	18°30'N	70°45'W
Azuaga, Spain (ä-thwä'gä)	128	38°15'N	5°42'W
Azucar, Presa de, res., Mex.	80	26°06'N	98°44'W
Azuero, Península de, pen., Pan.	87	7°30'N	80°34'W
Azufre, Cerro (Copiapó), mtn., Chile	102	27°10'S	69°00'W
Azul, Arg. (ä-sōōl')	102	36°46'S	59°51'W
Azul, Cordillera, mts., Peru	100	7°15'S	75°30'W
Azul, Sierra, mts., Mex.	88	23°20'N	98°28'W
Azusa, Ca., U.S. (á-zōō'sȧ)	75a	34°08'N	117°55'W
Az Zahrān (Dhahran), Sau. Ar.	154	26°13'N	50°00'E
Az Zaqāzīq, Egypt	185	30°36'N	31°36'E
Az Zarqā', Jord.	153a	32°03'N	36°07'E
Az Zāwiyah, Libya	184	32°28'N	11°55'E

B

PLACE (Pronunciation)	PAGE	LAT.	LONG.
Baadheere (Bardera), Som.	192a	2°13'N	42°24'E
Baal, Ger. (bäl)	127c	51°02'N	6°17'E
Baao, Phil. (bä'ō)	169a	13°27'N	123°22'E
Baarle-Hertog, Bel.	115a	51°26'N	4°57'E
Baarn, Neth.	115a	52°12'N	5°18'E
Babaeski, Tur. (bä'bä-ĕs'kĭ)	131	41°25'N	27°05'E
Babahoyo, Ec. (bä-bä-ō'yō)	100	1°56'S	79°24'W
Babana, Nig.	189	10°36'N	3°50'E
Babanango, S. Afr.	187c	28°24'S	31°11'E
Babanūsah, Sudan	185	11°30'N	27°55'E
Babar, Pulau, i., Indon. (bä'bär)	169	7°50'S	129°15'E
Bab-el-Mandeb *see* Mandeb, Bab-el-, strt.	154	13°17'N	42°49'E
Babelthuap, i., Palau	170b	7°30'N	134°36'E
Babia, Arroyo de la, r., Mex.	80	28°26'N	101°50'W
Babine, r., Can.	54	55°10'N	127°00'W
Babine Lake, l., Can. (băb'ēn)	52	54°45'N	126°00'W
Bābol, Iran	154	36°30'N	52°48'E
Babruysk, Bela.	136	53°07'N	29°13'E
Babushkin, Russia (bä'bŏŏsh-kĭn)	140	51°47'N	106°08'W
Babushkin, Russia	132	55°52'N	37°42'E
Babuyan Islands, is., Phil. (bä-bōō-yän')	168	19°30'N	122°38'E
Babyak, Blg. (bäb'zhák)	131	41°59'N	23°42'E
Babylon, N.Y., U.S. (băb'ĭ-lŏn)	68a	40°42'N	73°19'W
Babylon, hist., Iraq	154	32°15'N	45°23'E
Bacalar, Laguna de, l., Mex. (lä-gōō-nä-dĕ-bä-kä-lär')	90a	18°50'N	88°31'W
Bacan, Pulau, i., Indon.	169	0°30'S	127°00'E
Bacarra, Phil. (bä-kär'rä)	165	18°22'N	120°40'E
Bacău, Rom.	119	46°34'N	27°00'E
Baccarat, Fr. (bá-ká-rá')	127	48°29'N	6°42'E
Bacchus, Ut., U.S. (băk'ŭs)	75b	40°40'N	112°06'W
Bachajón, Mex. (bä-chä-hōn')	89	17°08'N	92°18'W
Bachu, China (bä-chōō)	160	39°50'N	78°23'E
Back, r., Can.	52	65°30'N	104°15'W
Bačka Palanka, Serb. (bäch'kä pälän-kä)	131	45°14'N	19°24'E
Bačka Topola, Serb. (bäch'kä tô'pô-lä')	131	45°48'N	19°38'E
Back Bay, India (băk)	159b	18°55'N	72°45'E
Backstairs Passage, strt., Austl. (băk-stârs')	174	35°50'S	138°15'E
Bac Lieu, Viet.	168	9°45'N	105°50'E
Bac Ninh, Viet. (bäk'nēn'')	165	21°10'N	106°02'E
Baco, Mount, mtn., Phil. (bä'kô)	169a	12°50'N	121°11'E
Bacoli, Italy (bä-kō-lē')	129c	40°33'N	14°05'E
Bacolod, Phil. (bä-kō'lŏd)	169	10°42'N	123°03'E
Bácsalmás, Hung. (bäch'ôl-mäs)	125	46°07'N	19°18'E
Bacup, Eng., U.K. (băk'ŭp)	114a	53°42'N	2°12'W
Bad, r., S.D., U.S. (băd)	70	44°04'N	100°58'W
Badajoz, Spain (bá-dhä-hōth')	118	38°52'N	6°56'W
Badalona, Spain (bä-dhä-lō'nä)	129	41°27'N	2°15'E
Badanah, Sau. Ar.	154	30°49'N	40°45'E
Bad Axe, Mi., U.S. (băd' ăks)	66	43°50'N	82°55'W
Bad Bramstedt, Ger. (bät bräm'shtĕt)	115c	53°55'N	9°53'E
Baden, Aus. (bä'dĕn)	124	48°00'N	16°14'E
Baden, Switz.	124	47°28'N	8°17'E
Baden-Baden, Ger. (bä'dĕn-bä'dĕn)	117	48°46'N	8°11'E
Bad Freienwalde, Ger. (bät frī'ĕn-väl'dĕ)	124	52°47'N	14°00'E
Bad Hersfeld, Ger. (bät hĕrsh'fĕlt)	124	50°53'N	9°43'E
Badīn, Pak.	158	24°47'N	69°51'E
Bad Ischl, Aus. (bät ĭsh''l)	124	47°46'N	13°37'E
Bad Kissingen, Ger. (bät kĭs'ĭng-ĕn)	124	50°12'N	10°05'E
Bad Kreuznach, Ger. (bät kroits'näk)	124	49°52'N	7°53'E
Badlands, reg., N.D., U.S. (băd' lănds)	70	46°43'N	103°22'W
Badlands, reg., S.D., U.S.	70	43°43'N	102°36'W
Badlands National Park, rec., S.D., U.S.	70	43°56'N	102°37'W
Badlāpur, India	159b	19°12'N	73°12'E
Badogo, Mali	188	11°02'N	8°13'W
Bad Oldesloe, Ger. (bät ōl'dĕs-lōē)	124	53°48'N	10°21'E
Bad Reichenhall, Ger. (bät rī'kĕn-häl)	124	47°43'N	12°53'E
Bad River Indian Reservation, I.R., Wi., U.S. (băd)	71	46°41'N	90°36'W
Bad Segeberg, Ger. (bät sĕ'gĕ-bōōrgh)	115c	53°56'N	10°18'E
Bad Tölz, Ger. (bät tûltz)	124	47°46'N	11°35'E
Badulla, Sri L.	159	6°55'N	81°07'E
Bad Vöslau, Aus.	115e	47°58'N	16°13'E
Badwater Creek, r., Wy., U.S. (băd'wô-tēr)	73	43°13'N	107°55'W
Baena, Spain (bä-ā'nä)	118	37°38'N	4°20'W
Baependi, Braz. (bä-å-pĕn'dĭ)	99a	21°57'S	44°51'W
Baffin Bay, b., N.A. (băf'ĭn)	49	72°00'N	65°00'W

224 Baffin Bay – Baranpauh PRONOUNCING INDEX

PLACE (Pronunciation)	PAGE	LAT.	LONG.
Baffin Bay, b., Tx., U.S.	81	27°11'N	97°35'W
Baffin Island, i., Can.	49	67°20'N	71°00'W
Bāfq, Iran (bäfk)	154	31°48'N	55°23'E
Bafra, Tur. (bäf'rä)	119	41°30'N	35°50'E
Bagabag, Phil. (bä-gä-bäg')	169a	16°38'N	121°16'E
Bāgalkot, India	159	16°14'N	75°40'E
Bagamoyo, Tan. (bä-gä-mō'yō)	187	6°26'S	38°54'E
Bagaryak, Russia (bȧ-gȧr-yäk')	142a	56°13'N	61°32'E
Bagbele, D.R.C.	191	4°21'N	29°17'E
Bagdad *see* Baghdād, Iraq	154	33°14'N	44°22'E
Baghdād, Iraq (bȧgh-dȧd') (băg'dăd)	154	33°14'N	44°22'E
Bagheria, Italy (bä-gå-rē'ä)	130	38°03'N	13°32'E
Bagley, Mn., U.S. (băg'lē)	70	47°31'N	95°24'W
Bagnara, Italy (bän-yä'rä)	130	38°17'N	15°52'E
Bagnell Dam, Mo., U.S. (băg'nĕl)	79	38°13'N	92°40'W
Bagnères-de-Bigorre, Fr. (bän-yâr'dē-bē-gor')	126	43°04'N	0°09'E
Bagnères-de-Luchon, Fr. (băn-yâr' dē-lu chôɴ')	126	42°46'N	0°36'E
Bagnols-sur-Ceze, Fr. (bä-nyôl')	126	44°09'N	4°37'E
Bago, Mya.	168	17°17'N	96°29'E
Bagoé, r., Mali (bȧ-gō'å)	184	12°22'N	6°34'W
Baguio, Phil. (bä-gē-ō')	168	16°24'N	120°36'E
Bagzane, Monts, mtn., Niger	184	18°40'N	8°40'E
Bahamas, nation, N.A. (bȧ-hā'mȧs)	87	26°15'N	76°00'W
Bahau, Malay.	153b	2°48'N	102°25'E
Bahāwalpur, Pak. (bŭ-hä'wŭl-po͞or)	155	29°29'N	71°41'E
Bahia, state, Braz.	101	11°05'S	43°00'W
Bahía, Islas de la, i., Hond. (ē's-läs-dĕ-lä-bä-ē'ä)	86	16°15'N	86°30'W
Bahia Blanca, Arg. (bä-ē'ä bläŋ'kä)	102	38°45'S	62°07'W
Bahía de Caráquez, Ec. (bä-e'ä dā kä-rä'kĕz)	100	0°45'S	80°29'W
Bahía Negra, Para. (bä-ē'ä nā'grä)	101	20°11'S	58°05'W
Bahi Swamp, sw., Tan.	191	6°05'S	35°10'E
Bahoruco, Sierra de, mts., Dom. Rep. (sē-ĕ'r-rä-dĕ-bä-ō-ro͞o'kō)	93	18°10'N	71°25'W
Bahrain, nation, Asia (bä-rān')	154	26°15'N	51°17'E
Baḩr al Ghazāl, hist. reg., Sudan (bär ĕl ghä-zäl')	185	7°56'N	27°15'E
Baḩrīyah, oasis, Egypt (bȧ-hȧ-rē'yä)	119	28°34'N	29°01'E
Baía dos Tigres, Ang.	190	16°36'S	11°43'E
Baia Mare, Rom. (bä'yä mä'rā)	119	47°40'N	23°35'E
Baidyabātī, India	158a	22°47'N	88°21'E
Baie-Comeau, Can.	60	49°13'N	68°10'W
Baie de Wasai, Mi., U.S. (bā dē wä-sä'ē)	75k	46°27'N	84°15'W
Baie-Saint Paul, Can. (bā'sănt-pôl')	51	47°27'N	70°30'W
Baigou, China (bī-gō)	162	39°08'N	116°02'E
Baihe, China (bī-hŭ)	164	32°30'N	110°15'E
Bai Hu, l., China (bī-ho͞o)	162	31°22'N	117°38'E
Baiju, China (bī-jyo͞o)	162	33°04'N	120°17'E
Baikal, Lake *see* Baykal, Ozero, l., Russia	135	53°00'N	109°28'E
Bailén, Spain (bā-ē-lān')	128	38°05'N	3°48'W
Băileşti, Rom. (bȧ-ĭ-lĕsh'tē)	131	44°01'N	23°21'E
Bainbridge, Ga., U.S. (bān'brĭj)	82	30°52'N	84°35'W
Bainbridge Island, i., Wa., U.S.	74a	47°39'N	122°32'W
Baipu, China (bī-po͞o)	162	32°15'N	120°47'E
Baiquan, China (bī-chyuäŋ)	164	47°22'N	126°00'E
Baird, Tx., U.S. (bârd)	80	32°22'N	99°28'W
Bairdford, Pa., U.S. (bârd'fōrd)	69e	40°37'N	79°53'W
Baird Mountains, mts., Ak., U.S.	63	67°35'N	160°10'W
Bairnsdale, Austl. (bârnz'dāl)	175	37°50'S	147°39'E
Baïse, r., Fr. (bä-ēz')	126	43°52'N	0°23'E
Baiyang Dian, l., China (bī-yäŋ-dĭĕn)	162	39°00'N	115°45'E
Baiyu Shan, mts., China (bī-yo͞o shän)	164	37°02'N	108°30'E
Baja, Hung. (bŏ'yŏ)	125	46°11'N	18°55'E
Baja California, state, Mex. (bä-hä)	86	30°15'N	117°25'W
Baja California, pen., Mex.	49	28°00'N	113°30'W
Baja California Sur, state, Mex.	86	26°00'N	113°30'W
Bajo, Canal, can., Spain	129a	40°36'N	3°41'W
Bakal, Russia (bä'kȧl)	142a	54°57'N	58°50'E
Baker, Mt., U.S. (bā'kēr)	73	46°21'N	104°12'W
Baker, Or., U.S.	64	44°46'N	117°52'W
Baker, i., Oc.	2	1°00'N	176°00'W
Baker, l., Can.	52	63°51'N	96°10'W
Baker, Mount, mtn., Wa., U.S.	64	48°46'N	121°52'W
Baker Creek, r., Il., U.S.	69a	41°13'N	87°47'W
Bakersfield, Ca., U.S. (bā'kērz-fēld)	64	35°23'N	119°00'W
Bakerstown, Pa., U.S. (bā'kerz-toun)	69e	40°39'N	79°56'W
Bakewell, Eng., U.K. (bāk'wĕl)	114a	53°12'N	1°40'W
Bakhchysarai, Ukr.	133	44°46'N	33°54'E
Bakhmach, Ukr. (bȧĸ-mäch')	133	51°09'N	32°47'E
Bakhtarān, Iran	154	34°01'N	47°00'E
Bakhtegan, Daryācheh-ye, l., Iran	154	29°29'N	54°31'E
Bakhteyevo, Russia	142b	55°35'N	38°32'E
Bako, Eth. (bä'kö)	185	5°47'N	36°39'E
Bakony, mts., Hung. (bȧ-kōn'y')	125	46°57'N	17°30'E
Bakoye, r., Afr. (bȧ-kō'ê)	184	12°47'N	9°35'W
Bakr Uzyak, Russia (bȧkr o͞oz'yȧk)	142a	52°59'N	58°43'E
Baku (Bakı), Azer. (bȧ-ko͞o')	134	40°28'N	49°45'E
Bakwanga *see* Mbuji-Mayi, D.R.C.	190	6°09'S	23°28'E
Balabac Island, i., Phil. (bä'lä-bäk)	168	8°00'N	116°28'E
Balabac Strait, strt., Asia	168	7°23'N	116°30'E
Ba'labakk, Leb.	153a	34°00'N	36°13'E
Balabanovo, Russia (bȧ-lȧ-bä'nô-vô)	142b	56°10'N	37°44'E
Balagansk, Russia (bä-lä-gänsk')	140	53°58'N	103°09'E
Balaguer, Spain (bä-lä-gĕr')	129	41°48'N	0°50'E
Balakhta, Russia (bȧ'lȧk-tȧ')	135	55°22'N	91°43'E
Balakliia, Ukr.	133	49°28'N	36°51'E
Balakovo, Russia (bȧ'lȧ-kô'vō)	137	52°00'N	47°40'E
Balancán, Mex. (bä-läŋ-kän')	89	17°47'N	91°32'W
Balanga, Phil. (bä-läŋ'gä)	169a	14°41'N	120°31'E
Ba Lang An, Mui, c., Viet.	165	15°18'N	109°10'E
Balashikha, Russia (bȧ-lä'shĭ-kȧ)	142b	55°48'N	37°58'E
Balashov, Russia (bȧ'lȧ-shôf)	137	51°30'N	43°00'E
Balasore, India (bä-lȧ-sōr')	155	21°38'N	86°59'E
Balassagyarmat, Hung. (bô'lôsh-shô-dyôr'môt)	125	48°04'N	19°19'E
Balaton Lake, l., Hung. (bô'lô-tôn)	119	46°47'N	17°55'E
Balayan, Phil. (bä-lä-yān')	169a	13°56'N	120°44'E
Balayan Bay, b., Phil.	169a	13°46'N	120°46'E
Balboa Heights, Pan. (bäl-bô'ä)	91	8°59'N	79°33'W
Balboa Mountain, mtn., Pan.	86a	9°05'N	79°44'W
Balcarce, Arg. (bäl-kär'sā)	102	37°49'S	58°17'W
Balchik, Blg.	131	43°24'N	28°13'E
Bald Eagle, Mn., U.S. (bôld ē'g'l)	75g	45°06'N	93°01'W
Bald Eagle Lake, l., Mn., U.S.	75g	45°08'N	93°03'W
Baldock Lake, l., Can.	57	56°33'N	97°57'W
Baldwin Park, Ca., U.S. (bôld'wĭn)	75a	34°05'N	117°58'W
Baldwinsville, N.Y., U.S. (bôld'wĭns-vĭl)	67	43°10'N	76°20'W
Baldy Mountain, mtn., Can.	57	51°28'N	100°44'W
Baldy Peak, mtn., Az., U.S. (bôl'dē)	64	33°55'N	109°35'W
Baldy Peak, mtn., Tx., U.S. (bôl'dē pēk)	80	30°38'N	104°11'W
Balearic Islands *see* Balears, Illes, is., Spain	112	39°25'N	1°28'E
Balearic Sea, sea, Spain (băl-ē-ăr'ĭk)	129	39°40'N	1°05'E
Balears, Illes, is., Spain	112	39°25'N	1°28'E
Baleine, Grande Rivière de la, r., Can.	53	55°00'N	75°30'W
Baler, Phil. (bä-lar')	169a	15°46'N	121°33'E
Baler Bay, b., Phil.	169a	15°51'N	121°40'E
Balesin, i., Phil.	169a	14°28'N	122°10'E
Baley, Russia (bȧl-yā')	141	51°29'N	116°12'E
Balfate, Hond. (bäl-fä'tē)	90	15°48'N	86°24'W
Balfour, S. Afr. (băl'fōr)	192c	26°41'S	28°37'E
Bali, i., Indon. (bä'lē)	168	8°00'S	115°22'E
Balıkeşir, Tur. (balĭk'ĭysĭr)	137	39°40'N	27°50'E
Balikpapan, Indon. (bä'lēk-pä'pän)	168	1°13'S	116°52'E
Balintang Channel, strt., Phil. (bā-lĭn-täng')	168	19°50'N	121°08'E
Balkan Mountains *see* Stara Planina, mts., Blg.	112	42°50'N	24°45'E
Balkh, Afg. (bälk)	155	36°48'N	66°50'E
Balkhash, Lake *see* Balqash köli, l., Kaz.	139	45°58'N	72°15'E
Ballancourt, Fr. (bä-äɴ-kōr')	127b	48°31'N	2°23'E
Ballarat, Austl. (băl'ȧ-răt)	175	37°37'S	144°00'E
Ballard, l., Austl. (băl'ȧrd)	174	29°15'S	120°45'E
Ballater, Scot., U.K. (bäl'ȧ-tēr)	120	57°05'N	3°06'W
Balleny Islands, is., Ant. (băl'ĕ nē)	178	67°00'S	164°00'E
Ballina, Austl. (băl-ĭ-nä')	176	28°50'S	153°35'E
Ballina, Ire.	120	54°06'N	9°05'W
Ballinasloe, Ire. (băl'ĭ-nȧ-slō')	120	53°20'N	8°09'W
Ballinger, Tx., U.S. (băl'ĭn-jēr)	80	31°45'N	99°58'W
Ballston Spa, N.Y., U.S. (bôls'tŭn spä')	67	43°05'N	73°50'W
Balmazújváros, Hung. (bŏl'mŏz-o͞o'y'vä'rōsh)	125	47°35'N	21°23'E
Balobe, D.R.C.	191	0°05'N	28°00'E
Balonne, r., Austl. (băl-ōn')	175	27°00'S	149°10'E
Bālotra, India	158	25°56'N	72°12'E
Balqash, Kaz.	139	46°58'N	75°00'E
Balqash köli, l., Kaz.	139	45°58'N	72°15'E
Balranald, Austl. (băl'-rȧn-ăld)	176	34°42'S	143°30'E
Balsam, l., Can. (bôl'sȧm)	59	44°30'N	78°50'W
Balsas, Braz. (bäl'säs)	101	7°09'S	46°04'W
Balsas, r., Mex.	86	18°00'N	101°00'W
Balta, Ukr. (bȧl'tȧ)	137	47°57'N	29°38'E
Bălţi, Mol.	137	47°47'N	27°57'E
Baltic Sea, sea, Eur. (bôl'tĭk)	112	55°20'N	16°50'E
Baltīm, Egypt (bȧl-tēm')	192b	31°33'N	31°04'E
Baltimore, Md., U.S. (bôl'tĭ-mōr)	65	39°20'N	76°38'W
Baltiysk, Russia (bäl-tēysk')	123	54°40'N	19°55'E
Baluarte, Río del, Mex. (rē'ō-dĕl-bä-lo͞o'r-tĕ)	88	23°09'N	105°42'W
Baluchistān, hist. reg., Asia (bȧ-lo͝o-chĭ-stän')	155	27°30'N	65°30'E
Balzac, Can. (bōl'zăk)	62e	51°10'N	114°01'W
Bama, Nig.	189	11°30'N	13°41'E
Bamako, Mali (bä-mā-kō')	184	12°39'N	8°00'W
Bambang, Phil. (bäm-bäng')	169a	16°24'N	121°08'E
Bambari, C.A.R. (bäm-bȧ-rē)	185	5°44'N	20°40'E
Bamberg, Ger. (bäm'bĕrgh)	117	49°53'N	10°52'E
Bamberg, S.C., U.S. (băm'bûrg)	83	33°17'N	81°04'W
Bamenda, Cam.	189	5°56'N	10°10'E
Bamingui, r., C.A.R.	189	7°35'N	19°45'E
Bampton, Eng., U.K. (băm'tŭn)	114b	51°42'N	1°33'W
Bampūr, Iran (bŭm-po͞or')	154	27°15'N	60°22'E
Bam Yanga, Ngao, mts., Cam.	189	8°20'N	14°40'E
Banahao, Mount, mtn., Phil. (bä-nä-hä'ō)	169a	14°04'N	121°45'E
Banalia, D.R.C.	191	1°33'N	25°20'E
Banamba, Mali	188	13°33'N	7°27'W
Bananal, Braz. (bä-nä-näl')	99a	22°42'S	44°17'W
Bananal, Ilha do, i., Braz. (ē'lä-dō-bä-nă-năl')	101	12°09'S	50°27'W
Banās, r., India (băn-äs')	155	25°20'N	75°20'E
Banās, Ra's, c., Egypt	185	23°48'N	36°39'E
Banat, reg., Rom. (bä-nät')	131	45°35'N	21°05'E
Bancroft, Can. (băn'krŏft)	51	45°05'N	77°55'W
Bancroft *see* Chililabombwe, Zam.	191	12°18'S	27°43'E
Bānda, India (bän'dă)	155	25°36'N	80°21'E
Banda, Kepulauan, is., Indon.	169	4°40'S	129°56'E
Banda, Laut (Banda Sea), sea, Indon.	169	6°05'S	127°28'E
Banda Aceh, Indon.	168	5°10'N	95°10'E
Banda Banda, Mount, mtn., Austl. (băn'dȧ băn'dȧ)	176	31°09'S	152°15'E
Bandama Blanc, r., C. Iv. (bän-dä'mă)	188	6°15'N	5°00'W
Bandar Beheshtī, Iran	154	25°18'N	60°45'E
Bandar-e 'Abbās, Iran (bän-där' ȧb-bäs')	154	27°04'N	56°22'E
Bandar-e Būshehr, Iran	154	28°48'N	50°53'E
Bandar-e Lengeh, Iran	154	26°44'N	54°47'E
Bandar-e Torkeman, Iran	154	37°05'N	54°08'E
Bandar Lampung, Indon.	168	5°16'S	105°06'E
Bandar Maharani, Malay. (bän-där' mä-hä-rä'nē)	153b	2°02'N	102°34'E
Bandar Seri Begawan, Bru.	168	5°00'N	114°59'E
Bande, Spain	128	42°02'N	7°58'W
Bandeira, Pico da, mtn., Braz. (pē'kō dā bän dā'rä)	101	20°27'S	41°47'W
Bandelier National Monument, rec., N.M., U.S. (băn-dĕ-lēr')	77	35°50'N	106°45'W
Banderas, Bahía de, b., Mex. (bä-ē'ä dĕ bän-dĕ'räs)	88	20°38'N	105°35'W
Bandirma, Tur. (bän-dĭr'mȧ)	119	40°25'N	27°50'E
Bandon, Or., U.S. (băn'dŭn)	72	43°06'N	124°25'W
Bāndra, India	159b	19°04'N	72°49'E
Bandundu, D.R.C.	186	3°18'S	17°20'E
Bandung, Indon.	168	7°00'S	107°22'E
Banes, Cuba (bä'nās)	93	21°00'N	75°45'W
Banff, Can. (bănf)	50	51°10'N	115°34'W
Banff, Scot., U.K.	120	57°39'N	2°37'W
Banff National Park, rec., Can.	52	51°38'N	116°22'W
Bánfield, Arg. (bȧ'n-fyĕ'ld)	102a	34°44'S	58°24'W
Banfora, Burkina	188	10°38'N	4°46'W
Bangalore, India (băŋ'gȧ'lôr)	155	13°03'N	77°39'E
Bangassou, C.A.R. (bäɴ-gȧ-so͞o')	185	4°47'N	22°49'E
Bangeta, Mount, mtn., Pap. N. Gui.	169	6°20'S	147°00'E
Banggai, Kepulauan, is., Indon. (bäng-gī')	169	1°05'S	123°45'E
Banggi, Pulau, i., Malay.	168	7°12'N	117°10'E
Banghāzī, Libya	185	32°07'N	20°04'E
Bangka, i., Indon. (bäŋ'kȧ)	168	2°24'S	106°55'E
Bangkalan, Indon. (bäng-kȧ-län')	168	6°07'S	112°50'E
Bangkok, Thai.	168	13°50'N	100°29'E
Bangladesh, nation, Asia	155	24°15'N	90°00'E
Bangong Co, l., Asia (bäŋ-gōŋ tswo)	158	33°40'N	79°30'E
Bangor, Wales, U.K. (băŋ'ŏr)	120	53°13'N	4°05'W
Bangor, Me., U.S. (băn'gēr)	65	44°47'N	68°47'W
Bangor, Mi., U.S.	66	42°20'N	86°05'W
Bangor, Pa., U.S.	67	40°55'N	75°10'W
Bangs, Mount, mtn., Az., U.S. (băngs)	77	36°45'N	113°50'W
Bangued, Phil. (bän-gād')	169a	17°36'N	120°38'E
Bangui, C.A.R. (bäɴ-gē')	185	4°22'N	18°35'E
Bangweulu, Lake, l., Zam. (băng-wē-o͞o'lo͞o)	186	10°55'S	30°10'E
Bangweulu Swamp, sw., Zam.	191	11°25'S	30°10'E
Bani, Dom. Rep. (bä'-nē)	93	18°15'N	70°25'W
Bani, Phil. (bä'nē)	169a	16°11'N	119°51'E
Bani, r., Mali	184	13°00'N	5°30'W
Bánica, Dom. Rep. (bä'-nē-kä)	93	19°00'N	71°35'W
Banī Mazār, Egypt	156	28°29'N	30°48'E
Banister, r., Va., U.S. (băn'ĭs-tēr)	83	36°45'N	79°17'W
Banī Suwayf, Egypt	185	29°05'N	31°06'E
Banja Luka, Bos. (bän-yä-lo͞o'kȧ)	119	44°45'N	17°11'E
Banjarmasin, Indon. (bän-jēr-mä'sēn)	168	3°18'S	114°32'E
Banjin, China (bän-jyĭn)	162	32°23'N	120°14'E
Banjul, Gam.	184	13°28'N	16°39'W
Bankberg, mts., S. Afr. (bȧŋk'bûrg)	187c	32°18'S	25°15'E
Banks, Or., U.S. (bănks)	74c	45°37'N	123°07'W
Banks, i., Austl.	175	10°10'S	143°08'E
Banks, Cape, c., Austl.	173b	34°01'S	151°17'E
Banks Island, i., Can.	49	73°00'N	123°00'W
Banks Island, i., Can.	54	53°25'N	130°10'W
Banks Islands, is., Vanuatu	175	13°38'S	168°23'E
Banks Peninsula, pen., N.Z.	177	43°45'S	172°20'E
Banks Strait, strt., Austl.	176	40°45'S	148°00'E
Bankstown, Austl.	173b	33°55'S	151°02'E
Bann, r., N. Ire., U.K. (băn)	120	54°50'N	6°29'W
Banning, Ca., U.S. (băn'ĭng)	75a	33°56'N	116°53'W
Bannockburn, Austl.	173a	38°03'S	144°11'E
Bannu, Pak.	158	33°03'N	70°39'E
Baños, Ec. (bä'-nyōs)	100	1°30'S	78°22'W
Banská Bystrica, Slvk. (bän'skä bē'strē-tzä)	117	48°46'N	19°10'E
Bansko, Blg. (bän'skō)	131	41°51'N	23°33'E
Banstead, Eng., U.K. (băn'stĕd)	114b	51°18'N	0°09'W
Banton, i., Phil. (bän-tōn')	169a	12°54'N	121°55'E
Bantry, Ire. (băn'trĭ)	120	51°39'N	9°30'W
Bantry Bay, b., Ire.	120	51°25'N	10°09'W
Banyak, Kepulauan, is., Indon.	168	2°08'N	97°15'E
Banyuwangi, Indon. (bän-jō-wäŋ'gē)	168	8°15'S	114°15'E
Baocheng, China (bou-chŭŋ)	164	33°15'N	106°58'E
Baodi, China (bou-dē)	164	39°44'N	117°19'E
Baoding, China (bou-dĭŋ)	161	38°52'N	115°31'E
Baoji, China (bou-jyē)	164	34°10'N	106°58'E
Baoshan, China (bou-shän)	160	25°14'N	99°03'E
Baoshan, China	162	31°25'N	121°29'E
Baotou, China (bou-tō)	161	40°28'N	110°10'E
Baoying, China (bou-yĭŋ)	164	33°14'N	119°20'E
Bapsfontein, S. Afr. (băps-fŏn-tān')	187b	26°01'S	28°26'E
Baqueroncito, Col. (bä-kĕ-rō'n-sē-tō)	100a	3°18'N	74°40'W
Baraawe, Som.	192a	1°20'N	44°00'E
Barabinsk, Russia (bȧ'rȧ-bĭnsk)	140	55°18'N	78°00'E
Baraboo, Wi., U.S. (bâr'ȧ-bo͞o)	71	43°29'N	89°44'W
Baracoa, Cuba (bä-rä-kō'ä)	93	20°20'N	74°25'W
Baracoa, Cuba	93a	23°03'N	82°34'W
Baradères, Baie des, b., Haiti (bä-rä-dâr')	93	18°35'N	73°35'W
Baradero, Arg. (bä-rä-dĕ'ŏ)	99c	33°50'S	59°30'W
Barahona, Dom. Rep. (bä-rä-ô'nä)	93	18°15'N	71°10'W
Barajas de Madrid, Spain (bä-rȧ'häs dā mä-drēdh')	129a	40°28'N	3°35'W
Baranagar, India	158	22°38'N	88°25'E
Baranavichy, Bela. (bä'rä-nō-vē'chē)	136	53°08'N	25°59'E
Baranco, Belize (bä-räŋ'kō)	90	16°01'N	88°55'W
Baranof, i., Ak., U.S. (bä-rä'nôf)	63	56°48'N	136°08'W
Baranpauh, Indon.	153b	0°40'N	103°28'E

ăt; finȧl; rāte; senåte; ärm; ȧsk; sofȧ; fâre; ch-choose; dh-as th in other; bē; ēvent; bĕt; recĕnt; cratēr; g-gō; gh-guttural g; bĭt; ĭ-short neutral; rīde; ĸ-guttural k as ch in German ich;

PRONOUNCING INDEX

Barão de Melgaço – Beatty 225

PLACE (Pronunciation)	PAGE	LAT.	LONG.
Barão de Melgaço, Braz. (bä-roun-dĕ-mĕl-gä'sō)	101	16°12'S	55°48'W
Bārāsat, India	158a	22°42'N	88°29'E
Barataria Bay, b., La., U.S.	81	29°13'N	89°50'W
Baraya, Col. (bä-rá'yä)	100a	3°10'N	75°04'W
Barbacena, Braz. (bär-bä-sā'ná)	101	21°15'S	43°46'W
Barbacoas, Col. (bär-bä-kō'äs)	100	1°39'N	78°12'W
Barbacoas, Ven. (bä-bä-kô'äs)	101b	9°30'N	66°58'W
Barbados, nation, N.A. (bär-bā'dōz)	87	13°30'N	59°00'W
Barbar, Sudan	185	18°11'N	34°00'E
Barbastro, Spain (bär-bäs'trō)	129	42°05'N	0°05'E
Barbeau, Mi., U.S. (bár-bō')	75k	46°17'N	84°16'W
Barberton, S. Afr.	186	25°48'S	31°04'E
Barberton, Oh., U.S. (bär'bĕr-tŭn)	69d	41°01'N	81°37'W
Barbezieux, Fr. (bärb'zyû')	126	45°30'N	0°11'W
Barbosa, Col. (bär-bō'-sá)	100a	6°26'N	75°19'W
Barboursville, W.V., U.S. (bär'bĕrs-vĭl)	66	38°20'N	82°20'W
Barbourville, Ky., U.S.	82	36°52'N	83°58'W
Barbuda, i., Antig. (bär-bōō'dá)	87	17°45'N	61°15'W
Barcaldine, Austl. (bär'kôl-dĭn)	175	23°33'S	145°17'E
Barcarrota, Spain (bär-kär-rō'tä)	128	38°31'N	6°50'W
Barcellona, Italy (bä-chĕl-lō'nä)	130	38°07'N	15°15'E
Barcelona, Spain (bär-thå-lō'nä)	110	41°25'N	2°08'E
Barcelona, Ven. (bär-så-lō'nä)	100	10°09'N	64°41'W
Barcelos, Braz. (bär-sĕ'lōs)	100	1°04'S	63°00'W
Barcelos, Port. (bär-thå'lōs)	128	41°34'N	8°39'W
Bardawīl, Sabkhat al, b., Egypt	153a	31°20'N	33°24'E
Bardejov, Czech Rep. (bär'dyĕ-yôf)	125	49°18'N	21°18'E
Bardsey Island, i., Wales, U.K. (bärd'sē)	120	52°45'N	4°50'W
Bardstown, Ky., U.S. (bärds'toun)	66	37°50'N	85°30'W
Bardwell, Ky., U.S. (bärd'wĕl)	82	36°51'N	88°57'W
Bareilly, India	155	28°21'N	79°25'E
Barents Sea, sea, Eur. (bä'rĕnts)	134	72°14'N	37°28'E
Barentu, Erit. (bä-rĕn'tōō)	185	15°06'N	37°39'E
Barfleur, Pointe de, c., Fr. (bár-flûr')	126	49°43'N	1°17'W
Barguzin, Russia (bär'gōō-zĭn)	135	53°44'N	109°28'E
Bar Harbor, Me., U.S. (bär här'bĕr)	60	44°22'N	68°13'W
Bari, Italy (bä'rē)	110	41°08'N	16°53'E
Barinas, Ven. (bä-rē'näs)	100	8°36'N	70°14'W
Baring, Cape, c., Can. (bâr'ĭng)	52	70°07'N	119°48'W
Barisan, Pegunungan, mts., Indon. (bä-rē-sän')	168	2°38'S	101°45'E
Barito, r., Indon. (bä-rē'tō)	168	2°10'S	114°38'E
Barka, r., Afr.	185	16°44'N	37°34'E
Barkley Sound, strt., Can.	54	48°53'N	125°20'W
Barkly East, S. Afr. (bärk'lē ēst)	187c	30°58'S	27°37'E
Barkly Tableland, plat., Austl. (bär'klē)	174	18°15'S	137°05'E
Barkol, China (bär-kůl)	160	43°43'N	92°50'E
Bârlad, Rom.	119	46°15'N	27°43'E
Bar-le-Duc, Fr. (băr-lē-dük')	127	48°47'N	5°05'E
Barlee, l., Austl. (bär-lē')	174	29°45'S	119°00'E
Barletta, Italy (bär-lĕt'tä)	119	41°19'N	16°20'E
Barmstedt, Ger. (bärm'shtĕt)	115c	53°47'N	9°46'E
Barnaul, Russia (bär-nä-ōl')	134	53°18'N	83°23'E
Barnesboro, Pa., U.S. (bärnz'bĕr-ō)	67	40°45'N	78°50'W
Barnesville, Ga., U.S. (bärnz'vĭl)	82	33°03'N	84°10'W
Barnesville, Mn., U.S.	70	46°38'N	96°25'W
Barnesville, Oh., U.S.	66	39°55'N	81°10'W
Barnet, Vt., U.S. (bär'nĕt)	67	44°20'N	72°00'W
Barnetby le Wold, Eng., U.K. (bär'nĕt-bī)	114a	53°34'N	0°26'W
Barnett Harbor, b., Bah.	92	25°40'N	79°20'W
Barnsdall, Ok., U.S. (bärnz'dôl)	79	36°38'N	96°14'W
Barnsley, Eng., U.K. (bärnz'lĭ)	114a	53°33'N	1°29'W
Barnsley, co., Eng., U.K.	114a	53°33'N	1°30'W
Barnstaple, Eng., U.K. (bärn'stā-p'l)	120	51°06'N	4°05'W
Barnwell, S.C., U.S. (bärn'wĕl)	83	33°14'N	81°23'W
Baro, Nig. (bä'rō)	184	8°37'N	6°25'E
Baroda, India (bär-rō'dä)	155	22°21'N	73°12'E
Barotse Plain, pl., Zam.	190	15°50'S	22°55'E
Barqah (Cyrenaica), hist. reg., Libya	185	31°09'N	21°45'E
Barquisimeto, Ven. (bär-kē-sē-mā'tō)	100	10°04'N	69°16'W
Barra, Braz. (bär'rä)	101	11°04'S	43°11'W
Barraba, Austl.	176	30°22'S	150°36'E
Barrackpore, India	158a	22°46'N	88°21'E
Barra do Corda, Braz. (bär'rä dō cōr-dä)	101	5°33'S	45°13'W
Barra Mansa, Braz. (bär'rä män'sä)	99a	22°35'S	44°09'W
Barrancabermeja, Col. (bär-rän'kä-bĕr-mā'hä)	100	7°06'N	73°49'W
Barranquilla, Col. (bär-rän-kēl'yä)	100	10°57'N	75°00'W
Barras, Braz. (bá'r-räs)	101	4°13'S	42°14'W
Barre, Vt., U.S. (băr'ē)	67	44°15'N	72°30'W
Barreiras, Braz. (bär-rā'räs)	101	12°13'S	44°59'W
Barreiro, Port. (bär-rĕ'ē-rō)	118	38°39'N	9°05'W
Barren, r., Ky., U.S.	82	37°00'N	86°20'W
Barren, Cape, c., Austl. (băr'ĕn)	175	40°20'S	149°00'E
Barren, Nosy, is., Madag.	187	18°18'S	43°57'E
Barren River Lake, res., Ky., U.S.	82	36°45'N	86°02'W
Barretos, Braz. (bär-rā'tōs)	101	20°40'S	48°36'W
Barrhead, Can. (băr'hĕd)	50	54°08'N	114°24'W
Barrie, Can. (băr'ĭ)	51	44°25'N	79°45'W
Barrington, Can. (bă-rĕng-tŏn)	62a	45°07'N	73°35'W
Barrington, Il., U.S.	69a	42°09'N	88°08'W
Barrington, R.I., U.S.	68b	41°44'N	71°16'W
Barrington Tops, mtn., Austl.	176	32°00'S	151°25'E
Bar River, Can. (bär)	75k	46°27'N	84°02'W
Barron, Wi., U.S. (băr'ŭn)	71	45°24'N	91°51'W
Barrow, Ak., U.S. (băr'ō)	64a	71°20'N	156°00'W
Barrow, i., Austl.	174	20°50'S	115°00'E
Barrow, r., Ire. (bă-rå)	120	52°35'N	7°05'W
Barrow, Point, c., Ak., U.S.	63	71°20'N	156°00'W
Barrow Creek, Austl.	174	21°23'S	133°55'E
Barrow-in-Furness, Eng., U.K.	116	54°10'N	3°15'W
Barstow, Ca., U.S. (bär'stō)	76	34°53'N	117°03'W
Barstow, Md., U.S.	68e	38°32'N	76°37'W
Barth, Ger. (bärt)	124	54°20'N	12°43'E
Bartholomew Bayou, r., U.S. (bär-thŏl'ō-mū bī-ōō')	79	33°53'N	91°45'W
Barthurst, Can. (bär-thŭrst')	51	47°38'N	65°40'W
Bartica, Guy. (bär'tĭ-ká)	101	6°23'N	58°32'W
Bartın, Tur. (bär'tĭn)	119	41°35'N	32°12'E
Bartle Frere, Mount, mtn., Austl. (bärt''l frēr')	175	17°30'S	145°46'E
Bartlesville, Ok., U.S. (bär'tlz-vil)	79	36°44'N	95°58'W
Bartlett, Il., U.S. (bärt'lĕt)	69a	41°59'N	88°11'W
Bartlett, Tx., U.S.	81	30°48'N	97°25'W
Barton, Vt., U.S. (bär'tŭn)	67	44°45'N	72°05'W
Barton-upon-Humber, Eng., U.K. (bär'tŭn-ŭp'ŏn-hŭm'bĕr)	114a	53°41'N	0°26'W
Bartoszyce, Pol. (bär-tô-shĭ'tsä)	125	54°15'N	20°50'E
Bartow, Fl., U.S. (bär'tō)	83a	27°51'N	81°50'W
Barvinkove, Ukr.	133	48°55'N	36°59'E
Barwon, r., Austl. (bär'wŭn)	175	30°00'S	147°30'E
Barwon Heads, Austl.	173a	38°17'S	144°29'E
Barycz, r., Pol. (bä'rĭch)	124	51°30'N	16°38'E
Barysaw, Bela.	136	54°16'N	28°33'E
Basankusu, D.R.C. (bä-sän-kōō'sōō)	185	1°14'N	19°45'E
Basbeck, Ger. (bäs'bĕk)	115c	53°40'N	9°11'E
Basdahl, Ger. (bäs'däl)	115c	53°27'N	9°00'E
Basehor, Ks., U.S. (bās'hôr)	75f	39°08'N	94°55'W
Basel, Switz. (bä'z'l)	117	47°32'N	7°35'E
Bashee, r., S. Afr. (bă-shē')	187c	31°47'S	28°25'E
Bashi Channel, strt., Asia (bäsh'ē)	161	21°20'N	120°22'E
Bashkortostan, prov., Russia	136	54°12'N	57°15'E
Bashtanka, Ukr. (bàsh-tän'kà)	133	47°32'N	32°31'E
Basilan Island, i., Phil.	168	6°37'N	122°07'E
Basildon, Eng., U.K.	121	51°35'N	0°25'E
Basilicata, hist. reg., Italy (bä-zē-lē-kä'tä)	130	40°30'N	15°55'E
Basin, Wy., U.S. (bā'sĭn)	73	44°22'N	108°02'W
Basingstoke, Eng., U.K. (bā'zĭng-stōk)	114b	51°14'N	1°06'W
Baška, Cro. (bäsh'ka)	130	44°58'N	14°44'E
Baskale, Tur. (bäsh-kä'lĕ)	137	38°10'N	44°00'E
Baskatong, Réservoir, res., Can.	59	46°50'N	75°50'W
Baskunchak, l., Russia	137	48°20'N	46°40'E
Basoko, D.R.C. (bá-sō'kō)	185	0°52'N	23°50'E
Basque Provinces, hist. reg., Spain	128	43°00'N	2°46'W
Basra *see* Al Başrah, Iraq	154	30°35'N	47°59'E
Bassano, Can. (bäs-sän'ō)	50	50°47'N	112°28'W
Bassano del Grappa, Italy	130	45°46'N	11°44'E
Bassari, Togo	188	9°15'N	0°47'E
Bassas da India, i., Reu. (bäs'säs dä ēn'dē-à)	187	21°23'S	39°42'E
Basse Terre, Guad. (bås' târ')	87	16°00'N	61°43'W
Basseterre, St. K./N.	91b	17°20'N	62°42'W
Basse Terre, i., Guad.	91b	16°10'N	62°14'W
Bassett, Va., U.S. (băs'sĕt)	83	36°45'N	81°58'W
Bass Islands, is., Oh., U.S. (băs)	66	41°40'N	82°50'W
Bass Strait, strt., Austl.	175	39°40'S	145°40'E
Basswood, l., N.A. (băs'wŏŏd)	71	48°10'N	91°36'W
Bästad, Swe. (bô'stät)	122	56°26'N	12°46'E
Bastia, Fr. (bäs'tē-ä)	117	42°43'N	9°27'E
Bastogne, Bel. (bäs-tôn'y')	121	50°02'N	5°45'E
Bastrop, La., U.S. (băs'trŭp)	81	32°47'N	91°55'W
Bastrop, Tx., U.S.	81	30°08'N	97°18'W
Bastrop Bayou, Tx., U.S.	81a	29°07'N	95°22'W
Bata, Eq. Gui. (bä'tä)	184	1°51'N	9°45'E
Batabano, Golfo de, b., Cuba (gôl-fō-dĕ-bä-tä-bá'nō)	92	22°10'N	83°05'W
Batāla, India	158	31°54'N	75°18'E
Batam, i., Indon. (bä-täm')	153b	1°03'N	104°00'E
Batang, China (bä-tän)	160	30°08'N	99°00'E
Batangas, Phil. (bä-tän'gäs)	168	13°45'N	121°04'E
Batan Islands, is., Phil. (bä-tän')	168	20°58'N	122°20'E
Bátaszék, Hung. (bä'tá-sĕk)	125	46°07'N	18°40'E
Batavia, Il., U.S. (bá-tā'vĭ-á)	69a	41°51'N	88°18'W
Batavia, N.Y., U.S.	67	43°00'N	78°15'W
Batavia, Oh., U.S.	69f	39°05'N	84°10'W
Bataysk, Russia (bá-tīsk')	137	47°08'N	39°44'E
Bătdâmbâng, Camb. (bát-tám-băng')	168	13°14'N	103°15'E
Batesburg, S.C., U.S. (bāts'bûrg)	83	33°53'N	81°34'W
Batesville, Ar., U.S. (bāts'vĭl)	79	35°46'N	91°39'W
Batesville, In., U.S.	66	39°15'N	85°15'W
Batesville, Ms., U.S.	82	34°17'N	89°55'W
Batetska, Russia (bá-tĕ'tská)	132	58°36'N	30°21'E
Bath, Can. (báth)	60	46°31'N	67°36'W
Bath, Eng., U.K.	117	51°24'N	2°20'W
Bath, Me., U.S.	60	43°54'N	69°50'W
Bath, N.Y., U.S.	67	42°25'N	77°20'W
Bath, Oh., U.S.	69d	41°11'N	81°38'W
Bathsheba, Barb.	91b	13°13'N	60°30'W
Bathurst, Austl. (băth'ŭrst)	175	33°28'S	149°30'E
Bathurst *see* Banjul, Gam.	184	13°28'N	16°39'W
Bathurst, S. Afr. (băt-hûrst)	187c	33°26'S	26°53'E
Bathurst, i., Austl.	174	11°19'S	130°13'E
Bathurst, Cape, c., Can. (bath'-ûrst)	52	70°33'N	127°55'W
Bathurst Inlet, b., Can.	52	68°10'N	108°00'W
Batia, Benin	188	10°54'N	1°29'E
Batley, Eng., U.K. (băt'lĭ)	114a	53°43'N	1°37'W
Batna, Alg. (bät'nä)	184	35°41'N	6°12'E
Baton Rouge, La., U.S. (băt'ŭn rōōzh')	65	30°28'N	91°10'W
Batticaloa, Sri L.	159	7°40'N	81°10'E
Battle, r., Can.	56	52°20'N	111°59'W
Battle Creek, Mi., U.S. (băt''l krēk')	65	42°20'N	85°15'W
Battle Ground, Wa., U.S. (băt''l ground)	74c	45°47'N	122°32'W
Battle Harbour, Can. (băt''l här'bĕr)	51	52°17'N	55°33'W
Battle Mountain, Nv., U.S.	72	40°40'N	116°56'W
Battonya, Hung. (bät-tô'nyä)	125	46°17'N	21°00'E
Batu, Kepulauan, is., Indon. (bä'tōō)	168	0°10'S	98°00'E
Batumi, Geor. (bŭ-tōō'mē)	134	41°40'N	41°30'E
Batu Pahat, Malay.	168	1°51'N	102°56'E
Batupanjang, Indon.	153b	1°42'N	101°35'E
Bauang, Phil. (bä'wäng)	169a	16°31'N	120°19'E
Bauchi, Nig. (bà-ōō'chē)	184	10°19'N	9°50'E
Bauld, Cape, c., Can.	53a	51°38'N	55°25'W
Bāuria, India	158a	22°29'N	88°08'E
Bauru, Braz. (bou-rōō')	101	22°21'S	48°57'W
Bauska, Lat. (bou'skà)	123	56°24'N	24°12'E
Bauta, Cuba (bá'ōō-tä)	93a	22°59'N	82°33'W
Bautzen, Ger. (bout'sĕn)	117	51°11'N	14°27'E
Bavaria *see* Bayern, hist. reg., Ger.	124	49°00'N	11°16'E
Baw Baw, Mount, mtn., Austl.	176	37°50'S	146°17'E
Bawean, Pulau, i., Indon. (bá'vē-än)	168	5°50'S	112°40'E
Bawtry, Eng., U.K. (bôtrĭ)	114a	53°26'N	1°01'W
Baxley, Ga., U.S. (băks'lĭ)	83	31°47'N	82°22'W
Baxter, Austl.	173a	38°12'S	145°10'E
Baxter Springs, Ks., U.S. (băks'tĕr sprĭngs')	79	37°01'N	94°44'W
Bay, Laguna de, l., Phil. (lă-gōō'nä dā bä'ē)	169a	14°24'N	121°13'E
Bayaguana, Dom. Rep. (bä-yä-gwä'nä)	93	18°45'N	69°40'W
Bay al Kabīr, Wādī, val., Libya	118	29°52'N	14°28'E
Bayambang, Phil. (bä-yäm-bäng')	169a	15°50'N	120°26'E
Bayamo, Cuba (bä-yä'mō)	92	20°25'N	76°35'W
Bayamón, P.R.	87b	18°27'N	66°13'W
Bayan, China (bä-yän)	164	46°00'N	127°20'E
Bayanaūyl, Kaz.	139	50°43'N	75°37'E
Bayard, Ne., U.S. (bā'ĕrd)	70	41°45'N	103°20'W
Bayard, N.M., U.S.	77	32°45'N	108°07'W
Bayard, W.V., U.S.	67	39°15'N	79°20'W
Bayburt, Tur. (bä'ĭ-bört)	137	40°15'N	40°10'E
Bay City, Mi., U.S. (bā)	65	43°35'N	83°55'W
Bay City, Tx., U.S.	81	28°59'N	95°58'W
Baydaratskaya Guba, b., Russia	136	69°20'N	66°10'E
Bay de Verde, Can.	61	48°05'N	52°54'W
Baydhabo (Baidoa), Som.	192a	3°19'N	44°20'E
Baydrag, r., Mong.	160	46°09'N	98°52'E
Bayern, state, Ger.	115d	48°05'N	11°30'E
Bayern (Bavaria), hist. reg., Ger. (bī'ĕrn) (bá-vâ-rĭ-á)	124	49°00'N	11°16'E
Bayeux, Fr. (bá-yû')	117	49°19'N	0°41'W
Bayfield, Wi., U.S. (bā'fēld)	71	46°48'N	90°51'W
Baykal, Ozero (Lake Baikal), l., Russia	135	53°00'N	109°28'E
Baykal'skiy Khrebet, mts., Russia	135	53°30'N	107°30'E
Baykit, Russia (bī-kēt')	135	61°43'N	96°39'E
Baymak, Russia (báy'mäk)	142a	52°35'N	58°21'E
Bay Mills, Mi., U.S. (bā mĭlls)	75k	46°27'N	84°36'W
Bay Mills Indian Reservation, I.R., Mi., U.S.	71	46°19'N	85°03'W
Bay Minette, Al., U.S. (bā'mĭn-ĕt')	82	30°52'N	87°44'W
Bayombong, Phil. (bä-yōm-bōng')	169a	16°28'N	121°09'E
Bayonne, Fr. (bä-yŏn')	110	43°28'N	1°30'W
Bayonne, N.J., U.S. (bā-yōn')	68a	40°40'N	74°07'W
Bayou Bodcau Reservoir, res., La., U.S. (bī'yōō bŏd'kō)	65	32°49'N	93°22'W
Bayport, Mn., U.S. (bā'pōrt)	75g	45°02'N	92°46'W
Bayqongyr, Kaz.	139	47°46'N	66°11'E
Bayramiç, Tur.	131	39°48'N	26°35'E
Bayreuth, Ger. (bī-roit')	124	49°56'N	11°35'E
Bay Roberts, Can. (bā rŏb'ĕrts)	61	47°36'N	53°16'W
Bays, Lake of, l., Can. (bās)	59	45°15'N	79°00'W
Bay Saint Louis, Ms., U.S. (bā' sânt lōō'ĭs)	82	30°19'N	89°20'W
Bay Shore, N.Y., U.S. (bā' shôr)	68a	40°44'N	73°15'W
Bayt Lahm, W.B. (bĕth'lē-hĕm)	153a	31°42'N	35°13'E
Baytown, Tx., U.S. (bā'town)	81a	29°44'N	95°01'W
Bayview, Al., U.S. (bā'vū)	68h	33°34'N	86°59'W
Bayview, Wa., U.S.	74a	48°29'N	122°28'W
Bay Village, Oh., U.S. (bā)	69d	41°29'N	81°56'W
Baza, Spain (bä'thä)	118	37°29'N	2°46'W
Baza, Sierra de, mts., Spain	128	37°19'N	2°48'W
Bazar-Dyuzi, mtn., Azer. (bä'zár-dyōōz'ĕ)	137	41°20'N	47°40'E
Bazaruto, Ilha do, i., Moz. (bä-zä-rōō'tō)	186	21°42'S	36°10'E
Baziège, Fr.	126	43°25'N	1°41'E
Be, Nosy, i., Madag.	187	13°14'S	47°28'E
Beach, N.D., U.S. (bēch)	70	46°55'N	104°00'W
Beachy Head, c., Eng., U.K. (bēchē hĕd)	121	50°40'N	0°25'E
Beacon, N.Y., U.S. (bē'kŭn)	67	41°30'N	73°55'W
Beaconsfield, Can. (bē'kŭnz-fēld)	62a	45°26'N	73°51'W
Beals Creek, r., Tx., U.S. (bēls)	80	32°10'N	101°14'W
Bear, r., Ut., U.S.	75b	41°28'N	112°10'W
Bear, r., U.S.	73	42°17'N	111°42'W
Bear Brook, r., Can.	62c	45°24'N	75°15'W
Bear Creek, Mt., U.S. (bâr krēk)	73	45°11'N	109°07'W
Bear Creek, r., Al., U.S. (bâr)	82	34°27'N	88°00'W
Bear Creek, r., Tx., U.S.	75c	32°56'N	97°09'W
Beardstown, Il., U.S. (bērds'toun)	79	40°01'N	90°26'W
Bearfort Mountain, mtn., N.J., U.S. (bē'fŏrt)	68a	41°08'N	74°23'W
Bearhead Mountain, mtn., Wa., U.S. (bâr'hĕd)	74a	47°01'N	121°49'W
Bear Lake, l., Can.	57	55°08'N	96°00'W
Bear Lake, l., Id., U.S.	73	41°56'N	111°10'W
Bear River Range, mts., U.S.	73	41°50'N	111°30'W
Beas de Segura, Spain (bā'äs dā sä-gōō'rä)	128	38°16'N	2°53'W
Beata, i., Dom. Rep. (bĕ-ä'tä)	93	17°40'N	71°40'W
Beata, Cabo, c., Dom. Rep. (ká'bô-bĕ-ä'tä)	93	17°40'N	71°20'W
Beatrice, Ne., U.S. (bē'á-trĭs)	64	40°16'N	96°45'W
Beatty, Nv., U.S. (bēt'ē)	76	36°58'N	116°48'W

226 Beattyville – Bergedorf PRONOUNCING INDEX

PLACE (Pronunciation)	PAGE	LAT.	LONG.
Beattyville, Ky., U.S. (bēt′ė-vĭl)	66	37°35'N	83°40'W
Beaucaire, Fr. (bō-kâr′)	126	43°49'N	4°37'E
Beaucourt, Fr. (bō-ko͞or′)	127	47°30'N	6°54'E
Beaufort, N.C., U.S. (bō′fr̄t)	83	34°43'N	76°40'W
Beaufort, S.C., U.S.	83	32°25'N	80°40'W
Beaufort Sea, sea, N.A.	63	70°30'N	138°40'W
Beaufort West, S. Afr.	186	32°20'S	22°45'E
Beauharnois, Can. (bō-är-nwä′)	59	45°23'N	73°52'W
Beaumont, Can.	62b	46°50'N	71°01'W
Beaumont, Can.	62g	53°22'N	113°18'W
Beaumont, Ca., U.S. (bō′mŏnt)	75a	33°57'N	116°57'W
Beaumont, Tx., U.S.	65	30°05'N	94°06'W
Beaune, Fr. (bōn)	126	47°02'N	4°49'E
Beauport, Can. (bō-pôr′)	62b	46°52'N	71°11'W
Beauséjour, Can.	50	50°04'N	96°33'W
Beauvais, Fr. (bō-vĕ′)	126	49°25'N	2°05'E
Beaver, Ok., U.S. (bē′vẽr)	78	36°46'N	100°31'W
Beaver, Pa., U.S.	69e	40°42'N	80°18'W
Beaver, Ut., U.S.	77	38°15'N	112°40'W
Beaver, i., Mi., U.S.	66	45°40'N	85°30'W
Beaver, r., Can.	52	54°20'N	111°10'W
Beaver City, Ne., U.S.	78	40°08'N	99°52'W
Beaver Creek, r., Co., U.S.	78	39°42'N	103°37'W
Beaver Creek, r., Ks., U.S.	78	39°44'N	101°05'W
Beaver Creek, r., Mt., U.S.	70	46°45'N	104°18'W
Beaver Creek, r., Wy., U.S.	70	43°46'N	104°25'W
Beaver Dam, Wi., U.S.	71	43°29'N	88°50'W
Beaverhead, r., Mt., U.S.	73	45°25'N	112°35'W
Beaverhead Mountains, mts., Mt., U.S. (bē′vẽr-hĕd)	73	44°33'N	112°59'W
Beaver Indian Reservation, I.R., Mi., U.S.	66	45°40'N	85°30'W
Beaverton, Or., U.S. (bē′vẽr-tŭn)	74c	45°29'N	122°49'W
Bebington, Eng., U.K. (bē′bĭng-tŭn)	114a	53°20'N	2°59'W
Bečej, Serb. (bĕ′chā)	131	45°36'N	20°03'E
Béchar, Alg.	184	31°39'N	2°14'W
Becharof, l., Ak., U.S. (bĕk-ȧ-rôf)	63	57°58'N	156°58'W
Becher Bay, b., Can. (bēch′ẽr)	74a	48°18'N	123°37'W
Beckley, W.V., U.S. (bĕk′lĭ)	66	37°40'N	81°15'W
Bédarieux, Fr. (bā-dȧ-ryû′)	126	43°36'N	3°11'E
Beddington Creek, r., Can. (bĕd′ĕng tŭn)	62e	51°14'N	114°13'W
Bedford, Can. (bĕd′fẽrd)	59	45°10'N	73°00'W
Bedford, S. Afr.	187c	32°43'S	26°19'E
Bedford, Eng., U.K.	117	52°10'N	0°25'W
Bedford, Ia., U.S.	71	40°40'N	94°41'W
Bedford, In., U.S.	66	38°50'N	86°30'W
Bedford, Ma., U.S.	61a	42°30'N	71°17'W
Bedford, N.Y., U.S.	68a	41°12'N	73°38'W
Bedford, Oh., U.S.	69d	41°23'N	81°32'W
Bedford, Pa., U.S.	67	40°05'N	78°20'W
Bedford, Va., U.S.	83	37°19'N	79°27'W
Bedford Hills, N.Y., U.S.	68a	41°14'N	73°41'W
Beebe, Ar., U.S. (bē′bė)	79	35°04'N	91°54'W
Beecher, Il., U.S. (bē′chŭr)	69a	41°20'N	87°38'W
Beechey Head, c., Can. (bē′chĭ hĕd)	74a	48°19'N	123°40'W
Beech Grove, In., U.S. (bēch grōv)	69g	39°43'N	86°05'W
Beecroft Head, c., Austl. (bē′krûft)	176	35°03'S	151°15'E
Beelitz, Ger. (bĕ′lētz)	115b	52°14'N	12°59'E
Be'er Sheva', Isr. (bēr-shē′bȧ)	153a	31°15'N	34°48'E
Be'er Sheva', r., Isr.	153a	31°23'N	34°30'E
Beestekraal, S. Afr.	192c	25°22'S	27°34'E
Beeston, Eng., U.K. (bēs′t'n)	114a	52°55'N	1°11'W
Beetz, r., Ger. (bĕtz)	115b	52°28'N	12°37'E
Beeville, Tx., U.S. (bē′vĭl)	81	28°24'N	97°44'W
Bega, Austl. (bā′gaă)	175	36°50'S	149°49'E
Beggs, Ok., U.S. (bĕgz)	79	35°46'N	96°06'W
Bégles, Fr. (bĕ′gl')	126	44°47'N	0°34'W
Begoro, Ghana	188	6°23'N	0°23'W
Behala, India	158a	22°31'N	88°19'E
Behbehān, Iran	157	30°35'N	50°14'E
Behm Canal, can., Ak., U.S.	54	55°41'N	131°35'W
Bei, r., China (bā)	163a	22°54'N	113°08'E
Bei'an, China (bā-än)	164	48°05'N	126°26'E
Beicai, China (bā-tsī)	163b	31°12'N	121°33'E
Beifei, r., China (bā-fā)	162	33°14'N	117°03'E
Beihai, China (bā-hī)	160	21°30'N	109°10'E
Beihuangcheng Dao, i., China (bā-hůäŋ-chŭŋ dou)	162	38°23'N	120°55'E
Beijing, China	161	39°55'N	116°23'E
Beijing Shi, prov., China (bā-jyĭŋ shr)	164	40°07'N	116°00'E
Beira, Moz. (bā′rȧ)	186	19°45'N	34°58'E
Beira, hist. reg., Port. (bĕ′y-rä)	128	40°38'N	8°00'W
Beirut, Leb. (bā-ro͞ot′)	154	33°53'N	35°30'E
Beja, Port. (bā′zhä)	118	38°03'N	7°53'W
Béja, Tun.	118	36°52'N	9°20'E
Bejaïa (Bougie), Alg.	184	36°46'N	5°00'E
Bejar, Spain	128	40°25'N	5°43'W
Bejestān, Iran	154	34°30'N	58°22'E
Bejucal, Cuba (bā-ho͞o-käl′)	92	22°56'N	82°23'W
Bejuco, Pan. (bĕ-ĸo͞o′kō)	91	8°37'N	79°54'W
Békés, Hung. (bā′kāsh)	125	46°45'N	21°08'E
Békéscsaba, Hung. (bā′kāsh-chô′bô)	119	46°39'N	21°06'E
Beketova, Russia (bĕkė-to′vȧ)	141	53°23'N	125°21'E
Bela Crkva, Serb. (bĕ′lä tsĕrk′vä)	131	44°53'N	21°25'E
Belalcázar, Spain (bāl-ȧ-kä′thär)	128	38°35'N	5°12'W
Belarus, nation, Eur.	134	53°30'N	25°33'E
Belau *see* Palau, nation, Oc.	3	7°15'N	134°30'E
Bela Vista de Goiás, Braz.	101	16°57'S	48°47'W
Belawan, Indon. (bå-lä′wȧn)	168	3°43'N	98°43'E
Belaya, r., Russia (byĕ′lĭ-yȧ)	137	52°30'N	56°15'E
Belcher Islands, is., Can. (bĕl′chẽr)	53	56°20'N	80°40'W
Belding, Mi., U.S. (bĕl′dĭng)	66	43°05'N	85°25'W
Belebey, Russia (byĕ′lĕ-bå′ĭ)	136	54°00'N	54°10'E
Belém, Braz. (bå-lĕɴ′)	101	1°18'S	48°27'W
Belén, Para. (bā-lān′)	102	23°30'S	57°09'W
Belen, N.M., U.S. (bĕ-lån′)	77	34°40'N	106°45'W
Bélep, Îles, is., N. Cal.	175	19°30'S	164°00'E
Belëv, Russia (byĕl′yĕf)	136	53°49'N	36°06'E
Belfair, Wa., U.S. (bĕl′far)	74a	47°27'N	122°50'W
Belfast, N. Ire., U.K.	110	54°36'N	5°45'W
Belfast, Me., U.S. (bĕl′fȧst)	60	44°25'N	69°01'W
Belfast, Lough, b., N. Ire., U.K. (lŏĸ bĕl′fȧst)	120	54°45'N	6°00'W
Belford Roxo, Braz.	102b	22°46'S	43°24'W
Belfort, Fr. (bā-fôr′)	117	47°40'N	7°50'E
Belgaum, India	155	15°57'N	74°32'E
Belgium, nation, Eur. (bĕl′jĭ-ŭm)	110	51°00'N	2°52'E
Belgorod, Russia (byĕl′gŭ-rut)	137	50°36'N	36°32'E
Belgorod, prov., Russia	133	50°40'N	36°42'E
Belgrade (Beograd), Serb.	110	44°48'N	20°32'E
Belhaven, N.C., U.S. (bĕl′hā-vĕn)	83	35°33'N	76°37'W
Belington, W.V., U.S. (bĕl′ĭng-tŭn)	67	39°00'N	79°55'W
Belitung, i., Indon.	168	3°30'S	107°30'E
Belize, nation, N.A.	86	17°00'N	88°40'W
Belize, r., Belize	90a	17°16'N	88°56'W
Belize City, Belize (bĕ-lēz′)	86	17°31'N	88°10'W
Bel'kovo, Russia (byĕl′kô-vô)	142b	56°15'N	38°49'E
Bel'kovskiy, i., Russia (byĕl-kôf′skĭ)	141	75°45'N	137°00'E
Bell, i., Can. (bĕl)	61	50°45'N	55°35'W
Bell, r., Can.	59	49°25'N	77°15'W
Bella Bella, Can.	54	52°10'N	128°07'W
Bella Coola, Can.	54	52°22'N	126°46'W
Bellaire, Oh., U.S. (bĕl-âr′)	66	40°00'N	80°45'W
Bellaire, Tx., U.S.	81a	29°43'N	95°28'W
Bellary, India (bĕl-lä′rė)	155	15°15'N	76°56'E
Bella Union, Ur. (bĕ′l-yȧ-o͞o-nyô′n)	102	30°18'S	57°26'W
Bella Vista, Arg. (bā′lyȧ vēs′tȧ)	102	27°07'S	65°14'W
Bella Vista, Arg.	102	28°35'S	58°53'W
Bella Vista, Arg.	102a	34°35'S	58°41'W
Bella Vista, Para.	101	22°16'S	56°14'W
Belle-Anse, Haiti	93	18°15'N	72°00'W
Belle Bay, b., Can. (bĕl)	61	47°35'N	55°15'W
Belle Chasse, La., U.S. (bĕl shäs′)	68d	29°52'N	90°00'W
Bellefontaine, Oh., U.S. (bel-fōn′tån)	66	40°25'N	83°50'W
Bellefontaine Neighbors, Mo., U.S.	75e	38°46'N	90°13'W
Belle Fourche, S.D., U.S. (bĕl′ fo͞orsh′)	70	44°28'N	103°50'W
Belle Fourche, r., Wy., U.S.	70	44°29'N	104°40'W
Belle Fourche Reservoir, res., S.D., U.S.	70	44°51'N	103°44'W
Bellegarde, Fr. (bĕl-gärd′)	127	46°06'N	5°50'E
Belle Glade, Fl., U.S. (bĕl glād)	83a	26°39'N	80°37'W
Belle-Île, i., Fr. (bĕlēl′)	117	47°15'N	3°30'W
Belle Isle, Strait of, strt., Can.	53	51°35'N	56°30'W
Belle Mead, N.J., U.S. (bĕl mēd)	68a	40°28'N	74°40'W
Belleoram, Can.	61	47°31'N	55°25'W
Belle Plaine, Ia., U.S. (bĕl plān′)	71	41°52'N	92°19'W
Belle Vernon, Pa., U.S. (bĕl vŭr′nŭn)	69e	40°08'N	79°52'W
Belleville, Can. (bĕl′vĭl)	59	44°15'N	77°25'W
Belleville, Il., U.S.	75e	38°31'N	89°59'W
Belleville, Ks., U.S.	79	39°49'N	97°37'W
Belleville, Mi., U.S.	69b	42°12'N	83°29'W
Belleville, N.J., U.S.	68a	40°47'N	74°09'W
Bellevue, Ia., U.S. (bĕl′vū)	71	42°14'N	90°26'W
Bellevue, Ky., U.S.	69f	39°06'N	84°29'W
Bellevue, Mi., U.S.	66	42°30'N	85°00'W
Bellevue, Oh., U.S.	66	41°15'N	82°45'W
Bellevue, Pa., U.S.	69e	40°30'N	80°04'W
Bellevue, Wa., U.S.	74a	47°37'N	122°12'W
Belley, Fr. (bĕ-lē′)	127	45°46'N	5°41'E
Bellflower, Ca., U.S. (bĕl-flou′ẽr)	75a	33°53'N	118°08'W
Bell Gardens, Ca., U.S.	75a	33°59'N	118°11'W
Bellingham, Ma., U.S. (bĕl′ĭng-hăm)	61a	42°05'N	71°28'W
Bellingham, Wa., U.S.	64	48°46'N	122°29'W
Bellingham Bay, b., Wa., U.S.	74d	48°44'N	122°34'W
Bellingshausen Sea, sea, Ant. (bĕl′ĭngz houz'n)	178	72°00'S	80°30'W
Bellinzona, Switz. (bĕl-ĭn-tsō′nä)	124	46°10'N	9°09'E
Bellmore, N.Y., U.S. (bĕl-mōr)	68a	40°40'N	73°31'W
Bello, Col. (bĕ′l-yō̤)	100	6°20'N	75°33'W
Bellow Falls, Vt., U.S. (bĕl′ōz fôls)	67	43°10'N	72°30'W
Bellpat, Pak.	158	29°08'N	68°00'E
Bell Peninsula, pen., Can.	53	63°50'N	81°16'W
Bells Corners, Can.	62c	45°20'N	75°49'W
Bells Mountain, mtn., Wa., U.S. (bĕls)	74c	45°50'N	122°21'W
Belluno, Italy (bĕl-lo͞o′nō)	130	46°08'N	12°14'E
Bell Ville, Arg. (bĕl vēl′)	102	32°33'S	62°36'W
Bellville, S. Afr.	186a	33°54'S	18°38'E
Bellville, Tx., U.S. (bĕl′vĭl)	81	29°57'N	96°15'W
Bélmez, Spain (bĕl′mĕth)	128	38°17'N	5°17'W
Belmond, Ia., U.S. (bĕl′mŏnd)	71	42°49'N	93°37'W
Belmont, Ca., U.S.	74b	37°34'N	122°18'W
Belmonte, Braz. (bĕl-mōn′tå)	101	15°58'S	38°47'W
Belmopan, Belize	86	17°15'N	88°47'W
Belogorsk, Russia	135	51°09'N	128°32'E
Belo Horizonte, Braz. (bĕ′lôre-sô′n-tĕ)	101	19°54'S	43°56'W
Beloit, Ks., U.S. (bė-loit′)	78	39°26'N	98°06'W
Beloit, Wi., U.S.	65	42°31'N	89°04'W
Belomorsk, Russia (byĕl-ȯ-môrsk′)	136	64°30'N	34°42'E
Beloretsk, Russia (byĕ′lō-rĕtsk)	136	53°58'N	58°25'E
Belosarayskaya, Kosa, c., Ukr.	133	46°43'N	37°18'E
Belovo, Russia (bvĕ′lŭ-vû)	140	54°25'N	86°18'E
Beloye, l., Russia	136	60°10'N	38°05'E
Belozersk, Russia (byĕ-lŭ-zyôrsk′)	136	60°00'N	38°00'E
Belper, Eng., U.K. (bĕl′pẽr)	114a	53°01'N	1°28'W
Belt, Mt., U.S. (bĕlt)	73	47°11'N	110°58'W
Belt Creek, r., Mt., U.S.	73	47°19'N	110°58'W
Belton, Tx., U.S. (bĕl′tŭn)	81	31°04'N	97°27'W
Belton Lake, l., Tx., U.S.	81	31°15'N	97°35'W
Beltsville, Md., U.S. (belts-vĭl)	68e	39°03'N	76°56'W
Belukha, Mount, mtn., Asia	134	49°47'N	86°23'E
Belvidere, Il., U.S. (bĕl-vĕ-dēr′)	71	42°14'N	88°52'W
Belvidere, N.J., U.S.	67	40°50'N	75°05'W
Belyando, r., Austl. (bĕl-yän′dō)	175	22°09'S	146°48'E
Belyanka, Russia (byĕl′yȧn-kȧ)	142a	56°04'N	59°16'E
Belyy, Russia (byĕ′lē)	136	55°52'N	32°58'E
Belyy, i., Russia	134	73°19'N	72°00'E
Belyye Stolby, Russia (byĕ′lĭ-ye stôl′bĭ)	142b	55°20'N	37°52'E
Belzig, Ger. (bĕl′tsēg)	115b	52°08'N	12°35'E
Belzoni, Ms., U.S. (bĕl-zō′nė)	82	33°09'N	90°30'W
Bembe, Ang. (bĕɴ′bĕ)	186	7°00'S	14°20'E
Bembézar, r., Spain (bĕm-bā-thär′)	128	38°00'N	5°18'W
Bemidji, Mn., U.S. (bė-mĭj′ĭ)	71	47°28'N	94°54'W
Bena Dibele, D.R.C. (bĕn′ȧ dē-bĕ′lĕ)	186	4°00'S	22°49'E
Benalla, Austl. (bĕn-ăl′ȧ)	175	36°30'S	146°00'E
Benares *see* Vārānasi, India	155	25°25'N	83°00'E
Benavente, Spain (bā-nä-vĕn′tā)	118	42°01'N	5°43'W
Benbrook, Tx., U.S. (bĕn′bro͝ok)	75c	32°41'N	97°27'W
Benbrook Reservoir, res., Tx., U.S.	75c	32°35'N	97°30'W
Bend, Or., U.S. (bĕnd)	64	44°04'N	121°17'W
Bendeleben, Mount, mtn., Ak., U.S. (bĕn-dĕl-bĕn)	63	65°18'N	163°45'W
Bender Beyla, Som.	192a	9°40'N	50°45'E
Bendigo, Austl. (bĕn′dĭ-gō)	175	36°39'S	144°20'E
Benedict, Md., U.S. (bĕnĕ′dĭct)	68e	38°31'N	76°41'W
Benešov, Czech Rep. (bĕn′ĕ-shôf)	124	49°48'N	14°40'E
Benevento, Italy (bā-nā-vĕn′tō)	118	41°08'N	14°46'E
Bengal, Bay of, b., Asia (bĕn-gôl′)	152	17°30'N	87°00'E
Bengamisa, D.R.C.	191	0°57'N	25°10'E
Bengbu, China (bŭŋ-bo͞o)	161	32°52'N	117°22'E
Benghazi *see* Banghāzī, Libya	184	32°07'N	20°04'E
Bengkalis, Indon. (bĕng-kä′lĭs)	168	1°29'N	102°06'E
Bengkulu, Indon.	168	3°46'S	102°18'E
Benguela, Ang. (bĕn-gĕl′ȧ)	186	12°35'S	13°25'E
Beni, r., Bol. (bā′nė)	100	13°41'S	67°30'W
Béni-Abbas, Alg. (bā′nė ä-bĕs′)	184	30°11'N	2°13'W
Benicia, Ca., U.S. (bė-nĭsh′ĭ-ȧ)	74b	38°03'N	122°09'W
Benin, nation, Afr.	184	8°00'N	2°00'E
Benin, r., Nig. (bĕn-ēn′)	189	5°55'N	5°15'E
Benin, Bight of, b., Afr.	184	5°30'N	3°00'E
Benin City, Nig.	184	6°19'N	5°41'E
Beni Saf, Alg. (bā′nė säf′)	184	35°23'N	1°20'W
Benito, r., Eq. Gui.	190	1°35'N	10°45'E
Benkelman, Ne., U.S. (bĕn-kĕl-mȧn)	78	40°05'N	101°35'W
Benkovac, Cro. (bĕn′kō-vȧts)	130	44°02'N	15°41'E
Bennettsville, S.C., U.S. (bĕn′ĕts vĭl)	83	34°35'N	79°41'W
Bennington, Vt., U.S. (bĕn′ĭng-tŭn)	67	42°55'N	73°15'W
Benns Church, Va., U.S. (bĕnz′ chûrch′)	68g	36°47'N	76°35'W
Benoni, S. Afr. (bĕ-nō′nī)	186	26°11'S	28°19'E
Benoy, Chad	189	8°59'N	16°19'E
Benque Viejo, Belize (bĕn-kĕ bĭĕ′hō)	90a	17°07'N	89°07'W
Bensberg, Ger.	127c	50°58'N	7°09'E
Bensenville, Il., U.S. (bĕn′sĕn-vĭl)	69a	41°57'N	87°56'W
Bensheim, Ger. (bĕns-hīm)	124	49°42'N	8°38'E
Benson, Az., U.S. (bĕn-sŭn)	77	32°00'N	110°20'W
Benson, Mn., U.S.	70	45°18'N	95°36'W
Bentiaba, Ang.	190	14°15'S	12°21'E
Bentleyville, Pa., U.S. (bent′lē vĭl)	69e	40°07'N	80°01'W
Benton, Can.	60	45°59'N	67°36'W
Benton, Ar., U.S. (bĕn′tŭn)	79	34°34'N	92°34'W
Benton, Ca., U.S.	76	37°44'N	118°22'W
Benton, Il., U.S.	66	38°00'N	88°55'W
Benton Harbor, Mi., U.S. (bĕn′tŭn här′bẽr)	66	42°05'N	86°30'W
Bentonville, Ar., U.S. (bĕn′tŭn-vĭl)	79	36°22'N	94°11'W
Benue, r., Afr. (bā′no͞o-å)	184	8°00'N	8°00'E
Benut, r., Malay.	153b	1°43'N	103°20'E
Benwood, W.V., U.S. (bĕn-wŏŏd)	66	39°55'N	80°45'W
Benxi, China (bŭn-shyē)	164	41°25'N	123°50'E
Beograd *see* Belgrade, Serb.	110	44°48'N	20°32'E
Beppu, Japan (bĕ′po͞o)	167	33°16'N	131°30'E
Bequia Island, i., St. Vin. (bĕk-ē′ä)	91b	13°00'N	61°08'W
Berakit, Tanjung, c., Indon.	153b	1°16'N	104°44'E
Berat, Alb. (bĕ-rät′)	131	40°43'N	19°59'E
Berau, Teluk, b., Indon.	169	2°22'S	131°40'E
Berazategui, Arg. (bĕ-rä-zȧ′tĕ-gē)	102a	34°46'S	58°14'W
Berbera, Som. (bûr′bûr-ȧ)	192a	10°25'N	45°05'E
Berbérati, C.A.R.	189	4°16'N	15°47'E
Berck, Fr. (bĕrk)	126	50°26'N	1°36'E
Berdians'k, Ukr.	137	46°45'N	36°47'E
Berdians'ka kosa, c., Ukr.	133	46°38'N	36°42'E
Berdyaush, Russia (bĕr′dyȧûsh)	142a	55°10'N	59°12'E
Berdychiv, Ukr.	134	49°53'N	28°32'E
Berea, Ky., U.S. (bė-rē′ȧ)	82	37°30'N	84°19'W
Berea, Oh., U.S.	69d	41°22'N	81°51'W
Berehove, Ukr.	125	48°13'N	22°40'E
Bereku, Tan.	191	4°27'S	35°44'E
Berens, r., Can. (beřėnz)	57	52°15'N	96°30'W
Berens Island, i., Can.	57	52°18'N	97°40'W
Berens River, Can.	50	52°22'N	97°02'W
Beresford, S.D., U.S. (bĕr′ĕs-fẽrd)	70	43°05'N	96°46'W
Berettyóújfalu, Hung. (bĕ′rĕt-tyō-o͞o′y'fô-lo͞o)	125	47°14'N	21°33'E
Berezhany, Ukr. (bĕr-yĕ′zhȧ-nė)	125	49°25'N	24°58'E
Berezivka, Ukr.	133	47°12'N	30°56'E
Berezna, Ukr. (bĕr-yôz′nȧ)	133	51°32'N	31°47'E
Bereznehuvate, Ukr.	133	47°19'N	32°58'E
Berezniki, Russia (bĕr-yôz′nyė-kė)	136	59°25'N	56°46'E
Berëzovka, Russia	142a	57°35'N	57°19'E
Berëzovo, Russia (bĭr-yô′zē-vû)	134	64°10'N	65°10'E
Berëzovskiy, Russia (bĕr-yô′zôf-skĭ)	142a	56°54'N	60°47'E
Berga, Spain (bĕr′gä)	129	42°05'N	1°52'E
Bergama, Tur. (bĕr′gä-mä)	154	39°08'N	27°09'E
Bergamo, Italy (bĕr′gä-mō)	118	45°43'N	9°41'E
Bergantin, Ven. (bĕr-gän-tē′n)	101b	10°04'N	64°23'W
Bergara, Spain	128	43°08'N	2°23'W
Bergedorf, Ger. (bĕr′gĕ-dôrf)	115c	53°29'N	10°12'E

ăt; finăl; rāte; senåte; ärm; ȧsk; sofȧ; fâre; ch-choose; dh-as th in other; bē; ėvent; bĕt; recĕnt; cratẽr; g-gō; gh-guttural g; bĭt; ĭ-short neutral; rīde; ĸ-guttural k as ch in German ich;

PLACE (Pronunciation)	PAGE	LAT.	LONG.
Bergen, Ger. (bĕr'gĕn)	124	54°26'N	13°26'E
Bergen, Nor.	110	60°24'N	5°20'E
Bergenfield, N.J., U.S.	68a	40°55'N	73°59'W
Bergen op Zoom, Neth.	121	51°29'N	4°16'E
Bergerac, Fr. (bĕr-zhĕ-ràk')	117	44°49'N	0°28'E
Bergisch Gladbach, Ger. (bĕrg'ĭsh-glät'bäk)	127c	50°59'N	7°08'E
Berglern, Ger. (bĕrgh'lĕrn)	115d	48°24'N	11°55'E
Bergneustadt, Ger.	127c	51°01'N	7°39'E
Bergville, S. Afr. (bĕrg'vĭl)	187c	28°46'S	29°22'E
Berhampur, India	155	19°19'N	84°48'E
Bering Sea, sea, (bē'rĭng)	194	58°00'N	175°00'W
Bering Strait, strt.,	64a	64°50'N	169°50'W
Berja, Spain (bĕr'hä)	128	36°50'N	2°56'W
Berkeley, Ca., U.S. (bûrk'lĭ)	64	37°52'N	122°17'W
Berkeley, Mo., U.S.	75e	38°45'N	90°20'W
Berkeley Springs, W.V., U.S. (bûrk'lĭ springz)	67	39°40'N	78°10'W
Berkhamsted, Eng., U.K. (bĕrk'hăm'stĕd)	114b	51°44'N	0°34'W
Berkley, Mi., U.S. (bûrk'lĭ)	69b	42°30'N	83°10'W
Berkovitsa, Blg. (bĕ-kō'vē-tsà)	131	43°14'N	23°08'E
Berkshire, hist. reg., Eng., U.K.	114b	51°23'N	1°07'W
Berland, r., Can.	55	54°00'N	117°10'W
Berlenga, is., Port. (bĕr-lĕn'gäzh)	128	39°25'N	9°33'W
Berlin, Ger. (bĕr-lēn')	110	52°31'N	13°28'E
Berlin, S. Afr. (bĕr-lĭn)	187c	32°53'S	27°36'E
Berlin, N.H., U.S. (bûr-lĭn)	67	44°25'N	71°10'W
Berlin, N.J., U.S.	68f	39°47'N	74°56'W
Berlin, Wi., U.S. (bûr-lĭn')	71	43°58'N	88°58'W
Bermejo, r., S.A. (bĕr-mā'hō)	102	25°05'S	61°00'W
Bermeo, Spain (bĕr-mā'yō)	128	43°23'N	2°43'W
Bermuda, dep., N.A.	87	32°20'N	65°45'W
Bern, Switz. (bĕrn)	110	46°55'N	7°25'E
Bernal, Arg. (bĕr-näl')	102a	34°43'S	58°17'W
Bernalillo, N.M., U.S. (bĕr-nä-lē'yō)	77	35°20'N	106°30'W
Bernard, l., Can. (bēr-närd')	67	45°45'N	79°25'W
Bernardsville, N.J., U.S. (bûr nårds'vĭl)	68a	40°43'N	74°34'W
Bernau, Ger. (bĕr'nou)	124	52°40'N	13°35'E
Bernburg, Ger. (bĕrn'bŏorgh)	124	51°48'N	11°43'E
Berndorf, Aus. (bĕrn'dôrf)	124	47°57'N	16°05'E
Berne, In., U.S. (bûrn)	66	40°40'N	84°55'W
Berner Alpen, mts., Switz.	124	46°29'N	7°30'E
Bernier, i., Austl. (bĕr-nēr')	174	24°58'S	113°15'E
Bernina, Pizzo, mtn., Eur.	124	46°23'N	9°58'E
Bero, r., Ang.	190	15°10'S	12°20'E
Beroun, Czech Rep. (bā'rōn)	124	49°57'N	14°03'E
Berounka, r., Czech Rep. (bĕ-rōn'kà)	124	49°53'N	13°40'E
Berowra, Austl.	173b	33°36'S	151°10'E
Berre, Étang de, l., Fr. (ā-tôN' dĕ bâr')	126a	43°27'N	5°07'E
Berre-l'Étang, Fr. (bâr'lā-tôN')	126a	43°28'N	5°11'E
Berriozabal, Mex. (bā'rēō-zä-bäl')	89	16°47'N	93°16'W
Berriyyane, Alg.	118	32°50'N	3°49'E
Berry Creek, r., Can.	56	51°15'N	111°40'W
Berryessa, r., Ca., U.S. (bĕ'rĭ ĕs'ȧ)	76	38°35'N	122°33'W
Berry Islands, is., Bah.	92	25°40'N	77°50'W
Berryville, Ar., U.S. (bĕr'ē-vĭl)	79	36°21'N	93°34'W
Bershad', Ukr. (byĕr'shät)	133	48°22'N	29°31'E
Berthier, Can.	62b	46°56'N	70°44'W
Bertrand, r., Wa., U.S. (bûr'trănd)	74d	48°58'N	122°31'W
Berwick, Pa., U.S. (bûr'wĭk)	67	41°05'N	76°10'W
Berwick-upon-Tweed, Eng., U.K. (bûr'ĭk)	116	55°45'N	2°01'W
Berwyn, Il., U.S. (bûr'wĭn)	69a	41°49'N	87°47'W
Beryslav, Ukr.	133	46°49'N	33°24'E
Besalampy, Madag. (bĕz-à-làm-pē')	187	16°48'S	44°40'E
Besançon, Fr. (bĕ-säN-sôN)	117	47°14'N	6°02'E
Besar, Gunong, mtn., Malay.	153b	2°31'N	103°09'E
Besed', r., Eur. (byĕ'syĕt)	132	52°58'N	31°36'E
Beskid Mountains, mts., Eur.	125	49°23'N	19°00'E
Beskra, Alg.	184	34°52'N	5°39'E
Beslan, Russia	138	43°12'N	44°33'E
Bessarabia, hist. reg., Mol.	133	47°00'N	28°30'E
Bességes, Fr. (bĕ-sĕzh')	126	44°20'N	4°07'E
Bessemer, Al., U.S. (bĕs'ē-mĕr)	68h	33°24'N	86°58'W
Bessemer, Mi., U.S.	71	46°29'N	90°04'W
Bessemer City, N.C., U.S.	83	35°16'N	81°17'W
Bestensee, Ger. (bĕs'tĕn-zā)	115b	52°15'N	13°39'E
Betanzos, Spain (bĕ-tän'thōs)	128	43°18'N	8°14'W
Betatakin Ruin, Az., U.S. (bĕt-ȧ-tăk'ĭn)	77	36°40'N	110°29'W
Bethal, S. Afr. (bĕth'ăl)	192c	26°27'S	29°28'E
Bethalto, Il., U.S. (bȧ-thăl'tō)	75e	38°54'N	90°03'W
Bethanien, Nmb.	186	26°20'S	16°10'E
Bethany, Mo., U.S.	79	40°15'N	94°04'W
Bethel, Ak., U.S. (bĕth'ĕl)	64a	60°50'N	161°50'W
Bethel, Ct., U.S.	68a	41°22'N	73°24'W
Bethel, Vt., U.S.	67	43°50'N	72°40'W
Bethel Park, Pa., U.S.	69e	40°19'N	80°02'W
Bethesda, Md., U.S. (bĕ-thĕs'dȧ)	68e	39°00'N	77°10'W
Bethlehem, S. Afr.	186	28°14'S	28°18'E
Bethlehem, Pa., U.S. (bĕth'lē-hĕm)	67	40°40'N	75°25'W
Bethlehem *see* Bayt Lahm, W.B.	153a	31°42'N	35°13'E
Béthune, Fr. (bā-tün')	126	50°32'N	2°37'E
Betroka, Madag. (bĕ-trōk'ȧ)	187	23°13'S	46°17'E
Bet She'an, Isr.	153a	32°30'N	35°30'E
Betsiamites, Can.	51	48°57'N	68°36'W
Betsiamites, r., Can.	60	49°11'N	69°20'W
Betsiboka, r., Madag. (bĕt-sĭ-bō'kȧ)	187	16°47'S	46°45'E
Bettles Field, Ak., U.S. (bĕt'ŭls)	63	66°58'N	151°48'W
Betwa, r., India (bĕt'wȧ)	155	25°00'N	78°00'E
Betz, Fr. (bĕ)	127b	49°09'N	2°58'E
Beveren, Bel.	115a	51°13'N	4°14'E
B. Everett Jordan Lake, res., N.C., U.S.	83	35°45'N	79°00'W
Beverly, Ma., U.S.	61a	42°34'N	70°53'W
Beverly, N.J., U.S.	68f	40°03'N	74°56'W
Beverly Hills, Ca., U.S.	75a	34°05'N	118°24'W
Bevier, Mo., U.S. (bē-vēr')	79	39°44'N	92°36'W
Bewdley, Eng., U.K. (būd'lĭ)	114a	52°22'N	2°19'W
Bexhill, Eng., U.K. (bĕks'hĭl)	121	50°49'N	0°25'E
Bexley, Eng., U.K. (bĕks'ly)	114b	51°26'N	0°09'E
Beyla, Gui. (bā'là)	184	8°41'N	8°37'W
Beylul, Erit.	185	13°15'N	42°21'E
Beypazari, Tur. (bā-pà-zä'rĭ)	119	40°10'N	31°40'E
Beyşehir, Tur.	137	38°00'N	31°45'E
Beysugskiy, Liman, b., Russia (lĭ-män' bĕy-sŏŏg'skĭ)	133	46°07'N	38°35'E
Bezhetsk, Russia (byĕ-zhĕtsk')	136	57°46'N	36°40'E
Bezhitsa, Russia (byĕ-zhĭ'tsȧ)	136	53°19'N	34°18'E
Béziers, Fr. (bā-zyā')	117	43°21'N	3°12'E
Bhadreswar, India	158a	22°49'N	88°22'E
Bhāgalpur, India (bä'gŭl-pòr)	155	25°15'N	86°59'E
Bhamo, Mya. (bŭ-mō')	155	24°00'N	96°15'E
Bhāngar, India	158a	22°30'N	88°36'E
Bharatpur, India (bĕrt'pòr)	155	27°21'N	77°33'E
Bhatinda, India (bŭ-tĭn-dȧ)	155	30°19'N	74°56'E
Bhātpāra, India	158a	22°52'N	88°24'E
Bhaunagar, India (bäv-nŭg'ŭr)	155	21°45'N	72°58'E
Bhayandar, India	159b	19°20'N	72°50'E
Bhilai, India	158	21°14'N	81°23'E
Bhīma, r., India (bē'mȧ)	155	18°00'N	74°45'E
Bhiwandi, India	159b	19°18'N	73°03'E
Bhiwāni, India	158	28°53'N	76°08'E
Bhopāl, India (bō-päl)	155	23°20'N	77°25'E
Bhubaneswar, India (bò-bû-nāsh'vûr)	155	20°21'N	85°53'E
Bhuj, India (bōōj)	155	23°22'N	69°39'E
Bhutan, nation, Asia (bōō-tän')	155	27°15'N	90°30'E
Biafra, Bight of, b., Afr.	184	4°05'N	7°10'E
Biak, i., Indon. (bē'äk)	169	1°00'S	136°00'E
Biała Podlaska, Pol. (byä'wä pōd-läs'kä)	125	52°01'N	23°08'E
Białograd, Pol.	124	54°00'N	16°01'E
Bialystok, Pol. (byä-wĭs'tŏk)	110	53°08'N	23°12'E
Biankouma, C. Iv.	188	7°44'N	7°37'W
Biarritz, Fr. (byä-rēts')	117	43°27'N	1°39'W
Bibb City, Ga., U.S. (bĭb' sĭ'tē)	82	32°31'N	84°56'W
Biberach, Ger. (bē'bĕräk)	124	48°06'N	9°49'E
Bibiani, Ghana	188	6°28'N	2°20'W
Bic, Can. (bĭk)	60	48°22'N	68°42'W
Bicknell, In., U.S. (bĭk'nĕl)	66	38°45'N	87°20'W
Bicske, Hung. (bĭsh'kē)	125	47°29'N	18°38'E
Bida, Nig. (bē'dä)	184	9°05'N	6°01'E
Biddeford, Me., U.S. (bĭd'ē-fērd)	60	43°29'N	70°29'W
Biddulph, Eng., U.K. (bĭd'ŭlf)	114a	53°07'N	2°10'W
Biebrza, r., Pol. (byĕb'zhȧ)	125	53°18'N	22°25'E
Biel, Switz. (bēl)	124	47°09'N	7°12'E
Bielefeld, Ger. (bē'lē-fĕlt)	117	52°01'N	8°35'E
Biella, Italy (byĕl'lä)	130	45°34'N	8°05'E
Bielsk Podlaski, Pol. (byĕlsk pŭd-lä'skĭ)	117	52°47'N	23°14'E
Bien Hoa, Viet.	168	10°59'N	106°49'E
Bienville, Lac, l., Can.	53	55°32'N	72°45'W
Biesenthal, Ger. (bē'sĕn-täl)	115b	52°46'N	13°38'E
Biferno, r., Italy (bē-fĕr'nō)	130	41°49'N	14°46'E
Bifoum, Gabon	190	0°22'S	10°23'E
Biga, Tur. (bē'ghȧ)	131	40°13'N	27°14'E
Big Bay de Noc, Mi., U.S. (bĭg bā dē nok')	71	45°48'N	86°41'W
Big Bayou, Ar., U.S. (bĭg'bī'yōō)	79	33°04'N	91°28'W
Big Bear City, Ca., U.S. (bĭg bâr)	75a	34°16'N	116°51'W
Big Belt Mountains, mts., Mt., U.S. (bĭg bĕlt)	64	46°53'N	111°43'W
Big Bend Dam, S.D., U.S. (bĭg bĕnd)	70	44°11'N	99°33'W
Big Bend National Park, rec., Tx., U.S.	64	29°15'N	103°15'W
Big Black, r., Ms., U.S. (bĭg blăk)	82	32°05'N	90°49'W
Big Blue, r., Ne., U.S. (bĭg blōō)	79	40°53'N	97°00'W
Big Canyon, Tx., U.S. (bĭg kăn'yŭn)	80	30°27'N	102°19'W
Big Cypress Indian Reservation, I.R., Fl., U.S.	83a	26°19'N	81°11'W
Big Cypress Swamp, sw., Fl., U.S. (bĭg sī'prĕs)	83a	26°02'N	81°20'W
Big Delta, Ak., U.S. (bĭg dĕl'tȧ)	63	64°08'N	145°48'W
Big Fork, r., Mn., U.S. (bĭg fôrk)	71	48°08'N	93°47'W
Biggar, Can.	50	52°04'N	108°00'W
Big Hole, r., Mt., U.S. (bĭg hōl)	73	45°53'N	113°15'W
Big Hole National Battlefield, Mt., U.S. (bĭg hōl băt''l-fēld)	73	45°44'N	113°35'W
Bighorn, r., U.S.	64	45°30'N	108°00'W
Bighorn Lake, res., Mt., U.S.	73	45°00'N	108°10'W
Bighorn Mountains, mts., U.S. (bĭg hôrn)	64	44°47'N	107°40'W
Big Island, i., Can.	57	49°10'N	94°40'W
Big Lake, Wa., U.S. (bĭg lăk)	74a	48°24'N	122°14'W
Big Lake, l., Can.	62g	53°35'N	113°47'W
Big Lake, l., Wa., U.S.	74a	48°24'N	122°14'W
Big Lost, r., Id., U.S. (lŏst)	73	43°56'N	113°38'W
Big Mossy Point, c., Can.	57	53°45'N	97°50'W
Big Muddy, r., Il., U.S.	66	37°50'N	89°00'W
Big Muddy Creek, r., Mt., U.S. (bĭg mud'ĭ)	73	48°53'N	105°02'W
Bignona, Sen.	188	12°49'N	16°14'W
Big Porcupine Creek, r., Mt., U.S. (pôr'kŭ-pīn)	73	46°38'N	107°04'W
Big Quill Lake, l., Can.	52	51°55'N	104°22'W
Big Rapids, Mi., U.S. (bĭg răp'ĭdz)	66	43°40'N	85°30'W
Big River, Can.	50	53°50'N	107°01'W
Big Sandy, r., Az., U.S. (bĭg sănd'ē)	77	34°59'N	113°36'W
Big Sandy, r., Ky., U.S.	66	38°15'N	82°35'W
Big Sandy, r., Wy., U.S.	73	42°08'N	109°35'W
Big Sandy Creek, r., Co., U.S.	78	39°08'N	103°36'W
Big Sandy Creek, r., Mt., U.S.	73	48°20'N	110°08'W
Bigsby Island, i., Can.	57	49°04'N	94°35'W
Big Sioux, r., U.S. (bĭg sōō)	70	44°34'N	97°00'W
Big Spring, Tx., U.S. (bĭg sprĭng)	80	32°15'N	101°28'W
Big Stone, l., Mn., U.S. (bĭg stōn)	70	45°29'N	96°40'W
Big Stone Gap, Va., U.S.	83	36°50'N	82°50'W
Big Sunflower, r., Ms., U.S. (sŭn-flou'ēr)	82	32°57'N	90°40'W
Big Timber, Mt., U.S. (bĭg'tĭm-bēr)	73	45°50'N	109°57'W
Big Wood, r., Id., U.S. (bĭg wŏd)	73	43°02'N	114°30'W
Bihār, state, India (bē-här')	155	25°30'N	87°00'E
Biharamulo, Tan. (bē-hä-rä-mōō'lō)	186	2°38'S	31°20'E
Bihorului, Munţii, mts., Rom.	125	46°37'N	22°37'E
Bijagós, Arquipélago dos, is., Gui.-B.	184	11°20'N	17°10'W
Bijāpur, India	159	16°53'N	75°42'E
Bijeljina, Bos.	131	44°44'N	19°15'E
Bijelo Polje, Serb. (bē'yĕ-lō pō'lyĕ)	131	43°02'N	19°48'E
Bijiang, China (bē-jyän)	163a	22°57'N	113°15'E
Bijie, China (bē-jyĕ)	165	27°20'N	105°18'E
Bijou Creek, r., Co., U.S. (bē'zhōō)	78	39°41'N	104°13'W
Bīkaner, India (bĭ-kă'nûr)	155	28°07'N	73°19'E
Bikin, Russia (bē-kēn')	166	46°41'N	134°29'E
Bikin, r., Russia	166	46°37'N	135°55'E
Bikoro, D.R.C. (bē-kō'rō)	186	0°45'S	18°07'E
Bikuar, Parque Nacional do, rec., Ang.	190	15°07'S	14°40'E
Bilāspur, India (bē-läs'pōōr)	155	22°08'N	82°12'E
Bila Tserkva, Ukr.	137	49°48'N	30°09'E
Bilauktaung, mts., Asia	168	14°40'N	98°50'E
Bilbao, Spain (bĭl-bä'ō)	110	43°12'N	2°48'W
Bilbays, Egypt	192b	30°26'N	31°37'E
Bileća, Bos. (bē'lĕ-chä)	131	42°52'N	18°26'E
Bilecik, Tur. (bē-lĕd-zhĕk')	119	40°10'N	29°58'E
Bílé Karpaty, mts., Eur.	125	48°53'N	17°35'E
Biłgoraj, Pol. (bēw-gō'rī)	125	50°31'N	22°43'E
Bilhorod-Dnistrovs'kyi, Ukr.	137	46°09'N	30°19'E
Bilimbay, Russia (bē'lĭm-bày)	142a	56°59'N	59°53'E
Billabong, r., Austl. (bĭl'ă-bŏng)	175	35°15'S	145°20'E
Billerica, Ma., U.S. (bĭl'rĭk-ȧ)	61a	42°33'N	71°16'W
Billericay, Eng., U.K.	114b	51°38'N	0°25'E
Billings, Mt., U.S. (bĭl'ĭngz)	64	45°47'N	108°29'W
Bill Williams, r., Az., U.S. (bĭl-wĭl'yumz)	77	34°10'N	113°50'W
Bilma, Niger (bēl'mä)	185	18°41'N	13°20'E
Bilopillia, Ukr.	137	51°10'N	34°19'E
Bilovods'k, Ukr.	133	49°12'N	39°36'E
Biloxi, Ms., U.S. (bĭ-lŏk'sĭ)	65	30°24'N	88°50'W
Bilqās Qism Awwal, Egypt	192b	31°14'N	31°25'E
Bimberi Peak, mtn., Austl. (bĭm'bĕrĭ)	176	35°45'S	148°50'E
Binalonan, Phil. (bē-nä-lō'nän)	169a	16°03'N	120°35'E
Bingen, Ger. (bĭn'gĕn)	124	49°57'N	7°54'E
Bingham, Eng., U.K. (bĭng'ăm)	114a	52°57'N	0°57'W
Bingham, Me., U.S.	60	45°03'N	69°51'W
Bingham Canyon, Ut., U.S.	75b	40°33'N	112°09'W
Binghamton, N.Y., U.S. (bĭng'ăm-tŭn)	65	42°05'N	75°55'W
Bingo-Nada, b., Japan (bĭn'gō nä-dä)	167	34°06'N	133°14'E
Binjai, Indon.	168	3°59'N	108°00'E
Binnaway, Austl. (bĭn'ă-wā)	176	31°42'S	149°22'E
Bintan, i., Indon. (bĭn'tän)	153b	1°09'N	104°43'E
Bintimani, mtn., S.L.	188	9°13'N	11°07'W
Bintulu, Malay. (bēn'tōō-lōō)	168	3°07'N	113°06'E
Binxian, China	164	45°40'N	127°20'E
Binxian, China (bĭn-shyän)	162	37°27'N	117°58'E
Bio Gorge, val., Ghana	188	8°30'N	2°05'W
Bioko (Fernando Póo), i., Eq. Gui.	184	3°35'N	7°45'E
Bira, Russia (bē'rá)	166	49°00'N	133°18'E
Bira, r., Russia	166	48°55'N	132°25'E
Birātnagar, Nepal (bĭ-rät'nŭ-gŭr)	158	26°35'N	87°18'E
Birbka, Ukr.	125	49°36'N	24°18'E
Birch Bay, Wa., U.S. (bûrch)	74d	48°55'N	122°45'W
Birch Bay, b., Wa., U.S.	74d	48°55'N	122°52'W
Birch Island, i., Can.	57	52°25'N	99°55'W
Birch Mountains, mts., Can.	52	57°36'N	113°10'W
Birch Point, c., Wa., U.S.	74d	48°57'N	122°50'W
Bird Island, i., S. Afr. (bĕrd)	187c	33°51'S	26°21'E
Bird Rock, i., Bah. (bûrd)	93	22°50'N	74°20'W
Birds Hill, Can. (bûrds)	62f	49°58'N	97°00'W
Birdsville, Austl. (bûrdz'vĭl)	174	25°50'S	139°31'E
Birdum, Austl. (bûrd'ŭm)	174	15°45'S	133°25'E
Birecik, Tur. (bē-rĕd-zhĕk')	119	37°10'N	37°50'E
Bir Gara, Chad	189	13°11'N	15°58'E
Birjand, Iran (bēr'jänd)	154	33°07'N	59°16'E
Birkenfeld, Or., U.S.	74c	45°59'N	123°20'W
Birkenhead, Eng., U.K. (bûr'kĕn-hĕd)	120	53°23'N	3°02'W
Birkenwerder, Ger. (bĕr'kĕn-vĕr-dĕr)	115b	52°41'N	13°22'E
Birmingham, Eng., U.K.	110	52°29'N	1°53'W
Birmingham, Al., U.S. (bûr'mĭng-häm)	65	33°31'N	86°49'W
Birmingham, Mi., U.S.	69b	42°32'N	83°13'W
Birmingham, Mo., U.S.	75f	39°10'N	94°22'W
Birmingham Canal, can., Eng., U.K.	114a	53°07'N	2°40'W
Bi'r Misāhah, Egypt	185	22°16'N	28°04'E
Birnin Kebbi, Nig.	184	12°32'N	4°12'E
Birobidzhan, Russia (bē'rō-bē-jän')	135	48°42'N	133°28'E
Birsk, Russia (bĭrsk)	134	55°25'N	55°30'E
Birstall, Eng., U.K. (bûr'stôl)	114a	53°44'N	1°39'W
Biryulëvo, Russia (bēr-yōōl'yô-vô)	142b	55°35'N	37°39'E
Biryusa, r., Russia (bēr-yōō'sä)	140	56°43'N	97°30'E
Bi'r Za'farānah, Egypt	153a	29°07'N	32°38'E
Biržai, Lith. (bēr-zhä'ē)	123	56°11'N	24°45'E
Bisbee, Az., U.S. (bĭz'bē)	64	31°30'N	109°55'W
Biscay, Bay of, b., Eur. (bĭs'kā')	112	45°19'N	3°51'W
Biscayne Bay, b., Fl., U.S. (bĭs-kān')	83a	25°22'N	80°15'W
Bischeim, Fr. (bĭsh'hĭm)	127	48°40'N	7°48'E
Biscotasi Lake, l., Can.	58	47°20'N	81°55'W
Biser, Russia (bē'sĕr)	142a	58°24'N	58°54'E
Biševo, is., Serb. (bē'shĕ-vō)	130	42°58'N	15°50'E

ng-sing; ŋ-baŋk; N-nasalized n; nŏd; cŏmmit; ōld; ôbey; ôrder; oi-boil; fōōd; ŏ-as oo in foot; ou-out; s-soft; sh-dish; th-thin; pūre; ûnite; ûrn; stŭd; circŭs; ü-as in French tu; '-indeterminate vowel.

PLACE (Pronunciation)	PAGE	LAT.	LONG.
Bishkek, Kyrg.	139	42°49'N	74°42'E
Bisho, S. Afr.	186	32°50'S	27°20'E
Bishop, Ca., U.S. (bĭsh'ŭp)	76	37°22'N	118°25'W
Bishop, Tx., U.S.	81	27°35'N	97°46'W
Bishop's Castle, Eng., U.K. (bĭsh'ŏps käs''l)	114a	52°29'N	2°57'W
Bishopville, S.C., U.S. (bĭsh'ŭp-vĭl)	83	34°11'N	80°13'W
Bismarck, N.D., U.S. (bĭz'märk)	64	46°48'N	100°46'W
Bismarck Archipelago, is., Pap. N. Gui.	169	3°15'S	150°45'E
Bismarck Range, mts., Pap. N. Gui.	169	5°15'S	144°15'E
Bissau, Gui.-B. (bė-sa'ōō)	188	11°51'N	15°35'W
Bissett, Can.	57	51°01'N	95°45'W
Bistineau, l., La., U.S. (bĭs-tĭ-nō')	81	32°19'N	93°45'W
Bistrita, Rom. (bĭs-trĭt-sȧ)	119	47°09'N	24°29'E
Bistrita, r., Rom.	125	47°08'N	25°47'E
Bitlis, Tur. (bĭt-lēs')	154	38°30'N	42°00'E
Bitola, Mac. (bē'tô-lä) (mô'nä-stēr)	130	41°02'N	21°22'E
Bitonto, Italy (bē-tôn'tō)	130	41°08'N	16°42'E
Bitter Creek, r., Wy., U.S. (bĭt'ēr)	73	41°36'N	108°29'W
Bitterfeld, Ger. (bĭt'ēr-fĕlt)	124	51°39'N	12°19'E
Bitterroot, r., Mt., U.S.	73	46°28'N	114°10'W
Bitterroot Range, mts., U.S. (bĭt'ēr-ōōt)	64	47°15'N	115°13'W
Bityug, r., Russia (bĭt'yōōg)	133	51°23'N	40°33'E
Biu, Nig.	189	10°35'N	12°13'E
Biwabik, Mn., U.S. (bē-wä'bĭk)	71	47°32'N	92°24'W
Biwa-ko, l., Japan (bē-wä'kō)	167	35°03'N	135°51'E
Biya, r., Russia (bĭ'yȧ)	140	52°22'N	87°28'E
Biysk, Russia (bēsk)	134	52°32'N	85°28'E
Bizana, S. Afr. (bĭz-änä)	187c	30°51'S	29°54'E
Bizerte, Tun. (bē-zērt')	184	37°23'N	9°52'E
Bjelovar, Cro. (byĕ-lō'vär)	130	45°54'N	16°53'E
Bjørnafjorden, b., Nor.	122	60°11'N	5°26'E
Bla, Mali	188	12°57'N	5°46'W
Black, l., Mi., U.S. (blăk)	66	45°25'N	84°15'W
Black, l., N.Y., U.S.	67	44°30'N	75°35'W
Black, r., Asia	168	21°00'N	103°30'E
Black, r., Can.	58	49°20'N	81°15'W
Black, r., Az., U.S.	77	33°35'N	109°35'W
Black, r., N.Y., U.S.	67	43°45'N	75°20'W
Black, r., S.C., U.S.	83	33°55'N	80°10'W
Black, r., Wi., U.S.	71	44°07'N	90°56'W
Black, r., U.S.	79	35°47'N	91°22'W
Blackall, Austl. (blăk'ăl)	175	24°23'S	145°37'E
Black Bay, b., Can. (blăk)	58	48°36'N	88°32'W
Blackburn, Eng., U.K. (blăk'bûrn)	120	53°45'N	2°28'W
Blackburn Mount, mtn., Ak., U.S.	63	61°50'N	143°12'W
Black Butte Lake, res., Ca., U.S.	76	39°45'N	122°20'W
Black Canyon of the Gunnison National Park, rec., Co., U.S.	77	38°34'N	107°43'W
Black Diamond, Wa., U.S. (dī'mŭnd)	74a	47°19'N	122°00'W
Black Down Hills, hills, Eng., U.K. (blăk'doun)	120	50°58'N	3°19'W
Blackduck, Mn., U.S. (blăk'dŭk)	71	47°41'N	94°33'W
Blackfeet Indian Reservation, I.R., Mt., U.S.	73	48°40'N	113°00'W
Blackfoot, Id., U.S. (blăk'fŏŏt)	73	43°11'N	112°23'W
Blackfoot, r., Mt., U.S.	73	46°53'N	113°33'W
Blackfoot Indian Reservation, I.R., Mt., U.S.	73	48°49'N	112°53'W
Blackfoot Indian Reserve, I.R., Can.	55	50°45'N	113°00'W
Blackfoot Reservoir, res., Id., U.S.	73	42°53'N	111°23'W
Black Forest *see* Schwarzwald, for., Ger.	124	47°54'N	7°57'E
Black Hills, mts., U.S.	64	44°08'N	103°47'W
Black Island, i., Can.	57	51°10'N	96°30'W
Black Lake, Can.	59	46°02'N	71°24'W
Black Mesa, Az., U.S. (blăk māsȧ)	77	36°33'N	110°40'W
Blackmud Creek, r., Can. (blăk'mŭd)	62g	53°28'N	113°34'W
Blackpool, Eng., U.K. (blăk'pōōl)	120	53°49'N	3°02'W
Black Range, mts., N.M., U.S.	64	33°15'N	107°55'W
Black River, Jam. (blăk')	92	18°00'N	77°50'W
Black River Falls, Wi., U.S.	71	44°18'N	90°51'W
Black Rock Desert, des., Nv., U.S. (rŏk)	72	40°55'N	119°00'W
Blacksburg, S.C., U.S. (blăks'bûrg)	83	35°09'N	81°30'W
Black Sea, sea,	113	43°01'N	32°16'E
Blackshear, Ga., U.S. (blăk'shĭr)	83	31°20'N	82°15'W
Blackstone, Va., U.S. (blăk'stōn)	83	37°04'N	78°00'W
Black Sturgeon, r., Can. (stû'jŭn)	58	49°12'N	88°41'W
Blacktown, Austl. (blăk'toun)	173b	33°47'S	150°55'E
Blackville, Can. (blăk'vĭl)	60	46°44'N	65°50'W
Blackville, S.C., U.S.	83	33°21'N	81°19'W
Black Volta (Volta Noire), r., Afr.	184	11°30'N	4°00'W
Black Warrior, r., Al., U.S. (blăk wŏr'ĭ-ēr)	82	32°37'N	87°42'W
Blackwater, r., Ire. (blăk-wô'tēr)	120	52°05'N	9°02'W
Blackwater, r., Mo., U.S.	79	38°53'N	93°22'W
Blackwater, r., Va., U.S.	83	37°07'N	77°10'W
Blackwell, Ok., U.S. (blăk'wĕl)	79	36°47'N	97°19'W
Bladel, Neth.	115a	51°22'N	5°15'E
Blagodarnoye, Russia (blä'gō-där-nō'yĕ)	137	45°00'N	43°30'E
Blagoevgrad, Blg.	131	42°01'N	23°06'E
Blagoveshchensk, Russia (blä'gŏ-vyĕsh'chĕnsk)	135	50°16'N	127°47'E
Blagoveshchensk, Russia	142a	55°03'N	56°00'E
Blaine, Mn., U.S. (blān)	75g	45°11'N	93°14'W
Blaine, Wa., U.S.	74d	48°59'N	122°49'W
Blaine, W.V., U.S.	67	39°25'N	79°10'W
Blair, Ne., U.S. (blâr)	70	41°33'N	96°09'W
Blairmore, Can.	55	49°38'N	114°25'W
Blairsville, Pa., U.S. (blârs'vĭl)	67	40°30'N	79°40'W
Blake, i., Wa., U.S. (blāk)	74a	47°37'N	122°28'W
Blakely, Ga., U.S. (blāk'lē)	82	31°22'N	84°55'W
Blanc, Cap, c., Afr.	184	20°39'N	18°08'W
Blanc, Mont, mtn., Eur. (môɴ bläɴ)	112	45°50'N	6°53'E
Blanca, Bahía, b., Arg. (bä-ē'ä-blȧn'kä)	102	39°30'S	61°00'W
Blanca Peak, mtn., Co., U.S. (blăŋ'kȧ)	64	37°36'N	105°22'W
Blanche, r., Can.	62c	45°34'N	75°38'W
Blanche, Lake, l., Austl. (blănch)	176	29°20'S	139°12'E
Blanchester, Oh., U.S. (blăn'chĕs-tēr)	69f	39°18'N	83°58'W
Blanco, r., Mex.	88	24°05'N	99°21'W
Blanco, r., Mex.	89	18°42'N	96°03'W
Blanco, Cabo, c., Arg. (bläŋ'kō)	102	47°08'S	65°47'W
Blanco, Cabo, c., C.R. (kȧ'bō-bläŋ'kō)	90	9°29'N	85°15'W
Blanco, Cape, c., Or., U.S. (blăn'kō)	72	42°53'N	124°38'W
Blancos, Cayo, i., Cuba (kä'yō-bläŋ'kōs)	92	23°15'N	80°55'W
Blanding, Ut., U.S.	77	37°40'N	109°31'W
Blankenfelde, Ger. (blän'kĕn-fĕl-dĕ)	115b	52°20'N	13°24'E
Blanquefort, Fr.	126	44°53'N	0°38'W
Blanquilla, Arrecife, i., Mex. (är-rĕ-sē'fē-blän-kē'l-yä)	89	21°32'N	97°14'W
Blantyre, Mwi. (blăn-tīyr)	186	15°47'S	35°00'E
Blasdell, N.Y., U.S. (blāz'dĕl)	69c	42°48'N	78°51'W
Blato, Cro. (blä'tō)	130	42°55'N	16°47'E
Blaye-et-Sainte Luce, Fr. (blā'ā-săɴt-lüs')	126	45°08'N	0°40'W
Blażowa, Pol. (bwä-zhō'vȧ)	125	49°51'N	22°05'E
Bleus, Monts, mts., D.R.C.	191	1°10'N	30°10'E
Blind River, Can. (blīnd)	51	46°10'N	83°09'W
Blissfield, Mi., U.S. (blĭs-fēld)	66	41°50'N	83°50'W
Blithe, r., Eng., U.K. (blĭth)	114a	52°22'N	1°49'W
Blitta, Togo	188	8°19'N	0°59'E
Block, i., R.I., U.S. (blŏk)	67	41°05'N	71°35'W
Bloedel, Can.	54	50°07'N	125°23'W
Bloemfontein, S. Afr. (blōōm'fŏn-tān)	186	29°09'S	26°16'E
Blois, Fr. (blwä)	117	47°36'N	1°21'E
Blood Indian Reserve, I.R., Can.	55	49°30'N	113°10'W
Bloomer, Wi., U.S. (blōōm'ēr)	71	45°07'N	91°30'W
Bloomfield, Ia., U.S.	71	40°44'N	92°21'W
Bloomfield, In., U.S. (blōōm'fēld)	66	39°00'N	86°55'W
Bloomfield, Mo., U.S.	79	36°54'N	89°55'W
Bloomfield, Ne., U.S.	70	42°36'N	97°40'W
Bloomfield, N.J., U.S.	68a	40°48'N	74°12'W
Bloomfield Hills, Mi., U.S.	69b	42°35'N	83°15'W
Blooming Prairie, Mn., U.S. (blōōm'ĭng prā'rĭ)	71	43°52'N	93°04'W
Bloomington, Ca., U.S. (blōōm'ĭng-tŭn)	75a	34°04'N	117°24'W
Bloomington, Il., U.S.	65	40°30'N	89°00'W
Bloomington, In., U.S.	66	39°10'N	86°35'W
Bloomington, Mn., U.S.	75g	44°50'N	93°18'W
Bloomsburg, Pa., U.S. (blōōmz'bûrg)	67	41°00'N	76°25'W
Blossburg, Al., U.S. (blŏs'bûrg)	68h	33°38'N	86°57'W
Blossburg, Pa., U.S.	67	41°45'N	77°00'W
Bloubergstrand, S. Afr.	186a	33°48'S	18°28'E
Blountstown, Fl., U.S. (blŭnts'tun)	82	30°24'N	85°02'W
Bludenz, Aus. (blōō-dĕnts')	124	47°09'N	9°50'E
Blue Ash, Oh., U.S. (blōō ăsh)	69f	39°14'N	84°23'W
Blue Earth, Mn., U.S. (blōō ûrth)	71	43°38'N	94°05'W
Blue Earth, r., Mn., U.S.	71	43°55'N	94°16'W
Bluefield, W.V., U.S. (blōō'fēld)	83	37°15'N	81°11'W
Bluefields, Nic. (blōō'fēldz)	87	12°03'N	83°45'W
Blue Island, Il., U.S.	69a	41°39'N	87°41'W
Blue Mesa Reservoir, res., Co., U.S.	77	38°25'N	107°00'W
Blue Mountain, mtn., Can.	61	50°28'N	57°11'W
Blue Mountains, mts., Austl.	175	33°35'S	149°00'E
Blue Mountains, mts., Jam.	92	18°05'N	76°35'W
Blue Mountains, mts., U.S.	64	45°15'N	118°50'W
Blue Mud Bay, b., Austl. (bōō mŭd)	174	13°20'S	136°45'E
Blue Nile, r., Afr.	185	12°30'N	34°00'E
Blue Rapids, Ks., U.S. (blōō răp'ĭdz)	79	39°40'N	96°41'W
Blue Ridge, mtn., U.S. (blōō rĭj)	65	35°30'N	82°50'W
Blue River, Can.	50	52°05'N	119°17'W
Blue River, r., Mo., U.S.	75f	38°55'N	94°33'W
Bluff, Ut., U.S.	77	37°18'N	109°34'W
Bluff Park, Al., U.S.	68h	33°24'N	86°52'W
Bluffton, In., U.S. (blŭf-tŭn)	66	40°40'N	85°15'W
Bluffton, Oh., U.S.	66	40°50'N	83°55'W
Blumenau, Braz. (blōō'mĕn-ou)	102	26°53'S	48°58'W
Blumut, Gunong, mtn., Malay.	153b	2°03'N	103°34'E
Blyth, Eng., U.K. (blīth)	120	55°03'N	1°34'W
Blythe, Ca., U.S.	77	33°37'N	114°37'W
Blytheville, Ar., U.S. (blīth'vĭl)	79	35°55'N	89°51'W
Bo, S.L.	188	7°56'N	11°21'W
Boac, Phil.	169a	13°26'N	121°50'E
Boaco, Nic. (bō-ä'kō)	90	12°24'N	85°41'W
Bo'ai, China (bwo-ī)	164	35°10'N	113°08'E
Boa Vista, i., C.V. (bō-ä-vēsh'tȧ)	184b	16°01'N	23°52'W
Boa Vista do Rio Branco, Braz.	101	2°46'N	60°45'W
Bobo Dioulasso, Burkina (bō'bō-dyōō-lȧs-sō')	184	11°12'N	4°18'W
Bobr, Bela. (bô'b'r)	132	54°19'N	29°11'E
Bóbr, r., Pol. (bŭ'br)	124	51°44'N	15°13'E
Bobrov, Russia (bŭb-rôf')	137	51°07'N	40°01'E
Bobrovyts'a, Ukr.	133	50°43'N	31°27'E
Bobrynets', Ukr.	133	48°04'N	32°10'E
Boca del Pozo, Ven. (bô-kä-dĕl-pô'zō)	101b	11°00'N	64°21'W
Boca de Uchire, Ven. (bô-kä-dĕ-ōō-chē'rĕ)	101b	10°09'N	65°27'W
Bocaina, Serra da, mtn., Braz. (sĕ'r-rä-dȧ-bō-kä'ē-nä)	99a	22°47'S	44°39'W
Bocas, Mex. (bō'käs)	88	22°29'N	101°03'W
Bocas del Toro, Pan. (bō'käs dĕl tō'rō)	91	9°24'N	82°15'W
Bochnia, Pol. (bōĸ'nyä)	125	49°58'N	20°28'E
Bocholt, Ger. (bō'kōlt)	127c	51°50'N	6°37'E
Bochum, Ger.	124	51°29'N	7°13'E
Bockum-Hövel, Ger. (bō'kōm-hû'fĕl)	127c	51°41'N	7°45'E
Bodalang, D.R.C.	190	3°14'N	22°14'E
Bodaybo, Russia (bō-dī'bō)	135	57°12'N	114°46'E
Bodele, depr., Chad (bō-dȧ-lā')	185	16°45'N	17°05'E
Boden, Swe.	116	65°51'N	21°29'E
Bodensee, l., Eur. (bō'dĕn zā)	112	47°48'N	9°22'E
Bodmin, Eng., U.K. (bŏd'mĭn)	120	50°29'N	4°45'W
Bodmin Moor, Eng., U.K. (bŏd'mĭn mŏr)	120	50°36'N	4°43'W
Bodrum, Tur.	137	37°10'N	27°07'E
Boende, D.R.C. (bô-ĕn'dȧ)	186	0°13'S	20°52'E
Boerne, Tx., U.S. (bō'ērn)	80	29°49'N	98°44'W
Boesmans, r., S. Afr.	187c	33°29'S	26°09'E
Boeuf, r., U.S. (bĕf)	81	32°23'N	91°57'W
Boffa, Gui. (bôf'ȧ)	184	10°10'N	14°02'W
Bōfu, Japan (bō'fōō)	167	34°03'N	131°35'E
Bogalusa, La., U.S. (bō-gȧ-lōō'sȧ)	81	30°48'N	89°52'W
Bogan, r., Austl. (bō'gĕn)	176	32°10'S	147°40'E
Bogense, Den. (bō'gĕn-sĕ)	122	55°34'N	10°09'E
Boggy Peak, mtn., Antig. (bŏg'ĭ-pēk)	91b	17°03'N	61°50'W
Bogong, Mount, mtn., Austl.	176	36°50'S	147°15'E
Bogor, Indon.	168	6°45'S	106°45'E
Bogoroditsk, Russia (bō-gō'rō-dĭtsk)	132	53°48'N	38°06'E
Bogorodsk, Russia	136	56°02'N	43°40'E
Bogorodskoye, Russia (bô-gô-rôd'skô-yĕ)	142a	56°43'N	56°53'E
Bogotá, Col.	100	4°36'N	74°05'W
Bogotol, Russia (bŏ'gō-tŏl)	135	56°15'N	89°45'E
Boguchar, Russia (bō'gō-chär)	137	49°40'N	41°00'E
Bogue Chitto, Ms., U.S. (nôr'fēld)	82	31°26'N	90°25'W
Boguete, Pan. (bō-gĕ'tē)	91	8°54'N	82°29'W
Bo Hai, b., China	161	38°30'N	120°00'E
Bohai Haixia, strt., China (bwo-hī hī-shyä)	164	38°05'N	121°40'E
Bohain-en-Vermandois, Fr. (bō-ăɴ-ŏɴ-vâr-mäɴ-dwä')	126	49°58'N	3°22'E
Bohemia *see* Čechy, hist. reg., Czech Rep.	124	49°51'N	13°55'E
Bohemian Forest, mts., Eur. (bō-hē'mĭ-ăn)	112	49°35'N	12°27'E
Bohodukhiv, Ukr.	137	50°10'N	35°31'E
Bohol, i., Phil. (bō-hôl')	169	9°28'N	124°35'E
Bohom, Mex. (bō-ô'm)	89	16°47'N	92°42'W
Bohuslav, Ukr.	133	49°34'N	30°51'W
Boiestown, Can. (boiz'toun)	60	46°27'N	66°25'W
Bois Blanc, i., Mi., U.S. (boi' blăŋk)	66	45°45'N	84°30'W
Boischâtel, Can. (bwä-shä-tĕl')	62b	46°54'N	71°08'W
Bois-des-Filion, Can. (bōō-ä'dĕ-fē-yôɴ')	62a	45°40'N	73°46'W
Boise, Id., U.S. (boi'zē)	64	43°38'N	116°12'W
Boise, r., Id., U.S.	72	43°43'N	116°30'W
Boise City, Ok., U.S.	78	36°42'N	102°30'W
Boissevain, Can. (bois'vān)	50	49°14'N	100°03'W
Bojador, Cabo, c., W. Sah.	184	26°21'N	16°08'W
Bojnūrd, Iran	154	37°29'N	57°13'E
Bokani, Nig.	189	9°26'N	5°13'E
Boknafjorden, b., Nor.	116	59°12'N	5°37'E
Boksburg, S. Afr. (bŏĸs'bûrgh)	187b	26°13'N	28°15'E
Bokungu, D.R.C.	190	0°41'S	22°19'E
Bol, Chad	189	13°28'N	14°43'E
Bolai I, C.A.R.	189	4°20'N	17°21'E
Bolama, Gui.-B. (bô-lä'mä)	184	11°34'S	15°41'W
Bolan, mtn., Pak. (bō-län')	158	30°13'N	67°09'E
Bolaños, Mex. (bō-län'yōs)	88	21°40'N	103°48'W
Bolaños, r., Mex.	88	21°26'N	103°54'W
Bolan Pass, p., Pak.	155	29°50'N	67°10'E
Bolbec, Fr. (bōl-bĕk')	126	49°37'N	0°26'E
Bole, Ghana (bō'lā)	184	9°02'N	2°29'W
Bolesławiec, Pol. (bō-lē-slä'vyĕts)	124	51°15'N	15°35'E
Bolgatanga, Ghana	188	10°46'N	0°52'W
Bolhrad, Ukr.	137	45°41'N	28°38'E
Boli, China (bwo-lē)	161	45°40'N	130°38'E
Bolinao, Phil. (bō-lē-nä'ō)	169a	16°24'N	119°53'E
Bolívar, Arg. (bō-lē'vär)	102	36°15'S	61°05'W
Bolívar, Col.	100	1°46'N	76°58'W
Bolivar, Mo., U.S. (bŏl'ĭ-vȧr)	79	37°37'N	93°22'W
Bolivar, Tn., U.S.	82	35°14'N	88°56'W
Bolívar, Pico, mtn., Ven.	100	8°44'N	70°54'W
Bolivar Peninsula, pen., Tx., U.S. (bŏl'ĭ-vȧr)	81a	29°25'N	94°40'W
Bolivia, nation, S.A. (bō-lĭv'ĭ-ȧ)	100	17°00'S	64°00'W
Bolkhov, Russia (bŏl-ĸôf')	136	53°27'N	35°59'E
Bollin, r., Eng., U.K. (bŏl'ĭn)	114a	53°18'N	2°11'W
Bollington, Eng., U.K. (bôl'ĭng-tŭn)	114a	53°18'N	2°06'W
Bollnäs, Swe. (bôl'nĕs)	122	61°22'N	16°20'E
Bolmen, l., Swe. (bôl'mĕn)	122	56°58'N	13°25'E
Bolobo, D.R.C. (bō'lō-bō)	186	2°14'S	16°18'E
Bologna, Italy (bō-lōn'yä)	110	44°30'N	11°18'E
Bologoye, Russia (bō-lō-gô'yĕ)	136	57°52'N	34°02'E
Bolonchenticul, Mex. (bō-lôn-chĕn-tē-kōō'l)	90a	20°03'N	89°47'W
Bolondrón, Cuba (bô-lôn-drōn')	92	22°45'N	81°25'W
Bolseno, Lago di, l., Italy (lä'gō-dē-bôl-sā'nō)	130	42°35'N	11°40'E
Bol'shaya Anyuy, r., Russia	141	67°58'N	161°15'E
Bol'shaya Chuya, r., Russia	141	58°15'N	111°40'E
Bol'shaya Kinel', r., Russia	136	53°20'N	52°40'E
Bol'she Ust'ikinskoye, Russia (bôl'she ôs-tyĭ-kēn'skô-yĕ)	142a	55°58'N	58°18'E
Bol'shoy Begichëv, i., Russia	135	74°30'N	114°40'E
Bol'shoye Ivonino, Russia (ĭ-vô'nĭ-nô)	142a	59°41'N	61°12'E
Bol'shoy Kuyash, Russia (bôl'-shôy kōō'yȧsh)	142a	55°52'N	61°07'E
Bolsover, Eng., U.K. (bōl'zō-vēr)	114a	53°14'N	1°17'W
Boltaña, Spain (bôl-tä'nä)	129	42°28'N	0°03'E
Bolton, Can. (bōl'tŭn)	62d	43°53'N	79°44'W
Bolton, Eng., U.K.	120	53°35'N	2°26'W

ăt; finăl; rāte; senāte; ärm; åsk; sofȧ; fâre; ch-choose; dh-as th in other; bē; ēvent; bĕt; recĕnt; cratēr; g-gō; gh-guttural g; bĭt; ĭ-short neutral; rīde; ĸ-guttural k as ch in German ich;

PRONOUNCING INDEX
Bolton-upon-Dearne – Bratsk 229

PLACE (Pronunciation)	PAGE	LAT.	LONG.
Bolton-upon-Dearne, Eng., U.K. (bōl'tŭn-ŭp'ŏn-dûrn)	114a	53°31′N	1°19′W
Bolu, Tur. (bō'lò)	119	40°45′N	31°45′E
Bolva, r., Russia (bôl'vä)	132	53°30′N	34°30′E
Bolvadin, Tur. (bōl-vä-dēn')	119	38°50′N	30°50′E
Bolzano, Italy (bôl-tsä'nō)	118	46°31′N	11°22′E
Boma, D.R.C. (bō'mä)	186	5°51′S	13°03′E
Bombala, Austl. (bŭm-bä'lă)	175	36°55′S	149°07′E
Bombay *see* Mumbai, India	155	18°58′N	72°50′E
Bombay Harbour, b., India	159b	18°55′N	72°52′E
Bomi Hills, Lib.	184	7°00′N	11°00′W
Bom Jardim, Braz. (bôn zhär-dēn')	99a	22°10′S	42°25′W
Bom Jesus do Itabapoana, Braz.	99a	21°08′S	41°51′W
Bømlo, i., Nor. (bûmlô)	122	59°47′N	4°57′E
Bomongo, D.R.C.	185	1°22′N	18°21′E
Bom Sucesso, Braz. (bôn-sōō-sĕ'sō)	99a	21°02′S	44°44′W
Bomu *see* Mbomou, r., Afr.	185	4°50′N	24°00′E
Bon, Cap, c., Tun. (bôn)	118	37°04′N	11°13′E
Bonaire, i., Neth. Ant. (bō-nâr')	100	12°10′N	68°15′W
Bonavista, Can. (bō-nȧ-vĭs'tȧ)	53a	48°39′N	53°07′W
Bonavista Bay, b., Can.	53a	48°45′N	53°20′W
Bond, Co., U.S. (bŏnd)	78	39°53′N	106°40′W
Bondo, D.R.C. (bôn'dô)	140	3°49′N	23°40′E
Bondoc Peninsula, pen., Phil. (bôn-dŏk')	169a	13°24′N	122°30′E
Bondoukou, C. Iv. (bŏn-dōō'kōō)	184	8°02′N	2°48′W
Bonds Cay, i., Bah. (bŏnds kē)	92	25°30′N	77°45′W
Bondy, Fr.	127b	48°54′N	2°28′E
Bône *see* Annaba, Alg.	184	36°57′N	7°39′E
Bone, Teluk, b., Indon.	168	4°09′S	121°00′E
Bonete, Cerro, mtn., Arg. (bô'nētĕh çĕrrô)	102	27°50′S	68°35′W
Bonfim, Braz. (bôn-fē'n)	99a	20°20′S	44°15′W
Bongor, Chad	189	10°17′N	15°22′E
Bonham, Tx., U.S. (bŏn'ăm)	79	33°35′N	96°09′W
Bonhomme, Pic, mtn., Haiti	93	19°10′N	72°20′W
Bonifacio, Fr. (bō-nē-fä'chō)	130	41°23′N	9°10′E
Bonifacio, Strait of, strt., Eur.	118	41°14′N	9°02′E
Bonifay, Fl., U.S. (bŏn-ĭ-fā')	82	30°46′N	85°40′W
Bonin Islands, is., Japan (bō'nĭn)	195	26°30′N	141°00′E
Bonn, Ger. (bôn)	110	50°44′N	7°06′E
Bonne Bay, b., Can. (bŏn)	61	49°33′N	57°55′W
Bonners Ferry, Id., U.S. (bonērz fĕr'ĭ)	72	48°41′N	116°19′W
Bonner Springs, Ks., U.S. (bŏn'ēr sprĭngz)	75f	39°04′N	94°52′W
Bonne Terre, Mo., U.S. (bŏn târ')	79	37°55′N	90°32′W
Bonnet Peak, mtn., Can. (bŏn'ĭt)	55	51°26′N	115°53′W
Bonneville Dam, dam, U.S. (bŏn'ē-vĭl)	72	45°37′N	121°57′W
Bonny, Nig. (bŏn'ē)	184	4°29′N	7°13′E
Bonny Lake, Wa., U.S. (bŏn'ē lăk)	74a	47°11′N	122°11′W
Bonnyville, Can. (bŏnē-vĭl)	55	54°16′N	110°44′W
Bonorva, Italy (bō-nôr'vä)	130	40°26′N	8°46′E
Bonthain, Indon. (bŏn-tīn')	168	5°30′S	119°52′E
Bonthe, S.L.	184	7°32′N	12°30′W
Bontoc, Phil. (bŏn-tŏk')	169a	17°10′N	121°01′E
Booby Rocks, is., Bah. (bōō'bĭ rŏks)	92	23°55′N	77°00′W
Booker T. Washington National Monument, rec., Va., U.S. (bŏŏk'ēr tē wŏsh'ĭng-tŭn)	83	37°07′N	79°45′W
Boom, Bel.	115a	51°05′N	4°22′E
Boone, Ia., U.S. (bōōn)	71	42°04′N	93°51′W
Booneville, Ar., U.S. (bōōn'vĭl)	79	35°09′N	93°54′W
Booneville, Ky., U.S.	66	37°25′N	83°40′W
Booneville, Ms., U.S.	82	34°37′N	88°35′W
Boons, S. Afr.	192c	25°59′S	27°15′E
Boonton, N.J., U.S. (bōōn'tŭn)	68a	40°54′N	74°24′W
Boonville, In., U.S.	66	38°00′N	87°15′W
Boonville, Mo., U.S.	79	38°57′N	92°44′W
Boorama, Som.	192a	10°05′N	43°08′E
Boosaaso, Som.	192a	11°19′N	49°10′E
Boothbay Harbor, Me., U.S. (bōōth'bā här'bēr)	60	43°51′N	69°39′W
Boothia, Gulf of, b., Can. (bōō'thĭ-ȧ)	53	69°04′N	86°04′W
Boothia Peninsula, pen., Can.	49	73°30′N	95°00′W
Bootle, Eng., U.K. (bōōt'l)	114a	53°29′N	3°02′W
Bor, Sudan (bôr)	185	6°13′N	31°35′E
Bor, Tur. (bōr)	137	37°50′N	34°40′E
Boraha, Nosy, i., Madag.	187	16°58′S	50°15′E
Borah Peak, mtn., Id., U.S. (bō'rä)	73	44°12′N	113°47′W
Borås, Swe. (bō'rōs)	116	57°43′N	12°55′E
Borāzjān, Iran (bō-räz-jän')	154	29°13′N	51°13′E
Borba, Braz. (bôr'bä)	101	4°23′S	59°31′W
Borborema, Planalto da, plat., Braz. (plä-nàl'tô-dä-bôr-bō-rĕ'mä)	101	7°35′S	36°40′W
Bordeaux, Fr. (bôr-dō')	110	44°50′N	0°37′W
Bordentown, N.J., U.S. (bôr'dĕn-toun)	67	40°05′N	74°40′W
Bordj-bou-Arréridj, Alg. (bôrj-bōō-à-rā-rēj')	118	36°03′N	4°48′E
Bordj Omar Idriss, Alg.	184	28°06′N	6°34′E
Borgarnes, Ice.	116	64°31′N	21°40′W
Borger, Tx., U.S. (bôr'gēr)	78	35°40′N	101°23′W
Borgholm, Swe. (bôrg-hŏlm')	122	56°52′N	16°40′E
Borgne, l., La., U.S. (bôrn'y')	81	30°03′N	89°36′W
Borgomanero, Italy (bôr'gō-mä-nâ'rō)	130	45°40′N	8°28′E
Borgo Val di Taro, Italy (bô'r-zhō-väl-dē-tá'rō)	130	44°29′N	9°44′E
Börili, Kaz.	142a	53°36′N	61°55′E
Boring, Or., U.S. (bōring)	74c	45°26′N	122°22′W
Borisoglebsk, Russia (bŏ-rē sŏ-glyĕpsk')	134	51°20′N	42°00′E
Borisovka, Russia (bŏ-rē-sôf'ká)	137	50°38′N	36°00′E
Borivli, India	159b	19°15′N	72°48′E
Borja, Spain (bôr'hä)	128	41°50′N	1°33′W
Borken, Ger. (bōr'kĕn)	127c	51°50′N	6°51′E
Borkou, reg., Chad (bôr-kōō')	185	18°11′N	18°28′E
Borkum, i., Ger. (bōr'kōōm)	124	53°31′N	6°50′E
Borlänge, Swe. (bŏr-lĕŋ'gĕ)	122	60°30′N	15°24′E
Borneo, i., Asia	168	0°25′N	112°39′E
Bornholm, i., Den. (bôrn-hôlm)	112	55°16′N	15°15′E
Boromlia, Ukr.	133	50°36′N	34°58′E
Boromo, Burkina	188	11°45′N	2°56′W
Borovan, Blg. (bŏ-rō-vän')	131	43°24′N	23°47′E
Borovichi, Russia (bŏ-rō-vē'chē)	134	58°22′N	33°56′E
Borovsk, Russia (bô'rŏvsk)	132	55°13′N	36°26′E
Borraan, Som.	192a	10°38′N	48°30′E
Borracha, Isla la, i., Ven. (ē's-lä-lä-bôr-rà'chä)	101b	10°18′N	64°44′W
Borriana, Spain	118	39°53′N	0°05′W
Borroloola, Austl. (bŏr-rô-lōō'lȧ)	174	16°15′S	136°19′E
Borshchiv, Ukr.	125	48°47′N	26°04′E
Bort-les-Orgues, Fr. (bôr-lā-zôrg)	126	45°26′N	2°26′E
Borūjerd, Iran	154	33°45′N	48°53′E
Boryslav, Ukr.	125	49°17′N	23°24′E
Boryspil', Ukr.	133	50°17′N	30°54′E
Borzna, Ukr. (bôrz'ná)	137	51°15′N	32°26′E
Borzya, Russia (bôrz'yȧ)	135	50°37′N	116°53′E
Bosa, Italy (bō'sä)	130	40°18′N	8°34′E
Bosanska Dubica, Bos. (bō'sän-skä dōō'bĭt-sä)	130	45°10′N	16°49′E
Bosanska Gradiška, Bos. (bō'sän-skä grä-dĭsh'kä)	131	45°08′N	17°15′E
Bosanski Novi, Bos. (bō's sän-skĭ nō'vē)	130	45°00′N	16°22′E
Bosanski Petrovac, Bos. (bō'sän-skĭ pĕt'rō-väts)	130	44°33′N	16°23′E
Bosanski Šamac, Bos. (bō'sän-skĭ shä'màts)	131	45°03′N	18°30′E
Boscobel, Wi., U.S. (bŏs'kō-bĕl)	71	43°08′N	90°44′W
Bose, China (bwo-sŭ)	165	24°00′N	106°38′E
Boshan, China (bwo-shan)	161	36°32′N	117°51′E
Boskoop, Neth.	115a	52°04′N	4°39′E
Boskovice, Czech Rep. (bŏs'kō-vē-tsĕ)	124	49°26′N	16°37′E
Bosna, r., Serb.	131	44°19′N	17°54′E
Bosnia and Herzegovina, nation, Eur.	131	44°15′N	17°30′E
Bosobolo, D.R.C.	190	4°11′N	19°54′E
Bosporus *see* İstanbul Boğazı, strt., Tur.	154	41°10′N	29°10′E
Bossangoa, C.A.R.	189	6°29′N	17°27′E
Bossier City, La., U.S. (bōsh'ēr)	81	32°31′N	93°42′W
Bosten Hu, l., China (bwo-stŭn hōō)	160	42°06′N	88°01′E
Boston, Ga., U.S. (bôs'tŭn)	82	30°47′N	83°47′W
Boston, Ma., U.S.	65	42°15′N	71°07′W
Boston Heights, Oh., U.S.	69d	41°15′N	81°30′W
Boston Mountains, mts., Ar., U.S.	65	35°46′N	93°32′W
Botany Bay, b., Austl. (bŏt'ȧ-nĭ)	175	33°58′S	151°11′E
Botevgrad, Blg.	131	42°54′N	23°41′E
Bothaville, S. Afr. (bō'tä-vĭl)	192c	27°24′S	26°38′E
Bothell, Wa., U.S. (bŏth'ĕl)	74a	47°46′N	122°12′W
Bothnia, Gulf of, b., Eur. (bŏth'nĭ-á)	112	63°40′N	21°30′E
Botoşani, Rom. (bŏ-tô-shàn'ĭ)	125	47°46′N	26°40′E
Botswana, nation, Afr. (bŏtswănă)	186	22°10′S	23°13′E
Bottineau, N.D., U.S. (bŏt-ĭ-nō')	70	48°48′N	100°28′W
Bottrop, Ger. (bŏt'trōp)	124	51°31′N	6°56′E
Botwood, Can. (bŏt'wŏŏd)	53a	49°08′N	55°21′W
Bouafle, C. Iv. (bō-á-flā')	184	6°59′N	5°45′W
Bouar, C.A.R. (bōō-är)	185	5°57′N	15°36′E
Bou Areg, Sebkha, Mor.	128	35°09′N	3°02′W
Boubandjidah, Parc National de, rec., Cam.	189	8°20′N	14°40′E
Boucherville, Can. (bōō-shä-vēl')	62a	45°37′N	73°27′W
Boudenib, Mor. (bōō-dĕ-nēb')	184	32°14′N	3°04′W
Boudette, Mn., U.S. (bōō-dĕt)	71	48°42′N	94°34′W
Boudouaou, Alg.	129	36°44′N	3°25′E
Boufarik, Alg. (bōō-fä-rēk')	129	36°35′N	2°55′E
Bougainville, i., Pap. N. Gui.	170e	6°00′S	155°00′E
Bougainville Trench, deep, (bōō-găn-vēl')	195	7°00′S	152°00′E
Bougie *see* Bejaïa, Alg.	184	36°46′N	5°00′E
Bougouni, Mali (bōō-gōō-nē')	184	11°27′N	7°30′W
Bouïra, Alg. (boo-ē'rà)	118	36°25′N	3°55′E
Bouïra-Sahary, Alg. (bwē-rà sà'ä-rē)	129	35°16′N	3°23′E
Bouka, r., Gui.	188	11°05′N	10°40′W
Boulder, Co., U.S.	64	40°02′N	105°19′W
Boulder, r., Mt., U.S.	73	46°10′N	112°07′W
Boulder City, Nv., U.S.	64	35°57′N	114°50′W
Boulder Peak, mtn., Id., U.S.	73	43°53′N	114°33′W
Boulogne-Billancourt, Fr. (bōō-lôn'y'-bē-yän-kōōr')	126	48°50′N	2°14′E
Boulogne-sur-Mer, Fr. (bōō-lôn'y-sür-mâr')	117	50°44′N	1°37′E
Boumba, r., Cam.	189	3°20′N	14°40′E
Bouna, C. Iv. (bōō-nä')	184	9°16′N	3°00′W
Bouna, Parc National de, rec., C. Iv.	188	9°20′N	3°35′W
Boundary Bay, b., N.A. (boun'dȧ-rĭ)	74d	49°03′N	122°59′W
Boundary Peak, mtn., Nv., U.S.	76	37°52′N	118°20′W
Bound Brook, N.J., U.S. (bound brŏk)	68a	40°34′N	74°32′W
Bountiful, Ut., U.S. (boun'tĭ-fŏŏl)	75b	40°55′N	111°53′W
Bountiful Peak, mtn., Ut., U.S. (boun'tĭ-fŏŏl)	75b	40°58′N	111°49′W
Bounty Islands, is., N.Z.	5	47°42′S	179°05′E
Bourail, N. Cal.	170f	21°34′S	165°30′E
Bourem, Mali (bōō-rĕm')	184	16°43′N	0°15′W
Bourg-en-Bresse, Fr. (bōōr-gĕn-brĕs')	117	46°12′N	5°13′E
Bourges, Fr. (bōōrzh)	117	47°06′N	2°22′E
Bourget, Can. (bōōr-zhĕ')	62c	45°26′N	75°09′W
Bourgoin, Fr. (bōōr-gwăn')	127	45°46′N	5°17′E
Bourke, Austl. (bŭrk)	175	30°10′S	146°00′E
Bourne, Eng., U.K. (bōrn)	114a	52°46′N	0°22′W
Bournemouth, Eng., U.K. (bôrn'mŭth)	120	50°44′N	1°55′W
Bou Saâda, Alg. (bōō-sä'dä)	118	35°13′N	4°17′E
Bousso, Chad (bōō-sô')	185	10°33′N	16°45′E
Boutilimit, Maur.	184	17°30′N	14°54′W
Bouvetøya, i., Ant.	3	55°00′S	3°00′E
Bow, r., Can. (bō)	52	50°35′N	112°15′W
Bowbells, N.D., U.S. (bō'bĕls)	70	48°50′N	102°16′W
Bowdle, S.D., U.S. (bōd''l)	70	45°28′N	99°42′W
Bowen, Austl. (bō'ĕn)	175	20°02′S	148°14′E
Bowie, Md., U.S. (bōō'ĭ) (bō'ē)	68e	38°59′N	76°47′W
Bowie, Tx., U.S.	79	33°34′N	97°50′W
Bowling Green, Ky., U.S. (bōlĭng grēn)	65	37°00′N	86°26′W
Bowling Green, Mo., U.S.	79	39°19′N	91°09′W
Bowling Green, Oh., U.S.	66	41°25′N	83°40′W
Bowman, N.D., U.S. (bō'măn)	70	46°11′N	103°23′W
Bowron, r., Can. (bō'rŭn)	55	53°20′N	121°10′W
Boxelder Creek, r., Mt., U.S. (bŏks'ĕl-dēr)	70	45°35′N	104°28′W
Box Elder Creek, r., Mt., U.S.	73	47°17′N	108°37′W
Box Hill, Austl.	173a	37°49′S	145°08′E
Boxian, China (bwo shyĕn)	164	33°52′N	115°47′E
Boxing, China (bwo-shyĭŋ)	162	37°09′N	118°08′E
Boxtel, Neth.	115a	51°40′N	5°21′E
Boyabo, D.R.C.	190	3°43′N	18°46′E
Boyang, China (bwo-yäŋ)	165	29°00′N	116°42′E
Boyer, r., Can. (boi'ēr)	62b	46°45′N	70°56′W
Boyer, r., Ia., U.S.	70	41°45′N	95°36′W
Boyle, Ire. (boil)	120	53°59′N	8°15′W
Boyne, r., Ire. (boin)	120	53°40′N	6°40′W
Boyne City, Mi., U.S.	66	45°15′N	85°05′W
Boyoma Falls, wtfl., D.R.C.	185	0°30′N	25°12′E
Boysen Reservoir, res., Wy., U.S.	73	43°19′N	108°11′W
Bozcaada, Tur. (bōz-cä'dä)	131	39°50′N	26°05′E
Bozca Ada, i., Tur.	131	39°50′N	26°00′E
Bozeman, Mt., U.S. (bōz'măn)	64	45°41′N	111°00′W
Bozene, D.R.C.	190	2°56′N	19°12′E
Bozhen, China (bwo-jŭn)	162	38°05′N	116°35′E
Bozoum, C.A.R.	189	6°19′N	16°23′E
Bra, Italy (brä)	130	44°41′N	7°52′E
Bracciano, Lago di, l., Italy (lä'gō-dē-brä-chä'nō)	130	42°05′N	12°00′E
Bracebridge, Can. (brās'brĭj)	59	45°05′N	79°20′W
Braceville, Il., U.S. (brās'vĭl)	69a	41°13′N	88°16′W
Bräcke, Swe. (brĕk'kĕ)	116	62°44′N	15°28′E
Brackenridge, Pa., U.S. (brăk'ĕn-rĭj)	69e	40°37′N	79°44′W
Brackettville, Tx., U.S. (brăk'ĕt-vĭl)	80	29°19′N	100°24′W
Braço Maior, mth., Braz.	101	11°00′S	51°00′W
Braço Menor, mth., Braz. (brä'zō-mē-nō'r)	101	11°38′S	50°00′W
Bradano, r., Italy (brä-dä'nō)	130	40°43′N	16°22′E
Bradenton, Fl., U.S. (brā'dĕn-tŭn)	83a	27°28′N	82°35′W
Bradfield, Eng., U.K. (brăd'fēld)	114b	51°25′N	1°08′W
Bradford, Eng., U.K. (brăd'fērd)	116	53°47′N	1°44′W
Bradford, Oh., U.S.	66	40°10′N	84°30′W
Bradford, Pa., U.S.	67	42°00′N	78°40′W
Bradley, Il., U.S. (brăd'lĭ)	69a	41°09′N	87°52′W
Bradner, Can. (brăd'nēr)	74d	49°05′N	122°26′W
Brady, Tx., U.S. (brā'dĭ)	80	31°09′N	99°21′W
Braga, Port. (brä'gä)	118	41°20′N	8°25′W
Bragado, Arg. (brä-gä'dō)	102	35°07′S	60°28′W
Bragança, Braz. (brä-gän'sä)	101	1°02′S	46°50′W
Bragança, Port.	128	41°48′N	6°46′W
Bragança Paulista, Braz. (brä-gän'sä-pä'ōō-lē's-tà)	102	22°58′S	46°31′W
Bragg Creek, Can. (brăg)	62e	50°57′N	114°35′W
Brahmaputra, r., Asia (brä'mà-pōō'trà)	155	26°45′N	92°45′E
Brāhui, mts., Pak.	155	28°32′N	66°15′E
Braidwood, Il., U.S. (brād'wŏŏd)	69a	41°16′N	88°13′W
Brăila, Rom. (brē'ē-là)	110	45°15′N	27°58′E
Brainerd, Mn., U.S. (brān'ērd)	71	46°20′N	94°09′W
Braintree, Ma., U.S. (brān'trē)	61a	42°14′N	71°00′W
Braithwaite, La., U.S. (brīth'wīt)	68d	29°52′N	89°57′W
Brakpan, S. Afr. (brăk'păn)	187b	26°15′S	28°22′E
Bralorne, Can. (brā'lôrn)	55	50°47′N	122°49′W
Bramalea, Can.	62d	43°48′N	79°41′W
Brampton, Can. (brămp'tŭn)	59	43°41′N	79°46′W
Branca, Pedra, mtn., Braz. (pĕ'drä-brä'n-kä)	102b	22°55′S	43°28′W
Branchville, N.J., U.S. (brȧnch'vĭl)	68a	41°09′N	74°44′W
Branchville, S.C., U.S.	83	33°17′N	80°48′W
Branco, r., Braz. (brăŋ'kō)	101	2°21′N	60°38′W
Brandberg, mtn., Nmb.	186	21°15′S	14°15′E
Brandenburg, Ger. (brän'dĕn-bŏŏrgh)	117	52°25′N	12°33′E
Brandenburg, state, Ger.	115b	52°15′N	13°00′E
Brandenburg, hist. reg., Ger.	124	52°12′N	13°31′E
Brandfort, S. Afr. (brăn'd-fôrt)	192c	28°42′S	26°29′E
Brandon, Can. (brăn'dŭn)	50	49°50′N	99°57′W
Brandon, Vt., U.S.	67	43°45′N	73°05′W
Brandon Mountain, mtn., Ire. (brăn-dŏn)	120	52°15′N	10°12′W
Brandywine, Md., U.S. (brăndĭ'wīn)	68e	38°42′N	76°51′W
Branford, Ct., U.S. (brăn'fērd)	67	41°15′N	72°50′W
Braniewo, Pol. (brä-nyĕ'vô)	125	54°23′N	19°50′E
Brańsk, Pol. (brän' sk)	125	52°44′N	22°51′E
Branson, Mo., U.S.	79	36°39′N	93°13′W
Brantford, Can. (brănt'fērd)	59	43°09′N	80°17′W
Bras d'Or Lake, l., Can. (brä-dôr')	61	45°52′N	60°50′W
Brasília, Braz. (brä-sē'lvä)	101	15°49′S	47°39′W
Brasília Legal, Braz.	101	3°45′S	55°46′W
Brasópolis, Braz. (brä-sô'pō-lēs)	99a	22°30′S	45°36′W
Braşov, Rom.	119	45°39′N	25°35′E
Brass, Nig. (brăs)	184	4°28′N	6°28′E
Brasschaat, Bel. (bräs'kät)	115a	51°19′N	4°30′E
Bratenahl, Oh., U.S. (bră'tĕn-ôl)	69d	41°34′N	81°36′W
Bratislava, Slvk. (brä'tĭs-lä-vä)	110	48°09′N	17°07′E
Bratsk, Russia (brätsk)	135	56°10′N	102°04′E

ng-sing; ŋ-baŋk; N-nasalized n; nŏd; cŏmmit; ōld; ôbey; ôrder; oi-boil; fōōd; ŏŏ-as oo in foot; ou-out; s-soft; sh-dish; th-thin; pūre; ûnite; ûrn; stŭd; circŭs; ü-as in French tu; '-indeterminate vowel.

230 Bratskoye Vodokhranilishche – Buenos Aires PRONOUNCING INDEX

PLACE (Pronunciation)	PAGE	LAT.	LONG.
Bratskoye Vodokhranilishche, res., Russia	135	56°10'N	102°05'E
Bratslav, Ukr. (brät'slåf)	133	48°48'N	28°59'E
Brattleboro, Vt., U.S. (brăt''l-bŭr-ô)	67	42°50'N	72°35'W
Braunau, Aus. (brou'nou)	124	48°15'N	13°05'E
Braunschweig, Ger. (broun'shvīgh)	117	52°16'N	10°32'E
Bråviken, r., Swe.	122	58°40'N	16°40'E
Brawley, Ca., U.S. (brô'lĭ)	64	32°59'N	115°32'W
Bray, Ire. (brā)	120	53°10'N	6°05'W
Braymer, Mo., U.S. (brā'mẽr)	79	39°34'N	93°47'W
Brays Bay, Tx., U.S. (brās'bī'yōō)	81a	29°41'N	95°33'W
Brazeau, r., Can.	55	52°55'N	116°10'W
Brazeau, Mount, mtn., Can. (brä-zō')	55	52°33'N	117°21'W
Brazil, In., U.S. (brá-zĭl')	66	39°30'N	87°00'W
Brazil, nation, S.A.	101	9°00'S	53°00'W
Brazilian Highlands, mts., Braz. (brá zĭl yăn hī-lándz)	97	14°00'S	48°00'W
Brazos, r., Tx., U.S. (brä'zōs)	64	33°10'N	98°50'W
Brazos, Clear Fork, r., Tx., U.S.	80	32°56'N	99°14'W
Brazos, Double Mountain Fork, r., Tx., U.S.	78	33°23'N	101°21'W
Brazos, Salt Fork, r., Tx., U.S. (sôlt fôrk)	78	33°20'N	101°57'W
Brazzaville, Congo (brá-zá-vēl')	186	4°16'S	15°17'E
Brčko, Bos. (běrch'kô)	131	44°54'N	18°46'E
Brda, r., Pol. (bẽr-dä)	125	53°18'N	17°55'E
Brea, Ca., U.S. (brē'á)	75a	33°55'N	117°54'W
Breakeyville, Can.	62b	46°40'N	71°13'W
Breckenridge, Mn., U.S. (brĕk'ĕn-rĭj)	70	46°17'N	96°35'W
Breckenridge, Tx., U.S.	80	32°46'N	98°53'W
Brecksville, Oh., U.S. (brĕks'vĭl)	69d	41°19'N	81°38'W
Břeclav, Czech Rep. (brzhĕl'láf)	124	48°46'N	16°54'E
Breda, Neth. (brā-dä')	121	51°35'N	4°47'E
Bredasdorp, S. Afr. (brā'das-dôrp)	186	34°15'S	20°00'E
Bredy, Russia (brĕ'dĭ)	142a	52°25'N	60°23'E
Bregenz, Aus. (brā'gĕnts)	124	47°30'N	9°46'E
Bregovo, Blg. (brĕ'gô-vô)	131	44°07'N	22°45'E
Breidafjördur, b., Ice.	116	65°15'N	22°50'W
Breidbach, S. Afr. (brēd'băk)	187c	32°54'S	27°26'E
Breil-sur-Roya, Fr. (brĕ'y')	127	43°57'N	7°36'E
Brejo, Braz. (brå'zhô)	101	3°33'S	42°46'W
Bremangerlandet, i., Nor.	122	61°51'N	4°25'E
Bremen, Ger. (brā-mĕn)	110	53°05'N	8°50'E
Bremen, In., U.S. (brē'mĕn)	66	41°25'N	86°05'W
Bremerhaven, Ger. (brām-ẽr-hä'fĕn)	116	53°33'N	8°38'E
Bremerton, Wa., U.S. (brĕm'ẽr-tŭn)	72	47°34'N	122°38'W
Bremervörde, Ger. (brĕ'mẽr-fŭr-dĕ)	115c	53°29'N	9°09'E
Bremner, Can. (brĕm'nẽr)	62g	53°34'N	113°14'W
Bremond, Tx., U.S. (brĕm'ŭnd)	81	31°11'N	96°40'W
Brenham, Tx., U.S. (brĕn'ăm)	81	30°10'N	96°24'W
Brenner Pass, p., Eur. (brĕn'ẽr)	117	47°00'N	11°30'E
Brentwood, Eng., U.K. (brĕnt'wŏd)	121	51°37'N	0°18'E
Brentwood, Md., U.S.	67	39°00'N	76°55'W
Brentwood, Mo., U.S.	75e	38°37'N	90°21'W
Brentwood, Pa., U.S.	69e	40°22'N	79°59'W
Brescia, Italy (brā'shä)	118	45°33'N	10°15'E
Bressanone, Italy (brĕs-sä-nō'nā)	130	46°42'N	11°40'E
Bressuire, Fr. (grĕ-swēr')	126	46°49'N	0°14'W
Brest, Bela.	134	52°06'N	23°43'E
Brest, Fr. (brĕst)	110	48°24'N	4°30'W
Brest, prov., Bela.	132	52°30'N	26°50'E
Bretagne, hist. reg., Fr. (brĕ-tän'yĕ)	126	48°00'N	3°00'W
Breton, Pertuis, strt., Fr. (pâr-twĕ'brĕ-tôn')	126	46°18'N	1°43'W
Breton Sound, strt., La., U.S. (brĕt'ŭn)	82	29°38'N	89°15'W
Breukelen, Neth.	115a	52°09'N	5°00'E
Brevard, N.C., U.S. (brĕ-värd')	83	35°14'N	82°45'W
Breves, Braz. (brā'vĕzh)	101	1°32'S	50°13'W
Brevik, Nor. (brĕ'vĕk)	122	59°04'N	9°39'E
Brewarrina, Austl. (brōō-ẽr-rē'ná)	175	29°54'S	146°50'E
Brewer, Me., U.S. (brōō'ẽr)	60	44°46'N	68°46'W
Brewerville, Lib.	188	6°26'N	10°47'W
Brewster, N.Y., U.S. (brōō'stẽr)	68a	41°23'N	73°38'W
Brewster, Cerro, mtn., Pan. (sĕ'r-rô-brōō'stẽr)	91	9°19'N	79°15'W
Brewton, Al., U.S. (brōō'tŭn)	82	31°06'N	87°04'W
Brežice, Slvn. (brĕ'zhē-tsĕ)	130	45°55'N	15°37'E
Breznik, Blg. (brĕs'nêk)	131	42°44'N	22°55'E
Briancon, Fr. (brē-än-sôn')	127	44°54'N	6°39'E
Briare, Fr. (brē-är')	126	47°40'N	2°46'E
Bridal Veil, Or., U.S. (brīd'ál vāl)	74c	45°33'N	122°10'W
Bridge Point, c., Bah. (brĭj)	92	25°35'N	76°40'W
Bridgeport, Al., U.S. (brĭj'pôrt)	82	34°55'N	85°42'W
Bridgeport, Ct., U.S.	65	41°12'N	73°12'W
Bridgeport, Il., U.S.	66	38°40'N	87°45'W
Bridgeport, Ne., U.S.	70	41°40'N	103°06'W
Bridgeport, Oh., U.S.	66	40°00'N	80°45'W
Bridgeport, Pa., U.S.	68f	40°06'N	75°21'W
Bridgeport, Tx., U.S.	79	33°13'N	97°46'W
Bridgeton, Al., U.S. (brĭj'tŭn)	68h	33°27'N	86°39'W
Bridgeton, Mo., U.S.	75e	38°45'N	90°23'W
Bridgeton, N.J., U.S.	67	39°30'N	75°15'W
Bridgetown, Barb. (brĭj' toun)	87	13°08'N	59°37'W
Bridgetown, Can.	60	44°51'N	65°18'W
Bridgeville, Pa., U.S. (brĭj'vĭl)	69e	40°22'N	80°07'W
Bridgewater, Austl. (brĭj'wô-tẽr)	176	42°50'S	147°28'E
Bridgewater, Can.	51	44°23'N	64°31'W
Bridgnorth, Eng., U.K. (brĭj'nôrth)	114a	52°32'N	2°25'W
Bridgton, Me., U.S. (brĭj'tŭn)	60	44°04'N	70°45'W
Bridlington, Eng., U.K. (brĭd'lĭng-tŭn)	120	54°06'N	0°10'W
Brie-Comte-Robert, Fr. (brē-kônt-ĕ-rô-bâr')	127b	48°42'N	2°37'E
Brielle, Neth.	115a	51°54'N	4°08'E
Brierfield, Eng., U.K. (brī'ẽr fēld)	114a	53°49'N	2°14'W
Brierfield, Al., U.S. (brī'ẽr-fēld)	82	33°01'N	86°55'W
Brier Island, i., Can. (brī'ẽr)	60	44°16'N	66°24'W
Brieselang, Ger. (brē'zĕ-läng)	115b	52°36'N	12°59'E
Briey, Fr. (brē-ĕ')	127	49°15'N	5°57'E
Brig, Switz. (brēg)	117	46°17'N	7°59'E
Brigg, Eng., U.K. (brĭg)	114a	53°33'N	0°29'W
Brigham City, Ut., U.S. (brĭg'ăm)	75b	41°31'N	112°01'W
Brighouse, Eng., U.K. (brĭg'hous)	114a	53°42'N	1°47'W
Bright, Austl. (brīt)	176	36°43'S	147°00'E
Bright, In., U.S. (brīt)	69f	39°13'N	84°51'W
Brightlingsea, Eng., U.K. (brī't-lĭng-sē)	114b	51°50'N	1°00'E
Brighton, Austl.	173a	37°55'S	145°00'E
Brighton, Eng., U.K.	117	50°47'N	0°07'W
Brighton, Al., U.S. (brī'tŭn)	68h	33°27'N	86°56'W
Brighton, Co., U.S.	78	39°58'N	104°49'W
Brighton, Ia., U.S.	71	41°11'N	91°47'W
Brighton, Il., U.S.	75e	39°03'N	90°08'W
Brighton Indian Reservation, I.R., Fl., U.S.	83a	27°05'N	81°25'W
Brihuega, Spain (brê-wä'gä)	128	40°32'N	2°52'W
Brimley, Mi., U.S. (brĭm'lē)	75k	46°24'N	84°34'W
Brindisi, Italy (brēn'dē-zē)	110	40°38'N	17°57'E
Brinje, Cro. (brēn'yĕ)	130	45°00'N	15°08'E
Brinkley, Ar., U.S. (brĭŋk'lĭ)	79	34°52'N	91°12'W
Brinnon, Wa., U.S. (brĭn'ŭn)	74a	47°41'N	122°54'W
Brion, i., Can. (brē-ôn')	61	47°47'N	61°29'W
Brioude, Fr. (brē-ōōd')	126	45°18'N	3°22'E
Brisbane, Austl. (brĭz'bån)	176	27°30'S	153°10'E
Bristol, Eng., U.K.	117	51°29'N	2°39'W
Bristol, Ct., U.S. (brĭs'tŭl)	67	41°40'N	72°55'W
Bristol, Pa., U.S.	68f	40°06'N	74°51'W
Bristol, R.I., U.S.	68b	41°41'N	71°14'W
Bristol, Tn., U.S.	65	36°35'N	82°10'W
Bristol, Va., U.S.	65	36°36'N	82°00'W
Bristol, Vt., U.S.	67	44°10'N	73°00'W
Bristol, Wi., U.S.	69a	42°32'N	88°04'W
Bristol Bay, b., Ak., U.S.	63	58°05'N	158°54'W
Bristol Channel, strt., Eng., U.K.	117	51°20'N	3°47'W
Bristow, Ok., U.S. (brĭs'tō)	79	35°50'N	96°25'W
British Columbia, prov., Can. (brĭt'ĭsh kŏl'ŭm-bĭ-á)	50	56°00'N	124°53'W
British Indian Ocean Territory, dep., Afr.	2	7°00'S	72°00'E
British Isles, is., Eur.	112	54°00'N	4°00'W
Brits, S. Afr.	192c	25°39'S	27°47'E
Britstown, S. Afr. (brĭts'toun)	186	30°30'S	23°40'E
Britt, Ia., U.S. (brĭt)	71	43°05'N	93°47'W
Brittany *see* Bretagne, hist. reg., Fr.	126	48°00'N	3°00'W
Britton, S.D., U.S. (brĭt'ŭn)	70	45°47'N	97°44'W
Brive-la-Gaillarde, Fr. (brēv-lä-gī-yärd'ĕ)	117	45°10'N	1°31'E
Briviesca, Spain (brē-vyās'ká)	128	42°34'N	3°21'W
Brno, Czech Rep. (b'r'nô)	110	49°18'N	16°37'E
Broa, Ensenada de la, b., Cuba	92	22°30'N	82°00'W
Broach, India	158	21°47'N	72°58'E
Broad, r., Ga., U.S. (brôd)	82	34°15'N	83°14'W
Broad, r., N.C., U.S.	83	35°38'N	82°40'W
Broadmeadows, Austl. (brôd'mĕd-ōz)	173a	37°40'S	144°53'E
Broadview Heights, Oh., U.S. (brôd'vū)	69d	41°18'N	81°41'W
Brockport, N.Y., U.S. (brŏk'pôrt)	67	43°15'N	77°55'W
Brockton, Ma., U.S. (brŏk'tŭn)	61a	42°04'N	71°01'W
Brockville, Can. (brŏk'vĭl)	51	44°35'N	75°40'W
Brockway, Mt., U.S. (brŏk'wā)	73	47°24'N	105°41'W
Brodnica, Pol. (brôd'nĭt-sá)	125	53°16'N	19°26'E
Brody, Ukr. (brô'dĭ)	137	50°05'N	25°10'E
Broken Arrow, Ok., U.S. (brō'kĕn ăr'ō)	79	36°03'N	95°48'W
Broken Bay, b., Austl.	176	33°34'S	151°20'E
Broken Bow, Ne., U.S. (brō'kĕn bō)	70	41°24'N	99°37'W
Broken Bow, Ok., U.S.	79	34°02'N	94°43'W
Broken Hill, Austl. (brōk'ĕn)	175	31°55'S	141°35'E
Broken Hill *see* Kabwe, Zam.	186	14°27'S	28°27'E
Bromley, Eng., U.K. (brŭm'lĭ)	114b	51°23'N	0°01'E
Bromptonville, Can. (brŭmp'tŭn-vĭl)	59	45°30'N	72°00'W
Brønderslev, Den. (brûn'dẽr-slĕv)	122	57°15'N	9°56'E
Bronkhorstspruit, S. Afr.	192c	25°50'S	28°48'E
Bronnitsy, Russia (brô-nyĭ'tsĭ)	132	55°26'N	38°16'E
Bronson, Mi., U.S. (brŏn'sŭn)	66	41°55'N	85°15'W
Bronte Creek, r., Can.	62d	43°25'N	79°53'W
Brood, r., S.C., U.S. (brōōd)	83	34°46'N	81°25'W
Brookfield, Il., U.S. (brŏok'fēld)	69a	41°49'N	87°51'W
Brookfield, Mo., U.S.	79	39°45'N	93°04'W
Brookhaven, Ga., U.S. (brŏok'hāv'n)	68c	33°52'N	84°21'W
Brookhaven, Ms., U.S.	82	31°35'N	90°26'W
Brookings, Or., U.S. (brŏok'ĭngs)	72	42°04'N	124°16'W
Brookings, S.D., U.S.	70	44°18'N	96°47'W
Brookline, Ma., U.S. (brŏok'lĭn)	61a	42°20'N	71°08'W
Brookline, N.H., U.S.	61a	42°44'N	71°37'W
Brooklyn, Oh., U.S. (brŏok'lĭn)	69d	41°26'N	81°44'W
Brooklyn Center, Mn., U.S.	75g	45°05'N	93°21'W
Brook Park, Oh., U.S. (brŏok)	69d	41°24'N	81°50'W
Brooks, Can.	55	50°35'N	111°53'W
Brooks Range, mts., Ak., U.S. (brŏoks)	64a	68°20'N	159°00'W
Brooksville, Fl., U.S. (brŏoks'vĭl)	83a	28°32'N	82°28'W
Brookville, In., U.S. (brŏok'vĭl)	66	39°20'N	85°00'W
Brookville, Pa., U.S.	67	41°10'N	79°00'W
Brookwood, Al., U.S. (brŏok'wŏod)	82	33°15'N	87°17'W
Broome, Austl. (brōōm)	174	18°00'S	122°15'E
Brossard, Can.	62a	45°26'N	73°28'W
Brothers, is., Bah. (brŭd'hẽrs)	92	26°05'N	79°00'W
Broumov, Czech Rep. (brōō'môf)	124	50°33'N	15°55'E
Brown Bank, bk.,	93	21°30'N	74°35'W
Brownfield, Tx., U.S. (broun'fēld)	78	33°11'N	102°16'W
Browning, Mt., U.S. (broun'ĭng)	73	48°37'N	113°05'W
Brownsboro, Ky., U.S. (brounz'bô-rô)	69h	38°22'N	85°30'W
Brownsburg, Can. (brouns'bûrg)	62a	45°40'N	74°24'W
Brownsburg, In., U.S.	69g	39°51'N	86°23'W
Brownsmead, Or., U.S. (brounz'-mēd)	74c	46°13'N	123°33'W
Brownstown, In., U.S. (brounz'toun)	66	38°50'N	86°00'W
Brownsville, Pa., U.S. (brounz'vĭl)	69e	40°01'N	79°53'W
Brownsville, Tn., U.S.	82	35°35'N	89°15'W
Brownsville, Tx., U.S.	64	25°55'N	97°30'W
Brownville Junction, Me., U.S. (broun'vĭl)	60	45°20'N	69°04'W
Brownwood, Tx., U.S. (broun'wŏod)	64	31°44'N	98°58'W
Brownwood, l., Tx., U.S.	80	31°55'N	99°15'W
Brozas, Spain (brō'thäs)	128	39°37'N	6°44'W
Bruce, Mount, mtn., Austl. (brōōs)	174	22°35'S	118°15'E
Bruce Peninsula, pen., Can.	58	44°50'N	81°20'W
Bruceton, Tn., U.S. (brōōs'tŭn)	82	36°02'N	88°14'W
Bruchsal, Ger. (brŏok'zäl)	124	49°08'N	8°34'E
Bruck, Aus. (brŏok)	124	47°25'N	15°14'E
Bruck, Aus.	124	48°01'N	16°47'E
Brück, Ger. (brük)	115b	52°12'N	12°45'E
Bruderheim, Can. (brōō'dẽr-hīm)	62g	53°47'N	112°56'W
Brugge, Bel.	117	51°13'N	3°05'E
Brühl, Ger. (brül)	127c	50°49'N	6°54'E
Bruneau, r., Id., U.S. (brōō-nō')	72	42°47'N	115°43'W
Brunei, nation, Asia (brŏo-nī')	168	4°52'N	113°38'E
Brünen, Ger. (brü'nĕn)	127c	51°43'N	6°41'E
Brunete, Spain (brōō-nā'tā)	129a	40°24'N	4°00'W
Brunette, i., Can. (brŏo-nĕt')	61	47°16'N	55°54'W
Brunn am Gebirge, Aus. (brōōn'äm gĕ-bĭr'gĕ)	115e	48°07'N	16°18'E
Brunsbüttel, Ger. (brŏons'büt-tĕl)	115c	53°58'N	9°10'E
Brunswick, Ga., U.S. (brŭnz'wĭk)	65	31°08'N	81°30'W
Brunswick, Md., U.S.	67	39°20'N	77°35'W
Brunswick, Me., U.S.	60	43°54'N	69°57'W
Brunswick, Mo., U.S.	79	39°25'N	93°07'W
Brunswick, Oh., U.S.	69d	41°14'N	81°50'W
Brunswick, Península de, pen., Chile	102	53°25'S	71°15'W
Bruny, i., Austl. (brōō'nē)	175	43°30'S	147°50'E
Brush, Co., U.S. (brŭsh)	78	40°14'N	103°40'W
Brusque, Braz. (brōō's-kōōĕ)	102	27°15'S	48°45'W
Brussels, Bel.	110	50°51'N	4°21'E
Brussels, Il., U.S. (brŭs'ĕls)	75e	38°57'N	90°36'W
Bruxelles *see* Brussels, Bel.	110	50°51'N	4°21'E
Bryan, Oh., U.S. (brī'ăn)	66	41°25'N	84°30'W
Bryan, Tx., U.S.	81	30°40'N	96°22'W
Bryansk, Russia	134	53°15'N	34°22'E
Bryansk, prov., Russia	132	52°43'N	32°25'E
Bryant, S.D., U.S. (brī'ănt)	70	44°35'N	97°29'W
Bryant, Wa., U.S.	74a	48°14'N	122°10'W
Bryce Canyon National Park, rec., Ut., U.S. (brīs)	64	37°35'N	112°15'W
Bryn Mawr, Pa., U.S. (brĭn mår')	68f	40°02'N	75°20'W
Bryson City, N.C., U.S. (brīs'ŭn)	82	35°25'N	83°25'W
Bryukhovetskaya, Russia (b'ryŭk'ô-vyĕt-skä'yä)	133	45°56'N	38°58'E
Buala, Sol. Is.	170e	8°08'S	159°35'E
Buatan, Indon.	153b	0°45'N	101°49'E
Buba, Gui.-B. (bōō'bá)	184	11°39'N	14°58'W
Bucaramanga, Col. (bōō-kä'rä-mäŋ'gä)	100	7°12'N	73°14'W
Buccaneer Archipelago, is., Austl. (bŭk-á-nēr')	174	16°05'S	122°00'E
Buchach, Ukr. (bŏ'chách)	125	49°04'N	25°25'E
Buchanan, Lib. (bŏŏ-kăn'ăn)	184	5°57'N	10°02'W
Buchanan, Mi., U.S.	66	41°50'N	86°25'W
Buchanan, l., Austl. (bû-kăn'nŏn)	175	21°40'S	145°00'E
Buchanan, l., Tx., U.S. (bû-kăn'ăn)	80	30°55'N	98°40'W
Buchans, Can.	61	48°49'N	56°52'W
Bucharest, Rom.	110	44°23'N	26°10'E
Buchholz, Ger. (bōōk'hōltz)	115c	53°19'N	9°53'E
Buck Creek, r., In., U.S. (bŭk)	69g	39°43'N	85°58'W
Buckhannon, W.V., U.S. (bŭk-hăn'ŭn)	66	39°00'N	80°10'W
Buckhaven, Scot., U.K. (bŭk-hā'v'n)	120	56°10'N	3°10'W
Buckie, Scot., U.K. (bŭk'ĭ)	120	57°40'N	2°50'W
Buckingham, Can. (bŭk'ĭng-ăm)	62c	45°35'N	75°25'W
Buckingham, can., India (bŭk'ĭng-ăm)	159	15°18'N	79°50'E
Buckinghamshire, co., Eng., U.K.	114b	51°45'N	0°48'W
Buckland, Can. (bŭk'lănd)	62b	46°37'N	70°33'W
Buckland Tableland, reg., Austl.	175	24°31'S	148°00'E
Buckley, Wa., U.S. (bŭk'lē)	74a	47°10'N	122°02'W
Bucksport, Me., U.S. (bŭks'pôrt)	60	44°35'N	68°47'W
Bouctouche, Can. (bŭk-tōōsh')	60	46°28'N	64°43'W
Bucun, China (bōō-tsòn)	162	36°38'N	117°26'E
București *see* Bucharest, Rom.	110	44°23'N	26°10'E
Bucyrus, Oh., U.S. (bū-sī'rŭs)	66	40°50'N	82°55'W
Budapest, Hung. (bōō'dá-pĕsht')	110	47°30'N	19°05'E
Budge Budge, India	158a	22°28'N	88°08'E
Budjala, D.R.C.	190	2°39'N	19°42'E
Budyonnovsk, Russia	138	44°46'N	44°09'E
Buea, Cam.	189	4°09'N	9°14'E
Buechel, Ky., U.S. (bĕ-chûl')	69h	38°12'N	85°38'W
Bueil, Fr. (bwā')	127b	48°55'N	1°27'E
Buena Park, Ca., U.S. (bwā'ná pärk)	75a	33°52'N	118°00'W
Buenaventura, Col. (bwā'nä-vĕn-tōō'rá)	100	3°46'N	77°09'W
Buenaventura, Cuba	93a	22°53'N	82°22'W
Buenaventura, Bahía de, b., Col.	100	3°45'N	79°23'W
Buena Vista, Co., U.S. (bū'ná vĭs'tá)	78	38°51'N	106°07'W
Buena Vista, Ga., U.S.	82	32°15'N	84°30'W
Buena Vista, Va., U.S.	67	37°45'N	79°20'W
Buena Vista, Bahía, b., Cuba (bä-ē'ä-bwĕ-nä-vē's-tä)	92	22°30'N	79°10'W
Buena Vista Lake Bed, l., Ca., U.S. (bū'ná vĭs'tá)	76	35°14'N	119°17'W
Buendia, Embalse de, res., Spain	128	40°30'N	2°45'W
Buenos Aires, Arg. (bwā'nôs ī'rās)	102	34°20'S	58°30'W
Buenos Aires, Col.	100a	3°01'N	76°34'W
Buenos Aires, C.R.	91	9°10'N	83°21'W
Buenos Aires, prov., Arg.	102	36°15'S	61°45'W

ăt; finăl; rāte; senâte; ärm; åsk; sofá; fâre; ch-choose; dh-as th in other; bē; êvent; bĕt; recĕnt; cratẽr; g-gō; gh-guttural g; bĭt; ĭ-short neutral; rīde; κ-guttural k as ch in German ich;

PRONOUNCING INDEX Buenos Aires – Cahama 231

PLACE (Pronunciation)	PAGE	LAT.	LONG.
Buenos Aires, l., S.A.	102	46°30′S	72°15′W
Buffalo, Mn., U.S. (buf′ȧ lō)	71	45°10′N	93°50′W
Buffalo, N.Y., U.S.	65	42°54′N	78°51′W
Buffalo, Tx., U.S.	81	31°28′N	96°04′W
Buffalo, Wy., U.S.	73	44°19′N	106°42′W
Buffalo, r., S. Afr.	187c	28°35′S	30°27′E
Buffalo, r., Ar., U.S.	79	35°56′N	92°58′W
Buffalo, r., Tn., U.S.	82	35°24′N	87°10′W
Buffalo Bayou, Tx., U.S.	81a	29°46′N	95°32′W
Buffalo Creek, r., Mn., U.S.	71	44°46′N	94°28′W
Buffalo Head Hills, hills, Can.	52	57°16′N	116°18′W
Buford, Can. (bū′fūrd)	62g	53°15′N	113°55′W
Buford, Ga., U.S. (bū′fẽrd)	82	34°05′N	84°00′W
Bug (Zakhidnyy Buh), r., Eur.	125	52°29′N	21°20′E
Buga, Col. (bo͞o′gä)	100	3°54′N	76°17′W
Buggenhout, Bel.	115a	51°01′N	4°10′E
Buglandsfjorden, l., Nor.	122	58°53′N	7°55′E
Bugojno, Bos. (bo̤-gō ĭ nô)	131	44°03′N	17°28′E
Bugul′ma, Russia (bo̤-go̤l′mȧ)	134	54°40′N	52°40′E
Buguruslan, Russia (bo̤-go̤-ro̤s-lȧn′)	134	53°30′N	52°32′E
Buhi, Phil. (bo͞o′ē)	169a	13°26′N	123°31′E
Buhl, Id., U.S. (būl)	73	42°36′N	114°45′W
Buhl, Mn., U.S.	71	47°28′N	92°49′W
Buin, Chile (bo̤-ēn′)	99b	33°44′S	70°44′W
Buinaksk, Russia (bo̤′ē-näksk)	137	42°40′N	47°20′E
Buir Nur, l., Asia (bo͞o-ēr no͞or)	161	47°50′N	117°00′E
Bujalance, Spain (bo͞o-hä-län′thā)	128	37°54′N	4°22′W
Bujumbura, Bdi.	191	3°23′S	29°22′E
Buka Island, i., Pap. N. Gui.	170e	5°15′S	154°35′E
Bukama, D.R.C. (bo͞o-kä′mä)	186	9°08′S	26°00′E
Bukavu, D.R.C.	186	2°30′S	28°52′E
Bukhara, Uzb. (bo̤-kä′rä)	139	39°31′N	64°22′E
Bukitbatu, Indon.	153b	1°25′N	101°58′E
Bukittinggi, Indon.	168	0°25′S	100°28′E
Bukoba, Tan.	186	1°20′S	31°49′E
Bukovina, hist. reg., Eur. (bo̤-kô′vĭ-nȧ)	125	48°06′N	25°20′E
Bula, Indon. (bo͞o′lä)	169	3°00′S	130°30′E
Bulalacao, Phil. (bo͞o-lä-lä′kä-ô)	169a	12°30′N	121°20′E
Bulawayo, Zimb. (bo͞o-lȧ-wä′yō)	186	20°12′S	28°43′E
Buldir, i., Ak., U.S. (bŭl dĭr)	63a	52°22′N	175°50′E
Bulgaria, nation, Eur. (bo̤l-gā′rĭ-ă)	110	42°12′N	24°13′E
Bulkley Ranges, mts., Can. (bŭlk′lē)	54	54°30′N	127°30′W
Bullaque, r., Spain (bo̤-lä′kā)	128	39°15′N	4°13′W
Bullas, Spain (bo͞ol′yäs)	128	38°07′N	1°48′W
Bullfrog Creek, r., Ut., U.S.	77	37°45′N	110°55′W
Bull Harbour, Can. (hăr′bẽr)	54	50°45′N	127°55′W
Bull Head, mtn., Jam.	92	18°10′N	77°15′W
Bull Run, r., Or., U.S. (bo̤l)	74c	45°26′N	122°11′W
Bull Run Reservoir, res., Or., U.S.	74c	45°29′N	122°11′W
Bull Shoals Reservoir, res., U.S. (bo̤l shōlz)	65	36°35′N	92°57′W
Bulpham, Eng., U.K. (bo̤l′făn)	114b	51°33′N	0°21′E
Bultfontein, S. Afr. (bo̤lt′fŏn-tān′)	192c	28°18′S	26°10′E
Bulun, Russia (bo͞o-lōn′)	135	70°48′N	127°27′E
Bulungu, D.R.C. (bo͞o-lo̤ŋ′go͞o)	190	6°04′S	21°54′E
Bulwer, S. Afr. (bo̤l-wẽr)	187c	29°49′S	29°48′E
Bumba, D.R.C. (bo̤m′bȧ)	185	2°11′N	22°28′E
Bumbire Island, i., Tan.	191	1°40′S	32°05′E
Buna, Pap. N. Gui. (bo͞o′nȧ)	169	8°58′S	148°38′E
Bunbury, Austl. (bŭn′bŭrĭ)	174	33°25′S	115°45′E
Bundaberg, Austl. (bŭn′dȧ-bûrg)	175	24°45′S	152°18′E
Bunguran Utara, Kepulauan, is., Indon.	168	3°22′N	108°00′E
Bunia, D.R.C.	191	1°34′N	30°15′E
Bunker Hill, Il., U.S. (bŭnk′ẽr hĭl)	75e	39°03′N	89°57′W
Bunkie, La., U.S. (bŭŋ′kĭ)	81	30°55′N	92°10′W
Bun Plains, pl., Kenya	191	0°55′N	40°35′E
Bununu Dass, Nig.	189	10°00′N	9°31′E
Buor-Khaya, Guba, b., Russia	141	71°45′N	131°00′E
Buor Khaya, Mys, c., Russia	135	71°47′N	133°22′E
Bura, Kenya	191	1°06′S	39°57′E
Buraydah, Sau. Ar.	154	26°23′N	44°14′E
Burbank, Ca., U.S. (bûr′bănk)	75a	34°11′N	118°19′W
Burco, Som.	192a	9°20′N	45°45′E
Burdekin, r., Austl. (bûr′dē-kĭn)	175	19°22′S	145°07′E
Burdur, Tur. (bo͞or-do̤r′)	119	37°50′N	30°15′E
Burdwān, India (bo̤d-wän′)	155	23°29′N	87°53′E
Bureinskiy, Khrebet, mts., Russia	135	51°15′N	133°30′E
Bureya, Russia (bo̤rā′ȧ)	135	49°55′N	130°00′E
Bureya, r., Russia (bo̤-rā′yä)	141	51°00′N	131°15′E
Burford, Eng., U.K. (bûr-fẽrd)	114b	51°46′N	1°38′W
Burgas, Blg. (bo̤r-gäs′)	119	42°29′N	27°30′E
Burgas, Gulf of, b., Blg.	119	42°30′N	27°40′E
Burgaw, N.C., U.S. (bûr′gô)	83	34°31′N	77°56′W
Burgdorf, Switz. (bo̤rg′dôrf)	124	47°04′N	7°37′E
Burgenland, state, Aus.	115e	47°58′N	16°57′E
Burgeo, Can.	61	47°36′N	57°34′W
Burgess, Va., U.S.	67	37°53′N	76°21′W
Burgo de Osma, Spain	128	41°35′N	3°02′W
Burgos, Mex. (bo̤r′gōs)	80	24°57′N	98°47′W
Burgos, Phil.	169a	16°03′N	119°52′E
Burgos, Spain (bo͞o′r-gôs)	118	42°20′N	3°44′W
Burgsvik, Swe. (bo̤rgs′vĭk)	122	57°04′N	18°18′E
Burhānpur, India (bo̤r′hän-po͞or)	155	21°26′N	76°08′E
Burias Island, i., Phil. (bo͞o′rē-äs)	169a	12°56′N	122°56′E
Burias Pass, strt., Phil. (bo͞o′rē-äs)	169a	13°04′N	123°11′E
Burica, Punta, c., N.A. (po͞o′n-tä-bo͞o′rē-kä)	91	8°02′N	83°12′W
Burien, Wa., U.S. (bū′rĭ-ĕn)	74a	47°28′N	122°20′W
Burin, Can. (bûr′ĭn)	53a	47°02′N	55°10′W
Burin Peninsula, pen., Can.	61	47°00′N	55°40′W
Burkburnett, Tx., U.S. (bûrk-bûr′nĕt)	78	34°04′N	98°35′W
Burke, Vt., U.S. (bûrk)	67	44°40′N	72°00′W
Burke Channel, strt., Can.	54	52°07′N	127°38′W
Burketown, Austl. (bûrk′toun)	174	17°50′S	139°30′E
Burkina Faso, nation, Afr.	184	13°00′N	2°00′W
Burley, Id., U.S. (bûr′lĭ)	73	42°31′N	113°48′W
Burley, Wa., U.S.	74a	47°25′N	122°38′W
Burlingame, Ca., U.S. (bûr′lĭn-găm)	74b	37°35′N	122°22′W
Burlingame, Ks., U.S.	79	38°45′N	95°49′W
Burlington, Can. (bûr′lĭng-tŭn)	59	43°19′N	79°48′W
Burlington, Co., U.S.	78	39°17′N	102°26′W
Burlington, Ia., U.S.	65	40°48′N	91°05′W
Burlington, Ks., U.S.	79	38°10′N	95°46′W
Burlington, Ky., U.S.	69f	39°01′N	84°44′W
Burlington, Ma., U.S.	61a	42°31′N	71°13′W
Burlington, N.C., U.S.	83	36°05′N	79°26′W
Burlington, N.J., U.S.	68f	40°04′N	74°52′W
Burlington, Vt., U.S.	65	44°30′N	73°15′W
Burlington, Wa., U.S.	74a	48°28′N	122°20′W
Burlington, Wi., U.S.	69a	42°41′N	88°16′W
Burma *see* Myanmar, nation, Asia	150	21°00′N	95°15′E
Burnaby, Can.	50	49°14′N	122°58′W
Burnet, Tx., U.S. (bûrn′ĕt)	80	30°46′N	98°14′W
Burnham on Crouch, Eng., U.K. (bûrn′ăm-ŏn-krouch)	114b	51°38′N	0°48′E
Burnie, Austl. (bûr′nē)	175	41°15′S	146°05′E
Burnley, Eng., U.K. (bûrn′lē)	120	53°47′N	2°19′W
Burns, Or., U.S. (bûrnz)	72	43°35′N	119°05′W
Burnside, Ky., U.S. (bûrn′sīd)	82	36°57′N	84°33′W
Burns Lake, Can. (bûrnz lăk)	50	54°14′N	125°46′W
Burnsville, Can. (bûrnz′vĭl)	60	47°44′N	65°07′W
Burnt, r., Or., U.S. (bûrnt)	72	44°26′N	117°53′W
Burntwood, r., Can.	57	55°53′N	97°30′W
Burrard Inlet, b., Can. (bûr′ȧrd)	74d	49°19′N	123°15′W
Burr Gaabo, Som.	187	1°14′N	51°47′E
Burro, Serranías del, mts., Mex. (sĕr-rä-nē′äs dĕl bo͞o′r-rô)	80	29°39′N	102°07′W
Bursa, Tur. (bo͞or′sȧ)	154	40°10′N	28°10′E
Būr Safājah, Egypt	185	26°57′N	33°56′E
Burscheid, Ger. (bo͞or′shīd)	127c	51°05′N	7°07′E
Būr Sūdān, Sudan (so͞o-dän′)	185	19°30′N	37°10′E
Burt, N.Y., U.S. (bŭrt)	69c	43°19′N	78°45′W
Burt, l., Mi., U.S. (bŭrt)	66	45°25′N	84°45′W
Burton, Wa., U.S. (bûr′tŭn)	74a	47°24′N	122°28′W
Burton, Lake, res., Ga., U.S.	82	34°46′N	83°40′W
Burtonsville, Md., U.S. (bûrtŏns-vĭl)	68e	39°07′N	76°57′W
Burton-upon-Trent, Eng., U.K. (bŭr′tŭn-ŭp′-ŏn-trĕnt)	120	52°48′N	1°37′W
Buru, i., Indon.	169	3°30′S	126°30′E
Burullus, l., Egypt	192b	31°20′N	30°58′E
Burundi, nation, Afr.	186	3°00′S	29°30′E
Burwell, Ne., U.S. (bûr′wĕl)	70	41°46′N	99°08′W
Bury, Eng., U.K. (bĕr′ĭ)	114a	53°36′N	2°17′W
Buryatia, prov., Russia	141	55°15′N	112°00′E
Bury Saint Edmunds, Eng., U.K. (bĕr′ĭ-sānt ĕd′mŭndz)	121	52°14′N	0°44′E
Burzaco, Arg. (bo͞or-zá′kô)	102a	34°50′S	58°23′W
Busanga Swamp, sw., Zam.	191	14°10′S	25°50′E
Būsh, Egypt (bo͞osh)	192b	29°13′N	31°08′E
Bushmanland, hist. reg., S. Afr. (bo̤sh-măn lănd)	186	29°15′S	18°45′E
Bushnell, Il., U.S. (bo̤sh′nĕl)	79	40°33′N	90°28′W
Businga, D.R.C. (bo̤-sĭŋ′gȧ)	185	3°20′N	20°53′E
Busira, r., D.R.C.	190	0°05′S	19°20′E
Bus′k, Ukr.	125	49°58′N	24°39′E
Busselton, Austl. (bŭs′l-tŭn)	174	33°40′S	115°30′E
Bussum, Neth.	115a	52°16′N	5°10′E
Bustamante, Mex. (bo͞os-tá-män′tā)	80	26°34′N	100°30′W
Busto Arsizio, Italy (bo͞os′tō är-sēd′zē-ō)	130	45°47′N	8°51′E
Busuanga, i., Phil. (bo͞o-swäŋ′gä)	169a	12°20′N	119°43′E
Buta, D.R.C. (bo͞o′tȧ)	185	2°48′N	24°44′E
Butha Buthe, Leso. (bo͞o-thȧ-bo͞o′thȧ)	187c	28°49′S	28°16′E
Butler, Al., U.S. (bŭt′lẽr)	82	32°05′N	88°10′W
Butler, In., U.S.	66	41°25′N	84°50′W
Butler, Md., U.S.	68e	39°32′N	76°46′W
Butler, N.J., U.S.	68a	41°00′N	74°20′W
Butler, Pa., U.S.	67	40°50′N	79°55′W
Butovo, Russia (bo̤-tô′vô)	142b	55°33′N	37°36′E
Butsha, D.R.C.	191	0°57′N	29°13′E
Buttahatchee, r., Al., U.S. (bŭt-ȧ-hăch′ē)	82	34°02′N	88°05′W
Butte, Mt., U.S. (būt)	64	46°00′N	112°31′W
Butterworth, S. Afr. (bŭ tẽr′wûrth)	187c	32°20′S	28°09′E
Butt of Lewis, c., Scot., U.K. (bŭt ŏv lū′ĭs)	120	58°34′N	6°15′W
Butuan, Phil. (bo͞o-to͞o′än)	169	8°40′N	125°33′E
Buturlinovka, Russia (bo͞o-to͞o′lē-nôf′kȧ)	137	50°47′N	40°35′E
Buuhoodle, Som.	192a	8°15′N	46°20′E
Buulo Berde, Som.	192a	3°53′N	45°30′E
Buxtehude, Ger.	115c	53°29′N	9°42′E
Buxton, Eng., U.K. (bŭks′t′n)	114a	53°15′N	1°55′W
Buxton, Or., U.S.	74c	45°41′N	123°11′W
Buy, Russia (bwē)	134	58°30′N	41°48′E
Büyükmenderes, r., Tur.	154	37°50′N	28°20′E
Buzău, Rom. (bo͞o-zĕ′o̤)	131	45°09′N	26°51′E
Buzău, r., Rom.	133	45°17′N	27°22′E
Buzaymah, Libya	185	25°14′N	22°13′E
Buzi, China (bo͞o-dz)	162	33°48′N	118°13′E
Buzuluk, Russia (bo̤-zo̤-lo̤k′)	134	52°50′N	52°10′E
Bwendi, D.R.C.	191	4°01′N	26°41′E
Byala, Blg.	131	43°26′N	25°44′E
Byala Slatina, Blg. (byä′lȧ slä′tēnä)	131	43°26′N	23°56′E
Byalynichy, Bela. (byĕl-ĭ-nĭ′chĭ)	132	54°02′N	29°42′E
Byarezina, r., Bela. (bĕr-yĕ′zē-nȧ)	132	53°20′N	29°05′E
Byaroza, Bela.	125	52°29′N	24°59′E
Byblos *see* Jubayl, Leb.	153a	34°07′N	35°38′E
Bydgoszcz, Pol. (bĭd′gôshch)	116	53°07′N	18°00′E
Byelorussia *see* Belarus, nation, Eur.	134	53°30′N	25°33′E
Byerazino, Bela. (bĕr-yä′zē-nô)	132	53°51′N	28°54′E
Byeshankovichy, Bela.	132	55°04′N	29°29′E
Byesville, Oh., U.S. (bīz-vĭl)	66	39°55′N	81°35′W
Bygdin, l., Nor. (bügh-dēn′)	122	61°24′N	8°31′E
Byglandsfjord, Nor. (bügh′länds-fyôr)	122	58°40′N	7°49′E
Bykhaw, Bela.	132	53°32′N	30°15′E
Bykovo, Russia (bĭ-kô′vô)	142b	55°38′N	38°05′E
Byrranga, Gory, mts., Russia	140	74°15′N	94°28′E
Bytantay, r., Russia (byän′täy)	141	68°15′N	132°15′E
Bytom, Pol. (bĭ′tŭm)	117	50°21′N	18°55′E
Bytosh′, Russia (bĭ-tôsh′)	132	53°48′N	34°06′E
Bytow, Pol. (bĭ′tŭf)	125	54°10′N	17°30′E

C

PLACE (Pronunciation)	PAGE	LAT.	LONG.
Cabagan, Phil. (kä-bä-gän′)	169a	17°27′N	121°50′E
Cabalete, i., Phil. (kä-bä-lā′tā)	169a	14°19′N	122°00′E
Caballones, Canal de, strt., Cuba (kä-nä′l-dĕ-kä-bäl-yô′nĕs)	92	20°45′N	79°20′W
Caballo Reservoir, res., N.M., U.S. (kä-bä-lyō′)	77	33°00′N	107°20′W
Cabanatuan, Phil. (kä-bä-nä-twän′)	169a	15°30′N	120°56′E
Cabano, Can. (kä-bä-nō′)	60	47°41′N	68°54′W
Cabarruyan, i., Phil. (kä-bä-ro͞o′yän)	169a	16°21′N	120°10′E
Cabedelo, Braz. (kä-bē-dā′lo̤)	101	6°58′S	34°49′W
Cabeza, Arrecife, i., Mex.	89	19°07′N	95°52′W
Cabeza del Buey, Spain (kȧ-bā′thä dĕl bwā′)	128	38°43′N	5°18′W
Cabimas, Ven. (kä-bē′mäs)	100	10°21′N	71°27′W
Cabinda, Ang.	186	5°33′S	12°12′E
Cabinda, hist. reg., Ang. (kä-bĭn′dȧ)	186	5°10′S	10°00′E
Cabinet Mountains, mts., Mt., U.S. (kăb′ĭ-nĕt)	72	48°13′N	115°52′W
Cabo Frio, Braz. (kä′bô-frē′ô)	99a	22°53′S	42°02′W
Cabo Frio, Ilha do, Braz. (ē′lä-dô-kä′bô frē′ô)	99a	23°01′S	42°00′W
Cabo Gracias a Dios, Hond. (kȧ′bô-grä-syäs-ä-dyô′s)	91	15°00′N	83°13′W
Cabonga, Réservoir, res., Can.	59	47°25′N	76°35′W
Cabora Bassa Reservoir, res., Moz.	186	15°45′S	32°00′E
Cabot Head, c., Can. (kăb′ŭt)	58	45°15′N	81°20′W
Cabot Strait, strt., Can. (kăb′ŭt)	53a	47°35′N	60°00′W
Cabra, Spain (käb′rä)	128	37°28′N	4°29′W
Cabra, i., Phil.	169a	13°55′N	119°55′E
Cabrera, Illa de, i., Spain	129	39°08′N	2°57′E
Cabrera, Sierra de la, mts., Spain	128	42°15′N	6°45′W
Cabriel, r., Spain (kä-brē-ĕl′)	128	39°25′N	1°20′W
Cabrillo National Monument, rec., Ca., U.S. (kä-brēl′yō)	76a	32°41′N	117°03′W
Cabuçu, r., Braz. (kä-bo͞o′-so͞o)	102b	22°57′S	43°36′W
Cabugao, Phil. (kä-bo͞o′gä-ô)	169a	17°48′N	120°28′E
Čačak, Serb. (chä′chȧk)	131	43°51′N	20°22′E
Caçapava, Braz. (kä′sä-pȧ′vä)	99a	23°05′S	45°52′W
Cáceres, Braz. (kȧ′sĕ-rĕs)	101	16°11′S	57°32′W
Cáceres, Spain (kȧ′thȧ-rās)	118	39°28′N	6°20′W
Cachapoal, r., Chile (kä-chä-pô-ȧ′l)	99b	34°23′S	70°19′W
Cache, r., Ar., U.S. (kȧsh)	79	35°24′N	91°12′W
Cache Creek, Can.	55	50°48′N	121°19′W
Cache Creek, r., Ca., U.S. (kăsh)	76	38°53′N	122°24′W
Cache la Poudre, r., Co., U.S. (kȧsh lȧ po͞od′r′)	78	40°43′N	105°39′W
Cachi, Nevados de, mtn., Arg. (nĕ-vȧ′dôs-dĕ-kȧ′chē)	102	25°05′S	66°40′W
Cachinal, Chile (kä-chē-näl′)	102	24°57′S	69°33′W
Cachoeira, Braz. (kä-shô-ā′rä)	101	12°32′S	38°47′W
Cachoeirá do Sul, Braz. (kä-shô-ā′rä-dô-so͞o′l)	102	30°02′S	52°49′W
Cachoeiras de Macacu, Braz. (kä-shô-ā′räs-dĕ-mä-kȧ′ko͞o)	99a	22°28′S	42°39′W
Cachoeiro de Itapemirim, Braz.	101	20°51′S	41°06′W
Cacólo, Ang.	190	10°07′S	19°17′E
Caconda, Ang. (kä-kōn′dä)	186	13°43′S	15°06′E
Cacouna, Can.	60	47°54′N	69°31′W
Cacula, Ang.	190	14°29′S	14°10′E
Cadale, Som.	192a	2°45′N	46°15′E
Caddo, l., La., U.S. (kăd′ō)	81	32°37′N	94°15′W
Cadereyta, Mex. (kä-dȧ-rā′tä)	88	20°42′N	99°47′W
Cadereyta Jimenez, Mex. (kä-dä-rā′tä hē-mā′nāz)	80	25°36′N	99°59′W
Cadi, Sierra de, mts., Spain (sē-ĕ′r-rä-dĕ-kä′dē)	129	42°17′N	1°34′E
Cadillac, Mi., U.S. (kăd′ĭ-lăk)	66	44°15′N	85°25′W
Cádiz, Spain (kȧ′dēz)	110	36°34′N	6°20′W
Cadiz, Ca., U.S. (kā′dĭz)	76	34°33′N	115°30′W
Cadiz, Oh., U.S.	66	40°15′N	81°00′W
Cádiz, Golfo de, b., Spain (gôl-fô-dĕ-kȧ′dēz)	118	36°50′N	7°00′W
Caen, Fr. (käN)	117	49°13′N	0°22′W
Caernarfon, Wales, U.K.	116	53°08′N	4°17′W
Caernarfon Bay, b., Wales, U.K.	120	53°09′N	4°56′W
Cagayan, Phil. (kä-gä-yän′)	169	8°13′N	124°30′E
Cagayan, r., Phil.	168	16°45′N	121°55′E
Cagayan Islands, is., Phil.	168	9°40′N	120°30′E
Cagayan Sulu, i., Phil. (kä-gä-yän so͞o′lo͞o)	168	7°00′N	118°30′E
Cagli, Italy (käl′yē)	130	43°35′N	12°40′E
Cagliari, Italy (käl′yä-rē)	110	39°16′N	9°08′E
Cagliari, Golfo di, b., Italy (gôl-fô-dē-käl′yä-rē)	118	39°08′N	9°12′E
Cagnes, Fr. (kän′y′)	127	43°40′N	7°14′E
Cagua, Ven. (kä′gwä)	101b	10°12′N	67°27′W
Caguas, P.R. (kä′gwäs)	87b	18°12′N	66°01′W
Cahaba, r., Al., U.S. (kȧ hä-bä)	82	32°50′N	87°15′W
Cahama, Ang. (kä-ä′mä)	186	16°17′S	14°19′E

232 Cahokia – Cantanhede PRONOUNCING INDEX

PLACE (Pronunciation)	PAGE	LAT.	LONG.
Cahokia, Il., U.S. (kȧ-hō′kĭ-ȧ)	75e	38°34′N	90°11′W
Cahora-Bassa, wtfl., Moz.	191	15°40′S	32°50′E
Cahors, Fr. (kȧ-ŏr′)	117	44°27′N	1°27′E
Cahuacán, Mex. (kä-wä-kä′n)	89a	19°38′N	99°25′W
Cahuita, Punta, c., C.R. (pōō′n-tä-kä-wē′tȧ)	91	9°47′N	82°41′W
Cahul, Mol.	133	45°49′N	28°17′E
Caibarién, Cuba (kī-bä-rē-ĕn′)	92	22°35′N	79°30′W
Caicedonia, Col. (kī-sĕ-dô-nēä)	100a	4°21′N	75°48′W
Caicos Bank, bk., (kī′kōs)	93	21°35′N	72°00′W
Caicos Islands, is., T./C. Is.	87	21°45′N	71°50′W
Caicos Passage, strt., N.A.	93	21°55′N	72°45′W
Caillou Bay, b., La., U.S. (kȧ-yōō′)	81	29°07′N	91°00′W
Caimanera, Cuba (kī-mä-nä′rä)	93	20°00′N	75°10′W
Caiman Point, c., Phil. (kī′man)	169a	15°56′N	119°33′E
Caimito, r., Pan. (kä-ē-mē′tô)	86a	8°50′N	79°45′W
Caimito del Guayabal, Cuba (kä-ē-mē′tō-dĕl-gwä-yä-bä′l)	93a	22°57′N	82°36′W
Cairns, Austl. (kârnz)	175	17°02′S	145°49′E
Cairo, C.R. (kī′rō)	91	10°06′N	83°47′W
Cairo, Egypt	185	30°00′N	31°17′E
Cairo, Ga., U.S. (kā′rō)	82	30°48′N	84°12′W
Cairo, Il., U.S.	65	36°59′N	89°11′W
Caistor, Eng., U.K. (kās′tẽr)	114a	53°30′N	0°20′W
Caiundo, Ang.	190	15°46′S	17°28′E
Caiyu, China (tsī-yōō)	162	39°39′N	116°36′E
Cajamarca, Col. (kä-кä-mä′r-kä)	100a	4°25′N	75°25′W
Cajamarca, Peru (kä-hä-mär′kä)	100	7°16′S	78°30′W
Čajniče, Bos. (chī′nĭ-chĕ)	131	43°32′N	19°04′E
Cajon, Ca., U.S. (kȧ-hōn′)	75a	34°18′N	117°28′W
Cajuru, Braz. (kä-zhōō′rōō)	99a	21°17′S	47°17′W
Čakovec, Cro. (chä′kō-vĕts)	130	46°23′N	16°27′E
Cala, S. Afr. (cä-lȧ)	187c	31°33′S	27°41′E
Calabar, Nig. (kăl-ȧ-bär′)	184	4°57′N	8°19′E
Calabazar, Cuba (kä-lä-bä-zä′r)	93a	23°02′N	82°25′W
Calabozo, Ven. (kä-lä-bô′zō)	100	8°48′N	67°27′W
Calabria, hist. reg., Italy (kä-lä′brē-ä)	130	39°26′N	16°23′E
Calafat, Rom. (kȧ-lȧ-fȧt′)	131	43°59′N	22°56′E
Calaguas Islands, is., Phil. (kä-läg′wäs)	169a	14°30′N	123°06′E
Calahoo, Can. (kă-lȧ-hōō′)	62g	53°42′N	113°58′W
Calahorra, Spain (kä-lä-ôr′rä)	118	42°18′N	1°58′W
Calais, Fr. (kȧ-lĕ′)	110	50°56′N	1°51′E
Calais, Me., U.S.	65	45°11′N	67°15′W
Calama, Chile (kä-lä′mä)	102	22°17′S	68°58′W
Calamar, Col. (kä-lä-mär′)	100	10°24′N	75°00′W
Calamar, Col.	100	1°55′N	72°33′W
Calamba, Phil. (kä-läm′bä)	169a	14°12′N	121°10′E
Calamian Group, is., Phil. (kä-lä-myän′)	168	12°14′N	118°38′E
Calañas, Spain (kä-län′yäs)	128	37°41′N	6°52′W
Calanda, Spain	129	40°53′N	0°20′W
Calapan, Phil. (kä-lä-pän′)	169a	13°25′N	121°11′E
Călăraşi, Rom. (kŭ-lŭ-rásh′ĭ)	119	44°09′N	27°20′E
Calatayud, Spain (kä-lä-tä-yōōdh′)	118	41°23′N	1°37′W
Calauag Bay, b., Phil.	169a	14°07′N	122°10′E
Calaveras Reservoir, res., Ca., U.S. (kăl-ȧ-vĕr′ăs)	74b	37°29′N	121°47′W
Calavite, Cape, c., Phil. (kä-lä-vē′tä)	169a	13°29′N	120°00′E
Calcasieu, r., La., U.S. (kăl′kȧ-shū)	81	30°22′N	93°08′W
Calcasieu Lake, l., La., U.S.	81	29°58′N	93°08′W
Calcutta see Kolkata, India (kăl-kŭt′ȧ)	155	22°32′N	88°22′E
Caldas, Col. (kä′l-däs)	100a	6°06′N	75°38′W
Caldas, dept., Col.	100a	5°20′N	75°38′W
Caldas da Rainha, Port. (käl′däs dä rīn′yȧ)	128	39°25′N	9°08′W
Calder, r., Eng., U.K. (kôl′dẽr)	114a	53°39′N	1°30′W
Caldera, Chile (käl-dā′rä)	102	27°02′S	70°53′W
Calder Canal, can., Eng., U.K.	114a	53°48′N	2°25′W
Caldwell, Id., U.S. (kôld′wĕl)	72	43°40′N	116°43′W
Caldwell, Ks., U.S.	79	37°04′N	97°36′W
Caldwell, Oh., U.S.	66	39°40′N	81°30′W
Caldwell, Tx., U.S.	81	30°30′N	96°40′W
Caledon, Can. (kăl′ē-dŏn)	62d	43°52′N	79°59′W
Caledonia, Mn., U.S. (kăl-ē-dō′nĭ-ȧ)	71	43°38′N	91°31′W
Calella, Spain (kä-lĕl′yä)	129	41°37′N	2°39′E
Calera Victor Rosales, Mex. (kä-lā′rä-vē′k-tôr-rō-sá′lĕs)	88	22°57′N	102°42′W
Calexico, Ca., U.S. (kȧ-lĕk′sĭ-kō)	64	32°41′N	115°30′W
Calgary, Can. (kăl′gȧ-rĭ)	50	51°03′N	114°05′W
Calhoun, Ga., U.S. (kăl-hōōn′)	82	34°30′N	84°56′W
Cali, Col. (kä′lē)	100	3°26′N	76°30′W
Caliente, Nv., U.S. (käl-yĕn′tä)	77	37°38′N	114°30′W
California, Mo., U.S. (kăl-ĭ-fôr′nĭ-ȧ)	79	38°38′N	92°38′W
California, Pa., U.S.	69e	40°03′N	79°53′W
California, state, U.S.	64	38°10′N	121°20′W
California, Golfo de, b., Mex. (gôl-fô-dĕ-kä-lē-fôr-nyä)	86	30°30′N	113°45′W
California Aqueduct, aq., Ca., U.S.	76	37°10′N	121°10′W
Călimani, Munţii, mts., Rom.	125	47°05′N	24°47′E
Calimere, Point, c., India	159	10°20′N	80°20′E
Calimesa, Ca., U.S. (kă-lĭ-mā′sȧ)	75a	34°00′N	117°04′W
Calipatria, Ca., U.S. (kăl-ĭ-pàt′rĭ-ȧ)	76	33°03′N	115°30′W
Calkini, Mex. (käl-kē-nē′)	89	20°21′N	90°06′W
Callabonna, Lake, l., Austl. (călă′bŏnȧ)	176	29°35′S	140°28′E
Callao, Peru (käl-yä′ō)	100	12°02′S	77°07′W
Calling, l., Can. (kôl′ĭng)	55	55°15′N	113°12′W
Calmar, Can. (kăl′mär)	62g	53°16′N	113°49′W
Calmar, Ia., U.S.	71	43°12′N	91°54′W
Caloosahatchee, r., Fl., U.S. (kȧ-loo-sȧ-hăch′ē)	83a	26°45′N	81°41′W
Calotmul, Mex. (kä-lôt-mōōl)	90a	20°58′N	88°11′W
Calpulalpan, Mex. (käl-pōō-läl′pän)	88	19°35′N	98°33′W
Caltagirone, Italy (käl-tä-jē-rō′nä)	118	37°14′N	14°32′E
Caltanissetta, Italy (käl-tä-nē-sĕt′tä)	118	37°30′N	14°02′E

PLACE (Pronunciation)	PAGE	LAT.	LONG.
Caluango, Ang.	190	8°21′S	19°40′E
Calucinga, Ang.	190	11°18′S	16°12′E
Calumet, Mi., U.S. (kă-lū-mĕt′)	71	47°15′N	88°29′W
Calumet, Lake, l., Il., U.S.	69a	41°43′N	87°36′W
Calumet City, Il., U.S.	69a	41°37′N	87°33′W
Calunda, Ang.	190	12°06′S	23°23′E
Caluquembe, Ang.	190	13°47′S	14°44′E
Caluula, Som.	192a	11°53′N	50°40′E
Calvert, Tx., U.S. (kăl′vẽrt)	81	30°59′N	96°41′W
Calvert Island, i., Can.	52	51°35′N	128°00′W
Calvi, Fr. (käl′vē)	130	42°33′N	8°35′E
Calvillo, Mex. (käl-vēl′yō)	89	21°51′N	102°44′E
Calvinia, S. Afr. (kăl-vĭn′ĭ-ȧ)	186	31°20′S	19°50′E
Cam, r., Eng., U.K. (kăm)	121	52°15′N	0°05′E
Camagüey, Cuba (kä-mä-gwā′)	87	21°25′N	78°00′W
Camagüey, prov., Cuba	92	21°30′N	78°10′W
Camajuani, Cuba (kä-mä-hwä′nė)	92	22°25′N	79°50′W
Camano, Wa., U.S. (kä-mä′no)	74a	48°10′N	122°32′W
Camano Island, i., Wa., U.S.	74a	48°11′N	122°29′W
Camargo, Mex. (kä-mär gō)	80	26°19′N	98°49′W
Camarón, Cabo, c., Hond. (kȧ′bô-kä-mä-rōn′)	90	16°06′N	85°05′W
Camas, Wa., U.S. (kăm′ăs)	74c	45°36′N	122°24′W
Camas Creek, r., Id., U.S.	73	44°10′N	112°09′W
Camatagua, Ven. (kä-mä-tá′gwä)	101b	9°49′N	66°55′W
Ca Mau, Mui, c., Viet.	168	8°36′N	104°43′E
Cambay, India (kăm-bā′)	158	22°22′N	72°39′E
Cambodia, nation, Asia	168	12°15′N	104°00′E
Cambonda, Serra, mts., Ang.	190	12°10′S	14°15′E
Camborne, Eng., U.K. (kăm′bôrn)	120	50°15′N	5°28′W
Cambrai, Fr. (käɴ-brĕ′)	117	50°10′N	3°15′E
Cambrian Mountains, mts., Wales, U.K. (kăm′brĭ-ăn)	120	52°05′N	4°05′W
Cambridge, Can.	59	43°22′N	80°19′W
Cambridge, Eng., U.K. (kām′brĭj)	117	52°12′N	0°11′E
Cambridge, Ma., U.S.	61a	42°23′N	71°07′W
Cambridge, Md., U.S.	67	38°35′N	76°10′W
Cambridge, Mn., U.S.	71	45°35′N	93°14′W
Cambridge, Ne., U.S.	78	40°17′N	100°10′W
Cambridge, Oh., U.S.	66	40°00′N	81°35′W
Cambridge Bay see Kaluktutiak, Can.	52	69°15′N	105°00′W
Cambridge City, In., U.S.	66	39°45′N	85°15′W
Cambridgeshire, co., Eng., U.K.	114a	52°26′N	0°19′W
Cambuci, Braz. (käm-bōō′sē)	99a	21°35′S	41°54′W
Cambundi-Catembo, Ang.	190	10°09′S	17°31′E
Camby, In., U.S. (kăm′bē)	69g	39°40′N	86°19′W
Camden, Austl.	173b	34°03′S	150°42′E
Camden, Al., U.S. (kăm′dĕn)	82	31°58′N	87°15′W
Camden, Ar., U.S.	79	33°36′N	92°49′W
Camden, Me., U.S.	60	44°11′N	69°05′W
Camden, N.J., U.S.	65	39°56′N	75°06′W
Camden, S.C., U.S.	83	34°14′N	80°37′W
Cameia, Parque Nacional da, rec., Ang.	190	11°40′S	21°20′E
Camenca, Mol.	133	48°02′N	28°43′E
Cameron, Mo., U.S. (kăm′ẽr-ŭn)	79	39°44′N	94°14′W
Cameron, Tx., U.S.	81	30°52′N	96°57′W
Cameron, W.V., U.S.	66	39°40′N	80°35′W
Cameron Hills, hills, Can.	52	60°13′N	120°20′W
Cameroon, nation, Afr.	184	5°48′N	11°00′E
Cameroon Mountain, mtn., Cam.	184	4°12′N	9°11′E
Camiling, Phil. (kä-mē-lĭng′)	169a	15°42′N	120°24′E
Camilla, Ga., U.S. (kȧ-mĭl′ȧ)	82	31°13′N	84°12′W
Caminha, Port. (kä-mĭn′yȧ)	128	41°52′N	8°44′W
Camoçim, Braz. (kä-mô-sēɴ′)	101	2°56′S	40°55′W
Camooweal, Austl.	174	20°00′S	138°13′E
Campana, Arg. (käm-pä′nä)	99c	34°10′S	58°58′W
Campana, i., Chile (käm-pän′yä)	102	48°20′S	75°15′W
Campanario, Spain (käm-pä-nä′rē-ō)	128	38°51′N	5°36′W
Campanella, Punta, c., Italy (pō′n-tä-käm-pä-nĕ′lä)	129c	40°20′N	14°21′E
Campanha, Braz. (käm-pän-yăɴ′)	99a	21°51′S	45°24′W
Campania, hist. reg., Italy (käm-pän′yä)	130	41°00′N	14°40′E
Campbell, Ca., U.S. (kăm′bĕl)	74b	37°17′N	121°57′W
Campbell, Mo., U.S.	79	36°29′N	90°04′W
Campbell, is., N.Z.	3	52°30′S	169°00′E
Campbellpore, Pak.	158	33°49′N	72°24′E
Campbell River, Can.	50	50°01′N	125°15′W
Campbellsville, Ky., U.S. (kăm′bĕlz-vĭl)	82	37°19′N	85°20′W
Campbellton, Can. (kăm′bĕl-tŭn)	51	48°00′N	66°40′W
Campbelltown, Austl. (kăm′bĕl-toun)	173b	34°04′S	150°49′E
Campbelltown, Scot., U.K. (kăm′b′l-toun)	120	55°25′N	5°50′W
Camp Dennison, Oh., U.S. (dĕ′nĭ-sŏn)	69f	39°12′N	84°17′W
Campeche, Mex. (käm-pā′chå)	86	19°51′N	90°32′W
Campeche, state, Mex.	86	18°55′N	90°20′W
Campeche, Bahía de, b., Mex. (bä-ē′ä-dĕ-käm-pā′chä)	86	19°30′N	93°40′W
Campechuela, Cuba (käm-på-chwā′lä)	92	20°15′N	77°15′W
Camperdown, S. Afr. (kăm′pẽr-doun)	187c	29°44′S	30°33′E
Câmpina, Rom.	131	45°08′N	25°47′E
Campina Grande, Braz. (käm-pē′nä grän′dĕ)	101	7°15′S	35°49′W
Campinas, Braz. (käm-pē′näs)	101	22°53′S	47°03′W
Camp Indian Reservation, I.R., Ca., U.S. (kămp)	76	32°39′N	116°26′W
Campo, Cam. (käm′pō)	184	2°22′N	9°49′E
Campoalegre, Col. (kä′m-pô-ălĕ′grĕ)	100	2°34′N	75°20′W
Campobasso, Italy (käm′pô-bäs′sō)	130	41°35′N	14°39′E
Campo Belo, Braz.	99a	20°52′S	45°15′W
Campo de Criptana, Spain (käm′pō dā krēp-tä′nä)	128	39°24′N	3°09′W

PLACE (Pronunciation)	PAGE	LAT.	LONG.
Campo Florido, Cuba (kä′m-pō flô-rē′dô)	93a	23°07′N	82°07′W
Campo Grande, Braz. (käm-pò grän′dĕ)	101	20°28′S	54°32′W
Campo Grande, Braz.	102b	22°54′S	43°33′W
Campo Maior, Braz. (käm-pò mä-yôr′)	101	4°48′S	42°12′W
Campo Maior, Port.	128	39°03′N	7°06′W
Campo Real, Spain (käm′pô rå-äl′)	129a	40°21′N	3°23′W
Campos, Braz. (kä′m-pôs)	101	21°46′S	41°19′W
Campos do Jordão, Braz. (kä′m-pôs-dô-zhôr-dou′ɴ)	99a	22°45′S	45°35′W
Campos Gerais, Braz. (kä′m-pôs-zhĕ-ràĕs)	99a	21°17′S	45°43′W
Camps Bay, S. Afr. (kămps)	186a	33°57′S	18°22′E
Camp Springs, Md., U.S. (cămp sprĭngz)	68e	38°48′N	76°55′W
Câmpulung, Rom.	119	45°15′N	25°03′E
Câmpulung Moldovenesc, Rom.	125	47°31′N	25°36′E
Camp Wood, Tx., U.S. (kămp wŏŏd)	80	29°39′N	100°02′W
Camrose, Can. (kăm-rōz)	50	53°01′N	112°50′W
Camu, r., Dom. Rep. (kä′mōō)	93	19°05′N	70°15′W
Canada, nation, N.A. (kăn′ȧ-dȧ)	50	50°00′N	100°00′W
Canada Bay, b., Can.	61	50°43′N	56°10′W
Cañada de Gómez, Arg. (kä-nyä′dä-dĕ-gô′mĕz)	102	32°49′S	61°24′W
Canadian, Tx., U.S. (kȧ-nā′dĭ-ăn)	78	35°54′N	100°24′W
Canadian, r., U.S.	64	35°30′N	102°30′W
Canajoharie, N.Y., U.S. (kăn-ȧ-jō-hăr′ē)	67	42°55′N	74°35′W
Çanakkale, Tur. (chä-näk-kä′lĕ)	119	40°10′N	26°26′E
Çanakkale Boğazi (Dardanelles), strt., Tur.	119	40°05′N	25°50′E
Canandaigua, N.Y., U.S. (kăn-ăn-dā′gwȧ)	67	42°55′N	77°20′W
Canandaigua, l., N.Y., U.S.	67	42°45′N	77°20′W
Cananea, Mex. (kä-nä-nĕ′ä)	86	31°00′N	110°20′W
Canarias, Islas (Canary Is.), is., Spain (ĕ′s-läs-kä-nä′ryäs)	183	29°15′N	16°30′W
Canarreos, Archipiélago de los, is., Cuba	92	21°35′N	82°20′W
Canary Islands see Canarias, Islas, is., Spain	183	29°15′N	16°30′W
Cañas, C.R. (kä′-nyäs)	90	10°26′N	85°06′W
Cañas, r., C.R.	90	10°20′N	85°21′W
Cañasgordas, Col. (kä′nyäs-gô′r-däs)	100a	6°44′N	76°01′W
Canastota, N.Y., U.S. (kăn-ȧs-tō′tȧ)	67	43°05′N	75°45′W
Canastra, Serra de, mts., Braz. (sĕ′r-rä-dĕ-kä-nä′s-trä)	101	19°53′S	46°57′W
Canatlán, Mex. (kä-nät-län′)	80	24°30′N	104°45′W
Canaveral, Cape, c., Fl., U.S.	65	28°30′N	80°23′W
Canavieiras, Braz. (kä-nä-vē-ā′räs)	101	15°40′S	38°49′W
Canberra, Austl. (kăn′bĕr-ȧ)	175	35°21′S	149°10′E
Canby, Mn., U.S. (kăn′bĭ)	70	44°43′N	96°15′W
Canchyuaya, Cerros de, mts., Peru (sĕ′r-rôs-dĕ-kän-chōō-ȧ′īä)	100	7°30′S	74°30′W
Cancuc, Mex. (kän-kōōk)	89	16°58′N	92°17′W
Cancún, Mex.	90a	21°25′N	86°50′W
Candelaria, Cuba (kän-dĕ-lä′ryä)	92	22°45′N	82°55′W
Candelaria, Phil. (kän-dȧ-lä′rē-ä)	169a	15°39′N	119°55′E
Candelaria, r., Mex. (kän-dĕ-lä-ryä)	89	18°25′N	91°21′W
Candeleda, Spain (kän-dhå-lā′dhä)	128	40°09′N	5°18′W
Candia see Irákleio, Grc.	110	35°20′N	25°10′E
Candle, Ak., U.S. (kăn′d′l)	63	65°00′N	162°04′W
Cando, N.D., U.S. (kăn′dō)	70	48°27′N	99°13′W
Candon, Phil. (kän-dōn′)	169a	17°13′N	120°26′E
Canelones, Ur. (kä-nĕ-lô-nĕs)	99c	34°32′S	56°19′W
Canelones, dept., Ur.	99c	34°34′S	56°15′W
Cañete, Peru (kän-yā′tå)	100	13°06′S	76°17′W
Caney, Cuba (kä-nā′) (kā′nĭ)	93	20°05′N	75°45′W
Caney, Ks., U.S. (kā′nĭ)	79	37°00′N	95°57′W
Caney Fork, r., Tn., U.S.	82	36°10′N	85°50′W
Cangamba, Ang.	186	13°40′S	19°54′E
Cangas, Spain (kän′gäs)	128	42°15′N	8°43′W
Cangas de Narcea, Spain (kä′n-gäs-dĕ-när-sĕ-ä)	128	43°08′N	6°36′W
Cangzhou, China (tsäŋ-jō)	164	38°21′N	116°53′E
Caniapiscau, l., Can.	53	54°10′N	71°13′E
Caniapiscau, r., Can.	53	57°00′N	68°45′W
Canicatti, Italy (kä-nė-kät′tē)	130	37°18′N	13°58′E
Cañitas, Mex. (kän-yē′täs)	88	23°38′N	102°44′W
Cannell, Can.	62g	53°35′N	113°38′W
Cannelton, In., U.S. (kăn′ĕl-tŭn)	66	37°55′N	86°45′W
Cannes, Fr. (kȧn)	117	43°34′N	7°05′E
Canning, Can. (kăn′ĭng)	60	45°09′N	64°25′W
Cannock, Eng., U.K. (kăn′ŭk)	114a	52°41′N	2°02′W
Cannock Chase, reg., Eng., U.K. (kăn′ŭk chās)	114a	52°43′N	1°54′W
Cannon, r., Mn., U.S. (kăn′ŭn)	71	44°18′N	93°24′W
Cannonball, r., N.D., U.S. (kăn′ŭn-bäl)	70	46°17′N	101°35′W
Caño, Isla de, i., C.R. (ē′s-lä-dĕ-kä′nō)	91	8°38′N	84°00′W
Canoga Park, Ca., U.S. (kȧ-nō′gȧ)	75a	34°07′N	118°36′W
Canoncito Indian Reservation, I.R., N.M., U.S.	77	35°00′N	107°05′W
Canon City, Co., U.S. (kăn′yŭn)	78	38°27′N	105°16′W
Canonsburg, Pa., U.S. (kăn′ŭnz-bûrg)	69e	40°16′N	80°11′W
Canoochee, r., Ga., U.S. (kȧ-nōō′chē)	83	32°25′N	82°11′W
Canora, Can. (kȧ-nōrȧ)	50	51°37′N	102°26′W
Canosa, Italy (kä-nō′sä)	130	41°14′N	16°03′E
Canouan, i., St. Vin.	91b	12°44′N	61°10′W
Cansahcab, Mex.	90a	21°11′N	89°05′W
Canso, Can. (kăn′sō)	61	45°20′N	61°00′W
Canso, Cape, c., Can.	61	45°21′N	60°46′W
Canso, Strait of, strt., Can.	61	45°37′N	61°25′W
Cantabrica, Cordillera, mts., Spain	112	43°05′N	6°05′W
Cantagalo, Braz. (kän-tä-gá′lo)	99a	21°59′S	42°22′W
Cantanhede, Port. (kän-täɴ-yā′då)	128	40°22′N	8°35′W

ăt; finăl; rāte; senåte; ärm; ȧsk; sofȧ; fâre; ch-choose; dh-as th in other; bē; ēvent; bĕt; recĕnt; cratẽr; g-gō; gh-guttural g; bĭt; ĭ-short neutral; rīde; ĸ-guttural k as ch in German ich;

ng-sing; ŋ-baŋk; N-nasalized n; nŏd; cŏmmit; ōld; ōbey; ôrder; oi-boil; fōōd; ȯ-as oo in foot; ou-out; s-soft; sh-dish; th-thin; pūre; ŭnite; ûrn; stŭd; circŭs; ü-as in French tu; '-indeterminate vowel.

234 Castelli – Chame PRONOUNCING INDEX

PLACE (Pronunciation)	PAGE	LAT.	LONG.
Castelli, Arg. (kás-tĕ'zhē)	99c	36°07'S	57°48'W
Castelló de la Plana, Spain	118	39°59'N	0°05'W
Castelnaudary, Fr. (kás'tĕl-nō-dá-rē')	126	43°20'N	1°57'E
Castelo, Braz. (käs-tĕ'lô)	99a	20°37'S	41°13'W
Castelo Branco, Port. (käs-tā'lô brän'kô)	118	39°48'N	7°37'W
Castelo de Vide, Port. (käs-tā'lô dĭ vē'dĭ)	128	39°25'N	7°25'W
Castelsarrasin, Fr. (kás'tĕl-sá-rá-zăn')	126	44°03'N	1°05'E
Castelvetrano, Italy (käs'tĕl-vĕ-trä'nō)	130	37°43'N	12°50'E
Castilla, Peru (kás-tē'l-yä)	100	5°18'S	80°40'W
Castilla La Nueva, hist. reg., Spain (käs-tē'lyä lä nwä'vä)	128	39°15'N	3°55'W
Castilla La Vieja, hist. reg., Spain (käs-tēl'yä lä vyä'hä)	128	40°48'N	4°24'W
Castillo de San Marcos National Monument, rec., Fl., U.S. (käs-tē'lyä de-sän mär-kōs)	83	29°55'N	81°25'W
Castle, i., Bah. (kás''l)	93	22°05'N	74°20'W
Castlebar, Ire. (kás''l-bär)	120	53°55'N	9°15'W
Castle Dale, Ut., U.S. (kás'l dāl)	77	39°15'N	111°00'W
Castle Donington, Eng., U.K. (dŏn'ĭng-tŭn)	114a	52°50'N	1°21'W
Castleford, Eng., U.K. (kás''l-fērd)	114a	53°43'N	1°21'W
Castlegar, Can. (kás''l-gär)	55	49°19'N	117°40'W
Castlemaine, Austl. (käs''l-mān)	176	37°05'S	144°10'E
Castle Peak, mtn., Co., U.S.	77	39°00'N	106°50'W
Castle Rock, Wa., U.S. (kás''l-rŏk)	72	46°17'N	122°53'W
Castle Rock Flowage, res., Wi., U.S.	71	44°03'N	89°48'W
Castle Shannon, Pa., U.S. (shăn'ŭn)	69e	40°22'N	80°02'W
Castleton, In., U.S. (kás''l-tŭn)	69g	39°54'N	86°03'W
Castor, r., Can. (kăs'tôr)	62c	45°16'N	75°14'W
Castor, r., Mo., U.S.	79	36°59'N	89°53'W
Castres, Fr. (kás'tr')	126	43°36'N	2°13'E
Castries, St. Luc. (kás-trē')	91b	14°01'N	61°00'W
Castro, Braz. (käs'trô)	101	24°56'S	50°00'W
Castro, Chile (käs'trô)	102	42°27'S	73°48'W
Castro Daire, Port. (käs'trô dīr'ĭ)	128	40°56'N	7°57'W
Castro del Río, Spain (käs-trô-dĕl rē'ō)	128	37°42'N	4°28'W
Castrop Rauxel, Ger. (käs'trōp rou'ksĕl)	127c	51°33'N	7°19'E
Castro-Urdiales, Spain	118	43°23'N	3°11'W
Castro Valley, Ca., U.S.	74b	37°42'N	122°05'W
Castro Verde, Port. (käs-trō vĕr'dĕ)	128	37°43'N	8°05'W
Castrovillari, Italy (käs'trō-vēl-lyä'rē)	130	39°48'N	16°11'E
Castuera, Spain (käs-tô-ā'rä)	128	38°43'N	5°33'W
Casula, Moz.	191	15°25'S	33°40'E
Cat, i., Bah.	93	24°30'N	75°30'W
Catacamas, Hond. (kä-tä-kä'mäs)	90	14°52'N	85°55'W
Cataguases, Braz. (kä-tä-gwá'sĕs)	99a	21°23'S	42°42'W
Catahoula, l., La., U.S. (kăt-ȧ-hó'lȧ)	81	31°35'N	92°20'W
Catalão, Braz. (kä-tä-louN')	101	18°09'S	47°42'W
Catalina, i., Dom. Rep. (kä-tä-lē'nä)	93	18°20'N	69°00'W
Catalunya, hist. reg., Spain	129	41°23'N	0°50'E
Catamarca, Arg. (kä-tä-mä'r-kä)	102	28°29'S	65°45'W
Catamarca, prov., Arg. (kä-tä-mär'kä)	102	27°15'S	67°15'W
Catanauan, Phil. (kä-tä-nä'wän)	169a	13°36'N	122°20'E
Catanduanes Island, i., Phil. (kä-tän-dwä'nĕs)	169	13°55'N	125°00'E
Catanduva, Braz. (kä-tän-dōō'vä)	101	21°12'S	48°47'W
Catania, Italy (kä-tä'nyä)	110	37°30'N	15°09'E
Catania, Golfo di, b., Italy (gōl-fō-dē-kä-tä'nyä)	130	37°24'N	15°28'E
Catanzaro, Italy (kä-tän-dzä'rō)	119	38°53'N	16°34'E
Catarroja, Spain (kä-tär-rō'hä)	129	39°24'N	0°25'W
Catawba, r., N.C., U.S. (kȧ-tô'bȧ)	83	35°25'N	80°55'W
Catbalogan, Phil. (kät-bä-lō'gän)	169	11°45'N	124°52'E
Catemaco, Mex. (kä-tä-mä'kō)	89	18°26'N	95°06'W
Catemaco, Lago, l., Mex. (lä'gô-kä-tä-mä'kō)	89	18°23'N	95°04'W
Caterham, Eng., U.K. (kā'tēr-ŭm)	114b	51°16'N	0°04'W
Catete, Ang. (kä-tĕ'tĕ)	186	9°06'S	13°43'E
Cathedral Mountain, mtn., Tx., U.S. (kȧ-thē'drăl)	80	30°09'N	103°46'W
Cathedral Peak, mtn., Afr. (kȧ-thē'drăl)	187c	28°53'S	29°04'E
Catherine, Lake, l., Ar., U.S. (kă-thēr-ĭn)	79	34°26'N	92°47'W
Cathkin Peak, mtn., Afr. (kăth'kĭn)	186	29°08'S	29°22'E
Cathlamet, Wa., U.S. (kăth-lăm'ĕt)	74c	46°12'N	123°22'W
Catlettsburg, Ky., U.S. (kăt'lĕts-bûrg)	66	38°20'N	82°35'W
Catoche, Cabo, c., Mex. (kä-tō'chĕ)	86	21°30'N	87°15'W
Catonsville, Md., U.S. (kā'tŭnz-vĭl)	68e	39°16'N	76°45'W
Catorce, Mex. (kä-tôr'sā)	88	23°41'N	100°51'W
Catskill, N.Y., U.S. (kăts'kĭl)	67	42°15'N	73°50'W
Catskill Mountains, mts., N.Y., U.S.	65	42°20'N	74°35'W
Cattaraugus Indian Reservation, I.R., N.Y., U.S. (kăt'tȧ-rȧ-gŭs)	67	42°30'N	79°05'W
Catu, Braz. (ká-tōō)	101	12°26'S	38°12'W
Catuala, Ang.	190	16°29'S	19°03'E
Catumbela, r., Ang. (kä'tŏm-bĕl'ȧ)	190	12°40'S	14°10'E
Cauayan, Phil. (kou-ä'yän)	169a	16°56'N	121°46'E
Cauca, r., Col. (kou'kä)	100	7°30'N	75°26'W
Caucagua, Ven. (käö-kä'gwä)	101b	10°17'N	66°22'W
Caucasus, mts.	134	43°20'N	42°00'E
Cauchon Lake, l., Can. (kô-shŏn')	57	55°25'N	96°30'W
Caughnawaga, Can.	62a	45°24'N	73°41'W
Caulfield, Austl.	173a	37°53'S	145°03'E
Caulonia, Italy (kou-lō'nyä)	130	38°24'N	16°22'E
Cauquenes, Chile (kou-kā'nās)	102	35°54'S	72°14'W
Caura, r., Ven. (kou'rä)	100	6°48'N	64°40'W
Causapscal, Can.	60	48°22'N	67°14'W
Caution, Cape, c., Can. (kô'shŭn)	54	51°10'N	127°47'W
Cauto, r., Cuba (kou'tō)	92	20°33'N	76°20'W
Cauvery, r., India	155	12°00'N	77°00'E
Cava, Braz. (kä'vä)	102b	22°41'S	43°26'W
Cava de' Tirreni, Italy (kä'vä-dĕ-tēr-rē'nē)	129c	40°27'N	14°43'E
Cávado, r., Port. (kä-vä'dō)	128	41°43'N	8°08'W
Cavalcante, Braz. (kä-väl-kän'tā)	101	13°45'S	47°33'W
Cavalier, N.D., U.S. (kăv-ȧ-lēr')	70	48°45'N	97°39'W
Cavally, r., Afr.	188	4°40'N	7°30'W
Cavan, Ire. (kăv'ăn)	120	54°01'N	7°00'W
Cavarzere, Italy (kä-vär'dzā-rā)	130	45°08'N	12°06'E
Cavendish, Vt., U.S. (kăv'ĕn-dĭsh)	67	43°25'N	72°35'W
Caviana, Ilha, i., Braz. (kä-vyä'nä)	101	0°45'N	49°33'W
Cavite, Phil. (kä-vē'tā)	169a	14°30'N	120°54'E
Caxambu, Braz. (kä-shá'm-bōō)	101	22°00'S	44°45'W
Caxias, Braz. (kä'shē-äzh)	101	4°48'S	43°16'W
Caxias do Sul, Braz. (kä'shē-äzh-dô-sōō'l)	102	29°13'S	51°03'W
Caxito, Ang. (kä-shē'tô)	186	8°33'S	13°36'E
Cayambe, Ec. (ká-īä'm-bĕ)	100	0°03'N	79°09'W
Cayenne, Fr. Gu. (kā-ĕn')	101	4°56'N	52°18'W
Cayetano Rubio, Mex. (kä-yĕ-tá-nô-rōō'byô)	88	20°37'N	100°21'W
Cayey, P.R.	87b	18°05'N	66°12'W
Cayman Brac, i., Cay. Is. (kī-män' bràk)	92	19°45'N	79°50'W
Cayman Islands, dep., N.A.	92	19°30'N	80°30'W
Cay Sal Bank, bk., (kē-sāl)	92	23°55'N	80°20'W
Cayuga, l., N.Y., U.S. (kā-yōō'gȧ)	67	42°35'N	76°35'W
Cazalla de la Sierra, Spain	128	37°55'N	5°48'W
Cazaux, Étang de, l., Fr. (ā-tăN' dĕ kä-zō')	126	44°32'N	0°59'W
Cazenovia, N.Y., U.S. (kăz-ē-nō'vĭ-ȧ)	67	42°55'N	75°50'W
Cazenovia Creek, r., N.Y., U.S.	69c	42°49'N	78°45'W
Čazma, Cro. (chäz'mä)	130	45°44'N	16°39'E
Cazombo, Ang. (kä-zô'm-bŏ)	186	11°54'S	22°52'E
Cazones, r., Mex. (kä-zō'nĕs)	89	20°37'N	97°28'W
Cazones, Ensenada de, b., Cuba (ĕn-sĕ-nä-dä-dĕ-kä-zō'nås)	92	22°05'N	81°30'W
Cazones, Golfo de, b., Cuba (gŏl-fô-dĕ-kä-zō'nås)	92	21°55'N	81°15'W
Cazorla, Spain (kä-thôr'lä)	128	37°55'N	2°58'W
Cea, r., Spain (thā'ä)	128	42°18'N	5°10'W
Ceará-Mirim, Braz. (sā-ä-rä'mē-rē'N)	101	6°00'S	35°13'W
Cebaco, Isla, i., Pan. (ĕ's-lä-sā-bä'kō)	91	7°27'N	81°08'W
Cebolla Creek, r., Co., U.S. (sē-bōl'yä)	77	38°15'N	107°10'W
Cebreros, Spain (sĕ-brĕ'rôs)	128	40°28'N	4°28'W
Cebu, Phil. (sā-bōō')	169	10°22'N	123°49'E
Čechy (Bohemia), hist. reg., Czech Rep.	124	49°51'N	13°55'E
Cecil, Pa., U.S. (sē'sĭl)	69e	40°20'N	80°10'W
Cedar, r., Ia., U.S.	71	42°23'N	92°07'W
Cedar, r., Wa., U.S.	74c	45°56'N	122°32'W
Cedar, West Fork, r., Ia., U.S.	71	42°49'N	93°10'W
Cedar Bayou, Tx., U.S.	81a	29°54'N	94°58'W
Cedar Breaks National Monument, rec., Ut., U.S.	77	37°35'N	112°55'W
Cedarburg, Wi., U.S. (sē'dēr bûrg)	71	43°23'N	88°00'W
Cedar City, Ut., U.S.	77	37°40'N	113°10'W
Cedar Creek, r., N.D., U.S.	70	46°05'N	102°10'W
Cedar Falls, Ia., U.S.	71	42°31'N	92°29'W
Cedar Keys, Fl., U.S.	82	29°06'N	83°03'W
Cedar Lake, In., U.S.	69a	41°22'N	87°27'W
Cedar Lake, l., In., U.S.	69a	41°23'N	87°25'W
Cedar Lake, res., Can.	52	53°10'N	100°00'W
Cedar Rapids, Ia., U.S.	65	42°00'N	91°43'W
Cedar Springs, Mi., U.S.	66	43°15'N	85°40'W
Cedartown, Ga., U.S. (sē'dēr-toun)	82	34°00'N	85°15'W
Cedarville, S. Afr. (cĕdȧr'vĭl)	187c	30°23'S	29°04'E
Cedral, Mex. (sā-dräl')	88	23°47'N	100°42'W
Cedros, Hond. (sā'drōs)	90	14°36'N	87°07'W
Cedros, i., Mex.	86	28°10'N	115°10'W
Ceduna, Austl. (sē-dô'nȧ)	174	32°15'S	133°55'E
Ceel Buur, Som.	192a	4°35'N	46°40'E
Cega, r., Spain (thā'gä)	128	41°25'N	4°27'W
Cegléd, Hung. (tsā'glād)	125	47°10'N	19°49'E
Ceglie, Italy (chĕ'lyĕ)	131	40°39'N	17°32'E
Cehegín, Spain (thā-å-hēn')	128	38°05'N	1°48'W
Ceiba del Agua, Cuba (sā'bä-dĕl-ä'gwä)	93a	22°53'N	82°38'W
Cekhira, Tun.	184	34°17'N	10°00'E
Celaya, Mex. (sā-lä'yä)	86	20°33'N	100°49'W
Celebes (Sulawesi), i., Indon.	168	2°15'S	120°30'E
Celebes Sea, sea, Asia	168	3°45'N	121°52'E
Celestún, Mex. (sĕ-lĕs-tōō'n)	90a	20°57'N	90°18'W
Celina, Oh., U.S. (sĕlī'na)	66	40°30'N	84°35'W
Celje, Slvn. (tsĕl'yĕ)	130	46°13'N	15°17'E
Celle, Ger. (tsĕl'ĕ)	117	52°37'N	10°05'E
Cement, Ok., U.S. (sē-mĕnt')	78	34°56'N	98°07'W
Cenderawasih, Teluk, b., Indon.	169	2°20'S	135°30'E
Ceniza, Pico, mtn., Ven. (pē'kô-sĕ-nē'zä)	101b	10°24'N	67°26'W
Center, Tx., U.S. (sĕn'tēr)	81	31°50'N	94°10'W
Center Hill Lake, res., Tn., U.S. (sĕn'tēr-hĭl)	82	36°02'N	86°00'W
Center Line, Mi., U.S. (sĕn'tēr līn)	69b	42°29'N	83°01'W
Centerville, Ia., U.S. (sĕn'tēr-vĭl)	71	40°44'N	92°48'W
Centerville, Mn., U.S.	75g	45°10'N	93°03'W
Centerville, Pa., U.S.	69e	40°02'N	79°58'W
Centerville, S.D., U.S.	70	43°07'N	96°56'W
Centerville, Ut., U.S.	75b	40°55'N	111°53'W
Central, Cordillera, mts., Bol. (kôr-dēl-yĕ'rä-sĕn-trá'l)	100	19°18'S	65°29'W
Central, Cordillera, mts., Col.	100a	3°58'N	75°55'W
Central, Cordillera, mts., Dom. Rep.	93	19°05'N	71°30'W
Central, Cordillera, mts., Phil. (kôr-dēl-yĕ'rä-sĕn-träl')	169a	17°05'N	120°55'E
Central African Republic, nation, Afr.	185	7°50'N	21°00'E
Central America, reg., N.A. (ä-mĕr'ĭ-kȧ)	86	10°45'N	87°15'W
Central City, Ky., U.S. (sĕn'trăl)	82	37°15'N	87°09'W
Central City, Ne., U.S. (sĕn'trȧl sĭ'tĭ)	70	41°07'N	98°00'W
Central Falls, R.I., U.S. (sĕn'trăl fôlz)	68b	41°54'N	71°23'W
Centralia, Il., U.S. (sĕn-trā'lĭ-ȧ)	66	38°35'N	89°05'W
Centralia, Mo., U.S.	79	39°11'N	92°07'W
Centralia, Wa., U.S.	72	46°42'N	122°58'W
Central Plateau, plat., Russia	136	55°00'N	33°30'E
Central Valley, N.Y., U.S.	68a	41°19'N	74°07'W
Centreville, Il., U.S. (sĕn'tēr-vĭl)	75e	38°33'N	90°06'W
Centreville, Md., U.S.	67	39°05'N	76°05'W
Century, Fl., U.S. (sĕn'tů-rĭ)	82	30°57'N	87°15'W
Ceram (Seram), i., Indon.	169	2°45'S	129°30'E
Céret, Fr.	126	42°29'N	2°47'E
Cerignola, Italy (chā-rē-nyô'lä)	130	41°16'N	15°55'E
Cerknica, Slvn. (tsĕr'knē-tsá)	130	45°48'N	14°21'E
Cern'achovsk, Russia (chĕr-nyä'kôfsk)	136	54°38'N	21°49'E
Cerralvo, Mex. (sĕr-räl'vō)	80	26°05'N	99°37'W
Cerralvo, i., Mex.	86	24°00'N	109°59'W
Cerrito, Col. (sĕr-rē'-tô)	100a	3°41'N	76°17'W
Cerritos, Mex. (sĕr-rē'tôs)	88	22°26'N	100°16'W
Cerro de Pasco, Peru (sĕr'rō dā päs'kō)	100	10°45'S	76°14'W
Cerro Gordo, Arroyo de, r., Mex. (är-rô-yō-dĕ-sĕ'r-rō-gôr-dō)	80	26°12'N	104°06'W
Certegui, Col. (sĕr-tĕ'gē)	100a	5°21'N	76°35'W
Cervantes, Phil. (sĕr-vän'tās)	169a	16°59'N	120°42'E
Cervera del Río Alhama, Spain	128	42°02'N	1°55'W
Cerveteri, Italy (chĕr-vĕ'tĕ-rē)	129d	42°00'N	12°06'E
Cesena, Italy (chĕ'sĕ-nä)	130	44°08'N	12°16'E
Cēsis, Lat. (sā'sĭs)	123	57°19'N	25°17'E
Česká Lípa, Czech Rep. (chĕs'kä lē'pa)	124	50°41'N	14°31'E
České Budějovice, Czech Rep. (chĕs'kā bōō'dyĕ-yô-vĕt-sĕ)	117	49°00'N	14°30'E
Českomoravská Vysočina, hills, Czech Rep.	124	49°21'N	15°40'E
Český Těšín, Czech Rep.	125	49°43'N	18°22'E
Çeşme, Tur. (chĕsh'mē)	131	38°20'N	26°20'E
Cessnock, Austl.	175	32°58'S	151°15'E
Cestos, r., Lib.	188	5°40'N	9°25'W
Cetinje, Serb. (tsĕt'ĭn-yĕ)	110	42°23'N	18°55'E
Ceuta, Sp. N. Afr. (thā-ōō'tä)	184	36°04'N	5°36'W
Cévennes, reg., Fr. (sā-vĕn')	117	44°20'N	3°48'E
Ceylon *see* Sri Lanka, nation, Asia	159	8°45'N	82°30'E
Chabot, Lake, l., Ca., U.S. (sha'bŏt)	74b	37°44'N	122°06'W
Chacabuco, Arg. (chä-kä-bōō'kō)	99c	34°37'S	60°27'W
Chacaltianguis, Mex. (chä-käl-tē-äŋ'gwĕs)	89	18°18'N	95°50'W
Chachapoyas, Peru (chä-chä-poi'yäs)	100	6°16'S	77°48'W
Chaco, prov., Arg. (chä'kō)	102	26°00'S	60°45'W
Chaco Culture National Historic Park, rec., N.M., U.S. (chä'kō)	77	36°05'N	108°00'W
Chad, Russia (chäd)	142a	56°33'N	57°11'E
Chad, nation, Afr.	185	17°48'N	19°00'E
Chad, Lake, l., Afr.	185	13°55'N	13°40'E
Chadbourn, N.C., U.S. (chăd'bŭrn)	83	34°19'N	78°55'W
Chadron, Ne., U.S. (chăd'rŭn)	64	42°50'N	103°10'W
Chafarinas, Islas, is., Sp. N. Afr.	128	35°08'N	2°20'W
Chaffee, Mo., U.S. (chăf'ē)	79	37°10'N	89°39'W
Chāgal Hills, hills, Afg.	154	29°15'N	63°28'E
Chagodoshcha, r., Russia (chä-gō-dôsh-chä)	132	59°08'N	35°13'E
Chagres, r., Pan. (chä'grĕs)	91	9°18'N	79°22'W
Chagrin, r., Oh., U.S. (shȧ'grĭn)	69d	41°34'N	81°24'W
Chagrin Falls, Oh., U.S. (shȧ'grĭn fôls)	69d	41°26'N	81°23'W
Chahar, hist. reg., China (chä-här)	161	44°25'N	115°00'E
Chake Chake, Tan.	191	5°15'S	39°46'E
Chalatenango, El Sal. (chäl-ä-tĕ-nän'gō)	90	14°04'N	88°54'W
Chalbi Desert, des., Kenya	191	3°40'N	36°50'E
Chalcatongo, Mex. (chäl-kä-tôŋ'gō)	89	17°04'N	97°41'W
Chalchihuites, Mex. (chäl-chē-wē'tås)	88	23°28'N	103°57'W
Chalchuapa, El Sal. (chäl-chwä'pä)	90	14°01'N	89°39'W
Chalco, Mex. (chäl-kō)	89a	19°15'N	98°54'W
Chaleur Bay, b., Can. (shá-lûr')	53	47°58'N	65°33'W
Chalgrove, Eng., U.K. (chăl'grôv)	114b	51°38'N	1°05'W
Chaling, China (chä'lĭng)	165	27°00'N	113°31'E
Chalkída, Grc.	119	38°28'N	23°38'E
Chalmette, La., U.S. (shăl-mĕt')	68d	29°57'N	89°57'W
Châlons-sur-Marne, Fr. (shä-lôN'sür-märn)	117	48°57'N	4°23'E
Chalon-sur-Saône, Fr.	117	46°47'N	4°54'E
Chaltel, Cerro (Monte Fitzroy), mtn., S.A. (sĕ'r-rô-chäl'tĕl)	102	48°10'S	73°18'W
Chālūs, Iran	157	36°38'N	51°26'E
Chama, Rio, r., N.M., U.S. (chä'mä)	77	36°19'N	106°31'W
Chama, Sierra de, mts., Guat. (sē-ĕ'r-rä-dĕ-chä-mä)	90	15°48'N	90°20'W
Chamama, Mwi.	191	12°55'S	33°43'E
Chaman, Pak. (chŭm-än')	155	30°58'N	66°21'E
Chambal, r., India (chŭm-bäl')	155	24°30'N	75°30'E
Chamberlain, S.D., U.S. (chām'bēr-lĭn)	70	43°48'N	99°21'W
Chamberlain, l., Me., U.S.	60	46°15'N	69°10'W
Chambersburg, Pa., U.S. (chām'bērz-bûrg)	67	40°00'N	77°40'W
Chambéry, Fr. (shäm-bā-rē')	117	45°35'N	5°54'E
Chambeshi, r., Zam.	191	10°35'S	31°20'E
Chamblee, Ga., U.S. (chăm-blē')	68c	33°55'N	84°18'W
Chambly, Can. (shäN-blē')	62a	45°27'N	73°17'W
Chambly, Fr.	127b	49°11'N	2°14'E
Chambord, Can.	51	48°22'N	72°01'W
Chame, Punta, c., Pan. (pô'n-tä-chä'mȧ)	91	8°41'N	79°27'W

PRONOUNCING INDEX
Chamelecón – Chesnokovka 235

PLACE (Pronunciation)	PAGE	LAT.	LONG.
Chamelecón, r., Hond. (chä-mĕ-lĕ-kō'n)	90	15°09'N	88°42'W
Chamo, l., Eth.	185	5°58'N	37°00'E
Chamonix-Mont-Blanc, Fr. (shȧ-mō-nē')	127	45°55'N	6°50'E
Champagne, reg., Fr. (shäm-pän'yē)	126	48°53'N	4°48'E
Champaign, Il., U.S. (shăm-pān')	65	40°10'N	88°15'W
Champdāni, India	158a	22°48'N	88°21'E
Champerico, Guat. (chăm-pā-rē'kō)	90	14°18'N	91°55'W
Champion, Mi., U.S. (chăm'pĭ-ŭn)	71	46°30'N	87°59'W
Champlain, Lake, l., N.A. (shăm-plān')	65	44°45'N	73°20'W
Champlitte-et-le-Prálot, Fr. (shäɴ-plēt')	127	47°38'N	5°28'E
Champotón, Mex. (chäm-pō-tōn')	89	19°21'N	90°43'W
Champotón, r., Mex.	89	19°19'N	90°15'W
Chañaral, Chile (chän-yä-räl')	102	26°20'S	70°46'W
Chances Peak, vol., Monts.	91b	16°43'N	62°10'W
Chandeleur Islands, is., La., U.S. (shăn-dē-lōōr')	82	29°53'N	88°35'W
Chandeleur Sound, strt., La., U.S.	82	29°47'N	89°08'W
Chandīgarh, India	155	30°51'N	77°13'E
Chandler, Can. (chăn'dlēr)	51	48°21'N	64°41'W
Chandler, Ok., U.S.	79	35°42'N	96°52'W
Chandrapur, India	155	19°58'N	79°21'E
Chang *see* Yangtze, r., China	161	30°30'N	117°25'E
Changane, r., Moz.	186	22°42'S	32°46'E
Changara, Moz.	191	16°54'S	33°14'E
Changchun, China (chäŋ-chòn)	161	43°55'N	125°25'E
Changdang Hu, l., China (chäŋ-däŋ hōō)	162	31°37'N	119°29'E
Changde, China (chäŋ-dŭ)	161	29°00'N	111°38'E
Changhua, Tai. (chäng'hwä')	165	24°02'N	120°32'E
Changjŏn, Kor., N. (chäng'jŭn')	166	38°40'N	128°05'E
Changli, China (chäŋ-lē)	164	39°46'N	119°10'E
Changning, China (chäŋ-nĭŋ)	160	24°34'N	99°49'E
Changping, China (chäŋ-pĭŋ)	164	40°12'N	116°10'E
Changqing, China (chäŋ-chyĭŋ)	162	36°33'N	116°42'E
Changsan Got, c., Kor., N.	166	38°06'N	124°50'E
Changsha, China (chäŋ-shä)	161	28°20'N	113°00'E
Changshan Qundao, is., China (chäŋ-shän chyòn-dou)	162	39°08'N	122°26'E
Changshu, China (chäŋ-shōō)	162	31°40'N	120°45'E
Changting, China	165	25°50'N	116°18'E
Changwu, China (chäng'wōō')	164	35°12'N	107°45'E
Changxindianzhen, China (chäŋ-shyĭn-dĭĕn-jŭn)	164a	39°49'N	116°12'E
Changxing Dao, i., China (chäŋ-shyĭŋ dou)	162	39°38'N	121°10'E
Changyi, China (chäŋ-yē)	162	36°51'N	119°23'E
Changyuan, China (chyäŋ-yuän)	162	35°10'N	114°41'E
Changzhi, China (chäŋ-jr)	164	35°58'N	112°58'E
Changzhou, China (chäŋ-jō)	161	31°47'N	119°56'E
Changzhuyuan, China (chäŋ-jōō-yuän)	162	31°33'N	115°17'E
Chanhassen, Mn., U.S. (shăn'hăs-sĕn)	75g	44°52'N	93°32'W
Chaniá, Grc.	118	35°31'N	24°01'E
Channel Islands, is., Eur. (chăn'ĕl)	112	49°15'N	3°30'W
Channel Islands, is., Ca., U.S.	76	33°30'N	119°15'W
Channel-Port-aux-Basques, Can.	51	47°35'N	59°11'W
Channelview, Tx., U.S. (chănĕlvū)	81a	29°46'N	95°07'W
Chantada, Spain (chän-tä'dä)	128	42°38'N	7°36'W
Chanthaburi, Thai.	168	12°37'N	102°04'E
Chantilly, Fr. (shäɴ-tē-yē')	127b	49°12'N	2°30'E
Chantilly, Va., U.S. (shăn'tĭlē)	68e	38°53'N	77°26'W
Chantrey Inlet, b., Can. (chăn-trē)	52	67°49'N	95°00'W
Chanute, Ks., U.S. (shȧ-nōōt')	65	37°41'N	95°27'W
Chany, l., Russia (chä'nē)	134	54°15'N	77°31'E
Chao'an, China (chou-än)	161	23°48'N	116°35'E
Chao Hu, l., China	165	31°45'N	116°59'E
Chao Phraya, r., Thai.	168	16°13'N	99°33'E
Chaor, r., China (chou-r)	164	47°20'N	121°40'E
Chaoshui, China (chou-shwā)	162	37°43'N	120°56'E
Chaoxian, China (chou shyĕn)	162	31°37'N	117°50'E
Chaoyang, China	161	41°32'N	120°20'E
Chaoyang, China (chou-yäŋ)	165	23°18'N	116°32'E
Chapada, Serra da, mts., Braz. (sĕ'r-rä-dä-shä-pä'dä)	101	14°57'S	54°34'W
Chapadão, Serra do, mtn., Braz. (sĕ'r-rä-dô-shä-pȧ-dou'ɴ)	99a	20°31'S	46°20'W
Chapala, Mex. (chä-pä'lä)	88	20°18'N	103°10'W
Chapala, Lago de, l., Mex. (lä'gô-dĕ-chä-pä'lä)	86	20°14'N	103°02'W
Chapalagana, r., Mex. (chä-pä-lä-gä'nä)	88	22°11'N	104°09'W
Chaparral, Col. (chä-pär-rá'l)	100	3°44'N	75°28'W
Chapayevsk, Russia (châ-pī'ĕfsk)	136	53°00'N	49°30'E
Chapel Hill, N.C., U.S. (chăp''l hĭl)	83	35°55'N	79°05'W
Chaplain, l., Wa., U.S. (chăp'lĭn)	74a	47°58'N	121°50'W
Chapleau, Can. (chăp-lō')	51	47°43'N	83°28'W
Chapman, Mount, mtn., Can. (chăp'măn)	55	51°50'N	118°20'W
Chapman's Bay, b., S. Afr. (chăp'măns bā)	186a	34°06'S	18°17'E
Chappell, Ne., U.S. (chă-pĕl')	70	41°06'N	102°29'W
Chapultenango, Mex. (chä-pòl-tē-näŋ'gō)	89	17°19'N	93°08'W
Chá Pungana, Ang.	190	13°44'S	18°39'E
Chār Borjak, Afg.	157	30°17'N	62°03'E
Charcas, Mex. (chär'käs)	88	23°09'N	101°09'W
Charco de Azul, Bahía, b., Pan.	91	8°14'N	82°45'W
Charente, r., Fr. (shȧ-räɴt')	126	45°48'N	0°28'W
Chari, r., Afr. (shä-rē')	189	12°45'N	14°55'E
Charing, Eng., U.K. (chă'rĭng)	114b	51°13'N	0°49'E
Chariton, Ia., U.S. (châr'ĭ-tŭn)	71	41°02'N	93°16'W
Chariton, r., Mo., U.S.	79	40°24'N	92°38'W

PLACE (Pronunciation)	PAGE	LAT.	LONG.
Charjew, Turkmen.	139	38°52'N	63°37'E
Charlemagne, Can. (shärl-màny')	62a	45°43'N	73°29'W
Charleroi, Bel. (shȧr-lē-rwä')	117	50°25'N	4°35'E
Charleroi, Pa., U.S. (shär'lē-roi)	69e	40°08'N	79°54'W
Charles, Cape, c., Va., U.S. (chärlz)	67	37°05'N	75°48'W
Charlesbourg, Can. (shȧrl-bōōr')	62b	46°51'N	71°16'W
Charles City, Ia., U.S. (chärlz)	71	43°03'N	92°40'W
Charleston, Il., U.S. (chärlz'tŭn)	66	39°30'N	88°10'W
Charleston, Mo., U.S.	79	36°53'N	89°20'W
Charleston, Ms., U.S.	82	34°00'N	90°02'W
Charleston, S.C., U.S.	65	32°47'N	79°56'W
Charleston, W.V., U.S.	65	38°20'N	81°35'W
Charlestown, St. K./N.	91b	17°10'N	62°32'W
Charlestown, In., U.S. (chärlz'toun)	69h	38°46'N	85°39'W
Charleville, Austl. (chär'lē-vĭl)	175	26°16'S	146°28'E
Charleville Mézières, Fr. (shärl-vēl')	126	49°48'N	4°41'E
Charlevoix, Mi., U.S. (shär'lē-voi)	66	45°20'N	85°15'W
Charlevoix, Lake, l., Mi., U.S.	71	45°17'N	85°43'W
Charlotte, Mi., U.S. (shär'lŏt)	66	42°35'N	84°50'W
Charlotte, N.C., U.S.	65	35°15'N	80°50'W
Charlotte Amalie, V.I.U.S. (shär-lŏt'ĕ ä-mä'lĭ-ȧ)	87	18°21'N	64°54'W
Charlotte Harbor, b., Fl., U.S.	83a	26°49'N	82°00'W
Charlotte Lake, l., Can.	54	52°07'N	125°30'W
Charlottenberg, Swe. (shär-lŭt'ĕn-bĕrg)	122	59°53'N	12°17'E
Charlottesville, Va., U.S. (shär'lŏtz-vĭl)	65	38°00'N	78°25'W
Charlottetown, Can. (shär'lŏt-toun)	51	46°14'N	63°08'W
Charlotte Waters, Austl. (shär'lŏt)	174	26°00'S	134°50'E
Charmes, Fr. (shärm)	127	48°23'N	6°19'E
Charnwood Forest, for., Eng., U.K. (chärn'wòd)	114a	52°42'N	1°15'W
Charny, Can. (shär-nē')	62b	46°43'N	71°16'W
Chars, Fr. (shär)	127b	49°09'N	1°57'E
Chārsadda, Pak. (chŭr-sä'dä)	155a	34°17'N	71°43'E
Charters Towers, Austl. (chär'tērz)	175	20°03'S	146°20'E
Chartres, Fr. (shärt'r')	117	48°26'N	1°29'E
Chascomús, Arg. (chäs-kō-mōōs')	102	35°32'S	58°01'W
Chase City, Va., U.S. (chās)	83	36°45'N	78°27'W
Chashniki, Bela. (chäsh'nyē-kē)	132	54°51'N	29°08'E
Chaska, Mn., U.S. (chăs'kȧ)	75g	44°48'N	93°36'W
Châteaudun, Fr. (shä-tō-dăɴ')	126	48°04'N	1°23'E
Château-Gontier, Fr. (chȧ-tō-gôɴ'tyā')	126	47°48'N	0°43'W
Châteauguay, Can. (chȧ-tō-gā)	62a	45°22'N	73°45'W
Châteauguay, r., N.A.	62a	45°13'N	73°51'W
Châteauneaut, Fr.	126a	43°23'N	5°11'E
Château-Renault, Fr. (shȧ-tō-rē-nō')	126	47°36'N	0°57'E
Château-Richer, Can. (shä-tō'rē-shā')	62b	47°00'N	71°01'W
Châteauroux, Fr. (shȧ-tō-rōō')	117	46°47'N	1°39'E
Château-Thierry, Fr. (shȧ-tō'ty-ĕr-rē')	126	49°03'N	3°22'E
Châtellerault, Fr. (shä-tĕl-rō')	117	46°48'N	0°31'E
Chatfield, Mn., U.S. (chăt'fēld)	71	43°50'N	92°10'W
Chatham, Can. (chăt'ăm)	51	42°25'N	82°10'W
Chatham, Can.	51	47°02'N	65°28'W
Chatham, Eng., U.K. (chăt'ŭm)	121	51°23'N	0°32'E
Chatham, N.J., U.S. (chăt'ăm)	68a	40°44'N	74°23'W
Chatham, Oh., U.S.	69d	41°06'N	82°01'W
Chatham Islands, is., N.Z.	2	44°00'S	178°00'W
Chatham Sound, strt., Can.	54	54°32'N	130°35'W
Chatham Strait, strt., Ak., U.S.	63	57°00'N	134°40'W
Chatsworth, Ca., U.S. (chătz'wûrth)	75a	34°16'N	118°36'W
Chatsworth Reservoir, res., Ca., U.S.	75a	34°15'N	118°41'W
Chattahoochee, Fl., U.S. (chăt-tȧ-hōō' cheē)	82	30°42'N	84°47'W
Chattahoochee, r., U.S.	65	32°00'N	85°10'W
Chattanooga, Tn., U.S. (chăt-ȧ-nōō'gȧ)	65	35°01'N	85°15'W
Chattooga, r., Ga., U.S. (chă-tōō'gȧ)	82	34°47'N	83°13'W
Chaudière, r., Can. (shō-dyĕr')	59	46°26'N	71°10'W
Chaumont, Fr. (shō-môɴ')	117	48°08'N	5°07'E
Chaunskaya Guba, b., Russia	141	69°15'N	170°00'E
Chauny, Fr. (shō-nē')	126	49°40'N	3°09'E
Chau-phu, Viet.	168	10°49'N	104°57'E
Chautauqua, l., N.Y., U.S. (shȧ-tô'kwȧ)	67	42°10'N	79°25'W
Chavaniga, Russia	136	66°02'N	37°50'E
Chaves, Port. (chä'vĕzh)	128	41°44'N	7°30'W
Chavinda, Mex. (chä-vē'n-dä)	88	20°01'N	102°27'W
Chavusi, Bela.	132	53°57'N	30°58'E
Chazumba, Mex. (chä-zòm'bä)	89	18°11'N	97°41'W
Cheadle, Eng., U.K. (chē'd'l)	114a	52°59'N	1°59'W
Cheat, W.V., U.S. (chēt)	67	39°35'N	79°40'W
Cheb, Czech Rep. (kĕb)	124	50°05'N	12°23'E
Chebarkul', Russia (chĕ-bȧr-kŭl')	142a	54°59'N	60°22'E
Cheboksary, Russia (chyĕ-bôk-sä'rē)	136	56°00'N	47°20'E
Cheboygan, Mi., U.S. (shē-boi'găn)	66	45°40'N	84°30'W
Chech, Erg, des., Alg.	184	24°45'N	2°07'W
Chechen', i., Russia (chyĕch'ĕn)	137	44°00'N	48°10'E
Chechnya, prov., Russia	138	43°30'N	45°50'E
Checotah, Ok., U.S. (chē-kō'tȧ)	79	35°27'N	95°32'W
Chedabucto Bay, b., Can. (chĕd-ȧ-bŭk-tō)	61	45°23'N	61°10'W
Cheduba Island, i., Mya.	168	18°45'N	93°01'E
Cheecham Hills, hills, Can. (chēē'hăm)	56	56°20'N	111°10'W
Cheektowaga, N.Y., U.S. (chēk-tō-wä'gȧ)	69c	42°54'N	78°46'W
Chefoo *see* Yantai, China	161	37°32'N	121°22'E
Chegutu, Zimb.	186	18°18'S	30°10'E
Chehalis, Wa., U.S. (chē-hā'lĭs)	72	46°39'N	122°58'W
Chehalis, r., Wa., U.S.	72	46°47'N	123°17'W
Cheju, Kor., S. (chĕ'jōō')	166	33°29'N	126°40'E
Cheju (Quelpart), i., Kor., S.	166	33°20'N	126°25'E
Chekalin, Russia (chĕ-kä'lĭn)	132	54°05'N	36°13'E
Chela, Serra da, mts., Ang. (sĕr'rȧ dä shā'lȧ)	186	15°30'S	13°30'E

PLACE (Pronunciation)	PAGE	LAT.	LONG.
Chelan, Wa., U.S. (chē-lăn')	72	47°51'N	119°59'W
Chelan, Lake, l., Wa., U.S.	72	48°09'N	120°20'W
Cheleiros, Port. (shĕ-la'rōzh)	129b	38°54'N	9°19'W
Chéliff, r., Alg. (shā-lĕf)	184	36°00'N	2°00'E
Chelles, Fr.	127b	48°53'N	2°36'E
Chełm, Pol. (kĕlm)	117	51°08'N	23°30'E
Chełmno, Pol. (kĕlm'nō)	125	53°20'N	18°25'E
Chelmsford, Can.	58	46°35'N	81°12'W
Chelmsford, Eng., U.K. (chĕlm's-fērd)	121	51°44'N	0°28'E
Chelmsford, Ma., U.S.	61a	42°36'N	71°21'W
Chelsea, Austl.	173a	38°05'S	145°08'E
Chelsea, Can.	62c	45°30'N	75°46'W
Chelsea, Al., U.S. (chĕl'sē)	68h	33°20'N	86°38'W
Chelsea, Ma., U.S.	61a	42°23'N	71°02'W
Chelsea, Mi., U.S.	66	42°20'N	84°00'W
Chelsea, Ok., U.S.	79	36°32'N	95°23'W
Cheltenham, Eng., U.K. (chĕlt'nŭm)	120	51°57'N	2°06'W
Cheltenham, Md., U.S. (chĕltĕn-hăm)	68e	38°45'N	76°50'W
Chelyabinsk, Russia (chĕl-yä-bĕnsk')	134	55°10'N	61°25'E
Chelyuskin, Mys, c., Russia (chĕl-yòs'-kĭn)	135	77°45'N	104°45'E
Chemba, Moz.	191	17°08'S	34°52'E
Chemnitz, Ger.	117	50°48'N	12°53'E
Chemung, r., N.Y., U.S. (shē-mŭng)	67	42°20'N	77°25'W
Chën, Gora, mtn., Russia	135	65°13'N	142°12'E
Chenāb, r., Asia (chē-näb)	155	30°30'N	71°30'E
Chenachane, Alg. (shē-nȧ-shän')	184	26°14'N	4°14'W
Chencun, China (chŭn-tsòn)	163a	22°58'N	113°14'E
Cheney, Wa., U.S. (chē'nȧ)	72	47°29'N	117°34'W
Chengde, China (chŭŋ-dŭ)	161	40°50'N	117°50'E
Chengdong Hu, l., China (chŭŋ-dôŋ hōō)	162	32°22'N	116°32'E
Chengdu, China (chŭŋ-dōō)	160	30°30'N	104°10'E
Chenggu, China (chŭŋ-gōō)	164	33°05'N	107°25'E
Chenghai, China (chŭŋ-hī)	165	23°22'N	116°40'E
Chengshan Jiao, c., China (jyou chŭŋ-shän)	164	37°28'N	122°40'E
Chengxi Hu, l., China (chŭŋ-shyē hōō)	162	32°31'N	116°04'E
Chennai (Madras), India	155	13°08'N	80°15'E
Chenxian, China (chŭn-shyĕn)	165	25°40'N	113°00'E
Chepén, Peru (chĕ-pĕ'n)	100	7°17'S	79°24'W
Chepo, Pan. (chā'pō)	91	9°12'N	79°06'W
Chepo, r., Pan.	91	9°10'N	78°36'W
Cher, r., Fr. (shâr)	117	47°14'N	1°34'E
Cherán, Mex. (chā-rän')	88	19°41'N	101°54'W
Cherangany Hills, hills, Kenya	191	1°25'N	35°20'E
Cheraw, S.C., U.S. (chē'rô)	83	34°40'N	79°52'W
Cherbourg, Fr. (shăr-bōr')	110	49°39'N	1°43'W
Cherdyn', Russia (chĕr-dyĕn')	134	60°25'N	56°32'E
Cheremkhovo, Russia (chĕr'yĕm-kô-vô)	135	52°58'N	103°18'E
Cherëmukhovo, Russia (chĕr-yĕ-mû-kô-vô)	142a	60°20'N	60°00'E
Cherepanovo, Russia (chĕr'yĕ pä-nô'vô)	134	54°13'N	83°22'E
Cherepovets, Russia (chĕr-yĕ-pô'vyĕtz)	134	59°08'N	37°59'E
Chereya, Bela. (chĕr-ā'yä)	132	54°38'N	29°16'E
Chergui, i., Tun.	118	34°50'N	11°40'E
Chergui, Chott ech, l., Alg. (chĕr gē)	118	34°12'N	0°10'W
Cherkasy, Ukr.	133	49°26'N	32°03'E
Cherkasy, prov., Ukr.	133	48°58'N	30°55'E
Cherkessk, Russia	138	44°14'N	42°04'E
Cherlak, Russia (chĭr-läk')	134	54°04'N	74°28'E
Chermoz, Russia (chĕr-môz')	136	58°47'N	56°08'E
Chern', Russia (chĕrn)	132	53°28'N	36°49'E
Chërnaya Kalitva, r., Russia (chôr'nȧ yȧ kȧ-lēt'vȧ)	133	50°15'N	39°16'E
Chernihiv, Ukr.	137	51°23'N	31°15'E
Chernihiv, prov., Ukr.	133	51°28'N	31°18'E
Chernihivka, Ukr.	133	47°08'N	36°20'E
Chernivtsi, Ukr.	134	48°18'N	25°56'E
Chernobyl' *see* Chornobai, Ukr.	132	51°17'N	30°14'E
Chernogorsk, Russia (chĕr-nŏ-gôrsk')	140	54°01'N	91°07'E
Chernoistochinsk, Russia (chĕr-nôy-stô'chĭnsk)	142a	57°44'N	59°55'E
Chernyanka, Russia (chĕrn-yäŋ'kä)	133	50°56'N	37°48'E
Cherokee, Ia., U.S. (chĕr-ô-kē')	70	42°43'N	95°33'W
Cherokee, Ks., U.S.	79	37°21'N	94°50'W
Cherokee, Ok., U.S.	78	36°44'N	98°22'W
Cherokee Lake, res., Tn., U.S.	82	36°22'N	83°22'W
Cherokees, Lake of the, res., Ok., U.S. (chĕr-ô-kēz')	65	36°32'N	95°14'W
Cherokee Sound, Bah.	92	26°15'N	76°55'W
Cherryfield, Me., U.S. (chĕr'ĭ-fēld)	60	44°37'N	67°56'W
Cherry Grove, Or., U.S.	74c	45°27'N	123°15'W
Cherryvale, Ks., U.S.	79	37°16'N	95°33'W
Cherryville, N.C., U.S. (chĕr'ĭ-vĭl)	83	35°32'N	81°22'W
Cherskogo, Khrebet, mts., Russia	135	67°15'N	140°00'E
Chertsey, Eng., U.K.	114b	51°24'N	0°30'W
Chervonoye, Vozyera, l., Bela. (chĕr-vô'nô-yĕ)	132	52°24'N	28°00'E
Chervyen', Bela. (chĕr'vyĕn)	132	53°43'N	28°26'E
Cherykaw, Bela.	132	53°34'N	31°22'E
Chesaning, Mi., U.S. (chĕs'ȧ-nĭng)	66	43°10'N	84°10'W
Chesapeake, Va., U.S. (chĕs'ȧ-pēk)	66g	36°48'N	76°16'W
Chesapeake Bay, b., U.S.	65	38°20'N	76°15'W
Chesapeake Beach, Md., U.S.	68e	38°42'N	76°33'W
Chesham, Eng., U.K. (chĕsh'ŭm)	114b	51°41'N	0°37'W
Cheshire, Mi., U.S. (chĕsh'ĭr)	66	42°25'N	86°00'W
Cheshire, co., Eng., U.K.	114a	53°16'N	2°30'W
Chëshskaya Guba, b., Russia	134	67°25'N	46°00'E
Cheshunt, Eng., U.K.	114b	51°43'N	0°02'W
Chesma, Russia (chĕs'mȧ)	142a	53°50'N	60°42'E
Chesnokovka, Russia (chĕs-nŏ-kôf'kȧ)	134	53°28'N	83°41'E

236 Chester – Chuuk (Truk) PRONOUNCING INDEX

PLACE (Pronunciation)	PAGE	LAT.	LONG.
Chester, Eng., U.K. (chĕs′tẽr)	120	53°12′N	2°53′W
Chester, Il., U.S.	79	37°54′N	89°48′W
Chester, Pa., U.S.	68f	39°51′N	75°22′W
Chester, S.C., U.S.	83	34°42′N	81°11′W
Chester, Va., U.S.	83	37°20′N	77°24′W
Chester, W.V., U.S.	66	40°35′N	80°30′W
Chesterfield, Eng., U.K. (chĕs′tẽr-fēld)	120	53°14′N	1°26′W
Chesterfield, Îles, is., N. Cal.	175	19°38′S	160°08′E
Chesterfield Inlet *see* Igluligaarjuk, Can.	52	63°19′N	91°11′W
Chesterfield Inlet, b., Can.	53	63°59′N	92°09′W
Chestermere Lake, l., Can. (chĕs′tē-mēr)	62e	51°03′N	113°45′W
Chesterton, In., U.S. (chĕs′tẽr-tŭn)	66	41°35′N	87°05′W
Chestertown, Md., U.S. (chĕs′tẽr-toun)	67	39°15′N	76°05′W
Chesuncook, l., Me., U.S. (chĕs′ŭn-ko͝ok)	60	46°03′N	69°40′W
Chetek, Wi., U.S. (chĕ′tĕk)	71	45°18′N	91°41′W
Chetumal, Bahía de, b., N.A. (bä-ē-ä dĕ chĕt-o͞o-mäl′)	86	18°07′N	88°05′W
Chevelon Creek, r., Az., U.S. (shĕv′á-lŏn)	77	34°35′N	111°00′W
Cheviot, Oh., U.S. (shĕv′ĭ-ŭt)	69f	39°10′N	84°37′W
Chevreuse, Fr. (shĕ-vrûz′)	127b	48°42′N	2°02′E
Chevy Chase, Md., U.S. (shĕvĭ chās)	68e	38°58′N	77°06′W
Chew Bahir, Afr. (stĕf-a-nē)	185	4°46′N	37°31′E
Chewelah, Wa., U.S. (chē-wē′lä)	72	48°17′N	117°42′W
Cheyenne, Wy., U.S. (shī-ĕn′)	64	41°10′N	104°49′W
Cheyenne, r., U.S.	64	44°20′N	102°15′W
Cheyenne River Indian Reservation, I.R., S.D., U.S.	70	45°07′N	100°46′W
Cheyenne Wells, Co., U.S.	78	38°46′N	102°21′W
Chhattisgarh, state, India	155	23°00′N	83°00′E
Chhindwāra, India	158	22°08′N	78°57′E
Chiai, Tai. (chī′ī′)	165	23°28′N	120°28′E
Chiange, Ang.	190	15°45′S	13°48′E
Chiang Mai, Thai.	168	18°38′N	98°44′E
Chiang Rai, Thai.	168	19°53′N	99°48′E
Chiapa, Río de, r., Mex.	90	16°00′N	92°20′W
Chiapa de Corzo, Mex. (chē-ä′pä dā kôr′zō)	89	16°44′N	93°01′W
Chiapas, state, Mex. (chē-ä′päs)	86	17°10′N	93°00′W
Chiapas, Cordillera de, mts., Mex. (kôr-dēl-yĕ′rä-dĕ-chyä′räs)	89	15°55′N	93°15′W
Chiari, Italy (kyä′rē)	130	45°31′N	9°57′E
Chiasso, Switz.	124	45°50′N	8°57′E
Chiatura, Geor.	138	42°17′N	43°17′E
Chiautla, Mex. (chyä-o͞ot′lä)	88	18°16′N	98°37′W
Chiavari, Italy (kyä-vä′rē)	130	44°18′N	9°21′E
Chiba, Japan (chē′bä)	161	35°37′N	140°08′E
Chiba, dept., Japan	167a	35°47′N	140°02′E
Chibougamau, Can. (chē-bo͞o′gä-mou)	51	49°57′N	74°23′W
Chibougamau, l., Can.	59	49°53′N	74°21′W
Chicago, Il., U.S. (shĭ-kô-gō) (chĭ-kä′gō)	65	41°49′N	87°37′W
Chicago Heights, Il., U.S.	69a	41°30′N	87°38′W
Chicapa, r., Afr. (chē-kä′pä)	186	7°45′S	20°25′E
Chicbul, Mex. (chēk-bo͞o′l)	89	18°45′N	90°56′W
Chic-Chocs, Monts, mts., Can.	53	48°38′N	66°37′W
Chichagof, i., Ak., U.S. (chē-chä′gôf)	63	57°50′N	137°00′W
Chichancanab, Lago de, l., Mex. (lä′gô-dĕ-chē-chän-kä-nä′b)	90a	19°50′N	88°28′W
Chichén Itzá, hist., Mex.	90a	20°40′N	88°35′W
Chichester, Eng., U.K. (chĭch′ĕs-tẽr)	120	50°50′N	0°55′W
Chichimilá, Mex. (chē-chē-mē′lä)	90a	20°36′N	88°14′W
Chichiriviche, Ven. (chē-chē-rē-vē-chĕ)	101b	10°56′N	68°17′W
Chickamauga, Ga., U.S. (chĭk-á-mô′gá)	82	34°50′N	85°15′W
Chickamauga Lake, res., Tn., U.S.	82	35°18′N	85°22′W
Chickasawhay, r., Ms., U.S. (chĭk-á-sô′wä)	82	31°45′N	88°45′W
Chickasha, Ok., U.S. (chĭk′á-shā)	64	35°04′N	97°56′W
Chiclana de la Frontera, Spain (chē-klä′nä)	128	36°25′N	6°09′W
Chiclayo, Peru (chē-klä′yō)	100	6°46′S	79°50′W
Chico, Ca., U.S. (chē′kō)	76	39°43′N	121°51′W
Chico, Wa., U.S.	74a	47°37′N	122°43′W
Chico, r., Arg.	102	44°30′S	66°00′W
Chico, r., Arg.	102	49°15′S	69°30′W
Chico, r., Phil.	169a	17°33′N	121°24′E
Chicoloapan, Mex. (chē-kō-lwä′pän)	89a	19°24′N	98°54′W
Chiconautla, Mex.	89a	19°39′N	99°01′W
Chicontepec, Mex. (chē-kōn′tĕ-pĕk′)	88	20°58′N	98°08′W
Chicopee, Ma., U.S. (chĭk′ō-pē)	67	42°10′N	72°35′W
Chicoutimi, Can. (shē-ko͞o′tē-mē′)	51	48°26′N	71°04′W
Chicxulub, Mex. (chēk-so͞o-lo͞o′b)	90a	21°10′N	89°30′W
Chiefland, Fl., U.S. (chēf′lănd)	83	29°30′N	82°50′W
Chiemsee, l., Ger. (kēm zā)	124	47°58′N	12°20′E
Chieri, Italy (kyâ′rē)	130	45°03′N	7°48′E
Chieti, Italy (kyĕ′tē)	118	42°22′N	14°22′E
Chifeng, China (chr-fŭŋ)	161	42°18′N	118°52′E
Chignanuapan, Mex. (chē′g-ŋä-nwä-pá′n)	88	19°49′N	98°02′W
Chignecto Bay, b., Can. (shĭg-nĕk′tō)	60	45°33′N	64°50′W
Chignik, Ak., U.S. (chĭg′nĭk)	63	56°14′N	158°12′W
Chignik Bay, b., Ak., U.S.	63	56°18′N	157°22′W
Chigu Co, l., China (chr-go͞o tswo)	158	28°55′N	91°47′E
Chigwell, Eng., U.K.	114b	51°38′N	0°05′E
Chihe, China (chr-hŭ)	162	32°32′N	117°57′E
Chihuahua, Mex. (chē-wä′wä)	86	28°37′N	106°06′W
Chihuahua, state, Mex.	86	29°00′N	107°30′W
Chikishlyar, Turkmen. (chē-kēsh-lyär′)	139	37°40′N	53°50′E
Chilanga, Zam.	191	15°34′S	28°17′E
Chilapa, Mex. (chē-lä′pä)	88	17°34′N	99°14′W
Chilchota, Mex. (chēl-chō′tä)	88	19°40′N	102°04′W
Chilcotin, r., Can. (chĭl-kō′tĭn)	54	52°20′N	124°15′W
Childress, Tx., U.S. (chĭld′rĕs)	78	34°26′N	100°11′W
Chile, nation, S.A. (chē′lā)	102	35°00′S	72°00′W
Chilecito, Arg. (chē-lā-sē′tō)	102	29°06′S	67°25′W
Chilengue, Serra do, mts., Ang.	190	13°20′S	15°00′E
Chilibre, Pan. (chē-lē′brĕ)	86a	9°09′N	79°37′W
Chililabombwe, Zam.	191	12°18′S	27°43′E
Chilka, l., India	158	19°26′N	85°42′E
Chilko, r., Can. (chĭl′kō)	54	51°53′N	123°53′W
Chilko Lake, l., Can.	54	51°20′N	124°05′W
Chillán, Chile (chēl-yän′)	102	36°44′S	72°06′W
Chillicothe, Il., U.S. (chĭl-ĭ-kŏth′ē)	66	41°55′N	89°30′W
Chillicothe, Mo., U.S.	79	39°46′N	93°32′W
Chillicothe, Oh., U.S.	66	39°20′N	83°00′W
Chilliwack, Can. (chĭl′ĭ-wăk)	50	49°10′N	121°57′W
Chiloé, Isla de, i., Chile	102	42°30′S	73°55′W
Chilpancingo de los Bravo, Mex.	86	17°32′N	99°30′W
Chilton, Wi., U.S. (chĭl′tŭn)	71	44°00′N	88°12′W
Chilung, Tai. (chī′lung)	161	25°02′N	121°48′E
Chilwa, Lake, l., Afr.	186	15°12′S	36°30′E
Chimacum, Wa., U.S. (chĭm′ä-kŭm)	74a	48°01′N	122°47′W
Chimalpa, Mex. (chē-mäl′pä)	89a	19°26′N	99°22′W
Chimaltenango, Guat. (chē-mäl-tå-näŋ′gō)	90	14°39′N	90°48′W
Chimaltitan, Mex. (chēmäl-tē-tän′)	88	21°36′N	103°50′W
Chimbay, Uzb. (chĭm-bī′)	139	43°00′N	59°44′E
Chimborazo, mtn., Ec. (chēm-bô-rä′zō)	100	1°35′S	78°45′W
Chimbote, Peru (chēm-bō′tå)	100	9°02′S	78°33′W
China, Mex. (chē′nä)	80	25°43′N	99°13′W
China, nation, Asia (chī′nȧ)	160	36°45′N	93°00′E
Chinameca, El Sal. (Chē-nä-mä′kä)	90	13°31′N	88°18′W
Chinandega, Nic. (chē-nän-dā′gä)	90	12°38′N	87°08′W
Chinati Peak, mtn., Tx., U.S. (chĭ-nä′tē)	80	29°56′N	104°29′W
Chincha Alta, Peru (chĭn′chä äl′tä)	100	13°24′S	76°04′W
Chinchas, Islas, is., Peru (ē′s-läs-chē′n-chäs)	100	11°27′S	79°05′W
Chinchilla, Austl. (chĭn-chĭl′ȧ)	176	26°44′S	150°36′E
Chinchorro, Banco, bk., Mex. (bä′n-kô-chēn-chô′r-rō)	90a	18°43′N	87°25′W
Chincilla de Monte Aragon, Spain	128	38°54′N	1°43′W
Chinde, Moz. (shĕn′dĕ)	186	17°39′S	36°34′E
Chin Do, i., Kor., S.	166	34°30′N	125°43′E
Chindwin, r., Mya. (chĭn-dwĭn)	155	23°30′N	94°34′E
Chingola, Zam. (chĭng-gōlä)	186	12°32′S	27°52′E
Chinguar, Ang. (chĭng-gär)	186	12°33′S	16°15′E
Chinguetti, Maur. (chēŋ-gĕt′ē)	184	20°34′N	12°34′W
Chinhoyi, Zimb.	186	17°22′S	30°12′E
Chinju, Kor., S. (chĭn′jo͞o)	166	35°13′N	128°10′E
Chinko, r., C.A.R. (shĭn′kô)	185	6°37′N	24°31′E
Chinmen *see* Quemoy, Tai.	165	24°30′N	118°20′E
Chino, Ca., U.S. (chē′nō)	75a	34°01′N	117°42′W
Chinon, Fr. (shē-nôn′)	126	47°09′N	0°13′E
Chinook, Mt., U.S. (shĭn-ók′)	73	48°35′N	109°15′W
Chinsali, Zam.	191	10°34′S	32°03′E
Chinteche, Mwi. (chĭn-tĕ′chē)	186	11°48′S	34°14′E
Chioggia, Italy (kyôd′jä)	130	45°12′N	12°17′E
Chíos, Grc. (kē′ôs)	119	38°23′N	26°09′E
Chíos, i., Grc.	119	38°20′N	25°45′E
Chipata, Zam.	186	13°39′S	32°40′E
Chipera, Moz. (zhē-pē′rä)	186	15°16′S	32°30′E
Chipley, Fl., U.S. (chĭp′lĭ)	82	30°45′N	85°33′W
Chipman, Can. (chĭp′mȧn)	60	46°11′N	65°53′W
Chipola, r., Fl., U.S. (chĭ-pō′lȧ)	82	30°40′N	85°14′W
Chippawa, Can. (chĭp′ē-wä)	69c	43°03′N	79°03′W
Chippewa, r., Mn., U.S. (chĭp′ē-wä)	70	45°07′N	95°41′W
Chippewa, r., Wi., U.S.	71	45°07′N	91°19′W
Chippewa Falls, Wi., U.S.	71	44°55′N	91°26′W
Chippewa Lake, Oh., U.S.	69d	41°04′N	81°54′W
Chiputneticook Lakes, l., N.A. (chĭ-pŏt-nĕt′ĭ-ko͝ok)	60	45°47′N	67°45′W
Chiquimula, Guat. (chē-kē-mo͞o′lä)	90	14°47′N	89°31′W
Chiquimulilla, Guat. (chē-kē-mo͞o-lē′l-yä)	90	14°08′N	90°23′W
Chiquinquira, Col. (chē-kēŋ′kē-rä′)	100	5°33′N	73°49′W
Chirala, India	159	15°52′N	80°22′E
Chirchik, Uzb. (chĭr-chĕk′)	139	41°28′N	69°18′E
Chire (Shire), r., Afr.	191	17°15′S	35°25′E
Chiricahua National Monument, rec., Az., U.S. (chĭ-rä-cä′hwä)	77	32°02′N	109°18′W
Chirikof, i., Ak., U.S. (chĭ′rĭ-kôf)	63	55°50′N	155°35′W
Chiriquí, Punta, c., Pan. (pò′n-tä-chē-rē-kē′)	91	9°13′N	81°39′W
Chiriquí Grande, Pan. (chē-rē-kē′ grän′dā)	91	8°57′N	82°08′W
Chiri San, mtn., Kor., S. (chĭ′rĭ-sän′)	166	35°20′N	127°39′E
Chiromo, Mwi.	186	16°34′S	35°13′E
Chirpan, Blg.	119	42°12′N	25°19′E
Chirripó, Río, r., C.R.	91	9°50′N	83°20′W
Chisasibi, Can.	51	53°40′N	78°58′W
Chisholm, Mn., U.S. (chĭz′ŭm)	71	47°28′N	92°53′W
Chişinău, Mol.	134	47°02′N	28°52′E
Chistopol′, Russia (chĭs-tô′pôl-y′)	134	55°21′N	50°37′E
Chita, Russia (chē-tá′)	135	52°09′N	113°39′E
Chitambo, Zam.	191	12°55′S	30°39′E
Chitato, Ang.	190	7°20′S	20°47′E
Chitembo, Ang.	190	13°34′S	16°40′E
Chitina, Ak., U.S. (chĭ-tē′ná)	63	61°28′N	144°35′W
Chitokoloki, Zam.	190	13°50′S	23°13′E
Chitorgarh, India	158	24°59′N	74°42′E
Chitrāl, Pak. (chē-träl′)	155	35°58′N	71°48′E
Chittagong, Bngl. (chĭt-á-gông′)	155	22°26′N	90°51′E
Chitungwiza, Zimb.	186	17°51′S	31°05′E
Chiumbe, r., Afr. (chē-ŏm′bå)	186	9°45′S	21°00′E
Chivasso, Italy (kē-väs′sō)	130	45°13′N	7°52′E
Chivhu, Zimb.	186	18°59′S	30°58′E
Chivilcoy, Arg. (chē-vēl-koi′)	102	34°51′S	60°03′W
Chixoy, r., Guat. (chē-ĸoi′)	90	15°40′N	90°35′W
Chizu, Japan (chē-zo͞o′)	167	35°16′N	134°15′E
Chloride, Az., U.S. (klō′rīd)	77	35°25′N	114°15′W
Chmielnik, Pol. (ĸmyĕl′nĕĸ)	125	50°36′N	20°46′E
Choapa, r., Chile (chô-á′pä)	99b	31°56′S	70°48′W
Choctawhatchee, r., Fl., U.S.	82	30°37′N	85°56′W
Choctawhatchee Bay, b., Fl., U.S. (chôk-tô-hăch′ē)	82	30°15′N	86°32′W
Chodziez, Pol. (ĸōj′yĕsh)	124	52°59′N	16°55′E
Choele Choel, Arg. (chô-ĕ′lĕ-chô-ĕ′l)	102	39°14′S	65°46′W
Chōfu, Japan (chō′fo͞o′)	167a	35°39′N	139°33′E
Chōgo, Japan (chō-gō)	167a	35°25′N	139°28′E
Choiseul, i., Sol. Is. (shwä-zûl′)	175	7°30′S	157°30′E
Choisy-le-Roi, Fr.	127b	48°46′N	2°25′E
Chojnice, Pol. (ĸōi-nē-tsĕ)	125	53°41′N	17°34′E
Cholet, Fr. (shô-lĕ′)	117	47°06′N	0°54′W
Cholula, Mex. (chō-lo͞o′lä)	88	19°04′N	98°19′W
Choluteca, Hond. (chō-lo͞o-tā′kä)	90	13°18′N	87°12′W
Choluteco, r., Hond.	90	13°34′N	86°59′W
Chomutov, Czech Rep. (kō′mò-tôf)	124	50°27′N	13°23′E
Chona, r., Russia (chō′nä)	141	60°45′N	109°15′E
Chone, Ec. (chô′nĕ)	100	0°48′S	80°06′W
Chŏngjin, Kor., N. (chŭng-jĭn′)	161	41°48′N	129°46′E
Chŏngju, Kor., S. (chŭng-jo͞o′)	166	36°35′N	127°30′E
Chongming Dao, i., China (chôŋ-mĭŋ dou)	165	31°40′N	122°30′E
Chongqing, China (chôŋ-chyĭŋ)	160	29°38′N	107°30′E
Chongqing, prov., China	160	30°00′N	108°00′E
Chŏnju, Kor., S. (chŭn-jo͞o′)	166	35°48′N	127°08′E
Chonos, Archipiélago de los, is., Chile	102	44°35′N	76°15′W
Chorley, Eng., U.K. (chôr′lĭ)	114a	53°40′N	2°38′W
Chornaya, neigh., Russia	142b	55°45′N	38°04′E
Chornobai, Ukr.	133	51°17′N	30°14′E
Chornobay, Ukr. (chĕr-nō-bī′)	133	49°41′N	32°24′E
Chornomors′ke, Ukr.	137	45°29′N	32°43′E
Chorrillos, Peru (chôr-rē′l-yōs)	100	12°17′S	76°55′W
Chortkiv, Ukr.	125	49°01′N	25°48′E
Chosan, Kor., N. (chō-sän′)	166	40°44′N	125°48′E
Chosen, Fl., U.S. (chō′z′n)	83a	26°41′N	80°41′W
Chōshi, Japan (chō′shē)	166	35°40′N	140°55′E
Choszczno, Pol. (chôsh′chnô)	124	53°10′N	15°25′E
Chota Nagpur, plat., India	158	23°40′N	82°50′E
Choteau, Mt., U.S. (shō′tō)	73	47°51′N	112°10′W
Chowan, r., N.C., U.S. (chô-wän′)	83	36°13′N	76°46′W
Chowilla Reservoir, res., Austl.	176	34°05′S	141°20′E
Chown, Mount, mtn., Can. (choun)	55	53°24′N	119°22′W
Choybalsan, Mong.	161	47°50′N	114°15′E
Christchurch, N.Z. (krīst′chûrch)	175a	43°30′S	172°38′E
Christian, i., Can. (krĭs′chăn)	59	44°50′N	80°00′W
Christiansburg, Va., U.S. (krĭs′chănz-bûrg)	83	37°08′N	80°25′W
Christiansted, V.I.U.S.	87b	17°45′N	64°44′W
Christmas Island, dep., Oc.	168	10°35′S	105°40′E
Christopher, Il., U.S. (krĭs′tô-fẽr)	79	37°58′N	89°04′W
Chrudim, Czech Rep. (ĸro͞o′dyĕm)	124	49°57′N	15°46′E
Chrzanów, Pol. (kzhä′nòf)	125	50°08′N	19°24′E
Chuansha, China (chůän-shä)	163b	31°12′N	121°41′E
Chubut, prov., Arg. (chò-bo͞ot′)	102	44°00′S	69°15′W
Chubut, r., Arg. (chò-bo͞ot′)	102	43°05′S	69°00′W
Chuckatuck, Va., U.S. (chŭck á-tŭck)	68g	36°51′N	76°35′W
Chucunaque, r., Pan. (cho͞o-ko͞o-nä′kå)	91	8°36′N	77°48′W
Chudovo, Russia (chò′dô-vô)	132	59°03′N	31°56′E
Chudskoye Ozero, l., Eur. (chòt′skô-yĕ)	136	58°43′N	26°45′E
Chuguchak, hist. reg., China (cho͞o′go͞o-chäk′)	160	46°09′N	83°58′E
Chuguyevka, Russia (chò-go͞o′yĕf-kå)	166	43°58′N	133°49′E
Chugwater Creek, r., Wy., U.S. (chŭg′wô-tẽr)	70	41°43′N	104°54′W
Chuhuïv, Ukr.	137	49°52′N	36°40′E
Chukotskiy Poluostrov, pen., Russia	135	66°12′N	175°00′W
Chukotskoye Nagor′ye, mts., Russia	135	66°00′N	166°00′E
Chula Vista, Ca., U.S. (cho͞o′lá vĭs′tá)	76a	32°38′N	117°05′W
Chulkovo, Russia (cho͞ol-kô vô)	142b	55°33′N	38°04′E
Chulucanas, Peru	100	5°13′S	80°13′W
Chulum, r., Russia	140	57°52′N	84°45′E
Chumikan, Russia (cho͞o-mē-kän′)	135	54°47′N	135°09′E
Chun′an, China (chòn-än)	165	29°38′N	119°00′E
Chunchŏn, Kor., S. (chòn-chŭn′)	166	37°51′N	127°46′E
Chungju, Kor., S. (chŭng′jo͞o′)	166	37°00′N	128°19′E
Chungking *see* Chongqing, China	160	29°38′N	107°30′E
Chunya, Tan.	191	8°32′S	33°25′E
Chunya, r., Russia (chòn′yä′)	140	61°45′N	101°28′E
Chuquicamata, Chile (cho͞o-kē-kä-mä′tä)	102	22°08′S	68°57′W
Chur, Switz. (ko͞or)	117	46°51′N	9°32′E
Churchill, Can. (chûrch′ĭl)	51	58°50′N	94°10′W
Churchill, r., Can.	52	58°00′N	95°00′W
Churchill, Cape, c., Can.	53	59°07′N	93°50′W
Churchill Falls, wtfl., Can.	53	53°35′N	64°27′W
Churchill Lake, l., Can.	56	56°12′N	108°40′W
Churchill Peak, mtn., Can.	52	58°10′N	125°14′W
Church Stretton, Eng., U.K. (chûrch strĕt′ŭn)	114a	52°32′N	2°49′W
Churchton, Md., U.S.	68e	38°49′N	76°33′W
Churu, India	158	28°22′N	75°00′E
Churumuco, Mex. (cho͞o-ro͞o-mo͞o′kō)	88	18°39′N	101°40′W
Chuska Mountains, mts., Az., U.S. (chŭs-kȧ)	77	36°21′N	109°11′W
Chusovaya, r., Russia (cho͞o-sô-vä′yá)	136	58°08′N	58°35′E
Chusovoy, Russia (cho͞o-sô-vôy′)	134	58°18′N	57°50′E
Chust, Uzb. (chòst)	139	41°05′N	71°28′E
Chuuk (Truk), is., Micron.	170c	7°25′N	151°47′E

ăt; finȧl; rāte; senåte; ärm; åsk; sofȧ; fâre; ch-choose; dh-as th in other; bē; ēvent; bĕt; recĕnt; cratẽr; g-gō; gh-guttural g; bĭt; ĭ-short neutral; rīde; ĸ-guttural k as ch in German ich;

PRONOUNCING INDEX

Chuvashia – Codogno 237

PLACE (Pronunciation)	PAGE	LAT.	LONG.
Chuvashia, prov., Russia	136	55°45'N	46°00'E
Chuviscar, r., Mex. (chōō-vēs-kär')	80	28°34'N	105°36'W
Chuwang, China (chōō-wäŋ)	162	36°08'N	114°53'E
Chuxian, China (chōō shyĕn)	164	32°19'N	118°19'E
Chuxiong, China (chōō-shyôŋ)	160	25°19'N	101°34'E
Chyhyryn, Ukr.	133	49°02'N	32°39'E
Cicero, Il., U.S. (sĭs'ẽr-ō)	69a	41°50'N	87°46'W
Cide, Tur. (jē'dĕ)	119	41°50'N	33°00'E
Ciechanów, Pol. (tsyĕ-kä'nóf)	125	52°52'N	20°39'E
Ciego de Avila, Cuba (syā'gō dā ä'vē-lä)	87	21°50'N	78°45'W
Ciego de Avila, prov., Cuba	92	22°00'N	78°40'W
Ciempozuelos, Spain (thyĕm-pō-thwā'lōs)	128	40°09'N	3°36'W
Ciénaga, Col. (syā'nä-gä)	100	11°01'N	74°15'W
Cienfuegos, Cuba (syĕn-fwā'gōs)	87	22°10'N	80°30'W
Cienfuegos, prov., Cuba	92	22°15'N	80°40'W
Cienfuegos, Bahía, b., Cuba (bä-ē'ä-syĕn-fwā'gōs)	92	22°00'N	80°35'W
Ciervo, Isla de la, i., Nic. (ē's-lä-dĕ-lä-syĕ'r-vô)	91	11°56'N	83°20'W
Cieszyn, Pol. (tsyĕ'shēn)	125	49°47'N	18°45'E
Cieza, Spain (thyā'thä)	128	38°13'N	1°25'W
Cigüela, r., Spain	128	39°53'N	2°54'W
Cihuatlán, Mex. (sē-wä-tlä'n)	88	19°13'N	104°36'W
Cihuatlán, r., Mex.	88	19°11'N	104°30'W
Cijara, Embalse de, res., Spain	128	39°25'N	5°00'W
Cilician Gates, p., Tur.	137	37°30'N	35°30'E
Cimarron, r., Co., U.S.	78	37°13'N	102°30'W
Cimarron, r., U.S. (sĭm-ȧ-rōn')	64	36°26'N	98°27'W
Cinca, r., Spain (thēŋ'kä)	129	42°09'N	0°08'E
Cincinnati, Oh., U.S. (sĭn-sĭ-nát'ĭ)	65	39°08'N	84°30'W
Cinco Balas, Cayos, is., Cuba (kä'yōs-thēŋ'kō bä'läs)	92	21°05'N	79°25'W
Cintalapa, Mex. (sēn-tä-lä'pä)	89	16°41'N	93°44'W
Cinto, Monte, mtn., Fr. (chēn'tō)	117	42°24'N	8°54'E
Circle, Ak., U.S. (sûr'k'l)	64a	65°49'N	144°22'W
Circleville, Oh., U.S. (sûr'k'lvĭl)	66	39°35'N	83°00'W
Cirebon, Indon.	168	6°50'S	108°33'E
Ciri Grande, r., Pan. (sē'rē-grá'n'dĕ)	86a	8°55'N	80°04'W
Cisco, Tx., U.S. (sĭs'kō)	80	32°23'N	98°57'W
Cisneros, Col. (sēs-nĕ'rôs)	100a	6°33'N	75°05'W
Cisterna di Latina, Italy (chēs-tĕ'r-nä-dē-lä-tē'nä)	129d	41°36'N	12°53'E
Cistierna, Spain (thēs-tyĕr'nä)	128	42°48'N	5°08'W
Citronelle, Al., U.S. (cĭt-rō'nĕl)	82	31°05'N	88°15'W
Cittadella, Italy (chēt-tä-dĕl'lä)	130	45°39'N	11°51'E
Città di Castello, Italy (chēt-tä'dē käs-tĕl'lō)	130	43°27'N	12°17'E
Ciudad Altamirano, Mex. (syōō-dä'd-äl-tä-mē-rä'nō)	88	18°24'N	100°38'W
Ciudad Bolívar, Ven. (syōō-dhädh' bô-lē'vär)	100	8°07'N	63°41'W
Ciudad Camargo, Mex.	86	27°42'N	105°10'W
Ciudad Chetumal, Mex.	86	18°30'N	88°17'W
Ciudad Darío, Nic. (syōō-dhädh'dä'rē-ō)	90	12°44'N	86°08'W
Ciudad de la Habana, prov., Cuba	92	23°20'N	82°10'W
Ciudad del Carmen, Mex. (syōō-dä'd-dĕl-kà'r-mĕn)	86	18°39'N	91°49'W
Ciudad del Maíz, Mex. (syōō-dhädh'del mä-ēz')	88	22°24'N	99°37'W
Ciudad Fernández, Mex. (syōō-dhädh'fĕr-nän'dĕz)	88	21°56'N	100°03'W
Ciudad García, Mex. (syōō-dhädh'gär-sē'ä)	86	22°39'N	103°02'W
Ciudad Guayana, Ven.	100	8°30'N	62°45'W
Ciudad Guzmán, Mex. (syōō-dhädh'góz-män)	86	19°40'N	103°29'W
Ciudad Hidalgo, Mex. (syōō-dä'd-ē-dä'l-gô)	88	19°41'N	100°35'W
Ciudad Juárez, Mex. (syōō-dhädh hwä'räz)	86	31°44'N	106°28'W
Ciudad Madero, Mex. (syōō-dä'd-mä-dĕ'rô)	89	22°16'N	97°52'W
Ciudad Mante, Mex. (syōō-dä'd-mán'tē)	86	22°34'N	98°58'W
Ciudad Manuel Doblado, Mex. (syōō-dä'd-män-wāl'dō-blä'dō)	88	20°43'N	101°57'W
Ciudad Obregón, Mex. (syōō-dhädh-ô-brĕ-gô'n)	86	27°40'N	109°58'W
Ciudad Real, Spain (thyōō-dhädh'rā-äl')	128	38°59'N	3°55'W
Ciudad Rodrigo, Spain (thyōō-dhädh'rô-drē'gō)	118	40°38'N	6°34'W
Ciudad Serdán, Mex. (syōō-dä'd-sĕr-dá'n)	89	18°58'N	97°26'W
Ciudad Victoria, Mex. (syōō-dhädh'vēk-tō'rē-ä)	86	23°43'N	99°09'W
Ciutadella, Spain	129	40°00'N	3°52'E
Civitavecchia, Italy (chē'vē-tä-vĕk'kyä)	130	42°06'N	11°49'E
Cixian, China (tsē shyĕn)	162	36°22'N	114°23'E
Clackamas, Or., U.S. (klăc-kȧ'măs)	74c	45°25'N	122°34'W
Claire, l., Can. (klâr)	52	58°33'N	113°16'W
Clair Engle Lake, l., Ca., U.S.	72	40°51'N	122°41'W
Clairton, Pa., U.S. (klârtŭn)	69e	40°17'N	79°53'W
Clanton, Al., U.S. (klăn'tŭn)	82	32°50'N	86°38'W
Clare, Mi., U.S. (klâr)	66	43°50'N	84°45'W
Clare Island, i., Ire.	120	53°46'N	10°00'W
Claremont, Ca., U.S. (klâr'mŏnt)	75a	34°06'N	117°43'W
Claremont, N.H., U.S. (klâr'mŏnt)	67	43°20'N	72°20'W
Claremont, W.V., U.S.	66	37°55'N	81°00'W
Claremore, Ok., U.S. (klâr'mōr)	79	36°16'N	95°37'W
Claremorris, Ire. (klâr-mŏr'ĭs)	120	53°46'N	9°05'W
Clarence Strait, strt., Austl. (klăr'ĕns)	174	12°15'S	130°05'E
Clarence Strait, strt., Ak., U.S.	54	55°25'N	132°00'W
Clarence Town, Bah.	93	23°05'N	75°00'W
Clarendon, Ar., U.S. (klâr'ĕn-dŭn)	79	34°42'N	91°17'W
Clarendon, Tx., U.S.	78	34°55'N	100°52'W
Clarens, S. Afr. (clâ-rēns)	187c	28°34'S	28°26'E
Claresholm, Can. (klâr'ĕs-hōlm)	50	50°02'N	113°35'W
Clarinda, Ia., U.S. (klȧ-rĭn'dȧ)	70	40°42'N	95°00'W
Clarines, Ven. (klä-rē'nĕs)	101b	9°57'N	65°10'W
Clarion, Ia., U.S. (klăr'ĭ-ŭn)	71	42°43'N	93°45'W
Clarion, Pa., U.S.	67	41°10'N	79°25'W
Clark, S.D., U.S. (klärk)	70	44°52'N	97°45'W
Clark, Point, c., Can.	58	44°05'N	81°50'W
Clarkdale, Az., U.S. (klärk-dāl)	77	34°45'N	112°05'W
Clarke City, Can.	51	50°12'N	66°38'W
Clarke Range, mts., Austl.	175	20°30'S	148°00'E
Clark Fork, r., Mt., U.S.	72	47°50'N	115°35'W
Clarksburg, W.V., U.S. (klärkz'bûrg)	65	39°15'N	80°20'W
Clarksdale, Ms., U.S. (klärks-dāl)	82	34°10'N	90°31'W
Clark's Harbour, Can. (klärks)	60	43°26'N	65°38'W
Clarks Hill Lake, res., U.S. (klärk-hĭl)	65	33°50'N	82°35'W
Clarkston, Ga., U.S. (klärks'tŭn)	68c	33°49'N	84°15'W
Clarkston, Wa., U.S.	72	46°24'N	117°01'W
Clarksville, Ar., U.S. (klärks-vĭl)	79	35°28'N	93°26'W
Clarksville, Tn., U.S.	82	36°30'N	87°23'W
Clarksville, Tx., U.S.	79	33°37'N	95°02'W
Clatskanie, Or., U.S.	74c	46°04'N	123°11'W
Clatskanie, r., Or., U.S. (klăt-skā'nē)	74c	46°06'N	123°11'W
Clatsop Spit, Or., U.S. (klăt-sŏp)	74c	46°13'N	124°04'W
Cláudio, Braz. (klou'-dēō)	99a	20°26'S	44°44'W
Claveria, Phil. (klä-vä-rē'ä)	165	18°38'N	121°08'E
Clawson, Mi., U.S. (klô's'n)	69b	42°32'N	83°09'W
Claxton, Ga., U.S. (klăks'tŭn)	83	32°07'N	81°54'W
Clay, Ky., U.S. (klā)	82	37°28'N	87°50'W
Clay Center, Ks., U.S. (klā sĕn'tẽr)	79	39°23'N	97°08'W
Clay City, Ky., U.S. (klā sĭ'tĭ)	66	37°50'N	83°55'W
Claycomo, Mo., U.S. (kla-kō'mo)	75f	39°12'N	94°30'W
Clay Cross, Eng., U.K. (klā krŏs)	114a	53°10'N	1°25'W
Claye-Souilly, Fr. (klĕ-sōō-yĕ')	127b	48°56'N	2°43'E
Claymont, De., U.S. (klā-mŏnt)	68f	39°48'N	75°28'W
Clayton, Eng., U.K.	114a	53°47'N	1°49'W
Clayton, Al., U.S. (klā'tŭn)	82	31°52'N	85°25'W
Clayton, Ca., U.S.	74b	37°56'N	121°56'W
Clayton, Mo., U.S.	75e	38°39'N	90°20'W
Clayton, N.C., U.S.	83	35°40'N	78°27'W
Clayton, N.M., U.S.	78	36°26'N	103°12'W
Clear, l., Ca., U.S.	76	39°05'N	122°50'W
Clear Boggy Creek, r., Ok., U.S. (klēr bŏg'ĭ krēk)	79	34°21'N	96°22'W
Clear Creek, r., Az., U.S.	77	34°40'N	111°05'W
Clear Creek, r., Tx., U.S.	81a	29°34'N	95°13'W
Clear Creek, r., Wy., U.S.	73	44°35'N	106°20'W
Clearfield, Pa., U.S. (klēr-fēld)	67	41°00'N	78°25'W
Clearfield, Ut., U.S.	75b	41°07'N	112°01'W
Clear Hills, Can.	50	57°11'N	119°20'W
Clear Lake, Ia., U.S.	71	43°09'N	93°23'W
Clear Lake, Wa., U.S.	74a	48°27'N	122°14'W
Clear Lake Reservoir, res., Ca., U.S.	72	41°53'N	121°00'W
Clearwater, Fl., U.S. (klēr-wô'tẽr)	83a	27°43'N	82°45'W
Clearwater, r., Can.	55	52°22'N	114°57'W
Clearwater, r., Can.	55	52°00'N	120°10'W
Clearwater, r., Can.	56	56°10'N	110°40'W
Clearwater, r., Id., U.S.	72	46°27'N	116°33'W
Clearwater, Middle Fork, r., Id., U.S.	72	46°10'N	115°48'W
Clearwater, North Fork, r., Id., U.S.	72	46°34'N	116°08'W
Clearwater, South Fork, r., Id., U.S.	72	45°46'N	115°53'W
Clearwater Mountains, mts., Id., U.S.	72	45°56'N	115°15'W
Cleburne, Tx., U.S. (klē'bŭrn)	64	32°21'N	97°23'W
Cle Elum, Wa., U.S. (klē ĕl'ŭm)	72	47°12'N	120°55'W
Clementon, N.J., U.S. (klē'mĕn-tŭn)	68f	39°49'N	75°00'W
Cleobury Mortimer, Eng., U.K. (klēō-bĕr'ĭ môr'tĭ-mẽr)	114a	52°22'N	2°29'W
Clermont, Austl. (klēr'mŏnt)	175	23°02'S	147°46'E
Clermont, Can.	59	47°45'N	70°20'W
Clermont-Ferrand, Fr. (klĕr-môn'fĕr-răn')	110	45°47'N	3°03'E
Cleveland, Ms., U.S. (klēv'lȧnd)	82	33°45'N	90°42'W
Cleveland, Oh., U.S.	65	41°30'N	81°42'W
Cleveland, Ok., U.S.	79	36°18'N	96°28'W
Cleveland, Tn., U.S.	82	35°09'N	84°52'W
Cleveland, Tx., U.S.	81	30°18'N	95°05'W
Cleveland Heights, Oh., U.S.	69d	41°30'N	81°35'W
Cleveland Peninsula, pen., Ak., U.S.	54	55°45'N	132°00'W
Cleves, Oh., U.S. (klē'vĕs)	69f	39°10'N	84°45'W
Clew Bay, b., Ire. (klōō)	120	53°47'N	9°45'W
Clewiston, Fl., U.S. (klē'wis-tŭn)	83a	26°44'N	80°55'W
Clichy, Fr. (klē-shē')	126	48°54'N	2°18'E
Clifden, Ire. (klĭf'dĕn)	120	53°31'N	10°04'W
Clifton, Az., U.S. (klĭf'tŭn)	77	33°05'N	109°20'W
Clifton, N.J., U.S.	68a	40°52'N	74°09'W
Clifton, S.C., U.S.	83	35°00'N	81°47'W
Clifton, Tx., U.S.	81	31°45'N	97°31'W
Clifton Forge, Va., U.S.	67	37°50'N	79°50'W
Clinch, r., Tn., U.S. (klĭnch)	82	36°30'N	83°19'W
Clingmans Dome, mtn., U.S. (klĭng'măns dōm)	82	35°37'N	83°26'W
Clinton, Can. (klĭn-'tŭn)	50	51°05'N	121°35'W
Clinton, Ia., U.S.	71	41°50'N	90°13'W
Clinton, Il., U.S.	66	40°10'N	88°55'W
Clinton, In., U.S.	66	39°40'N	87°25'W
Clinton, Ky., U.S.	82	36°39'N	88°56'W
Clinton, Ma., U.S.	61a	42°25'N	71°41'W
Clinton, Md., U.S.	68e	38°46'N	76°54'W
Clinton, Mo., U.S.	79	38°23'N	93°46'W
Clinton, N.C., U.S.	83	34°58'N	78°20'W
Clinton, Ok., U.S.	78	35°31'N	98°56'W
Clinton, S.C., U.S.	83	34°27'N	81°53'W
Clinton, Tn., U.S.	82	36°05'N	84°08'W
Clinton, Wa., U.S.	74a	47°59'N	122°22'W
Clinton, r., Mi., U.S.	69b	42°36'N	83°00'W
Clinton-Colden, l., Can.	52	63°58'N	106°34'W
Clintonville, Wi., U.S. (klĭn'tŭn-vĭl)	71	44°37'N	88°46'W
Clio, Mi., U.S. (klē'ō)	66	43°10'N	83°45'W
Cloates, Point, c., Austl. (klōts)	174	22°47'S	113°45'E
Clocolan, S. Afr.	192c	28°56'S	27°35'E
Clonakilty Bay, b., Ire. (klŏn-ȧ-kĭltē)	120	51°30'N	8°50'W
Cloncurry, Austl. (klŏn-kŭr'ē)	174	20°58'S	140°42'E
Clonmel, Ire. (klŏn-mĕl)	120	52°21'N	7°45'W
Cloquet, Mn., U.S. (klō-kā')	75h	46°42'N	92°28'W
Closter, N.J., U.S. (klōs'tẽr)	68a	40°58'N	73°57'W
Cloud Peak, mtn., Wy., U.S. (kloud)	64	44°23'N	107°11'W
Clover, S.C., U.S. (klō'vẽr)	83	35°08'N	81°08'W
Clover Bar, Can. (klō'vẽr bär)	62g	53°34'N	113°20'W
Cloverdale, Can.	74d	49°06'N	122°44'W
Cloverdale, Ca., U.S. (klō'vẽr-dāl)	76	38°47'N	123°03'W
Cloverport, Ky., U.S. (klō'vẽr pōrt)	66	37°50'N	86°35'W
Clovis, N.M., U.S. (klō'vĭs)	64	34°24'N	103°11'W
Cluj-Napoca, Rom.	110	46°46'N	23°34'E
Clun, r., Eng., U.K. (klŭn)	114a	52°25'N	2°56'W
Cluny, Fr. (klü-nē')	126	46°27'N	4°40'E
Clutha, r., N.Z. (klōō'thȧ)	175a	45°52'S	169°30'E
Clwyd, hist. reg., Wales, U.K.	114a	53°01'N	2°59'W
Clyde, Ks., U.S.	79	39°34'N	97°23'W
Clyde, Oh., U.S.	66	41°15'N	83°00'W
Clyde, r., Scot., U.K.	120	55°35'N	3°50'W
Clyde, Firth of, b., Scot., U.K. (fûrth ŏv klīd)	120	55°28'N	5°01'W
Côa, r., Port. (kō'ä)	128	40°28'N	6°55'W
Coacalco, Mex. (kō-ä-käl'kō)	89a	19°37'N	99°06'W
Coachella, Canal, can., Ca., U.S. (kō'chĕl-lȧ)	76	33°15'N	115°25'W
Coahuayana, Río de, r., Mex. (rē'ō-dĕ-kō-ä-wä-yá'nä)	88	19°00'N	103°33'W
Coahuayutla, Mex. (kō'ä-wī-yōōt'lä)	88	18°19'N	101°44'W
Coahuila, state, Mex. (kō-ä-wē'lä)	86	27°30'N	103°00'W
Coal City, Il., U.S. (kōl sĭ'tĭ)	69a	41°17'N	88°17'W
Coalcomán, Río de, r., Mex. (rē'ō-dĕ-kō-äl-kō-män')	88	18°45'N	103°15'W
Coalcomán, Sierra de, mts., Mex.	88	18°30'N	102°45'W
Coalcomán de Matamoros, Mex.	88	18°46'N	103°10'W
Coaldale, Can. (kōl'dāl)	55	49°43'N	112°37'W
Coalgate, Ok., U.S. (kōl'gāt)	79	34°44'N	96°13'W
Coal Grove, Oh., U.S. (kōl grōv)	66	38°20'N	82°40'W
Coalinga, Ca., U.S. (kō-ȧ-lĭŋ'gȧ)	76	36°09'N	120°23'W
Coalville, Eng., U.K. (kōl'vĭl)	114a	52°43'N	1°21'W
Coamo, P.R. (kō-ä'mō)	87b	18°05'N	66°21'W
Coari, Braz. (kō-är'ē)	100	4°06'S	63°10'W
Coast Mountains, mts., N.A. (kōst)	52	54°10'N	128°00'W
Coast Ranges, mts., U.S.	64	41°28'N	123°30'W
Coatepec, Mex. (kō-ä-tā-pĕk)	88	19°23'N	98°44'W
Coatepec, Mex.	89	19°26'N	96°56'W
Coatepec, Mex.	89a	19°08'N	99°25'W
Coatepeque, El Sal.	90	13°56'N	89°30'W
Coatepeque, Guat. (kō-ä-tā-pā'kā)	90	14°40'N	91°52'W
Coatesville, Pa., U.S. (kōts'vĭl)	67	40°00'N	75°50'W
Coatetelco, Mex. (kō-ä-tā-tĕl'kō)	88	18°43'N	99°17'W
Coaticook, Can. (kō'tĭ-kók)	59	45°10'N	71°55'W
Coatlinchán, Mex. (kô-ä-tlē'n-chä'n)	89a	19°26'N	98°52'W
Coats, i., Can. (kōts)	53	62°23'N	82°11'W
Coats Land, reg., Ant.	178	74°00'S	30°00'W
Coatzacoalcos, Mex.	86	18°09'N	94°26'W
Coatzacoalcos, r., Mex.	89	17°40'N	94°41'W
Coba, hist., Mex. (kô'bä)	90a	20°23'N	87°23'W
Cobalt, Can. (kō'bôlt)	51	47°21'N	79°40'W
Cobán, Guat. (kō-bän')	86	15°28'N	90°19'W
Cobar, Austl.	175	31°28'S	145°50'E
Cobberas, Mount, mtn., Austl. (cŏ-bẽr-ăs)	176	36°45'S	148°15'E
Cobequid Mountains, mts., Can.	60	45°35'N	64°10'W
Cobh, Ire. (kôv)	110	51°52'N	8°09'W
Cobija, Bol. (kô-bē'hä)	100	11°12'S	68°49'W
Cobourg, Can. (kō'bórgh)	51	43°55'N	78°05'W
Cobre, r., Jam. (kō'brä)	92	18°05'N	77°00'W
Coburg, Austl.	173a	37°45'S	144°58'E
Coburg, Ger. (kō'bōōrg)	124	50°16'N	10°57'E
Cocentaina, Spain (kō-thán-tä-ē'nȧ)	129	38°44'N	0°27'W
Cochabamba, Bol.	100	17°24'S	66°09'W
Cochinos, Bahía, b., Cuba (bä-ē'ä-kô-chē'nōs)	92	22°05'N	81°10'W
Cochinos Banks, bk.,	92	22°20'N	76°15'W
Cochiti Indian Reservation, I.R., N.M., U.S.	77	35°37'N	106°20'W
Cochran, Ga., U.S. (kŏk'răn)	82	32°23'N	83°23'W
Cochrane, Can. (kŏk'răn)	51	49°01'N	81°06'W
Cochrane, Can.	62e	51°11'N	114°28'W
Cockburn, i., Can. (kŏk-bûrn)	58	45°55'N	83°25'W
Cockeysville, Md., U.S. (kŏk'ĭz-vĭl)	68e	39°30'N	76°40'W
Cockrell Hill, Tx., U.S. (kŏk'rĕl)	75c	32°44'N	96°53'W
Coco, r., N.A.	87	14°55'N	83°45'W
Coco, Cayo, i., Cuba (kä'-yō-kô'kô)	92	22°30'S	78°30'W
Coco, Isla del, i., C.R. (ē's-lä-dĕl-kô-kô)	86	5°33'N	87°02'W
Cocoa, Fl., U.S. (kō'kō)	83a	28°21'N	80°44'W
Cocoa Beach, Fl., U.S.	83a	28°20'N	80°35'W
Cocoli, Pan. (kô-kô'lē)	86a	8°58'N	79°36'W
Coconino, Plateau, plat., Az., U.S. (kō kō nē'nō)	77	35°45'N	112°28'W
Cocos (Keeling) Islands, is., Oc. (kō'kōs) (kē'lĭng)	3	11°50'S	90°50'E
Coco Solito, Pan. (kô-kô-sō-lē'tō)	86a	9°21'N	79°53'W
Cocula, Mex. (kō-kōō'lä)	88	20°23'N	103°47'W
Cocula, r., Mex.	88	18°17'N	99°45'W
Cod, Cape, pen., Ma., U.S.	65	41°42'N	70°15'W
Codajás, Braz. (kō-dä-häzh')	100	3°44'S	62°09'W
Codera, Cabo, c., Ven. (kȧ'bô-kō-dĕ'rä)	101b	10°35'N	66°06'W
Codogno, Italy (kô-dō'nyō)	130	45°08'N	9°43'E

238 Codrington – Copano Bay
PRONOUNCING INDEX

PLACE (Pronunciation)	PAGE	LAT.	LONG.
Codrington, Antig. (kŏd′rĭng-tŭn)	91b	17°39′N	61°49′W
Cody, Wy., U.S. (kō′dĭ)	73	44°31′N	109°02′W
Coelho da Rocha, Braz.	102b	22°47′S	43°23′W
Coemba, Ang.	190	12°08′S	18°05′E
Coesfeld, Ger. (kûs′fĕld)	127c	51°56′N	7°10′E
Coeur d'Alene, Id., U.S. (kûr dȧ-lān′)	64	47°43′N	116°35′W
Coeur d'Alene, r., Id., U.S.	72	47°26′N	116°35′W
Coeur d'Alene Indian Reservation, I.R., Id., U.S.	72	47°18′N	116°45′W
Coeur d'Alene Lake, l., Id., U.S.	72	47°32′N	116°39′W
Coffeyville, Ks., U.S. (kŏf′ĭ-vĭl)	65	37°01′N	95°38′W
Coff's Harbour, Austl.	176	30°20′S	153°10′E
Cofimvaba, S. Afr. (căfĭm′vä-bȧ)	187c	32°01′S	27°37′E
Coghinas, r., Italy (kō′gē-nás)	130	40°31′N	9°00′E
Cognac, Fr. (kōn-yak′)	117	45°41′N	0°22′W
Cohasset, Ma., U.S. (kō-hăs′ĕt)	61a	42°14′N	70°48′W
Cohoes, N.Y., U.S. (kō-hōz′)	67	42°50′N	73°40′W
Coig, r., Arg. (kō′ĕk)	102	51°15′N	71°00′W
Coimbatore, India (kō-ēm-bȧ-tōr′)	155	11°03′N	76°56′E
Coimbra, Port. (kō-ēm′brä)	110	40°14′N	8°23′W
Coín, Spain (kō-ēn′)	128	36°40′N	4°45′W
Coina, Port. (kō-ē′nȧ)	129b	38°35′N	9°03′W
Coina, r., Port. (kō′y-nä)	129b	38°35′N	9°02′W
Coipasa, Salar de, pl., Bol. (sä-lä′r-dē-koi-pä′-sä)	100	19°12′S	69°13′W
Coixtlahuaca, Mex. (kō-ēks′tlä-wä′kä)	89	17°42′N	97°17′W
Cojedes, dept., Ven. (kō-kĕ′dĕs)	101b	9°50′N	68°21′W
Cojimar, Cuba (kō-hē-mär′)	93a	23°10′N	82°19′W
Cojutepeque, El Sal. (kō-hō-tĕ-pā′kā)	90	13°45′N	88°50′W
Cokato, Mn., U.S. (kō-kā′tō)	71	45°03′N	94°11′W
Cokeburg, Pa., U.S. (kōk bûgh)	69e	40°06′N	80°03′W
Colac, Austl. (kŏ′lȧc)	176	38°25′S	143°40′E
Colares, Port. (kō-lä′rĕs)	129b	38°47′N	9°27′W
Colatina, Braz. (kō-lä-tē′nä)	101	19°33′S	40°42′W
Colby, Ks., U.S. (kōl′bĭ)	78	39°23′N	101°04′W
Colchagua, prov., Chile (kōl-chä′gwä)	99b	34°42′S	71°24′W
Colchester, Eng., U.K. (kōl′chĕs-tēr)	121	51°52′N	0°50′E
Cold Lake, l., Can. (kōld)	56	54°33′N	110°05′W
Coldwater, Ks., U.S. (kōld′wô-tēr)	78	37°14′N	99°21′W
Coldwater, Mi., U.S.	66	41°55′N	85°00′W
Coldwater, r., Ms., U.S.	82	34°25′N	90°12′W
Coldwater Creek, r., Tx., U.S.	78	36°10′N	101°45′W
Coleman, Tx., U.S. (kōl′mȧn)	80	31°50′N	99°25′W
Colenso, S. Afr. (kō-lĕnz′ō)	187c	28°48′S	29°49′E
Coleraine, N. Ire., U.K.	120	55°08′N	6°40′W
Coleraine, Mn., U.S. (kōl-rān′)	71	47°16′N	93°29′W
Coleshill, Eng., U.K. (kōlz′hĭl)	114a	52°30′N	1°42′W
Colfax, Ia., U.S. (kōl′făks)	71	41°40′N	93°13′W
Colfax, La., U.S.	81	31°31′N	92°42′W
Colfax, Wa., U.S.	72	46°53′N	117°21′W
Colhué Huapi, l., Arg. (kōl-wā′ōȧ′pē)	102	45°30′S	68°45′W
Coligny, S. Afr.	192c	26°20′S	26°18′E
Colima, Mex. (kōlē′mä)	86	19°13′N	103°45′W
Colima, state, Mex.	88	19°10′N	104°00′W
Colima, Nevado de, mtn., Mex. (nĕ-vä′dō-dĕ-kō-lē′mä)	86	19°30′N	103°38′W
Coll, i., Scot., U.K. (kōl)	120	56°42′N	6°23′W
College, Ak., U.S.	63	64°43′N	147°50′W
College Park, Ga., U.S. (kŏl′ĕj)	68c	33°39′N	84°27′W
College Park, Md., U.S.	68e	38°59′N	76°58′W
Collegeville, Pa., U.S. (kŏl′ĕj-vĭl)	68f	40°11′N	75°27′W
Collie, Austl. (kŏl′ē)	174	33°20′S	116°20′E
Collier Bay, b., Austl. (kŏl-yēr)	174	15°30′S	123°30′E
Collingswood, N.J., U.S. (kŏl′ĭngz-wŏŏd)	68f	39°54′N	75°04′W
Collingwood, Can.	59	44°30′N	80°20′W
Collins, Ms., U.S. (kŏl′ĭns)	82	31°40′N	89°34′W
Collinsville, Il., U.S. (kŏl′ĭnz-vĭl)	75e	38°41′N	89°59′W
Collinsville, Ok., U.S.	79	36°21′N	95°50′W
Colmar, Fr. (kōl′mär)	117	48°03′N	7°25′E
Colmenar de Oreja, Spain (kōl-mā-när′dāōrā′hä)	128	40°06′N	3°25′W
Colmenar Viejo, Spain (kōl-mā-när′vyā′hō)	128	40°40′N	3°46′W
Cologne, Ger.	110	50°56′N	6°57′E
Colombia, Col. (kō-lŏm′bē-ä)	100a	3°23′N	74°48′W
Colombia, nation, S.A.	100	3°30′N	72°30′W
Colombo, Sri L. (kō-lŏm′bō)	159	6°58′N	79°52′E
Colón, Arg. (kō-lōn′)	99c	33°55′S	61°08′W
Colón, Cuba (kō-lō′n)	92	22°45′N	80°55′W
Colón, Mex. (kō-lōn′)	88	20°46′N	100°02′W
Colón, Pan. (kō-lō′n)	87	9°22′N	79°54′W
Colón, Archipiélago de, is., Ec.	100	0°10′S	87°45′W
Colón, Montañas de, mts., Hond. (môn-tä′n-yäs-dĕ-kō-lō′n)	91	14°58′N	84°39′W
Colonia, Ur. (kō-lō′nē-ä)	102	34°27′S	57°50′W
Colonia, dept., Ur.	99c	34°08′S	57°50′W
Colonia Suiza, Ur. (kō-lō′nēä-sōē′zä)	99c	34°17′S	57°15′W
Colonna, Capo, c., Italy	131	39°02′N	17°15′E
Colonsay, i., Scot., U.K. (kōl-ōn-sā′)	121	56°08′N	6°08′E
Coloradas, Lomas, Arg. (lō′mäs-kō-lō-rä′däs)	102	43°30′S	68°00′W
Colorado, state, U.S.	64	39°30′N	106°55′W
Colorado, r., Arg.	102	38°30′S	66°00′W
Colorado, r., N.A.	64	36°00′N	113°30′W
Colorado, r., Tx., U.S.	64	30°08′N	97°33′W
Colorado City, Tx., U.S. (kŏl-ō-rä′dō sĭ′tĭ)	80	32°24′N	100°50′W
Colorado National Monument, rec., Co., U.S.	77	39°00′N	108°40′W
Colorado Plateau, plat., U.S.	64	36°20′N	109°25′W
Colorado River Aqueduct, aq., Ca., U.S.	76	33°38′N	115°43′W
Colorado River Indian Reservation, I.R., Az., U.S.	77	34°03′N	114°02′W
Colorados, Archipiélago de los, is., Cuba	92	22°25′N	84°25′W
Colorado Springs, Co., U.S. (kŏl-ō-rä′dō)	64	38°49′N	104°48′W
Colotepec, r., Mex. (kō-lō′tĕ-pĕk)	89	15°56′N	96°57′W
Colotlán, Mex. (kō-lō-tlän′)	88	22°06′N	103°14′W
Colotlán, r., Mex.	88	22°09′N	103°17′W
Colquechaca, Bol. (kōl-kā-chä′kä)	100	18°47′S	66°02′W
Colstrip, Mt., U.S. (kōl′strip)	73	45°54′N	106°38′W
Colton, Ca., U.S. (kōl′tŭn)	75a	34°04′N	117°20′W
Columbia, Il., U.S. (kō-lŭm′bĭ-ȧ)	75e	38°26′N	90°12′W
Columbia, Ky., U.S.	82	37°06′N	85°15′W
Columbia, Md., U.S.	68e	39°15′N	76°51′W
Columbia, Mo., U.S.	65	38°55′N	92°19′W
Columbia, Ms., U.S.	82	31°15′N	89°49′W
Columbia, Pa., U.S.	67	40°00′N	76°25′W
Columbia, S.C., U.S.	65	34°00′N	81°00′W
Columbia, Tn., U.S.	82	35°36′N	87°02′W
Columbia, r., N.A.	52	46°00′N	120°00′W
Columbia, Mount, mtn., Can.	55	52°09′N	117°25′W
Columbia City, In., U.S.	66	41°10′N	85°30′W
Columbia City, Or., U.S.	74c	45°53′N	112°49′W
Columbia Heights, Mn., U.S.	75g	45°03′N	93°15′W
Columbia Icefield, ice, Can.	55	52°08′N	117°26′W
Columbia Mountains, mts., N.A.	55	51°30′N	118°30′W
Columbiana, Al., U.S. (kō-ŭm-bĭ-ă′nȧ)	82	33°10′N	86°35′W
Columbretes, is., Spain (kō-lōōm-brĕ′tĕs)	129	39°54′N	0°54′E
Columbus, Ga., U.S. (kō-lŭm′bŭs)	65	32°29′N	84°56′W
Columbus, In., U.S.	66	39°15′N	85°55′W
Columbus, Ks., U.S.	79	37°10′N	94°50′W
Columbus, Ms., U.S.	82	33°30′N	88°25′W
Columbus, Mt., U.S.	73	45°39′N	109°15′W
Columbus, Ne., U.S.	70	41°25′N	97°25′W
Columbus, N.M., U.S.	77	31°50′N	107°40′W
Columbus, Oh., U.S.	65	40°00′N	83°00′W
Columbus, Tx., U.S.	81	29°44′N	96°34′W
Columbus, Wi., U.S.	71	43°20′N	89°01′W
Columbus Bank, bk., (kō-lŭm′bŭs)	93	22°05′N	75°30′W
Columbus Grove, Oh., U.S.	66	40°55′N	84°05′W
Columbus Point, c., Bah.	93	24°10′N	75°15′W
Colusa, Ca., U.S. (kō-lū′sȧ)	76	39°12′N	122°01′W
Colville, Wa., U.S. (kŏl′vĭl)	72	48°33′N	117°53′W
Colville, r., Ak., U.S.	63	69°00′N	156°25′W
Colville Indian Reservation, I.R., Wa., U.S.	72	48°15′N	119°00′W
Colville R, Wa., U.S.	72	48°25′N	117°58′W
Colvos Passage, strt., Wa., U.S. (kōl′vōs)	74a	47°24′N	122°32′W
Colwood, Can. (kŏl′wŏŏd)	74a	48°26′N	123°30′W
Comacchio, Italy (kō-mäk′kyō)	130	44°42′N	12°12′E
Comala, Mex. (kō-mä-lä′)	88	19°22′N	103°47′W
Comalapa, Guat. (kō-mä-lä′-pä)	90	14°43′N	90°56′W
Comalcalco, Mex. (kō-mäl-käl′kō)	89	18°16′N	93°13′W
Comanche, Ok., U.S. (kō-măn′chē)	79	34°20′N	97°58′W
Comanche, Tx., U.S.	80	31°54′N	98°37′W
Comanche Creek, r., Tx., U.S.	80	31°02′N	102°47′W
Comayagua, Hond. (kō-mä-yä′gwä)	86	14°24′N	87°36′W
Combahee, r., S.C., U.S. (kŏm-bȧ-hē′)	83	32°42′N	80°40′W
Comer, Ga., U.S. (kŭm′ēr)	82	34°02′N	83°07′W
Comete, Cape, c., T./C. Is. (kō-mā′tȧ)	93	21°45′N	71°25′W
Comilla, Bngl. (kō-mĭl′ä)	155	23°33′N	91°17′E
Comino, Cape, c., Italy (kō-mē′nō)	130	40°30′N	9°48′E
Comitán, Mex. (kō-mē-tän′)	86	16°16′N	92°09′W
Commencement Bay, b., Wa., U.S. (kō-mĕns′mĕnt bā)	74a	47°17′N	122°21′W
Commentry, Fr. (kō-mäN-trē′)	126	46°16′N	2°44′E
Commerce, Ga., U.S. (kŏm′ērs)	82	34°10′N	83°27′W
Commerce, Ok., U.S.	79	36°57′N	94°54′W
Commerce, Tx., U.S.	79	33°15′N	95°52′W
Como, Italy (kō′mō)	118	45°48′N	9°03′E
Como, Lago di, l., Italy (lä′gō-dē-kō′mō)	118	46°00′N	9°30′E
Comodoro Rivadavia, Arg.	102	45°47′S	67°31′W
Como-Est, Can.	62a	45°27′N	74°08′W
Comonfort, Mex. (kō-mōn-fō′rt)	88	20°43′N	100°47′W
Comorin, Cape, c., India (kō′mō-rĭn)	159	8°05′N	78°05′E
Comoros, nation, Afr.	187	12°30′S	42°45′E
Comox, Can. (kō′mŏks)	54	49°40′N	124°55′W
Companario, Cerro, mtn., S.A. (sĕ′r-rō-kôm-pä-nä′ryō)	99b	35°54′S	70°23′W
Compiègne, Fr. (kôN-pyĕn′y′)	117	49°25′N	2°49′E
Comporta, Port. (kôm-pôr′tȧ)	129b	38°24′N	8°48′W
Compostela, Mex. (kōm-pō-stā′lä)	88	21°14′N	104°54′W
Compton, Ca., U.S. (kŏmpt′tŭn)	75a	33°54′N	118°14′W
Comrat, Mol. (kôm-rät′)	137	46°17′N	28°38′E
Conakry, Gui. (kō-nȧ-krē′)	184	9°31′N	13°43′W
Conanicut, i., R.I., U.S. (kō-nă′nĭ-kŭt)	68b	41°34′N	71°20′W
Conasauga, r., Ga., U.S. (kō-nȧ)	82	34°40′N	84°51′W
Concarneau, Fr. (kôN-kär-nō′)	126	47°54′N	3°52′W
Concepción, Bol. (kōn-sĕp′syōn′)	101	15°47′S	61°08′W
Concepción, Chile	102	36°51′S	72°59′W
Concepción, Pan.	79	8°31′N	82°38′W
Concepción, Para.	102	23°29′S	57°18′W
Concepcion, Phil.	169a	15°19′N	120°40′E
Concepción, vol., Nic.	90	11°36′N	85°43′W
Concepción, r., Mex.	86	30°25′N	112°20′W
Concepción del Mar, Guat. (kōn-sĕp-syōn′dĕl mär′)	90	14°07′N	91°23′W
Concepción del Oro, Mex. (kōn-sĕp-syōn′ dĕl ō′rō)	86	24°39′N	101°24′W
Concepción del Uruguay, Arg. (kōn-sĕp-syō′n-dĕl-ōō-rōō-gwī′)	102	32°31′S	58°10′W
Conception, i., Bah.	93	23°50′N	75°05′W
Conception, Point, c., Ca., U.S.	64	34°27′N	120°28′W
Conception Bay, b., Can. (kŏn-sĕp′shŭn)	61	47°50′N	52°50′W
Concho, r., Tx., U.S. (kŏn′chō)	80	31°34′N	100°00′W
Conchos, r., Mex.	86	29°30′N	105°00′W
Conchos, r., Mex. (kōn′chōs)	80	25°03′N	99°00′W
Concord, Ca., U.S. (kŏŋ′kŏrd)	74b	37°58′N	122°02′W
Concord, Ma., U.S.	61a	42°28′N	71°21′W
Concord, N.C., U.S.	83	35°23′N	80°11′W
Concord, N.H., U.S.	65	43°10′N	71°30′W
Concordia, Arg. (kŏn-kôr′dĭ-ȧ)	102	31°18′S	57°59′W
Concordia, Col.	100a	6°04′N	75°54′W
Concordia, Mex. (kōn-kō′r-dyä)	88	23°17′N	106°06′W
Concordia, Ks., U.S.	79	39°32′N	97°39′W
Concrete, Wa., U.S. (kŏn-′krēt)	72	48°33′N	121°44′W
Conde, Fr.	126	48°50′N	0°36′W
Conde, S.D., U.S. (kŏn-dē′)	70	45°10′N	98°06′W
Condega, Nic. (kōn-dĕ′gä)	90	13°20′N	86°27′W
Condeúba, Braz. (kōn-dā-ōō′bä)	101	14°47′S	41°44′W
Condom, Fr.	126	43°58′N	0°22′E
Condon, Or., U.S. (kŏn′dŭn)	72	45°14′N	120°10′W
Conecun, r., Al., U.S. (kō-nē′kŭ)	82	31°05′N	86°52′W
Conegliano, Italy (kō-nål-yä′nō)	130	45°59′N	12°17′E
Conejos, r., Co., U.S. (kō-nā′hōs)	77	37°07′N	106°19′W
Conemaugh, Pa., U.S. (kŏn′ē-mô)	67	40°25′N	78°50′W
Coney Island, i., N.Y., U.S. (kō′nĭ)	68a	40°34′N	73°27′W
Confolens, Fr. (kôN-fä-läN′)	126	46°01′N	0°41′E
Congaree, r., S.C., U.S. (kŏŋ-gȧ-rē′)	83	33°53′N	80°55′W
Conghua, China (tsôŋ-hwä)	165	23°30′N	113°40′E
Congleton, Eng., U.K. (kŏŋ′g′l-tŭn)	114a	53°10′N	2°13′W
Congo, nation, Afr. (kŏn′gō)	186	3°00′S	13°48′E
Congo (Zaire), r., Afr. (kŏn′gō)	183	2°00′S	17°00′E
Congo, Democratic Republic of the (Zaire), nation, Afr.	186	1°00′S	22°15′E
Congo, Serra do, mts., Ang.	190	6°25′S	13°30′E
Congo Basin, basin, D.R.C.	183	2°47′N	20°58′E
Conisbrough, Eng., U.K. (kŏn′ĭs-bŭr-ō)	114a	53°29′N	1°13′W
Coniston, Can.	59	46°29′N	80°51′W
Conklin, Can. (kŏŋk′lĭn)	55	55°38′N	111°05′W
Conley, Ga., U.S. (kŏn′lĭ)	68c	33°38′N	84°19′W
Conn, Lough, l., Ire. (lŏk kŏn)	120	53°56′N	9°25′W
Connacht, hist. reg., Ire. (cŏn′ät)	120	53°50′N	8°45′W
Conneaut, Oh., U.S. (kŏn-ē-ôt′)	66	41°55′N	80°35′W
Connecticut, state, U.S. (kŏ-nĕt′ĭ-kŭt)	65	41°40′N	73°10′W
Connecticut, r., U.S.	65	43°55′N	72°15′W
Connellsville, Pa., U.S. (kŏn′nĕlz-vĭl)	67	40°00′N	79°40′W
Connemara, mts., Ire. (kŏn-nē-mȧ′rȧ)	120	53°30′N	9°54′W
Connersville, In., U.S. (kŏn′ērz-vĭl)	66	39°35′N	85°10′W
Connors Range, mts., Austl. (kŏn′nŏrs)	175	22°15′S	149°00′E
Conrad, Mt., U.S. (kŏn′răd)	73	48°11′N	111°56′W
Conrich, Can. (kŏn′rĭch)	62e	51°06′N	113°51′W
Conroe, Tx., U.S. (kŏn′rō)	81	30°18′N	95°23′W
Conselheiro Lafaiete, Braz.	101	20°40′S	43°46′W
Conshohocken, Pa., U.S. (kŏn-shō-hŏk′ĕn)	68f	40°04′N	75°18′W
Consolación del Sur, Cuba (kōn-sō-lä-syōn′)	92	22°30′N	83°55′W
Con Son, is., Viet.	168	8°30′N	106°28′E
Constance, Mount, mtn., Wa., U.S. (kŏn′stăns)	74a	47°46′N	123°08′W
Constanța, Rom. (kōn-stán′tsá)	110	44°12′N	28°36′E
Constantina, Spain (kōn-stän-tē′nä)	128	37°52′N	5°39′W
Constantine, Alg. (kôN-stäN′tēn′)	184	36°28′N	6°38′E
Constantine, Mi., U.S. (kŏn′stăn-tēn)	66	41°50′N	85°40′W
Constitución, Chile (kōn′stē-tōō-syōn′)	102	35°24′S	72°25′W
Constitution, Ga., U.S. (kŏn-stĭ-tū′shŭn)	68c	33°41′N	84°20′W
Contagem, Braz. (kōn-tá′zhĕm)	99a	19°54′S	44°05′W
Contepec, Mex. (kōn-tē-pĕk′)	88	20°04′N	100°07′W
Contreras, Mex. (kôn-trĕ′räs)	89a	19°18′N	99°14′W
Contwoyto, l., Can.	52	65°42′N	110°50′W
Converse, Tx., U.S. (kŏn′vērs)	75d	29°31′N	98°17′W
Conway, Ar., U.S. (kŏn′wā)	79	35°06′N	92°27′W
Conway, N.H., U.S.	67	44°00′N	71°10′W
Conway, S.C., U.S.	83	33°49′N	79°01′W
Conway, Wa., U.S.	74a	48°20′N	122°20′W
Conyers, Ga., U.S. (kŏn′yŏrz)	82	33°41′N	84°01′W
Cooch Behār, India (kōōch bē-här′)	155	26°25′N	89°34′E
Cook, Cape, c., Can. (kŏŏk)	54	50°08′N	127°55′W
Cook, Mount *see* Aoraki, mtn., N.Z.	175a	43°27′S	170°13′E
Cookeville, Tn., U.S. (kŏŏk′vĭl)	82	36°07′N	85°30′W
Cooking Lake, Can. (kŏŏk′ĭng)	62g	53°25′N	113°08′W
Cooking Lake, l., Can.	62g	53°25′N	113°02′W
Cook Inlet, b., Ak., U.S.	63	60°50′N	151°38′W
Cook Islands, dep., Oc.	2	20°00′S	158°00′W
Cook Strait, strt., N.Z.	175a	40°37′S	174°15′E
Cooktown, Austl. (kŏŏk′toun)	175	15°40′S	145°20′E
Cooleemee, N.C., U.S. (kōō-lē′mē)	83	35°50′N	80°32′W
Coolgardie, Austl. (kōōl-gär′dē)	174	31°00′S	121°25′E
Cooma, Austl. (kōō′mȧ)	175	36°22′S	149°10′E
Coonamble, Austl. (kōō-năm′b'l)	175	31°00′S	148°30′E
Coonoor, India	159	10°22′N	76°15′E
Coon Rapids, Mn., U.S. (kōōn)	75g	45°09′N	93°17′W
Cooper, Tx., U.S. (kōōp′ēr)	79	33°23′N	95°40′W
Cooper Center, Ak., U.S.	63	61°54′N	15°30′W
Coopers Creek, r., Austl. (kōō′pērz)	175	27°32′N	141°19′E
Cooperstown, N.D., U.S.	70	47°26′N	98°07′W
Cooperstown, N.Y., U.S. (kōōp′ērs-toun)	67	42°45′N	74°55′W
Coosa, Al., U.S. (kōō′sȧ)	82	32°43′N	86°25′W
Coosa, r., U.S.	65	34°00′N	86°00′W
Coosawattee, r., Ga., U.S.	82	34°37′N	84°45′W
Coos Bay, Or., U.S. (kōōs)	72	43°21′N	124°12′W
Coos Bay, b., Or., U.S.	72	43°19′N	124°40′W
Cootamundra, Austl. (kŏŏtȧ-mŭnd′rȧ)	176	34°25′S	148°00′E
Copacabana, Braz. (kō′pä-kȧ-bá′nä)	102b	22°57′S	43°11′W
Copalita, r., Mex. (kō-pä-lē′tä)	89	15°55′N	96°06′W
Copán, hist., Hond. (kō-pän′)	90	14°50′N	89°10′W
Copano Bay, b., Tx., U.S. (kō-pän′ō)	81	28°08′N	97°25′W

ăt; finăl; rāte; senāte; ärm; ȧsk; sofȧ; fâre; ch-choose; dh-as th in other; bē; ēvent; bĕt; recĕnt; cratēr; g-gō; gh-guttural g; bĭt; ĭ-short neutral; rīde; K-guttural k as ch in German ich;

PRONOUNCING INDEX

Copenhagen – Crofton 239

PLACE (Pronunciation)	PAGE	LAT.	LONG.
Copenhagen (København), Den.	110	55°43'N	12°27'E
Copiapó, Chile (kō-pyä-pō')	102	27°16'S	70°28'W
Copley, Oh., U.S. (kŏp'lē)	69d	41°06'N	81°38'W
Copparo, Italy (kôp-pä'rō)	130	44°53'N	11°50'E
Coppell, Tx., U.S. (kŏp'pĕl)	75c	32°57'N	97°00'W
Copper, r., Ak., U.S. (kŏp'ẽr)	63	62°38'N	145°00'W
Copper Cliff, Can.	58	46°28'N	81°04'W
Copper Harbor, Mi., U.S.	71	47°27'N	87°53'W
Copperhill, Tn., U.S. (kŏp'ẽr hĭl)	82	35°00'N	84°22'W
Coppermine *see* Kugluktuk, Can.	52	67°46'N	115°19'W
Coppermine, r., Can.	52	66°48'N	114°59'W
Copper Mountain, mtn., Ak., U.S.	54	55°14'N	132°36'W
Copperton, Ut., U.S. (kŏp'ẽr-tŭn)	75b	40°34'N	112°06'W
Coquilee, Or., U.S. (kō-kēl')	72	43°11'N	124°11'W
Coquilhatville *see* Mbandaka, D.R.C.	186	0°04'N	18°16'E
Coquimbo, Chile (kō-kēm'bō)	102	29°58'S	71°31'W
Coquimbo, prov., Chile	99b	31°50'S	71°05'W
Coquitlam Lake, l., Can. (kō-kwĭt-lám)	74d	49°23'N	122°44'W
Corabia, Rom. (kō-rä'bĭ-á)	119	43°45'N	24°29'E
Coracora, Peru (kō'rä-kō'rä)	100	15°12'S	73°42'W
Coral Gables, Fl., U.S.	83a	25°43'N	80°14'W
Coral Rapids, Can. (kŏr'ăl)	51	50°18'N	81°49'W
Coral Sea, sea, Oc. (kŏr'ăl)	175	13°30'S	150°00'E
Coralville Reservoir, res., Ia., U.S.	71	41°45'N	91°50'W
Corangamite, Lake, l., Austl. (cŏr-ăng'á-mīt)	176	38°05'S	142°55'E
Coraopolis, Pa., U.S. (kō-rä-ŏp'ō-lĭs)	69e	40°30'N	80°09'W
Corato, Italy (kō'rä-tō)	130	41°08'N	16°28'E
Corbeil-Essonnes, Fr. (kôr-bā'yĕ-sŏn')	126	48°31'N	2°29'E
Corbett, Or., U.S. (kôr'bĕt)	74c	45°31'N	122°17'W
Corbie, Fr. (kôr-bē')	126	49°55'N	2°27'E
Corbin, Ky., U.S. (kôr'bĭn)	82	36°55'N	84°06'W
Corby, Eng., U.K. (kôr'bĭ)	114a	52°29'N	0°38'W
Corcovado, mtn., Braz. (kôr-kō-vä'dō)	102b	22°57'S	43°13'W
Corcovado, Golfo, b., Chile (kôr-kō-vä'dhō)	102	43°40'S	75°00'W
Cordeiro, Braz. (kôr-dā'rō)	99a	22°03'S	42°22'W
Cordele, Ga., U.S. (kôr-dēl')	82	31°55'N	83°50'W
Cordell, Ok., U.S. (kôr-dĕl')	78	35°19'N	98°58'W
Córdoba, Arg. (kôr'dô-vä)	102	30°20'S	64°03'W
Córdoba, Mex. (kô'r-dô-bä)	86	18°53'N	96°54'W
Córdoba, Spain (kô'r-dô-bä)	128	37°55'N	4°45'W
Córdoba, prov., Arg. (kôr'dô-vä)	102	32°00'S	64°00'W
Córdoba, Sierra de, mts., Arg.	102	31°15'S	64°30'W
Cordova, Ak., U.S. (kôr'dô-vä)	64a	60°34'N	145°38'W
Cordova, Al., U.S. (kôr'dô-á)	82	33°45'N	86°22'W
Cordova Bay, b., Ak., U.S.	54	54°55'N	132°35'W
Corfu *see* Kérkyra, i., Grc.	112	39°33'N	19°36'E
Corigliano, Italy (kō-rē-lyä'nō)	130	39°35'N	16°30'E
Corinth *see* Kórinthos, Grc.	110	37°56'N	22°54'E
Corinth, Ms., U.S. (kŏr'ĭnth)	82	34°55'N	88°30'W
Corinto, Braz. (kô-rē'n-tō)	101	18°20'S	44°16'W
Corinto, Col.	100a	3°09'N	76°12'W
Corinto, Nic. (kōr-ĭn'to)	90	12°30'N	87°12'W
Corio, Austl.	173a	38°05'S	144°22'E
Corio Bay, b., Austl.	173a	38°07'S	144°25'E
Corisco, Isla de, i., Eq. Gui.	190	0°50'N	8°40'E
Cork, Ire. (kôrk)	110	51°54'N	8°25'W
Cork Harbour, b., Ire.	120	51°44'N	8°15'W
Corleone, Italy (kôr-lā-ō'nā)	130	37°48'N	13°18'E
Cormorant Lake, l., Can.	57	54°13'N	100°47'W
Cornelia, Ga., U.S. (kôr-nē'lyá)	82	34°31'N	83°30'W
Cornelis, r., S. Afr. (kôr-nē'lĭs)	192c	27°48'S	29°15'E
Cornell, Ca., U.S. (kôr-nĕl')	75a	34°06'N	118°46'W
Cornell, Wi., U.S.	71	45°10'N	91°10'W
Corner Brook, Can. (kôr'nẽr)	51	48°57'N	57°57'W
Corner Inlet, b., Austl.	176	38°55'S	146°45'E
Corning, Ar., U.S. (kôr'nĭng)	79	36°26'N	90°35'W
Corning, Ia., U.S.	71	40°58'N	94°40'W
Corning, N.Y., U.S.	67	42°10'N	77°05'W
Corno, Monte, mtn., Italy (kôr'nō)	118	42°28'N	13°37'E
Cornwall, Bah.	92	25°55'N	77°15'W
Cornwall, Can. (kôrn'wôl)	59	45°05'N	74°35'W
Coro, Ven. (kō'rō)	100	11°22'N	69°43'W
Corocoro, Bol. (kō-rō-kō'rō)	100	17°15'S	68°21'W
Coromandel Coast, cst., India (kŏr-ō-man'dĕl)	155	13°30'N	80°30'E
Coromandel Peninsula, pen., N.Z.	177	36°50'S	176°00'E
Corona, Al., U.S. (kō-rō'ná)	82	33°42'N	87°28'W
Corona, Ca., U.S.	75a	33°52'N	117°34'W
Coronada, Bahía de, b., C.R. (bä-ē'ä-dĕ-kō-rō-nä'dō)	91	8°47'N	84°04'W
Corona del Mar, Ca., U.S. (kō-rō'ná dĕl mär)	75a	33°36'N	117°53'W
Coronado, Ca., U.S. (kŏr-ō-nä'dō)	76a	32°42'N	117°12'W
Coronation Gulf, b., Can. (kŏr-ō-nā'shŭn)	52	68°07'N	112°50'W
Coronel, Chile (kō-rō-nĕl')	102	37°00'S	73°10'W
Coronel Brandsen, Arg. (kŏ-rō-nĕl-brá'nd-sĕn)	99c	35°09'S	58°15'W
Coronel Dorrego, Arg. (kŏ-rō-nĕl-dôr-rĕ'gō)	102	38°43'S	61°16'W
Coronel Oviedo, Para. (kŏ-rō-nĕl-ō-vēĕ'dō)	102	25°28'S	56°22'W
Coronel Pringles, Arg. (kŏ-rō-nĕl-prēn'glĕs)	102	37°54'S	61°22'W
Coronel Suárez, Arg. (kŏ-rō-nĕl-swä'rās)	102	37°27'S	61°49'W
Corowa, Austl. (cŏr-ōwá)	176	36°02'S	146°23'E
Corozal, Belize (cŏr-ōth-äl')	90a	18°25'N	88°23'W
Corpus Christi, Tx., U.S. (kôr'pŭs krĭstē)	64	27°48'N	97°24'W
Corpus Christi Bay, b., Tx., U.S.	81	27°47'N	97°14'W
Corpus Christi Lake, l., Tx., U.S.	80	28°08'N	98°20'W
Corral, Chile (kō-räl')	102	39°57'S	73°15'W
Corral de Almaguer, Spain (kō-räl'dā äl-mä-gâr')	128	39°45'N	3°10'W
Corralillo, Cuba (kō-rä-lē-yō)	92	23°00'N	80°40'W
Corregidor Island, i., Phil. (kō-rā-hē-dōr')	169a	14°21'N	120°25'E
Correntina, Braz. (kō-rĕn-tē-ná)	101	13°18'S	44°33'W
Corrib, Lough, l., Ire. (lŏk kŏr'ĭb)	120	53°25'N	9°19'W
Corrientes, Arg. (kō-ryĕn'tās)	102	27°25'S	58°39'W
Corrientes, prov., Arg.	102	28°45'S	58°00'W
Corrientes, Cabo, c., Col. (ká'bô-kō-ryĕn'tās)	100	5°34'N	77°35'W
Corrientes, Cabo, c., Cuba (ká'bô-kôr-rē-ĕn'tĕs)	92	21°50'N	84°25'W
Corrientes, Cabo, c., Mex.	86	20°25'N	105°41'W
Corry, Pa., U.S. (kŏr'ĭ)	67	41°55'N	79°40'W
Corse, Cap, c., Fr. (kôrs)	117	42°59'N	9°19'E
Corsica, i., Fr. (kô'r-sē-kä)	112	42°10'N	8°55'E
Corsicana, Tx., U.S. (kôr-sĭ-kăn'á)	64	32°06'N	96°28'W
Cortazar, Mex. (kôr-tä-zär)	88	20°30'N	100°57'W
Corte, Fr. (kôr'tā)	130	42°18'N	9°10'E
Cortegana, Spain (kôr-tā-gä'nä)	128	37°54'N	6°48'W
Cortés, Ensendada de, b., Cuba (ĕn-sĕ-nä-dä-dĕ-kôr-tās')	92	22°05'N	83°45'W
Cortez, Co., U.S.	77	37°21'N	108°35'W
Cortland, N.Y., U.S. (kôrt'lănd)	67	42°35'N	76°10'W
Cortona, Italy (kôr-tō'nä)	130	43°16'N	12°00'E
Corubal, r., Gui.-B.	188	11°43'N	14°40'W
Coruche, Port. (kō-rōō'she)	128	38°58'N	8°34'W
Çoruh, r., Asia (chō-rōōk')	137	40°30'N	41°10'E
Çorum, Tur. (chô-rōōm')	154	40°34'N	34°45'E
Corunna, Mi., U.S. (kô-rŭn'á)	66	43°00'N	84°05'W
Coruripe, Braz. (kō-rò-rē'pĭ)	101	10°09'S	36°13'W
Corvallis, Or., U.S. (kŏr-văl'ĭs)	64	44°34'N	123°17'W
Corve, r., Eng., U.K. (kôr'vē)	114a	52°28'N	2°43'W
Corydon, Ia., U.S.	71	40°45'N	93°20'W
Corydon, In., U.S. (kŏr'ĭ-dŭn)	66	38°10'N	86°05'W
Corydon, Ky., U.S.	66	37°45'N	87°40'W
Cosamaloápan, Mex. (kô-sä-mä-lwä'pän)	89	18°21'N	95°48'W
Coscomatepec, Mex. (kôs'kōmä-tē-pĕk')	89	19°04'N	97°03'W
Cosenza, Italy (kō-zĕnt'sä)	119	39°18'N	16°15'E
Coshocton, Oh., U.S. (kō-shŏk'tŭn)	66	40°15'N	81°55'W
Cosigüina, vol., Nic.	90	12°59'N	87°35'W
Cosmoledo Group, is., Sey. (kŏs-mō-lā'dō)	187	9°42'S	47°45'E
Cosmopolis, Wa., U.S. (kŏz-mŏp'ō-lĭs)	72	46°58'N	123°47'W
Cosne-sur-Loire, Fr. (kōn-sür-lwär')	126	47°25'N	2°57'E
Cosoleacaque, Mex. (kō sō lā-ä-kä'kē)	89	18°01'N	94°38'W
Costa de Caparica, Port.	129b	38°40'N	9°12'W
Costa Mesa, Ca., U.S. (kŏs'tá mā'sá)	75a	33°39'N	118°54'W
Costa Rica, nation, N.A. (kŏs'tá rē'ká)	87	10°30'N	84°30'W
Cosumnes, r., Ca., U.S. (kō-sŭm'nĕz)	76	38°21'N	121°17'W
Cotabambas, Peru (kō-tä-bám'bäs)	100	13°49'S	72°17'W
Cotabato, Phil. (kō-tä-bä'tō)	169	7°06'N	124°13'E
Cotaxtla, Mex. (kō-täs'tlä)	89	18°49'N	96°22'W
Cotaxtla, r., Mex.	89	18°54'N	96°21'W
Coteau-du-Lac, Can. (cō-tō'dü-läk)	62a	45°17'N	74°11'W
Coteau-Landing, Can.	62a	45°15'N	74°13'W
Coteaux, Haiti	93	18°15'N	74°05'W
Cote d'Ivoire (Ivory Coast), nation, Afr.	184	7°43'N	6°30'W
Côte d'Or, reg., Fr.	126	47°02'N	4°35'E
Cotija de la Paz, Mex. (kô-tē'-kä-dĕ-lä-pá'z)	88	19°46'N	102°43'W
Cotonou, Benin (kō-tô-nōō')	184	6°21'N	2°26'E
Cotopaxi, mtn., Ec. (kō-tō-păk'sē)	100	0°40'S	78°26'W
Cotorro, Cuba (kō-tôr-rō)	93a	23°03'N	82°17'W
Cotswold Hills, hills, Eng., U.K. (kŭtz'wōld)	120	51°35'N	2°16'W
Cottage Grove, Mn., U.S. (kŏt'āj grōv)	75g	44°50'N	92°52'W
Cottage Grove, Or., U.S.	72	43°48'N	123°04'W
Cottbus, Ger. (kōtt'bōōs)	117	51°47'N	14°20'E
Cottonwood, r., Mn., U.S. (kŏt'ŭn-wŏd)	70	44°25'N	95°35'W
Cotulla, Tx., U.S. (kō-tŭl'lá)	80	28°26'N	99°14'W
Coubert, Fr. (kōō-bâr')	127b	48°40'N	2°43'E
Coudersport, Pa., U.S. (kou'dẽrz-port)	67	41°45'N	78°00'W
Coudres, Île aux, i., Can.	60	47°17'N	70°12'W
Coulommiers, Fr. (kōō-lô-myā')	127b	48°49'N	3°05'E
Coulto, Serra do, mts., Braz. (sĕ'r-rä-dô-kô-ō'tō)	102b	22°33'S	43°27'W
Council Bluffs, Ia., U.S. (koun'sĭl blŭf)	65	41°16'N	95°53'W
Council Grove, Ks., U.S. (koun'sĭl grōv)	79	38°39'N	96°30'W
Coupeville, Wa., U.S. (kōōp'vĭl)	74a	48°13'N	122°41'W
Courantyne, r., S.A. (kôr'ăntĭn)	101	4°28'N	57°42'W
Courtenay, Can. (cōōrt-nā')	50	49°41'N	125°00'W
Coushatta, La., U.S. (kou-shăt'á)	81	32°02'N	93°21'W
Coutras, Fr. (kōō-trá')	126	45°02'N	0°07'W
Covelo, Ang.	190	12°06'S	13°55'E
Coventry, Eng., U.K. (kŭv'ĕn-trĭ)	120	52°25'N	1°29'W
Covina, Ca., U.S. (kô-vē'ná)	75a	34°06'N	117°54'W
Covington, Ga., U.S. (kŭv'ĭng-tŭn)	82	33°36'N	83°50'W
Covington, In., U.S.	66	40°10'N	87°15'W
Covington, Ky., U.S.	65	39°05'N	84°31'W
Covington, La., U.S.	81	30°30'N	90°06'W
Covington, Oh., U.S.	66	40°10'N	84°20'W
Covington, Ok., U.S.	79	36°18'N	97°32'W
Covington, Tn., U.S.	82	35°33'N	89°40'W
Covington, Va., U.S.	66	37°50'N	80°00'W
Cowal, Lake, l., Austl. (kou'ăl)	176	33°30'S	147°10'E
Cowan, l., Austl. (kou'án)	174	32°00'S	122°30'E
Cowansville, Can.	59	45°13'N	72°47'W
Cow Creek, r., Or., U.S. (kou)	72	42°45'N	123°35'W
Cowes, Eng., U.K. (kouz)	120	50°43'N	1°25'W
Cowichan Lake, l., Can.	54	48°54'N	124°20'W
Cowlitz, r., Wa., U.S. (kou'lĭts)	72	46°30'N	122°45'W
Cowra, Austl. (kou'rá)	176	33°50'S	148°33'E
Coxim, Braz. (kō-shēn')	101	18°32'S	54°43'W
Coxquihui, Mex. (kōz-kē-wē')	89	20°10'N	97°34'W
Cox's Bāzār, Bngl.	158	21°32'N	92°00'E
Coyaima, Col. (kô-yáē'mä)	100a	3°48'N	75°11'W
Coyame, Mex. (kō-yä'mā)	80	29°26'N	105°05'W
Coyanosa Draw, Tx., U.S. (kō yà-nō'sä)	80	30°55'N	103°07'W
Coyoacán, Mex. (kô-yō-á-kän')	88	19°21'N	99°10'W
Coyote, r., Ca., U.S. (kī'ōt)	74b	37°37'N	121°57'W
Coyuca de Benítez, Mex. (kō-yōō'kä dā bā-nē'tāz)	88	17°04'N	100°06'W
Coyuca de Catalán, Mex. (kō-yōō'kä dā kä-tä-län')	88	18°19'N	100°41'W
Coyutla, Mex. (kō-yōō'tlä)	89	20°13'N	97°40'W
Cozad, Ne., U.S. (kō'zăd)	78	40°53'N	99°59'W
Cozaddale, Oh., U.S. (kō-zăd-dāl)	69f	39°16'N	84°09'W
Cozoyoapan, Mex. (kô-zō-yô-ä-pá'n)	88	16°45'N	98°17'W
Cozumel, Mex. (kō-zōō-mē'l)	90a	20°31'N	86°55'W
Cozumel, Isla de, i., Mex. (ē's-lä-dĕ-kō-zōō-mē'l)	86	20°26'N	87°10'W
Crab Creek, r., Wa., U.S.	72	47°21'N	119°09'W
Crab Creek, r., Wa., U.S. (krăb)	72	46°47'N	119°43'W
Cradock, S. Afr. (kră'dŭk)	186	32°12'S	25°38'E
Crafton, Pa., U.S. (krăf'tŭn)	69e	40°26'N	80°04'W
Craig, Co., U.S. (krāg)	73	40°32'N	107°31'W
Craiova, Rom. (krá-yō'vá)	119	44°18'N	23°50'E
Cranberry, l., N.Y., U.S. (krăn'bĕr-ĭ)	67	44°10'N	74°50'W
Cranbourne, Austl.	173a	38°07'S	145°16'E
Cranbrook, Can. (krăn'brŏk)	50	49°31'N	115°46'W
Cranbury, N.J., U.S. (krăn'bē-rĭ)	68a	40°19'N	74°31'W
Crandon, Wi., U.S. (krăn'dŭn)	71	45°35'N	88°55'W
Crane Prairie Reservoir, res., Or., U.S.	72	43°50'N	121°55'W
Cranston, R.I., U.S. (krăns'tŭn)	68b	41°46'N	71°25'W
Crater Lake, l., Or., U.S. (krā'tẽr)	72	43°00'N	122°08'W
Crater Lake National Park, rec., Or., U.S.	72	42°58'N	122°40'W
Craters of the Moon National Monument, rec., Id., U.S. (krā'tẽr)	73	43°28'N	113°15'W
Crateús, Braz. (krä-tā-ōōzh')	101	5°09'S	40°35'W
Crato, Braz. (krä'tō)	101	7°19'S	39°13'W
Crawford, Ne., U.S. (krô'fẽrd)	70	42°41'N	103°25'W
Crawford, Wa., U.S.	74c	45°49'N	122°24'W
Crawfordsville, In., U.S. (krô'fẽrdz-vĭl)	66	40°00'N	86°55'W
Crazy Mountains, mts., Mt., U.S. (krā'zĭ)	73	46°11'N	110°25'W
Crazy Woman Creek, r., Wy., U.S.	73	44°08'N	106°40'W
Crecy, S. Afr. (krĕ-sē)	192c	24°38'S	28°52'E
Crécy-en-Brie, Fr. (krā-sē'-ĕɴ-brē')	127b	48°52'N	2°55'E
Crécy-en-Ponthieu, Fr.	126	50°13'N	1°48'E
Credit, r., Can.	62d	43°41'N	79°55'W
Cree, l., Can. (krē)	52	57°35'N	107°52'W
Creighton, S. Afr. (cre-tŏn)	187c	30°02'S	29°52'E
Creighton, Ne., U.S. (krā'tŭn)	70	42°27'N	97°54'W
Creil, Fr. (krĕ'y')	126	49°18'N	2°28'E
Crema, Italy (krā'mä)	130	45°21'N	9°53'E
Cremona, Italy (krā-mō'nä)	118	45°09'N	10°02'E
Crépy-en-Valois, Fr. (krā-pē'ĕɴ-vä-lwä')	127b	49°14'N	2°53'E
Cres, Cro. (tsrĕs)	130	44°58'N	14°21'E
Crescent Beach, Can.	74d	49°03'N	122°58'W
Crescent City, Ca., U.S. (krĕs'ĕnt)	72	41°46'N	124°13'W
Crescent City, Fl., U.S.	83	29°26'N	81°35'W
Crescent Lake, l., Fl., U.S. (krĕs'ĕnt)	83	29°33'N	81°30'W
Crescent Lake, l., Or., U.S.	72	43°25'N	121°58'W
Cresco, Ia., U.S. (krĕs'kō)	71	43°23'N	92°07'W
Crested Butte, Co., U.S. (krĕst'ĕd būt)	77	38°50'N	107°00'W
Crestline, Ca., U.S. (krĕst-līn)	75a	34°15'N	117°17'W
Crestline, Oh., U.S.	66	40°50'N	82°40'W
Crestmore, Ca., U.S. (krĕst'môr)	75a	34°02'N	117°23'W
Creston, Can. (krĕs'tŭn)	50	49°06'N	116°31'W
Creston, Ia., U.S.	71	41°04'N	94°22'W
Creston, Oh., U.S.	69d	40°59'N	81°54'W
Crestview, Fl., U.S. (krĕst'vū)	82	30°44'N	86°35'W
Crestwood, Ky., U.S. (krĕst'wŏd)	69h	38°20'N	85°28'W
Crestwood, Mo., U.S.	75e	38°33'N	90°23'W
Crete, Il., U.S. (krēt)	69a	41°26'N	87°38'W
Crete, Ne., U.S.	79	40°38'N	96°56'W
Crete, i., Grc.	112	35°15'N	24°30'E
Creus, Cap de, c., Spain	129	42°16'N	3°18'E
Creuse, r., Fr. (krûz)	126	46°51'N	0°49'E
Creve Coeur, Mo., U.S. (krēv kôr)	75e	38°40'N	90°27'W
Crevillent, Spain	129	38°12'N	0°48'W
Crewe, Eng., U.K. (krōō)	120	53°06'N	2°27'W
Crewe, Va., U.S.	83	37°09'N	78°08'W
Crimean Peninsula *see* Kryms'kyi Pivostriv, pen., Ukr.	137	45°18'N	33°30'E
Crimmitschau, Ger. (krĭm'ĭt-shou)	124	50°49'N	12°22'E
Cripple Creek, Co., U.S. (krĭp''l)	78	38°44'N	105°12'W
Crisfield, Md., U.S. (krĭs-fēld)	67	38°00'N	75°50'W
Cristal, Monts de, mts., Gabon	190	0°50'N	10°30'E
Cristina, Braz. (krĕs-tē'-nä)	99a	22°13'S	45°15'W
Cristóbal Colón, Pico, mtn., Col. (pē'kô-krēs-tô'bäl-kō-lôn')	100	11°00'N	74°00'W
Crişul Alb, r., Rom. (krē'shōōl älb)	125	46°20'N	22°15'E
Crna, r., Serb. (ts'r'nà)	131	41°03'N	21°46'E
Crna Gora (Montenegro), state, Serb.	131	42°55'N	18°52'E
Črnomelj, Slvn. (ch'r'nō-māl')	130	45°35'N	15°11'E
Croatia, nation, Eur.	130	45°24'N	15°18'E
Crockett, Ca., U.S. (krŏk'ĕt)	74b	38°03'N	122°14'W
Crockett, Tx., U.S.	81	31°19'N	95°28'W
Crofton, Md., U.S.	68e	39°01'N	76°43'W

ng-sing; ŋ-baŋk; ɴ-nasalized n; nŏd; cŏmmit; ōld; ôbey; ôrder; oi-boil; fōōd; ŏ-as oo in foot; ou-out; s-soft; sh-dish; th-thin; pūre; ůnite; ûrn; stŭd; circŭs; ü-as in French tu; '-indeterminate vowel.

240 Crofton – Dall Island PRONOUNCING INDEX

PLACE (Pronunciation)	PAGE	LAT.	LONG.
Crofton, Ne., U.S.	70	42°44′N	97°32′W
Croix, Lac la, l., N.A. (läk lä kro͞o-ä′)	71	48°19′N	91°53′W
Croker, i., Austl. (krō′kȧ)	174	10°45′S	132°25′E
Cronulla, Austl. (krō-nŭl′ȧ)	173b	34°03′S	151°09′E
Crooked, i., Bah.	93	22°45′N	74°10′W
Crooked, l., Can.	61	48°25′N	56°05′W
Crooked, r., Can.	55	54°30′N	122°55′W
Crooked, r., Or., U.S.	72	44°07′N	120°30′W
Crooked Creek, r., Il., U.S. (kro͞ok′ĕd)	79	40°21′N	90°49′W
Crooked Island Passage, strt., Bah.	93	22°40′N	74°50′W
Crookston, Mn., U.S. (kro͝oks′tŭn)	70	47°44′N	96°35′W
Crooksville, Oh., U.S. (kro͝oks′vĭl)	66	39°45′N	82°05′W
Crosby, Eng., U.K.	114a	53°30′N	3°02′W
Crosby, Mn., U.S. (krôz′bĭ)	71	46°29′N	93°58′W
Crosby, N.D., U.S.	70	48°55′N	103°18′W
Crosby, Tx., U.S.	81a	29°55′N	95°04′W
Cross, l., La., U.S.	81	32°33′N	93°58′W
Cross, r., Nig.	189	5°35′N	8°05′E
Cross City, Fl., U.S.	82	29°55′N	83°25′W
Crossett, Ar., U.S. (krôs′ĕt)	79	33°08′N	92°00′W
Cross Lake, l., Can.	52	54°45′N	97°30′W
Cross River Reservoir, res., N.Y., U.S. (krôs)	68a	41°14′N	73°34′W
Cross Sound, strt., Ak., U.S. (krŏs)	63	58°12′N	137°20′W
Crosswell, Mi., U.S. (krŏz′wĕl)	66	43°15′N	82°35′W
Croswell, i., Serb.	130	44°50′N	14°31′E
Crotch, l., Can.	59	44°55′N	76°55′W
Crotone, Italy (krō-tō′nĕ)	131	39°05′N	17°08′E
Croton Falls Reservoir, res., N.Y., U.S. (krōtŭn)	68a	41°22′N	73°44′W
Croton-on-Hudson, N.Y., U.S. (krō′tŭn-ŏn hŭd′sŭn)	68a	41°12′N	73°53′W
Crow, l., Can.	71	49°13′N	93°29′W
Crow Agency, Mt., U.S.	73	45°36′N	107°27′W
Crow Creek, r., Co., U.S.	78	41°08′N	104°25′W
Crow Creek Indian Reservation, I.R., S.D., U.S.	70	44°17′N	99°17′W
Crow Indian Reservation, I.R., Mt., U.S. (krō)	73	45°26′N	108°12′W
Crowle, Eng., U.K. (kroul)	114a	53°36′N	0°49′W
Crowley, La., U.S. (krou′lē)	81	30°13′N	92°22′W
Crown Mountain, mtn., Can. (kroun)	74d	49°24′N	123°05′W
Crown Mountain, mtn., V.I.U.S.	87c	18°22′N	64°58′W
Crown Point, In., U.S. (kroun point′)	69a	41°25′N	87°22′W
Crown Point, N.Y., U.S.	67	44°00′N	73°25′W
Crowsnest Pass, p., Can.	55	49°39′N	114°45′W
Crow Wing, r., Mn., U.S. (krō)	71	44°50′N	94°01′W
Crow Wing, r., Mn., U.S.	71	46°42′N	94°48′W
Crow Wing, North Fork, r., Mn., U.S.	71	45°16′N	94°28′W
Crow Wing, South Fork, r., Mn., U.S.	71	44°59′N	94°42′W
Croydon, Austl. (kroi′dŭn)	175	18°15′S	142°15′E
Croydon, Austl.	173a	37°48′S	145°17′E
Croydon, Eng., U.K.	117	51°22′N	0°06′W
Croydon, Pa., U.S.	68f	40°05′N	74°55′W
Crozet, Îles, is., Afr. (krō-zĕ′)	3	46°20′S	51°30′E
Cruces, Cuba (kro͞o′sås)	92	22°20′N	80°20′W
Cruces, Arroyo de, r., Mex. (är-rŏ′yô-dĕ-kro͞o′sĕs)	80	26°17′N	104°32′W
Cruillas, Mex. (kro͞o-ēl′yäs)	80	24°45′N	98°31′W
Cruz, Cabo, c., Cuba (kȧ′-bō-kro͞oz)	87	19°50′N	77°45′W
Cruz, Cayo, i., Cuba (kä′yō-kro͞oz)	92	22°15′N	77°50′W
Cruz Alta, Braz. (kro͞oz äl′tä)	102	28°41′S	54°02′W
Cruz del Eje, Arg. (kro͞o′s-dĕl-ĕ-kĕ)	102	30°46′S	64°45′W
Cruzeiro, Braz. (kro͞o-zā′rō)	99a	22°36′S	44°57′W
Cruzeiro do Sul, Braz. (kro͞o-zā′rō dō so͞ol)	100	7°34′S	72°40′W
Crysler, Can.	62c	45°13′N	75°09′W
Crystal City, Tx., U.S. (krĭs′tăl sĭ′tĭ)	80	28°40′N	99°50′W
Crystal Falls, Mi., U.S. (krĭs′tăl fôls)	71	46°06′N	88°21′W
Crystal Lake, Il., U.S. (krĭs′tăl lăk)	69a	42°15′N	88°18′W
Crystal Springs, Ms., U.S. (krĭs′tăl sprĭngz)	82	31°58′N	90°20′W
Crystal Springs, oasis, Ca., U.S.	74b	37°31′N	122°26′W
Csongrád, Hung. (chōn′gräd)	125	46°42′N	20°09′E
Csorna, Hung. (chōr′nä)	125	47°39′N	17°11′E
Cúa, Ven. (ko͞o′ä)	101b	10°10′N	66°54′W
Cuajimalpa, Mex. (kwä-hē-mäl′pä)	89a	19°21′N	99°18′W
Cuale, Sierra del, mts., Mex. (sē-ĕ′r-rä-dĕl-kwȧ′lĕ)	88	20°20′N	104°58′W
Cuamato, Ang. (kwä-mä′tō)	190	17°05′S	15°09′E
Cuamba, Moz.	191	14°49′S	36°33′E
Cuando, Ang. (kwän′dō)	190	16°32′S	22°07′E
Cuando, r., Afr.	186	14°30′S	20°00′E
Cuangar, Ang.	190	17°36′S	18°39′E
Cuango, r., Afr.	186	9°00′S	18°00′E
Cuanza, r., Ang. (kwän′zä)	186	9°45′S	15°00′E
Cuarto, r., Arg.	102	33°00′S	63°25′W
Cuatro Caminos, Cuba (kwä′trō-kä-mē′nōs)	93a	23°01′N	82°13′W
Cuatro Ciénegas, Mex. (kwä′trō syā′nå-gäs)	80	26°59′N	102°03′W
Cuauhtemoc, Mex. (kwä-o͞o-tĕ-mŏk′)	89	15°43′N	91°57′W
Cuautepec, Mex. (kwä-o͞o-tĕ-pĕk)	88	16°41′N	99°04′W
Cuautepec, Mex.	88	20°01′N	98°19′W
Cuautitlán, Mex. (kwä-o͞o-tēt-län′)	89a	19°40′N	99°12′W
Cuautla, Mex. (kwä-o͞o′tlȧ)	88	18°47′N	98°57′W
Cuba, Port. (ko͞o′bȧ)	128	38°10′N	7°55′W
Cuba, nation, N.A. (kū′bȧ)	87	22°00′N	79°00′W
Cubagua, Isla, i., Ven. (ē′s-lä-ko͞o-bä′gwä)	101b	10°48′N	64°10′W
Cubango (Okavango), r., Afr. (ko͞o-bäŋ′gō)	186	17°10′S	18°20′E
Cub Hills, hills, Can. (kŭb)	56	54°20′N	104°30′W
Cucamonga, Ca., U.S. (ko͞o-kȧ-mŏŋ′gȧ)	75a	34°05′N	117°35′W
Cuchi, Ang.	186	14°40′S	16°50′E
Cuchillo Parado, Mex. (ko͞o-chē′lyō pä-rä′dō)	80	29°26′N	104°52′W
Cuchumatanes, Sierra de los, mts., Guat.	90	15°35′N	91°10′W
Cúcuta, Col. (ko͞o′ko͞o-tä)	100	7°56′N	72°30′W
Cudahy, Wi., U.S. (kŭd′ȧ-hī)	69a	42°57′N	87°52′W
Cuddalore, India (kŭd ȧ-lōr′)	155	11°49′N	79°46′E
Cuddapah, India (kŭd′ȧ-pä)	155	14°31′N	78°52′E
Cue, Austl. (kū)	174	27°30′S	118°10′E
Cuéllar, Spain (kwā′lyär′)	128	41°24′N	4°15′W
Cuenca, Ec. (kwĕn′kä)	100	2°52′S	78°54′W
Cuenca, Spain	118	40°05′N	2°07′W
Cuenca, Sierra de, mts., Spain (sē-ĕ′r-rä-dĕ-kwĕ′n-kä)	128	40°02′N	1°50′W
Cuencame, Mex. (kwĕn-kä-mā′)	80	24°52′N	103°42′W
Cuerámaro, Mex. (kwā-rä′mä-rō)	88	20°39′N	101°44′W
Cuernavaca, Mex. (kwĕr-nä-vä′kä)	86	18°55′N	99°15′W
Cuero, Tx., U.S. (kwā′rō)	81	29°05′N	97°16′W
Cuetzalá del Progreso, Mex. (kwĕt-zä-lä dĕl prō-grā′sō)	88	18°07′N	99°51′W
Cuetzalan del Progreso, Mex.	89	20°02′N	97°33′W
Cuevas del Almanzora, Spain (kwĕ′väs-dĕl-äl-män-zō-rä)	118	37°19′N	1°54′W
Cuglieri, Italy (ko͞o-lyā′rē)	130	40°11′N	8°37′E
Cuicatlán, Mex. (kwē-kä-tlän′)	89	17°46′N	96°57′W
Cuilapa, Guat. (ko͝o-ē-lä′pä)	90	14°16′N	90°20′W
Cuilo (Kwilu), r., Afr.	190	9°15′S	19°30′E
Cuito, r., Ang. (ko͞o-ē-′tō)	186	14°45′S	19°00′E
Cuitzeo, Mex. (kwēt′zā-ō)	88	19°57′N	101°11′W
Cuitzeo, Laguna de, l., Mex. (lä-o͝o′nä-dĕ-kwēt′zā-ō)	88	19°58′N	101°05′W
Cul de Sac, pl., Haiti (ko͞o′l-dĕ-sä′k)	93	18°35′N	72°05′W
Culebra, i., P.R. (ko͞o-lā′brä)	87b	18°19′N	65°32′W
Culebra, Sierra de la, mts., Spain (sē-ĕ′r-rä-dĕ-lä-ko͞o-lĕ-brä)	128	41°52′N	6°21′W
Culemborg, Neth.	115a	51°57′N	5°14′E
Culfa, Azer.	138	38°58′N	45°38′E
Culgoa, r., Austl. (kŭl-gō′ȧ)	175	29°21′S	147°00′E
Culiacán, Mex. (ko͞o-lyä-kȧ′n)	86	24°45′N	107°30′W
Culion, Phil. (ko͞o-lē-ōn′)	168	11°43′N	119°58′E
Cúllar de Baza, Spain (ko͞o′l-yär-dĕ-bä′zä)	128	37°36′N	2°35′W
Cullera, Spain (ko͞o-lyā′rä)	118	39°12′N	0°15′W
Cullinan, S. Afr. (ko͝o′lĭ-nȧn)	192c	25°41′S	28°32′E
Cullman, Al., U.S. (kŭl′măn)	82	34°10′N	86°50′W
Culpeper, Va., U.S. (kŭl′pĕp-ẽr)	67	38°30′N	77°55′W
Culross, Can. (kŭl′rôs)	62f	49°43′N	97°54′W
Culver, In., U.S. (kŭl′vẽr)	66	41°15′N	86°25′W
Culver City, Ca., U.S.	75a	34°00′N	118°23′W
Cumaná, Ven.	100	10°28′N	64°10′W
Cumberland, Can. (kŭm′bẽr-lănd)	62c	45°31′N	75°25′W
Cumberland, Md., U.S.	65	39°40′N	78°40′W
Cumberland, Wa., U.S.	74a	47°17′N	121°55′W
Cumberland, Wi., U.S.	71	45°31′N	92°01′W
Cumberland, r., U.S.	82	36°45′N	85°33′W
Cumberland, Lake, res., Ky., U.S.	65	36°55′N	85°20′W
Cumberland Islands, is., Austl.	175	20°20′S	149°46′E
Cumberland Peninsula, pen., Can.	53	65°59′N	64°05′W
Cumberland Plateau, plat., U.S.	82	35°25′N	85°30′W
Cumberland Sound, strt., Can.	53	65°27′N	65°44′W
Cundinamarca, dept., Col. (ko͞on-dē-nä-mä′r-kȧ)	100a	4°57′N	74°27′W
Cunduacán, Mex. (ko͝on-do͞o-ȧ-kän′)	89	18°04′N	93°23′W
Cunene (Kunene), r., Afr.	186	17°05′S	12°35′E
Cuneo, Italy (ko͞o′nā-ō)	130	44°24′N	7°31′E
Cunha, Braz. (ko͞o′nyȧ)	99a	23°05′S	44°56′W
Cunnamulla, Austl. (kŭn-ȧ-mŭl-ȧ)	175	28°00′S	145°55′E
Cupula, Pico, mtn., Mex. (pē′kō-ko͞o′po͞o-lä)	86	24°45′N	111°10′W
Cuquío, Mex. (ko͞o-kē′ō)	88	20°55′N	103°03′W
Curaçao, i., Neth. Ant. (ko͞o-rä-sä′ō)	100	12°12′N	68°58′W
Curacautín, Chile (kä-rä-käo͞o-tē′n)	102	38°25′S	71°53′W
Curaumilla, Punta, c., Chile (ko͞o-rou-mē′lyä)	99b	33°05′S	71°44′W
Curepto, Chile (ko͞o-rĕp-tō)	99b	35°06′S	72°02′W
Curitiba, Braz. (ko͞o-rē-tē′bȧ)	101	25°20′S	49°15′W
Curly Cut Cays, is., Bah.	92	23°40′N	77°40′W
Currais Novos, Braz. (ko͞or-rä′ēs nŏ-vōs)	101	6°02′S	36°39′W
Curran, Can. (kü-räɴ′)	62c	45°30′N	74°59′W
Current, i., Bah. (kŭ-rĕnt)	92	25°20′N	76°50′W
Current, r., Mo., U.S. (kûr′ĕnt)	79	37°18′N	91°21′W
Currie, Mount, mtn., S. Afr. (kŭ-rē)	187c	30°28′S	29°23′E
Currituck Sound, strt., N.C., U.S. (kûr′ĭ-tŭk)	83	36°27′N	75°42′W
Curtis, Ne., U.S. (kûr′tĭs)	78	40°36′N	100°29′W
Curtis, i., Austl.	175	23°38′S	151°43′E
Curtisville, Pa., U.S. (kûr′tĭs-vĭl)	69e	40°38′N	79°50′W
Čurug, Serb. (cho͞o′ro͝og)	131	45°27′N	20°03′E
Curunga, Ang.	190	12°51′S	21°12′E
Curupira, Serra, mts., S.A. (sĕr′rȧ ko͞o-ro͞o-pē′rȧ)	100	1°00′N	65°30′W
Cururupu, Braz. (ko͞o-ro͝o-ro͝o-po͞o′)	101	1°40′S	44°56′W
Curvelo, Braz. (ko͝or-vĕl′o͝o)	101	18°47′S	44°14′W
Cusco, Peru	100	13°36′S	71°52′W
Cushing, Ok., U.S. (kŭsh′ĭng)	79	35°58′N	96°46′W
Custer, S.D., U.S. (kŭs′tẽr)	70	43°46′N	103°36′W
Custer, Wa., U.S.	74d	48°55′N	122°39′W
Cut Bank, Mt., U.S. (kŭt băngk)	73	48°38′N	112°19′W
Cuthbert, Ga., U.S. (kŭth′bẽrt)	82	31°47′N	84°48′W
Cuttack, India (kŭ-tăk′)	155	20°38′N	85°53′E
Cutzamalá, r., Mex. (ko͞o-tzä-mä-lä′)	88	18°57′N	100°41′W
Cutzamalá de Pinzón, Mex. (ko͞o-tzä-mä-lä′dĕ-pēn-zô′n)	88	18°28′N	100°36′W
Cuvo, r., Ang. (ko͞o′vō)	186	11°00′S	14°30′E
Cuxhaven, Ger. (ko͝oks′hä-fĕn)	116	53°51′N	8°43′E
Cuyahoga, r., Oh., U.S. (kī-ȧ-hō′gȧ)	69d	41°22′N	81°38′W
Cuyahoga Falls, Oh., U.S.	69d	41°08′N	81°29′W
Cuyapaire Indian Reservation, I.R., Ca., U.S. (kū-yȧ-pâr)	76	32°46′N	116°20′W
Cuyo Islands, is., Phil. (ko͞o′yō)	168	10°54′N	120°08′E
Cuyotenango, Guat. (ko͞o-yō-tĕ-näŋ′gō)	90	14°30′N	91°35′W
Cuyuni, r., S.A. (ko͞o-yo͞o′nē)	101	6°40′N	60°44′W
Cuyutlán, Mex. (ko͞o-yo͞o-tlän′)	88	18°54′N	104°04′W
Cyclades *see* Kikládes, is., Grc.	112	37°30′N	24°45′E
Cynthiana, Ky., U.S. (sĭn-thĭ-ăn′ȧ)	66	38°20′N	84°20′W
Cypress, Ca., U.S. (sī′prĕs)	75a	33°50′N	118°03′W
Cypress Hills, hills, Can.	56	49°40′N	110°20′W
Cypress Lake, l., Can.	56	49°28′N	109°43′W
Cyprus, nation, Asia (sī′prŭs)	154	35°00′N	31°00′E
Cyrenaica *see* Barqah, hist. reg., Libya	185	31°09′N	21°45′E
Czech Republic, nation, Eur.	110	50°00′N	15°00′E
Czersk, Pol. (chĕrsk)	125	53°47′N	17°58′E
Częstochowa, Pol. (chăɴ-stō kō′vȧ)	117	50°49′N	19°10′E

D

PLACE (Pronunciation)	PAGE	LAT.	LONG.
Da'an, China (dä-än)	164	45°25′N	124°22′E
Dabakala, C. Iv. (dä-bä-kä′lä)	184	8°16′N	4°36′W
Daba Shan, mts., China (dä-bä shän)	160	32°25′N	108°20′E
Dabeiba, Col. (dȧ-bā′bä)	100a	7°01′N	76°16′W
Dabie Shan, mts., China (dä-bĭĕ shän)	161	31°40′N	114°50′E
Dabnou, Niger	189	14°09′N	5°22′E
Dabob Bay, b., Wa., U.S. (dā′bŏb)	74a	47°50′N	122°50′W
Dabola, Gui.	188	10°45′N	11°07′W
Dąbrowa Białostocka, Pol.	125	53°37′N	23°18′E
Dacca *see* Dhaka, Bngl.	155	23°45′N	90°29′E
Dachang, China (dä-chäŋ)	163b	31°18′N	121°25′E
Dachangshan Dao, i., China (dä-chäŋ-shän dou)	162	39°21′N	122°31′E
Dachau, Ger. (dä′κou)	124	48°16′N	11°26′E
Dacotah, Can. (dȧ-kō′tȧ)	62f	49°52′N	97°38′W
Dade City, Fl., U.S. (dād)	83a	28°22′N	82°09′W
Dadeville, Al., U.S. (dād′vĭl)	82	32°48′N	85°44′W
Dādra & Nagar Haveli, India	155	20°00′N	73°00′E
Dadu, r., China (dä-do͞o)	165	29°20′N	103°03′E
Daet, mtn., Phil. (dä′āt)	169a	14°07′N	122°59′E
Dafoe, r., Can.	57	55°50′N	95°50′W
Dafter, Mi., U.S. (dăf′tẽr)	75k	46°21′N	84°26′W
Dagana, Sen. (dä-gä′nä)	184	16°31′N	15°30′W
Dagana, reg., Chad	189	12°20′N	15°15′E
Dagang, China (dä-gäŋ)	163a	22°48′N	113°24′E
Dagda, Lat. (dȧg′dȧ)	123	56°04′N	27°30′E
Dagenham, Eng., U.K. (dȧg′ĕn-ăm)	114b	51°32′N	0°09′E
Dagestan, prov., Russia (dä-gĕs-tän′)	137	43°40′N	46°10′E
Daggett, Ca., U.S. (dăg′ĕt)	76	34°50′N	116°52′W
Dagu, China (dä-go͞o)	164	39°00′N	117°42′E
Dagu, r., China	162	36°29′N	120°06′W
Dagupan, Phil. (dä-go͞o′pän)	169a	16°02′N	120°20′E
Daheishan Dao, i., China (dä-hā-shän dou)	162	37°57′N	120°37′E
Dahl, Ger. (däl)	127c	51°18′N	7°33′E
Dahlak Archipelago, is., Erit.	185	15°45′N	40°30′E
Dahomey *see* Benin, nation, Afr.	184	8°00′N	2°00′E
Dahra, Libya	156	29°34′N	17°50′E
Daibu, China (dī-bo͞o)	162	31°22′N	119°29′E
Daigo, Japan (dī-gō)	167b	34°57′N	135°49′E
Daimiel Manzanares, Spain (dī-myĕl′män-zä-nä′rĕs)	128	39°05′N	3°36′W
Dairen *see* Dalian, China	161	38°54′N	121°35′E
Dairy, r., Or., U.S. (dâr′ĭ)	74c	45°33′N	123°04′W
Dai-Sen, mtn., Japan (dī′sĕn′)	167	35°22′N	133°35′E
Dai-Tenjo-dake, mtn., Japan (dī-tĕn′jō dä-κā)	167	36°21′N	137°38′E
Daiyun Shan, mtn., China (dī-yo͝on shän)	165	25°40′N	118°08′E
Dajabón, Dom. Rep. (dä-κä-bô′n)	93	19°35′N	71°40′W
Dajarra, Austl. (dȧ-jăr′ȧ)	174	21°45′S	139°30′E
Dakar, Sen. (dȧ-kär′)	184	14°40′N	17°26′W
Dakhla, W. Sah.	184	23°45′N	16°04′W
Dakouraoua, Niger	189	13°58′N	6°15′E
Dakovica, Serb.	131	42°33′N	20°28′E
Dalälven, r., Swe.	112	60°26′N	15°50′E
Dalby, Austl. (dôl′bē)	175	27°10′S	151°15′E
Dalcour, La., U.S. (dăl-kour)	68d	29°49′N	89°59′W
Dale, Nor. (dä′lĕ)	122	60°35′N	5°55′E
Dale Hollow Lake, res., Tn., U.S. (dāl hŏl′ō)	65	36°33′N	85°03′W
Dalemead, Can. (dä′lĕ-mēd)	62e	50°53′N	113°38′W
Dalen, Nor. (dä′lĕn)	122	59°28′N	8°01′E
Daleside, S. Afr. (dāl′sīd)	192c	26°30′S	28°03′E
Dalesville, Can. (dālz′vĭl)	62a	45°42′N	74°23′W
Daley Waters, Austl. (dā lē)	174	16°15′N	133°30′E
Dalhart, Tx., U.S. (dăl härt)	78	36°04′N	102°32′W
Dalhousie, Can. (dăl-ho͞o′zĕ)	60	48°04′N	66°23′W
Dali, China (dä-lĕ)	163a	23°07′N	113°06′E
Dali, China	160	26°00′N	100°08′E
Dali, China	160	35°00′N	109°38′E
Dalian, China (lů-dä)	161	38°54′N	121°35′E
Dalian Wan, b., China (dä-lĭĕn wän)	162	38°55′N	121°50′E
Dalías, Spain (dä-lĕ′ȧs)	128	36°49′N	2°50′W
Dall, i., Ak., U.S. (dăl)	63	54°50′N	133°10′W
Dallas, Or., U.S. (dăl′lȧs)	72	44°55′N	123°20′W
Dallas, S.D., U.S.	70	43°13′N	99°34′W
Dallas, Tx., U.S.	64	32°45′N	96°48′W
Dalles Dam, Or., U.S.	72	45°36′N	121°08′W
Dall Island, i., Ak., U.S.	54	54°50′N	132°55′W

PLACE (Pronunciation)	PAGE	LAT.	LONG.
Dalmacija, hist. reg., Serb. (däl-mä´tsĕ-yä)	130	43°25´N	16°37´E
Dalnerechensk, Russia	135	46°07´N	133°21´E
Daloa, C. Iv.	188	6°53´N	6°27´W
Dalroy, Can. (dăl´roi)	62e	51°07´N	113°39´W
Dalrymple, Mount, mtn., Austl. (dăl´rĭm-p'l)	175	21°14´S	148°46´E
Dalton, S. Afr. (dôl´tŏn)	187c	29°21´S	30°41´E
Dalton, Ga., U.S. (dôl´tŭn)	82	34°46´N	84°58´W
Daly, r., Austl. (dā´lĭ)	174	14°15´S	131°15´E
Daly City, Ca., U.S. (dā´lē)	74b	37°42´N	122°27´W
Damān, India	155	20°32´N	72°53´E
Damanhûr, Egypt (dä-män-hōōr´)	185	30°59´N	30°31´E
Damar, Pulau, i., Indon.	169	7°15´S	129°15´E
Damara, C.A.R.	189	4°58´N	18°42´E
Damaraland, hist. reg., Nmb. (dä´na-ra-länd)	186	22°15´S	16°15´E
Damas Cays, is., Bah. (dä´mäs)	92	23°50´N	79°50´W
Damascus, Syria	154	33°30´N	36°18´E
Damāvand, Qolleh-ye, mtn., Iran	154	36°05´N	52°05´E
Damba, Ang. (däm´bä)	186	6°41´S	15°08´E
Dâmbovița, r., Rom.	131	44°43´N	25°41´E
Dame Marie, Cap, c., Haiti (däm märē´)	93	18°35´N	74°50´W
Dāmghān, Iran (däm-gän´)	154	35°50´N	54°15´E
Daming, China (dä-mǐŋ)	164	36°15´N	115°09´E
Dammartin-en-Goële, Fr. (dän-mär-tăn-än-gŏ-ĕl´)	127b	49°03´N	2°40´E
Dampier, Selat, strt., Indon. (dăm´pēr)	169	0°40´S	131°15´E
Dampier Archipelago, is., Austl.	174	20°15´S	116°25´E
Dampier Land, reg., Austl.	174	17°30´S	122°25´E
Dan, r., N.C., U.S. (dăn)	83	36°26´N	79°40´W
Dana, Mount, mtn., Ca., U.S.	76	37°54´N	119°13´W
Da Nang, Viet.	168	16°08´N	108°22´E
Danbury, Eng., U.K.	114b	51°42´N	0°34´E
Danbury, Ct., U.S. (dăn´bĕr-ĭ)	68a	41°23´N	73°27´W
Danbury, Tx., U.S.	81a	29°14´N	95°22´W
Dandenong, Austl. (dăn´dĕ-nŏng)	176	37°59´S	145°13´E
Dandong, China (dän-dŏŋ)	161	40°10´N	124°30´E
Dane, r., Eng., U.K. (dān)	114a	53°11´N	2°14´W
Danea, Gui.	188	11°27´N	13°12´W
Danforth, Me., U.S.	60	45°38´N	67°53´W
Dan Gora, Nig.	189	11°30´N	8°09´E
Dangtu, China (dän-tōō)	165	31°35´N	118°28´E
Dani, Burkina	184	13°43´N	0°10´W
Dania, Fl., U.S. (dā´nĭ-à)	83a	26°01´N	80°10´W
Danilov, Russia	136	58°12´N	40°08´E
Danissa Hills, hills, Kenya	191	3°20´N	40°55´E
Dänizkänarı, Azer.	138	40°13´N	49°33´E
Dankov, Russia (dän´kôf)	136	53°17´N	39°09´E
Dannemora, N.Y., U.S.	67	44°45´N	73°45´W
Dannhauser, S. Afr. (dän´hou-zĕr)	187c	28°07´S	30°04´E
Dansville, N.Y., U.S. (dănz´vĭl)	67	42°30´N	77°40´W
Danube, r., Eur.	112	43°00´N	24°00´E
Danube, Mouths of the, mth., Rom. (dän´ub)	133	45°13´N	29°37´E
Danvers, Ma., U.S. (dăn´vērz)	61a	42°34´N	70°57´W
Danville, Ca., U.S. (dăn´vĭl)	74b	37°49´N	122°00´W
Danville, Il., U.S.	66	40°10´N	87°35´W
Danville, In., U.S.	66	39°45´N	86°30´W
Danville, Ky., U.S.	66	37°35´N	84°50´W
Danville, Pa., U.S.	67	41°00´N	76°35´W
Danville, Va., U.S.	65	36°35´N	79°24´W
Danxian, China (dän shyĕn)	165	19°30´N	109°38´E
Danyang, China (dän-yän)	162	32°01´N	119°32´E
Danzig see Gdańsk, Pol.	110	54°20´N	18°40´E
Danzig, Gulf of, b., Eur. (dăn´tsĭk)	116	54°41´N	19°01´E
Daoxian, China (dou shyĕn)	165	25°35´N	111°27´E
Dapango, Togo	188	10°52´N	0°12´E
Daphnae, hist., Egypt	153a	30°43´N	32°12´E
Daqin Dao, i., China (dä-chyĭn dou)	162	38°18´N	120°50´E
Darabani, Rom. (dä-rä-bän´ĭ)	125	48°13´N	26°38´E
Daraj, Libya	184	30°10´N	10°14´E
Darāw, Egypt (dá-rä´ōō)	192b	24°24´N	32°56´E
Darbhanga, India (dŭr-bŭn´gä)	155	26°03´N	85°09´E
Darby, Pa., U.S. (där´bĭ)	68f	39°55´N	75°16´W
Darby, i., Bah.	92	23°50´N	76°20´W
Dardanelles see Çanakkale Boğazi, strt., Tur.	119	40°05´N	25°50´E
Dar es Salaam, Tan. (där ĕs sä-läm´)	187	6°48´S	39°17´E
Dārfūr, hist. reg., Sudan (där-fōōr´)	185	13°21´N	23°46´E
Dargai, Pak. (dŭr-gä´ē)	158	34°35´N	72°00´E
Darien, Col. (dä-rĭ-ĕn´)	100a	3°55´N	76°30´W
Darien, Ct., U.S. (dâ-rē-ĕn´)	68a	41°04´N	73°28´W
Darién, Cordillera de, mts., Nic.	90	13°00´N	85°42´W
Darien, Serranía del, mts.,	91	8°13´N	77°28´W
Darjeeling, India (dŭr-jē´lĭng)	155	27°05´N	88°16´E
Darling, r., Austl.	175	31°35´S	143°20´E
Darling Downs, reg., Austl.	175	27°22´S	150°00´E
Darling Range, mts., Austl.	174	30°30´S	115°45´E
Darlington, Eng., U.K. (där´lĭng-tŭn)	120	54°32´N	1°35´W
Darlington, S.C., U.S.	83	34°15´N	79°52´W
Darlington, Wi., U.S.	71	42°41´N	90°06´W
Darłowo, Pol. (där-lô´vô)	124	54°26´N	16°23´E
Darmstadt, Ger. (därm´shtät)	117	49°53´N	8°40´E
Darnah, Libya	185	32°44´N	22°41´E
Darnley Bay, b., Ak., U.S. (därn´lē)	63	70°00´N	124°00´W
Daroca, Spain (dä-rō-kä)	128	41°08´N	1°24´W
Dartford, Eng., U.K.	114b	51°27´N	0°14´E
Dartmoor, for., Eng., U.K. (därt´mōōr)	120	50°35´N	4°05´W
Dartmouth, Can.	51	44°40´N	63°34´W
Dartmouth, Eng., U.K.	120	50°33´N	3°28´W
Daru, Pap. N. Gui. (dä´rōō)	169	9°04´S	143°21´E
Daruvar, Cro. (där´rōō-vär)	131	45°37´N	17°16´E
Darwen, Eng., U.K. (där´wĕn)	114a	53°42´N	2°28´W
Darwin, Austl. (där´wĭn)	174	12°25´S	131°00´E
Darwin, Cordillera, mts., Chile (kôr-dĕl-yĕ´rä-där´wĕn)	102	54°40´S	69°30´W
Dashhowuz, Turkmen.	139	41°50´N	59°45´E
Dash Point, Wa., U.S. (dăsh)	74a	47°19´N	122°25´W
Dasht, r., Pak. (dŭsht)	154	25°30´N	62°30´E
Dasol Bay, b., Phil. (dä-sōl´)	169a	15°53´N	119°40´E
Datian Ding, mtn., China (dä-tǐĕn dǐŋ)	165	22°25´N	111°20´E
Datong, China (dä-tŏŋ)	164	40°00´N	113°30´E
Dattapukur, India	158a	22°45´N	88°32´E
Datu, Tandjung, c., Asia	168	2°08´N	110°15´E
Datuan, China (dä-tŭän)	163b	30°57´N	121°43´E
Daugava (Zapadnaya Dvina), r., Eur.	123	56°40´N	24°40´E
Daugavpils, Lat. (dou´gäv-pēls)	136	55°52´N	26°32´E
Dauphin, Can. (dô´fĭn)	50	51°09´N	100°00´W
Dauphin Lake, l., Can.	57	51°17´N	99°48´W
Dāvangere, India	159	14°30´N	75°55´E
Davao, Phil. (dä´vä-ô)	169	7°05´N	125°30´E
Davao Gulf, b., Phil.	169	6°30´N	125°45´E
Davenport, Ia., U.S. (dăv´ĕn-pōrt)	65	41°34´N	90°38´W
Davenport, Wa., U.S.	72	47°39´N	118°07´W
David, Pan. (dä-vēdh´)	87	8°27´N	82°27´W
David City, Ne., U.S. (dā´vĭd)	70	41°15´N	97°10´W
David-Gorodok, Bela. (dä-vēt´ gồ-rồ´dồk)	137	52°02´N	27°14´E
Davis, Ok., U.S. (dā´vĭs)	79	34°34´N	97°08´W
Davis, W.V., U.S.	67	39°15´N	79°25´W
Davis Lake, l., Or., U.S.	72	43°38´N	121°43´W
Davis Mountains, mts., Tx., U.S.	80	30°45´N	104°17´W
Davis Strait, strt., N.A.	49	66°00´N	60°00´W
Davlekanovo, Russia	136	54°15´N	55°05´E
Davos, Switz. (dä´vōs)	124	46°47´N	9°50´E
Dawa, r., Afr.	185	4°30´N	40°30´E
Dawāsir, Wādī ad, val., Sau. Ar.	154	20°48´N	44°07´E
Dawei, Mya.	168	14°04´N	98°19´E
Dawen, r., China (dä-wŭn)	162	35°58´N	116°53´E
Dawley, Eng., U.K. (dô´lĭ)	114a	52°38´N	2°28´W
Dawna Range, mts., Mya. (dô´nä)	168	17°02´N	98°01´E
Dawson, Can. (dô´sŭn)	50	64°04´N	139°22´W
Dawson, Ga., U.S.	82	31°45´N	84°29´W
Dawson, Mn., U.S.	70	44°54´N	96°03´W
Dawson, r., Austl.	175	24°20´S	149°45´E
Dawson Bay, b., Can.	57	52°55´N	100°50´W
Dawson Creek, Can.	50	55°46´N	120°14´W
Dawson Range, mts., Can.	63	62°15´N	138°10´W
Dawson Springs, Ky., U.S.	82	37°10´N	87°40´W
Dawu, China (dä-wōō)	162	31°33´N	114°07´E
Dax, Fr. (däks)	117	43°42´N	1°06´W
Daxian, China (dä-shyĕn)	160	31°12´N	107°30´E
Daxing, China (dä-shyĭŋ)	164a	39°44´N	116°19´E
Dayiqiao, China (dä-yē-chyou)	162	31°43´N	120°40´E
Dayr az Zawr, Syria (dà-ĕrĕz-zôr´)	154	35°15´N	40°01´E
Dayton, Ky., U.S. (dā´tŭn)	69f	39°07´N	84°28´W
Dayton, N.M., U.S.	78	32°44´N	104°23´W
Dayton, Oh., U.S.	65	39°54´N	84°15´W
Dayton, Tn., U.S.	82	35°30´N	85°00´W
Dayton, Tx., U.S.	81	30°03´N	94°53´W
Dayton, Wa., U.S.	72	46°18´N	117°59´W
Daytona Beach, Fl., U.S. (dā-tō´nà)	65	29°11´N	81°02´W
Dayu, China (dä-yōō)	165	25°20´N	114°20´E
Da Yunhe (Grand Canal), can., China (dä yôn-hŭ)	161	35°00´N	117°00´E
Dayville, Ct., U.S. (dā´vĭl)	67	41°50´N	71°55´W
De Aar, S. Afr. (dē-är´)	186	30°45´S	24°05´E
Dead, l., Mn., U.S. (dĕd)	70	46°28´N	96°00´W
Dead Sea, l., Asia	154	31°30´N	35°30´E
Deadwood, S.D., U.S. (dĕd´wồd)	64	44°23´N	103°43´W
Deal Island, Md., U.S. (dēl-ī´lănd)	67	38°10´N	75°55´W
Dean, r., Can. (dēn)	54	52°45´N	125°30´W
Dean Channel, strt., Can.	54	52°33´N	127°13´W
Deán Funes, Arg. (dē-ä´n-fōō-nĕs)	102	30°26´S	64°12´W
Dearborn, Mi., U.S. (dēr´bŭrn)	69b	42°18´N	83°15´W
Dearg, Ben, mtn., Scot., U.K. (bĕn dŭrg)	120	57°48´N	4°59´W
Dease Strait, strt., Can. (dēz)	52	68°50´N	108°20´W
Death Valley, Ca., U.S.	76	36°18´N	116°26´W
Death Valley, val., Ca., U.S.	64	36°30´N	117°00´W
Death Valley National Park, rec., U.S.	76	36°34´N	117°00´W
Debal'tseve, Ukr.	133	48°23´N	38°29´E
Debao, China (dŭ-bou)	160	23°18´N	106°40´E
Debar, Mac. (dĕ´bär) (dä´brä)	131	41°31´N	20°32´E
Dęblin, Pol. (dăn´blĭn)	125	51°34´N	21°49´E
Dębno, Pol. (dĕb-nō´)	124	52°47´N	13°43´E
Debo, Lac, l., Mali	188	15°15´N	4°40´W
Debrecen, Hung. (dĕ´brĕ-tsĕn)	110	47°32´N	21°40´E
Debre Markos, Eth.	185	10°15´N	37°45´E
Debre Tabor, Eth.	185	11°57´N	38°09´E
Decatur, Al., U.S. (dĕ-kā´tŭr)	82	34°35´N	87°00´W
Decatur, Ga., U.S.	68c	33°47´N	84°18´W
Decatur, Il., U.S.	65	39°50´N	88°59´W
Decatur, In., U.S.	66	40°50´N	84°55´W
Decatur, Mi., U.S.	66	42°10´N	86°00´W
Decatur, Tx., U.S.	79	33°14´N	97°33´W
Decazeville, Fr. (dĕ-käz´vēl´)	117	44°33´N	2°16´E
Deccan, plat., India (dĕk´ăn)	155	19°05´N	76°40´E
Deception Lake, l., Can.	56	56°33´N	104°15´W
Deception Pass, p., Wa., U.S. (dĕ-sĕp´shŭn)	74a	48°24´N	122°44´W
Děčín, Czech Rep. (dyĕ´chēn)	124	50°47´N	14°14´E
Decorah, Ia., U.S. (dĕ-kō´rá)	71	43°18´N	91°48´W
Dedenevo, Russia (dyĕ-dyĕ´nyĕ-vô)	142b	56°14´N	37°31´E
Dedham, Ma., U.S. (dĕd´ăm)	61a	42°15´N	71°11´W
Dedo do Deus, mtn., Braz. (dĕ-dồ-dồ-dĕ´ồs)	102b	22°30´S	43°02´W
Dédougou, Burkina (dā-dồ-gōō´)	184	12°38´N	3°28´W
Dee, r., Scot., U.K.	120	57°05´N	2°25´W
Dee, r., U.K.	114a	53°15´N	3°05´E
Deep, r., N.C., U.S. (dēp)	83	35°36´N	79°32´W
Deep Fork, r., Ok., U.S.	79	35°35´N	96°42´W
Deep River, Can.	59	46°06´N	77°20´W
Deepwater, Mo., U.S. (dep-wô-tēr)	79	38°15´N	93°46´W
Deer, i., Me., U.S.	60	44°07´N	68°38´W
Deerfield, Il., U.S. (dēr´fēld)	69a	42°10´N	87°51´W
Deer Island, Or., U.S.	74c	45°56´N	122°51´W
Deer Lake, Can.	53a	49°10´N	57°25´W
Deer Lake, l., Can.	57	52°40´N	94°30´W
Deer Lodge, Mt., U.S. (dēr lŏj)	73	46°23´N	112°42´W
Deer Park, Oh., U.S.	69f	39°12´N	84°24´W
Deer Park, Wa., U.S.	72	47°58´N	117°28´W
Deer River, Mn., U.S.	71	47°20´N	93°49´W
Defiance, Oh., U.S. (dē-fī´áns)	66	41°15´N	84°20´W
DeFuniak Springs, Fl., U.S. (dē fū´nĭ-ăk)	82	30°42´N	86°06´W
Deganga, India	158a	22°41´N	88°41´E
Degeh Bur, Eth.	192a	8°10´N	43°25´E
Deggendorf, Ger. (dĕ´ghĕn-dôrf)	124	48°50´N	12°59´E
Degollado, Mex. (dā-gồ-lyä´dồ)	88	20°27´N	102°11´W
DeGrey, r., Austl. (dē grā´)	174	20°20´S	119°25´E
Degtyarsk, Russia (dĕg-ty´ärsk)	142a	56°42´N	60°05´E
Dehiwala-Mount Lavinia, Sri L.	159	6°47´N	79°55´E
Dehra Dūn, India (dā´rŭ)	155	30°09´N	78°07´E
Dehua, China (dŭ-hwä)	165	25°30´N	118°15´E
Dej, Rom. (dăzh)	119	47°09´N	23°53´E
De Kalb, Il., U.S. (dĕ kälb´)	66	41°54´N	88°46´W
Dekese, D.R.C.	190	3°27´S	21°24´E
Delacour, Can. (dĕ-lä-kōōr´)	62e	51°09´N	113°45´W
Delagua, Co., U.S. (dĕl-ä´gwä)	78	37°19´N	104°42´W
De Land, Fl., U.S. (dē länd´)	83	29°00´N	81°19´W
Delano, Ca., U.S. (dĕl´á-nō)	76	35°47´N	119°15´W
Delano Peak, mtn., Ut., U.S.	64	38°25´N	112°25´W
Delavan, Wi., U.S. (dĕl´á-văn)	71	42°39´N	88°38´W
Delaware, Oh., U.S. (dĕl´á-wâr)	66	40°15´N	83°05´W
Delaware, state, U.S.	65	38°40´N	75°30´W
Delaware, r., Ks., U.S.	79	39°45´N	95°47´W
Delaware, r., U.S.	67	41°50´N	75°20´W
Delaware Bay, b., U.S.	65	39°05´N	75°10´W
Delaware Reservoir, res., Oh., U.S.	67	40°30´N	83°05´E
Delémont, Switz. (dĕ-lä-môn´)	124	47°21´N	7°18´E
De Leon, Tx., U.S. (dē lē-ôn´)	80	32°06´N	98°33´W
Delft, Neth. (dĕlft)	121	52°01´N	4°20´E
Delfzijl, Neth.	121	53°20´N	6°50´E
Delgada, Punta, c., Arg. (pồô´n-tä-dĕl-gä´dä)	102	43°46´S	63°46´W
Delgado, Cabo, c., Moz. (kä´bồ-dĕl-gä´dồ)	187	10°40´S	40°35´E
Delhi, India	155	28°54´N	77°13´E
Delhi, Il., U.S. (dĕl´hī)	75e	39°03´N	90°16´W
Delhi, La., U.S.	81	32°26´N	91°29´W
Delhi, state, India	155	28°30´N	76°50´E
Delitzsch, Ger. (dā´lĭch)	124	51°32´N	12°18´E
Dellansjöarna, l., Swe.	122	61°57´N	16°25´E
Delles, Alg. (dĕ´lĕs´)	184	36°59´N	3°40´E
Dell Rapids, S.D., U.S. (dĕl)	70	43°50´N	96°43´W
Dellwood, Mn., U.S. (dĕl´wồd)	75g	45°05´N	92°58´W
Del Mar, Ca., U.S. (dĕl mär´)	76a	32°57´N	117°16´W
Delmas, S. Afr. (dĕl´más)	192c	26°08´S	28°42´E
Delmenhorst, Ger. (dĕl´mĕn-hôrst)	124	53°03´N	8°38´E
Del Norte, Co., U.S. (dĕl nôrt´)	77	37°40´N	106°25´W
De-Longa, i., Russia	135	76°21´N	148°56´E
De Long Mountains, mts., Ak., U.S. (dē´lŏng)	63	68°38´N	162°30´W
Deloraine, Austl. (dĕ-lú-rān´)	176	41°30´S	146°40´E
Delphi, In., U.S. (dĕl´fī)	66	40°35´N	86°40´W
Delphos, Oh., U.S. (dĕl´fŏs)	66	40°50´N	84°20´W
Delray Beach, Fl., U.S. (dĕl-rā´)	83a	26°27´N	80°05´W
Del Rio, Tx., U.S. (dĕl rē´ồ)	64	29°21´N	100°52´W
Delson, Can. (dĕl´sŭn)	62a	45°24´N	73°32´W
Delta, Co., U.S.	77	38°45´N	108°05´W
Delta, Ut., U.S.	77	39°20´N	112°35´W
Delta Beach, Can.	62f	50°10´N	98°20´W
Delvine, Alb. (dĕl´vē-nĕ)	131	39°58´N	20°10´E
Dëma, r., Russia (dyĕm´ä)	136	53°40´N	54°30´E
Demba, D.R.C.	190	5°30´S	22°16´E
Dembi Dolo, Eth.	185	8°46´N	34°46´E
Demidov, Russia (dzyĕ´mĕ-dồ´f)	132	55°16´N	31°32´E
Deming, N.M., U.S. (dĕm´ĭng)	64	32°15´N	107°45´W
Demmin, Ger. (dĕm´mĕn)	124	53°54´N	13°04´E
Demnat, Mor. (dĕm-nät)	184	31°58´N	7°03´W
Demopolis, Al., U.S. (dĕ-mŏp´ồ-lĭs)	82	32°30´N	87°50´W
Demotte, In., U.S. (dĕ´mŏt)	69a	41°12´N	87°13´W
Dempo, Gunung, mtn., Indon. (dĕm´pồ)	168	4°04´S	103°11´E
Dem'yanka, r., Russia (dyĕm-yän´kä)	140	59°07´N	72°58´E
Demyansk, Russia (dyĕm-yänsk´)	132	57°39´N	32°26´E
Denain, Fr. (dĕ-năn´)	126	50°23´N	3°21´E
Denakil Plain, pl., Eth.	185	12°45´N	41°01´E
Denali National Park, rec., Ak., U.S.	64a	63°48´N	153°02´W
Denbigh, Wales, U.K. (dĕn´bĭ)	120	53°15´N	3°25´W
Dendermonde, Bel.	115a	51°02´N	4°04´E
Dendron, Va., U.S. (dĕn´drŭn)	83	37°02´N	76°53´W
Denezhkin Kamen, Gora, mtn., Russia (dzyĕ-nĕ´zhkĕn kämĕn)	142a	60°26´N	59°35´E
Denham, Mount, mtn., Jam.	87	18°20´N	77°30´W
Den Helder, Neth. (dĕn hĕl´dĕr)	121	52°55´N	5°45´E
Dénia, Spain	129	38°48´N	0°06´E
Deniliquin, Austl. (dĕ-nĭl´ĭ-kwĭn)	175	35°20´S	144°52´E
Denison, Ia., U.S. (dĕn´ĭ-sŭn)	70	42°01´N	95°22´W
Denison, Tx., U.S.	64	33°45´N	97°02´W
Denizli, Tur. (dĕn-ĭz-lē´)	119	37°50´N	29°10´E
Denklingen, Ger. (dĕn´klĕn-gĕn)	127c	50°54´N	7°40´E
Denmark, S.C., U.S. (dĕn´märk)	83	33°18´N	81°09´W
Denmark, nation, Eur.	110	56°14´N	8°30´E
Denmark Strait, strt., Eur.	49	66°30´N	27°00´W

ng-sing; ŋ-baŋk; N-nasalized n; nŏd; cŏmmit; ōld; ồbey; ôrder; oi-boil; fōōd; ồ-as oo in foot; ou-out; s-soft; sh-dish; th-thin; pūre; ûnite; ûrn; stŭd; cirčus; ü-as in French tu; ´-indeterminate vowel.

PLACE (Pronunciation)	PAGE	LAT.	LONG.
Dennilton, S. Afr. (dĕn-ĭl-tŭn)	192c	25°18′s	29°13′E
Dennison, Oh., U.S. (dĕn′ĭ-sŭn)	66	40°25′N	81°20′W
Denpasar, Indon.	168	8°35′s	115°10′E
Denton, Eng., U.K. (dĕn′tŭn)	114a	53°27′N	2°07′W
Denton, Md., U.S.	67	38°55′N	75°50′W
Denton, Tx., U.S.	79	33°12′N	97°06′W
D'Entrecasteaux, Point, c., Austl. (dän-tr′-kás-tō′)	174	34°50′s	114°45′E
D'Entrecasteaux Islands, is., Pap. N. Gui. (dän-tr′-kás-tō′)	169	9°45′s	152°00′E
Denver, Co., U.S. (dĕn′vĕr)	64	39°44′N	104°59′W
Deoli, India	158	25°52′N	75°23′E
De Pere, Wi., U.S. (dĕ pĕr′)	71	44°25′N	88°04′W
Depew, N.Y., U.S. (dĕ-pū′)	69c	42°55′N	78°43′W
Deping, China (dŭ-pĭŋ)	162	37°28′N	116°57′E
Depue, Il., U.S. (dĕ pū)	66	41°15′N	89°55′W
De Queen, Ar., U.S. (dĕ kwēn′)	79	34°02′N	94°21′W
De Quincy, La., U.S. (dĕ kwĭn′sĭ)	81	30°27′N	93°27′W
Dera, Lach, r., Afr. (läk dä′rä)	192a	0°45′N	41°26′E
Dera, Lak, r., Afr.	185	0°45′N	41°30′E
Dera Ghāzi Khān, Pak. (dä′rū gä-zē′ кän)	155	30°09′N	70°39′E
Dera Ismāīl Khān, Pak. (dä′rū īs-mä-ēl′ кän)	158	31°55′N	70°51′E
Derbent, Russia (dĕr-bĕnt′)	137	42°00′N	48°10′E
Derby, Austl. (där′bē) (dûr′bē)	174	17°20′s	123°40′E
Derby, S. Afr. (där′bī)	192c	25°55′s	27°02′E
Derby, Eng., U.K. (där′bē)	117	52°55′N	1°29′W
Derby, Ct., U.S. (dûr′bē)	67	41°20′N	73°05′W
Derbyshire, co., Eng., U.K.	114a	53°11′N	1°30′W
Derdepoort, S. Afr.	192c	24°39′s	26°21′E
Derg, Lough, l., Ire. (lŏk dĕrg)	120	53°00′N	8°09′W
De Ridder, La., U.S. (dĕ rĭd′ēr)	81	30°50′N	93°18′W
Dermott, Ar., U.S. (dûr′mŏt)	79	33°32′N	91°24′W
Derry, N.H., U.S. (dār′ĭ)	61a	42°53′N	71°22′W
Derventa, Bos. (dĕr′wĕn-tá)	131	44°58′N	17°58′E
Derwent, r., Austl. (dēr′wĕnt)	176	42°21′s	146°30′E
Derwent, r., Eng., U.K.	114a	52°54′N	1°24′W
Des Arc, Ar., U.S. (dāz ärk′)	79	34°59′N	91°31′W
Descalvado, Braz. (dĕs-käl-vá-dô)	99a	21°55′s	47°37′W
Descartes, Fr.	126	46°58′N	0°42′E
Deschambault Lake, l., Can.	56	54°40′N	103°35′W
Deschênes, Can.	62c	45°23′N	75°47′W
Deschenes, Lake, l., Can.	62c	45°25′N	75°53′W
Deschutes, r., Or., U.S. (dā-shoot′)	72	44°25′N	121°21′W
Desdemona, Tx., U.S. (dĕz-dĕ-mō′ná)	80	32°16′N	98°33′W
Dese, Eth.	185	11°00′N	39°51′E
Deseado, r., Arg. (dā-sā-ä′dhô)	102	46°50′s	67°45′W
Desirade Island, i., Guad. (dā-zē-räs′)	91b	16°21′N	60°51′W
De Smet, S.D., U.S. (dĕ smĕt′)	70	44°23′N	97°33′W
Des Moines, Ia., U.S. (dĕ moin′)	65	41°35′N	93°37′W
Des Moines, N.M., U.S.	78	36°42′N	103°48′W
Des Moines, Wa., U.S.	74a	47°24′N	122°20′W
Des Moines, r., U.S.	65	42°30′N	94°20′W
Desna, r., Eur. (dyĕs-ná′)	137	51°55′N	31°45′E
Desolación, i., Chile	102	53°05′s	74°00′W
De Soto, Mo., U.S. (dĕ sō′tō)	79	38°07′N	90°32′W
Des Peres, Mo., U.S. (dĕs pĕr′ĕs)	75e	38°36′N	90°26′W
Des Plaines, Il., U.S. (dĕs plānz′)	69a	42°02′N	87°54′W
Des Plaines, r., U.S.	69a	41°39′N	87°56′W
Dessau, Ger. (dĕs′ou)	117	51°50′N	12°15′E
Detmold, Ger. (dĕt′mōld)	124	51°57′N	8°55′E
Detroit, Mi., U.S. (dĕ-troit′)	65	42°22′N	83°10′W
Detroit, Tx., U.S.	79	33°41′N	95°16′W
Detroit Lake, res., Or., U.S.	72	44°42′N	122°10′W
Detroit Lakes, Mn., U.S. (dĕ-troit′lăkz)	70	46°48′N	95°51′W
Detva, Slvk. (dyĕt′vá)	115e	48°32′N	19°21′E
Deurne, Bel.	115a	51°13′N	4°27′E
Deutsch Wagram, Aus.	115e	48°19′N	16°34′E
Deux-Montagnes, Can.	62a	45°33′N	73°53′W
Deux Montagnes, Lac des, l., Can.	62a	45°28′N	74°00′W
Deva, Rom. (dā′vä)	119	45°52′N	22°52′E
Dévaványa, Hung. (dā′vô-vän-yô)	125	47°01′N	20°58′E
Develi, Tur. (dĕ′vá-lē)	137	38°20′N	35°10′E
Deventer, Neth. (dĕv′ĕn-tēr)	121	52°14′N	6°07′E
Devils, r., Tx., U.S.	80	29°55′N	101°10′W
Devils Island see Diable, Île du, i., Fr. Gu.	101	5°15′N	52°40′W
Devils Lake, N.D., U.S.	64	48°10′N	98°55′W
Devils Lake, l., N.D., U.S. (dĕv′′lz)	70	47°57′N	99°04′W
Devils Lake Indian Reservation, I.R., N.D., U.S.	70	48°08′N	99°40′W
Devils Postpile National Monument, rec., Ca., U.S.	76	37°42′N	119°12′W
Devils Tower National Monument, rec., Wy., U.S.	73	44°38′N	105°07′W
Devoll, r., Alb.	131	40°55′N	20°10′E
Devon, Can.	62g	53°23′N	113°43′W
Devon, S. Afr. (dĕv′ŭn)	192c	26°23′s	28°47′E
Devonport, Austl. (dĕv′ŭn-pôrt)	175	41°20′s	146°30′E
Devonport, N.Z.	175a	36°50′s	174°45′E
Devore, Ca., U.S. (dĕ-vôr′)	75a	34°13′N	117°24′W
Dewatto, Wa., U.S. (dĕ-wát′ô)	74a	47°27′N	123°04′W
Dewey, Ok., U.S. (dū′ĭ)	79	36°48′N	95°55′W
De Witt, Ar., U.S. (dĕ wĭt′)	79	34°17′N	91°22′W
De Witt, Ia., U.S.	71	41°46′N	90°34′W
Dewsbury, Eng., U.K. (dūz′bĕr-ĭ)	114a	53°42′N	1°39′W
Dexter, Me., U.S. (dĕks′tēr)	60	45°01′N	69°19′W
Dexter, Mo., U.S.	79	36°46′N	89°56′W
Dezfūl, Iran	154	32°14′N	48°37′E
Dezhnëva, Mys, c., Russia (dyězh′nyĭf)	152	68°00′N	172°00′W
Dezhou, China (dŭ-jō)	164	37°28′N	116°17′E
Dhahran see Az Zahrān, Sau. Ar.	154	26°13′N	50°00′E
Dhaka, Bngl. (dä′kä) (dăk′ä)	155	23°45′N	90°29′E
Dharamtar Creek, r., India	159b	18°49′N	72°54′E
Dharmavaram, India	159	14°32′N	77°43′E
Dhawalāgiri, mtn., Nepal	155	28°42′N	83°31′E
Dhībān, Jord.	153a	31°30′N	35°46′E
Dhidhimótikhon, Grc.	131	41°20′N	26°27′E
Dhule, India	155	20°58′N	74°43′E
Dia, i., Grc. (dē′ä)	130a	35°27′N	25°17′E
Diable, Île du, i., Fr. Gu.	101	5°15′N	52°40′W
Diablo, Mount, mtn., Ca., U.S. (dyä′blô)	74b	37°52′N	121°55′W
Diablo Heights, Pan. (dyä′blô)	86a	8°58′N	79°34′W
Diablo Range, mts., Ca., U.S.	74b	37°47′N	121°50′W
Diablotins, Morne, mtn., Dom.	91b	15°31′N	61°24′W
Diaca, Moz.	191	11°30′s	39°59′E
Diaka, r., Mali	189	14°40′N	5°00′E
Diamantina, Braz.	101	18°14′s	43°32′W
Diamantina, r., Austl. (dī′man-tē′ná)	174	25°38′s	139°53′E
Diamantino, Braz. (dē-á-män-tē′no)	101	14°22′s	56°23′W
Diamond Peak, mtn., Or., U.S.	72	43°32′N	122°08′W
Diana Bank, bk., (dī′än′á)	93	22°30′N	74°45′W
Dianbai, China (dĕn-bī)	165	21°30′N	111°20′E
Dian Chi, l., China (dĕn chē)	160	24°58′N	103°18′E
Dickinson, N.D., U.S. (dĭk′ĭn-sŭn)	64	46°52′N	102°49′W
Dickinson, Tx., U.S. (dĭk′ĭn-sŭn)	81a	29°28′N	95°02′W
Dickinson Bayou, Tx., U.S.	81a	29°26′N	95°08′W
Dickson, Tn., U.S. (dĭk′sŭn)	82	36°03′N	87°24′W
Dickson City, Pa., U.S.	67	41°25′N	75°40′W
Didcot, Eng., U.K. (dĭd′cŏt)	114b	51°35′N	1°15′W
Didiéni, Mali	188	13°53′N	8°06′W
Die, Fr. (dē)	127	44°45′N	5°22′E
Diefenbaker, res., Can.	52	51°20′N	108°10′W
Diego de Ocampo, Pico, mtn., Dom. Rep. (dē-ā′gō dĕ-ō-käm′pô)	93	19°40′N	70°45′W
Diego Ramirez, Islas, is., Chile (dē ā′gō rä-mē′räz)	102	56°15′s	70°15′W
Diéma, Mali	188	14°32′N	9°12′W
Dien Bien Phu, Viet.	160	21°38′N	102°49′E
Dieppe, Can. (dē-ĕp′)	60	46°06′N	64°45′W
Dieppe, Fr.	117	49°54′N	1°05′E
Dierks, Ar., U.S. (dērks)	79	34°06′N	94°02′W
Diessen, Ger. (dēs′sĕn)	115d	47°57′N	11°06′E
Diest, Bel.	115a	50°59′N	5°05′E
Digby, Can. (dĭg′bī)	51	44°37′N	65°46′W
Dighton, Ma., U.S. (dī-tŭn)	68b	41°49′N	71°05′W
Digne, Fr. (dēn′y′)	127	44°07′N	6°16′E
Digoin, Fr. (dē-gwăn′)	126	46°28′N	4°06′E
Digul, r., Indon.	169	7°00′s	140°27′E
Dijohan Point, c., Phil. (dē-kō-än)	169a	16°24′N	122°25′E
Dijon, Fr. (dē-zhôn′)	110	47°21′N	5°02′E
Dikson, Russia (dĭk′sŏn)	134	73°30′N	80°35′E
Dikwa, Nig. (dē′kwä)	185	12°06′N	13°53′E
Dili, E. Timor (dĭl′ē)	169	8°35′s	125°35′E
Di Linosa Island, i., Italy (dē-lē-nō′sä)	118	36°01′N	12°43′E
Dilizhan, Arm.	137	40°45′N	45°00′E
Dillingham, Ak., U.S. (dĭl′ĕng-hăm)	64a	59°10′N	158°38′W
Dillon, Mt., U.S. (dĭl′ŭn)	73	45°12′N	112°40′W
Dillon, S.C., U.S.	83	34°24′N	79°28′W
Dillon Reservoir, res., Oh., U.S.	66	40°05′N	82°05′W
Dilolo, D.R.C. (dē-lō′lō)	186	10°19′s	22°23′E
Dimashq see Damascus, Syria	154	33°31′N	36°18′E
Dimbokro, C. Iv.	188	6°39′N	4°42′W
Dimitrovo see Pernik, Blg.	119	42°36′N	23°04′E
Dimlang, mtn., Nig.	189	8°24′N	11°47′E
Dimona, Isr.	153a	31°03′N	35°01′E
Dinagat Island, i., Phil.	169	10°15′N	126°15′E
Dinajpur, Bngl.	158	25°38′N	87°39′E
Dinan, Fr. (dē-nän′)	126	48°27′N	2°03′W
Dinant, Bel. (dē-nän′)	121	50°17′N	4°50′E
Dinara, mts., Serb. (dē′nä-rä)	119	43°50′N	16°15′E
Dinard, Fr.	126	48°38′N	2°04′W
Dindigul, India	159	10°25′N	78°03′E
Dingalan Bay, b., Phil. (dĭn-gä′län)	169a	15°19′N	121°33′E
Dingle, Ire. (dĭng′′l)	120	52°10′N	10°13′W
Dingle Bay, b., Ire.	117	52°02′N	10°15′W
Dingo, Austl. (dĭn′gō)	175	23°45′s	149°26′E
Dinguiraye, Gui.	188	11°18′N	10°43′W
Dingwall, Scot., U.K. (dĭng′wôl)	120	57°37′N	4°23′W
Dingxian, China (dĭn shyĕn)	164	38°30′N	115°00′E
Dingxing, China (dĭn-shyĭn)	164	39°18′N	115°50′E
Dingyuan, China (dĭn-yüän)	162	32°32′N	117°40′E
Dingzi Wan, b., China	162	36°33′N	121°06′E
Dinosaur National Monument, rec., Co., U.S. (dī′nô-sôr)	73	40°45′N	109°17′W
Dinslaken, Ger. (dĕns′lä-kĕn)	127c	51°33′N	6°44′E
Dinteloord, Neth.	115a	51°38′N	4°21′E
Dinuba, Ca., U.S. (dĭ-nū′bà)	76	36°33′N	119°29′W
Dios, Cayo de, i., Cuba (kä′yō-dĕ-dē-ōs′)	92	22°05′N	83°05′W
Diourbel, Sen. (dē-ōōr-bĕl′)	184	14°40′N	16°15′W
Diphu Pass, p., Asia (dī-pōō)	160	28°15′N	96°45′E
Diquis, r., C.R. (dē-kēs′)	91	8°59′N	83°24′W
Dire Dawa, Eth.	185	9°40′N	41°47′E
Diriamba, Nic. (dē-ryäm′bä)	90	11°52′N	86°15′W
Dirk Hartog, i., Austl.	174	26°25′s	113°15′E
Dirksland, Neth.	115a	51°45′N	4°04′E
Dirranbandi, Austl. (dĭ-rä-bän′dē)	175	28°24′s	148°29′E
Dirty Devil, r., Ut., U.S. (dûr′tĭ dĕv′′l)	77	38°20′N	110°30′W
Disappointment, l., Austl.	174	23°20′s	123°00′E
Disappointment, Cape, c., Wa., U.S. (dĭs ä-point′mĕnt)	74c	46°16′N	124°11′W
Discovery, S. Afr. (dĭs-kŭv′ēr-ĭ)	187b	26°10′s	27°53′E
Discovery, is., Can. (dĭs-kŭv′ēr-ē)	74a	48°25′N	123°13′W
Disko, i., Grnld. (dĭs′kō)	49	70°00′N	54°00′W
Disna, Bela. (dēs′nä)	136	55°34′N	28°15′E
Dispur, India	158	26°00′N	91°50′E
Disraëli, Can. (dĭs-rā′lĭ)	59	45°53′N	71°23′W
District of Columbia, state, U.S.	65	38°50′N	77°00′W
Distrito Federal, state, Braz. (dĕs-trē′tô-fĕ-dĕ-rä′l)	101	15°49′s	47°39′W
Distrito Federal, state, Mex.	88	19°14′N	99°08′W
Disūq, Egypt (dē-sōōk′)	192b	31°07′N	30°41′E
Diu, India (dē′ōō)	155	20°48′N	70°58′E
Divilacan Bay, b., Phil. (dē-vē-lä′kän)	169a	17°26′N	122°25′E
Divinópolis, Braz. (dē-vē-nô′pō-lês)	101	20°10′s	44°53′W
Divo, C. Iv.	188	5°50′N	5°22′W
Dixon, Il., U.S. (dĭks′ŭn)	71	41°50′N	89°30′W
Dixon Entrance, strt., N.A.	52	54°25′N	132°00′W
Diyarbakir, Tur. (dē-yär-bĕk′ĭr)	154	38°00′N	40°10′E
Dja, r., Afr.	185	2°30′N	14°00′E
Djambala, Congo	190	2°33′s	14°45′E
Djanet, Alg.	184	24°29′N	9°26′E
Djebobo, mtn., Ghana	188	8°20′N	0°37′E
Djedi, Oued, r., Alg.	118	34°18′N	4°39′E
Djember, Chad	189	10°25′N	17°50′E
Djerba, Île de, i., Tun.	118	33°53′N	11°26′E
Djerid, Chott, l., Tun. (jĕr′ĭd)	184	33°15′N	8°29′E
Djibasso, Burkina	188	13°07′N	4°10′W
Djibo, Burkina	188	14°06′N	1°38′W
Djibouti, Dji. (jē-bōō-tē′)	192a	11°34′N	43°00′E
Djibouti, nation, Afr.	192a	11°35′N	48°08′E
Djokoumatombi, Congo	190	0°47′N	15°22′E
Djokupunda, D.R.C.	186	5°27′s	20°58′E
Djoua, r., Afr.	190	1°25′N	13°40′E
Djursholm, Swe. (djōōrs′hōlm)	122	59°26′N	18°01′E
Dmitriyev-L'govskiy, Russia (d′mē′trī-yĕf l′gôf′skī)	132	52°07′N	35°05′E
Dmitrov, Russia (d′mē′trôf)	132	56°21′N	37°32′E
Dmitrovsk, Russia (d′mē′trôfsk)	132	52°30′N	35°10′E
Dmytrivka, Ukr.	133	47°57′N	38°56′E
Dnepropetrovsk see Dnipropetrovs'k, Ukr.	134	48°15′N	34°08′E
Dnieper (Dnipro), r., Eur.	137	46°45′N	33°40′E
Dniester, r., Eur.	137	48°21′N	28°10′E
Dniprodzerzhyns'k, Ukr.	137	48°32′N	34°38′E
Dniprodzerzhyns'ke vodoskhovyshche, res., Ukr.	134	49°00′N	34°10′E
Dnipropetrovs'k, Ukr.	134	48°15′N	34°08′E
Dnipropetrovs'k, prov., Ukr.	133	48°15′N	34°10′E
Dniprovs'kyi lyman, b., Ukr.	133	46°33′N	31°45′E
Dnistrovs'kyi lyman, l., Ukr.	133	46°13′N	29°50′E
Dno, Russia (d′nô′)	132	57°49′N	29°59′E
Do, Lac, l., Mali	188	15°50′N	2°20′W
Doba, Chad	189	8°39′N	16°51′E
Dobbs Ferry, N.Y., U.S. (dŏbz′fĕ′rĕ)	68a	41°01′N	73°53′W
Dobbyn, Austl. (dŏb′ĭn)	174	19°45′s	140°02′E
Dobele, Lat. (dô′bĕ-lĕ)	123	56°37′N	23°18′E
Doberai, Jazirah, pen., Indon.	169	1°25′s	133°15′E
Dobo, Indon.	169	6°00′s	134°18′E
Doboj, Bos. (dô′boi)	131	44°42′N	18°04′E
Dobrich, Blg.	119	43°33′N	27°52′E
Dobryanka, Russia (dôb-ryän′kà)	142a	58°27′N	56°26′E
Dobšiná, Slvk. (dŏp′shē-nä)	125	48°48′N	20°25′E
Doce, r., Braz. (dō′sä)	101	19°01′s	42°14′W
Doce, Canal Numero, can., Arg.	99c	36°47′s	59°00′W
Doce Leguas, Cayos de las, is., Cuba	92	20°55′N	79°05′W
Doctor Arroyo, Mex. (dôk-tōr′ är-rō′yō)	88	23°41′N	100°10′W
Doddington, Eng., U.K. (dŏd′dĭng-tŏn)	114b	51°17′N	0°47′E
Dodecanese see Dodekanisoy, is., Grc.	131	38°00′N	26°10′E
Dodekanisoy (Dodecanese), is., Grc.	131	38°00′N	26°10′E
Dodge City, Ks., U.S. (dŏj)	64	37°44′N	100°01′W
Dodgeville, Wi., U.S. (dŏj′vĭl)	71	42°58′N	90°07′W
Dodoma, Tan. (dō′dô-mä)	186	6°11′s	35°45′E
Dog, l., Can. (dŏg)	58	48°42′N	89°24′W
Dogger Bank, bk., (dŏg′gĕr)	121	55°07′N	2°25′E
Dogubayazit, Tur.	137	39°35′N	44°00′E
Doha see Ad Dawhah, Qatar	154	25°02′N	51°28′E
Dohad, India	158	22°52′N	74°18′E
Dokshytsy, Bela.	132	54°53′N	27°49′E
Dolbeau, Can.	51	48°52′N	72°16′W
Dole, Fr. (dōl)	117	47°07′N	5°28′E
Dolgaya, Kosa, c., Russia (kô′sä dôl-gä′yä)	133	46°42′N	37°42′E
Dolgeville, N.Y., U.S.	67	43°10′N	74°45′W
Dolgiy, i., Russia	136	69°20′N	59°20′E
Dolgoprudnyy, Russia	142b	55°57′N	37°33′E
Dolinsk, Russia (dá-lēnsk′)	141	47°29′N	142°31′E
Dollar Harbor, b., Bah.	92	25°30′N	79°15′W
Dolomite, Al., U.S. (dŏl′ô-mīt)	68h	33°28′N	86°57′W
Dolomiti, mts., Italy	130	46°16′N	11°43′E
Dolores, Arg. (dô-lō′rĕs)	102	36°20′s	57°42′W
Dolores, Col.	100a	3°33′s	74°54′W
Dolores, Ur.	99c	33°32′s	58°15′W
Dolores, Tx., U.S. (dô-lō′rĕs)	80	27°42′N	99°47′W
Dolores, r., Co., U.S.	77	38°35′N	108°50′W
Dolores Hidalgo, Mex. (dô-lō′rĕs-ē-däl′gō)	88	21°09′N	100°56′W
Dolphin and Union Strait, strt., Can.	52	69°22′N	117°10′W
Dolyna, Ukr.	125	48°57′N	24°01′E
Domažlice, Czech Rep. (dô′mäzh-lĕ-tsĕ)	124	49°27′N	12°55′E
Dombasle-sur-Meurthe, Fr. (dôn-bäl′)	127	48°38′N	6°18′E
Dombóvár, Hung. (dôm′bō-vär)	125	46°22′N	18°08′E
Domeyko, Cordillera, mts., Chile (kôr-dēl-yĕ′rä-dô-mā′kô)	100	20°50′s	69°02′W
Dominica, nation, N.A. (dô-mĭ-nē′kà)	87	15°30′N	60°45′W
Dominica Channel, strt., N.A.	91b	15°00′N	61°30′W
Dominican Republic, nation, N.A. (dô-mĭn′ĭ-kăn)	87	19°00′N	70°45′W
Dominion, Can. (dô-mĭn′yŭn)	61	46°13′N	60°01′W

ăt; fināl; rāte; senāte; ärm; ásk; sofá; fāre; ch-choose; dh-as th in other; bē; ēvent; bĕt; recĕnt; crātēr; g-gō; gh-guttural g; bĭt; ī-short neutral; rīde; к-guttural k as ch in German ich;

PLACE (Pronunciation)	PAGE	LAT.	LONG.
Dalmacija, hist. reg., Serb.			
(däl-mä′tsĕ-yä)	130	43°25′N	16°37′E
Dalnerechensk, Russia	135	46°07′N	133°21′E
Daloa, C. Iv.	188	6°53′N	6°27′W
Dalroy, Can. (dăl′roi)	62e	51°07′N	113°39′W
Dalrymple, Mount, mtn., Austl.	175	21°14′S	148°46′E
(dăl′rĭm-p′l)			
Dalton, S. Afr. (dôl′tŏn)	187c	29°21′S	30°41′E
Dalton, Ga., U.S. (dôl′tŭn)	82	34°46′N	84°58′W
Daly, r., Austl. (dā′lĭ)	174	14°15′S	131°15′E
Daly City, Ca., U.S. (dā′lē)	74b	37°42′N	122°27′W
Damān, India	155	20°32′N	72°53′E
Damanhûr, Egypt (dä-män-hōōr′)	185	30°59′N	30°31′E
Damar, Pulau, i., Indon.	169	7°15′S	129°15′E
Damara, C.A.R.	189	4°58′N	18°42′E
Damaraland, hist. reg., Nmb.			
(dä′na-rä-land)	186	22°15′S	16°15′E
Damas Cays, is., Bah. (dä′mäs)	92	23°50′N	79°50′W
Damascus, Syria	154	33°30′N	36°18′E
Damāvand, Qolleh-ye, mtn., Iran	154	36°05′N	52°05′E
Damba, Ang. (däm′bä)	186	6°41′S	15°08′E
Dâmboviţa, r., Rom.	131	44°43′N	25°41′E
Dame Marie, Cap, c., Haiti			
(däm märē′)	93	18°35′N	74°50′W
Dāmghān, Iran (däm-gän′)	154	35°50′N	54°15′E
Daming, China (dä-mĭŋ)	164	36°15′N	115°09′E
Dammartin-en-Goële, Fr.			
(däx-mär-tăn-än-gô-ĕl′)	127b	49°03′N	2°40′E
Dampier, Selat, strt., Indon. (däm′pēr)	169	0°40′S	131°15′E
Dampier Archipielago, is., Austl.	174	20°15′S	116°25′E
Dampier Land, reg., Austl.	174	17°30′S	122°25′E
Dan, r., N.C., U.S. (dăn)	83	36°26′N	79°40′W
Dana, Mount, mtn., Ca., U.S.	76	37°54′N	119°13′W
Da Nang, Viet.	168	16°08′N	108°22′E
Danbury, Eng., U.K.	114b	51°42′N	0°34′E
Danbury, Ct., U.S. (dăn′bĕr-ĭ)	68a	41°23′N	73°27′W
Danbury, Tx., U.S.	81a	29°14′N	95°22′W
Dandenong, Austl. (dăn′dĕ-nông)	176	37°59′S	145°13′E
Dandong, China	161	40°10′N	124°30′E
Dane, r., Eng., U.K. (dān)	114a	53°11′N	2°14′W
Danea, Gui.	188	11°27′N	13°12′W
Danforth, Me., U.S.	60	45°38′N	67°53′W
Dan Gora, Nig.	189	11°30′N	8°09′E
Dangtu, China (däŋ-tōō)	165	31°35′N	118°28′E
Dani, Burkina	184	13°43′N	0°10′W
Dania, Fl., U.S. (dā′nĭ-ä)	83a	26°01′N	80°10′W
Danilov, Russia (dá-nē-lôf)	136	58°12′N	40°08′E
Danissa Hills, hills, Kenya	191	3°20′N	40°55′E
Dänizkänari, Azer.	138	40°13′N	49°33′E
Dankov, Russia (dän′kôf)	136	53°17′N	39°09′E
Dannemora, N.Y., U.S. (dăn-ĕ-mō′rá)	67	44°45′N	73°45′W
Dannhauser, S. Afr. (dän′hou-zĕr)	187c	28°07′S	30°04′E
Dansville, N.Y., U.S. (dănz′vĭl)	67	42°30′N	77°40′W
Danube, r., Eur.	112	43°00′N	24°00′E
Danube, Mouths of the, mth., Rom.			
(dän′ub)	133	45°13′N	29°37′E
Danvers, Ma., U.S. (dăn′vērz)	61a	42°34′N	70°57′W
Danville, Ca., U.S. (dăn′vĭl)	74b	37°49′N	122°00′W
Danville, Il., U.S.	66	40°10′N	87°35′W
Danville, In., U.S.	66	39°45′N	86°30′W
Danville, Ky., U.S.	66	37°35′N	84°50′W
Danville, Pa., U.S.	67	41°00′N	76°35′W
Danville, Va., U.S.	65	36°35′N	79°24′W
Danxian, China (dän shyĕn)	165	19°30′N	109°38′E
Danyang, China (dän-yäŋ)	162	32°01′N	119°32′E
Danzig see Gdańsk, Pol.	110	54°20′N	18°40′E
Danzig, Gulf of, b., Eur. (dăn′tsĭk)	116	54°41′N	19°01′E
Daoxian, China (dou shyĕn)	165	25°35′N	111°27′E
Dapango, Togo	188	10°52′N	0°12′E
Daphnae, hist., Egypt	153a	30°43′N	32°12′E
Daqin Dao, i., China (dä-chyĭn dou)	162	38°18′N	120°50′E
Darabani, Rom. (dä-rä-bän′ĭ)	125	48°13′N	26°38′E
Daraj, Libya	184	30°12′N	10°14′E
Daráw, Egypt (dà-rä′ōō)	192b	24°24′N	32°56′E
Darbhanga, India (dŭr-bŭn′gä)	155	26°03′N	85°09′E
Darby, Pa., U.S. (där′bĭ)	68f	39°55′N	75°16′W
Darby, i., Bah.	92	23°50′N	76°20′W
Dardanelles see Çanakkale Boğazi,			
strt., Tur.	119	40°05′N	25°50′E
Dar es Salaam, Tan. (där ĕs sä-läm′)	187	6°48′S	39°17′E
Dârfûr, hist. reg., Sudan (där-fōōr′)	185	13°21′N	23°46′E
Dargai, Pak. (dŭr-gä′ē)	158	34°35′N	72°01′E
Darien, Col. (dä-rĭ-ĕn′)	100a	3°56′N	76°30′W
Darien, Ct., U.S. (dà-rē-ĕn′)	68a	41°04′N	73°28′W
Darién, Cordillera de, mts., Nic.	90	13°00′N	85°42′W
Darien, Serranía del, mts.	91	8°13′N	77°28′W
Darjeeling, India (dŭr-jē′lĭng)	155	27°05′N	88°16′E
Darling, r., Austl.	175	30°35′S	143°20′E
Darling Downs, reg., Austl.	175	27°22′S	150°00′E
Darling Range, mts., Austl.	174	30°30′S	115°45′E
Darlington, Eng., U.K. (där′lĭng-tŭn)	120	54°32′N	1°35′W
Darlington, S.C., U.S.	83	34°15′N	79°52′W
Darlington, Wi., U.S.	71	42°40′N	90°06′W
Darłowo, Pol. (där-lô′vô)	124	54°26′N	16°23′E
Darmstadt, Ger. (därm′shtät)	117	49°53′N	8°40′E
Darnah, Libya	185	32°44′N	22°41′E
Darnley Bay, b., Ak., U.S. (därn′lē)	63	70°00′N	124°00′W
Daroca, Spain (dä-rō-kä)	128	41°08′N	1°24′W
Dartford, Eng., U.K.	114b	51°27′N	0°14′E
Dartmoor, for., Eng., U.K.			
(därt′mōōr)	120	50°35′N	4°05′W
Dartmouth, Can. (därt′mŭth)	51	44°40′N	63°34′W
Dartmouth, Eng., U.K.	120	50°33′N	3°28′W
Daru, Pap. N. Gui. (dä′rōō)	169	9°04′S	143°21′E
Daruvar, Cro. (dä′rōō-vär′)	131	45°37′N	17°16′E
Darwen, Eng., U.K. (där′wĕn)	114a	53°42′N	2°28′W
Darwin, Austl. (där′wĭn)	174	12°25′S	131°00′E

PLACE (Pronunciation)	PAGE	LAT.	LONG.
Darwin, Cordillera, mts., Chile			
(kôr-dēl-yĕ′rä-där′wĕn)	102	54°40′S	69°30′W
Dashhowuz, Turkmen.	139	41°50′N	59°45′E
Dash Point, Wa., U.S. (dăsh)	74a	47°19′N	122°25′W
Dasht, r., Pak. (dŭsht)	154	25°30′N	62°30′E
Dasol Bay, b., Phil. (dä-sōl′)	169a	15°53′N	119°40′E
Datian Ding, mtn., China (dä-tiĕn dĭŋ)	165	22°25′N	111°20′E
Datong, China (dä-tôŋ)	164	40°00′N	113°30′E
Dattapukur, India	158a	22°45′N	88°32′E
Datteln, Ger. (dät′tĕln)	127c	51°39′N	7°20′E
Datu, Tandjung, c., Asia	168	2°08′N	110°15′E
Datuan, China (dä-tŭän)	163b	30°57′N	121°43′E
Daugava (Zapadnaya Dvina), r., Eur.	123	56°40′N	24°40′E
Daugavpils, Lat. (dou′gäv-pēls)	136	55°52′N	26°32′E
Dauphin, Can. (dô′fĭn)	50	51°09′N	100°00′W
Dauphin Lake, l., Can.	57	51°17′N	99°48′W
Dāvangere, India	159	14°30′N	75°55′E
Davao, Phil. (dä′vä-ô)	169	7°05′N	125°30′E
Davao Gulf, b., Phil.	169	6°30′N	125°45′E
Davenport, Ia., U.S. (dăv′ĕn-pôrt)	65	41°34′N	90°38′W
Davenport, Wa., U.S.	72	47°39′N	118°07′W
David, Pan. (dà-vēdh′)	87	8°27′N	82°27′W
David City, Ne., U.S. (dā′vĭd)	70	41°15′N	97°10′W
David-Gorodok, Bela.			
(dà-vét′ gô-rō′dŏk)	137	52°02′N	27°14′E
Davis, Ok., U.S. (dā′vĭs)	79	34°34′N	97°08′W
Davis, W.V., U.S.	67	39°15′N	79°25′W
Davis Lake, l., Or., U.S.	72	43°38′N	121°43′W
Davis Mountains, mts., Tx., U.S.	80	30°45′N	104°17′W
Davis Strait, strt., N.A.	49	66°00′N	60°00′W
Davlekanovo, Russia	136	54°15′N	55°05′E
Davos, Switz. (dä′vōs)	124	46°47′N	9°50′E
Dawa, r., Afr.	185	4°30′N	40°30′E
Dawāsir, Wādī ad, val., Sau. Ar.	154	20°48′N	44°07′E
Dawei, Mya.	168	14°04′N	98°19′E
Dawen, r., China (dä-wŭn)	162	35°58′N	116°53′E
Dawley, Eng., U.K. (dô′lĭ)	114a	52°38′N	2°28′W
Dawna Range, mts., Mya.	168	17°02′N	98°01′E
Dawson, r., Austl.	175	24°20′S	149°45′E
Dawson, Ga., U.S.	82	31°45′N	84°29′W
Dawson, Mn., U.S.	70	44°54′N	96°03′W
Dawson, r., Austl.	175	24°20′S	149°45′E
Dawson Bay, b., Can.	57	52°55′N	100°50′W
Dawson Creek, Can.	50	55°46′N	120°14′W
Dawson Range, mts., Can.	63	62°15′N	138°10′W
Dawson Springs, Ky., U.S.	82	37°10′N	87°40′W
Dawu, China (dä-wōō)	162	31°33′N	114°07′E
Dax, Fr. (däks)	117	43°42′N	1°06′W
Daxian, China (dä-shyĕn)	160	31°12′N	107°30′E
Daxing, China (dä-shyĭŋ)	164a	39°44′N	116°19′E
Dayiqiao, China (dä-ē-chyou)	162	31°43′N	120°40′E
Dayr az Zawr, Syria (dä-ĕrĕz-zôr′)	154	35°15′N	40°01′E
Dayton, Ky., U.S. (dā′tŭn)	69f	39°07′N	84°28′W
Dayton, N.M., U.S.	78	32°44′N	104°23′W
Dayton, Oh., U.S.	65	39°54′N	84°15′W
Dayton, Tn., U.S.	82	35°30′N	85°00′W
Dayton, Tx., U.S.	81	30°03′N	94°53′W
Dayton, Wa., U.S.	72	46°18′N	117°59′W
Daytona Beach, Fl., U.S. (dā-tō′ná)	65	29°11′N	81°02′W
Dayu, China (dä-yōō)	165	25°20′N	114°20′E
Da Yunhe (Grand Canal), can., China			
(dä yŏn-hŭ)	161	35°00′N	117°00′E
Dayville, Ct., U.S. (dā′vĭl)	68a	41°50′N	71°55′W
De Aar, S. Afr. (dē-är′)	186	30°45′S	24°05′E
Dead, l., Mn., U.S. (dĕd)	70	46°28′N	96°00′W
Dead Sea, l., Asia	154	31°30′N	35°30′E
Deadwood, S.D., U.S. (dĕd′wŏd)	64	44°23′N	103°43′W
Deal Island, Md., U.S. (dēl-ī′lănd)	67	38°10′N	75°55′W
Dean, r., Can. (dēn)	54	52°45′N	125°30′W
Dean Channel, strt., Can.	54	52°33′N	127°13′W
Deán Funes, Arg. (dĕ-à′n-fōō-nĕs)	102	30°26′S	64°12′W
Dearborn, Mi., U.S. (dēr′bŭrn)	69b	42°18′N	83°15′W
Dearg, Ben, mtn., Scot., U.K.			
(bĕn dŭrg)	120	57°48′N	4°59′W
Dease Strait, strt., Can. (dēz)	52	68°50′N	108°20′W
Death Valley, Ca., U.S.	76	36°18′N	116°26′W
Death Valley, val., Ca., U.S.	64	36°30′N	117°00′W
Death Valley National Park, rec.,			
U.S.	76	36°34′N	117°00′W
Debal'tseve, Ukr.	138	48°23′N	38°29′E
Debao, China (dü-bou)	160	23°18′N	106°40′E
Debar, Mac. (dĕ′bär) (dä′brä)	131	41°31′N	20°32′E
Dęblin, Pol. (dĕb′blĭn)	125	51°34′N	21°49′E
Dębno, Pol. (dĕb-nô′)	124	52°47′N	13°43′E
Debo, Lac, l., Mali	188	15°15′N	4°40′W
Debrecen, Hung. (dĕ′brĕ-tsĕn)	110	47°32′N	21°40′E
Debre Markos, Eth.	185	10°15′N	37°45′E
Debre Tabor, Eth.	185	11°57′N	38°09′E
Decatur, Al., U.S. (dĕ-kā′tŭr)	82	34°35′N	87°00′W
Decatur, Ga., U.S.	68c	33°47′N	84°18′W
Decatur, Il., U.S.	65	39°50′N	88°59′W
Decatur, In., U.S.	66	40°50′N	84°55′W
Decatur, Mi., U.S.	66	42°10′N	86°00′W
Decatur, Tx., U.S.	79	33°14′N	97°33′W
Decazeville, Fr. (dĕ-käz′vēl′)	117	44°33′N	2°16′E
Deccan, plat., India (dĕk′ăn)	155	19°05′N	76°40′E
Deception Lake, l., Can.	56	56°33′N	104°15′W
Deception Pass, p., Wa., U.S.			
(dĕ-sĕp′shŭn)	74a	48°24′N	122°44′W
Děčín, Czech Rep. (dĕ′chēn)	124	50°47′N	14°14′E
Decorah, Ia., U.S. (dĕ-kō′rá)	71	43°18′N	91°48′W
Dedenevo, Russia (dyĕ-dyĕ′nyĕ-vô)	142b	56°14′N	37°31′E
Dedham, Ma., U.S. (dĕd′ăm)	61a	42°15′N	71°11′W
Dedo do Deus, mtn., Braz.			
(dĕ-dô-dô-dĕ′ōōs)	102b	22°30′S	43°02′W
Dédougou, Burkina (dä-dô-gōō′)	184	12°38′N	3°28′W
Dee, r., Scot., U.K.	120	57°05′N	2°25′W

PLACE (Pronunciation)	PAGE	LAT.	LONG.
Dee, r., U.K.	114a	53°15′N	3°05′E
Deep, r., N.C., U.S. (dēp)	83	35°36′N	79°32′W
Deep Fork, r., Ok., U.S.	79	35°35′N	96°42′W
Deep River, Can.	59	46°06′N	77°20′W
Deepwater, Mo., U.S. (dep-wô-tĕr)	79	38°15′N	93°46′W
Deer, i., Me., U.S.	60	44°07′N	68°38′W
Deerfield, Il., U.S. (dēr′fĕld)	69a	42°10′N	87°51′W
Deer Island, Or., U.S.	74c	45°56′N	122°51′W
Deer Lake, Can.	53a	49°10′N	57°25′W
Deer Lake, l., Can.	57	52°40′N	94°30′W
Deer Lodge, Mt., U.S. (dēr lŏj)	73	46°23′N	112°42′W
Deer Park, Oh., U.S.	69f	39°12′N	84°24′W
Deer Park, Wa., U.S.	72	47°58′N	117°28′W
Deer River, Mn., U.S.	71	47°20′N	93°49′W
Defiance, Oh., U.S. (dē-fī′ăns)	66	41°15′N	84°20′W
DeFuniak Springs, Fl., U.S.			
(dē fū′nĭ-ăk)	82	30°42′N	86°06′W
Deganga, India	158a	22°41′N	88°41′E
Degeh Bur, Eth.	192a	8°10′N	43°25′E
Deggendorf, Ger. (dĕ′ghĕn-dôrf)	124	48°50′N	12°59′E
Degollado, Mex. (dā-gô-lyä′dō)	88	20°27′N	102°11′W
De Grey, r., Austl. (dē grā′)	174	20°20′S	119°25′E
Degtyarsk, Russia (dĕg-ty′arsk)	142a	56°42′N	60°05′E
Dehiwala-Mount Lavinia, Sri L.	159	6°47′N	79°55′E
Dehra Dūn, India (dā′rŭ)	155	30°09′N	78°07′E
Dehua, China (dü-hwä)	165	25°30′N	118°15′E
Dej, Rom. (dăzh)	119	47°09′N	23°53′E
De Kalb, Il., U.S. (dē kălb′)	66	41°54′N	88°46′W
Dekese, D.R.C.	190	3°27′S	21°24′E
Delacour, Can. (dē-lä-kōōr′)	62e	51°09′N	113°45′W
Delagua, Co., U.S. (dĕl-à′gwä)	78	37°19′N	104°42′W
De Land, Fl., U.S. (dē länd′)	83	29°00′N	81°19′W
Delano, Ca., U.S. (dĕl′á-nō)	76	35°47′N	119°15′W
Delano Peak, mtn., Ut., U.S.	64	38°25′N	112°25′W
Delavan, Wi., U.S. (dĕl′á-văn)	71	42°39′N	88°38′W
Delaware, Oh., U.S. (dĕl′á-wâr)	66	40°15′N	83°05′W
Delaware, state, U.S.	65	38°40′N	75°30′W
Delaware, r., Ks., U.S.	79	39°45′N	95°47′W
Delaware, r., U.S.	67	41°50′N	75°00′W
Delaware Bay, b., U.S.	65	39°05′N	75°10′W
Delaware Reservoir, res., Oh., U.S.	67	40°30′N	83°05′E
Delémont, Switz. (dĕ-lä-môn′)	124	47°21′N	7°18′E
De Leon, Tx., U.S. (dē lē-ŏn′)	80	32°06′N	98°33′W
Delft, Neth. (dĕlft)	121	52°01′N	4°20′E
Delfzijl, Neth.	121	53°20′N	6°50′E
Delgada, Punta, c., Arg.			
(pōō′n-tä-dĕl-gä′dä)	102	43°46′S	63°46′W
Delgado, Cabo, c., Moz.			
(kä′bô-dĕl-gä′dō)	187	10°40′S	40°35′E
Delhi, India	155	28°54′N	77°13′E
Delhi, Il., U.S. (dĕl′hī)	75e	39°03′N	90°16′W
Delhi, La., U.S.	81	32°26′N	91°29′W
Delhi, state, India	155	28°30′N	76°50′E
Delitzsch, Ger. (dā′lĭch)	124	51°32′N	12°18′E
Dellansjöarna, l., Swe.	122	61°57′N	16°25′E
Delles, Alg. (dĕ′lĕs′)	184	36°59′N	3°40′E
Dell Rapids, S.D., U.S. (dĕl)	70	43°50′N	96°43′W
Dellwood, Mn., U.S. (dĕl′wŏd)	75g	45°05′N	92°58′W
Del Mar, Ca., U.S. (dĕl mär′)	76a	32°57′N	117°16′W
Delmas, S. Afr. (dĕl′mȧs)	192c	26°08′S	28°43′E
Delmenhorst, Ger. (dĕl′mĕn-hôrst)	124	53°03′N	8°38′E
Del Norte, Co., U.S. (dĕl nôrt′)	77	37°40′N	106°25′W
De-Longa, i., Russia	135	76°21′N	148°56′E
De Long Mountains, mts., Ak., U.S.			
(dē′lŏŋg)	63	68°38′N	162°30′W
Deloraine, Austl. (dĕ-lṳ-rān)	176	41°30′S	146°40′E
Delphi, In., U.S. (dĕl′fī)	66	40°35′N	86°40′W
Delphos, Oh., U.S. (dĕl′fōs)	66	40°50′N	84°20′W
Delray Beach, Fl., U.S. (dĕl-rā′)	83a	26°27′N	80°05′W
Del Rio, Tx., U.S. (dĕl rē′ō)	64	29°21′N	100°52′W
Delson, Can. (dĕl′sŭn)	62a	45°24′N	73°32′W
Delta, Co., U.S.	77	38°45′N	108°05′W
Delta, Ut., U.S.	77	39°20′N	112°35′W
Delta Beach, Can.	62f	50°10′N	98°20′W
Delvine, Alb. (dĕl′vĕ-nä)	131	39°58′N	20°10′E
Dёma, r., Russia (dyĕm′ä)	136	53°40′N	54°30′E
Demba, D.R.C.	190	5°30′S	22°16′E
Dembi Dolo, Eth.	185	8°46′N	34°46′E
Demidov, Russia (dzyĕ′mĕ-dô′f)	132	55°16′N	31°32′E
Deming, N.M., U.S. (dĕm′ĭng)	64	32°15′N	107°45′W
Demmin, Ger. (dĕm′mĕn)	124	53°54′N	13°04′E
Demnat, Mor. (dĕm-nät)	184	31°58′N	7°03′W
Demopolis, Al., U.S. (dĕ-mŏp′ô-lĭs)	82	32°30′N	87°50′W
Demotte, In., U.S. (dē′mŏt)	69a	41°12′N	87°13′W
Dempo, Gunung, mtn., Indon.			
(dĕm′pō)	168	4°04′S	103°11′E
Dem′yanka, r., Russia (dyĕm-yän′kä)	140	59°07′N	72°58′E
Demyansk, Russia (dyĕm-yänsk′)	132	57°39′N	32°26′E
Denain, Fr. (dē-năn′)	126	50°23′N	3°21′E
Denakil Plain, pl., Eth.	185	12°45′N	41°01′E
Denali National Park, rec., Ak., U.S.	64a	63°48′N	153°02′W
Denbigh, Wales, U.K. (dĕn′bī)	120	53°15′N	3°25′W
Dendermonde, Bel.	115a	51°02′N	4°04′E
Dendron, Va., U.S. (dĕn′drŭn)	83	37°02′N	76°53′W
Denezhkin Kamen, Gora, mtn., Russia			
(dzyĕ-nĕ′zhkĕn kämĕn)	142a	60°26′N	59°35′E
Denham, Mount, mtn., Jam.	87	18°20′N	77°30′W
Den Helder, Neth. (dĕn hĕl′dĕr)	121	52°55′N	5°45′E
Dénia, Spain	129	38°48′N	0°06′E
Deniliquin, Austl. (dĕ-nĭl′ĭ-kwĭn)	175	35°20′S	144°52′E
Denison, Ia., U.S. (dĕn′ĭ-sŭn)	70	42°01′N	95°22′W
Denison, Tx., U.S.	64	33°45′N	97°02′W
Denizli, Tur. (dĕn′ĭz-lē)	119	37°40′N	29°10′E
Denklingen, Ger. (dĕn′klĕn-gĕn)	127c	50°54′N	7°40′E
Denmark, S.C., U.S. (dĕn′märk)	83	33°18′N	81°09′W
Denmark, nation, Eur.	110	56°14′N	8°30′E
Denmark Strait, strt., Eur.	49	66°30′N	27°00′W

PLACE (Pronunciation)	PAGE	LAT.	LONG.
Dennilton, S. Afr. (děn-ĭl-tŭn)	192c	25°18′s	29°13′E
Dennison, Oh., U.S. (děn´ĭ-sŭn)	66	40°25′N	81°20′W
Denpasar, Indon.	168	8°35′s	115°10′E
Denton, Eng., U.K. (děn´tŭn)	114a	53°27′N	2°07′W
Denton, Md., U.S.	67	38°55′N	75°50′W
Denton, Tx., U.S.	79	33°12′N	97°06′W
D'Entrecasteaux, Point, c., Austl. (dän-tr´-käs-tō´)	174	34°50′s	114°45′E
D'Entrecasteaux Islands, is., Pap. N. Gui. (dän-tr´-käs-tō´)	169	9°45′s	152°00′E
Denver, Co., U.S. (děn´vēr)	64	39°44′N	104°59′W
Deoli, India	158	25°52′N	75°23′E
De Pere, Wi., U.S. (dē pēr´)	71	44°25′N	88°04′W
Depew, N.Y., U.S. (dē-pū´)	69c	42°55′N	78°43′W
Deping, China (dŭ-pĭŋ)	162	37°28′N	116°57′E
Depue, Il., U.S. (dē pū)	66	41°15′N	89°55′W
De Queen, Ar., U.S. (dē kwēn´)	79	34°02′N	94°21′W
De Quincy, La., U.S. (dē kwĭn´sĭ)	81	30°27′N	93°27′W
Dera, Lach., r., Afr. (läk dä´rä)	192a	0°45′N	41°26′E
Dera, Lak, r., Afr.	185	0°45′N	41°30′E
Dera Ghāzi Khān, Pak. (dā´rŭ gä-zē´ kan´)	155	30°09′N	70°39′E
Dera Ismāil Khān, Pak. (dā´rŭ ĭs-mä-ēl´ kän´)	158	31°55′N	70°51′E
Derbent, Russia (děr-běnt´)	137	42°00′N	48°10′E
Derby, Austl. (där´bē) (dûr´bē)	174	17°20′s	123°40′E
Derby, S. Afr. (där´bĭ)	192c	25°55′s	27°02′E
Derby, Eng., U.K. (där´bē)	117	52°55′N	1°29′W
Derby, Ct., U.S. (dûr´bē)	67	41°20′N	73°05′W
Derbyshire, co., Eng., U.K.	114a	53°11′N	1°30′W
Derdepoort, S. Afr.	192c	24°39′s	26°21′E
Derg, Lough, l., Ire. (lŏk děrg)	120	53°00′N	8°09′W
De Ridder, La., U.S. (dē rĭd´ēr)	81	30°50′N	93°18′W
Dermott, Ar., U.S. (dûr´mŏt)	79	33°32′N	91°24′W
Derry, N.H., U.S. (dâr´ĭ)	61a	42°53′N	71°22′W
Derventa, Bos. (děr´ven-tä)	131	44°59′N	17°58′E
Derwent, r., Austl. (dûr´wěnt)	176	42°21′s	146°30′E
Derwent, r., Eng., U.K.	114a	53°11′N	1°24′W
Des Arc, Ar., U.S. (dăz ärk´)	79	34°59′N	91°31′W
Descalvado, Braz. (děs-käl-vá-dô)	99a	21°55′s	47°37′W
Descartes, Fr.	126	46°58′N	0°42′E
Deschambault Lake, l., Can.	56	54°40′N	103°35′W
Deschênes, Can.	62c	45°23′N	75°47′W
Deschenes, Lake, l., Can.	62c	45°25′N	75°53′W
Deschutes, r., Or., U.S. (dā-shōōt´)	72	44°25′N	121°21′W
Desdemona, Tx., U.S. (děz-dē-mō´ná)	80	32°16′N	98°33′W
Dese, Eth.	185	11°00′N	39°51′E
Deseado, r., Arg. (dā-sā-ä´dhō)	102	46°55′s	67°45′W
Desirade Island, i., Guad. (dā-zē-räs´)	91b	16°21′N	60°51′W
De Smet, S.D., U.S. (dē smět´)	70	44°23′N	97°33′W
Des Moines, Ia., U.S. (dē moin´)	65	41°35′N	93°37′W
Des Moines, N.M., U.S.	78	36°42′N	103°48′W
Des Moines, Wa., U.S.	74a	47°24′N	122°20′W
Des Moines, r., U.S.	65	42°30′N	94°20′W
Desna, r., Eur. (dyěs-ná´)	137	51°55′N	31°45′E
Desolación, i., Chile (dě-sō-lä-syō´n)	102	53°05′s	74°00′W
De Soto, Mo., U.S. (dē sō´tō)	79	38°07′N	90°32′W
Des Peres, Mo., U.S. (děs pěr´ěs)	75e	38°36′N	90°26′W
Des Plaines, Il., U.S. (děs plānz´)	69a	42°02′N	87°54′W
Des Plaines, r., U.S.	69a	41°39′N	87°56′W
Dessau, Ger. (děsŏu)	117	51°50′N	12°15′E
Detmold, Ger. (dět´mōld)	124	51°57′N	8°55′E
Detroit, Mi., U.S. (dē-troit´)	65	42°22′N	83°10′W
Detroit, Tx., U.S.	79	33°41′N	95°16′W
Detroit Lake, res., Or., U.S.	72	44°42′N	122°10′W
Detroit Lakes, Mn., U.S. (dē-troit´läkz)	70	46°48′N	95°51′W
Detva, Slvk. (dyět´vá)	125	48°32′N	19°21′E
Deurne, Bel.	115a	51°13′N	4°27′E
Deutsch Wagram, Aus.	115e	48°19′N	16°34′E
Deux-Montagnes, Can.	62a	45°33′N	73°53′W
Deux Montagnes, Lac des, l., Can.	62a	45°28′N	74°00′W
Deva, Rom. (dā´vä)	119	45°52′N	22°52′E
Dévaványa, Hung. (dā´vô-vän-yô)	125	47°01′N	20°58′E
Develi, Tur. (dě´vá-lē)	137	38°20′N	35°10′E
Deventer, Neth. (děv´ěn-tēr)	121	52°14′N	6°07′E
Devils, r., Tx., U.S.	80	29°55′N	101°10′W
Devils Island see Diable, Île du, i., Fr. Gu.	101	5°15′N	52°40′W
Devils Lake, N.D., U.S.	64	48°10′N	98°55′W
Devils Lake, l., N.D., U.S. (děv´lz)	70	47°57′N	99°04′W
Devils Lake Indian Reservation, I.R., N.D., U.S.	70	48°08′N	99°40′W
Devils Postpile National Monument, rec., Ca., U.S.	76	37°42′N	119°12′W
Devils Tower National Monument, rec., Wy., U.S.	73	44°38′N	105°07′W
Devoll, r., Alb.	131	40°55′N	20°15′E
Devon, Can.	62g	53°23′N	113°43′W
Devon, S. Afr. (dēv´ŭn)	192c	26°23′s	28°47′E
Devonport, Austl. (děv´ŭn-pôrt)	175	41°20′s	146°30′E
Devonport, N.Z.	175a	36°50′s	174°45′E
Devore, Ca., U.S. (dě-vôr´)	75a	34°13′N	117°24′W
Dewatto, Wa., U.S. (dē-wät´ō)	74a	47°27′N	123°04′W
Dewey, Ok., U.S. (dū´ĭ)	79	36°48′N	95°55′W
De Witt, Ar., U.S. (dē wĭt´)	79	34°17′N	91°22′W
De Witt, Ia., U.S.	71	41°46′N	90°34′W
Dewsbury, Eng., U.K. (dūz´bēr-ĭ)	114a	53°42′N	1°38′W
Dexter, Me., U.S. (děks´tēr)	60	45°01′N	69°19′W
Dexter, Mo., U.S.	79	36°46′N	89°56′W
Dezfūl, Iran	154	32°14′N	48°37′E
Dezhnëva, Mys, c., Russia (dyězh´nyĭf)	152	68°00′N	172°00′W
Dezhou, China (dŭ-jō)	164	37°28′N	116°17′E
Dhahran see Aẓ Ẓahrān, Sau. Ar.	154	26°13′N	50°00′E
Dhaka, Bngl. (dăk´ä) (dăk´a)	155	23°45′N	90°29′E
Dharamtar Creek, r., India	159b	18°49′N	72°54′E
Dharmavaram, India	159	14°32′N	77°43′E
Dhawalāgiri, mtn., Nepal	155	28°42′N	83°31′E
Dhībān, Jord.	153a	31°30′N	35°46′E
Dhidhimótikhon, Grc.	131	41°20′N	26°27′E
Dhule, India	155	20°58′N	74°43′E
Día, i., Grc. (dē´ä)	130a	35°27′N	25°17′E
Diable, Île du, i., Fr. Gu.	101	5°15′s	52°40′W
Diablo, Mount, mtn., Ca., U.S. (dyä´blō)	74b	37°52′N	121°55′W
Diablo Heights, Pan. (dyä´blō)	86a	8°58′N	79°34′W
Diablo Range, mts., Ca., U.S.	74b	37°47′N	121°50′W
Diablotins, Morne, mtn., Dom.	91b	15°31′N	61°24′W
Diaca, Moz.	191	11°30′s	39°59′E
Diaka, r., Mali	189	14°40′N	5°00′E
Diamantina, Braz.	101	18°14′s	43°32′W
Diamantina, r., Austl. (dī´man-tē´ná)	174	25°38′s	139°53′E
Diamantino, Braz. (dē-ä-män-tē´no)	101	14°22′s	56°23′W
Diamond Peak, mtn., Or., U.S.	72	43°32′N	122°08′W
Diana Bank, bk., (dī´än´á)	93	22°30′N	74°45′W
Dianbai, China (dēn-bī)	165	21°30′N	111°20′E
Dian Chi, l., China (dĭěn chē)	160	24°58′N	103°18′E
Dickinson, N.D., U.S. (dĭk´ĭn-sŭn)	64	46°52′N	102°49′W
Dickinson, Tx., U.S. (dĭk´ĭn-sŭn)	81a	29°28′N	95°02′W
Dickinson Bayou, Tx., U.S.	81a	29°26′N	95°08′W
Dickson, Tn., U.S. (dĭk´sŭn)	82	36°03′N	87°24′W
Dickson City, Pa., U.S.	67	41°25′N	75°40′W
Didcot, Eng., U.K. (dĭd´cŏt)	114b	51°35′N	1°15′W
Didiéni, Mali	188	13°53′N	8°06′W
Die, Fr. (dē)	127	44°45′N	5°22′E
Diefenbaker, res., Can.	52	51°20′N	108°10′W
Diego de Ocampo, Pico, mtn., Dom. Rep. (pē´-kô-dyē´gô-dē-ô-kä´m-pô)	93	19°40′N	70°45′W
Diego Ramírez, Islas, is., Chile (dē ä´gô rä-mē´räz)	102	56°15′s	70°15′W
Diéma, Mali	188	14°32′N	9°12′W
Dien Bien Phu, Viet.	160	21°38′N	102°49′E
Dieppe, Can. (dē-ěp´)	60	46°06′N	64°45′W
Dieppe, Fr.	117	49°54′N	1°05′E
Dierks, Ar., U.S. (dērks)	79	34°06′N	94°02′W
Diessen, Ger. (dēs´sěn)	115d	47°57′N	11°06′E
Diest, Bel.	115a	50°59′N	5°05′E
Digby, Ma., U.S. (dĭg´bĭ)	51	44°37′N	65°46′W
Dighton, Ma., U.S. (dī-tŭn)	68b	41°49′N	71°05′W
Digne, Fr. (dēn´y´)	127	44°07′N	6°16′E
Digoin, Fr. (dē-gwän´)	126	46°28′N	4°06′E
Digul, r., Indon.	169	7°00′s	140°27′E
Dijohan Point, c., Phil. (dē-kô-än)	169a	16°24′N	122°25′E
Dijon, Fr. (dē-zhôn´)	110	47°21′N	5°02′E
Dikson, Russia (dĭk´sŏn)	134	73°30′N	80°35′E
Dikwa, Nig. (dē´kwä)	185	12°06′N	13°53′E
Dili, E. Timor (dĭl´ē)	169	8°35′s	125°35′E
Di Linosa Island, i., Italy (dē-lē-nô´sä)	118	12°43′E	
Dilizhan, Arm.	137	40°45′N	45°00′E
Dillingham, Ak., U.S. (dĭl´ěng-hăm)	64a	59°10′N	158°38′W
Dillon, Mt., U.S. (dĭl´ŭn)	73	45°13′N	112°40′W
Dillon, S.C., U.S.	83	34°24′N	79°28′W
Dillon Reservoir, res., Oh., U.S.	66	40°05′N	82°05′W
Dilolo, D.R.C. (dē-lō´lō)	186	10°19′s	22°23′E
Dimashq see Damascus, Syria	154	33°31′N	36°18′E
Dimbokro, C. Iv.	188	6°39′N	4°42′W
Dimitrovo see Pernik, Blg.	119	42°36′N	23°04′E
Dimlang, mtn., Nig.	189	8°24′N	11°47′E
Dimona, Isr.	153a	31°03′N	35°01′E
Dinagat Island, i., Phil.	169	10°15′N	126°15′E
Dinājpur, Bngl.	158	25°38′N	87°39′E
Dinan, Fr. (dē-näv´)	126	48°27′N	2°03′W
Dinant, Bel. (dē-näv´)	121	50°17′N	4°50′E
Dinara, mts., Serb. (dē´nä-rä)	119	43°50′N	16°15′E
Dinard, Fr.	126	48°38′N	2°04′W
Dindigul, India	159	10°25′N	78°03′E
Dingalan Bay, b., Phil. (dĭŋ-gä´län)	169a	15°19′N	121°33′E
Dingle, Ire. (dĭng´´l)	120	52°10′N	10°13′W
Dingle Bay, b., Ire.	117	52°02′N	10°15′W
Dingo, Austl. (dĭn´gō)	175	23°45′s	149°26′E
Dinguiraye, Gui.	188	11°18′N	10°43′W
Dingwall, Scot., U.K. (dĭng´wŏl)	120	57°37′N	4°23′W
Dingxian, China (dĭŋ shyěn)	164	38°30′N	115°00′E
Dingxing, China (dĭŋ-shyĭŋ)	164	39°18′N	115°50′E
Dingyuan, China (dĭŋ-yŭän)	162	32°32′N	117°40′E
Dingzi Wan, b., China	162	36°33′N	121°06′E
Dinosaur National Monument, rec., Co., U.S. (dī´nō-sôr)	73	40°45′N	109°17′W
Dinslaken, Ger. (dēns´lä-kěn)	127c	51°33′N	6°44′E
Dinteloord, Neth.	115a	51°38′N	4°21′E
Dinuba, Ca., U.S. (dĭ-nū´bá)	76	36°33′N	119°29′W
Dios, Cayo de, i., Cuba (kä´yō-dē-dē-ōs´)	92	22°05′N	83°05′W
Diourbel, Sen. (dē-ōōr-běl´)	184	14°40′N	16°15′W
Diphu Pass, p., Asia (dĭ-pōō)	160	28°15′N	96°45′E
Diquis, r., C.R. (dē-kēs´)	91	8°59′N	83°24′W
Dire Dawa, Eth.	185	9°40′N	41°47′E
Diriamba, Nic. (dēr-yäm´bä)	90	11°52′N	86°15′W
Dirk Hartog, i., Austl.	174	26°25′s	113°15′E
Dirksland, Neth.	115a	51°45′N	4°04′E
Dirranbandi, Austl. (dĭ-rä-băn´dě)	175	28°24′s	148°29′E
Dirty Devil, r., Ut., U.S. (dûr´tĭ děv´´l)	77	38°20′N	110°30′W
Disappointment, l., Austl.	174	23°20′s	123°00′E
Disappointment, Cape, c., Wa., U.S. (dĭs´ä-point´ment)	74c	46°16′N	124°11′W
Discovery, S. Afr. (dĭs-kŭv´ěr-ĭ)	187b	26°10′s	27°53′E
Discovery, is., Can. (dĭs-kŭv´ěr-ē)	74a	48°25′s	123°13′W
Disko, i., Grnld. (dĭs´kō)	49	70°00′N	54°00′W
Disna, Bela. (dēs´ná)	136	55°34′N	28°15′E
Dispur, India	158	26°00′N	91°50′E
Disraëli, Can. (dĭs-rā´lĭ)	59	45°53′N	71°23′W
District of Columbia, state, U.S.	65	38°50′N	77°00′W
Distrito Federal, state, Braz. (dēs-trē´tô-fě-dē-rä´l)	101	15°49′s	47°39′W
Distrito Federal, state, Mex.	88	19°14′N	99°08′W
Disūq, Egypt (dē-sōōk´)	192b	31°07′N	30°41′E
Diu, India (dē´ōō)	155	20°48′N	70°58′E
Divilacan Bay, b., Phil. (dē-vē-lä´kän)	169a	17°26′N	122°25′E
Divinópolis, Braz. (dē-vē-nô´pō-lěs)	101	20°10′s	44°53′W
Divo, C. Iv.	188	5°50′N	5°22′W
Dixon, Il., U.S. (dĭks´ŭn)	71	41°50′N	89°30′W
Dixon Entrance, strt., N.A.	52	54°25′N	132°00′W
Diyarbakir, Tur. (dē-yär-běk´ĭr)	154	38°00′N	40°10′E
Dja, r., Afr.	185	2°30′N	14°00′E
Djambala, Congo	190	2°33′s	14°45′E
Djanet, Alg.	184	24°29′N	9°26′E
Djebobo, mtn., Ghana	188	8°20′N	0°37′E
Djedi, Oued, r., Alg.	118	34°18′N	4°39′E
Djember, Chad	189	10°25′N	17°50′E
Djerba, Île de, i., Tun.	118	33°53′N	11°26′E
Djerid, Chott, l., Tun. (jěr´īd)	184	33°15′N	8°29′E
Djibasso, Burkina	188	13°07′N	4°10′W
Djibo, Burkina	188	14°06′N	1°38′W
Djibouti, Dji. (jē-bōō-tē´)	192a	11°34′N	43°00′E
Djibouti, nation, Afr.	192a	11°35′N	48°08′E
Djokoumatombi, Congo	190	0°47′N	15°22′E
Djokupunda, D.R.C.	186	5°27′s	20°58′E
Djoua, r., Afr.	190	1°25′s	13°40′E
Djursholm, Swe. (djoors´hŏlm)	122	59°26′N	18°01′E
Dmitriyev-L'govskiy, Russia (d´mě´trī-yěf l´gôf´skī)	132	52°07′N	35°05′E
Dmitrov, Russia (d´mě´trôf)	132	56°21′N	37°32′E
Dmitrovsk, Russia (d´mě´trôfsk)	132	52°30′N	35°10′E
Dmytrivka, Ukr.	133	47°57′N	38°56′E
Dnepropetrovsk see Dnipropetrovs'k, Ukr.	134	48°15′N	34°40′E
Dnieper (Dnipro), r., Eur.	137	46°45′N	33°40′E
Dniester, r., Eur.	137	48°21′N	28°10′E
Dniprodzerzhyns'k, Ukr.	137	48°32′N	34°38′E
Dniprodzerzhyns'ke vodoskhovyshche, res., Ukr.	134	49°00′N	34°10′E
Dnipropetrovs'k, Ukr.	134	48°15′N	34°08′E
Dnipropetrovs'k, prov., Ukr.	133	48°30′N	34°10′E
Dniprovs'kyi lyman, b., Ukr.	133	46°33′N	31°45′E
Dnistrovs'kyi lyman, l., Ukr.	133	46°15′N	29°50′E
Dno, Russia (d´nô´)	132	57°49′N	29°59′E
Do, Lac, l., Mali	188	15°50′N	2°20′W
Doba, Chad	189	8°39′N	16°51′E
Dobbs Ferry, N.Y., U.S. (dŏbz´fē´rě)	68a	41°01′N	73°53′W
Dobbyn, Austl. (dŏb´ĭn)	174	19°45′s	140°02′E
Dobele, Lat. (dô´bě-lě)	123	56°37′N	23°18′E
Doberai, Jazirah, pen., Indon.	169	1°25′s	133°15′E
Dobo, Indon.	169	6°00′s	134°18′E
Doboj, Bos. (dô´boi)	131	44°42′N	18°04′E
Dobřich, Blg.	119	43°33′N	27°52′E
Dobryanka, Russia (dŏb-ryän´kà)	142a	58°27′N	56°26′E
Dobšina, Slvk. (dŏp´shě-nä)	125	48°48′N	20°25′E
Doce, r., Braz. (dô´sä)	101	19°01′s	42°14′W
Doce, Canal Numero, can., Arg.	99c	36°47′s	59°00′W
Doce Leguas, Cayos de las, is., Cuba	92	20°55′N	79°05′W
Doctor Arroyo, Mex. (dŏk-tōr´ är-rō´yō)	88	23°41′N	100°10′W
Doddington, Eng., U.K. (dŏd´dĭng-tŏn)	114b	51°17′N	0°47′E
Dodecanese see Dodekanisoy, is., Grc.	131	38°00′N	26°10′E
Dodekanisoy (Dodecanese), is., Grc.	131	38°00′N	26°10′E
Dodge City, Ks., U.S. (dŏj)	64	37°44′N	100°01′W
Dodgeville, Wi., U.S. (dŏj´vil)	71	42°58′N	90°07′W
Dodoma, Tan. (dō´dō-mä)	186	6°11′s	35°45′E
Dog, l., Can. (dŏg)	58	48°42′s	89°24′W
Dogger Bank, bk., (dŏg´gēr)	121	55°07′N	2°25′E
Dogubayazit, Tur.	137	39°35′N	44°00′E
Doha see Ad Dawhah, Qatar	154	25°02′N	51°28′E
Dohad, India	158	22°52′N	74°18′E
Dokshytsy, Bela. (dŏk-shětsě)	132	54°53′N	27°49′E
Dolbeau, Can.	51	48°52′N	72°16′W
Dole, Fr. (dōl)	117	47°07′N	5°28′E
Dolgaya, Kosa, c., Russia (kô´sä dôl-gä´yä)	133	46°42′N	37°42′E
Dolgeville, N.Y., U.S.	67	43°10′N	74°45′W
Dolgiy, i., Russia	136	69°20′N	59°20′E
Dolgoprudnyy, Russia	142b	55°57′N	37°33′E
Dolinsk, Russia (dá-lēnsk´)	141	47°29′N	142°31′E
Dollar Harbor, b., Bah.	92	25°30′N	79°15′W
Dolomite, Al., U.S. (dŏl´ō-mīt)	68h	33°28′N	86°57′W
Dolomites, mts., Italy	130	46°16′N	11°43′E
Dolores, Arg. (dō-lō´rěs)	102	36°20′s	57°42′W
Dolores, Col.	100a	3°33′s	74°54′W
Dolores, Ur.	99c	33°32′s	58°15′W
Dolores, Tx., U.S. (dō-lō´rěs)	80	27°42′N	99°47′W
Dolores, r., Co., U.S.	77	38°35′N	108°50′W
Dolores Hidalgo, Mex. (dō-lō´rěs-ē-däl´gō)	88	21°09′N	100°56′W
Dolphin and Union Strait, strt., Can. (dŏl´fĭn ūn´yŭn)	52	69°22′N	117°10′W
Dolyna, Ukr.	125	48°57′N	24°01′E
Domažlice, Czech Rep. (dō´mäzh-lě-tsě)	124	49°27′N	12°55′E
Dombasle-sur-Meurthe, Fr. (dôn-bäl´)	127	48°38′N	6°18′E
Dombóvár, Hung. (dôm´bô-vär)	125	46°22′N	18°08′E
Domeyko, Cordillera, mts., Chile (kôr-dēl-yě´rä-dō-mā´kô)	100	20°50′s	69°02′W
Dominica, nation, N.A. (dô-mĭ-nē´kà)	87	15°30′N	60°45′W
Dominica Channel, strt., N.A.	91b	15°00′N	61°30′W
Dominican Republic, nation, N.A. (dô-mĭn´ĭ-kăn)	87	19°00′N	70°45′W
Dominion, Can. (dō-mĭn´yŭn)	61	46°13′N	60°01′W

ăt; finăl; rāte; senăte; ärm; àsk; sofà; fāre; ch-choose; dh-as th in other; bē; ĕvent; bĕt; recēnt; crātēr; g-gō; gh-guttural g; bĭt; ī-short neutral; rīde; κ-guttural k as ch in German ich;

PLACE (Pronunciation)	PAGE	LAT.	LONG.
Domiongo, D.R.C.	190	4°37′s	21°15′e
Domodedovo, Russia (dŏ-mŏ-dyĕ′do-vô)	142b	55°27′N	37°45′e
Dom Silvério, Braz. (don-sĕl-vĕ′ryŏ)	99a	20°09′s	42°57′w
Don, r., Russia	134	49°50′N	41°30′e
Don, r., Eng., U.K.	114a	53°39′N	0°58′w
Don, r., Scot., U.K.	120	57°19′N	2°39′w
Donaldson, Mi., U.S.	75k	46°19′N	84°22′w
Donaldsonville, La., U.S. (dŏn′ăld-sŭn-vĭl)	81	30°05′N	90°58′w
Donalsonville, Ga., U.S.	82	31°02′N	84°50′w
Donawitz, Aus. (dŏ′nȧ-vĭts)	124	47°23′N	15°05′e
Don Benito, Spain (dōn′bȧ-nē′tō)	128	38°55′N	5°52′w
Doncaster, Austl. (dŏn′kȧs-tẽr)	173a	37°47′s	145°08′e
Doncaster, Eng., U.K. (dŏŋ′kȧs-tẽr)	120	53°32′N	1°07′w
Doncaster, co., Eng., U.K.	114a	53°35′N	1°10′w
Dondo, Ang. (dŏn′dō)	186	9°38′s	14°25′e
Dondo, Moz.	186	19°33′s	34°47′e
Dondra Head, c., Sri L.	159	5°52′N	80°52′e
Donegal, Ire. (dŏn-ê-gôl′)	120	54°44′N	8°05′w
Donegal Bay, b., Ire. (dŏn-ê-gôl′)	116	54°35′N	8°36′w
Donets Coal Basin, reg., Ukr. (dō-nyĕts′)	133	48°15′N	38°50′e
Donets′k, Ukr.	134	48°00′N	37°35′e
Donets′k, prov., Ukr.	133	47°55′N	37°40′e
Dong, r., China (dŏŋ)	161	24°13′N	115°08′e
Dongara, Austl. (dŏn-gä′rä)	174	29°15′s	115°00′e
Dongba, China (dŏŋ-bä)	162	31°40′N	119°02′e
Dong′e, China (dŏŋ-ŭ)	162	36°21′N	116°14′e
Dong′ezhen, China	164	36°11′N	116°16′e
Dongfang, China (dŏŋ-făŋ)	165	19°08′N	108°42′e
Donggala, Indon. (dŏn-gä′lä)	168	0°45′s	119°32′e
Dongguan, China (dŏŋ-gŭän)	163a	23°03′N	113°46′e
Dongguang, China (dŏŋ-gŭäŋ)	162	37°54′N	116°33′e
Donghai, China (dŏŋ-hī)	164	34°35′N	119°05′e
Dong Hoi, Viet. (dŏng-hô-ē′)	168	17°25′N	106°42′e
Dongila, Eth.	185	11°17′N	37°00′e
Dongming, China (dŏŋ-mĭŋ)	162	35°16′N	115°06′e
Dongo, Ang. (dŏŋ′gō)	186	14°45′s	15°30′e
Dongon Point, c., Phil. (dŏng-ŏn′)	169a	12°43′N	120°35′e
Dongou, Congo (dŏŋ-gōō′)	185	2°02′N	18°04′e
Dongping, China (dŏŋ-pĭŋ)	164	35°50′N	116°24′e
Dongping Hu, l., China (dŏŋ-pĭŋ hōō)	162	36°06′N	116°24′e
Dongshan, China (dŏŋ-shän)	162	31°05′N	120°24′e
Dongtai, China	162	32°51′N	120°20′e
Dongting Hu, l., China (dŏŋ-tĭŋ hōō)	161	29°10′N	112°30′e
Dongxiang, China (dŏŋ-shyän)	165	28°18′N	116°38′e
Doniphan, Mo., U.S. (dŏn′ĭ-făn)	79	36°37′N	90°50′w
Donji Vakuf, Bos. (dŏn′yĭ väk′ôf)	131	44°08′N	17°25′e
Don Martin, Presa de, res., Mex. (prĕ′sä-dĕ-dôn-mär-tē′n)	80	27°35′N	100°38′w
Donnacona, Can.	59	46°40′N	71°46′w
Donnemarie-en-Montois, Fr. (dôn-mä-rē′ĕn-môn-twä′)	127b	48°29′N	3°09′e
Donner und Blitzen, r., Or., U.S. (dŏn′ẽr ŏnt′blĭ′tsĕn)	72	42°45′N	118°57′w
Donnybrook, S. Afr. (dŏ-nĭ-brŏk′)	187c	29°56′s	29°54′e
Donora, Pa., U.S. (dŏ-nō′rȧ)	69e	40°10′N	79°51′w
Donostia-San Sebastián, Spain	110	43°19′N	1°59′w
Donoússa, i., Grc.	131	37°09′N	25°53′e
Doolow, Som.	192a	4°10′N	42°05′e
Doonerak, Mount, mtn., Ak., U.S. (dōō′nĕ-răk)	63	68°00′N	150°34′w
Doorn, Neth.	115a	52°02′N	5°21′e
Door Peninsula, pen., Wi., U.S. (dōr)	71	44°40′N	87°36′w
Dora Baltea, r., Italy (dō′rä bäl′tä-ä)	130	45°40′N	7°34′e
Doraville, Ga., U.S. (dō′rȧ-vĭl)	68c	33°54′N	84°17′w
Dorchester, Eng., U.K. (dôr′chĕs-tẽr)	120	50°45′N	2°34′w
Dordogne, r., Fr. (dôr-dôn′yĕ)	112	44°53′N	0°16′e
Dordrecht, Neth. (dôr′drĕkt)	121	51°48′N	4°39′e
Dordrecht, S. Afr. (dôr′drĕkt)	187c	31°24′s	27°06′e
Doré Lake, l., Can.	56	54°31′N	107°06′w
Dorgali, Italy (dôr′gä-lē)	130	40°18′N	9°37′e
Dörgön Nuur, l., Mong.	160	47°47′N	94°01′e
Dorion-Vaudreuil, Can. (dôr-yō)	62a	45°23′N	74°01′w
Dorking, Eng., U.K. (dôr′kĭŋg)	114b	51°12′N	0°20′w
Dormont, Pa., U.S. (dôr′mŏnt)	69e	40°24′N	80°02′w
Dornbirn, Aus. (dôrn′bĕrn)	124	47°24′N	9°45′e
Dornoch, Scot., U.K. (dôr′nŏk)	116	57°55′N	4°01′w
Dornoch Firth, b., Scot., U.K. (dôr′nŏk fẽrth)	120	57°55′N	3°55′w
Dorogobuzh, Russia (dôrŏgô′-bōō′zh)	132	54°57′N	33°18′e
Dorohoi, Rom. (dô-rô-hoi′)	125	47°57′N	26°28′e
Dorre Island, i., Austl. (dôr)	174	25°19′s	113°10′e
Dorsten, Ger.	127c	51°40′N	6°58′e
Dortmund, Ger. (dôrt′mŏnt)	117	51°31′N	7°28′e
Dortmund-Ems-Kanal, can., Ger. (dôrt′mōōnd-ĕms′kä-näl′)	127c	51°50′N	7°25′e
Dörtyol, Tur. (dûrt′yôl)	119	36°50′N	36°20′e
Dorval, Can. (dôr-vál′)	62a	45°26′N	73°44′w
Dos Bahías, Cabo, c., Arg. (kä′bŏ-dŏs-bä-ē′äs)	102	44°55′s	65°35′w
Dos Caminos, Ven. (dŏs-kä-mĕ′nŏs)	101b	9°38′N	67°17′w
Dosewallips, r., Wa., U.S. (dŏ′sĕ-wäl′lĭps)	74a	47°45′N	123°04′w
Dos Hermanas, Spain (dōsĕr-mä′näs)	128	37°17′N	5°56′w
Dosso, Niger (dŏs-ō′)	184	13°03′N	3°12′e
Dothan, Al., U.S. (dō′thăn)	65	31°13′N	85°23′w
Douai, Fr. (dōō-ā′)	117	50°23′N	3°04′e
Douala, Cam. (dōō-ä′lä)	184	4°03′N	9°42′e
Douarnenez, Fr. (dōō-àr nĕ-nĕs′)	126	48°06′N	4°18′w
Double Bayou, Tx., U.S. (dŭb′′l bī′yōō)	81a	29°40′N	94°38′w
Doubs, r., Eur.	127	46°15′N	5°50′e
Douentza, Mali	188	15°00′N	2°57′w
Douglas, I. of Man (dŭg′lȧs)	120	54°10′N	4°24′w
Douglas, Ak., U.S. (dŭg′lȧs)	63	58°18′N	134°35′w
Douglas, Az., U.S.	64	31°20′N	109°30′w
Douglas, Ga., U.S.	83	31°30′N	82°53′w
Douglas, Wy., U.S. (dŭg′lȧs)	73	42°45′N	105°21′w
Douglas, r., Eng., U.K. (dŭg′lȧs)	114a	53°38′N	2°48′w
Douglas Channel, strt., Can.	54	53°30′N	129°12′w
Douglas Lake, res., Tn., U.S. (dŭg′lȧs)	82	36°00′N	83°35′w
Douglas Lake Indian Reserve, I.R., Can.	55	50°10′N	120°49′w
Douglasville, Ga., U.S. (dŭg′lȧs-vĭl)	82	33°45′N	84°47′w
Dourada, Serra, mts., Braz. (sĕ′r-rä-dôō-rä′dä)	101	15°11′s	49°57′w
Dourdan, Fr. (dōōr-dän′)	127b	48°32′N	2°01′e
Douro, r., Port. (dō′ō-rô)	128	41°03′N	8°12′w
Dove, r., Eng., U.K. (dŭv)	114a	52°53′N	1°47′w
Dover, S. Afr.	192c	27°05′s	27°44′e
Dover, De., U.S. (dō vẽr)	65	39°10′N	75°30′w
Dover, N.H., U.S.	67	43°15′N	71°00′w
Dover, N.J., U.S.	68a	40°53′N	74°33′w
Dover, Oh., U.S.	66	40°35′N	81°30′w
Dover, Strait of, strt., Eur.	112	50°50′N	1°15′w
Dover-Foxcroft, Me., U.S. (dō′vẽr fŏks′krôft)	60	45°10′N	69°15′w
Dovre Fjell, mts., Nor. (dŏv′rĕ fyĕl′)	112	62°03′N	8°36′e
Dow, Il., U.S. (dou)	75e	39°01′N	90°20′w
Dowagiac, Mi., U.S. (dô-wŏ′jăk)	66	42°00′N	86°05′w
Downers Grove, Il., U.S. (dou′nẽrz grōv)	69a	41°48′N	88°00′w
Downey, Ca., U.S. (dou′nĭ)	75a	33°56′N	118°08′w
Downieville, Ca., U.S. (dou′nĭ-nĭl)	76	39°35′N	120°48′w
Downs, Ks., U.S. (dounz)	78	39°29′N	98°32′w
Doylestown, Oh., U.S.	69d	40°58′N	81°43′w
Doylestown, Pa., U.S. (doilz′toun)	68a	40°19′N	75°08′w
Drâa, Cap, c., Mor. (drä)	184	28°39′N	12°15′w
Drâa, Oued, r., Afr.	184	28°00′N	9°31′w
Drabiv, Ukr.	133	49°57′N	32°14′e
Drac, r., Fr. (dräk)	127	44°50′N	5°47′e
Dracut, Ma., U.S. (drä′kŭt)	61a	42°40′N	71°19′w
Draganovo, Blg. (drä-gä-nō′vô)	131	43°13′N	25°45′e
Drăgăşani, Rom. (drä-gä-shän′ĭ)	131	44°39′N	24°18′e
Draguignan, Fr. (drä-gēn-yän′)	127	43°33′N	6°28′e
Drahichyn, Bela.	125	52°10′N	25°11′e
Drakensberg, mts., Afr. (drä′kĕnz-bẽrgh)	186	29°15′s	29°07′e
Drake Passage, strt., (dräk päs′ĭj)	97	57°00′s	65°00′w
Dráma, Grc. (drä′mä)	119	41°09′N	24°10′e
Drammen, Nor. (dräm′ĕn)	116	59°45′N	10°15′e
Drau (Drava), r., Eur. (drou)	124	46°44′N	13°45′e
Drava, r., Eur. (drä′vä)	112	45°45′N	17°30′e
Dravograd, Slvn. (drä′vô-gräd′)	130	46°37′N	15°01′e
Drawsko Pomorskie, Pol. (dräv′skô pō-môr′skyĕ)	124	53°31′N	15°50′e
Drayton Harbor, b., Wa., U.S. (drä′tŭn)	74d	48°58′N	122°40′w
Drayton Plains, Mi., U.S.	69b	42°41′N	83°23′w
Drayton Valley, Can.	55	53°13′N	114°59′w
Drensteinfurt, Ger. (drĕn′shtĭn-fōōrt)	127c	51°47′N	7°44′e
Dresden, Ger. (dräs′dĕn)	110	51°05′N	13°45′e
Dreux, Fr. (drû)	126	48°44′N	1°24′e
Driefontein, S. Afr.	192c	25°53′s	29°10′e
Drin, r., Alb. (drēn)	131	42°13′N	20°13′e
Drina, r., Serb. (drē′nä)	119	44°09′N	19°17′e
Drinit, Pellg i, b., Alb.	131	41°42′N	19°17′e
Dr. Ir. W. J. van Blommestein Meer, res., Sur.	101	4°45′N	55°05′w
Drissa, r., Eur.	125	55°44′N	28°58′e
Driver, Va., U.S.	68g	36°50′N	76°30′w
Drøbak, Nor. (drû′bäk)	122	59°40′N	10°35′e
Drobeta-Turnu Severin, Rom.	119	43°54′N	24°49′e
Drogheda, Ire. (drŏ′hĕ-dȧ)	116	53°43′N	6°15′w
Drohobych, Ukr.	125	49°21′N	23°31′e
Drôme, r., Fr. (drōm)	126	44°42′N	4°53′e
Dronfield, Eng., U.K. (drŏn′fĕld)	114a	53°18′N	1°28′w
Drummond, i., Mi., U.S. (drŭm′ŭnd)	66	46°00′N	83°50′w
Drummondville, Can. (drŭm′ŭnd-vĭl)	51	45°53′N	72°33′w
Drumright, Ok., U.S. (drŭm′rĭt)	79	35°59′N	96°37′w
Drunen, Neth.	115a	51°41′N	5°10′e
Drut′, r., Bela. (drōōt)	132	53°40′N	29°45′e
Druya, Bela. (drŏ′yä)	132	55°45′N	27°26′e
Drweca, r., Pol. (dŭr-vān′tsä)	125	53°00′N	19°13′e
Dryden, Can. (drī-dĕn)	51	49°47′N	92°50′w
Drysdale, Austl.	173a	38°11′s	144°34′e
Dry Tortugas, is., Fl., U.S. (tôr-tōō′gäz)	83a	24°37′N	82°45′w
Dry Tortugas National Park, rec., Fl., U.S.	83a	24°42′N	83°02′w
Dschang, Cam. (dshäng)	184	5°34′N	10°09′e
Duabo, Lib.	188	5°40′N	8°05′w
Duagh, Can.	62g	53°43′N	113°24′w
Duarte, Pico, mtn., Dom. Rep. (dū′ärtĕh pĕcô)	87	19°00′N	71°00′w
Duas Barras, Braz. (dōō′äs-bä′r-räs)	99a	22°03′s	42°30′w
Dubai see Dubayy, U.A.E.	154	25°18′N	55°26′e
Dubăsari, Mol.	133	47°16′N	29°11′e
Dubawnt, l., Can. (dōō-bônt′)	52	63°27′N	103°00′w
Dubawnt, r., Can.	52	61°30′N	103°49′w
Dubayy, U.A.E.	154	25°18′N	55°26′e
Dubbo, Austl. (dŭb′ō)	175	32°20′s	148°42′e
Dubie, D.R.C.	191	8°33′s	28°32′e
Dublin, Ire.	110	53°20′N	6°15′w
Dublin, Ca., U.S. (dŭb′lĭn)	74b	37°42′N	121°56′w
Dublin, Ga., U.S.	83	32°33′N	82°55′w
Dublin, Tx., U.S.	80	32°05′N	98°20′w
Dubna, Russia	132	56°44′N	37°10′e
Dubno, Ukr. (dōō′b-nô)	125	50°24′N	25°44′e
Du Bois, Pa., U.S. (dŏ-bois′)	67	41°10′N	78°45′w
Dubovka, Russia (dò-bôf′kä)	137	49°00′N	44°50′e
Dubrovka, Russia (dōō-brôf′kä)	142c	59°51′N	30°56′e
Dubrovnik, Cro. (dò′brôv-nĕk) (rä-gōō′sä)	110	42°40′N	18°10′e
Dubrowna, Bela.	132	54°39′N	30°54′e
Dubuque, Ia., U.S. (dò-būk′)	65	42°30′N	90°43′w
Duchesne, Ut., U.S. (dò-shān′)	77	40°12′N	110°23′w
Duchesne, r., Ut., U.S.	77	40°20′N	110°50′w
Duchess, Austl. (dŭch′ĕs)	174	21°30′s	139°55′e
Ducie Island, i., Pit. (dū-sē′)	2	25°30′s	126°20′w
Duck, r., Tn., U.S.	82	35°55′N	87°40′w
Duckabush, r., Wa., U.S. (dŭk′ȧ-bòsh)	74a	47°41′N	123°09′w
Duck Lake, Can.	56	52°47′N	106°13′w
Duck Mountain, mtn., Can.	57	51°35′N	101°00′w
Ducktown, Tn., U.S. (dŭk′toun)	82	35°03′N	84°20′w
Duck Valley Indian Reservation, I.R., Id., U.S.	72	42°02′N	115°49′w
Duckwater Peak, mtn., Nv., U.S. (dŭk-wô-tẽr)	76	39°00′N	115°31′w
Duda, r., Col. (dōō′dä)	100a	3°25′N	74°23′w
Dudinka, Russia (dōō-dĭn′kä)	134	69°15′N	85°42′e
Dudley, Eng., U.K. (dŭd′lĭ)	117	52°28′N	2°07′e
Duero, r., Eur.	112	41°30′N	4°30′w
Dufourspitze, mtn., Eur.	124	45°55′N	7°52′e
Dugger, In., U.S. (dŭg′ẽr)	66	39°00′N	87°10′w
Dugi Otok, i., Serb. (dōō′gē o′tŏk)	130	44°03′N	14°40′e
Duisburg, Ger. (dōō′ĭs-bòrgh)	117	51°26′N	6°46′e
Dukhān, Qatar	157	25°25′N	50°48′e
Dukhovshchina, Russia (dōō-kôfsh-′chēnä)	132	55°13′N	32°26′e
Dukinfield, Eng., U.K. (dŭk′ĭn-fēld)	114a	53°28′N	2°05′w
Dukla Pass, p., Eur. (dò′klä)	117	49°25′N	21°44′e
Dulce, Golfo, b., C.R. (gōl′fô dōōl′sä)	87	8°25′N	83°13′w
Dülken, Ger. (dül′kĕn)	127c	51°15′N	6°21′e
Dülmen, Ger. (dül′mĕn)	127c	51°50′N	7°17′e
Duluth, Mn., U.S. (dò-lōōth′)	65	46°50′N	92°07′w
Dumai, Indon.	153b	1°39′N	101°30′e
Dumali Point, c., Phil. (dōō-mä′lē)	169a	13°07′N	121°42′e
Dumas, Tx., U.S.	78	35°52′N	101°58′w
Dumbarton, Scot., U.K. (dŭm′bär-tŭn)	120	56°00′N	4°35′w
Dum-Dum, India	158a	22°37′N	88°25′e
Dumfries, Scot., U.K. (dŭm-frēs′)	120	55°05′N	3°40′w
Dumjor, India	158a	22°37′N	88°14′e
Dumont, N.J., U.S. (dōō′mônt)	68a	40°56′N	74°00′w
Dumyât, Egypt	185	31°22′N	31°50′e
Dunaföldvár, Hung. (dó′nô-fûld′vär)	125	46°48′N	18°55′e
Dunăvtsi, Ukr.	133	48°52′N	26°51′e
Dunajec, r., Pol.	125	49°52′N	20°53′e
Dunaújváros, Hung.	125	46°57′N	18°55′e
Dunay, Russia (dōō′nī)	142c	59°59′N	30°57′e
Dunbar, W.V., U.S.	66	38°20′N	81°45′w
Duncan, Can. (dŭŋ′kȧn)	50	48°47′N	123°42′w
Duncan, Ok., U.S.	79	34°29′N	97°56′w
Duncan, r., Can.	55	50°30′N	116°45′w
Duncan Dam, dam, Can.	55	50°15′N	116°55′w
Duncan Lake, l., Can.	55	50°20′N	117°00′w
Duncansby Head, c., Scot., U.K. (dŭn′kănz-bī)	120	58°40′N	3°01′w
Duncanville, Tx., U.S. (dŭn′kȧn-vĭl)	75c	32°39′N	96°55′w
Dundalk, Ire. (dŭn′kôk)	116	54°00′N	6°18′w
Dundalk, Md., U.S.	68e	39°16′N	76°31′w
Dundalk Bay, b., Ire. (dŭn′dôk)	120	53°55′N	6°15′w
Dundas, Can. (dŭn′dȧs)	53	43°16′N	79°58′w
Dundas, I., Austl. (dŭn′dȧs)	174	32°15′s	122°00′e
Dundas Island, i., Can.	54	54°33′N	130°55′w
Dundas Strait, strt., Austl.	174	10°35′s	131°15′e
Dundedin, Fl., U.S. (dŭn-ē′dĭn)	83a	28°00′N	82°43′w
Dundee, S. Afr.	187c	28°14′s	30°16′e
Dundee, Scot., U.K.	110	56°30′N	2°55′w
Dundee, Il., U.S.	69a	42°06′N	88°17′w
Dundrum Bay, b., N. Ire., U.K. (dŭn-drŭm′)	120	54°13′N	5°47′w
Dunedin, N.Z.	175a	45°48′s	170°32′e
Dunellen, N.J., U.S. (dŭn-ĕl′l′n)	68a	40°36′N	74°28′w
Dunfermline, Scot., U.K. (dŭn-fẽrm′lĭn)	120	56°05′N	3°30′w
Dungarvan, Ire. (dŭn-gär′văn)	120	52°06′N	7°50′w
Dungeness, Wa., U.S. (dŭnj-nĕs′)	74a	48°09′N	123°07′w
Dungeness, r., Wa., U.S.	74a	48°03′N	123°10′w
Dungeness Spit, Wa., U.S.	74a	48°11′N	123°03′w
Dunhua, China (dŭn-hwä)	161	43°18′N	128°10′e
Dunkerque, Fr. (dŭn-kĕrk′)	117	51°02′N	2°37′e
Dunkirk, In., U.S. (dŭn′kûrk)	66	40°20′N	85°25′w
Dunkwa, Ghana	188	5°22′N	1°12′w
Dun Laoghaire, Ire. (dŭn-lā′rĕ)	116	53°16′N	6°09′w
Dunlap, Tn., U.S. (dŭn′lăp)	70	43°51′N	95°33′w
Dunlap, Tn., U.S.	82	35°23′N	85°23′w
Dunmore, Pa., U.S. (dŭn′mōr)	67	41°25′N	75°30′w
Dunn, N.C., U.S. (dŭn)	83	35°18′N	78°37′w
Dunnellon, Fl., U.S. (dŭn-ĕl′ŏn)	83	29°02′N	82°28′w
Dunnville, Can. (dŭn′vĭl)	59	42°55′N	79°40′w
Dunqulah, Sudan	185	19°21′N	30°19′e
Dunsmuir, Ca., U.S. (dŭnz′mūr)	72	41°08′N	122°17′w
Dunwoody, Ga., U.S. (dŭn-wòd′ĭ)	68c	33°57′N	84°20′w
Duolun, China (dwŏ-lōōn)	161	42°12′N	116°15′e
Du Page, East Branch, r., Il., U.S.	69a	41°41′N	88°11′w
Du Page, West Branch, r., Il., U.S.	69a	41°42′N	88°09′w
Dupax, Phil. (dōō′päks)	169a	16°16′N	121°06′e
Dupo, Il., U.S. (dū′pō)	75e	38°31′N	90°12′w
Duque de Caxias, Braz. (dōō′kĕ-dĕ-kä′shyäs)	99a	22°46′s	43°18′w
Duquesne, Pa., U.S. (dū-kān′)	69e	40°20′N	79°51′w
Du Quoin, Il., U.S. (dò-kwoin′)	79	38°01′N	89°14′w
Durance, r., Fr. (dü-räns′)	117	43°46′N	5°52′e
Durand, Mi., U.S. (dū-rănd′)	66	42°50′N	84°00′w
Durand, Wi., U.S.	71	44°37′N	91°58′w

244 Durango – El Cajon PRONOUNCING INDEX

PLACE (Pronunciation)	PAGE	LAT.	LONG.
Durango, Mex. (dōō-rä′n-gō)	86	24°02′N	104°42′W
Durango, Co., U.S. (dō-räŋ′gō)	77	37°15′N	107°55′W
Durango, state, Mex.	86	25°00′N	106°00′W
Durant, Ms., U.S. (dů-rănt′)	82	33°05′N	89°50′W
Durant, Ok., U.S.	79	33°59′N	96°23′W
Duratón, r., Spain (dōō-rä-tōn′)	128	41°30′N	3°55′W
Durazno, Ur. (dōō-räz′nō)	102	33°21′S	56°31′W
Durazno, dept., Ur.	99c	33°00′S	56°35′W
Durban, S. Afr. (dûr′băn)	186	29°48′S	31°00′E
Durbanville, S. Afr. (dûr-băn′vĭl)	186a	33°50′S	18°39′E
Durbe, Lat. (dōōr′bĕ)	123	56°36′N	21°24′E
Đurđevac, Cro.	119	46°03′N	17°03′E
Düren, Ger. (dü′rĕn)	127c	50°48′N	6°30′E
Durham, Eng., U.K. (dûr′ăm)	120	54°47′N	1°46′W
Durham, N.C., U.S.	65	36°00′N	78°55′W
Durham Downs, Austl.	176	27°30′S	141°55′E
Durrës, Alb. (dŏr′ĕs)	110	41°19′N	19°27′E
Duryea, Pa., U.S. (dōōr-yā′)	67	41°20′N	75°50′W
Dushan, China	162	31°38′N	116°16′E
Dushan, China (dōō-shän)	165	25°50′N	107°42′E
Dushanbe, Taj.	139	38°30′N	68°45′E
Düsseldorf, Ger. (düs′ĕl-dôrf)	117	51°14′N	6°47′E
Dussen, Neth.	115a	51°43′N	4°58′E
Dutalan Ula, mts., Mong.	164	49°25′N	112°40′E
Dutch Harbor, Ak., U.S. (dŭch här′bēr)	64a	53°58′N	166°30′W
Duvall, Wa., U.S. (dōō′vâl)	74a	47°44′N	121°59′W
Duwamish, r., Wa., U.S. (dōō-wăm′ĭsh)	74a	47°24′N	122°18′W
Duyun, China (dōō-yŏn)	160	26°18′N	107°40′E
Dvinskaya Guba, b., Russia	136	65°10′N	38°40′E
Dwārka, India	158	22°18′N	68°59′E
Dwight, Il., U.S. (dwīt)	66	41°00′N	88°20′W
Dworshak Res, Id., U.S.	72	46°45′N	115°50′W
Dyat′kovo, Russia (dyät′kō-vō)	132	53°36′N	34°19′E
Dyer, In., U.S. (dī′ēr)	69a	41°30′N	87°31′W
Dyersburg, Tn., U.S. (dī′ērz-bûrg)	82	36°02′N	89°23′W
Dyersville, Ia., U.S. (dī′ērz-vĭl)	71	42°28′N	91°09′W
Dyes Inlet, Wa., U.S. (dīz)	74a	47°37′N	122°45′W
Dykhtau, Gora, mtn., Russia	138	43°03′N	43°08′E
Dyment, Can. (dī′mĕnt)	57	49°37′N	92°19′W
Dzamïn Üüd, Mong.	161	44°38′N	111°32′E
Dzaoudzi, May. (dzou′dzĭ)	187	12°44′S	45°15′E
Dzavhan, r., Mong.	160	48°19′N	94°08′E
Dzerzhinsk, Russia	136	56°20′N	43°50′E
Dzerzhyns′k, Ukr.	133	48°26′N	37°50′E
Dzhalal-Abad, Kyrg. (jȧ-läl′ȧ-bät′)	139	40°56′N	73°00′E
Dzhambul *see* Zhambyl, Kaz.	139	42°51′N	71°29′E
Dzhankoi, Ukr.	137	45°43′N	34°22′E
Dzhizak, Uzb. (dzhė′zäk)	139	40°13′N	67°58′E
Dzhugdzhur Khrebet, mts., Russia (jŏg-jōōr′)	135	56°15′N	137°00′E
Działoszyce, Pol. (jyä-wō-shē′tsĕ)	125	50°21′N	20°22′E
Dzibalchén, Mex. (zē-bäl-chĕ′n)	90a	19°25′N	89°39′W
Dzidzantún, Mex. (zēd-zän-tōō′n)	90a	21°18′N	89°00′W
Dzierżoniów, Pol. (dzyĕr-zhôn′yŭf)	124	50°44′N	16°38′E
Dzilam González, Mex. (zē-lä′m-gôn-zä′lĕz)	90a	21°21′N	88°53′W
Dzitás, Mex. (zē-tá′s)	90a	20°47′N	88°32′W
Dzungaria, reg., China (dzŏŋ-gä′rĭ-ȧ)	160	44°39′N	86°13′E
Dzungarian Gate, p., Asia	160	45°00′N	88°00′E
Dzyarzhynsk, Bela.	132	53°41′N	27°14′E

E

PLACE (Pronunciation)	PAGE	LAT.	LONG.
Eagle, W.V., U.S.	66	38°10′N	81°20′W
Eagle, r., Co., U.S.	77	39°32′N	106°28′W
Eaglecliff, Wa., U.S. (ē′gl-klĭf)	74c	46°10′N	123°13′W
Eagle Creek, r., In., U.S.	69g	39°54′N	86°17′W
Eagle Grove, Ia., U.S.	71	42°39′N	93°55′W
Eagle Lake, Me., U.S.	60	47°03′N	68°38′W
Eagle Lake, Tx., U.S.	81	29°37′N	96°20′W
Eagle Lake, l., Ca., U.S.	72	40°45′N	120°52′W
Eagle Mountain, Ca., U.S.	76	33°49′N	115°27′W
Eagle Mountain L, Tx., U.S.	75c	32°56′N	97°27′W
Eagle Pass, Tx., U.S.	64	28°49′N	100°30′W
Eagle Pk, Ca., U.S.	72	41°18′N	120°11′W
Ealing, Eng., U.K. (ē′lĭng)	114b	51°29′N	0°19′W
Earle, Ar., U.S. (ûrl)	79	35°14′N	90°28′W
Earlington, Ky., U.S. (ûr′lĭng-tŭn)	82	37°15′N	87°31′W
Easley, S.C., U.S. (ēz′lĭ)	83	34°48′N	82°37′W
East, Mount, mtn., Pan.	86a	9°09′N	79°46′W
East Alton, Il., U.S. (ôl′tŭn)	75e	38°53′N	90°08′W
East Angus, Can. (ăŋ′gŭs)	59	45°35′N	71°40′W
East Aurora, N.Y., U.S. (ô-rō′rȧ)	69c	42°46′N	78°38′W
East Bay, b., Tx., U.S.	81a	29°30′N	94°41′W
East Bernstadt, Ky., U.S. (bûrn′stȧt)	82	37°09′N	84°08′W
Eastbourne, Eng., U.K. (ēst′bôrn)	121	50°48′N	0°16′E
East Caicos, i., T./C. Is. (kī′kōs)	93	21°40′N	71°35′W
East Cape, c., N.Z.	175a	37°37′S	178°33′E
East Cape *see* Dezhnëva, Mys, c., Russia	152	68°00′N	172°00′W
East Carondelet, Il., U.S. (kȧ-rŏn′dĕ-lĕt)	75e	38°33′N	90°14′W
East Cherokee Indian Reservation, I.R., N.C., U.S.	82	35°33′N	83°12′W
East Chicago, In., U.S. (shĭ-kô′gō)	69a	41°39′N	87°29′W
East China Sea, sea, Asia	161	30°28′N	125°52′E
East Cleveland, Oh., U.S. (klēv′lănd)	69d	41°33′N	81°35′W
East Cote Blanche Bay, b., La., U.S. (kōt blänsh′)	81	29°30′N	92°07′W
East Des Moines, r., Ia., U.S. (dē moin′)	71	42°57′N	94°17′W
East Detroit, Mi., U.S. (dė-troit′)	69b	42°28′N	82°57′W
Easter Island *see* Pascua, Isla de, i., Chile	195	26°50′S	109°00′W
Eastern Ghāts, mts., India	155	13°50′N	78°45′E
Eastern Turkestan, hist. reg., China (tŏr-kĕ-stän′)(tûr-kĕ-stăn′)	160	39°40′N	78°20′E
East Grand Forks, Mn., U.S. (grănd fôrks)	70	47°56′N	97°02′W
East Greenwich, R.I., U.S. (grĭn′ĭj)	68b	41°40′N	71°27′W
Easthampton, Ma., U.S. (ēst-hămp′tŭn)	67	42°15′N	72°45′W
East Hartford, Ct., U.S. (härt′fērd)	67	41°45′N	72°35′W
East Helena, Mt., U.S. (hĕ-hē′nȧ)	73	46°31′N	111°50′W
East Ilsley, Eng., U.K. (ĭl′slē)	114b	51°30′N	1°18′W
East Jordan, Mi., U.S. (jôr′dăn)	66	45°05′N	85°05′W
East Kansas City, Mo., U.S. (kăn′zȧs)	75f	39°09′N	94°30′W
Eastland, Tx., U.S. (ēst′lănd)	80	32°24′N	98°47′W
East Lansing, Mi., U.S. (lăn′sĭng)	66	42°45′N	84°30′W
Eastlawn, Mi., U.S.	69b	42°15′N	83°35′W
East Leavenworth, Mo., U.S. (lĕv′ĕn-wûrth)	75f	39°18′N	94°50′W
East Liverpool, Oh., U.S. (lĭv′ēr-pōōl)	66	40°40′N	80°35′W
East London, S. Afr. (lŭn′dŭn)	186	33°02′S	27°54′E
East Los Angeles, Ca., U.S. (lōs äŋ′hå-lās)	75a	34°01′N	118°09′W
Eastmain, r., Can. (ēst′mān)	53	52°12′N	73°19′W
Eastman, Ga., U.S. (ēst′măn)	82	32°10′N	83°11′W
East Millstone, N.J., U.S. (mĭl′stōn)	68a	40°30′N	74°35′W
East Moline, Il., U.S. (mō-lēn′)	71	41°31′N	90°28′W
East Nishnabotna, r., Ia., U.S. (nĭsh-nȧ-bŏt′nȧ)	70	40°53′N	95°23′W
Easton, Md., U.S. (ēs′tŭn)	67	38°45′N	76°05′W
Easton, Pa., U.S.	67	40°45′N	75°15′W
Easton L, Ct., U.S.	68a	41°18′N	73°17′W
East Orange, N.J., U.S. (ŏr′ĕnj)	68a	40°46′N	74°12′W
East Pakistan *see* Bangladesh, nation, Asia	155	24°15′N	90°00′E
East Palo Alto, Ca., U.S.	74b	37°27′N	122°07′W
East Peoria, Il., U.S. (pē-ō′rĭ-ȧ)	66	40°40′N	89°30′W
East Pittsburgh, Pa., U.S. (pĭts′bûrg)	69e	40°24′N	79°50′W
East Point, Ga., U.S.	68c	33°41′N	84°27′W
Eastport, Me., U.S. (ēst′pōrt)	60	44°53′N	67°01′W
East Providence, R.I., U.S. (prŏv′ĭ-dĕns)	68b	41°49′N	71°22′W
East Retford, Eng., U.K. (rĕt′fērd)	114a	53°19′N	0°56′W
East Riding of Yorkshire, co., Eng., U.K.	114a	53°45′N	0°40′W
East Rochester, N.Y., U.S. (rŏch′ĕs-tēr)	67	43°10′N	77°30′W
East Saint Louis, Il., U.S.	65	38°38′N	90°10′W
East Siberian Sea, sea, Russia (sī-bĭr′y′n)	135	73°00′N	153°28′E
Eastsound, Wa., U.S. (ēst-sound)	74d	48°42′N	122°42′W
East Stroudsburg, Pa., U.S. (stroudz′bûrg)	67	41°00′N	75°10′W
East Syracuse, N.Y., U.S. (sĭr′ȧ-kūs)	67	43°05′N	76°00′W
East Tavaputs Plateau, plat., Ut., U.S. (tă-vă′-pŭts)	77	39°25′N	109°45′W
East Tawas, Mi., U.S. (tô′wăs)	66	44°15′N	83°30′W
East Timor, nation, Asia	169	9°00′S	125°30′E
East Walker, r., U.S. (wôk′ēr)	76	38°36′N	119°02′W
Eaton, Co., U.S. (ē′tŭn)	78	40°31′N	104°42′W
Eaton, Oh., U.S.	66	39°45′N	84°40′W
Eaton Estates, Oh., U.S.	69d	41°19′N	82°01′W
Eaton Rapids, Mi., U.S. (răp′ĭdz)	66	42°30′N	84°40′W
Eatonton, Ga., U.S. (ētŭn-tŭn)	82	33°20′N	83°24′W
Eatontown, N.J., U.S. (ē′tŭn-toun)	68a	40°18′N	74°04′W
Eau Claire, Wi., U.S. (ō klâr′)	65	44°47′N	91°32′W
Ebeltoft, Den. (ĕ′bĕl-tŭft)	122	56°11′N	10°39′E
Ebensburg, Pa., U.S.	67	40°29′N	78°44′W
Ebersberg, Ger. (ĕ′bĕrs-bĕrgh)	115d	48°05′N	11°58′E
Ebingen, Ger. (ā′bĭng-ĕn)	124	48°13′N	9°04′E
Eboli, Italy (ĕb′ō-lē)	130	40°38′N	15°04′E
Ebolowa, Cam.	184	2°54′N	11°09′E
Ebreichsdorf, Aus.	115e	47°58′N	16°24′E
Ebrié, Lagune, b., C. Iv.	188	5°20′N	4°50′W
Ebro, r., Spain (ā′brō)	112	42°00′N	2°00′W
Eccles, Eng., U.K. (ĕk′′lz)	114a	53°29′N	2°20′W
Eccles, W.V., U.S.	66	37°45′N	81°10′W
Eccleshall, Eng., U.K.	114a	52°51′N	2°15′W
Eceabat, Tur.	131	40°10′N	26°21′E
Echague, Phil. (ā-chä′gwā)	169a	16°43′N	121°40′E
Echandi, Cerro, mtn., N.A. (sĕ′r-rō-ĕ-chä′nd)	91	9°05′N	82°51′W
Ech Cheliff, Alg.	184	36°14′N	1°32′E
Echimamish, r., Can.	57	54°15′N	97°30′W
Echmiadzin, Arm.	138	40°10′N	44°18′E
Echo Bay, Can. (ĕk′ō)	75k	46°29′N	84°04′W
Echoing, r., Can. (ĕk′ō-ĭng)	57	55°15′N	91°30′W
Echternach, Lux. (ĕk′tēr-näk)	127	49°48′N	6°25′E
Echuca, Austl. (ė-chō′kȧ)	175	36°10′S	144°47′E
Écija, Spain (ā′thė-hä)	118	37°20′N	5°07′W
Eckernförde, Ger.	124	54°27′N	9°51′E
Eclipse, Va., U.S. (ė-klĭps′)	68g	36°55′N	76°29′W
Ecorse, Mi., U.S. (ė-kôrs′)	69b	42°15′N	83°09′W
Ecuador, nation, S.A. (ĕk′wȧ-dôr)	100	0°00′N	78°30′W
Ed, Erit.	185	13°57′N	41°37′E
Eddyville, Ky., U.S. (ĕd′ĭ-vĭl)	82	37°03′N	88°03′W
Ede, Nig.	189	7°44′N	4°27′E
Edéa, Cam. (ĕ-dā′ä)	184	3°48′N	10°08′E
Eden, Tx., U.S.	80	31°13′N	99°51′W
Eden, Ut., U.S.	75b	41°18′N	111°49′W
Eden, r., Eng., U.K. (ē′dĕn)	120	54°40′N	2°35′W
Edenbridge, Eng., U.K. (ē′dĕn-brĭj)	114b	51°11′N	0°05′E
Edenham, Eng., U.K. (ē′d′n-ăm)	114a	52°46′N	0°25′W
Eden Prairie, Mn., U.S. (prâr′ĭ)	75g	44°51′N	93°29′W
Edenton, N.C., U.S. (ē′dĕn-tŭn)	83	36°02′N	76°37′W
Edenton, Oh., U.S.	69f	39°14′N	84°02′W
Edenvale, S. Afr. (ēd′ĕn-vāl)	187b	26°09′S	28°10′E
Edenville, S. Afr. (ē′d′n-vĭl)	192c	27°33′S	27°42′E
Eder, r., Ger. (ā′dēr)	124	51°05′N	8°52′E
Édessa, Grc.	119	40°48′N	22°04′E
Edgefield, S.C., U.S. (ĕj′fēld)	83	33°52′N	81°55′W
Edgeley, N.D., U.S. (ĕj′lĭ)	70	46°24′N	98°43′W
Edgemont, S.D., U.S. (ĕj′mŏnt)	70	43°19′N	103°50′W
Edgerton, Wi., U.S. (ĕj′ēr-tŭn)	71	42°49′N	89°06′W
Edgewater, Al., U.S. (ĕj-wô-tēr)	68h	33°31′N	86°52′W
Edgewater, Md., U.S.	68e	38°58′N	76°35′W
Edgewood, Can. (ĕj′wŏd)	55	49°47′N	118°08′W
Edina, Mn., U.S. (ė-dī′nȧ)	75g	44°55′N	93°20′W
Edina, Mo., U.S.	79	40°10′N	92°11′W
Edinburg, In., U.S. (ĕd′′n-bûrg)	66	39°20′N	85°55′W
Edinburg, Tx., U.S.	80	26°18′N	98°08′W
Edinburgh, Scot., U.K. (ĕd′′n-bŭr-ō)	110	55°57′N	3°10′W
Edirne, Tur.	131	41°41′N	26°35′E
Edisto, r., S.C., U.S. (ĕd′ĭs-tō)	83	33°10′N	80°50′W
Edisto, North Fork, r., S.C., U.S.	83	33°42′N	81°24′W
Edisto, South Fork, r., S.C., U.S.	83	33°43′N	81°35′W
Edisto Island, S.C., U.S.	83	32°32′N	80°20′W
Edmond, Ok., U.S. (ĕd′mŭnd)	79	35°39′N	97°29′W
Edmonds, Wa., U.S. (ĕd′mŭndz)	74a	47°49′N	122°23′W
Edmonton, Can.	50	53°33′N	113°28′W
Edmundston, Can. (ĕd′mŭn-stŭn)	51	47°22′N	68°20′W
Edna, Tx., U.S. (ĕd′nȧ)	81	28°59′N	96°39′W
Edremit, Tur. (ĕd-rĕ-mēt′)	119	39°35′N	27°00′E
Edremit Körfezi, b., Tur.	131	39°28′N	26°35′E
Edson, Can. (ĕd′sŭn)	50	53°35′N	116°26′W
Edward, i., Can. (ĕd′wērd)	58	48°21′N	88°29′W
Edward, l., Afr.	186	0°25′S	29°40′E
Edwardsville, Il., U.S. (ĕd′wērdz-vĭl)	75e	38°49′N	89°58′W
Edwardsville, In., U.S.	69h	38°17′N	85°53′W
Edwardsville, Ks., U.S.	75f	39°04′N	94°49′W
Eel, r., Ca., U.S. (ēl)	72	40°39′N	124°15′W
Eel, r., In., U.S.	66	40°50′N	85°55′W
Efate, i., Vanuatu (å-fä′tā)	175	18°02′S	168°29′E
Effigy Mounds National Monument, rec., Ia., U.S. (ĕf′ĭ-jŭ mounds)	71	43°04′N	91°15′W
Effingham, Il., U.S. (ĕf′ĭng-hăm)	66	39°05′N	88°30′W
Ega, r., Spain (ā′gä)	128	42°40′N	2°20′W
Egadi, Isole, is., Italy (ė′sō-lĕ-ĕ′gä-dē)	118	38°01′N	12°00′E
Egegik, Ak., U.S. (ĕg′ĕ-jĭt)	63	58°10′N	157°22′W
Eger, Hung. (ĕ gĕr)	125	47°53′N	20°24′E
Egersund, Nor. (ĕ′ghĕr-sŏn′)	116	58°29′N	6°01′E
Egg Harbor, N.J., U.S. (ĕg här′bēr)	67	39°30′N	74°35′W
Egham, Eng., U.K. (ĕg′ŭm)	114b	51°24′N	0°33′W
Egiyn, r., Mong.	160	49°41′N	100°40′E
Egmont, Cape, c., N.Z. (ĕg′mŏnt)	175a	39°18′S	173°49′E
Egypt, nation, Afr. (ē′jĭpt)	185	26°58′N	27°01′E
Eha-Amufu, Nig.	189	6°40′N	7°46′E
Eibar, Spain (ā′ė-bär)	128	43°12′N	2°20′W
Eichstätt, Ger. (īk′shtät)	124	48°54′N	11°14′E
Eichwalde, Ger. (īκ′väl-dĕ)	115b	52°22′N	13°37′E
Eidfjord, Nor. (ĕĭd′fyōr)	122	60°28′N	7°04′E
Eidsvoll, Nor. (īdhs′vôl)	116	60°19′N	11°15′E
Eifel, mts., Ger. (ī′fĕl)	124	50°08′N	6°30′E
Eighty Mile Beach, cst., Austl.	174	19°00′S	121°00′E
Eilenburg, Ger. (ī′lĕn-bŏrgh)	124	51°27′N	12°38′E
Einbeck, Ger. (īn′bĕk)	124	51°49′N	9°52′E
Eindhoven, Neth. (īnd′hō-vĕn)	121	51°29′N	5°20′E
Eisenach, Ger. (ī′zĕn-äκ)	117	50°58′N	10°18′E
Eisenhüttenstadt, Ger.	124	52°08′N	14°40′E
Eivissa, Spain	129	38°55′N	1°24′E
Eivissa, i., Spain	112	38°55′N	1°24′E
Ejea de los Caballeros, Spain	128	42°07′N	1°05′W
Ejura, Ghana	188	7°23′N	1°22′W
Ejutla de Crespo, Mex. (å-hōt′lä dā krās′pō)	89	16°34′N	96°44′W
Ekanga, D.R.C.	190	2°23′S	23°14′E
Ekenäs, Fin. (ĕ′kĕ-nâs)	123	59°59′N	23°25′E
Ekeren, Bel.	115a	51°17′N	4°27′E
Ekoli, D.R.C.	190	0°23′S	24°16′E
El Aaiún, W. Sah.	184	26°45′N	13°15′W
El Affroun, Alg. (ĕl ȧf-froun′)	129	36°28′N	2°38′E
Elands, r., S. Afr. (ėlånds)	187c	31°48′S	26°09′E
Elands, r., S. Afr.	192c	25°11′S	28°52′E
El Arahal, Spain (ĕl ä-rä-äl′)	128	37°17′N	5°32′W
El Arba, Alg.	129	36°35′N	3°10′E
Elat, Isr.	154	29°34′N	34°57′E
Elazığ, Tur. (ĕl-ä′zėz)	154	38°40′N	39°00′E
Elba, Al., U.S. (ĕl′bȧ)	82	31°25′N	86°01′W
Elba, Isola d′, i., Italy (ė-sō lä-d-ĕl′bȧ)	118	42°42′N	10°25′E
El Banco, Col. (ĕl băn′cō)	100	8°58′N	74°01′W
Elbansan, Alb. (ĕl-bä-sän′)	119	41°08′N	20°05′E
Elbe (Labe), r., Eur. (ĕl′bĕ)(lä′bĕ)	112	52°30′N	11°30′E
Elbert, Mount, mtn., Co., U.S. (ĕl′bērt)	64	39°05′N	106°25′W
Elberton, Ga., U.S. (ĕl′bēr-tŭn)	83	34°05′N	82°53′W
Elbeuf, Fr. (ĕl-bûf′)	117	49°16′N	0°59′E
El Beyadh, Alg.	118	33°42′N	1°06′E
Elbistan, Tur. (ēl-bē-stän′)	119	38°20′N	37°10′E
Elblag, Pol. (ĕl′bläŋg)	116	54°11′N	19°25′E
El Bonillo, Spain (ĕl bō-nēl′yō)	128	38°56′N	2°31′W
El Boulaïda, Alg.	184	36°33′N	2°45′E
Elbow, r., Can. (ĕl′bō)	62e	51°03′N	114°24′W
Elbow Cay, i., Bah.	92	26°25′N	76°55′W
Elbow Lake, Mn., U.S.	70	46°00′N	95°59′W
El′brus, Gora, mtn., Russia (ĕl′brŏs′)	134	43°20′N	42°25′E
Elbrus, Mount *see* El′brus, Gora, mtn., Russia	134	43°20′N	42°25′E
Elburz Mountains, mts., Iran (ĕl′bŏrz′)	154	36°30′N	51°00′E
El Cajon, Col. (ĕl-kä-kô′n)	100a	4°50′N	76°35′W

ăt; finăl; rāte; senåte; ärm; åsk; sofȧ; fâre; ch-choose; dh-as th in other; bē; ėvent; bĕt; recĕnt; cratẽr; g-gō; gh-guttural g; bĭt; ĭ-short neutral; rīde; κ-guttural k as ch in German ich;

PRONOUNCING INDEX

El Cajon – Erlanger 245

PLACE (Pronunciation)	PAGE	LAT.	LONG.
El Cajon, Ca., U.S.	76a	32°48'N	116°58'W
El Cambur, Ven. (käm-bōōr')	101b	10°24'N	68°06'W
El Campo, Tx., U.S. (kăm'pō)	81	29°13'N	96°17'W
El Carmen, Chile (kä'r-mĕn)	99b	34°14'S	71°23'W
El Carmen, Col. (kä'r-mĕn)	100	9°54'N	75°12'W
El Casco, Ca., U.S. (kăs'kô)	75a	33°59'N	117°08'W
El Centro, Ca., U.S. (sĕn'trô)	76	32°47'N	115°33'W
El Cerrito, Ca., U.S. (sĕr-rē'tō)	74b	37°55'N	122°19'W
El Cuyo, Mex.	90a	21°30'N	87°42'W
Elda, Spain (ĕl'dä)	129	38°28'N	0°44'W
El Djelfa, Alg.	184	34°40'N	3°17'E
El Djouf, des., Afr. (ĕl djōōf)	184	21°45'N	7°05'W
Eldon, Ia., U.S. (ĕl-dŭn)	71	40°55'N	92°15'W
Eldon, Mo., U.S.	79	38°21'N	92°36'W
Eldora, Ia., U.S. (ĕl-dō'rá)	71	42°21'N	93°08'W
El Dorado, Ar., U.S. (ĕl dô-rä'dō)	65	33°13'N	92°39'W
Eldorado, Il., U.S.	66	37°50'N	88°30'W
El Dorado, Ks., U.S.	79	37°49'N	96°51'W
Eldorado Springs, Mo., U.S. (sprĭngz)	79	37°51'N	94°02'W
Eldoret, Kenya (ĕl-dô-rĕt')	191	0°31'N	35°17'E
El Ebano, Mex. (ā-bä'nō)	88	22°13'N	98°26'W
Electra, Tx., U.S. (ê-lĕk'trá)	78	34°02'N	98°54'W
Electric Peak, mtn., Mt., U.S. (ê-lĕk'trĭk)	73	45°03'N	110°52'W
Elek, r.,	137	51°20'N	53°10'E
Elektrogorsk, Russia (ĕl-yĕk'trô-gôrsk)	142b	55°53'N	38°48'E
Elektrostal', Russia (ĕl-yĕk'trō-stál)	142b	55°47'N	38°27'E
Elektrougli, Russia	142b	55°43'N	38°13'E
Elephant Butte Reservoir, res., N.M., U.S. (ĕl'ê-fănt būt)	64	33°25'N	107°10'W
El Escorial, Spain (ĕl-ĕs-kô-ryä'l)	129a	40°38'N	4°08'W
El Espino, Nic. (ĕl-ĕs-pē'nō)	90	13°26'N	86°48'W
Eleuthera, i., Bah. (ê-lū'thēr-á)	87	25°05'N	76°10'W
Eleuthera Point, c., Bah.	92	24°35'N	76°05'W
Eleven Point, r., Mo., U.S. (ê-lĕv'ĕn)	79	36°53'N	91°39'W
Elgin, Scot., U.K.	120	57°40'N	3°30'W
Elgin, Il., U.S. (ĕl'jĭn)	69a	42°03'N	88°16'W
Elgin, Ne., U.S.	70	41°58'N	98°04'W
Elgin, Or., U.S.	72	45°34'N	117°58'W
Elgin, Tx., U.S.	81	30°21'N	97°22'W
Elgin, Wa., U.S.	74a	47°23'N	122°42'W
Elgon, Mount, mtn., Afr. (ĕl'gŏn)	185	1°00'N	34°25'E
El Grara, Alg.	118	32°50'N	4°26'E
El Grullo, Mex. (grōōl-yô)	88	19°46'N	104°10'W
El Guapo, Ven. (gwà'pô)	101b	10°07'N	66°00'W
El Hank, reg., Afr.	184	23°44'N	6°45'W
El Hatillo, Ven. (ä-tē'l-yô)	101b	10°08'N	65°13'W
Elie, Can. (ē'lē)	62f	49°55'N	97°45'W
Elila, r., D.R.C. (ê-lē'lá)	186	3°30'S	28°00'E
Elisa, i., Wa., U.S. (ê-lī'sà)	74d	48°43'N	122°37'W
Elisabethville see Lubumbashi, D.R.C.	186	11°40'S	27°28'E
Elisenvaara, Russia (ā-lē'sĕn-vä'rà)	123	61°25'N	29°46'E
Elizabeth, La., U.S. (ê-lĭz'á-bĕth)	81	30°50'N	92°47'W
Elizabeth, N.J., U.S.	68a	40°40'N	74°13'W
Elizabeth, Pa., U.S.	69e	40°16'N	79°53'W
Elizabeth City, N.C., U.S.	83	36°15'N	76°15'W
Elizabethton, Tn., U.S. (ê-lĭz-à-bĕth'tŭn)	83	36°19'N	82°12'W
Elizabethtown, Ky., U.S. (ê-lĭz'á-bĕth-toun)	66	37°40'N	85°55'W
El Jadida, Mor.	184	33°14'N	8°34'W
Elk, Pol.	116	53°53'N	22°23'E
Elk, r., Can.	55	50°00'N	115°00'W
Elk, r., Tn., U.S.	82	35°05'N	86°36'W
Elk, r., W.V., U.S.	66	38°30'N	81°05'W
El Kairouan, Tun. (kĕr-ō-än)	184	35°46'N	10°04'E
Elk City, Ok., U.S. (ĕlk)	78	35°23'N	99°23'W
El Kef, Tun. (xĕf')	118	36°14'N	8°42'E
Elkhart, In., U.S. (ĕlk'härt)	66	41°40'N	86°00'W
Elkhart, Ks., U.S.	78	37°00'N	101°54'W
Elkhart, Tx., U.S.	81	31°38'N	95°35'W
Elkhorn, Wi., U.S. (ĕlk'hôrn)	71	42°39'N	88°32'W
Elkhorn, r., Ne., U.S.	70	42°06'N	97°46'W
Elkin, N.C., U.S. (ĕl'kĭn)	83	36°15'N	80°50'W
Elk Island, i., Can.	57	50°45'N	96°32'W
Elk Island National Park, rec., Can. (ĕlk ī'lănd)	52	53°37'N	112°45'W
Elko, Nv., U.S. (ĕl'kō)	64	40°51'N	115°46'W
Elk Point, S.D., U.S.	70	42°41'N	96°41'W
Elk Rapids, Mi., U.S. (răp'ĭdz)	66	44°55'N	85°25'W
Elk River, Id., U.S. (rĭv'ēr)	72	46°47'N	116°11'W
Elk River, Mn., U.S.	71	45°17'N	93°33'W
Elkton, Ky., U.S. (ĕlk'tŭn)	82	36°47'N	87°08'W
Elkton, Md., U.S.	67	39°35'N	75°50'W
Elkton, S.D., U.S.	70	44°15'N	96°28'W
Elland, Eng., U.K. (ĕl'ănd)	114a	53°41'N	1°50'W
Ellen, Mount, mtn., Ut., U.S. (ĕl'ĕn)	77	38°05'N	110°50'W
Ellendale, N.D., U.S. (ĕl'ĕn-dāl)	70	46°01'N	98°33'W
Ellensburg, Wa., U.S. (ĕl'ĕnz-bûrg)	72	47°00'N	120°31'W
Ellenville, N.Y., U.S. (ĕl'ĕn-vĭl)	67	41°40'N	74°25'W
Ellerslie, Can. (ĕl'ērz-lē)	62g	53°25'N	113°30'W
Ellesmere, Eng., U.K. (ĕlz'mēr)	114a	52°55'N	2°54'W
Ellesmere Island, i., Can.	49	81°00'N	80°00'W
Ellesmere Port, Eng., U.K.	114a	53°17'N	2°54'W
Ellice Islands see Tuvalu, nation, Oc.	3	5°20'S	174°00'E
Ellicott City, is., Md., U.S. (ĕl'ĭ-kŏt sĭ'tē)	68e	39°16'N	76°48'W
Ellicott Creek, r., N.Y., U.S.	69c	43°00'N	78°46'W
Elliot, S. Afr.	187c	31°19'S	27°52'E
Elliot, Wa., U.S. (ĕl'ĭ-ŭt)	74a	47°28'N	122°08'W
Elliotdale, S. Afr. (ĕl-ĭ-ŏt'dāl)	187c	31°58'S	28°42'E
Elliot Lake, Can.	58	46°23'N	82°39'W
Ellis, Ks., U.S. (ĕl'ĭs)	78	38°56'N	99°34'W
Ellisville, Mo., U.S.	75e	38°35'N	90°35'W
Ellisville, Ms., U.S. (ĕl'ĭs-vĭl)	82	31°37'N	89°10'W
Ellsworth, Ks., U.S. (ĕlz'wûrth)	78	38°43'N	98°14'W
Ellsworth, Me., U.S.	60	44°33'N	68°26'W
Ellsworth Mountains, mts., Ant.	178	77°00'S	90°00'W
Ellwangen, Ger. (ĕl'väŋ-gĕn)	124	48°47'N	10°08'E
Elm, Ger. (ĕlm)	115c	53°31'N	9°13'E
Elm, r., S.D., U.S.	70	45°47'N	98°28'W
Elm, r., W.V., U.S.	66	38°30'N	81°05'W
Elma, Wa., U.S. (ĕl'mà)	72	47°02'N	123°20'W
El Mahdia, Tun. (mä-dēä)(mä'dē-á)	118	35°30'N	11°09'E
Elmendorf, Tx., U.S. (ĕl'mĕn-dôrf)	75d	29°16'N	98°20'W
El Menia, Alg.	184	30°39'N	2°52'E
Elm Fork, Tx., U.S. (ĕlm fôrk)	75c	32°55'N	96°56'W
Elmhurst, Il., U.S. (ĕlm'hûrst)	69a	41°54'N	87°56'W
El Miliyya, Alg. (mē'ä)	184	36°30'N	6°16'E
Elmira, N.Y., U.S. (ĕl-mī'rá)	67	42°05'N	76°50'W
Elmira Heights, N.Y., U.S.	67	42°10'N	76°50'W
El Modena, Ca., U.S. (mô-dē'nô)	75a	33°47'N	117°48'W
El Mohammadia, Alg.	129	35°35'N	0°05'E
El Monte, Ca., U.S. (mŏn'tâ)	75a	34°04'N	118°02'W
El Morro National Monument, rec., N.M., U.S.	77	35°05'N	108°20'W
Elmshorn, Ger. (ĕlms'hôrn)	124	53°45'N	9°39'E
Elmwood Place, Oh., U.S. (ĕlm'wŏŏd plās)	69f	39°11'N	84°30'W
Elokomin, r., Wa., U.S. (ê-lō'kô-mĭn)	74c	46°16'N	123°16'W
El Oro, Mex. (ô-rô)	88	19°49'N	100°04'W
El Pao, Ven. (ĕl pà'ō)	100	8°08'N	62°37'W
El Paraíso, Hond. (pä-rä-ē'sō)	90	13°55'N	86°35'W
El Pardo, Spain (pä'r-dô)	129a	40°31'N	3°47'W
El Paso, Tx., U.S. (pas'ō)	64	31°47'N	106°27'W
El Pilar, Ven. (pē-lä'r)	101b	9°56'N	64°48'W
El Porvenir, Pan. (pôr-vä-nēr')	91	9°34'N	78°55'W
El Puerto de Santa María, Spain	128	36°36'N	6°18'W
El Qala, Alg.	118	36°52'N	8°23'E
El Qoll, Alg.	184	37°02'N	6°29'E
El Real, Pan. (rā-äl)	91	8°07'N	77°43'W
El Reno, Ok., U.S. (rē'nō)	79	35°31'N	97°57'W
Elroy, Wi., U.S. (ĕl'roi)	71	43°44'N	90°17'W
Elsa, Can.	63	63°55'N	135°25'W
Elsah, Il., U.S. (ĕl'zá)	75e	38°57'N	90°22'W
El Salto, Mex. (säl'tō)	88	23°48'N	105°22'W
El Salvador, nation, N.A.	86	14°00'N	89°30'W
El Sauce, Nic. (ĕl-sá'ô-sē)	90	13°00'N	86°40'W
Elsberry, Mo., U.S. (ĕlz'bĕr-ĭ)	79	39°09'N	90°44'W
Elsdorf, Ger. (ĕls'dōrf)	127c	50°56'N	6°35'E
El Segundo, Ca., U.S. (sĕgŭn'dō)	75a	33°55'N	118°24'W
Elsinore, Ca., U.S. (ĕl'sĭ-nôr)	75a	33°40'N	117°19'W
Elsinore Lake, l., Ca., U.S.	75a	33°38'N	117°21'W
Elstorf, Ger. (ĕls'tôrf)	115c	53°25'N	9°48'E
Eltham, Austl. (ĕl'thăm)	173a	37°43'S	145°08'E
El Tigre, Ven. (tē'grĕ)	100	8°49'N	64°15'W
El'ton, l., Russia	137	49°10'N	47°00'E
El Toro, Ca., U.S. (tō'rō)	75a	33°37'N	117°42'W
El Triunfo, El Sal.	90	13°17'N	88°32'W
El Triunfo, Hond. (ĕl-trē-ōō'n-fô)	90	13°06'N	87°00'W
Elūru, India	155	16°44'N	80°09'E
El Vado Res, N.M., U.S.	77	36°37'N	106°30'W
Elvas, Port. (ĕl'väzh)	118	38°53'N	7°11'W
Elverum, Nor. (ĕl'vĕ-rŏŏm)	122	60°53'N	11°33'E
El Viejo, Nic. (ĕl-vyĕ'kŏ)	90	12°10'N	87°10'W
El Viejo, vol., Nic.	90	12°44'N	87°03'W
Elvins, Mo., U.S. (ĕl'vĭnz)	79	37°49'N	90°31'W
El Wad, Alg.	184	33°23'N	6°49'E
El Wak, Kenya (wäk')	185	3°00'N	41°00'E
Elwell, Lake, res., Mt., U.S.	73	48°22'N	111°17'W
Elwood, Il., U.S. (ĕl'wŏŏd)	69a	41°24'N	88°07'W
Elwood, In., U.S.	66	40°15'N	85°50'W
Elx, Spain	129	38°15'N	0°42'W
Ely, Eng., U.K. (ē'lĭ)	121	52°25'N	0°17'E
Ely, Mn., U.S.	71	47°54'N	91°53'W
Ely, Nv., U.S.	64	39°16'N	114°53'W
Elyria, Oh., U.S. (ê-lĭr'ĭ-á)	69d	41°22'N	82°07'W
Ema, r., Est. (å'mà)	123	58°25'N	27°00'E
Emāmshahr, Iran	154	36°25'N	55°01'E
Emån, r., Swe.	122	57°15'N	15°46'E
Embarrass, r., Il., U.S. (ĕm-băr'ăs)	66	39°15'N	88°05'W
Embrun, Can. (ĕm'brŭn)	62c	45°16'N	75°17'W
Embrun, Fr. (äɴ-brûɴ')	127	44°35'N	6°32'E
Embu, Kenya	191	0°32'S	37°27'E
Emden, Ger. (ĕm'dĕn)	124	53°21'N	7°15'E
Emerson, Can. (ĕm'ēr-sŭn)	50	49°00'N	97°12'W
Emeryville, Ca., U.S. (ĕm'ēr-ĭ-vĭl)	74b	37°50'N	122°17'W
Emi Koussi, mtn., Chad (ā'mê kōō-sē')	185	19°50'N	18°30'E
Emiliano Zapata, Mex. (ĕ-mē-lyà'nô-zä-pà'tà)	89	17°45'N	91°46'W
Emilia-Romagna, hist. reg., Italy (ē-mēl'yä rô-mä'n-yä)	130	44°35'N	10°48'E
Eminence, Ky., U.S. (ĕm'ĭ-nĕns)	66	38°25'N	85°15'W
Emira Island, i., Pap. N. Gui. (ā-mê-rä')	169	1°40'S	150°28'E
Emmen, Neth. (ĕm'ĕn)	121	52°48'N	6°55'E
Emmerich, Ger. (ĕm'ēr-ĭk)	127c	51°51'N	6°16'E
Emmetsburg, Ia., U.S. (ĕm'ĕts-bûrg)	71	43°07'N	94°41'W
Emmett, Id., U.S. (ĕm'ĕt)	72	43°53'N	116°30'W
Emmons, Mount, mtn., Ut., U.S. (ĕm'ŭnz)	64	40°43'N	110°20'W
Emory Peak, mtn., Tx., U.S. (ĕ'mô-rē pēk)	80	29°13'N	103°20'W
Empoli, Italy (ām'pô-lē)	130	43°43'N	10°55'E
Emporia, Ks., U.S. (ĕm-pō'rĭ-á)	64	38°24'N	96°11'W
Emporia, Va., U.S.	83	37°40'N	77°34'W
Emporium, Pa., U.S. (ĕm-pō'rĭ-ŭm)	67	41°30'N	78°15'W
Empty Quarter see Ar Rub' al Khālī, des., Asia	154	20°00'N	51°00'E
Ems, r., Ger. (ĕms)	124	52°52'N	7°16'E
Ems-Weser Kanal, can., Ger.	124	52°23'N	8°11'E
Enånger, Swe. (ĕn-ôŋ'gĕr)	122	61°36'N	16°55'E
Encantada, Cerro de la, mtn., Mex. (sĕ'r-rô-dĕ-lä-ĕn-kän-tä'dä)	86	31°58'N	115°15'W
Encanto, Cape, c., Phil. (ĕn-kän'tō)	169a	15°44'N	121°46'E
Encarnación, Para. (ĕn-kär-nä-syōn')	102	27°26'S	55°52'W
Encarnación de Díaz, Mex. (ĕn-kär-nä-syōn dà dē'äz)	88	21°34'N	102°15'W
Encinal, Tx., U.S. (ĕn'sĭ-nôl)	80	28°02'N	99°22'W
Encontrados, Ven. (ĕn-kōn-trä'dōs)	100	9°01'N	72°10'W
Encounter Bay, b., Austl. (ĕn-koun'tēr)	174	35°50'S	138°45'E
Endako, r., Can.	54	54°05'N	125°30'W
Endau, r., Malay.	153b	2°29'N	103°40'E
Enderbury, i., Kir. (ĕn'dēr-bŭrĭ)	194	2°00'S	171°00'W
Enderby Land, reg., Ant. (ĕn'dĕr bĭ)	178	72°00'S	52°00'E
Enderlin, N.D., U.S. (ĕn'dēr-lĭn)	70	46°38'N	97°37'W
Endicott, N.Y., U.S. (ĕn'dĭ-kŏt)	67	42°05'N	76°00'W
Endicott Mountains, mts., Ak., U.S.	63	67°30'N	153°45'W
Enez, Tur.	131	40°42'N	26°05'E
Enfer, Pointe d', c., Mart.	91b	14°21'N	60°48'W
Enfield, Eng., U.K.	114b	51°38'N	0°06'W
Enfield, Ct., U.S. (ĕn'fēld)	67	41°55'N	72°35'W
Enfield, N.C., U.S.	83	36°10'N	77°41'W
Engaño, Cabo, c., Dom. Rep. (kä'-bô- ĕn-gä-nô)	87	18°40'N	68°30'W
Engcobo, S. Afr. (ĕŋg-cô-bô)	187c	31°41'S	27°59'E
Engel's, Russia (ĕn'gĕls)	137	51°20'N	45°40'E
Engelskirchen, Ger. (ĕn'gĕls-kēr'kĕn)	127c	50°59'N	7°25'E
Enggano, Pulau, i., Indon. (ĕng-gä'nō)	168	5°22'S	102°18'E
England, Ar., U.S. (ĭŋ'glănd)	79	34°33'N	91°58'W
England, state, U.K. (ĭŋ'glănd)	110	51°35'N	1°40'W
Englewood, Co., U.S. (ĕn'g'l-wŏŏd)	78	39°39'N	105°00'W
Englewood, N.J., U.S.	68a	40°54'N	73°59'W
English, In., U.S. (ĭn'glĭsh)	66	38°15'N	86°25'W
English, r., Can.	53	50°31'N	94°12'W
English Channel, strt., Eur.	112	49°45'N	3°06'W
Enguera, Spain (ān'gārä)	129	38°58'N	0°42'W
Enid, Ok., U.S. (ē'nĭd)	64	36°25'N	97°52'W
Enid Lake, res., Ms., U.S.	82	34°13'N	89°47'W
Enkeldoring, S. Afr. (ĕŋ'k'l-dôr-ĭng)	192c	25°24'S	28°43'E
Enköping, Swe. (ĕn'kû-pĭng)	122	59°39'N	17°05'E
Ennedi, mts., Chad (ĕn-nĕd'ê)	185	16°45'N	22°45'E
Ennis, Ire. (ĕn'ĭs)	120	52°54'N	9°05'W
Ennis, Tx., U.S.	81	32°20'N	96°38'W
Enniscorthy, Ire. (ĕn-ĭs-kôr'thĭ)	120	52°33'N	6°27'W
Enniskillen, N. Ire., U.K. (ĕn-ĭs-kĭl'ĕn)	120	54°20'N	7°25'W
Ennis Lake, res., Mt., U.S.	73	45°15'N	111°30'W
Enns, r., Aus. (ĕns)	117	47°37'N	14°35'E
Enoree, S.C., U.S. (ê-nō'rē)	83	34°43'N	81°58'W
Enoree, r., S.C., U.S.	83	34°35'N	81°55'W
Enriquillo, Dom. Rep. (ĕn-rê-kê'l-yō)	93	17°55'N	71°15'W
Enriquillo, Lago, l., Dom. Rep. (lä'gô-ĕn-rê-kê'l-yô)	93	18°35'N	71°35'W
Enschede, Neth. (ĕns'kà-dĕ)	117	52°10'N	6°50'E
Enseñada, Arg.	99c	34°50'S	57°55'W
Ensenada, Mex. (ĕn-sĕ-nä'dä)	86	32°00'N	116°30'W
Enshi, China (ŭn-shr)	160	30°18'N	109°25'E
Enshū-Nada, b., Japan (ĕn'shōō nä-dä)	167	34°25'N	137°14'E
Entebbe, Ug.	185	0°04'N	32°28'E
Enterprise, Al., U.S. (ĕn'tēr-prīz)	82	31°20'N	85°50'W
Enterprise, Or., U.S.	72	45°25'N	117°16'W
Entiat, L, Wa., U.S.	72	45°43'N	120°11'W
Entraygues, Fr. (ĕɴ-trĕg')	126	44°39'N	2°33'E
Entre Rios, prov., Arg.	102	31°30'S	59°00'W
Enugu, Nig. (ĕ-nōō'gōō)	184	6°27'N	7°27'E
Enumclaw, Wa., U.S. (ĕn'ŭm-klô)	74a	47°12'N	121°59'W
Envigado, Col. (ĕn-vē-gá'dō)	100a	6°10'N	75°34'W
Eolie, Isole, is., Italy (ê'sō-lĕ-ĕ-ô'lyĕ)	118	38°43'N	14°43'E
Epe, Nig.	189	6°37'N	3°59'E
Épernay, Fr. (ā-pĕr-nĕ')	117	49°02'N	3°54'E
Épernon, Fr. (ā-pĕr-nôɴ')	127b	48°36'N	1°41'E
Ephraim, Ut., U.S. (ē'frà-ĭm)	77	39°20'N	111°40'W
Ephrata, Wa., U.S. (ĕfrā'tá)	72	47°18'N	119°35'W
Épi, Vanuatu (ā'pē)	175	16°59'S	168°29'E
Épila, Spain (ā'pē-lä)	128	41°38'N	1°15'W
Épinal, Fr. (ā-pē-nál')	117	48°11'N	6°27'E
Episkopi, Cyp.	153a	34°38'N	32°55'E
Epping, Eng., U.K. (ĕp'ĭng)	114b	51°41'N	0°06'E
Epsom, Eng., U.K.	114b	51°20'N	0°16'W
Epupa Falls, wtfl., Afr.	190	17°00'S	13°05'E
Epworth, Eng., U.K. (ĕp'wûrth)	114a	53°31'N	0°50'W
Equatorial Guinea, nation, Afr.	184	2°00'N	7°15'E
Équilles, Fr.	126a	43°34'N	5°21'E
Eramosa, r., Can. (ĕr-á-mō'sá)	62d	43°39'N	80°08'W
Erba, Jabal, mtn., Sudan (ĕr-bá)	185	20°53'N	36°45'E
Erciyeş Daği, mtn., Tur.	119	38°30'N	35°36'E
Erding, Ger. (ĕr'dĕng)	115d	48°19'N	11°54'E
Erechim, Braz. (ĕ-rĕ-shê'ɴ)	102	27°43'S	52°11'W
Ereğli, Tur. (ĕ-rā'ĭ-le)	119	37°40'N	34°00'E
Ereğli, Tur.	119	41°15'N	31°25'E
Erfurt, Ger. (ĕr'fŏŏrt)	117	50°59'N	11°04'E
Ergene, r., Tur. (ĕr'gĕ-nĕ)	131	41°17'N	26°50'E
Erges, r., Eur. (ĕr'-zhĕs)	128	39°45'N	7°01'W
Ergļi, Lat.	123	56°54'N	25°38'E
Eria, r., Spain (ā-rē'ä)	128	42°10'N	6°08'W
Erick, Ok., U.S. (âr'ĭk)	78	35°14'N	99°51'W
Erie, Ks., U.S. (ē'rĭ)	79	37°35'N	95°17'W
Erie, Pa., U.S.	65	42°05'N	80°05'W
Erie, Lake, l., N.A.	65	42°15'N	81°25'W
Erimo Saki, c., Japan (ā'rē-mō sä-kē)	161	41°53'N	143°20'E
Erin, Can. (ĕ'rĭn)	62d	43°46'N	80°04'W
Eritrea, nation, Afr. (ā-rê-trā'á)	185	16°15'N	38°30'E
Erlangen, Ger. (ĕr'läng-ĕn)	124	49°36'N	11°03'E
Erlanger, Ky., U.S. (ĕr'lăng-ĕr)	69f	39°01'N	84°36'W

246 Ermoúpoli – Farmers Branch PRONOUNCING INDEX

PLACE (Pronunciation)	PAGE	LAT.	LONG.
Ermoúpoli, Grc.	131	37°30'N	24°56'E
Ernākulam, India	155	9°58'N	76°23'E
Erne, Lower Lough, I., N. Ire., U.K.	120	54°30'N	7°40'W
Erne, Upper Lough, I., N. Ire., U.K. (lŏk ûrn)	120	54°20'N	7°24'W
Erode, India	159	11°20'N	77°45'E
Eromanga, i., Vanuatu	175	18°58'S	169°18'E
Eros, La., U.S. (ē'rōs)	81	32°23'N	92°22'W
Errego, Moz.	191	16°02'S	37°14'E
Errigal, mtn., Ire. (ĕr-ĭ-gôl')	120	55°02'N	8°07'W
Errol Heights, Or., U.S.	74c	45°29'N	122°38'W
Erstein, Fr. (ĕr'shtīn)	127	48°27'N	7°40'E
Erwin, N.C., U.S. (ûr'wĭn)	83	35°16'N	78°40'W
Erwin, Tn., U.S.	83	36°07'N	82°25'W
Erzgebirge, mts., Eur. (ĕrts'gĕ-bē'gĕ)	112	50°29'N	12°40'E
Erzincan, Tur. (ĕr-zĭn-jän')	154	39°50'N	39°30'E
Erzurum, Tur. (ĕrz'rōōm')	154	39°55'N	41°10'E
Esambo, D.R.C.	190	3°40'S	23°24'E
Esashi, Japan (ĕs'ä-shē)	161	41°50'N	140°10'E
Esbjerg, Den. (ĕs'byĕrgh)	116	55°29'N	8°25'E
Escalante, Ut., U.S. (ĕs-kȧ-lăn'tē)	77	37°50'N	111°40'W
Escalante, r., Ut., U.S.	77	37°40'N	111°20'W
Escalón, Mex.	80	26°45'N	104°20'W
Escambia, r., Fl., U.S. (ĕs-kăm'bĭ-ȧ)	82	30°38'N	87°20'W
Escanaba, Mi., U.S. (ĕs-kȧ-nô'bȧ)	65	45°44'N	87°05'W
Escanaba, r., Mi., U.S.	71	46°10'N	87°22'W
Escarpada Point, Phil.	168	18°40'N	122°45'E
Esch-sur-Alzette, Lux.	127	49°32'N	6°21'E
Eschwege, Ger. (ĕsh'vā-gē)	124	51°11'N	10°02'E
Eschweiler, Ger. (ĕsh'vī-lēr)	127c	50°49'N	6°15'E
Escondido, Ca., U.S. (ĕs-kŏn-dē'dō)	76	33°07'N	117°00'W
Escondido, r., Nic.	91	12°04'N	84°09'W
Escondido, Río, r., Mex. (rē'ō-ĕs-kōn-dē'dô)	80	28°30'N	100°45'W
Escudo de Veraguas, i., Pan. (ĕs-kōō'dä dā vä-rä'gwäs)	91	9°07'N	81°25'W
Escuinapa, Mex. (ĕs-kwē-nä'pä)	86	22°49'N	105°44'W
Escuintla, Guat. (ĕs-kwēn'tlä)	90	14°16'N	90°47'W
Ese, Cayos de, i., Col.	91	12°24'N	81°07'W
Eşfahān, Iran	154	32°38'N	51°30'E
Esgueva, r., Spain (ĕs-gĕ'vä)	128	41°48'N	4°10'W
Esher, Eng., U.K.	114b	51°23'N	0°22'W
Eshowe, S. Afr. (ĕsh'ō-wē)	187c	28°54'S	31°28'E
Esiama, Ghana	188	4°56'N	2°21'W
Eskdale, W.V., U.S. (ĕsk'dāl)	66	38°05'N	81°25'W
Eskifjördur, Ice. (ĕs'kē-fyür'dōōr)	110	65°04'N	14°01'W
Eskilstuna, Swe. (å'shĕl-stü-na)	116	59°23'N	16°28'E
Eskimo Lakes, l., Can. (es'kĭ-mō)	52	69°40'N	130°10'W
Eskişehir, Tur. (ĕs-kē-shĕ'h'r)	154	39°40'N	30°20'E
Esko, Mn., U.S. (ĕs'kō)	75h	46°27'N	92°22'W
Esla, r., Spain (ĕs-lä)	128	41°50'N	5°48'W
Eslöv, Swe. (ĕs'lûv)	122	55°50'N	13°17'E
Esmeraldas, Ec. (ĕs-mȧ-räl'däs)	100	0°58'N	79°45'W
Espanola, Can. (ĕs-pȧ-nō'lȧ)	51	46°11'N	81°59'W
Esparta, C.R. (ĕs-pär'tä)	91	9°59'N	84°40'W
Esperance, Austl. (ĕs'pē-rȧns)	174	33°45'S	122°07'E
Esperanza, Cuba (ĕs-pĕ-rä'n-zä)	92	22°30'N	80°10'W
Espichel, Cabo, c., Port. (ká'bō-ĕs-pē-shĕl')	128	38°25'N	9°13'W
Espinal, Col. (ĕs-pē-näl')	100	4°10'N	74°53'W
Espinhaço, Serra do, mts., Braz. (sĕ'r-rä-dô-ĕs-pē-nä-sô)	101	16°00'S	44°00'W
Espinillo, Punta, c., Ur. (pōō'n-tä-ĕs-pē-nē'l-yô)	99c	34°49'S	56°27'W
Espírito Santo, Braz. (ĕs-pē'rē-tô-sän'tô)	101	20°27'S	40°18'W
Espírito Santo, state, Braz.	101	19°57'S	40°58'W
Espiritu Santo, i., Vanuatu (ĕs-pē'rē-tōō sän'tō)	175	15°45'S	166°50'E
Espíritu Santo, Bahía del, b., Mex.	90a	19°25'N	87°28'W
Espita, Mex. (ĕs-pē'tä)	90a	20°57'N	88°22'W
Espoo, Fin.	123	60°13'N	24°41'E
Es Port de Pollença, Spain	129	39°50'N	3°00'E
Esposende, Port. (ĕs-pō-zĕn'dā)	128	41°33'N	8°45'W
Esquel, Arg. (ĕs-kĕ'l)	102	42°47'S	71°22'W
Esquimalt, Can. (ĕs-kwī'mŏlt)	54	48°26'N	123°24'W
Essaouira, Mor.	184	31°34'N	9°44'W
Essen, Bel.	115a	51°28'N	4°27'E
Essen, Ger. (ĕs'sĕn)	110	51°26'N	6°59'E
Essendon, Austl.	173a	37°46'S	144°55'E
Essequibo, r., Guy. (ĕs-ä-kē'bō)	101	4°26'N	58°17'W
Essex, Il., U.S.	69a	41°11'N	88°11'W
Essex, Ma., U.S.	61a	42°38'N	70°47'W
Essex, Md., U.S.	68e	39°19'N	76°29'W
Essex, Vt., U.S.	67	44°30'N	73°05'W
Essex Fells, N.J., U.S. (ĕs'ĕks fĕlz)	68a	40°50'N	74°16'W
Essexville, Mi., U.S. (ĕs'ĕks-vĭl)	66	43°35'N	83°50'W
Esslingen, Ger. (ĕs'slēn-gĕn)	124	48°45'N	9°19'E
Estacado, Llano, pl., U.S. (yä-nō ĕs-tȧcȧ-dō')	64	33°50'N	103°20'W
Estância, Braz. (ĕs-tän'sĭ-ä)	101	11°17'S	37°18'W
Estarreja, Port. (ĕ-tär-rä'zhä)	128	40°44'N	8°39'W
Estats, Pique d', mtn., Eur.	129	42°43'N	1°30'E
Estcourt, S. Afr. (ĕst-coort)	187c	29°04'S	29°53'E
Este, Italy (ĕs'tā)	130	45°13'N	11°40'E
Estella, Spain (ĕs-tāl'yä)	128	42°40'N	2°01'W
Estepa, Spain (ĕs-tā'pä)	128	37°18'N	4°54'W
Estepona, Spain (ĕs-tȧ-pō'nä)	128	36°26'N	5°08'W
Esterhazy, Can. (ĕs-tĕr-hä-zē)	57	50°40'N	102°08'W
Estero Bay, b., Ca., U.S. (ĕs-tā'rōs)	76	35°22'N	121°04'W
Estevan, Can. (ĕ-stē'vȧn)	50	49°07'N	103°05'W
Estevan Group, is., Can.	54	53°05'N	129°40'W
Estherville, Ia., U.S. (ĕs'tĕr-vĭl)	71	43°24'N	94°49'W
Estill, S.C., U.S. (ĕs'tĭl)	83	32°46'N	81°15'W
Eston, Can.	56	51°10'N	108°45'W
Estonia, nation, Eur.	134	59°10'N	25°00'E
Estoril, Port. (ĕs-tô-rēl')	129b	38°45'N	9°24'W
Estrêla, mtn., Port. (mäl-you'ĸ-dä-ĕs-trē'lä)	128	40°20'N	7°38'W
Estrêla, r., Braz. (ĕs-trē'lȧ)	102b	22°39'S	43°16'W
Estrêla, Serra da, mts., Port. (sĕr'rȧ dä ĕs-trā'lȧ)	128	40°25'N	7°45'W
Estremadura, hist. reg., Port. (ĕs-trä-mä-dōō'rȧ)	128	39°00'N	8°36'W
Estremoz, Port. (ĕs-trä-mōzh')	128	38°50'N	7°35'W
Estrondo, Serra do, mts., Braz. (sĕr'-rȧ dọ ĕs-trōn'-dō)	101	9°52'S	48°56'W
Esumba, Île, i., D.R.C.	190	2°00'N	21°12'E
Esztergom, Hung. (ĕs'tĕr-gōm)	125	47°46'N	18°45'E
Etah, Grnld. (ē'tȧ)	49	78°20'N	72°42'W
Étampes, Fr. (ā-tänp')	126	48°26'N	2°09'E
Étaples, Fr. (ā-täp'l')	126	50°32'N	1°38'E
Etchemin, r., Can. (ĕch'ĕ-mĭn)	62b	46°39'N	71°03'W
Ethiopa, nation, Afr. (ē-thē-ō'pē-ȧ)	185	7°53'N	37°55'E
Eticoga, Gui.-B.	188	11°09'N	16°08'W
Etiwanda, Ca., U.S. (ĕ-tĭ-wän'dȧ)	75a	34°07'N	117°31'W
Etna, Pa., U.S. (ĕt'nȧ)	69e	40°30'N	79°55'W
Etna, Mount, vol., Italy	112	37°48'N	15°00'E
Etobicoke Creek, r., Can.	62d	43°44'N	79°48'W
Etolin Strait, strt., Ak., U.S. (ĕt ō lĭn)	63	60°35'S	165°40'W
Etoshapan, pl., Nmb. (ĕtō'shä)	186	19°07'S	15°30'E
Etowah, Tn., U.S. (ĕt'ō-wä)	82	35°18'N	84°31'W
Etowah, r., Ga., U.S.	82	34°23'N	84°19'W
Étréchy, Fr. (ā-trā-shē')	127b	48°29'N	2°12'E
Etten-Leur, Neth.	115a	51°34'N	4°38'E
Etterbeek, Bel. (ĕt'ĕr-bāk)	115a	50°51'N	4°24'E
Etzatlán, Mex. (ĕt-zä-tlän')	88	20°44'N	104°04'W
Eucla, Austl. (ū'klȧ)	174	31°45'S	128°50'E
Euclid, Oh., U.S. (ū'klĭd)	69d	41°34'N	81°32'W
Eudora, Ar., U.S. (u-dō'rȧ)	79	33°07'N	91°16'W
Eufaula, Al., U.S. (û-fō'lȧ)	82	31°53'N	85°09'W
Eufaula, Ok., U.S.	79	35°16'N	95°35'W
Eufaula Reservoir, res., Ok., U.S.	79	35°00'N	94°45'W
Eugene, Or., U.S. (ů-jēn')	64	44°02'N	123°06'W
Euless, Tx., U.S. (ū'lĕs)	75c	32°50'N	97°05'W
Eunice, La., U.S. (ū'nĭs)	81	30°30'N	92°25'W
Eupen, Bel. (oi'pĕn)	121	50°39'N	6°05'E
Euphrates, r., Asia (û-frā'tēz)	154	36°00'N	40°00'E
Eure, r., Fr. (ûr)	126	49°03'N	1°22'E
Eureka, Ca., U.S. (û-rē'kȧ)	64	40°45'N	124°10'W
Eureka, Ks., U.S.	79	37°48'N	96°17'W
Eureka, Mt., U.S.	72	48°53'N	115°07'W
Eureka, Nv., U.S.	76	39°33'N	115°58'W
Eureka, S.D., U.S.	70	45°46'N	99°38'W
Eureka, Ut., U.S.	77	39°55'N	112°10'W
Eureka Springs, Ar., U.S.	79	36°24'N	93°43'W
Europe, cont. (ū'rŭp)	112	50°00'N	15°00'E
Eustis, Fl., U.S. (ūs'tĭs)	83	28°50'N	81°41'W
Eutaw, Al., U.S. (ū-tȧ)	82	32°48'N	87°50'W
Eutsuk Lake, l., Can. (ōōt'sŭk)	54	53°20'N	126°44'W
Evanston, Il., U.S. (ĕv'ăn-stŭn)	65	42°03'N	87°41'W
Evanston, Wy., U.S.	73	41°17'N	111°02'W
Evansville, In., U.S. (ĕv'ănz-vĭl)	65	38°00'N	87°30'W
Evansville, Wi., U.S.	71	42°46'N	89°19'W
Evart, Mi., U.S. (ĕv'ērt)	66	43°55'N	85°10'W
Evaton, S. Afr. (ĕv'ȧ-tŏn)	192c	26°32'S	27°53'E
Eveleth, Mn., U.S. (ĕv'ĕ-lĕth)	71	47°27'N	92°35'W
Everard, l., Austl. (ĕv'ĕr-ȧrd)	174	31°20'S	134°10'E
Everard Ranges, mts., Austl.	174	27°15'S	132°00'E
Everest, Mount, mtn., Asia (ĕv'ĕr-ĕst)	155	28°00'N	86°57'E
Everett, Ma., U.S. (ĕv'ēr-ĕt)	61a	42°24'N	71°03'W
Everett, Wa., U.S. (ĕv'ēr-ĕt)	64	47°59'N	122°11'W
Everett Mountains, mts., Can.	53	62°34'N	68°00'W
Everglades, The, sw., Fl., U.S.	83a	25°35'N	80°55'W
Everglades City, Fl., U.S. (ĕv'ēr-glādz)	83a	25°50'N	81°25'W
Everglades National Park, rec., Fl., U.S.	65	25°39'N	80°57'W
Evergreen, Al., U.S. (ĕv'ēr-grēn)	82	31°25'N	87°56'W
Evergreen Park, Il., U.S.	69a	41°44'N	87°42'W
Everman, Tx., U.S. (ĕv'ēr-măn)	75c	32°38'N	97°17'W
Everson, Wa., U.S. (ĕv'ēr-sŭn)	74d	48°55'N	122°21'W
Évora, Port. (ĕv'ō-rä)	118	38°35'N	7°54'W
Évreux, Fr. (ā-vrû')	117	49°02'N	1°11'E
Evrótas, r., Grc. (ĕv-rō'täs)	131	37°15'N	22°17'E
Évvoia, i., Grc.	119	38°38'N	23°45'E
'Ewa Beach, Hi., U.S. (ē'wä)	84a	21°17'N	158°03'W
Ewaso Ng'iro, r., Kenya	185	0°59'N	37°47'E
Excelsior, Mn., U.S. (ĕk-sel'sĭ-ōr)	75g	44°54'N	93°35'W
Excelsior Springs, Mo., U.S.	79	39°20'N	94°13'W
Exe, r., Eng., U.K. (ĕks)	120	50°57'N	3°37'W
Exeter, Eng., U.K.	117	50°45'N	3°33'W
Exeter, Ca., U.S. (ĕk'sē-tēr)	76	36°18'N	119°09'W
Exeter, N.H., U.S.	67	43°00'N	71°00'W
Exmoor, for., Eng., U.K. (ĕks'mōr)	120	51°10'N	3°55'W
Exmouth, Eng., U.K. (ĕks'mŭth)	120	50°40'N	3°20'W
Exmouth Gulf, b., Austl.	174	21°45'S	114°30'E
Exploits, r., Can. (ĕks-ploits')	61	48°50'N	56°15'W
Extórrax, r., Mex. (ĕx-tô'rȧx)	88	21°04'N	99°39'W
Extrema, Braz. (ĕsh-trĕ'mȧ)	99a	22°52'S	46°19'W
Extremadura, hist. reg., Spain (ĕks-trä-mä-doo'rä)	128	38°43'N	6°30'W
Exuma Sound, strt., Bah. (ĕk-sōō'mä)	92	24°20'N	76°20'W
Eyasi, Lake, l., Tan. (å-yä'sē)	186	3°25'S	34°55'E
Eyjafjördur, b., Ice.	116	66°21'N	18°20'W
Eyl, Som.	192a	7°53'N	49°45'E
Eyrarbakki, Ice.	116	63°51'N	20°52'W
Eyre, Austl. (âr)	174	32°15'S	126°20'E
Eyre, l., Austl.	174	28°43'S	137°50'E
Eyre Peninsula, pen., Austl.	174	33°30'S	136°00'E
Ezeiza, Arg. (ĕ-zā'zä)	102a	34°52'S	58°31'W
Ezine, Tur. (â'zĭ-nâ)	131	39°47'N	26°18'E

F

PLACE (Pronunciation)	PAGE	LAT.	LONG.
Faaborg, Den. (fō'bôrg)	122	55°06'N	10°19'E
Fabens, Tx., U.S. (fā'bĕnz)	80	31°30'N	106°07'W
Fabriano, Italy (fä-brē-ä'nō)	130	43°20'N	12°55'E
Fada, Chad (fä'dä)	185	17°06'N	21°18'E
Fada Ngourma, Burkina (fä'dä''n gōōr'mä)	184	12°04'N	0°21'E
Faddeya, i., Russia (fȧd-yä')	135	76°12'N	145°00'E
Faenza, Italy (fä-ĕnd'zä)	130	44°16'N	11°53'E
Fafe, Port. (fä'fā)	128	41°30'N	8°10'W
Fafen, r., Eth.	192a	8°15'N	42°40'E
Făgăras, Rom. (fȧ-gä'räsh)	131	45°50'N	24°55'E
Fagerness, Nor. (fä'ghĕr-nĕs)	116	61°00'N	9°10'E
Fagnano, l., S.A. (fäk-nä'nō)	102	54°35'S	68°20'W
Faguibine, Lac, l., Mali	188	16°50'N	4°20'W
Faial, i., Port. (fä-yä'l)	184a	38°40'N	29°19'W
Fā'id, Egypt (fä-yēd')	192d	30°19'N	32°18'E
Fairbanks, Ak., U.S. (fâr'bănks)	64a	64°50'N	147°48'W
Fairbury, Il., U.S. (fâr'bĕr-ĭ)	66	40°45'N	88°25'W
Fairbury, Ne., U.S.	79	40°09'N	97°11'W
Fairchild Creek, r., Can. (fâr'chīld)	62d	43°18'N	80°10'W
Fairfax, Mn., U.S. (fâr'făks)	71	44°29'N	94°44'W
Fairfax, S.C., U.S.	83	32°29'N	81°13'W
Fairfax, Va., U.S.	68e	38°51'N	77°20'W
Fairfield, Austl.	173b	33°52'S	150°57'E
Fairfield, Al., U.S. (fâr'fēld)	68h	33°30'N	86°50'W
Fairfield, Ct., U.S.	68a	41°08'N	73°22'W
Fairfield, Ia., U.S.	71	41°00'N	91°59'W
Fairfield, Il., U.S.	66	38°25'N	88°20'W
Fairfield, Me., U.S.	60	44°35'N	69°38'W
Fairhaven, Ma., U.S. (fâr-hā'vĕn)	67	41°35'N	70°55'W
Fair Haven, Vt., U.S.	67	43°35'N	73°15'W
Fair Island, i., Scot., U.K. (fâr)	120a	59°34'N	1°41'W
Fairmont, Mn., U.S. (fâr'mŏnt)	71	43°39'N	94°26'W
Fairmont, W.V., U.S.	66	39°30'N	80°10'W
Fairmont City, Il., U.S.	75e	38°39'N	90°05'W
Fairmount, In., U.S.	66	40°25'N	85°45'W
Fairmount, Ks., U.S.	75f	39°12'N	95°55'W
Fair Oaks, Ga., U.S. (fâr ōks)	68c	33°56'N	84°33'W
Fairport, N.Y., U.S. (fâr'pōrt)	67	43°05'N	77°30'W
Fairport Harbor, Oh., U.S.	66	41°45'N	81°15'W
Fairview, Ok., U.S. (fâr'vū)	78	36°16'N	98°28'W
Fairview, Or., U.S.	74c	45°32'N	112°26'W
Fairview, Ut., U.S.	77	39°35'N	111°30'W
Fairview Park, Oh., U.S.	69d	41°27'N	81°52'W
Fairweather, Mount, mtn., N.A. (fâr-wĕdh'ēr)	63	59°12'N	137°22'W
Faisalabad, Pak.	155	31°29'N	73°06'E
Faith, S.D., U.S. (fāth)	70	45°02'N	120°02'W
Faizābād, India	155	26°50'N	82°17'E
Fajardo, P.R.	87b	18°20'N	65°40'W
Fakfak, Indon.	169	2°56'S	132°25'E
Faku, China (fä-kōō)	164	42°28'N	123°20'E
Falcón, dept., Ven. (fäl-kō'n)	101b	11°00'N	68°28'W
Falconer, N.Y., U.S. (fô'k'n-ēr)	67	42°10'N	79°10'W
Falcon Heights, Mn., U.S. (fô'k'n)	75g	44°59'N	93°10'W
Falcon Reservoir, res., N.A. (fôk'n)	80	26°47'N	99°03'W
Fălești, Mol.	133	47°33'N	27°46'E
Falfurrias, Tx., U.S. (făl'fōō-rē'ás)	80	27°15'N	98°08'W
Falher, Can. (făl'ēr)	55	55°44'N	117°12'W
Falkenberg, Swe. (fäl'kĕn-bĕrgh)	122	56°54'N	12°25'E
Falkensee, Ger. (fäl'kĕn-zā)	115b	52°34'N	13°05'E
Falkenthal, Ger. (fäl'kĕn-täl)	115b	52°54'N	13°18'E
Falkirk, Scot., U.K. (fôl'kûrk)	120	55°59'N	3°55'W
Falkland Islands, dep., S.A. (fôk'lănd)	102	50°45'S	61°00'W
Falköping, Swe. (fäl'chûp-ĭng)	122	58°09'N	13°30'E
Fall City, Wa., U.S.	74a	47°34'N	121°53'W
Fall Creek, r., In., U.S. (fôl)	69g	39°52'N	86°04'W
Fallon, Nv., U.S. (făl'ŭn)	76	39°30'N	118°48'W
Fall River, Ma., U.S.	65	41°42'N	71°07'W
Falls Church, Va., U.S. (fâlz chûrch)	68e	38°53'N	77°10'W
Falls City, Ne., U.S.	79	40°04'N	95°37'W
Fallston, Md., U.S. (fâls'ton)	68e	39°32'N	76°26'W
Falmouth, Jam.	92	18°30'N	77°40'W
Falmouth, Eng., U.K. (făl'mŭth)	120	50°08'N	5°04'W
Falmouth, Ky., U.S.	66	38°40'N	84°20'W
False Divi Point, c., India	159	15°45'N	80°50'E
Falster, i., Den. (fäls'tĕr)	122	54°48'N	11°58'E
Fălticeni, Rom. (fŭl-tē-chän'y')	125	47°27'N	26°17'E
Falun, Swe. (fä-lōōn')	116	60°38'N	15°35'E
Famagusta, Cyp. (fä-mä-gōōs'tä)	119	35°08'N	33°59'E
Famatina, Sierra de, mts., Arg.	102	29°00'S	67°50'W
Fangxian, China (fäŋ-shyĕn)	164	32°05'N	110°45'E
Fanning, i., Can.	62f	49°45'N	97°46'W
Fano, Italy (fä'nō)	130	43°49'N	13°01'E
Fanø, i., Den. (fän'û)	122	55°24'N	8°10'E
Fan Si Pan, mtn., Viet.	165	22°25'N	103°50'E
Farafangana, Madag. (fä-rä-fäŋ-gä'nä)	187	23°18'S	47°59'E
Farāh, Afg. (fä-rä')	154	32°15'N	62°13'E
Farallón, Punta, c., Mex. (pô'n-tä-fä-rä-lōn)	88	19°21'N	105°03'W
Faranah, Gui. (fä-rä'nä)	184	10°02'N	10°44'W
Farasān, Jaza'ir, is., Sau. Ar.	154	16°45'N	41°08'E
Fārigh, Wādī al, r., Libya (wädē ĕl fä-rēg')	119	30°10'N	19°34'E
Farewell, Cape, c., N.Z. (fâr-wĕl')	175a	40°37'S	172°40'E
Fargo, N.D., U.S. (fär'gō)	64	46°53'N	96°48'W
Far Hills, N.J., U.S. (fär hĭlz)	68a	40°41'N	74°38'W
Faribault, Mn., U.S. (fâ'rĭ-bō)	71	44°19'N	93°16'W
Farilhões, is., Port. (fä-rē-lyônzh')	128	39°28'N	9°32'W
Faringdon, Eng., U.K. (fă'rĭng-dŏn)	114b	51°38'N	1°35'W
Fāriskūr, Egypt (fä-rēs-kōōr')	192b	31°19'N	31°46'E
Farit, Amba, mtn., Eth.	185	10°51'N	37°52'E
Farley, Mo., U.S. (fär'lē)	75f	39°16'N	94°49'W
Farmers Branch, Tx., U.S.	75c	32°56'N	96°53'W

ăt; finăl; rāte; senåte; ärm; åsk; sofȧ; fâre; ch-choose; dh-as th in other; bē; ēvent; bĕt; recĕnt; cratēr; g-gō; gh-guttural g; bĭt; ĭ-short neutral; rīde; ĸ-guttural k as ch in German ich;

PRONOUNCING INDEX
Farmersburg – Fonseca 247

PLACE (Pronunciation)	PAGE	LAT.	LONG.
Farmersburg, In., U.S. (fär'mẽrz-bûrg)	66	39°15'N	87°25'W
Farmersville, Tx., U.S. (fär'mẽrz-vĭl)	79	33°11'N	96°22'W
Farmingdale, N.J., U.S. (färm'ĕng-dāl)	68a	40°11'N	74°10'W
Farmingdale, N.Y., U.S.	68a	40°44'N	73°26'W
Farmingham, Ma., U.S. (färm-ĭng-hăm)	61a	42°17'N	71°25'W
Farmington, Il., U.S. (färm-ĭng-tŭn)	79	40°42'N	90°01'W
Farmington, Me., U.S.	60	44°40'N	70°10'W
Farmington, Mi., U.S.	69b	42°28'N	83°23'W
Farmington, Mo., U.S.	79	37°46'N	90°26'W
Farmington, N.M., U.S.	77	36°40'N	108°10'W
Farmington, Ut., U.S.	75b	40°59'N	111°53'W
Farmville, N.C., U.S. (färm-vĭl)	83	35°35'N	77°35'W
Farmville, Va., U.S.	83	37°15'N	78°23'W
Farnborough, Eng., U.K. (färn'bŭr-ô)	114b	51°15'N	0°45'W
Farne Islands, is., Eng., U.K. (färn)	120	55°40'N	1°32'W
Farnham, Can. (fär'năm)	67	45°15'N	72°55'W
Farningham, Eng., U.K. (fär'nĭng-ŭm)	114b	51°22'N	0°14'E
Farnworth, Eng., U.K. (färn'wŭrth)	114a	53°34'N	2°24'W
Faro, Braz. (fä'rò)	101	2°05'S	56°32'W
Faro, Port.	118	37°01'N	7°57'W
Farodofay, Madag.	187	24°59'S	46°58'E
Faroe Islands, is., Eur.	112	62°00'N	5°45'W
Fårön, i., Swe.	123	57°57'N	19°10'E
Farquhar, Cape, c., Austl. (fär'kwär)	174	23°50'S	112°55'E
Farrell, Pa., U.S. (făr'ĕl)	66	41°10'N	80°30'W
Farrukhābād, India (fŭ-rŏk-hä-bäd')	155	27°29'N	79°35'E
Fársala, Grc.	131	39°18'N	22°25'E
Farsund, Nor. (fär'sòn)	122	58°05'N	6°47'E
Fartak, Ra's, c., Yemen	154	15°43'N	52°17'E
Fartura, Serra da, mts., Braz. (sĕ'r-rä-dà-fär-tōō'rä)	102	26°40'S	53°15'W
Farvel, Kap, c., Grnld.	49	60°00'N	44°00'W
Farwell, Tx., U.S. (fär'wĕl)	78	34°24'N	103°03'W
Fasano, Italy (fä-zä'nō)	131	40°50'N	17°22'E
Fastiv, Ukr.	133	50°04'N	29°57'E
Fatëzh, Russia	132	52°06'N	35°51'E
Fatima, Port.	129	39°36'N	9°36'E
Fatsa, Tur. (fät'sä)	119	40°50'N	37°30'E
Faucilles, Monts, mts., Fr. (môn' fō-sēl')	127	48°07'N	6°13'E
Fauske, Nor.	116	67°15'N	15°24'E
Faust, Can. (foust)	55	55°19'N	115°38'W
Faustovo, Russia	142b	55°27'N	38°29'E
Faversham, Eng., U.K. (fă'vẽr-sh'm)	114b	51°19'N	0°54'E
Faxaflói, b., Ice.	116	64°33'N	22°40'W
Fayette, Al., U.S. (fā-yĕt')	82	33°40'N	87°54'W
Fayette, Ia., U.S.	71	42°49'N	91°49'W
Fayette, Mo., U.S.	79	39°09'N	92°41'W
Fayette, Ms., U.S.	82	31°43'N	91°00'W
Fayetteville, Ar., U.S. (fā-yĕt'vĭl)	79	36°03'N	94°08'W
Fayetteville, N.C., U.S.	83	35°02'N	78°54'W
Fayetteville, Tn., U.S.	82	35°10'N	86°33'W
Fazao, Forêt Classée du, for., Togo	188	8°50'N	0°40'E
Fazilka, India	158	30°30'N	74°02'E
Fazzān (Fezzan), hist. reg., Libya	185	26°45'N	13°01'E
Fdérik, Maur.	184	22°45'N	12°38'W
Fear, Cape, c., N.C., U.S. (fēr)	83	33°52'N	77°48'W
Feather, r., Ca., U.S. (fĕth'ẽr)	76	38°56'N	121°41'W
Feather, Middle Fork of, r., Ca., U.S.	76	39°49'N	121°10'W
Feather, North Fork of, r., Ca., U.S.	76	40°00'N	121°20'W
Featherstone, Eng., U.K. (fĕdh'ẽr stŭn)	114a	53°39'N	1°21'W
Fécamp, Fr. (fā-käɴ')	117	49°45'N	0°20'E
Federal, Distrito, dept., Ven. (dēs-trē'tô-fĕ-dĕ-rä'l)	101b	10°34'N	66°55'W
Federal Way, Wa., U.S.	74a	47°20'N	122°20'W
Fëdorovka, Russia (fyô'dô-rôf-kà)	142b	56°15'N	37°14'E
Fehmarn, i., Ger. (fā'märn)	124	54°28'N	11°15'E
Fehrbellin, Ger. (fĕr'bĕl-lēn)	115b	52°49'N	12°46'E
Feia, Logoa, l., Braz. (lô-gôä-fĕ'yä)	99a	21°54'S	41°15'W
Feicheng, China (fā-chŭŋ)	162	36°18'N	116°45'E
Feidong, China (fā-dôŋ)	162	31°53'N	117°28'E
Feira de Santana, Braz. (fĕ'ē-rä dä sänt-än'ä)	101	12°16'S	38°46'W
Feixian, China (fā-shyĕn)	162	35°17'N	117°59'E
Felanitx, Spain (fā-lä-nēch')	118	39°29'N	3°09'E
Feldkirch, Aus. (fĕlt'kĭrk)	124	47°15'N	9°36'E
Feldkirchen, Ger. (fĕld'kēr-kĕn)	115d	48°09'N	11°44'E
Felipe Carrillo Puerto, Mex.	90a	19°36'N	88°04'W
Feltre, Italy (fĕl'trā)	130	46°02'N	11°56'E
Femunden, l., Nor.	116	62°17'N	11°40'E
Fengcheng, China (fŭŋ-chŭŋ)	164	40°28'N	124°03'E
Fengcheng, China	163b	30°55'N	121°38'E
Fengdu, China (fŭŋ-dōō)	160	29°58'N	107°50'E
Fengjie, China (fŭŋ-jyĕ)	160	31°02'N	109°30'E
Fengming Dao, i., China (fŭŋ-mĭŋ dou)	162	39°19'N	121°15'E
Fengrun, China (fŭŋ-ròn)	162	39°51'N	118°06'E
Fengtai, China (fŭŋ-tī)	164a	39°51'N	116°19'E
Fengxian, China (fŭŋ-shyĕn)	163b	30°55'N	121°26'E
Fengxian, China	162	34°41'N	116°36'E
Fengxiang, China (fŭŋ-shyäŋ)	160	34°25'N	107°20'E
Fengyang, China (fŭng'yăng')	164	32°55'N	117°32'E
Fengzhen, China (fŭŋ-jŭn)	161	40°28'N	113°20'E
Fennimore Pass, strt., Ak., U.S. (fĕn-ĭ-mōr)	63a	51°40'N	175°38'W
Fenoarivo Atsinanana, Madag.	187	17°30'S	49°31'E
Fenton, Mi., U.S. (fĕn-tŭn)	66	42°50'N	83°40'W
Fenton, Mo., U.S.	75e	38°31'N	90°27'W
Fenyang, China	161	37°20'N	111°48'E
Feodosiia, Ukr.	137	45°02'N	35°21'E
Ferdows, Iran	154	34°00'N	58°13'E
Ferentino, Italy (fā-rĕn-tē'nō)	130	41°42'N	13°18'E
Fergana, Uzb.	139	40°23'N	71°46'E
Fergus Falls, Mn., U.S. (fûr'gŭs)	65	46°17'N	96°03'W
Ferguson, Mo., U.S. (fûr-gŭ-sŭn)	75e	38°45'N	90°18'W
Ferkéssédougou, C. Iv.	188	9°36'N	5°12'W
Fermo, Italy (fĕr'mō)	130	43°10'N	13°43'E
Fermoselle, Spain (fĕr-mō-sāl'yā)	128	41°20'N	6°23'W
Fermoy, Ire. (fûr-moi')	120	52°05'N	8°06'W
Fernandina Beach, Fl., U.S. (fûr-năn-dē'nȧ)	83	30°38'N	81°29'W
Fernando de Noronha, Arquipélago, is., Braz.	101	3°51'S	32°25'W
Fernando Póo *see* Bioko, i., Eq. Gui.	184	3°35'N	7°45'E
Fernán-Núñez, Spain (fĕr-nän'nōōn'yåth)	128	37°42'N	4°43'W
Fernão Veloso, Baia de, b., Moz.	191	14°20'S	40°55'E
Ferndale, Ca., U.S. (fûrn'dāl)	72	40°34'N	124°18'W
Ferndale, Mi., U.S.	69b	42°27'N	83°08'W
Ferndale, Wa., U.S.	74d	48°51'N	122°36'W
Fernie, Can. (fûr'nĭ)	50	49°30'N	115°03'W
Fern Prairie, Wa., U.S. (fûrn prâr'ĭ)	74c	45°38'N	122°25'W
Ferrara, Italy (fĕr-rä'rä)	118	44°50'N	11°37'E
Ferrat, Cap, c., Alg. (kăp fĕr-rät)	129	35°49'N	0°29'W
Ferreira do Alentejo, Port.	128	38°03'N	8°06'W
Ferreira do Zezere, Port. (fĕr-rĕ'ē-rä dò zā-zā'rĕ)	128	39°49'N	8°17'W
Ferrelview, Mo., U.S. (fĕr'rĕl-vū)	75f	39°18'N	94°40'W
Ferreñafe, Peru (fĕr-rĕn-yä'fĕ)	100	6°38'S	79°48'W
Ferriday, La., U.S. (fĕr'ĭ-dā)	81	31°38'N	91°33'W
Ferrol, Spain	110	43°30'N	8°12'W
Fershampenuaz, Russia (fĕr-shàm'pĕn-wäz)	142a	53°32'N	59°50'E
Fertile, Mn., U.S. (fûr'tĭl)	70	47°33'N	96°18'W
Fès, Mor. (fĕs)	184	34°08'N	5°00'W
Fessenden, N.D., U.S. (fĕs'ĕn-dĕn)	70	47°39'N	99°40'W
Festus, Mo., U.S. (fĕst'ŭs)	79	38°12'N	90°22'W
Fethiye, Tur. (fĕt-hē'yĕ)	119	36°40'N	29°05'E
Feuilles, Rivière aux, r., Can.	53	58°30'N	70°50'W
Ffestiniog, Wales, U.K.	120	52°59'N	3°58'W
Fianarantsoa, Madag. (fyȧ-nä-rȧn-tsō'ȧ)	187	21°21'S	47°15'E
Ficksburg, S. Afr. (fĭks'bûrg)	192c	28°53'S	27°53'E
Fidalgo Island, i., Wa., U.S. (fĭ-dăl'gō)	74a	48°28'N	122°39'W
Fieldbrook, Ca., U.S. (fēld'brók)	72	40°59'N	124°02'W
Fier, Alb. (fyĕr)	131	40°43'N	19°34'E
Fife Ness, c., Scot., U.K. (fīf'nes')	120	56°15'N	2°19'W
Fifth Cataract, wtfl., Sudan	185	18°27'N	33°38'E
Figeac, Fr. (fē-zhȧk')	126	44°37'N	2°02'E
Figeholm, Swe. (fē-ghĕ-hōlm)	122	57°24'N	16°33'E
Figueira da Foz, Port. (fē-gwĕy-rä-dȧ-fō'z)	128	40°10'N	8°50'W
Figuig, Mor.	184	32°20'N	1°30'W
Fiji, nation, Oc. (fē'jē)	3	18°40'S	175°00'E
Filadelfia, C.R. (fĭl-ȧ-dĕl'fĭ-ȧ)	90	10°26'N	85°37'W
Filatovskoye, Russia (fĭ-lä'tôf-skô-yĕ)	142a	56°49'N	62°20'E
Filchner Ice Shelf, ice, Ant. (fĭlk'nĕr)	178	80°00'S	35°00'W
Filicudi, i., Italy (fē'le-kōō'dē)	130	38°34'N	14°39'E
Filippovskoye, Russia (fĭ-lĭ-pôf'skô-yĕ)	142b	56°06'N	38°38'E
Filipstad, Swe. (fĭl'ĭps-städh)	122	59°44'N	14°09'E
Fillmore, Ut., U.S. (fĭl'môr)	77	39°00'N	112°20'W
Filsa, Nor.	122	60°35'N	12°03'E
Fimi, r., D.R.C.	186	2°43'S	17°50'E
Finch, Can. (fĭnch)	62c	45°09'N	75°06'W
Findlay, Oh., U.S. (fĭnd'lå)	66	41°05'N	83°40'W
Fingoe, Moz.	191	15°12'S	31°50'E
Finke, r., Austl. (fĭŋ'kē)	174	25°25'S	134°30'E
Finland, nation, Eur. (fĭn'lănd)	110	62°45'N	26°13'E
Finland, Gulf of, b., Eur. (fĭn'lănd)	112	59°35'N	23°35'E
Finlandia, Col. (fēn-lä'n-dēä)	100a	4°38'N	75°39'W
Finlay, r., Can. (fĭn'lā)	52	57°45'N	125°30'W
Finow, Ger. (fē'nōv)	115b	52°50'N	13°44'E
Finowfurt, Ger. (fē'nō-fōōrt)	115b	52°50'N	13°41'E
Fircrest, Wa., U.S. (fûr'krĕst)	74a	47°14'N	122°31'W
Firenze *see* Florence, Italy	110	43°47'N	11°15'E
Firenzuola, Italy (fē-rĕnt-swô'lä)	130	44°08'N	11°21'E
Firozpur, India	155	30°58'N	74°39'E
Fischa, r., Aus.	115e	48°04'N	16°33'E
Fischamend Markt, Aus.	115e	48°07'N	16°37'E
Fish, r., Nmb. (fĭsh)	186	28°00'S	17°30'E
Fish Cay, i., Bah.	93	22°30'N	74°20'W
Fish Creek, r., Can. (fĭsh)	62e	50°52'N	114°21'W
Fisher, La., U.S. (fĭsh'ẽr)	81	31°28'N	93°30'W
Fisher Bay, b., Can.	57	51°30'N	97°16'W
Fisher Channel, strt., Can.	54	52°10'N	127°42'W
Fisher Strait, strt., Can.	53	62°43'N	84°28'W
Fisterra, Cabo de, c., Spain	112	42°52'N	9°48'W
Fitchburg, Ma., U.S. (fĭch'bûrg)	67	42°35'N	71°48'W
Fitri, Lac, l., Chad	189	12°50'N	17°28'E
Fitzgerald, Ga., U.S. (fĭts-jĕr'ăld)	82	31°42'N	83°17'W
Fitz Hugh Sound, strt., Can. (fĭts hū)	54	51°40'N	127°57'W
Fitzroy, r., Austl. (fĭts-roi')	174	18°00'S	124°05'E
Fitzroy, r., Austl.	175	23°45'S	150°02'E
Fitzroy, Monte (Cerro Chaltel), mtn., S.A.	102	48°10'S	73°18'W
Fitzroy Crossing, Austl.	174	18°08'S	126°00'E
Fitzwilliam, i., Can. (fĭts-wĭl'yŭm)	58	45°30'N	81°45'W
Fiume *see* Rijeka, Cro.	118	45°22'N	14°24'E
Fiumicino, Italy (fyōō-mē-chē'nō)	129d	41°47'N	12°19'E
Fjällbacka, Swe. (fyĕl'bäk-ȧ)	122	58°37'N	11°17'E
Flagstaff, S. Afr. (flăg'stăf)	187c	31°06'S	29°31'E
Flagstaff, Az., U.S. (flăg-stăf)	64	35°15'N	111°40'W
Flagstaff, l., Me., U.S. (flăg-stăf)	67	45°05'N	70°30'W
Flåm, Nor. (flôm)	122	60°50'N	7°00'E
Flambeau, r., Wi., U.S. (flăm-bō')	71	45°32'N	91°05'W
Flaming Gorge Reservoir, res., U.S.	64	41°13'N	109°30'W
Flamingo, Fl., U.S. (flȧ-mĭŋ'gô)	83	25°10'N	80°55'W
Flamingo Cay, i., Bah. (flȧ-mĭŋ'gô)	93	22°50'N	75°50'W
Flamingo Point, c., V.I.U.S.	87c	18°19'N	65°00'W
Flanders, hist. reg., Fr. (flăn'dẽrz)	121	50°53'N	2°29'E
Flandreau, S.D., U.S. (flăn'drō)	70	44°02'N	96°35'W
Flathead, r., N.A.	55	49°30'N	114°30'W
Flathead, Middle Fork, r., Mt., U.S.	73	48°30'N	113°47'W
Flathead, North Fork, r., N.A.	73	48°45'N	114°20'W
Flathead, South Fork, r., Mt., U.S.	73	48°05'N	113°45'W
Flathead Indian Reservation, I.R., Mt., U.S.	73	47°30'N	114°25'W
Flathead Lake, l., Mt., U.S. (flăt'hĕd)	64	47°57'N	114°20'W
Flatow, Ger.	115b	52°44'N	12°58'E
Flat Rock, Mi., U.S. (flăt rŏk)	69b	42°06'N	83°17'W
Flattery, Cape, c., Wa., U.S. (flăt'ẽr-ĭ)	72	48°22'N	124°45'W
Flatwillow Creek, r., Mt., U.S. (flat wĭl'ô)	73	46°45'N	108°47'W
Flekkefjord, Nor. (flăk'kĕ-fyôr)	122	58°19'N	6°38'E
Flemingsburg, Ky., U.S. (flĕm'ĭngz-bûrg)	66	38°25'N	83°45'W
Flensburg, Ger. (flĕns'bòrgh)	116	54°48'N	9°27'E
Flers, Fr. (flĕr)	117	48°43'N	0°37'W
Fletcher, N.C., U.S.	83	35°26'N	82°30'W
Flinders, i., Austl.	175	39°35'S	148°10'E
Flinders, r., Austl.	175	18°48'S	141°07'E
Flinders, reg., Austl. (flĭn'dẽrz)	174	32°15'S	138°45'E
Flinders Reefs, rf., Austl.	175	17°30'S	149°02'E
Flin Flon, Can. (flĭn flŏn)	50	54°46'N	101°53'W
Flint, Wales, U.K.	114a	53°15'N	3°07'W
Flint, Mi., U.S.	65	43°00'N	83°45'W
Flint, r., Ga., U.S. (flĭnt)	65	31°25'N	84°15'W
Flintshire, co., Wales, U.K.	114a	53°13'N	3°00'W
Flora, Il., U.S. (flō'rȧ)	66	38°40'N	88°25'W
Flora, In., U.S.	66	40°25'N	86°30'W
Florala, Al., U.S. (flŏr-ăl'ȧ)	82	31°01'N	86°19'W
Floral Park, N.Y., U.S. (flŏr'ăl pärk)	68a	40°42'N	73°42'W
Florence, Italy	110	43°47'N	11°15'E
Florence, Al., U.S. (flŏr'ĕns)	65	34°46'N	87°40'W
Florence, Az., U.S.	77	33°00'N	111°25'W
Florence, Co., U.S.	78	38°23'N	105°08'W
Florence, Ks., U.S.	79	38°14'N	96°56'W
Florence, S.C., U.S.	83	34°10'N	79°45'W
Florence, Wa., U.S.	74a	48°13'N	122°21'W
Florencia, Col. (flō-rĕn'sē-ȧ)	100	1°31'N	75°13'W
Florencio Sánchez, Ur. (flō-rĕn-sēō-sá'n-chĕz)	99c	33°52'S	57°24'W
Florencio Varela, Arg. (flō-rĕn'sē-o vä-rā'lä)	102a	34°50'S	58°16'W
Flores, Braz. (flō'rĕzh)	101	7°57'S	37°48'W
Flores, Guat.	90a	16°53'N	89°54'W
Flores, dept., Ur.	99c	33°33'S	57°00'W
Flores, i., Indon.	168	8°14'S	121°08'E
Flores, r., Arg.	99c	36°13'S	60°28'W
Flores, Laut (Flores Sea), sea, Indon.	168	7°09'S	120°30'E
Floresville, Tx., U.S. (flō'rĕs-vĭl)	80	29°10'N	98°08'W
Floriano, Braz. (flō-rà-ä'nò)	101	6°17'S	42°58'W
Florianópolis, Braz. (flō-rē-ä-nō'pô-lēs)	102	27°30'S	48°30'W
Florida, Col. (flō-rē'dä)	100a	3°20'N	76°12'W
Florida, Cuba	92	22°10'N	79°50'W
Florida, S. Afr.	187b	26°11'S	27°56'E
Florida, Ur. (flō-rē-dhä)	102	34°06'S	56°14'W
Florida, N.Y., U.S. (flŏr'ĭ-dȧ)	68a	41°20'N	74°21'W
Florida, state, U.S. (flŏr'ĭ-dȧ)	65	30°30'N	84°40'W
Florida, dept., Ur. (flō-rē'dhä)	99c	33°48'S	56°15'W
Florida, i., Sol. Is.	175	8°56'S	159°45'E
Florida, Straits of, strt., N.A.	87	24°10'N	81°00'W
Florida Bay, b., Fl., U.S. (flŏr'ĭ-dȧ)	83a	24°55'N	80°55'W
Florida Keys, is., Fl., U.S.	65	24°33'N	81°20'W
Florida Mountains, mts., N.M., U.S.	77	32°10'N	107°35'W
Florido, Río, r., Mex. (flō-rē'dō)	80	27°21'N	104°48'W
Floridsdorf, Aus. (flō'rĭds-dôrf)	115e	48°16'N	16°25'E
Florina, Grc. (flō-rē'nä)	119	40°48'N	21°24'E
Florissant, Mo., U.S. (flŏr'ĭ-sănt)	75e	38°47'N	90°20'W
Floyd, r., Ia., U.S. (floid)	70	42°38'N	96°15'W
Floydada, Tx., U.S. (floi-dā'dȧ)	78	33°59'N	101°19'W
Floyds Fork, r., Ky., U.S. (floi-dz)	69h	38°08'N	85°30'W
Flumendosa, r., Italy	130	39°45'N	9°18'E
Flushing, Mi., U.S. (flŭsh'ĭng)	66	43°05'N	83°50'W
Fly, r., Pap. N. Gui. (flī)	169	8°00'S	141°45'E
Foča, Bos. (fō'chä)	131	43°29'N	18°48'E
Fochville, S. Afr. (fŏk'vĭl)	192c	26°29'S	27°29'E
Focşani, Rom. (fòk-shä'nê)	125	45°41'N	27°17'E
Fogang, China (fwo-gäŋ)	165	23°50'N	113°35'E
Foggia, Italy (fôd'jä)	119	41°30'N	15°34'E
Fogo, Can. (fō'gō)	61	49°43'N	54°17'W
Fogo, i., Can.	59	49°40'N	54°13'W
Fogo, i., C.V.	184b	14°46'N	24°51'W
Fohnsdorf, Aus. (fōns'dôrf)	124	47°13'N	14°40'E
Föhr, i., Ger. (fûr)	124	54°47'N	8°30'E
Foix, Fr. (fwä)	126	42°58'N	1°34'E
Fokku, Nig.	189	11°40'N	4°31'E
Folādī, Koh-e, mtn., Afg.	155	34°38'N	67°32'E
Folgares, Ang.	190	14°54'S	15°08'E
Foligno, Italy (fō-lēn'yō)	130	42°58'N	12°41'E
Folkeston, Eng., U.K.	121	51°05'N	1°18'E
Folkingham, Eng., U.K. (fō'kĭng-ăm)	114a	52°53'N	0°24'W
Folkston, Ga., U.S.	83	30°50'N	82°01'W
Folsom, Ca., U.S.	76	38°40'N	121°10'W
Folsom, N.M., U.S. (fōl'sŭm)	78	36°47'N	103°56'W
Fomento, Cuba (fō-mĕ'n-tō)	92	21°35'N	78°20'W
Fómeque, Col. (fō'mĕ-kĕ)	100a	4°29'N	73°52'W
Fonda, Ia., U.S. (fŏn'dȧ)	71	42°33'N	94°51'W
Fond du Lac, Wi., U.S. (fŏn dū lăk')	65	43°47'N	88°29'W
Fond du Lac Indian Reservation, I.R., Mn., U.S.	71	46°44'N	93°04'W
Fondi, Italy (fōn'dē)	130	41°23'N	13°25'E
Fonseca, Golfo de, b., N.A. (gôl-fô-dĕ-fōn-sä'kä)	86	13°09'N	87°55'W

ng-sing; ŋ-baŋk; ɴ-nasalized n; nŏd; cŏmmit; ōld; ôbey; ôrder; oi-boil; fōōd; ò-as oo in foot; ou-out; s-soft; sh-dish; th-thin; pūre; ûnite; ûrn; stŭd; circŭs; ü-as in French tu; '-indeterminate vowel.

248 Fontainebleau – Freienried PRONOUNCING INDEX

PLACE (Pronunciation)	PAGE	LAT.	LONG.
Fontainebleau, Fr. (fôn-tĕn-blō')	117	48°24'N	2°42'E
Fontana, Ca., U.S. (fŏn-tă'nȧ)	75a	34°06'N	117°27'W
Fonte Boa, Braz. (fōn'tȧ bō'ȧ)	100	2°32'S	66°05'W
Fontenay-le-Comte, Fr. (fônt-nĕ'lē-kônt')	126	46°28'N	0°53'W
Fontenay-Trésigny, Fr. (fôn-te-nā' tra-sēn-yē')	127b	48°43'N	2°53'E
Fontenelle Reservoir, res., Wy., U.S.	73	42°05'N	110°05'W
Fontera, Punta, c., Mex. (pōō'n-tä-fōn-tĕ'rä)	89	18°36'N	92°43'W
Fontibón, Col. (fōn-tē-bōn')	100a	4°42'N	74°09'W
Fontur, c., Ice.	112	66°21'N	14°02'W
Foothills, S. Afr. (fŏt-hĭls)	187b	25°55'S	27°36'E
Footscray, Austl.	173a	37°48'S	144°54'E
Foraker, Mount, mtn., Ak., U.S. (fôr'ȧ-kēr)	63	62°40'N	152°40'W
Forbach, Fr. (fōr'bäk)	127	49°12'N	6°54'E
Forbes, Austl. (fôrbz)	175	33°24'S	148°05'E
Forbes, Mount, mtn., Can.	55	51°52'N	116°56'W
Forchheim, Ger. (fōrк'hīm)	124	49°43'N	11°05'E
Fordyce, Ar., U.S. (fôr'dīs)	79	33°48'N	92°24'W
Forécariah, Gui. (fōr-kȧ-rē'ä')	184	9°26'N	13°06'W
Forel, Mont, mtn., Grnld.	49	65°50'N	37°41'W
Forest, Ms., U.S. (fŏr'ĕst)	82	32°22'N	89°29'W
Forest, r., N.D., U.S.	70	48°08'N	97°45'W
Forest City, Ia., U.S.	71	43°14'N	93°40'W
Forest City, N.C., U.S.	83	35°20'N	81°52'W
Forest City, Pa., U.S.	67	41°35'N	75°30'W
Forest Grove, Or., U.S. (grōv)	74c	45°31'N	123°07'W
Forest Hill, Md., U.S.	68e	39°35'N	76°26'W
Forest Hill, Tx., U.S.	75c	32°40'N	97°16'W
Forestville, Can. (fŏr'ĕst-vĭl)	60	48°45'N	69°06'W
Forestville, Md., U.S.	68e	38°51'N	76°55'W
Forez, Monts du, mts., Fr. (môn dü fô-rā')	126	44°55'N	3°43'E
Forfar, Scot., U.K. (fôr'fȧr)	120	57°10'N	2°55'W
Forillon, Parc National, rec., Can.	60	48°50'N	64°05'W
Forio, mtn., Italy (fô'ryō)	129c	40°29'N	13°55'E
Forked Creek, r., Il., U.S. (fôrk'd)	69a	41°16'N	88°01'W
Forked Deer, r., Tn., U.S.	82	35°53'N	89°29'W
Forli, Italy (fôr-lē')	118	44°13'N	12°03'E
Formby, Eng., U.K. (fôrm'bē)	114a	53°34'N	3°04'W
Formby Point, c., Eng., U.K.	114a	53°33'N	3°06'W
Formentera, Isla de, i., Spain (ē's-lä-dĕ-fôr-mĕn-tā'rä)	118	38°43'N	1°25'E
Formiga, Braz. (fôr-mē'gȧ)	101	20°27'S	45°25'W
Formigas Bank, bk., (fŏr-mē'gäs)	93	18°30'N	75°40'W
Formosa, Arg. (fôr-mō'sä)	102	27°25'S	58°12'W
Formosa, Braz.	101	15°32'S	47°10'W
Formosa, prov., Arg.	102	24°30'S	60°45'W
Formosa, Serra, mts., Braz. (sĕ'r-rä)	101	12°59'S	55°11'W
Formosa Bay, b., Kenya	191	2°45'S	40°30'E
Formosa Strait *see* Taiwan Strait, strt., Asia	161	24°30'N	120°00'E
Fornosovo, Russia (fôr-nô'sô vô)	142c	59°35'N	30°34'E
Forrest City, Ar., U.S. (fŏr'ĕst sĭ'tĭ)	79	35°00'N	90°46'W
Forsayth, Austl. (fôr-sīth')	175	18°33'S	143°42'E
Forshaga, Swe. (fôrs'hä'gä)	122	59°34'N	13°25'E
Forst, Ger. (fôrst)	117	51°45'N	14°38'E
Forsyth, Ga., U.S. (fôr-sīth')	82	33°02'N	83°56'W
Forsyth, Mt., U.S.	73	46°15'N	106°41'W
Fort Albany, Can. (fôrt ôl'bȧ nĭ)	51	52°20'N	81°30'W
Fort Alexander Indian Reserve, I.R., Can.	57	50°27'N	96°15'W
Fortaleza, Braz. (fôr'tä-lā'zȧ)	101	3°35'S	38°31'W
Fort Apache Indian Reservation, I.R., Az., U.S. (ȧ-păch'ē)	77	34°02'N	110°27'W
Fort Atkinson, Wi., U.S. (ăt'kĭn-sŭn)	71	42°55'N	88°46'W
Fort Beaufort, S. Afr. (bō'fōrt)	187c	32°47'S	26°39'E
Fort Belknap Indian Reservation, I.R., Mt., U.S.	73	48°16'N	108°38'W
Fort Bellefontaine, Mo., U.S. (bĕl-fōn-tān')	75f	38°50'N	90°15'W
Fort Benton, Mt., U.S. (bĕn'tŭn)	73	47°51'N	110°40'W
Fort Berthold Indian Reservation, I.R., N.D., U.S. (bĕrth'ôld)	70	47°47'N	103°28'W
Fort Bragg, Ca., U.S.	76	39°26'N	123°48'W
Fort Branch, In., U.S. (brȧnch)	66	38°15'N	87°35'W
Fort Chipewyan, Can.	50	58°46'N	111°15'W
Fort Cobb Reservoir, res., Ok., U.S.	78	35°12'N	98°28'W
Fort Collins, Co., U.S. (kŏl'ĭns)	64	40°36'N	105°04'W
Fort Crampel, C.A.R. (krȧm-pĕl')	185	6°59'N	19°11'E
Fort-de-France, Mart. (dē fräns)	87	14°37'N	61°06'W
Fort Deposit, Al., U.S. (dē-pŏz'ĭt)	82	31°58'N	86°35'W
Fort-de-Possel, C.A.R. (dē pô-sĕl')	185	5°03'N	19°11'E
Fort Dodge, Ia., U.S. (dŏj)	71	42°31'N	94°10'W
Fort Edward, N.Y., U.S. (wĕrd)	67	43°15'N	73°30'W
Fort Erie, Can. (ē'rĭ)	69c	42°55'N	78°56'W
Fortescue, r., Austl. (fôr'tĕs-kū)	174	21°25'S	116°50'E
Fort Fairfield, Me., U.S. (fâr'fēld)	60	46°46'N	67°53'W
Fort Fitzgerald, Can. (fĭts-jĕr'ăld)	50	59°48'N	111°50'W
Fort Frances, Can. (frăn'sĕs)	51	48°36'N	93°24'W
Fort Frederica National Monument, rec., Ga., U.S. (frĕd'ē-rĭ-kȧ)	82	31°13'N	85°25'W
Fort Gaines, Ga., U.S. (gānz)	82	31°35'N	85°03'W
Fort Gibson, Ok., U.S. (gĭb'sŭn)	79	35°50'N	95°13'W
Fort Good Hope, Can. (go͝od hōp)	50	66°19'N	128°52'W
Forth, Firth of, b., Scot., U.K. (fûrth ŏv fôrth)	112	56°04'N	3°03'W
Fort Hall, Kenya (hôl)	187	0°47'S	37°13'E
Fort Hall Indian Reservation, I.R., Id., U.S.	73	43°02'N	112°21'W
Fort Huachuca, Az., U.S. (wä-chōō'kä)	77	31°30'N	110°25'W
Fortier, Can. (fôr'tyā')	62f	49°56'N	97°55'W
Fort Kent, Me., U.S. (kĕnt)	60	47°14'N	68°37'W
Fort Langley, Can. (lăng'lĭ)	74d	49°10'N	122°35'W
Fort Lauderdale, Fl., U.S. (lô'dēr-dāl)	83a	26°07'N	80°09'W
Fort Lee, N.J., U.S.	68a	40°50'N	73°58'W
Fort Liard, Can.	50	60°16'N	123°34'W
Fort Loudoun Lake, res., Tn., U.S. (fôrt lou'dĕn)	82	35°52'N	84°10'W
Fort Lupton, Co., U.S. (lŭp'tŭn)	78	40°04'N	104°54'W
Fort Macleod, Can. (mȧ-kloud')	50	49°43'N	113°25'W
Fort Madison, Ia., U.S. (măd'ĭ-sŭn)	71	40°40'N	91°17'W
Fort Matanzas, Fl., U.S. (mä-tän'zäs)	83	29°39'N	81°17'W
Fort McDermitt Indian Reservation, I.R., Or., U.S. (măk dēr'mĭt)	72	42°04'N	118°07'W
Fort McMurray, Can. (măk-mŭr'ĭ)	50	56°44'N	111°23'W
Fort McPherson, Can. (măk-fŭr's'n)	50	67°37'N	134°59'W
Fort Meade, Fl., U.S. (mēd)	83a	27°45'N	81°48'W
Fort Mill, S.C., U.S. (mĭl)	83	35°03'N	80°57'W
Fort Mojave Indian Reservation, I.R., Ca., U.S. (mô-hä'vâ)	76	34°59'N	115°02'W
Fort Morgan, Co., U.S. (môr'gȧn)	78	40°14'N	103°49'W
Fort Myers, Fl., U.S. (mī'ērz)	83a	26°36'N	81°45'W
Fort Nelson, Can. (nĕl'sŭn)	50	58°57'N	122°30'W
Fort Nelson, r., Can. (nĕl'sŭn)	52	58°44'N	122°20'W
Fort Payne, Al., U.S. (pān)	82	34°26'N	85°41'W
Fort Peck, Mt., U.S. (pĕk)	73	47°58'N	106°30'W
Fort Peck Indian Reservation, I.R., Mt., U.S.	70	48°22'N	105°40'W
Fort Peck Lake, res., Mt., U.S.	64	47°52'N	106°59'W
Fort Pierce, Fl., U.S. (pērs)	83a	27°25'N	80°20'W
Fort Portal, Ug. (pôr'tál)	185	0°40'N	30°16'E
Fort Providence, Can. (prŏv'ĭ-dĕns)	50	61°27'N	117°59'W
Fort Pulaski National Monument, rec., Ga., U.S. (pu-lăs'kĭ)	83	31°59'N	80°56'W
Fort Qu'Appelle, Can.	56	50°46'N	103°55'W
Fort Randall Dam, dam, S.D., U.S.	70	42°48'N	98°35'W
Fort Resolution, Can. (rĕz'ō-lū'shŭn)	50	61°08'N	113°42'W
Fort Riley, Ks., U.S. (rī'lĭ)	79	39°05'N	96°46'W
Fort Saint James, Can. (fôrt sânt jāmz)	50	54°26'N	124°15'W
Fort Saint John, Can. (sānt jŏn)	50	56°15'N	120°51'W
Fort Sandeman, Pak. (săn'da-măn)	155	31°28'N	69°29'E
Fort Saskatchewan, Can. (săs-kăt'chōō-ân)	62g	53°43'N	113°13'W
Fort Scott, Ks., U.S. (skŏt)	65	37°50'N	94°43'W
Fort Severn, Can. (sĕv'ērn)	51	55°58'N	87°50'W
Fort-Shevchenko, Kaz. (shĕv-chĕn'kô)	139	44°30'N	50°18'E
Fort Sibut, C.A.R. (fôr sē-bü')	185	5°44'N	19°05'E
Fort Sill, Ok., U.S. (fôrt sĭl)	78	34°41'N	98°25'W
Fort Simpson, Can. (sĭmp'sŭn)	50	61°52'N	121°48'W
Fort Smith, Can.	50	60°09'N	112°08'W
Fort Smith, Ar., U.S. (smĭth)	65	35°23'N	94°24'W
Fort Stockton, Tx., U.S. (stŏk'tŭn)	80	30°54'N	102°51'W
Fort Sumner, N.M., U.S. (sŭm'nēr)	78	34°30'N	104°17'W
Fort Sumter National Monument, rec., S.C., U.S. (sŭm'tēr)	83	32°43'N	79°54'W
Fort Thomas, Ky., U.S. (tŏm'ăs)	69f	39°05'N	84°27'W
Fortuna, Ca., U.S. (fôr-tū'nȧ)	72	40°36'N	124°10'W
Fortune, Can. (fôr'tŭn)	61	47°04'N	55°51'W
Fortune, i., Bah.	93	22°35'N	74°20'W
Fortune Bay, b., Can.	53a	47°25'N	55°25'W
Fort Union National Monument, rec., N.M., U.S. (ūn'yŭn)	78	35°51'N	104°57'W
Fort Valley, Ga., U.S. (văl'ĭ)	82	32°33'N	83°53'W
Fort Vermilion, Can. (vēr-mĭl'yŭn)	50	58°23'N	115°50'W
Fort Victoria *see* Masvingo, Zimb.	186	20°07'S	30°47'E
Fort Wayne, In., U.S. (wān)	65	41°00'N	85°10'W
Fort William, Scot., U.K. (wĭl'yăm)	120	56°50'N	3°00'W
Fort William, Mount, mtn., Austl. (wĭ'ĭ-ăm)	176	24°45'S	151°15'E
Fort Worth, Tx., U.S. (wûrth)	64	32°45'N	97°20'W
Fort Yukon, Ak., U.S. (yōō'kŏn)	64a	66°30'N	145°00'W
Fort Yuma Indian Reservation, I.R., Ca., U.S. (yōō'mä)	77	32°54'N	114°47'W
Foshan, China	161	23°02'N	113°07'E
Fossano, Italy (fôs-sä'nō)	130	44°34'N	7°42'E
Fossil Creek, r., Tx., U.S. (fŏs-ĭl)	75c	32°53'N	97°19'W
Fossombrone, Italy (fôs-sôm-brō'nā)	130	43°41'N	12°48'E
Foss Res, Ok., U.S.	78	35°38'N	99°11'W
Fosston, Mn., U.S. (fŏs'tŭn)	70	47°34'N	95°44'W
Fosterburg, Il., U.S. (fŏs'tēr-bûrg)	75e	38°58'N	90°04'W
Fostoria, Oh., U.S. (fŏs-tō'rĭ-ȧ)	66	41°10'N	83°20'W
Fougéres, Fr. (fōō-zhâr')	117	48°23'N	1°14'W
Foula, i., Scot., U.K. (fou'lȧ)	120a	60°08'N	2°04'W
Foulwind, Cape, c., N.Z. (foul'wĭnd)	175a	41°45'S	171°00'E
Foumban, Cam. (fōōm-bän')	184	5°43'N	10°55'E
Fountain Creek, r., Co., U.S. (foun'tĭn)	78	38°36'N	104°37'W
Fountain Valley, Ca., U.S.	75a	33°42'N	117°57'W
Fourche la Fave, r., Ar., U.S. (fōōrsh lä fāv')	79	34°46'N	93°45'W
Fouriesburg, S. Afr. (fō'rēz-bûrg)	192c	28°38'S	28°13'E
Fourmies, Fr. (fōōr-mē')	126	50°01'N	4°01'E
Four Mountains, Islands of the, is., Ak., U.S.	63a	52°58'N	170°40'W
Fourth Cataract, wtfl., Sudan	185	18°52'N	32°07'E
Fouta Djallon, mts., Gui. (fōō'tä jä-lôn)	184	11°37'N	12°29'W
Foveaux Strait, strt., N.Z. (fō-vō')	175a	46°30'S	167°43'E
Fowler, Co., U.S. (foul'ēr)	78	38°04'N	104°02'W
Fowler, In., U.S.	66	40°35'N	87°20'W
Fowler, Point, c., Austl.	174	32°05'S	132°30'E
Fowlerton, Tx., U.S. (foul'ēr-tŭn)	80	28°26'N	98°48'W
Fox, i., Wa., U.S. (fŏks)	74a	47°15'N	122°08'W
Fox, r., Il., U.S.	71	41°35'N	88°43'W
Fox, r., Wi., U.S.	71	44°18'N	88°23'W
Foxboro, Ma., U.S. (fŏks'bŭrō)	61a	42°04'N	71°15'W
Foxe Basin, b., Can. (fŏks)	53	67°35'N	79°21'W
Foxe Channel, strt., Can.	53	64°30'N	79°23'W
Foxe Peninsula, pen., Can.	53	64°57'N	77°26'W
Fox Islands, is., Ak., U.S. (fŏks)	63a	53°04'N	167°30'W
Fox Lake, Il., U.S. (lăk)	69a	42°24'N	88°11'W
Fox Lake, l., Il., U.S.	69a	42°24'N	88°07'W
Fox Point, Wi., U.S.	69a	43°10'N	87°54'W
Foyle, Lough, b., Eur. (lŏk foil')	120	55°07'N	7°08'W
Foz do Cunene, Ang.	190	17°16'S	11°50'E
Fraga, Spain (frä'gä)	129	41°31'N	0°20'E
Fragoso, Cayo, i., Cuba (kä'yō-frä-gō'sō)	92	22°45'N	79°30'W
Framnes Mountains, mts., Ant.	178	67°50'S	62°35'E
Franca, Braz. (frä'n-kä)	101	20°28'S	47°20'W
Francavilla, Italy (fräŋ-kä-vēl'lä)	131	40°32'N	17°37'E
France, nation, Eur. (frȧns)	110	46°39'N	0°47'E
Frances, l., Can. (frăn'sĭs)	52	61°27'N	128°28'W
Frances, Cabo, c., Cuba (kä'bô-frän-sĕ's)	92	21°55'N	84°05'W
Frances, Punta, c., Cuba (pōō'n-tä-frän-sĕ's)	92	21°45'N	83°10'W
Francés Viejo, Cabo, c., Dom. Rep. (kä'bô-frän'sås vyā'hō)	93	19°40'N	69°35'W
Franceville, Gabon (fräns-vēl')	186	1°38'S	13°35'E
Francis Case, Lake, res., S.D., U.S. (frăn'sĭs)	64	43°15'N	99°00'W
Francisco Sales, Braz. (frän-sē's-kô-sá'lĕs)	99a	21°42'S	44°26'W
Francistown, Bots. (frăn'sĭs-toun)	186	21°17'S	27°28'E
Frankfort, S. Afr. (frănk'fŏrt)	187c	32°43'S	27°28'E
Frankfort, S. Afr.	192c	27°17'S	28°30'E
Frankfort, Il., U.S. (frăŋk'fŭrt)	69a	41°30'N	87°51'W
Frankfort, In., U.S.	66	40°15'N	86°30'W
Frankfort, Ks., U.S.	79	39°42'N	96°27'W
Frankfort, Ky., U.S.	65	38°10'N	84°55'W
Frankfort, Mi., U.S.	66	44°40'N	86°15'W
Frankfort, N.Y., U.S.	67	43°05'N	75°05'W
Frankfurt am Main, Ger.	110	50°07'N	8°40'E
Frankfurt an der Oder, Ger.	117	52°20'N	14°31'E
Franklin, S. Afr.	187c	30°19'S	29°28'E
Franklin, In., U.S. (frănk'lĭn)	66	39°25'N	86°00'W
Franklin, Ky., U.S.	82	36°42'N	86°34'W
Franklin, La., U.S.	81	29°47'N	91°31'W
Franklin, Ma., U.S.	61a	42°05'N	71°24'W
Franklin, Ne., U.S.	78	40°06'N	99°01'W
Franklin, N.H., U.S.	67	43°25'N	71°40'W
Franklin, N.J., U.S.	68a	41°08'N	74°35'W
Franklin, Oh., U.S.	66	39°30'N	84°20'W
Franklin, Pa., U.S.	67	41°25'N	79°50'W
Franklin, Tn., U.S.	82	35°54'N	86°54'W
Franklin, Va., U.S.	83	36°41'N	76°57'W
Franklin, l., Nv., U.S.	76	40°23'N	115°10'W
Franklin D. Roosevelt Lake, res., Wa., U.S.	72	48°12'N	118°43'W
Franklin Mountains, mts., Can.	52	65°36'N	125°55'W
Franklin Park, Il., U.S.	69a	41°56'N	87°53'W
Franklin Square, N.Y., U.S.	68a	40°43'N	73°40'W
Franklinton, La., U.S. (frăŋk'lĭn-tŭn)	81	30°49'N	90°09'W
Frankston, Austl.	173a	38°09'S	145°08'E
Franksville, Wi., U.S. (frănkz'vĭl)	69a	42°46'N	87°55'W
Fransta, Swe.	122	62°30'N	16°04'E
Franz Josef Land *see* Zemlya Frantsa-Iosifa, is., Russia	134	81°32'N	40°00'E
Frascati, Italy (fräs-kä'tē)	130	41°49'N	12°45'E
Fraser, Mi., U.S. (frā'zēr)	69b	42°32'N	82°57'W
Fraser, i., Austl.	175	25°12'S	153°00'E
Fraser, r., Can.	52	51°30'N	122°00'W
Fraserburgh, Scot., U.K. (frā'zēr-bûrg)	120	57°40'N	2°01'W
Fraser Plateau, plat., Can.	55	51°30'N	122°00'W
Frattamaggiore, Italy (frät-tä-mäg-zhyô'rē)	129c	40°41'N	14°16'E
Fray Bentos, Ur. (frī bĕn'tōs)	102	33°10'S	58°19'W
Frazee, Mn., U.S. (frā-zē')	70	46°42'N	95°43'W
Fraziers Hog Cay, i., Bah.	92	25°25'N	77°55'W
Frechen, Ger. (frĕ'кĕn)	127c	50°54'N	6°49'E
Fredericia, Den. (frĕdh-ĕ-rē'tsē-ȧ)	122	55°35'N	9°45'E
Frederick, Md., U.S. (frĕd'ēr-ĭk)	65	39°25'N	77°25'W
Frederick, Ok., U.S.	78	34°23'N	99°01'W
Frederick House, r., Can.	58	49°05'N	81°20'W
Fredericksburg, Tx., U.S. (frĕd'ēr-ĭkz-bûrg)	80	30°16'N	98°52'W
Fredericksburg, Va., U.S.	67	38°20'N	77°30'W
Fredericktown, Mo., U.S. (frĕd'ēr-ĭk-toun)	79	37°32'N	90°16'W
Fredericton, Can. (frĕd'-ēr-ĭk-tŭn)	51	45°48'N	66°39'W
Frederikshavn, Den. (frĕdh'ĕ-rĕks-houn)	116	57°27'N	10°31'E
Frederikssund, Den. (frĕdh'ĕ-rĕks-sōn)	122	55°51'N	12°04'E
Fredonia, Col. (frĕ-dō'nyȧ)	100a	5°55'N	75°40'W
Fredonia, Ks., U.S. (frē-dō'nĭ-ȧ)	79	36°31'N	95°50'W
Fredonia, N.Y., U.S.	67	42°25'N	79°20'W
Fredrikstad, Nor. (frădh'rĕks-städ)	116	59°14'N	10°58'E
Freeburg, Il., U.S. (frē'bûrg)	75e	38°26'N	89°59'W
Freehold, N.J., U.S. (frē'hōld)	68a	40°15'N	74°16'W
Freeland, Pa., U.S. (frē'lănd)	67	41°00'N	75°50'W
Freeland, Wa., U.S.	74a	48°01'N	122°32'W
Freels, Cape, c., Can. (frēlz)	61	46°37'N	53°45'W
Freelton, Can. (frēl'tŭn)	62d	43°24'N	80°02'W
Freeport, Bah.	92	26°30'N	78°45'W
Freeport, Il., U.S. (frē'pōrt)	65	42°19'N	89°30'W
Freeport, N.Y., U.S.	68a	40°39'N	73°35'W
Freeport, Tx., U.S.	81	28°56'N	95°21'W
Freetown, S.L. (frē'toun)	184	8°30'N	13°15'W
Fregenal de la Sierra, Spain (frā-hā-näl' dā lä syĕr'rä)	128	38°09'N	6°40'W
Fregene, Italy (frĕ-zhĕ'-nĕ)	129d	41°52'N	12°12'E
Freiberg, Ger. (frī'bĕrgh)	117	50°54'N	13°18'E
Freiburg, Ger.	117	48°00'N	7°50'E
Freienried, Ger. (frī'ĕn-rēd)	115d	48°20'N	11°08'E

ăt; fĭnăl; rāte; senâte; ärm; ȧsk; sofȧ; fâre; ch-choose; dh-as th in other; bē; ēvent; bĕt; recĕnt; cratēr; g-gō; gh-guttural g; bĭt; ĭ-short neutral; rīde; ĸ-guttural k as ch in German ich;

PRONOUNCING INDEX

Freirina – Gardner Canal 249

PLACE (Pronunciation)	PAGE	LAT.	LONG.
Freirina, Chile (frå-ī-rē′nä)	102	28°35′S	71°26′W
Freising, Ger. (frī′zĭng)	124	48°25′N	11°45′E
Fréjus, Fr. (frā-zhüs′)	127	43°28′N	6°46′E
Fremantle, Austl. (frē′măn-t'l)	174	32°03′S	116°05′E
Fremont, Ca., U.S. (frē-mŏnt′)	74b	37°33′N	122°00′W
Fremont, Mi., U.S.	66	43°25′N	85°55′W
Fremont, Ne., U.S.	70	41°26′N	96°30′W
Fremont, Oh., U.S.	66	41°20′N	83°05′W
Fremont, r., Ut., U.S.	77	38°20′N	111°30′W
Fremont Peak, mtn., Wy., U.S.	73	43°05′N	109°35′W
French Broad, r., Tn., U.S. (frĕnch brôd)	82	35°59′N	83°01′W
French Frigate Shoals, Hi., U.S.	84b	23°30′N	167°10′W
French Guiana, dep., S.A. (gē-ä′nä)	101	4°20′N	53°00′W
French Lick, In., U.S. (frĕnch lĭk)	66	38°35′N	86°35′W
Frenchman, r., N.A.	56	49°25′N	108°30′W
Frenchman Creek, r., Mt., U.S. (frĕnch-m*ă*n)	73	48°51′N	107°20′W
Frenchman Creek, r., Ne., U.S.	78	40°24′N	101°50′W
Frenchman Flat, Nv., U.S.	76	36°55′N	116°11′W
French Polynesia, dep., Oc.	2	15°00′S	140°00′W
French River, Mn., U.S.	75h	46°54′N	91°54′W
Freshfield, Mount, mtn., Can. (frĕsh′fēld)	55	51°44′N	116°57′W
Fresnillo, Mex. (frås-nēl′yō)	86	23°10′N	102°52′W
Fresno, Col. (frĕs′nô)	100a	5°10′N	75°01′W
Fresno, Ca., U.S.	64	36°44′N	119°46′W
Fresno, r., Ca., U.S. (frĕz′nō)	76	37°00′N	120°24′W
Fresno Slough, Ca., U.S.	76	36°39′N	120°12′W
Freudenstadt, Ger. (froi′den-shtät)	124	48°28′N	8°26′E
Freycinet Peninsula, pen., Austl. (frā-sē-nĕ′)	176	42°13′S	148°56′E
Fria, Gui.	188	10°05′N	13°32′W
Fria, r., Az., U.S. (frē-ä)	77	34°03′N	112°12′W
Fria, Cape, c., Nmb. (frī*á*)	186	18°15′S	12°10′E
Friant-Kern Canal, can., Ca., U.S. (kûrn)	76	36°57′N	119°37′W
Frias, Arg. (frē-äs)	102	28°43′S	65°03′W
Fribourg, Switz. (frē-bōōr′)	117	46°48′N	7°07′E
Fridley, Mn., U.S. (frĭd′lĭ)	75g	45°05′N	93°16′W
Friedberg, Ger. (frēd′bĕrgh)	115d	48°22′N	11°00′E
Friedland, Ger. (frēt′länt)	124	53°39′N	13°34′E
Friedrichshafen, Ger. (frē-drēks-häf′ĕn)	124	47°39′N	9°28′E
Friend, Ne., U.S. (frĕnd)	79	40°40′N	97°16′W
Friendswood, Tx., U.S. (frĕnds′wŏŏd)	81a	29°31′N	95°11′W
Fries, Va., U.S. (frēz)	83	36°42′N	80°59′W
Friesack, Ger. (frē′säk)	115b	52°44′N	12°35′E
Frio, Cabo, c., Braz. (kä′bō-frē′ō)	101	22°58′S	42°08′W
Frio, r., Tx., U.S.	80	29°00′N	99°15′W
Frisian Islands, is., Neth. (frē′zh*ă*n)	116	53°30′N	5°20′E
Friuli-Venezia Giulia, hist. reg., Italy	130	46°20′N	13°20′E
Frobisher Bay, b., Can.	53	62°49′N	66°41′W
Frobisher Lake, l., Can. (frŏb′ĭsh-ēr)	52	56°25′N	108°20′W
Frodsham, Eng., U.K. (frŏdz′*ă*m)	114a	53°18′N	2°48′W
Frohavet, b., Nor.	116	63°49′N	9°12′E
Frome, Lake, l., Austl. (frōōm)	174	30°40′S	140°13′E
Frontenac, Ks., U.S. (frŏn′tē-năk)	79	37°27′N	94°41′W
Frontera, Mex. (frôn-tā′rä)	89	18°34′N	92°38′W
Front Range, mts., Co., U.S. (frŭnt)	78	40°59′N	105°29′W
Front Royal, Va., U.S. (frŭnt)	67	38°55′N	78°10′W
Frosinone, Italy (frō-zē-nō′nā)	130	41°38′N	13°22′E
Frostburg, Md., U.S. (frôst′bûrg)	67	39°40′N	78°55′W
Fruita, Co., U.S. (frōōt-*á*)	77	39°10′N	108°45′W
Frunze *see* Bishkek, Kyrg.	139	42°49′N	74°42′E
Fryanovo, Russia (f'ryä′nô-vô)	142b	56°08′N	38°28′E
Fryazino, Russia (f'ryä′zĭ-nô)	142b	55°58′N	38°05′E
Frydlant, Czech Rep. (frēd′länt)	124	50°56′N	15°05′E
Fucheng, China (fōō-chŭŋ)	162	37°53′N	116°08′E
Fuchu, Japan (fōō′chōō)	167a	35°41′N	139°29′E
Fuchun, r., China (fōō-chòn)	165	29°50′N	120°00′E
Fuego, vol., Guat. (fwā′gō)	90	14°29′N	90°52′W
Fuencarral, Spain (fuān-kär-räl′)	129a	40°29′N	3°42′W
Fuensalida, Spain (fwān-sä-lē′dä)	128	40°04′N	4°15′W
Fuente, Mex. (fwĕ′n-tĕ′)	80	28°39′N	100°34′W
Fuente de Cantos, Spain (fwĕn′tå dā kän′tōs)	128	38°15′N	6°18′W
Fuente el Saz, Spain (fwĕn′tå ĕl säth′)	129a	40°39′N	3°30′W
Fuenteobejuna, Spain	128	38°15′N	5°30′W
Fuentesaúco, Spain (fwĕn-tå-sä-ōō′kō)	128	41°18′N	5°25′W
Fuerte, Río del, r., Mex. (rē′ō-dĕl-fōō-ĕ′r-tĕ)	86	26°15′N	108°50′W
Fuerte Olimpo, Para. (fwĕr′tå ō-lēm-pō)	102	21°10′S	57°49′W
Fuerteventura Island, i., Spain (fwĕr′tå-vĕn-tōō′rä)	184	28°24′N	13°21′W
Fuhai, China	160	47°01′N	87°07′E
Fuji, Japan (jōō′jē)	167	35°11′N	138°44′E
Fuji, r., Japan	167	35°20′N	138°23′E
Fujian, prov., China (fōō-jyĕn)	161	25°40′N	117°30′E
Fujidera, Japan	167b	34°34′N	135°37′E
Fujin, China (fōō-jyĭn)	161	47°13′N	132°11′E
Fuji San, mtn., Japan (fōō′jē sän)	161	35°23′N	138°44′E
Fujisawa, Japan (fōō′jē-sä′wa)	167a	35°20′N	139°29′E
Fujiyama *see* Fuji San, mtn., Japan	161	35°23′N	138°44′E
Fukuchiyama, Japan (fō′kō-chē-yä′mä)	167	35°18′N	135°07′E
Fukue, i., Japan (fō-kōō′ä)	166	32°40′N	129°02′E
Fukui, Japan (fōō′kōō-ē)	161	36°05′N	136°14′E
Fukuoka, Japan (fōō′kô-ō′kä)	161	33°35′N	130°23′E
Fukuoka, Japan	167a	35°52′N	139°31′E
Fukushima, Japan (fōō′kô-shē′mä)	166	37°45′N	140°29′E
Fukuyama, Japan (fōō′kô-yä′mä)	166	34°31′N	133°21′E
Fulda, Ger.	117	50°33′N	9°41′E
Fulda, r., Ger. (fōl′dä)	124	51°05′N	9°40′E
Fuling, China (fōō-lĭŋ)	160	29°40′N	107°30′E
Fullerton, Ca., U.S. (fŏŏl′ēr-t*ŭ*n)	75a	33°53′N	117°56′W
Fullerton, La., U.S.	81	31°00′N	93°00′W
Fullerton, Ne., U.S.	70	41°21′N	97°59′W
Fulton, Ky., U.S. (fŭl′tŭn)	82	36°30′N	88°53′W
Fulton, Mo., U.S.	79	38°51′N	91°56′W
Fulton, N.Y., U.S.	67	43°20′N	76°25′W
Fultondale, Al., U.S. (fŭl′tŭn-dāl)	68h	33°37′N	86°48′W
Funabashi, Japan (fōō′nä-bä′shē)	167	35°43′N	139°59′E
Funaya, Japan (fōō-nä′yä)	167b	34°45′N	135°52′E
Funchal, Port. (fòn-shäl′)	184	32°41′N	16°15′W
Fundación, Col. (fōōn-dä-syō′n)	100	10°43′N	74°13′W
Fundão, Port. (fòn-douɴ′)	128	40°08′N	7°32′W
Fundy, Bay of, b., Can. (fŭn′dĭ)	53	45°00′N	66°00′W
Fundy National Park, rec., Can.	53	45°38′N	65°00′W
Funing, China (fōō-nĭŋ)	164	33°55′N	119°54′E
Funing, China	162	39°55′N	119°16′E
Funing Wan, b., China	165	26°48′N	120°35′E
Funtua, Nig.	189	11°31′N	7°17′E
Furancungo, Moz.	191	14°55′S	33°35′E
Furbero, Mex. (fōōr-bĕ′rô)	89	20°21′N	97°32′W
Furgun, mtn., Iran	154	28°47′N	57°00′E
Furmanov, Russia (fûr-mä′nôf)	136	57°14′N	41°11′E
Furnas, Reprêsa de, res., Braz.	101	21°00′S	46°00′W
Furneaux Group, is., Austl. (fûr′nō)	175	40°15′S	146°27′E
Fürstenfeld, Aus. (für′stĕn-fĕlt)	124	47°02′N	16°03′E
Fürstenfeldbruck, Ger. (für′stĕn-fĕld′brŏŏk)	115d	48°11′N	11°16′E
Fürstenwalde, Ger. (für′stĕn-väl-dĕ)	124	52°21′N	14°04′E
Fürth, Ger. (fürt)	117	49°28′N	11°03′E
Furuichi, Japan (fōō′rò-ē′chē)	167b	34°33′N	135°37′E
Fusa, Japan (fōō′sä)	167a	35°52′N	140°08′E
Fuse, Japan	167b	34°40′N	135°33′E
Fushimi, Japan (fōō′shē-mē)	167b	34°57′N	135°47′E
Fushun, China (fōō′shōōn′)	161	41°50′N	124°00′E
Fusong, China (fōō-sòŋ)	164	42°12′N	127°12′E
Futtsu, Japan (fōō′tsōō′)	167a	35°19′N	139°49′E
Futtsu Misaki, c., Japan (fōōt′tsōō′ mē-sä′kē)	167a	35°19′N	139°46′E
Fuwah, Egypt (fōō′wä)	192b	31°13′N	30°35′E
Fuxian, China (fōō-shyĕn)	162	39°36′N	121°59′E
Fuxin, China (fōō-shyĭn)	164	42°05′N	121°40′E
Fuyang, China (fōō-yäŋ)	161	32°53′N	115°48′E
Fuyang, China	165	30°10′N	119°58′E
Fuyang, r., China (fōō-yäŋ)	162	36°59′N	114°48′E
Fuyu, China (fōō-yōō)	161	45°20′N	125°00′E
Fuzhou, China (fōō-jō)	161	26°02′N	119°18′E
Fuzhou, r., China	162	39°38′N	121°43′E
Fuzhoucheng, China (fōō-jō-chŭŋ)	162	39°46′N	121°44′E
Fyn, i., Den. (fü′n)	122	55°24′N	10°33′E
Fyne, Loch, l., Scot., U.K. (fīn)	120	56°14′N	5°10′W
Fyresvatn, l., Nor.	122	59°04′N	7°55′E

G

PLACE (Pronunciation)	PAGE	LAT.	LONG.
Gaalkacyo, Som.	192a	7°00′N	47°30′E
Gabela, Ang.	190	10°48′S	14°20′E
Gabès, Tun. (gä′bĕs)	184	33°51′N	10°04′E
Gabès, Golfe de, b., Tun.	184	32°22′N	10°59′E
Gabil, Chad	189	11°09′N	18°12′E
Gąbin, Pol. (gôɴ′bĕn)	125	52°23′N	19°47′E
Gabon, nation, Afr. (gà-bôɴ′)	186	0°30′S	10°45′E
Gaborone, Bots.	186	24°28′S	25°59′E
Gabriel, r., Tx., U.S. (gā′brĭ-ĕl)	81	30°38′N	97°15′W
Gabrovo, Blg. (gäb′rô-vō)	131	42°52′N	25°19′E
Gachsārān, Iran	157	30°12′N	50°47′E
Gacko, Bos. (gäts′kô)	131	43°10′N	18°34′E
Gadsden, Al., U.S. (gădz′d*ĕ*n)	65	34°00′N	86°00′W
Găeşti, Rom. (gä-yĕsh′tē)	131	44°43′N	25°21′E
Gaeta, Italy (gä-å′tä)	130	41°18′N	13°34′E
Gaffney, S.C., U.S. (găf′nĭ)	83	35°04′N	81°47′W
Gafsa, Tun. (gäf′sä)	184	34°16′N	8°37′E
Gagarin, Russia	132	55°32′N	34°58′E
Gagnoa, C. Iv.	188	6°08′N	5°56′W
Gagra, Geor.	138	43°20′N	40°15′E
Gaillac-sur-Tarn, Fr. (gà-yäk′sür-tärn′)	126	43°54′N	1°52′E
Gaillard Cut, reg., Pan. (gä-ēl-yä′rd)	86a	9°03′N	79°42′W
Gainesville, Fl., U.S. (gānz′vĭl)	65	29°40′N	82°20′W
Gainesville, Ga., U.S.	82	34°16′N	83°48′W
Gainesville, Tx., U.S.	79	33°38′N	97°08′W
Gainsborough, Eng., U.K. (gānz′b*ŭr*-ô)	114a	53°23′N	0°46′W
Gairdner, Lake, l., Austl. (gård′nēr)	174	32°20′S	136°30′E
Gaithersburg, Md., U.S. (gā′thērs′bûrg)	68e	39°08′N	77°13′W
Gaixian, China (gī-shyĕn)	164	40°25′N	122°20′E
Galana, r., Kenya	191	3°00′S	39°30′E
Galapagar, Spain (gä-lä-pä-gär′)	129a	40°36′N	4°00′W
Galapagos Islands *see* Colón, Archipiélago de, is., Ec.	100	0°10′S	87°45′W
Galaria, r., Italy	129d	41°58′N	12°21′E
Galashiels, Scot., U.K. (găl-*á*-shēlz)	120	55°40′N	2°57′W
Galaţi, Rom.	110	45°25′N	28°05′E
Galatina, Italy (gä-lä-tē′nä)	131	40°10′N	18°12′E
Galaxídi, Grc.	131	38°26′N	22°22′E
Galdhøpiggen, mtn., Nor.	122	61°37′N	8°17′E
Galeana, Mex. (gä-lå-ä′nä)	80	24°50′N	100°04′W
Galena, Il., U.S. (g*á*-lē′n*á*)	71	42°26′N	90°27′W
Galena, In., U.S.	69h	38°21′N	85°55′W
Galena Peak, mtn., Tx., U.S.	81a	29°44′N	95°14′W
Galera, Cerro, mtn., Pan. (sĕ′r-rô-gä-lĕ′rä)	86a	8°55′N	79°38′W
Galeras, vol., Col. (gä-lĕ′räs)	100	0°57′N	77°27′W
Gales, r., Or., U.S. (gālz)	74c	45°33′N	123°11′W
Galesburg, Il., U.S. (gālz′bûrg)	65	40°56′N	90°21′W
Galesville, Wi., U.S. (gālz′vĭl)	71	44°04′N	91°22′W
Galeton, Pa., U.S. (gāl′t*ŭ*n)	67	41°45′N	77°40′W
Galich, Russia (gál′ĭch)	136	58°20′N	42°38′E
Galicia, hist. reg., Pol. (gá-lĭsh′ĭ-á)	125	49°48′N	21°05′E
Galicia, hist. reg., Spain (gä-lē′thyä)	128	43°35′N	8°03′W
Galilee, l., Austl. (găl′ĭ-lē)	175	22°23′S	145°09′E
Galilee, Sea of, l., Isr.	153a	32°53′N	35°45′E
Galina Point, c., Jam. (gä-lē′nä)	92	18°25′N	76°50′W
Galion, Oh., U.S. (găl′ĭ-*ŭ*n)	66	40°45′N	82°50′W
Galisteo, N.M., U.S. (gä-lĭs-tå′ō)	78	35°20′N	106°00′W
Gallarate, Italy (gäl-lä-rä′tā)	130	45°37′N	8°48′E
Gallardon, Fr. (gä-lär-dôɴ′)	127b	48°31′N	1°40′E
Gallatin, Mo., U.S. (găl′*á*-tĭn)	79	39°55′N	93°58′W
Gallatin, Tn., U.S.	82	36°23′N	86°28′W
Gallatin, r., Mt., U.S.	73	45°12′N	111°10′W
Galle, Sri L. (gäl)	159	6°13′N	80°10′E
Gállego, r., Spain (gäl-yā′gō)	129	42°27′N	0°37′W
Gallinas, Punta de, c., Col. (gä-lyē′näs)	100	12°10′N	72°10′W
Gallipoli, Italy (gäl-lē′pô-lē)	131	40°03′N	17°58′E
Gallipoli *see* Gelibolu, Tur. (gäl-lē′pô-lē)	119	40°25′N	26°40′E
Gallipoli Peninsula, pen., Tur.	131	40°23′N	25°10′E
Gallipolis, Oh., U.S. (găl-ĭ-pô-lēs)	66	38°50′N	82°10′W
Gällivare, Swe. (yĕl-ĭ-vär′ĕ)	116	68°06′N	20°29′E
Gallo, r., Spain (gäl′yō)	128	40°43′N	1°42′W
Gallup, N.M., U.S. (găl′*ŭ*p)	64	35°30′N	108°45′W
Galty Mountains, mts., Ire.	120	52°19′N	8°20′W
Galva, Il., U.S. (găl′vȧ)	79	41°11′N	90°02′W
Galveston, Tx., U.S. (găl′vĕs-t*ŭ*n)	65	29°18′N	94°48′W
Galveston Bay, b., Tx., U.S.	65	29°39′N	94°45′W
Galveston I, Tx., U.S.	81a	29°12′N	94°53′W
Galway, Ire.	110	53°16′N	9°05′W
Galway Bay, b., Ire. (gôl′wā)	120	53°10′N	9°47′W
Gamba, China (gäm-bä)	158	28°23′N	89°42′E
Gambaga, Ghana (gäm-bä′gä)	184	10°32′N	0°26′W
Gambela, Eth. (găm-bā′l*á*)	185	8°15′N	34°33′E
Gambia (Gambie), r., Afr.	188	13°20′N	15°55′W
Gambia, The, nation, Afr.	184	13°38′N	19°38′W
Gambie, r., Afr.	184	12°30′N	13°00′W
Gamboma, Congo (gäm-bō′mä)	186	1°53′S	15°51′E
Gamleby, Swe. (gäm′lĕ-bü)	122	57°54′N	16°20′E
Gan, r., China (gän)	165	26°50′N	115°00′E
Gäncä, Azer.	136	40°40′N	46°22′E
Gandak, r., India	158	26°37′N	84°22′E
Gander, Can. (găn′dēr)	51	48°57′N	54°34′W
Gander, r., Can.	61	49°10′N	54°35′W
Gander Lake, l., Can.	61	48°55′N	55°40′W
Gandhinagar, India	158	23°30′N	72°47′E
Gandi, Nig.	189	12°55′N	5°49′E
Gandía, Spain (gän-dē′ä)	129	38°56′N	0°10′W
Gangdisê Shan (Trans Himalayas), mts., China	160	30°25′N	83°43′E
Ganges, r., Asia (găn′jēz)	155	24°00′N	89°30′E
Ganges, Mouths of the, mth., Asia (găn′jēz)	155	21°18′N	88°40′E
Gangi, Italy (gän′jē)	130	37°48′N	14°15′E
Gangtok, India	155	27°15′N	88°30′E
Gannan, China (gän-nän)	164	47°50′N	123°30′E
Gannett Peak, mtn., Wy., U.S. (găn′ĕt)	64	43°10′N	109°38′W
Gano, Oh., U.S. (g′nō)	69f	39°18′N	84°24′W
Gänserndorf, Aus.	115e	48°21′N	16°43′E
Gansu, prov., China (gän-sōō)	160	38°50′N	101°10′E
Ganwo, Nig.	189	11°13′N	4°42′E
Ganyu, China (gän-yōō)	162	34°52′N	119°07′E
Ganzhou, China (gän-jō)	161	25°50′N	114°30′E
Gao, Mali (gä′ō)	184	16°16′N	0°03′W
Gao'an, China (gou-än)	165	28°30′N	115°02′E
Gaomi, China (gou-mē)	162	36°23′N	119°46′E
Gaoqiao, China (gou-chyou)	163b	31°21′N	121°35′E
Gaoshun, China (gou-shòn)	162	31°22′N	118°50′E
Gaotang, China (gou-täŋ)	162	36°52′N	116°12′E
Gaoyao, China (gou-you)	165	23°08′N	112°25′E
Gaoyi, China (gou-yē)	162	37°37′N	114°39′E
Gaoyou, China (gou-yō)	164	32°46′N	119°26′E
Gaoyou Hu, l., China (kä′ō-yōō′hōō)	161	32°42′N	118°40′E
Gap, Fr. (gáp)	117	44°34′N	6°08′E
Gapan, Phil. (gä-pän)	169a	15°18′N	120°56′E
Gar, China	160	31°11′N	80°35′E
Garanhuns, Braz. (gä-răn-yònsh′)	101	8°49′S	36°28′W
Garber, Ok., U.S. (gár′bēr)	79	36°28′N	97°35′W
Garching, Ger. (gär′кĕng)	115d	48°15′N	11°39′E
Garcia, Mex. (gär-sē′ä)	80	25°50′N	100°37′W
García de la Cadena, Mex.	88	21°14′N	103°26′W
Garda, Lago di, l., Italy (lä-gō-dē-gär′dä)	118	45°43′N	10°26′E
Gardanne, Fr. (gȧr-dȧn′)	126a	43°28′N	5°29′E
Gardelegen, Ger. (gär-dē-lā′ghĕn)	124	52°32′N	11°22′E
Garden, i., Mi., U.S. (gär′d'n)	66	45°50′N	85°50′W
Gardena, Ca., U.S. (gär-dē′nȧ)	75a	33°53′N	118°19′W
Garden City, Ks., U.S.	78	37°58′N	100°52′W
Garden City, Mi., U.S.	69b	42°20′N	83°21′W
Garden Grove, Ca., U.S. (gär′d'n grōv)	75a	33°47′N	117°56′W
Garden Reach, India	158a	22°33′N	88°17′E
Garden River, Can.	75k	46°33′N	84°10′W
Gardeyz, Afg.	158	33°43′N	69°09′E
Gardiner, Me., U.S. (gärd′nēr)	60	44°12′N	69°46′W
Gardiner, Mt., U.S.	73	45°03′N	110°43′W
Gardiner, Wa., U.S.	74a	48°03′N	122°55′W
Gardiner Dam, dam, Can.	56	51°17′N	106°51′W
Gardner, Ma., U.S.	67	42°35′N	72°00′W
Gardner Canal, strt., Can.	54	53°28′N	128°15′W

ăt; fīnál; rāte; senåte; ärm; åsk; sofà; fåre; ch-choose; dh-as th in other; bē; ĕvent; bĕt; recĕnt; cratēr; g-gō; gh-guttural g; bĭt; ĭ-short neutral; rīde; ĸ-guttural k as ch in German ich;

PRONOUNCING INDEX

Gladstone – Grande 251

PLACE (Pronunciation)	PAGE	LAT.	LONG.
Gladstone, Austl.	174	33°15'S	138°20'E
Gladstone, Mi., U.S.	71	45°50'N	87°04'W
Gladstone, N.J., U.S.	68a	40°43'N	74°39'W
Gladstone, Or., U.S.	74c	45°23'N	122°36'W
Gladwin, Mi., U.S. (glăd'wĭn)	66	44°00'N	84°25'W
Glåma, r., Nor.	112	61°30'N	10°30'E
Glarus, Switz. (glä'rôs)	124	47°02'N	9°03'E
Glasgow, Scot., U.K. (glås'gō)	110	55°54'N	4°25'W
Glasgow, Ky., U.S.	82	37°00'N	85°55'W
Glasgow, Mo., U.S.	79	39°14'N	92°48'W
Glasgow, Mt., U.S.	73	48°14'N	106°39'W
Glassport, Pa., U.S. (glås'pōrt)	69e	40°19'N	79°53'W
Glauchau, Ger. (glou'κou)	124	50°51'N	12°28'E
Glazov, Russia (glä'zôf)	134	58°05'N	52°52'E
Glen, r., Eng., U.K. (glĕn)	114a	52°44'N	0°18'W
Glénan, Îles de, is., Fr. (ēl-dē-glā-näN')	126	47°43'N	4°42'W
Glen Burnie, Md., U.S. (bûr'nē)	68e	39°10'N	76°38'W
Glen Canyon, p., Ut., U.S.	77	37°10'N	110°50'W
Glen Canyon Dam, dam, Az., U.S. (glĕn kăn'yŭn)	64	36°57'N	111°25'W
Glen Canyon National Recreation Area, rec., U.S.	77	37°00'N	111°20'W
Glen Carbon, Il., U.S. (kär'bŏn)	75e	38°45'N	89°59'W
Glencoe, S. Afr. (glĕn-cō)	187c	28°14'S	30°09'E
Glencoe, Il., U.S.	69a	42°08'N	87°45'W
Glencoe, Mn., U.S. (glĕn'kō)	71	44°44'N	94°07'W
Glen Cove, N.Y., U.S. (kōv)	68a	40°51'N	73°38'W
Glendale, Az., U.S. (glĕn'dāl)	77	33°30'N	112°15'W
Glendale, Ca., U.S.	64	34°09'N	118°15'W
Glendale, Oh., U.S.	69f	31°16'N	84°22'W
Glendive, Mt., U.S. (glĕn'dīv)	64	47°08'N	104°41'W
Glendo, Wy., U.S.	73	42°32'N	104°54'W
Glendora, Ca., U.S. (glĕn-dō'rá)	75a	34°08'N	117°52'W
Glenelg, r., Austl.	176	37°20'S	141°30'E
Glen Ellyn, Il., U.S. (glĕn ĕl'-lĕn)	69a	41°53'N	88°04'W
Glen Innes, Austl. (ĭn'ĕs)	175	29°45'S	152°02'E
Glenns Ferry, Id., U.S. (fĕr'ĭ)	72	42°58'N	115°21'W
Glen Olden, Pa., U.S. (ōl'd'n)	68f	39°54'N	75°17'W
Glenomra, La., U.S. (glĕn-mō'rá)	81	30°58'N	92°36'W
Glenrock, Wy., U.S. (glĕn'rŏk)	73	42°50'N	105°53'W
Glens Falls, N.Y., U.S. (glĕnz fôlz)	67	43°20'N	73°40'W
Glenshaw, Pa., U.S. (glĕn'shô)	69e	40°33'N	79°57'W
Glen Valley, Can.	74d	49°09'N	122°30'W
Glenview, Il., U.S. (glĕn'vū)	69a	42°04'N	87°48'W
Glenville, Ga., U.S. (glĕn'vĭl)	83	31°55'N	81°56'W
Glenwood, Ia., U.S.	70	41°03'N	95°44'W
Glenwood, Mn., U.S.	70	45°39'N	95°23'W
Glenwood, N.M., U.S.	77	33°19'N	108°52'W
Glenwood Springs, Co., U.S.	77	39°35'N	107°20'W
Glienicke, Ger. (glē'nē-kē)	115b	52°38'N	13°19'E
Glinde, Ger. (glĕn'dĕ)	115c	53°32'N	10°13'E
Glittertinden, mtn., Nor.	122	61°39'N	8°33'E
Gliwice, Pol. (gwĭ-wĭt'sĕ)	117	50°18'N	18°40'E
Globe, Az., U.S. (glōb)	64	33°20'N	110°50'W
Głogów, Pol. (gwō'gōōv)	117	51°40'N	16°04'E
Glommen, r., Nor. (glôm'ĕn)	122	60°03'N	11°15'E
Glonn, Ger. (glônn)	115d	47°59'N	11°52'E
Glorieuses, Îles, is., Reu.	187	11°28'S	47°50'E
Glossop, Eng., U.K. (glŏs'ŭp)	114a	53°26'N	1°57'W
Gloster, Ms., U.S. (glŏs'tēr)	82	31°10'N	91°00'W
Gloucester, Eng., U.K. (glŏs'tēr)	117	51°54'N	2°11'W
Gloucester, Ma., U.S.	61a	42°37'N	70°40'W
Gloucester City, N.J., U.S.	68f	39°53'N	75°08'W
Glouster, Oh., U.S. (glŏs'tēr)	66	39°35'N	82°05'W
Glover Island, i., Can. (glŭv'ēr)	61	48°44'N	57°45'W
Gloversville, N.Y., U.S. (glŭv'ērz-vĭl)	67	43°05'N	74°20'W
Glovertown, Can. (glŭv'ēr-toun)	61	48°41'N	54°02'W
Glückstadt, Ger. (glük-shtät)	115c	53°47'N	9°25'E
Glushkovo, Russia (glōōsh'kō-vō)	133	51°21'N	34°43'E
Gmünden, Aus. (g'mōōn'dĕn)	124	47°57'N	13°47'E
Gniezno, Pol. (g'nyăz'nō)	117	52°32'N	17°34'E
Gnjilane, Serb. (gnyē'la-nĕ)	131	42°28'N	21°27'E
Goa, state, India (gō'á)	155	15°45'N	74°00'E
Goascorán, Hond. (gō-äs'kō-rän')	90	13°37'N	87°43'W
Goba, Eth. (gō'bä)	185	7°17'N	39°58'E
Gobabis, Nmb. (gō-bä'bĭs)	186	22°25'S	18°50'E
Gobi, des., Asia (gō'bē)	160	43°29'N	103°15'E
Goble, Or., U.S. (gō'b'l)	74c	46°01'N	122°53'W
Goch, Ger. (gōκ)	127c	51°35'N	6°10'E
Godāvari, r., India (gō-dä'vŭ-rē)	155	19°00'N	78°30'E
Goddards Soak, sw., Austl. (gŏd'árdz)	174	31°20'S	123°30'E
Goderich, Can. (gōd'rĭch)	58	43°45'N	81°45'W
Godfrey, Il., U.S. (gŏd'frē)	75e	38°57'N	90°12'W
Godhavn, Grnld. (gōdh'håvn)	49	69°15'N	53°30'W
Gods, r., Can. (gŏdz)	57	55°17'N	93°35'W
Gods Lake, Can.	51	54°40'N	94°09'W
Godthåb, Grnld. (gŏt'hŏb)	49	64°10'N	51°32'W
Goéland, Lac au, l., Can.	59	49°47'N	76°41'W
Goffs, Ca., U.S. (gôfs)	76	34°57'N	115°06'W
Gogebic, l., Mi., U.S. (gō-gē'bĭk)	71	46°24'N	89°25'W
Gogebic Range, mts., Mi., U.S.	71	46°37'N	89°48'W
Göggingen, Ger. (gûg'gĕn-gĕn)	115d	48°21'N	10°53'E
Gogland, i., Russia	123	60°04'N	26°55'E
Gogonou, Benin	189	10°50'N	2°50'E
Gogorrón, Mex. (gō-gō-rōn')	88	21°51'N	100°54'W
Goiânia, Braz. (gô-yá'nyä)	101	16°41'S	48°57'W
Goiás, Braz. (gô-yá's)	101	15°57'S	50°10'W
Goiás, state, Braz.	101	16°00'S	48°00'W
Goirle, Neth.	115a	51°31'N	5°06'E
Gökçeada, i., Tur.	131	40°10'N	25°27'E
Göksu, r., Tur. (gûk'sōō')	137	36°40'N	33°30'E
Gol, Nor. (gûl)	122	60°58'N	8°54'E
Golax, Va., U.S. (gō'lăks)	83	36°41'N	80°56'W
Golcar, Eng., U.K. (gōl'kár)	114a	53°38'N	1°52'W
Golconda, Il., U.S. (gŏl-kŏn'dá)	79	37°21'N	88°32'W
Gołdap, Pol. (gôl'dăp)	125	54°17'N	22°17'E
Golden, Can.	55	51°18'N	116°58'W
Golden, Co., U.S.	78	39°44'N	105°15'W
Goldendale, Wa., U.S. (gōl'dĕn-dāl)	72	45°49'N	120°48'W
Golden Gate, strt., Ca., U.S. (gōl'dĕn gāt)	74b	37°48'N	122°32'W
Golden Hinde, mtn., Can. (hīnd)	54	49°40'N	125°45'W
Golden's Bridge, N.Y., U.S.	68a	41°17'N	73°41'W
Golden Valley, Mn., U.S.	75g	44°58'N	93°23'W
Goldfield, Nv., U.S. (gōld'fēld)	76	37°42'N	117°15'W
Gold Hill, mtn., Pan.	86a	9°03'N	79°08'W
Gold Mountain, mtn., Wa., U.S. (gōld)	74a	47°33'N	122°48'W
Goldsboro, N.C., U.S. (gōldz-bûr'ō)	83	35°23'N	77°59'W
Goldthwaite, Tx., U.S. (gōld'thwāt)	80	31°27'N	98°34'W
Goleniów, Pol. (gô-lĕ-nyŭf')	124	53°33'N	14°51'E
Golets-Purpula, Gora, mtn., Russia	135	59°08'N	115°22'E
Golfito, C.R. (gôl-fē'tō)	91	8°40'N	83°12'W
Goliad, Tx., U.S. (gō-lĭ-ăd')	81	28°40'N	97°12'W
Golo, r., Fr.	130	42°28'N	9°18'E
Golo Island, i., Phil. (gō'lō)	169a	13°38'N	120°17'E
Golovchino, Russia (gō-lôf'chē-nō)	133	50°34'N	35°52'E
Golyamo Konare, Blg. (gō'lå-mō-kō'nä-rĕ)	131	42°16'N	24°33'E
Golzow, Ger. (gōl'tsōv)	115b	52°17'N	12°36'E
Gombe, Nig.	184	10°19'N	11°02'E
Gomera Island, i., Spain (gō-mā'rä)	184	28°00'N	18°01'W
Gomez Farias, Mex. (gō'māz fä-rē'ás)	88	24°59'N	101°02'W
Gómez Palacio, Mex. (pä-lä'syō)	86	25°35'N	103°30'W
Gonaïves, Haiti (gō-ná-ēv')	87	19°25'N	72°45'W
Gonaïves, Golfe des, b., Haiti (gō-ná-ēv')	93	19°20'N	73°20'W
Gonâve, Île de la, i., Haiti (gō-náv')	87	18°50'N	73°30'W
Gonda, India	158	27°13'N	82°00'E
Gondal, India	158	22°02'N	70°47'E
Gonder, Eth.	185	12°39'N	37°30'E
Gonesse, Fr. (gō-nĕs')	127b	48°59'N	2°28'E
Gongga Shan, mtn., China (gôŋ-gä shän)	160	29°16'N	101°46'E
Goniri, Nig.	189	11°30'N	12°20'E
Gonō, r., Japan (gō'nō)	167	35°00'N	132°25'E
Gonor, Can. (gō'nŏr)	62f	50°04'N	96°57'W
Gonubie, S. Afr. (gŏn'ōō-bē)	187c	32°56'S	28°02'E
Gonzales, Mex. (gōn-zá'lĕs)	88	22°47'N	98°26'W
Gonzales, Tx., U.S. (gŏn-zä'lĕz)	81	29°31'N	97°25'W
González Catán, Arg. (gōn-zá'lĕz-kä-tá'n)	102a	34°47'S	58°39'W
Good Hope, Cape of, c., S. Afr. (kāp ov gōōd hōp)	186	34°21'S	18°29'E
Good Hope Mountain, mtn., Can.	54	51°09'N	124°10'W
Gooding, Id., U.S. (gōd'ĭng)	73	42°55'N	114°43'W
Goodland, In., U.S. (gōd'lănd)	66	40°50'N	87°15'W
Goodland, Ks., U.S.	78	39°19'N	101°43'W
Goodwood, S. Afr. (gōd'wōd)	186a	33°54'S	18°33'E
Goole, Eng., U.K. (gōōl)	114a	53°42'N	0°52'W
Goose, r., N.D., U.S.	70	47°40'N	97°41'W
Gooseberry Creek, r., Wy., U.S. (gōōs-bĕr'ĭ)	73	44°04'N	108°35'W
Goose Creek, r., Id., U.S. (gōōs)	73	42°07'N	113°53'W
Goose Lake, l., Ca., U.S.	72	41°56'N	120°35'W
Gorakhpur, India (gō'rŭk-pōōr)	155	26°45'N	82°39'E
Gorda, Punta, c., Cuba (pōō'n-tä-gôr-dä)	92	22°25'N	82°10'W
Gorda Cay, i., Bah. (gôr'dä)	92	26°05'N	77°30'W
Gordon, Can. (gôr'dŭn)	62f	50°00'N	97°20'W
Gordon, Ne., U.S.	70	42°47'N	102°14'W
Gore, Eth. (gō'rē)	185	8°12'N	35°34'E
Gorgān, Iran	154	36°44'N	54°30'E
Gorgona, Isola di, Italy (gôr-gō'nä)	118	43°27'N	9°55'E
Gori, Geor. (gō'rē)	137	42°00'N	44°08'E
Gorinchem, Neth. (gō'rĭn-κĕm)	115a	51°50'N	4°59'E
Goring, Eng., U.K. (gôr'ĭng)	114b	51°30'N	1°08'W
Gorizia, Italy (gō-rē'tsē-yä)	130	45°56'N	13°40'E
Gor'kiy *see* Nizhniy Novgorod, Russia	134	56°15'N	44°05'E
Gor'kovskoye, res., Russia	134	56°38'N	43°40'E
Gorlice, Pol. (gôr-lē'tsĕ)	125	49°38'N	21°11'E
Görlitz, Ger. (gûr'lĭts)	117	51°10'N	15°01'E
Gorman, Tx., U.S. (gôr'măn)	80	32°13'N	98°40'W
Gorna Oryakhovitsa, Blg. (gôr'nä-ôr-yĕk'ō-vē-tsá)	131	43°08'N	25°40'E
Gornji Milanovac, Serb. (gôrn'yē-mē'la-nō-väts)	131	44°02'N	20°29'E
Gorno-Altay, prov., Russia	140	51°00'N	86°00'E
Gorno-Altaysk, Russia (gôr'nŭ'ŭl-tĭsk')	134	51°58'N	85°58'E
Gorodishche, Russia (gō-rô'dĭsh-chĕ)	142a	57°57'N	57°03'E
Gorodok, Russia	135	50°30'N	103°58'E
Gorontalo, Indon. (gō-rōn-tä'lo)	169	0°40'N	123°04'E
Gorzów Wielkopolski, Pol. (gō-zhōōv'vyĕl-ko-pōl'skē)	116	53°44'N	15°15'E
Gosely, Eng., U.K.	114a	52°33'N	2°10'W
Goshen, In., U.S. (gō'shĕn)	66	41°35'N	85°50'W
Goshen, Ky., U.S.	69h	38°24'N	85°34'W
Goshen, N.Y., U.S.	68a	41°24'N	74°19'W
Goshen, Oh., U.S.	69f	39°14'N	84°09'W
Goshute Indian Reservation, I.R., Ut., U.S. (gō-shōōt')	77	39°50'N	114°00'W
Goslar, Ger. (gôs'lär)	124	51°55'N	10°25'E
Gospa, r., Ven. (gôs-pä)	101b	9°43'N	64°23'W
Gostivar, Mac. (gos'tē-vär)	131	41°46'N	20°58'E
Gostynin, Pol. (gôs-tē'nĭn)	125	52°24'N	19°30'E
Göta, r., Swe. (gōētä)	122	58°11'N	12°03'E
Göta Kanal, can., Swe. (yû'tá)	122	58°35'N	15°24'E
Göteborg, Swe. (yû'tē-bôrgh)	110	57°39'N	11°56'E
Gotel Mountains, mts., Afr.	189	7°05'N	11°20'E
Gotera, El Sal. (gō-tä'rä)	90	13°41'N	88°06'W
Gotha, Ger. (gō'tá)	117	50°47'N	10°43'E
Gothenburg *see* Göteborg, Swe.	110	57°39'N	11°56'E
Gothenburg, Ne., U.S. (gŏth'ĕn-bûrg)	78	40°57'N	100°08'W
Gotland, i., Swe.	112	57°35'N	17°35'E
Gotska Sandön, i., Swe.	123	58°24'N	19°15'E
Göttingen, Ger. (gŭt'ĭng-ĕn)	124	51°32'N	9°57'E
Gouda, Neth. (gou'dä)	115a	52°00'N	4°42'E
Gough, i., St. Hel. (gŏf)	2	40°00'S	10°00'W
Gouin, Réservoir, res., Can.	53	48°15'N	74°15'W
Goukou, China (gō-kō)	161	48°45'N	121°42'E
Goulais, r., Can.	58	46°45'N	84°10'W
Goulburn, Austl. (gōl'bŭrn)	175	34°47'S	149°40'E
Goumbati, mtn., Sen.	188	13°08'N	12°06'W
Goumbou, Mali (gōōm-bōō')	184	14°59'N	7°27'W
Gouna, Cam.	189	8°32'N	13°34'E
Goundam, Mali (gōōn-däN')	184	16°29'N	3°37'W
Gouverneur, N.Y., U.S. (gŭv-ēr-nōōr')	67	44°20'N	75°25'W
Govenlock, Can. (gŭvĕn-lŏk)	50	49°15'N	109°48'W
Governador, Ilha do, i., Braz. (gô-vĕr-nä-dō-'r-ē-lä'dō)	102b	22°48'S	43°13'W
Governador Portela, Braz. (pōr-tē'lä)	102b	22°28'S	43°30'W
Governador Valadares, Braz. (vä-lä-dä'rĕs)	101	18°47'S	41°45'W
Governor's Harbour, Bah.	92	25°15'N	76°15'W
Gowanda, N.Y., U.S. (gō-wŏn'dá)	67	42°30'N	78°55'W
Goya, Arg. (gō'yä)	102	29°06'S	59°12'W
Göyçay, Azer. (gē-ôk'chī)	137	40°40'N	47°40'E
Goyt, r., Eng., U.K. (goit)	114a	53°19'N	2°03'W
Graaff-Reinet, S. Afr. (gräf'rī'nĕt)	186	32°10'S	24°40'E
Gračac, Cro. (grä'chäts)	130	44°16'N	15°50'E
Gračanica, Bos.	131	44°42'N	18°18'E
Graceville, Fl., U.S. (grās'vĭl)	82	30°57'N	85°30'W
Graceville, Mn., U.S.	70	45°33'N	96°25'W
Gracias, Hond. (grä'sē-äs)	90	14°35'N	88°37'W
Graciosa Island, i., Port. (grä-syō'sä)	184a	39°07'N	27°30'W
Gradačac, Bos. (gra-dä'chats)	119	44°50'N	18°28'E
Grado, Spain (grä'dō)	128	43°24'N	6°04'W
Gräfelfing, Ger. (grä'fĕl-fēng)	115d	48°07'N	11°27'E
Grafing bei München, Ger. (grä'fēng)	115d	48°03'N	11°58'E
Grafton, Austl. (graf'tŭn)	175	29°38'S	153°05'E
Grafton, Il., U.S.	75e	38°58'N	90°26'W
Grafton, Ma., U.S.	61a	42°13'N	71°41'W
Grafton, N.D., U.S.	70	48°24'N	97°25'W
Grafton, Oh., U.S.	69d	41°16'N	82°04'W
Grafton, W.V., U.S.	66	39°20'N	80°00'W
Gragnano, Italy (grän-yä'nō)	129c	40°27'N	14°32'E
Graham, N.C., U.S. (grā'ăm)	83	36°03'N	79°23'W
Graham, Tx., U.S.	78	33°07'N	98°34'W
Graham, Wa., U.S.	74a	47°03'N	122°18'W
Graham, i., Can.	52	53°50'N	132°40'W
Grahamstown, S. Afr. (grā'ăms'toun)	187c	33°19'S	26°33'E
Grajewo, Pol. (grä-yā'vo)	125	53°38'N	22°28'E
Grama, Serra de, mtn., Braz. (sĕ'r-rä-dĕ-grá'mà)	99a	20°42'S	42°28'W
Gramada, Blg. (grä'mä-dä)	131	43°46'N	22°41'E
Gramatneusiedl, Aus.	115e	48°02'N	16°29'E
Grampian Mountains, mts., Scot., U.K. (grăm'pĭ-ăn)	112	56°30'N	4°55'W
Granada, Nic. (grä-nä'dhä)	86	11°55'N	85°58'W
Granada, Spain (grä-nä'dä)	118	37°13'N	3°37'W
Gran Bajo, reg., Arg. (grán'bä'kō)	102	47°35'S	68°45'W
Granbury, Tx., U.S. (grăn'bĕr-ĭ)	81	32°26'N	97°45'W
Granby, Can. (grăn'bĭ)	51	45°30'N	72°40'W
Granby, Mo., U.S.	79	36°54'N	94°15'W
Granby, l., Co., U.S.	78	40°07'N	105°40'W
Gran Canaria Island, i., Spain (grän'kä-nä'rē-ä)	184	27°39'N	15°39'W
Gran Chaco, reg., S.A. (grán'chá'kō)	102	25°30'S	62°15'W
Grand, i., Mi., U.S.	71	46°37'N	86°38'W
Grand, l., Can.	60	45°59'N	66°15'W
Grand, l., Me., U.S.	60	45°17'N	67°42'W
Grand, r., Can.	59	43°45'N	80°20'W
Grand, r., Mi., U.S.	66	42°58'N	85°13'W
Grand, r., Mo., U.S.	79	39°50'N	93°52'W
Grand, r., S.D., U.S.	70	45°40'N	101°55'W
Grand, North Fork, r., U.S.	70	45°52'N	102°49'W
Grand, South Fork, r., S.D., U.S.	70	45°38'N	102°56'W
Grand Bahama, i., Bah.	87	26°35'N	78°30'W
Grand Bank, Can. (grănd băngk)	53a	47°06'N	55°47'W
Grand Bassam, C. Iv. (grän bá-säN')	184	5°12'N	3°44'W
Grand Bourg, Guad. (grän bōōr')	91b	15°54'N	61°20'W
Grand Caicos, i., T./C. Is. (grănd kä-ē'kōs)	93	21°45'N	71°50'W
Grand Canal *see* Da Yunhe, can., China	161	35°00'N	117°00'E
Grand Canal, can., Ire.	120	53°21'N	7°15'W
Grand Canyon, Az., U.S.	77	36°05'N	112°10'W
Grand Canyon, p., Az., U.S.	64	35°50'N	113°16'W
Grand Canyon National Park, rec., Az., U.S.	64	36°15'N	112°20'W
Grand Canyon-Parashant National Monument, rec., Az., U.S.	77	36°25'N	113°45'W
Grand Cayman, i., Cay. Is. (kā'măn)	87	19°15'N	81°15'W
Grand Coulee Dam, dam, Wa., U.S. (kōō'lē)	64	47°58'N	119°28'W
Grande, r., Arg.	99b	35°25'S	70°14'W
Grande, r., Bol.	100	16°49'S	63°19'W
Grande, r., Braz.	101	19°48'S	49°54'W
Grande, r., Mex.	89	17°37'N	96°41'W
Grande, r., Nic. (grän'dĕ)	91	13°01'N	84°21'W
Grande, r., Ur.	99c	33°19'S	57°15'W
Grande, Arroyo, r., Mex. (är-rō'yō-grä'n-dĕ)	88	23°30'N	98°45'W
Grande, Bahía, b., Arg. (bä-ē'ä-grán'dĕ)	102	50°45'S	68°00'W
Grande, Boca, mth., Ven. (bŏ'kä-grä'n-dĕ)	101	8°46'N	60°17'W

252 Grande – Gros Morne PRONOUNCING INDEX

PLACE (Pronunciation)	PAGE	LAT.	LONG.
Grande, Cuchilla, mts., Ur. (kōō-chē'l-yä)	102	33°00'S	55°15'W
Grande, Ilha, i., Braz. (grän'dě)	99a	23°11'S	44°14'W
Grande, Rio, r., N.A. (grän'dä)	64	26°50'N	99°10'W
Grande, Salinas, l., Arg. (sä-lē'näs)	102	29°45'S	65°00'W
Grande, Salto, wtfl., Braz. (säl-tô)	101	16°18'S	39°38'W
Grande Cayemite, Île, i., Haiti	93	18°45'N	73°45'W
Grande de Otoro, r., Hond. (grä'dą dā ō-tō'rô)	90	14°42'N	88°21'W
Grande de Santiago, Río, r., Mex. (rēô-grä'n-dě-dě-sän-tyá'gô)	86	20°30'N	104°00'W
Grande Pointe, Can. (grǎnd point')	62f	49°47'N	97°03'W
Grande Prairie, Can. (prâr'ĭ)	50	55°10'N	118°48'W
Grand Erg Occidental, des., Alg.	184	30°00'N	1°00'E
Grand Erg Oriental, des., Alg.	184	30°00'N	7°00'E
Grande Rivière du Nord, Haiti (rē-vyâr' dü nôr')	93	19°35'N	72°10'W
Grande Ronde, r., Or., U.S. (rônd')	72	45°32'N	117°52'W
Gran Desierto, des., Mex. (grän-dě-syě'r-tô)	77	32°14'N	114°28'W
Grande Terre, i., Guad.	91b	16°28'N	61°13'W
Grande Vigie, Pointe de la, c., Guad. (grǎNd vē-gē')	91b	16°32'N	61°25'W
Grand Falls, Can. (fôlz)	53a	48°56'N	55°40'W
Grandfather Mountain, mtn., N.C., U.S. (grǎnd-fä-thēr)	83	36°07'N	81°48'W
Grandfield, Ok., U.S. (grǎnd'fēld)	78	34°13'N	98°39'W
Grand Forks, Can. (fôrks)	50	49°02'N	118°27'W
Grand Forks, N.D., U.S.	64	47°55'N	97°05'W
Grand Haven, Mi., U.S. (hā'v'n)	66	43°05'N	86°15'W
Grand Island, Ne., U.S. (ī'lǎnd)	64	40°56'N	98°20'W
Grand Island, i., N.Y., U.S.	69c	43°03'N	78°58'W
Grand Junction, Co., U.S. (jüngk'shǔn)	64	39°05'N	108°35'W
Grand Lake, l., Can. (lāk)	53a	49°00'N	57°10'W
Grand Lake, l., La., U.S.	81	29°57'N	91°25'W
Grand Lake, l., Mn., U.S.	75h	46°54'N	92°26'W
Grand Ledge, Mi., U.S. (lěj)	66	42°45'N	84°50'W
Grand Lieu, Lac de, l., Fr. (grän'-lyû)	126	47°00'N	1°45'W
Grand Manan, i., Can. (mǎ-nǎn)	60	44°40'N	66°50'W
Grand Mère, Can. (grän mâr')	51	46°36'N	72°43'W
Grândola, Port. (grǎN'dô-lǎ)	128	38°10'N	8°36'W
Grand Portage Indian Reservation, I.R., Mn., U.S. (pōr'tĭj)	71	47°54'N	89°34'W
Grand Portage National Monument, rec., Mi., U.S.	71	47°59'N	89°47'W
Grand Prairie, Tx., U.S. (prě'rē)	75c	32°45'N	97°00'W
Grand Rapids, Can.	57	53°08'N	99°20'W
Grand Rapids, Mi., U.S. (rǎp'ĭdz)	65	43°00'N	85°45'W
Grand Rapids, Mn., U.S.	71	47°16'N	93°33'W
Grand-Riviere, Can.	60	48°26'N	64°30'W
Grand Staircase-Escalante National Monument, rec., Ut., U.S.	77	37°25'N	111°30'W
Grand Teton, mtn., Wy., U.S.	64	43°46'N	110°50'W
Grand Teton National Park, rec., Wy., U.S. (tē'tŏn)	73	43°54'N	110°15'W
Grand Traverse Bay, b., Mi., U.S. (trǎv'ērs)	66	45°00'N	85°30'W
Grand Turk, T./C. Is. (tûrk)	93	21°30'N	71°10'W
Grand Turk, i., T./C. Is.	93	21°30'N	71°10'W
Grandview, Mo., U.S. (grǎnd'vyōō)	75f	38°53'N	94°32'W
Granger, Wy., U.S. (grān'jēr)	73	41°37'N	109°58'W
Grangeville, Id., U.S. (grānj'vĭl)	72	45°56'N	116°08'W
Granite City, Il., U.S. (grǎn'ĭt sĭt'ĭ)	75e	38°42'N	90°09'W
Granite Falls, Mn., U.S. (fôlz)	70	44°46'N	95°34'W
Granite Falls, N.C., U.S.	83	35°49'N	81°25'W
Granite Falls, Wa., U.S.	74a	48°05'N	121°59'W
Granite Lake, l., Can.	61	48°01'N	57°00'W
Granite Peak, mtn., Mt., U.S.	64	45°13'N	109°48'W
Graniteville, S.C., U.S. (grǎn'ĭt-vĭl)	83	33°35'N	81°50'W
Granito, Braz. (grä-nē'tō)	101	7°39'S	39°34'W
Granma, prov., Cuba	92	20°10'N	76°50'W
Gränna, Swe. (grěn'à)	122	58°02'N	14°38'E
Granollers, Spain (grä-nôl-yěrs')	129	41°36'N	2°19'E
Gran Pajonal, reg., Peru (grä'n-pä-kô-näl')	100	11°14'S	71°45'W
Gran Paradiso, mtn., Italy	130	45°32'N	7°16'E
Gran Piedra, mtn., Cuba (grän-pyě'drä)	93	20°00'N	75°40'W
Grantham, Eng., U.K. (grǎn'tǎm)	120	52°54'N	0°38'W
Grant Park, Il., U.S. (grǎnt pärk)	69a	41°14'N	87°39'W
Grants Pass, Or., U.S. (grǎnts pǎs)	72	42°26'N	123°20'W
Granville, Fr. (grän-vēl')	117	48°52'N	1°35'W
Granville, N.Y., U.S. (grǎn'vĭl)	67	43°25'N	73°15'W
Granville, l., Can.	52	56°18'N	100°30'W
Grão Mogol, Braz. (groun' mò-gôl')	101	16°34'S	42°35'W
Grapevine, Tx., U.S. (grāp'vīn)	75c	32°56'N	97°05'W
Gräso, i., Swe.	122	60°30'N	18°35'E
Grass, r., N.Y., U.S.	67	44°45'N	75°10'W
Grass Cay, i., V.I.U.S.	87c	18°22'N	64°50'W
Grasse, Fr. (gräs)	127	43°39'N	6°57'E
Grass Mountain, mtn., Wa., U.S. (gràs)	74a	47°13'N	121°48'W
Grates Point, c., Can. (grāts)	61	48°09'N	52°57'W
Gravelbourg, Can. (grǎv'ěl-bôrg)	50	49°53'N	106°34'W
Gravesend, Eng., U.K. (grāvz'ěnd')	114b	51°26'N	0°22'E
Gravina, Italy (grä-vē'nä)	130	40°48'N	16°27'E
Gravois, Pointe à, c., Haiti (grà-vwä')	93	18°00'N	74°20'W
Gray, Fr. (grå)	127	47°26'N	5°35'E
Grayling, Mi., U.S. (grā'lĭng)	66	44°40'N	84°40'W
Grays Harbor, b., Wa., U.S. (grās)	64	46°55'N	124°23'W
Grayslake, Il., U.S. (grāz'lāk)	69a	42°20'N	88°20'W
Grays Peak, mtn., Co., U.S. (grāz)	78	39°29'N	105°52'W
Grays Thurrock, Eng., U.K. (thŭ'rŏk)	114b	51°28'N	0°19'E
Grayvoron, Russia (grà-ē'vô-rôn)	133	50°28'N	35°41'E
Graz, Aus. (gräts)	110	47°05'N	15°26'E
Great Abaco, i., Bah. (ä'bä-kō)	87	26°30'N	77°05'W
Great Artesian Basin, basin, Austl. (är-tēzh-ǎn bā-sĭn)	175	23°16'S	143°37'E
Great Australian Bight, b., Austl. (ôs-trā'lĭ-ǎn bĭt)	174	33°30'S	127°00'E
Great Bahama Bank, bk., (bà-hā'mà)	92	25°00'N	78°50'W
Great Barrier, i., N.Z. (bǎr'ĭ-ēr)	175a	36°10'S	175°30'E
Great Barrier Reef, rf., Austl. (bà-rĭ-ēr rēf)	175	16°43'S	146°34'E
Great Basin, basin, U.S. (grāt bā's'n)	64	40°08'N	117°10'W
Great Bear Lake, l., Can. (bâr)	52	66°10'N	119°53'W
Great Bend, Ks., U.S. (běnd)	78	38°41'N	98°46'W
Great Bitter Lake, l., Egypt	192b	30°24'N	32°27'E
Great Blasket Island, i., Ire. (blǎs'kět)	120	52°05'N	10°55'W
Great Corn Island, i., Nic.	91	12°10'N	82°54'W
Great Dismal Swamp, sw., U.S. (dĭz'mǎl)	83	36°35'N	76°34'W
Great Divide Basin, basin, Wy., U.S. (dĭ-vīd' bā's'n)	73	42°10'N	108°10'W
Great Dividing Range, mts., Austl. (dĭ-vī-dĭng rānj)	175	35°16'S	146°38'E
Great Duck, i., Can. (dŭk)	58	45°40'N	83°22'W
Greater Antilles, is., N.A.	87	20°30'N	79°15'W
Greater Khingan Range, mts., China (dä hĭŋ-gän lĭŋ)	161	46°30'N	120°00'E
Greater Leech Indian Reservation, I.R., Mn., U.S. (grāt'ēr lēch)	71	47°39'N	94°27'W
Greater Manchester, hist. reg., Eng., U.K.	114a	53°34'N	2°41'W
Greater Sunda Islands, is., Asia	168	4°00'S	108°00'E
Great Exuma, i., Bah. (ěk-sōō'mä)	92	23°35'N	76°00'W
Great Falls, Mt., U.S. (fôlz)	64	47°30'N	111°15'W
Great Falls, S.C., U.S.	83	34°32'N	80°53'W
Great Guana Cay, i., Bah. (gwä'nä)	92	24°00'N	76°20'W
Great Harbor Cay, i., Bah. (kē)	92	25°45'N	77°50'W
Great Inagua, i., Bah. (ē-nä'gwä)	87	21°00'N	73°15'W
Great Indian Desert, des., Asia	155	27°35'N	71°37'E
Great Isaac, i., Bah. (ī'zǎk)	92	26°05'N	79°05'W
Great Karroo, plat., S. Afr. (grāt kǎ'rōō)	186	32°45'S	22°00'E
Great Limpopo Transfrontier Park, rec., Afr.	186	22°00'S	31°30'E
Great Namaland, hist. reg., Nmb.	186	25°45'S	16°15'E
Great Neck, N.Y., U.S. (něk)	68a	40°48'N	73°44'W
Great Nicobar Island, i., India (nĭk-ô-bär')	168	7°00'N	94°18'E
Great Pedro Bluff, c., Jam.	92	17°50'N	78°05'W
Great Pee Dee, r., S.C., U.S. (pē-dē')	65	34°01'N	79°26'W
Great Plains, pl., N.A. (plāns)	49	45°00'N	104°00'W
Great Ragged, i., Bah.	93	22°10'N	75°45'W
Great Ruaha, r., Tan.	186	7°30'S	37°00'E
Great Salt Lake, l., Ut., U.S. (sôlt lāk)	64	41°19'N	112°48'W
Great Salt Lake Desert, des., Ut., U.S.	64	41°00'N	113°30'W
Great Salt Plains Reservoir, res., Ok., U.S.	78	36°56'N	98°14'W
Great Sand Dunes National Monument, rec., Co., U.S.	78	37°56'N	105°25'W
Great Sand Hills, hills, Can. (sǎnd)	56	50°35'N	109°05'W
Great Sandy Desert, des., Austl. (sǎn'dē)	174	21°50'S	123°10'E
Great Sandy Desert, des., Or., U.S. (sǎn'dĭ)	72	43°43'N	120°44'W
Great Sitkin, i., Ak., U.S. (sĭt-kĭn)	63a	52°18'N	176°22'W
Great Slave Lake, l., Can. (slāv)	52	61°37'N	114°58'W
Great Smoky Mountains National Park, rec., U.S. (smōk-ē)	65	35°43'N	83°20'W
Great Stirrup Cay, i., Bah. (stĭr-ŭp)	92	25°50'N	77°55'W
Great Victoria Desert, des., Austl. (vĭk-tō'rĭ-à)	174	29°45'S	124°30'E
Great Wall, hist., China	160	38°00'N	109°00'E
Great Waltham, Eng., U.K. (wôl'thŭm)	114b	51°47'N	0°27'E
Great Yarmouth, Eng., U.K. (yär-mŭth)	117	52°35'N	1°45'E
Grebbestad, Swe. (grěb-bě-städh)	122	58°42'N	11°15'E
Gréboun, Mont, mtn., Niger	184	20°00'N	8°35'E
Gredos, Sierra de, mts., Spain (syě'rä dā grā'dōs)	128	40°13'N	5°30'W
Greece, nation, Eur. (grēs)	110	39°00'N	21°30'E
Greeley, Co., U.S. (grē'lĭ)	64	40°25'N	104°41'W
Green, r., Ky., U.S. (grēn)	82	37°13'N	86°30'W
Green, r., N.D., U.S.	70	47°05'N	103°05'W
Green, r., Ut., U.S.	77	38°30'N	110°05'W
Green, r., Wa., U.S.	74a	47°17'N	121°57'W
Green, r., Wy., U.S.	73	41°08'N	110°27'W
Green, r., U.S.	64	38°30'N	110°10'W
Greenbank, Wa., U.S. (grēn'bǎnk)	74a	48°06'N	122°35'W
Green Bay, Wi., U.S.	65	44°30'N	88°04'W
Green Bay, b., U.S.	65	44°55'N	87°40'W
Green Bayou, Tx., U.S.	81a	29°53'N	95°13'W
Greenbelt, Md., U.S. (grēn'bělt)	68e	38°59'N	76°53'W
Greencastle, In., U.S. (grēn-kás''l)	66	39°40'N	86°50'W
Green Cay, i., Bah.	92	24°05'N	77°10'W
Green Cove Springs, Fl., U.S. (kōv)	83	29°56'N	81°42'W
Greendale, Wi., U.S. (grēn'dāl)	69a	42°56'N	87°59'W
Greenfield, Ia., U.S.	71	41°16'N	94°30'W
Greenfield, In., U.S. (grēn'fēld)	66	39°45'N	85°40'W
Greenfield, Ma., U.S.	67	42°35'N	72°35'W
Greenfield, Mo., U.S.	79	37°23'N	93°48'W
Greenfield, Oh., U.S.	66	39°15'N	83°25'W
Greenfield, Tn., U.S.	82	36°08'N	88°45'W
Greenfield Park, Can.	62a	45°29'N	73°29'W
Greenhills, Oh., U.S. (grēn-hĭls)	69f	39°16'N	84°31'W
Greenland, dep., N.A. (grēn'lǎnd)	49	74°00'N	40°00'W
Greenland Sea, sea,	198	77°00'N	1°00'W
Green Mountain, mtn., Or., U.S.	74c	45°52'N	123°24'W
Green Mountain Reservoir, res., Co., U.S.	77	39°50'N	106°20'W
Green Mountains, mts., N.A.	65	43°10'N	73°05'W
Greenock, Scot., U.K. (grēn'ŭk)	116	55°55'N	4°45'W
Green Peter Lake, res., Or., U.S.	72	44°28'N	122°30'W
Green Pond Mountain, mtn., N.J., U.S. (pŏnd)	68a	41°00'N	74°32'W
Greenport, N.Y., U.S.	67	41°06'N	72°22'W
Green River, Ut., U.S. (grēn rĭv'ēr)	77	39°00'N	110°05'W
Green River, Wy., U.S.	73	41°32'N	109°26'W
Green River Lake, res., Ky., U.S.	82	37°15'N	85°15'W
Greensboro, Al., U.S. (grēnz'bŭro)	82	32°42'N	87°36'W
Greensboro, Ga., U.S. (grēns-bûr'ô)	82	33°34'N	83°11'W
Greensboro, N.C., U.S.	65	36°04'N	79°45'W
Greensburg, In., U.S. (grēnz'bûrg)	66	39°20'N	85°30'W
Greensburg, Ks., U.S. (grēns-bûrg)	78	37°36'N	99°17'W
Greensburg, Pa., U.S.	67	40°20'N	79°30'W
Greenville, Lib.	184	5°01'N	9°03'W
Greenville, Al., U.S. (grēn'vĭl)	82	31°49'N	86°39'W
Greenville, Il., U.S.	79	38°52'N	89°22'W
Greenville, Ky., U.S.	82	37°11'N	87°11'W
Greenville, Me., U.S.	60	45°26'N	69°35'W
Greenville, Mi., U.S.	66	43°10'N	85°25'W
Greenville, Ms., U.S.	65	33°25'N	91°00'W
Greenville, N.C., U.S.	83	35°35'N	77°22'W
Greenville, Oh., U.S.	66	40°05'N	84°35'W
Greenville, Pa., U.S.	66	41°20'N	80°25'W
Greenville, S.C., U.S.	65	34°50'N	82°25'W
Greenville, Tn., U.S.	83	36°08'N	82°50'W
Greenville, Tx., U.S.	81	33°09'N	96°07'W
Greenwich, Eng., U.K.	114b	51°28'N	0°00'
Greenwich, Ct., U.S.	68a	41°01'N	73°37'W
Greenwood, Ar., U.S. (grēn-wŏŏd)	79	35°13'N	94°15'W
Greenwood, In., U.S.	69g	39°37'N	86°07'W
Greenwood, Ms., U.S.	82	33°30'N	90°09'W
Greenwood, S.C., U.S.	83	34°10'N	82°10'W
Greenwood, Lake, res., S.C., U.S.	83	34°17'N	81°55'W
Greenwood Lake, l., N.Y., U.S.	68a	41°13'N	74°20'W
Greer, S.C., U.S. (grēr)	83	34°55'N	81°56'W
Grefrath, Ger. (grěf'rät)	127c	51°20'N	6°21'E
Gregory, S.D., U.S. (grěg'ô-rĭ)	70	43°12'N	99°27'W
Gregory, Lake, l., Austl. (grěg'ô-rê)	174	28°47'S	139°15'E
Gregory Range, mts., Austl.	175	19°23'S	143°45'E
Greifenberg, Ger. (grī'fěn-běrgh)	115d	48°04'N	11°06'E
Greifswald, Ger. (grīfs'vält)	124	54°05'N	13°24'E
Greiz, Ger. (grīts)	124	50°39'N	12°14'E
Gremyachinsk, Russia (grâ'myá-chĭnsk)	142a	58°35'N	57°53'E
Grenada, Ms., U.S. (grê-nä'da)	82	33°45'N	89°47'W
Grenada, nation, N.A.	87	12°02'N	61°15'W
Grenada Lake, res., Ms., U.S.	82	33°52'N	89°30'W
Grenadines, The, is., N.A. (grěn'à-dēnz)	91b	12°37'N	61°35'W
Grenen, c., Den.	116	57°43'N	10°31'E
Grenoble, Fr. (grē-nô'bl')	117	45°14'N	5°45'E
Grenora, N.D., U.S. (grě-nō'rǎ)	70	48°38'N	103°55'W
Grenville, Can. (grěn'vĭl)	67	45°40'N	74°35'W
Grenville, Gren.	91b	12°07'N	61°38'W
Gresham, Or., U.S. (grěsh'ǎm)	74c	45°30'N	122°25'W
Gretna, La., U.S. (grět'nǎ)	68d	29°56'N	90°03'W
Grevelingen Krammer, r., Neth.	115a	51°42'N	4°03'E
Grevenbroich, Ger. (grě'fěn-broik)	127c	51°05'N	6°36'E
Grey, r., Can. (grā)	61	47°53'N	57°00'W
Grey, Point, c., Can.	74d	49°22'N	123°16'W
Greybull, Wy., U.S. (grā'bŏŏl)	73	44°28'N	108°05'W
Greybull, r., Wy., U.S.	73	44°13'N	108°43'W
Greylingstad, S. Afr. (grā-lĭng'shtát)	192c	26°40'S	29°13'E
Greymouth, N.Z. (grā'mouth)	175a	42°27'S	171°17'E
Grey Range, mts., Austl.	175	28°40'S	142°05'E
Greytown, S. Afr. (grā'toun)	187c	29°07'S	30°38'E
Grey Wolf Peak, mtn., Wa., U.S. (grā wŏolf)	74a	48°53'N	123°12'W
Gridley, Ca., U.S. (grĭd'lĭ)	76	39°22'N	121°43'W
Griffin, Ga., U.S. (grĭf'ĭn)	82	33°15'N	84°16'W
Griffith, Austl. (grĭf-ĭth)	176	34°16'S	146°10'E
Griffith, In., U.S.	69a	41°31'N	87°26'W
Grigoriopol', Mol. (grĭ'gor-ĭ-ŏ'pôl)	133	47°09'N	29°18'E
Grijalva, r., Mex. (grē-häl'vä)	89	17°25'N	93°23'W
Grim, Cape, c., Austl. (grĭm)	176	40°43'S	144°30'E
Grimma, Ger. (grĭm'à)	124	51°14'N	12°43'E
Grimsby, Can. (grĭmz'bĭ)	62d	43°11'N	79°33'W
Grimsby, Eng., U.K.	116	53°35'N	0°05'W
Grímsey, i., Ice. (grĭms'â)	116	66°30'N	17°50'W
Grimstad, Nor. (grĭm-städh)	116	58°21'N	8°30'E
Grindstone Island, Can.	61	47°25'N	61°51'W
Grinnel, Ia., U.S. (grĭ-něl')	71	41°44'N	92°44'W
Griswold, Ia., U.S. (grĭz'wŭld)	70	41°11'N	95°05'W
Groais Island, i., Can.	61	50°57'N	55°35'W
Grobina, Lat. (grō'bĭnĭa)	123	56°35'N	21°10'E
Groblersdal, S. Afr.	192c	25°11'S	29°25'E
Grodzisk, Pol. (grō'jēsk)	124	52°14'N	16°22'E
Grodzisk Masowiecki, Pol. (grō'jēsk mä-zō-vyěts'ke)	125	52°06'N	20°40'E
Groesbeck, Tx., U.S. (grōs'běk)	81	31°32'N	96°31'W
Groix, Île de, i., Fr. (ēl dē grwä')	126	47°39'N	3°28'W
Grójec, Pol. (grō'yěts)	125	51°53'N	20°52'E
Gronau, Ger. (grō'nou)	124	52°12'N	7°05'E
Groningen, Neth. (grō'nĭng-ěn)	116	53°13'N	6°30'E
Groote Eylandt, i., Austl. (grō'tē ī'länt)	174	13°50'S	137°30'E
Grootfontein, Nmb. (grōt'fŏn-tān')	186	19°30'S	18°15'E
Groot-Kei, r., Afr. (kē)	187c	32°17'S	27°30'E
Grootkop, mtn., S. Afr.	186a	34°11'S	18°23'E
Groot Marico, S. Afr.	192c	25°36'S	26°23'E
Groot Marico, r., Afr.	192c	25°13'S	26°20'E
Groot-Vis, r., S. Afr.	187c	33°04'S	26°08'E
Groot Vloer, pl., S. Afr. (grōt' vlōr')	186	30°00'S	21°00'E
Gros-Mécatina, i., Can.	61	50°50'N	58°33'W
Gros Morne, mtn., Can. (grō môrn')	61	49°36'N	57°48'W

ăt; finăl; rāte; senāte; ärm; åsk; sofà; fâre; ch-choose; dh-as th in other; bē; ēvent; bět; recěnt; cratēr; g-gō; gh-guttural g; bĭt; ĭ-short neutral; rīde; ĸ-guttural k as ch in German ich;

PRONOUNCING INDEX

Gros Morne National Park – Hadlock 253

PLACE (Pronunciation)	PAGE	LAT.	LONG.
Gros Morne National Park, rec., Can.	53a	49°45′N	59°15′W
Gros Pate, mtn., Can.	61	50°16′N	57°25′W
Grosse Island, i., Mi., U.S. (grōs)	69b	42°08′N	83°09′W
Grosse Isle, Can. (īl′)	62f	50°04′N	97°27′W
Grossenhain, Ger. (grōs′ĕn-hīn)	124	51°17′N	13°33′E
Gross-Enzersdorf, Aus.	115e	48°13′N	16°33′E
Grosse Pointe, Mi., U.S. (point′)	69b	42°23′N	82°54′W
Grosse Pointe Farms, Mi., U.S. (färm)	69b	42°25′N	82°53′W
Grosse Pointe Park, Mi., U.S. (pärk)	69b	42°23′N	82°55′W
Grosseto, Italy (grŏs-sā′tō)	130	42°46′N	11°09′E
Grossglockner, mtn., Aus.	117	47°06′N	12°45′E
Gross Höbach, Ger. (hû′bäk)	115d	48°21′N	11°36′E
Gross Kreutz, Ger. (kroitz)	115b	52°24′N	12°47′E
Gross Schönebeck, Ger. (shō′nĕ-bĕk)	115b	52°54′N	13°32′E
Gros Ventre, r., Wy., U.S. (grōvĕn′t′r)	73	43°38′N	110°34′W
Groton, Ct., U.S. (grŏt′ŭn)	67	41°20′N	72°00′W
Groton, Ma., U.S.	61a	42°37′N	71°34′W
Groton, S.D., U.S.	70	45°25′N	98°04′W
Grottaglie, Italy (grŏt-täl′yā)	131	40°32′N	17°26′E
Grouard Mission, Can.	50	55°31′N	116°09′W
Groveland, Ma., U.S. (grōv′land)	61a	42°25′N	71°02′W
Groveton, N.H., U.S. (grōv′tŭn)	67	44°35′N	71°30′W
Groveton, Tx., U.S.	81	31°04′N	95°07′W
Groznyy, Russia (grŏz′nĭ)	134	43°20′N	45°40′E
Grudziądz, Pol. (grō′jyŏnts)	116	53°30′N	18°48′E
Grues, Île aux, i., Can. (ō grü)	62b	47°05′N	70°32′W
Grundy Center, Ia., U.S. (grŭn′dĭ sĕn′tẽr)	71	42°22′N	92°45′W
Gruñidora, Mex. (grōō-nyĕ-dō′rō)	88	24°10′N	101°49′W
Grünwald, Ger. (grōōn′väld)	115d	48°04′N	11°34′E
Gryazi, Russia (gryä′zĭ)	132	52°31′N	39°59′E
Gryazovets, Russia (gryä′zō-vĕts)	136	58°52′N	40°14′E
Gryfice, Pol. (grĭ′fĭ-tsĕ)	124	53°55′N	15°11′E
Gryfino, Pol. (grĭ′fē-nō)	124	53°16′N	14°30′E
Guabito, Pan. (gwä-bē′tō)	91	9°30′N	82°33′W
Guacanayabo, Golfo de, b., Cuba (gôl-fō-dĕ-gwä-kä-nä-yä′bō)	92	20°30′N	77°40′W
Guacara, Ven. (gwä′kä-rä)	101b	10°16′N	67°48′W
Guadalajara, Mex. (gwä-dhä-lä-hä′rä)	86	20°41′N	103°21′W
Guadalajara, Spain (gwä-dä-lä-kä′rä)	118	40°37′N	3°10′W
Guadalcanal, Spain (gwä-dhäl-kä-näl′)	128	38°05′N	5°48′W
Guadalcanal, i., Sol. Is.	175	9°48′S	158°43′E
Guadalcázar, Mex. (gwä-dhäl-kä′zär)	88	22°38′N	100°24′W
Guadalete, r., Spain (gwä-dhä-lā′tā)	128	36°53′N	5°38′W
Guadalhorce, r., Spain (gwä-dhäl-ôr′thā)	128	37°05′N	4°50′W
Guadalimar, r., Spain (gwä-dhä-lē-mär′)	128	38°29′N	2°53′W
Guadalope, r., Spain (gwä-dä-lō-pĕ)	129	40°48′N	0°10′W
Guadalquivir, Río, r., Spain (rē′ō-gwä-dhäl-kē-vēr′)	112	37°30′N	5°00′W
Guadalupe, Mex.	80	31°23′N	106°06′W
Guadalupe, i., Mex.	86	29°00′N	118°45′W
Guadalupe, r., Tx., U.S. (gwä-dhä-lōō′pä)	80	29°54′N	99°03′W
Guadalupe, Sierra de, mts., Spain (syĕr′rä dā gwä-dhä-lōō′pä)	118	39°30′N	5°25′W
Guadalupe Mountains, mts., N.M., U.S.	80	32°00′N	104°55′W
Guadalupe Peak, mtn., Tx., U.S.	80	31°55′N	104°55′W
Guadarrama, r., Spain (gwä-dhär-rä′mä)	129a	40°34′N	3°58′W
Guadarrama, Sierra de, mts., Spain (gwä-dhär-rä′mä)	112	41°00′N	3°40′W
Guadatentin, r., Spain	128	37°43′N	1°58′W
Guadeloupe, dep., N.A. (gwä-dē-lōōp)	87	16°40′N	61°10′W
Guadeloupe Passage, strt., N.A.	91b	16°26′N	62°00′W
Guadiana, r., Eur. (gwä-dvä′nä)	112	39°00′N	6°00′W
Guadiana, Bahía de, b., Cuba (bä-ē′ä-dĕ-gwä-dhē-ä′nä)	92	22°10′N	84°35′W
Guadiana Alto, r., Spain (äl′tō)	128	39°02′N	2°52′W
Guadiana Menor, r., Spain (mā′nōr)	128	37°43′N	2°45′W
Guadiaro, r., Spain (gwä-dhē-ä′rō)	128	36°38′N	5°25′W
Guadiela, r., Spain (gwä-dhē-ā′lä)	128	40°27′N	2°05′W
Guadix, Spain (gwä-dēsh′)	128	37°18′N	3°09′W
Guaira, Braz. (gwä-ē-rä)	101	24°03′S	54°02′W
Guaire, r., Ven. (gwī′rĕ)	101b	10°25′N	66°43′W
Guajaba, Cayo, i., Cuba (kä′yō-gwä-hä′bä)	92	21°50′N	77°35′W
Guajará Mirim, Braz. (gwä-zhä-rä′mē-rēN′)	100	10°58′S	65°12′W
Guajira, Península de, pen., S.A.	100	12°35′N	73°00′W
Gualán, Guat. (gwä-län′)	90	15°08′N	89°21′W
Gualeguay, Arg. (gwä-lĕ-gwä′y)	102	33°10′S	59°20′W
Gualeguay, r., Arg.	102	32°49′S	59°05′W
Gualicho, Salina, l., Arg. (sä-lē′nä-gwä-lē′chō)	102	40°20′S	65°15′W
Guam, i., Oc. (gwäm)	3	14°00′N	143°20′E
Guamo, Col. (gwä′mō)	100a	4°02′N	74°58′W
Gu'an, China (gōō-än)	164a	39°25′N	116°18′E
Guan, r., China (gůän)	162	31°56′N	115°19′E
Guanabacoa, Cuba (gwä-nä-bä-kō′ä)	87	23°08′N	82°19′W
Guanabara, Baía de, b., Braz.	99a	22°44′S	43°09′W
Guanacaste, Cordillera, mts., C.R.	90	10°54′N	85°27′W
Guanacevi, Mex. (gwä-nä-sĕ-vē′)	86	25°30′N	105°45′W
Guanahacabibes, Península de, pen., Cuba	92	21°55′N	84°35′W
Guanajay, Cuba (gwänä-hī′)	92	22°55′N	82°40′W
Guanajuato, Mex. (gwä-nä-hwä′tō)	86	21°01′N	101°16′W
Guanajuato, state, Mex.	86	21°00′N	101°00′W
Guanape, Ven. (gwä-nä′pĕ)	101b	9°55′N	65°32′W
Guanape, r., Ven.	101b	9°52′N	65°20′W
Guanare, Ven. (gwä-nä′rä)	100	8°57′N	69°47′W
Guanduçu, r., Braz. (gwä′n-dōō′sōō)	102b	22°50′S	43°40′W
Guane, Cuba (gwä′nā)	92	22°10′N	84°05′W
Guangchang, China (gůäŋ-chäŋ)	165	26°50′N	116°18′E
Guangde, China (gůäŋ-dŭ)	165	30°40′N	119°20′E
Guangdong, prov., China (gůäŋ-dôŋ)	161	23°45′N	113°15′E
Guanglu Dao, i., China (gůäŋ-lōō dou)	162	39°13′N	122°21′E
Guangping, China (gůäŋ-pĭŋ)	162	36°30′N	114°57′E
Guangrao, China (gůaŋ-rou)	162	37°04′N	118°24′E
Guangshan, China (gůäŋ-shän)	162	32°02′N	114°53′E
Guangxi Zhuangzu, prov., China (gůäŋ-shyē)	160	24°00′N	108°30′E
Guangzhou (Canton), China	160	23°07′N	113°15′W
Guanhu, China (gůän-hōō)	162	34°26′N	117°59′E
Guannan, China (gůän-nän)	162	34°17′N	119°17′E
Guanta, Ven. (gwän′tä)	101b	10°15′N	64°35′W
Guantánamo, Cuba (gwän-tä′nä-mō)	93	20°10′N	75°10′W
Guantánamo, prov., Cuba	93	20°10′N	75°05′W
Guantánamo, Bahía de, b., Cuba	93	19°35′N	75°35′W
Guantao, China (gůän-tou)	162	36°39′N	115°25′E
Guanxian, China (gůän-shyĕn)	162	36°30′N	115°28′E
Guanyao, China (gůän-you)	163a	23°13′N	113°04′E
Guanyun, China (gůän-yōn)	162	34°28′N	119°16′E
Guapiles, C.R. (gwä-pē-lĕs)	91	10°05′N	83°54′W
Guapimirim, Braz. (gwä-pē-mē-rē′N)	102b	22°31′S	42°59′W
Guaporé, r., S.A. (gwä-pō-rā′)	100	12°11′S	63°47′W
Guaqui, Bol. (guä′kē)	100	16°42′S	68°47′W
Guara, Sierra de, mts., Spain (sē-ĕ′r-rä-dĕ-gwä′rä)	129	42°24′N	0°15′W
Guarabira, Braz. (gwä-rä-bē′rá)	101	6°49′S	35°27′W
Guaranda, Ec. (gwä-rán′dä)	100	1°39′S	78°57′W
Guarapari, Braz. (gwä-rä-pä′rĕ)	101	20°34′S	40°31′W
Guarapiranga, Represa do, res., Braz.	99a	23°45′S	46°44′W
Guarapuava, Braz. (gwä-rä-pwä′vá)	102	25°29′S	51°26′W
Guarda, Port. (gwär′dä)	128	40°32′N	7°17′W
Guardiato, r., Spain	128	38°10′N	5°05′W
Guarena, Spain (gwä-rā′nyä)	128	38°52′N	6°08′W
Guaribe, r., Ven. (gwä-rē′bĕ)	101b	9°48′N	65°17′W
Guárico, dept., Ven.	101b	9°42′N	67°25′W
Guarulhos, Braz. (gwä-rō′l-yōs)	99a	23°28′S	46°30′W
Guarus, Braz. (gwä′rōōs)	99a	21°44′S	41°19′W
Guasca, Col. (gwäs′kä)	100a	4°52′N	73°52′W
Guasipati, Ven. (gwä-sē-pä′tē)	101	7°26′N	61°57′W
Guastalla, Italy (gwäs-täl′lä)	130	44°53′N	10°39′E
Guasti, Ca., U.S. (gwäs′tĭ)	75a	34°04′N	117°35′W
Guatemala, Guat. (guä-tå-mä′lä)	86	14°37′N	90°32′W
Guatemala, nation, N.A.	86	15°45′N	91°45′W
Guatire, Ven. (gwä-tē′rĕ)	101b	10°28′N	66°34′W
Guaviare, r., Col.	100	3°35′N	69°28′W
Guayabal, Cuba (gwä-yä-bä′l)	92	20°40′N	77°40′W
Guayalejo, r., Mex. (gwä-yä-lĕ′hō)	88	23°24′N	99°09′W
Guayama, P.R. (gwä-yä′mä)	87b	18°00′N	66°08′W
Guayamouc, r., Haiti	93	19°05′N	72°00′W
Guayaquil, Ec. (gwī-ä-kēl′)	100	2°16′S	79°53′W
Guayaquil, Golfo de, b., Ec. (gôl-fō-dĕ)	100	3°03′S	82°12′W
Guaymas, Mex. (gwä′y-mäs)	86	27°49′N	110°58′W
Guayubin, Dom. Rep. (gwä-yōō-bē′n)	93	19°40′N	71°25′W
Guazacapán, Guat. (gwä-zä-kä-pän′)	90	14°04′N	90°26′W
Gubakha, Russia (gōō-bä′kå)	134	58°53′N	57°35′E
Gubbio, Italy (gōōb′byō)	130	43°23′N	12°36′E
Guben, Ger.	124	51°57′N	14°43′E
Gucheng, China (gōō-chŭŋ)	162	39°09′N	115°43′E
Gúdar, Sierra de, mts., Spain	129	40°28′N	0°47′W
Gudena, r., Den.	122	56°20′N	9°47′E
Gudermes, Russia	138	43°20′N	46°08′E
Gudvangen, Nor. (gōōdh′väŋ-gĕn)	122	60°52′N	6°45′E
Guebwiller, Fr. (gĕb-vē-lâr′)	127	47°53′N	7°10′E
Guédi, Mont, mtn., Chad	189	12°14′N	18°58′E
Guelma, Alg. (gwĕl′mä)	184	36°32′N	7°17′E
Guelph, Can. (gwĕlf)	59	43°33′N	80°15′W
Güere, r., Ven. (gwĕ′rĕ)	101b	9°39′N	65°00′W
Guéret, Fr. (gā-rĕ′)	126	46°09′N	1°52′E
Guernsey, dep., Eur.	126	49°28′N	2°35′W
Guernsey, i., Guern. (gûrn′zĭ)	117	49°27′N	2°36′W
Guerrero, Mex. (gĕr-rā′rō)	80	26°47′N	99°20′W
Guerrero, Mex.	80	28°20′N	100°24′W
Guerrero, state, Mex.	86	17°45′N	100°15′W
Gueydan, La., U.S. (gā′dăn)	81	30°01′N	92°31′W
Guia de Pacobaíba, Braz. (gwē′ä-dĕ-pä′kō-bī′bä)	102b	22°42′S	43°10′W
Guiana Highlands, mts., S.A.	97	3°20′N	60°00′W
Guichi, China (gwā-chr)	165	30°35′N	117°28′E
Guichicovi, Mex. (gwē-chē-kō′vē)	89	16°58′N	95°10′W
Guidonia, Italy (gwē-dō′nyä)	130	42°00′N	12°45′E
Guiglo, C. Iv.	188	6°33′N	7°29′W
Guignes-Rabutin, Fr. (gēN′yĕ)	127b	48°38′N	2°48′E
Güigüe, Ven. (gwē′gwĕ)	101b	10°05′N	67°48′W
Guija, Lago, l., N.A. (gē′hä)	90	14°16′N	89°21′W
Guildford, Eng., U.K. (gĭl′fērd)	120	51°13′N	0°34′W
Guilford, In., U.S. (gĭl′fērd)	69f	39°10′N	84°55′W
Guilin, China (gwā-lĭn)	161	25°18′N	110°22′E
Guimarães, Port. (gē-mä-rănsh′)	128	41°27′N	8°22′W
Guinea, nation, Afr. (gĭn′ē)	184	10°48′N	12°28′W
Guinea, Gulf of, b., Afr.	184	2°00′N	1°00′E
Guinea-Bissau, nation, Afr. (gĭn′ē)	184	12°00′N	20°00′W
Guingamp, Fr. (găN-găN′)	126	48°35′N	3°10′W
Guir, r., Mor.	118	31°55′N	2°48′W
Güira de Melena, Cuba (gwē′rä dā mä-lā′nä)	92	22°45′N	82°30′W
Güiria, Ven. (gwē-rē′ä)	100	10°43′N	62°16′W
Guise, Fr. (güēz)	126	49°54′N	3°37′E
Guisisil, vol., Nic. (gē-sē-sēl′)	90	12°40′N	86°11′W
Guiyang, China (gwā-yäŋ)	160	26°45′N	107°00′E
Guizhou, China (gwā-jō)	163a	22°46′N	113°15′E
Guizhou, prov., China	160	27°00′N	106°10′E
Gujānwāla, Pak. (gōj-rän′va-lá)	155	32°08′N	74°14′E
Gujarat, India	155	22°54′N	72°00′E
Gulbarga, India (gōl-bûr′gá)	155	17°25′N	76°52′E
Gulbene, Lat. (gōl-bă′nĕ)	123	57°09′N	26°49′E
Gulfport, Ms., U.S. (gŭlf′pōrt)	82	30°24′N	89°05′W
Gulja *see* Yining, China	160	43°58′N	80°40′E
Gull Lake, Can.	56	50°10′N	108°25′W
Gull Lake, l., Can.	55	52°35′N	114°00′W
Gulu, Ug.	191	2°47′N	32°18′E
Gumaca, Phil. (gōō-mä-kä′)	169a	13°55′N	122°06′E
Gumbeyka, r., Russia (gōm-bĕy′kå)	142a	53°20′N	59°42′E
Gumel, Nig.	184	12°39′N	9°22′E
Gummersbach, Ger. (gōm′ĕrs-bäk)	124	51°02′N	7°34′E
Gummi, Nig.	189	12°09′N	5°09′E
Gumpoldskirchen, Aus.	115e	48°04′N	16°15′E
Guna, India	158	24°44′N	77°17′E
Gunisao, r., Can. (gŭn-ĭ-sā′ō)	57	53°40′N	97°35′W
Gunisao Lake, l., Can.	57	53°35′N	96°10′W
Gunnedah, Austl. (gŭ′nē-dä)	176	31°00′S	150°10′E
Gunnison, Co., U.S. (gŭn′ĭ-sŭn)	77	38°33′N	106°56′W
Gunnison, Ut., U.S.	77	39°10′N	111°50′W
Gunnison, r., Co., U.S.	77	38°45′N	108°20′W
Guntersville, Al., U.S. (gŭn′tērz-vĭl)	82	34°20′N	86°19′W
Guntersville Lake, res., Al., U.S.	82	34°30′N	86°20′W
Guntramsdorf, Aus.	115e	48°04′N	16°19′E
Guntūr, India (gōn′tōōr)	155	16°22′N	80°29′E
Guoyang, China (gwō-yäŋ)	162	33°32′N	116°10′E
Gurdon, Ar., U.S. (gûr′dŭn)	79	33°56′N	93°10′W
Gurgueia, r., Braz.	101	8°12′S	43°49′W
Guri, Embalse, res., Ven.	100	7°30′N	63°00′W
Gurnee, Il., U.S. (gûr′nē)	69a	42°22′N	87°55′W
Gurskøy, i., Nor. (gōōrskûė)	122	62°18′N	5°20′E
Gurupi, Serra do, mts., Braz. (sĕ′r-rä-dō-gōō-rōō-pē′)	101	5°32′S	47°02′W
Guru Sikhar, mtn., India	158	29°42′N	72°50′E
Gur'yevsk, Russia (gōōr-yĭfsk′)	134	54°17′N	85°56′E
Gusau, Nig. (gōō-zä′ōō)	184	12°12′N	6°40′E
Gusev, Russia (gōō′sĕf)	123	54°35′N	22°15′E
Gushi, China (gōō-shr)	162	32°11′N	115°39′E
Gushiago, Ghana	188	9°55′N	0°12′W
Gusinje, Serb. (gōō-sēn′yĕ)	131	42°34′N	19°54′E
Gus'-Khrustal'nyy, Russia (gōōs-krōō-stäl′ny′)	136	55°39′N	40°41′E
Gustavo A. Madero, Mex. (gōōs-tä′vô-á-mä-dĕ′rō)	88	19°29′N	99°07′W
Güstrow, Ger. (güs′trō)	124	53°48′N	12°12′E
Gütersloh, Ger. (gü′tērs-lō)	124	51°54′N	8°22′E
Guthrie, Ok., U.S. (gŭth′rĭ)	79	35°52′N	97°26′W
Guthrie Center, Ia., U.S.	71	41°41′N	94°33′W
Gutiérrez Zamora, Mex. (gōō-tĭ-âr′räz zä-mō′rä)	89	20°27′N	97°17′W
Guttenberg, Ia., U.S. (gŭt′ĕn-bûrg)	71	42°48′N	91°09′W
Guyana, nation, S.A. (gŭy′ănă)	101	7°45′N	59°00′W
Guyang, China (gōō-yäŋ)	162	34°56′N	114°57′E
Guye, China (gōō-yŭ)	162	39°46′N	118°23′E
Guymon, Ok., U.S. (gī′mŏn)	78	36°41′N	101°29′W
Guysborough, Can. (gīz′bŭr-ô)	61	45°23′N	61°30′W
Guzhen, China (gōō-jŭn)	164	33°20′N	117°18′E
Gvardeysk, Russia (gvár-dĕysk′)	123	54°39′N	21°11′E
Gwadabawa, Nig.	189	13°20′N	5°15′E
Gwādar, Pak. (gwä′dŭr)	154	25°15′N	62°29′E
Gwalior, India	155	26°13′N	78°10′E
Gwane, D.R.C. (gwän)	185	4°43′N	25°50′E
Gwardafuy, Gees, c., Som.	192a	11°55′N	51°30′E
Gwda, r., Pol.	124	53°27′N	16°52′E
Gwembe, Zam.	191	16°30′S	27°35′E
Gweru, Zimb.	186	19°15′S	29°48′E
Gwinn, Mi., U.S. (gwĭn)	71	46°15′N	87°30′W
Gyaring Co, l., China	158	30°37′N	88°33′E
Gydan, Khrebet (Kolymskiy), mts., Russia	135	61°45′N	155°00′E
Gydanskiy Poluostrov, pen., Russia	134	70°42′N	76°03′E
Gympie, Austl. (gĭm′pē)	175	26°20′S	152°50′E
Gyöngyös, Hung. (dyûn′dyûsh)	119	47°47′N	19°55′E
Györ, Hung. (dyûr)	119	47°40′N	17°37′E
Gyōtoku, Japan (gyō′tō-kōō′)	167a	35°42′N	139°56′E
Gypsumville, Can. (jĭp′sŭm-vĭl)	50	51°45′N	98°35′W
Gýtheio, Grc.	131	36°50′N	22°37′E
Gyula, Hung. (dyō′lä)	125	46°38′N	21°18′E
Gyumri, Arm.	137	40°40′N	43°50′E
Gyzylarbat, Turkmen.	139	38°55′N	56°33′E

H

PLACE (Pronunciation)	PAGE	LAT.	LONG.
Haan, Ger. (hän)	127c	51°12′N	7°00′E
Haapamäki, Fin. (häp′ä-mĕ-kē)	123	62°16′N	24°20′E
Haapsalu, Est. (häp′sä-lō)	123	58°56′N	23°33′E
Haar, Ger. (här)	115d	48°06′N	11°44′E
Ha'Arava (Wādī al Jayb), val., Asia	153a	30°33′N	35°10′E
Haarlem, Neth. (här′lĕm)	121	52°22′N	4°37′E
Habana, prov., Cuba (hä-vä′nä)	92	22°45′N	82°25′W
Häbra, India	158a	22°49′N	88°38′E
Hachinohe, Japan (hä′chē-nō′hå)	166	40°29′N	141°40′E
Hachiōji, Japan (hä′chē-ō′jē)	166	35°39′N	139°18′E
Hackensack, N.J., U.S. (hăk′ĕn-săk)	68a	40°54′N	74°03′W
Hadd, Ra's al, c., Oman	154	22°29′N	59°46′E
Haddonfield, N.J., U.S. (hăd′ŭn-fēld)	68f	39°53′N	75°02′W
Haddon Heights, N.J., U.S. (hăd′ŭn hīts)	68f	39°53′N	75°03′W
Hadejia, Nig. (hä-dā′jä)	184	12°30′N	9°59′E
Hadejia, r., Nig.	184	12°15′N	10°00′E
Hadera, Isr. (κå-dĕ′rá)	153a	32°26′N	34°55′E
Haderslev, Den. (hä′dhĕrs-lĕv)	122	55°17′N	9°28′E
Hadiach, Ukr.	137	50°22′N	33°59′E
Hadīdū, Yemen	154	12°40′N	53°50′E
Hadlock, Wa., U.S. (hăd′lŏk)	74a	48°02′N	122°46′W

ng-sing; ŋ-baŋk; N-nasalized n; nŏd; cŏmmit; ōld; ōbey; ôrder; oi-boil; fōōd; ŏŏ-as oo in foot; ou-out; s-soft; sh-dish; th-thin; pūre; ûnite; ûrn; stŭd; circŭs; ü-as in French tu; ′-indeterminate vowel.

PLACE (Pronunciation)	PAGE	LAT.	LONG.
Haḑramawt, reg., Yemen	154	15°22'N	48°40'E
Hadūr Shu'ayb, mtn., Yemen	154	15°45'N	43°45'E
Haeju, Kor., N. (hä'ē-jŭ)	166	38°03'N	125°42'E
Hafnarfjördur, Ice.	116	64°02'N	21°32'W
Haft Gel, Iran	157	31°27'N	49°27'E
Hafun, Ras, c., Som. (hä-fōōn')	192a	10°15'N	51°35'E
Hageland, Mt., U.S. (hăge'lănd)	73	48°53'N	108°43'W
Hagen, Ger. (hä'gĕn)	124	51°21'N	7°29'E
Hagerstown, In., U.S. (hā'gērz-toun)	66	39°55'N	85°10'W
Hagerstown, Md., U.S.	65	39°40'N	77°45'W
Hagi, Japan (hä'gĭ)	167	34°25'N	131°25'E
Hague, Cap de la, c., Fr. (dē lä äg')	126	49°44'N	1°55'W
Haguenau, Fr. (åg'nō')	127	48°47'N	7°48'E
Hai'an, China (hī-än)	162	32°35'N	120°25'E
Haibara, Japan (hä'ē-bä'rä)	167	34°29'N	135°57'E
Haicheng, China (hī-chŭŋ)	164	40°58'N	122°45'E
Haidian, China (hī-dĭĕn)	162	39°59'N	116°17'E
Haifa, Isr. (hä'ē-fà)	154	32°48'N	35°00'E
Haifeng, China (hä'ē-fĕng')	165	23°00'N	115°20'E
Haifuzhen, China (hī-fōō-jŭn)	162	31°57'N	121°48'E
Haikou, China (hī-kō)	165	20°00'N	110°20'E
Ḩā'il, Sau. Ar.	154	27°30'N	41°47'E
Hailar, China	161	49°10'N	118°40'E
Hailey, Id., U.S. (hā'lĭ)	73	43°31'N	114°19'W
Haileybury, Can.	59	47°27'N	79°38'W
Haileyville, Ok., U.S. (hā'lĭ-vĭl)	79	34°51'N	95°34'W
Hailing Dao, i., China (hī-lĭŋ dou)	165	21°30'N	112°15'E
Hailong, China (hī-loŋ)	164	42°32'N	125°52'E
Hailun, China (hä'ē-lōōn')	161	47°18'N	126°50'E
Hainan, prov., China	160	19°00'N	109°30'E
Hainan Dao, i., China (hī-nän dou)	161	19°00'N	111°10'E
Hainburg, Aus.	124	48°09'N	16°57'E
Haines, Ak., U.S. (hānz)	63	59°10'N	135°38'W
Haines City, Fl., U.S.	83a	28°05'N	81°38'W
Hai Phong, Viet. (hī'fông')(hä'ēp-hŏng)	168	20°52'N	106°40'E
Haisyn, Ukr.	137	48°46'N	29°22'E
Haiti, nation, N.A. (hā'tĭ)	87	19°00'N	72°15'W
Haizhou, China	162	34°34'N	119°11'E
Haizhou Wan, b., China	164	34°49'N	120°35'E
Hajdúböszormény, Hung. (hôl'dò-bû'sûr-mān')	125	47°41'N	21°30'E
Hajdúhadház, Hung. (hô'ĭ-dò-hôd'häz)	125	47°32'N	21°32'E
Hajdúnánás, Hung. (hô'ĭ-dò-nä'näsh)	125	47°52'N	21°27'E
Hakodate, Japan (hä-kō-dä't å)	161	41°46'N	140°42'E
Haku-San, mtn., Japan (hä'kōō-sän')	166	36°11'N	136°45'E
Halā'ib, Egypt (hä-lä'ĕb)	185	22°10'N	36°40'E
Halbe, Ger. (häl'bĕ)	115b	52°07'N	13°43'E
Halberstadt, Ger. (häl'bĕr-shtät)	124	51°54'N	11°07'E
Halcon, Mount, mtn., Phil. (häl-kōn')	169a	13°19'N	120°55'E
Halden, Nor. (häl'dĕn)	116	59°10'N	11°21'E
Haldensleben, Ger.	124	52°18'N	11°23'E
Hale, Eng., U.K. (hāl)	114a	53°22'N	2°20'W
Haleakalā Crater, depr., Hi., U.S. (hä'lā-ä'kä-lä)	84a	20°44'N	156°15'W
Haleakalā National Park, rec., Hi., U.S.	84a	20°46'N	156°00'W
Hales Corners, Wi., U.S. (hālz kŏr'nērz)	69a	42°56'N	88°03'W
Halesowen, Eng., U.K. (hālz'ō-wĕn)	114a	52°26'N	2°03'W
Halethorpe, Md., U.S. (hāl-thôrp)	68e	39°15'N	76°40'W
Haleyville, Al., U.S. (hā'lĭ-vĭl)	82	34°11'N	87°36'W
Half Moon Bay, Ca., U.S. (hăf'mōōn)	74b	37°28'N	122°26'W
Halfway House, S. Afr. (häf-wā hous)	187b	26°00'S	28°08'E
Halfweg, Neth.	115a	52°23'N	4°45'E
Halifax, Can. (hăl'ĭ-făks)	51	44°39'N	63°36'W
Halifax, Eng., U.K.	120	53°44'N	1°52'W
Halifax Bay, b., Austl. (hăl'ĭ-făx)	175	18°56'S	147°07'E
Halifax Harbour, b., Can.	60	44°35'N	63°31'W
Halkett, Cape, c., Ak., U.S.	63	70°50'N	151°15'W
Hallam Peak, mtn., Can.	55	52°11'N	118°46'E
Halla San, mtn., Kor., S. (häl'là-sän)	166	33°20'N	126°37'E
Halle, Bel. (häl'lĕ)	115a	50°45'N	4°13'E
Halle, Ger.	117	51°30'N	11°59'E
Hallettsville, Tx., U.S. (hăl'ĕts-vĭl)	81	29°26'N	96°55'W
Hallock, Mn., U.S. (hăl'ŭk)	70	48°46'N	96°57'W
Hall Peninsula, pen., Can. (hôl)	53	63°14'N	65°40'W
Halls Bayou, Tx., U.S.	81a	29°55'N	95°23'W
Hallsberg, Swe. (häls'bĕrgh)	122	59°04'N	15°04'E
Halls Creek, Austl. (hôlz)	174	18°15'S	127°45'E
Halmahera, i., Indon. (häl-mä-hā'rä)	169	0°45'N	128°45'E
Halmahera, Laut, Indon.	169	1°00'S	129°00'E
Halmstad, Swe. (hälm'städ)	116	56°40'N	12°46'E
Halsafjorden, b., Nor. (häl'sĕ fyôrd)	122	63°03'N	8°23'E
Halstead, Ks., U.S. (hôl'stĕd)	79	38°02'N	97°36'W
Haltern, Ger. (häl'tĕrn)	127c	51°45'N	7°10'E
Haltom City, Tx., U.S. (hôl'tŏm)	75c	32°48'N	97°13'W
Halver, Ger.	127c	51°11'N	7°30'E
Hamada, Japan	166	34°53'N	132°05'E
Hamadān, Iran (hŭ-mŭ-dän')	154	34°45'N	48°07'E
Ḩamah, Syria (hä'mä)	154	35°08'N	36°53'E
Hamamatsu, Japan (hä'mä-mät'sò)	166	34°41'N	137°43'E
Hamar, Nor. (hä'mär)	116	60°49'N	11°05'E
Hamasaka, Japan (hä'mä-sä'kä)	167	35°57'N	134°27'E
Hamborn, Ger. (häm'bōrn)	127c	51°30'N	6°43'E
Hamburg, Ger. (häm'bòòrgh)	110	53°34'N	10°02'E
Hamburg, S. Afr. (hăm'bûrg)	187c	33°18'S	27°28'E
Hamburg, Ar., U.S. (hăm'bûrg)	79	33°15'N	91°49'W
Hamburg, N.J., U.S.	68a	41°09'N	74°35'W
Hamburg, N.Y., U.S.	69c	42°44'N	78°51'W
Hamden, Ct., U.S. (hăm'dĕn)	67	41°20'N	72°55'W
Hämeenlinna, Fin. (hĕ'mān-lĭn-nà)	116	61°00'N	24°29'E
Hameln, Ger. (hä'mĕln)	124	52°06'N	9°23'E
Hamelwörden, Ger. (hä'mĕl-vûr-dĕn)	115c	53°47'N	9°19'E

PLACE (Pronunciation)	PAGE	LAT.	LONG.
Hamersley Range, mts., Austl. (hăm'ērz-lē)	174	22°15'S	117°50'E
Hamhŭng, Kor., N. (häm'hòng')	161	39°57'N	127°35'E
Hami, China (hä-mē)	160	42°58'N	93°14'E
Hamilton, Austl. (hăm'ĭl-tŭn)	175	37°50'S	142°10'E
Hamilton, Can.	51	43°15'N	79°52'W
Hamilton, N.Z.	175a	37°45'S	175°28'E
Hamilton, Al., U.S.	82	34°09'N	88°01'W
Hamilton, Ma., U.S.	61a	42°37'N	70°52'W
Hamilton, Mo., U.S.	79	39°43'N	93°59'W
Hamilton, Mt., U.S.	73	46°15'N	114°09'W
Hamilton, Oh., U.S.	65	39°22'N	84°33'W
Hamilton, Tx., U.S.	80	31°42'N	98°07'W
Hamilton, Lake, l., Ar., U.S.	79	34°25'N	93°32'W
Hamilton Harbour, b., Can.	62d	43°17'N	79°50'W
Hamilton Inlet, b., Can.	53	54°20'N	56°57'W
Hamina, Fin. (hà'mē-nà)	123	60°34'N	27°15'E
Hamlet, N.C., U.S. (hăm'lĕt)	83	34°53'N	79°42'W
Hamlin, Tx., U.S. (hăm'lĭn)	78	32°54'N	100°08'W
Hamm, Ger. (häm)	124	51°40'N	7°48'E
Hammanskraal, S. Afr. (hä-mäns-kräl')	192c	25°24'S	28°17'E
Hamme, Bel.	115a	51°06'N	4°07'E
Hamme-Oste Kanal, can., Ger. (hä'mē-ōs'tē kä-näl)	115c	53°20'N	8°59'E
Hammerfest, Nor. (hä'mĕr-fĕst)	110	70°38'N	23°59'E
Hammond, In., U.S. (hăm'ŭnd)	65	41°37'N	87°31'W
Hammond, La., U.S.	81	30°30'N	90°28'W
Hammond, Or., U.S.	74c	46°12'N	123°57'W
Hammonton, N.J., U.S. (hăm'ŭn-tŭn)	67	39°40'N	74°45'W
Hampden, Me., U.S. (hăm'dĕn)	60	44°44'N	68°51'W
Hampstead, Md., U.S.	68e	39°36'N	76°54'W
Hampstead Norris, Eng., U.K. (hămp-stĕd nŏ'rĭs)	114b	51°27'N	1°14'W
Hampton, Can. (hămp'tŭn)	60	45°32'N	65°51'W
Hampton, Ia., U.S.	71	42°43'N	93°15'W
Hampton, Va., U.S.	67	37°02'N	76°21'W
Hampton Roads, b., Va., U.S.	68g	36°56'N	76°23'W
Hams Fork, r., Wy., U.S.	73	41°55'N	110°40'W
Hamtramck, Mi., U.S. (hăm-trăm'ĭk)	69b	42°24'N	83°03'W
Han, r., China (hän)	165	25°00'N	116°35'E
Han, r., China	161	31°40'N	112°04'E
Han, r., Kor., S.	166	37°10'N	127°40'E
Hāna, Hi., U.S. (hä'nä)	84a	20°43'N	155°59'W
Hanábana, r., Cuba (hä-nä-bä'nä)	92	22°30'N	80°55'W
Hanalei Bay, b., Hi., U.S. (hä-nä-lā'ē)	84a	22°15'N	159°40'W
Hanang, mtn., Tan.	191	4°26'S	35°24'E
Hanau, Ger. (hä'nou)	124	50°08'N	8°56'E
Hancock, Mi., U.S. (hăn'kŏk)	65	47°08'N	88°37'W
Handan, China (hän-dän)	162	36°37'N	114°30'E
Haney, Can. (hā-nē)	55	49°13'N	122°36'W
Hanford, Ca., U.S. (hăn'fērd)	76	36°20'N	119°38'W
Hangayn Nuruu, mts., Mong.	160	48°03'N	99°45'E
Hango, Fin. (häŋ'gû)	110	59°49'N	22°56'E
Hangzhou, China (häng'chō')	161	30°17'N	120°12'E
Hangzhou Wan, b., China (häŋ-jō wän)	165	30°20'N	121°25'E
Hankamer, Tx., U.S. (hăn'kà-mēr)	81a	29°52'N	94°42'W
Hankinson, N.D., U.S. (häŋ'kĭn-sŭn)	70	46°04'N	96°54'W
Hankou, China (hän-kō)	165	30°42'N	114°22'E
Hann, Mount, mtn., Austl. (hän)	174	16°05'S	126°07'E
Hanna, Can. (hăn'à)	50	51°38'N	111°54'W
Hanna, Wy., U.S.	73	41°51'N	106°34'W
Hannah, N.D., U.S.	70	48°58'N	98°42'W
Hannibal, Mo., U.S. (hăn'ĭ băl)	65	39°42'N	91°22'W
Hannover, Ger. (hän-ō'vĕr)	110	52°22'N	9°45'E
Hannover, hist. reg., Ger.	124	52°52'N	8°27'E
Hanöbukten, b., Swe.	122	55°54'N	14°55'E
Hanoi, Viet. (hä-noi')	168	21°04'N	105°50'E
Hanover, Can. (hăn'ō-vēr)	58	44°10'N	81°05'W
Hanover, Ma., U.S.	61a	42°07'N	70°49'W
Hanover, N.H., U.S.	67	43°45'N	72°15'W
Hanover, Pa., U.S.	67	39°50'N	77°00'W
Hanover, i., Chile	102	51°00'S	74°45'W
Hanshan, China (hän'shän')	162	31°43'N	118°06'E
Hans Lollick, i., V.I.U.S. (häns'lŏl'ĭk)	87c	18°24'N	64°55'W
Hanson, Ma., U.S. (hăn'sŭn)	61a	42°04'N	70°53'W
Hansville, Wa., U.S. (häns'-vĭl)	74a	47°55'N	122°33'W
Hantengri Feng, mtn., Asia (hän-tŭŋ-rē fŭŋ)	160	42°10'N	80°20'E
Hantsport, Can. (hänts'pōrt)	60	45°04'N	64°11'W
Hanyang, China (han'yäng')	161	30°30'N	114°10'E
Hanzhong, China (hän-jôŋ)	164	33°02'N	107°00'E
Haocheng, China (hou-chŭŋ)	162	33°19'N	117°33'E
Haparanda, Swe. (hä-pa-rän'dä)	116	65°54'N	23°57'E
Hapeville, Ga., U.S. (hāp'vĭl)	68c	33°39'N	84°25'W
Happy Camp, Ca., U.S.	72	41°47'N	123°22'W
Happy Valley-Goose Bay, Can.	51	53°19'N	60°33'W
Ḩaql, Sau. Ar.	153a	29°15'N	34°57'E
Har, Laga, r., Kenya	191	2°15'N	39°30'E
Haradok, Bela.	132	55°27'N	29°58'E
Harare, Zimb.	186	17°50'S	31°03'E
Harbin, China	161	45°40'N	126°30'E
Harbor Beach, Mi., U.S. (här'bēr bēch)	66	43°50'N	82°40'W
Harbor Springs, Mi., U.S.	66	45°25'N	85°05'W
Harbour Breton, Can. (brĕt'ŭn) (brē-tôn')	61	47°29'N	55°48'W
Harbour Grace, Can. (grās)	61	47°32'N	53°13'W
Harburg, Ger. (här-bòrgh)	115c	53°28'N	9°58'E
Hardangerfjorden, Nor. (här-däng'ēr fyôrd)	116	59°58'N	6°30'E
Hardin, Mt., U.S. (här'dĭn)	73	45°44'N	107°36'W
Harding, S. Afr. (här'dĭng)	186	30°34'S	29°54'E
Harding, Lake, res., U.S.	82	32°43'N	85°00'W
Hardwār, India (hŭr'dvär)	155	29°56'N	78°06'E
Hardy, r., Mex. (här'dĭ)	76	32°04'N	115°10'W

PLACE (Pronunciation)	PAGE	LAT.	LONG.
Hare Bay, b., Can. (hâr)	61	51°18'N	55°50'W
Harer, Eth.	185	9°43'N	42°10'E
Harerge, hist. reg., Eth.	185	8°15'N	41°00'E
Hargeysa, Som. (här-gā'ē-sà)	192a	9°20'N	43°57'E
Harghita, Munţii, mts., Rom.	125	46°25'N	25°40'E
Harima-Nada, b., Japan (hä'rē-mä nä-dä)	167	34°34'N	134°37'E
Haringvliet, r., Neth.	115a	51°49'N	4°03'E
Harīrūd, r., Asia	154	34°29'N	61°16'E
Harlan, Ia., U.S. (här'lăn)	79	41°40'N	95°10'W
Harlan, Ky., U.S.	82	36°50'N	83°19'W
Harlan County Reservoir, res., Ne., U.S.	78	40°03'N	99°51'W
Harlem, Mt., U.S. (hár'lĕm)	73	48°33'N	108°50'W
Harlingen, Neth. (här'lĭng-ĕn)	121	53°10'N	5°24'E
Harlingen, Tx., U.S.	64	26°12'N	97°42'W
Harlow, Eng., U.K. (här'lō)	114b	51°46'N	0°08'E
Harlowton, Mt., U.S. (hár'lō-tun)	73	46°26'N	109°50'W
Harmony, In., U.S. (här'mô-nĭ)	66	39°35'N	87°00'W
Harney Basin, Or., U.S. (här'nĭ)	72	43°26'N	120°19'W
Harney Lake, l., Or., U.S.	72	43°11'N	119°23'W
Harney Peak, mtn., S.D., U.S.	64	43°52'N	103°32'W
Härnosand, Swe. (hĕr-nû-sänd)	116	62°37'N	17°54'E
Haro, Spain (ä'rō)	128	42°35'N	2°49'W
Haro Strait, strt., N.A. (hă'rō)	74a	48°27'N	123°11'W
Harpenden, Eng., U.K. (här'pĕn-d'n)	114b	51°48'N	0°22'W
Harper, Lib.	184	4°25'N	7°43'W
Harper, Ks., U.S. (här'pēr)	78	37°17'N	98°02'W
Harper, Wa., U.S.	74a	47°31'N	122°32'W
Harpers Ferry, W.V., U.S. (här'pērz)	67	39°20'N	77°45'W
Harricana, r., Can.	59	50°10'N	78°50'W
Harriman, Tn., U.S. (hăr'ĭ-măn)	82	35°55'N	84°34'W
Harrington, De., U.S. (hăr'ĭng-tŭn)	67	38°55'N	75°35'W
Harris, i., Scot., U.K. (hăr'ĭs)	120	57°55'N	6°40'W
Harris, Lake, l., Fl., U.S.	83a	28°43'N	81°40'W
Harrisburg, Il., U.S. (hăr'ĭs-bûrg)	66	37°45'N	88°35'W
Harrisburg, Pa., U.S.	65	40°15'N	76°50'W
Harrismith, S. Afr. (hă-rĭs'mĭth)	192c	28°17'S	29°08'E
Harrison, Ar., U.S. (hăr'ĭ-sŭn)	79	36°13'N	93°06'W
Harrison, Oh., U.S.	69f	39°16'N	84°45'W
Harrisonburg, Va., U.S. (hăr'ĭ-sŭn-bûrg)	67	38°30'N	78°50'W
Harrison Lake, l., Can.	55	49°31'N	121°59'W
Harrisonville, Mo., U.S. (hăr-ĭ-sŭn-vĭl)	79	38°39'N	94°21'W
Harrisville, Ut., U.S. (hăr'ĭs-vĭl)	75b	41°17'N	112°00'W
Harrisville, W.V., U.S.	66	39°10'N	81°05'W
Harrodsburg, Ky., U.S. (hăr'ŭdz-bûrg)	66	37°45'N	84°50'W
Harrods Creek, r., Ky., U.S. (hăr'ŭdz)	69h	38°24'N	85°33'W
Harrow, Eng., U.K. (hăr'ō)	114b	51°34'N	0°21'W
Harsefeld, Ger. (här'zĕ-fĕld')	115c	53°27'N	9°30'E
Harstad, Nor. (här'städh)	116	68°49'N	16°10'E
Hart, Mi., U.S. (härt)	66	43°40'N	86°25'W
Hartbeesfontein, S. Afr.	192c	26°46'S	26°25'E
Hartbeespoortdam, res., S. Afr.	187b	25°47'S	27°43'E
Hartford, Al., U.S. (härt'fērd)	82	31°05'N	85°42'W
Hartford, Ar., U.S.	79	35°01'N	94°21'W
Hartford, Ct., U.S.	65	41°45'N	72°40'W
Hartford, Il., U.S.	75e	38°50'N	90°06'W
Hartford, Ky., U.S.	82	37°25'N	86°50'W
Hartford, Mi., U.S.	66	42°15'N	86°15'W
Hartford, Wi., U.S.	71	43°19'N	88°25'W
Hartford City, In., U.S.	66	40°35'N	85°25'W
Hartington, Eng., U.K. (härt'ĭng-tŭn)	114a	53°08'N	1°48'W
Hartington, Ne., U.S.	70	42°37'N	97°18'W
Hartland Point, c., Eng., U.K.	120	51°03'N	4°40'W
Hartlepool, Eng., U.K. (här't'l-pōōl)	116	54°40'N	1°12'W
Hartley, Ia., U.S. (härt'lĭ)	70	43°12'N	95°29'W
Hartley Bay, Can.	54	53°25'N	129°15'W
Hart Mountain, mtn., Can. (härt)	57	52°25'N	101°30'W
Hartsbeespoort, S. Afr.	187b	25°44'S	27°51'E
Hartselle, Al., U.S. (härt'sĕl)	82	34°24'N	86°55'W
Hartshorne, Ok., U.S. (härts'hôrn)	79	34°49'N	95°34'W
Hartsville, S.C., U.S. (härts'vĭl)	83	34°20'N	80°04'W
Hartwell, Ga., U.S. (härt'wĕl)	83	34°21'N	82°56'W
Hartwell Lake, res., U.S.	65	34°30'N	83°00'W
Hārua, India	158a	22°36'N	88°40'E
Harvard, Il., U.S. (här'vàrd)	71	42°25'N	88°39'W
Harvard, Ma., U.S.	61a	42°30'N	71°35'W
Harvard, Ne., U.S.	78	40°36'N	98°08'W
Harvard, Mount, mtn., Co., U.S.	77	38°55'N	106°20'W
Harvey, Can.	60	45°44'N	64°46'W
Harvey, Il., U.S.	69a	41°37'N	87°39'W
Harvey, La., U.S.	68d	29°54'N	90°05'W
Harvey, N.D., U.S.	70	47°46'N	99°55'W
Harwich, Eng., U.K. (här'wĭch)	121	51°53'N	1°13'E
Haryana, state, India	155	29°00'N	75°45'E
Harz Mountains, mts., Ger. (härts)	124	51°42'N	10°50'E
Hashimoto, Japan (hä'shē-mō'tō)	167	34°19'N	135°37'E
Haskell, Ok., U.S. (hăs'kĕl)	79	35°49'N	95°41'W
Haskell, Tx., U.S.	78	33°09'N	99°43'W
Haslingden, Eng., U.K. (hăz'lĭng dĕn)	114a	53°43'N	2°19'W
Hassi Messaoud, Alg.	184	31°17'N	6°13'E
Hässleholm, Swe. (hĕs'lĕ-hōlm)	122	56°10'N	13°44'E
Hastings, N.Z.	175a	39°33'S	176°53'E
Hastings, Eng., U.K. (hās'tĭngz)	117	50°52'N	0°28'E
Hastings, Mi., U.S.	66	42°40'N	85°20'W
Hastings, Mn., U.S.	75g	44°44'N	92°51'W
Hastings, Ne., U.S.	64	40°34'N	98°42'W
Hastings-on-Hudson, N.Y., U.S. (ŏn-hŭd'sŭn)	68a	40°59'N	75°53'W
Hatay, Tur.	154	36°20'N	36°10'E
Hatchie, r., Tn., U.S. (hăch'ē)	82	35°28'N	89°14'W
Haţeg, Rom. (kät-säg')	131	45°35'N	22°57'E
Hatfield Broad Oak, Eng., U.K. (hăt-fēld brôd ōk)	114b	51°50'N	0°14'E
Hatogaya, Japan (hä'tō-gä-yä)	167a	35°50'N	139°45'E

ăt; finăl; rāte; senåte; ärm; åsk; sofà; fâre; ch-choose; dh-as th in other; bē; ēvent; bĕt; recĕnt; cratēr; g-gō; gh-guttural g; bĭt; ĭ-short neutral; rīde; κ-guttural k as ch in German ich;

PRONOUNCING INDEX

Hatsukaichi – Hillsboro 255

PLACE (Pronunciation)	PAGE	LAT.	LONG.
Hatsukaichi, Japan (hät′sōō-kȧ′ê-chê)	167	34°22′N	132°19′E
Hatteras, Cape, c., N.C., U.S. (hăt′ēr-ȧs)	65	35°15′N	75°24′W
Hattiesburg, Ms., U.S. (hăt′ĭz-bûrg)	65	31°20′N	89°18′W
Hattingen, Ger. (hä′tĕn-gĕn)	127c	51°24′N	7°11′E
Hatvan, Hung. (hôt′vôn)	125	47°39′N	19°44′E
Hat Yai, Thai.	168	7°01′N	100°29′E
Haugesund, Nor. (hou′gē-soon′)	116	59°26′N	5°20′E
Haukivesi, l., Fin. (hou′kê-vĕ′sê)	123	62°02′N	29°02′E
Haultain, r., Can.	56	56°15′N	106°35′W
Hauptsrus, S. Afr.	192c	26°35′S	26°16′E
Hauraki Gulf, b., N.Z. (hä-ōō-rä′kê)	175a	36°30′S	175°00′E
Haut, Isle au, Me., U.S. (hō)	60	44°03′N	68°13′W
Haut Atlas, mts., Mor.	118	32°10′N	5°49′W
Hauterive, Can.	60	49°11′N	68°16′W
Hau'ula, Hi., U.S.	84a	21°37′N	157°45′W
Havana, Cuba	87	23°08′N	82°23′W
Havana, Il., U.S. (hȧ-vă′nȧ)	79	40°17′N	90°02′W
Havasu, Lake, res., U.S. (hăv′å-sōō)	77	34°26′N	114°09′W
Havel, r., Ger. (hä′fĕl)	124	53°09′N	13°10′E
Havel-Kanal, can., Ger.	115b	52°36′N	13°12′E
Haverhill, Ma., U.S. (hā′vēr-hĭl)	61a	42°46′N	71°05′W
Haverhill, N.H., U.S.	67	44°00′N	72°05′W
Haverstraw, N.Y., U.S. (hā′vẽr-strô)	68a	41°11′N	73°58′W
Havlíčkův Brod, Czech Rep.	117	49°38′N	15°34′E
Havre, Mt., U.S. (hăv′ēr)	64	48°34′N	109°42′W
Havre-Boucher, Can. (hăv′rȧ-bōō-shā′)	61	45°42′N	61°30′W
Havre de Grace, Md., U.S. (hăv′ēr dē gräs′)	67	39°35′N	76°05′W
Havre-Saint Pierre, Can.	60	50°15′N	63°36′W
Haw, r., N.C., U.S. (hô)	83	36°17′N	79°46′W
Hawaii, state, U.S.	64c	20°00′N	157°40′W
Hawai'i, i., Hi., U.S. (häw wī′ē)	64c	19°30′N	155°30′W
Hawai'ian Islands, is., Hi., U.S. (hä-wī′ȧn)	64c	22°00′N	158°00′W
Hawai'i Volcanoes National Park, rec., Hi., U.S.	64c	19°30′N	155°25′W
Hawarden, Ia., U.S. (hā′wȧr-dĕn)	70	43°00′N	96°28′W
Hawi, Hi., U.S. (hä′wē)	84a	20°16′N	155°48′W
Hawick, Scot., U.K. (hô′ĭk)	120	55°25′N	2°55′W
Hawke Bay, b., N.Z. (hôk)	175a	39°17′S	177°20′E
Hawker, Austl. (hô′kēr)	176	31°58′S	138°12′E
Hawkesbury, Can. (hôks′bĕr-ĭ)	59	45°35′N	74°35′W
Hawkinsville, Ga., U.S. (hô′kĭnz-vĭl)	82	32°15′N	83°30′W
Hawks Nest Point, c., Bah.	93	24°05′N	75°30′W
Hawley, Mn., U.S. (hô′lĭ)	70	46°52′N	96°18′W
Haworth, Eng., U.K. (hā′wûrth)	114a	53°50′N	1°57′W
Hawthorne, Ca., U.S. (hô′thôrn)	75a	33°55′N	118°22′W
Hawthorne, Nv., U.S.	76	38°33′N	118°39′W
Haxtun, Co., U.S. (hăks′tŭn)	78	40°39′N	102°38′W
Hay, r., Austl. (hā)	174	23°00′S	136°45′E
Hay, r., Can.	52	60°21′N	117°14′W
Hayama, Japan (hä-yä′mä)	167a	35°16′N	139°35′E
Hayashi, Japan (hä-yä′shē)	167a	35°13′N	139°38′E
Hayden, Az., U.S. (hā′dĕn)	77	33°00′N	110°50′W
Hayes, r., Can.	53	55°25′N	93°55′W
Hayes, Mount, mtn., Ak., U.S. (hāz)	63	63°32′N	146°40′W
Haynesville, La., U.S. (hānz′vĭl)	81	32°55′N	93°08′W
Hayrabolu, Tur.	131	41°14′N	27°05′E
Hay River, Can.	50	60°50′N	115°53′W
Hays, Ks., U.S. (hāz)	78	38°51′N	99°20′W
Haystack Mountain, mtn., Wa., U.S. (hā-stăk′)	74a	48°26′N	122°07′W
Hayward, Ca., U.S. (hā′wẽrd)	74b	37°40′N	122°06′W
Hayward, Wi., U.S.	71	46°01′N	91°31′W
Hazard, Ky., U.S. (hăz′ȧrd)	82	37°13′N	83°10′W
Hazelhurst, Ga., U.S. (hā′z'l-hûrst)	83	31°50′N	82°36′W
Hazelhurst, Ms., U.S.	82	31°52′N	90°23′W
Hazel Park, Mi., U.S.	69b	42°28′N	83°06′W
Hazelton, Can. (hā′z'l-tŭn)	50	55°15′N	127°40′W
Hazelton Mountains, mts., Can.	54	55°00′N	128°00′W
Hazleton, Pa., U.S.	67	41°00′N	76°00′W
Headland, Al., U.S. (hĕd′lănd)	82	31°22′N	85°20′W
Healdsburg, Ca., U.S. (hēldz′bûrg)	76	38°37′N	122°52′W
Healdton, Ok., U.S. (hēld′tŭn)	79	34°13′N	97°28′W
Heanor, Eng., U.K. (hēn′ŏr)	114a	53°01′N	1°22′W
Heard Island, i., Austl. (hûrd)	3	53°10′S	74°35′E
Hearne, Tx., U.S. (hûrn)	81	30°53′N	96°35′W
Hearst, Can. (hûrst)	51	49°36′N	83°40′W
Heart, r., N.D., U.S. (härt)	70	46°46′N	102°34′W
Heart Lake Indian Reserve, I.R., Can.	55	55°02′N	111°30′W
Heart's Content, Can. (härts kŏn′tĕnt)	61	47°52′N	53°22′W
Heavener, Ok., U.S. (hēv′nẽr)	79	34°52′N	94°36′W
Hebbronville, Tx., U.S. (hē′brŭn-vĭl)	80	27°18′N	98°40′W
Hebei, prov., China (hŭ-bā)	161	39°15′N	115°40′E
Heber City, Ut., U.S. (hē′bẽr)	77	40°30′N	111°25′W
Heber Springs, Ar., U.S.	79	35°28′N	91°59′W
Hebgen Lake, res., Mt., U.S. (hĕb′gĕn)	73	44°47′N	111°38′W
Hebrides, is., Scot., U.K.	112	57°00′N	6°30′W
Hebrides, Sea of the, sea, Scot., U.K.	120	57°00′N	7°00′W
Hebron, Can. (hēb′rŭn)	51	58°11′N	62°56′W
Hebron, In., U.S.	69a	41°19′N	87°13′W
Hebron, Ky., U.S.	69f	39°04′N	84°43′W
Hebron, N.D., U.S.	70	46°54′N	102°04′W
Hebron, Ne., U.S.	79	40°11′N	97°36′W
Hebron *see* Al Khalīl, W.B.	153a	31°31′N	35°07′E
Heby, Swe. (hĭ′bü)	122	59°56′N	16°48′E
Hecate Strait, strt., Can. (hĕk′ȧ-tē)	52	53°00′N	131°00′W
Hecelchakán, Mex. (ā-sĕl-chä-kän′)	89	20°10′N	90°09′W
Hechi, China (hŭ-chr)	165	24°50′N	108°18′E
Hechuan, China (hŭ-chyuän)	160	30°00′N	106°20′E
Hecla Island, i., Can.	57	51°08′N	96°45′W
Hedemora, Swe. (hĭ-dĕ-mō′rä)	122	60°16′N	15°55′E
Hedon, Eng., U.K. (hĕ-dŭn)	114a	53°44′N	0°12′W
Heemstede, Neth.	115a	52°20′N	4°36′E
Heerlen, Neth.	121	50°55′N	5°58′E

PLACE (Pronunciation)	PAGE	LAT.	LONG.
Hefei, China (hŭ-fā)	161	31°51′N	117°15′E
Heflin, Al., U.S. (hĕf′lĭn)	82	33°40′N	85°33′W
Heide, Ger. (hī′dē)	124	54°13′N	9°06′E
Heidelberg, Austl. (hī′dĕl-bûrg)	173a	37°45′S	145°04′E
Heidelberg, Ger. (hīdĕl-bĕrgh)	117	49°24′N	8°43′E
Heidelberg, S. Afr.	192c	26°32′S	28°22′E
Heidenheim, Ger. (hī′dĕn-hīm)	124	48°41′N	10°09′E
Heilbron, S. Afr. (hīl′brōn)	192c	27°17′S	27°58′E
Heilbronn, Ger. (hīl′brōn)	117	49°09′N	9°16′E
Heiligenhaus, Ger. (hī′lē-gĕn-houz)	127c	51°19′N	6°58′E
Heiligenstadt, Ger. (hī′lē-gĕn-shtät)	124	51°21′N	10°10′E
Heilongjiang, prov., China (hā-lôŋ-jyäŋ)	161	46°36′N	128°07′E
Heinola, Fin. (hå-nō′lȧ)	123	61°13′N	26°03′E
Heinsberg, Ger. (hīnz′bĕrgh)	127c	51°04′N	6°07′E
Heist-op-den-Berg, Bel.	115a	51°05′N	4°14′E
Hejaz *see* Al Ḩijāz, reg., Sau. Ar.	154	23°45′N	39°08′E
Hejian, China (hŭ-jyĕn)	164	38°28′N	116°05′E
Hekla, vol., Ice.	112	63°53′N	19°37′W
Hel, Pol. (hāl)	125	54°37′N	18°53′E
Helagsfjället, mtn., Swe.	116	62°54′N	12°24′E
Helan Shan, mts., China (hŭ-län shän)	160	38°02′N	105°20′E
Helena, Ar., U.S. (hē-lē′nȧ)	65	34°33′N	90°35′W
Helena, Mt., U.S. (hĕ-lĕ′nȧ)	64	46°35′N	112°01′W
Helensburgh, Austl. (hĕl′ĕnz-bŭr-ô)	173b	34°11′S	150°59′E
Helensburgh, Scot., U.K.	120	56°01′N	4°53′W
Helgoland, i., Ger. (hĕl′gô-länd)	124	54°13′N	7°30′E
Hellier, Ky., U.S. (hĕl′yẽr)	83	37°16′N	82°27′W
Hellín, Spain (ĕl-yén′)	118	38°30′N	1°40′W
Hells Canyon, p., U.S.	72	45°20′N	116°45′W
Helmand, r., Afg. (hĕl′mŭnd)	154	31°00′N	63°48′E
Hel'miaziv, Ukr.	133	49°49′N	31°54′E
Helmond, Neth. (hĕl′mōnt) (ĕl′môɴ′)	121	51°35′N	5°04′E
Helmstedt, Ger. (hĕlm′shtĕt)	124	52°14′N	11°03′E
Helotes, Tx., U.S. (hē′lōts)	75d	29°35′N	98°41′W
Helper, Ut., U.S. (hĕlp′ēr)	77	39°40′N	110°55′W
Helsingborg, Swe. (hĕl′sĭng-bôrgh)	116	56°04′N	12°40′E
Helsingfors *see* Helsinki, Fin.	110	60°10′N	24°53′E
Helsingør, Den. (hĕl-sĭng-ŭr′)	116	56°03′N	12°33′E
Helsinki, Fin. (hĕl′sên-kê)	110	60°10′N	24°53′E
Hemel Hempstead, Eng., U.K. (hĕm′ĕl hĕmp′stĕd)	114b	51°43′N	0°29′W
Hemer, Ger.	127c	51°22′N	7°46′E
Hemet, Ca., U.S. (hĕm′ĕt)	75a	33°45′N	116°57′W
Hemingford, Ne., U.S. (hĕm′ĭng-fẽrd)	70	42°21′N	103°30′W
Hemphill, Tx., U.S. (hĕmp′hĭl)	81	31°20′N	93°48′W
Hempstead, N.Y., U.S. (hĕmp′stĕd)	68a	40°42′N	73°37′W
Hempstead, Tx., U.S.	81	30°07′N	96°05′W
Hemse, Swe. (hĕm′sẽ)	122	57°15′N	18°25′E
Hemsön, i., Swe.	122	62°43′N	18°22′E
Henan, prov., China (hŭ-nän)	161	33°58′N	112°33′E
Henares, r., Spain (å-nä′rås)	128	40°50′N	2°55′W
Henderson, Ky., U.S. (hĕn′dẽr-sŭn)	66	37°50′N	87°30′W
Henderson, N.C., U.S.	83	36°18′N	78°24′W
Henderson, Nv., U.S.	76	36°09′N	115°04′W
Henderson, Tn., U.S.	82	35°25′N	88°40′W
Henderson, Tx., U.S.	81	32°09′N	94°48′W
Hendersonville, N.C., U.S. (hĕn′dẽr-sŭn-vĭl)	83	35°17′N	82°28′W
Hendersonville, Tn., U.S.	82	36°18′N	86°37′W
Hendon, Eng., U.K. (hĕn′dŭn)	114b	51°34′N	0°13′W
Hendrina, S. Afr. (hĕn-drē′nȧ)	192c	26°10′S	29°44′E
Hengch'un, Tai. (hĕng′chŭn′)	165	22°00′N	120°42′E
Hengelo, Neth. (hĕngẽ-lō)	121	52°20′N	6°45′E
Hengshan, China (hĕng′shän′)	165	27°20′N	112°40′E
Hengshui, China (hĕng′shōō-ē′)	162	37°43′N	115°42′E
Hengxian, China (hŭŋ shyĕn)	165	22°40′N	109°20′E
Hengyang, China	161	26°58′N	112°30′E
Heniches'k, Ukr.	137	46°11′N	34°47′E
Henley on Thames, Eng., U.K. (hĕn′lē ŏn tĕmz)	114b	51°31′N	0°54′W
Henlopen, Cape, c., De., U.S. (hĕn-lō′pĕn)	67	38°45′N	75°05′W
Hennebont, Fr. (ĕn-bôɴ′)	126	47°47′N	3°16′W
Hennenman, S. Afr.	192c	27°59′S	27°03′E
Hennessey, Ok., U.S. (hĕn′ĕ-sĭ)	79	36°04′N	97°53′W
Hennigsdorf, Ger. (hĕ′nēngz-dōrf))	115b	52°39′N	13°12′E
Hennops, r., S. Afr. (hĕn′ŏps)	187b	25°51′S	27°57′E
Hennopsrivier, S. Afr.	187b	25°50′S	27°59′E
Henrietta, Ok., U.S. (hĕn-rĭ-ĕt′ȧ)	79	35°25′N	95°58′W
Henrietta, Tx., U.S. (hen-rĭ-ĕ′tȧ)	78	33°47′N	98°11′W
Henrietta Maria, Cape, c., Can. (hĕn-rĭ-ĕt′ȧ)	53	55°10′N	82°20′W
Henry Mountains, mts., Ut., U.S. (hĕn′rĭ)	64	37°55′N	110°45′W
Henrys Fork, r., Id., U.S.	73	43°52′N	111°55′W
Henteyn Nuruu, mtn., Russia	164	49°40′N	111°00′E
Hentiyn Nuruu, mts., Mong.	160	49°25′N	107°51′E
Henzada, Mya.	155	17°38′N	95°28′E
Heppner, Or., U.S. (hĕp′nẽr)	72	45°21′N	119°33′W
Hepu, China (hŭ-pōō)	165	21°28′N	109°10′E
Herāt, Afg. (hĕ-rät′)	154	34°28′N	62°13′E
Hercules, Can.	62g	53°27′N	113°20′W
Herdecke, Ger. (hĕr′dĕ-kĕ)	127c	51°24′N	7°26′E
Heredia, C.R. (ā-rā′dhê-ä)	91	10°04′N	84°06′W
Hereford, Eng., U.K. (hĕrĕ′fẽrd)	120	52°05′N	2°44′W
Hereford, Md., U.S.	68e	39°35′N	76°42′W
Hereford, Tx., U.S. (hĕr′ĕ-fẽrd)	78	34°47′N	102°25′W
Hereford and Worcester, co., Eng., U.K.	114a	52°24′N	2°15′W
Herencia, Spain (å-rān′thê-ä)	128	39°23′N	3°22′W
Herentals, Bel.	115a	51°10′N	4°51′E
Herford, Ger. (hĕr′fôrt)	124	52°06′N	8°42′E
Herington, Ks., U.S. (hĕr′ĭng-tŭn)	79	38°41′N	96°57′W
Herisau, Switz. (hā′rê-zou)	124	47°23′N	9°18′E
Herk-de-Stad, Bel.	115a	50°56′N	5°13′E
Herkimer, N.Y., U.S. (hûr′kĭ-mẽr)	67	43°05′N	75°00′W

PLACE (Pronunciation)	PAGE	LAT.	LONG.
Hermansville, Mi., U.S. (hûr′măns-vĭl)	66	45°40′N	87°35′W
Hermantown, Mn., U.S. (hẽr′mȧn-toun)	75h	46°46′N	92°12′W
Hermanusdorings, S. Afr.	192c	24°08′S	27°46′E
Herminie, Pa., U.S. (hûr-mī′nē)	69e	40°16′N	79°45′W
Hermitage Bay, b., Can. (hûr′mĭ-tĕj)	61	47°35′N	56°05′W
Hermit Islands, is., Pap. N. Gui. (hûr′mĭt)	169	1°48′S	144°55′E
Hermosa Beach, Ca., U.S. (hẽr-mō′sȧ)	75a	33°51′N	118°24′W
Hermosillo, Mex. (ĕr-mô-sē′l-yŏ)	86	29°00′N	110°57′W
Herndon, Va., U.S. (hẽrn′don)	68e	38°58′N	77°22′W
Herne, Ger. (hĕr′nĕ)	127c	51°32′N	7°13′E
Herning, Den. (hĕr′nĭng)	116	56°08′N	8°55′E
Heron, l., Mn., U.S. (hĕr′ŭn)	70	43°42′N	95°23′W
Heron Lake, Mn., U.S.	70	43°48′N	95°20′W
Herrero, Punta, Mex. (pō′n-tä-ĕr-rĕ′rô)	90a	19°18′N	87°24′W
Herrin, Il., U.S. (hĕr′ĭn)	66	37°50′N	89°00′W
Herschel, S. Afr. (hẽr′-shĕl)	187c	30°37′S	27°12′E
Herscher, Il., U.S. (hēr′shēr)	69a	41°03′N	88°06′W
Herstal, Bel. (hĕr′stäl)	121	50°42′N	5°32′E
Hertford, Eng., U.K.	120	51°48′N	0°05′W
Hertford, N.C., U.S. (hŭrt′fẽrd)	83	36°10′N	76°30′W
Hertfordshire, co., Eng., U.K.	114b	51°46′N	0°05′W
Hertzberg, Ger. (hĕrtz′bĕrgh)	115b	52°54′N	12°58′E
Hervás, Spain	128	40°16′N	5°51′W
Herzliyya, Isr.	153a	32°10′N	34°49′E
Hessen, hist. reg., Ger. (hĕs′ĕn)	124	50°42′N	9°00′E
Hetch Hetchy Aqueduct, Ca., U.S. (hĕtch hĕt′chĭ ȧk′wê-dŭkt)	76	37°27′N	120°54′W
Hettinger, N.D., U.S. (hĕt′ĭn-jẽr)	70	45°58′N	102°36′W
Heuningspruit, S. Afr.	192c	27°28′S	27°26′E
Hexian, China (hŭ shyĕn)	165	24°20′N	111°28′E
Hexian, China	162	31°44′N	118°20′E
Heyang, China (hŭ-yäŋ)	164	35°18′N	110°18′E
Heystekrand, S. Afr.	192c	25°16′S	27°14′E
Heyuan, China (hŭ-yůän)	165	23°48′N	114°45′E
Heywood, Eng., U.K. (hā′wȯd)	114a	53°36′N	2°12′W
Heze, China (hŭ-dzŭ)	162	35°13′N	115°28′E
Hialeah, Fl., U.S. (hī-ȧ-lē′ăh)	83a	25°49′N	80°18′W
Hiawatha, Ks., U.S. (hī-ȧ-wô′thȧ)	79	39°50′N	95°33′W
Hiawatha, Ut., U.S.	77	39°25′N	111°05′W
Hibbing, Mn., U.S. (hĭb′ĭng)	65	47°26′N	92°58′W
Hickman, Ky., U.S. (hĭk′mȧn)	82	34°33′N	89°10′W
Hickory, N.C., U.S. (hĭk′ô-rĭ)	83	35°43′N	81°21′W
Hicksville, N.Y., U.S.	66	41°15′N	84°45′W
Hicksville, N.Y., U.S. (hĭks′vĭl)	68a	40°47′N	73°25′W
Hico, Tx., U.S. (hī′kō)	80	32°00′N	98°02′W
Hidalgo, Mex. (ê-dhäl′gō)	88	24°14′N	99°25′W
Hidalgo, Mex.	80	27°49′N	99°53′W
Hidalgo, state, Mex.	86	20°45′N	99°30′W
Hidalgo del Parral, Mex. (ê-dä′l-gō-dĕl-pär-rȧ′l)	86	26°55′N	105°40′W
Hidalgo Yalalag, Mex. (ê-dhäl′gō-yä-lä-läg)	89	17°12′N	96°11′W
Hierro Island, i., Spain (yĕ′r-rô)	184	27°37′N	18°29′W
Higashimurayama, Japan	167a	35°46′N	139°28′E
Higashiōsaka, Japan	167b	34°40′N	135°44′E
Higgins, l., Mi., U.S. (hĭg′ĭnz)	66	44°20′N	84°45′W
Higginsville, Mo., U.S. (hĭg′ĭnz-vĭl)	79	39°05′N	93°44′W
High, i., Mi., U.S.	66	45°45′N	85°45′W
High Bluff, Can.	62f	50°01′N	98°08′W
Highborne Cay, i., Bah. (hībôrn kē)	92	24°45′N	76°50′W
Highgrove, Ca., U.S. (hī′grōv)	75a	34°01′N	117°20′W
High Island, Tx., U.S.	81a	29°34′N	94°24′W
Highland, Ca., U.S. (hī′lănd)	75a	34°08′N	117°13′W
Highland, Il., U.S.	79	38°44′N	89°41′W
Highland, In., U.S.	69a	41°33′N	87°28′W
Highland, Mi., U.S.	69b	42°38′N	83°37′W
Highland Park, Il., U.S.	69a	42°11′N	87°47′W
Highland Park, Mi., U.S.	69b	42°24′N	83°06′W
Highland Park, N.J., U.S.	68a	40°30′N	74°25′W
Highland Park, Tx., U.S.	75c	32°49′N	96°48′W
Highlands, N.J., U.S. (hī-lăndz)	68a	40°24′N	73°59′W
Highlands, Tx., U.S.	81a	29°49′N	95°01′W
Highmore, S.D., U.S. (hī′mōr)	70	44°30′N	99°26′W
High Ongar, Eng., U.K. (on′gẽr)	114b	51°43′N	0°15′E
High Peak, mtn., Phil.	169a	15°38′N	120°05′E
High Point, N.C., U.S.	83	35°55′N	80°00′W
High Prairie, Can.	50	55°26′N	116°29′W
High Ridge, Mo., U.S.	75e	38°27′N	90°32′W
High River, Can.	50	50°35′N	113°52′W
High Rock Lake, res., N.C., U.S. (hī′-rŏk)	83	35°40′N	80°15′W
High Springs, Fl., U.S.	83	29°48′N	82°38′W
High Tatra Mountains, mts., Eur.	125	49°15′N	19°40′E
Hightstown, N.J., U.S. (hīts-toun)	68a	40°16′N	74°32′W
High Wycombe, Eng., U.K. (wĭ-kŭm)	120	51°36′N	0°45′W
Higuero, Punta, c., P.R.	87b	18°21′N	67°11′W
Higuerote, Ven. (ē-gĕ-rô′tĕ)	101b	10°29′N	66°06′W
Higüey, Dom. Rep. (ê-gwĕ′y)	93	18°40′N	68°45′W
Hiiumaa, i., Est. (hē′ȯm-ô)	136	58°47′N	22°05′E
Hikone, Japan (hē′kô-nĕ)	167	35°15′N	136°15′E
Hildburghausen, Ger. (hĭld′bȯrg hou-zĕn)	124	50°26′N	10°45′E
Hilden, Ger. (hēl′dĕn)	127c	51°10′N	6°56′E
Hildesheim, Ger. (hĭl′dĕs-hīm)	117	52°08′N	9°56′E
Hillaby, Mount, mtn., Barb. (hĭl′ȧ-bĭ)	91b	13°15′N	59°35′W
Hill City, Ks., U.S. (hĭl)	78	39°22′N	99°54′W
Hill City, Mn., U.S.	71	46°58′N	93°38′W
Hillegersberg, Neth.	115a	51°57′N	4°29′E
Hillerød, Den. (hê′lĕ-rûdh)	122	55°56′N	12°17′E
Hillsboro, Il., U.S. (hĭlz′bŭr-ō)	79	39°09′N	89°28′W
Hillsboro, Ks., U.S.	79	38°22′N	97°11′W
Hillsboro, N.D., U.S.	70	47°23′N	97°05′W

256 Hillsboro – Hrodna

PRONOUNCING INDEX

PLACE (Pronunciation)	PAGE	LAT.	LONG.
Hillsboro, N.H., U.S.	67	43°05'N	71°55'W
Hillsboro, Oh., U.S.	66	39°10'N	83°40'W
Hillsboro, Or., U.S.	74c	45°31'N	122°59'W
Hillsboro, Tx., U.S.	81	32°01'N	97°06'W
Hillsboro, Wi., U.S.	71	43°39'N	90°20'W
Hillsburgh, Can. (hĭlz´bûrg)	62d	43°48'N	80°09'W
Hills Creek Lake, res., Or., U.S.	72	43°41'N	122°26'W
Hillsdale, Mi., U.S. (hĭls-dāl)	77	41°55'N	84°35'W
Hilo, Hi., U.S. (hē´lō)	64c	19°44'N	155°01'W
Hilvarenbeek, Neth.	115a	51°29'N	5°10'E
Hilversum, Neth. (hĭl´vēr-sŭm)	115a	52°13'N	5°10'E
Himachal Pradesh, India	155	32°00'N	77°30'E
Himalayas, mts., Asia	155	29°30'N	85°02'E
Himeji, Japan (hē´må-jė)	166	34°50'N	134°42'E
Himmelpforten, Ger. (hē´mĕl-pfōr-tĕn)	115c	53°37'N	9°19'E
Ḩimş, Syria	154	34°44'N	36°43'E
Hinche, Haiti (hēn´chå) (ăɴsh)	93	19°10'N	72°05'W
Hinchinbrook, i., Austl. (hĭn-chĭn-brŏŏk)	174	18°23'S	146°57'W
Hinckley, Eng., U.K. (hĭnk´lĭ)	114a	52°32'N	1°21'W
Hindley, Eng., U.K. (hĭnd´lĭ)	114a	53°32'N	2°35'W
Hindu Kush, mts., Asia (hĭn´dōō kōōsh´)	155	35°15'N	68°44'E
Hindupur, India (hĭn´dōō-pōōr)	159	13°52'N	77°34'E
Hingham, Ma., U.S. (hĭng´ăm)	61a	42°14'N	70°53'W
Hinkley, Oh., U.S. (hĭnk´-lĭ)	69d	41°14'N	81°45'W
Hinojosa del Duque, Spain (ė-nō-kô´sä)	128	38°30'N	5°09'W
Hinsdale, Il., U.S. (hĭnz´dāl)	69a	41°48'N	87°56'W
Hinton, Can. (hĭn´tŭn)	55	53°25'N	117°34'W
Hinton, W.V., U.S. (hĭn´tŭn)	66	37°40'N	80°55'W
Hirado, i., Japan (hē´rä-dō)	166	33°19'N	129°18'E
Hirakata, Japan (hē´rä-kä´tä)	167b	34°49'N	135°40'E
Hirara, Japan	170d	24°48'N	125°17'E
Hiratsuka, Japan (hē-rät-sōō´kå)	167	35°20'N	139°19'E
Hirosaki, Japan (hē´rō-sä´kė)	161	40°31'N	140°38'E
Hirose, Japan (hē´rō-sā)	167	35°20'N	133°11'E
Hiroshima, Japan (hē-rō-shē´må)	161	34°22'N	132°25'E
Hirson, Fr. (ēr-sôɴ´)	126	49°54'N	4°00'E
Hisar, India	158	29°15'N	75°47'E
Hispaniola, i., N.A. (hĭ´spăn-ĭ-ō-là)	87	17°30'N	73°15'W
Hitachi, Japan (hē-tä´chē)	166	36°42'N	140°47'E
Hitchcock, Tx., U.S. (hĭch´kŏk)	81a	29°21'N	95°01'W
Hitoyoshi, Japan (hē´tō-yō´shė)	167	32°13'N	130°45'E
Hitra, i., Nor. (hĭträ)	116	63°34'N	7°37'E
Hittefeld, Ger. (hē´tĕ-fĕld)	115c	53°23'N	9°59'E
Hiwasa, Japan (hē´wå-sä)	167	33°44'N	134°31'E
Hiwassee, r., Tn., U.S. (hī-wôs´sē)	82	35°10'N	84°35'W
Hjälmaren, l., Swe.	116	59°07'N	16°05'E
Hjo, Swe. (yō)	122	58°19'N	14°11'E
Hjørring, Den. (jûr´ĭng)	116	57°27'N	9°59'E
Hlobyne, Ukr.	133	49°22'N	33°17'E
Hlohovec, Slvk. (hlō´ho-vĕts)	125	48°24'N	17°49'E
Hlukhiv, Ukr.	137	51°42'N	33°52'E
Hlybokaye, Bela.	136	55°08'N	27°44'E
Hobart, Austl. (hō´bȧrt)	175	43°00'S	147°30'E
Hobart, In., U.S.	69a	41°31'N	87°15'W
Hobart, Ok., U.S.	78	35°02'N	99°06'W
Hobart, Wa., U.S.	74a	47°25'N	121°58'W
Hobbs, N.M., U.S. (hŏbs)	78	32°41'N	103°15'W
Hoboken, Bel. (hō´bō-kĕn)	115a	51°11'N	4°20'E
Hoboken, N.J., U.S.	68a	40°43'N	74°03'W
Hobro, Den. (hō-brō´)	122	56°38'N	9°47'E
Hobson, Va., U.S. (hŏb´sŭn)	68g	36°54'N	76°31'W
Hobson's Bay, b., Austl. (hŏb´sŭnz)	173a	37°54'S	144°45'E
Hobyo, Som.	192a	5°24'N	48°28'E
Ho Chi Minh City, Viet.	168	10°46'N	106°34'E
Hockinson, Wa., U.S. (hŏk´ĭn-sŭn)	74c	45°44'N	122°29'W
Hoctún, Mex. (ôk-tōō´n)	90a	20°52'N	89°10'W
Hodgenville, Ky., U.S. (hŏj´ĕn-vĭl)	66	37°35'N	85°45'W
Hodges Hill, mtn., Can. (hŏj´ĕz)	61	49°04'N	55°53'W
Hódmezövásárhely, Hung. (hōd´mĕ-zû-vô´shôr-hĕl-y´)	125	46°24'N	20°21'E
Hodna, Chott el, l., Alg.	118	35°20'N	3°27'E
Hodonin, Czech Rep. (hē´dō-nén)	125	48°50'N	17°06'E
Hoegaarden, Bel.	115a	50°46'N	4°55'E
Hoek van Holland, Neth.	115a	51°59'N	4°05'E
Hoeryŏng, Kor., N. (hwĕr´yŭng)	166	42°28'N	129°39'E
Hof, Ger. (hōf)	124	50°19'N	11°55'E
Hofsjökull, ice, Ice. (hōfs´yü´kōōl)	116	64°55'N	18°40'W
Hog, i., Mi., U.S.	66	45°50'N	85°20'W
Hogansville, Ga., U.S. (hō´gănz-vĭl)	82	33°10'N	84°54'W
Hog Cay, i., Bah.	93	23°35'N	75°30'W
Hogsty Reef, rf., Bah.	93	21°45'N	73°50'W
Hohenbrunn, Ger. (hō´hĕn-brōōn)	115d	48°03'N	11°42'E
Hohenlimburg, Ger. (hō´hĕn lēm´bōōrg)	127c	51°20'N	7°35'E
Hohen Neuendorf, Ger. (hō´hēn noi´ĕn-dōrf)	115b	52°40'N	13°22'E
Hohe Tauern, mts., Aus. (hō´ĕ tou´ērn)	124	47°11'N	12°12'E
Hohhot, China (hŭ-hōō-tŭ)	161	41°05'N	111°50'E
Hohoe, Ghana	188	7°09'N	0°28'E
Hohokus, N.J., U.S. (hō-hō-kŭs)	68a	41°01'N	74°08'W
Hoi An, Viet.	165	15°48'N	108°30'E
Hoisington, Ks., U.S. (hoi´zĭng-tŭn)	78	38°30'N	98°46'W
Hojo, Japan (hō´jō)	167	33°58'N	132°50'E
Hokitika, N.Z. (hō-kĭ-tē´kä)	175a	42°43'S	170°59'E
Hokkaidō, i., Japan (hōk´kī-dō)	166	43°30'N	142°45'E
Holbaek, Den. (hōl´bĕk)	122	55°42'N	11°40'E
Holbox, Mex. (ôl-bŏ´x)	90a	21°33'N	87°19'W
Holbox, Isla, i., Mex. (ē´s-lä-ôl-bŏ´x)	90a	21°40'N	87°21'W
Holbrook, Az., U.S. (hōl´brook)	77	34°55'N	110°15'W
Holbrook, Ma., U.S.	61a	42°10'N	71°01'W
Holden, Ma., U.S. (hōl´dĕn)	61a	42°21'N	71°51'W
Holden, Mo., U.S.	79	38°42'N	94°00'W
Holden, W.V., U.S.	66	37°45'N	82°05'W
Holdenville, Ok., U.S. (hōl´dĕn-vĭl)	79	35°05'N	96°25'W
Holdrege, Ne., U.S. (hōl´drĕj)	78	40°25'N	99°28'W
Holguín, Cuba (ôl-gēn´)	87	20°55'N	76°15'W
Holguín, prov., Cuba	92	20°40'N	76°15'W
Holidaysburg, Pa., U.S. (hŏl´ĭ-dāz-bûrg)	67	40°30'N	78°30'W
Hollabrunn, Aus.	124	48°33'N	16°04'E
Holland, Mi., U.S. (hŏl´ănd)	66	42°45'N	86°10'W
Hollands Diep, strt., Neth.	115a	51°43'N	4°25'E
Hollenstedt, Ger. (hō´lĕn-shtĕt)	115c	53°22'N	9°43'E
Hollis, N.H., U.S. (hŏl´ĭs)	61a	42°30'N	71°29'W
Hollis, Ok., U.S.	78	34°39'N	99°56'W
Hollister, Ca., U.S. (hŏl´ĭs-tēr)	76	36°50'N	121°25'W
Holliston, Ma., U.S. (hŏl´ĭs-tŭn)	61a	42°12'N	71°25'W
Holly, Mi., U.S. (hŏl´ĭ)	66	42°45'N	83°30'W
Holly, Wa., U.S.	74a	47°34'N	122°58'W
Holly Springs, Ms., U.S. (hŏl´ĭ sprĭngz)	82	34°45'N	89°28'W
Hollywood, Ca., U.S. (hŏl´ė-wŏd)	75a	34°06'N	118°20'W
Hollywood, Fl., U.S.	83a	26°00'N	80°11'W
Holmes Reefs, rf., Austl. (hōmz)	175	16°33'S	148°43'E
Holmestrand, Nor. (hōl´mĕ-strän)	122	59°29'N	10°17'E
Holmsbu, Nor. (hōlms´bōō)	122	59°36'N	10°26'E
Holmsjön, l., Swe.	122	62°23'N	15°43'E
Holstebro, Den. (hōl´stĕ-brō)	116	56°22'N	8°39'E
Holstein, hist. reg., Ger.	124	54°10'N	9°40'E
Holston, r., Tn., U.S. (hōl´stŭn)	82	36°02'N	83°42'W
Holt, Eng., U.K. (hōlt)	114a	53°05'N	2°53'W
Holton, Ks., U.S. (hōl´tŭn)	79	39°27'N	95°43'W
Holy Cross, Ak., U.S. (hō´lĭ krôs)	63	62°10'N	159°40'W
Holyhead, Wales, U.K. (hŏl'ė-hĕd)	120	53°18'N	4°45'W
Holy Island, i., Eng., U.K.	120	55°43'N	1°48'W
Holy Island, i., Wales, U.K. (hō´lĭ)	120	53°15'N	4°45'W
Holyoke, Co., U.S. (hōl´yōk)	78	40°36'N	102°18'W
Holyoke, Ma., U.S.	67	42°10'N	72°40'W
Homano, Japan (hō-mä´nō)	167a	35°33'N	140°08'E
Homberg, Ger. (hōm´bĕrgh)	127c	51°27'N	6°42'E
Hombori, Mali	188	15°17'N	1°42'W
Home Gardens, Ca., U.S. (hōm gär´d'nz)	75a	33°53'N	117°32'W
Homeland, Ca., U.S. (hōm´lănd)	75a	33°44'N	117°07'W
Homer, Ak., U.S. (hō´mēr)	63	59°42'N	151°30'W
Homer, La., U.S.	81	32°46'N	93°05'W
Homer Youngs Peak, mtn., Mt., U.S.	73	45°19'N	113°41'W
Homestead, Fl., U.S. (hōm´stĕd)	83a	25°27'N	80°28'W
Homestead, Mi., U.S.	75k	46°20'N	84°07'W
Homestead, Pa., U.S.	69e	40°29'N	79°55'W
Homestead National Monument of America, rec., Ne., U.S.	79	40°16'N	96°51'W
Homewood, Al., U.S. (hōm´wŏd)	68h	33°28'N	86°48'W
Homewood, Il., U.S.	69a	41°34'N	87°40'W
Hominy, Ok., U.S. (hŏm´ĭ-nĭ)	79	36°25'N	96°24'W
Homochitto, r., Ms., U.S. (hō-mō-chĭt´ō)	82	31°23'N	91°15'W
Homyel', Bela.	136	52°20'N	31°03'E
Homyel', prov., Bela.	132	52°18'N	29°00'E
Honda, Col. (hōn´dȧ)	100	5°13'N	74°45'W
Honda, Bahía, b., Cuba (bä-ē´ä-ô´n-dä)	92	23°10'N	83°20'W
Hondo, Tx., U.S.	80	29°20'N	99°08'W
Hondo, r., N.M., U.S.	78	33°22'N	105°06'W
Hondo, Río, r., N.A. (hon-dō´)	90a	18°16'N	88°32'W
Honduras, nation, N.A. (hŏn-dōō´rȧs)	86	14°30'N	88°00'W
Honduras, Gulf of, b., N.A.	86	16°30'N	87°30'W
Honea Path, S.C., U.S. (hŭn´ĭ păth)	83	34°25'N	82°16'W
Hönefoss, Nor. (hē´nē-fōs)	116	60°10'N	10°15'E
Honesdale, Pa., U.S. (hōnz´dāl)	67	41°30'N	75°15'W
Honey Grove, Tx., U.S. (hŭn´ĭ grōv)	79	33°35'N	95°54'W
Honey Lake, l., Ca., U.S. (hŭn´ĭ)	76	40°11'N	120°34'W
Honfleur, Can. (ôɴ-flûr´)	62b	46°39'N	70°53'W
Honfleur, Fr. (ôɴ-flûr´)	126	49°26'N	0°13'E
Hon Gay, Viet.	165	20°58'N	107°10'E
Hong Kong (Xianggang), China	161	21°45'N	115°00'E
Hongshui, r., China (hōŋ-shwā)	160	24°30'N	105°00'E
Honguedo, Détroit d', strt., Can.	60	49°08'N	63°45'W
Hongze Hu, l., China	161	33°17'N	118°37'E
Honiara, Sol. Is.	175	9°26'S	159°57'E
Honiton, Eng., U.K. (hŏn´ĭ-tŏn)	120	50°49'N	3°10'W
Honolulu, Hi., U.S. (hŏn-ō-lōō´lōō)	64c	21°18'N	157°50'W
Honomu, Hi., U.S. (hŏn´ō-mōō)	84a	19°50'N	155°04'W
Honshū, i., Japan	161	36°00'N	138°00'E
Hood, Mount, mtn., Or., U.S.	64	45°20'N	121°43'W
Hood Canal, b., Wa., U.S. (hŏd)	74a	47°45'N	122°45'W
Hood River, Or., U.S.	64	45°42'N	121°30'W
Hoodsport, Wa., U.S. (hŏdz´pōrt)	74a	47°25'N	123°09'W
Hoogly, r., India (hōōg´lĭ)	155	21°35'N	87°50'E
Hoogstraten, Bel.	115a	51°24'N	4°46'E
Hooker, Ok., U.S. (hŏk´ēr)	78	36°49'N	101°13'W
Hool, Mex. (ōō´l)	90a	19°32'N	90°22'W
Hoonah, Ak., U.S. (hōō´nā)	63	58°05'N	135°25'W
Hoopa Valley Indian Reservation, I.R., Ca., U.S.	72	41°18'N	123°35'W
Hooper, Ne., U.S. (hŏp´ēr)	79	41°37'N	96°31'W
Hooper, Ut., U.S.	75b	41°10'N	112°08'W
Hooper Bay, Ak., U.S.	63	61°32'N	166°02'W
Hoopeston, Il., U.S. (hōōps´tŭn)	66	40°35'N	87°40'W
Hoosick Falls, N.Y., U.S. (hōō´sĭk)	67	42°55'N	73°15'W
Hoover Dam, Nv., U.S. (hōō´vēr)	76	36°00'N	115°06'W
Hoover Dam, dam, U.S.	64	36°00'N	114°27'W
Hopatcong, Lake, l., N.J., U.S. (hō-păt´kong)	68a	40°57'N	74°38'W
Hope, Ak., U.S. (hōp)	63	60°54'N	149°48'W
Hope, Ar., U.S.	79	33°41'N	93°35'W
Hope, N.D., U.S.	70	47°17'N	97°45'W
Hope, Ben, mtn., Scot., U.K. (bĕn hōp)	120	58°25'N	4°25'W
Hopedale, Can. (hōp´dāl)	51	55°26'N	60°11'W
Hopedale, Ma., U.S. (hōp´dāl)	61a	42°08'N	71°33'W
Hopelchén, Mex. (o-pĕl-chĕ´n)	90a	19°47'N	89°51'W
Hopes Advance, Cap, c., Can. (hōps ăd-vans´)	53	61°05'N	69°35'W
Hopetoun, Austl. (hōp´toun)	174	33°50'S	120°15'E
Hopetown, S. Afr. (hōp´toun)	186	29°35'S	24°10'E
Hopewell, Va., U.S. (hōp´wĕl)	83	37°14'N	77°15'W
Hopewell Culture National Historical Park, rec., Oh., U.S.	66	39°25'N	83°00'W
Hopi Indian Reservation, I.R., Az., U.S. (hō´pė)	77	36°20'N	110°30'W
Hopkins, Mn., U.S. (hŏp´kĭns)	75g	44°55'N	93°24'W
Hopkinsville, Ky., U.S. (hŏp´kĭns-vĭl)	65	36°50'N	87°28'W
Hopkinton, Ma., U.S. (hŏp´kĭn-tŭn)	61a	42°14'N	71°31'W
Hoquiam, Wa., U.S. (hō´kwĭ-ăm)	64	47°00'N	123°53'W
Horconcitos, Pan. (ŏr-kôn-sė´-tōs)	91	8°18'N	82°11'W
Horgen, Switz. (hôr´gĕn)	124	47°16'N	8°35'E
Horicon, Wi., U.S. (hŏr´ĭ-kŏn)	71	43°26'N	88°40'W
Horlivka, Ukr.	137	48°17'N	38°03'E
Hormuz, Strait of, strt., Asia (hôr´mŭz´)	154	26°30'N	56°30'E
Horn, i., Austl. (hôrn)	175	10°30'S	143°30'E
Horn, Cape *see* Hornos, Cabo de, c., Chile	102	56°00'S	67°00'W
Hornavan, l., Swe.	116	65°54'N	16°17'E
Horneburg, Ger. (hōr´nĕ-bŏrgh)	115c	53°30'N	9°35'E
Hornell, N.Y., U.S. (hōr-nĕl´)	67	42°20'N	77°40'W
Hornos, Cabo de, c., Chile	102	56°00'S	67°00'W
Horn Plateau, plat., Can.	52	62°12'N	120°29'W
Hornsby, Austl. (hônrz´bĭ)	173b	33°43'S	151°06'E
Horodenka, Ukr.	125	48°40'N	25°30'E
Horodnia, Ukr.	133	51°54'N	31°31'E
Horodok, Ukr.	125	49°47'N	23°39'E
Horqueta, Para. (ŏr-kĕ´tä)	102	23°20'S	57°00'W
Horse Creek, r., Co., U.S. (hôrs)	78	38°49'N	103°48'W
Horse Creek, r., Wy., U.S.	70	41°33'N	104°39'W
Horse Islands, is., Can.	61	50°11'N	55°45'W
Horsens, Den. (hôrs´ĕns)	122	55°50'N	9°49'E
Horseshoe Bay, Can. (hôrs-shōō)	74d	49°23'N	123°16'W
Horsforth, Eng., U.K. (hôrs´fŭrth)	114a	53°50'N	1°38'W
Horsham, Austl. (hôr´shăm) (hôrs´ăm)	175	36°42'S	142°17'E
Horst, Ger. (hôrst)	115c	53°49'N	9°37'E
Horten, Nor. (hōr´tĕn)	122	59°26'N	10°27'E
Horton, Ks., U.S. (hôr´tŭn)	79	39°38'N	95°32'W
Horton, r., Ak., U.S. (hôr´tŭn)	63	68°38'N	122°00'W
Horwich, Eng., U.K. (hŏr´ĭch)	114a	53°36'N	2°33'W
Horyn', r., Eur. (gō´rĕn´)	125	50°55'N	26°07'E
Hososhima, Japan (hō´sō-shē´mä)	166	32°25'N	131°40'E
Hoste, i., Chile (ŏs´tā)	102	55°20'S	70°45'W
Hostotipaquillo, Mex. (ōs-tō´tĭ-pä-kēl´yō)	88	21°09'N	104°05'W
Hota, Japan (hō´tä)	167a	35°08'N	139°50'E
Hotan, China (hwō-tän)	160	37°11'N	79°50'E
Hotan, r., China	160	39°09'N	81°08'E
Hoto Mayor, Dom. Rep. (ô-tô-mä-yō´r)	93	18°45'N	69°10'W
Hot Springs, Ak., U.S. (hŏt sprĭngs)	63	65°00'N	150°20'W
Hot Springs, Ar., U.S.	65	34°29'N	93°02'W
Hot Springs, S.D., U.S.	70	43°28'N	103°32'W
Hot Springs, Va., U.S.	67	38°00'N	79°55'W
Hot Springs National Park, rec., Ar., U.S.	65	34°30'N	93°00'W
Hotte, Massif de la, mts., Haiti	93	18°25'N	74°00'W
Hotville, Ca., U.S. (hŏt´vĭl)	76	32°50'N	115°24'W
Houdan, Fr. (ōō-däɴ´)	127b	48°47'N	1°36'E
Houghton, Mi., U.S. (hō´tŭn)	71	47°06'N	88°36'W
Houghton, l., Mi., U.S.	66	44°20'N	84°45'W
Houilles, Fr. (ōō-yēs´)	127b	48°55'N	2°11'E
Houjie, China (hwō-jyĕ)	163a	22°58'N	113°39'E
Houlton, Me., U.S. (hōl´tŭn)	60	46°07'N	67°50'W
Houma, La., U.S. (hōō´må)	81	29°36'N	90°43'W
Housatonic, r., U.S. (hōō-sȧ-tŏn´ĭk)	67	41°50'N	73°25'W
House Springs, Mo., U.S. (hous sprĭngs)	75e	38°24'N	90°34'W
Houston, Ms., U.S. (hūs´tŭn)	82	33°53'N	89°00'W
Houston, Tx., U.S.	65	29°46'N	95°21'W
Houston Ship Channel, strt., Tx., U.S.	81a	29°38'N	94°57'W
Houtbaai, S. Afr.	186a	34°03'S	18°22'E
Houtman Rocks, is., Austl. (hout´män)	174	28°15'S	112°45'E
Houzhen, China (hwō-jŭn)	162	36°59'N	118°59'E
Hovd, Mong.	160	48°08'N	91°40'E
Hovd Gol, r., Mong.	160	49°06'N	91°16'E
Hove, Eng., U.K. (hōv)	120	50°50'N	0°09'W
Hövsgöl Nuur, l., Mong.	160	51°11'N	99°11'E
Howard, Ks., U.S. (hou´ȧrd)	79	37°27'N	96°10'W
Howard, S.D., U.S.	70	44°01'N	97°31'W
Howden, Eng., U.K. (hou´dĕn)	114a	53°44'N	0°52'W
Howe, Cape, c., Austl. (hou)	175	37°30'S	150°40'E
Howell, Mi., U.S. (hou´ĕl)	66	42°40'N	84°00'W
Howe Sound, strt., Can.	54	49°22'N	123°18'W
Howick, Can. (hou´ĭk)	62a	45°11'N	73°51'W
Howick, S. Afr.	187c	29°29'S	30°16'E
Howland, i., Oc. (hou´lănd)	2	1°00'N	176°00'W
Howrah, India (hou´rä)	155	22°33'N	88°20'E
Howse Peak, mtn., Can.	55	51°30'N	116°40'W
Howson Peak, mtn., Can.	54	54°25'N	127°45'W
Hoxie, Ar., U.S. (kŏh´sĭ)	79	36°03'N	91°00'W
Hoy, i., Scot., U.K. (hoi)	120a	58°53'N	3°10'W
Hōya, Japan	167a	35°45'N	139°35'E
Hoylake, Eng., U.K. (hoi-lāk´)	114a	53°23'N	3°11'W
Hoyo, Sierra del, mts., Spain (sē-ĕ´r-rä-dĕl-ô´yō)	129a	40°39'N	3°56'W
Hradec Králové, Czech Rep.	117	50°12'N	15°50'E
Hradyz'k, Ukr.	133	49°12'N	33°06'E
Hranice, Czech Rep. (hrän´yė-tsĕ)	125	49°33'N	17°45'E
Hröby, Swe. (hûr´bü)	122	55°50'N	13°41'E
Hrodna, Bela.	136	53°40'N	23°49'E

ăt; finăl; rāte; senåte; ärm; åsk; sofȧ; fâre; ch-choose; dh-as th in other; bē; ėvent; bĕt; recĕnt; cratẽr; g-gō; gh-guttural g; bĭt; ĭ-short neutral; rīde; ĸ-guttural k as ch in German ich;

PLACE (Pronunciation)	PAGE	LAT.	LONG.
Hron, r., Slvk.	125	48°22'N	18°42'E
Hrubieszów, Pol. (hrōō-byä'shōōf)	125	50°48'N	23°54'E
Hsawnhsup, Mya.	160	24°29'N	94°45'E
Hsinchu, Tai. (hsĭn'chōō')	165	24°48'N	121°00'E
Huadian, China (hwä-dĭĕn)	164	42°38'N	126°45'E
Huai, r., China (hwī)	161	32°07'N	114°38'E
Huai'an, China (hwī-än)	164	33°31'N	119°11'E
Huailai, China	164	40°20'N	115°45'E
Huailin, China (hwī-lĭn)	162	31°27'N	117°36'E
Huainan, China	162	32°38'N	117°02'E
Huaiyang, China (hōōäī'yang)	164	33°45'N	114°54'E
Huaiyuan, China (hwī-yůän)	164	32°53'N	117°13'E
Huajicori, Mex. (wä-jē-kō'rē)	88	22°41'N	105°24'W
Huajuapan de León, Mex. (wäj-wä'päm dā lā-ón')	89	17°46'N	97°45'W
Hualapai Indian Reservation, I.R., Az., U.S. (wälåpī)	77	35°41'N	113°38'W
Hualapai Mountains, mts., Az., U.S.	77	34°53'N	113°54'W
Hualien, Tai. (hwä'lyĕn')	165	23°58'N	121°58'E
Huallaga, r., Peru (wäl-yä'gä)	100	8°12'S	76°34'W
Huamachuco, Peru (wä-mä-chōō'kō)	100	7°52'S	78°11'W
Huamantla, Mex. (wä-män'tlå)	89	19°18'N	97°54'W
Huambo, Ang.	186	12°44'S	15°47'E
Huamuxtitlán, Mex. (wä-mōōs-tē-tlän')	88	17°49'N	98°38'W
Huancavelica, Peru (wän'kä-vä-lē'kä)	100	12°47'S	75°02'W
Huancayo, Peru (wän-kä'yō)	100	12°09'S	75°04'W
Huanchaca, Bol. (wän-chä'kä)	100	20°09'S	66°40'W
Huang (Yellow), r., China (hůäŋ)	161	35°06'N	113°39'E
Huang, Old Beds of the, mth., China	160	40°28'N	106°34'E
Huang, Old Course of the, r., China	164	34°28'N	116°59'E
Huangchuan, China (hůäŋ-chůän)	164	32°07'N	115°01'E
Huanghua, China (hůäŋ-hwä)	162	38°28'N	117°18'E
Huanghuadian, China (hůäŋ-hwä-dĭĕn)	162	39°22'N	116°53'E
Huangli, China (hōōäng'lē)	162	31°39'N	119°42'E
Huangpu, China (hůäŋ-pōō)	163a	22°44'N	113°20'E
Huangpu, r., China	163b	30°56'N	121°16'E
Huangqiao, China (hůän-chyou)	162	32°15'N	120°13'E
Huangxian, China (hůäŋ shyĕn)	162	37°39'N	120°32'E
Huangyuan, China (hůäŋ-yůän)	160	37°00'N	101°01'E
Huanren, China (hůän-rŭn)	164	41°10'N	125°30'E
Huánuco, Peru (wä-nōō'kō)	100	9°50'S	76°17'W
Huánuni, Bol. (wä-nōō'nē)	100	18°11'S	66°43'W
Huaquechula, Mex. (wä-kĕ-chōō'lä)	88	18°44'N	98°37'W
Huaral, Peru (wä-rä'l)	100	11°28'S	77°11'W
Huarás, Peru (òä'rà's)	100	9°32'S	77°29'W
Huascarán, Nevados, mts., Peru (wäs-kä-rän')	100	9°05'S	77°50'W
Huasco, Chile (wäs'kō)	102	28°32'S	71°16'W
Huatla de Jiménez, Mex. (wà'tlä-dĕ-κĕ-mĕ'nĕz)	89	18°08'N	96°49'W
Huatlatlauch, Mex. (wä'tlä-tlä-ōō'ch)	88	18°40'N	98°04'W
Huatusco, Mex. (wä-tōōs'kō)	89	19°09'N	96°57'W
Huauchinango, Mex. (wä-ōō-chē-näŋ'gō)	88	20°09'N	98°03'W
Huaunta, Nic. (wä-ó'n-tä)	91	13°30'N	83°32'W
Huaunta, Laguna, l., Nic. (lä-gó'nä-wä-ó'n-tä)	91	13°35'N	83°46'W
Huautla, Mex. (wä-ōō'tlä)	88	21°04'N	98°13'W
Huaxian, China (hwä shyĕn)	164	35°34'N	114°32'E
Huaynamota, Río de, r., Mex. (rē'ō-dĕ-wäy-nä-mō'tä)	88	22°10'N	104°36'W
Huazolotitlán, Mex. (wäzô-lô-tlē-tlän')	89	16°18'N	97°55'W
Hubbard, N.H., U.S. (hŭb'ērd)	61a	42°53'N	71°12'W
Hubbard, Tx., U.S.	81	31°53'N	96°46'W
Hubbard, l., Mi., U.S.	66	44°45'N	83°30'W
Hubbard Creek Reservoir, res., Tx., U.S.	80	32°50'N	98°55'W
Hubei, prov., China (hōō-bā)	161	31°20'N	111°58'E
Hubli, India (hò'blē)	155	15°25'N	75°09'E
Hückeswagen, Ger. (hü'kĕs-vä'gĕn)	127c	51°09'N	7°20'E
Hucknall, Eng., U.K. (hŭk'năl)	114a	53°02'N	1°12'W
Huddersfield, Eng., U.K. (hŭd'ērz-fēld)	120	53°39'N	1°47'W
Hudiksvall, Swe. (hōō'dĭks-väl)	116	61°44'N	17°05'E
Hudson, Can. (hŭd'sŭn)	62a	45°26'N	74°08'W
Hudson, Ma., U.S.	61a	42°24'N	71°34'W
Hudson, Mi., U.S.	66	41°50'N	84°15'W
Hudson, N.Y., U.S.	67	42°15'N	73°45'W
Hudson, Oh., U.S.	69d	41°15'N	81°27'W
Hudson, Wi., U.S.	75g	44°59'N	92°45'W
Hudson, r., U.S.	65	42°30'N	73°55'W
Hudson Bay, Can.	57	52°52'N	102°25'W
Hudson Bay, b., Can.	53	60°15'N	85°30'W
Hudson Falls, N.Y., U.S.	67	43°20'N	73°30'W
Hudson Heights, Can.	62a	45°28'N	74°09'W
Hudson Strait, strt., Can.	53	63°25'N	74°05'W
Hue, Viet. (ü-ā')	168	16°28'N	107°42'E
Huebra, r., Spain (wĕ'brä)	128	40°44'N	6°17'W
Huehuetenango, Guat. (wā-wå-tå-näŋ'gō)	90	15°19'N	91°26'W
Huejotzingo, Mex. (wā-hō-tzĭŋ'gō)	88	19°09'N	98°24'W
Huejúcar, Mex. (wā-hōō'kär)	88	22°26'N	103°12'W
Huejuquilla el Alto, Mex. (wā-hōō-kēl'yä ĕl äl'tō)	88	22°42'N	103°54'W
Huejutla, Mex. (wā-hōō'tlä)	88	21°08'N	98°26'W
Huelma, Spain (wĕl'mä)	128	37°39'N	3°36'W
Huelva, Spain (wĕl'vä)	118	37°16'N	6°58'W
Huércal-Overa, Spain (wĕr-käl' ō-vā'rä)	128	37°12'N	1°58'W
Huerfano, r., Co., U.S. (wâr'fá-nō)	78	37°41'N	105°13'W
Huésca, Spain (wĕs-kä)	118	42°07'N	0°25'W
Huéscar, Spain (wās'kär)	128	37°50'N	2°34'W
Huetamo de Núñez, Mex.	88	18°34'N	100°53'W
Huete, Spain (wā'tå)	128	40°09'N	2°42'W
Hueycatenango, Mex. (wĕy-kà-tĕ-nä'n-gô)	88	17°31'N	99°10'W
Hueytlalpan, Mex. (wā'ī-tlăl'păn)	89	20°03'N	97°41'W
Hueytown, Al., U.S.	68h	33°28'N	86°59'W
Huffman, Al., U.S. (hŭf'măn)	68h	33°36'N	86°42'W
Hugh Butler, l., Ne., U.S.	78	40°21'N	100°40'W
Hughenden, Austl. (hū'ĕn-dĕn)	175	20°58'S	144°13'E
Hughes, Austl. (hūz)	174	30°45'S	129°30'E
Hughesville, Md., U.S.	68e	38°32'N	76°48'W
Hugo, Mn., U.S. (hū'gō)	75g	45°10'N	93°00'W
Hugo, Ok., U.S.	79	34°01'N	95°32'W
Hugoton, Ks., U.S. (hū'gō-tŭn)	78	37°10'N	101°28'W
Hugou, China (hōō-gō)	162	33°22'N	117°07'E
Huichapan, Mex. (wē-chä-pän')	88	20°22'N	99°39'W
Huila, dept., Col. (wē'lä)	100a	3°10'N	75°20'W
Huila, Nevado de, mtn., Col. (nĕ-vä-dô-de-wē'lä)	100a	2°59'N	76°01'W
Huilai, China	165	23°02'N	116°18'E
Huili, China	160	26°48'N	102°20'E
Huimanguillo, Mex. (wē-män-gēl'yō)	89	17°50'N	93°16'W
Huimin, China (hōōĭ mĭn)	161	37°29'N	117°32'E
Huitzilac, Mex. (ōē't-zē-lä'k)	89a	19°01'N	99°16'W
Huitzitzilingo, Mex. (wē-tzē-tzē-lē'n-go)	88	21°11'N	98°42'W
Huitzuco, Mex. (wē-tzōō'kō)	88	18°16'N	99°20'W
Huixquilucan, Mex. (ōē'x-kē-lōō-kä'n)	89a	19°21'N	99°22'W
Huiyang, China	165	23°05'N	114°25'E
Hukou, China (hōō-kō)	161	29°58'N	116°20'E
Hulan, China (hōō'län')	161	45°58'N	126°32'E
Hulan, r., China	164	47°20'N	126°30'E
Huliaipole, Ukr.	133	47°39'N	36°12'E
Hulin, China (hōō'lĭn')	166	45°45'N	133°25'E
Hull, Can. (hŭl)	51	45°26'N	75°43'W
Hull, Ma., U.S.	61a	42°18'N	70°54'W
Hull, r., Eng., U.K.	114a	53°47'N	0°20'W
Hulst, Neth. (hòlst)	115a	51°17'N	4°01'E
Huludao, China (hōō-lōō-dou)	161	40°40'N	120°55'E
Hulun Nur, l., China (hōō-lòn nòr)	161	48°50'N	116°45'E
Humacao, P.R. (ōō-mä-kä'ō)	87b	18°09'N	65°49'W
Humansdorp, S. Afr. (hōō'mäns-dòrp)	186	33°57'S	24°45'E
Humbe, Ang. (hòm'bå)	186	16°50'S	14°55'E
Humber, r., Can.	62d	43°53'N	79°40'W
Humber, r., Eng., U.K. (hŭm'bēr)	116	53°30'N	0°30'E
Humbermouth, Can. (hŭm'bēr-mŭth)	61	48°58'N	57°55'W
Humberside, hist. reg., Eng., U.K.	114a	53°47'N	0°36'W
Humble, Tx., U.S. (hŭm'b'l)	81	29°58'N	95°15'W
Humboldt, Can. (hŭm'bōlt)	50	52°12'N	105°07'W
Humboldt, Ia., U.S.	71	42°43'N	94°11'W
Humboldt, Ks., U.S.	79	37°48'N	95°26'W
Humboldt, Ne., U.S.	79	40°10'N	95°57'W
Humboldt, r., Nv., U.S.	64	40°30'N	116°50'W
Humboldt, East Fork, r., Nv., U.S.	72	40°59'N	115°21'W
Humboldt, North Fork, r., Nv., U.S.	72	41°25'N	115°45'W
Humboldt Bay, b., Ca., U.S.	72	40°48'N	124°25'W
Humboldt Range, mts., Nv., U.S.	76	40°12'N	118°16'W
Humbolt, Tn., U.S.	82	35°47'N	88°55'W
Humbolt Salt Marsh, Nv., U.S.	76	39°49'N	117°41'W
Humbolt Sink, Nv., U.S.	76	39°58'N	118°54'W
Humen, China (hōō-mŭn)	163a	22°49'N	113°39'E
Humphreys Peak, mtn., Az., U.S. (hŭm'frĭs)	64	35°20'N	111°40'W
Humpolec, Czech Rep. (hòm'pō-lĕts)	124	49°33'N	15°21'E
Humuya, r., Hond. (ōō-mōō'yä)	90	14°38'N	87°36'W
Hunaflói, b., Ice. (hōō'nä-flō'ī)	116	65°41'N	20°44'W
Hunan, prov., China (hōō'nän')	161	28°08'N	111°25'E
Hunchun, China (hòn-chůn)	161	42°53'N	130°34'E
Hunedoara, Rom. (κōō'nĕd-wä'rà)	131	45°45'N	22°54'E
Hungary, nation, Eur. (hŭŋ'gá-rĭ)	110	46°44'N	17°55'E
Hungerford, Austl. (hŭn'gēr-fērd)	175	28°50'S	144°32'E
Hungry Horse Reservoir, res., Mt., U.S. (hŭŋ'gá-rĭ hôrs)	73	48°11'N	113°30'W
Hunsrück, mts., Ger. (hōōns'rük)	124	49°43'N	7°12'E
Hunte, r., Ger. (hòn'tĕ)	124	52°45'N	8°26'E
Hunter Islands, is., Austl. (hŭn-tēr)	175	40°33'S	143°36'E
Huntingburg, In., U.S. (hŭnt'ĭng-bûrg)	66	38°15'N	86°55'W
Huntingdon, Can. (hŭnt'ĭng-dŭn)	59	45°10'N	74°05'W
Huntingdon, Can.	74d	49°00'N	122°16'W
Huntingdon, Tn., U.S.	82	36°00'N	88°23'W
Huntington, In., U.S.	66	40°55'N	85°30'W
Huntington, Pa., U.S.	67	40°30'N	78°00'W
Huntington, W.V., U.S.	65	38°25'N	82°25'W
Huntington Beach, Ca., U.S.	75a	33°39'N	118°00'W
Huntington Park, Ca., U.S.	75a	33°59'N	118°14'W
Huntington Station, N.Y., U.S.	68a	40°51'N	73°25'W
Huntley, Mt., U.S.	73	45°54'N	108°01'W
Huntsville, Can.	51	45°20'N	79°15'W
Huntsville, Al., U.S. (hŭnts'vĭl)	82	34°44'N	86°36'W
Huntsville, Mo., U.S.	79	39°24'N	92°32'W
Huntsville, Tx., U.S.	81	30°44'N	95°34'W
Huntsville, Ut., U.S.	75b	41°16'N	111°46'W
Huolu, China (hòŭ lōō)	162	38°05'N	114°20'E
Huon Gulf, b., Pap. N. Gui.	169	7°15'S	147°45'E
Huoqiu, China (hwò-chyò)	162	32°19'N	116°17'E
Huoshan, China (hwò-shän)	165	31°30'N	116°25'E
Ḩurayḑin, Wādī, r., Egypt	153a	30°55'N	34°12'E
Hurd, Cape, c., Can. (hûrd)	58	45°15'N	81°45'W
Hurdiyo, Som.	192a	10°43'N	51°05'E
Hurley, Wi., U.S. (hûr'lĭ)	71	46°26'N	90°11'W
Hurlingham, Arg. (ōō'r-lēn-gäm)	102a	34°36'S	58°38'W
Huron, Oh., U.S. (hū'rŏn)	66	41°20'N	82°35'W
Huron, S.D., U.S.	64	44°22'N	98°15'W
Huron, r., Mi., U.S.	69b	42°12'N	83°26'W
Huron, Lake, l., N.A. (hū'rŏn)	65	45°15'N	82°40'W
Huron Mountains, mts., Mi., U.S. (hū'rŏn)	71	46°47'N	87°52'W
Hurricane, Ak., U.S. (hŭr'ĭ-kān)	63	63°00'N	149°30'W
Hurricane, Ut., U.S.	77	37°10'N	113°20'W
Hurricane Flats, bk., (hŭ-rĭ-kán flăts)	92	23°35'N	78°30'W
Hurst, Tx., U.S.	75c	32°48'N	97°12'W
Húsavik, Ice.	116	66°00'N	17°10'W
Huşi, Rom. (kósh')	133	46°52'N	28°04'E
Huskvarna, Swe. (hòsk-vär'nà)	122	57°48'N	14°16'E
Husum, Ger. (hōō'zòm)	124	54°29'N	9°04'E
Hutchins, Tx., U.S. (hŭch'ĭnz)	75c	32°38'N	96°43'W
Hutchinson, Ks., U.S. (hŭch'ĭn-sŭn)	64	38°02'N	97°56'W
Hutchinson, Mn., U.S.	71	44°53'N	94°23'W
Hutuo, r., China	164	38°10'N	114°00'E
Huy, Bel. (ū-ē') (hů'ē)	121	50°33'N	5°14'E
Hvannadalshnúkur, mtn., Ice.	116	64°09'N	16°46'W
Hvar, i., Serb. (κhvär)	130	43°08'N	16°28'E
Hwange, Zimb.	186	18°22'S	26°29'E
Hwangju, Kor., N. (hwäng'jōō')	166	38°39'N	125°49'E
Hyargas Nuur, l., Mong.	160	48°00'N	92°32'E
Hyattsville, Md., U.S. (hī'ăt's-vĭl)	68e	38°57'N	76°58'W
Hyco Lake, res., N.C., U.S. (rōks' bŭr-ō)	83	36°22'N	78°58'W
Hydaburg, Ak., U.S. (hī-dă'bûrg)	63	55°12'N	132°49'W
Hyde, Eng., U.K. (hīd)	114a	53°27'N	2°05'W
Hyderābād, India (hī-dēr-å-băd')	155	17°29'N	78°28'E
Hyderābād, India	155	18°30'N	76°50'E
Hyderābād, Pak.	155	25°29'N	68°28'E
Hyéres, Fr. (ē-âr')	117	43°09'N	6°08'E
Hyéres, Îles d', is., Fr. (ēl'dyâr')	117	42°57'N	6°17'E
Hyesanjin, Kor., N. (hyĕ'sän-jĭn')	166	41°11'N	128°12'E
Hymera, In., U.S. (hī-mē'rá)	66	39°10'N	87°20'W
Hyndman Peak, mtn., Id., U.S. (hīnd'măn)	64	43°38'N	114°04'W
Hyōgo, dept., Japan (hĭ'yō'gō)	167b	34°54'N	135°15'E

I

PLACE (Pronunciation)	PAGE	LAT.	LONG.
Ia, r., Japan (ē'ä)	167b	34°54'N	135°34'E
Iahotyn, Ukr.	133	50°18'N	31°46'E
Ialomiţa, r., Rom.	131	44°37'N	26°42'E
Iaşi, Rom. (yä'shē)	110	47°10'N	27°40'E
Iasinia, Ukr.	125	48°17'N	24°21'E
Iavoriv, Ukr.	125	49°56'N	23°24'E
Iba, Phil. (ē'bä)	169a	15°20'N	119°59'E
Ibadan, Nig. (ē-bä'dän)	184	7°17'N	3°30'E
Ibagué, Col.	100	4°27'N	75°14'W
Ibar, r., Serb. (ē'bär)	131	43°22'N	20°35'E
Ibaraki, Japan (ē-bä'rä-gē)	167b	34°49'N	135°35'E
Ibarra, Ec. (ē-bär'rä)	100	0°19'N	78°08'W
Ibb, Yemen	157	14°01'N	44°10'E
Iberville, Can. (ē-bår-vēl') (ī'bēr-vĭl)	59	45°14'N	73°01'W
Ibi, Nig. (ē'bē)	184	8°12'N	9°45'E
Ibiapaba, Serra da, mts., Braz. (sĕ'r-rä-dä-ē-byä-på'bä)	101	3°30'S	40°55'W
Ibiza *see* Eivissa, i., Spain	112	38°55'N	1°24'E
Ibo, Moz. (ē'bō)	187	12°20'S	40°35'E
Ibrāhīm, Būr, b., Egypt	192d	29°57'N	32°33'E
Ibrahim, Jabal, mtn., Sau. Ar.	154	20°31'N	41°17'E
Ibwe Munyama, Zam.	191	16°09'S	28°34'E
Ica, Peru (ē'kä)	100	14°09'S	75°42'W
Icá (Putumayo), r., S.A.	100	3°00'S	69°00'W
Içana, Braz. (ē-sä'nä)	100	0°15'N	67°19'W
Ice Harbor Dam, Wa., U.S.	72	46°15'N	118°54'W
İçel, Tur.	154	37°00'N	34°40'E
Iceland, nation, Eur. (īs'lănd)	110	65°12'N	19°45'W
Ichibusayama, mtn., Japan (ē'chē-bōō'sà-yä'mä)	167	32°19'N	131°08'E
Ichihara, Japan	167a	35°31'N	140°05'E
Ichikawa, Japan (ē'chē-kä'wä)	167a	35°44'N	139°54'E
Ichinomiya, Japan (ē'chē-nō-mē'yä)	167	35°19'N	136°49'E
Ichinomoto, Japan (ē-chē'nō-mō-tō)	167b	34°37'N	135°50'E
Ichnia, Ukr.	137	50°47'N	32°23'E
Icy Cape, c., Ak., U.S. (ī'sĭ)	63	70°20'N	161°40'W
Idabel, Ok., U.S. (ī'dá-bĕl)	79	33°52'N	94°47'W
Idagrove, Ia., U.S. (ī'dá-grōv)	70	42°22'N	95°29'W
Idah, Nig. (ē'dä)	184	7°07'N	6°43'E
Idaho, state, U.S. (ī'dá-hō)	64	44°00'N	115°10'W
Idaho Falls, Id., U.S.	66	43°30'N	112°01'W
Idaho Springs, Co., U.S.	78	39°43'N	105°32'W
Idanha-a-Nova, Port. (ē-dän'yà-ä-nō'vá)	128	39°58'N	7°13'W
Ider, r., Mong.	160	48°58'N	98°38'E
Idi, Indon. (ē'dē)	168	4°58'N	97°47'E
Idkū Lake, l., Egypt	192b	31°13'N	30°22'E
Idle, r., Eng., U.K. (īd''l)	114a	53°22'N	0°56'W
Idlib, Syria	156	35°55'N	36°38'E
Idriaj, Slvn. (ē'drē-à)	130	46°01'N	14°01'E
Idutywa, S. Afr. (ē-dò-tī'wá)	187c	32°06'S	28°18'E
Ienakiieve, Ukr.	133	48°14'N	38°12'E
Ieper, Bel.	121	50°50'N	2°53'E
Ierápetra, Grc.	130a	35°01'N	25°48'E
Iesi, Italy (yā'sē)	130	43°37'N	13°20'E
Ievpatoriia, Ukr.	137	45°13'N	33°22'E
Ife, Nig.	184	7°30'N	4°30'E
Iferouâne, Niger (ēf'rōō-än')	184	19°04'N	8°24'E
Ifôghas, Adrar des, plat., Afr.	184	19°55'N	2°00'E
Igalula, Tan.	191	5°14'S	33°00'E
Igarka, Russia (ē-gär'kà)	134	67°22'N	86°16'E
Iglesias, Italy (ē-lē'syōs)	118	39°20'N	8°34'E
Igli, Alg. (ē-glē')	184	30°32'N	2°15'W
Igluligaarjuk (Chesterfield Inlet), Can.	51	63°19'N	91°11'W
Iglulik, Can.	51	69°33'N	81°18'W
Ignacio, Ca., U.S. (ĭg-nā'cĭ-ō)	74b	38°05'N	122°32'W
Iguaçu, r., Braz. (ē-gwä-sōō')	102b	22°42'S	43°19'W

PLACE (Pronunciation)	PAGE	LAT.	LONG.
Iguala, Mex. (ē-gwä'lä)	88	18°18'N	99°34'W
Igualada, Spain (ē-gwä-lä'dä)	129	41°35'N	1°38'E
Iguassu, r., S.A. (ē-gwä-sōō')	102	25°45'S	52°30'W
Iguassu Falls, wtfl., S.A.	101	25°40'S	54°16'W
Iguatama, Braz. (ē-gwä-tà'mä)	99a	20°13'S	45°40'W
Iguatu, Braz. (ē-gwä-tōō')	101	6°22'S	39°17'W
Iguidi, Erg, Afr.	184	26°22'N	6°53'W
Iguig, Phil. (ē-gēg')	169a	17°46'N	121°44'E
Iharana, Madag.	187	13°35'S	50°05'E
Ihiala, Nig.	189	5°51'N	6°51'E
Iida, Japan (ē'ē-dà)	167	35°39'N	137°53'E
Iijoki, r., Fin. (ē'yō'kĭ)	136	65°28'N	27°00'E
Iizuka, Japan (ē'ē-zò-kà)	167	33°39'N	130°39'E
Ijebu-Ode, Nig. (ē-jĕ'bōō ōdà)	184	6°50'N	3°56'E
IJmuiden, Neth.	115a	52°27'N	4°36'E
IJsselmeer, l., Neth. (ī'sĕl-mār)	121	52°46'N	5°14'E
Ikaalinen, Fin. (ē'kä-lĭ-nĕn)	123	61°47'N	22°55'E
Ikaría, i., Grc. (ē-kä'ryà)	131	37°43'N	26°07'E
Ikeda, Japan (ē'kà-dä)	167b	34°49'N	135°26'E
Ikerre, Nig.	189	7°31'N	5°14'E
Ikhtiman, Blg. (ėk'tē-män)	131	42°26'N	23°49'E
Iki, i., Japan (ē'kē)	166	33°46'N	129°44'E
Ikoma, Japan	167b	34°41'N	135°43'E
Ikoma, Tan. (ē-kō'mä)	186	2°08'S	34°47'E
Iksha, Russia (ĭk'shà)	142b	56°10'N	37°30'E
Ila, Nig.	189	8°01'N	4°55'E
Ilagan, Phil.	169a	17°09'N	121°52'E
Ilan, Tai. (ē'län')	165	24°50'N	121°42'E
Iława, Pol. (ē-lä'và)	125	53°35'N	19°36'E
Île-à-la-Crosse, Can.	56	55°34'N	108°00'W
Ilebo, D.R.C.	186	4°19'S	20°35'E
Ilek, Russia (ē'lyĕk)	137	51°30'N	53°10'E
Île-Perrot, Can. (yl-pĕ-rōt')	62a	45°21'N	73°54'W
Ilesha, Nig.	184	7°38'N	4°45'E
Ilford, Eng., U.K. (ĭl'fērd)	114b	51°33'N	0°06'E
Ilfracombe, Eng., U.K. (ĭl-frȧ-kōōm')	120	51°13'N	4°08'W
Ilhabela, Braz. (ē'lä-bĕ'lä)	99a	23°47'S	45°21'W
Ilha Grande, Baía de, b., Braz. (ēl'yȧ grän'dĕ)	99a	23°17'S	44°25'W
Ílhavo, Port. (ēl'yȧ-vō)	118	40°36'N	8°41'W
Ilhéus, Braz. (ē-lĕ'ōōs)	101	14°52'S	39°00'W
Ili, r., Asia	140	44°30'N	76°45'E
Iliamna, Ak., U.S. (ē-lē-ăm'nȧ)	63	59°45'N	155°05'W
Iliamna, Ak., U.S.	63	60°18'N	153°25'W
Iliamna, l., Ak., U.S.	63	59°25'N	155°30'W
Ilim, r., Russia (ē-lyêm')	140	57°28'N	103°00'E
Ilimsk, Russia (ē-lyêmsk')	135	56°47'N	103°43'E
Ilin Island, i., Phil. (ē-lyēn')	169a	12°16'N	120°57'E
Ilion, N.Y., U.S. (ĭl'ĭ-ŭn)	67	43°00'N	75°05'W
Ilkeston, Eng., U.K. (ĭl'kĕs-tŭn)	114a	52°58'N	1°19'W
Illampu, Nevado, mtn., Bol. (nĕ-vá'dō-ēl-yäm-pōō')	100	15°50'S	68°15'W
Illapel, Chile (ē-zhä-pĕ'l)	102	31°37'S	71°10'W
Iller, r., Ger. (ĭlēr)	124	47°52'N	10°06'E
Illimani, Nevado, mtn., Bol. (nĕ-vá'dō-ēl-yē-mä'nē)	100	16°50'S	67°38'W
Illinois, state, U.S. (ĭl-ĭ-noi') (ĭl-ĭ-noiz')	65	40°25'N	90°40'W
Illinois, r., Il., U.S.	65	39°00'N	90°30'W
Illintsi, Ukr.	133	49°07'N	29°13'E
Illizi, Alg.	184	26°35'N	8°24'E
Il'men, l., Russia (ô'zē-rô el''men'') (ĭl'mĕn)	136	58°18'N	32°00'E
Ilo, Peru	100	17°46'S	71°13'W
Ilobasco, El Sal. (ē-lô-bäs'kô)	90	13°57'N	88°46'W
Iloilo, Phil. (ē-lô-ē'lō)	168	10°49'N	122°33'E
Ilopango, Lago, l., El Sal. (ē-lô-päŋ'gō)	90	13°48'N	88°50'W
Ilorin, Nig. (ē-lô-rēn')	184	8°30'N	4°32'E
Ilūkste, Lat.	123	55°59'N	26°20'E
Ilwaco, Wa., U.S. (ĭl-wä'kō)	74c	46°19'N	124°02'W
Ilych, r., Russia (ē'l'ĭch)	136	62°30'N	57°30'E
Imabari, Japan (ē'mä-bä'rē)	166	34°05'N	132°58'E
Imai, Japan (ē-mī')	167b	34°30'N	135°47'E
Iman, r., Russia (ē-män')	166	45°40'N	134°31'E
Imandra, l., Russia (ē-män'drȧ)	136	67°40'N	32°30'E
Imbābah, Egypt (ēm-bä'bȧ)	192b	30°06'N	31°09'E
Imeni Morozova, Russia (ĭm-yĕ'nyĭ mô rô'zô vȧ)	142c	59°58'N	31°02'E
Imeni Moskvy, Kanal (Moscow Canal), can., Russia (kȧ-näl'ĭm-yȧ'nĭ mŏs-kvĭ)	132	56°33'N	37°15'E
Imeni Tsyurupy, Russia	142b	55°30'N	38°39'E
Imeni Vorovskogo, Russia	142b	55°43'N	38°21'E
Imlay City, Mi., U.S. (ĭm'lā)	66	43°00'N	83°15'W
Immenstadt, Ger. (ĭm'ĕn-shtät)	124	47°34'N	10°12'E
Immerpan, S. Afr. (ĭmĕr-pän)	192c	24°29'S	29°14'E
Imola, Italy (ē'mô-lä)	130	44°19'N	11°43'E
Imotski, Cro. (ē-môts'kē)	131	43°25'N	17°15'E
Impameri, Braz.	101	17°44'S	48°03'W
Impendle, S. Afr. (ĭm-pĕnd'lȧ)	187c	29°38'S	29°54'E
Imperia, Italy (ēm-pā'rē-ä)	118	43°52'N	8°00'E
Imperial, Pa., U.S. (ĭm-pē'rĭ-ăl)	69e	40°27'N	80°15'W
Imperial Beach, Ca., U.S.	76a	32°34'N	117°08'W
Imperial Valley, Ca., U.S.	76	33°00'N	115°22'W
Impfondo, Congo (ĭmp-fōn'dō)	185	1°37'N	18°04'E
Imphāl, India (ĭmp'hŭl)	155	24°42'N	94°00'E
Ina, r., Japan (ē-nä')	167b	34°56'N	135°21'E
Inaja Indian Reservation, I.R., Ca., U.S. (ē-nä'hä)	76	32°56'N	116°37'W
Inari, l., Fin.	116	69°02'N	26°22'E
Inca, Spain (ēŋ'kä)	129	39°43'N	2°53'E
Ince Burun, c., Tur. (ĭn'jȧ)	119	42°00'N	35°00'E
Inch'ŏn, Kor., S. (ĭn'chŭn)	161	37°26'N	126°46'E
Incudine, Monte, mtn., Fr. (ēn-kōō-dē'nȧ) (ăɴ-kü-dēn')	130	41°53'N	9°17'E
Indalsälven, r., Swe.	116	62°50'N	16°50'E
Independence, Ks., U.S. (ĭn-dē-pĕn'dĕns)	79	37°14'N	95°42'W
Independence, Mo., U.S.	75f	39°06'N	94°26'W
Independence, Oh., U.S.	69d	41°23'N	81°39'W
Independence, Or., U.S.	72	44°49'N	123°13'W
Independence Mountains, mts., Nv., U.S.	72	41°15'N	116°02'W
Inder köli, l., Kaz.	137	48°20'N	52°10'E
India, nation, Asia (ĭn'dĭ-ȧ)	155	23°00'N	77°30'E
Indian, l., Mi., U.S. (ĭn'dĭ-ăn)	71	46°04'N	86°34'W
Indian, r., N.Y., U.S.	67	44°05'N	75°45'W
Indiana, Pa., U.S. (ĭn-dĭ-än'ȧ)	67	40°40'N	79°10'W
Indiana, state, U.S.	65	39°50'N	86°45'W
Indianapolis, In., U.S. (ĭn-dĭ-ăn-ăp'ô-lĭs)	65	39°45'N	86°08'W
Indian Arm, b., Can. (ĭn'dĭ-ăn ärm)	74d	49°21'N	122°55'W
Indian Head, Can.	50	50°29'N	103°44'W
Indian Lake, l., Can.	58	47°00'N	82°00'W
Indian Ocean, o.	5	10°00'S	70°00'E
Indianola, Ia., U.S. (ĭn-dĭ-ăn-ō'lȧ)	71	41°22'N	93°33'W
Indianola, Ms., U.S.	82	33°29'N	90°35'W
Indigirka, r., Russia (ēn-dē-gēr'kȧ)	141	67°45'N	145°45'E
Indio, r., Pan. (ē'n-dyô)	86a	9°13'N	79°28'W
Indochina, reg., Asia (ĭn-dō-chī'nȧ)	168	17°22'N	105°18'E
Indonesia, nation, Asia (ĭn'dô-nē-zhȧ)	168	4°38'S	118°45'E
Indore, India (ĭn-dōr')	155	22°48'N	76°51'E
Indragiri, r., Indon. (ĭn-drȧ-jē'rē)	168	0°27'S	102°05'E
Indrāvati, r., India (ĭn-drŭ-vä'tē)	155	19°00'N	82°00'E
Indre, r., Fr. (ăɴ'dr')	126	47°13'N	0°29'E
Indus, Can. (ĭn'dŭs)	62e	50°55'N	113°45'W
Indus, r., Asia	155	26°43'N	67°41'E
Indwe, S. Afr. (ĭnd'wå)	187c	31°30'S	27°21'E
Inebolu, Tur. (ē-nå-bō'lōō)	119	41°50'N	33°40'E
Inego, Tur. (ē'nå-gŭ)	137	40°05'N	29°20'E
Infanta, Phil. (ēn-fän'tä)	169a	14°44'N	121°39'E
Infanta, Phil.	169a	15°50'N	119°53'E
Inferror, Laguna, l., Mex. (lä-gō'nä-ēn-fĕr-rôr)	89	16°18'N	94°40'W
Infiernillo, Presa de, res., Mex.	88	18°50'N	101°50'W
Infiesto, Spain (ēn-fyĕ's-tō)	128	43°21'N	5°24'W
I-n-Gall, Niger	189	16°47'N	6°56'E
Ingersoll, Can. (ĭn'gēr-sŏl)	58	43°05'N	81°00'W
Ingham, Austl. (ĭng'ăm)	175	18°45'S	146°14'E
Ingles, Cayos, is., Cuba (kä-yōs-ē'n-glĕ's)	92	21°55'N	82°35'W
Inglewood, Can.	62d	43°48'N	79°56'W
Inglewood, Ca., U.S. (ĭn'g'l-wŏŏd)	75a	33°57'N	118°22'W
Ingoda, r., Russia (ēn-gō'dȧ)	141	51°29'N	112°32'E
Ingolstadt, Ger. (ĭŋ'gōl-shtät)	124	48°46'N	11°27'E
Ingur, r., Geor. (ēn-gōr')	137	42°30'N	42°00'E
Ingushetia, prov., Russia	138	43°15'N	45°00'E
Inhambane, Moz. (ēn-äm-bä'-nĕ)	186	23°47'S	35°28'E
Inhambupe, Braz. (ēn-yäm-bōō'pä)	101	11°47'S	38°13'W
Inharrime, Moz. (ēn-yär-rē'mä)	186	24°17'S	35°07'E
Inhomirim, Braz. (ē-nô-mē-rē'ɴ)	102b	22°34'S	43°11'W
Inhul, r., Ukr.	133	47°22'N	32°52'E
Inhulets', r., Ukr.	133	47°12'N	33°12'E
Inírida, r., Col. (ē-nē-rē'dä)	100	2°25'N	70°38'W
Injune, Austl. (ĭn'jōn)	176	25°52'S	148°30'E
Inkeroinem, Fin. (ĭn'kĕr-oi-nĕn)	123	60°42'N	26°50'E
Inkster, Mi., U.S. (ĭngk'stēr)	69b	42°18'N	83°19'W
Inn, r., Eur. (ĭn)	117	48°00'N	12°00'E
Innamincka, Austl. (ĭnn-ȧ'mĭn-kȧ)	176	27°50'S	140°48'E
Inner Brass, i., V.I.U.S. (bräs)	87c	18°23'N	64°58'W
Inner Hebrides, is., Scot., U.K.	120	57°20'N	6°20'W
Inner Mongolia see Nei Monggol, prov., China	160	40°15'N	105°00'E
Innisfail, Can.	50	52°02'N	113°57'W
Innsbruck, Aus. (ĭns'brŏŏk)	117	47°15'N	11°25'E
Ino, Japan (ē'nō)	167	33°34'N	133°23'E
Inongo, D.R.C. (ē-nôŋ'gō)	186	1°57'S	18°16'E
Inowrocław, Pol. (ē-nô-vrōts'läf)	125	52°48'N	18°16'E
In Salah, Alg.	184	27°13'N	2°22'E
Inscription House Ruin, Az., U.S. (ĭn'skrĭp-shŭn hous rōō'ĭn)	77	36°45'N	110°47'W
International Falls, Mn., U.S. (ĭn'tēr-năsh'ŭn-ăl fôlz)	65	48°34'N	93°26'W
Inuvik, Can.	50	68°40'N	134°10'W
Inuyama, Japan (ē'nōō-yä'mä)	167	35°24'N	137°01'E
Invercargil, N.Z. (ĭn-vēr-kär'gĭl)	177	46°25'S	168°27'E
Inverell, Austl. (ĭn-vēr-el')	175	29°50'S	151°32'E
Invergrove Heights, Mn., U.S. (ĭn'vēr-grōv)	75g	44°51'N	93°01'W
Inverness, Can. (ĭn-vĕr-nĕs')	61	46°14'N	61°18'W
Inverness, Scot., U.K.	116	57°30'N	4°07'W
Inverness, Fl., U.S.	83	28°48'N	82°22'W
Investigator Strait, strt., Austl. (ĭn-vĕst'ĭ'gā-tôr)	176	35°33'S	137°00'E
Inyangani, mtn., Zimb. (ēn-yän-gä'nē)	186	18°06'S	32°37'E
Inyokern, Ca., U.S.	76	35°39'N	117°51'W
Inyo Mountains, mts., Ca., U.S. (ĭn'yō)	64	36°55'N	118°04'W
Inzer, r., Russia (ĭn'zĕr)	142a	54°24'N	57°17'E
Inzia, r., D.R.C.	190	5°55'S	17°50'E
Ioánnina, Grc. (yô-ä'nē-nȧ)	119	39°39'N	20°52'E
Ioco, Can.	74d	49°18'N	122°53'W
Iola, Ks., U.S. (ī-ō'lȧ)	79	37°55'N	95°23'W
Iôna, Parque Nacional do, rec., Ang.	190	16°35'S	12°00'E
Ionia, Mi., U.S. (ī-ō'nĭ-ȧ)	66	43°00'N	85°10'W
Ionian Islands, is., Grc. (ī-ō'nĭ-ăn)	119	39°10'N	20°05'E
Ionian Sea, sea, Eur.	112	38°59'N	18°48'E
Iori, r., Asia	138	41°03'N	46°17'E
Íos, i., Grc. (ī'ŏs)	131	36°48'N	25°25'E
Iowa, state, U.S. (ī'ô-wȧ)	65	42°05'N	94°20'W
Iowa, r., Ia., U.S.	71	41°55'N	92°20'W
Iowa City, Ia., U.S.	65	41°39'N	91°31'W
Iowa Falls, Ia., U.S.	71	42°32'N	93°16'W
Iowa Park, Tx., U.S.	78	33°57'N	98°39'W
Ipala, Tan.	191	4°30'S	32°53'E
Ipeirus, hist. reg., Grc.	131	39°35'N	20°45'E
Ipel', r., Eur. (ē'pĕl)	125	48°08'N	19°00'E
Ipiales, Col. (ē-pē-ä'lås)	100	0°48'N	77°45'W
Ipoh, Malay.	168	4°45'N	101°05'E
Ipswich, Austl. (ĭps'wĭch)	175	27°40'S	152°50'E
Ipswich, Eng., U.K.	117	52°05'N	1°05'E
Ipswich, Ma., U.S.	61a	42°41'N	70°50'W
Ipswich, S.D., U.S.	70	45°26'N	99°01'W
Ipu, Braz. (ē-pōō)	101	4°11'S	40°45'W
Iput', r., Eur. (ē-pòt')	137	52°53'N	31°57'E
Iqaluit, Can.	51	63°48'N	68°31'W
Iquique, Chile (ē-kē'kĕ)	100	20°16'S	70°07'W
Iquitos, Peru (ē-kē'tōs)	100	3°39'S	73°18'W
Irákleio, Grc.	110	35°20'N	25°10'E
Iran, nation, Asia (ē-rän')	154	31°15'N	53°30'E
Iran, Plateau of, plat., Iran	154	32°28'N	58°00'E
Iran Mountains, mts., Asia	168	2°30'N	114°30'E
Irapuato, Mex. (ē-rä-pwä'tō)	88	20°41'N	101°24'W
Iraq, nation, Asia (ē-räk')	154	32°00'N	42°30'E
Irazú, vol., C.R. (ē-rä-zōō')	91	9°58'N	83°54'W
Irbid, Jord. (ēr-bēd')	156	32°33'N	35°51'E
Irbit, Russia (ēr-bēt')	134	57°40'N	63°10'E
Irébou, D.R.C. (ē-rā'bōō)	186	0°40'S	17°48'E
Ireland, nation, Eur. (īr-lănd)	110	53°33'N	8°00'W
Iremel', Gora, mtn., Russia (gȧ-rä'ĭ-rĕ'mĕl)	142a	54°32'N	58°52'E
Irene, S. Afr. (ĭ-rē-nē)	187b	25°53'S	28°13'E
Irîgui, reg., Mali	188	16°45'N	5°35'W
Iriklinskoye Vodokhranilishche, res., Russia	137	52°20'N	58°50'E
Iringa, Tan. (ē-rĭŋ'gä)	186	7°46'S	35°42'E
Iriomote Jima, i., Japan (ērē'-ō-mō-tä)	161	24°20'N	123°30'E
Iriona, Hond. (ē-rē-ō'nä)	90	15°53'N	85°12'W
Irish Sea, sea, Eur. (ī'rĭsh)	112	53°55'N	5°25'W
Irkutsk, Russia (ĭr-kòtsk')	135	52°16'N	104°00'E
Irlam, Eng., U.K. (ûr'lăm)	114a	53°26'N	2°26'W
Irois, Cap des, c., Haiti	93	18°25'N	74°50'W
Iron Bottom Sound, strt., Sol. Is.	170e	9°15'S	160°00'E
Irondale, Al., U.S. (ī'ērn-dăl)	68h	33°32'N	86°43'W
Iron Gate, val., Eur.	131	44°43'N	22°32'E
Iron Knob, Austl. (ī-ăn nŏb)	176	32°47'S	137°10'E
Iron Mountain, Mi., U.S. (ī'ērn)	71	45°49'N	88°04'W
Iron River, Mi., U.S.	71	46°09'N	88°39'W
Ironton, Oh., U.S. (ī'ērn-tŭn)	66	38°30'N	82°45'W
Ironwood, Mi., U.S. (ī'ērn-wŏŏd)	71	46°28'N	90°10'W
Ironwood Forest National Monument, rec., Az., U.S.	77	32°30'N	111°25'W
Iroquois, r., Il., U.S. (ĭr'ô-kwoi)	66	40°55'N	87°20'W
Iroquois Falls, Can.	51	48°41'N	80°39'W
Irō-Saki, c., Japan (ē'rō sä'kē)	166	34°35'N	138°54'E
Irpin, r., Ukr.	133	50°13'N	29°55'E
Irrawaddy, r., Mya. (ĭr-ȧ-wäd'ē)	155	23°27'N	96°25'E
Irtysh, r., Asia (ĭr-tĭsh')	134	59°00'N	69°00'E
Irumu, D.R.C. (ē-rō'mōō)	185	1°30'N	29°52'E
Irun, Spain (ē-rōōn')	128	43°20'N	1°47'W
Irvine, Scot., U.K.	120	55°39'N	4°40'W
Irvine, Ca., U.S. (ûr'vĭn)	75a	33°40'N	117°45'W
Irvine, Ky., U.S.	66	37°40'N	84°00'W
Irving, Tx., U.S. (ûr'vēng)	75c	32°49'N	96°57'W
Irvington, N.J., U.S. (ûr'vēng-tŭn)	68a	40°43'N	74°15'W
Irwin, Pa., U.S. (ûr'wĭn)	69e	40°19'N	79°42'W
Is, Russia (ēs)	142a	58°48'N	59°44'E
Isa, Nig.	189	13°14'N	6°24'E
Isaacs, Mount, mtn., Pan. (ē-sä-ä'ks)	86a	9°22'N	79°31'W
Isabela, i., Ec. (ē-sä-bä'lä)	100	0°47'S	91°35'W
Isabela, Cabo, c., Dom. Rep. (kä'bô-ē-sä-bē'lä)	93	20°00'N	71°00'W
Isabella, Cordillera, mts., Nic. (kôr-dēl-yĕ'rä-ē-sä-bēlä)	90	13°20'N	85°37'W
Isabella Indian Reservation, I.R., Mi., U.S. (ĭs-ȧ-bĕl'-lȧ)	66	43°35'N	84°55'W
Isaccea, Rom. (ē-säk'chä)	133	45°16'N	28°26'E
Ísafjördur, Ice. (ēs'ä-fÿr-dòr)	116	66°09'N	22°39'W
Isangi, D.R.C. (ē-säŋ'gē)	160	0°46'N	24°15'E
Isar, r., Ger. (ē'zär)	117	48°30'N	12°30'E
Isarco, r., Italy (ē-sär'kō)	130	46°37'N	11°25'E
Isarog, Mount, mtn., Phil. (ē-sä-rō-g)	169a	13°40'N	123°23'E
Ischia, Italy (ēs'kyä)	129c	40°29'N	13°58'E
Ischia, Isola d', i., Italy (dē'sh-kyä)	118	40°26'N	13°55'E
Ise, Japan (ĭs'hē) (ŭ'gē-yä'mä'dȧ)	166	34°30'N	136°43'E
Iseo, Lago d', l., Italy (lä'-gō-dē-ē-zĕ'ō)	130	45°50'N	9°55'E
Isère, r., Fr. (ē-zâr')	117	45°15'N	5°15'E
Iserlohn, Ger. (ē'zĕr-lōn)	127c	51°22'N	7°42'E
Isernia, Italy (ē-zĕr'nyä)	130	41°35'N	14°14'E
Ise-Wan, b., Japan (ē'sĕ wän)	166	34°49'N	136°44'E
Iseyin, Nig.	184	7°58'N	3°36'E
Ishigaki, Japan	170d	24°20'N	124°09'E
Ishikari Wan, b., Japan (ē'shē-kä-rē wän)	166	43°30'N	141°05'E
Ishim, Russia (ĭsh-ĕm')	134	56°07'N	69°13'E
Ishim, r., Asia	134	53°17'N	67°45'E
Ishimbay, Russia (ē-shĕm-bī')	142a	53°28'N	56°02'E
Ishinomaki, Japan (ĭsh-nō-mä'kē)	161	38°22'N	141°22'E
Ishinomaki Wan, b., Japan (ē-shē-nō-mä'kē wän)	166	38°10'N	141°40'E
Ishly, Russia (ĭsh'lĭ)	142a	54°13'N	55°55'E
Ishlya, Russia (ĭsh'lyȧ)	142a	53°54'N	57°48'E
Ishmant, Egypt	192b	29°17'N	31°15'E
Ishpeming, Mi., U.S. (ĭsh'pē-mĭng)	71	46°28'N	87°42'W
Isipingo, S. Afr. (ĭs-ĭ-pĭng-gō)	187c	29°59'S	30°58'E
Isiro, D.R.C.	185	2°47'N	27°37'E
İskenderun, Tur. (ĭs-kĕn'dĕr-ōōn)	154	36°45'N	36°15'E
İskenderun Körfezi, b., Tur.	119	36°22'N	35°25'E
İskilip, Tur. (ēs'kĭ-lēp')	119	40°40'N	34°30'E

ăt; finăl; rāte; senāte; ärm; åsk; sofȧ; fâre; ch-choose; dh-as th in other; bē; ēvent; bĕt; recĕnt; cratēr; g-gō; gh-guttural g; bĭt; ĭ-short neutral; rīde; ɴ-guttural k as ch in German ich;

PLACE (Pronunciation)	PAGE	LAT.	LONG.
Iskūr, r., Blg. (ĭs'k'r)	131	43°05'N	23°37'E
Isla-Cristina, Spain (ī'lä-krē-stē'nä)	128	37°13'N	7°20'W
Islāmābād, Pak.	155	33°55'N	73°05'E
Isla Mujeres, Mex. (ē's-lä-mōō-kĕ'rĕs)	90a	21°25'N	86°53'W
Island Lake, l., Can.	53	53°47'N	94°25'W
Islands, Bay of, b., Can. (ī'lăndz)	61	49°10'N	58°15'W
Islay, i., Scot., U.K. (ī'lā)	116	55°55'N	6°35'W
Isle, r., Fr. (ēl)	126	45°02'N	0°29'E
Isle of Axholme, reg., Eng., U.K. (ăks'-hôm)	114a	53°33'N	0°48'W
Isle of Man, dep., Eur. (măn)	120	54°26'N	4°21'W
Isle Royale National Park, rec., Mi., U.S. (īl'roi-ăl')	65	47°57'N	88°37'W
Isleta, N.M., U.S. (ēs-lā'tȧ) (ī-lē'tȧ)	77	34°55'N	106°45'W
Isleta Indian Reservation, I.R., N.M., U.S.	77	34°55'N	106°45'W
Ismailia, Egypt (ēs-mā-ēl'ēȧ)	192b	30°35'N	32°17'E
Ismā'īlīyah Canal, can., Egypt	192b	30°25'N	31°45'E
Ismail Samani, pik, mtn., Taj.	139	38°57'N	72°01'E
Ismaning, Ger. (ēz'mä-nēng)	115d	48°14'N	11°41'E
Isparta, Tur. (ē-spär'tȧ)	154	37°50'N	30°40'E
Israel, nation, Asia	154	32°40'N	34°00'E
Issaquah, Wa., U.S. (ĭz'sȧ-kwăh)	74a	47°32'N	122°02'W
Isselburg, Ger. (ē'sĕl-bōōrg)	127c	51°50'N	6°28'E
Issoire, Fr. (ē-swȧr')	126	45°32'N	3°13'E
Issoudun, Fr. (ē-sōō-dăN')	126	46°56'N	2°00'E
Issum, Ger. (ē'sōōm)	127c	51°32'N	6°24'E
Issyk-Kul, Ozero, l., Kyrg.	139	42°13'N	76°12'E
İstanbul, Tur. (ē-stän-bōōl')	154	41°02'N	29°00'E
İstanbul Boğazı (Bosporus), strt., Tur.	154	41°10'N	29°10'E
Istiaía, Grc. (ĭs-tyī'yä)	131	38°58'N	23°11'E
Istmina, Col. (ēst-mē'nä)	100a	5°10'N	76°40'W
Istokpoga, Lake, l., Fl., U.S. (ĭs-tŏk-pō'gȧ)	83a	27°20'N	81°33'W
Istra, pen., Serb. (ē-strä)	130	45°18'N	13°48'E
Istranca Dağlari, mts., Eur. (ī-strän'jȧ)	131	41°50'N	27°25'E
Istres, Fr. (ēs'tr')	126a	43°30'N	5°00'E
Itabaiana, Braz. (ē-tä-bä-yȧ-nä)	101	10°42'S	37°17'W
Itabapoana, Braz. (ē-tä'-bä-pôä'nä)	99a	21°19'S	40°58'W
Itabapoana, r., Braz.	99a	21°11'S	41°18'W
Itabirito, Braz. (ē-tä-bē-rē'tô)	99a	20°15'S	43°46'W
Itabuna, Braz. (ē-tä-bōō'nȧ)	101	14°47'S	39°17'W
Itacoara, Braz. (ē-tä-kô'ä-rä)	99a	21°41'S	42°04'W
Itacoatiara, Braz. (ē-tȧ-kwä-tyä'rȧ)	101	3°03'S	58°18'W
Itagüí, Col. (ē-tä'gwê)	100a	6°11'N	75°36'W
Itagui, r., Braz.	102b	22°53'S	43°43'W
Itaipava, Braz. (ē-tī-pá'-vä)	102b	22°23'S	43°09'W
Itaipu, Braz. (ē-tī'pōō)	102b	22°58'S	43°02'W
Itaituba, Braz. (ē-tä'ī-tōō'bȧ)	101	4°12'S	56°00'W
Itajái, Braz. (ē-tä-zhī')	102	26°52'S	48°39'W
Italy, Tx., U.S.	81	32°11'N	96°51'W
Italy, nation, Eur. (ĭt'ȧ-lē)	110	43°58'N	11°14'E
Itambi, Braz. (ē-tä'm-bē)	102b	22°44'S	42°57'W
Itami, Japan (ē'tä'mē')	167b	34°47'N	135°25'E
Itapecerica, Braz. (ē-tä-pĕ-sĕ-rē'kȧ)	99a	20°29'S	45°08'W
Itapecuru-Mirim, Braz. (ē-tä-pĕ'kōō-rōō-mē-rēN')	101	3°17'S	44°15'W
Itaperuna, Braz. (ē-tȧ'pȧ-rōō'nä)	101	21°12'S	41°53'W
Itapetininga, Braz. (ē-tä-pē-tē-nē'N-gä)	101	23°37'S	48°03'W
Itapira, Braz. (ē-tȧ-pē'rȧ)	101	20°42'S	51°19'W
Itapira, Braz.	99a	22°27'S	46°47'W
Itarsi, India	158	22°43'N	77°45'E
Itasca, Tx., U.S. (ī-tăs'kȧ)	81	32°09'N	97°08'W
Itasca, l., Mn., U.S.	70	47°13'N	95°14'W
Itatiaia, Pico da, mtn., Braz. (pē'-kô-dȧ-ē-tä-tyȧ'ēä)	101	22°18'S	44°41'W
Itatiba, Braz. (ē-tä-tē'bä)	99a	23°01'S	46°48'W
Itaúna, Braz. (ē-tä-ōō'nä)	99a	20°05'S	44°35'W
Ithaca, Mi., U.S. (ĭth'ȧ-kȧ)	66	43°20'N	84°35'W
Ithaca, N.Y., U.S.	65	42°25'N	76°30'W
Itháka, i., Grc. (ē'thä-kē)	131	38°27'N	20°48'E
Itigi, Tan.	191	5°42'S	34°29'E
Itimbiri, r., D.R.C.	190	2°40'N	23°30'E
Itoko, D.R.C. (ē-tō'kō)	186	1°13'S	22°07'E
Itu, Braz. (ē-tōō')	99a	23°16'S	47°16'W
Ituango, Col. (ē-twäŋ'gō)	100	7°07'N	75°44'W
Ituiutaba, Braz. (ē-tōō-ēōō-tä'bä)	101	18°56'S	49°17'W
Itumirim, Braz. (ē-tōō-mē-rē'N)	99a	21°20'S	44°51'W
Itundujia Santa Cruz, Mex. (ē-tōōn-dōō-hē'ȧ sä'n-tä krōō'z)	89	16°50'N	97°43'W
Iturbide, Mex. (ē'tōōr-bē'dhȧ)	90a	19°38'N	89°31'W
Iturup, i., Russia (ē-tōō-rōōp')	141	45°35'N	147°15'E
Ituzaingo, Arg. (ē-tōō-zä-ē'n-gō)	102a	34°40'S	58°40'W
Itzehoe, Ger. (ē'tzĕ-hō)	124	53°55'N	9°31'E
Iuka, Ms., U.S. (ī-ū'kȧ)	82	34°47'N	88°10'W
Iúna, Braz. (ē-ōō'-nä)	99a	20°22'S	41°32'W
Ivanhoe, Austl. (ĭv'ăn-hō)	176	32°53'S	144°10'E
Ivanivka, Ukr.	132	46°43'N	34°33'E
Ivano-Frankivs'k, Ukr.	137	48°53'N	24°46'E
Ivanopil', Ukr.	133	49°51'N	28°11'E
Ivanovo, Russia (ē-vä'nō-vō)	132	57°02'N	41°54'E
Ivanovo, prov., Russia	132	56°55'N	40°30'E
Ivanteyevka, Russia (ē-vȧn-tyĕ'yĕf-kȧ)	142b	55°58'N	37°56'E
Ivdel', Russia (ĭv'dyĕl)	142a	60°42'N	60°27'E
Iviza *see* Eivissa, i., Spain	112	38°55'N	1°24'E
Ivohibé, Madag. (ē-vô-hē-bā')	187	22°28'S	46°59'E
Ivory Coast *see* Cote d'Ivoire, nation, Afr.	184	7°43'N	6°30'W
Ivrea, Italy (ē-vrĕ'ä)	118	45°25'N	7°54'E
Ivry-sur-Seine, Fr.	127b	48°49'N	2°23'E
Ivujivik, Can.	51	62°17'N	77°52'W
Ivvavik National Park, rec., Can.	63	69°10'N	139°30'W
Iwaki, Japan	166	37°03'N	140°57'E
Iwate Yama, mtn., Japan (ē-wä-tē-yä'mä)	166	39°50'N	140°56'E
Iwatsuki, Japan	167a	35°48'N	139°43'E
Iwaya, Japan (ē'wȧ-yȧ)	167b	34°35'N	135°01'E
Iwo, Nig.	184	7°38'N	4°11'E
Ixcateopán, Mex. (ēs-kä-tä-ō-pän')	88	18°29'N	99°49'W
Ixelles, Bel.	115a	50°49'N	4°23'E
Ixhautlán, Mex. (ēs-wȧt-län')	88	20°41'N	98°01'W
Ixhuatán, Mex. (ēs-hwä-tän')	89	16°19'N	94°30'W
Ixmiquilpan, Mex. (ēs-mē-kēl'pän)	88	20°30'N	99°12'W
Ixopo, S. Afr.	187c	30°10'S	30°04'E
Ixtacalco, Mex. (ēs-tȧ-käl'kō)	89a	19°23'N	99°07'W
Ixtaltepec, Mex. (ēs-täl-tĕ-pĕk')	89	16°33'N	95°04'W
Ixtapalapa, Mex. (ēs'tä-pä-lä'pä)	89a	19°21'N	99°06'W
Ixtapaluca, Mex. (ēs'tä-pä-lōō'kä)	89a	19°18'N	98°53'W
Ixtepec, Mex. (ēks-tĕ'pĕk)	89	16°37'N	95°09'W
Ixtlahuaca, Mex. (ēs-tlä-wä'kä)	88	19°34'N	99°46'W
Ixtlán de Juárez, Mex. (ēs-tlän' dā hwä'rȧz)	89	17°20'N	96°29'W
Ixtlán del Río, Mex. (ēs-tlän'dĕl rē'ō)	88	21°05'N	104°22'W
Iya, r., Russia	140	53°45'N	99°30'E
Iyo-Nada, b., Japan (ē'yō nä-dä)	167	33°33'N	132°07'E
Izabal, Guat. (ē'zä-bäl')	90	15°23'N	89°10'W
Izabal, Lago, l., Guat.	90	15°30'N	89°04'W
Izalco, El Sal. (ē-zäl'kō)	90	13°50'N	89°40'W
Izamal, Mex. (ē-zä-mä'l)	90a	20°55'N	89°00'W
Izberbash, Russia	138	42°33'N	47°52'E
Izhevsk, Russia (ē-zhyĕfsk')	134	56°50'N	53°15'E
Izhma, Russia (ĭzh'mȧ)	136	65°00'N	54°05'E
Izhma, r., Russia	136	64°00'N	53°00'E
Izhora, r., Russia (ēz'hô-rȧ)	142c	59°36'N	30°20'E
Izmaïl, Ukr.	137	45°00'N	28°49'E
İzmir, Tur. (ĭz-mēr')	154	38°25'N	27°05'E
İzmit, Tur. (ĭz-mēt')	119	40°45'N	29°45'E
Iznajar, Embalse de, res., Spain	128	37°15'N	4°30'W
Iztaccíhuatl, mtn., Mex.	88	19°10'N	98°38'W
Izuhara, Japan (ē'zōō-hä'rä)	167	34°11'N	129°18'E
Izumi-Ōtsu, Japan (ē'zōō-mōō ō'tsōō)	167b	34°30'N	135°24'E
Izumo, Japan (ē'zōō-mō)	167	35°22'N	132°45'E
Izu Shichitō, is., Japan	161	34°32'N	139°25'E

J

PLACE (Pronunciation)	PAGE	LAT.	LONG.
Jabal, Bahr al, r., Sudan	185	7°30'N	31°00'E
Jabalpur, India	155	23°18'N	79°59'E
Jablonec nad Nisou, Czech Rep. (yäb'lô-nyĕts)	124	50°43'N	15°12'E
Jablunkov Pass, p., Eur. (yäb'lōn-kôf)	125	49°31'N	18°35'E
Jaboatão, Braz. (zhä-bô-ȧ-touN)	101	8°14'S	35°08'W
Jaca, Spain (hä'kä)	129	42°35'N	0°30'W
Jacala, Mex. (hä-kȧ'lä)	88	21°01'N	99°11'W
Jacaltenango, Guat. (hä-kȧl-tĕ-nȧn'gō)	90	15°39'N	91°41'W
Jacarézinho, Braz. (zhä-kä-rĕ'zē-nyō)	101	23°13'S	49°58'W
Jachymov, Czech Rep. (yä'chĭ-môf)	124	50°22'N	12°51'E
Jacinto City, Tx., U.S. (hä-sēn'tō) (jȧ-sĭn'tō)	81a	29°45'N	95°14'W
Jacksboro, Tx., U.S. (jăks'bŭr-ō)	78	33°13'N	98°11'W
Jackson, Al., U.S. (jăk'sŭn)	82	31°31'N	87°52'W
Jackson, Ca., U.S.	76	38°22'N	120°47'W
Jackson, Ga., U.S.	82	33°19'N	83°55'W
Jackson, Ky., U.S.	82	37°32'N	83°17'W
Jackson, La., U.S.	81	30°50'N	91°13'W
Jackson, Mi., U.S.	65	42°15'N	84°25'W
Jackson, Mn., U.S.	70	43°37'N	95°00'W
Jackson, Mo., U.S.	79	37°23'N	89°40'W
Jackson, Ms., U.S.	65	32°17'N	90°10'W
Jackson, Oh., U.S.	66	39°00'N	82°40'W
Jackson, Tn., U.S.	65	35°37'N	88°49'W
Jackson, Port, b., Austl.	173b	33°50'S	151°18'E
Jackson Lake, l., Wy., U.S.	73	43°57'N	110°28'W
Jacksonville, Al., U.S. (jăk'sŭn-vĭl)	82	33°52'N	85°45'W
Jacksonville, Fl., U.S.	65	30°20'N	81°40'W
Jacksonville, Il., U.S.	65	39°43'N	90°12'W
Jacksonville, Tx., U.S.	81	31°58'N	95°18'W
Jacksonville Beach, Fl., U.S.	83	31°18'N	81°25'W
Jacmel, Haiti (zhȧk-mĕl')	93	18°15'N	72°30'W
Jaco, l., Mex. (hä'kō)	80	27°51'N	103°50'W
Jacobābad, Pak.	158	28°22'N	68°30'E
Jacobina, Braz. (zhä-kō-bē'nȧ)	101	11°13'S	40°30'W
Jacques-Cartier, r., Can.	62b	47°04'N	71°28'W
Jacques Cartier, Détroit de, strt., Can.	60	50°07'S	63°58'W
Jacques-Cartier, Mont, mtn., Can.	60	48°59'N	66°00'W
Jacquet River, Can. (zhȧ-kĕ') (jăk'ĕt)	60	47°55'N	66°00'W
Jacutinga, Braz. (zhȧ-kōō-tēn'gä)	99a	22°17'S	46°36'W
Jadebusen, b., Ger.	124	53°28'N	8°17'E
Jadotville *see* Likasi, D.R.C.	186	10°59'S	26°44'E
Jaén, Peru (Kä-ĕ'n)	100	5°38'S	78°49'W
Jaen, Spain	118	37°45'N	3°48'W
Jaffa, Cape, c., Austl. (jăf'ȧ)	174	36°58'S	139°29'E
Jaffna, Sri L. (jăf'nȧ)	159	9°44'N	80°09'E
Jagüey Grande, Cuba (hä'gwā grän'dā)	92	22°35'N	81°05'W
Jahore Strait, strt., Asia	153b	1°22'N	103°37'E
Jahrom, Iran	154	28°30'N	53°28'E
Jaibo, r., Cuba (hä-ē'bō)	93	20°10'N	75°20'W
Jaipur, India	155	27°00'N	75°50'E
Jaisalmer, India	158	27°00'N	70°54'E
Jajce, Bos. (yī'tsĕ)	131	44°20'N	17°19'E
Jajpur, India	155	20°49'N	86°37'E
Jakarta, Indon. (yä-kär'tä)	168	6°17'S	106°45'E
Jakobstad, Fin. (yȧ'kōb-städh)	116	63°33'N	22°31'E
Jalacingo, Mex. (hä-lä-sĭŋ'gō)	89	19°47'N	97°16'W
Jalālābād, Afg. (jŭ-lä-lȧ-bäd)	155a	34°25'N	70°27'E
Jalālah al Baḩrīyah, Jabal, mts., Egypt	192b	29°20'N	32°00'E
Jalapa, Guat. (hä-lä'pȧ)	90	14°38'N	89°58'W
Jalapa de Díaz, Mex.	89	18°06'N	96°33'W
Jalapa del Marqués, Mex. (dĕl mär-kās')	89	16°30'N	95°29'W
Jaleswar, Nepal	158	26°50'N	85°55'E
Jalgaon, India	158	21°08'N	75°33'E
Jalisco, Mex. (hä-lēs'kō)	88	21°27'N	104°54'W
Jalisco, state, Mex.	86	20°07'N	104°45'W
Jalón, r., Spain (hä-lōn')	128	41°22'N	1°46'W
Jalostotitlán, Mex. (hä-lōs-tē-tlȧn')	88	21°09'N	102°30'W
Jalpa, Mex. (häl'pä)	89	18°12'N	93°06'W
Jalpa, Mex. (häl'pä)	88	21°40'N	103°04'W
Jalpan, Mex. (häl'pän)	88	21°13'N	99°31'W
Jaltepec, Mex. (häl-tĕ-pĕk')	89	17°20'N	95°15'W
Jaltipan, Mex. (häl-tȧ-pän')	89	17°59'N	94°42'W
Jaltocan, Mex. (häl-tô-kän')	88	21°08'N	98°32'W
Jamaare, r., Nig.	189	11°50'N	10°10'E
Jamaica, nation, N.A.	87	17°45'N	78°00'W
Jamaica Cay, i., Bah.	93	22°45'N	75°55'W
Jamālpur, Bngl.	158	24°56'N	89°58'E
Jamay, Mex. (hä-mī')	88	20°16'N	102°43'W
Jambi, Indon. (mäm'bē)	168	1°45'S	103°28'E
James, r., Mo., U.S.	79	36°51'N	93°22'W
James, r., Va., U.S.	65	37°35'N	77°50'W
James, r., U.S.	64	46°25'N	98°55'W
James, Lake, res., N.C., U.S.	83	36°07'N	81°48'W
James Bay, b., Can. (jāmz)	53	53°53'N	80°40'W
Jamesburg, N.J., U.S. (jāmz'bûrg)	68a	40°21'N	74°26'W
James Point, c., Bah.	92	25°20'N	76°30'W
James Range, mts., Austl.	174	24°15'S	133°30'E
James Ross, i., Ant.	97	64°20'S	58°20'W
Jamestown, S. Afr.	187c	31°07'S	26°49'E
Jamestown, N.D., U.S.	64	46°54'N	98°42'W
Jamestown, N.Y., U.S. (jāmz'toun)	65	42°05'N	79°15'W
Jamestown, R.I., U.S.	68b	41°30'N	71°21'W
Jamestown Reservoir, res., N.D., U.S.	70	47°16'N	98°40'W
Jamiltepec, Mex. (hä-mĕl-tȧ-pĕk)	89	16°16'N	97°54'W
Jammerbugten, b., Den.	122	57°20'N	9°28'E
Jammu, India	155	32°50'N	74°52'E
Jammu and Kashmīr, state, India (kȧsh-mēr')	155	34°30'N	76°00'E
Jammu and Kashmīr, hist. reg., Asia (kȧsh-mēr')	155	39°10'N	75°05'E
Jāmnagar, India (jäm-nŭ'gŭr)	155	22°33'N	70°03'E
Jamshedpur, India (jäm'shād-pōōr)	155	22°52'N	86°11'E
Jándula, r., Spain (hän'dōō-lȧ)	128	38°28'N	3°52'W
Janesville, Wi., U.S. (jānz'vĭl)	71	42°41'N	89°03'W
Janin, W.B.	153a	32°27'N	35°19'E
Jan Mayen, i., Nor. (yän mī'ĕn)	116	70°59'N	8°05'W
Jánoshalma, Hung. (yä'nôsh-hôl-mô)	125	46°17'N	19°18'E
Janów Lubelski, Pol. (yä'nōōf lŭ-bĕl'skĭ)	125	50°40'N	22°25'E
Januária, Braz. (zhä-nwä'rē-ä)	101	15°31'S	44°17'W
Japan, nation, Asia (jȧ-păn')	161	36°30'N	133°30'E
Japan, Sea of, sea, Asia (jȧ-păn')	161	40°08'N	132°55'E
Japeri, Braz. (zhȧ-pĕ'rē)	102b	22°38'S	43°40'W
Japurá (Caquetá), r., S.A.	100	2°00'S	68°00'W
Jarabacoa, Dom. Rep. (Kä-rä-bä-kô'ä)	93	19°05'N	70°40'W
Jaral del Progreso, Mex. (hä-räl dĕl prô-grā'sō)	88	20°21'N	101°05'W
Jarama, r., Spain (hä-rä'mä)	128	40°33'N	3°30'W
Jarash, Jord.	153a	32°17'N	35°53'E
Jardines, Banco, bk., Cuba (bä'n-kô-här-dē'nȧs)	92	21°45'N	81°40'W
Jargalant, Mong.	164	46°28'N	115°10'E
Jari, r., Braz. (zhä-rē)	101	0°28'N	53°00'W
Jarocin, Pol. (yä-rō'tsyēn)	125	51°58'N	17°31'E
Jarosław, Pol. (yȧ-rôs-wȧf)	117	50°01'N	22°41'E
Jarud Qi, China (jya-lōō-tŭ shyē)	161	44°35'N	120°40'E
Jasin, Malay.	153b	2°19'N	102°26'E
Jašiūnai, Lith. (dzä-shōō-nä'yē)	123	54°27'N	25°25'E
Jāsk, Iran (jäsk)	154	25°46'N	57°48'E
Jasło, Pol. (yȧs'wō)	125	49°44'N	21°28'E
Jason Bay, b., Malay.	153b	1°53'N	104°14'E
Jasonville, In., U.S. (jā'sŭn-vĭl)	66	39°10'N	87°15'W
Jasper, Can.	50	52°53'N	118°05'W
Jasper, Al., U.S. (jăs'pēr)	82	33°50'N	87°17'W
Jasper, Fl., U.S.	83	30°30'N	82°56'W
Jasper, In., U.S.	66	38°20'N	86°55'W
Jasper, Mn., U.S.	70	43°51'N	96°22'W
Jasper, Tx., U.S.	81	30°55'N	93°59'W
Jasper National Park, rec., Can.	52	53°09'N	117°45'W
Jászapáti, Hung. (yäs'ô-pä-tē)	125	47°29'N	20°10'E
Jászberény, Hung.	125	47°30'N	19°56'E
Jatibonico, Cuba (hä-tē-bō-nē'kō)	92	22°00'N	79°15'W
Jauja, Peru (Kä-ō'K)	100	11°43'S	75°32'W
Jaumave, Mex. (hou-mä'vȧ)	88	23°23'N	99°24'W
Jaunjelgava, Lat. (youn'yĕl'gȧ-vȧ)	136	56°37'N	25°06'E
Java (Jawa), i., Indon.	168	8°35'S	111°11'E
Javari, r., S.A. (Kȧ-vä-rē)	100	4°25'S	72°07'W
Java Trench, deep,	168	9°45'S	107°30'E
Jawa, Laut (Java Sea), sea, Indon.	168	5°10'S	110°30'E
Jawor, Pol. (yä'vôr)	124	51°04'N	16°12'E
Jaworzno, Pol. (yä-vôzh'nô)	125	50°11'N	19°18'E
Jaya, Puncak, mtn., Indon.	169	4°00'S	137°00'E
Jayapura, Indon.	168	2°30'S	140°45'W
Jayb, Wādī al (Ha'Arava), val., Asia	153a	30°33'N	35°10'E
Jazzīn, Leb.	153a	33°34'N	35°37'E
Jeanerette, La., U.S. (jĕn-ĕr-et') (zhän-rĕt')	81	29°54'N	91°41'W
Jebba, Nig. (jĕb'ȧ)	184	9°07'N	4°46'E
Jeddore Lake, l., Can.	61	48°07'N	55°35'W
Jędrzejów, Pol. (yǎN-dzhā'yôf)	125	50°38'N	20°18'E
Jefferson, Ga., U.S. (jĕf'ĕr-sŭn)	82	34°05'N	83°35'W
Jefferson, Ia., U.S.	71	42°10'N	94°22'W

ng-sing; ŋ-baŋk; N-nasalized n; nŏd; cŏmmit; ōld; ôbey; ôrder; oi-boil; fōōd; ŏ-as oo in foot; ou-out; s-soft; sh-dish; th-thin; pūre; ûnite; ûrn; stŭd; circŭs; ü-as in French tu; '-indeterminate vowel.

260 Jefferson – Jylland

PRONOUNCING INDEX

PLACE (Pronunciation)	PAGE	LAT.	LONG.
Jefferson, La., U.S.	68d	29°57'N	90°04'W
Jefferson, Tx., U.S.	81	32°47'N	94°21'W
Jefferson, Wi., U.S.	71	42°59'N	88°45'W
Jefferson, r., Mt., U.S.	73	45°37'N	112°22'W
Jefferson, Mount, mtn., Or., U.S.	72	44°41'N	121°50'W
Jefferson City, Mo., U.S.	65	38°34'N	92°10'W
Jeffersontown, Ky., U.S. (jĕf'ẽr-sŭn-toun)	69h	38°11'N	85°34'W
Jeffersonville, In., U.S. (jĕf'ẽr-sŭn-vĭl)	69h	38°17'N	85°44'W
Jega, Nig.	189	12°15'N	4°23'E
Jehol, hist. reg., China (jĕ-hŏl)	161	42°31'N	118°12'E
Jēkabpils, Lat. (yĕk'ȧb-pĭls)	136	56°29'N	25°50'E
Jelenia Góra, Pol. (yĕ-lĕn'yȧ gŏ'rȧ)	124	50°53'N	15°43'E
Jelgava, Lat.	123	56°39'N	23°42'E
Jellico, Tn., U.S. (jĕl'ĭ-kō)	82	36°34'N	84°06'W
Jemez Indian Reservation, I.R., N.M., U.S.	77	35°35'N	106°45'W
Jena, Ger. (yā'nä)	117	50°55'N	11°37'E
Jenkins, Ky., U.S. (jĕŋ'kĭnz)	83	37°09'N	82°38'W
Jenkintown, Pa., U.S. (jĕŋ'kĭn-toun)	68f	40°06'N	75°08'W
Jennings, La., U.S. (jĕn'ĭngz)	81	30°14'N	92°40'W
Jennings, Mi., U.S.	66	44°20'N	85°20'W
Jennings, Mo., U.S.	75e	38°43'N	90°16'W
Jequitinhonha, r., Braz. (zhĕ-kē-tēŋ-ō'n-yä)	101	16°47'S	41°19'W
Jérémie, Haiti (zhā-rā-mē')	93	18°40'N	74°10'W
Jeremoabo, Braz. (zhĕ-rā-mō-ȧ'bō)	101	10°03'S	38°13'W
Jerez, Punta, c., Mex. (pōō'n-tä-κĕ-räz')	89	23°04'N	97°44'W
Jerez de la Frontera, Spain	118	36°42'N	6°09'W
Jerez de los Caballeros, Spain	128	38°20'N	6°45'W
Jericho, Austl. (jĕr'ĭ-kō)	175	23°38'S	146°24'E
Jericho, S. Afr. (jĕr-ĭkō)	192c	25°16'N	27°47'E
Jericho *see* Arīḥā, W.B.	153a	31°51'N	35°28'E
Jerome, Az., U.S. (jĕ-rōm')	64	34°45'N	112°10'W
Jerome, Id., U.S.	73	42°44'N	114°31'W
Jersey, dep., Eur.	126	49°15'N	2°10'W
Jersey, i., Jersey (jûr'zĭ)	117	49°13'N	2°07'W
Jersey City, N.J., U.S.	65	40°43'N	74°05'W
Jersey Shore, Pa., U.S.	67	41°10'N	77°15'W
Jerseyville, Il., U.S. (jẽr'zē-vĭl)	79	39°07'N	90°18'W
Jerusalem, Isr. (jē-rōō'sȧ-lĕm)	154	31°46'N	35°14'E
Jesup, Ga., U.S. (jĕs'ŭp)	83	31°36'N	81°53'W
Jesús Carranza, Mex. (hĕ-sōō's-kär-rä'n-zä)	89	17°26'N	95°01'W
Jewel, Or., U.S. (jū'ĕl)	74c	45°56'N	123°30'W
Jewel Cave National Monument, rec., S.D., U.S.	70	43°44'N	103°52'W
Jhālawār, India	155	24°30'N	76°00'E
Jhang Maghiāna, Pak.	158	31°21'N	72°19'E
Jhānsi, India (jän'sē)	155	25°29'N	78°32'E
Jharkhand, state, India	155	23°30'N	85°00'E
Jhārsuguda, India	158	22°51'N	84°13'E
Jhelum, Pak.	155	32°59'N	73°43'E
Jhelum, r., Asia (jā'lŭm)	155	31°40'N	71°51'E
Jiading, China (jyä-dĭŋ)	162	31°23'N	121°15'E
Jialing, r., China (jyä-lĭŋ)	160	32°30'N	105°30'E
Jiamusi, China	166	46°50'N	130°21'E
Ji'an, China (jyē-än)	161	27°15'N	115°10'E
Ji'an, China	164	41°00'N	126°04'E
Jianchangying, China (jyĕn-chäŋ-yĭŋ)	162	40°09'N	118°47'E
Jiangcun, China (jyäŋ-tsȯn)	163a	23°16'N	113°14'E
Jiangling, China (jyäŋ-lĭŋ)	161	30°30'N	112°10'E
Jiangshanzhen, China (jyäŋ-shän-jŭn)	162	36°39'N	120°31'E
Jiangsu, prov., China (jyäŋ-sōō)	161	33°45'N	120°30'E
Jiangwan, China (jyäŋ-wän)	163b	31°18'N	121°29'E
Jiangxi, prov., China (jyäŋ-shyē)	161	28°15'N	116°00'E
Jiangyin, China (jyäŋ-yĭn)	165	31°54'N	120°15'E
Jianli, China (jyĕn-lē)	165	29°50'N	112°52'E
Jianning, China (jyĕn-nĭŋ)	165	26°50'N	116°55'E
Jian'ou, China (jyĕn-ō)	165	27°10'N	118°18'E
Jianshi, China (jyĕn-shr)	165	30°40'N	109°45'E
Jiaohe, China	162	38°03'N	116°18'E
Jiaohe, China (jyou-hŭ)	164	43°40'N	127°20'E
Jiaoxian, China (jyou shyĕn)	161	36°18'N	120°01'E
Jiaozuo, China (jyou-dzwȯ)	162	35°15'N	113°18'E
Jiashan, China (jyä-shän)	162	32°41'N	118°00'E
Jiaxing, China (jyä-shyĭŋ)	161	30°45'N	120°50'E
Jiayu, China (jyä-yōō)	165	30°00'N	114°00'E
Jiazhou Wan, b., China (jyä-jō wän)	161	36°10'N	119°55'E
Jicarilla Apache Indian Reservation, I.R., N.M., U.S. (κē-kȧ-rēl'yȧ)	77	36°45'N	107°00'W
Jicarón, Isla, i., Pan. (κē-kä-rōn')	91	7°14'N	81°41'W
Jiddah, Sau. Ar.	154	21°30'N	39°15'E
Jieshou, China	162	33°17'N	115°20'E
Jieyang, China (jyē-yäŋ)	161	23°38'N	116°20'E
Jiggalong, Austl. (jĭg'ȧ-lông)	174	23°20'S	120°45'E
Jiguani, Cuba (κē-gwä-nē')	92	20°20'N	76°30'W
Jigüey, Bahía, b., Cuba (bä-ē'ä-κē'gwā)	92	22°15'N	78°10'W
Jihlava, Czech Rep. (yē'hlȧ-vȧ)	117	49°23'N	15°33'E
Jijel, Alg.	117	36°49'N	5°47'E
Jijia, r., Rom.	125	47°35'N	27°02'E
Jijiashi, China (jyē-jyä-shr)	162	32°10'N	120°17'E
Jijiga, Eth.	192a	9°15'N	42°48'E
Jilin, China (jyē-lĭn)	161	43°58'N	126°40'E
Jilin, prov., China	161	44°20'N	124°50'E
Jiloca, r., Spain (κē-lō'kȧ)	128	41°13'N	1°30'W
Jilotepeque, Guat. (κē-lō-tĕ-pĕ'kĕ)	90	14°39'N	89°36'W
Jima, Eth.	185	7°41'N	36°52'E
Jimbolia, Rom. (zhĭm-bô'lyä)	131	45°45'N	20°44'E
Jiménez, Mex. (κē-mä'nāz)	88	24°12'N	98°29'W
Jiménez, Mex.	80	29°03'N	100°42'W
Jiménez, Mex.	80	27°09'N	104°55'W
Jiménez del Téul, Mex. (tĕ-ōō'l)	88	21°28'N	103°51'W
Jimo, China (jyē-mwo)	164	36°22'N	120°28'E
Jim Thorpe, Pa., U.S. (jĭm' thôrp')	67	40°50'N	75°45'W
Jinan, China (jyē-nän)	161	36°40'N	117°01'E
Jincheng, China (jyĭn-chŭŋ)	164	35°30'N	112°50'E
Jindřichův Hradec, Czech Rep. (yĕn'd'r-zhĭ-kōōf hrä'dĕts)	124	49°09'N	15°02'E
Jing, r., China (jyĭŋ)	164	34°40'N	108°20'E
Jing'anji, China (jyĭŋ-än-jē)	162	34°30'N	116°55'E
Jingdezhen, China (jyĭŋ-dŭ-jŭn)	165	29°18'N	117°18'E
Jingjiang, China (jyĭŋ-jyäŋ)	162	32°02'N	120°15'E
Jingning, China (jyĭŋ-nĭŋ)	164	35°28'N	105°50'E
Jingpo Hu, l., China (jyĭŋ-pwo hōō)	164	44°10'N	129°00'E
Jingxian, China (jyĭŋ shyĕn)	165	26°32'N	109°45'E
Jingxian, China	162	37°43'N	116°17'E
Jingxing, China (jyĭŋ-shyĭŋ)	164	47°00'N	123°00'E
Jingzhi, China (jyĭŋ-jr)	162	36°19'N	119°23'E
Jinhua, China (jyĭn-hwä)	161	29°10'N	119°42'E
Jining, China (jyē-nĭŋ)	161	35°26'N	116°34'E
Jining, China	164	41°00'N	113°10'E
Jinja, Ug. (jĭn'jä)	185	0°26'N	33°12'E
Jinotega, Nic. (κē-nŏ-tä'gä)	90	13°07'N	86°00'W
Jinotepe, Nic. (κē-nŏ-tā'pā)	90	11°52'N	86°12'W
Jinqiao, China (jyĭn-chyou)	162	31°46'N	116°46'E
Jinshan, China (jyĭn-shän)	163b	30°53'N	121°09'E
Jinta, China (jyĭn-tä)	160	40°11'N	98°45'E
Jintan, China (jyĭn-tän)	162	31°47'N	119°34'E
Jin Xian, China (jyĭn shyĕn)	162	39°04'N	121°40'E
Jinxiang, China (jyĭn-shyäŋ)	162	35°03'N	116°20'E
Jinyun, China (jyĭn-yȯn)	165	28°40'N	120°08'E
Jinzhai, China (jyĭn-jī)	162	31°41'N	115°51'E
Jinzhou, China (jyĭn-jō)	161	41°00'N	121°00'E
Jinzhou Wan, b., China (jyĭn-jō wän)	162	39°07'N	121°17'E
Jinzū-Gawa, r., Japan (jēn'zōō gä'wä)	167	36°26'N	137°18'E
Jipijapa, Ec. (κē-pē-hä'pä)	100	1°36'S	80°52'W
Jiquilisco, El Sal. (κē-kē-lē's-kô)	90	13°18'N	88°32'W
Jiquilpan de Juárez, Mex. (κē-kēl'pän dā hwä'räz)	88	20°00'N	102°43'W
Jiquipilco, Mex. (hē-kē-pē'l-kō)	89a	19°32'N	99°37'W
Jitotol, Mex. (κē-tō-tōl')	89	17°03'N	92°54'W
Jiu, r., Rom.	131	44°45'N	23°17'E
Jiujiang, China (jyô-jyän)	163a	22°50'N	113°02'E
Jiujiang, China	161	29°43'N	116°00'E
Jiuquan, China (jyô-chyän)	160	39°46'N	98°26'E
Jiurongcheng, China (jyô-rôŋ-chŭŋ)	162	37°23'N	122°31'E
Jiushouzhang, China (jyô-shō-jäŋ)	162	35°59'N	115°52'E
Jiuwuqing, China (jyô-wōō-chyĭŋ)	164a	32°31'N	116°51'E
Jiuyongnian, China (jyô-yôŋ-nĭĕn)	162	36°41'N	114°46'E
Jixian, China (jyē shyĕn)	162	35°25'N	114°03'E
Jixian, China	162	37°37'N	115°33'E
Jixian, China	162	40°03'N	117°25'E
Jiyun, r., China (jyē-yōōm)	162	39°35'N	117°34'E
Joachimsthal, Ger.	115b	52°58'N	13°45'E
João Pessoa, Braz.	101	7°09'S	34°45'W
João Ribeiro, Braz. (zhô-uɴ-rē-bā'rō)	99a	20°42'S	44°03'W
Jobabo, r., Cuba (hō-bä'bä)	92	20°50'N	77°15'W
Jock, r., Can. (jŏk)	62c	45°08'N	75°51'W
Jocotepec, Mex. (jō-kō-tȧ-pĕk')	88	20°17'N	103°26'W
Jodar, Spain (hō'där)	128	37°54'N	3°20'W
Jodhpur, India (hōd'pōōr)	155	26°23'N	73°00'E
Joensuu, Fin. (yō-ĕn'sōō)	123	62°35'N	29°46'E
Joffre, Mount, mtn., Can. (jŏ'f'r)	55	50°32'N	115°13'W
Jõgeva, Est. (yû'gĕ-vȧ)	123	58°45'N	26°23'E
Joggins, Can. (jŏ'gĭnz)	60	45°42'N	64°27'W
Johannesburg, S. Afr. (yō-hän'ĕs-bōrgh)	186	26°08'S	27°54'E
John Day, r., Or., U.S. (jŏn'dā)	72	44°46'N	120°15'W
John Day, Middle Fork, r., Or., U.S.	72	44°53'N	119°04'W
John Day, North Fork, r., Or., U.S.	72	45°03'N	118°50'W
John Day Dam, Or., U.S.	72	45°40'N	120°15'W
John H. Kerr Reservoir, res., U.S.	65	36°30'N	78°38'W
John Martin Reservoir, res., Co., U.S. (jŏn mär'tĭn)	78	37°57'N	103°04'W
Johnson, r., Or., U.S. (jŏn'sŭn)	74c	45°27'N	122°20'W
Johnsonburg, Pa., U.S. (jŏn'sŭn-bûrg)	67	41°30'N	78°40'W
Johnson City, Il., U.S. (jŏn'sŭn)	66	37°50'N	88°55'W
Johnson City, N.Y., U.S.	67	42°10'N	76°00'W
Johnson City, Tn., U.S.	65	36°17'N	82°23'W
Johnston, i., Oc. (jŏn'stŭn)	2	17°00'N	168°00'W
Johnstone Strait, strt., Can.	54	50°25'N	126°00'W
Johnston Falls, wtfl., Afr.	191	10°35'S	28°50'E
Johnstown, N.Y., U.S. (jonz'toun)	67	43°00'N	74°20'W
Johnstown, Pa., U.S.	65	40°20'N	78°50'W
Johor, r., Malay. (jŭ-hōr')	153b	1°39'N	103°52'E
Johor Baharu, Malay.	168	1°28'N	103°46'E
Jõhvi, Est. (yû'vĭ)	123	59°21'N	27°21'E
Joigny, Fr. (zhwán-yē')	126	47°58'N	3°26'E
Joinville, Braz. (zhwăɴ-vēl')	102	26°18'S	48°47'W
Joinville, Fr.	127	48°28'N	5°05'E
Joinville, i., Ant.	97	63°00'S	53°30'W
Jojutla, Mex. (hō-hōō'tlä)	88	18°39'N	99°11'W
Jola, Mex. (kō'lä)	88	21°08'N	104°26'W
Joliet, Il., U.S. (jō-lĭ-ĕt')	69a	41°32'N	88°05'W
Joliette, Can. (zhō-lyĕt')	51	46°01'N	73°30'W
Jolo, Phil. (hō-lō)	168	5°59'N	121°05'E
Jolo Island, i., Phil.	168	5°55'N	121°15'E
Jomalig, i., Phil. (hō-mä'lēg)	169a	14°44'N	122°34'E
Jomulco, Mex. (hō-mōōl'kō)	88	21°08'N	104°24'W
Jonacatepec, Mex.	88	18°39'N	98°46'W
Jonava, Lith. (yō-nä'vȧ)	123	55°05'N	24°15'E
Jones, Phil. (jōnz)	169a	12°56'N	122°05'E
Jones, Phil.	169a	16°35'N	121°39'E
Jonesboro, Ar., U.S. (jōnz'bûro)	65	35°49'N	90°42'W
Jonesboro, La., U.S.	81	32°14'N	92°43'W
Jonesville, La., U.S. (jōnz'vĭl)	81	31°35'N	91°50'W
Jonesville, Mi., U.S.	66	42°00'N	84°45'W
Jong, r., S.L.	188	8°10'N	12°10'W
Joniškis, Lith. (yō'nĭsh-kĭs)	123	56°14'N	23°36'E
Jönköping, Swe. (yûn'chû-pĭng)	116	57°47'N	14°10'E
Jonquière, Can. (zhôɴ-kyâr')	51	48°25'N	71°15'W
Jonuta, Mex. (hō-nōō'tä)	89	18°07'N	92°09'W
Jonzac, Fr. (zhôɴ-zák')	126	45°27'N	0°27'W
Joplin, Mo., U.S. (jŏp'lĭn)	65	37°05'N	94°31'W
Jordan, nation, Asia (jôr'dăn)	154	30°15'N	38°00'E
Jordan, r., Asia	153a	32°05'N	35°35'E
Jordan, r., Ut., U.S.	75b	40°42'N	111°56'W
Jorhāt, India (jōr-hät')	155	26°43'N	94°16'E
Jorullo, Volcán de, vol., Mex. (vôl-kä'n-dĕ-hō-rōōl'yō)	88	18°54'N	101°38'W
José C. Paz, Arg.	102a	34°32'S	58°44'W
Joseph Bonaparte Gulf, b., Austl. (jō'sĕf bō'nȧ-pärt)	174	13°30'S	128°40'E
Josephburg, Can.	62g	53°45'N	113°06'W
Joseph Lake, l., Can. (jō'sĕf läk)	62g	53°18'N	113°06'W
Joshua Tree National Park, rec., Ca., U.S. (jō'shū-ȧ trē)	76	34°02'N	115°53'W
Jos Plateau, plat., Nig. (jŏs)	189	9°53'N	9°05'E
Jostedalsbreen, ice, Nor. (yôstĕ-däls-brēĕn)	116	61°40'N	6°55'E
Jotunheimen, mts., Nor.	116	61°44'N	8°11'E
Joulter's Cays, is., Bah. (jōl'tẽrz)	92	25°20'N	78°10'W
Jouy-le-Chatel, Fr. (zhwē-lĕ-shä-tĕl')	127b	48°40'N	3°07'E
Jovellanos, Cuba (hō-vĕl-yä'nōs)	92	22°50'N	81°10'W
J. Percy Priest Lake, res., Tn., U.S.	82	36°00'N	86°45'W
Juan Aldama, Mex. (kōä'n-äl-dä'mä)	88	24°16'N	103°21'W
Juan de Fuca, Strait of, strt., N.A. (hwän' dā fōō'kä)	52	48°25'N	124°37'W
Juan de Nova, Île, i., Reu.	187	17°18'S	43°07'E
Juan Diaz, r., Pan. (κōōä'n-dē'äz)	86a	9°05'N	79°30'W
Juan Fernández, Islas de, is., Chile	97	33°30'S	79°00'W
Juan L. Lacaze, Ur. (hōōä'n-ĕ'lĕ-lä-kä'zĕ)	99c	34°25'S	57°28'W
Juan Luis, Cayos de, is., Cuba (kä-yōs-dĕ-hwän lōō-ēs')	92	22°15'N	82°00'W
Juárez, Arg. (hōōá'rĕz)	102	37°42'S	59°46'W
Juázeiro, Braz. (zhōōá'zā'rō)	101	9°27'S	40°28'W
Juazeiro do Norte, Braz. (zhōōá'zā'rō-dō-nōr-tĕ)	101	7°16'S	38°57'W
Jubayl, Leb. (jōō-bīl')	153a	34°07'N	35°38'E
Jubba (Genale), r., Afr.	192a	1°30'N	42°25'E
Juby, Cap, c., Mor. (yōō'bē)	184	28°01'N	13°21'W
Júcar, r., Spain (hōō'kär)	118	39°10'N	1°22'W
Júcaro, Cuba (hōō'kä-rō)	92	21°40'N	78°50'W
Juchipila, Mex. (hōō-chē-pē'lȧ)	88	21°26'N	103°09'W
Juchitán, Mex. (hōō-chē-tän')	86	16°15'N	95°00'W
Juchitlán, Mex. (hōō-chē-tlän)	88	20°05'N	104°07'W
Jucuapa, El Sal. (κōō-kwä'pä)	90	13°30'N	88°24'W
Judenburg, Aus. (jōō'dĕn-bûrg)	124	47°10'N	14°40'E
Judith, r., Mt., U.S. (jōō'dĭth)	73	47°20'N	109°36'W
Juhua Dao, i., China (jyōō-hwä dou)	162	40°30'N	120°47'E
Juigalpa, Nic. (hwē-gäl'pä)	90	12°02'N	85°24'W
Juiz de Fora, Braz. (zhō-ēzh' dā fō'rä)	101	21°47'S	43°20'W
Jujuy, Arg. (hōō-hwē')	102	24°14'S	65°15'W
Jujuy, prov., Arg. (hōō-hwē')	102	23°00'S	65°45'W
Jukskei, r., S. Afr.	187b	25°58'S	27°58'E
Julesburg, Co., U.S. (jōōlz'bûrg)	78	40°59'N	102°16'W
Juliaca, Peru (hōō-lē-ä'kä)	100	15°26'S	70°12'W
Julian Alps, mts., Serb.	118	46°05'N	14°05'E
Julianehåb, Grnld.	49	60°07'N	46°20'W
Jülich, Ger. (yü'lĕk)	127c	50°55'N	6°22'E
Jullundur, India	155	31°29'N	75°39'E
Julpaiguri, India	158	26°35'N	88°48'E
Jumento Cays, is., Bah. (hōō-mĕn'tō)	93	23°05'N	75°40'W
Jumilla, Spain (hōō-mēl'yä)	128	38°28'N	1°20'W
Jump, r., Wi., U.S. (jŭmp)	71	45°18'N	90°53'W
Jumpingpound Creek, r., Can. (jŭmp-ĭng-pound)	62e	51°01'N	114°34'W
Jumrah, Indon.	153b	1°48'N	101°04'E
Junagādh, India (jŏ-nä'gŭd)	155	21°33'N	70°25'E
Junayfah, Egypt	192d	30°11'N	32°26'E
Junaynah, Ra's al, mtn., Egypt	153a	29°02'N	33°58'E
Junction, Tx., U.S. (jŭŋk'shŭn)	80	30°29'N	99°48'W
Junction City, Ks., U.S.	79	39°01'N	96°49'W
Jundiaí, Braz.	101	23°11'S	46°52'W
Juneau, Ak., U.S. (jōō'nō)	64a	58°25'N	134°30'W
Jungfrau, mtn., Switz. (yòng'frou)	124	46°30'N	7°59'E
Junín, Arg. (hōō-nē'n)	102	34°35'S	60°56'W
Junín, Col.	100a	4°47'N	73°39'W
Juniyah, Leb. (jōō-nē'ē)	153a	33°59'N	35°38'E
Jupiter, r., Can.	60	49°40'N	63°20'W
Jupiter, Mount, mtn., Wa., U.S.	74a	47°42'N	123°04'W
Jur, r., Sudan (jōr)	185	6°38'N	27°52'E
Jura, mts., Eur. (zhü-rá')	117	46°55'N	6°49'E
Jura, i., Scot., U.K. (jōō'rȧ)	120	56°09'N	6°45'W
Jura, Sound of, strt., Scot., U.K. (jōō'rȧ)	120	55°45'N	5°55'W
Jurbarkas, Lith. (yōōr-bär'kȧs)	123	55°06'N	22°50'E
Jūrmala, Lat.	123	56°57'N	23°37'E
Jurong, China (jyōō-rôŋ)	162	31°58'N	119°12'E
Juruá, r., S.A.	100	5°30'S	67°30'W
Juruena, r., Braz. (zhōō-rōōĕ'nä)	101	12°22'S	58°34'W
Jutiapa, Guat. (hōō-tē-ä'pä)	90	14°16'N	89°55'W
Juticalpa, Hond. (hōō-tē-käl'pä)	86	14°35'N	86°17'W
Jutland *see* Jylland, reg., Den.	116	56°04'N	9°00'E
Juventino Rosas, Mex.	88	20°38'N	101°02'W
Juventud, Isla de la, i., Cuba	87	21°40'N	82°45'W
Juxian, China (jyōō shyĕn)	164	35°35'N	118°50'E
Juxtlahuaca, Mex. (hōōs-tlȧ-hwä'kä)	88	17°20'N	98°02'W
Juye, China (jyōō-yŭ)	162	35°25'N	116°05'E
Južna Morava, r., Serb. (ŭ'zhnȧ mô'rȧ-vä)	131	42°30'N	22°00'E
Jylland, reg., Den.	116	56°04'N	9°00'E

ăt; finăl; rāte; senāte; ärm; ȧsk; sofȧ; fâre; ch-choose; dh-as th in other; bē; ēvent; bĕt; recĕnt; cratẽr; g-gō; gh-guttural g; bĭt; ĭ-short neutral; rīde; κ-guttural k as ch in German ich;

K

PLACE (Pronunciation)	PAGE	LAT.	LONG.
K2 (Qogir Feng), mtn., Asia	155	36°06'N	76°38'E
Kaabong, Ug.	191	3°31'N	34°08'E
Kaalfontein, S. Afr. (kärl-fŏn-tän)	187b	26°02'S	28°16'E
Kaappunt, c., S. Afr.	186a	34°21'S	18°30'E
Kabaena, Pulau, i., Indon. (kä-bȧ-ā'nȧ)	168	5°35'S	121°07'E
Kabala, S.L. (kȧ-bä'lȧ)	184	9°43'N	11°39'W
Kabale, Ug.	191	1°15'S	29°59'E
Kabalega Falls, wtfl., Ug.	185	2°15'N	31°41'E
Kabalo, D.R.C. (kä-bä'lō)	186	6°03'S	26°55'E
Kabambare, D.R.C. (kä-bäm-bä'rȧ)	186	4°47'S	27°45'E
Kabardino-Balkaria, prov., Russia	136	43°30'N	43°30'E
Kabba, Nig.	189	7°50'N	6°03'E
Kabe, Japan (kä'bā)	167	34°32'N	132°30'E
Kabinakagami, r., Can.	58	49°00'N	84°15'W
Kabinda, D.R.C. (kä-bēn'dä)	186	6°08'S	24°29'E
Kabompo, r., Zam. (kȧ-bôm'pō)	186	14°00'S	23°40'E
Kabongo, D.R.C. (kȧ-bông'ō)	186	7°58'S	25°10'E
Kabot, Gui.	188	10°48'N	14°57'W
Kaboudia, Ra's, c., Tun.	118	35°17'N	11°28'E
Kābul, Afg. (kä'bo͝ol)	155	34°39'N	69°14'E
Kabul, r., Asia (kä'bo͝ol)	155	34°44'N	69°43'E
Kabunda, D.R.C.	191	12°25'S	29°22'E
Kabwe, Zam.	186	14°27'S	28°27'E
Kachuga, Russia (kȧ-cho͞o-gȧ)	135	54°09'N	105°43'E
Kadei, r., Afr.	189	4°00'N	15°10'E
Kadnikov, Russia (käd'nē-kôf)	136	59°30'N	40°10'E
Kadoma, Japan	167b	34°43'N	135°36'E
Kadoma, Zimb.	186	18°21'S	29°55'E
Kaduna, Nig. (kä-do͞o'nä)	184	10°33'N	7°27'E
Kaduna, r., Nig.	189	9°30'N	6°00'E
Kaédi, Maur. (kä-ā-dē')	184	16°09'N	13°30'W
Ka'ena Point, c., Hi., U.S. (kä'ā-nä)	64d	21°33'N	158°19'W
Kaesŏng, Kor., N. (kä'ĕ-sŭng) (kī'jō)	161	38°00'N	126°35'E
Kafanchan, Nig.	189	9°36'N	8°17'E
Kafia Kingi, Sudan (kä'fē-ȧ kĭŋ'gē)	185	9°17'N	24°28'E
Kafue, Zam. (kä'fo͞o)	186	15°45'S	28°17'E
Kafue, r., Zam.	186	15°45'S	26°30'E
Kafue Flats, sw., Zam.	191	16°15'S	26°30'E
Kafue National Park, rec., Zam.	191	15°00'S	25°35'E
Kafwira, D.R.C.	191	12°10'S	27°33'E
Kagal'nik, r., Russia (kä-gäl''nēk)	133	46°58'N	39°25'E
Kagera, r., Afr. (kä-gä'rȧ)	186	1°10'S	31°10'E
Kagoshima, Japan (kä'gō-shē'mȧ)	161	31°35'N	130°31'E
Kagoshima-Wan, b., Japan (kä'gō-shē'mä wän)	166	31°24'N	130°39'E
Kahayan, r., Indon.	168	1°45'S	113°40'E
Kahemba, D.R.C.	190	7°17'S	19°00'E
Kahia, D.R.C.	191	6°21'S	28°24'E
Kahoka, Mo., U.S. (kȧ-hō'kȧ)	79	40°26'N	91°42'W
Kaho'olawe, i., Hi., U.S. (kä-ho͞o-lä'wē)	64c	20°28'N	156°48'W
Kahramanmaraş, Tur.	154	37°40'N	36°50'E
Kahshahpiwi, r., Can.	71	48°24'N	90°56'W
Kahuku Point, c., Hi., U.S. (kä-ho͞o'ko͞o)	64d	21°50'N	157°50'W
Kahului, Hi., U.S.	64c	20°53'N	156°28'W
Kai, Kepulauan, is., Indon.	169	5°35'S	132°45'E
Kaiang, Malay.	153b	3°00'N	101°47'E
Kaiashk, r., Can.	58	49°40'N	89°30'W
Kaibab Indian Reservation, I.R., Az., U.S. (kä'ē-bȧb)	77	36°55'N	112°45'W
Kaibab Plat, Az., U.S.	77	36°30'N	112°10'W
Kaidu, r., China (kī-do͞o)	160	42°35'N	84°04'E
Kaieteur Fall, wtfl., Guy. (kī-ĕ-to͞or')	101	4°48'N	59°24'W
Kaifeng, China (kī-fŭŋ)	161	34°48'N	114°22'E
Kai Kecil, i., Indon.	169	5°45'S	132°40'E
Kailua, Hi., U.S. (kä'ē-lo͞o'ä)	64c	21°18'N	157°43'W
Kailua Kona, Hi., U.S.	84a	19°49'N	155°59'W
Kaimana, Indon.	169	3°32'S	133°47'E
Kaimanawa Mountains, mts., N.Z.	177	39°10'S	176°00'E
Kainan, Japan (kä'ē-nän')	167	34°09'N	135°14'E
Kainji Lake, res., Nig.	184	10°25'N	4°50'E
Kaiserslautern, Ger. (kī-zĕrs-lou'tĕrn)	117	49°26'N	7°46'E
Kaitaia, N.Z. (kä-ē-tä'ē-ȧ)	175a	35°30'S	173°28'E
Kaiwi Channel, strt., Hi., U.S. (käē-wē)	64c	21°10'N	157°38'W
Kaiyuan, China (kū-yuän)	165	23°42'N	103°20'E
Kaiyuan, China	164	42°30'N	124°00'E
Kaiyuh Mountains, mts., Ak., U.S. (kī-yo͞o')	63	64°25'N	157°38'W
Kajaani, Fin. (kä'yȧ-nē)	116	64°15'N	27°16'E
Kajang, Gunong, mtn., Malay.	153b	2°47'N	104°05'E
Kajiki, Japan (kä'jē-kē)	166	31°44'N	130°41'E
Kakhovka, Ukr. (kä-kôf'kȧ)	133	46°46'N	33°32'E
Kakhovs'ke vodoskhovyshche, res., Ukr.	134	47°21'N	33°33'E
Kākināda, India	155	16°58'N	82°18'E
Kaktovik, Ak., U.S. (kăk-tō'vĭk)	63	70°08'N	143°51'W
Kakwa, r., Can. (kăk'wȧ)	55	54°00'N	118°55'W
Kalach, Russia (kȧ-läch')	137	50°15'N	40°55'E
Kaladan, r., Asia	160	21°07'N	93°04'E
Kalae, c., Hi., U.S.	84a	18°55'N	155°41'W
Kalahari Desert, des., Afr. (kä-lä-hä'rē)	186	23°00'S	22°03'E
Kalama, Wa., U.S. (kȧ-lăm'ȧ)	74c	46°01'N	122°50'W
Kalama, r., Wa., U.S.	74c	46°03'N	122°47'W
Kalamáta, Grc.	110	37°04'N	22°08'E
Kalamazoo, Mi., U.S. (kăl-ȧ-mȧ-zo͞o')	65	42°20'N	85°40'W
Kalamazoo, r., Mi., U.S.	66	42°35'N	86°00'W
Kalanchak, Ukr. (kä-län-chäk')	133	46°17'N	33°14'E
Kalandula, Ang. (do͞o'kȧ dȧ brä-gäN'sä)	186	9°06'S	15°57'E
Kalaotoa, Pulau, i., Indon.	168	7°22'S	122°30'E
Kalapana, Hi., U.S. (kä-lä-pä'nä)	84a	19°25'N	155°00'W
Kalar, mtn., Iran	154	31°43'N	51°41'E
Kalāt, Pak. (kŭ-lät')	155	29°05'N	66°36'E
Kalemie, D.R.C.	186	5°56'S	29°12'E
Kalgan *see* Zhangjiakou, China	161	40°45'N	114°58'E
Kalgoorlie-Boulder, Austl. (kăl-go͞or'lē)	174	30°45'S	121°35'E
Kaliakra, Nos, c., Blg.	119	43°25'N	28°42'E
Kalima, D.R.C.	191	2°34'S	26°37'E
Kaliningrad, Russia	134	54°42'N	20°32'E
Kaliningrad, Russia (kä-lē-nēn'grät)	142b	55°55'N	37°49'E
Kalinkavichy, Bela.	132	52°07'N	29°19'E
Kalispel Indian Reservation, I.R., Wa., U.S. (kăl-ĭ-spĕl')	72	48°25'N	117°30'W
Kalispell, Mt., U.S. (kăl'ĭ-spĕl)	64	48°12'N	114°18'W
Kalisz, Pol. (kä'lēsh)	117	51°45'N	18°05'E
Kaliua, Tan.	191	5°04'S	31°48'E
Kalixälven, r., Swe.	116	67°12'N	22°00'E
Kalmar, Swe. (käl'mär)	116	56°40'N	16°19'E
Kalmarsund, strt., Swe. (käl'mär)	122	56°30'N	16°17'E
Kal'mius, r., Ukr. (käl''myo͞os)	133	47°15'N	37°38'E
Kalmykia, prov., Russia	137	46°56'N	46°00'E
Kalocsa, Hung. (kä'lō-chä)	125	46°32'N	19°00'E
Kalohi Channel, strt., Hi., U.S. (kä-lō'hĭ)	84a	20°55'N	157°15'W
Kaloko, D.R.C.	191	6°47'S	25°48'E
Kalomo, Zam. (kä-lō'mō)	186	17°02'S	26°30'E
Kalsubai Mount, mtn., India	158	19°43'N	73°47'E
Kaltenkirchen, Ger. (käl'tĕn-kēr-kĕn)	115c	53°50'N	9°57'E
Kālu, r., India	159b	19°18'N	73°14'E
Kaluga, Russia (kȧ-lo͞o'gä)	134	54°29'N	36°12'E
Kaluga, prov., Russia	132	54°10'N	35°00'E
Kaluktutiak (Cambridge Bay), Can.	50	69°15'N	105°00'W
Kalundborg, Den. (kä-lo͝on''bôr')	122	55°42'N	11°07'E
Kalush, Ukr. (kä'lo͝osh)	125	49°02'N	24°24'E
Kalvarija, Lith. (käl-vä-rē'yȧ)	123	54°24'N	23°17'E
Kalwa, India	159b	19°12'N	72°59'E
Kal'ya, Russia (käl'yȧ)	142a	60°17'N	59°58'E
Kalyān, India	158	19°16'N	73°07'E
Kalyazin, Russia (käl-yä'zēn)	132	57°13'N	37°55'E
Kama, r., Russia (kä'mä)	134	56°10'N	53°50'E
Kamaishi, Japan (kä'mä-ē'shē)	166	39°16'N	142°03'E
Kamakura, Japan (kä'mä-ko͞o'rä)	167	35°19'N	139°33'E
Kamarān, i., Yemen	154	15°19'N	41°47'E
Kāmārhāti, India	158a	22°41'N	88°23'E
Kambove, D.R.C. (käm-bō'vē)	186	10°58'S	26°43'E
Kamchatka, r., Russia	141	54°15'N	158°38'E
Kamchatka, Poluostrov, pen., Russia	141	55°19'N	157°45'E
Kamen, Ger. (kä'mĕn)	127c	51°35'N	7°40'E
Kamenjak, Rt, c., Cro. (kä'mĕ-nyäk)	130	44°45'N	13°57'E
Kamen'-na-Obi, Russia (kä-mĭny'nŭ ô'bē)	134	53°43'N	81°28'E
Kamensk-Shakhtinskiy, Russia (kä'mĕnsk shäk'tĭn-skĭ)	133	48°17'N	40°16'E
Kamensk-Ural'skiy, Russia (kä'mĕnsk o͞o-räl'skĭ)	136	56°27'N	61°55'E
Kamenz, Ger. (kä'mĕnts)	124	51°16'N	14°05'E
Kameoka, Japan (kä'mā-ōkä)	167b	35°01'N	135°35'E
Kāmet, mtn., Asia	158	30°50'N	79°42'E
Kamianets'-Podil's'kyi, Ukr.	137	48°41'N	26°34'E
Kamianka-Buz'ka, Ukr.	125	50°06'N	24°20'E
Kamień Pomorski, Pol.	124	53°57'N	14°48'E
Kamikoma, Japan (kä'mē-kō'mä)	167b	34°45'N	135°50'E
Kamina, D.R.C.	186	8°44'S	25°00'E
Kaministikwia, r., Can. (kȧ-mĭ-nĭ-stĭk'wĭ-ȧ)	71	48°40'N	89°41'W
Kamituga, D.R.C.	191	3°04'S	28°11'E
Kamloops, Can. (kăm'lo͞ops)	50	50°40'N	120°20'W
Kamp, r., Aus. (kämp)	124	48°30'N	15°45'E
Kampala, Ug. (käm-pä'lä)	185	0°19'N	32°25'E
Kampar, r., Indon. (käm'pär)	168	0°30'N	101°30'E
Kampene, D.R.C.	191	3°36'S	26°40'E
Kampenhout, Bel.	115a	50°56'N	4°33'E
Kamp-Lintfort, Ger. (kämp-lĕnt'fōrt)	127c	51°30'N	6°34'E
Kâmpóng Saôm, Camb.	168	10°40'N	103°50'E
Kâmpóng Thum, Camb. (kŏm'pŏng-tŏm)	168	12°41'N	104°29'E
Kâmpôt, Camb. (käm'pōt)	168	10°41'N	104°07'E
Kampuchea *see* Cambodia, nation, Asia	168	12°15'N	104°00'E
Kamsack, Can. (kăm'săk)	50	51°34'N	101°54'W
Kamskoye, res., Russia	134	59°08'N	56°30'E
Kamudilo, D.R.C.	191	7°42'S	27°18'E
Kamuela, Hi., U.S.	84a	20°01'N	155°40'W
Kamui Misaki, c., Japan	166	43°25'N	139°35'E
Kámuk, Cerro, mtn., C.R. (sĕ'r-rō-kä-mo͞o'k)	91	9°18'N	83°02'W
Kamyshevatskaya, Russia	133	46°24'N	37°58'E
Kamyshin, Russia (kä-mwēsh'ĭn)	134	50°08'N	45°20'E
Kamyshlov, Russia (kä-mēsh'lôf)	134	56°50'N	62°32'E
Kan, r., Russia (kän)	140	56°30'N	94°17'E
Kanab, Ut., U.S. (kăn'ăb)	77	37°00'N	112°30'W
Kanabeki, Russia (kȧ-nä'byĕ-kĭ)	142a	57°48'N	57°16'E
Kanab Plateau, plat., Az., U.S.	77	36°31'N	112°55'W
Kanaga, i., Ak., U.S. (kă-nă'gă)	63a	52°02'N	177°38'W
Kanagawa, dept., Japan (kä'nä-gä'wä)	167a	35°29'N	139°32'E
Kanā'is, Ra's al, c., Egypt	119	31°14'N	28°08'E
Kanamachi, Japan (kä-nä-mä'chē)	167a	35°46'N	139°52'E
Kananga, D.R.C.	186	6°14'S	22°17'E
Kananikol'skoye, Russia	142a	52°48'N	57°29'E
Kanasín, Mex. (kä-nä-sē'n)	90a	20°54'N	89°31'W
Kanatak, Ak., U.S. (kă-nă'tŏk)	63	57°35'N	155°48'W
Kanawha, r., W.V., U.S. (kȧ-nô'wȧ)	65	37°55'N	81°50'W
Kanaya, Japan (kä'nä'yä)	167a	35°10'N	139°49'E
Kanazawa, Japan (kä'nȧ-zä'wȧ)	161	36°34'N	136°38'E
Kānchenjunga, mtn., Asia (kĭn-chĭn-jo͞on'gȧ)	155	27°30'N	88°18'E
Kānchipuram, India	155	12°55'N	79°43'E
Kandahār, Afg.	155	31°43'N	65°58'E
Kanda Kanda, D.R.C. (kän'dȧ kän'dȧ)	186	6°56'S	23°36'E
Kandalaksha, Russia (kȧn-dȧ-läk'shȧ)	134	67°10'N	33°05'E
Kandalakshskiy Zaliv, b., Russia	136	66°20'N	35°00'E
Kandava, Lat. (kän'dȧ-vȧ)	123	57°03'N	22°45'E
Kandi, Benin (käN-dē')	184	11°08'N	2°56'E
Kandiāro, Pak.	158	27°09'N	68°12'E
Kandla, India (kŭnd'lŭ)	158	23°00'N	70°20'E
Kandy, Sri L. (kän'dē)	159	7°18'N	80°42'E
Kane, Pa., U.S. (kān)	67	41°40'N	78°50'W
Kāne'ohe, Hi., U.S. (kä-nä-ō'hā)	84a	21°25'N	157°47'W
Kāne'ohe Bay, b., Hi., U.S.	64d	21°32'N	157°40'W
Kanevskaya, Russia (kȧ-nyĕf'skȧ)	133	46°07'N	38°58'E
Kangaroo, i., Austl. (kăŋ-gȧ-ro͞o')	174	36°05'S	137°05'E
Kangāvar, Iran (kŭŋ'gä-vär)	154	34°37'N	46°45'E
Kangean, Kepulauan, is., Indon. (käŋ'gē-än)	168	6°50'S	116°22'E
Kanggye, Kor., N. (käng'gyĕ)	161	40°55'N	126°40'E
Kanghwa, i., Kor., S. (käng'hwä)	166	37°38'N	126°00'E
Kangnŭng, Kor., S. (käng'nô ng)	166	37°42'N	128°50'E
Kango, Gabon (käN-gō)	186	0°09'N	10°08'E
Kangowa, D.R.C.	190	9°55'S	22°48'E
Kanin, Poluostrov, pen., Russia	134	68°00'N	45°00'E
Kaningo, Kenya	191	0°49'S	38°32'E
Kanin Nos, Mys, c., Russia	136	68°40'N	44°00'E
Kaniv, Ukr.	133	49°46'N	31°27'E
Kanivs'ke vodoskhovyshche, res., Ukr.	134	50°10'N	30°40'E
Kanjiža, Serb. (kä'nyē-zhä)	131	46°05'N	20°02'E
Kankakee, Il., U.S. (kăŋ-kȧ-kē')	66	41°07'N	87°53'W
Kankakee, r., Il., U.S.	66	41°15'N	88°15'W
Kankan, Gui. (käN-käN) (kän-kän')	184	10°23'N	9°18'W
Kannapolis, N.C., U.S. (kăn-ăp'ō-lĭs)	83	35°30'N	80°38'W
Kannoura, Japan (kä'nō-o͞o'rä)	167	33°34'N	134°18'E
Kano, Nig. (kä'nō)	184	12°00'N	8°30'E
Kanonkop, mtn., S. Afr.	186a	33°49'S	18°37'E
Kanopolis Reservoir, res., Ks., U.S. (kăn-ŏp'ō-lĭs)	78	38°44'N	98°01'W
Kānpur, India (kän'pûr)	158	26°30'N	80°10'E
Kansas, state, U.S. (kăn'zȧs)	64	38°30'N	99°40'W
Kansas, r., Ks., U.S.	79	39°08'N	95°52'W
Kansas City, Ks., U.S.	65	39°06'N	94°39'W
Kansas City, Mo., U.S.	65	39°05'N	94°35'W
Kansk, Russia	135	56°14'N	95°43'E
Kansŏng, Kor., S.	166	38°09'N	128°29'E
Kantang, Thai. (kän'täng')	168	7°26'N	99°28'E
Kantchari, Burkina	188	12°29'N	1°31'E
Kanton, i., Kir.	194	3°50'S	174°00'W
Kantunilkin, Mex. (kän-to͞o-nēl-kē'n)	90a	21°07'N	87°30'W
Kanzhakovskiy Kamen, Gora, mtn., Russia (kȧn-zhä'kôvs-kēē kämĭen)	142a	59°38'N	59°12'E
Kaohsiung, Tai. (kä-ō-syŏng')	161	22°35'N	120°25'E
Kaolack, Sen.	184	14°09'N	16°04'W
Kaouar, oasis, Niger	185	19°16'N	13°09'E
Kapaa, Hi., U.S.	84a	22°06'N	159°20'W
Kapanga, D.R.C.	190	8°21'S	22°35'E
Kapfenberg, Aus. (kăp'fĕn-bĕrgh)	124	47°27'N	15°16'E
Kapiri Mposhi, Zam.	191	13°58'S	28°41'E
Kapoeta, Sudan	185	4°45'N	33°35'E
Kaposvár, Hung. (kô'pôsh-vär)	125	46°21'N	17°45'E
Kapsan, Kor., N. (käp'sän')	166	40°59'N	128°22'E
Kapuskasing, Can.	51	49°28'N	82°22'W
Kapuskasing, r., Can.	58	48°55'N	82°55'W
Kapustin Yar, Russia (kä'pŏs-tēn yär')	137	48°30'N	45°40'E
Kaputar, Mount, mtn., Austl. (kă-pû-tăr)	176	30°11'S	150°11'E
Kapuvár, Hung. (kô'po͞o-vär)	125	47°35'N	17°02'E
Kara, Russia (kärȧ)	134	68°42'N	65°30'E
Kara, r., Russia	136	68°30'N	65°20'E
Karabalā', Iraq (kŭr'bȧ-lä)	154	32°31'N	43°58'E
Karabanovo, Russia (kä'rȧ-bȧ-nō-vō)	132	56°19'N	38°43'E
Karabash, Russia (kó-rȧ-bäsh')	142a	55°27'N	60°14'E
Kara-Bogaz-Gol, Zaliv, b., Turkmen. (kȧ-rä' bŭ-gäs')	139	41°30'N	53°40'E
Karachay-Cherkessia, prov., Russia	138	44°00'N	42°00'E
Karachev, Russia (kȧ-rȧ-chôf')	136	53°08'N	34°54'E
Karāchi, Pak.	155	24°59'N	68°56'E
Karaganda *see* Qaraghandy, Kaz.	139	49°42'N	73°18'E
Karaidel', Russia (kä'rī-dĕl)	142a	55°52'N	56°54'E
Karakoram Pass, p., Asia	155	35°35'N	77°45'E
Karakoram Range, mts., India (kä'rä kō'rŏm)	155	35°24'N	76°38'E
Karakorum, hist., Mong.	160	47°25'N	102°22'E
Kara-Kum, des., Turkmen.	139	40°00'N	57°00'E
Kara Kum Canal, can., Turkmen.	139	37°35'N	61°50'E
Karaman, Tur. (kä-rä-män')	119	37°10'N	33°00'E
Karamay, China (kär-äm-ā)	160	45°37'N	84°53'E
Karamea Bight, b., N.Z. (kȧ-rȧ-mē'ȧ bĭt)	175a	41°20'S	171°30'E
Kara Sea *see* Karskoye More, sea, Russia	134	74°00'N	68°00'E
Karashahr (Yanqui), China (kä-rä-shä-är) (yän-chyē)	160	42°14'N	86°28'E
Karatsu, Japan (kä'rȧ-tso͞o)	167	33°28'N	129°59'E
Karaul, Russia (kä-rä-ôl')	140	70°13'N	83°46'E
Karawanken, mts., Eur.	124	46°32'N	14°07'E
Karcag, Hung. (kär'tsäg)	125	47°18'N	20°58'E
Kárditsa, Grc.	131	39°23'N	21°57'E
Kärdla, Est. (kĕrd'lȧ)	123	58°59'N	22°44'E
Karelia, prov., Russia	140	62°30'N	32°35'E
Karema, Tan.	186	6°49'S	30°26'E
Kargat, Russia (kär-gät')	134	55°17'N	80°07'E
Karghalik *see* Yecheng, China	160	37°54'N	77°25'E
Kargopol', Russia (kär-gō-pōl'')	134	61°30'N	38°50'E
Kariba, Lake, res., Afr.	186	17°15'S	27°55'E
Karibib, Nmb. (kȧr'ȧ-bĭb)	186	21°55'S	15°50'E

262 Kārikāl – Key West PRONOUNCING INDEX

PLACE (Pronunciation)	PAGE	LAT.	LONG.
Kārikāl, India (kä-rē-käl′)	159	10°58′N	79°49′E
Karimata, Kepulauan, is., Indon. (kä-rē-mä′tà)	168	1°08′S	108°10′E
Karimata, Selat, strt., Indon.	168	1°00′S	107°10′E
Karimun Besar, i., Indon.	153b	1°10′N	103°28′E
Karimunjawa, Kepulauan, is., Indon. (kä′rē-mo͞on-yä′vä)	168	5°36′S	110°15′E
Karin, Som. (kär′ĭn)	192a	10°43′N	45°50′E
Karkar Island, i., Pap. N. Gui. (kär′kär)	169	4°50′S	146°45′E
Karkheh, r., Iran	154	32°45′N	47°50′E
Karkinits′ka zatoka, b., Ukr.	133	45°50′N	32°45′E
Karkūk, Iraq	154	35°28′N	44°22′E
Karlivka, Ukr.	133	49°26′N	35°08′E
Karlobag, Cro. (kär-lō-bäg′)	130	44°30′N	15°03′E
Karlovac, Cro. (kär′lô-väts)	119	45°29′N	15°16′E
Karlovo, Blg. (kär′lô-vō)	131	42°39′N	24°48′E
Karlovy Vary, Czech Rep. (kär′lô-vē vä′rē)	117	50°13′N	12°53′E
Karlshamn, Swe. (kärls′häm)	122	56°11′N	14°50′E
Karlskrona, Swe. (kärls′krō-nä)	116	56°10′N	15°33′E
Karlsruhe, Ger. (kärls′ro͞o-ĕ)	117	49°00′N	8°23′E
Karlstad, Swe. (kärl′städ)	110	59°25′N	13°28′E
Karluk, Ak., U.S. (kär′lŭk)	63	57°30′N	154°22′W
Karmøy, i., Nor. (kärm-ûe)	122	59°14′N	5°00′E
Karnataka, state, India	155	14°55′N	75°00′E
Karnobat, Blg. (kär-nô′bät)	131	42°39′N	26°59′E
Karonga, Mwi. (kȧ-rōŋ′gȧ)	186	9°52′S	33°57′E
Kárpathos, i., Grc.	119	35°34′N	27°26′E
Karpinsk, Russia (kär′pĭnsk)	142a	59°46′N	60°00′E
Kars, Tur. (kärs)	154	40°35′N	43°00′E
Kārsava, Lat. (kär′sä-vä)	123	56°46′N	27°39′E
Karshi, Uzb. (kär′shē)	139	38°30′N	66°08′E
Karskiye Vorota, Proliv, strt., Russia	134	70°30′N	58°07′E
Karskoye More (Kara Sea), sea, Russia	134	74°00′N	68°00′E
Kartaly, Russia (kár′tá lē)	134	53°05′N	60°40′E
Karunagapalli, India	159	9°09′N	76°34′E
Karvina, Czech Rep.	125	49°50′N	18°30′E
Kasai (Cassai), r., Afr.	186	3°45′S	19°10′E
Kasama, Zam. (kȧ-sä′mȧ)	186	10°13′S	31°12′E
Kasanga, Tan. (kȧ-säŋ′gȧ)	186	8°28′S	31°09′E
Kasaoka, Japan (kä′sä-ō′kȧ)	167	34°33′N	133°29′E
Kasba-Tadla, Mor. (käs′bȧ-täd′lȧ)	184	32°37′N	5°57′W
Kasempa, Zam. (kȧ-sĕm′pȧ)	186	13°27′S	25°50′E
Kasenga, D.R.C. (kȧ-seŋ′gȧ)	186	10°22′S	28°38′E
Kasese, D.R.C.	191	1°38′S	27°07′E
Kasese, Ug.	191	0°10′N	30°05′E
Kāshān, Iran (kä-shän′)	154	33°52′N	51°15′E
Kashgar *see* Kashi, China	160	39°29′N	76°00′E
Kashi (Kashgar), China (kä-shr) (käsh-gär)	160	39°29′N	76°00′E
Kashihara, Japan (kä′shē-hä′rä)	167b	34°31′N	135°48′E
Kashiji Plain, pl., Zam.	190	13°25′S	22°30′E
Kashin, Russia (kä-shēn′)	132	57°20′N	37°38′E
Kashira, Russia (kä-shē′rȧ)	132	54°49′N	38°11′E
Kashiwa, Japan (kä′shē-wä)	167a	35°51′N	139°58′E
Kashiwara, Japan	167b	34°35′N	135°38′E
Kashiwazaki, Japan (kä′shē-wä-zä′kē)	166	37°06′N	138°17′E
Kāshmar, Iran	157	35°12′N	58°27′E
Kashmīr *see* Jammu and Kashmīr, state, India	155	34°30′N	76°00′E
Kashmor, Pak.	158	28°33′N	69°34′E
Kashtak, Russia (käsh′tȧk)	142a	55°18′N	61°25′E
Kasimov, Russia (kȧ-sē′môf)	136	54°56′N	41°23′E
Kaskanak, Ak., U.S. (kăs-kă′năk)	63	60°00′N	158°00′W
Kaskaskia, r., Il., U.S. (kăs-kăs′kĭ-ȧ)	66	39°10′N	88°50′W
Kaskattama, r., Can. (kăs-kȧ-tă′mȧ)	57	56°28′N	90°55′W
Kaskö (Kaskinen), Fin. (käs′kû) (käs′kē-nĕn)	123	62°24′N	21°18′E
Kasli, Russia (käs′lĭ)	136	55°53′N	60°46′E
Kasongo, D.R.C. (kä-sôŋ′gō)	186	4°31′S	26°42′E
Kásos, i., Grc.	119	35°20′N	26°55′E
Kaspiysk, Russia	138	42°52′N	47°38′E
Kassándras, Kólpos, b., Grc.	131	40°10′N	23°35′E
Kassel, Ger. (käs′ĕl)	117	51°19′N	9°30′E
Kasson, Mn., U.S. (kăs′ŭn)	71	44°01′N	92°45′W
Kastamonu, Tur. (kä-stȧ-mō′no͞o)	154	41°20′N	33°50′E
Kastoría, Grc. (kás-tō′rĭ-ȧ)	119	40°28′N	21°17′E
Kasūr, Pak.	158	31°10′N	74°29′E
Kataba, Zam.	191	16°05′S	25°10′E
Katahdin, Mount, mtn., Me., U.S. (kȧ-tä′dĭn)	60	45°56′N	68°57′W
Katanga, hist. reg., D.R.C. (kȧ-täŋ′gȧ)	186	8°30′S	25°00′E
Katanning, Austl. (kȧ-tăn′ĭng)	174	33°45′S	117°45′E
Katav-Ivanovsk, Russia (kä′tȧf ĭ-vä′nôfsk)	142a	54°46′N	58°13′E
Kateninskiy, Russia (kátyĕ′nĭs-kĭ)	142a	53°12′N	61°05′E
Kateríni, Grc.	131	40°18′N	22°36′E
Katete, Zam.	191	14°05′S	32°07′E
Katherine, Austl. (kăth′ẽr-ĭn)	174	14°15′S	132°20′E
Kāthiāwār, pen., India (kä′tyȧ-wär′)	155	22°10′N	70°20′E
Kathmandu, Nepal (kăt-măn-do͞o′)	155	27°49′N	85°21′E
Kathryn, Can. (kăth′rĭn)	62e	51°13′N	113°42′W
Kathryn, Ca., U.S.	75a	33°42′N	117°45′W
Katihār, India	158	25°39′N	87°39′E
Katiola, C. Iv.	188	8°08′N	5°06′W
Katmai National Park, rec., Ak., U.S. (kăt′mī)	64a	58°38′N	155°00′W
Katompi, D.R.C.	191	6°11′S	26°20′E
Katopa, D.R.C.	191	2°45′S	25°06′E
Katowice, Pol.	110	50°15′N	19°00′E
Katrineholm, Swe. (kȧ-trē′nĕ-hôlm)	122	59°01′N	16°10′E
Katsbakhskiy, Russia (kȧts-băk′skĭ)	142a	52°57′N	59°37′E
Katsina, Nig. (kät′sē-nȧ)	184	13°00′N	7°32′E
Katsina Ala, Nig.	184	7°10′N	9°17′E
Katsura, r., Japan (kä′tso͞o-rä)	167b	34°55′N	135°43′E
Katta-Kurgan, Uzb. (kȧ-tȧ-kŏr-gän′)	139	39°45′N	66°42′E
Kattegat, strt., Eur. (kăt′ĕ-gät)	112	56°57′N	11°25′E
Katumba, D.R.C.	191	7°45′S	25°18′E
Katun′, r., Russia (kȧ-tōn′)	140	51°30′N	86°18′E
Katwijk aan Zee, Neth.	115a	52°12′N	4°23′E
Kaua′i, i., Hi., U.S.	64c	22°09′N	159°15′W
Kauai Channel, strt., Hi., U.S. (kä-o͞o-ä′ē)	64c	21°35′N	158°52′W
Kaufbeuren, Ger. (kouf′boi-rĕn)	124	47°52′N	10°38′E
Kaufman, Tx., U.S. (kôf′măn)	81	32°36′N	96°18′W
Kaukauna, Wi., U.S. (kô-kô′nȧ)	71	44°17′N	88°15′W
Kaulakahi Channel, strt., Hi., U.S. (kä′o͞o-lȧ-kä′hē)	84a	22°00′N	159°55′W
Kaunakakai, Hi., U.S. (kä′o͞o-nȧ-kä′kī)	84a	21°06′N	156°59′W
Kaunas, Lith. (kou′nȧs) (kôv′nô)	134	54°42′N	23°54′E
Kaura Namoda, Nig.	184	12°35′N	6°35′E
Kavála, Grc. (kä-vä′lä)	119	40°55′N	24°24′E
Kavieng, Pap. N. Gui. (kä-vē-ĕng′)	169	2°44′S	151°02′E
Kavīr, Dasht-e, des., Iran (dŭsht-ē-ka-vēr′)	154	34°41′N	53°30′E
Kawagoe, Japan (kä-wä-gō′ȧ)	167	35°55′N	139°29′E
Kawaguchi, Japan (kä-wä-go͞o-chē)	167a	35°48′N	139°44′E
Kawaikini, mtn., Hi., U.S. (kä-wä′ē-kĭ-nī)	84a	22°05′N	159°33′W
Kawanishi, Japan (kä-wä′nē-shē)	167b	34°49′N	135°26′E
Kawasaki, Japan (kä-wä-sä′kē)	166	35°32′N	139°43′E
Kaxgar, r., China	160	39°30′N	75°00′E
Kaya, Burkina (kä′yä)	184	13°05′N	1°05′W
Kayan, r., Indon.	168	1°45′N	115°38′E
Kaycee, Wy., U.S. (kā-sē′)	73	43°43′N	106°38′W
Kayes, Mali (kāz)	184	14°27′N	11°26′W
Kayseri, Tur. (kī′sĕ-rē)	154	38°45′N	35°20′E
Kazach′ye, Russia	135	70°46′N	135°47′E
Kazakhstan, nation, Asia	134	48°45′N	59°00′E
Kazan′, Russia (kȧ-zän′)	134	55°50′N	49°18′E
Kazanka, Ukr. (kȧ-zän′kȧ)	133	47°49′N	32°50′E
Kazanlŭk, Blg. (kȧ′zȧn-lĕk)	131	42°47′N	25°23′E
Kazbek, Gora, mtn. (kȧz-bĕk′)	137	42°42′N	44°31′E
Kāzerūn, Iran	154	29°37′N	51°44′E
Kazincbarcika, Hung. (kô′zĭnts-bôr-tsĭ-ko)	125	48°15′N	20°39′E
Kazungula, Zam.	191	17°45′S	25°20′E
Kazusa Kameyama, Japan (kä-zo͞o-sä kä-mâ′yä-mä)	167a	35°14′N	140°06′E
Kazym, r., Russia (kä-zēm′)	140	63°30′N	67°41′E
Kéa, i., Grc.	131	37°36′N	24°13′E
Kealaikahiki Channel, strt., Hi., U.S. (kā-ä′lä-ē-kä-hē′kē)	84a	20°38′N	157°00′W
Keansburg, N.J., U.S. (kēnz′bûrg)	68a	40°26′N	74°08′W
Kearney, Ne., U.S. (kär′nĭ)	70	40°42′N	99°05′W
Kearny, N.J., U.S.	68a	40°46′N	74°09′W
Keasey, Or., U.S. (kēz′ĭ)	74c	45°51′N	123°20′W
Kebnekaise, mtn., Swe. (kĕp′nĕ-kȧ-ĕs′ĕ)	112	67°53′N	18°10′E
Kecskemét, Hung. (kĕch′kĕ-māt)	119	46°52′N	19°42′E
Kedah, hist. reg., Malay. (kä′dä)	168	6°00′N	100°31′E
Kėdainiai, Lith. (kē-dī′nĭ-ī)	123	55°16′N	23°58′E
Kedgwick, Can. (kĕdj′wĭk)	60	47°39′N	67°21′W
Keenbrook, Ca., U.S. (kēn′bro͝ok)	75a	34°16′N	117°29′W
Keene, N.H., U.S. (kēn)	67	42°55′N	72°15′W
Keetmanshoop, Nmb. (kāt′mȧns-hōp)	186	26°30′S	18°05′E
Keet Seel Ruin, Az., U.S. (kēt sĕl)	77	36°46′N	110°32′W
Keewatin, Mn., U.S. (kē-wä′tĭn)	71	47°24′N	93°03′W
Kefallonía, i., Grc.	119	38°08′N	20°58′E
Keffi, Nig. (kĕf′ē)	184	8°51′N	7°52′E
Ke Ga, Mui, c., Viet.	168	12°58′N	109°50′E
Kei, r., Afr. (kā)	187c	32°57′S	26°50′E
Keila, Est. (kā′lȧ)	123	59°19′N	24°25′E
Keilor, Austl.	173a	37°43′S	144°50′E
Kei Mouth, S. Afr.	187c	32°40′S	28°23′E
Keiskammahoek, S. Afr. (kās′kämä-ho͝ok′)	187c	32°42′S	27°11′E
Kéita, Bahr, r., Chad	189	9°30′N	19°17′E
Keitele, l., Fin. (kā′tĕ-lĕ)	123	62°50′N	25°40′E
Kekaha, Hi., U.S.	84a	21°57′N	159°42′W
Kelafo, Eth.	192a	5°40′N	44°00′E
Kelang, Malay.	168	3°20′N	101°27′E
Kelang, r., Malay.	153b	3°00′N	101°40′E
Kelkit, r., Tur.	119	40°38′N	37°03′E
Keller, Tx., U.S. (kĕl′ẽr)	75c	32°56′N	97°15′W
Kellinghusen, Ger. (kĕ′lĕng-ho͞o-zĕn)	115c	53°57′N	9°43′E
Kellogg, Id., U.S. (kĕl′ŏg)	72	47°32′N	116°07′W
Kelme′, Lith. (kĕl-mȧ)	123	55°36′N	22°53′E
Kélo, Chad	189	9°19′N	15°48′E
Kelowna, Can.	50	49°53′N	119°29′W
Kelsey Bay, Can. (kĕl′sē)	54	50°24′N	125°57′W
Kelso, Wa., U.S.	74c	46°09′N	122°54′W
Keluang, Malay.	153b	2°01′N	103°19′E
Kem′, Russia (kĕm)	134	65°00′N	34°48′E
Kemah, Tx., U.S. (kē′mȧ)	81a	29°32′N	95°01′W
Kemerovo, Russia	134	55°31′N	86°05′E
Kemi, Fin. (kā′mē)	116	65°48′N	24°38′E
Kemi, r., Fin.	116	67°02′N	27°50′E
Kemigawa, Japan (kĕ′mē-gä′wä)	167a	35°38′N	140°07′E
Kemijarvi, Fin. (kā′mē-yĕr-vē)	116	66°48′N	27°21′E
Kemi-joki, l., Fin.	116	66°37′N	28°13′E
Kemmerer, Wy., U.S. (kĕm′ẽr-ẽr)	73	41°48′N	110°36′W
Kemp, l., Tx., U.S. (kĕmp)	78	33°55′N	99°22′W
Kempen, Ger. (kĕm′pĕn)	127c	51°22′N	6°25′E
Kempsey, Austl. (kĕmp′sē)	175	30°59′S	152°50′E
Kempt, l., Can. (kĕmpt)	59	47°28′N	74°00′W
Kempten, Ger. (kĕmp′tĕn)	117	47°44′N	10°17′E
Kempton Park, S. Afr. (kĕmp′tŏn pärk)	192c	26°07′S	28°29′E
Ken, r., India	158	25°00′N	79°55′E
Kenai, Ak., U.S. (kĕ-nī′)	63	60°38′N	151°18′W
Kenai Fjords National Park, rec., Ak., U.S.	63	59°45′N	150°00′W
Kenai Mountains, mts., Ak., U.S.	63	60°00′N	150°00′W
Kenai Pen, Ak., U.S.	63	64°40′N	150°18′W
Kendal, S. Afr.	192c	26°03′S	28°58′E
Kendal, Eng., U.K. (kĕn′dȧl)	120	54°20′N	1°48′W
Kendallville, In., U.S. (kĕn′dăl-vĭl)	66	41°25′N	85°20′W
Kenedy, Tx., U.S. (kĕn′ĕ-dĭ)	81	28°49′N	97°50′W
Kenema, S.L.	188	7°52′N	11°12′W
Kenitra, Mor. (kĕ-nē′trȧ)	118	34°21′N	6°34′W
Kenmare, N.D., U.S. (kĕn-mâr′)	70	48°41′N	102°05′W
Kenmore, N.Y., U.S. (kĕn′mōr)	69c	42°58′N	78°53′W
Kennebec, r., Me., U.S. (kĕn-ē-bĕk′)	60	44°23′N	69°48′W
Kennebunk, Me., U.S. (kĕn-ē-buŋk′)	60	43°24′N	70°33′W
Kennedale, Tx., U.S. (kĕn′-ē-dāl)	75c	32°38′N	97°13′W
Kennedy, Cape *see* Canaveral, Cape, c., Fl., U.S.	65	28°30′N	80°23′W
Kennedy, Mount, mtn., Can.	63	60°25′N	138°50′W
Kenner, La., U.S. (kĕn′ẽr)	81	29°58′N	90°15′W
Kennett, Mo., U.S. (kĕn′ĕt)	79	36°14′N	90°01′W
Kennewick, Wa., U.S. (kĕn′ē-wĭk)	72	46°12′N	119°06′W
Kenney Dam, dam, Can.	54	53°37′N	124°58′W
Kennydale, Wa., U.S. (kĕn-nē′dāl)	74a	47°31′N	122°12′W
Kénogami, Can. (kĕn-ô′gä-mē)	51	48°26′N	71°14′W
Kenogamissi Lake, l., Can.	58	48°15′N	81°31′W
Keno Hill, Can.	63	63°58′N	135°18′W
Kenora, Can. (kē-nō′rȧ)	51	49°47′N	94°29′W
Kenosha, Wi., U.S. (kē-nō′shȧ)	65	42°34′N	87°50′W
Kenova, W.V., U.S. (kē-nō′vȧ)	66	38°20′N	82°35′W
Kensico Reservoir, res., N.Y., U.S. (kĕn′sĭ-kō)	68a	41°08′N	73°45′W
Kent, Oh., U.S. (kĕnt)	66	41°05′N	81°20′W
Kent, Wa., U.S.	74a	47°23′N	122°14′W
Kentani, S. Afr. (kĕnt-änĭ′)	187c	32°31′S	28°19′E
Kentland, In., U.S. (kĕnt′lȧnd)	66	40°50′N	87°25′W
Kenton, Oh., U.S. (kĕn′tŭn)	66	40°40′N	83°35′W
Kent Peninsula, pen., Can.	52	68°28′N	108°10′W
Kentucky, state, U.S. (kĕn-tŭk′ĭ)	65	37°30′N	87°35′W
Kentucky, res., U.S.	65	36°20′N	88°50′W
Kentucky, r., Ky., U.S.	65	38°15′N	85°01′W
Kentwood, La., U.S. (kĕnt′wo͝od)	81	30°56′N	90°31′W
Kenya, nation, Afr. (kĕn′yȧ)	186	1°00′N	36°53′E
Kenya, Mount (Kirinyaga), mtn., Kenya	187	0°10′S	37°20′E
Kenyon, Mn., U.S. (kĕn′yŭn)	71	44°15′N	92°58′W
Keokuk, Ia., U.S. (kē′ô-kŭk)	65	40°24′N	91°34′W
Keoma, Can. (kē-ō′mȧ)	62e	51°13′N	113°39′W
Kepenkeck Lake, l., Can.	61	48°13′N	54°45′W
Kępno, Pol. (kăɴ′pnō)	125	51°17′N	17°59′E
Kerala, state, India	155	16°38′N	76°00′E
Kerang, Austl. (kē-răng′)	175	35°32′S	143°58′E
Kerch, Ukr.	134	45°20′N	36°26′E
Kerchenskiy Proliv, strt., Eur. (kĕr-chĕn′skĭ prô′lĭf)	133	45°08′N	36°35′E
Kerempe Burun, c., Tur.	119	42°00′N	33°20′E
Keren, Erit.	185	15°46′N	38°28′E
Kerguélen, Îles, is., Afr. (kĕr′gȧ-lĕn)	3	49°50′S	69°30′E
Kericho, Kenya	191	0°22′S	35°17′E
Kerinci, Gunung, mtn., Indon.	168	1°45′S	101°18′E
Keriya *see* Yutian, China	160	36°55′N	81°39′E
Keriya, r., China (kē′rē-yä)	160	37°13′N	81°59′E
Kerkebet, Erit.	156	16°18′N	37°24′E
Kerkenna, Îles, i., Tun. (kĕr′kĕn-nä)	184	34°49′N	11°37′E
Kerki, Turkmen. (kĕr′kē)	139	37°52′N	65°15′E
Kérkyra, Grc.	119	39°36′N	19°56′E
Kérkyra, i., Grc.	118	39°33′N	19°36′E
Kermadec Islands, is., N.Z. (kẽr-măd′ĕk)	3	30°30′S	177°00′E
Kermān, Iran (kẽr-män′)	154	30°23′N	57°08′E
Kermānshāh *see* Bakhtarān, Iran	154	34°01′N	47°00′E
Kern, r., Ca., U.S.	76	35°31′N	118°37′W
Kern, South Fork, r., Ca., U.S.	76	35°40′N	118°15′W
Kerpen, Ger. (kĕr′pĕn)	127c	50°52′N	6°42′E
Kerrobert, Can.	56	51°53′N	109°13′W
Kerrville, Tx., U.S. (kûr′vĭl)	80	30°02′N	99°07′W
Kerulen, r., Asia (kĕr′o͞o-lĕn)	161	47°52′N	113°22′E
Kesagami Lake, l., Can.	59	50°23′N	80°15′W
Keşan, Tur. (kĕ′shȧn)	131	40°50′N	26°37′E
Keshan, China (kŭ-shän)	161	48°00′N	126°30′E
Kesour, Monts des, mts., Alg.	118	32°51′N	0°30′W
Kestell, S. Afr. (kĕs′tĕl)	192c	28°19′N	28°43′E
Keszthely, Hung. (kĕst′hĕl-lĭ)	125	46°46′N	17°12′E
Ket′, r., Russia (kyĕt)	140	58°30′N	84°15′E
Keta, Ghana	184	6°00′N	1°00′E
Ketamputih, Indon.	153b	1°25′N	102°19′E
Ketapang, Indon. (kĕ-tȧ-päng′)	168	2°00′S	109°57′E
Ketchikan, Ak., U.S. (kĕch-ĭ-kăn′)	64a	55°21′N	131°35′W
Kętrzyn, Pol. (kăɴ′t′r-zĭn)	125	54°04′N	21°24′E
Kettering, Eng., U.K. (kĕt′ẽr-ĭng)	114a	52°23′N	0°43′W
Kettering, Oh., U.S.	66	39°40′N	84°15′W
Kettle, r., Can.	55	49°40′N	119°00′W
Kettle, r., Mn., U.S. (kĕt′′l)	71	46°20′N	92°57′W
Kettwig, Ger. (kĕt′vēg)	127c	51°22′N	6°56′E
Kęty, Pol. (kăɴ tĭ)	125	49°54′N	19°16′E
Ketzin, Ger. (kĕ′tzēn)	115b	52°29′N	12°51′E
Keuka, l., N.Y., U.S. (kē-ū′kȧ)	67	42°30′N	77°10′W
Kevelaer, Ger. (kĕ′fĕ-lär)	127c	51°35′N	6°15′E
Kew, Austl.	173a	37°49′S	145°02′E
Kewanee, Il., U.S. (kē-wä′nē)	71	41°15′N	89°55′W
Kewaunee, Wi., U.S. (kē-wô′nē)	71	44°27′N	87°33′W
Keweenaw Bay, b., Mi., U.S. (kē′wē-nô)	71	46°59′N	88°15′W
Keweenaw Peninsula, pen., Mi., U.S.	71	47°28′N	88°12′W
Keya Paha, r., S.D., U.S. (kē-yȧ pä′hä)	70	43°11′N	100°10′W
Key Largo, i., Fl., U.S.	83a	25°11′N	80°15′W
Keyport, N.J., U.S. (kē′pōrt)	68a	40°26′N	74°12′W
Keyport, Wa., U.S.	74a	47°42′N	122°38′W
Keyser, W.V., U.S. (kī′sẽr)	67	39°25′N	79°00′W
Key West, Fl., U.S. (kē wĕst′)	65	24°31′N	81°47′W

ăt; finăl; rāte; senāte; ärm; àsk; sofȧ; fâre; ch-choose; dh-as th in other; bē; ēvent; bĕt; recĕnt; cratẽr; g-gō; gh-guttural g; bĭt; ĭ-short neutral; rīde; ĸ-guttural k as ch in German ich;

PLACE (Pronunciation)	PAGE	LAT.	LONG.
Kežmarok, Slvk. (kĕzh′má-rôk)	125	49°10′N	20°27′E
Khabarovo, Russia (kŭ-bár-ôvŏ)	134	69°31′N	60°41′E
Khabarovsk, Russia (kä-bä′rôfsk)	135	48°35′N	135°12′E
Khakassia, prov., Russia	140	52°32′N	89°33′E
Khālāpur, India	159b	18°48′N	73°17′E
Khalkidhiki, pen., Grc.	131	40°30′N	23°18′E
Khal′mer-Yu, Russia (kŭl-myĕr′-yōō′)	134	67°52′N	64°25′E
Khalturin, Russia (käl′tōō-rĕn)	136	58°28′N	49°00′E
Khambhāt, Gulf of, b., India	155	21°20′N	72°27′E
Khammam, India	159	17°09′N	80°13′E
Khānābād, Afg.	158	36°43′N	69°11′E
Khandwa, India	158	21°53′N	76°22′E
Khaníon, Kólpos, b., Grc.	130a	35°35′N	23°55′E
Khanka, l., Asia (kän′ká)	135	45°09′N	133°28′E
Khānpur, Pak.	158	28°42′N	70°42′E
Khanty-Mansiysk, Russia (kŭn-te′mŭn-sēsk′)	134	61°02′N	69°01′E
Khān Yūnus, Gaza	153a	31°21′N	34°19′E
Kharagpur, India (kŭ-rŭg′pòr)	155	22°26′N	87°21′E
Kharkiv, Ukr.	134	50°00′N	36°10′E
Kharkiv, prov., Ukr.	133	49°33′N	35°55′E
Kharkov *see* Kharkiv, Ukr.	134	50°00′N	36°10′E
Kharlovka, Russia	136	68°47′N	37°20′E
Kharmanli, Blg. (kár-män′lē)	131	41°54′N	25°55′E
Khartoum, Sudan	185	15°34′N	32°36′E
Khasavyurt, Russia	138	43°15′N	46°37′E
Khāsh, Iran	154	28°08′N	61°08′E
Khāsh, r., Afg.	154	32°30′N	64°27′E
Khasi Hills, hills, India	155	25°38′N	91°55′E
Khaskovo, Blg. (kás′kô-vô)	119	41°56′N	25°32′E
Khatanga, Russia (ká-tän′gá)	135	71°48′N	101°47′E
Khatangskiy Zaliv, b., Russia (kä-täŋ′g-skē)	135	73°45′N	108°30′E
Khaybār, Sau. Ar.	154	25°45′N	39°28′E
Kherson, Ukr. (kĕr-sôn′)	137	46°38′N	32°34′E
Kherson, prov., Ukr.	133	46°32′N	32°55′E
Khiitola, Russia (khē′tō-lá)	123	61°14′N	29°40′E
Khimki, Russia (kēm′kĭ)	142b	55°54′N	37°27′E
Khmel′nyts′kyi, Ukr.	137	49°29′N	26°54′E
Khmel′nyts′kyy, prov., Ukr.	133	49°27′N	26°30′E
Khmil′nyk, Ukr.	133	49°34′N	27°58′E
Kholm, Russia (kôlm)	132	57°09′N	31°07′E
Kholmsk, Russia (kŭlmsk)	135	47°09′N	142°33′E
Khomeynīshahr, Iran	157	32°41′N	51°31′E
Khon Kaen, Thai.	168	16°37′N	102°41′E
Khopër, r., Russia (kô′pēr)	137	52°00′N	43°00′E
Khor, Russia (kôr′)	166	47°50′N	134°52′E
Khor, r., Russia	166	47°23′N	135°20′E
Khóra Sfakíon, Grc.	130a	35°12′N	24°10′E
Khorog, Taj.	139	37°30′N	71°36′E
Khorol, Ukr. (kô′rôl)	133	49°48′N	33°17′E
Khorol, r., Ukr.	133	49°50′N	33°21′E
Khorramābād, Iran	157	33°30′N	48°20′E
Khorramshahr, Iran (kô-ram′shär)	154	30°36′N	48°15′E
Khot′kovo, Russia	142b	56°15′N	38°00′E
Khotyn, Ukr.	137	48°29′N	26°32′E
Khoyniki, Bela.	133	51°54′N	30°00′E
Khudzhand, Taj.	139	40°17′N	69°37′E
Khulna, Bngl.	155	22°50′N	89°38′E
Khūryān Mūryān, is., Oman	154	17°27′N	56°02′E
Khust, Ukr. (kōōst)	125	48°10′N	23°18′E
Khvalynsk, Russia (kvá-lĭnsk′)	137	52°30′N	48°00′E
Khvoy, Iran	154	38°32′N	45°01′E
Khyber Pass, p., Asia (kī′bēr)	155	34°28′N	71°18′E
Kialwe, D.R.C.	191	9°22′S	27°08′E
Kiambi, D.R.C. (kyäm′bē)	186	7°20′S	28°01′E
Kiamichi, r., Ok., U.S. (kyá-mē′chē)	79	34°31′N	95°34′W
Kianta, l., Fin. (kyán′tá)	136	65°00′N	28°15′E
Kibenga, D.R.C.	190	7°55′S	17°35′E
Kibiti, Tan.	191	7°44′S	38°57′E
Kibombo, D.R.C.	191	3°54′S	25°55′E
Kibondo, Tan.	191	3°35′S	30°42′E
Kičevo, Mac. (kē′chĕ-vô)	131	41°30′N	20°59′E
Kickapoo, r., Wi., U.S. (kĭk′á-pōō)	71	43°20′N	90°55′W
Kicking Horse Pass, p., Can.	55	51°25′N	116°10′W
Kidal, Mali (kē-dál′)	184	18°33′N	1°00′E
Kidderminster, Eng., U.K. (kĭd′ēr-mĭn-stēr)	114a	52°23′N	2°14′W
Kidd's Beach, S. Afr. (kĭdz)	187c	33°09′S	27°43′E
Kidsgrove, Eng., U.K. (kĭdz′grōv)	114a	53°05′N	2°15′W
Kiel, Ger. (kēl)	110	54°19′N	10°08′E
Kiel, Wi., U.S.	71	43°52′N	88°04′W
Kiel Bay, b., Ger.	124	54°33′N	10°19′E
Kiel Canal *see* Nord-Ostsee Kanal, can., Ger.	124	54°03′N	9°23′E
Kielce, Pol. (kyĕl′tsĕ)	125	50°50′N	20°41′E
Kieldrecht, Bel. (kēl′drĕkt)	115a	51°17′N	4°09′E
Kiev (Kyiv), Ukr.	134	50°27′N	30°30′E
Kiffa, Maur. (kēf′á)	184	16°37′N	11°24′W
Kigali, Rw. (kē-gä′lē)	186	1°59′S	30°05′E
Kigoma, Tan. (kē-gō′mä)	186	4°57′S	29°38′E
Kii-Suido, strt., Japan (kē sōō-ē′dō)	166	33°53′N	134°55′E
Kikaiga, i., Japan	166	28°25′N	130°10′E
Kikinda, Serb. (kē′kēn-dä)	131	45°49′N	20°30′E
Kikládes, is., Grc.	118	37°30′N	24°45′E
Kikwit, D.R.C. (kē′kwēt)	186	5°02′S	18°49′E
Kil, Swe. (kēl)	122	59°30′N	13°15′E
Kilauea, Hi., U.S. (kē-lä-ōō-ā′ä)	84a	22°12′N	159°25′W
Kilauea Crater, depr., Hi., U.S.	84a	19°28′N	155°18′W
Kilbuck Mountains, mts., Ak., U.S. (kĭl-bŭk)	63	60°05′N	160°00′W
Kilchu, Kor., N. (kĭl′chò)	166	40°59′N	129°23′E
Kildare, Ire. (kĭl-dār′)	120	53°09′N	7°05′W
Kilembe, D.R.C.	190	5°42′S	19°55′E
Kilgore, Tx., U.S.	81	32°23′N	94°53′W
Kilia, Ukr.	133	45°28′N	29°17′E
Kilifi, Kenya	191	3°38′S	39°51′E
Kilimanjaro, mtn., Tan. (kyl-ē-män-jä′rô)	187	3°09′S	37°19′E
Kilimatinde, Tan. (kĭl-ē-mä-tĭn′dā)	186	5°48′S	34°58′E
Kilindoni, Tan.	191	7°55′S	39°39′E
Kilingi-Nõmme, Est. (kē′lĭn-gē-nôm′mĕ)	123	58°08′N	25°03′E
Kilis, Tur. (kē′lēs)	119	36°50′N	37°20′E
Kilkenny, Ire. (kĭl-kĕn-ĭ)	117	52°40′N	7°30′W
Kilkis, Grc. (kĭl′kĭs)	131	40°59′N	22°51′E
Killala, Ire. (kĭ-lä′lá)	120	54°11′N	9°10′W
Killarney, Ire.	120	52°03′N	9°05′W
Killdeer, N.D., U.S. (kĭl′dēr)	70	47°22′N	102°45′W
Killiniq Island, i., Can.	53	60°32′N	63°56′W
Kilmarnock, Scot., U.K. (kĭl-mär′nŭk)	120	55°38′N	4°25′W
Kilrush, Ire. (kĭl′rŭsh)	120	52°40′N	9°16′W
Kilwa Kisiwani, Tan.	191	8°58′S	39°30′E
Kilwa Kivinje, Tan.	187	8°43′S	39°18′E
Kim, r., Cam.	189	5°40′N	11°17′E
Kimamba, Tan.	191	6°47′S	37°08′E
Kimba, Austl. (kĭm′bá)	176	33°08′S	136°25′E
Kimball, Ne., U.S. (kĭm-bál)	70	41°14′N	103°41′W
Kimball, S.D., U.S.	70	43°44′N	98°58′W
Kimberley, Can. (kĭm′bēr-lĭ)	50	49°41′N	115°59′W
Kimberley, S. Afr.	186	28°40′S	24°50′E
Kimi, Cam.	189	6°05′N	11°30′E
Kimmirut (Lake Harbour), Can.	51	62°43′N	69°40′W
Kímolos, i., Grc. (kē′mô-lôs)	131	36°52′N	24°20′E
Kimry, Russia (kĭm′rē)	136	56°53′N	37°24′E
Kimvula, D.R.C.	190	5°44′S	15°58′E
Kinabalu, Gunong, mtn., Malay.	168	5°45′N	115°26′E
Kincardine, Can. (kĭn-kär′dĭn)	51	44°10′N	81°15′W
Kinda, D.R.C.	191	9°18′S	25°04′E
Kindanba, Congo	190	3°44′S	14°31′E
Kinder, La., U.S. (kĭn′dēr)	81	30°30′N	92°50′W
Kindersley, Can. (kĭn′dērz-lē)	50	51°27′N	109°10′W
Kindia, Gui. (kĭn′dē-á)	184	10°04′N	12°51′W
Kindu, D.R.C.	186	2°57′S	25°56′E
Kinel′-Cherkassy, Russia	136	53°32′N	51°32′E
Kineshma, Russia (kē-nĕsh′má)	136	57°27′N	41°02′E
King, i., Austl. (kĭng)	175	39°35′S	143°40′E
Kingaroy, Austl. (kĭn′gă-roi)	176	26°37′S	151°50′E
King City, Can.	62d	43°56′N	79°32′W
King City, Ca., U.S. (kĭng sĭ′tĭ)	76	36°12′N	121°08′W
Kingcome Inlet, b., Can. (kĭng′kŭm)	54	50°50′N	126°10′W
Kingfisher, Ok., U.S. (kĭng′fĭsh-ēr)	79	35°51′N	97°55′W
King George Sound, strt., Austl. (jôrj)	174	35°17′S	118°30′E
Kingisepp, Russia (kĭŋ-gē-sep′)	136	59°22′N	28°38′E
King Leopold Ranges, mts., Austl. (lē′ō-pōld)	174	16°25′S	125°00′E
Kingman, Az., U.S. (kĭng′măn)	77	35°10′N	114°05′W
Kingman, Ks., U.S. (kĭng′măn)	78	37°38′N	98°07′W
Kings, r., Ca., U.S.	76	36°28′N	119°43′W
Kings Canyon National Park, rec., Ca., U.S. (kăn′yŭn)	64	36°52′N	118°53′W
Kingsclere, Eng., U.K. (kings-clēr)	114b	51°18′N	1°15′W
Kingscote, Austl. (kĭngz′kŭt)	176	35°45′S	137°32′E
King's Lynn, Eng., U.K. (kĭngz lĭn′)	121	52°45′N	0°20′E
Kings Mountain, N.C., U.S.	83	35°13′N	81°30′W
Kings Norton, Eng., U.K. (nôr′tŭn)	114a	52°25′N	1°54′W
King Sound, strt., Austl.	174	16°50′S	123°35′E
Kings Park, N.Y., U.S. (kĭngz pärk)	68a	40°53′N	73°16′W
Kings Peak, mtn., Ut., U.S.	64	40°46′N	110°20′W
Kingsport, Tn., U.S. (kĭngz′pōrt)	83	36°33′N	82°36′W
Kingston, Austl. (kĭngz′tŭn)	174	37°52′S	139°52′E
Kingston, Can.	51	44°15′N	76°30′W
Kingston, Jam.	87	18°00′N	76°45′W
Kingston, N.Y., U.S.	65	42°00′N	74°00′W
Kingston, Pa., U.S.	67	41°15′N	75°50′W
Kingston, Wa., U.S.	74a	47°04′N	122°29′W
Kingston upon Hull, Eng., U.K.	110	53°45′N	0°25′W
Kingstown, St. Vin. (kĭngz′toun)	87	13°10′N	61°14′W
Kingstree, S.C., U.S. (kĭngz′trē)	83	33°30′N	79°50′W
Kingsville, Tx., U.S. (kĭngz′vĭl)	81	27°32′N	97°52′W
King William Island, i., Can. (kĭng wĭl′yăm)	52	69°25′N	97°00′W
King William's Town, S. Afr. (kĭng-wĭl′-yŭmz-toun)	187c	32°53′S	27°24′E
Kinira, r., S. Afr.	187c	30°37′S	28°52′E
Kinloch, Mo., U.S. (kĭn-lŏk)	75e	38°44′N	90°19′W
Kinnaird, Can. (kĭn-ärd′)	55	49°17′N	117°39′W
Kinnairds Head, c., Scot., U.K. (kĭn-ârds′hĕd)	116	57°42′N	3°55′W
Kinomoto, Japan (kē′nō-mōtō)	167	33°53′N	136°07′E
Kinosaki, Japan (kē′nō-sä′kē)	167	35°38′N	134°47′E
Kinshasa, D.R.C.	186	4°18′S	15°18′E
Kinsley, Ks., U.S. (kĭnz′lĭ)	78	37°55′N	99°24′W
Kinston, N.C., U.S. (kĭnz′tŭn)	83	35°15′N	77°35′W
Kintampo, Ghana (kĕn-täm′pō)	184	8°03′N	1°43′W
Kintyre, pen., Scot., U.K.	120	55°50′N	5°40′W
Kiowa, Ks., U.S. (kī′ô-wá)	78	37°01′N	98°30′W
Kiowa, Ok., U.S.	79	34°42′N	95°53′W
Kipawa, Lac, l., Can.	59	46°55′N	79°00′W
Kipembawe, Tan. (kē-pĕm-bä′wå)	186	7°39′S	33°24′E
Kipengere Range, mts., Tan.	191	9°10′S	34°00′E
Kipili, Tan.	191	7°26′S	30°36′E
Kipushi, D.R.C.	191	11°46′S	27°14′E
Kirakira, Sol. Is.	170e	10°27′S	161°55′E
Kirby, Tx., U.S. (kŭr′bĭ)	75d	29°29′N	98°23′W
Kirbyville, Tx., U.S. (kŭr′bĭ-vĭl)	81	30°39′N	93°54′W
Kirenga, r., Russia (kē-rĕŋ′gá)	141	56°30′N	108°18′E
Kirensk, Russia (kē-rĕnsk′)	135	57°47′N	108°22′E
Kirgiz Range, mts., Asia	139	42°30′N	74°00′E
Kiri, D.R.C.	190	1°27′S	19°00′E
Kiribati, nation, Oc.	3	1°30′S	173°00′E
Kirin *see* Chilung, Tai.	161	25°02′N	121°48′E
Kiritimati, i., Kir.	2	2°20′N	157°40′W
Kirkby, Eng., U.K.	114a	53°29′N	2°54′W
Kirkby-in-Ashfield, Eng., U.K. (kûrk′bē-ĭn-ăsh′fēld)	114a	53°06′N	1°16′W
Kirkcaldy, Scot., U.K. (kēr-kô′dĭ)	120	56°06′N	3°15′W
Kirkenes, Nor.	116	69°40′N	30°03′E
Kirkham, Eng., U.K. (kûrk′ăm)	114a	53°47′N	2°53′W
Kirkland, Wa., U.S. (kûrk′lănd)	74a	47°41′N	122°12′W
Kirklareli, Tur. (kērk′lär-ĕ′lē)	119	41°44′N	27°15′E
Kirksville, Mo., U.S. (kûrks′vĭl)	65	40°12′N	92°35′W
Kirkwall, Scot., U.K. (kûrk′wôl)	116	58°58′N	2°59′W
Kirkwood, S. Afr.	187c	33°26′S	25°24′E
Kirkwood, Mo., U.S. (kûrk′wŏd)	75e	38°35′N	90°24′W
Kirn, Ger. (kērn)	124	49°47′N	7°23′E
Kirov, Russia	132	54°04′N	34°19′E
Kirov, Russia	134	58°35′N	49°35′E
Kirovakan, Arm.	138	40°48′N	44°30′E
Kirovgrad, Russia (kē′rŭ-vŭ-grad)	142a	57°26′N	60°03′E
Kirovohrad, Ukr.	137	48°33′N	32°17′E
Kirovohrad, prov., Ukr.	133	48°23′N	31°10′E
Kirovsk, Russia (kē-rôfsk′)	142c	59°52′N	30°59′E
Kirovsk, Russia	134	67°40′N	33°58′E
Kirsanov, Russia (kēr-sá′nôf)	137	52°40′N	42°40′E
Kırşehir, Tur. (kēr-shĕ′hēr)	154	39°10′N	34°00′E
Kirtachi Seybou, Niger	189	12°48′N	2°29′E
Kīrthar Range, mts., Pak. (kĭr-tŭr)	155	27°00′N	67°10′E
Kirton, Eng., U.K. (kûr′tŭn)	114a	53°29′N	0°35′W
Kiruna, Swe. (kē-rōō′nä)	116	67°49′N	20°08′E
Kirundu, D.R.C.	191	0°44′S	25°32′E
Kirwin Reservoir, res., Ks., U.S. (kûr′wĭn)	78	39°34′N	99°04′W
Kiryū, Japan	166	36°24′N	139°20′E
Kirzhach, Russia (kēr-zhàk′)	132	56°08′N	38°53′E
Kisaki, Tan. (kē-sá′kē)	187	7°37′S	37°43′E
Kisangani, D.R.C.	185	0°30′N	25°12′E
Kisarazu, Japan (kē′sá-rá′zōō)	167a	35°23′N	139°55′E
Kiselëvsk, Russia (kē-sĭ-lyôfsk′)	134	54°00′N	86°39′E
Kishinev *see* Chişinău, Mol.	134	47°02′N	28°52′E
Kishiwada, Japan (kē′shē-wä′dä)	166	34°25′N	135°18′E
Kishkino, Russia (kēsh′kĭ-nô)	142b	55°15′N	38°04′E
Kisiwani, Tan.	191	4°08′S	37°57′E
Kiska, i., Ak., U.S. (kĭs′kä)	64b	52°08′N	177°10′E
Kiskatinaw, r., Can.	55	55°10′N	120°20′W
Kiskittogisu Lake, l., Can.	57	54°05′N	99°00′W
Kiskitto Lake, l., Can. (kĭs-kĭ′tō)	57	54°16′N	98°34′W
Kiskunfélegyháza, Hung. (kĭsh′kòn-fā′lĕd-y′hä′zô)	125	46°42′N	19°52′E
Kiskunhalas, Hung. (kĭsh′kòn-hô′lôsh)	125	46°24′N	19°26′E
Kiskunmajsa, Hung. (kĭsh′kòn-mī′shô)	125	46°29′N	19°42′E
Kislovodsk, Russia	138	43°55′N	42°44′E
Kismaayo, Som.	187	0°18′S	42°30′E
Kiso-Gawa, r., Japan (kē′sō-gä′wä)	167	35°29′N	137°12′E
Kiso-Sammyaku, mts., Japan (kē′sō säm′myá-kōō)	167	35°47′N	137°39′E
Kíssamos, Grc.	130a	35°13′N	23°35′E
Kissidougou, Gui. (kē′sē-dōō′gōō)	184	9°11′N	10°06′W
Kissimmee, Fl., U.S. (kĭ-sĭm′ē)	83a	28°17′N	81°25′W
Kissimmee, r., Fl., U.S.	83a	27°45′N	81°07′W
Kissimmee, Lake, l., Fl., U.S.	83a	27°58′N	81°17′W
Kisujszállás, Hung.	125	47°12′N	20°47′E
Kisumu, Kenya (kē′sōō-mōō)	186	0°06′S	34°45′E
Kita, Mali (kē′tá)	184	13°03′N	9°29′W
Kitakami Gawa, r., Japan	166	39°20′N	141°10′E
Kitakyūshū, Japan	161	33°53′N	130°50′E
Kitale, Kenya	191	1°01′N	35°00′E
Kit Carson, Co., U.S.	78	38°46′N	102°48′W
Kitchener, Can. (kĭch′ĕ-nēr)	51	43°25′N	80°35′W
Kitenda, D.R.C.	190	6°53′S	17°21′E
Kitgum, Ug. (kĭt′gòm)	185	3°29′N	33°04′E
Kitimat, Can. (kĭ′tĭ-măt)	50	54°03′N	128°33′W
Kitimat, r., Can.	54	53°50′N	129°00′W
Kitimat Ranges, mts., Can.	54	53°30′N	128°50′W
Kitlope, r., Can. (kĭt′lōp)	54	53°00′N	128°00′W
Kitsuki, Japan (kēt′sò-kē)	167	33°24′N	131°35′E
Kittanning, Pa., U.S. (kĭ-tăn′ĭng)	67	40°50′N	79°30′W
Kittatinny Mountains, mts., N.J., U.S. (kĭ-tă-tĭ′nē)	68a	41°16′N	74°44′W
Kittery, Me., U.S. (kĭt′ēr-ĭ)	60	43°07′N	70°45′W
Kittsee, Aus.	115e	48°05′N	17°05′E
Kitty Hawk, N.C., U.S. (kĭt′ē hôk)	83	36°04′N	75°42′W
Kitunda, Tan.	191	6°48′S	33°13′E
Kitwe, Zam.	191	12°49′S	28°13′E
Kitzingen, Ger. (kĭt′zĭng-ĕn)	124	49°44′N	10°08′E
Kiunga, Kenya	191	1°45′S	41°29′E
Kivu, Lac, l., Afr.	186	1°45′S	28°55′E
Kīyose, Japan	167a	35°47′N	139°32′E
Kizel, Russia (kē′zĕl)	136	59°05′N	57°42′E
Kızıl, r., Tur.	154	40°00′N	34°00′E
Kizil′skoye, Russia (kĭz′ĭl-skô-yĕ)	142a	52°43′N	58°53′E
Kizlyar, Russia (kĭz-lyär′)	137	44°00′N	46°50′E
Kizlyarskiy Zaliv, b., Russia	138	44°33′N	46°55′E
Kizu, Japan (kē′zōō)	167	34°43′N	135°49′E
Klaas Smits, r., S. Afr.	187c	31°45′S	26°33′E
Klaaswaal, Neth.	115a	51°46′N	4°25′E
Kladno, Czech Rep. (kläd′nō)	124	50°10′N	14°05′E
Klagenfurt, Aus. (klä′gĕn-fòrt)	117	46°38′N	14°19′E
Klaipėda, Lith. (klī′pà-dá)	136	55°43′N	21°10′E
Klamath, r., U.S.	72	41°40′N	123°25′W
Klamath Falls, Or., U.S.	64	42°13′N	121°49′W
Klamath Mountains, mts., Ca., U.S.	72	42°00′N	123°25′W
Klarälven, r., Swe.	116	60°40′N	13°00′E
Klaskanine, r., Or., U.S. (klăs′kă-nīn)	74c	46°02′N	123°43′W
Klatovy, Czech Rep. (klä′tô-vē)	117	49°23′N	13°18′E
Klawock, Ak., U.S. (klă′wăk)	63	55°32′N	133°10′W
Kleinmachnow, Ger. (klīn-mäk′nō)	115b	52°22′N	13°12′E
Klerksdorp, S. Afr. (klērks′dôrp)	192c	26°52′S	26°40′E
Klerksraal, S. Afr. (klērks′käl)	192c	26°15′S	27°10′E
Kletnya, Russia (klyĕt′nyá)	132	53°19′N	33°14′E
Kleve, Ger. (klĕ′fĕ)	124	51°47′N	6°09′E

ng-sing; ŋ-baŋk; ɴ-nasalized n; nŏd; cŏmmit; ōld; ŏbey; ôrder; oi-boil; fōōd; ŏŏ-as oo in foot; ou-out; s-soft; sh-dish; th-thin; pūre; ůnite; ûrn; stŭd; circŭs; ü-as in French tu; ′-indeterminate vowel.

264 Klickitat – Krasnousol'skiy PRONOUNCING INDEX

PLACE (Pronunciation)	PAGE	LAT.	LONG.
Klickitat, r., Wa., U.S.	72	46°01'N	121°07'W
Klimovichi, Bela. (klē-mô-vē'chē)	132	53°37'N	31°21'E
Klimovsk, Russia (klĭ'môfsk)	142b	55°21'N	37°32'E
Klin, Russia (klēn)	132	56°18'N	36°43'E
Klintehamn, Swe. (klēn'tē-häm)	122	57°24'N	18°14'E
Klintsy, Russia (klĭn'tsĭ)	137	52°46'N	32°14'E
Klip, r., S. Afr. (klĭp)	192c	27°18'N	29°25'E
Klipgat, S. Afr.	192c	25°26'S	27°57'E
Klippan, Swe. (klyp'pan)	122	56°08'N	13°09'E
Kłodzko, Pol. (klôd'skô)	124	50°26'N	16°38'E
Klondike Region, hist. reg., N.A. (klŏn'dīk)	50	64°12'N	142°38'W
Klosterfelde, Ger. (klōs'tĕr-fĕl-dĕ)	115b	52°47'N	13°29'E
Klosterneuburg, Aus. (klōs-tĕr-noi'bŏorgh)	115e	48°19'N	16°20'E
Kluane, l., Can.	52	61°15'N	138°40'W
Kluane National Park, rec., Can.	52	60°25'N	137°53'W
Kluczbork, Pol. (klŏoch'bôrk)	125	50°59'N	18°15'E
Klyaz'ma, r., Russia (klyäz'må)	132	55°49'N	39°19'E
Klyetsk, Bela. (klĕtsk)	132	53°04'N	26°43'E
Klyuchevskaya, vol., Russia (klyŏo-chĕfskä'yä)	135	56°13'N	160°00'E
Klyuchi, Russia (klyōō'chĭ)	142a	57°03'N	57°20'E
Knezha, Blg. (knyā'zhà)	119	43°27'N	24°03'E
Knife, r., N.D., U.S. (nīf)	70	47°06'N	102°33'W
Knight Inlet, b., Can. (nīt)	54	50°41'N	125°40'W
Knightstown, In., U.S. (nīts'toun)	66	39°45'N	85°30'W
Knin, Cro. (knēn)	130	44°02'N	16°14'E
Knittelfeld, Aus.	117	47°13'N	14°50'E
Knob Peak, mtn., Phil. (nŏb)	169a	12°30'N	121°20'E
Knottingley, Eng., U.K. (nŏt'ĭng-lĭ)	114a	53°42'N	1°14'W
Knox, In., U.S. (nŏks)	66	41°15'N	86°40'W
Knox, Cape, c., Can.	54	54°12'N	133°20'W
Knoxville, Ia., U.S. (nŏks'vĭl)	71	41°19'N	93°05'W
Knoxville, Tn., U.S.	65	35°58'N	83°55'W
Knutsford, Eng., U.K. (nŭts'fẽrd)	114a	53°18'N	2°22'W
Knyszyn, Pol. (knĭ'shĭn)	125	53°16'N	22°59'E
Kobayashi, Japan (kō'bà-yä'shē)	167	31°58'N	130°59'E
Kōbe, Japan (kō'bē)	161	34°30'N	135°10'E
Kobeliaky, Ukr.	137	49°11'N	34°12'E
København see Copenhagen, Den.	110	55°43'N	12°27'E
Koblenz, Ger. (kō'blĕntz)	117	50°18'N	7°36'E
Kobozha, r., Russia (kô-bō'zhà)	132	58°55'N	35°18'E
Kobrinskoye, Russia (kô-brĭn'skô-yĕ)	142c	59°25'N	30°07'E
Kobryn, Bela. (kō'brĕn')	137	52°13'N	24°23'E
Kobuk, r., Ak., U.S. (kō'bŭk)	63	66°58'N	158°48'W
Kobuk Valley National Park, rec., Ak., U.S.	63	67°20'N	159°00'W
Kobuleti, Geor. (kô-bô-lyā'tē)	137	41°50'N	41°40'E
Kočani, Mac. (kō'chä-nē)	131	41°54'N	22°25'E
Kočevje, Slvn. (kō'chāv-ye)	130	45°38'N	14°51'E
Kocher, r., Ger. (kôk'ẽr)	124	49°00'N	9°52'E
Kochi, India	159	9°58'N	76°19'E
Kōchi, Japan (kō'chē)	161	33°35'N	133°32'E
Kodaira, Japan	167a	35°43'N	139°29'E
Kodiak, Ak., U.S. (kō'dyăk)	64a	57°50'N	152°30'W
Kodiak Island, i., Ak., U.S.	64a	57°24'N	153°32'W
Kodok, Sudan (ko'dŏk)	185	9°57'N	32°08'E
Koforidua, Ghana (kō fō-rĭ-dōō'à)	184	6°03'N	0°17'W
Kōfu, Japan (kō'fōō')	166	35°41'N	138°34'E
Koga, Japan (kō'gà)	167	36°13'N	139°40'E
Kogan, r., Gui.	188	11°30'N	14°05'W
Kogane, Japan (kō'gä-nå)	167a	35°50'N	139°56'E
Koganei, Japan (kō'gä-nā)	167a	35°42'N	139°31'E
Køge, Den. (kû'gĕ)	122	55°27'N	12°09'E
Køge Bugt, b., Den.	122	55°30'N	12°25'E
Kogoni, Mali	188	14°44'N	6°02'W
Kohīma, India (kô-ē'mà)	155	25°45'N	94°41'E
Kohyl'nyk, r., Eur.	133	46°08'N	29°10'E
Koito, r., Japan (kô'ē-tō)	167a	35°19'N	139°58'E
Kōje, i., Kor., S. (kû'jĕ)	166	34°53'N	129°00'E
Kokand, Uzb. (kô-känt')	139	40°27'N	71°07'E
Kokemäenjoki, r., Fin.	123	61°23'N	22°03'E
Kokhma, Russia (kôk'mà)	132	56°57'N	41°08'E
Kokkola, Fin. (kô'kô-lä)	116	63°47'N	22°58'E
Kokomo, In., U.S. (kō'kô-mō)	66	40°30'N	86°20'W
Koko Nor (Qinghai Hu), l., China (kō'kô nor) (chyĭŋ-hī hōō)	160	37°26'N	98°30'E
Kokopo, Pap. N. Gui. (kô-kō'pō)	169	4°25'S	152°27'E
Kökshetaū, Kaz.	139	53°15'N	69°13'E
Koksoak, r., Can. (kôk'sô-ăk)	53	57°42'N	69°50'W
Kokstad, S. Afr. (kôk'shtät)	187c	30°33'S	29°27'E
Kokubu, Japan (kō'kōō-bōō)	167	31°42'N	130°46'E
Kokuou, Japan (kô'kōō-ô'ōō)	167b	34°34'N	135°39'E
Kola Peninsula see Kol'skiy Poluostrov, pen., Russia	134	67°15'N	37°40'E
Kolār (Kolār Gold Fields), India (kōl-är')	155	13°39'N	78°33'E
Kolárvo, Slvk. (kōl-ärōvō)	125	47°54'N	17°59'E
Kolbio, Kenya	191	1°10'S	41°15'E
Kol'chugino, Russia (kôl-chò'gē-nô)	132	56°19'N	39°29'E
Kolda, Sen.	188	12°53'N	14°57'W
Kolding, Den. (kŭl'dĭng)	122	55°29'N	9°24'E
Kole, D.R.C. (kō'lâ)	186	3°19'S	22°46'E
Kolguyev, i., Russia (kôl-gò'yĕf)	134	69°00'N	49°00'E
Kolhāpur, India	159	16°48'N	74°15'E
Kolín, Czech Rep. (kō'lēn)	124	50°01'N	15°11'E
Kolkasrags, c., Lat. (kôl-käs'rágz)	123	57°46'N	22°39'E
Kolkata (Calcutta), India	155	22°32'N	88°22'E
Köln see Cologne, Ger.	127c	50°56'N	6°57'E
Kolno, Pol. (kôw'nô)	125	53°23'N	21°56'E
Koło, Pol. (kô'wô)	125	52°11'N	18°37'E
Kołobrzeg, Pol. (kô-lôb'zhĕk)	116	54°10'N	15°35'E
Kolomna, Russia (kál-ôm'ná)	136	55°06'N	38°47'E
Kolomyia, Ukr.	125	48°32'N	25°04'E
Kolp', r., Russia (kôlp)	132	59°18'N	35°32'E
Kolpashevo, Russia (kŭl på shô'vá)	134	58°16'N	82°43'E
Kolpino, Russia (kôl'pē-nô)	136	59°45'N	30°37'E
Kolpny, Russia (kôlp'nyē)	132	52°14'N	36°54'E
Kol'skiy Poluostrov, pen., Russia	134	67°15'N	37°40'E
Kolva, r., Russia	136	61°00'N	57°00'E
Kolwezi, D.R.C. (kōl-wē'zē)	186	10°43'S	25°28'E
Kolyberovo, Russia (kô-lĭ-byâ'rô-vô)	142b	55°16'N	38°45'E
Kolyma, r., Russia	135	66°30'N	151°45'E
Kolymskiy Mountains see Gydan, Khrebet, mts., Russia	135	61°45'N	155°00'E
Kom, r., Afr.	190	2°15'N	12°05'E
Komadugu Gana, r., Nig.	189	12°15'N	11°10'E
Komae, Japan	167a	35°37'N	139°35'E
Komandorskiye Ostrova, is., Russia	153	55°40'N	167°13'E
Komárno, Slvk. (kô'mär-nô)	125	47°46'N	18°08'E
Komarno, Ukr.	125	49°38'N	23°42'E
Komárom, Hung. (kô'mä-rôm)	125	47°45'N	18°06'E
Komatipoort, S. Afr. (kō-mä'tē-pôrt)	186	25°21'S	32°00'E
Komatsu, Japan (kō-mät'sōō)	166	36°23'N	136°26'E
Komatsushima, Japan (kō-mät'sōō-shē'mä)	167	34°04'N	134°32'E
Komeshia, D.R.C.	191	8°01'S	27°07'E
Komga, S. Afr. (kōm'gà)	187c	32°36'S	27°54'E
Komi, prov., Russia (kômē)	140	63°00'N	55°00'E
Kommetijie, S. Afr.	186a	34°09'S	18°19'E
Komoé, r., C. Iv.	188	5°40'N	3°40'W
Komsomolets, Kaz.	142a	53°45'N	62°04'E
Komsomol'sk-na-Amure, Russia	135	50°46'N	137°14'E
Kona, Mali	188	14°57'N	3°53'W
Konda, r., Russia (kôn'dà)	136	60°50'N	64°00'E
Kondas, r., Russia (kôn'dás)	142a	59°30'N	56°28'E
Kondoa, Tan. (kôn-dō'à)	186	4°52'S	36°00'E
Kondolole, D.R.C.	191	1°20'N	25°58'E
Koné, N. Cal.	170f	21°04'S	164°52'E
Kong, C. Iv. (kông)	184	9°05'N	4°41'W
Kongbo, C.A.R.	190	4°44'N	21°23'E
Kongolo, D.R.C. (kôŋ'gō'lō)	186	5°23'S	27°00'E
Kongsberg, Nor. (kŭngs'bĕrg)	122	59°40'N	9°36'E
Kongsvinger, Nor. (kŭngs'vĭŋ-gẽr)	122	60°12'N	12°00'E
Koni, D.R.C. (kō'nē)	186	10°32'S	27°27'E
Königsberg see Kaliningrad, Russia	134	54°42'N	20°32'E
Königsbrunn, Ger. (kû'nēgs-brōōn)	115d	48°16'N	10°53'E
Königs Wusterhausen, Ger. (kû'nēgs vōōs'tĕr-hou-zĕn)	115b	52°18'N	13°38'E
Konin, Pol. (kô'nyĕn)	117	52°11'N	18°17'E
Kónitsa, Grc. (kô'nyē'tsà)	131	40°03'N	20°46'E
Konjic, Bos. (kôn'yēts)	131	43°38'N	17°59'E
Konju, Kor., S.	166	36°21'N	127°05'E
Konnagar, India	158a	22°41'N	88°22'E
Konotop, Ukr. (kô-nô-tôp')	137	51°13'N	33°14'E
Konpienga, r., Burkina	188	11°15'N	0°35'E
Konqi, r., China (kôn-chyē)	160	41°09'N	87°46'E
Końskie, Pol. (koin''skyĕ)	125	51°12'N	20°26'E
Konstanz, Ger. (kôn'shtänts)	124	47°39'N	9°10'E
Kontagora, Nig. (kōn-tà-gō'rä)	184	10°24'N	5°28'E
Konya, Tur. (kōn'yà)	154	36°55'N	32°25'E
Koocanusa, Lake, res., N.A.	72	49°00'N	115°10'W
Kootenay (Kootenai), r., N.A.	55	49°45'N	117°05'W
Kootenay Lake, l., Can.	55	49°35'N	116°50'W
Kootenay National Park, rec., Can. (kōō'tē-nâ)	50	51°06'N	117°02'W
Kōō-zan, mtn., Japan (kōō'zän)	167b	34°53'N	135°32'E
Kopervik, Nor. (kô'pĕr-vêk)	122	59°18'N	5°20'E
Kopeysk, Russia (kô-pāsk')	140	55°07'N	61°37'E
Köping, Swe. (chû'pĭng)	122	59°32'N	15°58'E
Kopparberg, Swe. (kōp'pár-bĕrgh)	122	59°53'N	15°00'E
Koppeh Dāgh, mts., Asia	154	37°28'N	58°29'E
Koppies, S. Afr.	192c	27°15'S	27°35'E
Koprivnica, Cro. (kô'prēv-nē'tsà)	130	46°10'N	16°48'E
Kopychyntsi, Ukr.	125	49°06'N	25°55'E
Korčula, i., Serb. (kôr'chōō-là)	131	42°50'N	17°05'E
Korea, North, nation, Asia	161	40°00'N	127°00'E
Korea, South, nation, Asia	161	36°30'N	128°00'E
Korea Bay, b., Asia	164	39°18'N	123°50'E
Korean Archipelago, is., Kor., S.	161	34°05'N	125°35'E
Korea Strait, strt., Asia	161	33°30'N	128°30'E
Korets', Ukr.	125	50°35'N	27°13'E
Korhogo, C. Iv. (kôr-hō'gō)	184	9°27'N	5°38'W
Korinthiakós Kólpos, b., Grc.	119	38°15'N	22°33'E
Kórinthos, Grc. (kô-rēn'thôs) (kŏr'ĭnth)	110	37°56'N	22°54'E
Koriukivka, Ukr.	133	51°44'N	32°24'E
Kōriyama, Japan (kō'rē-yä'mä)	166	37°18'N	140°25'E
Korkino, Russia (kôr'kē-nŭ)	142a	54°53'N	61°25'E
Korla, China (kōr-lä)	160	41°37'N	86°03'E
Körmend, Hung. (kûr'mĕnt)	124	47°02'N	16°36'E
Kornat, i., Serb. (kôr-nät')	130	43°46'N	15°10'E
Korneuburg, Aus. (kôr'noi-bòrgh)	115e	48°22'N	16°21'E
Koro, Mali	188	14°04'N	3°05'W
Korocha, Russia (kô-rō'chà)	133	50°50'N	37°13'E
Korop, Ukr. (kō'rôp)	133	51°33'N	32°54'E
Koro Sea, sea, Fiji	170g	18°00'S	179°50'E
Korosten', Ukr. (kô'rôs-tĕn)	137	50°51'N	28°39'E
Korostyshiv, Ukr.	133	50°19'N	29°05'E
Koro Toro, Chad	189	16°05'N	18°30'E
Korotoyak, Russia (kô'rô-tô-yák')	133	51°00'N	39°06'E
Korsakov, Russia (kôr'sà-kôf')	135	46°42'N	143°16'E
Korsnäs, Fin. (kôrs'nĕs)	123	62°51'N	21°17'E
Korsør, Den. (kôrs'ûr')	122	55°19'N	11°08'E
Kortrijk, Bel.	121	50°49'N	3°10'E
Koryakskiy Khrebet, mts., Russia	135	62°00'N	168°45'E
Kosa Byriuchyi ostriv, i., Ukr.	133	46°07'N	35°12'E
Kościan, Pol. (kŭsh'tsyàn)	124	52°05'N	16°38'E
Kościerzyna, Pol. (kŭsh-tsyĕ-zhē'nà)	125	54°08'N	17°59'E
Kosciusko, Ms., U.S. (kŏs-ĭ-ŭs'kō)	82	33°04'N	89°35'W
Kosciuszko, Mount, mtn., Austl.	175	36°26'S	148°20'E
Kosha, Sudan	185	20°49'N	30°27'E
Koshigaya, Japan (kō'shē-gä'yä)	167a	35°53'N	139°48'E
Kōshim, r., Kaz.	137	50°30'N	50°40'E
Kosi, r., India (kô'sē)	158	26°00'N	86°20'E
Košice, Slvk. (kō'shē-tsĕ')	111	48°43'N	21°17'E
Kosmos, S. Afr. (kŏz'mŏs)	187b	25°45'S	27°51'E
Kosobrodskiy, Russia (kä-sô'brôd-skĭ)	142a	54°14'N	60°53'E
Kosovo, hist. reg., Serb.	131	42°35'N	21°00'E
Kosovska Mitrovica, Serb. (kô'sôv-skä' mē'trô-vē-tsä')	131	42°51'N	20°50'E
Kostajnica, Cro. (kôs'tä-ē-nē'tsà)	130	45°14'N	16°32'E
Koster, S. Afr.	192c	25°52'S	26°52'E
Kostiantynivka, Ukr.	133	48°33'N	37°42'E
Kostino, Russia (kôs'tĭ-nô)	142b	55°54'N	37°51'E
Kostroma, Russia (kôs-trô-má')	134	57°46'N	40°55'E
Kostroma, prov., Russia	132	57°50'N	41°10'E
Kostrzyn, Pol. (kôst'chĕn)	117	52°35'N	14°38'E
Kos'va, r., Russia (kôs'và)	142a	58°44'N	57°08'E
Koszalin, Pol. (kô-shä'lĭn)	116	54°12'N	16°10'E
Kőszeg, Hung. (kû'sĕg)	124	47°21'N	16°32'E
Kota, India	155	25°17'N	75°49'E
Kota Baharu, Malay. (kō'tä bä'rōō)	168	6°15'N	102°23'E
Kotabaru, Indon.	168	3°22'S	116°15'E
Kota Kinabalu, Malay.	168	5°55'N	116°05'E
Kota Tinggi, Malay.	153b	1°43'N	103°54'E
Kotel, Blg. (kō-tĕl')	131	42°54'N	26°28'E
Kotel'nich, Russia (kô-tyĕl'nēch)	136	58°15'N	48°20'E
Kotel'nyy, i., Russia (kô-tyĕl'nē)	135	74°51'N	134°09'E
Kotka, Fin. (kŏt'kà)	116	60°28'N	26°56'E
Kotlas, Russia (kôt'lás)	136	61°10'N	46°50'E
Kotlin, Ostrov, i., Russia (ôs-trôf' kôt'lĭn)	142c	60°02'N	29°49'E
Kotor, Serb.	131	42°25'N	18°46'E
Kotorosl', r., Russia (kô-tô'rôsl)	132	57°18'N	39°08'E
Kotovs'k, Ukr.	133	47°49'N	29°31'E
Kotto, r., C.A.R.	185	5°17'N	22°04'E
Kotuy, r., Russia (kô-tōō')	140	71°00'N	103°15'E
Kotzebue, Ak., U.S. (kôt'sĕ-bōō)	64a	66°48'N	162°42'W
Kotzebue Sound, strt., Ak., U.S.	63	67°00'N	164°28'W
Kouchibouguac National Park, rec., Can.	60	46°53'N	65°35'W
Koudougou, Burkina (kōō-dōō'gōō)	184	12°15'N	2°22'W
Kouilou, r., Congo	186	4°30'S	12°00'E
Koula-Moutou, Gabon	190	1°08'S	12°29'E
Koulikoro, Mali (kōō-lē-kō'rō)	184	12°53'N	7°33'W
Koulouguidi, Mali	189	13°27'N	17°33'E
Koumac, N. Cal.	170f	20°33'S	164°17'E
Koumra, Chad	189	8°55'N	17°33'E
Koundara, Gui.	188	12°29'N	13°18'W
Kouroussa, Gui. (kōō-rōō'sä)	184	10°39'N	9°53'W
Koutiala, Mali (kōō-tē-ä'lä)	184	12°29'N	5°29'W
Kouvola, Fin. (kō'ô-vô-lä)	123	60°51'N	26°40'E
Kouzhen, China (kô-jŭn)	162	36°19'N	117°37'E
Kovda, l., Russia (kôv'dà)	136	66°45'N	32°00'E
Kovel', Ukr. (kō'vĕl)	137	51°13'N	24°45'E
Kovno see Kaunas, Lith.	134	54°42'N	23°54'E
Kovrov, Russia (kôv-rôf')	136	56°23'N	41°21'E
Koyuk, Ak., U.S. (kô-yōōk')	63	65°00'N	161°18'W
Koyukuk, r., Ak., U.S. (kô-yōō'kôk)	63	66°25'N	153°50'W
Kozáni, Grc.	119	40°16'N	21°51'E
Kozelets', Ukr. (kôzĕ-lyĕts)	133	50°53'N	31°07'E
Kozel'sk, Russia (kô-zĕlsk')	132	54°01'N	35°49'E
Kozhikode, India	155	11°19'N	75°49'E
Koziatyn, Ukr.	137	49°43'N	28°50'E
Kozienice, Pol. (kō-zyĕ-nē'tsĕ)	125	51°34'N	21°35'E
Koźle, Pol. (kôzh'lĕ)	125	50°19'N	18°10'E
Kozloduy, Blg. (kŭz'lô-dwē)	131	43°45'N	23°42'E
Kōzu, i., Japan (kō'zōō)	167	34°16'N	139°03'E
Kra, Isthmus of, isth., Asia	168	9°30'S	99°45'E
Kraai, r., S. Afr. (krā'ē)	187c	30°50'S	27°03'E
Krabbendijke, Neth.	115a	51°26'N	4°05'E
Krâchéh, Camb.	168	12°28'N	106°06'E
Kragujevac, Serb. (krä'gōō'yĕ-vàts)	119	44°01'N	20°55'E
Kraków, Pol. (krä'kôf)	110	50°05'N	20°00'E
Kraljevo, Serb. (kräl'ye-vô)	119	43°39'N	20°48'E
Kramators'k, Ukr.	133	48°43'N	37°32'E
Kramfors, Swe. (kräm'fôrs)	122	62°54'N	17°49'E
Kranj, Slvn. (krän')	118	46°16'N	14°23'E
Kranskop, S. Afr. (kränz'kŏp)	187c	28°57'S	30°54'E
Krāslava, Lat. (kräs'lä-vä)	123	55°53'N	27°12'E
Kraslice, Czech Rep. (kräs'lē-tsĕ)	124	50°19'N	12°30'E
Krasnaya Gorka, Russia	142a	55°12'N	56°40'E
Krasnaya Sloboda, Russia	137	48°25'N	44°35'E
Kraśnik, Pol. (kräsh'nĭk)	125	50°53'N	22°15'E
Krasnoarmeysk, Russia (kräs'nô-ár-maśk')	142b	56°06'N	38°09'E
Krasnoarmiis'k, Ukr.	133	48°19'N	37°04'E
Krasnodar, Russia (kräs'nô-dár)	134	45°03'N	38°55'E
Krasnodarskiy, prov., Russia (kräs-nô-där'skĭ ôb'låst)	133	45°25'N	38°10'E
Krasnogorsk, Russia	142b	55°49'N	37°20'E
Krasnogorskiy, Russia (kräs-nô-gôr'skĭ)	142a	54°36'N	61°15'E
Krasnogvardeyskiy, Russia (krå'sno-gvăr-dzyē ĕs-kēē)	142a	57°17'N	62°05'E
Krasnohrad, Ukr.	133	49°23'N	35°26'E
Krasnokamsk, Russia (kräs-nô-kämsk')	136	58°00'N	55°45'E
Krasnokuts'k, Ukr.	133	50°03'N	35°05'E
Krasnoslobodsk, Russia (kräs'nô-slôbôtsk')	136	54°20'N	43°50'E
Krasnotur'insk, Russia (krŭs-nŭ-tōō-rensk')	134	59°47'N	60°15'E
Krasnoufimsk, Russia (krŭs-nŭ-ōō-fēmsk')	134	56°38'N	57°46'E
Krasnoural'sk, Russia (kräs'nô-ōō-rälsk')	136	58°21'N	60°05'E
Krasnousol'skiy, Russia (kräs-nô-ŏ-sôl'skĭ)	142a	53°54'N	56°27'E

ăt; finăl; rāte; senâte; ärm; ȧsk; sofá; fâre; ch-choose; dh-as th in other; bē; ēvent; bĕt; recĕnt; cratẽr; g-gō; gh-guttural g; bĭt; ĭ-short neutral; rīde; ĸ-guttural k as ch in German ich;

266 La Dorado – Larch Mountain PRONOUNCING INDEX

PLACE (Pronunciation)	PAGE	LAT.	LONG.
La Dorado, Col. (lä dô-rä'dä)	100	5°28'N	74°42'W
Ladozhskoye Ozero, l., Russia (lá-dôsh'skô-yĕ ô'zĕ-rô)	134	60°59'N	31°30'E
La Durantaye, Can. (lä dü-räN-tā')	62b	46°51'N	70°51'W
Lady Frere, S. Afr. (lā-dē frâ'r')	187c	31°48'S	27°16'E
Lady Grey, S. Afr.	187c	30°44'S	27°17'E
Ladysmith, Can. (lā'dĭ-smĭth)	54	48°58'N	123°49'W
Ladysmith, S. Afr.	186	28°38'S	29°48'E
Ladysmith, Wi., U.S.	71	45°27'N	91°07'W
Lae, Pap. N. Gui. (lä'å)	169	6°15'S	146°57'E
Laerdalsøyri, Nor.	122	61°08'N	7°26'E
La Esperanza, Hond. (lä ĕs-på-rän'zä)	90	14°20'N	88°21'W
Lafayette, Al., U.S.	82	32°52'N	85°25'W
Lafayette, Ca., U.S.	74b	37°53'N	122°07'W
Lafayette, Ga., U.S. (lă-fā-yĕt')	82	34°41'N	85°19'W
Lafayette, In., U.S.	65	40°25'N	86°55'W
Lafayette, La., U.S.	65	30°15'N	92°02'W
La Fayette, R.I., U.S.	68b	41°34'N	71°29'W
La Ferté-Alais, Fr. (lä-fĕr-tā'ä-lā')	127b	48°29'N	2°19'E
La Ferté-sous-Jouarre, Fr. (lá fĕr-tā'sōō-zhōō-är')	127b	48°56'N	3°07'E
Lafia, Nig.	189	8°30'N	8°30'E
Lafiagi, Nig.	189	8°52'N	5°25'E
La Flèche, Fr. (lá flâsh')	126	47°43'N	0°03'W
La Follete, Tn., U.S. (lä-fŏl'ĕt)	82	36°23'N	84°07'W
Lafourche, Bayou, r., La., U.S. (bä-yōō'lå-fōōrsh')	81	29°25'N	90°15'W
La Gaiba, Braz. (lä-gī'bä)	101	17°54'S	57°32'W
La Galite, i., Tun. (gä-lēt)	118	37°36'N	8°03'E
Lågan, r., Nor. (lô'ghĕn)	112	61°00'N	10°00'E
Lagan, r., Swe.	122	56°34'N	13°25'E
Lagan, r., N. Ire., U.K. (lā'găn)	120	54°30'N	6°00'W
Lagarto, r., Pan. (lä-gä'r-tô)	86a	9°08'N	80°05'W
Lagartos, l., Mex. (lä-gä'r-tôs)	90a	21°32'N	88°15'W
Laghouat, Alg. (lä-gwät')	184	33°45'N	2°49'E
Lagkadás, Grc.	131	40°44'N	23°10'E
Lagny, Fr. (län-yē')	127b	48°53'N	2°41'E
Lagoa da Prata, Braz. (lä-gô'ä-dä-prä'tä)	99a	20°04'S	45°33'W
Lagoa Dourada, Braz. (lä-gô'ä-dô-rä'dä)	99a	20°55'S	44°03'W
Lagogne, Fr. (laN-gōn'y')	126	44°43'N	3°50'E
Lagonay, Phil.	169a	13°44'N	123°31'E
Lagos, Nig. (lä'gōs)	184	6°27'N	3°24'E
Lagos, Port. (lä'gōzh)	128	37°08'N	8°43'W
Lagos de Moreno, Mex. (lä'gōs dā mô-rā'nō)	86	21°21'N	101°55'W
La Grand' Combe, Fr. (lå gräN kôNb')	126	44°12'N	4°03'E
La Grande, Or., U.S. (lå grănd')	64	45°20'N	118°06'W
La Grande, r., Can.	53	53°55'N	77°30'W
La Grange, Austl. (lä grānj)	174	18°40'S	122°00'E
La Grange, Ga., U.S. (lå-grānj')	65	33°01'N	85°00'W
La Grange, Il., U.S.	69a	41°49'N	87°53'W
Lagrange, In., U.S.	66	41°40'N	85°25'W
La Grange, Ky., U.S.	66	38°20'N	85°25'W
La Grange, Mo., U.S.	79	40°04'N	91°30'W
Lagrange, Oh., U.S.	69d	41°14'N	82°07'W
Lagrange, Tx., U.S.	81	29°55'N	96°50'W
La Grita, Ven. (lä grē'tä)	100	8°02'N	71°59'W
La Guaira, Ven. (lä gwä'ē-rä)	100	10°36'N	66°54'W
La Guardia, Spain (lä gwär'dē-ä)	128	41°55'N	8°48'W
Laguna, Braz. (lä-gōō'nä)	102	28°19'S	48°42'W
Laguna, Cayos, is., Cuba (kä'yōs-lä-gò'nä)	92	22°15'N	82°45'W
Laguna Indian Reservation, I.R., N.M., U.S.	77	35°00'N	107°30'W
Lagunillas, Bol. (lä-gōō-nēl'yäs)	100	19°42'S	63°38'W
Lagunillas, Mex. (lä-gōō-nē'l-yäs)	88	21°34'N	99°41'W
La Habana *see* Havana, Cuba	87	23°08'N	82°23'W
La Habra, Ca., U.S. (lá hàb'rà)	75a	34°56'N	117°57'W
Lahaina, Hi., U.S. (lä-hä'ē-nä)	84a	20°52'N	156°39'W
Lāhījān, Iran	157	37°12'N	50°01'E
Laholm, Swe. (lä'hôlm)	122	56°30'N	13°00'E
La Honda, Ca., U.S. (lä hôn'dä)	74b	37°20'N	122°16'W
Lahore, Pak. (lä-hōr')	155	32°00'N	74°18'E
Lahr, Ger. (lär)	124	48°19'N	7°52'E
Lahti, Fin. (lä'tē)	116	60°59'N	27°39'E
Lai, Chad	185	9°29'N	16°18'E
Lai'an, China (lī-än)	162	32°27'N	118°25'E
Laibin, China (lī-bĭn)	165	23°42'N	109°20'E
L'Aigle, Fr. (lĕ'gl')	126	48°45'N	0°37'E
Laisamis, Kenya	191	1°36'N	37°48'E
Laiyang, China (läī'yäng)	164	36°59'N	120°42'E
Laizhou Wan, b., China (lī-jō wän)	161	37°22'N	119°19'E
Laja, Río de la, r., Mex. (rē'ō-dē-lä-lä'κä)	88	21°17'N	100°57'W
Lajas, Cuba (lä'häs)	92	22°25'N	80°20'W
Lajeado, Braz. (lä-zhĕä'dô)	102	29°24'S	51°46'W
Lajes, Braz. (lå'zhĕs)	102	27°47'S	50°17'W
Lajinha, Braz. (lä-zhē'nyä)	99a	20°08'S	41°36'W
La Jolla, Ca., U.S. (lå hoi'yå)	76a	32°51'N	117°16'W
La Jolla Indian Reservation, I.R., Ca., U.S.	76	33°19'N	116°21'W
La Junta, Co., U.S. (lå hōōn'tå)	78	37°59'N	103°35'W
Lake Arthur, La., U.S. (är'thŭr)	81	30°06'N	92°40'W
Lake Barkley, res., U.S.	82	36°45'N	88°00'W
Lake Benton, Mn., U.S. (bĕn'tŭn)	70	44°15'N	96°17'W
Lake Bluff, Il., U.S. (blŭf)	69a	42°17'N	87°50'W
Lake Brown, Austl. (broun)	174	31°03'S	118°30'E
Lake Charles, La., U.S. (chärlz')	65	30°15'N	93°14'W
Lake City, Fl., U.S.	83	30°09'N	82°40'W
Lake City, Ia., U.S.	71	42°14'N	94°43'W
Lake City, Mn., U.S.	71	44°28'N	92°19'W
Lake City, S.C., U.S.	83	33°57'N	79°45'W
Lake Clark National Park, rec., Ak., U.S.	63	60°30'N	153°15'W
Lake Cowichan, Can. (kou'ĭ-chăn)	54	48°50'N	124°03'W

PLACE (Pronunciation)	PAGE	LAT.	LONG.
Lake Crystal, Mn., U.S. (krĭs'tăl)	71	44°05'N	94°12'W
Lake District, reg., Eng., U.K. (lāk)	120	54°25'N	3°20'W
Lake Elmo, Mn., U.S. (ĕlmō)	75g	45°00'N	92°53'W
Lake Forest, Il., U.S. (fŏr'ĕst)	69a	42°16'N	87°50'W
Lake Fork, r., Ut., U.S.	77	40°30'N	110°25'W
Lake Geneva, Wi., U.S. (jĕ-nē'vå)	71	42°36'N	88°28'W
Lake Havasu City, Az., U.S.	77	34°27'N	114°22'W
Lake June, Tx., U.S. (jōōn)	75c	32°43'N	96°45'W
Lakeland, Fl., U.S. (lāk'lănd)	65	28°02'N	81°58'W
Lakeland, Ga., U.S.	82	31°02'N	83°02'W
Lakeland, Mn., U.S.	75g	44°57'N	92°47'W
Lake Linden, Mi., U.S. (lĭn'dĕn)	71	47°11'N	88°26'W
Lake Louise, Can. (lōō-ēz')	55	51°26'N	116°11'W
Lake Mead National Recreation Area, rec., U.S.	77	36°00'N	114°30'W
Lake Mills, Ia., U.S. (mĭlz')	71	43°25'N	93°32'W
Lakemore, Oh., U.S. (lāk-mōr)	69d	41°01'N	81°24'W
Lake Odessa, Mi., U.S.	66	42°50'N	85°15'W
Lake Oswego, Or., U.S. (ŏs-wē'go)	74c	45°25'N	122°40'W
Lake Placid, N.Y., U.S.	67	44°17'N	73°59'W
Lake Point, Ut., U.S.	75b	40°41'N	112°16'W
Lakeport, Ca., U.S. (lāk'pōrt)	76	39°03'N	122°54'W
Lake Preston, S.D., U.S. (prĕs'tŭn)	70	44°21'N	97°23'W
Lake Providence, La., U.S. (prŏv'ĭ-dĕns)	81	32°48'N	91°12'W
Lake Red Rock, res., Ia., U.S.	71	41°30'N	93°15'W
Lake Sharpe, res., S.D., U.S.	70	44°30'N	100°00'W
Lakeside, Ca., U.S. (lāk'sīd)	76a	32°52'N	116°55'W
Lake Station, In., U.S.	69a	41°34'N	87°15'W
Lake Stevens, Wa., U.S.	74a	48°01'N	122°04'W
Lake Success, N.Y., U.S. (sŭk-sĕs')	68a	40°46'N	73°43'W
Lakeview, Or., U.S.	72	42°11'N	120°21'W
Lake Village, Ar., U.S.	79	33°20'N	91°17'W
Lake Wales, Fl., U.S. (wālz')	83a	27°54'N	81°35'W
Lakewood, Ca., U.S. (lāk'wŏŏd)	75a	33°50'N	118°09'W
Lakewood, Co., U.S.	78	39°44'N	105°06'W
Lakewood, Oh., U.S.	65	41°29'N	81°48'W
Lakewood, Pa., U.S.	67	40°05'N	74°10'W
Lakewood, Wa., U.S.	74a	48°09'N	122°13'W
Lakewood Center, Wa., U.S.	74a	47°10'N	122°31'W
Lake Worth, Fl., U.S. (wûrth')	83a	26°37'N	80°04'W
Lake Worth Village, Tx., U.S.	75c	32°49'N	97°26'W
Lake Zurich, Il., U.S. (tsü'rĭk)	69a	42°11'N	88°05'W
Lakhdenpokh'ya, Russia (l'äk-dīe'npōkyå)	123	61°33'N	30°10'E
Lakhtinskiy, Russia (låk-tĭn'skĭ)	142c	59°59'N	30°10'E
Lakota, N.D., U.S. (lå-kō'tå)	70	48°04'N	98°21'W
Lakshadweep, state, India	155	10°10'N	72°50'E
Lakshadweep, is., India	155	11°00'N	73°02'E
La Libertad, El Sal.	90	13°29'N	89°20'W
La Libertad, Guat. (lä lē-bĕr-tädh')	90	15°31'N	91°44'W
La Libertad, Guat.	90a	16°46'N	90°12'W
La Ligua, Chile (lä lē'gwä)	99b	32°21'S	71°13'W
Lalín, Spain (lä-lē'n)	128	42°40'N	8°05'W
La Línea, Spain (lä lē'nå-ä)	118	36°11'N	5°22'W
Lalitpur, Nepal	155	27°23'N	85°24'E
La Louviere, Bel. (lå lōō-vyår')	121	50°30'N	4°10'E
La Luz, Mex. (lä lōōz')	88	21°04'N	101°19'W
Lama-Kara, Togo	188	9°33'N	1°12'E
La Malbaie, Can. (lä mäl-bå')	51	47°39'N	70°10'W
La Mancha, reg., Spain (lä män'chä)	128	38°55'N	4°20'W
Lamar, Co., U.S. (lå-mär')	78	38°04'N	102°44'W
Lamar, Mo., U.S.	79	37°28'N	94°15'W
La Marmora, Punta, mtn., Italy (lä-mä'r-mô-rä)	118	40°00'N	9°28'E
La Marque, Tx., U.S. (lå-märk)	81a	29°23'N	94°58'W
Lamas, Peru (lä'mås)	100	6°24'S	76°41'W
Lamballe, Fr. (läN-bäl')	126	48°29'N	2°36'W
Lambari, Braz. (läm-bà'rē)	99a	21°58'S	45°22'W
Lambasa, Fiji	170g	16°26'S	179°24'E
Lambayeque, Peru (läm-bä-yā'kå)	100	6°41'S	79°58'W
Lambert, Ms., U.S. (lăm'bērt)	82	34°10'N	90°16'W
Lambertville, N.J., U.S. (lăm'bērt-vĭl)	67	40°20'N	75°00'W
Lame Deer, Mt., U.S. (lām dēr')	73	45°36'N	106°40'W
Lamego, Port. (lä-mā'gō)	128	41°07'N	7°47'W
La Mesa, Col.	100a	4°38'N	74°27'W
La Mesa, Ca., U.S. (lä mā'sä)	76a	32°46'N	117°01'W
Lamesa, Tx., U.S.	78	32°44'N	101°54'W
Lamía, Grc. (lå-mē'å)	119	38°54'N	22°25'E
Lamon Bay, b., Phil. (lä-mōn')	168	14°35'N	121°52'E
La Mora, Chile (lä-mō'rä)	99b	32°28'S	70°56'W
La Moure, N.D., U.S. (lå mōōr')	70	46°23'N	98°17'W
Lampa, r., Chile (lá'm-pä)	99b	33°15'S	70°55'W
Lampasas, Tx., U.S. (lăm-păs'ås)	80	31°06'N	98°10'W
Lampasas, r., Tx., U.S.	80	31°18'N	98°08'W
Lampazos, Mex. (läm-pä'zōs)	86	27°03'N	100°30'W
Lampedusa, i., Italy (läm-på-dōō'sä)	118	35°29'N	12°58'E
Lamstedt, Ger. (läm'shtĕt)	115c	53°38'N	9°06'E
Lamu, Kenya (lä'mōō)	187	2°16'S	40°54'E
Lamu Island, i., Kenya	191	2°25'S	40°50'E
La Mure, Fr. (lå mür')	127	44°55'N	5°50'E
Lan', r., Bela. (lån')	132	52°38'N	27°05'E
Lāna'i, i., Hi., U.S. (lä-nä'ē)	64c	20°48'N	157°06'W
Lanai City, Hi., U.S.	84a	20°50'N	156°56'W
Lanak La, p., China	160	34°40'N	79°50'E
Lanark, Scot., U.K. (lăn'ărk)	120	55°40'N	3°50'W
Lancashire, co., Eng., U.K. (lăŋ'kå-shĭr)	114a	53°49'N	2°42'W
Lancaster, Eng., U.K.	116	54°04'N	2°55'W
Lancaster, Ky., U.S.	66	37°35'N	84°30'W
Lancaster, Ma., U.S.	61a	42°28'N	71°40'W
Lancaster, N.H., U.S.	67	44°25'N	71°30'W
Lancaster, N.Y., U.S.	69c	42°54'N	78°42'W
Lancaster, Oh., U.S.	66	39°40'N	82°35'W
Lancaster, Pa., U.S.	65	40°05'N	76°20'W
Lancaster, Tx., U.S.	75c	32°36'N	96°45'W
Lancaster, Wi., U.S.	71	42°51'N	90°44'W

PLACE (Pronunciation)	PAGE	LAT.	LONG.
Lândana, Ang. (län-dä'nä)	186	5°15'S	12°07'E
Landau, Ger. (län'dou)	124	49°13'N	8°07'E
Lander, Wy., U.S. (lăn'dēr)	73	42°49'N	108°24'W
Landerneau, Fr. (läN-dĕr-nō')	126	48°28'N	4°14'W
Landes, reg., Fr. (läNd)	126	44°22'N	0°52'W
Landsberg, Ger. (länds'bōōrgh)	124	48°03'N	10°53'E
Lands End, c., Eng., U.K.	112	50°03'N	5°45'W
Landshut, Ger. (länts'hōōt)	117	48°32'N	12°09'E
Landskrona, Swe. (låns-krô'nå)	122	55°51'N	12°47'E
Lanett, Al., U.S. (lå-nĕt')	82	32°52'N	85°13'W
Langat, r., Malay.	153b	2°46'N	101°33'E
Langdon, Can. (lăng'dŭn)	62e	50°58'N	113°40'W
Langdon, Mn., U.S.	75g	44°49'N	92°56'W
L'Ange-Gardien, Can. (läNzh gär-dyăN')	62b	46°55'N	71°06'W
Langeland, i., Den.	122	54°52'N	10°46'E
Langenzersdorf, Aus.	115e	48°30'N	16°22'E
Langesund, Nor. (läng'ĕ-sōn')	122	58°59'N	9°38'E
Langfjorden, b., Nor.	122	62°40'N	7°45'E
Langhorne, Pa., U.S. (lăng'hôrn)	68f	40°10'N	74°55'W
Langia Mountains, mts., Ug.	191	3°35'N	33°35'E
Langjökull, ice, Ice. (läng-yû'kōōl)	116	64°40'N	20°31'W
Langla Co, l., China (läŋ-lä tswo)	158	30°42'N	80°40'E
Langley, Can. (lăng'lĭ)	55	49°06'N	122°39'W
Langley, S.C., U.S.	83	33°32'N	81°52'W
Langley, Wa., U.S.	74a	48°02'N	122°25'W
Langley Indian Reserve, I.R., Can.	74d	49°12'N	122°31'W
Langnau, Switz. (läng'nou)	124	46°56'N	7°46'E
Langon, Fr. (läN-gôN')	126	44°34'N	0°16'W
Langres, Fr. (läN'gr')	127	47°53'N	5°20'E
Langres, Plateau de, plat., Fr. (plä-tō'dĕ-läN'grĕ)	126	47°39'N	5°00'E
Langsa, Indon. (läng'så)	168	4°33'N	97°52'E
Lang Son, Viet. (läng'sŏn')	168	21°52'N	106°42'E
L'Anguille, r., Ar., U.S. (läN-gē'y')	79	35°23'N	90°52'W
Langxi, China (läŋ-shyē)	162	31°10'N	119°09'E
Langzhong, China (läŋ-jôŋ)	160	31°40'N	106°05'E
Lanham, Md., U.S. (lăn'ăm)	68e	38°58'N	76°54'W
Lanigan, Can. (lăn'ĭ-găn)	50	51°52'N	105°02'W
Länkäran, Azer. (lĕn-kô-rän')	134	38°52'N	48°58'E
Lankoviri, Nig.	189	9°00'N	11°25'E
Lansdale, Pa., U.S. (lănz'dāl)	67	40°20'N	75°15'W
Lansdowne, Pa., U.S. (lănz'doun)	68f	39°57'N	75°17'W
L'Anse, Mi., U.S. (låns)	71	46°43'N	88°28'W
L'Anse and Vieux Desert Indian Reservation, I.R., Mi., U.S.	71	46°41'N	88°12'W
Lansford, Pa., U.S. (lănz'fērd)	67	40°50'N	75°50'W
Lansing, Ia., U.S.	71	43°22'N	91°16'W
Lansing, Il., U.S.	69a	41°34'N	87°33'W
Lansing, Ks., U.S.	75f	39°15'N	94°53'W
Lansing, Mi., U.S.	65	42°45'N	84°35'W
Lanús, Arg. (lä-nōōs')	102a	34°42'S	58°24'W
Lanusei, Italy (lä-nōō-sĕ'y)	130	39°51'N	9°34'E
Lanúvio, Italy (lä-nōō'vyō)	129d	41°41'N	12°42'E
Lanzarote Island, i., Spain (län-zä-rō'tä)	184	29°04'N	13°03'W
Lanzhou, China (län-jō)	160	35°55'N	103°55'E
Laoag, Phil. (lä-wäg')	168	18°13'N	120°38'E
Laon, Fr. (läN)	126	49°36'N	3°35'E
La Oroya, Peru (lä-ô-rô'yä)	100	11°30'S	76°00'W
Laos, nation, Asia (lā-ōs) (lá-ōs')	168	20°15'N	102°00'E
Laoshan Wan, b., China (lou-shän wän)	162	36°21'N	120°48'E
La Palma, Pan. (lä-päl'mä)	91	8°25'N	78°07'W
La Palma, Spain	128	37°24'N	6°36'W
La Palma Island, i., Spain	184	28°42'N	19°03'W
La Pampa, prov., Arg.	102	37°25'S	67°00'W
Lapa Rio Negro, Braz. (lä-pä-rē'ō-nĕ'grô)	102	26°12'S	49°56'W
La Paz, Arg. (lä päz')	102	30°48'S	59°47'W
La Paz, Bol.	100	16°31'S	68°03'W
La Paz, Hond.	90	14°15'N	87°40'W
La Paz, Mex. (lä-pà'z)	88	23°39'N	100°44'W
La Paz, Mex.	86	24°00'N	110°15'W
Lapeer, Mi., U.S. (lå-pēr')	66	43°05'N	83°15'W
La-Penne-sur-Huveaune, Fr. (la-pĕn'sür-ü-vōn')	126a	43°18'N	5°33'E
La Perouse, Austl.	173b	33°59'S	151°14'E
La Piedad Cabadas, Mex. (lä pyå-dhädh' kä-bä'dhäs)	88	20°20'N	102°04'W
Lapland, hist. reg., Eur. (lăp'lănd)	110	68°20'N	22°00'E
La Plata, Arg. (lä plä'tä)	102	34°54'S	57°57'W
La Plata, Mo., U.S. (lå plā'tå)	79	40°03'N	92°28'W
La Plata Peak, mtn., Co., U.S.	77	39°00'N	106°25'W
La Pocatière, Can. (lå pô-kà-tyår')	59	47°24'N	70°01'W
La Poile Bay, b., Can. (lå pwäl')	61	47°38'N	58°20'W
La Porte, In., U.S. (lå pōrt')	66	41°35'N	86°45'W
Laporte, Oh., U.S.	69d	41°19'N	82°05'W
La Porte, Tx., U.S.	81a	29°40'N	95°01'W
La Porte City, Ia., U.S.	71	42°20'N	92°10'W
Lappeenranta, Fin. (lä'pĕn-rän'tä)	123	61°04'N	28°08'E
La Prairie, Can. (lå-prå-rē')	62a	45°24'N	73°30'W
Lâpseki, Tur. (läp'sá-kē)	131	40°20'N	26°41'E
Laptev Sea, sea, Russia (läp'tyĭf)	135	75°39'N	120°00'E
La Puebla de Montalbán, Spain	128	39°54'N	4°21'W
La Puente, Ca., U.S. (pwĕn'tē)	75a	34°01'N	117°57'W
Lapuşul, r., Rom. (lä'pōō-shōōl)	125	47°29'N	23°46'E
La Quiaca, Arg. (lä kē-ä'kä)	102	22°15'S	65°44'W
L'Aquila, Italy (lá'kē-lä)	118	42°22'N	13°24'E
Lār, Iran (lär)	154	27°31'N	54°12'E
Lara, Austl.	173a	38°02'S	144°24'E
Larache, Mor. (lä-räsh')	184	35°15'N	6°09'W
Laramie, Wy., U.S. (lăr'å-mĭ)	64	41°20'N	105°40'W
Laramie, r., Co., U.S.	78	40°56'N	105°55'W
Larchmont, N.Y., U.S. (lärch'mŏnt)	68a	40°56'N	73°46'W
Larch Mountain, mtn., Or., U.S. (lärch)	74c	45°32'N	122°06'W

PLACE (Pronunciation)	PAGE	LAT.	LONG.
Laredo, Spain (lȧ-rā′dhō)	128	43°24′N	3°24′W
Laredo, Tx., U.S.	64	27°31′N	99°29′W
La Réole, Fr. (lȧ rå-ōl′)	126	44°37′N	0°03′W
Largeau, Chad (lär-zhō′)	185	17°55′N	19°07′E
Largo, Cayo, Cuba (kä′yō-lär′gō)	92	21°40′N	81°30′W
Larimore, N.D., U.S. (lăr′ĭ-môr)	70	47°53′N	97°38′W
Larino, Italy (lä-rē′nō̊)	130	41°48′N	14°54′E
La Rioja, Arg. (lä rē̊-ōhä)	102	29°18′S	67°42′W
La Rioja, prov., Arg. (lä-rē̊-ô′kä)	102	28°45′S	68°00′W
Lárisa, Grc. (lä′rē̊-sȧ)	119	39°38′N	22°25′E
Lārkāna, Pak.	158	27°40′N	68°12′E
Larnaka, Cyp.	119	34°55′N	33°37′E
Lárnakos, Kólpos, b., Cyp.	153a	36°50′N	33°45′E
Larned, Ks., U.S. (lär′nĕd)	78	38°09′N	99°07′W
La Robla, Spain (lä rōb′lä)	128	42°48′N	5°36′W
La Rochelle, Fr. (lȧ rō̊-shĕl′)	110	46°10′N	1°09′W
La Roche-sur-Yon, Fr. (lȧ rō̊sh′sûr-yôɴ′)	117	46°39′N	1°27′W
La Roda, Spain (lä rō′dä)	128	39°13′N	2°08′W
La Romona, Dom. Rep. (lä-rä-mô′nä)	93	18°25′N	69°00′W
Larrey Point, c., Austl. (lăr′ē̊)	174	19°15′S	118°15′E
Laruns, Fr. (lȧ-răɴs′)	126	42°58′N	0°28′W
Larvik, Nor. (lär′vēk)	116	59°06′N	10°03′E
La Sabana, Ven. (lä-sä-bȧ′nä)	101b	10°38′N	66°24′W
La Sabina, Cuba (lä-sä-bē′nä)	93a	22°51′N	82°05′W
La Sagra, mtn., Spain (lä sä′grä)	118	37°56′N	2°35′W
La Sal, Ut., U.S. (lȧ sål′)	77	38°10′N	109°20′W
La Salle, Can. (l*ȧ* săl′)	69b	42°14′N	83°06′W
La Salle, Can.	62a	45°26′N	73°39′W
La Salle, Can.	62f	49°41′N	97°16′W
La Salle, Il., U.S.	66	41°20′N	89°05′W
Las Animas, Co., U.S. (lȧs ă′nĭ-m*ȧ*s)	78	38°03′N	103°16′W
La Sarre, Can.	51	48°43′N	79°12′W
Lascahobas, Haiti (läs-kä-ō′bȧs)	93	19°00′N	71°55′W
Las Cruces, Mex. (läs-kro͞o′sĕs)	89	16°37′N	93°54′W
Las Cruces, N.M., U.S.	64	32°20′N	106°50′W
La Selle, Massif de, mtn., Haiti (lä′sĕl′)	93	18°25′N	72°05′W
La Serena, Chile (lä-sĕ-rĕ′nä)	102	29°55′S	71°24′W
La Seyne, Fr. (lä-sån′)	117	43°07′N	5°52′E
Las Flores, Arg. (läs flo′rē̊s)	102	36°01′S	59°07′W
Lashio, Mya. (lăsh′ē̊-ō)	160	22°58′N	98°03′E
Las Juntas, C.R. (läs-ᴋo͞o′n-täs)	90	10°15′N	85°00′W
Las Maismas, sw., Spain (läs-mī′s-mäs)	128	37°05′N	6°25′W
La Solana, Spain (lä-sō̊-lä-nä)	128	38°56′N	3°13′W
Las Palmas, Pan.	91	8°08′N	81°30′W
Las Palmas de Gran Canaria, Spain (läs päl′mäs)	184	28°07′N	15°28′W
La Spezia, Italy (lä-spĕ′zyä)	110	44°07′N	9°48′E
Las Piedras, Ur. (läs-pyĕ′drȧs)	99c	34°42′S	56°08′W
Las Pilas, vol., Nic. (läs-pē′läs)	90	12°32′N	86°43′W
Las Rosas, Mex. (läs rō thäs)	89	16°24′N	92°23′W
Las Rozas de Madrid, Spain (läs rō′thas dā mä-dhrēd′)	129a	40°29′N	3°53′W
Lassee, Aus.	115e	48°14′N	16°50′E
Lassen Peak, mtn., Ca., U.S. (lăs′ĕn)	64	40°30′N	121°32′W
Lassen Volcanic National Park, rec., Ca., U.S.	64	40°43′N	121°35′W
L'Assomption, Can. (lȧs-sôm-syôɴ)	62a	45°50′N	73°25′W
Lass Qoray, Som.	192a	11°13′N	48°19′E
Las Tablas, Pan. (läs tä′bläs)	91	7°48′N	80°16′W
Last Mountain, l., Can. (lȧst moun′tĭn)	52	51°05′N	105°10′W
Lastoursville, Gabon (lȧs-to͞or-vēl′)	186	1°00′S	12°49′E
Las Tres Vírgenes, Volcán, vol., Mex. (vē′r-hĕ-nĕs)	86	26°00′N	111°45′W
Las Tunas, prov., Cuba	92	21°05′N	77°00′W
Las Vacas, Mex. (läs-vȧ′käs)	89	16°24′N	95°48′W
Las Vegas, Chile (läs-vĕ′gäs)	99b	32°50′S	70°59′W
Las Vegas, N.M., U.S.	64	35°36′N	105°13′W
Las Vegas, Nv., U.S. (läs vā′gäs)	64	36°12′N	115°10′W
Las Vegas, Ven. (läs-vĕ′gäs)	101b	10°26′N	64°08′W
Las Vigas, Mex.	89	19°38′N	97°03′W
Las Vizcachas, Meseta de, plat., Arg.	102	49°35′S	71°00′W
Latacunga, Ec. (lä-tä-kȯŋ′gä)	100	1°02′S	78°33′W
Latakia *see* Al Lādhiqīyah, Syria	154	35°32′N	35°51′E
La Teste-de-Buch, Fr. (lä-tĕst-d*ĕ*-büsh)	126	44°38′N	1°11′W
Lathrop, Mo., U.S. (lā′thr*ŭ*p)	79	39°32′N	94°21′W
La Tortuga, Isla, i., Ven. (ē̊′s-lä-lä-tôr-to͞o′gä)	100	10°55′N	65°18′W
Latorytsia, r., Eur.	125	48°27′N	22°30′E
Latourell, Or., U.S. (lȧ-tou′rĕl)	74c	45°32′N	122°13′W
La Tremblade, Fr. (lä-trĕɴ-bläd′)	126	45°45′N	1°12′W
Latrobe, Pa., U.S. (lȧ-trōb′)	67	40°25′N	79°15′W
La Tuque, Can. (lȧ′tük′)	51	47°27′N	72°49′W
Lātūr, India (lä-to͞or′)	158	18°20′N	76°35′E
Latvia, nation, Eur.	134	57°28′N	24°29′E
Lau Group, is., Fiji	170g	18°20′S	178°30′W
Launceston, Austl. (lôn′sĕs-t*ŭ*n)	175	41°35′S	147°22′E
Launceston, Eng., U.K. (lôrn′st*ŏ*n)	120	50°38′N	4°26′W
La Unión, Chile (lä-o͞o-nyô′n)	102	40°15′S	73°04′W
La Unión, El Sal.	90	13°18′N	87°51′W
La Unión, Mex. (lä o͞on-nyōn′)	88	17°59′N	101°48′W
La Unión, Spain	118	37°38′N	0°50′W
Laura, Austl. (lôr*ȧ*)	175	15°40′S	144°45′E
Laurel, De., U.S. (lô′r*ĕ*l)	67	38°30′N	75°40′W
Laurel, Md., U.S.	68e	39°06′N	76°51′W
Laurel, Ms., U.S.	65	31°42′N	89°07′W
Laurel, Mt., U.S.	73	45°41′N	108°45′W
Laurel, Wa., U.S.	74d	48°52′N	122°29′W
Laurelwood, Or., U.S. (lô′rĕl-wȯd)	74c	45°25′N	123°05′W
Laurens, S.C., U.S. (lô′r*ĕ*nz)	83	34°29′N	82°03′W
Laurentian Highlands, hills, Can. (lô′r*ĕ*n-tĭ-ȧn)	49	49°00′N	74°50′W
Laurentides, Can. (lô′r*ĕ*n-tīdz)	62a	45°51′N	73°46′W
Lauria, Italy (lou′rē̊-ä)	119	40°03′N	15°02′E
Laurinburg, N.C., U.S. (lô′rĭn-bûrg)	83	34°45′N	79°27′W
Laurium, Mi., U.S. (lô′rĭ-*ŭ*m)	71	47°13′N	88°28′W
Lausanne, Switz. (lō-zȧn′)	110	46°32′N	6°35′E
Laut, Pulau, i., Indon.	168	3°39′S	116°07′E
Lautaro, Chile (lou-tä′rō̊)	102	38°40′S	72°24′W
Laut Kecil, Kepulauan, is., Indon.	168	4°44′S	115°43′E
Lautoka, Fiji	170g	17°37′S	177°27′E
Lauzon, Can. (lō-zōɴ′)	62b	46°50′N	71°10′W
Lava Beds National Monument, rec., Ca., U.S. (lä′v*ȧ* bĕds)	72	41°38′N	121°44′W
Lavaca, r., Tx., U.S. (lȧ-vȧk′*ȧ*)	81	29°05′N	96°50′W
Lava Hot Springs, Id., U.S.	73	42°37′N	111°58′W
Laval, Can.	51	45°31′N	73°44′W
Laval, Fr. (lä-väl′)	117	48°05′N	0°47′W
La Vecilla de Curueño, Spain	128	42°53′N	5°18′W
La Vega, Dom. Rep. (lä-vĕ′gä)	93	19°15′N	70°35′W
Lavello, Italy (lä-vĕl′lō̊)	130	41°05′N	15°50′E
La Verne, Ca., U.S. (lȧ vûrn′)	75a	34°06′N	117°46′W
Laverton, Austl. (lā′vẽr-t*ŭ*n)	174	28°45′S	122°30′E
La Victoria, Ven. (lä vē̊k-tō′rē̊-ä)	100	10°14′N	67°20′W
La Vila Joiosa, Spain	129	38°30′N	0°14′W
Lavonia, Ga., U.S. (l*ȧ*-vō′nĭ-*ȧ*)	82	34°26′N	83°05′W
Lavon Reservoir, res., Tx., U.S.	81	33°06′N	96°20′W
Lavras, Braz. (lä′vräzh)	99a	21°15′S	44°59′W
Lávrio, Grc.	131	37°44′N	24°05′E
Lavry, Russia (lou′rȧ)	132	57°35′N	27°28′E
Lawndale, Ca., U.S. (lôn′dāl)	75a	33°54′N	118°22′W
Lawra, Ghana	188	10°39′N	2°52′W
Lawrence, In., U.S. (lô′r*ĕ*ns)	69g	39°59′N	86°01′W
Lawrence, Ks., U.S.	65	38°57′N	95°13′W
Lawrence, Ma., U.S.	61a	42°42′N	71°09′W
Lawrence, Pa., U.S.	69e	40°18′N	80°07′W
Lawrenceburg, In., U.S. (lô′r*ĕ*ns-bûrg)	69f	39°06′N	84°47′W
Lawrenceburg, Ky., U.S.	66	38°00′N	85°00′W
Lawrenceburg, Tn., U.S.	82	35°13′N	87°20′W
Lawrenceville, Ga., U.S. (lô-r*ĕ*ns-vĭl)	82	33°56′N	83°57′W
Lawrenceville, Il., U.S.	66	38°45′N	87°45′W
Lawrenceville, N.J., U.S.	68a	40°17′N	74°44′W
Lawrenceville, Va., U.S.	83	36°43′N	77°52′W
Lawsonia, Md., U.S. (lô-sō′nĭ-*ȧ*)	67	38°00′N	75°50′W
Lawton, Ok., U.S. (lô′t*ŭ*n)	64	34°36′N	98°25′W
Lawz, Jabal al, mtn., Sau. Ar.	154	28°46′N	35°37′E
Layang Layang, Malay. (lä-yäng′ lä-yäng′)	153b	1°49′N	103°28′E
Laysan, i., Hi., U.S.	84b	26°00′N	171°00′W
Layton, Ut., U.S. (lā′t*ŭ*n)	75b	41°04′N	111°58′W
Lazdijai, Lith. (läzh′dē-yī′)	123	54°12′N	23°35′E
Lazio (Latium), hist. reg., Italy	130	42°05′N	12°25′E
Lead, S.D., U.S. (lēd)	64	44°22′N	103°47′W
Leader, Can.	56	50°55′N	109°32′W
Leadville, Co., U.S. (lĕd′vĭl)	78	39°14′N	106°18′W
Leaf, r., Ms., U.S. (lēf)	82	31°43′N	89°20′W
League City, Tx., U.S. (lēg)	81a	29°31′N	95°05′W
Leamington, Can. (lĕm′ĭng-t*ŭ*n)	58	42°05′N	82°35′W
Leamington, Eng., U.K. (lĕ′mĭng-t*ŭ*n)	120	52°17′N	1°25′W
Leatherhead, Eng., U.K. (lĕdh′ẽr-hĕd′)	114b	51°17′N	0°20′W
Leavenworth, Ks., U.S. (lĕv′*ĕ*n-wûrth)	65	39°19′N	94°54′W
Leavenworth, Wa., U.S.	72	47°35′N	120°39′W
Leawood, Ks., U.S. (lē′wȯd)	75f	38°58′N	94°37′W
Łeba, Pol. (lā′bä)	125	54°45′N	17°34′E
Lebam, r., Malay.	153b	1°35′N	104°09′E
Lebango, Congo	190	0°22′N	14°49′E
Lebanon, Il., U.S. (lĕb′*ȧ*-nŭn)	75e	38°36′N	89°49′W
Lebanon, In., U.S.	66	40°00′N	86°30′W
Lebanon, Ky., U.S.	82	37°32′N	85°15′W
Lebanon, Mo., U.S.	79	37°40′N	92°43′W
Lebanon, N.H., U.S.	67	43°40′N	72°15′W
Lebanon, Oh., U.S.	66	39°25′N	84°10′W
Lebanon, Or., U.S.	72	44°31′N	122°53′W
Lebanon, Pa., U.S.	67	40°20′N	76°20′W
Lebanon, Tn., U.S.	82	36°10′N	86°16′W
Lebanon, nation, Asia	154	34°00′N	34°00′E
Lebedyan', Russia (lyĕ′bĕ-dyän′)	136	53°03′N	39°08′E
Lebedyn, Ukr.	137	50°34′N	34°27′E
Le Blanc, Fr. (lĕ-bläɴ′)	126	46°38′N	0°59′E
Le Borgne, Haiti (lĕ bōrn′y′)	93	19°50′N	72°30′W
Lębork, Pol. (lăɴ-bȯrk′)	125	54°33′N	17°46′E
Lebrija, Spain (lå-brē′hä)	128	36°55′N	6°06′W
Lecce, Italy (lĕt′chā)	119	40°22′N	18°11′E
Lecco, Italy (lĕk′kō)	130	45°52′N	9°28′E
Lech, r., Ger. (lĕk)	124	47°41′N	10°52′E
Le Châtelet-en-Brie, Fr. (lĕ-shä-t*ĕ*-lā′ĕɴ-brē′)	127b	48°29′N	2°50′E
Leche, Laguna de, l., Cuba (lä-gȯ′nä-dĕ-lĕ′chĕ)	92	22°10′N	78°30′W
Leche, Laguna de la, l., Mex.	80	27°16′N	102°45′W
Lecompte, La., U.S.	81	31°06′N	92°25′W
Le Creusot, Fr. (lẽkrû-zō̊)	117	46°48′N	4°23′E
Ledesma, Spain (lå-dĕs′mä)	128	41°05′N	5°59′W
Leduc, Can. (l*ĕ*-do͞ok′)	55	53°16′N	113°33′W
Leech, l., Mn., U.S. (lēch)	71	47°06′N	94°16′W
Leeds, Eng., U.K.	110	53°48′N	1°33′W
Leeds, Al., U.S. (lēdz)	68h	33°33′N	86°33′W
Leeds, N.D., U.S.	70	48°18′N	99°24′W
Leeds, co., Eng., U.K.	114a	53°50′N	1°30′W
Leeds and Liverpool Canal, can., Eng., U.K. (lĭv′ẽr-po͞ol)	114a	53°36′N	2°38′W
Leegebruch, Ger. (lĕh′gĕn-bro͞oᴋ)	115b	52°43′N	13°12′E
Leek, Eng., U.K. (lēk)	114a	53°06′N	2°01′W
Leer, Ger. (lār)	124	53°14′N	7°27′E
Leesburg, Fl., U.S. (lēz′bûrg)	83	28°49′N	81°53′W
Leesburg, Va., U.S.	67	39°10′N	77°30′W
Lees Summit, Mo., U.S.	75f	38°55′N	94°23′W
Lee Stocking, i., Bah.	92	23°45′N	76°05′W
Leesville, La., U.S. (lēz′vĭl)	81	31°09′N	93°17′W
Leetonia, Oh., U.S. (lē̊-tō′nĭ-*ȧ*)	66	40°50′N	80°45′W
Leeuwarden, Neth. (lā′wär-dĕn)	117	52°12′N	5°50′E
Leeuwin, Cape, c., Austl. (lo͞o′wĭn)	174	34°15′S	114°30′E
Leeward Islands, is., N.A. (lē′wẽrd)	87	17°00′N	62°15′W
Lefkáda, Grc.	131	38°49′N	20°43′E
Lefkáda, i., Grc.	119	38°42′N	20°22′E
Le François, Mart.	91b	14°37′N	60°55′W
Lefroy, l., Austl. (lē-froi′)	174	31°30′S	122°00′E
Leganés, Spain (lå-gä′nås)	129a	40°20′N	3°46′W
Legazpi, Phil. (lå-gäs′pē̊)	169	13°09′N	123°44′E
Legge Peak, mtn., Austl. (lĕg)	176	41°33′S	148°10′E
Leggett, Ca., U.S.	76	39°51′N	123°42′W
Leghorn *see* Livorno, Italy	110	43°32′N	11°18′E
Legnano, Italy (lå-nyä′nō)	130	45°35′N	8°53′E
Legnica, Pol. (lĕk-nĭt′sȧ)	117	51°13′N	16°10′E
Leh, India (lā)	158	34°10′N	77°40′E
Le Havre, Fr. (lẽ ȧv′r′)	110	49°31′N	0°07′E
Lehi, Ut., U.S. (lē′hī)	77	40°25′N	111°55′W
Lehman Caves National Monument, rec., Nv., U.S. (lē′m*ă*n)	77	38°54′N	114°08′W
Lehnin, Ger. (lĕh′nēn)	115b	52°19′N	12°45′E
Leicester, Eng., U.K. (lĕs′tẽr)	110	52°37′N	1°08′W
Leicestershire, co., Eng., U.K.	114a	52°40′N	1°12′W
Leichhardt, r., Austl. (līk′härt)	174	18°30′S	139°45′E
Leiden, Neth. (lī′d*ĕ*n)	121	52°09′N	4°29′E
Leigh Creek, Austl. (lē krēk)	176	30°33′S	138°30′E
Leikanger, Nor. (lī′käŋ′gẽr)	122	61°11′N	6°51′E
Leimuiden, Neth.	115a	52°13′N	4°40′E
Leine, r., Ger. (lī′nĕ)	124	51°58′N	9°56′E
Leinster, hist. reg., Ire. (lĕn-stẽr)	120	52°45′N	7°19′W
Leipsic, Oh., U.S. (līp′sĭk)	66	41°05′N	84°00′W
Leipzig, Ger. (līp′tsĭk)	110	51°20′N	12°24′E
Leiria, Port. (lā-rē′ä)	128	39°45′N	8°50′W
Leitchfield, Ky., U.S. (lēch′fēld)	82	37°28′N	86°20′W
Leitha, r., Aus.	115e	48°04′N	16°57′E
Leitrim, Can.	62c	45°20′N	75°36′W
Leivádia, Grc.	131	38°25′N	22°51′E
Leizhou Bandao, pen., China (lā-jō bän-dou)	160	20°42′N	109°10′E
Leksand, Swe. (lĕk′sänd)	122	60°45′N	14°56′E
Leland, Wa., U.S. (lē′l*ă*nd)	74a	47°54′N	122°53′W
Leliu, China (lŭ-lĭō̊)	163a	22°52′N	113°09′E
Le Locle, Switz. (lẽ lō̊′kl′)	124	47°03′N	6°43′E
Le Maire, Estrecho de, strt., Arg. (ĕs-trĕ′chō̊-dĕ-lĕ-mī′rĕ)	102	55°15′S	65°30′W
Le Mans, Fr. (lẽ mäɴ′)	117	48°01′N	0°12′E
Le Marin, Mart.	91b	14°28′N	60°55′W
Le Mars, Ia., U.S. (lē̊ märz′)	70	42°46′N	96°09′W
Lemay, Mo., U.S.	75e	38°32′N	90°17′W
Lemdiyya, Alg.	184	36°18′N	2°40′E
Lemery, Phil. (lā-mā-rē′)	169a	13°51′S	120°55′E
Lemhi, r., Id., U.S.	73	44°40′N	113°27′W
Lemhi Range, mts., Id., U.S. (lĕm′hī)	73	44°35′N	113°33′W
Lemmon, S.D., U.S. (lĕm′*ŭ*n)	70	45°55′N	102°10′W
Le Môle, Haiti (lẽ mōl′)	93	19°50′N	73°20′W
Lemon Grove, Ca., U.S. (lĕm′*ŭ*n-grōv)	76a	32°44′N	117°02′W
Le Moule, Guad. (lẽ mo͞ol′)	91b	16°19′N	61°22′W
Lempa, r., N.A. (lĕm′pä)	90	13°20′N	88°46′W
Lemvig, Den. (lĕm′vēgh)	122	56°33′N	8°16′E
Lena, r., Russia	135	68°00′N	123°00′E
Lençóes Paulista, Braz. (lĕɴ-sôɴs′ pou-lēs′tȧ)	102	22°30′S	48°45′W
Lençóis, Braz. (lĕn-sóis)	101	12°38′S	41°28′W
Lenexa, Ks., U.S. (lē′nĕx-*ă*)	75f	38°58′N	99°44′W
Lengyandong, China (lŭŋ-yän-dō̊ŋ)	163a	23°12′N	113°21′E
Lenik, r., Malay.	153b	1°59′N	102°51′E
Leningrad *see* Saint Petersburg, Russia	134	59°57′N	30°20′E
Leningrad, prov., Russia	132	59°15′N	30°30′E
Leningradskaya, Russia (lyĕ-nĭn-gräd′skȧ-yȧ)	133	46°19′N	39°23′E
Lenino, Russia (lyĕ′nĭ-nô)	142b	55°37′N	37°41′E
Leninogorsk, Kaz.	139	50°29′N	83°25′E
Lenīnsk, Kaz.	139	45°39′N	63°19′E
Leninsk, Russia (lyĕ-nēnsk′)	137	48°40′N	45°10′E
Leninsk-Kuznetski, Russia (lyĕ-nē̊nsk′ko͞oz-nyĕt′skĭ)	134	54°28′N	86°48′E
Lennox, S.D., U.S. (lĕn′*ŭ*ks)	70	43°22′N	96°53′W
Lenoir, N.C., U.S. (lē̊-nōr′)	83	35°54′N	81°35′W
Lenoir City, Tn., U.S.	82	35°47′N	84°16′W
Lenox, Ia., U.S.	71	40°51′N	94°29′W
Léo, Burkina	188	11°06′N	2°06′W
Leoben, Aus. (lå-ō′b*ĕ*n)	124	47°22′N	15°09′E
Léogane, Haiti (lā-ō̊-gan′)	93	18°30′N	72°35′W
Leola, S.D., U.S. (lē̊-ō′l*ȧ*)	70	45°43′N	99°55′W
Leominster, Ma., U.S. (lĕm′ĭn-stẽr)	67	42°32′N	71°45′W
León, Mex. (lå-ōn′)	86	21°08′N	101°41′W
León, Nic. (lĕ-ô′n)	86	12°28′N	86°53′W
León, Spain (lĕ-ô′n)	118	42°38′N	5°33′W
Leon, Ia., U.S. (lē′ŏn)	71	40°43′N	93°44′W
León, hist. reg., Spain	128	41°18′N	5°50′W
Leon, r., Tx., U.S. (lē′ōn)	80	31°54′N	98°20′W
Leonforte, Italy (lā-ō̊n-fō̊r′tā)	130	37°40′N	14°27′E
Leopold II, Lac *see* Mai-Ndombe, Lac, l., D.R.C.	186	2°16′S	19°00′E
Leopoldina, Braz. (lā-ō̊-pō̊l-dē′nä)	99a	21°32′S	42°38′W
Leopoldsburg, Bel.	115a	51°07′N	5°18′E
Leopoldsdorf im Marchfelde, Aus. (lā′ō̊-pō̊lts-dō̊rf′)	115e	48°14′N	16°42′E
Léopoldville *see* Kinshasa, D.R.C.	186	4°18′S	15°18′E
Leova, Mol.	133	46°30′N	28°16′E
Lepe, Spain (lā′pā)	128	37°15′N	7°12′W
Leping, China (lŭ-pĭŋ)	165	29°02′N	117°12′E
L'Épiphanie, Can. (lā-pē̊-fä-nē′)	62a	45°51′N	73°29′W
Le Plessis-Belleville, Fr. (lĕ-plĕ-sē′bĕl-vēl′)	127b	49°05′N	2°46′E
Lepreau, Can. (lē̊-prō′)	60	45°10′N	66°28′W
Le Puy, Fr. (lē pwē′)	117	45°02′N	3°54′E
Lercara Friddi, Italy (lĕr-kä′rä)	130	37°47′N	13°36′E
Lerdo, Mex. (lĕr′dō)	86	25°31′N	103°30′W

ng-sing; ŋ-baŋk; ɴ-nasalized n; nŏd; c*ŏ*mmit; ōld; ō̊bey; ôrder; oi-boil; fo͞od; ȯ-as oo in foot; ou-out; s-soft; sh-dish; th-thin; pūre; ůnite; ûrn; stŭd; circ*ŭ*s; ü-as in French tu; ′-indeterminate vowel.

ăt; fināl; rāte; senāte; ärm; ȧsk; sofá; fāre; ch-choose; dh-as th in other; bē; ĕvent; bĕt; recĕnt; cratẽr; g-gō; gh-guttural g; bĭt; ĭ-short neutral; rīde; ĸ-guttural k as ch in German ich;

PRONOUNCING INDEX Liski – Loosdrechtsche Plassen 269

PLACE (Pronunciation)	PAGE	LAT.	LONG.
Liski, Russia (lyēs′kė)	133	50°56′N	39°28′E
Lisle, Il., U.S. (līl)	69a	41°48′N	88°04′W
L'Isle-Adam, Fr. (lēl-ädäN′)	127b	49°05′N	2°13′E
Lismore, Austl. (līz′môr)	175	28°48′S	153°18′E
Litani, r., Leb.	153a	33°28′N	35°42′E
Litchfield, Il., U.S. (lĭch′fēld)	79	39°10′N	89°38′W
Litchfield, Mn., U.S.	71	45°08′N	94°34′W
Litchfield, Oh., U.S.	69d	41°10′N	82°01′W
Lithgow, Austl. (lĭth′gō)	175	33°23′S	149°31′E
Lithinon, Akra, c., Grc.	130a	34°59′N	24°35′E
Lithonia, Ga., U.S. (lĭ-thō′nĭ-à)	68c	33°43′N	84°07′W
Lithuania, nation, Eur. (lĭth-ū-ā′nĭ-à)	134	55°42′N	23°30′E
Litóchoro, Grc.	131	40°05′N	22°29′E
Litoko, D.R.C.	190	1°13′S	24°47′E
Litoměřice, Czech Rep. (lē′tô-myĕr′zhĭ-tsĕ)	124	50°33′N	14°10′E
Litomyšl, Czech Rep. (lē′tô-mêsh′l)	124	49°52′N	16°14′E
Litoo, Tan.	191	9°45′S	38°24′E
Little, r., Austl.	173a	37°54′S	144°27′E
Little, r., Tn., U.S.	82	36°28′N	89°39′W
Little, r., Tx., U.S.	81	30°48′N	96°50′W
Little Abaco, i., Bah. (ä′bä-kō)	92	26°55′N	77°45′W
Little Abitibi, r., Can.	58	50°15′N	81°30′W
Little America, sci., Ant.	178	78°30′S	161°30′W
Little Andaman, i., India (ăn-dà-măn′)	168	10°39′N	93°08′E
Little Bahama Bank, bk., (bà-hā′mà)	92	26°55′N	78°40′W
Little Belt Mountains, mts., Mt., U.S. (bĕlt)	64	47°00′N	110°50′W
Little Bighorn, r., Mt., U.S. (bĭg-hôrn)	73	45°08′N	107°30′W
Little Bighorn Battlefield National Monument, rec., Mt., U.S. (bĭg-hôrn băt′'l-fēld)	73	45°44′N	107°15′W
Little Bitter Lake, l., Egypt	192b	30°10′N	32°36′E
Little Bitterroot, r., Mt., U.S. (bĭt′ēr-ōōt)	73	47°45′N	114°45′W
Little Blue, r., Ia., U.S. (blōō)	75f	38°52′N	94°25′W
Little Blue, r., Ne., U.S.	78	40°15′N	98°01′W
Littleborough, Eng., U.K. (lĭt′'l-bŭr-ô)	114a	53°39′N	2°06′W
Little Calumet, r., Il., U.S. (kăl-ū-mĕt′)	69a	41°38′N	87°38′W
Little Cayman, i., Cay. Is. (kā′măn)	92	19°40′N	80°05′W
Little Colorado, r., Az., U.S. (kŏl-ô-rä′dō)	64	36°05′N	111°35′W
Little Compton, R.I., U.S. (kŏmp′tŭn)	68b	41°31′N	71°07′W
Little Corn Island, i., Nic.	91	12°19′N	82°50′W
Little Exuma, i., Bah. (ĕk-sōō′mä)	93	23°25′N	75°40′W
Little Falls, Mn., U.S. (fôlz)	71	45°58′N	94°23′W
Little Falls, N.Y., U.S.	67	43°05′N	74°55′W
Littlefield, Tx., U.S. (lĭt′'l-fēld)	78	33°55′N	102°17′W
Little Fork, r., Mn., U.S. (fôrk)	71	48°24′N	93°30′W
Little Goose Dam, dam, Wa., U.S.	72	46°35′N	118°02′W
Little Hans Lollick, i., V.I.U.S. (häns lŏl′lĭk)	87c	18°25′N	64°54′W
Little Humboldt, r., Nv., U.S. (hŭm′bōlt)	72	41°10′N	117°40′W
Little Inagua, i., Bah. (ē-nä′gwä)	93	21°30′N	73°00′W
Little Isaac, i., Bah. (ī′zăk)	92	25°55′N	79°00′W
Little Kanawha, r., W.V., U.S. (kà-nô′wà)	66	39°05′N	81°30′W
Little Karroo, plat., S. Afr. (kä-rōō)	186	33°50′S	21°02′E
Little Mecatina, r., Can. (mĕ cà tĭ nå)	53	52°40′N	62°21′W
Little Miami, r., Oh., U.S. (mī-ăm′ĭ)	69f	39°19′N	84°15′W
Little Minch, strt., Scot., U.K.	120	57°35′N	6°45′W
Little Missouri, r., Ar., U.S. (mĭ-sōō′rĭ)	79	34°15′N	93°54′W
Little Missouri, r., U.S.	64	46°00′N	104°00′W
Little Pee Dee, r., S.C., U.S. (pē-dē′)	83	34°35′N	79°21′W
Little Powder, r., Wy., U.S. (pou′dēr)	73	44°51′N	105°20′W
Little Red, r., Ar., U.S. (rĕd)	79	35°25′N	91°55′W
Little Red, r., Ok., U.S.	79	33°53′N	94°38′W
Little Rock, Ar., U.S. (rôk)	65	34°42′N	92°16′W
Little Sachigo Lake, l., Can. (săchĭ-gō)	57	54°09′N	92°11′W
Little Salt Lake, l., Ut., U.S.	77	37°55′N	112°53′W
Little San Salvador, i., Bah. (săn săl′vä-dôr)	93	24°35′N	75°55′W
Little Satilla, r., Ga., U.S. (sà-tĭl′à)	83	31°43′N	82°47′W
Little Sioux, r., Ia., U.S. (sōō)	70	42°22′N	95°47′W
Little Smoky, r., Can. (smōk′ĭ)	55	55°10′N	116°55′W
Little Snake, r., Co., U.S. (snāk)	73	40°40′N	108°21′W
Little Tallapoosa, r., Al., U.S. (tăl-à-pōō′sä)	82	32°25′N	85°28′W
Little Tennessee, r., Tn., U.S. (tĕn-ĕ-sē′)	82	35°36′N	84°05′W
Littleton, Co., U.S. (lĭt′'l-tŭn)	78	39°34′N	105°01′W
Littleton, Ma., U.S.	61a	42°32′N	71°29′W
Littleton, N.H., U.S.	67	44°15′N	71°45′W
Little Wabash, r., Il., U.S. (wô′băsh)	66	38°50′N	88°30′W
Little Wood, r., Id., U.S. (wŏŏd)	73	43°00′N	114°08′W
Lityn, Ukr.	133	49°16′N	28°11′E
Liubar, Ukr.	133	49°56′N	27°44′E
Liuhe, China	164	42°10′N	125°38′E
Liuli, Tan.	191	11°05′S	34°38′E
Liupan Shan, mts., China	164	36°20′N	105°30′E
Liuwa Plain, pl., Zam.	190	14°30′S	22°40′E
Liuyang, China (lyōō′yäng′)	165	28°10′N	113°35′E
Liuyuan, China (liō-yüän)	162	36°09′N	114°37′E
Liuzhou, China (liō-jō)	160	24°25′N	109°30′E
Līvāni, Lat. (lē′vá-nė)	123	56°24′N	26°12′E
Lively, Can.	58	46°26′N	81°09′W
Livengood, Ak., U.S. (lĭv′ĕn-gŏŏd)	63	65°30′N	148°35′W
Live Oak, Fl., U.S. (līv′ōk)	82	30°15′N	83°00′W
Livermore, Ca., U.S. (lĭv′ēr-mōr)	74b	37°41′N	121°46′W
Livermore, Ky., U.S.	66	37°30′N	87°05′W
Liverpool, Austl. (lĭv′ēr-pōōl)	173b	33°55′S	150°56′E
Liverpool, Can.	51	44°02′N	64°41′W
Liverpool, Eng., U.K.	110	53°25′N	2°52′W
Liverpool, Tx., U.S.	81a	29°18′N	95°17′W
Liverpool Bay, b., Can.	63	69°45′N	130°00′W
Liverpool Range, mts., Austl.	175	31°47′S	151°00′E
Livindo, r., Afr.	185	1°09′N	13°30′E
Livingston, Guat.	90	15°50′N	88°45′W
Livingston, Al., U.S. (lĭv′ĭng-stŭn)	82	32°35′N	88°09′W
Livingston, Il., U.S.	75e	38°58′N	89°51′W
Livingston, Mt., U.S.	64	45°40′N	110°35′W
Livingston, N.J., U.S.	68a	40°47′N	74°20′W
Livingston, Tn., U.S.	82	36°23′N	85°20′W
Livingstone, Zam. (lĭv-ĭng-stôn)	186	17°50′S	25°53′E
Livingstone, Chutes de, wtfl., Afr.	190	4°50′S	14°30′E
Livingstonia, Mwi. (lĭv-ĭng-stō′nĭ-à)	186	10°36′S	34°07′E
Livno, Bos. (lēv′nō)	119	43°50′N	17°03′E
Livny, Russia (lēv′nė)	137	52°28′N	37°36′E
Livonia, Mi., U.S. (lĭ-vō-nĭ-à)	69b	42°25′N	83°23′W
Livorno, Italy (lė-vôr′nō) (lĕg′hôrn)	110	43°32′N	11°18′E
Livramento, Braz. (lē-vrä-mĕ′n-tô)	102	30°46′S	55°21′W
Lixian, China (lē shyĕn)	165	29°42′N	111°40′E
Lixian, China	162	38°30′N	115°38′E
Liyang, China (lē′yäng′)	165	31°30′N	119°29′E
Lizard Point, c., Eng., U.K. (lĭz′ård)	117	49°55′N	5°09′W
Lizy-sur-Ourcq, Fr. (lēk-sē′sür-ōōrk′)	127b	49°01′N	3°02′E
Ljubljana, Slvn. (lyōō′blyä′na)	110	46°04′N	14°29′E
Ljubuški, Bos. (lyōō′bōsh-kė)	131	43°11′N	17°29′E
Ljungan, r., Swe.	122	62°50′N	13°45′E
Ljungby, Swe. (lyŏng′bü)	122	56°49′N	13°56′E
Ljusdal, Swe. (lyōōs′däl)	122	61°50′N	16°11′E
Ljusnan, r., Swe.	116	61°55′N	15°33′E
Llandudno, Wales, U.K. (lăn-düd′nō)	120	53°20′N	3°46′W
Llanelli, Wales, U.K. (là-nĕl′ĭ)	117	51°44′N	4°09′W
Llanes, Spain (lyä′nås)	118	43°25′N	4°41′W
Llano, Tx., U.S. (lä′nō) (lyä′nō)	80	30°45′N	98°41′W
Llano, r., Tx., U.S.	80	30°38′N	99°04′W
Llanos, reg., S.A. (lyà′nōs)	100	4°00′N	71°15′W
Lleida, Spain	118	41°38′N	0°37′E
Llera, Mex. (lyā′rä)	88	23°16′N	99°03′W
Llerena, Spain (lyā-rā′nä)	128	38°14′N	6°02′W
Lliria, Spain	129	39°35′N	0°34′W
Llobregat, r., Spain (lyô-brē-gät′)	129	41°55′N	1°55′E
Lloyd Lake, l., Can. (loid)	62e	50°52′N	114°13′W
Lloydminster, Can.	50	53°17′N	110°00′W
Llucena, Spain	129	40°08′N	0°18′W
Llucmajor, Spain	129	39°28′N	2°53′E
Llullaillaco, Volcán, vol., S.A. (lyōō-lyī-lyä′kō)	102	24°50′S	68°30′W
Loange, r., Afr. (lō-äŋ′gä)	186	5°00′S	20°15′E
Lobamba, Swaz.	186	26°27′S	31°12′E
Lobatse, Bots. (lō-bä′tsē)	186	25°13′S	25°35′E
Lobería, Arg. (lô-bĕ′rē′ä)	102	38°13′S	58°48′W
Lobito, Ang. (lō-bē′tō)	186	12°30′S	13°34′E
Lobnya, Russia (lôb′nyà)	142b	56°01′N	37°29′E
Lobo, Phil.	169a	13°39′N	121°14′E
Lobos, Arg. (lō′bōs)	99c	35°10′S	59°08′W
Lobos, Cayo, i., Bah. (lō′bōs)	92	22°25′N	77°40′W
Lobos, Isla de, i., Mex. (ē′s-lä-dĕ-lô′bōs)	89	21°24′N	97°11′W
Lobos de Tierra, i., Peru (lô′bō-dĕ-tyĕ′r-rä)	100	6°29′S	80°55′W
Lobva, Russia (lôb′vå)	142a	59°12′N	60°28′E
Lobva, r., Russia	142a	59°14′N	60°17′E
Locarno, Switz. (lō-kär′nō)	124	46°10′N	8°43′E
Loches, Fr. (lôsh)	126	47°08′N	0°56′E
Loch Raven Reservoir, res., Md., U.S.	68e	39°28′N	76°38′W
Lockeport, Can.	60	43°42′N	65°07′W
Lockhart, S.C., U.S. (lŏk′härt)	83	34°47′N	81°30′W
Lockhart, Tx., U.S.	81	29°54′N	97°40′W
Lock Haven, Pa., U.S. (lŏk′hā-vĕn)	67	41°05′N	77°30′W
Lockland, Oh., U.S. (lŏk′lănd)	69f	39°14′N	84°27′W
Lockport, Il., U.S.	69a	41°35′N	88°04′W
Lockport, N.Y., U.S.	67	43°11′N	78°43′W
Loc Ninh, Viet. (lōk′nĭng′)	168	12°00′N	106°30′E
Lod, Isr. (lôd)	153a	31°57′N	34°55′E
Lodève, Fr. (lō-dĕv′)	126	43°43′N	3°18′E
Lodeynoye Pole, Russia (lô-dĕy-nô′yĕ)	136	60°43′N	33°24′E
Lodge Creek, r., N.A. (lŏj)	73	49°20′N	110°20′W
Lodge Creek, r., Mt., U.S.	73	48°51′N	109°30′W
Lodgepole Creek, r., Wy., U.S. (lŏj′pōl)	70	41°22′N	104°48′W
Lodhran, Pak.	158	29°40′N	71°39′E
Lodi, Italy (lô′dē)	130	45°18′N	9°30′E
Lodi, Ca., U.S. (lō′dī)	76	38°07′N	121°17′W
Lodi, Oh., U.S. (lō′dī)	69d	41°02′N	82°01′W
Lodosa, Spain (lō-dō′sä)	128	42°27′N	2°04′W
Lodwar, Kenya	191	3°07′N	35°36′E
Łódź, Pol.	110	51°46′N	19°30′E
Loeches, Spain (lō-āch′ĕs)	129a	40°22′N	3°25′W
Loffa, r., Afr.	188	7°10′N	10°35′W
Lofoten, is., Nor. (lô′fō-tĕn)	112	68°26′N	13°42′E
Logan, Oh., U.S. (lō′găn)	66	39°35′N	82°25′W
Logan, Ut., U.S.	64	41°46′N	111°51′W
Logan, W.V., U.S.	66	37°50′N	82°00′W
Logan, Mount, mtn., Can.	52	60°54′N	140°33′W
Logansport, In., U.S. (lō′gănz-pōrt)	65	40°45′N	86°25′W
Logone, r., Afr. (lō-gō′nä) (lō-gōn′)	185	10°20′N	15°30′E
Logroño, Spain (lō-grō′nyō)	118	42°28′N	2°25′W
Logrosán, Spain (lō-grō-sän′)	128	39°22′N	5°29′W
Løgstør, Den. (lügh-stûr′)	122	56°56′N	9°15′E
Loir, r., Fr. (lwår)	126	47°40′N	0°07′E
Loire, r., Fr.	112	47°30′N	2°00′E
Loja, Ec. (lō′hä)	100	3°49′S	79°13′W
Loja, Spain (lō′-kä)	128	37°10′N	4°11′W
Loka, D.R.C.	190	0°20′N	17°57′E
Lokala Drift, Bots. (lō′kä-là drĭft)	192c	24°00′S	26°38′E
Lokandu, D.R.C.	191	2°31′S	25°47′E
Lokhvytsia, Ukr.	137	50°21′N	33°16′E
Lokichar, Kenya	191	2°23′N	35°39′E
Lokitaung, Kenya	191	4°16′N	35°45′E
Lokofa-Bokolongo, D.R.C.	190	0°12′N	19°22′E
Lokoja, Nig. (lō-kō′yä)	184	7°47′N	6°45′E
Lokolama, D.R.C.	190	2°34′S	19°53′E
Lokosso, Burkina	188	10°19′N	3°40′W
Lol, r., Sudan (lōl)	185	9°06′N	28°09′E
Loliondo, Tan.	191	2°03′S	35°37′E
Lolland, i., Den. (lôl′än′)	122	54°41′N	11°00′E
Lolo, Mt., U.S.	73	46°45′N	114°05′W
Lom, Blg. (lōm)	119	43°48′N	23°15′E
Loma Linda, Ca., U.S. (lō′mà lĭn′dà)	75a	34°04′N	117°16′W
Lomami, r., D.R.C.	186	0°50′S	24°40′E
Lomas de Zamora, Arg. (lō′mäs dā zä-mō′rä)	99c	34°46′S	58°24′W
Lombard, Il., U.S. (lŏm-bärd)	69a	41°53′N	88°01′W
Lombardia, hist. reg., Italy (lôm-bär-dē′ä)	130	45°20′N	9°30′E
Lomblen, Pulau, i., Indon. (lŏm-blĕn′)	169	8°08′S	123°45′E
Lombok, i., Indon. (lŏm-bŏk′)	168	9°15′S	116°15′E
Lomé, Togo	184	6°08′N	1°13′E
Lomela, D.R.C. (lō-mā′là)	186	2°19′S	23°33′E
Lomela, r., D.R.C.	186	0°35′S	21°20′E
Lometa, Tx., U.S. (lō-mē′tà)	80	31°10′N	98°25′W
Lomié, Cam. (lō-mė-ā′)	189	3°10′N	13°37′E
Lomita, Ca., U.S. (lō-mē′tà)	75a	33°48′N	118°20′W
Lommel, Bel.	115a	51°14′N	5°21′E
Lommond, Loch, l., Scot., U.K. (lŏk lō′mŭnd)	120	56°15′N	4°40′W
Lomonosov, Russia (lō-mô′nô-sof)	142c	59°54′N	29°47′E
Lompoc, Ca., U.S. (lŏm-pōk′)	76	34°39′N	120°30′W
Łomża, Pol. (lôm′zhà)	125	53°11′N	22°04′E
Lonaconing, Md., U.S. (lō-nà-kō′nĭng)	67	39°35′N	78°55′W
London, Can. (lŭn′dŭn)	51	43°00′N	81°20′W
London, Eng., U.K.	110	51°30′N	0°07′W
London, Ky., U.S.	82	37°07′N	84°06′W
London, Oh., U.S.	66	39°50′N	83°30′W
Londonderry, Can. (lŭn′dŭn-dĕr-ĭ)	60	45°29′N	63°36′W
Londonderry, N. Ire., U.K.	116	55°00′N	7°19′W
Londonderry, Cape, c., Austl.	174	13°30′S	127°00′E
Londrina, Braz. (lôn-drē′nä)	101	21°53′S	51°17′W
Lonely, i., Can. (lōn′lĭ)	53	45°35′N	81°30′W
Lone Pine, Ca., U.S.	76	36°36′N	118°03′W
Lone Star, Nic.	91	13°58′N	84°25′W
Long, i., Bah.	87	23°25′N	75°10′W
Long, i., Can.	60	44°21′S	66°25′W
Long, l., N.D., U.S.	70	46°47′N	100°14′W
Long, l., Wa., U.S.	74a	47°29′N	122°36′W
Longa, r., Ang. (lôŋ′gà)	186	10°20′S	15°15′E
Long Bay, b., S.C., U.S.	83	33°30′N	78°54′W
Long Beach, Ca., U.S. (lông bēch)	64	33°46′N	118°12′W
Long Beach, N.Y., U.S.	68a	40°35′N	73°38′W
Long Branch, N.J., U.S. (lông brånch)	68a	40°18′N	73°59′W
Longdon, N.D., U.S. (lông′-dŭn)	70	48°45′N	98°23′W
Long Eaton, Eng., U.K. (ē′tŭn)	114a	52°54′N	1°16′W
Longford, Ire. (lŏng′fērd)	120	53°43′N	7°40′W
Longgu, China (lôŋ-gōō)	162	34°52′N	116°48′E
Longhorn, Tx., U.S. (lông-hôrn)	75d	29°33′N	98°23′W
Longido, Tan.	191	2°44′S	36°41′E
Long Island, i., Pap. N. Gui.	169	5°10′S	147°30′E
Long Island, i., Ak., U.S.	54	54°54′N	132°45′W
Long Island, i., N.Y., U.S. (lông)	65	40°50′N	72°50′W
Long Island Sound, strt., U.S. (lông ī′lănd)	65	41°05′N	72°45′W
Longjumeau, Fr. (lôN-zhü-mō′)	127b	48°42′N	2°17′E
Longkou, China (lôŋ-kō)	162	37°39′N	120°21′E
Longlac, Can. (lông′låk)	51	49°41′N	86°28′W
Longlake, S.D., U.S. (lông-lāk)	70	45°52′N	99°06′W
Long Lake, l., Can.	58	49°10′N	86°45′W
Longmont, Co., U.S. (lông′mŏnt)	78	40°11′N	105°07′W
Longnor, Eng., U.K. (lông′nôr)	114a	53°11′N	1°52′W
Long Pine, Ne., U.S. (lông pīn)	70	42°31′N	99°42′W
Long Point, c., Can.	57	53°02′N	98°40′W
Long Point, c., Can.	61	48°48′N	58°46′W
Long Point, c., Can.	59	42°35′N	80°05′W
Long Point Bay, b., Can.	59	42°40′N	80°10′W
Long Range Mountains, mts., Can.	53a	48°00′N	58°30′W
Longreach, Austl. (lông′rēch)	175	23°32′S	144°17′E
Long Reach, r., Can.	60	45°26′N	66°05′W
Long Reef, c., Austl.	173b	33°45′S	151°22′E
Longridge, Eng., U.K. (lông′rĭj)	114a	53°51′N	2°37′W
Longs Peak, mtn., Co., U.S. (lôngz)	64	40°17′N	105°37′W
Longtansi, China (lôŋ-tä-sz)	162	32°12′N	115°53′E
Longton, Eng., U.K. (lông′tŭn)	114a	52°59′N	2°08′W
Longueuil, Can. (lôN-gû′y′)	59	45°32′N	73°30′W
Longview, Tx., U.S.	81	32°29′N	94°44′W
Longview, Wa., U.S. (lông-vū)	72	46°06′N	123°02′W
Longville, La., U.S. (lông-vĭl)	81	30°36′N	93°14′W
Longwy, Fr. (lôN-wē′)	127	49°32′N	6°14′E
Longxi, China (lôŋ-shyē)	160	35°00′N	104°40′E
Long Xuyen, Viet. (loung′ sōō′yĕn)	168	10°31′N	105°28′E
Longzhou, China (lôŋ-jō)	160	22°20′N	107°02′E
Lonoke, Ar., U.S. (lō′nōk)	79	34°48′N	91°52′W
Lons-le-Saunier, Fr. (lôN-lē-sō-nyā′)	127	46°40′N	5°33′E
Lontue, r., Chile (lōn-tōē′)	99b	35°20′S	70°45′W
Looc, Phil. (lō-ōk′)	169a	12°16′N	121°59′E
Loogootee, In., U.S.	66	38°40′N	86°55′W
Lookout, Cape, c., N.C., U.S. (lŏŏk′out)	83	34°34′N	76°38′W
Lookout Point Lake, res., Or., U.S.	72	43°51′N	122°38′W
Loolmalasin, mtn., Tan.	191	3°03′S	35°46′E
Looma, Can. (ō′mä)	62g	53°22′N	113°15′W
Loop Head, c., Ire. (lōōp)	120	52°32′N	9°59′W
Loosahatchie, r., Tn., U.S. (lōz-à-hă′chē)	82	35°20′N	89°45′W
Loosdrechtsche Plassen, l., Neth.	115a	52°11′N	5°09′E

PLACE (Pronunciation)	PAGE	LAT.	LONG.
Lopatka, Mys, c., Russia (lô-pät'kà)	153	51°00'N	156°52'E
Lopez, Cap, c., Gabon	190	0°37'N	8°43'E
Lopez Bay, b., Phil. (lō'pāz)	169a	14°04'N	122°00'E
Lopez I, Wa., U.S.	74a	48°25'N	122°53'W
Lopori, r., D.R.C. (lō-pō'rē)	185	1°35'N	20°43'E
Lora, Spain (lō'rä)	128	37°40'N	5°31'W
Lorain, Oh., U.S. (lô-rān')	69d	41°28'N	82°10'W
Loralai, Pak. (lō-rŭ-lī')	155	30°31'N	68°35'E
Lorca, Spain (lôr'kä)	118	37°39'N	1°40'W
Lord Howe, i., Austl. (lôrd hou)	174	31°44'S	157°56'W
Lordsburg, N.M., U.S. (lôrdz'bûrg)	77	32°20'N	108°45'W
Lorena, Braz. (lô-rā'nà)	99a	22°45'S	45°07'W
Loreto, Braz. (lô-rā'tō)	101	7°09'S	45°10'W
Loretteville, Can. (lô-rĕt-vēl')	62b	46°51'N	71°21'W
Lorica, Col. (lô-rē'kä)	100	9°14'N	75°54'W
Lorient, Fr. (lô-rē'äN')	117	47°45'N	3°22'W
Lorn, Firth of, b., Scot., U.K. (fûrth ŏv lôrn')	120	56°10'N	6°09'W
Lörrach, Ger. (lûr'äk)	124	47°36'N	7°38'E
Lorraine, hist. reg., Fr.	127	49°00'N	6°00'E
Los Alamitos, Ca., U.S. (lōs àl-à-mē'tōs)	75a	33°48'N	118°04'W
Los Alamos, N.M., U.S. (àl-à-môs')	77	35°53'N	106°20'W
Los Altos, Ca., U.S. (àl-tôs')	74b	37°23'N	122°06'W
Los Andes, Chile (àn'dĕs)	99b	32°44'S	70°36'W
Los Angeles, Chile (äŋ'hå-lās)	102	37°27'S	72°15'W
Los Angeles, Ca., U.S.	64	34°03'N	118°14'W
Los Angeles, Ca., U.S.	76	34°03'N	118°14'W
Los Angeles, r., Ca., U.S.	75a	33°50'N	118°13'W
Los Angeles Aqueduct, Ca., U.S.	76	35°12'N	118°02'W
Los Bronces, Chile (lôs brō'n-sĕs)	99b	33°09'S	70°18'W
Loscha, r., Id., U.S. (lŏs'chă)	72	46°20'N	115°11'W
Los Estados, Isla de, i., Arg. (ē's-lä dĕ lôs ĕs-dôs)	102	54°45'S	64°25'W
Los Gatos, Ca., U.S. (gä'tôs)	76	37°13'N	121°59'W
Los Herreras, Mex. (ĕr-rā-räs)	80	25°55'N	99°23'W
Los Indios, Cayos de, is., Cuba (kä'yōs dĕ lôs ē'n-dyô's)	92	21°50'N	83°10'W
Los Llanos, Dom. Rep. (lôs ē-lä'nōs)	93	18°35'N	69°30'W
Lošinj, i., Serb.	130	44°35'N	14°34'E
Losino Petrovskiy, Russia	142b	55°52'N	38°12'E
Los Nietos, Ca., U.S. (nyā'tōs)	75a	33°57'N	118°05'W
Los Palacios, Cuba	92	22°35'N	83°15'W
Los Pinos, r., Co., U.S. (pē'nōs)	77	36°58'N	107°35'W
Los Reyes, Mex.	86	19°35'N	102°29'W
Los Reyes, Mex.	89a	19°21'N	98°58'W
Los Santos, Pan. (săn'tōs)	91	7°57'N	80°24'W
Los Santos de Maimona, Spain (sän'tōs)	128	38°38'N	6°30'W
Lost, r., Or., U.S.	72	42°07'N	121°30'W
Los Teques, Ven. (tĕ'kĕs)	100	10°22'N	67°04'W
Lost River Range, mts., Id., U.S. (rĭ'vēr)	73	44°23'N	113°48'W
Los Vilos, Chile (vē'lôs)	102	31°56'S	71°29'W
Lot, r., Fr. (lôt)	117	44°30'N	1°30'E
Lota, Chile (lō'tä)	102	37°11'S	73°14'W
Lothian, Md., U.S. (lōth'ĭăn)	68e	38°50'N	76°38'W
Lotikipi Plain, pl., Afr.	191	4°25'N	34°55'E
Lötschberg Tunnel, trans., Switz.	124	46°26'N	7°54'E
Louangphrabang, Laos (lōō-ang'prä-bäng')	168	19°47'N	102°15'E
Loudon, Tn., U.S. (lou'dŭn)	82	35°43'N	84°20'W
Loudonville, Oh., U.S. (lou'dŭn-vĭl)	66	40°40'N	82°15'W
Loudun, Fr.	126	47°03'N	0°00'
Loughborough, Eng., U.K. (lŭf'bŭr-ô)	114a	52°46'N	1°12'W
Louisa, Ky., U.S. (lōō'ēz-à)	66	38°05'N	82°40'W
Louisade Archipelago, is., Pap. N. Gui.	175	10°44'S	153°58'E
Louisberg, N.C., U.S. (lōō'ĭs-bûrg)	83	36°05'N	79°19'W
Louisburg, Can. (lōō'ĭs-bourg)	61	45°55'N	59°58'W
Louiseville, Can.	59	46°17'N	72°58'W
Louisiana, Mo., U.S. (lōō-ē-zē-ăn'à)	79	39°24'N	91°03'W
Louisiana, state, U.S.	65	30°50'N	92°50'W
Louis Trichardt, S. Afr. (lōō'ĭs trĭchårt)	186	22°52'S	29°53'E
Louisville, Co., U.S. (lōō'ĭs-vĭl) (lōō'ē-vĭl)	78	39°58'N	105°08'W
Louisville, Ga., U.S.	83	33°00'N	82°25'W
Louisville, Ky., U.S.	65	38°15'N	85°45'W
Louisville, Ms., U.S.	82	33°07'N	89°02'W
Louis XIV, Pointe, c., Can.	53	54°35'N	79°51'W
Louny, Czech Rep. (lō'nė)	124	50°20'N	13°47'E
Loup, r., Ne., U.S. (lōōp)	70	41°17'N	97°58'W
Loup City, Ne., U.S.	70	41°15'N	98°59'W
Lourdes, Fr. (lōōrd)	117	43°06'N	0°03'W
Lourenço Marques *see* Maputo, Moz.	186	26°50'S	32°30'E
Loures, Port. (lō'rĕzh)	129b	38°49'N	9°10'W
Lousa, Port. (lō'zà)	128	40°05'N	8°12'W
Louth, Eng., U.K. (louth)	120	53°27'N	0°02'W
Louvain *see* Leuven, Bel.	121	50°53'N	4°42'E
Louviers, Fr. (lōō-vyā')	126	49°13'N	1°11'E
Lovech, Blg. (lō'vĕts)	131	43°10'N	24°40'E
Loveland, Co., U.S. (lŭv'lănd)	78	40°24'N	105°04'W
Loveland, Oh., U.S.	69f	39°16'N	84°15'W
Lovell, Wy., U.S. (lŭv'ĕl)	73	44°50'N	108°23'W
Lovelock, Nv., U.S. (lŭv'lŏk)	76	40°10'N	118°37'W
Lovick, Al., U.S. (lŭ'vĭk)	68h	33°34'N	86°38'W
Loviisa, Fin. (lô'vē-sà)	123	60°28'N	26°10'E
Low, Cape, c., Can. (lō)	53	62°58'N	86°50'W
Lowa, r., D.R.C. (lō'wà)	186	1°30'S	27°18'E
Lowell, In., U.S.	69a	41°17'N	87°26'W
Lowell, Ma., U.S.	65	42°38'N	71°18'W
Lowell, Mi., U.S.	66	42°55'N	85°20'W
Löwenberg, Ger. (lû'vĕn-bĕrgh)	115b	52°53'N	13°09'E
Lower Brule Indian Reservation, I.R., S.D., U.S. (brü'lā)	70	44°15'N	100°21'W
Lower California *see* Baja California, pen., Mex.	49	28°00'N	113°30'W
Lower Granite Dam, dam, Wa., U.S.	72	46°40'N	117°26'W
Lower Hutt, N.Z. (hŭt)	175a	41°10'S	174°55'E
Lower Klamath Lake, l., Ca., U.S. (klăm'ăth)	72	41°55'N	121°50'W
Lower Lake, l., Ca., U.S.	72	41°21'N	119°53'W
Lower Marlboro, Md., U.S. (lō'ēr märl'bōrō)	68e	38°40'N	76°42'W
Lower Monumental Dam, dam, Wa., U.S.	72	46°34'N	118°32'W
Lower Otay Lake, res., Ca., U.S. (ō'tā)	76a	32°37'N	116°46'W
Lower Red Lake, l., Mn., U.S.	71	47°58'N	94°31'W
Lower Saxony *see* Niedersachsen, state, Ger.	115c	53°30'N	9°30'E
Lowestoft, Eng., U.K. (lō'stŏf)	121	52°31'N	1°45'E
Łowicz, Pol. (lô'vĭch)	125	52°06'N	19°57'E
Lowville, N.Y., U.S. (lou'vĭl)	67	43°45'N	75°30'W
Loxicha, Mex.	89	16°03'N	96°46'W
Loxton, Austl. (lôks'tŭn)	176	34°25'S	140°38'E
Loyauté, Îles, is., N. Cal.	175	21°00'S	167°00'E
Loznica, Serb. (lōz'nē-tsà)	119	44°31'N	19°16'E
Lozova, Ukr.	137	48°53'N	36°23'E
Luama, r., D.R.C. (lōō'ä-mà)	186	4°17'S	27°45'E
Lu'an, China (lōō-än)	165	31°45'N	116°29'E
Luan, r., China	161	41°25'N	117°15'E
Luanda, Ang. (lōō-än'dä)	186	8°48'S	13°14'E
Luanguinga, r., Afr. (lōō-ä-gĭŋ'gà)	186	14°00'S	20°45'E
Luanshya, Zam.	191	13°08'S	28°24'E
Luanxian, China (luän shyĕn)	162	39°47'N	118°40'E
Luao, Ang.	190	10°42'S	22°12'E
Luarca, Spain (lwä'kä)	118	43°33'N	6°30'W
Lubaczów, Pol. (lōō-bä'chôf)	125	50°08'N	23°10'E
Lubán, Pol. (lōō'bän')	124	51°08'N	15°17'E
Lubānas Ezers, l., Lat. (lōō-bä'nás ä'zĕrs)	123	56°48'N	26°30'E
Lubang, Phil. (lōō-bäng')	169a	13°49'N	120°07'E
Lubang Islands, is., Phil.	168	13°47'N	119°56'E
Lubango, Ang.	186	14°55'S	13°30'E
Lubartów, Pol. (lōō-bär'tôf)	125	51°27'N	22°37'E
Lubawa, Pol. (lōō-bä'vä)	125	53°31'N	19°47'E
Lübben, Ger. (lüb'ĕn)	124	51°56'N	13°53'E
Lubbock, Tx., U.S.	64	33°35'N	101°50'E
Lubec, Me., U.S. (lū'bĕk)	60	44°49'N	67°01'W
Lübeck, Ger. (lü'bĕk)	110	53°53'N	10°42'E
Lübecker Bucht, b., Ger. (lü'bĕ-kĕr bōōĸt)	116	54°10'N	11°20'E
Lubilash, r., D.R.C. (lōō-bē-lāsh')	186	7°35'S	23°55'E
Lubin, Pol. (lyō'bĭn)	124	51°24'N	16°14'E
Lublin, Pol. (lyō'blēn')	110	51°14'N	22°33'E
Lubny, Ukr. (lōb'nė)	137	50°01'N	33°02'E
Lubuagan, Phil. (lō-bwä-gä'n)	169a	17°24'N	121°11'E
Lubudi, D.R.C.	191	9°57'S	25°58'E
Lubudi, r., D.R.C. (lō-bō'dē)	186	10°00'S	24°30'E
Lubumbashi, D.R.C.	186	11°40'S	27°28'E
Lucano, Ang.	190	11°16'S	21°38'E
Lucca, Italy (lōōk'kä)	118	43°51'N	10°29'E
Lucea, Jam.	92	18°25'N	78°10'W
Luce Bay, b., Scot., U.K. (lūs)	120	54°45'N	4°45'W
Lucena, Phil. (lōō-sā'nä)	169a	13°55'N	121°36'E
Lucena, Spain (lōō-thā'nä)	118	37°25'N	4°28'W
Lučenec, Slvk. (lōō'châ-nyĕts)	117	48°19'N	19°41'E
Lucera, Italy (lōō-châ'rä)	130	41°31'N	15°22'E
Luchi, China	165	28°18'N	110°10'E
Lucin, Ut., U.S. (lů-sēn')	73	41°23'N	113°59'W
Lucipara, Kepulauan, is., Indon.	169	5°45'S	128°15'E
Luckenwalde, Ger.	124	52°05'N	13°10'E
Lucknow, India (lŭk'nou)	155	26°54'N	80°58'E
Lucky Peak Lake, res., Id., U.S.	72	43°33'N	116°00'W
Luçon, Fr. (lü-sôN')	126	46°27'N	1°12'W
Lucrecia, Cabo, c., Cuba	93	21°05'N	75°30'W
Luda Kamchiya, r., Blg.	131	42°46'N	27°13'E
Lüdenscheid, Ger. (lü'dĕn-shīt)	127c	51°13'N	7°38'E
Lüderitz, Nmb. (lü'dĕr-ĭts) (lü'dĕ-rĭts)	186	26°35'S	15°15'E
Lüderitz Bucht, b., Nmb.	186	26°35'S	14°30'E
Ludhiāna, India	155	31°00'N	75°52'E
Lüdinghausen, Ger.	127c	51°46'N	7°27'E
Ludington, Mi., U.S. (lŭd'ĭng-tŭn)	66	44°00'N	86°25'W
Ludlow, Eng., U.K. (lŭd'lō)	114a	52°22'N	2°43'W
Ludlow, Ky., U.S.	69f	39°05'N	84°33'W
Ludvika, Swe. (loodh-vē'kà)	122	60°10'N	15°09'E
Ludwigsburg, Ger.	124	48°53'N	9°14'E
Ludwigsfelde, Ger.	115b	52°18'N	13°16'E
Ludwigshafen, Ger.	124	49°29'N	8°26'E
Ludwigslust, Ger.	124	53°18'N	11°31'E
Ludza, Lat. (lōōd'zà)	123	56°33'N	27°45'E
Luebo, D.R.C. (lōō-ā'bô)	186	5°15'S	21°22'E
Luena, Ang.	186	11°45'S	19°55'E
Luena, D.R.C.	191	9°27'S	25°47'E
Lufira, r., D.R.C. (lōō-fē'rà)	186	9°32'S	27°15'E
Lufkin, Tx., U.S. (lŭf'kĭn)	81	31°21'N	94°43'W
Luga, Russia (lōō'gà)	136	58°43'N	29°52'E
Luga, r., Russia	132	59°00'N	29°25'E
Lugano, Switz. (lōō-gä'nō)	124	46°01'N	8°52'E
Lugenda, r., Moz.	187	12°05'S	38°15'E
Lugo, Italy (lōō'gô)	130	44°28'N	11°57'E
Lugo, Spain (lōō'gō)	118	43°01'N	7°32'W
Lugoj, Rom.	119	45°51'N	21°56'E
Luhans'k, Ukr.	134	48°34'N	39°18'E
Luhans'k, prov., Ukr.	133	49°30'N	38°35'E
Luhe, China (lōō-hŭ)	162	32°22'N	118°50'E
Luiana, Ang.	190	17°23'S	23°03'E
Luilaka, r., D.R.C. (lōō-ē-lä'kà)	186	2°18'S	21°15'E
Luis Moya, Mex. (lōōē's-mô-yä)	88	22°26'N	102°14'W
Luján, Arg. (lōō-hän')	99c	34°36'S	59°07'W
Luján, r., Arg.	99c	34°33'S	58°59'W
Lujia, China (lōō-jyä)	162	31°17'N	120°54'W
Lukanga Swamp, sw., Zam. (lōō-käŋ'gà)	186	14°30'S	27°25'E
Lukenie, r., D.R.C. (lōō-kā'nyà)	186	3°10'S	19°05'E
Lukolela, D.R.C.	186	1°03'S	17°01'E
Lukovit, Blg. (lōō'kō-vēt')	131	43°13'N	24°07'E
Łuków, Pol. (wō'kôf)	125	51°57'N	22°25'E
Lukuga, r., D.R.C. (lōō-kōō'gà)	186	5°50'S	27°35'E
Lüleburgaz, Tur. (lü'lĕ-bŏr-gäs')	131	41°25'N	27°23'E
Luling, Tx., U.S. (lū'lĭng)	81	29°41'N	97°38'W
Lulong, China (lōō-lôŋ)	161	39°54'N	118°53'E
Lulonga, r., D.R.C.	190	1°00'N	18°37'E
Luluabourg *see* Kananga, D.R.C.	186	6°14'S	22°17'E
Lulu Island, i., Can.	74d	49°09'N	123°05'W
Lulu Island, i., Ak., U.S.	54	55°28'N	133°30'W
Lumajangdong Co, l., China	158	34°00'N	81°47'E
Lumber, r., N.C., U.S. (lŭm'bēr)	83	34°45'N	79°10'W
Lumberton, Ms., U.S. (lŭm'bēr-tŭn)	82	31°00'N	89°25'W
Lumberton, N.C., U.S.	83	34°47'N	79°00'W
Luminárias, Braz. (lōō-mē-nä'ryäs)	99a	21°32'S	44°53'W
Lummi, i., Wa., U.S.	74d	48°42'N	122°43'W
Lummi Bay, b., Wa., U.S. (lŭm'ĭ)	74d	48°47'N	122°44'W
Lummi Island, Wa., U.S.	74d	48°44'N	122°42'W
Lumwana, Zam.	191	11°50'S	25°10'E
Lün, Mong.	160	47°58'N	104°52'E
Luna, Phil. (lōō'nä)	169a	16°51'N	120°22'E
Lund, Swe. (lŭnd)	116	55°42'N	13°10'E
Lundy, i., Eng., U.K. (lŭn'dē)	120	51°12'N	4°50'W
Lüneburg, Ger. (lü'nē-bôrgh)	124	53°16'N	10°25'E
Lunel, Fr. (lü-nĕl')	126	43°41'N	4°07'E
Lünen, Ger. (lü'nĕn)	127c	51°36'N	7°30'E
Lunenburg, Can. (lōō'nĕn-bûrg)	51	44°23'N	64°19'W
Lunenburg, Ma., U.S.	61a	42°36'N	71°44'W
Lunéville, Fr. (lü-nà-vel')	127	48°37'N	6°29'E
Lunga, Ang.	190	14°42'S	18°32'E
Lungué-Bungo, r., Afr.	186	13°00'S	20°30'E
Lunsar, S.L.	188	8°41'N	12°32'W
Luodian, China (lwô-diĕn)	162	31°25'N	121°20'E
Luoding, China (lwô-dĭŋ)	165	23°42'N	111°35'E
Luohe, China (lwô-hŭ)	162	33°35'N	114°02'E
Luoyang, China (lwô-yäŋ)	161	34°45'N	112°32'E
Luozhen, China (lwô-jŭn)	162	37°45'N	118°29'E
Luque, Para. (loo'kā)	102	25°18'S	57°17'W
Luray, Va., U.S. (lū-rā')	67	38°40'N	78°25'W
Lurgan, N. Ire., U.K. (lûr'găn)	116	54°27'N	6°28'W
Lúrio, Moz. (lōō'rē-ô)	187	13°17'S	40°29'E
Lúrio, Moz.	187	14°00'S	38°45'E
Lusaka, D.R.C.	191	7°10'S	29°27'E
Lusaka, Zam. (lô-sä'kà)	186	15°25'S	28°17'E
Lusambo, D.R.C. (lōō-säm'bō)	186	4°58'S	23°27'E
Lusanga, D.R.C.	186	5°13'S	18°43'E
Lusangi, D.R.C.	191	4°37'S	27°08'E
Lushan, China	164	33°45'N	113°00'E
Lushiko, r., Afr.	190	6°35'S	19°45'E
Lushoto, Tan. (lōō-shō'tō)	187	4°47'S	38°17'E
Lüshun, China (lü-shŭn)	161	38°49'N	121°15'E
Lusikisiki, S. Afr. (lōō-sē-kē-sē'kē)	187c	31°22'S	29°37'E
Lusk, Wy., U.S. (lŭsk)	70	42°46'N	104°27'W
Lūt, Dasht-e, des., Iran (dă'sht-ē-lōōt)	154	31°47'N	58°38'E
Lutcher, La., U.S. (lŭch'ēr)	81	30°03'N	90°43'W
Luton, Eng., U.K. (lū'tŭn)	120	51°55'N	0°28'W
Luts'k, Ukr.	137	50°45'N	25°20'E
Luuq, Som.	192a	3°38'N	42°35'E
Luverne, Al., U.S. (lū-vûn')	82	31°42'N	86°15'W
Luverne, Mn., U.S.	70	43°40'N	96°13'W
Luwingu, Zam.	191	10°15'S	29°55'E
Luxapallila Creek, r., U.S. (lŭk-sà-pôl'ĭ-là)	82	33°36'N	88°08'W
Luxembourg, Lux.	110	49°38'N	6°30'E
Luxembourg, nation, Eur.	110	49°30'N	6°22'E
Luxeuil-les-Baines, Fr.	127	47°49'N	6°19'E
Luxomni, Ga., U.S. (lŭx'ôm-nī)	68c	33°54'N	84°07'W
Luxor *see* Al Uqşur, Egypt	185	25°38'N	32°59'E
Luya Shan, mtn., China	164	38°50'N	111°40'E
Luyi, China (lōō-yē)	162	33°52'N	115°32'E
Luzern, Switz. (lô-tsĕrn)	117	47°03'N	8°18'E
Luzhou, China (lōō-jō)	160	28°58'N	105°25'E
Luziânia, Braz. (lōō-zyà'nēà)	101	16°17'S	47°44'W
Luzon, i., Phil. (lōō-zŏn')	168	17°10'N	119°45'E
Luzon Strait, strt., Asia	165	20°40'N	121°00'E
L'viv, Ukr.	134	49°50'N	24°00'E
L'vov *see* L'viv, Ukr.	134	49°50'N	24°00'E
Lyalta, Can.	62e	51°07'N	113°36'W
Lyalya, r., Russia (lyä'lyä)	142a	58°58'N	60°17'E
Lyaskovets, Blg.	131	43°07'N	25°41'E
Lydenburg, S. Afr. (lī'dĕn-bûrg)	186	25°06'S	30°21'E
Lyell, Mount, mtn., Ca., U.S. (lī'ĕl)	76	37°44'N	119°22'W
Lyepye', Bela. (lyĕ-pĕl')	132	54°52'N	28°41'E
Lykens, Pa., U.S. (lī'kĕnz)	67	40°35'N	76°45'W
Lykhivka, Ukr.	133	48°52'N	33°57'E
Lyna, r., Eur. (lĭn'à)	125	53°56'N	20°30'E
Lynch, Ky., U.S. (lĭnch)	83	36°56'N	82°55'W
Lynchburg, Va., U.S. (lĭnch'bûrg)	65	37°23'N	79°08'W
Lynch Cove, Wa., U.S. (lĭnch)	74a	47°26'N	122°54'W
Lynden, Can. (lĭn'dĕn)	62d	43°14'N	80°08'W
Lynden, Wa., U.S.	74d	48°56'N	122°27'W
Lyndhurst, Austl.	173a	38°03'S	145°14'E
Lyndon, Ky., U.S. (lĭn'dŭn)	69h	38°15'N	85°36'W
Lyndonville, Vt., U.S. (lĭn'dŭn-vĭl)	67	44°35'N	72°00'W
Lynn, Ma., U.S. (lĭn)	65	42°28'N	70°57'W
Lynn Lake, Can. (lăk)	50	56°51'N	101°05'W
Lynwood, Ca., U.S. (lĭn'wŏŏd)	75a	33°56'N	118°13'W
Lyon, Fr. (lē-ôN')	110	45°44'N	4°52'E
Lyons, Ga., U.S. (lī'ŭnz)	83	32°08'N	82°19'W
Lyons, Ks., U.S.	78	38°20'N	98°11'W
Lyons, Ne., U.S.	70	41°57'N	96°28'W
Lyons, N.J., U.S.	68a	40°41'N	74°33'W

ăt; finăl; rāte; senåte; ärm; àsk; sofà; fâre; ch-choose; dh-as th in other; bē; ėvent; bĕt; recĕnt; cratēr; g-gō; gh-guttural g; bĭt; ĭ-short neutral; rīde; ĸ-guttural k as ch in German ich;

PRONOUNCING INDEX Lyons – Maldon 271

PLACE (Pronunciation)	PAGE	LAT.	LONG.
Lyons, N.Y., U.S.	67	43°05'N	77°00'W
Lyptsi, Ukr.	133	50°11'N	36°25'E
Lysefjorden, b., Nor.	122	58°59'N	6°35'E
Lysekil, Swe. (lü'sĕ-kėl)	122	58°17'N	11°22'E
Lys'va, Russia (lĭs'vȧ)	136	58°07'N	57°47'E
Lytham, Eng., U.K. (lĭth'ăm)	114a	53°44'N	2°58'W
Lytkarino, Russia	142b	55°35'N	37°55'E
Lyttelton, S. Afr. (lĭt'l'ton)	187b	25°51'S	28°13'E
Lyuban', Russia (lyōō'bȧn)	132	59°21'N	31°15'E
Lyubertsy, Russia (lyōō'bĕr-tsė)	132	55°40'N	37°55'E
Lyubim, Russia (lyōō-bėm')	132	58°24'N	40°39'E
Lyublino, Russia (lyōōb'lĭ-nȯ)	142b	55°41'N	37°45'E
Lyudinovo, Russia (lū-dē'novō)	132	53°52'N	34°28'E

M

PLACE (Pronunciation)	PAGE	LAT.	LONG.
Ma'ān, Jord. (mä-än')	154	30°12'N	35°45'E
Maartensdijk, Neth.	115a	52°09'N	5°10'E
Maas (Meuse), r., Eur.	121	51°50'N	5°40'E
Maastricht, Neth. (mäs'trĭKt)	121	50°51'N	5°35'E
Mabaia, Ang.	190	7°13'S	14°03'E
Mabana, Wa., U.S. (mä-bä-nä)	74a	48°06'N	122°25'W
Mabank, Tx., U.S. (mā'bănk)	81	32°21'N	96°05'W
Mabeskraal, S. Afr.	192c	25°12'S	26°47'E
Mableton, Ga., U.S. (mā'b'l-tŭn)	68c	33°49'N	84°34'W
Mabrouk, Mali	184	19°27'N	1°16'W
Mabula, S. Afr. (mă'bōō-la)	192c	24°49'S	27°59'E
Macalelon, Phil. (mä-kä-lā-lōn')	169a	13°46'N	122°09'E
Macau, Braz. (mä-kȧ'ō)	101	5°12'S	36°34'W
Macau, China	161	22°00'N	113°00'E
Macaya, Pico de, mtn., Haiti	93	18°25'N	74°00'W
Macclesfield, Eng., U.K. (măk''lz-fēld)	114a	53°15'N	2°07'W
Macclesfield Canal, can., Eng., U.K. (măk''lz-fēld)	114a	53°14'N	2°07'W
Macdona, Tx., U.S. (măk-dō'nā)	75d	29°20'N	98°42'W
Macdonald, l., Austl. (măk-dŏn'ăld)	174	23°40'S	127°40'E
Macdonnell Ranges, mts., Austl. (măk-dŏn'ĕl)	174	23°40'S	131°30'E
MacDowell Lake, l., Can. (măk-dou ĕl)	57	52°15'N	92°45'W
Macdui, Ben, mtn., Scot., U.K. (bĕn măk-dōō'ė)	116	57°06'N	3°45'W
Macedonia, Oh., U.S. (măs-ė-dō'nĭ-ȧ)	69d	41°19'N	81°30'E
Macedonia, nation, Eur.	131	41°50'N	22°00'E
Macedonia, hist. reg., Eur. (măs-ė-dō'nĭ-ȧ)	119	41°05'N	22°15'E
Maceió, Braz.	101	9°40'S	35°43'W
Macerata, Italy (mä-chȧ-rä'tä)	130	43°18'N	13°28'E
Macfarlane, Lake, l., Austl. (mȧc'fär-lān)	176	32°10'S	137°00'E
Machache, mtn., Leso.	187c	29°22'S	27°53'E
Machado, Braz. (mä-shȧ-dô)	99a	21°42'S	45°55'W
Machakos, Kenya	191	1°31'S	37°16'E
Machala, Ec. (mä-chȧ'lä)	100	3°18'S	78°54'W
Machens, Mo., U.S. (măk'ĕns)	75e	38°54'N	90°20'W
Machias, Me., U.S. (mȧ-chī'ȧs)	60	44°22'N	67°29'W
Machida, Japan (mä-chė'dä)	167a	35°32'N	139°28'E
Machilīpatnam, India	155	16°22'N	81°10'E
Machu Picchu, Peru (mȧ'chȯ-pė'k-chȯ)	100	13°07'S	72°34'W
Măcin, Rom. (mȧ-chēn')	133	45°15'N	28°09'E
Macina, reg., Mali	188	14°50'N	4°40'W
Mackay, Austl. (mă-kī')	175	21°15'S	149°08'E
Mackay, Id., U.S. (măk-kā')	73	43°55'N	113°38'W
Mackay, l., Austl. (mă-kī')	174	22°30'S	127°45'E
MacKay, l., Can. (măk-kā')	52	64°10'N	112°35'W
Mackenzie, r., Can.	52	63°38'N	124°23'W
Mackenzie Bay, b., Can.	63	69°20'N	137°10'W
Mackenzie Mountains, mts., Can. (mȧ-kĕn'zĭ)	52	63°41'N	129°27'W
Mackinaw, r., Il., U.S.	66	40°35'N	89°25'W
Mackinaw City, Mi., U.S. (măk'ĭ-nô)	66	45°45'N	84°45'W
Mackinnon Road, Kenya	191	3°44'S	39°03'E
Macleantown, S. Afr. (măk-lăn'toun)	187c	32°48'S	27°48'E
Maclear, S. Afr. (mȧ-klēr')	186	31°06'S	28°23'E
Macomb, Il., U.S. (mȧ-kōōm')	79	40°27'N	90°40'W
Mâcon, Fr. (mä-kôN)	117	46°19'N	4°51'E
Macon, Ga., U.S. (mā'kŏn)	65	32°49'N	83°39'W
Macon, Mo., U.S.	79	39°42'N	92°29'W
Macon, Ms., U.S.	82	32°07'N	88°31'W
Macquarie, r., Austl.	175	31°43'S	148°04'E
Macquarie Islands, is., Austl. (mȧ-kwŏr'ė)	3	54°36'S	158°45'E
Macuelizo, Hond. (mä-kwĕ-lē'zô)	90	15°22'N	88°32'W
Mad, r., Ca., U.S. (măd)	72	40°38'N	123°37'W
Madagascar, nation, Afr. (măd-ȧ-găs'kȧr)	187	18°05'S	43°12'E
Madame, i., Can. (mȧ-dȧm')	61	45°33'N	61°02'W
Madanapalle, India	159	13°06'N	78°09'E
Madang, Pap. N. Gui. (mä-däng')	169	5°15'S	145°45'E
Madaoua, Niger (mȧ-dou'ȧ)	184	14°04'N	6°03'E
Madawaska, r., Can. (mäd-ȧ-wôs'kȧ)	59	45°20'N	77°25'W
Madeira, r., S.A.	100	6°48'S	62°43'W
Madeira, Arquipélago da, is., Port.	183	33°26'N	16°44'W
Madeira, Ilha da, i., Port. (mä-dā'rä)	184	32°41'N	16°15'W
Madeleine, Îles de la, is., Can.	53	47°30'N	61°45'W
Madelia, Mn., U.S. (mȧ-dē'lĭ-ȧ)	71	44°03'N	94°23'W
Madeline, i., Wi., U.S. (măd'ĕ-lĭn)	71	46°47'N	91°30'W
Madera, Ca., U.S. (mȧ-dā'rȧ)	76	36°57'N	120°04'W
Madera, vol., Nic.	90	11°27'N	85°30'W
Madgaon, India	159	15°09'N	73°58'E
Madhya Pradesh, state, India (mŭd'vŭ prŭ-dāsh')	155	22°04'N	77°48'E
Madill, Ok., U.S. (mȧ-dĭl')	79	34°04'N	96°45'W
Madīnat ash Sha'b, Yemen	154	12°45'N	44°00'E
Madingo, Congo	190	4°07'S	11°22'E
Madingou, Congo	190	4°09'S	13°34'E
Madison, Fl., U.S. (măd'ĭ-sŭn)	82	30°28'N	83°25'W
Madison, Ga., U.S.	82	33°34'N	83°29'W
Madison, Il., U.S.	75e	38°40'N	90°09'W
Madison, In., U.S.	66	38°45'N	85°25'W
Madison, Ks., U.S.	79	38°08'N	96°07'W
Madison, Me., U.S.	60	44°47'N	69°52'W
Madison, Mn., U.S.	70	44°59'N	96°13'W
Madison, N.C., U.S.	83	36°22'N	79°59'W
Madison, Ne., U.S.	70	41°49'N	97°27'W
Madison, N.J., U.S.	68a	40°46'N	74°25'W
Madison, S.D., U.S.	70	44°01'N	97°08'W
Madison, Wi., U.S.	65	43°05'N	89°23'W
Madison Res, Mt., U.S.	73	45°25'N	111°28'W
Madisonville, Ky., U.S. (măd'ĭ-sŭn-vĭl)	66	37°20'N	87°30'W
Madisonville, La., U.S.	81	30°22'N	90°10'W
Madisonville, Tx., U.S.	81	30°57'N	95°55'W
Madjori, Burkina	188	11°26'N	1°15'E
Mado Gashi, Kenya	191	0°44'N	39°10'E
Madona, Lat. (mȧ'dō'nȧ)	123	56°50'N	26°14'E
Madrakah, Ra's al, c., Oman	154	18°53'N	57°48'E
Madras *see* Chennai, India	155	13°08'N	80°15'E
Madre, Laguna, l., Mex. (lă-gōō'nă mă'drȧ)	81	25°08'N	97°41'W
Madre, Sierra, mts., N.A. (sē-ĕ'r-rä-mȧ'drĕ)	89	15°55'N	92°40'W
Madre, Sierra, mts., Phil.	169a	16°40'N	122°10'E
Madre de Dios, r., S.A. (mä'drȧ dā dė-ōs')	100	12°07'S	68°02'W
Madre de Dios, Archipiélago, is., Chile (mȧ'drȧ dā dė-ōs')	102	50°40'S	76°30'W
Madre del Sur, Sierra, mts., Mex. (sē-ĕ'r-rä-mä'drȧ dĕlsōōr')	86	17°35'N	100°35'W
Madre Occidental, Sierra, mts., Mex.	86	29°30'N	107°30'W
Madre Oriental, Sierra, mts., Mex.	86	25°30'N	100°45'W
Madrid, Spain (mä-drė'd)	110	40°26'N	3°42'W
Madrid, Ia., U.S. (măd'rĭd)	71	41°51'N	93°48'W
Madridejos, Spain (mä-dhrė-dhā'hōs)	128	39°29'N	3°32'W
Madura, i., Indon. (mȧ-dōō'rä)	168	6°45'S	113°30'E
Madurai, India (mä-dōō'rä)	155	9°57'N	78°04'E
Madureira, Serra do, mtn., Braz. (sĕ'r-rä-dȯ-mä-dōō-rā'rȧ)	102b	22°49'S	43°30'W
Maebashi, Japan (mä-ĕ-bä'shė)	161	36°26'N	139°04'E
Maestra, Sierra, mts., Cuba (sē-ĕ'r-rȧ-mä-ās'trä)	87	20°05'N	77°05'W
Maewo, i., Vanuatu	175	15°17'S	168°16'E
Mafeking, S. Afr. (mȧf'ê'kĭng)	186	25°46'S	24°45'E
Mafra, Braz. (mä'frä)	102	26°21'N	49°59'W
Mafra, Port. (mäf'rȧ)	129b	38°56'N	9°20'W
Magadan, Russia (mȧ-gȧ-dän')	135	59°39'N	150°43'E
Magadan Oblast, Russia	141	65°00'N	160°00'E
Magadi, Kenya	191	1°54'S	36°17'E
Magalies, r., S. Afr. (mä-gä'lyĕs)	187b	25°51'S	27°42'E
Magaliesberg, mts., S. Afr.	187b	25°45'S	27°43'E
Magaliesburg, S. Afr.	192c	26°01'S	27°32'E
Magallanes, Estrecho de, strt., S.A.	102	52°30'S	68°45'W
Magat, r., Phil. (mä-gät')	169a	16°45'N	121°16'E
Magdalena, Arg. (mäg-dä-lā'nä)	99c	35°05'S	57°32'W
Magdalena, Bol.	100	13°17'S	63°57'W
Magdalena, Mex.	64	30°34'N	110°50'W
Magdalena, N.M., U.S.	77	34°10'N	107°45'W
Magdalena, i., Chile	102	44°45'S	73°15'W
Magdalena, r., Col.	100	7°45'N	74°04'W
Magdalena, Bahía, b., Mex. (bä-ē'ä-mäg-dä-lā'nä)	86	24°30'N	114°00'W
Magdeburg, Ger. (mäg'dĕ-bȯrgh)	110	52°07'N	11°39'E
Magellan, Strait of *see* Magallanes, Estrecho de, strt., S.A.	102	52°30'S	68°45'W
Magenta, Italy (mȧ-jĕn'tȧ)	130	45°26'N	8°53'E
Magerøya, i., Nor.	116	71°10'N	24°11'E
Maggiore, Lago, l., Italy	118	46°03'N	8°25'E
Maghāghah, Egypt	192b	28°38'N	30°50'W
Maghniyya, Alg.	118	34°52'N	1°40'W
Magiscatzin, Mex. (mä-kės-kät-zēn')	88	22°48'N	98°42'W
Maglaj, Bos. (mȧ'glä-ė)	131	44°34'N	18°12'E
Maglie, Italy (mäl'yā)	131	40°06'N	18°20'E
Magna, Ut., U.S. (măg'nȧ)	75b	40°43'N	112°06'W
Magnitogorsk, Russia (mȧg-nyē'tȯ-gȯrsk)	134	53°26'N	59°05'E
Magnolia, Ar., U.S. (măg-nō'lĭ-ȧ)	79	33°16'N	93°13'W
Magnolia, Ms., U.S.	82	31°08'N	90°27'W
Magny-en-Vexin, Fr. (mä-nyē'ĕN-vĕ-sȧN')	127b	49°09'N	1°45'E
Magog, Can. (mȧ-gŏg')	59	45°15'N	72°10'W
Magpie, r., Can.	60	50°40'N	64°30'W
Magpie, r., Can.	58	48°13'N	84°50'W
Magpie, Lac, l., Can.	60	50°55'N	64°39'W
Magrath, Can.	50	49°25'N	112°52'W
Magude, Moz. (mä-gōō'dȧ)	186	24°58'S	32°39'E
Magwe, Mya. (mŭg-wā')	155	20°19'N	94°57'E
Mahābād, Iran	157	36°55'N	45°50'E
Mahahi Port, D.R.C. (mä-hä'gė)	185	2°14'N	31°12'E
Mahajanga, Madag.	187	15°12'S	46°26'E
Mahakam, r., Indon.	168	0°30'S	116°15'E
Mahali Mountains, mts., Tan.	191	6°20'S	30°00'E
Mahaly, Madag. (mȧ-hȧl-ē')	187	24°09'S	46°20'E
Mahanoro, Madag. (mȧ-hȧ-nȯ'rō)	187	19°57'S	48°47'E
Mahanoy City, Pa., U.S. (mä-hȧ-noi')	67	40°50'N	76°10'W
Maḩaţţat al Qaţrānah, Jord.	153a	31°15'N	36°04'E
Maḩaţţat 'Aqabat al Ḩijāzīyah, Jord.	153a	29°45'N	35°55'E
Maḩaţţat ar Ramlah, Jord.	153a	29°31'N	35°57'E
Maḩaţţat Jurf ad Darāwīsh, Jord.	153a	30°41'N	35°51'E
Mahd adh-Dhahab, Sau. Ar.	157	23°30'N	40°52'E
Mahe, India (mä-ā')	155	11°42'N	75°39'E
Mahenge, Tan. (mä-hĕŋ'gȧ)	186	7°38'S	36°16'E
Mahi, r., India	158	23°16'N	73°20'E
Mahilyow, Bela.	136	53°53'N	30°22'E
Mahilyow, prov., Bela.	132	53°28'N	30°15'E
Māhīm Bay, b., India	159b	19°03'N	72°45'E
Mahlabatini, S. Afr. (mä'lȧ-bȧ-tē'nė)	187c	28°15'S	31°29'E
Mahlow, Ger. (mä'lōv)	115b	52°23'N	13°24'E
Mahnomen, Mn., U.S. (mô-nō'mĕn)	70	47°18'N	95°58'W
Mahone Bay, Can. (mȧ-hōn')	60	44°27'N	64°23'W
Mahone Bay, b., Can.	60	44°30'N	64°15'W
Mahopac, Lake, l., N.Y., U.S. (mā-hō'păk)	68a	41°24'N	73°45'W
Mahwah, N.J., U.S. (mȧ-wă')	68a	41°05'N	74°09'W
Maidenhead, Eng., U.K. (mād'ĕn-hĕd)	114b	51°30'N	0°44'W
Maidstone, Eng., U.K.	121	51°17'N	0°32'E
Maiduguri, Nig. (mä'ē-dȧ-gōō'rė)	185	11°51'N	13°10'E
Maigualida, Sierra, mts., Ven. (sē-ĕ'r-rȧ-mī-gwä'lē-dĕ)	100	6°30'N	65°50'W
Maijdi, Bngl.	158	22°59'N	91°08'E
Maikop *see* Maykop, Russia	134	44°35'N	40°07'E
Main, r., Ger. (mīn)	124	49°49'N	9°20'E
Main Barrier Range, mts., Austl. (băr''ĕr)	175	31°25'S	141°40'E
Mai-Ndombe, Lac, l., D.R.C.	186	2°16'S	19°00'E
Maine, state, U.S. (mān)	65	45°25'N	69°50'W
Mainland, i., Scot., U.K. (mān-lănd)	116	60°19'N	2°40'W
Maintenon, Fr. (mäN-tĕ-nŏN')	127b	48°35'N	1°35'E
Maintirano, Madag. (mä'ėn-tė-rä'nō)	187	18°05'S	44°08'E
Mainz, Ger. (mīnts)	110	49°59'N	8°16'E
Maio, i., C.V. (mä'yo)	184b	15°15'N	22°50'W
Maipo, S.A.	102	34°08'S	69°51'W
Maipo, r., Chile (mī'pȯ)	99b	33°45'S	71°08'W
Maiquetía, Ven. (mī-kĕ-tė'ä)	100	10°37'N	66°56'W
Maison-Rouge, Fr. (mȧ-zŏN-rōōzh')	127b	48°34'N	3°09'E
Maisons-Laffitte, Fr.	127b	48°57'N	2°09'E
Maitland, Austl. (māt'lănd)	175	32°45'S	151°40'E
Maizuru, Japan (mä-ĭ'zōō-rōō)	167	35°26'N	135°15'E
Majene, Indon.	168	3°34'S	119°00'E
Maji, Eth.	185	6°14'N	35°34'E
Majorca *see* Mallorca, i., Spain	112	39°18'N	2°22'E
Makah Indian Reservation, I.R., Wa., U.S.	72	48°17'N	124°52'W
Makanya, Tan. (mä-kän'yä)	187	4°15'S	37°49'E
Makanza, D.R.C.	185	1°42'N	19°08'E
Makarakomburu, Mount, mtn., Sol. Is.	170e	9°43'S	160°02'E
Makarska, Cro. (mȧ'kär-skȧ)	131	43°17'N	17°05'E
Makar'yev, Russia	136	57°50'N	43°48'E
Makasar *see* Ujungpandang, Indon.	168	5°08'S	119°28'E
Makasar, Selat (Makassar Strait), strt., Indon.	168	2°00'S	118°07'E
Makaw, D.R.C.	190	3°29'S	18°19'E
Make, i., Japan (mä'kȧ)	167	30°43'N	130°49'E
Makeni, S.L.	184	8°53'N	12°03'W
Makgadikgadi Pans, pl., Bots.	186	20°38'S	21°31'E
Makhachkala, Russia (mäK'äch-kä'lä)	137	43°00'N	47°40'E
Makhaleng, r., Leso.	187c	29°53'S	27°33'E
Makiïvka, Ukr.	137	48°03'N	38°00'E
Makindu, Kenya	191	2°17'S	37°49'E
Makkah *see* Mecca, Sau. Ar.	154	21°27'N	39°45'E
Makkovik, Can.	51	55°01'N	59°10'W
Makokou, Gabon (mȧ-kȯ-kōō')	184	0°34'N	12°52'E
Maków Mazowiecki, Pol. (mä'kōov mä-zō-vyĕts'kē)	125	52°51'N	21°07'E
Makuhari, Japan (mä-kōō-hä'rē)	167a	35°39'N	140°04'E
Makurazaki, Japan (mä'kȯ-rä-zä'kė)	167	31°16'N	130°18'E
Makurdi, Nig.	184	7°45'N	8°32'E
Makushin, Ak., U.S. (mȧ-kȯ'shĭn)	63	53°57'N	166°28'W
Makushino, Russia (mȧ-kȯ-shėn'ȯ)	134	55°03'N	67°43'E
Mala, Punta, c., Pan. (pȯ'n-tä-mä'lä)	91	7°32'N	79°44'W
Malabar Coast, cst., India (măl'ȧ-bär)	159	11°19'N	75°33'E
Malabar Point, c., India	159b	18°57'N	72°47'E
Malabo, Eq. Gui.	184	3°45'N	8°47'E
Malabon, Phil.	169a	14°39'N	120°57'E
Malacca, Strait of, strt., Asia (mȧ-lăk'ȧ)	168	4°15'N	99°44'E
Malad City, Id., U.S. (mȧ-lăd')	73	42°11'N	112°15'W
Maladzyecha, Bela.	136	54°18'N	26°57'E
Málaga, Col. (mȧ'lä-gȧ)	100	6°41'N	72°46'W
Málaga, Spain	110	36°45'N	4°25'W
Malagón, Spain (mä-lä-gōn')	128	39°12'N	3°52'W
Malaita, i., Sol. Is. (mȧ-lä'ė-tȧ)	175	8°38'S	161°15'E
Malakāl, Sudan (mȧ-lȧ-käl')	185	9°46'N	31°54'E
Malakhovka, Russia (mȧ-läk'ôf-kȧ)	142b	55°38'N	38°01'E
Malang, Indon.	168	8°06'S	112°50'E
Malanje, Ang. (mä-läŋ-gȧ)	186	9°32'S	16°20'E
Malanville, Benin	184	12°04'N	3°09'E
Mälaren, l., Swe.	116	59°38'N	16°55'E
Malartic, Can.	51	48°07'N	78°11'W
Malatya, Tur. (mȧ-lä'tyȧ)	154	38°30'N	38°15'E
Malawi, nation, Afr.	186	11°15'S	33°45'E
Malawi, Lake *see* Nyasa, Lake, l., Afr.	186	10°45'S	34°30'E
Malaya Vishera, Russia (vė-shā'rä)	134	58°51'N	32°13'E
Malay Peninsula, pen., Asia (mȧ-lā') (mā'lā)	168	6°00'N	101°00'E
Malaysia, nation, Asia (mȧ-lā'zhȧ)	168	4°10'N	101°22'E
Malbon, Austl. (măl'bŭn)	174	21°15'S	140°30'E
Malbork, Pol. (mäl'bȯrk)	116	54°02'N	19°04'E
Malcabran, r., Port. (mäl-kä-brän')	129b	38°47'N	8°46'W
Malden, Ma., U.S. (môl'dĕn)	61a	42°26'N	71°04'W
Malden, Mo., U.S.	79	36°32'N	89°56'W
Malden, i., Kir.	2	4°20'S	154°30'W
Maldives, nation, Asia	150	4°30'N	71°30'E
Maldon, Eng., U.K. (môrl'dŏn)	114b	51°44'N	0°39'E

272 Maldonado – Margate PRONOUNCING INDEX

PLACE (Pronunciation)	PAGE	LAT.	LONG.
Maldonado, Ur. (mäl-dō-nä′dȯ)	102	34°54′S	54°57′W
Maldonado, Punta, c., Mex. (po͞o′n-tä)	88	16°18′N	98°34′W
Maléas, Ákra, c., Grc.	119	36°31′N	23°13′E
Mālegaon, India	158	20°35′N	74°30′E
Malé Karpaty, mts., Slvk.	125	48°31′N	17°15′E
Malekula, i., Vanuatu (mä-lā-ko͞o′lä)	175	16°44′S	167°45′E
Malema, Moz.	191	14°57′S	37°20′E
Malheur, r., Or., U.S. (mȧ-lo͞or′)	72	43°45′N	117°41′W
Malheur Lake, l., Or., U.S. (mȧ-lo͞or′)	72	43°16′N	118°37′W
Mali, nation, Afr.	184	15°45′N	0°15′W
Malibu, Ca., U.S. (mă′lĭ-bo͞o)	75a	34°03′N	118°38′W
Malik, Wādī al, r., Sudan	185	16°48′N	29°30′E
Malimba, Monts, mts., D.R.C.	191	7°45′S	29°15′E
Malinalco, Mex. (mä-lē-näl′kō)	88	18°54′N	99°31′W
Malinaltepec, Mex. (mä-lē-näl-tȧ-pĕk′)	88	17°01′N	98°41′W
Malindi, Kenya (mä-lēn′dē)	187	3°14′S	40°04′E
Malin Head, c., Ire.	116	55°23′N	7°24′W
Malino, Russia (mä′lĭ-nô)	142b	55°07′N	38°12′E
Malkara, Tur. (mäl′kȧ-rȧ)	131	40°51′N	26°52′E
Malko Tŭrnovo, Blg. (mäl′kō-t′r′nô-vȧ)	131	41°59′N	27°28′E
Mallaig, Scot., U.K.	120	56°59′N	5°55′W
Mallet Creek, Oh., U.S. (măl′ĕt)	69d	41°10′N	81°55′W
Mallorca, i., Spain	112	39°30′N	3°00′E
Mallow, Ire. (măl′ō)	120	52°07′N	9°04′W
Malmédy, Bel. (mȧl-mā-dē′)	121	50°25′N	6°01′E
Malmesbury, S. Afr. (mämz′bēr-ĭ)	186	33°30′S	18°35′E
Malmköping, Swe. (mälm′chû′pĭng)	122	59°09′N	16°39′E
Malmö, Swe.	110	55°36′N	13°00′E
Malmyzh, Russia (mȧl-mēzh′)	135	49°58′N	137°07′E
Malmyzh, Russia	136	56°30′N	50°48′E
Maloarkhangelsk, Russia (mä′lō-är-ĸän′gĕlsk)	132	52°26′N	36°29′E
Malolos, Phil. (mä-lō′lōs)	169a	14°51′N	120°49′E
Malomal′sk, Russia (mȧ-lō-mälsk′′)	142a	58°47′N	59°55′E
Malone, N.Y., U.S. (mȧ-lōn′)	67	44°50′N	74°20′W
Malonga, D.R.C.	190	10°24′S	23°10′E
Maloti Mountains, mts., Leso.	187c	29°00′S	28°29′E
Maloyaroslavets, Russia (mä′lō-yä-rō-slä-vyĕts)	132	55°01′N	36°25′E
Malozemel′skaya Tundra, reg., Russia	136	67°30′N	50°00′E
Malpas, Eng., U.K. (măl′păz)	114a	53°01′N	2°46′W
Malpelo, Isla de, i., Col. (mäl-pā′lō)	100	3°55′N	81°30′W
Malpeque Bay, b., Can. (môl-pĕk′)	60	46°30′N	63°47′W
Malta, Mt., U.S. (môl′tȧ)	73	48°20′N	107°50′W
Malta, nation, Eur.	110	35°52′N	13°30′E
Maltahöhe, Nmb. (mäl′tä-hō′ē)	186	24°45′S	16°45′E
Maltrata, Mex. (mäl-trä′tä)	89	18°48′N	97°16′W
Maluku (Moluccas), is., Indon.	169	2°22′S	128°25′E
Maluku, Laut (Molucca Sea), sea, Indon.	169	0°15′N	125°41′E
Malūţ, Sudan	185	10°30′N	32°17′E
Mālvan, India	159	16°08′N	73°32′E
Malvern, Ar., U.S. (măl′vĕrn)	79	34°21′N	92°47′W
Malyn, Ukr.	133	50°44′N	29°15′E
Malynivka, Ukr.	133	49°50′N	36°43′E
Malyy Anyuy, r., Russia	141	67°52′N	164°30′E
Malyy Tamir, i., Russia	141	78°10′N	107°30′E
Mamantel, Mex. (mä-mȧn-tĕl′)	89	18°36′N	91°06′W
Mamaroneck, N.Y., U.S. (măm′ȧ-rō-nĕk)	68a	40°57′N	73°44′W
Mambasa, D.R.C.	191	1°21′N	29°03′E
Mamburao, Phil. (mäm-bo͞o′rä-ō)	169a	13°14′N	120°35′E
Mamfe, Cam. (mäm′fĕ)	184	5°46′N	9°17′E
Mamihara, Japan (mä′mē-hä-rä)	167	32°41′N	131°12′E
Mammoth Cave, Ky., U.S. (măm′ŏth)	82	37°10′N	86°04′W
Mammoth Cave National Park, rec., Ky., U.S.	65	37°20′N	86°21′W
Mammoth Hot Springs, Wy., U.S. (măm′ŭth hŏt sprĭngz)	73	44°55′N	110°50′W
Mamnoli, India	159b	19°17′N	73°15′E
Mamoré, r., S.A.	100	13°00′S	65°20′W
Mamou, Gui.	184	10°26′N	12°07′W
Mampong, Ghana	188	7°04′N	1°24′W
Mamry, Jezioro, l., Pol. (mäm′rĭ)	125	54°10′N	21°28′E
Man, C. Iv.	188	7°24′N	7°33′W
Manacor, Spain (mä-nä-kôr′)	129	39°35′N	3°15′E
Manado, Indon.	169	1°29′N	124°50′E
Managua, Cuba (mä-nä′gwä)	93a	22°58′N	82°17′W
Managua, Nic.	86	12°10′N	86°16′W
Managua, Lago de, l., Nic. (lä′gō-dē)	90	12°28′N	86°10′W
Manakara, Madag. (mä-nä-kä′rŭ)	187	22°17′S	48°06′E
Manama see Al Manāmah, Bahr.	154	26°01′N	50°33′E
Mananara, r., Madag. (mä-nä-nä′rŭ)	187	23°15′S	48°15′E
Mananjary, Madag. (mä-nän-zhä′rē)	187	20°16′S	48°13′E
Manas, China	160	44°30′N	86°00′E
Manassas, Va., U.S. (mȧ-năs′ȧs)	67	38°45′N	77°30′W
Manaus, Braz. (mä-nä′o͞ozh)	101	3°01′S	60°00′W
Mancelona, Mi., U.S. (măn-sē-lō′nȧ)	66	44°50′N	85°05′W
Mancha Real, Spain (män′chä rā-äl′)	128	37°48′N	3°37′W
Manchazh, Russia (män′chȧsh)	142a	56°30′N	58°10′E
Manchester, Eng., U.K.	110	53°28′N	2°14′W
Manchester, Ct., U.S. (măn′chĕs-tẽr)	67	41°45′N	72°30′W
Manchester, Ga., U.S.	82	32°50′N	84°37′W
Manchester, Ia., U.S.	71	42°30′N	91°30′W
Manchester, Ma., U.S.	61a	42°35′N	70°47′W
Manchester, Mo., U.S.	75e	38°36′N	90°31′W
Manchester, N.H., U.S.	65	43°00′N	71°30′W
Manchester, Oh., U.S.	66	38°40′N	83°35′W
Manchester Ship Canal, Eng., U.K.	114a	53°20′N	2°40′W
Manchuria, hist. reg., China (măn-cho͞o′rē-ȧ)	161	48°00′N	124°58′E
Mandal, Nor. (män′däl)	122	58°03′N	7°28′E
Mandalay, Mya. (măn′dȧ-lā)	155	22°00′N	96°08′E
Mandalselva, r., Nor.	122	58°25′N	7°30′E
Mandan, N.D., U.S. (măn′dăn)	64	46°49′N	100°54′W
Mandara Mountains, mts., Afr. (män-dä′rä)	185	10°15′N	13°23′E
Mandau Siak, r., Indon.	153b	1°03′N	101°25′E
Mandeb, Bab-el-, strt., (băb′ĕl măn-dĕb′)	154	13°17′N	42°49′E
Mandimba, Moz.	191	14°21′S	35°39′E
Mandinga, Pan. (män-dĭŋ′gä)	91	9°32′N	79°04′W
Mandla, India	158	22°43′N	80°23′E
Mándra, Grc. (män′drä)	131	38°06′N	23°32′E
Mandritsara, Madag. (mȧn-drēt-sä′rȧ)	187	15°49′S	48°47′E
Manduria, Italy (män-do͞o′rē-ä)	131	40°23′N	17°41′E
Mandve, India	159b	18°47′N	72°52′E
Māndvi, India (mŭnd′vē)	159b	19°29′N	72°53′E
Māndvi, India (mŭnd′vē)	155	22°54′N	69°23′E
Mandya, India	159	12°40′N	77°00′E
Manfredonia, Italy (män-frȧ-dô′nyä)	130	41°39′N	15°55′E
Manfredónia, Golfo di, b., Italy (gôl-fô-dē)	130	41°34′N	16°05′E
Mangabeiras, Chapada das, pl., Braz.	101	8°05′S	47°32′W
Mangalore, India (mŭŋ-gŭ-lōr′)	155	12°53′N	74°52′E
Mangaratiba, Braz. (män-gä-rä-tē′bȧ)	99a	22°56′S	44°03′W
Mangatarem, Phil. (män′gȧ-tä′rĕm)	169a	15°48′N	120°18′E
Mange, D.R.C.	190	0°54′N	20°30′E
Mangkalihat, Tanjung, c., Indon.	168	1°25′N	119°55′E
Mangles, Islas de, Cuba (ĕ′s-läs-dĕ-män′glās) (măn′g′lz)	92	22°05′N	82°50′W
Mangoche, Mwi.	186	14°16′S	35°14′E
Mangoky, r., Madag. (män-gō′kē)	187	22°02′S	44°11′E
Mangole, Pulau, i., Indon.	169	1°35′S	126°22′E
Mangualde, Port. (män-gwäl′dĕ)	128	40°38′N	7°44′W
Mangueira, Lagoa da, l., Braz.	102	33°15′S	52°45′W
Mangum, Ok., U.S. (măŋ′gŭm)	78	34°52′N	99°31′W
Mangzhangdian, China (mäŋ-jäŋ-dĭĕn)	162	32°07′N	114°44′E
Manhattan, Il., U.S.	69a	41°25′N	87°29′W
Manhattan, Ks., U.S. (măn-hăt′ăn)	64	39°11′N	96°34′W
Manhattan Beach, Ca., U.S.	75a	33°53′N	118°24′W
Manhuaçu, Braz. (män-o͝oà′so͞o)	99a	20°17′S	42°01′W
Manhumirim, Braz. (män-o͞o-mē-rē′ɴ)	99a	22°30′S	41°57′W
Manicouagane, r., Can.	53	50°00′N	68°35′W
Manicouagane, Lac, res., Can.	53	51°30′N	68°19′W
Manicuare, Ven. (mä-nē-kwä′rĕ)	101b	10°35′N	64°10′W
Manihiki Islands, is., Cook Is. (mä′nē-hē′kē)	195	9°40′S	158°00′W
Manila, Phil.	168	14°37′N	121°00′E
Manila Bay, b., Phil. (mȧ-nĭl′ȧ)	169a	14°38′N	120°46′E
Manisa, Tur. (mä′nē-sȧ)	119	38°40′N	27°30′E
Manistee, Mi., U.S. (măn-ĭs-tē′)	66	44°15′N	86°20′W
Manistee, r., Mi., U.S.	66	44°25′N	85°45′W
Manistique, Mi., U.S. (măn-ĭs-tēk′)	71	45°58′N	86°16′W
Manistique, l., Mi., U.S.	71	46°14′N	85°30′W
Manistique, r., Mi., U.S.	71	46°05′N	86°09′W
Manitoba, prov., Can. (măn-ĭ-tō′bȧ)	50	55°12′N	97°29′W
Manitoba, Lake, l., Can.	52	51°00′N	98°45′W
Manito Lake, l., Can. (măn′ĭ-tō)	56	52°45′N	109°45′W
Manitou, i., Mi., U.S. (măn′ĭ-to͞o)	71	47°21′N	87°33′W
Manitou, l., Can.	71	49°21′N	93°01′W
Manitou Islands, is., Mi., U.S.	66	45°05′N	86°00′W
Manitoulin Island, i., Can. (măn-ĭ-to͞o′lĭn)	53	45°45′N	81°30′W
Manitou Springs, Co., U.S.	78	38°51′N	104°58′W
Manitowoc, Wi., U.S. (măn-ĭ-tō-wŏk′)	71	44°05′N	87°42′W
Manitqueira, Serra da, mts., Braz.	99a	22°40′S	45°12′W
Maniwaki, Can.	59	46°23′N	76°00′W
Manizales, Col. (mä-nē-zä′läs)	100	5°05′N	75°31′W
Manjacaze, Moz. (man′yä-kä′zĕ)	186	24°37′S	33°49′E
Mankato, Ks., U.S. (măn-kā′tō)	78	39°45′N	98°12′W
Mankato, Mn., U.S.	65	44°10′N	93°59′W
Mankim, Cam.	189	5°01′N	12°00′E
Manlléu, Spain (män-lyā′o͞o)	129	42°00′N	2°16′E
Mannar, Sri L. (mȧ-när′)	159	9°48′N	80°03′E
Mannar, Gulf of, b., Asia	155	8°47′N	78°33′E
Mannheim, Ger. (män′hīm)	117	49°30′N	8°31′E
Manning, Ia., U.S. (măn′ĭng)	70	41°53′N	95°04′W
Manning, S.C., U.S.	83	33°41′N	80°12′W
Mannington, W.V., U.S. (măn′ĭng-tŭn)	66	39°30′N	80°55′W
Mano, r., Afr.	188	7°00′N	11°25′W
Man of War Bay, b., Bah.	93	21°05′N	74°05′W
Man of War Channel, strt., Bah.	92	22°45′N	76°10′W
Manokwari, Indon. (mȧ-nŏk-wä′rē)	169	0°56′S	134°10′E
Manono, D.R.C.	191	7°18′S	27°25′E
Manor, Can. (măn′ẽr)	57	49°36′N	102°05′W
Manor, Wa., U.S.	74c	45°45′N	122°36′W
Manori, neigh., India	159b	19°13′N	72°43′E
Manosque, Fr. (mȧ-nôsh′)	127	43°51′N	5°48′E
Manotick, Can.	62c	45°13′N	75°41′W
Manouane, r., Can.	59	50°15′N	70°30′W
Manouane, Lac, l., Can. (mä-no͞o′ȧn)	60	50°36′N	70°50′W
Manresa, Spain (män-rā′sä)	118	41°44′N	1°52′E
Mansa, Zam.	186	11°12′S	28°53′E
Mansel, i., Can. (măn′sĕl)	53	61°56′N	81°10′W
Manseriche, Pongo de, reg., Peru (pŏ′n-gō-dĕ-män-sĕ-rē′chĕ)	100	4°15′S	77°45′W
Mansfield, Eng., U.K. (mănz′fēld)	114a	53°08′N	1°12′W
Mansfield, La., U.S.	81	32°02′N	93°43′W
Mansfield, Oh., U.S.	66	40°45′N	82°30′W
Mansfield, Wa., U.S.	72	47°48′N	119°39′W
Mansfield, Mount, mtn., Vt., U.S.	67	44°30′N	72°45′W
Mansfield Woodhouse, Eng., U.K. (wŏd-hous)	114a	53°08′N	1°12′W
Manta, Ec. (män′tä)	100	1°03′S	80°16′W
Manteno, Il., U.S. (măn-tē-nō)	69a	41°15′N	87°50′W
Manteo, N.C., U.S.	83	35°55′N	75°40′W
Mantes-la-Jolie, Fr. (mänt-ĕ-lä-zhō-lē′)	126	48°59′N	1°42′E
Manti, Ut., U.S. (măn′tī)	77	39°15′N	11°40′W
Mantova, Italy (män′tô-vä) (măn′tū-ȧ)	118	45°09′N	10°47′E
Mantua, Cuba (män-to͞o′ȧ)	92	22°20′N	84°15′W
Mantua see Mantova, Italy	118	45°09′N	10°47′E
Mantua, Ut., U.S. (măn′tū-ȧ)	75b	41°30′N	111°57′W
Manua Islands, is., Am. Sam.	170a	14°13′S	169°35′W
Manui, Pulau, i., Indon. (mä-no͞o′ē)	169	3°35′S	123°38′E
Manus Island, i., Pap. N. Gui. (mä′no͞os)	169	2°22′S	146°22′E
Manvel, Tx., U.S. (măn′vel)	81a	29°28′N	95°22′W
Manville, N.J., U.S. (măn′vĭl)	68a	40°33′N	74°36′W
Manville, R.I., U.S.	68b	41°57′N	71°27′W
Manzala Lake, l., Egypt	192b	31°14′N	32°04′E
Manzanares, Col. (män-sä-nä′rĕs)	100a	5°15′N	75°09′W
Manzanares, r., Spain (mänz-nä′rĕs)	129a	40°36′N	3°48′W
Manzanares, Canal del, Spain (kä-nä′l-dĕl-män-thä-nä′rĕs)	129a	40°20′N	3°38′W
Manzanillo, Cuba (män′zä-nēl′yō)	87	20°20′N	77°05′W
Manzanillo, Mex.	86	19°02′N	104°21′W
Manzanillo, Bahía de, b., Mex. (bä-ē′ä-dĕ-män-zä-nē′l-yō)	88	19°00′N	104°38′W
Manzanillo, Bahía de, b., N.A.	93	19°55′N	71°50′W
Manzanillo, Punta, c., Pan.	91	9°40′N	79°33′W
Manzhouli, China (män-jō-lē)	161	49°25′N	117°15′E
Manzovka, Russia (mȧn-zhô′f-kȧ)	166	44°16′N	132°13′E
Mao, Chad (mä′ō)	185	14°07′N	15°19′E
Mao, Dom. Rep.	93	19°35′N	71°10′W
Maó, Spain	118	39°52′N	4°15′E
Maoke, Pegunungan, mts., Indon.	169	4°00′S	138°00′E
Maoming, China	161	21°55′N	110°40′E
Maoniu Shan, mtn., China (mou-nĭō shän)	164	32°45′N	104°09′E
Mapastepec, Mex. (ma-päs-tȧ-pĕk′)	89	15°24′N	92°52′W
Mapia, Kepulauan, i., Indon.	169	0°57′N	134°22′E
Mapimí, Mex. (mä-pē-mē′)	80	25°50′N	103°50′W
Mapimí, Bolsón de, des., Mex. (bôl-sŏ′n-dĕ-mä-pē′mē)	80	27°27′N	103°20′W
Maple Creek, Can. (mā′p′l) (crēk)	50	49°55′N	109°27′W
Maple Grove, Can. (grōv)	62a	45°19′N	73°51′W
Maple Heights, Oh., U.S.	69d	41°25′N	81°34′W
Maple Shade, N.J., U.S. (shād)	68f	39°57′N	75°01′W
Maple Valley, Wa., U.S. (văl′ē)	74a	47°24′N	122°02′W
Maplewood, Mn., U.S. (wŏd)	75g	45°00′N	93°03′W
Maplewood, Mo., U.S.	75e	38°37′N	90°20′W
Mapumulo, S. Afr. (mä-pä-mo͞o′lō)	187c	29°12′S	31°05′E
Maputo, Moz.	186	26°50′S	32°30′E
Maquela do Zombo, Ang. (mȧ-kā′lȧ dō zŏm′bō)	186	6°08′S	15°15′E
Maquoketa, Ia., U.S. (mȧ-kō-kē-tȧ)	71	42°04′N	90°42′W
Maquoketa, r., Ia., U.S.	71	42°08′N	90°40′W
Mar, Serra do, mts., Braz. (sĕr′rȧ dō mär′)	102	26°30′S	49°15′W
Maracaibo, Ven. (mä-rä-kī′bō)	100	10°38′N	71°45′W
Maracaibo, Lago de, l., Ven. (lä′gô-dĕ-mä-rä-kī′bō)	100	9°55′N	72°13′W
Maracay, Ven. (mä-rä-kāy′)	100	10°15′N	67°35′W
Marādah, Libya	185	29°10′N	19°07′E
Maradi, Niger (mȧ-rȧ-dē′)	184	13°29′N	7°06′E
Marāgheh, Iran	157	37°20′N	46°10′E
Maraisburg, S. Afr.	187b	26°12′S	27°57′E
Marais des Cygnes, r., Ks., U.S.	79	38°30′N	95°30′W
Marajó, Ilha de, i., Braz.	101	1°00′S	49°30′W
Maralal, Kenya	191	1°06′N	36°42′E
Marali, C.A.R.	189	6°01′N	18°24′E
Marand, Iran	157	38°26′N	45°46′E
Maranguape, Braz. (mä-rȧŋ-gwä′pĕ)	101	3°48′S	38°38′W
Maranhão, state, Braz. (mä-rän-youɴ)	101	5°15′S	45°52′W
Maranoa, r., Austl. (mä-rä-nō′ä)	175	27°01′S	148°03′E
Marano di Napoli, Italy (mä-rä′nô-dē-nä′pô-lē)	129c	40°39′N	14°12′E
Marañón, r., Peru (mä-rä-nyōn′)	100	4°26′S	75°08′W
Marapanim, Braz. (mä-rä-pä-nē′ɴ)	101	0°45′S	47°42′W
Marathon, Can.	51	48°50′N	86°10′W
Marathon, Fl., U.S. (măr′ȧ-thŏn)	83a	24°41′N	81°06′W
Marathon, Oh., U.S.	69f	39°09′N	83°59′W
Maravatío, Mex. (mä-rȧ-vä′tē-ō)	88	19°54′N	100°25′W
Marawi, Sudan	185	18°07′N	31°57′E
Marble Bar, Austl. (märb′′l bär)	174	21°15′S	119°15′E
Marble Canal, can., Az., U.S. (mär′b′l)	77	36°21′N	111°48′W
Marblehead, Ma., U.S. (mär′b′l-hĕd)	61a	42°30′N	70°51′W
Marburg an der Lahn, Ger.	124	50°49′N	8°46′E
Marca, Ponta da, c., Ang.	190	16°31′S	11°42′E
Marcala, Hond. (mär-kä-lä)	90	14°08′N	88°01′W
Marceline, Mo., U.S. (mär-sē-lēn′)	79	39°42′N	92°56′W
Marche, hist. reg., Italy (mär′kä)	130	43°35′N	12°33′E
Marchegg, Aus.	115e	48°18′N	16°55′E
Marchena, Spain (mär-chā′nä)	118	37°20′N	5°25′W
Marchena, i., Ec. (ĕ′s-lä-mär-chĕ′nä)	100	0°29′N	90°31′W
Marchfeld, reg., Aus.	115e	48°14′N	16°37′E
Mar Chiquita, Laguna, l., Arg. (lä-go͞o′nä-mär-chē-kē′tä)	99c	34°25′S	61°10′W
Marcos Paz, Arg. (mär-kōs′ päz)	99c	34°49′S	58°51′W
Marcus, i., Japan (mär′kŭs)	195	24°00′N	155°00′E
Marcus Hook, Pa., U.S. (mär′kŭs hŏk)	68f	39°49′N	75°25′W
Marcy, Mount, mtn., N.Y., U.S. (mär′sē)	67	44°10′N	73°55′W
Mar de Espanha, Braz. (mär-dĕ-ĕs-pä′nyȧ)	99a	21°53′S	43°00′W
Mar del Plata, Arg. (mär dĕl- plä′ta)	102	37°59′S	57°35′W
Mardin, Tur. (mär-dēn′)	154	37°25′N	40°40′E
Maré, i., N. Cal. (mȧ-rā′)	175	21°53′S	168°30′E
Maree, Loch, b., Scot., U.K. (mä-rē′)	120	57°40′N	5°44′W
Marengo, Ia., U.S. (mȧ-rĕŋ′gō)	71	41°47′N	92°04′W
Marennes, Fr. (mȧ-rĕn′)	126	45°49′N	1°08′W
Marfa, Tx., U.S. (mär′fȧ)	80	30°19′N	104°01′W
Margarita, Pan. (mär-go͞o-rē′tä)	86a	9°20′N	79°55′W
Margarita, Isla de, i., Ven. (mä-gȧ-rē′tȧ)	100	11°00′N	64°15′W
Margate, S. Afr. (mä-gāt′)	187c	30°52′S	30°21′E

ăt; finăl; rāte; senȧte; ärm; ȧsk; sofȧ; fâre; ch-choose; dh-as th in other; bē; ēvent; bĕt; recĕnt; cratẽr; g-gō; gh-guttural g; bĭt; ĭ-short neutral; rīde; ĸ-guttural k as ch in German ich;

PLACE (Pronunciation)	PAGE	LAT.	LONG.
Margate, Eng., U.K. (mär'gāt)	121	51°21'N	1°17'E
Margherita Peak, mtn., Afr.	185	0°22'N	29°51'E
Marguerite, r., Can.	60	50°39'N	66°42'W
Marhanets', Ukr.	133	47°41'N	34°33'E
Maria, Can. (mȧ-rē'ȧ)	60	48°10'N	66°04'W
Mariager, Den. (mä-rē̇-ägh'ẽr)	122	56°38'N	10°00'E
Mariana, Braz. (mä-ryä'nä)	99a	20°23'S	43°24'W
Mariana Islands, is., Oc.	5	16°00'N	145°30'E
Marianao, Cuba (mä-rē̇-ä-nä'ō)	87	23°05'N	82°26'W
Mariana Trench, deep,	195	12°00'N	144°00'E
Marianna, Ar., U.S. (mä-rĭ-ăn'ȧ)	79	34°45'N	90°45'W
Marianna, Fl., U.S.	82	30°46'N	85°14'W
Marianna, Pa., U.S.	69e	40°01'N	80°05'W
Mariano Acosta, Arg. (mä-rē̇ä'nō-ȧ-kōs'tä)	102a	34°28'S	58°48'W
Mariánské Lázně, Czech Rep. (mär'yȧn-skĕ'läz'nyĕ)	124	49°58'N	12°42'E
Marias, r., Mt., U.S. (mȧ-rī'ȧz)	73	48°15'N	110°50'W
Marias, Islas, is., Mex. (mä-rē'äs)	86	21°30'N	106°40'W
Mariato, Punta, c., Pan.	91	7°17'N	81°09'W
Maribo, Den. (mä'rē̇-bō)	122	54°46'N	11°29'E
Maribor, Slvn. (mä're-bôr)	110	46°33'N	15°37'E
Maricaban, i., Phil. (mä-rē̇-kä-bän')	169a	13°40'N	120°44'E
Mariefred, Swe. (mä-rē'ĕ-frĭd)	122	59°17'N	17°09'E
Marie Galante, i., Guad. (mȧ-rē' gȧ-läNt')	91b	15°58'N	61°05'W
Mariehamn, Fin. (mä-rē'ē-hăm''n)	123	60°07'N	19°57'E
Mari El, prov., Russia	136	56°30'N	48°00'E
Mariestad, Swe. (mä-rē'ĕ-städ')	122	58°43'N	13°45'E
Marietta, Ga., U.S. (mā-rĭ-ĕt'ȧ)	68c	33°57'N	84°33'W
Marietta, Oh., U.S.	66	39°25'N	81°30'W
Marietta, Ok., U.S.	79	33°53'N	97°07'W
Marietta, Wa., U.S.	74d	48°48'N	122°35'W
Mariinsk, Russia (mȧ-rē'ĭnsk)	140	56°15'N	87°28'E
Marijampole, Lith. (mä-rē̇-yäm-pô'lē)	123	54°33'N	23°26'E
Marikana, S. Afr. (mä'-rĭ-kä-nȧ)	192c	25°40'S	27°28'E
Marília, Braz. (mä-rē'lyȧ)	101	22°02'S	49°48'W
Marimba, Ang.	190	8°28'S	17°08'E
Marín, Spain	128	42°24'N	8°40'W
Marinduque Island, i., Phil. (mä-rēn-dōō'kä)	169a	13°14'N	121°45'E
Marine, Il., U.S. (mȧ-rēn')	75e	38°48'N	89°47'W
Marine City, Mi., U.S.	66	42°45'N	82°30'W
Marine Lake, l., Mn., U.S.	75g	45°13'N	92°55'W
Marine on Saint Croix, Mn., U.S.	75g	45°11'N	92°47'W
Marinette, Wi., U.S. (mär-ĭ-nĕt')	65	45°04'N	87°40'W
Maringa, r., D.R.C. (mȧ-rĭŋ'gä)	185	0°30'N	21°00'E
Marinha Grande, Port. (mä-rēn'yȧ grän'dē)	128	39°49'N	8°53'W
Marion, Al., U.S. (măr'ĭ-ŭn)	82	32°36'N	87°19'W
Marion, Ia., U.S.	71	42°01'N	91°39'W
Marion, Il., U.S.	66	37°40'N	88°55'W
Marion, In., U.S.	65	40°35'N	85°45'W
Marion, Ks., U.S.	79	38°21'N	97°02'W
Marion, Ky., U.S.	82	37°19'N	88°05'W
Marion, N.C., U.S.	83	35°40'N	82°00'W
Marion, N.D., U.S.	70	46°37'N	98°20'W
Marion, Oh., U.S.	66	40°35'N	83°10'W
Marion, S.C., U.S.	83	34°08'N	79°23'W
Marion, Va., U.S.	83	36°48'N	81°33'W
Marion, Lake, res., S.C., U.S.	83	33°25'N	80°35'W
Marion Reef, rf., Austl.	175	18°57'S	151°31'E
Mariposa, Chile (mä-rē̇-pō'sä)	99b	35°33'S	71°21'W
Mariposa Creek, r., Ca., U.S.	76	37°14'N	120°30'W
Mariquita, Col. (mä-rē-kē'tä)	100a	5°13'N	74°52'W
Mariscal Estigarribia, Para.	102	22°03'S	60°28'W
Marisco, Ponta do, c., Braz. (pô'n-tä-dō-mä-rē's-kö)	102b	23°01'S	43°17'W
Maritime Alps, mts., Eur. (mȧ'rĭ-tīm ălps)	117	44°20'N	7°02'E
Mariupol', Ukr.	134	47°07'N	37°32'E
Mariveles, Phil.	169a	14°27'N	120°29'E
Marj Uyan, Leb.	153a	33°21'N	35°36'E
Marka, Som.	192a	1°45'N	44°47'E
Markaryd, Swe. (mär'kä-rüd)	122	56°30'N	13°34'E
Marked Tree, Ar., U.S. (märkt trē)	79	35°31'N	90°26'W
Marken, i., Neth.	115a	52°26'N	5°08'E
Market Bosworth, Eng., U.K. (bŏz'wûrth)	114a	52°37'N	1°23'W
Market Deeping, Eng., U.K. (dēp'ĭng)	114a	52°40'N	0°19'W
Market Drayton, Eng., U.K. (drā'tŭn)	114a	52°54'N	2°29'W
Market Harborough, Eng., U.K. (här'bŭr-ō)	114a	52°28'N	0°55'W
Market Rasen, Eng., U.K. (rā'zĕn)	114a	53°23'N	0°21'W
Markham, Can. (märk'ȧm)	59	43°53'N	79°15'W
Markham, Mount, mtn., Ant.	178	82°59'S	159°30'E
Markivka, Ukr.	133	49°32'N	39°34'E
Markovo, Russia (mär'kô-vô)	135	64°46'N	170°48'E
Markrāna, India	158	27°08'N	74°43'E
Marks, Russia	137	51°42'N	46°46'E
Marksville, La., U.S. (märks'vĭl)	81	31°09'N	92°05'W
Markt Indersdorf, Ger. (märkt ēn'dĕrs-dōrf)	115d	48°22'N	11°23'E
Marktredwitz, Ger. (märk-rĕd'vēts)	124	50°02'N	12°05'E
Markt Schwaben, Ger. (märkt shvä'bĕn)	115d	48°12'N	11°52'E
Marl, Ger. (märl)	127c	51°40'N	7°05'E
Marlboro, N.J., U.S.	68a	40°18'N	74°15'W
Marlborough, Ma., U.S.	61a	42°21'N	71°33'W
Marlette, Mi., U.S. (mär-lĕt')	66	43°25'N	83°05'W
Marlin, Tx., U.S. (mär'lĭn)	81	31°18'N	96°52'W
Marlinton, W.V., U.S. (mär'lĭn-tŭn)	66	38°15'N	80°10'W
Marlow, Eng., U.K. (mär'lō)	114b	51°33'N	0°46'W
Marlow, Ok., U.S.	79	34°38'N	97°56'W
Marls, The, b., Bah. (märls)	92	26°30'N	77°15'W
Marmande, Fr. (mȧr-mäNd')	126	44°30'N	0°10'E
Marmara Denizi, sea, Tur.	154	40°40'N	28°00'E
Marmarth, N.D., U.S. (mär'mȧrth)	70	46°19'N	103°57'W
Mar Muerto, l., Mex. (mär-mōĕ'r-tô)	89	16°13'N	94°22'W
Marne, Ger. (mär'nĕ)	115c	53°57'N	9°01'E
Marne, r., Fr. (märn)	117	49°00'N	4°30'E
Maroa, Ven. (mä-rō'ä)	100	2°43'N	67°37'W
Maroantsetra, Madag. (mȧ-rō-äŋ-tsā'trȧ)	187	15°18'S	49°48'E
Maro Jarapeto, mtn., Col. (mȧ-rô-hä-rä-pĕ'tô)	100a	6°29'N	76°39'W
Maromokotro, mtn., Madag.	187	14°00'S	49°11'E
Marondera, Zimb.	186	18°10'S	31°36'E
Maroni, r., S.A. (mȧ-rō'nē̇)	101	3°02'N	53°54'W
Maro Reef, rf., Hi., U.S.	84b	25°15'N	170°00'W
Maroua, Cam. (mär'wä)	185	10°36'N	14°20'E
Marple, Eng., U.K. (mär'p'l)	114a	53°24'N	2°04'W
Marquard, S. Afr.	192c	28°41'S	27°26'E
Marquesas Islands, is., Fr. Poly. (mär-kĕ'säs)	2	8°50'S	141°00'W
Marquesas Keys, is., Fl., U.S. (mär-kē'zȧs)	83a	24°37'N	82°15'W
Marquês de Valença, Braz. (mȧr-kĕ's-dĕ-vä-lĕ'n-sȧ)	99a	22°16'S	43°42'W
Marquette, Can. (mȧr-kĕt')	62f	50°04'N	97°43'W
Marquette, Mi., U.S.	65	46°32'N	87°25'W
Marquez, Tx., U.S. (mär-kāz')	81	31°14'N	96°15'W
Marra, Jabal, mtn., Sudan (jĕb'ĕl mär'ä)	185	13°00'N	23°47'E
Marrakech, Mor. (mȧr-rä'kĕsh)	184	31°38'N	8°00'W
Marree, Austl. (măr'rē)	174	29°38'S	137°55'E
Marrero, La., U.S.	68d	29°55'N	90°06'W
Marrupa, Moz.	191	13°08'S	37°30'E
Mars, Pa., U.S. (märz)	69e	40°42'N	80°01'W
Marsabit, Kenya	191	2°20'N	37°59'E
Marsala, Italy (mär-sä'lä)	118	37°48'N	12°28'E
Marsden, Eng., U.K. (märz'dĕn)	114a	53°36'N	1°55'W
Marseille, Fr. (mȧr-sâ'y')	110	43°18'N	5°25'E
Marseilles, Il., U.S. (mär-sĕlz')	66	41°20'N	88°40'W
Marshall, Il., U.S. (mär'shȧl)	66	39°20'N	87°40'W
Marshall, Mi., U.S.	66	42°20'N	84°55'W
Marshall, Mn., U.S.	70	44°28'N	95°49'W
Marshall, Mo., U.S.	79	39°07'N	93°12'W
Marshall, Tx., U.S.	65	32°33'N	94°22'W
Marshall Islands, nation, Oc.	3	10°00'N	165°00'E
Marshalltown, Ia., U.S. (mär'shȧl-toun)	71	42°02'N	92°55'W
Marshallville, Ga., U.S. (mär'shȧl-vĭl)	82	32°29'N	83°55'W
Marshfield, Ma., U.S. (märsh'fēld)	61a	42°06'N	70°43'W
Marshfield, Mo., U.S.	79	37°20'N	92°53'W
Marshfield, Wi., U.S.	71	44°40'N	90°10'W
Marsh Harbour, Bah.	92	26°30'N	77°00'W
Mars Hill, In., U.S. (märz'hĭl')	69g	39°43'N	86°15'W
Mars Hill, Me., U.S.	60	46°34'N	67°54'W
Marstrand, Swe. (mär'stränd)	122	57°54'N	11°33'E
Marsyaty, Russia (märs'yȧ-tĭ)	142a	60°03'N	60°28'E
Mart, Tx., U.S. (märt)	81	31°32'N	96°49'W
Martaban, Gulf of, b., Mya. (mär-tŭ-bän')	168	16°34'N	96°58'E
Martapura, Indon.	168	3°19'S	114°45'E
Martha's Vineyard, i., Ma., U.S. (mär'thȧz vĭn'yȧrd)	67	41°25'N	70°35'W
Martigny, Switz. (mȧr-tē̇-nyē')	124	46°06'N	7°00'E
Martigues, Fr.	127	43°24'N	5°05'E
Martin, Tn., U.S. (mär'tĭn)	82	36°20'N	88°45'W
Martina Franca, Italy (mär-tē'nä fräŋ'kä)	131	40°43'N	17°21'E
Martinez, Ca., U.S. (mär-tē'nĕz)	74b	38°01'N	122°08'W
Martinez, Tx., U.S.	75d	29°25'N	98°20'W
Martinique, dep., N.A. (mȧr-tē̇-nēk')	87	14°50'N	60°40'W
Martin Lake, res., Al., U.S.	82	32°40'N	86°05'W
Martin Point, c., Ak., U.S.	63	70°10'N	142°00'W
Martinsburg, W.V., U.S. (mär'tĭnz-bûrg)	67	39°30'N	78°00'W
Martins Ferry, Oh., U.S. (mär'tĭnz)	66	40°05'N	80°45'W
Martinsville, In., U.S. (mär'tĭnz-vĭl)	66	39°25'N	86°25'W
Martinsville, Va., U.S.	83	36°40'N	79°53'W
Martos, Spain (mär'tōs)	128	37°43'N	3°58'W
Martre, Lac la, l., Can. (läk la märtr)	52	63°24'N	119°58'W
Marugame, Japan (mä'rōō-gä'mä)	167	34°19'N	133°48'E
Marungu, mts., D.R.C.	191	7°50'S	29°50'E
Marve, neigh., India	159b	19°12'N	72°43'E
Mary, Turkmen. (mä'rē̇)	139	37°45'N	61°47'E
Mar'yanskaya, Russia (mȧr-yän'skȧ-yȧ)	133	45°04'N	38°39'E
Maryborough, Austl. (mă'rĭ-bŭr-ō)	175	25°35'S	152°40'E
Maryborough, Austl.	175	37°00'S	143°50'E
Maryland, state, U.S. (mĕr'ĭ-lănd)	65	39°10'N	76°25'W
Marys, r., Nv., U.S. (mā'rĭz)	72	41°25'N	115°10'W
Marystown, Can. (mâr'ĭz-toun)	61	47°11'N	55°10'W
Marysville, Can.	60	45°59'N	66°35'W
Marysville, Ca., U.S.	76	39°09'N	121°37'W
Marysville, Oh., U.S.	66	40°15'N	83°25'W
Marysville, Wa., U.S.	74a	48°03'N	122°11'W
Maryville, Il., U.S. (mā'rĭ-vĭl)	75e	38°44'N	89°57'W
Maryville, Mo., U.S.	79	40°21'N	94°51'W
Maryville, Tn., U.S.	82	35°44'N	83°59'W
Mārzuq, Libya	185	26°00'N	14°09'E
Marzūq, Idehan, des., Libya	184	24°30'N	13°00'E
Masai Steppe, plat., Tan.	191	4°30'S	36°40'E
Masaka, Ug.	191	0°20'S	31°44'E
Masalasef, Chad	189	11°43'N	17°08'E
Masalembo-Besar, i., Indon.	168	5°40'S	114°28'E
Masan, Kor., S. (mä-sän')	161	35°10'N	128°31'E
Masangwe, Tan.	191	5°28'S	30°05'E
Masasi, Tan. (mä-sä'sē̇)	187	10°43'S	38°48'E
Masatepe, Nic. (mä-sä-tĕ'pĕ)	90	11°57'N	86°10'W
Masaya, Nic. (mä-sä'yä)	90	11°58'N	86°05'W
Masbate, Phil. (mäs-bä'tä)	169a	12°21'N	123°38'E
Masbate, i., Phil.	169	12°19'N	123°03'E
Mascarene Islands, is., Afr.	5	20°20'S	56°40'E
Mascot, Tn., U.S. (măs'kŏt)	82	36°04'N	83°45'W
Mascota, Mex. (mäs-kō'tä)	88	20°33'N	104°45'W
Mascota, r., Mex.	88	20°33'N	104°52'W
Mascouche, Can. (mȧs-kōōsh')	62a	45°45'N	73°36'W
Mascouche, r., Can.	62a	45°44'N	73°45'W
Mascoutah, Il., U.S. (măs-kū'tä)	75e	38°29'N	89°48'W
Maseru, Leso. (măz'ĕr-ōō)	186	29°09'S	27°11'E
Mashhad, Iran	154	36°17'N	59°30'E
Māshkel, Hāmūn-i-, l., Asia (hä-mōōn'ē māsh-kĕl')	154	28°28'N	64°13'E
Mashra'ar Raqq, Sudan	185	8°28'N	29°15'E
Masi-Manimba, D.R.C.	190	4°46'S	17°55'E
Masindi, Ug. (mä-sēn'dē̇)	185	1°44'N	31°43'E
Masjed Soleymān, Iran	154	31°45'N	49°17'E
Mask, Lough, b., Ire. (lŏk mȧsk)	120	53°35'N	9°23'W
Maslovo, Russia (mäs'lô-vô)	142a	60°08'N	60°28'E
Mason, Mi., U.S. (mā'sŭn)	66	42°35'N	84°25'W
Mason, Oh., U.S.	69f	39°22'N	84°18'W
Mason, Tx., U.S.	80	30°46'N	99°14'W
Mason City, Ia., U.S.	65	43°08'N	93°14'W
Massa, Italy (mäs'sä)	130	44°02'N	10°08'E
Massachusetts, state, U.S. (măs-ȧ-chōō'sĕts)	65	42°20'N	72°30'W
Massachusetts Bay, b., Ma., U.S.	60	42°26'N	70°20'W
Massafra, Italy (mäs-sä'frä)	131	40°35'N	17°05'E
Massa Marittima, Italy	130	43°03'N	10°55'E
Massapequa, N.Y., U.S.	68a	40°41'N	73°28'W
Massaua *see* Mitsiwa, Erit.	185	15°40'N	39°19'E
Massena, N.Y., U.S. (mă-sē'nȧ)	67	44°55'N	74°55'W
Masset, Can. (măs'ĕt)	50	54°02'N	132°09'W
Masset Inlet, b., Can.	55	53°42'N	132°20'E
Massif Central, Fr. (mȧ-sēf' säN-trȧl')	110	45°12'N	3°02'E
Massillon, Oh., U.S. (măs'ĭ-lŏn)	66	40°50'N	81°35'W
Massinga, Moz. (mä-sĭn'gä)	186	23°18'S	35°18'E
Massive, Mount, mtn., Co., U.S. (măs'ĭv)	64	39°05'N	106°30'W
Masson, Can. (măs-sŭn)	62c	45°33'N	75°25'W
Masuda, Japan (mä-sōō'dȧ)	167	34°42'N	131°53'E
Masuria, reg., Pol.	125	53°40'N	21°10'E
Masvingo, Zimb.	186	20°07'S	30°47'E
Matadi, D.R.C. (mȧ-tä'dē̇)	186	5°49'S	13°27'E
Matagalpa, Nic. (mä-tä-gäl'pä)	86	12°52'N	85°57'W
Matagami, l., Can. (mȧ-tä-gä'mē̇)	53	50°10'N	78°28'W
Matagorda Bay, b., Tx., U.S. (măt-ȧ-gôr'dȧ)	81	28°32'N	96°13'W
Matagorda Island, i., Tx., U.S.	81	28°13'N	96°27'W
Matam, Sen. (mä-täm')	184	15°40'N	13°15'W
Matamoros, Mex. (mä-tä-mō'rōs)	80	25°32'N	103°13'W
Matamoros, Mex.	86	25°52'N	97°30'W
Matane, Can. (mȧ-tăn')	51	48°51'N	67°32'W
Matanzas, Cuba (mä-tän'zäs)	87	23°05'N	81°35'W
Matanzas, prov., Cuba	92	22°45'N	81°20'W
Matanzas, Bahía, b., Cuba (bä-ē'ä)	92	23°10'N	81°30'W
Matapalo, Cabo, c., C.R. (kä'bō-mä-tä-pä'lō)	91	8°22'N	83°25'W
Matapédia, Can. (mă-tȧ-pā'dē-ȧ)	60	47°58'N	66°56'W
Matapédia, l., Can.	60	48°33'N	67°32'W
Matapédia, r., Can.	60	48°10'N	67°10'W
Mataquito, r., Chile (mä-tä-kē'tô)	99b	35°08'S	71°35'W
Matara, Sri L. (mä-tä'rä)	159	5°59'N	80°35'E
Mataram, Indon.	168	8°45'S	116°15'E
Matatiele, S. Afr. (mä-tä-tyā'lä)	187c	30°21'S	28°49'E
Matawan, N.J., U.S.	68a	40°24'N	74°13'W
Matehuala, Mex. (mä-tä-wä'lä)	86	23°38'N	100°39'W
Matera, Italy (mä-tâ'rä)	130	40°42'N	16°37'E
Mateur, Tun. (mȧ-tûr')	118	37°09'N	9°43'E
Mātherān, India	159b	18°58'N	73°16'E
Matheson, Can.	59	48°35'N	80°33'W
Mathews, Lake, l., Ca., U.S. (măth ūz)	75a	33°50'N	117°24'W
Mathura, India (mŭ-tŏŏ'rŭ)	155	27°39'N	77°39'E
Matias Barbosa, Braz. (mä-tē'äs-bär-bô-sä)	99a	21°53'S	43°19'W
Matillas, Laguna, l., Mex. (lä-gô'nä-mä-tē'l-yäs)	89	18°02'N	92°36'W
Matina, C.R. (mä-tē'nä)	91	10°06'N	83°20'W
Matiši, Lat. (mä'tē̇-sē̇)	123	57°43'N	25°09'E
Matlalcueyetl, Cerro, mtn., Mex. (sĕ'r-rä-mä-tläl-kwĕ'yĕtl)	88	19°13'N	98°02'W
Matlock, Eng., U.K. (măt'lŏk)	114a	53°08'N	1°33'W
Matochkin Shar, Russia (mä'tŏch-kĭn)	134	73°57'N	56°16'E
Mato Grosso, Braz. (mät'ō grō's'ō)	101	15°04'S	59°58'W
Mato Grosso, state, Braz.	101	14°38'S	55°36'W
Mato Grosso, Chapada de, hills, Braz. (shä-pä'dä-dē)	101	13°39'S	55°42'W
Mato Grosso do Sul, state, Braz.	101	20°00'S	56°00'W
Matosinhos, Port.	128	41°10'N	8°48'W
Maṭraḥ, Oman (mȧ-trä')	154	23°36'N	58°27'E
Matsubara, Japan	167b	34°34'N	135°34'E
Matsudo, Japan (mät'sō-dō)	167a	35°48'N	139°55'E
Matsue, Japan (mät'sō-ĕ)	161	35°29'N	133°04'E
Matsumoto, Japan (mät'sō-mō'tō)	166	36°15'N	137°59'E
Matsuyama, Japan (mät'sō-yä'mä)	161	33°48'N	132°45'E
Matsuzaka, Japan (mät'sō-zä'kä)	167	34°35'N	136°34'E
Mattamuskeet, Lake, l., N.C., U.S. (măt-tȧ-mŭs'kēt)	83	35°34'N	76°03'W
Mattaponi, r., Va., U.S. (măt'ȧ-poni')	67	37°45'N	77°00'W
Mattawa, Can. (măt'ȧ-wȧ)	51	46°15'N	78°49'W
Matterhorn, mtn., Eur. (mät'ẽr-hôrn)	124	45°57'N	7°36'E
Matteson, Il., U.S. (mătt'ē-sŭn)	69a	41°30'N	87°42'W
Matthew Town, Bah. (măth'ū toun)	93	21°00'N	73°40'W
Mattoon, Il., U.S. (mă-tōōn')	65	39°30'N	88°20'W
Maturín, Ven. (mä-tōō-rēn')	100	9°48'N	63°16'W
Maúa, Moz.	191	13°51'S	37°10'E
Mauban, Phil. (mä'ōō-bän')	169a	14°11'N	121°44'E

ng-sing; ŋ-baŋk; N-nasalized n; nŏd; cŏmmit; ōld; ôbey; ôrder; oi-boil; fōōd; ŏŏ-as oo in foot; ou-out; s-soft; sh-dish; th-thin; pūre; ŭnite; ûrn; stŭd; circŭs; ü-as in French tu; '-indeterminate vowel.

274 Maubeuge – Mendota PRONOUNCING INDEX

PLACE (Pronunciation)	PAGE	LAT.	LONG.
Maubeuge, Fr. (mô-bûzh')	126	50°18'N	3°57'E
Maud, Oh., U.S. (môd)	69f	39°21'N	84°23'W
Mauer, Aus. (mou'ēr)	115e	48°09'N	16°16'E
Maués, Braz. (mà-wĕ's)	101	3°34'S	57°30'W
Mau Escarpment, clf., Kenya	191	0°45'S	35°50'E
Maui, i., Hi., U.S. (mä'ōō-ē)	64c	20°52'N	156°02'W
Maule, r., Chile (má'ō-lĕ)	99b	35°45'S	70°50'W
Maumee, Oh., U.S. (mô-mē')	66	41°30'N	83°40'W
Maumee, r., In., U.S.	66	41°10'N	84°50'W
Maumee Bay, b., Oh., U.S.	66	41°50'N	83°20'W
Maun, Bots. (mä-ōn')	186	19°52'S	23°40'E
Mauna Kea, mtn., Hi., U.S. (mä'ŏ-nä-kā'ä)	64c	19°52'N	155°30'W
Mauna Loa, mtn., Hi., U.S. (mä'ŏ-nälō'ä)	64c	19°28'N	155°38'W
Maurepas Lake, l., La., U.S. (mō-rē-pä')	81	30°18'N	90°40'W
Mauricie, Parc National de la, rec., Can.	59	46°46'N	73°00'W
Mauritania, nation, Afr. (mô-rē-tā'nĭ-à)	184	19°38'N	13°30'W
Mauritius, nation, Afr. (mô-rĭsh'ĭ-ŭs)	3	20°18'S	57°36'E
Maury, Wa., U.S. (mô'rĭ)	74a	47°22'N	122°23'W
Mauston, Wi., U.S. (môs'tŭn)	71	43°46'N	90°05'W
Maverick, r., Az., U.S. (mă-vûr'ĭk)	77	33°40'N	109°30'W
Mavinga, Ang.	190	15°50'S	20°21'E
Mawlamyine, Mya.	168	16°30'N	97°39'E
Maxville, Can. (măks'vĭl)	62c	45°17'N	74°52'W
Maxville, Mo., U.S.	75e	38°26'N	90°24'W
Maya, r., Russia (mä'yä)	141	58°00'N	135°45'E
Mayaguana, i., Bah.	93	22°25'N	73°00'W
Mayaguana Passage, strt., Bah.	93	22°20'N	73°25'W
Mayagüez, P.R. (mä-yä-gwāz')	87	18°12'N	67°10'W
Mayari, r., Cuba	93	20°25'N	75°35'W
Mayas, Montañas, mts., N.A. (mŏntăṇ'ăs mä'äs)	90a	16°43'N	89°00'W
Mayd, i., Som.	192a	11°24'N	46°38'E
Mayen, Ger. (mī'ĕn)	124	50°19'N	7°14'E
Mayenne, r., Fr. (mà-yĕn)	126	48°14'N	0°45'W
Mayfield, Ky., U.S. (mā'fēld)	82	36°44'N	88°19'W
Mayfield Creek, r., Ky., U.S.	82	36°54'N	88°47'W
Mayfield Heights, Oh., U.S.	69d	41°31'N	81°26'W
Mayfield Lake, res., Wa., U.S.	72	46°31'N	122°34'W
Maykop, Russia	134	44°35'N	40°07'E
Maykor, Russia (mī-kôr')	142a	59°01'N	55°52'E
Maymyo, Mya. (mī'myō)	160	22°14'N	96°32'E
Maynard, Ma., U.S. (mā'nàrd)	61a	42°25'N	71°27'W
Mayne, Can. (mān)	74d	48°51'N	123°18'W
Mayne, i., Can.	74d	48°52'N	123°14'W
Mayo, Can. (mā-yō')	50	63°40'N	135°51'W
Mayo, Fl., U.S.	82	30°02'N	83°08'W
Mayo, Md., U.S.	68e	38°54'N	76°31'W
Mayodan, N.C., U.S. (mā-yō'dăn)	83	36°25'N	79°59'W
Mayon Volcano, vol., Phil. (mä-yōn')	169a	13°21'N	123°43'E
Mayotte, dep., Afr. (mà-yôt')	187	13°07'S	45°32'E
May Pen, Jam.	92	18°00'N	77°25'W
Mayraira Point, c., Phil.	165	18°40'N	120°45'E
Mayran, Laguna de, l., Mex. (lä-ŏ'nä-dĕ-mī-rän')	86	25°40'N	102°35'W
Mayskiy, Russia	138	43°38'N	44°04'E
Maysville, Ky., U.S. (māz'vĭl)	66	38°35'N	83°45'W
Mayumba, Gabon	186	3°25'S	10°39'E
Mayville, N.D., U.S.	70	47°30'N	97°20'W
Mayville, N.Y., U.S. (mā'vĭl)	67	42°15'N	79°30'W
Mayville, Wi., U.S.	71	43°30'N	88°45'W
Maywood, Ca., U.S. (mā'wŏŏd)	75a	33°59'N	118°11'W
Maywood, Il., U.S.	69a	41°53'N	87°51'W
Mazabuka, Zam. (mä-zä-bōō'kä)	186	15°51'S	27°46'E
Mazagão, Braz. (mä-zà-gou'n)	101	0°05'S	51°27'W
Mazapil, Mex. (mä-zä-pēl')	80	24°40'N	101°30'W
Mazara del Vallo, Italy (mät-sä'rä dĕl väl'lō)	130	37°40'N	12°37'E
Mazār-i-Sharīf, Afg. (mà-zär'-ē-shà-rēf')	155	36°48'N	67°12'E
Mazarrón, Spain (mä-zär-rō'n)	128	37°37'N	1°29'W
Mazatenango, Guat. (mä-zä-tā-näŋ'gō)	86	14°30'N	91°30'W
Mazatla, Mex.	89a	19°30'N	99°24'W
Mazatlán, Mex.	86	23°14'N	106°27'W
Mazatlán (San Juan), Mex. (mä-zä-tlän') (sän hwän')	89	17°05'N	95°26'W
Mažeikiai, Lith. (mà-zhā'kē-ĭ)	123	56°19'N	22°24'E
Mazhafah, Jabal, mtn., Sau. Ar.	153a	28°56'N	35°05'E
Mazyr, Bela.	137	52°03'N	29°14'E
Mbabane, Swaz. (m'bä-bä'nĕ)	186	26°18'S	31°14'E
Mbaiki, C.A.R. (m'bà-ē'kē)	185	3°53'N	18°00'E
Mbakana, Montagne de, mts., Cam.	189	7°55'N	14°40'E
Mbakaou, Barrage de, dam, Cam.	189	6°10'N	12°55'E
Mbala, Zam.	186	8°50'S	31°22'E
Mbale, Ug.	191	1°05'N	34°10'E
Mbamba Bay, Tan.	191	11°17'S	34°46'E
Mbandaka, D.R.C.	186	0°04'N	18°16'E
M'banza Congo, Ang.	186	6°30'S	14°10'E
Mbanza-Ngungu, D.R.C.	186	5°20'S	10°55'E
Mbarara, Ug.	191	0°37'S	30°39'E
Mbasay, Chad	189	7°39'N	15°40'E
Mbigou, Gabon (m-bē-gōō')	186	2°07'S	11°30'E
Mbinda, Congo	190	2°00'S	12°55'E
Mbogo, Tan.	191	7°26'S	33°26'E
Mbomou (Bomu), r., Afr. (m'bō'mōō)	185	4°50'N	24°00'E
Mbout, Maur. (m'bōō')	184	16°03'N	12°31'W
Mbuji-Mayi, D.R.C.	190	6°09'S	23°38'E
McAdam, Can. (măk-ăd'ăm)	60	45°36'N	67°20'W
McAfee, N.J., U.S. (măk-à'fē)	68a	41°10'N	74°32'W
McAlester, Ok., U.S. (măk ăl'ĕs-tēr)	65	34°55'N	95°45'W
McAllen, Tx., U.S. (măk-ăl'ĕn)	80	26°12'N	98°14'W
McBride, Can. (măk-brīd')	50	53°18'N	120°10'W
McCalla, Al., U.S. (măk-kăl'lā)	68h	33°20'N	87°00'W
McCamey, Tx., U.S. (mă-kā'mĭ)	80	31°08'N	102°13'W
McColl, S.C., U.S. (mà-kŏl')	83	34°40'N	79°34'W
McComb, Ms., U.S. (mà-kōm')	82	31°14'N	90°27'W
McConaughy, Lake, l., Ne., U.S. (măk kō'nō ĭ')	70	41°24'N	101°40'W
McCook, Ne., U.S. (mà-kŏk')	78	40°13'N	100°37'W
McCormick, S.C., U.S. (mà-kôr'mĭk)	83	33°56'N	82°20'W
McDonald, Pa., U.S. (măk-dŏn'ăld)	69e	40°22'N	80°13'W
McDonald Island, i., Austl.	178	53°00'S	72°45'E
McDonald Lake, l., Can. (măk-dŏn-ăld)	62e	51°12'N	113°53'W
McGehee, Ar., U.S. (mà-gē')	79	33°39'N	91°22'W
McGill, Nv., U.S. (mà-gĭl')	77	39°25'N	114°47'W
McGowan, Wa., U.S. (măk-gou'ăn)	74c	46°15'N	123°55'W
McGrath, Ak., U.S. (măk'gràth)	64a	62°58'N	155°20'W
McGregor, Can. (măk-grĕg'ēr)	69b	42°08'N	82°58'W
McGregor, Ia., U.S.	71	42°58'N	91°12'W
McGregor, Tx., U.S.	81	31°26'N	97°23'W
McGregor, r., Can.	55	54°10'N	121°00'W
McGregor Lake, l., Can. (măk-grĕg'ēr)	62c	45°38'N	75°44'W
McHenry, Il., U.S. (măk-hĕn'rĭ)	69a	42°21'N	88°16'W
Mchinji, Mwi.	186	13°42'S	32°50'E
McIntosh, S.D., U.S. (măk'ĭn-tŏsh)	70	45°54'N	101°22'W
McKay, r., Or., U.S.	74c	45°43'N	123°00'W
McKeesport, Pa., U.S. (mà-kez'pōrt)	69e	40°21'N	79°51'W
McKees Rocks, Pa., U.S. (mà-kēz' rŏks)	69e	40°29'N	80°05'W
McKenzie, Tn., U.S. (mà-kĕn'zĭ)	82	36°07'N	88°30'W
McKenzie, r., Or., U.S.	72	44°07'N	122°20'W
McKinley, Mount, mtn., Ak., U.S. (mà-kĭn'lĭ)	64a	63°00'N	151°02'W
McKinney, Tx., U.S. (mà-kĭn'ĭ)	79	33°12'N	96°35'W
McLaughlin, S.D., U.S. (măk-lŏf'lĭn)	70	45°48'N	100°45'W
McLean, Va., U.S. (măc'lān)	68e	38°56'N	77°11'W
McLeansboro, Il., U.S. (mà-klānz'bŭr-ō)	66	38°10'N	88°35'W
McLennan, Can. (măk-lĭn'năn)	50	55°42'N	116°54'W
McLeod, r., Can.	55	53°45'N	115°55'W
McLeod Lake, Can.	54	54°59'N	123°02'W
McLoughlin, Mount, mtn., Or., U.S. (mák-lŏk'lĭn)	72	42°27'N	122°20'W
McMillan Lake, l., Tx., U.S. (măk-mĭl'ăn)	80	32°40'N	104°09'W
McMillin, Wa., U.S. (măk-mĭl'ĭn)	74a	47°08'N	122°14'W
McMinnville, Or., U.S. (măk-mĭn'vĭl)	72	45°13'N	123°13'W
McMinnville, Tn., U.S.	82	35°41'N	85°47'W
McMurray, Wa., U.S. (măk-mŭr'ĭ)	74a	48°19'N	122°15'W
McNary, Az., U.S. (măk-nâr'ē)	77	34°10'N	109°55'W
McNary, La., U.S.	81	30°58'N	92°32'W
McNary Dam, Or., U.S.	72	45°57'N	119°15'W
McPherson, Ks., U.S. (măk-fŭr's'n)	79	38°21'N	97°41'W
McRae, Ga., U.S. (măk-rā')	83	32°02'N	82°55'W
McRoberts, Ky., U.S. (măk-rŏb'ērts)	83	37°12'N	82°40'W
Mead, Ks., U.S. (mēd)	78	37°17'N	100°21'W
Mead, Lake, l., U.S.	64	36°20'N	114°14'W
Meade Peak, mtn., Id., U.S.	73	42°19'N	111°16'W
Meadow Lake, Can. (mĕd'ō lăk)	50	54°08'N	108°26'W
Meadows, Can. (mĕd'ōz)	62f	50°02'N	97°35'W
Meadville, Pa., U.S. (mēd'vĭl)	66	41°40'N	80°10'W
Meaford, Can. (mē'fērd)	59	44°35'N	80°40'W
Mealy Mountains, mts., Can. (mē'lē)	53	53°32'N	57°58'W
Meandarra, Austl. (mē-ăn-dä'rà)	176	27°47'S	149°40'E
Meaux, Fr. (mō)	126	48°58'N	2°53'E
Mecapalapa, Mex. (mā-kä-pä-lä'pä)	89	20°32'N	97°52'W
Mecatina, r., Can. (mā-kà-tē'nà)	61	50°50'N	59°45'W
Mecca (Makkah), Sau. Ar. (mĕk'à)	154	21°27'N	39°45'E
Mechanic Falls, Me., U.S. (mē-kăn'ĭk)	60	44°05'N	70°23'W
Mechanicsburg, Pa., U.S. (mē-kăn'ĭks-bûrg)	67	40°15'N	77°00'W
Mechanicsville, Md., U.S. (mē-kăn'ĭks-vĭl)	68e	38°27'N	76°45'W
Mechanicville, N.Y., U.S. (mē-kăn'ĭk-vĭl)	67	42°55'N	73°45'W
Mechelen, Bel.	121	51°01'N	4°28'E
Mechriyya, Alg.	118	33°30'N	0°13'W
Mecklenburg, hist. reg., Ger.	124	53°30'N	13°00'E
Medan, Indon. (mā-dän')	168	3°35'N	98°35'E
Medanosa, Punta, c., Arg. (pōō'n-tä-mĕ-dä-nō'sä)	102	47°50'S	65°53'W
Medden, r., Eng., U.K. (mĕd'ĕn)	114a	53°14'N	1°05'W
Medellín, Col. (må-dhĕl-yēn')	100	6°15'N	75°34'W
Medellín, Mex. (mĕ-dĕl-yĕ'n)	89	19°03'N	96°08'W
Medenine, Tun. (mā-dĕ-nēn')	118	33°22'N	10°33'E
Medfeld, Ma., U.S. (mĕd'fēld)	61a	42°11'N	71°19'W
Medford, Ma., U.S. (mĕd'fērd)	61a	42°25'N	71°07'W
Medford, N.J., U.S.	68f	39°54'N	74°50'W
Medford, Ok., U.S.	79	36°47'N	97°44'W
Medford, Or., U.S.	64	42°19'N	122°52'W
Medford, Wi., U.S.	71	45°09'N	90°22'W
Media, Pa., U.S. (mē'dĭ-à)	68f	39°55'N	75°24'W
Mediaş, Rom. (mĕd-yäsh')	125	46°09'N	24°21'E
Medical Lake, Wa., U.S. (mĕd'ĭ-kăl)	72	47°34'N	117°40'W
Medicine Bow, r., Wy., U.S.	73	41°58'N	106°30'W
Medicine Bow Range, mts., Co., U.S. (mĕd'ĭ-sĭn bō)	78	40°55'N	106°02'W
Medicine Hat, Can. (mĕd'ĭ-sĭn hăt)	50	50°03'N	110°40'W
Medicine Lake, l., Mt., U.S. (mĕd'ĭ-sĭn)	73	48°24'N	104°15'W
Medicine Lodge, Ks., U.S.	78	37°17'N	98°37'W
Medicine Lodge, r., Ks., U.S.	78	37°20'N	98°57'W
Medina *see* Al Madīnah, Sau. Ar.	154	24°26'N	39°42'E
Medina, N.Y., U.S. (mē-dī'nà)	67	43°15'N	78°20'W
Medina, Oh., U.S.	69d	41°08'N	81°52'W
Medina, r., Tx., U.S.	80	29°45'N	99°13'W
Medina del Campo, Spain (mā-dē'nä dĕl käm'pō)	118	41°18'N	4°54'W
Medina de Ríoseco, Spain (mā-dē'nä dā rē-ō-sā'kō)	128	41°53'N	5°05'W
Medina Lake, l., Tx., U.S.	80	29°36'N	98°47'W
Medina Sidonia, Spain	128	36°28'N	5°58'W
Mediterranean Sea, sea (mĕd-ĭ-tēr-ā'nē-ăn)	118	36°22'N	13°25'E
Medjerda, Oued, r., Afr.	118	36°43'N	9°54'E
Mednogorsk, Russia	134	51°27'N	57°22'E
Medveditsa, r., Russia (mĕd-vyĕ'dē tsà)	137	50°10'N	43°40'E
Medvezhegorsk, Russia (mĕd-vyĕzh'yē-gôrsk')	136	63°00'N	34°20'E
Medway, Ma., U.S. (mĕd'wā)	61a	42°08'N	71°23'W
Medway Towns, co., Eng., U.K.	114b	51°27'N	0°30'E
Medyn', Russia (mĕ-dēn')	132	54°58'N	35°53'E
Medzhybizh, Ukr.	133	49°23'N	27°29'E
Meekatharra, Austl. (mē-kà-thăr'à)	174	26°30'S	118°38'E
Meeker, Co., U.S. (mēk'ēr)	77	40°00'N	107°55'W
Meelpaeg Lake, l., Can. (mēl'pà-ĕg)	61	48°22'N	56°52'W
Meerane, Ger. (mā-rä'nĕ)	124	50°51'N	12°27'E
Meerbusch, Ger.	127c	51°15'N	6°41'E
Meerut, India (mē'rŏŏt)	155	28°59'N	77°43'E
Megalópoli, Grc.	131	37°22'N	22°08'E
Mégara, Grc. (mĕg'à-rà)	131	37°59'N	23°21'E
Megget, S.C., U.S. (mĕg'ĕt)	83	32°44'N	80°15'W
Megler, Wa., U.S. (mĕg'lēr)	74c	46°15'N	123°52'W
Mehanom, Mys, c., Ukr.	133	44°48'N	35°17'E
Meherrin, r., Va., U.S. (mē-hĕr'ĭn)	83	36°40'N	77°49'W
Mehlville, Mo., U.S.	75e	38°30'N	90°19'W
Mehsāna, India	158	23°42'N	72°23'E
Mehun-sur-Yèvre, Fr. (mē-ŭn-sür-yĕvr')	126	47°11'N	2°14'E
Meiling Pass, p., China (mā'lĭng')	161	25°22'N	115°00'E
Meinerzhagen, Ger. (mī'nĕrts-hä-gĕn)	127c	51°06'N	7°39'E
Meiningen, Ger. (mī'nĭng-ĕn)	124	50°35'N	10°25'E
Meiringen, Switz.	124	46°45'N	8°11'E
Meissen, Ger.	124	51°11'N	13°28'E
Meizhu, China (mā-jōō)	162	31°17'N	119°12'E
Mejillones, Chile (mĕ-kē-lyō'nās)	102	23°07'S	70°31'W
Mekambo, Gabon	190	1°01'N	13°56'E
Mekele, Eth.	185	13°31'N	39°19'E
Meknès, Mor. (mĕk'nĕs) (mĕk-nĕs')	184	33°56'N	5°44'W
Mekong, r., Asia	168	18°00'N	104°30'E
Melaka, Malay.	168	2°11'N	102°15'E
Melaka, state, Malay.	153b	2°19'N	102°09'E
Melanesia, is., Oc.	194	13°00'S	164°00'E
Melbourne, Austl. (mĕl'bŭrn)	175	37°52'S	145°08'E
Melbourne, Eng., U.K.	114a	52°49'N	1°26'W
Melbourne, Fl., U.S.	83a	28°05'N	80°37'W
Melbourne, Ky., U.S.	69f	39°02'N	84°22'W
Melcher, Ia., U.S. (mĕl'chēr)	71	41°13'N	93°11'W
Melekess, Russia (mĕl-yĕk-ĕs)	136	54°14'N	49°39'E
Melenki, Russia (mĕ-lyĕŋ'kē)	136	55°25'N	41°34'E
Melfort, Can. (mĕl'fôrt)	50	52°52'N	104°36'W
Melghir, Chott, l., Alg.	184	33°52'N	5°22'E
Melilla, Sp. N. Afr. (mā-lēl'yä)	184	35°24'N	3°30'W
Melipilla, Chile (må-lē-pē'lyä)	102	33°40'S	71°12'W
Melita, Can.	57	49°11'N	101°09'W
Melitopol', Ukr. (mā-lē-tô'pôl-y')	137	46°49'N	35°19'E
Melívoia, Grc.	131	39°42'N	22°47'E
Melkrivier, S. Afr.	192c	24°01'S	28°23'E
Mellen, Wi., U.S. (mĕl'ĕn)	71	46°20'N	90°40'W
Mellerud, Swe. (mål'ē-rōōdh)	122	58°43'N	12°25'E
Melmoth, S. Afr.	187c	28°38'S	31°26'E
Melo, Ur. (mā'lō)	102	32°18'S	54°07'W
Melocheville, Can. (mē-lôsh-vēl')	62a	45°24'N	73°56'W
Melozha, r., Russia (myĕ'lô-zhà)	142b	56°06'N	38°34'E
Melrose, Ma., U.S. (mĕl'rōz)	61a	42°29'N	71°06'W
Melrose, Mn., U.S.	71	45°39'N	94°49'W
Melrose Park, Il., U.S.	69a	41°54'N	87°52'W
Meltham, Eng., U.K. (mĕl'thăm)	114a	53°35'N	1°51'W
Melton, Austl. (mĕl'tŭn)	173a	37°41'S	144°35'E
Melton Mowbray, Eng., U.K. (mō'brâ)	114a	52°45'N	0°52'W
Melúli, r., Moz.	191	16°10'S	39°30'E
Melun, Fr. (mē-lŭn')	117	48°32'N	2°40'E
Melunga, Ang.	190	17°16'S	16°24'E
Melville, Can. (mĕl'vĭl)	50	50°55'N	102°48'W
Melville, La., U.S.	81	30°39'N	91°45'W
Melville, i., Austl.	174	11°30'S	131°12'E
Melville, l., Can.	53	53°46'N	59°31'W
Melville, Cape, c., Austl.	175	14°15'S	145°50'E
Melville Hills, hills, Can.	52	69°18'N	124°57'W
Melville Peninsula, pen., Can.	53	67°44'N	84°09'W
Melvindale, Mi., U.S. (mĕl'vĭn-dāl)	69b	42°17'N	83°11'W
Melyana, Alg.	117	36°19'N	1°56'E
Mélykút, Hung. (mā'l'kōōt)	125	46°14'N	19°21'E
Memba, Moz. (mĕm'bà)	187	14°12'S	40°35'E
Memel *see* Klaipėda, Lith.	136	55°43'N	21°10'E
Memel, S. Afr. (mĕ'mĕl)	192c	27°42'S	29°35'E
Memmingen, Ger. (mĕm'ĭng-ĕn)	124	47°59'N	10°10'E
Memo, r., Ven. (mĕ'mō)	101b	9°32'N	66°30'W
Memphis, Mo., U.S. (mĕm'fĭs)	79	40°27'N	92°11'W
Memphis, Tn., U.S. (mĕm'fĭs)	65	35°07'N	90°03'W
Memphis, Tx., U.S.	78	34°42'N	100°33'W
Memphis, hist., Egypt	192b	29°50'N	31°12'E
Mena, Ukr. (mĕ-nà')	133	51°31'N	32°14'E
Mena, Ar., U.S. (mē'nà)	79	34°35'N	94°09'W
Menangle, Austl.	173b	34°08'S	150°48'E
Menard, Tx., U.S. (mē-närd')	80	30°56'N	99°48'W
Menasha, Wi., U.S. (mē-năsh'à)	71	44°12'N	88°29'W
Mende, Fr. (mänd)	126	44°31'N	3°30'E
Menden, Ger. (mĕn'dĕn)	127c	51°26'N	7°47'E
Mendes, Braz. (mĕ'n-dĕs)	102b	22°32'S	43°44'W
Mendocino, Ca., U.S.	76	39°18'N	123°47'W
Mendocino, Cape, c., Ca., U.S. (mĕn'dô-sē'nō)	65	40°25'N	12°42'W
Mendota, Il., U.S. (mĕn-dō'tà)	71	41°34'N	89°06'W
Mendota, l., Wi., U.S.	71	43°09'N	89°41'W

ăt; finăl; rāte; senåte; ärm; åsk; sofà; fâre; ch-choose; dh-as th in other; bē; êvent; bĕt; recĕnt; cratēr; g-gō; gh-guttural g; bĭt; ĭ-short neutral; rīde; κ-guttural k as ch in German ich;

PRONOUNCING INDEX
Mendoza – Millicent 275

PLACE (Pronunciation) PAGE LAT. LONG.

Mendoza, Arg. (mĕn-dō′sä) 102 32°48′S 68°45′W
Mendoza, prov., Arg. 102 35°10′S 69°00′W
Mengcheng, China (mŭŋ-chŭŋ) 162 33°15′N 116°34′E
Meng Shan, mts., China (mŭŋ shän) 162 35°47′N 117°23′E
Mengzi, China 160 23°22′N 103°20′E
Menindee, Austl. (mĕ-nĭn-dē) 176 32°23′S 142°30′E
Menlo Park, Ca., U.S. (mĕn′lō pärk) 74b 37°27′N 122°11′W
Menno, S.D., U.S. (mĕn′ō) 70 43°14′N 97°34′W
Menominee, Mi., U.S. (mē-nŏm′ĭ-nē) 71 45°08′N 87°40′W
Menominee, r., Mi., U.S. 71 45°37′N 87°54′W
Menominee Falls, Wi., U.S. (fôls) 69a 43°11′N 88°06′W
Menominee Range, Mi., U.S. 71 46°07′N 88°53′W
Menomonee, r., Wi., U.S. 69a 43°09′N 88°06′W
Menomonie, Wi., U.S. 71 44°53′N 91°55′W
Menongue, Ang. 190 14°36′S 17°48′E
Menorca (Minorca), i., Spain (mĕ-nō′r-kä) 112 40°05′N 3°58′E
Mentana, Italy (mĕn-tà′nä) 129d 42°02′N 12°40′E
Mentawai, Kepulauan, is., Indon. (mĕn-tä-vī′) 168 1°08′S 98°10′E
Menton, Fr. (mäN-tôN′) 127 43°46′N 7°37′E
Mentone, Ca., U.S. (mĕn′tōne) 75a 34°05′N 117°08′W
Mentz, l., S. Afr. (mĕnts) 187c 33°13′S 25°15′E
Menzel Bourguiba, Tun. 118 37°12′N 9°51′E
Menzelinsk, Russia (mĕn′zyĕ-lĕnsk′) 136 55°40′N 53°15′E
Menzies, Austl. (mĕn′zēz) 174 29°45′S 122°15′E
Meogui, Mex. (må-ō′gē) 80 28°17′N 105°28′W
Meppel, Neth. (mĕp′ĕl) 121 52°41′N 6°08′E
Meppen, Ger. (mĕp′ĕn) 124 52°40′N 7°18′E
Merabéllou, Kólpos, b., Grc. 130a 35°16′N 25°55′E
Meramec, r., Mo., U.S. (mĕr′ȧ-mĕk) 79 38°06′N 91°06′W
Merano, Italy (må-rä′nō) 118 46°39′N 11°10′E
Merasheen, i., Can. (mē′rà-shēn) 61 47°30′N 54°15′W
Merauke, Indon. (må-rou′kä) 169 8°32′S 140°17′E
Meraux, La., U.S. (mē-ro′) 68d 29°56′N 89°56′W
Mercato San Severino, Italy 129c 40°34′N 14°38′E
Merced, Ca., U.S. (mĕr-sĕd′) 76 37°17′N 120°30′W
Merced, r., Ca., U.S. 76 37°25′N 120°31′W
Mercedario, Cerro, mtn., Arg. (mĕr-så-dhä′rê-ō) 102 31°58′S 70°07′W
Mercedes, Arg. 99c 34°41′S 59°26′W
Mercedes, Arg. (mĕr-sā′dhās) 102 29°04′S 58°01′W
Mercedes, Ur. 102 33°17′S 58°04′W
Mercedes, Tx., U.S. 81 26°09′N 97°55′W
Mercedita, Chile (mĕr-sē-dē′tä) 99b 33°51′S 71°10′W
Mercer Island, Wa., U.S. (mûr′sẽr) 74a 47°35′N 122°15′W
Mercês, Braz. (mĕ-sē′s) 99a 21°13′S 43°20′W
Merchtem, Bel. 115a 50°57′N 4°13′E
Mercier, Can. 62a 45°19′N 73°45′W
Mercy, Cape, c., Can. 53 64°48′N 63°22′W
Meredith, N.H., U.S. (mĕr′ē-dĭth) 67 43°35′N 71°35′W
Merefa, Ukr. (må-rĕf′à) 133 49°49′N 36°04′E
Merendón, Serranía de, mts., Hond. 90 15°01′N 89°05′W
Mereworth, Eng., U.K. (mĕ-rĕ wûrth) 114b 51°15′N 0°23′E
Mergui, Mya. (mĕr-gē′) 168 12°29′N 98°39′E
Mergui Archipelago, is., Mya. 168 12°04′N 97°02′E
Meric (Maritsa), r., Eur. 123 40°43′N 26°19′E
Mérida, Mex. 86 20°58′N 89°37′W
Mérida, Ven. 100 8°30′N 71°15′W
Mérida, Cordillera de, mts., Ven. (mĕ′rē-dhä) 100 8°30′N 70°45′W
Meriden, Ct., U.S. (mĕr′ĭ-dĕn) 67 41°30′N 72°50′W
Meridian, Ms., U.S. (mė-rĭd-ĭ-ăn) 65 32°21′N 88°41′W
Meridian, Tx., U.S. 81 31°56′N 97°37′W
Mérignac, Fr. 126 44°50′N 0°40′W
Merikarvia, Fin. (mā′rē-kàr′vē-ä) 123 61°51′N 21°30′E
Mering, Ger. (mē′rĕng) 115d 48°16′N 11°00′E
Merkel, Tx., U.S. (mûr′kĕl) 80 32°26′N 100°02′W
Merkinė, Lith. 123 54°10′N 24°10′E
Merksem, Bel. 115a 51°15′N 4°27′E
Merkys, r., Lith. (măr′kĭs) 125 54°23′N 25°00′E
Merlo, Arg. (mĕr-lô) 102a 34°40′S 58°44′W
Meron, Hare, mtn., Isr. 153a 32°58′N 35°25′E
Merriam, Ks., U.S. (mĕr-rĭ-yàm) 75f 39°01′N 94°42′W
Merriam, Mn., U.S. 75g 44°44′N 93°36′W
Merrick, N.Y., U.S. (mĕr′ĭk) 68a 40°40′N 73°33′W
Merrifield, Va., U.S. (mĕr′ĭ-fēld) 68e 38°50′N 77°12′W
Merrill, Wi., U.S. (mĕr′ĭl) 71 45°11′N 89°42′W
Merrimac, Ma., U.S. (mĕr′ĭ-măk) 61a 45°20′N 71°00′W
Merrimack, N.H., U.S. 61a 42°51′N 71°25′W
Merrimack, r., Ma., U.S. (mĕr′ĭ-măk) 67 43°10′N 71°30′W
Merritt, Can. (mĕr′ĭt) 50 50°07′N 120°47′W
Merryville, La., U.S. (mĕr′ĭ-vĭl) 81 30°46′N 93°34′W
Mersa Fatma, Erit. 185 14°54′N 40°14′E
Merseburg, Ger. (mĕr′zĕ-bo͝orgh) 124 51°21′N 11°59′E
Mersey, r., Eng., U.K. (mûr′zė) 114a 53°20′N 2°55′W
Merseyside, hist. reg., Eng., U.K. 114a 53°29′N 2°59′W
Mersing, Malay. 153b 2°25′N 103°51′E
Merta Road, India (mār′tŭ rōd) 158 26°50′N 73°54′E
Merthyr Tydfil, Wales, U.K. (mûr′thẽr tĭd′vĭl) 120 51°46′N 3°30′W
Mértola Almodóvar, Port. (mĕr-tô-lȧ-äl-mō-dō′vär) 128 37°39′N 8°04′W
Méru, Fr. (mā-rü′) 126 49°14′N 2°08′E
Meru, Kenya (mā′ro͞o) 185 0°01′N 37°45′E
Meru, Mount, mtn., Tan. 191 3°15′S 36°43′E
Merume Mountains, mts., Guy. (mĕr-ü′mĕ) 101 5°45′N 60°15′W
Merwede Kanaal, can., Neth. 115a 52°15′N 5°01′E
Merwin, l., Wa., U.S. (mĕr′wĭn) 74c 45°58′N 122°27′W
Merzifon, Tur. (mĕr′ze-fōn) 154 40°50′N 35°30′E
Mesa, Az., U.S. (mā′sȧ) 77 33°25′N 111°50′W
Mesabi Range, mts., Mn., U.S. (mā-sŏb′bē) 71 47°17′N 93°04′W
Mesagne, Italy (mā-sän′yā) 131 40°34′N 17°51′E
Mesa Verde National Park, rec., Co., U.S. (vĕr′dē) 64 37°22′N 108°27′W
Mescalero Apache Indian Reservation, I.R., N.M., U.S. (mĕs-kä-lā′rō) 77 33°10′N 105°45′W
Meshchovsk, Russia (myĕsh′chĕfsk) 132 54°17′N 35°19′E
Mesilla, N.M., U.S. (må-sē′yä) 77 32°15′N 106°45′W
Meskine, Chad 189 11°25′N 15°21′E
Mesolóngi, Grc. 131 38°23′N 21°28′E
Mesopotamia, hist. reg., Asia 157 34°00′N 44°00′E
Mesquita, Braz. 102b 22°48′S 43°26′W
Messina, Italy (mĕ-sē′nȧ) 110 38°11′N 15°34′E
Messina, S. Afr. 186 22°17′S 30°13′E
Messina, Stretto di, strt., Italy (stĕ′t-tō dē) 119 38°10′N 15°34′E
Messíni, Grc. 131 37°05′N 22°00′E
Mestaganem, Alg. 184 36°04′N 0°11′E
Mestre, Italy (mĕs′trā) 130 45°29′N 12°15′E
Meta, dept., Col. (mĕ′tä) 100a 3°28′N 74°07′W
Meta, r., S.A. 100 4°33′N 72°09′W
Métabetchouane, r., Can. (mĕ-tà-bĕt-cho͞o-än′) 59 47°45′N 72°00′W
Metairie, La., U.S. 81 30°00′N 90°11′W
Metán, Arg. (mĕ-tà′n) 102 25°32′S 64°51′W
Metangula, Moz. 186 12°42′S 34°48′E
Metapán, El Sal. (må-täpän′) 90 14°21′N 89°26′W
Metcalfe, Can. (mĕt-käf) 62c 45°14′N 75°27′W
Metchosin, Can. 74a 48°22′N 123°33′W
Metepec, Mex. (må-tĕ-pĕk′) 88 18°56′N 98°31′W
Metepec, Mex. 88 19°15′N 99°36′W
Methow, r., Wa., U.S. (mĕt′hou) (mĕt hou′) 72 48°26′N 120°15′W
Methuen, Ma., U.S. (mē-thū′ĕn) 61a 42°44′N 71°11′W
Metković, Cro. (mĕt′kō-vĭch) 131 43°02′N 17°40′E
Metlakatla, Ak., U.S. (mĕt-lȧ-kăt′lȧ) 63 55°08′N 131°35′W
Metropolis, Il., U.S. (mē-trŏp′ō-lĭs) 79 37°09′N 88°46′W
Metter, Ga., U.S. (mĕt′ẽr) 83 32°21′N 82°05′W
Mettmann, Ger. (mĕt′män) 127c 51°15′N 6°58′E
Metuchen, N.J., U.S. (mĕ-tŭ′chĕn) 68a 40°32′N 74°21′W
Metz, Fr. (mĕtz) 117 49°08′N 6°10′E
Metztitlán, Mex. (mĕtz-tĕt-län) 88 20°36′N 98°45′W
Meuban, Cam. 189 2°27′N 12°41′E
Meuse (Maas), r., Eur. (mûz) (müz) 121 50°32′N 5°22′E
Mexborough, Eng., U.K. (mĕks′bŭr-ô) 114a 53°30′N 1°17′W
Mexia, Tx., U.S. (må-hē′ä) 81 31°32′N 96°29′W
Mexian, China 161 24°20′N 116°10′E
Mexicalcingo, Mex. (mĕ-kē-käl-sēn′go) 89a 19°13′N 99°34′W
Mexicali, Mex. (mäk-sē-kä′lē) 86 32°28′N 115°29′W
Mexicana, Altiplanicie, plat., Mex. 88 22°38′N 102°33′W
Mexican Hat, Ut., U.S. (mĕk′sĭ-kăn hăt) 77 37°10′N 109°55′W
Mexico, Me., U.S. (mĕk′sĭ-kō) 60 44°34′N 70°33′W
Mexico, Mo., U.S. 79 39°09′N 91°51′W
Mexico, nation, N.A. 86 23°45′N 104°00′W
Mexico, Gulf of, b., N.A. 86 25°15′N 93°45′W
Mexico City, Mex. (mĕk′sĭ-kō) 86 19°28′N 99°09′W
Mexticacán, Mex. (mĕs′tê-kä-kän′) 88 21°12′N 102°43′W
Meyersdale, Pa., U.S. (mī′ẽrz-dāl) 67 39°55′N 79°00′W
Meyerton, S. Afr. (mī′ẽr-tŭn) 192c 26°35′S 28°01′E
Meymaneh, Afg. 154 35°53′N 64°38′E
Mezen′, Russia 134 65°50′N 44°05′E
Mezen′, r., Russia 136 65°20′N 44°45′E
Mézenc, Mont, mtn., Fr. (mōN-mä-zĕN′) 126 44°55′N 4°12′E
Mezha, r., Eur. (myā′zhà) 132 55°53′N 31°44′E
Mézieres-sur-Seine, Fr. (mā-zyâr′sür-sân′) 127b 48°58′N 1°49′E
Mezőkövesd, Hung. (mĕ′zû-kû′vĕsht) 125 47°49′N 20°36′E
Mezötur, Hung. (mĕ′zû-to͞or) 125 47°00′N 20°36′E
Mezquital, Mex. (måz-kė-täl′) 88 23°30′N 104°20′W
Mezquitic, Mex. (måz-kė-tēk′) 88 22°25′N 103°43′W
Mezquitic, r., Mex. 88 22°25′N 103°45′W
Mfangano Island, i., Kenya 191 0°28′S 33°35′E
Mga, Russia (m′gȧ) 142c 59°45′N 31°04′E
Mglin, Russia (m′glēn′) 132 53°03′N 32°52′W
Mia, Oued, r., Alg. 118 29°26′N 3°15′E
Miacatlán, Mex. (mė′ä-kä-tlän′) 88 18°42′N 99°17′W
Miahuatlán, Mex. (mė′ä-wä-tlän′) 89 16°20′N 96°38′W
Miajadas, Spain (mė-ä-hä′däs) 128 39°10′N 5°53′W
Miami, Az., U.S. 64 33°20′N 110°55′W
Miami, Fl., U.S. 65 25°45′N 80°11′W
Miami, Ok., U.S. 79 36°51′N 94°51′W
Miami, Tx., U.S. 78 35°41′N 100°39′W
Miami Beach, Fl., U.S. 83a 25°47′N 80°07′W
Miamisburg, Oh., U.S. (mī-ăm′ĭz-bûrg) 66 39°40′N 84°20′W
Miamitown, Oh., U.S. (mī-ăm′ĭ-toun) 69f 39°13′N 84°43′W
Mīāneh, Iran 154 37°15′N 47°13′E
Miangas, Pulau, i., Indon. 169 5°30′N 127°00′E
Miaoli, Tai. (mē-ou′lĭ) 165 24°30′N 120°48′E
Miaozhen, China (miou-jŭn) 162 31°44′N 121°28′E
Miass, Russia (mī-äs′) 140 54°59′N 60°06′E
Miastko, Pol. (myäst′kô) 124 54°01′N 17°00′E
Miccosukee Indian Reservation, I.R., Fl., U.S. 83a 26°10′N 80°50′W
Michalovce, Slvk. (mē′kä-lôf′tsĕ) 125 48°44′N 21°56′E
Michel Peak, mtn., Can. 54 53°35′N 126°25′W
Michelson, Mount, mtn., Ak., U.S. (mĭch′ĕl-sŭn) 63 69°11′N 144°12′W
Michendorf, Ger. (mē′kĕn-dōrf) 115b 52°19′N 13°02′E
Miches, Dom. Rep. (mē′chĕs) 93 19°00′N 69°05′W
Michigan, state, U.S. (mĭsh-ĭ-găn) 65 45°55′N 87°00′W
Michigan, Lake, l., U.S. 65 43°20′N 87°10′W
Michigan City, In., U.S. 66 41°40′N 86°55′W
Michipicoten, r., Can. 71 47°56′N 84°42′W
Michipicoten Harbour, Can. 71 47°58′N 84°58′W
Michurinsk, Russia (mĭ-cho͞o-rĭnsk′) 137 52°53′N 40°32′E
Mico, Punta, c., Nic. (po͞o′n-tä-mē′kô) 91 11°38′N 83°24′W
Micronesia, is., Oc. 194 11°00′N 159°00′E
Micronesia, Federated States of, nation, Oc. 3 5°00′N 152°00′E
Midas, Nv., U.S. (mī′dȧs) 72 41°15′N 116°50′W
Middelfart, Den. (mēd′'l-fărt) 122 55°30′N 9°45′E
Middle, r., Can. 54 55°00′N 125°50′W
Middle Andaman, i., India (ăn-dȧ-măn′) 168 12°44′N 93°21′E
Middle Bayou, Tx., U.S. 81a 29°38′N 95°06′W
Middleburg, S. Afr. (mĭd′ĕl-bûrg) 186 31°30′S 25°00′E
Middleburg, S. Afr. 192c 25°47′S 29°30′E
Middlebury, Vt., U.S. (mĭd′'l-bĕr-ĭ) 67 44°00′N 73°10′W
Middle Concho, Tx., U.S. (kōn′chō) 80 31°21′N 100°50′W
Middle River, Md., U.S. 68e 39°20′N 76°27′W
Middlesboro, Ky., U.S. (mĭd′'lz-bŭr-ô) 82 36°36′N 83°42′W
Middlesbrough, Eng., U.K. (mĭd′'lz-brŭ) 116 54°35′N 1°18′W
Middlesex, N.J., U.S. (mĭd′'l-sĕks) 68a 40°34′N 74°30′W
Middleton, Can. (mĭd′'l-tŭn) 60 44°57′N 65°04′W
Middleton, Eng., U.K. 114a 53°34′N 2°12′W
Middletown, Ct., U.S. 67 41°35′N 72°40′W
Middletown, De., U.S. 67 39°30′N 75°40′W
Middletown, Ma., U.S. 61a 42°35′N 71°01′W
Middletown, N.Y., U.S. 67 41°26′N 74°25′W
Middletown, Oh., U.S. 66 39°30′N 84°25′W
Middlewich, Eng., U.K. (mĭd′'l-wĭch) 114a 53°11′N 2°27′W
Middlewit, S. Afr. (mĭd′l′wĭt) 192c 24°50′S 27°00′E
Midfield, Al., U.S. 68h 33°28′N 86°54′W
Midi, Canal du, Fr. (kä-näl-dü-mē-dē′) 117 43°22′N 1°35′E
Mid Illovo, S. Afr. (mĭd ĭl′ô-vō) 187c 29°59′S 30°32′E
Midland, Can. (mĭd′lȧnd) 51 44°45′N 79°50′W
Midland, Mi., U.S. 66 43°40′N 84°20′W
Midland, Tx., U.S. 80 32°05′N 102°05′W
Midvale, Ut., U.S. (mĭd′vāl) 75b 40°37′N 111°54′W
Midway, Al., U.S. (mĭd′wā) 82 32°03′N 85°30′W
Midway Islands, is., Oc. 2 28°00′N 179°00′W
Midwest, Wy., U.S. (mĭd-wĕst′) 73 43°25′N 106°15′W
Midye, Tur. (mēd′yĕ) 137 41°35′N 28°10′E
Międzyrzecz, Pol. (myăN-dzû′zhĕch) 124 52°26′N 15°35′E
Mielec, Pol. (myĕ′lĕts) 125 50°17′N 21°27′E
Mier, Mex. (myâr) 80 26°26′N 99°08′W
Mieres, Spain (myā′rās) 128 43°14′N 5°45′W
Mier y Noriega, Mex. (myâr′ē nō-rē-ā′gä) 88 23°28′N 100°08′W
Miguel Auza, Mex. 88 24°17′N 103°27′W
Miguel Pereira, Braz. 102b 22°27′S 43°28′W
Mijares, r., Spain 129 39°55′N 0°01′W
Mikage, Japan (mē′kà-gå) 167b 34°42′N 135°15′E
Mikawa-Wan, b., Japan (mē′kä-wä wän) 167 34°43′N 137°09′E
Mikhaylov, Russia (mē-käy′lôf) 136 54°14′N 39°03′E
Mikhaylovka, Russia 142a 55°35′N 57°57′E
Mikhaylovka, Russia 142c 59°20′N 30°21′E
Mikhaylovka, Russia 137 50°05′N 43°10′E
Mikhnëvo, Russia (mĭk-nyô′vô) 142b 55°08′N 37°57′E
Miki, Japan (mē′kē) 167b 34°47′N 134°59′E
Mikindani, Tan. (mē-kēn-dä′nē) 187 10°17′S 40°07′E
Mikkeli, Fin. (mēk′ĕ-lĭ) 116 61°42′N 27°14′E
Mikulov, Czech Rep. (mĭ′ko͞o-lôf) 124 48°47′N 16°39′E
Mikumi, Tan. 191 7°24′S 36°59′E
Mikuni, Japan (mē′ko͞o-nē) 167 36°09′N 136°14′E
Mikuni-Sammyaku, mts., Japan (säm′myä-ko͞o) 167 36°51′N 138°38′E
Mikura, i., Japan (mē′ko͞o-rà) 167 33°53′N 139°26′E
Milaca, Milaca, Mn., U.S. (mė-lăk′ȧ) 71 45°45′N 93°41′W
Milan (Milano), Italy (mė-lä′nō) 130 45°29′N 9°12′E
Milan, Mi., U.S. (mĭl′ăn) 66 42°05′N 83°40′W
Milan, Mo., U.S. 79 40°13′N 93°07′W
Milan, Tn., U.S. 82 35°54′N 88°47′W
Milâs, Tur. (mē′läs) 119 37°10′N 27°25′E
Milazzo, Italy 130 38°13′N 15°17′E
Milbank, S.D., U.S. (mĭl′băŋk) 70 45°13′N 96°38′W
Mildura, Austl. (mĭl-dū′rȧ) 175 34°10′S 142°18′E
Miles City, Mt., U.S. (mīlz) 64 46°24′N 105°50′W
Milford, Ct., U.S. (mĭl′fẽrd) 67 41°15′N 73°05′W
Milford, De., U.S. 67 38°55′N 75°25′W
Milford, Ma., U.S. 61a 42°09′N 71°31′W
Milford, Mi., U.S. 69b 42°35′N 83°36′W
Milford, N.H., U.S. 67 42°50′N 71°40′W
Milford, Oh., U.S. 69f 39°11′N 84°18′W
Milford, Ut., U.S. 77 38°20′N 113°05′W
Milford Sound, strt., N.Z. 177 44°35′S 167°47′E
Miling, Austl. (mĭl′ĭng) 174 30°30′S 116°25′E
Milipitas, Ca., U.S. (mĭl-ĭ-pĭ′tăs) 74b 37°26′N 121°54′W
Milk, r., N.A. 64 48°30′N 107°00′W
Millau, Fr. (mē-yō′) 117 44°06′N 3°04′E
Millbrae, Ca., U.S. (mĭl′brā) 74b 37°36′N 122°23′W
Millbury, Ma., U.S. (mĭl′bĕr-ĭ) 61a 42°12′N 71°46′W
Mill Creek, r., Can. (mĭl) 62g 53°28′N 113°25′W
Mill Creek, r., Ca., U.S. 76 40°07′N 121°55′W
Milledgeville, Ga., U.S. (mĭl′ĕj-vĭl) 82 33°05′N 83°15′W
Mille Îles, Rivière des, r., Can. (rê-vyâr′ dā mēl′ēl′) 62a 45°41′N 73°40′W
Mille Lac Indian Reservation, I.R., Mn., U.S. (mĭl lăk′) 71 46°14′N 94°13′W
Mille Lacs, l., Mn., U.S. 71 46°25′N 93°22′W
Mille Lacs, Lac des, l., Can. (läk dĕ mēl läks) 58 48°52′N 90°53′W
Millen, Ga., U.S. (mĭl′ĕn) 83 32°47′N 81°55′W
Miller, S.D., U.S. (mĭl′ẽr) 70 44°31′N 99°00′W
Millerovo, Russia (mĭl′ĕ-rō-vō) 137 48°58′N 40°27′E
Millersburg, Ky., U.S. (mĭl′ẽrz-bûrg) 66 38°15′N 84°10′W
Millersburg, Oh., U.S. 66 40°35′N 81°55′W
Millersburg, Pa., U.S. 67 40°35′N 76°55′W
Millerton, Can. (mĭl′ẽr-tŭn) 60 46°56′N 65°40′W
Millertown, Can. (mĭl′ẽr-toun) 61 48°49′N 56°32′W
Millicent, Austl. (mĭl-ĭ-sĕnt) 176 37°30′S 140°20′E

276 Millinocket – Molody Tud PRONOUNCING INDEX

PLACE (Pronunciation)	PAGE	LAT.	LONG.
Millinocket, Me., U.S. (mĭl-ĭ-nŏk′ĕt)	60	45°40′N	68°44′W
Millis, Ma., U.S. (mĭl-ĭs)	61a	42°10′N	71°22′W
Millstadt, Il., U.S. (mĭl′stăt)	75e	38°27′N	90°06′W
Millstone, r., N.J., U.S. (mĭl′stōn)	68a	40°27′N	74°38′W
Millstream, Austl. (mĭl′strēm)	174	21°45′S	117°10′E
Milltown, Can. (mĭl′toun)	60	45°13′N	67°19′W
Mill Valley, Ca., U.S. (mĭl)	74b	37°54′N	122°32′W
Millwood Reservoir, res., Ar., U.S.	79	33°00′N	94°00′W
Milly-la-Forêt, Fr. (mē-yē′-la-fō-rĕ′)	127b	48°24′N	2°28′E
Milnerton, S. Afr. (mĭl′nẽr-tŭn)	186a	33°52′S	18°30′E
Milnor, N.D., U.S. (mĭl′nẽr)	70	46°17′N	97°29′W
Milo, Me., U.S.	60	44°16′N	69°01′W
Milos, i., Grc. (mē′lŏs)	119	36°45′N	24°35′E
Milpa Alta, Mex. (mē′l-pä-ä′l-tä)	89a	19°11′N	99°01′W
Milton, Can.	62d	43°31′N	79°53′W
Milton, Fl., U.S. (mĭl′tŭn)	82	30°37′N	87°02′W
Milton, Pa., U.S.	67	41°00′N	76°50′W
Milton, Ut., U.S.	75b	41°04′N	111°44′W
Milton, Wa., U.S.	74a	47°15′N	122°20′W
Milton, Wi., U.S.	71	42°45′N	89°00′W
Milton-Freewater, Or., U.S.	72	45°57′N	118°25′W
Milvale, Pa., U.S. (mĭl′văl)	69e	40°29′N	79°58′W
Milville, N.J., U.S. (mĭl′vĭl)	67	39°25′N	75°00′W
Milwaukee, Wi., U.S.	65	43°03′N	87°55′W
Milwaukee, r., Wi., U.S.	69a	43°10′N	87°56′W
Milwaukie, Or., U.S. (mĭl-wô′kē)	72	45°27′N	122°38′W
Mimiapan, Mex. (mē-myä-pan′)	89a	19°26′N	99°28′W
Mimoso do Sul, Braz. (mē-mō′sō-dō-sōō′l)	99a	21°03′S	41°21′W
Min, r., China (mēn)	161	26°03′N	118°30′E
Min, r., China	165	29°30′N	104°00′E
Mina, r., Alg. (mē′nȧ)	129	35°24′N	0°51′E
Minago, r., Can. (mĭ-nä′gō)	57	54°25′N	98°45′W
Minakuchi, Japan (mē′nȧ-kōō′chē)	167	34°59′N	136°06′E
Minas, Cuba (mē′näs)	92	21°30′N	77°35′W
Minas, Indon.	153b	0°52′N	101°29′E
Minas, Ur. (mē′näs)	102	34°18′S	55°12′W
Minas, Sierra de las, mts., Guat. (syĕr′rä dä läs mē′näs)	90	15°08′N	90°25′W
Minas Basin, b., Can. (mĭ′nȧs)	60	45°20′N	64°00′W
Minas Channel, strt., Can.	60	45°15′N	64°45′W
Minas de Oro, Hond. (mē′näs-dĕ-dĕ-ō-rô)	90	14°52′N	87°19′W
Minas de Riotinto, Spain (mē′näs dä rē-ō-tēn′tō)	128	37°43′N	6°35′W
Minas Novas, Braz. (mē′näzh nō′väzh)	101	17°20′S	42°19′W
Minatare, l., Ne., U.S. (mĭn′ȧ-târ)	70	41°56′N	103°07′W
Minatitlán, Mex. (mē-nä-tē-tlän′)	86	17°59′N	94°33′W
Minatitlán, Mex.	88	19°21′N	104°02′W
Minato, Japan (mē′nȧ-tō)	167	35°13′N	139°52′E
Minch, The, strt., Scot., U.K.	112	58°04′N	6°04′W
Mindanao, i., Phil.	169	8°00′N	125°00′E
Mindanao Sea, sea, Phil.	169	8°55′N	124°00′E
Minden, Ger. (mĭn′dĕn)	124	52°17′N	8°58′E
Minden, La., U.S.	81	32°36′N	93°19′W
Minden, Ne., U.S.	78	40°30′N	98°54′W
Mindoro, i., Phil.	168	12°50′N	121°05′E
Mindoro Strait, strt., Phil.	169a	12°28′N	120°33′E
Mindyak, Russia (mēn′dyák)	142a	54°01′N	58°48′E
Mineola, N.Y., U.S. (mĭn-ē-ō′lȧ)	68a	40°43′N	73°38′W
Mineola, Tx., U.S.	81	32°39′N	95°31′W
Mineral del Chico, Mex. (mē-nä-räl′dĕl chē′kō)	88	20°13′N	98°46′W
Mineral del Monte, Mex. (mē-nä-räl′ dĕl mōn′tā)	88	20°18′N	98°39′W
Mineral'nyye Vody, Russia	137	44°10′N	43°15′E
Mineral Point, Wi., U.S. (mĭn′ẽr-ăl)	71	42°50′N	90°10′W
Mineral Wells, Tx., U.S. (mĭn′ẽr-ăl wĕlz)	80	32°48′N	98°06′W
Minerva, Oh., U.S. (mĭ-nur′vȧ)	66	40°45′N	81°10′W
Minervino, Italy (mē-nĕr-vē′nō)	130	41°07′N	16°05′E
Mineyama, Japan (mē-nĕ-yä′mä)	167	35°38′N	135°05′E
Mingäçevir, Azer.	138	40°45′N	47°03′E
Mingäçevir su anbarı, res., Azer.	138	40°50′N	46°50′E
Mingan, Can.	51	50°18′N	64°02′W
Mingenew, Austl. (mĭn′gĕ-nû)	174	29°15′S	115°45′E
Mingo Junction, Oh., U.S. (mĭŋ′gō)	66	40°15′N	80°40′W
Minho, hist. reg., Port. (mēn yō)	128	41°32′N	8°13′W
Minho (Miño), r., Eur. (mē′n-yō)	128	41°28′N	9°05′W
Ministik Lake, l., Can. (mĭ-nĭs′tĭk)	62g	53°23′N	113°05′W
Minna, Nig. (mĭn′ȧ)	184	9°37′N	6°33′E
Minneapolis, Ks., U.S. (mĭn-ē-ăp′ō-lĭs)	79	39°07′N	97°41′W
Minneapolis, Mn., U.S.	65	44°58′N	93°15′W
Minnedosa, Can. (mĭn-ē-dō′sȧ)	50	50°14′N	99°51′W
Minneota, Mn., U.S. (mĭn-ē-ō′tȧ)	70	44°34′N	95°59′W
Minnesota, state, U.S. (mĭn-ē-sō′tȧ)	65	46°10′N	90°20′W
Minnesota, r., Mn., U.S.	65	44°30′N	95°00′W
Minnetonka, l., Mn., U.S. (mĭn-ē-tôŋ′kȧ)	71	44°52′N	93°34′W
Minnitaki Lake, l., Can. (mĭ′nĭ-tä′kē)	57	49°58′N	92°00′W
Mino, r., Japan	167b	34°56′N	135°06′E
Minonk, Il., U.S. (mĭ′nŏnk)	66	40°55′N	89°00′W
Minooka, Il., U.S. (mĭ-nōō′kȧ)	69a	41°27′N	88°15′W
Minot, N.D., U.S.	64	48°13′N	101°17′W
Minsk, Bela. (mēnsk)	134	53°54′N	27°35′E
Minsk, prov., Bela.	132	53°50′N	27°43′E
Mińsk Mazowiecki, Pol. (mēn′sk mä-zō-vyĕt′skĭ)	125	52°10′N	21°35′E
Minsterley, Eng., U.K. (mĭnstẽr-lē)	114a	52°38′N	2°55′W
Minto, Can.	60	46°05′N	66°05′W
Minto, l., Can.	53	57°18′N	75°50′W
Minturno, Italy (mēn-tōōr′nō)	130	41°17′N	13°44′E
Minūf, Egypt (mē-nōōf′)	192b	30°26′N	30°55′E
Minusinsk, Russia (mē-nō-sēnsk′)	135	53°47′N	91°45′E
Min'yar, Russia	142a	55°06′N	57°33′E
Miquelon Lake, l., Can. (mĭ′kē-lôn)	62g	53°16′N	112°55′W
Miquihuana, Mex. (mē-kē-wä′nä)	88	23°36′N	99°45′W
Mir, Bela. (mēr)	125	53°27′N	26°25′E
Miracema, Braz. (mē-rä-sĕ′mä)	99a	21°24′S	42°10′W
Miracema do Tocantins, Braz.	101	9°34′S	48°24′W
Mirador, Braz. (mē-rȧ-dōr′)	101	6°19′S	44°12′W
Miraflores, Col. (mē-rä-flō′räs)	100	5°10′N	73°13′W
Miraflores, Peru	100	16°19′S	71°20′W
Miraflores Locks, trans., Pan.	86a	9°00′N	79°35′W
Miragoâne, Haiti (mē-rȧ-gwän′)	93	18°25′N	73°05′W
Mira Loma, Ca., U.S. (mĭ′rȧ lō′mȧ)	75a	34°01′N	117°32′W
Miramar, Ca., U.S. (mĭr′ȧ-mär)	76a	32°53′N	117°08′W
Miramas, Fr.	126	43°35′N	5°00′E
Miramichi Bay, b., Can. (mĭr′ȧ-mē′shē)	60	47°08′N	65°08′W
Miranda, Col. (mē-rä′n-dä)	100a	3°14′N	76°11′W
Miranda, Ca., U.S.	76	40°14′N	123°49′W
Miranda, Ven.	101b	10°09′N	68°24′W
Miranda, dept., Ven.	101b	10°17′N	66°41′W
Miranda de Ebro, Spain (mē-rä′n-dä-dĕ-ĕ′brō)	128	42°42′N	2°59′W
Miranda do Douro, Port. (mē-rän′dä dō-dwĕ′rō)	128	41°30′N	6°17′W
Mirandela, Port. (mē-rän-dä′lȧ)	128	41°28′N	7°10′W
Mirando City, Tx., U.S. (mĭr-ăn′dō)	80	27°25′N	99°03′W
Mira Por Vos Islets, is., Bah. (mē′rä pôr vōs)	93	22°05′N	74°30′W
Mira Por Vos Pass, strt., Bah.	93	22°10′N	74°35′W
Mirbāţ, Oman	154	16°58′N	54°42′E
Mirebalais, Haiti (mēr-bȧ-lē′)	93	18°50′N	72°05′W
Mirecourt, Fr. (mēr-kōōr′)	127	48°20′N	6°08′E
Mirfield, Eng., U.K. (mûr′fēld)	114a	53°41′N	1°42′W
Miri, Malay. (mē′rē)	168	4°13′N	113°56′E
Mirim, Lagoa, l., S.A. (mē-rēn′)	102	33°00′S	53°15′W
Miropol'ye, Ukr. (mē-rō-pôl′yĕ)	133	51°02′N	35°13′E
Mīrpur Khās, Pak. (mēr′pōōr kȧs)	158	25°36′N	69°10′E
Mīrzāpur, India (mēr′zä-pōōr)	155	25°12′N	82°38′E
Misantla, Mex. (mē-sän′tlä)	89	19°55′N	96°49′W
Miscou, i., Can. (mĭs′kō)	60	47°58′N	64°35′W
Miscou Point, c., Can.	60	48°04′N	64°32′W
Miseno, Cape, c., Italy (mē-zĕ′nō)	129c	40°33′N	14°12′E
Misery, Mount, mtn., St. K./N. (mĭz′rē-ĭ)	91b	17°28′N	62°47′W
Mishan, China (mē′shän)	166	45°32′N	132°19′E
Mishawaka, In., U.S. (mĭsh-ȧ-wôk′ȧ)	66	41°45′N	86°15′W
Mishina, Japan (mē′shē-mä)	167	35°09′N	138°56′E
Misiones, prov., Arg. (mē-syō′näs)	102	27°00′S	54°30′W
Miskito, Cayos, is., Nic.	91	14°34′N	82°30′W
Miskolc, Hung. (mĭsh′kôlts)	110	48°07′N	20°50′E
Misool, Pulau, i., Indon. (mē-sōl′)	169	2°00′S	130°05′E
Misquah Hills, Mn., U.S. (mĭs-kwä′ hĭlz)	71	47°50′N	90°30′W
Mişr al Jadīdah, Egypt	192b	30°06′N	31°35′E
Misrātah, Libya	185	32°23′N	14°58′E
Missinaibi, r., Can. (mĭs′ĭn-ä′ē-bē)	53	50°27′N	83°01′W
Missinaibi Lake, l., Can.	58	48°23′N	83°40′W
Mission, Ks., U.S. (mĭsh′ŭn)	75f	39°02′N	94°39′W
Mission, Tx., U.S.	80	26°14′N	98°19′W
Mission City, Can. (sĭ′tĭ)	55	49°08′N	112°18′W
Missisagi, r., Can.	58	46°35′N	83°30′W
Mississauga, Can.	59	43°34′N	79°37′W
Mississippi, state, U.S. (mĭs-ĭ-sĭp′ē)	65	32°30′N	89°45′W
Mississippi, l., Can.	59	45°05′N	76°15′W
Mississippi, r., U.S.	65	32°00′N	91°30′W
Mississippi Sound, strt., Ms., U.S.	82	34°16′N	89°10′W
Missoula, Mt., U.S. (mĭ-zōō′lȧ)	64	46°55′N	114°00′W
Missouri, state, U.S. (mĭ-sōō′rē)	65	38°00′N	93°40′W
Missouri, r., U.S.	64	40°40′N	96°00′W
Missouri City, Tx., U.S.	81a	29°37′N	95°32′W
Missouri Coteau, hills, U.S.	64	47°30′N	101°00′W
Missouri Valley, Ia., U.S.	70	41°35′N	95°53′W
Mist, Or., U.S. (mĭst)	74c	46°00′N	123°15′W
Mistassini, Can. (mĭs-tȧ-sĭ′nē)	59	48°56′N	71°55′W
Mistassini, l., Can. (mĭs-tȧ-sĭ′nē)	53	50°48′N	73°30′W
Mistelbach, Aus. (mĭs′tĕl-bäk)	124	48°34′N	16°33′E
Misteriosa, Lago, l., Mex. (mēs-tē-ryō′sä)	90a	18°05′N	90°15′W
Misti, Volcán, vol., Peru	100	16°04′S	71°20′W
Mistretta, Italy (mē-strĕt′tä)	130	37°54′N	14°22′E
Misty Fjords National Monument, rec., Ak., U.S.	63	51°00′N	131°00′W
Mita, Punta de, c., Mex. (pōō′n-tä-dĕ-mē′tä)	88	20°44′N	105°34′W
Mitaka, Japan (mē′tä-kä)	167a	35°42′N	139°34′E
Mitchell, Il., U.S. (mĭch′ĕl)	75e	38°46′N	90°05′W
Mitchell, In., U.S.	66	38°45′N	86°25′W
Mitchell, Ne., U.S.	70	41°56′N	103°49′W
Mitchell, S.D., U.S.	64	43°42′N	98°01′W
Mitchell, Mount, mtn., N.C., U.S.	65	35°47′N	82°15′W
Mīt Ghamr, Egypt	192b	30°43′N	31°20′E
Mitla Pass, p., Egypt	153a	30°03′N	32°40′E
Mito, Japan (mē′tō)	166	36°20′N	140°23′E
Mitsiwa, Erit.	185	15°40′N	39°19′E
Mitsu, Japan (mēt′sò)	167	34°21′N	132°49′E
Mittelland Kanal, can., Ger. (mĭt′ĕl-länd)	124	52°18′N	10°42′E
Mittenwalde, Ger. (mē′tĕn-väl-dĕ)	115b	52°16′N	13°33′E
Mittweida, Ger. (mĭt-vī′dä)	124	50°59′N	12°58′E
Mitumba, Monts, mts., D.R.C.	191	10°50′S	27°00′E
Mityayevo, Russia (mĭt-yä′yĕ-vô)	142a	60°17′N	61°02′E
Miura, Japan	167a	35°08′N	139°37′E
Miwa, Japan (mē′wä)	167b	34°32′N	135°51′E
Mixico, Guat. (mēs′kō)	90	14°37′N	90°37′W
Mixquiahuala, Mex. (mēs-kē-wä′lä)	88	20°12′N	99°13′W
Mixteco, r., Mex. (mēs-tä′kō)	88	17°45′N	98°10′W
Miyake, Japan (mē′yä-kȧ)	167b	34°35′N	135°34′E
Miyake, i., Japan (mē′yä-kȧ)	167	34°06′N	139°21′E
Miyakonojō, Japan	166	31°44′N	131°04′E
Miyazaki, Japan (mē′yä-zä′kē)	166	31°55′N	131°27′E
Miyoshi, Japan (mē-yō′shē′)	166	34°48′N	132°49′E
Mizdah, Libya (mĕz′dä)	156	31°29′N	13°09′E
Mizil, Rom. (mē′zēl)	131	45°01′N	26°30′E
Mizoram, state, India	155	23°25′N	92°45′E
Mjölby, Swe. (myûl′bü)	122	58°20′N	15°09′E
Mjörn, l., Swe.	122	57°55′N	12°22′E
Mjösa, l., Nor. (myûsä)	116	60°41′N	11°25′E
Mkalama, Tan.	186	4°07′S	34°38′E
Mkushi, Zam.	191	13°40′S	29°20′E
Mkwaja, Tan.	191	5°47′S	38°51′E
Mladá Boleslav, Czech Rep. (mlä′dä bô′lĕ-slȧf)	124	50°26′N	14°52′E
Mlala Hills, hills, Tan.	191	6°47′S	31°45′E
Mlanje Mountains, mts., Mwi.	191	15°55′S	35°30′E
Mława, Pol. (mwä′vȧ)	116	53°07′N	20°25′E
Mmabatho, S. Afr.	186	25°42′S	25°43′E
Moa, r., Afr.	188	7°40′N	11°15′W
Moa, Pulau, i., Indon.	169	8°30′S	128°30′E
Moab, Ut., U.S. (mō′ăb)	77	38°35′N	109°35′W
Moanda, Gabon	186	1°37′S	13°09′E
Moar Lake, l., Can. (môr)	57	52°00′N	95°09′W
Moba, D.R.C.	186	7°12′S	29°39′E
Mobaye, C.A.R. (mō-bä′y′)	185	4°19′N	21°11′E
Mobayi-Mbongo, D.R.C.	185	4°14′N	21°11′E
Moberly, Mo., U.S. (mō′bẽr-lĭ)	65	39°24′N	92°25′W
Mobile, Al., U.S. (mō-bēl′)	65	30°42′N	88°03′W
Mobile, r., Al., U.S.	82	31°15′N	88°00′W
Mobile Bay, b., Al., U.S.	65	30°26′N	87°56′W
Mobridge, S.D., U.S. (mō′brĭj)	70	45°32′N	100°26′W
Moca, Dom. Rep. (mô′kä)	93	19°25′N	70°35′W
Moçambique, Moz. (mō-säŋ-bē′kĕ)	191	15°03′S	40°42′E
Moçâmedes, Ang. (mō-zȧ-mĕ-dĕs)	186	15°10′S	12°09′E
Moçâmedes, hist. reg., Ang.	186	16°00′S	12°15′E
Mochitlán, Mex. (mō-chē-tlän′)	88	17°10′N	99°19′W
Mochudi, Bots. (mō-chōō′dē)	186	24°13′S	26°07′E
Mocímboa da Praia, Moz. (mō-sē′ēm-bō-ȧ prä′ēȧ)	187	11°20′S	40°21′E
Moclips, Wa., U.S.	72	47°14′N	124°13′W
Môco, Serra do, mtn., Ang.	190	12°25′S	15°10′E
Mococa, Braz. (mō-kō′kä)	99a	21°29′S	46°58′W
Moctezuma, Mex. (mōk′tȧ-zōō′mä)	88	22°44′N	101°06′W
Mocuba, Moz.	191	16°50′S	36°59′E
Modderfontein, S. Afr.	187b	26°06′S	28°10′E
Modena, Italy (mô′dĕ-nä)	118	44°38′N	10°54′E
Modesto, Ca., U.S. (mō-dĕs′tō)	76	37°39′N	121°00′W
Mödling, Aus. (mûd′lĭng)	115e	48°06′N	16°17′E
Moelv, Nor.	122	60°55′N	10°40′E
Moengo, Sur.	101	5°43′N	54°19′W
Moenkopi, Az., U.S.	77	36°07′N	111°13′W
Moers, Ger. (mûrs)	127c	51°27′N	6°38′E
Moffat Tunnel, trans., Co., U.S. (mŏf′ăt)	78	39°52′N	106°20′W
Mogadishu (Muqdisho), Som.	192a	2°08′N	45°22′E
Mogadore, Oh., U.S. (mŏg-ȧ-dōr′)	69d	41°04′N	81°23′E
Mogaung, Mya. (mō-gä′ông)	155	25°30′N	96°52′E
Mogi das Cruzes, Braz. (mō-gē-däs-krōō′sĕs)	101	23°33′S	46°10′W
Mogi-Guaçu, r., Braz. (mō-gē-gwȧ′sōō)	99a	22°06′S	47°12′W
Mogilno, Pol. (mō-gēl′nō)	124	52°38′N	17°58′W
Mogi-Mirim, Braz. (mō-gē-mē-rē′N)	99a	22°26′S	46°57′W
Mogok, Mya. (mō-gōk′)	155	23°14′N	96°38′E
Mogol, r., S. Afr. (mō-gōl)	192c	24°12′S	27°55′E
Mogollon Plateau, plat., Az., U.S.	64	34°15′N	110°45′W
Mogollon Rim, clf., Az., U.S. (mō-gō-yōn′)	77	34°26′N	111°17′W
Moguer, Spain (mō-gĕr′)	128	37°15′N	6°50′W
Mohács, Hung. (mō′häch)	125	45°59′N	18°38′E
Mohale's Hoek, Leso.	187c	30°09′S	27°28′E
Mohall, N.D., U.S. (mō′hôl)	70	48°46′N	101°29′W
Mohave, l., Nv., U.S. (mō-hä′vä)	77	35°23′N	114°40′W
Mohe, China (mwo-hŭ)	161	53°33′N	122°30′E
Mohenjo-Dero, hist., Pak.	155	27°20′N	68°10′E
Mohyliv-Podil's'kyi, Ukr.	137	48°27′N	27°51′E
Mõisaküla, Est. (mẽ′sȧ-kü′lä)	123	58°07′N	25°12′E
Moissac, Fr. (mwä-săk′)	126	44°07′N	1°05′E
Moita, Port. (mō-ē′tȧ)	129b	38°39′N	9°00′W
Mojave, Ca., U.S.	76	35°06′N	118°09′W
Mojave, r., Ca., U.S. (mō-hä′vä)	76	34°46′N	117°24′W
Mojave Desert, Ca., U.S.	76	35°05′N	117°30′W
Mojave Desert, des., Ca., U.S.	64	35°00′N	117°00′W
Mokhotlong, Leso.	187c	29°18′S	29°06′E
Mokp'o, Kor., S. (mōk′pō′)	161	34°50′N	126°30′E
Mol, Bel.	115a	51°21′N	5°09′E
Moldavia *see* Moldova, nation, Eur.	134	48°00′N	28°00′E
Moldavia, hist. reg., Rom.	125	47°20′N	27°12′E
Molde, Nor. (môl′dĕ)	116	62°44′N	7°15′E
Moldova, nation, Eur.	134	48°00′N	28°00′E
Moldova, r., Rom.	125	47°17′N	26°27′E
Moldoveanu, Vârful, mtn., Rom.	131	45°33′N	24°38′E
Molepolole, Bots. (mō-lȧ-pō-lō′lȧ)	186	24°15′S	25°33′W
Molfetta, Italy (môl-fĕt′tä)	119	41°11′N	16°38′E
Molina, Chile (mō-lē′nä)	99b	35°07′S	71°17′W
Molina de Aragón, Spain (mô-lē′nä dĕ ä-rä-gô′n)	128	40°40′N	1°54′W
Molina de Segura, Spain (mô-lē′nä dĕ sĕ-gōō′rä)	128	38°03′N	1°07′W
Moline, Il., U.S. (mô-lēn′)	79	41°31′N	90°34′W
Moliro, D.R.C.	186	8°13′S	30°34′E
Moliterno, Italy (mōl-ē-tĕr′nō)	130	40°13′N	15°54′W
Mollendo, Peru (mô-lyĕn′dō)	100	17°02′S	71°59′W
Moller, Port, Ak., U.S. (pōrt mōl′ẽr)	63	56°18′N	161°30′W
Mölndal, Swe. (mûln′däl)	122	57°39′N	12°01′E
Molochna, r., Ukr.	133	47°05′N	35°22′E
Molochnyĭ lyman, l., Ukr.	133	46°35′N	35°32′E
Molody Tud, Russia (mō-lō-dō′ē tōō′d)	142b	55°17′N	37°31′E

ăt; finăl; rāte; senāte; ärm; ȧsk; sofȧ; fâre; ch-choose; dh-as th in other; bē; ēvent; bĕt; recĕnt; cratẽr; g-gō; gh-guttural g; bĭt; ĭ-short neutral; rīde; ĸ-guttural k as ch in German ich;

PRONOUNCING INDEX Moloka'i – Moreton Bay 277

PLACE (Pronunciation)	PAGE	LAT.	LONG.
Moloka'i, i., Hi., U.S. (mō-lō kä'ē)	64c	21°15'N	157°05'W
Molokcha, r., Russia (mô'lôk-chä)	142b	56°15'N	38°29'E
Molopo, r., Afr. (mō-lō-pô)	186	27°45'S	20°45'E
Molson Lake, l., Can. (mōl'sŭn)	57	54°12'N	96°45'W
Molteno, S. Afr. (môl-tā'nô)	187c	31°24'S	26°23'E
Moluccas see Maluku, is., Indon.	169	2°22'S	128°25'E
Moma, Moz.	191	16°44'S	39°14'E
Mombasa, Kenya (mōm-bä'sä)	187	4°03'S	39°40'E
Mombetsu, Japan (môm'bĕt-sōō')	166	44°21'N	142°48'E
Momence, Il., U.S. (mō-mĕns')	69a	41°09'N	87°40'W
Momostenango, Guat. (mō-mōs-tā-näŋ'gō)	90	15°02'N	91°25'W
Momotombo, Nic.	90	12°25'N	86°43'W
Mompog Pass, strt., Phil. (mōm-pōg')	169a	13°35'N	122°09'E
Mompos, Col. (mōm-pōs')	100	9°05'N	74°30'W
Momtblanc, Spain	129	41°20'N	1°08'E
Møn, i., Den. (mûn)	122	54°54'N	12°30'E
Monaca, Pa., U.S. (mō-nă'kō)	69e	40°41'N	80°17'W
Monaco, nation, Eur. (mŏn'ȧ-kō)	110	43°43'N	7°47'E
Monaghan, Ire. (mŏ'ā-găn)	120	54°16'N	7°20'W
Mona Passage, strt., N.A. (mō'nä)	87	18°00'N	68°10'W
Monarch Mountain, mtn., Can. (mŏn'ērk)	54	51°41'N	125°53'W
Monashee Mountains, mts., Can. (mŏ-nă'shē)	55	50°30'N	118°30'W
Monastir see Bitola, Mac.	130	41°02'N	21°22'E
Monastir, Tun. (mōn-ås-tēr')	118	35°49'N	10°56'E
Monastyrshchina, Russia (mō-nås-tērsh'chĭ-nå)	132	54°19'N	31°49'E
Monastyryshche, Ukr.	133	48°57'N	29°53'E
Monção, Braz. (mon-soun')	101	3°39'S	45°23'W
Moncayo, mtn., Spain (mōn-kä'yō)	128	41°44'N	1°48'W
Monchegorsk, Russia (mōn'chĕ-gôrsk)	136	69°00'N	33°35'E
Mönchengladbach, Ger. (mûn'kĕn gläd'bäk)	124	51°12'N	6°28'E
Moncique, Serra de, mts., Port. (sĕr'rä dā mōn-chē'kĕ)	128	37°22'N	8°37'W
Monclova, Mex. (mōn-klō'vä)	86	26°53'N	101°25'W
Moncton, Can. (mŭŋk'tŭn)	51	46°06'N	64°47'W
Mondêgo, r., Port. (mōn-dĕ'gō)	128	40°10'N	8°36'W
Mondego, Cabo, c., Port. (kå'bō mōn-dā'gō)	128	40°12'N	8°55'W
Mondombe, D.R.C. (mōn-dôm'bå)	186	0°45'S	23°06'E
Mondoñedo, Spain (mōn-dô-nyā'dō)	128	43°35'N	7°18'W
Mondovi, Wi., U.S. (mŏn-dō'vĭ)	71	44°35'N	91°42'W
Monee, Il., U.S. (mō-nĭ)	69a	41°25'N	87°45'W
Monessen, Pa., U.S. (mō-nĕs'sen)	69e	40°09'N	79°53'W
Monett, Mo., U.S. (mō-nĕt')	79	36°55'N	93°55'W
Monfalcone, Italy	130	45°49'N	13°30'E
Monforte de Lemos, Spain (mōn-fôr'tā dĕ lĕ'mōs)	128	42°30'N	7°30'W
Mongala, r., D.R.C. (mŏn-gål'ȧ)	185	3°20'N	21°30'E
Mongalla, Sudan	185	5°11'N	31°46'E
Monghyr, India (mōn-gēr')	155	25°23'N	86°34'E
Mongo, r., Afr.	188	9°50'N	11°50'W
Mongolia, nation, Asia (mōŋ-gō'lĭ-ȧ)	160	46°00'N	100°00'E
Mongos, Chaîne des, mts., C.A.R.	185	8°04'N	21°59'E
Mongoumba, C.A.R. (mōŋ-gōōm'bȧ)	185	3°38'N	18°36'E
Mongu, Zam. (mōŋ-gōō')	186	15°15'S	23°09'E
Monkey Bay, Mwi.	191	14°05'S	34°55'E
Monkey River, Belize (mŭŋ'kĭ)	90a	16°22'N	88°33'W
Monkland, Can. (mŭngk-lănd)	62c	45°12'N	74°52'W
Monkoto, D.R.C. (mōn-kō'tō)	186	1°38'S	20°39'E
Monmouth, Il., U.S. (mŏn'mŭth) (mŏn'mouth)	79	40°54'N	90°38'W
Monmouth Junction, N.J., U.S. (mŏn'mouth jŭngk'shŭn)	68a	40°23'N	74°33'W
Monmouth Mountain, mtn., Can. (mŏn'mŭth)	54	51°00'N	123°47'W
Mono, r., Afr.	188	7°20'N	1°25'E
Mono Lake, l., Ca., U.S. (mō'nō)	76	38°04'N	119°00'W
Monon, In., U.S. (mō'nŏn)	66	40°55'N	86°55'W
Monongah, W.V., U.S. (mō-nŏŋ'gȧ)	66	39°25'N	80°10'W
Monongahela, Pa., U.S. (mō-nŏn-gȧ-hē'lȧ)	69a	40°11'N	79°55'W
Monongahela, r., W.V., U.S.	66	39°30'N	80°10'W
Monopoli, Italy (mō-nô'pô-lē)	131	40°55'N	17°17'E
Monóvar, Spain (mō-nō'vär)	129	38°26'N	0°50'W
Monreale, Italy (mōn-rå-ä'lā)	130	38°04'N	13°15'E
Monroe, Ga., U.S. (mŭn-rō')	82	33°47'N	83°43'W
Monroe, La., U.S.	65	32°30'N	92°06'W
Monroe, Mi., U.S.	66	41°55'N	83°25'W
Monroe, N.C., U.S.	83	34°58'N	80°34'W
Monroe, N.Y., U.S.	68a	41°19'N	74°11'W
Monroe, Ut., U.S.	77	38°35'N	112°10'W
Monroe, Wa., U.S.	74a	47°52'N	121°58'W
Monroe, Wi., U.S.	71	42°35'N	89°40'W
Monroe, Lake, l., Fl., U.S.	83	28°50'N	81°15'W
Monroe City, Mo., U.S.	79	39°38'N	91°41'W
Monroeville, Al., U.S. (mŭn-rō'vĭl)	82	31°33'N	87°19'W
Monroeville, Pa., U.S.	69e	40°26'N	79°46'W
Monrovia, Lib.	184	6°18'N	10°47'W
Monrovia, Ca., U.S. (mōn-rō'vĭ-ȧ)	75a	34°09'N	118°00'W
Mons, Bel. (môns')	117	50°29'N	3°55'E
Monson, Me., U.S. (mŏn'sŭn)	60	45°17'N	69°28'W
Mönsterås, Swe. (mŭn'stĕr-ōs)	122	57°04'N	16°24'E
Montagne Tremblant Provincial Park, rec., Can.	65	46°30'N	75°51'W
Montague, Can. (mŏn'tȧ-gū)	61	46°10'N	62°39'W
Montague, Mi., U.S.	66	43°30'N	86°25'W
Montague, i., Ak., U.S.	63	60°10'N	147°00'W
Montalbán, Ven. (mōnt-äl-bän)	101b	10°14'N	68°19'W
Montalegre, Port. (mōn-tä-lā'grĕ)	128	41°49'N	7°48'W
Montana, state, U.S. (mŏn-tăn'ȧ)	64	47°10'N	111°50'W
Montánchez, Spain (mōn-tän'chāth)	128	39°18'N	6°09'W
Montargis, Fr. (môn-tȧr-zhē')	117	47°59'N	2°42'E
Montataire, Fr. (môn-tȧ-târ)	127b	49°15'N	2°26'E
Montauban, Fr. (môn-tō-bän')	117	44°01'N	1°22'E
Montauk, N.Y., U.S.	67	41°03'N	71°57'W
Montauk Point, c., N.Y., U.S. (mŏn-tôk')	67	41°05'N	71°55'W
Montbard, Fr. (môn-bȧr')	126	47°40'N	4°19'E
Montbéliard, Fr. (môn-bā-lyȧr')	127	47°32'N	6°45'E
Mont Belvieu, Tx., U.S. (mŏnt bĕl'vū)	81a	29°51'N	94°53'W
Montbrison, Fr. (môn-brē-zon')	126	45°38'N	4°06'E
Montceau, Fr. (môn-sō')	126	46°39'N	4°22'E
Montclair, N.J., U.S. (mŏnt-klâr')	68a	40°49'N	74°13'W
Mont-de-Marsan, Fr. (môn-dē-mȧr-sän')	117	43°54'N	0°32'W
Montdidier, Fr. (môn-dē-dyā')	126	49°42'N	2°33'E
Monte, Arg. (mō'n-tĕ)	99c	35°25'S	58°49'W
Monteagudo, Bol. (mōn'tā-ä-gōō'dhō)	100	19°49'S	63°48'W
Montebello, Can.	62c	45°40'N	74°56'W
Montebello, Ca., U.S. (mŏn-tē-bĕl'ō)	75a	34°01'N	118°06'W
Monte Bello Islands, is., Austl.	174	20°30'S	114°10'E
Monte Caseros, Arg. (mō'n-tĕ-kä-sĕ'rōs)	102	30°16'S	57°39'W
Montecillos, Cordillera de, mts., Hond.	90	14°19'N	87°52'W
Monte Cristi, Dom. Rep. (mō'n-tĕ-krē's-tē)	93	19°50'N	71°40'W
Montecristo, Isola di, i., Italy (mōn'tå-krēs'tō)	130	42°20'N	10°19'E
Monte Escobedo, Mex. (mōn'tā ĕs-kō-bā'dhō)	88	22°18'N	103°34'W
Monteforte Irpino, Italy (mōn-tĕ-fō'r-tĕ ē'r-pē'nō)	129c	40°39'N	14°42'E
Montefrío, Spain (mōn-tå-frē'ō)	128	37°20'N	4°02'W
Montego Bay, Jam. (mŏn-tē'gō)	87	18°30'N	77°55'W
Montelavar, Port. (mōn-tĕ-lȧ-vär')	129b	38°51'N	9°20'W
Montélimar, Fr. (môn-tā-lē-mär')	117	44°33'N	4°47'E
Montellano, Spain (mōn-tå-lyä'nō)	128	37°00'N	5°34'W
Montello, Wi., U.S. (mŏn-tĕl'ō)	71	43°47'N	89°20'W
Montemorelos, Mex. (mōn'tā-mō-rā'lōs)	86	25°14'N	99°50'W
Montemor-o-Novo, Port. (mōn-tĕ-mōr'ō-nō'vō)	128	38°39'N	8°11'W
Montenegro see Crna Gora, state, Serb.	131	42°55'N	18°52'E
Montenegro, reg., Moz.	191	13°07'S	39°00'E
Montepulciano, Italy (mōn'tå-pōōl-chä'nō)	130	43°05'N	11°48'E
Montereau-faut-Yonne, Fr. (môn-t'rō'fō-yōn')	126	48°24'N	2°57'E
Monterey, Ca., U.S. (mŏn-tĕ-rā')	64	36°36'N	121°53'W
Monterey, Tn., U.S.	82	36°06'N	85°15'W
Monterey Bay, b., Ca., U.S.	64	36°48'N	122°01'W
Monterey Park, Ca., U.S.	75a	34°04'N	118°08'W
Montería, Col. (mōn-tå-rā'ä)	100	8°47'N	75°57'W
Monteros, Arg. (mōn-tĕ'rōs)	102	27°14'S	65°29'W
Monterotondo, Italy (mōn-tĕ-rō-tô'n-dō)	129d	42°03'N	12°39'E
Monterrey, Mex. (mōn-tĕr-rā')	86	25°43'N	100°19'W
Montesano, Wa., U.S. (mŏn-tē-sä'nō)	72	46°59'N	123°35'W
Monte Sant'Angelo, Italy (mō'n-tĕ sän ä'n-gzhĕ-lō)	119	41°43'N	15°59'E
Montes Claros, Braz. (môn-tĕs-klä'rōs)	101	16°44'S	43°41'W
Montevallo, Al., U.S. (mŏn-tē-văl'ō)	82	33°05'N	86°49'W
Montevarchi, Italy (mōn-tå-vär'kē)	130	43°30'N	11°45'E
Montevideo, Ur. (mōn'tå-vē-dhā'ō)	102	34°50'S	56°10'W
Montevideo, Mn., U.S. (mŏn'tē-vĭd'ē-ō)	70	44°56'N	95°42'W
Monte Vista, Co., U.S. (mŏn'tē vĭs'tȧ)	77	37°35'N	106°10'W
Montezuma, Ga., U.S. (mŏn-tē-zōō'mȧ)	82	32°17'N	84°00'W
Montezuma Castle National Monument, rec., Az., U.S.	77	34°38'N	111°50'W
Montfoort, Neth.	115a	52°02'N	4°56'E
Montfor-l'Amaury, Fr. (môn-fōr'lä-mō-rē')	127b	48°47'N	1°49'E
Montfort, Fr. (mŏn-fōr)	126	48°09'N	1°58'W
Montgomery, Al., U.S. (mŏnt-gŭm'ēr-ĭ)	65	32°23'N	86°17'W
Montgomery, W.V., U.S.	66	38°10'N	81°25'W
Montgomery City, Mo., U.S.	79	38°58'N	91°29'W
Monticello, Ar., U.S. (mŏn-tĭ-sĕl'ō)	79	33°38'N	91°47'W
Monticello, Fl., U.S.	82	30°32'N	83°53'W
Monticello, Ga., U.S.	82	33°00'N	83°11'W
Monticello, Ia., U.S.	71	42°14'N	91°13'W
Monticello, Il., U.S.	66	40°05'N	88°35'W
Monticello, In., U.S.	66	40°40'N	86°50'W
Monticello, Ky., U.S.	82	36°47'N	84°50'W
Monticello, Me., U.S.	60	46°19'N	67°53'W
Monticello, Mn., U.S.	71	45°18'N	93°48'W
Monticello, N.Y., U.S.	67	41°35'N	74°40'W
Monticello, Ut., U.S.	77	37°55'N	109°25'W
Montijo, Port. (mōn-tē'zhō)	129b	38°42'N	8°58'W
Montijo, Spain (mōn-tē'hō)	128	38°55'N	6°35'W
Montijo, Bahía, b., Pan. (bä-ē'ä mōn-tē'hō)	87	7°36'N	81°11'W
Mont-Joli, Can. (môn zhō-lē')	51	48°35'N	68°11'W
Montluçon, Fr. (môn-lü-sôn')	117	46°20'N	2°35'E
Montmagny, Can. (môn-mán-yē')	59	46°59'N	70°33'W
Montmorency, Fr. (môn'mō-rän-sē')	127b	48°59'N	2°19'E
Montmorency, r., Can. (mŏnt-mō-rĕn'sĭ)	62b	47°03'N	71°10'W
Montmorillon, Fr. (môn'mō-rē-yôn')	126	46°26'N	0°50'E
Montone, r., Italy (mōn-tō'nĕ)	130	44°03'N	11°45'E
Montoro, Spain (mōn-tō'rō)	128	38°01'N	4°22'W
Montpelier, Id., U.S.	73	42°19'N	111°19'W
Montpelier, In., U.S. (mŏnt-pēl'yēr)	66	40°35'N	85°20'W
Montpelier, Oh., U.S.	66	41°35'N	84°35'W
Montpelier, Vt., U.S.	65	44°20'N	72°35'W
Montpellier, Fr. (môn-pĕ-lyā')	117	43°38'N	3°53'E
Montréal, Can. (mŏn-trē-ôl')	51	45°30'N	73°35'W
Montreal, r., Can.	59	47°50'N	80°30'W
Montreal, r., Can.	58	47°15'N	84°20'W
Montreal Lake, l., Can.	56	54°20'N	105°40'W
Montréal-Nord, Can.	62a	45°36'N	73°38'W
Montreuil, Fr.	127b	48°52'N	2°27'E
Montreux, Switz. (môn-trû')	124	46°26'N	6°52'E
Montrose, Scot., U.K.	120	56°45'N	2°25'W
Montrose, Ca., U.S. (mŏnt-rōz)	75a	34°13'N	118°13'W
Montrose, Co., U.S. (mŏn-trōz')	77	38°30'N	107°55'W
Montrose, Oh., U.S.	69d	41°08'N	81°38'W
Montrose, Pa., U.S. (mŏnt-rōz')	67	41°50'N	75°50'W
Montrouge, Fr.	127b	48°49'N	2°19'E
Mont-Royal, Can.	62a	47°31'N	73°39'W
Monts, Pointe des, c., Can. (pwănt' dā môn')	60	49°19'N	67°22'W
Mont Saint Martin, Fr. (môn sân mär-tăn')	127	49°34'N	6°13'E
Montserrat, dep., N.A. (mŏnt-sĕ-răt')	87	16°48'N	63°15'W
Montvale, N.J., U.S. (mŏnt-văl')	68a	41°02'N	74°01'W
Monywa, Mya. (mŏn'yōō-wȧ)	155	22°02'N	95°16'E
Monza, Italy (mōn'tsä)	130	45°34'N	9°17'E
Monzón, Spain (mōn-thōn')	129	41°54'N	0°09'E
Moody, Tx., U.S. (mōō'dĭ)	81	31°18'N	97°20'W
Mooi, r., S. Afr. (mōō'ĭ)	192c	26°34'S	27°03'E
Mooi, r., S. Afr.	187c	29°00'S	30°15'E
Mooirivier, S. Afr.	187c	29°14'S	29°59'E
Moolap, Austl.	173a	38°11'S	144°26'E
Moonta, Austl. (mōōn'tȧ)	174	34°05'S	137°42'E
Moora, Austl. (mōr'ȧ)	174	30°35'S	116°12'E
Moorabbin, Austl.	173a	37°56'S	145°02'E
Moore, l., Austl. (mōr)	174	29°50'S	118°12'E
Moorenweis, Ger. (mō'rĕn-vīz)	115d	48°10'N	11°05'E
Moore Reservoir, res., Vt., U.S.	67	44°20'N	72°10'W
Moorestown, N.J., U.S. (morz'toun)	68f	39°58'N	74°56'W
Mooresville, In., U.S. (mōrz'vĭl)	69g	39°37'N	86°22'W
Mooresville, N.C., U.S.	83	35°34'N	80°48'W
Moorhead, Mn., U.S. (mōr'hĕd)	70	46°52'N	96°44'W
Moorhead, Ms., U.S.	82	33°25'N	90°30'W
Moose, r., Can.	53	51°01'N	80°42'W
Moose Creek, Can.	62c	45°16'N	74°58'W
Moosehead, Me., U.S. (mōōs'hĕd)	60	45°37'N	69°15'W
Moose Island, i., Can.	57	51°50'N	97°09'W
Moose Jaw, Can. (mōōs jô)	50	50°23'N	105°32'W
Moose Jaw, r., Can.	56	50°34'N	105°17'W
Moose Lake, Can.	57	53°40'N	100°28'W
Moose Mountain, mtn., Can.	57	49°45'N	102°37'W
Moose Mountain Creek, r., Can.	57	49°12'N	102°10'W
Moosilauke, mtn., N.H., U.S. (mōō-sĭ-lȧ'kē)	67	44°00'N	71°50'W
Moosinning, Ger. (mō'zē-nēng)	115d	48°17'N	11°51'E
Moosomin, Can. (mōō'sō-mĭn)	57	50°07'N	101°40'W
Moosonee, Can. (mōō'sô-nē)	51	51°20'N	80°44'W
Mopti, Mali (mŏp'tē)	184	14°30'N	4°12'W
Moquegua, Peru (mō-kā'gwä)	100	17°15'S	70°54'W
Mór, Hung. (mōr)	125	47°25'N	18°14'E
Mora, India	159b	18°54'N	72°56'E
Mora, Spain (mō-rä)	128	39°42'N	3°45'W
Mora, Swe. (mō'rä)	122	61°00'N	14°29'E
Mora, Mn., U.S. (mō'rȧ)	71	45°52'N	93°18'W
Mora, N.M., U.S.	78	35°58'N	105°17'W
Morādābād, India (mō-rä-dä-bäd')	155	28°57'N	78°48'E
Morales, Guat. (mō-rä'lĕs)	90	15°29'N	88°46'W
Moramanga, Madag. (mō-rä-män'gä)	187	18°48'S	48°09'E
Morant Point, c., Jam. (mō-rănt')	92	17°55'N	76°10'W
Morata de Tajuña, Spain (mō-rä'tä dā tä-hōō'nyä)	129a	40°14'N	3°27'W
Moratuwa, Sri L.	159	6°35'N	79°59'E
Morava (Moravia), hist. reg., Czech Rep.	124	49°21'N	16°57'E
Morava, r., Eur.	117	49°00'N	17°30'E
Moravia see Morava, hist. reg., Czech Rep.	124	49°21'N	16°57'E
Morawhanna, Guy. (mō-rȧ-hwä'nȧ)	101	8°12'N	59°33'W
Moray Firth, b., Scot., U.K. (mŭr'ā)	112	57°41'N	3°55'W
Mörbylånga, Swe. (mûr'bü-lôŋ'gȧ)	122	56°32'N	16°23'E
Morden, Can. (môr'dĕn)	50	49°11'N	98°05'W
Mordialloc, Austl. (môr-dĭ-ăl'ŏk)	173a	38°00'S	145°05'E
Mordvinia, prov., Russia	136	54°18'N	43°50'E
More, Ben, mtn., Scot., U.K. (bĕn môr)	120	58°09'N	5°01'W
Moreau, r., S.D., U.S. (mō-rō')	70	45°13'N	102°22'W
Moree, Austl. (mō'rē)	175	29°20'S	149°50'E
Morehead, Ky., U.S.	66	38°10'N	83°25'W
Morehead City, N.C., U.S. (mōr'hĕd)	83	34°43'N	76°43'W
Morehouse, Mo., U.S. (mōr'hous)	79	36°49'N	89°41'W
Morelia, Mex. (mō-rā'lyä)	86	19°43'N	101°12'W
Morella, Spain (mō-rāl'yä)	129	40°38'N	0°07'W
Morelos, Mex. (mō-rā'lōs)	88	22°46'N	102°36'W
Morelos, Mex.	89a	19°41'N	99°29'W
Morelos, Mex.	80	28°24'N	100°51'W
Morelos, r., Mex.	80	25°27'N	99°35'W
Morena, Sierra, mtn., Ca., U.S. (syĕr'rä mō-rā'nä)	74b	37°24'N	122°19'W
Morena, Sierra, mts., Spain (syĕr'rä mō-rā'nä)	112	38°15'N	5°45'W
Morenci, Az., U.S. (mō-rĕn'sĭ)	77	33°05'N	109°25'W
Morenci, Mi., U.S.	66	41°50'N	84°50'W
Moreno, Arg. (mō-rĕ'nō)	102a	34°39'S	58°47'W
Moreno, Ca., U.S.	75a	33°55'N	117°09'W
Mores, i., Bah. (mōrz)	92	26°20'N	77°35'W
Moresby, i., Can. (mōrz'bĭ)	74d	48°43'N	123°15'W
Moresby Island, i., Can.	52	52°50'N	131°55'W
Moreton, i., Austl. (môr'tŭn)	176	26°53'S	152°42'E
Moreton Bay, b., Austl. (môr'tŭn)	176	27°12'S	153°10'E

ng-sing; ŋ-baŋk; n-nasalized n; nŏd; cŏmmit; ōld; ôbey; ôrder; oi-boil; fōōd; ŏŏ-as oo in foot; ou-out; s-soft; sh-dish; th-thin; pūre; ûnite; ûrn; stŭd; circŭs; ü-as in French tu; '-indeterminate vowel.

278 Morewood - Muqui PRONOUNCING INDEX

PLACE (Pronunciation)	PAGE	LAT.	LONG.
Morewood, Can. (mōr′wo͝od)	62c	45°11′N	75°17′W
Morgan, Mt., U.S. (môr′gǎn)	73	48°55′N	107°56′W
Morgan, Ut., U.S.	73	41°04′N	111°42′W
Morgan City, La., U.S.	81	29°41′N	91°11′W
Morganfield, Ky., U.S. (môr′gǎn-fēld)	66	37°40′N	87°55′W
Morgan's Bay, S. Afr.	187c	32°42′S	28°19′E
Morganton, N.C., U.S. (môr′gǎn-tǔn)	83	35°44′N	81°42′W
Morgantown, W.V., U.S. (môr′gǎn-toun)	67	39°40′N	79°55′W
Morga Range, mts., Afg.	155a	34°02′N	70°38′E
Morgenzon, S. Afr. (môr′gänt-sŏn)	192c	26°44′S	29°39′E
Moriac, Austl.	173a	38°15′S	144°20′E
Morice Lake, l., Can.	54	54°00′N	127°37′W
Moriguchi, Japan (mō′rē-go͞o′chē)	167b	34°44′N	135°34′E
Morinville, Can. (mō′rĭn-vĭl)	62g	53°48′N	113°39′W
Morioka, Japan (mō′rē-ō′kà)	161	39°40′N	141°21′E
Morkoka, r., Russia (môr-kô′kà)	141	65°35′N	111°00′E
Morlaix, Fr. (mŏr-lĕ′)	117	48°36′N	3°48′W
Morley, Can. (mŏr′lē)	62e	51°10′N	114°51′W
Mormant, Fr.	127b	48°35′N	2°54′E
Morne Gimie, St. Luc. (môrn′ zhē-mē′)	91b	13°53′N	61°03′W
Mornington, Austl.	173a	38°13′S	145°02′E
Morobe, Pap. N. Gui.	169	8°03′S	147°45′E
Morocco, nation, Afr. (mô-rŏk′ō)	184	32°00′N	7°00′W
Morogoro, Tan. (mō-rō-gō′rō)	187	6°49′S	37°40′E
Moroleón, Mex. (mô-rō-lā-ōn′)	88	20°07′N	101°15′W
Morombe, Madag. (mo͞o-ro͞om′bā)	187	21°39′S	43°34′E
Morón, Arg. (mo-rō′n)	99c	34°39′S	58°37′W
Morón, Cuba (mô-rōn′)	92	22°05′N	78°35′W
Morón, Ven. (mô-rō′n)	101b	10°29′N	68°11′W
Morondava, Madag. (mô-rōn-dä′vä)	187	20°17′S	44°18′E
Morón de la Frontera, Spain (mô-rōn′dā läf rôn-tā′rä)	128	37°08′N	5°20′W
Morongo Indian Reservation, I.R., Ca., U.S. (mō-rôŋ′gō)	76	33°54′N	116°47′W
Moroni, Com.	187	11°41′S	43°16′E
Moroni, Ut., U.S. (mō-rō′nī)	77	39°30′N	111°40′W
Morotai, i., Indon. (mō-rō-tä′ē)	169	2°12′N	128°30′E
Moroto, Ug.	191	2°32′N	34°39′E
Morozovsk, Russia	137	48°20′N	41°50′E
Morrill, Ne., U.S. (mŏr′ĭl)	70	41°59′N	103°54′W
Morrilton, Ar., U.S. (mŏr′ĭl-tǔn)	79	35°09′N	92°42′W
Morrinhos, Braz. (mô-rēn′yōzh)	101	17°45′S	48°56′W
Morris, Can. (mŏr′ĭs)	50	49°21′N	97°22′W
Morris, Il., U.S.	66	41°20′N	88°25′W
Morris, Mn., U.S.	70	45°35′N	95°53′W
Morris, r., Can.	57	49°30′N	97°30′W
Morrison, Il., U.S. (mŏr′ĭ-sǔn)	71	41°48′N	89°58′W
Morris Reservoir, res., Ca., U.S.	75a	34°11′N	117°49′W
Morristown, N.J., U.S. (mŏr′rĭs-toun)	68a	40°48′N	74°29′W
Morristown, Tn., U.S.	82	36°10′N	83°18′W
Morrisville, Pa., U.S. (mŏr′ĭs-vĭl)	68f	40°12′N	74°46′W
Morro do Chapéu, Braz. (môr-ò dò-shä-pĕ′o͞o)	101	11°34′S	41°03′W
Morrow, Oh., U.S. (mŏr′ō)	69f	39°21′N	84°07′W
Mors, i., Den.	122	56°46′N	8°38′E
Morshansk, Russia (môr-shänsk′)	136	53°25′N	41°35′E
Mortara, Italy (môr-tä′rä)	130	45°13′N	8°47′E
Morteros, Arg. (môr-tĕ′tôs)	102	30°47′S	62°00′W
Mortes, Rio das, r., Braz. (rēō-däs-mô′r-tĕs)	99a	21°04′S	44°29′W
Morton Indian Reservation, I.R., Mn., U.S. (mŏr′tǔn)	71	44°35′N	94°48′W
Mortsel, Bel. (môr-sĕl′)	115a	51°10′N	4°28′E
Morvan, mts., Fr. (mōr-väɴ′)	126	47°11′N	4°10′E
Morzhovets, i., Russia (mŏr′zhô-vyĕts′)	136	66°40′N	42°30′E
Mosal'sk, Russia (mō-zàlsk′)	132	54°27′N	34°57′E
Moscavide, Port.	129b	38°47′N	9°06′W
Moscow (Moskva), Russia	134	55°45′N	37°37′E
Moscow, Id., U.S. (mŏs′kō)	64	46°44′N	116°57′W
Mosel (Moselle), r., Eur. (mō′sĕl) (mô-zĕl)	124	49°49′N	7°00′E
Moses, r., S. Afr.	192c	25°17′S	29°04′E
Moses Lake, Wa., U.S.	72	47°08′N	119°15′W
Moses Lake, l., Wa., U.S. (mō′zĕz)	72	47°09′N	119°30′W
Moshchnyy, is., Russia (môsh′chnĭ)	123	59°56′N	28°07′E
Moshi, Tan. (mō′shē)	187	3°21′S	37°20′E
Mosjøen, Nor.	116	65°50′N	13°10′E
Moskva *see* Moscow, Russia	134	55°45′N	37°37′E
Moskva, prov., Russia	132	55°38′N	36°48′E
Moskva, r., Russia	136	55°30′N	37°05′E
Mosonmagyaróvár, Hung.	125	47°51′N	17°16′E
Mosquitos, Costa de, cst., Nic. (kŏs-tä-dĕ-mōs-kē′tō)	91	12°05′N	83°49′W
Mosquitos, Gulfo de los, b., Pan. (go͞o′l-fô-dĕ-lôs-mōs-kē′tôs)	87	9°17′N	80°59′W
Moss, Nor. (môs)	116	59°29′N	10°39′E
Moss Beach, Ca., U.S. (môs bēch)	74b	37°32′N	122°31′W
Mosselbaai, S. Afr. (mō′sǔl bā)	186	34°06′S	22°23′E
Mossendjo, Congo	190	2°57′S	12°44′E
Mossley, Eng., U.K. (môs′lĭ)	114a	53°31′N	2°02′W
Moss Point, Ms., U.S. (môs)	82	30°25′N	88°32′W
Most, Czech Rep. (môst)	124	50°32′N	13°37′E
Mostar, Bos. (môs′tär)	119	43°20′N	17°51′E
Móstoles, Spain (mōs-tō′läs)	129a	40°19′N	3°52′W
Mostoos Hills, hills, Can. (mŏs′to͞os)	56	54°50′N	108°45′W
Mosvatnet, l., Nor.	122	59°55′N	7°50′E
Motagua, r., N.A. (mô-tä′gwä)	90	15°29′N	88°39′W
Motala, Swe. (mô-tô′lä)	122	58°34′N	15°00′E
Motherwell, Scot., U.K. (mŭdh′ẽr-wĕl)	116	55°45′N	4°05′W
Motril, Spain (mô-trēl′)	118	36°44′N	3°32′W
Motul, Mex. (mō-to͞o′l)	90a	21°07′N	89°14′W
Mouaskar, Alg.	184	35°25′N	0°08′E
Mouchoir Bank, bk., (mo͞o-shwàr′)	93	21°35′N	70°40′W
Mouchoir Passage, strt., T./C. Is.	93	21°05′N	71°05′W
Moudjéria, Maur.	188	17°53′N	12°20′W
Mouila, Gabon	190	1°52′S	11°01′E
Mouille Point, c., S. Afr.	186a	33°54′S	18°19′E
Moulins, Fr. (mo͞o-lăɴ′)	117	46°34′N	3°19′E
Moulouya, Oued, r., Mor. (mo͞o-lo͞o′yà)	184	34°00′N	4°00′W
Moultrie, Ga., U.S. (mōl′trĭ)	82	31°10′N	83°48′W
Moultrie, Lake, l., S.C., U.S.	83	33°12′N	80°00′W
Mound City, Il., U.S.	79	37°06′N	89°13′W
Mound City, Mo., U.S.	79	40°08′N	95°13′W
Moundou, Chad	189	8°34′N	16°05′E
Moundsville, W.V., U.S. (moundz′vĭl)	66	39°50′N	80°50′W
Mount, Cape, c., Lib.	188	6°47′N	11°20′W
Mountain Brook, Al., U.S. (moun′tĭn bro͝ok)	68h	33°30′N	86°45′W
Mountain Creek Lake, l., Tx., U.S.	75c	32°43′N	97°03′W
Mountain Grove, Mo., U.S. (grōv)	79	37°07′N	92°16′W
Mountain Home, Id., U.S. (hōm)	72	43°08′N	115°43′W
Mountain Park, Can. (pärk)	50	52°55′N	117°14′W
Mountain View, Ca., U.S. (moun′tĭn vū)	74b	37°25′N	122°07′W
Mountain View, Mo., U.S.	79	36°59′N	91°46′W
Mount Airy, N.C., U.S. (âr′ĭ)	83	36°28′N	80°37′W
Mount Ayliff, S. Afr. (ā′lĭf)	187c	30°48′S	29°24′E
Mount Ayr, Ia., U.S. (âr)	71	40°43′N	94°06′W
Mount Carmel, Il., U.S. (kär′mĕl)	66	38°25′N	87°45′W
Mount Carmel, Pa., U.S.	67	40°50′N	76°25′W
Mount Carooll, Il., U.S.	71	42°05′N	89°55′W
Mount Clemens, Mi., U.S. (klĕm′ĕnz)	69b	42°36′N	82°52′W
Mount Desert, i., Me., U.S. (dĕ-zûrt′)	60	44°15′N	68°08′W
Mount Dora, Fl., U.S. (dō′rà)	83a	28°45′N	81°38′W
Mount Duneed, Austl.	173a	38°15′S	144°20′E
Mount Eliza, Austl.	173a	38°11′S	145°05′E
Mount Fletcher, S. Afr. (flĕ′chẽr)	187c	30°42′S	28°32′E
Mount Forest, Can. (fŏr′ĕst)	59	44°00′N	80°45′W
Mount Frere, S. Afr. (frâr′)	187c	30°54′S	29°02′E
Mount Gambier, Austl. (găm′bēr)	174	37°30′S	140°53′E
Mount Gilead, Oh., U.S. (gĭl′ēăd)	66	40°30′N	82°50′W
Mount Healthy, Oh., U.S. (hĕlth′ē)	69f	39°14′N	84°32′W
Mount Holly, N.J., U.S. (hŏl′ĭ)	68f	39°59′N	74°47′W
Mount Hope, Can.	62d	43°09′N	79°55′W
Mount Hope, N.J., U.S. (hōp)	68a	40°55′N	74°32′W
Mount Hope, W.V., U.S.	66	37°55′N	81°10′W
Mount Isa, Austl. (ī′zà)	174	21°00′S	139°45′E
Mount Kisco, N.Y., U.S. (kĭs′kō)	68a	41°12′N	73°44′W
Mountlake Terrace, Wa., U.S. (mount lāk tĕr′ĭs)	74a	47°48′N	122°19′W
Mount Lebanon, Pa., U.S. (lĕb′á-nŭn)	69e	40°22′N	80°03′W
Mount Magnet, Austl. (măg-nĕt)	174	28°00′S	118°00′E
Mount Martha, Austl.	173a	38°17′S	145°01′E
Mount Morgan, Austl. (môr-gǎn)	175	23°42′S	150°45′E
Mount Moriac, Austl.	173a	38°13′S	144°12′E
Mount Morris, Mi., U.S. (mŏr′ĭs)	66	43°10′N	83°45′W
Mount Morris, N.Y., U.S.	67	42°45′N	77°50′W
Mount Nimba National Park, rec., C. Iv.	188	7°35′N	8°10′W
Mount Olive, N.C., U.S. (ŏl′ĭv)	83	35°11′N	78°05′W
Mount Peale, Ut., U.S.	77	38°26′N	109°16′W
Mount Pleasant, Ia., U.S. (plĕz′ănt)	71	40°59′N	91°34′W
Mount Pleasant, Mi., U.S.	66	43°35′N	84°45′W
Mount Pleasant, S.C., U.S.	83	32°46′N	79°51′W
Mount Pleasant, Tn., U.S.	82	35°31′N	87°12′W
Mount Pleasant, Tx., U.S.	81	33°10′N	94°56′W
Mount Pleasant, Ut., U.S.	77	39°35′N	111°20′W
Mount Prospect, Il., U.S. (prŏs′pĕkt)	69a	42°03′N	87°56′W
Mount Rainier National Park, rec., Wa., U.S. (rå-nēr′)	64	46°47′N	121°17′W
Mount Revelstoke National Park, rec., Can. (rĕv′ĕl-stōk)	50	51°22′N	120°15′W
Mount Savage, Md., U.S. (săv′ăj)	67	39°45′N	78°55′W
Mount Shasta, Ca., U.S. (shăs′tà)	72	41°18′N	122°17′W
Mount Sterling, Il., U.S. (stûr′lĭng)	79	39°59′N	90°44′W
Mount Sterling, Ky., U.S.	66	38°05′N	84°00′W
Mount Stewart, Can. (stū′ărt)	61	46°22′N	62°52′W
Mount Union, Pa., U.S. (ūn′yǔn)	67	40°25′N	77°50′W
Mount Vernon, Il., U.S. (vûr′nǔn)	66	38°20′N	88°50′W
Mount Vernon, In., U.S.	66	37°55′N	87°50′W
Mount Vernon, Mo., U.S.	79	37°09′N	93°48′W
Mount Vernon, N.Y., U.S.	68a	40°55′N	73°51′W
Mount Vernon, Oh., U.S.	66	40°25′N	82°30′W
Mount Vernon, Va., U.S.	68e	38°43′N	77°06′W
Mount Vernon, Wa., U.S.	72	48°25′N	122°20′W
Moura, Braz. (mō′rà)	101	1°33′S	61°38′W
Moura, Port.	128	38°08′N	7°28′W
Mourne Mountains, mts., N. Ire., U.K. (mōrn)	120	54°10′N	6°09′W
Moussoro, Chad	189	13°39′N	16°29′E
Moûtiers, Fr. (mo͞o-tyâr′)	127	45°31′N	6°34′E
Mowbullan, Mount, mtn., Austl.	176	26°50′S	151°34′E
Moyahua, Mex. (mô-yä′wä)	88	21°16′N	103°10′W
Moyale, Kenya (mô-yä′lä)	185	3°28′N	39°04′E
Moyamba, S.L. (mô-yäm′bä)	184	8°10′N	12°26′W
Moyen Atlas, mts., Mor.	118	32°49′N	5°28′W
Moyeuvre-Grande, Fr.	127	49°15′N	6°26′E
Moyie, r., Id., U.S. (moi′yē)	72	38°50′N	116°10′W
Moyobamba, Peru (mō-yō-bäm′bä)	100	6°12′S	76°56′W
Moyuta, Guat. (mô-ē-o͞o′tä)	90	14°01′N	90°05′W
Moyyero, r., Russia	140	67°15′N	104°10′E
Moyynqum, des., Kaz.	139	44°30′N	70°00′E
Mozambique, nation, Afr. (mō-zǎm-bēk′)	186	20°15′S	33°53′E
Mozambique Channel, strt., Afr. (mō-zǎm-bek′)	187	24°00′S	38°00′E
Mozdok, Russia (môz-dôk′)	137	43°45′N	44°35′E
Mozhaysk, Russia (mô-zhäysk′)	132	55°31′N	36°02′E
Mozhayskiy, Russia (mô-zhày′skĭ)	142c	59°42′N	30°08′E
Mpanda, Tan.	191	6°22′S	31°02′E
Mpika, Zam.	191	11°54′S	31°26′E
Mpimbe, Mwi.	191	15°18′S	35°04′E
Mporokoso, Zam. (′m-pō-rō-kō′sō)	186	9°23′S	30°05′E
Mpwapwa, Tan. (′m-pwä′pwä)	186	6°21′S	36°29′E
Mqanduli, S. Afr. (′m-kän′do͞o-lē)	187c	31°50′S	28°42′E
Mrągowo, Pol. (mräŋ′gô-vô)	125	53°52′N	21°18′E
M'Sila, Alg. (m′sē′lä)	184	35°47′N	4°34′E
Msta, r., Russia (m′stá′)	136	58°30′N	33°00′E
Mstsislaw, Bela.	132	54°01′N	31°42′E
Mtakataka, Mwi.	191	14°12′S	34°32′E
Mtamvuna, r., Afr.	187c	30°43′S	29°53′E
Mtata, r., S. Afr.	187c	31°48′S	29°03′E
Mtsensk, Russia (m′tsĕnsk)	136	53°17′N	36°33′E
Mtwara, Tan.	191	10°16′S	40°11′E
Muar, r., Malay.	153b	2°18′N	102°43′E
Mubende, Ug.	191	0°35′N	31°23′E
Mubi, Nig.	189	10°18′N	13°20′E
Mucacata, Moz.	191	13°20′S	39°59′E
Much, Ger. (mo͞oκ)	127c	50°54′N	7°24′E
Muchinga Mountains, mts., Zam.	191	12°40′S	30°50′E
Much Wenlock, Eng., U.K. (mŭch wĕn′lŏk)	114a	52°35′N	2°33′W
Muckalee Creek, r., Ga., U.S. (mŭk′ă lē)	82	31°55′N	84°10′W
Muckleshoot Indian Reservation, I.R., Wa., U.S. (mŭck′′l-sho͞ot)	74a	47°21′N	122°04′W
Mucubela, Moz.	191	16°55′S	37°52′E
Mud, l., Mi., U.S. (mŭd)	71	46°12′N	84°32′W
Mudan, r., China (mo͞o-dän)	164	45°30′N	129°40′E
Mudanjiang, China (mo͞o-dän-jyäŋ)	164	44°28′N	129°38′E
Muddy, r., Nv., U.S. (mŭd′ĭ)	77	36°56′N	114°42′W
Muddy Boggy Creek, r., Ok., U.S. (mud′ĭ bôg′ĭ)	79	34°42′N	96°11′W
Muddy Creek, r., Ut., U.S. (mŭd′ĭ)	77	38°45′N	111°10′W
Mudgee, Austl. (mŭ-jē)	176	32°47′S	149°10′E
Mudjatik, r., Can.	56	56°23′N	107°40′W
Mufulira, Zam.	191	12°33′S	28°14′E
Muğla, Tur. (mo͞og′lä)	154	37°10′N	28°20′E
Mühldorf, Ger. (mül-dôrf)	124	48°15′N	12°33′E
Mühlhausen, Ger. (mül′hou-zĕn)	124	51°13′N	10°25′E
Muhu, i., Est. (mo͞o′ho͞o)	123	58°41′N	22°55′E
Muir Woods National Monument, rec., Ca., U.S. (mūr)	76	37°54′N	123°22′W
Muizenberg, S. Afr. (mwĭz-ĕn-bûrg′)	186a	34°07′S	18°28′E
Mukacheve, Ukr.	125	48°25′N	22°43′E
Mukden *see* Shenyang, China	161	41°45′N	123°22′E
Mukhtuya, Russia (mòk-to͞o′yà)	135	61°00′N	113°00′E
Mukilteo, Wa., U.S. (mū-kĭl-tā′ō)	74a	47°57′N	122°18′W
Muko, Japan (mo͞o′kō)	167b	34°57′N	135°43′E
Muko, r., Japan (mo͞o′kō)	167b	34°52′N	135°17′E
Mukutawa, r., Can.	57	53°10′N	97°28′W
Mukwonago, Wi., U.S. (mū-kwō-ná′gō)	69a	42°52′N	88°19′W
Mula, Spain (mo͞o′lä)	128	38°05′N	1°12′W
Mula, Al., U.S. (mŭl′gȧ)	68h	33°33′N	86°59′W
Mulde, r., Ger. (mòl′dĕ)	124	50°30′N	12°30′E
Muleros, Mex. (mo͞o-lā′rōs)	88	23°44′N	104°00′W
Muleshoe, Tx., U.S. (mūl′shō)	78	34°13′N	102°43′W
Mulgrave, Can. (mŭl′grāv)	61	45°37′N	61°23′W
Mulhacén, mtn., Spain	118	37°04′N	3°18′W
Mülheim, Ger. (mül′hīm)	127c	51°25′N	6°53′E
Mulhouse, Fr. (mü-lo͞oz′)	117	47°46′N	7°20′E
Muling, China (mo͞o-lĭŋ)	164	44°32′N	130°18′E
Muling, r., China	164	44°40′N	130°30′E
Mull, Island of, i., Scot., U.K. (mŭl)	120	56°40′N	6°19′W
Mullan, Id., U.S. (mŭl′ǎn)	72	47°26′N	115°50′W
Müller, Pegunungan, mts., Indon. (mül′ẽr)	168	0°22′N	113°05′E
Mullingar, Ire. (mŭl-ĭn-gär)	120	53°31′N	7°26′W
Mullins, S.C., U.S. (mŭl′ĭnz)	83	34°11′N	79°13′W
Mullins River, Belize	90a	17°08′N	88°18′W
Multān, Pak. (mò-tän′)	155	30°17′N	71°13′E
Multnomah Channel, strt., Or., U.S. (mŭl nō mà)	74c	45°41′N	122°53′W
Mulumbe, Monts, mts., D.R.C.	191	8°47′S	27°20′E
Mulvane, Ks., U.S. (mŭl-vān′)	79	37°30′N	97°13′W
Mumbai (Bombay), India	155	18°58′N	72°50′E
Mumbwa, Zam. (mòm′bwä)	186	14°59′S	27°04′E
Mumias, Kenya	191	0°20′N	34°29′E
Muna, Mex. (mo͞o′nà)	90a	20°28′N	89°42′W
München *see* Munich, Ger.	110	48°08′N	11°35′E
Muncie, In., U.S. (mŭn′sĭ)	65	40°10′N	85°30′W
Mundelein, Il., U.S. (mŭn-dĕ-lĭn′)	69a	42°16′N	88°00′W
Mundonueva, Pico de, mtn., Col. (pē′kō-dĕ-mo͞o′n-dō-nwĕ′vä)	100a	4°18′N	74°12′W
Muneco, Cerro, mtn., Mex. (sĕ′r-rō-mo͞o-nĕ′kō)	89a	19°13′N	99°20′W
Mungana, Austl. (mŭn-găn′ȧ)	175	17°15′S	144°18′E
Mungbere, D.R.C.	191	2°38′N	28°30′E
Munger, Mn., U.S. (mŭn′gẽr)	75h	46°48′N	92°20′W
Mungindi, Austl. (mŭn-gĭn′dē)	175	29°00′S	148°45′E
Munhall, Pa., U.S. (mŭn′hôl)	69e	40°24′N	79°53′W
Munhango, Ang. (mòn-häŋ′gà)	186	12°15′S	18°55′E
Munich, Ger.	110	48°08′N	11°35′E
Munising, Mi., U.S. (mū′nĭ-sĭng)	71	46°24′N	86°41′W
Muniz Freire, Braz.	99a	20°29′S	41°25′W
Munku Sardyk, mtn., Asia (mòn′kò sär-dĭk′)	135	51°45′N	100°30′E
Muñoz, Phil. (mo͞on-nyōth′)	169a	15°44′N	120°53′E
Münster, Ger. (mün′stẽr)	117	51°57′N	7°38′E
Munster, In., U.S. (mŭn′stẽr)	69a	41°34′N	87°31′W
Munster, hist. reg., Ire. (mŭn-stẽr)	120	52°30′N	9°24′W
Muntok, Indon. (mòn-tŏk′)	168	2°05′S	105°11′E
Muong Sing, Laos (mo͞o′ông-sĭng′)	168	21°06′N	101°17′E
Muping, China (mo͞o-pĭŋ)	162	37°23′N	121°36′E
Muqui, Braz. (mo͞o-kòē)	99a	20°56′S	41°20′W

ăt; finăl; rāte; senåte; ärm; åsk; sofȧ; fâre; ch-choose; dh-as th in other; bē; ēvent; bĕt; recĕnt; cratẽr; g-gō; gh-guttural g; bĭt; ĭ-short neutral; rīde; κ-guttural k as ch in German ich;

PLACE (Pronunciation)	PAGE	LAT.	LONG.
Mur, r., Eur. (mo͞or)	117	47°00'N	15°00'E
Muradiye, Tur. (mo͞o-rä'dê-yĕ)	137	39°00'N	43°40'E
Murat, Fr. (mü-rä')	126	45°05'N	2°56'E
Murat, r., Tur. (mo͞o-rät')	154	39°00'N	42°00'E
Murchison, r., Austl. (mûr'chĭ-sŭn)	174	26°45'S	116°15'E
Murcia, Spain (mo͞or'thyä)	110	38°00'N	1°10'W
Murcia, hist. reg., Spain	128	38°35'N	1°51'W
Murdo, S.D., U.S. (mûr'dô)	70	43°53'N	100°42'W
Mureş, r., Rom. (mo͞o'rĕsh)	119	46°02'N	21°50'E
Muret, Fr. (mü-rĕ')	126	43°28'N	1°17'E
Murfreesboro, Tn., U.S. (mûr'frēz-bŭr-ô)	82	35°50'N	86°19'W
Murgab, Taj.	139	38°10'N	73°59'E
Murgab, r., Asia (mo͞or-gäb')	154	37°07'N	62°32'E
Muriaé, r., Braz.	99a	21°20'S	41°40'W
Murino, Russia (mo͞o'rĭ-nô)	142c	60°03'N	30°28'E
Müritz, l., Ger. (mür'its)	124	53°20'N	12°33'E
Murmansk, Russia (mo͞or-mänsk')	134	69°00'N	33°20'E
Murom, Russia (mo͞o'rôm)	134	55°30'N	42°00'W
Muroran, Japan (mo͞o'rô-rän)	161	42°21'N	141°05'E
Muros, Spain (mo͞o'rōs)	128	42°48'N	9°00'W
Muroto-Zaki, c., Japan (mo͞o'rô-tō zä'kê)	166	33°14'N	134°12'E
Murphy, Mo., U.S. (mŭr'fĭ)	75e	38°29'N	90°29'W
Murphy, N.C., U.S.	82	35°05'N	84°00'W
Murphysboro, Il., U.S. (mŭr'fĭz-bŭr-ô)	79	37°46'N	89°21'W
Murray, Ky., U.S. (mŭr'ĭ)	82	36°39'N	88°17'W
Murray, Ut., U.S.	75b	40°40'N	111°53'W
Murray, r., Austl.	174	34°20'S	140°00'E
Murray, r., Can.	55	55°00'N	121°00'W
Murray, Lake, res., S.C., U.S. (mŭr'ĭ)	83	34°07'N	81°18'W
Murray Bridge, Austl.	174	35°10'S	139°35'E
Murray Harbour, Can.	61	46°00'N	62°31'W
Murray Region, reg., Austl. (mŭ'rē)	175	33°20'S	142°30'E
Murrumbidgee, r., Austl. (mŭr-ŭm-bĭd'jē)	175	34°30'S	145°20'E
Murrupula, Moz.	191	15°27'S	38°47'E
Murshidābād, India (mòr'shē-dä-bäd')	158	24°08'N	88°11'E
Murska Sobota, Slvn. (mo͞or'skä sô'bô-tä)	130	46°40'N	16°14'E
Muruasigar, mtn., Kenya	191	3°08'N	35°02'E
Murwāra, India	155	23°54'N	80°23'E
Murwillumbah, Austl. (mŭr-wĭl'lŭm-bŭ)	176	28°15'S	153°30'E
Mürz, r., Aus. (mürts)	124	47°30'N	15°21'E
Mürzzuschlag, Aus. (mürts'tso͞o-shlägh)	124	47°37'N	15°41'E
Mus, Tur. (mo͞osh)	137	38°55'N	41°30'E
Musala, mtn., Blg.	131	42°05'N	23°24'E
Musan, Kor., N. (mo͞o'sän)	161	41°11'N	129°10'E
Musashino, Japan (mo͞o-sä'shē-nō)	167a	35°43'N	139°35'E
Muscat, Oman (mŭs-kăt')	154	23°23'N	58°30'E
Muscat and Oman *see* Oman, nation, Asia	154	20°00'N	57°45'E
Muscatine, Ia., U.S. (mŭs-kȧ-tēn)	71	41°26'N	91°00'W
Muscle Shoals, Al., U.S. (mŭs''l shōlz)	82	34°44'N	87°38'W
Musgrave Ranges, mts., Austl. (mŭs'grāv)	174	26°15'S	131°15'E
Mushie, D.R.C. (mŭsh'ē)	186	3°04'S	16°50'E
Mushin, Nig.	189	6°32'N	3°22'E
Musi, r., Indon. (mo͞o'sē)	168	2°40'S	103°42'E
Musinga, Alto, mtn., Col. (ä'l-tô-mo͞o-sē'n-gä)	100a	6°40'N	76°13'W
Muskego Lake, l., Wi., U.S. (mŭs-kē'gō)	69a	42°53'N	88°10'W
Muskegon, Mi., U.S. (mŭs-kē'gŭn)	65	43°15'N	86°20'W
Muskegon, r., Mi., U.S.	66	43°20'N	85°55'W
Muskegon Heights, Mi., U.S.	66	43°10'N	86°20'W
Muskingum, r., Oh., U.S. (mŭs-kĭŋ'gŭm)	66	39°45'N	81°55'W
Muskogee, Ok., U.S. (mŭs-kō'gê)	65	35°44'N	95°21'W
Muskoka, l., Can. (mŭs-kō'kȧ)	59	45°00'N	79°30'W
Musoma, Tan.	191	1°30'S	33°48'E
Mussau Island, i., Pap. N. Gui. (mo͞o-sä'o͞o)	169	1°30'S	149°32'E
Musselshell, r., Mt., U.S. (mŭs''l-shĕl)	73	46°25'N	108°20'W
Mussende, Ang.	190	10°32'S	16°05'E
Mussuma, Ang.	190	14°14'S	21°59'E
Mustafakemalpaşa, Tur.	119	40°05'N	28°30'E
Mustang Bayou, Tx., U.S.	81a	29°22'N	95°12'W
Mustang Creek, r., Tx., U.S. (mŭs'tăng)	78	36°22'N	102°46'W
Mustang Island, i., Tx., U.S.	81	27°43'N	97°00'W
Mustique, i., St. Vin. (mŭs-tēk')	91b	12°53'N	61°03'W
Mustvee, Est. (mo͞ost'vĕ-ĕ)	123	58°50'N	26°54'E
Musu Dan, c., Kor., N. (mo͞o'so͞o dän)	161	40°51'N	130°00'E
Muswellbrook, Austl. (mŭs'wŭnl-brók)	176	32°15'S	150°50'E
Mutare, Zimb.	186	18°49'S	32°39'E
Mutombo Mukulu, D.R.C. (mo͞o-tôm'bō mo͞o-ko͞o'lo͞o)	186	8°12'S	23°56'E
Mutsu Wan, b., Japan (mo͞ot'so͞o wän)	166	41°20'N	140°55'E
Mutton Bay, Can. (mŭt''n)	61	50°48'N	59°02'W
Mutum, Braz. (mo͞o-to͞o'm)	99a	19°48'S	41°24'W
Muzaffargarh, Pak.	158	30°09'N	71°15'E
Muzaffarpur, India	158	26°13'N	85°20'E
Muzon, Cape, c., Ak., U.S.	54	54°41'N	132°44'W
Muzquiz, Mex. (mo͞oz'kēz)	80	27°53'N	101°31'W
Muztagata, mtn., China	160	38°20'N	75°28'E
Mvomero, Tan.	191	6°20'S	37°25'E
Mvoti, r., S. Afr.	187c	29°18'S	30°52'E
Mwali, i., Com.	187	12°15'S	43°45'E
Mwanza, Tan. (mwän'zä)	186	2°31'S	32°54'E
Mwaya, Tan. (mwä'yä)	186	9°19'S	33°51'E
Mwenga, D.R.C.	191	3°02'S	28°26'E
Mweru, l., Afr.	186	8°50'S	28°50'E
Mwingi, Kenya	191	0°56'S	38°04'E
Myanmar (Burma), nation, Asia	150	21°00'N	95°15'E
Myingyan, Mya. (myĭng-yŭn')	155	21°37'N	95°26'E
Myitkyina, Mya. (myĭ'chē-nȧ)	155	25°33'N	97°25'E
Myjava, Slvk. (mûê'yȧ-vä)	125	48°45'N	17°33'E
Mykhailivka, Ukr.	133	47°16'N	35°12'E
Mykolaïv, Ukr.	134	46°58'N	32°02'E
Mykolaïv, prov., Ukr.	133	47°27'N	31°25'E
Mýkonos, i., Grc.	131	37°26'N	25°30'E
Mymensingh, Bngl.	155	24°48'N	90°28'E
Mynämäki, Fin.	123	60°41'N	21°58'E
Myohyang San, mtn., Kor., N. (myō'hyang)	166	40°00'N	126°12'E
Mýrdalsjökull, ice, Ice. (mür'däls-yû'kôl)	116	63°34'N	18°04'W
Myrhorod, Ukr.	137	49°56'N	33°36'E
Mýrina, Grc.	131	39°52'N	25°01'E
Myrtle Beach, S.C., U.S. (mûr't'l)	83	33°42'N	78°53'W
Myrtle Point, Or., U.S.	72	43°04'N	124°08'W
Mysen, Nor.	122	59°32'N	11°16'E
Myshikino, Russia (mêsh'kê-nô)	132	57°48'N	38°21'E
Mysore, India (mī-sōr')	155	12°31'N	76°42'E
Mysovka, Russia (mê' sôf-kȧ)	123	55°11'N	21°17'E
Mystic, Ia., U.S. (mĭs'tĭk)	71	40°47'N	92°54'W
Mytilíni, Grc.	119	39°09'N	26°35'E
Mytishchi, Russia (mê-tēsh'chi)	142b	55°55'N	37°46'E
Mziha, Tan.	191	5°54'S	37°47'E
Mzimba, Mwi. ('m-zĭm'bä)	186	11°52'S	33°34'E
Mzimkulu, r., Afr.	187c	30°12'S	29°57'E
Mzimvubu, r., S. Afr.	187c	31°22'S	29°20'E
Mzuzu, Mwi.	191	11°30'S	34°10'E

N

PLACE (Pronunciation)	PAGE	LAT.	LONG.
Naab, r., Ger. (näp)	124	49°38'N	12°15'E
Naaldwijk, Neth.	115a	52°00'N	4°11'E
Nā'ālehu, Hi., U.S.	84a	19°00'N	155°35'W
Naantali, Fin. (nän'tȧ-lē)	123	60°29'N	22°03'E
Nabberu, l., Austl. (năb'ēr-o͞o)	174	26°05'S	120°35'E
Naberezhnyye Chelny, Russia	134	55°42'N	52°19'E
Nabeul, Tun. (nā-bûl')	184	36°34'N	10°45'E
Nabiswera, Ug.	191	1°28'N	32°16'E
Naboomspruit, S. Afr.	192c	24°32'S	28°43'E
Nābulus, W.B.	153a	32°13'N	35°16'E
Nacala, Moz. (nă-kȧ'lä)	187	14°34'S	40°41'E
Nacaome, Hond. (nä-kä-ō'mȧ)	90	13°32'N	87°28'W
Na Cham, Viet. (nä chäm')	165	22°02'N	106°30'E
Naches, r., Wa., U.S. (năch'ĕz)	72	46°51'N	121°03'W
Náchod, Czech Rep. (näk'ôt)	124	50°25'N	16°08'E
Nacimiento, Lake, res., Ca., U.S. (nȧ-sĭ-myĕn'tō)	76	35°50'N	121°00'W
Nacogdoches, Tx., U.S. (năk'ô-dō'chĕz)	81	31°36'N	94°40'W
Nadadores, Mex. (nä-dä-dō'räs)	80	27°04'N	101°36'W
Nadiād, India	158	22°45'N	72°51'E
Nadir, V.I.U.S.	87c	18°19'N	64°53'W
Nădlac, Rom.	131	46°09'N	20°52'E
Nadvirna, Ukr.	125	48°37'N	24°35'E
Nadym, r., Russia (nä'dĭm)	140	64°30'N	72°48'E
Naestved, Den. (nĕst'vĭdh)	116	55°14'N	11°46'E
Nafada, Nig.	189	11°08'N	11°20'E
Nafishah, Egypt	192d	30°34'N	32°15'E
Náfplio, Grc.	131	37°33'N	22°46'E
Nafūd ad Daḩy, des., Sau. Ar.	154	22°15'N	44°15'E
Nag, Co, l., China	158	31°38'N	91°18'E
Naga, Phil. (nä'gä)	169	13°37'N	123°12'E
Naga, i., Japan	167	32°09'N	130°16'E
Nagahama, Japan (nä'gä-hä'mä)	167	33°32'N	132°29'E
Nagahama, Japan	167	35°23'N	136°16'E
Nagaland, India	155	25°47'N	94°15'E
Nagano, Japan (nä'gä-nô)	161	36°42'N	138°12'E
Nagaoka, Japan (nä'gȧ-ō'kȧ)	161	37°22'N	138°49'E
Nagaoka, Japan	167b	34°54'N	135°42'E
Nāgappattinam, India	155	10°48'N	79°51'E
Nagarote, Nic. (nä-gä-rō'tĕ)	90	12°17'N	86°35'W
Nagasaki, Japan (nä'gȧ-sä'kē)	161	32°48'N	129°53'E
Nāgaur, India	158	27°19'N	73°41'E
Nagaybakskiy, Russia (nȧ-gày-băk'skĭ)	142a	53°33'N	59°33'E
Nagcarlan, Phil. (näg-kär-län')	169a	14°07'N	121°24'E
Nāgercoil, India	159	8°15'N	77°29'E
Nagorno Karabakh, hist. reg., Azer. (nu-gôr'nŭ-kŭ-rŭ-bäk')	137	40°10'N	46°50'E
Nagoya, Japan	161	35°09'N	136°53'E
Nāgpur, India (näg'po͞or)	155	21°12'N	79°09'E
Nagua, Dom. Rep. (nä'gwä)	93	19°20'N	69°40'W
Nagykanizsa, Hung. (nôd'y'kô'nē-shô)	119	46°27'N	17°00'E
Nagykőrös, Hung. (nôd'y'kŭ-rŭsh)	125	47°02'N	19°46'E
Naha, Japan (nä'hä)	161	26°02'N	127°43'E
Nahanni National Park, rec., Can.	52	62°10'N	125°15'W
Nahant, Ma., U.S. (nȧ-hănt)	61a	42°26'N	70°55'W
Nahariyya, Isr.	153a	33°01'N	35°06'E
Nahuel Huapi, l., Arg. (nä'wĕl wä'pē)	102	41°00'S	71°30'W
Nahuizalco, El Sal. (nä-wē-zäl'kō)	90	13°50'N	89°43'W
Naic, Phil. (nä-ēk)	169a	14°20'N	120°46'E
Naica, Mex. (nä-ē'kä)	80	27°53'N	105°30'W
Naiguata, Pico, mtn., Ven. (pē'kô)	101b	10°32'N	66°44'W
Nain, Can. (nīn)	51	56°29'N	61°52'W
Nā'īn, Iran	157	32°52'N	53°05'E
Nairn, Scot., U.K. (nârn)	120	57°35'N	3°54'W
Nairobi, Kenya (nī-rō'bē)	186	1°17'S	36°49'E
Naivasha, Kenya (nī-vä'shȧ)	186	0°47'S	36°29'E
Najd, hist. reg., Sau. Ar.	154	25°18'N	42°38'E
Najin, Kor., N. (nä'jĭn)	161	42°04'N	130°35'E
Najran, des., Sau. Ar. (nŭj-rän')	154	17°29'N	45°30'E
Naju, Kor., S. (nä'jo͞o')	166	35°02'N	126°42'E
Najusa, r., Cuba (nä-ho͞o'sä)	92	20°55'N	77°55'W
Nakatsu, Japan (nä'käts-o͞o)	166	33°34'N	131°10'E
Nakhodka, Russia (nŭ-kôt'kŭ)	135	43°03'N	133°08'E
Nakhon Ratchasima, Thai.	168	14°56'N	102°14'E
Nakhon Sawan, Thai.	168	15°42'N	100°06'E
Nakhon Si Thammarat, Thai.	168	8°27'N	99°58'E
Nakło nad Notecia, Pol.	125	53°10'N	17°35'E
Nakskov, Den. (näk'skou)	116	54°51'N	11°06'E
Naktong, r., Kor., S. (näk'tŭng)	166	36°10'N	128°30'E
Nal'chik, Russia (nȧl-chēk')	137	43°30'N	43°35'E
Nalón, r., Spain (nä-lōn')	128	43°15'N	5°38'W
Nālūt, Libya (nä-lo͞ot')	184	31°51'N	10°49'E
Namak, Daryacheh-ye, l., Iran	154	34°58'N	51°33'E
Namakan, l., Mn., U.S. (nă'mȧ-kăn)	71	48°20'N	92°43'W
Namangan, Uzb. (nȧ-män-gän')	139	41°08'N	71°59'E
Namao, Can.	62g	53°43'N	113°30'W
Namatanai, Pap. N. Gui. (nä'mä-tȧ-nä'ē)	169	3°43'S	152°26'E
Nambour, Austl. (năm'bòr)	176	26°48'S	153°00'E
Nam Co, l., China (näm tswo)	160	30°30'N	91°10'E
Nam Dinh, Viet. (näm dēnĸ')	168	20°30'N	106°10'E
Nametil, Moz.	191	15°43'S	39°21'E
Namhae, i., Kor., S. (näm'hī')	166	34°23'N	128°05'E
Namib Desert, des., Nmb. (nä-mēb')	186	18°45'S	12°45'E
Namibia, nation, Afr.	186	19°30'S	16°13'E
Namoi, r., Austl. (nămoi)	175	30°10'S	148°43'E
Namous, Oued en, r., Alg. (nȧ-mo͞os')	118	31°48'N	0°19'W
Nampa, Id., U.S. (năm'pȧ)	64	43°35'N	116°35'W
Namp'o, Kor., N.	161	38°47'N	125°28'E
Nampuecha, Moz.	191	13°59'S	40°18'E
Nampula, Moz.	191	15°07'S	39°15'E
Namsos, Nor. (näm'sôs)	116	64°28'N	11°14'E
Namu, Can.	54	51°53'N	127°50'W
Namuli, Serra, mts., Moz.	191	15°05'S	37°05'E
Namur, Bel. (nȧ-mür')	117	50°29'N	4°55'E
Namutoni, Nmb. (nȧ-mo͞o-tō'nê)	186	18°45'S	17°00'E
Nan, r., Thai.	168	18°11'N	100°29'E
Nanacamilpa, Mex. (nä-nä-kä-mē'l-pä)	89a	19°30'N	98°33'W
Nanaimo, Can. (nȧ-nī'mō)	50	49°10'N	123°56'W
Nanam, Kor., N. (nä'nän')	166	41°38'N	129°37'E
Nanao, Japan (nä'nä-ō)	166	37°03'N	136°59'E
Nan'ao Dao, i., China (nän-ou dou)	165	23°30'N	117°30'E
Nanchang, China (nän'chäng')	161	28°38'N	115°48'E
Nanchangshan Dao, i., China (nän-chăŋ-shän dou)	162	37°56'N	120°42'E
Nancheng, China (nän-chŭŋ)	161	26°50'N	116°40'E
Nanchong, China (nän-chôŋ)	160	30°45'N	106°05'E
Nancy, Fr. (näɴ-sē')	117	48°42'N	6°11'E
Nancy Creek, r., Ga., U.S. (năn'cē)	68c	33°51'N	84°25'W
Nanda Devi, mtn., India (nän'dä dā'vē)	155	30°30'N	80°25'E
Nānded, India	158	19°13'N	77°21'E
Nandurbār, India	158	21°29'N	74°13'E
Nandyāl, India	159	15°54'N	78°09'E
Nanga Parbat, mtn., Pak.	158	35°20'N	74°35'E
Nangi, India	158a	22°30'N	88°14'E
Nangis, Fr. (näɴ-zhē')	127b	48°33'N	3°01'E
Nangong, China (nän-gôŋ)	164	37°22'N	115°22'E
Nangweshi, Zam.	190	16°26'S	23°17'E
Nanhuangcheng Dao, i., China (nän-hůäŋ-chŭŋ dou)	162	38°22'N	120°54'E
Nanhui, China	162	31°03'N	121°45'E
Nanjing, China (nän-jyĭŋ)	161	32°04'N	118°46'E
Nanjuma, r., China (nän-jyo͞o-mä)	162	39°37'N	115°45'E
Nanking *see* Nanjing, China	161	32°04'N	118°46'E
Nanle, China (nän-lŭ)	162	36°03'N	115°13'E
Nan Ling, mts., China	161	25°15'N	111°40'E
Nanliu, r., China (nän-lĭō)	165	22°00'N	109°18'E
Nannine, Austl. (nă-nēn')	174	25°50'S	118°30'E
Nanning, China (nän'nĭng')	160	22°56'N	108°10'E
Nanpan, r., China (nän-pän)	165	24°50'N	105°30'E
Nanping, China (nän-pĭŋ)	161	26°40'N	118°05'E
Nansei-shotō, is., Japan	161	27°30'N	127°00'E
Nansemond, Va., U.S. (năn'sē-mŭnd)	68g	36°46'N	76°32'W
Nantai Zan, mtn., Japan (nän-täē zän)	166	36°47'N	139°28'E
Nantes, Fr. (näɴt')	110	47°13'N	1°37'W
Nanteuil-le-Haudouin, Fr. (näɴ-tû-lĕ-ō-dwăɴ')	127b	49°08'N	2°49'E
Nanticoke, Pa., U.S. (năn'tĭ-kōk)	67	41°10'N	76°00'W
Nantong, China (nän-tôŋ)	162	32°02'N	120°51'E
Nantong, China	162	32°08'N	121°06'E
Nantucket, i., Ma., U.S. (năn-tŭk'ĕt)	65	41°15'N	70°05'W
Nantwich, Eng., U.K. (nănt'wĭch)	114a	53°04'N	2°31'W
Nanxiang, China (nän-shyäŋ)	162	31°17'N	121°17'E
Nanxiong, China (nän-shôŋ)	165	25°10'N	114°20'E
Nanyang, China	161	33°00'N	112°42'E
Nanyang Hu, l., China (nän-yäŋ ho͞o)	162	35°14'N	116°24'E
Nanyuan, China (nän-yůän)	164a	39°48'N	116°24'E
Naolinco, Mex. (nä-o-lēŋ'kō)	89	19°39'N	96°50'W
Náousa, Grc. (nä'o͞o-sȧ)	131	40°38'N	22°05'E
Naozhou Dao, i., China (nou-jō dou)	165	20°58'N	110°58'E
Napa, Ca., U.S. (năp'ȧ)	64	38°20'N	122°17'W
Napanee, Can. (năp'ȧ-nē)	59	44°15'N	77°00'W
Naperville, Il., U.S. (nā'pēr-vĭl)	69a	41°46'N	88°09'W
Napier, N.Z. (nā'pĭ-ēr)	175a	39°30'S	177°00'E
Napierville, Can. (nā'pĭ-ē-vĭl)	62a	45°11'N	73°24'W
Naples (Napoli), Italy	110	40°37'N	14°12'E
Naples, Fl., U.S. (nā'p'lz)	83a	26°07'N	81°46'W
Napo, r., S.A. (nä'pō)	100	1°49'S	74°20'W

PLACE (Pronunciation)	PAGE	LAT.	LONG.
Napoleon, Oh., U.S. (nȧ-pō'lē-ŭn)	66	41°20'N	84°10'W
Napoleonville, La., U.S. (nȧ-pō'lē-ŭn-vĭl)	81	29°56'N	91°03'W
Napoli see Naples, Italy	110	40°37'N	14°12'E
Napoli, Golfo di, b., Italy	118	40°29'N	14°08'E
Nappanee, In., U.S. (năp'ȧ-nē)	66	41°30'N	86°00'W
Nara, Japan (nä'rä)	161	34°41'N	135°50'E
Nara, Mali	184	15°09'N	7°27'W
Nara, dept., Japan	167b	34°36'N	135°49'E
Nara, r., Russia	132	55°05'N	37°16'E
Narach, Vozyera, l., Bela.	132	54°51'N	27°00'E
Naracoorte, Austl. (nȧ-rȧ-kōōn'tē)	174	36°50'S	140°50'E
Narashino, Japan	167a	35°41'N	140°01'E
Naraspur, India	159	16°32'N	81°43'E
Narberth, Pa., U.S. (när'bûrth)	68f	40°01'N	75°17'W
Narbonne, Fr. (nȧr-bôn')	117	43°12'N	3°00'E
Nare, Col. (nä'rē)	100a	6°12'N	74°37'W
Narew, r., Pol. (när'ĕf)	125	52°43'N	21°19'E
Narmada, r., India	155	22°30'N	75°30'E
Narodnaya, Gora, mtn., Russia (nȧ-rôd'nȧ-yȧ)	134	65°10'N	60°10'E
Naro-Fominsk, Russia (nä'rô-mēnsk')	136	55°23'N	36°43'E
Narrabeen, Austl. (năr-ȧ-bĭn)	173b	33°44'S	151°18'E
Narragansett, R.I., U.S. (năr-ă-găn'sĕt)	68b	41°26'N	71°27'W
Narragansett Bay, b., R.I., U.S.	67	41°20'N	71°15'W
Narrandera, Austl. (nȧ-rȧn-dē'rȧ)	175	34°40'S	146°40'E
Narrogin, Austl. (năr'ô-gĭn)	174	33°00'S	117°15'E
Narva, Est. (nȧr'vȧ)	136	59°24'N	28°12'E
Narvacan, Phil. (när-vä-kän')	169a	17°27'N	120°29'E
Narva Jõesuu, Est. (nȧr'vȧ ò-ô-ă'sōō-ò)	123	59°26'N	28°02'E
Narvik, Nor. (när'vêk)	110	68°21'N	17°18'E
Narvskiy Zaliv, b., Eur. (nȧr'vskĭ zä'lĭf)	123	59°35'N	27°25'E
Narvskoye, res., Eur.	123	59°18'N	28°14'E
Nar'yan-Mar, Russia (när'yȧn mär')	134	67°42'N	53°30'E
Naryilco, Austl. (năr-ĭl'kō)	176	28°40'S	141°50'E
Narym, Russia (nä-rēm')	134	58°47'N	82°05'E
Naryn, r., Asia (nŭ-rĭn')	140	41°20'N	76°00'E
Naseby, Eng., U.K. (nāz'bĭ)	114a	52°23'N	0°59'W
Nashua, Mo., U.S. (năsh'ū-ȧ)	75f	39°18'N	94°34'W
Nashua, N.H., U.S.	65	42°47'N	71°23'W
Nashville, Ar., U.S. (năsh'vĭl)	79	33°56'N	93°50'W
Nashville, Ga., U.S.	82	31°12'N	83°15'W
Nashville, Il., U.S.	79	38°21'N	89°42'W
Nashville, Mi., U.S.	66	42°35'N	85°50'W
Nashville, Tn., U.S.	65	36°10'N	86°48'W
Nashwauk, Mn., U.S. (năsh'wôk)	71	47°21'N	93°12'W
Näsi, l., Fin.	116	61°42'N	24°05'E
Našice, Cro. (nä'shē-tsĕ)	119	45°29'N	18°06'E
Nasielsk, Pol. (nä'syĕlsk)	125	52°35'N	20°50'E
Nāsik, India (nä'sĭk)	155	20°02'N	73°49'E
Nāşir, Sudan (nä-zēr')	185	8°30'N	33°06'E
Nasirabād, India	158	26°13'N	74°48'E
Naskaupi, r., Can. (năs'kô-pĭ)	53	53°59'N	61°10'W
Nasondoye, D.R.C.	191	10°22'S	25°06'E
Nass, r., Can. (năs)	54	55°00'N	129°30'W
Nassau, Bah. (năs'ô)	87	25°05'N	77°20'W
Nassenheide, Ger. (nä'sĕn-hī-dē)	115b	52°49'N	13°13'E
Nasser, Lake, res., Egypt	185	23°50'N	32°50'E
Nasugbu, Phil. (nȧ-sòg-bōō')	169a	14°05'N	120°37'E
Nasworthy Lake, l., Tx., U.S. (năz'wûr-thē)	80	31°17'N	100°30'W
Natagaima, Col. (nä-tä-gī'mä)	100a	3°38'N	75°07'W
Natal, Braz. (nä-täl')	101	6°00'S	35°13'W
Natashquan, Can. (nä-tăsh'kwän)	51	50°11'N	61°49'W
Natashquan, r., Can.	61	50°35'N	61°35'W
Natchez, Ms., U.S. (năch'ĕz)	65	31°35'N	91°20'W
Natchitoches, La., U.S. (năk'ĭ-tŏsh)(nȧch-ĭ-tōsh')	81	31°46'N	93°06'W
Natick, Ma., U.S. (nā'tĭk)	61a	42°17'N	71°21'W
National Bison Range, I.R., Mt., U.S. (năsh'ŭn-ăl bī's'n)	73	47°18'N	113°58'W
National City, Ca., U.S.	76a	32°38'N	117°01'W
Natitingou, Benin	184	10°19'N	1°22'E
Natividade, Braz. (nä-tē-vē-dä'dē)	101	11°43'S	47°34'W
Natron, Lake, l., Tan. (nā'trŏn)	186	2°17'S	36°10'E
Natrona Heights, Pa., U.S. (nā'trō nȧ)	69e	40°38'N	79°43'W
Naţrūn, Wādī an, val., Egypt	192b	30°33'N	30°12'E
Natuna Besar, i., Indon.	168	4°00'N	106°50'E
Natural Bridges National Monument, rec., Ut., U.S. (năt'ū-răl brĭj'ĕs)	77	37°20'N	110°20'W
Naturaliste, Cape, c., Austl. (năt-ū-rȧ-lĭst')	174	33°30'S	115°10'E
Nau, Cap de la, c., Spain	112	38°43'N	0°14'E
Naucalpan de Juárez, Mex.	89a	19°28'N	99°14'W
Nauchampatepetl, mtn., Mex. (näōō-chäm-pä-tē'pĕtl)	89	19°32'N	97°09'W
Nauen, Ger. (nou'ĕn)	115b	52°36'N	12°53'E
Naugatuck, Ct., U.S. (nô'gȧ-tŭk)	67	41°25'N	73°05'W
Naujan, Phil. (nä-ò-hän')	169a	13°19'N	121°17'E
Naumburg, Ger. (noum'bòrgh)	124	51°10'N	11°50'E
Nauru, nation, Oc.	3	0°30'S	167°00'E
Nautla, Mex. (nä-ōōt'lä)	86	20°14'N	96°44'W
Nava, Mex. (nä'vä)	80	28°25'N	100°44'W
Nava del Rey, Spain (nä-vä dĕl rā'ē)	128	41°22'N	5°04'W
Navahermosa, Spain (nä-vä-ĕr-mō'sä)	128	39°39'N	4°28'W
Navajas, Cuba (nä-vä-häs')	92	22°40'N	81°20'W
Navajo Hopi Joint Use Area, I.R., Az., U.S.	77	36°15'N	110°30'W
Navajo Indian Reservation, I.R., U.S. (năv'ȧ-hō)	77	36°31'N	109°24'W
Navajo National Monument, rec., Az., U.S.	77	36°43'N	110°39'W
Navajo Reservoir, res., N.M., U.S.	77	36°57'N	107°26'W
Navalcarnero, Spain (nä-väl'kär-nā'rō)	129a	40°17'N	4°05'W
Navalmoral de la Mata, Spain	128	39°53'N	5°32'W
Navan, Can. (nă'vȧn)	62c	45°25'N	75°26'W
Navarino, i., Chile (nȧ-vä-rē'nō)	102	55°30'S	68°15'W
Navarra, hist. reg., Spain (nä-vär'rä)	128	42°40'N	1°35'W
Navarro, Arg. (nä-vá'r-rō)	99c	35°00'S	59°16'W
Navasota, Tx., U.S. (năv-ȧ-sō'tȧ)	81	30°24'N	96°05'W
Navasota, r., Tx., U.S.	81	31°03'N	96°11'W
Navassa, i., N.A. (nȧ-văs'ȧ)	93	18°25'N	75°15'W
Navia, r., Spain (nä-vē'ä)	128	43°10'N	6°45'W
Navidad, Chile (nä-vē-dä'd)	99b	33°57'S	71°51'W
Navidad Bank, bk., (nä-vē-dädh')	93	20°05'N	69°00'W
Navidade do Carangola, Braz. (nä-vē-dä'dô-kä-rän-gô'la)	99a	21°04'S	41°58'W
Navojoa, Mex. (nä-vò-kō'ä)	86	27°00'N	109°40'W
Nawābshāh, Pak. (nȧ-wäb'shä)	158	26°20'N	68°30'E
Naxçıvan, Azer.	137	39°10'N	45°30'E
Naxçıvan Muxtar, state, Azer.	138	39°20'N	45°30'E
Náxos, i., Grc. (nák'sôs)	119	37°15'N	25°20'E
Nayarit, state, Mex. (nä-yä-rēt')	86	22°00'N	105°15'W
Nayarit, Sierra de, mts., Mex. (sē-ĕ'r-rä-dē)	88	23°20'N	105°07'W
Naye, Sen.	188	14°25'N	12°12'W
Naylor, Md., U.S. (nā'lŏr)	68e	38°43'N	76°46'W
Nazaré da Mata, Braz. (dä-mä-tä)	101	7°46'S	35°13'W
Nazas, Mex. (nä'zäs)	80	25°14'N	104°08'W
Nazas, r., Mex.	86	25°30'N	104°40'W
Nazerat, Isr.	153a	32°43'N	35°19'E
Nazilli, Tur. (nȧ-zĭ-lē')	137	37°40'N	28°10'E
Naziya, r., Russia (nȧ-zē'yȧ)	142c	59°48'N	31°18'E
Nazko, r., Can.	54	52°35'N	123°10'W
N'dalatando, Ang.	190	9°18'S	14°54'E
Ndali, Benin	189	9°51'N	2°43'E
Ndikiniméki, Cam.	189	4°46'N	10°50'E
N'Djamena, Chad	185	12°07'N	15°03'E
Ndola, Zam. (n'dō'lä)	186	12°58'S	28°38'E
Ndoto Mountains, mts., Kenya	191	1°55'N	37°05'E
Ndrhamcha, Sebkha de, l., Maur.	188	18°50'N	15°15'W
Nduye, D.R.C.	191	1°50'N	29°01'E
Neagh, Lough, l., N. Ire., U.K. (lŏk nā)	116	54°40'N	6°47'W
Néa Páfos, Cyp.	153a	34°46'N	32°27'E
Neapean, r., Austl.	173b	33°40'S	150°39'E
Neápoli, Grc.	131	36°35'N	23°08'E
Neápolis, Grc.	130a	35°17'N	25°37'E
Near Islands, is., Ak., U.S. (nēr)	63a	52°20'N	172°40'E
Neath, Wales, U.K. (nēth)	120	51°41'N	3°50'W
Nebine Creek, r., Austl. (nĕ-bēnē')	176	27°50'S	147°00'E
Nebitdag, Turkmen.	139	39°30'N	54°20'E
Nebraska, state, U.S. (nē-brăs'kȧ)	64	41°45'N	101°30'W
Nebraska City, Ne., U.S.	79	40°40'N	95°50'W
Nechako, r., Can.	54	53°45'N	124°55'W
Nechako Plateau, plat., Can. (nĭ-chă'kō)	54	54°00'N	124°30'W
Nechako Range, mts., Can.	54	53°20'N	124°30'W
Nechako Reservoir, res., Can.	54	53°25'N	125°10'W
Neches, r., Tx., U.S. (nĕch'ĕz)	81	31°03'N	94°40'W
Neckar, r., Ger. (nĕk'är)	124	49°16'N	9°06'E
Necker Island, i., Hi., U.S.	84b	24°00'N	164°00'W
Necochea, Arg. (nā-kô-chā'ä)	102	38°30'S	58°45'W
Nedryhailiv, Ukr.	133	50°49'N	33°52'E
Needham, Ma., U.S. (nēd'ăm)	61a	42°17'N	71°14'W
Needles, Ca., U.S. (nē'd'lz)	77	34°51'N	114°39'W
Neenah, Wi., U.S. (nē'nȧ)	71	44°10'N	88°30'W
Neepawa, Can.	50	50°13'N	99°29'W
Nee Reservoir, res., Co., U.S. (nee)	78	38°26'N	102°56'W
Negareyama, Japan (nå'gä-rä-yä'mä)	167a	35°52'N	139°54'E
Negaunee, Mi., U.S. (nē-gô'nē)	71	46°30'N	87°37'W
Negeri Sembilan, state, Malay. (nā'grē-sĕm-bē-län')	153b	2°46'N	101°54'E
Negev, des., Isr. (nĕ'gĕv)	153a	30°34'N	34°43'E
Negombo, Sri L.	159	7°39'N	79°49'E
Negotin, Serb. (nĕ'gō-tēn)	131	44°13'N	22°33'E
Negro, r., Arg.	102	39°50'S	65°00'W
Negro, r., N.A.	90	13°01'N	87°10'W
Negro, r., S.A.	99c	33°17'S	58°18'W
Negro, r., S.A. (nā'grō)	100	0°18'S	63°21'W
Negro, Cerro, mtn., Pan. (sĕ'-rrô-nā'grô)	91	8°44'N	80°37'W
Negros, i., Phil. (nā'grōs)	168	9°50'N	121°45'E
Nehalem, r., Or., U.S. (nê-hăl'ĕm)	72	45°52'N	123°37'W
Nehaus an der Oste, Ger. (noi'houz)(ōz'tē)	115c	53°48'N	9°02'E
Nehbandān, Iran	157	31°32'N	60°02'E
Nehe, China (nŭ-hŭ)	164	48°23'N	124°58'E
Neheim-Hüsten, Ger. (nĕ'hĭm)	127c	51°28'N	7°58'E
Neiba, Dom. Rep. (nå-ē'bä)	93	18°30'N	71°20'W
Neiba, Bahía de, b., Dom. Rep.	93	18°10'N	71°00'W
Neiba, Sierra de, mts., Dom. Rep. (sē-ĕr'rä-dē)	93	18°40'N	71°40'W
Neihart, Mt., U.S. (nī'hart)	73	46°54'N	110°39'W
Neijiang, China (nā-jyäŋ)	165	29°38'N	105°01'E
Neillsville, Wi., U.S. (nēlz'vĭl)	71	44°35'N	90°37'W
Nei Monggol (Inner Mongolia), prov., China	160	40°15'N	105°00'E
Neiqiu, China (nā-chyō)	162	37°17'N	114°32'E
Neira, Col. (nā'rä)	100a	5°10'N	75°32'W
Neisse, r., Eur. (nēs)	124	51°30'N	15°00'E
Neiva, Col. (nå-ē'vä)(nā'vä)	100	2°55'N	75°16'W
Neixiang, China (nā-shyäŋ)	164	33°00'N	111°38'E
Nekemte, Eth.	185	9°09'N	36°29'E
Nekoosa, Wi., U.S. (nê-kōō'sȧ)	71	44°19'N	89°54'W
Neligh, Ne., U.S. (nē'-lē)	70	42°06'N	98°02'W
Nel'kan, Russia (nĕl-kän')	135	57°45'N	136°36'E
Nellore, India (nĕl-lōr')	155	14°28'N	79°59'E
Nel'ma, Russia (nĕl-mä')	166	47°34'N	139°05'E
Nelson, Can. (nĕl'sŭn)	50	49°29'N	117°17'W
Nelson, N.Z.	175a	41°15'S	173°22'E
Nelson, Eng., U.K.	114a	53°50'N	2°13'W
Nelson, i., Ak., U.S.	63	60°38'N	164°42'W
Nelson, r., Can.	57	56°50'N	93°40'W
Nelson, Cape, c., Austl.	176	38°29'S	141°20'E
Nelsonville, Oh., U.S. (nĕl'sŭn-vĭl)	66	39°30'N	82°15'W
Néma, Maur. (nā'mä)	184	16°37'N	7°15'W
Nemadji, r., Wi., U.S. (nĕ-măd'jē)	75h	46°33'N	92°16'W
Neman, Russia (nĕ'-mȧn)	123	55°02'N	22°01'E
Neman, r., Eur.	136	53°28'N	24°45'E
Nembe, Nig.	189	4°35'N	6°26'E
Nemeiben Lake, l., Can. (nĕ-mē'bȧn)	56	55°20'N	105°20'W
Nemours, Fr.	126	48°16'N	2°41'E
Nemuro, Japan (nā'mò-rō)	161	43°13'N	145°10'E
Nemuro Strait, strt., Asia	166	43°07'N	145°10'E
Nemyriv, Ukr.	133	48°56'N	28°51'E
Nen, r., China (nŭn)	161	47°07'N	123°28'E
Nen, r., Eng., U.K. (nĕn)	114a	52°32'N	0°19'W
Nenagh, Ire. (nē'nȧ)	120	52°50'N	8°05'W
Nenana, Ak., U.S. (nå-nä'nȧ)	63	64°28'N	149°18'W
Nenikyul', Russia (nĕ-nyĕ'kyûl)	142c	59°26'N	30°40'E
Nenjiang, China (nŭn-jyäŋ)	161	49°02'N	125°15'E
Neodesha, Ks., U.S. (nē-ô-dĕ-shô')	79	37°24'N	95°41'W
Neosho, Mo., U.S.	79	36°51'N	94°22'W
Neosho, r., Ks., U.S. (nē-ō'shō)	79	38°07'N	95°40'W
Nepal, nation, Asia (nĕ-pôl')	155	28°45'N	83°00'E
Nephi, Ut., U.S. (nē'fī)	77	39°40'N	111°50'W
Nepomuceno, Braz. (nĕ-pô-mōō-sĕ'no)	99a	21°15'S	45°13'W
Nera, r., Italy (nā'rä)	130	42°45'N	12°54'E
Nérac, Fr. (nā-råk')	126	44°08'N	0°19'E
Nerchinsk, Russia (nyĕr'chĕnsk)	135	51°47'N	116°17'E
Nerchinskiy Khrebet, mts., Russia	135	50°30'N	118°30'E
Nerchinskiy Zavod, Russia (nyĕr'chĕn-skĭzȧ-vôt')	135	51°35'N	119°46'E
Nerekhta, Russia (nyĕ-rĕκ'tȧ)	132	57°29'N	40°34'E
Neretva, r., Serb. (nĕ'rĕt-vȧ)	131	43°08'N	17°50'E
Nerja, Spain (nĕr'hä)	128	36°45'N	3°53'W
Nerl', r., Russia (nyĕrl)	132	56°59'N	37°57'E
Nerskaya, r., Russia (nyĕr'skȧ-yȧ)	142b	55°31'N	38°46'E
Nerussa, r., Russia (nyå-rōō'sȧ)	132	52°24'N	34°20'E
Ness, Loch, l., Scot., U.K. (lŏκ nĕs)	120	57°23'N	4°20'W
Ness City, Ks., U.S. (nĕs)	78	38°27'N	99°55'W
Nesterov, Russia (nyĕs-tă'rôf)	123	54°39'N	22°38'E
Néstos (Mesta), r., Eur. (nās'tōs)	131	41°25'N	24°12'E
Netanya, Isr.	153a	32°19'N	34°52'E
Netcong, N.J., U.S. (nĕt'cŏnj)	68a	40°54'N	74°42'W
Netherlands, nation, Eur. (nĕdh'ēr-lăndz)	110	53°01'N	3°57'E
Netherlands Guiana see Suriname, nation, S.A.	101	4°00'N	56°00'W
Nettilling, l., Can.	53	66°30'N	70°40'W
Nett Lake Indian Reservation, I.R., Mn., U.S. (nĕt lāk)	71	48°23'N	93°19'W
Nettuno, Italy (nĕt-tōō'nô)	129d	41°28'N	12°40'E
Neubeckum, Ger. (noi'bĕ-kōōm)	127c	51°48'N	8°01'E
Neubrandenburg, Ger. (noi-brän'dĕn-bòrgh)	124	53°33'N	13°16'E
Neuburg, Ger. (noi'bòrgh)	124	48°43'N	11°12'E
Neuchâtel, Switz. (nû-shȧ-tĕl')	117	47°00'N	6°52'E
Neuchâtel, Lac de, l., Switz.	124	46°48'N	6°53'E
Neuenhagen, Ger. (noi'ĕn-hä-gĕn)	115b	52°31'N	13°41'E
Neuenrade, Ger. (noi'ĕn-rä-dĕ)	127c	51°17'N	7°47'E
Neufchâtel-en-Bray, Fr. (nû-shä-tĕl'ĕn-brā')	126	49°43'N	1°25'E
Neulengbach, Aus.	115e	48°13'N	15°55'E
Neumarkt, Ger. (noi'märkt)	124	49°17'N	11°30'E
Neumünster, Ger. (noi'münstēr)	116	54°04'N	10°00'E
Neunkirchen, Aus. (noin'kĭrκ-ĕn)	124	47°43'N	16°05'E
Neuquén, Arg. (nĕ-ò-kān')	102	38°52'S	68°12'W
Neuquén, prov., Arg.	102	39°40'S	70°45'W
Neuquén, r., Arg.	102	38°45'S	69°00'W
Neuruppin, Ger. (noi'rōō-pēn)	124	52°55'N	12°48'E
Neuse, r., N.C., U.S. (nūz)	83	36°12'N	78°50'W
Neusiedler See, l., Eur. (noi-zēd'lĕr)	124	47°54'N	16°31'E
Neuss, Ger. (nois)	127c	51°12'N	6°41'E
Neustadt, Ger. (noi'shtät)	124	49°21'N	8°08'E
Neustadt bei Coburg, Ger. (bī kō'bōōrgh)	124	50°20'N	11°09'E
Neustadt in Holstein, Ger.	124	54°06'N	10°50'E
Neustrelitz, Ger. (noi-strā'lĭts)	124	53°21'N	13°05'E
Neutral Hills, hills, Can. (nū'trăl)	56	52°10'N	110°50'W
Neu Ulm, Ger. (noi ò lm')	124	48°23'N	10°01'E
Neuville, Can. (nū'vĭl)	62b	46°39'N	71°35'W
Neuwied, Ger. (noi'vēdt)	124	50°26'N	7°28'E
Neva, r., Russia (nyĕ-vä')	132	59°49'N	30°54'E
Nevada, Ia., U.S. (nê-vä'dȧ)	71	42°01'N	93°27'W
Nevada, Mo., U.S.	79	37°49'N	94°21'W
Nevada, state, U.S. (nē vá'dȧ)	64	39°30'N	117°00'W
Nevada, Sierra, mts., Spain (syĕr'rä nä-vä'dhä)	112	37°01'N	3°28'W
Nevada, Sierra, mts., U.S. (sē-ĕ'r-rä nê-vä'dȧ)	64	39°20'N	120°05'W
Nevado, Cerro el, mtn., Col. (sĕ'r-rô-ĕl-nĕ-vä'dô)	100a	4°02'N	74°08'W
Neva Stantsiya, Russia (nyĕ-vä' stän'tsĭ-yȧ)	142c	59°53'N	30°30'E
Neve, Serra da, mts., Ang.	190	13°40'S	13°20'E
Nevel', Russia (nyĕ'vĕl)	136	56°03'N	29°57'E
Neveri, r., Ven. (nĕ-vĕ-rē)	101b	10°13'N	64°18'W
Nevers, Fr. (nĕ-vâr')	117	47°00'N	3°10'E
Neves, Braz.	102b	22°51'S	43°06'W
Nevesinje, Bos. (nĕ-vĕ'sēn-yĕ)	131	43°15'N	18°08'E
Nevinnomyssk, Russia	138	44°38'N	41°56'E
Nevis, i., St. K./N. (nē'vĭs)	87	17°05'N	62°38'W
Nevis, Ben, mtn., Scot., U.K. (bĕn)	116	56°47'N	5°00'W

ăt; finăl; rāte; senȧte; ärm; ȧsk; sofȧ; fâre; ch-choose; dh-as th in other; bē; êvent; bĕt; recĕnt; cratēr; g-gō; gh-guttural g; bĭt; ĭ-short neutral; rīde; κ-guttural k as ch in German ich;

PRONOUNCING INDEX

Nevis Peak – Nipissing 281

PLACE (Pronunciation)	PAGE	LAT.	LONG.
Nevis Peak, mtn., St. K./N.	91b	17°11'N	62°33'W
Nevşehir, Tur. (něv-shě'hěr)	119	38°40'N	34°35'E
Nev'yansk, Russia (něv-yänsk')	134	57°29'N	60°14'E
New, r., Va., U.S. (nū)	83	37°20'N	80°35'W
Newala, Tan.	191	10°56'S	39°18'E
New Albany, In., U.S. (nû ôl'bá-nĭ)	69h	38°17'N	85°49'W
New Albany, Ms., U.S.	83	34°28'N	39°00'W
New Amsterdam, Guy. (ăm'stēr-dăm)	101	6°14'N	57°30'W
Newark, Eng., U.K. (nū'ērk)	114a	53°04'N	0°49'W
Newark, Ca., U.S. (nū'ērk)	74b	37°32'N	122°02'W
Newark, De., U.S. (nōō'ärk)	67	39°40'N	75°45'W
Newark, N.J., U.S. (nōō'ûrk)	65	40°44'N	74°10'W
Newark, N.Y., U.S. (nū'ērk)	67	43°05'N	77°10'W
Newark, Oh., U.S.	66	40°05'N	82°25'W
Newaygo, Mi., U.S. (nū'wā-go)	66	43°25'N	85°50'W
New Bedford, Ma., U.S. (běd'fērd)	65	41°35'N	70°55'W
Newberg, Or., U.S. (nū'bûrg)	66	45°17'N	122°58'W
New Bern, N.C., U.S. (bûrn)	65	35°05'N	77°05'W
Newbern, Tn., U.S.	82	36°05'N	89°12'W
Newberry, Mi., U.S. (nū'běr-ĭ)	71	46°22'N	85°31'W
Newberry, S.C., U.S.	83	34°15'N	81°40'W
New Boston, Mi., U.S. (bôs'tŭn)	69b	42°10'N	83°24'W
New Boston, Oh., U.S.	66	38°45'N	82°55'W
New Braunfels, Tx., U.S. (nū broun'fěls)	80	29°43'N	98°07'W
New Brighton, Mn., U.S. (brī'tŭn)	75g	45°04'N	93°12'W
New Brighton, Pa., U.S.	69e	40°34'N	80°18'W
New Britain, Ct., U.S. (brĭt''n)	67	41°40'N	72°45'W
New Britain, i., Pap. N. Gui.	169	6°45'S	149°38'E
New Brunswick, N.J., U.S. (brŭnz'wĭk)	68a	40°29'N	74°27'W
New Brunswick, prov., Can.	51	47°14'N	66°30'W
Newburg, In., U.S.	66	38°00'N	87°25'W
Newburg, Mo., U.S.	79	37°54'N	91°53'W
Newburgh, N.Y., U.S.	67	41°30'N	74°00'W
Newburgh Heights, Oh., U.S.	69d	41°27'N	81°40'W
Newbury, Eng., U.K. (nū'běr-ĭ)	120	51°24'N	1°26'W
Newbury, Ma., U.S.	61a	42°48'N	70°52'W
Newbury, co., Eng., U.K.	114b	51°25'N	1°15'W
Newburyport, Ma., U.S. (nū'bēr-ĭ-pōrt)	61a	42°48'N	70°53'W
New Caledonia, dep., Oc.	175	21°28'S	164°40'E
New Canaan, Ct., U.S. (kā-nán)	68a	41°06'N	73°30'W
New Carlisle, Can. (kär-līl')	51	48°01'N	65°20'W
Newcastle, Austl. (nū-kás''l)	176	33°00'S	151°55'E
Newcastle, Can.	51	47°00'N	65°34'W
New Castle, De., U.S.	67	39°40'N	75°35'W
New Castle, In., U.S.	66	39°55'N	85°25'W
New Castle, Oh., U.S.	66	40°20'N	82°10'W
New Castle, Pa., U.S.	66	41°00'N	80°25'W
Newcastle, Tx., U.S.	78	33°13'N	98°44'W
Newcastle, Wy., U.S.	70	43°51'N	104°11'W
Newcastle under Lyme, Eng., U.K. (nů-kás''l) (nū-käs''l)	114a	53°01'N	2°14'W
Newcastle, Eng., U.K.	110	55°00'N	1°45'W
Newcastle Waters, Austl. (wô'tērz)	174	17°10'S	133°25'E
Newcomerstown, Oh., U.S. (nū'kŭm-ērz-toun)	66	40°15'N	81°40'W
New Croton Reservoir, res., N.Y., U.S. (krō'tŏn)	68a	41°15'N	73°47'W
New Delhi, India (děl'hī)	155	28°43'N	77°18'E
Newell, S.D., U.S. (nū'ěl)	70	44°43'N	103°26'W
New England Range, mts., Austl. (nū ĭŋ'glănd)	175	29°32'S	152°30'E
Newenham, Cape, c., Ak., U.S. (nū-ěn-hăm)	63	58°40'N	162°32'W
Newfane, N.Y., U.S. (nū-fān)	69c	43°17'N	78°44'W
Newfoundland, i., Can.	53a	48°30'N	56°00'W
Newfoundland and Labrador, prov., Can.	51	48°15'N	56°53'W
Newgate, Can. (nů'gāt)	55	49°01'N	115°10'W
New Georgia, i., Sol. Is. (jôr'jĭ-á)	175	8°08'S	158°00'E
New Georgia Group, is., Sol. Is.	170e	8°30'S	157°20'E
New Georgia Sound, strt., Sol. Is.	170e	8°00'S	158°10'E
New Glasgow, Can. (glás'gō)	51	45°35'N	62°36'W
New Guinea, i., (gĭne)	169	5°45'S	140°00'E
Newhalem, Wa., U.S. (nū hā'lŭm)	72	48°44'N	121°11'W
New Hampshire, state, U.S. (hămp'shĭr)	65	43°55'N	71°40'W
New Hampton, Ia., U.S. (hămp'tŭn)	71	43°03'N	92°20'W
New Hanover, S. Afr. (hăn'ôvēr)	187c	29°23'S	30°32'E
New Hanover, i., Pap. N. Gui.	169	2°37'S	150°15'E
New Harmony, In., U.S. (nū här'mō-nĭ)	66	38°10'N	87°55'W
New Haven, Ct., U.S. (hā'vĕn)	65	41°20'N	72°55'W
New Haven, In., U.S. (nū hāv''n)	66	41°05'N	85°00'W
New Hebrides, is., Vanuatu	175	16°00'S	167°00'E
New Holland, Eng., U.K. (hŏl'ănd)	114a	53°42'N	0°21'W
New Holland, N.C., U.S.	83	35°27'N	76°14'W
New Hope Mountain, mtn., Al., U.S. (hōp)	68h	33°23'N	86°45'W
New Hudson, Mi., U.S. (hŭd'sŭn)	69b	42°30'N	83°36'W
New Iberia, La., U.S. (ī-bē'rĭ-á)	81	30°00'N	91°50'W
Newington, Can. (nū'ĕng-tŏn)	62c	45°07'N	75°00'W
New Ireland, i., Pap. N. Gui. (īr'lănd)	169	3°15'S	152°30'E
New Jersey, state, U.S. (jûr'zĭ)	65	40°30'N	74°50'W
New Kensington, Pa., U.S. (kĕn'zĭng-tŭn)	69e	40°34'N	79°35'W
Newkirk, Ok., U.S. (nū'kûrk)	79	36°52'N	97°03'W
New Lenox, Il., U.S. (lĕn'ŭk)	69a	41°31'N	87°58'W
New Lexington, Oh., U.S. (lĕk'sĭng-tŭn)	66	39°40'N	82°10'W
New Lisbon, Wi., U.S. (lĭz'bŭn)	71	43°52'N	90°11'W
New Liskeard, Can.	59	47°30'N	79°40'W
New London, Ct., U.S. (lŭn'dŭn)	67	41°20'N	72°05'W
New London, Wi., U.S.	71	44°24'N	88°45'W
New Madrid, Mo., U.S. (măd'rĭd)	79	36°34'N	89°31'W

PLACE (Pronunciation)	PAGE	LAT.	LONG.
Newman's Grove, Ne., U.S. (nū'măn grōv)	70	41°46'N	97°44'W
Newmarket, Can. (nū'mär-kĕt)	59	44°00'N	79°30'W
New Martinsville, W.V., U.S. (mär'tĭnz-vĭl)	66	39°35'N	80°50'W
New Meadows, Id., U.S.	72	44°58'N	116°20'W
New Mexico, state, U.S. (mĕk'sĭ-kō)	64	34°30'N	107°10'W
New Mills, Eng., U.K. (mĭlz)	114a	53°22'N	2°00'W
New Munster, Wi., U.S. (mŭn'stēr)	69a	42°35'N	88°13'W
Newnan, Ga., U.S. (nū'năn)	82	33°22'N	84°47'W
New Norfolk, Austl. (nôr'fŏk)	175	42°50'S	147°17'E
New Orleans, La., U.S. (ôr'lê-ănz)	65	30°00'N	90°05'W
New Philadelphia, Oh., U.S. (fĭl-á-dĕl'fĭ-á)	66	40°30'N	81°30'W
New Plymouth, N.Z. (plĭm'ŭth)	175a	39°04'S	174°13'E
Newport, Austl.	173b	33°39'S	151°19'E
Newport, Eng., U.K. (nū-pôrt)	120	50°41'N	1°25'W
Newport, Eng., U.K.	114a	52°46'N	2°22'W
Newport, Wales, U.K.	117	51°36'N	3°05'W
Newport, Ar., U.S. (nū'pōrt)	79	35°35'N	91°16'W
Newport, Ky., U.S.	65	39°05'N	84°30'W
Newport, Me., U.S.	60	44°49'N	69°20'W
Newport, Mn., U.S.	75g	44°52'N	92°59'W
Newport, N.H., U.S.	67	43°20'N	72°10'W
Newport, Or., U.S.	72	44°39'N	124°02'W
Newport, R.I., U.S.	67	41°29'N	71°16'W
Newport, Tn., U.S.	82	35°55'N	83°12'W
Newport, Vt., U.S.	67	44°55'N	72°15'W
Newport, Wa., U.S.	72	48°12'N	117°01'W
Newport Beach, Ca., U.S. (bēch)	75a	33°36'N	117°55'W
Newport News, Va., U.S.	65	36°59'N	76°24'W
New Prague, Mn., U.S. (nū prāg)	71	44°33'N	93°35'W
New Providence, i., Bah. (prŏv'ĭ-dĕns)	92	25°00'N	77°25'W
New Richmond, Oh., U.S. (rĭch'mŭnd)	66	38°55'N	84°15'W
New Richmond, Wi., U.S.	71	45°07'N	92°34'W
New Roads, La., U.S. (rōds)	81	30°42'N	91°26'W
New Rochelle, N.Y., U.S. (rū-shĕl')	68a	40°55'N	73°47'W
New Rockford, N.D., U.S. (rŏk'fōrd)	70	47°40'N	99°08'W
New Ross, Ire. (rôs)	120	52°25'N	6°55'W
New Sarepta, Can.	62g	53°17'N	113°09'W
New Siberian Islands *see* Novosibirskiye Ostrova, is., Russia	135	74°00'N	140°30'E
New Smyrna Beach, Fl., U.S. (smûr'ná)	83	29°00'N	80°57'W
New South Wales, state, Austl. (wālz)	175	32°45'S	146°14'E
Newton, Can. (nū'tŭn)	62f	49°56'N	98°04'W
Newton, Eng., U.K.	114a	53°27'N	2°37'W
Newton, Ia., U.S.	71	41°42'N	93°04'W
Newton, Il., U.S.	66	39°00'N	88°10'W
Newton, Ks., U.S.	79	38°03'N	97°22'W
Newton, Ma., U.S.	61a	42°21'N	71°13'W
Newton, Ms., U.S.	82	32°18'N	89°10'W
Newton, N.C., U.S.	83	35°40'N	81°19'W
Newton, N.J., U.S.	68a	41°03'N	74°45'W
Newton, Tx., U.S.	81	30°47'N	93°45'W
Newtonsville, Oh., U.S. (nū'tŭnz-vĭl)	69f	39°11'N	84°04'W
Newtown, N.D., U.S. (nū'toun)	70	47°57'N	102°25'W
Newtown, Oh., U.S.	69f	39°08'N	84°22'W
Newtown, Pa., U.S.	68f	40°13'N	74°56'W
Newtownards, N. Ire., U.K. (nu-t'n-ardz')	120	54°35'N	5°39'W
New Ulm, Mn., U.S. (ŭlm)	71	44°18'N	94°27'W
New Waterford, Can. (wô'tēr-fērd)	51	46°15'N	60°05'W
New Westminster, Can. (wĕst'mĭn-stēr)	55	49°12'N	122°55'W
New York, N.Y., U.S. (yôrk)	65	40°40'N	73°58'W
New York, state, U.S.	65	42°45'N	78°05'W
New Zealand, nation, Oc. (zē'lănd)	175a	42°00'S	175°00'E
Nexapa, r., Mex. (nĕks-ä'pä)	88	18°32'N	98°29'W
Neya-gawa, Japan (nā'yä gä'wä)	167b	34°47'N	135°38'E
Neyshābūr, Iran	154	36°06'N	58°45'E
Neyva, r., Russia (nĕy'vá)	142a	57°39'N	60°37'E
Nezahualcóyotl, Mex.	89a	19°27'N	99°03'W
Nez Perce, Id., U.S. (nĕz' pûrs')	72	46°16'N	116°15'W
Nez Perce Indian Reservation, I.R., Id., U.S.	72	46°20'N	116°30'W
Ngami, l., Bots. (n'gä'mē)	186	20°56'S	22°31'E
Ngangerabeli Plain, pl., Kenya	191	1°20'S	40°10'E
Ngangla Ringco, l., China (ŋäŋ-lä rĭŋ-tswo)	158	31°42'N	82°53'E
Ngarimbi, Tan.	191	8°28'S	38°36'E
Ngoko, r., Afr.	190	1°55'N	15°53'E
Ngol-Kedju Hill, mtn., Cam.	189	6°20'N	9°45'E
Ngong, Kenya ('n-gông)	186	1°27'S	36°39'E
Ngounié, r., Gabon	190	1°15'S	10°43'E
Ngoywa, Tan.	191	5°56'S	32°48'E
Ngqeleni, S. Afr. ('ng-kĕ-lā'nĕ)	187c	31°41'S	29°04'E
Nguigmi, Niger ('n-gĕg'mē)	185	14°15'N	13°07'E
Ngurore, Nig.	189	9°18'N	12°14'E
Nguru, Nig. ('n-gōō'rōō)	184	12°53'N	10°26'E
Nguru Mountains, mts., Tan.	191	6°10'S	37°35'E
Nha Trang, Viet. (nyä-träng')	168	12°08'N	108°56'E
Niafounke, Mali	184	16°03'N	4°17'W
Niagara, Wi., U.S. (nī-ăg'á-rá)	71	45°45'N	88°05'W
Niagara, r., N.A.	69c	43°12'N	79°03'W
Niagara Falls, Can.	69c	43°05'N	79°05'W
Niagara Falls, N.Y., U.S.	65	43°06'N	79°02'W
Niagara-on-the-Lake, Can.	62d	43°16'N	79°05'W
Niakaramandougou, C. Iv.	188	8°40'N	5°17'W
Niamey, Niger (nē-ä-mâ')	184	13°31'N	2°07'E
Niamtougou, Togo	188	9°46'N	1°06'E
Niangara, D.R.C. (nē-äŋ-gá'rà)	185	3°42'N	27°52'E
Niangua, r., Mo., U.S. (nī-ăŋ'gwá)	79	37°30'N	93°05'W
Nias, Pulau, i., Indon. (nē'äs')	168	0°58'N	97°43'E
Nibe, Den. (nē'bĕ)	122	56°57'N	9°36'E

PLACE (Pronunciation)	PAGE	LAT.	LONG.
Nicaragua, nation, N.A. (nĭk-á-rä'gwá)	86	12°45'N	86°15'W
Nicaragua, Lago de, l., Nic. (lä'gô dĕ)	86	11°45'N	85°28'W
Nicastro, Italy (nê-käs'trō)	119	38°39'N	16°15'E
Nicchehabin, Punta, c., Mex. (pōō'n-tä-nĕk-chĕ-ä-bē'n)	90a	19°50'N	87°20'W
Nice, Fr. (nēs)	110	43°42'N	7°21'E
Nicheng, China (nē-chŭŋ)	163b	30°54'N	121°48'E
Nichicun, l., Can. (nĭch'ĭ-kŭn)	53	53°07'N	72°10'W
Nicholas Channel, strt., N.A. (nĭk'ô-lás)	92	23°30'N	80°20'W
Nicholasville, Ky., U.S. (nĭk'ô-lás-vĭl)	66	37°55'N	84°35'W
Nicobar Islands, is., India (nĭk-ô-bär')	168	8°28'N	94°04'E
Nicolai Mountain, mtn., Or., U.S. (nē-cō lī')	74c	46°05'N	123°27'W
Nicolás Romero, Mex. (nē-kô-lá's rô-mĕ'rô)	89a	19°38'N	99°20'W
Nicolet, Lake, l., Mi., U.S. (nĭ'kŏ-lĕt)	75k	46°22'N	84°14'W
Nicolls Town, Bah.	92	25°10'N	78°00'W
Nicols, Mn., U.S. (nĭk'ĕls)	75g	44°50'N	93°12'W
Nicomeki, r., Can.	74d	49°04'N	122°47'W
Nicosia, Cyp. (nē-kô-sē'á)	154	35°10'N	33°22'E
Nicoya, C.R. (nē-kō'yä)	90	10°08'N	85°27'W
Nicoya, Golfo de, b., C.R. (gôl-fô-dĕ)	90	10°03'N	85°04'W
Nicoya, Península de, pen., C.R.	90	10°05'N	86°00'W
Nidzica, Pol. (nĕ-jĕt'sá)	125	53°21'N	20°30'E
Niedere Tauern, mts., Aus.	124	47°15'N	13°41'E
Niederkrüchten, Ger. (nĕ'dĕr-krük-tĕn)	127c	51°12'N	6°14'E
Niederösterreich, state, Aus.	115e	48°24'N	16°20'E
Niedersachsen (Lower Saxony), state, Ger. (nĕ'dĕr-zäk-sĕn)	115c	53°30'N	9°30'E
Niellim, Chad	189	9°42'N	17°49'E
Nienburg, Ger. (nē'ĕn-bôrgh)	124	52°40'N	9°15'E
Nietverdiend, S. Afr.	192c	25°02'S	26°10'E
Nieuw Nickerie, Sur. (nē-nĕ'kĕ-rē')	101	5°51'N	57°00'W
Nieves, Mex. (nyā'vås)	88	24°00'N	102°57'W
Niğde, Tur. (nĭg'dĕ)	119	37°55'N	34°40'E
Nigel, S. Afr. (nī'jĕl)	192c	26°26'S	28°27'E
Niger, nation, Afr. (nī'jēr)	184	18°02'N	8°30'E
Niger, r., Afr.	184	8°00'N	6°00'E
Niger Delta, d., Nig.	189	4°45'N	5°20'E
Nigeria, nation, Afr. (nī-jē'rĭ-á)	184	8°57'N	6°30'E
Nihoa, i., Hi., U.S.	84b	23°15'N	161°30'W
Nii, i., Japan (nē)	167	34°26'N	139°23'E
Niigata, Japan (nē'ê-gä'tà)	161	37°47'N	139°04'E
Ni'ihau, i., Hi., U.S. (nē'ê-ha'ōō)	64c	21°50'N	160°05'W
Niimi, Japan (nē'mê)	167	34°59'N	133°28'E
Niiza, Japan	167a	35°48'N	139°34'E
Nijmegen, Neth. (nī'mâ-gĕn)	121	51°50'N	5°52'E
Nikitinka, Russia (nĕ-kĭ'tĭn-kà)	132	55°33'N	33°19'E
Nikolayevka, Russia (nē-kô-lä'yĕf-ká)	142c	59°29'N	29°48'E
Nikolayevka, Russia	166	48°37'N	134°09'E
Nikolayevskiy, Russia	137	50°00'N	45°30'E
Nikolayevsk-na-Amure, Russia	135	53°18'N	140°49'E
Nikol'sk, Russia (nê-kôlsk')	134	59°30'N	45°40'E
Nikol'skoye, Russia (nê-kôl'skô-yĕ)	142c	59°27'N	30°00'E
Nikopol, Blg. (nē'kô-pōl')	119	43°41'N	24°52'E
Nikopol', Ukr.	137	47°36'N	34°24'E
Nilahue, r., Chile (nē-lá'wĕ)	99b	34°36'S	71°50'W
Nile, r., Afr. (nīl)	185	27°30'N	31°00'E
Niles, Mi., U.S. (nīlz)	66	41°50'N	86°15'W
Niles, Oh., U.S.	66	41°15'N	80°45'W
Nileshwar, India	159	12°08'N	74°14'E
Nilgiri Hills, hills, India	159	12°05'N	76°22'E
Nilópolis, Braz. (nē-lô'pō-lēs)	99a	22°48'S	43°25'W
Nīmach, India	158	24°32'N	74°51'E
Nimba, Mont, mtn., Afr. (nĭm'bà)	184	7°40'N	8°33'W
Nimba Mountains, mts., Afr.	188	7°30'N	8°35'W
Nîmes, Fr. (nēm)	110	43°49'N	4°22'E
Nimrod Reservoir, res., Ar., U.S. (nĭm'rŏd)	79	34°58'N	93°46'W
Nimule, Sudan (nē-mōō'lā)	185	3°38'N	32°12'E
Ninda, Ang.	190	14°47'S	21°24'E
Nine Mile Creek, r., Ut., U.S. (mīn'ĭmōd')	77	39°50'N	110°30'W
Ninety Mile Beach, cst., Austl.	175	38°20'S	147°30'E
Nineveh, Iraq (nĭn'ê-vá)	154	36°30'N	43°10'E
Ning'an, China (nĭŋ-än)	161	44°20'N	129°20'E
Ningbo, China (nĭŋ-bwo)	161	29°56'N	121°30'E
Ningde, China (nĭŋ-dŭ)	161	26°38'N	119°33'E
Ninghai, China (nĭng'hī')	165	29°20'N	121°20'E
Ninghe, China (nĭŋ-hŭ)	162	39°20'N	117°50'E
Ningjin, China (nĭŋ-jyĭn)	162	37°39'N	116°47'E
Ningjin, China	162	37°37'N	114°55'E
Ningming, China	165	22°22'N	107°06'E
Ningwu, China (nĭng'wōō')	161	39°00'N	112°12'E
Ningxia Huizu, prov., China (nĭŋ-shyä)	160	37°10'N	106°00'E
Ningyang, China (nĭng'yăng')	162	35°46'N	116°48'E
Ninh Binh, Viet. (nēn bĕnk')	168	20°22'N	106°00'E
Ninigo Group, is., Pap. N. Gui.	169	1°15'S	143°30'E
Ninnescah, r., Ks., U.S. (nĭn'ĕs-kä)	78	37°37'N	98°31'W
Nioaque, Braz. (nēō-á'-kĕ)	101	21°14'S	55°41'W
Niobrara, r., U.S. (nī-ō-brâr'á)	64	42°46'N	98°46'W
Niokolo Koba, Parc National du, rec., Sen.	188	13°05'N	13°00'W
Nioro du Sahel, Mali (nê-ō'rō)	184	15°15'N	9°35'W
Nipawin, Can.	50	53°22'N	104°00'W
Nipe, Bahía de, b., Cuba (bä-ē'ä-dĕ-nē'pā)	93	20°50'N	75°30'W
Nipe, Sierra de, mts., Cuba (sē-ĕ'r-rä-dĕ)	93	20°20'N	75°50'W
Nipigon, Can. (nĭp'ĭ-gŏn)	51	48°58'N	88°17'W
Nipigon, l., Can.	53	49°37'N	89°55'W
Nipigon Bay, b., Can.	58	48°56'N	88°00'W
Nipisiguit, r., Can. (nĭ-pĭ'sĭ-kwĭt)	60	47°26'N	66°15'W
Nipissing, l., Can. (nĭp'ĭ-sĭng)	53	45°59'N	80°19'W

ng-sing; ŋ-baŋk; N-nasalized n; nŏd; cŏmmit; ōld; ôbey; ôrder; oi-boil; fōōd; ŏ-as oo in foot; ou-out; s-soft; sh-dish; th-thin; pūre; ûnite; ûrn; stŭd; circŭs; ü-as in French tu; '-indeterminate vowel.

ăt; fīnăl; rāte; senâte; ärm; àsk; sofà; fâre; ch-choose; dh-as th in other; bē; ěvent; bĕt; recĕnt; cratẽr; g-gō; gh-guttural g; bĭt; ī-short neutral; rīde; ᴋ-guttural k as ch in German ich;

PRONOUNCING INDEX

North Yorkshire - Oceanside 283

PLACE (Pronunciation)	PAGE	LAT.	LONG.
North Yorkshire, co., Eng., U.K.	114a	53°50'N	1°10'W
Norton, Ks., U.S. (nôr'tŭn)	78	39°40'N	99°54'W
Norton, Ma., U.S.	68b	41°58'N	71°08'W
Norton, Va., U.S.	83	36°54'N	82°36'W
Norton Bay, b., Ak., U.S.	63	64°22'N	162°18'W
Norton Reservoir, res., Ma., U.S.	68b	42°01'N	71°07'W
Norton Sound, strt., Ak., U.S.	63	63°48'N	164°50'W
Norval, Can. (nôr'vål)	62d	43°39'N	79°52'W
Norwalk, Ca., U.S. (nôr'wôk)	75a	33°54'N	118°05'W
Norwalk, Ct., U.S.	68a	41°06'N	73°25'W
Norwalk, Oh., U.S.	66	41°15'N	82°35'W
Norway, Me., U.S.	60	44°11'N	70°35'W
Norway, Mi., U.S.	71	45°47'N	87°55'W
Norway, nation, Eur. (nôr'wā)	110	63°48'N	11°17'E
Norway House, Can.	50	53°59'N	97°50'W
Norwegian Sea, sea, Eur. (nôr-wē'jăn)	116	66°54'N	1°43'E
Norwell, Ma., U.S. (nôr'wĕl)	61a	42°10'N	70°47'W
Norwich, Eng., U.K.	117	52°40'N	1°15'E
Norwich, Ct., U.S. (nôr'wĭch)	67	41°20'N	72°00'W
Norwich, N.Y., U.S.	67	42°35'N	75°30'W
Norwood, Ma., U.S. (nôr'wo͝od)	61a	42°11'N	71°13'W
Norwood, N.C., U.S.	83	35°15'N	80°08'W
Norwood, Oh., U.S.	69f	39°10'N	84°27'W
Nose Creek, r., Can. (nōz)	62e	51°09'N	114°02'W
Noshiro, Japan (nō'shē-rō)	166	40°09'N	140°02'E
Nosivka, Ukr. (nô'sôf-kȧ)	133	50°54'N	31°35'E
Nossob, r., Afr. (nô'sŏb)	186	24°15'S	19°10'E
Noteć, r., Pol. (nô'tĕcn)	124	52°50'N	16°19'E
Notodden, Nor. (nōt'ôd'n)	122	59°35'N	9°15'E
Notre Dame, Monts, mts., Can.	60	46°35'N	70°35'W
Notre Dame Bay, b., Can. (nō't'r dàm')	53a	49°45'N	55°15'W
Notre-Dame-du-Lac, Can.	60	47°37'N	68°51'W
Nottawasaga Bay, b., Can. (nŏt'ȧ-wȧ-sä'gȧ)	59	44°45'N	80°35'W
Nottaway, r., Can. (nŏt'ȧ-wā)	53	50°58'N	78°02'W
Nottingham, Eng., U.K. (nŏt'ĭng-ăm)	117	52°58'N	1°09'W
Nottingham Island, i., Can.	53	62°58'N	78°53'W
Nottinghamshire, co., Eng., U.K.	114a	53°03'N	1°05'W
Nottoway, r., Va., U.S. (nŏt'ȧ-wā)	83	36°53'N	77°47'W
Notukeu Creek, r., Can.	56	49°55'N	106°30'W
Nouadhibou, Maur.	184	21°02'N	17°09'W
Nouakchott, Maur.	184	18°06'N	15°57'W
Nouamrhar, Maur.	184	19°22'N	16°31'W
Nouméa, N. Cal. (no͞o-mā'ä)	175	22°16'S	166°27'E
Nouvelle, Can. (no͞o-vĕl')	60	48°09'N	66°22'W
Nouvelle-France, Cap de, c., Can.	53	62°03'N	74°00'W
Nouzonville, Fr. (no͞o-zôN-vēl')	126	49°51'N	4°43'E
Nova Cruz, Braz. (nō'vȧ-kro͞o'z)	101	6°22'S	35°20'W
Nova Friburgo, Braz. (frê-bo͞or'gò)	101	22°18'S	42°31'W
Nova Iguaçu, Braz. (nō'vä-ē-gwä-so͞o')	101	22°45'S	43°27'W
Nova Lima, Braz. (lē'mä)	99a	19°59'S	43°51'W
Nova Lisboa *see* Huambo, Ang.	186	12°44'S	15°47'E
Nova Mambone, Moz. (nō'vȧ-mäm-bō'nĕ)	186	21°04'S	35°13'E
Nova Odesa, Ukr.	133	47°18'N	31°48'E
Nova Praha, Ukr.	133	48°34'N	32°54'E
Novara, Italy (nô-vä'rä)	118	45°24'N	8°38'E
Nova Resende, Braz.	99a	21°12'S	46°25'W
Nova Scotia, prov., Can. (skō'shȧ)	51	44°28'N	65°00'W
Nova Vodolaha, Ukr.	133	49°43'N	35°51'E
Novaya Ladoga, Russia (nô'vȧ-ya lȧ-dô-gȧ)	123	60°06'N	32°16'E
Novaya Lyalya, Russia (lyȧ'lyȧ)	142a	59°03'N	60°36'E
Novaya Sibir, i., Russia (sē-bēr')	135	75°00'N	149°00'E
Novaya Zemlya, i., Russia (zĕm-lyȧ')	134	72°00'N	54°46'E
Nova Zagora, Blg. (zä'gō-rȧ)	131	42°30'N	26°01'E
Novelda, Spain (nō-vĕl'dȧ)	129	38°22'N	0°46'W
Nové Mesto nad Váhom, Slvk. (nô'vĕ myĕs'tō)	125	48°44'N	17°47'E
Nové Zámky, Slvk. (zȧm'kĕ)	117	47°58'N	18°10'E
Novgorod, Russia (nôv'gō-rŏt)	136	58°32'N	31°16'E
Novgorod, prov., Russia	132	58°27'N	31°55'E
Novhorod-Sivers'kyi, Ukr.	137	52°01'N	33°14'E
Novi, Mi., U.S. (nō'vĭ)	69b	42°29'N	83°28'W
Novigrad, Cro. (nō'vĭ grȧd)	130	44°09'N	15°34'E
Novi Ligure, Italy (nô'vē)	130	44°43'N	8°48'E
Novinger, Mo., U.S. (nŏv'ĭn-jĕr)	79	40°14'N	92°43'W
Novi Pazar, Blg. (pä-zär')	131	43°22'N	27°26'E
Novi Pazar, Serb. (pȧ-zär')	119	43°08'N	20°30'E
Novi Sad, Serb. (säd')	110	45°15'N	19°53'E
Novoaidar, Ukr.	133	48°57'N	39°01'E
Novoasbest, Russia (nô-vô-äs-bĕst')	142a	57°43'N	60°14'E
Novocherkassk, Russia (nô'vô-chĕr-kȧsk')	137	47°25'N	40°04'E
Novokuznetsk, Russia (nō'vô-kŏ'z-nyĕ'tsk) (stȧ'lênsk)	134	53°43'N	86°59'E
Novo-Ladozhskiy Kanal, can., Russia (nô-vô-lä'dôzh-skĭ kȧ-näl')	123	59°54'N	31°19'E
Novo Mesto, Slvn. (nôvô mās'tô)	130	45°48'N	15°13'E
Novomoskovsk, Russia (nô'vô-môs-kôfsk')	134	54°06'N	38°08'E
Novomoskovs'k, Ukr.	137	48°37'N	35°12'E
Novomyrhorod, Ukr.	133	48°46'N	31°44'E
Novonikol'skiy, Russia (nô'vô-nyĭ-kôl'skĭ)	142a	52°28'N	57°12'E
Novorossiysk, Russia (nô'vô-rô-sēsk')	134	44°43'N	37°48'E
Novorzhev, Russia (nô'vô-rzhêv')	132	57°01'N	29°17'E
Novo-Selo, Blg. (nō'vô-sĕ'lô)	131	44°09'N	22°46'E
Novosibirsk, Russia (nô'vô-sē-bērsk')	134	55°09'N	82°58'E
Novosibirskiye Ostrova (New Siberian Islands), is., Russia	135	74°00'N	140°30'E
Novosil', Russia (nô'vô-sĭl)	132	52°58'N	37°03'E
Novosokol'niki, Russia (nô'vô-sô-kôl'nē-kē)	132	56°18'N	30°07'E

PLACE (Pronunciation)	PAGE	LAT.	LONG.
Novotatishchevskiy, Russia (nô'vô-tȧ-tyĭsh'chĕv-skĭ)	142a	53°22'N	60°24'E
Novoukraïnka, Ukr.	137	48°18'N	31°33'E
Novouzensk, Russia (nô-vô-ô-zĕnsk')	137	50°40'N	48°08'E
Novozybkov, Russia (nô'vô-zĕp'kôf)	137	52°31'N	31°54'E
Novyi Buh, Ukr.	133	47°43'N	32°33'E
Nový Jičín, Czech Rep. (nô'vē yē'chēn)	125	49°36'N	18°02'E
Novyy Oskol, Russia (ôs-kôl')	133	50°46'N	37°53'E
Novyy Port, Russia (nô'vē)	134	67°19'N	72°28'E
Nowa Sól, Pol. (nō'vȧ sŭl')	124	51°49'N	15°41'E
Nowata, Ok., U.S. (nô-wä'tȧ)	79	36°42'N	95°38'W
Nowood Creek, r., Wy., U.S.	73	44°02'N	107°37'W
Nowra, Austl. (nou'rȧ)	176	34°55'S	150°45'E
Nowy Dwór Mazowiecki, Pol. (nō'vĭ dvo͞or mä-zo-vyĕts'ke)	125	52°26'N	20°46'E
Nowy Sącz, Pol. (nō'vē sônch')	125	49°36'N	20°42'E
Nowy Targ, Pol. (tärk')	125	49°29'N	20°02'E
Noxon Reservoir, res., Mt., U.S.	72	47°50'N	115°40'W
Noxubee, r., Ms., U.S. (nŏks'û-bē)	82	33°20'N	88°55'W
Noyes Island, i., Ak., U.S. (noiz)	54	55°30'N	133°40'W
Nozaki, Japan (nō'zä-kē)	167b	34°43'N	135°39'E
Nqamakwe, S. Afr. ('n-gä-mä'kwȧ)	187c	32°13'S	27°57'E
Nqutu, S. Afr. ('n-ko͞o'to͞o)	187c	28°17'S	30°41'E
Nsawam, Ghana	188	5°50'N	0°20'W
Ntshoni, mtn., S. Afr.	187c	29°34'S	30°03'E
Ntwetwe Pan, pl., Bots.	186	20°00'S	24°18'E
Nubah, Jibāl an, mts., Sudan	185	12°22'N	30°39'E
Nubian Desert, des., Sudan (no͞o'bĭ-ăn)	185	21°13'N	33°09'E
Nudo Coropuna, mtn., Peru (no͞o'dô kō-rō-po͞o'nä)	100	15°53'S	72°04'W
Nudo de Pasco, mtn., Peru (dē pȧs'kô)	100	10°34'S	76°12'W
Nueces, r., Tx., U.S. (nů-ā'sȧs)	64	28°20'N	98°08'W
Nueltin, l., Can. (nwĕl'tin)	52	60°14'N	101°00'W
Nueva Armenia, Hond. (nwā'vä är-mä'nê-ȧ)	90	15°47'N	86°32'W
Nueva Esparta, dept., Ven. (nwĕ'vä ĕs-pä'r-tä)	101b	10°50'N	64°35'W
Nueva Gerona, Cuba (kĕ-rō'nä)	92	21°55'N	82°45'W
Nueva Palmira, Ur. (päl-mē'rä)	99c	33°53'S	58°23'W
Nueva Rosita, Mex. (nòĕ'vä rô-sē'tä)	64	27°55'N	101°10'W
Nueva San Salvador, El Sal.	90	13°41'N	89°16'W
Nueve, Canal Numero, can., Arg.	99c	36°22'S	58°19'W
Nueve de Julio, Arg. (nwā'vȧ dā ho͞o'lyô)	102	35°26'S	60°51'W
Nuevitas, Cuba (nwä-vē'täs)	87	21°35'N	77°15'W
Nuevitas, Bahía de, b., Cuba (bä-ē'ä dĕ nwä-vē'täs)	92	21°30'N	77°05'W
Nuevo, Ca., U.S. (nwā'vō)	75a	33°48'N	117°09'W
Nuevo Laredo, Mex. (lä-rā'dhō)	86	27°29'N	99°30'W
Nuevo León, state, Mex. (lȧ-ōn')	86	26°00'N	100°00'W
Nuevo San Juan, Pan. (nwĕ'vô sän ko͞o-ä'n)	86a	9°14'N	79°43'W
Nugumanovo, Russia (nû-gû-mä'nô-vô)	142a	55°28'N	61°50'E
Nulato, Ak., U.S. (no͞o-lä'tō)	63	64°40'N	158°18'W
Nullagine, Austl. (nŭ-lă'jēn)	174	22°00'S	120°07'E
Nullarbor Plain, pl., Austl. (nŭ-lär'bôr)	174	31°45'S	126°30'E
Numabin Bay, b., Can. (no͞o-mä'bĭn)	56	56°30'N	103°08'W
Numansdorp, Neth.	115a	51°43'N	4°25'E
Numazu, Japan (no͞o'mä-zo͞o)	166	35°06'N	138°55'E
Numfoor, Pulau, i., Indon.	169	1°20'S	134°48'E
Nun, r., Nig.	189	5°05'N	6°10'E
Nunavut, ter., Can.	50	70°00'N	95°00'W
Nunawading, Austl.	173a	37°49'S	145°10'E
Nuneaton, Eng., U.K. (nūn'ē-tŭn)	120	52°31'N	1°28'W
Nunivak, i., Ak., U.S. (no͞o'nĭ-văk)	64a	60°25'N	167°42'W
Nunyama, Russia (nŭn-yä'mȧ)	63	65°49'N	170°32'W
Nuoro, Italy (nwô'rō)	130	40°29'N	9°20'E
Nura, r., Kaz.	140	49°48'N	73°54'E
Nurata, Uzb. (no͞or'ät'ä)	139	40°33'N	65°28'E
Nuremberg *see* Nürnberg, Ger.	110	49°28'N	11°07'E
Nürnberg, Ger. (nürn'bĕrgh)	110	49°28'N	11°07'E
Nurse Cay, i., Bah.	93	22°30'N	75°50'W
Nusabyin, Tur. (no͞o'sĭ-bĕn)	137	37°05'N	41°10'E
Nushagak, r., Ak., U.S. (nū-shă-găk')	63	59°28'N	157°40'W
Nushan Hu, l., China	162	32°50'N	117°59'E
Nushki, Pak. (nŭsh'kē)	155	29°30'N	66°02'E
Nuthe, r., Ger. (no͞o'tē)	115b	52°15'N	13°11'E
Nutley, N.J., U.S. (nŭt'lē)	68a	40°49'N	74°09'W
Nutter Fort, W.V., U.S. (nŭt'ēr fôrt)	66	39°15'N	80°15'W
Nutwood, Il., U.S. (nŭt'wòd)	75e	39°05'N	90°34'W
Nuwaybi 'al Muzayyinah, Egypt	153a	28°59'N	34°40'E
Nuweland, S. Afr.	186a	33°58'S	18°28'E
Nyack, N.Y., U.S. (nī'ăk)	68a	41°05'N	73°55'W
Nyainqêntanglha Shan, mts., China (nyä-ĭn-chyŭn-tăn-lä shän)	160	29°55'N	88°08'E
Nyakanazi, Tan.	191	3°00'S	31°15'E
Nyala, Sudan	185	12°00'N	24°52'E
Nyanga, r., Gabon	190	2°45'S	10°30'E
Nyanza, Rw.	191	2°21'S	29°45'E
Nyasa, Lake, l., Afr. (nyä'sä)	186	10°45'S	34°30'E
Nyasvizh, Bela. (nyĕs'vēsh)	132	53°13'N	26°44'E
Nyazepetrovsk, Russia (nyä'zĕ-pĕ-trôvsk')	142a	56°04'N	59°38'E
Nyborg, Den. (nü'bôr'')	122	55°20'N	10°45'E
Nybro, Swe. (nü'brô)	122	56°44'N	15°56'E
Nyeri, Kenya	191	0°25'S	36°57'E
Nyika Plateau, plat., Mwi.	191	10°30'S	35°50'E
Nyíregyháza, Hung. (nyē'rĕd-y'hä'zä)	119	47°58'N	21°45'E
Nykøbing, Den. (nü'kû-bĭng)	116	56°46'N	8°47'E
Nykøbing, Den.	122	54°45'N	11°54'E
Nykøbing Sjaelland, Den.	122	55°55'N	11°37'E
Nyköping, Swe. (nü'chû-pĭng)	116	58°46'N	16°58'E
Nylstroom, S. Afr. (nĭl'strōm)	186	24°42'S	28°25'E

PLACE (Pronunciation)	PAGE	LAT.	LONG.
Nymagee, Austl. (nī-mȧ-gē')	175	32°17'S	146°18'E
Nymburk, Czech Rep. (nĕm'bòrk)	117	50°12'N	15°03'E
Nynäshamn, Swe. (nü-nĕs-hȧm'n)	122	58°53'N	17°55'E
Nyngan, Austl. (nĭŋ'gȧn)	175	31°31'S	147°25'E
Nyong, r., Cam. (nyông)	184	4°00'N	12°00'E
Nyou, Burkina	188	12°46'N	1°56'W
Nýřany, Czech Rep. (nēr-zhä'nē)	124	49°43'N	13°13'E
Nysa, Pol. (nē'sȧ)	125	50°29'N	17°20'E
Nytva, Russia	136	58°00'N	55°10'E
Nyungwe, Mwi.	191	10°16'S	34°07'E
Nyunzu, D.R.C.	191	5°57'S	28°01'E
Nyuya, r., Russia (nyo͞o'yȧ)	141	60°30'N	111°45'E
Nyzhni Sirohozy, Ukr.	133	46°51'N	34°25'E
Nzega, Tan.	191	4°13'S	33°11'E
N'zeto, Ang.	186	7°14'S	12°52'E
Nzi, r., C. Iv.	188	7°00'N	4°27'W
Nzwani, i., Com. (äN-zhwäN)	187	12°14'S	44°47'E

O

PLACE (Pronunciation)	PAGE	LAT.	LONG.
Oahe, Lake, res., U.S.	64	45°20'N	100°00'W
O'ahu, i., Hi., U.S. (ō-ä'ho͞o) (ō-ä'hü)	64c	21°38'N	157°48'W
Oak Bay, Can.	54	48°27'N	123°18'W
Oak Bluff, Can. (ōk blŭf)	62f	49°47'N	97°21'W
Oak Creek, Co., U.S. (ōk krēk')	73	40°20'N	106°50'W
Oakdale, Ca., U.S. (ōk'dāl)	76	37°45'N	120°52'W
Oakdale, Ky., U.S.	66	38°15'N	85°50'W
Oakdale, La., U.S.	81	30°49'N	92°40'W
Oakdale, Pa., U.S.	69e	40°24'N	80°11'W
Oakengates, Eng., U.K. (ōk'ĕn-gāts)	114a	52°41'N	2°27'W
Oakes, N.D., U.S. (ōks)	70	46°10'N	98°50'W
Oakfield, Me., U.S. (ōk'fēld)	60	46°08'N	68°10'W
Oakford, Pa., U.S. (ōk'fōrd)	68f	40°08'N	74°58'W
Oak Grove, Or., U.S. (grōv)	74c	45°25'N	122°38'W
Oakham, Eng., U.K. (ōk'ăm)	114a	52°40'N	0°38'W
Oak Harbor, Oh., U.S. (ōk'här'bēr)	66	41°30'N	83°05'W
Oak Harbor, Wa., U.S.	74a	48°18'N	122°39'W
Oakland, Ca., U.S. (ōk'lănd)	64	37°48'N	122°16'W
Oakland, Ne., U.S.	70	41°50'N	96°28'W
Oakland City, In., U.S.	66	38°20'N	87°20'W
Oak Lawn, Il., U.S.	69a	41°43'N	87°45'W
Oakleigh, Austl. (ōk'lȧ)	173a	37°54'S	145°05'E
Oakley, Id., U.S. (ōk'lĭ)	72	42°15'N	135°53'W
Oakley, Ks., U.S.	78	39°08'N	100°49'W
Oakman, Al., U.S. (ōk'măn)	82	33°42'N	87°20'W
Oakmont, Pa., U.S. (ōk'mōnt)	69e	40°31'N	79°50'W
Oak Mountain, mtn., Al., U.S.	68h	33°22'N	86°42'W
Oak Park, Il., U.S. (pärk)	69a	41°53'N	87°48'W
Oak Point, Wa., U.S.	74c	46°11'N	123°11'W
Oak Ridge, Tn., U.S. (rĭj)	82	36°01'N	84°15'W
Oakville, Can. (ōk'vĭl)	59	43°27'N	79°40'W
Oakville, Can.	62f	49°56'N	97°58'W
Oakville, Mo., U.S.	75e	38°27'N	90°18'W
Oakville Creek, r., Can.	62d	43°34'N	79°54'W
Oakwood, Tx., U.S. (ōk'wòd)	81	31°36'N	95°48'W
Oatman, Az., U.S. (ōt'măn)	77	34°00'N	114°25'W
Oaxaca, Mex.	86	17°03'N	96°42'W
Oaxaca, state, Mex. (wä-hä'kä)	86	16°45'N	97°00'W
Oaxaca, Sierra de, mts., Mex. (sē-ĕ'r-rä dē)	89	16°15'N	97°25'W
Ob', r., Russia	134	62°15'N	67°00'E
Oba, Can. (ō'bȧ)	51	48°58'N	84°09'W
Obama, Japan (ō'bȧ-mä)	167	35°29'N	135°44'E
Oban, Scot., U.K. (ō'băn)	120	56°25'N	5°35'W
Oban Hills, hills, Nig.	189	5°35'N	8°30'E
O'Bannon, Ky., U.S. (ō-băn'nŏn)	69h	38°17'N	85°30'W
O Barco de Valdeorras, Spain	128	42°26'N	6°58'W
Obatogamau, l., Can. (ō-bȧ-tō'găm-ô)	59	49°38'N	74°10'W
Oberhausen, Ger. (ō'bĕr-hou'zĕn)	127c	51°27'N	6°51'E
Oberlin, Ks., U.S. (o'bĕr-lĭn)	78	39°49'N	100°30'W
Oberlin, Oh., U.S.	66	41°15'N	82°15'W
Oberroth, Ger. (ō'bĕr-rōt)	115d	48°19'N	11°20'E
Obi, Kepulauan, is., Indon. (ō'bē)	169	1°25'S	128°15'E
Obi, Pulau, i., Indon.	169	1°30'S	127°45'E
Óbidos, Braz. (ō'bē-dòzh)	101	1°57'S	55°32'W
Obihiro, Japan (ō'bē-hē'rō)	166	42°55'N	142°50'E
Obion, r., Tn., U.S.	82	36°10'N	89°25'W
Obion, North Fork, r., Tn., U.S. (ō-bī'ŏn)	82	35°49'N	89°06'W
Obitsu, r., Japan (ō'bēt'so͞o)	167a	35°19'N	140°03'E
Obock, Dji. (ō-bŏk')	192a	11°55'N	43°15'E
Obol', r., Bela. (ō-bōl')	132	55°24'N	29°24'E
Oboyan', Russia (ô-bô-yän')	137	51°14'N	36°16'E
Obskaya Guba, b., Russia	134	67°13'N	73°45'E
Obuasi, Ghana	188	6°14'N	1°39'W
Obukhiv, Ukr.	133	50°07'N	30°36'E
Obukhovo, Russia	142b	55°50'N	38°17'E
Obytichna kosa, spit, Ukr.	133	46°32'N	36°07'E
Ocala, Fl., U.S. (ō-kä'lȧ)	83	29°11'N	82°09'W
Ocampo, Mex. (ō-käm'pō)	88	22°49'N	99°23'W
Ocaña, Col. (ō-kän'yä)	100	8°15'N	73°37'W
Ocaña, Spain (ō-kä'n-yä)	128	39°58'N	3°31'W
Occidental, Cordillera, mts., Col.	100a	5°05'N	76°04'W
Occidental, Cordillera, mts., Peru	100	10°12'S	76°58'W
Ocean Beach, Ca., U.S. (ō'shăn bēch)	76a	32°44'N	117°14'W
Ocean Bight, b., Bah.	93	21°15'N	73°15'W
Ocean City, Md., U.S.	67	38°20'N	75°10'W
Ocean City, N.J., U.S.	67	39°15'N	74°35'W
Ocean Falls, Can. (Fôls)	50	52°21'N	127°40'W
Ocean Grove, Austl.	173a	38°16'S	144°32'E
Ocean Grove, N.J., U.S. (grōv)	67	40°10'N	74°00'W
Oceanside, Ca., U.S. (ō'shăn-sīd)	76	33°11'N	117°22'W

284 Oceanside – Opava PRONOUNCING INDEX

PLACE (Pronunciation)	PAGE	LAT.	LONG.
Oceanside, N.Y., U.S.	68a	40°38'N	73°39'W
Ocean Springs, Ms., U.S. (sprĭngs)	82	30°25'N	88°49'W
Ochakiv, Ukr.	133	46°38'N	31°33'E
Ochamchira, Geor.	138	42°44'N	41°28'E
Ochlockonee, r., Fl., U.S. (ŏk-lô-kō'nē)	82	30°10'N	84°38'W
Ocilla, Ga., U.S. (ô-sĭl'ȧ)	82	31°36'N	83°15'W
Ockelbo, Swe. (ôk'ĕl-bô)	122	60°54'N	16°35'E
Ocklawaha, Lake, res., Fl., U.S.	83	29°30'N	81°50'W
Ocmulgee, r., Ga., U.S.	82	32°25'N	83°30'W
Ocmulgee National Monument, rec., Ga., U.S. (ôk-mŭl'gē)	82	32°45'N	83°28'W
Ocoa, Bahía de, b., Dom. Rep.	93	18°20'N	70°40'W
Ococingo, Mex. (ô-kô-sē'n-gô)	89	17°03'N	92°18'W
Ocom, Lago, l., Mex. (ô-kô'm)	90a	19°26'N	88°18'W
Oconee, r., Ga., U.S. (ô-kō'nē)	65	32°45'N	83°00'W
Oconee, Lake, res., Ga., U.S.	82	33°30'N	83°15'W
Oconomowoc, Wi., U.S. (ô-kŏn'ô-mô-wŏk')	71	43°06'N	88°24'W
Oconto, Wi., U.S. (ô-kŏn'tō)	71	44°54'N	87°55'W
Oconto, r., Wi., U.S.	71	45°08'N	88°24'W
Oconto Falls, Wi., U.S.	71	44°53'N	88°11'W
Ocós, Guat. (ô-kōs')	90	14°31'N	92°12'W
Ocotal, Nic. (ō-kô-täl')	90	13°36'N	86°31'W
Ocotepeque, Hond. (ô-kō-tā-pā'kā)	90	14°25'N	89°13'W
Ocotlán, Mex. (ō-kô-tlän')	88	20°19'N	102°44'W
Ocotlán de Morelos, Mex. (dā mô-rā'lōs)	89	16°46'N	96°41'W
Ocozocoautla, Mex. (ô-kō'zô-kwä-ōō'tlä)	89	16°44'N	93°22'W
Ocumare del Tuy, Ven. (ō-kōō-mä'ra del twē')	100	10°07'N	66°47'W
Oda, Ghana	188	5°55'N	0°59'W
Odawara, Japan (ō'dá-wä'rä)	167	35°15'N	139°10'E
Odda, Nor. (ôdh-ȧ)	122	60°04'N	6°30'E
Odebolt, Ia., U.S. (ō'dē-bōlt)	70	42°20'N	95°14'W
Odemira, Port. (ō-dā-mē'rȧ)	128	37°35'N	8°40'W
Ödemiş, Tur. (û'dĕ-mēsh)	119	38°12'N	28°00'E
Odendaalsrus, S. Afr. (ō'dĕn-däls-rûs')	192c	27°52'S	26°41'E
Odense, Den. (ō'dhĕn-sĕ)	116	55°24'N	10°20'E
Odenton, Md., U.S. (ō'dĕn-tŭn)	68e	39°05'N	76°43'W
Odenwald, for., Ger. (ō'dĕn-väld)	124	49°39'N	8°55'E
Oder, r., Eur. (ō'dēr)	112	52°40'N	14°19'E
Oderhaff, l., Eur.	124	53°47'N	14°02'E
Odesa, Ukr.	134	46°28'N	30°44'E
Odesa, prov., Ukr.	133	46°05'N	29°48'E
Odessa, Tx., U.S. (ô-dĕs'ȧ)	80	31°52'N	102°21'W
Odessa, Wa., U.S.	72	47°20'N	118°42'W
Odiel, r., Spain (ō-dē-ĕl')	128	37°47'N	6°42'W
Odiham, Eng., U.K. (ŏd'ē-ăm)	114b	51°14'N	0°56'W
Odintsovo, Russia (ô-dēn'tsô-vô)	142b	55°40'N	37°16'E
Odiongan, Phil. (ō-dē-ôŋ'gän)	169a	12°24'N	121°59'E
Odivelas, Port. (ō-dē-vā'lyäs)	129b	38°47'N	9°11'W
Odobeşti, Rom. (ō-dô-bĕsh't')	125	45°46'N	27°08'E
O'Donnell, Tx., U.S. (ō-dŏn'ĕl)	78	32°59'N	101°51'W
Odorhei, Rom. (ō-dôr-hā')	125	46°18'N	25°17'E
Odra *see* Oder, r., Eur. (ô'drȧ)	112	52°40'N	14°19'E
Oeiras, Braz. (wä-ē-räzh')	101	7°05'S	42°01'W
Oeirás, Port. (ô-ĕ'y-rȧ's)	129b	38°42'N	9°18'W
Oelwein, Ia., U.S. (ōl'wīn)	71	42°40'N	91°56'W
O'Fallon, Il., U.S. (ō-făl'ŭn)	75e	38°36'N	89°55'W
O'Fallon Creek, r., Mt., U.S.	73	46°25'N	104°47'W
Ofanto, r., Italy (ô-fän'tō)	130	41°08'N	15°33'E
Offa, Nig.	189	8°09'N	4°44'E
Offenbach, Ger. (ôf'ĕn-bäk)	124	50°06'N	8°50'E
Offenburg, Ger. (ôf'ĕn-bŏorgh)	124	48°28'N	7°57'E
Ofuna, Japan (ō'fōō-nä)	167a	35°21'N	139°32'E
Ogaden Plateau, plat., Eth.	192a	6°45'N	44°53'E
Ogaki, Japan	166	35°21'N	136°36'E
Ogallala, Ne., U.S. (ō-gä-lä'lä)	70	41°08'N	101°44'W
Ogbomosho, Nig. (ŏg-bô-mō'shō)	184	8°08'N	4°15'E
Ogden, Ia., U.S. (ŏg'dĕn)	71	42°10'N	94°20'W
Ogden, Ut., U.S.	64	41°14'N	111°58'W
Ogden, r., Ut., U.S.	75b	41°16'N	111°54'W
Ogden Peak, mtn., Ut., U.S.	75b	41°11'N	111°51'W
Ogdensburg, N.J., U.S. (ŏg'dĕnz-bûrg)	68a	41°05'N	74°36'W
Ogdensburg, N.Y., U.S.	65	44°40'N	75°30'W
Ogeechee, r., Ga., U.S. (ô-gē'chē)	83	32°35'N	81°50'W
Ogies, S. Afr.	192c	26°03'S	29°04'E
Ogilvie Mountains, mts., Can. (ō'g'l-vĭ)	52	64°45'N	138°10'W
Oglesby, Il., U.S. (ō'g'lz-bĭ)	66	41°20'N	89°00'W
Oglio, r., Italy (ôl'yō)	130	45°15'N	10°19'E
Ogo, Japan (ō'gō)	167b	34°49'N	135°06'E
Ogou, r., Togo	188	8°05'N	1°30'E
Ogudnëvo, Russia (ôg-ŏŏd-nyô'vô)	142b	56°04'N	38°17'E
Ogulin, Cro. (ō-gōō-lēn')	130	45°17'N	15°11'E
Ogwashi-Uku, Nig.	189	6°10'N	6°31'E
O'Higgins, prov., Chile (ô-kē'gēns)	99b	34°17'S	70°52'W
Ohio, state, U.S. (ô'hī'ō)	65	40°30'N	83°15'W
Ohio, r., U.S.	65	37°25'N	88°05'W
Ohoopee, r., Ga., U.S. (ô-hōō'pe-mc)	83	32°32'N	82°38'W
Ohře, r., Eur. (ōr'zhĕ)	124	50°08'N	12°45'E
Ohrid, Mac. (ō'krēd)	131	41°08'N	20°46'E
Ohrid, Lake, l., Eur.	131	40°58'N	20°35'E
Ōi, Japan (oi')	167a	35°51'N	139°31'E
Oi-Gawa, r., Japan (ō'ē-gä'wȧ)	167	35°09'N	138°05'E
Oil City, Pa., U.S. (oil sĭ'tĭ)	67	41°25'N	79°40'W
Oirschot, Neth.	115a	51°30'N	5°20'E
Oise, r., Fr. (wäz)	117	49°30'N	2°56'E
Oisterwijk, Neth.	115a	51°34'N	5°13'E
Oita, Japan (ō'ē-tä)	166	33°14'N	131°38'E
Oji, Japan (ō'jē)	167b	34°36'N	135°43'E
Ojinaga, Mex. (ō-kē-nä'gä)	86	29°34'N	104°26'W
Ojitlán, Mex. (ōkē-tlän') (sän-lōō'käs)	89	18°04'N	96°23'W
Ojo Caliente, Mex. (ōkō käl-yĕn'tȧ)	88	21°50'N	100°43'W
Ojocaliente, Mex. (ô-kô-kä-lyĕ'n-tĕ)	88	22°39'N	102°15'W
Ojo del Toro, Pico, mtn., Cuba (pē'kô-ô-kō-dĕl-tô'rō)	92	19°55'N	77°25'W
Oka, Can. (ō-kä)	62a	45°28'N	74°05'W
Oka, r., Russia (ô-kä')	136	55°10'N	42°10'E
Oka, r., Russia (ō-kä')	140	53°28'N	101°09'E
Oka, r., Russia (ō-kä')	137	52°10'N	35°20'E
Okahandja, Nmb.	186	21°50'S	16°45'E
Okanagan (Okanogan), r., N.A. (ō'kȧ-näg'ăn)	55	49°06'N	119°43'W
Okanagan Lake, l., Can.	52	50°00'N	119°28'W
Okano, r., Gabon (ō'kä'nō)	184	0°15'N	11°08'E
Okanogan, Wa., U.S.	72	48°20'N	119°34'W
Okanogan, r., Wa., U.S.	72	48°36'N	119°33'W
Okatibbee, r., Ms., U.S. (ō'kä-tĭb'ē)	82	32°37'N	88°54'W
Okatoma Creek, r., Ms., U.S. (ô-kȧ-tō'mä)	82	31°43'N	89°34'W
Okavango (Cubango), r., Afr.	186	18°00'S	20°00'E
Okavango Swamp, sw., Bots.	186	19°30'S	23°02'E
Okaya, Japan (ō'kȧ-yȧ)	167	36°04'N	138°01'E
Okayama, Japan (ō'kȧ-yä'mȧ)	161	34°39'N	133°54'E
Okazaki, Japan (ō'kȧ-zä'kē)	166	34°58'N	137°09'E
Okeechobee, Fl., U.S. (ō-kē-chō'bē)	83	27°15'N	80°50'W
Okeechobee, Lake, l., Fl., U.S.	65	27°00'N	80°49'W
Okeene, Ok., U.S. (ô-kēn')	78	36°06'N	98°19'W
Okefenokee Swamp, sw., U.S. (ō'kē-fē-nō'kē)	83	30°54'N	82°20'W
Okemah, Ok., U.S. (ô-kē'mä)	79	35°26'N	96°18'W
Okene, Nig.	189	7°33'N	6°15'E
Okha, Russia (ŭ-kä')	135	53°44'N	143°12'E
Okhotino, Russia (ô-kô'tĭ-nô)	142b	56°14'N	38°24'E
Okhotsk, Russia (ô-kôtsk')	135	59°28'N	143°32'E
Okhotsk, Sea of, sea, Asia (ô-kôtsk')	135	56°45'N	146°00'E
Okhtyrka, Ukr.	137	50°18'N	34°53'E
Okinawa, i., Japan	161	26°30'N	128°00'E
Okino, i., Japan (ō'kē-nô)	167	36°22'N	133°27'E
Okino Erabu, i., Japan (ō-kē'nô-ȧ-rä'bōō)	166	27°18'N	129°00'E
Oklahoma, state, U.S. (ô-klȧ-hō'mȧ)	64	36°00'N	98°20'W
Oklahoma City, Ok., U.S.	64	35°27'N	97°32'W
Oklawaha, r., Fl., U.S. (ŏk-lȧ-wô'hô)	83	29°13'N	82°00'W
Okmulgee, Ok., U.S. (ôk-mŭl'gē)	79	35°37'N	95°58'W
Okolona, Ky., U.S. (ō-kô-lō'nȧ)	69h	38°08'N	85°41'W
Okolona, Ms., U.S.	82	33°59'N	88°43'W
Oktemberyan, Arm.	138	40°09'N	44°02'E
Okushiri, i., Japan (ō'koo-shē'rē)	166	42°12'N	139°30'E
Okuta, Nig.	189	9°14'N	3°15'E
Olalla, Wa., U.S. (ō-lä'lȧ)	74a	47°26'N	122°33'W
Olanchito, Hond. (ō'län-chē'tō)	90	15°28'N	86°35'W
Öland, i., Swe. (û-lănd')	112	57°03'N	17°15'E
Olathe, Ks., U.S. (ō-lā'thē)	75f	38°53'N	94°49'W
Olavarría, Arg. (ō-lä-vär-rē'ä)	102	36°49'N	60°15'W
Oława, Pol. (ô-lä'vȧ)	125	50°57'N	17°18'E
Olazoago, Arg. (ô-läz-kôä'gô)	99c	35°14'S	60°37'W
Olbia, Italy (ô'l-byä)	130	40°55'N	9°28'E
Olching, Ger. (ōl'kēng)	115d	48°13'N	11°21'E
Old Bahama Channel, strt., N.A. (bȧ-hā'mȧ)	92	22°45'N	78°30'W
Old Bight, Bah.	93	24°15'N	75°20'W
Old Bridge, N.J., U.S. (brĭj)	68a	40°24'N	74°22'W
Old Crow, Can. (crō)	50	67°51'N	139°58'W
Oldenburg, Ger. (ôl'dĕn-bŏorgh)	116	53°09'N	8°13'E
Old Forge, Pa., U.S. (fôrj)	67	41°20'N	75°50'W
Oldham, Eng., U.K. (ōld'ăm)	120	53°32'N	2°07'W
Oldham, co., Eng., U.K.	114a	53°35'N	2°05'W
Old Harbor, Ak., U.S. (här'bēr)	63	57°18'N	153°20'W
Old Head of Kinsale, c., Ire. (ōld hĕd ŏv kĭn-sāl)	120	51°35'N	8°35'W
Old R, Tx., U.S.	81a	29°54'N	94°52'W
Olds, Can. (ōldz)	50	51°47'N	114°06'W
Old Tate, Bots.	186	21°18'S	27°43'E
Old Town, Me., U.S. (toun)	60	44°55'N	68°42'W
Old Wives Lake, l., Can. (wīvz)	56	50°05'N	106°00'W
Olean, N.Y., U.S. (ō-lē-ăn')	65	42°05'N	78°25'W
Olecko, Pol. (ô-lĕt'skô)	125	54°02'N	22°29'E
Olekma, r., Russia (ô-lyĕk-mȧ')	141	55°41'N	120°33'E
Olëkminsk, Russia (ô-lyĕk-mĕnsk')	135	60°39'N	120°40'E
Oleksandriia, Ukr.	132	48°40'N	33°07'E
Olenëk, r., Russia (ô-lyĕ-nyôk')	135	68°00'N	113°00'E
Oléron, Île, d', i., Fr. (ēl' dō lā-rôn')	117	45°52'N	1°58'W
Oleśnica, Pol. (ô-lĕsh-nĭ'tsȧ)	117	51°13'N	17°24'E
Olfen, Ger. (ōl'fĕn)	127c	51°43'N	7°22'E
Ol'ga, Russia (ôl'gȧ)	135	43°48'N	135°44'E
Ol'gi, Zaliv, b., Russia (zä'lĭf ôl'gĭ)	166	43°43'N	135°25'E
Olhão, Port. (ôl-youn')	118	37°02'N	7°54'W
Ol'hopil', Ukr.	133	48°11'N	29°28'E
Olievenhoutpoort, S. Afr.	187b	25°58'S	27°55'E
Ólimbos, mtn., Cyp.	153a	34°56'N	32°52'E
Olinda, Braz. (ô-lē'n-dä)	101	8°00'S	34°58'W
Olinda, Braz.	102b	22°49'S	43°25'W
Oliva, Spain (ô-lē'vä)	129	38°54'N	0°07'W
Oliva de la Frontera, Spain (ô-lē'vä dä)	128	38°33'N	6°55'W
Olive Hill, Ky., U.S. (ŏl'ĭv)	66	38°15'N	83°10'W
Oliveira, Braz. (ô-lē-vā'rä)	99a	20°42'S	44°49'W
Olivenza, Spain (ô-lē-vĕn'thä)	128	38°42'N	7°06'W
Oliver, Can. (ŏ'lĭ-vēr)	50	49°11'N	119°33'W
Oliver, Can.	62g	53°38'N	113°21'W
Oliver, Wi., U.S. (ŏ'lĭvēr)	75h	46°39'N	92°12'W
Oliver Lake, l., Can.	62g	53°19'N	113°00'W
Olivia, Mn., U.S. (ō-lĭv'ē-ȧ)	70	44°46'N	95°00'W
Olivos, Arg. (ōlē'vōs)	102a	34°30'S	58°29'W
Ollagüe, Chile (ô-lyä'gȧ)	100	21°17'S	68°17'W
Ollerton, Eng., U.K. (ŏl'ēr-tŭn)	114a	53°12'N	1°02'W
Olmos Park, Tx., U.S. (ōl'mŭs pärk')	75d	29°27'N	98°32'W
Olney, Il., U.S. (ŏl'nĭ)	66	38°45'N	88°05'W
Olney, Or., U.S. (ŏl'nē)	74c	46°06'N	123°45'W
Olney, Tx., U.S.	78	33°24'N	98°43'W
Olomane, r., Can. (ō'lō mä'nē)	61	51°05'N	60°50'W
Olomouc, Czech Rep. (ō'lô-mōts)	117	49°37'N	17°15'E
Olonets, Russia (ô-lô'nĕts)	123	60°58'N	32°54'E
Olongapo, Phil.	168	14°49'S	120°17'E
Oloron, Gave d', r., Fr. (gäv-dō-lō-rôn')	126	43°21'N	0°44'W
Oloron-Sainte Marie, Fr. (ō-lô-rônt'sănt mȧ-rē')	126	43°11'N	1°37'W
Olot, Spain (ô-lōt')	118	42°09'N	2°30'E
Olpe, Ger. (ōl'pĕ)	127c	51°02'N	7°51'E
Olsnitz, Ger. (ōlz'nētz)	124	50°25'N	12°11'E
Olsztyn, Pol. (ôl'shtĭn)	116	53°47'N	20°28'E
Olt, r., Rom.	119	44°09'N	24°40'E
Olten, Switz. (ōl'tĕn)	124	47°20'N	7°53'E
Olteniţa, Rom. (ôl-tā'nĭ-tsȧ)	131	44°05'N	26°39'E
Olvera, Spain (ôl-vĕ'rä)	128	36°55'N	5°16'W
Olympia, Wa., U.S. (ô-lĭm'pĭ-ȧ)	64	47°02'N	122°52'W
Olympic Mountains, mts., Wa., U.S.	72	47°54'N	123°58'W
Olympic National Park, rec., Wa., U.S. (ô-lĭm'pĭk)	64	47°54'N	123°00'W
Ólympos, mtn., Grc.	118	40°05'N	22°21'E
Olympus, Mount, mtn., Wa., U.S. (ô-lĭm'pŭs)	72	47°43'N	123°30'W
Olyphant, Pa., U.S. (ŏl'ĭ-fănt)	67	41°30'N	75°40'W
Olyutorskiy, Mys, c., Russia (ŭl-yōō'tŏr-skē)	135	59°49'N	167°16'E
Omae-Zaki, c., Japan (ō'mä-ā zä'kē)	167	34°37'N	138°15'E
Omagh, N. Ire., U.K. (ō'mä)	120	54°35'N	7°25'W
Omaha, Ne., U.S. (ō'mȧ-hä)	65	41°18'N	95°57'W
Omaha Indian Reservation, I.R., Ne., U.S.	70	42°09'N	96°08'W
Oman, nation, Asia	154	20°00'N	57°45'E
Oman, Gulf of, b., Asia	154	24°24'N	58°58'E
Omaruru, Nmb. (ō-mä-rōō'rōō)	186	21°25'S	16°50'E
Ombrone, r., Italy (ôm-brō'nā)	130	42°48'N	11°18'E
Omdurman, Sudan	185	15°45'N	32°30'E
Omealca, Mex. (ōmā-äl'kō)	89	18°44'N	96°45'W
Ometepec, Mex. (ô-mā-tȧ-pĕk')	88	16°41'N	98°27'W
Om Hajer, Eth.	185	14°06'N	36°46'E
Omineca, r., Can. (ō-mĭ-nĕk'ȧ)	54	55°50'N	125°45'W
Omineca Mountains, mts., Can.	54	56°00'N	125°00'W
Ōmiya, Japan (ō'mē-yȧ)	167	35°54'S	139°38'E
Omo, r., Eth. (ō'mō)	185	5°54'N	36°09'E
Omoa, Hond. (ô-mō'rä)	90	15°43'N	88°03'W
Omoko, Nig.	189	5°20'N	6°39'E
Omolon, r., Russia (ō'mō)	141	67°43'N	159°15'E
Ōmori, Japan (ō'mô-rē)	167a	35°50'N	140°09'E
Omotepe, Isla de, i., Nic. (ē's-lä-dē-ō-mô-tā'pȧ)	90	11°32'N	85°30'W
Omro, Wi., U.S. (ŏm'rō)	71	44°01'N	89°46'W
Omsk, Russia (ômsk)	134	55°12'N	73°19'E
Ōmura, Japan (ō'mōō-rȧ)	167	32°56'N	129°57'E
Ōmuta, Japan (ō-mô-tȧ)	167	33°02'N	130°28'E
Omutninsk, Russia (ō'mōō-tnēnsk)	136	58°38'N	52°10'E
Onawa, Ia., U.S. (ŏn-ȧ-wȧ)	70	42°02'N	96°05'W
Onaway, Mi., U.S.	66	45°25'N	84°10'W
Oncócua, Ang.	190	16°34'S	13°28'E
Onda, Spain (ōn'dä)	129	39°58'N	0°13'W
Ondava, r., Slvk. (ōn'dȧ-vȧ)	125	48°51'N	21°40'E
Ondo, Nig.	189	7°04'N	4°47'E
Öndörhaan, Mong.	161	47°20'N	110°40'E
Onega, Russia (ô-nyĕ'gȧ)	134	63°50'N	38°08'E
Onega, r., Russia	136	63°20'N	39°20'E
Onega, Lake *see* Onezhskoye Ozero, l., Russia	136	62°02'N	34°35'E
Oneida, N.Y., U.S. (ô-nī'dȧ)	67	43°05'N	75°40'W
Oneida, l., N.Y., U.S.	67	43°10'N	76°00'W
O'Neill, Ne., U.S. (ō-nēl')	70	42°28'N	98°38'W
Oneonta, N.Y., U.S. (ō-nē-ŏn'tȧ)	67	42°25'N	75°05'W
Onezhskaja Guba, b., Russia	136	64°30'N	36°00'E
Onezhskiy, Poluostrov, pen., Russia	136	64°30'N	37°40'E
Onezhskoye Ozero, Russia (ô-năsh'skô-yĕ ô'zĕ-rô)	136	62°02'N	34°35'E
Ongiin Hiid, Mong.	160	46°00'N	102°46'E
Ongole, India	159	15°36'N	80°03'E
Onilahy, r., Madag.	187	23°41'S	45°00'E
Onitsha, Nig. (ô-nĭt'shȧ)	184	6°09'N	6°47'W
Onomichi, Japan (ō'nô-mē'chē)	166	34°27'N	133°12'E
Onon, r., Asia (ô'nôn)	135	49°00'N	112°00'E
Onoto, Ven. (ô-nō'tō)	101b	9°38'N	65°03'W
Onslow, Austl. (ŏnz'lō)	174	21°53'S	115°00'E
Onslow, b., N.C., U.S. (ŏnz'lō)	83	34°22'N	77°35'W
Ontake San, mtn., Japan (ōn'tä-kȧ sän)	166	35°55'N	137°29'E
Ontario, Ca., U.S. (ŏn-tā'rĭ-ō)	75a	34°04'N	117°39'E
Ontario, Or., U.S.	72	44°02'N	116°57'W
Ontario, prov., Can.	51	50°47'N	88°50'W
Ontario, Lake, l., N.A.	65	43°35'N	79°05'W
Ontinyent, Spain	129	38°48'N	0°35'W
Ontonagon, Mi., U.S. (ŏn-tō-năg'ŏn)	71	46°50'N	89°20'W
Ōnuki, Japan (ō'nōō-kē)	167a	35°17'N	139°51'E
Oodnadatta, Austl. (ōōd'nȧ-dȧ'tȧ)	174	27°38'S	135°40'E
Ooldea Station, Austl. (ōō-dā'ä)	174	30°35'S	132°08'E
Oologah Reservoir, res., Ok., U.S.	65	36°43'N	95°32'W
Ooltgensplaat, Neth.	115a	51°41'N	4°19'E
Oostanaula, r., Ga., U.S. (ōō-stȧ-nô'lȧ)	82	34°25'N	85°10'W
Oostende, Bel. (ōst-ĕn'dē)	117	51°14'N	2°55'E
Oosterhout, Neth.	115a	51°38'N	4°52'E
Ooster Schelde, r., Neth.	115a	51°40'N	3°40'E
Ootsa Lake, l., Can.	54	53°49'N	126°18'W
Opalaca, Sierra de, mts., Hond. (sē-sē'r-rä-dĕ-ô-pä-lä'kä)	90	14°30'N	88°29'W
Opasquia, Can. (ō-pás'kwē-ȧ)	57	53°16'N	93°53'W
Opatów, Pol. (ô-pä'tôf)	125	50°47'N	21°25'E
Opava, Czech Rep. (ō'pä-vä)	125	49°56'N	17°52'E

ăt; finăl; rāte; senâte; ärm; åsk; sofȧ; fâre; ch-choose; dh-as th in other; bē; ēvent; bĕt; recĕnt; cratēr; g-gō; gh-guttural g; bĭt; ĭ-short neutral; rīde; κ-guttural k as ch in German ich;

PRONOUNCING INDEX Opelika – Ouro Fino 285

PLACE (Pronunciation)	PAGE	LAT.	LONG.
Opelika, Al., U.S. (ŏp-ė-lī'kȧ)	82	32°39'N	85°23'W
Opelousas, La., U.S. (ŏp-ē-lōō'sȧs)	81	30°33'N	92°04'W
Opeongo, l., Can. (ŏp-ė-ŏŋ'gō)	59	45°40'N	78°20'W
Opheim, Mt., U.S. (ō-fīm')	73	48°51'N	106°19'W
Ophir, Ak., U.S. (ō'fēr)	63	63°10'N	156°28'W
Ophir, Mount, mtn., Malay.	153b	2°22'N	102°37'E
Opico, El Sal. (ō-pē'kō)	90	13°50'N	89°23'W
Opinaca, r., Can. (ŏp-ĭ-nä'kȧ)	53	52°28'N	77°40'W
Opishnia, Ukr.	133	49°57'N	34°34'E
Opladen, Ger. (ŏp'lä-dĕn)	127c	51°04'N	7°00'E
Opobo, Nig.	189	4°34'N	7°27'E
Opochka, Russia (ō-pôch'kȧ)	136	56°43'N	28°39'E
Opoczno, Pol. (ō-pôch'nō)	125	51°22'N	20°18'E
Opole, Pol. (ō-pôl'ä)	117	50°42'N	17°55'E
Opole Lubelskie, Pol. (ō-pō'lă lōō-bĕl'skyĕ)	125	51°09'N	21°58'E
Opp, Al., U.S. (ŏp)	82	31°18'N	86°15'W
Oppdal, Nor. (ôp'däl)	122	62°37'N	9°41'E
Opportunity, Wa., U.S. (ŏp-ŏr tū'nĭ tĭ)	72	47°37'N	117°20'W
Oquirrh Mountains, mts., Ut., U.S. (ō'kwēr)	75b	40°38'N	112°11'W
Oradea, Rom. (ō-räd'yä)	110	47°02'N	21°55'E
Oral, Kaz.	139	51°14'N	51°22'E
Oran, Alg. (ō-rän)(ō-räN')	184	35°46'N	0°45'W
Orán, Arg. (ō-rá'n)	102	23°13'S	64°17'W
Oran, Mo., U.S. (ôr'ăn)	79	37°05'N	89°39'W
Oran, Sebkha d', l., Alg.	129	35°28'N	0°28'W
Orange, Austl. (ŏr'ĕnj)	175	33°15'S	149°08'E
Orange, Fr. (ō-räNzh')	117	44°08'N	4°48'E
Orange, Ca., U.S.	75a	33°48'N	117°51'W
Orange, Ct., U.S.	67	41°15'N	73°00'W
Orange, N.J., U.S.	68a	40°46'N	74°14'W
Orange, Tx., U.S.	79	30°07'N	93°44'W
Orange, r., Afr.	186	29°15'S	17°30'E
Orange, Cabo, c., Braz. (kȧ-bô-rȧ'n-zhĕ)	101	4°25'N	51°30'W
Orangeburg, S.C., U.S. (ŏr'ĕnj-bûrg)	83	33°30'N	80°50'W
Orange Cay, i., Bah. (ŏr-ĕnj kē)	92	24°55'N	79°05'W
Orange City, Ia., U.S.	70	43°01'N	96°06'W
Orange Lake, l., Fl., U.S.	83	29°30'N	82°12'W
Orangeville, Can. (ŏr'ĕnj-vĭl)	59	43°55'N	80°06'W
Orangeville, S. Afr.	192c	27°05'S	28°13'E
Orange Walk, Belize (wôl''k)	90a	18°09'N	88°32'W
Orani, Phil. (ō-rä'nē)	169a	14°47'N	120°32'E
Oranienburg, Ger. (ō-rä'nē-ĕn-bŏrgh)	124	52°45'N	13°14'E
Oranjemund, Nmb.	186	28°33'S	16°20'E
Orăştie, Rom. (ō-rûsh'tyȧ)	131	45°50'N	23°14'E
Orbetello, Italy (ôr-bȧ-tĕl'lō)	130	42°27'N	11°15'E
Orbigo, r., Spain (ôr-bē'gō)	128	42°30'N	5°55'W
Orbost, Austl. (ŏr'bŭst)	176	37°43'S	148°20'E
Orcas, i., Wa., U.S. (ôr'kȧs)	74d	48°43'N	122°52'W
Orchard Farm, Mo., U.S. (ôr'chērd färm)	75e	38°53'N	90°27'W
Orchard Park, N.Y., U.S.	69c	42°46'N	78°46'W
Orchards, Wa., U.S. (ôr'chĕdz)	74c	45°40'N	122°33'W
Orchila, Isla, i., Ven.	100	11°47'N	66°34'W
Ord, Ne., U.S. (ôrd)	70	41°35'N	98°57'W
Ord, r., Austl.	174	17°30'S	128°40'E
Ord, Mount, mtn., Az., U.S.	77	33°55'N	109°40'W
Orda, Kaz. (ŏr'dȧ)	137	48°50'N	47°30'E
Orda, Russia (ôr'dȧ)	142a	57°10'N	57°12'E
Ordes, Spain	128	43°00'N	8°24'W
Ordos Desert, des., China	160	39°12'N	108°10'E
Ordu, Tur. (ôr'dŏ)	119	41°00'N	37°50'E
Ordway, Co., U.S. (ôrd'wā)	78	38°11'N	103°46'W
Örebro, Swe. (û'rē-brō)	116	59°16'N	15°11'E
Oredezh, r., Russia (ô'rĕ-dĕzh)	142c	59°23'N	30°21'E
Oregon, Il., U.S.	71	42°01'N	89°21'W
Oregon, state, U.S.	64	43°40'N	121°50'W
Oregon Caves National Monument, rec., Or., U.S. (cāvz)	72	42°05'N	123°13'W
Oregon City, Or., U.S.	74c	45°21'N	122°36'W
Öregrund, Swe. (û-rē-grŏnd)	122	60°20'N	18°26'E
Orekhovo, Blg.	131	43°43'N	23°59'E
Orekhovo-Zuyevo, Russia (ôr-yĕ'kô-vô zô'yĕ-vô)	134	55°46'N	39°00'E
Orël, Russia (ôr-yôl')	134	52°59'N	36°05'E
Orël, prov., Russia	132	52°35'N	36°08'E
Orem, Ut., U.S. (ō'rĕm)	77	40°15'N	111°50'W
Ore Mountains *see* Erzgebirge, mts., Eur.	112	50°29'N	12°40'E
Orenburg, Russia (ō'rĕn-bōōrg)	134	51°50'N	55°05'E
Øresund, strt., Eur.	122	55°50'N	12°40'E
Órganos, Sierra de los, mts., Cuba (sē-ĕ'r-rä-dĕ-lôs-ŏ'r-gä-nôs)	92	22°20'N	84°10'W
Organ Pipe Cactus National Monument, rec., Az., U.S. (ôr'găn pīp kăk'tŭs)	77	32°14'N	113°05'W
Orgãos, Serra das, mtn., Braz. (sĕ'r-rä-däs-ôr-gouN's)	99a	22°30'S	43°01'W
Orhei, Mol.	137	47°27'N	28°49'E
Orhon, r., Mong.	160	48°33'N	103°07'E
Oriental, Cordillera, mts., Col. (kôr-dēl-yĕ'rä)	100a	3°30'N	74°27'W
Oriental, Cordillera, mts., Dom. Rep. (kôr-dēl-yĕ'rä-ō-ryĕ'n-täl)	93	18°55'N	69°40'W
Oriental, Cordillera, mts., S.A. (kôr-dēl-yĕ'rä ō-rē-ĕn-täl')	100	14°00'S	68°33'W
Orikhiv, Ukr.	133	47°34'N	35°51'E
Oril', r., Ukr.	133	49°08'N	34°55'E
Orillia, Can. (ō-rĭl'ĭ-ȧ)	51	44°35'N	79°25'W
Orin, Wy., U.S.	73	42°40'N	105°10'W
Orinda, Ca., U.S.	74b	37°53'N	122°11'W
Orinoco, r., Ven. (ō-rĭ-nō'kō)	100	8°32'N	63°13'W
Oriola, Spain	129	38°04'N	0°55'W
Orion, Phil. (ō-rē-ōn')	169a	14°37'N	120°34'E
Orissa, state, India (ō-rĭs'ȧ)	155	25°09'N	83°50'E
Oristano, Italy (ō-rēs-tä'nō)	118	39°53'N	8°38'E
Oristano, Golfo di, b., Italy (gôl-fō-dē-ō-rēs-tä'nō)	130	39°53'N	8°12'E
Orituco, r., Ven. (ō-rē-tōō'kō)	101b	9°37'N	66°25'W
Oriuco, r., Ven. (ō-rēōō'kō)	101b	9°36'N	66°25'W
Orivesi, l., Fin.	123	62°15'N	29°55'E
Orizaba, Mex. (ō-rē-zä'bä)	87	18°52'N	97°05'E
Orizaba, Pico de, vol., Mex.	86	19°04'N	97°14'W
Orkanger, Nor.	122	63°19'N	9°54'W
Orkla, r., Nor. (ôr'klȧ)	122	62°55'N	9°50'E
Orkney, S. Afr. (ôrk'nĭ)	192c	26°58'S	26°39'E
Orkney Islands, is., Scot., U.K.	112	59°01'N	2°08'W
Orlando, S. Afr. (ôr-lăn-dō)	187b	26°15'S	27°56'E
Orlando, Fl., U.S. (ôr-lăn'dō)	65	28°32'N	81°22'W
Orland Park, Il., U.S. (ôr-lăn')	69a	41°38'N	87°52'W
Orleans, Can. (ôr-lā-äN')	62c	45°28'N	75°31'W
Orléans, Fr. (ŏr-lā-äN')	110	47°55'N	1°56'E
Orleans, In., U.S. (ôr-lēnz')	66	38°40'N	86°25'W
Orléans, Île d', i., Can.	59	46°56'N	70°57'W
Orly, Fr.	127b	48°45'N	2°24'E
Ormond Beach, Fl., U.S. (ôr'mŏnd)	83	29°15'N	81°05'W
Ormskirk, Eng., U.K. (ôrms'kērk)	114a	53°34'N	2°53'W
Ormstown, Can. (ôrms'toun)	62a	45°07'N	74°00'W
Orneta, Pol. (ôr-nyĕ'tä)	125	54°07'N	20°10'E
Örnsköldsvik, Swe. (ûrn'skŏlts-vĕk)	116	63°10'N	18°32'E
Oro, Río del, r., Mex. (rē'ō dĕl ō'rō)	88	18°04'N	100°59'W
Oro, Río del, r., Mex.	80	26°04'N	105°40'W
Orobie, Alpi, mts., Italy (äl'pē-ô-rô'byĕ)	130	46°05'N	9°47'E
Oron, Nig.	189	4°48'N	8°14'E
Orosei, Golfo di, b., Italy (gôl-fō-dē-ō-rô-sā'ē)	130	40°12'N	9°45'E
Orosháza, Hung. (ō-rôsh-hä'sô)	125	46°33'N	20°31'E
Orosi, vol., C.R. (ō-rō'sē)	90	11°00'N	85°30'W
Oroville, Ca., U.S. (ŏr'ō-vĭl)	76	39°29'N	121°34'W
Oroville, Wa., U.S.	72	48°55'N	119°25'W
Oroville, Lake, res., Ca., U.S.	76	39°32'N	121°25'W
Orreagal, Spain	128	43°00'N	1°17'W
Orrville, Oh., U.S. (ôr'vĭl)	66	40°45'N	81°50'W
Orsa, Swe. (ōr'sä)	122	61°08'N	14°35'E
Orsha, Bela. (ôr'shȧ)	136	54°29'N	30°28'E
Orsk, Russia (ôrsk)	134	51°15'N	58°50'E
Orşova, Rom. (ôr'shō-vä)	131	44°43'N	22°26'E
Ortega, Col. (ôr-tĕ'gä)	100a	3°56'N	75°12'W
Ortegal, Cabo, c., Spain (kȧ'bô-ôr-tȧ-gäl')	118	43°46'N	8°15'W
Orth, Aus.	115e	48°09'N	16°42'E
Orthez, Fr. (ŏr-tĕz')	127	43°29'N	0°43'W
Órthrys, Óros, mtn., Grc.	131	39°00'N	22°15'E
Ortigueira, Spain (ôr-tē-gā'ē-rä)	118	43°40'N	7°50'W
Orting, Wa., U.S. (ôrt'ĭng)	74a	47°06'N	122°12'W
Ortona, Italy (ôr-tō'nä)	130	42°22'N	14°22'E
Ortonville, Mn., U.S. (ôr-tŭn-vĭl)	70	45°18'N	96°26'W
Orūmīyeh, Iran	154	37°30'N	45°15'E
Orūmīyeh, Daryacheh-ye, l., Iran	154	38°01'N	45°17'E
Oruro, Bol. (ō-rōō'rō)	100	17°57'S	66°59'W
Orvieto, Italy (ôr-vyâ'tō)	130	42°43'N	12°08'E
Osa, Russia (ō'sȧ)	136	57°18'N	55°25'E
Osa, Península de, pen., C.R. (ō'sä)	91	8°30'N	83°25'W
Osage, Ia., U.S. (ō'sāj)	71	43°16'N	92°49'W
Osage, r., Mo., U.S.	79	38°10'N	93°12'W
Osage City, Ks., U.S. (ō'sāj sĭ'tĭ)	79	38°28'N	95°53'W
Ōsaka, Japan (ō'sä-kä)	161	34°40'N	135°27'E
Ōsaka, dept., Japan	167b	34°45'N	135°36'E
Ōsaka-Wan, b., Japan (wän)	166	34°34'N	135°16'E
Osakis, Mn., U.S. (ō-sā'kĭs)	70	45°51'N	95°09'W
Osakis, l., Mn., U.S.	71	45°55'N	94°55'W
Osawatomie, Ks., U.S. (ŏs-ȧ-wăt'ō-mē)	79	38°29'N	94°57'W
Osborne, Ks., U.S. (ŏz'bŭrn)	78	39°25'N	98°42'W
Osceola, Ar., U.S. (ŏs-ē-ō'lȧ)	79	35°42'N	89°58'W
Osceola, Ia., U.S.	71	41°04'N	93°45'W
Osceola, Mo., U.S.	79	38°02'N	93°41'W
Osceola, Ne., U.S.	70	41°11'N	97°34'W
Oscoda, Mi., U.S. (ŏs-kō'dȧ)	66	44°25'N	83°20'W
Osëtr, r., Russia (ō'sĕt'r)	132	54°27'N	38°15'E
Osgood, In., U.S. (ŏz'gŏd)	66	39°10'N	85°20'W
Osgoode, Can.	62c	45°09'N	75°37'W
Osh, Kyrg. (ôsh)	139	40°33'N	72°48'E
Oshawa, Can. (ŏsh'ȧ-wȧ)	51	43°50'N	78°50'W
Ōshima, i., Japan (ō'shē'mä)	167	34°47'N	139°35'E
Oshkosh, Ne., U.S. (ŏsh'kŏsh)	70	41°24'N	102°22'W
Oshkosh, Wi., U.S.	65	44°01'N	88°35'W
Oshogbo, Nig.	184	7°47'N	4°34'E
Osijek, Cro. (ôs'ĭ-yĕk)	119	45°33'N	18°48'E
Osinniki, Russia (ŭ-sē'nyĭ-kē)	140	53°37'N	87°21'E
Oskaloosa, Ia., U.S. (ŏs-kȧ-lōō'sȧ)	71	41°16'N	92°40'W
Oskarshamn, Swe. (ôs'kärs-häm'n)	122	57°16'N	16°24'E
Oskarström, Swe. (ôs'kärs-strûm)	122	56°48'N	12°55'E
Öskemen, Kaz.	139	49°58'N	82°38'E
Oskil, r., Eur.	137	51°00'N	37°41'E
Oslo, Nor. (ôs'lō)	110	59°56'N	10°41'E
Oslofjorden, b., Nor.	122	59°03'N	10°35'E
Osmaniye, Tur.	119	37°10'N	36°30'E
Osnabrück, Ger. (ôs-nä-brük')	124	52°16'N	8°05'E
Osorno, Chile (ō-sō'r-nō)	102	40°42'S	73°13'W
Osøyra, Nor.	122	60°24'N	5°22'E
Osprey Reef, rf., Austl. (ŏs'prā)	175	14°00'S	146°45'E
Ossa, Mount, mtn., Austl. (ŏsȧ)	175	41°45'S	146°05'E
Osseo, Mn., U.S. (ŏs'sē-ō)	75g	45°07'N	93°24'W
Ossining, N.Y., U.S. (ŏs'ĭ-nĭng)	68a	41°09'N	73°51'W
Ossipee, N.H., U.S. (ŏs'ĭ-pē)	60	43°42'N	71°08'W
Ossjøen, l., Nor. (ôs-syûĕn)	122	61°20'N	12°00'E
Ostashkov, Russia (ôs-täsh'kôf)	136	57°07'N	33°04'E
Oster, Ukr. (ōs'tĕr)	133	50°55'N	30°52'E
Osterdalälven, r., Swe.	116	61°40'N	13°00'E
Osterfjord, b., Nor. (ûs'tĕr fyôr')	122	60°40'N	5°25'E
Östersund, Swe. (ûs'tĕr-sōōnd)	116	63°09'N	14°49'E
Östhammar, Swe. (ûst'häm'är)	122	60°16'N	18°21'E
Ostrava, Czech Rep.	110	49°51'N	18°18'E
Ostróda, Pol. (ôs'trŏt-ȧ)	125	53°41'N	19°58'E
Ostrogozhsk, Russia (ôs-tr-gôzhk')	137	50°53'N	39°03'E
Ostroh, Ukr.	137	50°21'N	26°40'E
Ostrołęka, Pol. (ôs-trō-woN'kȧ)	125	53°04'N	21°35'E
Ostrov, Russia (ôs-trôf')	136	57°21'N	28°22'E
Ostrowiec Świętokrzyski, Pol. (ôs-trō'vyĕts shvyĕN-tō-kzhĭ'ske)	117	50°55'N	21°24'E
Ostrów Lubelski, Pol. (ôs'trôf lōō'bĕl-skĭ)	125	51°32'N	22°49'E
Ostrów Mazowiecka, Pol. (mä-zô-vyĕt'skä)	117	52°47'N	21°54'E
Ostrów Wielkopolski, Pol. (ôs'trōōf vyĕl-kō-pōl'skē)	117	51°38'N	17°49'E
Ostrzeszów, Pol. (ôs-tzhā'shôf)	125	51°26'N	17°56'E
Ostuni, Italy (ôs-tōō'nē)	131	40°44'N	17°35'E
Osum, r., Alb. (ō'sòm)	131	40°37'N	20°00'E
Osuna, Spain (ō-sōō'nä)	128	37°18'N	5°05'W
Osveya, Bela. (ôs'vē-yȧ)	132	56°00'N	28°08'E
Oswaldtwistle, Eng., U.K. (ŏz-wăld-twĭs''l)	114a	53°44'N	2°23'W
Oswegatchie, r., N.Y., U.S. (ŏs-wē-găch'ĭ)	67	44°15'N	75°20'W
Oswego, Ks., U.S. (ŏs-wē'gō)	79	37°10'N	95°08'W
Oswego, N.Y., U.S.	65	43°25'N	76°30'W
Oświęcim, Pol. (ôsh-vyăN'tsyĭm)	125	50°02'N	19°17'E
Otaru, Japan (ō'tȧ-rŏ)	161	43°07'N	141°00'E
Otavalo, Ec. (ōtä-vä'lō)	100	0°14'N	78°16'W
Otavi, Nmb. (ō-tä'vē)	186	19°35'S	17°20'E
Otay, Ca., U.S. (ō'tā)	76a	32°36'N	117°04'W
Otepää, Est.	123	58°03'N	26°30'E
Oti, r., Afr.	188	9°00'N	0°10'E
Otish, Monts, mts., Can. (ō-tĭsh')	53	52°15'N	70°20'W
Otjiwarongo, Nmb. (ŏt-jē-wȧ-rôn'gō)	186	20°20'S	16°25'E
Otočac, Cro. (ō'tô-chȧts)	130	44°53'N	15°15'E
Otra, r., Nor.	122	59°13'N	7°20'E
Otra, r., Russia (ŏt'rȧ)	142b	55°22'N	38°20'E
Otradnoye, Russia (ô-trȧ'd-nôyĕ)	142c	59°46'N	30°50'E
Otranto, Italy (ō'trän-tô) (ô-trän'tō)	131	40°07'N	18°30'E
Otranto, Strait of, strt., Eur.	112	40°30'N	18°45'E
Otsego, Mi., U.S. (ŏt-sē'gō)	66	42°25'N	85°45'W
Otsu, Japan (ō'tsŏ)	166	35°00'N	135°54'E
Otta, l., Nor. (ŏt'tȧ)	122	61°53'N	8°40'E
Ottawa, Can. (ŏt'ȧ-wȧ)	51	45°25'N	75°43'W
Ottawa, Il., U.S.	66	41°20'N	88°50'W
Ottawa, Ks., U.S.	79	38°37'N	95°16'W
Ottawa, Oh., U.S.	66	41°00'N	84°00'W
Ottawa, r., Can.	53	46°05'N	77°20'W
Otter Creek, r., Ut., U.S. (ŏt'ĕr)	77	38°20'N	111°55'W
Otter Creek, r., Vt., U.S.	67	44°05'N	73°15'W
Otter Point, c., Can.	74a	48°21'N	123°50'W
Otter Tail, l., Mn., U.S.	70	46°21'N	95°52'W
Otterville, Il., U.S. (ŏt'ēr-vĭl)	75e	39°03'N	90°24'W
Ottery, S. Afr. (ŏt'ĕr-ĭ)	186a	34°02'S	18°31'E
Ottumwa, Ia., U.S. (ō-tŭm'wȧ)	65	41°00'N	92°26'W
Otukpa, Nig.	189	7°09'N	7°41'E
Otumba, Mex. (ō-tŭm'bä)	88	19°41'N	98°46'W
Otway, Cape, c., Austl. (ŏt'wā)	175	38°55'S	153°40'E
Otway, Seno, b., Chile (sĕ'nō-ō't-wä'y)	102	53°00'S	73°00'W
Otwock, Pol. (ôt'vôtsk)	125	52°05'N	21°18'E
Ouachita, r., U.S.	65	33°25'N	92°30'W
Ouachita Mountains, mts., U.S. (wŏsh'ĭ-tô)	65	34°29'N	95°01'W
Ouagadougou, Burkina (wä'gȧ-dōō'gōō)	184	12°22'N	1°31'W
Ouahigouya, Burkina (wä-ē-gōō'yä)	184	13°35'N	2°25'W
Oualâta, Maur. (wä-lä'tä)	184	17°11'N	6°50'W
Ouallene, Alg. (wäl-lân')	184	24°43'N	1°15'E
Ouanaminthe, Haiti	93	19°35'N	71°45'W
Ouarane, reg., Maur.	184	20°44'N	10°27'W
Ouarkoye, Burkina	188	12°05'N	3°40'W
Ouassel, r., Alg.	129	35°30'N	1°55'E
Oubangui (Ubangi), r., Afr. (ōō-băŋ'gē)	190	4°30'N	20°35'E
Oude Rijn, r., Neth.	115a	52°09'N	4°33'E
Oudewater, Neth.	115a	52°01'N	4°52'E
Oud-Gastel, Neth.	115a	51°35'N	4°27'E
Oudtshoorn, S. Afr. (outs'hōrn)	186	33°33'S	23°36'E
Oued Rhiou, Alg.	129	35°55'N	0°57'E
Oued Tlelat, Alg.	129	35°33'N	0°28'W
Oued-Zem, Mor. (wĕd-zĕm')	184	33°05'N	5°49'W
Ouessant, Island d', i., Fr. (ēl-dwĕ-säN')	117	48°28'N	5°00'W
Ouesso, Congo	185	1°37'N	16°04'E
Ouest, Point, c., Haiti	93	19°00'N	73°25'W
Ouezzane, Mor. (wĕ-zan')	184	34°48'N	5°40'W
Ouham, r., Afr.	189	8°30'N	17°50'E
Ouidah, Benin (wē-dä')	184	6°25'N	2°05'E
Oujda, Mor.	184	34°41'N	1°45'W
Oulins, Fr. (ōō-lăN')	127b	48°52'N	1°27'E
Oullins, Fr. (ōō-lăN')	126	45°44'N	4°46'E
Oulu, Fin. (ō'lŏ)	110	64°58'N	25°43'E
Oulujärvi, l., Fin.	116	64°20'N	25°48'E
Oum Chalouba, Chad (ōōm shä-lōō'bä)	185	15°48'N	20°30'E
Oum Hadjer, Chad	189	13°18'N	19°41'E
Ounas, r., Fin. (ō'nȧs)	116	67°46'N	24°40'E
Oundle, Eng., U.K. (ōn'd'l)	114a	52°28'N	0°28'W
Ounianga Kébir, Chad (ōō-nē-äŋ'gȧ kē-bēr')	185	19°04'N	20°22'E
Ouray, Co., U.S. (ōō-rā')	78	38°00'N	107°40'W
Ourense, Spain	128	42°20'N	7°52'W
Ourinhos, Braz. (ôō-rē'nyōs)	101	23°04'S	49°45'W
Ourique, Port. (ō-rē'kĕ)	128	37°39'N	8°10'W
Ouro Fino, Braz. (ōū-rô-fē'nō)	99a	22°18'S	46°21'W

ng-sing; ŋ-baŋk; N-nasalized n; nŏd; cŏmmit; ōld; ôbey; ôrder; oi-boil; fōōd; ò-as oo in foot; ou-out; s-soft; sh-dish; th-thin; pūre; ûnite; ûrn; stŭd; circŭs; ü-as in French tu; '-indeterminate vowel.

286 Ouro Prêto – Paracale PRONOUNCING INDEX

PLACE (Pronunciation)	PAGE	LAT.	LONG.
Ouro Prêto, Braz. (ō′rò prā′tò)	102	20°24′S	43°30′W
Outardes, Rivière aux, r., Can.	53	50°53′N	68°50′W
Outer, i., Wi., U.S. (out′ēr)	71	47°03′N	90°20′W
Outer Brass, i., V.I.U.S. (bräs)	87c	18°24′N	64°58′W
Outer Hebrides, is., Scot., U.K.	120	57°20′N	7°50′W
Outjo, Nmb. (ōt′yō)	186	20°05′S	17°10′E
Outlook, Can.	56	51°31′N	107°05′W
Outremont, Can. (o͞o-trē-môN′)	62a	45°31′N	73°36′W
Ouvéa, i., N. Cal.	175	20°43′S	166°48′E
Ouyen, Austl. (o͞o-ĕn)	176	35°05′S	142°10′E
Ovalle, Chile (ō-väl′yä)	102	30°43′S	71°16′W
Ovando, Bahía de, b., Cuba (bä-ē′ä-dĕ-ō-vä′n-dō)	93	20°10′N	74°05′W
Ovar, Port. (ô-vär′)	128	40°52′N	8°38′W
Overijse, Bel.	115a	50°46′N	4°32′E
Overland, Mo., U.S. (ō-vēr-lănd)	75e	38°42′N	90°22′W
Overland Park, Ks., U.S.	75f	38°59′N	94°40′W
Overlea, Md., U.S. (ō′vēr-lā)(ō′vēr-lē)	68e	39°21′N	76°31′W
Övertornea, Swe.	116	66°19′N	23°31′E
Ovidiopol′, Ukr.	133	46°15′N	30°28′E
Oviedo, Dom. Rep. (ô-vyĕ′dō)	93	17°50′N	71°25′W
Oviedo, Spain (ō-vē-ā′dhō)	110	43°22′N	5°50′W
Ovruch, Ukr.	133	51°19′N	28°51′E
Owada, Japan (ō′wä-dä)	167a	35°49′N	139°33′E
Owambo, hist. reg., Nmb.	186	18°10′S	15°00′E
Owando, Congo	186	0°29′S	15°55′E
Owasco, l., N.Y., U.S. (ō-wăsk′kō)	67	42°50′N	76°30′W
Owase, Japan (ō′wä-shĕ)	167	34°03′N	136°12′E
Owego, N.Y., U.S. (ō-wē′gō)	67	42°05′N	76°15′W
Owen, Wi., U.S. (ō′ĕn)	71	44°56′N	90°35′W
Owensboro, Ky., U.S. (ō′ĕnz-bŭr-ô)	65	37°45′N	87°05′W
Owens Lake, l., Ca., U.S.	76	37°13′N	118°20′W
Owen Sound, Can. (ō′ĕn)	51	44°30′N	80°55′W
Owen Stanley Range, mts., Pap. N. Gui. (stăn′lē)	169	9°00′S	147°30′E
Owensville, In., U.S. (ō′ĕnz-vĭl)	66	38°15′N	87°40′W
Owensville, Mo., U.S.	79	38°20′N	91°29′W
Owensville, Oh., U.S.	69f	39°08′N	84°07′W
Owenton, Ky., U.S. (ō′ĕn-tŭn)	66	38°35′N	84°55′W
Owerri, Nig. (ô-wĕr′ē)	184	5°26′N	7°04′E
Owings Mill, Md., U.S. (ōwĭngz mĭl)	68e	39°25′N	76°50′W
Owl Creek, r., Wy., U.S. (oul)	73	43°45′N	108°46′W
Owo, Nig.	189	7°15′N	5°37′E
Owosso, Mi., U.S. (ô-wŏs′ō)	66	43°00′N	84°15′W
Owyhee, r., U.S.	64	43°04′N	117°45′W
Owyhee, Lake, res., Or., U.S.	64	43°27′N	117°30′W
Owyhee, South Fork, r., Id., U.S.	72	42°07′N	116°43′W
Owyhee Mountains, mts., Id., U.S. (ô-wī′hē)	64	43°15′N	116°48′W
Oxbow, Can.	57	49°12′N	102°11′W
Oxchuc, Mex. (ôs-cho͞ok′)	89	16°47′N	92°24′W
Oxford, Can. (ŏks′fērd)	60	45°44′N	63°52′W
Oxford, Eng., U.K.	117	51°43′N	1°16′W
Oxford, Al., U.S. (ŏks′fērd)	83	33°38′N	80°46′W
Oxford, Ma., U.S.	61a	42°07′N	71°52′W
Oxford, Mi., U.S.	66	42°50′N	83°15′W
Oxford, Ms., U.S.	82	34°22′N	89°30′W
Oxford, N.C., U.S.	83	36°17′N	78°35′W
Oxford, Oh., U.S.	66	39°30′N	84°45′W
Oxford Lake, l., Can.	57	54°51′N	95°37′W
Oxfordshire, co., Eng., U.K.	114b	51°36′N	1°30′W
Oxkutzcab, Mex. (ôx-ko͞o′tz-käb)	90a	20°18′N	89°22′W
Oxmoor, Al., U.S. (ŏks′mòr)	68h	33°25′N	86°52′W
Oxnard, Ca., U.S. (ŏks′närd)	76	34°08′N	119°12′W
Oxon Hill, Md., U.S. (ŏks′ōn hĭl)	68e	38°48′N	77°00′W
Oyapock, r., S.A. (ō-yä-pŏk′)	101	2°45′N	52°15′W
Oyem, Gabon	184	1°37′N	11°35′E
Øyeren, l., Nor. (ûĭērĕn)	122	59°50′N	11°25′E
Oymyakon, Russia (oi-myŭ-kôn′)	135	63°14′N	142°58′E
Oyo, Nig. (ō′yō)	184	7°51′N	3°56′E
Oyonnax, Fr. (ô-yō-näks′)	127	46°16′N	5°40′E
Oyster Bay, N.Y., U.S.	68a	40°52′N	73°32′W
Oyster Bayou, Tx., U.S.	81a	29°41′N	94°33′W
Oyster Creek, r., Tx., U.S. (ois′tēr)	81a	29°13′N	95°29′W
Oyyl, r., Kaz.	137	49°30′N	55°10′E
Ozama, r., Dom. Rep. (ô-zä′mä)	93	18°45′N	69°55′W
Ozamiz, Phil. (ô-zä′mēz)	169	8°06′N	123°43′E
Ozark, Al., U.S. (ō′zärk)	82	31°28′N	85°28′W
Ozark, Ar., U.S.	79	35°29′N	93°49′W
Ozark Plateau, plat., U.S.	65	36°37′N	93°56′W
Ozarks, Lake of the, l., Mo., U.S. (ō′zärksz)	65	38°06′N	93°26′W
Ozëry, Russia (ô-zyô′rê)	132	54°53′N	38°31′E
Ozieri, Italy	118	40°38′N	8°53′E
Ozorków, Pol. (ô-zôr′kôf)	125	51°58′N	19°20′E
Ozuluama, Mex.	89	21°34′N	97°52′W
Ozumba, Mex.	89a	19°02′N	98°48′W
Ozurgeti, Geor.	138	41°56′N	42°00′E
P			
Paarl, S. Afr. (pärl)	186	33°45′S	18°55′E
Pa′auilo, Hi., U.S. (pä-ä-o͞o′ē-lō)	84a	20°03′N	155°25′W
Pabianice, Pol. (pä-byä-nē′tsĕ)	125	51°40′N	19°29′E
Pacaás Novos, Massiço de, mts., Braz.	100	11°03′S	64°02′W
Pacaraima, Serra, mts., S.A. (sĕr′rá pä-kä-rä-ē′mä)	100	3°45′N	62°30′W
Pacasmayo, Peru (pä-käs-mä′yō)	100	7°24′S	79°30′W
Pachuca, Mex. (pä-cho͞o′kä)	86	20°07′N	98°43′W
Pacific, Wa., U.S. (pá-sĭf′ĭk)	74a	47°16′N	122°15′W
Pacifica, Ca., U.S. (pá-sĭf′ĭ-kä)	74b	37°38′N	122°29′W
Pacific Beach, Ca., U.S.	76a	32°47′N	117°22′W
Pacific Grove, Ca., U.S.	76	36°37′N	121°54′W
Pacific Islands, Trust Territory of the *see* Palau, nation, Oc.	3	7°15′N	134°30′E
Pacific Ocean, o.	2	0°00′	170°00′W
Pacific Ranges, mts., Can.	54	51°00′N	125°30′W
Pacific Rim National Park, rec., Can.	54	49°00′N	126°00′W
Pacolet, r., S.C., U.S. (pă′cō-lĕt)	83	34°55′N	81°49′W
Pacy-sur-Eure, Fr. (pä-sē-sür-ûr′)	127b	49°01′N	1°24′E
Padang, Indon. (pä-däng′)	168	1°01′S	100°28′E
Padang, i., Indon.	153b	1°12′N	102°21′E
Padang Endau, Malay.	153b	2°39′N	103°38′E
Paden City, W.V., U.S. (pā′dĕn)	66	39°30′N	80°55′W
Paderborn, Ger. (pä-dĕr-bôrn′)	124	51°43′N	8°46′E
Padibe, Ug.	191	3°28′N	32°50′E
Padiham, Eng., U.K. (păd′ĭ-hăm)	114a	53°48′N	2°19′W
Padilla, Mex. (pä-dēl′yä)	88	24°00′N	98°45′W
Padilla Bay, b., Wa., U.S. (pä-dēl′lá)	74a	48°31′N	122°34′W
Padova, Italy (pä′dô-vä)(păd′û-á)	118	45°24′N	11°53′E
Padre Island, i., Tx., U.S. (pä′drā)	81	27°09′N	97°15′W
Padua *see* Padova, Italy	118	45°24′N	11°53′E
Paducah, Ky., U.S.	65	37°05′N	88°36′W
Paducah, Tx., U.S.	78	34°01′N	100°18′W
Paektu-san, mtn., Asia (pâk′to͞o-sän′)	166	42°00′N	128°03′E
Pag, i., Serb. (päg)	130	44°30′N	14°48′E
Pagai Selatan, Pulau, i., Indon.	168	2°48′S	100°22′E
Pagai Utara, Pulau, i., Indon.	168	2°45′S	100°02′E
Pagasitikós Kólpos, b., Grc.	131	39°15′N	23°00′E
Page, Az., U.S.	77	36°57′N	111°27′W
Pago Pago, Am. Sam.	170a	14°16′S	170°42′W
Pagosa Springs, Co., U.S. (pá-gō′sá)	78	37°15′N	107°05′W
Pāhala, Hi., U.S. (pä-hä′lä)	84a	19°11′N	155°28′W
Pahang, state, Malay.	153b	3°02′N	102°57′E
Pahang, r., Malay.	168	3°39′N	102°41′E
Pahokee, Fl., U.S. (pá-hō′kē)	83a	26°45′N	80°40′W
Paide, Est. (pī′dĕ)	123	58°54′N	25°30′E
Päijänne, l., Fin. (pĕ′ē-yĕn-nĕ)	116	61°38′N	25°05′E
Pailolo Channel, strt., Hi., U.S. (pä-ē-lō′lō)	84a	21°05′N	156°41′W
Paine, Chile (pī′nĕ)	99b	33°49′S	70°44′W
Painesville, Oh., U.S. (pānz′vĭl)	66	41°40′N	81°15′W
Painted Desert, des., Az., U.S. (pānt′ĕd)	78	36°15′N	111°35′W
Painted Rock Reservoir, res., Az., U.S.	77	33°00′N	113°05′W
Paintsville, Ky., U.S. (pānts′vĭl)	66	37°50′N	82°50′W
Paisley, Scot., U.K. (pāz′lĭ)	116	55°50′N	4°30′W
Paita, Peru (pä-ē′tä)	100	5°11′S	81°12′W
Pai T′ou Shan, mts., Kor., N.	161	40°30′N	127°20′E
Paiute Indian Reservation, I.R., Ut., U.S.	77	38°17′N	113°50′W
Pajápan, Mex. (pä-hä′pän)	89	18°16′N	94°41′W
Pakanbaru, Indon.	168	0°43′N	101°15′E
Pakhra, r., Russia (päĸ′rá)	142b	55°29′N	37°51′E
Pakistan, nation, Asia	155	28°00′N	67°30′E
Pakokku, Mya. (pá-kŏk′kò)	160	21°29′N	95°00′E
Paks, Hung. (pôksh)	125	46°38′N	18°53′E
Pala, Chad	189	9°22′N	14°54′E
Palacios, Tx., U.S. (pá-lä′syōs)	81	28°42′N	96°12′W
Palagruža, Otoci, is., Cro.	130	42°20′N	16°23′E
Palaiseau, Fr. (pá-lĕ-zō′)	127b	48°44′N	2°16′E
Palana, Russia	135	59°07′N	159°58′E
Palanan Bay, b., Phil. (pä-lä′nän)	169a	17°14′N	122°35′E
Palanan Point, c., Phil.	169a	17°12′N	122°40′E
Pālanpur, India (pä′lŭn-po͞or)	155	24°08′N	73°29′E
Palapye, Bots. (pá-läp′yĕ)	186	22°34′S	27°28′E
Palatine, Il., U.S. (păl′á-tīn)	69a	42°07′N	88°03′W
Palatka, Fl., U.S. (pá-lăt′ká)	83	29°39′N	81°40′W
Palau (Belau), nation, Oc. (pä-lä′ō)	3	7°15′N	134°30′E
Palauig, Phil. (pá-lou′ēg)	169a	15°27′N	119°54′E
Palawan, i., Phil. (pä-lä′wán)	168	9°50′N	117°38′E
Pālayankottai, India	159	8°50′N	77°50′E
Paldiski, Est. (päl′dĭ-skĭ)	123	59°22′N	24°04′E
Palembang, Indon. (pä-lĕm-bäng′)	168	2°57′S	104°40′E
Palencia, Guat. (pä-lĕn′sē-á)	90	14°40′N	90°22′W
Palencia, Spain (pä-lĕ′n-syä)	118	42°02′N	4°32′W
Palenque, Mex. (pä-lĕŋ′ká)	89	17°34′N	91°58′W
Palenque, Punta, c., Dom. Rep. (po͞o′n-tä)	93	18°10′N	70°10′W
Palermo, Col. (pä-lĕr′mô)	100a	2°53′N	75°26′W
Palermo, Italy	110	38°08′N	13°24′E
Palestine, Tx., U.S.	65	31°46′N	95°38′W
Palestine, hist. reg., Asia (păl′ĕs-tīn)	153a	31°33′N	35°00′E
Paletwa, Mya. (pŭ-lĕt′wä)	155	21°19′N	92°52′E
Palghāt, India	159	10°49′N	76°40′E
Pāli, India	158	25°53′N	73°18′E
Palín, Guat. (pä-lēn′)	90	14°42′N	90°42′W
Palizada, Mex. (pä-lē-zä′dä)	89	18°17′N	92°04′W
Palk Strait, strt., Asia (pôk)	155	10°00′N	79°23′E
Palma, Braz. (päl′mä)	99a	21°23′S	42°18′W
Palma, Spain	110	39°35′N	2°38′E
Palma, Bahía de, b., Spain	129	39°24′N	2°37′E
Palma del Río, Spain	128	37°43′N	5°19′W
Palmares, Braz. (päl-mä′rĕs)	101	8°46′S	35°28′W
Palmas, Braz. (päl′mäs)	102	26°20′S	51°56′W
Palmas, Braz.	101	10°08′S	48°18′W
Palmas, Cape, c., Lib.	184	4°22′N	7°44′W
Palma Soriano, Cuba (sō-ré-ä′nō)	92	20°15′N	76°00′W
Palm Beach, Fl., U.S. (päm bēch′)	83a	26°43′N	80°03′W
Palmeira dos Índios, Braz. (pä-mā′rä-dôs-ē′n-dyôs)	101	9°26′S	36°33′W
Palmeirinhas, Ponta das, c., Ang.	190	9°05′S	13°00′E
Palmela, Port. (päl-mā′lä)	128	38°34′N	8°54′W
Palmer, Ak., U.S. (päm′ēr)	63	61°38′N	149°15′W
Palmer, Wa., U.S.	74a	47°19′N	121°53′W
Palmerston North, N.Z. (päm′ēr-stŭn)	175a	40°20′S	175°35′E
Palmerville, Austl. (päm′ēr-vĭl)	175	16°08′S	144°15′E
Palmetto, Fl., U.S. (pál-mĕt′ō)	83a	27°32′N	82°34′W
Palmetto Point, c., Bah.	93	21°15′N	73°25′W
Palmi, Italy (päl′mē)	130	38°21′N	15°54′E
Palmira, Col. (päl-mē′rä)	100	3°33′N	76°17′W
Palmira, Cuba	92	22°15′N	80°25′W
Palmyra, Mo., U.S. (păl-mī′rá)	79	39°45′N	91°32′W
Palmyra, N.J., U.S.	68f	40°01′N	75°00′W
Palmyra, i., Oc.	2	6°00′N	162°20′W
Palmyra, hist., Syria	154	34°25′N	38°28′E
Palmyras Point, c., India	158	20°42′N	87°45′E
Palo Alto, Ca., U.S. (pă′lō ăl′tō)	74b	37°27′N	122°09′W
Paloduro Creek, r., Tx., U.S. (pă-lō-do͞o′rô)	78	36°16′N	101°12′W
Paloh, Malay.	153b	2°11′N	103°12′E
Paloma, l., Mex. (pä-lō′mä)	80	26°53′N	104°02′W
Palomo, Cerro el, mtn., Chile (sĕ′r-rô-ĕl-pä-lō′mô)	99b	34°36′S	70°20′W
Palos, Cabo de, c., Spain (ká′bô-dĕ-pä′lôs)	118	39°38′N	0°43′W
Palos Verdes Estates, Ca., U.S. (pă′lŭs vûr′dĭs)	75a	33°48′N	118°24′W
Palouse, Wa., U.S. (pá-lo͞oz′)	72	46°54′N	117°04′W
Palouse, r., Wa., U.S.	72	47°02′N	117°35′W
Palu, Tur. (pä-lo͞o′)	137	38°55′N	40°10′E
Paluan, Phil. (pä-lo͞o′än)	169a	13°25′N	120°29′E
Pamiers, Fr. (pá-myā′)	117	43°07′N	1°34′E
Pamirs, mts., Asia	155	38°14′N	72°27′E
Pamlico, r., N.C., U.S. (păm′lĭ-kō)	83	35°25′N	76°59′W
Pamlico Sound, strt., N.C., U.S.	65	35°10′N	76°10′W
Pampa, Tx., U.S. (păm′pá)	64	35°32′N	100°56′W
Pampa de Castillo, pl., Arg. (pä′m-pä-dĕ-käs-tē′l-yô)	102	45°30′S	67°30′W
Pampana, r., S.L.	188	8°35′N	11°55′W
Pampanga, r., Phil. (päm-päŋ′gä)	169a	15°20′N	120°48′E
Pampas, reg., Arg. (päm′päs)	102	37°00′S	64°30′W
Pampilhosa do Botão, Port. (päm-pē-lyō′sá-dô-bō-touN)	128	40°21′N	8°32′W
Pamplona, Col. (päm-plō′nä)	100	7°19′N	72°41′W
Pamplona, Spain (päm-plō′nä)	118	42°49′N	1°39′W
Pamunkey, r., Va., U.S. (pá-mŭŋ′kĭ)	67	37°40′N	77°20′W
Pana, Il., U.S. (pā′ná)	66	39°25′N	89°05′W
Panagyurishte, Blg. (pá-ná-gyo͞o′rēsh-tĕ)	131	42°30′N	24°11′E
Panaji (Panjim), India	155	15°33′N	73°52′E
Panamá, Pan.	87	8°58′N	79°32′W
Panama, nation, N.A.	87	9°00′N	80°00′W
Panamá, Istmo de, isth., Pan.	87	9°00′N	80°00′W
Panama Canal, can., Pan.	86a	9°20′N	79°55′W
Panama City, Fl., U.S. (păn-á mä′ sĭ′tĭ)	82	30°08′N	85°39′W
Panamint Range, mts., Ca., U.S. (păn-á-mĭnt′)	76	36°40′N	117°30′W
Panarea, i., Italy (pä-nä′rē-ä)	130	38°37′N	15°05′E
Panaro, r., Italy (pä-nä′rô)	130	44°47′N	11°06′E
Panay, i., Phil. (pä-nī′)	168	11°15′N	121°38′E
Pančevo, Serb. (pán′chĕ-vô)	119	44°52′N	20°42′E
Panchor, Malay.	153b	2°11′N	102°43′E
Pānchur, India	158a	22°31′N	88°17′E
Panda, D.R.C. (pän′dä′)	186	10°59′S	27°24′E
Pan de Guajaibon, mtn., Cuba (pän dä gwä-já-bōn′)	92	22°50′N	83°20′W
Panevėžys, Lith. (pä′nyĕ-vāzh′ēs)	136	55°44′N	24°21′E
Panga, D.R.C. (päŋ′gä)	185	1°51′N	26°25′E
Pangani, Tan. (pän-gä′nē)	187	5°28′S	38°58′E
Pangani, r., Tan.	191	4°40′S	37°45′E
Pangkalpinang, Indon. (päng-käl′pē-näng′)	168	2°11′S	106°04′E
Pangnirtung, Can.	51	66°08′N	65°26′W
Panguitch, Ut., U.S. (pän′gwĭch)	77	37°50′N	112°30′W
Panié, Mont, mtn., N. Cal.	170f	20°36′S	164°46′E
Pānihāti, India	158a	22°42′N	88°23′E
Panimávida, Chile (pä-nē-má′vē-dä)	99b	35°44′S	71°26′W
Panshi, China (pän-shē)	164	42°50′N	126°48′E
Pantar, Pulau, i., Indon. (pän′tär)	169	8°40′N	123°45′E
Pantelleria, i., Italy (pän-tĕl-lå-rē′ä)	118	36°43′N	11°59′E
Pantepec, Mex. (pän-tå-pĕk′)	89	17°11′N	93°04′W
Panuco, Mex. (pä′no͞o-kô)	88	22°04′N	98°11′W
Pánuco, Mex. (pä′no͞o-kô)	88	23°25′N	105°55′W
Panuco, r., Mex.	86	21°59′N	98°20′W
Pánuco de Coronado, Mex. (pä′no͞o-kô dā kō-rô-nä′dhō)	80	24°33′N	104°20′W
Panvel, India	159b	18°59′N	73°06′E
Panyu, China (pä-yo͞o)	163a	22°56′N	113°22′E
Panzós, Guat. (pä-zós′)	90	15°26′N	89°40′W
Pao, r., Ven. (pá′ō)	101b	9°52′N	67°57′W
Paola, Ks., U.S. (pā-ō′lá)	79	38°34′N	94°51′W
Paoli, In., U.S. (pā-ō′lī)	66	38°35′N	86°30′W
Paoli, Pa., U.S.	68f	40°03′N	75°29′W
Paonia, Co., U.S. (pā-ō′nyá)	77	38°50′N	107°40′W
Pápa, Hung. (pä′pô)	119	47°18′N	17°27′E
Papagayo, r., Mex. (pä-pä-gä′yō)	88	16°52′N	99°41′W
Papagayo, Golfo del, b., C.R. (gôl-fô-dĕl-pä-pä-gá′yō)	90	10°44′N	85°56′W
Papagayo, Laguna, l., Mex. (lä-ō-nä)	88	16°44′N	99°44′W
Papantla de Olarte, Mex. (pä-pän′tlä dā-ō-lä′r-tĕ)	86	20°30′N	97°15′W
Papatoapan, r., Mex. (pä-pä-tô-ä-pá′n)	89	18°00′N	96°22′W
Papenburg, Ger. (päp′ĕn-bòrgh)	124	53°05′N	7°23′E
Papinas, Arg. (pä-pē′näs)	99c	35°30′S	57°19′W
Papineauville, Can. (pä-pē-nō′vĕl)	62c	45°38′N	75°01′W
Papua, Gulf of, b., Pap. N. Gui. (păp-o͞o-á)	169	8°20′S	144°45′E
Papua New Guinea, nation, Oc. (păp-o͞o-á)(gĭne)	169	7°00′S	142°15′E
Papudo, Chile (pä-po͞o′dô)	99b	32°30′S	71°25′W
Paquequer Pequeno, Braz. (pä-kĕ-kĕ′r-pĕ-kĕ′nô)	102b	22°19′S	43°02′W
Para, r., Russia	132	53°45′N	40°58′E
Paracale, Phil. (pä-rä-kä′lå)	169a	14°17′N	122°47′E

PRONOUNCING INDEX

Paracambi – Pelee Island 287

PLACE (Pronunciation)	PAGE	LAT.	LONG.
Paracambi, Braz.	102b	22°36'S	43°43'W
Paracatu, Braz. (pä-rä-kä-to͞o')	101	17°17'S	46°43'W
Paracel Islands, is., Asia	168	16°40'N	113°00'E
Paraćin, Serb. (pá'rä-chėn)	119	43°51'N	21°26'E
Para de Minas, Braz. (pä-rä-dě-mē'näs)	101	19°52'S	44°37'W
Paradise, i., Bah.	92	25°05'N	77°20'W
Paradise Valley, Nv., U.S. (păr'á-dīs)	72	41°28'N	117°32'W
Parados, Cerro de los, mtn., Col. (sě'r-rô-dě-lôs-pä-rä'dōs)	100a	5°44'N	75°13'W
Paragould, Ar., U.S. (păr'ȧ-go͞old)	79	36°03'N	90°29'W
Paraguaçu, r., Braz. (pä-rä-gwä-zo͞o')	101	12°25'S	39°46'W
Paraguay, nation, S.A. (păr'ȧ-gwā)	102	24°00'S	57°00'W
Paraguay, r., S.A. (pä-rä-gwä'y)	102	21°12'S	57°31'W
Paraíba, state, Braz. (pä-rä-ē'bä)	101	7°11'S	37°05'W
Paraíba, r., Braz.	99a	23°02'S	45°43'W
Paraíba do Sul, Braz. (dô-so͞o'l)	99a	22°10'S	43°18'W
Paraibuna, Braz. (pä-räē-bo͞o'nä)	99a	23°23'S	45°38'W
Paraíso, C.R.	91	9°50'N	83°53'W
Paraíso, Mex.	89	18°24'N	93°11'W
Paraíso, Pan. (pä-rä-ē'sō)	86a	9°02'N	79°38'W
Paraisópolis, Braz. (pä-räē-sô'pō-lēs)	99a	22°35'S	45°45'W
Paraitinga, r., Braz. (pä-rä-ē-tē'n-gä)	99a	23°15'S	45°24'W
Parakou, Benin (pȧ-rȧ-ko͞o')	184	9°21'N	2°37'E
Paramaribo, Sur. (pä-rȧ-mä'rē-bō)	101	5°50'N	55°15'W
Paramatta, Austl. (păr-ȧ-măt'ȧ)	173b	33°49'S	150°59'E
Paramillo, mtn., Col. (pä-rä-mē'l-yō)	100a	7°06'N	75°55'W
Paramus, N.J., U.S.	68a	40°56'N	74°04'W
Paran, r., Asia	153a	30°05'N	34°50'E
Paraná, Arg.	102	31°44'S	60°32'W
Paraná, r., S.A.	102	24°00'S	54°00'W
Paranaíba, Braz. (pä-rä-nȧ-ē'bȧ)	101	19°43'S	51°13'W
Paranaíba, r., Braz.	101	18°58'S	50°44'W
Paraná Ibicuy, r., Arg.	99c	33°27'S	59°26'W
Paranam, Sur.	101	5°39'N	55°13'W
Paranápanema, r., Braz. (pä-rä-nä'pä-ně-mä)	101	22°28'S	52°15'W
Paraopeda, r., Braz. (pä-rä-o-pě'dä)	99a	20°09'S	44°14'W
Parapara, Ven. (pä-rä-pä-rä)	101b	9°44'N	67°17'W
Parati, Braz. (pä-rätē)	99a	23°14'S	44°43'W
Paray-le-Monial, Fr. (pȧ-rě'lē-mô-nyäl')	126	46°27'N	4°14'E
Pārbati, r., India	158	24°50'N	76°44'E
Parchim, Ger. (par'kĭm)	124	53°25'N	11°52'E
Parczew, Pol. (pär'chĕf)	125	51°38'N	22°53'E
Pardo, r., Braz. (pär'dō)	101	15°25'S	39°40'W
Pardo, r., Braz.	99a	21°32'S	46°40'W
Pardubice, Czech Rep. (pär'do̤-bĭt-sĕ)	124	50°02'N	15°47'E
Parecis, Serra dos, mts., Braz. (sĕr'rȧ dōs pä-rȧ-sēzh')	101	13°45'S	59°28'W
Paredes de Nava, Spain (pä-rā'dås dä nä'vä)	128	42°10'N	4°41'W
Paredón, Mex.	80	25°56'N	100°58'W
Parent, Can.	51	47°59'N	74°30'W
Parent, Lac, l., Can.	59	48°40'N	77°00'W
Parepare, Indon.	168	4°01'S	119°38'E
Pargolovo, Russia (pár-gô'lô vô)	142c	60°04'N	30°18'E
Paria, r., Az., U.S.	77	37°07'N	111°51'W
Paria, Golfo de, b., (gôl-fô-dě-br-pä-rē-ä)	100	10°33'N	62°14'W
Paricutín, Volcán, vol., Mex.	88	19°27'N	102°14'W
Parida, Río de la, r., Mex. (rē'ō-dě-lä-pä-rē'dä)	80	26°23'N	104°40'W
Parima, Serra, mts., S.A. (sěr'rȧ pä-rē'mȧ)	100	3°45'N	64°00'W
Pariñas, Punta, c., Peru (po͞o'n-tä-pä-rē'n-yäs)	100	4°30'S	81°23'W
Parintins, Braz. (pä-rĭn-tĭnzh')	101	2°34'S	56°30'W
Paris, Can.	59	43°15'N	80°23'W
Paris, Fr. (pȧ-rē')	110	48°51'N	2°20'E
Paris, Ar., U.S. (păr'ĭs)	79	35°17'N	93°43'W
Paris, Il., U.S.	66	39°35'N	87°40'W
Paris, Ky., U.S.	66	38°15'N	84°15'W
Paris, Mo., U.S.	79	39°27'N	91°59'W
Paris, Tn., U.S.	82	36°16'N	88°20'W
Paris, Tx., U.S.	65	33°39'N	95°33'W
Parita, Golfo de, b., Pan. (gôl-fô-dě-pä-rē'tä)	91	8°06'N	80°10'W
Park City, Ut., U.S.	73	40°39'N	111°33'W
Parker, S.D., U.S. (pär'kēr)	70	43°24'N	97°10'W
Parker Dam, dam, U.S.	64	34°20'N	114°00'W
Parkersburg, W.V., U.S. (pär'kērz-bûrg)	65	39°15'N	81°35'W
Parkes, Austl. (pärks)	176	33°10'S	148°10'E
Park Falls, Wi., U.S. (pärk)	71	45°55'N	90°29'W
Park Forest, Il., U.S.	69a	41°29'N	87°41'W
Parkland, Wa., U.S. (pärk'lănd)	74a	47°09'N	122°26'W
Park Range, mts., Co., U.S.	73	40°54'N	106°40'W
Park Rapids, Mn., U.S.	70	46°53'N	95°05'W
Park Ridge, Il., U.S.	69a	42°00'N	87°50'W
Park River, N.D., U.S.	70	48°22'N	97°43'W
Parkrose, Or., U.S. (pärk'rōz)	74c	45°33'N	122°33'W
Park Rynie, S. Afr.	187c	30°22'S	30°43'E
Parkston, S.D., U.S. (pärks'tŭn)	70	43°22'N	97°59'W
Parkville, Md., U.S.	68e	39°22'N	76°32'W
Parkville, Mo., U.S.	75f	39°12'N	94°41'W
Parla, Spain (pär'lä)	129a	40°14'N	3°46'W
Parma, Italy (pär'mä)	118	44°48'N	10°20'E
Parma, Oh., U.S.	69d	41°23'N	81°44'W
Parma Heights, Oh., U.S.	69d	41°23'N	81°36'W
Parnaíba, Braz. (pär-nä-ē'bä)	101	3°00'S	41°42'W
Parnaíba, r., Braz.	101	3°57'S	42°30'W
Parnassós, mtn., Grc.	131	38°36'N	22°35'E
Parndorf, Aus.	115e	48°00'N	16°52'E
Pärnu, Est. (pĕr'no͞o)	136	58°24'N	24°29'E
Pärnu, r., Est.	123	58°40'N	25°05'E
Pärnu Laht, b., Est. (läkt)	123	58°15'N	24°17'E

PLACE (Pronunciation)	PAGE	LAT.	LONG.
Paro, Bhu. (pä'rô)	158	27°30'N	89°30'E
Paroo, r., Austl. (pä'ro͞o)	175	30°00'S	144°00'E
Páros, Grc. (pä'rôs) (pā'rŏs)	131	37°05'N	25°14'E
Páros, i., Grc.	119	37°11'N	25°00'E
Parow, S. Afr. (pă'rô)	186a	33°54'S	18°36'E
Parowan, Ut., U.S. (păr'ô-wăn)	77	37°50'N	112°50'W
Parral, Chile (pär-rä'l)	102	36°07'S	71°47'W
Parral, r., Mex.	80	27°25'N	105°08'W
Parramatta, r., Austl. (păr-ȧ-măt'ȧ)	173b	33°42'S	150°58'E
Parras, Mex. (pär-räs')	80	25°28'N	102°08'W
Parrita, C.R. (pär-rē'tä)	91	9°32'N	84°17'W
Parrsboro, Can. (pärz'bŭr-ô)	60	45°24'N	64°20'W
Parry, i., Can. (păr'ĭ)	59	45°15'N	80°00'W
Parry, Mount, mtn., Can.	54	52°53'N	128°45'W
Parry Islands, is., Can.	49	75°30'N	110°00'W
Parry Sound, Can.	51	45°20'N	80°00'W
Parsnip, r., Can. (pärs'nĭp)	55	54°45'N	122°20'W
Parsons, Ks., U.S. (pär's'nz)	65	37°20'N	95°16'W
Parsons, W.V., U.S.	67	39°05'N	79°40'W
Parthenay, Fr. (pȧr-t'nĕ')	126	46°39'N	0°16'W
Partinico, Italy (pär-tē'nē-kô)	130	38°02'N	13°11'E
Partizansk, Russia	135	43°15'N	133°19'E
Parys, S. Afr. (pȧ-rīs')	192c	26°53'S	27°28'E
Pasadena, Ca., U.S. (păs-ȧ-dē'nȧ)	64	34°09'N	118°09'W
Pasadena, Md., U.S.	68e	39°06'N	76°35'W
Pasadena, Tx., U.S.	81a	29°43'N	95°13'W
Pascagoula, Ms., U.S. (păs-kȧ-go͞o'lȧ)	82	30°22'N	88°33'W
Pascagoula, r., Ms., U.S.	82	30°52'N	88°48'W
Paşcani, Rom. (päsh-kän')	125	47°46'N	26°42'E
Pasco, Wa., U.S. (păs'kō)	72	46°13'N	119°04'W
Pascua, Isla de (Easter Island), i., Chile	195	26°50'S	109°00'W
Pasewalk, Ger. (pä'zē-välk)	124	53°31'N	14°01'E
Pashiya, Russia (pä'shĭ-yȧ)	142a	58°27'N	58°17'E
Pashkovo, Russia (pȧsh-kô'vô)	166	48°52'N	131°09'E
Pashkovskaya, Russia (pȧsh-kôf'skȧ-yȧ)	133	45°00'N	39°04'E
Pasig, Phil.	169a	14°34'N	121°05'E
Pasión, Río de la, r., Guat. (rē'ō-dě-lä-pä-syōn')	90a	16°31'N	90°11'W
Paso de los Libres, Arg. (pä-sô-dě-lôs-lē'brĕs)	102	29°33'S	57°05'W
Paso de los Toros, Ur. (tŏ'rôs)	99c	32°43'S	56°33'W
Paso Robles, Ca., U.S. (pă'sō rō'blĕs)	76	35°38'N	120°44'W
Pasquia Hills, hills, Can. (păs'kwē-ȧ)	57	53°13'N	102°37'W
Passaic, N.J., U.S. (pă-sā'ĭk)	68a	40°52'N	74°08'W
Passaic, r., N.J., U.S.	68a	40°42'N	74°26'W
Passamaquoddy Bay, b., N.A. (păs'ȧ-mȧ-kwŏd'ĭ)	60	45°06'N	66°59'W
Passa Tempo, Braz. (pä's-sä-tĕ'm-pô)	99a	20°40'S	44°29'W
Passau, Ger. (päsou)	117	48°34'N	13°27'E
Pass Christian, Ms., U.S. (pȧs krĭs'tyĕn)	82	30°20'N	89°15'W
Passero, Cape, c., Italy (päs-sĕ'rô)	112	36°34'N	15°13'E
Passo Fundo, Braz. (pä'sō fōn'dō)	102	28°16'S	52°13'W
Passos, Braz. (pä's-sōs)	101	20°45'S	46°37'W
Pastaza, r., S.A. (päs-tä'zä)	100	3°05'S	76°18'W
Pasto, Col. (päs'tō)	100	1°15'N	77°19'W
Pastora, Mex. (päs-tô-rä)	88	22°08'N	100°04'W
Pasuruan, Indon.	168	7°45'S	112°50'E
Pasvalys, Lith. (päs-vä-lēs')	123	56°04'N	24°23'E
Patagonia, reg., Arg. (păt-ȧ-gō'nĭ-ȧ)	102	46°45'S	69°30'W
Pātālganga, r., India	159b	18°52'N	73°08'E
Patapsco, r., Md., U.S. (pȧ-tăps'kō)	68e	39°12'N	76°30'W
Pateros, Lake, res., Wa., U.S.	72	48°05'N	119°45'W
Paterson, N.J., U.S. (păt'ēr-sŭn)	68a	40°55'N	74°10'W
Pathein, Mya.	155	16°46'N	94°47'E
Pathfinder Reservoir, res., Wy., U.S. (păth'fīn-dēr)	73	42°22'N	107°10'W
Patiāla, India (pŭt-ē-ä'lŭ)	155	30°25'N	76°28'E
Pati do Alferes, Braz. (pä-tē-dô-ál-fĕ'rĕs)	102b	22°25'S	43°25'W
Patna, India (pŭt'nŭ)	155	25°33'N	85°18'E
Patnanongan, i., Phil. (pät-nä-nôn'gän)	169a	14°50'N	122°25'E
Patoka, r., In., U.S. (pȧ-tō'kȧ)	66	38°25'N	87°25'W
Patom Plateau, plat., Russia	135	59°30'N	115°00'E
Patos, Braz. (pä'tōzh)	101	7°03'S	37°14'W
Patos, Wa., U.S. (pä'tōs)	74d	48°47'N	122°57'W
Patos, Lagoa dos, l., Braz. (lä'gô-ȧ dozh pä'tōzh)	102	31°15'S	51°30'W
Patos de Minas, Braz. (dě-mē'näzh)	101	18°39'S	46°31'W
Pátra, Grc.	119	38°15'N	21°48'E
Patraïkós Kólpos, b., Grc.	131	38°16'N	21°19'E
Patras *see* Pátra, Grc.	119	38°15'N	21°48'E
Patrocínio, Braz. (pä-trō-sē'nē-ò)	101	18°48'S	46°47'W
Pattani, Thai. (pät'ȧ-nē)	168	6°56'N	101°13'E
Patten, Me., U.S. (păt''n)	60	45°59'N	68°27'W
Patterson, La., U.S. (păt'ēr-sŭn)	81	29°41'N	91°20'W
Patterson, i., Can.	58	48°38'N	87°14'W
Patton, Pa., U.S.	67	40°40'N	78°45'W
Patuca, r., Hond.	91	15°22'N	84°31'W
Patuca, Punta, c., Hond. (po͞o'n-tä-pä-to͞o'kä)	91	15°55'N	84°05'W
Patuxent, r., Md., U.S. (pȧ-tŭk'sĕnt)	67	39°10'N	77°10'W
Pátzcuaro, Mex. (päts'kwä-rō)	88	19°30'N	101°36'W
Pátzcuaro, Lago de, l., Mex. (lä'gô-dě)	88	19°36'N	101°38'W
Patzicia, Guat. (pät-zē'syä)	90	14°36'N	90°57'W
Patzún, Guat. (pät-zo͞on')	90	14°40'N	91°00'W
Pau, Fr. (pō)	117	43°18'N	0°23'W
Pau, Gave de, r., Fr. (gäv-dĕ)	126	43°33'N	0°51'W
Paulding, Oh., U.S. (pôl'dĭng)	66	41°05'N	84°35'W
Paulinenaue, Ger. (pou'lē-nĕ-nou-ĕ)	115b	52°40'N	12°43'E
Paulistano, Braz. (pá'ó-lēs-tà-nä)	101	8°13'S	41°06'W
Paulo Afonso, Salto, wtfl., Braz. (säl-tô-pou'lô äf-fôn'sô)	101	9°33'S	38°32'W
Paul Roux, S. Afr. (pôrl ro͞o)	192c	28°18'S	27°57'E

PLACE (Pronunciation)	PAGE	LAT.	LONG.
Paulsboro, N.J., U.S. (pôlz'bē-rô)	68f	39°50'N	75°16'W
Pauls Valley, Ok., U.S. (pôlz văl'ē)	79	34°43'N	97°13'W
Pavarandocito, Col. (pä-vä-rän-dô-sē'tô)	100a	7°18'N	76°32'W
Pavda, Russia (päv'da)	142a	59°16'N	59°32'E
Pavia, Italy (pä-vē'ä)	118	45°12'N	9°11'E
Pavlodar, Kaz. (pȧv-lô-dár')	139	52°17'N	77°23'E
Pavlof Bay, b., Ak., U.S. (păv-lŏf)	63	55°20'N	161°20'W
Pavlohrad, Ukr.	137	48°32'N	35°52'E
Pavlovsk, Russia (pȧv-lôfsk')	133	50°28'N	40°05'E
Pavlovsk, Russia	142c	59°41'N	30°27'E
Pavlovskiy Posad, Russia (pȧv-lôf'skĭ pô-sȧt')	136	55°47'N	38°39'E
Pavuna, Braz. (pä-vo͞o'nȧ)	102b	22°48'S	43°21'W
Päwesin, Ger. (pă'vĕ-zēn)	115b	52°31'N	12°44'E
Pawhuska, Ok., U.S. (pô-hŭs'kȧ)	79	36°41'N	96°20'W
Pawnee, Ok., U.S. (pô-nē')	79	36°20'N	96°47'W
Pawnee, r., Ks., U.S.	78	38°18'N	99°42'W
Pawnee City, Ne., U.S.	79	40°08'N	96°09'W
Paw Paw, Mi., U.S. (pô'pô)	66	42°15'N	85°55'W
Paw Paw, r., Mi., U.S.	71	42°14'N	86°21'W
Pawtucket, R.I., U.S. (pô-tŭk'ĕt)	67	41°53'N	71°23'W
Paxoi, i., Grc.	131	39°14'N	20°15'E
Paxton, Il., U.S. (păks'tŭn)	66	40°35'N	88°00'W
Payette, Id., U.S. (pȧ-ĕt')	72	44°05'N	116°55'W
Payette, r., Id., U.S.	72	43°57'N	116°26'W
Payette, North Fork, r., Id., U.S.	72	44°10'N	116°10'W
Payette, South Fork, r., Id., U.S.	72	44°07'N	115°43'W
Pay-Khoy, Khrebet, mts., Russia	136	68°08'N	63°04'E
Payne, l., Can. (pān)	53	59°22'N	73°16'W
Paynesville, Mn., U.S. (pānz'vĭl)	71	45°23'N	94°43'W
Paysandú, Ur. (pī-sän-do͞o')	102	32°16'S	57°55'W
Payson, Ut., U.S. (pā's'n)	77	40°05'N	111°45'W
Pazardzhik, Blg. (pä-zár-dzhek')	119	42°10'N	24°22'E
Pazin, Cro. (pä'zēn)	130	45°14'N	13°57'E
Peabody, Ks., U.S. (pē'bŏd-ĭ)	79	38°09'N	97°09'W
Peabody, Ma., U.S.	61a	42°32'N	70°56'W
Peace, r., Can.	52	57°30'N	117°30'W
Peace Creek, r., Fl., U.S. (pēs)	83a	27°16'N	81°53'W
Peace Dale, R.I., U.S. (dāl)	68b	41°27'N	71°30'W
Peace River, Can. (rĭv'ēr)	50	56°14'N	117°17'W
Peacock Hills, hills, Can. (pē-kŏk' hĭlz)	52	66°08'N	109°55'W
Peak Hill, Austl.	174	25°38'S	118°50'E
Pearl, r., U.S. (pûrl)	65	30°30'N	89°45'W
Pearland, Tx., U.S. (pûrl'ănd)	81a	29°34'N	95°17'W
Pearl Harbor, Hi., U.S.	84a	21°20'N	157°53'W
Pearl Harbor, b., Hi., U.S.	64d	21°22'N	157°58'W
Pearsall, Tx., U.S. (pēr'sôl)	80	28°53'N	99°06'W
Pearse Island, i., Can. (pērs)	54	54°51'N	130°21'W
Pearston, S. Afr. (pē'ērstŏn)	187c	32°36'S	25°09'E
Peary Land, reg., Grnld. (pēr'ĭ)	198	82°00'N	40°00'W
Pease, r., Tx., U.S. (pēz)	78	34°07'N	99°53'W
Peason, La., U.S. (pēz''n)	81	31°25'N	93°19'W
Pebane, Moz. (pĕ-bá'nē)	187	17°10'S	38°08'E
Pecan Bay, Tx., U.S. (pē-kăn')	80	32°04'N	99°15'W
Peçanha, Braz. (pā-kän'yȧ)	101	18°37'S	42°26'W
Pecatonica, r., Il., U.S. (pĕk-ȧ-tŏn-ĭ-kȧ)	71	42°21'N	89°28'W
Pechenga, Russia (pyĕ'chĕŋ-gȧ)	136	69°30'N	31°10'E
Pechora, r., Russia	134	66°00'N	54°00'E
Pechora Basin, Russia (pyĕ-chô'rȧ)	134	67°55'N	58°37'E
Pechori, Russia (pĕt'sĕ-rĕ)	132	57°48'N	27°33'E
Pecos, N.M., U.S. (pā'kōs)	77	35°29'N	105°41'W
Pecos, Tx., U.S.	80	31°26'N	103°30'W
Pecos, r., U.S.	64	31°10'N	103°10'W
Pécs, Hung. (pāch)	119	46°04'N	18°15'E
Peddie, S. Afr.	187c	33°13'S	27°09'E
Pedley, Ca., U.S. (pĕd'lē)	75a	33°59'N	117°29'W
Pedra Azul, Braz. (pā'drä-zo͞o'l)	101	16°03'S	41°13'W
Pedreiras, Braz. (pĕ-drā'räs)	101	4°30'S	44°31'W
Pedro, Point, c., Sri L. (pē'drô)	159	9°50'N	80°14'E
Pedro Antonio Santos, Mex.	90a	18°55'N	88°13'W
Pedro Betancourt, Cuba (bā-tăŋ-kōrt')	92	22°40'N	81°15'W
Pedro de Valdivia, Chile (pĕ'drô-dĕ-väl-dē'vē-ä)	102	22°32'S	69°55'W
Pedro do Rio, Braz. (dô-rē'rô)	102b	22°20'S	43°09'W
Pedro II, Braz. (pā'drô sĕ-gôn'dô)	101	4°20'S	41°27'W
Pedro Juan Caballero, Para. (hòä'n-kä-bäl-yĕ'rō)	102	22°40'S	55°42'W
Pedro Miguel, Pan. (mē-gäl')	86a	9°01'N	79°36'W
Pedro Miguel Locks, trans., Pan. (mē-gäl')	86a	9°01'N	79°36'W
Peebinga, Austl. (pē-bĭng'ȧ)	175	34°43'S	140°55'E
Peebles, Scot., U.K. (pē'b'lz)	120	55°40'N	3°15'W
Peekskill, N.Y., U.S. (pēks'kĭl)	68a	41°17'N	73°55'W
Pegasus Bay, b., N.Z. (pĕg'ȧ-sŭs)	175a	43°18'S	173°25'E
Pegnitz, r., Ger. (pĕgh-nēts)	124	49°38'N	11°40'E
Pego, Spain (pā'gō)	129	38°50'N	0°09'W
Peguis Indian Reserve, I.R., Can.	57	51°20'N	97°35'W
Pegu Yoma, mts., Mya. (pĕ-go͞o'yō'mä)	155	19°16'N	95°59'E
Pehčevo, Mac. (pĕk'chĕ-vô)	131	41°42'N	22°57'E
Peigan Indian Reserve, I.R., Can.	55	49°35'N	113°40'W
Peipus, Lake *see* Chudskoye Ozero, l., Eur.	136	58°43'N	26°45'E
Peiraiás, Grc.	119	37°57'N	23°38'E
Pekin, Il., U.S. (pē'kĭn)	66	40°35'N	89°30'W
Peking *see* Beijing, China	161	39°55'N	116°23'E
Pelagie, Isole, is., Italy	118	35°46'N	12°32'E
Pélagos, i., Grc.	131	39°17'N	24°05'E
Pelahatchie, Ms., U.S. (pĕl-ȧ-hăch'ē)	82	32°17'N	89°48'W
Pelat, Mont, mtn., Fr. (pē-lȧ')	117	44°16'N	6°43'E
Peleduy, Russia (pyĕl-yĭ-do͞o'ē)	135	59°50'N	112°47'E
Pelée, Mont, mtn., Mart. (pē-lā')	91b	14°49'N	61°10'W
Pelee, Point, c., Can.	58	41°55'N	82°30'W
Pelee Island, i., Can. (pē'lē)	58	41°45'N	82°30'W

ng-sing; ŋ-baŋk; N-nasalized n; nŏd; cŏmmit; ōld; ôbey; ôrder; oi-boil; fo͞od; ò-as oo in foot; ou-out; s-soft; sh-dish; th-thin; pūre; ûnite; ûrn; stŭd; circŭs; ü-as in French tu; '-indeterminate vowel.

288 Pelequén – Piedmont PRONOUNCING INDEX

PLACE (Pronunciation)	PAGE	LAT.	LONG.
Pelequén, Chile (pĕ-lĕ-kĕ´n)	99b	34°26′S	71°52′W
Pelham, Ga., U.S. (pĕl´hăm)	82	31°07′N	84°10′W
Pelham, N.H., U.S.	61a	42°43′N	71°22′W
Pelican, l., Mn., U.S.	71	46°36′N	94°00′W
Pelican Bay, b., Can.	57	52°45′N	100°20′W
Pelican Harbor, b., Bah. (pĕl´ĭ-kăn)	92	26°20′N	76°45′W
Pelican Rapids, Mn., U.S. (pĕl´ĭ-kăn)	70	46°34′N	96°05′W
Pella, Ia., U.S. (pĕl´á)	71	41°25′N	92°50′W
Pellworm, i., Ger. (pĕl´vōrm)	124	54°33′N	8°25′E
Pelly, l., Can.	52	66°08′N	102°57′W
Pelly, r., Can.	52	62°20′N	133°00′W
Pelly Bay, b., Can. (pĕl´ĭ)	53	68°57′N	91°05′W
Pelly Crossing, Can.	63	62°50′N	136°50′W
Pelly Mountains, mts., Can.	52	61°50′N	133°05′W
Peloncillo Mountains, mts., Az., U.S. (pĕl-ôn-sĭl´lō)	77	32°40′N	109°20′W
Peloponnisos, pen., Grc.	131	37°28′N	22°14′E
Pelotas, Braz. (pā-lō´tăzh)	102	31°45′S	52°18′W
Pelton, Can. (pĕl´tŭn)	69b	42°15′N	82°57′W
Pelym, r., Russia	136	60°20′N	63°05′E
Pelzer, S.C., U.S. (pĕl´zēr)	83	34°38′N	82°30′W
Pemanggil, i., Malay.	153b	2°37′N	104°41′E
Pematangsiantar, Indon.	168	2°58′N	99°03′E
Pemba, Moz. (pĕm´bá)	187	12°58′S	40°30′E
Pemba, Zam.	186	15°29′S	27°22′E
Pemba Channel, strt., Afr.	191	5°10′S	39°30′E
Pemba Island, i., Tan.	191	5°20′S	39°57′E
Pembina, N.D., U.S. (pĕm´bĭ-ná)	70	48°58′N	97°15′W
Pembina, r., Can.	55	53°05′N	114°30′W
Pembina, r., N.A.	57	49°08′N	98°20′W
Pembroke, Can. (pĕm´ brōk)	51	45°50′N	77°00′W
Pembroke, Wales, U.K.	120	51°40′N	5°00′W
Pembroke, Ma., U.S. (pĕm´brōk)	61a	42°05′N	70°49′W
Pen, India	159b	18°44′N	73°06′E
Penafiel, Port. (pā-ná-fyĕl´)	128	41°12′N	8°19′W
Peñafiel, Spain (pā-nyá-fyĕl´)	128	41°38′N	4°08′W
Peñalara, mtn., Spain (pā-nyä-lä´rä)	118	40°52′N	3°57′W
Pena Nevada, Cerro, Mex.	88	23°47′N	99°52′W
Peñaranda de Bracamonte, Spain	128	40°54′N	5°11′W
Peñarroya-Pueblonuevo, Spain (pĕn-yär-rō´yä-pwĕ´blō-nwĕ´vō)	128	38°18′N	5°18′W
Peñas, Cabo de, c., Spain (kä´bō-dĕ-pā´nyäs)	128	43°42′N	6°12′W
Penas, Golfo de, b., Chile (gôl-fô-dĕ-pĕ´n-äs)	102	47°15′S	77°30′W
Penasco, r., Tx., U.S. (pā-näs´kō)	80	32°50′N	104°45′W
Pendembu, S.L. (pĕn-dĕm´bōō)	184	8°06′N	10°42′W
Pender, Ne., U.S. (pĕn´dēr)	70	42°08′N	96°43′W
Penderisco, r., Col. (pĕn-dĕ-rē´s-kō)	100a	6°30′N	76°21′W
Pendjari, Parc National de la, rec., Benin	188	11°25′N	1°30′E
Pendleton, Or., U.S. (pĕn´d'l-tŭn)	64	45°41′N	118°47′W
Pend Oreille, r., Wa., U.S.	72	48°44′N	117°20′W
Pend Oreille, Lake, l., Id., U.S. (pŏn-dô-rā´) (pĕn-dô-rĕl´)	64	48°09′N	116°38′W
Penedo, Braz. (pâ-nā´dò)	101	10°17′S	36°28′W
Penetanguishene, Can. (pĕn´ē-tăŋ-gĭ-shēn´)	59	44°45′N	79°55′W
Pengcheng, China (pŭŋ-chŭŋ)	162	36°24′N	114°11′E
Penglai, China (pŭŋ-lī)	164	37°49′N	120°45′E
Peniche, Port. (pĕ-nē´chā)	128	39°22′N	9°24′W
Peninsula, Oh., U.S. (pĕn-ĭn´sū-lá)	69d	41°14′N	81°32′W
Penistone, Eng., U.K. (pĕn´ĭ-stŭn)	114a	53°31′N	1°38′W
Penjamillo, Mex. (pĕn-hä-mēl´yō)	88	20°06′N	101°56′W
Pénjamo, Mex. (pān´hä-mō)	88	20°27′N	101°43′W
Penk, r., Eng., U.K. (pĕnk)	114a	52°41′N	2°10′W
Penkridge, Eng., U.K. (pĕnk´rĭj)	114a	52°43′N	2°07′W
Penne, Italy (pĕn´nā)	130	42°28′N	13°57′E
Penner, r., India (pĕn´ēr)	155	14°43′N	79°09′E
Pennines, hills, Eng., U.K. (pĕn-īn')	120	54°30′N	2°10′W
Pennines, Alpes, mts., Eur.	124	46°02′N	7°07′E
Pennsboro, W.V., U.S. (pĕnz´bŭr-ô)	66	39°10′N	81°00′W
Penns Grove, N.J., U.S. (pĕnz grōv)	68f	39°44′N	75°28′W
Pennsylvania, state, U.S. (pĕn-sĭl-vā´nĭ-á)	65	41°00′N	78°10′W
Penn Yan, N.Y., U.S. (pĕn yăn´)	67	42°40′N	77°00′W
Pennycutaway, r., Can.	57	56°10′N	93°25′W
Peno, l., Russia (pā´nô)	132	56°55′N	32°28′E
Penobscot, r., Me., U.S.	65	45°00′N	68°36′W
Penobscot Bay, b., Me., U.S. (pē-nŏb´skŏt)	60	44°20′N	69°00′W
Penong, Austl. (pē-nông´)	174	32°00′S	133°00′E
Penrith, Austl.	173b	33°45′S	150°42′E
Pensacola, Fl., U.S. (pĕn-sá-kō´lá)	65	30°25′N	87°13′W
Pensacola Dam, Ok., U.S.	79	36°27′N	95°02′W
Pensilvania, Col. (pĕn-sēl-vá´nyä)	100a	5°31′N	75°05′W
Pentecost, i., Vanuatu (pĕn´tē-kōst)	175	16°05′S	168°28′E
Penticton, Can.	50	49°30′N	119°35′W
Pentland Firth, strt., Scot., U.K. (pĕnt´lănd)	120	58°44′N	3°25′W
Penza, Russia (pĕn´zá)	134	53°10′N	45°00′E
Penzance, Eng., U.K. (pĕn-zăns´)	120	50°07′N	5°40′W
Penzberg, Ger. (pĕnts´bĕrgh)	124	47°43′N	11°21′E
Penzhina, r., Russia (pyĭn-zē-nŭ)	141	62°15′N	166°30′E
Penzhino, Russia	135	63°42′N	168°00′E
Penzhinskaya Guba, b., Russia	141	60°30′N	161°30′E
Peoria, Il., U.S. (pē-ō´rĭ-á)	65	40°45′N	89°35′W
Peotillos, Mex. (pā-ō-tel´yōs)	88	22°30′N	100°39′W
Peotone, Il., U.S. (pē´ō-tōn)	69a	41°20′N	87°47′W
Pepacton Reservoir, res., N.Y., U.S. (pĕp-ăc´tŭn)	67	42°05′N	74°40′W
Pepe, Cabo, c., Cuba (kä´bō-pĕ´pĕ)	92	21°30′N	83°10′W
Pepperell, Ma., U.S. (pĕp´ēr-ĕl)	61a	42°40′N	71°36′W
Peqin, Alb. (pĕ-kēn´)	131	41°03′N	19°48′E
Perales, r., Spain (pā-rä´lās)	129a	40°24′N	4°07′W
Perales de Tajuña, Spain (dā tä-hōō´nyä)	129a	40°14′N	3°22′W
Perche, Collines du, hills, Fr.	126	48°25′N	0°40′E
Perchtoldsdorf, Aus. (pĕrk´tōlts-dôrf)	115e	48°07′N	16°17′E
Perdekop, S. Afr.	192c	27°11′S	29°38′E
Perdido, r., Al., U.S. (pĕr-dī´dō)	82	30°45′N	87°38′W
Perdido, Monte, mtn., Spain (pĕr-dē´dō)	129	42°40′N	0°00′
Perdões, Braz. (pĕr-dô´ĕs)	99a	21°05′S	45°05′W
Pereiaslav-Khmel'nyts'kyi, Ukr.	137	50°05′N	31°25′E
Pereira, Col. (pā-rā´rä)	100	4°49′N	75°42′W
Pere Marquette, Mi., U.S.	66	43°55′N	86°10′W
Pereshchepyne, Ukr.	133	49°02′N	35°19′E
Pereslavl'-Zalesskiy, Russia (pâ-rå-slàv´'l zà-lyĕs´kĭ)	136	56°43′N	38°52′E
Pergamino, Arg. (pĕr-gä-mē´nō)	102	33°53′S	60°36′W
Perham, Mn., U.S. (pēr´hăm)	70	46°37′N	95°35′W
Peribonca, r., Can. (pĕr-ĭ-bŏŋ´kȧ)	53	50°30′N	71°00′W
Périgueux, Fr. (pā-rē-gû´)	117	45°12′N	0°43′E
Perija, Sierra de, mts., Col. (sē-ĕ´r-rä-dĕ-pĕ-rē´kä)	100	9°25′N	73°30′W
Perkam, Tanjung, c., Indon.	169	1°20′S	138°45′E
Perkins, Can. (pĕr´kĕns)	62c	45°37′N	75°37′W
Perlas, Archipiélago de las, is., Pan.	91	8°29′N	79°15′W
Perlas, Laguna las, l., Nic. (lä-gó´nä-dĕ-läs)	91	12°34′N	83°19′W
Perleberg, Ger. (pĕr´lĕ-bĕrg)	124	53°06′N	11°51′E
Perm', Russia (pĕrm)	134	58°00′N	56°15′E
Pernambuco *see* Recife, Braz.	101	8°09′S	34°59′W
Pernambuco, state, Braz. (pĕr-năm-bōō´kō)	101	8°08′S	38°54′W
Pernik, Blg. (pĕr-nēk´)	119	42°36′N	23°04′E
Péronne, Fr. (pā-rŏn´)	126	49°57′N	2°49′E
Perote, Mex. (pĕ-rō´tĕ)	89	19°33′N	97°13′W
Perovo, Russia (pâ´rô-vô)	142b	55°43′N	37°47′E
Perpignan, Fr. (pĕr-pē-nyäN´)	117	42°42′N	2°48′E
Perris, Ca., U.S. (pĕr´ĭs)	75a	33°46′N	117°14′W
Perros, Bahía, b., Cuba (bä-ē´ä-pā´rōs)	92	22°25′N	78°35′W
Perrot, Île, i., Can.	62a	45°23′N	73°57′W
Perry, Fl., U.S. (pĕr´ĭ)	82	30°06′N	83°35′W
Perry, Ga., U.S.	82	32°27′N	83°44′W
Perry, Ia., U.S.	71	41°49′N	94°40′W
Perry, N.Y., U.S.	67	42°45′N	78°00′W
Perry, Ok., U.S.	79	36°17′N	97°18′W
Perry, Ut., U.S.	75b	41°27′N	112°02′W
Perry Hall, Md., U.S.	68e	39°24′N	76°29′W
Perryopolis, Pa., U.S. (pē-rē-ŏ´pō-lĭs)	69e	40°05′N	79°45′W
Perrysburg, Oh., U.S. (pĕr´ĭz-bûrg)	66	41°35′N	83°35′W
Perryton, Tx., U.S. (pĕr´ĭ-tŭn)	78	36°23′N	100°48′W
Perryville, Ak., U.S. (pĕr-ĭ-vĭl)	63	55°58′N	159°28′W
Perryville, Mo., U.S.	79	37°41′N	89°52′W
Persan, Fr. (pĕr-säN´)	127b	49°09′N	2°15′E
Persepolis, hist., Iran (pĕr-sĕpō-lĭs)	154	30°15′N	53°08′E
Persian Gulf, b., Asia (pûr´zhăn)	154	27°38′N	50°30′E
Perth, Austl. (pûrth)	174	31°50′S	116°10′E
Perth, Can.	59	44°40′N	76°15′W
Perth, Scot., U.K.	116	56°24′N	3°25′W
Perth Amboy, N.J., U.S. (ăm´boi)	68a	40°31′N	74°16′W
Pertuis, Fr. (pĕr-tüē´)	127	43°43′N	5°29′E
Peru, Il., U.S. (pē-rōō´)	66	41°20′N	89°10′W
Peru, In., U.S.	66	40°45′N	86°00′W
Peru, nation, S.A.	100	10°00′S	75°00′W
Peru-Chile Trench, deep,	97	25°00′S	71°30′W
Perugia, Italy (pā-rōō´jä)	118	43°08′N	12°24′E
Peruque, Mo., U.S. (pĕ rō´kĕ)	75e	38°52′N	90°36′W
Pervomais'k, Ukr.	137	48°04′N	30°52′E
Pervoural'sk, Russia (pĕr-vô-ô-rálsk´)	142a	56°54′N	59°58′E
Pesaro, Italy (pā´zä-rō)	118	43°54′N	12°55′E
Pescado, r., Ven. (pĕs-kä´dō)	101b	9°33′N	65°32′W
Pescara, Italy (pās-kä´rä)	130	42°26′N	14°15′E
Pescara, r., Italy	130	42°18′N	13°22′E
Peschanyy müyisi, c., Kaz.	137	43°10′N	51°20′E
Pescia, Italy (pä´shä)	130	43°53′N	11°42′E
Peshāwar, Pak. (pĕ-shä´wŭr)	155	34°01′N	71°34′E
Peshtera, Blg.	131	42°03′N	24°19′E
Peshtigo, Wi., U.S. (pĕsh´tē-gō)	71	45°03′N	87°46′W
Peshtigo, r., Wi., U.S.	71	45°15′N	88°14′W
Peski, Russia (pyâs´kĭ)	142b	55°13′N	38°48′E
Pêso da Régua, Port. (pā-sò-dä-rā´gwä)	128	41°09′N	7°47′W
Pespire, Hond. (pås-pē´rå)	90	13°35′N	87°20′W
Pesqueria, r., Mex. (pås-kå-rē´á)	80	25°55′N	100°25′W
Pessac, Fr.	126	44°48′N	0°38′W
Petacalco, Bahía de, b., Mex. (bä-ē´ä-dĕ-pĕ-tä-käl´kô)	88	17°55′N	102°00′W
Petah Tiqwa, Isr.	153a	32°05′N	34°53′E
Petaluma, Ca., U.S. (pĕt-á-lò´má)	76	38°15′N	122°38′W
Petare, Ven. (pĕ-tä´rĕ)	101b	10°28′N	66°48′W
Petatlán, Mex. (pā-tä-tlän´)	88	17°31′N	101°17′W
Petawawa, Can.	59	45°54′N	77°17′W
Petén, Laguna de, l., Guat. (lä-gò´nä-dĕ-pâ-tān´)	90a	17°05′N	89°54′W
Petenwell Reservoir, res., Wi., U.S.	71	44°10′N	89°55′W
Peterborough, Austl.	174	32°53′S	138°58′E
Peterborough, Can. (pē´tēr-bŭr-ô)	51	44°20′N	78°20′W
Peterborough, Eng., U.K.	120	52°35′N	0°14′W
Peterhead, Scot., U.K. (pē-tēr-hĕd´)	120	57°36′N	3°47′W
Peter Pond Lake, l., Can. (pŏnd)	52	55°55′N	108°44′W
Petersburg, Ak., U.S. (pē´tērz-bûrg)	63	56°52′N	133°10′W
Petersburg, Il., U.S.	79	40°01′N	89°51′W
Petersburg, In., U.S.	66	38°30′N	87°15′W
Petersburg, Ky., U.S.	69f	39°04′N	84°52′W
Petersburg, Va., U.S.	65	37°12′N	77°30′W
Petershagen, Ger. (pĕ´tĕrs-hä-gĕn)	115b	52°32′N	13°46′E
Petershausen, Ger. (pĕ´tĕrs-hou-zĕn)	115d	48°25′N	11°29′E
Pétionville, Haiti	93	18°30′N	72°20′W
Petitcodiac, Can. (pē-tē-kō-dyák´)	60	45°56′N	65°10′W
Petite Terre, i., Guad. (pē-tēt´târ´)	91b	16°12′N	61°00′W
Petit Goâve, Haiti (pē-tē´ gô-áv´)	93	18°25′N	72°50′W
Petit Jean Creek, r., Ar., U.S. (pē-tē´zhäN´)	79	35°05′N	93°55′W
Petit Loango, Gabon	190	2°16′S	9°35′E
Petlalcingo, Mex. (pĕ-tläl-sĕŋ´gô)	89	18°05′N	97°53′W
Peto, Mex. (pĕ´tô)	90a	20°07′N	88°49′W
Petorca, Chile (pā-tōr´kà)	99b	32°14′S	70°55′W
Petoskey, Mi., U.S. (pê-tŏs-kĭ)	66	45°25′N	84°55′W
Petra, hist., Jord.	153a	30°21′N	35°25′E
Petra Velikogo, Zaliv, b., Russia	166	42°40′N	131°50′E
Petre, Point, c., Can.	59	43°50′N	77°00′W
Petrich, Blg. (pā´trĭch)	119	41°24′N	23°13′E
Petrified Forest National Park, rec., Az., U.S. (pĕt´rĭ-fīd fôr´ĕst)	77	34°58′N	109°35′W
Petrinja, Cro. (pā´trēn-yá)	130	45°25′N	16°17′E
Petrodvorets, Russia (pyĕ-trô-dvô-ryĕts´)	142c	59°53′N	29°55′E
Petrokrepost', Russia (pyĕ´trô-krĕ-pôst)	136	59°56′N	31°03′E
Petrolia, Can. (pê-trō´lĭ-á)	58	42°50′N	82°10′W
Petrolina, Braz. (pĕ-trō-lē´ná)	101	9°18′S	40°28′W
Petronell, Aus.	115e	48°07′N	16°52′E
Petropavlivka, Ukr.	133	48°24′N	36°23′E
Petropavlovka, Russia	142a	54°10′N	59°50′E
Petropavlovsk, Kaz.	139	54°44′N	69°07′E
Petropavlovsk-Kamchatskiy, Russia (käm-chät´skĭ)	135	53°13′N	158°56′E
Petrópolis, Braz. (pá-trô-pô-lēzh´)	101	22°31′S	43°10′W
Petroşani, Rom.	131	45°24′N	23°24′E
Petrovsk, Russia (pyĕ-trôfsk´)	137	52°20′N	45°15′E
Petrovskaya, Russia (pyĕ-trôf´skà-yá)	133	45°25′N	37°50′E
Petrovskoye, Russia	137	45°20′N	43°00′E
Petrovsk-Zabaykal'skiy, Russia (pyĕ-trôfskzä-bī-käl´skĭ)	135	51°13′N	109°08′E
Petrozavodsk, Russia (pyā´trô-zà-vôtsk´)	134	61°46′N	34°25′E
Petrus Steyn, S. Afr.	192c	27°40′S	28°09′E
Petrykivka, Ukr.	133	48°43′N	34°29′E
Pewaukee, Wi., U.S. (pĭ-wô´kê)	69a	43°05′N	88°15′W
Pewaukee Lake, l., Wi., U.S.	69a	43°03′N	88°18′W
Pewee Valley, Ky., U.S. (pe wē)	69h	38°19′N	85°29′W
Peza, r., Russia (pyā´zá)	136	65°35′N	46°50′E
Pézenas, Fr. (pā-zĕ-nä´)	126	43°26′N	3°24′E
Pforzheim, Ger. (pfôrts´hīm)	117	48°52′N	8°43′E
Phalodi, India	158	27°13′N	72°22′E
Phan Thiet, Viet. (p'hän´)	168	11°30′N	108°43′E
Phelps Lake, l., N.C., U.S.	83	35°46′N	76°27′W
Phenix City, Al., U.S. (fē´nĭks)	82	32°29′N	85°00′W
Philadelphia, Ms., U.S. (fĭl-á-dĕl´phĭ-á)	82	32°45′N	89°07′W
Philadelphia, Pa., U.S.	65	40°00′N	75°13′W
Philip, S.D., U.S. (fĭl´ĭp)	70	44°03′N	101°35′W
Philippeville *see* Skikda, Alg.	184	36°58′N	6°51′E
Philippines, nation, Asia (fĭl´ĭ-pēnz)	169	14°25′N	125°00′E
Philippine Sea, sea, (fĭl´ĭ-pēn)	195	16°00′N	133°00′E
Philippine Trench, deep,	169	10°30′N	127°15′E
Philipsburg, Pa., U.S. (fĭl´lĭps-bĕrg)	67	40°55′N	78°10′W
Philipsburg, Wy., U.S.	73	46°19′N	113°19′W
Phillip, i., Austl. (fĭl´ĭp)	176	38°32′S	145°10′E
Phillip Channel, strt., Indon.	153b	1°04′N	103°40′E
Phillipi, W.V., U.S. (fĭ-lĭp´ĭ)	66	39°10′N	80°00′W
Phillips, Wi., U.S. (fĭl´ĭps)	71	45°41′N	90°24′W
Phillipsburg, Ks., U.S. (fĭl´lĭps-bĕrg)	78	39°44′N	99°19′W
Phillipsburg, N.J., U.S.	67	40°45′N	75°10′W
Phitsanulok, Thai.	168	16°51′N	100°15′E
Phnom Penh (Phnum Pénh), Camb. (nŏm´pĕn´)	168	11°39′N	104°53′E
Phnum Pénh *see* Phnom Penh, Camb.	168	11°39′N	104°53′E
Phoenix, Az., U.S. (fē´nĭks)	64	33°30′N	112°00′W
Phoenix, Md., U.S.	68e	39°31′N	76°40′W
Phoenix Islands, is., Kir.	2	4°00′S	174°00′W
Phoenixville, Pa., U.S. (fē´nĭks-vĭl)	68f	40°08′N	75°31′W
Phou Bia, mtn., Laos	168	19°36′N	103°00′E
Phra Nakhon Si Ayutthaya, Thai.	168	14°16′N	100°37′E
Phuket, Thai.	168	7°57′N	98°19′E
Phu Quoc, Dao, i., Viet.	168	10°13′N	104°00′E
Pi, r., China (bē)	162	32°06′N	116°31′E
Piacenza, Italy (pyä-chĕnt´sä)	118	45°02′N	9°42′E
Pianosa, i., Italy (pyä-nō´sä)	130	42°13′N	15°45′E
Piave, r., Italy (pyä´vā)	130	45°45′N	12°15′E
Piazza Armerina, Italy (pyät´sá är-må-rē´nä)	130	37°23′N	14°26′E
Pibor, r., Sudan (pē´bôr)	185	7°21′N	32°54′E
Pic, r., Can. (pĕk)	58	48°48′N	86°28′W
Picara Point, c., V.I.U.S. (pē-kä´rä)	87c	18°23′N	64°57′W
Picayune, Ms., U.S. (pĭk´á yōōn)	82	30°32′N	89°41′W
Picher, Ok., U.S. (pĭch´ēr)	79	36°58′N	94°49′W
Pichilemu, Chile (pē-chē-lĕ´mōō)	99b	34°22′S	72°01′W
Pichucalco, Mex. (pē-chōō-käl´kō)	89	17°34′N	93°06′W
Pickerel, l., Can. (pĭk´ēr-ĕl)	58	48°35′N	91°10′W
Pickwick Lake, res., U.S. (pĭk´wĭck)	82	35°04′N	88°05′W
Pico, Ca., U.S. (pē´kō)	75a	34°01′N	118°05′W
Pico Island, i., Port. (pē´kō)	184a	38°16′N	28°49′W
Pico Riviera, Ca., U.S.	75a	34°01′N	118°05′W
Picos, Braz. (pē´kōzh)	101	7°13′S	41°23′W
Picton, Austl. (pĭk´tŭn)	173b	34°11′S	150°37′E
Picton, Can.	59	44°00′N	77°15′W
Pictou, Can. (pĭk-tōō´)	61	45°41′N	62°43′W
Pidálion, Akrotírion, c., Cyp.	153a	34°50′N	34°05′E
Pidurutalagala, mtn., Sri L. (pē´dò-rò-tä´lá-gä´lá)	159	7°00′N	80°46′E
Pidvolochys'k, Ukr.	133	49°32′N	26°16′E
Pie, i., Can. (pī)	58	48°10′N	89°07′W
Piedade, Braz. (pyā-dä´dĕ)	99a	23°42′S	47°25′W
Piedmont, Al., U.S. (pēd´mŏnt)	82	33°54′N	85°36′W
Piedmont, Ca., U.S.	74b	37°50′N	122°14′W
Piedmont, Mo., U.S.	79	37°09′N	90°42′W

PRONOUNCING INDEX

Piedmont – Pointe-des-Cascades 289

PLACE (Pronunciation)	PAGE	LAT.	LONG.
Piedmont, S.C., U.S.	83	34°40'N	82°27'W
Piedmont, W.V., U.S.	67	39°30'N	79°05'W
Piedrabuena, Spain (pyā-drä-bwā'nä)	128	39°01'N	4°10'W
Piedras, Punta, c., Arg. (pōō'n-tä-pyĕ'dräs)	99c	35°25'S	57°10'W
Piedras Negras, Mex. (pyā'dräs nā'gräs)	86	28°41'N	100°33'W
Pieksämäki, Fin. (pyĕk'sĕ-mĕ-kē)	123	62°18'N	27°14'E
Piemonte, hist. reg., Italy (pyĕ-mō'n-tĕ)	130	44°30'N	7°42'E
Pienaars, r., S. Afr.	192c	25°13'S	28°05'E
Pienaarsrivier, S. Afr.	192c	25°12'S	28°18'E
Pierce, Ne., U.S. (pērs)	70	42°11'N	97°33'W
Pierce, W.V., U.S.	67	39°15'N	79°30'W
Piermont, N.Y., U.S. (pēr'mŏnt)	68a	41°03'N	73°55'W
Pierre, S.D., U.S. (pēr)	64	44°22'N	100°20'W
Pierrefonds, Can.	62a	45°29'N	73°52'W
Piešt'any, Slvk.	125	48°36'N	17°48'E
Pietermaritzburg, S. Afr. (pē-tēr-mȧ-rĭts-bûrg)	186	29°36'S	30°23'E
Pietersburg, S. Afr. (pē'tērz-bûrg)	186	23°56'S	29°30'E
Piet Retief, S. Afr. (pēt rĕ-tēf')	186	27°00'S	30°58'E
Pietrosu, Vârful, mtn., Rom.	125	47°35'N	24°49'E
Pieve di Cadore, Italy (pyā'vā dē kä-dō'rå)	118	46°26'N	12°22'E
Pigeon, r., N.A. (pĭj'ŭn)	71	48°05'N	90°13'W
Pigeon Lake, Can.	62f	49°57'N	97°36'W
Pigeon Lake, l., Can.	55	53°00'N	114°00'W
Piggott, Ar., U.S. (pĭg-ŭt)	79	36°22'N	90°10'W
Pijijiapan, Mex. (pēkē-kē-ä'pän)	89	15°40'N	93°12'W
Pijnacker, Neth.	115a	52°01'N	4°25'E
Pikes Peak, mtn., Co., U.S. (pīks)	64	38°49'N	105°03'W
Pikeville, Ky., U.S. (pīk'vĭl)	66	37°28'N	82°31'W
Pikou, China (pē-kō)	164	39°25'N	122°19'E
Pikwitonei, Can. (pĭk'wĭ-tōn)	57	55°35'N	97°09'W
Piła, Pol. (pē'lȧ)	124	53°09'N	16°44'E
Pilansberg, mtn., S. Afr. (pē'ăns'bûrg)	192c	25°08'S	26°55'E
Pilar, Arg. (pē'lär)	99c	34°27'S	58°55'W
Pilar, Para.	102	27°00'S	58°15'W
Pilar de Goiás, Braz. (dĕ-gô'yà's)	101	14°47'S	49°33'W
Pilchuck, r., Wa., U.S.	74a	48°03'N	121°58'W
Pilchuck Creek, r., Wa., U.S. (pĭl'chŭck)	74a	48°19'N	122°11'W
Pilchuck Mountain, mtn., Wa., U.S.	74a	48°03'N	121°48'W
Pilcomayo, r., S.A. (pēl-cō-mī'ō)	102	24°45'S	59°15'W
Pili, Phil. (pē'lē)	169a	13°34'N	123°17'E
Pilica, r., Pol. (pē-lēt'sä)	125	51°00'N	19°48'E
Pillar Point, c., Wa., U.S. (pĭl'ăr)	74a	48°14'N	124°06'W
Pillar Rock, Wa., U.S.	74c	46°16'N	123°35'W
Pilón, r., Mex. (pē-lōn')	88	24°13'N	99°03'W
Pilot Point, Tx., U.S. (pī'lŭt)	79	33°24'N	97°00'W
Pilsen see Plzeň, Czech Rep.	110	49°45'N	13°23'E
Piltene, Lat. (pĭl'tĕ-nĕ)	123	57°17'N	21°40'E
Pimal, Cerra, mtn., Mex. (sĕ'r-rä-pē-mäl')	88	22°58'N	104°19'W
Pimba, Austl. (pĭm'bȧ)	174	31°15'S	137°50'E
Pimville, neigh., S. Afr. (pĭm'vĭl)	187b	26°17'S	27°54'E
Pinacate, Cerro, mtn., Mex. (sĕ'r-rô-pē-nä-kä'tĕ)	86	31°45'N	113°30'W
Pinamalayan, Phil. (pē-nä-mä-lä'yän)	169a	13°04'N	121°31'E
Pinang see George Town, Malay.	168	5°21'N	100°09'E
Pınarbaşı, Tur. (pē'när-bä'shĭ)	119	38°50'N	36°10'E
Pinar del Río, Cuba (pē-när' dĕl rē'ō)	87	22°25'N	83°35'W
Pinar del Río, prov., Cuba	92	22°45'N	83°25'W
Pinatubo, mtn., Phil. (pē-nä-tōō'bō)	169a	15°09'N	120°19'E
Pincher Creek, Can. (pĭn'chēr krēk)	55	49°29'N	113°57'W
Pinckneyville, Il., U.S. (pĭnk'nĭ-vĭl)	79	38°06'N	89°22'W
Pińczów, Pol. (pĕn''chôf)	125	50°32'N	20°33'E
Pindamonhangaba, Braz. (pē'n-dä-mōnyä'n-gä-bä)	99a	22°56'S	45°26'W
Pinder Point, c., Bah.	92	26°35'N	78°35'W
Pindiga, Nig.	189	9°59'N	10°54'E
Píndos Óros, mts., Grc.	112	39°48'N	21°19'E
Pine, r., Can. (pīn)	55	55°30'N	122°20'W
Pine, r., Wi., U.S.	71	45°50'N	88°37'W
Pine Bluff, Ar., U.S. (pīn blŭf)	65	34°13'N	92°01'W
Pine City, Mn., U.S. (pīn)	71	45°50'N	93°01'W
Pine Creek, Austl.	174	13°45'S	132°00'E
Pine Creek, r., Nv., U.S.	76	40°15'N	116°17'W
Pine Falls, Can.	57	50°35'N	96°15'W
Pine Flat Lake, res., Ca., U.S.	76	36°52'N	119°18'W
Pine Forest Range, mts., Nv., U.S.	72	41°35'N	118°45'W
Pinega, Russia (pē-nyĕ'gȧ)	134	64°40'N	43°30'E
Pinega, r., Russia	136	64°10'N	42°30'E
Pine Hill, N.J., U.S. (pīn hĭl)	68f	39°47'N	74°59'W
Pineiós, r., Grc.	131	39°30'N	21°40'E
Pine Island Sound, strt., Fl., U.S.	83a	26°32'N	82°30'W
Pine Lake Estates, Ga., U.S. (lāk ĕs-tāts')	68c	33°47'N	84°13'W
Pinelands, S. Afr. (pīn'lănds)	186a	33°57'S	18°30'E
Pine Lawn, Mo., U.S. (lôn)	75e	38°42'N	90°17'W
Pine Pass, p., Can.	55	55°22'N	122°40'W
Pinerolo, Italy (pē-nå-rô'lō)	130	44°47'N	7°18'E
Pines, Lake o' the, Tx., U.S.	81	32°50'N	94°40'W
Pinetown, S. Afr. (pīn'toun)	187c	29°47'S	30°52'E
Pine View Reservoir, res., Ut., U.S. (vū)	75b	41°17'N	111°54'W
Pineville, Ky., U.S. (pīn'vĭl)	82	36°48'N	83°43'W
Pineville, La., U.S.	81	31°20'N	92°25'W
Ping, r., Thai.	168	17°54'N	98°29'E
Pingding, China (pĭŋ-dĭŋ)	164	37°50'N	113°30'E
Pingdu, China (pĭŋ-dōō)	164	36°46'N	119°57'E
Pinggir, Indon.	153b	1°05'N	101°12'E
Pinghe, China (pĭŋ-hŭ)	165	24°30'N	117°02'E
Pingle, China (pĭŋ-lŭ)	165	24°30'N	110°22'E
Pingliang, China (pĭng'lyäng')	160	35°12'N	106°50'E
Pingquan, China (pĭŋ-chyůän)	164	40°58'N	118°40'E
Pingtan, China (pĭŋ-tän)	165	25°30'N	119°45'E
Pingtan Dao, i., China (pĭŋ-tän dou)	165	25°40'N	119°45'E
P'ingtung, Tai.	165	22°40'N	120°35'E
Pingwu, China (pĭŋ-wōō)	164	32°20'N	104°40'E
Pingxiang, China (pĭŋ-shyäŋ)	165	27°40'N	113°50'E
Pingyi, China (pĭŋ-yē)	162	35°30'N	117°38'E
Pingyuan, China (pĭŋ-yůän)	162	37°11'N	116°26'E
Pingzhou, China (pĭŋ-jō)	163a	23°01'N	113°11'E
Pinhal, Braz. (pē-nyä'l)	99a	22°11'S	46°43'W
Pinhal Novo, Port. (nô vô)	129b	38°38'N	8°54'W
Pinhel, Port. (pēn-yĕl')	128	40°45'N	7°03'W
Pini, Pulau, i., Indon.	168	0°07'S	98°38'E
Pinnacles National Monument, rec., Ca., U.S. (pĭn'ȧ-k'lz)	76	36°30'N	121°00'W
Pinneberg, Ger. (pĭn'ĕ-bĕrg)	115c	53°40'N	9°48'E
Pinole, Ca., U.S. (pĭ-nō'lē)	74b	38°01'N	122°17'W
Pinos-Puente, Spain (pwān'tā)	128	37°15'N	3°43'W
Pinotepa Nacional, Mex. (pē-nô-tā'pä nä-syô-näl')	88	16°21'N	98°04'W
Pins, Île des, i., N. Cal.	175	22°44'S	167°44'E
Pinsk, Bela. (pēn'sk)	134	52°07'N	26°05'E
Pinta, i., Ec.	100	0°41'N	90°47'W
Pintendre, Can. (pĕN-täNdr')	62b	46°45'N	71°07'W
Pinto, Spain (pēn'tō)	129a	40°14'N	3°42'W
Pinto Butte, Can. (pĭn'tō)	56	49°22'N	107°25'W
Pioche, Nv., U.S. (pī-ō'chē)	77	37°56'N	114°28'W
Piombino, Italy (pyôm-bē'nō)	118	42°56'N	10°33'E
Pioneer Mountains, mts., Mt., U.S. (pī'ô-nēr')	73	45°23'N	112°51'W
Piotrków Trybunalski, Pol. (pyōtr'kōōv trĭ-bōō-nal'skē)	117	51°23'N	19°44'E
Piper, Al., U.S. (pī'pēr)	82	33°04'N	87°00'W
Piper, Ks., U.S.	75f	39°09'N	94°51'W
Pipe Spring National Monument, rec., Az., U.S. (pīp sprĭng)	77	36°50'N	112°45'W
Pipestone, Mn., U.S. (pīp'stōn)	70	44°00'N	96°19'W
Pipestone National Monument, rec., Mn., U.S.	70	44°03'N	96°24'W
Pipmuacan, Réservoir, res., Can. (pĭp-mä-kän')	59	49°45'N	70°00'W
Piqua, Oh., U.S. (pĭk'wȧ)	66	40°10'N	84°15'W
Piracaia, Braz. (pē-rä-kà'yä)	99a	23°04'S	46°20'W
Piracicaba, Braz. (pē-rä-sē-kä'bä)	101	22°43'S	47°39'W
Piraíba, r., Braz. (pä-rä-ē'bȧ)	99a	21°38'S	41°29'W
Piramida, mtn., Russia	135	54°00'N	96°00'E
Piran, Slvn. (pē-rå'n)	130	45°31'N	13°34'E
Piranga, Braz. (pē-rä'n-gä)	99a	20°41'S	43°17'W
Pirapetinga, Braz. (pē-rä-pĕ-tē'n-gä)	99a	21°40'S	42°20'W
Pirapora, Braz. (pē-rȧ-pō'rȧ)	101	17°39'S	44°54'W
Pirassununga, Braz. (pē-rä-sōō-nōō'n-gä)	99a	22°00'S	47°24'W
Pirenópolis, Braz. (pē-rĕ-nô'pō-lēs)	101	15°56'S	48°49'W
Piritu, Laguna de, l., Ven. (lä-gō'nä-dĕ-pē-rē'tōō)	101b	10°00'N	64°57'W
Pirmasens, Ger. (pĭr-mä-zĕns')	124	49°12'N	7°34'E
Pirna, Ger. (pĭr'nä)	124	50°57'N	13°56'E
Pirot, Serb. (pē'rōt)	119	43°09'N	22°35'E
Pirtleville, Az., U.S. (pûr't'l-vĭl)	77	31°25'N	109°35'W
Piru, Indon. (pē-rōō')	169	3°15'S	128°25'E
Pisa, Italy (pē'sä)	118	43°52'N	10°24'E
Pisagua, Chile (pē-sä'gwä)	100	19°43'S	70°12'W
Piscataway, Md., U.S. (pĭs-kă-tă-wā)	68e	38°42'N	76°59'W
Piscataway, N.J., U.S.	68a	40°35'N	74°27'W
Pisco, Peru (pēs'kō)	100	13°43'S	76°07'W
Pisco, Bahía de, b., Peru	100	13°43'S	77°48'W
Piseco, l., N.Y., U.S. (pĭ-sā'kô)	67	43°25'N	74°35'W
Pisek, Czech Rep. (pē'sĕk)	117	49°18'N	14°08'E
Pisticci, Italy (pēs-tē'chē)	130	40°24'N	16°34'E
Pistoia, Italy (pês-tô'yä)	118	43°57'N	11°54'E
Pisuerga, r., Spain (pê-swĕr'gä)	128	41°48'N	4°28'W
Pit, r., Ca., U.S. (pĭt)	72	40°58'N	121°42'W
Pitalito, Col. (pē-tä-lē'tō)	100	1°45'N	75°09'W
Pitcairn, dep., Oc.	2	25°04'S	130°05'W
Piteälven, r., Swe.	116	66°08'N	18°51'E
Pitești, Rom. (pē-tĕsht'')	131	44°51'N	24°51'E
Pithara, Austl. (pĭt'ărȧ)	174	30°27'S	116°45'E
Pithiviers, Fr. (pē-tē-vyā')	126	48°12'N	2°14'E
Pitman, N.J., U.S. (pĭt'măn)	68f	39°44'N	75°08'W
Pitseng, Leso.	187c	29°03'S	28°13'E
Pitt, r., Can.	74d	49°19'N	122°39'W
Pitt Island, i., Can.	54	53°35'N	129°45'W
Pittsburg, Ca., U.S. (pĭts'bûrg)	74b	38°01'N	121°52'W
Pittsburg, Ks., U.S.	65	37°25'N	94°43'W
Pittsburg, Tx., U.S.	79	32°00'N	94°57'W
Pittsburgh, Pa., U.S.	65	40°26'N	80°01'W
Pittsfield, Il., U.S. (pĭts'fēld)	79	39°37'N	90°47'W
Pittsfield, Ma., U.S.	67	42°25'N	73°15'W
Pittsfield, Me., U.S.	60	44°45'N	69°44'W
Pittston, Pa., U.S. (pĭts'tŭn)	67	41°20'N	75°50'W
Piũi, Braz. (pē-ōō'ē)	99a	20°27'S	45°57'W
Piura, Peru (pē-ōō'rä)	100	5°13'S	80°46'W
Pivdennyi Buh, r., Ukr.	137	48°12'N	30°13'E
Piya, Russia (pē'yȧ)	142a	58°34'N	61°12'E
Placentia, Can.	61	47°15'N	53°58'W
Placentia, Ca., U.S. (plă-sĕn'shĭ-ȧ)	75a	33°52'N	117°50'W
Placentia Bay, b., Can.	53a	47°14'N	54°30'W
Placerville, Ca., U.S. (plăs'ēr-vĭl)	76	38°43'N	120°47'W
Placetas, Cuba (plä-thā'täs)	92	22°10'N	79°40'W
Placid, l., N.Y., U.S. (plăs'ĭd)	67	44°20'N	74°00'W
Plain City, Ut., U.S. (plān)	75b	41°18'N	112°06'W
Plainfield, Il., U.S. (plān'fēld)	69a	41°37'N	88°12'W
Plainfield, In., U.S.	69g	39°42'N	86°23'W
Plainfield, N.J., U.S.	68a	40°38'N	74°25'W
Plainview, Ar., U.S. (plān'vū)	79	34°59'N	93°15'W
Plainview, Mn., U.S.	71	44°09'N	93°12'W
Plainview, Ne., U.S.	70	42°20'N	97°47'W
Plainview, Tx., U.S.	78	34°11'N	101°42'W
Plainwell, Mi., U.S. (plan'wĕl)	66	42°25'N	85°40'W
Plaisance, Can. (plĕ-zäNs')	62c	45°37'N	75°07'W
Plana or Flat Cays, is., Bah. (plä'nä)	93	22°35'N	73°35'W
Planegg, Ger. (plä'nĕg)	115d	48°06'N	11°27'E
Plano, Tx., U.S. (plā'nō)	79	33°01'N	96°42'W
Plantagenet, Can. (plän-täzh-nĕ')	62c	45°33'N	75°00'W
Plant City, Fl., U.S. (plȧnt sĭ'tĭ)	83a	28°00'N	82°07'W
Plaquemine, La., U.S. (plăk'mēn')	81	30°17'N	91°14'W
Plasencia, Spain (plä-sĕn'thē-ä)	128	40°02'N	6°07'W
Plast, Russia (plást)	136	54°22'N	60°48'E
Plaster Rock, Can. (plås'tēr rŏk)	60	46°54'N	67°24'W
Plastun, Russia (plås-tōōn')	166	44°41'N	136°08'E
Plata, Río de la, est., S.A. (dālä plä'tä)	102	34°35'S	58°15'W
Platani, r., Italy (plä-tä'nē)	130	37°26'N	13°28'E
Plateforme, Pointe, c., Haiti	93	19°35'N	73°50'W
Platinum, Ak., U.S. (plăt'ĭ-nŭm)	63	59°00'N	161°27'W
Plato, Col. (plä'tō)	100	9°49'N	74°48'W
Platón Sánchez, Mex. (plä-tōn' sän'chĕz)	88	21°14'N	98°20'W
Platte, S.D., U.S. (plăt)	70	43°22'N	98°51'W
Platte, r., Mo., U.S.	79	40°09'N	94°40'W
Platte, r., Ne., U.S.	64	40°50'N	100°40'W
Platteville, Wi., U.S. (plăt'vĭl)	71	42°44'N	90°31'W
Plattsburg, Mo., U.S. (plăts'bûrg)	79	39°33'N	94°26'W
Plattsburg, N.Y., U.S.	67	44°40'N	73°30'W
Plattsmouth, Ne., U.S. (plăts'mŭth)	70	41°00'N	95°53'W
Plauen, Ger. (plou'ĕn)	117	50°30'N	12°08'E
Playa de Guanabo, Cuba (plä-yä-dĕ-gwä-nä'bô)	93a	23°10'N	82°07'W
Playa de Santa Fe, Cuba	93a	23°05'N	82°31'W
Playas Lake, l., N.M., U.S. (plä'yȧs)	77	31°50'N	108°30'W
Playa Vicente, Mex. (vē-sĕn'tå)	89	17°49'N	95°49'W
Playa Vicente, r., Mex.	89	17°36'N	96°13'W
Playgreen Lake, l., Can. (plā'grēn)	57	54°00'N	98°10'W
Pleasant, l., N.Y., U.S. (plĕz'ănt)	67	43°25'N	74°25'W
Pleasant Grove, Al., U.S.	68h	33°29'N	86°57'W
Pleasant Hill, Ca., U.S.	74b	37°57'N	122°04'W
Pleasant Hill, Mo., U.S.	79	38°46'N	94°18'W
Pleasanton, Ca., U.S. (plĕz'ăn-tŭn)	74b	37°40'N	121°53'W
Pleasanton, Ks., U.S.	79	38°10'N	94°41'W
Pleasanton, Tx., U.S.	80	28°58'N	98°30'W
Pleasant Plain, Oh., U.S. (plĕz'ănt)	69f	39°17'N	84°06'W
Pleasant Ridge, Mi., U.S.	69b	42°28'N	83°09'W
Pleasant View, Ut., U.S. (plĕz'ănt vū)	75b	41°20'N	112°02'W
Pleasantville, N.Y., U.S. (plĕz'ănt-vĭl)	68a	41°08'N	73°47'W
Pleasure Ridge Park, Ky., U.S. (plĕzh'ēr rĭj)	69h	38°09'N	85°49'W
Plenty, Bay of, b., N.Z. (plĕn'tē)	175a	37°30'S	177°10'E
Plentywood, Mt., U.S. (plĕn'tē-wŏd)	73	48°47'N	104°38'W
Ples, Russia (plyĕs)	132	57°26'N	41°29'E
Pleshcheyevo, l., Russia (plĕsh-chā'yĕ-vô)	132	56°50'N	38°22'E
Plessisville, Can. (plĕ-sē-vēl')	59	46°12'N	71°47'W
Pleszew, Pol. (plĕ'zhĕf)	125	51°54'N	17°48'E
Plettenberg, Ger. (plĕ'tĕn-bĕrgh)	127c	51°13'N	7°53'E
Pleven, Blg. (plĕ'vĕn)	119	43°24'N	24°26'E
Pljevlja, Serb. (plĕv'lyä)	119	43°20'N	19°21'E
Płock, Pol. (pwŏtsk)	117	52°32'N	19°44'E
Ploërmel, Fr. (plô-ĕr-mĕl')	126	47°56'N	2°25'W
Ploiești, Rom. (plô-yĕsht'')	110	44°56'N	26°01'E
Plomári, Grc.	131	38°51'N	26°24'E
Plomb du Cantal, mtn., Fr. (plôN'dükän-täl')	117	45°30'N	2°49'E
Plonge, Lac la, l., Can. (plōnzh)	56	55°08'N	107°25'W
Plovdiv, Blg. (plôv'dĭf) (fĭl-ĭp-ŏp'ô-lĭs)	110	42°09'N	24°43'E
Pluma Hidalgo, Mex. (plōō'mä ē-däl'gō)	89	15°54'N	96°23'W
Plunge, Lith. (plōn'gā)	123	55°56'N	21°45'E
Plymouth, Monts.	91b	16°43'N	62°12'W
Plymouth, Eng., U.K. (plĭm'ŭth)	117	50°25'N	4°14'W
Plymouth, In., U.S.	66	41°20'N	86°20'W
Plymouth, Ma., U.S.	67	42°00'N	70°45'W
Plymouth, Mi., U.S.	69b	42°23'N	83°27'W
Plymouth, N.C., U.S.	83	35°50'N	76°44'W
Plymouth, N.H., U.S.	67	43°50'N	71°40'W
Plymouth, Pa., U.S.	67	41°15'N	75°55'W
Plymouth, Wi., U.S.	71	43°45'N	87°59'W
Plyussa, r., Russia (plyōō'sȧ)	132	58°33'N	28°30'E
Plzeň, Czech Rep.	110	49°45'N	13°23'E
Po, r., Italy	112	45°10'N	11°00'E
Pocahontas, Ar., U.S. (pō-kȧ-hŏn'tȧs)	79	36°15'N	91°01'W
Pocahontas, Ia., U.S.	71	42°43'N	94°41'W
Pocatello, Id., U.S. (pō-kȧ-tĕl'ō)	64	42°54'N	112°30'W
Pochëp, Russia (pô-chĕp')	137	52°56'N	33°27'E
Pochinok, Russia (pô-chē'nôk)	132	54°14'N	32°27'E
Pochinski, Russia	136	54°40'N	44°50'E
Pochotitán, Mex. (pô-chô-tē-tá'n)	88	21°37'N	104°33'W
Pochutla, Mex.	89	15°46'N	96°28'W
Pocomoke City, Md., U.S. (pō-kō-mōk')	67	38°05'N	75°35'W
Pocono Mountains, mts., Pa., U.S. (pō-cō'nō)	67	41°10'N	75°30'W
Poços de Caldas, Braz. (pō-sôs-dĕ-käl'dás)	101	21°48'S	46°34'W
Poder, Sen. (pō-dôr')	184	16°35'N	15°04'W
Podgorica, Serb.	131	42°25'N	19°15'E
Podkamennaya Tunguska, r., Russia	135	61°43'N	93°45'E
Podol'sk, Russia (pô-dôl''sk)	136	55°26'N	37°33'E
Poggibonsi, Italy (pôd-jē-bôn'sē)	130	43°27'N	11°12'E
Pogodino, Bela. (pô-gô'dē-nô)	136	54°17'N	31°00'E
P'ohangdong, Kor., S.	166	35°57'N	129°23'E
Pointe-à-Pitre, Guad. (pwăNt' á pē-tr')	87	16°15'N	61°32'W
Pointe-aux-Trembles, Can. (pōō-ăNt' ō-träNbl)	62a	45°39'N	73°30'W
Pointe Claire, Can. (pōō-ăNt' klĕr)	62a	45°27'N	73°48'W
Pointe-des-Cascades, Can. (käs-kädz')	62a	45°19'N	73°58'W

ng-sing; ŋ-baŋk; N-nasalized n; nŏd; cŏmmit; ōld; ôbey; ôrder; oi-boil; fōōd; ŏŏ-as oo in foot; ou-out; s-soft; sh-dish; th-thin; pūre; ůnite; ûrn; stŭd; circŭs; ü-as in French tu; '-indeterminate vowel.

290 Pointe Fortune – Poygan PRONOUNCING INDEX

PLACE (Pronunciation)	PAGE	LAT.	LONG.
Pointe Fortune, Can. (fôr′tūn)	62a	45°34′N	74°23′W
Pointe-Gatineau, Can. (po͞o-äɴt′gä-tē-nō′)	62c	45°28′N	75°42′W
Pointe Noire, Congo	186	4°48′S	11°51′E
Point Hope, Ak., U.S. (hōp)	63	68°18′N	166°38′W
Point Pleasant, W.V., U.S. (plĕz′ănt)	66	38°50′N	82°10′W
Point Roberts, Wa., U.S. (rŏb′ērts)	74d	48°59′N	123°04′W
Poissy, Fr. (pwȧ-sē′)	127b	48°55′N	2°02′E
Poitiers, Fr. (pwȧ-tyā′)	117	46°35′N	0°18′E
Pokaran, India (pō′kŭr-ŭn)	158	27°00′N	72°05′E
Pokrov, Russia (pỏ-krôf)	132	55°56′N	39°09′E
Pokrovskoye, Russia (pỏ-krôf′skỏ-yĕ)	133	47°27′N	38°54′E
Pola, r., Russia (pō′lä)	132	57°44′N	31°53′E
Pola de Laviana, Spain (dĕ-lä-vyä′nä)	128	43°15′N	5°29′W
Pola de Siero, Spain	128	43°24′N	5°39′W
Poland, nation, Eur. (pō′lănd)	110	52°37′N	17°01′E
Polangui, Phil. (pô-läŋ′gē)	169a	13°18′N	123°29′E
Polatsk, Bela.	136	55°30′N	28°48′E
Polazna, Russia (pô′láz-nȧ)	142a	58°18′N	56°25′E
Polessk, Russia (pô′lĕsk)	123	54°50′N	21°14′E
Polevskoy, Russia (pô-lĕ′vs-kô′ê)	142a	56°28′N	60°14′E
Polgár, Hung. (pỏl′gär)	125	47°54′N	21°10′E
Policastro, Golfo di, b., Italy	130	40°00′N	13°23′E
Polichnítos, Grc.	131	39°05′N	26°11′E
Poligny, Fr. (pō-lē-nyē′)	127	46°48′N	5°42′E
Polillo, Phil. (pỏ-lēl′yō)	169a	14°42′N	121°56′W
Polillo Islands, is., Phil.	155	15°05′N	122°15′E
Polillo Strait, strt., Phil.	169a	15°02′N	121°40′E
Polist′, r., Russia (pô′lĭst)	132	57°42′N	31°02′E
Polistena, Italy (pō-lės-tā′nä)	130	38°25′N	16°05′E
Polkan, Gora, mtn., Russia	135	60°18′N	92°08′E
Polochic, r., Guat. (pō-lỏ-chēk′)	90	15°19′N	89°45′W
Polonne, Ukr.	133	50°07′N	27°31′E
Polpaico, Chile (pỏl-pȧ′y-kỏ)	99b	33°10′S	70°53′W
Polson, Mt., U.S. (pōl′sŭn)	73	47°40′N	114°10′W
Poltava, Ukr. (pỏl-tä′vä)	134	49°35′N	34°33′E
Poltava, prov., Ukr.	133	49°53′N	32°58′E
Põltsamaa, Est. (pŏlt′sȧ-mä)	123	58°39′N	26°00′E
Polunochnoye, Russia (pô-lo͞o-nỏ′ch-nô′yĕ)	142a	60°52′N	60°27′E
Poluy, r., Russia (pôl′wė)	140	65°45′N	68°15′E
Polyakovka, Russia (pŭl-yä′kỏv-kȧ)	142a	54°38′N	59°42′E
Polyarnyy, Russia (pŭl-yär′nē)	134	69°10′N	33°30′E
Polygyros, Grc.	131	40°23′N	23°27′E
Polynesia, is., Oc.	194	4°00′S	156°00′W
Pomba, r., Braz. (pô′m-bȧ)	99a	21°28′S	42°28′W
Pomerania, hist. reg., Pol. (pŏm-ė-rā′nĭ-ȧ)	124	53°50′N	15°20′E
Pomeroy, S. Afr. (pŏm′ēr-roi)	187c	28°36′S	30°26′E
Pomeroy, Wa., U.S. (pŏm′ēr-oi)	72	46°28′N	117°35′W
Pomezia, Italy (pỏ-mĕ′t-zyä)	129d	41°41′N	12°31′E
Pomigliano d′Arco, Italy (pỏ-mē-lyȧ′nỏ-d-ä′r-kỏ)	129c	40°39′N	14°23′E
Pomme de Terre, Mn., U.S. (pỏm dē tĕr′)	70	45°22′N	95°52′W
Pomona, Ca., U.S. (pỏ-mō′nȧ)	64	34°04′N	117°45′W
Pomorie, Blg.	119	42°24′N	27°41′E
Pompano Beach, Fl., U.S. (pŏm′pă-nỏ)	83a	26°12′N	80°07′W
Pompeii Ruins, hist., Italy	129c	40°31′N	14°29′E
Pompton Lakes, N.J., U.S. (pŏmp′tŏn)	68a	41°01′N	74°16′W
Pomuch, Mex. (pô-mo͞o′ch)	90a	20°12′N	90°10′W
Ponca, Ne., U.S. (pŏn′kȧ)	70	42°34′N	96°43′W
Ponca City, Ok., U.S.	79	36°42′N	97°07′W
Ponce, P.R. (pōn′sā)	87	18°01′N	66°43′W
Pondicherry, India	155	11°58′N	79°48′E
Pondicherry, state, India	155	11°50′N	74°50′E
Ponferrada, Spain (pỏn-fĕr-rä′dhä)	118	42°33′N	6°38′W
Ponoka, Can. (pŏ-nō′kȧ)	50	52°42′N	113°35′W
Ponoy, Russia	136	66°58′N	41°00′E
Ponoy, r., Russia	136	67°00′N	39°00′E
Ponta Delgada, Port. (pỏn′tȧ dĕl-gä′dȧ)	184a	37°40′N	25°45′W
Ponta Grossa, Braz. (grō′sȧ)	101	25°09′S	50°05′W
Pont-à-Mousson, Fr. (pôɴ′tȧ-mo͞osôɴ′)	127	48°55′N	6°02′E
Pontarlier, Fr. (pôɴ′tȧr-lyā′)	127	46°53′N	6°22′E
Pont-Audemer, Fr. (pôɴ′tōd′mâr′)	126	49°23′N	0°28′E
Pontchartrain Lake, l., La., U.S. (pôɴ-shȧr-trăn′)	81	30°10′N	90°10′W
Ponteareas, Spain	128	42°09′N	8°23′W
Pontedera, Italy (pōn-tȧ-dā′rä)	130	43°37′N	10°37′E
Ponte de Sor, Port.	128	39°14′N	8°03′W
Pontefract, Eng., U.K. (pŏn′tė-frăkt)	114a	53°41′N	1°18′W
Ponte Nova, Braz. (pô′n-tĕ-nô′vȧ)	101	20°26′S	42°52′W
Pontevedra, Spain (pỏn-tĕ-vĕ-drä)	118	42°28′N	8°38′W
Ponthierville *see* Ubundi, D.R.C.	186	0°21′S	25°29′E
Pontiac, Il., U.S. (pŏn′tĭ-ăk)	66	40°55′N	88°35′W
Pontiac, Mi., U.S.	65	42°37′N	83°17′W
Pontianak, Indon. (pỏn-tė-ä′nȧk)	168	0°04′S	109°20′E
Pontian Kechil, Malay.	153b	1°29′N	103°24′E
Pontic Mountains, mts., Tur.	137	41°20′N	34°30′E
Pontivy, Fr. (pôɴ-tė-vē′)	126	48°05′N	2°57′W
Pontoise, Fr. (pôɴ-twȧz′)	126	49°03′N	2°05′E
Pontonnyy, Russia (pôn′tôn-nyĭ)	142c	59°47′N	30°39′E
Pontotoc, Ms., U.S. (pŏn-tỏ-tŏk′)	82	34°11′N	88°59′W
Pontremoli, Italy (pỏn-trĕm′ô-lē)	130	44°21′N	9°50′E
Ponziane, Isole, i., Italy (ė′sō-lĕ)	118	40°55′N	12°58′E
Poole, Eng., U.K. (po͞ol)	120	50°43′N	2°00′W
Poolesville, Md., U.S. (poolĕs-vĭl)	68e	39°08′N	77°26′W
Pooley Island, i., Can. (po͞o′lē)	54	52°44′N	128°16′W
Poopó, Lago de, l., Bol.	100	18°45′S	67°07′W
Popayán, Col. (pō-pä-yän′)	100	2°21′N	76°43′W
Poplar, Mt., U.S. (pŏp′lēr)	73	48°08′N	105°10′W
Poplar, r., Mt., U.S.	73	48°34′N	105°20′W
Poplar, West Fork, r., Mt., U.S.	73	48°59′N	106°06′W
Poplar Bluff, Mo., U.S. (blŭf)	79	36°43′N	90°22′W
Poplar Plains, Ky., U.S. (plāns)	66	38°20′N	83°40′W
Poplar Point, Can.	62f	50°04′N	97°57′W
Poplarville, Ms., U.S. (pŏp′lēr-vĭl)	82	30°50′N	89°33′W
Popocatépetl Volcán, Mex. (pỏ-pỏ-kä-tā′pĕ′t′l)	86	19°01′N	98°38′W
Popokabaka, D.R.C. (pō′pỏ-kȧ-bä′kȧ)	186	5°42′S	16°35′E
Popovo, Blg. (pô′pỏ-vō)	131	43°23′N	26°17′E
Porbandar, India (pōr-bŭn′dŭr)	155	21°44′N	69°40′E
Porce, r., Col. (pôr-sĕ)	100a	7°11′N	74°55′W
Porcher Island, i., Can. (pôr′kēr)	54	53°57′N	130°30′W
Porcuna, Spain (pôr-ko͞o′nä)	128	37°54′N	4°10′W
Porcupine, r., N.A.	63	67°38′N	140°07′W
Porcupine Creek, r., Mt., U.S.	73	48°27′N	106°24′W
Porcupine Hills, hills, Can.	57	52°30′N	101°45′W
Pordenone, Italy (pōr-dȧ-nō′nȧ)	130	45°58′N	12°38′E
Pori, Fin. (pô′rė)	116	61°29′N	21°45′E
Poriúncula, Braz.	99a	20°58′S	42°02′W
Porkhov, Russia (pôr′κôf)	136	57°46′N	29°33′E
Porlamar, Ven. (pỏr-lä-mär′)	100	11°00′N	63°55′W
Pornic, Fr. (pỏr-nēk′)	126	47°08′N	2°07′W
Poronaysk, Russia (pô′rỏ-nīsk)	135	49°21′N	143°23′E
Porrentruy, Switz. (pỏ-räɴ-trüē′)	124	47°25′N	7°02′E
Porsgrunn, Nor. (pỏrs′grỏn′)	122	59°09′N	9°36′E
Portachuelo, Bol. (pỏrt-ä-chwā′lỏ)	100	17°20′S	63°12′W
Portage, Pa., U.S. (pôr′tȧj)	67	40°25′N	78°35′W
Portage, Wi., U.S.	71	43°33′N	89°29′W
Portage Des Sioux, Mo., U.S. (dē so͞o)	75e	38°56′N	90°21′W
Portage la Prairie, Can. (lä-prā′rĭ)	50	49°57′N	98°25′W
Port Alberni, Can. (pōr äl-bĕr-nē′)	50	49°14′N	124°48′W
Portalegre, Port. (pôr-tä-lā′grĕ)	118	39°18′N	7°26′W
Portales, N.M., U.S. (pôr-tä′lĕs)	78	34°10′N	103°11′W
Port Alfred, S. Afr.	186	33°36′S	26°55′E
Port Alice, Can. (ăl′ĭs)	50	50°23′N	127°27′W
Port Allegany, Pa., U.S. (ăl-ė-gā′nĭ)	67	41°50′N	78°10′W
Port Angeles, Wa., U.S. (ăn′jē-lĕs)	64	48°07′N	123°26′W
Port Antonio, Jam.	87	18°10′N	76°25′W
Portarlington, Austl.	173a	38°07′S	144°39′E
Port Arthur, Tx., U.S.	65	29°52′N	93°59′W
Port Augusta, Austl. (ô-gŭs′tă)	176	32°28′S	137°50′E
Port au Port Bay, b., Can. (pỏr′tō pōr′)	61	48°41′N	58°45′W
Port-au-Prince, Haiti (prăɴs′)	87	18°35′N	72°20′W
Port Austin, Mi., U.S. (ôs′tĭn)	66	44°00′N	83°00′W
Port Blair, India (blâr)	168	12°07′N	92°45′E
Port Bolivar, Tx., U.S. (bŏl′ĭ-vȧr)	81a	29°22′N	94°46′W
Port Borden, Can. (bôr′dĕn)	60	46°15′N	63°42′W
Port-Bouët, C. Iv.	184	5°24′N	3°56′W
Port-Cartier, Can.	60	50°01′N	66°53′W
Port Chester, N.Y., U.S. (chĕs′tēr)	68a	40°59′N	73°40′W
Port Chicago, Ca., U.S. (shĭ-kô′gō)	74b	38°03′N	122°01′W
Port Clinton, Oh., U.S. (klĭn′tŭn)	66	41°30′N	83°00′W
Port Colborne, Can.	59	42°53′N	79°13′W
Port Coquitlam, Can. (kỏ-kwĭt′lȧm)	55	49°16′N	122°46′W
Port Credit, Can. (krĕd′ĭt)	62d	43°33′N	79°35′W
Port-de-Bouc, Fr. (pỏr-dē-bo͝ok′)	126a	43°24′N	5°00′E
Port de Paix, Haiti (pĕ)	93	19°55′N	72°50′W
Port Dickson, Malay. (dĭk′sŭn)	153b	2°33′N	101°49′E
Port Discovery, b., Wa., U.S. (dĭs-kŭv′ēr-ĭ)	74a	48°05′N	122°55′W
Port Edward, S. Afr. (ĕd′wērd)	187c	31°04′S	30°14′E
Port Elgin, Can. (ĕl′jĭn)	60	46°03′N	64°05′W
Port Elizabeth, S. Afr. (ė-lĭz′ȧ-bĕth)	186	33°57′S	25°37′E
Porterdale, Ga., U.S. (pōr′tēr-dāl)	82	33°34′N	83°53′W
Porterville, Ca., U.S. (pōr′tēr-vĭl)	76	36°03′N	119°05′W
Port Francqui *see* Ilebo, D.R.C.	186	4°19′S	20°35′E
Port Gamble, Wa., U.S. (găm′bŭl)	74a	47°52′N	122°36′W
Port Gamble Indian Reservation, I.R., Wa., U.S.	74a	47°54′N	122°33′W
Port-Gentil, Gabon (zhäɴ-tē′)	186	0°43′S	8°47′E
Port Gibson, Ms., U.S.	82	31°56′N	90°57′W
Port Harcourt, Nig. (här′kŭrt)	184	4°43′N	7°05′E
Port Hardy, Can. (här′dĭ)	54	50°43′N	127°29′W
Port Hawkesbury, Can.	61	45°37′N	61°21′W
Port Hedland, Austl. (hĕd′lănd)	174	20°30′S	118°30′E
Porthill, Id., U.S.	72	49°00′N	116°30′W
Port Hood, Can. (hỏd)	61	46°01′N	61°32′W
Port Hope, Can. (hōp)	59	43°55′N	78°10′W
Port Huron, Mi., U.S. (hū′rŏn)	65	43°00′N	82°30′W
Portici, Italy (pôr′tė-chė)	129c	40°34′N	14°20′E
Portillo, Chile (pôr-tē′l-yỏ)	99b	32°51′S	70°09′W
Portimão, Port. (pôr-tē-moůɴ)	128	37°09′N	8°34′W
Port Jervis, N.Y., U.S. (jûr′vĭs)	68a	41°22′N	74°41′W
Portland, Austl. (pōrt′lănd)	175	38°20′S	142°40′E
Portland, In., U.S.	66	40°25′N	85°00′W
Portland, Me., U.S.	65	43°40′N	70°16′W
Portland, Mi., U.S.	66	42°50′N	85°00′W
Portland, Or., U.S.	64	45°31′N	122°41′W
Portland, Tx., U.S.	81	27°53′N	97°20′W
Portland Bight, b., Jam.	92	17°45′N	77°05′W
Portland Canal, can., Ak., U.S.	54	55°10′N	130°08′W
Portland Inlet, b., Can.	54	54°50′N	130°15′W
Portland Point, c., Jam.	92	17°40′N	77°20′W
Port Lavaca, Tx., U.S. (lȧ-vä′kȧ)	81	28°36′N	96°38′W
Port Lincoln, Austl. (lĭŋ-kŭn)	174	34°39′S	135°50′E
Port Ludlow, Wa., U.S. (lŭd′lō)	74a	47°26′N	122°41′W
Port Macquarie, Austl. (mȧ-kwŏ′rĭ)	175	31°25′S	152°45′E
Port Madison Indian Reservation, I.R., Wa., U.S. (măd′ĭ-sŭn)	74a	47°46′N	122°38′W
Port Maria, Jam. (mȧ-rī′ȧ)	92	18°20′N	76°55′W
Port Moody, Can. (mo͞od′ĭ)	55	49°17′N	122°51′W
Port Moresby, Pap. N. Gui. (mōrz′bė)	169	9°34′S	147°20′E
Port Neches, Tx., U.S. (nĕch′ĕz)	81	29°59′N	93°57′W
Port Nelson, Can. (nĕl′sŭn)	57	57°03′N	92°36′W
Portneuf-Sur-Mer, Can. (pỏr-nûf′sür mĕr)	60	48°36′N	69°06′W
Port Nolloth, S. Afr. (nŏl′ŏth)	186	29°10′S	17°00′E
Porto (Oporto), Port. (pōr′tỏ)	110	41°10′N	8°38′W
Porto Acre, Braz. (ä′krĕ)	100	9°38′S	67°34′W
Porto Alegre, Braz. (ä-lā′grĕ)	102	29°58′S	51°11′W
Porto Amboim, Ang.	186	11°01′S	13°45′E
Portobelo, Pan. (pôr′tỏ-bā′lỏ)	87	9°32′N	79°40′W
Pôrto de Pedras, Braz. (pā′dräzh)	101	9°09′S	35°20′W
Pôrto Feliz, Braz. (fĕ-lē′s)	99a	23°12′S	47°30′W
Portoferraio, Italy (pōr′tỏ-fĕr-rä′yō)	130	42°47′N	10°20′E
Port of Spain, Trin. (spān)	101	10°44′N	61°24′W
Portogruaro, Italy (pōr′tỏ-grỏ-ä′rō)	130	45°48′N	12°49′E
Portola, Ca., U.S. (pôr′tō-lȧ)	76	39°47′N	120°29′W
Porto Mendes, Braz. (mĕ′n-dĕs)	101	24°41′S	54°13′W
Porto Murtinho, Braz. (mỏr-tēn′yỏ)	101	21°43′S	57°43′W
Porto Nacional, Braz. (nȧ-syỏ-näl′)	101	10°43′S	48°14′W
Porto Novo, Benin (pōr′tỏ-nō′vō)	184	6°29′N	2°37′E
Port Orchard, Wa., U.S. (ôr′chērd)	74a	47°32′N	122°38′W
Port Orchard, b., Wa., U.S.	74a	47°40′N	122°39′W
Porto Santo, Ilha de, i., Port. (säɴ′tỏ)	184	32°41′N	16°15′W
Porto Seguro, Braz. (sā-go͞o′rỏ)	101	16°26′S	38°59′W
Porto Torres, Italy (tôr′rĕs)	130	40°49′N	8°25′E
Porto-Vecchio, Fr. (vĕk′ê-ỏ)	130	41°36′N	9°17′E
Porto Velho, Braz. (vāl′yỏ)	100	8°45′S	63°43′W
Portoviejo, Ec. (pôr-tō-vyā′hō)	100	1°11′S	80°28′W
Port Phillip Bay, b., Austl. (fĭl′ĭp)	175	37°57′S	144°50′E
Port Pirie, Austl. (pĭ′rė)	174	33°10′S	138°00′E
Port Royal, b., Jam. (roi′ăl)	92	17°50′N	76°45′W
Port Said, Egypt	192d	31°15′N	32°19′E
Port Saint Johns, S. Afr. (sȧnt jŏnz)	186	31°37′S	29°32′E
Port Saint Lucie, Fl., U.S.	83a	27°20′N	80°20′W
Port Shepstone, S. Afr. (shĕps′tŭn)	186	30°45′S	30°23′E
Portsmouth, Dom.	91b	15°33′N	61°28′W
Portsmouth, Eng., U.K. (pôrts′mŭth)	110	50°45′N	1°03′W
Portsmouth, N.H., U.S.	65	43°05′N	70°50′W
Portsmouth, Oh., U.S.	65	38°45′N	83°00′W
Portsmouth, Va., U.S.	65	36°50′N	76°19′W
Port Sulphur, La., U.S. (sŭl′fēr)	82	29°28′N	89°41′W
Port Susan, b., Wa., U.S. (sū-zȧn′)	74a	48°11′N	122°25′W
Port Townsend, Wa., U.S. (tounz′ĕnd)	74a	48°07′N	122°46′W
Port Townsend, b., Wa., U.S.	74a	48°05′N	122°47′W
Portugal, nation, Eur. (pōr′tu-găl)	110	38°15′N	8°08′W
Portugalete, Spain (pōr-to͞o-gä-lā′tā)	128	43°18′N	3°05′W
Portuguese West Africa *see* Angola, nation, Ang.	186	14°15′S	16°00′E
Port Vendres, Fr.	126	42°32′N	3°07′E
Port Vila, Vanuatu	175	17°44′S	168°19′E
Port Wakefield, Austl. (wāk′fēld)	174	34°12′S	138°10′E
Port Washington, N.Y., U.S. (wôsh′ĭng-tŭn)	68a	40°49′N	73°42′W
Port Washington, Wi., U.S.	71	43°24′N	87°52′W
Posadas, Arg. (pō-sä′dhäs)	102	27°32′S	55°56′W
Posadas, Spain (pỏ-sä-däs)	128	37°48′N	5°09′W
Poshekhon′ye Volodarsk, Russia (pô-shyĕ′κŏn-yĕ vôl′ỏ-dȧrsk)	132	58°31′N	39°07′E
Poso, Danau, l., Indon. (pō′sō)	168	2°00′S	119°40′E
Pospelokova, Russia (pỏs-pyĕl′kô-vȧ)	142a	59°25′N	60°50′E
Possession Sound, strt., Wa., U.S. (pŏ-zĕsh-ŭn)	74a	47°59′N	122°17′W
Possum Kingdom Reservoir, res., Tx., U.S. (pŏs′ŭm kĭng′dŭm)	80	32°58′N	98°12′W
Post, Tx., U.S. (pōst)	78	33°12′N	101°21′W
Postojna, Slvn. (pōs-tōynȧ)	130	45°45′N	14°13′E
Pos′yet, Russia (pos-yĕt′)	166	42°27′N	130°47′E
Potawatomi Indian Reservation, I.R., Ks., U.S. (pŏt-ȧ-wā′tō mĕ)	79	39°30′N	96°11′W
Potchefstroom, S. Afr. (pŏch′ĕf-strōm)	186	26°42′S	27°06′E
Poteau, Ok., U.S. (pỏ-tō′)	79	35°03′N	94°37′W
Poteet, Tx., U.S. (pỏ-tēt)	80	29°05′N	98°35′W
Potenza, Italy (pỏ-tĕnt′sä)	119	40°39′N	15°49′E
Potenza, r., Italy	130	43°09′N	13°00′E
Potgietersrus, S. Afr. (pỏt-κē′tērs-rûs)	186	24°09′S	29°04′E
Potholes Reservoir, res., Wa., U.S.	72	47°00′N	119°20′W
Poti, Geor. (pô′tė)	137	42°10′N	41°40′E
Potiskum, Nig.	184	11°43′N	11°05′E
Potomac, Md., U.S. (pỏ-tō′măk)	68e	39°01′N	77°13′W
Potomac, r., U.S. (pỏ-tō′măk)	65	38°15′N	76°55′W
Potosí, Bol.	100	19°35′S	65°45′W
Potosi, Mo., U.S. (pŏ-tō′sĭ)	79	37°56′N	90°46′W
Potosi, r., Mex. (pō-tỏ-sē′)	80	25°04′N	99°36′W
Potrerillos, Hond. (pō-trȧ-rēl′yỏs)	90	15°13′N	87°58′W
Potsdam, Ger. (pỏts′däm)	117	52°24′N	13°04′E
Potsdam, N.Y., U.S. (pŏts′dăm)	67	44°40′N	75°00′W
Pottenstein, Aus.	115e	47°58′N	16°06′E
Potters Bar, Eng., U.K. (pŏt′ĕz bär)	114b	51°41′N	0°12′W
Pottstown, Pa., U.S. (pŏts′toun)	67	40°15′N	75°40′W
Pottsville, Pa., U.S. (pôts′vĭl)	67	40°40′N	76°15′W
Poughkeepsie, N.Y., U.S. (pŏ-kĭp′sė)	65	41°45′N	73°55′W
Poulsbo, Wa., U.S. (pōlz′bỏ)	74a	47°44′N	122°38′W
Poulton-le-Fylde, Eng., U.K. (pōl′tŭn-lē-fīld′)	114a	53°52′N	2°59′W
Pouso Alegre, Braz. (pō′zỏ ä-lā′grĕ)	101	22°13′S	45°56′W
Póvoa de Varzim, Port. (pō-vō′ȧ dā vär′zēɴ)	118	41°23′N	8°44′W
Powder, r., Or., U.S.	72	44°55′N	117°35′W
Powder, r., U.S. (pou′dēr)	64	45°18′N	105°37′W
Powder, South Fork, r., Wy., U.S.	73	43°13′N	106°54′W
Powder River, Wy., U.S.	73	43°06′N	106°55′W
Powell, Wy., U.S. (pou′ĕl)	73	44°44′N	108°44′W
Powell, Lake, res., U.S.	64	37°26′N	110°25′W
Powell Lake, l., Can.	54	50°10′N	124°13′W
Powell Point, c., Bah.	92	24°50′N	76°20′W
Powell Reservoir, res., Ky., U.S.	82	36°30′N	83°35′W
Powell River, Can.	50	49°52′N	124°33′W
Poyang Hu, l., China	161	29°20′N	116°28′E
Poygan, r., Wi., U.S. (poi′găn)	71	44°10′N	89°05′W

PLACE (Pronunciation)	PAGE	LAT.	LONG.
Požarevac, Serb. (pô´zhȧ´rĕ-vȧts)	131	44°38′N	21°12′E
Poza Rica, Mex. (pô-zô-rē´kä)	89	20°32′N	97°25′W
Poznań, Pol.	110	52°25′N	16°55′E
Pozoblanco, Spain (pȯ-thō-bläŋ´kō)	128	38°23′N	4°50′W
Pozos, Mex. (pȯ´zōs)	88	22°05′N	100°50′W
Pozuelo de Alarcón, Spain (pȯ-thwā´lō dā ä-lär-kōn´)	129a	40°27′N	3°49′W
Pozzuoli, Italy (pȯt-swô´lē)	130	40°34′N	14°08′E
Pra, r., Ghana (prȧ)	188	5°45′N	1°35′W
Pra, r., Russia	132	55°00′N	40°13′E
Prachin Buri, Thai. (prä´chēn)	168	13°59′N	101°15′E
Pradera, Col. (prä-dĕ´rä)	100a	3°24′N	76°13′W
Prades, Fr. (prȧd)	126	42°37′N	2°23′E
Prado, Col. (prädȯ)	100a	3°44′N	74°55′W
Prado Reservoir, res., Ca., U.S. (prä´dō)	75a	33°45′N	117°40′W
Prados, Braz. (prȧ´dȯs)	99a	21°05′S	44°04′W
Prague, Czech Rep.	124	50°05′N	14°26′E
Praha *see* Prague, Czech Rep.	110	50°05′N	14°26′E
Praia, C.V. (prä´yȧ)	184b	15°00′N	23°30′W
Praia Funda, Ponta da, c., Braz. (pôn´tä-dȧ-prä´yȧ-fo͞o´n-dä)	102b	23°04′S	43°34′W
Prairie du Chien, Wi., U.S. (prā´rĭ dȯ shēn´)	71	43°02′N	91°10′W
Prairie Grove, Can. (prā´rĭ grōv)	62f	49°48′N	96°57′W
Prairie Island Indian Reservation, I.R., Mn., U.S.	71	44°42′N	92°32′W
Prairies, Rivière des, r., Can. (rē-vyâr´ dā prȧ-rē´)	62a	45°40′N	73°34′W
Pratas Island, i., Asia	165	20°40′N	116°30′E
Prato, Italy (prä´tō)	130	43°53′N	11°03′E
Pratt, Ks., U.S. (prăt)	78	37°37′N	98°43′W
Prattville, Al., U.S. (prăt´vĭl)	82	32°28′N	86°27′W
Pravdinsk, Russia	123	54°26′N	21°00′E
Pravdinskiy, Russia (prȧv-dĕn´skĭ)	142b	56°03′N	37°52′E
Pravia, Spain (prä´vē-ä)	128	43°30′N	6°08′W
Pregolya, r., Russia (prĕ-gô´lă)	123	54°37′N	20°50′E
Premont, Tx., U.S. (prē-mŏnt´)	80	27°20′N	98°07′W
Prenzlau, Ger. (prĕnts´lou)	124	53°19′N	13°52′E
Přerov, Czech Rep. (przhĕ´rôf)	117	49°28′N	17°28′E
Prescot, Eng., U.K. (prĕs´kŭt)	114a	53°25′N	2°48′W
Prescott, Can. (prĕs´kŭt)	67	44°45′N	75°35′W
Prescott, Ar., U.S.	79	33°47′N	93°23′W
Prescott, Az., U.S. (prĕs´kŏt)	64	34°30′N	112°30′W
Prescott, Wi., U.S. (prĕs´kŏt)	75g	44°45′N	92°48′W
Presho, S.D., U.S. (prĕsh´ō)	70	43°56′N	100°04′W
Presidencia Rogue Sáenz Peña, Arg.	102	26°52′S	60°15′W
Presidente Epitácio, Braz. (prā-sē-dĕn´tĕ å-pē-tä´syȯ)	101	21°56′S	52°01′W
Presidio, Tx., U.S. (prĕ-sĭ´dĭ-ō)	80	29°33′N	104°23′W
Presidio, Río del, r., Mex. (rē´ō-dĕl-prĕ-sē´dyô)	88	23°54′N	105°44′W
Prešov, Slvk. (prĕ´shôf)	117	49°00′N	21°18′E
Prespa, Lake, l., Eur. (prĕs´pä)	131	40°49′N	20°50′E
Prespuntal, r., Ven.	101b	9°55′N	64°32′W
Presque Isle, Me., U.S. (prĕsk'ēl´)	60	46°41′N	68°03′W
Pressbaum, Aus.	115e	48°12′N	16°06′E
Prestea, Ghana	188	5°27′N	2°08′W
Preston, Austl.	173a	37°45′S	145°01′E
Preston, Eng., U.K. (prĕs´tŭn)	120	53°46′N	2°42′W
Preston, Id., U.S. (pres´tŭn)	73	42°05′N	111°54′W
Preston, Mn., U.S. (prĕs´tŭn)	71	43°42′N	92°06′W
Preston, Wa., U.S.	74a	47°31′N	121°56′W
Prestonburg, Ky., U.S. (prĕs´tŭn-bûrg)	66	37°35′N	82°50′W
Prestwich, Eng., U.K. (prĕst´wĭch)	114a	53°32′N	2°17′W
Pretoria, S. Afr. (prē-tō´rĭ-ȧ)	186	25°43′S	28°16′E
Pretoria North, S. Afr. (prĕ-tô´rĭ-ȧ no͞ord)	192c	25°41′S	28°11′E
Préveza, Grc. (prĕ´vå-zä)	131	38°58′N	20°44′E
Pribilof Islands, is., Ak., U.S. (prĭ´bĭ-lof)	63	57°00′N	169°20′W
Priboj, Serb. (prē´boi)	131	43°33′N	19°33′E
Price, Ut., U.S. (prīs)	77	39°35′N	110°50′W
Price, r., Ut., U.S.	77	39°21′N	110°35′W
Prichard, Al., U.S. (prĭt´chârd)	82	30°44′N	88°04′W
Priddis, Can. (prĭd´dĭs)	62e	50°53′N	114°20′W
Priddis Creek, r., Can.	62e	50°56′N	114°32′W
Priego, Spain (prē-ā´gō)	128	37°27′N	4°13′W
Prienai, Lith. (prē-ĕn´ī)	123	54°38′N	23°56′E
Prieska, S. Afr. (prē-ĕs´kȧ)	186	29°40′S	22°50′E
Priest Lake, l., Id., U.S. (prēst)	72	48°30′N	116°43′W
Priest Rapids Dam, Wa., U.S.	72	46°39′N	119°55′W
Priest Rapids Lake, res., Wa., U.S.	72	46°42′N	119°58′W
Priiskovaya, Russia (prĭ-ēs´kô-vȧ-yȧ)	142a	60°50′N	58°55′E
Prijedor, Bos. (prē´yĕ-dȯr)	130	44°58′N	16°43′E
Prijepolje, Serb. (prē´yĕ-pȯ´lyĕ)	131	43°22′N	19°41′E
Prilep, Mac. (prē´lĕp)	119	41°20′N	21°35′E
Primorsk, Russia (prē-môrsk´)	123	60°24′N	28°35′E
Primorsko-Akhtarskaya, Russia (prē-môr´skô äκ-tär´skĭ-ē)	137	46°03′N	38°09′E
Primrose, S. Afr.	187b	26°11′S	28°11′E
Primrose Lake, l., Can.	56	54°55′N	109°45′W
Prince Albert, Can. (prĭns ăl´bẽrt)	50	53°12′N	105°46′W
Prince Albert National Park, rec., Can.	52	54°10′N	105°25′W
Prince Albert Sound, strt., Can.	52	70°23′N	116°57′W
Prince Charles Island, i., Can. (chärlz)	53	67°41′N	74°10′W
Prince Edward Island, prov., Can.	51	46°45′N	63°10′W
Prince Edward Islands, is., S. Afr.	178	46°36′S	37°57′E
Prince Edward National Park, rec., Can. (ĕd´wẽrd)	53	46°33′N	63°35′W
Prince Edward Peninsula, pen., Can.	67	44°00′N	77°15′W
Prince Frederick, Md., U.S. (prĭnce frĕdẽrĭk)	68e	38°33′N	76°35′W
Prince George, Can. (jôrj)	50	53°51′N	122°57′W
Prince of Wales, i., Austl.	175	10°47′S	142°15′E
Prince of Wales, i., Ak., U.S.	63	55°47′N	132°50′W
Prince of Wales, Cape, c., Ak., U.S. (wālz)	63	65°48′N	169°08′W
Prince Rupert, Can. (roo´pẽrt)	50	54°19′N	130°19′W
Princes Risborough, Eng., U.K. (prĭns´ĕz rĭz´brŭ)	114b	51°41′N	0°51′W
Princess Charlotte Bay, b., Austl. (shär´lŏt)	175	13°45′S	144°15′E
Princess Royal Channel, strt., Can. (roi´ăl)	54	53°10′N	128°37′W
Princess Royal Island, i., Can.	54	52°57′N	128°49′W
Princeton, Can. (prĭns´tŭn)	50	49°27′N	120°31′W
Princeton, Il., U.S.	66	41°20′N	89°25′W
Princeton, In., U.S.	66	38°20′N	87°35′W
Princeton, Ky., U.S.	82	37°07′N	87°52′W
Princeton, Mi., U.S.	71	46°16′N	87°33′W
Princeton, Mn., U.S.	71	45°34′N	93°36′W
Princeton, Mo., U.S.	79	40°23′N	93°34′W
Princeton, N.J., U.S.	67	40°21′N	74°40′W
Princeton, Wi., U.S.	71	43°50′N	89°09′W
Princeton, W.V., U.S.	83	37°21′N	81°05′W
Prince William Sound, strt., Ak., U.S. (wĭl´yăm)	63	60°40′N	147°10′W
Príncipe, i., S. Tom./P. (prēn´sē-pĕ)	184	1°37′N	7°25′E
Principe Channel, strt., Can. (prĭn´sĭ-pē)	54	53°28′N	129°45′W
Prineville, Or., U.S. (prīn´vĭl)	72	44°17′N	120°48′W
Prineville Reservoir, res., Or., U.S.	72	44°07′N	120°45′W
Prinzapolca, Nic. (prēn-zä-pōl´kä)	91	13°18′N	83°35′W
Prinzapolca, r., Nic.	91	13°23′N	84°23′W
Prior Lake, Mn., U.S. (prī´ẽr)	75g	44°43′N	93°26′W
Priozërsk, Russia (prĭ-ô´zĕrsk)	123	61°03′N	30°08′E
Pripet, r., Eur.	137	51°50′N	29°45′E
Pripet Marshes, sw., Eur.	137	52°10′N	27°30′E
Priština, Serb. (prēsh´tĭ-nä)	119	42°39′N	21°12′E
Pritzwalk, Ger. (prēts´välk)	124	53°09′N	12°12′E
Privas, Fr. (prē-väs´)	126	44°44′N	4°37′E
Prizren, Serb. (prē´zrĕn)	119	42°11′N	20°45′E
Procida, Italy (prô´chē-dă)	129c	40°31′N	14°02′E
Procida, Isola di, i., Italy	129c	40°32′N	13°57′E
Proctor, Mn., U.S. (prŏk´tẽr)	75h	46°45′N	92°14′W
Proctor, Vt., U.S.	67	43°40′N	73°00′W
Proebstel, Wa., U.S. (prōb´stĕl)	74c	45°40′N	122°29′W
Proenca-a-Nova, Port. (prȯ-ān´sȧ-ä-nō´vȧ)	128	39°44′N	7°55′W
Progreso, Hond. (prȯ-grĕ´sȯ)	90	15°28′N	87°49′W
Progreso, Mex. (prȯ-grā´sō)	86	21°14′N	89°39′W
Progreso, Mex.	80	27°29′N	101°05′W
Prokhladnyy, Russia	138	43°46′N	44°00′E
Prokop'yevsk, Russia	140	53°53′N	86°45′E
Prokuplje, Serb. (prô´kȯp'l-yĕ)	131	43°16′N	21°40′E
Prome, Mya.	168	18°46′N	95°15′E
Pronya, r., Bela. (prô´nyä)	132	54°08′N	30°58′E
Pronya, r., Russia	132	54°08′N	39°30′E
Prospect, Ky., U.S. (prŏs´pĕkt)	69h	38°21′N	85°36′W
Prospect Park, Pa., U.S. (prŏs´pĕkt pärk)	68f	39°53′N	75°18′W
Prosser, Wa., U.S. (prŏs´ẽr)	72	46°10′N	119°46′W
Prostějov, Czech Rep. (prôs´tyĕ-yôf)	125	49°28′N	17°08′E
Protection, i., Wa., U.S. (prȯ-tĕk´shŭn)	74a	48°07′N	122°56′W
Protoka, r., Russia (prôt´ō-kȧ)	132	55°00′N	36°42′E
Provadiya, Blg. (prō-väd´ē-yȧ)	131	43°13′N	27°28′E
Providence, Ky., U.S. (prŏv´ĭ-dĕns)	66	37°25′N	87°45′W
Providence, R.I., U.S.	65	41°50′N	71°23′W
Providence, Ut., U.S.	73	41°42′N	111°50′W
Providencia, Isla de, i., Col.	91	13°21′N	80°55′W
Providenciales, i., T./C. Is.	93	21°50′N	72°15′W
Provideniya, Russia (prȯ-vĭ-dă´nĭ-yȧ)	63	64°30′N	172°54′W
Provincetown, Ma., U.S.	67	42°03′N	70°11′W
Provo, Ut., U.S. (prō´vō)	64	40°15′N	111°40′W
Prozor, Bos. (prȯ´zôr)	131	43°48′N	17°59′E
Prudence Island, i., R.I., U.S. (pro͞o´dĕns)	68b	41°38′N	71°20′W
Prudhoe Bay, b., Ak., U.S.	63	70°40′N	147°25′W
Prudnik, Pol. (prȯd´nĭk)	125	50°19′N	17°34′E
Prussia, hist. reg., Eur. (prŭsh´ȧ)	124	50°43′N	8°35′E
Pruszków, Pol. (prȯsh´kȯf)	125	52°09′N	20°50′E
Prut, r., Eur. (pro͞ot)	112	48°05′N	27°07′E
Pryluky, Ukr.	137	50°36′N	32°21′E
Prymors'k, Ukr.	133	46°43′N	36°21′E
Pryor, Ok., U.S. (prī´ẽr)	79	36°16′N	95°19′W
Pryvil'ne, Ukr.	133	47°30′N	32°21′E
Przedbórz, Pol.	125	51°05′N	19°53′E
Przemyśl, Pol. (pzhĕ´mĭsh'l)	110	49°47′N	22°45′E
Przheval'sk, Kyrg. (p'r-zhĭ-välsk´)	139	42°29′N	78°24′E
Psel, r., Eur.	137	49°45′N	33°42′E
Pskov, Russia (pskôf)	134	57°48′N	28°19′E
Pskov, prov., Russia	132	57°33′N	29°05′E
Pskovskoye Ozero, l., Eur. (p'skôv´skô´yĕ ôzē-rô)	136	58°05′N	28°15′E
Ptich', r., Bela. (p'tēch)	136	53°17′N	28°16′E
Ptuj, Slvn. (pto͞o´ē)	130	46°24′N	15°54′E
Pucheng, China (po͞o´chĕng´)	165	28°02′N	118°25′E
Pucheng, China (po͞o-chŭŋ)	162	35°43′N	115°22′E
Puck, Pol. (pȯtsk)	125	54°43′N	18°23′E
Pudozh, Russia (po͞o´dôzh)	136	61°50′N	36°50′E
Puebla, Mex. (pwā´blä)	86	19°02′N	98°11′W
Puebla, state, Mex.	89	19°00′N	97°45′W
Puebla de Don Fadrique, Spain	128	37°55′N	2°55′W
Pueblo, Co., U.S. (pwā´blō)	64	38°15′N	104°36′W
Pueblo Nuevo, Mex. (nwā´vȯ)	88	23°23′N	105°21′W
Pueblo Viejo, Mex. (vyā´hȯ)	89	17°23′N	93°46′W
Puente Alto, Chile (pwĕ´n-tĕ ȧl´tȯ)	99b	33°36′S	70°34′W
Puentedeume, Spain (pwĕn-tå-dhå-o͞o´må)	128	43°28′N	8°09′W
Puente-Genil, Spain (pwĕn´tå-hå-nēl´)	128	37°25′N	4°18′W
Puerco, Rio, r., N.M., U.S. (pwĕr´kō)	77	35°15′N	107°05′W
Puerto Aisén, Chile (pwĕ´r-tō ä´y-sĕ´n)	102	45°28′S	72°44′W
Puerto Angel, Mex. (pwĕ´r-tō äŋ´hål)	89	15°42′N	96°32′W
Puerto Armuelles, Pan. (pwe´r-tȯ är-mo͞o-ā´lyās)	91	8°18′N	82°52′W
Puerto Barrios, Guat. (pwĕ´r-tȯ bär´rē-ȯs)	86	15°43′N	88°36′W
Puerto Bermúdez, Peru (pwĕ´r-tō bĕr-mo͞o´dāz)	100	10°17′S	74°57′W
Puerto Berrío, Col. (pwĕ´r-tō bĕr-rē´ō)	100	6°29′N	74°27′W
Puerto Cabello, Ven. (pwĕ´r-tō kä-bĕl´yō)	100	10°28′N	68°01′W
Puerto Cabezas, Nic. (pwĕ´r-tȯ kä-bā´zäs)	91	14°01′N	83°26′W
Puerto Casado, Para. (pwĕ´r-tō kä-sä´dō)	102	22°16′S	57°57′W
Puerto Castilla, Hond. (pwĕ´r-tȯ käs-tēl´yō)	90	16°01′N	86°01′W
Puerto Chicama, Peru (pwĕ´r-tō chē-kä´mä)	100	7°46′S	79°18′W
Puerto Colombia, Col. (pwĕr´tȯ kȯ-lȯm´bē-ȧ)	100	11°08′N	75°09′W
Puerto Cortés, C.R. (pwĕ´r-tȯ kȯr-tās´)	91	9°00′N	83°37′W
Puerto Cortés, Hond. (pwĕ´r-tȯ kȯr-tās´)	86	15°48′N	87°57′W
Puerto Cumarebo, Ven. (pwĕ´r-tō ko͞o-mä-rĕ´bȯ)	100	11°25′N	69°17′W
Puerto de Luna, N.M., U.S. (pwĕr´tȯ då lo͞o´nä)	78	34°49′N	104°36′W
Puerto de Nutrias, Ven. (pwĕ´r-tō dĕ no͞o-trē-äs´)	100	8°02′N	69°19′W
Puerto Deseado, Arg. (pwĕ´r-tō dā-så-ä´dhō)	102	47°38′S	66°00′W
Puerto de Somport, p., Eur.	129	42°51′N	0°25′W
Puerto Eten, Peru (pwĕ´r-tō ĕ-tĕ´n)	100	6°59′S	79°51′W
Puerto Jiménez, C.R. (pwĕ´r-tȯ κē-mĕ´nĕz)	91	8°35′N	83°23′W
Puerto La Cruz, Ven. (pwĕ´r-tō lä kro͞o´z)	100	10°14′N	64°38′W
Puertollano, Spain (pwĕ-tȯl-yä´nō)	118	38°41′N	4°05′W
Puerto Madryn, Arg. (pwĕ´r-tō mä-drēn´)	102	42°45′S	65°01′W
Puerto Maldonado, Peru (pwĕ´r-tō mäl-dō-nä´dȯ)	100	12°43′S	69°01′W
Puerto Miniso, Mex. (pwĕ´r-tō mē-nē´sȯ)	88	16°06′N	98°02′W
Puerto Montt, Chile (pwĕ´r-tō mȯ´nt)	102	41°29′S	73°00′W
Puerto Natales, Chile (pwĕ´r-tō nä-tȧ´lĕs)	102	51°48′S	72°01′W
Puerto Niño, Col. (pwĕ´r-tō nē´n-yô)	100a	5°57′N	74°36′W
Puerto Padre, Cuba (pwĕ´r-tō pä´drā)	92	21°10′N	76°40′W
Puerto Peñasco, Mex. (pwĕ´r-tȯ pĕn-yä´s-kȯ)	86	31°39′N	113°15′W
Puerto Pinasco, Para. (pwĕ´r-tō pē-nȧ´s-kȯ)	102	22°31′S	57°50′W
Puerto Píritu, Ven. (pwĕ´r-tō pē´rē-to͞o)	101b	10°05′N	65°04′W
Puerto Plata, Dom. Rep. (pwĕ´r-tȯ plä´tä)	87	19°50′N	70°40′W
Puerto Princesa, Phil. (pwĕr-tȯ prēn-sā´sä)	168	9°45′N	118°41′E
Puerto Rico, dep., N.A. (pwĕr´tȯ rē´kō)	87	18°16′N	66°50′W
Puerto Rico Trench, deep,	87	19°45′N	66°30′W
Puerto Salgar, Col. (pwĕ´r-tō säl-gär´)	100a	5°30′N	74°39′W
Puerto Santa Cruz, Arg. (pwĕ´r-tō sän´tä kro͞oz´)	102	50°04′S	68°32′W
Puerto Suárez, Bol. (pwĕ´r-tō swä´råz)	101	18°55′S	57°39′W
Puerto Tejada, Col. (pwĕ´r-tō tĕ-κä´dä)	100	3°13′N	76°23′W
Puerto Vallarta, Mex. (pwĕ´r-tō väl-yär´tä)	88	20°36′N	105°13′W
Puerto Varas, Chile (pwĕ´r-tō vä´räs)	102	41°16′S	73°03′W
Puerto Wilches, Col. (pwĕ´r-tō vēl´c-hĕs)	100	7°19′N	73°54′W
Pugachëv, Russia (po͞o´gȧ-chyôf)	137	52°00′N	48°40′E
Puget, Wa., U.S. (pū´jĕt)	74c	46°10′N	123°23′W
Puget Sound, strt., Wa., U.S.	72	47°49′N	122°26′W
Puglia (Apulia), hist. reg., Italy (po͞o´lyä) (ä-po͞o´lyä)	130	41°13′N	16°10′E
Pukaskwa National Park, rec., Can.	53	48°22′N	85°55′W
Pukeashun Mountain, mtn., Can.	55	51°12′N	119°14′W
Pukin, r., Malay.	153b	2°53′N	102°54′E
Pula, Cro. (po͞o´lä)	118	44°52′N	13°55′E
Pulacayo, Bol. (po͞o-lä-kä´yō)	100	20°12′N	66°33′W
Pulaski, Tn., U.S. (pů-lăs´kĭ)	82	35°11′N	87°03′W
Pulaski, Va., U.S.	83	37°00′N	81°45′W
Puławy, Pol. (pȯ-wä´vē)	125	51°24′N	21°59′E
Pulicat, r., India	159	13°58′N	79°52′E
Pullman, Wa., U.S. (pȯl´măn)	72	46°44′N	117°10′W
Pulog, Mount, mtn., Phil. (po͞o´lôg)	169a	16°38′N	120°53′E
Puma Yumco, l., China (po͞o-mä yo͞om-tswo)	158	28°30′N	90°10′E
Pumpkin Creek, r., Mt., U.S. (pŭmp´kĭn)	73	45°47′N	105°35′W
Punakha, Bhu. (po͞o-nŭk´ŭ)	155	27°45′N	89°59′E
Punata, Bol. (po͞o-nä´tä)	100	17°43′S	65°43′W
Pune, India	155	18°38′N	73°53′E
Punjab, state, India (pŭn´jäb´)	155	31°00′N	75°30′E
Puno, Peru (po͞o´nȯ)	100	15°58′S	70°02′W
Punta Arenas, Chile (po͞o´n-tä-rĕ´näs)	102	53°09′S	70°48′W
Punta de Piedras, Ven. (po͞o´n-tä dĕ pyĕ´dräs)	101b	10°54′N	64°06′W
Punta Gorda, Belize (pȯn´tä gôr´dä)	90	16°07′N	88°50′W
Punta Gorda, Fl., U.S. (pŭn´tȧ gôr´dȧ)	83a	26°55′N	82°02′W
Punta Gorda, Río, r., Nic. (po͞o´n-tä gô´r-dä)	91	11°34′N	84°13′W

ăt; finăl; rāte; senāte; ärm; ȧsk; sofà; fâre; ch-choose; dh-as th in other; bē; êvent; bĕt; recĕnt; cratĕr; g-gō; gh-guttural g; bĭt; ī-short neutral; rīde; κ-guttural k as ch in German ich;

ng-sing; ŋ-baŋk; N-nasalized n; nŏd; cŏmmit; ōld; ȯbey; ôrder; oi-boil; fōōd; ȯ-as oo in foot; ou-out; s-soft; sh-dish; th-thin; pūre; ûnite; ûrn; stŭd; circŭs; ü-as in French tu; ´-indeterminate vowel.

294 Resolution Island – Rockdale PRONOUNCING INDEX

PLACE (Pronunciation)	PAGE	LAT.	LONG.
Resolution Island, i., N.Z. (rĕz-ŏl-ūshŭn)	175a	45°43'S	166°20'E
Restigouche, r., Can.	60	47°35'N	67°35'W
Restrepo, Col. (rĕs-trĕ'pô)	100a	3°49'N	76°31'W
Restrepo, Col.	100a	4°16'N	73°32'W
Retalhuleu, Guat. (rā-täl-ōō-lān')	90	14°31'N	91°41'W
Rethel, Fr. (r-tl')	126	49°34'N	4°20'E
Réthimnon, Grc.	130a	35°21'N	24°30'E
Retie, Bel.	115a	51°16'N	5°08'E
Retsil, Wa., U.S. (rĕt'sĭl)	74a	47°33'N	122°37'W
Réunion, dep., Afr. (rā-ü-nyôn')	3	21°06'S	55°36'E
Reus, Spain (rā'ōōs)	118	41°08'N	1°05'E
Reutlingen, Ger. (roit'lĭng-ĕn)	124	48°29'N	9°14'E
Reutov, Russia (rĕ-ōō'ôf)	142b	55°45'N	37°52'E
Revda, Russia (ryâv'dá)	142a	56°48'N	59°57'E
Revelstoke, Can. (rĕv'ĕl-stōk)	50	51°00'N	118°12'W
Reventazón, Río, r., C.R. (rā-vĕn-tä-zōn')	91	10°10'N	83°30'W
Revere, Ma., U.S. (rē-vēr')	61a	42°24'N	71°01'W
Revillagigedo, Islas, is., Mex. (ĕ's-läs-rē-vēl-yä-hĕ'gĕ-dō)	86	18°45'N	111°00'W
Revillagigedo Chan, Ak., U.S. (rĕ-vĭl'á-gĭ-gē'dō)	54	55°10'N	131°13'W
Revillagigedo Island, i., Ak., U.S.	54	55°35'N	131°23'W
Revin, Fr. (rē-văn)	126	49°56'N	4°34'E
Rewa, India (rā'wä)	155	24°41'N	81°11'E
Rewāri, India	158	28°19'N	76°39'E
Rexburg, Id., U.S. (rĕks'bûrg)	73	43°50'N	111°48'W
Rey, Iran	157	35°35'N	51°25'E
Rey, l., Mex. (rā)	80	27°00'N	103°33'W
Rey, Isla del, i., Pan. (ĕ's-lä-dĕl-rā'ē)	91	8°20'N	78°40'W
Reyes, Bol. (rā'yĕs)	100	14°19'S	67°16'W
Reyes, Point, c., Ca., U.S.	76	38°00'N	123°00'W
Reykjanes, c., Ice. (rā'kyá-nĕs)	112	63°37'N	24°33'W
Reykjavík, Ice. (rā'kyá-vēk)	110	64°09'N	21°39'W
Reynosa, Mex. (rā-ē-nō'sä)	80	26°05'N	98°21'W
Rēzekne, Lat. (rå'zĕk-nĕ)	136	56°31'N	27°19'E
Rezh, Russia (rĕzh')	142a	57°22'N	61°23'E
Rezina, Mol. (ryĕzh'ĕ-nĭ)	133	47°44'N	28°56'E
Rhaetian Alps, mts., Eur.	124	46°30'N	10°00'E
Rhaetien Alps, mts., Eur.	130	46°22'N	10°33'E
Rheinberg, Ger. (rīn'bĕrgh)	127c	51°33'N	6°37'E
Rheine, Ger. (rī'nĕ)	124	52°16'N	7°26'E
Rheinkamp, Ger.	127c	51°30'N	6°37'E
Rheinland, hist. reg., Ger.	124	50°05'N	6°40'E
Rheydt, Ger. (rĕ'yt)	127c	51°10'N	6°28'E
Rhin, r., Ger. (rēn)	115b	52°52'N	12°49'E
Rhine, r., Eur.	112	50°34'N	7°21'E
Rhinelander, Wi., U.S. (rīn'lăn-dēr)	71	45°39'N	89°25'W
Rhin Kanal, can., Ger. (rēn kä-näl')	115b	52°47'N	12°40'E
Rhiou, r., Alg.	129	35°45'N	1°18'E
Rhode Island, state, U.S. (rōd ī'lănd)	65	41°35'N	71°40'W
Rhode Island, i., R.I., U.S.	68b	41°31'N	71°14'W
Rhodes, S. Afr. (rōdz)	187c	30°48'S	27°56'E
Rhodes *see* Ródos, i., Grc.	112	36°00'N	28°29'E
Rhodesia *see* Zimbabwe, nation, Afr.	186	17°50'S	29°30'E
Rhodope Mountains, mts., Eur. (rô'dō-pĕ)	112	42°00'N	24°08'E
Rhondda, Wales, U.K. (rŏn'dhá)	120	51°40'N	3°40'W
Rhône, r., Fr. (rōn)	112	44°30'N	4°45'E
Rhoon, Neth.	115a	51°52'N	4°24'E
Rhum, i., Scot., U.K. (rŭm)	120	57°00'N	6°20'W
Riachão, Braz. (rē-ä-choun')	101	7°15'S	46°30'W
Rialto, Ca., U.S. (rē-ăl'tō)	75a	34°06'N	117°23'W
Riau, prov., Indon.	153b	0°56'N	101°25'E
Riau, Kepulauan, i., Indon.	168	0°30'N	104°55'E
Riau, Selat, strt., Indon.	153b	0°40'N	104°27'E
Riaza, r., Spain (rē-ä'thä)	128	41°25'N	3°25'W
Ribadavia, Spain (rē-bä-dhä'vē-ä)	128	42°18'N	8°06'W
Ribadeo, Spain (rē-bä-dhā'ō)	128	43°32'N	7°05'W
Ribadesella, Spain (rē'bä-dā-sāl'yä)	128	43°30'N	5°02'W
Ribe, Den. (rē'bĕ)	122	55°20'N	8°45'E
Ribeirão Prêto, Braz. (rē-bā-roun-prĕ'tô)	101	21°11'S	47°47'W
Ribera, N.M., U.S. (rē-bĕ'rä)	78	35°23'N	105°27'W
Riberalta, Bol. (rē-bå-räl'tä)	100	11°06'S	66°02'W
Rib Lake, Wi., U.S. (rĭb lāk)	71	45°20'N	90°11'W
Rîbniţa, Mol.	133	47°45'N	29°02'E
Rice, l., Can.	59	44°05'N	78°10'W
Rice Lake, Wi., U.S.	71	45°30'N	91°44'W
Rice Lake, l., Mn., U.S.	75g	45°10'N	93°09'W
Richards Island, i., Can. (rĭch'ĕrds)	63	69°45'N	135°30'W
Richards Landing, Can. (lănd'ĭng)	75k	46°18'N	84°02'W
Richardson, Tx., U.S. (rĭch'ĕrd-sŭn)	75c	32°56'N	96°44'W
Richardson, Wa., U.S.	74a	48°27'N	122°54'W
Richardson Mountains, mts., Can.	52	66°58'N	136°19'W
Richardson Mountains, mts., N.Z.	177	44°50'S	168°30'E
Richardson Park, De., U.S. (pärk)	67	39°45'N	75°35'W
Richelieu, r., Can. (rēsh'lyû')	59	45°05'N	73°25'W
Richfield, Mn., U.S.	75g	44°53'N	93°17'W
Richfield, Oh., U.S.	69d	41°14'N	81°38'W
Richfield, Ut., U.S.	77	38°45'N	112°05'W
Richford, Vt., U.S. (rĭch'fērd)	67	45°00'N	72°35'W
Rich Hill, Mo., U.S. (rĭch hĭl)	79	38°05'N	94°21'W
Richibucto, Can. (rĭ-chĭ-bŭk'tō)	51	46°41'N	64°52'W
Richland, Ga., U.S. (rĭch'lănd)	82	32°05'N	84°40'W
Richland, Wa., U.S.	72	46°17'N	119°19'W
Richland Center, Wi., U.S. (sĕn'tēr)	71	43°20'N	90°25'W
Richmond, Austl. (rĭch'mŭnd)	175	20°47'S	143°14'E
Richmond, Austl.	173b	33°36'S	150°45'E
Richmond, Can.	62c	45°12'N	75°49'W
Richmond, Can.	59	45°40'N	72°07'W
Richmond, S. Afr.	187c	29°52'S	30°17'E
Richmond, Il., U.S.	69a	42°29'N	88°18'W
Richmond, In., U.S.	66	39°50'N	85°00'W
Richmond, Ky., U.S.	66	37°45'N	84°20'W
Richmond, Mo., U.S.	79	39°16'N	93°58'W
Richmond, Tx., U.S.	81	29°35'N	95°45'W
Richmond, Ut., U.S.	73	41°55'N	111°50'W
Richmond, Va., U.S.	65	37°35'N	77°30'W
Richmond Beach, Wa., U.S.	74a	47°47'N	122°23'W
Richmond Heights, Mo., U.S.	75e	38°38'N	90°20'W
Richmond Highlands, Wa., U.S.	74a	47°46'N	122°22'W
Richmond Hill, Can. (hĭl)	59	43°53'N	79°26'W
Richton, Ms., U.S. (rĭch'tŭn)	82	31°20'N	89°54'W
Richwood, W.V., U.S. (rĭch'wŏŏd)	66	38°10'N	80°30'W
Ridderkerk, Neth.	115a	51°52'N	4°35'E
Rideau, r., Can.	62c	45°17'N	75°41'W
Rideau Lake, l., Can. (rē-dō')	59	44°40'N	76°20'W
Ridgefield, Ct., U.S. (rĭj'fēld)	68a	41°16'N	73°30'W
Ridgefield, Wa., U.S.	74c	45°49'N	122°40'W
Ridgeway, Can. (rĭj'wā)	69c	42°53'N	79°02'W
Ridgewood, N.J., U.S. (rĭdj'wŏŏd)	68a	40°59'N	74°08'W
Ridgway, Pa., U.S.	67	41°25'N	78°40'W
Riding Mountain, mtn., Can. (rīd'ĭng)	57	50°37'N	99°37'W
Riding Mountain National Park, rec., Can. (rīd'ĭng)	52	50°59'N	99°19'W
Riding Rocks, is., Bah.	92	25°20'N	79°10'W
Riebeek-Oos, S. Afr.	187c	33°14'S	26°09'E
Ried, Aus. (rēd)	124	48°13'N	13°30'E
Riesa, Ger. (rē'zá)	124	51°17'N	13°17'E
Rieti, Italy (rē-ā'tē)	118	42°25'N	12°51'E
Rievleidam, res., S. Afr.	187b	25°52'S	28°18'E
Riffe Lake, res., Wa., U.S.	72	46°20'N	122°10'W
Rifle, Co., U.S. (rī'f'l)	77	39°35'N	107°50'W
Riga, Lat. (rē'gá)	134	56°55'N	24°05'E
Riga, Gulf of, b., Eur.	136	57°56'N	23°05'E
Rīgān, Iran	154	28°45'N	58°55'E
Rigaud, Can. (rē-gō')	62a	45°29'N	74°18'W
Rigby, Id., U.S. (rĭg'bē)	73	43°40'N	111°55'W
Rigeley, W.V., U.S. (rĭj'lē)	67	39°40'N	78°45'W
Rīgestān, des., Afg.	154	30°53'N	64°42'E
Rigolet, Can. (rĭg-ō-lā')	51	54°10'N	58°40'W
Riihimäki, Fin.	123	60°44'N	24°44'E
Rijeka, Cro. (rĭ-yĕ'kä)	118	45°22'N	14°24'E
Rijkevorsel, Bel.	115a	51°21'N	4°46'E
Rijswijk, Neth.	115a	52°03'N	4°19'E
Rika, r., Ukr. (rē'ká)	125	48°21'N	23°37'E
Rima, r., Nig.	189	13°30'N	5°50'E
Rimavska Sobota, Slvk. (rĭ'máf-ská sô'bō-tá)	125	48°25'N	20°01'E
Rimbo, Swe. (rĕm'bō)	122	59°45'N	18°22'E
Rimini, Italy (rē'mē-nē)	118	44°03'N	12°33'E
Rimouski, Can. (rē-mōōs'kē)	51	48°27'N	68°32'W
Rincón de Romos, Mex. (rēn-kōn dā rô-mōs')	88	22°13'N	102°21'W
Ringkøbing, Den. (rĭng'kûb-ĭng)	116	56°06'N	8°14'E
Ringkøbing Fjord, b., Den.	122	55°55'N	8°04'E
Ringsted, Den. (rĭng'stĕdh)	122	55°27'N	11°49'E
Ringvassøya, i., Nor. (rĭng'väs-ûē)	116	69°58'N	16°43'E
Ringwood, Austl.	173a	37°49'S	145°14'E
Rinjani, Gunung, mtn., Indon.	168	8°39'S	116°22'E
Río Abajo, Pan. (rē'ō-ä-bä'κō)	86a	9°01'N	78°30'W
Río Balsas, Mex. (rē'ō-bäl-säs)	88	17°59'N	99°45'W
Riobamba, Ec. (rē'ō-bäm-bä)	100	1°45'S	78°37'W
Rio Bonito, Braz. (rē'ō bō-nē'tō)	99a	22°44'S	42°38'W
Rio Branco, Braz. (rē'ō brän'kō)	100	9°57'S	67°50'W
Rio Branco, Ur. (rīō brănċō)	102	32°33'S	53°29'W
Rio Casca, Braz. (rē'ō-ká's-kä)	99a	20°15'S	42°39'W
Río Chico, Ven. (rē'ō chē'kō)	101b	10°20'N	65°58'W
Rio Claro, Braz. (rē'ō klä'rō)	101	22°25'S	47°33'W
Río Cuarto, Arg. (rē'ō kwär'tō)	102	33°05'S	64°15'W
Rio das Flores, Braz. (rē'ō-däs-flō-rĕs)	99a	22°10'S	43°35'W
Rio de Janeiro, Braz. (rē'ō dā zhä-nä'ē-rō)	102b	22°50'S	43°20'W
Rio de Janeiro, state, Braz.	101	22°27'S	42°43'W
Río de Jesús, Pan.	91	7°54'N	80°59'W
Río Frío, Mex. (rē'ō-frē'ō)	89a	19°21'N	98°40'W
Río Gallegos, Arg. (rē'ō gä-lā'gōs)	102	51°43'S	69°15'W
Rio Grande, Braz. (rē'ō grän'dĕ)	102	31°04'S	52°14'W
Río Grande, Mex. (rē'ō grän'dā)	88	23°51'N	102°59'W
Riogrande, Tx., U.S. (rē'ō grän-dā)	80	26°23'N	98°48'W
Rio Grande do Norte, state, Braz.	101	5°26'S	37°20'W
Rio Grande do Sul, state, Braz. (rē'ō grän'dĕ-dō-sōō'l)	102	29°00'S	54°00'W
Ríohacha, Col. (rē'ō-ä'chä)	100	11°30'N	72°54'W
Río Hato, Pan. (rē-ō-ä'tō)	91	8°19'N	80°11'W
Riom, Fr. (rē-ôn')	126	45°54'N	3°08'E
Rio Muni, hist. reg., Eq. Gui. (rē'ō mōō'nē)	184	1°47'N	8°33'E
Rionegro, Col. (rē'ō-nĕ'grō)	100a	6°09'N	75°22'W
Río Negro, prov., Arg. (rē'ō nā'grō)	102	40°15'S	68°15'W
Río Negro, dept., Ur. (rē'ō-nĕ'grō)	99c	32°48'S	57°45'W
Río Negro, Embalse del, res., Ur.	102	32°45'S	55°50'W
Rionero, Italy (rē-ō-nā'rō)	130	40°55'N	15°42'E
Rioni, r., Geor.	138	42°08'N	41°39'E
Rio Novo, Braz. (rē'ō-nō'vō)	99a	21°30'S	43°08'W
Rio Pardo de Minas, Braz. (rē'ō pär'dō-dĕ-mē'näs)	101	15°43'S	42°24'W
Rio Pombo, Braz. (rē'ō pôm'bä)	99a	21°17'S	43°09'W
Rio Sorocaba, Represa do, res., Braz.	99a	23°37'S	47°19'W
Ríosucio, Col. (rē'ō-sōō'syō)	100a	5°25'N	75°41'W
Río Tercero, Arg. (rē'ō dĕr-sĕ'rō)	102	32°12'S	63°59'W
Rio Verde, Braz. (vĕr'dĕ)	101	17°47'S	50°49'W
Ríoverde, Mex. (rē'ō-vĕr'dā)	86	21°54'N	99°59'W
Ripley, Eng., U.K. (rĭp'lē)	114a	53°03'N	1°24'W
Ripley, Ms., U.S.	82	34°44'N	88°55'W
Ripley, Tn., U.S.	82	35°44'N	89°34'W
Ripoll, Spain (rē-pōl'')	129	42°10'N	2°10'E
Ripon, Wi., U.S. (rĭp'ŏn)	71	43°49'N	88°50'W
Ripon, i., Austl.	174	20°05'S	118°10'E
Ripon Falls, wtfl., Ug.	186	0°38'N	33°02'E
Risaralda, dept., Col.	100a	5°15'N	76°00'W
Risdon, Austl. (rĭz'dŭn)	175	42°37'S	147°32'E
Rishiri, i., Japan (rē-shē'rē)	166	45°10'N	141°08'E
Rishon le Ziyyon, Isr.	153a	31°57'N	34°48'E
Rishra, India	158a	22°42'N	88°22'E
Rising Sun, In., U.S. (rīz'ĭng sŭn)	66	38°55'N	84°55'W
Risor, Nor. (rēs'ûr)	116	58°44'N	9°10'E
Ritacuva, Alto, mtn., Col. (ä'l-tō-rē-tä-kōō'vä)	100	6°22'N	72°13'W
Rittman, Oh., U.S. (rĭt'năn)	69d	40°58'N	81°47'W
Ritzville, Wa., U.S. (rĭts'vĭl)	72	47°08'N	118°23'W
Riva, Dom. Rep. (rē'vä)	93	19°10'N	69°55'W
Riva, Italy (rē'vä)	130	45°54'N	10°49'E
Riva, Md., U.S. (rī'vä)	68e	38°57'N	76°36'W
Rivas, Nic. (rē'väs)	90	11°25'N	85°51'W
Rive-de-Gier, Fr. (rēv-dē-zhē-ā')	126	45°32'N	4°37'E
Rivera, Ur. (rē-vā'rä)	102	30°52'S	55°32'W
River Cess, Lib. (rĭv'ēr sĕs)	184	5°46'N	9°52'W
Riverdale, Il., U.S. (rĭv'ēr dāl)	69a	41°38'N	87°36'W
Riverdale, Ut., U.S.	75b	41°11'N	112°00'W
River Falls, Al., U.S.	82	31°20'N	86°25'W
River Falls, Wi., U.S.	71	44°48'N	92°38'W
Riverhead, N.Y., U.S. (rĭv'ēr hĕd)	67	40°55'N	72°40'W
Riverina, reg., Austl. (rĭv-ēr-ē'nä)	175	34°55'S	144°30'E
River Jordan, Can. (jôr'dăn)	74a	48°25'N	124°03'W
River Oaks, Tx., U.S. (ōkz)	75c	32°47'N	97°24'W
River Rouge, Mi., U.S. (rōōzh)	69b	42°16'N	83°09'W
Rivers, Can.	57	50°01'N	100°15'W
Riverside, Ca., U.S. (rĭv'ēr-sīd)	64	33°59'N	117°21'W
Riverside, N.J., U.S.	68f	40°02'N	74°58'W
Rivers Inlet, Can.	54	51°45'N	127°15'W
Riverstone, Austl.	173b	33°41'S	150°52'E
Riverton, Va., U.S.	67	39°00'N	78°15'W
Riverton, Wy., U.S.	73	43°02'N	108°24'W
Rivesaltes, Fr. (rēv'zält')	126	42°48'N	2°48'E
Riviera Beach, Fl., U.S. (rĭv-ĭ-ĕr'á bēch)	83a	26°46'N	80°04'W
Riviera Beach, Md., U.S.	68e	39°10'N	76°32'W
Rivière-Beaudette, Can.	62a	45°14'N	74°20'W
Rivière-du-Loup, Can. (rē-vyâr' dü lōō')	51	47°50'N	69°32'W
Rivière Qui Barre, Can. (rēv-yĕr' kē-bär)	62g	53°47'N	113°51'W
Rivière-Trois-Pistoles, Can. (trwä'pēs-tōl')	60	48°07'N	69°10'W
Rivne, Ukr.	133	48°11'N	31°46'E
Rivne, Ukr.	137	50°37'N	26°17'E
Rivne, prov., Ukr.	133	50°55'N	27°00'E
Riyadh, Sau. Ar.	154	24°31'N	46°47'E
Rize, Tur. (rē'zĕ)	119	41°00'N	40°30'E
Rizhao, China (rē-jou)	164	35°27'N	119°28'E
Rizzuto, Cape, c., Italy (rēt-sōō'tō)	131	38°53'N	17°05'E
Rjukan, Nor. (ryōō'kăn)	116	59°53'N	8°30'E
Roanne, Fr. (rō-än')	117	46°02'N	4°04'E
Roanoke, Al., U.S. (rō'á-nōk)	82	33°08'N	85°21'W
Roanoke, Va., U.S.	65	37°16'N	79°55'W
Roanoke, r., U.S.	65	36°17'N	77°22'W
Roanoke Rapids, N.C., U.S.	83	36°25'N	77°40'W
Roanoke Rapids Lake, res., N.C., U.S.	83	36°28'N	77°37'W
Roan Plateau, plat., Co., U.S. (rōn)	77	39°25'N	110°00'W
Roatan, Hond. (rō-ä-tän')	90	16°18'N	86°33'W
Roatán, i., Hond.	90	16°19'N	86°46'W
Robbeneiland, i., S. Afr.	186a	33°48'S	18°22'E
Robbins, Il., U.S. (rŏb'ĭnz)	69a	41°39'N	87°42'W
Robbinsdale, Mn., U.S. (rŏb'ĭnz-dāl)	75g	45°03'N	93°22'W
Robe, Wa., U.S. (rōb)	74a	48°06'N	121°50'W
Roberts, Mount, mtn., Austl. (rŏb'ērts)	175	28°05'S	152°30'E
Roberts, Point, c., Wa., U.S. (rŏb'ērts)	74d	48°58'N	123°05'W
Robertson, Lac, l., Can.	61	51°00'N	59°10'W
Robertsport, Lib. (rŏb'ērts-pōrt)	184	6°45'N	11°22'W
Roberval, Can. (rŏb'ēr-văl) (rō-bĕr-vål')	51	48°32'N	72°15'W
Robinson, Can.	61	48°16'N	58°50'W
Robinson, Il., U.S. (rŏb'ĭn-sŭn)	66	39°00'N	87°45'W
Robinvale, Austl. (rŏb-ĭn'vál)	176	34°45'S	142°45'E
Roblin, Can.	57	51°15'N	101°25'W
Robson, Mount, mtn., Can. (rŏb'sŭn)	55	53°07'N	119°09'W
Robstown, Tx., U.S. (rŏbz'toun)	81	27°46'N	97°41'W
Roca, Cabo da, c., Port. (ká'bō-dä-rô'kă)	128	38°47'N	9°30'W
Rocas, Atol das, atoll, Braz. (ä-tôl-däs-rō'kás)	101	3°50'S	33°46'W
Rocha, Ur. (rō'chás)	102	34°26'S	54°14'W
Rochdale, Eng., U.K. (rŏch'dāl)	120	53°37'N	2°09'W
Roche à Bateau, Haiti (rôsh á bá-tō')	93	18°10'N	74°00'W
Rochefort, Fr. (rôsh-fōr')	117	45°55'N	0°57'W
Rochelle, Il., U.S. (rō-shĕl')	71	41°53'N	89°06'W
Rochester, Eng., U.K.	114a	51°24'N	0°30'E
Rochester, In., U.S. (rŏch'ĕs-tēr)	66	41°05'N	86°20'W
Rochester, Mi., U.S.	69b	42°41'N	83°09'W
Rochester, Mn., U.S.	65	44°01'N	92°30'W
Rochester, N.H., U.S.	67	43°20'N	71°00'W
Rochester, N.Y., U.S.	65	43°15'N	77°35'W
Rochester, Pa., U.S.	69e	40°42'N	80°16'W
Rock, r., la., U.S.	70	43°17'N	96°13'W
Rock, r., Or., U.S.	74c	45°34'N	122°52'W
Rock, r., Or., U.S.	74c	45°52'N	123°14'W
Rock, r., U.S.	65	41°40'N	90°00'W
Rockaway, N.J., U.S. (rŏck'á-wā)	68a	40°54'N	74°30'W
Rockbank, Austl.	173a	37°44'S	144°40'E
Rockcliffe Park, Can. (rok'klĭf pärk)	62c	45°27'N	75°40'W
Rock Creek, r., Can. (rŏk)	73	49°01'N	107°00'W
Rock Creek, r., Il., U.S.	69a	41°16'N	87°54'W
Rock Creek, r., Mt., U.S.	73	46°25'N	113°40'W
Rock Creek, r., Or., U.S.	72	45°30'N	120°06'W
Rock Creek, r., Wa., U.S.	72	47°09'N	117°50'W
Rockdale, Austl.	173b	33°57'S	151°08'E

ăt; finăl; rāte; senåte; ärm; åsk; sofá; fâre; ch-choose; dh-as th in other; bē; ēvent; bĕt; recĕnt; cratēr; g-gō; gh-guttural g; bĭt; ĭ-short neutral; rīde; κ-guttural k as ch in German ich;

PLACE (Pronunciation)	PAGE	LAT.	LONG.
Rockdale, Md., U.S.	68e	39°22'N	76°49'W
Rockdale, Tx., U.S. (rŏk'dāl)	81	30°39'N	97°00'W
Rock Falls, Il., U.S. (rŏk fôlz)	71	41°45'N	89°42'W
Rockford, Il., U.S. (rŏk'fẽrd)	65	42°16'N	89°07'W
Rockhampton, Austl. (rŏk-hămp'tŭn)	175	23°26'S	150°29'E
Rock Hill, S.C., U.S. (rŏk'hĭl)	65	34°55'N	81°01'W
Rockingham, N.C., U.S. (rŏk'ĭng-hăm)	83	34°54'N	79°45'W
Rockingham Forest, for., Eng., U.K. (rok'ĭng-hăm)	114a	52°29'N	0°43'W
Rock Island, Il., U.S.	65	41°31'N	90°37'W
Rock Island Dam, Wa., U.S. (ī lănd)	72	47°17'N	120°33'W
Rockland, Can. (rŏk'lănd)	62c	45°33'N	75°17'W
Rockland, Ma., U.S.	61a	42°07'N	70°55'W
Rockland, Me., U.S.	60	44°06'N	69°09'W
Rockland Reservoir, res., Austl.	176	36°55'S	142°20'E
Rockmart, Ga., U.S. (rŏk'märt)	82	33°58'N	85°00'W
Rockmont, Wi., U.S. (rŏk'mŏnt)	75h	46°34'N	91°54'W
Rockport, In., U.S. (rŏk'pōrt)	66	38°20'N	87°00'W
Rockport, Ma., U.S.	61a	42°39'N	70°37'W
Rockport, Mo., U.S.	79	40°25'N	95°30'W
Rockport, Tx., U.S.	81	28°03'N	97°03'W
Rock Rapids, Ia., U.S. (răp'ĭdz)	70	43°26'N	96°10'W
Rock Sound, strt., Bah.	92	24°50'N	76°05'W
Rocksprings, Tx., U.S. (rŏk sprĭngs)	80	30°02'N	100°12'W
Rock Springs, Wy., U.S.	64	41°35'N	109°13'W
Rockstone, Guy. (rŏk'stôn)	101	5°55'N	57°27'W
Rock Valley, Ia., U.S. (văl'ĭ)	70	43°13'N	96°17'W
Rockville, In., U.S. (rŏk'vĭl)	66	39°45'N	87°15'W
Rockville, Md., U.S.	68e	39°05'N	77°11'W
Rockville Centre, N.Y., U.S. (sĕn'tẽr)	68a	40°39'N	73°39'W
Rockwall, Tx., U.S. (rŏk'wôl)	79	32°55'N	96°23'W
Rockwell City, Ia., U.S. (rŏk'wĕl)	71	42°22'N	94°37'W
Rockwood, Can. (rŏk-wòd)	62d	43°37'N	80°08'W
Rockwood, Me., U.S.	60	45°39'N	69°45'W
Rockwood, Tn., U.S.	82	35°51'N	84°41'W
Rocky, East Branch, r., Oh., U.S.	69d	41°13'N	81°43'W
Rocky, West Branch, r., Oh., U.S.	69d	41°17'N	81°54'W
Rocky Boys Indian Reservation, I.R., Mt., U.S.	73	48°08'N	109°34'W
Rocky Ford, Co., U.S.	78	38°02'N	103°43'W
Rocky Hill, N.J., U.S. (hĭl)	68a	40°24'N	74°38'W
Rocky Island Lake, l., Can.	58	46°56'N	83°04'W
Rocky Mount, N.C., U.S.	83	35°55'N	77°47'W
Rocky Mountain House, Can.	55	52°22'N	114°55'W
Rocky Mountain National Park, rec., Co., U.S.	64	40°29'N	106°06'W
Rocky Mountains, mts., N.A.	49	50°00'N	114°00'W
Rocky River, Oh., U.S.	69d	41°29'N	81°51'W
Rodas, Cuba (rō'dhàs)	92	22°20'N	80°35'W
Roden, r., Eng., U.K. (rō'dĕn)	114a	52°49'N	2°38'W
Rodeo, Mex. (rô-dā'ō)	80	25°12'N	104°34'W
Rodeo, Ca., U.S. (rō'dēō)	74b	38°02'N	122°16'W
Roderick Island, i., Can. (rŏd'ē-rĭk)	54	52°40'N	128°22'W
Rodez, Fr. (rô-dĕz')	117	44°22'N	2°34'E
Rodnei, Munţii, mts., Rom.	125	47°41'N	24°05'E
Rodniki, Russia (rôd'nē-kē)	136	57°08'N	41°48'E
Rodonit, Kep I, c., Alb.	131	41°38'N	19°01'E
Ródos, Grc.	119	36°24'N	28°15'E
Ródos, i., Grc.	119	36°00'N	28°29'E
Roebling, N.J., U.S. (rōb'lĭng)	68f	40°07'N	74°48'W
Roebourne, Austl. (rō'bŭrn)	174	20°50'S	117°15'E
Roebuck Bay, b., Austl. (rō'bŭck)	174	18°15'S	121°10'E
Roedtan, S. Afr.	192c	24°37'S	29°08'E
Roeselare, Bel.	121	50°55'N	3°05'E
Roesiger, l., Wa., U.S. (rōz'ĭ-gẽr)	74a	47°59'N	121°56'W
Roes Welcome Sound, strt., Can. (rōz)	53	64°10'N	87°23'W
Rogatica, Bos. (rô-gä'tē-tsä)	131	43°46'N	19°00'E
Rogers, Ar., U.S. (rŏj-ẽrz)	79	36°19'N	94°07'W
Rogers City, Mi., U.S.	66	45°30'N	83°50'W
Rogersville, Tn., U.S.	82	36°21'N	83°00'W
Rognac, Fr. (rŏn-yäk')	126a	43°29'N	5°15'E
Rogoaguado, l., Bol. (rō'gō-ä-gwä-dō)	100	12°42'S	66°46'W
Rogovskaya, Russia (rô-gôf'ská-yá)	133	45°43'N	38°42'E
Rogózno, Pol. (rô'gôzh-nō)	124	52°44'N	16°53'E
Rogue, r., Or., U.S. (rōg)	72	42°32'N	124°13'W
Rohatyn, Ukr.	125	49°22'N	24°37'E
Rojas, Arg. (rō'häs)	99c	34°11'S	60°42'W
Rojo, Cabo, c., Mex. (rō'hō)	89	21°35'N	97°16'W
Rojo, Cabo, c., P.R. (rō'hō)	87b	17°55'N	67°14'W
Rokel, r., S.L.	188	9°00'N	11°55'W
Rokkō-Zan, mtn., Japan (rôk'kō zän)	167b	34°46'N	135°16'E
Rokycany, Czech Rep. (rô'kĭ'tsä-nĭ)	124	49°44'N	13°37'E
Roldanillo, Col. (rôl-dä-nē'l-yō)	100a	4°24'N	76°09'W
Rolla, Mo., U.S.	79	37°56'N	91°45'W
Rolla, N.D., U.S.	70	48°52'N	99°32'W
Rolleville, Bah.	92	23°40'N	76°00'W
Roma, Austl. (rō'mȧ)	175	26°30'S	148°48'E
Roma *see* Rome, Italy	110	41°52'N	12°37'E
Roma, Leso.	187c	29°28'S	27°43'E
Romaine, r., Can. (rô-mĕn')	53	51°22'N	63°23'W
Roman, Rom. (rō'män)	125	46°56'N	26°57'E
Romania, nation, Eur. (rō-mā'nē-ȧ)	110	46°18'N	22°53'E
Romano, Cape, c., Fl., U.S. (rō-mä'nō)	83a	25°48'N	82°00'W
Romano, Cayo, i., Cuba (kä'yō-rô-mä'nô)	92	22°15'N	78°00'W
Romanovo, Russia (rô-mä'nô-vô)	142a	59°09'N	61°24'E
Romans, Fr. (rô-mäN')	126	45°04'N	4°49'E
Romblon, Phil. (rōm-blōn')	169a	12°34'N	122°16'E
Romblon Island, i., Phil.	169a	12°33'N	122°17'E
Rome (Roma), Italy	110	41°52'N	12°37'E
Rome, Ga., U.S. (rōm)	65	34°14'N	85°10'W
Rome, N.Y., U.S.	67	43°15'N	75°25'W
Romeo, Mi., U.S. (rō'mē-ō)	66	42°50'N	83°00'W
Romford, Eng., U.K. (rŭm'fẽrd)	114b	51°35'N	0°11'E
Romilly-sur-Seine, Fr. (rô-mē-yē'sür-sān')	126	48°32'N	3°41'E
Romita, Mex. (rō-mē'tä)	88	20°53'N	101°32'W
Romny, Ukr. (rôm'nĭ)	137	50°46'N	33°31'E
Rømø, i., Den. (rûm'û)	122	55°08'N	8°17'E
Romoland, Ca., U.S. (rō'mō'lănd)	75a	33°44'N	117°11'W
Romorantin-Lanthenay, Fr. (rô-mô-räN-tăN')	126	47°24'N	1°46'E
Rompin, Malay.	153b	2°42'N	102°30'E
Rompin, r., Malay.	153b	2°54'N	103°10'E
Romsdalsfjorden, Nor.	122	62°40'N	7°05'W
Romulus, Mi., U.S. (rom'ū lŭs)	69b	42°14'N	83°24'W
Ron, Mui, c., Viet.	165	18°05'N	106°45'E
Ronan, Mt., U.S. (rō'năn)	73	47°28'N	114°03'W
Roncador, Serra do, mts., Braz. (sĕr'rä dò rôn-kä-dôr')	101	12°44'S	52°19'W
Ronceverte, W.V., U.S. (rŏn'sē-vûrt)	66	37°45'N	80°30'W
Ronda, Spain (rōn'dä)	118	36°45'N	5°10'W
Ronda, Sierra de, mts., Spain	128	36°35'N	5°03'W
Rondônia, state, Braz.	100	10°15'S	63°07'W
Ronge, Lac la, l., Can. (rōnzh)	52	55°10'N	105°00'W
Rongjiang, China (rôŋ-jyäŋ)	165	25°52'N	108°45'E
Rongxian, China	165	22°50'N	110°32'E
Rønne, Den. (rûn'ĕ)	116	55°08'N	14°46'E
Ronneby, Swe. (rōn'ĕ-bü)	122	56°13'N	15°17'E
Ronne Ice Shelf, ice, Ant.	178	77°30'S	38°00'W
Roodepoort, S. Afr. (rō'dĕ-pōrt)	187b	26°10'S	27°52'E
Roodhouse, Il., U.S. (rōōd'hous)	79	39°29'N	90°21'W
Rooiberg, S. Afr.	192c	24°46'S	27°42'E
Roosendaal, Neth. (rō'zĕn-däl)	115a	51°32'N	4°27'E
Roosevelt, Ut., U.S. (rōz''vĕlt)	77	40°20'N	110°00'W
Roosevelt, r., Braz. (rô'sĕ-vĕlt)	101	9°22'S	60°28'W
Roosevelt Island, i., Ant.	178	79°30'S	168°00'W
Root, r., Wi., U.S.	69a	42°49'N	87°54'W
Roper, r., Austl. (rōp'ẽr)	174	14°50'S	134°00'E
Ropsha, Russia (rôp'shá)	142c	59°44'N	29°53'E
Roque Pérez, Arg. (rô'kĕ-pĕ'rĕz)	99c	35°23'S	59°22'W
Roques, Islas los, is., Ven.	100	12°25'N	67°40'W
Roraima, state, Braz. (rô'rīy-mă)	100	2°00'N	62°15'W
Roraima, Mount, mtn., S.A. (rô-rä-ē'mä)	101	5°12'N	60°52'W
Røros, Nor. (rûr'ôs)	116	62°36'N	11°25'E
Ros', r., Ukr. (rôs)	133	49°40'N	30°22'E
Rosa, Monte, mtn., Italy (mōn'tä rō'zä)	118	45°56'N	7°51'E
Rosales, Mex. (rō-zä'läs)	80	28°15'N	100°43'W
Rosales, Phil. (rô-sä'lĕs)	169a	15°54'N	120°38'E
Rosamorada, Mex. (rō'zä-mō-rä'dhä)	88	22°06'N	105°16'W
Rosaria, Laguna, l., Mex. (lä-gò'nä-rô-sä'ryä)	89	17°50'N	93°51'W
Rosario, Arg. (rô-zä'rē-ō)	102	32°58'S	60°42'W
Rosario, Braz. (rô-zä'rē-ò)	101	2°49'S	44°15'W
Rosario, Mex.	80	26°31'N	105°40'W
Rosario, Mex.	88	22°58'N	105°54'W
Rosario, Phil.	169a	13°49'N	121°13'W
Rosario, Ur.	99c	34°19'S	57°24'E
Rosario, Cayo, i., Cuba (kä'yō-rô-sä'ryō)	92	21°40'N	81°55'W
Rosário do Sul, Braz. (rô-zä'rē-ò-dò-sōō'l)	102	30°17'S	54°52'W
Rosário Oeste, Braz. (ō'ĕst'ĕ)	101	14°47'S	56°20'W
Rosario Strait, strt., Wa., U.S.	74a	48°27'N	122°45'W
Rosbach, Ger. (rōz'bäk)	127c	50°47'N	7°38'E
Roscoe, Tx., U.S. (rôs'kō)	80	32°26'N	100°38'W
Roseau, Dom.	91b	15°17'N	61°23'W
Roseau, Mn., U.S. (rô-zō')	70	48°52'N	95°47'W
Roseau, r., Mn., U.S.	70	48°52'N	96°11'W
Roseberg, Or., U.S. (rōz'bûrg)	64	43°13'N	123°30'W
Rosebud, r., Can. (rōz'bŭd)	55	51°20'N	112°20'W
Rosebud Creek, r., Mt., U.S.	73	45°48'N	106°34'W
Rosebud Indian Reservation, I.R., S.D., U.S.	70	43°13'N	100°42'W
Rosedale, Ms., U.S.	82	33°49'N	90°56'W
Rosedale, Wa., U.S.	74a	47°20'N	122°39'W
Roseires Reservoir, res., Sudan	185	11°15'N	34°45'E
Roselle, Il., U.S. (rō-zĕl')	69a	41°59'N	88°05'W
Rosemère, Can. (rōz'mĕr)	62a	45°38'N	73°48'W
Rosemount, Mn., U.S. (rōz'mount)	75g	44°44'N	93°08'W
Rosendal, S. Afr. (rô-sĕn'täl)	192c	28°32'S	27°56'E
Rosenheim, Ger. (rō'zĕn-hīm)	117	47°52'N	12°06'E
Roses, Golf de, b., Spain	129	42°10'N	3°20'E
Rosetown, Can. (rōz'toun)	50	51°33'N	108°00'W
Rosetta *see* Rashīd, Egypt	156	31°22'N	30°25'E
Rosettenville, neigh., S. Afr.	187b	26°15'S	28°04'E
Roseville, Ca., U.S. (rōz'vĭl)	76	38°44'N	121°19'W
Roseville, Mi., U.S.	69b	42°30'N	82°55'W
Roseville, Mn., U.S.	75g	45°01'N	93°10'W
Rosiclare, Il., U.S. (rōz'y-klâr)	66	37°30'N	88°15'W
Rosignol, Guy. (rŏs-ĭg-nŏl)	101	6°16'N	57°37'W
Roşiori de Vede, Rom. (rô-shôr'ĕ dĕ vĕ-dĕ)	131	44°06'N	25°00'E
Roskilde, Den. (rôs'kĕl-dĕ)	122	55°39'N	12°04'E
Roslavl', Russia (rôs'läv'l)	136	53°56'N	32°52'E
Roslyn, Wa., U.S. (rŏz'lĭn)	72	47°14'N	121°00'W
Rösrath, Ger. (rûz'rät)	127c	50°53'N	7°11'E
Ross, Oh., U.S. (rôs)	69f	39°19'N	84°39'W
Rossano, Italy (rô-sä'nō)	119	39°34'N	16°38'E
Rossan Point, c., Ire.	120	54°45'N	8°30'W
Ross Creek, r., Can.	62g	53°40'N	113°08'W
Rosseau, l., Can. (rôs-sō')	59	45°15'N	79°30'W
Rossel, i., Pap. N. Gui. (rô-sĕl')	175	11°31'S	154°00'E
Rosser, Can. (rôs'sẽr)	62f	49°59'N	97°27'W
Ross Ice Shelf, ice, Ant.	178	81°30'S	175°00'W
Rossignol, Lake, l., Can.	60	44°10'N	65°10'W
Ross Island, i., Can.	57	54°14'N	97°45'W
Ross Lake, res., Wa., U.S.	72	48°40'N	121°07'W
Rossland, Can. (rôs'lănd)	50	49°05'N	118°48'W
Rossosh', Russia (rôs'sŭsh)	137	50°12'N	39°32'E
Rossouw, S. Afr.	187c	31°12'S	27°18'E
Ross Sea, sea, Ant.	178	76°00'S	178°00'W
Rossvatnet, l., Nor.	116	65°36'N	13°08'E
Rossville, Ga., U.S. (rôs'vĭl)	82	34°57'N	85°22'W
Rosthern, Can.	56	52°41'N	106°25'W
Rostock, Ger. (rôs'tŭk)	116	54°04'N	12°06'E
Rostov, Russia	136	57°13'N	39°23'E
Rostov, prov., Russia	133	47°38'N	39°15'E
Rostov-na-Donu, Russia (rôstôv-nȧ-dô-nōō)	134	47°16'N	39°47'E
Roswell, Ga., U.S. (rŏz'wĕl)	82	34°02'N	84°21'W
Roswell, N.M., U.S.	64	33°23'N	104°32'W
Rotan, Tx., U.S. (rô-tăn')	78	32°51'N	100°27'W
Rothenburg, Ger.	124	49°20'N	10°10'E
Rotherham, Eng., U.K. (rŏdh'ẽr-ăm)	114a	53°26'N	1°21'W
Rotherham, co., Eng., U.K.	114a	53°52'N	1°45'W
Rothesay, Can. (rôth'sȧ)	60	45°23'N	66°00'W
Rothesay, Scot., U.K.	120	55°50'N	3°14'W
Rothwell, Eng., U.K.	114a	53°44'N	1°30'W
Roti, Pulau, i., Indon. (rō'tē)	168	10°30'S	122°52'E
Roto, Austl. (rō'tô)	176	33°07'S	145°30'E
Rotorua, N.Z.	177	38°07'S	176°17'E
Rotterdam, Neth. (rôt'ẽr-däm')	110	51°55'N	4°27'E
Rottweil, Ger. (rōt'vīl)	124	48°10'N	8°36'E
Roubaix, Fr. (rōō-bĕ')	126	50°42'N	3°10'E
Rouen, Fr. (rōō-äN')	110	49°25'N	1°05'E
Rouge, r., Can. (rōōzh)	62d	43°53'N	79°21'W
Rouge, r., Can.	59	46°40'N	74°50'W
Rouge, r., Mi., U.S.	69b	42°30'N	83°15'W
Rough River Reservoir, res., Ky., U.S.	66	37°45'N	86°10'W
Round Lake, Il., U.S.	69a	42°21'N	88°05'W
Round Pond, l., Can.	61	48°15'N	55°57'W
Round Rock, Tx., U.S.	81	30°31'N	97°41'W
Round Top, mtn., Or., U.S. (tŏp)	74c	45°41'N	123°22'W
Roundup, Mt., U.S. (round'ŭp)	73	46°25'N	108°35'W
Rousay, i., Scot., U.K. (rōō'zȧ)	120a	59°10'N	3°04'W
Rouyn, Can. (rōōn)	51	48°22'N	79°03'W
Rovaniemi, Fin. (rô'vȧ-nyĕ'mĭ)	116	66°29'N	25°45'E
Rovato, Italy (rô-vä'tō)	130	45°33'N	10°00'E
Roven'ki, Russia	133	49°54'N	38°54'E
Roven'ky, Ukr.	133	48°06'N	39°44'E
Rovereto, Italy (rō-vā-rā'tô)	130	45°53'N	11°05'E
Rovigo, Italy (rô-vē'gô)	130	45°05'N	11°48'E
Rovinj, Cro. (rô'ēn')	130	45°05'N	13°40'E
Rovira, Col. (rô-vē'rä)	100a	4°14'N	75°13'W
Rovuma (Ruvuma), r., Afr.	191	10°50'S	39°50'E
Rowley, Ma., U.S. (rou'lē)	61a	42°43'N	70°53'W
Roxana, Il., U.S. (rŏks'ăn-nȧ)	75e	38°51'N	90°05'W
Roxas, Phil. (rô-xäs)	168	11°30'N	122°47'E
Roxo, Cap, c., Sen.	188	12°20'N	16°43'W
Roy, N.M., U.S. (roi)	78	35°54'N	104°09'W
Roy, Ut., U.S.	75b	41°10'N	112°02'W
Royal, i., Bah.	92	25°30'N	76°50'W
Royal Canal, can., Ire. (roi-ăl)	120	53°28'N	6°45'W
Royal Natal National Park, rec., S. Afr.	187c	28°35'S	28°54'E
Royal Oak, Can. (roi'ăl ōk)	74a	48°30'N	123°24'W
Royal Oak, Mi., U.S.	69b	42°29'N	83°09'W
Royalton, Mi., U.S. (roi'ăl-tŭn)	66	42°00'N	86°25'W
Royan, Fr. (rwä-yäN')	126	45°40'N	1°02'W
Roye, Fr. (rwä)	126	49°43'N	2°40'E
Royersford, Pa., U.S. (rô' yẽrz-fẽrd)	68f	40°11'N	75°32'W
Royston, Ga., U.S. (roiz'tŭn)	82	34°15'N	83°06'W
Royton, Eng., U.K. (roi'tŭn)	114a	53°34'N	2°07'W
Rozay-en-Brie, Fr. (rô-zā-ĕN-brē')	127b	48°41'N	2°57'E
Rozdil'na, Ukr.	133	46°47'N	30°08'E
Rozhaya, r., Russia (rô'zhá-yȧ)	142b	55°20'N	37°37'E
Rozivka, Ukr.	133	47°14'N	36°35'E
Rožňava, Slvk. (rôzh'nyá-vȧ)	125	48°39'N	20°32'E
Rtishchevo, Russia ('r-tĭsh'chĕ-vô)	137	52°15'N	43°40'E
Ru, r., China (rōō)	162	33°07'N	114°18'E
Ruacana Falls, wtfl., Afr.	186	17°15'S	14°45'E
Ruaha National Park, rec., Tan.	191	7°15'S	34°50'E
Ruapehu, vol., N.Z. (rò-ȧ-pā'hōō)	175a	39°15'S	175°37'E
Rub' al Khali *see* Ar Rub' al Khālī, des., Asia	154	20°00'N	51°00'E
Rubeho Mountains, mts., Tan.	191	6°45'S	36°15'E
Rubidoux, Ca., U.S.	75a	33°59'N	117°24'W
Rubizhne, Ukr.	133	48°53'N	38°29'E
Rubondo Island, i., Tan.	191	2°10'S	31°55'E
Rubtsovsk, Russia	134	51°31'N	81°17'E
Ruby, Ak., U.S. (rōō'bē)	64a	64°38'N	155°22'W
Ruby, l., Nv., U.S.	76	40°11'N	115°20'W
Ruby, r., Mt., U.S.	73	45°06'N	112°10'W
Ruby Mountains, mts., Nv., U.S.	76	40°11'N	115°36'W
Rudkøbing, Den. (rōōdh'kûb-ĭng)	122	54°56'N	10°44'E
Rüdnitz, Ger. (rüd'nētz)	115b	52°44'N	13°38'E
Rudolf, Lake, l., Afr. (rōō'dŏlf)	185	3°30'N	36°05'E
Rufā'ah, Sudan (rōō-fä'ä)	185	14°52'N	33°30'E
Ruffec, Fr. (rü-fĕk')	126	46°03'N	0°11'E
Rufiji, r., Tan. (rōō-fē'jē)	187	8°00'S	38°00'E
Rufisque, Sen. (rü-fēsk')	184	14°43'N	17°17'W
Rufunsa, Zam.	191	15°05'S	29°40'E
Rufus Woods, Wa., U.S.	72	48°02'N	119°33'W
Rugao, China (rōō-gou)	164	32°24'N	120°33'E
Rugby, Eng., U.K. (rŭg'bē)	114a	52°22'N	1°15'W
Rugby, N.D., U.S.	70	48°22'N	100°00'W
Rugeley, Eng., U.K. (rōōj'lē)	114a	52°46'N	1°56'W
Rügen, i., Ger. (rü'ghĕn)	112	54°28'N	13°47'E
Ruhnu-Saar, i., Est. (rōōnò-sä'är)	123	57°46'N	23°15'E
Ruhr, r., Ger. (rōr)	124	51°18'N	8°17'E
Rui'an, China (rwā-än)	165	27°48'N	120°40'E
Ruiz, Mex. (rōē'z)	88	21°55'N	105°09'W
Ruiz, Nevado del, vol., Col. (nĕ-vä'dô-dĕl-rōōē'z)	100a	4°52'N	75°20'W
Rūjiena, Lat. (rò'yĭ-ă-nȧ)	123	57°54'N	25°19'E
Ruki, r., D.R.C.	190	0°05'S	18°55'E
Rukwa, Lake, l., Tan. (rōōk-wä')	186	8°00'S	32°25'E

ng-sing; ŋ-baŋk; N-nasalized n; nŏd; cŏmmit; ōld; ôbey; ôrder; oi-boil; fōōd; ò-as oo in foot; ou-out; s-soft; sh-dish; th-thin; pūre; ûnite; ûrn; stŭd; circŭs; ü-as in French tu; '-indeterminate vowel.

296 Rum – Saint George PRONOUNCING INDEX

PLACE (Pronunciation)	PAGE	LAT.	LONG.
Rum, r., Mn., U.S. (rŭm)	71	45°52'N	93°45'W
Ruma, Serb. (rōō'mà)	131	45°00'N	19°53'E
Rumbek, Sudan (rŭm'bĕk)	185	6°52'N	29°43'E
Rum Cay, i., Bah.	93	23°40'N	74°50'W
Rumford, Me., U.S. (rŭm'fẽrd)	60	44°32'N	70°35'W
Rummah, Wādī ar, val., Sau. Ar.	154	26°17'N	41°45'E
Rummānah, Egypt	153a	31°01'N	32°39'E
Runan, China (rōō-nän)	164	32°59'N	114°22'E
Runcorn, Eng., U.K. (rŭŋ'kôrn)	114a	53°20'N	2°44'W
Ruo, r., China (rwô)	160	41°15'N	100°46'E
Rupat, i., Indon. (rōō'pät)	153b	1°55'N	101°35'E
Rupat, Selat, strt., Indon.	153b	1°55'N	101°17'E
Rupert, Id., U.S. (rōō'pẽrt)	73	42°36'N	113°41'W
Rupert, Rivière de, r., Can.	53	51°35'N	76°30'W
Ruse, Blg. (rōō'sĕ) (rò'sĕ)	110	43°50'N	25°59'E
Rushan, China (rōō-shän)	162	36°54'N	121°31'E
Rush City, Mn., U.S.	71	45°40'N	92°59'W
Rushville, Il., U.S. (rŭsh'vĭl)	79	40°08'N	90°34'W
Rushville, In., U.S.	66	39°35'N	85°30'W
Rushville, Ne., U.S.	70	42°43'N	102°27'W
Rusizi, r., Afr.	191	3°00'S	29°05'E
Rusk, Tx., U.S. (rŭsk)	81	31°49'N	95°09'W
Ruskin, Can. (rŭs'kĭn)	74d	49°10'N	122°25'W
Russ, r., Aus.	115e	48°12'N	16°55'E
Russas, Braz. (rōō's-säs)	101	4°48'S	37°50'W
Russell, Can. (rŭs'ĕl)	50	50°47'N	101°15'W
Russell, Can.	62c	45°15'N	75°22'W
Russell, Ca., U.S.	74b	37°39'N	122°08'W
Russell, Ks., U.S.	78	38°51'N	98°51'W
Russell, Ky., U.S.	66	38°30'N	82°45'W
Russel Lake, l., Can.	57	56°15'N	101°30'W
Russell Islands, is., Sol. Is.	175	9°16'S	158°30'E
Russellville, Al., U.S. (rŭs'ĕl-vĭl)	82	34°29'N	87°44'W
Russellville, Ar., U.S.	79	35°16'N	93°08'W
Russelville, Ky., U.S.	82	36°48'N	86°51'W
Russia, nation, Eur., Asia	134	61°00'N	60°00'E
Russian, r., Ca., U.S. (rŭsh'ăn)	76	38°59'N	123°10'W
Rustavi, Geor.	138	41°33'N	45°02'E
Rustenburg, S. Afr. (rŭs'tĕn-bûrg)	192c	25°40'S	27°15'E
Ruston, La., U.S. (rŭs'tŭn)	81	32°32'N	92°39'W
Ruston, Wa., U.S.	74a	47°18'N	122°30'W
Rute, Spain (rōō'tā)	128	38°20'N	4°34'W
Ruth, Nv., U.S. (rōōth)	76	39°17'N	115°00'W
Ruthenia, hist. reg., Ukr.	125	48°25'N	23°00'E
Rutherfordton, N.C., U.S. (rŭdh'ẽr-fẽrd-tŭn)	83	35°23'N	81°58'W
Rutland, Vt., U.S.	67	43°35'N	72°55'W
Rutledge, Md., U.S. (rŭt'lĕdj)	68e	39°34'N	76°33'W
Rutog, China	160	33°29'N	79°26'E
Rutshuru, D.R.C. (rōōt-shōō'rōō)	186	1°11'S	29°27'E
Ruvo, Italy (rōō'vô)	130	41°07'N	16°32'E
Ruvuma, r., Afr.	186	11°30'S	37°00'E
Ruza, Russia (rōō'zà)	132	55°42'N	36°12'E
Ruzhany, Bela. (rò-zhän'ĭ)	125	52°49'N	24°54'E
Rwanda, nation, Afr.	186	2°10'S	29°37'E
Ryabovo, Russia (ryä'bô-vô)	142c	59°24'N	31°08'E
Ryazan', Russia (ryä-zän'')	134	54°37'N	39°43'E
Ryazan', prov., Russia	132	54°10'N	39°37'E
Ryazhsk, Russia (ryäzh'sk')	136	53°43'N	40°04'E
Rybachiy, Poluostrov, pen., Russia	136	69°50'N	33°20'E
Rybatskoye, Russia	142c	59°50'N	30°31'E
Rybinsk, Russia	134	58°02'N	38°52'E
Rybinskoye, res., Russia	134	58°23'N	38°15'E
Rybnik, Pol. (rĭb'nĕk)	125	50°06'N	18°37'E
Ryde, Eng., U.K. (rīd)	120	50°43'N	1°16'W
Rye, N.Y., U.S. (rī)	68a	40°58'N	73°42'W
Ryl'sk, Russia (rĕl''sk)	137	51°33'N	34°42'E
Ryōtsu, Japan (ryōt'sōō)	166	38°02'N	138°23'E
Rypin, Pol. (rĭ'pĕn)	125	53°04'N	19°25'E
Rysy, mtn., Eur.	125	49°12'N	20°04'E
Ryukyu Islands *see* Nansei-shotō, is., Japan	161	27°30'N	127°00'E
Rzeszów, Pol. (zhà-shoof)	117	50°02'N	22°00'E
Rzhev, Russia ('r-zhĕf)	134	56°16'N	34°17'E
Rzhyshchiv, Ukr.	133	49°58'N	31°05'E

S

PLACE (Pronunciation)	PAGE	LAT.	LONG.
Saale, r., Ger. (sä-lĕ)	124	51°14'N	11°52'E
Saalfeld, Ger. (säl'fĕlt)	124	50°38'N	11°20'E
Saarbrücken, Ger. (zähr'brü-kĕn)	117	49°15'N	7°01'E
Saaremaa, i., Est.	136	58°25'N	22°30'E
Saavedra, Arg. (sä-ä-vā'drä)	102	37°45'S	62°23'W
Saba, i., Neth. Ant. (sä'bä)	91b	17°39'N	63°20'W
Šabac, Serb. (shä'bàts)	119	44°45'N	19°49'E
Sabadell, Spain (sä-bä-dhäl')	118	41°32'N	2°07'E
Sabah, hist. reg., Malay.	168	5°10'N	116°25'E
Sabana, Archipiélago de, is., Cuba	92	23°05'N	80°00'W
Sabana, Río, r., Pan. (sä-bä'nä)	91	8°40'N	78°02'W
Sabana de la Mar, Dom. Rep. (sä-bä'nä dā lä mär')	93	19°05'N	69°30'W
Sabana de Uchire, Ven. (sä-bà'nä dĕ ōō-chē'rĕ)	101b	10°02'N	65°32'W
Sabanagrande, Hond. (sä-bä'nä-grä'n-dĕ)	90	13°47'N	87°16'W
Sabanalarga, Col. (sä-bà'nä-lär'gä)	100	10°38'N	75°02'W
Sabanas Páramo, mtn., Col. (sä-bà'näs pà'rä-mô)	100a	6°28'N	76°08'W
Sabancuy, Mex. (sä-bäŋ-kwē')	89	18°58'N	91°09'W
Sabang, Indon. (sä'bäng)	168	5°52'N	95°26'E
Sabaudia, Italy (sà-bou'dē-ä)	130	41°19'N	13°00'E
Sabetha, Ks., U.S. (sà-bĕth'à)	79	39°54'N	95°49'W
Sabi (Rio Save), r., Afr. (sä'bē)	186	20°18'S	32°07'E
Sabile, Lat. (sà'bē-lĕ)	123	57°03'N	22°34'E
Sabinal, Tx., U.S. (sà-bī'nàl)	80	29°19'N	99°27'W
Sabinal, Cayo, i., Cuba (kä'yō sä-bē-näl')	92	21°40'N	77°20'W
Sabinas, Mex.	86	28°05'N	101°30'W
Sabinas, r., Mex. (sä-bē'näs)	80	26°37'N	99°52'W
Sabinas, Río, r., Mex. (rē'ō sä-bē'näs)	80	27°25'N	100°33'W
Sabinas Hidalgo, Mex. (ē-däl'gô)	80	26°30'N	100°10'W
Sabine, Tx., U.S. (sà-bēn')	81	29°44'N	93°54'W
Sabine, r., U.S.	65	32°00'N	94°30'W
Sabine, Mount, mtn., Ant.	178	72°05'S	169°10'E
Sabine Lake, l., La., U.S.	81	29°53'N	93°41'W
Sablayan, Phil. (säb-lä-yän')	169a	12°49'N	120°47'E
Sable, Cape, c., Can. (sā'b'l)	53	43°25'N	65°24'W
Sable, Cape, c., Fl., U.S.	65	25°12'N	81°10'W
Sables, Rivière aux, r., Can.	59	49°00'N	70°20'W
Sablé-sur-Sarthe, Fr. (säb-lā-sür-särt')	126	47°50'N	0°17'W
Sablya, Gora, mtn., Russia	136	64°50'N	59°00'E
Sàbor, r., Port. (sä-bōr')	128	41°18'N	6°54'W
Sabunchu, Azer.	138	40°26'N	49°56'E
Sabzevār, Iran	157	36°13'N	57°42'E
Sac, r., Mo., U.S. (sôk)	79	38°11'N	93°45'W
Sacandaga Reservoir, res., N.Y., U.S. (sä-kăn-dâ'gà)	67	43°10'N	74°15'W
Sacavém, Port. (sä-kä-vĕn')	129b	38°47'N	9°06'W
Sacavém, r., Port.	129b	38°52'N	9°06'W
Sac City, Ia., U.S. (sôk)	70	42°25'N	95°00'W
Sachigo Lake, l., Can. (săch'ĭ-gō)	57	53°49'N	92°08'W
Sachsen, hist. reg., Ger. (zäk'sĕn)	124	50°45'N	12°17'E
Sacketts Harbor, N.Y., U.S. (săk'ĕts)	67	43°55'N	76°05'W
Sackville, Can. (săk'vĭl)	60	45°54'N	64°22'W
Saco, Me., U.S. (sô'kô)	60	43°30'N	70°28'W
Saco, r., Braz. (sä'kô)	102b	22°20'S	43°26'W
Saco, r., Me., U.S.	60	43°53'N	70°46'W
Sacramento, Mex.	80	25°45'N	103°22'W
Sacramento, Mex.	80	27°05'N	101°45'W
Sacramento, Ca., U.S. (săk-rà-mĕn'tō)	64	38°35'N	121°30'W
Sacramento, r., Ca., U.S.	76	40°20'N	122°07'W
Şa'dah, Yemen	154	16°50'N	43°45'E
Saddle Lake Indian Reserve, I.R., Can.	55	54°00'N	111°40'W
Saddle Mountain, mtn., Or., U.S. (săd''l)	74c	45°58'N	123°40'W
Sadiya, India (sŭ-dē'yä)	155	27°53'N	95°35'E
Sado, i., Japan (sä'dō)	161	38°05'N	138°26'E
Sado, r., Port. (sä'dò)	128	38°15'N	8°20'W
Saeby, Den. (sĕ'bü)	122	57°21'N	10°29'E
Saeki, Japan (sä'å-kĕ)	166	32°56'N	131°51'E
Säffle, Swe.	122	59°10'N	12°55'E
Safford, Az., U.S. (săf'fẽrd)	77	32°50'N	109°45'W
Safi, Mor. (sä'fē) (äs'fē)	184	32°24'N	9°09'W
Safīd Koh, Selseleh-ye, mts., Afg.	154	34°45'N	63°58'E
Saga, Japan (sä'gä)	167	33°15'N	130°18'E
Sagami-Nada, b., Japan (sä'gä'mē nä-dä)	167	35°06'N	139°24'E
Sagamore Hills, Oh., U.S. (săg'à-môr hĭlz)	69d	41°19'N	81°34'W
Saganaga, l., N.A. (să-gà-nă'gà)	71	48°13'N	91°17'W
Sāgar, India	155	23°55'N	78°45'E
Saghyz, r., Kaz.	137	48°30'N	56°10'E
Saginaw, Mi., U.S. (săg'ĭ-nô)	65	43°25'N	84°00'W
Saginaw, Mn., U.S.	75h	46°51'N	92°26'W
Saginaw, Tx., U.S.	75c	32°52'N	97°22'W
Saginaw Bay, b., Mi., U.S.	65	43°50'N	83°40'W
Saguache, Co., U.S. (sà-wăch')	77	38°05'N	106°10'W
Saguache Creek, r., Co., U.S.	66	38°05'N	106°40'W
Sagua de Tánamo, Cuba (sä-gwä dĕ tà'nä-mō)	93	20°40'N	75°15'W
Sagua la Grande, Cuba (sä-gwä lä grä'n-dĕ)	92	22°45'N	80°05'W
Saguaro National Park, rec., Az., U.S. (säg-wä'rō)	77	32°12'N	110°40'W
Saguenay, r., Can. (săg-ĕ-nā')	53	48°20'N	70°15'W
Sagunt, Spain	129	38°58'N	1°29'E
Sagunto, Spain (sä-gòn'tō)	118	39°40'N	0°17'W
Sahara, des., Afr. (sà-hä'rà)	184	23°44'N	1°40'W
Saharan Atlas, mts., Afr.	118	32°51'N	1°02'W
Sahāranpur, India (sŭ-hä'rŭn-pōōr')	155	29°58'N	77°41'E
Sahara Village, Ut., U.S. (sà-hä'rà)	75b	41°06'N	111°58'W
Sahel *see* Sudan, reg., Afr.	184	15°00'N	7°00'E
Sāhiwāl, Pak.	158	30°43'N	73°04'E
Sahuayo de Dias, Mex.	88	20°03'N	102°43'W
Saigon *see* Ho Chi Minh City, Viet.	168	10°46'N	106°34'E
Saijō, Japan (sä'ē-jō)	167	33°55'N	133°13'E
Saimaa, l., Fin. (sä'ĭ-mä)	116	61°24'N	28°45'E
Sain Alto, Mex. (sä-ēn' äl'tō)	88	23°35'N	103°13'W
Saint Adolphe, Can. (sănt a'dôlf) (săN' tà-dôlf')	62f	49°40'N	97°07'W
Saint Afrique, Fr. (săN' tà-frēk')	126	43°58'N	2°52'E
Saint Albans, Austl. (sânt ôl'bănz)	173a	37°44'S	144°47'E
Saint Albans, Eng., U.K.	120	51°44'N	0°20'W
Saint Albans, Vt., U.S.	67	44°50'N	73°05'W
Saint Albans, W.V., U.S.	66	38°20'N	81°50'W
Saint Albert, Can. (sănt ăl'bẽrt)	55	53°38'N	113°38'W
Saint Amand-Mont Rond, Fr. (săN't à-mäN' môN-rôN')	126	46°44'N	2°28'E
Saint André-Est, Can.	62a	45°33'N	74°19'W
Saint Andrews, Can.	51	45°05'N	67°03'W
Saint Andrews, Scot., U.K.	120	56°20'N	2°40'W
Saint Andrew's Channel, strt., Can.	61	46°06'N	60°28'W
Saint Anicet, Can. (sĕNt ä-nē-sĕ')	62a	45°07'N	74°23'W
Saint Ann, Mo., U.S. (sănt ăn')	75e	38°44'N	90°23'W
Sainte Anne, Guad.	91b	16°15'N	61°23'W
Saint Anne, Il., U.S.	69a	41°01'N	87°44'W
Sainte Anne, r., Can. (săNt àn') (sănt ăn')	59	46°55'N	71°46'W
Sainte-Anne, r., Can.	62b	47°07'N	70°50'W
Sainte Anne-des-Plaines, Can. (dā plĕN)	62a	45°46'N	73°49'W
Saint Ann's Bay, Jam.	92	18°25'N	77°15'W
Saint Anns Bay, b., Can. (ănz)	61	46°20'N	60°30'W
Saint Anselme, Can. (săN' täN-sĕlm')	62b	46°37'N	70°58'W
Saint Anthony, Can. (săN ăn'thô-nē)	51	51°24'N	55°35'W
Saint Anthony, Id., U.S. (sănt ăn'thô-nē)	73	43°59'N	111°42'W
Saint Antoine-de-Tilly, Can.	62b	46°40'N	71°31'W
Saint Apollinaire, Can. (săN' tà-pôl-ē-nâr')	62b	46°36'N	71°30'W
Saint Arnoult-en-Yvelines, Fr. (săN-tär-nōō'ĕN-nĕv-lĕn')	127b	48°33'N	1°55'E
Saint Augustin-de-Québec, Can. (sĕn tō-güs-tĕn')	62b	46°45'N	71°27'W
Saint Augustin-Deux-Montagnes, Can.	62a	45°38'N	73°59'W
Saint Augustine, Fl., U.S. (sânt ô'gŭs-tēn)	65	29°53'N	81°21'W
Sainte Barbe, Can. (sănt bärb')	62a	45°14'N	74°12'W
Saint Barthélemy, i., Guad.	91b	17°55'N	62°32'W
Saint Bees Head, c., Eng., U.K. (sănt bēz' hĕd)	120	54°30'N	3°40'W
Saint Benoît, Can. (sĕN bĕ-nōō-ä')	62a	45°34'N	74°05'W
Saint Bernard, La., U.S. (bẽr-närd')	68d	29°52'N	89°52'W
Saint Bernard, Oh., U.S.	69f	39°10'N	84°30'W
Saint Bride, Mount, mtn., Can. (sănt brīd)	55	51°30'N	115°57'W
Saint Brieuc, Fr. (săN' brēs')	117	48°32'N	2°47'W
Saint Bruno, Can. (brū'nō)	62a	45°31'N	73°20'W
Saint Canut, Can. (săN' kà-nü')	62a	45°43'N	74°04'W
Saint Casimir, Can. (kà-zē-mēr')	59	46°45'N	72°34'W
Saint Catharines, Can. (kăth'à-rĭnz)	51	43°10'N	79°14'W
Saint Catherine, Mount, mtn., Gren.	91b	12°10'N	61°42'W
Saint Chamas, Fr. (săN-shä-mä')	126a	43°32'N	5°03'E
Saint Chamond, Fr. (săN' shà-môN')	117	45°30'N	4°17'E
Saint Charles, Can. (săN' shärlz')	62b	46°47'N	70°57'W
Saint Charles, Il., U.S. (sânt chärlz')	69a	41°55'N	88°19'W
Saint Charles, Mi., U.S.	66	43°20'N	84°10'W
Saint Charles, Mn., U.S.	71	43°56'N	92°05'W
Saint Charles, Mo., U.S.	75e	38°47'N	90°29'W
Saint Charles, Lac, l., Can.	62b	46°56'N	71°21'W
Saint Christopher-Nevis *see* Saint Kitts and Nevis, nation, N.A.	87	17°24'N	63°30'W
Saint Clair, Mi., U.S. (sânt klâr)	66	42°55'N	82°30'W
Saint Clair, l., Can.	65	42°25'N	82°30'W
Saint Clair, r., Can.	58	42°45'N	82°25'W
Sainte Claire, Can.	62b	46°36'N	70°52'W
Saint Clair Shores, Mi., U.S.	69b	42°30'N	82°54'W
Saint Claude, Fr. (săN' klōd')	127	46°24'N	5°53'E
Saint Clet, Can. (săNt' klā')	62a	45°22'N	74°21'W
Saint Cloud, Fl., U.S. (sânt kloud')	83a	28°13'N	81°17'W
Saint Cloud, Mn., U.S.	65	45°33'N	94°08'W
Saint Constant, Can. (kŏn'stănt)	62a	45°23'N	73°34'W
Saint Croix, i., V.I.U.S. (sânt kroi')	87	17°40'N	64°43'W
Saint Croix, r., N.A. (kroi')	60	45°28'N	67°32'W
Saint Croix, r., U.S. (sânt kroi')	65	45°45'N	93°00'W
Saint Croix Indian Reservation, I.R., Wi., U.S.	71	45°40'N	92°21'W
Saint Croix Island, i., S. Afr. (săN krwä)	187c	33°48'S	25°45'E
Saint Damien-de-Buckland, Can. (sânt dā'mē-ĕn)	62b	46°37'N	70°39'W
Saint David, Can. (dā'vĭd)	62b	46°47'N	71°11'W
Saint David's Head, c., Wales, U.K.	120	51°54'N	5°25'W
Saint-Denis, Fr. (săN'dē-nē')	117	48°26'N	2°22'E
Saint Dizier, Fr. (dē-zyā')	117	48°49'N	4°55'E
Saint Dominique, Can. (sĕN dō-mē-nēk')	62a	45°19'N	74°09'W
Saint Edouard-de-Napierville, Can. (sĕN-tĕ-dōō-är')	62a	45°14'N	73°31'W
Saint Elias, Mount, mtn., N.A. (sânt ē-lī'ăs)	52	60°25'N	141°00'W
Saint Étienne, Fr.	117	45°26'N	4°22'E
Saint Etienne-de-Lauzon, Can. (săN' tā-tyĕn')	62b	46°39'N	71°19'W
Sainte Euphémie, Can. (sĕNt û-fĕ-mē')	62b	46°47'N	70°27'W
Saint Eustache, Can. (săN' tû-stásh')	62a	45°34'N	73°54'W
Saint Eustache, Can.	62f	49°58'N	97°47'W
Sainte Famille, Can. (săN't fà-mē'y')	62b	46°58'N	70°58'W
Saint Félicien, Can. (săN fā-lē-syăN')	51	48°39'N	72°28'W
Sainte Felicite, Can.	60	48°54'N	67°20'W
Saint Féréol, Can. (fa-rā-ôl')	62b	47°07'N	70°52'W
Saint Florent-sur-Cher, Fr. (săN' flô-räN'sür-shâr')	126	46°58'N	2°15'E
Saint Flour, Fr. (săN flōōr')	126	45°02'N	3°09'E
Sainte Foy, Can. (săNt fwä)	59	46°47'N	71°18'W
Saint Francis, r., Ar., U.S.	79	35°56'N	90°27'W
Saint Francis Lake, l., Can. (săN frăn'sĭs)	59	45°00'N	74°20'W
Saint François, Can. (săN'fräN-swä')	62b	47°01'N	70°49'W
Saint François de Boundji, Congo	190	1°03'S	15°22'E
Saint Francois Xavier, Can.	62f	49°55'N	97°32'W
Saint Gaudens, Fr. (gō-däNs')	126	43°07'N	0°43'E
Sainte Genevieve, Mo., U.S. (sânt jĕn'ē-vēv)	79	37°58'N	90°02'W
Saint George, Austl. (sânt jôrj')	175	28°02'S	148°40'E
Saint George, Can. (săN jôrj')	51	45°08'N	66°49'W
Saint George, Can. (săN'zhôrzh')	62d	43°14'N	80°15'W
Saint George, S.C., U.S. (sânt jôrj')	83	33°11'N	80°35'W
Saint George, Ut., U.S.	77	37°05'N	113°40'W
Saint George, i., Ak., U.S.	63	56°30'N	169°40'W
Saint George, Cape, c., Can.	53a	48°28'N	59°15'W

ăt; finăl; rāte; senāte; ärm; åsk; sofá; fâre; ch-choose; dh-as th in other; bē; ēvent; bĕt; recĕnt; cratẽr; g-gō; gh-guttural g; bĭt; ĭ-short neutral; rīde; K-guttural k as ch in German ich;

PRONOUNCING INDEX

Saint George – Salina 297

PLACE (Pronunciation)	PAGE	LAT.	LONG.
Saint George, Cape, c., Fl., U.S.	82	29°30′N	85°20′W
Saint George's, Can. (jôrj′ĕs)	51	48°26′N	58°29′W
Saint Georges, Fr. Gu.	101	3°48′N	51°47′W
Saint George's, Gren.	91b	12°02′N	61°57′W
Saint George's Bay, b., Can.	53a	48°20′N	59°00′W
Saint Georges Bay, b., Can.	61	45°49′N	61°45′W
Saint George's Channel, strt., Eur. (jôr-jĕz)	112	51°45′N	6°30′W
Saint Germain-en-Laye, Fr. (săN′ zhĕr-măN-äN-lā′)	126	48°53′N	2°05′E
Saint Gervais, Can. (zhĕr-vĕ′)	62b	46°43′N	70°53′W
Saint Girons, Fr. (zhē-rôN′)	126	42°58′N	1°08′E
Saint Gotthard Pass, p., Switz.	124	46°33′N	8°34′E
Saint Gregory, Mount, mtn., Can. (sånt grĕg′ẽr-ė)	61	49°19′N	58°13′W
Saint Helena, i., St. Hel.	183	16°01′S	5°16′W
Saint Helenabaai, b., S. Afr.	186	32°25′S	17°15′E
Saint Helens, Eng., U.K. (sånt hĕl′ĕnz)	114a	53°27′N	2°44′W
Saint Helens, Or., U.S. (hĕl′ĕnz)	74c	45°52′N	122°49′W
Saint Helens, Mount, vol., Wa., U.S.	72	46°13′N	122°10′W
Saint Helier, Jersey (hyĕl′yẽr)	126	49°12′N	2°06′W
Saint Henri, Can. (săN′ hĕn′rė)	62b	46°41′N	71°04′W
Saint Hubert, Can.	62a	45°29′N	73°24′W
Saint Hyacinthe, Can.	51	45°35′N	72°55′W
Saint Ignace, Mi., U.S. (sånt ĭg′nås)	71	45°51′N	84°39′W
Saint Ignace, i., Can. (săN′ ĭg′nås)	58	48°47′N	88°14′W
Saint Irenee, Can. (săN′ tē-rå-nā′)	59	47°34′N	70°15′W
Saint Isidore-de-Laprairie, Can.	62a	45°18′N	73°41′W
Saint Isidore-de-Prescott, Can. (săN′ ĭz′ĭ-dôr-prĕs-kŏt)	62c	45°23′N	74°54′W
Saint Isidore-Dorchester, Can. (dŏr-chĕs′tẽr)	62b	46°35′N	71°05′W
Saint Jacob, Il., U.S. (jā-kŏb)	75e	38°43′N	89°46′W
Saint James, Mn., U.S. (sånt jāmz′)	71	43°58′N	94°37′W
Saint James, Mo., U.S.	79	37°59′N	91°37′W
Saint James, Cape, c., Can.	54	51°58′N	131°00′W
Saint Janvier, Can. (săN′ zhän-vyā′)	62a	45°43′N	73°56′W
Saint Jean, Can. (săN′ zhäN′)	51	45°20′N	73°15′W
Saint Jean, Can.	62b	46°55′N	70°54′W
Saint Jean, Lac, l., Can.	53	48°35′N	72°00′W
Saint Jean-Chrysostome, Can. (krĭ-zōs-tōm′)	62b	46°43′N	71°12′W
Saint Jean-d'Angely, Fr. (däN-zhå-lē′)	126	45°56′N	0°33′W
Saint Jean-de-Luz, Fr. (dĕ lüz′)	126	43°23′N	1°40′W
Saint Jérôme, Can. (sånt jĕ-rōm′) (săN zhā-rōm′)	62a	45°47′N	74°00′W
Saint Joachim-de-Montmorency, Can. (sånt jō′*a*-kĭm)	62b	47°04′N	70°51′W
Saint John, Can. (sånt jŏn)	51	45°16′N	66°03′W
Saint John, In., U.S.	69a	41°27′N	87°29′W
Saint John, Ks., U.S.	78	37°59′N	98°44′W
Saint John, N.D., U.S.	70	48°57′N	99°42′W
Saint John, i., V.I.U.S.	87b	18°16′N	64°48′W
Saint John, r., N.A.	53	47°00′N	68°00′W
Saint John, Cape, c., Can.	61	50°00′N	55°32′W
Saint Johns, Antig.	91b	17°07′N	61°50′W
Saint John's, Can. (jŏns)	53a	47°34′N	52°43′W
Saint Johns, Az., U.S. (jŏnz)	77	34°30′N	109°25′W
Saint Johns, Mi., U.S.	66	43°05′N	84°35′W
Saint Johns, r., Fl., U.S.	65	29°54′N	81°32′W
Saint Johnsbury, Vt., U.S. (jŏnz′bĕr-ė)	67	44°25′N	72°00′W
Saint Joseph, Dom.	91b	15°25′N	61°26′W
Saint Joseph, Mi., U.S.	66	42°05′N	86°30′W
Saint Joseph, Mo., U.S. (sånt jỏ-sĕf)	65	39°44′N	94°49′W
Saint Joseph, i., Can.	66	46°15′N	83°55′W
Saint Joseph, l., Can. (jō′zhûf)	53	51°31′N	90°40′W
Saint Joseph, r., Mi., U.S. (sånt jỏ′sĕf)	66	41°45′N	85°50′W
Saint Joseph Bay, b., Fl., U.S. (jō′zhûf)	82	29°48′N	85°26′W
Saint Joseph-de-Beauce, Can. (sĕN zhō-zĕf′dĕ bōs)	59	46°18′N	70°52′W
Saint Joseph-du-Lac, Can. (sĕN zhō-zĕf′ dü läk)	62a	45°32′N	74°00′W
Saint Joseph Island, i., Tx., U.S. (sånt jỏ-sĕf)	81	27°58′N	96°50′W
Saint Junien, Fr. (săN′zhü-nyăN′)	126	45°53′N	0°54′E
Sainte Justine-de-Newton, Can. (sånt jŭs-tēn′)	62a	45°22′N	74°22′W
Saint Kilda, Austl.	173a	37°52′S	144°59′E
Saint Kilda, i., Scot., U.K. (kĭl′d*a*)	120	57°50′N	8°32′W
Saint Kitts, i., St. K./N. (sānt kĭtts)	87	17°24′N	63°30′W
Saint Kitts and Nevis, nation, N.A.	87	17°24′N	63°30′W
Saint Lambert, Can.	67	45°29′N	73°29′W
Saint Lambert-de-Lévis, Can.	62b	46°35′N	71°12′W
Saint Laurent, Can. (săN′lō-rän)	62a	45°31′N	73°41′W
Saint Laurent, Fr. Gu.	101	5°27′N	53°56′W
Saint Laurent-d'Orleans, Can.	62b	46°52′N	71°00′W
Saint Lawrence, Can. (sānt lô′rĕns)	61	46°55′N	55°23′W
Saint Lawrence, i., Ak., U.S. (sånt lô′rĕns)	64a	63°10′N	172°12′W
Saint Lawrence, r., N.A.	53	48°24′N	69°30′W
Saint Lawrence, Gulf of, b., Can.	53	48°00′N	62°00′W
Saint Lazare, Can. (săN′lȧ-zär′)	62b	46°39′N	70°48′W
Saint Lazare-de-Vaudreuil, Can.	62a	45°24′N	74°08′W
Saint Léger-en-Yvelines, Fr. (sȧN-lā-zhĕ′ĕN-nēv-lēn′)	127b	48°43′N	1°45′E
Saint Leonard, Can. (sånt lĕn′ȧrd)	60	47°10′N	67°56′W
Saint Léonard, Can.	62a	45°36′N	73°35′W
Saint Leonard, Md., U.S.	68e	38°29′N	76°31′W
Saint Lô, Fr.	117	49°07′N	1°05′W
Saint-Louis, Sen.	184	16°02′N	16°30′W
Saint Louis, Mi., U.S. (sånt lo͞o′ĭs)	66	43°25′N	84°35′W
Saint Louis, Mo., U.S. (sånt lo͞o′ĭs) (lo͞o′ė)	65	38°39′N	90°15′W
Saint Louis, r., Mn., U.S. (sånt lo͞o′ĭs)	71	46°57′N	92°58′W
Saint Louis, Lac, l., Can. (săN′ lo͞o-ē′)	62a	45°24′N	73°51′W
Saint Louis-de-Gonzague, Can. (săN′ lo͞o ē′)	62a	45°13′N	74°00′W
Saint Louis Park, Mn., U.S.	75g	44°56′N	93°21′W
Saint Lucia, nation, N.A.	87	13°54′N	60°40′W
Saint Lucia Channel, strt., N.A. (lū′shĭ-*a*)	91b	14°15′N	61°00′W
Saint Lucie Canal, can., Fl., U.S. (lū′sė)	83a	26°57′N	80°25′W
Saint Magnus Bay, b., Scot., U.K. (măg′nŭs)	120a	60°25′N	2°09′W
Saint Malo, Fr. (săN′ mȧ-lỏ′)	117	48°40′N	2°02′W
Saint Malo, Golfe de, b., Fr. (gôlf-dĕ-sȧN-mä-lō′)	117	48°50′N	2°49′W
Saint Marc, Haiti (săN′ mȧrk′)	93	19°10′N	72°40′W
Saint-Marc, Canal de, strt., Haiti	93	19°05′N	73°15′W
Saint Marcellin, Fr. (mär-sĕ-lăN′)	127	45°08′N	5°15′E
Saint Margarets, Md., U.S.	68e	39°02′N	76°30′W
Sainte Marie, Cap, c., Madag.	187	25°31′S	45°00′E
Sainte-Marie-aux-Mines, Fr. (sȧN′tĕ-mä-rē′ō-mēn′)	127	48°14′N	7°08′E
Sainte Marie-Beauce, Can. (sănt′mȧ-rē′)	59	46°27′N	71°03′W
Saint Maries, Id., U.S. (sånt mā′rēs)	72	47°18′N	116°34′W
Saint Martin, i., N.A. (mär′tĭn)	91b	18°06′N	62°54′W
Sainte Martine, Can.	62a	45°14′N	73°37′W
Saint Martins, Can. (mär′tĭnz)	60	45°21′N	65°32′W
Saint Martinville, La., U.S. (mär′tĭn-vĭl)	81	30°08′N	91°50′W
Saint Mary, r., Can. (mā′rė)	55	49°25′N	113°00′W
Saint Mary, Cape, c., Gam.	188	13°28′N	16°40′W
Saint Mary Reservoir, res., Can.	55	49°30′N	113°00′W
Saint Marys, Austl. (mā′rėz)	176	41°40′S	148°10′E
Saint Marys, Can.	58	43°15′N	81°10′W
Saint Marys, Ga., U.S.	83	30°43′N	81°35′W
Saint Mary's, Ks., U.S.	79	39°12′N	96°03′W
Saint Mary's, Oh., U.S.	66	40°30′N	84°25′W
Saint Marys, Pa., U.S.	67	41°25′N	78°30′W
Saint Marys, W.V., U.S.	66	39°20′N	81°15′W
Saint Marys, r., N.A.	75k	46°27′N	84°33′W
Saint Marys, r., U.S.	83	30°37′N	82°05′W
Saint Mary's Bay, b., Can.	61	46°50′N	53°47′W
Saint Mary's Bay, b., Can.	60	44°20′N	66°10′W
Saint Mathew, S.C., U.S. (măth′ū)	83	33°40′N	80°46′W
Saint Matthew, i., Ak., U.S.	63	60°25′N	172°10′W
Saint Matthews, Ky., U.S. (măth′ūz)	69h	38°15′N	85°39′W
Saint Maur-des-Fossés, Fr.	127b	48°48′N	2°29′E
Saint Maurice, r., Can. (săN′ mỏ-rēs′) (sånt mô′rĭs)	53	47°20′N	72°55′W
Saint Michael, Ak., U.S. (sånt mī′kĕl)	63	63°22′N	162°20′W
Saint Michel, Can. (săN′mė-shĕl′)	62b	46°52′N	70°54′W
Saint Michel, Bras, r., Can.	62b	46°47′N	70°51′W
Saint Michel-de-l'Atalaye, Haiti	93	19°25′N	72°20′W
Saint Michel-de-Napierville, Can.	62a	45°14′N	73°34′W
Saint Mihiel, Fr. (săN′ mē-yĕl′)	127	48°53′N	5°30′E
Saint Nazaire, Fr. (săN′nȧ-zâr′)	110	47°18′N	2°13′W
Saint Nérée, Can. (nā-rā′)	62b	46°43′N	70°43′W
Saint Nicolas, Can. (ne-kỏ-lä′)	62b	46°42′N	71°22′W
Saint Nicolas, Cap, c., Haiti	93	19°45′N	73°35′W
Saint Omer, Fr. (săN′tỏ-mâr′)	126	50°44′N	2°16′E
Saint Pascal, Can. (sĕN pā-skăl′)	60	47°32′N	69°48′W
Saint Paul, Can. (sånt pôl′)	50	53°59′N	111°17′W
Saint Paul, Mn., U.S.	65	44°57′N	93°05′W
Saint Paul, Ne., U.S.	70	41°13′N	98°28′W
Saint Paul, i., Can.	61	47°15′N	60°10′W
Saint Paul, i., Ak., U.S.	63	57°10′N	170°20′W
Saint Paul, r., Lib.	188	7°10′N	10°00′W
Saint Paul, Île, i., Afr.	3	38°43′S	77°31′E
Saint Paul Park, Mn., U.S. (pärk)	75g	44°51′N	93°00′W
Saint Pauls, N.C., U.S. (pôls)	83	34°47′N	78°57′W
Saint Peter, Mn., U.S. (pē tẽr)	71	44°20′N	93°56′W
Saint Peter Port, Guern.	126	49°27′N	2°35′W
Saint Petersburg (Sankt-Peterburg) (Leningrad), Russia	134	59°57′N	30°20′E
Saint Petersburg, Fl., U.S. (pē′tẽrz-bûrg)	65	27°47′N	82°38′W
Sainte Pétronille, Can. (sĕNt pĕt-rō-nēl′)	62b	46°51′N	71°08′W
Saint Philémon, Can. (sĕN fēl-mōN′)	62b	46°41′N	70°28′W
Saint Philippe-d'Argenteuil, Can. (săn′fe-lēp′)	62a	45°38′N	74°25′W
Saint Philippe-de-Laprairie, Can.	62a	45°20′N	73°28′W
Saint Pierre, Mart. (săn′pyâr′)	91b	14°45′N	61°12′W
Saint Pierre, St. P./M.	61	46°47′N	56°11′W
Saint Pierre, i., St. P./M.	61	46°47′N	56°11′W
Saint Pierre, Lac, l., Can.	59	46°07′N	72°45′W
Saint Pierre and Miquelon, dep., N.A.	53a	46°53′N	56°40′W
Saint Pierre-d'Orléans, Can.	62b	46°53′N	71°04′W
Saint Pierre-Montmagny, Can.	62b	46°55′N	70°37′W
Saint Placide, Can. (plăs′ĭd)	62a	45°32′N	74°11′W
Saint Pol-de-Léon, Fr. (săN-pô′dĕ-lā-ôN′)	126	48°41′N	4°00′W
Saint Quentin, Fr. (săN′kän-tăN′)	117	49°52′N	3°16′E
Saint Raphaël, Can. (rä-fä-él′)	62b	46°48′N	70°46′W
Saint Raymond, Can.	59	46°50′N	71°51′W
Saint Rédempteur, Can. (săN rā-dāNp-tûr′)	62b	46°42′N	71°18′W
Saint Rémi, Can. (sĕN rĕ-mē′)	62a	45°15′N	73°36′W
Saint Romuald-d'Etchemin, Can. (sĕN rŏ′mo͞o-äl)	59	46°45′N	71°14′W
Sainte Rose, Guad.	91b	16°19′N	61°45′W
Saintes, Fr.	126	45°44′N	0°41′W
Sainte Scholastique, Can. (skỏ-lȧs-tēk′)	62a	45°39′N	74°05′W
Saint Siméon, Can.	59	47°51′N	69°55′W
Saint Stanislas-de-Kostka, Can.	62a	45°11′N	74°08′W
Saint Stephen, Can. (stē′vĕn)	51	45°12′N	66°17′W
Saint Sulpice, Can.	62a	45°50′N	73°21′W
Saint Thérèse-de-Blainville, Can. (tĕ-rĕz′ dĕ blĕN-vēl′)	59	45°38′N	73°51′W
Saint Thomas, Can. (tŏm′*a*s)	51	42°45′N	81°15′W
Saint Thomas, i., V.I.U.S.	87	18°22′N	64°57′W
Saint Thomas Harbor, b., V.I.U.S. (tŏm′*a*s)	87c	18°19′N	64°56′W
Saint Timothée, Can. (tē-mô-tā′)	62a	45°17′N	74°03′W
Saint Tropez, Fr. (trỏ-pĕ′)	127	43°15′N	6°42′E
Saint Valentin, Can. (văl-ĕn-tīn)	62a	45°07′N	73°19′W
Saint Valéry-sur-Somme, Fr. (vȧ-lā-rē′)	126	50°10′N	1°39′E
Saint Vallier, Can. (väl-yā′)	62b	46°54′N	70°49′W
Saint Victor, Can. (vĭk′tẽr)	59	46°09′N	70°56′W
Saint Vincent, Gulf, b., Austl. (vĭn′sĕnt)	176	34°55′S	138°00′E
Saint Vincent and the Grenadines, nation, N.A.	87	13°20′N	60°50′W
Saint Vincent Passage, strt., N.A.	91b	13°35′N	61°10′W
Saint Walburg, Can.	50	53°39′N	109°12′W
Saint Yrieix-la-Perche, Fr. (ē-rė-ĕ)	126	45°30′N	1°08′E
Saitama, dept., Japan (sī′tä-mä)	167a	35°52′N	139°40′E
Saitbaba, Russia (sá-ĕt′bá-bȧ)	142a	54°06′N	56°42′E
Sajama, Nevada, mtn., Bol. (nĕ-vȧ′dä-sä-hȧ′mä)	100	18°13′S	68°53′W
Sakai, Japan (sä′kä-ė)	166	34°34′N	135°28′E
Sakaiminato, Japan	167	35°33′N	133°15′E
Sakākah, Sau. Ar.	154	29°58′N	40°03′E
Sakakawea, Lake, res., N.D., U.S.	64	47°49′N	101°58′W
Sakania, D.R.C. (sȧ-kä′nĭ-ȧ)	186	12°45′S	28°34′E
Sakarya, r., Tur. (sȧ-kär′yȧ)	154	40°10′N	31°00′E
Sakata, Japan (sä′kä-tä)	161	38°56′N	139°57′E
Sakchu, Kor., N. (säk′chò)	166	40°29′N	125°09′E
Sakha (Yakutia), prov., Russia	141	65°21′N	117°13′E
Sakhalin, i., Russia (sȧ-kȧ-lēn′)	135	52°00′N	143°00′E
Šakiai, Lith. (shä′kĭ-ī)	123	54°59′N	23°05′E
Sakishima-guntō, is., Japan (sä′kė-shē′ma gòn′tō′)	161	24°25′N	125°00′E
Sakmara, r., Russia	137	52°00′N	56°10′E
Sakomet, r., R.I., U.S. (sä-kō′mĕt)	68b	41°32′N	71°11′W
Sakurai, Japan	167b	34°31′N	135°51′E
Sakwaso Lake, l., Can. (s*a*-kw*a*′sō)	57	53°01′N	91°55′W
Sal, i., C.V. (säal)	184b	16°45′N	22°39′W
Sal, r., Russia (sȧl)	137	47°30′N	43°00′E
Sal, Cay, i., Bah. (kē săl)	92	23°45′N	80°25′W
Sala, Swe. (sô′lä)	122	59°56′N	16°34′E
Sala Consilina, Italy (sä′lä kỏn-sė-lē′nä)	130	40°24′N	15°38′E
Salada, Laguna, l., Mex. (lä-gò′nä-sä-lä′dä)	76	32°34′N	115°45′W
Saladillo, Arg. (sä-lä-dēl′yỏ)	102	35°38′S	59°48′W
Salado, Hond. (sä-lä′dhō)	90	15°44′N	87°03′W
Salado, r., Arg.	99c	35°53′S	58°12′W
Salado, r., Arg.	102	37°00′S	67°00′W
Salado, r., Arg. (sä-lä′dỏ)	102	26°05′S	63°35′W
Salado, r., Mex.	86	28°00′N	102°00′W
Salado, r., Mex. (sä-lä′dô)	89	18°30′N	97°29′W
Salado Creek, r., Tx., U.S.	75d	29°23′N	98°25′W
Salado de los Nadadores, Río, r., Mex. (dĕ-lỏs-nä-dä-dỏ′rĕs)	80	27°26′N	101°35′W
Salal, Chad	189	14°51′N	17°13′E
Salamanca, Chile (sä-lä-mä′n-kä)	99b	31°48′S	70°57′W
Salamanca, Mex.	86	20°36′N	101°10′W
Salamanca, Spain (sä-lä-mä′n-kȧ)	110	40°54′N	5°42′W
Salamanca, N.Y., U.S. (săl-ȧ-măŋ′kȧ)	67	42°10′N	78°45′W
Salamat, Bahr, r., Chad (bär sä-lä-mät′)	185	10°06′N	19°16′E
Salamina, Col. (sä-lä-mē′-nä)	100a	5°25′N	75°29′W
Salamína, Grc.	131	37°58′N	23°30′E
Salat-la-Canada, Fr.	126	44°52′N	1°13′E
Salaverry, Peru (sä-lä-vā′rė)	100	8°16′S	78°54′W
Salawati, i., Indon. (sä-lä-wä′tė)	169	1°07′S	130°52′E
Salawe, Tan.	191	3°19′S	32°52′E
Sala y Gómez, Isla, i., Chile	195	26°50′S	105°50′W
Salcedo, Dom. Rep. (säl-sā′dō)	93	19°25′N	70°30′W
Saldaña, r., Col. (säl-dȧ′n-yä)	100a	3°42′N	75°16′W
Saldanha, S. Afr.	186	32°55′S	18°05′E
Saldus, Lat. (sȧl′dòs)	123	56°39′N	22°30′E
Sale, Austl. (sāl)	176	38°10′S	147°07′E
Sale, Eng., U.K.	114a	53°24′N	2°20′W
Sale, r., Can. (sȧl′rė-vyâr′)	62f	49°44′N	97°11′W
Salekhard, Russia (sŭ-lyĭ-kärt)	136	66°35′N	66°50′E
Salem, India	155	11°39′N	78°11′E
Salem, S. Afr.	187c	33°29′S	26°30′E
Salem, Il., U.S. (sā′lĕm)	66	38°40′N	89°00′W
Salem, In., U.S.	66	38°35′N	86°00′W
Salem, Ma., U.S.	61a	42°31′N	70°54′W
Salem, Mo., U.S.	79	37°36′N	91°33′W
Salem, N.H., U.S.	61a	42°46′N	71°16′W
Salem, N.J., U.S.	67	39°35′N	75°30′W
Salem, Oh., U.S.	66	40°55′N	80°50′W
Salem, Or., U.S.	64	44°55′N	123°03′W
Salem, S.D., U.S.	70	43°43′N	97°23′W
Salem, Va., U.S.	83	37°16′N	80°05′W
Salem, W.V., U.S.	66	39°15′N	80°35′W
Salemi, Italy (sä-lå′mē)	130	37°49′N	12°48′E
Salerno, Italy (sä-lĕr′nỏ)	118	40°27′N	14°46′E
Salerno, Golfo di, b., Italy (gỏl-fỏ-dē)	118	40°30′N	14°40′E
Salford, Eng., U.K. (săl′fẽrd)	120	53°26′N	2°19′W
Salgótarján, Hung. (shôl′gỏ-tôr-yän)	125	48°06′N	19°50′E
Salhyr, r., Ukr.	133	45°25′N	34°22′E
Salida, Co., U.S. (sȧ-lī′dȧ)	78	38°31′N	106°01′W
Salies-de-Béan, Fr.	126	43°27′N	0°58′W
Salima, Mwi.	191	13°47′S	34°26′E
Salina, Ks., U.S. (s*a*-lī′n*a*)	64	38°50′N	97°37′W
Salina, Ut., U.S.	77	39°00′N	111°55′W

ng-sing; ŋ-baŋk; N-nasalized n; nŏd; cŏmmit; ōld; ỏbey; ôrder; oi-boil; fo͞od; ȯ-as oo in foot; ou-out; s-soft; sh-dish; th-thin; pūre; ủnite; ûrn; stŭd; circŭs; ü-as in French tu; ′-indeterminate vowel.

298 Salina - San Francisco PRONOUNCING INDEX

PLACE (Pronunciation)	PAGE	LAT.	LONG.
Salina, i., Italy (sä-lē'nä)	130	38°35'N	14°48'E
Salina Cruz, Mex. (sä-lē'nä krōōz')	86	16°10'N	95°12'W
Salina Point, c., Bah.	93	22°10'N	74°20'W
Salinas, Mex.	86	22°38'N	101°42'W
Salinas, P.R.	87b	17°58'N	66°16'W
Salinas, Ca., U.S. (sȧ-lē'nȧs)	76	36°41'N	121°40'W
Salinas, r., Mex. (sä-lē'näs)	89	16°15'N	90°31'W
Salinas, r., Ca., U.S.	76	36°33'N	121°29'W
Salinas, Bahía de, b., N.A. (bä-ē'ä-dĕ-sȧ-lē'nȧs)	90	11°05'N	85°55'W
Salinas National Monument, rec., N.M., U.S.	77	34°10'N	106°05'W
Salinas Victoria, Mex. (sä-lē'näs vēk-tō'rē-ä)	80	25°59'N	100°19'W
Saline, r., Ar., U.S. (sȧ-lēn')	79	34°06'N	92°30'W
Saline, r., Ks., U.S.	78	39°05'N	99°43'W
Salins-les-Bains, Fr. (sȧ-lăn'-lā-băn')	127	46°55'N	5°54'E
Salisbury, Can.	60	46°03'N	65°05'W
Salisbury, Eng., U.K. (sôlz'bē-rē)	117	50°35'N	1°51'W
Salisbury, Md., U.S.	67	38°20'N	75°40'W
Salisbury, Mo., U.S.	79	39°24'N	92°47'W
Salisbury, N.C., U.S.	83	35°40'N	80°29'W
Salisbury *see* Harare, Zimb.	186	17°50'S	31°03'E
Salisbury Island, i., Can.	53	63°36'N	76°20'W
Salisbury Plain, pl., Eng., U.K.	120	51°15'N	1°52'W
Salkehatchie, r., S.C., U.S. (sô-kē-hăch'ē)	83	33°09'N	81°10'W
Sallisaw, Ok., U.S. (săl'ĭ-sô)	79	35°27'N	94°48'W
Salmon, Id., U.S. (săm'ŭn)	73	45°11'N	113°54'W
Salmon, r., Can.	54	54°00'N	123°50'W
Salmon, r., Can.	60	46°19'N	65°36'W
Salmon, r., Id., U.S.	64	45°30'N	115°45'W
Salmon, r., N.Y., U.S.	67	44°35'N	74°15'W
Salmon, r., Wa., U.S.	74c	45°44'N	122°36'W
Salmon, Middle Fork, r., Id., U.S.	72	44°50'N	114°52'W
Salmon Arm, Can.	55	50°42'N	119°16'W
Salmon Falls Creek, r., Id., U.S.	73	42°22'N	114°53'W
Salmon Gums, Austl. (gŭmz)	174	33°00'S	122°00'E
Salmon River Mountains, mts., Id., U.S.	64	44°15'N	115°44'W
Salon-de-Provence, Fr. (sȧ-lôn-dē-prō-văns')	127	43°48'N	5°09'E
Salonika *see* Thessaloníki, Grc.	110	40°38'N	22°59'E
Salonta, Rom. (sä-lôn'tä)	125	46°46'N	21°38'E
Saloum, r., Sen.	188	14°10'N	15°45'W
Salsette Island, i., India	159b	19°12'N	72°52'E
Sal'sk, Russia (sälsk)	137	46°30'N	41°20'E
Salt, r., Az., U.S. (sôlt)	64	33°28'N	111°35'W
Salt, r., Mo., U.S.	79	39°54'N	92°11'W
Salta, Arg. (säl'tä)	102	24°50'S	65°16'W
Salta, prov., Arg.	102	25°15'S	65°00'W
Saltair, Ut., U.S. (sôlt'âr)	75b	40°46'N	112°09'W
Salt Cay, i., T./C. Is.	93	21°20'N	71°15'W
Salt Creek, r., Il., U.S. (sôlt)	69a	42°01'N	88°01'W
Saltillo, Mex. (säl-tēl'yō)	86	25°24'N	100°59'W
Salt Lake City, Ut., U.S. (sôlt lāk sĭ'tĭ)	64	40°45'N	111°52'W
Salto, Arg. (säl'tō)	99c	34°17'S	60°15'W
Salto, Ur.	102	31°18'S	57°45'W
Salto, r., Mex.	88	22°16'N	99°18'W
Salto, Serra do, mtn., Braz. (sĕ'r-rä-dō)	99a	20°26'S	43°28'W
Salto Grande, Braz. (grän'dä)	101	22°57'S	49°58'W
Salton Sea, Ca., U.S. (sôlt'ŭn)	76	33°28'N	115°43'W
Salton Sea, l., Ca., U.S.	64	33°19'N	115°50'W
Saltpond, Ghana	184	5°16'N	1°07'W
Salt River Indian Reservation, I.R., Az., U.S. (sôlt rĭv'ẽr)	77	33°40'N	112°01'W
Saltsjöbaden, Swe. (sält'shû-bäd'ĕn)	122	59°15'N	18°20'E
Saltspring Island, i., Can. (sält'sprĭng)	54	48°47'N	123°30'W
Saltville, Va., U.S. (sôlt'vĭl)	83	36°50'N	81°45'W
Saltykovka, Russia (sȧl-tē'kôf-kȧ)	142b	55°45'N	37°56'E
Salud, Mount, mtn., Pan. (sä-lōō'th)	86a	9°14'N	79°42'W
Saluda, S.C., U.S. (sȧ-lōō'dȧ)	83	34°02'N	81°46'W
Saluda, r., S.C., U.S.	83	34°07'N	81°48'W
Saluzzo, Italy (sä-lōōt'sō)	130	44°39'N	7°31'E
Salvador, Braz. (säl-vä-dōr') (bä-ē'ȧ)	101	12°59'S	38°27'W
Salvador Lake, l., La., U.S.	81	29°45'N	90°20'W
Salvador Point, c., Bah.	92	24°30'N	77°45'W
Salvatierra, Mex. (säl-vä-tyĕr'rä)	88	20°13'N	100°52'W
Salween, r., Asia	152	21°00'N	98°00'E
Salyan, Azer.	137	39°40'N	49°10'E
Salzburg, Aus. (sälts'bòrgh)	117	47°48'N	13°04'E
Salzwedel, Ger. (sälts-vä'dĕl)	124	52°51'N	11°10'E
Samālūt, Egypt (sä-mä-lōōt')	156	28°17'N	30°43'E
Samanä, Cabo, c., Dom. Rep.	87	19°20'N	69°00'W
Samana or Atwood Cay, i., Bah.	93	23°05'N	73°45'W
Samar, i., Phil. (sä'mär)	169	11°30'N	126°07'E
Samara (Kuybyshev), Russia	136	53°10'N	50°05'E
Samara, r., Russia	137	52°50'N	50°35'E
Samara, r., Ukr. (sȧ-mä'rȧ)	133	48°47'N	35°30'E
Samarai, Pap. N. Gui. (sä-mä-rä'ē)	169	10°45'S	150°49'E
Samarinda, Indon.	168	0°30'S	117°10'E
Samarkand, Uzb. (sȧ-mȧr-känt')	139	39°42'N	67°00'E
Şamaxı, Azer.	137	40°35'N	48°40'E
Samba, D.R.C.	191	4°38'S	26°22'E
Sambalpur, India (sŭm'bŭl-pòr)	155	21°30'N	84°05'E
Sāmbhar, r., India	158	27°00'N	74°58'E
Sambir, Ukr.	125	49°31'N	23°12'E
Samborombón, r., Arg.	99c	35°20'S	57°52'W
Samborombón, Bahía, b., Arg. (bä-ē'ä-säm-bô-rôm-bô'n)	99c	35°57'S	57°05'W
Sambre, r., Eur. (säN'br')	121	50°20'N	4°15'E
Sambungo, Ang.	190	8°39'S	20°43'E
Sammamish, r., Wa., U.S.	74a	47°43'N	122°08'W
Sammamish, Lake, l., Wa., U.S. (sȧ-măm'ĭsh)	74a	47°35'N	122°02'W
Samoa, nation, Oc.	2	14°30'S	172°00'W
Samoa Islands, is., Oc.	170a	14°00'S	171°00'W
Samokov, Blg. (sä'mô-kôf)	131	42°20'N	23°33'E
Samora Correia, Port. (sä-mō'rä-kôr-rĕ'yä)	129b	38°55'N	8°52'W
Samorovo, Russia (sȧ-mȧ-rô'vô)	140	60°47'N	69°13'E
Sámos, i., Grc. (să'mŏs)	119	37°53'N	26°35'E
Samothráki, i., Grc.	119	40°23'N	25°10'E
Sampaloc Point, c., Phil. (säm-pä'lôk)	169a	14°43'N	119°56'E
Sam Rayburn Reservoir, res., Tx., U.S.	81	31°10'N	94°15'W
Samson, Al., U.S. (săm'sŭn)	82	31°06'N	86°02'W
Samsu, Kor., N. (säm'sōō')	166	41°12'N	128°00'E
Samsun, Tur. (sȧm'sōōn')	154	41°20'N	36°05'E
Samtredia, Geor. (sȧm'trĕ-dē)	137	42°18'N	42°25'E
Samuel, i., Can. (săm'ū-ĕl)	74d	48°50'N	123°10'W
Samur, r., (sä-mōōr')	137	41°40'N	47°20'E
San, Mali (sän)	184	13°18'N	4°54'W
San, r., Eur.	117	50°33'N	22°12'E
Şan'ā', Yemen (sän'ä)	154	15°17'N	44°05'E
Sanaga, r., Cam. (sä-nä'gä)	184	4°30'N	12°00'E
San Ambrosio, Isla, i., Chile (ē's-lä-dĕ-sän äm-brō'zē-ō)	97	26°40'S	80°00'W
Sanana, Pulau, i., Indon.	169	2°15'S	126°38'E
Sanandaj, Iran	154	36°44'N	46°43'E
San Andreas, Ca., U.S. (săn ăn'drē-ăs)	76	38°10'N	120°42'W
San Andreas, l., Ca., U.S.	74b	37°36'N	122°26'W
San Andrés, Col. (sän-än-drĕ's)	100a	6°57'N	75°41'W
San Andrés, Mex. (sän än-drās')	89a	19°15'N	99°10'W
San Andrés, i., Col.	91	12°32'N	81°34'W
San Andrés, Laguna de, l., Mex.	89	22°40'N	97°50'W
San Andres Mountains, mts., N.M., U.S. (săn ăn'drē-ăs)	64	33°00'N	106°40'W
San Andrés Tuxtla, Mex. (sän-än-drā's-tōōs'tlä)	86	18°27'N	95°12'W
San Angelo, Tx., U.S. (săn ăn-jĕ-lō)	64	31°28'N	100°22'W
San Antioco, Isola di, i., Italy (ē'sō-lä-dē-sän-än-tyŏ'kō)	130	39°00'N	8°25'E
San Antonio, Chile (sän-ȧn-tō'nyō)	102	33°34'S	71°36'W
San Antonio, Col.	100a	2°57'N	75°06'W
San Antonio, Col.	100a	3°55'N	75°28'W
San Antonio, Phil.	169a	14°57'N	120°05'E
San Antonio, Tx., U.S. (săn ăn-tō'nĕ-ō)	64	29°25'N	98°30'W
San Antonio, r., Tx., U.S.	81	29°00'N	97°58'W
San Antonio, Cabo, c., Cuba (kä'bō-sän-än-tō'nyō)	87	21°55'N	84°55'W
San Antonio, Lake, res., Ca., U.S.	76	36°00'N	121°13'W
San Antonio Bay, b., Tx., U.S.	81	28°20'N	97°08'W
San Antonio de Areco, Arg. (dā ä-rā'kō)	99c	34°16'S	59°30'W
San Antonio de las Vegas, Cuba	93a	22°51'N	82°23'W
San Antonio de los Baños, Cuba (dā lōs bän'yōs)	92	22°54'N	82°30'W
San Antonio de los Cobres, Arg. (dā lōs kō'brās)	102	24°15'S	66°29'W
San Antônio de Pádua, Braz. (dĕ-pȧ'dwä)	99a	21°32'S	42°09'W
San Antonio de Tamanaco, Ven.	101b	9°42'N	66°03'W
San Antonio Oeste, Arg. (sän-nä-tō'nyō ō-ĕs'tā)	102	40°49'S	64°56'W
San Antonio Peak, mtn., Ca., U.S. (săn ăn-tō'nĭ-ō)	75a	34°17'N	117°39'W
Sanarate, Guat. (sä-nä-rä'tĕ)	90	14°47'N	90°12'W
San Augustine, Tx., U.S. (săn ô'gŭs-tēn)	81	31°33'N	94°08'W
San Bartolo, Mex. (sän bär-tō'lô)	89a	19°36'N	99°43'W
San Bartolo, Mex.	80	24°43'N	103°12'W
San Bartolomeo, Italy (bär-tô-lô-mā'ō)	130	41°25'N	15°04'E
San Benedetto del Tronto, Italy (bā'nȧ-dĕt'tō dĕl trōn'tō)	130	42°58'N	13°54'E
San Benito, Tx., U.S. (săn bĕ-nē'tō)	81	26°07'N	97°37'W
San Benito, r., Ca., U.S.	76	36°40'N	121°20'W
San Bernardino, Ca., U.S. (bûr-när-dē'nō)	64	34°07'N	117°19'W
San Bernardino Mountains, mts., Ca., U.S.	76	34°05'N	116°23'W
San Bernardo, Chile (sän bĕr-när'dō)	99b	33°35'S	70°42'W
San Blas, Mex. (sän bläs')	86	21°33'N	105°19'W
San Blas, Cape, c., Fl., U.S.	65	29°38'N	85°38'W
San Blas, Cordillera de, mts., Pan.	91	9°17'N	78°20'W
San Blas, Golfo de, b., Pan.	91	9°33'N	78°42'W
San Blas, Punta, c., Pan.	91	9°35'N	78°55'W
San Bruno, Ca., U.S. (săn brū-nō)	74b	37°38'N	122°25'W
San Buenaventura, Mex. (bwā'nȧ-vĕn-tōō'rä)	80	27°07'N	101°30'W
San Carlos, Chile (sän-kȧ'r-lōs)	102	36°23'S	71°58'W
San Carlos, Col.	100a	6°11'N	74°58'W
San Carlos, Eq. Gui.	190	3°27'N	8°33'E
San Carlos, Mex. (sän kär'lōs)	89	17°49'N	92°33'W
San Carlos, Mex.	80	24°36'N	98°52'W
San Carlos, Nic. (sän-kä'r-lōs)	91	11°08'N	84°48'W
San Carlos, Phil.	169a	15°56'N	120°20'E
San Carlos, Ca., U.S. (săn kär'lōs)	74b	37°30'N	122°15'W
San Carlos, Ven.	100	9°36'N	68°35'W
San Carlos, r., C.R.	91	10°36'N	84°18'W
San Carlos de Bariloche, Arg.	102	41°15'S	71°26'W
San Carlos Indian Reservation, I.R., Az., U.S. (săn kär'lōs)	77	33°27'N	110°15'W
San Carlos Lake, res., Az., U.S.	77	33°05'N	110°29'W
San Casimiro, Ven. (kä-sē-mē'rō)	101b	10°01'N	67°02'W
San Cataldo, Italy (kä-täl'dō)	130	37°30'N	13°59'E
Sánchez, Dom. Rep. (sän'chĕz)	87	19°15'N	69°40'W
Sanchez, Río de los, r., Mex. (rē'ō-dĕ-lōs)	88	20°31'N	102°29'W
Sánchez Román, Mex. (rō-mȧ'n)	88	21°48'N	103°20'W
San Clemente, Spain (sän klȧ-mĕn'tȧ)	128	39°25'N	2°24'W
San Clemente Island, i., Ca., U.S.	64	32°54'N	118°29'W
San Cristóbal, Dom. Rep. (krēs-tô'bäl)	93	18°25'N	70°05'W
San Cristóbal, Guat.	90	15°22'N	90°26'W
San Cristóbal, Ven.	100	7°43'N	72°15'W
San Cristobal, i., Sol. Is.	175	10°47'S	162°17'E
San Cristóbal de las Casas, Mex.	86	16°44'N	92°39'W
Sancti Spíritus, Cuba (säŋk'tē spē'rē-tōōs)	87	21°55'N	79°25'W
Sancti Spíritus, prov., Cuba	92	22°05'N	79°20'W
Sancy, Puy de, mtn., Fr. (pwē-dĕ-sȧn-sē')	117	45°30'N	2°53'E
Sand, i., Or., U.S. (sănd)	74c	46°16'N	124°01'W
Sand, i., Wi., U.S.	71	46°03'N	91°09'W
Sand, r., S. Afr.	187c	28°30'S	29°30'E
Sand, r., S. Afr.	192c	28°09'S	26°46'E
Sanda, Japan (sän'dä)	167	34°53'N	135°14'E
Sandakan, Malay. (sȧn-dä'kȧn)	168	5°51'N	118°03'E
Sanday, i., Scot., U.K. (sănd'ā)	120a	59°17'N	2°25'W
Sandbach, Eng., U.K. (sănd'băch)	114a	53°08'N	2°22'W
Sandefjord, Nor. (sän'dĕ-fyôr')	122	59°09'N	10°14'E
San de Fuca, Wa., U.S. (de-fōō-cä)	74a	48°14'N	122°44'W
Sanders, Az., U.S.	77	35°13'N	109°20'W
Sanderson, Tx., U.S. (săn'dẽr-sŭn)	80	30°09'N	102°24'W
Sandersville, Ga., U.S. (săn'dẽrz-vĭl)	83	32°57'N	82°50'W
Sandhammaren, c., Swe. (sänt'häm-mȧr)	116	55°24'N	14°37'E
Sand Hills, reg., Ne., U.S. (sănd)	70	41°57'N	101°29'W
Sand Hook, N.J., U.S. (sănd hòk)	68a	40°29'N	74°05'W
Sandhurst, Eng., U.K. (sănd'hûrst)	114b	51°20'N	0°48'W
Sandia Indian Reservation, I.R., N.M., U.S.	77	35°15'N	106°30'W
San Diego, Ca., U.S. (săn dē-ā'gō)	64	32°43'N	117°10'W
San Diego, Tx., U.S.	78	27°47'N	98°13'W
San Diego, r., Ca., U.S.	76	32°53'N	116°57'W
San Diego de la Unión, Mex. (sän dē-ā-gō dā lä ōō-nyōn')	88	21°27'N	100°52'W
Sandies Creek, r., Tx., U.S. (sănd'ēz)	81	29°13'N	97°34'W
San Dimas, Mex. (dē-mäs')	88	24°08'N	105°57'W
San Dimas, Ca., U.S. (săn dē-mȧs)	75a	34°07'N	117°49'W
Sandnes, Nor. (sänd'nĕs)	122	58°52'N	5°44'E
Sandoa, D.R.C. (sän-dō'ä)	186	9°39'S	23°00'E
Sandomierz, Pol. (sän-dô'myĕzh)	125	50°39'N	21°45'E
San Doná di Piave, Italy (sän dô nä' dē pyä'vĕ)	130	45°38'N	12°34'E
Sandoway, Mya. (sän-dô-wī')	155	18°24'N	94°28'E
Sandpoint, Id., U.S. (sănd point)	72	48°17'N	116°34'W
Sandringham, Austl. (săn'drĭng-ăm)	173a	37°57'S	145°01'E
Sandrio, Italy (sä'n-dryô)	130	46°11'N	9°53'E
Sand Springs, Ok., U.S. (sănd sprĭnz)	79	36°08'N	96°06'W
Sandstone, Austl. (sănd'stōn)	174	28°00'S	119°25'E
Sandstone, Mn., U.S.	71	46°08'N	92°53'W
Sanduo, China (sän-dwô)	162	32°49'N	119°39'E
Sandusky, Al., U.S. (săn-dŭs'kē)	68h	33°32'N	86°50'W
Sandusky, Mi., U.S.	66	43°25'N	82°50'W
Sandusky, Oh., U.S.	65	41°25'N	82°45'W
Sandusky, r., Oh., U.S.	66	41°10'N	83°20'W
Sandwich, Il., U.S. (sănd'wĭch)	66	42°35'N	88°53'W
Sandy, Or., U.S. (sănd'ē)	74c	45°24'N	122°16'W
Sandy, Ut., U.S.	75b	40°36'N	111°53'W
Sandy, r., Or., U.S.	74c	45°28'N	122°17'W
Sandy Cape, c., Austl.	175	24°25'S	153°10'E
Sandy Hook, Ct., U.S. (hòk)	68a	41°25'N	73°17'W
Sandy Lake, l., Can.	62g	53°46'N	113°58'W
Sandy Lake, l., Can.	61	49°16'N	57°00'W
Sandy Lake, l., Can.	57	53°00'N	93°07'W
Sandy Point, Tx., U.S.	81a	29°22'N	95°27'W
Sandy Point, c., Wa., U.S.	74d	48°48'N	122°42'W
Sandy Springs, Ga., U.S. (springz)	68c	33°55'N	84°23'W
San Estanislao, Para. (ĕs-tä-nēs-lä'ô)	102	24°38'S	56°20'W
San Esteban, Hond. (ĕs-tĕ'bän)	90	15°13'N	85°53'W
San Fabian, Phil. (fä-byä'n)	169a	16°14'N	120°28'E
San Felipe, Chile (fĕ-lē'pä)	102	32°45'S	70°43'W
San Felipe, Mex. (fĕ-lē'pĕ)	88	21°29'N	101°13'W
San Felipe, Mex.	88	22°21'N	105°26'W
San Felipe, Ven. (fĕ-lē'pĕ)	100	10°13'N	68°45'W
San Felipe, Cayos de, is., Cuba (kä'yōs-dĕ-sän-fĕ-lē'pĕ)	92	22°00'N	83°30'W
San Felipe Creek, r., Ca., U.S. (săn fē-lēp'ā)	76	33°10'N	116°03'W
San Felipe Indian Reservation, I.R., N.M., U.S.	77	35°26'N	106°26'W
San Félix, Isla, i., Chile (ē's-lä-dĕ-sän fā-lēks')	97	26°20'S	80°10'W
San Fernanda, Spain (fĕr-nä'n-dä)	128	36°28'N	6°13'W
San Fernando, Arg. (fĕr-nä'n-dō)	99c	34°26'S	58°34'W
San Fernando, Chile	99b	35°36'S	70°58'W
San Fernando, Mex. (fĕr-nän'dō)	80	24°52'N	98°10'W
San Fernando, Phil. (sän fĕr-nä'n-dō)	168	16°38'N	120°19'E
San Fernando, Ca., U.S. (fĕr-nän'dō)	75a	34°17'N	118°27'W
San Fernando, r., Mex. (sän fĕr-nän'dō)	80	25°07'N	98°25'W
San Fernando de Apure, Ven. (sän-fĕr-nä'n-dō-dĕ-ä-pōō'rä)	100	7°46'N	67°29'W
San Fernando de Atabapo, Ven. (dĕ-ä-tä-bä'pô)	100	3°58'N	67°41'W
San Fernando de Henares, Spain (dĕ-ä-nä'räs)	129a	40°23'N	3°31'W
Sånfjället, mtn., Swe.	116	62°19'N	13°30'E
Sanford, Can. (săn'fẽrd)	62f	49°41'N	97°27'W
Sanford, Fl., U.S. (săn'fôrd)	65	28°46'N	81°18'W
Sanford, Me., U.S. (săn'fẽrd)	60	43°26'N	70°47'W
Sanford, N.C., U.S.	83	35°26'N	79°10'W
San Francisco, Arg. (săn frăn'sĭs'kô)	102	31°23'S	62°09'W
San Francisco, El Sal.	90	13°48'N	88°11'W

PRONOUNCING INDEX San Francisco – Santa Clara 299

PLACE (Pronunciation)	PAGE	LAT.	LONG.
San Francisco, Ca., U.S.	64	37°45'N	122°26'W
San Francisco, r., N.M., U.S.	77	33°35'N	108°55'W
San Francisco Bay, b., Ca., U.S. (săn frăn'sĭs'kô)	76	37°45'N	122°21'W
San Francisco del Oro, Mex. (dĕl ō'rō)	86	27°00'N	106°37'W
San Francisco del Rincón, Mex. (dĕl rēn-kōn')	88	21°01'N	101°51'W
San Francisco de Macaira, Ven. (dĕ-mä-kī'rä)	101b	9°58'N	66°17'W
San Francisco de Macoris, Dom. Rep. (dä-mä-kō'rēs)	93	19°20'N	70°15'W
San Francisco de Paula, Cuba (dā pou'lä)	93a	23°04'N	82°18'W
San Gabriel, Ca., U.S. (săn gä-brē-ĕl') (gä'brē-ĕl)	75a	34°06'N	118°06'W
San Gabriel, r., Ca., U.S.	75a	33°47'N	118°06'W
San Gabriel Chilac, Mex. (sän-gä-brē-ĕl-chē-läk')	89	18°19'N	97°22'W
San Gabriel Mts, Ca., U.S.	75a	34°17'N	118°03'W
San Gabriel Reservoir, res., Ca., U.S.	75a	34°14'N	117°48'W
Sangamon, r., Il., U.S. (săn'gá-mŭn)	79	40°08'N	90°08'W
Sanger, Ca., U.S. (săng'ẽr)	76	36°42'N	119°33'W
Sangerhausen, Ger. (säng'ẽr-hou-zĕn)	124	51°28'N	11°17'E
Sangha, r., Afr.	185	2°40'N	16°10'E
Sangihe, Pulau, i., Indon.	169	3°30'N	125°30'E
San Gil, Col. (sän-kē'l)	100	6°32'N	73°13'W
San Giovanni in Fiore, Italy (sän jô-vän'nē ēn fyō'rā)	130	39°15'N	16°40'E
San Giuseppe Vesuviano, Italy	129c	40°36'N	14°31'E
Sangju, Kor., S. (säng'jōō')	166	36°20'N	128°07'E
Sāngli, India	155	16°56'N	74°38'E
Sangmélima, Cam.	189	2°56'N	11°59'E
San Gorgonio Mountain, mtn., Ca., U.S. (săn gôr-gō'nĭ-ō)	75a	34°06'N	116°50'W
Sangre de Cristo Mountains, mts., U.S.	64	37°45'N	105°50'W
San Gregoria, Ca., U.S. (săn grĕ-gôr'ā)	74b	37°20'N	122°23'W
Sangro, r., Italy (säŋ'grō)	130	41°38'N	13°56'E
Sangüesa, Spain (sän-gwĕ'sä)	128	42°36'N	1°15'W
Sanhe, China (sän-hŭ)	162	39°59'N	117°06'E
Sanibel Island, i., Fl., U.S. (săn'ĭ-bĕl)	83a	26°26'N	82°15'W
San Ignacio, Belize	90a	17°11'N	89°04'W
San Ildefonso, Cape, c., Phil. (sän-ēl-dĕ-fōn-sō)	169a	16°03'N	122°10'E
San Ildefonso o la Granja, Spain (ō lä grän'khä)	128	40°54'N	4°02'W
San Isidro, Arg. (ē-sē'drō)	99c	34°28'S	58°31'W
San Isidro, C.R.	91	9°24'N	83°43'W
San Jacinto, Phil. (sän hä-sēn'tō)	169a	12°33'N	123°43'E
San Jacinto, Ca., U.S. (săn já-sĭn'tô)	75a	33°47'N	116°57'W
San Jacinto, r., Ca., U.S. (săn já-sĭn'tô)	75a	33°44'N	117°14'W
San Jacinto, r., Tx., U.S.	81	30°25'N	95°05'W
San Jacinto, West Fork, r., Tx., U.S.	81	30°35'N	95°37'W
San Javier, Chile (sän-hà-vē'ĕr)	99b	35°35'S	71°43'W
San Jerónimo, Mex.	89a	19°31'N	98°46'W
San Jerónimo de Juárez, Mex. (hå-rō'nē-mô dā hwä'râz)	88	17°08'N	100°30'W
San Joaquin, Ven.	101b	10°16'N	67°47'W
San Joaquin, r., Ca., U.S. (săn hwä-kēn')	76	37°10'N	120°51'W
San Joaquin Valley, Ca., U.S.	76	36°45'N	120°30'W
San Jorge, Golfo, b., Arg. (gôl-fô-sän-kô'r-kĕ)	102	46°15'S	66°45'W
San José, C.R. (sän hô-sā')	87	9°57'N	84°05'W
San Jose, Phil.	169a	12°22'N	121°04'E
San Jose, Phil.	169a	15°49'N	120°57'E
San Jose, Ca., U.S. (săn hô-zā')	64	37°20'N	121°54'W
San José, i., Mex. (kô-sĕ')	86	25°00'N	110°35'W
San José, Isla de, i., Pan. (ē's-lä-dĕ-sän hô-sā')	91	8°17'N	79°20'W
San Jose, Rio, r., N.M., U.S. (săn hô-zā')	77	35°15'N	108°10'W
San José de Feliciano, Arg. (dā lē ĕs-kĕ'nä)	102	30°26'S	58°44'W
San José de Gauribe, Ven. (sän-hô-sĕ'dĕ-gàōō-rē'bĕ)	101b	9°51'N	65°49'W
San José de las Lajas, Cuba (sän-kô-sĕ'dĕ-läs-lá'käs)	93a	22°58'N	82°10'W
San José Iturbide, Mex. (ē-tōōr-bē'dĕ)	88	21°00'N	100°24'W
San Juan, Arg. (hwän')	102	31°36'S	68°29'W
San Juan, Col. (hòá'n)	100a	3°23'N	73°48'W
San Juan, Dom. Rep. (sän hwän')	93	18°50'N	71°15'W
San Juan, Phil.	169a	16°41'N	120°20'E
San Juan, P.R. (sän hwän')	87	18°30'N	66°10'W
San Juan, prov., Arg.	102	31°00'S	69°30'W
San Juan, r., Mex. (sän-hōō-än')	89	18°10'N	95°23'W
San Juan, r., N.A.	87	10°58'N	84°18'W
San Juan, r., U.S.	64	36°30'N	109°00'W
San Juan, Cabezas de, c., P.R.	87b	18°29'N	65°30'W
San Juan, Cabo, c., Eq. Gui.	190	1°08'N	9°23'E
San Juan, Pico, mtn., Cuba (pē'kô-sän-kòá'n)	92	21°55'N	80°00'W
San Juan, Río, r., Mex. (rē'ō-sän-hwän)	80	25°35'N	99°15'W
San Juan Bautista, Para. (sän hwän' bou-tēs'tä)	102	26°48'S	57°09'W
San Juan Capistrano, Mex. (sän-hōō-än' kä-pēs-trä'nō)	88	22°41'N	104°07'W
San Juan Creek, r., Ca., U.S. (săn hwän')	76	35°24'N	120°12'W
San Juan de Guadalupe, Mex. (sän hwan dā gwä-dhä-lōō'pä)	80	24°37'N	102°43'W
San Juan del Norte, Nic.	91	10°55'N	83°44'W
San Juan del Norte, Bahía de, b., Nic.	91	11°12'N	83°40'W
San Juan de los Lagos, Mex. (sän-hōō-än'dā los lä'gōs)	88	21°15'N	102°18'W
San Juan de los Lagos, r., Mex. (dā lōs lä'gōs)	88	21°13'N	102°12'W
San Juan de los Morros, Ven. (dĕ-lôs-mô'r-rôs)	101b	9°54'N	67°22'W
San Juan del Río, Mex.	88	20°21'N	99°59'W
San Juan del Río, Mex. (sän hwän del rē'ō)	80	24°47'N	104°29'W
San Juan del Sur, Nic. (dĕl sōōr)	86	11°15'N	85°53'W
San Juan Evangelista, Mex. (sän-hōō-ä'n-å-väŋ-kå-lēs'ta')	89	17°57'N	95°08'W
San Juan Island, i., Wa., U.S.	74a	48°28'N	123°08'W
San Juan Islands, is., Can. (sän hwän)	54	48°49'N	123°14'W
San Juan Islands, is., Wa., U.S.	142a	48°36'N	122°50'W
San Juan Ixtenco, Mex. (ēx-tĕ'n-kō)	89	19°14'N	97°52'W
San Juan Martínez, Cuba	92	22°15'N	83°50'W
San Juan Mountains, mts., Co., U.S. (san hwán')	64	37°50'N	107°30'W
San Julián, Arg. (sän hōō-lyä'n)	102	49°17'S	68°02'W
San Justo, Arg. (hōōs'tô)	102a	34°40'S	58°33'W
Sankanbiriwa, mtn., S.L.	188	8°56'N	10°48'W
Sankarani, r., Afr. (sän'kä-rä'nē)	184	11°10'N	8°35'W
Sankt Gallen, Switz.	117	47°25'N	9°22'E
Sankt Moritz, Switz. (sånt mō'rĭts) (zäŋkt mō'rēts)	124	46°31'N	9°50'E
Sankt Pölten, Aus. (zäŋkt-pûl'tĕn)	124	48°12'N	15°38'E
Sankt Veit, Aus. (zäŋkt vīt')	124	46°46'N	14°20'E
Sankuru, r., D.R.C. (sän-kōō'rōō)	186	4°00'S	22°35'E
San Lázaro, Cabo, c., Mex. (sän-lá'zä-rō)	86	24°58'N	113°30'W
San Leandro, Ca., U.S. (săn lē-ăn'drô)	74b	37°43'N	122°10'W
Şanlıurfa, Tur.	154	37°20'N	38°45'E
San Lorenzo, Arg. (sän lô-rĕn'zô)	102	32°46'S	60°44'W
San Lorenzo, Hond. (sän lô-rĕn'zô)	90	13°24'N	87°24'W
San Lorenzo, Ca., U.S. (săn lô-rĕn'zô)	74b	37°41'N	122°08'W
San Lorenzo de El Escorial, Spain	128	40°36'N	4°09'W
Sanlúcar de Barrameda, Spain (sän-lōō'kär)	118	36°46'N	6°21'W
San Lucas, Bol. (lōō'kàs)	100	20°12'S	65°06'W
San Lucas, Cabo, c., Mex.	86	22°45'N	109°45'W
San Luis, Arg. (lò-ēs')	102	33°16'S	66°15'W
San Luis, Col. (lòē's)	100a	6°03'N	74°57'W
San Luis, Cuba	93	20°15'N	75°50'W
San Luis, Guat.	90	14°38'N	89°42'W
San Luis, prov., Arg.	102	32°45'S	66°00'W
San Luis de la Paz, Mex. (dā lä päz')	88	21°17'N	100°32'W
San Luis del Cordero, Mex. (dĕl kôr-dā'rō)	80	25°25'N	104°20'W
San Luis Obispo, Ca., U.S. (ô-bĭs'pō)	64	35°18'N	120°40'W
San Luis Obispo Bay, b., Ca., U.S.	76	35°07'N	121°05'W
San Luis Potosí, Mex.	86	22°08'N	100°58'W
San Luis Potosí, state, Mex.	86	22°45'N	101°45'W
San Luis Rey, r., Ca., U.S. (rā'ē)	76	33°22'N	117°06'W
San Manuel, Az., U.S. (sän măn'ū-ĕl)	77	32°30'N	110°45'W
San Marcial, N.M., U.S. (săn màr-shäl')	77	33°40'N	107°00'W
San Marco, Italy (sän mär'kô)	130	41°53'N	15°50'E
San Marcos, Guat. (mär'kōs)	90	14°57'N	91°49'W
San Marcos, Mex.	88	16°46'N	99°23'W
San Marcos, Tx., U.S. (săn mär'kŏs)	81	29°53'N	97°56'W
San Marcos, r., Tx., U.S.	80	30°08'N	98°15'W
San Marcos de Colón, Hond. (sän-má'r-kōs-dĕ-kô-lô'n)	90	13°17'N	86°50'W
San Maria di Léuca, Cape, c., Italy (dē-lĕ'ōō-kä)	119	39°47'N	18°20'E
San Marino, S. Mar. (sän mä-rē'nô)	130	44°55'N	12°26'E
San Marino, Ca., U.S. (săn mẽr-ē'nô)	75a	34°07'N	118°06'W
San Marino, nation, Eur.	110	43°40'N	13°00'E
San Martín, Col. (sän mär-tē'n)	100a	3°42'N	73°44'W
San Martín, vol., Mex. (mär-tē'n)	89	18°36'N	95°11'W
San Martín, l., S.A.	102	48°15'S	72°30'W
San Martín Chalchicuautla, Mex.	88	21°22'N	98°39'W
San Martin de la Vega, Spain (sän mär ten' dā lä vā'gä)	129a	40°12'N	3°34'W
San Martín Hidalgo, Mex. (sän mär-tē'n-ē-däl'gô)	88	20°27'N	103°55'W
San Mateo, Mex.	89	16°59'N	97°04'W
San Mateo, Ca., U.S. (săn mä-tā'ô)	74b	37°34'N	122°20'W
San Mateo, Ven. (sän mà-tĕ'ô)	101b	9°45'N	64°34'W
San Matías, Golfo, b., Arg. (sän mä-tē'äs)	102	41°30'S	63°45'W
Sanmen Wan, b., China	165	29°00'N	122°15'E
San Miguel, El Sal. (sän mē-gål')	86	13°28'N	88°11'W
San Miguel, Mex. (sän mē-gål')	89	18°18'N	97°09'W
San Miguel, Pan.	91	8°26'N	78°55'W
San Miguel, Phil. (sän mē-gē'l)	169a	15°09'N	120°56'E
San Miguel, Ven. (sän mē-gē'l)	101b	9°56'N	64°58'W
San Miguel, vol., El Sal.	90	13°27'N	88°17'W
San Miguel, i., Ca., U.S.	76	34°03'N	120°23'W
San Miguel, r., Bol. (sän-mē-gĕl')	100	13°34'S	63°58'W
San Miguel, r., N.A. (sän mē-gål')	89	15°27'N	92°00'W
San Miguel, r., Co., U.S. (săn mē-gĕl')	77	38°15'N	108°40'W
San Miguel, Bahía, b., Pan. (bä-ē'ä-sän mē-gål')	91	8°17'N	78°26'W
San Miguel Bay, b., Phil.	169a	13°55'N	123°12'E
San Miguel de Allende, Mex. (dā ä-lyĕn'då)	88	20°54'N	100°44'W
San Miguel el Alto, Mex. (ĕl äl'tō)	88	21°03'N	102°26'W
Sannār, Sudan	185	14°25'N	33°30'E
San Narciso, Phil. (sän när-sē'sō)	169a	15°01'N	120°05'E
San Narciso, Phil.	169a	13°34'N	122°33'E
San Nicolás, Arg. (sän nē-kô-lä's)	102	33°20'S	60°14'W
San Nicolas, Phil. (nē-kô-läs')	169a	16°05'N	120°45'E
San Nicolas, i., Ca., U.S. (săn nĭ'kō-lá)	76	33°14'N	119°10'W
San Nicolás, r., Mex.	88	19°40'N	105°08'W
Sanniquellie, Lib.	188	7°22'N	8°43'W
Sannūr, Wādī, Egypt	192b	28°48'N	31°12'E
Sanok, Pol. (sä'nôk)	125	49°31'N	22°13'E
San Pablo, Phil. (sän-pä-blō)	169a	14°05'N	121°20'E
San Pablo, Ca., U.S. (săn păb'lō)	74b	37°58'N	122°21'W
San Pablo, Ven. (sän-pá'blō)	101b	9°46'N	65°04'W
San Pablo, r., Pan. (sän päb'lō)	91	8°12'N	81°12'W
San Pablo Bay, b., Ca., U.S. (săn păb'lō)	74b	38°04'N	122°25'W
San Pablo Res, Ca., U.S.	74b	37°55'N	122°12'W
San Pascual, Phil. (päs-kwäl')	169a	13°08'N	122°59'E
San Pedro, Arg. (sän pā'drō)	102	24°15'S	64°15'W
San Pedro, Arg.	99c	33°41'S	59°42'W
San Pedro, Chile (sän pĕ'drō)	99b	33°54'S	71°27'W
San Pedro, El Sal. (sän pā'drō)	90	13°49'N	88°58'W
San Pedro, Mex. (sän pā'drō)	89	18°38'N	92°25'W
San Pedro, Para. (sän-pĕ'drō)	102	24°13'S	57°00'W
San Pedro, Ca., U.S. (săn pē'drô)	75a	33°44'N	118°17'W
San Pedro, r., Cuba (sän-pĕ'drō)	92	21°05'N	78°15'W
San Pedro, r., Mex. (sän pā'drō)	88	22°08'N	104°59'W
San Pedro, r., Mex.	80	27°56'N	105°50'W
San Pedro, r., Az., U.S.	77	32°48'N	110°37'W
San Pedro, Río de, r., Mex.	88	21°51'N	102°24'W
San Pedro, Río de, r., N.A.	89	18°23'N	92°13'W
San Pedro Bay, b., Ca., U.S. (săn pē'drô)	75a	33°42'N	118°12'W
San Pedro de las Colonias, Mex. (dĕ-läs-kô-lô'nyäs)	80	25°47'N	102°58'W
San Pedro de Macorís, Dom. Rep. (sän-pĕ'drō-då mä-kô-rēs')	93	18°30'N	69°30'W
San Pedro Lagunillas, Mex. (sän pā'drō lä-gōō-nēl'yäs)	88	21°12'N	104°47'W
San Pedro Sula, Hond. (sän pā'drô sōō'lä)	90	15°29'N	88°01'W
San Pietro, Isola di, i., Italy (ē'sō-lä-dē-sän pyā'trô)	130	39°09'N	8°15'E
San Quentin, Ca., U.S. (săn kwĕn-tĕn')	74b	37°57'N	122°29'W
San Quintin, Phil. (sän kēn-tēn')	169a	15°59'N	120°47'E
San Rafael, Arg. (sän rä-fä-āl')	102	34°30'S	68°13'W
San Rafael, Col. (sän-rä-fä-ĕ'l)	100a	6°18'N	75°02'W
San Rafael, Ca., U.S. (săn rá-fĕl)	74b	37°58'N	122°31'W
San Rafael, r., Ut., U.S. (săn rá-fĕl')	77	39°05'N	110°50'W
San Rafael, Cabo, c., Dom. Rep. (ká'bô)	93	19°00'N	68°50'W
San Ramón, C.R.	91	10°07'N	84°30'W
San Ramon, Ca., U.S. (săn rä-mōn')	74b	37°47'N	122°59'W
San Remo, Italy (sän rā'mô)	130	43°48'N	7°46'E
San Roque, Col. (sän-rô'kĕ)	100a	6°29'N	75°00'W
San Roque, Spain	128	36°13'N	5°23'W
San Saba, Tx., U.S. (săn sä'bá)	80	31°12'N	98°43'W
San Saba, r., Tx., U.S.	80	30°58'N	99°12'W
San Salvador, El Sal. (sän säl-vä-dōr')	86	13°45'N	89°11'W
San Salvador (Watling), i., Bah. (săn săl'và-dôr)	93	24°05'N	74°30'W
San Salvador, i., Ec.	100	0°14'S	90°50'W
San Salvador, r., Ur. (sän-säl-vä-dô'r)	99c	33°42'S	58°04'W
Sansanné-Mango, Togo (sän-sà-nā' män'gō)	184	10°21'N	0°28'E
San Sebastian, Spain (săn sā-bås-tyän')	184	28°09'N	17°11'W
San Sebastián *see* Donostia-San Sebastián, Spain	110	43°19'N	1°59'W
San Sebastian, Ven. (sän-sĕ-bäs-tyá'n)	101b	9°58'N	67°11'W
San Sebastiàn de los Reyes, Spain	129a	40°33'N	3°38'W
San Severo, Italy (sän sĕ-vā'rō)	119	41°43'N	15°24'E
Sanshui, China (sän-shwā)	161	23°14'N	112°51'E
San Simon Creek, r., Az., U.S. (săn sī-mōn')	77	32°45'N	109°30'W
Santa Ana, El Sal.	86	14°02'N	89°35'W
Santa Ana, Mex. (sän'tä ä'nä)	88	19°18'N	98°10'W
Santa Ana, Ca., U.S. (săn'tá ăn'á)	64	33°45'N	117°52'W
Santa Ana, r., Ca., U.S.	75a	33°41'N	117°57'W
Santa Ana Mountains, mts., Ca., U.S.	75a	33°44'N	117°36'W
Santa Anna, Tx., U.S.	80	31°44'N	99°18'W
Santa Antão, i., C.V. (sä-tä-ä'n-zhĕ-lō)	184b	17°20'N	26°05'W
Santa Bárbara, Braz. (sän-tä-bá'r-bä-rä)	101	19°57'S	43°25'W
Santa Bárbara, Hond.	90	14°52'N	88°20'W
Santa Bárbara, Mex.	80	26°48'N	105°50'W
Santa Barbara, Ca., U.S.	64	34°26'N	119°43'W
Santa Barbara, i., Ca., U.S.	76	33°30'N	118°44'W
Santa Barbara Channel, strt., Ca., U.S.	76	34°15'N	120°00'W
Santa Branca, Braz. (sän-tä-brä'N-kä)	99a	23°25'S	45°52'W
Santa Catalina, i., Ca., U.S.	64	33°29'N	118°37'W
Santa Catalina, Cerro de, mtn., Pan.	91	8°39'N	81°36'W
Santa Catalina, Gulf of, b., Ca., U.S. (săn'tá kă-tá-lē'ná)	76	33°00'N	117°58'W
Santa Catarina, Mex. (sän'tä kä-tä-rē'nä)	80	25°41'N	100°27'W
Santa Catarina, state, Braz. (sän-tä-kä-tä-rē'nä)	102	27°15'S	50°30'W
Santa Catarina, r., Mex.	88	16°31'N	98°39'W
Santa Clara, Cuba (sän'tä klä'rä)	87	22°25'N	80°00'W
Santa Clara, Mex.	80	24°29'N	103°22'W
Santa Clara, Ur.	102	32°46'S	54°51'W
Santa Clara, Ca., U.S. (săn'tá klâr'á)	72	37°21'N	121°56'W
Santa Clara, vol., Nic.	90	12°44'N	87°00'W
Santa Clara, r., Ca., U.S. (săn'tá klă'rá)	76	34°22'N	118°53'W
Santa Clara, Bahía de, b., Cuba (bä-ē'ä-dĕ-sän-tä-klä-rä)	92	23°05'N	80°50'W
Santa Clara, Sierra, mts., Mex. (sē-ĕ'r-rä-sän'tä klä'rä)	86	27°30'N	113°50'W

ng-sing; ŋ-baŋk; N-nasalized n; nŏd; cŏmmit; ōld; ôbey; ôrder; oi-boil; fōōd; ŏ-as oo in foot; ou-out; s-soft; sh-dish; th-thin; pūre; ûnite; ûrn; stŭd; circŭs; ü-as in French tu; '-indeterminate vowel.

300 Santa Clara Indian Reservation – Sarnia PRONOUNCING INDEX

PLACE (Pronunciation)	PAGE	LAT.	LONG.
Santa Clara Indian Reservation, I.R., N.M., U.S.	77	35°59'N	106°10'W
Santa Cruz, Bol. (sän'tä krōōz')	100	17°45'S	63°03'W
Santa Cruz, Braz. (sän-tä-krōō's)	102	29°43'S	52°15'W
Santa Cruz, Braz.	102b	22°55'S	43°41'W
Santa Cruz, Chile	99b	34°38'S	71°21'W
Santa Cruz, C.R.	90	10°16'N	85°37'W
Santa Cruz, Mex.	80	25°50'N	105°25'W
Santa Cruz, Phil.	169a	13°28'N	122°02'E
Santa Cruz, Phil.	169a	14°17'N	121°25'E
Santa Cruz, Phil.	169a	15°46'N	119°53'E
Santa Cruz, Ca., U.S.	64	36°59'N	122°02'W
Santa Cruz, prov., Arg.	102	48°00'S	70°00'W
Santa Cruz, i., Ec. (sän-tä-krōō'z)	100	0°38'S	90°20'W
Santa Cruz, r., Arg. (sän'tä krōōz')	102	50°05'S	71°00'W
Santa Cruz, r., Az., U.S. (săn'tȧ krōōz')	77	32°30'N	111°30'W
Santa Cruz Barillas, Guat. (sän-tä-krōō'z-bä-rē'l-yäs)	90	15°47'N	91°22'W
Santa Cruz del Sur, Cuba (sän-tä-krōō's-dĕl-sōō'r)	92	20°45'N	78°00'W
Santa Cruz de Tenerife, Spain (sän'tä krōōz dā tā-nå-rē'fā)	182	28°07'N	15°27'W
Santa Cruz Islands, is., Sol. Is.	175	10°58'S	166°47'E
Santa Cruz Mountains, mts., Ca., U.S. (săn'tȧ krōōz')	74b	37°30'N	122°19'W
Santa Domingo, Cay, i., Bah.	93	21°50'N	75°45'W
Santa Fe, Arg. (sän'tä fā')	102	31°33'S	60°45'W
Santa Fé, Cuba (sän-tä-fē')	92	21°45'N	82°40'W
Santa Fe, Spain (sän'tä-fā')	128	37°12'N	3°43'W
Santa Fe, N.M., U.S. (săn'tȧ fā')	64	35°40'N	106°00'W
Santa Fe, prov., Arg. (sän'tä fā')	102	32°00'S	61°15'W
Santa Fe de Bogotá see Bogotá, Col.	100	4°36'N	74°05'W
Santa Filomena, Braz. (sän-tä-fē-lō-mĕ'nä)	101	9°09'S	44°45'W
Santa Genoveva, mtn., Mex. (sän-tä-hĕ-nō-vĕ'vä)	86	23°30'N	110°00'W
Santai, China (san-tī)	160	31°02'N	105°02'E
Santa Inés, Ven. (sän'tä ē-nĕ's)	101b	9°54'N	64°21'W
Santa Inés, i., Chile (sän'tä ē-nās')	102	53°45'S	74°15'W
Santa Isabel, i., Sol. Is.	175	7°57'S	159°28'E
Santa Isabel, Pico de, mtn., Eq. Gui.	189	3°35'N	8°46'E
Santa Lucía, Cuba (sän'tä lōō-sē'ä)	92	21°15'N	77°30'W
Santa Lucía, Ur. (sän-tä-lōō-sē'ä)	102	34°27'S	56°23'W
Santa Lucía, Ven.	101b	10°18'N	66°40'W
Santa Lucía, r., Ur.	99c	34°19'S	56°13'W
Santa Lucia Bay, b., Cuba (sän'tä lōō-sē'ä)	92	22°55'N	84°20'W
Santa Margarita, i., Mex. (sän'tä mär-gä-rē'tä)	86	24°15'N	112°00'W
Santa Maria, Braz. (sän'tä mä-rē'ä)	102	29°40'S	54°00'W
Santa Maria, Italy (sän-tä mä-rē'ä)	130	41°05'N	14°15'E
Santa María, Phil. (sän-tä-mä-rē'ä)	169a	14°48'N	120°57'E
Santa Maria, Ca., U.S. (săn-tȧ mȧ-rē'ȧ)	76	34°57'N	120°28'W
Santa María, vol., Guat.	90	14°45'N	91°33'W
Santa María, r., Mex. (sän'tä mä-rē'ä)	88	21°33'N	100°17'W
Santa Maria, Cabo de, c., Port. (kä'bō-dĕ-sän-tä-mä-rē'ä)	128	36°58'N	7°54'W
Santa Maria, Cape, c., Bah.	93	23°45'N	75°30'W
Santa María, Cayo, i., Cuba	92	22°40'N	79°00'W
Santa María del Oro, Mex. (sän'tä-mä-rē'ä-dĕl-ô-rō)	88	21°21'N	104°35'W
Santa María de los Ángeles, Mex. (dĕ-lôs-ä'n-hĕ-lĕs)	88	22°10'N	103°34'W
Santa María del Río, Mex.	88	21°46'N	100°43'W
Santa María de Ocotán, Mex.	88	22°56'N	104°30'W
Santa Maria Island, i., Port. (sän-tä-mä-rē'ä)	184a	37°09'N	26°02'W
Santa Maria Madalena, Braz.	99a	22°00'S	42°00'W
Santa Marta, Col. (sän'tä mär'tä)	100	11°15'N	74°13'W
Santa Marta, Cabo de, c., Ang.	190	13°52'S	12°25'E
Santa Monica, Ca., U.S. (săn'tȧ mŏn'ĭ-kȧ)	64	34°01'N	118°29'W
Santa Monica Mountains, mts., Ca., U.S.	75a	34°08'N	118°38'W
Santana, r., Braz. (sän-tä'nä)	102b	22°33'S	43°37'W
Santander, Col. (sän-tän-dĕr')	100a	3°00'N	76°25'W
Santander, Spain (sän-tän-där')	110	43°27'N	3°50'W
Sant Antoni de Portmany, Spain	129	38°59'N	1°17'E
Santa Paula, Ca., U.S. (săn'tȧ pô'lȧ)	76	34°24'N	119°05'W
Santarém, Braz. (sän-tä-rĕn')	101	2°28'S	54°37'W
Santarém, Port.	128	39°18'N	8°48'W
Santaren Channel, strt., Bah. (sän-tȧ-rĕn')	92	24°15'N	79°30'W
Santa Rita do Sapucai, Braz. (sä-pō-kä'ē)	99a	22°15'S	45°41'W
Santa Rosa, Arg. (sän-tä-rō-sä)	102	36°45'S	64°10'W
Santa Rosa, Col. (sän-tä-rō-sä)	100a	6°38'N	75°26'W
Santa Rosa, Ec.	100	3°29'S	79°55'W
Santa Rosa, Guat. (sän'tä rō'sȧ)	90	14°21'N	90°16'W
Santa Rosa, Hond.	90	14°45'N	88°51'W
Santa Rosa, Ca., U.S. (săn'tȧ rō'zȧ)	64	38°27'N	122°42'W
Santa Rosa, N.M., U.S. (săn'tȧ rō'sȧ)	78	34°55'N	104°41'W
Santa Rosa, Ven. (sän-tä-rō-sä)	101b	9°37'N	64°10'W
Santa Rosa de Cabal, Col. (sän-tä-rō-sä-dĕ-kä-bä'l)	100a	4°53'N	75°38'W
Santa Rosa de Viterbo, Braz. (sän-tä-rō-sä-dĕ-vē-tĕr'-bō)	99a	21°30'S	47°21'W
Santa Rosa Indian Reservation, I.R., Ca., U.S. (săn'tȧ rō'zȧ')	76	33°28'N	116°50'W
Santa Rosalía, Mex. (sän'tä rō-zä'lē-ä)	86	27°13'N	112°15'W
Santa Rosa Range, mts., Nv., U.S. (săn'tȧ rō'zȧ)	72	41°33'N	117°50'W
Santa Susana, Ca., U.S. (săn'tȧ sōō-zä'nȧ)	75a	34°16'N	118°42'W
Santa Teresa, Arg. (sän-tä-tĕ-rĕ'sä)	99c	33°27'S	60°47'W
Santa Teresa, Ven.	101b	10°14'N	66°40'W
Santa Uxia, Spain	128	42°34'N	8°55'W
Santa Vitória do Palmar, Braz. (sän-tä-vē-tō'ryä-dō-päl-mär)	102	33°30'S	53°16'W
Santa Ynez, r., Ca., U.S. (săn'tȧ ē-nĕz')	76	34°40'N	120°20'W
Santa Ysabel Indian Reservation, I.R., Ca., U.S. (săn-tȧ ĭ-zȧ-bĕl')	76	33°05'N	116°46'W
Santee, Ca., U.S. (săn tē')	76a	32°50'N	116°58'W
Santee, r., S.C., U.S.	65	33°00'N	79°45'W
Sant' Eufemia, Golfo di, b., Italy (gôl-fō-dē-sän-tĕ'ōō-fĕ'myä)	130	38°53'N	15°53'E
Sant Feliu de Guixols, Spain	129	41°45'N	3°01'E
Santiago, Braz. (sän-tyá'gô)	102	29°05'S	54°46'W
Santiago, Chile (sän-tē-ä'gō)	102	33°26'S	70°40'W
Santiago, Pan.	87	8°07'N	80°58'W
Santiago, Phil. (sän-tyä'gō)	169a	16°42'N	121°33'E
Santiago, prov., Chile (sän-tyá'gō)	99b	33°28'S	70°55'W
Santiago, i., Phil.	169a	16°29'N	120°03'E
Santiago de Compostela, Spain	118	42°52'N	8°32'W
Santiago de Cuba, Cuba (sän-tyá'gô-dā kōō'bȧ)	87	20°00'N	75°50'W
Santiago de Cuba, prov., Cuba	92	20°20'N	76°05'W
Santiago de las Vegas, Cuba (sän-tyá'gô-dĕ-läs-vĕ'gäs)	93a	22°58'N	82°23'W
Santiago del Estero, Arg.	102	27°50'S	64°14'W
Santiago del Estero, prov., Arg. (sän-tē-á'gō-dĕl ĕs-tā-rō)	102	27°15'S	63°30'W
Santiago de los Caballeros, Dom. Rep.	87	19°30'N	70°45'W
Santiago Mountains, mts., Tx., U.S. (sän-tē-ä'gō)	64	30°00'N	103°30'W
Santiago Reservoir, res., Ca., U.S.	75a	33°47'N	117°42'W
Santiago Rodríguez, Dom. Rep. (sän-tyá'gô-rô-drē'gĕz)	93	19°30'N	71°25'W
Santiago Tuxtla, Mex. (sän-tyá'gô-tōō'x-tlä)	89	18°28'N	95°18'W
Santiaguillo, Laguna de, l., Mex. (lä-oō'nä-dĕ-sän-tē-ä-gēl'yō)	80	24°51'N	104°43'W
Santisteban del Puerto, Spain (sän'tē stå-bän'dĕl pwĕr'tō)	128	38°15'N	3°12'W
Sant Mateu, Spain	129	40°26'N	0°09'E
Santo Amaro, Braz. (sän'tō ä-mä'rō)	101	12°32'S	38°33'W
Santo Amaro de Campos, Braz.	99a	22°01'S	41°05'W
Santo André, Braz.	99a	23°40'S	46°31'W
Santo Angelo, Braz. (sän-tô-ä'n-zhĕ-lô)	102	28°16'S	53°59'W
Santo Antônio do Monte, Braz. (sän-tō-än-tō'nyō-dō-môn'tĕ)	99a	20°06'S	45°18'W
Santo Domingo, Cuba (sän'tô-dōmĭn'gô)	92	22°35'N	80°20'W
Santo Domingo, Dom. Rep. (sän'tô dô-mĭn'gô)	87	18°30'N	69°55'W
Santo Domingo, Nic. (sän-tô-dô-mē'n-gō)	91	12°15'N	84°56'W
Santo Domingo de la Calzada, Spain (dā lä käl-thä'dä)	128	42°27'N	2°55'W
Santoña, Spain (sän-tō'nyä)	128	43°25'N	3°27'W
Santos, Braz. (sän'tozh)	101	23°58'S	46°20'W
Santos Dumont, Braz. (sän'tôs-dō-mô'nt)	101	21°28'S	43°33'W
Sanuki, Japan (sä'nōō-kē)	167a	35°16'N	139°53'E
San Urbano, Arg. (sän-ōr-bä'nō)	99c	33°39'S	61°28'W
San Valentin, Monte, mtn., Chile (sän-vä-lĕn-tē'n)	102	46°41'S	73°30'W
San Vicente, Arg. (sän-vē-sĕn'tĕ)	99c	35°00'S	58°26'W
San Vicente, Chile	99b	34°25'S	71°06'W
San Vicente, El Sal. (sän vē-sĕn'tå)	90	13°41'N	88°43'W
San Vicente de Alcántara, Spain	128	39°24'N	7°08'W
San Vito al Tagliamento, Italy (san vē'tô)	130	45°53'N	12°52'E
San Xavier Indian Reservation, I.R., Az., U.S. (x-ä'vĭēr)	77	32°07'N	111°12'W
San Ysidro, Ca., U.S. (săn ysĭ-drō')	76a	32°33'N	117°02'W
Sanyuanli, China (sän-yûän-lē)	163a	23°11'N	113°16'E
São Bernardo do Campo, Braz. (soun-bĕr-nár'dō-dō-kä'm-pô)	99a	23°44'S	46°33'W
São Borja, Braz. (soun-bôr-zhä)	102	28°44'S	55°59'W
São Carlos, Braz. (soun kär'lōzh)	101	22°02'S	47°54'W
São Cristovão, Braz. (soun-krēs-tō-voun)	101	11°04'S	37°11'W
São Fidélis, Braz. (soun-fē-dĕ'lĕs)	99a	21°41'S	41°45'W
São Francisco, Braz. (soun frän-sēsh'kō)	101	15°59'S	44°42'W
São Francisco, r., Braz. (sän-frän-sē's-kō)	101	8°56'S	40°20'W
São Francisco do Sul, Braz. (soun frän-sēsh'kō-dō-sōō'l)	102	26°15'S	48°42'W
São Gabriel, Braz. (soun'gä-brē-ĕl')	102	30°28'S	54°11'W
São Geraldo, Braz. (soun-zhĕ-rä'l-dō)	99a	21°01'S	42°49'W
São Gonçalo, Braz. (soun'gōŋ-sä'lō)	99a	22°55'S	43°04'W
Sao Hill, Tan.	191	8°20'S	35°12'E
São João, Gui.-B.	188	11°32'N	15°26'W
São João da Barra, Braz. (soun-zhōun-dä-bà'rä)	99a	21°40'S	41°03'W
São João da Boa Vista, Braz. (soun-zhōun-dä-bôä-vē's-tä)	99a	21°58'S	46°45'W
São João del Rei, Braz. (soun zhô-oun'dĕl-rā)	102	21°08'S	44°14'W
São João de Meriti, Braz. (soun-zhōun-dĕ-mĕ-rē-tē)	102b	22°47'S	43°22'W
São João do Araguaia, Braz. (soun zhô-oun'dô-ä-rä-gwä'yä)	101	5°29'S	48°44'W
São João dos Lampas, Port. (soun' zhô-oun' dōzh län-päzh')	129b	38°52'N	9°24'W
São João Nepomuceno, Braz. (soun-zhōun-nĕ-pô-mōō-sĕ-nō)	99a	21°33'S	43°00'W
São Jorge Island, i., Port. (soun zhôr'zhĕ)	184a	38°28'N	27°34'W
São José do Rio Pardo, Braz. (soun-zhô-sĕ'dô-rē'ō-pá'r-dō)	99a	21°36'S	46°50'W
São José do Rio Prêto, Braz. (soun zhô-zĕ'dô-re'ō-prĕ-tō)	101	20°57'S	49°12'W
São José dos Campos, Braz. (soun zhô-zā'dōzh kän pōzh')	99a	23°12'S	45°53'W
São Leopoldo, Braz. (soun-lĕ-ô-pôl'dō)	102	29°46'S	51°09'W
São Luis, Braz.	101	2°31'S	43°14'W
São Luis do Paraitinga, Braz. (soun-lōōē's-dō-pä-rä-ē-tē'n-gä)	99a	23°15'S	45°18'W
São Manuel, r., Braz.	101	8°28'S	57°07'E
São Mateus, Braz. (soun mä-tā'ōzh)	101	18°44'S	39°45'W
São Mateus, Braz.	102b	22°49'S	43°23'W
São Miguel Arcanjo, Braz. (soun-mē-gĕ'l-ár-kän-zhō)	99a	23°54'S	47°59'W
São Miguel Island, i., Port.	184a	37°59'N	26°38'W
Saona, i., Dom. Rep. (sä-ō'nä)	93	18°10'N	68°55'W
Saône, r., Fr. (sōn)	112	47°00'N	5°30'E
São Nicolau, i., C.V. (soun' nĕ-kô-loun')	184b	16°19'N	25°19'W
São Paulo, Braz. (soun' pou'lō)	101	23°34'S	46°38'W
São Paulo, state, Braz. (soun pou'lō)	101	21°45'S	50°47'W
São Paulo de Olivença, Braz. (soun'pou'lōdā ô-lē-vĕn'sȧ)	100	3°32'S	68°46'W
São Pedro, Braz. (soun-pĕ'drō)	99a	22°34'S	47°54'W
São Pedro de Aldeia, Braz. (soun-pĕ'drō-dĕ-äl-dĕ'yä)	99a	22°50'S	42°04'W
São Pedro e São Paulo, Rocedos, rocks, Braz.	97	1°50'N	30°00'W
São Raimundo Nonato, Braz. (soun' rī-mô'n-do nō-nä'tō)	101	9°09'S	42°32'W
São Roque, Braz. (soun' rō'kĕ)	99a	23°32'S	47°08'W
São Roque, Cabo de, c., Braz. (kä'bō-dĕ-soun' rō'kĕ)	101	5°06'S	35°11'W
São Sebastião, Braz. (soun så-bäs-tē-oun')	99a	23°48'S	45°25'W
São Sebastião, Ilha de, i., Braz.	99a	23°52'S	45°22'W
São Sebastião do Paraíso, Braz.	99a	20°54'S	46°58'W
São Simão, Braz. (soun-sē-moun)	99a	21°30'S	47°33'W
São Tiago, i., C.V. (soun tē-ä'gō)	184b	15°09'N	24°45'W
São Tomé, S. Tom./P.	184	0°20'N	6°44'E
Sao Tome and Principe, nation, Afr. (prēn'sē-pē)	184	1°00'N	6°00'E
Saoura, Oued, r., Alg.	184	29°39'N	1°42'W
São Vicente, Braz. (soun ve-se'n-tē)	101	23°57'S	46°25'W
São Vicente, i., C.V. (soun vē-sĕn'tå)	184b	16°51'N	24°35'W
São Vicente, Cabo de, c., Port. (kà'bō-dĕ-sän-vē-sĕ'n-tĕ)	112	37°03'N	9°31'W
Sapele, Nig. (sä-pā'lā)	184	5°54'N	5°41'E
Sapitwa, mtn., Mwi.	191	15°58'S	35°38'E
Sa Pobla, Spain	129	39°46'N	3°02'E
Sapozhok, Russia (sä-pô-zhôk')	132	53°58'N	40°44'E
Sapporo, Japan (säp-pô'rō)	161	43°02'N	141°29'E
Sapronovo, Russia (sap-rô'nō-vō)	142b	55°13'N	38°25'E
Sapucaí, r., Braz. (sä-pōō-kȧ-ē')	99a	22°20'S	45°53'W
Sapucaia, Braz. (sä-pōō-kä'yȧ)	99a	22°01'S	42°54'W
Sapucaí Mirim, r., Braz. (sä-pōō-kȧ-ē'mē-rēn)	99a	21°06'S	47°03'W
Sapulpa, Ok., U.S. (sȧ-pŭl'pȧ)	79	36°01'N	96°05'W
Saqqez, Iran	157	36°14'N	46°16'E
Saquarema, Braz. (sä-kwä-rĕ-mä)	99a	22°56'S	42°32'W
Sara, Wa., U.S. (sä'rä)	74c	45°45'N	122°42'W
Sara, Bahr, r., Chad (bär)	185	8°19'N	17°44'E
Sarajevo, Bos. (sä-rà-yĕv'ô) (sá-rá'ya-vô)	110	43°50'N	18°26'E
Sarakhs, Iran	157	36°32'N	61°11'E
Sarana, Russia (sá-rä'nȧ)	142a	56°31'N	57°44'E
Saranac Lake, N.Y., U.S.	67	44°20'N	74°05'W
Saranac Lake, l., N.Y., U.S. (săr'ȧ-năk)	67	44°15'N	74°20'W
Sarandi, Arg. (sä-rän'dē)	102a	34°41'S	58°21'W
Sarandi Grande, Ur. (sä-rän'dē-grän'dĕ)	99c	33°42'S	56°21'W
Saranley, Som.	192a	2°28'N	42°15'E
Saransk, Russia (sá-ränsk')	134	54°10'N	45°10'E
Sarany, Russia (sá-rä'nĭ)	142a	58°33'N	58°48'E
Sara Peak, mtn., Nig.	189	9°37'N	9°25'E
Sarapul, Russia (sä-räpōl')	136	56°28'N	53°50'E
Sarasota, Fl., U.S. (săr-ȧ-sōtȧ)	83a	27°27'N	82°30'W
Saratoga, Tx., U.S. (săr-ȧ-tō'gȧ)	81	30°17'N	94°31'W
Saratoga, Wa., U.S.	74a	48°04'N	122°29'W
Saratoga Pass, Wa., U.S.	74a	48°09'N	122°33'W
Saratoga Springs, N.Y., U.S. (sprĭngz)	67	43°05'N	74°50'W
Saratov, Russia (sá rä'tôf)	134	51°30'N	45°30'E
Saravane, Laos	165	15°48'N	106°40'E
Sarawak, hist. reg., Malay. (sá-rä'wäk)	168	2°30'N	112°45'E
Sárbogárd, Hung. (shär'bō-gärd)	125	46°53'N	18°38'E
Sarcee Indian Reserve, I.R., Can. (sär'sē)	62e	50°58'N	114°23'W
Sarcelles, Fr.	127b	49°00'N	2°23'E
Sardalas, Libya	184	25°59'N	10°33'E
Sardinia, i., Italy (sär-dĭn'ĭȧ)	112	40°08'N	9°05'E
Sardis, Ms., U.S. (sär'dĭs)	82	34°26'N	89°55'W
Sardis Lake, res., Ms., U.S.	82	34°27'N	89°43'W
Sargent, Ne., U.S. (sär'jĕnt)	70	41°40'N	99°38'W
Sarh, Chad (är-chan-bô')	185	9°09'N	18°23'E
Sarikamis, Tur.	137	40°30'N	42°40'E
Sariñena, Spain (sä-rēn-yĕ'nä)	129	41°46'N	0°11'W
Sark, i., Guern. (särk)	126	49°28'N	2°22'W
Şarköy, Tur. (shär'kû-ē)	131	40°39'N	27°07'E
Sarmiento, Monte, mtn., Chile (mô'n-tĕ-sär-myĕn'tō)	102	54°28'S	70°40'W
Sarnia, Can. (sär'nē-ȧ)	51	43°00'N	82°25'W

PRONOUNCING INDEX
Sarno – Semichi Islands 301

PLACE (Pronunciation)	PAGE	LAT.	LONG.
Sarno, Italy (sä'r-nô)	129c	40°35'N	14°38'E
Sarny, Ukr. (sär'nē)	137	51°17'N	26°39'E
Saronikós Kólpos, b., Grc.	131	37°51'N	23°30'E
Saros Körfezi, b., Tur. (sä'rôs)	131	40°30'N	26°20'E
Sárospatak, Hung. (shä'rôsh-pô'tôk)	125	48°19'N	21°35'E
Šar Planina, mts., Serb. (shär plä'nē-na)	131	42°07'N	21°54'E
Sarpsborg, Nor. (särps'bôrg)	122	59°17'N	11°07'E
Sarrebourg, Fr. (sär-bōōr')	127	48°44'N	7°02'E
Sarreguemines, Fr. (sär-gē-mēn')	117	49°06'N	7°05'E
Sarria, Spain (sär'ē-ä)	118	42°14'N	7°17'W
Sarstun, r., N.A. (särs-tōō'n)	90	15°50'N	89°26'W
Sartène, Fr. (sär-tĕn')	130	41°36'N	8°59'E
Sarthe, r., Fr. (särt)	117	47°44'N	0°32'W
Şärur, Azer.	138	39°33'N	44°58'E
Sárvár, Hung. (shär'vär)	124	47°14'N	16°55'E
Sarych, Mys, c., Ukr. (mĭs sá-rēch')	137	44°25'N	33°00'E
Saryesik-Atyraū, des., Kaz.	139	45°30'N	76°00'E
Sary-Ishikotrau, Peski, des., Kyrg. (sä'rē ē' shĕk-ō'trou)	139	46°12'N	75°30'E
Sarysū, r., Kaz. (sä'rē-sōō)	139	47°47'N	69°14'E
Sasarām, India (sŭs-ŭ-räm')	155	25°00'N	84°00'E
Sasayama, Japan (sä'sä-yä'mä)	167	35°05'N	135°14'E
Sasebo, Japan (sä'sâ-bô)	161	33°12'N	129°43'E
Saskatchewan, prov., Can.	50	54°46'N	107°40'W
Saskatchewan, r., Can. (săs-kăch'ē-wän)	52	53°45'N	103°20'W
Saskatoon, Can. (săs-kȧ-tōōn')	50	52°07'N	106°38'W
Sasolburg, S. Afr.	192c	26°52'S	27°47'E
Sasovo, Russia (sás'ô-vô)	136	54°20'N	42°00'E
Saspamco, Tx., U.S. (săs-păm'cō)	75d	29°13'N	98°18'W
Sassandra, C. Iv.	188	4°58'N	6°05'W
Sassandra, r., C. Iv. (sȧs-sän'drȧ)	184	5°35'N	6°25'W
Sassari, Italy (säs'sä-rē)	118	40°44'N	8°33'E
Sassnitz, Ger. (säs'nēts)	124	54°31'N	13°37'E
Satadougou, Mali (sä-tá-dōō-gōō')	188	12°21'N	12°07'W
Säter, Swe. (sĕ'tĕr)	122	60°21'N	15°50'E
Satilla, r., Ga., U.S. (sȧ-tĭl'ȧ)	83	31°15'N	82°13'W
Satka, Russia (sät'kȧ)	136	55°03'N	59°02'E
Sátoraljaujhely, Hung. (shä'tô-rô-lyô-ōō'yĕl')	125	48°24'N	21°40'E
Satu Mare, Rom. (sȧ'tōō-mȧ'rē)	119	47°50'N	22°53'E
Saturna, Can. (să-tûr'nȧ)	74d	48°48'N	123°12'W
Saturna, i., Can.	74d	48°47'N	123°03'W
Sauda, Nor.	116	59°40'N	6°21'E
Saudárkrókur, Ice.	110	65°41'N	19°38'W
Saudi Arabia, nation, Asia (sȧ-ô'dĭ ȧ-rā'bĭ-ȧ)	154	22°40'N	46°00'E
Sauerlach, Ger. (zou'ĕr-läĸ)	115d	47°58'N	11°39'E
Saugatuck, Mi., U.S. (sô'gȧ-tŭk)	66	42°40'N	86°10'W
Saugeen, r., Can.	58	44°20'N	81°20'W
Saugerties, N.Y., U.S. (sô'gĕr-tēz)	67	42°05'N	73°55'W
Saugus, Ma., U.S. (sô'gŭs)	61a	42°28'N	71°01'W
Sauk, r., Mn., U.S. (sôk)	71	45°30'N	94°45'W
Sauk Centre, Mn., U.S.	71	45°43'N	94°58'W
Sauk City, Wi., U.S.	71	43°16'N	89°45'W
Sauk Rapids, Mn., U.S. (răp'ĭd)	71	45°35'N	94°08'W
Sault Sainte Marie, Can.	51	46°31'N	84°20'W
Sault Sainte Marie, Mi., U.S. (sōō sānt mȧ-rē')	65	46°29'N	84°21'W
Saumatre, Étang, l., Haiti	93	18°40'N	72°10'W
Saunders Lake, l., Can. (sän'dĕrs)	62g	53°18'N	113°25'W
Saurimo, Ang.	186	9°39'S	20°24'E
Sausalito, Ca., U.S. (sô-sȧ-lē'tô)	74b	37°51'N	122°29'W
Sausset-les-Pins, Fr. (sō-sĕ'lā-păɴ')	126a	43°20'N	5°08'E
Saútar, Ang.	190	11°06'S	18°27'E
Sauvie Island, i., Or., U.S. (sô'vē)	74c	45°43'N	123°49'W
Sava, r., Serb. (sä'vä)	112	44°50'N	18°30'E
Savage, Md., U.S. (săv'ĕj)	68e	39°07'N	76°49'W
Savage, Mn., U.S.	75g	44°47'N	93°20'W
Savai'i, i., Samoa	170a	13°35'S	172°25'W
Savalen, l., Nor.	122	62°19'N	10°15'E
Savalou, Benin	184	7°56'N	1°58'E
Savanna, Il., U.S. (sȧ-văn'ȧ)	71	42°05'N	90°09'W
Savannah, Ga., U.S. (sȧ-văn'ȧ)	65	32°04'N	81°07'W
Savannah, Mo., U.S.	79	39°58'N	94°49'W
Savannah, Tn., U.S.	82	35°13'N	88°14'W
Savannah, r., U.S.	65	33°11'N	81°51'W
Savannakhét, Laos	168	16°33'N	104°45'E
Savanna la Mar, Jam. (sȧ-vän'ȧ lȧ mär')	92	18°10'N	78°10'W
Save, r., Fr.	126	43°32'N	0°50'E
Save, Rio (Sabi), r., Afr. (rē'ō-sä'vĕ)	186	21°28'S	34°14'E
Säveh, Iran	157	35°01'N	50°20'E
Saverne, Fr. (sȧ-vĕrn')	127	48°40'N	7°22'E
Savigliano, Italy (sä-vēl-yä'nô)	130	44°38'N	7°42'E
Savigny-sur-Orge, Fr.	127b	48°41'N	2°22'E
Savona, Italy (sä-nō'nä)	118	44°19'N	8°28'E
Savonlinna, Fin. (sȧ'vôn-lēn'nȧ)	123	61°53'N	28°49'E
Savran', Ukr. (säv-rän')	133	48°07'N	30°09'E
Sawahlunto, Indon.	168	0°37'S	100°50'E
Sawākin, Sudan	185	19°02'N	37°19'E
Sawda, Jabal as, mts., Libya	185	28°14'N	13°46'E
Sawhāj, Egypt	185	26°34'N	31°40'E
Sawknah, Libya	185	29°04'N	15°53'E
Sawu, Laut (Savu Sea), sea, Asia	168	9°15'S	122°15'E
Sawyer, l., Wa., U.S. (sô'yĕr)	74a	47°20'N	122°02'W
Saxony see Sachsen, hist. reg., Ger.	124	50°45'N	12°17'E
Say, Niger (sä'ē)	184	13°09'N	2°16'E
Sayan Khrebet, mts., Russia (sŭ-yän')	135	51°30'N	90°00'E
Sayhūt, Yemen	154	15°23'N	51°28'E
Sayre, Ok., U.S. (sā'ĕr)	78	35°19'N	99°40'W
Sayre, Pa., U.S.	67	41°55'N	76°30'W
Sayreton, Al., U.S. (sā'ĕr-tŭn)	68h	33°34'N	86°51'W
Sayreville, N.J., U.S. (sâr'vĭl)	68a	40°28'N	74°21'W
Sayr Usa, Mong.	160	44°15'N	107°00'E
Sayula, Mex. (sä-yōō'lä)	89	17°51'N	94°56'W
Sayula, Mex.	88	19°50'N	103°33'W
Sayula, Laguna de, l., Mex. (lä-gô'nä-dē)	88	20°00'N	103°33'W
Say'un, Yemen	154	16°00'N	48°59'E
Sayville, N.Y., U.S. (sā'vĭl)	67	40°45'N	73°10'W
Sazanit, i., Alb.	119	40°30'N	19°17'E
Sázava, r., Czech Rep.	124	49°36'N	15°24'E
Sazhino, Russia (sàz-hē'nô)	142a	56°20'N	58°15'E
Scandinavian Peninsula, pen., Eur.	152	62°00'N	14°00'E
Scanlon, Mn., U.S. (skăn'lôn)	75h	46°27'N	92°26'W
Scappoose, Or., U.S. (skă-pōōs')	74c	45°46'N	122°53'W
Scappoose, r., Or., U.S.	74c	45°47'N	122°57'W
Scarborough, Eng., U.K. (skär'bŭr-ô)	120	54°16'N	0°19'W
Scarsdale, N.Y., U.S. (skärz'dāl)	68a	41°01'N	73°47'W
Scatari I, Can. (skăt'ȧ-rē)	61	46°00'N	59°44'W
Schaerbeek, Bel. (skär'bāk)	115a	50°50'N	4°23'E
Schaffhausen, Switz. (shäf'hou-zĕn)	117	47°42'N	8°38'E
Schefferville, Can.	51	54°52'N	67°01'W
Schelde,, r., Eur.	121	51°04'N	3°55'E
Schenectady, N.Y., U.S. (skĕ-nĕk'tá-dē)	65	42°50'N	73°55'W
Scheveningen, Neth.	115a	52°06'N	4°15'E
Schiedam, Neth.	115a	51°55'N	4°23'E
Schiltigheim, Fr. (shĕl'tegh-hĭm)	127	48°48'N	7°47'E
Schio, Italy (skē'ô)	130	45°43'N	11°23'E
Schleswig, Ger. (shlĕs'vĕgh)	116	54°32'N	9°32'E
Schleswig, hist. reg., Ger. (shlĕs'vĕgh)	124	54°40'N	9°10'E
Schleswig-Holstein, state, Ger. (shlĕs'vĕgh-hōl'shtīn)	115c	53°40'N	9°45'E
Schmalkalden, Ger. (shmäl'käl-dĕn)	124	50°41'N	10°25'E
Schneider, In., U.S. (schnīd'ĕr)	69a	41°12'N	87°26'W
Schofield, Wi., U.S. (skō'fēld)	71	44°52'N	89°37'W
Schönebeck, Ger. (shû'nē-bergh)	124	52°01'N	11°44'E
Schoonhoven, Neth.	115a	51°56'N	4°51'E
Schramberg, Ger. (shräm'bĕrgh)	124	48°14'N	8°24'E
Schreiber, Can.	58	48°50'N	87°10'W
Schroon, l., N.Y., U.S. (skrōōn)	67	43°50'N	73°50'W
Schultzendorf, Ger. (shōōl'tzĕn-dōrf)	115b	52°21'N	13°55'E
Schumacher, Can.	58	48°30'N	81°30'W
Schuyler, Ne., U.S. (slī'lĕr)	70	41°28'N	97°05'W
Schuylkill, r., Pa., U.S. (skōōl'kĭl)	68f	40°10'N	75°31'W
Schuylkill-Haven, Pa., U.S. (skōōl'kĭl hā-vĕn)	67	40°35'N	76°10'W
Schwabach, Ger. (shvä'bäk)	124	49°19'N	11°02'E
Schwäbische Alb, mts., Ger. (shvă'bē-shĕ älb)	124	48°11'N	9°09'E
Schwäbisch Gmünd, Ger. (shvă'bĕsh gmünd)	124	48°47'N	9°49'E
Schwäbisch Hall, Ger. (häl)	124	49°08'N	9°44'E
Schwandorf, Ger. (shvän'dôrf)	124	49°19'N	12°08'E
Schwaner, Pegunungan, mts., Indon. (skvän'ĕr)	168	1°05'S	112°30'E
Schwarzwald, for., Ger. (shvärts'väld)	124	47°54'N	7°57'E
Schwaz, Aus.	124	47°20'N	11°45'E
Schwechat, Aus. (shvĕk'ät)	124	48°09'N	16°29'E
Schwedt, Ger. (shvĕt)	124	53°04'N	14°17'E
Schweinfurt, Ger. (shvīn'fōrt)	124	50°03'N	10°14'E
Schwelm, Ger. (shvĕlm)	127c	51°17'N	7°18'E
Schwerin, Ger. (shvĕ-rēn')	124	53°36'N	11°25'E
Schweriner See, l., Ger. (shvĕ'rē-nĕr zā)	124	53°40'N	11°06'E
Schwerte, Ger. (shvĕr'tē)	127c	51°26'N	7°34'E
Schwielowsee, l., Ger. (shvē'lōv zā)	115b	52°20'N	12°52'E
Schwyz, Switz. (schēts)	124	47°01'N	8°38'E
Sciacca, Italy (shē-äk'kä)	130	37°30'N	13°09'E
Scilly, Isles of, is., Eng., U.K. (sĭl'ē)	112	49°56'N	6°50'W
Scioto, r., Oh., U.S. (sī-ō'tō)	65	39°10'N	82°55'W
Scituate, Ma., U.S. (sĭt'ū-āt)	61a	42°12'N	70°45'W
Scobey, Mt., U.S. (skō'bē)	73	48°48'N	105°29'W
Scoggin, Or., U.S. (skō'gĭn)	74c	45°28'N	123°14'W
Scotch, r., Can. (skŏch)	62c	45°21'N	74°56'W
Scotia, Ca., U.S. (skō'shȧ)	72	40°29'N	124°06'W
Scotland, S.D., U.S.	70	43°08'N	97°43'W
Scotland, state, U.K. (skŏt'lȧnd)	110	57°05'N	5°10'W
Scotland Neck, N.C., U.S. (nĕk)	83	36°06'N	77°25'W
Scotstown, Can. (skŏts'toun)	67	45°35'N	71°15'W
Scott, r., Ca., U.S.	72	41°20'N	122°55'W
Scott, Cape, c., Can. (skŏt)	52	50°47'N	128°26'W
Scott, Mount, mtn., Or., U.S.	74c	45°27'N	122°33'W
Scott, Mount, mtn., Or., U.S.	72	42°55'N	122°00'W
Scott Air Force Base, Il., U.S.	75e	38°33'N	89°52'W
Scottburgh, S. Afr. (skŏt'bŭr-ô)	186	30°18'S	30°42'E
Scott City, Ks., U.S.	78	38°28'N	100°54'W
Scottdale, Ga., U.S. (skŏt'dāl)	68c	33°47'N	84°16'W
Scott Islands, is., Ant.	178	67°00'S	178°00'E
Scottsbluff, Ne., U.S. (skŏts'blŭf)	70	41°52'N	103°40'W
Scottsboro, Al., U.S. (skŏts'bŭro)	82	34°40'N	86°03'W
Scottsburg, In., U.S. (skŏts'bûrg)	66	38°40'N	85°50'W
Scottsdale, Austl. (skŏts'dāl)	176	41°12'S	147°37'E
Scottsville, Ky., U.S. (skŏts'vĭl)	82	36°45'N	86°10'W
Scottville, Mi., U.S.	66	44°00'N	86°20'W
Scranton, Pa., U.S. (skrăn'tŭn)	65	41°15'N	75°45'W
Scugog, l., Can. (skū'gôg)	59	44°05'N	78°55'W
Scunthorpe, Eng., U.K. (skŭn'thôrp)	114a	53°36'N	0°38'W
Scutari see Shkodër, Alb.	110	42°04'N	19°30'E
Scutari, Lake, l., Eur. (skōō'tä-rē)	119	42°14'N	19°33'E
Seabeck, Wa., U.S. (sē'bĕck)	74a	47°38'N	122°50'W
Sea Bright, N.J., U.S. (sē brīt)	68a	40°22'N	73°58'W
Seabrook, Tx., U.S. (sē'brŏk)	81	29°34'N	95°01'W
Seaford, De., U.S. (sē'fērd)	67	38°35'N	75°40'W
Seagraves, Tx., U.S. (sē'grāvs)	78	32°51'N	102°38'W
Sea Islands, is., Ga., U.S. (sē)	83	31°21'N	81°05'W
Seal, r., Can.	52	59°08'N	96°37'W
Seal Beach, Ca., U.S.	75a	33°44'N	118°06'W
Seal Cays, is., Bah.	93	22°40'N	75°55'W
Seal Cays, is., T./C. Is.	93	21°10'N	71°45'W
Seal Island, i., S. Afr. (sēl)	186a	34°07'S	18°36'E
Sealy, Tx., U.S. (sē'lē)	81	29°46'N	96°10'W
Searcy, Ar., U.S. (sûr'sē)	79	35°13'N	91°43'W
Searles, l., Ca., U.S. (sûrl's)	76	35°44'N	117°22'W
Searsport, Me., U.S. (sērz'pōrt)	60	44°28'N	68°55'W
Seaside, Or., U.S. (sē'sīd)	72	45°59'N	123°55'W
Seattle, Wa., U.S. (sē-ăt''l)	64	47°36'N	122°20'W
Sebaco, Nic. (sē-bä'kō)	90	12°50'N	86°03'W
Sebago, Me., U.S. (sē-bā'gō)	60	43°52'N	70°20'W
Sebastián Vizcaíno, Bahía, b., Mex.	86	28°45'N	115°15'W
Sebastopol, Ca., U.S. (sē-bȧs'tō-pōl)	76	38°27'N	122°50'W
Sebderat, Erit.	185	15°30'N	36°45'E
Sebewaing, Mi., U.S. (sē'bē-wăng)	66	43°45'N	83°25'W
Sebezh, Russia (syĕ'bĕzh)	132	56°16'N	28°29'E
Sebinkarahisar, Tur.	119	40°15'N	38°10'E
Sebnitz, Ger. (zĕb'nĕts)	124	51°01'N	14°16'E
Sebou, Oued, r., Mor.	184	34°23'N	5°18'W
Sebree, Ky., U.S. (sē-brē')	66	37°35'N	87°30'W
Sebring, Fl., U.S. (sē'brĭng)	83a	27°30'N	81°26'W
Sebring, Oh., U.S.	66	40°55'N	81°05'W
Secchia, r., Italy (sĕ'kyä)	130	44°25'N	10°25'E
Seco, r., Mex. (sĕ'kō)	89	18°11'N	93°18'W
Sedalia, Mo., U.S.	65	38°42'N	93°12'W
Sedan, Fr. (sē-däɴ)	117	49°49'N	4°55'E
Sedan, Ks., U.S. (sē-dăn')	79	37°07'N	96°08'W
Sedom, Isr.	153a	31°04'N	35°24'E
Sedro Woolley, Wa., U.S. (sē'drô-wòl'ē)	74a	48°30'N	122°14'W
Šeduva, Lith. (shē'dò-vá)	123	55°46'N	23°45'E
Seestall, Ger. (zā'shtäl)	115d	47°58'N	10°52'E
Sefrou, Mor. (sē-frōō')	118	33°49'N	4°46'W
Seg, l., Russia (syĕgh)	136	63°20'N	33°30'E
Segamat, Malay. (sā'gá-māt)	153b	2°30'N	102°49'E
Segang, China (sŭ-gäŋ)	162	31°59'N	114°13'E
Segbana, Benin	189	10°56'N	3°42'E
Segorbe, Spain (sē-gôr-bē)	129	39°50'N	0°30'W
Ségou, Mali (sā-gōō')	184	13°27'N	6°16'W
Segovia, Col. (sē-gô'vēä)	100a	7°08'N	74°42'W
Segovia, Spain (sā-gō'vē-ä)	118	40°58'N	4°05'W
Segre, r., Spain (sā'grā)	129	41°54'N	1°10'E
Seguam, i., Ak., U.S. (sē'gwäm)	63a	52°16'N	172°10'W
Seguam Passage, strt., Ak., U.S.	63a	52°20'N	173°00'W
Séguédine, Niger	189	20°12'N	12°59'E
Séguéla, C. Iv. (sā-gā-lā')	184	7°57'N	6°40'W
Seguin, Tx., U.S. (sĕ-gēn')	81	29°35'N	97°58'W
Segula, i., Ak., U.S. (sē-gū'lá)	63a	52°08'N	178°35'E
Segura, r., Spain	118	38°24'N	2°12'W
Segura, Sierra de, mts., Spain (sē-ĕ'r-rä-dē)	128	38°05'N	2°45'W
Sehwān, Pak.	158	26°33'N	67°51'E
Seibo, Dom. Rep. (sĕ'y-bō)	93	18°45'N	69°05'W
Seiling, Ok., U.S.	78	36°09'N	98°56'W
Seim, r., Eur.	137	51°23'N	33°22'E
Seinäjoki, Fin. (sā'ē-nĕ-yō'kē)	123	62°47'N	22°50'E
Seine, r., Can. (sån)	62f	49°48'N	97°03'W
Seine, r., Can. (sån)	58	49°04'N	91°00'W
Seine, r., Fr.	112	48°00'N	4°30'E
Seine, Baie de la, b., Fr. (bī dĕ lä sån)	126	49°37'N	0°53'W
Seio do Venus, mtn., Braz. (sĕ-yô-dô-vĕ'nōōs)	102b	22°28'S	43°12'W
Seixal, Port. (sā-ē-shäl')	129b	38°38'N	9°06'W
Sekenke, Tan.	191	4°16'S	34°10'E
Şeki, Azer.	138	41°12'N	47°12'E
Sekondi-Takoradi, Ghana (sĕ-kôn'dē tä-kō-rä'dē)	184	4°59'N	1°43'W
Sekota, Eth.	185	12°47'N	38°59'E
Selangor, state, Malay. (sā-län'gōr)	153b	2°53'N	101°29'E
Selanovtsi, Blg. (sāl'á-nôv-tsĭ)	131	43°42'N	24°05'E
Selaru, Pulau, i., Indon.	169	8°30'S	130°30'E
Selatan, Tanjung, c., Indon. (sȧ-lä'tän)	168	4°09'S	114°40'E
Selawik, Ak., U.S. (sē-lá-wĭk)	63	66°30'N	160°09'W
Selayar, Pulau, i., Indon.	168	6°15'S	121°15'E
Selbusjøen, l., Nor. (sĕl'bōō)	122	63°18'N	11°55'E
Selby, Eng., U.K. (sĕl'bē)	114a	53°47'N	1°03'W
Seldovia, Ak., U.S. (sĕl-dō'vē-ȧ)	63	59°26'N	151°42'W
Selemdzha, r., Russia (sā-lĕmt-zhä')	141	52°28'N	131°50'E
Selenga (Selenge), r., Asia (sĕ lĕŋ gä')	135	49°00'N	102°00'E
Selenge, r., Asia	160	49°04'N	102°23'E
Selennyakh, r., Russia (sĕl-yĭn-yäk)	141	67°42'N	141°45'E
Sélestat, Fr. (sē-lē-stä')	127	48°16'N	7°27'E
Sélibaby, Maur. (sā-lē-bȧ-bē')	184	15°21'N	12°11'W
Seliger, l., Russia (sĕl'lē-gĕr)	136	57°14'N	33°18'E
Selizharovo, Russia (sā'lē-zhä'rô-vô)	132	56°51'N	33°28'E
Selkirk, Can. (sĕl'kûrk)	50	50°09'N	96°52'W
Selkirk Mountains, mts., Can.	52	51°00'N	117°40'W
Selleck, Wa., U.S. (sĕl'ĕck)	74a	47°22'N	121°52'W
Sellersburg, In., U.S. (sĕl'ĕrs-bûrg)	69h	38°25'N	85°45'W
Sellya Khskaya, Guba, b., Russia (sĕl-yäk'skȧ-yȧ)	141	72°30'N	136°00'E
Selma, Al., U.S. (sĕl'mȧ)	65	32°25'N	87°00'W
Selma, Ca., U.S.	76	36°34'N	119°37'W
Selma, N.C., U.S.	83	35°33'N	78°16'W
Selma, Tx., U.S.	75d	29°33'N	98°19'W
Selmer, Tn., U.S.	82	35°11'N	88°36'W
Selsingen, Ger. (zĕl'zēn-gĕn)	115c	53°22'N	9°13'E
Selway, r., Id., U.S. (sĕl'wȧ)	72	46°07'N	115°12'W
Selwyn, l., Can. (sĕl'wĭn)	52	59°41'N	104°30'W
Seman, r., Alb.	131	40°48'N	19°53'E
Semarang, Indon. (sē-mä'räng)	168	7°03'S	110°27'E
Semenivka, Ukr.	137	52°10'N	32°34'E
Semeru, Gunung, mtn., Indon.	168	8°06'S	112°55'E
Semey (Semipalatinsk), Kaz.	139	50°28'N	80°29'E
Semiahmoo Indian Reserve, I.R., Can.	74d	49°01'N	122°43'W
Semiahmoo Spit, Wa., U.S. (sĕm'ĭ-ȧ-mōō)	74d	48°59'N	122°52'W
Semichi Islands, is., Ak., U.S. (sē-mē'chī)	63a	52°40'N	174°50'E

ng-sing; ŋ-baŋk; ɴ-nasalized n; nŏd; cŏmmit; ōld; ôbey; ôrder; oi-boil; fōōd; ò-as oo in foot; ou-out; s-soft; sh-dish; th-thin; pūre; ûnite; ûrn; stŭd; circŭs; ü-as in French tu; '-indeterminate vowel.

302 Seminoe Reservoir – Shiawassee PRONOUNCING INDEX

PLACE (Pronunciation)	PAGE	LAT.	LONG.
Seminoe Reservoir, res., Wy., U.S. (sĕm'ĭ nō)	73	42°08'N	107°10'W
Seminole, Ok., U.S. (sĕm'ĭ-nōl)	79	35°13'N	96°41'W
Seminole, Tx., U.S.	80	32°43'N	102°39'W
Seminole, Lake, res., U.S.	82	30°57'N	84°46'W
Semipalatinsk *see* Semey, Kaz.	139	50°28'N	80°29'E
Semisopochnoi, i., Ak., U.S. (sĕ-mē-sá-pōsh' noi)	63a	51°45'N	179°25'E
Semliki, r., Afr. (sĕm'lē-kē)	185	0°45'N	29°36'E
Semmering Pass, p., Aus. (sĕm'ĕr-ĭng)	124	47°39'N	15°50'E
Senador Pompeu, Braz. (sĕ-nä-dōr-pôm-pĕ'ō)	101	5°34'S	39°18'W
Senaki, Geor.	138	42°17'N	42°04'E
Senatobia, Ms., U.S. (sĕ-nà-tō'bē-*á*)	82	34°36'N	89°56'W
Sendai, Japan (sĕn-dī')	161	38°18'N	141°02'E
Seneca, Ks., U.S. (sĕn'ĕ-k*á*)	79	39°49'N	96°03'W
Seneca, Md., U.S.	68e	39°04'N	77°20'W
Seneca, S.C., U.S.	83	34°40'N	82°58'W
Seneca, l., N.Y., U.S.	67	42°30'N	76°55'W
Seneca Falls, N.Y., U.S.	67	42°55'N	76°55'W
Sénégal, nation, Afr. (sĕn-ĕ-gôl')	184	14°53'N	14°58'W
Sénégal, r., Afr.	184	16°00'N	14°00'W
Senekal, S. Afr. (sĕn'ĕ-k*ă*l)	192c	28°20'S	27°37'E
Senftenberg, Ger. (zĕnf'tĕn-bĕrgh)	124	51°32'N	14°00'E
Sengunyane, r., Leso.	187c	29°35'S	28°08'E
Senhor do Bonfim, Braz. (sĕn-yôr dō bôɴ-fē'ɴ)	101	10°21'S	40°09'W
Senigallia, Italy (så-nē-gäl'lyä)	130	43°42'N	13°16'E
Senj, Cro. (sĕn')	130	44°58'N	14°55'E
Senja, i., Nor. (sĕnyä)	116	69°28'N	16°10'E
Senlis, Fr. (säɴ-lēs')	127b	49°13'N	2°35'E
Sennar Dam, dam, Sudan	185	13°38'N	33°38'E
Senneterre, Can.	51	48°20'N	77°22'W
Sens, Fr. (säɴs)	126	48°05'N	3°18'E
Sensuntepeque, El Sal. (sĕn-sōōn-tå-pā'kå)	90	13°53'N	88°34'W
Senta, Serb. (sĕn'tä)	119	45°54'N	20°05'E
Senzaki, Japan (sĕn'zä-kē)	167	34°22'N	131°09'E
Seoul (Sŏul), Kor., S.	161	37°35'N	127°03'E
Sepang, Malay.	153b	2°43'N	101°45'E
Sepetiba, Baía de, b., Braz. (bäē'ä dĕ så-på-tē'b*á*)	102b	23°01'S	43°42'W
Sepik, r., (sĕp-ēk')	169	4°07'S	142°40'E
Septentrional, Cordillera, mts., Dom. Rep.	93	19°50'N	71°15'W
Septeuil, Fr. (sĕ-tû')	127b	48°53'N	1°40'E
Sept-Îles, Can. (sĕ-tēl')	60	50°12'N	66°23'W
Sequatchie, r., Tn., U.S. (sē-kwăch'ē)	82	35°33'N	85°14'W
Sequim, Wa., U.S. (sē'kwĭm)	74a	48°05'N	123°07'W
Sequim Bay, b., Wa., U.S.	74a	48°04'N	122°58'W
Sequoia National Park, rec., Ca., U.S. (sē-kwoi'*á*)	64	36°34'N	118°37'W
Seraing, Bel. (sĕ-răɴ')	121	50°38'N	5°28'E
Serāmpore, India	158a	22°44'N	88°21'E
Serang, Indon. (så-räng')	168	6°13'S	106°10'E
Seranggung, Indon.	153b	0°49'N	104°11'E
Serbia and Montenegro (Yugoslavia), nation, Eur.	110	44°00'N	21°00'E
Serbia *see* Srbija, hist. reg., Serb.	131	44°05'N	20°35'E
Serdobsk, Russia (sĕr-dôpsk')	137	52°30'N	44°20'E
Sered', Slvk.	125	48°17'N	17°43'E
Seredyna-Buda, Ukr.	132	52°11'N	34°03'E
Seremban, Malay. (sĕr-ĕm-bän')	153b	2°44'N	101°57'E
Serengeti National Park, rec., Tan.	191	2°20'S	34°50'E
Serengeti Plain, pl., Tan.	191	2°40'S	34°55'E
Serenje, Zam. (sē-rĕn'yē)	186	13°12'S	30°49'E
Seret, r., Ukr. (sĕr'ĕt)	125	49°45'N	25°30'E
Sergeya Kirova, i., Russia (sĕr-gyĕ'yá kĕ'rô-vå)	140	77°30'N	86°10'E
Sergipe, state, Braz. (sĕr-zhĕ'pĕ)	101	10°27'S	37°04'W
Sergiyev Posad, Russia	142b	56°18'N	38°08'E
Sergiyevsk, Russia	136	53°58'N	51°00'E
Sérifos, Grc.	131	37°10'N	24°32'E
Sérifos, i., Grc.	131	37°42'N	24°17'E
Serodino, Arg. (sĕ-rô-dē'nō)	99c	32°36'S	60°56'W
Seropédica, Braz. (sĕ-rô-pĕ'dē-kä)	102b	22°44'S	43°43'W
Serov, Russia (syĕ-rôf')	140	59°36'N	60°30'E
Serowe, Bots. (sĕ-rō'wĕ)	186	22°18'S	26°39'E
Serpa, Port. (sĕr-pä)	128	37°56'N	7°38'W
Serpukhov, Russia (syĕr'pô-κôf)	134	54°53'N	37°27'E
Sérres, Grc. (sĕr'rĕ) (sĕr'ēs)	119	41°06'N	23°36'E
Serrinha, Braz. (sĕr-rēn'yá)	101	11°43'S	38°49'W
Serta, Port. (sĕr'tá)	128	39°48'N	8°01'W
Sertânia, Braz. (sĕr-tá'nyä)	101	8°28'S	37°13'W
Sertãozinho, Braz. (sĕr-touɴ-zē'n-yō)	99a	21°10'S	47°58'W
Serting, r., Malay.	153b	3°01'N	102°32'E
Sese Islands, is., Ug.	191	0°30'S	32°30'E
Sesia, r., Italy (sâz'yä)	130	45°33'N	8°25'E
Sesimbra, Port. (sĕ-sē'm-brä)	129b	38°27'N	9°06'W
Sesmyl, r., S. Afr.	187b	25°51'S	28°06'E
Ses Salines, Cap de, c., Spain	129	39°16'N	3°03'E
Sestri Levante, Italy (sĕs'trē lå-vän'tå)	130	44°15'N	9°24'E
Sestroretsk, Russia (sĕs-trô-rĕtsk)	136	60°06'N	29°58'E
Sestroretskiy Razliv, Ozero, l., Russia	142c	60°05'N	30°07'E
Seta, Japan (sĕ'tä)	167b	34°58'N	135°56'W
Sète, Fr. (sĕt)	117	43°24'N	3°42'E
Sete Lagoas, Braz. (sĕ-tĕ lä-gō'äs)	101	19°23'S	43°58'W
Sete Pontes, Braz.	102b	22°51'S	43°05'W
Seto, Japan (sĕ'tō)	167	35°11'N	137°07'E
Seto-Naikai, sea, Japan (sĕ'tō nī'kī)	167	33°50'N	132°25'E
Settat, Mor. (sĕt-ät') (sē-tá')	184	33°02'N	7°30'W
Sette-Cama, Gabon (sĕ-tĕ-kä-mä')	186	2°29'S	9°40'E
Settlement Point, c., Bah. (sĕt'l-m*ĕ*nt)	92	26°40'N	79°00'W
Settlers, S. Afr. (sĕt'lĕrs)	192c	24°57'S	28°33'E
Settsu, Japan	167b	34°46'N	135°33'E
Setúbal, Port. (så-tōō'bäl)	118	30°32'N	8°54'W

PLACE (Pronunciation)	PAGE	LAT.	LONG.
Setúbal, Baía de, b., Port.	128	38°27'N	9°08'W
Seul, Lac, l., Can. (lák sûl)	53	50°20'N	92°30'W
Sevan, l., Arm. (syĭ-vän')	137	40°10'N	45°20'E
Sevastopol', Ukr. (syĕ-väs-tô'pôl'')	134	44°34'N	33°34'E
Sevenoaks, Eng., U.K. (sē-vĕn-ōks')	114b	51°16'N	0°12'E
Severka, r., Russia (sâ'vĕr-kå)	142b	55°11'N	38°41'E
Severn, r., Can. (sĕv'ĕrn)	53	55°21'N	88°42'W
Severn, r., U.K.	120	51°50'N	2°25'W
Severna Park, Md., U.S. (sĕv'ĕrn-á)	68e	39°04'N	76°33'W
Severnaya Dvina, r., Russia	134	63°00'N	42°40'E
Severnaya Zemlya (Northern Land), is., Russia (sĕ-vyĭr-nīŭ zĭ-m'lyä')	135	79°33'N	101°15'E
Severoural'sk, Russia (sĕ-vyĭ-rŭ-ōō-rälsk')	140	60°08'N	59°53'E
Sevier, r., Ut., U.S.	64	39°25'N	112°20'W
Sevier, East Fork, r., Ut., U.S.	77	37°45'N	112°10'W
Sevier Lake, l., Ut., U.S. (sē-vēr')	77	38°55'N	113°10'W
Sevilla, Col. (sĕ-vē'l-yä)	100a	4°16'N	75°56'W
Sevilla, Spain (så-vēl'yä)	110	37°29'N	5°58'W
Seville, Oh., U.S. (sē'vĭl)	69d	41°01'N	81°45'W
Sevlievo, Blg. (sĕv'lyĕ-vô)	119	43°02'N	25°05'E
Sevsk, Russia (syĕfsk)	132	52°08'N	34°28'E
Seward, Ak., U.S. (sū'ård)	64a	60°18'N	149°28'W
Seward, Ne., U.S.	79	40°55'N	97°06'W
Seward Peninsula, pen., Ak., U.S.	63	65°40'N	164°00'W
Sewell, Chile (sĕ'ō-ĕl)	102	34°01'S	70°18'W
Sewickley, Pa., U.S. (sē-wĭk'lē)	69e	40°33'N	80°11'W
Seybaplaya, Mex. (sā-ē-bä-plä'yä)	89	19°38'N	90°40'W
Seychelles, nation, Afr. (sā-shĕl')	3	5°20'S	55°10'E
Seydisfjördur, Ice. (sā'dēs-fyûr-dôr)	116	65°21'N	14°08'W
Seyhan, r., Tur.	119	37°28'N	35°40'E
Seylac, Som.	192a	11°19'N	43°20'E
Seymour, S. Afr. (sē'môr)	187c	32°33'S	26°48'E
Seymour, Ia., U.S.	71	40°41'N	93°03'W
Seymour, In., U.S. (sē'mōr)	66	38°55'N	85°55'W
Seymour, Tx., U.S.	78	33°35'N	99°16'W
Sezela, S. Afr.	187c	30°33'S	30°37'W
Sezze, Italy (sĕt'så)	130	41°32'N	13°00'E
Sfântu Gheorghe, Rom.	119	45°53'N	25°49'E
Sfax, Tun. (sfäks)	184	34°51'N	10°45'E
's-Gravenhage *see* The Hague, Neth. ('s krä'vĕn-hä'κĕ) (hāg)	110	52°05'N	4°16'E
Sha, r., China (shä)	161	33°33'N	114°30'E
Shaanxi, prov., China (shän-shyē)	160	35°30'N	109°10'E
Shabeelle (Shebele), r., Afr.	192a	1°38'N	43°50'E
Shache, China (shä-chŭ)	160	38°15'N	77°15'E
Shackleton Ice Shelf, ice, Ant. (shăk''l-tŭn)	178	65°00'S	100°00'E
Shades Creek, r., Al., U.S. (shādz)	68h	33°20'N	86°55'W
Shades Mountain, mtn., Al., U.S.	68h	33°22'N	86°51'W
Shagamu, Nig.	189	6°51'N	3°39'E
Shāhdād, Namakzār-e, l., Iran (nŭ-mŭk-zär')	154	31°00'N	58°30'E
Shāhjahānpur, India (shä-jŭ-hän'pōōr)	155	27°58'N	79°58'E
Shajing, China (shä-jyĭŋ)	163a	22°44'N	113°48'E
Shaker Heights, Oh., U.S. (shā'kĕr)	69d	41°28'N	81°34'W
Shakhty, Russia (shäκ'tē)	134	47°41'N	40°11'E
Shaki, Nig.	189	8°39'N	3°25'E
Shakopee, Mn., U.S. (shăk'ō-pe)	75g	44°48'N	93°31'W
Shala Lake, l., Eth. (shä'lå)	185	7°34'N	39°00'E
Shalqar, Kaz.	139	47°52'N	59°41'E
Shalqar köli, l., Kaz.	137	50°30'N	51°30'E
Shām, Jabal ash, mtn., Oman	154	23°01'N	57°45'E
Shambe, Sudan (shäm'bå)	185	7°08'N	30°46'E
Shammar, Jabal, mts., Sau. Ar. (jĕb'*ĕ*l shŭm'âr)	154	27°13'N	40°16'E
Shamokin, Pa., U.S. (shá-mō'kĭn)	67	40°45'N	76°30'W
Shamrock, Tx., U.S. (shăm'rŏk)	78	35°14'N	100°12'W
Shamva, Zimb. (shäm'vá)	186	17°18'S	31°35'E
Shandon, Oh., U.S. (shăn-dŭn)	69f	39°20'N	84°13'W
Shandong, prov., China (shän-dôŋ)	161	36°08'N	117°09'E
Shandong Bandao, pen., China (shän-dôŋ bän-dou)	161	37°00'N	120°10'E
Shangcai, China (shäŋ-tsī)	162	33°16'N	114°16'E
Shangcheng, China (shäŋ-chŭŋ)	162	31°47'N	115°22'E
Shangdu, China (shäŋ-dōō)	164	41°38'N	113°22'E
Shanghai, China (shäng'hī')	161	31°14'N	121°27'E
Shanghai Shi, prov., China (shäŋ-hī shr)	161	31°30'N	121°45'E
Shanghe, China (shäŋ-hŭ)	162	37°18'N	117°10'E
Shanglin, China (shäŋ-lĭn)	162	38°20'N	116°05'E
Shangqiu, China (shäŋ-chyō)	164	34°24'N	115°39'E
Shangrao, China (shäŋ-rou)	165	28°25'N	117°58'E
Shangzhi, China (shäŋ-jr)	164	45°18'N	127°52'E
Shanhaiguan, China	164	40°01'N	119°45'E
Shannon, Al., U.S. (shăn'*ŭ*n)	68h	33°23'N	86°52'W
Shannon, r., Ire. (shăn'*ŏ*n)	117	52°30'N	10°15'W
Shanshan, China (shän'shän')	160	42°51'N	89°53'E
Shantar, i., Russia (shän'tär)	141	55°13'N	138°42'E
Shantou, China (shän-tō)	161	23°20'N	116°40'E
Shanxi, prov., China (shän-shyē)	161	37°30'N	112°00'E
Shan Xian, China (shän shyĕn)	162	34°47'N	116°04'E
Shaobo, China (shou-bwo)	164	32°33'N	119°30'E
Shaobo Hu, l., China (shou-bwo hōō)	162	32°47'N	119°13'E
Shaoguan, China (shou-gŭän)	161	24°58'N	113°42'E
Shaoxing, China (shou-shyĭŋ)	161	30°00'N	120°40'E
Shaoyang, China	161	27°15'N	111°28'E
Shapki, Russia (shäp'kĭ)	142c	59°36'N	31°11'E
Shark Bay, b., Austl. (shärk)	174	25°30'S	113°00'E
Sharon, Ma., U.S. (shăr'*ŏ*n)	61a	42°07'N	71°11'W
Sharon, Pa., U.S.	66	41°15'N	80°30'W
Sharon Springs, Ks., U.S.	78	38°51'N	101°45'W
Sharonville, Oh., U.S. (shăr'*ŏ*n vĭl)	69f	39°16'N	84°24'W
Sharpsburg, Pa., U.S. (shärps'bûrg)	69e	40°30'N	79°54'W
Sharr, Jabal, mtn., Sau. Ar.	154	28°00'N	36°07'E
Shashi, China (shä-shē)	161	30°20'N	112°18'E
Shasta, Mount, mtn., Ca., U.S.	64	41°35'N	122°12'W

PLACE (Pronunciation)	PAGE	LAT.	LONG.
Shasta Lake, res., Ca., U.S. (shăs't*á*)	64	40°51'N	122°32'W
Shatsk, Russia (shátsk)	136	54°00'N	41°40'E
Shattuck, Ok., U.S. (shăt'ŭk)	78	36°16'N	99°53'W
Shaunavon, Can.	50	49°40'N	108°25'W
Shaw, Ms., U.S. (shô)	82	33°36'N	90°44'W
Shawano, Wi., U.S. (sh*á*-wô'nô)	71	44°41'N	88°13'W
Shawinigan, Can.	51	46°32'N	72°46'W
Shawnee, Ks., U.S. (shô-nē')	75f	39°01'N	94°43'W
Shawnee, Ok., U.S.	64	35°20'N	96°54'W
Shawneetown, Il., U.S. (shô'nē-toun)	66	37°40'N	88°05'W
Shayang, China	165	31°00'N	112°38'E
Shchara, r., Bela. (sh-chá'rå)	125	53°17'N	25°12'E
Shchëlkovo, Russia (shchĕl'kô-vô)	132	55°55'N	38°00'E
Shchigry, Russia (shchē'grē)	133	51°52'N	36°54'E
Shchors, Ukr. (shchôrs)	133	51°38'N	31°58'E
Shchuch'ye Ozero, Russia (shchōōch'yĕ ô'zĕ-rō)	142a	56°31'N	56°35'E
Sheakhala, India	158a	22°47'N	88°10'E
Shebele (Shabeelle), r., Afr. (shā'bå-lē)	192a	6°07'N	43°10'E
Sheboygan, Wi., U.S. (shē-boi'g*ă*n)	65	43°45'N	87°44'W
Sheboygan Falls, Wi., U.S.	71	43°43'N	87°51'W
Shechem, hist., W.B.	153a	32°15'N	35°22'E
Shedandoah, Pa., U.S.	67	40°50'N	76°15'W
Shediac, Can. (shĕ'dē-ăk)	60	46°13'N	64°32'W
Shedin Peak, mtn., Can. (shĕd'*ĭ*n)	54	55°55'N	127°32'W
Sheerness, Eng., U.K. (shēr'nĕs)	114b	51°26'N	0°46'E
Sheffield, Can.	62d	43°20'N	80°13'W
Sheffield, Eng., U.K.	116	53°23'N	1°28'W
Sheffield, Al., U.S. (shĕf'fēld)	82	35°42'N	87°42'W
Sheffield, Oh., U.S.	69d	41°26'N	82°05'W
Sheffield, co., Eng., U.K.	114a	53°52'N	1°35'W
Sheffield Lake, Oh., U.S.	69d	41°30'N	82°03'W
Sheksna, r., Russia (shĕks'nà)	136	59°50'N	38°40'E
Shelagskiy, Mys, c., Russia (shĭ-läg'skē)	135	70°08'N	170°52'E
Shelbina, Ar., U.S. (shĕl-bī'nà)	79	39°41'N	92°03'W
Shelburn, In., U.S. (shĕl'b*ŭ*rn)	66	39°10'N	87°30'W
Shelburne, Can.	51	43°46'N	65°19'W
Shelburne, Can.	59	44°04'N	80°12'W
Shelby, In., U.S. (shĕl'bē)	69a	41°12'N	87°21'W
Shelby, Mi., U.S.	66	43°35'N	86°20'W
Shelby, Ms., U.S.	82	33°56'N	90°44'W
Shelby, Mt., U.S.	73	48°35'N	111°55'W
Shelby, N.C., U.S.	83	35°16'N	81°35'W
Shelby, Oh., U.S.	66	40°50'N	82°40'W
Shelbyville, Il., U.S. (shĕl'bē-vĭl)	66	39°20'N	88°45'W
Shelbyville, In., U.S.	66	39°30'N	85°45'W
Shelbyville, Ky., U.S.	66	38°10'N	85°15'W
Shelbyville, Tn., U.S.	82	35°30'N	86°28'W
Shelbyville Reservoir, res., Il., U.S.	66	39°30'N	88°45'W
Sheldon, Ia., U.S. (shĕl'd*ŭ*n)	70	43°10'N	95°50'W
Sheldon, Tx., U.S.	81a	29°52'N	95°07'W
Shelekhova, Zaliv, b., Russia	135	60°00'N	156°00'E
Shelikof Strait, strt., Ak., U.S. (shĕ'lē-kôf)	63	57°56'N	154°20'W
Shellbrook, Can.	56	53°15'N	106°22'W
Shelley, Id., U.S. (shĕl'lē)	73	43°24'N	112°06'W
Shellrock, r., Ia., U.S. (shĕl'rŏk)	71	43°25'N	93°19'W
Shelon', r., Russia (shâ'lôn)	132	57°50'N	29°40'E
Shelton, Ct., U.S. (shĕl't*ŭ*n)	67	41°15'N	73°05'W
Shelton, Ne., U.S.	78	40°46'N	98°41'W
Shelton, Wa., U.S.	72	47°14'N	123°05'W
Shemakha, Russia (shĕ-má-kä')	142a	56°16'N	59°19'E
Shenandoah, Ia., U.S. (shĕn-*ă*n-dō'*á*)	79	40°46'N	95°23'W
Shenandoah, Va., U.S.	67	38°30'N	78°30'W
Shenandoah, r., Va., U.S.	67	38°55'N	78°05'W
Shenandoah National Park, rec., Va., U.S.	65	38°35'N	78°25'W
Shendam, Nig.	189	8°53'N	9°32'E
Shengfang, China (shengfäng)	162	39°05'N	116°40'E
Shenkursk, Russia (shĕn-kōōrsk')	134	62°10'N	43°08'E
Shenmu, China	164	38°55'N	110°35'E
Shenqiu, China	164	33°11'N	115°06'E
Shenxian, China (shŭn shyän)	162	38°02'N	115°33'E
Shenxian, China (shŭn shyĕn)	162	36°14'N	115°38'E
Shenyang, China (shŭn-yäŋ)	161	41°45'N	123°22'E
Shenze, China (shŭn-dzŭ)	162	38°12'N	115°12'E
Shenzhen, China	165	22°32'N	114°08'E
Sheopur, India	155	25°37'N	77°10'E
Shepard, Can. (shĕ'pârd)	62e	50°57'N	113°55'W
Shepetivka, Ukr.	137	50°10'N	27°01'E
Shepparton, Austl. (shĕp'âr-t*ŭ*n)	176	36°15'S	145°25'E
Sherborn, Ma., U.S. (shûr'b*ŭ*rn)	61a	42°15'N	71°22'W
Sherbrooke, Can.	51	45°24'N	71°54'W
Sherburn, Eng., U.K. (shûr'b*ŭ*rn)	114a	53°47'N	1°15'W
Shereshevo, Bela. (shĕ-rĕ-shĕ-vô)	125	52°31'N	24°08'E
Sheridan, Ar., U.S. (shĕr'ĭ-d*ă*n)	79	34°19'N	92°21'W
Sheridan, Or., U.S.	72	45°06'N	123°22'W
Sheridan, Wy., U.S.	64	44°48'N	106°56'W
Sherman, Tx., U.S. (shĕr'măn)	64	33°39'N	96°37'W
Sherna, r., Russia (shĕr'nà)	142b	56°08'N	38°45'E
Sherridon, Can.	57	55°10'N	101°10'W
's Hertogenbosch, Neth. (sĕr-tō'ghĕn-bôs)	121	51°41'N	5°19'E
Sherwood, Or., U.S.	74c	45°21'N	122°50'W
Sherwood Forest, for., Eng., U.K.	114a	53°11'N	1°07'W
Sherwood Park, Can.	55	53°31'N	113°19'W
Shetland Islands, is., Scot., U.K. (shĕt'l*ă*nd)	112	60°35'N	2°10'W
Shewa Gimira, Eth.	185	7°13'N	35°49'E
Shexian, China (shŭ shyĕn)	162	36°34'N	113°42'E
Sheyang, r., China (shĕ-yäŋ)	162	33°42'N	119°40'E
Sheyenne, r., N.D., U.S. (shī-ĕn')	70	46°42'N	97°52'W
Shi, r., China (shr)	162	31°58'N	115°50'E
Shi, r., China	162	32°09'N	114°11'E
Shiawassee, r., Mi., U.S. (shī-á-wôs'ē)	66	43°15'N	84°05'W

PRONOUNCING INDEX Shibām – Siracusa 303

PLACE (Pronunciation)	PAGE	LAT.	LONG.
Shibām, Yemen (shē'bäm)	154	16°02'N	48°40'E
Shibīn al Kawn, Egypt (shē-bēn'ĕl kōm')	192b	30°31'N	31°01'E
Shibīn al Qanāṭir, Egypt (kà-nä'tēr)	192b	30°18'N	31°21'E
Shicun, China (shr-tsòn)	162	33°47'N	117°18'E
Shields, r., Mt., U.S. (shēldz)	73	45°54'N	110°40'W
Shifnal, Eng., U.K. (shĭf'năl)	114a	52°40'N	2°22'W
Shijian, China (shr-jyĕn)	162	31°27'N	117°51'E
Shijiazhuang, China (shr-jyä-jůäŋ)	161	38°04'N	114°31'E
Shijiu Hu, l., China (shr-jyō hōō)	162	31°29'N	119°07'E
Shikārpur, Pak.	155	27°51'N	68°52'E
Shiki, Japan (shē'kē)	167a	35°50'N	139°35'E
Shikoku, i., Japan (shē'kō'kōō)	161	33°43'N	133°33'E
Shilka, r., Russia (shĭl'kà)	141	53°00'N	118°45'E
Shilla, mtn., India	158	32°18'N	78°17'E
Shillong, India (shĕl-lông')	155	25°39'N	91°58'E
Shiloh, Il., U.S. (shī'lō)	75e	38°34'N	89°54'W
Shilong, China (shr-lôŋ)	165	23°05'N	113°58'E
Shilou, China	163a	22°58'N	113°29'E
Shimabara, Japan (shē'mä-bä'rä)	167	32°46'N	130°22'E
Shimada, Japan (shē'mä-dä)	167	34°49'N	138°13'E
Shimbiris, mtn., Som.	192a	10°40'N	47°23'E
Shimizu, Japan (shē'mē-zōō)	166	35°00'N	138°29'E
Shimminato, Japan (shēm'mē'nä-tô)	167	36°47'N	137°05'E
Shimoda, Japan (shē'mô-dà)	167	34°41'N	138°58'E
Shimoga, India	159	13°59'N	75°38'E
Shimoni, Kenya	191	4°39'S	39°23'E
Shimonoseki, Japan	161	33°58'N	130°55'E
Shimo-Saga, Japan (shē'mô sä'gä)	167b	35°01'N	135°41'E
Shin, Loch, l., Scot., U.K. (lŏk shĭn)	120	58°08'N	4°02'W
Shinagawa-Wan, b., Japan (shē'nä-gä'wä wän)	167a	35°37'N	139°49'E
Shinano-Gawa, r., Japan (shē-nä'nō gä'wä)	167	36°43'N	138°22'E
Shīndand, Afg.	157	33°18'N	62°08'E
Shinji, l., Japan (shĭn'jē)	167	35°23'N	133°05'E
Shinkolobwe, D.R.C.	191	11°02'S	26°35'E
Shinyanga, Tan. (shĭn-yäŋ'gä)	186	3°40'S	33°26'E
Shiono Misaki, c., Japan (shē-ō'nō mē'sä-kē)	166	33°20'N	136°10'E
Shipai, China (shr-pī)	163a	23°07'N	113°23'E
Ship Channel Cay, i., Bah. (shĭp chă-nĕl kē)	92	24°50'N	76°50'W
Shipley, Eng., U.K. (shĭp'lē)	114a	53°50'N	1°47'W
Shippegan, Can. (shĭ'pē-găn)	60	47°45'N	64°42'W
Shippegan Island, i., Can.	60	47°50'N	64°38'W
Shippenburg, Pa., U.S. (shĭp'ĕn bûrg)	67	40°00'N	77°30'W
Shipshaw, r., Can. (shĭp'shô)	59	48°50'N	71°03'W
Shiqma, r., Isr.	153a	31°31'N	34°40'E
Shirane-san, mtn., Japan (shē'rä'nå-sän')	167	35°44'N	138°14'E
Shirati, Tan. (shē-rä'tē)	186	1°15'S	34°02'E
Shīrāz, Iran (shē-räz')	154	29°32'N	52°27'E
Shire, r., Afr. (shē'rå)	186	15°00'S	35°00'E
Shiriya Saki, c., Japan (shē'rē sä'kē)	166	41°25'N	142°10'E
Shirley, Ma., U.S. (shûr'lē)	61a	42°33'N	71°39'W
Shishaldin Volcano, vol., Ak., U.S. (shī-shăl'dĭn)	63a	54°48'N	164°00'W
Shively, Ky., U.S. (shīv'lē)	69h	38°11'N	85°47'W
Shivpuri, India	155	25°31'N	77°46'E
Shivta, Horvot, hist., Isr.	153a	30°54'N	34°36'E
Shivwits Plateau, plat., Az., U.S.	77	36°13'N	113°42'W
Shiwan, China (shr-wän)	163a	23°01'N	113°04'E
Shiwan Dashan, mts., China (shr-wän dä-shän)	165	22°10'N	107°30'E
Shizuki, Japan (shĭ'zōō-kē)	167	34°29'N	134°51'E
Shizuoka, Japan (shē'zōō'ōkä)	166	34°58'N	138°24'E
Shklow, Bela.	132	54°11'N	30°23'E
Shkodër, Alb. (shkô'dûr) (shkōō'tárĕ)	110	42°04'N	19°30'E
Shkotovo, Russia (shkô'tô-vô)	166	43°15'N	132°21'E
Shoal Creek, r., Il., U.S. (shōl)	79	38°37'N	89°25'W
Shoal Lake, l., Can.	57	49°32'N	95°00'W
Shoals, In., U.S. (shōlz)	66	38°40'N	86°45'W
Shōdo, i., Japan (shō'dō)	167	34°27'N	134°27'E
Sholāpur, India (shō'lä-pōōr)	155	17°42'N	75°51'E
Shorewood, Wi., U.S. (shōr'wòd)	69a	43°05'N	87°54'W
Shoshone, Id., U.S. (shò-shōn'tē)	73	42°56'N	114°24'W
Shoshone, r., Wy., U.S.	73	44°35'N	108°50'W
Shoshone Lake, l., Wy., U.S.	73	44°17'N	110°50'W
Shoshoni, Wy., U.S.	73	43°14'N	108°05'W
Shostka, Ukr. (shôst'kà)	133	51°51'N	33°31'E
Shouguang, China (shō-gůäŋ)	162	36°53'N	118°45'E
Shouxian, China (shō shyĕn)	162	32°36'N	116°45'E
Shpola, Ukr. (shpô'là)	137	49°01'N	31°36'E
Shreveport, La., U.S. (shrēv'pôrt)	65	32°30'N	93°46'W
Shrewsbury, Eng., U.K. (shrōōz'bēr-ĭ)	120	52°43'N	2°44'W
Shrewsbury, Ma., U.S.	61a	42°18'N	71°43'W
Shropshire, co., Eng., U.K.	114a	52°36'N	2°45'W
Shroud Cay, i., Bah.	92	24°20'N	76°40'W
Shuangcheng, China (shůäŋ-chŭŋ)	164	45°18'N	126°18'E
Shuanghe, China (shůäŋ-hŭ)	162	31°33'N	116°48'E
Shuangliao, China	161	43°37'N	123°30'E
Shuangyang, China	164	43°28'N	125°45'E
Shuhedun, China (shōō-hŭ-dòn)	162	31°33'N	117°01'E
Shuiye, China (shwā-yŭ)	162	36°08'N	114°07'E
Shule, r., China (shōō-lŭ)	160	40°53'N	94°55'E
Shullsburg, Wi., U.S. (shŭlz'bûrg)	71	42°35'N	90°16'W
Shumagin, is., Ak., U.S. (shōō'mà-gĕn)	63	55°22'N	159°20'W
Shumen, Blg.	119	43°15'N	26°54'E
Shunde, China (shòn-dŭ)	163a	22°50'N	113°15'E
Shungnak, Ak., U.S. (shŭng'nák)	63	66°55'N	157°20'W
Shunut, Gora, mtn., Russia (gà-rä shōō'nòt)	142a	56°33'N	59°45'E
Shunyi, China (shòn-yē)	162	40°09'N	116°38'E
Shuqrah, Yemen	154	13°32'N	46°02'E
Shūrāb, r., Iran (shōō räb)	154	31°08'N	55°30'E
Shuri, Japan (shōō'rē)	166	26°10'N	127°48'E
Shurugwi, Zimb.	186	19°34'S	30°03'E
Shūshtar, Iran (shōōsh'tŭr)	154	31°50'N	48°46'E
Shuswap Lake, l., Can. (shōōs'wŏp)	55	50°57'N	119°15'W
Shuya, Russia (shōō'yà)	134	56°52'N	41°23'E
Shuyang, China (shōō yäng)	162	34°09'N	118°47'E
Shweba, Mya.	155	22°23'N	96°13'E
Shymkent, Kaz.	139	42°17'N	69°42'E
Shyroke, Ukr.	133	47°40'N	33°18'E
Siak Kecil, r., Indon.	153b	1°01'N	101°45'E
Siaksriinderapura, Indon. (sē-äks'rĭ ĕn'drà-pōō'rä)	153b	0°48'N	102°05'E
Siālkot, Pak. (sē-äl'kōt)	155	32°39'N	74°30'E
Siátista, Grc. (syä'tĭs-ta)	131	40°15'N	21°32'E
Siau, Pulau, i., Indon.	169	2°40'N	126°00'E
Šiauliai, Lith. (shē-ou'lē-ī)	136	55°57'N	23°19'E
Sibay, Russia (sē'báy)	142a	52°41'N	58°40'E
Šibenik, Cro. (shē-bā'nĕk)	119	43°44'N	15°55'E
Siberia, reg., Russia	152	57°00'N	97°00'E
Siberut, Pulau, i., Indon. (sē'bå-rōōt)	168	1°22'S	99°45'E
Sibiti, Congo (sē-bē-tē')	186	3°41'S	13°21'E
Sibiu, Rom. (sē-bĭ-ōō')	119	45°47'N	24°09'E
Sibley, Ia., U.S. (sĭb'lē)	70	43°24'N	95°33'W
Sibolga, Indon. (sē-bō'gä)	168	1°45'N	98°45'E
Sibsāgar, India (sēb-sŭ'gŭr)	155	26°47'N	94°45'E
Sibutu Island, i., Phil.	168	4°40'N	119°30'E
Sibuyan, i., Phil. (sē-bōō-yän')	169a	12°19'N	122°25'E
Sibuyan Sea, sea, Phil.	168	12°43'N	122°38'E
Sichuan, prov., China (sz-chůän)	160	31°20'N	103°00'E
Sicily, i., Italy (sĭs'ĭ-lē)	112	37°38'N	13°30'E
Sico, r., Hond. (sē-kō)	90	15°32'N	85°42'W
Sidamo, hist. reg., Eth. (sē-dä'mô)	185	5°08'N	37°45'E
Siderno Marina, Italy (sē-dĕr'nô mä-rē'nä)	130	38°18'N	16°19'E
Sídheros, Ákra, c., Grc.	130a	35°19'N	26°20'E
Sidi Aïssa, Alg.	129	35°53'N	3°44'E
Sidi bel Abbès, Alg. (sē'dē-bĕl á-bĕs')	184	35°15'N	0°43'W
Sidi Ifni, Mor. (ēf'nē)	184	29°22'N	10°15'W
Sidirókastro, Grc.	131	41°13'N	23°27'E
Sidley, Mount, mtn., Ant. (sĭd'lē)	178	77°25'S	129°00'W
Sidney, Can.	54	48°39'N	123°24'W
Sidney, Mt., U.S. (sĭd'nē)	73	47°43'N	104°07'W
Sidney, Ne., U.S.	70	41°10'N	103°00'W
Sidney, Oh., U.S.	66	40°20'N	84°10'W
Sidney Lanier, Lake, res., Ga., U.S. (lăn'yēr)	65	34°27'N	83°56'W
Sido, Mali	188	11°40'N	7°36'W
Sidon *see* Saydā, Leb.	154	33°34'N	35°23'E
Sidr, Wādī, r., Egypt	153a	29°43'N	32°58'E
Sidra, Gulf of *see* Surt, Khalīj, b., Libya	185	31°30'N	18°28'E
Siedlce, Pol. (syĕd''l-tsĕ)	125	52°09'N	22°20'E
Siegburg, Ger. (zēg'bōōrgh)	124	50°48'N	7°13'E
Siegen, Ger. (zē'ghĕn)	124	50°52'N	8°01'E
Sieghartskirchen, Aus.	115e	48°16'N	16°00'E
Siemiatycze, Pol. (syĕm'yä'tē-chĕ)	125	52°26'N	22°52'E
Siemionówka, Pol. (sĕĕ-mēō'nôf-kä)	125	52°53'N	23°50'E
Siem Reap, Camb. (syĕm'rā'áp)	168	13°32'N	103°54'E
Siena, Italy (sē-ĕn'ä)	118	43°19'N	11°21'E
Sieradz, Pol. (syĕ'rädz)	125	51°35'N	18°45'E
Sierpc, Pol. (syĕrpts)	125	52°51'N	19°42'E
Sierra Blanca, Tx., U.S. (sē-ĕ'rá blaŋ-kä)	80	31°10'N	105°20'W
Sierra Blanca Peak, mtn., N.M., U.S. (blăn'ká)	64	33°25'N	105°50'W
Sierra Leone, nation, Afr. (sē-ĕr'rä lā-ō'nå)	184	8°48'N	12°30'W
Sierra Madre, Ca., U.S. (mä'drē)	75a	34°10'N	118°03'W
Sierra Mojada, Mex. (sē-ĕ'r-rä-mô-kä'dä)	80	27°22'N	103°42'W
Sífnos, i., Grc.	131	36°58'N	24°30'E
Sigean, Fr. (sē-zhŏn')	126	43°02'N	2°56'E
Sigeurney, Ia., U.S. (sē-gûr-nĭ)	71	41°16'N	92°10'W
Sighetu Marmaţiei, Rom.	125	47°57'N	23°55'E
Sighişoara, Rom. (sē-gē-shwä'rà)	125	46°11'N	24°48'E
Siglufjördur, Ice.	116	66°06'N	18°45'W
Signakhi, Geor.	137	41°45'N	45°50'E
Signal Hill, Ca., U.S. (sĭg'năl hĭl)	75a	33°48'N	118°11'W
Sigsig, Ec. (sēg-sēg')	100	3°04'S	78°44'W
Sigtuna, Swe. (sēgh-tōō'nà)	122	59°40'N	17°39'E
Siguanea, Ensenada de la, b., Cuba	92	21°45'N	83°15'W
Siguatepeque, Hond. (sē-gwà'tĕ-pē-kĕ)	90	14°33'N	87°51'W
Sigüenza, Spain (sē-gwĕ'n-zä)	118	41°03'N	2°38'W
Siguiri, Gui. (sē-gē-rē')	184	11°25'N	9°10'W
Sihong, China (sz-hôŋ)	162	33°25'N	118°13'E
Siirt, Tur. (sĭ-ērt')	137	38°00'N	42°00'E
Sikalongo, Zam.	191	16°46'S	27°07'E
Sikasso, Mali (sē-käs'sō)	184	11°19'N	5°40'W
Sikeston, Mo., U.S. (sīks'tŭn)	79	36°50'N	89°35'W
Sikhote Alin', Khrebet, mts., Russia (se-kô'ta a-lēn')	135	45°00'N	135°45'E
Sikinos, i., Grc. (sĭ'kĭ-nōs)	131	36°45'N	24°55'E
Sikkim, state, India	155	27°42'N	88°25'E
Siklós, Hung. (sĭ'klōsh)	125	45°51'N	18°18'E
Sil, r., Spain (sē'l)	128	42°20'N	7°13'W
Silang, Phil. (sē-läng')	169a	14°14'N	120°58'E
Silao, Mex. (sē-lä'ō)	88	20°56'N	101°25'W
Silchar, India (sĭl-chär')	155	24°52'N	92°50'E
Silent Valley, S. Afr. (sī'lĕnt vă'lē)	192c	24°32'S	26°40'E
Siler City, N.C., U.S. (sī'lēr)	83	35°45'N	79°29'W
Silesia, hist. reg., Pol. (sĭ-lē'shà)	124	50°58'N	16°53'E
Silifke, Tur.	119	36°20'N	34°00'E
Siling Co, l., China	160	32°05'N	89°10'E
Silistra, Blg. (sē-lēs'trà)	119	44°01'N	27°13'E
Siljan, l., Swe. (sēl'yän)	116	60°48'N	14°28'E
Silkeborg, Den. (sĭl'kĕ-bôr')	122	56°10'N	9°33'E
Sillery, Can. (sēl'-re')	62b	46°46'N	71°15'W
Siloam Springs, Ar., U.S. (sī-lōm)	79	36°10'N	94°32'W
Siloana Plains, pl., Zam.	190	16°55'S	23°10'E
Silocayoápan, Mex. (sē-lô-kä-yô-ä'pän)	88	17°29'N	98°09'W
Silsbee, Tx., U.S. (sĭlz' bē)	81	30°19'N	94°09'W
Šilutė, Lith.	123	55°21'N	21°29'E
Silva Jardim, Braz. (sē'l-vä-zhär-dēn)	99a	22°40'N	42°24'W
Silvana, Wa., U.S. (sĭ-vän'á)	74a	48°12'N	122°16'W
Silvânia, Braz. (sēl-vá'nyä)	101	16°43'S	48°33'W
Silvassa, India	158	20°10'N	73°00'E
Silver, l., Mo., U.S.	79	39°38'N	93°12'W
Silverado, Ca., U.S. (sĭl-vēr-ä'dō)	75a	33°45'N	117°40'W
Silver Bank, bk.,	93	20°40'N	69°40'W
Silver Bank Passage, strt., N.A.	93	20°40'N	70°20'W
Silver Bay, Mn., U.S.	71	47°24'N	91°07'W
Silver City, Pan.	91	9°20'N	79°54'W
Silver City, N.M., U.S. (sĭl'vēr sĭ'tĭ)	77	32°45'N	108°20'W
Silver Creek, N.Y., U.S. (crēk)	67	42°35'N	79°10'W
Silver Creek, r., Az., U.S.	77	34°30'N	110°05'W
Silver Creek, r., In., U.S.	69h	38°20'N	85°45'W
Silver Creek, Muddy Fork, r., In., U.S.	69h	38°26'N	85°52'W
Silverdale, Wa., U.S. (sĭl'vēr-dāl)	74a	49°39'N	122°42'W
Silver Lake, Wi., U.S. (lāk)	69a	42°33'N	88°10'W
Silver Lake, l., Wi., U.S.	69a	42°35'N	88°08'W
Silver Spring, Md., U.S. (sprĭng)	68e	39°00'N	77°00'W
Silver Star Mountain, mtn., Wa., U.S.	74c	45°45'N	122°15'W
Silverthrone Mountain, mtn., Can. (sĭl'vēr-thrōn)	54	51°31'N	126°06'W
Silverton, S. Afr.	192c	25°45'S	28°13'E
Silverton, Co., U.S. (sĭl'vēr-tŭn)	77	37°50'N	107°40'W
Silverton, Oh., U.S.	69f	39°12'N	84°24'W
Silverton, Or., U.S.	72	45°02'N	122°46'W
Silves, Port. (sēl'vĕzh)	118	37°15'N	8°24'W
Silvies, r., Or., U.S. (sĭl'vēz)	72	43°44'N	119°15'W
Sim, Russia (sĭm)	142a	55°00'N	57°42'E
Sim, r., Russia	142a	54°50'N	56°50'E
Simao, China (sz-mou)	160	22°56'N	101°07'E
Simard, Lac, l., Can.	59	47°38'N	78°40'W
Simba, D.R.C.	190	0°36'N	22°55'E
Simcoe, Can. (sĭm'kō)	120	42°50'N	80°20'W
Simcoe, l., Can.	53	44°30'N	79°20'W
Simeulue, Pulau, i., Indon.	168	2°27'N	95°30'E
Simferopol', Ukr.	134	44°58'N	34°04'E
Similk Beach, Wa., U.S. (sē'mĭlk)	74a	48°27'N	122°35'W
Simla, India (sĭm'là)	155	31°09'N	77°15'E
Şimleu Silvaniei, Rom.	119	47°14'N	22°46'E
Simms Point, c., Bah.	92	25°00'N	77°40'W
Simojovel, Mex. (sē-mō-hŏ-vĕl')	89	17°12'N	92°43'W
Simonésia, Braz. (sē-mô-nĕ'syä)	99a	20°04'S	41°53'W
Simonette, r., Can. (sī-mŏn-ĕt')	55	54°15'N	118°00'W
Simonstad, S. Afr.	186a	34°11'S	18°25'E
Simood Sound, Can.	54	50°45'N	126°25'W
Simplon Pass, p., Switz. (sĭm'plŏn) (săn-plôn')	124	46°13'N	7°53'E
Simpson, i., Can.	71	48°43'N	87°44'W
Simpson Desert, des., Austl. (sĭmp-sŭn)	174	24°40'S	136°40'E
Simrishamn, Swe. (sēm'rĕs-häm'n)	122	55°35'N	14°19'E
Sims Bayou, Tx., U.S. (sĭmz bī-yōō')	81a	29°37'N	95°23'W
Simushir, i., Russia (se-mōō'shēr)	161	47°15'N	150°47'E
Sinaia, Rom. (sĭ-nä'yà)	131	45°20'N	25°30'E
Sinai Peninsula, pen., Egypt (sī'nī)	185	29°24'N	33°29'E
Sinaloa, state, Mex. (sē-nä-lô-ä)	86	25°15'N	107°45'W
Sinan, China (sz-nän)	160	27°50'N	108°30'E
Sinanju, Kor., N. (sĭ'nän-jŏ')	166	39°39'N	125°41'E
Sincelejo, Col. (sēn-så-lā'hō)	100	9°12'N	75°30'W
Sinclair Inlet, Wa., U.S. (sĭn-klâr')	74a	47°31'N	122°41'W
Sinclair Mills, Can.	55	54°02'N	121°41'W
Sindi, Est. (sēn'dē)	123	58°20'N	24°40'E
Sines, Port. (sē'năzh)	128	37°57'N	8°50'W
Singapore, Sing. (sĭn'gá-pōr')	168	1°18'N	103°52'E
Singapore, nation, Asia	168	1°22'N	103°45'E
Singapore Strait, strt., Asia	153b	1°14'N	104°20'E
Singu, Mya. (sĭn'gŭ)	160	22°37'N	96°04'E
Siniye Lipyagi, Russia (sēn'ê lēp'yä-gê)	133	51°24'N	38°29'E
Sinj, Cro. (sēn')	130	43°42'N	16°39'E
Sinjah, Sudan	185	13°09'N	33°52'E
Sinkāt, Sudan	156	18°50'N	36°50'E
Sinkiang *see* Xinjiang, prov., China	160	40°15'N	82°15'E
Sin'kovo, Russia (sĭn-kô'vô)	142b	56°23'N	37°19'E
Sinnamary, Fr. Gu.	101	5°15'N	52°52'W
Sinni, r., Italy (sēn'nē)	130	40°05'N	16°15'E
Sinnūris, Egypt	192b	29°25'N	30°52'E
Sino, Pedra de, mtn., Braz. (pĕ'drä-dô-sē'nô)	102b	22°27'S	43°02'W
Sinop, Tur.	154	42°00'N	35°05'E
Sint Eustatius, i., Neth. Ant.	91b	17°32'N	62°45'W
Sint Niklaas, Bel.	115a	51°10'N	4°07'E
Sinton, Tx., U.S. (sĭn'tŭn)	81	28°03'N	97°30'W
Sintra, Port. (sēn'trá)	128	38°48'N	9°23'W
Sint Truiden, Bel.	115a	50°49'N	5°14'E
Sinŭiju, Kor., N. (sĭ'nòĭ-jōō)	161	40°04'N	124°33'E
Sinyavino, Russia (sĭn-yä'vĭ-nô)	142c	59°50'N	31°07'E
Sinyaya, r., Eur. (sēn'yá-yá)	132	56°40'N	28°20'E
Sion, Switz. (sē'ôn')	124	46°15'N	7°17'E
Sioux City, Ia., U.S. (sōō sĭ'tĭ)	65	42°30'N	96°25'W
Sioux Falls, S.D., U.S. (fôlz)	64	43°33'N	96°43'W
Sioux Lookout, Can.	51	50°06'N	91°55'W
Siping, China (sz-pĭŋ)	161	43°05'N	124°24'E
Sipiwesk, Can.	50	55°27'N	97°24'W
Sipsey, r., Al., U.S. (sĭp'sē)	82	33°26'N	87°42'W
Sipura, Pulau, i., Indon.	168	2°15'S	99°33'E
Siqueros, Mex. (sē-kā'rōs)	88	23°19'N	106°14'W
Siquia, Río, r., Nic. (sē-kē'ä)	91	12°23'N	84°36'W
Siracusa, Italy (sē-rä-koo'sä)	119	37°02'N	15°19'E

ng-sing; ŋ-baŋk; ɴ-nasalized n; nŏd; cŏmmit; ōld; ôbey; ôrder; oi-boil; fōōd; ò-as oo in foot; ou-out; s-soft; sh-dish; th-thin; pūre; ûnite; ûrn; stŭd; circŭs; ü-as in French tu; '-indeterminate vowel.

304 Sirājganj – Songwe PRONOUNCING INDEX

PLACE (Pronunciation)	PAGE	LAT.	LONG.
Sirājganj, Bngl. (sĭ-räj´gŭnj)	155	24°23'N	89°43'E
Sirama, El Sal. (Sē-rä-mä)	90	13°23'N	87°55'W
Sir Douglas, Mount, mtn., Can. (sûr dŭg´l*ă*s)	55	50°44'N	115°20'W
Sir Edward Pellew Group, is., Austl. (pĕl´ū)	174	15°15'S	137°15'E
Siret, Rom.	125	47°58'N	26°01'E
Siret, r., Eur.	119	47°00'N	27°00'E
Sirhān, Wādī, depr., Sau. Ar.	154	31°02'N	37°16'E
Sirsa, India	158	29°39'N	75°02'E
Sir Sandford, Mount, mtn., Can. (sûr sănd´fẽrd)	55	51°40'N	117°52'W
Sirvintos, Lith. (shēr´vĭn-tôs)	123	55°02'N	24°59'E
Sir Wilfrid Laurier, Mount, mtn., Can. (sûr wĭl´frĭd lôr´yēr)	55	52°47'N	119°45'W
Sisak, Cro. (sē´sȧk)	119	45°29'N	16°20'E
Sisal, Mex. (sē-säl´)	86	21°09'N	90°03'W
Sishui, China (sz-shwā)	162	35°40'N	117°17'E
Sisquoc, r., Ca., U.S. (sĭs´kwŏk)	76	34°47'N	120°13'W
Sisseton, S.D., U.S. (sĭs´t*ŭ*n)	70	45°39'N	97°04'W
Sistān, Daryacheh-ye, l., Asia	154	31°45'N	61°15'E
Sisteron, Fr. (sēst´rôɴ´)	127	44°10'N	5°55'E
Sisterville, W.V., U.S. (sĭs´tẽr-vĭl)	66	39°30'N	81°00'W
Sitía, Grc. (sē´tĭ-ȧ)	130a	35°09'N	26°10'E
Sitka, Ak., U.S. (sĭt´kȧ)	64a	57°08'N	135°18'W
Sittingbourne, Eng., U.K. (sĭt-ĭng-bôrn)	114b	51°20'N	0°44'E
Sittwe, Mya.	155	20°09'N	92°54'E
Sivas, Tur. (sē´väs)	154	39°50'N	36°50'E
Siverek, Tur. (sē´vĕ-rĕk)	154	37°50'N	39°20'E
Siverskaya, Russia (sē´vĕr-skȧ-yȧ)	123	59°17'N	30°03'E
Sivers'kyi Donets', r., Eur.	133	48°48'N	38°42'E
Sīwah, Egypt	156	29°12'N	25°31'E
Siwah, oasis, Egypt (sē´wä)	185	29°33'N	25°11'E
Sixaola, r., C.R.	91	9°31'N	83°07'W
Sixian, China (sz shyĕn)	162	33°37'N	117°51'E
Sixth Cataract, wtfl., Sudan	185	16°26'N	32°44'E
Siyang, China (sz-yäŋ)	162	33°43'N	118°42'E
Sjaelland, i., Den. (shĕl´lăn´)	122	55°34'N	11°35'E
Sjenica, Serb. (syĕ´nē-tsȧ)	131	43°15'N	20°02'E
Skadovs'k, Ukr.	133	46°08'N	32°54'E
Skagen, Den. (skä´ghĕn)	122	57°43'N	10°32'E
Skagerrak, strt., Eur. (skä-ghĕ-räk´)	112	57°43'N	8°28'E
Skagit, r., Wa., U.S.	72	48°29'N	121°52'W
Skagit Bay, b., Wa., U.S. (skăg´ĭt)	74a	48°20'N	122°32'W
Skagway, Ak., U.S. (skăg-wā)	64a	59°30'N	135°28'W
Skälderviken, b., Swe.	122	56°20'N	12°25'E
Skalistyy, Golets, mtn., Russia	135	57°28'N	119°48'E
Skalistyy Khrebet, mts., Russia	138	43°15'N	43°00'E
Skamania, Wa., U.S. (sk*ȧ*-mā´nĭ-*ȧ*)	74c	45°37'N	112°03'W
Skamokawa, Wa., U.S.	74c	46°16'N	123°27'W
Skanderborg, Den. (skän-ẽr-bôr´)	122	56°04'N	9°55'E
Skaneateles, N.Y., U.S. (skăn-ē-ăt´lĕs)	67	42°55'N	76°25'W
Skaneateles, l., N.Y., U.S.	67	42°50'N	76°20'W
Skänninge, Swe. (shĕn´ĭng-ĕ)	122	58°24'N	15°02'E
Skanör-Falseterbo, Swe. (skän´ûr)	122	55°24'N	12°49'E
Skara, Swe. (skä´rȧ)	122	58°25'N	13°24'E
Skeena, r., Can. (skē´nȧ)	52	54°30'N	129°00'W
Skeena Mountains, mts., Can.	54	56°00'N	128°00'W
Skeerpoort, S. Afr.	187b	25°49'S	27°45'E
Skeerpoort, r., S. Afr.	187b	25°58'S	27°41'E
Skeldon, Guy. (skĕl´d*ŭ*n)	101	5°49'N	57°15'W
Skellefteå, Swe. (shĕl´ĕf-tē-a´)	116	64°47'N	20°48'E
Skellefteälven, r., Swe.	116	65°15'N	19°30'E
Skhodnya, Russia (skôd´nyȧ)	142b	55°57'N	37°21'E
Skhodnya, r., Russia	142b	55°55'N	37°16'E
Skíathos, i., Grc. (skē´ȧ-thôs)	131	39°15'N	23°25'E
Skibbereen, Ire. (skĭb´ẽr-ēn)	120	51°32'N	9°25'W
Skidegate, b., Can. (skĭ´-dē-gāt´)	54	53°15'N	132°00'W
Skidmore, Tx., U.S. (skĭd´môr)	81	28°16'N	97°40'W
Skien, Nor. (skē´*ĕ*n)	116	59°13'N	9°35'E
Skierniewice, Pol. (skyĕr-nyē-vēt´sē)	125	51°58'N	20°13'E
Skihist Mountain, mtn., Can.	55	50°11'N	121°54'W
Skikda, Alg.	184	36°58'N	6°51'E
Skilpadfontein, S. Afr.	192c	25°02'S	28°50'E
Skive, Den. (skē´vĕ)	122	56°34'N	8°56'E
Skjálfandafljót, r., Ice. (skyäl´fänd-ô)	116	65°24'N	16°40'W
Skjerstad, Nor. (skyĕr-städ)	116	67°12'N	15°37'E
Škofja Loka, Slvn. (shkôf´yȧ lō´kȧ)	130	46°10'N	14°20'E
Skokie, Il., U.S. (skō´kē)	69a	42°02'N	87°45'W
Skokomish Indian Reservation, I.R., Wa., U.S. (Skō-kō´mĭsh)	74a	47°22'N	123°07'W
Skole, Ukr. (skô´lĕ)	125	49°03'N	23°32'E
Skópelos, i., Grc. (skô´pȧ-lôs)	131	39°04'N	23°31'E
Skopin, Russia (skô´pēn)	136	53°49'N	39°35'E
Skopje, Mac. (skôp´yĕ)	130	42°02'N	21°26'E
Skövde, Swe. (shûv´d*ĕ*)	116	58°25'N	13°48'E
Skovorodino, Russia (skô´vô-rô´dĭ-nô)	135	53°53'N	123°56'E
Skowhegan, Me., U.S. (skou-hē´g*ă*n)	60	44°45'N	69°27'W
Skradin, Cro. (skrä´dēn)	131	43°49'N	17°58'E
Skreia, Nor. (skrā´ȧ)	122	60°40'N	10°55'E
Skudeneshavn, Nor. (skōō´dĕ-nes-houn´)	122	59°10'N	5°19'E
Skull Valley Indian Reservation, I.R., Ut., U.S. (skŭl)	77	40°25'N	112°50'W
Skuna, r., Ms., U.S. (skû´nȧ)	82	33°57'N	89°36'W
Skunk, r., Ia., U.S. (skŭnk)	71	41°12'N	92°14'W
Skuodas, Lith. (skwô´dȧs)	123	56°16'N	21°32'E
Skurup, Swe. (skû´rŏp)	122	55°29'N	13°27'E
Skvyra, Ukr.	137	49°43'N	29°41'E
Skwierzyna, Pol. (skvē-ĕr´zhĭ-nȧ)	124	52°35'N	15°30'E
Skye, Island of, i., Scot., U.K. (skī)	116	57°25'N	6°17'W
Skykomish, r., Wa., U.S. (skī´kō-mĭsh)	74a	47°50'N	121°55'W
Skyring, Seno de, b., Chile (sĕ´nō-s-krĕ´ng)	102	52°35'S	72°30'W
Skýros, Grc.	131	38°53'N	24°32'E
Skýros, i., Grc.	119	38°50'N	24°43'E
Slagese, Den.	122	55°25'N	11°19'E
Slamet, Gunung, mtn., Indon. (slä´mĕt)	168	7°15'S	109°15'E
Slănic, Rom. (slŭ´nĕk)	131	45°13'N	25°56'E
Slater, Mo., U.S. (slāt´ẽr)	79	39°13'N	93°03'W
Slatina, Rom. (slä´tē-nä)	131	44°26'N	24°21'E
Slaton, Tx., U.S. (slā´t*ŭ*n)	78	33°26'N	101°38'W
Slave, r., Can. (slāv)	52	59°40'N	111°21'W
Slavgorod, Russia (slăf´gô-rôt)	134	52°58'N	78°43'E
Slavonija, hist. reg., Serb. (slä-vô´nē-yä)	131	45°29'N	17°31'E
Slavonska Požega, Cro. (slä-vôn´skä pô´zhĕ-gä)	131	45°18'N	17°42'E
Slavonski Brod, Cro. (skä-vôn´skē brôd)	119	45°10'N	18°01'E
Slavuta, Ukr. (slȧ-vōō´tȧ)	133	50°18'N	27°01'E
Slavyanskaya, Russia (slȧv-yȧn´skȧ-yȧ)	133	45°14'N	38°09'E
Sławno, Pol. (swav´nō)	124	54°21'N	16°38'E
Slayton, Mn., U.S. (slā´t*ŭ*n)	70	44°00'N	95°44'W
Sleaford, Eng., U.K. (slē´fẽrd)	114a	53°00'N	0°25'W
Sleepy Eye, Mn., U.S. (slēp´ī ī)	71	44°17'N	94°44'W
Slidell, La., U.S. (slī-dĕl´)	81	30°17'N	89°47'W
Sliedrecht, Neth.	115a	51°49'N	4°46'E
Sligo, Ire. (slī´gō)	116	54°17'N	8°19'W
Slite, Swe. (slē´tĕ)	122	57°41'N	18°47'E
Sliven, Blg. (slē´vĕn)	119	42°41'N	26°20'E
Sloatsburg, N.Y., U.S. (slōts´bûrg)	68a	41°09'N	74°11'W
Slonim, Bela. (swō´nĕm)	125	53°05'N	25°19'E
Slough, Eng., U.K. (slou)	114b	51°29'N	0°36'W
Slovakia, nation, Eur.	125	48°50'N	20°00'E
Slovenia, nation, Eur.	130	45°58'N	14°43'E
Slovians'k, Ukr.	137	48°52'N	37°34'E
Sluch, r., Ukr.	137	50°56'N	26°48'E
Slunj, Cro. (slōn´)	130	45°08'N	15°46'E
Słupsk, Pol. (swòpsk)	116	54°28'N	17°02'E
Slutsk, Bela. (slòtsk)	136	53°02'N	27°34'E
Slyne Head, c., Ire. (slīn)	116	53°25'N	10°05'W
Smackover, Ar., U.S. (smăk´ô-vẽr)	79	33°22'N	92°42'W
Smederevo, Serb.	131	44°39'N	20°54'E
Smederevska Palanka, Serb. (smĕ-dĕ-rĕv´skä pä-län´kä)	131	44°21'N	21°00'E
Smedjebacken, Swe. (smĭ´tyĕ-bä-kĕn)	122	60°09'N	15°19'E
Smethport, Pa., U.S. (smĕth´pōrt)	67	41°50'N	78°25'W
Smethwick, Eng., U.K.	120	52°31'N	2°04'W
Smila, Ukr.	137	49°14'N	31°52'E
Smile, Ukr.	133	50°55'N	33°36'E
Smiltene, Lat. (smēl´tĕ-nĕ)	123	57°26'N	25°57'E
Smith, Can. (smĭth)	50	55°10'N	114°02'W
Smith, i., Wa., U.S.	74a	48°20'N	122°53'W
Smith, r., Mt., U.S.	73	47°00'N	111°20'W
Smith Center, Ks., U.S. (sĕn´tẽr)	78	39°45'N	98°46'W
Smithers, Can. (smĭth´ẽrs)	50	54°47'N	127°10'W
Smithfield, N.C., U.S. (smĭth´fēld)	83	35°30'N	78°21'W
Smithfield, Ut., U.S.	73	41°50'N	111°49'W
Smithland, Ky., U.S. (smĭth´l*ă*nd)	66	37°10'N	88°25'W
Smith Mountain Lake, res., Va., U.S.	83	37°00'N	79°45'W
Smith Point, Tx., U.S.	81a	29°32'N	94°45'W
Smiths Falls, Can. (smĭths)	51	44°55'N	76°05'W
Smithton, Austl. (smĭth´t*ŭ*n)	176	40°55'S	145°12'E
Smithton, Il., U.S.	75e	38°24'N	89°59'W
Smithville, Tx., U.S. (smĭth´vĭl)	81	30°00'N	97°08'W
Smitswinkelvlakte, pl., S. Afr.	186a	34°16'S	18°25'E
Smoke Creek Desert, des., Nv., U.S. (smōk crēk)	76	40°28'N	119°40'W
Smoky, r., Can. (smōk´ĭ)	55	55°30'N	117°30'W
Smoky Hill, r., U.S. (smōk´ĭ hĭl)	64	38°40'N	100°00'W
Smøla, i., Nor. (smûlä)	116	63°16'N	7°40'E
Smolensk, Russia (smô-lyĕnsk´)	134	54°46'N	32°03'E
Smolensk, prov., Russia	132	55°00'N	32°18'E
Smyadovo, Blg.	131	43°04'N	27°00'E
Smyrna *see* İzmir, Tur.	154	38°25'N	27°05'E
Smyrna, De., U.S. (smûr´n*ȧ*)	67	39°20'N	75°35'W
Smyrna, Ga., U.S.	68c	33°53'N	84°31'W
Snag, Can. (snăg)	63	62°18'N	140°30'W
Snake, r., Mn., U.S. (snāk)	71	45°58'N	93°20'W
Snake, r., U.S.	64	45°30'N	117°00'W
Snake Range, mts., Nv., U.S.	77	39°20'N	114°15'W
Snake River Plain, pl., Id., U.S.	73	43°08'N	114°46'W
Snap Point, c., Bah.	92	23°45'N	77°30'W
Sneffels, Mount, mtn., Co., U.S. (snĕf´*ĕ*lz)	77	38°00'N	107°50'W
Snelgrove, Can. (snĕl´grōv)	62d	43°44'N	79°50'W
Sniardwy, Jezioro, l., Pol. (snyärt´vĭ)	125	53°46'N	21°59'E
Snöhetta, mtn., Nor. (snû-hĕttä)	116	62°18'N	9°12'E
Snohomish, Wa., U.S. (snô-hō´mĭsh)	74a	47°55'N	122°05'W
Snohomish, r., Wa., U.S.	74a	47°53'N	122°04'W
Snoqualmie, Wa., U.S. (snō qwâl´mē)	74a	47°32'N	121°50'W
Snoqualmie, r., Wa., U.S.	72	47°32'N	121°53'W
Snov, r., Eur. (snôf)	133	51°38'N	31°38'E
Snowdon, mtn., Wales, U.K.	120	53°05'N	4°04'W
Snow Hill, Md., U.S. (hĭl)	67	38°15'N	75°20'W
Snow Lake, Can.	57	54°50'N	100°10'W
Snowy Mountains, mts., Austl. (snō´ē)	175	36°17'S	148°30'E
Snyder, Ok., U.S. (snī´dẽr)	78	34°40'N	98°57'W
Snyder, Tx., U.S.	80	32°48'N	100°53'W
Soar, r., Eng., U.K. (sōr)	114a	52°44'N	1°09'W
Sobat, r., Sudan (sō´bät)	185	9°04'N	32°02'E
Sobinka, Russia (sô-bĭŋ´kȧ)	132	55°59'N	40°02'E
Sobo Zan, mtn., Japan (sō´bô zän)	166	32°47'N	131°27'E
Sobral, Braz. (sô-brä´l)	101	3°39'S	40°16'W
Sochaczew, Pol. (sô-ĸä´chĕf)	125	52°14'N	20°18'E
Sochi, Russia (sôch´ĭ)	134	43°35'N	39°50'E
Society Islands, is., Fr. Poly. (sô-sī´*ĕ*-tē)	195	15°00'S	157°30'W
Socoltenango, Mex. (sô-kôl-tē-näŋ´gō)	89	16°17'N	92°20'W
Socorro, Braz. (sō-kô´r-rō)	99a	22°35'S	46°32'W
Socorro, Col. (sô-kôr´rō)	100	6°23'N	73°19'W
Socorro, N.M., U.S.	77	34°05'N	106°55'W
Socuéllamos, Spain (sô-kōō-āl´yä-môs)	128	39°18'N	2°48'W
Soda, l., Ca., U.S. (sō´d*ȧ*)	76	35°12'N	116°25'W
Soda Peak, mtn., Wa., U.S.	74c	45°53'N	122°04'W
Soda Springs, Id., U.S. (sprĭngz)	73	42°39'N	111°37'W
Söderhamn, Swe. (sû-dẽr-häm´'n)	116	61°20'N	17°00'E
Söderköping, Swe.	122	58°30'N	16°14'E
Södertälje, Swe. (sû-dĕr-tĕl´yĕ)	116	59°12'N	17°35'E
Sodo, Eth.	185	7°03'N	37°46'E
Soest, Ger. (zōst)	124	51°35'N	8°05'E
Sofia (Sofiya), Blg. (sô´fē-yȧ) (sô´fē-ȧ)	110	42°43'N	23°20'E
Sofiïvka, Ukr.	133	48°03'N	33°53'E
Sofiya *see* Sofia, Blg.	110	42°43'N	23°20'E
Soga, Japan (sō´gä)	167a	35°35'N	140°08'E
Sogamoso, Col. (sô-gä-mô´sō)	100	5°42'N	72°51'W
Sognafjorden, b., Nor.	112	61°09'N	5°30'E
Sogozha, r., Russia (sô´gô-zhȧ)	132	58°35'N	39°08'E
Sohano, Pap. N. Gui.	170e	5°27'S	154°40'E
Soissons, Fr. (swä-sôɴ´)	126	49°23'N	3°17'E
Sōka, Japan (sō´kä)	167a	35°50'N	139°49'E
Sokal', Ukr. (sô´käl´)	125	50°28'N	24°20'E
Söke, Tur. (sû´kĕ)	119	37°40'N	27°10'E
Sokólka, Pol. (sô-kòl´kȧ)	125	53°23'N	23°30'E
Sokolo, Mali (sô-kô-lō´)	184	14°51'N	6°09'W
Sokołów Podlaski, Pol. (sô-kô-wòf´ pŭd-lä´skĭ)	125	52°24'N	22°15'E
Sokone, Sen.	188	13°53'N	16°22'W
Sokoto, Nig. (sō´kô-tō)	184	13°04'N	5°16'E
Sola de Vega, Mex.	89	16°31'N	96°58'W
Solander, Cape, c., Austl.	173b	34°03'S	151°16'E
Solano, Phil. (sō-lä´nō)	169a	16°31'N	121°11'E
Soledad, Col. (sō-lē-dä´d)	100	10°47'N	75°00'W
Soledad Díez Gutiérrez, Mex.	88	22°19'N	100°54'W
Soleduck, r., Wa., U.S. (sōl´dŭk)	72	47°59'N	124°28'W
Solentiname, Islas de, is., Nic. (ĕ´s-läs-dĕ-sô-lĕn-tē-nä´mȧ)	90	11°15'N	85°16'W
Solihull, Eng., U.K. (sō´lĭ-hŭl)	114a	52°25'N	1°46'W
Solihull, co., Eng., U.K.	114a	52°25'N	1°42'W
Solikamsk, Russia (sô-lē-kȧmsk´)	136	59°38'N	56°48'E
Sol'-Iletsk, Russia	134	51°10'N	55°05'E
Solimões *see* Amazon, r., Braz.	100	2°45'S	67°44'W
Solingen, Ger. (zō´lĭng-*ĕ*n)	124	51°10'N	7°05'E
Sóller, Spain (sō´lyĕr)	129	39°45'N	2°40'E
Sologne, reg., Fr. (sō-lōn´y*ĕ*)	126	47°36'N	1°53'E
Sololá, Guat. (sô-lō´lä)	90	14°45'N	91°12'W
Solomon, r., Ks., U.S.	78	39°24'N	98°19'W
Solomon, North Fork, r., Ks., U.S.	78	39°34'N	99°52'W
Solomon, South Fork, r., Ks., U.S.	78	39°19'N	99°52'W
Solomon Islands, nation, Oc. (sŏ´lō-mŭn)	3	7°00'S	160°00'E
Solon, China (swo-lōōn)	161	46°32'N	121°18'E
Solon, Oh., U.S. (sō´l*ŭ*n)	69d	41°23'N	81°26'W
Solothurn, Switz. (zō´lō-thōōrn)	124	47°13'N	7°30'E
Solovetskiye Ostrova, is., Russia	136	65°10'N	35°40'E
Šolta, i., Serb. (shôl´tä)	130	43°20'N	16°15'E
Soltau, Ger. (sôl´tou)	124	53°00'N	9°50'E
Sol'tsy, Russia (sôl´tsĕ)	132	58°04'N	30°13'E
Solvay, N.Y., U.S. (sŏl´vā)	67	43°05'N	76°10'W
Sölvesborg, Swe. (sûl´vĕs-bôrg)	122	56°04'N	14°35'E
Sol'vychegodsk, Russia (sôl´vē-chĕ-gôtsk´)	136	61°18'N	46°58'E
Solway Firth, b., U.K. (sŏl´wāfûrth´)	116	54°42'N	3°55'W
Solwezi, Zam.	191	12°11'S	26°25'E
Soly, Bela.	122	54°31'N	26°11'E
Somalia, nation, Afr. (sō-mä´lē-*ȧ*)	192a	3°28'N	44°47'E
Somanga, Tan.	191	8°24'S	39°17'E
Sombor, Serb. (sôm´bôr)	119	45°45'N	19°10'E
Sombrerete, Mex. (sôm-brȧ-rā´tȧ)	88	23°38'N	103°37'W
Sombrero, Cayo, i., Ven. (kä-yô-sôm-brĕ´rō)	101b	10°52'N	68°12'W
Somerset, Ky., U.S. (sŭm´ẽr-sĕt)	82	37°05'N	84°35'W
Somerset, Ma., U.S.	68b	41°46'N	71°05'W
Somerset, Pa., U.S.	67	40°00'N	79°05'W
Somerset, Tx., U.S.	75d	29°13'N	98°39'W
Somerset East, S. Afr.	187c	32°44'S	25°36'E
Somersworth, N.H., U.S. (sŭm´ẽrz-wûrth)	60	43°16'N	70°53'W
Somerton, Az., U.S. (sŭm´ẽr-t*ŭ*n)	77	32°36'N	114°43'W
Somerville, Ma., U.S. (sŭm´ẽr-vĭl)	61a	42°23'N	71°06'W
Somerville, N.J., U.S.	68a	40°34'N	74°37'W
Somerville, Tn., U.S.	82	35°14'N	89°21'W
Somerville, Tx., U.S.	81	30°21'N	96°31'W
Someș, r., Eur.	125	47°43'N	23°09'E
Somma Vesuviana, Italy (sôm´mä vā-zōō-vē-ä´nä)	129c	40°38'N	14°27'E
Somme, r., Fr. (sŏm)	126	50°02'N	2°04'E
Sommerfeld, Ger. (zō´mĕr-fĕld)	115b	52°48'N	13°02'E
Sommerville, Austl.	173a	38°14'S	145°10'E
Somoto, Nic. (sô-mō´tō)	90	13°28'N	86°37'W
Son, r., India (sōn)	155	24°40'N	82°35'E
Sŏnchŏn, Kor., N. (sŭn´shŭn)	166	39°49'N	124°56'E
Sondags, r., S. Afr.	187c	33°17'S	25°14'E
Sønderborg, Den. (sûn´'er-bôrgh)	116	54°55'N	9°47'E
Sondershausen, Ger. (zôn´dẽrz-hou´zĕn)	124	51°17'N	10°45'E
Song Ca, r., Viet.	165	19°15'N	105°00'E
Songea, Tan. (sôn-gā´ȧ)	186	10°41'S	35°39'E
Songjiang, China	161	31°01'N	121°14'E
Sŏngjin, Kor., N. (sŭng´jĭn´)	166	40°38'N	129°10'E
Songkhla, Thai. (sông´klä´)	168	7°09'N	100°34'E
Songwe, D.R.C.	191	12°25'S	29°40'E

PRONOUNCING INDEX Sonneberg – Sri Jayewardenepura Kotte 305

PLACE (Pronunciation)	PAGE	LAT.	LONG.
Sonneberg, Ger. (sŏn′ē-bĕrgh)	124	50°20′N	11°14′E
Sonora, Ca., U.S. (sō-nō′rȧ)	76	37°58′N	120°22′W
Sonora, Tx., U.S.	80	30°33′N	100°38′W
Sonora, state, Mex.	86	29°45′N	111°15′W
Sonora, r., Mex.	86	28°45′N	111°35′W
Sonora Peak, mtn., Ca., U.S.	64	38°22′N	119°39′W
Sonseca, Spain (sôn-sā′kä)	128	39°41′N	3°56′W
Sonsón, Col. (sôn-sôn′)	100	5°42′N	75°28′W
Sonsonate, El Sal. (sōn-sō-nä′tå)	90	13°46′N	89°43′W
Sonsorol Islands, is., Palau (sŏn-sŏ-rōl′)	169	5°03′N	132°33′E
Sooke Basin, b., Can. (sŏk)	74a	48°21′N	123°47′W
Soo Locks, trans., Mi., U.S. (sōō lŏks)	75a	46°30′N	84°30′W
Sopetrán, Col. (sō-pě-trä′n)	100a	6°30′N	75°44′W
Sopot, Pol. (sô′pôt)	125	54°26′N	18°25′E
Sopron, Hung. (shŏp′rôn)	119	47°41′N	16°36′E
Sora, Italy (sō′rä)	130	41°43′N	13°37′E
Sorbas, Spain (sôr′bäs)	128	37°05′N	2°07′W
Sordo, r., Mex. (sô′r-dō)	89	16°39′N	97°33′W
Sorel, Can. (sō-rĕl′)	51	46°01′N	73°07′W
Sorell, Cape, c., Austl.	176	42°10′S	144°50′E
Soresina, Italy (sō-rå-zē′nä)	130	45°17′N	9°51′E
Soria, Spain (sō′rē-ä)	118	41°46′N	2°28′W
Soriano, dept., Ur. (sō-rēä′nō)	99c	33°25′S	58°00′W
Soroca, Mol.	137	48°09′N	28°17′E
Sorocaba, Braz. (sō-rō-kä′bȧ)	101	23°29′S	47°27′W
Sorong, Indon. (sō-rông′)	169	1°00′S	131°20′E
Sorot′, r., Russia (sō-rō′tzh)	132	57°08′N	29°23′E
Soroti, Ug. (sō-rō′tē)	185	1°43′N	33°37′E
Sørøya, i., Nor.	116	70°37′N	20°58′E
Sorraia, r., Port. (sôr-rī′ȧ)	128	38°55′N	8°42′W
Sorrento, Italy (sôr-rĕn′tō)	130	40°23′N	14°23′E
Sorsogon, Phil. (sŏr-sōgōn′)	169	12°51′N	124°02′E
Sortavala, Russia (sôr′tä-vä-lä)	134	61°43′N	30°40′E
Sosna, r., Russia (sôs′nȧ)	133	50°33′N	38°15′E
Sosnogorsk, Russia	134	63°13′N	54°09′E
Sosnowiec, Pol. (sôs-nō′vyĕts)	125	50°17′N	19°10′E
Sosnytsia, Ukr.	133	51°30′N	32°29′E
Sosunova, Mys, c., Russia (mĭs sô′sŏ-nôf′ȧ)	166	46°28′N	138°06′E
Sos′va, r., Russia (sôs′vȧ)	142a	59°55′N	60°40′E
Sos′va, r., Russia (sôs′vȧ)	136	63°10′N	63°30′E
Sota, r., Benin	189	11°10′N	3°20′E
Sota la Marina, Mex. (sō-tä-lä-mä-rē′nä)	88	23°45′N	98°11′W
Soteapan, Mex. (sō-tå-ä′pän)	89	18°14′N	94°51′W
Soto la Marina, Río, r., Mex. (rē′ō-so′tō lä mä-rē′nä)	88	23°55′N	98°30′W
Sotuta, Mex. (sō-tōō′tä)	90a	20°35′N	89°00′W
Soublette, Ven. (sō-ōō-blĕ′tĕ)	101b	9°55′N	66°06′W
Souflí, Grc.	131	41°12′N	26°17′E
Soufrière, St. Luc. (sōō-frē-âr′)	91b	13°50′N	61°03′W
Soufrière, mtn., St. Vin.	91b	13°19′N	61°12′W
Soufrière, vol., Guad. (sōō-frē-âr′)	91b	16°06′N	61°42′W
Sŏul *see* Seoul, Kor., S.	161	37°35′N	127°03′E
Sounding Creek, r., Can. (soun′dĭng)	56	51°35′N	111°00′W
Souq Ahras, Alg.	117	36°23′N	8°00′E
Sources, Mount aux, mtn., Afr. (môn′tō sôrs′)	186	28°47′S	29°04′E
Soure, Port. (sōr-ĕ)	128	40°04′N	8°37′W
Souris, Can. (sōō′rē′)	61	46°20′N	62°17′W
Souris, Can.	50	49°38′N	100°15′W
Souris, r., N.A.	52	48°30′N	101°30′W
Sourlake, Tx., U.S. (sour′lāk)	81	30°09′N	94°24′W
Sousse, Tun. (sōōs)	184	36°00′N	10°39′E
South, r., Ga., U.S.	68c	33°40′N	84°15′W
South, r., N.C., U.S.	83	34°49′N	78°33′W
South Africa, nation, Afr.	186	28°00′S	24°50′E
South Amboy, N.J., U.S. (south′ăm′boi)	68a	40°28′N	74°17′W
South America, cont.	97	15°00′S	60°00′W
Southampton, Eng., U.K. (south-ămp′tŭn)	110	50°54′N	1°30′W
Southampton, N.Y., U.S.	67	40°53′N	72°24′W
Southampton Island, i., Can.	53	64°38′N	84°00′W
South Andaman Island, i., India (ăn-dȧ-măn′)	168	11°57′N	93°24′E
South Australia, state, Austl. (ôs-trā′lĭ-ȧ)	174	29°45′S	132°00′E
South Bay, b., Bah.	93	20°55′N	73°35′W
South Bend, In., U.S. (bĕnd)	65	41°40′N	86°20′W
South Bend, Wa., U.S. (bĕnd)	72	46°39′N	123°48′W
South Bight, b., Bah.	92	24°20′N	77°35′W
South Bimini, i., Bah. (bē′mē-nē)	92	25°40′N	79°20′W
Southborough, Ma., U.S. (south′bŭr-ō)	61a	42°18′N	71°33′W
South Boston, Va., U.S. (bôs′tŭn)	83	36°41′N	78°55′W
Southbridge, Ma., U.S. (south′brĭj)	67	42°05′N	72°00′W
South Caicos, i., T./C. Is. (kī′kōs)	93	21°30′N	71°35′W
South Carolina, state, U.S. (kăr-ō-lī′nȧ)	65	34°15′N	81°10′W
South Cave, Eng., U.K. (cāv)	114a	53°45′N	0°35′W
South Charleston, W.V., U.S.	66	38°20′N	81°40′W
South China Sea, sea, Asia (chī′nȧ)	168	15°23′N	114°12′E
South Creek, r., Austl.	173b	33°43′S	150°50′E
South Dakota, state, U.S. (dȧ-kō′tȧ)	64	44°20′N	101°55′W
South Downs, Eng., U.K. (dounz)	120	50°55′N	1°13′W
South Dum-Dum, India	158a	22°36′N	88°25′E
South East Cape, c., Austl.	175	43°47′S	146°03′E
Southend-on-Sea, Eng., U.K. (south-ĕnd′)	121	51°33′N	0°41′E
Southern Alps, mts., N.Z. (sŭ-thûrn ălps)	175a	43°35′S	170°00′E
Southern Cross, Austl.	174	31°13′S	119°30′E
Southern Indian, l., Can. (sŭth′ẽrn ĭn′dĭ-ăn)	52	56°46′N	98°57′W
Southern Pines, N.C., U.S. (sŭth′ẽrn pīnz)	83	35°10′N	79°23′W
Southern Ute Indian Reservation, I.R., Co., U.S. (ūt)	77	37°05′N	108°23′W
South Euclid, Oh., U.S. (ū′klĭd)	69d	41°30′N	81°34′W
South Fox, i., Mi., U.S. (fŏks)	66	45°25′N	85°55′W
South Gate, Ca., U.S. (gāt)	75a	33°57′N	118°13′W
South Georgia, i., S. Geor. (jôr′jȧ)	97	54°00′S	37°00′W
South Haven, Mi., U.S. (hāv′′n)	66	42°25′N	86°15′W
South Hill, Va., U.S.	83	36°44′N	78°08′W
South Holston Lake, res., U.S.	83	36°35′N	82°00′W
South Indian Lake, Can.	57	56°50′N	99°00′W
Southington, Ct., U.S. (sŭdh′ĭng-tŭn)	67	41°35′N	72°55′W
South Island, i., N.Z.	175a	42°40′S	169°00′E
South Loup, r., Ne., U.S. (lōōp)	70	41°21′N	100°08′W
South Magnetic Pole, pt. of. i.,	178	65°18′S	139°30′E
South Merrimack, N.H., U.S. (mĕr′ĭ-măk)	61a	42°47′N	71°36′W
South Milwaukee, Wi., U.S. (mĭl-wô′kē)	69a	42°55′N	87°52′W
South Moose Lake, l., Can.	57	53°51′N	100°20′W
South Nation, r., Can.	59	45°00′N	75°25′W
South Negril Point, c., Jam. (nå-grĕl′)	92	18°15′N	78°25′W
South Ogden, Ut., U.S. (ŏg′dĕn)	75b	41°12′N	111°58′W
South Orkney Islands, is., Ant.	97	57°00′S	45°00′W
South Ossetia, hist. reg., Geor.	138	42°20′N	44°00′E
South Paris, Me., U.S. (păr′ĭs)	60	44°13′N	70°32′W
South Park, Ky., U.S. (pärk)	69h	38°06′N	85°43′W
South Pasadena, Ca., U.S. (păs-å-dē′nȧ)	75a	34°06′N	118°08′W
South Pease, r., Tx., U.S. (pēz)	78	33°54′N	100°45′W
South Pender, i., Can. (pĕn′dẽr)	74d	48°45′N	123°09′W
South Pittsburg, Tn., U.S. (pĭs′bûrg)	82	35°00′N	85°42′W
South Platte, r., U.S. (plăt)	64	40°40′N	102°40′W
South Point, c., Barb.	91b	13°00′N	59°43′W
South Point, c., Mi., U.S.	66	44°50′N	83°20′W
South Pole, pt. of. i., Ant.	178	90°00′S	0°00′
South Porcupine, Can.	58	48°28′N	81°13′W
Southport, Austl. (south′pōrt)	175	27°57′S	153°27′E
Southport, Eng., U.K. (south′pôrt)	120	53°38′N	3°00′W
Southport, In., U.S.	69g	39°40′N	86°07′W
Southport, N.C., U.S.	83	35°55′N	78°02′W
South Portland, Me., U.S. (pōrt-lănd)	60	43°37′N	70°15′W
South Prairie, Wa., U.S. (prā′rĭ)	74a	47°08′N	122°06′W
South Range, Wi., U.S. (rănj)	75h	46°37′N	91°59′W
South River, N.J., U.S. (rĭv′ẽr)	68a	40°27′N	74°23′W
South Ronaldsay, i., Scot., U.K. (rŏn′ăld-s′ā)	120a	58°48′N	2°55′W
South Saint Paul, Mn., U.S.	75g	44°54′N	93°02′W
South Salt Lake, Ut., U.S. (sôlt lāk)	75b	40°44′N	111°53′W
South Sandwich Islands, is., S. Geor. (sănd′wĭch)	97	58°00′S	27°00′W
South Sandwich Trench, deep,	97	55°00′S	27°00′W
South San Francisco, Ca., U.S. (săn frăn-sĭs′kō)	74b	37°39′N	122°24′W
South Saskatchewan, r., Can. (sås-kach′ĕ-wän)	52	50°30′N	110°30′W
South Shetland Islands, is., Ant.	97	62°00′S	70°00′W
South Shields, Eng., U.K. (shēldz)	116	55°00′N	1°22′W
South Sioux City, Ne., U.S. (sōō sĭt′ē)	70	42°48′N	96°26′W
South Taranaki Bight, b., N.Z. (tä-rä-nä′kē)	175a	39°35′S	173°50′E
South Thompson, r., Can. (tŏmp′sŭn)	55	50°41′N	120°21′W
Southton, Tx., U.S. (south′tŭn)	75d	29°18′N	98°26′W
South Uist, i., Scot., U.K. (ū′ĭst)	120	57°15′N	7°24′W
South Umpqua, r., Or., U.S. (ŭmp′kwȧ)	72	43°00′N	122°54′W
Southwell, Eng., U.K. (south′wĕl)	114a	53°04′N	0°56′W
South West Africa *see* Namibia, nation, Afr.	186	19°30′S	16°13′E
Southwest Miramichi, r., Can. (mĭr ȧ-mē′shē)	60	46°35′N	66°17′W
Southwest Point, c., Bah.	92	25°50′N	77°10′W
Southwest Point, c., Bah.	93	23°55′N	74°30′W
South Yorkshire, hist. reg., Eng., U.K.	114a	53°29′N	1°35′W
Sovetsk, Russia (sō-vyĕtsk′)	136	55°04′N	21°54′E
Sovetskaya Gavan′, Russia (sŭ-vyĕt′skī-u gä′vŭn′)	135	48°59′N	140°14′E
Sow, r., Eng., U.K. (sou)	114a	52°45′N	2°12′W
Soya Kaikyō, strt., Asia	166	45°45′N	141°38′E
Sōya Misaki, c., Japan (sō′yȧ mē′sä-kē)	166	45°35′N	141°25′E
Soyo, Ang.	186	6°10′S	12°25′E
Sozh, r., Eur. (sôzh)	137	52°50′N	31°00′E
Sozopol, Blg. (sôz′ō-pôl′)	131	42°18′N	27°50′E
Spa, Bel. (spä)	121	50°30′N	5°50′E
Spain, nation, Eur. (spān)	110	40°15′N	4°30′W
Spalding, Ne., U.S. (spôl′dĭng)	70	41°43′N	98°23′W
Spanaway, Wa., U.S. (spăn′ȧ-wā)	74a	47°06′N	122°26′W
Spangler, Pa., U.S. (spăng′lẽr)	67	40°40′N	78°50′W
Spanish Fork, Ut., U.S. (spăn′ĭsh fôrk)	77	40°10′N	111°40′W
Spanish Town, Jam.	87	18°00′N	76°55′W
Sparks, Nv., U.S. (spärks)	76	39°34′N	119°45′W
Sparrows Point, Md., U.S. (spăr′ōz)	68e	39°13′N	76°29′W
Sparta *see* Spárti, Grc.	131	37°07′N	22°28′E
Sparta, Ga., U.S. (spär′tȧ)	83	33°16′N	82°59′W
Sparta, Il., U.S.	79	38°07′N	89°42′W
Sparta, Mi., U.S.	66	43°10′N	85°45′W
Sparta, Tn., U.S.	82	35°54′N	85°26′W
Sparta, Wi., U.S.	71	43°56′N	90°50′W
Sparta Mountains, mts., N.J., U.S.	68a	41°00′N	74°38′W
Spartanburg, S.C., U.S. (spär′tăn-bûrg)	65	34°57′N	82°13′W
Spartel, Cap, c., Mor. (spär-tĕl′)	128	35°48′N	5°50′W
Spárti (Sparta), Grc.	131	37°07′N	22°28′E
Spartivento, Cape, c., Italy (spär-tē-vĕn′tô)	130	37°55′N	16°09′E
Spartivento, Cape, c., Italy	112	38°54′N	8°52′E
Spas-Demensk, Russia (spȧs dyĕ-mĕnsk′)	132	54°24′N	34°02′E
Spas-Klepiki, Russia (spȧs klĕp′ē-kē)	132	55°09′N	40°11′E
Spassk-Ryazanskiy, Russia (ryä-zän′skĭ)	132	54°24′N	40°21′E
Spassk-Dal′niy, Russia (spŭsk′dăl′nyē)	135	44°30′N	133°00′E
Spátha, Ákra, c., Grc.	130a	35°42′N	23°45′E
Spaulding, Al., U.S. (spôl′dĭng)	68h	33°27′N	86°50′W
Spear, Cape, c., Can. (spēr)	61	47°32′N	52°32′W
Spearfish, S.D., U.S. (spēr′fĭsh)	70	44°28′N	103°52′W
Speed, In., U.S. (spēd)	69h	38°25′N	85°45′W
Speedway, In., U.S. (spēd′wā)	69g	39°47′N	86°14′W
Speichersee, l., Ger.	115d	48°12′N	11°47′E
Spencer, Ia., U.S.	70	43°09′N	95°08′W
Spencer, In., U.S. (spĕn′sẽr)	66	39°15′N	86°45′W
Spencer, N.C., U.S.	83	35°43′N	80°25′W
Spencer, W.V., U.S.	66	38°55′N	81°20′W
Spencer Gulf, b., Austl. (spĕn′sẽr)	174	34°20′S	136°55′E
Sperenberg, Ger. (shpĕ′rĕn-bĕrgh)	115b	52°09′N	13°22′E
Spey, l., Scot., U.K. (spā)	120	57°25′N	3°29′W
Speyer, Ger. (shpī′ẽr)	124	49°18′N	8°26′E
Sphinx, hist., Egypt (sfĭnks)	192b	29°57′N	31°08′E
Spijkenisse, Neth.	115a	51°51′N	4°18′E
Spinazzola, Italy (spē-nät′zō-lä)	130	40°58′N	16°05′E
Spirit Lake, Ia., U.S. (lāk)	70	43°25′N	95°08′W
Spirit Lake, Id., U.S. (spĭr′ĭt)	72	47°58′N	116°51′W
Spišská Nová Ves, Slvk. (spēsh′skä nō′vä vĕs)	117	48°56′N	20°35′E
Spitsbergen *see* Svalbard, dep., Nor.	134	77°00′N	20°00′E
Split, Cro. (splēt)	110	43°30′N	16°28′E
Split Lake, l., Can.	57	56°08′N	96°15′W
Spokane, Wa., U.S. (spōkăn′)	64	47°39′N	117°25′W
Spokane, r., Wa., U.S.	72	47°47′N	118°00′W
Spokane Indian Reservation, I.R., Wa., U.S.	72	47°55′N	118°00′W
Spoleto, Italy (spō-lā′tō)	130	42°44′N	12°44′E
Spoon, r., Il., U.S. (spōōn)	79	40°36′N	90°22′W
Spooner, Wi., U.S. (spōōn′ẽr)	71	45°50′N	91°53′W
Spotswood, N.J., U.S. (spŏtz′wŏŏd)	68a	40°23′N	74°22′W
Sprague, r., Or., U.S. (sprāg)	72	42°30′N	121°42′W
Spratly, i., Asia (sprăt′lē)	168	8°38′N	111°54′E
Spray, N.C., U.S. (sprā)	83	36°30′N	79°44′W
Spree, r., Ger. (shprā)	124	51°53′N	14°08′E
Spremberg, Ger. (shprĕm′bĕrgh)	124	51°35′N	14°23′E
Spring, r., Ar., U.S.	79	36°25′N	91°35′W
Springbok, S. Afr. (sprĭng′bŏk)	186	29°35′S	17°55′E
Spring Creek, r., Nv., U.S. (sprĭng)	76	40°18′N	117°45′W
Spring Creek, r., Tx., U.S.	81	30°03′N	95°43′W
Spring Creek, r., Tx., U.S.	80	31°08′N	100°50′W
Springdale, Can.	61	49°30′N	56°05′W
Springdale, Ar., U.S. (sprĭng′dāl)	79	36°10′N	94°07′W
Springdale, Pa., U.S.	69e	40°33′N	79°46′W
Springer, N.M., U.S. (sprĭng′ẽr)	78	36°21′N	104°37′W
Springerville, Az., U.S.	77	34°08′N	109°17′W
Springfield, Co., U.S. (sprĭng′fēld)	78	37°24′N	102°04′W
Springfield, Il., U.S.	65	39°46′N	89°37′W
Springfield, Ky., U.S.	66	37°35′N	85°10′W
Springfield, Ma., U.S.	65	42°05′N	72°35′W
Springfield, Mn., U.S.	71	44°14′N	94°59′W
Springfield, Mo., U.S.	65	37°13′N	93°17′W
Springfield, Oh., U.S.	65	39°55′N	83°50′W
Springfield, Or., U.S.	72	44°01′N	123°02′W
Springfield, Tn., U.S.	82	36°30′N	86°53′W
Springfield, Vt., U.S.	67	43°20′N	72°35′W
Springfontein, S. Afr. (sprĭng′fōn-tĭn)	186	30°16′S	25°45′E
Springhill, Can. (sprĭng-hĭl′)	51	45°39′N	64°03′W
Spring Mountains, mts., Nv., U.S.	76	36°18′N	115°49′W
Springs, S. Afr. (sprĭngs)	192c	26°16′S	28°27′E
Springstein, Can. (sprĭng′stīn)	62f	49°49′N	97°29′W
Springton Reservoir, res., Pa., U.S. (sprĭng-tŭn)	68f	39°57′N	75°26′W
Springvale, Austl.	173a	37°57′N	145°09′E
Spring Valley, Ca., U.S.	76a	32°46′N	117°01′W
Springvalley, Il., U.S. (sprĭng-văl′ĭ)	66	41°20′N	89°15′W
Spring Valley, Mn., U.S.	71	43°41′N	92°26′W
Spring Valley, N.Y., U.S.	68a	41°07′N	74°03′W
Springville, Ut., U.S. (sprĭng-vĭl)	77	40°10′N	111°40′W
Springwood, Austl.	173b	33°42′S	150°34′E
Spruce Grove, Can. (sprōōs grōv)	62g	53°32′N	113°55′W
Spur, Tx., U.S. (spûr)	78	33°29′N	100°51′W
Squam, l., N.H., U.S. (skwôm)	67	43°45′N	71°30′W
Squamish, Can. (skwŏ′mĭsh)	54	49°42′N	123°09′W
Squamish, r., Can.	54	50°10′N	123°30′W
Squillace, Golfo di, b., Italy (gōō′l-fō-dē skwĕl-lä′chå)	130	38°44′N	16°47′E
Srbija (Serbia), hist. reg., Serb. (sr bē-yä) (sĕr′bē-ä)	131	44°05′N	20°35′E
Srbobran, Serb. (s′r′bō-brän′)	131	45°32′N	19°50′E
Sredne-Kolymsk, Russia (s′rĕd′nyē kō-lēmsk′)	135	67°49′N	154°55′E
Sredne Rogatka, Russia (s′red′nȧ-ya) (rô gär′tkȧ)	142c	59°49′N	30°20′E
Sredniy Ik, r., Russia (srĕd′nĭ ĭk)	142a	55°46′N	58°50′E
Sredniy Ural, mts., Russia (ŏ′rál)	142a	57°47′N	59°00′E
Srem, Pol. (shrĕm)	125	52°06′N	17°01′E
Sremska Karlovci, Serb. (srĕm′skē kär′lov-tsē)	131	45°10′N	19°57′E
Sremska Mitrovica, Serb. (srĕm′skä mē′trō-vē-tsä′)	131	44°59′N	19°39′E
Sretensk, Russia (s′rĕ′tĕnsk)	135	52°13′N	117°39′E
Sri Jayewardenepura Kotte, Sri L.	159	6°50′N	80°05′E

ng-sing; ŋ-baŋk; ɴ-nasalized n; nŏd; cŏmmit; ōld; ôbey; ôrder; oi-boil; fōōd; ŏ-as oo in foot; ou-out; s-soft; sh-dish; th-thin; pūre; ůnite; ûrn; stŭd; circŭs; ü-as in French tu; ′-indeterminate vowel.

306 Sri Lanka – Suide PRONOUNCING INDEX

PLACE (Pronunciation)	PAGE	LAT.	LONG.
Sri Lanka, nation, Asia	159	8°45'N	82°30'E
Srinagar, India (srē-nŭg'ŭr)	155	34°11'N	74°49'E
Środa, Pol. (shrô'dà)	125	52°14'N	17°17'E
Stabroek, Bel.	115a	51°20'N	4°21'E
Stade, Ger. (shtä'dĕ)	124	53°36'N	9°28'E
Städjan, mtn., Swe. (stĕd'yän)	122	61°53'N	12°50'E
Stafford, Eng., U.K. (stăf'fērd)	120	52°48'N	2°06'W
Stafford, Ks., U.S.	78	37°58'N	98°37'W
Staffordshire, co., Eng., U.K.	114a	52°45'N	2°00'W
Stahnsdorf, Ger. (shtäns'dôrf)	115b	52°22'N	13°10'E
Staines, Eng., U.K.	114b	51°26'N	0°13'W
Stakhanov, Ukr.	137	48°34'N	38°37'E
Stalingrad see Volgograd, Russia	134	48°40'N	42°20'E
Stalybridge, Eng., U.K.	114a	53°29'N	2°03'W
Stambaugh, Mi., U.S. (stăm'bô)	71	46°03'N	88°38'W
Stamford, Eng., U.K.	114a	52°39'N	0°28'W
Stamford, Ct., U.S. (stăm'fērd)	68a	41°03'N	73°32'W
Stamford, Tx., U.S.	78	32°57'N	99°48'W
Stammersdorf, Aus. (shtäm'ērs-dôrf)	115e	48°19'N	16°25'E
Stamps, Ar., U.S. (stămps)	79	33°22'N	93°31'W
Stanberry, Mo., U.S. (stan'bĕr-ē)	79	40°12'N	94°34'W
Standerton, S. Afr. (stăn'dēr-tŭn)	186	26°57'S	29°17'E
Standing Rock Indian Reservation, I.R., N.D., U.S. (stănd'ĭng rŏk)	70	47°07'N	101°05'W
Standish, Eng., U.K. (stăn'dĭsh)	114a	53°36'N	2°39'W
Stanford, Ky., U.S. (stăn'fērd)	82	37°29'N	84°40'W
Stanger, S. Afr. (stăŋ-ger)	187c	29°22'S	31°18'E
Staniard Creek, Bah.	92	24°50'N	77°55'W
Stanislaus, r., Ca., U.S. (stăn'ĭs-lô)	76	38°10'N	120°16'W
Stanley, Can. (stăn'lē)	60	46°17'N	66°44'W
Stanley, Falk. Is.	102	51°46'S	57°59'W
Stanley, N.D., U.S.	70	48°20'N	102°25'W
Stanley, Wi., U.S.	71	44°56'N	90°56'W
Stanley Pool, l., Afr.	186	4°07'S	15°40'E
Stanley Reservoir, res., India (stăn'lē)	159	12°07'N	77°27'E
Stanleyville see Kisangani, D.R.C.	185	0°30'S	25°12'E
Stann Creek, Belize (stăn krēk)	90a	17°01'N	88°14'W
Stanovoy Khrebet, mts., Russia (stŭn-à-voi')	135	56°12'N	127°12'E
Stanton, Ca., U.S. (stăn'tŭn)	75a	33°48'N	118°00'W
Stanton, Ne., U.S.	70	41°57'N	97°15'W
Stanton, Tx., U.S.	80	32°08'N	101°46'W
Stanwood, Wa., U.S. (stăn'wŏŏd)	74a	48°14'N	122°23'W
Staples, Mn., U.S. (stā'p'lz)	71	46°21'N	94°48'W
Stapleton, Al., U.S.	82	30°45'N	87°48'W
Stara Planina, mts., Blg.	112	42°50'N	24°45'E
Staraya Kupavna, Russia (stä'ra-ya kû-päf'nà)	142b	55°48'N	38°10'E
Staraya Russa, Russia (stä'rà-yà rōō'sà)	136	57°58'N	31°21'E
Stara Zagora, Blg. (zä'gô-rà)	119	42°26'N	25°37'E
Starbuck, Can. (stär'bŭk)	62f	49°46'N	97°36'W
Stargard Szczeciński, Pol. (shtär'gärt shchĕ-chyn'skē)	116	53°19'N	15°03'E
Staritsa, Russia (stä'rē-tsà)	132	56°29'N	34°58'E
Starke, Fl., U.S. (stärk)	83	29°55'N	82°07'W
Starkville, Co., U.S. (stärk'vĭl)	78	37°06'N	104°34'W
Starkville, Ms., U.S.	82	33°27'N	88°47'W
Starnberg, Ger. (shtärn-bĕrgh)	115d	47°59'N	11°20'E
Starnberger See, l., Ger.	124	47°58'N	11°30'E
Starobil's'k, Ukr.	137	49°19'N	38°57'E
Starodub, Russia (stä-rô-drŏop')	132	52°25'N	32°49'E
Starograd Gdański, Pol. (stä'rō-grad gdĕn'skē)	116	53°58'N	18°33'E
Starokostiantyniv, Ukr.	137	49°45'N	27°12'E
Staro-Minskaya, Russia (stä'rô mĭn'skà-yà)	137	46°19'N	38°51'E
Staro-Shcherbinovskaya, Russia	133	46°38'N	38°38'E
Staro-Subkhangulovo, Russia (stäro-sōōb-kan-gōō'lōvō)	142a	53°08'N	57°24'E
Staroutkinsk, Russia (stá-rô-ōōt'kĭnsk)	142a	57°14'N	59°21'E
Starovirivka, Ukr.	133	49°31'N	35°48'E
Start Point, c., Eng., U.K. (stärt)	117	50°14'N	3°34'W
Staryi Ostropil', Ukr.	133	49°48'N	27°32'E
Stary Sącz, Pol. (stä-rē sônch')	125	49°32'N	20°36'E
Staryy Oskol, Russia (stä'rē ôs-kôl')	137	51°18'N	37°51'E
Stassfurt, Ger. (shtäs'fōōrt)	124	51°52'N	11°35'E
Staszów, Pol. (stä'shôf)	125	50°32'N	21°13'E
State College, Pa., U.S. (stāt kŏl'ĕj)	67	40°50'N	77°55'W
State Line, Mn., U.S. (līn)	75h	46°36'N	92°18'W
Staten Island, i., N.Y., U.S. (stăt'ĕn)	68a	40°35'N	74°10'W
Statesboro, Ga., U.S. (stāts'bŭr-ô)	83	32°26'N	81°47'W
Statesville, N.C., U.S. (stās'vĭl)	83	34°45'N	80°54'W
Staunton, Il., U.S. (stôn'tŭn)	75e	39°01'N	89°47'W
Staunton, Va., U.S.	67	38°10'N	79°05'W
Stavanger, Nor. (stä'väng'ēr)	110	58°59'N	5°44'E
Stave, r., Can. (stāv)	74d	49°12'N	122°24'W
Staveley, Eng., U.K. (stāv'lē)	114a	53°17'N	1°21'W
Stavenisse, Neth.	115a	51°35'N	3°59'E
Stavropol', Russia	134	45°05'N	41°50'E
Steamboat Springs, Co., U.S. (stēm'bōt')	78	40°30'N	106°48'W
Stebliv, Ukr.	133	49°23'N	31°03'E
Steel, r., Can. (stēl)	58	49°08'N	86°55'W
Steelton, Pa., U.S. (stēl'tŭn)	67	40°15'N	76°45'W
Steenbergen, Neth.	115a	51°35'N	4°18'E
Steens Mountain, mts., Or., U.S. (stēnz)	72	42°15'N	118°52'W
Steep Point, c., Austl. (stēp)	174	26°15'N	112°05'E
Stefanie, Lake see Chew Bahir, l., Afr.	185	4°46'N	37°31'E
Steinbach, Can.	50	49°32'N	96°41'W
Steinkjer, Nor. (stĕīn-kyĕr)	116	64°00'N	11°19'E
Stella, Wa., U.S. (stĕl'à)	74c	46°11'N	123°12'W
Stellarton, Can. (stĕl'ár-tŭn)	51	45°34'N	62°40'W
Stendal, Ger. (shtĕn'däl)	124	52°37'N	11°51'E
Stepanakert see Xankändi, Azer.	136	39°50'N	46°40'E
Stephens, Port, b., Austl. (stē'fĕns)	176	32°43'N	152°55'E
Stephenville, Can. (stē'vĕn-vĭl)	53a	48°33'N	58°35'W
Stepnogorsk, Kaz.	139	52°20'N	72°05'E
Sterkrade, Ger. (shtĕr'krädĕ)	127c	51°31'N	6°51'E
Sterkstroom, S. Afr.	187c	31°33'S	26°36'E
Sterling, Co., U.S. (stûr'lĭng)	64	40°38'N	103°14'W
Sterling, Il., U.S.	66	41°48'N	89°42'W
Sterling, Ks., U.S.	78	38°11'N	98°11'W
Sterling, Ma., U.S.	61a	42°26'N	71°41'W
Sterling, Tx., U.S.	80	31°53'N	100°58'W
Sterlitamak, Russia (styĕr'lē-ta-mäk')	134	53°38'N	55°56'E
Šternberk, Czech Rep. (shtĕrn'bĕrk)	125	49°44'N	17°18'E
Stettin see Szczecin, Pol.	110	53°25'N	14°35'E
Stettler, Can.	50	52°19'N	112°43'W
Steubenville, Oh., U.S. (stū'bĕn-vĭl)	66	40°20'N	80°40'W
Stevens, l., Wa., U.S. (stē'vĕnz)	74a	47°59'N	122°06'W
Stevens Point, Wi., U.S.	71	44°30'N	89°35'W
Stevensville, Mt., U.S. (stē'vĕnz-vĭl)	73	46°31'N	114°03'E
Stewart, r., Can. (stū'ērt)	52	63°27'N	138°48'W
Stewart Island, i., N.Z.	175a	46°56'S	167°40'E
Stewiacke, Can. (stū'wē-āk)	60	45°08'N	63°21'W
Steynsrus, S. Afr. (stīns'rōōs)	192c	27°58'S	27°33'E
Steyr, Aus. (shtīr)	117	48°03'N	14°24'E
Stif, Alg.	184	36°18'N	5°21'E
Stikine, r., Can. (stĭ-kēn')	52	58°17'N	130°10'W
Stikine Ranges, Can.	50	59°05'N	130°00'W
Stillaguamish, r., Wa., U.S.	74a	48°11'N	122°18'W
Stillaguamish, South Fork, r., Wa., U.S. (stĭl-ă-gwä'mĭsh)	74a	48°05'N	121°59'W
Stillwater, Mn., U.S. (stĭl'wô-tēr)	75g	45°04'N	92°48'W
Stillwater, Mt., U.S.	73	45°23'N	109°45'W
Stillwater, Ok., U.S.	79	36°06'N	97°03'W
Stillwater, r., Mt., U.S.	73	48°47'N	114°40'W
Stillwater Range, mts., Nv., U.S.	76	39°43'N	118°11'W
Štip, Mac. (shtĭp)	131	41°43'N	22°07'E
Stirling, Scot., U.K. (stûr'lĭng)	120	56°05'N	3°59'W
Stittsville, Can. (stĭts'vĭl)	62c	45°15'N	75°54'W
Stizef, Alg. (mĕr-syā' lá-kôNb)	129	35°18'N	0°11'W
Stjördalshalsen, Nor. (styûr-däls-hälsĕn)	122	63°26'N	11°00'E
Stockbridge Munsee Indian Reservation, I.R., Wi., U.S. (stŏk'brĭdj mŭn-sē)	71	44°49'N	89°00'W
Stockerau, Aus. (shtô'kĕ-rou)	124	48°24'N	16°13'E
Stockholm, Swe. (stôk'hôlm)	110	59°23'N	18°00'E
Stockholm, Me., U.S. (stŏk'hōlm)	60	47°05'N	68°08'W
Stockport, Eng., U.K. (stŏk'pôrt)	120	53°24'N	2°09'W
Stockton, Eng., U.K.	120	54°35'N	1°25'W
Stockton, Ca., U.S. (stŏk'tŭn)	64	37°56'N	121°16'W
Stockton, Ks., U.S.	78	39°26'N	99°16'W
Stockton, i., Wi., U.S.	71	46°56'N	90°25'W
Stockton Plateau, plat., Tx., U.S.	64	30°34'N	102°35'W
Stockton Reservoir, res., Mo., U.S.	79	37°40'N	93°45'W
Stöde, Swe. (stû'dĕ)	122	62°26'N	16°35'E
Stoeng Trêng, Camb. (stông'trĕng')	168	13°36'N	106°00'E
Stoke-on-Trent, Eng., U.K. (stōk-ŏn-trĕnt)	116	53°01'N	2°12'W
Stokhid, r., Ukr.	125	51°24'N	25°20'E
Stolac, Bos. (stô'lâts)	131	43°03'N	17°59'E
Stolbovoy, is., Russia (stôl-bô-voi')	141	74°05'N	136°00'E
Stolin, Bela. (stô'lēn)	125	51°54'N	26°52'E
Stömstad, Swe.	122	58°58'N	11°09'E
Stone, Eng., U.K.	114a	52°54'N	2°09'W
Stoneham, Can. (stōn'ăm)	62b	46°59'N	71°22'W
Stoneham, Ma., U.S.	61a	42°30'N	71°05'W
Stonehaven, Scot., U.K. (stōn'hā-v'n)	120	56°57'N	2°09'W
Stone Mountain, Ga., U.S. (stōn)	68c	33°49'N	84°10'W
Stonewall, Can. (stōn'wôl)	62f	50°09'N	97°21'W
Stonewall, Ms., U.S.	82	32°08'N	88°44'W
Stoney Creek, Can. (stō'nē)	62d	43°13'N	79°45'W
Stonington, Ct., U.S. (stōn'ĭng-tŭn)	67	41°20'N	71°55'W
Stony Indian Reserve, I.R., Can.	62e	51°10'N	114°45'W
Stony Mountain, Can.	62f	50°05'N	97°13'W
Stony Plain, Can. (stō'nē plān)	62g	53°32'N	114°00'W
Stony Plain Indian Reserve, I.R., Can.	62g	53°29'N	113°48'W
Stony Point, N.Y., U.S.	68a	41°13'N	73°58'W
Stora Sotra, i., Nor.	122	60°24'N	4°35'E
Stord, i., Nor. (stôrd)	122	59°54'N	5°15'E
Store Baelt, strt., Den.	122	55°25'N	10°50'E
Storfjorden, b., Nor.	122	62°17'N	6°19'E
Stormberg, mts., S. Afr. (stôrm'bûrg)	187c	31°28'S	26°35'E
Storm Lake, Ia., U.S.	70	42°39'N	95°12'W
Stormy Point, c., V.I.U.S. (stôr'mē)	87c	18°22'N	65°01'W
Stornoway, Scot., U.K. (stôr'nō-wā)	116	58°13'N	6°21'W
Storozhynets', Ukr.	125	48°10'N	25°44'E
Störsjo, Swe. (stôr'shû)	122	62°49'N	13°08'E
Störsjoen, l., Nor. (stôr-syûĕn)	122	61°32'N	11°30'E
Störsjon, l., Swe.	116	63°06'N	14°00'E
Storvik, Swe.	122	60°37'N	16°31'E
Stoughton, Wi., U.S.	71	42°54'N	89°15'W
Stour, r., Eng., U.K. (stour)	121	52°09'N	0°29'E
Stourbridge, Eng., U.K. (stour'brĭj)	114a	52°27'N	2°08'W
Stow, Ma., U.S. (stō)	61a	42°56'N	71°31'W
Stow, Oh., U.S.	69d	41°09'N	81°26'W
Straatsdrif, S. Afr.	192c	25°19'S	26°22'E
Strabane, N. Ire., U.K. (stră-băn')	120	54°59'N	7°27'W
Straelen, Ger. (shtră'lĕn)	127c	51°26'N	6°16'E
Strahan, Austl. (strā'ăn)	175	42°08'S	145°28'E
Strakonice, Czech Rep. (strä'kô-nyē-tsĕ)	124	49°18'N	13°52'E
Straldzha, Blg. (sträl'dzhà)	131	42°37'N	26°44'E
Stralsund, Ger. (shräl'sŏŏnt)	116	54°18'N	13°04'E
Strangford Lough, l., N. Ire., U.K.	120	54°30'N	5°34'W
Strängnäs, Swe. (strĕng'nĕs)	122	59°23'N	16°59'E
Stranraer, Scot., U.K. (străn-rär')	120	54°55'N	5°05'W
Strasbourg, Fr. (strás-bōōr')	110	48°36'N	7°49'E
Stratford, Can. (străt'fērd)	58	43°20'N	81°05'W
Stratford, Ct., U.S.	67	41°10'N	73°05'W
Stratford, Wi., U.S.	71	44°16'N	90°02'W
Stratford-upon-Avon, Eng., U.K.	120	52°13'N	1°41'W
Straubing, Ger. (strou'bĭng)	124	48°52'N	12°36'E
Strausberg, Ger. (strous'bĕrgh)	124	52°35'N	13°50'E
Strawberry, r., Ut., U.S.	77	40°05'N	110°55'W
Strawn, Tx., U.S. (strôn)	80	32°38'N	98°28'W
Streator, Il., U.S. (strē'tēr)	66	41°05'N	88°50'W
Streeter, N.D., U.S.	70	46°40'N	99°22'W
Streetsville, Can. (strēts'vĭl)	62d	43°34'N	79°43'W
Strehaia, Rom. (strē-kä'yà)	131	44°37'N	23°13'E
Strel'na, Russia (strĕl'nà)	142c	59°52'N	30°01'E
Stretford, Eng., U.K. (strĕt'fērd)	114a	53°25'N	2°19'W
Strickland, r., Pap. N. Gui. (strĭk'lănd)	169	6°15'S	142°00'E
Strijen, Neth.	115a	51°44'N	4°32'E
Stromboli, Italy (strôm'bô-lē)	119	38°46'N	15°16'E
Stromyn, Russia (strô'mĭn)	142b	56°02'N	38°29'E
Strong, r., Ms., U.S. (strông)	82	32°03'N	89°42'W
Strongsville, Oh., U.S. (strôngz'vĭl)	69d	41°19'N	81°50'W
Stronsay, i., Scot., U.K. (strŏn'sā)	120a	59°09'N	2°35'W
Stroudsburg, Pa., U.S. (stroudz'bûrg)	67	41°00'N	75°15'W
Struer, Den.	122	56°29'N	8°34'E
Strugi Krasnyye, Russia (strōō'gĭ krä's-ny'yĕ)	132	58°14'N	29°10'E
Struma, r., Eur. (strōō'mà)	131	41°55'N	23°05'E
Strumica, Mac. (strōō'mĭ-tsà)	131	41°26'N	22°38'E
Strunino, Russia	142b	56°23'N	38°34'E
Struthers, Oh., U.S. (strŭdh'ērz)	66	41°00'N	80°35'W
Struvenhütten, Ger. (shtrōō'vĕn-hü-tĕn)	115c	53°52'N	10°04'E
Strydpoortberge, mts., S. Afr.	192c	24°08'N	29°18'E
Stryi, Ukr.	125	49°16'N	23°51'E
Strzelce Opolskie, Pol. (stzhĕl'tsĕ o-pōl'skyĕ)	125	50°31'N	18°20'E
Strzelin, Pol. (stzhĕ-lĭn)	125	50°48'N	17°06'E
Strzelno, Pol. (stzhäl'nô)	125	52°37'N	18°10'E
Stuart, Fl., U.S. (stū'ērt)	83a	27°10'N	80°14'W
Stuart, Ia., U.S.	71	41°31'N	94°20'W
Stuart, i., Ak., U.S.	63	63°25'N	162°45'W
Stuart, i., Wa., U.S.	74d	48°42'N	123°10'W
Stuart Lake, l., Can.	54	54°32'N	124°35'W
Stuart Range, mts., Austl.	174	29°00'S	134°30'E
Sturgeon, r., Can.	62g	53°41'N	113°46'W
Sturgeon, r., Mi., U.S.	71	46°43'N	88°43'W
Sturgeon Bay, Wi., U.S.	71	44°50'N	87°22'W
Sturgeon Bay, b., Can.	57	52°00'N	98°00'W
Sturgeon Falls, Can.	51	46°19'N	79°49'W
Sturgis, Ky., U.S.	66	37°35'N	88°00'W
Sturgis, Mi., U.S.	66	41°45'N	85°25'W
Sturgis, S.D., U.S.	70	44°25'N	103°31'W
Sturt Creek, r., Austl.	174	19°40'S	127°40'E
Sturtevant, Wi., U.S. (stûr'tē-vănt)	69a	42°42'N	87°54'W
Stutterheim, S. Afr. (stürt'ēr-hīm)	187c	32°34'S	27°27'E
Stuttgart, Ger. (shtōōt'gärt)	110	48°48'N	9°15'E
Stuttgart, Ar., U.S. (stŭt'gärt)	79	34°30'N	91°33'W
Stykkishólmur, Ice.	116	65°00'N	21°48'W
Styr', r., Eur. (stēr)	125	51°44'N	26°07'E
Suao, Tai. (sōōou)	165	24°35'N	121°45'E
Subarnarekha, r., India	158	22°38'N	86°26'E
Subata, Lat. (sò'bá-tá)	123	56°02'N	25°54'E
Subic, Phil. (sōō'bĭk)	169a	14°52'N	120°15'E
Subic Bay, b., Phil.	169a	14°41'N	120°11'E
Subotica, Serb. (sōō'bô'tē-tsà)	110	46°06'N	19°41'E
Subugo, mtn., Kenya	191	1°40'S	35°49'E
Succasunna, N.J., U.S. (sŭk'kȧ-sŭn'nȧ)	68a	40°52'N	74°37'W
Suceava, Rom. (sōō-chȧ-ä'và)	125	47°39'N	26°17'E
Suceava, r., Rom.	125	47°45'N	26°10'E
Sucha, Pol. (sōō'kȧ)	125	49°44'N	19°40'E
Suchiapa, Mex. (sōō-chē-ä'pä)	89	16°38'N	93°08'W
Suchiapa, r., Mex.	89	16°27'N	93°26'W
Suchitoto, El Sal. (sōō-chē-tô'tô)	90	13°58'N	89°03'W
Sucio, r., Col. (sōō'syô)	100a	6°55'N	76°15'W
Suck, r., Ire. (sŭk)	120	53°34'N	8°16'W
Sucre, Bol. (sōō'krā)	100	19°06'S	65°16'W
Sucre, dept., Ven. (sōō'krē)	101b	10°18'N	64°12'W
Sud, Canal du, strt., Haiti	93	18°40'N	73°15'W
Sud, Rivière du, r., Can. (rē-vyâr'dü süd')	62b	46°56'N	70°35'W
Suda, Russia (sò'dȧ)	142a	56°58'N	56°45'E
Suda, r., Russia (sò'dȧ)	132	59°24'N	36°40'E
Sudair, Sau. Ar. (sŭ-dä'ēr)	154	25°48'N	46°28'E
Sudalsvatnet, l., Nor.	122	59°35'N	6°59'E
Sudan, nation, Afr.	185	14°00'N	28°00'E
Sudan, reg., Afr. (sōō-dän')	184	15°00'N	7°00'E
Sudbury, Can. (sŭd'bĕr-ē)	51	46°28'N	81°00'W
Sudbury, Ma., U.S.	61a	42°23'N	71°25'W
Sudetes, mts., Eur.	112	50°41'N	15°37'E
Sudogda, Russia (sò'dôk-dà)	132	55°57'N	40°29'E
Sudost', r., Eur. (sò-dôst')	132	52°43'N	33°13'E
Sudzha, Russia (sòd'zhȧ)	133	51°14'N	35°11'E
Sueca, Spain (swā'kä)	129	39°12'N	0°18'W
Suez, Egypt	185	29°58'N	32°34'E
Suez, Gulf of, b., Egypt (sōō-ĕz')	185	29°53'N	32°33'E
Suez Canal, can., Egypt	185	30°53'N	32°21'E
Suffern, N.Y., U.S. (sŭf'fērn)	68a	41°07'N	74°09'W
Suffolk, Va., U.S. (sŭf'ŭk)	68g	36°43'N	76°35'W
Sugar City, Co., U.S.	78	38°12'N	103°42'W
Sugar Creek, Mo., U.S.	75f	39°07'N	94°27'W
Sugar Creek, r., Il., U.S. (shòg'ēr)	79	40°14'N	89°28'W
Sugar Creek, r., In., U.S.	66	39°55'N	87°10'W
Sugar Island, i., Mi., U.S.	75k	46°31'N	84°12'W
Sugarloaf Point, c., Austl. (sògēr'lōf)	176	32°19'S	153°04'E
Suggi Lake, l., Can.	57	54°22'N	102°47'W
Sühbaatar, Mong.	160	50°18'N	106°31'E
Suhl, Ger. (zōōl)	124	50°37'N	10°41'E
Suichuan, mtn., China	165	26°25'N	114°10'E
Suide, China (swā-dü)	164	37°32'N	110°12'E

ăt; finăl; rāte; senāte; ärm; ȧsk; sofȧ; fâre; ch-choose; dh-as th in other; bē; ēvent; bĕt; recĕnt; cratēr; g-gō; gh-guttural g; bĭt; ĭ-short neutral; rīde; K-guttural k as ch in German ich;

PRONOUNCING INDEX Suifenhe – Tablas 307

PLACE (Pronunciation)	PAGE	LAT.	LONG.
Suifenhe, China (swā-fŭn-hŭ)	161	44°47'N	131°13'E
Suihua, China	161	46°38'N	126°50'E
Suining, China (so͞o'ē-nĭng')	162	33°54'N	117°57'E
Suipacha, Arg. (swē-pä'chä)	99c	34°45'S	59°43'W
Suiping, China (swā-pĭŋ)	162	33°09'N	113°58'E
Suir, r., Ire. (sūr)	120	52°20'N	7°32'W
Suisun Bay, b., Ca., U.S. (so͞oē-so͞on')	74b	38°07'N	122°02'W
Suita, Japan (so͞o'ē-tä)	167b	34°45'N	135°32'E
Suitland, Md., U.S. (so͞ot'lănd)	68e	38°51'N	76°57'W
Suixian, China (swā shyĕn)	165	31°42'N	113°20'E
Suiyüan, hist. reg., China (swā-yüĕn)	160	41°31'N	107°04'E
Suizhong, China (swā-jôŋ)	164	40°22'N	120°20'E
Sukabumi, Indon.	168	6°52'S	106°56'E
Sukadana, Indon.	168	1°15'S	110°30'E
Sukagawa, Japan (so͞o'kä-gä'wä)	167	37°08'N	140°07'E
Sukhinichi, Russia (so͞o'kē-nē-chē)	136	54°07'N	35°18'E
Sukhona, r., Russia (so͞o-kô'nà)	136	59°30'N	42°20'E
Sukhoy Log, Russia (so͞o'kôy lôg)	142a	56°55'N	62°03'E
Sukhumi, Geor. (so͞o-ko͞om')	137	43°00'N	41°00'E
Sukkur, Pak. (sŭk'ŭr)	155	27°49'N	68°50'E
Sukkwan Island, i., Ak., U.S.	54	55°05'N	132°45'W
Suksun, Russia (so͞ok'so͞on)	142a	57°08'N	57°22'E
Sukumo, Japan (so͞o'ko͞o-mō)	167	32°58'N	132°45'E
Sukunka, r., Can.	55	55°00'N	121°50'W
Sula, r., Ukr. (so͞o-lä')	133	50°36'N	33°13'E
Sula, Kepulauan, is., Indon.	169	2°20'S	125°20'E
Sulaco, r., Hond. (so͞o-lä'kō)	90	14°55'N	87°31'W
Sulaimān Range, mts., Pak. (so͞o-lä-ē-män')	155	29°47'N	69°10'E
Sulak, r., Russia (so͞o-lák')	137	43°30'N	47°00'E
Sulfeld, Ger. (zo͞ol'fĕld)	115c	53°48'N	10°13'E
Sulina, Rom. (so͞o-lē'nà)	119	45°08'N	29°38'E
Sulitelma, mtn., Eur. (so͞o-lē-tyĕl'mà)	116	67°03'N	16°35'E
Sullana, Peru (so͞o-lyä'nä)	100	4°57'S	80°47'W
Sulligent, Al., U.S. (sŭl'ĭ-jĕnt)	82	33°52'N	88°06'W
Sullivan, Il., U.S. (sŭl'ĭ-văn)	66	41°35'N	88°35'W
Sullivan, In., U.S.	66	39°05'N	87°20'W
Sullivan, Mo., U.S.	79	38°13'N	91°09'W
Sulmona, Italy (so͞ol-mō'nä)	130	42°02'N	13°58'E
Sulphur, Ok., U.S. (sŭl'fŭr)	79	34°31'N	96°58'W
Sulphur, r., Tx., U.S.	79	33°26'N	95°06'W
Sulphur Springs, Tx., U.S. (sprĭngz)	79	33°09'N	95°36'W
Sultan, Wa., U.S. (sŭl'tăn)	74a	47°52'N	121°49'W
Sultan, r., Wa., U.S.	74a	47°55'N	121°49'W
Sultepec, Mex. (so͞ol-tå-pĕk')	88	18°50'N	99°51'W
Sulu Archipelago, is., Phil. (so͞o'lo͞o)	168	5°52'N	122°00'E
Suluntah, Libya	119	32°39'N	21°49'E
Sulūq, Libya	185	31°39'N	20°15'E
Sulu Sea, sea, Asia	168	8°25'N	119°00'E
Suma, Japan (so͞o'mä)	167b	34°39'N	135°08'E
Sumas, Wa., U.S. (sū'màs)	74d	49°00'N	122°16'W
Sumatera, i., Indon. (so͞o-mä-trà)	168	2°06'N	99°40'E
Sumatra see Sumatera, i., Indon.	168	2°06'N	99°40'E
Sumba, i., Indon. (sŭm'bà)	168	9°52'S	119°00'E
Sumba, Île, i., D.R.C.	190	1°44'N	19°32'E
Sumbawa, i., Indon. (so͞om-bä'wä)	168	9°00'S	118°18'E
Sumbawa-Besar, Indon.	168	8°32'S	117°20'E
Sumbawanga, Tan.	191	7°58'S	31°37'E
Sumbe, Ang.	186	11°13'S	13°50'E
Sümeg, Hung. (shü'mĕg)	125	46°59'N	17°19'E
Sumida, r., Japan (so͞o'mē-dä)	167	36°01'N	139°24'E
Sumidouro, Braz. (so͞o-mē-dō'rò)	99a	22°04'S	42°41'W
Sumiyoshi, Japan (so͞o'mē-yō'shē)	167b	34°43'N	135°16'E
Summer Lake, l., Or., U.S. (sŭm'ẽr)	72	42°50'N	120°35'W
Summerland, Can. (sŭ'mẽr-lănd)	55	49°39'N	119°40'W
Summerside, Can. (sŭm'ẽr-sīd)	51	46°25'N	63°47'W
Summerton, S.C., U.S. (sŭm'ẽr-tŭn)	83	33°37'N	80°22'W
Summerville, S.C., U.S. (sŭm'ẽr-vĭl)	83	33°00'N	80°10'W
Summit, Il., U.S. (sŭm'mĭt)	69a	41°47'N	87°48'W
Summit, N.J., U.S.	68a	40°43'N	74°21'W
Summit Lake Indian Reservation, I.R., Nv., U.S.	72	41°35'N	119°30'W
Summit Peak, mtn., Co., U.S.	77	37°20'N	106°40'W
Sumner, Wa., U.S. (sŭm'nẽr)	74a	47°12'N	122°14'W
Šumperk, Czech Rep. (shŏom'pĕrk)	125	49°57'N	17°02'E
Sumqayıt, Azer.	138	40°36'N	49°38'E
Sumrall, Ms., U.S. (sŭm'rôl)	82	31°25'N	89°34'W
Sumter, S.C., U.S. (sŭm'tẽr)	83	33°55'N	80°21'W
Sumy, Ukr. (so͞o'mĭ)	134	50°54'N	34°47'E
Sumy, prov., Ukr.	133	51°02'N	34°05'E
Sun, r., Mt., U.S. (sŭn)	73	47°34'N	111°53'W
Sunburst, Mt., U.S.	73	48°53'N	111°55'W
Sunda, Selat, strt., Indon.	168	5°45'S	106°15'E
Sundance, Wy., U.S. (sŭn'dăns)	73	44°24'N	104°27'W
Sundarbans, sw., Asia (so͞on'dẽr-bŭns)	155	21°50'N	89°00'E
Sunday Strait, strt., Austl. (sŭn'dā)	174	15°50'S	122°45'E
Sundbyberg, Swe. (so͞on'bü-bĕrgh)	122	59°24'N	17°56'E
Sunderland, Eng., U.K. (sŭn'dẽr-lănd)	116	54°55'N	1°25'W
Sunderland, Md., U.S.	68e	38°41'N	76°36'W
Sundsvall, Swe. (so͞onds'väl)	110	62°24'N	19°19'E
Sungari (Songhua), r., China	161	46°09'N	127°53'E
Sungari Reservoir, res., China	164	42°55'N	127°50'E
Sungurlu, Tur. (so͞on'go͞or-lo͞o')	119	40°08'N	34°20'E
Sun Kosi, r., Nepal	158	27°13'N	85°52'E
Sunland, Ca., U.S. (sŭn-lănd)	75a	34°16'N	118°18'W
Sunne, Swe. (so͞on'ĕ)	122	59°51'N	13°07'E
Sunninghill, Eng., U.K. (sŭnĭng'hĭl)	114b	51°23'N	0°40'W
Sunnymead, Ca., U.S. (sŭn'ĭ-mēd)	75a	33°56'N	117°15'W
Sunnyside, Ut., U.S.	77	39°35'N	110°20'W
Sunnyside, Wa., U.S.	72	46°19'N	120°00'W
Sunnyvale, Ca., U.S. (sŭn-nē-vāl)	74b	37°23'N	122°02'W
Sunol, Ca., U.S. (so͞o'nŭl)	74b	37°36'N	122°53'W
Sunset, Ut., U.S. (sŭn-sĕt)	75b	41°08'N	112°02'W
Sunset Crater National Monument, rec., Az., U.S. (krā'tẽr)	77	35°20'N	111°30'W
Sunshine, Austl.	173a	37°47'S	144°50'E
Suntar, Russia (so͞on-tär')	135	62°14'N	117°49'E
Sunyani, Ghana	188	7°20'N	2°20'W
Suoyarvi, Russia (so͞o'ô-yĕr'vē)	136	62°12'N	32°29'E
Superior, Az., U.S. (su-pē'rĭ-ẽr)	77	33°15'N	111°10'W
Superior, Ne., U.S.	78	40°04'N	98°05'W
Superior, Wi., U.S.	65	46°44'N	92°06'W
Superior, Wy., U.S.	73	41°45'N	108°57'W
Superior, Laguna, l., Mex. (lä-go͞o'nä so͞o-pā-rē-ōr')	89	16°20'N	94°55'W
Superior, Lake, l., N.A.	65	47°38'N	89°20'W
Superior Village, Wi., U.S.	75h	46°38'N	92°07'W
Sup'ung Reservoir, res., Asia (so͞o'po͞ong)	166	40°35'N	126°00'E
Suqian, China (so͞o-chyĕn)	162	33°57'N	118°17'E
Suquamish, Wa., U.S. (so͞o-gwä'mĭsh)	74a	47°44'N	122°34'W
Suquṭrā (Socotra), i., Yemen (sô-kō'trà)	154	13°00'N	52°30'E
Şūr, Leb. (so͞or) (tĭr)	153a	33°16'N	35°13'E
Şūr, Oman	154	22°23'N	59°28'E
Surabaya, Indon.	168	7°23'S	112°45'E
Surakarta, Indon.	168	7°35'S	110°45'E
Šurany, Slvk. (sho͞o'rà-nû')	125	48°05'N	18°11'E
Surat, Austl. (sū-răt)	176	27°18'S	149°00'E
Surat, India (so͞o'rŭt)	155	21°08'N	73°22'E
Surat Thani, Thai.	168	8°59'N	99°14'E
Surazh, Bela.	132	55°24'N	30°46'E
Surazh, Russia (so͞o-rázh')	132	53°02'N	32°27'E
Surgères, Fr. (sür-zhâr')	126	46°06'N	0°51'W
Surgut, Russia (so͞or-go͞ot')	134	61°18'N	73°38'E
Suriname, nation, S.A. (so͞o-rē-näm')	101	4°00'N	56°00'W
Sūrmaq, Iran	157	31°03'N	52°48'E
Surt, Libya	185	31°14'N	16°37'E
Surt, Khalīj, b., Libya	185	31°30'N	18°28'E
Suruga-Wan, b., Japan (so͞o'ro͞o-gä wän)	166	34°52'N	138°36'E
Susa, Japan	167	34°40'N	131°39'E
Sušak, i., Serb.	130	42°45'N	16°30'E
Susak, Otok, i., Serb.	130	44°31'N	14°15'E
Susaki, Japan (so͞o'sä-kē)	167	33°23'N	133°16'E
Sušice, Czech Rep.	124	49°14'N	13°31'E
Susitna, Ak., U.S. (so͞o-sĭt'nà)	63	61°28'N	150°28'W
Susitna, r., Ak., U.S.	63	62°00'N	150°28'W
Susong, China (so͞o-sôŋ)	165	30°18'N	116°08'E
Susquehanna, Pa., U.S. (sŭs'kwē-hăn'à)	67	41°55'N	73°55'W
Susquehanna, r., U.S.	67	39°50'N	76°20'W
Sussex, Can. (sŭs'ĕks)	51	45°43'N	65°31'W
Sussex, N.J., U.S.	68a	41°12'N	74°36'W
Sussex, Wi., U.S.	69a	43°08'N	88°12'W
Sutherland, Austl. (sŭdh'ẽr-lănd)	173b	34°02'S	151°04'E
Sutherland, S. Afr. (sŭ'thẽr-lănd)	186	32°25'S	20°40'E
Sutlej, r., Asia (sŭt'lĕj)	155	30°15'N	73°00'E
Sutton, Eng., U.K. (sŭt''n)	114b	51°21'N	0°12'W
Sutton, Ma., U.S.	61a	42°09'N	71°46'W
Sutton Coldfield, Eng., U.K. (kōld'fēld)	114a	52°34'N	1°49'W
Sutton-in-Ashfield, Eng., U.K. (ĭn-ăsh'fēld)	114a	53°07'N	1°15'W
Suurberge, mts., S. Afr.	187c	33°15'S	25°32'E
Suva, Fiji	170g	18°08'S	178°25'E
Suwa, Japan (so͞o'wä)	167	36°03'N	138°08'E
Suwałki, Pol. (so͞o-vou'kē)	125	54°05'N	22°58'E
Suwanee Lake, l., Can.	57	56°08'N	100°10'W
Suwannee, r., U.S. (so͞o-wô'nē)	65	29°42'N	83°00'W
Suways al Ḥulwah, Tur' at As, can., Egypt	192b	30°15'N	32°20'E
Suxian, China (so͞o shyĕn)	164	33°29'N	117°51'E
Suzdal', Russia (so͞oz'dál)	132	56°26'N	40°29'E
Suzhou, China (so͞o-jō)	161	31°19'N	120°37'E
Suzu Misaki, c., Japan (so͞o'zo͞o mē'sä-kē)	166	37°30'N	137°35'E
Svalbard (Spitsbergen), dep., Nor. (sväl'bärt) (spĭts'bûr-gĕn)	134	77°00'N	20°00'E
Svaneke, Den. (svä'nĕ-kĕ)	122	55°08'N	15°07'E
Svatove, Ukr.	137	49°23'N	38°10'E
Svedala, Swe. (svĕ'dä-lä)	122	55°29'N	13°11'E
Sveg, Swe.	122	62°03'N	14°22'E
Svelvik, Nor. (svĕl'vĕk)	122	59°37'N	10°18'E
Svenčionys, Lith.	123	55°09'N	26°09'E
Svendborg, Den. (svĕn'bôrgh)	122	55°05'N	10°35'E
Svensen, Or., U.S. (svĕn'sĕn)	74c	46°10'N	123°39'W
Sverdlovsk see Yekaterinburg, Russia	134	56°51'N	60°36'E
Svetlaya, Russia (svyĕt'lá-yá)	166	46°09'N	137°53'E
Svicha, r., Ukr.	125	49°09'N	24°10'E
Svilajnac, Serb. (svē'lá-ē-náts)	131	44°12'N	21°14'E
Svilengrad, Blg. (svĕl'ĕn-grát)	131	41°44'N	26°11'E
Svir', r., Russia	136	60°55'N	33°40'E
Svir Kanal, can., Russia (kà-näl')	123	60°10'N	32°40'E
Svishtov, Blg. (svēsh'tôf)	119	43°36'N	25°21'E
Svisloch', r., Bela. (svēs'lôk)	132	53°38'N	28°10'E
Svitavy, Czech Rep.	124	49°46'N	16°28'E
Svobodnyy, Russia (svô-bôd'nĭ)	135	51°28'N	128°28'E
Svolvaer, Nor. (svôl'vĕr)	116	68°15'N	14°29'E
Svyatoy Nos, Mys, c., Russia (svyŭ'toi nôs)	135	72°18'N	139°28'E
Swadlincote, Eng., U.K. (swŏd'lĭn-kōt)	114a	52°46'N	1°33'W
Swain Reefs, rf., Austl. (swān)	175	22°12'S	152°08'E
Swainsboro, Ga., U.S. (swānz'bŭr-ô)	83	32°37'N	82°21'W
Swakopmund, Nmb. (svä'kôp-mo͝ont) (swá'kôp-mŏnd)	186	22°40'S	14°30'E
Swallowfield, Eng., U.K. (swŏl'ô-fēld)	114b	51°21'N	0°58'W
Swampscott, Ma., U.S. (swômp'skŏt)	61a	42°28'N	70°55'W
Swan, r., Austl.	174	31°30'S	116°30'E
Swan, r., Can.	57	51°58'N	101°45'W
Swan, r., Mt., U.S.	73	47°50'N	113°40'W
Swan Hill, Austl.	175	35°20'S	143°30'E
Swan Hills, Can. (hĭlz)	50	54°52'N	115°45'W
Swan Island, i., Austl. (swŏn)	173a	38°15'S	144°41'E
Swan Lake, l., Can.	57	52°30'N	100°45'W
Swanland, reg., Austl. (swŏn'lănd)	174	31°45'S	119°15'E
Swan Range, mts., Mt., U.S.	73	47°50'N	113°40'W
Swan River, Can. (swŏn rĭv'ẽr)	50	52°06'N	101°16'W
Swansea, Wales, U.K.	117	51°37'N	3°59'W
Swansea, Il., U.S. (swŏn'sē)	75e	38°32'N	89°59'W
Swansea, Ma., U.S.	68b	41°45'N	71°09'W
Swanson Reservoir, res., Ne., U.S. (swŏn'sŭn)	78	40°13'N	101°30'W
Swartberg, mtn., Afr.	187c	30°08'S	29°34'E
Swartkop, mtn., S. Afr.	186a	34°13'S	18°27'E
Swartruggens, S. Afr.	192c	25°40'S	26°40'E
Swartspruit, S. Afr.	187b	25°44'S	28°01'E
Swatow see Shantou, China	161	23°20'N	116°40'E
Swaziland, nation, Afr. (Swä'zē-lănd)	186	26°45'S	31°30'E
Sweden, nation, Eur. (swē'dĕn)	110	60°10'N	14°10'E
Swedesboro, N.J., U.S. (swēdz'bē-rô)	68f	39°45'N	75°22'W
Sweetwater, Tn., U.S. (swēt'wô-tẽr)	82	35°36'N	84°29'W
Sweetwater, Tx., U.S.	64	32°28'N	100°25'W
Sweetwater, l., N.D., U.S.	70	48°15'N	98°35'W
Sweetwater, r., Wy., U.S.	73	42°19'N	108°35'W
Sweetwater Reservoir, res., Ca., U.S.	76a	32°42'N	116°54'W
Świdnica, Pol. (shvĭd-nē'tsä)	124	50°50'N	16°30'E
Świdwin, Pol. (shvĭd'vĭn)	124	53°46'N	15°48'E
Świebodzice, Pol.	124	50°51'N	16°17'E
Świebodzin, Pol. (shvyĕn-bo'jĕts)	124	52°16'N	15°36'E
Świecie, Pol. (shvyän'tsyĕ)	125	53°23'N	18°26'E
Świętokrzyskie, Góry, mts., Pol. (shvyĕn-tō-kzhĭ'skyĕ go͞o'rĭ)	125	50°57'N	21°02'E
Swift, r., Eng., U.K.	114a	52°26'N	1°08'W
Swift, r., Me., U.S. (swĭft)	61	44°42'N	70°40'E
Swift Creek Reservoir, res., Wa., U.S.	72	46°03'N	122°10'W
Swift Current, Can. (swĭft kûr'ĕnt)	50	50°17'N	107°50'W
Swindle Island, i., Can.	54	52°32'N	128°35'W
Swindon, Eng., U.K. (swĭn'dŭn)	120	51°35'N	1°55'W
Swinomish Indian Reservation, I.R., Wa., U.S. (swĭ-nō'mĭsh)	74a	48°25'N	122°27'W
Świnoujście, Pol. (shvĭ-nĭ-ô-wēsh'chyĕ)	124	53°56'N	14°14'E
Swinton, Eng., U.K. (swĭn'tŭn)	114a	53°30'N	1°19'W
Swissvale, Pa., U.S. (swĭs'vāl)	69e	40°25'N	79°53'W
Switzerland, nation, Eur. (swĭt'zẽr-lănd)	110	46°30'N	7°43'E
Syanno, Bela. (syĕ'nô)	132	54°48'N	29°43'E
Syas', r., Russia (syäs)	132	59°28'N	33°24'E
Sycamore, Il., U.S. (sĭk'à-mōr)	71	42°00'N	88°42'W
Sycan, r., Or., U.S.	72	42°45'N	121°00'W
Sychëvka, Russia (sē-chôf'kà)	132	55°52'N	34°18'E
Sydney, Austl. (sĭd'nē)	175	33°55'S	151°17'E
Sydney, Can.	51	46°09'N	60°11'W
Sydney Mines, Can.	51	46°14'N	60°14'W
Syktyvkar, Russia (sük-tüf'kär)	134	61°35'N	50°40'E
Sylacauga, Al., U.S. (sĭl-à-kô'gà)	82	33°10'N	86°15'W
Sylarna, mtn., Eur.	122	63°00'N	12°10'E
Sylt, i., Ger. (sĭlt)	124	54°55'N	8°30'E
Sylvania, Ga., U.S. (sĭl-vā'nĭ-à)	83	32°44'N	81°40'W
Sylvester, Ga., U.S. (sĭl-vĕs'tẽr)	82	31°32'N	83°50'W
Sými, i., Grc.	119	36°27'N	27°41'E
Synel'nykove, Ukr.	137	48°19'N	35°33'E
Syracuse, Ks., U.S. (sĭr'à-kūs)	78	37°59'N	101°44'W
Syracuse, N.Y., U.S.	65	43°05'N	76°10'W
Syracuse, Ut., U.S.	75b	41°06'N	112°04'W
Syr Darya, r., Asia	134	44°15'N	65°45'E
Syria, nation, Asia (sĭr'ĭ-à)	154	35°00'N	37°15'E
Syrian Desert, des., Asia	154	32°00'N	40°00'E
Sýros, i., Grc.	119	37°23'N	24°55'E
Sysert', Russia (sĕ'sĕrt)	142a	56°30'N	60°48'E
Sysola, r., Russia	136	60°50'N	50°40'E
Syvash, zatoka, b., Ukr.	133	45°55'N	34°42'E
Syzran', Russia (sĕz-rän')	134	53°09'N	48°27'E
Szamotuły, Pol. (shá-mô-to͞o'wē)	124	52°36'N	16°34'E
Szarvas, Hung. (sôr'vôsh)	125	46°51'N	20°36'E
Szczebrzeszyn, Pol. (shchĕ-bzhä'shĕn)	125	50°41'N	22°58'E
Szczecin, Pol. (shchĕ'tsĭn)	110	53°25'N	14°35'E
Szczecinek, Pol. (shchĕ-tsĭ-nĕk)	116	53°41'N	16°42'E
Szczuczyn, Pol. (shcho͞o'chĕn)	125	53°32'N	22°17'E
Szczytno, Pol. (shchĭt'nô)	125	53°33'N	21°00'E
Szechwan Basin, basin, China	160	30°45'N	104°40'E
Szeged, Hung. (sĕ'gĕd)	110	46°15'N	20°12'E
Székesfehérvár, Hung. (sā'kĕsh-fĕ'hâr-vär)	119	47°12'N	18°26'E
Szekszárd, Hung. (sĕk'särd)	119	46°19'N	18°42'E
Szentendre, Hung. (sĕnt'ĕn-drĕ)	125	47°40'N	19°07'E
Szentes, Hung. (sĕn'tĕsh)	125	46°38'N	20°18'E
Szigetvar, Hung. (sĕ'gĕt-vär)	125	46°05'N	17°50'E
Szolnok, Hung.	125	47°11'N	20°12'E
Szombathely, Hung. (sôm'bôt-hĕl')	119	47°13'N	16°35'E
Szprotawa, Pol. (shprō-tä'vä)	124	51°34'N	15°29'E
Szydłowiec, Pol. (shĭd-wô'vyets)	125	51°13'N	20°53'E

T

PLACE (Pronunciation)	PAGE	LAT.	LONG.
Taal, l., Phil. (tä-äl')	169a	13°58'N	121°06'E
Tabaco, Phil. (tä-bä'kō)	169a	13°27'N	123°40'E
Tabankulu, S. Afr. (tā-băn-ko͞o'la)	187c	30°56'S	29°19'E
Tabasará, Serranía de, mts., Pan.	91	8°29'N	81°22'W
Tabasco, Mex. (tä-bäs'kō)	88	21°47'N	103°04'W
Tabasco, state, Mex.	86	18°10'N	93°00'W
Taber, Can.	50	49°47'N	112°08'W
Tablas, i., Phil. (tä'bläs)	169a	12°26'N	122°00'E

308 Tablas Strait – Tarboro PRONOUNCING INDEX

PLACE (Pronunciation)	PAGE	LAT.	LONG.
Tablas Strait, strt., Phil.	169a	12°17'N	121°41'E
Table Bay, b., S. Afr. (tā'b'l)	186a	33°41'S	18°27'E
Table Mountain, mtn., S. Afr.	186a	33°58'S	18°26'E
Table Rock Lake, Mo., U.S.	79	36°37'N	93°29'W
Tabligbo, Togo	188	6°35'N	1°30'E
Taboga, i., Pan. (tä-bō'gä)	86a	8°48'N	79°35'W
Taboguilla, i., Pan. (tä-bō-gē'l-yä)	86a	8°48'N	79°31'W
Tábor, Czech Rep. (tä'bôr)	124	49°25'N	14°40'E
Tabora, Tan. (tä-bō'rä)	186	5°01'S	32°48'E
Tabou, C. Iv. (tȧ-bōō')	184	4°25'N	7°21'W
Tabrīz, Iran (tȧ-brēz')	154	38°00'N	46°13'E
Tabuaeran, i., Kir.	2	3°52'N	159°20'W
Tabwémasana, Mont, mtn., Vanuatu	170f	15°20'S	166°44'E
Tacámbaro, r., Mex. (tä-käm'bä-rō)	88	18°55'N	101°25'W
Tacámbaro de Codallos, Mex.	88	19°12'N	101°28'W
Tacarigua, Laguna de la, l., Ven.	101b	10°18'N	65°43'W
Tacheng, China (tä-chŭŋ)	160	46°50'N	83°24'E
Tachie, r., Can.	54	54°30'N	125°00'W
Tacloban, Phil. (tä-klō'bän)	169	11°06'N	124°58'E
Tacna, Peru (täk'nä)	100	18°34'S	70°16'W
Tacoma, Wa., U.S. (tȧ-kō'mȧ)	64	47°14'N	122°27'W
Taconic Range, mts., N.Y., U.S. (tȧ-kŏn'ĭk)	67	41°55'N	73°40'W
Tacotalpa, Mex. (tä-kō-täl'pä)	89	17°37'N	92°51'W
Tacotalpa, r., Mex.	89	17°24'N	92°38'W
Tademaït, Plateau du, plat., Alg. (tä-dē-mä'ēt)	184	28°00'N	2°15'E
Tadio, Lagune, b., C. Iv.	188	5°20'N	5°25'W
Tadjoura, Dji. (tȧd-zhōō'rȧ)	192a	11°48'N	42°54'E
Tadley, Eng., U.K. (tăd'lē)	114b	51°19'N	1°08'W
Tadotsu, Japan (tä'dō-tsō)	167	34°14'N	133°43'E
Tadoussac, Can. (tȧ-dōō-sȧk')	59	48°09'N	69°43'W
Tadzhikistan *see* Tajikistan, nation, Asia	134	39°22'N	69°30'E
Taebaek Sanmaek, mts., Asia (tī-bīk' sän-mīk')	166	37°20'N	128°50'E
Taedong, r., Kor., N. (tī-dŏng)	166	38°38'N	124°32'E
Taegu, Kor., S. (tī'gōō')	161	35°49'N	128°41'E
Taejŏn, Kor., S.	166	36°20'N	127°26'E
Tafalla, Spain (tä-fäl'yä)	128	42°30'N	1°42'W
Tafna, r., Alg. (täf'nä)	128	35°28'N	1°00'W
Taft, Ca., U.S. (tȧft)	76	35°09'N	119°27'W
Tagama, reg., Niger	189	15°50'N	6°30'E
Taganrog, Russia (tȧ-gȧn-rôk')	137	47°12'N	38°56'E
Taganrogskiy Zaliv, b., Eur. (tȧ-gȧn-rôk'skĭ zä'lĭf)	137	46°55'N	38°17'E
Tagula, i., Pap. N. Gui. (tä'gōō-lȧ)	175	11°45'S	153°46'E
Tagus (Tajo), r., Eur. (tā'gŭs)	112	39°40'N	5°07'W
Tahan, Gunong, mtn., Malay.	168	4°33'N	101°52'E
Tahat, mtn., Alg. (tä-hät')	184	23°22'N	5°21'E
Tahiti, i., Fr. Poly. (tä-hē'tē) (tä'ē-tā')	2	17°30'S	149°30'W
Tahkuna Nina, c., Est. (tȧh-kōō'nȧ nē'nȧ)	123	59°08'N	22°03'E
Tahlequah, Ok., U.S. (tä-lĕ-kwä')	79	35°54'N	94°58'W
Tahoe, l., U.S. (tä'hō)	64	39°09'N	120°18'W
Tahoua, Niger (tä'ōō-ä)	184	14°54'N	5°16'E
Tahtsa Lake, l., Can.	54	53°33'N	127°47'W
Tahuya, Wa., U.S. (tȧ-hū-yă')	74a	47°23'N	123°03'W
Tahuya, r., Wa., U.S.	74a	47°28'N	122°55'W
Tai'an, China (tī-än)	164	36°13'N	117°08'E
Taibai Shan, mtn., China (tī-bī shän)	164	33°42'N	107°25'E
Taibus Qi, China (tī-bōō-sz chyē)	164	41°52'N	115°25'E
Taicang, China (tī-tsäŋ)	162	31°26'N	121°06'E
T'aichung, Tai. (tī'chŏng)	161	24°10'N	120°42'E
Tai'erzhuang, China (tī-är-jüäŋ)	162	34°34'N	117°44'E
Taigu, China (tī-gōō)	164	37°25'N	112°35'E
Taihang Shan, mts., China (tī-häŋ shän)	164	35°45'N	112°00'E
Taihe, China (tī-hŭ)	162	33°10'N	115°38'E
Tai Hu, l., China (tī hōō)	161	31°13'N	120°00'E
Tailagoin, reg., Mong. (tī'lȧ-gän' kä'rä)	160	43°39'N	105°54'E
Tailai, China (tī-lī)	164	46°20'N	123°10'E
Tailem Bend, Austl. (tā-lĕm)	176	35°15'S	139°30'E
T'ainan, Tai. (tī'nan')	161	23°08'N	120°18'E
Taínaro, c., Grc.	118	37°45'N	22°00'E
Taining, China (tī'nĭng')	165	26°58'N	117°15'E
T'aipei, Tai. (tī'pā')	161	25°02'N	121°38'E
Taiping, pt. of. i., Malay.	168	4°56'N	100°39'E
Taiping Ling, mtn., China	164	47°03'N	120°30'E
Taisha, Japan (tī'shä)	167	35°23'N	132°40'E
Taishan, China (tī-shän)	165	22°15'N	112°50'E
Tai Shan, mts., China (tī shän)	164	36°16'N	117°05'E
Taitao, Península de, pen., Chile	102	46°20'S	77°15'W
T'aitung, Tai. (tī'tōōng')	165	22°45'N	121°02'E
Taiwan, nation, Asia (tī-wän) (fôr-mō'sȧ)	161	23°30'N	122°20'E
Taiwan Strait, strt., Asia	161	24°30'N	120°00'E
Taixian, China (tī shyĕn)	162	32°31'N	119°54'E
Taixing, China (tī-shyĭŋ)	162	32°12'N	119°58'E
Taiyuan, China (tī-yüän)	161	37°32'N	112°38'E
Taizhou, China (tī-jō)	162	32°23'N	119°41'E
Ta'Izz, Yemen	157	13°38'N	44°04'E
Tajano de Morais, Braz. (tĕ-zhä'nō-dĕ-mô-rä'ēs)	99a	22°05'S	42°04'W
Tajikistan, nation, Asia	134	39°22'N	69°30'E
Tajumulco, vol., Guat. (tä-hōō-mōōl'kō)	90	15°03'N	91°53'W
Tajuña, r., Spain (tä-ĸōō'n-yä)	128	40°23'N	2°36'W
Tājūrā', Libya	118	32°56'N	13°24'W
Tak, Thai.	168	16°57'N	99°12'E
Taka, i., Japan (tä'kä)	167	30°47'N	130°23'E
Takada, Japan (tä'kȧ-dä)	166	37°08'N	138°30'E
Takahashi, Japan (tä'kä'hä-shĭ)	167	34°47'N	133°35'E
Takaishi, Japan	167b	34°32'N	135°27'E
Takamatsu, Japan (tä'kä'mä-tsōō')	161	34°20'N	134°02'E
Takamori, Japan (tä'kä'mô-rē')	167	32°50'N	131°08'E
Takaoka, Japan (ta'kä'ô-kä')	166	36°45'N	136°59'E
Takapuna, N.Z.	177	36°48'S	174°47'E
Takarazuka, Japan (tä'kä-rä-zōō'kä)	167b	34°48'N	135°22'E
Takasaki, Japan (tä'kät'sōō-kē')	166	36°20'N	139°00'E
Takatsu, Japan (tä-kät'sōō) (mē'zō-nō-kō'chē)	167a	35°36'N	139°37'E
Takatsuki, Japan (tä'kät'sōō-kē')	167b	34°51'N	135°38'E
Takayama, Japan (tä'kä'yä'mä)	167	36°11'N	137°16'E
Takefu, Japan (tä'kē-fōō)	166	35°57'N	136°09'E
Take-shima, is., Asia	166	37°15'N	131°51'E
Takla Lake, l., Can.	52	55°25'N	125°53'W
Takla Makan, des., China (mä-kän')	160	39°22'N	82°34'E
Takoma Park, Md., U.S. (tă'kōmă pärk)	68e	38°59'N	77°00'W
Takum, Nig.	189	7°17'N	9°59'E
Tala, Mex. (tä'lä)	88	20°39'N	103°42'W
Talagante, Chile (tä-lä-gä'n-tĕ)	99b	33°39'S	70°54'W
Talamanca, Cordillera de, mts., C.R.	91	9°37'N	83°55'W
Talanga, Hond. (tä-lä'n-gä)	90	14°21'N	87°09'W
Talara, Peru (tä-lä'rä)	100	4°32'S	81°17'W
Talasea, Pap. N. Gui. (tä-lä-sā'ä)	169	5°20'S	150°00'E
Talata Mafara, Nig.	189	12°35'N	6°04'E
Talaud, Kepulauan, is., Indon. (tä-lout')	169	4°17'N	127°30'E
Talavera de la Reina, Spain	118	39°58'N	4°51'W
Talca, Chile (täl'kä)	102	35°25'S	71°39'W
Talca, prov., Chile	99b	35°23'S	71°15'W
Talca, Punta, c., Chile (pōō'n-tä-täl'kä)	99b	33°25'S	71°42'W
Talcahuano, Chile (täl-kä-wä'nō)	102	36°41'S	73°05'W
Taldom, Russia (täl-dôm)	132	56°44'N	37°33'E
Taldyqorghan, Kaz.	139	45°03'N	77°18'E
Talea de Castro, Mex. (tä'lå-ä dā käs'trō)	89	17°22'N	96°14'W
Talibu, Pulau, i., Indon.	169	1°30'S	125°00'E
Talim, i., Phil. (tä-lēm')	169a	14°21'N	121°14'E
Talisay, Phil. (tä-lē'sī)	169a	14°08'N	122°56'E
Talkeetna, Ak., U.S. (tál-kēt'nȧ)	63	62°18'N	150°02'W
Talladega, Al., U.S. (tăl-ȧ-dē'gȧ)	82	33°25'N	86°06'W
Tallahassee, Fl., U.S. (tăl-ȧ-hăs'ē)	65	30°25'N	84°17'W
Tallahatchie, r., Ms., U.S. (tal-ȧ hăch'ē)	82	34°21'N	90°03'W
Tallapoosa, Ga., U.S. (tăl-ȧ-pōō'sȧ)	82	33°44'N	85°15'W
Tallapoosa, r., Al., U.S.	82	32°22'N	86°08'W
Tallassee, Al., U.S. (tăl'ȧ-sē)	82	32°30'N	85°54'W
Tallinn, Est. (tál'lĕn) (rā'väl)	134	59°26'N	24°44'E
Tallmadge, Oh., U.S. (tăl'mĭj)	69d	41°06'N	81°26'W
Tallulah, La., U.S. (tă-lōō'lȧ)	81	32°25'N	91°13'W
Tal'ne, Ukr.	133	48°52'N	30°43'E
Talo, mtn., Eth.	185	10°45'N	37°55'E
Taloje Budrukh, India	159b	19°05'N	73°05'E
Talpa de Allende, Mex. (täl'pä dā äl-yēn'dȧ)	88	20°25'N	104°48'W
Talquin, Lake, res., Fl., U.S.	82	30°26'N	84°33'W
Talsi, Lat. (tal'sĭ)	123	57°16'N	22°35'E
Taltal, Chile (täl-täl')	102	25°26'S	70°32'W
Taly, Russia (täl'ĭ)	133	49°51'N	40°07'E
Tama, Ia., U.S. (tā'mȧ)	71	41°57'N	92°36'W
Tama, r., Japan	167a	35°38'N	139°35'E
Tamale, Ghana (tä-mä'lā)	184	9°25'N	0°50'W
Taman', Russia (tȧ-män'')	133	45°13'N	36°46'E
Tamanaco, r., Ven. (tä-mä-nä'kō)	101b	9°32'N	66°00'W
Tamaqua, Pa., U.S. (tȧ-mô'kwȧ)	67	40°45'N	75°50'W
Tamar, r., Eng., U.K. (tā'mär)	120	50°35'N	4°15'W
Tamarite de Litera, Spain (tä-mä-rē'tä)	129	41°52'N	0°24'E
Tamaulipas, state, Mex. (tä-mä-ōō-lē'päs')	86	23°45'N	98°30'W
Tamazula de Gordiano, Mex.	88	19°44'N	103°09'W
Tamazulapan del Progreso, Mex.	89	17°41'N	97°34'W
Tamazunchale, Mex. (tä-mä-zòn-chä'lä)	88	21°16'N	98°46'W
Tambacounda, Sen. (täm-bä-kōōn'dä)	184	13°47'N	13°40'W
Tambador, Serra do, mts., Braz. (sĕ'r-rä-dô-täm'bä-dōr)	101	10°33'S	41°16'W
Tambelan, Kepulauan, is., Indon. (täm-bå-län')	168	0°38'N	107°38'E
Tambo, Austl. (tăm'bō)	175	24°50'S	146°15'E
Tambov, Russia (tám-bôf')	134	52°45'N	41°10'E
Tambov, prov., Russia	132	52°50'N	40°42'E
Tambre, r., Spain (täm'brā)	128	42°59'N	8°33'W
Tambura, Sudan (täm-bōō'rä)	185	5°34'N	27°30'E
Tame, r., Eng., U.K. (tām)	114a	52°41'N	1°42'W
Tâmega, r., Port. (tȧ-mā'gȧ)	128	41°30'N	7°45'W
Tamenghest, Alg.	184	22°34'N	5°34'E
Tamenghest, Oued, r., Alg.	184	22°15'N	2°51'E
Tamgak, Monts, mtn., Niger (tam-gäk')	184	18°40'N	8°40'E
Tamgué, Massif du, mtn., Gui.	184	12°15'N	12°35'W
Tamiahua, Mex. (tä-myä-wä)	89	21°17'N	97°26'W
Tamiahua, Laguna, l., Mex. (lä-gō'nä-tä-myä-wä)	89	21°38'N	97°33'W
Tamiami Canal, can., Fl., U.S. (tă-mī-ăm'ĭ)	83a	25°52'N	80°08'W
Tamil Nadu, state, India	155	11°30'N	78°00'E
Tampa, Fl., U.S. (tăm'pȧ)	65	27°57'N	82°25'W
Tampa Bay, b., Fl., U.S.	65	27°35'N	82°38'W
Tampere, Fin. (tȧm'pĕ-rĕ)	116	61°21'N	23°39'E
Tampico, Mex. (täm-pē'kō)	86	22°14'N	97°51'W
Tampico Alto, Mex. (täm-pē'kō äl'tō)	89	22°07'N	97°48'W
Tampin, Malay.	153b	2°28'N	102°15'E
Tam Quan, Viet.	165	14°20'N	109°10'E
Tamuín, Mex. (tä-mōō-ē'n)	88	22°04'N	98°47'W
Tamworth, Austl. (tăm'wûrth)	175	31°01'S	151°00'E
Tamworth, Eng., U.K.	114a	52°38'N	1°41'W
Tana, i., Vanuatu	175	19°32'S	169°27'E
Tana, r., Kenya (tä'nä)	187	0°30'S	39°30'E
Tanabe, Japan (tä-nä'bā)	166	33°45'N	135°21'E
Tanabe, Japan	167b	34°49'N	135°46'E
Tanacross, Ak., U.S. (tă'nȧ-crōs)	63	63°20'N	143°30'W
Tanaga, i., Ak., U.S. (tä-nä'gä)	63a	51°28'N	178°10'W
Tanahbala, Pulau, i., Indon. (tȧ-nä-bä'lä)	168	0°30'S	98°22'E
Tanahmasa, Pulau, i., Indon. (tȧ-nä-mä'sä)	168	0°03'S	97°30'E
Tanakpur, India (tăn'ăk-pŏr)	158	29°10'N	80°07'E
Tana Lake, l., Eth.	185	12°09'N	36°41'E
Tanami, Austl. (tȧ-nă'mē)	174	19°45'S	129°50'E
Tanana, Ak., U.S. (tă'nȧ-nô)	63	65°18'N	152°20'W
Tanana, r., Ak., U.S.	63	64°26'N	148°40'W
Tanaro, r., Italy (tä-nä'rô)	130	44°45'N	8°02'E
Tanashi, Japan	167a	35°44'N	139°34'E
Tanbu, China (tän-bōō)	163a	23°20'N	113°06'E
Tancheng, China (tän-chŭŋ)	164	34°37'N	118°22'E
Tanchŏn, Kor., N. (tän'chŭn)	166	40°29'N	128°50'E
Tancítaro, Mex. (tän-sē'tä-rō)	88	19°16'N	102°24'W
Tancítaro, Cerro de, mtn., Mex. (sĕ'r-rō-dē)	88	19°24'N	102°19'W
Tancoco, Mex. (tän-kō'kō)	89	21°16'N	97°45'W
Tandil, Arg. (tän-dēl')	102	36°16'S	59°01'W
Tandil, Sierra del, mts., Arg.	102	38°40'S	59°40'W
Tanega, i., Japan (tä'nȧ-gä')	161	30°36'N	131°11'E
Tanezrouft, reg., Alg. (tä'nĕz-rôft)	184	24°17'N	0°30'W
Tang, r., China (täŋ)	162	33°38'N	117°29'E
Tang, r., China	162	39°13'N	114°45'E
Tanga, Tan. (täŋ'gȧ)	187	5°04'S	39°06'E
Tangancícuaro, Mex. (täŋ-gän-sē'kwa-rô)	88	19°52'N	102°13'W
Tanganyika, Lake, l., Afr.	186	5°15'S	29°40'E
Tanger, Mor. (tăn-jēr')	184	35°52'N	5°55'W
Tangermünde, Ger. (täŋ'ĕr-mün'de)	124	52°33'N	11°58'E
Tanggu, China (täŋ-gōō)	162	39°04'N	117°41'E
Tanggula Shan, mts., China (täŋ-gōō-lä shän)	160	33°15'N	89°07'E
Tanghe, China	164	32°40'N	112°50'E
Tangier *see* Tanger, Mor.	184	35°52'N	5°55'W
Tangipahoa, r., La., U.S. (tăn'jē-pȧ-hō'ȧ)	81	30°48'N	90°28'W
Tangra Yumco, l., China (täŋ-rä yōōm-tswo)	158	30°50'N	85°40'E
T'angshan, China	164	39°38'N	118°11'E
Tangxian, China (täŋ shyĕn)	162	38°49'N	115°00'E
Tangzha, China (täŋ-jä)	162	32°06'N	120°48'E
Tanimbar, Kepulauan, is., Indon.	169	8°00'S	132°00'E
Tanjong Piai, c., Malay.	153b	1°16'N	103°11'E
Tanjong Ramunia, c., Malay.	153b	1°27'N	104°44'E
Tanjungbalai, Indon. (tän'jŏng-bä'lȧ)	153b	1°00'N	103°26'E
Tanjungpandan, Indon.	168	2°47'S	107°51'E
Tanjungpinang, Indon. (tän'jŏng-pē'näng)	153b	0°55'N	104°29'E
Tannu-Ola, mts., Asia	135	51°00'N	94°00'E
Tannūrah, Ra's at, c., Sau. Ar.	154	26°45'N	49°59'E
Tano, r., Afr.	188	5°40'N	2°30'W
Tanquijo, Arrecife, i., Mex. (är-rē-sē'fē-tän-kē'kō)	89	21°07'N	97°16'W
Ṭanṭā, Egypt	185	30°47'N	31°00'E
Tantoyuca, Mex. (tän-tō-yōō'kä)	88	21°22'N	98°13'W
Tanyang, Kor., S.	166	36°53'N	128°20'E
Tanzania, nation, Afr.	186	6°48'S	33°58'E
Tao, r., China (tou)	164	35°30'N	103°40'E
Tao'an, China (tou-än)	161	45°15'N	122°45'E
Tao'er, r., China (tou-är)	161	45°40'N	122°00'E
Taormina, Italy (tä-ôr-mē'nä)	130	37°53'N	15°18'E
Taos, N.M., U.S. (tä'ōs)	77	36°25'N	105°35'W
Taoudenni, Mali (tä'ōō-dē-nē')	184	22°57'N	3°37'W
Taoussa, Mali	188	16°55'N	0°35'W
Taoyuan, China (tou-yüän)	165	29°00'N	111°15'E
Tapa, Est. (tá'pȧ)	123	59°16'N	25°56'E
Tapachula, Mex.	90	14°55'N	92°20'W
Tapajós, r., Braz. (tä-pä-zhô's)	101	3°27'S	55°33'W
Tapalque, Arg. (tä-päl-kē')	99c	36°22'S	60°05'W
Tapanatepec, Mex. (tä-pä-nä-tē-pĕk)	89	16°22'N	94°19'W
Tāpi, r., India	155	21°00'N	76°30'E
Tappi Saki, c., Japan (täp'pē sä'kē)	166	41°05'N	139°40'E
Tapps, l., Wa., U.S. (tăpz)	74a	47°20'N	122°12'W
Taquara, Serra de, mts., Braz. (sĕ'r-rä-dē-tä-kwä'rä)	101	15°28'S	54°33'W
Taquari, r., Braz. (tä-kwä'rĭ)	101	18°35'S	56°50'W
Tar, r., N.C., U.S. (tär)	83	35°58'N	78°06'W
Tara, Russia (tä'rȧ)	134	56°58'N	74°13'E
Tara, i., Phil. (tä'rä)	169a	12°18'N	120°28'E
Tara, r., Russia (tä'rȧ)	140	56°32'N	76°13'E
Ṭarābulus, Leb. (tȧ-rä'bō-lōōs)	154	34°25'N	35°50'E
Tarābulus (Tripolitania), hist. reg., Libya	184	31°00'N	12°26'E
Tarakan, Indon.	168	3°17'N	118°04'E
Taranaki, Mount, vol., N.Z.	177	39°18'S	174°04'E
Tarancón, Spain (tä-rän-kōn')	128	40°01'N	3°00'W
Taranto, Italy (tä'rän-tô)	119	40°30'N	17°15'E
Taranto, Golfo di, b., Italy (gôl-fô-dē tä'rän-tō)	112	40°03'N	17°10'E
Tarapoto, Peru (tä-rä-pô'tō)	100	6°29'S	76°26'W
Tarare, Fr. (tȧ-rär')	126	45°55'N	4°23'E
Tarascon, Fr. (tä-rȧs-kôN')	126	42°53'N	1°35'E
Tarascon, Fr. (tä-räs-kôN)	126	43°47'N	4°41'E
Tarashcha, Ukr. (tä'rásh-chá)	133	49°34'N	30°52'E
Tarata, Bol. (tä-rä'tä)	100	17°43'S	66°00'W
Taravo, r., Fr.	130	41°54'N	8°58'E
Tarazit, Massif de, mts., Niger	189	20°05'N	7°35'E
Tarazona, Spain (tä-rä-thō'nä)	128	41°54'N	1°45'W
Tarazona de la Mancha, Spain (tä-rä-zō'nä-dĕ-lä-mä'n-chä)	128	39°13'N	1°50'W
Tarbes, Fr. (tärb)	117	43°04'N	0°05'E
Tarboro, N.C., U.S. (tär'bŭr-ō)	83	35°53'N	77°34'W

PRONOUNCING INDEX Taree – Testa del Gargano 309

PLACE (Pronunciation)	PAGE	LAT.	LONG.
Taree, Austl. (tă-rē′)	176	31°52′S	152°21′E
Tarentum, Pa., U.S. (tȧ-rĕn′tŭm)	69e	40°36′N	79°44′W
Tarfa, Wādī at, val., Egypt	192b	28°14′N	31°00′E
Târgovişte, Rom.	119	44°54′N	25°29′E
Târgu Jiu, Rom.	119	45°02′N	23°17′E
Târgu Mureş, Rom.	119	46°33′N	24°33′E
Târgu Neamţ, Rom.	125	47°14′N	26°23′E
Târgu Ocna, Rom.	125	46°18′N	26°38′E
Târgu Secuiesc, Rom.	125	46°04′N	26°06′E
Tarhūnah, Libya	156	32°26′N	13°38′E
Tarija, Bol. (tär-rē′hä)	100	21°42′S	64°52′W
Tarim, Yemen (tȧ-rĭm′)	154	16°13′N	49°08′E
Tarim, r., China (tȧ-rĭm′)	160	40°45′N	85°39′E
Tarim Basin, basin, China (tȧ-rĭm′)	160	39°52′N	82°34′E
Tarka, r., S. Afr. (tä′kȧ)	187c	32°15′S	26°00′E
Tarkastad, S. Afr.	187c	32°01′S	26°18′E
Tarkhankut, Mys, c., Ukr. (mĭs tȧr-κän′kŏŏt)	137	45°21′N	32°30′E
Tarkio, Mo., U.S. (tär′kĭ-ō)	79	40°27′N	95°22′W
Tarkwa, Ghana (tärk′wä)	184	5°19′N	1°59′W
Tarlac, Phil. (tär′läk)	168	15°29′N	120°36′E
Tarlton, S. Afr. (tärl′tŭn)	187b	26°05′S	27°38′E
Tarma, Peru (tär′mä)	100	11°26′S	75°40′W
Tarn, r., Fr. (tärn)	117	43°45′N	2°00′E
Târnăveni, Rom.	125	46°19′N	24°18′E
Tarnów, Pol. (tär′nōf)	117	50°02′N	21°00′E
Taro, r., Italy (tä′rō)	130	44°41′N	10°03′E
Taroudant, Mor. (tȧ-rōō-dänt′)	184	30°39′N	8°52′W
Tarpon Springs, Fl., U.S. (tär′pŏn)	83a	28°07′N	82°44′W
Tarporley, Eng., U.K. (tär′pēr-lē)	114a	53°09′N	2°40′W
Tarpum Bay, b., Bah. (tär′pŭm)	92	25°05′N	76°20′W
Tarquinia, Italy (tär-kwē′nē-ä)	130	42°16′N	11°46′E
Tarragona, Spain (tär-rä-gō′nä)	110	41°05′N	1°15′E
Tarrant, Al., U.S. (tăr′ănt)	68h	33°35′N	86°46′W
Tárrega, Spain (tä rå-gä)	129	41°40′N	1°09′E
Tarrejón de Ardoz, Spain (tär-rĕ-κō′n-dĕ-är-dôz)	129a	40°28′N	3°29′W
Tarrytown, N.Y., U.S. (tăr′ĭ-toun)	68a	41°04′N	73°52′W
Tarsus, Tur. (tȧr′sŏŏs) (tär′sŭs)	154	37°00′N	34°50′E
Tartagal, Arg. (tär-tä-gȧ′l)	102	23°31′S	63°47′W
Tartu, Est. (tär′tōō) (dôr′pät)	134	58°23′N	26°44′E
Ţarţūs, Syria	156	34°54′N	35°59′E
Tarumi, Japan (tä′rōō-mē)	167b	34°38′N	135°04′E
Tarusa, Russia (tä-rōōs′ȧ)	132	54°43′N	37°11′E
Tarzana, Ca., U.S. (tär-ză′ȧ)	75a	34°10′N	118°32′W
Tashkent, Uzb. (tȧsh′kĕnt)	139	41°23′N	69°04′E
Tasman Bay, b., N.Z. (tăz′măn)	175a	40°50′S	173°20′E
Tasmania, state, Austl.	175	41°28′S	142°30′E
Tasman Peninsula, pen., Austl.	176	43°00′S	148°30′E
Tasman Sea, sea, Oc.	195	29°30′S	155°00′E
Tasquillo, Mex. (täs-kē′lyō)	88	20°34′N	99°21′W
Tatarsk, Russia (tȧ-tärsk′)	134	55°13′N	75°58′E
Tatarstan, prov., Russia	136	55°00′N	51°00′E
Tatar Strait, strt., Russia	135	51°00′N	141°45′E
Tater Hill, mtn., Or., U.S. (tāt′ēr hĭl)	74c	45°47′N	123°02′W
Tateyama, Japan (tä′tĕ-yä′mä)	167	35°04′N	139°52′E
Tatlow, Mount, mtn., Can.	54	51°23′N	123°52′W
Tau, Nor.	122	59°05′N	5°59′E
Tauern Tunnel, trans., Aus.	124	47°12′N	13°17′E
Taung, S. Afr. (tä′ŏng)	186	27°25′S	24°47′E
Taunton, Ma., U.S. (tän′tŭn)	67	41°54′N	71°03′W
Taunton, r., R.I., U.S.	68b	41°50′N	71°02′W
Taupo, Lake, l., N.Z. (tä′ōō-pō)	175a	38°42′S	175°55′E
Taurage, Lith. (tou′rȧ-gā)	123	55°15′N	22°18′E
Taurus Mountains *see* Toros Dağları, mts., Tur.	154	37°00′N	32°40′E
Tauste, Spain (tä-ōōs′tå)	128	41°55′N	1°15′W
Tavda, Russia (tȧv-dȧ′)	134	58°00′N	64°44′E
Tavda, r., Russia	140	58°30′N	64°15′E
Taverny, Fr. (tȧ-vĕr-nē′)	127b	49°02′N	2°13′E
Taviche, Mex. (tä-vē′chĕ)	89	16°43′N	96°35′W
Tavira, Port. (tä-vē′rȧ)	128	37°09′N	7°42′W
Tavşanlı, Tur. (tȧv′shän-lĭ)	137	39°30′N	29°30′E
Tawakoni, l., Tx., U.S.	81	32°51′N	95°59′W
Tawaramoto, Japan (tä′wä-rä-mô-tō)	167b	34°33′N	135°48′E
Tawas City, Mi., U.S.	66	44°15′N	83°30′W
Tawas Point, c., Mi., U.S. (tô′wȧs)	66	44°15′N	83°25′W
Tawitawi Group, is., Phil. (tä′wē-tä′wē)	168	4°52′N	120°35′E
Tawkar, Sudan	185	18°28′N	37°46′E
Taxco de Alarcón, Mex. (täs′kō dĕ ä-lär-kō′n)	88	18°34′N	99°37′W
Tay, r., Scot., U.K.	120	56°35′N	3°37′W
Tay, Loch, l., Scot., U.K.	120	56°25′N	4°07′W
Tayabas Bay, b., Phil. (tä-yä′bäs)	169a	13°44′N	121°40′E
Tayga, Russia (tī′gä)	140	56°12′N	85°47′E
Taygonos, Mys, c., Russia	135	60°37′N	160°17′E
Taylor, Tx., U.S.	81	30°35′N	97°25′W
Taylor, Mount, mtn., N.M., U.S.	64	35°20′N	107°40′W
Taylorville, Il., U.S. (tā′lēr-vĭl)	66	39°30′N	89°20′W
Taymyr, l., Russia (tī-mĭr′)	135	74°13′N	100°45′E
Taymyr, Poluostrov, pen., Russia	135	75°15′N	95°00′E
Tayshet, Russia (tī-shĕt′)	135	56°09′N	97°49′E
Tayug, Phil.	169a	16°01′N	120°45′E
Taz, r., Russia (tȧz)	140	67°15′N	80°45′E
Taza, Mor. (tä′zä)	184	34°08′N	4°00′W
Tazovskoye, Russia	134	66°58′N	78°28′E
Tbessa, Alg.	184	35°27′N	8°13′E
Tbilisi, Geor. (′tbĭl-yē′sē)	137	41°40′N	44°45′E
Tchentlo Lake, l., Can.	54	55°11′N	125°00′W
Tchibanga, Gabon (chē-băŋ′gä)	186	2°51′S	11°02′E
Tchien, Lib.	188	6°04′N	8°08′W
Tchigai, Plateau du, plat., Afr.	189	21°20′N	14°50′E
Tczew, Pol. (t′chĕf′)	116	54°06′N	18°48′E
Teabo, Mex. (tĕ-ä′bō)	90a	20°25′N	89°14′W
Teague, Tx., U.S.	81	31°39′N	96°16′W
Teapa, Mex. (tä-ä′pä)	89	17°35′N	92°56′W
Tebing Tinggi, i., Indon. (teb′ing-tĭng′gä)	153b	0°54′N	102°39′E
Tecalitlán, Mex. (tā-kä-lē-tlän′)	88	19°28′N	103°17′W
Techiman, Ghana	188	7°35′N	1°56′W
Tecoanapa, Mex. (tāk-wä-nä-pä′)	88	16°33′N	98°46′W
Tecoh, Mex. (tĕ-kô)	90a	20°46′N	89°27′W
Tecolotlán, Mex. (tā-kô-lô-tlän′)	88	20°13′N	103°57′W
Tecolutla, Mex. (tā-kô-lōō′tlä)	89	20°33′N	97°00′W
Tecolutla, r., Mex.	89	20°16′N	97°14′W
Tecomán, Mex. (tā-kô-män′)	88	18°53′N	103°53′W
Tecómitl, Mex. (tĕ-kō′mētl)	89a	19°13′N	98°59′W
Tecozautla, Mex. (tā′kô-zä-ōō′tlä)	88	20°33′N	99°38′W
Tecpan de Galeana, Mex. (tĕk-pän′ dā gä-lā-ä′nä)	88	17°13′N	100°41′W
Tecpatán, Mex. (tĕk-pä-tä′n)	89	17°08′N	93°18′W
Tecuala, Mex. (tĕ-kwä-lä)	88	22°24′N	105°29′W
Tecuci, Rom. (tȧ-kōch′)	119	45°51′N	27°30′E
Tecumseh, Can. (tē-kŭm′sĕ)	69b	42°19′N	82°53′W
Tecumseh, Mi., U.S.	66	42°00′N	84°00′W
Tecumseh, Ne., U.S.	79	40°21′N	96°09′W
Tecumseh, Ok., U.S.	79	35°18′N	96°55′W
Tees, r., Eng., U.K. (tēz)	120	54°40′N	2°10′W
Teganuna, l., Japan (tā′gä-nōō′nä)	167a	35°50′N	140°02′E
Tegucigalpa, Hond. (tā-gōō-sē-gäl′pä)	86	14°08′N	87°15′W
Tehachapi Mountains, mts., Ca., U.S. (tē-hȧ-shä′pĭ)	76	34°50′N	118°55′W
Tehrān, Iran (tĕ-hrän′)	154	35°45′N	51°30′E
Tehuacán, Mex. (tā-wä-kän′)	86	18°27′N	97°23′W
Tehuantepec, Mex.	86	16°20′N	95°14′W
Tehuantepec, r., Mex.	89	16°30′N	95°23′W
Tehuantepec, Golfo de, b., Mex. (gôl-fô dĕ)	86	15°45′N	95°00′W
Tehuantepec, Istmo de, isth., Mex. (ē′st-mô dĕ)	89	17°55′N	94°35′W
Tehuehuetla, Arroyo, r., Mex. (tē-wĕ-wĕ′tlä ȧr-rô-yō)	88	17°54′N	100°26′W
Tehuitzingo, Mex. (tā-wē-tzĭŋ′gō)	88	18°21′N	98°16′W
Tejeda, Sierra de, mts., Spain (sē-ĕ′r-rä dĕ tĕ-kĕ′dä)	128	36°55′N	4°00′W
Tejúpan, Mex. (tĕ-κōō-pä′n) (sän-tyȧ′gō)	89	17°39′N	97°34′W
Tejúpan, Punta, c., Mex.	88	18°19′N	103°30′W
Tejupilco de Hidalgo, Mex. (tȧ-hōō-pēl′kō dā ē-dhäl′gō)	88	18°52′N	100°07′W
Tekamah, Ne., U.S. (tē-kȧ′mȧ)	70	41°46′N	96°13′W
Tekax de Alvaro Obregon, Mex.	90a	20°12′N	89°11′W
Tekeze, r., Afr.	185	13°38′N	38°00′E
Tekit, Mex. (tĕ-kē′t)	90a	20°35′N	89°18′W
Tekoa, Wa., U.S. (tē-kō′ȧ)	72	47°15′N	117°03′W
Tela, Hond. (tā′lä)	86	15°45′N	87°25′W
Tela, Bahía de, b., Hond.	90	15°53′N	87°29′W
Telapa Burok, Gunong, mtn., Malay.	153b	2°51′N	102°04′E
Telavi, Geor.	137	42°00′N	45°20′E
Tel Aviv-Yafo, Isr. (tĕl-ä-vēv′jä′jȧ′fȧ)	154	32°03′N	34°46′E
Telegraph Creek, Can. (tĕl′ē-grȧf)	50	57°59′N	131°22′W
Teleneşti, Mol.	133	47°31′N	28°22′E
Telescope Peak, mtn., Ca., U.S. (tĕl′ē skōp)	64	36°12′N	117°05′W
Telesung, Indon.	153b	1°07′N	102°53′E
Telica, vol., Nic. (tā-lē′kä)	90	12°38′N	86°52′W
Tell City, In., U.S. (tĕl)	66	38°00′N	86°45′W
Teller, Ak., U.S. (tĕl′ẽr)	63	65°17′N	166°28′W
Tello, Col. (tĕ′l-yō)	100a	3°05′N	75°08′W
Telluride, Co., U.S. (tĕl′ū-rīd)	77	37°55′N	107°50′W
Telok Datok, Malay.	153b	2°51′N	101°33′E
Teloloapan, Mex. (tā′lô-lô-ä′pän)	88	18°19′N	99°54′W
Tel′pos-Iz, Gora, mtn., Russia (tyĕl′pôs-ēz′)	134	63°50′N	59°20′E
Telšiai, Lith. (tĕl′shä′ē)	123	55°59′N	22°17′E
Teltow, Ger. (tĕl′tō)	115b	52°24′N	13°12′E
Teluklecak, Indon.	153b	1°53′N	101°45′E
Tema, Ghana	188	5°38′N	0°01′E
Temascalcingo, Mex. (tā′mäs-käl-sĭŋ′gō)	88	19°55′N	100°00′W
Temascaltepec, Mex. (tā′mäs-käl-tȧ pĕk)	88	19°00′N	100°03′W
Temax, Mex. (tĕ′mäx)	86	21°10′N	88°51′W
Temir, Kaz.	139	49°10′N	57°15′E
Temirtau, Kaz.	139	50°08′N	73°13′E
Temiscouata, l., Can. (tĕ′mĭs-kô-ä′tä)	60	47°40′N	68°50′W
Témiskaming, Can. (tē-mĭs′kȧ-mĭng)	51	46°41′N	79°01′W
Temoaya, Mex. (tĕ-mô-ȧ-um-yä)	89a	19°28′N	99°36′W
Tempe, Az., U.S.	77	33°24′N	111°54′W
Temperley, Arg. (tĕ′m-pĕr-lā)	102a	34°47′S	58°24′W
Tempio Pausania, Italy (tĕm′pē-ō pou-sä′nē-ä)	130	40°55′N	9°05′E
Temple, Tx., U.S. (tĕm′p′l)	81	31°06′N	97°20′W
Temple City, Ca., U.S.	75a	34°07′N	118°02′W
Templeton, Can. (tĕm′p′l-tŭn)	62c	45°29′N	75°37′W
Templin, Ger. (tĕm-plēn′)	124	53°08′N	13°30′E
Tempoal, r., Mex. (tĕm-pô-ä′l)	88	21°38′N	98°23′W
Temryuk, Russia (tyĕm-ryôk′)	137	45°17′N	37°21′E
Temuco, Chile (tȧ-mōō′kō)	102	38°46′S	72°38′W
Temyasovo, Russia (tĕm-yä′sô-vô)	142a	53°00′N	58°06′E
Tenāli, India	159	16°10′N	80°32′E
Tenamaxtlán, Mex. (tā′nä-mäs-tlän′)	88	20°13′N	104°06′W
Tenancingo, Mex. (tȧ-nän-sĕŋ′gō)	88	18°54′N	99°36′W
Tenango, Mex. (tȧ-näŋ′gō)	89a	19°09′N	98°51′W
Tenasserim, Mya. (tĕn-ăs′ẽr-ĭm)	168	12°09′N	99°01′E
Tendrivs′ka Kosa, ostriv, i., Ukr.	133	46°12′N	31°17′E
Tenerife Island, i., Spain (tå-nå-rē′fä) (tĕn-ẽr-ĭf′)	184	28°41′N	17°02′W
Ténès, Alg. (tā-nĕs′)	117	36°28′N	1°22′E
Tengiz köli, l., Kaz.	139	50°45′N	68°39′E
Tengxian, China (tŭŋ shyĕn)	164	35°07′N	117°08′E
Tenjin, Japan (tĕn′jĕn)	167b	34°54′N	135°04′E
Tenke, D.R.C. (tĕŋ′kȧ)	186	11°26′S	26°45′E
Tenkiller Ferry Reservoir, res., Ok., U.S. (tĕn-kĭl′ẽr)	79	35°42′N	94°47′W
Tenkodogo, Burkina (tĕŋ-kô-dō′gô)	184	11°47′N	0°22′W
Tenmile, r., Wa., U.S. (tĕn mīl)	74d	48°52′N	122°32′W
Tennant Creek, Austl. (tĕn′ănt)	174	19°45′S	134°00′E
Tennessee, state, U.S. (tĕn-ĕ-sē′)	65	35°50′N	88°00′W
Tennessee, r., U.S.	65	35°35′N	88°20′W
Tennille, Ga., U.S. (tĕn′ĭl)	82	32°55′N	86°50′W
Teno, r., Chile (tĕ′nô)	99b	34°55′S	71°00′W
Tenora, Austl. (tĕn-ŏrȧ)	176	34°23′S	147°33′E
Tenosique, Mex. (tā-nô-sē′kå)	89	17°27′N	91°25′W
Tenri, Japan	167b	34°36′N	135°50′E
Tenryū-Gawa, r., Japan (tĕn′ryōō′gä′wä)	167	35°16′N	137°54′E
Tensas, r., La., U.S. (tĕn′sô)	81	31°54′N	91°30′W
Tensaw, r., Al., U.S. (tĕn′sô)	82	30°45′N	87°52′W
Tenterfield, Austl. (tĕn′tẽr-fēld)	175	29°00′S	152°06′E
Ten Thousand, Islands, is., Fl., U.S. (tĕn thou′zănd)	83a	25°45′N	81°35′W
Teocaltiche, Mex. (tā′ô-käl-tē′chå)	88	21°27′N	102°38′W
Teocelo, Mex. (tā-ô-sā′lō)	89	19°22′N	96°57′W
Teocuitatlán de Corona, Mex.	88	20°06′N	103°22′W
Teófilo Otoni, Braz. (tē-ô′fē-lō-tô′nē)	101	17°49′S	41°18′W
Teoloyucan, Mex. (tā′ô-lô-yōō′kän)	88	19°43′N	99°12′W
Teopisca, Mex. (tā-ô-pēs′kä)	89	16°30′N	92°33′W
Teotihuacán, Mex. (tĕ-ô-tē-wä-kä′n)	89a	19°40′N	98°52′W
Teotitlán del Camino, Mex. (tā-ô-tē-tlän′ dĕl kä-mē′nô)	89	18°07′N	97°04′W
Tepalcatepec, Mex. (tā′päl-kä-tā′pĕk)	88	19°11′N	102°51′W
Tepalcatepec, r., Mex.	88	18°54′N	102°25′W
Tepalcingo, Mex. (tā-päl-sĕŋ′gô)	88	18°34′N	98°49′W
Tepatitlán de Morelos, Mex. (tā-pä-tē-tlän′ dā mô-rā′los)	86	20°55′N	102°47′W
Tepeaca, Mex. (tā-pā-ä′kä)	89	18°57′N	97°54′W
Tepecoacuilco de Trujano, Mex.	88	18°15′N	99°29′W
Tepeji del Río, Mex. (tā-pā-κe′ dĕl rē′ō)	88	19°55′N	99°22′W
Tepelmeme, Mex. (tā′pĕl-mā′må)	89	17°51′N	97°23′W
Tepetlaoxtoc, Mex. (tā′pā-tlä′ôs-tōk′)	88	19°34′N	98°49′W
Tepezala, Mex. (tā-pā-zä-lä′)	88	22°12′N	102°12′W
Tepic, Mex. (tā-pēk′)	86	21°32′N	104°53′W
Tëplaya Gora, Russia (tyôp′lȧ-yȧ gô-rȧ)	142a	58°32′N	59°08′W
Teplice, Czech Rep.	117	50°39′N	13°50′E
Teposcolula, Mex.	89	17°33′N	97°29′W
Tequendama, Salto de, wtfl., Col. (sä′l-tô dĕ tĕ-kĕn-dä′mä)	100	4°34′N	74°18′W
Tequila, Mex. (tå-kē′lä)	88	20°53′N	103°48′W
Tequisistlán, r., Mex. (tĕ-kē-sēs-tlä′n)	89	16°20′N	95°40′W
Tequisquiapan, Mex. (tå-kēs-kē-ä′pän)	88	20°33′N	99°57′W
Ter, r., Spain (tĕr)	129	42°04′N	2°52′E
Téra, Niger	188	14°01′N	0°45′E
Tera, r., Spain (tā′rä)	128	42°05′N	6°24′W
Teramo, Italy (tā′rä-mô)	130	42°40′N	13°41′E
Terborg, Neth. (tĕr-bōrg)	127c	51°55′N	6°23′E
Tercan, Tur. (tĕr′jän)	137	39°40′N	40°12′E
Terceira Island, i., Port. (tĕr-sā′rä)	184a	38°49′N	26°36′W
Terebovlia, Ukr.	125	49°18′N	25°43′E
Terek, r., Russia	137	43°30′N	45°10′E
Terenkul′, Russia (tĕ-rĕn′kŏl)	142a	55°38′N	62°18′E
Teresina, Braz. (tĕr-å-sē′nȧ)	101	5°04′S	42°42′W
Teresópolis, Braz. (tĕr-ā-sô′pō-lēzh)	99a	22°25′S	42°59′W
Teriberka, Russia (tyĕr-ē-byôr′kȧ)	136	69°00′N	35°15′E
Terme, Tur. (tĕr′mē)	137	41°05′N	37°00′E
Termez, Uzb. (tyĕr′mĕz)	139	37°19′N	67°20′E
Termini, Italy (tĕr′mē-nē)	130	37°58′N	13°39′E
Términos, Laguna de, l., Mex. (lä-gō′nä dĕ ĕ′r-mē-nôs)	86	18°37′N	91°32′W
Termoli, Italy (tĕr′mô-lē)	130	42°00′N	15°01′E
Tern, r., Eng., U.K. (tûrn)	114a	52°49′N	2°31′W
Ternate, Indon. (tĕr-nä′tā)	169	0°52′N	127°25′E
Terni, Italy (tĕr′nē)	118	42°38′N	12°41′E
Ternopil′, Ukr.	137	49°32′N	25°36′E
Terpeniya, Mys, c., Russia	135	48°44′N	144°42′E
Terpeniya, Zaliv, b., Russia (zä′lĭf tĕr-pä′nĭ-yȧ)	166	49°10′N	143°05′E
Terrace, Can. (tĕr′ĭs)	50	54°31′N	128°35′W
Terracina, Italy (tĕr-rä-chē′nä)	118	41°18′N	13°14′E
Terra Nova National Park, rec., Can.	53a	48°37′N	54°15′W
Terrassa, Spain	129	41°34′N	2°01′E
Terrebonne, Can. (tĕr-bŏn′)	67	45°42′N	73°38′W
Terrebonne Bay, b., La., U.S.	81	28°55′N	90°30′W
Terre Haute, In., U.S. (tĕr-ē hōt′)	65	39°25′N	87°25′W
Terrell, Tx., U.S. (tĕr′ĕl)	81	32°44′N	96°15′W
Terrell, Wa., U.S.	74d	48°53′N	122°44′W
Terrell Hills, Tx., U.S. (tĕr′ĕl hĭlz)	75d	29°28′N	98°27′W
Terschelling, i., Neth. (tĕr-skĕl′ĭng)	121	53°25′N	5°12′E
Teruel, Spain (tā-rōō-ĕl′)	118	40°20′N	1°05′W
Tešanj, Bos. (tĕ′shän′)	131	44°36′N	17°59′E
Teschendorf, Ger. (tĕ′shĕn-dōrf)	115b	52°51′N	13°10′E
Tesecheacan, Mex. (tĕ-sĕ-chĕ-ä-kä′n)	89	18°10′N	95°41′W
Teshekpuk, l., Ak., U.S. (tē-shĕk′pŭk)	63	70°18′N	152°36′W
Teshio Dake, mtn., Japan (tĕsh′ē-ō-dä′kā)	166	44°00′N	142°50′E
Teshio Gawa, r., Japan (tĕsh′ē-ô gä′wä)	166	44°53′N	144°55′E
Tesiyn, r., Asia	160	49°45′N	96°00′E
Teslin, Can. (tĕs-lĭn)	63	60°10′N	132°30′W
Teslin, l., Can.	52	60°12′N	132°08′W
Teslin, r., Can.	52	61°18′N	134°14′W
Tessaoua, Niger (tĕs-sä′ō-ä)	184	13°53′N	7°53′E
Tessenderlo, Bel.	115a	51°04′N	5°08′E
Test, r., Eng., U.K. (tĕst)	120	51°10′N	1°30′W
Testa del Gargano, c., Italy (tās′tä dĕl gär-gä′nō)	130	41°48′N	16°13′E

310 Tetachuck Lake – Tlalchapa

PRONOUNCING INDEX

PLACE (Pronunciation)	PAGE	LAT.	LONG.
Tetachuck Lake, l., Can.	54	53°20'N	125°50'W
Tete, Moz. (tā'tĕ)	186	16°13'S	33°35'E
Tête Jaune Cache, Can. (tĕt'zhōn-kăsh)	55	52°57'N	119°26'W
Teteriv, r., Ukr.	137	51°05'N	29°30'E
Teterow, Ger. (tā'tĕ-rō)	124	53°46'N	12°33'E
Teteven, Blg. (tĕt'ĕ-ven')	131	42°57'N	24°15'E
Teton, r., Mt., U.S. (tē'tŏn)	73	47°54'N	111°37'W
Tétouan, Mor.	184	35°42'N	5°34'W
Tetovo, Mac. (tā'tō-vō)	131	42°01'N	21°00'E
Tetyukhe-Pristan, Russia (tĕt-yōō'kĕ prĭ-stän')	166	44°21'N	135°44'E
Tetyushi, Russia (tyĭt-yô'shĭ)	136	54°57'N	48°50'E
Teupitz, Ger. (toi'pētz)	115b	42°08'N	13°37'E
Tevere, r., Italy	118	42°30'N	12°14'E
Teverya, Isr.	153a	32°48'N	35°32'E
Tewksbury, Ma., U.S. (tūks'bĕr-ĭ)	61a	42°37'N	71°14'W
Texada Island, i., Can.	54	49°40'N	124°24'W
Texarkana, Ar., U.S. (tĕk-sär-kăn'ȧ)	65	33°26'N	94°02'W
Texarkana, Tx., U.S.	65	33°26'N	94°04'W
Texas, state, U.S.	64	31°00'N	101°00'W
Texas City, Tx., U.S.	81	29°23'N	94°54'W
Texcaltitlán, Mex. (tās-käl'tē-tlän')	88	18°54'N	99°51'W
Texcoco, Mex. (tās-kō'kō)	88	19°31'N	98°53'W
Texcoco, Lago de, l., Mex.	89a	19°30'N	99°00'W
Texel, i., Neth. (tĕk'sĕl)	121	53°10'N	4°45'E
Texistepec, Mex. (tĕk-sēs-tā-pĕk')	89	17°51'N	94°46'W
Texmelucan, Mex.	88	19°17'N	98°26'W
Texoma, Lake, res., U.S. (tĕk'ō-mă)	64	34°03'N	96°28'W
Texontepec, Mex. (tā-zōn-tā-pĕk')	88	19°52'N	98°48'W
Texontepec de Aldama, Mex. (dā äl-dä'mä)	88	20°19'N	99°19'W
Teyateyaneng, Leso.	187c	29°11'S	27°43'E
Teykovo, Russia (tĕy-kô-vô)	136	56°52'N	40°34'E
Teziutlán, Mex. (tā-zē-ōō-tlän')	89	19°48'N	97°21'W
Tezpur, India	158	26°42'N	92°52'E
Tha-anne, r., Can.	52	60°50'N	96°56'W
Thabana Ntlenyana, mtn., Leso.	187c	29°28'S	29°17'E
Thabazimbi, S. Afr.	192c	24°36'S	27°22'E
Thailand, nation, Asia	168	16°30'N	101°00'E
Thailand, Gulf of, b., Asia	168	11°37'N	100°46'E
Thale Luang, l., Thai.	168	7°51'N	99°39'E
Thame, Eng., U.K. (tām)	114b	51°43'N	0°59'W
Thames, r., Can. (tĕmz)	58	42°40'N	81°45'W
Thames, r., Eng., U.K.	112	51°30'N	1°30'W
Thāmit, Wādī, r., Libya	119	30°39'N	16°23'E
Thāna, India (thä'nŭ)	158	19°13'N	72°58'E
Thāna Creek, r., India	159b	19°03'N	72°58'E
Thanh Hoa, Viet. (tän'hō'ä)	168	19°46'N	105°42'E
Thanjāvūr, India	155	10°51'N	79°11'E
Thann, Fr. (tän)	127	47°49'N	7°05'E
Thaon-les-Vosges, Fr. (tä-ŏn-lā-vōzh')	127	48°16'N	6°24'E
Thargomindah, Austl. (thär'gō-mĭn'dȧ)	175	27°58'S	143°57'E
Thásos, i., Grc. (thă'sôs)	119	40°41'N	24°53'E
Thatch Cay, i., V.I.U.S. (thăch)	87c	18°22'N	64°53'W
Thaya, r., Eur. (tā'yȧ)	124	48°48'N	15°40'E
Thayer, Mo., U.S. (thâ'ẽr)	79	36°30'N	91°34'W
Thebes see Thíva, Grc.	119	38°20'N	23°18'E
Thebes, hist., Egypt (thēbz)	185	25°47'N	32°39'E
The Brothers, mtn., Wa., U.S.	74a	47°39'N	123°08'W
The Coorong, l., Austl. (kō'rŏng)	176	36°07'S	139°45'E
The Coteau, hills, Can.	56	51°10'N	107°30'W
The Dalles, Or., U.S. (dălz)	64	45°36'N	121°10'W
The Father, mtn., Pap. N. Gui.	169	5°05'S	151°30'E
The Hague ('s-Gravenhage), Neth.	110	52°05'N	4°16'E
The Oaks, Austl.	173b	34°04'S	150°36'E
Theodore, Austl. (thēō'dôr)	176	24°51'S	150°09'E
Theodore Roosevelt Dam, dam, Az., U.S. (thē-ō-dōr rōō-sȧ-vĕlt)	77	33°46'N	111°25'W
Theodore Roosevelt Lake, res., Az., U.S.	77	33°45'N	111°00'W
Theodore Roosevelt National Park, rec., N.D., U.S.	70	47°20'N	103°42'W
Theológos, Grc.	131	40°37'N	24°41'E
The Pas, Can. (pä)	50	53°50'N	101°15'W
Thermopolis, Wy., U.S. (thẽr-mŏp'ō-lĭs)	73	43°38'N	108°11'W
The Round Mountain, mtn., Austl.	176	30°17'S	152°19'E
Thessalía, hist. reg., Grc.	131	39°50'N	22°09'E
Thessalon, Can.	51	46°11'N	83°37'W
Thessaloníki, Grc. (thĕs-sȧ-lô-nē'kē)	110	40°38'N	22°59'E
Thetford Mines, Can. (thĕt'fẽrd mīns)	59	46°05'N	71°20'W
The Twins, mtn., Afr. (twĭnz)	187c	30°09'S	28°29'E
Theunissen, S. Afr.	192c	28°25'S	26°44'E
The Wrekin, co., Eng., U.K.	114a	53°43'N	2°30'W
Thibaudeau, Can. (tĭ'bō-dō')	57	57°05'N	94°08'W
Thibodaux, La., U.S. (tē-bō-dō')	81	29°48'N	90°48'W
Thief, l., Mn., U.S. (thēf)	70	48°32'N	95°46'W
Thief, r., Mn., U.S.	71	48°18'N	96°07'E
Thief Rivers Falls, Mn., U.S. (thēf rĭv'ẽr fôlz)	70	48°07'N	96°11'W
Thiers, Fr. (tyâr)	126	45°51'N	3°32'E
Thiès, Sen. (tē-ĕs')	184	14°48'N	16°56'W
Thika, Kenya	191	1°03'S	37°05'E
Thimphu, Bhu.	155	27°33'N	89°42'E
Thingvallavatn, l., Ice.	116	64°12'N	20°22'W
Thio, N. Cal.	170f	21°37'S	166°14'E
Thionville, Fr. (tyôn-vēl')	117	49°23'N	6°31'E
Third Cataract, wtfl., Sudan	185	19°53'N	30°11'E
Thiruvananthapuram, India	159	8°34'N	76°58'E
Thisted, Den. (tēs'tĕdh)	122	56°57'N	8°38'E
Thistilfjördur, b., Ice.	116	66°29'N	14°59'W
Thistle, i., Austl. (thĭs''l)	176	34°55'S	136°11'E
Thíva (Thebes), Grc.	119	38°20'N	23°18'E
Thjórsá, r., Ice. (tyûr'sȧ)	116	64°23'N	19°18'W
Thohoyandou, S. Afr.	186	23°00'S	30°29'E
Tholen, Neth.	115a	51°32'N	4°11'E
Thomas, Ok., U.S. (tŏm'ăs)	78	35°44'N	98°43'W
Thomas, W.V., U.S.	67	39°15'N	79°30'W
Thomaston, Ga., U.S. (tŏm'ăs-tŭn)	82	32°51'N	84°17'W
Thomasville, Al., U.S. (tŏm'ăs-vĭl)	82	31°55'N	87°43'W
Thomasville, N.C., U.S.	83	35°52'N	80°05'W
Thomlinson, Mount, mtn., Can.	54	55°33'N	127°29'W
Thompson, Can.	50	55°48'N	97°59'W
Thompson, r., Can.	55	50°15'N	121°20'W
Thompson, r., Mo., U.S.	79	40°32'N	93°49'W
Thompson Falls, Mt., U.S.	72	47°35'N	115°20'W
Thomson, r., Austl. (tŏm-sŏn)	175	24°30'S	143°07'E
Thomson's Falls, Kenya	191	0°02'N	36°22'E
Thonon-les-Bains, Fr. (tô-nôn'lā-băn')	127	46°22'N	6°27'E
Thorne, Eng., U.K. (thôrn)	114a	53°37'N	0°58'W
Thorntown, In., U.S. (thôrn'tŭn)	66	40°05'N	86°35'W
Thorold, Can. (thō'rōld)	59	43°13'N	79°12'W
Thouars, Fr. (tōō-är')	126	47°00'N	0°17'W
Thousand Islands, is., N.Y., U.S. (thou'zănd)	67	44°15'N	76°10'W
Thrace, hist. reg., (thrās)	131	41°20'N	26°07'E
Thrapston, Eng., U.K. (thrăp'stŭn)	114a	52°23'N	0°32'W
Three Forks, Mt., U.S. (thrē fôrks)	73	45°56'N	111°35'W
Three Oaks, Mi., U.S. (thrē ōks)	66	41°50'N	86°40'W
Three Points, Cape, c., Ghana	184	4°45'N	2°06'W
Three Rivers, Mi., U.S.	66	42°00'N	83°40'W
Thule, Grnld.	49	76°34'N	68°47'W
Thun, Switz. (tōōn)	124	46°46'N	7°34'E
Thunder Bay, Can.	58	48°28'N	89°12'W
Thunder Bay, b., Can. (thŭn'dẽr)	58	48°29'N	88°52'W
Thunder Hills, hills, Can.	56	54°30'N	106°00'W
Thunersee, l., Switz.	124	46°40'N	7°30'E
Thurber, Tx., U.S. (thûr'bẽr)	80	32°30'N	98°23'W
Thüringen (Thuringia), hist. reg., Ger. (tü'rĭng-ĕn)	124	51°07'N	10°45'E
Thurles, Ire. (thûrlz)	120	52°44'N	7°45'W
Thurrock, co., Eng., U.K.	114b	51°30'N	0°22'E
Thursday, i., Austl. (thûrz-dā)	175	10°17'S	142°23'E
Thurso, Can. (thŭn'sô)	62c	45°36'N	75°15'W
Thurso, Scot., U.K.	120	58°35'N	3°40'W
Thurston Island, i., Ant. (thûrs'tŭn)	178	71°20'S	98°00'W
Tiachiv, Ukr.	125	48°01'N	23°42'E
Tiandong, China (tĭĕn-dôŋ)	165	23°32'N	107°10'E
Tianjin, China	161	39°08'N	117°14'E
Tianjin Shi, prov., China (tĭĕn-jyĭn shr)	164	39°30'N	117°13'E
Tianmen, China (tĭĕn-mŭn)	165	30°40'N	113°10'E
Tianshui, China (tĭĕn-shwā)	164	34°25'N	105°40'E
Tiasmyn, r., Ukr.	133	49°14'N	32°23'E
Tibagi, Braz. (tē'bȧ-zhē)	101	24°40'S	50°35'W
Tibasti, Sarir, des., Libya	185	24°00'N	16°30'E
Tibati, Cam.	189	6°27'N	12°38'E
Tiber see Tevere, r., Italy	118	42°30'N	12°14'E
Tibesti, mts., Chad	185	20°40'N	17°48'E
Tibet see Xizang, prov., China (tĭ-bĕt')	160	32°22'N	83°30'E
Tibnīn, Leb.	153a	33°12'N	35°23'E
Tiburon, Haiti	93	18°35'N	74°25'W
Tiburon, Ca., U.S. (tē-bōō-rōn')	74b	37°53'N	122°27'W
Tiburón, i., Mex.	86	28°45'N	113°10'W
Tiburón, Cabo, c., (kȧ'bô)	91	8°42'N	77°19'W
Tiburon Island, i., Ca., U.S.	74b	37°52'N	122°26'W
Ticao Island, i., Phil. (tē-kä'ō)	169a	12°40'N	123°30'E
Tickhill, Eng., U.K. (tĭk'ĭl)	114a	53°26'N	1°06'W
Ticonderoga, N.Y., U.S. (tī-kŏn-dẽr-ō'gȧ)	67	43°50'N	73°30'W
Ticul, Mex. (tē-kōō'l)	90a	20°22'N	89°32'W
Tidaholm, Swe. (tē'dȧ-hōlm)	122	58°11'N	13°53'E
Tideswell, Eng., U.K. (tīdz'wĕl)	114a	53°17'N	1°47'W
Tidikelt, reg., Alg. (tē-dē-kĕlt')	184	25°53'N	2°11'E
Tidjikdja, Maur. (tē-jĭk'jä)	184	18°33'N	11°25'W
Tidra, Ile, i., Maur.	188	19°50'N	16°45'W
Tieling, China (tĭĕ-lĭŋ)	161	42°18'N	123°50'E
Tielmes, Spain (tyâl-màs')	129a	40°15'N	3°20'W
Tienen, Bel.	115a	50°49'N	4°58'E
Tien Shan, mts., Asia	160	42°00'N	78°46'E
Tientsin see Tianjin, China	161	39°08'N	117°14'E
Tierp, Swe. (tyĕrp)	122	60°21'N	17°28'E
Tierpoort, S. Afr.	187b	25°53'S	28°26'E
Tierra Blanca, Mex. (tyĕ'rä-blän'kä)	89	18°28'N	96°19'W
Tierra del Fuego, i., S.A. (tyĕr'rä dĕl fwā'gō)	102	53°50'S	68°45'W
Tiétar, r., Spain (tē-ā'tär)	128	39°56'N	5°44'W
Tiffin, Oh., U.S. (tĭf'ĭn)	66	41°10'N	83°15'W
Tifton, Ga., U.S. (tĭf'tŭn)	82	31°25'N	83°34'W
Tigard, Or., U.S. (tī'gärd)	74c	45°25'N	122°46'W
Tighina, Mol.	137	46°49'N	29°29'E
Tignish, Can. (tĭg'nĭsh)	60	46°57'N	64°02'W
Tigoda, r., Russia (tē'gô-dȧ)	142c	59°29'N	31°15'E
Tigre, r., Peru	100	2°20'S	75°41'W
Tigres, Península dos, pen., Ang. (pĕ-nē'ŋ-sōō-lä-dôs-tē'grĕs)	186	16°30'S	11°45'E
Tigris, r., Asia	154	34°45'N	44°10'E
Tih, Jabal at, mts., Egypt	153a	29°23'N	34°05'E
Tihert, Alg.	184	35°28'N	1°15'E
Tihuatlán, Mex. (tē-wä-tlän')	89	20°43'N	97°34'W
Tijuana, Mex. (tē-hwä'nä)	86	32°32'N	117°02'W
Tijuca, Pico da, mtn., Braz. (pē'kō-dä-tē-zhōō'kȧ)	102b	22°56'S	43°17'W
Tikal, hist., Guat. (tē-käl')	90a	17°16'N	89°49'W
Tikhoretsk, Russia (tē-kôr-yĕtsk')	137	45°55'N	40°05'E
Tikhvin, Russia (tēk-vēn')	134	59°36'N	33°38'E
Tikrit, Iraq	154	34°36'N	43°31'E
Tiksi, Russia (tēk-sē')	135	71°42'N	128°32'E
Tilburg, Neth. (tĭl'bûrg)	117	51°33'N	5°05'E
Tilbury, Eng., U.K.	114b	51°28'N	0°23'E
Tilemsi, Vallée du, val., Mali	188	17°50'N	0°25'E
Tilichiki, Russia (tyĭ-le-chĭ-kē)	135	60°49'N	166°14'E
Tilimsen, Alg.	184	34°53'N	1°21'W
Tillabéry, Niger (tē-yȧ-bā-rē')	184	14°14'N	1°30'E
Tillamook, Or., U.S. (tĭl'ȧ-mo͝ok)	72	45°27'N	123°50'W
Tillamook Bay, b., Or., U.S.	72	45°32'N	124°26'W
Tillberga, Swe. (tēl-bĕr'ghȧ)	122	59°40'N	16°34'E
Tillsonburg, Can. (tĭl'sŭn-bûrg)	59	42°50'N	80°50'W
Tim, Russia (tēm)	133	51°39'N	37°07'E
Timaru, N.Z. (tĭm'ȧ-rōō)	175a	44°26'S	171°17'E
Timashevskaya, Russia (tēmȧ-shĕfs-kȧ'yȧ)	137	45°47'N	38°57'E
Timbalier Bay, b., La., U.S. (tĭm'bȧ-lēr)	81	28°55'N	90°14'W
Timber, Or., U.S. (tĭm'bẽr)	74c	45°43'N	123°17'W
Timbo, Gui. (tĭm'bō)	184	10°41'N	11°51'W
Timbuktu see Tombouctou, Mali	184	16°46'N	3°01'W
Timétrine Monts, mts., Mali	188	19°50'N	0°30'W
Timimoun, Alg. (tē-mē-mōōn')	184	29°14'N	0°22'E
Timiris, Cap, c., Maur.	184	19°23'N	16°32'W
Timiş, r., Eur.	131	45°28'N	21°06'E
Timişoara, Rom.	119	45°44'N	21°21'E
Timmins, Can. (tĭm'ĭnz)	51	48°25'N	81°22'W
Timmonsville, S.C., U.S. (tĭm'ŭnz-vĭl)	83	34°09'N	79°55'W
Timok, r., Eur.	131	43°35'N	22°13'E
Timor, i., Asia (tē-mōr')	169	10°08'S	125°00'E
Timor Sea, sea,	174	12°40'S	125°00'E
Timpanogos Cave National Monument, rec., Ut., U.S. (tĭ-măn'ō-gŏz)	77	40°25'N	111°45'W
Timpson, Tx., U.S. (tĭmp'sŭn)	81	31°55'N	94°24'W
Timsâh, l., Egypt (tĭm'sä)	192b	30°34'N	32°22'E
Tina, r., S. Afr. (tē'nȧ)	187c	30°50'S	28°44'E
Tina, Monte, mtn., Dom. Rep. (mô'n-tē-tē'nȧ)	93	18°50'N	70°40'W
Tinaguillo, Ven. (tē-nä-gē'l-yō)	101b	9°55'N	68°18'W
Tinah, Khalij at, b., Egypt	153a	31°06'N	32°42'E
Tindouf, Alg. (tēn-dōōf')	184	27°43'N	7°44'W
Tinggi, i., Malay.	153b	2°16'N	104°16'E
Tinghert, Plateau du, plat., Alg.	184	27°30'N	7°30'E
Tingi Mountains, mts., S.L.	188	9°00'N	10°50'W
Tinglin, China	163b	30°53'N	121°18'E
Tingo María, Peru (tē'ngō-mä-rē'ä)	100	9°15'S	76°04'W
Tingréla, C. Iv.	188	10°29'N	6°24'W
Tingsryd, Swe. (tĭngs'rüd)	122	56°32'N	14°58'E
Tinguindio, Mex.	88	19°38'N	102°02'W
Tinguiririca, r., Chile (tē'n-gē-rē-rē'kä)	99b	34°48'S	70°45'W
Tinley Park, Il., U.S. (tĭn'lē)	69a	41°34'N	87°47'W
Tinnoset, Nor. (tĕn'nôs'sĕt)	122	59°44'N	9°00'E
Tinogasta, Arg. (tē-nō-gäs'tä)	102	28°07'S	67°30'W
Tínos, i., Grc.	119	37°45'N	25°12'E
Tinsukia, India (tin-sōō'kĭ-ȧ)	154	27°18'N	95°29'W
Tintic, Ut., U.S. (tĭn'tĭk)	77	39°55'N	112°15'W
Tio, Pic de, mtn., Gui.	188	8°55'N	8°55'W
Tioman, i., Malay.	153b	2°50'N	104°15'E
Tipitapa, Nic. (tē-pē-tä'pä)	90	12°14'N	86°05'W
Tipitapa, r., Nic.	90	12°13'N	85°57'W
Tippah Creek, r., Ms., U.S. (tĭp'pȧ)	82	34°43'N	88°15'W
Tippecanoe, r., In., U.S. (tĭp-ē-kȧ-nōō')	66	40°55'N	86°45'W
Tipperary, Ire. (tĭ-pē-râ'rē)	117	52°28'N	8°13'W
Tippo Bay, Ms., U.S. (tĭp'ō bīōō')	79	33°35'N	90°06'W
Tipton, Ia., U.S.	71	41°46'N	91°10'W
Tipton, In., U.S.	66	40°15'N	86°00'W
Tiranë, Alb. (tē-rä'nä)	110	41°48'N	19°50'E
Tirano, Italy (tē-rä'nō)	130	46°12'N	10°09'E
Tiraspol, Mol.	137	46°52'N	29°38'E
Tire, Tur. (tē'rĕ)	119	38°05'N	27°48'E
Tiree, i., Scot., U.K. (tī-rē')	116	56°34'N	6°30'W
Tirlyanskiy, Russia (tĭr-lyän'skĭ)	142a	54°13'N	58°37'E
Tiruchchirāppalli, India (tĭr'ō-chĭ-rä'pȧ-lĭ)	155	10°49'N	78°48'E
Tirunelveli, India	159	8°53'N	77°43'E
Tiruppur, India	159	11°11'N	77°08'E
Tisdale, Can. (tĭz'dāl)	50	52°51'N	104°04'W
Tista, r., Asia	158	26°00'N	89°30'E
Tisza, r., Eur. (tē'sä)	112	47°30'N	21°00'E
Titāgarh, India	158a	22°44'N	88°23'E
Titicaca, Lago, l., S.A. (lä'gô-tē-tē-kä'kä)	100	16°12'S	70°33'W
Titiribi, Col. (tē-tē-rē-bē')	100a	6°05'N	75°47'W
Tito, Lagh, r., Kenya	191	2°25'N	39°05'E
Titov Veles, Mac. (tē'tôv vĕ'lĕs)	131	41°42'N	21°50'E
Titterstone Clee Hill, hill, Eng., U.K. (klē)	114a	52°24'N	2°37'W
Titule, D.R.C.	191	3°17'N	25°32'E
Titusville, Fl., U.S. (tī'tŭs-vĭl)	83a	28°37'N	80°44'W
Titusville, Pa., U.S.	67	40°40'N	79°40'W
Titz, Ger. (tētz)	127c	51°00'N	6°26'E
Tiverton, R.I., U.S. (tĭv'ẽr-tun)	68b	41°38'N	71°11'W
Tivoli, Italy (tē'vô-lē)	118	41°38'N	12°48'E
Tixkokob, Mex. (tēx-kō-kō'b)	90a	21°01'N	89°23'W
Tixtla de Guerrero, Mex. (tē'x-tlä-dĕ-gĕr-rĕ'rô)	88	17°36'N	99°24'W
Tizard Bank and Reef, rf., Asia (tĭz'ȧrd)	168	10°51'N	113°20'E
Tizimín, Mex. (tē-zē-mē'n)	90a	21°08'N	88°10'W
Tizi-Ouzou, Alg. (tē'zē-ōō-zōō')	184	36°44'N	4°04'E
Tiznados, r., Ven. (tēz-nä'dôs)	101b	9°53'N	67°49'W
Tiznit, Mor. (tēz-nēt)	184	29°52'N	9°39'W
Tkvarcheli, Geor.	138	42°15'N	41°41'E
Tlacolula de Matamoros, Mex.	89	16°56'N	96°29'W
Tlacotálpan, Mex. (tlä-kō-täl'pän)	89	18°39'N	95°40'W
Tlacotepec, Mex. (tlä-kō-tȧ-pĕ'k)	88	17°46'N	99°57'W
Tlacotepec, Mex.	88	19°11'N	99°41'W
Tlacotepec, Mex.	89	18°41'N	97°40'W
Tláhuac, Mex. (tlä-wäk')	89a	19°16'N	99°00'W
Tlajomulco de Zúñiga, Mex. (tlä-hô-mōō'l-ko-dĕ-zōō'n-yē-gä)	88	20°30'N	103°27'W
Tlalchapa, Mex. (tläl-chä'pä)	88	18°26'N	100°29'W

PRONOUNCING INDEX Tlalixcoyán – Travnik 311

PLACE (Pronunciation)	PAGE	LAT.	LONG.
Tlalixcoyán, Mex. (tlä-lēs'kō-yän')	89	18°53'N	96°04'W
Tlalmanalco, Mex. (tläl-mä-nä'l-kô)	89a	19°12'N	98°48'W
Tlalnepantla, Mex.	89a	19°32'N	99°13'W
Tlalnepantla, Mex. (tläl-nå-pän'tlä)	89a	18°59'N	99°01'W
Tlalpan, Mex. (tläl-pä'n)	88	19°17'N	99°10'W
Tlalpujahua, Mex. (tläl-pōō-kä'wä)	88	19°50'N	100°10'W
Tlapa, Mex. (tlä'pä)	88	17°30'N	98°30'W
Tlapacoyán, Mex. (tlä-pä-kô-yä'n)	89	19°57'N	97°11'W
Tlapehuala, Mex. (tlä-på-wä'lä)	88	18°17'N	100°30'W
Tlaquepaque, Mex. (tlä-kē-pä'kē)	88	20°39'N	103°17'W
Tlatlaya, Mex. (tlä-tlä'yä)	88	18°36'N	100°14'W
Tlaxcala, Mex. (tläs-kä'lä)	86	19°16'N	98°14'W
Tlaxcala, state, Mex.	88	19°30'N	98°15'W
Tlaxco, Mex. (tläs'kō)	88	19°37'N	98°06'W
Tlaxiaco Santa María Asunción, Mex.	89	17°16'N	97°41'W
Tlayacapán, Mex. (tlä-yä-kä-pä'n)	89a	18°57'N	99°00'W
Tlevak Strait, strt., Ak., U.S.	54	53°03'N	132°58'W
Tlumach, Ukr. (t'lû-mäch')	125	48°47'N	25°00'E
Toa, r., Cuba (tō'ä)	93	20°25'N	74°35'W
Toamasina, Madag.	187	18°14'S	49°25'E
Toar, Cuchillas de, mts., Cuba (kōō-chē'l-lyäs-dĕ-tô-ä'r)	93	20°20'N	74°50'W
Tobago, i., Trin. (tô-bā'gō)	87	11°15'N	60°30'W
Toba Inlet, b., Can.	54	50°20'N	124°50'W
Tobarra, Spain (tô-bär'rä)	128	38°37'N	1°42'W
Tobol (Tobyl), r., Asia	140	56°00'N	66°30'E
Tobol'sk, Russia (tô-bôlsk')	140	58°09'N	68°28'E
Tobyl *see* Tobol, r., Asia	140	52°00'N	62°00'E
Tocaima, Col. (tô-kä'y-mä)	100a	4°28'N	74°38'W
Tocantinópolis, Braz. (tō-kän-tē-nŏ'pō-lēs)	101	6°27'S	47°18'W
Tocantins, state, Braz.	101	10°00'S	48°00'W
Tocantins, r., Braz. (tō-kän-tēns')	101	3°28'S	49°22'W
Toccoa, Ga., U.S. (tŏk'ō-ȧ)	82	34°35'N	83°20'W
Toccoa, r., Ga., U.S.	82	34°53'N	84°24'W
Tochigi, Japan (tō'chē-gĭ)	167	36°25'N	139°45'E
Tocoa, Hond. (tō-kō'ä)	90	15°37'N	86°01'W
Tocopilla, Chile (tō-kô-pēl'yä)	102	22°03'S	70°08'W
Tocuyo de la Costa, Ven. (tô-kōō'yō-dĕ-lä-kôs'tä)	101b	11°03'N	68°24'W
Toda, Japan	167a	35°48'N	139°42'E
Todmorden, Eng., U.K. (tŏd'môr-dĕn)	114a	53°43'N	2°05'W
Tofino, Can. (tō-fē'nō)	54	49°09'N	125°54'W
Töfsingdalens National Park, rec., Swe.	122	62°09'N	13°05'E
Tōgane, Japan (tō'gä-nä)	167	35°29'N	140°16'E
Togian, Kepulauan, is., Indon.	168	0°20'S	122°00'E
Togo, nation, Afr. (tō'gō)	184	8°00'N	0°52'E
Toguzak, r., Russia (tô'gô-zák)	142a	53°40'N	61°42'E
Tohono O'odham Indian Reservation, I.R., Az., U.S.	77	32°33'N	112°12'W
Tohopekaliga, Lake, l., Fl., U.S. (tō'hō-pē'kȧ-lī'gȧ)	83a	28°16'N	81°09'W
Tohor, Tanjong, c., Malay.	153b	1°53'N	102°29'E
Toijala, Fin. (toi'yȧ-lȧ)	123	61°11'N	23°46'E
Toi-Misaki, c., Japan (toi mē'sä-kē)	166	31°20'N	131°20'E
Toiyabe, Nv., U.S. (toi'yă-bē)	76	38°59'N	117°22'W
Tokachi Gawa, r., Japan (tō-kä'chē gä'wä)	166	43°10'N	142°30'E
Tokaj, Hung. (tô'kô-ē)	125	48°06'N	21°24'E
Tokat, Tur. (tô-kät')	154	40°20'N	36°30'E
Tokelau, dep., Oc. (tō-kē-lä'ō)	2	8°00'S	176°00'W
Tokmak, Kyrg. (tôk'mák)	139	42°44'N	75°41'E
Tokmak, Ukr.	133	47°17'N	35°48'E
Tokorozawa, Japan (tō'kô-rô-zä'wä)	167a	35°47'N	139°29'E
Tok-to, atoll, Asia	166	37°15'N	131°51'E
Tokuno, i., Japan (tô-kōō'nō)	161	27°42'N	129°25'E
Tokushima, Japan (tō'kô'shē-mä)	161	34°06'N	134°31'E
Tokuyama, Japan (tō'kô'yä-mä)	167	34°04'N	131°49'E
Tōkyō, Japan	161	35°42'N	139°46'E
Tōkyō-Wan, b., Japan (tō'kyō wän)	167	35°56'N	139°56'E
Tolcayuca, Mex. (tôl-kä-yōō'kä)	88	19°55'N	98°54'W
Toledo, Spain (tô-lĕ'dô)	118	39°53'N	4°02'W
Toledo, Ia., U.S. (tô-lē'dō)	71	41°59'N	92°35'W
Toledo, Oh., U.S.	65	41°40'N	83°35'W
Toledo, Or., U.S.	72	44°37'N	123°58'W
Toledo, Montes de, mts., Spain (mô'n-tĕs-dĕ-tô-lĕ'dô)	128	39°33'N	4°40'W
Toledo Bend Reservoir, res., U.S.	65	31°30'N	93°30'W
Toliara, Madag.	187	23°16'S	43°44'E
Tolima, dept., Col. (tô-lē'mä)	100a	4°07'N	75°20'W
Tolima, Nevado del, mtn., Col. (nĕ-vä-dô-dĕl-tô-lē'mä)	100a	4°40'N	75°20'W
Tolimán, Mex. (tô-lē-män')	88	20°54'N	99°54'W
Tollesbury, Eng., U.K. (tōl'z-bĕrĭ)	114b	51°46'N	0°49'E
Tolmezzo, Italy (tôl-mĕt'zō)	130	46°25'N	13°03'E
Tolmin, Slvn. (tôl'mēn)	130	46°12'N	13°45'E
Tolna, Hung. (tôl'nô)	125	46°25'N	18°47'E
Tolo, Teluk, b., Indon. (tō'lō)	168	2°00'S	122°06'E
Tolosa, Spain (tô-lō'sä)	118	43°10'N	2°05'W
Tolt, r., Wa., U.S. (tōlt)	74a	47°13'N	121°49'W
Toluca, Mex. (tô-lōō'kä)	86	19°17'N	99°40'W
Toluca, Il., U.S. (tô-lōō'kȧ)	66	41°00'N	89°10'W
Toluca, Nevado de, mtn., Mex. (nĕ-vä-dô-dĕ-tô-lōō'kä)	86	19°09'N	99°42'W
Tolyatti, Russia	136	53°30'N	49°10'E
Tom', r., Russia	140	55°33'N	85°00'E
Tomah, Wi., U.S. (tō'mȧ)	71	43°58'N	90°31'W
Tomahawk, Wi., U.S. (tŏm'ȧ-hôk)	71	45°27'N	89°44'W
Tomakivka, Ukr.	133	47°49'N	34°43'E
Tomanivi, mtn., Fiji	170g	17°37'S	178°01'E
Tomar, Port. (tō-mär')	128	39°36'N	8°26'W
Tomashovka, Bela.	125	51°34'N	23°37'E
Tomaszów Lubelski, Pol. (tô-mä'shôf lōō-bĕl'skĭ)	125	50°20'N	23°27'E

PLACE (Pronunciation)	PAGE	LAT.	LONG.
Tomaszów Mazowiecki, Pol. (tô-mä'shôf mä-zō'vyĕt-skĭ)	125	51°33'N	20°00'E
Tomatlán, Mex. (tô-mä-tlä'n)	88	19°54'N	105°14'W
Tombadonkéa, Gui.	188	11°00'N	14°23'W
Tombador, Serra do, mts., Braz. (sĕr'rȧ dò tōm-bä-dôr')	101	11°31'S	57°33'W
Tombigbee, r., U.S. (tŏm-bĭg'bē)	65	33°00'N	88°30'W
Tombos, Braz. (tô'm-bōs)	99a	20°53'S	42°00'W
Tombouctou, Mali	184	16°46'N	3°01'W
Tombstone, Az., U.S. (tōōm'stōn)	77	31°40'N	110°00'W
Tombua, Ang. (ȧ-lē-zhän'drē)	186	15°49'S	11°53'E
Tomelilla, Swe. (tô'mĕ-lēl-lä)	122	55°34'N	13°55'E
Tomelloso, Spain (tō-mål-lyō'sō)	128	39°09'N	3°02'W
Tommot, Russia (tŏm-mŏt')	135	59°13'N	126°22'E
Tomsk, Russia (tŏmsk)	134	56°29'N	84°57'E
Tonala, Mex.	88	20°38'N	103°14'W
Tonalá, r., Mex.	89	18°05'N	94°08'W
Tonawanda, N.Y., U.S. (tŏn-ȧ-wŏn'dȧ)	69c	43°01'N	78°53'W
Tonawanda Creek, r., N.Y., U.S.	69c	43°05'N	78°43'W
Tonbridge, Eng., U.K. (tŭn-brij)	114b	51°11'N	0°17'E
Tonda, Japan (tôn'dä)	167b	34°51'N	135°38'E
Tondabayashi, Japan (tôn-dä-bä'yä-shē)	167b	34°29'N	135°36'E
Tondano, Indon. (tôn-dä'nō)	169	1°15'N	124°50'E
Tønder, Den. (tûn'nĕr)	122	54°47'N	8°49'E
Tone-Gawa, r., Japan (tô'nĕ gä'wa)	167	36°12'N	139°19'E
Tonga, nation, Oc. (tŏŋ'gȧ)	194	18°50'S	175°20'W
Tong'an, China (tôŋ-än)	165	24°48'N	118°02'E
Tonga Trench, deep,	194	23°00'S	172°30'W
Tongbei, China (tôŋ-bā)	161	48°00'N	126°48'E
Tongguan, China (tôŋ-gūän)	161	34°48'N	110°25'E
Tonghe, China (tôŋ-hŭ)	164	45°58'N	128°40'E
Tonghua, China (tôŋ-hwä)	161	41°43'N	125°50'E
Tongjiang, China (tôŋ-jyäŋ)	161	47°38'N	132°54'E
Tongliao, China (tôŋ-lĭou)	164	43°30'N	122°15'E
Tongo, Cam.	189	5°11'N	14°00'E
Tongoy, Chile (tōn-goi')	102	30°16'S	71°29'W
Tongren, China (tôŋ-rŭn)	160	27°45'N	109°12'E
Tongshan, China (tôŋ-shän)	162	34°27'N	116°27'E
Tongtian, r., China (tôŋ-tĭĕn)	160	33°00'N	97°00'E
Tongue, r., Mt., U.S. (tŭng)	73	45°08'N	106°40'W
Tongxian, China (tôŋ shyĕn)	162	39°55'N	116°40'E
Tonj, r., Sudan (tônj)	185	6°18'N	28°33'E
Tonk, India (Tŏŋk)	155	26°13'N	75°45'E
Tonkawa, Ok., U.S. (tŏŋ kȧ-wô)	79	36°42'N	97°19'W
Tonkin, Gulf of, b., Asia (tôn-kăn')	168	20°30'N	108°10'E
Tonle Sap, l., Camb. (tôn'lā säp')	168	13°03'N	102°49'E
Tonneins, Fr. (tô-näN')	126	44°24'N	0°18'E
Tönning, Ger. (tû'nĕng)	124	54°20'N	8°55'E
Tonopah, Nv., U.S. (tō-nô-pä')	64	38°04'N	117°15'W
Tönsberg, Nor. (tûns'bĕrgh)	116	59°19'N	10°25'E
Tonto, r., Mex. (tōn'tō)	89	18°15'N	96°13'W
Tonto Creek, r., Az., U.S.	77	34°05'N	111°15'W
Tonto National Monument, rec., Az., U.S. (tôn'tô)	77	33°33'N	111°08'W
Tooele, Ut., U.S. (tô-ĕl ē)	75b	40°33'N	112°17'W
Toowoomba, Austl. (tô wōōm'bȧ)	175	27°32'S	152°10'E
Topanga, Ca., U.S. (tō'păn-gȧ)	75a	34°05'N	118°36'W
Topeka, Ks., U.S. (tô-pē'kȧ)	65	39°02'N	95°41'W
Topilejo, Mex. (tô-pē-lĕ'hô)	89a	19°12'N	99°09'W
Topock, Az., U.S.	77	34°40'N	114°20'W
Topol'čany, Slvk. (tô-pôl'chä-nü)	125	48°38'N	18°10'E
Topolobampo, Mex. (tô-pō-lô-bä'm-pô)	86	25°45'N	109°00'W
Topolovgrad, Blg.	131	42°05'N	26°19'E
Toppenish, Wa., U.S. (tôp'ĕn-ĭsh)	72	46°22'N	120°00'W
Torbat-e Ḩeydarīyeh, Iran	157	35°16'N	59°13'E
Torbat-e Jām, Iran	157	35°14'N	60°36'E
Torbay, Can. (tôr-bā')	61	47°40'N	52°43'W
Torbay *see* Torquay, Eng., U.K.	120	50°30'N	3°26'W
Torbreck, Mount, mtn., Austl. (tôr-brĕk)	176	37°05'S	146°55'E
Torch, l., Mi., U.S. (tôrch)	66	45°00'N	85°30'W
Töreboda, Swe. (tū'rĕ-bō'dä)	122	58°44'N	14°04'E
Torhout, Bel.	121	51°01'N	3°04'E
Toribío, Col. (tô-rē-bē'ō)	100a	2°58'N	76°14'W
Toride, Japan (tô'rē-dä)	167a	35°54'N	104°04'E
Torino *see* Turin, Italy	110	45°05'N	7°44'E
Tormes, r., Spain (tōr'mās)	128	41°12'N	6°15'W
Torneälven, r., Eur.	112	67°00'N	22°30'E
Torneträsk, l., Swe. (tôr'nĕ trĕsk)	116	68°10'N	20°36'E
Torngat Mountains, mts., Can.	53	59°18'N	64°35'W
Tornio, Fin. (tôr'nĭ-ō)	110	65°55'N	24°09'E
Toro, Lac, l., Can.	59	46°53'N	73°46'W
Toronto, Can. (tô-rŏn'tō)	51	43°40'N	79°23'W
Toronto, Oh., U.S.	66	40°30'N	80°35'W
Toronto, res., Mex.	80	27°35'N	105°37'W
Toropets, Russia (tô'rô-pyĕts)	136	56°31'N	31°37'E
Toros Dağları, mts., Tur. (tô'rŭs)	154	37°00'N	32°40'E
Torote, r., Spain (tô-rō'tä)	129a	40°36'N	3°24'W
Torquay, Eng., U.K. (tôr-kē')	120	50°30'N	3°26'W
Torra, Cerro, mtn., Col. (sĕ'r-rô-tō'r-rä)	100a	4°41'N	76°22'W
Torrance, Ca., U.S. (tŏr'rănc)	75a	33°50'N	118°20'W
Torre Annunziata, Italy (tôr'rä ä-nōōn-tsē-ä'tä)	129c	40°31'N	14°27'E
Torreblanca, Spain	129	40°18'N	0°12'E
Torre del Greco, Italy (tôr'rå dĕl grĕ'kô)	130	40°32'N	14°23'E
Torrejoncillo, Spain (tôr'rā-hōn-thē'lyō)	128	39°54'N	6°26'W
Torrelavega, Spain (tôr-rā'lä-vä'gä)	128	43°22'N	4°02'W
Torre Maggiore, Italy (tôr'rā mäd-jō'rā)	130	41°41'N	15°18'E
Torrens, Lake, l., Austl. (tŏr-ĕns)	174	30°07'S	137°40'E
Torrent, Spain	129	39°25'N	0°28'W

PLACE (Pronunciation)	PAGE	LAT.	LONG.
Torreón, Mex. (tôr-rå-ōn')	86	25°32'N	103°26'W
Torres Islands, is., Vanuatu (tôr'rĕs) (tŏr'ĕz)	175	13°18'N	165°59'E
Torres Martinez Indian Reservation, I.R., Ca., U.S. (tôr'ĕz mär-tē'nĕz)	76	33°33'N	116°21'W
Torres Novas, Port. (tôr'rĕzh nō'väzh)	128	39°28'N	8°37'W
Torres Strait, strt., Austl. (tôr'rĕs)	175	10°30'S	141°30'E
Torres Vedras, Port. (tôr'rĕsh vă'dräzh)	128	39°08'N	9°18'W
Torrevieja, Spain (tôr-rā-vyā'hä)	129	37°58'N	0°40'W
Torrijos, Phil. (tôr-rē'hōs)	169a	13°19'N	122°06'E
Torrington, Ct., U.S. (tŏr'ĭng-tŭn)	67	41°50'N	73°10'W
Torrington, Wy., U.S.	70	42°04'N	104°11'W
Torro, Spain (tô'r-rō)	128	41°27'N	5°23'W
Torsby, Swe. (tôrs'bü)	122	60°07'N	12°56'E
Torshälla, Swe. (tôrs'hĕl-ä)	122	59°26'N	16°21'E
Tórshavn, Far. Is. (tôrs-houn')	110	62°00'N	6°55'W
Tortola, i., Br. Vir. Is. (tôr-tō'lä)	87b	18°34'N	64°40'W
Tortona, Italy (tôr-tō'nä)	130	44°52'N	8°52'W
Tortosa, Spain (tôr-tō'sä)	110	40°59'N	0°33'E
Tortosa, Cap de, c., Spain	129	40°42'N	0°55'E
Tortue, Canal de la, strt., Haiti (tôr-tü')	93	20°05'N	73°20'W
Tortue, Île de la, i., Haiti	93	20°10'N	73°00'W
Tortue, Rivière de la, r., Can. (lä tōr-tü')	62a	45°12'N	73°32'W
Toruń, Pol.	110	53°02'N	18°35'E
Tõrva, Est. (t'r'vȧ)	123	58°02'N	25°56'E
Torzhok, Russia (tôr'zhôk)	136	57°03'N	34°53'E
Toscana, hist. reg., Italy (tôs-kä'nä)	130	43°23'N	11°08'E
Tosna, r., Russia	142c	59°28'N	30°53'E
Tosno, Russia (tôs'nô)	132	59°32'N	30°52'E
Tostado, Arg. (tôs-tä'dô)	102	29°10'S	61°43'W
Tosya, Tur. (tôz'yȧ)	119	41°00'N	34°00'E
Totana, Spain (tô-tä-nä)	128	37°45'N	1°28'W
Tot'ma, Russia (tôt'mȧ)	136	60°00'N	42°20'E
Totness, Sur.	101	5°51'N	56°17'W
Totonicapán, Guat. (tōtō-nē-kä'pän)	86	14°55'N	91°20'W
Totoras, Arg. (tô-tō'räs)	99c	32°33'S	61°13'W
Totsuka, Japan (tōt'sōō-kä)	167a	35°24'N	139°32'E
Tottenham, Eng., U.K. (tŏt'ĕn-ȧm)	114b	51°35'N	0°06'W
Tottori, Japan (tô'tô-rē)	161	35°30'N	134°15'E
Touba, C. Iv.	188	8°17'N	7°41'W
Touba, Sen.	188	14°51'N	15°53'W
Toubkal, Jebel, mtn., Mor.	184	31°15'N	7°46'W
Tougan, Burkina	188	13°04'N	3°04'W
Touggourt, Alg. (tō-gōōrt') (tōō-gōōr')	184	33°09'N	6°07'E
Touil, Oued, r., Alg. (tōō-ēl')	118	34°42'N	2°16'E
Toul, Fr. (tōōl)	117	48°39'N	5°51'E
Toulon, Fr. (tōō-lôN')	110	43°09'N	5°54'E
Toulouse, Fr. (tōō-lōōz')	110	43°37'N	1°27'E
Toungoo, Mya. (tŏ-ôŋ-gōō')	168	19°00'N	96°29'E
Tourcoing, Fr. (tōr-kwăN')	117	50°44'N	3°06'E
Tournan-en-Brie, Fr. (tōōr-näN-ĕN-brē')	127b	48°45'N	2°47'E
Tours, Fr. (tōōr)	110	47°23'N	0°39'E
Touside, Pic, mtn., Chad (tōō-sē-dā')	185	21°10'N	16°30'E
Tovdalselva, r., Nor. (tôv-däls-ĕlvå)	122	58°23'N	8°16'E
Towanda, Pa., U.S. (tô-wän'dȧ)	67	41°45'N	76°30'W
Town Bluff Lake, l., Tx., U.S.	81	30°52'N	94°30'W
Towner, N.D., U.S. (tou'nĕr)	70	48°21'N	100°24'W
Townsend, Ma., U.S. (toun'zĕnd)	61a	42°41'N	71°42'W
Townsend, Mt., U.S.	73	46°19'N	111°35'W
Townsend, Mount, mtn., Wa., U.S.	74a	47°52'N	123°03'W
Townsville, Austl. (tounz'vĭl)	175	19°18'S	146°50'E
Towson, Md., U.S. (tou'sŭn)	68e	39°24'N	76°36'W
Towuti, Danau, l., Indon. (tō-wōō'tē)	168	3°00'S	121°45'E
Toxkan, r., China	160	40°34'N	77°15'E
Toyah, Tx., U.S. (tô'yȧ)	80	31°19'N	103°46'W
Toyama, Japan (tō'yä-mä)	161	36°42'N	137°14'E
Toyama-Wan, b., Japan	167	36°58'N	137°16'E
Toyohashi, Japan (tō'yō-hä'shē)	166	34°44'N	137°21'E
Toyonaka, Japan (tō'yō-nä'kä)	167b	34°47'N	135°28'E
Tozeur, Tun. (tō-zûr')	118	33°59'N	8°11'E
Trabzon, Tur. (tráb'zôn)	154	41°00'N	39°45'E
Tracy, Can.	59	46°00'N	73°13'W
Tracy, Ca., U.S. (trā'sē)	76	37°45'N	121°27'W
Tracy, Mn., U.S.	70	44°13'N	95°37'W
Tracy City, Tn., U.S.	82	35°15'N	85°44'W
Trafalgar, Cabo, c., Spain (kȧ'bô-trä-fäl-gä'r)	128	36°10'N	6°02'W
Trafonomby, mtn., Madag.	187	24°32'S	46°35'E
Trail, Can. (trāl)	50	49°06'N	117°42'W
Traisen, r., Aus.	115e	48°15'N	15°55'E
Traiskirchen, Aus.	115e	48°01'N	16°18'E
Trakai, Lith. (trȧ-kāy)	123	54°38'N	24°59'E
Trakiszki, Pol. (trȧ-kē'-sh-kē)	125	54°16'N	23°07'E
Tralee, Ire. (trȧ-lē')	117	52°16'N	9°20'W
Tranås, Swe. (trän'ôs)	122	58°03'N	14°56'E
Trancoso, Port. (träŋ-kō'sō)	128	40°46'N	7°23'W
Trangan, Pulau, i., Indon. (träŋ'gän)	169	6°52'S	133°30'E
Trani, Italy (trä'nē)	130	41°15'N	16°25'E
Transylvania, hist. reg., Rom. (trăn-sĭl-vā'nĭ-ȧ)	125	46°30'N	22°35'E
Trapani, Italy	118	38°01'N	12°31'E
Trappes, Fr. (tràp)	127b	48°47'N	2°01'E
Traralgon, Austl. (tră'răl-gŏn)	176	38°15'S	146°33'E
Trarza, reg., Maur.	188	17°35'N	15°15'W
Trasimeno, Lago, l., Italy (lä'gō trä-sē-mā'nō)	130	43°00'N	12°12'E
Trás-os-Montes, hist. reg., Port. (träzh'ōzh mōn'tāzh)	118	41°33'N	7°13'W
Traun, r., Aus. (troun)	124	48°10'N	14°15'E
Traunstein, Ger. (troun'stīn)	124	47°52'N	12°38'E
Traverse, Lake, l., Mn., U.S. (trăv'ĕrs)	70	45°46'N	96°53'W
Traverse City, Mi., U.S.	66	44°45'N	85°40'W
Travnik, Bos. (träv'nēk)	131	44°13'N	17°43'E

ng-sing; ŋ-baŋk; N-nasalized n; nŏd; cŏmmit; ōld; ôbey; ôrder; oi-boil; fōōd; ŏ-as oo in foot; ou-out; s-soft; sh-dish; th-thin; pūre; ûnite; ûrn; stŭd; circŭs; ü-as in French tu; '-indeterminate vowel.

312 Treasure Island – Turnov PRONOUNCING INDEX

PLACE (Pronunciation)	PAGE	LAT.	LONG.
Treasure Island, i., Ca., U.S. (trĕzh′ēr)	74b	37°49′N	122°22′W
Trebbin, Ger. (trĕ′bēn)	115b	52°13′N	13°13′E
Trebinje, Bos. (trå′bēn-yĕ)	131	42°43′N	18°21′E
Trebišov, Slvk. (trĕ′bẻ-shỏf)	125	48°36′N	21°32′E
Tregrosse Islands, is., Austl. (trẻ-grōs′)	175	18°08′S	150°53′E
Treinta y Tres, Ur. (trå-ēn′tä ē träs′)	102	33°14′S	54°17′W
Trelew, Arg. (trē′lū)	102	43°15′S	65°25′W
Trelleborg, Swe.	122	55°24′N	13°07′E
Tremiti, Isole, is., Italy (ẻ′sō-lĕ trā-mē′tē)	130	42°07′N	16°33′E
Trenčín, Czech Rep. (trĕn′chēn)	117	48°52′N	18°02′E
Trenque Lauquén, Arg. (trĕn′kĕ-lá′ò-kĕ′n)	102	35°50′S	62°44′W
Trent, r., Can. (trĕnt)	59	44°15′N	77°55′W
Trent, r., Eng., U.K.	114a	53°25′N	0°45′W
Trent and Mersey Canal, can., Eng., U.K. (trĕnt) (mûr zē)	114a	53°11′N	2°24′W
Trentino-Alto Adige, hist. reg., Italy	130	46°16′N	10°47′E
Trento, Italy (trĕn′tô)	118	46°04′N	11°07′E
Trenton, Can. (trĕn′tŭn)	51	44°05′N	77°35′W
Trenton, Can.	61	45°37′N	62°38′W
Trenton, Mi., U.S.	69b	42°08′N	83°12′W
Trenton, Mo., U.S.	79	40°05′N	93°36′W
Trenton, N.J., U.S.	65	40°13′N	74°46′W
Trenton, Tn., U.S.	82	35°57′N	88°55′W
Trepassey, Can. (trẻ-păs′ẻ)	61	46°44′N	53°22′W
Trepassey Bay, b., Can.	61	46°40′N	53°20′W
Tres Arroyos, Arg. (trās′är-rō′yōs)	102	38°18′S	60°16′W
Três Corações, Braz. (trĕ′s kō-rä-zô′ĕs)	99a	21°41′S	45°14′W
Tres Cumbres, Mex. (trĕ′s ko͞o′m-brĕs)	89a	19°03′N	99°14′W
Três Lagoas, Braz. (trĕ′s lä-gô′ás)	101	20°48′S	51°42′W
Três Marias, Represa, res., Braz.	101	18°15′S	45°30′W
Tres Morros, Alto de, mtn., Col. (á′l-tō dĕ trĕ′s mô′r-rôs)	100a	7°08′N	76°10′W
Três Pontas, Braz. (trĕ′pô′n-täs)	99a	21°22′S	45°30′W
Três Pontas, Cabo das, c., Ang.	190	10°23′S	13°32′E
Três Rios, Braz. (trĕ′s rẻ′ōs)	99a	22°07′S	43°13′W
Très-Saint Rédempteur, Can. (săɴ rā-dăɴp-tûr′)	62a	45°26′N	74°23′W
Treuenbrietzen, Ger. (troi′ĕn-brē-tzĕn)	115b	52°06′N	12°52′E
Treviglio, Italy (trā-vē′lyô)	130	45°30′N	9°34′E
Treviso, Italy (trĕ-vẻ′sō)	118	45°39′N	12°15′E
Trichardt, S. Afr. (trĭ-ᴋärt′)	192c	26°32′N	29°16′E
Trier, Ger.	117	49°45′N	6°38′E
Trieste, Italy (trẻ-ĕs′tā)	110	45°39′N	13°48′E
Triglav, mtn., Slvn.	130	46°23′N	13°50′E
Trigueros, Spain (trẻ-gā′rōs)	128	37°23′N	6°50′W
Tríkala, Grc.	119	39°33′N	21°49′E
Trikora, Puncak, mtn., Indon.	169	4°15′S	138°45′E
Trim Creek, r., Il., U.S. (trĭm)	69a	41°19′N	87°39′W
Trincomalee, Sri L. (trĭŋ-kỏ-mȧ-lē′)	159	8°39′N	81°12′E
Tring, Eng., U.K. (trĭng)	114b	51°46′N	0°40′W
Trinidad, Bol. (trē-nẻ-dhädh′)	100	14°48′S	64°43′W
Trinidad, Cuba (trē-nẻ-dhädh′)	87	21°50′N	80°00′W
Trinidad, Ur.	102	33°29′S	56°55′W
Trinidad, Co., U.S. (trĭn′ĭdăd)	64	37°11′N	104°31′W
Trinidad, i., Trin. (trĭn′ĭ-dăd)	101	10°00′N	61°00′W
Trinidad, r., Pan.	86a	8°55′N	80°01′W
Trinidad, Sierra de, mts., Cuba (sē-ĕ′r-rá dĕ trẻ-nẻ-dä′d)	92	21°50′N	79°55′W
Trinidad and Tobago, nation, N.A. (trĭn′ĭ-dăd) (tỏ-bā′gō)	87	11°00′N	61°00′W
Trinitaria, Mex. (trē-nē-tä′ryä)	89	16°09′N	92°04′W
Trinity, Can. (trĭn′ĭ-tẻ)	61	48°59′N	53°55′W
Trinity, Tx., U.S.	81	30°52′N	95°27′W
Trinity, is., Ak., U.S.	63	56°25′N	153°15′W
Trinity, r., Ca., U.S.	72	40°50′N	123°20′W
Trinity, r., Tx., U.S.	65	30°50′N	95°09′W
Trinity, East Fork, r., Tx., U.S.	79	33°24′N	96°42′W
Trinity, West Fork, r., Tx., U.S.	78	33°22′N	98°26′W
Trinity Bay, b., Can.	53	48°00′N	53°40′W
Trino, Italy (trē′nô)	130	45°11′N	8°16′E
Trion, Ga., U.S. (trī′ŏn)	82	34°32′N	85°18′W
Trípoli, Grc.	119	37°32′N	22°32′E
Tripoli (Tarābulus), Libya	185	32°50′N	13°13′E
Tripolitania *see* Tarābulus, hist. reg., Libya	184	31°00′N	12°26′E
Tripura, state, India	155	24°00′N	92°00′E
Tristan da Cunha Islands, is., St. Hel. (trẻs-tän′dä ko͞on′yȧ)	2	35°30′S	12°15′W
Triste, Golfo, b., Ven. (gỏl-fô trĕ′s-tĕ)	101b	10°40′N	68°05′W
Triticus Reservoir, res., N.Y., U.S. (trī tĭ-cŭs)	68a	41°20′N	73°36′W
Trnava, Slvk. (t′r′nȧ-vȧ)	125	48°22′N	17°34′E
Trobriand Islands, is., Pap. N. Gui. (trō-brẻ-ănd′)	169	8°25′S	151°45′E
Trogir, Cro. (trỏ′gēr)	130	43°32′N	16°17′E
Trois Fourches, Cap des, c., Mor.	128	35°28′N	2°58′W
Trois-Rivières, Can. (trwä′rẻ-vyâ′)	51	46°21′N	72°35′W
Troitsk, Russia (trô′ẻtsk)	140	54°06′N	61°35′E
Troits′ke, Ukr.	133	47°39′N	30°16′E
Troitsko-Pechorsk, Russia (trô′ĭtsk-ỏ-pyĕ-chôrsk′)	134	62°18′N	56°07′E
Trollhättan, Swe. (trôl′hĕt-ĕn)	116	58°17′N	12°17′E
Trollheimen, mts., Nor. (trỏll-hĕĭm)	122	62°48′N	9°05′E
Trona, Ca., U.S. (trō′nȧ)	76	35°49′N	117°20′W
Tronador, Cerro, mtn., S.A. (sĕ′r-rỏ trō-nä′dỏr)	102	41°17′S	71°56′W
Troncoso, Mex. (trỏn-kỏ′sō)	88	22°43′N	102°22′W
Trondheim, Nor. (trôn′hâm)	110	63°25′N	11°35′E
Trosa, Swe. (trỏ′sä)	122	58°54′N	17°25′E
Trout, l., Can.	53	51°16′N	92°46′W
Trout, l., Can.	52	61°10′N	121°30′W
Trout Creek, r., Or., U.S.	72	42°18′N	118°31′W
Troutdale, Or., U.S. (trout′dāl)	74c	45°32′N	122°23′W
Trout Lake, Mi., U.S.	71	46°20′N	85°02′W
Trouville, Fr. (tro͞o-vēl′)	126	49°23′N	0°05′E
Troy, Al., U.S. (troi)	82	31°47′N	85°46′W
Troy, Il., U.S.	75e	38°44′N	89°53′W
Troy, Ks., U.S.	79	39°46′N	95°07′W
Troy, Mo., U.S.	78	38°56′N	99°57′W
Troy, Mt., U.S.	72	48°28′N	115°56′W
Troy, N.C., U.S.	83	35°21′N	79°58′W
Troy, N.Y., U.S.	65	42°45′N	73°45′W
Troy, Oh., U.S.	66	40°00′N	84°10′W
Troy, hist., Tur.	154	39°59′N	26°14′E
Troyes, Fr. (trwä)	117	48°18′N	4°03′E
Trstenik, Serb. (t′r′stĕ-nẻk)	119	43°36′N	21°00′E
Trubchëvsk, Russia (tròp′chĕfsk)	137	52°36′N	33°46′E
Trucial States *see* United Arab Emirates, nation, Asia	154	24°00′N	54°00′E
Truckee, Ca., U.S. (trŭk′ẻ)	76	39°20′N	120°12′W
Truckee, r., Ca., U.S.	76	39°25′N	120°07′W
Truganina, Austl.	173a	37°49′N	144°44′E
Trujillo, Col. (trò-ᴋẻ′l-yō)	100a	4°10′N	76°20′W
Trujillo, Peru	100	8°08′S	79°00′W
Trujillo, Spain (tro͞o-ᴋẻ′l-yỏ)	118	39°27′N	5°50′W
Trujillo, Ven.	100	9°15′N	70°28′W
Trujillo, r., Mex.	88	23°12′N	103°10′W
Trujín, Lago, l., Dom. Rep. (tro͞o-ᴋēn′)	93	17°45′N	71°25′W
Truk *see* Chuuk, is., Micron.	170c	7°25′N	151°47′E
Trumann, Ar., U.S. (tro͞o′măn)	79	35°41′N	90°31′W
Trŭn, Blg. (trŭn)	131	42°49′N	22°39′E
Truro, Can. (tro͞o′rō)	51	45°22′N	63°16′W
Truro, Eng., U.K.	120	50°17′N	5°05′W
Trussville, Al., U.S. (trŭs′vĭl)	68h	33°37′N	86°37′W
Truth or Consequences, N.M., U.S. (tro͞oth ŏr kŏn′sẻ-kwĕn-sĭs)	77	33°10′N	107°20′W
Trutnov, Czech Rep. (tròt′nỏf)	124	50°36′N	15°36′E
Trzcianka, Pol. (tchyän′kȧ)	124	53°02′N	16°27′E
Trzebiatów, Pol. (tchĕ-byä′tò-v)	124	54°03′N	15°16′E
Tsaidam Basin, basin, China (tsī-däm)	160	37°19′N	94°08′E
Tsala Apopka Lake, r., Fl., U.S. (tsä′lȧ ă-pŏp′kȧ)	83	28°57′N	82°11′W
Tsast Bogd, mtn., Mong.	160	46°44′N	92°34′E
Tsavo National Park, rec., Kenya	191	2°35′S	38°45′E
Tsawwassen Indian Reserve, I.R., Can.	74d	49°03′N	123°11′W
Tsentral′nyy-Kospashskiy, Russia (tsĕn-träl′nyĭ-kôs-pȧsh′skĭ)	142a	59°03′N	57°48′E
Tshela, D.R.C. (tshā′lȧ)	186	4°59′S	12°56′E
Tshikapa, D.R.C. (tshẻ-kä′pä)	186	6°25′S	20°48′E
Tshofa, D.R.C.	191	5°14′S	25°15′E
Tshuapa, r., D.R.C.	186	0°30′S	22°00′E
Tsiafajovona, mtn., Madag.	187	19°17′S	47°27′E
Tsiribihina, r., Madag. (tsē′rẻ-bē-hẻ-nä′)	187	19°45′S	43°30′E
Tsitsa, r., S. Afr. (tsẻ′tsȧ)	187c	31°28′S	28°53′E
Tskhinvali, Geor.	138	42°13′N	43°56′E
Tsolo, S. Afr. (tsō′lō)	187c	31°19′S	28°47′E
Tsomo, S. Afr.	187c	32°03′S	27°49′E
Tsomo, r., S. Afr.	187c	31°53′S	27°48′E
Tsu, Japan (tso͞o)	166	34°42′N	136°31′E
Tsuchiura, Japan (tso͞o′chẻ-o͞o-rä)	167	36°04′N	140°09′E
Tsuda, Japan (tso͞o′dä)	167b	34°48′N	135°43′E
Tsugaru Kaikyō, strt., Japan	161	41°25′N	140°20′E
Tsumeb, Nmb. (tso͞o′mĕb)	186	19°10′S	17°45′E
Tsunashima, Japan (tso͞o′nä-shẻ′mä)	167a	35°32′N	139°37′E
Tsuruga, Japan (tso͞o′rò-gä)	166	35°39′N	136°04′E
Tsurugi San, mtn., Japan (tso͞o′rò-gẻ sän)	166	33°52′N	134°07′E
Tsuruoka, Japan (tso͞o′rò-ō′kä)	166	38°43′N	139°51′E
Tsurusaki, Japan (tso͞o′rò-sä′kẻ)	167	33°15′N	131°42′E
Tsu Shima, is., Japan (tso͞o shẻ′mä)	161	34°28′N	129°30′E
Tsushima Strait, strt., Asia	161	34°00′N	129°00′E
Tsuwano, Japan (tso͞o′wȧ-nỏ′)	167	34°28′N	131°47′E
Tsuyama, Japan (tso͞o′yä-mä′)	166	35°05′N	134°00′E
Tua, r., Port. (to͞o′ä)	128	41°23′N	7°18′W
Tualatin, r., Or., U.S. (to͞o′ȧ-lā-tĭn)	74c	45°25′N	122°54′W
Tuamoto, Îles, Fr. Poly. (to͞o-ä-mō′to͞o)	195	19°00′S	141°20′W
Tuapse, Russia (tò′áp-sĕ)	137	44°00′N	39°10′E
Tuareg, hist. reg., Alg.	184	21°26′N	2°51′E
Tubarão, Braz. (to͞o-bä-rouɴ′)	102	28°23′N	48°56′W
Tübingen, Ger. (tü′bĭng-ĕn)	124	48°33′N	9°05′E
Tubinskiy, Russia (tû bĭn′skĭ)	142a	52°53′N	58°15′E
Tubruq, Libya	185	32°03′N	24°04′E
Tucacas, Ven. (to͞o-kä′käs)	100	10°48′N	68°20′W
Tucker, Ga., U.S. (tŭk′ēr)	68c	33°51′N	84°13′W
Tucson, Az., U.S. (to͞o-sŏn′)	64	32°15′N	111°00′W
Tucumán, Arg. (to͞o-ko͞o-män′)	102	26°52′S	65°08′W
Tucumán, prov., Arg.	102	26°30′S	65°30′W
Tucumcari, N.M., U.S. (tò′kŭm-kâr-ẻ)	78	35°11′N	103°43′W
Tucupita, Ven. (to͞o-ko͞o-pẻ′tä)	100	9°00′N	62°09′W
Tudela, Spain (to͞o-dhā′lä)	118	42°03′N	1°37′W
Tugaloo, r., Ga., U.S. (tŭg′ȧ-lo͞o)	82	34°35′N	83°05′W
Tugela, r., S. Afr. (to͞o-gel′ȧ)	187c	28°50′S	30°52′E
Tugela Ferry, S. Afr.	187c	28°44′S	30°27′E
Tug Fork, r., U.S. (tŭg)	66	37°50′N	82°30′W
Tuguegarao, Phil. (to͞o-gā-gȧ-rä′ō)	168	17°37′N	121°44′E
Tuhai, r., China (to͞o-hī)	162	37°05′N	116°56′E
Tui, Slvn.	128	42°03′N	8°38′W
Tuinplaas, S. Afr.	192c	24°54′S	28°46′E
Tujunga, Ca., U.S. (to͞o-jŭŋ′gȧ)	75a	34°15′N	118°16′W
Tukan, Russia (to͞o′kȧn)	142a	53°52′N	57°25′E
Tukangbesi, Kepulauan, is., Indon.	169	6°00′S	124°15′E
Tūkrah, Libya	185	32°34′N	20°47′E
Tuktoyaktuk, Can.	50	69°32′N	132°37′W
Tuktut Nogait National Park, rec., Can.	52	69°00′N	122°00′W
Tukums, Lat. (tò′kòms)	136	56°57′N	23°09′E
Tukuyu, Tan. (to͞o-ko͞o′yȧ)	186	9°13′S	33°43′E
Tukwila, Wa., U.S. (tŭk′wĭ-lȧ)	74a	47°28′N	122°16′W
Tula, Mex. (to͞o′lä)	88	20°04′N	99°22′W
Tula, Russia (to͞o′lȧ)	136	54°12′N	37°37′E
Tula, prov., Russia	132	53°45′N	37°19′E
Tula, r., Mex. (to͞o′lä)	88	20°40′N	99°27′W
Tulagai, i., Sol. Is. (to͞o-lä′gē)	175	9°15′S	160°17′E
Tulaghi, Sol. Is.	170e	9°06′S	160°09′E
Tulalip, Wa., U.S. (tū-lă′lĭp)	74a	48°04′N	122°18′W
Tulalip Indian Reservation, I.R., Wa., U.S.	74a	48°06′N	122°16′W
Tulancingo, Mex. (to͞o-län-sĭŋ′gō)	86	20°04′N	98°24′W
Tulangbawang, r., Indon.	168	4°17′S	105°00′E
Tulare, Ca., U.S. (to͞o-lä′rȧ) (tul-âr′)	76	36°12′N	119°22′W
Tulare Lake Bed, l., Ca., U.S.	76	35°57′N	120°18′W
Tularosa, N.M., U.S. (to͞o-lȧ-rō′zȧ)	77	33°05′N	106°05′W
Tulcán, Ec. (to͞ol-kän′)	100	0°44′N	77°52′W
Tulcea, Rom. (tòl′chȧ)	119	45°10′N	28°47′E
Tul′chyn, Ukr.	137	48°42′N	28°53′E
Tulcingo, Mex. (to͞ol-sĭŋ′gō)	88	18°03′N	98°27′W
Tule, r., Ca., U.S. (to͞ol′lå)	76	36°08′N	118°50′W
Tule River Indian Reservation, I.R., Ca., U.S. (to͞o′lå)	76	36°00′N	118°40′W
Tuli, Zimb. (to͞o′lẻ)	186	20°58′S	29°12′E
Tulia, Tx., U.S. (to͞o′lĭ-ȧ)	78	34°32′N	101°46′W
Tulik Volcano, vol., Ak., U.S. (tò′lĭk)	63a	53°28′N	168°10′W
Tūlkarm, W.B. (to͞ol kärm)	153a	32°19′N	35°02′E
Tullahoma, Tn., U.S. (tŭl-ȧ-hō′mȧ)	82	35°21′N	86°12′W
Tullamore, Ire. (tŭl-ȧ-mōr′)	120	53°15′N	7°29′W
Tulle, Fr. (tül)	126	45°15′N	1°45′E
Tulln, Aus. (tòln)	124	48°21′N	16°04′E
Tullner Feld, reg., Aus.	115e	48°20′N	15°59′E
Tulpetlac, Mex. (to͞ol-på-tlȧk′)	89a	19°33′N	99°04′W
Tulsa, Ok., U.S. (tŭl′sȧ)	65	36°08′N	95°58′W
Tulum, Mex. (to͞o-lò′m)	90a	20°17′N	87°26′W
Tulun, Russia (tò-lo͞on′)	135	54°29′N	100°43′E
Tuma, r., Nic. (to͞o′mä)	90	13°07′N	85°32′W
Tumba, Lac, l., D.R.C. (tòm′bä)	186	0°50′S	17°45′E
Tumbes, Peru (to͞o′m-bĕs)	100	3°39′S	80°27′W
Tumbiscatío, Mex. (to͞om-bẻ-skȧ-tē′ō)	88	18°32′N	102°23′W
Tumbo, i., Can.	74d	48°49′N	123°04′W
Tumacocori National Monument, rec., Az., U.S. (to͞o-mä-kä′kå-rē)	77	31°36′N	110°20′W
Tumen, China (to͞o-mŭn)	164	43°00′N	129°50′E
Tumen, r., Asia	166	42°08′N	128°40′E
Tumeremo, Ven. (to͞o-må-rā′mō)	101	7°15′N	61°28′W
Tumkūr, India	159	13°22′N	77°05′E
Tumuc-Humac Mountains, mts., S.A. (to͞o-mòk′o͞o-mäk′)	101	2°15′N	54°50′W
Tunas de Zaza, Cuba (to͞o′näs dā zä′zä)	92	21°40′N	79°35′W
Tunbridge Wells, Eng., U.K. (tŭn′brĭj welz′)	121	51°05′N	0°09′E
Tunduru, Tan.	191	11°07′S	37°21′E
Tungabhadra Reservoir, res., India	159	15°26′N	75°57′E
Tuni, India	159	17°29′N	82°38′E
Tunica, Ms., U.S. (tū′nĭ-kȧ)	82	34°41′N	90°23′W
Tunis, Tun. (tū′nĭs)	184	36°59′N	10°06′E
Tunis, Golfe de, b., Tun.	118	37°06′N	10°43′E
Tunisia, nation, Afr. (tu-nĭzh′ẻ-ȧ)	184	35°00′N	10°11′E
Tunja, Col. (to͞o′n-hȧ)	100	5°32′N	73°19′W
Tunkhannock, Pa., U.S. (tŭnk-hăn′ŭk)	67	41°35′N	75°55′W
Tunnel, r., Wa., U.S. (tŭn′ĕl)	74a	47°48′N	123°04′W
Tuoji Dao, i., China (twỏ-jyē dou)	162	38°11′N	120°45′E
Tuolumne, r., Ca., U.S. (twô-lŭm′nẻ)	76	37°35′N	120°37′W
Tuostakh, r., Russia	141	67°09′N	137°30′E
Tupelo, Ms., U.S. (tū′pẻ-lō)	82	34°14′N	88°43′W
Tupinambaranas, Ilha, i., Braz.	101	3°04′S	58°09′W
Tupiza, Bol. (to͞o-pē′zä)	100	21°26′S	65°43′W
Tupper Lake, N.Y., U.S. (tŭp′ēr)	67	44°15′N	74°25′W
Tüpqaraghan tübegi, pen., Kaz.	137	44°30′N	50°40′E
Tupungato, Cerro, vol., S.A.	102	33°30′S	69°52′W
Tuquerres, Col. (to͞o-kĕ′r-rĕs)	100	1°12′N	77°44′W
Tura, Russia (tòr′ȧ)	135	64°08′N	99°58′E
Turbio, r., Mex. (to͞or-byỏ)	88	20°28′N	101°40′W
Turbo, Col. (to͞o′bō)	100	8°02′N	76°43′W
Turda, Rom. (tòr′dȧ)	125	46°35′N	23°47′E
Turfan Depression, depr., China	160	42°16′N	90°00′E
Turffontein, neigh., S. Afr.	187b	26°15′S	28°02′E
Tŭrgovishte, Blg.	131	43°14′N	26°36′E
Turgutlu, Tur.	137	38°30′N	27°20′E
Türi, Est. (tü′rĭ)	123	58°49′N	25°29′E
Turia, r., Spain (to͞o′ryä)	128	40°12′N	1°18′W
Turicato, Mex. (to͞o-rẻ-kä′tō)	88	19°03′N	101°24′W
Turiguano, i., Cuba (to͞o-rẻ-gwä′nō)	92	22°20′N	78°35′W
Turin, Italy	110	45°05′N	7°44′E
Turiya, r., Ukr.	125	51°18′N	24°55′E
Turka, Ukr. (tòr′kȧ)	125	49°10′N	23°02′E
Turkestan, hist. reg., Asia	134	43°27′N	62°14′E
Turkey, nation, Asia	111	38°45′N	32°00′E
Turkey, r., Ia., U.S. (tûrk′ẻ)	71	43°20′N	92°16′W
Türkistan, Kaz.	139	44°00′N	68°00′E
Turkmenbashy, Turkmen.	139	40°00′N	52°50′E
Turkmenistan, nation, Asia	134	40°46′N	56°01′E
Turks, is., T./C. Is. (tûrks)	87	21°40′N	71°45′W
Turks Island Passage, strt., T./C. Is.	93	21°15′N	71°25′W
Turku, Fin. (tòrgokò)	110	60°28′N	22°12′E
Turlock, Ca., U.S. (tûr′lŏk)	76	37°30′N	120°51′W
Turneffe, i., Belize	86	17°25′N	87°43′W
Turner, Ks., U.S. (tûr′nēr)	75f	39°05′N	94°42′W
Turner Sound, strt., Bah.	92	24°20′N	78°05′W
Turners Peninsula, pen., S.L.	188	7°20′N	12°40′W
Turnhout, Bel. (tŭrn-hout′)	121	51°19′N	4°58′E
Turnov, Czech Rep. (tòr′nỏf)	124	50°36′N	15°12′E

ăt; finăl; rāte; senåte; ärm; ȧsk; sofȧ; fâre; ch-choose; dh-as th in other; bē; ẻvent; bĕt; recĕnt; cratēr; g-gō; gh-guttural g; bĭt; ĭ-short neutral; rīde; ᴋ-guttural k as ch in German ich;

PRONOUNCING INDEX

Turnu Măgurele – Urdaneta 313

PLACE (Pronunciation)	PAGE	LAT.	LONG.
Turnu Măgurele, Rom.	119	43°54′N	24°49′E
Turpan, China (to͞o-är-pän)	160	43°06′N	88°41′E
Turquino, Pico, mtn., Cuba (pē′kō dā to͞or-kē′nō)	92	20°00′N	76°50′W
Turrialba, C.R. (to͞or-ryä′l-bä)	91	9°54′N	83°41′W
Turtkul′, Uzb. (tort-kol′)	139	41°28′N	61°02′E
Turtle, r., Can.	57	49°20′N	92°30′W
Turtle Bay, b., Tx., U.S.	81a	29°48′N	94°38′W
Turtle Creek, r., S.D., U.S.	70	44°40′N	98°53′W
Turtle Mountain Indian Reservation, I.R., N.D., U.S.	70	48°45′N	99°57′W
Turtle Mountains, mts., N.D., U.S.	70	48°57′N	100°11′W
Turukhansk, Russia (to͞o-ro͞o-känsk′)	134	66°03′N	88°39′E
Tuscaloosa, Al., U.S. (tŭs-kȧ-lo͞o′sȧ)	65	33°10′N	87°35′W
Tuscarora, Nv., U.S. (tŭs-kȧ-rō′rȧ)	72	41°18′N	116°15′W
Tuscarora Indian Reservation, I.R., N.Y., U.S.	69c	43°10′N	78°51′W
Tuscola, Il., U.S. (tŭs-kō-lȧ)	66	39°50′N	88°20′W
Tuscumbia, Al., U.S. (tŭs-kŭm′bĭ-ȧ)	82	34°41′N	87°42′W
Tushino, Russia (to͞o′shĭ-nô)	142b	55°51′N	37°24′E
Tuskegee, Al., U.S. (tŭs-kē′gė)	82	32°25′N	85°40′W
Tustin, Ca., U.S. (tŭs′tĭn)	75a	33°44′N	117°49′W
Tutayev, Russia (to͞o-tȧ-yĕf′)	136	57°53′N	39°34′E
Tutbury, Eng., U.K. (tŭt′bĕr-ė)	114a	52°52′N	1°51′W
Tuticorin, India (to͞o-tė-kō-rĭn′)	159	8°51′N	78°09′E
Tutitlán, Mex. (to͞o-tē-tlä′n)	89a	19°38′N	99°10′W
Tutóia, Braz. (to͞o-tō′yȧ)	101	2°42′S	42°21′W
Tutrakan, Blg.	119	44°02′N	26°36′E
Tuttle Creek Reservoir, res., Ks., U.S.	79	39°30′N	96°38′W
Tuttlingen, Ger. (tot′lĭng-ĕn)	124	47°58′N	8°50′E
Tutuila, i., Am. Sam.	170a	14°18′S	170°42′W
Tutwiler, Ms., U.S. (tŭt′wĭ-lẽr)	82	34°01′N	90°25′W
Tuva, prov., Russia	140	51°15′N	90°45′E
Tuvalu, nation, Oc.	3	5°20′S	174°00′E
Tuwayq, Jabal, mts., Sau. Ar.	154	20°45′N	46°30′E
Tuxedo Park, N.Y., U.S. (tŭk-sē′dō pärk)	68a	41°11′N	74°11′W
Tuxford, Eng., U.K. (tŭks′fẽrd)	114a	53°14′N	0°54′W
Túxpan, Mex. (to͞os′pän)	88	19°34′N	103°22′W
Túxpan, Mex.	86	20°57′N	97°26′W
Túxpan, r., Mex. (to͞os′pän)	89	20°55′N	97°52′W
Túxpan, Arrecife, i., Mex. (är-rĕ-sė′fē-to͞o′x-pä′n)	89	21°01′N	97°12′W
Tuxtepec, Mex. (tos-tå-pĕk′)	89	18°06′N	96°09′W
Tuxtla Gutiérrez, Mex. (tos′tlä go͞o-tyår′rĕs)	86	16°44′N	93°08′W
Tuy, r., Ven. (to͞o′ė)	101b	10°15′N	66°03′W
Tuyra, r., Pan. (to͞o-ē′rȧ)	91	7°55′N	77°37′W
Tuz Gölü, l., Tur.	136	38°45′N	33°25′E
Tuzigoot National Monument, rec., Az., U.S.	77	34°40′N	111°52′W
Tuzla, Bos. (toz′lȧ)	119	44°33′N	18°46′E
Tvedestrand, Nor. (tvĭ′dhĕ-stränd)	122	58°39′N	8°54′E
Tveitsund, Nor. (tvåt′sond)	122	59°03′N	8°29′E
Tver′, Russia	134	56°52′N	35°57′E
Tver′, prov., Russia	132	56°50′N	33°08′E
Tvertsa, r., Russia (tvĕr′tsȧ)	132	56°58′N	35°22′E
Tweed, r., U.K. (twēd)	120	55°32′N	2°35′W
Tweeling, S. Afr. (twē′lĭng)	192c	27°34′S	28°31′E
Twenty Mile Creek, r., Can. (twĕn′tĭ mīl)	62d	43°09′N	79°49′W
Twickenham, Eng., U.K. (twĭk′′n-ăm)	114b	51°26′N	0°20′W
Twillingate, Can. (twĭl′ĭn-gāt)	53a	49°39′N	54°46′W
Twin Bridges, Mt., U.S. (twĭn brĭ-jĕz)	73	45°34′N	112°17′W
Twin Falls, Id., U.S. (fôls)	64	42°33′N	114°29′W
Twinsburg, Oh., U.S. (twĭnz′bŭrg)	69d	41°19′N	81°26′W
Twitchell Reservoir, res., Ca., U.S.	76	34°50′N	120°10′W
Two Butte Creek, r., Co., U.S. (to͞o būt)	78	37°39′N	102°45′W
Two Harbors, Mn., U.S.	71	47°00′N	91°42′W
Two Prairie Bay, Ar., U.S. (prā′rĭ bī o͞o′)	79	34°48′N	92°07′W
Two Rivers, Wi., U.S. (rĭv′ẽrz)	71	44°09′N	87°36′W
Tyabb, Austl.	173a	38°16′S	145°11′E
Tylden, S. Afr. (tĭl-dĕn)	187c	32°08′S	27°06′E
Tyldesley, Eng., U.K. (tĭldz′lė)	114a	53°32′N	2°28′W
Tyler, Mn., U.S. (tī′lẽr)	70	44°18′N	96°08′W
Tyler, Tx., U.S.	65	32°21′N	95°19′W
Tylertown, Ms., U.S. (tī′lẽr-toun)	82	31°08′N	90°06′W
Tylihul, r., Ukr.	133	47°25′N	30°27′E
Tyndall, S.D., U.S. (tĭn′dȧl)	70	42°58′N	97°52′W
Tyndinskiy, Russia	135	55°22′N	124°45′E
Tyne, r., Eng., U.K. (tīn)	120	54°59′N	1°56′W
Tynemouth, Eng., U.K. (tīn′mŭth)	116	55°04′N	1°39′W
Tyngsboro, Ma., U.S. (tĭnj-bûr′ō)	61a	42°40′N	71°27′W
Tynset, Nor. (tün′sĕt)	116	62°17′N	10°45′E
Tyre *see* Şūr, Leb.	153a	33°16′N	35°13′E
Tyrifjorden, l., Nor.	122	60°03′N	10°25′E
Tyrnavos, Grc.	131	39°50′N	22°14′E
Tyrone, Pa., U.S.	67	40°40′N	78°15′W
Tyrrell, Lake, l., Austl. (tir′ĕll)	176	35°12′S	143°00′E
Tyrrhenian Sea, sea, Italy (tĭr-rē′nĭ-ȧn)	112	40°10′N	12°15′E
Tyukalinsk, Russia (tyo-kȧ-lĭnsk′)	134	56°03′N	71°43′E
Tyukyan, r., Russia (tyok′yȧn)	141	65°42′N	116°09′E
Tyuleniy, i., Russia	137	44°30′N	48°00′E
Tyumen′, Russia (tyo͞o-mĕn′)	134	57°02′N	65°28′E
Tzucacab, Mex. (tzo͞o-kä-kä′b)	90a	20°06′N	89°03′W

U

PLACE (Pronunciation)	PAGE	LAT.	LONG.
Uaupés, Braz. (wä-o͞o′pās)	100	0°02′S	67°03′W
Ubangi, r., Afr. (o͞o-bäŋ′gė)	185	3°00′N	18°00′E
Ubatuba, Braz. (o͞o-bå-to͞o′bȧ)	99a	23°25′S	45°06′W
Ubeda, Spain (o͞o′bå-dä)	128	38°01′N	3°23′W
Uberaba, Braz. (o͞o-bå-rä′bȧ)	101	19°47′S	47°47′W
Uberlândia, Braz. (o͞o-bĕr-lȧ′n-dyä)	101	18°54′S	48°11′W
Ubombo, S. Afr. (o͞o-bôm′bō)	186	27°33′S	32°13′E
Ubon Ratchathani, Thai. (o͞o′bŭn rä′chätá-nē)	168	15°15′N	104°52′E
Ubort′, r., Eur. (o͞o-bôrt′)	133	51°18′N	27°43′E
Ubrique, Spain (o͞o-brē′kå)	128	36°43′N	5°36′W
Ubundu, D.R.C.	186	0°21′S	25°29′E
Ucayali, r., Peru (o͞o′kä-yä′lē)	100	8°58′S	74°13′W
Uccle, Bel. (ū′kl′)	115a	50°48′N	4°17′E
Uchaly, Russia (û-chä′lĭ)	142a	54°22′N	59°28′E
Uchiko, Japan (o͞o′chė-kō)	167	33°30′N	132°39′E
Uchinoura, Japan (o͞o′chė-nō-o͞o′rȧ)	167	31°16′N	131°03′E
Uchinskoye Vodokhranilishche, res., Russia	142b	56°08′N	37°44′E
Uchiura-Wan, b., Japan (o͞o′chė-o͞o′rä wän)	166	42°20′N	140°44′E
Uchur, r., Russia (o-chor′)	141	57°25′N	130°35′E
Uda, r., Russia	141	53°54′N	131°29′E
Uda, r., Russia (o′dȧ)	141	52°28′N	110°51′E
Udai, r., Ukr.	133	50°45′N	32°13′E
Udaipur, India (o-dŭ′ė-po͞or)	158	24°41′N	73°41′E
Uddevalla, Swe. (o͞od′dĕ-väl-ȧ)	116	58°21′N	11°55′E
Udine, Italy (o͞o′dė-nå)	118	46°05′N	13°14′E
Udmurtia, prov., Russia	136	57°00′N	53°00′E
Udon Thani, Thai.	168	17°31′N	102°51′E
Udskaya Guba, b., Russia	135	55°00′N	136°30′E
Ueckermünde, Ger.	124	53°43′N	14°01′E
Ueda, Japan (wā′dä)	166	36°26′N	138°16′E
Uele, r., D.R.C. (wā′lå)	185	3°55′N	23°30′E
Uelzen, Ger. (ült′sĕn)	124	52°58′N	10°34′E
Ufa, Russia (o′fa)	134	54°45′N	55°57′E
Ufa, r., Russia	136	56°00′N	57°05′E
Ugab, r., Nmb. (o͞o′gäb)	186	21°10′S	14°00′E
Ugalla, r., Tan. (o͞o-gä′lä)	186	6°15′S	32°30′E
Uganda, nation, Afr. (o͞o-gän′dä) (û-găn′dȧ)	185	2°00′N	32°28′E
Ugashik Lake, l., Ak., U.S. (o͞o′gȧ-shėk)	63	57°36′N	157°10′W
Ugie, S. Afr. (o′jē)	187c	31°13′S	28°14′E
Uglegorsk, Russia (o͞o-glĕ-gôrsk)	135	49°00′N	142°31′E
Ugleural′sk, Russia (og-lĕ-o-rálsk′)	142a	58°58′N	57°35′E
Uglich, Russia (o͞og-lėch′)	132	57°33′N	38°19′E
Uglitskiy, Russia (og-lĭt′skĭ)	142a	53°50′N	60°18′E
Uglovka, Russia (o͞og-lôf′kȧ)	132	58°14′N	33°24′E
Ugra, r., Russia (o͞og′rȧ)	136	54°43′N	34°20′E
Ugŭrchin, Blg.	131	43°06′N	24°23′E
Uhrichsville, Oh., U.S. (ū′rĭks-vĭl)	66	40°25′N	81°20′W
Uíge, Ang.	186	7°37′S	15°03′E
Uiju, Kor., N. (o′ėjo͞o)	161	40°09′N	124°33′E
Uinkaret Plateau, plat., Az., U.S. (ů-ĭn′kâr-ēt)	77	36°43′N	113°15′W
Uinskoye, Russia (o-ĭn′skô-yĕ)	142a	56°53′N	56°25′E
Uinta, r., Ut., U.S. (ů-ĭn′tȧ)	77	40°25′N	109°55′W
Uintah and Ouray Indian Reservation, I.R., Ut., U.S.	77	40°20′N	110°20′W
Uinta Mountains, mts., Ut., U.S.	64	40°35′N	111°00′W
Uitenhage, S. Afr.	186	33°46′S	25°26′E
Uithoorn, Neth.	115a	52°13′N	4°49′E
Uji, Japan (o͞o′jē)	167b	34°53′N	135°49′E
Ujiji, Tan. (o͞o-jē′jė)	186	4°55′S	29°41′E
Ujjain, India (o͞o-jŭėn)	155	23°18′N	75°37′E
Ujungpandang, Indon.	168	5°08′S	119°28′E
Ukerewe Island, i., Tan.	191	2°00′S	32°40′E
Ukhta, Russia (o͞ok′tȧ)	136	65°22′N	31°30′E
Ukhta, Russia	136	63°08′N	53°42′E
Ukiah, Ca., U.S. (ū-kī′ȧ)	76	39°09′N	122°12′W
Ukmerge, Lith. (ok′mĕr-ghå)	136	55°16′N	24°45′E
Ukraine, nation, Eur.	134	49°15′N	30°15′E
Uku, i., Japan (o͞ok′o͞o)	167	33°18′N	129°02′E
Ulaangom, Mong.	160	50°23′N	92°14′E
Ulan Bator (Ulaanbaatar), Mong.	160	47°56′N	107°00′E
Ulan-Ude, Russia (o͞o′län o͞o′då)	135	51°59′N	107°41′E
Ulchin, Kor., S. (o͞ol′chėn′)	166	36°57′N	129°26′E
Ulcinj, Serb. (o͞ol′tsėn′)	119	41°56′N	19°15′E
Ulhās, r., India	159b	19°13′N	73°03′E
Ulhāsnagar, India	158	19°10′N	73°07′E
Uliastay, Mong.	160	47°49′N	97°00′E
Ulindi, r., D.R.C. (o͞o-lĭn′dė)	186	1°55′S	26°17′E
Ulla, Bela. (ol′ȧ)	132	55°14′N	29°15′E
Ulla, r., Bela.	132	54°58′N	29°03′E
Ulla, r., Spain (o͞o′lä)	128	42°45′N	8°33′W
Ullŭng, i., Kor., S. (o͞ol′long′)	166	37°29′N	130°50′E
Ulm, Ger. (olm)	117	48°24′N	9°59′E
Ulmer, Mount, mtn., Ant. (ŭl′mûr′)	178	77°30′S	86°00′W
Ulricehamn, Swe. (ol-rē′sĕ-häm)	122	57°49′N	13°23′E
Ulsan, Kor., S. (o͞ol′sän′)	166	35°35′N	129°22′E
Ulster, hist. reg., Eur. (ŭl′stẽr)	120	54°41′N	7°10′W
Ulua, r., Hond. (o͞o-lo͞o′ȧ)	90	15°49′N	87°45′W
Ulubāria, India	158a	22°27′N	88°09′E
Ulukışla, Tur. (o͞o-lo͞o-kēsh′lȧ)	119	36°40′N	34°30′E
Ulunga, Russia (o-lo͞on′gȧ)	166	46°16′N	136°29′E
Ulungur, r., China (o͞o-lo͞on-gůr)	160	46°31′N	88°00′E
Uluru (Ayers Rock), mtn., Austl.	174	25°23′S	131°05′E
Ulu-Telyak, Russia (o͞o lo′tĕlyăk)	142a	54°54′N	57°01′E
Ulverstone, Austl. (ŭl′vẽr-stŭn)	175	41°20′S	146°22′E
Ul′yanovka, Russia	142c	59°38′N	30°47′E
Ul′yanovsk, Russia (o͞o-lyä′nôfsk)	134	54°20′N	48°24′E
Ulysses, Ks., U.S. (ū-lĭs′ēz)	78	37°34′N	101°25′W
Umán, Mex. (o͞o-män′)	90a	20°52′N	89°44′W
Uman′, Ukr. (o-mȧn′)	137	48°44′N	30°13′E
Umatilla Indian Reservation, I.R., Or., U.S. (ū-mȧ-tĭl′ȧ)	72	45°38′N	118°35′W
Umberpāda, India	159b	19°28′N	73°04′E
Umbria, hist. reg., Italy (ŭm′brĭ-ȧ)	130	42°53′N	12°22′E
Umeälven, r., Swe.	112	64°57′N	18°51′E
Umhlatuzi, r., S. Afr. (om′hlȧ-to͞o′zĭ)	187c	28°47′S	31°17′E
Umiat, Ak., U.S. (o͞o′mĭ-ăt)	64a	69°20′N	152°28′W
Umkomaas, S. Afr. (om-kō′mäs)	187c	30°12′S	30°48′E
Umnak, i., Ak., U.S. (o͞om′nȧk)	64b	53°10′N	169°08′W
Umnak Pass, Ak., U.S.	63a	53°10′N	168°04′W
Umniati, r., Zimb.	186	17°08′S	29°11′E
Umpqua, r., Or., U.S. (ŭmp′kwȧ)	72	43°42′N	123°50′W
Umtata, S. Afr. (om-tä′tä)	186	31°36′S	28°47′E
Umtentweni, S. Afr.	187c	30°41′S	30°29′E
Umzimkulu, S. Afr. (om-zėm-ko͞o′lo͞o)	187c	30°12′S	29°53′E
Umzinto, S. Afr. (om-zĭn′tō)	187c	30°19′S	30°41′E
Una, r., Serb. (o͞o′nȧ)	130	44°38′N	16°10′E
Unalakleet, Ak., U.S. (ū-nȧ-lȧk′lēt)	63	63°50′N	160°42′W
Unalaska, Ak., U.S. (ū-nȧ-lȧs′kȧ)	63a	53°30′N	166°20′W
Unare, r., Ven.	101b	9°45′N	65°12′W
Unare, Laguna de, l., Ven. (lä-go′nä-de-o͞o-nȧ′rĕ)	101b	10°07′N	65°23′W
Unayzah, Sau. Ar.	154	25°50′N	44°02′E
Uncas, Can. (ŭŋ′kȧs)	62g	53°30′N	113°02′W
Uncia, Bol. (o͞on′sē-ä)	100	18°28′S	66°32′W
Uncompahgre, r., Co., U.S.	77	38°20′N	107°45′W
Uncompahgre Peak, mtn., Co., U.S. (ŭn-kŭm-pä′grė)	77	38°00′N	107°30′W
Uncompahgre Plateau, plat., Co., U.S.	77	38°40′N	108°40′W
Underberg, S. Afr. (ŭn′dẽr-bûrg)	187c	29°51′S	29°32′E
Unecha, Russia (o-nĕ′chȧ)	132	52°51′N	32°44′E
Ungava, Péninsule d′, pen., Can.	53	59°55′N	74°00′W
Ungava Bay, b., Can. (ŭŋ-gȧ′vȧ)	53	59°46′N	67°18′W
União da Vitória, Braz. (o͞o-nė-ouɴ′ dä vē-tô′ryä)	102	26°17′S	51°13′W
Unije, i., Serb. (o͞o′nė-yĕ)	130	44°39′N	14°10′E
Unimak, i., Ak., U.S. (o͞o-nė-mȧk′)	63	54°30′N	163°35′W
Unimak Pass, Ak., U.S.	63a	54°22′N	165°22′W
Union, Mo., U.S.	79	38°28′N	90°59′W
Union, Ms., U.S. (ūn′yŭn)	82	32°35′N	89°07′W
Union, N.C., U.S.	83	34°42′N	81°40′W
Union, Or., U.S.	72	45°13′N	117°52′W
Union City, Ca., U.S.	74b	37°36′N	122°01′W
Union City, In., U.S.	66	40°10′N	85°00′W
Union City, Mi., U.S.	66	42°00′N	85°10′W
Union City, Pa., U.S.	67	41°50′N	79°50′W
Union City, Tn., U.S.	82	36°25′N	89°04′W
Unión de Reyes, Cuba	92	22°45′N	81°30′W
Unión de San Antonio, Mex.	88	21°07′N	101°56′W
Unión de Tula, Mex.	88	19°57′N	104°14′W
Union Grove, Wi., U.S. (ūn-yŭn grōv)	69a	42°41′N	88°03′W
Unión Hidalgo, Mex. (ė-dä′lgō)	89	16°29′N	94°51′W
Union Point, Ga., U.S.	82	33°37′N	83°08′W
Union Springs, Al., U.S. (sprĭngz)	82	32°08′N	85°43′W
Uniontown, Al., U.S. (ūn′yŭn-toun)	82	32°26′N	87°30′W
Uniontown, Oh., U.S.	69d	40°58′N	81°25′W
Uniontown, Pa., U.S.	67	39°55′N	79°45′W
Unionville, Mo., U.S. (ūn′yŭn-vĭl)	79	40°28′N	92°58′W
Unisan, Phil. (o͞o-nē′sän)	169a	13°50′N	121°59′E
United Arab Emirates, nation, Asia	154	24°00′N	54°00′E
United Kingdom, nation, Eur.	110	56°30′N	1°40′W
United States, nation, N.A.	64	38°00′N	110°00′W
Unity, Can.	56	52°27′N	109°10′W
Universal, In., U.S. (ū-nĭ-vûr′sȧl)	66	39°35′N	87°30′W
University City, Mo., U.S. (ū′nĭ-vûr′sĭ-tĭ)	75e	38°40′N	90°19′W
University Park, Tx., U.S.	75c	32°51′N	96°48′W
Unna, Ger. (o͞o′nä)	127c	51°32′N	7°41′E
Uno, Canal Numero, can., Arg.	99c	36°43′S	58°14′W
Unterhaching, Ger. (o͞on′tĕr-hä-kēng)	115d	48°03′N	11°38′E
Ünye, Tur. (ün′yĕ)	119	41°00′N	37°10′E
Unzha, r., Russia (on′zhȧ)	136	57°45′N	44°10′E
Upa, r., Russia (o′pȧ)	132	53°54′N	36°48′E
Upata, Ven. (o͞o-pä′tä)	100	7°58′N	62°27′W
Upemba, Parc National de l′, rec., D.R.C.	191	9°10′S	26°15′E
Upington, S. Afr. (ŭp′ĭng-tŭn)	186	28°25′S	21°15′E
Upland, Ca., U.S. (ŭp′lănd)	75a	34°06′N	117°38′W
Upolu, i., Samoa	170a	13°55′S	171°45′W
Upolu Point, c., Hi., U.S. (o͞o-pô′lo͞o)	84a	20°15′N	155°48′W
Upper Arrow Lake, l., Can. (ăr′ō)	55	50°30′N	117°55′W
Upper Darby, Pa., U.S. (där′bĭ)	68f	39°58′N	75°16′W
Upper des Lacs, l., N.A. (dĕ läk)	70	48°58′N	101°55′W
Upper Kapuas Mountains, mts., Asia	168	1°45′N	112°06′E
Upper Klamath Lake, l., Or., U.S.	72	42°23′N	122°55′W
Upper Lake, l., Nv., U.S. (ŭp′ẽr)	72	41°42′N	119°59′W
Upper Marlboro, Md., U.S. (ûpẽr mărl′bōrō)	68e	38°49′N	76°46′W
Upper Mill, Wa., U.S. (mĭl)	74a	47°11′N	121°55′W
Upper Red Lake, l., Mn., U.S. (rĕd)	71	48°14′N	94°53′W
Upper Sandusky, Oh., U.S. (săn-dŭs′kė)	66	40°50′N	83°20′W
Upper San Leandro Reservoir, res., Ca., U.S. (ŭp′ẽr săn lė-ăn′drō)	74b	37°47′N	122°04′W
Upper Volta *see* Burkina Faso, nation, Afr.	184	13°00′N	2°00′W
Uppingham, Eng., U.K. (ŭp′ĭng-ăm)	114a	52°35′N	0°43′W
Uppsala, Swe. (o͞op′sȧ-lä)	110	59°53′N	17°39′E
Uptown, Ma., U.S. (ŭp′toun)	61a	42°10′N	71°36′W
Uraga, Japan (o͞o′rȧ-gȧ′)	167a	35°15′N	139°43′E
Ural, r., (o-räl′′) (ū-rôl)	134	48°00′N	51°00′E
Urals, mts., Russia	134	56°28′N	58°13′E
Uran, India (o͞o-rän′)	159b	18°53′N	72°46′E
Uranium City, Can.	50	59°34′N	108°59′W
Urawa, Japan (o͞o′rä-wä′)	166	35°52′N	139°39′E
Urayasu, Japan (o͞o′rä-yä′so͞o)	167a	35°40′N	139°54′W
Urazovo, Russia (o-rȧ′zō-vō)	133	50°08′N	38°03′E
Urbana, Il., U.S. (ûr-băn′ȧ)	66	40°10′N	88°15′W
Urbana, Oh., U.S.	66	40°05′N	83°50′W
Urbino, Italy (o͞or-bē′nō)	130	43°43′N	12°37′E
Urdaneta, Phil. (o͞or-dä-nā′tä)	169a	15°59′N	120°34′E

314 Urdinarrain – Vega PRONOUNCING INDEX

PLACE (Pronunciation)	PAGE	LAT.	LONG.
Urdinarrain, Arg. (o͞or-dē-när-rȧē′n)	99c	32°43′S	58°53′W
Uritsk, Russia (o͞o′rĭtsk)	142c	59°50′N	30°11′E
Urla, Tur. (ŏŏr′lȧ)	131	38°20′N	26°44′E
Urman, Russia (ŏŏr′mȧn)	142a	54°53′N	56°52′E
Urmi, r., Russia (ŏŏr′mė)	166	48°50′N	134°00′E
Uromi, Nig.	189	6°44′N	6°18′E
Urrao, Col. (o͞or-rȧ′ô)	100	6°19′N	76°11′W
Urshel′skiy, Russia (o͞or-shĕl′skēė)	132	55°50′N	40°11′E
Ursus, Pol.	125	52°12′N	20°53′E
Urubamba, r., Peru (o͞o-ro͞o-bäm′bä)	100	11°48′S	72°34′W
Uruguaiana, Braz.	102	29°45′S	57°00′W
Uruguay, nation, S.A. (o͞o-ro͞o-gwī′) (ū′ro͞o-gwā)	102	32°45′S	56°00′W
Uruguay, r., S.A. (o͞o-ro͞o-gwī′)	102	27°05′S	55°15′W
Ürümqi, China (ů-rům-chyē)	160	43°49′N	87°43′E
Urup, i., Russia (ŏŏ′rŏŏp′)	161	46°00′N	150°00′E
Uryupinsk, Russia (ŏŏr′yŏŏ-pēn-sk′)	137	50°50′N	42°00′E
Ürzhar, Kaz.	139	47°28′N	82°00′E
Urziceni, Rom. (ŏŏ-zē-chĕn′′)	131	44°45′N	26°42′E
Usa, Japan	166	33°31′N	131°22′E
Usa, r., Russia (ŏŏ′sȧ)	136	66°00′N	58°20′E
Uşak, Tur. (o͞o′shȧk)	119	38°45′N	29°15′E
Usakos, Nmb. (o͞o-sä′kōs)	186	22°00′S	15°40′E
Usambara Mountains, mts., Tan.	191	4°40′S	38°25′E
Usangu Flats, sw., Tan.	191	8°10′S	34°00′E
Ushaki, Russia (o͞o′shá-kĭ)	142c	59°28′N	31°00′E
Ushakovskoye, Russia (ŏŏ-shá-kôv′skô-yĕ)	142a	56°18′N	62°23′E
Ushashi, Tan.	191	2°00′S	33°57′E
Ushiku, Japan (o͞o′shė-ko͞o)	167a	35°24′N	140°09′E
Ushimado, Japan (o͞o′shė-mä′dō)	167	34°37′N	134°09′E
Ushuaia, Arg. (o͞o-sho͞o-ī′ä)	102	54°46′S	68°24′W
Usman′, Russia (o͞os-mȧn′)	137	52°03′N	39°40′E
Usol′ye, Russia (ŏŏ-sô′lyĕ)	142a	59°24′N	56°40′E
Usol′ye-Sibirskoye, Russia (ŏŏ-sô′lyĕsĭ′ bėr′skô-yĕ)	140	52°44′N	103°46′E
Uspallata Pass, p., S.A. (o͞os-pä-lyä′tä)	102	32°47′S	70°08′W
Uspanapa, r., Mex. (o͞os-pä-nä′pä)	89	17°43′N	94°14′W
Ussel, Fr. (üs′ĕl)	126	45°33′N	2°17′E
Ussuri, r., Asia (o͞o-so͞o′rė)	141	47°30′N	134°00′E
Ussuriysk, Russia	135	43°48′N	132°09′E
Ust′-Bol′sheretsk, Russia	135	52°41′N	157°00′E
Ustica, Isola di, i., Italy	130	38°43′N	12°11′E
Ústí nad Labem, Czech Rep.	124	50°40′N	14°02′E
Ust′-Izhora, Russia (ŏŏst-ēz′hô-rȧ)	142c	59°49′N	30°35′E
Ustka, Pol. (o͞ost′kȧ)	124	54°34′N	16°52′E
Ust′-Kamchatsk, Russia	135	56°13′N	162°18′E
Ust′-Katav, Russia (ŏŏst kȧ′táf)	142a	54°55′N	58°12′E
Ust′-Kishert′, Russia (ŏŏst kē′shĕrt)	142a	57°21′N	57°13′E
Ust′-Kulom, Russia (kŏŏ′lŭm)	134	61°38′N	54°00′E
Ust′-Maya, Russia (mȧ′yȧ)	135	60°33′N	134°43′E
Ust′ Olenëk, Russia	135	72°52′N	120°15′E
Ust-Ordynskiy, Russia (ŏŏst-ôr-dyēnsk′ĭ)	140	52°47′N	104°39′E
Ust′ Penzhino, Russia	141	63°00′N	165°10′E
Ust′ Port, Russia (ŏŏst′pôrt′)	134	69°20′N	83°41′E
Ust′-Tsil′ma, Russia (tsĭl′mȧ)	134	65°25′N	52°10′E
Ust′-Tyrma, Russia (tur′mȧ)	135	50°27′N	131°17′E
Ust′ Uls, Russia	142a	60°35′N	58°32′E
Ust-Urt, Plateau, plat., Asia	134	44°03′N	54°58′E
Ustynivka, Ukr.	133	47°59′N	32°31′E
Ustyuzhna, Russia (yo͞ozh′nȧ)	136	58°49′N	36°19′E
Usu, China (ů-so͞o)	160	44°28′N	84°07′E
Usuki, Japan (o͞o′so͞o-kė′)	167	33°06′N	131°47′E
Usulutan, El Sal. (o͞o-so͞o-lä-tän′)	90	13°22′N	88°25′W
Usumacinta, r., N.A. (o͞o′so͞o-mä-sēn′tō)	89	18°24′N	92°30′W
Us′va, Russia (o͞os′vȧ)	142a	58°41′N	57°38′E
Utah, state, U.S. (ū′tô)	64	39°25′N	112°40′W
Utah Lake, l., Ut., U.S.	77	40°10′N	111°55′W
Utan, India	159b	19°17′N	72°43′E
Ute Mountain Indian Reservation, I.R., N.M., U.S.	77	36°57′N	108°34′W
Utena, Lith. (o͞o′tā-nä)	123	55°32′N	25°40′E
Utete, Tan. (o͞o-tā′tå)	187	8°05′S	38°47′E
Utica, In., U.S. (ū′tĭ-kȧ)	69h	38°20′N	85°39′W
Utica, N.Y., U.S.	65	43°05′N	75°10′W
Utiel, Spain (o͞o-tyål′)	128	39°34′N	1°13′W
Utika, Mi., U.S. (ū′tĭ-kȧ)	69b	42°37′N	83°02′W
Utik Lake, l., Can.	57	55°16′N	96°00′W
Utikuma Lake, l., Can.	55	55°50′N	115°25′W
Utila, i., Hond. (o͞o-tē′lä)	90	16°07′N	87°05′W
Uto, Japan (o͞o′tō′)	166	32°43′N	130°39′E
Utrecht, Neth. (ü′trĕĸt) (ū′trĕkt)	117	52°05′N	5°06′E
Utrera, Spain (o͞o-trā′rä)	118	37°12′N	5°48′W
Utsunomiya, Japan (o͞ot′sŏŏ-nŏ-mē-yȧ′)	161	36°35′N	139°52′E
Uttaradit, Thai.	168	17°47′N	100°10′E
Uttaranchal, state, India	155	29°30′N	78°30′E
Uttarpara-Kotrung, India	158a	22°40′N	88°21′E
Uttar Pradesh, state, India (ŏŏt-tär-prä-dĕsh)	155	27°00′N	80°00′E
Uttoxeter, Eng., U.K. (ŭt-tôk′sė-tẽr)	114a	52°54′N	1°52′W
Utuado, P.R. (o͞o-to͞o-ä′dhō)	87b	18°16′N	66°40′W
Uusikaupunki, Fin.	123	60°48′N	21°24′E
Uvalde, Tx., U.S. (ū-văl′dė)	80	29°14′N	99°47′W
Uvel′skiy, Russia (ŏŏ-vyĕl′skĭ)	142a	54°27′N	61°22′E
Uvinza, Tan.	191	5°06′S	30°22′E
Uvira, D.R.C. (o͞o-vē′rä)	186	3°28′S	29°03′E
Uvod′, r., Russia (ŏŏ-vôd′)	132	56°40′N	41°10′E
Uvongo Beach, S. Afr.	187c	30°49′S	30°23′E
Uvs Nuur, l., Asia	160	50°29′N	93°32′E
Uwajima, Japan (o͞o-wä′jė-mä)	166	33°12′N	132°35′E
Uxbridge, Ma., U.S. (ŭks′brĭj)	61a	42°05′N	71°38′W
Uxmal, hist., Mex. (o͞o′x-mä′l)	90a	20°22′N	89°44′W
Uy, r., Russia (o͞oy)	142a	54°05′N	62°11′E
Uyskoye, Russia (ûy′skô-yĕ)	142a	54°22′N	60°01′E
Uyuni, Bol. (o͞o-yo͞o′nė)	100	20°28′S	66°45′W
Uyuni, Salar de, pl., Bol. (sä-lär-dĕ)	100	20°58′S	67°09′W
Uzbekistan, nation, Asia	134	42°42′N	60°00′E
Uzh, r., Ukr. (ŏŏzh)	133	51°07′N	29°05′E
Uzhhorod, Ukr.	125	48°38′N	22°18′E
Užice, Serb. (o͞o′zhė-tsĕ)	131	43°51′N	19°53′E
Uzunköprü, Tur.	131	41°17′N	26°42′E

V

PLACE (Pronunciation)	PAGE	LAT.	LONG.
Vaal, r., S. Afr. (väl)	186	28°15′S	24°30′E
Vaaldam, res., S. Afr.	192c	26°58′S	28°37′E
Vaalplaas, S. Afr.	192c	25°39′S	28°56′E
Vaalwater, S. Afr.	192c	24°17′S	28°08′E
Vaasa, Fin. (vä′sȧ)	110	63°06′N	21°39′E
Vác, Hung. (väts)	125	47°46′N	19°10′E
Vache, Île à, i., Haiti	93	18°05′N	73°40′W
Vadstena, Swe. (väd′stĭ′nȧ)	122	58°27′N	14°53′E
Vaduz, Liech. (vä′dŏŏts)	124	47°10′N	9°32′E
Vaga, r., Russia (va′gȧ)	136	61°55′N	42°30′E
Vah, r., Slvk. (väĸ)	117	48°07′N	17°52′E
Vaigai, r., India	159	10°20′N	78°13′E
Vakh, r., Russia (vȧĸ)	140	61°30′N	81°33′E
Valachia, hist. reg., Rom.	131	44°45′N	24°17′E
Valcartier-Village, Can. (väl-kärt-yĕ′vē-läzh′)	62b	46°56′N	71°28′W
Valdai Hills, hills, Russia (vȧl-dī′ gô′rĭ)	136	57°50′N	32°35′E
Valday, Russia (vȧl-dī′)	136	57°58′N	33°13′E
Valdecañas, Embalse de, res., Spain	128	39°45′N	5°30′W
Valdemārpils, Lat.	123	57°22′N	22°34′E
Valdemorillo, Spain (väl-då-mŏ-rēl′yō)	129a	40°30′N	4°04′W
Valdepeñas, Spain (väl-då-pān′yäs)	118	38°46′N	3°22′W
Valderaduey, r., Spain (väl-dĕ-rä-dwĕ′y)	128	41°39′N	5°35′W
Valdés, Península, pen., Arg. (vȧl-dĕ′s)	102	42°15′S	63°15′W
Valdez, Ak., U.S. (văl′dĕz)	63	61°10′N	146°18′W
Valdilecha, Spain (väl-dė-lā′chä)	129a	40°17′N	3°19′W
Valdivia, Chile (väl-dė′vä)	102	39°47′S	73°13′W
Valdivia, Col. (väl-dė′vēä)	100a	7°10′N	75°26′W
Val-d′Or, Can.	51	48°03′N	77°50′W
Valdosta, Ga., U.S. (văl-dŏs′tȧ)	65	30°50′N	83°18′W
Vale, Or., U.S. (vāl)	72	43°59′N	117°14′W
Valença, Braz. (vä-lĕɴ′sȧ)	101	13°43′S	38°58′W
Valença, Port.	128	42°03′N	8°36′W
Valence, Fr. (vä-lĕɴɴs)	117	44°56′N	4°54′E
Valencia, Spain	110	39°26′N	0°23′W
Valencia, Ven. (vä-lĕn′syä)	100	10°11′N	68°00′W
Valencia, hist. reg., Spain	129	39°08′N	0°43′W
Valencia, Golf de, b., Spain	129	39°50′N	0°30′E
Valencia, Lago de, l., Ven.	101b	10°11′N	67°45′W
Valencia de Alcántara, Spain	128	39°34′N	7°13′W
Valenciennes, Fr. (vȧ-läɴ-syĕn′)	126	50°24′N	3°36′E
Valentine, Ne., U.S. (vȧ läɴ-tė-nyē′)	64	42°52′N	100°34′W
Valera, Ven. (vä-lĕ′rä)	100	9°12′N	70°45′W
Valerianovsk, Russia (vȧ-lĕ-rĭ-ä′nôvsk)	142a	58°47′N	59°34′E
Valga, Est. (väl′gȧ)	136	57°47′N	26°03′E
Valhalla, S. Afr. (vȧl-hăl-ȧ)	187b	25°49′S	28°09′E
Valier, Mt., U.S. (vă-lēr′)	73	48°17′N	112°14′W
Valjevo, Serb. (väl′yå-vŏ)	131	44°17′N	19°57′E
Valky, Ukr.	133	49°49′N	35°40′E
Valladolid, Mex. (väl-yä-dhŏ-lēdh′)	86	20°39′N	88°13′W
Valladolid, Spain (väl-yä-dhŏ-lēdh′)	110	41°41′N	4°41′W
Valle, Arroyo del, Ca., U.S. (ä-rō′yō dĕl väl′yå)	76	37°36′N	121°43′W
Vallecas, Spain (väl-yā′käs)	129a	40°23′N	3°37′W
Valle de Allende, Mex. (väl′yā dā äl-yĕn′då)	80	26°55′N	105°25′W
Valle de Bravo, Mex. (brä′vŏ)	88	19°12′N	100°07′W
Valle de Guanape, Ven. (vä′l-yĕ-dĕ-gwä-nä′pĕ)	101b	9°54′N	65°41′W
Valle de la Pascua, Ven. (lä-pä′s-ko͞oä)	100	9°12′N	65°08′W
Valle del Cauca, dept., Col. (vä′l-yĕ del kou′kä)	100a	4°03′N	76°13′W
Valle de Santiago, Mex. (sän-tė-ä′gŏ)	88	20°23′N	101°11′W
Valledupar, Col. (do͞o-pär′)	100	10°13′N	73°39′W
Valle Grande, Bol. (grän′dā)	100	18°27′S	64°03′W
Vallejo, Ca., U.S. (vä-yā′hō) (vă-lā′hō)	64	38°06′N	122°15′W
Vallejo, Sierra de, mts., Mex. (sē-ĕ′r-rä-dĕ-väl-yĕ′ĸŏ)	88	21°00′N	105°10′W
Vallenar, Chile (väl-yå-när′)	102	28°39′S	70°52′W
Valles, Mex.	86	21°59′N	99°02′W
Valletta, Malta (väl-lĕt′ä)	118	35°50′N	14°29′E
Valle Vista, Ca., U.S. (väl′yā vĭs′tȧ)	75a	33°45′N	116°53′W
Valley City, N.D., U.S.	64	46°55′N	97°59′W
Valley City, Oh., U.S. (văl′ĭ)	69d	41°14′N	81°56′W
Valley Falls, Ks., U.S.	79	39°25′N	95°26′W
Valleyfield, Can. (văl′ė-fēld)	51	45°16′N	74°09′W
Valley Park, Mo., U.S. (văl′ė pärk)	75e	38°33′N	90°30′W
Valley Stream, N.Y., U.S. (văl′ĭ strēm)	68a	40°39′N	73°42′W
Valli di Comácchio, l., Italy (vä′lē-dē-kô-mȧ′chyŏ)	130	44°38′N	12°15′E
Vallière, Haiti (väl-yâr′)	93	19°30′N	71°55′W
Vallimanca, r., Arg. (väl-yē-mä′n-kä)	99c	36°21′S	60°55′W
Valls, Spain (väls)	118	41°15′N	1°15′E
Valmiera, Lat. (vȧl′myĕ-rȧ)	136	57°34′N	25°54′E
Valognes, Fr. (vȧ-lŏn′y′)	126	49°32′N	1°30′W
Valona *see* Vlorë, Alb.	119	40°28′N	19°31′E
Valozhyn, Bela.	132	54°04′N	26°38′E
Valparaíso, Chile (väl′pä-rä-ē′sŏ)	102	33°02′S	71°32′W
Valparaíso, Mex.	88	22°49′N	103°33′W
Valparaiso, In., U.S. (văl-pȧ-rā′zŏ)	66	41°25′N	87°05′W
Valpariso, prov., Chile	99b	32°58′S	71°23′W
Valréas, Fr. (vȧl-rȧ-ä′)	126	44°25′N	4°56′E
Vals, r., S. Afr.	192c	27°32′S	26°51′E
Vals, Tanjung, c., Indon.	169	8°30′S	137°15′E
Valsbaai, b., S. Afr.	186a	34°14′S	18°35′E
Valuyevo, Russia (vȧ-lo͞o′yĕ-vô)	142b	55°34′N	37°21′E
Valuyki, Russia (vȧ-lŏŏ-ē′kė)	137	50°14′N	38°04′E
Valverde del Camino, Spain (väl-vĕr-dĕ-dĕl-kä-mė′nō)	128	37°34′N	6°44′W
Vammala, Fin.	123	61°19′N	22°51′E
Van, Tur. (vȧn)	154	38°04′N	43°10′E
Van Buren, Ar., U.S. (văn bū′rĕn)	79	35°26′N	94°20′W
Van Buren, Me., U.S.	60	47°09′N	67°58′W
Vanceburg, Ky., U.S. (văns′bûrg)	66	38°35′N	83°20′W
Vancouver, Can. (văn-ko͞o′vẽr)	50	49°16′N	123°06′W
Vancouver, Wa., U.S.	64	45°37′N	122°40′W
Vancouver Island, i., Can.	52	49°50′N	125°05′W
Vancouver Island Ranges, mts., Can.	54	49°25′N	125°25′W
Vandalia, Il., U.S. (văn-dā′lĭ-ȧ)	66	39°00′N	89°00′W
Vandalia, Mo., U.S.	79	39°19′N	91°30′W
Vanderbijlpark, S. Afr.	192c	26°43′S	27°50′E
Vanderhoof, Can.	50	54°01′N	124°01′W
Van Diemen, Cape, c., Austl. (văndē′mĕn)	174	11°05′S	130°15′E
Van Diemen Gulf, b., Austl.	174	11°50′S	131°30′E
Vanegas, Mex. (vä-nĕ′gäs)	86	23°54′N	100°54′W
Vänern, l., Swe.	112	58°52′N	13°17′E
Vänersborg, Swe. (vĕ′nẽrs-bôr′)	116	58°24′N	12°15′E
Vanga, Kenya (väŋ′gä)	187	4°38′S	39°10′E
Vangani, India	159b	19°07′N	73°15′E
Van Gölü, l., Tur.	136	38°33′N	42°46′E
Van Horn, Tx., U.S.	80	31°03′N	104°50′W
Vanier, Can.	62c	45°27′N	75°39′W
Van Lear, Ky., U.S. (văn lēr′)	66	37°45′N	82°50′W
Vannes, Fr. (vȧn)	117	47°42′N	2°46′W
Van Nuys, Ca., U.S. (văn nīz′)	75a	34°11′N	118°27′W
Van Rees, Pegunungan, mts., Indon.	169	2°30′S	138°45′E
Vantaan, r., Fin.	123	60°25′N	24°43′E
Vanua Levu, i., Fiji	170g	16°33′S	179°15′E
Vanuatu, nation, Oc.	175	16°02′S	169°15′E
Van Wert, Oh., U.S. (văn wûrt′)	66	40°50′N	84°35′W
Vara, Swe. (vä′rä)	122	58°17′N	12°55′E
Varaklāni, Lat.	123	56°38′N	26°46′E
Varallo, Italy (vä-räl′lô)	130	45°44′N	8°14′E
Vārānasi (Benares), India	155	25°25′N	83°00′E
Varangerfjorden, b., Nor.	113	70°05′N	30°20′E
Varano, Lago di, l., Italy (lä′gō-dē-vä-rä′nŏ)	130	41°52′N	15°55′E
Varaždin, Cro. (vä′räzh′dėn)	119	46°17′N	16°20′E
Varazze, Italy (vä-rät′sā)	130	44°23′N	8°34′E
Varberg, Swe. (vär′bĕrg)	122	57°06′N	12°16′E
Vardar, r., Serb. (vär′där)	131	41°40′N	21°50′E
Varėna, Lith. (vä-rä′nȧ)	123	54°16′N	24°35′E
Varennes, Can. (vȧ-rĕn′)	62a	45°41′N	73°27′W
Vareš, Bos. (vä′rĕsh)	131	44°10′N	18°20′E
Varese, Italy (vä-rā′så)	130	45°45′N	8°49′E
Varginha, Braz. (vär-zhė′n-yä)	101	21°33′S	45°25′W
Varkaus, Fin. (vär′kous)	123	62°19′N	27°51′E
Varlamovo, Russia (vár-lȧ′mô-vô)	142a	54°37′N	60°41′E
Varna, Blg. (vär′nȧ)	110	43°14′N	27°58′E
Varna, Russia	142a	53°22′N	60°59′E
Värnamo, Swe. (vĕr′nȧ-mŏ)	122	57°11′N	13°45′E
Varnsdorf, Czech Rep. (värns′dôrf)	124	50°54′N	14°36′E
Varnville, S.C., U.S. (värn′vĭl)	83	32°49′N	81°05′W
Vasa, India	159b	19°20′N	72°47′E
Vascongadas *see* Basque Provinces, hist. reg., Spain	128	43°00′N	2°46′W
Vashka, r., Russia	136	64°00′N	48°00′E
Vashon, Wa., U.S. (văsh′ŭn)	74a	47°27′N	122°28′W
Vashon Heights, Wa., U.S. (hītz)	74a	47°30′N	122°28′W
Vashon Island, i., Wa., U.S.	74a	47°27′N	122°27′W
Vaslui, Rom. (vȧs-lo͞o′ė)	125	46°39′N	27°49′E
Vassar, Mi., U.S. (văs′ẽr)	66	43°25′N	83°35′W
Vassouras, Braz. (väs-sō′räzh)	99a	22°25′S	43°40′W
Västerås, Swe. (vĕs′tĕr-ŏs)	116	59°39′N	16°30′E
Västerdalälven, r., Swe.	116	61°06′N	13°10′E
Västervik, Swe. (vĕs′tĕr-vēk)	116	57°45′N	16°35′E
Vasto, Italy (väs′tô)	118	42°06′N	12°42′E
Vasyl′kiv, Ukr.	137	50°10′N	30°22′E
Vasyugan, r., Russia (vȧs-yo͞o-gȧn′)	140	58°52′N	77°30′E
Vatican City, nation, Eur.	130	41°54′N	12°22′E
Vaticano, Cape, c., Italy (vä-tė-kä′nô)	130	38°38′N	15°52′E
Vatnajökull, ice, Ice. (vät′nȧ-yû-kŏŏl)	116	64°34′N	16°41′W
Vatomandry, Madag.	187	18°53′S	48°13′E
Vatra Dornei, Rom. (vät′rȧ dôr′nå′)	125	47°22′N	25°20′E
Vättern, l., Swe.	112	58°15′N	14°24′E
Vattholma, Swe.	122	60°01′N	17°40′E
Vaudreuil, Can. (vŏ-drû′y′)	62a	45°24′N	74°02′W
Vaugh, Wa., U.S. (vôn)	74a	47°21′N	122°47′W
Vaughan, Can.	62d	43°47′N	79°36′W
Vaughn, N.M., U.S.	78	34°37′N	105°13′W
Vaupés, r., S.A. (vȧ′o͞o-pĕ′s)	100	1°18′N	71°14′W
Vawkavysk, Bela. (vôl-kŏ-vėsk′)	125	53°11′N	24°29′E
Vaxholm, Swe. (väks′hŏlm)	122	59°26′N	18°19′E
Växjo, Swe. (vĕks′shû)	116	56°53′N	14°46′E
Vaygach, i., Russia (vī-gȧch′)	134	70°00′N	59°00′E
Veadeiros, Chapadas dos, hills, Braz. (shä-pä′däs-dŏs-vĕ-ȧ-dā′rōs)	101	14°00′S	47°00′W
Vedea, r., Rom. (vå′dyȧ)	131	44°25′N	24°45′E
Vedia, Arg. (vĕ′dyä)	99c	34°29′S	61°30′W
Veedersburg, In., U.S. (vē′dẽrz-bûrg)	66	40°05′N	87°15′W
Vega, i., Nor.	116	65°38′N	10°51′E

ăt; finăl; rāte; senåte; ärm; ȧsk; sofȧ; fâre; ch-choose; dh-as th in other; bē; ėvent; bĕt; recĕnt; cratẽr; g-gō; gh-guttural g; bĭt; ĭ-short neutral; rīde; ĸ-guttural k as ch in German ich;

PRONOUNCING INDEX Vega de Alatorre – Villa Hidalgo 315

PLACE (Pronunciation)	PAGE	LAT.	LONG.
Vega de Alatorre, Mex. (vā'gä dā ä-lä-tōr'rā)	89	20°02'N	96°39'W
Vega Real, reg., Dom. Rep. (vĕ'gä-rĕ-ä'l)	93	19°30'N	71°05'W
Vegreville, Can.	50	53°30'N	112°03'W
Vehār Lake, l., India	159b	19°11'N	72°52'E
Veinticinco de Mayo, Arg.	99c	35°26'S	60°09'W
Vejer de la Frontera, Spain	128	36°15'N	5°58'W
Vejle, Den. (vī'lĕ)	116	55°41'N	9°29'E
Velbert, Ger. (fĕl'bĕrt)	127c	51°20'N	7°03'E
Velebit, mts., Serb. (vā'lĕ-bĕt)	119	44°25'N	15°23'E
Velen, Ger. (fĕ'lĕn)	127c	51°54'N	7°00'E
Vélez-Málaga, Spain (vā'lāth-mä'lä-gä)	128	36°48'N	4°05'W
Vélez-Rubio, Spain (rōō'bē-ō)	128	37°38'N	2°05'W
Velika Kapela, mts., Serb. (vĕ'lē-kä kä-pĕ'lä)	119	45°03'N	15°20'E
Velika Morava, r., Serb. (mō'rä-vä)	119	44°00'N	21°30'E
Velikaya, r., Russia (vå-lē'kä-yà)	132	57°25'N	28°07'E
Velikiye Luki, Russia (vyē-lē'-kyĕ lōō'ke)	134	56°19'N	30°32'E
Velikiy Ustyug, Russia (vå-lē'kĭ ōōs-tyòg')	134	60°45'N	46°38'E
Veliko Tŭrnovo, Blg.	119	43°06'N	25°38'E
Velikoye, Russia (vå-lē'kô-yĕ)	132	57°21'N	39°45'E
Velikoye, l., Russia	132	57°00'N	36°53'E
Veli Lošinj, Cro. (lô'shēn')	130	44°30'N	14°29'E
Velizh, Russia (vå'lēzh)	136	55°37'N	31°11'E
Vella Lavella, i., Sol. Is.	175	8°00'S	156°42'E
Velletri, Italy (vĕl-lā'trē)	130	41°42'N	12°48'E
Vellore, India (vĕl-lōr')	155	12°57'N	79°09'E
Vels, Russia (vĕls)	142a	60°35'N	58°47'E
Vel'sk, Russia (vĕlsk)	134	61°00'N	42°18'E
Velten, Ger. (fel'tĕn)	115b	52°41'N	13°11'E
Velya, r., Russia (vĕl'yà)	142b	56°23'N	37°54'E
Velyka Lepetykha, Ukr.	133	47°11'N	33°58'E
Velykyi Bychkiv, Ukr.	125	47°59'N	24°01'E
Venadillo, Col. (vĕ-nä-dē'l-yō)	100a	4°43'N	74°55'W
Venado, Mex. (vå-mä'dō)	88	22°54'N	101°07'W
Venado Tuerto, Arg. (vĕ-nä'dô-tōōĕ'r-tô)	102	33°28'S	61°47'W
Vendôme, Fr. (väN-dōm')	126	47°46'N	1°05'E
Veneto, hist. reg., Italy (vĕ-nĕ'tô)	130	45°58'N	11°24'E
Venëv, Russia (vĕn-ĕf')	136	54°19'N	38°14'E
Venezia *see* Venice, Italy	110	45°25'N	12°18'E
Venezuela, nation, S.A. (vĕn-ĕ-zwē'lá)	100	8°00'N	65°00'W
Venezuela, Golfo de, b., S.A. (gôl-fô-dĕ)	100	11°34'N	71°02'W
Veniaminof, Mount, mtn., Ak., U.S.	63	56°12'N	159°20'W
Venice, Italy	110	45°25'N	12°18'E
Venice, Ca., U.S. (vĕn'ĭs)	75a	33°59'N	118°28'W
Venice, Il., U.S.	75e	38°40'N	90°10'W
Venice, Gulf of, b., Italy	118	45°23'N	13°00'E
Venlo, Neth.	127c	51°22'N	6°11'E
Venta, r., Eur. (vĕn'tà)	123	57°05'N	21°45'E
Ventana, Sierra de la, mts., Arg. (sē-ĕ-rá-dĕ-lä-vĕn-tä'nä)	102	38°00'S	63°00'W
Ventersburg, S. Afr. (vĕn-tĕrs'bûrg)	192c	28°06'S	27°10'E
Ventersdorp, S. Afr. (vĕn-tĕrs'dôrp)	192c	26°20'S	26°48'E
Ventimiglia, Italy (vĕn-tē-mēl'yä)	130	43°46'N	7°37'E
Ventnor, N.J., U.S. (vĕnt'nēr)	67	39°20'N	74°25'W
Ventspils, Lat. (vĕnt'spēls)	136	57°24'N	21°41'E
Ventuari, r., Ven. (vĕn-tōōä'rē)	100	4°47'N	65°56'W
Ventura, Ca., U.S. (vĕn-tōō'rá)	76	34°18'N	119°18'W
Venukovsky, Russia (vĕ-nōō'kôv-skĭ)	142b	55°10'N	37°26'E
Venustiano Carranza, Mex. (vĕ-nōōs-tyä'nō-kär-rä'n-zä)	88	19°44'N	103°48'W
Venustiano Carranzo, Mex. (kär-rä'n-zô)	89	16°21'N	92°36'W
Vera, Arg. (vĕ-rä)	102	29°22'S	60°09'W
Vera, Spain (vā'rä)	128	37°18'N	1°53'W
Veracruz, Mex.	86	19°13'N	96°07'W
Veracruz, state, Mex. (vā-rä-krōōz')	86	20°30'N	97°15'W
Verāval, India (vĕr'vŭ-väl)	155	20°59'N	70°49'E
Vercelli, Italy (vĕr-chĕl'lē)	130	45°18'N	8°27'E
Verchères, Can. (vĕr-shâr')	62a	45°46'N	73°21'W
Verde, i., Phil. (vĕr'dā)	169a	13°34'N	121°11'E
Verde, r., Mex.	88	21°48'N	99°50'W
Verde, r., Mex.	88	20°50'N	103°00'W
Verde, r., Mex.	89	16°05'N	97°44'W
Verde, r., Az., U.S. (vûrd)	77	34°04'N	111°40'W
Verde, Cap, c., Bah.	93	22°50'N	75°00'W
Verde, Cay, i., Bah.	93	22°00'N	75°05'W
Verde Island Passage, strt., Phil. (vĕr'dē)	169a	13°36'N	120°39'E
Verdemont, Ca., U.S. (vûr'dē-mŏnt)	75a	34°12'N	117°22'W
Verden, Ger. (fĕr'dĕn)	124	52°55'N	9°15'E
Verdigris, r., Ok., U.S. (vûr'dē-grēs)	79	36°50'N	95°29'W
Verdun, Can. (vĕr'dŭN')	59	45°27'N	73°34'W
Verdun, Fr. (vâr-dûN')	117	49°09'N	5°21'E
Verdun, Fr.	127	43°48'N	1°10'E
Vereeniging, S. Afr. (vĕ-rā'nĭ-gĭng)	192c	26°40'S	27°56'E
Verena, S. Afr. (vĕr-ĕn á)	192c	25°30'S	29°02'E
Vereya, Russia (vĕ-rā'yà)	132	55°21'N	36°08'E
Verín, Spain (vå-rēn')	128	41°56'N	7°26'W
Verkhne-Kamchatsk, Russia (vyĕrk'nyĕ kàm-chatsk')	135	54°42'N	158°41'E
Verkhne Neyvinskiy, Russia (nā-vĭn'skĭ)	142a	57°17'N	60°10'E
Verkhne Ural'sk, Russia (ò-ralsk')	134	53°53'N	59°13'E
Verkhniy Avzyan, Russia (vyĕrk'nyĕ àv-zyàn')	142a	53°32'N	57°30'E
Verkhniye Kigi, Russia (vyĕrk'nĭ-yĕ kĭ'gĭ)	142a	55°23'N	58°37'E
Verkhniy Ufaley, Russia (ò-fà'lā)	142a	56°04'N	60°15'E
Verkhnyaya Pyshma, Russia (vyĕrk'nyä-yä pōōsh'mà)	142a	56°57'N	60°37'E
Verkhnyaya Salda, Russia (säl'dà)	142a	58°03'N	60°33'E
Verkhnyaya Tunguska (Angara), r., Russia (tòn-gòs'kà)	140	58°13'N	97°00'E
Verkhnyaya Tura, Russia (tò'rà)	142a	58°22'N	59°51'E
Verkhnyaya Yayva, Russia (yäy'và)	142a	59°28'N	57°38'E
Verkhotur'ye, Russia (vyĕr-kô-tōōr'yĕ)	142a	58°52'N	60°47'E
Verkhoyansk, Russia (vyĕr-kô-yänsk')	135	67°43'N	133°33'E
Verkhoyanskiy Khrebet, mts., Russia (vyĕr-kô-yänskĭ)	135	67°45'N	128°00'E
Vermilion, Can. (vĕr-mĭl'yŭn)	50	53°22'N	110°51'W
Vermilion, l., Mn., U.S.	71	47°49'N	92°35'W
Vermilion, r., Can.	59	47°30'N	73°15'W
Vermilion, r., Can.	56	53°30'N	111°00'W
Vermilion, r., Il., U.S.	66	41°05'N	89°00'W
Vermilion, r., Mn., U.S.	71	48°09'N	92°31'W
Vermilion Hills, hills, Can.	56	50°43'N	106°50'W
Vermilion Range, mts., Mn., U.S.	71	47°55'N	91°59'W
Vermillion, S.D., U.S.	70	42°46'N	96°56'W
Vermillion, r., S.D., U.S.	70	43°54'N	97°14'W
Vermillion Bay, b., La., U.S.	81	29°47'N	92°00'W
Vermont, state, U.S. (vĕr-mŏnt')	65	43°50'N	72°50'W
Vernal, Ut., U.S. (vûr'nál)	73	40°29'N	109°40'W
Verneuk Pan, pl., S. Afr. (vĕr-nŭk')	186	30°10'S	21°46'E
Vernon, Can. (vĕr-nôN')	50	50°18'N	119°15'W
Vernon, Can.	62c	45°10'N	75°27'W
Vernon, Ca., U.S. (vûr'nŭn)	75a	34°01'N	118°12'W
Vernon, In., U.S. (vûr'nŭn)	66	39°00'N	85°40'W
Vernon, N.J., U.S.	68a	39°00'N	85°40'W
Vernon, Tx., U.S.	78	34°09'N	99°16'W
Vernonia, Or., U.S. (vûr-nō'nyá)	74c	45°52'N	123°12'W
Vero Beach, Fl., U.S. (vē'rô)	83a	27°36'N	80°25'W
Véroia, Grc.	131	40°30'N	22°13'E
Verona, Italy (vā-rō'nä)	118	45°28'N	11°02'E
Versailles, Fr. (vĕr-sī'y')	117	48°48'N	2°07'E
Versailles, Ky., U.S. (vĕr-sālz')	66	38°05'N	84°45'W
Versailles, Mo., U.S.	79	38°27'N	92°52'W
Vert, Cap, c., Sen.	184	14°43'N	17°30'W
Verulam, S. Afr. (vĕ-rōō-lăm)	187c	29°39'S	31°08'E
Verviers, Bel. (vĕr-vyā')	121	50°35'N	5°57'E
Vesele, Ukr.	133	46°59'N	34°56'E
Vesijärvi, l., Fin.	123	61°09'N	25°10'E
Vesoul, Fr. (vē-sōōl')	127	47°38'N	6°11'E
Vestavia Hills, Al., U.S.	68h	33°26'N	86°46'W
Vesterålen, is., Nor. (vĕs'tĕr ô'lĕn)	116	68°54'N	14°03'E
Vestfjord, b., Nor.	112	67°33'N	12°59'E
Vestmannaeyjar, Ice. (vĕst'män-ä-ā'yär)	116	63°12'N	20°17'W
Vesuvio, vol., Italy (vĕ-sōō'vyä)	112	40°35'N	14°26'E
Ves'yegonsk, Russia (vĕ-syĕ-gônsk')	132	58°42'N	37°09'E
Veszprem, Hung. (vĕs'prām)	125	47°05'N	17°53'E
Vészto, Hung. (vĕs'tû)	125	46°55'N	21°18'E
Vet, r., S. Afr. (vĕt)	192c	28°25'S	26°37'E
Vetlanda, Swe. (vĕt-län'dä)	122	57°26'N	15°05'E
Vetluga, Russia (vyĕt-lōō'gà)	136	57°50'N	45°42'E
Vetluga, r., Russia	136	56°50'N	45°50'E
Vetovo, Blg. (vā'tô-vô)	131	43°42'N	26°18'E
Vetren, Blg. (vĕt'rĕn')	131	42°16'N	24°04'E
Vevay, In., U.S. (vē'vā)	66	38°45'N	85°05'W
Veynes, Fr. (vän'')	127	44°31'N	5°47'E
Vézère, r., Fr. (vā-zer')	126	45°01'N	1°00'E
Viacha, Bol. (vēä'chà)	100	16°43'S	68°16'W
Viadana, Italy (vē-ä-dä'nä)	130	44°55'N	10°30'E
Vian, Ok., U.S. (vī'ăn)	79	35°30'N	95°00'W
Viana, Braz. (vē-ä'nä)	101	3°09'S	44°44'W
Viana do Alentejo, Port. (vē-ä'ná dō ä-lĕn-tā'hô)	128	38°20'N	8°02'W
Viana do Bolo, Spain	128	42°10'N	7°07'W
Viana do Castelo, Port. (dò käs-tā'lò)	118	41°41'N	8°45'W
Viangchan, Laos	168	18°07'N	102°33'E
Viar, r., Spain (vē-ä'rä)	128	38°15'N	6°08'W
Viareggio, Italy (vē-ä-rĕd'jô)	130	43°52'N	10°14'E
Viborg, Den. (vē'bôr)	122	56°27'N	9°22'E
Vibo Valentia, Italy (vē'bô-vä-lĕ'n-tyä)	130	38°47'N	16°06'E
Vic, Spain	129	41°55'N	2°14'E
Vicálvaro, Spain	129a	40°25'N	3°37'W
Vicente López, Arg. (vē-sĕ'n-tĕ-lŏ'pĕz)	102a	34°31'S	58°29'W
Vicenza, Italy (vē-chĕnt'sä)	118	45°33'N	11°33'E
Vichuga, Russia (vē-chōō'gà)	136	57°13'N	41°58'E
Vichy, Fr. (vē-shē')	117	46°06'N	3°28'E
Vickersund, Nor.	122	60°00'N	9°59'E
Vicksburg, Mi., U.S. (vĭks'bûrg)	66	42°10'N	85°30'W
Vicksburg, Ms., U.S.	65	32°20'N	90°50'W
Viçosa, Braz. (vē-sô'sä)	99a	20°46'S	42°51'W
Victoria, Arg. (vēk-tô'rēä)	102	32°36'S	60°09'W
Victoria, Can. (vĭk-tō'rĭ-á)	50	48°26'N	123°23'W
Victoria, Chile (vēk-tô-rēä)	102	38°15'S	72°16'W
Victoria, Col. (vēk-tô'rēä)	100a	5°19'N	74°54'W
Victoria, Phil. (vēk-tô-ryä)	169a	15°34'N	120°41'E
Victoria, Tx., U.S. (vĭk-tō'rĭ-á)	81	28°48'N	97°00'W
Victoria, Va., U.S.	83	36°57'N	78°13'W
Victoria, state, Austl.	175	36°46'S	143°15'E
Victoria, l., Afr.	186	0°50'S	32°50'E
Victoria, r., Austl.	174	17°25'S	130°50'E
Victoria, Mount, mtn., Mya.	155	21°26'N	93°59'E
Victoria, Mount, mtn., Pap. N. Gui.	169	9°35'S	147°45'E
Victoria de las Tunas, Cuba (vēk-tō'rē-ä dā läs tōō'näs)	92	20°55'N	77°05'W
Victoria Falls, wtfl., Afr.	186	17°55'S	25°51'E
Victoria Island, i., Can.	49	70°13'N	107°45'W
Victoria Lake, l., Can.	61	48°20'N	57°40'W
Victoria Land, reg., Ant.	178	75°00'S	160°00'E
Victoria Nile, r., Ug.	191	2°20'N	31°35'E
Victoria Peak, mtn., Belize (vēk-tōrĭ'á)	90a	16°47'N	88°40'W
Victoria Peak, mtn., Can.	54	50°03'N	126°06'W
Victoria River Downs, Austl. (vĭc-tōr'ĭá)	174	16°30'S	131°10'E
Victoria Strait, strt., Can. (vĭk-tō'rĭ-á)	52	69°10'N	100°58'W
Victoriaville, Can. (vĭk-tō'rĭ-á-vĭl)	51	46°04'N	71°59'W
Victoria West, S. Afr. (wĕst)	186	31°25'S	23°10'E
Vidalia, Ga., U.S. (vĭ-dā'lĭ-á)	83	32°10'N	82°26'W
Vidalia, La., U.S.	81	31°33'N	91°28'W
Vidin, Blg. (vĭ'dĭn)	119	44°00'N	22°53'E
Vidnoye, Russia	142b	55°33'N	37°41'E
Vidzy, Bela. (vē'dzĭ)	132	55°23'N	26°46'E
Viedma, Arg. (vyād'mä)	102	40°55'S	63°03'W
Viedma, l., Arg.	102	49°40'S	72°35'W
Viejo, r., Nic. (vyā'hō)	90	12°45'N	86°19'W
Vienna (Wien), Aus.	110	48°13'N	16°22'E
Vienna, Ga., U.S. (vē-ĕn'á)	82	32°03'N	83°50'W
Vienna, Il., U.S.	79	37°24'N	88°50'W
Vienna, Va., U.S.	68e	38°54'N	77°16'W
Vienne, Fr. (vyĕn')	117	45°31'N	4°54'E
Vienne, r., Fr.	126	47°06'N	0°20'E
Vientiane *see* Viangchan, Laos	168	18°07'N	102°33'E
Vieques, P.R. (vyā'kås)	87b	18°09'N	65°27'W
Vieques, i., P.R. (vyā'kås)	87b	18°05'N	65°28'W
Vierfontein, S. Afr. (vĕr'fōn-tān)	192c	27°06'S	26°45'E
Viersen, Ger. (fēr'zĕn)	127c	51°15'N	6°24'E
Vierwaldstätter See, l., Switz.	124	46°54'N	8°36'E
Vierzon, Fr. (vyâr-zôN')	117	47°14'N	2°04'E
Viesca, Mex. (vē-ās'kä)	80	25°21'N	102°47'W
Viesca, Laguna de, l., Mex. (lä-ò'nä-dĕ)	80	25°30'N	102°40'W
Vieste, Italy (vyĕs'tå)	130	41°52'N	16°10'E
Vietnam, nation, Asia (vyĕt'näm')	168	18°00'N	107°00'E
Vigan, Phil. (vēgän)	168	17°36'N	120°22'E
Vigevano, Italy (vē-jå-vä'nô)	130	45°18'N	8°52'E
Vigny, Fr. (vēn-y'ē')	127b	49°05'N	1°54'E
Vigo, Spain (vē'gō)	110	42°18'N	8°42'W
Vihti, Fin. (vē'tĭ)	123	60°27'N	24°18'E
Vijayawāda, India	155	16°31'N	80°37'E
Viksøyri, Nor.	122	61°06'N	6°35'E
Vila Caldas Xavier, Moz.	191	15°59'S	34°12'E
Vila de Manica, Moz. (vē'lä dā mä-nē'kà)	186	18°48'S	32°49'E
Vila de Rei, Port. (vē'lá dā rā'ĭ)	128	39°42'N	8°03'W
Vila do Conde, Port. (vē'lä dò kôn'dē)	128	41°21'N	8°44'W
Vilafranca del Penedès, Spain	129	41°20'N	1°40'E
Vilafranca de Xira, Port. (fräŋ'kà dā shē'rà)	128	38°58'N	8°59'W
Vilaine, r., Fr. (vē-lân')	126	47°34'N	2°15'W
Vilalba, Spain	128	43°18'N	7°43'W
Vilanculos, Moz. (vē-län-kōō'lôs)	186	22°03'S	35°13'E
Vilāni, Lat. (vē'lá-nĭ)	123	56°31'N	27°00'E
Vila Nova de Foz Côa, Port. (nō'và dā fôz-kō'á)	128	41°08'N	7°11'W
Vila Nova de Gaia, Port. (vē'lä nō'và dā gä'yä)	128	41°08'N	8°40'W
Vila Nova de Milfontes, Port. (nō'và dā mēl-fôn'tāzh)	128	37°44'N	8°48'W
Vila Real, Port. (rā-äl')	118	41°18'N	7°48'W
Vila-real, Spain	129	39°55'N	0°07'W
Vila Real de Santo Antonio, Port.	128	37°14'N	7°25'W
Vila Viçosa, Port. (vē-sō'zà)	128	38°47'N	7°24'W
Vileyka, Bela. (vē-lā'ē-kà)	132	54°19'N	26°58'E
Vilhelmina, Swe.	116	64°37'N	16°30'E
Viljandi, Est. (vēl'yän-dē)	136	58°24'N	25°34'E
Viljoenskroon, S. Afr.	192c	27°13'S	26°58'E
Vilkaviškis, Lith. (vēl-kà-vēsh'kēs)	123	54°40'N	23°08'E
Vil'kitskogo, i., Russia (vyl-kēts-kōgō)	140	73°25'N	76°00'E
Villa Acuña, Mex. (vēl'yä-kōō'n-yä)	80	29°20'N	100°56'W
Villa Ahumada, Mex. (ä-ōō-mä'dä)	80	30°43'N	106°30'W
Villa Alta, Mex. (äl'tä)(sän ēl-då-fōn'sō)	89	17°20'N	96°08'W
Villa Angela, Arg. (vē'l-yä á'n-ĸē-lä)	102	27°31'S	60°42'W
Villa Ballester, Arg. (vē'l-yä-bál-yĕs-tĕr)	102a	34°33'S	58°33'W
Villa Bella, Bol. (bē'l-yä)	100	10°25'S	65°22'W
Villablino, Spain (vēl-yä-blē'nō)	128	42°58'N	6°18'W
Villacañas, Spain (vēl-yä-kän'yäs)	128	39°39'N	3°20'W
Villacarrillo, Spain (vēl-yä-kä-rēl'yō)	128	38°09'N	3°07'W
Villach, Aus. (fē'läĸ)	117	46°38'N	13°50'E
Villacidro, Italy (vē-lä-chē'drô)	130	39°28'N	8°41'E
Villa Clara, prov., Cuba	92	22°40'N	80°10'W
Villa Constitución, Arg. (kōn-stē-tōō-syōn')	99c	33°15'S	60°19'W
Villa Coronado, Mex. (kō-rō-nä'dhō)	80	26°45'N	105°10'W
Villa Cuauhtémoc, Mex. (vēl'yä-kōō-äò-tē'mōk)	89	22°11'N	97°50'W
Villa de Allende, Mex. (vēl'yä'dā äl-yĕn'dā)	80	25°18'N	100°01'W
Villa de Alvarez, Mex. (vēl'yä-dĕ-ä'l-vä-rĕz)	88	19°17'N	103°44'W
Villa de Cura, Ven. (dĕ-kōō'rä)	101b	10°03'N	67°29'W
Villa de Guadalupe, Mex. (dĕ-gwä-dhä-lōō'pā)	88	23°22'N	100°44'W
Villa de Mayo, Arg.	102a	34°31'S	58°41'W
Villa Dolores, Arg. (vēl'yä dô-lō'rās)	102	31°50'S	65°05'W
Villa Escalante, Mex. (vēl'yä-ĕs-kä-lán'tĕ)	88	19°24'N	101°36'W
Villa Flores, Mex. (vēl'yä-flō'rās)	89	16°13'N	93°17'W
Villafranca, Italy (vēl-lä-fräŋ'kä)	130	45°22'N	10°53'E
Villafranca del Bierzo, Spain	128	42°37'N	6°49'W
Villafranca de los Barros, Spain	128	38°34'N	6°22'W
Villafranche-de-Rouergue, Fr. (dē-rōō-ĕrg')	126	44°21'N	2°02'E
Villa García, Mex. (gär-sē'ä)	88	22°07'N	101°55'W
Villagarcía, Spain	128	42°38'N	8°43'W
Villagrán, Mex.	80	24°28'N	99°30'W
Villa Grove, Il., U.S. (vĭl'á grōv')	66	39°55'N	88°15'W
Villaguay, Arg. (vē'l-yä-gwī)	102	31°47'S	58°53'W
Villa Hayes, Para. (vēl'yä äyås)(hāz)	102	25°07'S	57°31'W
Villahermosa, Mex. (vēl'yä-ĕr-mō'sä)	86	17°59'N	92°56'W
Villa Hidalgo, Mex. (vēl'yäē-däl'gō)	88	21°39'N	102°41'W

316 Villaldama – Waipahu PRONOUNCING INDEX

PLACE (Pronunciation)	PAGE	LAT.	LONG.
Villaldama, Mex. (vēl-yäl-dä′mä)	86	26°30′N	100°26′W
Villa Lopez, Mex. (vēl′yä lō′pĕz)	80	27°00′N	105°02′W
Villalpando, Spain (vēl-yäl-pän′dō)	128	41°54′N	5°24′W
Villa María, Arg. (vē′l-yä-mä-rē′ä)	102	32°17′S	63°08′W
Villamatín, Spain (vēl-yä-mä-tē′n)	128	36°50′N	5°38′W
Villa Mercedes, Arg. (mĕr-sā′dås)	102	33°38′S	65°16′W
Villa Montes, Bol. (vē′l-yä-mô′n-tĕs)	100	21°13′S	63°26′W
Villa Morelos, Mex. (mô-rĕ′lomcs)	88	20°01′N	101°24′W
Villanueva, Col. (vē′l-yä-nòĕ′vä)	100	10°44′N	73°08′W
Villanueva, Hond. (vēl′yä-nwä′vä)	90	15°19′N	88°02′W
Villanueva, Mex. (vēl′yä-nòĕ′vä)	88	22°25′N	102°53′W
Villanueva de Córdoba, Spain (vēl-yä-nwĕ′vä-dā kôr′dô-bä)	128	38°18′N	4°38′W
Villanueva de la Serena, Spain (lä sā-rā′nä)	128	38°59′N	5°56′W
Villa Obregón, Mex. (vē′l-yä-ô-brĕ-gô′n)	89a	19°21′N	99°11′W
Villa Ocampo, Mex. (ô-käm′pō)	80	26°26′N	105°30′W
Villa Pedro Montoya, Mex. (vēl′yä-pĕ′drô-môn-tô′yá)	88	21°38′N	99°51′W
Villard-Bonnot, Fr. (vēl-yär′bôn-nō′)	127	45°15′N	5°53′E
Villarrica, Para. (vēl-yä-rē′kä)	102	25°55′S	56°23′W
Villarrobledo, Spain (vēl-yär-rô-blā′dhō)	118	39°15′N	2°37′W
Villa Unión, Mex. (vēl′yä-ōō-nyōn′)	88	23°10′N	106°14′W
Villavicencio, Col. (vē′l-yä-vē-sĕ′n-syō)	100	4°09′N	73°38′W
Villaviciosa de Odón, Spain	129a	40°22′N	3°38′W
Villavieja, Col. (vē′l-yä-vē-ĕ′kà)	100a	3°13′N	75°13′W
Villazón, Bol. (vē′l-yä-zô′n)	100	22°02′S	65°42′W
Villefranche, Fr.	117	45°59′N	4°43′E
Villejuif, Fr. (vēl′zhüst′)	127b	48°48′N	2°22′E
Ville-Marie, Can.	51	47°18′N	79°22′W
Villena, Spain (vē-lyā′nà)	118	38°37′N	0°52′W
Villeneuve, Can. (vēl′nûv′)	62g	53°40′N	113°49′W
Villeneuve-Saint Georges, Fr. (săn-zhôrzh′)	127b	48°43′N	2°27′E
Villeneuve-sur-Lot, Fr. (sür-lō′)	126	44°25′N	0°41′E
Ville Platte, La., U.S. (vēl plát′)	81	30°41′N	92°17′W
Villers Cotterêts, Fr. (vē-âr′kô-trā′)	127b	49°15′N	3°05′E
Villerupt, Fr. (vēl′rüp′)	127	49°28′N	6°16′E
Ville-Saint Georges, Can. (vīl-sĕn-zhôrzh′)	59	46°07′N	70°40′W
Villeta, Col. (vē′l-yĕ′tä)	100a	5°02′N	74°29′W
Villeurbanne, Fr. (vēl-ûr-bän′)	117	45°43′N	4°55′E
Villiers, S. Afr. (vĭl′ĭ-ērs)	192c	27°03′S	28°38′E
Villingen-Schwenningen, Ger.	124	48°04′N	8°33′E
Villisca, Ia., U.S. (vĭ′lĭs′kà)	71	40°56′N	94°56′W
Villupuram, India	159	11°59′N	79°33′E
Vilnius, Lith. (vĭl′nē-ōs)	134	54°40′N	25°26′E
Vilppula, Fin. (vĭl′pŭ-là)	123	62°01′N	24°24′E
Vil′shanka, Ukr.	133	48°14′N	30°52′E
Vil′shany, Ukr.	133	50°02′N	35°54′E
Vilvoorde, Bel.	115a	50°56′N	4°25′E
Vilyuy, r., Russia (vēl′yĭ)	135	63°00′N	121°00′E
Vilyuysk, Russia (vē-lyōō′ĭsk′)	135	63°41′N	121°47′E
Vimmerby, Swe. (vĭm′ĕr-bü)	122	57°41′N	15°51′E
Vimperk, Czech Rep. (vĭm-pĕrk′)	124	49°04′N	13°41′E
Viña del Mar, Chile (vē′nyä dĕl mär′)	102	33°00′S	71°33′W
Vinalhaven, Me., U.S. (vī-năl-hā′vĕn)	60	44°03′N	68°49′W
Vinaròs, Spain	129	40°29′N	0°27′E
Vincennes, Fr. (văn-sĕn′)	127b	48°51′N	2°27′E
Vincennes, In., U.S. (vĭn-zĕnz′)	65	38°40′N	87°30′W
Vincent, Al., U.S. (vĭn′sĕnt)	82	33°21′N	86°25′W
Vindelälven, r., Swe.	116	65°02′N	18°30′E
Vindeln, Swe. (vĭn′dĕln)	116	64°10′N	19°52′E
Vindhya Range, mts., India (vĭnd′yä)	155	22°30′N	75°50′E
Vineland, N.J., U.S. (vīn′lănd)	67	39°30′N	75°00′W
Vinh, Viet. (vēn′y′)	168	18°38′N	105°42′E
Vinhais, Port. (vēn-yä′ēzh)	128	41°51′N	7°00′W
Vinings, Ga., U.S. (vī′nĭngz)	68c	33°52′N	84°28′W
Vinita, Ok., U.S. (vĭ-nē′tà)	79	36°38′N	95°09′W
Vinkovci, Cro. (vēn′kôv-tsē)	131	45°17′N	18°47′E
Vinnytsia, Ukr.	134	49°13′N	28°31′E
Vinnytsya, prov., Ukr.	133	48°45′N	28°01′E
Vinogradovo, Russia (vĭ-nô-grä′do-vô)	142b	55°25′N	38°33′E
Vinson Massif, mtn., Ant.	178	77°40′S	87°00′W
Vinton, Ia., U.S. (vĭn′tŭn)	71	42°08′N	92°01′W
Vinton, La., U.S.	81	30°12′N	93°35′W
Violet, La., U.S. (vī′ô-lĕt)	68d	29°54′N	89°54′W
Virac, Phil. (vē-räk′)	165	13°38′N	124°20′E
Virbalis, Lith. (vēr′bá-lēs)	123	54°38′N	22°55′E
Virden, Can. (vûr′dĕn)	50	49°51′N	101°55′W
Virden, Il., U.S.	79	39°28′N	89°46′W
Virgin, r., U.S.	77	36°51′N	113°50′W
Virginia, S. Afr.	192c	28°07′S	26°54′E
Virginia, Mn., U.S. (vĕr-jĭn′yà)	65	47°32′N	92°36′W
Virginia, state, U.S.	65	37°00′N	80°45′W
Virginia Beach, Va., U.S.	67	36°50′N	75°58′W
Virginia City, Nv., U.S.	76	39°18′N	119°40′W
Virgin Islands, is., N.A. (vûr′jĭn)	87	18°15′N	64°00′W
Viroqua, Wi., U.S. (vĭ-rō′kwà)	71	43°33′N	90°54′W
Virovitica, Cro. (vē-rô-vē′tē-tsä)	131	45°50′N	17°24′E
Virpazar, Serb. (vēr′pä-zär′)	131	42°16′N	19°06′E
Virrat, Fin. (vĭr′ät)	123	62°15′N	23°45′E
Virserum, Swe. (vĭr′sĕ-rôm)	122	57°22′N	15°35′E
Vis, Cro. (vēs)	130	43°03′N	16°11′E
Vis, i., Serb.	119	43°00′N	16°10′E
Visalia, Ca., U.S. (vĭ-sā′lĭ-à)	76	36°20′N	119°18′W
Visby, Swe. (vĭs′bü)	110	57°39′N	18°19′E
Viscount Melville Sound, strt., Can.	49	74°00′N	110°00′W
Višegrad, Bos. (vē′shĕ-gräd)	131	43°48′N	19°17′E
Vishākhapatnam, India	155	17°48′N	83°21′E
Vishera, r., Russia (vĭ′shĕ-rà)	142a	60°40′N	58°46′E
Vishnyakovo, Russia	142b	55°44′N	38°10′E
Vishoek, S. Afr.	186a	34°13′S	18°26′E

PLACE (Pronunciation)	PAGE	LAT.	LONG.
Visim, Russia (vē′sĭm)	142a	57°38′N	59°32′E
Viskan, r., Swe.	122	57°20′N	12°25′E
Višķi, Lat. (vēs′kĭ)	123	56°02′N	26°47′E
Visoko, Bos. (vē′sô-kô)	131	43°59′N	18°10′E
Vistula *see* Wisła, r., Pol.	112	52°30′N	20°00′E
Vitebsk, prov., Bela.	132	55°05′N	29°18′E
Viterbo, Italy (vē-tĕr′bō)	118	42°24′N	12°08′E
Viti Levu, i., Fiji	170g	18°00′S	178°00′E
Vitim, Russia (vē′tĕm)	135	59°22′N	112°43′E
Vitim, r., Russia (vē′tĕm)	135	54°00′N	115°00′E
Vitino, Russia (vē′tĭ-nô)	142c	59°40′N	29°51′E
Vitória, Braz. (vē-tō′rē-ä)	101	20°09′S	40°17′W
Vitoria, Spain (vē-tô-ryä)	118	42°43′N	2°43′W
Vitória de Conquista, Braz. (vē-tô′rē-ä-dä-kōn-kwē′s-tä)	101	14°51′S	40°44′W
Vitry-le-François, Fr. (vē-trē′lē-frän-swá′)	126	48°44′N	4°34′E
Vitsyebsk, Bela. (vē′tyĕpsk)	136	55°12′N	30°16′E
Vittorio, Italy (vē-tô′rē-ô)	130	45°59′N	12°17′E
Viveiro, Spain	128	43°39′N	7°37′W
Vivian, La., U.S. (vĭv′ĭ-ăn)	81	32°51′N	93°59′W
Vizianagaram, India	155	18°10′N	83°29′E
Vlaardingen, Neth. (vlär′dĭng-ĕn)	121	51°54′N	4°20′E
Vladikavkaz, Russia	137	43°05′N	44°35′E
Vladimir, Russia (vlå-dyē′mēr)	134	56°08′N	40°24′E
Vladimir, prov., Russia (vlä-dyē′mēr)	132	56°08′N	39°53′E
Vladimiro-Aleksandrovskoye, Russia	166	42°50′N	133°00′E
Vladivostok, Russia (vlá-dē-vôs-tök′)	135	43°06′N	131°47′E
Vlasenica, Bos. (vlä′sĕ-nĕt′sà)	131	44°11′N	18°58′E
Vlasotince, Serb. (vlä′sô-tēn-tsĕ)	131	42°58′N	22°08′E
Vlieland, i., Neth. (vlē′länt)	121	53°19′N	4°55′E
Vlissingen, Neth. (vlĭs′sĭng-ĕn)	121	51°30′N	3°34′E
Vlorë, Alb.	119	40°27′N	19°30′E
Vltava, r., Czech Rep.	124	49°24′N	14°18′E
Vodl, l., Russia (vôd′l)	136	62°20′N	37°20′E
Voerde, Ger.	127c	51°35′N	6°41′E
Voghera, Italy (vô-gā′rä)	130	44°58′N	9°02′E
Voight, r., Wa., U.S.	74a	47°03′N	122°08′W
Voinjama, Lib.	188	8°25′N	9°45′W
Voiron, Fr. (vwá-rôn′)	127	45°23′N	5°48′E
Voisin, Lac, l., Can. (vwŏ′-zĭn)	56	54°13′N	107°15′W
Volchansk, Ukr. (vôl-chänsk′)	137	50°18′N	36°56′E
Volga, r., Russia (vôl′gä)	134	47°30′N	46°20′E
Volga, Mouths of the, mth.	137	46°00′N	49°10′E
Volgograd, Russia (vŏl-gō-grä′t)	134	48°40′N	42°20′E
Volgogradskoye, res., Russia (vôl-gô-grad′skô-yĕ)	134	51°10′N	45°10′E
Volkhov, Russia (vôl′kôf)	123	59°54′N	32°21′E
Volkhov, r., Russia	136	58°45′N	31°40′E
Volodarskiy, Russia (vô-lô-där′skĭ)	142c	59°49′N	30°06′E
Volodymyr-Volyns′kyi, Ukr.	125	50°50′N	24°20′E
Vologda, Russia (vô′lôg-dà)	134	59°12′N	39°52′E
Vologda, prov., Russia	132	59°00′N	37°26′E
Volokolamsk, Russia (vô-lô-kô′lamsk)	132	56°02′N	35°58′E
Volokonovka, Russia (vô-lô-kô′nôf-kå)	133	50°28′N	37°52′E
Vol′sk, Russia (vôl′sk)	137	52°02′N	47°23′E
Volta, r., Ghana	188	6°05′N	0°30′E
Volta, Lake, res., Ghana	184	7°10′N	0°30′W
Volta Blanche (White Volta), r., Afr.	188	11°30′N	0°40′W
Volta Noire *see* Black Volta, r., Afr.	184	11°30′N	4°00′W
Volta Redonda, Braz. (vôl′tä-rā-dôn′dä)	101	22°32′S	44°05′W
Volterra, Italy (vôl-tĕr′rä)	130	43°22′N	10°51′E
Voltri, Italy (vôl′trē)	130	44°25′N	8°45′E
Volturno, r., Italy (vôl-tōōr′nô)	130	41°12′N	14°20′E
Vólvi, Límni, l., Grc.	131	40°41′N	23°23′E
Volzhskoye, l., Russia (vôl′sh-skô-yĕ)	132	56°43′N	36°18′E
Von Ormy, Tx., U.S. (vŏn ôr′mē)	75d	29°18′N	98°36′W
Vööpsu, Est. (vōōp′sò)	123	58°06′N	27°30′E
Voorburg, Neth.	115a	52°04′N	4°21′E
Voortrekkerhoogte, S. Afr.	187b	25°48′S	28°10′E
Vop′, r., Russia (vôp)	132	55°20′N	32°55′E
Vopnafjördur, Ice.	116	65°43′N	14°58′W
Vordingborg, Den. (vôr′dĭng-bôr)	122	55°10′N	11°55′E
Vóreioi Sporades, is., Grc.	131	38°55′N	24°05′E
Vóreios Evvoïkós Kólpos, b., Grc.	131	38°48′N	23°02′E
Vorkuta, Russia (vôr-kōō′tá)	134	67°28′N	63°40′E
Vormsi, i., Est. (vôrm′sĭ)	123	59°06′N	23°05′E
Vorona, r., Russia (vô-rô′na)	137	51°50′N	42°00′E
Voronava, Bela.	125	54°07′N	25°16′E
Voronezh, Russia (vô-rô′nyĕzh)	134	51°39′N	39°11′E
Voronezh, prov., Russia	133	51°10′N	39°13′E
Voronezh, r., Russia	137	52°17′N	39°32′E
Vorontsovka, Russia (vô-rônt′sôv-kà)	142a	59°40′N	60°14′E
Voron′ya, r., Russia (vô-rônyá)	136	68°20′N	35°20′E
Võrts-Järv, l., Est. (vôrts yärv)	123	58°15′N	26°12′E
Võru, Est. (vô′rû)	136	57°50′N	26°58′E
Vorya, r., Russia (vôr′yá)	142b	55°55′N	38°15′E
Vosges, mts., Fr. (vōzh)	117	48°09′N	6°57′E
Voskresensk, Russia (vôs-krĕ-sĕnsk′)	142b	55°20′N	38°42′E
Voss, Nor. (vôs)	116	60°40′N	6°24′E
Vostryakovo, Russia	142b	55°23′N	37°49′E
Votkinsk, Russia (vôt-kēnsk′)	136	57°00′N	54°00′E
Votkinskoye Vodokhranilishche, res., Russia	136	57°30′N	55°00′E
Vouga, r., Port. (vō′gà)	128	40°43′N	7°51′W
Vouziers, Fr. (vōō-zyā′)	126	49°25′N	4°40′E
Voxnan, r., Swe.	122	61°30′N	15°24′E
Voyageurs National Park, rec., Mn., U.S.	71	48°30′N	92°40′W
Vozhe, l., Russia (vôzh′yĕ)	136	60°40′N	39°00′E
Voznesens′k, Ukr.	137	47°34′N	31°22′E
Vradiïvka, Ukr.	133	47°51′N	30°38′E
Vrangelya (Wrangel), i., Russia	134	71°25′N	178°30′W
Vranje, Serb. (vrän′yĕ)	131	42°33′N	21°55′E
Vratsa, Blg. (vrät′tsá)	119	43°12′N	23°31′E

PLACE (Pronunciation)	PAGE	LAT.	LONG.
Vrbas, Serb. (v′r′bäs)	131	45°34′N	19°43′E
Vrbas, r., Serb.	131	44°25′N	17°17′E
Vrchlabí, Czech Rep. (v′r′chlä-bē)	124	50°32′N	15°51′E
Vrede, S. Afr. (vrĭ′dĕ)(vrĕd)	192c	27°25′S	29°11′E
Vredefort, S. Afr. (vrĭ′dĕ-fōrt)(vrĕd′fōrt)	192c	27°00′S	27°21′E
Vreeswijk, Neth.	115a	52°00′N	5°06′E
Vršac, Serb. (v′r′shats)	119	45°08′N	21°18′E
Vrutky, Slvk. (vrōōt′kē)	125	49°09′N	18°55′E
Vryburg, S. Afr. (vrī′bûrg)	186	26°55′S	24°45′E
Vryheid, S. Afr. (vrī′hīt)	186	27°43′S	30°58′E
Vsetín, Czech Rep. (fsĕt′yēn)	125	49°21′N	18°01′E
Vsevolozhskiy, Russia (vsyĕ′vôlô′zh-skĕĕ)	142c	60°01′N	30°41′E
Vuelta Abajo, reg., Cuba (vwĕl′tä ä-bä′hō)	92	22°20′N	83°45′W
Vught, Neth.	115a	51°38′N	5°18′E
Vukovar, Cro. (vò′kô-vär)	131	45°20′N	19°00′E
Vulcan, Mi., U.S. (vŭl′kăn)	66	45°45′N	87°50′W
Vulcano, i., Italy (vōōl-kä′nô)	130	38°23′N	15°00′E
Vûlchedrŭma, Blg.	131	43°43′N	23°29′E
Vuntut National Park, rec., Can.	52	68°27′N	139°58′W
Vyartsilya, Russia (vyár-tsē′lyá)	123	62°10′N	30°40′E
Vyatka, r., Russia (vyát′kà)	136	59°20′N	51°25′E
Vyazemskiy, Russia (vyà-zĕm′skĭ)	166	47°29′N	134°39′E
Vyaz′ma, Russia (vyàz′mà)	136	55°12′N	34°17′E
Vyazniki, Russia (vyáz′nē-kē)	136	56°10′N	42°10′E
Vyborg, Russia (vwē′bôrk)	134	60°43′N	28°46′E
Vychegda, r., Russia (vē′chĕg-dà)	136	61°40′N	48°00′E
Vyerkhnyadzvinsk, Bela.	132	55°48′N	27°59′E
Vyetka, Bela. (vyĕt′kà)	132	52°36′N	31°05′E
Vylkove, Ukr.	137	45°24′N	29°36′E
Vym, r., Russia (vwĕm)	136	63°15′N	51°20′E
Vyritsa, Russia (vē′rĭ-tsá)	142c	59°24′N	30°20′E
Vyshnevolotskoye, l., Russia (vŭy′sh-ŋĕ′vôlôt′s-kô′yĕ)	132	57°30′N	34°27′E
Vyshniy Volochëk, Russia (vĕsh′nyĭ vôl-ô-chĕk′)	134	57°34′N	34°35′E
Vyškov, Czech Rep. (vēsh′kôf)	124	49°17′N	16°58′E
Vysoké Mýto, Czech Rep. (vû′sô-kā mû′tô)	124	49°58′N	16°07′E
Vysokovsk, Russia (vĭ-sô′kôfsk)	132	56°16′N	36°32′E
Vytegra, Russia (vû′tĕg-rà)	134	61°00′N	36°20′E
Vyzhnytsia, Ukr.	125	48°16′N	25°12′E

W

PLACE (Pronunciation)	PAGE	LAT.	LONG.
W, Parcs Nationaux du, rec., Niger	189	12°20′N	2°40′E
Waal, r., Neth. (väl)	121	51°46′N	5°00′E
Waalwijk, Neth.	115a	51°41′N	5°05′E
Wabamun, Grc.	119	39°23′N	22°56′E
Wabamuno, Can. (wŏ′bă-mŭn)	55	53°33′N	114°28′W
Wabasca, Can. (wŏ-bás′kă)	55	56°00′N	113°53′W
Wabash, In., U.S. (wô′băsh)	66	40°45′N	85°50′W
Wabash, r., U.S.	65	38°00′N	88°00′W
Wabasha, Mn., U.S. (wä′bà-shô)	71	44°24′N	92°04′W
Wabe Gestro, r., Eth.	185	6°25′N	41°21′E
Wabowden, Can. (wä-bō′d′n)	57	54°55′N	98°38′W
Wąbrzeźno, Pol. (vôŋ-bzĕzh′nô)	125	53°17′N	18°59′E
Wabu Hu, l., China (wä-bōō hōō)	162	32°25′N	116°35′E
W. A. C. Bennett Dam, dam, Can.	55	56°01′N	122°10′W
Waccamaw, r., S.C., U.S. (wăk′à-mô)	83	33°47′N	78°55′W
Waccasassa Bay, b., Fl., U.S. (wă-kă-să′sà)	82	29°02′N	83°10′W
Wachow, Ger. (vä′kōv)	115b	52°32′N	12°46′E
Waco, Tx., U.S. (wā′kō)	64	31°35′N	97°06′W
Waconda Lake, res., Ks., U.S.	78	39°45′N	98°15′W
Wadayama, Japan (wä′dä′yä-mä)	167	35°19′N	134°49′E
Waddenzee, sea, Neth.	121	53°00′N	4°50′E
Waddington, Mount, mtn., Can. (wŏd′dĭng-tŭn)	52	51°23′N	125°15′W
Wadena, Can.	56	51°57′N	103°50′W
Wadena, Mn., U.S. (wŏ-dē′nà)	70	46°26′N	95°09′W
Wadesboro, N.C., U.S. (wādz′bŭr-ô)	83	34°57′N	80°05′W
Wadley, Ga., U.S. (wŭd′lē)	83	32°54′N	82°25′W
Wad Madani, Sudan (wäd mĕ-dä′nē)	185	14°27′N	33°31′E
Wadowice, Pol. (vá-dô′vēt-sĕ)	125	49°53′N	19°31′E
Wadsworth, Oh., U.S. (wŏdz′wûrth)	69d	41°01′N	81°44′W
Wager Bay, b., Can. (wā′jēr)	53	65°48′N	88°19′W
Wagga Wagga, Austl. (wôg′à wôg′à)	175	35°10′S	147°30′E
Wagoner, Ok., U.S. (wăg′ŭn-ēr)	79	35°58′N	95°22′W
Wagon Mound, N.M., U.S. (wăg′ŭn mound)	78	35°59′N	104°45′W
Wągrowiec, Pol. (vôŋ-grô′vyĕts)	125	52°47′N	17°14′E
Waha, Libya	156	28°16′N	19°54′E
Wahiawā, Hi., U.S.	64d	21°30′N	158°03′W
Wahoo, Ne., U.S. (wä-hōō′)	70	41°14′N	96°39′W
Wahpeton, N.D., U.S. (wô′pē-tŭn)	70	46°17′N	96°38′W
Waialua, Hi., U.S. (wä′ē-ä-lōō′ä)	84a	21°33′N	158°08′W
Wai′anae, Hi., U.S. (wä′ē-á-nä′ā)	84a	21°25′N	158°11′W
Waidhofen, Aus. (vīd′hôf-ĕn)	124	47°58′N	14°46′E
Waigeo, Pulau, i., Indon. (wä-ē-gā′ō)	169	0°07′N	131°00′E
Waikato, r., N.Z. (wä′ē-kä′to)	175a	38°10′S	175°35′E
Waikerie, Austl. (wā′kēr-ē)	176	34°15′S	140°00′E
Wailuku, Hi., U.S. (wä′ē-lōō′kōō)	84c	20°55′N	156°30′W
Waimānalo, Hi., U.S. (wä-ē-mä′nä-lo)	84a	21°19′N	157°43′W
Waimea, Hi., U.S. (wä-ē-mā′ä)	84a	21°56′N	159°38′W
Wainganga, r., India (wä-ēn-gŭŋ′gä)	155	20°30′N	80°15′E
Waingapu, Indon.	168	9°32′S	120°00′E
Wainwright, Can.	50	52°49′N	110°52′W
Wainwright, Ak., U.S. (wān-rīt)	63	74°40′N	159°00′W
Waipahu, Hi., U.S. (wä′ē-pä′hōō)	64d	21°20′N	158°02′W

PRONOUNCING INDEX

Waiska – Webster Springs 317

PLACE (Pronunciation)	PAGE	LAT.	LONG.
Waiska, r., Mi., U.S. (wȧ-ĭz-kȧ)	75k	46°20'N	84°38'W
Waitsburg, Wa., U.S. (wāts'bûrg)	72	46°17'N	118°08'W
Wajima, Japan (wä'jê-mä)	167	37°23'N	136°56'E
Wajir, Kenya	191	1°45'N	40°04'E
Wakami, r., Can.	58	47°43'N	82°22'W
Wakasa-Wan, b., Japan (wä'kä-sä wän)	166	35°43'N	135°39'E
Wakatipu, l., N.Z. (wä-kä-tē'pōō)	175a	45°04'S	168°30'E
Wakayama, Japan (wä-kä'yä-mä)	161	34°14'N	135°11'E
Wake, i., Oc. (wāk)	3	19°25'N	167°00'E
Wa Keeney, Ks., U.S. (wô-kē'nê)	78	39°01'N	99°53'W
Wakefield, Can. (wāk-fēld)	62c	45°39'N	75°55'W
Wakefield, Eng., U.K.	120	53°41'N	1°25'W
Wakefield, Ma., U.S.	61a	42°31'N	71°05'W
Wakefield, Mi., U.S.	71	46°28'N	89°55'W
Wakefield, Ne., U.S.	70	42°15'N	96°52'W
Wakefield, R.I., U.S.	68b	41°26'N	71°30'W
Wakefield, co., Eng., U.K.	114a	53°12'N	1°25'W
Wake Forest, N.C., U.S. (wāk fŏr'ĕst)	83	35°58'N	78°31'W
Waki, Japan (wä'kê)	167	34°05'N	134°10'E
Wakkanai, Japan (wä'kä-nä'ê)	161	45°19'N	141°43'E
Wakkerstroom, S. Afr. (vȧk'ēr-strōm)(wăk'ēr-strōōm)	186	27°19'S	30°04'E
Wakonassin, r., Can.	58	46°35'N	82°10'W
Waku Kundo, Ang.	186	11°25'S	15°07'E
Wałbrzych, Pol. (väl'bzhûk)	124	50°46'N	16°16'E
Walcott, Lake, res., Id., U.S.	73	42°40'N	113°23'W
Wałcz, Pol. (välch)	124	53°11'N	16°30'E
Waldoboro, Me., U.S. (wôl'dô-bŭr-ô)	60	44°06'N	69°22'W
Waldo Lake, l., Or., U.S. (wôl'dō)	72	43°46'N	122°10'W
Waldorf, Md., U.S. (wäl'dôrf)	68e	38°37'N	76°57'W
Waldron, Mo., U.S.	75f	39°14'N	94°47'W
Waldron, i., Wa., U.S.	74d	48°42'N	123°02'W
Wales, Ak., U.S. (wālz)	63	65°35'N	168°14'W
Wales, state, U.K.	110	52°12'N	3°40'W
Walewale, Ghana	188	10°21'N	0°48'W
Walgett, Austl. (wôl'gĕt)	175	30°00'S	148°10'E
Walhalla, S.C., U.S. (wŭl-hăl'ȧ)	82	34°45'N	83°04'W
Walikale, D.R.C.	191	1°25'S	28°03'E
Walkden, Eng., U.K.	114a	53°32'N	2°24'W
Walker, Mn., U.S. (wôk'ēr)	71	47°06'N	94°37'W
Walker, r., Nv., U.S.	76	39°07'N	119°10'W
Walker, Mount, mtn., Wa., U.S.	74a	47°47'N	122°54'W
Walker Lake, l., Can.	57	54°42'N	96°57'W
Walker Lake, l., Nv., U.S.	76	38°46'N	118°30'W
Walker River Indian Reservation, I.R., Nv., U.S.	76	39°06'N	118°20'W
Walkerville, Mt., U.S. (wôk'ēr-vĭl)	73	46°20'N	112°32'W
Wallace, Id., U.S. (wŏl'ås)	72	47°27'N	115°55'W
Wallaceburg, Can.	58	42°39'N	82°25'W
Wallacia, Austl.	173b	33°52'S	150°40'E
Wallaroo, Austl. (wŏl-ȧ-rōō)	174	33°52'S	137°45'E
Wallasey, Eng., U.K. (wŏl'ȧ-sê)	114a	53°25'N	3°03'W
Walla Walla, Wa., U.S. (wŏl'ȧ wŏl'ȧ)	64	46°03'N	118°20'W
Walled Lake, Mi., U.S. (wôl'd lāk)	69b	42°32'N	83°29'W
Wallel, Tulu, mtn., Eth.	185	9°00'N	34°52'E
Wallingford, Eng., U.K. (wŏl'ĭng-fērd)	114b	51°34'N	1°08'W
Wallingford, Vt., U.S.	67	43°30'N	72°55'W
Wallis and Futuna Islands, dep., Oc.	194	13°00'S	176°10'E
Wallisville, Tx., U.S. (wŏl'ĭs-vĭl)	81a	29°50'N	94°44'W
Wallowa, Or., U.S. (wôl'ô-wȧ)	72	45°34'N	117°32'W
Wallowa, r., Or., U.S.	72	45°28'N	117°28'W
Wallowa Mountains, mts., Or., U.S.	72	45°10'N	117°22'W
Wallula, Wa., U.S.	72	46°08'N	118°55'W
Walnut, Ca., U.S. (wôl'nŭt)	75a	34°00'N	117°51'W
Walnut, r., Ks., U.S.	79	37°28'N	97°06'W
Walnut Canyon National Mon, rec., Az., U.S.	77	35°10'N	111°30'W
Walnut Creek, Ca., U.S.	74b	37°54'N	122°04'W
Walnut Creek, r., Tx., U.S.	75c	32°37'N	97°03'W
Walnut Ridge, Ar., U.S. (rĭj)	79	36°04'N	90°56'W
Walpole, Ma., U.S. (wôl'pōl)	61a	42°09'N	71°15'W
Walpole, N.H., U.S.	67	43°05'N	72°25'W
Walsall, Eng., U.K. (wôl-sôl)	120	52°35'N	1°58'W
Walsenburg, Co., U.S. (wôl'sĕn-bûrg)	78	37°38'N	104°46'W
Walsum, Ger.	127c	51°32'N	6°41'E
Walter F. George Reservoir, res., U.S.	82	32°00'N	85°00'W
Walters, Ok., U.S. (wôl'tērz)	78	34°21'N	98°19'W
Waltham, Ma., U.S. (wôl'thăm)	61a	42°22'N	71°14'W
Walthamstow, Eng., U.K. (wôl'tăm-stō)	114b	51°34'N	0°01'W
Walton, N.Y., U.S.	67	42°10'N	75°05'W
Walton-le-Dale, Eng., U.K. (lē-dāl')	114a	53°44'N	2°40'W
Walvis Bay, Nmb. (wôl'vĭs)	186	22°50'S	14°30'E
Walworth, Wi., U.S. (wôl'wŭrth)	71	42°33'N	88°39'W
Wama, Ang.	190	12°14'S	15°33'E
Wamba, r., D.R.C.	186	7°00'S	18°00'E
Wamego, Ks., U.S. (wŏ-mē'gō)	79	39°13'N	96°17'W
Wami, r., Tan. (wä'mē)	187	6°31'S	37°17'E
Wanapitei Lake, l., Can.	59	46°45'N	80°45'W
Wanaque, N.J., U.S. (wŏn'ȧ-kū)	68a	41°03'N	74°16'W
Wanaque Reservoir, res., N.J., U.S.	68a	41°06'N	74°20'W
Wanda Shan, mts., China (wän-dä shän)	161	45°54'N	131°45'E
Wandoan, Austl.	176	26°09'S	149°51'E
Wandsbek, Ger. (vänds'bĕk)	115c	53°34'N	10°07'E
Wandsworth, Eng., U.K. (wôndz'wûrth)	114b	51°26'N	0°12'W
Wanganui, N.Z. (wôŋ'gȧ-nōō'ê)	175a	39°53'S	175°01'E
Wangaratta, Austl. (wŏŋ'gȧ-răt'ȧ)	176	36°23'S	146°18'E
Wangeroog, i., Ger. (vän'gē-rōg)	124	53°49'N	7°57'E
Wangqingtuo, China (wäŋ-chyĭŋ-twô)	162	39°14'N	116°56'E
Wangsi, China (wäŋ-sē)	162	37°59'N	116°57'E
Wantage, Eng., U.K. (wŏn'tȧj)	114b	51°33'N	1°26'W
Wantagh, N.Y., U.S.	68a	40°41'N	73°30'W
Wanxian, China (wän shyĕn)	162	38°51'N	115°10'E
Wanxian, China (wän-shyĕn)	160	30°48'N	108°22'E
Wanzai, China (wän-dzī)	165	28°05'N	114°25'E
Wanzhi, China (wän-jr)	162	31°11'N	118°31'E
Wapakoneta, Oh., U.S. (wä'pȧ-kô-nĕt'ȧ)	66	40°35'N	84°10'W
Wapawekka Hills, hills, Can. (wŏ'pă-wē'kă-hĭlz)	56	54°45'N	104°20'W
Wapawekka Lake, l., Can.	56	54°55'N	104°40'W
Wapello, Ia., U.S. (wŏ-pĕl'ō)	71	41°10'N	91°11'W
Wappapello Reservoir, res., Mo., U.S. (wă'pȧ-pĕl-lō)	65	37°07'N	90°10'W
Wappingers Falls, N.Y., U.S. (wŏp'ĭn-jērz)	67	41°35'N	73°55'W
Wapsipinicon, r., Ia., U.S. (wŏp'sĭ-pĭn'ĭ-kŏn)	71	42°16'N	91°35'W
Wapusk National Park, rec., Can.	52	58°00'N	94°15'W
Warabi, Japan (wä'rä-bê)	167a	35°50'N	139°41'E
Warangal, India (wŭ'răŋ-gȧl)	155	18°03'N	79°45'E
Warburton, The, r., Austl. (wŏr'bûr-tŭn)	174	27°30'S	138°45'E
Wardān, Wādī, r., Egypt	153a	29°22'N	33°00'E
Ward Cove, Ak., U.S.	54	55°24'N	131°43'W
Warden, S. Afr. (wôr'dĕn)	192c	27°52'S	28°59'E
Wardha, India (wŭr'dä)	155	20°46'N	78°42'E
War Eagle, W.V., U.S. (wôr ē'g'l)	66	37°30'N	81°50'W
Waren, Ger. (vä'rĕn)	124	53°32'N	12°43'E
Warendorf, Ger. (vä'rĕn-dōrf)	127c	51°57'N	7°59'E
Wargla, Alg.	184	32°00'N	5°18'E
Warialda, Austl.	176	29°32'S	150°34'E
Warmbad, Nmb. (värm'bäd)	186	28°25'S	18°45'E
Warmbad, S. Afr.	192c	24°52'S	28°18'E
Warm Beach, Wa., U.S. (wôrm)	74a	48°10'N	122°22'W
Warm Springs Indian Reservation, I.R., Or., U.S. (wôrm sprĭnz)	72	44°55'N	121°30'W
Warm Springs Reservoir, res., Or., U.S.	72	43°42'N	118°40'W
Warner Mountains, mts., Ca., U.S.	64	41°30'N	120°17'W
Warner Robins, Ga., U.S.	82	32°37'N	83°36'W
Warnow, r., Ger. (vär'nō)	124	53°51'N	11°55'E
Warracknabeal, Austl.	176	36°20'S	142°28'E
Warragamba Reservoir, res., Austl.	176	33°40'S	150°00'E
Warrego, r., Austl. (wôr'ê-gô)	175	27°13'S	145°58'E
Warren, Can.	62f	50°08'N	97°32'W
Warren, Ar., U.S. (wŏr'ĕn)	79	33°37'N	92°03'W
Warren, In., U.S.	66	40°40'N	85°25'W
Warren, Mi., U.S.	69b	42°33'N	83°03'W
Warren, Mn., U.S.	70	48°11'N	96°44'W
Warren, Oh., U.S.	66	41°15'N	80°50'W
Warren, Or., U.S.	74c	45°49'N	122°51'W
Warren, Pa., U.S.	67	41°50'N	79°10'W
Warren, R.I., U.S.	68b	41°44'N	71°14'W
Warrendale, Pa., U.S. (wŏr'ĕn-dāl)	69e	40°39'N	80°04'W
Warrensburg, Mo., U.S. (wŏr'ĕnz-bûrg)	79	38°45'N	93°42'W
Warrenton, Ga., U.S. (wŏr'ĕn-tŭn)	83	33°26'N	82°37'W
Warrenton, Or., U.S.	74c	46°10'N	123°56'W
Warrenton, Va., U.S.	67	38°45'N	77°50'W
Warri, Nig. (wär'ê)	184	5°33'N	5°43'E
Warrington, Eng., U.K.	114a	53°22'N	2°30'W
Warrington, Fl., U.S. (wŏ'ĭng-tŭn)	82	30°21'N	87°15'W
Warrnambool, Austl. (wôr'năm-bōōl)	175	38°20'S	142°28'E
Warroad, Mn., U.S. (wôr'rōd)	70	48°55'N	95°20'W
Warrumbungle Range, mts., Austl. (wôr'ŭm-bŭŋ-g'l)	175	31°18'S	150°00'E
Warsaw, Pol.	110	52°15'N	21°05'E
Warsaw, Il., U.S. (wôr'sô)	79	40°21'N	91°26'W
Warsaw, In., U.S.	66	41°15'N	85°50'W
Warsaw, N.Y., U.S.	67	42°45'N	78°10'W
Warsaw, NC, N.C., U.S.	83	35°00'N	78°07'W
Warsop, Eng., U.K. (wôr'sŭp)	114a	53°13'N	1°05'W
Warszawa *see* Warsaw, Pol.	110	52°15'N	21°05'E
Warta, r., Pol. (vär'tȧ)	117	52°30'N	16°00'E
Wartburg, S. Afr.	187c	29°26'S	30°39'E
Warwick, Austl. (wŏr'ĭk)	175	28°05'S	152°10'E
Warwick, Can.	59	45°58'N	71°57'W
Warwick, Eng., U.K.	120	52°19'N	1°46'W
Warwick, N.Y., U.S.	68a	41°15'N	74°22'W
Warwick, R.I., U.S.	67	41°42'N	71°27'W
Warwickshire, co., Eng., U.K.	114a	52°30'N	1°35'W
Wasatch Mountains, mts., Ut., U.S. (wô'săch)	75b	40°45'N	111°46'W
Wasatch Plateau, plat., Ut., U.S.	77	38°55'N	111°40'W
Wasatch Range, mts., U.S.	64	39°10'N	111°30'W
Wasbank, S. Afr.	187c	28°27'S	30°09'E
Wasco, Or., U.S. (wăs'kō)	72	45°36'N	120°42'W
Waseca, Mn., U.S. (wô-sē'kȧ)	71	44°04'N	93°31'W
Wash, The, Eng., U.K. (wŏsh)	116	53°00'N	0°20'E
Washburn, Me., U.S. (wŏsh'bŭrn)	60	46°46'N	68°10'W
Washburn, Wi., U.S.	71	46°41'N	90°55'W
Washburn, Mount, mtn., Wy., U.S.	73	44°55'N	110°10'W
Washington, D.C., U.S. (wŏsh'ĭng-tŭn)	65	38°50'N	77°00'W
Washington, Ga., U.S.	83	33°43'N	82°46'W
Washington, Ia., U.S.	71	41°17'N	91°42'W
Washington, In., U.S.	66	38°40'N	87°10'W
Washington, Ks., U.S.	79	39°48'N	97°04'W
Washington, Mo., U.S.	79	38°33'N	91°00'W
Washington, N.C., U.S.	83	35°32'N	77°01'W
Washington, Pa., U.S.	66	40°10'N	80°14'W
Washington, state, U.S.	64	47°30'N	121°10'W
Washington, i., Wi., U.S.	71	45°18'N	86°42'W
Washington, Lake, l., Wa., U.S.	74a	47°34'N	122°12'W
Washington, Mount, mtn., N.H., U.S.	65	44°15'N	71°15'W
Washington Court House, Oh., U.S.	66	39°30'N	83°25'W
Washington Park, Il., U.S.	75e	38°38'N	90°06'W
Washita, r., Ok., U.S. (wŏsh'ĭ-tô)	78	35°33'N	99°16'W
Washougal, Wa., U.S. (wŏ-shōō'gȧl)	74c	45°35'N	122°21'W
Washougal, r., Wa., U.S.	74c	45°38'N	122°17'W
Wasilków, Pol. (vȧ-sēl'kôf)	125	53°12'N	23°13'E
Waskaiowaka Lake, l., Can. (wŏ'skă-yō'wō-kă)	57	56°30'N	96°20'W
Wassenberg, Ger. (vä'sĕn-bĕrgh)	127c	51°06'N	6°07'E
Wassuk Range, mts., Nv., U.S. (wȧs'sŭk)	76	38°58'N	119°00'W
Waswanipi, Lac, l., Can.	59	49°35'N	76°15'W
Water, i., V.I.U.S. (wô'tēr)	87c	18°20'N	64°57'W
Waterberge, mts., S. Afr. (wôrtēr'bûrg)	192c	24°25'S	27°53'E
Waterboro, S.C., U.S. (wô'tēr-bûr-ō)	83	32°50'N	80°40'W
Waterbury, Ct., U.S. (wô'tēr-bĕr-ê)	67	41°30'N	73°00'W
Water Cay, i., Bah.	93	22°55'N	75°50'W
Waterdown, Can. (wô'tēr-doun)	62d	43°20'N	79°54'W
Wateree Lake, res., S.C., U.S. (wô'tēr-ē)	83	34°40'N	80°48'W
Waterford, Ire. (wô'tēr-fērd)	117	52°20'N	7°03'W
Waterford, Wi., U.S.	69a	42°46'N	88°13'W
Waterloo, Bel.	115a	50°44'N	4°24'E
Waterloo, Can. (wô-tēr-lōō')	59	43°30'N	80°40'W
Waterloo, Can.	59	45°25'N	72°30'W
Waterloo, Ia., U.S.	65	42°30'N	92°22'W
Waterloo, Il., U.S.	79	38°19'N	90°08'W
Waterloo, Md., U.S.	68e	39°11'N	76°50'W
Waterloo, N.Y., U.S.	67	42°55'N	76°50'W
Waterton-Glacier International Peace Park, rec., N.A. (wô'ter-tŭn-glā'shûr)	64	48°55'N	114°10'W
Waterton Lakes National Park, rec., Can.	55	49°05'N	113°50'W
Watertown, Ma., U.S. (wô'tēr-toun)	61a	42°22'N	71°11'W
Watertown, N.Y., U.S.	65	44°00'N	75°55'W
Watertown, S.D., U.S.	64	44°53'N	97°07'W
Watertown, Wi., U.S.	71	43°13'N	88°40'W
Water Valley, Ms., U.S. (văl'ê)	82	34°08'N	89°38'W
Waterville, Me., U.S.	60	44°34'N	69°37'W
Waterville, Mn., U.S.	71	44°10'N	93°35'W
Waterville, Wa., U.S.	72	47°38'N	120°04'W
Watervliet, N.Y., U.S. (wô'tēr-vlēt')	67	42°45'N	73°54'W
Watford, Eng., U.K. (wŏt'fōrd)	120	51°38'N	0°24'W
Wathaman Lake, l., Can.	56	56°55'N	103°43'W
Watlington, Eng., U.K.	114b	51°37'N	1°01'W
Watonga, Ok., U.S. (wŏ-tôŋ'gȧ)	79	35°50'N	98°26'E
Watsa, D.R.C. (wät'sä)	185	3°03'N	29°32'E
Watseka, Il., U.S. (wŏt-sē'kȧ)	66	40°45'N	87°45'W
Watson, In., U.S. (wŏt'sŭn)	69h	38°21'N	85°42'W
Watson Lake, Can.	50	60°18'N	128°50'W
Watsonville, Ca., U.S. (wŏt'sŭn-vĭl)	76	36°55'N	121°46'W
Wattenscheid, Ger. (vä'tĕn-shīd)	127c	51°30'N	7°07'E
Watts, Ca., U.S. (wŏts)	75a	33°56'N	118°15'W
Watts Bar Lake, res., Tn., U.S. (bär)	82	35°45'N	84°49'W
Waubay, S.D., U.S. (wô'bā)	70	45°19'N	97°18'W
Wauchula, Fl., U.S. (wô-chōō'lȧ)	83a	27°32'N	81°48'W
Wauconda, Il., U.S. (wô-kŏn'dȧ)	69a	42°15'N	88°08'W
Waukegan, Il., U.S. (wŏ-kē'găn)	65	42°22'N	87°51'W
Waukesha, Wi., U.S. (wô'kē-shô)	69a	43°01'N	88°13'W
Waukon, Ia., U.S. (wô kŏn)	71	43°15'N	91°30'W
Waupaca, Wi., U.S. (wô-păk'ȧ)	71	44°22'N	89°06'W
Waupun, Wi., U.S. (wô-pŭn')	71	43°37'N	88°45'W
Waurika, Ok., U.S. (wô-rē'kȧ)	79	34°09'N	97°59'W
Wausau, Wi., U.S. (wô'sô)	65	44°58'N	89°40'W
Wausaukee, Wi., U.S. (wô-sô'kê)	71	45°22'N	87°58'W
Wauseon, Oh., U.S. (wô'sê-ŏn)	66	41°30'N	84°10'W
Wautoma, Wi., U.S. (wô-tō'mȧ)	71	44°04'N	89°11'W
Wauwatosa, Wi., U.S. (wô-wȧ-t'ō'sȧ)	69a	43°03'N	88°00'W
Waveney, r., Eng., U.K. (wāv'nê)	121	52°27'N	1°17'E
Waverly, S. Afr.	187c	31°54'S	26°29'E
Waverly, Ia., U.S. (wā'vēr-lê)	71	42°43'N	92°29'W
Waverly, Tn., U.S.	82	36°04'N	87°46'W
Wāw, Sudan	185	7°41'N	28°00'E
Wawa, Can.	58	47°59'N	84°47'W
Wāw al-Kabīr, Libya	185	25°23'N	16°52'E
Wawanesa, Can. (wŏ'wō-nē'sȧ)	57	49°36'N	99°41'W
Wawasee, l., In., U.S. (wô-wô-sē')	66	41°25'N	85°45'W
Waxahachie, Tx., U.S. (wăk-sȧ-hăch'ê)	81	32°23'N	96°50'W
Wayland, Ky., U.S. (wā'lănd)	83	37°25'N	82°47'W
Wayland, Ma., U.S.	61a	42°23'N	71°22'W
Wayne, Mi., U.S.	69b	42°17'N	83°23'W
Wayne, Ne., U.S.	70	42°13'N	97°03'W
Wayne, N.J., U.S.	68a	40°56'N	74°16'W
Wayne, Pa., U.S.	68f	40°03'N	75°22'W
Waynesboro, Ga., U.S. (wānz'bŭr-ô)	83	33°05'N	82°02'W
Waynesboro, Pa., U.S.	67	39°45'N	77°35'W
Waynesboro, Va., U.S.	67	38°05'N	78°50'W
Waynesburg, Pa., U.S. (wānz'bûrg)	66	39°55'N	80°10'W
Waynesville, N.C., U.S. (wānz'vĭl)	83	35°28'N	82°58'W
Waynoka, Ok., U.S. (wā-nō'kȧ)	78	36°34'N	98°52'W
Wayzata, Mn., U.S. (wā-zä-tȧ)	75g	44°58'N	93°31'W
Wazīrbad, Pak.	158	32°39'N	74°11'E
Weagamow Lake, l., Can. (wē'ăg-ă-mou)	57	52°53'N	91°22'W
Weald, The, reg., Eng., U.K. (wēld)	120	50°58'N	0°15'W
Weatherford, Ok., U.S. (wĕ-dhēr-fērd)	78	85°32'N	98°41'W
Weatherford, Tx., U.S.	81	32°45'N	97°46'W
Weaver, r., Eng., U.K. (wē'vēr)	114a	53°09'N	2°31'W
Weaverville, Ca., U.S. (wē'vēr-vĭl)	72	40°44'N	122°55'W
Webb City, Mo., U.S.	79	37°10'N	94°26'W
Weber, r., Ut., U.S.	75b	41°13'N	112°07'W
Webster, Ma., U.S.	61a	42°04'N	71°52'W
Webster, S.D., U.S.	70	45°19'N	97°30'W
Webster City, Ia., U.S.	71	42°28'N	93°49'W
Webster Groves, Mo., U.S. (grōvz)	75e	38°36'N	90°22'W
Webster Springs, W.V., U.S. (sprĭngz)	66	38°30'N	80°20'W

ăt; fĭnăl; rāte; senāte; ärm; ásk; sofà; fâre; ch-choose; dh-as th in other; bē; ĕvent; bĕt; recĕnt; cratēr; g-gō; gh-guttural g; bĭt; ĭ-short neutral; rīde; ᴋ-guttural k as ch in German ich;

PRONOUNCING INDEX

Whiterock Reservoir – Wolkersdorf 319

PLACE (Pronunciation)	PAGE	LAT.	LONG.
Whiterock Reservoir, res., Tx., U.S. (hwīt′rŏk)	75c	32°51′N	96°40′W
White Russia *see* Belarus, nation, Eur.	134	53°30′N	25°33′E
Whitesail Lake, l., Can. (whīt′sāl)	54	53°30′N	127°00′W
White Sands National Monument, rec., N.M., U.S.	77	32°50′N	106°20′W
White Sea, sea, Russia	134	66°00′N	40°00′E
White Settlement, Tx., U.S.	75c	32°45′N	97°28′W
White Sulphur Springs, Mt., U.S.	73	46°32′N	110°49′W
White Umfolzi, r., S. Afr. (ŭm-fō-lō′zē)	187c	28°12′S	30°55′E
Whiteville, N.C., U.S. (hwīt′vĭl)	83	34°18′N	78°45′W
White Volta (Volta Blanche), r., Afr.	188	9°40′N	1°10′W
Whitewater, Wi., U.S. (whīt-wŏt′ẽr)	71	42°49′N	88°40′W
Whitewater, l., Can.	57	49°14′N	100°39′W
Whitewater, r., In., U.S.	69f	39°19′N	84°55′W
Whitewater Bay, b., Fl., U.S.	83a	25°16′N	80°21′W
Whitewater Creek, r., Mt., U.S.	73	48°50′N	107°50′W
Whitewell, Tn., U.S. (hwīt′wĕl)	82	35°11′N	85°31′W
Whitewright, Tx., U.S. (hwīt′rīt)	79	33°33′N	96°25′W
Whitham, r., Eng., U.K. (wĭth′ăm)	114a	53°08′N	0°15′W
Whiting, In., U.S. (hwīt′ĭng)	69a	41°41′N	87°30′W
Whitinsville, Ma., U.S. (hwīt′ĕns-vĭl)	61a	42°06′N	71°40′W
Whitman, Ma., U.S. (hwĭt′măn)	61a	42°05′N	70°57′W
Whitmire, S.C., U.S. (hwĭt′mīr)	83	34°30′N	81°40′W
Whitney, Mount, mtn., Ca., U.S.	64	36°34′N	118°18′W
Whitney Lake, l., Tx., U.S. (hwĭt′nē)	81	32°02′N	97°36′W
Whitstable, Eng., U.K. (wĭt′stăb′l)	114b	51°22′N	1°03′E
Whitsunday, i., Austl. (hwĭt′s′n-dā)	175	20°16′S	149°00′E
Whittier, Ca., U.S. (hwĭt′ĭ-ẽr)	75a	33°58′N	118°02′W
Whittlesea, S. Afr. (wĭt′l′sē)	187c	32°11′S	26°51′E
Whitworth, Eng., U.K. (hwĭt′wŭrth)	114a	53°40′N	2°10′W
Whyalla, Austl. (hwī-ăl′ă)	174	33°00′S	137°32′E
Whymper, Mount, mtn., Can. (wĭm′pẽr)	54	48°57′N	124°10′W
Wiarton, Can. (wī′ăr-tŭn)	51	44°45′N	80°45′W
Wichita, Ks., U.S. (wĭch′ĭ-tô)	64	37°42′N	97°21′W
Wichita, r., Tx., U.S.	78	33°50′N	99°38′W
Wichita Falls, Tx., U.S. (fôls)	64	33°54′N	98°29′W
Wichita Mountains, mts., Ok., U.S.	64	34°48′N	98°43′W
Wick, Scot., U.K. (wĭk)	116	58°25′N	3°05′W
Wickatunk, N.J., U.S. (wĭk′à-tŭnk)	68a	40°21′N	74°15′W
Wickenburg, Az., U.S.	77	33°58′N	112°44′W
Wickiup Reservoir, res., Or., U.S.	72	43°40′N	121°43′W
Wickliffe, Oh., U.S. (wĭk′klĭf)	69d	41°37′N	81°29′W
Wicklow, Ire.	120	52°59′N	6°06′W
Wicklow Mountains, mts., Ire. (wĭk′lō)	120	52°49′N	6°20′W
Wickup Mountain, mtn., Or., U.S. (wĭk′ŭp)	74c	46°06′N	123°35′W
Wiconisco, Pa., U.S. (wĭ-kŏn′ĭs-kō)	67	43°35′N	76°45′W
Widen, W.V., U.S. (wī′dĕn)	66	38°25′N	80°55′W
Widnes, Eng., U.K. (wĭd′nĕs)	114a	53°21′N	2°44′W
Wieliczka, Pol. (vyĕ-lēch′kà)	125	49°58′N	20°06′E
Wien *see* Vienna, Aus.	110	48°13′N	16°22′E
Wien, state, Aus.	115e	48°11′N	16°23′E
Wiener Neustadt, Aus. (vē′nẽr noi′shtät)	117	47°48′N	16°15′E
Wiener Wald, for., Aus.	115e	48°09′N	16°05′E
Wieprz, r., Pol. (vyĕpzh)	125	51°25′N	22°45′E
Wiergate, Tx., U.S. (wẽr′gāt)	81	31°00′N	93°42′W
Wiesbaden, Ger. (vēs′bä-dĕn)	117	50°05′N	8°15′E
Wigan, Eng., U.K. (wĭg′ăn)	120	53°33′N	2°37′W
Wiggins, Ms., U.S. (wĭg′ĭnz)	82	30°51′N	89°05′W
Wight, Isle of, i., Eng., U.K. (wīt)	120	50°44′N	1°17′W
Wilber, Ne., U.S. (wĭl′bẽr)	79	40°29′N	96°57′W
Wilburton, Ok., U.S. (wĭl′bẽr-tŭn)	79	34°54′N	95°18′W
Wilcannia, Austl. (wĭl-căn-ĭă)	175	31°30′S	143°30′E
Wildau, Ger. (vēl′dou)	115b	52°20′N	13°39′E
Wildberg, Ger. (vēl′bĕrgh)	115b	52°52′N	12°39′E
Wildcat Hill, hill, Can. (wīld′kăt)	57	53°17′N	102°30′W
Wildhay, r., Can. (wīld′hā)	55	53°15′N	117°20′W
Wildomar, Ca., U.S. (wĭl′dô-mär)	75a	33°35′N	117°17′W
Wild Rice, r., Mn., U.S.	70	47°10′N	96°40′W
Wild Rice, r., N.D., U.S.	70	46°10′N	97°12′W
Wild Rice Lake, l., Mn., U.S.	75h	46°54′N	92°10′W
Wildspitze, mtn., Aus.	124	46°55′N	10°50′E
Wildwood, N.J., U.S.	67	39°00′N	74°50′W
Wiley, Co., U.S. (wī′lē)	78	38°08′N	102°41′W
Wilge, r., S. Afr. (wĭl′jē)	192c	25°38′S	29°09′E
Wilge, r., S. Afr.	192c	27°27′S	28°46′E
Wilhelm, Mount, mtn., Pap. N. Gui.	169	5°58′S	144°58′E
Wilhelmina Gebergte, mts., Sur.	101	4°30′N	57°00′W
Wilhelmina Kanaal, can., Neth.	115a	51°37′N	4°55′E
Wilhelmshaven, Ger. (vēl-hĕlms-hä′fĕn)	116	53°30′N	8°10′E
Wilkes-Barre, Pa., U.S. (wĭlks′băr-ē)	65	41°15′N	75°50′W
Wilkes Land, reg., Ant.	178	71°00′S	126°00′E
Wilkeson, Wa., U.S. (wĭl-kē′sŭn)	74a	47°06′N	122°03′W
Wilkie, Can. (wĭlk′ē)	50	52°25′N	108°43′W
Wilkinsburg, Pa., U.S. (wĭl′kĭnz-bûrg)	69e	40°26′N	79°53′W
Willamette, r., Or., U.S.	64	45°00′N	123°00′W
Willapa Bay, b., Wa., U.S.	72	46°37′N	124°00′W
Willard, Oh., U.S. (wĭl′ărd)	66	41°00′N	82°50′W
Willard, Ut., U.S.	75b	41°24′N	112°02′W
Willcox, Az., U.S. (wĭl′kŏks)	77	32°15′N	109°50′W
Willcox Playa, l., Az., U.S.	77	32°08′N	109°51′W
Willemstad, Neth. Ant.	100	12°12′N	68°58′W
Willesden, Eng., U.K. (wĭlz′dĕn)	114b	51°31′N	0°17′W
William "Bill" Dannelly Reservoir, res., Al., U.S.	82	32°10′N	87°15′W
William Creek, Austl. (wĭl′yăm)	174	28°45′S	136°20′E
Williams, Az., U.S. (wĭl′yămz)	77	35°15′N	112°15′W
Williams, i., Bah.	92	24°30′N	78°30′W
Williamsburg, Ky., U.S. (wĭl′yămz-bûrg)	82	36°42′N	84°09′W
Williamsburg, Oh., U.S.	69f	39°04′N	84°02′W

PLACE (Pronunciation)	PAGE	LAT.	LONG.
Williamsburg, Va., U.S.	83	37°15′N	76°41′W
Williams Lake, Can.	55	52°08′N	122°09′W
Williamson, W.V., U.S. (wĭl′yăm-sŭn)	66	37°40′N	82°15′W
Williamsport, Md., U.S.	67	39°35′N	77°45′W
Williamsport, Pa., U.S.	67	41°15′N	77°05′W
Williamston, N.C., U.S. (wĭl′yămz-tŭn)	83	35°50′N	77°04′W
Williamston, S.C., U.S.	83	34°36′N	82°30′W
Williamstown, Austl.	173a	37°52′S	144°54′E
Williamstown, W.V., U.S. (wĭl′yămz-toun)	66	39°20′N	81°30′W
Williamsville, N.Y., U.S. (wĭl′yăm-vĭl)	69c	42°58′N	78°46′W
Willimantic, Ct., U.S. (wĭl-ĭ-măn′tĭk)	67	41°40′N	72°10′W
Willis, Tx., U.S. (wĭl′ĭs)	81	30°24′N	95°29′W
Willis Islands, is., Austl.	175	16°15′S	150°30′E
Williston, N.D., U.S. (wĭl′ĭs-tŭn)	64	48°08′N	103°38′W
Williston, Lake, l., Can.	52	55°40′N	123°40′W
Willmar, Mn., U.S. (wĭl′mär)	70	45°07′N	95°05′W
Willoughby, Oh., U.S. (wĭl′ô-bē)	69d	41°39′N	81°25′W
Willow, Ak., U.S.	63	61°50′N	150°00′W
Willow Creek, r., Or., U.S.	72	44°21′N	117°34′W
Willow Grove, Pa., U.S.	68f	40°07′N	75°07′W
Willowick, Oh., U.S. (wĭl′ô-wĭk)	69d	41°39′N	81°28′W
Willowmore, S. Afr. (wĭl′ô-môr)	186	33°15′S	23°37′E
Willow Run, Mi., U.S. (wĭl′ô rŭn)	69b	42°16′N	83°34′W
Willows, Ca., U.S. (wĭl′ōz)	76	39°32′N	122°11′W
Willow Springs, Mo., U.S. (sprĭngz)	79	36°59′N	91°56′W
Willowvale, S. Afr. (wĭ-lô′văl)	187c	32°17′S	28°32′E
Wills Point, Tx., U.S. (wĭlz point)	81	32°42′N	96°02′W
Wilmer, Tx., U.S. (wĭl′mẽr)	75c	32°35′N	96°40′W
Wilmette, Il., U.S. (wĭl-mĕt′)	69a	42°04′N	87°42′W
Wilmington, Austl.	176	32°39′S	138°07′E
Wilmington, Ca., U.S. (wĭl′mĭng-tŭn)	75a	33°46′N	118°16′W
Wilmington, De., U.S.	65	39°45′N	75°33′W
Wilmington, Il., U.S.	69a	41°19′N	88°09′W
Wilmington, Ma., U.S.	61a	42°34′N	71°10′W
Wilmington, N.C., U.S.	65	34°12′N	77°56′W
Wilmington, Oh., U.S.	66	39°20′N	83°50′W
Wilmore, Ky., U.S. (wĭl′mōr)	66	37°50′N	84°35′W
Wilmslow, Eng., U.K. (wĭlmz′lō)	114a	53°19′N	2°14′W
Wilno *see* Vilnius, Lith.	134	54°40′N	25°26′E
Wilpoort, S. Afr.	192c	26°57′S	26°17′E
Wilson, Ar., U.S. (wĭl′sŭn)	79	35°35′N	90°02′W
Wilson, N.C., U.S.	83	35°42′N	77°55′W
Wilson, Ok., U.S.	79	34°09′N	97°27′W
Wilson, r., Al., U.S.	82	34°53′N	87°28′W
Wilson, Mount, mtn., Ca., U.S.	75a	34°15′N	118°06′W
Wilson, Point, c., Austl.	173a	38°05′S	144°31′E
Wilson Lake, res., Al., U.S.	65	34°45′N	87°30′W
Wilson's Promontory, pen., Austl. (wĭl′sŭnz)	175	39°05′S	146°50′E
Wilsonville, Il., U.S. (wĭl′sŭn-vĭl)	75e	39°04′N	89°52′W
Wilstedt, Ger. (vēl′shtĕt)	115c	53°45′N	10°04′E
Wilster, Ger. (vēl′stẽr)	115c	53°55′N	9°23′E
Wilton, Ct., U.S. (wĭl′tŭn)	68a	41°11′N	73°25′W
Wilton, N.D., U.S.	70	47°09′N	100°47′W
Wiluna, Austl. (wĭ-lōō′nă)	174	26°35′S	120°25′E
Winamac, In., U.S. (wĭn′ă măk)	66	41°05′N	86°40′W
Winburg, S. Afr. (wĭm-bûrg)	192c	28°31′S	27°02′E
Winchester, Eng., U.K.	120	51°04′N	1°20′W
Winchester, Ca., U.S. (wĭn′chĕs-tẽr)	75a	33°41′N	117°06′W
Winchester, Id., U.S.	72	46°14′N	116°39′W
Winchester, In., U.S.	66	40°10′N	84°50′W
Winchester, Ky., U.S.	66	38°00′N	84°15′W
Winchester, Ma., U.S.	61a	42°28′N	71°09′W
Winchester, N.H., U.S.	67	42°45′N	72°25′W
Winchester, Tn., U.S.	82	35°11′N	86°06′W
Winchester, Va., U.S.	67	39°10′N	78°10′W
Wind, r., Wy., U.S.	73	43°17′N	109°02′W
Windber, Pa., U.S. (wĭnd′bẽr)	67	40°15′N	78°45′W
Wind Cave National Park, rec., S.D., U.S.	70	43°36′N	103°53′W
Winder, Ga., U.S. (wīn′dẽr)	82	33°58′N	83°43′W
Windermere, Eng., U.K. (wĭn′dẽr-mẽr)	120	54°25′N	2°59′W
Windham, Ct., U.S. (wĭnd′ăm)	67	41°45′N	72°05′W
Windham, N.H., U.S.	61a	42°49′N	71°21′W
Windhoek, Nmb. (vĭnt′hŏk)	186	22°05′S	17°10′E
Wind Lake, l., Wi., U.S.	69a	42°49′N	88°06′W
Wind Mountain, mtn., N.M., U.S.	80	32°02′N	105°30′W
Windom, Mn., U.S. (wĭn′dŭm)	70	43°50′N	95°04′W
Windorah, Austl. (wĭn-dō′ră)	175	25°15′S	142°50′E
Wind River Indian Reservation, I.R., Wy., U.S.	73	43°26′N	109°00′W
Wind River Range, mts., Wy., U.S.	64	43°19′N	109°47′W
Windsor, Austl. (wĭn′zẽr)	173b	33°37′S	150°49′E
Windsor, Can.	51	42°19′N	83°00′W
Windsor, Can.	53a	48°57′N	55°40′W
Windsor, Can.	51	44°59′N	64°08′W
Windsor, Eng., U.K.	120	51°27′N	0°37′W
Windsor, Co., U.S.	78	40°27′N	104°51′W
Windsor, Mo., U.S.	79	38°32′N	93°31′W
Windsor, N.C., U.S.	83	35°58′N	76°57′W
Windsor, Vt., U.S.	67	43°30′N	72°25′W
Windward Islands, is., N.A. (wĭnd′wẽrd)	87	12°45′N	61°40′W
Windward Passage, strt., N.A.	87	19°30′N	74°20′W
Winefred Lake, l., Can.	56	55°30′N	110°35′W
Winfield, Ks., U.S.	79	37°14′N	97°00′W
Winifred, Mt., U.S. (wĭn ĭ frĕd)	73	47°35′N	109°20′W
Winisk, r., Can.	53	54°30′N	86°30′W
Wink, Tx., U.S. (wĭŋk)	80	31°48′N	103°06′W
Winkler, Can. (wĭnk′lẽr)	57	49°11′N	97°56′W
Winneba, Ghana (wĭn′ē-bà)	188	5°25′N	0°36′W
Winnebago, Mn., U.S. (wĭn′ē-bā′gō)	71	43°45′N	94°08′W
Winnebago, Lake, l., Wi., U.S.	71	44°09′N	88°10′W

PLACE (Pronunciation)	PAGE	LAT.	LONG.
Winnebago Indian Reservation, I.R., Ne., U.S.	70	42°15′N	96°06′W
Winnemucca, Nv., U.S. (wĭn-ē-mŭk′ă)	64	40°59′N	117°43′W
Winnemucca, l., Nv., U.S.	76	40°06′N	119°07′W
Winner, S.D., U.S. (wĭn′ẽr)	70	43°22′N	99°50′W
Winnetka, Il., U.S. (wĭ-nĕt′kȧ)	69a	42°07′N	87°44′W
Winnett, Mt., U.S. (wĭn′ĕt)	73	47°01′N	108°20′W
Winnfield, La., U.S. (wĭn′fēld)	81	31°56′N	92°39′W
Winnibigoshish, l., Mn., U.S. (wĭn′ĭ-bĭ-gō′shĭsh)	71	47°30′N	93°45′W
Winnipeg, Can. (wĭn′ĭ-pĕg)	50	49°53′N	97°09′W
Winnipeg, r., Can.	52	50°30′N	95°00′W
Winnipeg, Lake, l., Can.	52	52°00′N	97°00′W
Winnipegosis, Can. (wĭn′ĭ-pē-gō′sĭs)	50	51°39′N	99°56′W
Winnipegosis, l., Can.	52	52°30′N	100°00′W
Winnipesaukee, l., N.H., U.S. (wĭn′ē-pē-sô′kē)	67	43°40′N	71°20′W
Winnsboro, La., U.S. (wĭnz′bŭr′ô)	81	32°09′N	91°42′W
Winnsboro, S.C., U.S.	83	34°29′N	81°05′W
Winnsboro, Tx., U.S.	79	32°56′N	95°15′W
Winona, Can. (wĭ-nō′nȧ)	62d	43°13′N	79°39′W
Winona, Mn., U.S.	65	44°03′N	91°40′W
Winona, Ms., U.S.	82	33°29′N	89°43′W
Winooski, Vt., U.S. (wĭ-nōōs′kē)	67	44°30′N	73°10′W
Winsen, Ger. (vēn′zĕn)	115c	53°22′N	10°13′E
Winsford, Eng., U.K. (wĭnz′fẽrd)	114a	53°11′N	2°30′W
Winslow, Az., U.S. (wĭnz′lō)	77	35°00′N	110°45′W
Winslow, Wa., U.S.	74a	47°38′N	122°31′W
Winsted, Ct., U.S. (wĭn′stĕd)	67	41°55′N	73°05′W
Winster, Eng., U.K. (wĭn′stẽr)	114a	53°08′N	1°38′W
Winston-Salem, N.C., U.S. (wĭn stŭn-sā′lĕm)	65	36°05′N	80°15′W
Winterberge, mts., Afr.	187c	32°18′S	26°25′E
Winter Garden, Fl., U.S. (wĭn′tẽr gär′d′n)	83a	28°32′N	81°35′W
Winter Haven, Fl., U.S. (hā′vĕn)	83a	28°01′N	81°38′W
Winter Park, Fl., U.S. (pärk)	83a	28°35′N	81°21′W
Winters, Tx., U.S. (wĭn′tẽrz)	80	31°59′N	99°58′W
Winterset, Ia., U.S. (wĭn′tẽr-sĕt)	71	41°19′N	94°03′W
Winterswijk, Neth.	127c	51°58′N	6°44′E
Winterthur, Switz. (vĭn′tĕr-tōōr)	124	47°30′N	8°32′E
Winterton, S. Afr.	187c	28°51′S	29°33′E
Winthrop, Ma., U.S.	61a	42°23′N	70°59′W
Winthrop, Me., U.S. (wĭn′thrŭp)	60	44°19′N	70°00′W
Winthrop, Mn., U.S.	71	44°31′N	94°20′W
Winton, Austl. (wĭn-tŭn)	175	22°17′S	143°08′E
Wipperfürth, Ger. (vē′pĕr-fürt)	127c	51°07′N	7°23′E
Wirksworth, Eng., U.K. (wûrks′wûrth)	114a	53°05′N	1°35′W
Wisconsin, state, U.S. (wĭs-kŏn′sĭn)	65	44°30′N	91°00′W
Wisconsin, r., Wi., U.S.	65	43°14′N	90°34′W
Wisconsin Dells, Wi., U.S.	71	43°38′N	89°46′W
Wisconsin Rapids, Wi., U.S.	71	44°24′N	89°50′W
Wishek, N.D., U.S. (wĭsh′ĕk)	70	46°15′N	99°34′W
Wisła, r., Pol. (vēs′wà)	112	52°30′N	20°00′E
Wisłoka, r., Pol. (vēs-wô′kà)	125	49°55′N	21°26′E
Wismar, Ger. (vĭs′mär)	116	53°53′N	11°28′E
Wismar, Guy. (wĭs′măr)	101	5°58′N	58°15′W
Wisner, Ne., U.S. (wĭz′nẽr)	70	42°00′N	96°55′W
Wissembourg, Fr. (vē-sän-bōōr′)	127	49°03′N	7°58′E
Wister, Lake, l., Ok., U.S. (vĭs′tẽr)	79	35°02′N	94°52′W
Witbank, S. Afr. (wĭt-băŋk)	192c	25°53′S	29°14′E
Witberg, mtn., Afr.	187c	30°32′S	27°18′E
Witham, Eng., U.K. (wĭdh′ăm)	114b	51°48′N	0°37′E
Witham, r., Eng., U.K.	114a	53°11′N	0°20′W
Withamsville, Oh., U.S. (wĭdh′ămz-vĭl)	69f	39°04′N	84°16′W
Withlacoochee, r., Fl., U.S. (wĭth-là-kōō′chē)	83a	28°58′N	82°30′W
Withlacoochee, r., Ga., U.S.	82	31°15′N	83°30′W
Withrow, Mn., U.S. (wĭdh′rō)	75g	45°08′N	92°54′W
Witney, Eng., U.K. (wĭt′nē)	114b	51°45′N	1°30′W
Witt, Il., U.S. (vĭt)	66	39°10′N	89°15′W
Witten, Ger. (vē′tĕn)	127c	51°26′N	7°19′E
Wittenberg, Ger. (vē′tĕn-bĕrgh)	124	51°53′N	12°40′E
Wittenberge, Ger. (vĭt-ĕn-bĕr′gē)	124	52°59′N	11°45′E
Wittlich, Ger. (vĭt′lĭk)	124	49°58′N	6°54′E
Witu, Kenya (wē′tōō)	187	2°18′S	40°28′E
Witu Islands, is., Pap. N. Gui.	169	4°45′S	149°50′E
Witwatersberg, mts., S. Afr. (wĭt-wôr-tẽrz-bûrg)	187b	25°58′S	27°53′E
Witwatersrand, mtn., S. Afr. (wĭt-wôr′tẽrs-rănd)	192c	25°55′S	26°27′E
Wkra, r., Pol. (f′krà)	125	52°40′N	20°35′E
Włocławek, Pol. (vwô-tswä′vĕk)	125	52°38′N	19°08′E
Włodawa, Pol. (vwô-dä′vä)	125	51°33′N	23°33′E
Włoszczowa, Pol. (vwôsh-chô′vä)	125	50°51′N	19°58′E
Woburn, Ma., U.S. (wô′bŭrn) (wō′bŭrn)	61a	42°29′N	71°10′W
Woerden, Neth.	115a	52°05′N	4°52′E
Woking, Eng., U.K.	114b	51°18′N	0°33′W
Wokingham, Eng., U.K. (wô′kĭng-hăm)	114b	51°23′N	0°50′W
Wolcott, Ks., U.S. (wōl′kŏt)	75f	39°12′N	94°47′W
Wolf, i., Can. (wŏlf)	59	44°10′N	76°25′W
Wolf, r., Ms., U.S.	82	30°45′N	89°36′W
Wolf, r., Wi., U.S.	71	45°14′N	88°45′W
Wolfenbüttel, Ger. (vôl′fĕn-büt-ĕl)	124	52°10′N	10°32′E
Wolf Lake, l., Il., U.S.	69a	41°39′N	87°33′W
Wolf Point, Mt., U.S. (wŏlf point)	73	48°07′N	105°40′W
Wolfratshausen, Ger. (vōlf′räts-hou-zĕn)	115d	47°55′N	11°25′E
Wolfsburg, Ger. (vôlfs′bōōrgh)	124	52°30′N	10°37′E
Wolfville, Can. (wŏlf′vĭl)	60	45°05′N	64°22′W
Wolgast, Ger. (vôl′găst)	124	54°04′N	13°46′E
Wolhuterskop, S. Afr.	187b	25°41′S	27°40′E
Wolkersdorf, Aus.	115e	48°24′N	16°31′E

ăt; finăl; rāte; senåte; ärm; åsk; sofȧ; fâre; ch-choose; dh-as th in other; bē; ĕvent; bĕt; recĕnt; cratẽr; g-gō; gh-guttural g; bĭt; ĭ-short neutral; rīde; ᴋ-guttural k as ch in German ich;

PRONOUNCING INDEX Yandongi – Zaltbommel 321

PLACE (Pronunciation)	PAGE	LAT.	LONG.
Yandongi, D.R.C.	190	2°51'N	22°16'E
Yangcheng Hu, l., China (yäŋ-chŭŋ hōō)	162	31°30'N	120°31'E
Yangchun, China (yäŋ-chòn)	165	22°08'N	111°48'E
Yang'erzhuang, China (yäŋ-är-jůäŋ)	162	38°18'N	117°31'E
Yanggezhuang, China (yäŋ-gŭ-jůäŋ)	164a	40°10'N	116°48'E
Yanggu, China (yäŋ-gōō)	162	36°06'N	115°46'E
Yanghe, China (yäŋ-hŭ)	162	33°48'N	118°23'E
Yangjiang, China (yäŋ-jyäŋ)	165	21°52'N	111°58'E
Yangjiaogou, China (yäŋ-jyou-gō)	162	37°17'N	118°53'E
Yangon *see* Rangoon, Mya.	155	16°46'N	96°09'E
Yangquan, China (yäŋ-chyůän)	162	37°52'N	113°36'E
Yangtze (Chang), r., China (yäng'tse) (chäŋ)	161	30°30'N	117°25'E
Yangxin, China (yäŋ-shyĭn)	162	37°39'N	117°34'E
Yangyang, Kor., S. (yäng'yäng')	166	38°02'N	128°38'E
Yangzhou, China (yäŋ-jō)	161	32°24'N	119°24'E
Yanji, China (yän-jyē)	161	42°55'N	129°35'E
Yanjiahe, China (yän-jyä-hŭ)	162	31°55'N	114°47'E
Yanjin, China (yän-jyĭn)	162	35°09'N	114°13'E
Yankton, S.D., U.S. (yănk'tŭn)	64	42°51'N	97°24'W
Yanling, China (yän-lĭŋ)	162	34°07'N	114°12'E
Yanshan, China (yän-shän)	164	38°05'N	117°15'E
Yanshou, China (yän-shō)	164	45°25'N	128°43'E
Yantai, China	161	37°32'N	121°22'E
Yanychi, Russia (yä-nĭ-chĭ)	142a	57°42'N	56°24'E
Yanzhou, China (yäŋ-jō)	161	35°35'N	116°50'E
Yanzhuang, China (yän-jůäŋ)	162	36°08'N	117°47'E
Yao, Chad (yä'ō)	185	13°00'N	17°38'E
Yao, Japan	167b	34°37'N	135°37'E
Yaoundé, Cam.	184	3°52'N	11°31'E
Yap, i., Micron. (yăp)	3	11°00'N	138°00'E
Yapen, Pulau, i., Indon.	169	1°30'S	136°15'E
Yaque del Norte, r., Dom. Rep. (yä'kå dĕl nôr'tå)	87	19°40'N	71°25'W
Yaque del Sur, r., Dom. Rep. (yä-kĕ-dĕl-sōō'r)	93	18°35'N	71°05'W
Yaqui, r., Mex. (yä'kē)	86	28°15'N	109°40'W
Yaracuy, dept., Ven. (yä-rä-kōō'ē)	101b	10°10'N	68°31'W
Yaraka, Austl. (yå-răk'ȧ)	175	24°50'S	144°08'E
Yaransk, Russia (yä-ränsk')	134	57°18'N	48°05'E
Yarda, oasis, Chad (yär'dȧ)	185	18°29'N	19°13'E
Yare, r., Eng., U.K.	121	52°40'N	1°32'E
Yarkand *see* Shache, China	160	38°15'N	77°15'E
Yarmouth, Can. (yär'mŭth)	60	43°50'N	66°07'W
Yaroslavka, Russia (yå-rô-släv'kå)	142a	55°52'N	57°59'E
Yaroslavl', Russia (yå-rô-släv''l)	134	57°37'N	39°54'E
Yaroslavl', prov., Russia	132	58°05'N	38°05'E
Yarra, r., Austl.	173a	37°51'S	144°54'E
Yarro-to, l., Russia (yá'rô-tô')	136	67°55'N	71°35'E
Yartsevo, Russia (yär'tsyê-vô)	136	55°04'N	32°38'E
Yartsevo, Russia	135	60°13'N	89°52'E
Yarumal, Col. (yä-rōō-mäl')	100	6°57'N	75°24'W
Yasawa Group, is., Fiji	170g	17°00'S	177°23'E
Yasel'da, r., Bela. (yä-syŭl'dá)	125	52°13'N	25°53'E
Yateras, Cuba (yä-tā'räs)	93	20°00'N	75°00'W
Yates Center, Ks., U.S. (yāts)	79	37°53'N	95°44'W
Yathkyed, l., Can. (yàth-kī-ĕd')	52	62°41'N	98°00'W
Yatsuga-take, mtn., Japan (yät'sōō-gä dä'kä)	167	36°01'N	138°21'W
Yatsushiro, Japan (yät'sōō'shē-rô)	167	32°30'N	130°35'E
Yatta Plateau, plat., Kenya	191	1°55'S	38°10'E
Yautepec, Mex. (yä-ōō-tå-pĕk')	88	18°53'N	99°04'W
Yawata, Japan (yä'wä-tä)	167	34°52'N	135°43'E
Yawatahama, Japan (yä'wä'tä'hä-mä)	167	33°24'N	132°25'E
Yaxian, China (yä shyĕn)	165	18°10'N	109°32'E
Yayama, D.R.C.	190	1°16'S	23°07'E
Yayao, China (yä-you)	163a	23°10'N	113°40'E
Yazd, Iran	154	31°59'N	54°03'E
Yazoo, r., Ms., U.S. (yă'zōō)	65	32°32'N	90°40'W
Yazoo City, Ms., U.S.	82	32°50'N	90°18'W
Ydra, i., Grc.	131	37°20'N	23°30'E
Ye, Mya. (yā)	168	15°13'N	97°52'E
Yeadon, Pa., U.S. (yē'dŭn)	68f	39°56'N	75°16'W
Yecla, Spain (yā'klä)	128	38°35'N	1°09'W
Yefremov, Russia (yĕ-frā'môf)	132	53°08'N	38°04'E
Yegor'yevsk, Russia (yĕ-gôr'yĕfsk)	136	55°23'N	38°59'E
Yeji, China (yŭ-jyē)	162	31°52'N	115°57'E
Yekaterinburg, Russia	134	56°51'N	60°36'E
Yelabuga, Russia (yĕ-lä'bò-gá)	136	55°50'N	52°18'E
Yelan, Russia	137	50°50'N	44°00'E
Yelets, Russia (yĕ-lyĕts')	134	52°35'N	38°28'E
Yelizavetpol'skiy, Russia (yĕ'lĭ-za-vĕt-pôl-skĭ)	142a	52°51'N	60°38'E
Yelizavety, Mys, c., Russia (yĕ-lyē-sá-vyĕ'tĭ)	135	54°28'N	142°59'E
Yell, i., Scot., U.K. (yĕl)	120a	60°35'N	1°27'W
Yellow *see* Huang, r., China	161	35°06'N	113°39'E
Yellow, r., Fl., U.S. (yĕl'ō)	82	30°33'N	86°53'W
Yellowhead Pass, p., Can. (yĕl'ō-hĕd)	55	52°52'N	118°35'W
Yellowknife, Can. (yĕl'ō-nīf)	50	62°29'N	114°38'W
Yellow Sea, sea, Asia	161	35°20'N	122°15'E
Yellowstone, r., U.S.	64	46°00'N	108°00'W
Yellowstone, Clarks Fork, r., U.S.	73	44°55'N	109°05'W
Yellowstone Lake, l., Wy., U.S.	64	44°27'N	110°03'W
Yellowstone National Park, rec., U.S. (yĕl'ō-stōn)	64	44°45'N	110°35'W
Yel'nya, Russia (yĕl'nyá)	132	54°34'N	33°12'E
Yemanzhelinsk, Russia (yĕ-mán-zhâ'lĭnsk)	142a	54°47'N	61°24'E
Yemen, nation, Asia (yĕm'ĕn)	154	15°00'N	47°00'E
Yemetsk, Russia	136	63°28'N	41°28'E
Yenangyaung, Mya. (yā'nän-d oung)	155	20°27'N	94°59'E
Yencheng, China	160	37°30'N	79°26'E
Yendi, Ghana (yĕn'dē)	184	9°26'N	0°01'W
Yengisar, China (yŭn-gē-sär)	160	39°01'N	75°29'E
Yenice, r., Tur.	137	41°10'N	33°00'E
Yenisey, r., Russia (yĕ-nê-sĕ'ê)	134	71°00'N	82°00'E
Yeniseysk, Russia (yĕ-nĭĕsă'ĭsk)	135	58°27'N	90°28'E
Yeo, l., Austl. (yō)	174	28°15'S	124°00'E
Yerevan, Arm. (yĕ-rĕ-vän')	137	40°10'N	44°30'E
Yerington, Nv., U.S. (yê'rĭng-tŭn)	76	38°59'N	119°10'W
Yermak, i., Russia	136	66°45'N	71°30'E
Yeste, Spain (yĕs'tå)	128	38°23'N	2°19'W
Yeu, Île d', i., Fr. (ēl dyû)	117	46°43'N	2°45'W
Yevlax, Azer.	138	40°36'N	47°09'E
Yexian, China (yŭ-shyĕn)	162	37°09'N	119°57'E
Yeya, r., Russia (yă'yȧ)	133	46°25'N	39°17'E
Yeysk, Russia (yĕysk)	137	46°41'N	38°13'E
Yi, r., China	162	34°38'N	118°07'E
Yibin, China (yē-bĭn)	160	28°50'N	104°40'E
Yichang, China (yē-chäŋ)	161	30°38'N	111°22'E
Yidu, China (yē-dōō)	164	36°42'N	118°30'E
Yilan, China (yē-län)	161	46°10'N	129°40'E
Yinchuan, China (yĭn-chůän)	160	38°22'N	106°22'E
Yingkou, China (yĭŋ-kō)	161	40°35'N	122°10'E
Yining, China (yē-nĭŋ)	160	43°58'N	80°40'E
Yin Shan, mts., China (yĭng'shän')	164	40°50'N	110°30'E
Yishan, China (yē-shän)	160	24°32'N	108°42'E
Yishui, China (yē-shwā)	162	35°49'N	118°40'E
Yitong, China (yē-tôŋ)	161	43°15'N	125°10'E
Yixian, China (yē shyĕn)	164	41°30'N	121°15'E
Yixing, China	162	31°26'N	119°57'E
Yiyang, China (yē-yäŋ)	165	28°52'N	112°12'E
Yoakum, Tx., U.S. (yō'kŭm)	81	29°18'N	97°09'W
Yockanookany, r., Ms., U.S. (yŏk'á-nōō-kă-nĭ)	82	32°47'N	89°38'W
Yodo-Gawa, strt., Japan (yō'dō'gä-wä)	167b	34°46'N	135°35'E
Yog Point, c., Phil. (yōg)	165	14°00'N	124°30'E
Yogyakarta, Indon. (yŏg-yá-kär'tá)	168	7°50'S	110°20'E
Yoho National Park, rec., Can. (yō'hō)	50	51°26'N	116°30'W
Yojoa, Lago de, l., Hond. (lä'gô dĕ yô-hō'ä)	90	14°49'N	87°53'W
Yokkaichi, Japan (yō'kä'ē-chê)	166	34°58'N	136°35'E
Yokohama, Japan (yō'kō-hä'mạ)	161	35°37'N	139°40'E
Yokosuka, Japan (yō-kō'sò-kä)	166	35°17'N	139°40'E
Yokota, Japan (yō-kō'tä)	167a	35°23'N	140°02'E
Yola, Nig. (yō'lä)	184	9°13'N	12°27'E
Yolaina, Cordillera de, mts., Nic.	91	11°34'N	84°34'W
Yomou, Gui.	188	7°34'N	9°16'W
Yonago, Japan (yō'nä-gō)	166	35°27'N	133°19'E
Yonezawa, Japan (yō'nĕ'zá-wä)	166	37°50'N	140°07'E
Yong'an, China (yôŋ-än)	165	26°00'N	117°22'E
Yongding, r., China (yôŋ-dĭŋ)	164	40°25'N	115°00'E
Yŏngdŏk, Kor., S. (yŭng'dŭk')	166	36°28'N	129°25'E
Yŏnghŭng, Kor., N. (yŭng'hŏng')	166	39°31'N	127°11'E
Yŏnghŭng Man, b., Kor., N.	166	39°10'N	128°00'E
Yongnian, China (yôŋ-nĭĕn)	164	36°47'N	114°32'E
Yongqing, China (yôŋ-chyĭŋ)	164a	39°18'N	116°27'E
Yongshun, China (yôŋ-shòn)	160	29°05'N	109°58'E
Yonkers, N.Y., U.S. (yŏŋ'kĕrz)	68a	40°57'N	73°54'W
Yonne, r., Fr. (yôn)	126	48°18'N	3°15'E
Yono, Japan (yō'nō)	167a	35°53'N	139°36'E
Yorba Linda, Ca., U.S. (yôr'bá lĭn'dá)	75a	33°55'N	117°51'W
York, Austl.	174	32°00'S	117°00'E
York, Eng., U.K.	116	53°58'N	1°10'W
York, Al., U.S. (yôrk)	82	32°33'N	88°16'W
York, Ne., U.S.	79	40°52'N	97°36'W
York, Pa., U.S.	65	40°00'N	76°40'W
York, S.C., U.S.	83	34°59'N	81°14'W
York, Cape, c., Austl.	175	10°45'S	142°35'E
York, Kap, c., Grnld.	49	75°30'N	73°00'W
Yorke Peninsula, pen., Austl.	176	34°24'S	137°20'E
Yorketown, Austl.	176	35°00'S	137°28'E
York Factory, Can.	57	57°05'N	92°18'W
Yorkshire Wolds, Eng., U.K. (yôrk'shĭr)	120	54°00'N	0°35'W
Yorkton, Can. (yôrk'tŭn)	50	51°13'N	102°28'W
Yorktown, Tx., U.S. (yôrk'toun)	81	28°57'N	97°30'W
Yorktown, Va., U.S.	83	37°12'N	76°31'W
Yoro, Hond. (yō'rō)	90	15°09'N	87°05'W
Yoron, i., Japan	166	26°48'N	128°40'E
Yosemite National Park, rec., Ca., U.S. (yô-sĕm'ĭ-tê)	64	38°03'N	119°36'W
Yoshida, Japan (yō'shê-dȧ)	167	34°39'N	132°41'E
Yoshikawa, Japan (yō-shê'kä'wä')	167a	35°53'N	139°51'E
Yoshino, r., Japan (yō'shē-nō)	167	34°04'N	133°57'E
Yoshkar-Ola, Russia (yôsh-kär'ō-lä')	136	56°35'N	48°05'E
Yos Sudarsa, Pulau, i., Indon.	169	7°20'S	138°30'E
Yŏsu, Kor., S. (yŭ'sōō')	166	34°42'N	127°42'W
You, r., China (yō)	165	23°55'N	106°50'E
Youghal, Ire. (yōō'ôl) (yôl)	121	51°58'N	7°57'E
Youghal Bay, b., Ire.	120	51°52'N	7°46'W
Young, Austl. (yŭng)	176	34°15'S	148°18'E
Young, Ur. (yô-ōō'ng)	99c	32°42'S	57°38'W
Youngs, l., Wa., U.S. (yŭngz)	74a	47°25'N	122°08'W
Youngstown, N.Y., U.S.	69c	43°15'N	79°02'W
Youngstown, Oh., U.S.	66	41°05'N	80°40'W
Yozgat, Tur. (yōz'gȧd)	154	39°50'N	34°50'E
Ypsilanti, Mi., U.S. (ĭp-sĭ-lăn'tĭ)	69b	42°15'N	83°37'W
Yreka, Ca., U.S. (wī-rē'kȧ)	72	41°43'N	122°36'W
Yrghyz, Kaz.	139	48°30'N	61°17'E
Yrghyz, r., Kaz.	112	49°30'N	60°32'E
Ysleta, Tx., U.S. (ēz-lĕ'tȧ)	80	31°42'N	106°18'W
Yssingeaux, Fr. (ē-săN-zhō)	126	45°09'N	4°08'E
Ystad, Swe.	116	55°25'N	13°49'E
Ystādeh-ye Moqor, Āb-e, l., Afg.	158	32°35'N	68°00'E
Yu'alliq, Jabal, mts., Egypt	153a	30°12'N	33°42'E
Yuan, r., China (yůän)	161	28°50'N	110°50'E
Yuan'an, China (yůän-än)	165	31°08'N	111°28'E
Yuanling, China (yůän-lĭŋ)	165	28°30'N	110°18'E
Yuanshi, China (yůän-shr)	164	37°45'N	114°32'E
Yuasa, Japan	167	34°02'N	135°10'E
Yuba City, Ca., U.S. (yōō'bȧ)	76	39°08'N	121°38'W
Yucaipa, Ca., U.S. (yŭ-kȧ-ē'pȧ)	75a	34°02'N	117°02'W
Yucatán, state, Mex. (yōō-kä-tän')	86	20°45'N	89°00'W
Yucatán Channel, strt., N.A.	86	22°30'N	87°00'W
Yucatán Peninsula, pen., N.A.	90	19°30'N	89°00'W
Yucheng, China (yōō-chŭŋ)	162	34°31'N	115°54'E
Yucheng, China	164	36°55'N	116°39'E
Yuci, China (yōō-tsz)	164	37°32'N	112°40'E
Yudoma, r., Russia (yōō-dō'mȧ)	141	59°13'N	137°00'E
Yueqing, China (yůĕ-chyĭn)	165	28°02'N	120°40'E
Yueyang, China (yůĕ-yäŋ)	161	29°25'N	113°05'E
Yuezhuang, China (yůĕ-jůäŋ)	162	36°13'N	118°17'E
Yug, r., Russia (yôg)	136	59°50'N	45°55'E
Yugoslavia *see* Serbia and Montenegro, nation, Eur. (yōō-gô-slä-vĭ-ȧ)	110	44°00'N	21°00'E
Yukhnov, Russia (yôk'nof)	132	54°44'N	35°15'E
Yukon, ter., Can. (yōō'kŏn)	50	63°16'N	135°30'W
Yukon, r., N.A.	64a	64°00'N	159°30'W
Yukutat Bay, b., Ak., U.S. (yōō-kū tăt')	63	59°34'N	140°50'W
Yuldybayevo, Russia (yòld'bä'yĕ-vô)	142a	52°20'N	57°52'E
Yulin, China (yōō-lĭn)	165	22°38'N	110°10'E
Yulin, China	160	38°18'N	109°45'E
Yuma, Az., U.S. (yōō'mä)	64	32°40'N	114°40'W
Yuma, Co., U.S.	78	40°08'N	102°50'W
Yuma, r., Dom. Rep.	93	19°05'N	70°05'W
Yumbi, D.R.C.	191	1°14'S	26°14'E
Yumen, China (yōō-mŭn)	160	40°14'N	96°56'E
Yuncheng, China (yòn-chŭŋ)	164	35°00'N	110°40'E
Yunnan, prov., China (yun'nän')	160	24°23'N	101°03'E
Yunnan Plat, plat., China (yò-nän)	160	26°03'N	101°26'E
Yunxian, China (yòn shyĕn)	161	32°50'N	110°55'E
Yunxiao, China (yòn-shyou)	165	24°00'N	117°20'E
Yura, Japan (yōō'rä)	167	34°18'N	134°54'E
Yurécuaro, Mex. (yōō-rā'kwä-rô)	88	20°21'N	102°16'W
Yurimaguas, Peru (yōō-rê-mä'gwäs)	100	5°59'S	76°12'W
Yuriria, Mex. (yōō'rê-rē'ä)	88	20°11'N	101°08'W
Yurovo, Russia	142b	55°30'N	38°24'E
Yur'yevets, Russia	136	57°15'N	43°08'E
Yuscarán, Hond. (yōōs-kä-rän')	90	13°57'N	86°48'W
Yushan, China (yōō-shän)	165	28°42'N	118°20'E
Yü Shan, mtn., Tai.	161	23°38'N	121°05'E
Yushu, China (yōō-shōō)	164	44°58'N	126°32'E
Yutian, China (yōō-tĭĕn)	164	39°54'N	117°45'E
Yutian, China (yōō-tĭĕn) (kŭ-r-yä)	160	36°55'N	81°39'E
Yuty, Para. (yōō-tē')	102	26°45'S	56°13'W
Yuwangcheng, China (yü'wäng'chĕng)	162	31°32'N	114°26'E
Yuxian, China (yōō shyĕn)	164	39°40'N	114°38'E
Yuzha, Russia (yōō'zhȧ)	136	56°38'N	42°20'E
Yuzhno-Sakhalinsk, Russia (yōōzh'nô-sä-kä-lĭnsk')	135	47°11'N	143°04'E
Yuzhnoural'skiy, Russia (yōōzh-nô-ò-rál'skĭ)	142a	54°26'N	61°17'E
Yuzhnyy Ural, mts., Russia (yōō'zhnĭ ò-räl')	142a	52°51'N	57°48'E
Yverdon, Switz. (ê-vĕr-dôn)	124	46°46'N	6°35'E
Yvetot, Fr. (ēv-tō')	126	49°39'N	0°45'E

Z

PLACE (Pronunciation)	PAGE	LAT.	LONG.
Za, r., Mor.	118	34°19'N	2°23'W
Zaachila, Mex. (sä-ä-chē'lä)	89	16°56'N	96°45'W
Zaandam, Neth. (zän'dám)	121	52°25'N	4°49'E
Ząbkowice Śląskie, Pol.	124	50°35'N	16°48'E
Zabrze, Pol. (zäb'zhĕ)	117	50°18'N	18°48'E
Zacapa, Guat. (sä-kä'pä)	90	14°56'N	89°30'W
Zacapoaxtla, Mex. (sä-kä-pō-äs'tlä)	89	19°51'N	97°34'W
Zacatecas, Mex. (sä-kä-tā'käs)	86	22°44'N	102°32'W
Zacatecas, state, Mex.	86	24°00'N	102°45'W
Zacatecoluca, El Sal. (sä-kä-tå-kô-lōō'kä)	90	13°31'N	88°50'W
Zacatelco, Mex.	88	19°12'N	98°12'W
Zacatepec, Mex. (sä-kä-tå-pĕk') (sän-tê-ä'gô)	89	17°10'N	95°53'W
Zacatlán, Mex. (sä-kä-tlän')	89	19°55'N	97°57'W
Zacoalco de Torres, Mex. (sä-kô-äl'kô dä tōr'rĕs)	88	20°12'N	103°33'W
Zacualpan, Mex. (sä-kò-äl-pän')	88	18°43'N	99°46'W
Zacualtipan, Mex. (sá-kò-äl-tē-pän')	88	20°38'N	98°39'W
Zadar, Cro. (zä'där)	110	44°08'N	15°16'E
Zadonsk, Russia (zä-dônsk')	132	52°22'N	38°55'E
Žagare, Lat. (zhȧgȧrē)	123	56°21'N	23°14'E
Zagarolo, Italy (tzä-gä-rō'lô)	129d	41°51'N	12°53'E
Zaghouan, Tun. (zá-gwän')	184	36°30'N	10°04'E
Zagreb, Cro. (zä'grĕb)	110	45°50'N	15°58'E
Zagros Mountains, mts., Iran	154	33°30'N	46°30'E
Zāhedān, Iran (zä'hå-dān)	154	29°37'N	60°31'E
Zahlah, Leb. (zä'lā')	153a	33°50'N	35°54'E
Zaire *see* Congo, Democratic Republic of the, nation, Afr.	186	1°00'S	22°15'E
Zaječar, Serb. (zä'yĕ-chär')	131	43°54'N	22°16'E
Zakhidnyi Buh (Bug), r., Eur.	124	52°29'N	21°20'E
Zakopane, Pol. (zá-kô-pä'nĕ)	125	49°18'N	19°57'E
Zakouma, Parc National de, rec., Chad	189	10°50'N	19°20'E
Zákynthos, Grc.	131	37°48'N	20°55'E
Zákynthos, i., Grc.	119	37°45'N	20°32'E
Zalaegerszeg, Hung. (zŏ'lô-ĕ'gĕr-sĕg)	124	46°50'N	16°50'E
Zalău, Rom. (zȧ-lŭ'ò)	125	47°11'N	23°06'E
Zaltan, Libya	185	28°20'N	19°40'E
Zaltbommel, Neth.	115a	51°48'N	5°15'E

322 Zambezi – Zyryanovsk PRONOUNCING INDEX

PLACE (Pronunciation)	PAGE	LAT.	LONG.
Zambezi, r., Afr. (zȧm-bā´zē)	186	16°00′S	29°45′E
Zambia, nation, Afr. (zăm´bē-ȧ)	186	14°23′S	24°15′E
Zamboanga, Phil. (säm-bō-aŋ´gä)	168	6°58′N	122°02′E
Zambrów, Pol. (zäm´brōf)	125	52°29′N	22°17′E
Zamora, Mex. (sä-mō´rä)	86	19°59′N	102°16′W
Zamora, Spain (thä-mō´rä)	118	41°32′N	5°43′W
Zanatepec, Mex.	89	16°30′N	94°22′W
Zandvoort, Neth.	115a	52°22′N	4°30′E
Zanesville, Oh., U.S. (zānz´vĭl)	66	39°55′N	82°00′W
Zangasso, Mali	188	12°09′N	5°37′W
Zanjān, Iran	154	36°26′N	48°24′E
Zanzibar, Tan. (zăn´zĭ-bär)	187	6°10′S	39°11′E
Zanzibar, i., Tan.	187	6°20′S	39°37′E
Zanzibar Channel, strt., Tan.	191	6°05′S	39°00′E
Zaozhuang, China (dzou-jůäŋ)	162	34°51′N	117°34′E
Zapadnaya Dvina *see* Western Dvina, r., Eur.	123	55°30′N	28°27′E
Zapala, Arg. (zä-pä´lä)	102	38°53′S	70°02′W
Zapata, Tx., U.S. (sä-pä´tä)	80	26°52′N	99°18′W
Zapata, Ciénaga de, sw., Cuba (syĕ´nä-gä-dĕ-zä-pȧ´tä)	92	22°30′N	81°20′W
Zapata, Península de, pen., Cuba (pĕ-nē´n-so͞o-lä-dĕ-zä-pȧ´tä)	92	22°20′N	81°30′W
Zapatera, Isla, i., Nic. (ē´s-lä-sä-pä-tā´rō)	90	11°45′N	85°45′W
Zapopan, Mex. (sä-pō´pän)	88	20°42′N	103°23′W
Zaporizhzhia, Ukr.	134	47°50′N	35°10′E
Zaporizhzhia, prov., Ukr.	133	47°20′N	35°05′E
Zaporoshskoye, Russia (zȧ-pō-rôsh´skō-yĕ)	123	60°36′N	30°31′E
Zapotiltic, Mex. (sä-pō-tēl-tēk´)	88	19°37′N	103°25′W
Zapotitlán, Mex. (sä-pō-tē-tlän´)	88	17°13′N	98°58′W
Zapotitlán, Punta, c., Mex.	89	18°34′N	94°48′W
Zapotlanejo, Mex. (sä-pō-tlä-nā´hō)	88	20°38′N	103°05′W
Zaragoza, Mex. (sä-rä-gō´sä)	88	23°59′N	99°45′W
Zaragoza, Mex.	88	22°02′N	100°45′W
Zaragoza, Spain (thä-rä-gō´thä)	110	41°39′N	0°53′W
Zarand, Munţii, mts., Rom.	125	46°07′N	22°21′E
Zaranda Hill, mtn., Nig.	189	10°15′N	9°35′E
Zaranj, Afg.	157	31°06′N	61°53′E
Zarasai, Lith. (zä-rä-sī´)	123	55°45′N	26°18′E
Zárate, Arg. (zä-rä´tā)	102	34°05′S	59°05′W
Zaraysk, Russia (zä-rä´ēsk)	136	54°46′N	38°53′E
Zaria, Nig. (zä´rē-ä)	184	11°07′N	7°44′E
Zarqā', r., Jord.	153a	32°13′N	35°43′E
Zarzal, Col. (zär-zȧ´l)	100a	4°23′N	76°04′W
Zashiversk, Russia (zȧ´shĭ-vĕrsk´)	135	67°08′N	144°02′E
Zastavna, Ukr. (zȧs-täf´nȧ)	125	48°32′N	25°50′E
Zastron, S. Afr. (zȧs´trŭn)	187c	30°19′S	27°07′E
Žatec, Czech Rep. (zhä´tĕts)	124	50°19′N	13°32′E
Zavitinsk, Russia	141	50°12′N	129°44′E
Zawiercie, Pol. (zȧ-vyĕr´tsyĕ)	125	50°28′N	19°25′E
Zāwiyat al-Baydā', Libya	185	32°49′N	21°46′E
Zāyandeh, r., Iran	154	32°15′N	51°00′E
Zaysan, Kaz. (zī´sȧn)	139	47°43′N	84°44′E
Zaza, r., Cuba (zȧ´zȧ)	92	21°40′N	79°25′W
Zbarazh, Ukr. (zbä-räzh´)	125	49°39′N	25°48′E
Zbruch, r., Ukr. (zbrōch)	125	48°56′N	26°18′E
Zdolbuniv, Ukr.	125	50°31′N	26°17′E
Zduńska Wola, Pol. (zdo͞on´'skä vō´lä)	125	51°36′N	18°27′E
Zebediela, S. Afr.	192c	24°19′S	29°21′E
Zeeland, Mi., U.S. (zē´lănd)	66	42°50′N	86°00′W
Zefat, Isr.	153a	32°58′N	35°30′E
Zehdenick, Ger. (tsā´dē-nēk)	124	52°59′N	13°20′E
Zehlendorf, Ger. (tsā´lĕn-dôrf)	115b	52°47′N	13°23′E
Zeist, Neth.	115a	52°05′N	5°14′E
Zelenogorsk, Russia (zĕ-lă´nō-gōrsk)	123	60°13′N	29°39′E
Zella-Mehlis, Ger. (tsâl´ȧ-mā´lēs)	124	50°40′N	10°38′E
Zémio, C.A.R. (za-myō´)	185	5°03′N	25°11′E
Zemlya Frantsa-Iosifa (Franz Josef Land), is., Russia	134	81°32′N	40°00′E
Zempoala, Punta, c., Mex. (po͞o´n-tä-sĕm-pō-ä´lä)	89	19°30′N	96°18′W
Zempoatlépetl, mtn., Mex. (sĕm-pō-ä-tlā'pĕt´l)	89	17°13′N	95°59′W
Zemun, Serb. (zĕ´mo͞on) (sĕm´lĭn)	119	44°50′N	20°25′E
Zengcheng, China (dzŭŋ-chŭŋ)	163a	23°18′N	113°49′E
Zenica, Bos. (zĕ´nēt-sä)	131	44°10′N	17°54′E
Zeni-Su, is., Japan (zĕ´nē so͞o)	167	33°55′N	138°55′E
Žepče, Bos. (zhĕp´chĕ)	133	44°26′N	18°01′E
Zepernick, Ger. (tsĕ´pĕr-nēk)	115b	52°39′N	13°32′E
Zerbst, Ger. (tsĕrbst)	124	51°58′N	12°03′E
Zerpenschleuse, Ger. (tsĕr´pĕn-shloi-zĕ)	115b	52°51′N	13°30′E
Zeuthen, Ger. (tsoi´tĕn)	115b	52°21′N	13°38′E
Zevenaar, Neth.	127c	51°56′N	6°06′E
Zevenbergen, Neth.	115a	51°38′N	4°36′E
Zeya, Russia (zå´yȧ)	135	53°43′N	127°29′E
Zeya, r., Russia	141	52°31′N	128°30′E
Zeytun, Tur. (zā-to͞on´)	137	38°00′N	36°40′E
Zezere, r., Port. (zĕ´zå-rĕ)	128	39°54′N	8°12′W
Zgierz, Pol. (zgyĕzh)	125	51°51′N	19°26′E
Zhambyl, Kaz.	139	42°51′N	71°29′E
Zhangaqazaly, Kaz.	139	45°47′N	62°00′E
Zhangbei, China (jäŋ-bā)	161	41°12′N	114°50′E
Zhanggezhuang, China (jäŋ-gŭ-jůäŋ)	162	40°09′N	116°56′E
Zhangguangcai Ling, mts., China (jäŋ-gůäŋ-tsī lĭŋ)	164	43°50′N	127°55′E
Zhangjiakou, China	161	40°45′N	114°58′E
Zhangqiu, China (jäŋ-chyō)	162	36°50′N	117°29′E
Zhangye, China (jäŋ-yu)	160	38°46′N	101°00′E
Zhangzhou, China (jäŋ-jō)	161	24°35′N	117°45′E
Zhangzi Dao, i., China (jäŋ-dz dou)	162	39°02′N	122°44′E
Zhanhua, China (jän-hwä)	162	37°42′N	117°49′E
Zhanjiang, China (jän-jyäŋ)	161	21°20′N	110°28′E
Zhanyu, China (jän-yo͞o)	164	44°30′N	122°30′E
Zhao'an, China (jou-än)	165	23°48′N	117°10′E
Zhaodong, China (jou-dōŋ)	164	45°58′N	126°00′E
Zhaotong, China (jou-tōŋ)	160	27°18′N	103°50′E
Zhaoxian, China (jou shyĕn)	162	37°46′N	114°48′E
Zhaoyuan, China (jou-yuän)	162	37°22′N	120°23′E
Zharkent, Kaz.	139	44°12′N	79°58′E
Zhaysang köli, l., Kaz.	139	48°16′N	84°05′E
Zhecheng, China (jŭ-chŭŋ)	164	34°05′N	115°19′E
Zhegao, China (jŭ-gou)	162	31°47′N	117°44′E
Zhejiang, prov., China (jŭ-jyäŋ)	161	29°30′N	120°00′E
Zhelaniya, Mys, c., Russia (zhĕ´lä-nĭ-yȧ)	134	75°43′N	69°10′E
Zhem, r., Kaz.	137	46°50′N	54°10′E
Zhengding, China (jŭŋ-dĭŋ)	164	38°10′N	114°35′E
Zhengyang, China (jŭŋ-yäŋ)	162	32°34′N	114°22′E
Zhengzhou, China (jŭŋ-jō)	161	34°46′N	113°42′E
Zhenjiang, China (jŭn-jyäŋ)	161	32°13′N	119°24′E
Zhenyuan, China (jŭn-yůän)	165	27°08′N	108°30′E
Zhetïqara, Kaz.	139	52°12′N	61°18′E
Zhigalovo, Russia (zhē-gä´lō-vō)	135	54°52′N	105°05′E
Zhigansk, Russia (zhē-gȧnsk´)	135	66°45′N	123°20′E
Zhijiang, China (jr-jyäŋ)	165	27°25′N	109°45′E
Zhizdra, Russia (zhēz´drȧ)	132	53°47′N	34°41′E
Zhizhitskoye, l., Russia (zhē-zhēt´skō-yĕ)	132	56°08′N	31°34′E
Zhmerynka, Ukr.	137	49°02′N	28°09′E
Zhongwei, China (jōŋ-wā)	160	37°32′N	105°10′E
Zhongxian, China (jōŋ shyĕn)	160	30°20′N	108°00′E
Zhongxin, China (jōŋ-shyĭn)	163a	23°16′N	113°38′E
Zhoucun, China (jō-tso͞on)	164	36°49′N	117°52′E
Zhoukouzhen, China (jō-kō-jŭn)	162	33°39′N	114°40′E
Zhoupu, China (jō-po͞o)	162	31°07′N	121°33′E
Zhoushan Qundao, is., China (jō-shän-chyōn-dou)	161	30°00′N	123°00′E
Zhouxian, China (jō shyĕn)	164	39°30′N	115°59′E
Zhovkva, Ukr.	125	50°03′N	23°58′E
Zhu, r., China (jo͞o)	163a	22°48′N	113°36′E
Zhuanghe, China (jůäŋ-hŭ)	164	39°40′N	123°00′E
Zhuanqiao, China (jůäŋ-chyou)	163b	31°02′N	121°24′E
Zhucheng, China (jo͞o-chŭŋ)	164	36°01′N	119°24′E
Zhuji, China (jo͞o-jyē)	165	29°58′N	120°10′E
Zhujiang Kou, b., Asia (jo͞o-jyäŋ kō)	165	22°00′N	114°00′E
Zhukovskiy, Russia (zhō-kôf´skĭ)	142b	55°33′N	38°09′E
Zhurivka, Ukr.	133	50°31′N	31°43′E
Zhytomyr, Ukr.	134	50°15′N	28°40′E
Zhytomyr, prov., Ukr.	133	50°40′N	28°07′E
Zi, r., China (dzē)	165	26°50′N	111°00′E
Zia Indian Reservation, I.R., N.M., U.S.	77	35°30′N	106°43′W
Zibo, China (dzē-bwo)	162	36°48′N	118°04′E
Ziel, Mount, mtn., Austl. (zēl)	174	23°15′S	132°45′E
Zielona Góra, Pol. (zhyĕ-lô´nä go͞o´rä)	124	51°56′N	15°30′E
Zigazinskiy, Russia (zĭ-gazinskēē)	142a	53°50′N	57°18′E
Ziguinchor, Sen.	184	12°35′N	16°16′W
Zile, Tur. (zē-lĕ´)	119	40°20′N	35°50′E
Žilina, Slvk. (zhē´lĭ-nä)	117	49°14′N	18°45′E
Zillah, Libya	185	28°26′N	17°52′E
Zima, Russia (zē´mȧ)	140	53°58′N	102°08′E
Zimapan, Mex. (sē-mä´pän)	88	20°43′N	99°23′W
Zimatlán de Alvarez, Mex.	89	16°52′N	96°47′W
Zimba, Zam.	191	17°19′S	26°13′E
Zimbabwe, nation, Afr. (rō-dē´zhĭ-ȧ)	186	17°50′S	29°30′E
Zimnicea, Rom. (zēm-nē´chȧ)	131	43°39′N	25°22′E
Zin, r., Isr.	153a	30°50′N	35°12′E
Zinacatepec, Mex. (zē-nä-kä-tĕ´pĕk)	89	18°19′N	97°15′W
Zinapécuaro, Mex. (sē-nä-pā´kwȧ-rō)	88	19°50′N	100°49′W
Zinder, Niger (zĭn´dẽr)	184	13°48′N	8°59′E
Zin'kiv, Ukr.	133	50°13′N	34°23′E
Zion, Il., U.S. (zī´ŭn)	69a	42°27′N	87°50′W
Zion National Park, rec., Ut., U.S.	64	37°20′N	113°00′W
Zionsville, In., U.S. (zīŭnz-vĭl)	69g	39°57′N	86°15′W
Zirandaro, Mex. (sē-rän-dä´rō)	88	18°28′N	101°02′W
Zitacuaro, Mex. (sē-tȧ-kwä´rō)	88	19°25′N	100°22′W
Zitlala, Mex. (sē-tlä´lä)	88	17°38′N	99°09′W
Zittau, Ger. (tsē´tou)	124	50°55′N	14°48′E
Ziway, l., Eth.	185	8°08′N	39°11′E
Ziya, r., China (dzē-yä)	162	38°38′N	116°31′E
Zlatograd, Blg.	131	41°24′N	25°05′E
Zlatoust, Russia (zlȧ-tō-ōst´)	134	55°13′N	59°39′E
Zlītan, Libya	185	32°27′N	14°33′E
Złoczew, Pol. (zwô´chĕf)	125	51°23′N	18°34′E
Zlynka, Russia (zlēŋ´kȧ)	132	52°28′N	31°39′E
Znamensk, Russia (znä´mĕnsk)	123	54°37′N	21°13′E
Znamianka, Ukr.	133	48°43′N	32°35′E
Znojmo, Czech Rep. (znoi´mô)	117	48°52′N	16°03′E
Zoetermeer, Neth.	115a	52°08′N	4°29′E
Zoeterwoude, Neth.	115a	52°08′N	4°29′E
Zolochiv, Ukr.	125	49°48′N	24°55′E
Zolotonosha, Ukr. (zō´lō-tō-nô´shȧ)	137	49°41′N	32°03′E
Zolotoy, Mys, c., Russia (mĭs zō-lō-tôy´)	166	47°24′N	139°10′E
Zomba, Mwi. (zōm´bȧ)	186	15°23′S	35°18′E
Zongo, D.R.C. (zŏŋ´gō)	185	4°19′N	18°36′E
Zonguldak, Tur. (zōn´go͞ol´dȧk)	154	41°25′N	31°50′E
Zonhoven, Bel.	115a	50°59′N	5°24′E
Zoquitlán, Mex. (sō-kēt-län´)	89	18°09′N	97°02′W
Zorita, Spain (thō-rē´tä)	128	39°18′N	5°41′W
Zossen, Ger. (tsō´sĕn)	115b	52°13′N	13°27′E
Zouar, Chad	189	20°27′N	16°32′E
Zouxian, China (dzō shyĕn)	164	35°24′N	116°54′E
Zubtsov, Russia (zōp-tsôf´)	132	56°13′N	34°34′E
Zuera, Spain (thwā´rä)	129	41°40′N	0°48′W
Zugdidi, Geor.	138	42°30′N	41°53′E
Zuger See, l., Switz. (tso͞og)	124	47°10′N	8°40′E
Zugspitze, mtn., Eur.	124	47°25′N	11°00′E
Zuidelijk Flevoland, reg., Neth.	115a	52°22′N	5°20′E
Zújar, r., Spain (zo͞o´ĸär)	128	38°55′N	5°05′W
Zújar, Embalse del, res., Spain	128	38°50′N	5°20′W
Zulueta, Cuba (zo͞o-lō-ĕ´tä)	92	22°20′N	79°35′W
Zumbo, Moz. (zo͞om´bō)	186	15°36′S	30°25′E
Zumbro, r., Mn., U.S. (zŭm´brō)	71	44°18′N	92°14′W
Zumbrota, Mn., U.S. (zŭm-brō´tȧ)	71	44°16′N	92°39′W
Zumpango, Mex. (sōm-päŋ-gō)	88	19°48′N	99°06′W
Zundert, Neth.	115a	51°28′N	4°39′E
Zungeru, Nig. (zōŋ-gā´ro͞o)	184	9°48′N	6°09′E
Zunhua, China (dzōn-hwä)	164	40°12′N	117°55′E
Zuni, r., Az., U.S.	77	34°40′N	109°30′W
Zuni Indian Reservation, I.R., N.M., U.S. (zo͞o´nē)	77	35°10′N	108°40′W
Zuni Mountains, mts., N.M., U.S.	77	35°10′N	108°10′W
Zunyi, China	160	27°58′N	106°40′E
Zürich, Switz. (tsü´rĭk)	110	47°22′N	8°32′E
Zürichsee, l., Switz.	124	47°18′N	8°47′E
Zushi, Japan (zo͞o´shē)	167a	35°17′N	139°35′E
Zuwārah, Libya	184	32°58′N	12°07′E
Zuwayzā, Jord.	153a	31°42′N	35°55′E
Zvenigorod, Russia (zvā-nē´gô-rôt)	132	55°46′N	36°54′E
Zvenyhorodka, Ukr.	137	49°07′N	30°59′E
Zvishavane, Zimb.	186	20°15′S	30°28′E
Zvolen, Slvk. (zvô´lĕn)	125	48°35′N	19°10′E
Zvornik, Bos. (zvôr´nēk)	131	44°24′N	19°08′E
Zweibrücken, Ger. (tsvī-brük´ĕn)	124	49°16′N	7°20′E
Zwickau, Ger. (tsvĭkou)	117	50°43′N	12°30′E
Zwolle, Neth. (zvōl´ĕ)	117	52°33′N	6°05′E
Żyradów, Pol. (zhē-rär´dōf)	125	52°04′N	20°28′E
Zyryanka, Russia (zē-ryän´kȧ)	135	65°45′N	151°15′E
Zyryanovsk, Kaz.	139	49°43′N	84°20′E

ăt; finăl; rāte; senåte; ärm; åsk; sofȧ; fâre; ch-choose; dh-as th in other; bē; ēvent; bĕt; recĕnt; cratẽr; g-gō; gh-guttural g; bĭt; ĭ-short neutral; rīde; ĸ-guttural k as ch in German ich;

Listed below are major topics covered by the thematic maps, graphs and/or statistics.
Page citations are for world, continent and country maps and for world tables.

SOURCES

The following sources have been consulted during the process of creating and updating the thematic maps and statistics for the 21st Edition.

Air Carrier Traffic at Canadian Airports, Statistics Canada
Annual Coal Report, U.S. Dept. of Energy, Energy Information Administration
Armed Conflicts Report, Project Ploughshares
Atlas of Canada, Natural Resources Canada
Canadian Minerals Yearbook, Statistics Canada
Census of Canada, Statistics Canada
Census of Population, U.S. Census Bureau
Chromium Industry Directory, International Chromium Development Association
Coal Fields of the Conterminous United States, U.S. Geological Survey
Coal Quality and Resources of the Former Soviet Union, U.S. Geological Survey
Coal-Bearing Regions and Structural Sedimentary Basins of China and Adjacent Seas, U.S. Geological Survey
Commercial Service Airports in the United States with Percent Boardings Change, Federal Aviation Administration (FAA)
Completed Peacekeeping Operations, Center for Defense Information
Conventional Arms Transfers to Developing Nations, Library of Congress, Congressional Research Service
Current Status of the World's Major Episodes of Political Violence: Hot Wars and Hot Spots, Center for Systemic Peace
Dependencies and Areas of Special Sovereignty, U.S. Dept. of State, Bureau of Intelligence and Research
Earth's Seasons—Equinoxes, Solstices, Perihelion, and Aphelion, U.S. Naval Observatory
EarthTrends: The Environmental Information Portal, World Resources Institute and World Conservation Monitoring Centre 2003. Available at http://earthtrends.wri.org/ Washington, D.C.: World Resources Institute
Economic Census, U.S. Census Bureau
Employment, Hours, and Earnings from the Current Employment Statistics Survey, U.S. Dept. of Labor, Bureau of Labor Statistics
Energy Statistics Yearbook, United Nations Dept. of Economic and Social Affairs
Epidemiological Fact Sheets by Country, Joint United Nations Program on HIV/AIDS (UNAIDS), World Health Organization, United Nations Children's Fund (UNICEF)
Estimated Water Use in the United States, U.S. Geological Survey
Estimates of Health Personnel, World Health Organization
FAO Food Balance Sheet, Food and Agriculture Organization of the United Nations (FAO)
FAO Statistical Databases (FAOSTAT), Food and Agriculture Organization of the United Nations (FAO)
Fishstat Plus, Food and Agriculture Organization of the United Nations (FAO)
Geothermal Resources Council Bulletin, Geothermal Resources Bulletin
Geothermal Resources in China, Bob Lawrence and Associates, Inc.
Global Alcohol Database, World Health Organization
Global Forest Resources Assessment, Food and Agriculture Organization of the United Nations (FAO), Forest Resources Assessment Programme
Great Lakes Factsheet Number 1, U.S. Environmental Protection Agency
The Hop Atlas, Joh. Barth & Sohn GmbH & Co. KG
Human Development Report 2003, United Nations Development Programme, © 2003 by United Nations Development Programme. Used by permission of Oxford University Press, Inc.
Installed Generating Capacity, International Geothermal Association
International Database, U.S. Census Bureau
International Energy Annual, U.S. Dept. of Energy, Energy Information Administration
International Journal on Hydropower and Dams, International Commission on Large Dams
International Petroleum Encyclopedia, PennWell Publishing Co.
International Sugar and Sweetener Report, F.O. Licht, Licht Interactive Data
International Trade Statistics, World Trade Organization
International Water Power and Dam Construction Yearbook, Wilmington Publishing
Iron and Steel Statistics, U.S. Geological Survey, Thomas D. Kelly and Michael D. Fenton
Lakes at a Glance, LakeNet
Land Scan Global Population Database, U.S. Dept. of Energy, Oak Ridge National Laboratory (© 2003 UT-Battelle, LLC. All rights reserved. Notice: These data were produced by UT-Battelle, LLC under Contract No. DE-AC05-00OR22725 with the Department of Energy. The Government has certain rights in this data. Neither UT-Battelle, LLC nor the United States Department of Energy, nor any of their employees, makes any warranty, express or implied, or assumes any legal liability or responsibility for the accuracy, completeness, or usefulness of any data, apparatus, product, or process disclosed, or represents that its use would not infringe privately owned rights.)
Largest Rivers in the United States, U.S. Geological Survey
Lengths of the Major Rivers, U.S. Geological Survey
Likely Nuclear Arsenals Under the Strategic Offensive Reductions Treaty, Center for Defense Information
Major Episodes of Political Violence, Center for Systemic Peace
Maps of Nuclear Power Reactors, International Nuclear Safety Center
Mineral Commodity Summaries, U.S. Geological Survey, Bureau of Mines
Mineral Industry Surveys, U.S. Geological Survey, Bureau of Mines
Minerals Yearbook, U.S. Geological Survey, Bureau of Mines
National Priorities List, U.S. Environmental Protection Agency
National Tobacco Information Online System (NATIONS), U.S. Dept. of Health and Human Services, Centers for Disease Control and Prevention (CDC)
Natural Gas Annual, U.S. Dept. of Energy, Energy Information Administration
New and Recent Conflicts of the World, The History Guy
Nuclear Power Reactors in the World, International Atomic Energy Agency
Oil and Gas Journal DataBook, PennWell Publishing Co.
Oil and Gas Resources of the World, Oilfield Publications, Ltd.
Petroleum Supply Annual, U.S. Dept. of Energy, Energy Information Administration
Population of Capital Cities and Cities of 100,000 and More Inhabitants, United Nations Dept. of Economic and Social Affairs
Preliminary Estimate of the Mineral Production of Canada, Natural Resources Canada
Red List of Threatened Species, International Union for Conservation and Natural Resources
Significant Earthquakes of the World, U.S. Geological Survey
State of Food Insecurity in the World, Food and Agriculture Organization of the United Nations (FAO)
State of the World's Children, United Nations Children's Fund (UNICEF)
Statistical Abstract of the United States, U.S. Census Bureau
Statistics on Asylum-Seekers, Refugees and Others of Concern to UNHCR, United Nations High Commissioner for Refugees (UNHCR)
Survey of Energy Resources, World Energy Council
Tables of Nuclear Weapons Stockpiles, Natural Resources Defense Council
TeleGeography Research, PriMetrica, Inc. (www.primetrica.com)
Tobacco Atlas, World Health Organization
Tobacco Control Country Profiles, World Health Organization
Transportation in Canada, Minister of Public Works and Government Services, Transport Canada
UNESCO Statistical Tables, United Nations Educational, Scientific and Cultural Organization (UNESCO)
United Nations Commodity Trade Statistics (COMTRADE), United Nations Dept. of Economic and Social Affairs
United Nations Peacekeeping in the Service of Peace, United Nations Dept. of Peacekeeping Operations
United Nations Peacekeeping Operations, United Nations Dept. of Peacekeeping Operations
Uranium: Resources, Production and Demand, United Nations Organization for Economic Co-operation and Development (OECD)
Volcanoes of the World, Smithsonian National Museum of Natural History
Water Account for Australia, Australian Bureau of Statistics
Women in National Parliaments, Inter-Parliamentary Union
Women's Suffrage, Inter-Parliamentary Union
The World at War, Center for Defense Information, The Defense Monitor
The World at War, Federation of American Scientists, Military Analysis Network
World Conflict List, National Defense Council Foundation
World Contraceptive Use, United Nations Dept. of Economic and Social Affairs
The World Factbook, U.S. Dept. of State, Central Intelligence Agency (CIA)
World Facts and Maps, Rand McNally
World Lakes Database, International Lake Environment Committee
World Population Prospects, United Nations Dept. of Economic and Social Affairs
World Urbanization Prospects, United Nations Dept. of Economic and Social Affairs
World Water Resources and Their Use, State Hydrological Institute of Russia/UNESCO
The World's Nuclear Arsenal, Center for Defense Information

Special Acknowledgements

The American Geographical Society, for permission to use the Miller cylindrical projection.
The Association of American Geographers, for permission to use R. Murphy's landforms map.
The McGraw-Hill Book Company, for permission to use G. Trewartha's climatic regions map.
The University of Chicago Press, for permission to use Goode's Homolosine equal-area projection.